TENNESSEE

Teacher's Edition

McDougal Littell

World Cultures and GEOGRAPHY

Tennessee Highlights

- Guide to Understanding Tennessee's Social Studies Curriculum Standards
- Guide to the Tennessee Comprehensive Assessment Program (TCAP) Preparation and Practice
- Tennessee Social Studies Curriculum Standards at point-of-use for easy planning

Senior Consultants

Sarah Witham Bednarz is associate professor of geography at Texas A&M University, where she has taught since 1988. She earned a Ph.D. in Educational Curriculum and Instruction specializing in geography in 1992 from Texas A&M University and has written extensively about geography literacy and education. Dr. Bednarz was an author of Geography for Life: National Geography Standards, 1994. In 2005 she received the prestigious George J. Miller Award from the National Council for Geographic Education.

Marci Smith Deal is the K–12 Social Studies Curriculum Coordinator for Hurst-Euless-Bedford Independent School District in Texas. She received the 2000 Distinguished Geographer Award for the State of Texas, and was one of the honorees of the 2001 National Council for Geographic Education Distinguished Teacher Award. She has served as president for the Texas Council for Social Studies Supervisors and as vice-president for the Texas Council for Social Studies. She currently serves as a teacher consultant for National Geographic Society.

Inés M. Miyares is professor of geography at Hunter College-City University of New York. Born in Havana, Cuba, and fluent in Spanish, Dr. Miyares has focused much of her scholarship on Latin America, immigration and refugee policy, and urban ethnic geography. She holds a Ph.D. in geography from Arizona State University. In 1999 Dr. Miyares was the recipient of the Hunter College Performance Excellence Award for excellence in teaching, research, scholarly writing, and service. She has just published a co-edited book on contemporary ethnic geographies in the United States.

Donna Ogle is professor of Reading and Language at National-Louis University in Chicago, Illinois, and is a specialist in reading in the content areas, with an interest in social studies. She is past president of the International Reading Association and a former social studies teacher. Dr. Ogle is currently directing two content literacy projects in Chicago schools and is Senior Consultant to the Striving Readers Research Project. She continues to explore applications of the K-W-L strategy she developed and is adding a Partner Reading component—PRC2 (Partner Reading and Content 2).

Charles S. White is associate professor in the School of Education at Boston University, where he teaches methods of instruction in social studies. Dr. White has written and spoken extensively on the importance of civic education, both in the United States and overseas. He has worked in Russia since 1997 on curriculum reform and teacher preparation in education for democracy. Dr. White has received numerous awards for his scholarship, including the 1995 Federal Design Achievement Award from the National Endowment for the Arts, for the Teaching with Historic Places project. In 1997, Dr. White taught his Models of Teaching doctoral course at the Universidad San Francisco de Quito, Ecuador.

Middle School Teacher Consultant

Judith K. Bock earned a B.A. in Elementary Education and an M.A. in Geography and Environmental Sciences from Northeastern Illinois University. She taught as a middle school educator in gifted education for 33 years and is currently an adjunct geography professor in college geography departments in metropolitan Chicago. Ms. Bock serves on several local, state, and national geography education committees, writes geography lesson plans and curriculum, and facilitates geography workshops and institutes. She received the Distinguished Teaching Award from NCGE and was awarded the Outstanding Geographer in Illinois–2004 from the Illinois Geographical Society.

Acknowledgments for copyrighted material are on pages R109–R116 and constitute an extension of this page. **Teacher Reviewers** *Erik Branch* Photo by Mara Branch; *Ann Christianson* Photo by Nick Snoeyenbos; *Jim Easton* Photo by Gayle Haab; *Ed Felten* Photo by Edward J. Felten; *Greta Frensley* Photo courtesy of Greta Frensley; *Todd Harrison* Photo by Mary Ann Gilchrist; *Esther Howse* Photo by Dennis Howse; *Amber McVey* Photo by Marcia M. Hoffman; *Ken Metz* Photo by Kenneth M. Metz; *Suzanne Moen* Photo by Barb Johanningmeier; *Tim Mortensen* Photo by Tim Mortensen; *Matt Parker* Photo by Mark T. Riggs; *Terry Rhodes* Photo by Melissa Rhodes; *Matthew Scheidler* Photo by Kristin Peterson; *Don Stringfellow* Photo by Don Stringfellow.

ISBN-13: 978-0-618-88980-8 ISBN-10: 0-618-88980-9

Printed in the United States of America

01 02 03 04 05 06 07 08 09—CKI—12 11 10 09 08 07

Internet Web Site: http://www.mcdougallittell.com

Your time is valuable.

That's why McDougal Littell *World Cultures and Geography* gives you the freedom and resources to:

- **Connect your students to the world**
- **Teach the way you want to teach**
- **Use assessment as a teaching tool**

Connect your students to the world

Integrated print and technology capture the imagination and help your students connect to the essential questions of geography.

With the click of a mouse, **Animated Geography** gives your students a close-up view of important details and concepts.

Animated GEOGRAPHY

Animated Geography engages your students with eye-opening, interactive visuals that bring geography to life.

Easily accessed through:

- Pupil's Edition
- eEdition DVD-ROM or Online
- Power Presentations DVD-ROM
- ClassZone.com

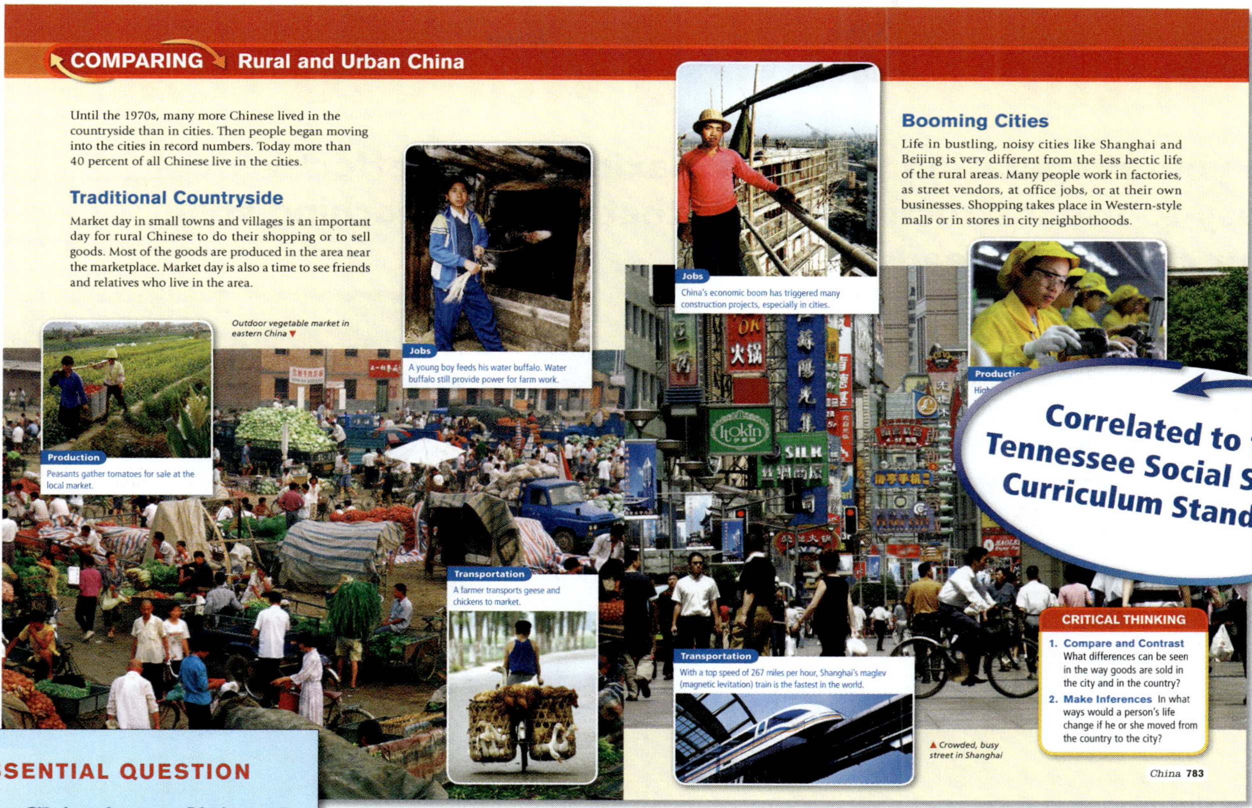

Correlated to the Tennessee Social Studies Curriculum Standards

ESSENTIAL QUESTION

How is China's traditional way of life changing in response to global trends?

The **Essential Question** connects the people, places, and events required in the Tennessee Social Studies Curriculum Standards in every chapter.

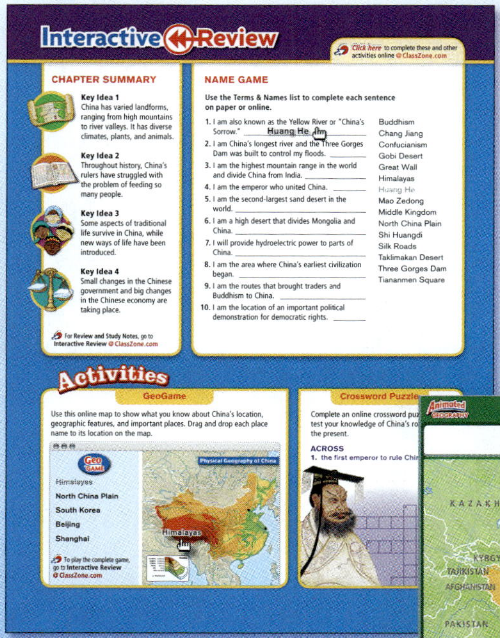

Interactive Review

Interactive Review provides online games and graphic organizers designed to help your students review and master the standards.

Available through:

- Pupil's Edition
- eEdition DVD-ROM or Online
- ClassZone.com
- Power Presentations DVD-ROM

Teach the way you want to teach

Reduce your preparation time and maximize your effectiveness with easy-to-use instructional tools that fit your teaching style and your classroom needs.

The **Presentation Toolkit** gives you fun and exciting ways to connect to your students through activities, classroom presentations, and more.

Resources2Go™ puts your complete print resources on one USB drive, giving you a fast and easy way to access and edit your lessons and resources.

Instant access to:
- Complete Teacher's Edition
- All print ancillaries for the program
- Best Practices Toolkit

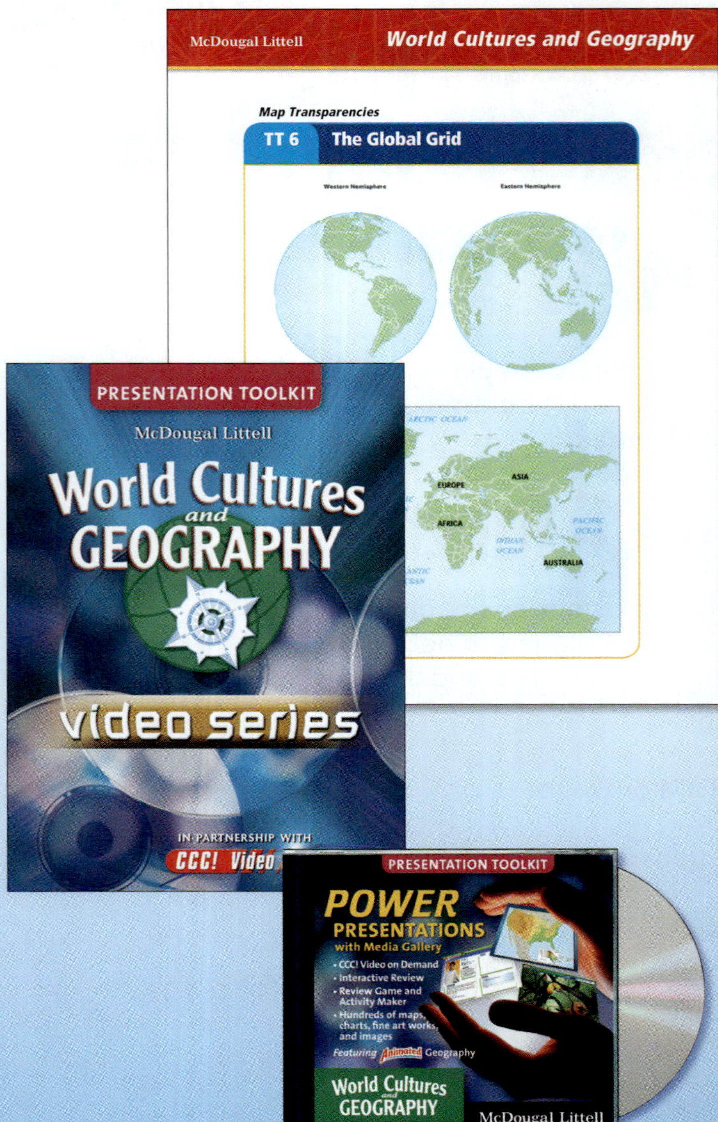

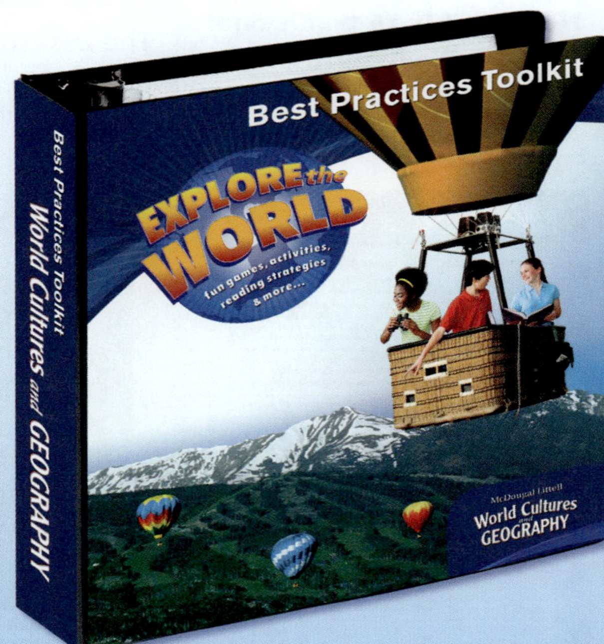

The **Best Practices Toolkit** helps you chart a path to success for every student with a variety of games, projects, reading strategies, and hands-on, differentiated activities designed specifically for middle school students.

Use assessment as a teaching tool

Easy-to-use testing and remediation tools help you save time as you keep every student on track with the standards.

Review and Practice for TCAP

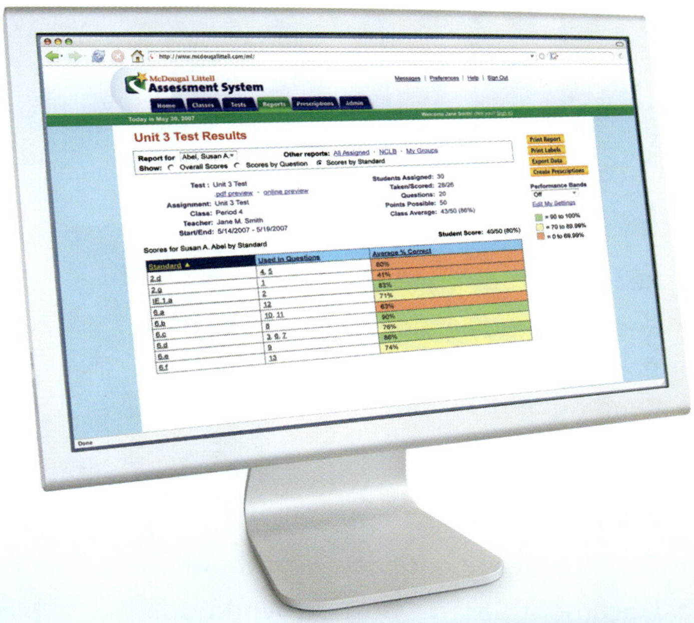

McDougal Littell Assessment System

The **McDougal Littell Assessment System** is a flexible, web-based program that allows you to use assessment as a teaching tool.

Test
Unique testing is custom-built to your state standards.

Score
Automatic scoring gives you results in minutes.

Report
Diagnostic reports show you what standards were missed.

Reteach
Personalized remediation helps you target reteaching.

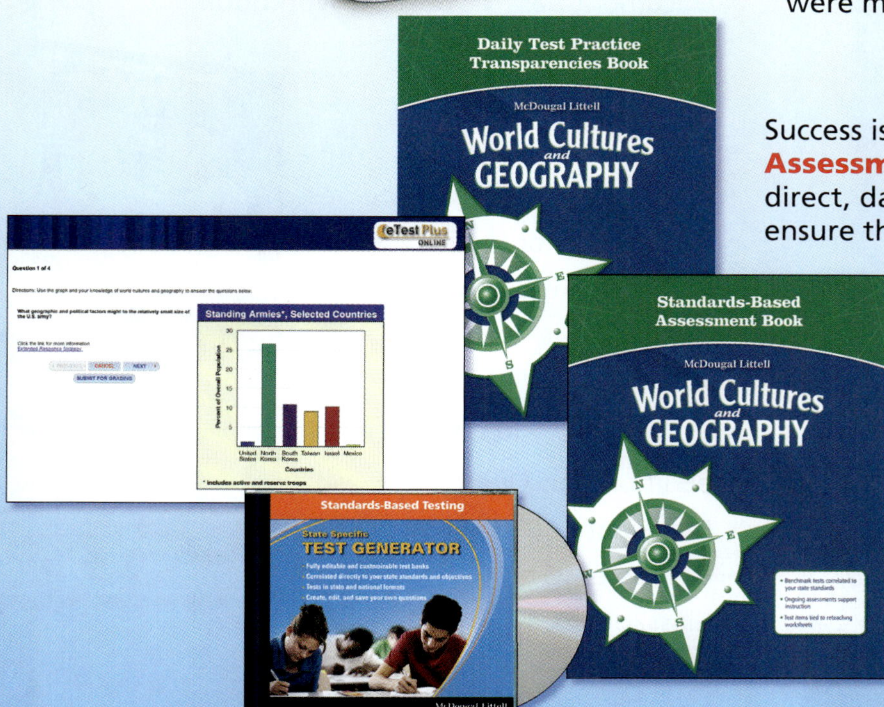

Success is as easy as 1–2–3 with the **Assessment and Reteaching Package**, giving you the direct, daily, and differentiated resources you need to ensure that your students master important concepts.

1. **DIRECT**
   Test preparation strategy instruction right in the Pupil's Edition
2. **DAILY**
   Test practice transparencies for every section of the Pupil's Edition
3. **DIFFERENTIATED**
   Standards-based options and online practice for all learners

Where Great Lessons Begin

McDougal Littell World Cultures and GEOGRAPHY

Pupil's Edition
Teacher's Edition

Resource Manager

- Lesson Plans
- Unit Resource Books

Presentation Toolkit

- Transparencies Books
- Power Presentations DVD-ROM
- *World Cultures and Geography* Video DVD

Assessment and Reteaching Package

- McDougal Littell Assessment System
- McDougal Littell Test Generator
- Standards-Based Assessment Book
- Daily Test Practice Transparencies

Student Workbooks

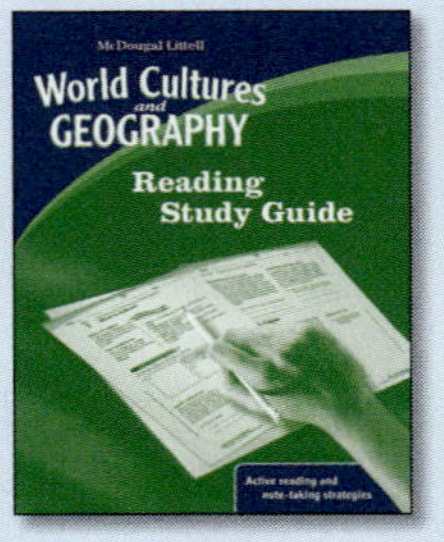

- Reading Study Guide
- Reading Study Guide with Additional Support
- Reading Study Guide in Spanish
- Reading Study Guide with Additional Support in Spanish
- Test Practice and Review Workbook
- Document-Based Question Workbook

Best Practices Toolkit

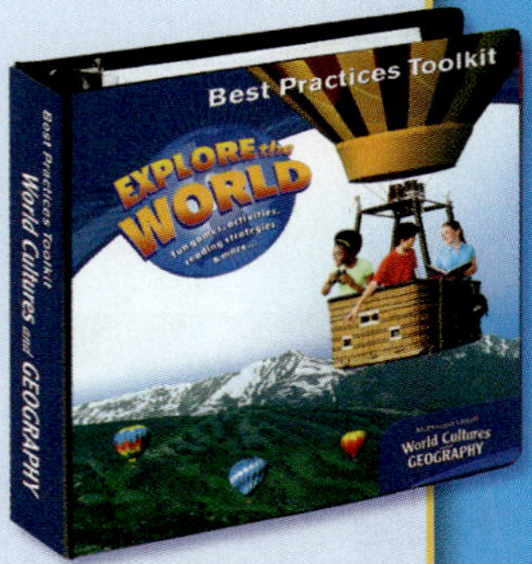

- Games, Projects, and Activities
- Interactive Reading Strategies
- Differentiation Activity Bank
- What the Research Tells Us

Technology

- eEdition DVD-ROM
- eEdition Plus Online
- EasyPlanner DVD-ROM
- Resources2Go™
- ClassZone.com
- Audio Book in English
- Audio Book in Spanish

TENNESSEE

OVERVIEW

Tennessee Teacher Edition

Smoky Mountains in Tennessee © Terrance Klassen/Alamy

TENNESSEE CONTENTS

Tennessee Social Studies Curriculum Standards: Process Standards and World Geography Seventh Grade (Standards, Learning Expectations, and Performance Indicators)

correlated to

McDougal Littell *World Cultures and Geography*

World Cultures and Geography	Tennessee Social Studies Curriculum Standards: Process Standards and World Geography Seventh Grade (Standards, Learning Expectations, and Performance Indicators)
Unit 1 Introduction to World Geography (pp. 2–103)	
Chapter 1 (pp. 2–23) Understanding the Earth and Its Peoples	
Themes and Elements of Geography pp. 4–9	**Process Standards** P1, P2, P6, P7, P8, P14, P19, P20, P21, P36 **Learning Expectations** 3.02, 3.03, 3.04, 3.05, 3.06, 3.07, 3.08, 3.09, 6.03 **Performance Indicators** 7.3.spi.6., 7.3.spi.10., 7.3.spi.11., 7.3.spi.13., 7.3.spi.18., 7.3.spi.19., 7.3.spi.20., 7.3.tpi.5., 7.5.spi.2., 7.5.tpi.3.
Technology Tools for Geographers pp. 10–15	**Process Standards** P1, P2, P6, P7, P8, P11, P12, P16, P19, P20, P21, P23, P32, P33, P36, P37 **Learning Expectations** 1.04 **Performance Indicators** 7.3.spi.6., 7.3.tpi.7.
Careers in Geography pp. 16–20	**Process Standards** P1, P2, P6, P7, P8, P10, P19, P20, P21, P22, P26, P36, P37 **Performance Indicators** 7.3.tpi.7., 7.3.tpi.18.

World Cultures and Geography	Tennessee Social Studies Curriculum Standards: Process Standards and World Geography Seventh Grade (Standards, Learning Expectations, and Performance Indicators)
Chapter 2 (pp. 24–55) Earth's Interlocking Systems	
The Earth and Its Forces pp. 26–33	**Process Standards** P1, P2, P3, P6, P7, P8, P15, P19, P20, P21, P33, P36 **Learning Expectations** 3.06, 3.07 **Performance Indicators** 7.3.spi.2., 7.3.spi.11., 7.3.tpi.16., 7.5.tpi.1., 7.5.tpi.2., 7.5.tpi.3., 7.5.tpi.4.
Bodies of Water and Landforms pp. 34–39	**Process Standards** P1, P2, P3, P6, P7, P8, P14, P15, P16, P19, P20, P21, P36 **Performance Indicators** 7.3.spi.9.
Climate and Vegetation pp. 40–47	**Process Standards** P1, P2, P3, P6, P7, P8, P14, P15, P19, P20, P21, P28, P33, P36 **Learning Expectations** 3.04, 3.05, 3.06, 3.08 **Performance Indicators** 7.3.spi.6., 7.3.spi.10., 7.3.spi.11., 7.3.spi.13., 7.3.spi.18., 7.3.tpi.5.
Environmental Challenges pp. 48–52	**Process Standards** P1, P2, P3, P4, P5, P6, P7, P8, P10, P13, P14, P15, P16, P17, P18, P19, P20, P21, P22, P23, P32, P33, P36, P37 **Learning Expectations** 3.08, 5.02, 6.01, 6.02 **Performance Indicators** 7.3.spi.13., 7.3.spi.18., 7.3.tpi.14., 7.5.spi.4., 7.5.tpi.4., 7.6.spi.1., 7.6.spi.2., 7.6.spi.3., 7.6.tpi.4.
Chapter 3 (pp. 56–85) Human Geography	
The Geography of Population pp. 58–65	**Process Standards** P1, P2, P3, P6, P7, P8, P10, P14, P15, P16, P17, P18, P19, P20, P21, P22, P23, P32, P33, P36, P37 **Learning Expectations** 1.04, 3.09, 4.02, 5.01 **Performance Indicators** 7.1.spi.2., 7.2.spi.3., 7.2.tpi.3., 7.3.spi.4., 7.3.spi.8., 7.3.spi.19., 7.3.spi.20., 7.3.tpi.17., 7.5.spi.2.
Why People Move pp. 66–71	**Process Standards** P1, P2, P3, P6, P7, P8, P10, P14, P15, P19, P20, P22, P33, P36 **Learning Expectations** 1.01, 1.03, 1.04, 3.09, 5.01, 6.01, 6.02 **Performance Indicators** 7.3.spi.4., 7.3.spi.15., 7.3.spi.19., 7.3.spi.20., 7.5.spi.2., 7.6.spi.1., 7.6.spi.2., 7.6.spi.3.

World Cultures and Geography	Tennessee Social Studies Curriculum Standards: Process Standards and World Geography Seventh Grade (Standards, Learning Expectations, and Performance Indicators)
Resources and Economics pp. 72–77	**Process Standards** P1, P2, P3, P6, P7, P8, P10, P14, P19, P20, P21, P22, P23, P32, P33, P36, P37 **Learning Expectations** 2.01, 2.02, 2.03, 3.07, 4.02 **Performance Indicators** 7.2.spi.1., 7.2.spi.2., 7.2.spi.4., 7.2.spi.5., 7.2.tpi.1., 7.2.tpi.2., 7.2.tpi.3., 7.2.tpi.6., 7.2.tpi.7., 7.2.tpi.8., 7.2.tpi.10., 7.3.spi.5., 7.5.tpi.3., 7.6.tpi.4.
Why We Need Government pp. 78–82	**Process Standards** P1, P2, P5, P6, P7, P8, P13, P15, P19, P20, P21, P26, P33, P36 **Learning Expectations** 4.01, 6.01, 6.02 **Performance Indicators** 7.1.spi.1., 7.1.tpi.3., 7.4.spi.1., 7.4.spi.4., 7.4.tpi.1., 7.4.tpi.3., 7.4.tpi.4., 7.6.spi.1., 7.6.spi.2., 7.6.spi.3.
Chapter 4 (pp. 86–103) People and Culture	
What is Culture? pp. 88–95	**Process Standards** P1, P2, P3, P5, P6, P7, P8, P13, P14, P15, P19, P20, P21, P23, P32, P33, P36 **Learning Expectations** 1.01, 1.02, 3.05, 6.01, 6.02 **Performance Indicators** 7.1.spi.1., 7.1.spi.2., 7.1.spi.3., 7.1.tpi.3., 7.3.spi.10., 7.6.spi.1., 7.6.spi.2., 7.6.spi.3., 7.6.tpi.5., 7.6.tpi.7.
How Does Culture Change? pp. 96–100	**Process Standards** P1, P2, P3, P6, P7, P8, P15, P19, P20, P21, P33, P36 **Learning Expectations** 1.01, 1.03, 6.01, 6.02 **Performance Indicators** 7.3.spi.16., 7.6.spi.1., 7.6.spi.2., 7.6.spi.3., 7.6.tpi.5.

World Cultures and Geography	Tennessee Social Studies Curriculum Standards: Process Standards and World Geography Seventh Grade (Standards, Learning Expectations, and Performance Indicators)
Unit 2 The United States and Canada (pp. 104–183)	
Unit 2 Regional Atlas: The United States and Canada pp. 106–119	**Process Standards** P1, P6, P7, P8, P14, P19, P33, P36 **Learning Expectations** 1.01, 2.02, 3.07, 3.09, 5.01 **Performance Indicators** 7.1.spi.1., 7.1.spi.2., 7.1.tpi.3., 7.2.spi.4., 7.2.spi.5., 7.2.tpi.1., 7.2.tpi.7., 7.2.tpi.10., 7.3.spi.5., 7.3.spi.19., 7.3.spi.20., 7.5.spi.2., 7.5.tpi.3.
Chapter 5 (pp. 120–153) The United States	
Geography: From Coast to Coast pp. 122–129	**Process Standards** P1, P2, P3, P6, P7, P8, P10, P14, P15, P16, P17, P18, P19, P20, P21, P22, P33, P36, P37 **Learning Expectations** 3.06, 3.08, 3.09 **Performance Indicators** 7.3.spi.9., 7.3.spi.11., 7.3.spi.13., 7.3.spi.15., 7.3.spi.18., 7.3.spi.19., 7.3.spi.20., 7.5.spi.2., 7.5.tpi.4.
History: Creating a Nation pp. 130–137	**Process Standards** P1, P2, P3, P6, P7, P8, P14, P15, P16, P19, P20, P21, P23, P26, P32, P33, P36, P37 **Learning Expectations** 3.07, 3.08, 3.09, 4.02, 5.02, 6.01, 6.02 **Performance Indicators** 7.3.spi.18., 7.3.spi.19., 7.3.spi.20., 7.5.spi.2., 7.5.spi.4., 7.5.tpi.2., 7.5.tpi.3., 7.6.spi.1., 7.6.spi.2., 7.6.spi.3., 7.6.tpi.7.
Culture: A Diverse Culture pp. 138–145	**Process Standards** P1, P2, P3, P6, P7, P8, P14, P15, P19, P20, P21, P33, P36, P37 **Learning Expectations** 1.01, 1.02, 5.01, 5.02 **Performance Indicators** 7.1.spi.1., 7.1.spi.2., 7.1.tpi.3., 7.3.tpi.17., 7.4.tpi.6., 7.5.spi.4., 7.6.tpi.10.
Government & Economics: Democracy and Free Enterprise pp. 146–150	**Process Standards** P1, P2, P6, P7, P8, P14, P15, P16, P19, P20, P21, P33, P36 **Learning Expectations** 2.01, 2.02, 2.03, 3.07, 3.09, 4.01, 4.02 **Performance Indicators** 7.1.spi.1., 7.1.tpi.3., 7.2.spi.1., 7.2.spi.2., 7.2.spi.4., 7.2.spi.5., 7.2.tpi.1., 7.2.tpi.10., 7.3.spi.5., 7.3.spi.19., 7.3.spi.20., 7.4.spi.1., 7.4.tpi.3., 7.5.spi.2., 7.5.tpi.3.

World Cultures and Geography	Tennessee Social Studies Curriculum Standards: Process Standards and World Geography Seventh Grade (Standards, Learning Expectations, and Performance Indicators)
Chapter 6 (pp. 154–183) Canada	
Geography: Mountains, Prairies, and Coastlands pp. 156–163	**Process Standards** P1, P2, P3, P6, P7, P7, P8, P10, P14, P15, P19, P20, P21, P22, P33, P36, P37 **Learning Expectations** 3.02, 3.03, 3.06, 3.07, 5.02 **Performance Indicators** 7.3.spi.9., 7.3.spi.11., 7.3.spi.15., 7.5.spi.4., 7.5.tpi.3., 7.5.tpi.4.
History: First Peoples to a Modern Nation pp. 164–169	**Process Standards** P1, P2, P3, P4, P6, P7, P8, P14, P15, P19, P20, P21, P23, P32, P34, P36, P37 **Learning Expectations** 1.01, 1.03, 3.08, 3.09, 4.02, 5.01, 6.01, 6.02 **Performance Indicators** 7.2.tpi.11., 7.3.spi.13., 7.3.spi.18., 7.3.spi.19., 7.3.spi.20., 7.5.spi.2., 7.6.spi.1., 7.6.spi.2., 7.6.spi.3.
Culture: A Nation of Immigrants pp. 170–175	**Process Standards** P1, P2, P3, P6, P7, P8, P14, P15, P19, P20, P21, P33, P36 **Learning Expectations** 1.01, 1.02, 1.04, 3.07, 3.09, 5.01, 5.02 **Performance Indicators** 7.1.spi.1., 7.1.tpi.3., 7.3.spi.19., 7.3.spi.20., 7.3.tpi.17., 7.5.spi.2., 7.5.spi.4., 7.5.tpi.3., 7.6.tpi.10.
Government & Economics: Rich Resources in a Vast Land pp. 176–180	**Process Standards** P1, P2, P6, P7, P8, P19, P20, P21, P33, P36, P37 **Learning Expectations** 2.01, 2.02, 2.03, 3.07, 4.01, 4.02 **Performance Indicators** 7.1.spi.1., 7.1.tpi.3., 7.1.tpi.5., 7.2.spi.2., 7.2.spi.4., 7.2.tpi.5., 7.2.tpi.10., 7.2.tpi.11., 7.3.tpi.12., 7.4.spi.1., 7.4.spi.4., 7.4.tpi.3., 7.4.tpi.6., 7.5.spi.1., 7.5.tpi.3.

World Cultures and Geography	Tennessee Social Studies Curriculum Standards: Process Standards and World Geography Seventh Grade (Standards, Learning Expectations, and Performance Indicators)
Unit 3 Latin America (pp. 184–311)	
Unit 3 Regional Atlas: Latin America pp. 186–195	**Process Standards** P1, P7, P8, P14, P19, P33, P36 **Learning Expectations** 1.01, 2.03, 3.06, 3.07, 3.09, 5.01 **Performance Indicators** 7.1.spi.1., 7.1.spi.2., 7.1.tpi.3., 7.2.spi.2., 7.2.spi.4., 7.3.spi.11., 7.3.spi.19., 7.3.spi.20., 7.3.tpi.9., 7.5.spi.2., 7.5.tpi.3.
Chapter 7 (pp. 196–227) Latin America: Physical Geography and History	
Physical Geography of Mexico, Central America, and the Caribbean pp. 198–203	**Process Standards** P1, P2, P3, P6, P7, P8, P14, P15, P19, P20, P21, P33, P36 **Learning Expectations** 3.02, 3.03, 3.06, 3.07, 3.08, 5.02 **Performance Indicators** 7.2.tpi.3., 7.3.spi.9., 7.3.spi.11., 7.3.spi.13., 7.3.spi.15., 7.3.spi.18., 7.5.spi.4., 7.5.tpi.2., 7.5.tpi.3., 7.5.tpi.4.
Physical Geography of South America pp. 204–211	**Process Standards** P1, P2, P3, P6, P7, P8, P14, P15, P19, P20, P33, P36 **Learning Expectations** 3.02, 3.03, 3.06, 3.07 **Performance Indicators** 7.3.spi.9., 7.3.spi.11., 7.5.tpi.3.
History: Ancient Civilizations pp. 212–219	**Process Standards** P1, P2, P3, P4, P5, P6, P7, P8, P13, P14, P15, P16, P19, P20, P23, P32, P33, P36, P37 **Learning Expectations** 1.03, 3.08, 6.01, 6.02 **Performance Indicators** 7.3.spi.13., 7.3.spi.18., 7.6.spi.1., 7.6.spi.2., 7.6.spi.3.
History: From Colonization to Independence pp. 220–224	**Process Standards** P1, P2, P3, P6, P7, P8, P10, P15, P19, P20, P21, P22, P34, P36, P37 **Learning Expectations** 3.08, 3.09, 4.02, 5.01, 5.02, 6.01, 6.02 **Performance Indicators** 7.3.spi.13., 7.3.spi.18., 7.3.spi.19., 7.3.spi.20., 7.5.spi.2., 7.5.spi.4., 7.6.spi.1., 7.6.spi.2., 7.6.spi.3., 7.6.tpi.7.

World Cultures and Geography	Tennessee Social Studies Curriculum Standards: Process Standards and World Geography Seventh Grade (Standards, Learning Expectations, and Performance Indicators)
Chapter 8 (pp. 228–251) Mexico	
History & Government: Struggle Toward Democracy pp. 230–235	**Process Standards** P1, P2, P3, P6, P7, P8, P10, P11, P14, P15, P16, P19, P20, P21, P22, P33, P36, P37 **Learning Expectations** 3.02, 3.03, 4.01, 4.02, 5.02, 6.01, 6.02 **Performance Indicators** 7.1.spi.1., 7.1.tpi.3., 7.4.spi.1., 7.4.spi.4., 7.4.tpi.3., 7.4.tpi.6., 7.5.spi.4., 7.6.spi.1., 7.6.spi.2., 7.6.spi.3., 7.6.tpi.7.
Culture: A Blend of Traditions pp. 236–243	**Process Standards** P1, P2, P3, P6, P7, P8, P9, P10, P14, P15, P19, P20, P21, P22, P34, P36, P37 **Learning Expectations** 1.01, 1.02, 1.03, 3.09, 5.01 **Performance Indicators** 7.1.spi.1., 7.1.tpi.3., 7.3.spi.19., 7.3.spi.20., 7.5.spi.2., 7.6.tpi.7., 7.6.tpi.10., 7.6.tpi.11.
Economics: Creating a New Economy pp. 244–248	**Process Standards** P1, P2, P3, P4, P5, P6, P7, P8, P10, P12, P13, P15, P16, P17, P18, P19, P20, P21, P22, P23, P32, P33, P36, P37 **Learning Expectations** 1.01, 2.01, 2.02, 2.03, 5.01 **Performance Indicators** 7.2.spi.1., 7.2.spi.2., 7.2.tpi.3., 7.2.tpi.10., 7.2.tpi.11., 7.3.tpi.12., 7.4.spi.4., 7.5.spi.1.
Chapter 9 (pp. 252–287) Middle America and Spanish-Speaking South America	
Central America pp. 254–259	**Process Standards** P1, P2, P3, P6, P7, P8, P10, P11, P12, P15, P19, P20, P21, P22, P29, P33, P36, P37 **Learning Expectations** 1.01, 1.04, 2.02, 2.03, 3.07, 3.08, 4.02, 5.01, 5.02, 6.01, 6.02 **Performance Indicators** 7.1.tpi.5., 7.2.spi.2., 7.2.tpi.3., 7.2.tpi.10., 7.2.tpi.11., 7.3.spi.13., 7.3.spi.18., 7.3.tpi.15., 7.5.spi.4., 7.5.tpi.3., 7.5.tpi.4., 7.6.spi.1., 7.6.spi.2., 7.6.spi.3., 7.6.tpi.10.
The Caribbean pp. 260–265	**Process Standards** P1, P2, P3, P6, P7, P8, P14, P15, P16, P17, P18, P19, P20, P21, P33, P36 **Learning Expectations** 1.01, 1.03, 1.04, 2.03, 3.09, 4.01, 5.01, 5.02, 6.01, 6.02 **Performance Indicators** 7.1.spi.1., 7.1.tpi.3., 7.1.tpi.5., 7.2.spi.1., 7.2.spi.2., 7.2.tpi.5., 7.2.tpi.11., 7.3.spi.19., 7.3.spi.20., 7.4.spi.1., 7.4.tpi.3., 7.5.spi.2., 7.5.spi.4., 7.6.spi.1., 7.6.spi.2., 7.6.spi.3., 7.6.tpi.10.

World Cultures and Geography	Tennessee Social Studies Curriculum Standards: Process Standards and World Geography Seventh Grade (Standards, Learning Expectations, and Performance Indicators)
Venezuela and Colombia pp. 266–271	**Process Standards** P1, P2, P3, P4, P5, P6, P7, P8, P13, P14, P15, P17, P19, P20, P21, P23, P32, P36, P37 **Learning Expectations** 1.01, 1.02, 2.03, 3.08, 5.01, 5.02, 6.01, 6.02 **Performance Indicators** 7.1.spi.1., 7.1.tpi.3., 7.2.spi.2., 7.2.tpi.3., 7.3.spi.18., 7.5.spi.4., 7.6.spi.1., 7.6.spi.2., 7.6.spi.3., 7.6.tpi.8., 7.6.tpi.10.
Peru, Bolivia, and Ecuador pp. 272–279	**Process Standards** P1, P2, P3, P6, P7, P8, P10, P14, P15, P16, P17, P19, P20, P21, P22, P33, P36, P37 **Learning Expectations** 1.01, 1.03, 2.02, 2.03, 3.06, 3.07, 3.08, 3.09, 5.01, 6.01, 6.02 **Performance Indicators** 7.1.spi.1., 7.1.tpi.3., 7.1.tpi.5., 7.2.spi.2., 7.2.tpi.3., 7.2.tpi.8., 7.2.tpi.10., 7.3.spi.11., 7.3.spi.13., 7.3.spi.18., 7.3.spi.19., 7.3.spi.20., 7.5.spi.2., 7.5.tpi.3., 7.6.spi.1., 7.6.spi.2., 7.6.spi.3., 7.6.tpi.10.
The Southern Cone pp. 280–284	**Process Standards** P1, P2, P3, P6, P7, P8, P10, P14, P15, P16, P19, P20, P21, P22, P33, P36, P37 **Learning Expectations** 1.01, 1.03, 2.01, 2.02, 2.03, 3.07, 3.09, 5.01, 5.02, 6.01, 6.02 **Performance Indicators** 7.1.spi.1., 7.1.tpi.3., 7.2.spi.1., 7.2.spi.2., 7.2.tpi.3., 7.2.tpi.8., 7.2.tpi.10., 7.3.spi.19., 7.3.spi.20., 7.5.spi.2., 7.5.spi.4., 7.5.tpi.3., 7.6.spi.1., 7.6.spi.2., 7.6.spi.3., 7.6.tpi.10.
Chapter 10 (pp. 288–311) Brazil	
History and Government: From Portuguese Colony to Modern Giant pp. 290–295	**Process Standards** P1, P2, P3, P6, P7, P8, P10, P15, P16, P19, P20, P21, P22, P33, P34, P36, P37 **Learning Expectations** 3.09, 4.01, 4.02, 5.01, 5.02, 6.01, 6.02 **Performance Indicators** 7.1.spi.1., 7.1.spi.2., 7.1.tpi.3., 7.3.spi.19., 7.3.spi.20., 7.3.tpi.12., 7.3.tpi.17., 7.4.spi.1., 7.4.tpi.1., 7.4.tpi.2., 7.4.tpi.3., 7.4.tpi.6., 7.5.spi.1., 7.5.spi.2., 7.5.spi.4., 7.5.tpi.4., 7.6.spi.1., 7.6.spi.2., 7.6.spi.3.
Culture: A Multicultural Society pp. 296–303	**Process Standards** P1, P2, P3, P5, P6, P7, P8, P10, P13, P14, P15, P19, P20, P21, P22, P23, P32, P33, P36, P37 **Learning Expectations** 1.01, 1.02, 1.03, 3.09, 5.01, 5.02 **Performance Indicators** 7.1.spi.1., 7.1.tpi.3., 7.3.spi.19., 7.3.spi.20., 7.5.spi.2., 7.5.spi.4., 7.6.tpi.10.
Economics: Developing an Abundant Land pp. 304–308	**Process Standards** P1, P2, P3, P6, P7, P8, P15, P16, P17, P18, P19, P20, P21, P31, P36 **Learning Expectations** 1.03, 2.02, 2.03, 3.07, 3.08, 5.02, 6.01, 6.02 **Performance Indicators** 7.2.spi.2., 7.2.tpi.3., 7.2.tpi.10., 7.3.spi.13., 7.3.spi.18., 7.3.tpi.14., 7.5.spi.4., 7.5.tpi.3., 7.5.tpi.4., 7.6.spi.1., 7.6.spi.2., 7.6.spi.3., 7.6.tpi.8.

World Cultures and Geography	Tennessee Social Studies Curriculum Standards: Process Standards and World Geography Seventh Grade (Standards, Learning Expectations, and Performance Indicators)
Unit 4 Europe (pp. 312–447)	
Unit 4 Regional Atlas: Europe pp. 314–323	**Process Standards** P1, P7, P8, P14, P19, P33, P36 **Learning Expectations** 3.06, 3.07, 3.09, 5.01 **Performance Indicators** 7.1.spi.1., 7.1.spi.2., 7.1.tpi.3., 7.2.spi.4., 7.3.spi.11., 7.3.spi.19., 7.3.spi.20., 7.5.spi.2., 7.5.tpi.3.
Chapter 11 (pp. 324–359) Europe: Physical Geography and History	
Geography: Europe's Dramatic Landscape pp. 326–333	**Process Standards** P1, P2, P3, P6, P7, P8, P10, P14, P15, P16, P19, P20, P21, P22, P26, P33, P36 **Learning Expectations** 3.02, 3.03, 3.06, 3.07, 3.09, 5.02, 6.01, 6.02 **Performance Indicators** 7.2.spi.1., 7.2.spi.2., 7.2.tpi.3., 7.3.spi.9., 7.3.spi.11., 7.3.spi.19., 7.3.spi.20., 7.3.tpi.9., 7.5.spi.2., 7.5.spi.4., 7.5.tpi.3., 7.6.spi.1., 7.6.spi.2., 7.6.spi.3.
History: Classical Greece and Rome pp. 334–341	**Process Standards** P1, P2, P3, P4, P5, P6, P7, P8, P13, P14, P15, P16, P17, P18, P19, P20, P21, P23, P32, P33, P36, P37 **Learning Expectations** 3.08, 4.01, 4.02, 6.01, 6.02 **Performance Indicators** 7.3.spi.13., 7.3.spi.15., 7.3.spi.18., 7.4.spi.1., 7.4.tpi.3., 7.6.spi.1., 7.6.spi.2., 7.6.spi.3., 7.6.tpi.7.
History: The Middle Ages and Renaissance pp. 342–349	**Process Standards** P1, P2, P3, P6, P7, P8, P9, P10, P14, P15, P19, P20, P21, P22, P24, P35, P36, P37 **Learning Expectations** 3.09, 4.01, 6.01, 6.02 **Performance Indicators** 7.3.spi.19., 7.3.spi.20., 7.3.spi.21., 7.4.spi.1., 7.4.tpi.3., 7.5.spi.2., 7.6.spi.1., 7.6.spi.2., 7.6.spi.3.
History: Modern European History pp. 350–356	**Process Standards** P1, P2, P3, P6, P7, P8, P14, P15, P16, P17, P19, P20, P21, P24, P33, P34, P36 **Learning Expectations** 1.04, 2.02, 3.02, 3.03, 4.02, 5.01, 5.02, 6.01, 6.02 **Performance Indicators** 7.2.tpi.10., 7.2.tpi.11., 7.3.tpi.12., 7.4.tpi.5., 7.5.spi.1., 7.5.spi.4., 7.6.spi.1., 7.6.spi.2., 7.6.spi.3.

World Cultures and Geography	Tennessee Social Studies Curriculum Standards: Process Standards and World Geography Seventh Grade (Standards, Learning Expectations, and Performance Indicators)
Chapter 12 (pp. 360–397) Western Europe	
Greece and Italy pp. 362–369	**Process Standards** P1, P2, P3, P6, P7, P8, P14, P15, P16, P19, P20, P21, P23, P32, P33, P36 **Learning Expectations** 1.01, 1.02, 2.02, 2.03, 3.07, 5.01, 5.02, 6.01, 6.02 **Performance Indicators** 7.1.spi.2., 7.1.tpi.3., 7.2.spi.2., 7.2.tpi.3., 7.2.tpi.8., 7.2.tpi.10., 7.2.tpi.11., 7.5.spi.4., 7.5.tpi.3., 7.6.spi.1., 7.6.spi.2., 7.6.spi.3., 7.6.tpi.7.
Spain and Portugal pp. 370–375	**Process Standards** P1, P2, P3, P6, P7, P8, P9, P14, P15, P19, P20, P21, P26, P33, P36 **Learning Expectations** 1.01, 1.02, 1.03, 2.02, 2.03, 3.09, 5.01, 5.02, 6.01, 6.02 **Performance Indicators** 7.1.tpi.1., 7.2.spi.2., 7.2.tpi.3., 7.2.tpi.10., 7.2.tpi.11., 7.3.spi.19., 7.3.spi.20., 7.3.tpi.12., 7.5.spi.1., 7.5.spi.2., 7.5.spi.4., 7.6.spi.1., 7.6.spi.2., 7.6.spi.3., 7.6.tpi.10.
France and the Benelux Countries pp. 376–381	**Process Standards** P1, P2, P3, P6, P7, P8, P15, P16, P17, P19, P20, P21, P31, P34, P36 **Learning Expectations** 1.01, 1.02, 2.02, 2.03, 3.07, 3.08, 5.01, 5.02, 6.01, 6.02 **Performance Indicators** 7.1.tpi.1., 7.2.spi.2., 7.2.tpi.10., 7.2.tpi.11., 7.3.spi.13., 7.3.spi.18., 7.5.spi.4., 7.5.tpi.2., 7.5.tpi.3., 7.6.spi.1., 7.6.spi.2., 7.6.spi.3.
Germany and the Alpine Countries pp. 382–387	**Process Standards** P1, P2, P3, P5, P6, P7, P8, P10, P11, P12, P13, P15, P16, P19, P20, P21, P22, P23, P24, P32, P34, P36, P37 **Learning Expectations** 1.01, 1.02, 1.03, 2.02, 5.01, 5.02, 6.01, 6.02 **Performance Indicators** 7.2.spi.1., 7.2.tpi.10., 7.4.spi.4., 7.5.spi.4., 7.6.spi.1., 7.6.spi.2., 7.6.spi.3., 7.6.tpi.7., 7.6.tpi.9., 7.6.tpi.10.
The Nordic Countries pp. 388–394	**Process Standards** P1, P2, P3, P6, P7, P8, P10, P11, P12, P14, P15, P16, P17, P18, P19, P20, P21, P22, P24, P25, P33, P36, P37 **Learning Expectations** 1.01, 1.02, 1.03, 2.02, 2.03, 3.07, 3.09, 5.01, 5.02, 6.01, 6.02 **Performance Indicators** 7.1.spi.1., 7.1.tpi.3., 7.2.spi.2., 7.2.tpi.10., 7.3.spi.15., 7.3.spi.19., 7.3.spi.20., 7.4.tpi.5., 7.5.spi.2., 7.5.spi.4., 7.5.tpi.3., 7.6.spi.1., 7.6.spi.2., 7.6.spi.3., 7.6.tpi.10.

World Cultures and Geography	Tennessee Social Studies Curriculum Standards: Process Standards and World Geography Seventh Grade (Standards, Learning Expectations, and Performance Indicators)
Chapter 13 (pp. 398–421) United Kingdom	
History: Building a British Empire pp. 400–407	**Process Standards** P1, P2, P3, P6, P7, P8, P10, P11, P12, P15, P19, P20, P21, P22, P23, P29, P32, P36, P37 **Learning Expectations** 1.01, 1.03, 1.04, 3.07, 3.09, 4.01, 4.02, 5.01, 5.02, 6.01, 6.02 **Performance Indicators** 7.3.spi.19., 7.3.spi.20., 7.3.tpi.12., 7.4.spi.1., 7.4.tpi.3., 7.5.spi.1., 7.5.spi.2., 7.5.spi.4., 7.5.tpi.3., 7.6.spi.1., 7.6.spi.2., 7.6.spi.3., 7.6.tpi.7.
Culture: From Shakespeare to J. K. Rowling pp. 408–413	**Process Standards** P1, P2, P3, P6, P7, P8, P15, P19, P20, P21, P36 **Learning Expectations** 1.01, 1.02, 3.09, 5.01, 5.02 **Performance Indicators** 7.1.spi.1., 7.1.tpi.3., 7.3.spi.19., 7.3.spi.20., 7.3.tpi.12., 7.3.tpi.17., 7.5.spi.1., 7.5.spi.2., 7.5.spi.4., 7.6.tpi.10.
Government & Economics: Parliament and Free Enterprise pp. 414–418	**Process Standards** P1, P2, P3, P6, P7, P8, P10, P11, P12, P14, P15, P16, P19, P20, P21, P22, P26, P33, P36, P37 **Learning Expectations** 1.03, 2.02, 2.03, 3.07, 3.09, 4.01, 5.02 **Performance Indicators** 7.1.spi.1., 7.1.tpi.3., 7.2.spi.1., 7.2.spi.2., 7.2.spi.4., 7.2.tpi.10., 7.2.tpi.11., 7.3.spi.19., 7.3.spi.20., 7.4.spi.1., 7.4.spi.4., 7.4.tpi.1., 7.4.tpi.2., 7.4.tpi.3., 7.5.spi.2., 7.5.spi.4., 7.5.tpi.3.
Chapter 14 (pp. 422–447) Eastern Europe	
Poland, Ukraine, and the Baltic States pp. 424–431	**Process Standards** P1, P2, P3, P4, P5, P6, P7, P8, P10, P13, P14, P15, P16, P19, P20, P21, P22, P23, P24, P32, P33, P36, P37 **Learning Expectations** 1.01, 1.02, 2.01, 2.02, 2.03, 3.07, 3.08, 3.09, 4.02, 5.01, 5.02, 6.01, 6.02 **Performance Indicators** 7.1.spi.1., 7.1.tpi.3., 7.1.tpi.5., 7.2.spi.1., 7.2.spi.2., 7.2.tpi.8., 7.2.tpi.10., 7.2.tpi.11., 7.3.spi.13., 7.3.spi.16., 7.3.spi.18., 7.3.spi.19., 7.3.spi.20., 7.3.tpi.12., 7.5.spi.1., 7.5.spi.2., 7.5.spi.4., 7.5.tpi.3., 7.6.spi.1., 7.6.spi.2., 7.6.spi.3., 7.6.tpi.7., 7.6.tpi.10.
Hungary and the Czech Republic pp. 432–439	**Process Standards** P1, P2, P3, P6, P7, P8, P14, P15, P19, P20, P21, P26, P33, P36 **Learning Expectations** 1.01, 1.02, 2.02, 4.02, 5.01, 5.02, 6.01, 6.02 **Performance Indicators** 7.1.spi.1., 7.1.spi.2., 7.1.tpi.3., 7.2.tpi.3., 7.2.tpi.10., 7.3.tpi.12., 7.5.spi.1., 7.5.spi.4., 7.6.spi.1., 7.6.spi.2., 7.6.spi.3., 7.6.tpi.7., 7.6.tpi.10.

World Cultures and Geography	Tennessee Social Studies Curriculum Standards: Process Standards and World Geography Seventh Grade (Standards, Learning Expectations, and Performance Indicators)
The Balkans pp. 440–444	**Process Standards** P1, P2, P3, P6, P7, P8, P14, P15, P17, P19, P20, P21, P33, P36 **Learning Expectations** 1.01, 1.02, 2.02, 5.01, 5.02, 6.01, 6.02 **Performance Indicators** 7.1.spi.1., 7.1.spi.2., 7.1.tpi.3., 7.2.tpi.10., 7.2.tpi.11., 7.3.tpi.17., 7.5.spi.4., 7.6.spi.1., 7.6.spi.2., 7.6.spi.3.
Unit 5 Russia and the Eurasian Republics (pp. 448–519)	
Unit 5 Regional Atlas: Russia and the Eurasian Republics pp. 450–457	**Process Standards** P1, P7, P8, P14, P19, P33, P36 **Learning Expectations** 3.06, 3.07, 3.09, 5.01 **Performance Indicators** 7.1.spi.1., 7.1.spi.2., 7.1.tpi.3., 7.2.spi.4., 7.3.spi.11., 7.3.spi.19., 7.3.spi.20., 7.5.spi.2., 7.5.tpi.3.
Chapter 15 (pp. 458–493) Russia	
Geography: Sweeping Across Eurasia pp. 460–467	**Process Standards** P1, P2, P3, P6, P7, P8, P14, P15, P16, P19, P20, P21, P33, P36 **Learning Expectations** 1.03, 1.04, 2.03, 3.02, 3.03, 3.07, 3.08 **Performance Indicators** 7.2.spi.2., 7.2.tpi.3., 7.3.spi.9., 7.3.spi.13., 7.3.spi.18., 7.3.tpi.10., 7.3.tpi.14., 7.5.tpi.3.
History: Governing a Vast Land pp. 468–475	**Process Standards** P1, P2, P3, P6, P7, P8, P14, P15, P19, P20, P21, P24, P27, P31, P33, P36 **Learning Expectations** 1.03, 2.03, 3.02, 3.09, 4.02, 5.01, 5.02, 6.01, 6.02 **Performance Indicators** 7.3.spi.19., 7.3.spi.20., 7.5.spi.2., 7.5.spi.4., 7.5.tpi.2., 7.6.spi.1., 7.6.spi.2., 7.6.spi.3., 7.6.tpi.7.
Culture: Blending Europe and Asia pp. 476–485	**Process Standards** P1, P2, P3, P5, P6, P7, P8, P13, P14, P15, P19, P20, P21, P23, P32, P36, P37 **Learning Expectations** 1.01, 1.02, 3.07, 5.01, 5.02 **Performance Indicators** 7.1.spi.1., 7.1.tpi.3., 7.3.spi.16., 7.3.tpi.15., 7.5.spi.4., 7.5.tpi.3., 7.6.tpi.10.
Government & Economics: The Struggle for Reform pp. 486–490	**Process Standards** P1, P2, P3, P6, P7, P8, P10, P12, P14, P15, P19, P20, P21, P22, P33, P36, P37 **Learning Expectations** 3.07, 4.02, 5.02 **Performance Indicators** 7.1.spi.1., 7.1.tpi.3., 7.1.tpi.5., 7.2.spi.1., 7.3.tpi.12., 7.4.spi.4., 7.4.tpi.1., 7.4.tpi.2., 7.5.spi.1., 7.5.spi.4., 7.5.tpi.3.

World Cultures and Geography	Tennessee Social Studies Curriculum Standards: Process Standards and World Geography Seventh Grade (Standards, Learning Expectations, and Performance Indicators)
Chapter 16 (pp. 494–515) The Eurasian Republics	
Geography: Center of a Landmass pp. 496–503	**Process Standards** P1, P2, P3, P6, P7, P8, P10, P11, P12, P14, P15, P17, P18, P19, P20, P21, P22, P26, P32, P33, P36, P37 **Learning Expectations** 2.03, 3.02, 3.03, 3.06, 3.07, 3.08, 3.09, 4.02, 5.02, 6.01, 6.02 **Performance Indicators** 7.2.spi.2., 7.3.spi.9., 7.3.spi.11., 7.3.spi.13., 7.3.spi.18., 7.3.spi.19., 7.3.spi.20., 7.3.tpi.10., 7.3.tpi.12., 7.4.tpi.6., 7.5.spi.1., 7.5.spi.2., 7.5.spi.4., 7.5.tpi.3., 7.5.tpi.4., 7.6.spi.1., 7.6.spi.2., 7.6.spi.3., 7.6.tpi.8.
History & Culture: Historic Crossroads pp. 504–511	**Process Standards** P1, P2, P3, P6, P7, P8, P10, P14, P15, P19, P20, P21, P22, P23, P36, P37 **Learning Expectations** 1.01, 1.02, 1.03, 4.02, 5.01, 5.02, 6.01, 6.02 **Performance Indicators** 7.1.spi.1., 7.1.spi.2., 7.1.tpi.3., 7.2.tpi.3., 7.3.tpi.10., 7.3.tpi.12., 7.3.tpi.17., 7.5.spi.1., 7.5.spi.4., 7.6.spi.1., 7.6.spi.2., 7.6.spi.3., 7.6.tpi.10.
Government & Economics: The Challenge of Independence pp. 512–516	**Process Standards** P1, P2, P3, P4, P5, P6, P7, P8, P10, P13, P14, P15, P16, P17, P18, P19, P20, P21, P22, P23, P32, P36, P37 **Learning Expectations** 2.02, 2.03, 3.07, 4.02, 5.02, 6.01, 6.02 **Performance Indicators** 7.2.spi.1., 7.2.spi.2., 7.2.tpi.3., 7.2.tpi.10., 7.4.spi.4., 7.5.spi.4., 7.5.tpi.3., 7.6.spi.1., 7.6.spi.2., 7.6.spi.3., 7.6.tpi.7.
Unit 6 **Africa** **(pp. 520–639)**	
Unit 6 Regional Atlas: Africa pp. 522–533	**Process Standards** P1, P6, P7, P8, P14, P19, P33, P36 **Learning Expectations** 3.07, 3.09, 5.01 **Performance Indicators** 7.1.spi.1., 7.1.spi.2., 7.1.tpi.3., 7.2.spi.4., 7.3.spi.19., 7.3.spi.20., 7.3.tpi.9., 7.5.spi.2., 7.5.tpi.3.
Chapter 17 (pp. 534–551) Africa: Physical Geography and Early History	
Africa's Geographic Extremes pp. 536–543	**Process Standards** P1, P2, P3, P6, P7, P8, P15, P19, P20, P21, P33, P36 **Learning Expectations** 1.01, 1.03, 2.03, 3.02, 3.03, 3.06, 3.07, 3.08, 3.09, 4.02, 5.01, 5.02, 6.01, 6.02 **Performance Indicators** 7.2.spi.2., 7.3.spi.9., 7.3.spi.11., 7.3.spi.13., 7.3.spi.18., 7.3.spi.19., 7.3.spi.20., 7.3.tpi.17., 7.5.spi.2., 7.5.spi.4., 7.5.tpi.2., 7.5.tpi.3., 7.5.tpi.4., 7.6.spi.1., 7.6.spi.2., 7.6.spi.3.

World Cultures and Geography	Tennessee Social Studies Curriculum Standards: Process Standards and World Geography Seventh Grade (Standards, Learning Expectations, and Performance Indicators)
Human beginnings in Africa pp. 544–548	**Process Standards** P1, P2, P4, P5, P6, P7, P8, P13, P14, P19, P20, P21, P23, P32, P34, P36, P37 **Learning Expectations** 1.01
Chapter 18 (pp. 552–577) Egypt and North Africa	
History: Ancient Egypt—Pyramids and Pharaohs pp. 554–561	**Process Standards** P1, P2, P3, P6, P7, P8, P10, P15, P16, P19, P20, P21, P22, P28, P33, P36, P37 **Learning Expectations** 1.01, 1.03, 3.07, 3.08, 3.09, 5.01, 5.02, 6.01, 6.02 **Performance Indicators** 7.3.spi.13., 7.3.spi.18., 7.3.spi.19., 7.3.spi.20., 7.3.spi.21., 7.3.tpi.10., 7.3.tpi.15., 7.5.spi.2., 7.5.spi.3., 7.5.spi.4., 7.5.tpi.2., 7.5.tpi.3., 7.5.tpi.4., 7.6.spi.1., 7.6.spi.2., 7.6.spi.3., 7.6.tpi.4.
Modern Egypt & Sudan pp. 562–567	**Process Standards** P1, P2, P3, P6, P7, P8, P10, P11, P12, P14, P15, P19, P20, P21, P22, P33, P34, P36 **Learning Expectations** 1.01, 1.02, 3.07, 3.08, 4.02, 5.01, 5.02, 6.01, 6.02 **Performance Indicators** 7.1.spi.1., 7.2.tpi.3., 7.2.tpi.6., 7.3.spi.13., 7.3.spi.18., 7.5.spi.4., 7.5.tpi.3., 7.5.tpi.4., 7.6.spi.1., 7.6.spi.2., 7.6.spi.3., 7.6.tpi.7., 7.6.tpi.10.
Other Nations of North Africa pp. 568–574	**Process Standards** P1, P2, P3, P6, P7, P8, P10, P14, P15, P19, P20, P21, P22, P23, P32, P35, P36, P37 **Learning Expectations** 1.01, 1.02, 2.03, 3.09, 4.02, 5.01, 5.02, 6.01, 6.02 **Performance Indicators** 7.1.spi.1., 7.1.tpi.3., 7.2.spi.2., 7.2.tpi.8., 7.3.spi.19., 7.3.spi.20., 7.3.tpi.12., 7.5.spi.1., 7.5.spi.2., 7.5.spi.4., 7.6.spi.1., 7.6.spi.2., 7.6.spi.3., 7.6.tpi.9., 7.6.tpi.10.

World Cultures and Geography	Tennessee Social Studies Curriculum Standards: Process Standards and World Geography Seventh Grade (Standards, Learning Expectations, and Performance Indicators)
Chapter 19 (pp. 578–603) West Africa	
History: Three Trading Empires pp. 580–585	**Process Standards** P1, P2, P3, P6, P7, P8, P10, P14, P15, P19, P20, P21, P22, P33, P36 **Learning Expectations** 1.03, 1.04, 3.07, 3.08, 3.09, 4.02, 5.01, 5.02, 6.01, 6.02 **Performance Indicators** 7.1.spi.1., 7.1.tpi.3., 7.3.spi.13., 7.3.spi.18., 7.3.spi.19., 7.3.spi.20., 7.3.tpi.15., 7.5.spi.2., 7.5.spi.4., 7.5.tpi.3., 7.6.spi.1., 7.6.spi.2., 7.6.spi.3., 7.6.tpi.7., 7.6.tpi.10.
Nigeria and Ghana pp. 586–593	**Process Standards** P1, P2, P3, P4, P5, P6, P7, P8, P10, P13, P14, P15, P16, P19, P20, P21, P22, P23, P32, P33, P36, P37 **Learning Expectations** 1.01, 1.02, 1.03, 2.02, 2.03, 3.07, 3.08, 4.01, 5.02, 6.01, 6.02 **Performance Indicators** 7.2.spi.2., 7.2.tpi.3., 7.2.tpi.10., 7.3.spi.13., 7.3.spi.18., 7.4.spi.1., 7.4.tpi.3., 7.5.spi.4., 7.5.tpi.3., 7.6.spi.1., 7.6.spi.2., 7.6.spi.3., 7.6.tpi.7.
Mali and Côte d'Ivoire pp. 594–600	**Process Standards** P1, P2, P3, P6, P7, P8, P10, P14, P15, P16, P19, P20, P21, P22, P29, P36, P37 **Learning Expectations** 1.01, 1.02, 2.03, 3.09, 5.01, 5.02, 6.01, 6.02 **Performance Indicators** 7.1.tpi.5., 7.2.spi.1., 7.2.spi.2., 7.2.tpi.3., 7.3.spi.19., 7.3.spi.20., 7.3.tpi.12., 7.5.spi.1., 7.5.spi.2., 7.5.spi.4., 7.5.tpi.2., 7.6.spi.1., 7.6.spi.2., 7.6.spi.3., 7.6.tpi.10.

World Cultures and Geography	Tennessee Social Studies Curriculum Standards: Process Standards and World Geography Seventh Grade (Standards, Learning Expectations, and Performance Indicators)
Chapter 20 (pp. 604–639) East, Central, and Southern Africa	
Ethiopia pp. 606–611	**Process Standards** P1, P2, P3, P6, P7, P8, P14, P15, P19, P20, P21, P33, P36, P37 **Learning Expectations** 1.01, 1.02, 1.03, 2.02, 3.02, 3.03, 3.09, 5.01, 5.02, 6.01, 6.02 **Performance Indicators** 7.2.tpi.5., 7.2.tpi.10., 7.3.spi.15., 7.3.spi.19., 7.3.spi.20., 7.3.tpi.12., 7.5.spi.1., 7.5.spi.2., 7.5.spi.3., 7.5.spi.4., 7.5.tpi.2., 7.6.spi.1., 7.6.spi.2., 7.6.spi.3., 7.6.tpi.10.
Kenya pp. 612–617	**Process Standards** P1, P2, P3, P4, P5, P6, P7, P8, P13, P15, P16, P19, P20, P21, P23, P32, P33, P36, P37 **Learning Expectations** 1.01, 1.02, 1.03, 2.02, 2.03, 3.07, 4.02, 5.01, 5.02, 6.01, 6.02 **Performance Indicators** 7.1.spi.1., 7.1.tpi.3., 7.2.spi.2., 7.2.tpi.10., 7.5.spi.4., 7.5.tpi.3., 7.6.spi.1., 7.6.spi.2., 7.6.spi.3., 7.6.tpi.7., 7.6.tpi.10.
Democratic Republic of the Congo pp. 618–623	**Process Standards** P1, P2, P3, P6, P7, P8, P14, P15, P16, P19, P20, P21, P33, P36 **Learning Expectations** 1.01, 1.02, 1.03, 2.02, 3.07, 3.09, 5.01, 5.02, 6.01, 6.02 **Performance Indicators** 7.1.spi.2., 7.2.tpi.4., 7.2.tpi.10., 7.3.spi.15., 7.3.spi.19., 7.3.spi.20., 7.4.spi.4., 7.5.spi.2., 7.5.spi.4., 7.5.tpi.3., 7.6.spi.1., 7.6.spi.2., 7.6.spi.3., 7.6.tpi.7., 7.6.tpi.9., 7.6.tpi.10.
South Africa pp. 624–629	**Process Standards** P1, P2, P3, P6, P7, P8, P10, P14, P15, P19, P20, P21, P22, P24, P27, P31, P33, P34, P36, P37 **Learning Expectations** 1.01, 1.02, 2.02, 2.03, 3.07, 3.08, 3.09, 4.01, 4.02, 5.01, 5.02, 6.01, 6.02 **Performance Indicators** 7.1.tpi.3., 7.2.spi.2., 7.2.spi.4., 7.2.tpi.3., 7.2.tpi.10., 7.3.spi.4., 7.3.spi.13., 7.3.spi.18., 7.3.spi.19., 7.3.spi.20., 7.3.tpi.12., 7.4.spi.1., 7.4.spi.4., 7.4.tpi.3., 7.5.spi.1., 7.5.spi.2., 7.5.spi.4., 7.5.tpi.3., 7.6.spi.1., 7.6.spi.2., 7.6.spi.3., 7.6.tpi.7., 7.6.tpi.10.
Zimbabwe, Botswana, and Angola pp. 630–636	**Process Standards** P1, P2, P3, P6, P7, P8, P10, P12, P14, P15, P19, P20, P21, P22, P33, P36, P37 **Learning Expectations** 1.01, 1.02, 2.02, 2.03, 3.07, 3.08, 3.09, 4.02, 5.01, 5.02, 6.01, 6.02 **Performance Indicators** 7.1.spi.1., 7.1.tpi.3., 7.2.spi.2., 7.2.tpi.6., 7.2.tpi.10., 7.3.spi.13., 7.3.spi.18., 7.3.spi.19., 7.3.spi.20., 7.4.spi.1., 7.4.tpi.3., 7.5.spi.2., 7.5.spi.4., 7.5.tpi.3., 7.6.spi.1., 7.6.spi.2., 7.6.spi.3., 7.6.tpi.7., 7.6.tpi.10.

World Cultures and Geography	Tennessee Social Studies Curriculum Standards: Process Standards and World Geography Seventh Grade (Standards, Learning Expectations, and Performance Indicators)
Unit 7 Southwest Asia and South Asia (pp. 640–743)	
Unit 7 Regional Atlas: Southwest Asia and South Asia pp. 642–649	**Process Standards** P1, P6, P7, P8, P14, P19, P33 **Learning Expectations** 3.06, 3.07, 3.09, 5.01 **Performance Indicators** 7.1.spi.1., 7.1.spi.2., 7.1.tpi.3., 7.2.spi.4., 7.3.spi.11., 7.3.spi.19., 7.3.spi.20., 7.5.spi.2., 7.5.tpi.3.
Chapter 21 (pp. 650–673) Southwest Asia and South Asia: Physical Geography and History	
Physical Geography of Southwest Asia pp. 652–657	**Process Standards** P1, P2, P3, P6, P7, P8, P10, P14, P15, P16, P19, P20, P21, P22, P23, P32, P33, P36, P37 **Learning Expectations** 1.03, 2.03, 3.06, 3.07, 3.08, 3.09, 4.02, 6.01, 6.02 **Performance Indicators** 7.2.spi.2., 7.3.spi.9., 7.3.spi.11., 7.3.spi.13., 7.3.spi.18., 7.3.spi.19., 7.3.spi.20., 7.3.tpi.9., 7.3.tpi.10., 7.5.spi.2., 7.5.tpi.2., 7.5.tpi.3., 7.5.tpi.4., 7.6.spi.1., 7.6.spi.2., 7.6.spi.3.
Physical Geography of South Asia pp. 658–665	**Process Standards** P1, P2, P3, P5, P6, P7, P8, P13, P14, P15, P16, P17, P18, P19, P20, P21, P23, P26, P31, P32, P33, P36, P37 **Learning Expectations** 2.03, 3.02, 3.03, 3.06, 3.07 **Performance Indicators** 7.2.tpi.3., 7.2.tpi.5., 7.3.spi.9., 7.3.spi.11., 7.3.spi.15., 7.3.tpi.10., 7.5.tpi.2., 7.5.tpi.3., 7.5.tpi.4., 7.6.tpi.8.
History: River Valley Civilizations pp. 666–670	**Process Standards** P1, P2, P3, P6, P7, P8, P9, P10, P12, P14, P15, P19, P20, P21, P22, P30, P36, P37 **Learning Expectations** 3.08, 5.01, 5.02, 6.01, 6.02 **Performance Indicators** 7.3.spi.18., 7.3.tpi.15., 7.5.spi.3., 7.5.spi.4., 7.5.tpi.2., 7.5.tpi.3., 7.6.spi.1., 7.6.spi.2., 7.6.spi.3., 7.6.tpi.7.

World Cultures and Geography	Tennessee Social Studies Curriculum Standards: Process Standards and World Geography Seventh Grade (Standards, Learning Expectations, and Performance Indicators)
Chapter 22 (pp. 674–695) Southwest Asia	
Saudi Arabia, Iraq, and Iran pp. 676–681	**Process Standards** P1, P2, P3, P6, P7, P8, P14, P15, P16, P19, P20, P21, P24, P27, P33, P36, P37 **Learning Expectations** 1.01, 1.02, 1.03, 2.02, 2.03, 3.07, 3.09, 4.01, 4.02, 5.01, 5.02, 6.01, 6.02 **Performance Indicators** 7.1.spi.1., 7.1.spi.2., 7.1.tpi.3., 7.2.spi.2., 7.2.tpi.3., 7.2.tpi.5., 7.2.tpi.10., 7.2.tpi.11., 7.3.spi.15., 7.3.spi.19., 7.3.spi.20., 7.3.tpi.17., 7.4.spi.1., 7.4.tpi.2., 7.4.tpi.3., 7.5.spi.2., 7.5.spi.4., 7.5.tpi.3., 7.6.spi.1., 7.6.spi.2., 7.6.spi.3., 7.6.tpi.7., 7.6.tpi.8., 7.6.tpi.10.
Israel and the Palestinian Territories pp. 682–687	**Process Standards** P1, P2, P3, P4, P5, P6, P7, P8, P10, P11, P12, P13, P14, P15, P19, P20, P21, P22, P23, P30, P32, P33, P36, P37 **Learning Expectations** 1.01, 1.02, 1.04, 2.03, 3.02, 3.03, 3.08, 5.01, 5.02, 6.01, 6.02 **Performance Indicators** 7.1.spi.1., 7.1.spi.2., 7.1.tpi.3., 7.2.spi.2., 7.2.tpi.3., 7.3.spi.13., 7.3.spi.16., 7.3.spi.18., 7.5.spi.4., 7.6.spi.1., 7.6.spi.2., 7.6.spi.3., 7.6.tpi.8.
Turkey, Syria, and Lebanon pp. 688–692	**Process Standards** P1, P2, P3, P6, P7, P8, P14, P15, P19, P20, P22, P33, P34, P36 **Learning Expectations** 1.01, 3.02, 3.03, 3.07, 5.01, 5.02, 6.01, 6.02 **Performance Indicators** 7.1.spi.1., 7.1.tpi.1., 7.1.tpi.3., 7.2.tpi.11., 7.5.spi.4., 7.5.tpi.3., 7.6.spi.1., 7.6.spi.2., 7.6.spi.3., 7.6.tpi.10.
Chapter 23 (pp. 696–725) India	
History: Empires and Independence pp. 698–705	**Process Standards** P1, P2, P3, P4, P6, P7, P8, P14, P15, P19, P20, P21, P23, P26, P32, P33, P36, P37 **Learning Expectations** 3.02, 3.09, 4.02, 5.01, 5.02, 6.01, 6.02 **Performance Indicators** 7.3.spi.19., 7.3.spi.20., 7.5.spi.2., 7.5.spi.4., 7.6.spi.1., 7.6.spi.2., 7.6.spi.3.
Culture: A Rich and Ancient Legacy pp. 706–715	**Process Standards** P1, P2, P3, P6, P7, P8, P10, P14, P15, P16, P17, P19, P20, P21, P22, P33, P36 **Learning Expectations** 1.01, 1.02, 3.09, 5.01, 5.02 **Performance Indicators** 7.1.spi.1., 7.1.spi.2., 7.1.spi.3., 7.1.tpi.3., 7.2.tpi.3., 7.2.tpi.5., 7.3.spi.16., 7.3.spi.19., 7.3.spi.20., 7.3.tpi.12., 7.5.spi.1., 7.5.spi.2., 7.5.spi.4., 7.6.tpi.9., 7.6.tpi.10.

World Cultures and Geography	Tennessee Social Studies Curriculum Standards: Process Standards and World Geography Seventh Grade (Standards, Learning Expectations, and Performance Indicators)
Government & Economics: The World's Largest Democracy pp. 716–722	**Process Standards** P1, P2, P3, P6, P7, P8, P10, P12, P14, P15, P19, P20, P21, P22, P33, P36 **Learning Expectations** 2.02, 2.03, 3.07, 3.08, 4.01, 5.01, 5.02, 6.01, 6.02 **Performance Indicators** 7.1.spi.1., 7.1.tpi.3., 7.2.spi.2., 7.2.spi.4., 7.2.tpi.5., 7.2.tpi.10., 7.3.spi.13., 7.3.spi.18., 7.3.tpi.17., 7.4.spi.1., 7.4.tpi.1., 7.4.tpi.2., 7.4.tpi.3., 7.5.spi.4., 7.5.tpi.3., 7.6.spi.1., 7.6.spi.2., 7.6.spi.3.
Chapter 24 (pp. 726–743) South Asian Neighbors	
Afghanistan, Pakistan, and Bangladesh pp. 728–735	**Process Standards** P1, P2, P3, P4, P6, P7, P8, P14, P15, P16, P19, P20, P21, P23, P26, P32, P33, P36, P37 **Learning Expectations** 1.01, 1.02, 2.03, 3.07, 3.08, 3.09, 5.01, 5.02, 6.01, 6.02 **Performance Indicators** 7.1.spi.1., 7.1.tpi.3., 7.2.spi.2., 7.3.spi.13., 7.3.spi.18., 7.3.spi.19., 7.3.spi.20., 7.3.tpi.10., 7.5.spi.2., 7.5.spi.4., 7.5.tpi.2., 7.5.tpi.3., 7.6.spi.1., 7.6.spi.2., 7.6.spi.3., 7.6.tpi.7., 7.6.tpi.10.
Mountain Kingdoms and Island Nations pp. 736–740	**Process Standards** P1, P2, P3, P6, P7, P8, P14, P15, P19, P20, P21, P33, P36, P37 **Learning Expectations** 1.01, 1.03, 2.02, 3.07, 3.08, 5.02 **Performance Indicators** 7.1.spi.1., 7.1.spi.2., 7.1.tpi.3., 7.2.tpi.5., 7.2.tpi.10., 7.3.spi.13., 7.3.spi.18., 7.3.tpi.10., 7.3.tpi.15., 7.5.spi.4., 7.5.tpi.2., 7.5.tpi.3., 7.6.tpi.10.
Unit 8 East Asia and Southeast Asia (pp. 744–871)	
Unit 8 Regional Atlas: East Asia and Southeast Asia pp. 746–753	**Process Standards** P1, P6, P7, P8, P14, P19, P33, P36 **Learning Expectations** 3.06, 3.07, 3.09, 5.01 **Performance Indicators** 7.1.spi.1., 7.1.spi.2., 7.1.tpi.3., 7.2.spi.4., 7.3.spi.11., 7.3.spi.19., 7.3.spi.20., 7.5.spi.2., 7.5.tpi.3.

World Cultures and Geography	Tennessee Social Studies Curriculum Standards: Process Standards and World Geography Seventh Grade (Standards, Learning Expectations, and Performance Indicators)
Chapter 25 (pp. 754–793) China	
Geography: Land of Three Rivers pp. 756–763	**Process Standards** P1, P2, P3, P6, P7, P8, P14, P15, P19, P20, P21, P33, P33, P36 **Learning Expectations** 1.03, 3.02, 3.03, 3.07, 5.01 **Performance Indicators** 7.1.tpi.1., 7.2.tpi.3., 7.2.tpi.10., 7.3.spi.9., 7.3.tpi.10., 7.5.spi.4., 7.5.tpi.2., 7.5.tpi.3., 7.5.tpi.4.
History: Middle Kingdom to Communist Power pp. 764–773	**Process Standards** P1, P2, P3, P6, P7, P8, P15, P19, P20, P21, P23, P24, P32, P33, P36, P37 **Learning Expectations** 1.01, 3.09, 4.01, 5.01, 5.02, 6.01, 6.02 **Performance Indicators** 7.3.spi.15., 7.3.spi.19., 7.3.spi.20., 7.3.tpi.15., 7.4.spi.1., 7.4.tpi.3., 7.5.spi.2., 7.5.spi.4., 7.5.tpi.4., 7.6.spi.1., 7.6.spi.2., 7.6.spi.3., 7.6.tpi.7.
Culture: Changing Chinese Culture pp. 774–783	**Process Standards** P1, P2, P3, P6, P7, P8, P10, P14, P15, P16, P19, P20, P21, P22, P33, P36, P37 **Learning Expectations** 1.01, 1.02, 3.09, 5.01, 5.02 **Performance Indicators** 7.1.spi.1., 7.1.spi.2., 7.1.spi.3., 7.1.tpi.3., 7.2.tpi.3., 7.3.spi.19., 7.3.spi.20., 7.3.tpi.12., 7.3.tpi.17., 7.5.spi.1., 7.5.spi.2., 7.5.spi.4., 7.6.tpi.9., 7.6.tpi.10.
Government & Economics: A Rising Power pp. 784–790	**Process Standards** P1, P2, P3, P6, P7, P8, P14, P15, P16, P17, P18, P19, P20, P21, P33, P36 **Learning Expectations** 1.01, 1.03, 2.02, 3.08, 3.09, 4.01, 5.02, 6.01, 6.02 **Performance Indicators** 7.1.spi.1., 7.1.tpi.3., 7.2.spi.1., 7.2.tpi.10., 7.3.spi.13., 7.3.spi.18., 7.3.spi.19., 7.3.spi.20., 7.3.tpi.10., 7.4.spi.1., 7.4.tpi.2., 7.4.tpi.3., 7.4.tpi.5., 7.4.tpi.6., 7.5.spi.2., 7.5.spi.4., 7.6.spi.1., 7.6.spi.2., 7.6.spi.3., 7.6.tpi.10.
Chapter 26 (pp. 794–821) The Korean Peninsula	
Geography: A Rugged Peninsula pp. 796–803	**Process Standards** P1, P2, P3, P6, P7, P8, P10, P12, P14, P15, P19, P20, P21, P22, P26, P33, P36, P37 **Learning Expectations** 1.01, 1.03, 2.03, 3.07, 3.08, 3.09, 5.01 **Performance Indicators** 7.2.spi.2., 7.2.tpi.3., 7.3.spi.9., 7.3.spi.13., 7.3.spi.18., 7.3.spi.19., 7.3.spi.20., 7.3.tpi.14., 7.5.spi.2., 7.5.tpi.2., 7.5.tpi.3.

World Cultures and Geography	Tennessee Social Studies Curriculum Standards: Process Standards and World Geography Seventh Grade (Standards, Learning Expectations, and Performance Indicators)
History & Culture: Strong Traditions, Modern Innovations pp. 804–811	**Process Standards** P1, P2, P3, P4, P6, P7, P8, P10, P14, P15, P19, P20, P21, P22, P23, P30, P32, P33, P36, P37 **Learning Expectations** 1.01, 1.03, 3.08, 5.01, 5.02, 6.01, 6.02 **Performance Indicators** 7.1.spi.1., 7.1.tpi.3., 7.1.tpi.5., 7.3.spi.13., 7.3.spi.15., 7.3.spi.16., 7.3.spi.18., 7.3.tpi.15., 7.5.spi.4., 7.6.spi.1., 7.6.spi.2., 7.6.spi.3., 7.6.tpi.10.
Government & Economics: Contrasting Countries pp. 812–818	**Process Standards** P1, P2, P3, P4, P5, P6, P7, P8, P10, P13, P14, P15, P16, P17, P19, P20, P21, P22, P23, P32, P33, P36, P37 **Learning Expectations** 2.01, 2.03, 4.01, 5.02, 6.01, 6.02 **Performance Indicators** 7.1.spi.1., 7.1.tpi.3., 7.2.spi.1., 7.2.spi.2., 7.2.tpi.3., 7.4.spi.1., 7.4.spi.4., 7.4.tpi.2., 7.4.tpi.3., 7.4.tpi.5., 7.5.spi.4., 7.6.spi.1., 7.6.spi.2., 7.6.spi.3., 7.6.tpi.7., 7.6.tpi.10.
Chapter 27 (pp.822–849) Japan	
Geography: Land of the Rising Sun pp. 824–829	**Process Standards** P1, P2, P3, P6, P7, P8, P15, P19, P20, P21, P33, P36 **Learning Expectations** 1.01, 1.03, 2.03, 3.06, 3.07, 5.01 **Performance Indicators** 7.2.spi.2., 7.3.spi.9., 7.3.spi.11., 7.3.tpi.10., 7.3.tpi.12., 7.5.spi.1., 7.5.tpi.2., 7.5.tpi.3.
History & Culture: Samurai to Animé pp. 830–839	**Process Standards** P1, P2, P3, P6, P7, P8, P10, P14, P15, P19, P20, P21, P22, P27, P33, P34, P36, P37 **Learning Expectations** 1.01, 1.02, 1.04, 3.09, 5.01, 5.02, 6.01, 6.02 **Performance Indicators** 7.1.spi.1., 7.1.tpi.3., 7.3.spi.19., 7.3.spi.20., 7.3.tpi.12., 7.3.tpi.15., 7.5.spi.1., 7.5.spi.2., 7.5.spi.4., 7.6.spi.1., 7.6.spi.2., 7.6.spi.3., 7.6.tpi.9., 7.6.tpi.10.
Government & Economics: Economic Giant pp. 840–846	**Process Standards** P1, P2, P3, P4, P6, P7, P8, P10, P11, P12, P14, P15, P19, P20, P21, P22, P23, P26, P32, P33, P36, P37 **Learning Expectations** 2.02, 2.03, 3.07, 4.01, 5.02, 6.01, 6.02 **Performance Indicators** 7.1.spi.1., 7.1.tpi.3., 7.2.spi.1., 7.2.spi.2., 7.2.tpi.3., 7.2.tpi.10., 7.4.spi.1., 7.4.spi.4., 7.4.tpi.2., 7.4.tpi.3., 7.4.tpi.5., 7.5.spi.4., 7.5.tpi.3., 7.6.spi.1., 7.6.spi.2., 7.6.spi.3.

World Cultures and Geography	Tennessee Social Studies Curriculum Standards: Process Standards and World Geography Seventh Grade (Standards, Learning Expectations, and Performance Indicators)
Chapter 28 (pp. 850–871) Southeast Asia	
Physical Geography of Southeast Asia pp. 852–857	**Process Standards** P1, P2, P3, P6, P7, P8, P10, P14, P15, P16, P17, P18, P19, P20, P21, P22, P33, P36, P37 **Learning Expectations** 2.03, 3.06, 3.07 **Performance Indicators** 7.2.spi.2., 7.3.spi.9., 7.3.spi.11., 7.3.tpi.9., 7.3.tpi.15., 7.4.tpi.6., 7.5.tpi.2., 7.5.tpi.3., 7.5.tpi.4.
The Mainland Countries pp. 858–863	**Process Standards** P1, P2, P3, P6, P7, P8, P10, P14, P15, P19, P20, P21, P22, P33, P36 **Learning Expectations** 1.01, 1.02, 2.01, 2.02, 3.02, 3.03, 3.07, 3.09, 4.01, 5.01, 5.02, 6.01, 6.02 **Performance Indicators** 7.2.tpi.10., 7.3.spi.19., 7.3.spi.20., 7.4.tpi.3., 7.5.spi.2., 7.5.spi.4., 7.5.tpi.3., 7.6.spi.1., 7.6.spi.2., 7.6.spi.3., 7.6.tpi.7., 7.6.tpi.10.
Indonesia and the Philippines pp. 864–868	**Process Standards** P1, P2, P3, P4, P6, P7, P8, P10, P14, P15, P16, P17, P18, P19, P20, P21, P22, P23, P24, P32, P33, P36, P37 **Learning Expectations** 1.01, 2.03, 3.07, 3.09, 5.01, 5.02, 6.01, 6.02 **Performance Indicators** 7.2.spi.2., 7.3.spi.19., 7.3.spi.20., 7.3.tpi.17., 7.5.spi.2., 7.5.spi.4., 7.5.tpi.3., 7.6.spi.1., 7.6.spi.2., 7.6.spi.3., 7.6.tpi.9., 7.6.tpi.10.
Unit 9 **Oceania and Antarctica** **(pp. 872–927)**	
Unit 9 Regional Atlas: Oceania and Antarctica pp. 874–881	**Process Standards** P1, P6, P7, P8, P14, P19, P33, P36 **Learning Expectations** 3.06, 3.07, 3.09, 5.01 **Performance Indicators** 7.1.spi.1., 7.1.spi.2., 7.1.tpi.3., 7.2.spi.4., 7.3.spi.11., 7.3.spi.19., 7.3.spi.20., 7.5.spi.2., 7.5.tpi.3.
Chapter 29 (pp. 882–909) Australia and New Zealand	
Geography: Unique Lands pp. 884–891	**Process Standards** P1, P2, P3, P6, P7, P8, P14, P15, P16, P19, P20, P21, P33, P36, P37 **Learning Expectations** 3.07, 3.08, 5.02, 6.01, 6.02 **Performance Indicators** 7.3.spi.9., 7.3.spi.13., 7.3.spi.15., 7.3.spi.18., 7.5.spi.4., 7.5.tpi.3., 7.6.spi.1., 7.6.spi.2., 7.6.spi.3.

World Cultures and Geography	Tennessee Social Studies Curriculum Standards: Process Standards and World Geography Seventh Grade (Standards, Learning Expectations, and Performance Indicators)
Australia pp. 892–899	**Process Standards** P1, P2, P3, P6, P7, P8, P10, P14, P15, P19, P20, P21, P22, P36, P37 **Learning Expectations** 1.01, 1.03, 2.02, 2.03, 3.07, 3.09, 4.01, 4.02, 5.01, 5.02, 6.01, 6.02 **Performance Indicators** 7.1.spi.1., 7.1.tpi.3., 7.2.spi.2., 7.2.tpi.3., 7.2.tpi.8., 7.2.tpi.10., 7.3.spi.19., 7.3.spi.20., 7.4.spi.1., 7.4.tpi.2., 7.4.tpi.3., 7.5.spi.2., 7.5.spi.4., 7.5.tpi.3., 7.6.spi.1., 7.6.spi.2., 7.6.spi.3., 7.6.tpi.7., 7.6.tpi.10.
New Zealand pp. 900–906	**Process Standards** P1, P2, P3, P4, P6, P7, P8, P15, P19, P20, P21, P23, P32, P36, P37 **Learning Expectations** 1.01, 1.03, 2.03, 3.02, 3.03, 3.07, 3.09, 4.01, 5.01, 5.02, 6.01, 6.02 **Performance Indicators** 7.1.tpi.5., 7.2.spi.2., 7.3.spi.19., 7.3.spi.20., 7.3.tpi.15., 7.4.spi.1., 7.4.tpi.3., 7.4.tpi.5., 7.5.spi.2., 7.5.spi.4., 7.5.tpi.3., 7.6.spi.1., 7.6.spi.2., 7.6.spi.3., 7.6.tpi.10.
Chapter 30 (pp. 910–927) Oceania and Antarctica	
Oceania: The Pacific World pp. 912–919	**Process Standards** P1, P2, P3, P6, P7, P8, P14, P15, P19, P20, P21, P33, P36 **Learning Expectations** 1.01, 3.06, 3.09, 5.02, 6.01, 6.02 **Performance Indicators** 7.1.tpi.1., 7.2.tpi.3., 7.3.spi.9., 7.3.spi.11., 7.3.spi.19., 7.3.spi.20., 7.5.spi.2., 7.5.spi.4., 7.5.tpi.2., 7.5.tpi.4., 7.6.spi.1., 7.6.spi.2., 7.6.spi.3., 7.6.tpi.10.
Antarctica: A Frozen Desert pp. 920–924	**Process Standards** P1, P2, P3, P6, P7, P8, P10, P14, P15, P16, P17, P19, P20, P21, P22, P23, P32, P33, P36, P37 **Learning Expectations** 5.01 **Performance Indicators** 7.3.spi.6., 7.3.spi.9.

TENNESSEE EDITION

McDougal Littell

World Cultures and GEOGRAPHY

Sarah Bednarz

Marci Smith Deal

Inéz Miyares

Donna Ogle

Charles White

Senior Consultants

Sarah Witham Bednarz is associate professor of geography at Texas A&M University, where she has taught since 1988. She earned a Ph.D. in Educational Curriculum and Instruction specializing in geography in 1992 from Texas A&M University and has written extensively about geography literacy and education. Dr. Bednarz was an author of *Geography for Life: National Geography Standards, 1994*. In 2005 she received the prestigious George J. Miller Award from the National Council for Geographic Education.

Marci Smith Deal is the K–12 Social Studies Curriculum Coordinator for Hurst-Euless-Bedford Independent School District in Texas. She received the 2000 Distinguished Geographer Award for the State of Texas, and was one of the honorees of the 2001 National Council for Geographic Education Distinguished Teacher Award. She has served as president for the Texas Council for Social Studies Supervisors and as vice-president for the Texas Council for Social Studies. She currently serves as a teacher consultant for National Geographic Society.

Inés M. Miyares is professor of geography at Hunter College-City University of New York. Born in Havana, Cuba, and fluent in Spanish, Dr. Miyares has focused much of her scholarship on Latin America, immigration and refugee policy, and urban ethnic geography. She holds a Ph.D. in geography from Arizona State University. In 1999 Dr. Miyares was the recipient of the Hunter College Performance Excellence Award for excellence in teaching, research, scholarly writing, and service. She has just published a co-edited book on contemporary ethnic geographies in the United States.

Acknowledgments for copyrighted material are on pages R109–R116 and constitute an extension of this page.

ISBN-13: 978-0-618-88960-0 ISBN-10: 0-618-88960-4

Printed in the United States of America

01 02 03 04 05 06 07 08 09—CKI—12 11 10 09 08 07

Internet Web Site: http://www.mcdougallittell.com

TN2

Senior Consultants

Donna Ogle is professor of Reading and Language at National-Louis University in Chicago, Illinois, and is a specialist in reading in the content areas, with an interest in social studies. She is past president of the International Reading Association and a former social studies teacher. Dr. Ogle is currently directing two content literacy projects in Chicago schools and is Senior Consultant to the Striving Readers Research Project. She continues to explore applications of the K-W-L strategy she developed and is adding a Partner Reading component—PRC2 (Partner Reading and Content 2).

Charles S. White is associate professor in the School of Education at Boston University, where he teaches methods of instruction in social studies. Dr. White has written and spoken extensively on the importance of civic education, both in the United States and overseas. He has worked in Russia since 1997 on curriculum reform and teacher preparation in education for democracy. Dr. White has received numerous awards for his scholarship, including the 1995 Federal Design Achievement Award from the National Endowment for the Arts, for the Teaching with Historic Places project. In 1997, Dr. White taught his Models of Teaching doctoral course at the Universidad San Francisco de Quito, Ecuador.

Middle School Teacher Consultant

Judith K. Bock earned a B.A. in Elementary Education and an M.A. in Geography and Environmental Sciences from Northeastern Illinois University. She taught as a middle school educator in gifted education for 33 years and is currently an adjunct geography professor in college geography departments in metropolitan Chicago. Ms. Bock serves on several local, state, and national geography education committees, writes geography lesson plans and curriculum, and facilitates geography workshops and institutes. She received the Distinguished Teaching Award from NCGE and was awarded the Outstanding Geographer in Illinois–2004 from the Illinois Geographical Society.

TN3

Tennessee Reviewers

Sandra Miller
Portland Middle School
Portland, TN

Theresa Peacock
Blackman Middle School
Murfreesboro, TN

David Rector
Fred J. Page Middle School
Franklin, TN

Regional Reviewers

Edwin Bryant
Professor of Hinduism
Rutgers University
New Brunswick, New Jersey

Gary S. Elbow
Associate Dean of the Honors College and Professor of Geography
Texas Tech University
Lubbock, Texas

Ibipo Johnston-Anumonwo
Professor of Geography
State University of New York College at Cortland
Cortland, New York

James A. Millward
Associate Professor of History
Georgetown University
Washington, D.C.

Mark Peterson
Associate Professor of Korean
Brigham Young University
Provo, Utah

TN4

Teacher Reviewers

Erik Branch
James C. Wright
Middle School
Madison, Wisconsin

Ann Christianson
John Muir Middle School
Wausau, Wisconsin

Jim Easton
Roscoe Middle School
Roscoe, Illinois

Ed Felton
Coopersville
Middle School
Coopersville, Michigan

Greta Frensley
Knox Doss Middle School
Gallatin, Tennessee

Todd Harrison
Hardin County
Middle School
Savannah, Tennessee

Esther Howse
Byron Center West Middle School
Byron Center, Michigan

Amber McVey
Antioch Middle School
Gladstone, Missouri

Ken Metz
Glacier Creek
Middle School
Cross Plains, Wisconsin

Suzanne Moen
DeForest Middle School
DeForest, Wisconsin

Tim Mortenson
Patrick Marsh Middle School
Sun Prairie, Wisconsin

Matt Parker
Eastgate Middle School
Kansas City, Missouri

Terry Rhodes
McNair Middle School
Fayetteville, Arkansas

Matthew J. Scheidler
Wayzata East Middle School
Plymouth, Minnesota

Don Stringfellow
Mobile City Public
School System
Mobile, Alabama

TN5

STUDENT PANEL

The following Middle School Student Panel reviewed textbook materials and technology products for this program.

Jessica Baker
Nelly Benitez
Ameer Cannon
Katie Conley
Murad Dajani
Will DiFrancesca
Tom Foydel
Philippa Gillette
Jenny Gorelick
Michael Grassle
Danielle Jackson
DeJauna Jackson
Mark Johnson
David Lenz
Andrew Mack
Madelaine Martin
Jabari McIntyre
Victoria Meliska
Sarah Peters
Brianna Ransom
Andrés Rivera-Thompson
Simone Samuels
Ben Shoaf
Gabriel Siegal
Kyle Siegal
Brock Snider
Hank Strickler
Jasmine Wright
Hannah Wyler

TN6

Note from the Editors:

We were determined from the start to make this book as lively and engaging as possible, and called upon the students you see on these pages to help us. They contributed many helpful, positive suggestions for both the textbook and the technology products. Working with these students was one of the best parts of the project. We appreciate their insights, enthusiasm, and creativity.

Thanks, everyone!

TN7

TENNESSEE

OVERVIEW
Tennessee Student Edition

Smoky Mountains in Tennessee © *Terrance Klassen/Alamy*

TN9

UNIT 1 TENNESSEE

Introduction to World Geography

Online Activities @ ClassZone.com

Animated GEOGRAPHY
Amazon Rain Forest

Explore life in a rain forest—the Yanomamo people, plants, animals, and the scientists who study them.

Interactive Review

Solve a crossword of key terms and names related to Earth's physical geography.

Online Test Practice

Review test-taking strategies and practice for your test at the end of every chapter.

TN10

The United States and Canada

Online Activities
@ ClassZone.com

Worldster

Kirima's Web Page

Visit Kirima's Web page to experience the sights and sounds of her world.

Interactive Review

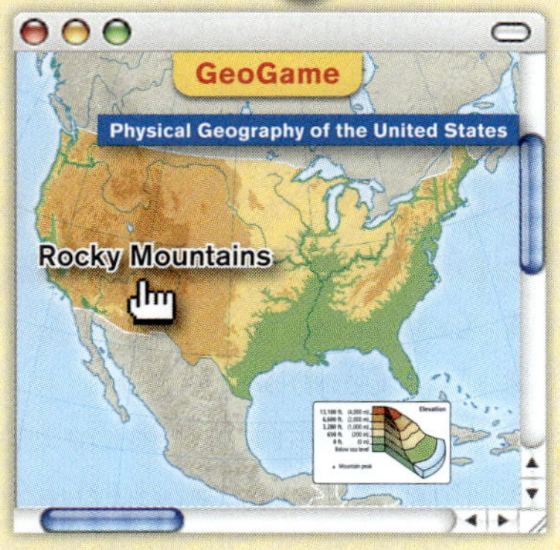

Play the GeoGame to test your knowledge of the geography of the United States.

Online Test Practice

Review test-taking strategies and practice for your test at the end of every chapter.

TN11

UNIT 3 TENNESSEE

Latin America

Online Activities
@ ClassZone.com

Worldster
Pedro's Web Page

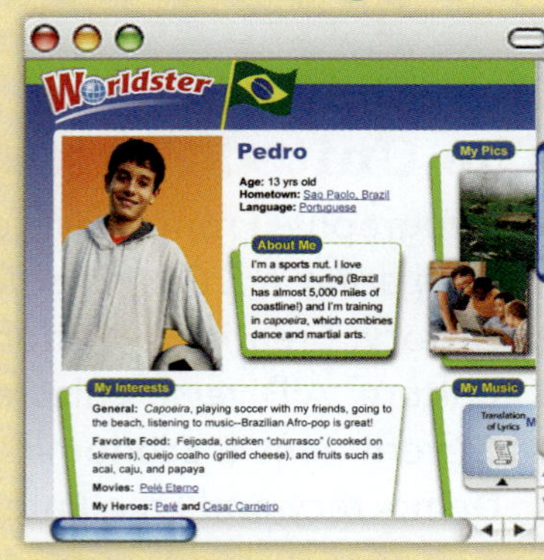

Visit Pedro's Web page to experience the sights and sounds of his world.

Interactive Review

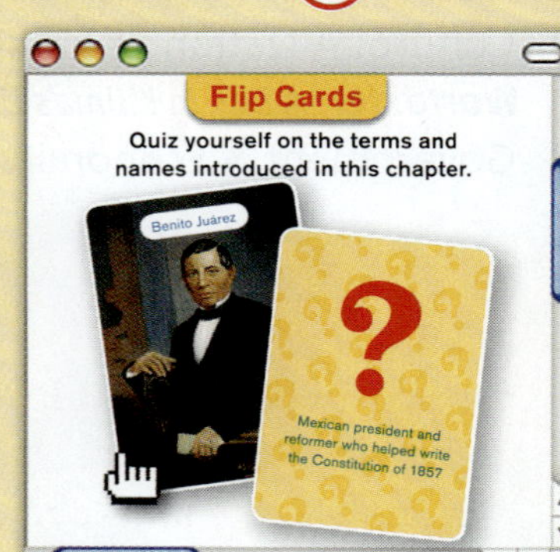

Use interactive Flip Cards to review key terms and names related to Mexico.

Online Test Practice

Review test-taking strategies and practice for your test at the end of every chapter.

TN12

UNIT 4 TENNESSEE

Europe

Online Activities
@ ClassZone.com

Animated GEOGRAPHY
Victorian London

Enter the world of Victorian London and visit homes, shops, parks, and an underground railway.

Interactive Review

Play the GeoGame to test your knowledge of the geography of Western Europe.

Online Test Practice

Review test-taking strategies and practice for your test at the end of every chapter.

TN13

Russia and the Eurasian Republics

Online Activities
@ ClassZone.com

Worldster
Zuhura's Web Page

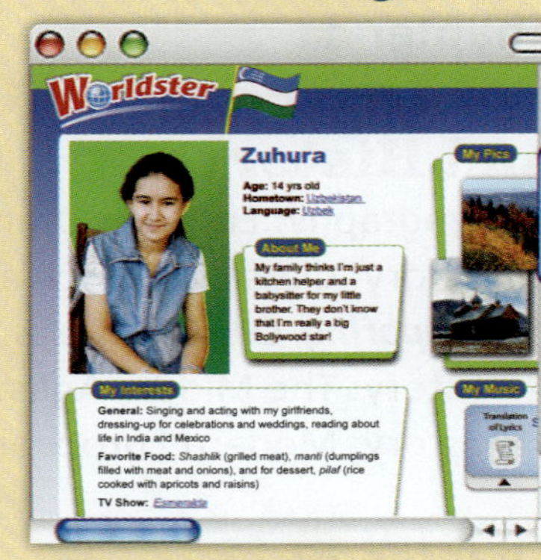

Visit Zuhura's Web page to experience the sights and sounds of her world.

Interactive Review

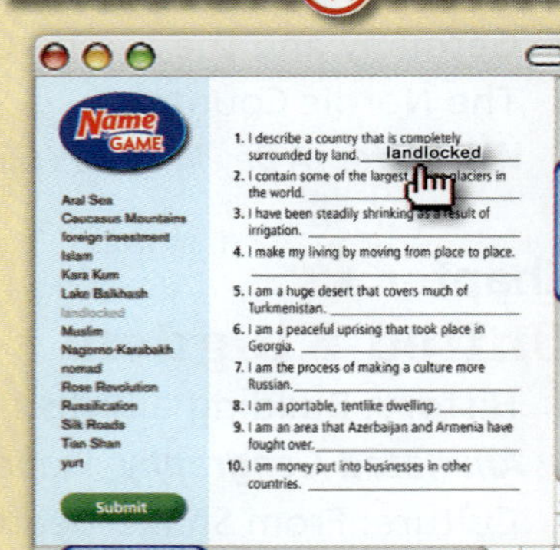

Play the Name Game to test your knowledge of the Eurasian Republics.

Online Test Practice

Review test-taking strategies and practice for your test at the end of every chapter.

TN14

UNIT 6 TENNESSEE

Africa

Online Activities
@ ClassZone.com

Animated GEOGRAPHY
The Great Pyramid

Enter the world of ancient Egypt to see how the pyramids were built.

Interactive Review

Play the GeoGame to test your knowledge of Egypt and North Africa.

Review test-taking strategies and practice for your test at the end of every chapter.

TN15

UNIT 7

TENNESSEE

Southwest Asia and South Asia

Online Activities
@ ClassZone.com

Animated GEOGRAPHY
Monsoons

Discover the life-giving and life-destroying effects of a monsoon. Hear news accounts of flooded villages.

Interactive Review

Solve a crossword of key terms and names related to Southwest Asia.

Online Test Practice

Review test-taking strategies and practice for your test at the end of every chapter.

TN16

East Asia and Southeast Asia

Online Activities
@ ClassZone.com

Worldster

Cheng's Web Page

Visit Cheng's Web page to experience the sights and sounds of his world.

Interactive Review

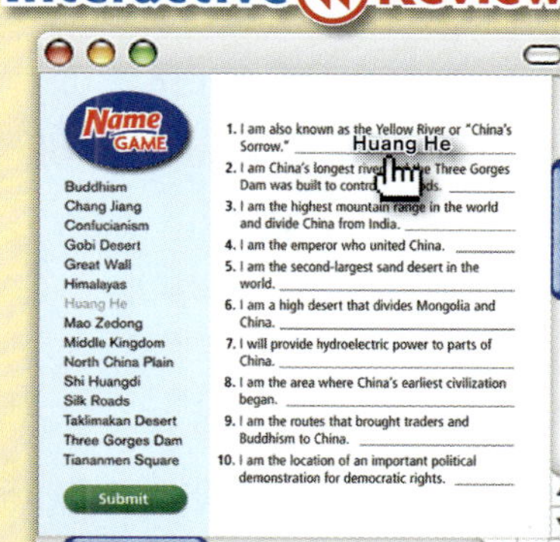

Play the Name Game to test your knowledge of China.

Online Test Practice

Review test-taking strategies and practice for your test at the end of every chapter.

TN17

UNIT 9 TENNESSEE

Oceania and Antarctica

Online Activities
@ ClassZone.com

Animated GEOGRAPHY
The Great Barrier Reef

Dive into Australia's waters and explore life on the reef.

Interactive Review

Use interactive Flip Cards to review key terms and names related to Oceania and Antarctica.

Online Test Practice

Review test-taking strategies and practice for your test at the end of every chapter.

TN18

Animated GEOGRAPHY

CONNECT

GEOGRAPHY

HISTORY

CULTURE

ECONOMICS

Activities

TN19

FEATURES

ANALYZING Primary Sources

HISTORY MAKERS

Fun Facts!

TN20

FEATURES

Worldster

COMPARING

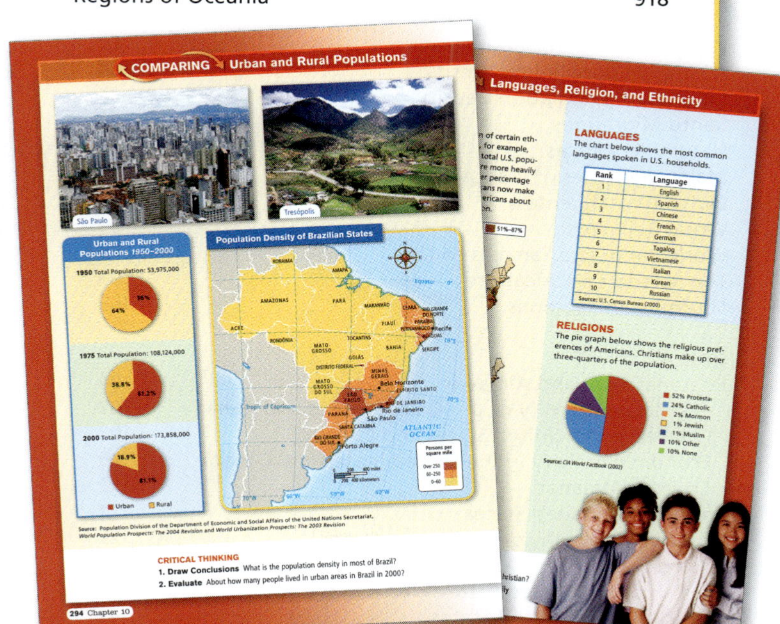

TN21

MAPS

MAPS

TN22

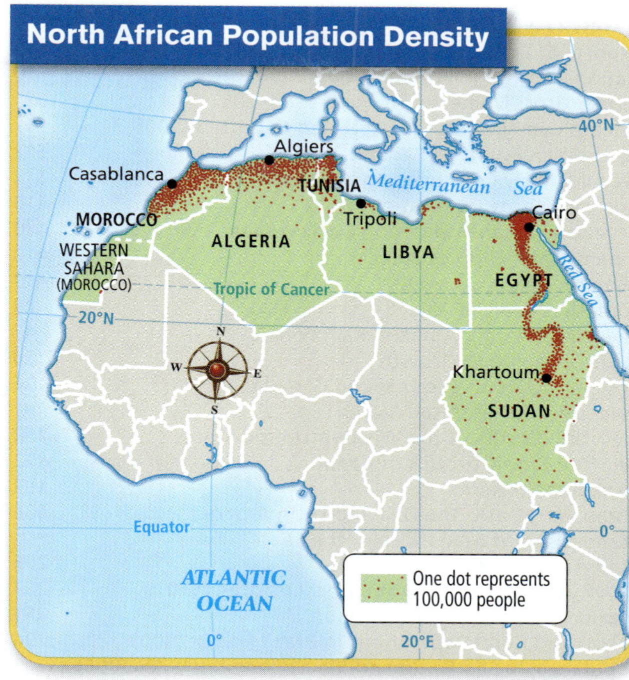
North African Population Density
Algiers
Casablanca
TUNISIA
Mediterranean Sea
Tripoli
Cairo
MOROCCO
WESTERN SAHARA (MOROCCO)
ALGERIA
LIBYA
EGYPT
Red Sea
Tropic of Cancer
20°N
40°N
N
S
E
W
Khartoum
SUDAN
Equator
0°
ATLANTIC OCEAN
One dot represents 100,000 people
0°
20°E

TN23

GRAPHS, TABLES, AND CHARTS

GRAPHS

Economic Activities of Kenya *2004*

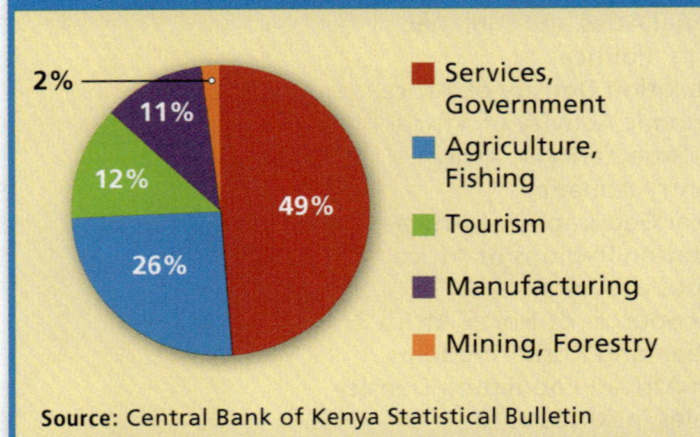

Source: Central Bank of Kenya Statistical Bulletin

"Say cheetah" A photographer gets a close-up view of a cheetah at Masai Mara National Reserve.

TABLES AND CHARTS

TN24

INFOGRAPHICS

Four Steps to Being a Strategic Reader

These pages explain how *World Cultures and Geography* chapters are organized. By using the four key strategies below, you'll become a more successful reader of geography, and more knowledgeable about your state's standards.

1. Set a Purpose for Reading
2. Build Your Social Studies Vocabulary
3. Use Active Reading Strategies
4. Check Your Understanding

Activity

Can you find it?

Find the following items on this and the next three pages.

- **one** chapter Essential Question
- **two** places where important words are defined
- **three** online games
- **four** key strategies for reading

1 Set a Purpose for Reading

Key features at the beginning of each chapter and section help you set a purpose for reading.

A Essential Question This key question sets the main purpose for reading.

B Connect Geography & History This feature helps you to consider the relationship between geography and history.

C Animated Geography identifies the region or country you'll be studying.

D Before, You Learned and **Now You Will Learn** This information helps you to connect what you've studied before to what you'll study next.

E Key Question Each topic covered in the chapter is followed by a Key Question that sets your purpose for reading about that topic.

2 Build Your Social Studies Vocabulary

The Reading for Understanding pages provide three important ways to build your vocabulary.

A Terms & Names cover the most important events, people, places, and social studies concepts in the section.

B Background Vocabulary lists words you need to know in order to understand the basic concepts and ideas discussed in the section.

C Visual Vocabulary features provide visual support for some definitions.

D Terms & Names and **Background Vocabulary** are highlighted and defined in the main text so that you'll understand them as you read.

TN26

Rice fields, Bali, Indonesia

CHAPTER 2

Earth's Interlocking Systems

1 GEOGRAPHY The Earth and Its Forces

2 GEOGRAPHY Bodies of Water and Landforms

3 GEOGRAPHY Climate and Vegetation

4 GEOGRAPHY

A

ESSENTIAL QUESTION

How do Earth's physical systems make life on Earth possible?

C

Animated GEOGRAPHY

Earth's Tectonic Plates

Click here to explore Earth and its systems @ClassZone.com

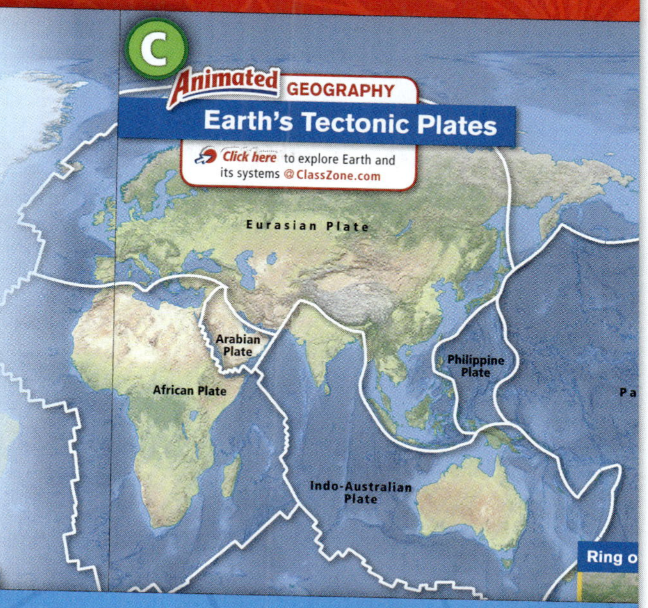

B

CONNECT Geography & History

Use the map and the time line to answer the following questions.

1. On which plate does most of the United States sit

2. Which event on the time line is supported by the

E

External Forces Shaping the Earth

KEY QUESTION What external forces shape the Earth?

D

External forces also reshape the Earth's surface. The two main exter-nal forces are weathering and erosion. **Weathering** is the gradual physi-cal and chemical breakdown of rocks near or on the Earth's surface. **Erosion** is the wearing away and movement of weathered materials from one place to another by the action of water, wind, or ice. As you can see, weathering and erosion work together to shape the Earth.

Weathering Weathering occurs slowly, over many years or even centuries. The two types of weathering are mechanical weathering and chemical weathering. Mechanical weathering is a process in which rocks are broken down into smaller pieces by physical means. It takes place when ice, extremes of hot and cold, or even tree roots cause rocks to split apart. It also occurs when hard objects, such as other rocks or sand, scrape or rub against a rock, and pieces of th rock break off.

Chemical weathering is caused by chemical reactions betwee the minerals in the rock and elements in the air or water. This pro ess changes the make-up of the rock itself. For example, most roc contain iron. When iron comes in contact with water, it rusts, whi helps to break down the rock. Water and elements in the air can cau other minerals in rocks to dissolve.

The Grand Canyon The Grand Canyon is located on the Colorado River in Arizona. It is an example of both weathering and erosion caused by wind and water.

SECTION 2

Reading for Understanding

Key Ideas

D

BEFORE, YOU LEARNED

Internal and external forces shape the surface of the Earth.

NOW YOU WILL LEARN

Interaction between landform of water makes life on Earth p

Vocabulary

A

TERMS & NAMES

drainage basin the area drained by a major river

ground water water found beneath the Earth's surface

hydrologic cycle the circulation of water between the Earth, the oceans, and the atmosphere

landform a feature on the Earth's surface formed by physical force

plateau a broad, flat area of land higher than the surrounding land

relief the difference in the elevation of a landform from its lowest point to its highest point

continental shelf the submerg at the edge of a continent

B

BACKGROUND VOCABULA

atmosphere the layer of gases surround the Earth

C

Visual Vocabulary landform

Reading Strategy

Re-create the web diagram shown at right. As you read and respond to the **KEY QUESTIONS**, use the diagram to organize important details about the Earth's landforms and bodies of water.

See Skillbuilder Handbook, page R4

FIND MAIN IDEAS

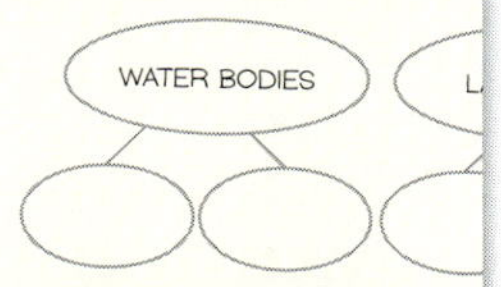

GRAPHIC ORGANIZERS Go to Interactive Review @ClassZon

TN27

READING FOR UNDERSTANDING

3 Use Active Reading Strategies

Active reading strategies help you note the most important information in each section.

A **Reading Strategy** Each Reading for Understanding page contains a Reading Strategy diagram to help you track and organize the information you read.

B **Skillbuilder Handbook** Every Reading Strategy is supported by a corresponding lesson in the Skillbuilder Handbook section at the back of this book.

C **Active Reading Strategies** in the Skillbuilder Handbook will help you to read and study *World Cultures and Geography*.

drainage basin the area drained by a major river

ground water water found beneath the Earth's surface

hydrologic cycle the circulation of water between the Earth, the oceans, and the atmosphere

landform a feature on the Earth's surface formed by physical force

plateau a broad, flat area of land higher than the surrounding land

relief the difference in the elevation of a landform from its lowest point to its highest point

continental shelf the sub… at the edge of a continent

BACKGROUND VOCAB…

atmosphere the layer of ga… surround the Earth

Visual Vocabulary landform

Reading Strategy **A**

Re-create the web diagram shown at right. As you read and respond to the **KEY QUESTIONS**, use the diagram to organize important details about the Earth's landforms and bodies of water.

See Skillbuilder Handbook, page R4

FIND MAIN IDEAS

WATER BODIES

GRAPHIC ORGANIZERS
Go to Interactive Review @ Class…

Skillbuilder Handbook

C **Contents**

Reading and Critical Thinking Skills

B **1.2 Finding Main Ideas**

Defining the Skill

A **main idea** is a statement that summarizes the subject of a speech, an article, a section of a book, or a paragraph. Main ideas can be stated or unstated. The main idea of a paragraph is often stated in the first or last sentence. If it is the first sentence, it is followed by sentences that support that main idea. If it is the last sentence, the details build up to the main idea. To find an unstated idea, use the details of the paragraph as clues.

Applying the Skill

The following paragraph provides reasons why Japan and Australia make such good trading partners. Use the strategies listed below to help you identify the main idea.

How to Find the Main Idea

Strategy 1 Identify what you think may be the stated main idea. Check the first and last sentences of the paragraph to see if either could be the stated main idea.

Strategy 2 Identify details that support the main idea. Some details explain that idea. Others give examples of what is stated in the main idea.

TRADING PARTNERS

Australia is an island. Japan is also an island, though much smaller than Australia. Australia is not as densely populated as Japan. 2 In Japan, an average of 881 people live in each square mile. Australia averages 7 people per square mile. 2 Australia has wide-open lands available for agriculture, ranching, and mining. 2 Japan buys wool from Australian ranches, wheat from Australian farms, and iron ore from Australian mines. To provide jobs for its many workers, 2 Japan developed industries. Those industries 2 sell electronics and cars to Australia. 1 Australia and Japan are trading partners because each has something the other needs.

Make a Chart

Making a chart can help you identify the main idea and details in a passage or paragraph. The chart below identifies the main idea and details in the paragraph you just read.

Main Idea: Australia and Japan are good trading partners because each supplies something the other needs.
Detail: Japan's population density is 881 people per square mile; Australia's is 7 per square mile.
Detail: Australia has land, natural resources, and agricultural products.
Detail: Japan buys wool, wheat, and iron ore from Australia.
Detail: Japan has many industries.
Detail: Australia buys electronics and cars from Japanese industries.

TN28

4 Check Your Understanding

One of the most important things you'll do as you study *World Cultures and Geography* is to check your understanding of events, people, places, and issues as you read.

A **Interactive Review** includes a Name Game and provides online activities to test your knowledge of the geography you just studied.

B **Section Assessment** reviews the section Terms & Names, revisits your Reading Strategy notes, and provides key questions about the section.

Interactive Review

Click here for activities

A CHAPTER SUMMARY

Key Idea 1
The Earth is composed of many layers, and its surface continually changes because of the drifting of the continents.

Key Idea 2
Interaction between landforms and bodies of water makes life on Earth possible.

Key Idea 3
The Earth's rotation and revolution influence weather, climate, and living conditions on Earth.

Key Idea 4
Human interference with physical systems causes problems with the environment.

For Review and Study Notes, go to Interactive Review @ ClassZone.com

NAME GAME

Use the Terms & Names list to complete each sentence on paper or online.

1. I am the hot metal center of the Earth. core
2. I am a naturally formed feature on the Earth's surface. ______
3. I fall in the form of rain, snow, sleet, or hail. ______
4. I am an increase in the Earth's temperature. ______
5. I move weathered materials from one place to another. ______
6. I am the trapping of the sun's heat by gases in the atmosphere. ______
7. I am a large rigid piece of the Earth's crust that is in motion. ______
8. I am typical weather conditions over a period of time. ______
9. I circulate water between the Earth, oceans, and atmosphere. ______
10. I am plants that grow in a region. ______

Activities

Flip Cards

Use the online flip cards to quiz yourself on terms and names introduced in this chapter.

Crossword Pu

Complete an online crosswor your knowledge of Earth's pl

ACROSS

1. a naturally formed featu Earth's surface

Erosion New landforms and new soil are formed by erosion. It occurs when materials loosened by weathering are moved by water, wind, or ice to new locations. Currents in streams and rivers pick up loose materials and deposit them downstream or carry them out to sea. These tiny pieces of rock, deposited by water, wind, or ice are called **sediment**. Sediment can be sand, stone, or finely ground particles called silt.

Wave action along coastlines carries rocks and sand from one place to another. Waves also pound boulders into smaller rocks. Wind erosion lifts particles from the Earth's surface and blows them great distances. The wind's actions can reshape rock s Arizona's Grand Canyon is a result of both wind and water

Another type of erosion is caused by glaciers. **Glaciers** a slow-moving masses of ice. They grind rocks and boulders un the ice and leave behind the rock when the ice melts. Pa central United States have been shaped by glacial erosion.

SYNTHESIZE Explain how external forces shape the Eart surface. Weathering wears down existing rock formations; its the loose materials in a different place to build another l

B Section 1 Assessment

ONLINE QUIZ
For test practice, go to Interactive Review @ ClassZone.com

TERMS & NAMES

1. **Explain the importance of**
   - continent
   - tectonic plate
   - weathering
   - erosion

USE YOUR READING NOTES

2. **Categorize** Use your completed chart to answer the following question:

   Are external or internal forces responsible for volcanoes? Explain your answer.

INTERNAL FORCES	EXTERNAL FORCES
1.	1.
2.	2.
3.	3.
4.	4.

KEY IDEAS

3. What are the five layers that make up the Earth's interior and exterior?
4. How were the continents formed?
5. What are the two major external forces reshaping the Earth?

CRITICAL THINKING

6. **Draw Conclusions** How does the movement of wind, water, or ice reshape the Earth's surface?
7. **Analyze Causes and Effects** What is the relationship between plate movement, volcanoes, and earthquakes?
8. CONNECT to Today In which parts of the United States are external forces shaping the landscape?
9. ART **Create a Puzzle** Make a copy of a map of the world. Cut out the continents. Use the continents as puzzle pieces to form the continent of Pangaea. When you have finished putting the pieces together, draw an outline around the entire supercontinent.

33

TN29

Exploring Geography Online

World Cultures and Geography provides a variety of tools to help you explore geography online. See geography come to life in the Animation Center. Find help for your research projects in the Research and Writing Center. Review for tests with the Interactive Review, or create your own activities in the Activity Center. Go to ClassZone.com to make ***World Cultures and Geography*** interactive!

ClassZone.com

is your gateway to exploring geography. Explore the different ClassZone Centers to help you study and have fun with geography.

A Interactive Review provides you with flip cards, a crossword puzzle, section quizzes, drag-and-drop map activities, and more.

B Activity Center lets you explore world cultures through a **Worldster** community and participate in **WebQuests** and geography games.

C Activity Maker lets you create your own activities so that you can focus on what *you* need to review.

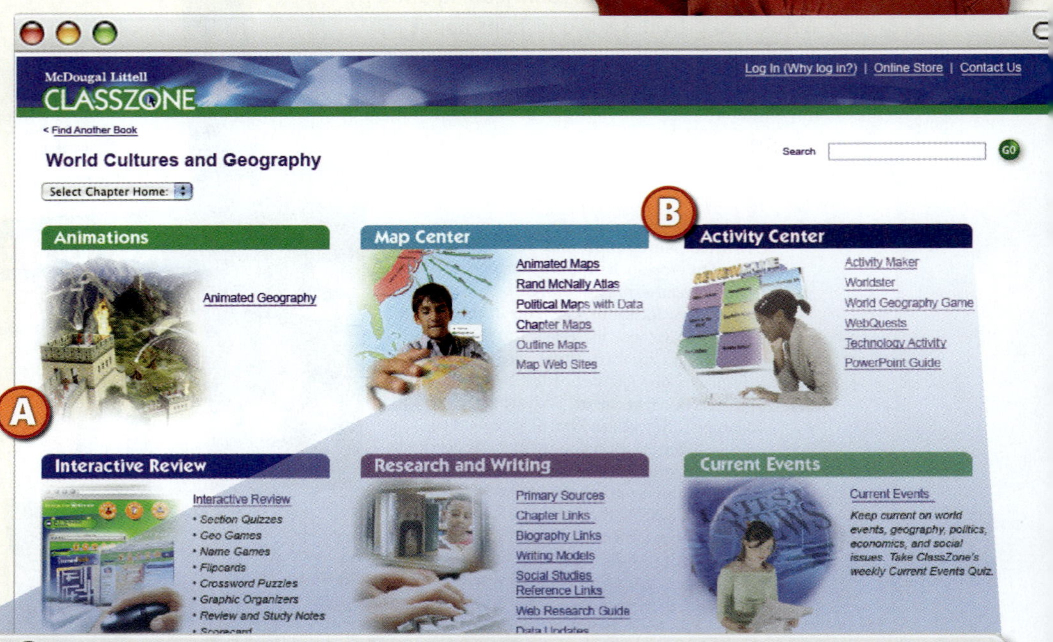

TN30

Animation Center

A rich collection of interactive features and maps on a wide variety of geographical topics at **ClassZone.com**

A Roll-overs
Explore the illustration by clicking on areas you'd like to know more about. This Amazon Rain Forest animation includes features about biodiversity, life in the rain forest, and more.

B Video and Photo Gallery
Explore the rain forest and the scientists who study it.

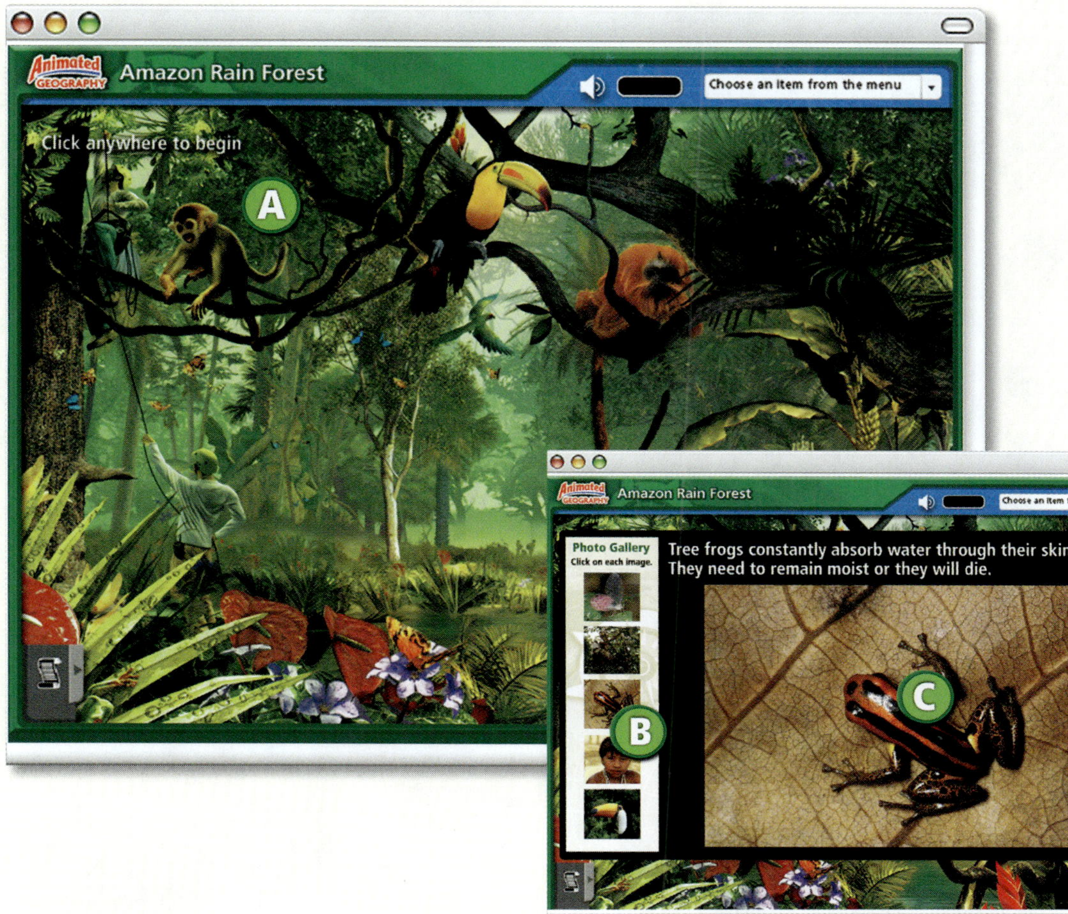

Review Game

- Create your own review game to study geography *your* way in the Activity Maker at **ClassZone.com**
- Select your own topics from any chapter to help you focus on specific regions, people, or events to review.
- Help your friends to explore geography online. Challenge them to play a review game that you create and modify!

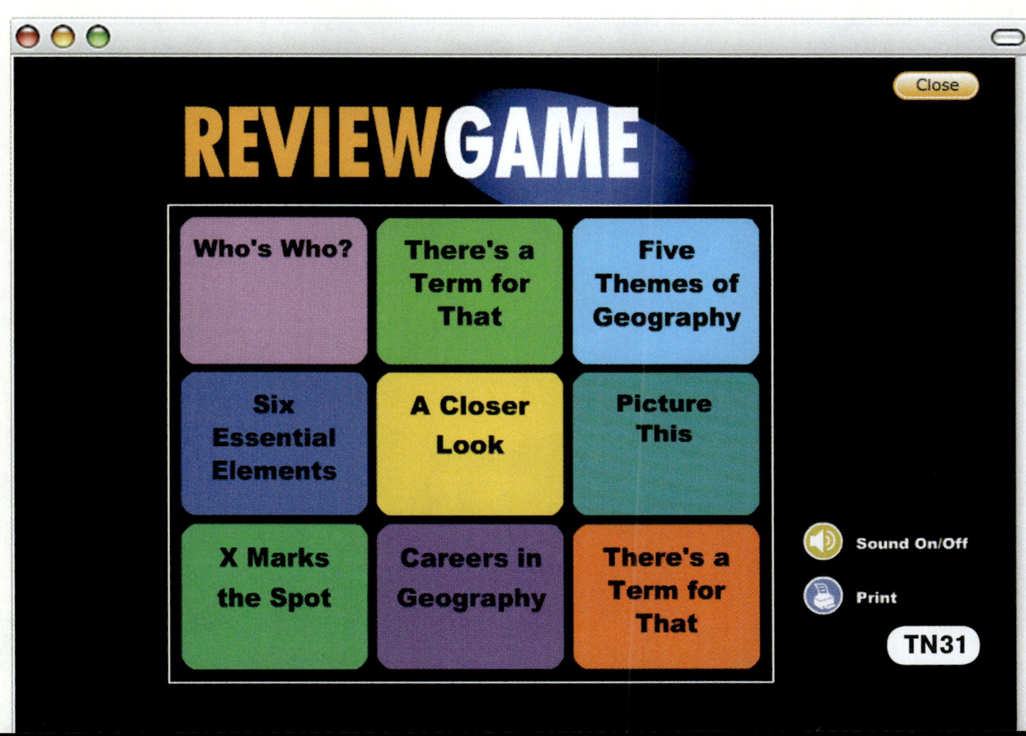

TENNESSEE

STEPS TO SUCCESS

- Tennessee's Social Studies Curriculum Standards for World Geography are organized into six content standards, and are divided into learning expectations and performance indicators.

- The TCAP is designed to measure your understanding and mastery of the content and objectives of the curriculum.

For a complete list of the Tennessee Social Studies Curriculum Standards and Blueprint for Learning, see page R117.

Smoky Mountains in Tennessee © Terrance Klassen/Alamy

TN33

Why Study Geography?

When you study geography, you will see how the past informs your everyday life. Features like "Connecting to Your World" will tie what you are learning in your textbook to your life.

This textbook will focus and motivate you to learn more about people who have lives very different from your own.

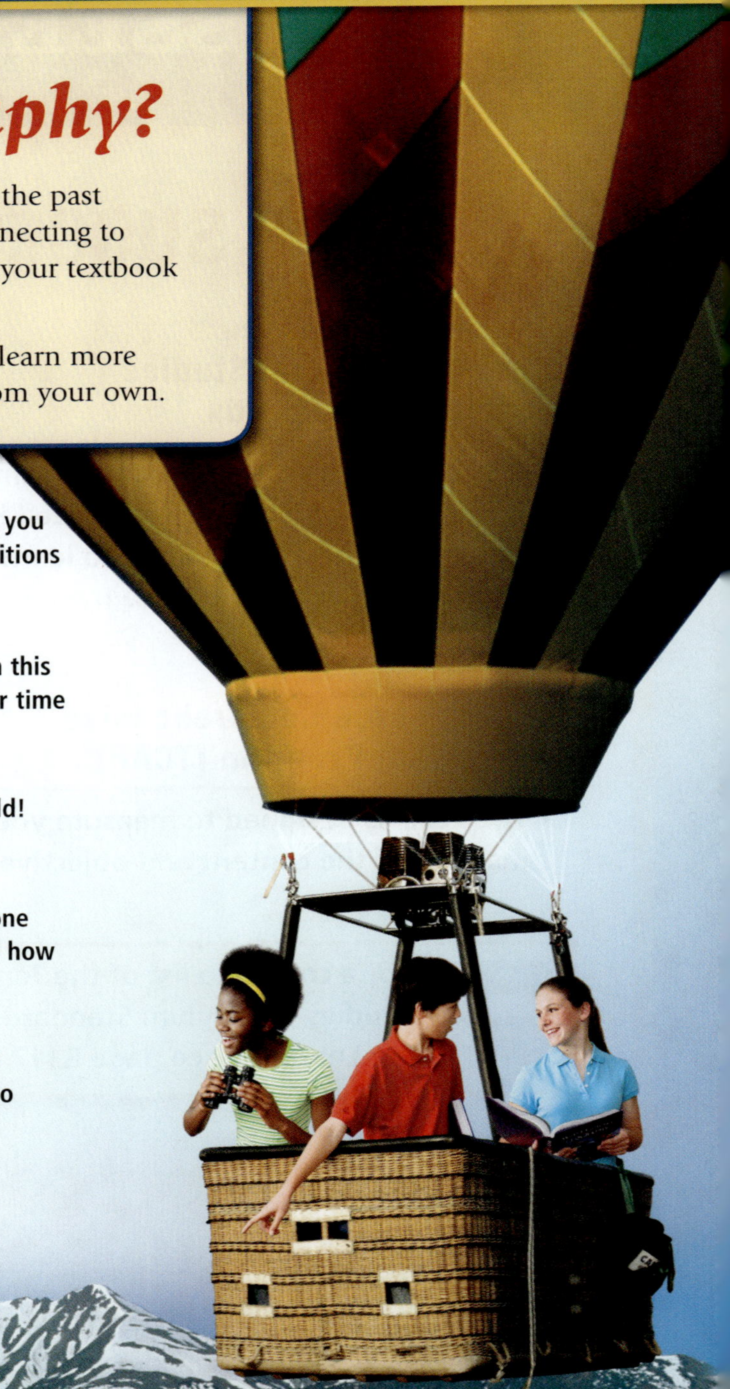

Explore Cultures **Your textbook will transport you to cultures around the world, and show you the traditions and beliefs that made those cultures unique.**

Geography allows you to view different societies to broaden your understanding of people and places. In this class, you will learn how societies have changed over time and the contributions each has made.

Improve Your Map Skills **Focus on the world! Your textbook will help you appreciate how much geography affects our lives every day.**

You will learn how to read all kinds of maps: time zone maps, elevation maps, and so on. You will even learn how to make a map on your own!

Become a Better Critical Thinker **Many questions throughout this book will encourage you to form your own opinion.**

For example, how would your life be different if you lived in China?

TN34

What Will I Learn?

Guide to Understanding Tennessee's Social Studies Curriculum Standards

Tennessee's Social Studies Curriculum Standards for World Geography are divided into six content standards:

- **Content Standard 1.0: Culture**
- **Content Standard 2.0: Economics**
- **Content Standard 3.0: Geography**
- **Content Standard 4.0: Governance and Civics**
- **Content Standard 5.0: History**
- **Content Standard 6.0: Individuals, Groups, and Interactions**

Each content standard is further divided into learning expectations. Each learning expectation is documented through performance indicators: these are the skills you need to succeed on the TCAP.

CONTENT STANDARD	LEARNING EXPECTATIONS	PERFORMANCE INDICATORS
1.0: Culture Culture encompasses similarities and differences among people including their beliefs, knowledge, changes, values, and traditions. Students will explore these elements of society to develop an appreciation and respect for the variety of human cultures.	**1.01** Understand the complex nature of culture. **1.02** Recognize the role of major religions.	*At Level I, the student is able to* **7.1.spi.1.** recognize cultural definitions.

CONNECT *To Your Tennessee Test*

TCAP Blueprint for Learning

At the end of the eighth grade, you will take a test called the **TCAP: Tennessee Comprehensive Assessment Program.**

This multiple-choice test is designed to measure your understanding and mastery of the content and objectives of the curriculum.

The objectives of the curriculum are explained in ***Tennessee's Blueprint for Learning***. The complete Blueprint for Learning for seventh grade is in the back of the book.

TENNESSEE CURRICULUM STANDARDS

TN35

Overview

Tennessee Social Studies Curriculum Standards

The charts that follow provide a brief overview of each content standard and its performance indicators. The performance indicators are the skills and knowledge you need to succeed on the TCAP. Look for them at the beginning of each section.

For a complete list of the Tennessee Social Studies Curriculum Standards and Blueprint for Learning, see page R117.

CONTENT STANDARD 1.0: CULTURE

Performance Indicators			What It Means to *You*
Level 1	**7.1.spi.1.**	recognize cultural definitions (i.e., language, religion, customs, political system, economic system).	Exploring different cultures will help you understand how people live in different parts of the world. By learning about other cultures, you will have a better sense of your own culture.
Level 2	**7.1.spi.2.**	locate cultural information on a thematic map (i.e., languages, political systems, economic systems, religions).	
Level 3	**7.1.spi.3.**	compare and contrast the tenets of the five major world religions (i.e., Christianity, Buddhism, Islam, Hinduism, Judaism).	

CONTENT STANDARD 2.0: ECONOMICS

Performance Indicators			What It Means to *You*
Level 1	**7.2.spi.1.**	recognize basic economic concepts (i.e., imports, exports, barter system, tariffs, closed and emerging markets, supply and demand, inflation, recession, depression).	You will learn how geography impacts economic systems around the world. By learning how to use visuals to interpret economic facts, you will be able to analyze the role economics plays in your life.
Level 2	**7.2.spi.4.**	interpret economic issues as expressed with maps, tables, diagrams, and charts.	
	7.2.spi.5.	select the major resources, industrial, and agricultural products for the three grand divisions from a map of Tennessee.	

CONTENT STANDARD 3.0: GEOGRAPHY

Performance Indicators			What It Means to *You*
Level 1	**7.3.spi.2.**	locate the Earth's major physical characteristics (i.e., 7 continents, 4 oceans).	Examining the different physical characteristics that define regions around the world will help you better understand the world. You will learn about the consequences of humans changing the physical environment around them.
Level 2	**7.3.spi.10.**	identify the characteristics that define a region geographically.	
Level 3	**7.3.spi.18.**	analyze the environmental consequences of humans changing their physical environment (i.e., air and water pollution, mining, deforestation, global warming).	

TN36

CONTENT STANDARD 4.0: GOVERNANCE AND CIVICS

Performance Indicators			What It Means to *You*
Level 1	7.4.spi.1.	define the different types of governments (i.e., democracy, autocracy, oligarchy, monarchy, dictatorship).	You will compare and contrast how different governments work and how government impacts your life. You will research current issues on the Internet to understand different perspectives on an issue.
Level 2	7.4.spi.3.	recognize how the boundaries of Congressional districts change in the state of Tennessee (i.e., statutory requirements, population shifts, political power shifts).	
	7.4.spi.4.	identify political leaders from selected contemporary settings (i.e., United States, India, Canada, Mexico, Great Britain, Russia, China).	

CONTENT STANDARD 5.0: HISTORY

Performance Indicators			What It Means to *You*
Level 1	7.5.spi.1.	identify the causes and consequences of urbanization (i.e., industrial development, education, health care, cultural opportunities, poverty, overcrowding, disease, pollution, crime).	You will examine how geography and geographic events have shaped the world you live in. You will analyze geographic and historic events to make predictions for future problems.
Level 2	7.5.spi.2.	identify reasons why people choose to settle in different places (i.e., occupation, family, climate, natural resources).	
Level 3	7.5.spi.4.	analyze the causes and effects of change in a place over time from a written passage.	

CONTENT STANDARD 6.0: INDIVIDUALS, GROUPS, AND INTERACTIONS

Performance Indicators			What It Means to *You*
Level 1	7.6.spi.1.	identify ways family, groups, and community influence daily life and personal choices.	You will analyze current geographic issues and how they might be resolved. You and your classmates will take the role of an international organization and try to solve a current global issue.
Level 2	7.6.spi.2.	differentiate between the rights, roles and state of the individual in relation to the general welfare in various regions of the world.	
Level 3	7.6.spi.3.	recognize the causes, consequences and possible solutions applied by governing bodies to persistent global issues using a narrative (i.e., health, security, resource allocation, economic development, environmental quality).	

TN37

N
E
W
S

TENNESSEE

Find online test practice at www.ClassZone.com

Guide to

The Tennessee Comprehensive Assessment Program (TCAP)

The chart below provides a guide to the test-taking strategies that will help prepare you for the **TCAP**.

- **Learn** each strategy by reviewing the numbered steps on the page listed in the column.
- **Practice** the strategy on the following page.
- **Apply** the strategies you learned at the end of each chapter.

STRATEGY	LEARN	PRACTICE	APPLY
Multiple Choice	p. S2	p. S3	end of every section and chapter
Primary and Secondary Sources	p. S4	p. S5	p. 153, Chapter 5 p. 603, Chapter 19
Political Cartoons	p. S6	p. S7	www.ClassZone.com
Charts	p. S8	p. S9	p. 55, Chapter 2 p. 821, Chapter 26
Line and Bar Graphs	p. S10	p. S11	p. 227, Chapter 7 p. 447, Chapter 14
Pie Graphs	p. S12	p. S13	p. 183, Chapter 6 p. 695, Chapter 22
Political Maps	p. S14	p. S15	p. 23, Chapter 1
Thematic Maps	p. S16	p. S17	p. 257, Chapter 8 p. 493, Chapter 15
Time Lines	p. S18	p. S19	beginning of every chapter
Reading Visuals	pp. S20–S21	pp. S22–S23	p. 519, Chapter 16
Document-Based Questions	pp. S24–S25	pp. S26–S27	p. 85, Chapter 3 p. 673, Chapter 21

S1

USING STRATEGIES FOR . . .

Multiple Choice

Multiple-choice questions test the ability to choose the correct answer from several choices. The test taker reads each stem and decides which is the correct answer. Test takers are often asked to mark their answers on an answer sheet.

Explain to students that they will do best on test questions by thinking them through carefully and by applying test-taking strategies, such as the following:

1. Pay close attention to modifiers, such as the word *mostly* in question 1. This word affects the meaning of the question and can provide some clues that can help you narrow down your options for correct answers. The correct answer is (A).
2. In question 2, words such as *every* (B) and *all* (D) make those particular answers highly unlikely and give you the confidence to eliminate them. The correct answer is (H).
3. In question 3, one of the possible answers is "all of the above." Make sure that the other choices are all correct before selecting this choice. In this case, (D) is the correct answer.
4. Be careful with questions that contain negatives. Identify all the choices in question 4 that are in southern Africa before making your final selection. The correct answer is (G).

General Test-Taking Tips

Share these tips with your students.

- The night before a test make sure you get at least eight hours of sleep.
- Have a healthy breakfast or lunch before taking your test.
- Wear clothes that make you feel comfortable.
- Relax and enjoy the challenge!

TCAP STRATEGIES

Multiple Choice

A multiple-choice question is a question or incomplete sentence and a set of choices. One of the choices correctly answers the question or completes the sentence.

1. Read the question or incomplete sentence carefully. Try to answer the question before looking at the choices.
2. Look for key words in the question. They may help you figure out the correct answer.
3. Read each choice with the question. Don't decide on your final answer until you have read all the choices.
4. Rule out any choices that you know are wrong.
5. Watch answers with words like *all, never,* and *only.* These answers are often incorrect.
6. Sometimes the last choice is *all of the above.* Make sure that the other choices are all correct before you pick this answer.
7. Be careful with questions that include the word *not.*

answers: 1 (A); 2 (H); 3 (D); 4 (G)

❶ **1 The Sahara is mostly**

❸ choices

A sand, rocks, and gravel
B boulders and sand
C cliffs and gulleys
D grasses and bushes

❷ Words like *mostly* or *partly* are key words in multiple choice. Look for answers that are mostly true or partly true about the subject.

❹ You know that if the Sahara is a desert, **D** is incorrect. A desert cannot be mostly covered with grass and bushes.

2 Over hundreds of years, Bantu-speaking people migrated from West Africa to

F South and Southwest Asia
G every continent on earth
H East and South Africa
J all of North Africa and Arabia

❺ Watch for answers that have words like *all, never, always, every,* and *only.* These answers are often incorrect.

3 The people of West Africa passed on their history by

A painting pictures
B telling stories
C creating dances
❻ D all of the above

4 Which of the following is *not* one of the nations in southern Africa?

F Zimbabwe
G Nigeria
H Mozambique
J Namibia

❼ First rule out all the answers that name southern African countries. The answer that remains is the correct choice.

S2

DIFFERENTIATING INSTRUCTION

English Learners

Make sure students understand the following terms and concepts in the sample questions on these pages.

Strategy Page

Question 1 *Sahara:* a desert

Question 2 *Bantu:* a group of related languages spoken in Africa

migrated: moved

continent: one of the seven large land masses in the world

Practice Page

Question 1 *bubonic plague:* a deadly disease that spread through Europe in the mid-1300s

prestige: high standing among others

Question 2 *reform:* effort to improve or correct flaws

Directions: Read each question carefully. Choose the best answer from the four choices.

1 **Which of the following was *not* a result of the bubonic plague?**

A Cities worked together during the plague.
B Europe lost one-third of its population.
C The Church lost its prestige among the people.
D The economies of many countries were ruined.

2 **Martin Luther started a reform movement when he**

F published the New Testament in German
G criticized some of the Church's practices
H wrote his 95 Theses and made them public
J all of the above

3 **The Ottoman Empire reached its greatest size and glory under the rule of**

A Mehmet II
B Selim the Grim
C Suleiman the Lawgiver
D Timur the Lame

4 **During the 1700s, England controlled which of the following?**

F the sugar trade
G the Atlantic slave trade
H the cotton trade
J the coconut trade

SKILL TESTED IN THE ITEMS

STRATEGY ITEMS		PRACTICE ITEMS	
Item Number	**Skill Tested**	**Item Number**	**Skill Tested**
1.	Categorizing	1.	Analyzing Causes and Effects
2.	Finding Main Ideas	2.	Analyzing Causes and Effects
3.	Summarizing	3.	Summarizing
4.	Categorizing	4.	Finding Main Ideas

PRACTICE QUESTIONS

Thinking It Through

Share the following explanations with students as they discuss the strategies they used to answer the practice questions.

1. This item is an example of a *not* question. Students should recognize that the bubonic plague affected Europe in every way except that described in (A), which is the correct answer.
2. This question is an *all of the above* question. Students should first look to see if any of the answers given are false. Since (F), (G), and (H) are all true, the correct answer is (J).
3. In this question, the key word is *greatest*. This word indicates that the correct answer will not be just a successful or powerful ruler, but the most successful of all. In this case, the correct answer is (C), Suleiman the Lawgiver.
4. *1700s* and *England* are key words in this question. In 1700, England controlled the Atlantic slave trade, so the correct answer is (G).

USING STRATEGIES FOR . . .

Primary and Secondary Sources

Students can learn about the past by studying primary and secondary sources. As students read these sources, have them try to answer the following questions: Who is it about? What is it about? When did it happen? Where did it happen? Why did it happen? Reading the source line will help students decide whether the source is primary or secondary and how reliable it is.

1. Review the primary source when you answer the questions. Question 1 asks you what Confucius is giving advice about in the passage. Notice that the title of this passage is *Good Government*. Then observe that both excerpts give advice on training people as subjects and workers. These clues support choosing (B), which is the correct answer.
2. The key word in Question 2 is *best*. Students can recognize that in this type of question, several answers may seem to be correct in some way. They must look for the choice that represents the *whole* passage. In this case, only (J) accurately summarizes all the information in the passage. It is the correct answer.

General Test-Taking Tips

Share these tips with your students.

- Read the directions carefully before you begin to answer the questions.
- Plan the time you are given to take the test.
- Check your answers.
- Believe in yourself.

TCAP STRATEGIES

Primary and Secondary Sources

Sometimes you will need to study a document to answer a question. Documents can be either primary sources or secondary sources. Primary sources are written or created by people who either saw an event or were actually part of the event. A primary source can be a photograph, letter, diary, speech, or autobiography.

A secondary source is an account of events by a person who did not actually experience them. Newspaper articles and history books are some examples of secondary sources.

1. Look at the source line to learn about the document and its author. If the author is well known and has been quoted often, the information is likely true.
2. Skim the article to get an idea of what it is about.
3. Note special punctuation. For example, ellipses (. . .) indicate that words have been left out.
4. Ask yourself questions about the document as you read.
5. Review the questions to see what information you will need to find. Then reread the document.

answers: 1 (B); 2 (J)

Good Government

Chap. 2.20 Lord Ji Kang asked, "What should I do in order to make the people respectful, loyal, and zealous [hard-working]?" The Master said: "Approach them with dignity and they will be respectful. Be yourself a good son and kind father, and they will be loyal. Raise the good and train the incompetent, and they will be zealous."

Chap. 13.2 Ran Yong . . . asked about government. The Master said: "Guide the officials. Forgive small mistakes. Promote [people] of talent." "How does one recognize that a [person] has talent and deserves to be promoted?" The Master said: "Promote those you know. Those whom you do not know will hardly remain ignored."

—*The Analects of Confucius*

The *Analects* is a book of thoughts and ideas by Confucius. He was a scholar and teacher in ancient China.

1 Confucius is giving advice on how to

A be a gentleman
B be a good ruler
C become wealthy
D raise a good family

2 Which sentence **best** expresses the idea of these paragraphs?

F The wise ruler governs people through fear.
G People should obey their rulers no matter what.
H A good ruler gives a lot of orders to people.
J If rulers do things well, people will follow them.

DIFFERENTIATING INSTRUCTION

English Learners

Make sure students understand the following terms and concepts in the sample questions on these pages.

Strategy Page: Primary Source

respectful: showing respect and appreciation for others

dignity: worthy of respect

incompetent: inadequate for a particular purpose

promote: to advance someone to a more responsible position

Practice Page: Secondary Source

allies: partners

fight on its side: fight along with, oppose a common enemy

two fronts: the areas of contact between opposing combat forces

neutral: does not choose sides, uninvolved

Question 2 *alliance:* agreement to advance common interests

Directions: Read this passage. Use the passage and your knowledge of world cultures and geography to answer questions 1 and 2.

Before World War I

In 1892, France and Russia had become military allies. Later, Germany signed an agreement to protect Austria. If any nation attacked Austria, Germany would fight on its side. France and Russia had to support each other as well. For instance, if France got into a war with Germany, Russia had to fight Germany, too. This meant that in any war, Germany would have to fight on two fronts: France on the west and Russia on the east.

If a war broke out, what part would Great Britain play? No one knew. It might remain neutral, like Belgium. It might, if given a reason, fight against Germany.

1 **If Russia and Germany went to war, which country had to help Russia?**

A Great Britain
B Belgium
C Austria
D France

2 **What was a result of the military alliance formed between France and Russia?**

F Belgium would have to fight a war on the side of Germany.
G Great Britain would remain neutral.
H Germany would have to fight a war on two fronts.
J France and Russia would support Germany in a war.

SKILL TESTED IN THE ITEMS

STRATEGY ITEMS		PRACTICE ITEMS	
Item Number	Skill Tested	Item Number	Skill Tested
1.	Summarizing	1.	Drawing Conclusions
2.	Finding Main Ideas	2.	Analyzing Causes and Effects

PRACTICE QUESTIONS

Thinking It Through

Share the following explanations with students as they discuss the strategies they used to answer the practice questions.

1. This question asks you to draw conclusions from the information in the passage about the military alliances before World War I. A key word is *had.* In other words, which country was required to support Russia? It might be helpful to make a T-chart to list the countries that supported one another. France and Russia would be in one column and Germany and Austria in the other. Belgium and Great Britain were neutral or unknown. This information would allow you to see that France was Russia's ally. The correct answer is (D).
2. To answer this question you must understand the cause and effect relationships described in the passage. The alliance between France and Russia is the cause. Evaluate each possible answer to see which one would be an effect of that alliance. You can eliminate (F) and (G) because the alliance had no effect on Belgium or Great Britain. Germany was not an ally of France and Russia, therefore (J) is incorrect. A key phrase in the passage is *this meant,* which is followed by the effect that is stated in (H), the correct answer.

USING STRATEGIES FOR . . .

Political Cartoons

Political cartoons are one kind of primary source. Remind students to analyze the political cartoon before reading the questions. They should identify the subject, note important symbols and details, interpret the message, and analyze the point of view. Then they should read the question to identify the information they need to find.

1. Question 1 asks you to draw on your knowledge of World War II and the swastika as a symbol. Since the swastika was the symbol of Nazi Germany, the correct answer is (B).
2. To answer question 2, analyze the cartoonist's point of view. Look at how the cartoonist uses caricature in the cartoon. The swastika, which represents Nazi Germany, is huge and appears to be turning like a wheel, about to roll down onto Poland and crush it. The title also gives a clue to the correct answer, which is (F).

General Test-Taking Tips

Share these tips with your students.

- Ask questions before the test begins.
- Know how to fill in the answer forms.
- Read and listen to directions carefully.

TCAP STRATEGIES

Political Cartoons

Cartoonists who draw political cartoons use both words and art to express opinions about political issues.

1. Try to figure out what the cartoon is about. Titles and captions may give clues.
2. Use labels to help identify the people, places, and events represented in the cartoon.
3. Note when and where the cartoon was published.
4. Look for symbols—that is, people, places, or objects that stand for something else.
5. The cartoonist often exaggerates the physical features of people and objects. This technique will give you clues as to how the cartoonist feels about the subject.
6. Try to figure out the cartoonist's message and summarize it in a sentence.

answers: 1 (B); 2 (F)

1 NEXT!

4 The cartoonist uses the swastika, a symbol used during World War II.

5 The swastika looks like a huge, frightening machine. It can easily crush Poland.

2 The label "Poland" tells which country is the subject of the cartoon's title.

Daniel Fitzpatrick / *St. Louis Post-Dispatch*, August 24, 1939.

3 The date is a clue that the cartoon refers to the beginning of World War II.

1 What does the swastika in the cartoon stand for?

A the Soviet Union
B Nazi Germany
C the Polish army
D Great Britain

6 **2 Which sentence best summarizes the cartoonist's message?**

F Germany will attack Poland next.
G Poland should stop Germany.
H Germany will lose this battle.
J Poland will fight a civil war.

DIFFERENTIATING INSTRUCTION

English Learners

Make sure students understand the following terms and concepts in the sample questions on these pages.

Strategy Page

Question 1 *stand for:* serve as a symbol of

Practice Page

Question 2 *Count Dracula:* fictional monster in literature, said to be a vampire who lives off the blood of others

Darth Vader: a fictional villain in the *Star Wars* films who appears in a suit of armor

Frankenstein: fictional creator of a monster made out of body parts from dead people

Directions: Study this cartoon. Use the cartoon and your knowledge of world cultures and geography to answer questions 1 and 2.

Steve Sack, *Minneapolis Star-Tribune*, January 1, 1992.

1 **How has the cartoonist drawn Saddam Hussein?**

A as Count Dracula

B as Darth Vader

C as Frankenstein's monster

D as a military leader

2 **Notice all the "MADE IN" labels on Saddam Hussein. The best title for the cartoon would be**

F "A Monster Never Sleeps"

G "The Monster of Iraq"

H "They've Made a Mistake"

J "They've Created a Monster"

SKILL TESTED IN THE ITEMS

STRATEGY ITEMS		PRACTICE ITEMS	
Item Number	**Skill Tested**	**Item Number**	**Skill Tested**
1.	Analyzing Political Cartoons	1.	Clarifying
2.	Finding Main Ideas	2.	Finding Main Ideas

PRACTICE QUESTIONS

Thinking It Through

Share the following explanations with students as they discuss the strategies they used to answer the practice questions.

1. To correctly answer this question, you must use your own knowledge. The caricature doesn't look like a military leader, so you can eliminate (D). The character in the cartoon is a caricature of Frankenstein's monster because it suggests that it's made from parts from different countries, similar to how Frankenstein's monster was made from different body parts. The correct answer is (C).

2. To answer this question, you need to analyze the cartoon to interpret its main idea. The "MADE IN" labels indicate that all the different countries created this monster, so the correct answer is (J).

USING STRATEGIES FOR . . .

Charts

Explain to students that they will do best on test questions by thinking them through carefully and by applying test-taking strategies, such as the following.

1. Question 1 asks you to compare and contrast the number of immigrants, shown in column 3, from row to row for the countries shown in the chart. The key word is *most.* The correct answer is (C) because the United States had the greatest number of immigrants.
2. To answer Question 2 correctly, you should focus first on column 2 and determine the earliest date shown, 1821. Then see which countries in column one relate to that year. The correct answer is (G) because all three of those countries received immigrants starting in 1821.

General Test-Taking Tips

Share these tips with your students.

- Use practice tests, such as the one you are taking now, to learn about your test-taking habits and weaknesses.
- Use this information to practice strategies that will help you become a successful test-taker.

flow chart

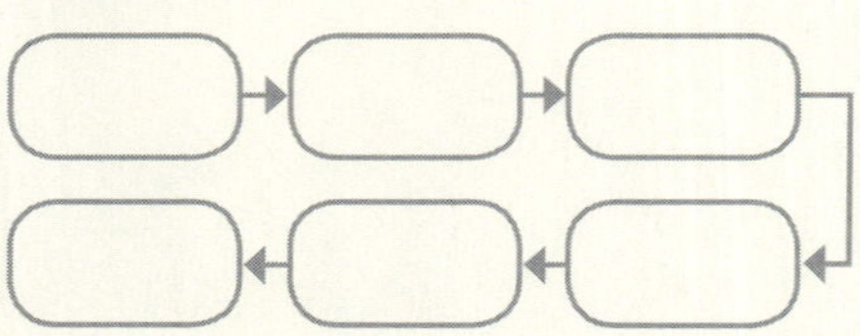

Venn diagram

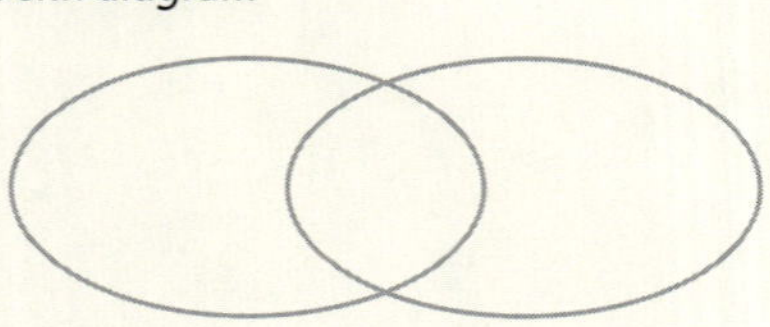

concept web

TCAP STRATEGIES

Charts

Charts present facts in a visual form. History textbooks use several different types of charts. The chart that is most often found on standardized tests is the table. A table organizes information in columns and rows.

1. Read the title of the chart to find out what information is represented.
2. Read the headings at the top of each column. Then read the headings at the left of each row.
3. Notice how the information in the chart is organized.
4. Notice that information from different years is used for different countries.
5. Of all the countries listed, six took in the most immigrants. Think about what these countries have in common.
6. Read the questions and then study the chart again.

1 This chart is about the number of people who immigrated to different countries.

4 Notice that information from different years is used for different countries.

Immigration to Selected Countries

Country	Years	Number of Immigrants
Argentina	1856–1932	6,405,000
Australia	1861–1932	2,913,000
Brazil	1821–1932	4,431,000
British West Indies	1836–1932	1,587,000
Canada	1821–1932	5,206,000
Cuba	1901–1932	857,000
Mexico	1911–1931	226,000
New Zealand	1851–1932	594,000
South Africa	1881–1932	852,000
United States	1821–1932	34,244,000
Uruguay	1836–1932	713,000

Source: Alfred W. Crosby, Jr. *The Columbian Exchange: Biological and Cultural Consequences of 1492*

3 This chart lists countries in alphabetical order. Other charts organize information by years or by numbers.

5 Of all the countries listed, six took in the most immigrants. Think about what these countries have in common.

1 The country that received the most immigrants was

A Canada
B the British West Indies
C the United States
D Brazil

2 Different countries received immigrants in different years. According to the chart, which countries received immigrants the earliest?

F Argentina, New Zealand, and Canada
G Canada, Brazil, and United States
H Mexico, United States, and British West Indies
J Brazil, South Africa, and Cuba

answers: 1 (C); 2 (G)

S8

DIFFERENTIATING INSTRUCTION

English Learners

Make sure students understand the following terms and concepts in the sample charts and strategies on these pages.

Strategy Page: Chart

column: reads down

row: reads across

flow chart: example at left

Venn diagram: example at left

concept web: example at left

immigrants: people who leave one country to settle in another

Practice Page: Chart

civilization: an advanced form of culture

B.C.: before Christ, dates counted backwards from the year Christ was believed to have been born

specialized workers: those with specific jobs based on specific skills

institutions: practices or groups of people with a specific purpose

Directions: Read the chart carefully. Use the chart and your knowledge of world cultures and geography to answer questions 1 and 2.

Ancient Civilizations				
Feature	**China**	**Egypt**	**Indus Valley**	**Mesopotamia**
Location	River valley	River valley	River valley	River valley
Period	2000 B.C.–400 B.C.	3100 B.C.–332 B.C.	2500 B.C.–1500 B.C.	3500 B.C.–500 B.C.
Specialized workers	Priests; government workers, soldiers; workers in bronze, silk; farmers	Priests; government workers, scribes, soldiers; workers in pottery, stone; farmers	Priests; government officials; workers in pottery, bricks; farmers	Priests; government officials, scribes, soldiers; workers in pottery, textiles; farmers
Institutions	Walled cities; oracle-bone reading	Ruling class of priests, nobles; education system	Ruling class of priests	Ruling class of priests and nobles; education for scribes
Record keeping	Pictographic writing	Hieroglyphic writing	Pictographic writing	Cuneiform writing
Advanced technology and artifacts	Writing; making bronze and silk; irrigation systems	Papyrus; mathematics; astronomy, engineering; pyramids; mummification; medicine	Irrigation systems; indoor plumbing; seals	Wheel; plow; sailboat; bronze weapons

1 **Which civilization appeared first?**

A China

B Egypt

C Indus Valley

D Mesopotamia

2 **The Indus Valley civilization did not have**

F irrigation systems

G walled cities

H government officials

J indoor plumbing

TENNESSEE TCAP

SKILL TESTED IN THE ITEMS

STRATEGY ITEMS		PRACTICE ITEMS	
Item Number	**Skill Tested**	**Item Number**	**Skill Tested**
1.	Comparing and Contrasting	1.	Sequencing Events
2.	Sequencing Events	2.	Categorizing

TCAP STRATEGIES AND PRACTICE

PRACTICE QUESTIONS

Thinking It Through

Share the following explanations with students as they discuss the strategies they used to answer the practice questions.

1. *First* is the key word. Look across the "Period" row to determine which civilization is oldest. Remember that during the B.C. period higher numbers indicate earlier dates because you are counting backwards. Mesopotamia appeared in 3500 B.C. and is the earliest civilization. The correct answer is (D).
2. This item is an example of a *not* question. First read down the Indus Valley column to see which three things the Indus Valley *did* have, namely (F), (H), and (J). There were no walled cities in the Indus Valley. Therefore, the correct answer is (G).

TCAP STRATEGIES AND PRACTICE

USING STRATEGIES FOR . . .

Line and Bar Graphs

Remind students that the vertical axis goes up and down and is normally shown on the left side of the graph. The horizontal axis runs across the bottom of the graph.

Explain to students that they will do best on test questions by thinking them through carefully and by applying test-taking strategies, such as the following.

1. To answer this question, you must test the information in each possible answer against the information in the graph. Recognize the two categories of information that are represented by the different colored lines on the graph. In this case, you will find that the only answer that is accurate is (C), which is the correct answer.
2. *Second largest* is a key phrase in question 2. To answer the question, you need to study the information in the bar graph to find the country with the second tallest bar, which represents the second largest amount of debt compared to the other countries' bars. Note that in this case the bars are arranged in order from lowest to highest, left to right. The correct answer is (H).

General Test-Taking Tips

Share these tips with your students.

- Do not spend too much time on one question.
- Skip a question you are having problems with. Go back to it later, if you have time.
- If you skip a question, be sure to skip the answer space for the same number on your answer sheet.

TCAP STRATEGIES

Line and Bar Graphs

Graphs are often used to show numbers. Line graphs often show changes over time. Bar graphs make it easy to compare numbers.

1. Read the title of the graph to find out what information is represented.
2. Study the labels on the graph.
3. Look at the source line that tells where the graph is from. Decide whether you can depend on the source to provide reliable information.
4. See if you can make any generalizations about the information in the graph. Note whether the numbers change over time.
5. Read the questions carefully and then study the graph again.

1 Exports of English Manufactured Goods, 1699–1774

4 Notice that both lines rise over time. This means the value of all exports increased.

3 Facts from scholarly journals are usually reliable.

2

5 **Source:** R. Davis, "English Foreign Trade, 1700–1774," *Economic History Review* (1962)

1 Which of the following is a true statement?

- **A** Exports to the Atlantic economy declined over time.
- **B** Total exports stayed the same over time.
- **C** Total exports rose sharply after 1724.
- **D** Exports to the Atlantic economy fell sharply after 1754.

1 Nations with High Foreign Debt, 2003

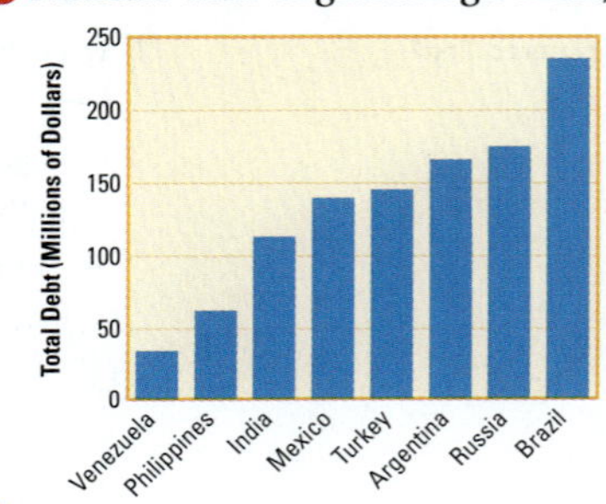

2 **Source:** The World Bank

3 Facts from major organizations, such as the World Bank, usually are reliable.

5 **2** The nation with the second largest foreign debt is

- **F** Brazil
- **G** Argentina
- **H** Russia
- **J** Mexico

Line graph adapted from "Exports of English Manufactured Goods, 1700–1774," from *A History of World Societies, Fifth Edition* by John P. McKay, Bennett D. Hill, John Buckler, and Patricia Buckley Ebrey. Used by permission.

answers: 1 (C); 2 (H)

DIFFERENTIATING INSTRUCTION

English Learners

Make sure students understand the following terms and concepts in the graphs and sample questions on these pages.

Strategy Page

Line graph Point out that the line graph covers years of discovery, exploration, and colonization (1699–1774), which explains why one land is called "Spanish America," a phrase not used now.

Practice Page

Question 1 *increased:* grew, as shown by upward movement on the graph

Directions: Study the graphs carefully. Use the graphs and your knowledge of world cultures and geography to answer questions 1 and 2.

Japan: Gross Domestic Product, 1984–2002

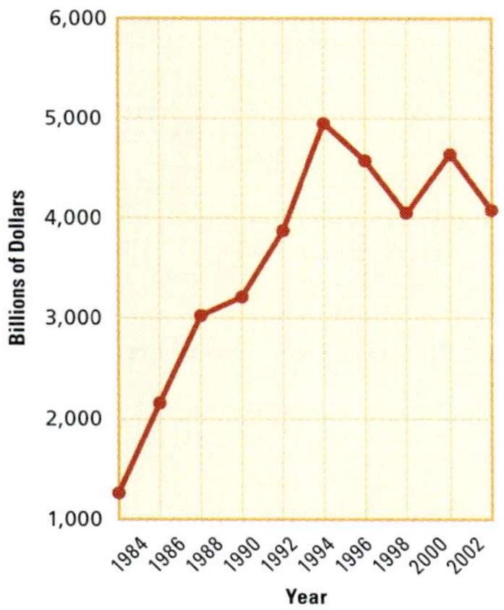

Source: Annual Report on National Accounts 2005, Cabinet Office of the Government of Japan

Unemployment Rates for Selected Countries, 2005

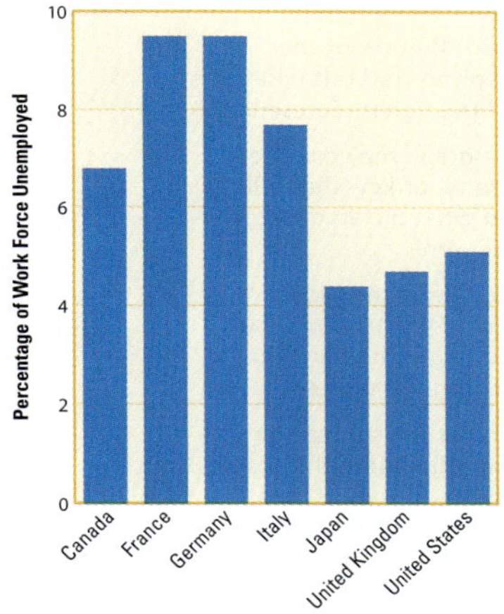

Source: Organization for Economic Cooperation and Development

1 **Japan's gross domestic product increased the most between**

A 1986 and 1990

B 1990 and 1994

C 1994 and 1998

D 1998 and 2002

2 **Which country had the lowest unemployment rate in 2005?**

F Japan

G Germany

H United States

J United Kingdom

PRACTICE QUESTIONS

Thinking It Through

Share the following explanations with students as they discuss the strategies they used to answer the practice questions.

1. *Increased the most* is the key phrase in question 1. To determine the answer, you need to compare and contrast how the line graph changes between specific dates. Examine the graph between each of the sets of dates shown in the answer choices. Note that the answer choices all cover four-year periods, so you will be looking at changes that cover three points on the graph. It is clear that the line rose most sharply between 1990 and 1994, when gross domestic product increased by almost $2,000 billion. The correct answer is (B).
2. To answer this item, you need to study the information in the bar graph to find the country with the shortest bar, or lowest unemployment rate. Note that the bars are arranged in the countries' alphabetical order, not by ranking of unemployment rate. The correct answer is (F).

SKILL TESTED IN THE ITEMS

STRATEGY ITEMS		PRACTICE ITEMS	
Item Number	**Skill Tested**	**Item Number**	**Skill Tested**
1.	Interpreting Graphs	1.	Interpreting Graphs
2.	Interpreting Graphs Comparing and Contrasting	2.	Interpreting Graphs Comparing and Contrasting

USING STRATEGIES FOR . . .

Pie Graphs

Explain to students that they will do best on test questions by thinking them through carefully and by applying test-taking strategies, such as the following.

1. The key words in Question 1 are *nearly two-thirds.* You need to recall that two-thirds is equal to about 67 percent. Find the slice of the pie that is closest to 67 percent. Then read the legend to see which region is represented by that slice. The correct answer is (D).
2. Use the pie graph and legend to answer this question. Compare the population percentage for each region in each answer choice to find the region that has a smaller share of the world's population than Latin America and the Caribbean. Of the four alternatives, only North America has a smaller share than Latin America and the Caribbean. The correct answer, therefore, is (H).

General Test-Taking Tips

Share these tips with your students.

- Read the question and each answer choice before answering.
- Many items include choices that may seem right at first glance, but are actually wrong.

TCAP STRATEGIES

Pie Graphs

A pie, or circle, graph shows the relationship among parts of a whole. These parts look like slices of a pie. Each slice is shown as a percentage of the whole pie.

1. Read the title of the graph to find out what information is represented.
2. The graph may provide a legend, or key, that tells you what different slices represent.
3. The size of the slice is related to the percentage. The larger the percentage, the larger the slice.
4. Look at the source line that tells where the graph is from. Ask yourself if you can depend on this source to provide reliable information.
5. Read the questions carefully, and study the graph again.

1 **World Population by Region, 2006**

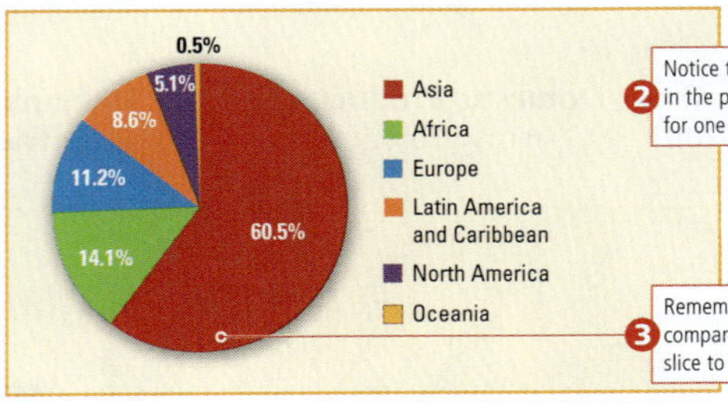

2 Notice that each color in the pie graph stands for one of the regions.

3 Remember, the numbers compare the size of each slice to the entire pie.

Source: Population Reference Bureau

4 The Population Reference Bureau studies population data for the United States and other countries.

5

1 **Which region accounts for nearly two-thirds of the world's population?**

A Africa
B North America
C Europe
D Asia

2 **A greater share of the world's population lives in Latin America and the Caribbean than lives in**

F Africa
G Europe
H North America
J Asia

For this question, find the "pie slices" for each of the regions listed in the alternatives. Compare each one to the "pie slice" for Latin America and the Caribbean.

answers: 1 (D); 2 (H)

S12

DIFFERENTIATING INSTRUCTION

English Learners

Make sure students understand the following terms and concepts in the graphs and sample questions on these pages.

Strategy Page

Question 1 *accounts for:* has

Practice Page

Pie Graph *energy consumption:* energy use

Directions: Study the pie graph. Use the graph and your knowledge of world cultures and geography to answer questions 1 and 2.

World Energy Consumption by Region

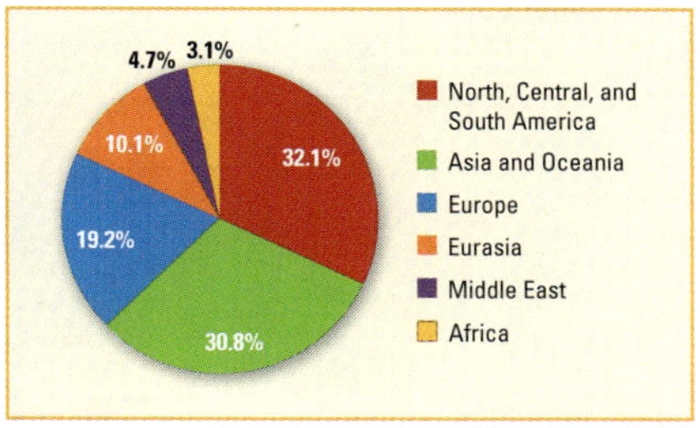

Source: Energy Information Administration/Annual Energy Review 2005

1 **Which region uses the least energy?**

A Europe

B Asia and Oceania

C Africa

D Middle East

2 **North, Central, and South America are grouped together because**

F they are in the same part of the world

G all three have about the same number of people

H all three are roughly the same size

J they use the same power sources

SKILL TESTED IN THE ITEMS

STRATEGY ITEMS		PRACTICE ITEMS	
Item Number	**Skill Tested**	**Item Number**	**Skill Tested**
1.	Interpreting Graphs	1.	Interpreting Graphs Comparing and Contrasting
2.	Interpreting Graphs Comparing and Contrasting	2.	Making Inferences

PRACTICE QUESTIONS

Thinking It Through

Share the following explanations with students as they discuss the strategies they used to answer the practice questions.

1. The key word is *least.* Compare the slices of the pie. At 3.1 percent, Africa has the lowest percentage of world energy use. The correct answer is (C).
2. To answer this question, you need to use your knowledge of world cultures and geography to know which of the statements in the answer choices are true. The title of the graph gives you a clue because it says that the pie is divided by regions. North, Central, and South America have different numbers of people, are of different sizes, and use different power sources. They are grouped together because they are in the same part of the world. The correct answer, therefore, is (F).

USING STRATEGIES FOR . . .

Political Maps

Explain to students that they will do best on test questions by thinking them through carefully and by applying test-taking strategies, such as the following.

1. To answer question 1, use the compass rose on the map to see which direction is east. Of the answer choices, only (B) and (C) are eastern states. By comparing the location of Maine and Georgia on the map, you can see that Maine extends farther east. The correct answer, therefore, is (B).
2. To answer question 2, you will need to use the map scale to estimate the distance between the Great Lakes and the Pacific Ocean. The Great Lakes are not labeled so you will also need to use your knowledge of geography to identify them. The question provides a clue: that they are located on the United States–Canada border. Use a ruler or your fingers to estimate the distance and compare it to the scale. The distance from the western end of the Great Lakes to the Pacific Ocean is about 4 times the length of the scale, or 1,600 miles. The correct answer is (G), which is closest to that distance.

General Test-Taking Tips

Share these tips with your students.

- Try to answer every question on the test.
- If you are not sure of an answer, make an educated guess.
- First eliminate the choices you are sure are not correct. Then choose from the choices that remain.

TCAP STRATEGIES

Political Maps

Political maps show the divisions within countries. A country may be divided into states, provinces, or other kinds of segments. The maps also show where major cities are. They may also show mountains, oceans, seas, lakes, and rivers.

1. Read the title of the map. This will give you the subject and purpose of the map.
2. Read the labels on the map. They also give information about the map's subject and purpose.
3. Study the key or legend to help you understand the symbols on the map.
4. Use the scale to estimate distances between places shown on the map. Maps usually show the distance in both miles and kilometers.
5. Use the compass rose to figure out directions on the map.
6. Read the questions. Carefully study the map to find the answers.

1 Present-Day United States

3 The legend gives symbols for the nation's capital and the states' capitals.

2 The labels identify the states of the United States.

6

1 What state extends farthest east?

A Oregon
B Maine
C Georgia
D Alaska

2 About how many miles is it from the Great Lakes to the Pacific Ocean along the United States–Canada border?

F 1,000
G 1,500
H 2,000
J 2,500

answers: 1 (B); 2 (G)

S14

DIFFERENTIATING INSTRUCTION

English Learners

Make sure students understand the following terms and concepts in the sample questions and strategies on these pages.

Strategy Page

political: type of map that shows borders of countries, states, cities, and the like

Strategy 4 *scale:* ratio between distances on a map and the corresponding distances on Earth

estimate: calculate approximately the distance, amount, or some other measure

kilometer: a measure of distance in the metric system equal to 1,000 meters, or roughly three-fifths of a mile

Question 1 *extends:* reaches

Practice Page

Question 1 *borders:* is right next to

Directions: Study the map carefully. Use the map and your knowledge of world cultures and geography to answer questions 1 and 2.

Present-Day Europe

1 What body of water borders Poland on the north?

A North Sea
B Norwegian Sea
C Baltic Sea
D Adriatic Sea

2 The most northern country in Europe is

F Ireland
G Denmark
H Iceland
J Norway

PRACTICE QUESTIONS

Thinking It Through

Share the following explanations with students as they discuss the strategies they used to answer the practice questions.

1. To answer question 1, use the compass rose on the map to see which direction is north. Then locate Poland and look at the northern edge of the country. The Baltic Sea is on Poland's northern border. The correct answer, therefore, is (C).
2. The key words in Question 2 are *most northern.* First, identify where north is on the map using the compass rose. Then, check each answer to see which country in Europe is most northern. The answer is (J), Norway.

SKILL TESTED IN THE ITEMS

STRATEGY ITEMS		PRACTICE ITEMS	
Item Number	Skill Tested	Item Number	Skill Tested
1.	Interpreting Maps Comparing and Contrasting	1.	Interpreting Maps
2.	Interpreting Maps	2.	Interpreting Maps Comparing and Contrasting

USING STRATEGIES FOR . . .

Thematic Maps

Explain to students that they will do best on test questions by thinking them through carefully and by applying test-taking strategies, such as the following.

1. To answer question 1, use the map labels and the legend. The shaded area shows you where Buddhism originated. The correct answer is (B).
2. To answer question 2, use the arrows and dates to trace the spread of Buddhism, which indicates it went from China to Korea and Japan. The correct answer is (H).

General Test-Taking Tips

Share these tips with your students.

- Think positively.
- Tell yourself that you can do it!
- If you have studied for the test, you are prepared to succeed.

TCAP STRATEGIES

Thematic Maps

Thematic maps focus on special topics. For example, a thematic map might show a country's natural resources or major battles in a war.

1. Read the title of the map. This will give you the subject and purpose of the map.
2. Read the labels on the map. They give information about the map's subject and purpose.
3. Study the key or legend to help you understand the symbols on the map.
4. Ask yourself whether the symbols show a pattern.
5. Read the questions. Carefully study the map to find the answers.

1 The Spread of Buddhism

2 The labels name the major areas of South and East Asia. The dates show when Buddhism first came to each area.

3 The arrows show where Buddhism spread.

4 Notice that the spread of Buddhism took several centuries.

5

1 Where did Buddhism start?

A Japan
B India
C China
D Afghanistan

2 Buddhism spread from China to

F Japan and Tibet
G Tibet and Korea
H Korea and Japan
J all of the above

answers: 1 (B); 2 (H)

S16

DIFFERENTIATING INSTRUCTION

English Learners

Make sure students understand the following terms and concepts in the maps and sample questions on these pages.

Strategy Page: Map

Buddhism: one of the world's major religions, which originated in India

Directions: Study the map carefully. Use the map and your knowledge of world cultures and geography to answer questions 1 and 2.

The Roman Empire, A.D. 400

1 **Which area was part of the Eastern Roman Empire?**

- **A** Spain
- **B** Gaul
- **C** Anatolia
- **D** all of the above

2 **The most northern region in the Western Roman Empire was**

- **F** Syria
- **G** Gaul
- **H** Spain
- **J** Britain

TENNESSEE TCAP

PRACTICE QUESTIONS

Thinking It Through

Share the following explanations with students as they discuss the strategies they used to answer the practice questions.

1. To answer question 1, use the key to identify the Eastern Roman Empire on the map. Check the locations listed in the answer choices. Anatolia is the only area located in the Eastern Roman Empire. Therefore, the correct answer is (C).
2. The key words in this question are *most northern.* Use the compass rose on the map to see which direction is north. Then, identify the color on the map that indicates the Western Roman Empire. Finally, check each answer to see which place within the Western Roman Empire is most northern. The answer is (J), Britain.

SKILL TESTED IN THE ITEMS

STRATEGY ITEMS		PRACTICE ITEMS	
Item Number	**Skill Tested**	**Item Number**	**Skill Tested**
1.	Interpreting Maps	1.	Interpreting Maps
2.	Interpreting Maps	2.	Interpreting Maps Comparing and Contrasting

USING STRATEGIES FOR . . .

Time Lines

Explain to students that they will do best on test questions by thinking them through carefully and by applying test-taking strategies, such as the following.

1. To answer question 1, read the captions for all the dates between 1961 and 1966, and count the number of countries listed that gained independence. The correct answer is (C), 11.
2. To answer question 2, you will need to draw on your knowledge of world history, cultures, and geography. Consider how World War II affected the nations of Europe, which controlled colonies in Africa. Because European nations were weaker after the war, they would not be able to manage their colonies as well. As a result, African nations became stronger and won their independence. The correct answer is (F).

General Test-Taking Tips

Share these tips with your students.

- Relax during the test.
- Several times during the test, take a few seconds to relax and breathe deeply.
- Occasional deep breaths will help relieve anxiety and keep you focused.

TCAP STRATEGIES

Time Lines

A time line is a chart that lists events in the order in which they occurred. Time lines can be vertical or horizontal.

1. Read the title to learn what period of time the time line covers.
2. Note the dates when the time line begins and ends.
3. Read the events in the order they occurred.
4. Think about what else was going on in the world on these dates. Try to make connections.
5. Read the questions. Then carefully study the time line to find the answers.

1 Dates of Independence in African Countries

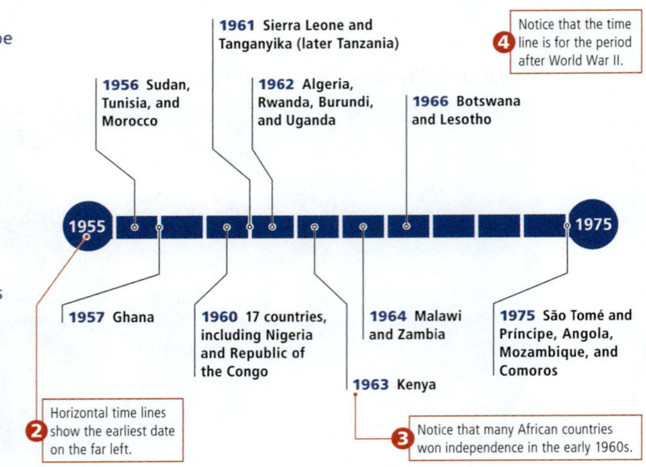

5

1 How many countries gained independence from 1961 to 1966?

A 7

B 9

C 11

D 13

2 Why do you think so many countries won their independence after World War II?

F European nations were weaker after the war.

G All Europeans in Africa moved back to Europe.

H Europe no longer wanted to own colonies.

J Europe gave each colony its own army after the war.

answers: 1 (C), 2 (F)

S18

DIFFERENTIATING INSTRUCTION

English Learners

Make sure students understand the following terms and concepts in the time lines and sample questions on these pages.

Strategy Page

Time Line and Questions 1 and 2
independence: freedom from outside control

Practice Page

Time Line and Question 2 *hardliners:* Communists who were unwilling to make compromises with reformers

Directions: Study the time line. Use the information shown and your knowledge of world cultures and geography to answer questions 1 and 2.

The Breakup of the Soviet Union

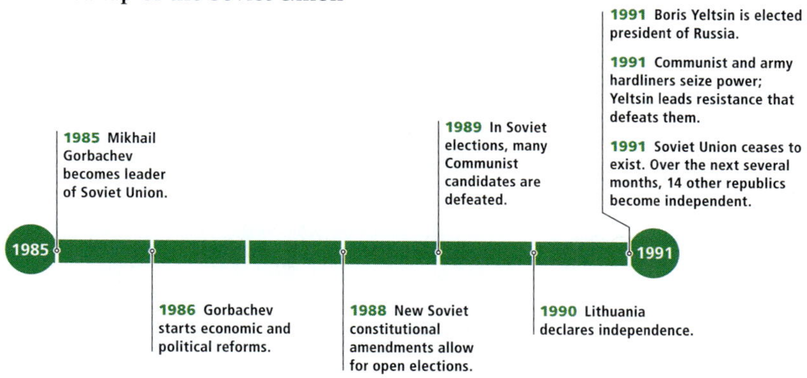

1 What happened after Lithuania declared independence?

- **A** Gorbachev started economic and political reforms.
- **B** Many other republics became independent.
- **C** New constitutional amendments allowed for open elections.
- **D** Gorbachev defeated Yeltsin in a new election.

2 In which year did Communist and army hardliners try to seize power?

- **F** 1985
- **G** 1988
- **H** 1990
- **J** 1991

TENNESSEE TCAP

PRACTICE QUESTIONS

Thinking It Through

Share the following explanations with students as they discuss the strategies they used to answer the practice questions.

1. The key word in Question 1 is *after.* First, find the name *Lithuania* on the time line. Then, compare the statements in the answer choices with the events that occurred in 1991, which were after Lithuania declared independence. The only choice that correctly reflects the information on the time line is (B), which is the correct answer.
2. Look for the phrase *Communist and army hardliners* on the time line. Notice the date that is connected with them. The correct answer is (J), 1991.

SKILL TESTED IN THE ITEMS

STRATEGY ITEMS		PRACTICE ITEMS	
Item Number	Skill Tested	Item Number	Skill Tested
1.	Summarizing	1.	Sequencing Events
2.	Making Inferences	2.	Summarizing

USING STRATEGIES FOR . . .

Reading Visuals

Remind students of the following:

- Useful information may be found in a title, a caption, or a source line as well as in the document itself.
- In the questions on this page, you need to use the document to answer the first two questions.
- Some questions do not include a document. Instead, all the questions may require you to use your knowledge of world cultures and geography to answer the questions.

General Test-Taking Tips

Share these tips with your students.

- Be sure to answer all parts of the reading visuals questions.

TCAP STRATEGIES

Reading Visuals

Questions based on reading visual texts focus on various kinds of documents, such as a photograph, cartoon, chart, graph, or time line. Each document is accompanied by one or more multiple choice questions. Sometimes, you can find the answer in the document. Other times, you will use what you already know about a subject to answer questions.

1. Read the title of the document to get an idea of what it is about.
2. Study the document (including any caption, if it is an image).
3. Read the questions carefully. Study the document again to find the answers.
4. Circle the correct answer.

answers: 1 (A), 2 (H)

1 The Salt March

Mohandas Gandhi and poet Sarojini Naidu lead Indians in a march down the west coast of India. They are protesting the Salt Acts of 1930.

2 This document is a photograph showing Mohandas K. Gandhi leading a demonstration.

1 Mohandas Gandhi was an important leader in what country? 3

4 **A** India
B Pakistan
C China
D Great Britain

2 What principle did Gandhi follow to win independence for India?

F violence
G anger
H non-violence
J acceptance

S20

DIFFERENTIATING INSTRUCTION

English Learners

Make sure students understand the following terms and concepts in the sample questions on these pages.

Strategy Page

Document *Mohandas Gandhi:* Make sure students know who Gandhi was and why he protested against British rule in India.
Question 3 *principle:* a rule or standard

Practice Page

Document and Questions Students may not understand every word in the document, but they should focus on understanding the main ideas. Help them understand key words: *Balkans, Bosnia and Herzegovina, Sarajevo, intercom, cellar, cope*

TCAP STRATEGIES

Read the column heads carefully. They offer important clues about the subject of the chart. For instance, the column head "Impact" is a clue about why these inventions were so important.

1 Inventions of the Industrial Revolution 3

Invention	Impact
Flying shuttle, spinning jenny, water frame, spinning mule, power loom	Spun thread and wove cloth faster; more factories were built and more people were hired
Cotton gin	Cleaned seeds from cotton faster; companies produced more cotton
Macadam road, steamboat, locomotive	Made travel over land and water faster; could carry larger, heavier loads; railroads needed more coal and iron
Mechanical reaper	Made harvesting easier; increased wheat production

2 **Directions:** Read the list of inventions and their impact. Then answer the questions.

1 What was the impact of the cotton gin?

4 **A** It allowed cotton to be picked much faster than by hand.
B It helped cotton pickers earn more money.
C It made cotton more expensive.
D It made cotton shirts more popular.

2 What crop was the mechanical reaper used to harvest?

F cotton
G tobacco
H grapes
J wheat

For many questions based on visuals, it is important to read the text in the visual. Read the information in charts and tables. Read the title and captions for pictures. They often contain important information.

1. Read the title of the document to get an idea of what it is about.
2. Carefully read directions and questions.
3. Study the document.
4. Circle the correct answer.

answers: 1 (A), 2 (J)

TENNESSEE TCAP

S21

TCAP STRATEGIES AND PRACTICE

USING STRATEGIES FOR . . .

Reading Visuals

Remind students of the following:

- Read all the questions that go with a document before beginning to answer any questions. Look for words that tell you how to organize your answers.
- In this case, you are to use a chart to record the impact of inventions of the Industrial Revolution and then answer the questions. You need to apply your knowledge of world cultures and geography to answer correctly.

DIFFERENTIATING INSTRUCTION

English Learners

Make sure students understand the following terms and concepts in the sample questions on these pages.

Strategy Page: Document

Explain that the Industrial Revolution was not a war like the American Revolution, but, rather, the term describes radical changes in making products and farming, notably through the use of machines.

Practice Page: Document

smallpox: a serious, often deadly, contagious disease, caused by a poxvirus

typhoid fever: a serious, contagious, bacterial disease

measles: a serious, sometimes deadly, contagious viral disease

PRACTICE QUESTIONS

Thinking It Through

Share the following explanations with students as they discuss the strategies they used to answer the practice questions.

1. To correctly answer this question, reread the passage, paying special attention the information about where Zlata's family is. The passage states, "The cellar is ugly, dark, smelly." The correct answer is (D).
2. To answer this question, reread the passage, while you look for information about Sarajevo. Zlata writes, "Today was truly, absolutely the worst day ever in Sarajevo. The shooting started around noon." The correct answer is (G).

TCAP PRACTICE

Directions: Read the following passage from *Zlata's Diary,* a diary kept by Zlata, a 12-year-old girl. Use the passage and your knowledge of world cultures and geography to answer the questions.

Saturday, May 2, 1992

Dear Mimmy,

Today was truly, absolutely the worst day ever in Sarajevo. The shooting started around noon. Mommy and I moved into the hall. Daddy was in his office, under our apartment, at the time. We told him on the intercom to run quickly to the downstairs lobby where we'd meet him . . . The gunfire was getting worse, and we couldn't get over the wall to the Bobars', so we ran down to our own cellar.

The cellar is ugly, dark, smelly. Mommy, who's terrified of mice, had two fears to cope with. The three of us were in the same corner as the other day. We listened to the pounding shells, the shooting, the thundering noise overhead. We even heard planes. At one moment I realized that this awful cellar was the only place that could save our lives. Suddenly, it started to look almost warm and nice. It was the only way we could defend ourselves against all this terrible shooting. We heard glass shattering in our street. Horrible. I put my fingers in my ears to block out the terrible sounds. I was worried about Cicko. We had left him behind in the lobby. Would he catch cold there? Would something hit him? I was terribly hungry and thirsty. We had left our half-cooked lunch in the kitchen.

—Zlata Filipovic, *Zlata's Diary: A Child's Life in Sarajevo* (1994)

1 **Where does Zlata's family hide?**

- **A** the bedroom
- **B** her father's office
- **C** the kitchen
- **D** the cellar

2 **What does Zlata say is happening in the city of Sarajevo?**

- **F** a fair
- **G** a war with shooting
- **H** a celebration
- **J** an election

SKILL TESTED IN THE ITEMS

STRATEGY ITEMS		PRACTICE ITEMS	
Item Number	**Skill Tested**	**Item Number**	**Skill Tested**
1.	Finding Main Ideas	1.	Making Inferences
2.	Drawing Conclusions	2.	Summarizing
3.	Summarizing	3.	Analyzing Causes and Effects

Directions: Use the drawing and passage below and your knowledge of world cultures and geography to answer the questions.

Smallpox Spreads Among the Aztecs

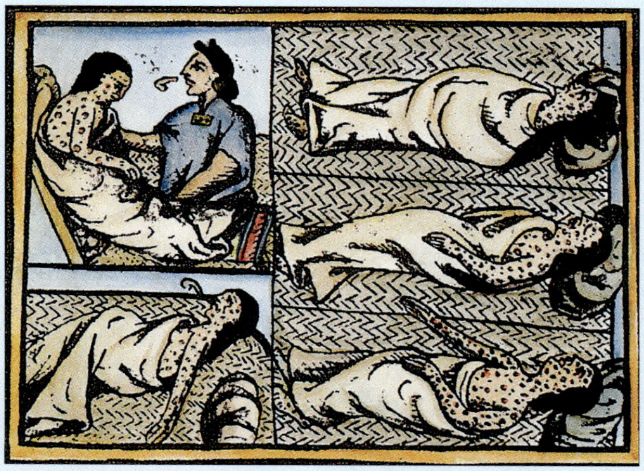

The Granger Collection, New York.

European diseases were like a second "army" of conquerors. Native people had no way to treat diseases like smallpox, typhoid fever, or measles. This "army" was more deadly than swords or guns.

—Based on P. M. Ashburn, *The Ranks of Death* (1947)

1 **Why were European diseases like a "second army"?**

- **A** The spread of disease killed more native people than swords or guns.
- **B** Guns were not reliable on the battlefield.
- **C** Native people had many ways to treat diseases like smallpox.
- **D** Europeans used the spread of disease to defeat native people.

2 **What treatments did the Aztecs have that could cure European diseases?**

- **F** medicine
- **G** herbs
- **H** plants
- **J** none

SKILL TESTED IN THE ITEMS

STRATEGY ITEMS		PRACTICE ITEMS	
Item Number	Skill Tested	Item Number	Skill Tested
1.	Analyzing Causes and Effects	1.	Drawing Conclusions Analyzing Causes and Effects
2.	Analyzing Causes and Effects Summarizing		

PRACTICE QUESTIONS

Thinking It Through

Share the following explanations with students as they discuss the strategies they used to answer the practice questions.

1. To correctly answer this question, you must determine the meaning behind "second army". Compare European diseases to an army, using the passage as a guide. The correct answer is (A).
2. To answer this question, reread the passage. The passage does not mention any cures. The correct answer is (J).

USING STRATEGIES FOR . . .

Document-Based Questions

Explain to students that they will do best on test questions by thinking them through carefully and by applying test-taking strategies, such as the following.

1. To correctly answer this question, you must use your own knowledge, and the introductory text above the document. The correct answer is (A), leader of the Mongols.
2. To answer this question, you need to analyze the introductory text, as well as the depiction of Genghis Khan in the document. The correct answer is (J), merciless.

Remind students of the following:

- Document-based questions are designed to help you work like a historian. You are given several documents from a variety of sources that you must analyze, evaluate, and synthesize in order to answer the questions correctly.
- Use information in the "Introduction" to help you answer the questions. The "Historical Context" gives you the focus of the document-based question. The document-based question here focuses on the conquests of the Mongol Empire.

Document-Based Questions

To answer a document-based question, read the document and any background information carefully. You should also use what you know about world cultures and geography.

1. Read the "Historical Context" section. It will give you an idea of the topic that will be covered in the questions.
2. Read the title of the document to get an idea of what it is about.
3. Study the document.
4. Read the questions carefully. Study the document again to find the answers.
5. Circle the correct answer.

answers: 1 (A), 2 (J)

1 Introduction

Historical Context: By the 8th century B.C., Mongol nomads lived in different tribes. They sometimes fought among themselves. In the early 1200s, a new leader—Genghis Khan—united these tribes. He turned the Mongols into a powerful fighting army.

2 Document 1: Mongol Warrior

3 

Victoria & Albert Museum, London/Art Resource, New York.

4 **1 Who was Genghis Khan?**

5 **A** leader of the Mongols

B leader of the Persians

C leader of the Egyptians

D leader of the Aztecs

2 What was a characteristic of a Mongol warrior?

F merciful

G passive

H nonviolent

J merciless

DIFFERENTIATING INSTRUCTION

English Learners

Make sure students understand the following terms and concepts in the sample questions on these pages.

Strategy Pages

Historical Context *nomads:* people who make a living by herding animals and moving from place to place as the seasons change

Document 1 *characteristics:* most distinctive features

Document 3 *mint:* place where money is made by government authority

TCAP STRATEGIES AND PRACTICE

Document 2: The Mongol Empire

1 **What trade route linked the Mongol Empire to Europe?**

- **A** Tibetan Path
- **B** Khan Road
- **C** India Way
- **D** Silk Road

Document 3: The Great Khan's Wealth

. . . All those who have gems and pearls and gold and silver must bring them to the Great Khan's mint . . . By this means the Great Khan acquires all the gold and silver and pearls and precious stones of all his territories [lands] . . .

. . . The Great Khan must have, as indeed he has, more treasure than anyone else in the world . . . All the world's great [rulers] put together have not such riches as belong to the Great Khan alone.

—Marco Polo, *The Travels of Marco Polo* (c. 1300)

1 **Why might Marco Polo's travels make Europeans want to see East Asia?**

- **F** Europeans thought they could conquer East Asia.
- **G** Europeans were interested in the treasures of East Asia.
- **H** Europeans wanted a route to the Americas.
- **J** Europeans thought the Great Khan could rule European countries.

answers: 1 (D), 1 (G)

TENNESSEE TCAP

USING STRATEGIES FOR . . .

Document-Based Questions

Explain to students that they will do best on test questions by thinking them through carefully and by applying test-taking strategies, such as the following.

1. To correctly answer this question, follow the trade route between Europe and the Mongol Empire. This trade route is labeled "Silk Road". The correct answer is (D).
2. To answer this question, reread the document from Marco Polo. The document describes the treasure of the Great Khan. The correct answer is (G).

SKILL TESTED IN THE ITEMS

STRATEGY ITEMS	
Item Number	**Skill Tested**
Document 1	Summarizing
Document 2	Interpreting Maps
Document 3	Making Inferences
Part 2: Essay	Drawing Conclusions Analyzing Causes and Effects

PRACTICE QUESTIONS

Thinking It Through

Share the following explanations with students as they discuss the strategies they used to answer the practice questions.

1. The key word in this question is *poor.* How are the poor depicted in the document? The correct answer is (C).
2. To answer this question correctly, reread the Historical Context above the document. The correct answer is (G).

Introduction

Historical Context: For many centuries, kings and queens ruled the countries of Europe. Their power was supported by nobles and armies. Then European society began to change. In the late 1700s, those changes produced a violent revolution in France.

Document 1: Social Classes in Pre-Revolutionary France

1 What does the cartoon say about the lives of the poor before the French Revolution?

- **A** The poor were helped by the Church.
- **B** The poor were helped by the nobility.
- **C** The poor were under the control of the nobility and church.
- **D** The poor led easy lives.

2 When did a violent revolution occur in France?

- **F** the early 1700s
- **G** the late 1700s
- **H** the early 1800s
- **J** the late 1800s

Le Grand Abus. Engraving of a cartoon held in the collection of M. de baron de Vinck d'Orp of Brussels/Mary Evans Picture Library, London.

S26

DIFFERENTIATING INSTRUCTION

English Learners

Make sure students understand the following terms and concepts in the sample questions on these pages.

Practice Pages

Introduction *revolution:* overthrow of the government

Task *radical:* extreme

Document 1 *peasant:* a member of a class of farmers who are small landowners or agricultural laborers

Document 2 *preservation of:* the keeping of *resistance to oppression:* fight against the loss of freedoms

Document 3 *Bastille:* government prison in Paris

feudalism: a political system in which lords gave land to vassals in exchange for services

Document 2: A Declaration of Rights

1. Men are born and remain free and equal in rights . . .
2. The aim of all political association is the preservation of the natural and [unlimited] rights of man. These rights are liberty, property, security, and resistance to oppression . . .

—*Declaration of the Rights of Man and of the Citizen* (1789)

1 **According to the "Declaration of Rights," how is man born?**

A free

B happy

C unequal

D unhappy

Document 3: The French Revolution—Major Events

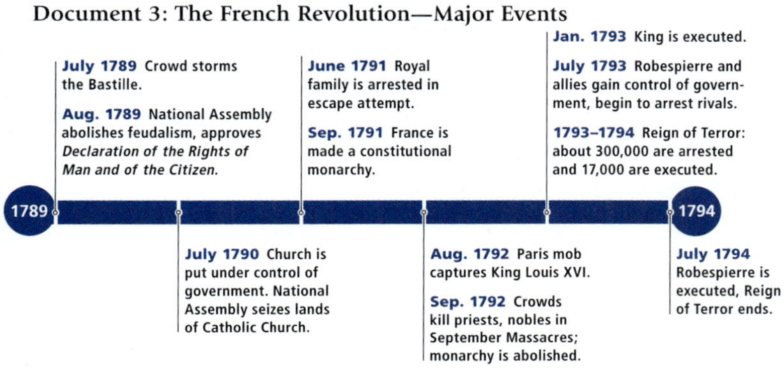

1 **When did the Reign of Terror end?**

F July 1790

G August 1792

H September 1792

J July 1794

SKILL TESTED IN THE ITEMS

STRATEGY ITEMS	
Item Number	**Skill Tested**
Document 1	Making Inferences
Document 2	Summarizing
Document 3	Drawing Conclusions
Part 2: Essay	Analyzing Causes and Effects Synthesizing

PRACTICE QUESTIONS

Thinking It Through

Share the following explanations with students as they discuss the strategies they used to answer the practice questions.

1. The key word in this question is *born.* Reread the document, paying special attention to #1. The correct answer is (A).
2. To answer this question correctly, read each entry on the timeline, looking for "Reign of Terror". The correct answer is (J).

1 Focus & Motivate

Objective

- Identify the different parts of a map and use them to locate information on maps.

Springboard Explain that maps are tools that help us study various aspects of geography. Ask students for examples of times they have used maps as tools. *(Possible answers: to plan a trip with their families, to locate an unfamiliar place in their community, to visually give directions to someone, to locate a place they've read about in a book or seen in a TV show or movie)*

2 Teach

Map Basics

Think, Pair, Share

- How has mapmaking changed over time? *(Photographs from airplanes and satellites have been added to data collected on land and sea.)*
- Why is every place on Earth located in two hemispheres? *(because its location can be stated in the northern or the southern hemisphere and in the eastern or the western hemisphere)*
- What are the advantages and disadvantages of a map compared to a globe? *(Possible answer: A map is flat and can be carried more easily than a globe. A globe is more accurate because it is a sphere like the Earth.)*
- **Compare** What are the similarities and differences in the two maps of Africa? *(Possible answer: Both show the shape of Africa and roughly the same areas of different land use. The map on A1 provides more detailed information about land use and resources as well as political boundaries and the names of different countries and cities.)*

Map Basics

Maps are an important tool for studying the use of space on Earth. This handbook covers the basic map skills and information that geographers rely on as they investigate the world—and the skills you will need as you study geography.

Mapmaking depends on surveying, or measuring and recording the features of Earth's surface. Until recently, this could be undertaken only on land or sea. Today, aerial photography and satellite imaging are the most popular ways to gather data.

Satellite Imagery A satellite image is a photograph taken from a satellite. This image shows Africa. The color of the land indicates whether it is desert, forest, farmland, or mountains.

Reading a Map Most maps have these parts, which help you to read and understand the information presented.

- **A Title** The title indicates the subject of the map and tells you what information it contains.
- **B Symbols** Symbols may stand for capital cities, economic activities, or natural resources. Check the map legend for more details.
- **C Labels** Labels are words or phrases that name features on the map.
- **D Colors** Colors show a variety of information on a map. The map legend tells what the colors mean.
- **E Lines of Longitude** These are imaginary lines that show distances east or west of the prime meridian.
- **F Lines of Latitude** These are imaginary lines that show distances north or south of the equator.
- **G Compass Rose** The compass rose shows you north (N), south (S), east (E), and west (W) on the map. Sometimes only north is shown.
- **H Locator Globe** The box shows where the area on the map is located in the world.
- **I Scale** A scale compares a unit of length on the map and a unit of distance on Earth.
- **J Legend** A legend, or key, lists and explains the symbols and colors used on the map.

DIFFERENTIATING INSTRUCTION

English Learners

Vocabulary: Word Relationships

Help students understand that on this page *image* and *photograph* are used as synonyms. Photographs are one kind of image. Other examples are drawings and paintings. Draw a word web on the board with *image* in the center. Have students find related words in the text, such as *imagery, imaging, photograph,* and *photography.*

Struggling Readers

Relate Text and Visuals

Have pairs of students match each of the parts of the map as described on this page with the reference on page A1. Note that the visual reference for (H) *Locator Globe* is right above the text description. One partner should read the text describing a map part while the other locates it on the economic activity map. The partner who locates the map part should then describe what that map part shows in his or her own words.

Economic Activity of Africa

Reading a Map This map shows the land use and natural resources that contribute to the economy of Africa. Point out that the thin white lines on the map show borders between the different African countries.

Point to the following items on the map and ask students to identify each one and describe its purpose in their own words.

- Compass rose *(shows the four directions on the map)*
- Scale *(shows how distances on the map relate to distances on Earth)*
- Legend *(shows what the colors and symbols on the map stand for)*

Have students use the compass rose, scale, and legend to answer the following questions.

- If you traveled from Ethiopia to Kenya what direction would you be traveling? *(south)*
- About how far is it from Johannesburg to Cape Town in South Africa? *(about 750 miles)*
- How is most of the land in northern Africa used? *(nomadic herding and limited agriculture)*
- **Make Inferences** Why do you think the land in this region is used in this way? Hint: Look at the satellite image for clues. *(Most of the land is desert.)*

CONNECT to Language Arts

Write a Paragraph

Assign students one country or region of Africa to examine more closely or allow students to choose an area. Suggest that they use the map to help them imagine what it would be like to live in that area. Remind them that economic activity refers to the way people earn a living. Then ask them to write a paragraph about life in the region based on their inferences from reading the map.

DIFFERENTIATING INSTRUCTION

Inclusion

Read Maps in Pairs

Pair students with visual impairments with a partner without impairments. Direct the students without impairments to describe the legend to their partners. Have partners work together to locate the different colors and symbols on the map and discuss the economic activity of different regions of Africa. Encourage the students with visual impairments to create mental maps of the various regions.

Gifted & Talented

Write a Dialogue

Encourage students to think about the differences between mapmakers who relied only on data collected by land and sea and those who now have satellite imagery to help them make maps. Invite students to work in pairs to write an imaginary dialogue between mapmakers from two different time periods in which they describe the process of making maps. Allow students to present their dialogues in small groups or to the class.

The Geographic Grid

Reading a Map Explain that latitude and longitude lines together form a grid that can be used to specify any point on the Earth's surface. Latitude lines are called parallels because they are always an equal distance apart. Longitude lines meet at the North and South poles.

- Where is 0° latitude? *(at the equator)*
- What is 0° longitude? *(the prime meridian)*

Large Scale: Washington, D.C.

Reading a Map Point out that the entire area shown on this map is only about 3 miles by 2 miles.

- About how far is it from the Lincoln Memorial to the Washington Monument? *(3/4 of a mile)*
- Why is a large scale commonly used on city maps? *(Possible answer: because they can cover a small area in great detail)*

Small Scale: Eastern United States

Reading a Map Explain that the use of different scales allows mapmakers to show vastly different areas in the same space on paper. This map shows an area approximately 1,500 miles by 1,000 miles.

- About how far is it between the cities that are farthest east and farthest west on the map? *(about 750 miles between Boston and Chicago)*
- How might both maps be useful if you were planning a trip from Atlanta to Washington, D.C.? *(Possible answer: The small scale map would help you determine the distance between the two cities and the large scale map would help you find specific sites once you reached Washington, D.C.)*

The Geographic Grid

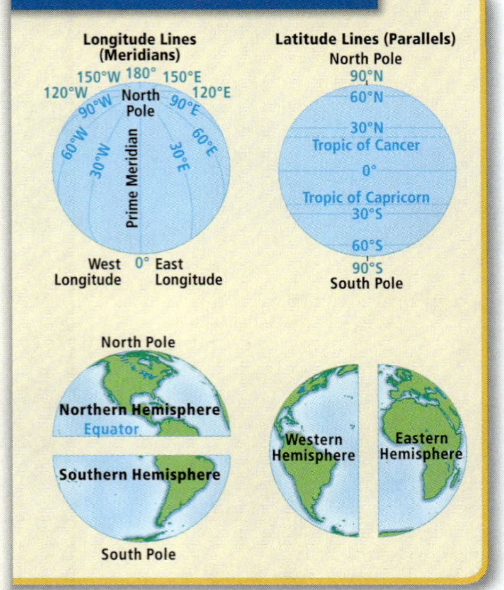

Longitude and Latitude Lines These lines appear together on a map and pinpoint the absolute locations of cities and other geographic features. You express these locations as coordinates of intersecting lines. These are measured in degrees.

Longitude Lines These imaginary lines run north and south; they are also known as meridians. They show distances in degrees east or west of the prime meridian. The prime meridian is a longitude line that runs from the North Pole to the South Pole through Greenwich, England. It marks 0° longitude.

Latitude Lines These imaginary lines run east to west around the globe; they are also known as parallels. They show distances in degrees north or south of the equator. The equator is a latitude line that circles Earth halfway between the North and South poles. It marks 0° latitude. The tropics of Cancer and Capricorn are parallels that form the boundaries of the tropics, a region that does not have a change of seasons.

Hemisphere The word *hemisphere* is a term for half the globe. The globe can be divided into northern and southern hemispheres (separated by the equator) or into eastern and western hemispheres. The United States is located in the northern and western hemispheres.

Scale What scale to use depends on how much detail is to be shown. If the area is small and many details are needed, a large scale is used. If the area is large and fewer details are needed, a small scale is used.

Large Scale: Washington, D.C.

Small Scale: Eastern United States

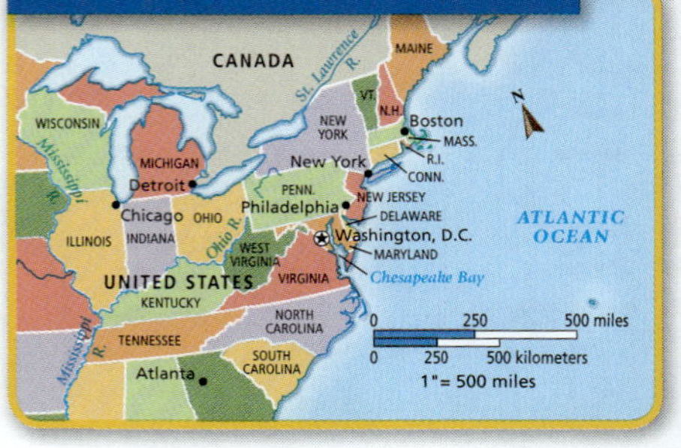

A2

DIFFERENTIATING INSTRUCTION: TIERED ACTIVITIES

OBJECTIVE Use hemispheres, latitude, and longitude in a geography game.

Basic

Have teams of five choose five cities. Direct each team to use a globe to fill in the blanks for each city: This city is in the ___ and ___ hemispheres. It is north of ___ and south of ___ (fill in latitudes). It is east of ___ and west of ___ (fill in longitudes). Have teams take turns identifying one another's city.

On Level

Create teams of five. Provide each team with a world atlas or map. Have each team choose five cities. Have the team members work together to write clues about each city in terms of its relative location to two other cities. Continue as in the Basic activity.

Challenge

Provide teams of five with a world map. Have each team choose five cities. Have the team members work together to write step-by-step clues for each city that help the other teams find each city's absolute location. (For example, "This city is east of Casablanca and west of Cairo.") Continue as in the Basic activity.

Projections A projection is a way of showing the curved surface of Earth on a flat map. Flat maps cannot show sizes, shapes, and directions with total accuracy. As a result, all projections distort some aspect of Earth's surface. Below are four projections.

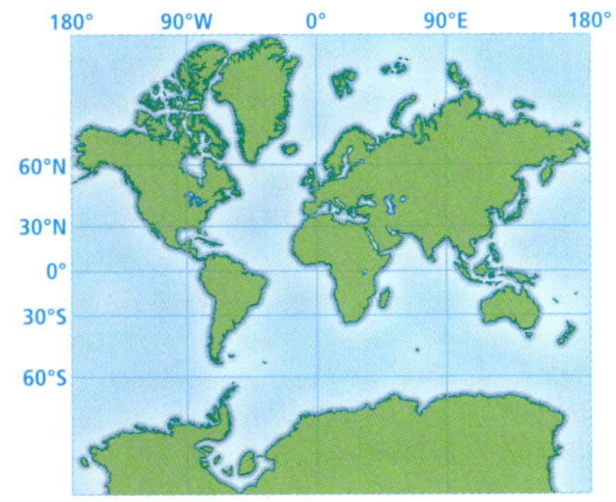

Mercator Projection The Mercator projection shows most of the continents as they look on a globe. However, the projection stretches out the lands near the North and South poles. The Mercator projection is used for navigation.

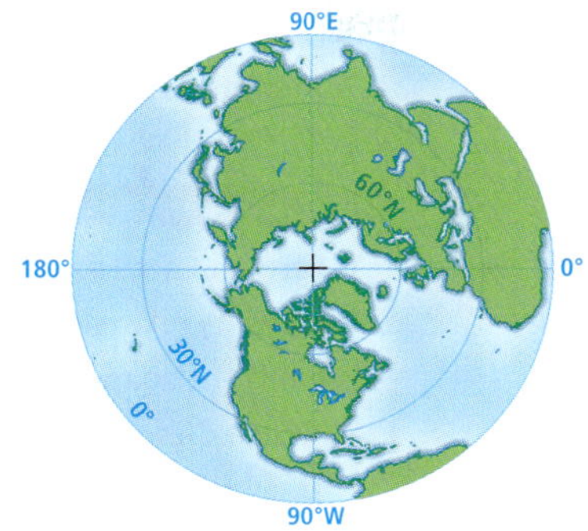

Azimuthal Projection An azimuthal projection shows Earth so that a straight line from the central point to any other point on the map corresponds to the shortest distance between the two points. Sizes and shapes of the continents are distorted.

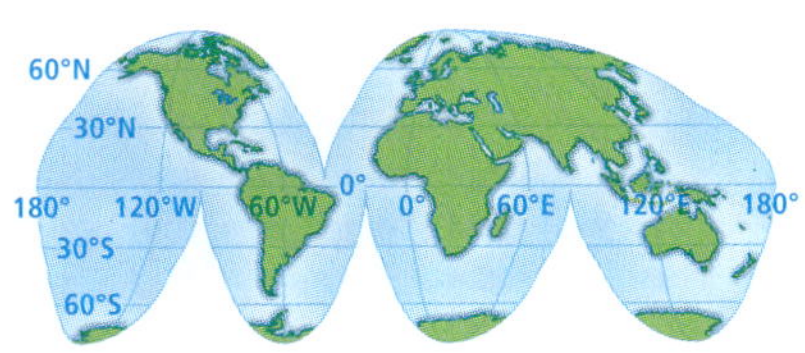

Homolosine Projection This projection shows landmasses' shapes and sizes accurately, but distances are not correct.

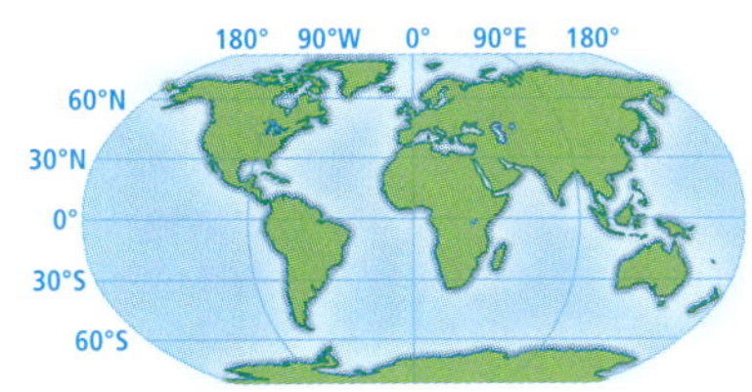

Robinson Projection For textbook maps, the Robinson projection is commonly used. It shows the entire Earth, with continents and oceans having nearly their true sizes and shapes. However, the landmasses near the poles appear flattened.

Map Practice

Use pages A1–A3 to help you answer these questions.

MAIN IDEA

1. What information is provided by the legend on page A1?
2. What do latitude and longitude lines show?

CRITICAL THINKING

3. Why do you think latitude and longitude are important to sailors?
4. How do the depictions of Antarctica in the Mercator and Robinson projections compare?

MAP PRACTICE ANSWERS

MAIN IDEA

1. land use and resources
2. Latitude lines show distances north and south of the equator; longitude lines show distances east and west of the prime meridian.

CRITICAL THINKING

3. Possible answer: Sailors rely on longitude and latitude because they have no visual landmarks to use.
4. On the Mercator projection, Antarctica looks stretched out and large; on the Robinson projection, Antarctica looks thin and narrow.

More About . . .

Azimuthal Projections

Only a globe can accurately represent the shape of the Earth and relationships among the Earth's surface features. An azimuthal projection depicts a part of the Earth as a flattened disk viewed from a certain point on Earth. For example, in a polar projection, such as the map shown here, the center point is the North Pole.

More About . . .

Mercator Projections

Flemish geographer Gerhardus Mercator created the Mercator projection in 1569. He was the first cartographer to create a navigational chart that included a grid in which lines of latitude and longitude are straight and cross at right angles. In a Mercator projection, parallels of latitude, which are equally spaced on a globe, are drawn farther and farther apart as their distances from the equator increases. This preserves the shapes of land areas, but distorts their sizes.

3 Wrap-Up

Assess Have students answer the Map Practice questions.

Reteach

Apply Show an overhead transparency of a map of a continent other than Africa. Read the terms describing the parts of a map and call on volunteers to identify each one on the map. Ask students to locate the farthest ranges of latitude and longitude shown on the map. In which hemispheres is this continent located? *(Answers will vary depending on the map shown, but should reflect an understanding of how latitude and longitude are related to the hemispheres.)*

1 Focus & Motivate

Objective

- Recognize and read different types of maps.

Springboard There are different kinds of maps for different purposes. What are some possible purposes of maps? *(Possible answers: to show travel routes, population changes, political divisions, physical features, historical landmarks, and events)*

2 Teach

Different Types of Maps

Reader, Recorder, Reporter

- How do physical, political, and thematic maps differ? *(Physical show physical features; political show political divisions; thematic: show information on a specific topic.)*
- How is color used differently in physical and political maps? *(Physical maps show different elevations, and political maps show different political divisions.)*

Eurasia: Physical

Reading a Map Explain that color on this map shows elevation. Point out that rivers usually run from higher to lower elevations and empty into a larger body of water. Direct students to refer to the map to answer the Think Like a Geographer questions.

THINK LIKE A GEOGRAPHER ANSWERS

Possible answers:

- between about 40°E and 90°E longitude and between about 55°N and 35°N latitude
- east of Eastern Europe and north of Southwest Asia
- a small strip between the Black and Caspian Seas and a rectangle east of the Caspian Sea
- north and south; transportation is easiest in the directions the rivers flow
- Caucasus, Ural, and Pamir mountains; Kara Kum and Kyzyl Kum deserts. These areas may be difficult to live in and provide a natural barrier that limits travel or invasion.

Different Types of Maps

Physical Maps Physical maps help you see the landforms and bodies of water in specific areas. By studying a physical map, you can learn the relative locations and characteristics of places in a region.

On a physical map, color, shading, or contour lines are used to show elevations or altitudes, also called relief.

THINK LIKE A GEOGRAPHER

- Where on Earth's surface is this area located?
- What is its relative location?
- What is the shape of the region?
- In which directions do the rivers flow? How might the directions of flow affect transportation in the region?
- Are there mountains or deserts? How might they affect the people living in the area?

Eurasia: Physical

DIFFERENTIATING INSTRUCTION

English Learners

Vocabulary: Multiple Meaning Words

Explain that some words used to describe different types of maps on pages A4 and A5 have more than one meaning. For example, *physical* can mean "relating to material things" or "relating to the body." Write both definitions on the board, and ask which definition applies. Follow a similar procedure with the words *relative, relief,* and *political.*

Inclusion

Compare Two Maps

Help students understand that the maps on pages A4 and A5 show the same region. Ask students to point out common features that can help orient them, such as the Black Sea and the same degrees of latitude and longitude. Make overhead transparencies of the maps and place one on top of the other. Then have students look again at the two maps individually. They should notice that physical features cross political boundaries.

Political Maps Political maps show features that humans have created on Earth's surface. Cities, states, provinces, territories, and countries are included on a political map.

THINK LIKE A GEOGRAPHER

- Where on Earth's surface is this area located?
- What is its relative location? How might a country's location affect its economy and its relationships with other countries?
- What is the shape and size of the country? How might its shape and size affect the people living in the country?
- Who are the region's, country's, state's, or city's neighbors?
- How populated does the area seem to be? How might that affect activities there?

Eurasia: Political

Eurasia: Political

Reading a Map Point out that color on this map is used to easily see the size and boundaries of the different countries.

- What is the capital of Uzbekistan? *(Tashkent)*
- What physical features are included on this map? *(bodies of water)*
- **Make Inferences** Why do you think they are included? *(Possible answers: They are important resources that affect life in different countries. They help describe the countries' relative locations.)*

Have students refer to the map to answer the Think Like a Geographer questions.

THINK LIKE A GEOGRAPHER ANSWERS

Possible answers:

- between about 40°E and 90°E longitude and between about 55°N and 35°N latitude
- east of Eastern Europe and north of Southwest Asia; location influences trading partners, allies, and potential enemies
- Most countries in the region are small and may be crowded or have few resources. Kazakhstan is large and may be harder to govern.
- The region's neighbors include Russia, Turkey, Iran, Afghanistan, and China.
- There are few cities shown on the map, which suggests smaller populations that may live in rural areas.

CONNECT to Math

Graph Populations

Have students use classroom resources to find information about the population of each of the countries shown on the map. Direct them to use the information to create bar graphs and then compare the size of the population of the countries in the region. Remind them to give their graphs a title and include labels as needed. Invite volunteers to share their graphs with the class and explain which countries have the largest and smallest populations. Discuss how the bar graph added to their knowledge of Eurasia.

DIFFERENTIATING INSTRUCTION: TIERED ACTIVITIES

OBJECTIVE Identify and analyze physical and political maps.

Basic

Assign pairs of students to find an example of either a physical or a political map in their textbooks. Have partners identify the different parts of the map and label them on a map they create. They may use the map on page A1 as a model. Ask students to write a caption explaining what kind of map it is and what it shows.

On Level

Have students find an example of both a physical and political map of the same state, country, or region. They may use their textbooks or other classroom resources. Have students complete a Venn diagram that shows the similarities and differences in the two maps. Allow students to share their Venn diagrams.

Challenge

Direct students to locate examples of both a physical and a political map of Africa. Have them compare it to the economic activity map on page A1. Ask them to write one or two paragraphs summarizing how the two additional maps add to their understanding of the economic activity of Africa.

Major Earthquakes in Japan

Reading a Map Point out that the title of the map is "Major Earthquakes in Japan *1986–2005*." Even the smallest dot shown represents a strong earthquake. An earthquake with a magnitude above 6.0 can cause much damage if it occurs in or near a city. An earthquake with a magnitude of 7.8 is ten times stronger than one with a magnitude of 6.8.

- What is the magnitude of the Kobe earthquake? *(6.8 to 7.2)*
- **Draw Conclusions** Why do you think this particular earthquake is labeled and dated? *(Possible answer: because it occurred in a large city and caused much damage)*

Have students refer to the maps to answer the Think Like a Geographer questions.

THINK LIKE A GEOGRAPHER ANSWERS

Possible answers:

- It suggests that the map is about the biggest earthquakes in Japan over a 20-year period.
- They show that they occurred in the Pacific Ocean off the northeast coast.

North Sea Oil and Gas Fields

Reading a Map Point out that the gray lines and labels show where each country's area of control is located in the North Sea.

- Which country controls no oil fields? *(Germany)*
- Which country controls the largest amount of oil and gas in the region? *(United Kingdom)*

THINK LIKE A GEOGRAPHER ANSWERS

Possible answers:

- They both indicate that the theme is the location of oil and gas fields.
- Oil and natural gas are a source of energy for the countries and could also be a resource they export.

Thematic Maps Geographers also rely on thematic maps, which focus on specific topics. For example, in this textbook you will see the thematic maps that show climates, types of vegetation, natural resources, population densities, and economic activities. Some thematic maps show historical trends; others focus on movements of people or ideas. Thematic maps may be presented in a variety of ways.

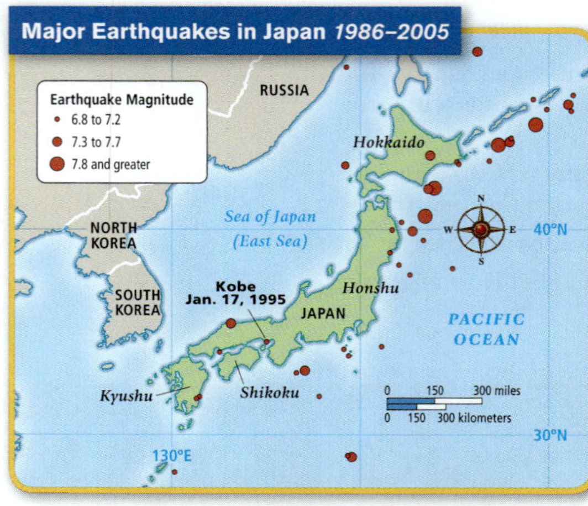

Proportional Symbol Map On a proportional symbol map, the size of each symbol indicates the quantity of something at a specific location. The map shown here depicts the magnitude of earthquakes that occurred throughout Japan.

THINK LIKE A GEOGRAPHER

- What does the map title suggest about the theme of the map?
- What do the symbols tell you about where the largest magnitude earthquakes occurred?

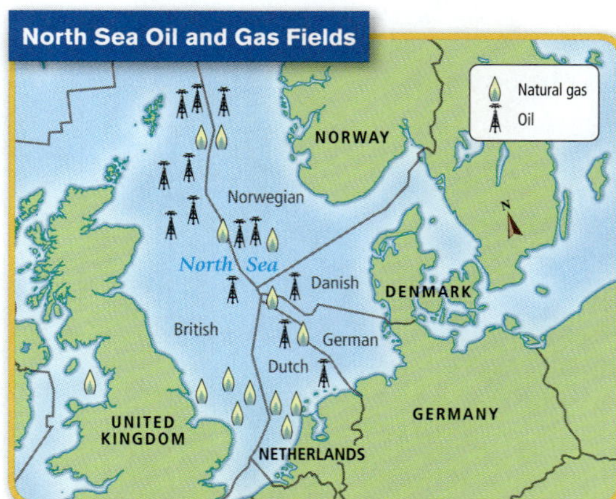

Point-Symbol Map On a point-symbol map, a symbol is used to show the location of a particular feature. This map shows the locations of oil and gas fields in the North Sea.

THINK LIKE A GEOGRAPHER

- How do the map title and legend help you know the theme of the map?
- How might the presence of oil and gas fields affect the countries located near them?

A6

DIFFERENTIATING INSTRUCTION

Struggling Readers

Read in Pairs

Some of the terminology used to describe thematic maps may be unfamiliar or difficult for students to comprehend. Have students work in pairs to read page A6. Suggest that partners take turns reading paragraphs aloud and identify any unfamiliar words or phrases. Encourage them to figure out the meaning from context clues and by using a dictionary. Ask volunteers to summarize each paragraph in their own words.

Gifted & Talented

Create a Display of Thematic Maps

Invite interested students to find additional examples of thematic maps, such as the kinds described in the first paragraph on page A6. Encourage them to use their textbooks or other classroom resources. Allow students to use their examples to create a bulletin board display about thematic maps. Students may then serve as "experts" to present the maps to their classmates and answer any questions.

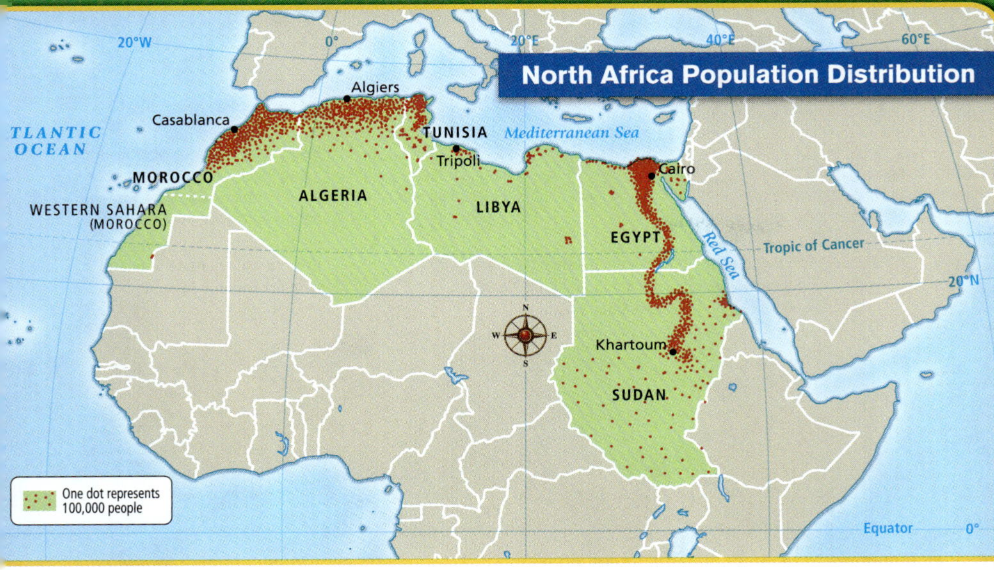

Dot Maps These kinds of maps use dots to show how something is distributed in a certain area. This map shows how the population of North Africa is distributed.

THINK LIKE A GEOGRAPHER

- How do the title and legend of the map identify the data being presented?
- What does each dot on the map represent?
- How does this kind of map help you to see how densely populated different parts of a region are?

Map Practice

Use pages A4–A7 to help you answer these questions.

MAIN IDEA

1. Which desert is located in Turkmenistan?
2. Baku is the capital of which Eurasian republic?

CRITICAL THINKING

3. What country in North Africa is the least densely populated?
4. How are thematic maps an effective way of presenting information?

MAP PRACTICE ANSWERS

MAIN IDEA

1. Kara Kum Desert
2. Azerbaijan

CRITICAL THINKING

3. Libya
4. Possible answer: Thematic maps focus on specific information.

North Africa Population Distribution

Reading a Map Explain that the lack of dots in an area does not mean there are no people there, only that the population is less than 100,000 people.

- What does the map show about the population of Sudan? *(Possible answer: The county has a large population. There is a strip of very densely populated area along the Nile and many other areas that have 100,000 people.)*
- **Categorize** What do the most densely populated areas on the map have in common? *(Possible answer: They are all located near water, either near the coast or along the Nile River.)*

Have students refer to the map to answer the Think Like a Geographer questions.

THINK LIKE A GEOGRAPHER ANSWERS

Possible answers:

- They show that the map is about where population is concentrated in northern Africa.
- 100,000 people
- The closer the dots are to each other, the denser the population.

3 Wrap-Up

Assess Have students answer the Map Practice questions.

Reteach

Draw Conclusions Organize students into three groups. Assign each group to review one of the three main types of maps discussed in the section. Encourage group members to list the main ideas about their topic. Call on volunteers to share their group's main ideas with the rest of the class. Clarify understanding as needed.

1 Focus & Motivate

Objective

- Recognize, understand, and use many geographical terms.

Springboard Like many other fields of study, geography has its own vocabulary. Some of these terms are common and easy to understand, such as *island*. Others are less common, such as *cataract*. This dictionary focuses on land and waterways.

- With their books closed, ask students to brainstorm a list of geographical terms for landforms and waterways. You might get them started by suggesting some common terms, such as *mountain* or *ocean*. *(Possible answers: canyon, cliff, delta)*

2 Teach

Geographic Dictionary

Think, Pair, Share

- **Compare and Contrast** How are buttes and mesas similar and different? *(Possible answers: Both are raised, flat areas of land with steep cliffs, but mesas are larger.)* What do swamps and marshes have in common? *(Possible answer: Both are watery areas.)*
- **Make Inferences** Why would people want to live near floodplains? Why wouldn't they want to live there? *(Possible answers: Deposits made by the river are often good for farming. It could be dangerous to live there because it may flood easily.)*
- What feature or features are most noticeable where you live? *(Accept answers describing landforms and waterways in your area.)*

More About . . .

Delta

The ancient Greeks first used the term *delta* to describe the triangular deposit of land at the mouth of the Nile River because its shape resembled the triangle-shaped letter *delta* of their alphabet. Both the delta and the river valley gave rise to the ancient civilization of Egypt. The Nile Delta stretches about 100 miles from Cairo to the Mediterranean and is about 120 miles wide along its outer edge.

Geographic Dictionary

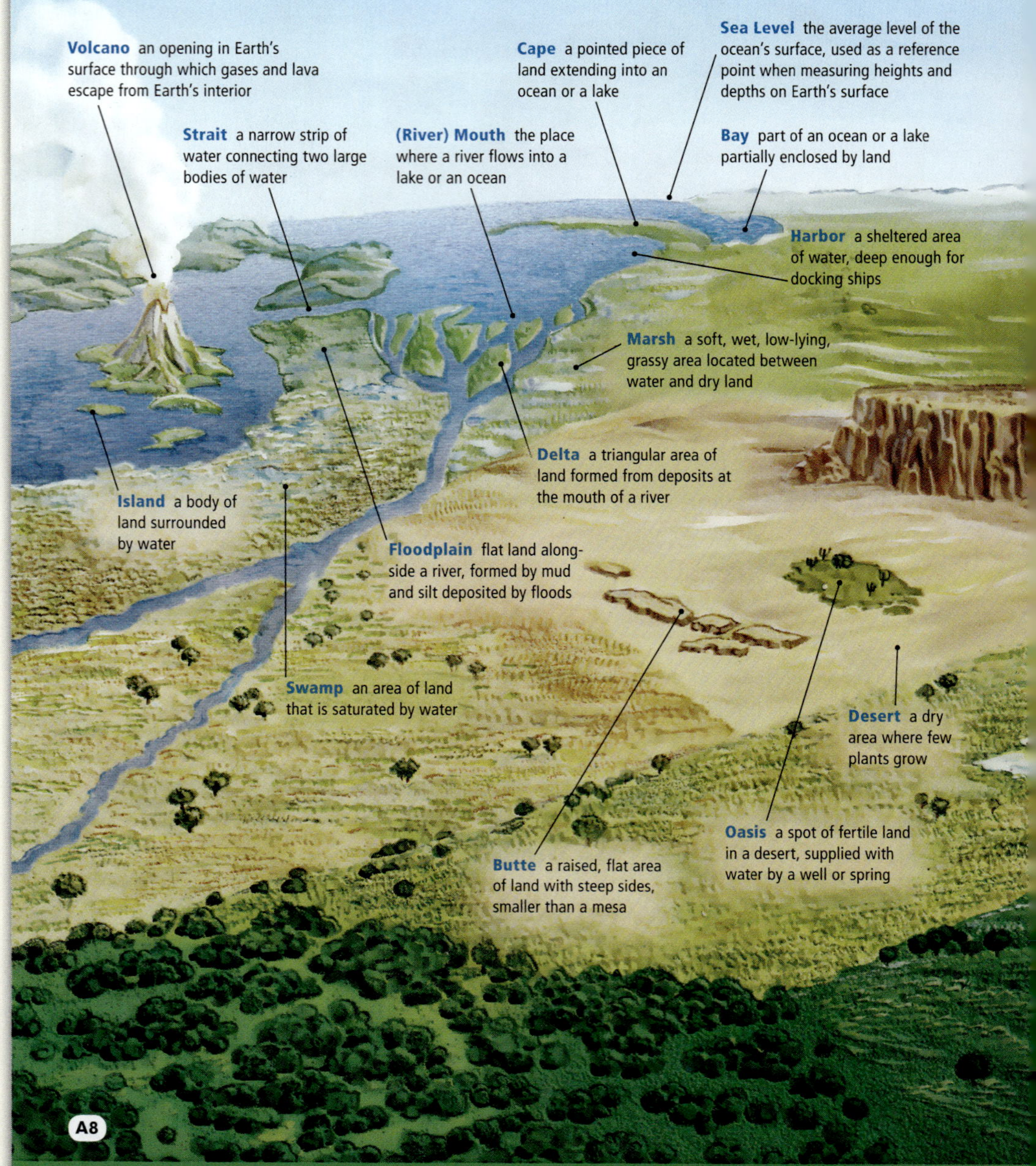

DIFFERENTIATING INSTRUCTION

Struggling Readers

Use Prior Knowledge and Visual Clues

First, ask students to scan the visuals on the pages to see if they recognize any landforms. Next, direct them to scan for any terms that look familiar. Have them match the term to the picture. Finally, have them identify terms that are not familiar. Suggest that they create a vocabulary card for each of these terms that include the term, its definition, and an illustration.

Inclusion

Create an Audio Dictionary

Allow students who learn best orally to record an audio dictionary of the geographic terms on pages A8 and A9. Have students work in pairs to read the terms and definitions into an audiotape or digital recorder. Then encourage them to play back the recording and pause after each term to find the related visual in the textbook. Encourage students to check their understanding by quizzing one another.

Comparing Regions

Deserts in Africa and Eurasia

The Sahara covers more than 3.5 million square miles of northern Africa and is the largest desert in the world. Rainfall varies by region from 1 to 5 inches per year. The Sahara is a hot desert with temperatures reaching as high as 136° F. About two million people live in the desert, mostly as farmers or nomadic herders.

The deserts of Eurasia are less than one-tenth the size of the Sahara, but they cover large areas of the region. The Kara Kum *(black sand)* Desert covers about three-quarters of Turkmenistan and the Kyzyl Kum *(red sand)* Desert occupies much of Uzbekistan.

3 Wrap-Up

Assess Use the following questions to check student understanding of geographical terms. Have students write their answers.

- How is a canyon related to a valley? *(A canyon is a deep, narrow valley.)*
- What are the similarities and differences between a floodplain and a delta? *(Both are formed by deposits from a river; a floodplain is flat land along the side of the river and a delta is a triangular area at the mouth of a river.)*
- Why might a desert community form around an oasis? *(An oasis is a rare source of water in a very dry region.)*

Reteach

Apply Assign each student to create two or three cards for a matching game. Have them write a geographic term on an index card and its definition on another card of a different color. Collect all the cards. Spread out the term cards face up. Place the definition cards face down in a pile. Organize students into two teams. Allow teams to take turns choosing a definition card and trying to match it with the correct term. The other team may challenge if they think an incorrect match has been made.

DIFFERENTIATING INSTRUCTION: TIERED ACTIVITIES

OBJECTIVE Expand geographic vocabulary by adding to the Geographic Dictionary.

Basic

Assign pairs of students three or four terms from the Geographic Dictionary. Provide index cards to each pair. Instruct each pair of students to write their assigned term and the definition on one side of the card. Have them illustrate the term on the other side.

On Level

To extend the Basic activity, assign pairs of students the following list: *archipelago, dune, estuary, fjord, isthmus, hill, levee,* and *peninsula.* Have each pair find a definition of each term. Instruct them to write the definition on one side of an index card and illustrate the term on the other side.

Challenge

Have pairs of students brainstorm a list of geographical terms not included in the Geographic Dictionary. Encourage students to go beyond the obvious (such as *ocean* or *continent*). Have them use resources to find the meaning of each term, then complete the activity as directed in the On Level activity.

1 Focus & Motivate

Objective

- Recognize and interpret different types of graphs.

Springboard Explain that geographers use graphs to present data about geographic topics in visual forms. Ask students to look at each of the graphs shown on page A10 and to state what kind of information each type of graph presents. *(Possible answers: bar graph—compares amount of nuclear energy produced; line graph—how population changed over time; pie graph—what part of exports went to different countries; climograph—precipitation and temperature changes by month)*

2 Teach

Graphs

Reader, Recorder, Reporter

- What do all the countries shown on the bar graph have in common? *(They all produce at least 40 percent of their energy from nuclear power.)*
- How long did it take the Aztec and Incan populations to decline to less than 5 million each? *(about 50 years)*
- Which Asian country is Japan's largest trading partner? *(China)*
- **Compare** How does the climate in Dakar in January compare to the climate in August? *(January is warm and dry; August is hotter and rainy.)*

3 Wrap-Up

Assess Use the following questions to check student understanding of graphs. Have students write their responses.

- How does France's production of nuclear energy compare to Slovenia's? *(France produces about twice as much of its energy from nuclear power.)*
- Why does the climograph use two different vertical axes? *(because it is graphing two different variables)*
- What type of graph is used to show the parts of a whole? *(pie graph)*

Reteach Call on volunteers to describe why each kind of graph was used to show a particular type of geographic information.

Graphs

Graphs are often used to show numbers in visual form. Bar graphs make it easy to compare numbers. Line graphs show changes over time. Pie graphs show the relationship among parts of a whole. Climographs show the monthly average temperature and precipitation in a place.

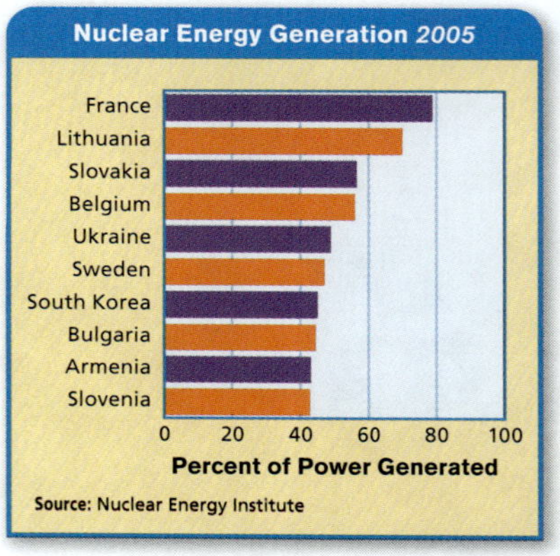

Bar Graph This bar graph compares ten countries' use of nuclear power to produce energy. The graph shows what percent of the energy each country used was from nuclear power.

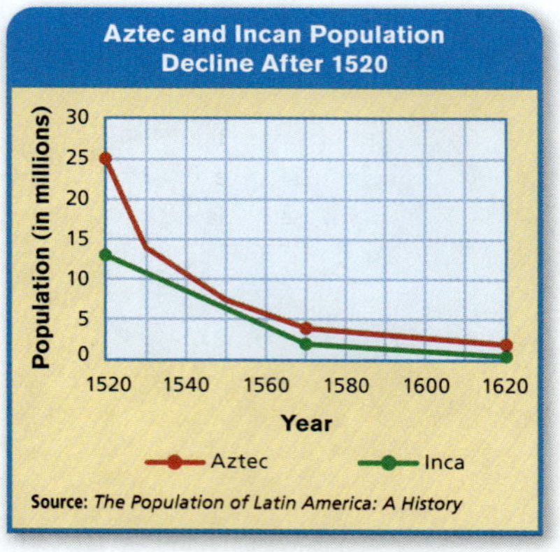

Line Graph This line graph shows how the population of the Aztec and Inca changed between 1520 and 1620.

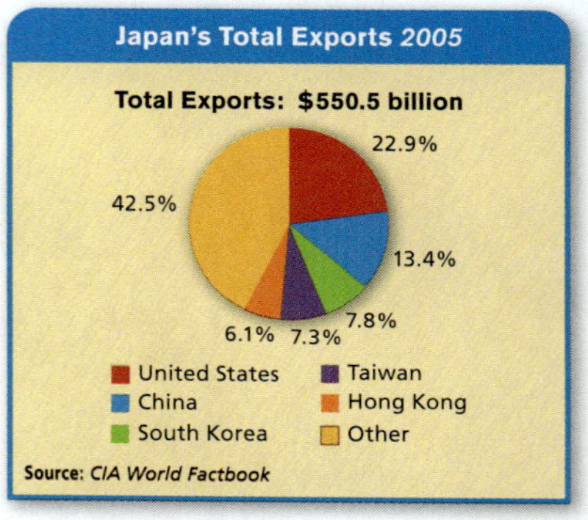

Pie Graph This pie graph shows how Japan's total exports are distributed. The graph shows what percentages of all the exports are sent to specific countries.

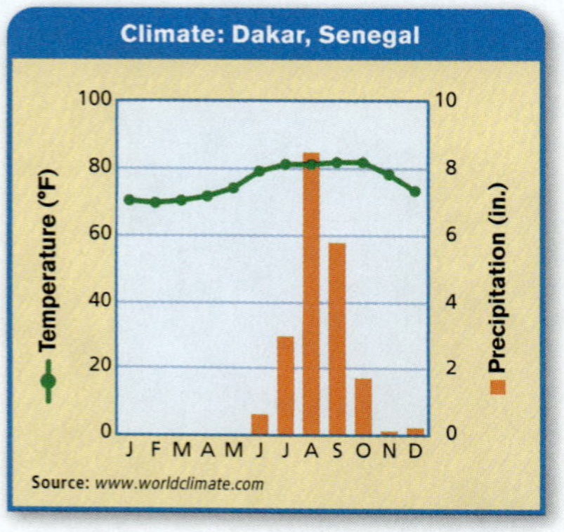

Climograph This climograph shows the climate of Dakar, Senegal. Bars represent the average monthly precipitation, and a line represents the average monthly temperature in the city.

DIFFERENTIATING INSTRUCTION

English Learners

Key Academic Vocabulary

Explain that the word *percent* means "per 100." Point out that the % symbol means the same thing as the word. Ask students to identify the two graphs that show percent *(bar graph and pie graph)*. Model sentences that can be used to describe the information on the graphs. For example, "France gets 80 percent of its energy from nuclear power." Have students work in pairs to practice similar statements.

Inclusion

Enlarge the Graphs

Students with visual impairments may benefit from working with larger versions of the graphs, which are available in the eEdition at **ClassZone.com**. Have students work in pairs to interpret the information shown on each of the graphs. Encourage them to take turns quizzing one another about the data from the graphs.

Geography Standards

To help you study the world, geographers have set up 18 geography standards. These standards provide guidelines for what you will need to be informed about the people, places, and environments of Earth.

THE WORLD IN SPATIAL TERMS

1. How to use maps and other tools
2. How to use mental maps to organize information
3. How to analyze the spatial organization of people, places, and environments

PLACES AND REGIONS

4. The physical and human characteristics of places
5. How people create regions to interpret Earth
6. How culture and experience influence people's perceptions of places and regions

PHYSICAL SYSTEMS

7. The physical processes that shape Earth's surface
8. The distribution of ecosystems on Earth

HUMAN SYSTEMS

9. The characteristics, distribution, and migration of human populations
10. The complexity of Earth's cultural mosaics
11. The patterns and networks of economic interdependence on Earth
12. The patterns of human settlement
13. The forces of cooperation and conflict

ENVIRONMENT AND SOCIETY

14. How human actions modify the physical environment
15. How physical systems affect human systems
16. The distribution and meaning of resources

THE USES OF GEOGRAPHY

17. How to apply geography to interpret the past
18. How to apply geography to interpret the present and plan for the future

DIFFERENTIATING INSTRUCTION

Struggling Readers

Use Jigsaw Reading

Organize students into groups of six and assign each student a number from one to six. Have students form new groups based on their numbers. Assign each new group one of the main categories of standards. Ask students to develop a summary statement that describes why the standards in their assigned category belong together. Have students return to their original groups and present their summaries to the other group members.

Gifted & Talented

Write an Essay

Have students choose a current issue that relates to one or more of the 18 standards and to identify the appropriate standards. The issue may be something that is an issue on a global level, such as climate change, or one that relates to the local community, such as land use for a new shopping center or park. Have students write a brief essay in which they explain how the current issue applies to a particular standard.

1 Focus & Motivate

Objective

- Explain how the geography standards serve as guidelines for learning about the people, places, and environment of Earth.

Springboard Ask students to study the three small photographs and choose one of the broad categories of standards that might apply to each. *(Possible answers: top—world in spatial terms; middle—places and regions; bottom—environment and society)*

2 Teach

Geography Standards

Talk About It

- Why do so many standards relate to human activity? *(Possible answers: because people are the ones who study geography, and human actions have the most impact on changing the environment)*
- How does the creation of regions help people interpret the Earth? *(Possible answer: It allows people to see the similarities that exist over a wide area.)*
- What are some of the physical and human characteristics of your community, state, and region? *(Answers will vary, but should reflect an understanding of what distinguishes each of these places.)*

3 Wrap-Up

Assess Use the following questions to check student understanding of geography standards. Have students write their responses.

- What are two ways humans interact with their environment? *(Humans are affected by their environment and also modify it to meet their needs.)*
- Why is it important to learn to use maps? *(Maps are important tools for understanding the spatial organization of people, places, and environment.)*

Reteach

Apply Write each of the standards on an index card. Write the six main categories of standards on the board. Call on volunteers to choose an index card and place it in the correct category. Encourage students to look for key words in the standards that can help them identify the category.

RAND McNALLY

World Atlas

CONTENTS

A12

Legend

Complete Legend for Physical and Political Maps

Water Features

Symbol	Meaning
ATLANTIC OCEAN	Ocean or sea
	Lake
	Salt lake
	Seasonal lake
Mississippi	River
Niagara Falls	Waterfall
	Ice pack

Land Features

Symbol	Meaning
Mt. Mitchell 6,684 ft. 2,037 m. △	Mountain peak
Mt. McKinley 20,320 ft. 6,194 m. ▲	Highest mountain peak
Great Basin	Physical feature (mountain range, desert, plateau, etc.)
Nantucket Island	Island

Cultural Features

Symbol	Meaning
(thick red line)	International boundary
(thin red line)	State boundary
CANADA	Country
KANSAS	State

Population Centers

National capital	State capital	Town	Population
✪	⊛	■	Over 1,000,000
✪	⊛	▣	250,000 – 1,000,000
✪	⊛	•	Under 250,000

Land Elevations and Ocean Depths

Land elevation

3,000 meters	9,840 feet
2,000 meters	6,560 feet
1,000 meters	3,280 feet
500 meters	1,640 feet
200 meters	656 feet
0 Sea level	0 Sea level

Water depth

0 Sea level	0 Sea level
200 meters	656 feet
2,000 meters	6,560 feet

RAND McNALLY

A13

World Physical
National capital
Major city
International boundary
Land elevation
3,000 meters
2,000 meters
1,000 meters
500 meters
200 meters
0 Sea level
9,840 feet
6,560 feet
3,280 feet
1,640 feet
656 feet
0 Sea level
Water depth
0 Sea level
200 meters
2,000 meters
0 Sea level
656 feet
6,560 feet
0 1000 2000 Miles
0 1000 2000 3000 Kilometers
180° 165° 150° 135° 120° 105° 90° 75° 60° 45° 30° 15° 0°
75° 60° 45° 30° 15° 0° 15° 30° 45° 60°
ARCTIC OCEAN
Arctic Circle
Tropic of Cancer
Equator
Tropic of Capricorn
Antarctic Circle
PACIFIC
OCEAN
ATLANTIC
OCEAN
NORTH
AMERICA
SOUTH
AMERICA
N
S
E
W
Greenland
Baffin Island
Baffin Bay
Hudson Bay
Jan Mayen
Iceland
Faroe Is.
British Isles
London
Mt. McKinley
20,320 ft.
6,194 m.
Yukon
Mackenzie
Canadian Shield
Aleutian Islands
Vancouver
Rocky Mountains
Great Plains
Newfoundland
St. Lawrence
Appalachian Mts.
New York
Cape Hatteras
Los Angeles
Colorado
Mississippi
Azores
Iberian Peninsula
Atlas
Canary Islands
Midway Is.
Baja California
Gulf of Mexico
Hawai'ian Islands
Yucatan Peninsula
Cuba
Hispaniola
Puerto Rico
Jamaica
Caribbean Sea
Cape Verde Islands
Cap Vert
Niger
A
Trinidad
Orinoco
Palmyra Atoll
Kiribati
Line Islands
Galapagos Islands
Amazon Basin
Amazon
Cabo de São Roque
Andes
Planalto do Mato Grosso
Marquesas Is.
Samoa Islands
Society Islands
Tonga Islands
Cook Islands
St. Helena
Rio de Janeiro
Paraná
Cerro Aconcagua
22,831 ft.
6,959 m.
Buenos Aires
Chatham Is.
Archipiélago de los Chonos
Patagonia
Falkland Is.
South Georgia
Tierra del Fuego
Cape Horn
South Sandwich Is.
South Orkney Is.
South Shetland Is.
Antarctic Peninsula
Weddell Sea
Marie Byrd Land
Vinson Massif
16,066 ft.
4,897 m.

A14

15° 30° 45° 60° 75° 90° 105° 120° 135° 150° 165° 180°
ARCTIC OCEAN
Spitsbergen
Franz Josef Land
Nordkapp
Novaya Zemlya
Scandinavia
North Sea
EUROPE
Moscow
Volga
Don
Ural Mts.
Ob
Yenisey
Lena
Siberia
Sea of Okhotsk
Bering Sea
Kamchatka Peninsula
Sakhalin
Amur
Altai Mts.
ASIA
Alps
Balkan Peninsula
Black Sea
Caucasus
Gora El'brus 18,510 ft. 5,642 m.
Aral Sea
Sardinia
Sicily
Crete
Cyprus
Mediterranean Sea
Mts.
Pamir
Gobi Desert
Beijing
Hokkaidō
Honshū
Sea of Japan
Huang
Plateau of Tibet
Himalayas
Indus
Ganges
Mt. Everest 29,028 ft. 8,848 m.
Yangtze
East China Sea
Kyūshū
Cairo
Nile
Zagros Mts.
Sahara
Arabian Peninsula
Red Sea
AFRICA
Sahel
Mumbai (Bombay)
Deccan
Arabian Sea
Bay of Bengal
Taiwan
PACIFIC
Tropic of Cancer
Mariana Islands
Wake Island
Hainan Dao
South China Sea
Mekong
Luzon
Guam
Socotra
Lakshadweep
Sri Lanka
Mindanao
Palau Islands
OCEAN
Ethiopian Plateau
Gulf of Guinea
Congo
Congo Basin
Rift Valley
Kilimanjaro 19,340 ft. 5,895 m.
Maldive Islands
Malay Peninsula
Caroline Islands
Marshall Islands
Borneo
Celebes
Equator
0°
Seychelles
Sumatra
New Guinea
Solomon Islands
Java
INDIAN
Cocos Islands
Zambezi
Madagascar
Mauritius
Reunion
Kalahari Desert
Johannesburg
OCEAN
Great Sandy Desert
Coral Sea
New Hebrides
New Caledonia
Fiji Is.
AUSTRALIA
Tropic of Capricorn
Great Dividing Range
Darling
Sydney
Cape of Good Hope
Cape Leeuwin
North Island
Aoraki (Mt. Cook) 12,316 ft. 3,754 m.
Tasmania
South Island
Îles Kerguélen
SOUTHERN
OCEAN
Antarctic Circle
Enderby Land
Wilkes Land
Queen Maud Land
Victoria Land
ANTARCTICA
75° 60° 45° 30° 15° 15° 30° 45° 60° 75°

A15

World Political
National capital
Major city
International boundary
0 1000 2000 Miles
0 1000 2000 3000 Kilometers
ARCTIC OCEAN
PACIFIC OCEAN
ATLANTIC OCEAN
SOUTHERN OCEAN
CANADA
UNITED STATES
MEXICO
BRAZIL
GREENLAND (Denmark)
ALASKA (U.S.)
RUSSIA
ICELAND
UNITED KINGDOM
IRELAND
PORTUGAL
SPAIN
MOROCCO
W. SAHARA
MAURITANIA
MALI
SENEGAL
GUINEA-BISSAU
GUINEA
SIERRA LEONE
LIBERIA
COTE D'IVOIRE
GHANA
BURKINA FASO
CAPE VERDE
BAHAMAS
CUBA
HAITI
DOM. REP.
PUERTO RICO (U.S.)
JAMAICA
BELIZE
GUAT.
HOND.
EL SAL.
NIC.
COSTA RICA
PANAMA
TRINIDAD AND TOBAGO
VENEZUELA
GUYANA
SURINAME
FRENCH GUIANA (Fr.)
COLOMBIA
ECUADOR
PERU
BOLIVIA
PARAGUAY
CHILE
ARGENTINA
URUGUAY
FALKLAND IS. (U.K.)
ST. HELENA (U.K.)
MIDWAY IS. (U.S.)
Hawai'ian Islands (U.S.)
KIRIBATI
SAMOA
AMERICAN SAMOA (U.S.)
COOK ISLANDS (N.Z.)
TONGA
FRENCH POLYNESIA (Fr.)
Easter Island (Chile)
Galapagos Islands (Ecuador)
Azores (Port.)
Canary Islands (Sp.)
FAROE IS. (Den.)
South Georgia (U.K.)
South Orkney Is. (U.K.)
South Shetland Is. (U.K.)
Weddell Sea
Anchorage
Vancouver
Montréal
Ottawa
Chicago
New York
Washington D.C.
Los Angeles
Houston
Mexico City
Caracas
Lima
Brasília
Rio de Janeiro
Santiago
Buenos Aires
London
Madrid
Casablanca
Aleutian Islands
Yukon
Hudson Bay
Baffin Bay
Newfoundland
Colorado
Mississippi
Gulf of Mexico
Caribbean Sea
Amazon
Niger
Arctic
Tropic of Cancer
Equator
Tropic of Capricorn
Antarctic Circle
International Date Line
N
S
E
W
180° 165° 150° 135° 120° 105° 90° 75° 60° 45° 30° 15° 0°
75° 60° 45° 30° 15° 0° 15° 30° 45° 60° 75°

A16

15° 30° 45° 60° 75° 90° 105° 120° 135° 150° 165° 180°
ARCTIC OCEAN
Franz Josef Land
Spitsbergen (Nor.)
Novaya Zemlya
Circle
NORWAY
FINLAND
SWEDEN
North Sea
DEN.
EST.
LAT.
LITH.
NETH.
GERMANY
POLAND
BELARUS
BEL.
CZ.
SLVK.
UKRAINE
FRANCE
SWITZ.
AUS.
HUNG.
MOLD.
ROM.
ITALY
CRO.
BOS.
SERB.
BUL.
ALB.
MA.
GREECE
Rome
Moscow
Volga
RUSSIA
Ob'
Yenisey
Lena
Novosibirsk
Sea of Okhotsk
Bering Sea
KAZAKHSTAN
MONGOLIA
Black Sea
GEO.
ARM.
AZER.
UZBEKISTAN
TURKMENISTAN
KYRG.
TAJIK.
TURKEY
Mediterranean Sea
Crete
CYPRUS
LEB.
SYRIA
ISRAEL
IRAQ
JORDAN
Tehrān
IRAN
AFGHANISTAN
PAKISTAN
KUWAIT
TUNISIA
ALGERIA
LIBYA
Cairo
EGYPT
SAUDI ARABIA
QATAR
U.A.E.
OMAN
Karāchi
NEPAL
BHU.
BNGL.
Ganges
CHINA
Beijing
Yangtze
Shanghai
NORTH KOREA
SOUTH KOREA
Seoul
Sea of Japan
JAPAN
Tōkyō
International Date Line
PACIFIC OCEAN
Tropic of Cancer
TAIWAN
Hong Kong
Kolkata (Calcutta)
INDIA
Mumbai (Bombay)
Arabian Sea
Bay of Bengal
MYANMAR
LAOS
THAILAND
Bangkok
CAMBODIA
VIETNAM
South China Sea
PHILIPPINES
Manila
NORTHERN MARIANA ISLANDS (U.S.)
WAKE ISLAND (U.S.)
GUAM (U.S.)
NIGER
CHAD
SUDAN
Nile
Red Sea
ERITREA
YEMEN
DJIBOUTI
Addis Ababa
ETHIOPIA
SOMALIA
BENIN
NIGERIA
Lagos
CENTRAL AFRICAN REPUBLIC
CAMEROON
EQUATORIAL GUINEA
GABON
CONGO
Congo
UGANDA
KENYA
RWANDA
DEM. REP. OF THE CONGO
BURUNDI
TANZANIA
SRI LANKA
MALDIVES
SEYCHELLES
COMOROS
BRUNEI
MALAYSIA
SINGAPORE
Borneo
Sumatra
Jakarta
Java
INDONESIA
EAST TIMOR
New Guinea
PAPUA NEW GUINEA
PALAU
FED. STATES OF MICRONESIA
MARSHALL ISLANDS
SOLOMON ISLANDS
Equator
ANGOLA
ZAMBIA
MALAWI
MOZAMBIQUE
ZIMBABWE
NAMIBIA
BOTSWANA
MADAGASCAR
MAURITIUS
REUNION (Fr.)
Johannesburg
SWAZILAND
LESOTHO
SOUTH AFRICA
INDIAN OCEAN
Coral Sea
VANUATU
NEW CALEDONIA (Fr.)
FIJI
Tropic of Capricorn
AUSTRALIA
Brisbane
Darling
Perth
Sydney
Melbourne
Auckland
NEW ZEALAND
Tasmania
Îles Kerguélen (Fr.)
SOUTHERN OCEAN
Antarctic Circle
ANTARCTICA
0° 15° 30° 45° 60° 75°
© Rand McNally & Co.
Made in U.S.A.
N-MDG10000-P1-1-1-1-1

A17

BRITISH COLUMBIA
ALBERTA
SASKATCHEWAN
MANITOBA
C A
WASHINGTON
Seattle
Mt. Olympus 7,969 ft. 2,429 m.
Mt. Rainier 14,411 ft. 4,392 m.
Mt. Saint Helens 8,364 ft. 2,549 m.
Mt. Hood 11,239 ft. 3,426 m.
Cape Flattery
Cape Blanco
Cape Mendocino
OREGON
Harney Basin
Goose Lake
Blue Mts.
IDAHO
Borah Peak 12,662 ft. 3,859 m.
Salmon River Mountains
Bitterroot Range
Flathead Lake
MONTANA
Fort Peck Lake
NORTH DAKOTA
SOUTH DAKOTA
Black Hills
Harney Peak 7,242 ft. 2,207 m.
WYOMING
Gannett Pk. 13,804 ft. 4,207 m.
Grand Teton 13,770 ft. 4,197 m.
Cloud Peak 13,167 ft. 4,013 m.
Yellowstone Lake
Great Divide Basin
Rocky Mountains
Great Plains
NEBRASKA
KANSAS
OKLAHO
TEXAS
Dallas
Edwards Plateau
Stockton Plateau
Llano Estacado
Guadalupe Pk. 8,749 ft. 2,667 m.
Emory Peak 7,825 ft. 2,385 m.
NEW MEXICO
Wheeler Peak 13,161 ft. 4,011 m.
Mt. Taylor 11,301 ft. 3,445 m.
Baldy Peak 11,403 ft. 3,476 m.
Sacramento Mts.
Rio Grande
COLORADO
Denver
Longs Peak 14,255 ft. 4,345 m.
Mt. Elbert 14,433 ft. 4,399 m.
Pikes Pk. 14,110 ft. 4,301 m.
San Juan Mts.
Front Range
UTAH
Great Salt Lake
Utah Lake
Wasatch Range
Uinta Mts.
Kings Peak 13,528 ft. 4,123 m.
Flaming Gorge Res.
Colorado Plateau
Lake Powell
ARIZONA
Phoenix
Grand Canyon
Humphreys Peak 12,633 ft. 3,851 m.
Painted Desert
NEVADA
Great Basin
Wheeler Peak 13,063 ft. 3,982 m.
Pyramid Lake
Lake Tahoe
CALIFORNIA
San Francisco
Monterey Bay
Los Angeles
Mt. Shasta 14,162 ft. 4,317 m.
Mt. Whitney 14,494 ft. 4,418 m.
Telescope Peak 11,049 ft. 3,368 m.
Death Valley
Mojave Desert
Sierra Nevada
Coast Ranges
Cascade Range
San Joaquin Valley
Point Arguello
Channel Islands
PACIFIC OCEAN
MEXICO
ARCTIC OCEAN
Beaufort Sea
Barrow
Brooks Range
ALASKA
Nome
Fairbanks
Alaska Range
Mt. McKinley 20,320 ft. 6,194 m.
Anchorage
Juneau
Gulf of Alaska
YUKON TERRITORY
NORTHWEST TERRITORY
CANADA
RUSSIA
PACIFIC OCEAN
HAWAII (U.S.)
Kaua'i
O'ahu
Honolulu
Moloka'i
Maui
Lāna'i
Hawai'i
Mauna Kea 13,796 ft. 4,205 m.
Mauna Loa 13,677 ft. 4,169 m.

United States Physical
National capitals
State capitals
Towns
Population
Over 1,000,000
250,000 – 1,000,000
International boundary
State boundary
Land elevation
3,000 meters — 9,840 feet
2,000 meters — 6,560 feet
1,000 meters — 3,280 feet
500 meters — 1,640 feet
200 meters — 656 feet
0 Sea level — 0 Sea level
Water depth
0 Sea level — 0 Sea level
200 meters — 656 feet
2,000 meters — 6,560 feet
0 100 200 300 Miles
0 100 200 300 400 Kilometers
ATLANTIC OCEAN
GULF OF MEXICO
Gulf of Maine
ONTARIO
QUÉBEC
NEW BRUNSWICK
Great Lakes
Lake Superior
Lake Michigan
Lake Huron
Lake Erie
Lake Ontario
Lake Winnipeg
Lake of the Woods
Lake Nipigon
Georgian Bay
Isle Royale
Keweenaw Peninsula
Whitefish Point
Upper Peninsula
Lower Peninsula
MINNESOTA
WISCONSIN
MICHIGAN
IOWA
ILLINOIS
INDIANA
OHIO
MISSOURI
KENTUCKY
TENNESSEE
ARKANSAS
MISSISSIPPI
ALABAMA
GEORGIA
FLORIDA
LOUISIANA
SOUTH CAROLINA
NORTH CAROLINA
VIRGINIA
WEST VIRGINIA
MARYLAND
DELAWARE
NEW JERSEY
PENNSYLVANIA
NEW YORK
CONN.
R.I.
MASS.
VERMONT
NEW HAMPSHIRE
MAINE
Minneapolis
Chicago
Detroit
Cleveland
St. Louis
Atlanta
New Orleans
Houston
Miami
Washington D.C.
Philadelphia
New York
Boston
Toronto
Ottawa
Montréal
Appalachian Mountains
Allegheny Plateau
Allegheny Mts.
Blue Ridge
Piedmont
Coastal Plain
Cumberland Plateau
Ozark Plateau
Ouachita Mts.
Boston Mts.
Adirondack Mountains
Catskill Mts.
Green Mts.
White Mts.
Mt. Mitchell 6,684 ft. 2,037 m.
Mt. Washington 6,288 ft. 1,917 m.
Mt. Katahdin 5,268 ft. 1,606 m.
Cape Cod
Cape Hatteras
Cape Lookout
Cape Fear
Cape Canaveral
Cape Sable
Cape San Blas
Chesapeake Bay
Delaware Bay
Albemarle Sound
Pamlico Sound
Long Island
Martha's Vineyard
Nantucket Island
Sea Islands
Florida Keys
The Everglades
Lake Okeechobee
Tampa Bay
Apalachee Bay
Atchafalaya Bay
Mississippi Delta
Lake Pontchartrain
Toledo Bend Res.
Sam Rayburn Res.
J. Strom Thurmond Reservoir
Kentucky Lake
Lake Barkley
Lake Cumberland
Lake of the Ozarks
Lake Winnebago
Lake Champlain
Moosehead Lake
Niagara Falls
Flint Hills
Mississippi
Ohio
Missouri
Tennessee
Cumberland
St. Lawrence
Ottawa
Hudson
Susquehanna
Potomac
James
Savannah
© Rand McNally & Co.

A19

RAND McNALLY

A20

RAND McNALLY

North America Physical
National capitals
Towns
Population
Over 1,000,000
250,000 – 1,000,000
International boundary
Land elevation
3,000 meters — 9,840 feet
2,000 meters — 6,560 feet
1,000 meters — 3,280 feet
500 meters — 1,640 feet
200 meters — 656 feet
0 Sea level — 0 Sea level
Water depth
0 Sea level — 0 Sea level
200 meters — 656 feet
2,000 meters — 6,560 feet
0 200 400 600 800 1000 Miles
0 300 600 900 1200 1500 Kilometers
ASIA
RUSSIA
North Pole
ARCTIC OCEAN
Bering Sea
Bering Strait
Point Hope
Point Barrow
Beaufort Sea
Cape Bathurst
Queen Elizabeth Islands
Melville Island
Banks Island
Victoria Island
Devon Island
Ellesmere Island
Baffin Bay
Cape Adair
Baffin Island
Davis Strait
GREENLAND (Denmark)
Kap Brewster
Kap Farvel
ICELAND
Norwegian Sea
Arctic Circle
Aleutian Islands
Alaska Peninsula
Brooks Range
U.S.
Yukon
Kuskokwim
Mt. McKinley 20,320 ft. 6,194 m.
Alaska Range
Anchorage
Gulf of Alaska
Mt. Logan 19,551 ft. 5,959 m.
Mackenzie
Great Bear Lake
Great Slave Lake
Foxe Basin
Cape Mercy
Labrador Sea
Péninsule d'Ungava
Hudson Bay
James Bay
CANADA
Canadian Shield
Lake Athabasca
Peace
Nelson
Saskatchewan
Lake Winnipeg
Albany
Smallwood Reservoir
Newfoundland
Gulf of St. Lawrence
Les Laurentides
Coast Mountains
Queen Charlotte Islands
Vancouver Island
Vancouver
Fraser
Rocky Mountains
PACIFIC OCEAN
Columbia
Cascade Range
Cape Blanco
Cape Mendocino
Coast Ranges
Sierra Nevada
Snake
Great Salt Lake
Great Basin
Mt. Whitney 14,494 ft. 4,418 m.
Colorado
Colorado Plateau
Los Angeles
Great Plains
Missouri
Lake Superior
Great Lakes
Lake Michigan
Lake Huron
Lake Erie
Lake Ontario
Niagara Falls
Montréal
Ottawa
Toronto
Chicago
New York
Cape Cod
Washington D.C.
UNITED STATES
Appalachian Mountains
Arkansas
Ozark Plateau
Ohio
Mississippi
Red
Coastal Plain
Cape Hatteras
BERMUDA (U.K.)
ATLANTIC OCEAN
Cape Canaveral
The Everglades
Miami
BAHAMAS
Tropic of Cancer
Pecos
Rio Grande
Houston
GULF OF MEXICO
Gulf of California
Baja California
Cabo San Lucas
MEXICO
Sierra Madre Occidental
Sierra Madre Oriental
Bahía de Campeche
Yucatan Peninsula
Mexico City
Volcán Pico de Orizaba 18,406 ft. 5,610 m.
Havana
CUBA
DOMINICAN REPUBLIC
PUERTO RICO (U.S.)
HAITI
Hispaniola
JAMAICA
BELIZE
GUATEMALA
HONDURAS
EL SALVADOR
NICARAGUA
Lago de Nicaragua
COSTA RICA
PANAMA
Golfo de Panamá
CARIBBEAN SEA
VENEZUELA
COLOMBIA
SOUTH AMERICA
BRAZ.
Equator
N
S
E
W
© Rand McNally & Co.
Made in U.S.A.

A22

North America Political
National capitals
Towns
Population
Over 1,000,000
250,000 – 1,000,000
Under 250,000
International boundary
0 200 400 600 800 1000 Miles
0 300 600 900 1200 1500 Kilometers
ASIA
RUSSIA
North Pole
ARCTIC OCEAN
Bering Sea
Bering Strait
Arctic Circle
Aleutian Islands
Beaufort Sea
Queen Elizabeth Islands
Ellesmere Island
Devon Island
Banks Island
Victoria Island
Baffin Bay
Baffin Island
GREENLAND (Denmark)
Godthåb
ICELAND
U.S.
Anchorage
Fairbanks
Yukon
Gulf of Alaska
Inuvik
Great Bear Lake
Mackenzie
Great Slave Lake
Yellowknife
Whitehorse
Juneau
Iqaluit
Hudson Bay
Churchill
Nelson
Peace
CANADA
Edmonton
Calgary
Saskatoon
Regina
Winnipeg
Lake Winnipeg
Vancouver
Victoria
Seattle
Thunder Bay
Québec
Montréal
Ottawa
Toronto
Gulf of St. Lawrence
Newfoundland
St. John's
Saint John
Halifax
PACIFIC OCEAN
Columbia
Spokane
Portland
Missouri
Billings
Minneapolis
Lake Superior
Milwaukee
Chicago
Detroit
Cleveland
Boston
New York
Philadelphia
Washington D.C.
Pittsburgh
Sacramento
San Francisco
Great Salt Lake
Omaha
UNITED STATES
Denver
Las Vegas
Colorado
Kansas City
Arkansas
Indianapolis
Cincinnati
St. Louis
Ohio
Nashville
Charlotte
Los Angeles
San Diego
Tijuana
Phoenix
Albuquerque
Oklahoma City
Red
Memphis
Atlanta
Mississippi
Dallas
Houston
San Antonio
New Orleans
Jacksonville
Tampa
Miami
Ciudad Juárez
Hermosillo
Chihuahua
Gulf of California
MEXICO
Rio Grande
Culiacán
Torreón
Monterrey
San Luis Potosí
Guadalajara
León
Mexico City
Veracruz
Puebla
Acapulco
Mérida
Cancún
GULF OF MEXICO
BERMUDA (U.K.)
ATLANTIC OCEAN
BAHAMAS
Tropic of Cancer
Havana
CUBA
DOMINICAN REPUBLIC
PUERTO RICO (U.S.)
HAITI
Port-au-Prince
Santo Domingo
Kingston
JAMAICA
CARIBBEAN SEA
BELIZE
Belmopan
GUATEMALA
Guatemala
HONDURAS
Tegucigalpa
San Salvador
EL SALVADOR
NICARAGUA
Lago de Nicaragua
Managua
COSTA RICA
San José
Panamá
PANAMA
Golfo de Panamá
Caracas
VENEZUELA
COLOMBIA
Bogotá
SOUTH AMERICA
BRAZ.
Equator
PACIFIC OCEAN
N
S
E
W
© Rand McNally & Co.
Made in U.S.A.

A23

South America Physical
National capitals
Towns
Population
Over 1,000,000
250,000 – 1,000,000
International boundary
Land elevation
3,000 meters — 9,840 feet
2,000 meters — 6,560 feet
1,000 meters — 3,280 feet
500 meters — 1,640 feet
200 meters — 656 feet
0 Sea level — 0 Sea level
Water depth
0 Sea level — 0 Sea level
200 meters — 656 feet
2,000 meters — 6,560 feet
0 200 400 600 800 1000 Miles
0 300 600 900 1200 1500 Kilometers
GULF OF MEXICO
Tropic of Cancer
NORTH AMERICA
MEXICO
BELIZE
GUATEMALA
HONDURAS
EL SALVADOR
NICARAGUA
COSTA RICA
PANAMA
CUBA
Yucatan Channel
Greater Antilles
JAMAICA
HAITI
DOMINICAN REPUBLIC
PUERTO RICO (U.S.)
Lesser Antilles
CARIBBEAN SEA
ATLANTIC OCEAN
Punta Gallinas
Pico Cristóbal Colón 18,947 ft. 5,775 m.
Caracas
TRINIDAD AND TOBAGO
Boca Grande
Lago de Maracaibo
Orinoco
Llanos
VENEZUELA
GUYANA
FRENCH GUIANA (FR.)
SURINAME
Cabo Orange
Golfo de Panamá
Bogotá
COLOMBIA
Nev. del Huila 18,865 ft. 5,750 m.
Punta Galera
Galapagos Islands (Ec.)
Equator
ECUADOR
Chimborazo 20,702 ft. 6,310 m.
Putumayo
Japurá
Negro
Manaus
Amazon
Amazon Basin
Ilha de Marajó
Xingu
Tocantins
Tapajós
Madeira
Juruá
Selvas
BRAZIL
Cabo de São Roque
Recife
Represa de Sobradinho
Punta Pariñas
Andes
PERU
Ucayali
Nev. Huascarán 22,133 ft. 6,746 m.
Lima
Planalto do Mato Grosso
Punta Carreta
Lago Titicaca
Nev. Illampu 21,066 ft. 6,421 m.
La Paz
BOLIVIA
Cordillera Real
Brasília
São Francisco
Serra do Espinhaço
PACIFIC OCEAN
Nev. Sajama 21,463 ft. 6,542 m.
Gran Chaco
PARAGUAY
Paraná
São Paulo
Rio de Janeiro
Cabo de São Tomé
Tropic of Capricorn
Atacama Desert
Isla San Ambrosio (Chile)
Isla San Felix (Chile)
Nev. Ojos del Salado 22,615 ft. 6,893 m.
Uruguay
Lagoa dos Patos
URUGUAY
Lagoa Mirim
Archipiélago Juan Fernández (Chile)
Cerro Aconcagua 22,831 ft. 6,959 m.
Santiago
Buenos Aires
Río de la Plata
Pampas
ARGENTINA
CHILE
Punta Lavapié
N
W
E
S
Golfo San Matías
Península Valdés
Isla Grande de Chiloé
Archipiélago de los Chonos
Patagonia
Cabo Dos Bahías
Golfo San Jorge
Cabo Tres Puntas
Isla Wellington
Bahía Grande
FALKLAND ISLANDS (U.K.)
West Falkland
East Falkland
Strait of Magellan
Tierra del Fuego
Isla Santa Inés
Cape Horn
South Georgia (U.K.)
Drake Passage
South Shetland Islands (U.K.)
South Orkney Islands (U.K.)
South Sandwich Islands (U.K.)
© Rand McNally & Co.
Made in U.S.A.
90° 80° 70° 60° 50° 40° 30° 20° 10°
20° 10° 0° 10° 20° 30° 40° 50° 60°

A24

RAND McNALLY

A25

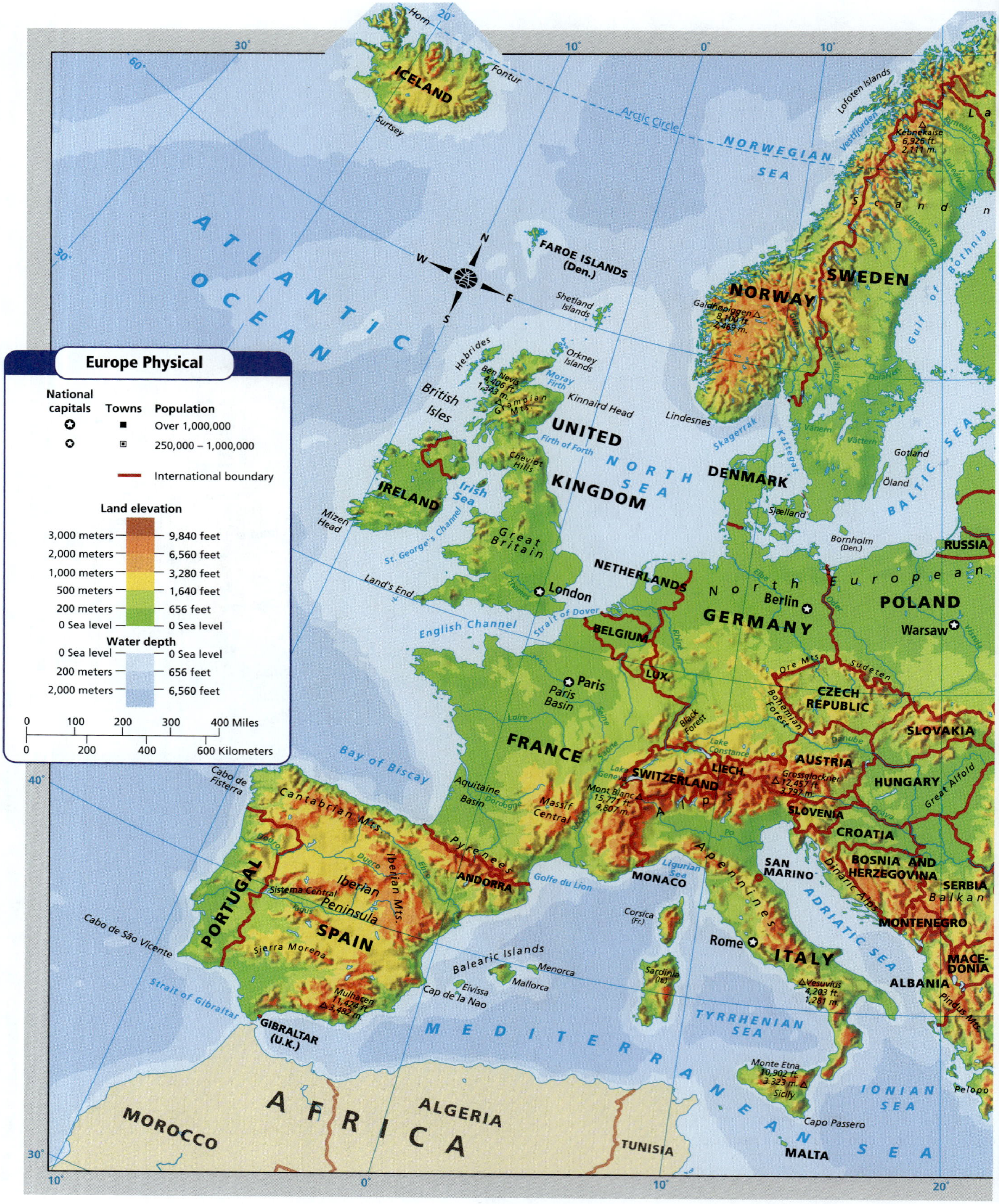
Europe Physical
National capitals
Towns
Population
Over 1,000,000
250,000 – 1,000,000
International boundary
Land elevation
3,000 meters — 9,840 feet
2,000 meters — 6,560 feet
1,000 meters — 3,280 feet
500 meters — 1,640 feet
200 meters — 656 feet
0 Sea level — 0 Sea level
Water depth
0 Sea level — 0 Sea level
200 meters — 656 feet
2,000 meters — 6,560 feet
0 100 200 300 400 Miles
0 200 400 600 Kilometers
ATLANTIC OCEAN
ICELAND
Surtsey
Fontur
Horn
Arctic Circle
NORWEGIAN SEA
FAROE ISLANDS (Den.)
Shetland Islands
Orkney Islands
Hebrides
British Isles
Ben Nevis 4,406 ft. 1,343 m.
Grampian Mts.
Moray Firth
Kinnaird Head
Firth of Forth
Cheviot Hills
UNITED KINGDOM
NORTH SEA
IRELAND
Irish Sea
Mizen Head
St. George's Channel
Great Britain
Land's End
London
English Channel
Strait of Dover
NORWAY
SWEDEN
Galdhøpiggen 8,100 ft. 2,469 m.
Kebnekaise 6,926 ft. 2,111 m.
Lofoten Islands
Vestfjorden
Lindesnes
Skagerrak
Kattegat
DENMARK
Sjælland
Bornholm (Den.)
Gotland
Öland
Vänern
Vättern
Gulf of Bothnia
BALTIC SEA
RUSSIA
NETHERLANDS
BELGIUM
LUX.
GERMANY
North European
Berlin
POLAND
Warsaw
Ore Mts.
Sudeten
CZECH REPUBLIC
Bohemian Forest
Black Forest
SLOVAKIA
AUSTRIA
Grossglockner 12,457 ft. 3,797 m.
HUNGARY
Great Alfold
Danube
LIECH.
SWITZERLAND
Lake Constance
Lake Geneva
Alps
Mont Blanc 15,771 ft. 4,807 m.
SLOVENIA
CROATIA
BOSNIA AND HERZEGOVINA
SERBIA
Balkan
MONTENEGRO
MACEDONIA
ALBANIA
Pindus Mts.
Dinaric Alps
FRANCE
Paris
Paris Basin
Loire
Seine
Rhine
Aquitaine Basin
Massif Central
Bay of Biscay
Pyrenees
ANDORRA
MONACO
Golfe du Lion
Ligurian Sea
SAN MARINO
Apennines
Po
Rome
ITALY
Vesuvius 4,203 ft. 1,281 m.
Corsica (Fr.)
Sardinia (It.)
TYRRHENIAN SEA
ADRIATIC SEA
IONIAN SEA
Monte Etna 10,902 ft. 3,323 m.
Sicily
Capo Passero
MALTA
Cabo de Fisterra
Cantabrian Mts.
Duero
Iberian Mts.
Iberian Peninsula
Sistema Central
Tagus
Ebro
PORTUGAL
SPAIN
Sierra Morena
Cabo de São Vicente
Mulhacén 11,424 ft. 3,482 m.
Strait of Gibraltar
GIBRALTAR (U.K.)
Balearic Islands
Menorca
Mallorca
Eivissa
Cap de la Nao
MEDITERRANEAN SEA
AFRICA
MOROCCO
ALGERIA
TUNISIA

A26

Murmansk
Kola Peninsula
Ponoy
WHITE SEA
Timan Ridge
Pechora
Gora Narodnaya 6,214 ft. 1,894 m.
Mezen'
Ural Mountains
Ob'
Irtysh
Severnaya Dvina
Onega
FINLAND
Lake Onega
Sukhona
Severnyye Uvaly (Hills)
Kama
Kama Reservoir
Lake Ladoga
Gulf of Finland
RUSSIA
Rybinsk Res.
Gorki Res.
Kuybyshev Res.
ESTONIA
Lake Peipus
Volga
Valdai Hills
Moscow
Oka
LATVIA
LITHUANIA
Central Russian Upland
Oka-Don Plain
Volga Upland
Khoper
Don
Neman
BELARUS
Volgograd Res.
Ural
KAZAKHSTAN
Caspian Depression
Pryp'yat'
Dnieper Lowland
Kiev
Donets Basin
Tsymlyansk Res.
UKRAINE
Dniester
Dnieper
Aral Sea
UZBEKISTAN
Amu Darya
ASIA
CASPIAN SEA
Carpathian Mts.
MOLDOVA
ROMANIA
Transylvanian Alps
Sea of Azov
Crimean Peninsula
Caucasus
Gora El'brus 18,510 ft. 5,642 m.
GEORGIA
TURKMENISTAN
BLACK SEA
Danube
ARMENIA
AZERBAIJAN
AZER.
BULGARIA
Rhodope Mts.
Istanbul
Sea of Marmara
TURKEY
IRAN
Mt. Olympus 9,570 ft. 2,917 m.
AEGEAN SEA
GREECE
Tigris
IRAQ
SYRIA
Euphrates
Rhodes
CYPRUS
LEBANON
Sea of Crete
Crete
© Rand McNally & Co.
Made in U.S.A.
N-MDG50000-A1- -1-1-1

A27

RAND McNALLY

A28

EUROPE: POLITICAL

RAND McNALLY

A29

Africa Physical
National capitals
Towns
Population
Over 1,000,000
250,000 – 1,000,000
International boundary
Land elevation
3,000 meters — 9,840 feet
2,000 meters — 6,560 feet
1,000 meters — 3,280 feet
500 meters — 1,640 feet
200 meters — 656 feet
0 Sea level — 0 Sea level
Water depth
0 Sea level — 0 Sea level
200 meters — 656 feet
2,000 meters — 6,560 feet
0 200 400 600 800 1000 Miles
0 300 600 900 1200 1500 Kilometers
ATLANTIC OCEAN
INDIAN OCEAN
EUROPE
ASIA
Mediterranean Sea
Black Sea
Caspian Sea
Aral Sea
Red Sea
Persian Gulf
Gulf of Aden
Gulf of Guinea
Gulf of Sidra
Strait of Gibraltar
Mozambique Channel
Tropic of Cancer
Equator
Tropic of Capricorn
FRANCE
AUS.
HUNG.
UKRAINE
RUSSIA
KAZ.
UZBEKISTAN
ROMANIA
BOS.
SERB.
MONT.
BUL.
ALB.
ITALY
PORTUGAL
SPAIN
GREECE
MALTA
TURKEY
GEORGIA
ARM.
AZER.
TURKMENISTAN
CYPRUS
SYRIA
LEBANON
ISRAEL
IRAQ
IRAN
JORDAN
KUWAIT
QATAR
U.A.E.
SAUDI ARABIA
OMAN
YEMEN
Socotra (Yem.)
Azores (Port.)
Madeira Islands (Port.)
Canary Islands (Spain)
Casablanca
Algiers
MOROCCO
Atlas Mountains
Jebel Toubkal 13,665 ft. 4,165 m.
Grand Erg Occidental
Grand Erg Oriental
TUNISIA
ALGERIA
Erg Chech
Tahat 9,541 ft. 2,908 m.
Ahaggar
LIBYA
Libyan Desert
Qattara Depression
Cairo
EGYPT
Nile
Lake Nasser
Nubian Desert
WESTERN SAHARA
Râs Nouâdhibou
MAURITANIA
Ijâfene
Sahara
Tibesti
Emi Koussi 11,204 ft. 3,415 m.
Ennedi
Ténéré
CAPE VERDE
Cap Vert
Dakar
SENEGAL
THE GAMBIA
GUINEA-BISSAU
GUINEA
SIERRA LEONE
LIBERIA
Growa Point
COTE D'IVOIRE
MALI
Sahel
BURKINA FASO
GHANA
Lake Volta
TOGO
BENIN
Accra
Lagos
NIGER
Niger
NIGERIA
Jos Plateau
Benue
Lake Chad
CHAD
SUDAN
Khartoum
Blue Nile
White Nile
As Sudd
ERITREA
Ras Dejen 15,158 ft. 4,620 m.
Lake Tana
DJIBOUTI
Ethiopian Plateau
Rift Valley
ETHIOPIA
SOMALIA
Gees Gwardafuy
CAMEROON
Cameroon Mtn. 13,451 ft. 4,100 m.
Bioko
EQUATORIAL GUINEA
SAO TOME AND PRINCIPE
GABON
CONGO
Congo Basin
Congo (Zaire)
Stanley Falls
Uele
Ubangi
CENTRAL AFRICAN REPUBLIC
DEM. REP. OF THE CONGO
Kinshasa
Lualaba
UGANDA
Lake Rudolf
KENYA
Lake Victoria
Kirinyaga 17,058 ft. 5,199 m.
Nairobi
RWANDA
BURUNDI
Serengeti Plain
Kilimanjaro 19,340 ft. 5,895 m.
Lake Tanganyika
Masai Steppe
Zanzibar
TANZANIA
SEYCHELLES
Ascension (St. Helena)
ST. HELENA (U.K.)
Luanda
Cuanza
Kasai
ANGOLA
Cunene
Okavango
ZAMBIA
Zambezi
Victoria Falls
Lake Kariba
MALAWI
Lake Nyasa
Cabo Delgado
COMOROS
MOZAMBIQUE
Tanjona Bobaomby
Maromokotro 9,436 ft. 2,876 m.
MADAGASCAR
MAURITIUS
REUNION (Fr.)
Tanjona Vohimena
NAMIBIA
Namib Desert
BOTSWANA
Kalahari Desert
ZIMBABWE
Limpopo
Ponta da Barra
Johannesburg
SWAZILAND
Vaal
Orange
LESOTHO
Drakensberg
SOUTH AFRICA
Cape of Good Hope
Kaap Agulhas
Prince Edward Islands (S. Af.)
Îles Crozet (Fr.)
N
S
E
W
© Rand McNally & Co.
Made in U.S.A.

A30

Africa Political
EUROPE
ASIA
ATLANTIC OCEAN
INDIAN OCEAN
Mediterranean Sea
Black Sea
Caspian Sea
Red Sea
Persian Gulf
Gulf of Aden
Mozambique Channel
Strait of Gibraltar
Gulf of Sidra
Tropic of Cancer
Equator
Tropic of Capricorn
FRANCE
PORTUGAL
SPAIN
ITALY
AUS.
HUNG.
ROMANIA
BOS.
SERB.
MONT.
BUL.
ALB.
GREECE
UKRAINE
RUSSIA
KAZ.
UZBEKISTAN
TURKMENISTAN
GEORGIA
ARM.
AZER.
TURKEY
CYPRUS
SYRIA
LEBANON
ISRAEL
JORDAN
IRAQ
IRAN
KUWAIT
QATAR
U.A.E.
SAUDI ARABIA
OMAN
YEMEN
Madrid
Rome
Athens
Riyadh
Azores (Port.)
Madeira Islands (Port.)
Canary Islands (Spain)
MOROCCO
Rabat
Casablanca
Marrakech
WESTERN SAHARA
Laayoune
ALGERIA
Algiers
Wahran
Qacentina
Ghardaïa
Tamenghest
TUNISIA
Tunis
MALTA
LIBYA
Tripoli
Banghāzī
Sabhā
EGYPT
Cairo
Alexandria
Suez
Asyût
Aswân
Lake Nasser
MAURITANIA
Nouakchott
CAPE VERDE
MALI
Tombouctou
Gao
Bamako
SENEGAL
Dakar
THE GAMBIA
GUINEA-BISSAU
GUINEA
Conakry
SIERRA LEONE
Freetown
LIBERIA
Monrovia
COTE D'IVOIRE
Abidjan
BURKINA FASO
Ouagadougou
GHANA
Accra
Lake Volta
TOGO
BENIN
Cotonou
NIGER
Niamey
Zinder
NIGERIA
Kano
Abuja
Ibadan
Lagos
CHAD
N'Djamena
Abéché
Lake Chad
SUDAN
Port Sudan
Omdurman
Khartoum
Juba
Blue Nile
White Nile
ERITREA
Asmara
DJIBOUTI
Djibouti
ETHIOPIA
Addis Ababa
Dire Dawa
Lake Tana
SOMALIA
Mogadishu
Socotra (Yem.)
CAMEROON
Douala
Yaoundé
Malabo
EQUATORIAL GUINEA
SAO TOME AND PRINCIPE
CENTRAL AFRICAN REPUBLIC
Bangui
GABON
Libreville
CONGO
Brazzaville
DEM. REP. OF THE CONGO
Kinshasa
Kisangani
Mbuji-Mayi
Kolwezi
Lubumbashi
Congo (Zaire)
UGANDA
Kampala
KENYA
Nairobi
Mombasa
Lake Victoria
RWANDA
Kigali
BURUNDI
Bujumbura
Lake Tanganyika
TANZANIA
Dodoma
Dar es Salaam
Lake Nyasa
SEYCHELLES
COMOROS
MAYOTTE (FR.)
Antsiranana
MADAGASCAR
Antananarivo
Fianarantsoa
MAURITIUS
REUNION (FR.)
ANGOLA
Luanda
Lobito
Huambo
ZAMBIA
Lusaka
Ndola
Zambezi
MALAWI
Lilongwe
MOZAMBIQUE
Beira
Maputo
ZIMBABWE
Harare
Bulawayo
NAMIBIA
Windhoek
Walvis Bay
BOTSWANA
Gaborone
Limpopo
SOUTH AFRICA
Pretoria
Johannesburg
Durban
Cape Town
Port Elizabeth
Orange
SWAZILAND
LESOTHO
Maseru
Ascension (St. Helena)
ST. HELENA (U.K.)
Prince Edward Islands (S. Af.)
Îles Crozet (Fr.)
Capitals not shown on map
Country | Capital
Benin | Porto-Novo
Cape Verde | Praia
Comoros | Moroni
Cote d'Ivoire | Yamoussoukro
Guinea-Bissau | Bissau
Mauritius | Port Louis
Mayotte (Fr.) | Mamoutzou
Réunion (Fr.) | Saint-Denis
St. Helena (U.K.) | Jamestown
Sao Tome and Principe | Sao Tome
Seychelles | Victoria
Swaziland | Lobamba, Mbabane
The Gambia | Banjul
Togo | Lomé
National capitals
Towns
Population
Over 1,000,000
250,000 – 1,000,000
Under 250,000
International boundary
0 200 400 600 800 1000 Miles
0 300 600 900 1200 1500 Kilometers
N
S
E
W
© Rand McNally & Co.
Made in U.S.A.

A31

ATLANTIC OCEAN
ARCTIC OCEAN
ICELAND
IRELAND
UNITED KINGDOM
FAROE ISLANDS (Den.)
Arctic Circle
Barents Sea
North Sea
DENMARK
NORWAY
SWEDEN
FINLAND
Severnaya Zemlya
Novaya Zemlya
Kara Sea
Yamal Pen.
PORTUGAL
SPAIN
GIBRALTAR (U.K.)
ANDORRA
FRANCE
MONACO
NETH.
BEL.
GERMANY
SWITZ.
AUSTRIA
CZECH REP.
POLAND
RUS.
LITH.
LATVIA
ESTONIA
SLOVAKIA
HUNGARY
SLO.
CROATIA
BELARUS
UKRAINE
MOLD.
ROMANIA
BOS.
SERB.
MONT.
ITALY
ALB.
MACE.
BULGARIA
GREECE
MOROCCO
ALGERIA
TUNISIA
Mediterranean Sea
Black Sea
Volga
Ural Mountains
West
Siberian
Lowland
Novosibirsk
Caspian Depression
Caucasus
Ankara
TURKEY
Mount Ararat 16,940 Ft. 5,165m
GEORGIA
ARM.
AZER.
Caspian Sea
KAZAKHSTAN
Astana
Aral Sea
Ust-Urt Plateau
Syr Darya
Lake Balkhash
UZBEKISTAN
Kara Kum (Desert)
TURKMENISTAN
KYRGYZSTAN
Tian Shan
TAJIKISTAN
Pamirs
Tarim Basin
K2 (Qogir Feng) 28,250 Ft. 8,611m
Altun Shan
Kunlun Mts.
Hindu Kush
Himalaya Mts.
LIBYA
CYPRUS
LEBANON
SYRIA
ISRAEL
JORDAN
Sinai Pen.
IRAQ
Tigris
Euphrates
Tehran
Dasht-e Kavir
IRAN
Zagros Mts.
KUWAIT
AFGHANISTAN
PAKISTAN
CHAD
EGYPT
Nile
An-Nafud
SAUDI ARABIA
Arabian Peninsula
Persian Gulf
BAHRAIN
QATAR
U.A.E.
Gulf of Oman
New Delhi
NEPAL
Mt. Everest 29,035 Ft. 8,850m
Great Indian Desert
Ganges
Red Sea
SUDAN
ERITREA
Rub Al-Khali
OMAN
YEMEN
INDIA
DJIBOUTI
ETHIOPIA
Gulf of Aden
Mumbai (Bombay)
Godavari
Deccan Plateau
Western Ghats
Eastern Ghats
Bay of Bengal
DEM. REP. OF THE CONGO (ZAIRE)
UGANDA
Socotra (Yem.)
Arabian Sea
RWANDA
KENYA
SOMALIA
BURUNDI
Lakshadweep (India)
SRI LANKA
TANZANIA
MALDIVES
INDIAN OCEAN
ZAMBIA
MALAWI
MOZAMBIQUE
N
S
E
W
20°
30°
40°
50°
60°
70°
80°
10°
0°

A32

Asia Physical
National capitals
Towns
Population
Over 1,000,000
250,000 – 1,000,000
International boundary
Land elevation
3,000 meters — 9,840 feet
2,000 meters — 6,560 feet
1,000 meters — 3,280 feet
500 meters — 1,640 feet
200 meters — 656 feet
0 Sea level — 0 Sea level
Water depth
0 Sea level — 0 Sea level
200 meters — 656 feet
2,000 meters — 6,560 feet
0 200 400 600 800 Miles
0 200 400 600 800 1000 Kilometers
80°
70°
60°
50°
40°
30°
90°
100°
110°
120°
130°
140°
150°
10°
0°
10°
20°
160°
170°
RUSSIA
Siberia
Central Siberian Uplands
Taymyr Peninsula
Laptev Sea
New Siberian Islands
East Siberian Sea
Arctic Circle
Bering Sea
Aleutian Islands (U.S.)
Verkhoyansk Mts.
Kamchatka Peninsula
Sea of Okhotsk
Sakhalin
Kuril Islands
Stanovoy Range
Lake Baikal
Sayan Mountains
Greater Khingan Range
Sikhote-Alin Mts.
Tatar Strait
Hokkaido
Honshu
Sea of Japan
Tōkyō
JAPAN
Mt. Fuji 12,388 ft. 3,776 m
Shikoku
Kyushu
MONGOLIA
Altai Mts.
Gobi Desert
Beijing
NORTH KOREA
SOUTH KOREA
Yellow Sea
Shanghai
East China Sea
Qilian Shan
CHINA
Qinling Shandi
Chang (Yangtze)
Huang
Xi
TAIWAN
Taiwan Strait
PACIFIC OCEAN
Tropic of Cancer
NORTHERN MARIANA ISLANDS (U.S.)
GUAM (U.S.)
FEDERATED STATES OF MICRONESIA
Philippine Sea
Luzon Strait
Luzon
PHILIPPINES
Manila
Mindanao
PALAU
Equator
BHUTAN
Brahmaputra
BNGL.
MYANMAR
LAOS
THAILAND
Bangkok
CAMBODIA
VIETNAM
Mekong
Red
Gulf of Tonkin
Hainan Island
South China Sea
Gulf of Thailand
Andaman Sea
Andaman Islands (India)
Nicobar Islands (India)
MALAY PENINSULA
MALAYSIA
Str. of Malacca
Singapore
Sumatra
BRUNEI
Borneo
Sulu Sea
Celebes Sea
Celebes
Moluccas
Ceram
Banda Sea
New Guinea
PAPUA NEW GUINEA
INDONESIA
Greater Sunda Islands
Java Sea
Java
Jakarta
EAST TIMOR
Timor
Timor Sea
Arafura Sea
Gulf of Carpentaria
AUSTRALIA
Coral Sea
© Rand McNally & Co.
Made in U.S.A.
N-MDG60000-A1

A33

RAND McNALLY

A34

RAND McNALLY

A35

AUSTRALIA AND OCEANIA: PHYSICAL

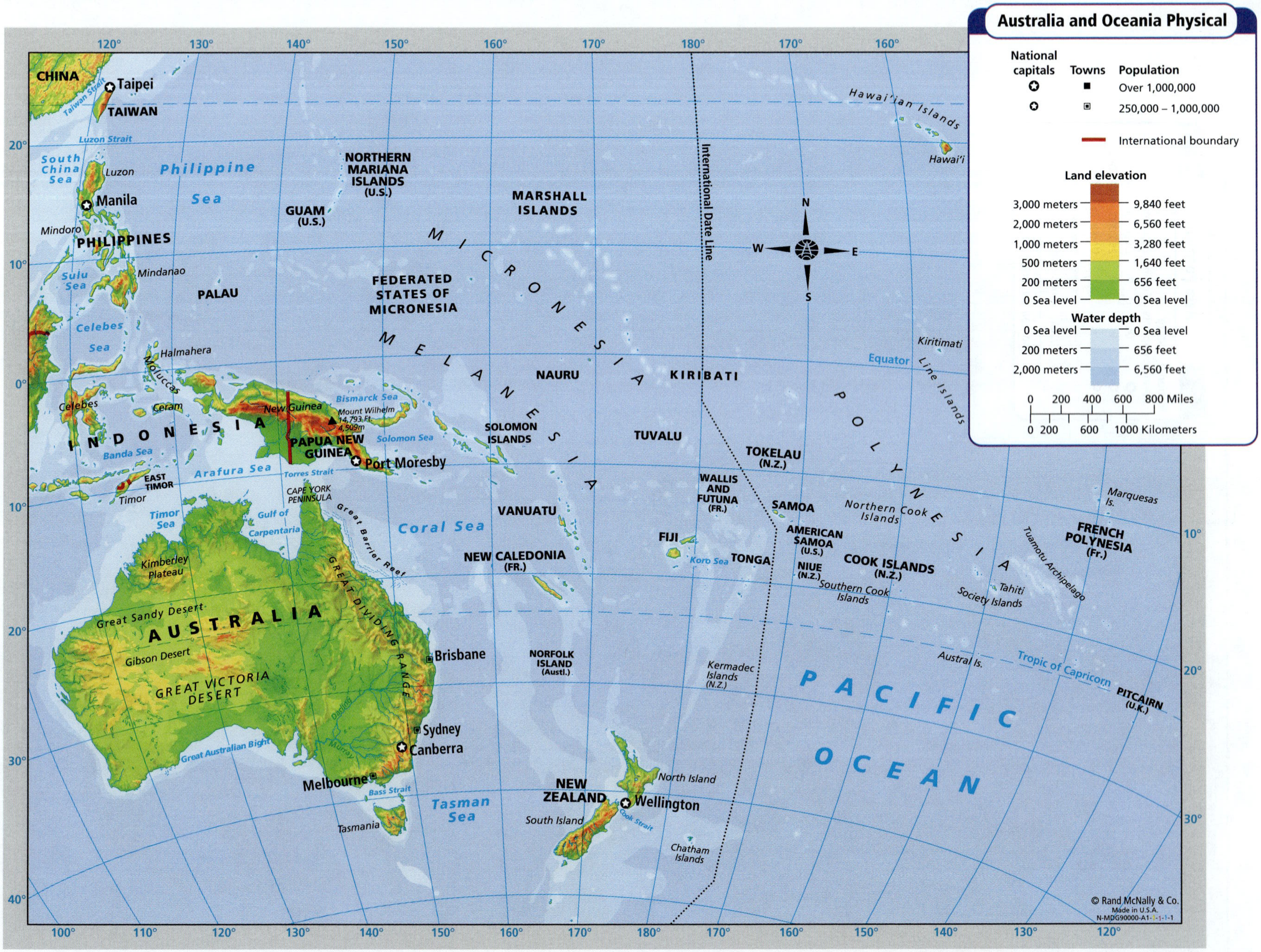

North Pole

- ✪ National capital
- • Town
- International boundary

0 200 400 600 800 1000 Miles

0 400 800 1200 1600 Kilometers

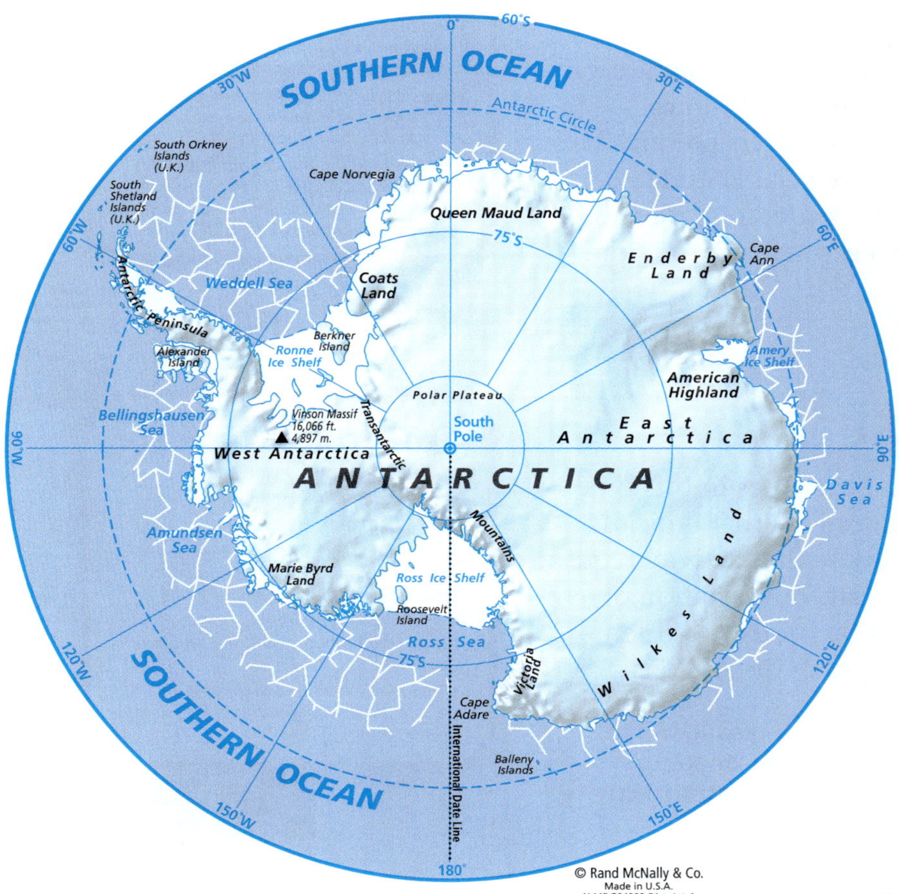

South Pole

0 200 400 600 800 1000 Miles

0 400 800 1200 1600 Kilometers

A37

Reading Strategies in the Teacher's Edition

Look for these cooperative learning reading strategies throughout the Teacher's Edition.

Think, Pair, Share

In this type of activity, students first think about how they would respond to the questions and then discuss their ideas with a partner. Once students have finished their discussion, the pairs share their answers with the class.

Teach
World Religions and Culture

Think, Pair, Share

- What continent was the original home of all of the world's major religions? *(Asia)*
- How does immigration affect religion in an area? *(Immigrants often bring their own religion with them, so the number of religions practiced often increases.)*
- **Find Main Ideas** What beliefs separate the three main types of religions? *(belief in one god, belief in more than one god, and belief in divine forces in nature)*

Reader, Recorder, Reporter

Three to six students take part in this kind of activity. One or two students serve as the reader and search for answers to the questions in the text. While the group discusses each question and answer, a recorder takes notes. Finally, one or two reporters present the group's answers to the class. Students may also appoint a leader to guide the proceedings and keep track of the time.

Teach
World Languages and Culture

Reader, Recorder, Reporter

- How do languages change? *(People borrow words from other languages and need new words for new objects or activities.)*
- How did Indo-European languages spread over such a wide area? *(Explorers and colonists brought them to all areas of the world.)*
- **Find Main Ideas** How can a common language help unify a culture? *(People can communicate easily and feel connected to each other, helping to create a cultural identity.)*

Talk About It

The entire class participates in this type of activity. The questions that typically accompany a Talk About It activity require students to synthesize information or form and support opinions and often spark lively large-group discussions.

3 Teach
Culture: A Way of Life

Talk About It

- What makes cultures unique? *(the ways they use the elements of culture—language, food, religion, and so on—to meet their needs)*
- What things do people within an ethnic group share? *(language, customs, a common heritage)*
- **Find Main Ideas** How do people learn culture? *(by observing others and from direct teaching)*

For more interactive strategies: Use the "Before-, During-, and After-Reading" strategies and the "Differentiation Activity Bank" in the *Best Practices Toolkit*.

UNIT 1 Introduction to World Geography

Why It Matters:

You live on a unique planet in the Sun's planetary system. It is the only planet capable of supporting a wide variety of life forms. As human beings we adapt and alter the environments on Earth.

Landsat satellite

CHAPTER 1 Understanding the Earth and Its Peoples

Fiery volcanic eruption

CHAPTER 2 Earth's Interlocking Systems

Indonesian rice fields

CHAPTER 3 Human Geography

Cave of Hands, Santa Cruz, Argentina

CHAPTER 4 People and Culture

Chapter 1 Understanding the Earth and Its Peoples PLANNING GUIDE

Complete print resources all on one USB drive allow you to customize lesson plans for all students.

Section Titles & Key Ideas	Unit Resource Book Introduction to World Geography	Assessment Resources
1 Themes and Elements of Geography **KEY IDEA** Geographers have specialized ways to view and interpret information about the world.	• **Reading Study Guide,** pp. 1–2 • **RSG with Additional Support,** pp. 7–8 • **Vocabulary Practice,** p. 25 • **Outline Map Activities,** p. 29 • **Interdisciplinary Projects,** pp. 34–35	**Unit Resource Book** • Section Quiz, p. 38 • Reteaching Activity, p. 41 **Test Generator CD-ROM** **Daily Test Practice Transparency,** TT1
2 Technology Tools for Geographers **KEY IDEA** Geographers use technological tools to help them understand both Earth's physical processes and the activities of people on Earth.	• **Reading Study Guide,** pp. 3–4 • **RSG with Additional Support,** pp. 9–10 • **Map and Graph Skills,** p. 27 • **History Makers: Ptolemy,** p. 31	**Unit Resource Book** • Section Quiz, p. 39 • Reteaching Activity, p. 42 **Test Generator CD-ROM** **Daily Test Practice Transparency,** TT2
3 Careers in Geography **KEY IDEA** Geographers do many different kinds of jobs as they gather data and analyze and interpret it.	• **Reading Study Guide,** pp. 5–6 • **RSG with Additional Support,** pp. 11–12 • **History Makers:The Urban Planner,** p. 32 • **Connect to Your World,** p. 36	**Unit Resource Book** • Section Quiz, p. 40 • Reteaching Activity, p. 43 • Chapter Test (Forms A, B, & C), pp. 45–56 **Test Generator CD-ROM** **Daily Test Practice Transparency,** TT3

McDougal Littell Assessment System

TEST SCORE REPORT RETEACH

No Time? To focus students on important content in the chapter, use the **"Reading Study Guide with Additional Support"** in ***Resources2Go.***

Teacher Presentation Options

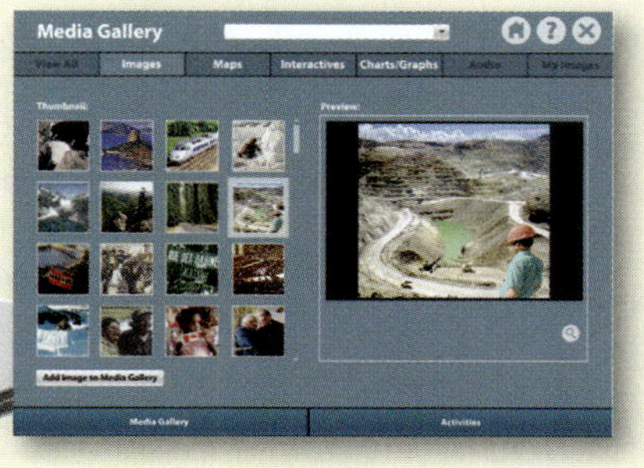

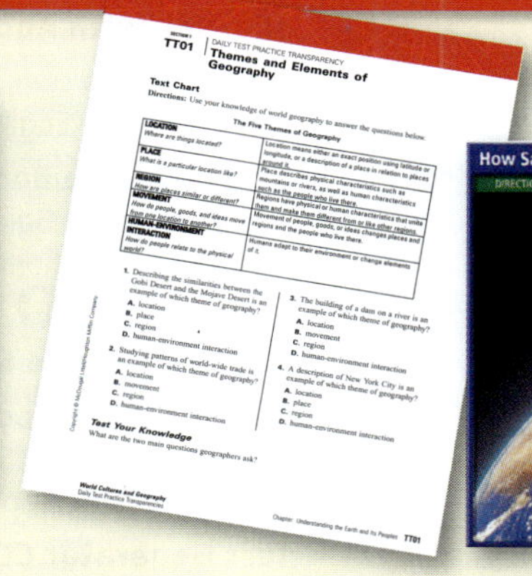

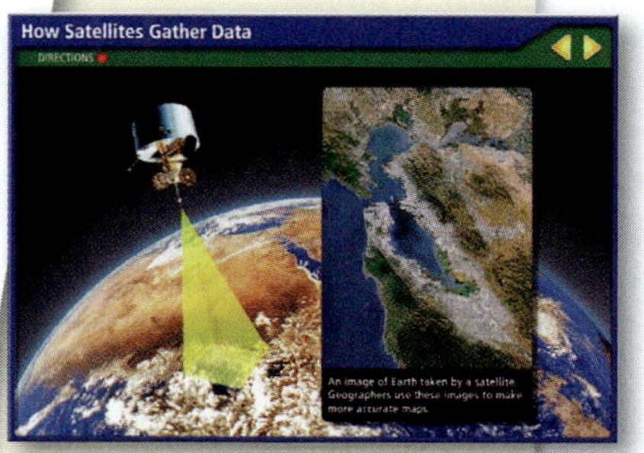

Electronic Books

eEdition DVD-ROM

eEdition Online

Presentation Toolkit

Video/DVD

- World Cultures and Geography

Power Presentations DVD-ROM

- Discussion Starters
- Media Gallery
- Review Game
- Interactive Review
- Animated Features

Unit Transparency Book

- 3-Minute Warm-Ups
- Essential Question
- Critical Thinking
- Maps
- Cultures

Daily Test Practice Transparencies

Animated Geography

- Satellite View of the Earth
- Landsat Satellite

OnlineActivities at ClassZone.com

Animated Geography

- Videos and Photos
- 3-D Animations
- Interactive Maps

Interactive Review

- Section Quizzes
- Review and Study Notes
- Vocabulary Flip Cards
- Graphic Organizers
- Name Game and Crossword Puzzles
- GeoGame

Online Test Practice

Map Center

- Atlas Maps
- Political Maps with Data
- Links to Mapping Web Sites

Activity Center

- Activity Maker—Review Games, Puzzles, Flip Cards, and Graphic Organizers
- Worldster

Research & Writing Center

- Web Sites
- Data Updates
- Primary Sources
- Writing Models

Current Events

Esther Howse
Byron Center West Middle School
Byron Center, Michigan

Make It Fun — Teacher-Tested Activities

As you teach this chapter, look for my classroom-proven ideas for:

- making the lesson fun
- focusing students' attention and comprehension
- integrating technology with traditional teaching

Teaching Bell to Bell When you see this timing guide, have students discuss in pairs, in groups of 3–6, or as a class.

- **Think, Pair, Share** (pairs)
- **Reader, Recorder, Reporter** (3–6)
- **Talk About It** (any size)

CHAPTER 1 RESOURCES FOR DIFFERENTIATED INSTRUCTION

Struggling Readers

eEdition with Audio DVD-ROM

Lesson Plans
- Strategies for Inclusion

Unit Resource Book
- Reading Study Guide with Additional Support, pp. 7–12 (B)
- Reteaching Activity, pp. 41–44
- Chapter Test (Form A), pp. 45–48 (A)

Best Practices Toolkit (C)
- Interactive Reading Strategies, pp. 90–145
- Differentiation Activity Bank: Struggling Readers, pp. 146–149

Teacher's Edition Activities
- Categorize, Sec. 3
- Tiered Activities, Sec. 1–3

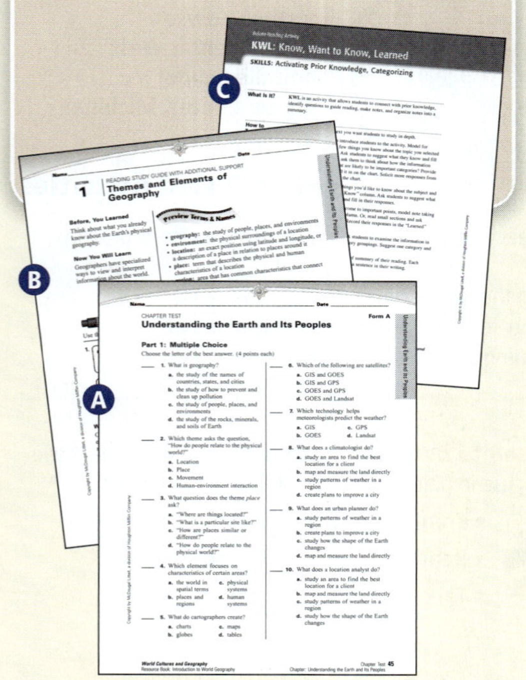

English Learners

Pupil's Edition in Spanish

eEdition with Audio DVD-ROM

Lesson Plans
- Strategies for English Learners

Unit Resource Book
- Reading Study Guide (Spanish), pp. 13–18
- RSG with Additional Support (Spanish), pp. 19–24 (A) (B)

Multi-Language Glossary (C)

Best Practices Toolkit
- Differentiation Activity Bank: English Learners, pp. 165–194

Test Generator CD-ROM
- Section Quizzes in Spanish
- Chapter Test in Spanish

Teacher's Edition Activities
- Pronounce and Preview, Sec. 1–3
- Key Academic Vocabulary, Sec. 1
- Comprehension: Self Monitor, Sec. 2
- Tiered Activities, Sec. 1–3

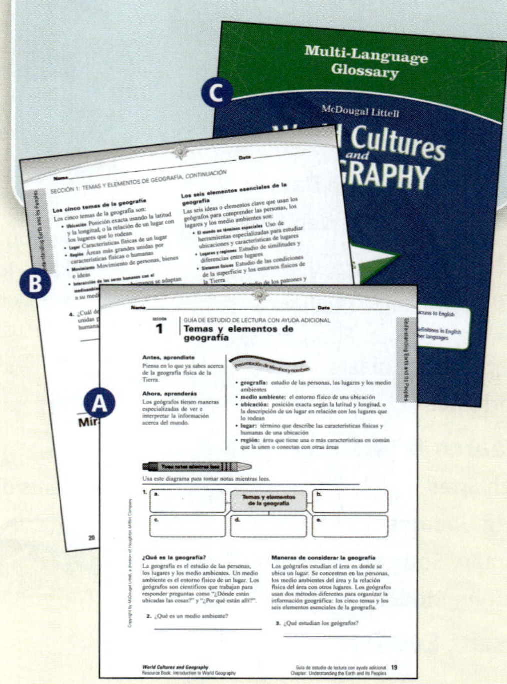

Inclusion

eEdition with Audio DVD-ROM

Lesson Plans
- Strategies for Inclusion

Unit Resource Book
- Reading Study Guide with Additional Support, pp. 19–24 (A) (B)
- Reteaching Activity, pp. 41–44
- Chapter Test (Form A), pp. 45–48

Best Practices Toolkit (C)
- Interactive Reading Strategies, pp. 90–145
- Differentiation Activity Bank: Struggling Readers, pp. 146–149

Teacher's Edition Activities
- Oral Terms and Names Definitions, Sec. 1
- Match Terms and Definitions, Sec. 2
- Spell and Define Vocabulary, Sec. 3
- Tiered Activities, Sec. 1–3

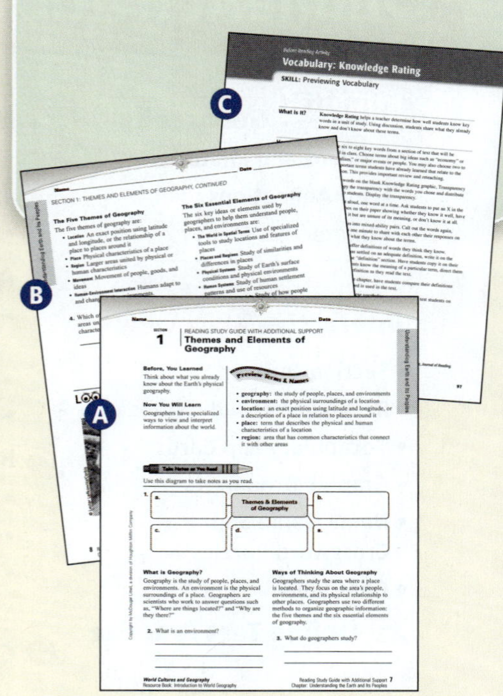

Interdisciplinary Activities

Gifted & Talented Students

Lesson Plans
- Core Lesson Plan

Unit Resource Book
- Map and Graph Skills, p. 27 (A)
- Outline Map Activities, p. 29
- History Maker, p. 31 (B)
- Interdisciplinary Projects, pp. 33–35 (C)
- Chapter Test (Form C), pp. 53–56

Best Practices Toolkit
- Student Products Rubrics, pp. 1–14
- Writing Templates, pp. 78–79

Teacher's Edition Activities
- Create a Board Game, Sec. 1
- Tiered Activities, Sec. 1–3

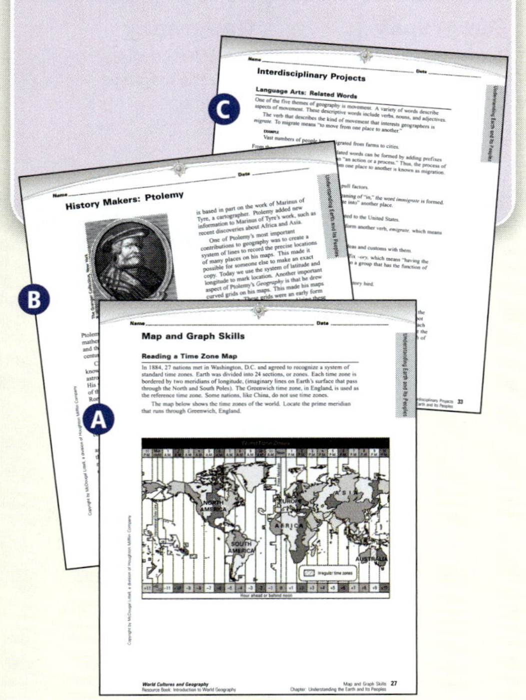

Pre-AP*

Lesson Plans
- Core Lesson Plan

Unit Resource Book
- Interdisciplinary Projects, pp. 33–35 (A)
- Connect to Your World, p. 36
- Primary Source, p. 37 (B)
- Chapter Test (Form C), pp. 53–56

Best Practices Toolkit (C)
- Differentiation Activity Bank: Pre-AP Students, pp. 195–200
- Writing Templates, pp. 78–79

* Pre-AP is a registered trademark of the College Entrance Examination Board, which was not involved in the production of and does not endorse this product.

Teacher's Edition Activities
- Create a Map Scrapbook, Sec. 2
- Tiered Activities, Sec. 1–3

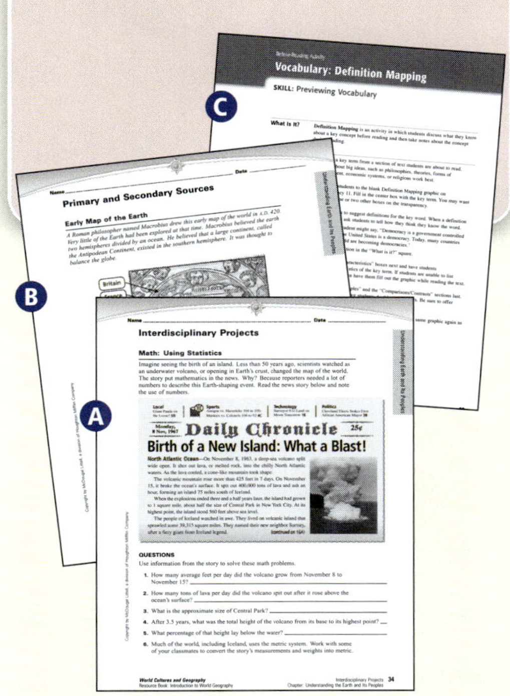

CONNECT to Math

Graphing Mountains

CLASS TIME 20 minutes

TASK Create a bar graph display showing the heights of the ten tallest mountains in the world and the five tallest mountains in the United States.

SUPPLIES
- data on the heights of the ten tallest mountain peaks in the world, and the five tallest mountain peaks in the United States
- posterboard and construction paper
- glue and rulers

ACTIVITY Have students re-read the Fun Facts in Section 2. Discuss Mt. Everest's height compared to other mountains, including mountains in the United States. Have students work in small groups to construct bar graph posters showing the data. Remind them to create the correct scale for the graphs and to use one color of paper for mountains throughout the world and a different color for U.S. mountains. Display the completed bar graph posters.

Good for Spatial Learners

CONNECT to Language Arts

Be a Location Analyst

CLASS TIME 20 minutes

TASK Write a journal entry from the point of view of a location analyst.

SUPPLIES
- pens and paper

ACTIVITY Have students re-read the first two paragraphs under the heading "Advising Businesses and Government" in Section 3. Tell students to imagine that they are a location analyst who has been asked to find a location for a shopping center in their city or town. Have students write a journal entry for the location analyst explaining the search for the location and the technologies used in the search. Journal entries should reflect an understanding of the factors location analysts consider in their work.

Good for Visual and Verbal Learners

Introduce the ESSENTIAL QUESTION

- Explain that geography is the study of people, places, and environments.
- Point out that geographers are interested in how human activities have shaped different places on Earth.
- Describe the differences between physical geographers and human geographers.

Key Ideas

Section 1: Geographers have specialized ways to view and interpret information about the world.

Section 2: Geographers use technological tools to help them understand both Earth's physical processes and the activities of people on Earth.

Section 3: Geographers do many different kinds of jobs as they gather data and analyze and interpret it.

CONNECT to the Essential Question

You may wish to revisit the Essential Question periodically with students as they work through the chapter.

Suggestions for reconnecting are provided at an appropriate point in the *Teacher's Edition* for each section.

Indonesian rice fields

CHAPTER 1 Understanding the Earth and Its Peoples

1 GEOGRAPHY **Themes and Elements of Geography**

2 GEOGRAPHY **Technology Tools for Geographers**

3 CULTURE **Careers in Geography**

ESSENTIAL QUESTION

In what ways does geography help us understand our world?

CONNECT Geography & History

Use the satellite image and the time line to answer the following questions.

1. The large continent in the center of the image is Africa. How would you describe the land?
2. Which of the events listed on the time line made this image possible?

A.D. 1

Culture
◄ **c. A.D. 1st century** Strabo describes the world known to the Greeks and Romans in his 17-volume *Geography.*

Geography
c. A.D. 2nd century Greek geographer Ptolemy writes his 8-volume *Geography* on mapmaking.

Geography
▲ **1100** Chinese begin using the magnetic compass.

TIME LINE DISCUSSION

Invite students to identify the items on the time line that helped physical geographers do their work better. *(first Landsat satellite, GIS development, GPS becomes operational)*

- Which events on the time line feature geographers writing about their discoveries in books? *(Strabo's 17-volume* Geography *in the 1st century and Ptolemy's 8-volume* Geography *in the 2nd century.)*
- Which events on the time line point out the development or use of important tools by geographers? *(Chinese use of the compass in 1100; John Hadley's design for the sextant in 1730; satellite launch in 1972; development of GIS and GPS)*

Animated GEOGRAPHY

Satellite View of the Earth

Click here to explore Earth and its people @ClassZone.com

A

Amazon Rain Forest The Amazon rain forest in South America is the largest rain forest in the world.

B

Sahara Desert The Sahara is the largest desert in the world.

History
1730 John Hadley creates the basic design for a sextant. ▼

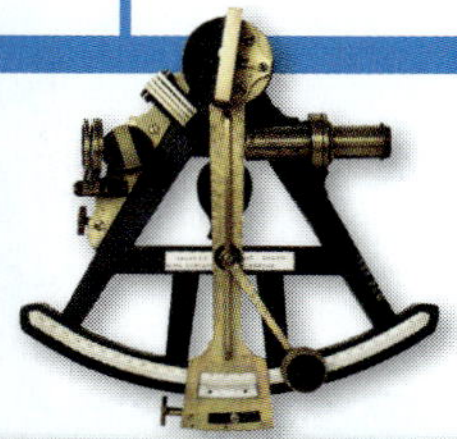

History
1960s Geographic Information Systems (GIS) development begins.

History
1972 First Landsat satellite is launched.

Geography
1983 The Global Positioning System (GPS) becomes operational. ►

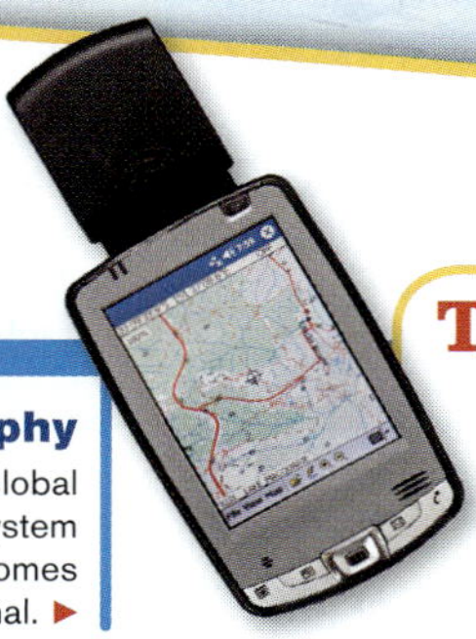

Today

3

TEACHER MEDIA FAVORITES

Books for the Teacher

Diamond, Jared. *Guns, Germs, and Steel.* W.W. Norton, 1999. Focuses on how different environments shaped different peoples throughout history.

Fortey, Richard. *Earth: An Intimate History,* Knopf, 2004. Explains how physical geography of Earth has changed over time.

Video / DVD

National Geographic's Strange Days on Planet Earth. National Geographic, 2005. Four-part series on how patterns of human behavior affect the environment.

World in the Balance: The Population Paradox. PBS, 2004. Episode of the PBS series *The People Paradox* about population trends throughout the world.

Internet

Go to the *Research and Writing Center* at **ClassZone.com** to link to:

- data on the region
- a variety of chapter-specific, content-reviewed sites
- primary sources

Animated GEOGRAPHY

Satellite View of the Earth

CLASSZONE.COM

Satellite photos help geographers gather information about different parts of Earth's surface. Physical characteristics such as deserts and rain forests can be studied for changes over time. Satellite photos also help geographers create maps and compare the characteristics of different places.

- About how much of the South American continent is covered by rain forest? *(Between one third and one half appears to be rain forest.)*
- What does the photo reveal about the size of Africa compared to the size of South America? *(Africa is larger.)*
- Besides rain forest and desert, what other major physical characteristic does the photo show? *(water or oceans)*

CONNECT GEOGRAPHY & HISTORY ANSWERS

1. some desert, some very green land
2. the 1972 Landsat satellite launch

Presentation Options

Use the **Power Presentations** DVD-ROM to project the animated map.

- Prompt students to use the pop-up images and audio from the map to discover what they already know about the Earth and its people.
- Challenge students to answer the map questions individually, in small groups, or as a class activity. Use all of the interactive material to help preview chapter content.

1 Plan & Prepare

Objectives

- Identify different ways of thinking about geography
- Explain the five themes of geography
- Describe the six essential elements of geography

Read for the Essential Question

Help students read for a purpose by reminding them of the Essential Question: "In what ways does geography help us understand our world?" Point out that they can use text features such as Key Questions, special features, images, and maps to form answers to the question.

Vocabulary

Best Practices Toolkit

Use the ***Best Practices Toolkit*** to model strategies for vocabulary notetaking. Vary strategies throughout the year. Choose from: *Knowledge Rating, Predicting ABCs, Definition Mapping, Word Sort, Word Wheel, Frayer Model (Word Squares), Magnet Words,* and *Student VOC.*

Vocabulary Strategies, TT9–TT16

Reading Strategy

Best Practices Toolkit

Display the Web Diagram transparency.

- Remind students that finding main ideas means identifying important ideas in the text.
- Model filling in some additional entries for the first head, starting with Ways of Thinking about Geography.
- Have students complete the organizer and display it for review.

Web Diagram, TT28

Section 1 Reading for Understanding

Key Ideas

BEFORE YOU READ

Think about what you already know about the Earth's physical geography.

NOW YOU WILL LEARN

Geographers have specialized ways to view and interpret information about the world.

Vocabulary

TERMS & NAMES

geography the study of people, places, and environments

environment the physical surroundings of a location

spatial where a place is located and its physical relationship to other places, people, or environments

location an exact position using latitude and longitude, or a description of a place in relation to places around it

place a geographical term that describes the physical and human characteristics of a location

region an area that has one or more common characteristics that unite or connect it with other areas

BACKGROUND VOCABULARY

three-dimensional an image in which there is a sense of depth and perspective

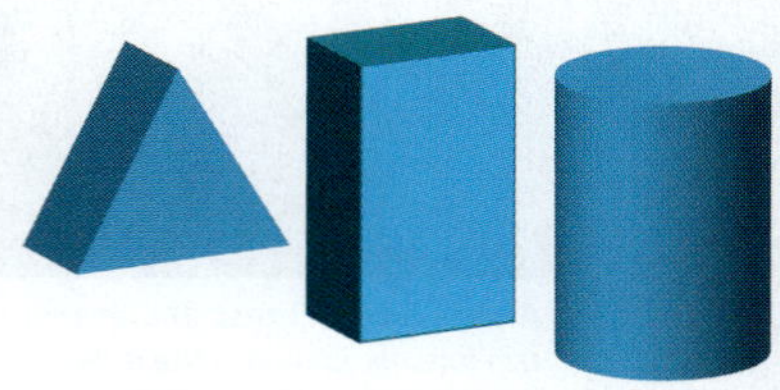

Visual Vocabulary Three-dimensional shapes

Reading Strategy

Re-create the web diagram shown at right. As you read and respond to the **KEY QUESTIONS**, use the diagram to help you find main ideas about the themes and elements of geography.

Skillbuilder Handbook, page R4

FIND MAIN IDEAS

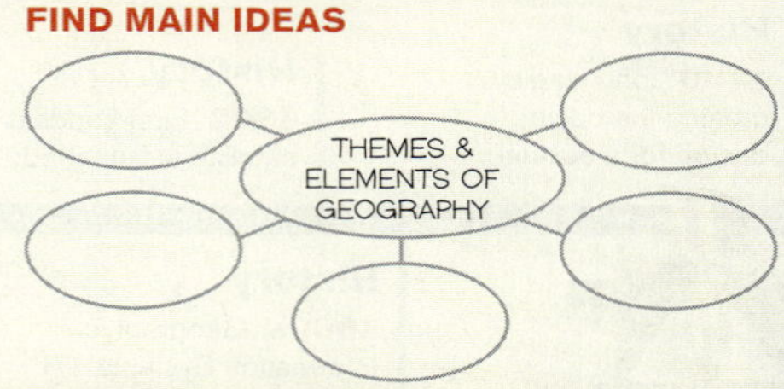

GRAPHIC ORGANIZERS
Go to **Interactive Review** @ClassZone.com

PRETEACHING VOCABULARY

English Learners

Pronounce and Preview

Pronounce each term for students. Review the meanings of words within definitions, such as *latitude, longitude,* and *characteristics.*

- To modify vocabulary learning, have students complete worksheets as they read, instead of afterward.

Unit Resource Book: Introduction to World Geography
- Vocabulary Practice, p. 25

Inclusion

Oral Terms and Names Definitions

Have a volunteer read each term and its definition aloud. Ask listeners to define each term orally with different words and then use it in a sentence. As a group, discuss challenging words or those that are hard to define.

Budapest, Hungary

A group of friends

SECTION 1 GEOGRAPHY

Themes and Elements of Geography

Connecting to Your World

Have you ever drawn a map to show someone how to get to your house? Or have you described your hometown to someone who doesn't live there? If you answered yes to these questions, you were doing what geographers do. Geographers try to answer the questions, "Where are things located?" and "Why are they there?" Basic questions like these form the framework for the subject called geography. **Geography** is the study of people, places, and environments. An **environment** is the physical surroundings of a location.

Ways of Thinking About Geography

KEY QUESTION What are the themes and elements of geography?

Geographers study the world in **spatial** terms. This means they look at the space where a place is located and its physical relationship to other places, people, and environments. Geographers—and students of geography—use two different methods to organize geographic information: the five themes and the six essential elements of geography. The categories vary slightly, but the graphic on the next two pages will help you learn how to apply these ideas as you read this text.

Three-Dimensional Model This computer-generated model is used to study geographic conditions in the Los Angeles region.

5

2 Focus & Motivate

3-Minute Warm-Up

Write on the board or display the transparency:

- Look at the image at the bottom of this page. What can you tell about the geographic conditions in the Los Angeles region? *(The region is mountainous.)*

Unit Transparency Book
- 3-Minute Warm-Ups, TT1

Connecting To Your World

More About. . .

California's Geography

Southern California sometimes has violent earthquakes caused by movement along a crack in the Earth's crust known as the San Andreas fault.

3 Teach

Ways of Thinking About Geography

Think, Pair, Share

- What two methods do geographers use to organize information? *(the five themes and six essential elements of geography)*
- How does studying the world in spatial terms help geographers? *(Physical relationships between people, places, and environments can be seen.)*
- **Find Main Ideas** How do the themes and elements influence the study of geography? *(They help make connections between places, processes, and human activities on Earth.)*

SECTION 1 PROGRAM RESOURCES

Unit Resource Book: Introduction to World Geography

ON LEVEL

Unit Resource Book
- Reading Study Guide, p. 1
- Outline Map Activities, p. 29
- Section Quiz, p. 38

STRUGGLING READERS

Unit Resource Book
- Reading Study Guide with Additional Support, p. 7
- Skillbuilder Practice, p. 26
- Vocabulary Practice, p. 25
- Section Quiz, p. 38
- Reteaching Activity, p. 41

eEdition with Audio DVD-ROM

ENGLISH LEARNERS

Pupil Edition in Spanish
eEdition with Audio DVD-ROM
Unit Resource Book
- RSG (Spanish), p. 13
- RSG with Additional Support (Spanish), p. 19

Multi-Language Glossary
Test Generator
- Section Quiz in Spanish

INCLUSION

Unit Resource Book
- RSG with Additional Support, p. 7
- Section Quiz, p. 38
- Reteaching Activity, p. 41

GIFTED & TALENTED

Unit Resource Book
- Section Quiz, p. 38

PRE-AP

Unit Resource Book
- Primary Sources p. 37
- Section Quiz, p. 38

TECHNOLOGY

Unit Transparency Book
- 3-Minute Warm-Ups, TT1
- Essential Question, TT2
- Critical Thinking, TT3
- Cultures, TT8

Daily Test Practice Transparencies
- Section 1, TT1

Power Presentations
ClassZone.com
World Cultures and Geography Video on DVD

Teach

The Five Themes of Geography

Reader, Recorder, Reporter

- What does the photo of Rio de Janeiro tell you about its location? *(It is both mountainous and located on a large body of water.)*
- How is the theme of region different from the theme of place? *(Regions are made up of many different places united by similar characteristics.)*
- **Find Main Ideas** What does the theme of human-environment interaction reveal about the study of geography? *(Possible answer: People are highly dependent on their environment for their needs.)*

More About . . .

Gobi Desert

The region of the Gobi Desert covers more than 1,000 miles from east to west in Mongolia and China. The western part of the desert is made up entirely of sand, while the eastern region has small grassy areas. These areas support nomadic peoples who herd sheep and goats. Fierce dust and sand storms are common in the Gobi, and can be seen on satellite photos from above the Earth. Many discoveries of prehistoric tools, mammals, and dinosaur eggs have been made in this desert. Ask students to compare this region to the region they live in.

The Five Themes of Geography

The world is a big place and studying it is a complicated task. You can make that job easier by learning five core themes of geography. These themes can help you answer geographic questions.

1 LOCATION

Where are things located?

Location means either an exact position using latitude or longitude, or a description of a place in relation to places around it.

Rio de Janeiro, Brazil

2 PLACE

What is a particular location like?

Place describes physical characteristics such as mountains or rivers, as well as human characteristics such as the people who live there.

Beijing, China

3 REGION

How are places similar or different?

Regions have physical or human characteristics that unite them and make them different from or similar to other regions.

Gobi Desert

4 MOVEMENT

How do people, goods, and ideas move from one location to another?

Movement of people, goods, and ideas changes places and regions and the people who live there.

Interstate Cloverleaf

5 HUMAN-ENVIRONMENT INTERACTION

How do people relate to the physical world?

Humans adapt to their environment and change elements of it.

Pacific Ocean Windsurfing

INTERDISCIPLINARY ACTIVITIES

CONNECT to Art

Make a Geography Collage

Have students work in pairs to collect or draw more images that illustrate the five themes of geography. Students should complete the project by adding a title, a caption, or both. Display the collages in the classroom.

CONNECT to Language Arts

Write a Paragraph

Have students write a paragraph describing the place or region they live in. Ask them to include examples of movement they are aware of in the region, as well examples of human-environment interaction.

Unit Resource Book

- Interdisciplinary Project: Related Words, p. 33

Six Essential Elements

Geographers use six key ideas or elements to help them understand people, places, and environments on the Earth.

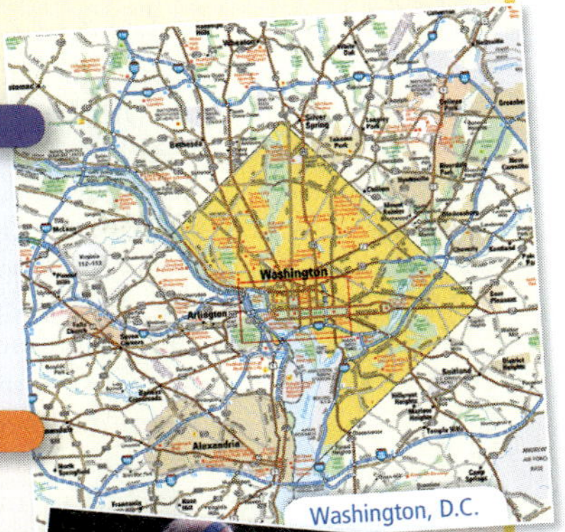
Washington, D.C.

1 THE WORLD IN SPATIAL TERMS

Geographers study the locations of places and distributions or patterns of features by using maps, data, and other geographic tools. Knowing about the world in spatial terms helps geographers understand physical and human patterns.

2 PLACES AND REGIONS

Geographers look for characteristics of places and then compare their similarities and differences.

Mount Everest

Volcano Specialist

3 PHYSICAL SYSTEMS

Geographers study changes in the Earth's surface. Where and why people choose to live in certain locations may depend on Earth's surface conditions.

South Asian Teapickers

4 HUMAN SYSTEMS

Geographers study human settlement patterns and use of resources. This information helps explain human interactions and lifestyles.

5 ENVIRONMENT AND SOCIETY

Geographers study how people interact with the environment and how they use resources.

Surveyor

6 THE USES OF GEOGRAPHY

Geographers study patterns and processes in the world. This information helps people understand the past and plan for the future.

Teach

Six Essential Elements

Talk About It

- How does a road map help a geographer understand the world in spatial terms? *(Possible Answer: It shows how different roads connect places within a region and provides approximate distances between locations.)*
- Which of the essential elements do you think would be most important for a volcano specialist? Why? *(Places and Regions, Physical Systems; volcano specialists might want to compare volcanoes in different parts of the world and how they change over time.)*
- **Find Main Ideas** How do the six essential elements differ from the five themes of geography? *(Possible Answer: The six elements are slightly more specific, and give geographers more choices for organizing information.)*

More About . . .

Tea Picking in South Asia

Most of the tea in the world is grown on plantations located in South Asia. Tea pickers are an important part of the tea harvesting process because tea leaves are delicate and need to be picked by hand. Tea pickers carry wicker baskets for their leaves, and must wear cotton gloves to pick certain expensive teas because the oils on human hands can affect the tea's flavor. Once leaves are picked, they are dried in either machines or the sun before being sold. A geographer in this region might study the human settlement patterns, the use of resources related to tea, and the movement of tea to other parts of the world.

DIFFERENTIATING INSTRUCTION

English Learners

Key Academic Vocabulary

To help students studying English as a second language understand words used in the text above, pair an English speaker with a non-English speaker. Have the English speaker explain the following terms: *physical, resource, interact, processes, environment.*

Gifted & Talented

Create a Board Game

Have students create a board game using the six essential elements of geography. Participants in the game will attempt to travel around the world in search of examples of the six elements while facing volcanoes and other challenges. The students should design a board, create the rules, and demonstrate play.

More About . . .

Water and Settlement

One of the largest dams in the world is Hoover Dam on the Colorado River between Nevada and Arizona. It helps with flood control and navigation on the river, and also generates hydroelectric power throughout the Southwest. Built during the Great Depression of the early 1930s, the dam is named for President Herbert Hoover. The dam holds back the waters of Lake Mead, the largest reservoir in the United States.

CONNECT to Geography

Chicago

Tell students that Chicago has more than 20 miles of shoreline along Lake Michigan. Its metropolitan area extends to the borders of Wisconsin to the north and Indiana to the south and east. Many interstate highways also intersect in Chicago.

CRITICAL THINKING ANSWER

Left caption—Theme(s): place, movement; Element(s): physical systems, human systems

Middle caption—Theme(s): human-environment interaction; Element(s): human systems, environment and society

Right caption—Theme(s): location, region; Element(s): the world in spatial terms

Five Themes of Geography Now that you have seen the geographic themes and elements side by side, let's look more closely at an example of one of the themes and elements as it applies to a particular place.

The theme of human-environment interaction is a good place to get an idea of how a geographer thinks. For thousands of years, people have found it valuable to settle by rivers. A river can provide food, water, transportation, and other needs of daily life. However, rivers can flood, destroying homes and villages, and taking human life. So, humans began to alter their environment by building walls called levees to protect the land from floods. Sometimes they created dams to control the flow of water and to save some water for times when they needed it. A geographer who asks the questions "Where do people choose to live?" and "Why here?" will answer that the river provides many needs for a group of people. So, people will likely be found in areas that have rivers as a resource. As it turns out, we know that early civilizations such as those in Egypt, Southwest Asia, India, and China began in river valleys.

Six Essential Elements Using the six elements helps geographers make sense both of physical processes on the Earth and of human systems devised by the people who live there.

Let's look at the element of physical systems. Geographers want to know how these systems work to reshape the Earth's surface and what impact these changes have on plants, animals, and people. Volcanoes

CONNECT to Geography

Chicago The skyline of Chicago towers over the shore of Lake Michigan. One of the largest cities in the United States, Chicago has long been the economic and cultural center of the region.

CRITICAL THINKING

Find Main Ideas Look at the image of Chicago at right. Which of the five themes and six elements of geography do the captions reflect?

High-rise buildings maximize the use of valuable land.

Chicago is a major hub for Great Lakes water transportation and a main railroad headquarters.

Chicago is located in the upper Midwest of the United States.

DIFFERENTIATING INSTRUCTION: TIERED ACTIVITIES

OBJECTIVE Gain a deeper understanding of the five themes and six essential elements of geography.

Basic

Ask students to write the five themes on one side of a sheet of paper and the six elements on the other side. Then have them write an example of each theme or element different from the ones in the text.

On Level

Have students study the five themes and the six elements. Ask them to identify similarities and differences between the categories. Then have them write a brief essay (at least two paragraphs) pointing out the similarities and differences.

Challenge

Ask students to imagine they are a geographer from another country visiting the United States. What examples of the themes and elements of geography might they find? What might they see that is different from their own country? Have students record their impressions in a journal entry.

are an example of a physical force that changes the shape of the Earth's surface and may have a dramatic effect on human populations. A volcanic eruption may kill people, plants, and animals living in the area. Flows of lava may change the landscape, burn forests or crops, and possibly alter the course of rivers. Islands in the Pacific Ocean have been created by volcanic eruptions, and still others have disappeared between the waves when they were blown apart by eruptions. Volcanoes can also trigger earthquakes. Geographers studying physical systems point out that many volcanoes take place in certain areas of the world. Studying this pattern of volcanic action helps explain where and why people live in certain locations.

Using the five themes of geography and the six essential elements will help you to think like a geographer. The themes and elements will help you to think about particular places and the physical processes and human activities that shaped those places in the past—and continue to do so. They will also enable you to look for patterns and connections in geographic information. You will be better able to answer the two main geographic questions, "Where are things located?" and "Why are they there?" You will learn about the tools used to record and analyze geographic information in the next section.

Themes are location, place, region, movement, and human-environment interaction; elements are the world in spatial terms, places and regions, physical systems, human systems, environment and society, and uses of geography. ◄

FIND MAIN IDEAS Identify the five themes and six essential elements of geography.

ONLINE QUIZ For test practice, go to Interactive Review @ClassZone.com

Section 1 Assessment

TERMS & NAMES

1. **Explain the importance of**
   - geography
   - environment
   - spatial

USE YOUR READING NOTES

2. **Find Main Ideas** Use your completed web diagram to answer the following question:

   What are the five themes of geography?

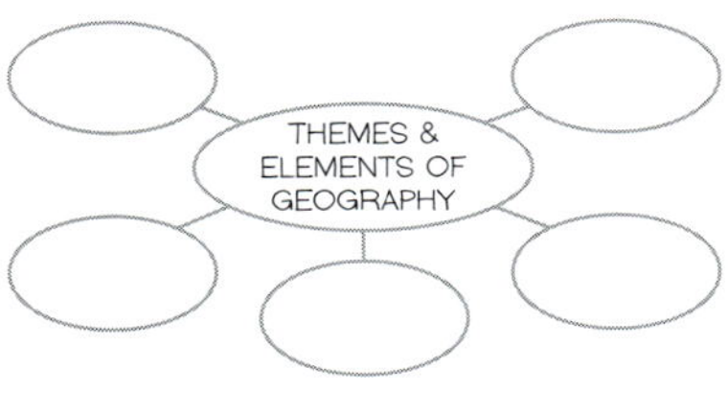

KEY IDEAS

3. What two questions do geographers try to answer?
4. How do geographers use the five themes?
5. What are the six elements that geographers use to look at the world?

CRITICAL THINKING

6. **Summarize** What does it mean to study the world in spatial terms?
7. **Compare and Contrast** How do you think the study of geography differs from that of history?
8. **CONNECT to Today** How does studying geography help you understand the world in which you live?
9. **ART** **Make a Poster** Create a poster that lists the five themes and six elements of geography. For each theme or element, include the definition and a photograph or drawing to illustrate it.

CONNECT to the Essential Question

In what ways does geography help us understand the world?

Ask students what they have learned so far that can help them answer this question. Students might mention:

- Geography is the study of people, places, and environments.
- Geographers try to answer the questions, "Where are things located?" and "Why are they there?"
- Geographers use five themes and six essential elements to organize their information.

4 Assess & Reteach

Assess Have students complete the Section Assessment.

Unit Resource Book
- Section Quiz, p. 38

Interactive Review @ClassZone.com
- Section Quiz

Power Presentations

Test Generator

Reteach Have small groups of students work to create panel discussions. One group presents a discussion about the five themes of geography. Another presents one on the six essential elements of geography, while a third presents one on similarities and differences between the two.

Unit Resource Book
- Reteaching Activity, p. 41

SECTION 1 ASSESSMENT ANSWERS

Terms & Names

1. geography: the study of people, places, and environments; environment: the physical surroundings of a location; spatial: location and its relationship to other people and places

Use Your Reading Notes

2. location, place, region, movement, human-environment interaction

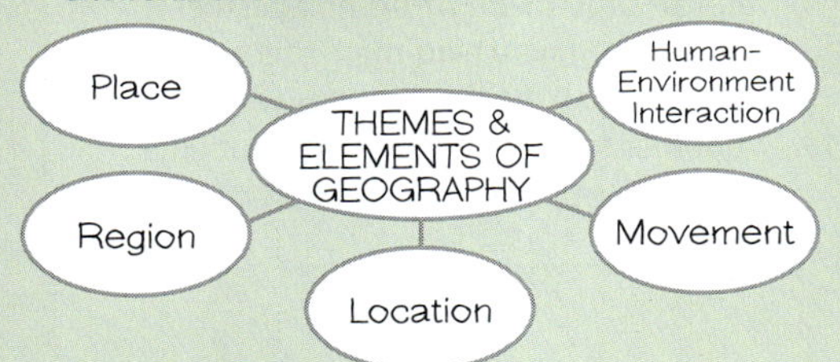

Key Ideas

3. "Where are things located?" and "Why are they there?"
4. to think about the world's people and its environment
5. spatial terms, places and regions, physical systems, human systems, environment and society, and the uses of geography

Critical Thinking

6. to look at place location and its relation to other places, people, and environments
7. Geography studies relationships between people and places; history studies them in time.
8. It can help you know the nature of your world and your place in it.
9. Use the rubric to score the poster.

Art Rubric

	Content	Accuracy
4	excellent; addresses all important points	no errors
3	good; addresses some important points	few/minor errors
2	fair; addresses few important points	several errors
1	poor; addresses no important points	many errors

1 Plan & Prepare

Objectives

- Describe the current science of mapmaking
- Explain how Geographic Information Systems work

Read for the Essential Question

Help students read for a purpose by reminding them of the Essential Question: "In what ways does geography help us understand our world?" Point out that they can use text features such as Key Questions, special features, images, and maps to form answers to the question.

Vocabulary

Best Practices Toolkit

Use the ***Best Practices Toolkit*** to model strategies for vocabulary notetaking. Vary strategies throughout the year. Choose from: *Knowledge Rating, Predicting ABCs, Definition Mapping, Word Sort, Word Wheel, Frayer Model (Word Squares), Magnet Words,* and *Student VOC.*

Vocabulary Strategies, TT9–TT16

Reading Strategy

Best Practices Toolkit

Display the Web Diagram transparency.

- Review the notion that a summary consists of main ideas and important details found in a text.
- Model adding details to the diagram and explain how the diagram can help students organize ideas about geographers' technological tools.
- Have students suggest entries and display the diagram for them to use in summarizing ideas.

Web Diagram, TT28

SECTION 2 Reading for Understanding

Key Ideas

BEFORE, YOU LEARNED

Geography is the study of Earth's physical features and the interaction of people with the environment and with each other.

NOW YOU WILL LEARN

Geographers use technological tools to help them understand both Earth's physical processes and the activities of people on Earth.

Vocabulary

TERMS & NAMES

globe a model of the earth in the shape of a sphere

map a representation of a part of the Earth

cartographer (kahr•TAHG•ruh•fur) a geographer who creates maps

surveyor a person who measures the land

remote sensing obtaining information about a site by using an instrument that is not physically in contact with the site

Landsat a series of information-gathering satellites that orbit above Earth

Global Positioning System (GPS) a system that uses a network of earth-orbiting satellites to pinpoint location

Geographic Information Systems (GIS) a computer or Internet-based mapping technology

BACKGROUND VOCABULARY

database a collection of information that can be analyzed

debris (duh•BREE) the scattered remains of something broken or destroyed

Reading Strategy

Re-create the web diagram shown at right. As you read and respond to the **KEY QUESTIONS**, use the diagram to summarize ideas about geographers' technological tools.

Skillbuilder Handbook, page R5

SUMMARIZE

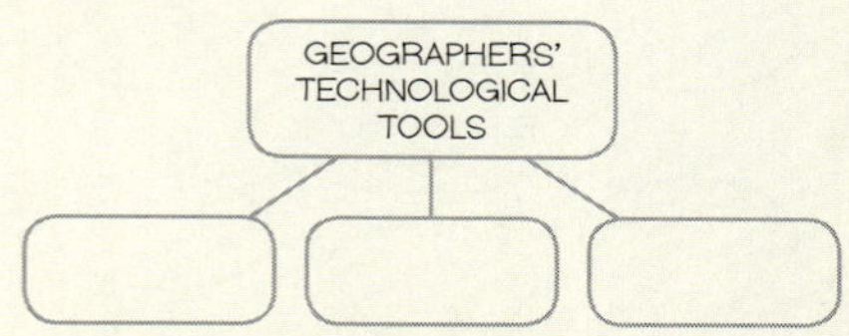

GRAPHIC ORGANIZERS
Go to Interactive Review @ClassZone.com

PRETEACHING VOCABULARY

English Learners

Pronounce and Preview

Pronounce each term for students. Review the meanings of words within definitions, such as *sphere, representation,* and *pinpoint.*

- To modify vocabulary learning, have students complete worksheets as they read, instead of afterward.

Unit Resource Book: Introduction to World Geography
- Vocabulary Practice, p. 25

Inclusion

Match Terms and Definitions

Have student pairs write the vocabulary words and definitions on index cards. Then have them play a matching game. Ask one student to shuffle the cards and arrange them. After their partner has finished matching the terms and definitions, have the students work together to correct any errors. Students should then switch roles.

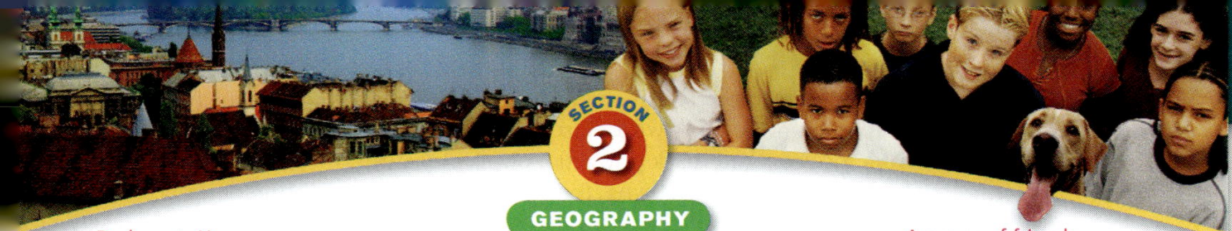

SECTION 2
GEOGRAPHY

Budapest, Hungary

A group of friends

Technology Tools for Geographers

Connecting to Your World

When you were a much younger student, you probably used paper and pencil to do your schoolwork. Now, when you have an assignment to complete, you most likely use a computer and the Internet. Today's geographers and other scientists use high-tech instruments and advanced computer software to create maps and databases. A **database** is a collection of information that can be analyzed. Geographers use these tools and their analysis to answer geographic questions.

Gerardus Mercator
A Flemish cartographer, Mercator developed a type of map still used today.

The Science of Mapmaking

KEY QUESTION How has technology changed mapmaking?

In their work, geographers use photographs, graphs, globes, and maps. A **globe** is a model of the Earth in the shape of a sphere. It shows the actual shape of the Earth. But you can only see half at any one time, and it is not easy to carry around. So, geographers use maps. A **map** is a representation of a part of the Earth. Maps can help geographers see patterns in the way human or physical processes occur. **Cartographers** (kahr•TAGH•ruh•furs) are geographers who create maps.

Animated GEOGRAPHY
Landsat Satellite
This satellite provides visible and infrared views of the Earth.

Click here to see how satellites gather data @ClassZone.com

2 Focus & Motivate

3-Minute Warm-Up

Write on the board or display the transparency:

- What do the images in this section suggest about modern maps? *(There are a variety.)*

Unit Transparency Book
- 3-Minute Warm-Ups, TT1

Connecting to Your World

More About . . .

Landsat Satellites

The Landsat I was launched in 1972, and the latest in 1999. Stations in South Dakota and Australia receive satellite information.

3 Teach

The Science of Mapmaking

Reader, Recorder, Reporter

- How does remote sensing equipment help create maps? *(It allows mapping of places that are difficult to reach.)*
- **Summarize** Describe the types of information that satellites collect. *(Landsat: images and data about Earth; GOES: weather information; GPS: information on locations)*

Animated GEOGRAPHY
Landsat Satellite
CLASSZONE.COM

Make Inferences How does a satellite image help us understand Earth's geography? *(It provides information for mapmaking.)*

SECTION 2 PROGRAM RESOURCES

Unit Resource Book: Introduction to World Geography

ON LEVEL
Unit Resource Book
- Reading Study Guide, p. 3
- Section Quiz, p. 39

STRUGGLING READERS
Unit Resource Book
- Reading Study Guide with Additional Support, p. 9
- Skillbuilder Practice, p. 26
- Vocabulary Practice, p. 25
- Section Quiz, p. 39
- Reteaching Activity, p. 42

eEdition with Audio DVD-ROM

ENGLISH LEARNERS
Pupil Edition in Spanish
eEdition with Audio DVD-ROM
Unit Resource Book
- RSG (Spanish), p. 15
- RSG with Additional Support (Spanish), p. 21

Multi-Language Glossary
Test Generator
- Section Quiz in Spanish

INCLUSION
Unit Resource Book
- RSG with Additional Support, p. 9
- Section Quiz, p. 39
- Reteaching Activity, p. 42

GIFTED & TALENTED
Unit Resource Book
- Map and Graph Skills, p. 27
- History Makers: Ptolemy, p. 31
- Section Quiz, p. 39

PRE-AP
Unit Resource Book
- Section Quiz, p. 39

TECHNOLOGY
Unit Transparency Book
- 3-Minute Warm-Ups, TT1
- Essential Question, TT2
- Critical Thinking, TT4
- Map: The Global Grid, TT6
- Map: Map Projections, TT7

Daily Test Practice Transparencies
- Section 2, TT2

Power Presentations
ClassZone.com
World Cultures and Geography Video on DVD

RESEARCH & WRITING CENTER

Analyzing Primary Sources

Most of Strabo's other works have been lost. He based *Geography* on his own observations of the world and the works of earlier writers, including Homer. Of the seventeen volumes, eight describe peoples and places in Europe, six describe Asia, and one focuses on Africa. Two introductory volumes define the concept of geography and outline its goals.

- What was the purpose of *Geography?* *(to provide information about the known world)*
- Why did Strabo include a map in his book? *(Possible Answer: It helped his readers locate the places he was describing.)*

DOCUMENT-BASED QUESTION ANSWER

Asia, Europe

Unit Resource Book: Introduction to World Geography

- Primary Sources: Early Map of the Earth, p. 37

More About . . .

Weather Satellites

Satellites such as GOES are important to weather forecasters trying to predict severe storms. Weather satellites can track conditions such as tornadoes, flash floods, hail storms, and hurricanes. In addition to following these weather conditions, satellites can estimate how much rain will fall in a specific place during a thunderstorm or hurricane. Weather satellites also help forecasters issue warnings about winter storms such as blizzards before they arrive.

Cartographers create maps from data collected in surveys. **Surveyors** are people who map and measure the land. They go out to a location and mark down the physical features they see, such as rivers, mountains, or towns. Today, cartographers use technologically advanced tools that provide a much more detailed and accurate picture of the world.

To create modern maps, geographers often use remote sensing equipment. **Remote sensing** means obtaining information about a site by using an instrument that is not physically in contact with the site. Generally, these instruments are cameras mounted on airplanes or Earth-orbiting satellites.

Satellites Two of the best-known satellites are Landsat and GOES. **Landsat** is actually a series of information-gathering satellites that orbit more than 100 miles above the Earth. Each has a variety of sensing devices to collect images and data. Each time a satellite makes an orbit, it gathers information from an area about 115 miles wide. Landsat can scan the entire Earth in 18 days.

GOES, or Geostationary Operational Environmental Satellite, is a weather satellite. This satellite flies in an orbit at the same speed as the

ONLINE PRIMARY SOURCE To see more works of Strabo, go to the **Research & Writing Center** @ClassZone.com

ANALYZING Primary Sources

Strabo (c. 64 B.C.–A.D. 23) was a Greek who wrote books about geography and history. His 17-volume *Geography* is the main source for information about the world known to the ancient Greeks and Romans. Strabo drew the map below for his geography book.

Strabo

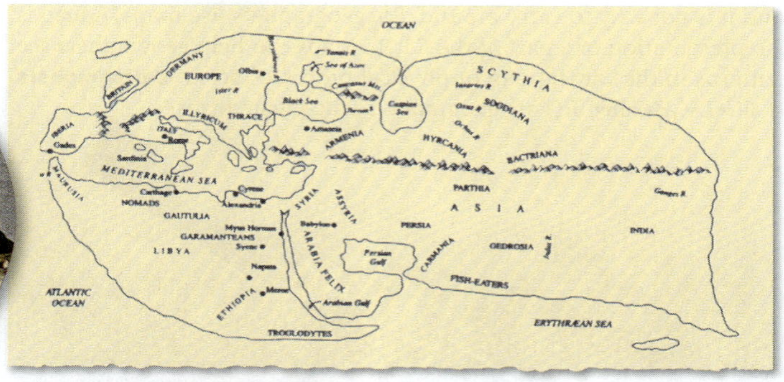

DOCUMENT-BASED QUESTION

Which two continents labeled on the map were known to the ancient Greeks and Romans?

12 Chapter 1

INTERDISCIPLINARY ACTIVITIES

CONNECT to Science

Make A Chart

Have students work in pairs to collect images from the Landsat or GOES satellites. These are available on the Internet or through the library. Students should choose an area of the world and make a compare-and-contrast chart of images taken at different times of the year. The chart should include the location of the images, what is shown, and include at least two statements comparing and contrasting the images.

CONNECT to Science

Be a Surveyor

Have small groups of students survey one of the following: the classroom, school, block, or neighborhood. The survey should include measurements of the location and annotations about physical features on a map of the area. As a class, discuss the difficulties of surveying an area without modern technology.

Unit Resource Book

- Map and Graph Skills, p. 27

Earth's rotation. In this way it remains "stationary" above a fixed area. It gathers images of conditions that are used to forecast the weather. In 2006, there were two GOES satellites. One provided images from the eastern United States and the other from the west. You see GOES images when you watch a TV weather forecast.

Global Positioning System The U.S. Department of Defense developed technology to help American military forces know exactly where they were. The **Global Positioning System (GPS)** employs a network of Earth-orbiting satellites to collect information about the location of a receiver. The satellites beam the receiver's exact position—latitude, longitude, elevation, and time—to Earth. This information is displayed on the receiver.

A GPS receiver can be small enough to fit in your hand. It has an electronic position locator that sends a beam from where you and the device are to an orbiting satellite. The satellite measures where the GPS device is and beams back your exact position.

GPS can be used from any point on the Earth and in any type of weather. You can use its data to help you figure out "Where am I?" and "Where am I going?" GPS data can be used to determine location, aid in navigating from place to place, create maps, and track the movement of people and things. Animal biologists, for example, use GPS devices to track animals and learn about their habits.

DRAW CONCLUSIONS Explain how technology has changed mapmaking. Maps can be more accurate and include more data.

COMPARING Mapping Styles: Washington, D.C.

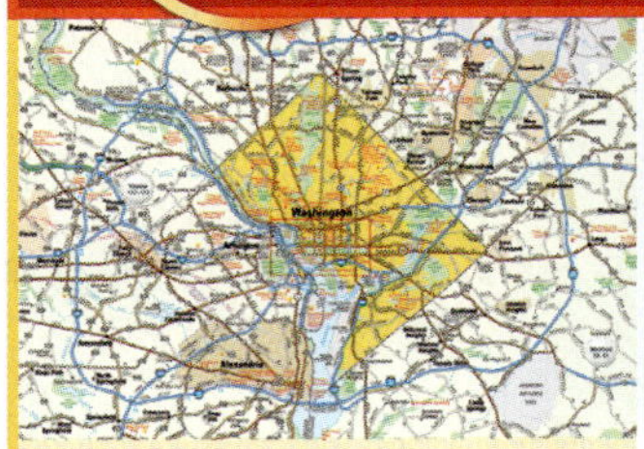

Road Maps Road maps are created from information including aerial photographs, road surveys made with hand-held digitizers, and maps showing the positions of such features as lakes and mountains.

Satellite Images Satellite images are produced by equipment that records information in a digital format. The information is then converted to images that look like photographs.

Infrared Images Infrared images measure the radiation emitted by water bodies, vegetation, and buildings. In this type of image, the warm areas appear in light blue, areas with vegetation appear red, and water is black.

CRITICAL THINKING

Evaluate Which of these mapping styles would be the most valuable for determining where earthquake damage has occurred?

DIFFERENTIATING INSTRUCTION

English Learners

Comprehension: Self-Monitor

The explanation of infrared images may be difficult for students learning English. Point out that the visual on this page illustrates infrared images. Tell students to read the explanation, pausing periodically to check the visual for the same information.

Pre-AP

Create a Map Scrapbook

Have students create a map scrapbook by collecting a variety of maps used in newspapers or magazines, and writing descriptions of maps used on TV or on the Internet. Organize into a scrapbook.

More About . . .

GPS

The U.S. Department of Defense developed GPS in the 1970s for use in military operations. In the 1980s, President Reagan decided to make the system available to the public. In order to prevent terrorists and other criminals from using this technology in attacks against the United States, the government installed a system called selective availability, which made non-military GPS receivers slightly less accurate by distorting their satellite signal.

Selective availability was discontinued in May 2000 in order to make GPS more useful to people and companies all over the world. Today, GPS has many uses in different fields. For example, many police and fire departments use GPS to determine the location of the police car or fire truck closest to an emergency.

Unit Transparency Book
- Map: The Global Grid, TT6

COMPARING

Mapping Styles: Washington, D.C.

Tell students that GOES satellites send back infrared images. The different colors in these images allow weather forecasters to quickly notice unusual features, such as high clouds with a large amount of water vapor. This can signal the beginning of a major storm.

Satellite images can be used to monitor changes in parts of the Earth's surface over time. For example, scientists studying icecaps in Antarctica rely heavily on satellite images. Road maps, while not created from satellite or infrared images, provide useful information to drivers throughout the world.

CRITICAL THINKING ANSWER

Evaluate satellite image

Unit Transparency Book
- Map: Map Projections, TT7

Teach

Geographic Information Systems

Think, Pair, Share

- How did the development of GIS affect the use of satellite images? *(They were integrated with other forms of information.)*
- For what types of projects does the federal government use GIS? *(hurricane evacuations, spread of disease, debris recovery)*
- **Summarize** In what ways did life for geographers change as a result of GIS technology? *(Possible answer: It made it easier for them to do their jobs.)*

CONNECT Geography & Technology

The development of GIS began in the 1960s. At first, the technology ran on powerful supercomputers. Today, however, many desktop computers are able to use GIS technology. Scanners can be used to create new digital maps from original ones. These maps can then be stored in the computer's memory.

Geographers need special training to use GIS because it is a complex technology with many features. Cartographers especially like using GIS, however, since it makes many of the tasks they do easier.

CRITICAL THINKING ANSWER

Summarize A terrain map, a land use map, and a base map are combined to examine possible sites for the airport.

Geographic Information Systems

KEY QUESTION How do Geographic Information Systems work?

A very technologically advanced tool geographers use is the **Geographic Information Systems (GIS)**. GIS is a computer-based mapping technology. The complete system is able to gather, store, analyze, and display spatial information about places. It combines information from a variety of sources into digital databases.

GIS can integrate geographic information, such as maps, aerial photographs, and satellite images. It can also include information such as population figures, economic statistics, or temperature readings. Someone using GIS selects the information needed to answer a geographic question. Then GIS combines layers of information to give the user a better understanding of how the data works together. It can display the information in different ways, such as on a map, design, chart, or graph. The diagram below shows how GIS works.

CONNECT Geography & Technology

GIS GIS is a tool that allows geographers to solve problems by combining geographic information about a location from several sources.

1 A question or problem is posed. An example is, "In what general area near this town might an airport be located?" A section of land is identified for problem solving.

2 Computer databases hold geographic information about the location. The user selects layers of information that answer the question, "What geographic characteristics are important for a good airport site?"

3 A terrain map is selected to identify all areas flat enough for landing airplanes.

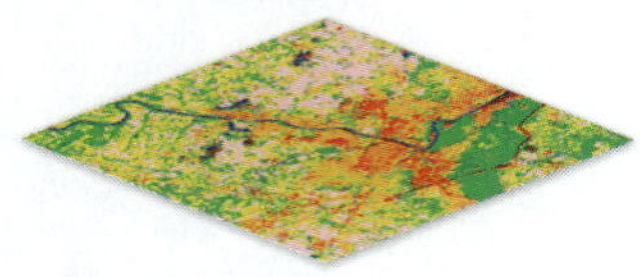

4 A land use map shows areas that have few homes.

5 A base map shows where roads are located so that the airport can be reached and safety concerns are addressed.

6 The layers of information are combined to create a composite map showing possible sites for the airport.

CRITICAL THINKING

Summarize What geographic information is combined in the final map?

DIFFERENTIATING INSTRUCTION: TIERED ACTIVITIES

OBJECTIVE Demonstrate an understanding of GIS and how it works.

Basic

Have students work in small groups. Each group should create a poster summarizing the steps that go into making a GIS map, design, chart, or graph. Display the posters on the bulletin board.

On Level

Provide student groups with data sheets about GIS and GPS. Students should decide which they think would be more useful for planning a car trip across the United States. One student from each group should report to the class.

Challenge

Have students write a letter from the point of view of a cartographer describing the use of GIS in his/her work. Students should include specific information about how GIS can help with simple site questions and complex global problems.

GIS projects can range from simple, specific site questions to more complex global problems. For example, you could use GIS to determine the quickest and safest path to walk to your school. The federal government was able to use GIS to predict the location of **debris** (duh•BREE) from the space shuttle *Columbia*, which broke up upon reentry into the Earth's atmosphere in 2003. GIS can be used to plan for hurricane evacuations or to monitor the possible spread of avian flu. Urban planners use GIS to determine where to place a park or where to relocate a dangerous highway intersection. Private companies use GIS to decide where to drill for oil or even where to place a new fast-food restaurant.

GIS makes it possible for geographers to answer geographic questions quickly and accurately. They are better able to see relationships between data, to understand the past and present, and to predict future situations.

In the next section, you will learn about the many different kinds of jobs geographers perform.

SUMMARIZE Explain how GIS works. The system integrates geographic information such as maps, photographs, satellite images, and statistical information to give a better understanding of a geographic question or pattern.

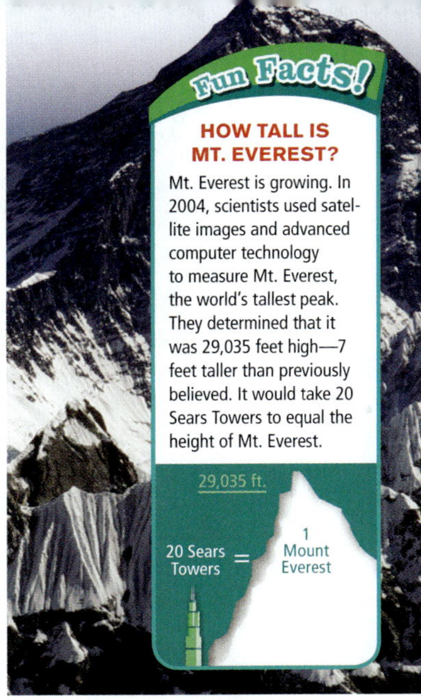

Fun Facts!

HOW TALL IS MT. EVEREST?

Mt. Everest is growing. In 2004, scientists used satellite images and advanced computer technology to measure Mt. Everest, the world's tallest peak. They determined that it was 29,035 feet high—7 feet taller than previously believed. It would take 20 Sears Towers to equal the height of Mt. Everest.

Section 2 Assessment

ONLINE QUIZ For test practice, go to **Interactive Review @ClassZone.com**

TERMS & NAMES

1. **Explain the importance of**
   - map
   - cartographer
   - Geographic Information Systems (GIS)

USE YOUR READING NOTES

2. **Summarize** Use your completed web diagram to answer the following question:

   What are some geographers' tools besides maps and globes?

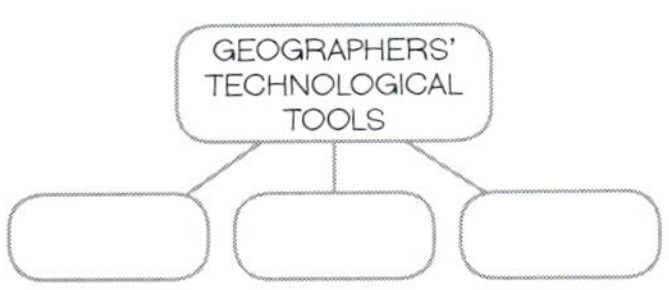

KEY IDEAS

3. What were two early means of showing the Earth's surface?
4. How did remote sensing change the way geographic data were obtained?
5. What are some ways GIS can be used?

CRITICAL THINKING

6. **Make Inferences** How were early geographers limited in gathering geographic information?
7. **Draw Conclusions** How does technology help geographers?
8. **CONNECT to Today** In what ways do you think that new geography technology might aid military forces in modern warfare?
9. **TECHNOLOGY** **Make a Multimedia Presentation** Use an Internet-based GIS to demonstrate the uses of this geographic tool. Give examples of the different tasks a GIS can do.

CONNECT to the Essential Question

In what ways does geography help us understand the world?

Ask students what they have learned so far that can help them answer this question. Students might mention:

- Technology helps cartographers create more accurate maps.
- GIS provides a better understanding of geographic questions or patterns.

4 Assess & Reteach

Assess Have students complete the Section Assessment.

Unit Resource Book
- Section Quiz, p. 39

Interactive Review @ClassZone.com
- Section Quiz

Power Presentations

Test Generator

Reteach Using the headings found in the text and pieces of information for each heading, students should create quiz questions. The pair can exchange questions with another pair and answer the questions.

Unit Resource Book
- Reteaching Activity, p. 42

SECTION 2 ASSESSMENT ANSWERS

Terms & Names

1. map: a representation of a part of the Earth; cartographer: a geographer who creates maps; GIS: a computer or Internet-based mapping technology

Use Your Reading Notes

2. Satellite images, GPS, and GIS

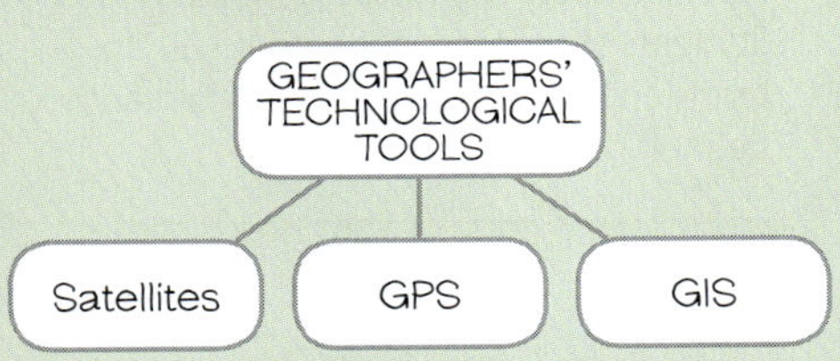

Key Ideas

3. globes and maps
4. Today, data is obtained by an instrument not in contact with the land.
5. to help with evacuations, locate parks, project paths of epidemics, suggest business locations

Critical Thinking

6. They only had their own experiences and those of others to use in preparing geographic information.
7. New technology uses remote sensing and computers to provide geographic data.
8. It provides their location and that of enemy forces and could provide them with data on physical features and climate.
9. Use the rubric to score the multimedia presentation.

Presentation Rubric

	Content	Accuracy
4	excellent; addresses all important points	no errors
3	good; addresses some important points	few/minor errors
2	fair; addresses few important points	several errors
1	poor; addresses no important points	many errors

1 Plan & Prepare

Objectives

- Identify the jobs that process geographic data
- Describe how geographers advise business and government
- Explain the available jobs in physical and human geography

Read for the Essential Question

Help students read for a purpose by reminding them of the Essential Question: "In what ways does geography help us understand our world?" Point out that they can use text features such as Key Questions, special features, images, and maps to form answers to the question.

Vocabulary

Best Practices Toolkit

Use the ***Best Practices Toolkit*** to model strategies for vocabulary notetaking. Vary strategies throughout the year. Choose from: *Knowledge Rating, Predicting ABCs, Definition Mapping, Word Sort, Word Wheel, Frayer Model (Word Squares), Magnet Words*, and *Student VOC.*

Vocabulary Strategies, TT9–TT16

Reading Strategy

Best Practices Toolkit

Display the Web Diagram transparency.

- Point out to students that while categorizing they must decide how the information in this section can be sorted into the categories.
- Tell students that categorizing information can help them remember and use information about a topic.
- Model filling in categories on the diagram, using a career mentioned in the section.

Web Diagram, TT28

SECTION 3 Reading for Understanding

Key Ideas

BEFORE, YOU LEARNED

Geographers use technology to help them do their jobs of finding information about selected areas.

NOW YOU WILL LEARN

Geographers do many different kinds of jobs as they gather data and analyze and interpret it.

Vocabulary

TERMS & NAMES

location analyst a person who studies an area to find the best location for a client

climatologist a geographer who studies climates

Visual Vocabulary Climatologist

urban planner a person who creates plans for developing and improving parts of a city

geomorphology (JEE•oh•mawr•FAHL•uh•jee) the study of how the shape of the Earth changes

REVIEW VOCABULARY

surveyor a person who maps and measures the land

Geographic Information Systems (GIS) a computer-based mapping technology

urban having to do with a city

Reading Strategy

Re-create the web diagram shown at right. As you read and respond to the **KEY QUESTIONS**, use the web diagram to categorize details about careers in geography.

Skillbuilder Handbook, page R7

CATEGORIZE

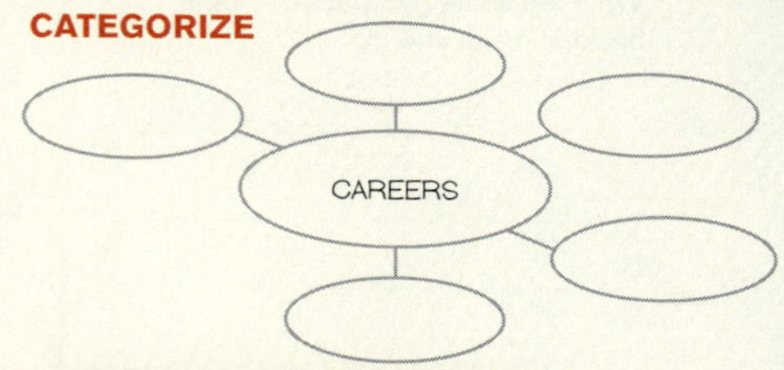

GRAPHIC ORGANIZERS
Go to Interactive Review @ ClassZone.com

PRETEACHING VOCABULARY

English Learners

Pronounce and Preview

Pronounce each term for students. Review the meanings of words within definitions, such as *client, climates,* and *developing.*

- To modify vocabulary learning, have students complete worksheets as they read, instead of afterward.

Unit Resource Book: Introduction to World Geography

- Vocabulary Practice, p. 25

Inclusion

Spell and Define Vocabulary

Have student pairs use letter tiles to spell out all the vocabulary terms. When a person spells out a term, his or her partner should say the definition out loud. Students should trade roles until all the terms and names have been spelled and defined.

Budapest, Hungary

SECTION 3 CULTURE

A group of friends

Careers in Geography

Connecting to Your World

"What are you going to be when you grow up?" is a question that you may have been asked. You may not even know there are a variety of geography-related jobs. The Association of American Geographers lists nearly 150 different geography jobs. So, if you are interested in people, places, and environments, consider a job in geography. Your work will not be limited to maps—it might range from analyzing data to planning projects, or making decisions about the environment.

Modern Cartographer A cartographer works at updating a street map.

Processing Geographic Data

KEY QUESTION What is a geographer's main activity?

A geographer's main activity is analyzing geographic information to answer geographic questions. Jobs processing geographic data begin, of course, with collecting the information. One on-the-ground job in data collection is that of a **surveyor**. Surveyors map and measure the land directly. They may mark boundaries, study the shape of the land, or even help find sewer and water systems beneath the Earth.

Surveyor A surveyor in Dhaka, Bangladesh, gathers data for a project.

2 Focus & Motivate

3-Minute Warm-Up

Write on the board or display the transparency:

- Look at the photograph at the bottom of this page. How does the equipment the man is using help you understand his job? *(The person in the bottom photo is a surveyor using an instrument called a theodolite to collect data about a work site.)*

Unit Transparency Book
- 3-Minute Warm-Ups, TT1

Connecting to Your World

More About . . .

Geography and Jobs

Some other different job titles that geographers can have include Aerial Photo Interpreter, Medical Geographer, Environmental Manager, and Meteorologist.

3 Teach

Processing Geographic Data

Talk About It

- How might a planner make a neighborhood a better place to live? *(Possible Answer: determining the location of new development such as parks, schools, and highways)*
- **Categorize** What jobs involve collecting geographic data? *(surveyor, remote sensing specialist, GIS analyst)*

SECTION 3 PROGRAM RESOURCES

Unit Resource Book: Introduction to World Geography

ON LEVEL

Unit Resource Book
- Reading Study Guide, p. 5
- Connect To Your World, p. 36
- Section Quiz, p. 40

STRUGGLING READERS

Unit 3 Resource Book
- Reading Study Guide with Additional Support, p. 11
- Skillbuilder Practice, p. 26
- Vocabulary Practice, p. 25
- Section Quiz, p. 40
- Reteaching Activity, p. 43

eEdition with Audio DVD-ROM

ENGLISH LEARNERS

Pupil Edition in Spanish

eEdition with Audio DVD-ROM

Unit Resource Book
- RSG (Spanish), p. 17
- RSG with Additional Support (Spanish), p. 23

Multi-Language Glossary

Test Generator
- Section Quiz in Spanish

INCLUSION

Unit Resource Book
- RSG with Additional Support, p. 11
- Section Quiz, p. 40
- Reteaching Activity, p. 43

GIFTED & TALENTED

Unit Resource Book
- History Makers p. 32
- Section Quiz, p. 40

PRE-AP

Unit Resource Book
- Interdisciplinary Project, pp. 34, 35

TECHNOLOGY

Unit Transparency Book
- 3-Minute Warm-Ups, TT1
- Essential Question, TT2
- Critical Thinking, TT5

Daily Test Practice Transparencies
- Section 3, TT3

Power Presentations

ClassZone.com

World Cultures and Geography Video on DVD

Teach

Advising Business and Government

Think, Pair, Share

- In which job do geographers find locations for airports and public transportation? *(urban planner)*
- Why might a location analyst read a GIS report? *(to help a client decide on the best location for a business)*
- **Categorize** What data do climatologists collect to help farmers? *(Long-term climate data is used to project future weather patterns that might affect crops.)*

CONNECT Geography & Culture

Cancún Then and Now

Located on the Yucatán Peninsula, which extends into the Gulf of Mexico, Cancún's location makes it vulnerable to hurricanes. In 2005, Cancún was struck by the 140 mile per hour winds of Hurricane Wilma. There was extensive damage to homes, hotels, and beaches. However, by 2006, much of the resort area had been rebuilt.

Teacher-Tested Activities

Esther Howse,
Byron Center West Middle School,
Byron Center, Michigan

To help students envision careers in geography, I have my students play, "And the Oscar Goes to . . . " This game gives students an opportunity to explore ideas about what geographers do in the real world. Here's how it works.

- After the section is read, assign groups of two to three students to "act out" silently the six careers mentioned in the section.
- Randomly select a group to present their skit. After the skit is presented the class should write down which of the careers was portrayed.
- Review answers as a class.

High-tech information-gathering jobs include working with **Geographic Information Systems (GIS)** data. Some examples of these jobs include remote sensing specialists and GIS analysts. Take a look back at the GIS feature in Section 2 to get an idea of what a job using GIS would be like. Data analysis jobs require the ability to think critically, high-level computer skills, and a college education.

Once data have been processed, a geographer may study the information to use in planning projects such as a new urban area, a disaster evacuation plan, or the placement of a new highway. Planners can also help determine how to make a neighborhood a better place to live. These jobs, too, require good critical thinking, writing, and computer skills, as well as a college education. Planners are valuable to the success of a community.

FIND MAIN IDEAS Explain the main job of a geographer. to analyze information to answer geographic questions

Advising Businesses and Government

KEY QUESTION How do geographers help businesses and government?

About half of jobs using geography are in business and government. All kinds of businesses use geographic information to help build and expand their operations. A **location analyst** studies an area to find the best location for a client. The client might be a large retail store chain that wants to know which location would be best for opening a new store. The location analyst can study GIS reports on such elements as transportation networks or population in an area and give the business owners the positive and negative points about a location being considered.

In 1967, the Mexican government was looking for a location to create a new international tourist resort. They used location analysts to find an area that had good beaches and was easy to reach from the United States. The result was Cancún, today one of the world's most desirable vacation sites.

Businesses connected with natural resources such as forests also rely on geographers. Geographers help them understand the relationship between their business and the environment where their business is located. **Climatologists** are geographers who study

CONNECT Geography & Culture

Cancún Then and Now

In 1967, Cancún was a small, swampy island on Mexico's Caribbean coast. It had white sand beaches, many birds and mangrove trees, but few people. After it was selected as a resort site, it was quickly transformed. Today, Cancún has more than 100 hotels and 500,000 permanent residents. Many work in the tourist industry that serves the millions of visitors who come each year from all over the world.

DIFFERENTIATING INSTRUCTION

Struggling Readers

Categorize

To help students differentiate among careers in geography, have them complete a chart like this one. After they have correctly identified the responsibilities of each job, have them re-read and respond to the **Key Questions** in this section.

Job	Responsibilities
location analyst	
climatologist	
urban planner	

climates. They are used by businesses that need information about climate to conduct their operations. For example, coffee-growers in Brazil must have an idea if the weather during the next year will be helpful or harmful to their crop. To determine this, a climatologist might study the long-term climate data about the region to project future weather patterns.

City governments often use an **urban planner** to create plans for improving parts of the city. Planners may help locate and design residential or business areas, or parks and recreational spaces. They may find a location for an airport, mass transit routes, or sewer and water lines.

FIND MAIN IDEAS Explain how geographers help business and government. They analyze information to answer geographic questions.

Urban Planners Planners help city governments locate projects being built in a city. **What questions might a planner ask to help a city find the right location for a project?** They may look at land use, transportation, population patterns, or public services.

CONNECT to Your World

Geography Competitions

Across the United States and in foreign countries, students in grades four through eight compete in geography contests. Students are quizzed on all types of geography, including physical geography—such as locations of places and land and water features—and human geography.

Students compete in Washington, D.C., for the right to represent the United States in international competition for the world championship.

CRITICAL THINKING

Journal Entry Start a page in your journal with questions you could use in a geography contest at your school.

19

More About . . .

Urban Planning

The Chinese city of Shanghai has a museum devoted to urban planning. The Shanghai Urban Planning Exhibition Hall features exhibits that show urban planning achievements, most of which have taken place since the 1990s. The theme of the museum is "city, man, environment, and development." The museum also focuses on the future of the city.

Unit Resource Book: Introduction to World Geography
- History Makers: The Urban Planner, p. 32

CONNECT to Your World

Geography Competitions

The one way to prepare for a geography competition is to know map or atlas details. But it is also important to understand information about places and people. For example, a sample question might ask: "Which country has the world's largest Muslim population—Indonesia or Mexico?" Knowing about world religions is the key to the correct answer, which is Indonesia.

CRITICAL THINKING ANSWER

Check that students' journals include different types of geography questions.

Unit Resource Book: Introduction to World Geography
- Connect to Your World: Geography Competitions, p. 36

CONNECT to the Essential Question

In what ways does geography help us understand the world?

Ask students what they have learned so far that can help them answer this question. Students might mention:
- Geographers gather data to understand the world or improve planning.
- Many geographers offer advice or possible solutions to problems posed by government agencies.

DIFFERENTIATING INSTRUCTION: TIERED ACTIVITIES

OBJECTIVE Brainstorm and create a plan for a recreation area.

Basic

Give pairs of students graph paper. Ask each pair to plan a playground. Have students think about the kinds of people who use playgrounds and what they like to do there. Show students how to use the graph paper to plan their space. Encourage them to label the parts of their plan.

On Level

Have small groups of students brainstorm the needs of a playground or recreation area. Remind them to consider how people will use the area and to address safety. Have each group draw and label a plan. In addition, the group should write a one-page description of the advantages of their plan.

Challenge

Have small groups plan a new recreation area or redesign an existing one. Students' plans should include a detailed map and a proposal describing the advantages of the plan. Information about population using the area, types of uses, safety, and approximate cost should be included.

Teach

Physical and Human Geography

Reader, Recorder, Reporter

- How are earth scientists different from human geographers? *(Earth scientists study how Earth's shape changes; human geographers analyze social and political issues.)*
- **Categorize** What kinds of people can help human geographers analyze a problem? *(Possible Answers: political scientists, economists, sociologists)*

4 Assess & Reteach

Assess Have students complete the Section Assessment.

Unit Resource Book: Introduction to World Geography
- Section Quiz, p. 40

Interactive Review @ClassZone.com
- Section Quiz

Power Presentations

Test Generator

Reteach Have students write a job description for the jobs in this section. Have them explain the responsibilities and who a geographer might work for.

Unit Resource Book: Introduction to World Geography
- Reteaching Activity, p. 43

Surveying the Public Human geographers ask residents about problems in the neighborhood. **Why would geographers ask residents questions about their neighborhood?**

They are gathering data to help them analyze a geographic question for a project.

Physical and Human Geography

KEY QUESTION What jobs are available related to physical and human geography?

Physical geographers are sometimes called earth scientists. Some study such topics as **geomorphology** (JEE•oh•mawr•FAHL•uh•jee), that is, the study of how the shape of the Earth changes. Others study weather and climate. Still others study water, the oceans, soils, or ecology. Jobs in these fields require special scientific training.

Some geographers study social, political, and economic issues as they relate to place or region. Human geographers are usually hired by government agencies to analyze a specific problem. These geographers work closely with political scientists, economists, and sociologists. Together, they provide possible solutions to problems from many different aspects of life in an area. And, of course, geographers teach the subject at all levels of education, from elementary schools to universities.

But no matter what geography jobs people might hold, they are always trying to answer the basic geographic questions: "Where are things located?" and "Why are they there?"

SUMMARIZE Name some physical and human geography jobs.
Possible answer: Jobs are available in earth sciences such as geomorphology, weather, climate, ecology, and in social sciences.

ONLINE QUIZ For test practice, go to Interactive Review @ClassZone.com

Section 3 Assessment

TERMS & NAMES

1. **Explain the importance of**
   - location analyst
   - climatologist
   - urban planner

USE YOUR READING NOTES

2. **Categorize** Use your completed web diagram to answer the following question:
   In what ways could a business use the skills of a geographer?

KEY IDEAS

3. What are three requirements for most careers in geography?
4. Who are the two major employers of geographers?
5. What questions do all types of geographers ask?

CRITICAL THINKING

6. **Find Main Ideas** How important are geographers to businesses?
7. **Summarize** How do you train for a geographer's job in business and government?
8. **CONNECT to Today** How do planners help governments with public projects?
9. **WRITING Write a Job Description** Select a specific job in the field of geography and research it on the Internet. Then write a description of the job's skill requirements and its responsibilities. If possible find a person doing that job and do an interview or have that person speak to the class.

SECTION 3 ASSESSMENT ANSWERS

Terms & Names

1. location analyst: studies an area to find the best location for a client ; climatologist: studies climates; urban planner: creates plans for developing and improving parts of a city

Use Your Reading Notes

2. explain relationships between the environment and humans living in the area

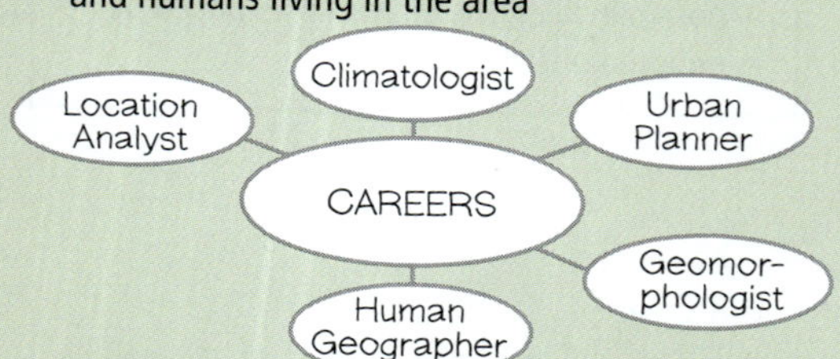

Key Ideas

3. the ability to think critically, good computer skills, and a college education
4. businesses and government
5. "Where are things located?" and "Why are they there?"

Critical Thinking

6. Geographers provide answers to business questions about location, weather, or environment.
7. get a college education in geography, develop critical thinking, writing, and computer skills
8. They use technology to analyze data to find the best locations, and work with social scientists to consider the social issues that affect the projects.
9. Use the rubric to score the job description.

Writing Rubric

	Content	Accuracy
4	excellent; addresses all important points	no errors
3	good; addresses some important points	few/minor errors
2	fair; addresses few important points	several errors
1	poor; addresses no important points	many errors

Interactive Review

Click here to complete these and other activities online @ ClassZone.com

CHAPTER SUMMARY

Key Idea 1
Geographers have specialized ways to view and interpret information about the world.

Key Idea 2
Geographers use technical tools to help them understand both the Earth's physical processes and the activities of people on Earth.

Key Idea 3
Geographers do many different kinds of jobs as they gather data and analyze and interpret it.

For **Review and Study Notes**, go to **Interactive Review** @ ClassZone.com

NAME GAME

Use the Terms & Names list to complete each sentence on paper or online.

1. I am a model of the Earth. globe
2. I am in orbit 100 miles above the Earth. ______
3. I am the study of people, places, and things. ______
4. I create plans for improving parts of a city. ______
5. I am an advanced technology tool used by geographers. ______
6. I am a representation of a part of the Earth. ______
7. I am a person who creates maps. ______
8. I am a group of places that have something in common. ______
9. I am a description of the physical and human characteristics of a location. ______
10. I am a person who maps and measures the land. ______

cartographer
climatologist
environment
Geographic Information Systems
geography
globe
Global Positioning System
Landsat
location analyst
map
spatial
surveyor
place
region
urban planner

Activities

Flip Cards

Use the online flip cards to quiz yourself on terms and names introduced in this chapter.

Crossword Puzzle

Complete an online crossword puzzle to test your knowledge of basic geographic terms.

ACROSS
1. a model of the Earth

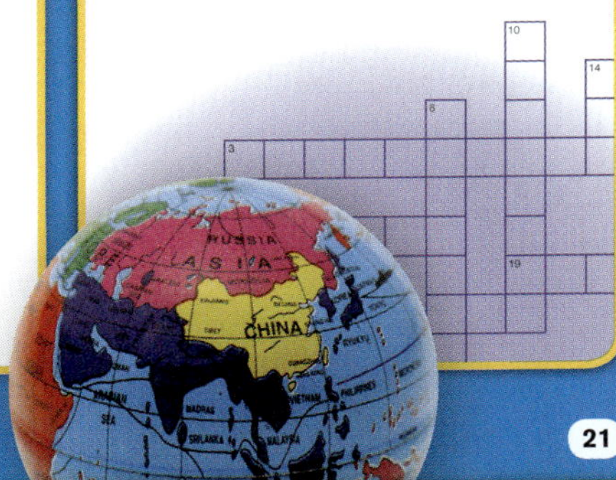

21

ONLINE AND ON-SCREEN ACTIVITIES

Flipcards

Vocabulary Flip Cards are virtual flash cards at **ClassZone.com.**

Students can quiz each other or work independently to study the flip cards for Chapter 1.

- They can quiz themselves on key terms and names by reviewing definitions.
- Demonstrate how to advance and shuffle the deck for another challenge.

Crossword Puzzle

Students can review important concepts of "Understanding the Earth and Its Peoples" by completing the online crossword puzzle at **ClassZone.com.**

- As an extension, or for more challenge, have students use the *Activity Maker* at **ClassZone.com** to make their own crossword puzzle for a section of the chapter.
- Have students trade puzzles and solve answers for each other's sections.

Chapter Summary

As a follow-up to the chapter, encourage students to pair up and go over the **Review and Study Notes** at **ClassZone.com.**

Alternatively, student pairs can review their **Reading Study Guide** notes, making sure that they can add important details to each section's "Key Idea."

Name Game

Teaching Tips

ON PAPER Have students work in pairs to complete this activity. As an extension, they can write questions for the terms that are not covered in the items shown here.

ONLINE Make sure that students understand how to drag and drop terms to fill in blanks. Ask them to print out their results and compare them with those of a classmate.

ANSWERS

1. globe
2. Landsat
3. geography
4. urban planner
5. Geographic Information Systems
6. map
7. cartographer
8. region
9. place
10. surveyor

Make It Fun

Teacher-Tested Activities

Esther Howse,
Byron Center West Middle School,
Byron Center, Michigan

Try this activity, "The Whole World in Your Hand," to review the five themes of geography. Have students trace their hand on a piece of paper. Next they should draw an image of earth from space in the palm space. Then they should fill in the five themes—one on each of the five fingers. (location, place, region, movement, human-environment interaction).

Vocabulary ANSWERS

1. Geographers view the world in spatial terms.
2. Remote sensing means finding information about Earth using instruments not in contact with the site.
3. The environment includes physical and human conditions on the earth.
4. A database provides information for answering geographic questions.
5. Geography is the study of the interaction between the earth's physical and human systems.
6. Both are ways to represent features found on Earth.
7. A cartographer is a geographer who creates maps.
8. A region is made up of places that are similar.
9. They are all technology tools.
10. They are all geography jobs.

Key Ideas ANSWERS

11. people, places, and environments
12. because it studies both physical processes and people
13. Location answers the question "Where is it located?" while place describes "What is it like?"
14. physical processes and human activities
15. obtaining information about a site using equipment that is not in contact with it
16. They collect images and data as they orbit the earth.
17. "Where am I?" and "Where am I going?"
18. Geographic Information Systems (GIS)
19. analyzing geographic information
20. studies geographic information to provide a client with positive and negative points about a location
21. create plans for improving parts of a city
22. geomorphology, studies of oceans, soils, weather, climate, and ecology

Critical Thinking ANSWERS

23. Table should show: Location: exact and relative; Place: physical and human characteristics; Region: common characteristics and political units where appropriate; Movement: people, things and/or ideas move and bring change; Human-envirnoment interaction: adapt or change environment

CHAPTER 1 Assessment

VOCABULARY

Explain the significance of each of the following.

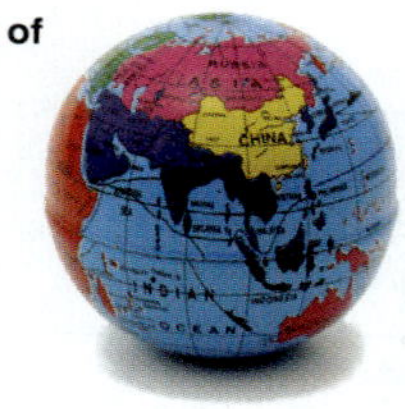

1. spatial
2. remote sensing
3. environment
4. database
5. geography

Explain how the terms and names in each group are related.

6. map, globe
7. map, cartographer
8. place, region
9. remote sensing, Landsat, Global Positioning System, Geographic Information System
10. cartographer, surveyor, climatologist, urban planner

KEY IDEAS

1 Themes and Elements of Geography

11. What are the three topics you study in geography?
12. Why is geography considered both science and social studies?
13. What is the difference between location and place?
14. What forces continually change Earth's surface?

2 Technology Tools for Geographers

15. What is remote sensing?
16. How do satellites aid in mapmaking?
17. What two basic geography questions does the Global Positioning System help you answer?
18. What is an example of a technologically advanced geographer's tool?

3 Careers in Geography

19. What is the main activity of a geographer's job?
20. What does a location analyst do?
21. What does an urban planner do?
22. What types of geography are included in physical geography?

CRITICAL THINKING

23. **Categorize** Create a table that lists the five themes of geography and shows two examples of geographic information that would be included each theme's description.

FIVE THEMES	GEOGRAPHIC INFORMATION

24. **Find Main Ideas** How do people adapt to their physical world?
25. **Summarize** How does modern technology help geographers?
26. **Connect to Economics** What role do geographers play in business operations?
27. **Connect Geography & History** How does the study of geographic patterns help us to understand past events?
28. **Five Themes: Human-Environment Interaction** What can be learned about people by studying their interaction with the environment?

Answer the ESSENTIAL QUESTION

In what ways does geography help us understand our world?

Written Response Write a two- or three-paragraph response to the Essential Question. Be sure to consider the key ideas of each section as well as specific ideas about how geographers answer geographic questions. Use the rubric below to guide your thinking.

Response Rubric

A strong response will:

- discuss the two basic geographic questions
- identify and describe the five themes and six essential elements of geography

ESSENTIAL QUESTION REVIEW

Before students begin their response, have them review the key chapter elements that will help them answer the Essential Question. Direct students' attention to significant features in the chapter, such as:

- Comparing Mapping Styles: Washington, D.C., Section 2
- Connect Geography and Technology, Section 2
- Connect To Your World, Section 3

Written Response Rubric

	Coverage of Key Ideas	Explanation	Historical Accuracy
4	all	clear, concise	no errors
3	most	good	few/minor errors
2	some	fair	several errors
1	two or fewer	poor	many errors

STANDARDS-BASED ASSESSMENT

Test Practice
- Online Test Practice @ClassZone.com
- Test-Taking Strategies and Practice at the front of this book

POLITICAL MAP

Use the map to answer questions 1 and 2 on your paper.

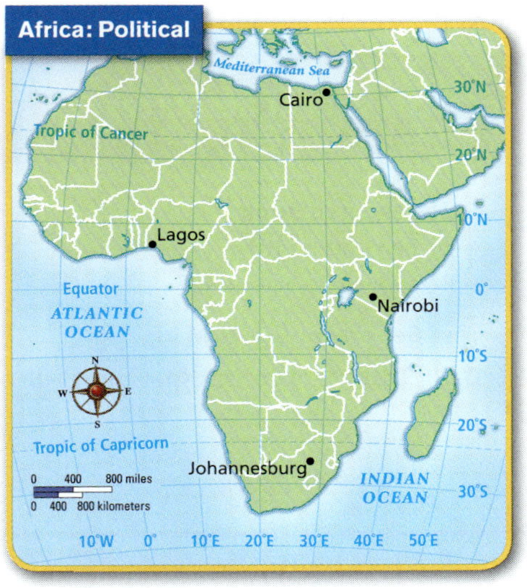

1. Which of the cities shown on the map is located south of the Tropic of Capricorn?

A. Cairo
B. Johannesburg
C. Nairobi
D. Lagos

2. Which of the cities shown on the map is located at approximately 30° N and 31° E?

A. Cairo
B. Johannesburg
C. Nairobi
D. Lagos

CLIMATE GRAPH

Examine the climate graph below. Use the information in the graph to answer question 3 on your paper.

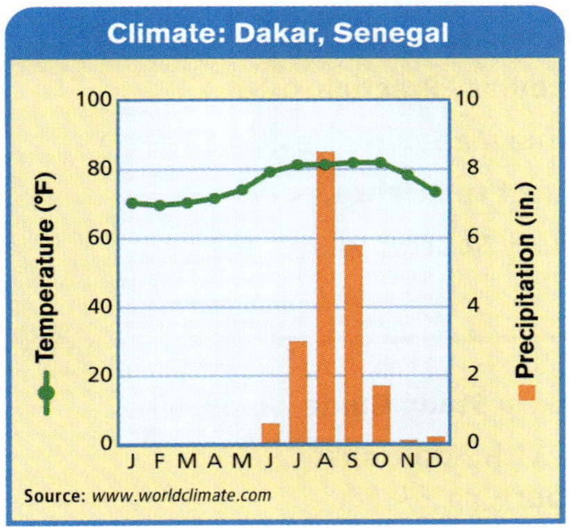

3. In which months does Dakar get the most rainfall?

GeoActivity

1. INTERDISCIPLINARY ACTIVITY–TECHNOLOGY

With a small group, review new geographic technology. Choose one new tool to research further and create a museum-style display. Be sure your display shows the tool and how it works. Also, explain how the tool has changed the way geographic information is gathered.

2. WRITING FOR SOCIAL STUDIES

Imagine that you are an urban planner and have been asked to develop a plan for placement of a park in your city. Write one paragraph describing the location you chose and the reasons you selected it.

3. MENTAL MAPPING

Create an outline map of your neighborhood, city, or town and locate and label any of the following that are present in the area you map:

- physical features
- your home
- your school
- fire station
- hospital
- shopping area

DIFFERENTIATING CHAPTER ASSESSMENT

Unit Resource Book

- **Chapter Test A,** pp. 45–48.
  Use for Struggling Readers, Beginning-to-Intermediate-Level English Learners, and for Inclusion of students whose IEP suggests supported test-taking.
- **Chapter Test B,** pp. 49–52.
  Use for On-Level Learners and ESOL students with Advanced-Level English.
- **Chapter Test C,** pp. 53–56.
  Use for Gifted and Talented students or students assigned Pre-AP work.

Test Generator

- Chapter Tests A, B, and C, in English and Spanish

McDougal Littell Assessment System

TEST **Test Generator** Use three levels of tests in English and Spanish tied to state standards. Plus, choose questions to customize tests for all populations.

SCORE Scan and Score plain paper student tests in minutes.

REPORT Create instant progress reports, sortable by student, by standard, or other criteria.

RETEACH Respond immediately to students' needs by using a wealth of reteaching resources.

24. People can change the environment to make it more livable, or they can choose to move to a different and better environment.

25. Modern technology that uses remote sensing equipment and computers that can store and organize all types of data. This helps geographers to gather more accurate information and helps them obtain it more rapidly.

26. Possible answer: Modern business can call on geographers to help determine where to locate their operations and how to solve problems that relate to the locations.

27. Possible answer: Studying geography to find patterns can help to explain why people acted as they did under certain circumstances and what events occurred when they did.

28. Possible answer: Studying how people interact with their environment can show how human activities are similar or different in various parts of the world, such as in warm or cold climates or in cities or less populated areas.

Standards-Based Assessment ANSWERS

1. B
2. A
3. August and September

GeoActivity

1. INTERDISCIPLINARY ACTIVITY RUBRIC

The display should
- show an example of a tool
- clearly demonstrate how the tool works
- accurately describe the tool's impact

2. WRITING FOR SOCIAL STUDIES RUBRIC

The paragraph should
- clearly describe the location chosen
- explain how the location was selected

3. MENTAL MAPPING RUBRIC

The outline map should
- represent the approximate shape of the student's neighborhood, city, or town
- include all geographic features listed in their correct location

Earth's Interlocking Systems
PLANNING GUIDE

Complete print resources all on one USB drive allow you to customize lesson plans for all students.

Section Titles & Key Ideas	Unit Resource Book Introduction to World Geography	Assessment Resources
1 The Earth and Its Forces **KEY IDEA** The Earth is composed of many layers. Its surface continually changes because of the drifting of its plates.	• **Reading Study Guide,** pp. 57–58 • **RSG with Additional Support,** pp. 65–66 • **Vocabulary Practice,** p. 89 • **Outline Map Activities,** pp. 93–94 • **Connect to Science,** p. 100 • **Primary Source,** pp. 101–102	**Unit Resource Book** • Section Quiz, p. 103 • Reteaching Activity, p. 107 **Test Generator CD-ROM** **Daily Test Practice Transparency,** TT9
2 Bodies of Water and Landforms **KEY IDEA** Interaction between landforms and bodies of water makes life on Earth possible.	• **Reading Study Guide,** pp. 59–60 • **RSG with Additional Support,** pp. 67–68 • **Outline Map Activities,** pp. 93–94 • **Primary Source,** pp. 101–102	**Unit Resource Book** • Section Quiz, p. 104 • Reteaching Activity, p. 108 **Test Generator CD-ROM** **Daily Test Practice Transparency,** TT10
3 Climate and Vegetation **KEY IDEA** The Earth's rotation and revolution influence weather, climate, and living conditions on Earth.	• **Reading Study Guide,** pp. 61–62 • **RSG with Additional Support,** pp. 69–70 • **Skillbuilder Practice,** p. 90 • **Interdisciplinary Projects,** pp. 96–98	**Unit Resource Book** • Section Quiz, p. 105 • Reteaching Activity, p. 109 **Test Generator CD-ROM** **Daily Test Practice Transparency,** TT11
4 Environmental Challenges **KEY IDEA** Human interference with physical systems can cause problems with the environment.	• **Reading Study Guide,** pp. 63–64 • **RSG with Additional Support,** pp. 71–72 • **Economics in the World,** p. 99 • **History Makers: Wangari Maathai,** p. 95	**Unit Resource Book** • Section Quiz, p. 106 • Reteaching Activity, p. 110 • Chapter Test (Forms A, B, & C), pp. 111–122 **Test Generator CD-ROM** **Daily Test Practice Transparency,** TT12

McDougal Littell Assessment System
TEST SCORE REPORT RETEACH

No Time? To focus students on important content in the chapter, use the **"Reading Study Guide with Additional Support"** in ***Resources2Go.***

Teacher Presentation Options

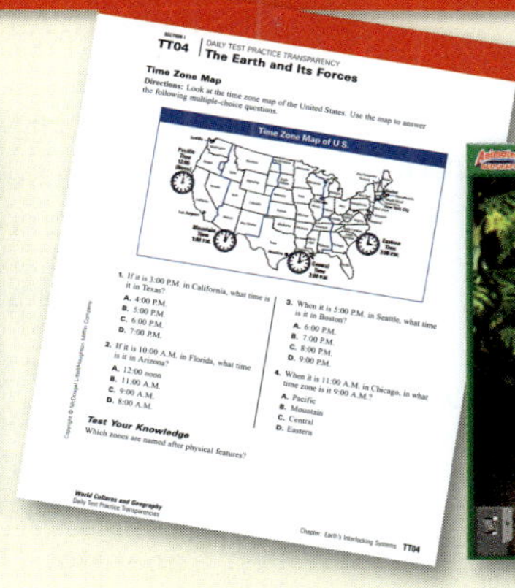

Electronic Books
eEdition DVD-ROM
eEdition Online

Presentation Toolkit
Video/DVD
- World Cultures and Geography

Power Presentations DVD-ROM
- Discussion Starters
- Media Gallery
- Review Game
- Interactive Review
- Animated Features

Unit Transparency Book
- 3-Minute Warm-Ups
- Essential Question
- Critical Thinking
- Maps
- Cultures

Daily Test Practice Transparencies

Animated Geography
- Earth's Tectonic Plates
- Continental Drift Theory
- Amazon Rain Forest

OnlineActivities at ClassZone.com

Animated Geography
- Videos and Photos
- 3-D Animations
- Interactive Maps

Interactive Review
- Section Quizzes
- Review and Study Notes
- Vocabulary Flip Cards
- Graphic Organizers
- Name Game and Crossword Puzzles
- GeoGame

Online Test Practice

Map Center
- Atlas Maps
- Political Maps with Data
- Links to Mapping Web Sites

Activity Center
- Activity Maker—Review Games, Puzzles, Flip Cards, and Graphic Organizers
- Worldster

Research & Writing Center
- Web Sites
- Data Updates
- Primary Sources
- Writing Models

Current Events

Eric Branch
Wright Middle School
Madison, Wisconsin

Make It Fun — Teacher-Tested Activities

As you teach this chapter, look for my classroom-proven ideas for:
- making the lesson fun
- focusing students' attention and comprehension
- integrating technology with traditional teaching

Teaching Bell to Bell When you see this timing guide, have students discuss in pairs, in groups of 3–6, or as a class.
- **Think, Pair, Share** (pairs)
- **Reader, Recorder, Reporter** (3–6)
- **Talk About It** (any size)

CHAPTER 2 RESOURCES FOR DIFFERENTIATED INSTRUCTION

Struggling Readers

eEdition with Audio DVD-ROM

Lesson Plans
- Strategies for Inclusion

Unit Resource Book
- Reading Study Guide with Additional Support, pp. 65–72 (B)
- Reteaching Activity, pp. 107–110
- Chapter Test (Form A), pp. 111–114 (A)

Best Practices Toolkit (C)
- Interactive Reading Strategies, pp. 90–145
- Differentiation Activity Bank: Struggling Readers, pp. 146–149

Teacher's Edition Activities
- Find Locations of Bodies of Water, Sec. 2
- Create an Illustration, Sec. 3
- Compare and Contrast, Sec. 4
- Tiered Activities, Sec. 1–4

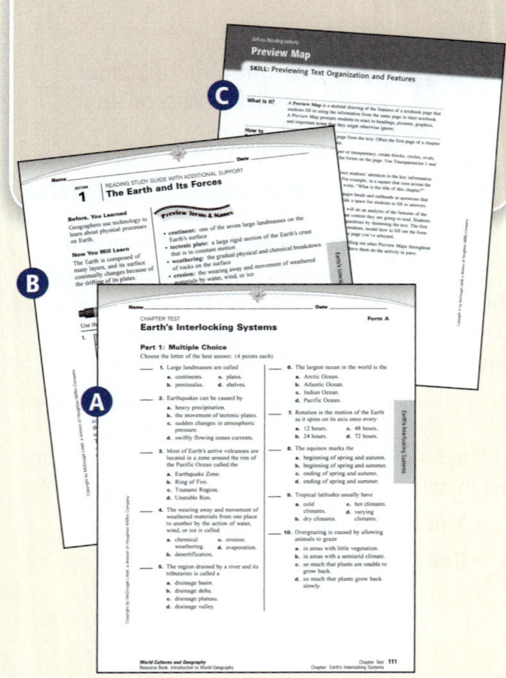

English Learners

Pupil's Edition in Spanish

eEdition with Audio DVD-ROM

Lesson Plans
- Strategies for English Learners

Unit Resource Book
- Reading Study Guide (Spanish), pp. 73–80
- RSG with Additional Support (Spanish), pp. 81–88 (A) (B)

Multi-Language Glossary (C)

Best Practices Toolkit
- Differentiation Activity Bank: English Learners, pp. 165–194

Test Generator CD-ROM
- Section Quizzes in Spanish
- Chapter Test in Spanish

Teacher's Edition Activities
- Pronounce and Preview, Sec. 1–4
- Comprehension: Self Monitor, Sec. 1
- Vocabulary: Multiple-Meaning Words, Sec. 1
- Key Academic Vocabulary, Sec. 2, 3
- Tiered Activities, Sec. 1–4

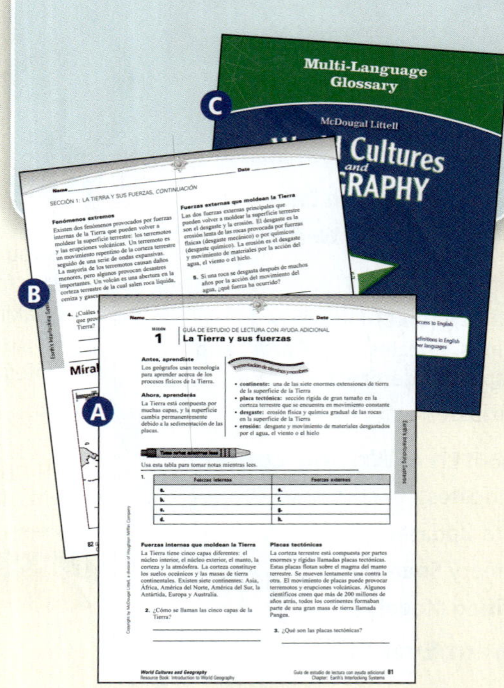

Inclusion

eEdition with Audio DVD-ROM

Lesson Plans
- Strategies for Inclusion

Unit Resource Book
- Reading Study Guide with Additional Support, pp. 62–72 (A) (B)
- Reteaching Activity, pp. 107–110
- Chapter Test (Form A), pp. 111–114

Best Practices Toolkit (C)
- Interactive Reading Strategies, pp. 90–145
- Differentiation Activity Bank: Struggling Readers, pp. 146–149

Teacher's Edition Activities
- Match Terms and Definitions, Sec. 1, 2
- Create Audio or Visual Vocaulary, Sec. 3
- Tiered Activities, Sec. 1–4

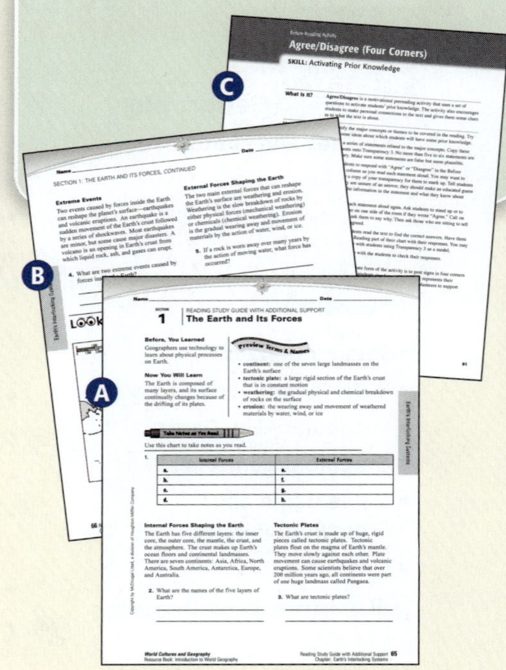

Interdisciplinary Activities

Gifted & Talented Students

Lesson Plans
- Core Lesson Plan

Unit Resource Book
- Map and Graph Skills, p. 91 (A)
- Outline Map Activities, pp. 93–94
- History Maker, p. 95 (B)
- Interdisciplinary Projects, pp. 96–98 (C)
- Economics in the World, p. 99
- Chapter Test (Form C), pp. 119–122

Best Practices Toolkit
- Student Products Rubrics, pp. 1–14
- Writing Templates, pp. 78–89

Teacher's Edition Activities
- Create a Board Game, Sec. 1
- Tiered Activities, Sec. 1–4

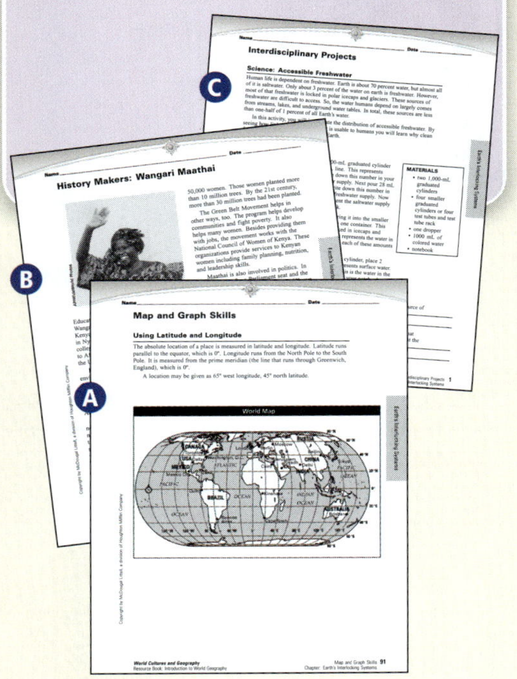

Pre-AP*

Lesson Plans
- Core Lesson Plan

Unit Resource Book
- Outline Map Activities, p. 94 (A)
- History Maker, p. 95
- Economics in the World, p. 99 (B)
- Connect to Science, p. 100
- Primary Source, pp. 101–102 (C)
- Chapter Test (Form C), pp. 119–122

Best Practices Toolkit
- Differentiation Activity Bank: Pre-AP Students, pp. 195–200
- Writing Templates, pp. 78–89

* Pre-AP is a registered trademark of the College Entrance Examination Board, which was not involved in the production of and does not endorse this product.

Teacher's Edition Activities
- Complie a List of Earthquakes and Volcanic Eruptions, Sec. 1
- Tiered Activities, Sec. 1–4

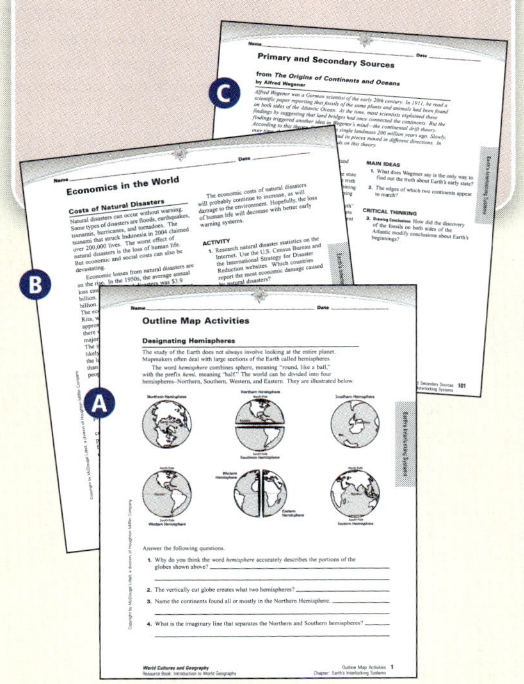

CONNECT to Science

Mapping Vegetation Regions

CLASS TIME 20 minutes

TASK Show vegetation regions on a map of the world.

SUPPLIES
- blank physical map of the world
- colored pencils or markers

ACTIVITY Have students review the information on Vegetation Regions in Section 3. Have the class identify the four basic vegetation regions and write the information on the board. Then have them study the Comparing World Climate Regions feature in Section 3. Using the chart information, have students color and label their blank maps to show the different vegetation regions. Ask them to identify as many different kinds of vegetation as they can in each region.

Good for Visual and Verbal Learners

CONNECT to Math

Bar Graph of Greenhouse Gas Emissions

CLASS TIME 30 minutes

TASK Create a bar graph that shows the top ten countries for greenhouse gas emissions.

SUPPLIES
- recent data on greenhouse gas emissions
- paper, colored markers or pencils

ACTIVITY Have students re-read the global warming subsection in Section 4. Supply copies of (or have students research) greenhouse gas emission data by country or state for a recent year. Then have them work in pairs to create bar graph posters showing the top ten countries or states for greenhouse gas emissions. Remind students to use the ruler to create the correct scale for the graphs. Suggest that they use different colors or patterns for each country or state. Display the completed bar graph posters in the classroom.

Good for Visual and Mathematical Learners

Introduce the ESSENTIAL QUESTION

- Explain how internal and external forces shape Earth.
- Point out the importance of water to plant and animal life on Earth.
- Describe how weather and climate affect vegetation regions on Earth.

Key Ideas

Section 1: The Earth is composed of many layers, and its surface continually changes because of the drifting of its plates.

Section 2: Interaction between landforms and bodies of water makes life on Earth possible.

Section 3: The Earth's rotation and revolution influence weather, climate, and living conditions on Earth.

Section 4: Human interference with physical systems can cause problems with the environment.

CONNECT to the Essential Question

You may wish to revisit the Essential Question periodically with students as they work through the chapter.

Suggestions for reconnecting are provided at an appropriate point in the *Teacher's Edition* for each section.

Rice fields, Bali, Indonesia

CHAPTER 2

Earth's Interlocking Systems

ESSENTIAL QUESTION

How do Earth's physical systems make life on Earth possible?

CONNECT Geography & History

Use the map and the time line to answer the following questions.

1. On which plate does most of the United States sit?
2. Which event on the time line is supported by the evidence on this map?

1500

Geography
1620 English philosopher Francis Bacon suggests the continents were once a supercontinent.

Geography
1883 Volcanic eruption destroys two-thirds of Krakatoa Island, Indonesia.

TIME LINE DISCUSSION

Tell students that even small movements of the Earth's tectonic plates can lead to major events such as volcanic eruptions and earthquakes.

- Which events on the time line highlight volcanic eruptions? *(1883—Volcanic eruption destroys two-thirds of Krakatoa Island, Indonesia; 1980—Mount St. Helens erupts in Washington and causes the world's largest landslide.)*
- Which event on the time line highlights an earthquake? *(2004—Indian Ocean earthquake and tsunami kill more than 280,000.)*
- In what way do you think the events highlighted for 1620 and 1912 are related? *(Possible Answer: Both involve theories proposed to explain Earth's landforms. The development of the second theory might have been affected by the first theory.)*

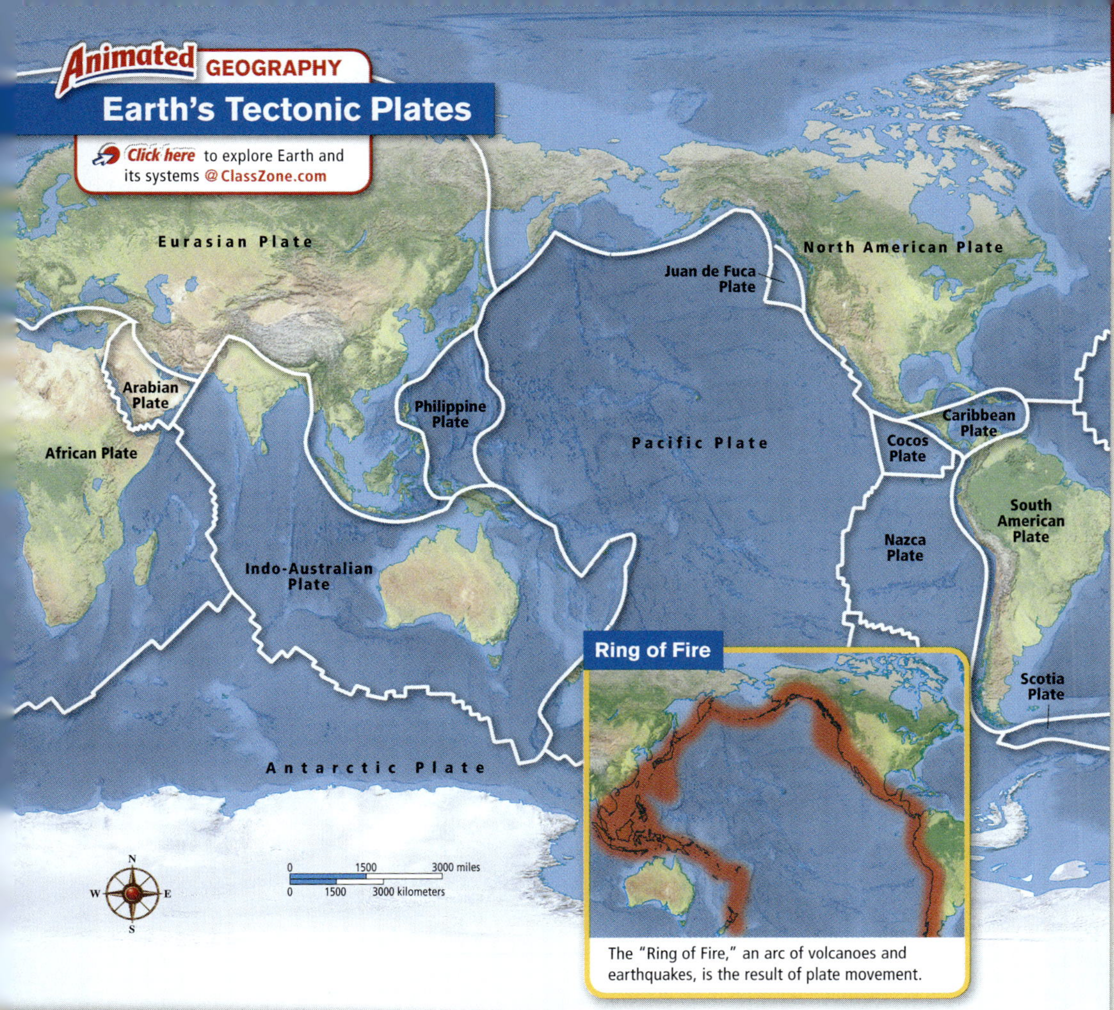

The "Ring of Fire," an arc of volcanoes and earthquakes, is the result of plate movement.

25

TEACHER MEDIA FAVORITES

Books for the Teacher

Albert, Toni. *The Remarkable Rainforest: An Active-Learning Book for Kids.* Trickle Creek Books, 2003. Provides a rich variety of activities to help students learn about rain forests and the environment.

Barr, Linda. *Volcano! When a Mountain Explodes.* Capstone Press, 2004. Audio book on the causes of volcanic eruptions, including the eruption of Mount St. Helens in 1980.

Van Rose, Susanna. *Volcano and Earthquake.* DK Publishers, Inc., 2000. Informative guide to volcanic eruptions and earthquakes that occur on Earth.

Video / DVD

Global Warming: The Signs and the Science. (DVD) PBS Home Video, 2005. TV documentary on the people and places in the world affected by global warming. (Narrated by Alanis Morrisette.)

Tropical Rainforest. (DVD) Vista Point Entertainment, 2005. Documentary about daily life in the rain forest. Narrated by Geoffrey Holder.

Internet

Go to the *Research and Writing Center* at **ClassZone.com** to link to:

- data on the region
- a variety of chapter-specific content-reviewed sites
- "History Makers" biographies
- primary sources

Animated GEOGRAPHY

Earth's Tectonic Plates

CLASSZONE.COM

Point out that this is a physical map of the Earth showing different tectonic plates. Explain why the arc of volcanoes and earthquakes displayed as an inset is called the "Ring of Fire," and point out its location on an expanded world map.

- According to the time line, what historic event took place on the North American plate? *(the eruption of Mt. St. Helens)*
- Why do you think the Eurasian tectonic plate is the largest? *(Possible Answer: because it covers two continents—Europe and Asia)*
- Compare this map to the Continental Drift Theory diagram in Section 1. What was different about Earth's landforms in the past? *(The continents were joined together as one supercontinent.)*

CONNECT GEOGRAPHY & HISTORY ANSWERS

1. North American Plate
2. 1960s: Scientists prove that the Earth's crust is broken into moving plates.

Presentation Options

Use the **Power Presentations** DVD-ROM to project the animated map.

- Prompt students to use the pop-up images and audio from the map to discover what they already know about the geography, culture, history, economics, or governments of the region.
- Challenge students to answer the map questions individually, in small groups, or as a class activity. Use all of the interactive material to help preview chapter content.

1 Plan & Prepare

Objectives

- Identify the internal forces shaping the Earth
- Describe two extreme events caused by internal forces
- Summarize the external forces shaping the Earth

Read for the Essential Question

Help students read for a purpose by reminding them of the Essential Question: "How do Earth's physical systems make life on Earth possible?" Point out that they can use text features such as Key Questions, special features, images, and maps to form answers to the question.

Vocabulary

Best Practices Toolkit

Use the ***Best Practices Toolkit*** to model strategies for vocabulary notetaking. Vary strategies throughout the year. Choose from: *Knowledge Rating, Predicting ABCs, Definition Mapping, Word Sort, Frayer Model (Word Squares), Magnet Words, Word Wheel,* and *Student VOC.*

Vocabulary Strategies, TT9–TT16

Reading Strategy

Best Practices Toolkit

Display the T Chart/Two-Column Chart transparency.

- Explain to students that in categorizing they need to determine how the information in the section can be sorted into the categories shown in the chart.
- Model this strategy by filling in an internal force and an external force from the section.
- Have students suggest entries using the categories in the chart.
- Display the graphic organizer for students to review.

T Chart/Two-Column Chart, TT20

SECTION 1 Reading for Understanding

Key Ideas

BEFORE, YOU LEARNED

Geographers use technology to learn about physical processes on Earth.

NOW YOU WILL LEARN

The Earth is composed of many layers. Its surface continually changes because of the drifting of its plates.

Vocabulary

TERMS & NAMES

magma molten rock

continent one of seven large landmasses on the Earth's surface

tectonic plate a large rigid section of the Earth's crust that is in constant motion

earthquake a sudden movement of the Earth's crust followed by a series of shocks

Ring of Fire a zone of volcanoes around the Pacific Ocean

volcano an opening in the Earth's crust from which molten rock, ash, and hot gases flow or are thrown out

weathering the gradual physical and chemical breakdown of rocks on the Earth's surface

erosion the wearing away and movement of weathered materials by water, wind, or ice

sediment pieces of rock in the form of sand, stone, or silt deposited by wind, water, or ice

glacier a large, slow-moving mass of ice

Visual Vocabulary glacier

Reading Strategy

Re-create the chart shown at right. As you read and respond to the **KEY QUESTIONS**, use the chart to organize important details about the external and internal forces shaping the Earth.

Skillbuilder Handbook, page R7

CATEGORIZE

INTERNAL FORCES	EXTERNAL FORCES
1.	1.
2.	2.
3.	3.
4.	4.

GRAPHIC ORGANIZERS Go to Interactive Review @ClassZone.com

PRETEACHING VOCABULARY

English Learners

Pronounce and Preview

Pronounce each term for students. Review the meanings of words within definitions, such as *rigid, molten,* and *silt.*

- To modify vocabulary learning, have students complete worksheets as they read, instead of afterward.

Unit Resource Book: Introduction to World Geography
- Vocabulary Practice, p. 89

Inclusion

Match Terms and Definitions

Have student pairs write the vocabulary words and definitions on index cards. Then have them cut out the words and definitions to create a matching game. Ask one student to shuffle the cards and arrange them. After their partner has finished matching the terms and definitions, have the students work together to correct any errors. Students should then switch roles.

Big Sur, California

Peruvian Rainbow Boa in the rain forest

SECTION 1 GEOGRAPHY

The Earth and Its Forces

Connecting to Your World

Have you ever experienced an earthquake or the eruption of a volcano? Probably not. But you may have seen these events on television. Perhaps you saw the coverage of the great earthquake in Indonesia on December 26, 2004. Hundreds of thousands of people were killed or displaced by this earthquake and the tsunami, or great wave, it caused. Earthquakes and volcanoes are just two of the many forces that change the Earth's surface.

Internal Forces Shaping the Earth

KEY QUESTION How is the Earth's surface shaped by internal forces?

The Earth is one of eight planets that orbit the sun. It is located about 93 million miles from the sun. The Earth's circumference, or distance around, is 24,900 miles. Its diameter, or distance through the center of Earth, is about 7,900 miles. Earth is unique in the solar system—the sun and its planets—because it supports life. This is because the Earth has lots of breathable air and usable water. In fact, the Earth's surface is about three-fourths water and one-fourth land. The Earth appears to be a solid ball, but it is actually made of different layers that are like shells set inside each other.

San Francisco Earthquake The earthquake in 1989 was the worst quake since 1906.

2 Focus & Motivate

3-Minute Warm-Up

Write on the board or display the transparency:

- How would you describe a volcanic eruption? An earthquake? *(Volcanic eruption: an explosion of lava and smoke; Earthquake: a great shaking of the Earth)*

Unit Transparency Book
- 3-Minute Warm-Ups, TT9

Connecting to Your World

More About . . .

The San Francisco Earthquakes

The 1906 earthquake is considered one of the worst natural disasters in U.S. history. Between 700 and 3,000 people were killed. The 1989 earthquake caused almost six billion dollars in damage in San Francisco.

3 Teach

Internal Forces Shaping the Earth

Talk About It

- Which of Earth's layers contains the continents? *(crust)*
- Why do tectonic plates cause earthquakes? *(because they move against or separate from each other)*
- **Categorize** What are the different kinds of tectonic plate movements? *(spreading, diving, colliding, and sliding)*

SECTION 1 PROGRAM RESOURCES

Unit Resource Book: Introduction to World Geography

ON LEVEL

Unit Resource Book
- Reading Study Guide, p. 57
- Connect to Science, p. 100
- Section Quiz, p. 103

STRUGGLING READERS

Unit Resource Book
- Reading Study Guide with Additional Support, p. 65
- Skillbuilder Practice, p. 90
- Vocabulary Practice, p. 89
- Section Quiz, p. 103
- Reteaching Activity, p. 107

eEdition with Audio DVD-ROM

ENGLISH LEARNERS

Pupil Edition in Spanish

eEdition with Audio DVD-ROM

Unit Resource Book
- RSG (Spanish), p. 73
- RSG with Additional Support (Spanish), p. 81

Multi-Language Glossary

Test Generator
- Section Quiz in Spanish

INCLUSION

Unit Resource Book
- RSG with Additional Support, p. 65
- Section Quiz, p. 103
- Reteaching Activity, p. 107

GIFTED & TALENTED

Unit Resource Book
- Map and Graph Skills, p. 91
- Section Quiz, p. 103

PRE-AP

Unit Resource Book
- Outline Map Activities, p. 94
- Section Quiz, p. 103

TECHNOLOGY

Unit Transparency Book
- 3-Minute Warm-Ups, TT9
- Essential Question, TT10
- Critical Thinking, TT11
- Map: World Physical, TT15

Daily Test Practice Transparencies
- Section 1, TT4

Power Presentations

ClassZone.com

World Cultures and Geography Video on DVD

More About . . .

Earth's Atmosphere

The Earth's atmosphere also contains several layers. The first five to ten miles above the Earth contain the troposphere. Air constantly moves in the troposphere, causing wind. Above the troposphere is the stratosphere, a zone of dry, thin air. Within the stratosphere is the ozone layer, a barrier that prevents most of the sun's powerful ultraviolet rays from reaching the Earth's surface.

Comparing Regions

Asia and Africa

Asia and Africa are the two largest continents in the world. Over three billion people live in Asia, or almost three fifths of the world's total population. Africa has less than a billion people, but has more countries. Asia's population is densely packed into areas such as the Ganges plains in India and the Yangtze river valley in northern China, where monsoon rains and fertile soil encourage farming. Africa has relatively little productive farming compared to Asia.

Earth's Layers

It is about 3,900 miles to the center of the Earth. The Earth has five different layers.

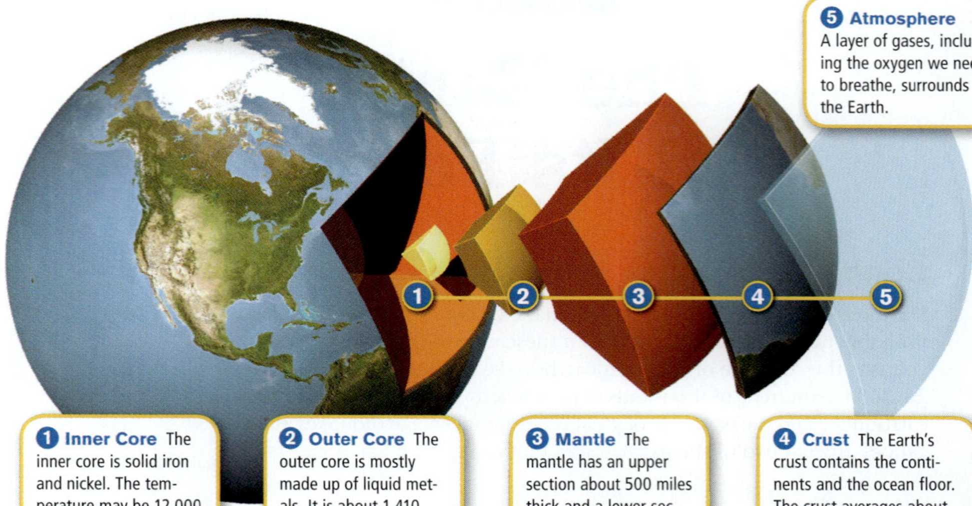

5 Atmosphere A layer of gases, including the oxygen we need to breathe, surrounds the Earth.

1 Inner Core The inner core is solid iron and nickel. The temperature may be 12,000 degrees Fahrenheit.

2 Outer Core The outer core is mostly made up of liquid metals. It is about 1,410 miles thick.

3 Mantle The mantle has an upper section about 500 miles thick and a lower section 1,330 miles thick.

4 Crust The Earth's crust contains the continents and the ocean floor. The crust averages about 20 miles thick.

Earth's Molten Interior The center of the Earth's interior is a hot metal core made up of one layer of iron and one of nickel. The inner core is solid. The outer core is liquid, because the metal has melted. Lying just above the core is the mantle. The mantle is a soft layer of hot rock, some of which is molten, or melted. It is the largest of the Earth's layers. The molten rock of the mantle is called **magma**.

The crust is the Earth's thin outer layer, or shell. It is the solid, rocky surface of the Earth that forms the ocean floors and the large landmasses called **continents**. The crust is the part of the Earth on which we live. It floats on top of the mantle. The crust is only about five miles thick under the oceans but averages about 22 miles thick under the continents. The Earth's layers are illustrated above.

Geographers have identified seven continents on the Earth's surface. The continents, in order of their size, are Asia, Africa, North America, South America, Antarctica, Europe, and Australia. Europe and Asia are actually one great landmass that is sometimes referred to as Eurasia or the Eurasian continent. Antarctica is a continent because it has a landmass beneath its icy surface. The Arctic does not, and so it is not a continent.

DIFFERENTIATING INSTRUCTION

English Learners

Comprehension: Self-Monitor

Students learning English may have difficulty understanding the explanation of tectonic plate movements on the next page. Point out that the visuals at the bottom of the next page illustrate the different kind of plate movements. As students read each explanation, tell them to pause periodically to check the visual to better understand the information.

Gifted & Talented

Create a Board Game

Have students create a board game using Earth's layers. Participants in the game will attempt to travel from the atmosphere to the inner core while facing earthquakes, volcanic eruptions, and other challenges. The students should design a game board, create the rules, and demonstrate how to play.

Tectonic Plates The surface of the Earth is constantly moving and changing, even as you read this sentence. Geographers use technological tools to observe and measure forces deep inside the Earth and on the surface that reshape the Earth's crust.

Plate movement, earthquakes, and the activity of volcanoes are all internal forces that change the landscape. The Earth's crust is divided into a number of large rigid pieces called **tectonic plates**. These plates are shown on the map at the beginning of the chapter. The continents and oceans are located on these plates, which float on the magma of the Earth's mantle. Heated magma cools as it reaches the crust and then sinks downward. This process causes the magma to act like a conveyor belt under the plates. The plates move slowly against each other, at a rate of up to four inches a year. The plate movement can cause earthquakes and volcanic eruptions. There are four types of tectonic plate movements that are shown in the diagrams below. Each of the movements causes changes in the shape of Earth's crust.

COMPARING Tectonic Plate Movements

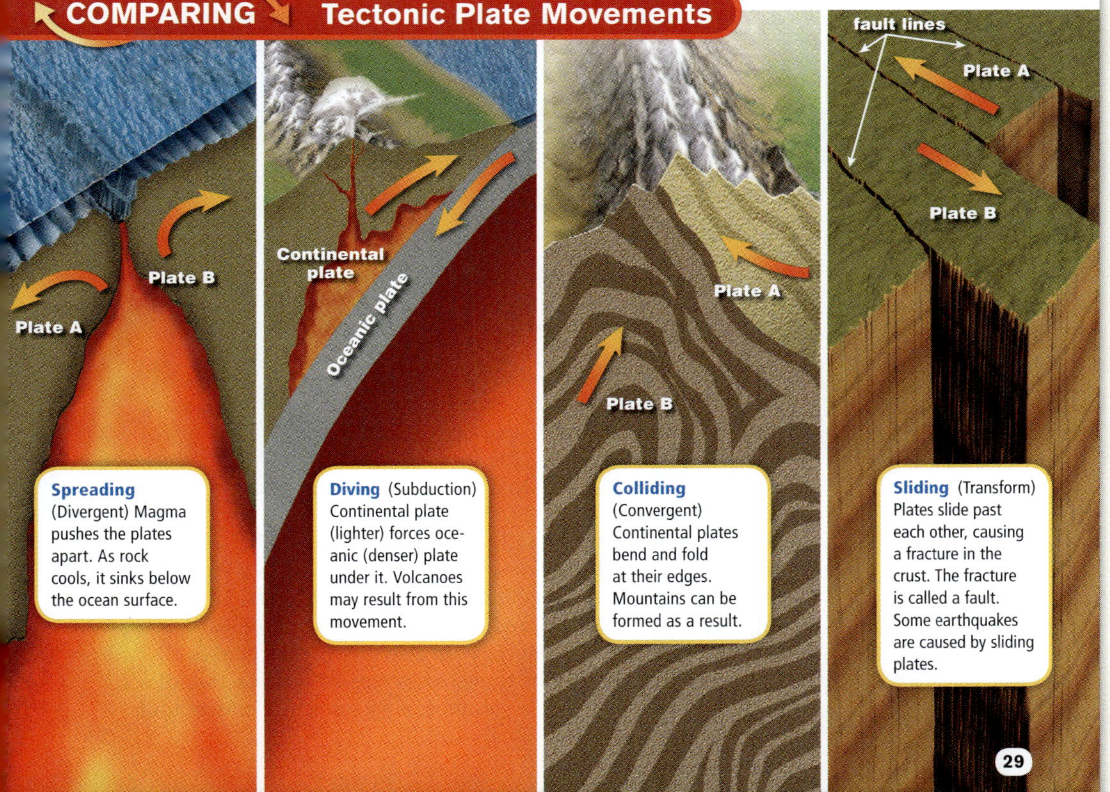

More About . . .

Tectonic Plate Movements

Point out that plate movement leads to the creation of new plate material as well as reshaping of the crust. The spreading that pushes plates apart allows magma to create new ridges of crust. The spreading beneath the Atlantic Ocean adds about 15 miles of new crust every one million years or so. This means that the Atlantic, once a small bay between Africa, Europe, and the Americas, is now more than 3,000 miles wide.

COMPARING

Tectonic Plate Movements

Diving and colliding plates can push up or sideways to create mountains. Earthquakes and volcanoes are often associated with these movements.

- How would you explain the difference between diving and colliding plates? *(Possible Answer: Diving involves a lighter plate forcing a heavier plate under it, resulting in volcanoes; colliding involves plates bending and folding at their edges, resulting in mountains.)*

Unit Resource Book: Introduction to World Geography
- Outline Map Activities: Designating Hemispheres, p. 94

INTERDISCIPLINARY ACTIVITIES

CONNECT to Science

Chart How Tectonic Plates Move

Have students make charts, such as the one at right, to help them better understand the movements of tectonic plates. Have them use their textbooks to complete their charts. As students finish each section of this chapter, have them update their charts.

Movement	Action	Result
Spreading *(Divergent)*	*(plates pushed apart)*	
Diving *(Subduction)*	*(heavier plate forced under lighter plate)*	
Colliding *(Convergent)*	*(plate edges bend and fold)*	
Sliding *(Transform)*	*(plates slide past each other)*	

Make It Fun

Teacher-Tested Activities

Eric Branch, Wright Middle School, Madison, Wisconsin

To help students visualize tectonic movement provide pairs of students with pieces of cardboard (from shoeboxes, lightweight boxes, or corrugated pieces). Instruct students to push the two pieces toward each other. Record what happens. Some may slide over the other, while others will fold. Repeat several times with different weights of cardboard. Have the students look at the illustrations on this page and match the results of their tests with the illustrations.

Animated GEOGRAPHY

Continental Drift Theory

CLASSZONE.COM

CONNECT Geography & History

Point out that the second map shows the continents 65 million years ago. The continents that are clearly connected look different from how they look now. Have students study both maps and notice the different shapes of the continents and how they have changed over time.

ANSWER

Draw Conclusions South America, Antarctica and Australia; Africa, Asia and North America

Unit Resource Book: Introduction to World Geography

- Skillbuilder Practice, p. 90
- Map and Graph Skills: Using Latitude and Longitude, p. 91

Teach

Extreme Events

Reader, Recorder, Reporter

- What two major events are caused by internal forces? *(earthquakes and volcanic eruptions)* What do they have in common? *(Both are destructive and can reshape the Earth's surface.)*
- Describe how volcanoes erupt. *(Molten rock rises to the Earth's mantle and is stored in a chamber beneath the crust; pressure builds and forces the molten rock to erupt.)*
- **Categorize** What are some significant effects of earthquakes? *(collapse of buildings, destruction of cities, many lives lost, fires, floods, landslides, avalanches)*

Animated GEOGRAPHY

Continental Drift Theory

Click here to see the movement of the continents @ ClassZone.com

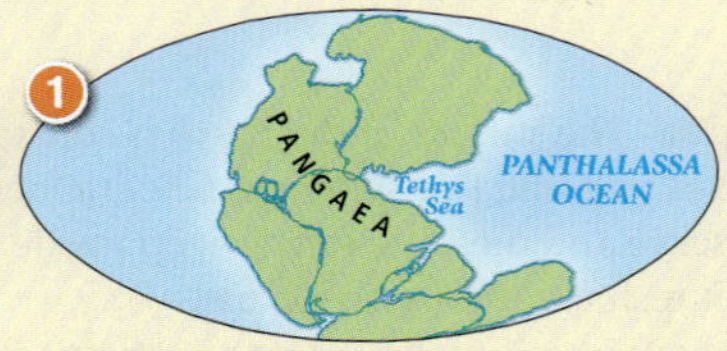

1 The world map at left shows the supercontinent of Pangaea about 200 million years ago as it begins to break apart.

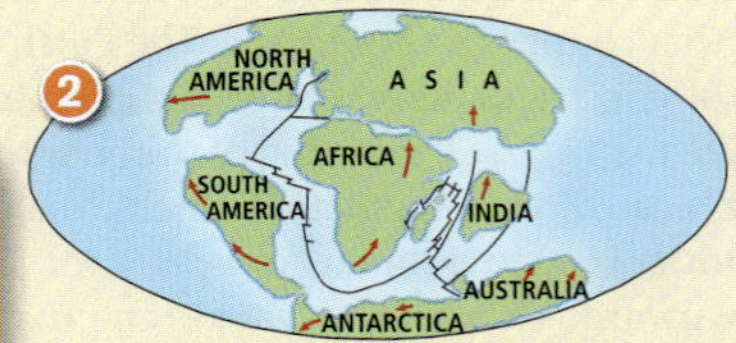

2 This world map shows the Earth about 65 million years ago. Some of the modern-day continents are visible.

CONNECT Geography & History

Draw Conclusions Which continents were connected 65 million years ago?

Continental Drift If you look at the continents, they look like giant pieces of a jigsaw puzzle waiting to be fit together. Alfred Wegener, a German meteorologist, noticed this pattern. He believed that the continents were once part of a supercontinent called Pangaea (pan•JEE•uh). He suggested that Pangaea divided and drifted apart about 200 million years ago. It contained almost all of the landmasses on Earth and was surrounded by one vast ocean. When it began to break apart, its pieces slowly moved in different directions, and some formed the continents we now know. To this day, the continents continue to move.

SUMMARIZE Explain how internal forces shape the Earth's surface. Cooling magma acts like a conveyer belt, moving tectonic plates and causing earthquakes and volcanoes.

Extreme Events

KEY QUESTION How are earthquakes and volcanoes connected?

Two events caused by internal forces may dramatically reshape the Earth's surface—earthquakes and volcanic eruptions. Both can be deadly, and both have powerful effects on human life.

Earthquakes An **earthquake** is a sudden movement of the Earth's crust that is followed by a series of shocks. Huge rocks along a line where faults are located slide apart and break up, causing that area of the Earth's surface to shake. The stress in the rocks builds for years. Then, energy is swiftly released outward through the ground in vibrations called shock waves.

Earthquakes occur constantly. Some we may not feel, but scientific instruments record them. Some produce slight shock waves. Other earthquakes and their aftershocks cause major disasters. Buildings collapse, cities are destroyed, and thousands of lives may be lost. Fires, floods, landslides, and avalanches can also follow earthquakes.

Earthquakes may also be triggered by the explosive action of a volcano. Most of Earth's active volcanoes are located in a zone around

DIFFERENTIATING INSTRUCTION

English Learners

Vocabulary: Multiple-Meaning Words

Examine the term *faults* in the sentence "Huge rocks along a line where faults are located slide apart and break up. . . ." Explain that *fault* is a homonym (words that have the same sound and/or spelling, but different meanings). Have students write the geologic meaning of *faults* as used in their textbooks. Then have them use a dictionary to find other meanings for it and use each in a sentence.

Pre-AP

Compile a List of Earthquakes and Volcanic Eruptions

Have students compile a list of major earthquakes and volcanic eruptions that have happened in the world since 1900. Assign pairs of students to a decade or region. Then have them list the year, the area where the event occurred, and the kinds of damage it caused, including after-effects, such as fires and avalanches. Compile the individual lists into a unified list, or class time line.

the rim of the Pacific Ocean called the **Ring of Fire**. A number of tectonic plates meet in this zone, and many of the largest earthquakes in the world have occurred there.

One of the biggest earthquakes ever recorded took place near the coast of Indonesia in 2004. It occurred on the floor of the Indian Ocean where two tectonic plates meet. The quake produced a giant ocean wave called a tsunami (tsu•NAH•mee). Within hours, the tsunami had devastated the coasts of 12 countries in southern Asia and eastern Africa. Whole villages were swept away, and more than 280,000 people were killed.

Volcanoes A **volcano** is an opening in the Earth's crust from which molten rock, ash, and hot gases flow. Molten rock rises from the Earth's mantle and is stored in a chamber beneath the crust. Pressure builds and forces the molten rock, or lava, to erupt onto the Earth's surface, burning or burying anything it touches. The eruption may also throw large clouds of gases and ash into the atmosphere.

Some volcanoes erupt constantly. Others are inactive, but could erupt at some future time. The most destructive eruption in recorded history happened in 1883, when two-thirds of the Indonesian island of Krakatoa (KRAK•uh•TOH•uh) was blown apart. The most active volcano in the United States is Mount St. Helens in Washington. It blew its top off in 1980. Activity occurred again in 2005.

UNDERSTAND CAUSES Explain how earthquakes and volcanoes are connected.
Both are related to the movement of magma under the Earth's crust.

CONNECT to Science

In a volcanic eruption, red-hot magma rises through cracks in the Earth's crust. Gases are released, causing a violent explosion of liquid rock, hot ash, and fiery gases into the air.

Kilauea, Hawaii

Activity
Make a Model of a Volcano

Materials
- small drink bottle (1/2 pint or 12 oz.)
- 1/4 cup of water
- a few drops of dishwashing detergent
- orange food coloring
- 1/4 cup vinegar
- 1 tablespoon baking soda
- a small square of tissue

1. Place the water, soap, food coloring, and vinegar in the bottle.
2. Wrap the baking soda in the tissue and drop the baking soda packet in the bottle. Watch the eruption.

INTERDISCIPLINARY ACTIVITIES

CONNECT to Math

Calculate Volcanic Eruptions in the United States

About 500 volcanoes are known to have erupted on Earth since the beginning of history. Supply students with this number and the number of volcanoes known to have erupted in the United States given in the Connect to Science side note above (65). Have students calculate the percentage of volcanic eruptions that have occurred in the United States. *(13%)* Offer them the following suggestions:

- Convert the number of volcanic eruptions in the United States in the Connect to Science side note above to a decimal. *(65/500, or 13%, becomes 0.13)*
- Multiply the total number of volcanic eruptions (500) by this decimal (0.13) to obtain the number of volcanic eruptions in the United States.

To extend the activity, have students research the number of volcanic eruptions in one of the following countries: Guatemala, Mexico, Japan. Then they should calculate the percentage of eruptions for each country.

More About . . .

Tsunamis

Tsunamis are actually a series of waves. The waves travel in all directions from the center of a disturbance and can move as fast as 500 miles-per-hour. On the open ocean, tsunamis are not generally large, making them hard to detect. However, as they approach shallow coastlines, they slow down and may grow as tall as 100 feet before hitting land. During the 2004 Indian Ocean tsunami, waves reached 30 feet tall.

CONNECT to Science

Tell students that the study of volcanoes is called volcanology. The term *volcano* refers to both the crack in the Earth's crust through which heated gases, lava, and ash rise, and the cone-shaped mountain built up around the vent by eruptions. Volcanoes are described as *active, dormant*, or *extinct*, and are known to have existed on the moon and on other planets as well. More than 500 volcanoes are known to have erupted on Earth, with 65 of those occurring in the United States.

Unit Resource Book: Introduction to World Geography
- Connect to Science: Make a Model of a Volcano, p. 100

Teach

External Forces Shaping the Earth

Think, Pair, Share

- What is the difference between mechanical weathering and chemical weathering? *(Mechanical weathering occurs when physical forces, such as ice, split rocks apart; chemical weathering occurs when chemical reactions break down rocks, such as iron in rocks coming in contact with water.)*
- How does erosion help build new landforms? *(by moving materials to new locations and creating sediment)*
- **Categorize** What are the three main types of erosion? *(water erosion, wind erosion, ice or glacial erosion)* Which type of erosion has affected parts of the central United States? *(glacial erosion)*

More About . . .

The Grand Canyon

The Grand Canyon is considered one of the natural wonders of the world. At around a mile deep, between 4 and 15 miles wide, and 277 miles long, the Grand Canyon is so large that it has several distinct areas. The North Rim, for example, contains lush spruce, pine, and fir tree forests. The South Rim, which is lower, tends to be warmer and drier, and to contain different kinds of trees and plants.

External Forces Shaping the Earth

KEY QUESTION What external forces shape the Earth?

External forces also reshape the Earth's surface. The two main external forces are weathering and erosion. **Weathering** is the gradual physical and chemical breakdown of rocks near or on the Earth's surface. **Erosion** is the wearing away and movement of weathered materials from one place to another by the action of water, wind, or ice. As you can see, weathering and erosion work together to shape the Earth.

Weathering Weathering occurs slowly, over many years or even centuries. The two types of weathering are mechanical weathering and chemical weathering. Mechanical weathering is a process in which rocks are broken down into smaller pieces by physical means. It takes place when ice, extremes of hot and cold, or even tree roots cause rocks to split apart. It also occurs when hard objects, such as other rocks or sand, scrape or rub against a rock, and pieces of the rock break off.

Chemical weathering is caused by chemical reactions between the minerals in the rock and elements in the air or water. This process changes the make-up of the rock itself. For example, most rocks contain iron. When iron comes in contact with water, it rusts, which helps to break down the rock. Water and elements in the air can cause other minerals in rocks to dissolve.

The Grand Canyon The Grand Canyon is located on the Colorado River in Arizona. It is an example of both weathering and erosion caused by wind and water.

DIFFERENTIATING INSTRUCTION: TIERED ACTIVITIES

OBJECTIVE Demonstrate an understanding of the impact of weathering and erosion on the Grand Canyon.

Basic

Have students work in small groups. Have each group create a poster with illustrations and labels showing the different impacts of weathering and erosion on the Grand Canyon. Ask each group to present their posters to the class.

On Level

Have students work in small groups to create a trivia-type game that tests knowledge of mechanical weathering, chemical weathering, and erosion, along with the impact of each on the Grand Canyon. Have each group present its game to the class.

Challenge

Have students write a journal entry from the point of view of an explorer seeing the Grand Canyon for the first time. The entry should describe the impact of weathering and erosion on the Grand Canyon and the differences between them.

Erosion New landforms and new soil are formed by erosion. It occurs when materials loosened by weathering are moved by water, wind, or ice to new locations. Currents in streams and rivers pick up loose materials and deposit them downstream or carry them out to sea. These tiny pieces of rock, deposited by water, wind, or ice are called **sediment**. Sediment can be sand, stone, or finely ground particles called silt.

Wave action along coastlines carries rocks and sand from one place to another. Waves also pound boulders into smaller rocks. Wind erosion lifts particles from the Earth's surface and blows them great distances. The wind's actions can reshape rock surfaces. Arizona's Grand Canyon is a result of both wind and water erosion.

Another type of erosion is caused by glaciers. **Glaciers** are large, slow-moving masses of ice. They grind rocks and boulders underneath the ice and leave behind the rock when the ice melts. Parts of the central United States have been shaped by glacial erosion.

Beach Erosion Ocean waves and tides eroded this section of beach in California.

SYNTHESIZE Explain how external forces shape the Earth's surface. Weathering wears down existing rock formations; erosion deposits the loose materials in a different place to build another landform.

Section 1 Assessment

ONLINE QUIZ For test practice, go to **Interactive Review** @ClassZone.com

TERMS & NAMES

1. **Explain the importance of**
   - continent
   - tectonic plate
   - weathering
   - erosion

USE YOUR READING NOTES

2. **Categorize** Use your completed chart to answer the following question:

   Are external or internal forces responsible for volcanoes? Explain your answer.

INTERNAL FORCES	EXTERNAL FORCES
1.	1.
2.	2.
3.	3.
4.	4.

KEY IDEAS

3. What are the five layers that make up the Earth's interior and exterior?
4. How were the continents formed?
5. What are the two major external forces reshaping the Earth?

CRITICAL THINKING

6. **Draw Conclusions** How does the movement of wind, water, or ice reshape the Earth's surface?
7. **Analyze Causes and Effects** What is the relationship between plate movement, volcanoes, and earthquakes?
8. **CONNECT to Today** In which parts of the United States are external forces shaping the landscape?
9. **ART** **Create a Puzzle** Make a copy of a map of the world. Cut out the continents. Use the continents as puzzle pieces to form the continent of Pangaea. When you have finished putting the pieces together, draw an outline around the entire supercontinent.

33

CONNECT to the Essential Question

How do Earth's physical systems make life on Earth possible?

Ask students what they have learned so far that can help them answer this question. Students might mention:

- Earth's atmosphere makes breathing possible.
- Earth's continents cover only one-fourth of its surface, leaving three-fourths covered by water.
- Powerful internal and external forces continually reshape Earth's surface.

4 Assess & Reteach

Assess Have students complete the Section Assessment.

Unit Resource Book: Introduction to World Geography
- Section Quiz, p. 103

Interactive Review @ClassZone.com
- Section Quiz

Power Presentations

Test Generator

Reteach Have pairs of students create five "True or False" statements about "The Earth and Its Forces." Then have each pair exchange their statements with another pair and evaluate the statements. For each false statement, have students correct it, using their textbooks.

Unit Resource Book: Introduction to World Geography
- Reteaching Activity, p. 107

SECTION 1 ASSESSMENT ANSWERS

Terms & Names

1. continent: one of seven large landmasses on Earth; tectonic plate: a rigid section of Earth that is in constant motion; weathering: the gradual breakdown of rocks on Earth's surface; erosion: the wearing away and movement of weathered materials by water, wind, or ice

Use Your Reading Notes

2. Internal forces push magma through openings in the Earth's surface, creating volcanoes.

INTERNAL FORCES	EXTERNAL FORCES
1. Tectonic plate movement 2. Earthquakes 3. Volcanoes	1. Mechanical weathering 2. Chemical weathering 3. Erosion

Key Ideas

3. inner core, outer core, mantle, crust, atmosphere
4. The continents were formed when the supercontinent Pangaea broke into pieces that drifted apart.
5. weathering and erosion

Critical Thinking

6. They transport sediments from one place to another, creating new rock surfaces, changing beaches, and reshaping the land.
7. Certain movements of the tectonic plates cause earthquakes and volcanoes.
8. Coastal areas of the United States are affected by erosion and almost any area of the country is affected by weathering.
9. Use this rubric to score the puzzle.

Puzzle Rubric

	Continents	Outline
4	all present and accurate	correctly reflects the supercontinent
3	all present and accurate	slightly incorrect in the orientation
2	missing one continent	several continents misplaced
1	missing more than one continent	completely incorrect orientation

1 Plan & Prepare

Objectives

- Identify the different bodies of water found on Earth
- Explain the hydrologic cycle
- Describe how landforms are created

Read for the Essential Question

Help students read for a purpose by reminding them of the Essential Question: "How do Earth's physical systems make life on Earth possible?" Point out that they can use text features such as Key Questions, special features, images, and maps to form answers to the question.

Vocabulary

Best Practices Toolkit

Use the ***Best Practices Toolkit*** to model strategies for vocabulary notetaking. Vary strategies throughout the year. Choose from: *Knowledge Rating, Predicting ABCs, Definition Mapping, Word Sort, Frayer Model (Word Squares), Magnet Words, Word Wheel,* and *Student VOC.*

Vocabulary Strategies, TT9–TT16

Reading Strategy

Best Practices Toolkit

Display the Web Diagram transparency.

- Remind students that main ideas are important ideas in a block of text.
- Model filling in main ideas on a graphic organizer using the subheads in this section.
- Have students suggest entries. Remind students that main ideas in each block of text in the section will help answer a Key Question.

Web Diagram, TT28

Section 2 Reading for Understanding

Key Ideas

BEFORE, YOU LEARNED

Internal and external forces shape the surface of the Earth.

NOW YOU WILL LEARN

Interaction between landforms and bodies of water makes life on Earth possible.

Vocabulary

TERMS & NAMES

drainage basin the area drained by a major river

ground water water found beneath the Earth's surface

hydrologic cycle the circulation of water between the Earth, the oceans, and the atmosphere

landform a feature on the Earth's surface formed by physical force

plateau a broad, flat area of land higher than the surrounding land

relief the difference in the elevation of a landform from its lowest point to its highest point

continental shelf the submerged land at the edge of a continent

BACKGROUND VOCABULARY

atmosphere the layer of gases that surround the Earth

Visual Vocabulary landform

Reading Strategy

Re-create the web diagram shown at right. As you read and respond to the **KEY QUESTIONS**, use the diagram to organize important details about the Earth's landforms and bodies of water.

See Skillbuilder Handbook, page R4

FIND MAIN IDEAS

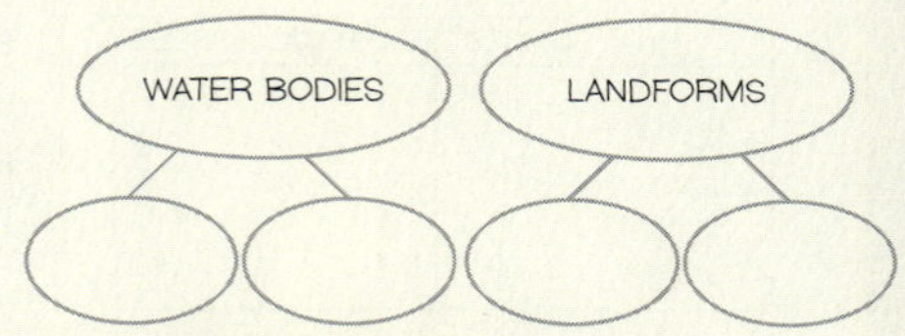

GRAPHIC ORGANIZERS
Go to **Interactive Review** @ClassZone.com

PRETEACHING VOCABULARY

English Learners

Pronounce and Preview

Pronounce each term for students. Review the meanings of words within definitions, such as *circulation, broad,* and *submerged.*

- To modify vocabulary learning, have students complete worksheets as they read, instead of afterward.

Unit Resource Book: Introduction to World Geography
- Vocabulary Practice, p. 89

Inclusion

Match Terms and Definitions

Have volunteers read each term and its definition into an audio cassette recorder. As students listen to the recordings play back, have them follow along with the words in their textbooks. Then have students use each term in a sentence.

SECTION 2 GEOGRAPHY

Big Sur, California

Peruvian Rainbow Boa in the rain forest

Bodies of Water and Landforms

Connecting to Your World

How important is water to your life and to life on Earth? Without enough usable water, there would be no life. The Earth is able to support plant and animal life because of its abundance of water. It appears to be the only planet in our solar system able to do so. The Earth is sometimes called the "blue planet" because bodies of water cover so much of its surface.

The Earth from Space

Bodies of Water

KEY QUESTION What are the two types of water supply found on Earth?

Almost three-fourths of the surface of the Earth is covered by water. Most of the water—more than 97 percent of it—is salt water. This is the water in the oceans and seas. Only about 2.5 percent of the Earth's water is fresh water—that is, water containing little or no salt.

Fresh Water Most fresh water is locked in frozen form in ice caps or glaciers. Much of the rest is found in rivers, streams, and lakes.

Iguacu Falls This series of 275 falls is located between Brazil and Argentina.

2 Focus & Motivate

3-Minute Warm-Up

Write on the board or display the transparency:

- Look at the map and photograph of Iguacu Falls. How do these help you understand the movement of fresh water? *(Possible Answer: They show how fresh water flows downhill and toward the ocean.)*

Unit Transparency Book
- 3-Minute Warm-Ups, TT9

Connecting to Your World

More About . . .

Iguacu Falls

Iguacu Falls is actually several hundred waterfalls separated by rocky islands. The highest waterfall is 210 feet tall.

3 Teach

Bodies of Water

Think, Pair, Share

- Which type of water supply uses a drainage basin? *(fresh water)* What is the largest drainage basin in the world? *(Amazon river system)*
- What is a sea? *(A sea contains salt water and is partly or completely surrounded by land.)*
- **Find Main Ideas** Why is there more salt water than freshwater on Earth's surface? *(because salt water is found in oceans and seas, which are much larger than freshwater sources such as rivers and lakes)*

SECTION 2 PROGRAM RESOURCES

Unit Resource Book: Introduction to World Geography

ON LEVEL

Unit Resource Book
- Reading Study Guide p. 59
- Section Quiz, p. 104

STRUGGLING READERS

Unit Resource Book
- Reading Study Guide with Additional Support, p. 67
- Vocabulary Practice, p. 89
- Section Quiz, p. 104
- Reteaching Activity, p. 108

eEdition with Audio DVD-ROM

ENGLISH LEARNERS

Pupil Edition in Spanish
eEdition with Audio DVD-ROM
Unit Resource Book
- RSG (Spanish), p. 75
- RSG with Additional Support (Spanish), p. 83

Multi-Language Glossary
Test Generator
- Section Quiz in Spanish

INCLUSION

Unit Resource Book
- RSG with Additional Support, p. 67
- Section Quiz, p. 104
- Reteaching Activity, p. 108

GIFTED & TALENTED

Unit Resource Book
- Outline Map Activities, p. 93
- Section Quiz, p. 104

PRE-AP

Unit Resource Book
- Primary Sources p. 101
- Section Quiz, p. 104

TECHNOLOGY

Unit Transparency Book
- 3-Minute Warm-Ups, TT9
- Essential Question, TT10
- Critical Thinking, TT12
- Map: Oceans in Relief, TT16

Daily Test Practice Transparencies
- Section 2, TT5

Power Presentations
ClassZone.com
World Cultures and Geography Video on DVD

CONNECT Geography & History

The Great Salt Lake

As one of a few bodies of water in a desert region, many different groups, including Native Americans, have used the Great Salt Lake's freshwater marshes and streams for hunting and fishing. The Lake itself has few if any fish because of its salt level, but it does have many shrimp. In 1847, a group of Mormon pioneers from Illinois settled near the lake. The settlement eventually became the city of Salt Lake City, Utah.

- What kinds of activities could people use the Great Salt Lake for? *(Possible Answers: shrimping, boating, sailing, other forms of recreation)*

CRITICAL THINKING ANSWER

Summarize When the water in Lake Bonneville began to evaporate, the salt it contained remained and became part of the body of water called the Great Salt Lake.

More About . . .

The Mediterranean Sea

The Mediterranean Sea contains a number of large islands, including Sicily, Corsica, Crete, and Malta. Its climate tends to be warm and mild with plenty of sunshine, though earthquakes and volcanic eruptions are common. Many ancient civilizations, including the Greek and Roman civilizations, flourished in the regions around the Mediterranean.

CONNECT Geography & History

The Great Salt Lake This huge lake in northern Utah covers about 1,700 square miles. It is a remnant, or remainder, of a huge ancient fresh water lake called Lake Bonneville. That lake existed around 14,000 years ago and was about ten times as large as the Great Salt Lake. As the climate became drier and warmer over the centuries, Lake Bonneville's waters began to evaporate. But the salt it contained did not. That's why the Great Salt Lake is one of the saltiest bodies of water in the world.

CRITICAL THINKING

Summarize How did freshwater Lake Bonneville become the Great Salt Lake?

Russia's Lake Baikal, for example, is the world's largest lake and contains 20 percent of all Earth's fresh water. Rivers and streams move water downhill to or from larger bodies of water. Smaller streams and rivers that flow into a major river are called tributaries. The region drained by a river and its tributaries is called a **drainage basin**. The Amazon River system in South America is the world's largest drainage basin.

Some fresh water, called **ground water**, is found beneath the Earth's surface. This water is held in the pores and cracks of rocks and can be pumped from the ground.

Salt Water The water in the Earth's oceans and seas is called salt water because it contains a small percentage of dissolved minerals and chemical compounds called salts. Actually, all of the oceans and seas are part of the same body of water, which is divided by the continents. Geographers gave names to the different areas of the oceans.

The Earth's oceans are the Pacific, Atlantic, Indian, and Arctic, and the Southern Ocean, which is the body of water around Antarctica. The Pacific Ocean is the largest and covers almost one-third of the Earth. A body of salt water that is completely or partly enclosed by land is called a sea. An example is the Mediterranean Sea. Oceans and seas are sources of food, energy, and minerals and are used for transportation and recreation. They also help to distribute Earth's heat.

COMPARE Compare the two types of water on the Earth.
The earth has fresh and salt water. Almost all of the water is salt water. Freshwater is found in lakes, rivers, streams, and ice.

DIFFERENTIATING INSTRUCTION

English Learners

Key Academic Vocabulary

Read this sentence from the textbook: "They also help to distribute Earth's heat." Point out that the word *distribute* is a transitive verb that signals an action. Ask: What is being distributed, or moved? *(heat)* Have students define the term *distribute*. Then have them use the term in a sentence to show something else being distributed.

Struggling Readers

Find Locations of Bodies of Water

Some students may not know where the Great Salt Lake is located in the United States. Ask them to identify the location on a map of the United States. Then ask them to identify the locations of the Pacific, Atlantic, Indian, Arctic, and Southern oceans on a globe or world map to help them understand how much of the Earth is covered by salt water.

Hydrologic Cycle

The Earth's water is renewed by a never-ending process called the hydrologic cycle, or water cycle. The **hydrologic cycle** is the circulation of water between the Earth, the oceans, and the atmosphere. Approximately 119,000 cubic miles of water evaporates into the atmosphere each year. The **atmosphere** is the layer of gases that surround the Earth.

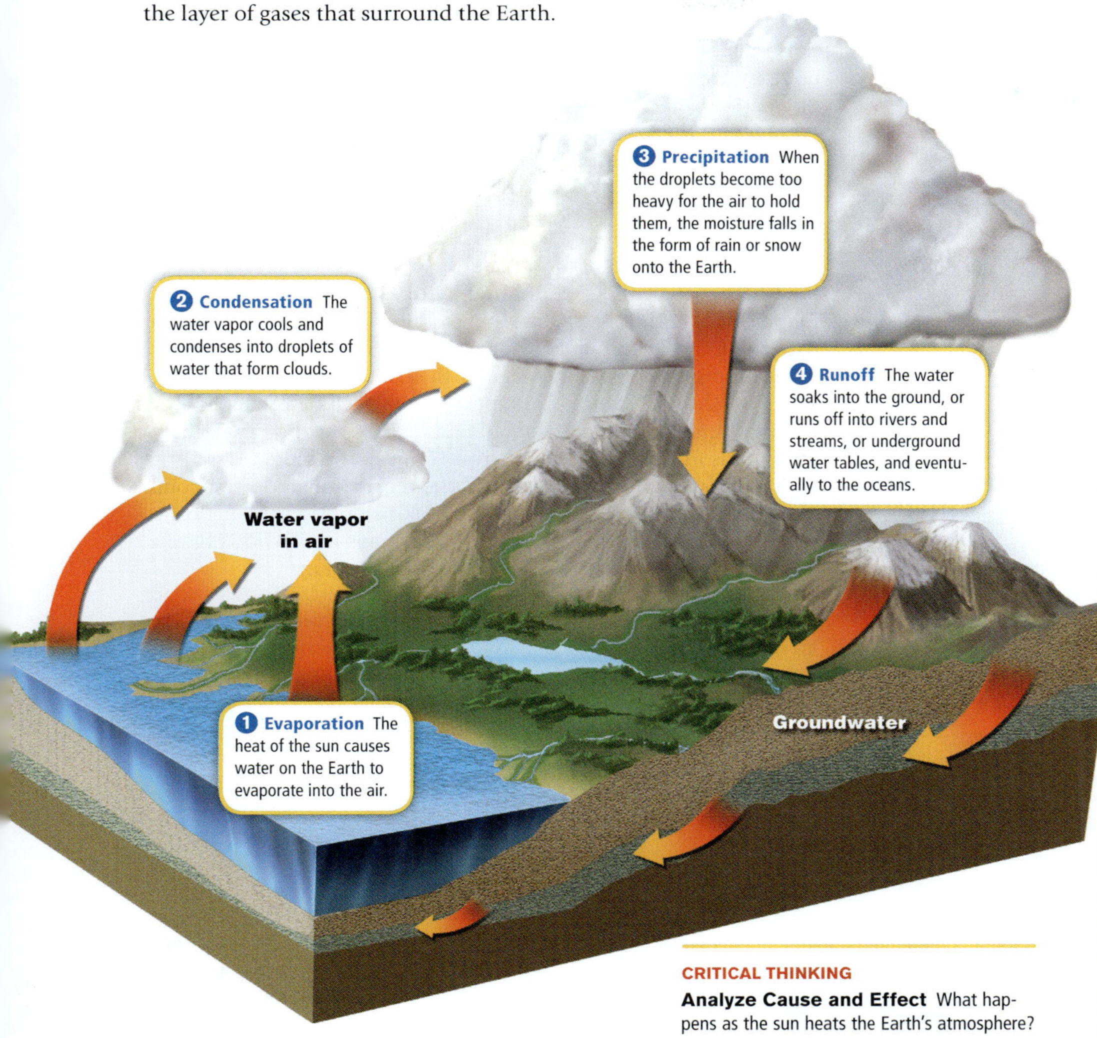

CRITICAL THINKING

Analyze Cause and Effect What happens as the sun heats the Earth's atmosphere?

Hydrologic Cycle

- What are the four steps in the hydrologic cycle? *(evaporation, condensation, precipitation, runoff)*
- How does the hydrologic cycle affect rain and snowfall totals? *(It returns water from the atmosphere as rain or snow. As more water condenses, more or bigger clouds are formed.)*

CRITICAL THINKING ANSWER

Analyze Cause and Effect Water on the Earth condenses, evaporates, and eventually returns to the Earth as precipitation.

More About . . .

Water Tables

Despite its name, a water table is rarely flat. It is the upper part of a ground water supply. Below the water table, the cracks and pores, or spaces between soil and sediment, in rocks are always filled with water. The pores in the soil above the water table are not always filled. These spaces may fill after a heavy rain, but during a drought can become dry. The water in these spaces, called soil water, is often used by plants.

INTERDISCIPLINARY ACTIVITIES

CONNECT to Science

Chart How the Hydrologic Cycle Works

Have students make charts, such as the one to the right, to help them better understand how the hydrologic cycle works. Have them use their textbooks to complete their charts.

Step	Description	Area(s)
Evaporation		
Condensation		
Precipitation		*Earth*
	water soaks into ground; eventually reaches oceans	

Teach
Landforms

Talk About It

- Describe how external forces affect landforms. *(External forces erode mountains and other rock formations. The materials move to other locations, where they form new landforms.)*
- How do oceanic landforms become islands? *(Possible Answer: They are pushed up by internal forces such as volcanic eruptions or formed from coral reefs.)*
- **Find Main Ideas** Why are the same types of landforms found on dry land and under water? *(because the internal and external forces affecting both are the same tectonic forces)*

Unit Resource Book: Introduction to World Geography
- Outline Map Activities: Recognizing Continents and Oceans, p. 93
- Primary and Secondary Sources: from *The Origins of Continents and Oceans*, pp. 101–102

Comparing Regions

Rocky Mountains and Appalachian Mountains

The Rocky Mountains and the Appalachian Mountains are two of the major mountain ranges in North America. The Rockies are located in the western United States, while the Appalachians are primarily in the eastern United States. In general, the peaks of the Rockies are much taller than the peaks of the Appalachians, with the highest, Mt. Elbert in Colorado, at 14,433 feet. The highest peak in the Appalachians, Mt. Mitchell in North Carolina, is only 6,684 feet.

The Rockies run from Alaska, through Canada, to New Mexico, while the Appalachians run from Newfoundland, Canada, south to Alabama. Both chains have been affected by erosion that has been occurring for millions of years.

Landforms

KEY QUESTION How are landforms created?

Features on the Earth's surface formed by physical forces are called **landforms**. Landforms are produced by the internal and external forces that reshape the Earth. Internal forces push, move, and raise up parts of the Earth's crust. The result is the creation of new rock formations, such as mountains. External forces wear down these formations and transport the eroded materials to other locations. The eroded materials then become new landforms. These processes take a long time, but they are constantly at work. The location and size of landforms often affect where people choose to live.

Many of the same landforms found on dry land are also found under water. Those on the land are called continental landforms. Those on the sea floor are called oceanic landforms.

Continental Landforms The major continental landforms are mountains, hills, plains, and plateaus. A **plateau** is a broad area of land higher than the surrounding land. The same landforms are found on all of the continents. In fact, satellite photographs show a pattern on most continents: wide plains in the center and a narrow belt of mountains near the edge of the continent, where tectonic plates collide. For example, the landscape of the United States has the Rocky Mountains and coastal mountains in the west, the Appalachian Mountains in the east, and the Great Plains in the center.

The difference in the elevation of a landform from its lowest point to its highest point is called **relief**. Mountains show great relief compared to plains and plateaus. Many of the maps in this book have a relief indicator to show these differences in elevation.

Oceanic Landforms The landforms on the ocean floor are like an invisible landscape. Most cannot be seen from the surface of the water. But high mountains, vast plains, deep valleys, and coral reefs are present under the water's surface. Some are the result of the same tectonic forces that shape the continental landforms. The submerged

Monument Valley, Arizona In this photograph, you can see dramatic examples of relief between the floor of the valley and the tops of the landforms.

DIFFERENTIATING INSTRUCTION: TIERED ACTIVITIES

OBJECTIVE Create a chart showing how landforms are created.

Basic

Have student pairs create a two-column chart, labeled *Continental Landforms and Oceanic Landforms.* Have them use their textbooks to fill in the types of continental landforms *(mountains, hills, plains, plateaus)* and oceanic landforms *(volcanoes, mountain chains, plains, trenches).*

On Level

Have students work individually to create a two-column chart, labeled *Continental Landforms and Oceanic Landforms.* Have students work individually, using their textbooks, to complete their charts. For each type of landform, have them also list one example of that type.

Challenge

In addition to the On Level activity, have students write a summary of how external and internal physical forces help to create each type of landform. Have students explain the formation of at least one landform, such as the Hawaiian islands.

land at the edge of a continent is called the **continental shelf**. It slopes downward and then descends to the deep part of the ocean. On the deep ocean floor are volcanoes, mountain chains, plains, and trenches. These landforms are created by the movement of tectonic plates, the same forces as those on the continent.

Some of the oceanic landforms are pushed up above the water and become islands. Many islands, such as the Hawaiian Islands, are formed by volcanic action. Other islands are formed from coral reefs. However, all islands are subject to the same external forces that wear down landforms on the continent. These forces include the weathering and erosion caused by wind, water, and ice.

Indian Ocean Reef
A diver views plant and animal life on a coral reef.

SUMMARIZE Explain how landforms are created.
Both continental and oceanic landforms are formed by and subject to external and internal physical forces such as movement of tectonic plates, weathering, and erosion.

Section 2 Assessment

ONLINE QUIZ
For test practice, go to **Interactive Review @ClassZone.com**

TERMS & NAMES

1. **Explain the importance of**
   - hydrologic cycle
   - landform
   - relief
   - continental shelf

USE YOUR READING NOTES

2. **Find Main Ideas** Use your completed web diagram to answer the following question:
   How is it possible for the oceans to have landforms?

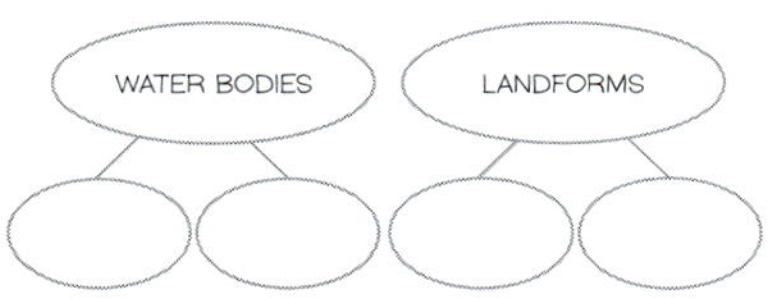

KEY IDEAS

3. How much water on the Earth's surface is fresh water?
4. What are the four components that keep the hydrologic cycle going?
5. What are the major continental landforms?

CRITICAL THINKING

6. **Find Main Ideas** Why are oceans important to life on Earth?
7. **Compare and Contrast** How are continental and oceanic landforms the same? different?
8. **CONNECT to Today** In what ways does water affect your daily life?
9. **TECHNOLOGY** **Make a Multimedia Presentation** Plan a slide show about the hydrologic cycle. Sketch each step in the cycle and explain what is happening.

CONNECT to the Essential Question

How do Earth's physical systems make life on Earth possible?

Ask students what they have learned so far that can help them answer this question. Students might mention:

- Earth's abundant water supply helps support plant and animal life.
- The hydrologic cycle continually circulates water among the Earth, its oceans, and its atmosphere.
- Continental and oceanic landforms provide habitats for plant and animal life.

4 Assess & Reteach

Assess Have students complete the Section Assessment.

Unit Resource Book: Introduction to World Geography
- Section Quiz, p. 104

Interactive Review @ClassZone.com
- Section Quiz

Power Presentations

Test Generator

Reteach Have students work in pairs to list three main ideas about "Bodies of Waters and Landforms" along with three supporting details for each main idea. Ask volunteers to share one of their main ideas, with the rest of the class filling in the supporting details, and then comparing them to the volunteers' list. Repeat until each pair has had a turn.

Unit Resource Book: Introduction to World Geography
- Reteaching Activity, p. 108

SECTION 2 ASSESSMENT ANSWERS

Terms & Names

1. hydrologic cycle: the circulation of water between the Earth, the oceans, and the atmosphere; landform: a feature on the Earth's surface formed by physical force; relief: the difference in the elevation of a landform from its lowest to its highest point; continental shelf: the submerged land at the edge of a continent

Use Your Reading Notes

2. The floor of the ocean is a part of the Earth's crust. Its landforms are mostly created by internal forces and are located under water.
   Diagram *Water Bodies: freshwater—rivers lakes, icecaps; salt water—oceans, seas; Landforms: continental—hills, mountains, plateaus, plains; oceanic—mountains, plains, trenches*

Key Ideas

3. Only 2.5 percent is fresh water.
4. sun, water, air, and gravity
5. mountains, hills, plains, and plateaus

Critical Thinking

6. Oceans are part of the hydrologic cycle that renews the Earth's waters. They also provide food, minerals, and energy, and are used for transportation and recreation.
7. The same landforms are found both places and the same external and internal forces reshape them. The major difference is that the oceanic landforms are under water.
8. Answers will vary, but should mention that water is necessary for life, sanitation, and the economy.
9. Use the rubric to score presentations.

Presentation Rubric

	Text and Illustrations	Use of Technology
4	clear and informative, covers all steps	excellent use of technical tools
3	adequate to explain the ideas	good use of technical tools
2	confusing	fair use of technical tools
1	unclear	poor use of technical tools

1 Plan & Prepare

Objectives

- Describe the Earth's rotation and revolution
- Explain the difference between weather and climate
- Compare and contrast the Earth's climate and vegetation regions

Read for the Essential Question

Help students read for a purpose by reminding them of the Essential Question: "How do Earth's physical systems make life on Earth possible?" Point out that they can use text features such as Key Questions, special features, images, and maps to form answers to the question.

Vocabulary

Best Practices Toolkit

Use the ***Best Practices Toolkit*** to model strategies for vocabulary notetaking. Vary strategies throughout the year. Choose from: *Knowledge Rating, Predicting ABCs, Definition Mapping, Word Sort, Frayer Model (Word Squares), Magnet Words, Word Wheel,* and *Student VOC.*

Vocabulary Strategies, TT9–TT16

Reading Strategy

Best Practices Toolkit

Display the Web Diagram transparency.

- Point out to students that when summarizing, they restate a paragraph, a passage, or a section in fewer words.
- Model summarizing the subsection, "Climate and Vegetation Regions."
- Have students suggest entries and practice summarizing using the other subheadings within the section.
- Display the graphic organizer for students to review.

Web Diagram, TT28

SECTION 3 Reading for Understanding

Key Ideas

BEFORE, YOU LEARNED

The Earth's surface is covered with both continental and oceanic landforms. The hydrologic cycle circulates the water.

NOW YOU WILL LEARN

The Earth's rotation and revolution influence weather, climate, and living conditions on Earth.

Vocabulary

TERMS & NAMES

solstice the time during the year when the sun reaches the farthest northern or southern point in the sky

equinox one of the two times a year when the sun's rays are over the equator and days and night around the world are equal in length

weather the condition of the Earth's atmosphere at a given time and place

climate the typical weather conditions of a region over a long period of time

precipitation falling water droplets in the form of rain, snow, sleet, or hail

vegetation region an area that has similar plants

savanna a vegetation region with a mix of grassland and scattered trees

desert a region with plants specially adapted to dry conditions

Visual Vocabulary desert

Reading Strategy

Re-create the diagram shown at right. As you read and respond to the **KEY QUESTONS**, use the diagram to help you summarize information about the world's climate and vegetation.

See Skillbuilder Handbook, page R5

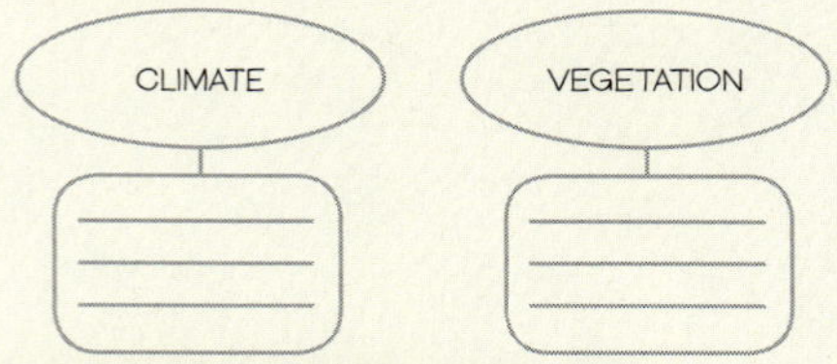

GRAPHIC ORGANIZERS Go to **Interactive Review @ClassZone.com**

PRETEACHING VOCABULARY

English Learners

Pronounce and Preview

Pronounce each term for students. Review the meanings of words within definitions, such as *equator* and *typical.*

- To modify vocabulary learning, have students complete worksheets as they read, instead of afterward.

Unit Resource Book: Introduction to World Geography
- Vocabulary Practice, p. 89

Inclusion

Word Sort/Word Splash

Have student pairs fill out the graphic organizer for the terms in the section.

- Use the transparency to model the strategy.
- Have volunteers suggest entries for Sections 1, 2, and 3.
- Then distribute copies of the organizer for students to complete with a partner.

Best Practices Toolkit
- Vocabulary: Word Sort/Word Wheel, TT13

Big Sur, California

Peruvian Rainbow Boa in the rain forest

SECTION 3 GEOGRAPHY

Climate and Vegetation

Connecting to Your World

Every moment in the day, weather and climate are a part of your life. They affect what clothes you wear and how you get to school. You might walk or ride your bike if it is not too cold or too wet. But if it's raining or snowing you may go in a car or by bus. Some school activities, like sports, depend on weather and climate, too. In fact, weather and climate affect plant and animal life and nearly every human activity.

Inupiat Woman The harsh climate of the Arctic makes wearing a fur parka necessary.

The Earth's Rotation and Revolution

KEY QUESTION How does Earth's revolution affect seasons?

The Earth rotates as it revolves around the sun. Rotation is the motion of the Earth as it spins on its axis once every 24 hours. Revolution is the motion of the Earth as it circles, or makes a year-long orbit, of the sun.

Earth's Movement The Earth is tilted at a 23.5° angle. The Earth's revolution around the sun affects patterns of Earth's weather and climate. The Earth's tilt stays the same as it revolves around the sun. As a result, different parts of the Earth get direct rays from the sun for more hours of the day at certain times of the year. This causes the changing seasons.

Midnight Sun in the Arctic Multiple exposures show the position of the sun over a 24-hour period in the Arctic summer.

2 Focus & Motivate

3-Minute Warm-Up

Write on the board or display the transparency:

- Look at the photograph, "Midnight Sun in the Arctic." What does it suggest about the position of the sun during the Arctic summer? *(Possible Answer: It never sets; it's at its highest point of the year.)*

Unit Transparency Book
- 3-Minute Warm-Ups, TT9

Connecting to Your World

More About . . .

The Inupiat

The Inupiat are a native people who live in northern Alaska near the Arctic Circle. Their isolated location prevented contact with the rest of the world until they encountered European explorers in the late 1700s.

3 Teach

The Earth's Rotation and Revolution

Think, Pair, Share

- Which type of Earth's movement takes 24 hours to complete? *(rotation)*
- Why do different parts of the Earth get more sun at certain times of the year? *(because of the Earth's tilt)*
- **Summarize** How are the summer and winter solstices related to the seasons? *(They indicate the longest and shortest days of the year and begin the summer and winter seasons.)*

SECTION 3 PROGRAM RESOURCES

Unit Resource Book: Introduction to World Geography

ON LEVEL
Unit Resource Book
- Reading Study Guide, p. 61
- Section Quiz, p. 105

STRUGGLING READERS
Unit Resource Book
- Reading Study Guide with Additional Support, p. 69
- Vocabulary Practice, p. 89
- Section Quiz, p. 105
- Reteaching Activity, p. 109

eEdition with Audio DVD-ROM

ENGLISH LEARNERS
Pupil Edition in Spanish
eEdition with Audio DVD-ROM
Unit Resource Book
- RSG (Spanish), p. 77
- RSG with Additional Support (Spanish), p. 85

Multi-Language Glossary
Test Generator
- Section Quiz in Spanish

INCLUSION
Unit Resource Book
- RSG with Additional Support, p. 69
- Section Quiz, p. 105
- Reteaching Activity, p. 109

GIFTED & TALENTED
Unit Resource Book
- Interdisciplinary Projects, pp. 96, 97, 98

PRE-AP
Unit Resource Book
- Interdisciplinary Projects, pp. 96, 97, 98

TECHNOLOGY
Unit Transparency Book
- 3-Minute Warm-Ups, TT9
- Essential Question, TT10
- Critical Thinking, TT13
- Cultures, TT17

Daily Test Practice Transparencies
- Section 3, TT6

Power Presentations
ClassZone.com
World Cultures and Geography Video on DVD

Seasons: Northern Hemisphere

Discuss with students how they could revise the graphic to make it of the Southern Hemisphere. *(The seasons would be reversed: fall becomes spring and so on.)*

- When are the sun's rays strongest at the Tropic of Capricorn? *(when it's summer solstice in the Southern Hemisphere, or December 22)*

CRITICAL THINKING ANSWER

Compare and Contrast Areas around the equator experience little change in seasons.

Comparing Regions

North Pole and South Pole

The Northern Hemisphere is the area between the North Pole and the equator, while the Southern Hemisphere is the area between the South Pole and the equator. When the North Pole is tilted away from the sun as far as possible, the farthest point that the sun can reach is the Arctic Circle. When the South Pole is tilted away from the sun as far as possible, the Antarctic Circle is the farthest point. Depending on the angle of Earth's tilt at different times of the year, the sun does not rise—or set—at either the North Pole or the South Pole.

Teach

Weather and Climate

Reader, Recorder, Reporter

- What is the most important factor affecting weather? *(solar energy)* Name two other factors that affect weather. *(Possible Answers: wind, landforms, bodies of water, water vapor, cloud cover, elevation)*
- Why do weather conditions tend to change more during summer? *(because the atmosphere receives more solar energy)*
- **Summarize** How does the sun affect weather and climate differently? *(Possible Answer: It affects weather in terms of seasons—how much solar energy a location gets at certain times of the year. It affects climate in terms of a place's location on Earth.)*

Seasons: Northern Hemisphere

Earth's Revolution
The seasons are related to the Earth's tilt and revolution around the sun.

CRITICAL THINKING

Compare and Contrast Which part of the Earth's surface doesn't experience seasons?

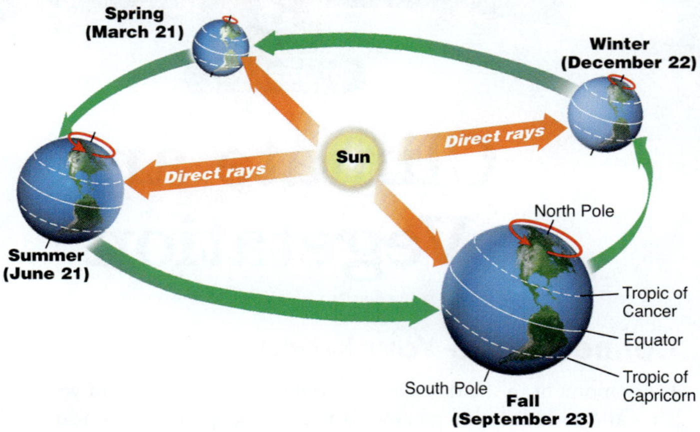

Seasons The term **solstice** is used to describe the time during the year when the sun reaches the farthest northern or southern point in the sky. In the Northern Hemisphere, the summer solstice is the longest day of the year and begins summer. The winter solstice is the shortest day of the year and begins winter. These dates are reversed in the Southern Hemisphere. The beginning of spring and autumn start on the **equinox**. On these two days, the sun's rays are directly over the equator, and days and nights around the world are equal in length. The Earth's revolution brings the temperature and weather changes we call seasons to many parts of the Earth. But in some regions there is little change. The illustration above shows the position of the Earth at the start of the four seasons in the Northern Hemisphere.

SUMMARIZE What causes seasons on Earth? Revolution around the sun and the Earth's tilt cause the seasons.

Weather and Climate

KEY QUESTION What is the difference between weather and climate?

People often confuse weather and climate. **Weather** is the condition of the Earth's atmosphere at a given time and place. For example, today may be sunny and warm. **Climate** is the term for the typical weather conditions of a certain region over a long period of time.

Causes of Weather Several factors interact to cause weather at a particular location. They include solar energy, wind, landforms,

DIFFERENTIATING INSTRUCTION

Struggling Readers

Create an Illustration

Have students use the diagram on this page to create a four-panel illustration showing seasons in the Northern Hemisphere. Each panel should have a short caption explaining the positions of the sun and earth and other differences between seasons.

English Learners

Key Academic Vocabulary

Review with students some of the key academic vocabulary words that appear on this page. You may want to focus on *region* ("Climate is the term for the typical weather conditions of a certain region over a long period of time.") and *revolution* ("The Earth's revolution brings the temperature and weather changes we call seasons to many parts of the Earth.").

bodies of water, water vapor, cloud cover, and elevation. The combination of all these factors varies from location to location, creating local weather conditions.

The most important weather factor is the amount of energy, in the form of heat from the sun, that a location receives. This is why, for example, the time of year influences the weather. In summer, much more solar energy is found in the atmosphere, and weather conditions change. Winds move the solar energy and moisture that air holds across the Earth. Also, land heats and cools more quickly than bodies of water do. So land located near a body of water has a different weather pattern from locations further inland.

Clouds and water vapor are connected to each other in weather patterns. Clouds hold water vapor in the atmosphere. Water vapor determines whether there will be **precipitation**, which is falling water droplets in the form of rain, snow, sleet, or hail.

Finally, as elevation above sea level rises, air becomes thinner and loses its ability to hold moisture, so it becomes cooler. The temperature drops by about 3.5°F for every 1,000-foot increase in elevation. So, you can find ice and snow on the tops of mountains even at the equator.

Causes of Climate There are many different climates around the world. A place's location on the Earth, especially its latitude, is important in determining climate. For example, climates are warmer near the equator and colder near the poles.

Wind and ocean currents help distribute the sun's heat from one part of the world to another. Ocean currents are like rivers flowing through the ocean. They move warm waters away from the equator and cold water from the poles. Air currents blowing over the ocean waters pick up heat and moisture and move them to other parts of the Earth.

COMPARE Explain the difference between weather and climate.
Weather is a condition of the Earth's atmosphere at a specific time and place, but climate is the pattern of these weather conditions over time.

Climate and Vegetation Regions

KEY QUESTION What are climate and vegetation regions?

To categorize climate regions, geographers divide the Earth into three general zones of latitude: tropical, middle, and high. These zones are found on both sides of the equator. Tropical latitudes are found between the equator and Tropic of Cancer and between the equator and the Tropic of Capricorn where it is usually hot. Middle latitudes extend from the tropic lines to the lines of the Arctic and Antarctic circles. High latitude refers to the cold areas around the North and South Poles.

Fun Facts!

IT'S RAINING FROGS!

Yes, it's true—there are recorded instances of frogs raining down on the Earth, most recently in Serbia in 2005. The most logical explanation is a strong wind that can suck light objects out of the water and later deposit them somewhere else.

In 1981, a city in southern Greece experienced a rain of frogs that landed in trees and in the streets. The species of frog was native to North Africa. That's one strong wind!

Fun Facts!

It's Raining Frogs!

A number of other creatures, including fish, snails, lizards, and worms have been known to rain from the sky during unusual storms. The most likely explanation for these occurrences is water spouts. A water spout is similar to a tornado, except that it occurs over water. Water spouts tend to suck up anything close to the water's surface, including animals. For example, in 1894, a rain of jellyfish was reported in Bath, England.

Teach

Climate and Vegetation Regions

Talk About It

- What role does latitude play in determining climate? *(Possible Answer: It helps geographers know whether a climate is warmer or colder based on how close the region is to the equator.)*
- Which climate regions can have hot weather? *(tropical, dry, mid-latitude)*
- **Summarize** What is the difference between climate regions and vegetation regions? *(Climate regions are areas of the Earth with similar climates, or weather patterns, usually based on latitude. Vegetation regions are areas of Earth that have similar plants. Vegetation regions and climate regions tend to overlap.)*

INTERDISCIPLINARY ACTIVITIES

CONNECT to Language Arts

Write a Paragraph

Have students think about their access to clean water. Remind them that they have easy access to water for drinking and cooking. Have them write a paragraph describing changes in their life if they had to collect water for their family daily.

Unit Resource Book: Introduction to World Geography
- Interdisciplinary Projects: Accessible Water, pp. 97–98

CONNECT to Math

Make a Precipitation Graph

Have students use the Internet to find the precipitation amounts in three selected cities. Then, they can construct a bar graph showing the months of the year and the amount of precipitation for the selected cities.

Unit Resource Book: Introduction to World Geography
- Interdisciplinary Projects: Rainfall Around the World, pp. 96, 98

COMPARING

World Climate Regions

- Identify some of the climate regions found in the continental United States. *(Dry: Desert, Semiarid; Mid-Latitude: Mediterranean, Marine West Coast, Humid Subtropical, Humid Continental; Highland; Tropical: Tropical Wet and Dry)*
- How is the combination of climate regions found in Africa different from the combination of climate regions in South America? *(Possible Answer: Africa has larger Dry/Desert and Semiarid areas; South America has larger Tropical/Tropical Wet and Tropical Wet and Dry areas.)*
- Considering the number of different climate regions, how would you explain the fact that most of northern Europe and northern Canada have the same climate? *(Possible Answer: Both are close to the Arctic Circle, far from the equator, leading to a High Latitude: Subarctic climate.)*

ANSWERS

1. **Region** Tropical Wet
2. **Place** Answers will vary.

Comparing Regions

Climates in the United States

Here is how some places compare in terms of climate according to National Oceanic & Atmospheric Administration:

- Albany, New York—average temperature in January, 22.2 degrees; average yearly snowfall, 64.4 inches
- Austin, Texas—average temperature in January, 50.2 degrees; average yearly snowfall, 0.9 inches
- Miami, Florida— average temperature in January, 68.1 degrees; average yearly rainfall, 58.5 inches
- Los Angeles, California— average temperature in January, 57.1 degrees; average yearly rainfall, 13.1 inches
- Seattle, Washington—average temperature in January, 40.1 degrees; average number of days with rainfall each year, 155

COMPARING World Climate Regions

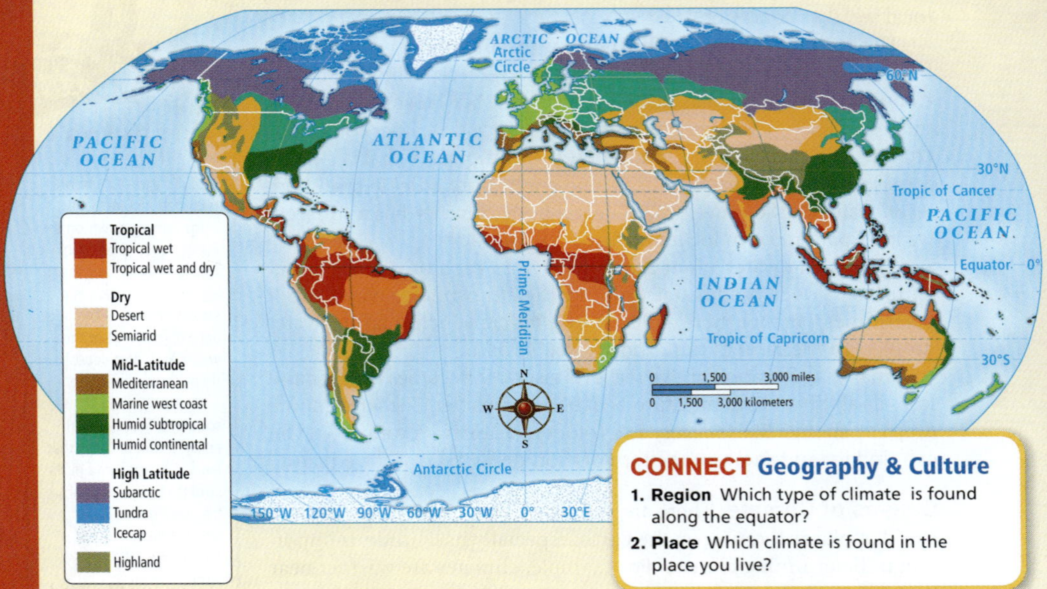

CONNECT Geography & Culture

1. **Region** Which type of climate is found along the equator?
2. **Place** Which climate is found in the place you live?

	TYPES	CHARACTERISTICS	VEGETATION
TROPICAL	Tropical Wet	rainy all year round	broadleaf trees
	Tropical Wet and Dry	has a rainy season	grasslands and savanna
DRY	Desert	less than 10 inches of precipitation	specially adapted plants
	Semiarid	up to 20 inches of rain	grasslands called steppe
MID-LATITUDE	Mediterranean	wet winter	plants that have adapted to a dry season in the summer
	Marine West Coast	rainy year round	mixed forests of broadleaf and needleleaf trees or only needleleaf trees
	Humid Subtropical	long, hot, humid summers	
	Humid Continental	wide range of summer and winter temperatures	grasslands and savanna
HIGH LATITUDE	Subarctic	cold dry climate	specially adapted plants
	Tundra	permanent ice and snow	specially adapted plants
	Icecap	permanent ice and snow	none
HIGHLAND	Varies with latitude, elevation, and continental location		

DIFFERENTIATING INSTRUCTION: TIERED ACTIVITIES

OBJECTIVE Create a glossary of terms related to world climate regions.

Basic

Provide student pairs with a list of climate categories and sub-categories. They should use the list to prepare a glossary of the definitions of climate categories. Glossaries should be in alphabetical order and include illustrations. Use the glossaries to review world climate regions.

On Level

Have students work independently to complete the Basic activity. In addition, have students add examples of one U.S. state and one country (not the United States) for each climate region. Use the completed glossaries to review world climate regions.

Challenge

In addition to the On Level activity, have students write a summary of the climates of the United States and of another country. Students should use at least one category and four sub-categories from their glossaries. Use the glossaries and summaries for review.

Climate Regions As you can see on the map on the opposite page, the Earth has five general climate regions: tropical, mid-latitude, high latitude, dry, and highland. Tropical climates are always hot and can be rainy most of the year or only during one season. The middle latitudes have the greatest variety of climates, ranging from hot and humid to cool and fairly dry. Climates along the oceans are also included in this category. High latitude climates are cool to cold all year long. Dry climates can be found in every latitude region. Highland climates are based on the elevation of a particular place. So, for example, as you go up a mountain, the climate may change from warm to cooler to cold.

Vegetation Regions The term **vegetation region** refers to an area that has similar plants. A vegetation region is named for the types of trees, grasslands, and specially adapted plants found there. The four basic types of vegetation are: forest, savanna, grasslands, and desert. Forests can be cold, tropical, or temperate. **Savanna** is a mix of grasslands and trees. Grasslands can have short or tall grasses, depending on the amount of rain. Finally, a **desert**—which can be hot or cold—has plants specially adapted to very dry conditions.

SUMMARIZE Identify the five main climate regions. tropical, dry, mid-latitude, high latitude, highland

Section 3 Assessment

ONLINE QUIZ For test practice, go to Interactive Review @ClassZone.com

TERMS & NAMES

1. **Explain the importance of**
   - weather
   - climate
   - precipitation
   - vegetation region

USE YOUR READING NOTES

2. **Summarize** Use your completed chart to answer the following question:
   What are the basic causes of weather and which factor is the most important?

CLIMATE | VEGETATION

KEY IDEAS

3. What causes the changing seasons?
4. What are the causes of climate?
5. How are vegetation regions named?

CRITICAL THINKING

6. **Analyze Causes and Effects** Why are the seasons reversed in the Northern and Southern Hemispheres?
7. **Draw Conclusions** How does location affect climate and vegetation?
8. **CONNECT to Today** What weather conditions have caused problems in the United States recently?
9. **WRITING** **Write a Description** Determine what climate and vegetation region you live in. Then write a paragraph describing the climate features and types of vegetation.

CONNECT to the Essential Question

How do Earth's physical systems make life on Earth possible?

Ask students what they have learned so far that can help them answer this question. Students might mention:

- The Earth's rotation and its revolution around the sun affects the Earth's weather and climate.
- Precipitation returns water to Earth in the form of rain, snow, sleet, or hail.
- Wind and ocean currents distribute solar energy throughout the world.

4 Assess & Reteach

Assess Have students complete the Section Assessment.

Unit Resource Book: Introduction to World Geography
- Section Quiz, p. 105

Interactive Review @ClassZone.com
- Section Quiz

Power Presentations

Test Generator

Reteach Have students prepare a slip of paper, containing the name of one climate zone or type, or one vegetation region. Place the slips in an open container. Call on a volunteer to select a strip, read it out loud, and then briefly tell what he or she knows about the climate zone or type, or vegetation region. Repeat the process until all the slips have been discussed.

Unit Resource Book: Introduction to World Geography
- Reteaching Activity, p. 109

SECTION 3 ASSESSMENT ANSWERS

Terms & Names

1. weather: the daily condition of Earth's atmosphere; climate: the weather conditions over a long period of time; precipitation: falling water droplets in the form of rain, snow, sleet, or hail; vegetation region: an area that has similar plants

Use Your Reading Notes

2. heat from the sun, water vapor, cloud cover, landforms, bodies of water, elevation, and wind; Heat from the sun is the most important.

CLIMATE: Earth's rotation and revolution, seasons

VEGETATION: vegetation regions, climate regions

Key Ideas

3. the earth's tilt and revolution
4. location, wind and ocean currents
5. The regions are named for the types of trees, grasslands, and plants found there.

Critical Thinking

6. The seasons are reversed because of the tilt of the Earth since one hemisphere gets more sun at one time than the other.
7. Locations closer to the equator produce warmer climates and the vegetation linked to it; locations near the poles produce colder climates and related vegetation.
8. hurricanes in the south, drought in the west, flooding in other parts of the country
9. Use the rubric to evaluate the paragraph.

Paragraph Rubric

	Identification of Regions	Paragraph Construction
4	clear and very complete	clear topic sentence, good support
3	clear	clear topic sentence, complete sentences
2	missing several regions	topic sentence, some fragments
1	missing many regions	no topic sentence, many fragments

1 Plan & Prepare

Objectives: Identify why the Amazon rain forest is an important physical feature, and how it affects the Earth.

Visual Vocabulary: Ask students to describe what they see in the illustration. Call out familiar items listed below, asking students to locate each one on the page. Then tell them the precise term, as listed in boldface. Add terms to a class vocabulary list, illustrated poster, or Word Wall.

- brightly colored bird **(toucan)**
- native people **(Yanamamö)**
- communal village hut **(shabono)**
- tree tops **(canopy)**
- purple flowers **(orchids)**

2 Teach

Engage . . .

DISPLAY the "Amazon Rain Forest" feature from the **Power Presentations** DVD-ROM or the *Animations Center* at **ClassZone.com.** As the animation loads on screen, ask students:

- Why are the scientists climbing a tree?
- What challenges do people living in the rain forest face?
- How is a shabono organized?
- How does the rain forest produce oxygen?

Encourage all answers.

Explore . . .

ROLL OVER sections of the illustration or click on items in the menu. Ask students to choose which areas they would like to explore further.

- Then return to the discussion questions. Have any answers changed?
- Have students jot down a list of at least three new questions that arise from watching the rollover features.

Animated GEOGRAPHY
Amazon Rain Forest

Click here to enter the rain forest @ClassZone.com

AMAZON RAIN FOREST

The Amazon rain forest is one of the world's most important physical features. It acts as the "lungs of the planet" by producing oxygen, and is the home of millions of plants and animals.

Click here to learn more about this frog and the amazing diversity of plants and animals in the forest.

Click here to see Yanamamö village life in the rain forest and learn about how the Yanamamö interact with the forest.

Click here to see the methods scientists are using to study plants and animals in the rain forest.

DIFFERENTIATING INSTRUCTION: TIERED ACTIVITIES

OBJECTIVE Discuss the importance of plant and animal diversity in the rain forest.

Basic

List the Animals and Plants of the Amazon Rain Forest

Have students take a sheet of paper and draw a line down the center. They should label one side *Animals of the Amazon Rain Forest* and the other side *Plants of the Amazon Rain Forest.* Using their textbooks and other classroom resources, have students complete their charts. To extend this activity, have students illustrate at least one animal and one plant, and write a caption for each.

GeoActivity

Plan a Scientific Study
With a small group, plan the scientific study of one of the plants or animals in the picture. Identify the subject you wish to study. Talk about what information you want to find. Then write questions that would help you track down that information.

47

On Level

List the Importance of the Rain Forest for Plants and Animals

Have students work in pairs. Using their textbooks and other classrooom resources have students create a list of plants and animals unique to the Amazon rain forest. Some of the listed plants and animals should include species that can be used as medicine. As a class, discuss the importance of plant and animal diversity in the rain forest for humans.

Challenge

Report on Rain Forest Development

Have students work in pairs, using library or Internet sources, to find out why the Amazon rain forest is disappearing. The report should cover development in the rain forest, and the effects of loggers and builders. The report should also explain how the disappearance of rain forest plants and animals affects the Earth. Have students present their reports to the class.

Explain . . .

CHOOSE topics to help students answer their questions, such as:

- **Rain Forest Animals:** Scientists discover new species of animals each year in the rain forest. Many animals live at the top of the canopy and cannot be seen from the ground. Others live about half way to the top, also making them difficult to study.
- **Yanamamö:** The Yanamamö have little if any contact with the modern world. They live in small groups, and their daily activities include farming, hunting, and making crafts. A typical Yanamamö village is built in the shape of a giant circular hut. Each section of the village has a roof and back wall, but no front or side walls. Individual family huts are joined together in a circle.
- **Plants:** More than half of the world's plants are found in the rain forest. Plants use solar energy to convert carbon dioxide (one of the gases in the atmosphere) and water into oxygen for breathing. This process is called photosynthesis.

Elaborate . . .

ASSIGN students into four teams to compete in a Geography Bee.

- Have each team write a five-question quiz on the Amazon rain forest. Each team will then answer the quiz questions of the other three teams. Only one team member may give the answer, but all team members may help determine the answer. The team with the most correct answers wins.

Evaluate . . .

Plan a Scientific Study

Activity Rubric

4 Excellent The plan identifies many plants and animals; has many good questions.

3 Strong The plan identifies some plants and animals; has some good questions.

2 Fair The plan identifies a few plants and animals; has a few good questions.

1 Poor The plan does not identify plants and animals; it fails to include questions.

1 Plan & Prepare

Objectives

- Explain the relationship between global warming and the greenhouse effect, and their effects on the Earth's climate
- Identify the causes and effects of desertification

Read for the Essential Question

Help students read for a purpose by reminding them of the Essential Question: "How do Earth's physical systems make life on Earth possible?" Point out that they can use text features such as Key Questions, special features, images, and maps to form answers to the question.

Vocabulary

Best Practices Toolkit

Use the ***Best Practices Toolkit*** to model strategies for vocabulary notetaking. Vary strategies throughout the year. Choose from: *Knowledge Rating, Predicting ABCs, Definition Mapping, Word Sort, Frayer Model (Word Squares), Magnet Words, Word Wheel,* and *Student VOC.*

Vocabulary Strategies, TT9–TT16

Reading Strategy

Best Practices Toolkit

Display the T Chart/Two-Column Chart transparency.

- Remind students that organizing information into categories can help in making effective comparisons and contrasts.
- Model filling in an example or detail for global warming and for desertification.
- Have students suggest entries. Have them revise their charts as they complete the section.

T Chart/Two-Column Chart, TT20

SECTION 4

Reading for Understanding

Key Ideas

BEFORE, YOU LEARNED

Many different physical systems influence the way we live on Earth.

NOW YOU WILL LEARN

Human interference with physical systems can cause problems with the environment.

Vocabulary

TERMS & NAMES

global warming an increase in the average temperature of the Earth's atmosphere

greenhouse effect the trapping of the sun's heat by gases in the Earth's atmosphere

greenhouse gas any gas in the atmosphere that contributes to the greenhouse effect

fossil fuels fuels such as coal, oil, and natural gas

desertification the process in which farmland becomes less productive because the land is degraded

sustainable using natural resources in a way that they exist for future generations

BACKGROUND VOCABULARY

carbon dioxide a gas composed of carbon and oxygen

emissions substances discharged into the air

degraded of lower quality

Visual Vocabulary desertification

Reading Strategy

Re-create the chart shown at right. As you read and respond to the **KEY QUESTIONS**, use the chart to compare and contrast details about environmental challenges the world faces.

Skillbuilder Handbook, page R9

COMPARE AND CONTRAST

GLOBAL WARMING	DESERTIFICATION

GRAPHIC ORGANIZERS Go to **Interactive Review** @ClassZone.com

PRETEACHING VOCABULARY

English Learners

Pronounce and Preview

Pronounce each term for students. Review the meanings of words within definitions, such as *degraded, generations,* and *discharged.*

- To modify vocabulary learning, have students complete worksheets as they read, instead of afterward.

Unit Resource Book: Introduction to World Geography
- Vocabulary Practice, p. 89

Inclusion

Create Audio or Visual Vocabulary

Have students work in pairs. Students with strong visual/reading skills can read the vocabulary terms and definitions aloud to their partners. Students should then work together to create either a visual review of the vocabulary terms, such as an illustrated glossary, or an audio review, such as a jingle or rhyming phrase about each term.

Big Sur, California

SECTION 4 GEOGRAPHY

Peruvian Rainbow Boa in the rain forest

Environmental Challenges

Connecting to Your World

Have your parents or other adults ever told you that the climate seems to be changing? Maybe they said something like, "We never had hurricanes or tornados like these when I was a kid." They could be right—climates do change. Some changes take place naturally over many years, such as the build-up of ice during ice ages. But recently, scientists have noticed rapid climate changes that some believe is the result of human activity.

Global Warming

KEY QUESTION How are global warming and the greenhouse effect related?

Global warming and desertification are two possible threats to the environment. **Global warming** is an increase in the average temperature of the Earth's atmosphere. It refers to an increase large enough to cause changes in the Earth's overall climate. The Earth's average temperature has risen between 0.9 and 1.3°F since the late 1800s. Many scientists think that temperatures may rise another 2.5 to 10.4°F by the end of the century. Some scientists believe that this warming is part of the larger cycle of warm and cold periods in the Earth's history. Others suspect that it is caused by humans.

Dori, Burkina Faso Changes in the climate have caused drought in this region, leading to fewer crops.

49

2 Focus & Motivate

3-Minute Warm-Up

Write on the board or display the transparency:

- Look at the Greenhouse Effect diagram in this section. What does it tell you about global warming? *(Possible Answer: Many sources of gases from burning fossil fuels contribute to trapping heat in the atmosphere.)*

Unit Transparency Book
- 3-Minute Warm-Ups, TT9

Connecting to Your World

More About . . .

Burkina Faso

This republic in West Africa generally receives little rainfall each year. With poor soil and a large semi-desert area, farming is difficult—much of the land isn't suited for farming without irrigation.

3 Teach

Global Warming

Think, Pair, Share

- How does the greenhouse effect contribute to raising Earth's average temperature? *(It traps the sun's heat in the atmosphere.)*
- **Compare and Contrast** How are greenhouse gases and fossil fuels similar? How are they different? *(Similar: Both contribute to the greenhouse effect. Different: Greenhouse gases, including water vapor, are all the gases that contribute to the greenhouse effect. Fossil fuels release certain greenhouse gases.)*

SECTION 4 PROGRAM RESOURCES

Unit Resource Book: Introduction to World Geography

ON LEVEL

Unit Resource Book
- Reading Study Guide, p. 63
- Section Quiz, p. 106

STRUGGLING READERS

Unit Resource Book
- Reading Study Guide with Additional Support, p. 71
- Vocabulary Practice, p. 89
- Section Quiz, p. 106
- Reteaching Activity, p. 110

eEdition with Audio DVD-ROM

ENGLISH LEARNERS

Pupil Edition in Spanish
eEdition with Audio DVD-ROM
Unit Resource Book
- RSG (Spanish), p. 79
- RSG with Additional Support (Spanish), p. 87

Multi-Language Glossary
Test Generator
- Section Quiz in Spanish

INCLUSION

Unit Resource Book
- RSG with Additional Support, p. 71
- Section Quiz, p. 106
- Reteaching Activity, p. 110

GIFTED & TALENTED

Unit Resource Book
- History Makers: Wangari Maathai, p. 95
- Section Quiz, p. 106

PRE-AP

Unit Resource Book
- Economics in the World, p. 99

TECHNOLOGY

Unit Transparency Book
- 3-Minute Warm-Ups, TT9
- Essential Question, TT10
- Critical Thinking, TT14

Daily Test Practice Transparencies
- Section 4, TT7

Power Presentations
ClassZone.com
World Cultures and Geography Video on DVD

Greenhouse Effect

Have students identify other sources of greenhouse gases, such as heating and cooling homes and offices, that they could add to this diagram.

- What makes up the heat trapping layer? *(greenhouse gases)*

More About . . .

Kyoto Protocol

More than 165 countries ratified the Kyoto Protocol by 2006. However, the United States was not one of them. Instead, the United States, Australia, China, and several other large nations entered into a different agreement called the Asia-Pacific Partnership on Clean Development and Climate. This partnership proposes to reduce emissions from member countries through the development and use of new technologies that do not increase carbon dioxide in the atmosphere. The partnership held its first meeting in 2006.

Comparing Regions

United States and Switzerland

The United States and Switzerland are very different when it comes to greenhouse gas emissions. Both countries have high standards of living and successful economies, but in the United States, approximately 23.35 tons of greenhouse gases are emitted per person each year—one of the highest rates in the world. In Switzerland, approximately 7.19 tons per person are emitted each year—one of the lowest rates in the world.

Many of the emissions come from motor vehicles. In the United States, people tend to drive longer distances. In Switzerland, a much smaller country in terms of population and area, people tend to rely more on public transportation.

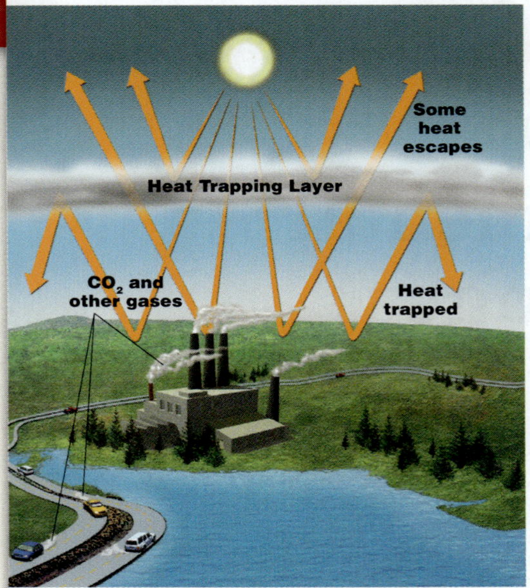

Greenhouse Effect Trapped gases and heat combine to change the Earth's temperature. **How do humans add to the greenhouse effect?**

The burning of fossil fuels adds to the greenhouse effect by adding more carbon dioxide to the atmosphere.

The Greenhouse Effect The trapping of the sun's heat by gases in the Earth's atmosphere is called the **greenhouse effect**. These gases act like the glass roof of a greenhouse. They let in solar energy, which heats up the planet, but they trap much of the heat that rises from the Earth's surface. So, the Earth becomes warmer. Some greenhouse effect is necessary. Without it, you would be living on a freezing cold planet. The temperature would be about zero degrees Fahenheit.

Any gas in the atmosphere that contributes to the greenhouse effect is called a **greenhouse gas**. Greenhouse gases include water vapor and **carbon dioxide**, a gas composed of carbon and oxygen. The burning of **fossil fuels**, such as coal, oil, and natural gas, has caused an increase in carbon dioxide in the atmosphere. As this gas builds in the atmosphere, the atmosphere becomes warmer and speeds up the heating effect.

The Impact on the Climate In theory, a more intense greenhouse effect could change the Earth's climate. Warmer temperatures could cause the ice caps and glaciers around the world to melt and sea levels to rise. Flooding could occur along coastal regions. Global land-use patterns would change. Some crops would no longer grow in certain areas. Some areas would become hotter and drier, with extreme heat waves, droughts, and more forest fires. These changes would alter the fragile relationships between living things and the environment.

Solutions Lowering the levels of greenhouse gases is a complex, worldwide goal. Some nations generate huge amounts of greenhouse gases. These gases affect not only those nations, but also the whole planet. So, humans need to take steps to reduce the levels of greenhouse gases. One solution might be to build more energy-efficient cars and factories. Another could be to use alternative energy sources such as energy produced by the sun, or the internal heat of the Earth.

In 1997, nations from around the world gathered at Kyoto, Japan, to discuss plans to reduce greenhouse gases. The result of their meeting was an agreement to cut emissions of carbon dioxide and other greenhouse gases. (**Emissions** are substances discharged into the air.) The agreement, called the Kyoto Protocol, went into effect in 2005.

SUMMARIZE Explain how global warming takes place. Greenhouse gases trap heat. The trapped heat increases the average world temperature.

DIFFERENTIATING INSTRUCTION

Struggling Readers

Compare and Contrast

To help students differentiate among the vocabulary terms on this page, have them complete a chart such as the one at right. After they have correctly identified each term and provided an example, have them re-read and respond to the Key Question in this subsection.

Term	Definition	Example
greenhouse effect		
greenhouse gas		
carbon dioxide		

Desertification

KEY QUESTION What causes desertification?

Desertification is the process in which farmland becomes less productive because the land is **degraded**. The land becomes more desert-like, that is, dry and unproductive. Desertification is a serious problem because it turns arid and semiarid areas into nonproductive wasteland. Each year, about 25,000 square miles of land—an area the size of West Virginia—is degraded. This process is happening in many parts of the world, including Africa, China, and the American West.

Causes and Effects In desertification, natural vegetation is removed or destroyed, and soil is exposed to wind. Without shade from the sun, the moisture in the soil evaporates more quickly. The dry top layers of soil particles then blow away. The soil becomes less able to support plant life. The loss of moisture and plants may in itself cause more desertification. But destructive practices in arid and semi-arid regions have speeded up the process. Some of these practices are

overgrazing allowing animals to graze so much that plants are unable to grow back

cultivation of marginal land planting crops on fragile soil

deforestation cutting down trees and not replanting new trees

ONLINE PRIMARY SOURCE To read more of Wangari Maathai's writing, go to the **Research & Writing Center** @ClassZone.com

ANALYZING Primary Sources

Wangari Maathai (born 1940) won the 2004 Nobel Peace Prize for her work fighting deforestation in Africa. To combat desertification, she founded the Green Belt Movement, which has planted 30 million trees across Africa, including her native Kenya.

> [The Green Belt Movement] encourages women to create jobs, prevent soil loss, slow the process of desertification and [to] plant and to eat indigenous [local] food crops.
>
> Source: Speech to the 4th United Nations World Women's Conference, Beijing, China, 1995

DOCUMENT-BASED QUESTION

What is the goal of the Green Belt Movement?

51

DIFFERENTIATING INSTRUCTION: TIERED ACTIVITIES

OBJECTIVE Demonstrate an understanding of the causes and effects of desertification.

Basic

Have students work in small groups. Each group should create a poster or other visual display, illustrating the causes and effects of desertification. Have students present their posters and displays, or post them in your classroom.

On Level

Have students write a letter from the point of view of a farmer in the American West describing the causes and effects of desertification on the farmer's land and crops.

Challenge

Provide student groups with a data sheet about desertification in Africa. Students should then create a plan, including visuals, on ways to slow desertification and improve soil quality in Africa in addition to planting trees. Each group should present their plan to the class.

Teach

Desertification

Reader, Recorder, Reporter

- Explain the process of desertification. *(Farmland becomes desert-like as natural vegetation is removed or destroyed, making it unusable.)*
- What steps can be taken to reduce destructive practices that cause desertification? *(building sand fences, using straw mats to protect plants, planting tree fences or grass belts, and use of solar ovens)*
- **Compare and Contrast** How are the practices of overgrazing, cultivation of marginal land, and deforestation similar? *(Possible Answer: They are similar because all three speed up desertification. Overgrazing and deforestation involve loss of plants that are not replaced, while cultivation of marginal land involves planting in weak soil.)*

Unit Resource Book: Introduction to World Geography

- Economics in the World: Costs of Natural Disasters, p. 99

RESEARCH & WRITING CENTER

ANALYZING Primary Sources

Wangari Maathai was the first woman from East and Central Africa to earn a doctorate degree in 1971. She introduced the idea of planting trees in 1976, and founded the Green Belt Movement in 1977. In 2002, Maathai was elected to the Kenyan parliament with 98% of the vote.

- How does the Green Belt Movement slow desertification? *(Possible Answer: By planting new trees in areas that have been deforested or overgrazed, soil loss is slowed or stopped.)*

DOCUMENT-BASED QUESTION ANSWER

The goal of the Green Belt movement is to create jobs, to work to prevent soil loss and slow desertification, and to replace local crops.

Unit Resource Book: Introduction to World Geography

- History Makers: Wangari Maathai, p. 95

CONNECT to the Essential Question

How do Earth's physical systems make life on Earth possible?

Ask students what they have learned so far that can help them answer this question. Students might mention:

- The greenhouse effect keeps Earth's atmosphere warm, but also contributes to global warming.
- Earth's soil supports plant life, but can be damaged by desertification.

Unit Transparency Book: Introduction to World Geography
- Essential Question, TT10

4 Assess & Reteach

Assess Have students complete the Section Assessment.

Unit Resource Book: Introduction to World Geography
- Section Quiz, p. 106

Interactive Review @ClassZone.com
- Section Quiz

Power Presentations

Test Generator

Reteach Have students work in small groups and create one graphic organizer on global warming and one on desertification. Have each group present their graphic organizer to the class. Students should revise their organizers as necessary after each group has presented.

Unit Resource Book: Introduction to World Geography
- Reteaching Activity, p. 110

CONNECT Geography & Culture

Green Belt Movement

The Green Belt Movement, founded by Wangari Maathai in Kenya in 1977, is an example of a program of sustainable development. It started out as a jobs program to pay rural and urban women to plant trees. But it soon became a movement to improve the environment, slow deforestation, and halt desertification.

Solutions There are different solutions to the growing problem of desertification. Each depends on the underlying cause. Some simple solutions are to build sand fences that interrupt the wind, or to use straw mats to reduce evaporation so young plants can take root. Still another is to use solar ovens in place of open fires that require firewood. Planting tree fences and grass belts also reduces the spread of sandy areas. This practice is being used in China and Africa today.

Solutions like the ones above are examples of sustainable practices. **Sustainable** means that these practices use natural resources in such a way as to ensure that they exist for future generations. Sustainable practices work with the environment to protect the land, preserve wildlife, and repair the damage that has been done to it. The practices allow people to live a better life and ensure that resources will be available both now and in the future.

DRAW CONCLUSIONS Explain what causes desertification.
a combination of activities by humans, including overgrazing, marginal land use, and deforestation

ONLINE QUIZ For test practice, go to **Interactive Review @ClassZone.com**

Section 4 Assessment

TERMS & NAMES

1. Explain the importance of
- global warming
- greenhouse effect
- desertification
- sustainable

USE YOUR READING NOTES

2. Compare and Contrast Use your completed chart to answer the following question:

How does the greenhouse effect contribute to global warming?

GLOBAL WARMING	DESERTIFICATION

KEY IDEAS

3. Why is global warming a problem?

4. What are greenhouse gases?

5. What are three simple solutions to controlling desertification?

CRITICAL THINKING

6. Evaluate Which environmental problem, global warming or desertification, is a greater threat to the Earth? Why?

7. Summarize How does deforestation cause desertification?

8. CONNECT to Today What might happen if more gasoline-powered motor vehicles were used around the world?

9. MATH **Make a Chart** Use the Internet to find information about desertification. Then make a chart that shows locations and the percentage of land that has been degraded.

SECTION 4 ASSESSMENT ANSWERS

Terms & Names

1. global warming: an increase in the average temperature of Earth's atmosphere; greenhouse effect: the trapping of the sun's heat by gases in the Earth's atmosphere; desertification: the process in which farmland becomes less productive because land is degraded; sustainable: using natural resources in a way that they exist for future generations

Use Your Reading Notes

2. Greenhouse gases trap the sun's heat in the atmosphere resulting a warmer climate. **Diagram** *Global Warming: Earth temperature up, more greenhouse gases; Desertification: less productive land, destructive practices.*

Key Ideas

3. because increases in the Earth's average temperature cause climate and weather changes that can severely affect life

4. gases in the atmosphere that cause the greenhouse effect

5. build sand fences, use straw mats to reduce evaporation, use solar ovens, plant tree fences and grass belts

Critical Thinking

6. Answers will vary. Global warming choice should indicate the severe effects of increased temperatures; desertification should indicate the consequences of the loss of productive lands.

7. When the trees are cut down, the soil is more exposed to wind, moisture is lost, and the soil is not able to support plant life.

8. More motor vehicles will cause more greenhouse gases, and this will contribute to global warming.

9. Use the rubric to score the chart.

Chart Rubric

	List of Locations	Presentation of Information
4	complete	complete
3	covers adequately	missing one site
2	barely explains	missing two sites
1	unclear or missing	missing more than two sites

Interactive Review

Click here to complete these and other activities online @ClassZone.com

CHAPTER SUMMARY

Key Idea 1
The Earth is composed of many layers, and its surface continually changes because of the drifting of the continents.

Key Idea 2
Interaction between landforms and bodies of water makes life on Earth possible.

Key Idea 3
The Earth's rotation and revolution influence weather, climate, and living conditions on Earth.

Key Idea 4
Human interference with physical systems causes problems with the environment.

For Review and Study Notes, go to Interactive Review @ClassZone.com

NAME GAME

Use the Terms & Names list to complete each sentence on paper or online.

1. I am the hot metal center of the Earth. core
2. I am a naturally formed feature on the Earth's surface. ______
3. I fall in the form of rain, snow, sleet, or hail. ______
4. I am an increase in the Earth's temperature. ______
5. I move weathered materials from one place to another. ______
6. I am the trapping of the sun's heat by gases in the atmosphere. ______
7. I am a large rigid piece of the Earth's crust that is in motion. ______
8. I am typical weather conditions over a period of time. ______
9. I circulate water between the Earth, oceans, and atmosphere. ______
10. I am plants that grow in a region. ______

climate
core
desertification
erosion
global warming
greenhouse effect
hydrologic cycle
landform
magma
precipitation
relief
sediment
tectonic plate
vegetation
weather

Activities

Flip Cards

Use the online flip cards to quiz yourself on terms and names introduced in this chapter.

Crossword Puzzle

Complete an online crossword puzzle to test your knowledge of Earth's physical systems.

ACROSS

1. a naturally formed feature on the Earth's surface

53

ONLINE AND ON-SCREEN ACTIVITIES

Flip Cards

Vocabulary Flip Cards are virtual flash cards at **ClassZone.com**. Students can quiz each other or work independently to study the flip cards for the chapter, "Earth's Interlocking Systems."

- They can quiz themselves on key terms and names by reviewing definitions.
- Demonstrate how to advance and shuffle the deck for another challenge.

Crossword Puzzle

Students can review important elements of Earth's Interlocking Systems by completing this crossword puzzle online at **ClassZone.com**.

- As an extension of this activity, have students use the *Activity Maker* at **ClassZone.com** to make their own crossword puzzle for one section of the chapter.
- Then have students trade puzzles with classmates to solve each other's puzzles.

Chapter Summary

As a follow-up to the chapter, encourage students to pair up and go over the **Review and Study Notes** at **ClassZone.com.**

Alternatively, students pairs can review their Reading Study Guide notes, making sure that they can add important details to each section's "Key Idea."

Name Game

Teaching Tips

ON PAPER Have students work in pairs to complete this activity. As an extension, they can write questions for the terms that are not covered in the items shown here.

ONLINE Make sure that students understand how to drag and drop terms to fill in blanks. Ask them to print out their results and compare them with those of a classmate.

ANSWERS

1. core
2. landform
3. precipitation
4. global warming
5. erosion
6. greenhouse effect
7. tectonic plate
8. climate
9. hydrologic cycle
10. vegetation

Make It Fun

Teacher-Tested Activities

Eric Branch, Wright Middle School, Madison, Wisconsin

Here's a way to review the chapter material using the Five Themes.

- First make a list of geographic features or forces covered in the chapter. Around the room, place five signs, each with one theme of geography.
- Model the activity for the students using volcanoes. For example, for Place, say that a site could be identifiable because of a nearby volcano. For Region, talk about the Ring of Fire. For Location, provide the volcano's latitude and longitude.
- Next call out a feature or force covered in the text. Students physically move to one of the themes' areas that they think they could apply to the item called out.
- Once in groups students should discuss one or two reasons for identifying with the theme.

Vocabulary ANSWERS

1. Tectonic plate movement causes volcanoes, earthquakes, and the reshaping of the Earth.
2. Greenhouse effect is linked to global warming.
3. The hydrologic cycle is the source of all water on Earth.
4. Global warming threatens to change life on Earth.
5. Desertification is a very large environmental threat in Africa and China.
6. They are all part of the Earth's structure.
7. They are all processes that reshape the Earth's structure.
8. They are all part of the way the Earth receives water.
9. Greenhouse gases are believed to contribute to global warming.
10. Vegetation is dependent on climate.

Key Ideas ANSWERS

11. that the seven continents were once part of a supercontinent that broke into pieces and drifted apart
12. spreading, diving, colliding, and sliding
13. Weathering wears down surface rock, and erosion takes the loosened materials and deposits them in other places.
14. The Earth appears blue from above because almost 75% of its surface is covered by water.
15. Atlantic, Pacific, Indian, Arctic, and Southern
16. a wide plains area in the center and narrow belts of mountains on the continent's edge
17. rotation and revolution
18. wind currents, ocean currents, latitude, and elevation
19. forest, savanna, grasslands, and desert
20. It is trapped by gases in the Earth's atmosphere.
21. energy from the sun or from internal heat of Earth
22. Africa, China, and the American West

Critical Thinking ANSWERS

23. Possible answers:
    temperature increases
    melting of ice caps and glaciers
    flooding along coastal regions
    people forced to move
    land-use patterns change
    more uninhabitable areas
    droughts
    forest fires

CHAPTER 2 Assessment

VOCABULARY

Explain the significance of each of the following.

1. tectonic plate
2. greenhouse effect
3. hydrologic cycle
4. global warming
5. desertification

Explain how the terms and names in each group are related.

6. magma, crust, and continent
7. weathering, erosion, and desertification
8. hydrologic cycle, precipitation, and atmosphere
9. greenhouse gas and global warming
10. vegetation and climate

KEY IDEAS

1 The Earth and Its Forces

11. What is the continental drift theory?
12. What are the four types of tectonic plate movements?
13. How do weathering and erosion reshape the Earth's surface?

2 Bodies of Water and Landforms

14. Why is the Earth sometimes called the "blue planet"?
15. What are the names of the world's five oceans?
16. What two general features do most continents have in common?

3 Climate and Vegetation

17. What are the two motions of the Earth in relation to the sun?
18. What are the four factors that influence climate?
19. What are the basic types of vegetation?

4 Environmental Challenges

20. What happens to the sun's heat in the greenhouse effect?
21. What are two alternative energy sources?
22. What are three areas of the world that have a serious problem with desertification?

CRITICAL THINKING

23. **Analyze Cause and Effect** Create a web diagram to show the effects of global warming. List at least five effects caused by a more intense greenhouse effect.

24. **Summarize** Why are so many earthquakes and volcanoes found near the Ring of Fire ?
25. **Draw Conclusions** Why is the surface of the Earth constantly changing?
26. **Connect to Economics** What are the economic costs of earthquakes or volcanoes?
27. **Five Themes: Human-Environment Interaction** What steps have been taken to control global warming?
28. **Connect Geography & Culture** Why might the location and size of landforms affect where people live?

Answer the ESSENTIAL QUESTION

How do Earth's physical systems make life on Earth possible?

Written Response Write a two- or three-paragraph response to the Essential Question. Be sure to consider the key ideas of each section as well as ways in which the systems interlock. Use the rubric below to guide your thinking.

Response Rubric
A strong response will:
- discuss how the hydrologic cycle makes human occupation of the Earth possible
- describe the causes of weather and climate
- explain how the above systems affect humans

ESSENTIAL QUESTION REVIEW

Before students begin their essays, have them review the key chapter elements that will help them answer the Essential Question. Direct them to significant features in the chapter, such as:

- Comparing Tectonic Plates Movements, Section 1
- Hydrologic Cycle, Section 2
- Comparing World Climate Regions, Section 3
- Animated Geography: Amazon Rain Forest, Section 3
- Greenhouse Effect, Section 4

Written Response Rubric

	Synthesis of Key Ideas	Paragraph Construction
4	clear, complete	clear topic sentence good supporting material
3	includes all key ideas	topic sentence complete sentence
2	missing one key idea	poor topic sentence some fragments
1	missing key ideas	lacks topic sentence many fragment

STANDARDS-BASED ASSESSMENT

Test Practice
- Online Test Practice @ClassZone.com
- Test-Taking Strategies and Practice at the front of this book

THEMATIC MAP

Use the map to answer questions 1 and 2 on your paper.

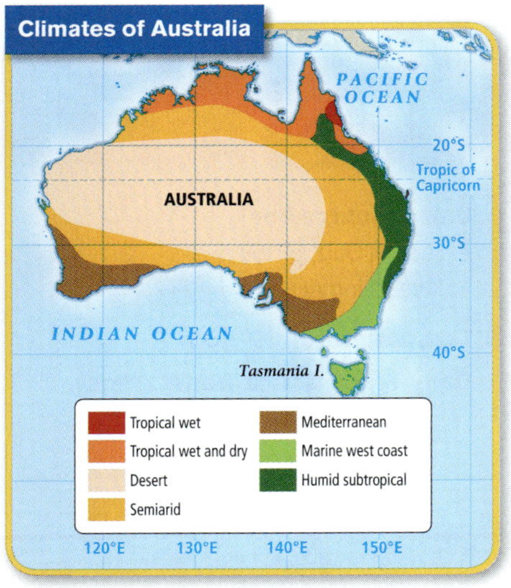

1. Which climate takes up the largest portion of Australia?

A. Desert
B. Mediterranean
C. Semiarid
D. Tropical wet and dry

2. In which part of Australia would you find a tropical wet climate?

A. northwestern
B. northeastern
C. southwestern
D. southeastern

CHART

Study the chart below. Use the information in the chart to answer questions 3 and 4.

Deadliest Earthquakes *1975* to *2005*

Country	Year	Deaths
Indonesia	2004	283,106
China	1976	255,000
Pakistan	2005	80,361
Iran	1990	40,000
Iran	2003	26,200
Armenia	1988	25,000
Guatemala	1976	23,000
India	2001	20,230
Turkey	1999	17,119

Source: USGS Earthquake Hazards Program

3. Which country has experienced the greatest number of deadly earthquakes?

4. In what year did the single most deadly earthquake strike?

GeoActivity

1. INTERDISCIPLINARY ACTIVITY–SCIENCE

Select one of the Earth's major mountain chains and illustrate its creation on a poster. Be sure the poster shows the plates involved, the direction of the collision, and the name of the mountains that were formed.

2. WRITING FOR SOCIAL STUDIES

Review the illustration of the Earth's interior in Section 1. Write a one-paragraph description that would help a younger student understand the layers of the Earth's interior.

3. MENTAL MAPPING

Create an outline map of the world and label the following:
- the seven continents
- the five oceans

DIFFERENTIATING CHAPTER ASSESSMENT

Unit Resource Book

- **Chapter Test A,** pp. 111–114.
  Use for Struggling Readers, Beginning-to-Intermediate-Level English Learners, and for Inclusion of students whose IEP suggests supported test-taking.
- **Chapter Test B,** pp. 115–118.
  Use for On-Level Learners and ESOL students with Advanced-Level English.
- **Chapter Test C,** pp. 119–122.
  Use for Gifted and Talented students or students assigned Pre-AP work.

Test Generator

- Chapter Tests A, B, and C, in English and Spanish

McDougal Littell Assessment System

TEST **Test Generator** Use three levels of tests in English and Spanish tied to state standards. Plus, choose questions to customize tests for all populations.

SCORE Scan and Score plain paper student tests in minutes.

REPORT Create instant progress reports, sortable by student, by standard, or other criteria.

RETEACH Respond immediately to students' needs by using a wealth of reteaching resources.

24. The Ring of Fire produces so many earthquakes and volcanoes because it is an area where several tectonic plates meet and move against one another.

25. The surface of the Earth is constantly changing because internal and external forces are creating new landforms; global warming and desertification are changing the landscape; human activity is altering physical systems.

26. Possible answer: Volcanoes and earthquakes can cause property damage, damage to the environment, loss of work hours, and injury or even death.

27. Possible answer: Wealthy nations who produce the most greenhouse gases are being asked to build more efficient cars and factories and use different energy sources. Also, an international treaty to reduce greenhouse gases was signed at Kyoto, Japan, by many of the world's nations.

28. Possible answer: People might choose to not live on or near some landforms that might make life difficult. But they could also choose to live somewhere because the landform offers certain advantages.

Standards-Based Assessment ANSWERS

1. A
2. B
3. Iran
4. 2004

GeoActivity

1. INTERDISCIPLINARY ACTIVITY RUBRIC

The poster should
- identify the mountain chain and show its creation
- show the plates and direction of movement
- clearly show the creation process

2. WRITING FOR SOCIAL STUDIES RUBRIC

The paragraph should
- correctly explain Earth's layers
- use language suitable for younger children

3. MENTAL MAPPING RUBRIC

The outline map should
- show seven continents correctly located and labeled
- show five oceans correctly located and labeled

CHAPTER 3 Human Geography PLANNING GUIDE

Complete print resources all on one USB drive allow you to customize lesson plans for all students.

Section Titles & Key Ideas	Unit Resource Book Introduction to World Geography	Assessment Resources
1 The Geography of Population **KEY IDEA** People are not equally distributed on the Earth's surface.	• **Reading Study Guide,** pp. 123–124 • **RSG with Additional Support,** pp. 131–132 • **Vocabulary Practice,** p. 155 • **Map and Graph Skills,** p. 157	**Unit Resource Book** • Section Quiz, p. 169 • Reteaching Activity, p. 173 **Test Generator CD-ROM** **Daily Test Practice Transparency,** TT18
2 Why People Move **KEY IDEA** People move from one place to another to meet their needs.	• **Reading Study Guide,** pp. 125–126 • **RSG with Additional Support,** pp. 133–134	**Unit Resource Book** • Section Quiz, p. 170 • Reteaching Activity, p. 174 **Test Generator CD-ROM** **Daily Test Practice Transparency,** TT19
3 Resources and Economics **KEY IDEA** Economic activities in an area depend on the presence of natural resources.	• **Reading Study Guide,** pp. 127–128 • **RSG with Additional Support,** pp. 135–136 • **Skillbuilder Practice,** p. 156 • **Interdisciplinary Projects,** pp. 163–164 • **Economics in the World,** p. 166	**Unit Resource Book** • Section Quiz, p. 171 • Reteaching Activity, p. 175 **Test Generator CD-ROM** **Daily Test Practice Transparency,** TT20
4 Why We Need Government **KEY IDEA** The world is divided into many political regions and organizations.	• **Reading Study Guide,** pp. 129–130 • **RSG with Additional Support,** pp. 137–138 • **Outline Map Activities,** p. 159 • **History Makers,** pp. 161–162	**Unit Resource Book** • Section Quiz, p. 172 • Reteaching Activity, p. 176 • Chapter Test (Forms A, B, & C), pp. 177–188 **Test Generator CD-ROM** **Daily Test Practice Transparency,** TT21

McDougal Littell Assessment System

TEST SCORE REPORT RETEACH

No Time? To focus students on important content in the chapter, use the **"Reading Study Guide with Additional Support"** in ***Resources2Go.***

Teacher Presentation Options

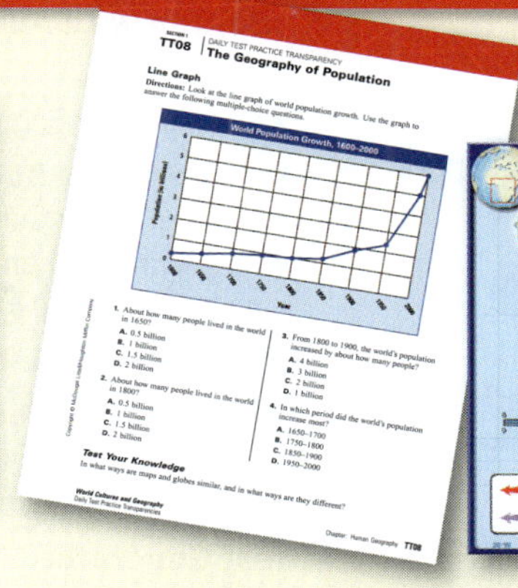

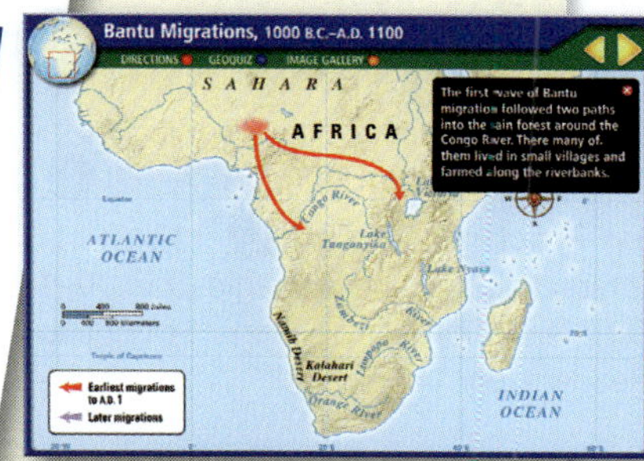

Electronic Books

eEdition DVD-ROM

eEdition Online

Presentation Toolkit

Video/DVD

- World Cultures and Geography

Power Presentations DVD-ROM

- Discussion Starters
- Media Gallery
- Review Game
- Interactive Review
- Animated Features

Unit Transparency Book

- 3-Minute Warm-Ups
- Essential Question
- Critical Thinking
- Maps
- Cultures

Daily Test Practice Transparencies

Animated Geography

- Earth at Night
- Bantu Migrations Map

OnlineActivities at ClassZone.com

Animated Geography

- Videos and Photos
- 3-D Animations
- Interactive Maps

Interactive Review

- Section Quizzes
- Review and Study Notes
- Vocabulary Flip Cards
- Graphic Organizers
- Name Game and Crossword Puzzles
- GeoGame

Online Test Practice

Map Center

- Atlas Maps
- Political Maps with Data
- Links to Mapping Web Sites

Activity Center

- Activity Maker—Review Games, Puzzles, Flip Cards, and Graphic Organizers
- Worldster

Research & Writing Center

- Web Sites
- Data Updates
- Primary Sources
- Writing Models

Current Events

Amber I. McVey
Antioch Middle School
Antioch, Illinois

Make It Fun — Teacher-Tested Activities

As you teach this chapter, look for my classroom-proven ideas for:

- making the lesson fun
- focusing students' attention and comprehension
- integrating technology with traditional teaching

Teaching Bell to Bell When you see this timing guide, have students discuss in pairs, in groups of 3–6, or as a class.

- **Think, Pair, Share** (pairs)
- **Reader, Recorder, Reporter** (3–6)
- **Talk About It** (any size)

CHAPTER 3

RESOURCES FOR DIFFERENTIATED INSTRUCTION

Struggling Readers

eEdition with Audio DVD-ROM

Lesson Plans
- Strategies for Inclusion

Unit Resource Book
- Reading Study Guide with Additional Support, pp. 131–138 B
- Reteaching Activity, pp. 173–176
- Chapter Test (Form A), pp. 177–180 A

Best Practices Toolkit C
- Interactive Reading Strategies, pp. 90–145
- Differentiation Activity Bank: Struggling Readers, pp. 146–149

Teacher's Edition Activities
- List Local and Regional Resources, Sec. 3
- Create Symbols of Government, Sec. 4
- Tiered Activities, Sec. 1–4

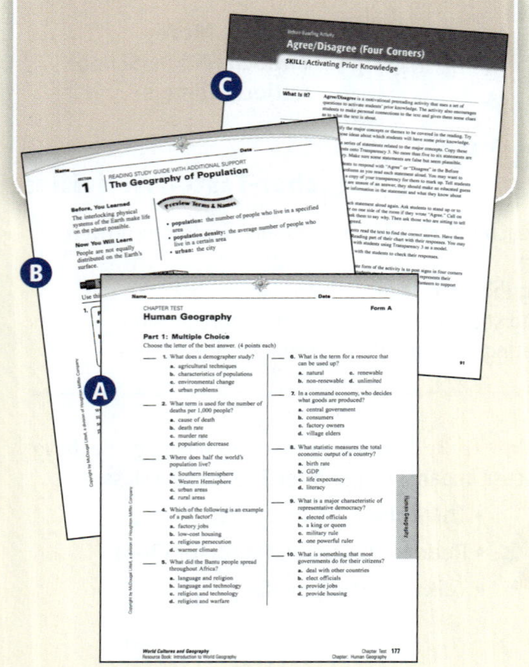

English Learners

Pupil's Edition in Spanish

eEdition with Audio DVD-ROM

Lesson Plans
- Strategies for English Learners

Unit Resource Book
- Reading Study Guide (Spanish), pp. 139–146
- RSG with Additional Support (Spanish), pp. 147–154 A B

Multi-Language Glossary C

Best Practices Toolkit
- Differentiation Activity Bank: English Learners, pp. 165–194

Test Generator CD-ROM
- Section Quizzes in Spanish
- Chapter Test in Spanish

Teacher's Edition Activities
- Pronounce and Preview, Sec. 1–4
- Vocabulary: Key Academic Vocabulary, Sec. 1
- Vocabulary: Suffixes, Sec. 2
- Tiered Activities, Sec. 1–4

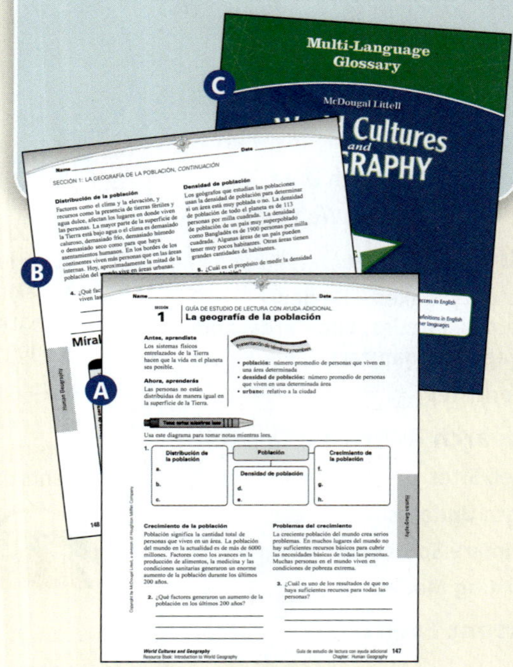

Inclusion

eEdition with Audio DVD-ROM

Lesson Plans
- Strategies for Inclusion

Unit Resource Book
- Reading Study Guide with Additional Support, pp. 131–138 A B
- Reteaching Activity, pp. 173–176
- Chapter Test (Form A), pp. 177–180

Best Practices Toolkit C
- Interactive Reading Strategies, pp. 90–145
- Differentiation Activity Bank: Struggling Readers, pp. 146–149

Teacher's Edition Activities
- Vocabulary: Definition Mapping, Sec. 1
- Reading and Defining Terms Verbally, Sec. 2
- Illustrate Terms and Names, Sec. 3
- Vocabulary; Student VOC Strategy, Sec. 4
- Tiered Activities, Sec. 1–4

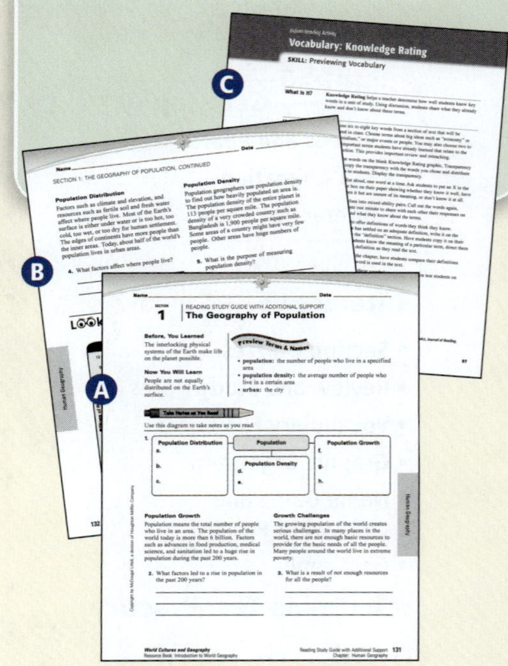

Gifted & Talented Students

Lesson Plans
- Core Lesson Plan

Unit Resource Book
- Outline Map Activities, p. 159
- History Makers, pp. 161–162 Ⓐ
- Interdisciplinary Projects, pp. 163–165 Ⓑ Ⓒ
- Economics in the World, p. 166
- Chapter Test (Form C), pp. 185–188

Best Practices Toolkit
- Student Products Rubrics, pp. 1–14
- Writing Templates, pp. 78–89

Teacher's Edition Activities
- Illustrate Habitable Land, Sec. 1
- Tiered Activities, Sec. 1–4

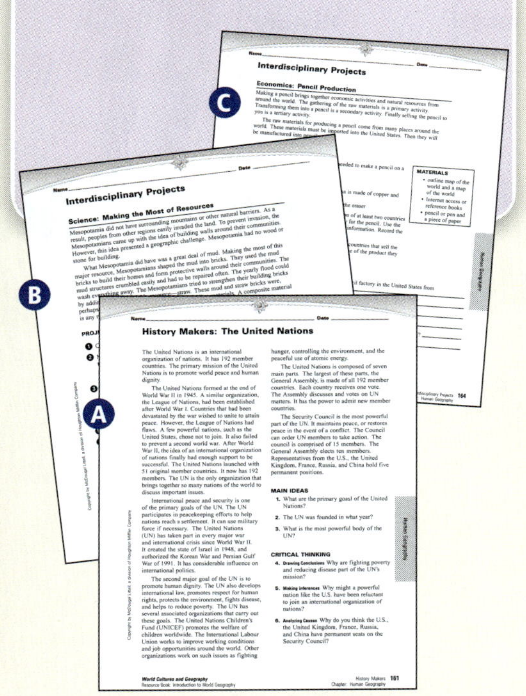

Pre-AP*

Lesson Plans
- Core Lesson Plan

Unit Resource Book
- Vocabulary Practice, p. 155
- Map and Graph Skills, pp. 157–158
- Economics in the World, p. 166 Ⓐ
- Connect to Math, p. 167 Ⓑ
- Primary Source, p. 168 Ⓒ
- Chapter Test (Form C), pp. 185–188

Best Practices Toolkit
- Writing Templates, pp. 78–89
- Differentiation Activity Bank: Pre-AP Students, pp. 195–200

* Pre-AP is a registered trademark of the College Entrance Examination Board, which was not involved in the production of and does not endorse this product.

Teacher's Edition Activities
- List Examples of Push and Pull Factors, Sec. 2
- Report on Differences in Monarchies, Sec 4
- Tiered Activities, Sec. 1–4

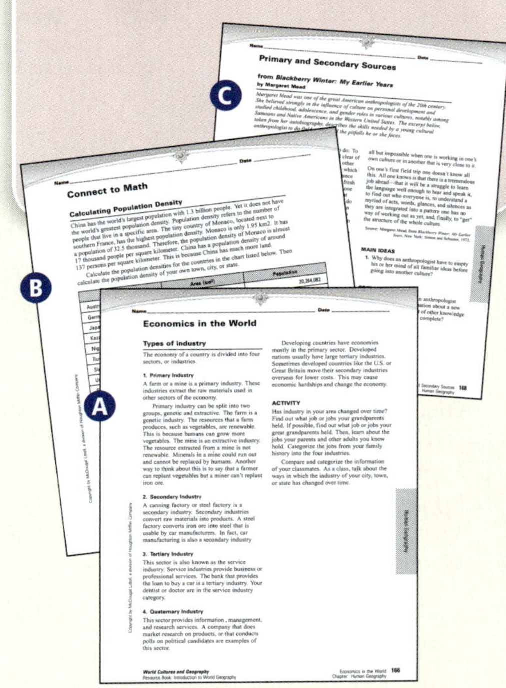

Interdisciplinary Activities

CONNECT to Science

Chart Renewable and Non-Renewable Resources: Past, Present, and Future

CLASS TIME 30 minutes

TASK Prepare a chart that compares the importance of renewable and non-renewable resources in the past, present, and future.

SUPPLIES
- art paper, colored pencils or markers

ACTIVITY Have students review the information in Section 3 on natural resources. Then ask small groups of students to prepare a chart, listing some renewable and non-renewable resources and assessing their importance in the past, present, and future. Suggest they consider the role of a resource in creating a past industry such as railroads, in boosting today's economy such as computers, and in reducing global warming in the future.

Good for Visual and Verbal Learners

CONNECT to Math

Bar Graphs Comparing Population and Population Densities of U.S. States

CLASS TIME 30 minutes

TASK Create a bar graph that compares the population and population densities of various states, including your home state.

SUPPLIES
- recent data on population densities
- art paper, colored markers or pencils

ACTIVITY Have students work in pairs to create two bar graphs that show total population and population density for at least five states, including their home state. They might use the following from the 2000 U.S. Census:

California—33,871,648; 217 persons per square mile;

Rhode Island—1,048,319; 1,003 persons per square mile;

Texas—20,851,820; 80 persons per square mile;

Illinois—12,419,293; 223 persons per square mile;

Alaska—626,932; 1 person per square mile.

Good for Visual and Mathematical/Logical Learners

Introduce the ESSENTIAL QUESTION

- Summarize the reasons people live in different places.
- Point out that people have moved since the earliest times and continue to move today.
- Explain how people manage natural resources and how natural resources contribute to a country's economy.
- Identify why people form governments and other organizations.

Key Ideas

Section 1: People are not equally distributed on the Earth's surface.

Section 2: People move from one place to another to meet their needs.

Section 3: Economic activities in an area depend on the presence of natural resources.

Section 4: The world is divided into many political regions and organizations.

CONNECT to the Essential Question

You may wish to revisit the Essential Question periodically with students as they work through the chapter.

Suggestions for reconnecting are provided at an appropriate point in the *Teacher's Edition* for each section.

Women in Mumbai, India

CHAPTER 3

Human Geography

1 GEOGRAPHY **The Geography of Population**

2 CULTURE **Why People Move**

3 ECONOMICS **Resources and Economics**

4 GOVERNMENT **Why We Need Government**

ESSENTIAL QUESTION

How do natural resources affect a country's population distribution and economy?

CONNECT Geography & History

Use the satellite image and the time line to answer the following questions.

1. How much did the world's population grow between 1 A.D. and 2000?
2. How different would this image have appeared when the population of the world hit one billion?

1000 B.C.

History ◀ **1000 B.C.** Bantu migrations in Africa

Geography **1 A.D.** Total world population hits 300 million.

History **180** Roman Empire at its peak ▶

TIME LINE DISCUSSION

Note that the time line covers a span of about 3,000 years and that the Earth's population has grown during this time, although not at a steady pace.

- Which event on the time line highlights people's movement from one place to another? *(1000 B.C.—Bantu migrations in Africa)*
- How much did the world population grow from A.D. 1 to 1804? *(700 million people)* From 1804 to 1999? *(5 billion people)*
- Why do you think the formation of the European Union is noted under "Economics" rather than "Government?" *(Possible Answer: It was formed to improve economic conditions in Europe, not to change the governments of its member nations.)*

Animated GEOGRAPHY

Earth at Night

Click here to explore human geography @ ClassZone.com

Geography
1804 Total population hits 1 billion.

Government
◄ **1788** United States Constitution is ratified. (Independence Hall)

Economics
◄ **1993** European Union established

Geography
1999 Total population hits 6 billion.

Today

TEACHER MEDIA FAVORITES

Books for the Teacher

Gallo, Donald R., editor. *First Crossing: Stories about Teen Immigrants.* Candlewick Press, 2004. Tales about young immigrants, written by well-known authors.

Perry, Michael. *Population: 485—Meeting Your Neighbors One Siren at a Time.* Harper Perennial, 2003. First-person account of the author's return home. (available on audio)

Video / DVD

Commanding Heights: The Battle for the World Economy. (DVD) WGBH Boston, 2002. Engaging documentary on the global economy, filmed on five continents.

The Middle Passage. (DVD) HBO Home Video, 2003. A look at the forced migration of Africans to the Americas.

The Population Paradox. (DVD) WGBH Boston, 2004. Examines world population trends through vivid personal stories.

Internet

Go to the *Research and Writing Center* at **ClassZone.com** to link to:

- data on the region
- a variety of chapter-specific content-reviewed sites
- "History Makers" biographies
- primary sources

Animated GEOGRAPHY

Earth at Night

CLASSZONE.COM

Point out that this image is actually a number of images—taken by satellite at night—that were put together. Explain that the concentration of lights (as indicated by the gold color) shows where people live. Conversely, the areas without lights indicate no people or few people.

- Judging from this map, which areas of the United States are most densely populated? *(the Northeast and upper Midwest)*
- Judging from this map, how would you describe the distribution of the population in Europe? *(It is densely populated, because most of its area shows a concentration of lights.)*
- Compare this map to the World Population Density map in Section 1. How would you describe population density in the areas where there is little or no vegetation? *(very low density)*

CONNECT GEOGRAPHY & HISTORY ANSWERS

1. 5,700,000,000 people
2. There likely would have been very little gold color since there were many fewer people, limited lighting at night, and no electric lights yet.

Presentation Options

Use the **Power Presentations** DVD-ROM to project the animated map.

- Prompt students to use the pop-up images and audio from the map to discover what they already know about the geography, culture, history, economics, or governments of the world.
- Challenge students to answer the map questions individually, in small groups, or as a class activity. Use all of the interactive material to help preview chapter content.

1 Plan & Prepare

Objectives

- Explain population growth, how it is measured, and the challenges it presents
- Identify the factors that influence population distribution and density

Read for the Essential Question

Help students read for a purpose by reminding them of the Essential Question: "How do natural resources affect a country's population distribution and economy?" Point out that they can use text features such as Key Questions, special features, images, and maps to form answers to the question.

Vocabulary

Best Practices Toolkit

Use the ***Best Practices Toolkit*** to model strategies for vocabulary notetaking. Vary strategies throughout the year. Choose from: *Knowledge Rating, Predicting ABCs, Definition Mapping, Word Sort, Frayer Model (Word Squares), Magnet Words, Word Wheel,* and *Student VOC*.

Vocabulary Strategies, TT9–TT16

Reading Strategy

Best Practices Toolkit

Display the Web Diagram transparency.

- Remind students that comparing means finding similarities and that contrasting means finding differences.
- Have students suggest basic definitions for population growth, population distribution, and population density. Then model filling in a few entries to compare and contrast the three terms.
- Have students suggest entries.
- Display the graphic organizer for students to review.

Web Diagram, TT28

SECTION 1 Reading for Understanding

Key Ideas

BEFORE, YOU LEARNED

The interlocking physical systems of the Earth make life on the planet possible.

NOW YOU WILL LEARN

People are not equally distributed on the Earth's surface.

Vocabulary

TERMS & NAMES

population the number of people who live in a specified area

birth rate the number of births per 1,000 people per year

death rate the number of deaths per 1,000 people per year

rate of natural increase the death rate subtracted from the birth rate

population density the average number of people who live in a certain area

urbanization the process of city development

demographer a geographer who studies the characteristics of human populations

BACKGROUND VOCABULARY

habitable lands lands suitable for human living

urban relating to, or located in, a city

rural relating to the country or farming

Visual Vocabulary Urbanization (Los Angeles, California)

Reading Strategy

Re-create the web diagram shown at right. As you read and respond to the **KEY QUESTIONS**, use the diagram to help you identify important details to compare and contrast population growth, distribution, and density.

See Skillbuilder Handbook, page R9

COMPARE AND CONTRAST

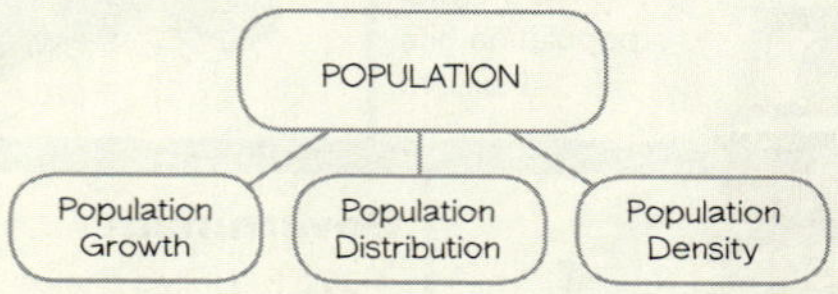

GRAPHIC ORGANIZERS
Go to **Interactive Review** @ClassZone.com

PRETEACHING VOCABULARY

English Learners

Pronounce and Preview

Pronounce each term for students. Review the meanings of words within definitions, such as *specified, development,* and *characteristics.*

- To modify vocabulary learning, have students complete worksheets as they read, instead of afterward.

Unit Resource Book: Introduction to World Geography
- Vocabulary Practice, p. 155

Inclusion

Vocabulary: Definition Mapping

Have students work in pairs to fill out the graphic organizer for the terms in the section.

- Use the transparency to model the strategy.
- Have volunteers suggest entries.
- Then distribute copies of the organizer for students to complete with a partner.

Best Practices Toolkit
- Vocabulary: Definition Mapping, TT11

SECTION 1 GEOGRAPHY

Chicago, Illinois, United States of America

A street scene in Turpan, China

The Geography of Population

Connecting to Your World

Do you know how many people there are in the world? About 6.5 billion! In the time it takes you to read this paragraph, another 140 people will be born somewhere in the world. Where do the 6.5 billion people live? More than half of them live in Asia. In fact, over 2 billion are located in just two countries—China and India. The world's third most populous country is the United States, which has over 300 million people.

Population Growth

KEY QUESTION What challenges does rapid population growth cause?

Geographers use the term **population** to mean the total number of people who live in a specified area. The population of the world today is more than 6 billion. It did not reach one billion until about 1804. Yet, over the last 200 years, the number has jumped by some 5 billion. What factors are responsible? The most important factors were increases in food production, discoveries in medical science, and improvements in sanitation. As a result, more babies survived and people lived longer, healthier lives.

Shanghai, China China's largest city is home to more than 12 million people.

2 Focus & Motivate

3-Minute Warm-Up

Write on the board or display the transparency:

- Look at the Population Density map in Section 1. Why do you think the densities are similar north of 60° N latitude? *(Possible Answer: Temperatures are cold and climates harsh, so few people live there.)*

Unit Transparency Book: Introduction to World Geography

- 3-Minute Warm-Ups, TT18

Connecting to Your World

More About . . .

Factoring U.S. Population Growth

The U.S. Census Bureau estimates that there is a birth in the United States every 7 seconds, a death every 13 seconds, a gain of one person by immigration every 31 seconds—resulting in the U.S. population "growing" by one person every 11 seconds.

3 Teach

Population Growth

Think, Pair, Share

- How is the rate of natural increase figured? *(by subtracting the death rate from the birth rate)*
- **Compare and Contrast** How does the rate of natural increase compare around the world? *(In most of Asia, Africa, and South America the rate is higher compared to most countries of North America and Europe, and Japan.)*

SECTION 1 PROGRAM RESOURCES

Unit Resource Book: Introduction to World Geography

ON LEVEL

Unit Resource Book
- Reading Study Guide, p. 123
- Map and Graph Skills, p. 157
- Section Quiz, p. 169

STRUGGLING READERS

Unit Resource Book
- Reading Study Guide with Additional Support, p. 131
- Skillbuilder Practice: Finding Main Ideas, p. 156
- Section Quiz, p. 169
- Reteaching Activity, p. 173

ENGLISH LEARNERS

Pupil Edition in Spanish
Pupil Edition Audio DVD
- RSG (Spanish), p. 139
- RSG with Additional Support (Spanish), p. 147

Multi-Language Glossary
Test Generator
- Section Quiz in Spanish

INCLUSION

Unit Resource Book
- RSG with Additional Support, p. 131
- Section Quiz, p. 169
- Reteaching Activity, p. 173

GIFTED & TALENTED

Unit Resource Book
- Interdisciplinary Projects, pp. 163–165

PRE-AP

Unit Resource Book
- Connect to Math, p. 167
- Primary Sources, p. 168

TECHNOLOGY

Unit Transparency Book
- 3-Minute Warm-Ups, TT18
- Essential Question, TT19
- Critical Thinking, TT20
- Map: World Population, TT25

Daily Test Practice Transparencies
- Section 1, TT8

Power Presentations
ClassZone.com
World Cultures and Geography Video on DVD

World Population Growth

Ask students:

- How long did it take for the world's population to double from one billion to two billion people? *(126 years)*
- How long did it take for the world's population to double from two billion to four billion people? *(45 years)*

Point out that just two countries—India and China—are predicted to add a billion people to the world by 2050. In contrast, the U.S. will likely remain the third largest country, adding about 100 million people.

More About . . .

Squatter Settlements

In some areas of South America and Africa, settlements of poor people without legal right to the property—called *squatters* in English—have been common since the 1920s. For example, the *favelas* (squatter settlements) around Rio de Janeiro are homes for many families who have lived there for generations.

Currently, about 33% of the 12 million people in Rio are thought to live in *favelas*. In other cities in Africa, South America, and Asia, the percent of people living in substandard housing may be as high as 60%–70%.

More About . . .

Clean Water and Sanitation

Although world organizations like the World Health Organization and UNICEF push for better living conditions for poor people around the world, they report that population growth continues to present huge challenges. More than one billion people—about one in every six—lack access to safe drinking water. And more than twice as many—2.6 billion—do not have access to basic sanitation (i.e., toilet facilities).

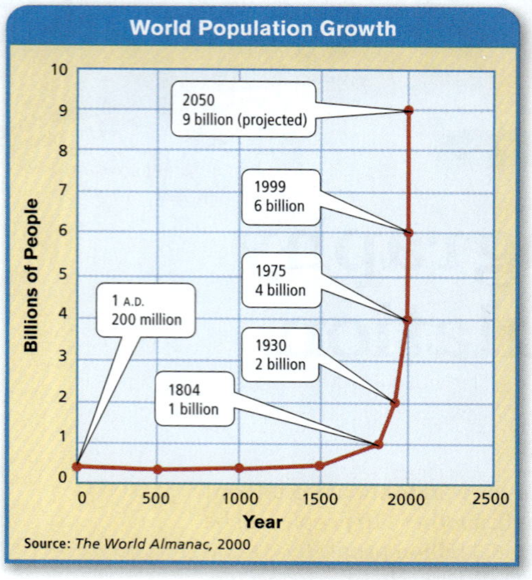

Measuring Growth Geographers measure population growth by figuring out how many people have been born and how many have died, how many have moved into and how many have moved out of a specific area. The **birth rate** measures the number of births per 1,000 people per year. The **death rate** measures the number of deaths per 1,000 people per year. To find out the **rate of natural increase**, the death rate is subtracted from the birth rate. This is the population growth that results from natural processes of birth and death.

Today, in most of Asia, Africa, and South America, the rate of natural increase is very high. Farming is a major way of life in these regions, and families need many children to help with the farm work. In most countries of Europe and North America, and also in Japan, the population growth rate is much lower. In these countries, most people live in cities and have few children, making their rate of natural increase much lower than in regions with a higher rural population.

Growth Challenges The expanding population creates serious challenges. The Earth's resources are limited and not evenly distributed throughout the world. In many countries, it is difficult just to provide the basic needs such as food, clean water, and housing. Many people move from rural areas to cities to make a better life for themselves and their families. If they cannot find housing, they often build houses on the outskirts of cities with whatever materials they can find. Some of these squatter settlements become very large, even as large as the city itself, but they often lack clean water, sewers, or paved roads.

Rapid popula tion growth puts pressure on the limited resources and makes it diffi- cult for peopl to obtain food, housing and jobs.

SUMMARIZE Explain the challenges of rapid population growth.

Rio de Janeiro, Brazil
These houses are part of squatter settlements on the outskirts of the main city.

60 Chapter 3

INTERDISCIPLINARY ACTIVITIES

CONNECT to Math

Compare Rates of Natural Increase

Ask students to calculate the rates of natural increase for the two countries: United States: 2004 birth rate—14 per 1,000; death rate—8/1,000 *(rate of natural increase = 6 per 1,000)*; India: 2003 birth rate—24 per 1,000; death rate—8/1,000 *(rate of natural increase = 16 per 1,000)*. Using the rates have them compute the estimated increase in population for one year, using 295 million for U.S. *(about 1.8 million)* and 1.065 billion for India *(about 17 million)*.

CONNECT to Science

Describe Local Sources of Clean Water

Have students use library resources or contact local officials for information about the source of their local water supply (where they are located, how they are kept safe, and so on). They should present their findings in a brief written report.

Unit Resource Book: Introduction to World Geography

- Interdisciplinary Projects: pp. 163–165

Population Distribution and Density

KEY QUESTION What are the factors that influence where people choose to live?

As you read earlier, people are not distributed equally around the world. Where they choose to live is partly affected by climate, elevation, and resources such as fertile soil and fresh water. Today, the largest populations are found in what are called **habitable lands**.

Population Distribution Only a small portion of the Earth's surface is suitable for humans to settle. Almost 75 percent of the Earth's surface is water. In addition, between 35 and 40 percent of the Earth's land is too hot, too cold, too wet, or too dry for large-scale settlement. Most people live in the Northern Hemisphere, between 20° North and 60° North latitude. Fewer people live in the Southern Hemisphere because there is less land available. The edges of continents are more heavily populated than interior lands. Many people choose to live in coastal lands and in river valleys because these locations offer people opportunities to earn a living. About two-thirds of the world's population is found within 300 miles of ocean waters.

COMPARING Urban Populations

More than 400 cities have populations of one million or more. Many cities in Africa and Asia, such as Lagos, Nigeria, and Mumbai, India, are expected to grow rapidly in the 21st century.

Mumbai, India

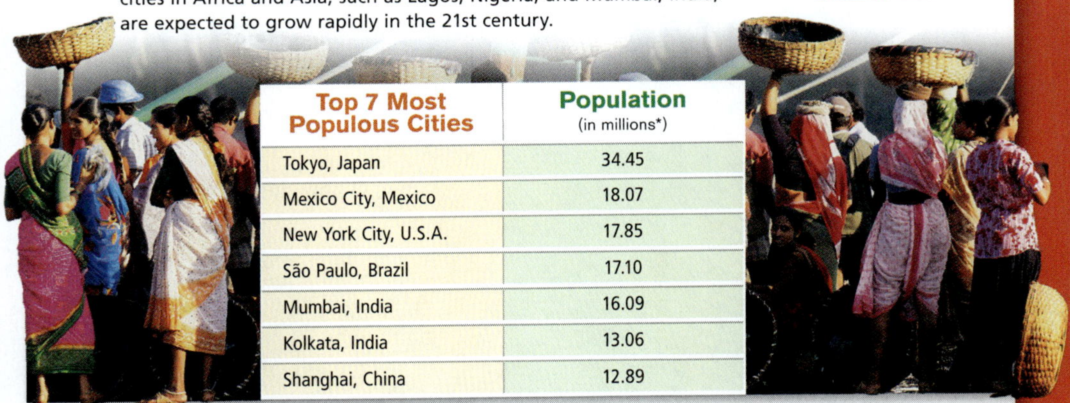

Top 7 Most Populous Cities	Population (in millions*)
Tokyo, Japan	34.45
Mexico City, Mexico	18.07
New York City, U.S.A.	17.85
São Paulo, Brazil	17.10
Mumbai, India	16.09
Kolkata, India	13.06
Shanghai, China	12.89

* Population shown is for an entire urban area, not just the city itself.

Source: *The World Almanac and Book of Facts, 2006*

CRITICAL THINKING

Compare and Contrast What is the total population of the Indian cities listed? How does that compare with the population of Tokyo?

DIFFERENTIATING INSTRUCTION

English Learners

Key Academic Vocabulary

Note the word *portion* in the sentence "Only a small portion of the Earth's surface is suitable. . . ." Read the sentence out loud and ask students to suggest definitions or synonyms for portion. *(a part, a share)* Then ask students to look up the word in a dictionary. Discuss other common uses of the word or demonstrate its meaning, such as by giving students a portion of a whole, such as a piece from a ball of clay.

Gifted & Talented

Illustrate Habitable Land

Have students review the first three sentences of the subsection "Population Distribution." Then ask them to use graph paper to show the amount of habitable land compared to the entire surface of the Earth. They might use blue to indicate water, brown for land areas not suitable for humans, and green for areas that are suitable. *(About 75% should be water, 25% land, and 40% of the land is not suitable.)*

Teach

Population Distribution and Density

Reader, Recorder, Reporter

- How has the distribution of the world's population changed over the last 100 years? *(It has changed from most people living in rural areas to about half now living in urban areas.)*
- Why can a country's population density be deceiving? *(because it is an average, and in large countries, the density can vary greatly from one area to another)*
- **Compare and Contrast** How is the population near the edges of continents different from that of interior lands? *(About two-thirds of all people live within 300 miles of the ocean.)*

Unit Resource Book: Introduction to World Geography
- Skillbuilder Practice: Finding Main Ideas, p. 156
- Primary and Secondary Sources: from *Blackberry Winter: My Earlier Years* by Margaret Mead, p. 168

COMPARING

Urban Populations

Point out that about 80% of the U.S. population lives in a metropolitan area. There are eight metropolitan areas in the United States with a population of over five million.

- What U.S. metropolitan areas would you predict rank second and third behind New York City? *(Los Angeles followed by Chicago)*

CRITICAL THINKING ANSWER

Compare and Contrast about 29 million; about 5 million less than Tokyo

More About . . .

The World's Largest Cities

London was the world's largest city in 1900 with a population over six million, which would rank it only about 20th in the world today. In 1900, Tokyo, Japan, and St. Petersburg, Russia, were the only top ten cities located outside of central Europe or the United States. Today, New York City is the only U.S. city among the top 20 in the world—and London the only European city.

CONNECT to Math

Discuss how population density can be used. For example, urban planners use population density when planning public transportation and other development.

- Based on the map, which areas have the highest population densities? *(Asian countries, such as India, Bangladesh, China, Japan, and Indonesia)*
- Based on the map, which country in North America is the least densely populated? *(Canada)*
- Why are such large areas of Africa and Australia so lightly populated? *(These are desert areas.)*

ACTIVITY ANSWERS

Afghanistan 47.96; Brazil 22.1; Chad 7.74; Finland 15.47; France 111.28; Iraq 61.28; Thailand 125.74; United States 30.99

Unit Resource Book: Introduction to World Geography
- Map and Graph Skills: Reading a Bar Graph, pp. 157–158
- Connect to Math: Calculating Population Density, p. 167

More About . . .

Largest and Smallest Population Densities

Some very small countries in terms of landmass have very high population densities. For example, Monaco has only about 32,000 people, but its area is only about three-quarters of a square mile. Thus, its density is the highest in the world—almost 42,000 people per square mile. Excluding the tiny "microstates," Bangladesh is considered the most densely populated country with nearly 2,600 people per square mile. Mongolia has the lowest population density, with less than three million people over 600,000 square miles (five people per square mile).

CONNECT to Math

Population density is different from population distribution. Population distribution shows where people live. **Population density** shows on average how many people are living in a specific size area such as a square mile or square kilometer. The density number helps explain how crowded an area is.

To find the population density number, add up the total number of people living in an area and divide by the total amount of land they occupy. Some areas of Earth are very lightly populated and others are quite densely populated.

Activity

Calculating Population Density

Materials
- a calculator
- paper and pencil

1. Review the paragraph at left to see how population density is calculated.
2. Use the information below to calculate the population density for eight countries.
3. After you have completed your calculations, create a bar graph showing the population densities of the eight nations.

Country	Area (km²)	Population
Afghanistan	647,500	31,056,997
Brazil	8,511,965	188,078,227
Chad	1,284,000	9,944,200
Finland	338,145	5,231,372
France	547,030	60,876,136
Iraq	437,072	26,783,383
Thailand	514,000	64,631,595
United States	9,631,420	298,444,215

Source: *CIA World Factbook*, 2006

World Population Density

DIFFERENTIATING INSTRUCTION: TIERED ACTIVITIES

OBJECTIVE Prepare and present a visual display on world population.

Basic

Have students work in small groups to create posters that illustrate population growth. Have them select three facts from their textbooks, such as from the World Population Growth feature, create illustrations for each fact, and write a title for their posters. Ask each group to present its poster to the class.

On Level

In addition to the Basic activity, have students include a summary statement that ties together the facts they've selected for their posters. Ask each group to present its poster to the class.

Challenge

Have students work in small groups to prepare posters that illustrate the challenges of the growing world population and its shift from rural to urban areas. Have them include an illustration and caption for each challenge, and list the ways these challenges can be overcome. Ask each group to present its poster to the class.

Rural vs. Urban Today, about half the world's population lives in **urban** areas, such as cities and their suburbs. This is a big change from a hundred years ago. Then, most people lived in **rural** areas, on farms and in small villages. Only about 14 percent lived in cities. By 2030, population experts believe that 60 percent of all people will live in urban areas. This process of development from small settlements to large ones is called **urbanization**.

Population Density Geographers who study the characteristics of human populations are called **demographers**. They use population density to find out how heavily populated an area is. One of the most densely populated countries is Bangladesh in South Asia, with over 1,900 people per square mile. The average population density of the entire planet is 113 people per square mile.

Statistics can be deceiving because the number is an average. Some areas of a country might be lightly populated, while others are heavily populated. In the United States, the population density is about 80 people per square mile. However, New Jersey has 1,134 people per square mile and Alaska has only 1.1 people per square mile.

UNDERSTAND CAUSES Identify the factors that influence where people choose to live. Where people live is affected by whether the land is habitable and where they can best make a living.

Section 1 Assessment

ONLINE QUIZ For test practice, go to **Interactive Review @ClassZone.com**

TERMS & NAMES

1. **Explain the importance of**
   - birth rate
   - death rate
   - population density
   - urbanization

USE YOUR READING NOTES

2. **Compare and Contrast** Use your completed web diagram to answer the following question:
   How is population density different from population distribution?

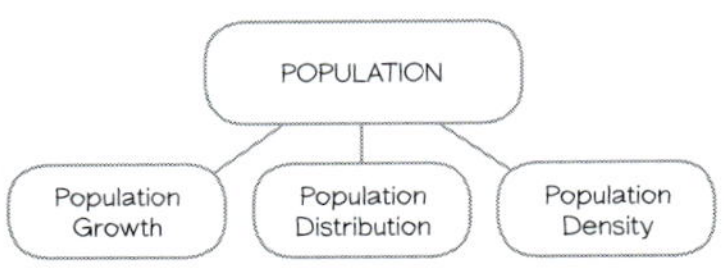

KEY IDEAS

3. What are three factors that have caused the world's population to grow so rapidly?
4. What is urbanization?
5. How is population density determined?

CRITICAL THINKING

6. **Make Inferences** Why are people unevenly distributed on the Earth's surface?
7. **Compare and Contrast** Why do some countries have much lower growth rates than others?
8. **CONNECT to Today** What do you think are the major problems that the world faces as the population continues to grow?
9. **MATH** **Create a Population Density Table** Use the Internet to find statistics for population and an average population density for each continent. Create a table and a bar chart showing these statistics. Continents should be listed in order from highest density to lowest density.

CONNECT to the Essential Question

How do natural resources affect a country's population distribution and economy?

Ask students what they have learned so far that can help them answer this question. Students might mention:

- The availability of natural resources allows some countries to meet basic needs, but makes it a challenge for others.
- Economic opportunities attract people to coastal areas and river valleys.
- More habitable land results in more people living in the Northern Hemisphere than in the Southern.

4 Assess & Reteach

Assess Have students complete the Section Assessment.

Unit Resource Book: Introduction to World Geography
- Section Quiz, p. 169

Interactive Review @ClassZone.com
- Section Quiz

Power Presentations

Test Generator

Reteach Have student pairs create multiple-choice questions and answers for the section. Pairs exchange their questions and answer them, using their textbooks for help if necessary.

Unit Resource Book: Introduction to World Geography
- Reteaching Activity, p. 173

SECTION 1 ASSESSMENT ANSWERS

Terms & Names

1. birth rate: number born in a year; death rate: number died in a year; population density: the average that explains the level of crowdedness in a specific region; urbanization: process of city development

Use Your Reading Notes

2. Distribution indicates where people live; density indicates how many people live in a specific area.

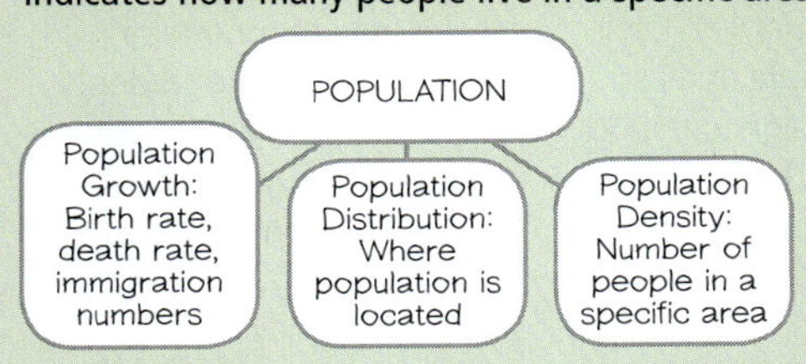

Key Ideas

3. increases in food production, discoveries in medical science, and improvements in sanitation
4. the process of city development from small settlements to large cities
5. by dividing the number of people in an area by the total amount of land they occupy

Critical Thinking

6. Possible answer: because large areas of the Earth are not suitable for humans to live
7. The birth rate is lower in more developed countries because fewer children are needed for physical labor than in less developed countries.
8. Answers will vary but should mention using up resources, overcrowding in urban areas, and inability to provide food, housing, and jobs for all the people.
9. Use this rubric to score the population density chart.

Chart Rubric

	Accuracy of information	Completeness
4	complete and correct	continents listed
3	correct	missing a continent
2	incorrect information	missing two continents
1	much incorrect	missing many continents

1 Focus & Motivate

Objective

Compare and contrast Australia's population density and population distribution with that of Indonesia.

Springboard Tell students to review the maps of Australia and Indonesia, including the scales and legends. Then ask students to estimate the size of Australia compared to the size of Indonesia. Which is larger? *(Australia)* How much larger? *(about four times as large).*

2 Teach

Population Distribution

Talk About It

- How would you describe the Australian Outback from looking at the pictures? *(Possible Answers: dry, dusty, very few people)*
- How does the climate and land affect Australia's population distribution? *(The large area of very dry land makes it difficult for people to live in central Australia.)*
- **Compare and Contrast** How is the location of Jakarta similar to the location of cities in Australia? *(Jakarta and Australian cities are all located in coastal areas with access to the ocean.)*

More About . . .

The Environment of the Outback

Although most of the Outback—an area about six times the size of Great Britain and two-and-a-half times that of Texas—is dry, there are wetlands and waterfalls. As in many wild areas, economic concerns and environmental concerns often come into conflict. For example, cattle grazing can pose a threat to wild bird and mammal habitats. Mining is a significant economic activity in the southern and western Outback, and it poses threats to surface lands. Australia's people and government are working to balance these interests.

COMPARING Population Distribution

Population density is an average figure for a specific area. However, it does not account for the distribution of population in an area. Notice in these two examples how the pattern of population density varies even within a country.

Australia

Australia is ranked 54th in the world for total population, but 191st in population density. This is because Australia has a great deal of land but a small population. It has a lot of open space sometimes called the Outback, shown below. The heaviest population is found in the coastal cities.

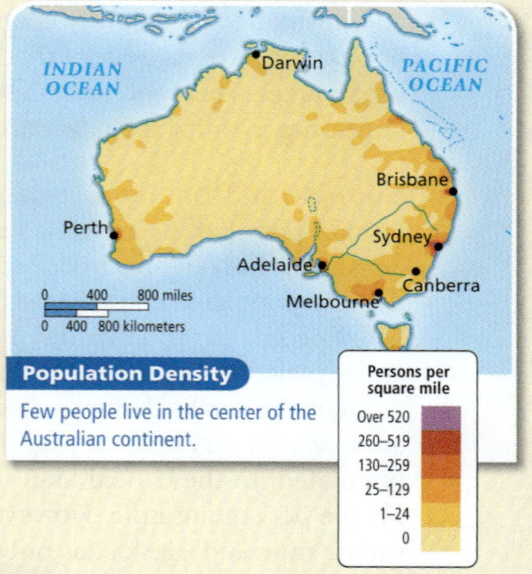

Population Density
Few people live in the center of the Australian continent.

Work
Australia's wide open spaces makes cattle ranching one of the biggest businesses.

Transportation
A lone vehicle travels a dusty Outback road.

DIFFERENTIATING INSTRUCTION: TIERED ACTIVITIES

OBJECTIVE Create a travel agency advertisement inviting people to visit or move to either Indonesia or Australia.

Basic

As a class, discuss what would draw visitors or immigrants to each country. Then have pairs of students create a newspaper or magazine ad for one country with an appropriate illustration and slogan. Have students present or post their ads.

On Level

Adapt the Basic activity, so that pairs of students create a four-page travel brochure for one of the countries. Each page should include at least one engaging image and block of text. Have students present or post their brochures.

Challenge

Adapt the Basic activity, so that pairs of students create a one-minute TV or radio advertisement for one of the countries. Have students write a script, make or obtain suitable props, and rehearse their ad. Students may enjoy recording their ad or presenting it "live."

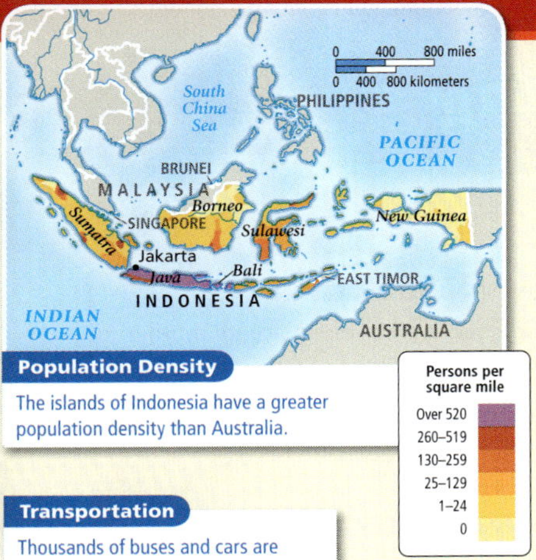

Population Density

The islands of Indonesia have a greater population density than Australia.

Transportation

Thousands of buses and cars are needed to move Jakarta's population.

Indonesia

Indonesia is ranked 4th in the world for total population, but 60th in population density. This is because two Indonesian islands, Bali and Java, and the capital, Jakarta, are heavily populated. Other parts of the island nation are not as densely populated. Jakarta is pictured below.

Work

Manufacturing takes place in the larger cities of Indonesia.

CRITICAL THINKING

1. **Compare and Contrast** How are the patterns of density different in the two countries?
2. **Draw Conclusions** In what ways do the images illustrate how density may affect ways of living in the two countries?

65

INTERDISCIPLINARY ACTIVITIES

CONNECT to Language Arts

Daily Journal Entries

Have students use the photos of either Australia or Indonesia to create a journal entry. Students should assume that they live in one of the countries in describing their activities for a day. Remind students to write in complete sentences and use vivid details.

CONNECT to Economics

Challenges Facing Indonesia and Australia

Working in small groups, have students list the economic challenges they think Jakarta and Australia face because of their population densities. Encourage students to think about economic issues such as jobs, housing, transportation, and the use of natural resources and the environment. Then as a class, discuss the similarities and differences between the two countries.

Comparing Regions

Asia and North America

Asia contains three of the four countries with the largest population in the world (China, India, and Indonesia) and has by far the greatest density of any continent—203 people per square mile. This is true even though it also contains the country with the lowest density—Mongolia, with just 5 people per square mile. North America has a population density of about 32 people per square mile. Canada has 9 people per square mile, while the United States has around 83 per square mile.

3 Wrap Up

Encourage students to review their Reading Notes chart for Section 1 and add additional details or explanations based on their reading of the Comparing Population Densities feature. To extend this activity, ask for volunteers to share some of the additions they made to their charts.

CRITICAL THINKING ANSWERS

1. **Compare and Contrast** Indonesia has entire islands with high population densities; Australia generally has a low density, but its densest population is in urban areas along its coast.
2. **Draw Conclusions** The areas of Indonesia with high population densities allow for manufacturing, but also result in crowded housing and transportation; the large, sparsely populated areas of Australia allow for ranching, but long-distance travel may be more challenging.

1 Plan & Prepare

Objectives

- Describe push and pull factors that encourage people to migrate
- Define internal and external migration, and give examples of each
- Summarize the cultural, economic, and political effects of migration

Read for the Essential Question

Help students read for a purpose by reminding them of the Essential Question: "How do natural resources affect a country's population distribution and economy?" Point out that they can use text features such as Key Questions, special features, images, and maps to form answers to the question.

Vocabulary

Best Practices Toolkit

Use the ***Best Practices Toolkit*** to model strategies for vocabulary notetaking. Vary strategies throughout the year. Choose from: *Knowledge Rating, Predicting ABCs, Definition Mapping, Word Sort, Frayer Model (Word Squares), Magnet Words, Word Wheel,* and *Student VOC.*

Vocabulary Strategies, TT9–TT16

Reading Strategy

Best Practices Toolkit

Display the T Chart/Two Column-Chart transparency.

- Remind students that comparing and contrasting means looking for similarities (What is the same?), while contrasting means looking for differences (What is different?).
- Model filling in details from the text about push factors in migration.
- Have students suggest entries for pull factors in migration.
- Display the graphic organizer for students to review.

T Chart/Two Column-Chart, TT20

Section 2 Reading for Understanding

Key Ideas

BEFORE, YOU LEARNED

Population patterns differ by place and region and may change over time.

NOW YOU WILL LEARN

People move from one place to another to meet their needs.

Vocabulary

TERMS & NAMES

migration the process of relocating to a new region

immigrant a person who leaves one area to settle in another

push factor a reason that causes people to leave an area

pull factor a reason that attracts people to another area

culture the shared attitudes, knowledge, and behaviors of a group

diversity having many different ways to think or to do something, or a variety of people

discrimination actions that might be hurtful to an individual or a group

BACKGROUND VOCABULARY

refugee a person who flees a place to find safety

persecution cruel treatment on the basis of religion, race, ethnic group, nationality, political views, gender, or class

Visual Vocabulary diversity

Reading Strategy

Re-create the chart shown at right. As you read and respond to the **KEY QUESTIONS**, use the chart to compare and contrast important details about the reasons people move.

See Skillbuilder Handbook, page R9

COMPARE AND CONTRAST

PUSH FACTORS	PULL FACTORS
1.	1.
2.	2.

GRAPHIC ORGANIZERS Go to Interactive Review @ClassZone.com

66 Chapter 3

PRETEACHING VOCABULARY

English Learners

Pronounce and Preview

Pronounce each term for students. Review the meanings of words within definitions, such as *relocating, behaviors,* and *gender.*

- To modify vocabulary learning, have students complete worksheets as they read, instead of afterward.

Unit Resource Book: Introduction to World Geography

- Vocabulary Practice, p. 155

Inclusion

Reading and Defining Terms Verbally

Have volunteers read one term and its definition aloud. Ask others to define the term in their own words verbally, and then use it in a sentence. Repeat for each term. As a group, discuss any words that are challenging to define or use in a sentence.

Chicago, Illinois, United States of America

A street scene in Turpan, China

Why People Move

Connecting to Your World

Have you ever moved from one part of the city or town where you live to another? Maybe you moved to be closer to your school or to where a parent works. It may be that one of your parents took a job in a different city. Perhaps your family had to move to a different climate for health reasons. Whatever the reason, moving to a new, unfamiliar area could change your life in many ways.

Moving Day About 112 million Americans relocate every year.

Causes of Migration

KEY QUESTION Why do people move from place to place?

From earliest times, people have moved to new locations. This process of relocating to a new region is called **migration**. A person who leaves one country to settle in another is called an **immigrant**. Population geographers often talk about push-pull factors when they study migration. The **push factors** are the reasons that cause people to leave an area. **Pull factors** are the reasons that attract people to another area. For example, a group may decide to move to a better location after their crops fail for several years. The crop failure is the "push" factor, and the better location is the "pull" factor. Together, they give the group a reason for migrating. Generally, the causes of migration are environmental, economic, cultural, or political.

Ellis Island, New York Millions of people entered the United States here between 1892 and 1954.

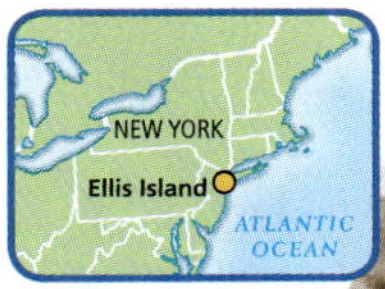

2 Focus & Motivate

3-Minute Warm-Up

Write on the board or display the transparency:

- Why is the United States often called "a nation of immigrants"? *(With the exception of Native Americans, everyone's ancestors immigrated from another country.)*

Unit Transparency Book
- 3-Minute Warm-Ups, TT18

Connecting to Your World

More About . . .

Moving Within the United States

Many Americans move, but stay within the same state. However, by 2025 over 11 million are expected to move into the South from other areas—about seven million from the Northeast, three-and-a-half million from the Midwest, and 350,000 from the West.

3 Teach

Causes of Migration

Think, Pair, Share

- What are some examples of pull factors? *(Economic: desire for land or jobs; Cultural: to return to a homeland or live near a holy site)*
- How can political actions push people to migrate? *(Possible Answers: They're seeking a safe place from the war.)*
- **Compare and Contrast** What is the basic difference between push factors and pull factors? *(Push factors are reasons people leave an area; Pull factors are reasons people go to another area.)*

SECTION 2 PROGRAM RESOURCES

Unit Resource Book: Introduction to World Geography

ON LEVEL

Unit Resource Book
- Reading Study Guide, p. 125
- Section Quiz, p. 170

STRUGGLING READERS

Unit Resource Book
- Reading Study Guide with Additional Support, p. 133
- Skillbuilder Practice: Finding Main Ideas, p. 156
- Section Quiz, p. 170
- Reteaching Activity, p. 174

ENGLISH LEARNERS

Pupil Edition in Spanish
Pupil Edition Audio DVD
- RSG (Spanish), p. 141
- RSG with Additional Support (Spanish), p. 149

Multi-Language Glossary

Test Generator
- Section Quiz in Spanish

INCLUSION

Unit Resource Book
- RSG with Additional Support, p. 133
- Section Quiz, p. 170
- Reteaching Activity, p. 174

GIFTED & TALENTED

Unit Resource Book
- Section Quiz, p. 170

PRE-AP

Unit Resource Book
- Section Quiz, p. 170

TECHNOLOGY

Unit Transparency Book
- 3-Minute Warm-Ups, TT18
- Essential Question, TT19
- Critical Thinking, TT21

Daily Test Practice Transparencies
- Section 2, TT9

Power Presentations
ClassZone.com
World Cultures and Geography Video on DVD

More About . . .

The Earliest Migrations

Not all scientists agree on the earliest migrations that spread the human race across the globe. New technology may make it possible to get a more definitive answer to these migrations. In a large new study, scientists will gather 100,000 DNA samples from around the world and compare them to get a more definitive answer as to where people migrated to and from, and which ethnic groups are most closely related to each other.

COMPARING

Migration Factors

Point out that these factors have affected different people and areas at different times. Ask students to think of examples they may know from their prior knowledge, such as the economic pull for immigrants to America today or the push factor of famine in parts of Africa.

- Imagine that a person from another country decided to move to the United States because he or she could open a store in Chicago. Would you describe the move as motivated by a push or a pull factor, or both? *(Possible Answer: both, because lack of economic opportunities pushed the person to leave his or her country and the chance to open a store pulled him or her to the United States)*

CRITICAL THINKING ANSWER

Make Inferences Possible Answers: Political—wars make it impossible to live in an area; environmental—changes can make an area unlivable; economic—everyone needs food and shelter; cultural—the draw of a homeland or religion is very strong. Accept all answers that are supported with appropriate reasons.

Push Factors In early times, a change in environment was a major cause of migration. For example, climate changes thousands of years ago brought on an ice age. People in northern Europe moved south in order to survive. Sometimes, an environment is unable to support growing numbers of people. So some part of the group needs to move to a place that would better support them.

Political actions may push a person or group to migrate to a new area. **Refugees** might flee a place to find safety from war. For example, beginning in 2003, an ethnic war in the Darfur region of Sudan caused about 2 million people to flee to refugee camps. Governments can force people to relocate even if they do not want to go. Cruel treatment, or **persecution**, of a particular group could also cause people to leave. People have sometimes been persecuted for their religion, race, nationality, political views, or membership in an ethnic group.

Pull Factors The desire for land has pulled people to new regions for thousands of years. During the 1800s, millions immigrated to the United States in search of land to farm or ranch. Economic opportunities are still a major reason for migration. Today, people move to find a job or to get a better job.

Sometimes people move for cultural reasons. For example, they may want to return to an area they consider the homeland of their people. Another pull factor may be that the land or the region has religious significance. Israel is an example of both reasons. Jews from all over the world immigrate to Israel because they consider it their homeland, and because it also contains many Jewish holy sites.

SUMMARIZE What are the push-pull factors of migration? They are the reasons people leave one area and are attracted to another area.

COMPARING Migration Factors

PUSH FACTORS	PULL FACTORS
Environmental • lack of resources to support an entire group • change in climate or vegetation	**Economic** • availability of land • job opportunities
Political • escape from war or persecution • forced removal	**Cultural** • return to a homeland • desire to live near a holy site

CRITICAL THINKING

Make Inferences Which factor do you think is most often the cause of migration?

68

DIFFERENTIATING INSTRUCTION

English Learners

Vocabulary: Suffixes

Note the word *environment* in the sentence "In early times, a change in environment was a major cause of migration." Point out that this word combines the root word *environ* meaning "surround" with the suffix *-ment* meaning "product, action, or state." Thus, environment can be explained as meaning "the state or condition of surroundings." Have students think of other words that have the suffix *-ment* and use them in sentences. *(treatment, confinement, excitement)*

Pre-AP

List Examples of Push and Pull Factors

Have students prepare a list or chart of push and full factors in immigration as discussed in their textbooks and provide specific examples of each, either from the text, prior knowledge, or research.

Where People Migrate

KEY QUESTION What are the two kinds of migration?

Geographers identify two different types of migration. Internal migration occurs when people move within a country. Moving across a continent or even to another continent is called external migration.

Internal Migration Internal migration happens when people move from one place to another but stay within the same country. Someone who does this is called a migrant. If you move from Pennsylvania to California, you have moved a very long distance, but you have stayed within the United States, so you are an internal migrant. Two of the most common forms of internal migration are moving from rural areas or small towns to cities, and from cities to suburbs.

The world is becoming more urban. In many countries, more than 70 percent of the population lives in urban areas. Pull factors attract people to cities. If migrants can't find affordable housing in the city, they often build houses on the outskirts. This kind of growth on the outskirts of a city is sometimes called urban sprawl. In developed countries such as the United States, sprawl happens as new suburbs are built around the edges of a city.

External Migration Migration across parts of a continent may take place quickly, but sometimes it takes hundreds or even thousands of years. For example, Bantu-speaking people of Africa slowly spread across the southern half of the continent from 1000 B.C. to A.D. 1100. The push factor was environmental. The numbers of Bantu people were increasing, and they needed more land for farming and herding.

Migration also takes place from one continent to another continent. The countries of the Western Hemisphere as well as Australia and New Zealand, are filled with immigrants from Europe, Africa, and Asia.

COMPARE Explain the difference between internal and external migration.
Internal occurs within a country; external occurs across a continent or from one continent to another.

Kuba Mask The Kuba are a Bantu-speaking people.

Animated GEOGRAPHY

Bantu Migrations

Click here to learn more about the migrations of Bantu people @ ClassZone.com

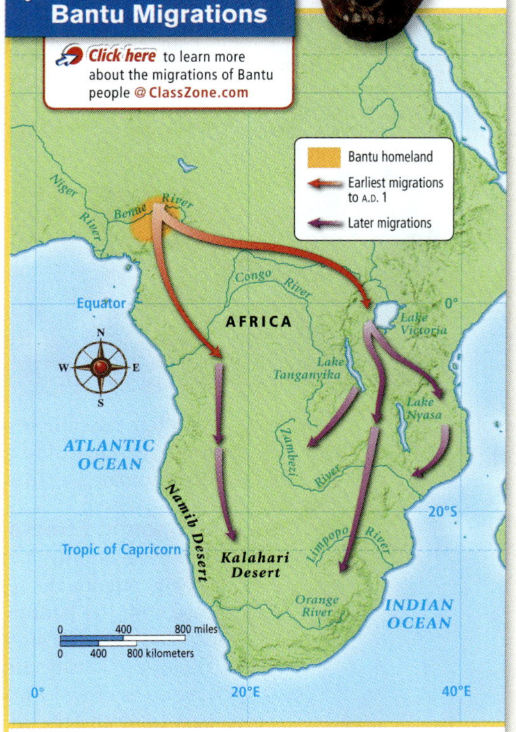

CONNECT Geography & Culture

Movement Toward which body of water did the eastern branch of the early migrations move?

INTERDISCIPLINARY ACTIVITIES

CONNECT to Art

Illustrate Internal Migration

Have students draw a map of the United States or another country, and illustrate at least two internal migrations (for example, African-American migrations to the North in the 1920s or migrations west during the Depression). Students should add short explanations of each move, identifying the push and pull factors. Explain that their illustrations may cover historical movements or ones they create.

CONNECT to Language Arts

Write a Journal Entry for a New Immigrant's First Day

Ask students to visualize the first day of a new immigrant who is entering the United States or another country. Then have them write a journal entry about that first day, describing some of the new sights, sounds, clothing, and so on, that would make that first day memorable.

Teach

Where People Migrate

Talk About It

- What are two common forms of internal migration? *(people moving from rural areas to cities and from cities to suburbs)*
- What is the definition of external migration? *(moving across a continent or to another continent)*
- **Compare and Contrast** Which person moves a greater distance—an internal or external migrant? *(Possible Answer: It could be either. Both an external and an internal migrant might move only a few miles—or thousands of miles across a country or to a new continent.)*

Animated GEOGRAPHY

Bantu Migrations

CLASSZONE.COM

CONNECT Geography & Culture

Note the time span of the migration as detailed in the text (over 2,000 years). Compare this external migration to the external migration that brought diverse people to the United States in a period of less than 400 years.

- What was the general direction of the Bantu migration in Africa? *(Bantu people generally migrated from west-central Africa to southern Africa.)*
- Why do you think the western branch of the Bantu migration stopped where it did? *(The Kalahari and Namib deserts prevented further migration.)*

ANSWER

Movement Earliest migrants to the east moved toward Lake Victoria; later migrants toward Lake Nyasa, Lake Tanganyika, the Zambezi and Limpopo rivers, and the Indian Ocean.

Presentation Options

Use the **Power Presentations** DVD-ROM to project the animated map.

- Click to animate the map.
- Listen to the audio and examine the images to learn more about Bantu migrations.

Teach

The Effects of Migration

Think, Pair, Share

- How does diversity change a culture? *(Possible Answer: It can provide new foods to eat, clothes to wear, and ideas and technology to use.)*
- What are some effects of discrimination? *(People who are discriminated against may not be able to get good jobs or good places to live.)*
- **Compare and Contrast** How can immigrants help or hurt a region's economy? *(Help: by bringing in needed workers, skills, or technology; Hurt: if jobs are scarce, or if they have limited resources or skills)*

Unit Resource Book: Introduction to World Geography
- Skillbuilder Practice: Finding Main Ideas, p. 156

CONNECT Geography & Culture

Bantu Languages

Many of the African languages developed from a single parent tongue, Proto-Bantu. Most Bantu speakers live in a region from south of the Sahara to the tip of Africa. A Bantu language is the first language of nearly one-third of all Africans.

Swahili is the most commonly spoken language in Africa after Arabic. Swahili uses Bantu basics along with Arabic and Persian words. It probably developed as people of East Africa interacted with Indian Ocean traders.

- Name a Bantu language and where its speakers live. *(Swahili; central and east Africa)*

More About . . .

Refugees

A 2003 world survey estimated that more than seven million people in the world had been restricted to refugee camps or segregated settlements for 10 or more years.

The Effects of Migration

KEY QUESTION How does migration affect people and lands?

People bring the customs and traditions of their culture with them when they relocate. **Culture** is the shared attitudes, knowledge, and behaviors of a group. As a result of migration, the cultures of both the immigrants and the people living in an area may change. Migration also has economic and political effects, both positive and negative.

Cultural Effects When different groups in an area interact, they learn about each other—what they eat, wear, and believe. If people accept these different ideas and behaviors, it adds to the diversity of the group. **Diversity** means having many different ways to think about or to do something, or it may refer to a variety of people. Many people believe that having diversity in a group makes it stronger.

CONNECT Geography & Culture

Bantu Languages

The Bantu spread their languages when they migrated across Africa. Today, about 240 million Africans speak one of the hundreds of Bantu languages as their first language. Some 50 million of them in central and east Africa speak Swahili (swah•HEE•lee), also known as Kiswahili (KEY•swah•HEE•lee).

Earlier you learned about the Bantu-speaking people and how they spread across southern Africa. Wherever they settled, the Bantu brought their language and iron-making skills to the people in the region. Even if the Bantu moved on, their language and iron-working techniques remained. So migration changed the existing culture in several ways.

Migration does not always benefit the people who move. This is especially true if a group did not want to move. For example, refugees who flee war may be forced into overcrowded camps with little hope of returning home. Another possible result of migration is **discrimination**, actions that might be hurtful to an individual or group. For example, the group discriminated against may not be able to get jobs or housing.

Economic Effects The arrival of a new group can help or hurt a region's economy. If more workers are needed or the new workers have special talents, the region's economy may be improved. This was the case with the Bantu who brought iron-making skills with them.

The arrival of large numbers of people sometimes strains a region's resources. This occurs when, for example, war refugees with limited resources crowd into refugee camps. The living conditions may be very poor and the additional numbers make life in the camps miserable. In 2005, many thousands of people fled their homes as warfare swept Sudan in Africa. The refugees found little shelter, water, or firewood in the camps. In fact, violence often broke out over the

DIFFERENTIATING INSTRUCTION: TIERED ACTIVITIES

OBJECTIVE Create a narrative work from an immigrant's point of view.

Basic

Each student will assume the role of an external immigrant and create an illustration of his or her experiences, from the first-person point-of-view. In their illustration have students identify one push factor and one pull factor that motivated immigration. Create a display or classroom library of them.

On Level

Adapt the Basic level activity, so that students also create a short story of their experiences, from the first-person point-of-view. In their stories, have students describe one push factor and one pull factor that motivated immigration. Encourage students to read their stories aloud to the class.

Challenge

Adapt the Basic level activity, so that students also create a narrative poem or song that describes push factors and pull factors that motivated immigration. Encourage students to read their poems aloud or perform their songs for the class.

available water. Security and protection for the people in the camps can also become a problem. Governments may not be able to adequately provide for the refugees or may need to ask other nations to help care for the people.

Political Effects The policies of a country or region can be affected by the arrival of immigrants. Sometimes immigrants may be viewed as unwanted or dangerous. The government of a country might then support actions to remove the immigrants or allow them to be treated badly in hopes that they will leave. The immigrants may worry about their personal safety and ability to provide for themselves and their families.

In the best of times, the new immigrants make contributions to the country and are viewed as assets. For example, many immigrant groups brought valuable skills to the United States and were given the opportunity to become citizens.

Water Cans at a Refugee Camp Water is scarce in this African refugee camp. Each day people line up in hopes of getting a supply of water.

DRAW CONCLUSIONS Identify positive and negative effects of migration. Changes may improve the area or the people already there, or the land may suffer.

Section 2 Assessment

ONLINE QUIZ For test practice, go to Interactive Review @ClassZone.com

TERMS & NAMES

1. **Explain the importance of**
   - migration
   - push factor
   - pull factor
   - culture

USE YOUR READING NOTES

2. **Compare and Contrast** Use your completed chart to answer the following question:
   Which factors do you think are more powerful in encouraging migration?

PUSH FACTORS	PULL FACTORS
1.	1.
2.	2.

KEY IDEAS

3. What has been a major push factor in migration since earliest times?
4. Which form of internal migration has affected most countries?
5. What are the three ways that migration affects people and lands?

CRITICAL THINKING

6. **Analyze Causes and Effects** How do push-pull factors work together?
7. **Summarize** How is a culture affected by migration?
8. **CONNECT to Today** What are the pull factors that would attract people to move to your community?
9. **HISTORY** **Create a Push-Pull Poster** Choose two immigrant groups that came to the United States at any time in its history. Create a poster with images to show where they came from, what pushed them from their homeland, and what pulled them to the United States.

CONNECT to the Essential Question

How do natural resources affect a country's population distribution and economy?

Ask students what they have learned so far that can help them answer this question. Students might mention:

- Natural resources can be a pull factor, which attracts immigrants. Lack of resources can be a push factor, which encourages people to leave.
- Immigrants can add labor and skills to an economy to help it grow. Too many immigrants can strain an economy and its resources.

4 Assess & Reteach

Assess Have students complete the Section Assessment.

Unit Resource Book: Introduction to World Geography
- Section Quiz, p. 170

Interactive Review @ClassZone.com
- Section Quiz

Power Presentations

Test Generator

Reteach Assign groups of students one of the subsections of Section 2, "Why People Move." Have them list the important events and ideas covered in the assigned subsection, and then present their lists to the class.

Unit Resource Book: Introduction to World Geography
- Reteaching Activity, p. 174

SECTION 2 ASSESSMENT ANSWERS

Terms & Names

1. migration: process of relocating to a new region; push factor: a reason for people to leave an area; pull factor: reason for attracting people to a new area; culture: the attitudes, knowledge, and behaviors of a group

Use Your Reading Notes

2. Possible answer: Push factors are likely more powerful since they often involve basic survival.
   Diagram Push Factors: *Environmental: lack of resources, change in climate; Political: war/persecution, forced removal.* Pull Factors: *Economic: available land, better jobs; Cultural: return to homeland, desire to be near a holy site*

Key Ideas

3. Since earliest times, people have migrated because of changes in their environment.
4. people moving from rural areas to cities
5. culturally, economically, and politically

Critical Thinking

6. Possible answer: People are pushed to leave an area by certain factors, and other factors pull them to choose a particular place to relocate.
7. People bring their culture with them when they migrate to another land, and usually their culture mixes with the culture of the people who already live there.
8. Answers will vary but should show an understanding of pull factors.
9. Use the rubric to score the poster.

Poster Rubric

	Information	Concept Clarity
4	clear and precise	enables good understanding of the push/pull factors
3	adequate	mostly clear on factors
2	barely explains	marginally clear on factors
1	unclear	fails to clarify factors

1 Plan & Prepare

Objectives

- Identify the different types of natural resources and their importance
- Explain how the four basic economic systems work
- Summarize the ways economic development is measured and the differences between developing and developed nations
- Describe the need for world trade and how it can lead to interdependence and competition

Read for the Essential Question

Help students read for a purpose by reminding them of the Essential Question: "How do natural resources affect a country's population distribution and economy?" Point out that they can use text features such as Key Questions, special features, images, and maps to form answers to the question.

Vocabulary

Best Practices Toolkit

Use the ***Best Practices Toolkit*** to model strategies for vocabulary notetaking. Vary strategies throughout the year. Choose from: *Knowledge Rating, Predicting ABCs, Definition Mapping, Word Sort, Frayer Model (Word Squares), Magnet Words, Word Wheel,* and *Student VOC.*

Vocabulary Strategies, TT9–TT16

Reading Strategy

Best Practices Toolkit

Display the Three-Column Chart transparency.

- Explain that in categorizing, students need to determine how information can be sorted into the categories in the chart.
- Model filling in categories on a chart using the information in the subsection "Economic Systems."
- Have students suggest entries.
- Display the graphic organizer for students to review.

Three-Column Chart, TT21

SECTION 3

Reading for Understanding

Key Ideas

BEFORE, YOU LEARNED

Migration changes places and regions by introducing new people and cultures.

NOW YOU WILL LEARN

Economic activities in an area depend on the presence of natural resources.

Vocabulary

TERMS & NAMES

natural resource something that is found in nature that is necessary or useful to humans

economy a system for producing and exchanging goods and services among a group of people

economic system a way people use resources to make and exchange goods

command economy an economic system in which the production of goods and services is decided by a central government

market economy an economic system in which the production of goods and services is decided by supply and the demand of consumers

Gross Domestic Product (GDP) the total value of all the goods and services produced in a country in a year

export a product or resource sold to another country

import a product or resource that comes into a country

specialization a focus on producing a limited number of a specific products

BACKGROUND VOCABULARY

raw material an unprocessed natural resource that will be converted to a finished product

Reading Strategy

Re-create the chart shown at right. As you read and respond to the **KEY QUESTIONS**, use the chart to categorize important details about types of economic systems.

Skillbuilder Handbook, page R7

CATEGORIZE

ECONOMIC SYSTEMS			
Traditional	Market	Command	Mixed
1.	1.	1.	1.
2.	2.	2.	2.
3.	3.	3.	3.

GRAPHIC ORGANIZERS Go to **Interactive Review** @ClassZone.com

PRETEACHING VOCABULARY

English Learners

Pronounce and Preview

Pronounce each term for students. Review the meanings of words within definitions, such as *necessary, exchange, consumers,* and *unprocessed.*

- To modify vocabulary learning, have students complete worksheets as they read, instead of afterward.

Unit Resource Book: Introduction to World Geography
- Vocabulary Practice, p. 155

Inclusion

Illustrate Terms and Names

Read the definition of each vocabulary term and name aloud, and answer any questions students might have. Then have students choose one term to illustrate or create a diagram for (or assign terms). Ask volunteers to share their illustrations or diagrams and have the other students match the correct terms to them.

SECTION 3 ECONOMICS

Chicago, Illinois, United States of America

A street scene in Turpan, China

Resources and Economics

Connecting to Your World

What natural resources are found in your area? A **natural resource** is something found in nature that is necessary or useful to humans. Forests, mineral deposits, and fresh water are examples of natural resources. Often, the presence of natural resources attracts people to a particular area. Think about how important natural resources are to your life, to your community, and to your country. How are they being used, and how long will they last?

Natural Resources

KEY QUESTION What are the different types of natural resources?

Natural resources are essential for economic development, but resources are not equally distributed around the world. People learn to use the resources in their own areas to their best advantage. However, just because a natural resource is present does not mean it can or will be used. People in some countries may not have the technology to take the resource and turn it into usable products. For example, iron ore is useless until technology turns it into iron and steel products. Technology changes over time, making the value of resources change as well.

Copper Mine This copper mine in Spain has been producing copper since Roman times.

2 Focus & Motivate

3-Minute Warm-Up

Write on the board or display the transparency:

- View the Natural Resources map in Section 3. Are there any resources listed which are not available in North America? *(No: All are available.)* South America? *(Yes: Coal and uranium are not available.)* Which type of resources are shown on the map? *(non-renewable resources)*

Unit Transparency Book: Introduction to World Geography
- 3-Minute Warm-Ups, TT18

Connecting to Your World

More About . . .

U.S. Coins and Copper

Surprisingly, only about 3 percent of a U.S. penny is copper. The U.S. dime, quarter, and half dollar each contain over 90 percent copper.

3 Teach

Natural Resouces

Talk About It

- Why is technology important to an economy? *(Technology is required to make resources into usable products.)*
- What is the difference between renewable and unlimited resources? *(Unlimited are always available no matter how much is used.)*
- **Categorize** What are some unlimited resources? *(sunlight, wind, geothermal heat, tides)*

SECTION 3 PROGRAM RESOURCES

Unit Resource Book: Introduction to World Geography

ON LEVEL

Unit Resource Book
- Reading Study Guide, p. 127
- Section Quiz, p. 171

STRUGGLING READERS

Unit Resource Book
- Reading Study Guide with Additional Support, p. 135
- Skillbuilder Practice: Finding Main Ideas, p. 156
- Section Quiz, p. 171
- Reteaching Activity, p. 175

ENGLISH LEARNERS

Pupil Edition in Spanish

Pupil Edition Audio DVD
- RSG (Spanish), p. 143
- RSG with Additional Support (Spanish), p. 151

Multi-Language Glossary

Test Generator
- Section Quiz in Spanish

INCLUSION

Unit Resource Book
- RSG with Additional Support, p. 135
- Section Quiz, p. 171
- Reteaching Activity, p. 175

GIFTED & TALENTED

Unit Resource Book
- Interdisciplinary Projects, pp. 163, 165
- Section Quiz, p. 171

PRE-AP

Unit Resource Book
- Economics in the World, p. 166

TECHNOLOGY

Unit Transparency Book
- 3-Minute Warm-Ups, TT18
- Essential Question, TT19
- Critical Thinking, TT22
- Cultures, TT26

Daily Test Practice Transparencies
- Section 3, TT10

Power Presentations

ClassZone.com

World Cultures and Geography Video on DVD

CONNECT Geography & Economics

Natural Resources

Note that the importance of different resources has changed over time. For example:

- Early civilizations valued salt from Africa and spices from Asia so much that kingdoms became rich and powerful trading them.
- Uranium was considered unimportant until the beginning of the nuclear age in the 20th century.
- Bauxite only became important in the late 1800s, when a low-cost method of using it to produce aluminum was discovered.
- Even petroleum was thought to have limited importance, until the introduction of internal combustion engines in the early 1900s. Today, petroleum—sometimes called "black gold"—is considered by many experts to be the world's most important economic resource.

CRITICAL THINKING ANSWER

Draw Conclusions Possible Answer: Non-renewable resources are the most desirable since their supply is limited.

Unit Resource Book: Introduction to World Geography

- Economics in the World: Types of Industry, p. 166

Comparing Regions

Southwest Asia and North America

Although scientist disagree on many aspects of estimated oil reserves, it is clear that certain regions have much greater proven reserves than others. Some estimates suggest that the relatively small region of Southwest Asia has more than 65 percent of the world's oil reserves, while the much larger region, North America, has only about five percent. Use of oil is uneven as well, with North America consuming over 30 percent of the world's oil production, while Southwest Asia only consumes about seven percent.

Geographers divide natural resources into three main categories: renewable, non-renewable, and unlimited. Renewable resources are those that nature can replace, such as trees or plants. Unlimited resources are things such as sunlight and wind—these never run out and often are used to produce energy. Non-renewable resources can't be easily replaced, so when they are used up, there aren't additional resources. Minerals and fuels like coal and oil fall into this category.

CATEGORIZE Identify three basic types of resources.
renewable, non-renewable, unlimited

CONNECT Geography & Economics

Natural Resources The map below shows the location of some major non-renewable resources. When supplies of these resources are gone, countries relying on them for their economies will have to change their economic focus.

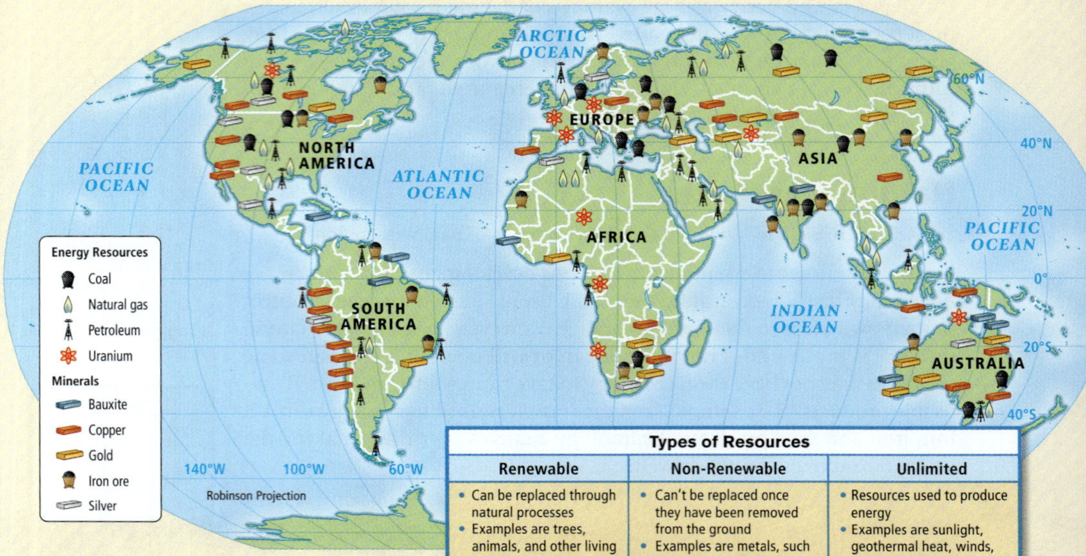

Types of Resources		
Renewable	**Non-Renewable**	**Unlimited**
• Can be replaced through natural processes • Examples are trees, animals, and other living things. • Can be lost if they are used faster than they can be replaced	• Can't be replaced once they have been removed from the ground • Examples are metals, such as gold and iron, and fossil fuels, such as oil and coal. • Once they are used up, these resources are gone forever.	• Resources used to produce energy • Examples are sunlight, geothermal heat, winds, and tides. • Always available no matter what amount is used

CRITICAL THINKING

Draw Conclusions Which of the three categories of natural resources are likely to be most desirable?

DIFFERENTIATING INSTRUCTION

Struggling Readers

List Local and Regional Resources

Discuss the definitions of the various types of resources, and then discuss local or regional resources that are important to families and local businesses. Help students fill in a chart (such as the one to the right) and discuss the reasons students categorized their entries the way they did.

Local and Regional Resources		
Renewable	**Non-renewable**	**Unlimited**
• *Trees* • *Fertile land*	• *Coal*	• *Sunlight*

Economic Systems

KEY QUESTION What are the four basic economic systems?

An **economy** consists of the production and exchange of goods and services among a group of people. Economies exist at the local, regional, national, and international levels. **Economic systems** are different ways that people use resources to make and exchange goods and services.

Four basic economic systems are traditional, command, market, and mixed.

Traditional economy Goods and services are traded, but money is rarely exchanged. This process is called "barter" and is the oldest economic system. It is not used much today.

Command economy A **command economy** is one in which production of goods and services is decided by a central government. The government usually owns most of the resources and businesses that make the products or provide the services. This type of economy is also called a planned economy.

Market economy When the production of goods and services is determined by the supply and the demand of consumers it is called a **market economy**. It is also known as a demand economy, or capitalism. This is the type of economy found in the United States.

Mixed economy In this economy, a combination of command and market economies provides goods and services.

Economic activities are all the different ways that people make a living under these economic systems, including manufacturing, agriculture, fishing, and providing services. Some countries have a wide mix of economic activities while others may have only one or two main economic activities.

SUMMARIZE Name the four basic types of economic systems.
traditional, command, market, mixed

Food Market in Vietnam Supply and demand help determine which products will be sold. **How is this market different from most food markets in the United States?**

◀ Most U.S. markets are indoors. Food producers sell to a huge distribution network, not to individuals.

Teach

Economic Systems

Think, Pair, Share

- Who controls a command economy? *(a central government)*
- How is the production of goods and services in a market economy determined? *(by supply and demand)*
- **Categorize** What type of economy is seldom used today? Why do you think that is true? *(traditional economy; modern supply and demand makes bartering very hard to accomplish)*

Unit Resource Book: Introduction to World Geography
- Skillbuilder Practice: Finding Main Ideas, p. 156

More About . . .

A Barter Economy in Argentina

As Argentina's economic and political organization neared collapse in 2002, people in Argentina organized an extensive barter economy. With about half of all Argentineans living below the poverty level, people began trading for goods and services. It was estimated that nearly three million people were participating in about 5,000 different barter markets. People could trade for goods and services, such as bread, clothing, school tutors, mental health counselors, and auto repairs.

INTERDISCIPLINARY ACTIVITIES

CONNECT to Economics

Roleplay a Barter Economy

Help students prepare to roleplay a barter economy by giving each student a piece of paper that stands for a particular number and type of item, such as one CD, one pair of sneakers, one lunch ticket, and so on. Allow students to trade for a set time, and then discuss the advantages and disadvantages of this type of economy.

CONNECT to Art

Illustrate Relationships in a Command or Market Economy

Ask students to select either a command or market economy, and then create a poster that illustrates the relationships among government, companies, and consumers. Encourage students to consider issues that affect these relationships. Display all posters for use as part of a class discussion about the pros and cons of each system.

Teach

Measuring Economic Development

Reader, Recorder, Reporter

- Why do you think the literacy rate helps identify developing and developed nations? *(Possible Answer: A literate work force helps nations develop technology and use natural resources more effectively.)*
- **Categorize** Besides literacy what are some ways to measure a country's economic development? *(GDP, infant mortality, life expectancy, health information)*

COMPARING

Economic Development

Point out that the United States has a larger GDP and GDP/Person than Japan or Germany, but a higher infant mortality rate and lower life expectancy.

- Why might Japan and Germany lead the United States in infant mortality and life expectancy? *(Possible Answer: They may devote a greater portion of GDP toward public health.)*

CRITICAL THINKING ANSWER

Compare and Contrast Uruguay's GDP, life expectancy, and literacy rate is higher and its infant mortality lower than Burkina Faso.

Unit Resource Book: Introduction to World Geography

- Interdisciplinary Projects: Making the Most of Resources, pp. 163, 165

CONNECT to the Essential Question

How do natural resources affect a country's population distribution and economy?

Ask students what they have learned so far that can help them answer this question. Students might mention:

- The presence of natural resources and the technology to use them are key to an economy.
- Developing nations often trade raw materials while developed countries specialize in processing them.

Measuring Economic Development

KEY QUESTION How is economic development measured?

To measure economic development, geographers may look at such figures as literacy, health information, life expectancy, or the value of a country's economy. One of the most important measures used is Gross Domestic Product (GDP). The **Gross Domestic Product (GDP)** measures the total value of all the goods and services produced in a country in a year. Countries are divided into two categories based on economic development: developing or developed nations.

Developing Nations These nations have low GDP and few economic activities. Many people raise food or animals to survive and have little or no machinery or advanced technology to do the work.

Developed Nations These nations have high GDP and many economic activities, especially business and information processing, and a lot of high-level technology. Food is grown on commercial farms, and most people work in offices and factories.

COMPARE Identify information used to measure development.
GDP, life expectancy, literacy, health information, infant mortality

COMPARING Economic Development

Geographers use many different measurements to look at the development level of a country's economy. Here is a comparison of some developing nations and some developed nations.

◄ *Brazilian coffee bean worker*

	Status	GDP	GDP/ Person In US Dollars*	Infant Mortality (1000 Live Births)	Life Expectancy (At Birth)	Literacy Rate (Percent)
Burkina Faso	developing	5.4 billion	1,300	91.4	48.9	26.6
India	developing	719.8 billion	3,400	54.6	64.7	59.5
Uruguay	developing	13.2 billion	9,600	11.6	76.3	98.0
Germany	developed	2.8 trillion	30,400	4.1	78.8	99.0
Japan	developed	4.9 trillion	31,500	3.2	81.3	99.0
United States	developed	12.5 trillion	42,000	6.4	77.9	99.0

* official exchange rate

Source: *CIA World Factbook, 2006*

CRITICAL THINKING

Compare and Contrast How can you tell Uruguay has developed more than Burkina Faso?

DIFFERENTIATING INSTRUCTION: TIERED ACTIVITIES

OBJECTIVE Categorize economic development based on exports.

Basic

Point out that developing countries often trade raw materials to developed countries. Assign small groups of students one of the regional atlases in the text. Then have groups categorize developing nations and developed nations based on exports listed on the country cards.

On Level

In addition to the Basic activity, have groups choose a second region to categorize. After they have categorized both regions, the group should prepare a comparison of the two regions in terms of economic development. Each group should present its findings to the class.

Challenge

In addition to the On Level activity, have groups brainstorm other measures that could be used to measure a country's economic development. Using that list and research, groups should categorize the countries in one region. Groups can present and explain their rankings to the class.

World Trade

KEY QUESTION Why do countries trade with one another?

Early trade networks started because people who did not have certain resources, such as salt, wanted them. For trade to happen, nations usually have to give up some of their resources in exchange. Products or resources sold to other countries are called **exports**. Products or resources that come into a country are called **imports**. For example, U.S.-made mining equipment sold to Brazil is an export of the United States. Brazilian coffee sold to the United States is a U.S. import.

A country may choose to focus on producing only one or two products or resources and exclude other economic activities. This practice is called **specialization**. Countries specialize because it allows them to trade for products they can't produce themselves. For example, a country may sell cocoa beans or wheat or iron ore in exchange for machinery, or chemicals, or electronic goods. In general, developing nations specialize in **raw materials** or low-cost items, while developed nations sell high-level technology goods or services. This focus leads to interdependence between countries. When countries produce the same trade items, competition results.

FIND MAIN IDEAS Explain why nations trade. Nations trade with one another to get products that they want and do not have.

Fun Facts!

OSTRICH MEAT EXPORTS

Chile is known for its major exports of copper, fruit, and paper. But it has added a new export—ostrich meat—that is bringing in more than 17 billion dollars per year. This nontraditional export is bringing money and jobs to Chile.

Section 3 Assessment

ONLINE QUIZ For test practice, go to **Interactive Review** @ClassZone.com

TERMS & NAMES

1. **Explain the importance of**
   - economic system
   - Gross Domestic Product (GDP)
   - export
   - import

USE YOUR READING NOTES

2. **Categorize** Use your completed chart to answer the following question:

   In which economic system are the production of goods and services determined by consumer demand?

ECONOMIC SYSTEMS			
Traditional	Market	Command	Mixed
1.	1.	1.	1.
2.	2.	2.	2.
3.	3.	3.	3.

KEY IDEAS

3. What is a natural resource?
4. What are the four basic economic systems?
5. What is the difference between an export and an import?

CRITICAL THINKING

6. **Summarize** How do geographers divide countries into developed and developing nations?
7. **Compare and Contrast** How are command and market economic systems different?
8. **CONNECT to Today** What natural resources are located in your area and how are they used?
9. **TECHNOLOGY** **Create a Multimedia Presentation** Use the Internet to study the imports and exports of the United States or another country. Plan a slide show about major exports and what countries buy them and major imports and what countries sell them.

Teach

World Trade

Think, Pair, Share

- Which do you think a nation would rather have—a high level of imports or exports? *(exports, because they bring money into the economy while imports cost money)*
- Why might specialization be a problem? *(if the need for a specific product decreased or its costs increased greatly)*
- **Categorize** What type of goods do developed/developing countries specialize in? *(developed: those requiring high-level technology; developing: raw materials or low-cost items)*

4 Assess & Reteach

Assess Have students complete the Section Assessment.

Unit Resource Book: Introduction to World Geography
- Section Quiz, p. 171

Interactive Review @ClassZone.com
- Section Quiz

Power Presentations

Test Generator

Reteach Each student prepares one question and answer about natural resources or economic systems. Then, discuss each question.

Unit Resource Book: Introduction to World Geography
- Reteaching Activity, p. 175

SECTION 3 ASSESSMENT ANSWERS

Terms & Names

1. economic system: a way people use resources to make and exchange goods; Gross Domestic Product: the total value of all the goods and services produced in a country in a year; export: a product or resource sold to another country; import: a product or resource that comes into a country

Use Your Reading Notes

2. Possible answer: Market economy, also called demand economy or capitalism. **Diagram** *Traditional: barter, not used much; Command: central government decides on production; Market: Supply and the demand of consumers determine production; Mixed: a combination of command and market economies*

Key Ideas

3. It is something found in nature that is necessary or useful to humans.
4. traditional, command, market, and mixed
5. An export is something that a country sells to another country, while an import is something a country buys from another country.

Critical Thinking

6. Geographers look at a country's economic activities and economic statistics to determine a level of development.
7. The government controls the economy in a command system, but consumer demand directs a market economy.
8. Answers will vary.
9. Use the rubric to score the multimedia presentation.

Presentation Rubric

	Text	Illustration
4	clear and informative	very helpful illustrations
3	adequate to explain the ideas	adequate illustration
2	confusing	poor illustration
1	unclear	unclear or missing

1 Plan & Prepare

Objectives

- Identify the main types of government and their characteristics
- Summarize the role of an active citizen in a democracy
- Describe some international organizations and their goals

Read for the Essential Question

Help students read for a purpose by reminding them of the Essential Question: "How do natural resources affect a country's population distribution and economy?" Point out that they can use text features such as Key Questions, special features, images, and maps to form answers to the question.

Vocabulary

Best Practices Toolkit

Use the ***Best Practices Toolkit*** to model strategies for vocabulary notetaking. Vary strategies throughout the year. Choose from: *Knowledge Rating, Predicting ABCs, Definition Mapping, Word Sort, Frayer Model (Word Squares), Magnet Words, Word Wheel,* and *Student VOC*.

Vocabulary Strategies, TT9–TT16

Reading Strategy

Best Practices Toolkit

Display the Three-Column Chart transparency.

- Explain to students that in categorizing they need to determine how the information in the section can be sorted into the categories shown in the chart.
- Model filling in categories on the chart using the "Democracy" row.
- Have students suggest entries, using the other categories in the chart.
- Display the graphic organizer for students to review.

Three-Column Chart, TT21

Section 4 Reading for Understanding

Key Ideas

BEFORE, YOU LEARNED

People have different ways to use and trade the Earth's natural resources.

NOW YOU WILL LEARN

The world is divided into many political regions and organizations.

Vocabulary

TERMS & NAMES

government an organization set up to make and enforce rules for a society

citizen a person who owes loyalty to a country and receives its protection

representative democracy a type of government in which citizens hold political power through elected representatives

monarchy a type of government in which a ruling family headed by a king or queen holds political power

oligarchy (AHL•ih•GAHR•kee) a type of government in which a small group of people holds power

dictatorship a type of government in which an individual holds complete political power

communism a type of government in which the Communist Party holds all political power and controls the economy

Visual Vocabulary monarchy (The queen of Denmark)

Reading Strategy

Re-create the chart shown at right. As you read and respond to the **KEY QUESTIONS**, use the chart to categorize important details about types of government.

See Skillbuilder Handbook, page R7

CATEGORIZE

	RULER	BASIS OF RULE
Democracy		
Monarchy		
Dictatorship		
Oligarchy		
Communism		

GRAPHIC ORGANIZERS Go to **Interactive Review** @ClassZone.com

PRETEACHING VOCABULARY

English Learners

Pronounce and Preview

Pronounce each term for students. Review the meanings of words within definitions, such as *loyalty, protection,* and *representatives.*

- To modify vocabulary learning, have students complete worksheets as they read, instead of afterward.

Unit Resource Book: Introduction

- Vocabulary Practice, p. 155

Inclusion

Vocabulary: Student VOC Strategy

Have students work in pairs to fill out the graphic organizer for the terms and names in the section.

- Use the transparency to model the strategy.
- Have volunteers suggest entries.
- Then distribute copies of the organizer for students to work on with a partner.

Best Practices Toolkit

- Vocabulary: Student VOC Strategy, TT16

SECTION 4
GOVERNMENT

Chicago, Illinois, United States of America

A street scene in Turpan, China

Why We Need Government

Passport An official document that allows a person to travel abroad.

Connecting to Your World

You live in one of nearly 200 countries in the world. Some countries are tiny when compared to the United States; others are larger in physical size or in population. All countries have one thing in common—a government. A **government** is an organization set up to make and enforce rules for a group of people. It has authority over the land within its boundaries.

Types of Government

KEY QUESTION What types of government operate around the world?

Government is needed to provide security, make and enforce the laws, furnish the services that keep a country running, and protect the rights of citizens. A **citizen** is a person who owes loyalty to a country and receives its protection. The government also acts on behalf of the people in the country when it deals with other countries. As you have read, all countries have some type of government. The types differ mainly over how much power the people have. Some countries are ruled by a single person, others by a small group, and still others are ruled by many people.

U.S. Capitol Building This is the official seat of government for the country.

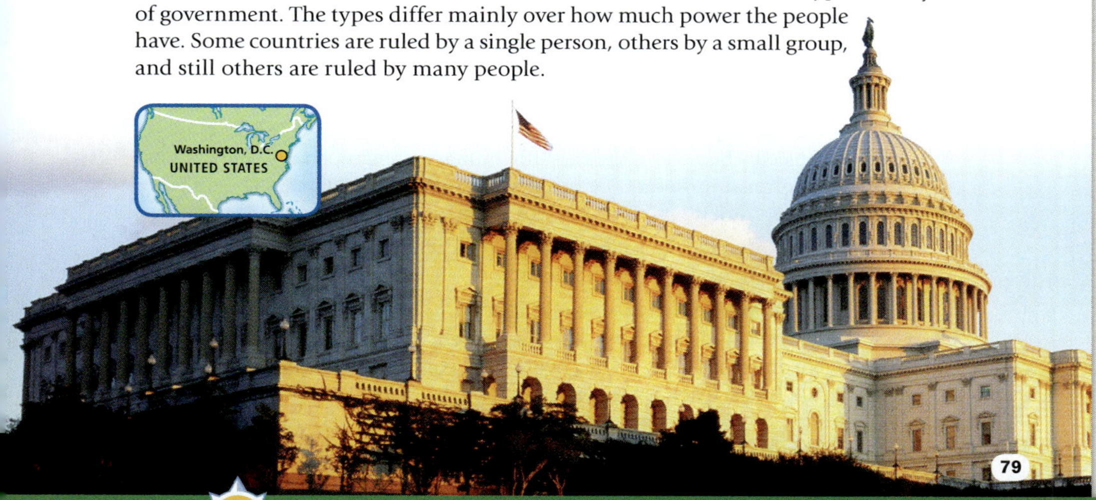

2 Focus & Motivate

3-Minute Warm-Up

Write on the board or display the transparency:

- People in every country have the right to vote for their leaders. True or False? *(False—only people in democracies are guaranteed the right to vote.)*

Unit Transparency Book: Introduction to World Geography
- 3-Minute Warm-Ups, TT18

Connecting to Your World

More About . . .

U.S. Passports

The U.S. Department of State began issuing passports in 1789. E-Passports became available in 2006. E-passports have a chip that stores the bearer's biographic information and a digital photograph.

3 Teach

Types of Government

Think, Pair, Share

- How is the monarchy of the United Kingdom different from Saudi Arabia's? *(The U.K. monarch's power is limited by a constitution.)*
- Who holds political power in a Communist country? *(The government is controlled by the Communist party.)*
- **Categorize** In which types of government do the smallest number of people hold power? *(A single person may hold power in a dictatorship or a monarchy.)*

SECTION 4 PROGRAM RESOURCES

Unit Resource Book: Introduction to World Geography

ON LEVEL

Unit Resource Book
- Reading Study Guide, p. 129
- Outline Map Activities, p. 159
- Section Quiz, p. 172

STRUGGLING READERS

Unit Resource Book
- Reading Study Guide with Additional Support, p. 137
- Section Quiz, p. 172
- Reteaching Activity, p. 176

ENGLISH LEARNERS

Pupil Edition in Spanish

Pupil Edition Audio DVD
- RSG (Spanish), p. 145
- RSG with Additional Support (Spanish), p. 153

Multi-Language Glossary

Test Generator
- Section Quiz in Spanish

INCLUSION

Unit Resource Book
- RSG with Additional Support, p. 137
- Section Quiz, p. 172
- Reteaching Activity, p. 176

GIFTED & TALENTED

Unit Resource Book
- History Makers: The United Nations, p. 161

PRE-AP

Unit Resource Book
- History Makers: The European Union, p. 162

TECHNOLOGY

Unit Transparency Book
- 3-Minute Warm-Ups, TT18
- Essential Question, TT19
- Critical Thinking, TT23
- Map: World Political, TT24

Daily Test Practice Transparencies
- Section 4, TT11

Power Presentations

ClassZone.com

World Cultures and Geography Video on DVD

COMPARING

Governments

Point out to students that countries sometimes change their type of government. For example, France was a monarchy until 1789, and struggled to form a lasting democracy, before finally abolishing the monarchy in 1875.

- How do you think people in a dictatorship, monarchy, or Communist country can change their government? *(Possible Answer: through widespread demonstrations, strikes, protests, and boycotts)*

CRITICAL THINKING ANSWER

Evaluate democracy

Unit Resource Book: Introduction to World Geography

- Outline Map Activities: World: Political, pp. 159–160

Teacher-Tested Activities

Amber I. McVey, Antioch Middle School, Antioch, Illinois

To help introduce the types of governments that countries generally have, I have my students play "That's Me!"

- Students are divided into five groups and each group is given a type of government.
- Each group reads about the characteristics of their assigned government and the group becomes "experts" in that type of government.
- Then, I read off characteristics such as : "This type of government holds complete power."
- If that characteristic describes their government, the group shouts out, "That's Me!"

It is quite possible for more than one group to shout out, "That's Me!" and this becomes an opportunity for class discussion or debate.

COMPARING Governments

TYPES OF GOVERNMENT

DEMOCRACY
- Rule by citizens through elected officials
- Rule is based on citizenship.
- Majority rules

DICTATORSHIP
- Rule by a single individual
- Ruler controls military.
- Citizens have little power to change government.

OLIGARCHY
- Rule by a small group of citizens
- Rule is based on wealth or privilege.
- Ruling group controls military.

MONARCHY
- Rule by a king or queen
- Rule is hereditary.
- May share power through a constitution

COMMUNISM
- Rule by the Communist Party on behalf of the people
- Government owns all economic goods and services.
- Citizens have little power to change government.

CRITICAL THINKING

Evaluate In which type of government do citizens have the most power?

Generally, the type of government a country has falls into one of the following categories:

Representative Democracy Citizens hold political power and rule through elected representatives. In a **representative democracy**, such as the United States, representatives create laws for all the people. If the people object to the laws, they can work to change the laws or change the representatives through elections.

Monarchy In a **monarchy**, a ruling family headed by a king, queen, emperor, or sultan holds political power. Power may or may not be shared with citizens. Saudi Arabia is an example of a traditional monarchy, in which the monarch has complete power. The United Kingdom is a constitutional monarchy, in which the monarch's power is limited by a constitution.

Oligarchy A government where a small group of people holds power, usually because of their wealth, military strength, family connections, political influence, or privilege, is called an **oligarchy**. The military government of the country of Burma is an oligarchy.

Dictatorship In a **dictatorship**, an individual holds complete political power. North Korea is an example.

Communism In **communism**, all political power and control of the economy is held by the government, which is controlled by the Communist Party. The government controls all economic goods and services. Cuba is an example.

DIFFERENTIATING INSTRUCTION

Struggling Readers

Create Symbols of Government

To help students better categorize the different types of government, have small groups of students work together to create a symbol (or series of symbols) for each one. For example, symbols might reflect who holds political power—and who doesn't—or items typically associated with a type of government, such as a crown for a monarchy. Have students present their symbols or create a classroom display.

Pre-AP

Report on Differences in Monarchies

Ask students to review the text on "Monarchy" on this page. Discuss how political power might be used if a monarch has complete power or if that power is limited by a constitution. Then ask students to research and report on the governments of Saudi Arabia and the United Kingdom, comparing and contrasting a traditional and a constitutional monarchy.

There are also different levels of government. The national government oversees the entire country. Countries often have smaller governmental units like state or provincial governments and local governments. Each of these political units deals with specific aspects of life at the state or local level.

DRAW CONCLUSIONS What are the major types of government? democracy, monarchy, oligarchy, dictatorship, and communism

Being an Active Citizen

KEY QUESTION What are your most important responsibilities as a citizen?

As a citizen in a democracy, you have important responsibilities. These responsibilities fall into two categories—personal and community. Personal responsibilities involve your personal behavior and relationships with others. They include taking care of yourself, helping your family, knowing right from wrong, and behaving in a respectful way. Community responsibilities involve the government and your community. They include obeying the law, voting, paying taxes, serving on a jury, and defending your country.

Being an informed citizen means you need to make yourself aware of the issues in an election and the positions held by the candidates running for office. You can also make elected officials aware of your concerns.

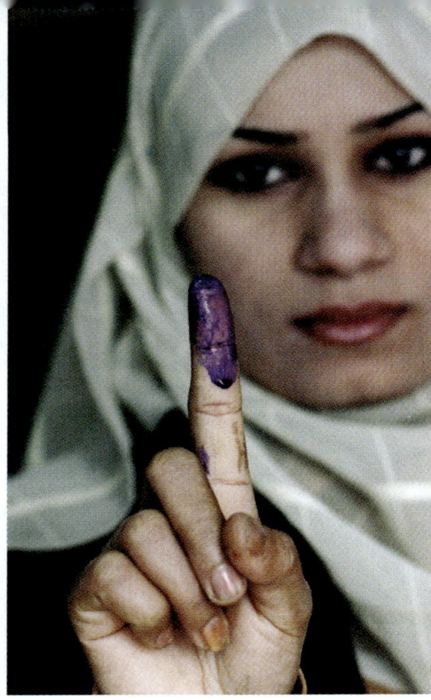

Iraqi Woman Voter This woman shows a purple finger, the sign she has voted. The 2005 election was the first free election in Iraq in 50 years. **How might having a democratic government change this woman's life?** She may have the chance to change the conditions in the country and improve life there.

EVALUATE Which responsibilities of a citizen apply to you? Answers may vary, but should include items from the under-18 and all ages categories.

Responsibilities of a U.S. Citizen

As a young person, you can be a good citizen in a number of ways. Notice that some responsibilities are especially for people under 18, and some are specifically for those over 18. And all citizens have some responsibilities in common.

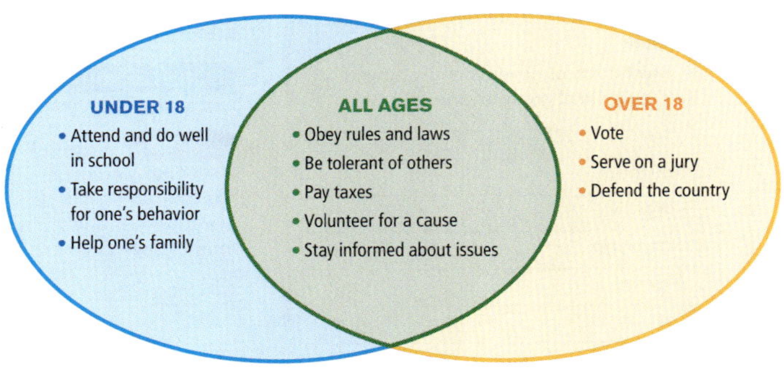

CRITICAL THINKING

Draw Conclusions Which of the responsibilities shown are community responsibilities?

DIFFERENTIATING INSTRUCTION: TIERED ACTIVITIES

OBJECTIVE Describe the responsibilities of citizenship in a democracy and their importance.

Basic

Review with students the subsection, "Being an Active Citizen." Then have students select one personal and one community responsibility, and write two or more sentences that describe each responsibility, how the student might fulfill it, and why it is important for families and society. Ask for volunteers to share their descriptions.

On Level

Modify the Basic activity, so that students write a paragraph about each responsibility they choose. Ask for volunteers to share their descriptions.

Challenge

In addition to the On Level activity, have students write an additional paragraph describing how each responsibility would be affected under a different type of government, such as a dictatorship. Ask for volunteers to share their descriptions.

Teach

Being an Active Citizen

Reader, Recorder, Reporter

- Why is it important to be an informed citizen? *(so people can vote for representatives who reflect their views and make their opinions known to officials)*
- **Categorize** How are voting, serving on a jury, and paying taxes similar in a democracy? *(They are all community responsibilities of citizens.)*

Responsibilities of a U.S. Citizen

Note that until 1971, only those 21 and older could vote in the United States. That year, the 26th Amendment was ratified, giving those 18 and older the right to vote.

- Why do you think some responsibilities are limited to those 18 and over? *(Possible Answers: Defending the country, voting, and serving on a jury all require greater maturity.)*

CRITICAL THINKING ANSWER

Draw Conclusions obeying rules/laws, paying taxes, volunteering, staying informed, voting, serving on a jury, defending the country

CONNECT to the Essential Question

How do natural resources affect a country's population distribution and economy?

Ask students what they have learned so far that can help them answer this question. Students might mention:

- Governments provide security and enforce laws to support the economy.
- International organizations help nations work together on common goals.

Unit Transparency Book: Introduction to World Geography
- Essential Question, TT19

Teach

International Organizations

Talk About It

- **Categorize** What is the largest international organization in the world? *(the United Nations)* Which two organizations mentioned in the text are concerned with economic issues? *(the UN and ECOWAS)*

Unit Resource Book: Introduction to World Geography
- History Makers: The United Nations, p. 161
- History Makers: The European Union, p. 162

4 Assess & Reteach

Assess Have students complete the Section Assessment.

Unit Resource Book: Introduction to World Geography
- Section Quiz, p. 172

Interactive Review @ClassZone.com
- Section Quiz

Power Presentations

Test Generator

Reteach Have small groups of students create one graphic organizer on types of government and one on citizen responsibilities. Have each group present its organizers, then revise them.

Unit Resource Book: Introduction to World Geography
- Reteaching Activity, p. 176

United Nations Security Council This part of the UN deals with political situations that pose a threat to its members. **Why might a nation want to join an organization like the UN?**

to improve its own security or that of the region it is located in

International Organizations

KEY QUESTION Why are international organizations formed?

Countries may join with other nations to form organizations to promote common goals. These organizations might have military, economic, or political goals. Sometimes, countries in a region will form an organization. One example, the Tsunami Warning System (TWS), is a group of 26 nations with coasts or territories on the Pacific Ocean. The TWS was organized to gather information about and send out warnings of these dangerous oceanic events. Other groups are organized to promote economic development, such as the Economic Community of West African States (ECOWAS).

The largest international organization in the world is the United Nations (UN). Nearly 200 countries belong to this organization. Its members work to improve political, cultural, educational, health, and economic conditions around the world.

SUMMARIZE Explain why nations form international organizations.
to solve political, economic, and cultural problems

Section 4 Assessment

ONLINE QUIZ For test practice, go to **Interactive Review @ClassZone.com**

TERMS & NAMES

1. **Explain the importance of**
   - government
   - citizen
   - representative democracy
   - dictatorship

USE YOUR READING NOTES

2. **Compare and Contrast** Use your completed chart to answer the following question:
   In which form of government is power held by a small group of people?

	RULER	BASIS OF RULE
Democracy		
Monarchy		
Dictatorship		
Oligarchy		
Communism		

KEY IDEAS

3. What are the five types of government?
4. What civic responsibilities does a citizen have?
5. What is the main reason for creating a regional or international organization?

CRITICAL THINKING

6. **Draw Conclusions** Why do people form governments?
7. **Make Inferences** Why do local governments exist?
8. **CONNECT to Today** Identify the name of your state and local governments, and explain what each does.
9. **WRITING** **Write a Web Log Entry** Imagine that someone living under a different system of government has been critical of democracy on a web log. Write a response telling the reasons why you support your form of government.

SECTION 4 ASSESSMENT ANSWERS

Terms & Names

1. government: provide protection and services to the people living within a country; citizen: a person who owes loyalty to a country and receives its protections; representative democracy: a government in which citizens have much control over government functions; dictatorship: a government in which citizens have little control over government functions

Use Your Reading Notes

2. oligarchy. **Diagram** *Ruler: Elected officials, king or queen, single individual, small group of citizens, Communist Party; Basis of Rule: majority rules, heredity, control of the military, wealth or privilege, Communist Party*

Key Ideas

3. democracy, monarchy, oligarchy, dictatorship, and communism
4. obey the law, vote, pay taxes, and defend the country
5. to work together on problems or issues that cannot be handled by one country alone

Critical Thinking

6. People form governments because they want order and protection.
7. because national governments usually do not deal with local issues
8. Students should name state and local governments.
9. Use the rubric to evaluate the web log entry.

Web Log Entry Rubric

	Reasons	Paragraph Construction
4	clear and very complete	clear topic sentence, good supporting sentences
3	clear	clear topic sentence, complete sentences
2	unclear, few reasons listed	topic sentence, some fragments
1	unclear, no reasons listed	no topic sentence, fragments

Interactive Review

Click here to complete these and other activities online @ClassZone.com

CHAPTER SUMMARY

Key Idea 1
People are not equally distributed on the Earth's surface.

Key Idea 2
People move from one place to another to meet their needs for certain resources.

Key Idea 3
Economic activities in an area depend on the presence of natural resources.

Key Idea 4
The world is divided into many political regions and organizations.

For **Review and Study Notes**, go to **Interactive Review** @ClassZone.com

NAME GAME

Use the Terms & Names list to complete each sentence on paper or online.

1. I am the average number of people who live in a specific area. **population density**
2. I am something that is found in nature that is necessary or useful to humans. ________
3. I am the process of relocating to a new region. ________
4. I am an organization set up to make and enforce rules for society. ________
5. I am the growth in the number of people living in urban areas. ________
6. I am the way people use resources to make and exchange goods. ________
7. I am a reason that attracts people to another area. ________
8. I am a type of government in which citizens hold political power. ________
9. I am a product or resource sold to another country. ________
10. I am a person who owes loyalty to a country and receives its protection. ________

citizen
democracy
export
economic system
government
Gross Domestic Product
immigrant
import
migration
natural resource
population
population density
pull factor
push factor
urbanization

Activities

Flip Cards

Use the online flip cards to quiz yourself on the terms and names introduced in this chapter.

Crossword Puzzle

Complete an online crossword puzzle to test your knowledge of human geography.

ACROSS
1. something found in nature that is useful to humans

83

ONLINE AND ON-SCREEN ACTIVITIES

Flip Cards

Vocabulary Flip Cards are virtual flash cards at **ClassZone.com.** Students can quiz each other or work independently to study the flip cards for the chapter, "Human Geography."

- They can quiz themselves on key terms and names by reviewing definitions.
- Demonstrate how to advance and shuffle the deck for another challenge.

Crossword Puzzle

Students can review important elements of Human Geography by completing this crossword puzzle online at **ClassZone.com.**

- As an extension of this activity, have students use the *Activity Maker* at **ClassZone.com** to make their own crossword puzzle for one section of the chapter.
- Then have students trade puzzles with classmates to solve each other's puzzles.

Chapter Summary

As a follow-up to the chapter, encourage students to pair up and go over the **Review and Study Notes at ClassZone.com.**

Alternatively, students pairs can review their Reading Study Guide notes, making sure that they can add important details to each section's "Key Idea."

Name Game

Teaching Tips

ON PAPER Have students work in pairs to complete this activity. As an extension, they can write questions for the terms that are not covered in the items shown here.

ONLINE Make sure that students understand how to drag and drop terms to fill in blanks. Ask them to print out their results and compare them with those of a classmate.

ANSWERS

1. population density	**6.** economic system
2. natural resource	**7.** pull factor
3. migration	**8.** democracy
4. government	**9.** export
5. urbanization	**10.** citizen

Make It Fun

Teacher-Tested Activities

Amber I. McVey, Antioch Middle School, Antioch, Illinois

I use this activity the day after we have discussed the three types of natural resources. This activity serves as a nice warm-up or review. To help students classify natural resources as renewable, nonrenewable, or unlimited, I give each student entering the room, an index card with a natural resource written on it.

- The room is divided into thirds—one-third for each of the types of resources.
- Students are asked to categorize themselves based on the natural resource on their index card.
- Students can help each other; the teacher only acts as an observer.
- When they are satisfied with their position, students sit at a desk in that section.
- Finally we identify the resources and discuss why they fit in the selected category.

Vocabulary ANSWERS

1. Urbanization has pulled many of the world's people into cities from rural areas.
2. Population density is a way to measure how crowded a place is.
3. An economic system determines how the goods and services of a place are produced and distributed.
4. Gross Domestic Product is one way to measure the development level of an economy.
5. All are related to the process of migration.
6. All are types of governments.
7. A government functions to protect and support the citizens of a country.

Key Ideas ANSWERS

8. in the Northern Hemisphere, between 20° North and 60° North latitude
9. climate, elevation, and access to water or resources
10. Settlements grow larger and more diverse as people move them from rural areas or other parts of the city.
11. Possible answer: People are pushed out by war or changes in the environment and are pulled by the desire for new land or better jobs.
12. In internal migration, people move within their country, but in external migration, people move out of the country across a continent or to another continent.
13. diversity and discrimination
14. Renewable resources are resources that can be replaced through natural processes, such as trees, animals, and other living things. Non-renewable resources are those that cannot be replaced once they have been used, such as oil and coal.
15. A mixed economy combines features of both market and command—some government direction of the economy with some consumer demand.
16. developed nations and developing nations
17. It is an organization set up to make and enforce rules for a group of people.
18. rule by citizens through elected officials, rule based on citizenship, majority rules
19. the United Nations; to improve economic, political, and cultural conditions around the world

CHAPTER 3 Assessment

VOCABULARY

Explain the significance of each of the following.

1. urbanization
2. population density
3. economic system
4. Gross Domestic Product (GDP)

Explain how the terms and names in each group are related.

5. immigrant, push factor, and pull factor
6. democracy, dictatorship, and communism
7. government and citizen

KEY IDEAS

1 The Geography of Population

8. Where do most of the world's people live?
9. What are three factors that influence where people choose to live?
10. How does urbanization occur?

2 Why People Move

11. What are two main push factors and two main pull factors of migration?
12. How do internal and external migrations differ?
13. What are two cultural effects of migration?

3 Resources and Economics

14. What types of resources are renewable? Non-renewable?
15. In what ways is a mixed economy like both market and command economies?
16. What two categories do geographers use to refer to a country's level of economic development?

4 How Governments Work

17. What is a government?
18. What are three characteristics of a democracy?
19. What is the largest international organization in the world, and why was it formed?

CRITICAL THINKING

20. **Compare and Contrast** Create a table to compare and contrast who holds political power in the five systems of government.

TYPE OF GOVERNMENT	WHO HOLDS POWER

21. **Analyze Causes and Effects** How might migration affect population density?
22. **Compare and Contrast** What is the difference between a democracy and a dictatorship?
23. **Connect to Economics** Why do countries need to engage in trade?
24. **Connect Geography & History** Why have people needed to form governments?
25. **Five Themes: Movement** What might happen when a group of people brings their culture to a new area?

Answer the ESSENTIAL QUESTION

How do natural resources affect a country's population distribution and economy?

Written Response Write a two- or three-paragraph response to the Essential Question. Be sure to consider the key ideas of each section as well as specific ideas about population distribution and economics. Use the rubric below to guide your thinking.

Response Rubric

A strong response will:
- explain why natural resources influence population distribution
- discuss how natural resources impact an economy

ESSENTIAL QUESTION REVIEW

Before students begin their essays, have them review the key chapter elements that will help them answer the Essential Question. Direct them to significant features in the chapter, such as:

- World Population Density map, Section 1
- Comparing Population Densities, Section 1
- Comparing Migration Factors, Section 2
- Animated Geography: Bantu Migrations, Section 2
- Connect Geography & Economics: Natural Resources, Section 3

Written Response Rubric

	Reasons	Paragraph Construction
4	clear and very complete	clear topic sentence and support
3	clear	clear topic sentence
2	unclear, few reasons	topic sentence, some fragments
1	unclear, no reasons	no topic sentence, fragments

STANDARDS-BASED ASSESSMENT

Test Practice
- Online Test Practice @ClassZone.com
- Test-Taking Strategies and Practice at the front of this book

THEMATIC MAP

Use the map to answer questions 1 and 2 on your paper.

1. **Which of the industries shown on the map is not present at Lyon?**
   A. aerospace
   B. chemicals
   C. electronics
   D. vehicles

2. **In which part of France are the fewest industries located?**
   A. northern
   B. central
   C. eastern
   D. southern

CHART

Use the information in the chart to briefly answer questions 3 and 4 on your paper.

Birth Rates

Country	Rank Out of 226	Birth Rate per 1000
Niger	1	50.73
Tajikistan	49	32.65
Ecuador	90	22.29
Vietnam	131	16.86
United States	155	14.14
Italy	222	8.72

Source: *CIA World Factbook*, 2006

3. **How does the birth rate of Niger compare with that of Italy?**
4. **How does the United States' rate compare with that of Vietnam?**

GeoActivity

1. INTERDISCIPLINARY ACTIVITY–ECONOMICS

With a small group, come up with a list of 10 products that you use daily, such as a computer. Research these products to find out where they are produced. Create a slide show that displays the product and the country or countries that produce it.

2. WRITING FOR SOCIAL STUDIES

Reread the part of Section 1 that discusses population geography. Write a series of newspaper headlines that tell the story of the growth of the world's population. Arrange them in chronological order on a poster.

3. MENTAL MAPPING

Create an outline map of your state. Label the sites of the following:
- the state government
- your county government
- your community government

DIFFERENTIATING CHAPTER ASSESSMENT

Unit Resource Book
- **Chapter Test A,** pp. 177–180. *Use for Struggling Readers, Beginning-to-Intermediate-Level English Learners, and for Inclusion of students whose IEP suggests supported test-taking.*
- **Chapter Test B,** pp. 181–184. *Use for On-Level Learners and ESOL students with Advanced-Level English.*
- **Chapter Test C,** pp. 185–188. *Use for Gifted and Talented students or students assigned Pre-AP work.*

Test Generator
- Chapter Tests A, B, and C, in English and Spanish

McDougal Littell Assessment System

TEST **Test Generator** Use three levels of tests in English and Spanish tied to state standards. Plus, choose questions to customize tests for all populations.

SCORE Scan and Score plain paper student tests in minutes.

REPORT Create instant progress reports, sortable by student, by standard, or other criteria.

RETEACH Respond immediately to students' needs by using a wealth of reteaching resources.

Critical Thinking ANSWERS

20. democracy/people; monarchy/king or queen; oligarchy/small group; dictatorship/single individual; communism/Communist Party

21. Possible answer: The population density of the region being left will be lowered, while the population density of the new region will increase.

22. Possible answer: In a country that is a democracy, the people rule through elected representatives. In a country that has a dictatorship, the people have no say in the government.

23. Countries engage in trade to sell goods and services that they produce and to buy those that they do not have.

24. People form governments to provide security, enforce laws, protect rights, and provide services to the citizens.

25. Possible answer: Often, the culture of the new people is accepted by the people already there and there is diversity. Sometimes the culture of the new people is not accepted and discrimination results.

Standards-Based Assessment ANSWERS

1. A

2. B

3. Niger is about 6 times greater or 5.82 more births per thousand.

4. The U.S. is smaller, but only by 1.19 births per thousand.

GeoActivity

1. INTERDISCIPLINARY ACTIVITY RUBRIC

The slide show should
- include ten products and their locations
- have images or illustrations of the product
- demonstrate good use of technical tools

2. WRITING FOR SOCIAL STUDIES RUBRIC

The newspaper headlines should
- have clear and informative language
- show correct chronological order

3. MENTAL MAPPING RUBRIC

The outline map should:
- represent the approximate shape of the state
- include the sites of the state capital, county seat, and local government buildings

CHAPTER 4

People and Culture PLANNING GUIDE

Complete print resources all on one USB drive allow you to customize lesson plans for all students.

Section Titles & Key Ideas	Unit Resource Book Introduction to World Geography	Assessment Resources
1 What Is Culture? **KEY IDEA** Human beings are members of social groups that have shared and unique behaviors and attitudes.	• **Reading Study Guide,** pp. 189–190 • **RSG with Additional Support,** pp. 193–194 • **Vocabulary Practice,** p. 205 • **Map and Graph Skills,** p. 207 • **History Makers,** pp. 211–212	**Unit Resource Book** • Section Quiz, p. 218 • Reteaching Activity, p. 220 **Test Generator CD-ROM** **Daily Test Practice Transparency,** TT12
2 How Does Culture Change? **KEY IDEA** Cultures do not remain the same but change over time.	• **Reading Study Guide,** pp. 191–192 • **RSG with Additional Support,** pp. 195–196 • **Outline Map Activities,** p. 209 • **Connect Geography and Culture,** p. 216	**Unit Resource Book** • Section Quiz, p. 219 • Reteaching Activity, p. 221 • Chapter Test (Forms A, B, & C), pp. 222–236 **Test Generator CD-ROM** **Daily Test Practice Transparency,** TT13

McDougal Littell Assessment System

TEST SCORE REPORT RETEACH

INTEGRATED TECHNOLOGY

No Time? To focus students on important content in the chapter, use the **"Reading Study Guide with Additional Support"** in ***Resources2GO.***

Teacher Presentation Options

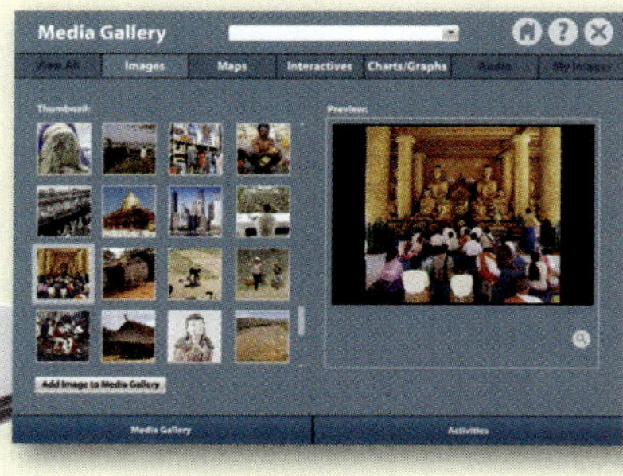

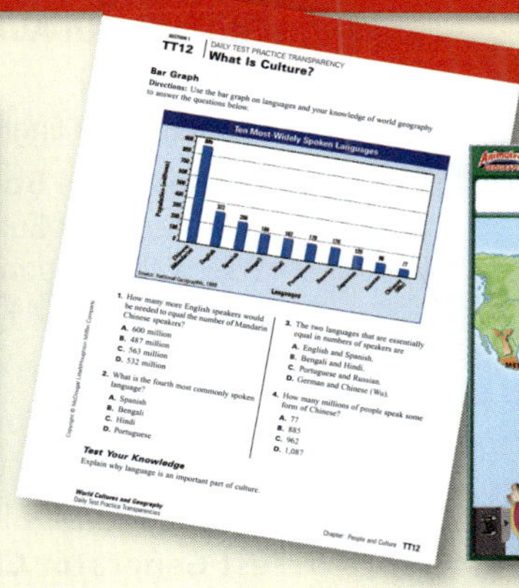

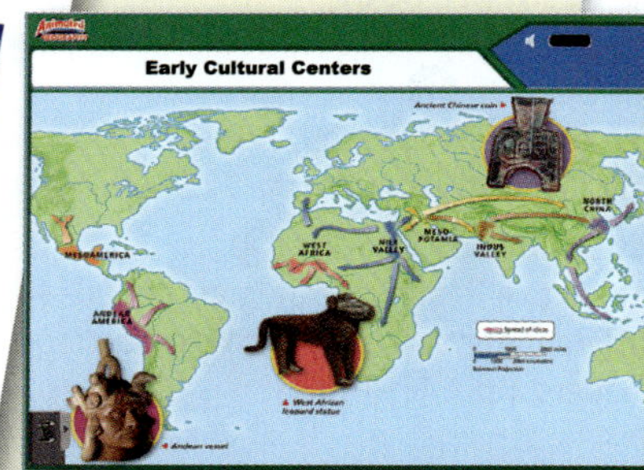

Electronic Books

eEdition DVD-ROM
eEdition Online

Presentation Toolkit

Video/DVD
- World Cultures and Geography

Power Presentations DVD-ROM
- Discussion Starters
- Media Gallery
- Review Game
- Interactive Review
- Animated Features

Unit Transparency Book
- 3-Minute Warm-Ups
- Essential Question
- Critical Thinking
- Maps
- Cultures

Daily Test Practice Transparencies

Animated Geography

- Early Cultural Centers Map
- Worldster: Culture and Your World

OnlineActivities at ClassZone.com

Animated Geography

- Videos and Photos
- 3-D Animations
- Interactive Maps

Interactive Review

- Section Quizzes
- Review and Study Notes
- Vocabulary Flip Cards
- Graphic Organizers
- Name Game and Crossword Puzzles
- GeoGame

Online Test Practice

Map Center

- Atlas Maps
- Political Maps with Data
- Links to Mapping Web Sites

Activity Center

- Activity Maker—Review Games, Puzzles, Flip Cards, and Graphic Organizers
- Worldster

Research & Writing Center

- Web Sites
- Data Updates
- Primary Sources
- Writing Models

Current Events

Ann Christianson
John Muir Middle School
Wausau, Wisconsin

Make It Fun — Teacher-Tested Activities

As you teach this chapter, look for my classroom-proven ideas for:
- making the lesson fun
- focusing students' attention and comprehension
- integrating technology with traditional teaching

Teaching Bell to Bell When you see this timing guide, have students discuss in pairs, in groups of 3–6, or as a class.

- **Think, Pair, Share** (pairs)
- **Reader, Recorder, Reporter** (3–6)
- **Talk About It** (any size)

RESOURCES FOR DIFFERENTIATED INSTRUCTION

CHAPTER 4

Struggling Readers

eEdition with Audio DVD-ROM

Lesson Plans
- Strategies for Inclusion

Unit Resource Book
- Reading Study Guide with Additional Support, pp. 193–196 B
- Reteaching Activity, pp. 220–221
- Chapter Test (Form A), pp. 222–226 A

Best Practices Toolkit C
- Interactive Reading Strategies, pp. 90–145
- Differentiation Activity Bank: Struggling Readers, pp. 146–149

Teacher's Edition Activities
- Define Terms Used to Explain Religion, Sec. 1
- Chart Changes in Technology, Sec. 2
- Tiered Activities, Sec. 1–2

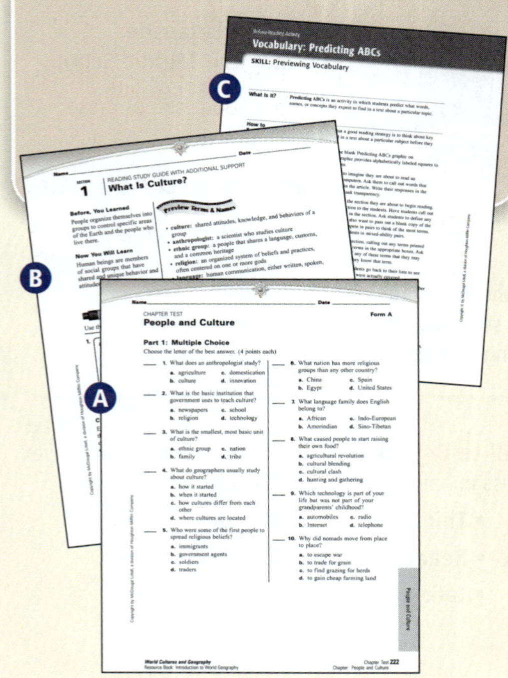

English Learners

Pupil's Edition in Spanish

eEdition with Audio DVD-ROM

Lesson Plans
- Strategies for English Learners

Unit Resource Book
- Reading Study Guide (Spanish), pp. 197–200
- RSG with Additional Support (Spanish), pp. 201–204 A B

Multi-Language Glossary C

Best Practices Toolkit
- Differentiation Activity Bank: English Learners, pp. 165–194

Test Generator CD-ROM
- Section Quizzes in Spanish
- Chapter Test in Spanish

Teacher's Edition Activities
- Pronounce and Preview, Sec. 1–2
- Vocabulary: Key Academic Vocabulary, Sec. 1
- Describe an Invention That Changed Lives, Sec. 2
- Tiered Activities, Sec. 1–2

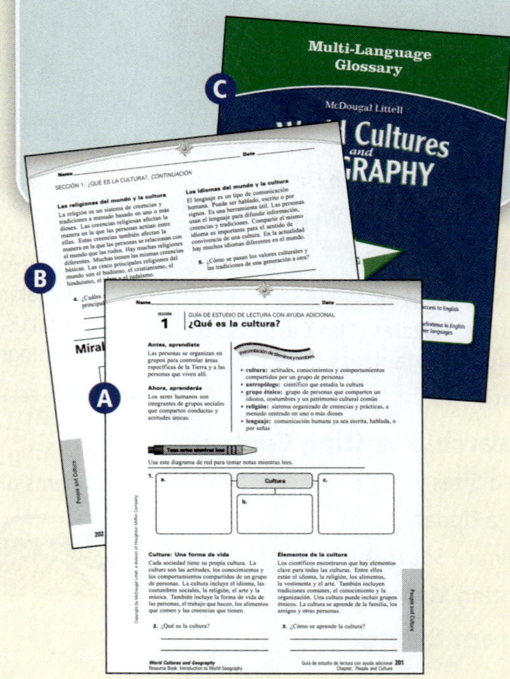

Inclusion

eEdition with Audio DVD-ROM

Lesson Plans
- Strategies for Inclusion

Unit Resource Book
- Reading Study Guide with Additional Support, pp. 193–196 A B
- Reteaching Activity, pp. 220–221
- Chapter Test (Form A), pp. 222–226

Best Practices Toolkit C
- Interactive Reading Strategies, pp. 90–145
- Differentiation Activity Bank: Struggling Readers, pp. 146–149

Teacher's Edition Activities
- Student VOC Strategy, Sec. 1
- Reading and Defining Terms Verbally, Sec. 2
- Tiered Activities, Sec. 1–2

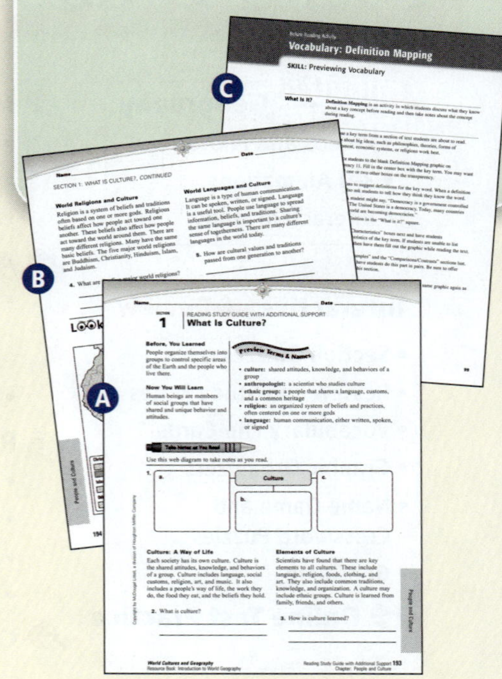

Interdisciplinary Activities

Gifted & Talented Students

Lesson Plans
- Core Lesson Plan

Unit Resource Book
- Outline Map Activities, p. 209
- History Makers, p. 211 (A)
- Interdisciplinary Projects, pp. 213–215 (B) (C)
- Primary Sources, p. 217
- Chapter Test (Form C), pp. 232–236

Best Practices Toolkit
- Student Products Rubrics, pp. 1–14
- Writing Templates, pp. 78–89

Teacher's Edition Activities
- Prepare a Display on the Five Major Religions, Sec. 1
- Tiered Activities, Sec. 1–2

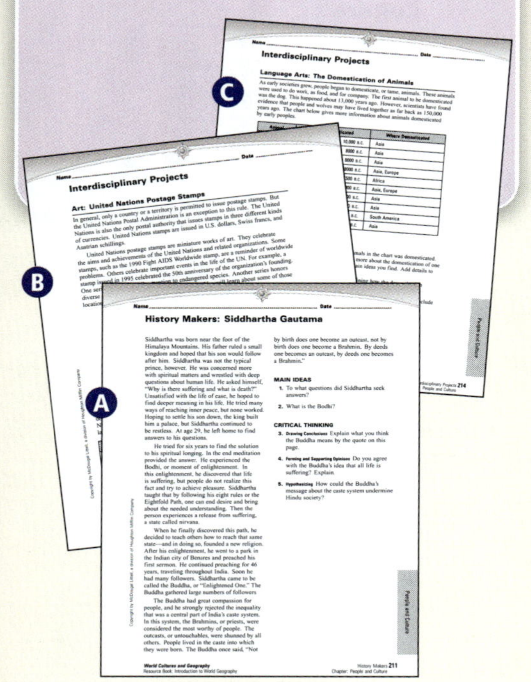

Pre-AP*

Lesson Plans
- Core Lesson Plan

Unit Resource Book
- Map and Graph Skills, p. 207
- History Makers, p. 212 (A)
- Connect Geography and Culture, p. 216 (B)
- Primary Source, p. 217 (C)
- Chapter Test (Form C), pp. 232–236

Best Practices Toolkit
- Writing Templates, pp. 78–89
- Differentiation Activity Bank: Pre-AP Students, pp. 195–200

* Pre-AP is a registered trademark of the College Entrance Examination Board, which was not involved in the production of and does not endorse this product.

Teacher's Edition Activities
- Chart Elements of Local Culture, Sec. 1
- Tiered Activities, Sec. 1–2

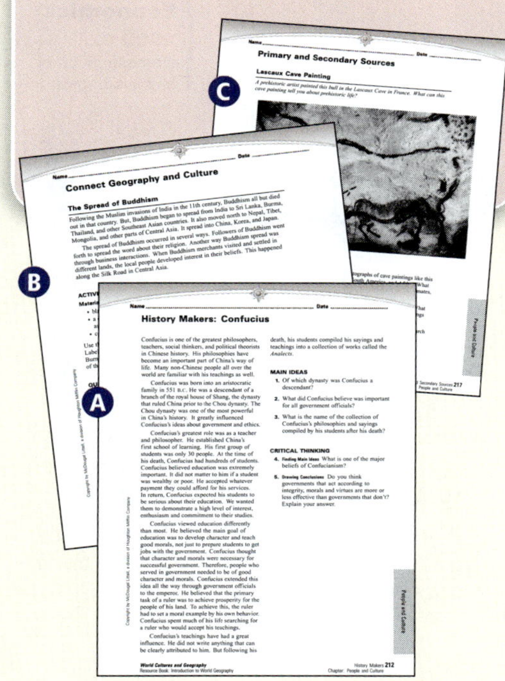

CONNECT to Civics

Report on a Community Cultural Organization

CLASS TIME 30 minutes

TASK Prepare and deliver an oral report on a community cultural organization and how it supports local culture.

SUPPLIES
- computer, or paper and pencil/pen
- list of local community cultural organizations

ACTIVITY Provide students with a list of local community cultural organizations, including a contact person, if possible. Have pairs of students contact one of these organizations (by phone or email) and conduct a short interview. Then have students prepare and present a short oral report on the organization, its history, and how it supports local culture.

Good for Interpersonal Learners

CONNECT to Language Arts

CLASS TIME 30 minutes

TASK Write a poem or short story on one aspect of American culture that they value.

SUPPLIES
- computer, or paper and pencil/pen

ACTIVITY Have students think about cultural activities that they enjoy in the United States, such as a celebration, food dish, style of music, and so on. Then have students create a poem or short story about one of these activities. The theme of their work should show how this activity is important to them or to American society as a whole. Ask volunteers to read their work to the class and answer any questions.

Good for Verbal Learners

Introduce the ESSENTIAL QUESTION

- Point out that all societies develop their own culture—a way of life held in common.
- Discuss the reasons that different cultures might meet their needs in different ways.
- Predict some ways that ideas spread from one place to another.

Key Ideas

Section 1: Human beings are members of social groups that have shared and unique behaviors and attitudes.

Section 2: Cultures do not remain the same but change over time.

CONNECT to the Essential Question

You may wish to revisit the Essential Question periodically with students as they work through the chapter.

Suggestions for reconnecting are provided at an appropriate point in the *Teacher's Edition* for each section.

Cave of Hands, Santa Cruz, Argentina

CHAPTER 4

People and Culture

1 CULTURE What Is Culture?

2 CULTURE How Does Culture Change?

ESSENTIAL QUESTION

How does culture develop and how does it shape our lives?

CONNECT Geography & History

Use the map and the time line to answer the following questions.

1. Which early cultural centers are found in Africa?
2. One of the world's oldest religions developed in the Indus River valley. What is the religion?

9000 B.C.

Economics
8000 B.C.
Development of agriculture
(Early farm tools)

Economics
9000 B.C.
Domestication of animals begins.

Culture
5000 B.C.
First cities develop.
(Home in early city)

86 Chapter 4

TIME LINE DISCUSSION

The domestication of animals and the development of agriculture led to significant changes in culture. Since then, cultures have continued to develop and change and continue that process today.

- About how many years after the domestication of animals did the development of agriculture start? *(1,000 years; 9000 B.C. to 8000 B.C.)*
- About when did people begin to live in larger settlements? *(around 5000 B.C., when the first cities developed)*
- How do you think the development of writing around 3000 B.C. might have influenced the spread of ideas? *(Possible Answer: It was possible for specific ideas, laws, and codes to be written down so others could read them, rather than relying on people's memories and oral descriptions.)*

Animated GEOGRAPHY

Early Cultural Centers

Click here to explore elements of human culture @ClassZone.com

Government
1792 B.C. Hammurabi's code of law develops in the Babylonian Empire.

1000 B.C.

Culture
3000 B.C. First system of writing is in use. (Cuneiform tablet)

Culture
1500 B.C. One of the world's oldest major religions, Hinduism, begins to develop. (Ganesha, a Hindu deity)

TEACHER MEDIA FAVORITES

Books for the Student

Freeman, Dena. *How People Live*. DK Children, 2003. Illustrated guide to more than 80 cultures, highlighting common bonds and unique cultural events and artifacts.

Pelta, Kathy. *Rediscovering Easter Island*. Lerner Publishing Group, 2001. Recounts the theories and science used to better understand this mysterious, ancient civilization.

Video / DVD

In the Shadow of the Himalayas: How People Live in Nepal. (DVD) Smudge Productions, 2004. Overview of Nepal's unique culture.

Hispanic Culture. (DVD) Video Knowledge, Inc., 2004. A look at Hispanic history and culture.

New Zealand Maori Culture, Traditions and History. (DVD) CustomFlix, 2006. A study of Maori culture.

Internet

Go to the *Research and Writing Center* at **ClassZone.com** to link to:

- data on human geography
- a variety of chapter-specific, content-reviewed sites
- "History Makers" biographies
- primary sources

Animated GEOGRAPHY

Early Culture Centers

CLASSZONE.COM

Note that some of these cultural centers existed around the same time, but because of the difficulty of travel, had little or no contact with each other. Archaeologists use sophisticated scientific techniques to establish the dates of artifacts such as statues, tools, and coins.

- In which general direction did the ideas from Mesoamerica travel? *(north)*
- Which cultural centers had ideas that spread to other cultural centers? *(Possible Answers: The Nile Valley center spread to West Africa and Mesopotamia; Mesopotamia spread to the Indus Valley and North China; Indus Valley spread to North China.)*
- Ideas from which cultural center appear to have spread the longest way? *(Mesopotamia's ideas to North China)*

CONNECT GEOGRAPHY & HISTORY ANSWERS

1. West Africa, Nile Valley
2. Hinduism

Unit Resource Book: Introduction to World Geography

- Outline Map Activities: Early Cultural Centers, pp. 209–210

Presentation Options

Use the **Power Presentations** DVD-ROM to project the animated map.

- Prompt students to use the pop-up images and audio from the map to discover what they already know about early cultural centers.
- Challenge students to answer the map questions individually, in small groups, or as a class activity. Use all of the interactive material to help preview chapter content.

1 Plan & Prepare

Objectives

- Describe culture and its elements
- Identify the major religions in the world and how they spread
- Summarize the roles language plays in culture and identify some families of languages and how they have spread

Read for the Essential Question

Help students read for a purpose by reminding them of the Essential Question: "How does culture develop and how does it shape our lives?" Point out that they can use text features such as Key Questions, special features, images, and maps to form answers to the question.

Vocabulary

Best Practices Toolkit

Use the ***Best Practices Toolkit*** to model strategies for vocabulary notetaking. Vary strategies throughout the year. Choose from: *Knowledge Rating, Predicting ABCs, Definition Mapping, Word Sort, Frayer Model (Word Squares), Magnet Words, Word Wheel,* and *Student VOC.*

Vocabulary Strategies, TT9–TT16

Reading Strategy

Best Practices Toolkit

Display the Web Diagram transparency.

- Remind students that the main idea in each chunk of text will be an answer to a "Key Question."
- Model filling in a main idea from the "Culture: A Way of Life" subsection.
- Have students suggest entries for other sections.
- Display the completed diagram for review.

Web Diagram, TT28

Section 1 Reading for Understanding

Key Ideas

BEFORE, YOU LEARNED

People organize themselves into groups to control specific areas of the Earth and the people who live there.

NOW YOU WILL LEARN

Human beings are members of social groups that have shared and unique behaviors and attitudes.

Vocabulary

TERMS & NAMES

culture shared attitudes, knowledge, and behaviors of a group

anthropologist (AN•thruh•PAHL•uh•jihst) a scientist who studies culture

ethnic group a people that shares a language, customs, and a common heritage

religion an organized system of beliefs and practices, often centered on one or more gods

language human communication, either written, spoken, or signed

language family a group of languages that have a common origin

BACKGROUND VOCABULARY

missionary a person sent to do religious work in another land

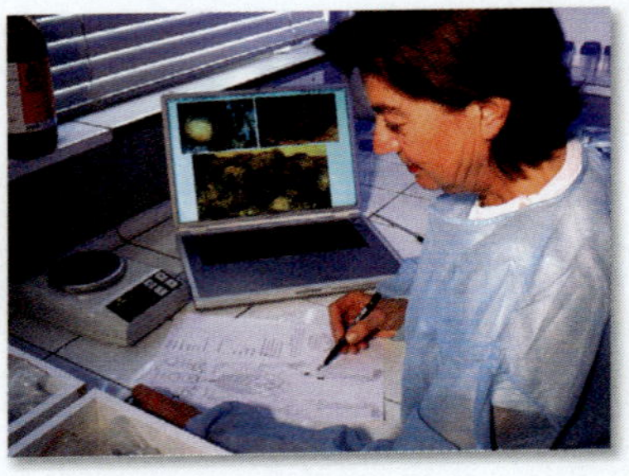

Visual Vocabulary anthropologist

Reading Strategy

Re-create the web diagram shown at right. As you read and respond to the **KEY QUESTIONS**, use the diagram to find main ideas about culture.

See Skillbuilder Handbook, page R4

FIND MAIN IDEAS

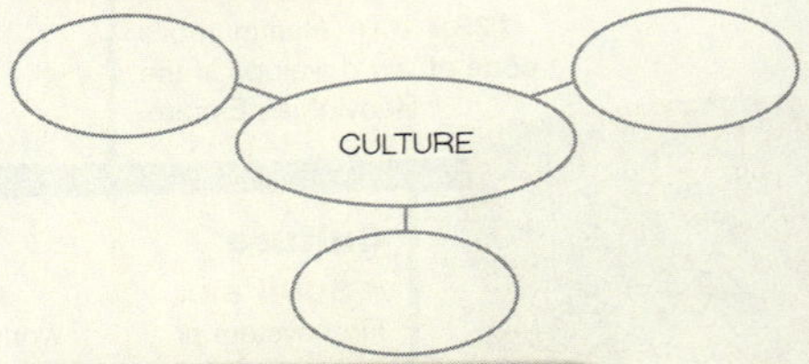

GRAPHIC ORGANIZERS
Go to **Interactive Review @ClassZone.com**

88 Chapter 4

PRETEACHING VOCABULARY

English Learners

Pronounce and Preview

Pronounce each term for students. Review the meanings of words within definitions, such as *knowledge*, *heritage*, and *signed*.

- To modify vocabulary learning, have students complete worksheets as they read, instead of afterward.

Unit Resource Book: Introduction to World Geography

- Vocabulary Practice, p. 205

Inclusion

Student VOC Strategy

Have student pairs fill out the graphic organizer for the section terms.

- Use the transparency to model the strategy.
- Have volunteers suggest entries.
- Distribute copies of the organizer for students to complete.

Best Practices Toolkit

- Vocabulary: Student VOC Strategy, TT16

Bush plane, Alaska

Hindu celebration, India

SECTION 1 CULTURE

What Is Culture?

Connecting to Your World

How does your daily life compare to the lives of students in other parts of the world? Maybe you get up, have cereal for breakfast, and then walk to school. A young person your age in a rural area in Africa may have to work in the fields. When your school day ends, you may have sports or club activities before you go home. If you lived in China's crowded capital of Beijing, you might go home to your family's one-room apartment, where you would study or watch television.

Culture: A Way of Life

KEY QUESTION What is culture?

People meet their basic needs in many different ways. This is because each society has its own culture. **Culture** is the shared attitudes, knowledge, and behaviors of a group. It is the total way of life held in common by a specific group of people. Culture includes language, religion, art, and music. It also includes how a group of people live, what work they do, what food they eat, what beliefs they hold, and how they use the environment to meet their needs. Cultures also create social customs and technologies to solve problems. A culture unites people by helping them to understand their world and to relate to others in the group or those outside the group. Culture is passed from generation to generation.

Newspaper Rock, Utah For perhaps 2,000 years, native peoples and passersby have carved symbols into this sandstone rock.

2 Focus & Motivate

3-Minute Warm-Up

Write on the board or display the transparency:

- Compare the World Religions map to the World Language Families Today map. How are they alike? How do they differ? *(Similar: Many different areas share the same religion and language; Different: some regions, such as Europe and Asia, appear more diverse in terms of religion.)*

Unit Transparency Book
- 3-Minute Warm-Ups, TT27

Connecting to Your World

More About . . .

Newspaper Rock

Newspaper Rock is one of many early Native American petroglyphs (pictures carved into stone) found throughout Utah. In the Navajo language, this rock is named *Tse' Hane*—the rock that tells a story.

3 Teach

Culture: A Way of Life

Talk About It

- What makes cultures unique? *(the ways they use the elements of culture—language, food, religion, and so on—to meet their needs)*
- What things do people within an ethnic group share? *(language, customs, a common heritage)*
- **Find Main Ideas** How do people learn culture? *(by observing others and from direct teaching)*

SECTION 1 PROGRAM RESOURCES

Unit Resource Book: Introduction to World Geography

ON LEVEL

Unit Resource Book
- Reading Study Guide, p. 189
- Map and Graph Skills, p. 207
- Section Quiz, p. 218

STRUGGLING READERS

Unit Resource Book
- Reading Study Guide with Additional Support, p. 193
- Skillbuilder Practice: Summarizing, p. 206
- Section Quiz, p. 218
- Reteaching Activity, p. 220

eEdition with Audio DVD-ROM

ENGLISH LEARNERS

Pupil Edition in Spanish
eEdition with Audio DVD-ROM
Unit Resource Book
- RSG (Spanish), p. 197
- RSG with Additional Support (Spanish), p. 201

Multi-Language Glossary
Test Generator
- Section Quiz in Spanish

INCLUSION

Unit Resource Book
- RSG with Additional Support, p. 193
- Section Quiz, p. 218
- Reteaching Activity, p. 220

GIFTED & TALENTED

Unit Resource Book
- History Makers, pp. 211–212
- Interdisciplinary Projects, pp. 213, 215

PRE-AP

Unit Resource Book
- Primary Sources, p. 217

TECHNOLOGY

Unit Transparency Book
- 3-Minute Warm-Ups, TT27
- Essential Question, TT28
- Critical Thinking, TT29
- Map: World Languages, TT32
- Cultures, TT33

Daily Test Practice Transparencies
- Section 1, TT12

Power Presentations
ClassZone.com
World Cultures and Geography Video on DVD

More About . . .

Ethnic Groups and Political Boundaries

Only in a very few countries today, such as Iceland and Japan, does the population consist of mainly one ethnic group. In most countries—largely due to immigration—many ethnic groups exist side-by-side. In some cases, one group has dominated the government and economy to the exclusion of others. These divisions can result in civil wars and movements to secede from a country. The conflicts in Darfur, Sudan and in Iraq are examples of areas with ethnic conflicts.

ANALYZING Primary Sources

The Negritude movement was created in the 1930s by three men: Aimé Césaire, who was born in Martinique, in the West Indies, Leopold Sedar Senghor from Senegal, and Léon-Gontran Damas from French Guiana. All were poets and writers who met in Paris, discussing the conflicts they saw between their African culture and the culture brought by the French, who had colonized their regions.

- How would you describe Césaire's view of culture? *(Possible Answer: He thought it involved individual and personal characteristics, as well as larger, public elements, such as literature.)*

DOCUMENT-BASED QUESTION ANSWER

Possible Answer: Césaire is saying that culture influences everything that a person does.

Unit Resource Book: Introduction to World Geography
- Primary and Secondary Sources: Lascaux Cave Painting, p. 217

Elements of Culture Scientists called **anthropologists** study culture. They have found that there are basic elements for all cultures. These include language, religion, certain foods and clothing, arts and crafts, technology, and government. Cultural elements also include a group's common practices, its shared understandings, and its social organization. The way a group uses these elements is what makes its culture unique, or one of a kind. Geographers study where different cultures are located and how they interact with their environment.

Every culture contains smaller social groups. The family is the smallest and most basic unit of a culture. Sometimes a culture includes ethnic groups. An **ethnic group** is a people that shares a language, customs, and a common heritage. Mexican Americans and Korean Americans are examples of ethnic groups that are part of the larger culture region that is the United States.

Learning Culture People are not born with cultural knowledge—they learn it from family, friends, and others. Generally, people learn culture in two ways: by observing others in their culture, and from direct teaching. Think about how you learned to speak. At first, you learned from listening to others and by imitating them. Later, when you went to school, you were directly taught the language, so that you could not only speak it, but write and read it too.

FIND MAIN IDEAS Identify the elements of culture. Culture is the total way of life held in common by a specific group of people. It includes language, religion, food, clothing, arts, crafts, technology, and government.

ANALYZING Primary Sources

Aimé Césaire (born 1913), a poet and political leader, was born in Martinique, a French island in the West Indies. He helped to found the Negritude movement. Its purpose was to glorify traditional African culture and identity.

> Culture is everything. Culture is the way we dress, the way we carry our heads, the way we walk, the way we tie our ties—it is not only the fact of writing books or building houses.
>
> Source: Aimé Césaire, speech to the World Congress of Black Writers and Artists, Paris, France, 1956

DOCUMENT-BASED QUESTION

What point is Césaire making about the role of culture in life?

90

DIFFERENTIATING INSTRUCTION

English Learners

Key Academic Vocabulary

Note the word *interact* in the last sentence of the first paragraph. Read the sentence out loud and ask students to suggest definitions. Then ask students to look up *interact* in a dictionary. You may want to discuss other words with the prefix *inter-* such as *international*, *interstate*, and *interchange*.

Pre-AP

Chart Elements of Local Culture

Ask students to review the "Elements of Culture" section on this page. Then ask each student to create a chart of the elements listed, giving an example or detail of each as a way to illustrate their own culture. Discuss elements such as "common practices" and "social organization," so that students can distinguish among the terms and cite an appropriate example.

Worldster

Culture and Your World

To learn more about how culture is learned, go to the **Activity Center** @ClassZone.com

How Culture Is Learned We learn about common practices, shared understandings, and social organization from direct teaching and from observing cultural practices. Family and friends, school, the media, the government, and religious institutions all help us learn our culture.

Family, Friends, and School Family and friends teach us social customs, values, religious and political beliefs, and the basics of living with others. At school, students learn about their culture and the cultures of others.

Media Media, such as television, the Internet, music, books, magazines, and newspapers, help communicate what is happening in our society and in the world around us.

Government Some of the most directed cultural learning comes from the government. It provides schools to instruct young people in the customs and traditions of their culture.

Religious Institutions Personal values and religious beliefs help people learn to live with others.

CONNECT to Your Life

Journal Entry Think about the culture in which you live. What objects might best represent parts of your culture? Record your ideas in your journal.

DIFFERENTIATING INSTRUCTION: TIERED ACTIVITIES

OBJECTIVE Prepare and present a visual display on an area of local American culture.

Basic

Invite students to work individually to create a visual display, such as a poster, that shows one favorite activity of their culture. Suggest they consider a leisure activity, meal, or family celebration. Ask volunteers to explain the significance of their chosen activity.

On Level

In addition to the Basic activity, have students include a description of how this part of their culture has been learned—such as from parents or other family members. Ask volunteers to explain their display.

Challenge

In addition to the On Level activity, have students include a comparison of their activity to another they have experienced or learned about, such as a traditional meal in another country. Ask volunteers to explain their display.

More About . . .

The Media and Youth Culture

Perhaps one of the biggest changes in youth culture today is how fast "what's cool" and "what's not" changes. Thanks to scores of cable television channels and web sites, "IM-ing" (instant messaging), and the ubiquitousness of cell phones, news about "coolness" travels fast—and far. It also gets shaped and transformed by young people themselves, more so today than by traditional media sources.

Comparing Regions

Family and Culture

In the United States, we generally think of family as the nuclear family—that is, the parents and their children. In many parts of Africa, the family is an extended family, including several generations all living very close to each other and possibly in the same dwelling. These family groups often are headed by the eldest member of the group. Depending on the culture, that person may be a male or a female. Decisions for the group may be made by a council of the eldest members. All members of the family are expected to abide by the decisions made by this group.

Presentation Options

Go to the *Activity Center* at **ClassZone.com** to project Worldster.

- Click on the multimedia features of the Culture and Your World Web page.
- Ask students which areas of culture they would like to explore. Listen to music, examine photographs, learn simple phrases in other languages, and more.

CONNECT to Your Life

Journal Entry

Remind students that photos can illustrate ideas not directly covered by the text. Tell students to review the photos on this page to help them with their own entries.

Teach

World Religions and Culture

Think, Pair, Share

- What continent was the original home of all of the world's major religions? *(Asia)*
- How does immigration affect religion in an area? *(Immigrants often bring their own religion with them, so the number of religions practiced often increases.)*
- **Find Main Ideas** What beliefs separate the three main types of religions? *(belief in one god, belief in more than one god, and belief in divine forces in nature)*

Unit Resource Book: Introduction to World Geography
- History Makers: Confucius, p. 212
- World Literature: How Thunder and Earthquake Made Ocean: A Creation Myth Told by Joseph Bruchac, pp. 239–240

COMPARING

World Religions

Religions classified as within the same family may still experience conflicts, sometimes violent ones. Differences between Sunni and Shi'ite groups have continued a split in Islam for nearly 1,500 years, leading to conflicts in Iraq and other areas of the Middle East. Roman Catholics and Protestant Christians in Northern Ireland are also struggling to resolve conflicts over religious, political, and economic issues.

- Which major religion has the most followers worldwide? *(Christianity)* The least? *(Judaism)*
- Which religion has no one leader? *(Hinduism)*

Unit Resource Book: Introduction to World Geography
- History Makers: Siddhartha Gautama, p. 211

World Religions and Culture

KEY QUESTION What role does religion have in a culture?

Because religion has such an influence on people's lives, it is an important element in most cultures. **Religion** is an organized system of beliefs and practices, often centered on one or more gods. Religion establishes beliefs and values. These beliefs and values guide people's behaviors toward each other and toward the environment.

Types of Religions There are thousands of religions in the world, but many religions have common elements. These elements may include specific behaviors to be practiced, important dates and rituals, holy books, and standards of proper behavior. Religions are often divided into three types—those with a belief in one god, those with a belief in more than one god, and those with a belief in divine forces in nature. The five major world religions are Buddhism, Christianity, Hinduism, Islam, and Judaism, as described in the chart below. You will learn more about them in other chapters of this book and in the World Religions handbook at the back of this text.

COMPARING World Religions

RELIGION	BASIC BELIEFS	TEACHER OR LEADER	FOLLOWERS
Buddhism	Followers can achieve enlightenment by understanding the Four Noble Truths and following the Eightfold Path.	Siddhartha Gautama, the Buddha	379 million
Christianity	There is only one God. Jesus is the Son of God. Jesus' death and resurrection made eternal life possible.	Jesus of Nazareth	2.1 billion
Hinduism	The soul never dies but is continually reborn until it becomes enlightened. Enlightenment comes after people free themselves from earthly desires.	no one leader	860 million
Islam	There is only one God. Persons achieve salvation by following the Five Pillars of Islam and living a just life.	Muhammad	1.3 billion
Judaism	There is only one God. According to believers, God loves and protects his people, but then holds them accountable for their sins.	Abraham	15.1 million

◄ A Chinese-Buddhist Enlightened Being

DIFFERENTIATING INSTRUCTION

Struggling Readers

Define Terms Used to Explain Religion

Ask students to review the information in the Comparing World Religions chart on this page. Discuss terms such as *enlightenment*, *resurrection*, *salvation*, and other terms that may be unfamiliar or challenging. Have students write definitions for these terms in their own words. Remind students to use the World Religions handbook at the back of their textbooks as a handy reference.

Gifted & Talented

Prepare a Display on the Five Major Religions

Using the information in the chart above and that found in the World Religions handbook, have students prepare a museum style display on the major religions of the world. The goal of the presentation is to help others better understand the religions.

Spread of Religion Over the centuries, religions have spread from their points of origin to the rest of the world. All of the world's major religions began in Asia and moved to other continents. At first, religious beliefs were carried to different places by followers of the religion or traders. Later, **missionaries**, people sent to do religious work in other lands, spread their faiths. For example, Christianity began in Southwest Asia and was spread throughout the world by missionaries. Still later, immigrants brought their religious beliefs with them as they moved to other countries.

In some lands, traditional religions have been practiced for as long as people have lived in a culture group. In areas where there has been little immigration, most of the people have the same religion. In countries where there has been much immigration, several religions might be practiced. This is the case in the United States, where all the world's major religions and many others are practiced. In fact, the United States has more religious groups than any other country in the world.

SUMMARIZE Explain the role religion has in culture. Religion is an important element of culture because it has such influence on people's lives.

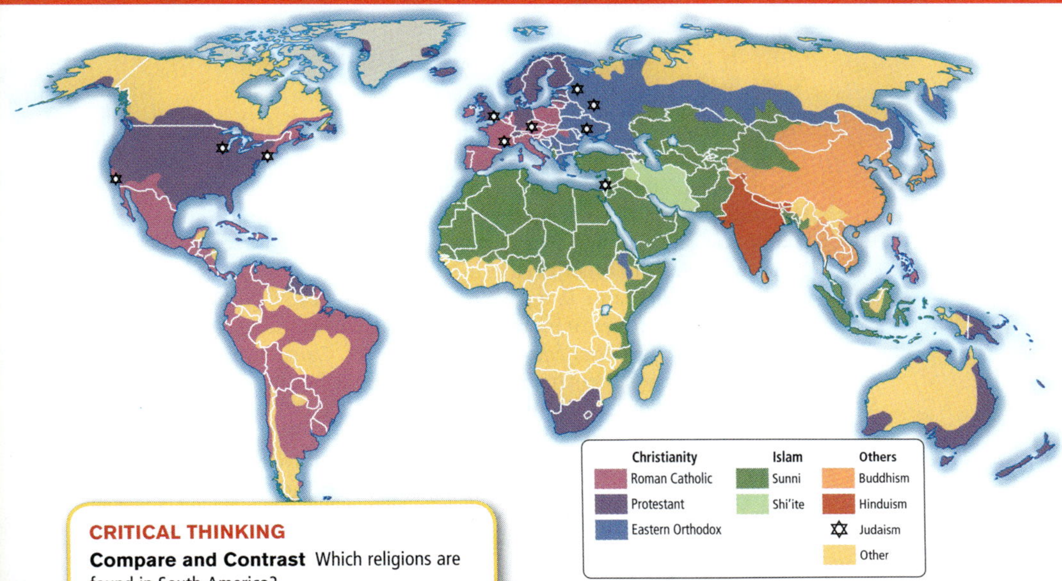

CRITICAL THINKING

Compare and Contrast Which religions are found in South America?

More About . . .

U.S. Immigration and Religion

Because of its constitutional guarantee of the separation of church and state, the U.S. government does not ask the religion of immigrants. Therefore, it cannot keep any statistics that detail the religion of immigrants. For the same reason, the U.S. Census also does not collect information on religion. Thus, statistics on religion must be compiled by private sources.

More About . . .

Religion in the United States

Current estimates count the percentages of U.S. citizens in particular religions as follows: Protestant—52%, Roman Catholic—24%, Mormon—2%, Jewish—1%, Muslim—1%, and other religions at about 10%. Also, 10% of the American people describe themselves as belonging to no religion.

COMPARING

World Religions

Point out the large areas on the map that indicate "other." Discuss with students why the religion in these areas is generally not one of the major ones. *(Possible Answers: These areas are remote, likely have few people, and would be difficult for missionaries to visit.)*

- What is the major religion in India? (*Hinduism*) In Mexico? (*Roman Catholic*)

CRITICAL THINKING ANSWER

Compare and Contrast Roman Catholic, Protestant, and Other

Unit Resource Book: Introduction to World Geography

- Skillbuilder Practice: Summarizing, p. 206

INTERDISCIPLINARY ACTIVITIES

CONNECT to Math

Graph Religions in the United States

Ask students to prepare a bar graph of the number of people of various religions in the United States. They should use the figures in the side column, More About, Religion in the United States.

Unit Resource Book: Introduction to World Geography

- Map and Graph Skills: Reading a Line Graph, p. 207

Teach

World Languages and Culture

Reader, Recorder, Reporter

- How do languages change? *(People borrow words from other languages and need new words for new objects or activities.)*
- How did Indo-European languages spread over such a wide area? *(Explorers and colonists brought them to all areas of the world.)*
- **Find Main Ideas** How can a common language help unify a culture? *(People can communicate easily and feel connected to each other, helping to create a cultural identity.)*

Unit Resource Book: Introduction to World Geography
- World Literature: from A Memoir by John Glenn, pp. 237–240

World Language Families Today

CONNECT Geography & Culture

Point out that sign language is not universal. Every language needs a distinct sign language, since words and forms of speech vary so much. For example, Swiss-German Sign Language is used in Switzerland while American Sign Language is used in the United States. Just as spoken languages encompass a variety of regional dialects, so do sign languages. Swiss-German Sign Language has five distinct dialects, while people in Great Britain can tell which part of the country someone is from by how they sign certain words.

- Which language families are most common in the far northern regions of the world? *(Amerindian, Ural-Asiatic)*

ANSWER

Place Indo-European, Amerindian

Unit Resource Book: Introduction to World Geography
- Interdisciplinary Projects: United Nations Postage Stamps, pp. 213, 215

World Language Families Today

CONNECT Geography & Culture

Place Which language families are found in North America?

World Languages and Culture

KEY QUESTION Why is language important?

Language is the way cultural values and traditions are passed from one generation to another. **Language** is human communication either written, spoken, or signed, such as American Sign Language. Your family and teachers use language to help you understand your world and how to live in it. Because language relates to all aspects of life, it helps a people to establish a cultural identity.

Sharing the same language is also important to a culture's sense of unity. Sometimes, if more than one language is spoken in an area, people don't feel connected to each other. In Canada, for example, both English and French are spoken. As a result, at times English- and French-speaking Canadians experience conflict.

Language Families Geographers believe that there are between 3,000 and 6,500 languages in the world today. India, for example, has 18 official languages, and more than 800 other languages are spoken there. Some of the world's languages are spoken by only a few

DIFFERENTIATING INSTRUCTION: TIERED ACTIVITIES

OBJECTIVE Plan a time capsule that displays local culture.

Basic

Have student groups devise a list of 10 objects to include in a time capsule to be opened 50 years from now. Each group should write a sentence for each one, explaining why they chose it. Discuss which objects express the more lasting parts of culture, rather than "fads."

On Level

In addition to the Basic activity, students should include a list of 10 local people they think would be most important to learn about 50 years from now. For each person, students should write a sentence, explaining his or her significance.

Challenge

In addition to the On Level activity, students should include a summary that gives an overview of their local culture.

thousand people. Other languages have millions of speakers. The language with the largest number of native speakers is Mandarin Chinese, spoken by an estimated 885 million people, mostly in China. English is the most widespread language in the world.

Scholars have arranged the world's languages into 11 main language families. A **language family** is a group of languages that have a common origin. English is in the Indo-European language family, the most widespread language family. An estimated one-half of the world's population speaks an Indo-European language.

Spread and Change of Language Geographers study how languages are distributed throughout the world as a way to learn more about cultures. Like religion and other elements of culture, language spreads in many ways. People bring their language and their culture with them when they move from place to place. Indo-European languages, such as English, Dutch, Portuguese, Spanish, and French, were carried to all parts of the world by European explorers and colonists.

Language not only spreads, it also changes. Language changes when people interact and borrow words from one another. Change also happens when people need new words to express new ideas or to represent new objects or activities, such as *weblog*, or *blog*.

FIND MAIN IDEAS Explain why language is important to culture. Because language relates to all aspects of life, it helps establish cultural identity.

Hello
Buon Giorno
Zdravstvui
Dzien Dobry
Dia Dhuit
Namasté
Gei Sou

Fun Facts!

INDO-EUROPEAN LANGUAGES

All of the above greetings come to you from the same language family—Indo-European. These words are variations of *hello* in English, Italian, Russian, Polish, Hindi, Greek, and Irish. These seven languages are a part of more than 400 in this language family.

Section 1 Assessment

ONLINE QUIZ For test practice, go to **Interactive Review @ClassZone.com**

TERMS & NAMES

1. **Explain the importance of**
   - culture
   - ethnic group
   - religion
   - language

USE YOUR READING NOTES

2. **Summarize** Use your completed chart to answer the following question:
   Why is language important in a culture?

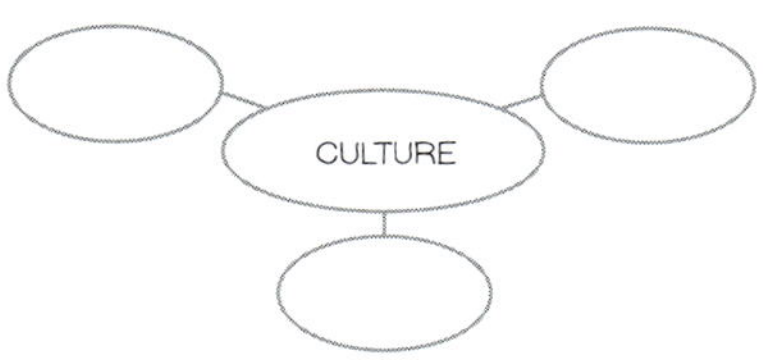

KEY IDEAS

3. What are the basic elements of culture?
4. What are the five major religions of the world?
5. How are the world's languages organized?

CRITICAL THINKING

6. **Draw Conclusions** Why is religion an important part of a group's culture?
7. **Sequence Events** How does culture pass from one generation to the next?
8. **CONNECT to Today** What are the most influential elements in the culture of the United States?
9. **WRITING** **Write a Brief Description** Choose a cultural group. Write a description that includes information on cultural elements, such as religion, language, government, technology, arts and crafts, and food and clothing.

CONNECT to the Essential Question

How does culture develop and how does it shape our lives?

Ask students what they have learned so far that can help them answer this question. Students might mention:

- Culture develops over many years and is made up of shared elements, such as language, religion, and food.
- Religions can guide people's beliefs and values and affect what choices they make.
- Language can help unify a culture and reinforce its unique aspects.

4 Assess & Reteach

Assess Have students complete the Section Assessment.

Unit Resource Book: Introduction to World Geography
- Section Quiz, p. 218

Interactive Review @ClassZone.com
- Section Quiz

Power Presentations

Test Generator

Reteach Have pairs of students create one multiple-choice question about each of the three "What Is Culture?" subsections. Then have each pair exchange their questions with another pair and answer each other's questions. For each incorrect answer, have students correct it, using their textbooks.

Unit Resource Book: Introduction to World Geography
- Reteaching Activity, p. 220

SECTION 1 ASSESSMENT ANSWERS

Terms & Names

1. culture: shared attitudes, knowledge, and behaviors of a group; ethnic group: a people who share a language, customs, and a common heritage; religion: an organized system of beliefs and practices, often centered on one or more gods; language: human speech, either written, spoken, or signed

Use Your Reading Notes

2. Language is one of the main vehicles for explaining aspects of a culture and teaching it to others. **Diagram** *way of life: all aspects of a society; religion: guides behavior; language: communicates culture*

Key Ideas

3. language, religion, foods and clothing, arts and crafts, technology, government
4. Hinduism, Buddhism, Judaism, Christianity, Islam
5. into 11 main language families

Critical Thinking

6. Possible answer: Religious beliefs guide the way a person acts.
7. Possible answer: An older generation teaches the younger generation.
8. Answers will vary but should include one or more of the basic elements of culture—language, religion, government, technology, arts and crafts, and food and clothing.
9. Use the rubric to score the brief descriptions.

Brief Description Rubric

	Cultural Elements	Paragraph Construction
4	very complete	clearly and correctly written
3	adequate coverage	clearly written some errors
2	missing several aspects	topic sentence, some fragments
1	missing many aspects	no topic sentence, fragments

1 Plan & Prepare

Objectives

- Describe the ways that cultures change and how cultures affect each other
- Summarize some responses to cultural change

Read for the Essential Question

Help students read for a purpose by reminding them of the Essential Question: "How does culture develop and how does it shape our lives?" Point out that they can use text features such as Key Questions, special features, images, and maps to form answers to the question.

Vocabulary

Best Practices Toolkit

Use the ***Best Practices Toolkit*** to model strategies for vocabulary notetaking. Vary strategies throughout the year. Choose from: *Knowledge Rating, Predicting ABCs, Definition Mapping, Word Sort, Frayer Model (Word Squares), Magnet Words, Word Wheel,* and *Student VOC.*

Vocabulary Strategies, TT9–TT16

Reading Strategy

Best Practices Toolkit

Display the Cause-and-Effect Diagram (Multiple Causes) transparency.

- Remind students that causes and effects are often signaled by words surrounding them. Have students brainstorm some signal words and phrases, such as *due to, as a result, since, because, so, which,* and *cause.*
- Model filling in causes in the "Innovation" box.
- Have students suggest entries for the "Diffusion" box.
- Display the completed diagram for review.

Cause-and-Effect Diagram (Multiple Causes), TT24

SECTION 2 Reading for Understanding

Key Ideas

BEFORE, YOU LEARNED

People belong to specific groups that share a common culture.

NOW YOU WILL LEARN

Cultures do not remain the same but change over time.

Vocabulary

TERMS & NAMES

agricultural revolution the shift from gathering food to raising food

innovation something new that is introduced for the first time

technology people's application of knowledge, tools, and inventions to meet their needs

diffusion the spread of ideas, inventions, and patterns of behavior from one group to another

cultural hearth an area where a culture originated and spread to other areas

BACKGROUND VOCABULARY

domestication the raising and tending of a plant or animal to be of use to humans

nomad a person who has no set home but moves from place to place in search of food for animals

Visual Vocabulary nomad group

Reading Strategy

Re-create the diagram at right. As you read and respond to the **KEY QUESTIONS**, use the diagram to analyze the causes and effects of cultural change.

Skillbuilder Handbook, page R8

ANALYZE CAUSE AND EFFECT

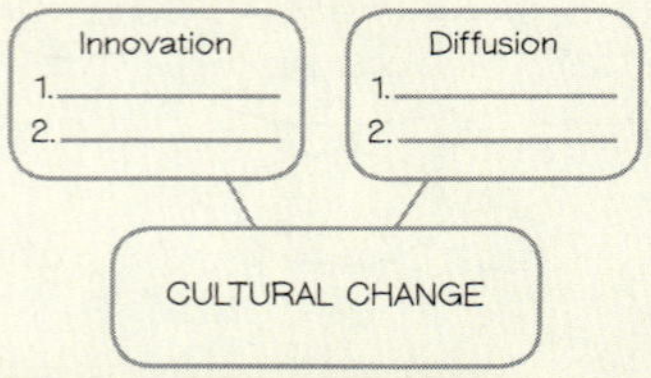

GRAPHIC ORGANIZERS
Go to **Interactive Review @ClassZone.com**

PRETEACHING VOCABULARY

English Learners

Pronounce and Preview

Pronounce each term for students. Review the meanings of words within definitions, such as *introduced, application,* and *inventions.*

- To modify vocabulary learning, have students complete worksheets as they read, instead of afterward.

Unit Resource Book: Introduction to World Geography

- Vocabulary Practice, p. 205

Inclusion

Reading and Defining Terms Verbally

Have volunteers read one term and its definition aloud. Ask others to define the term in their own words verbally, and then use it in a sentence. Repeat for each term. As a group, discuss any words that are challenging to define or difficult to use in a sentence.

SECTION 2 CULTURE

Bush plane, Alaska

Hindu celebration, India

How Does Culture Change?

Connecting to Your World

Think about all the information and communication tools you have in your daily life. Did you include cell phones, mp3 players, DVDs, digital cameras, or laptop computers? When your parents were your age, none of these devices were available to them. These improvements in technology have changed the ways many people live their lives—and this change has taken place in only one generation. Cultural change happens much more rapidly now than it did in the past.

Zulu Girl, South Africa, with Cell Phone

Culture Change and Exchange

KEY QUESTION How does culture change?

Culture changes over time. The changes may be very slow or quite rapid. Some changes are simple, such as changes in clothing styles. Other changes are more complex, such as the agricultural revolution. In the **agricultural revolution**, which happened thousands of years ago, humans shifted from gathering food to raising food. Because the food source was more reliable, fewer people were needed for raising food, and ways of life changed. Two ways that culture changes are through innovation and diffusion.

Camels Go to Pasture Traditional ways are replaced by modern ones.

97

2 Focus & Motivate

3-Minute Warm-Up

Write on the board or display the transparency:

- View the photographs of the Zulu girl, camels, and Amish transportation in this section. How are these examples of cultural changes? *(They combine the traditional with the modern.)*

Unit Transparency Book: Introduction
- 3-Minute Warm-Ups, TT27

Connecting to Your World

More About . . .

Music Technology Changes

In the 1980s, portable "boom boxes"—some weighing more than 20 pounds—let people listen to cassette tapes on the go. Today's portable music players weigh less than a pound and store thousands of songs.

3 Teach

Culture Change and Exchange

Think, Pair, Share

- What is an example of an innovation? *(Possible Answers: computers, domestication of plants, watering crops, storing food)*
- How did traders affect cultural diffusion? *(They brought new ideas along trade routes.)*
- **Analyze Cause and Effect** How have new types of mass communication encouraged cultural diffusion? *(Possible Answer: They allow contact between cultures, which may have been prevented by geographic barriers.)*

SECTION 2 PROGRAM RESOURCES

Unit Resource Book: Introduction to World Geography

ON LEVEL

Unit Resource Book
- Reading Study Guide, p. 191
- Outline Map Activities, p. 209
- Section Quiz, p. 219

STRUGGLING READERS

Unit Resource Book
- Reading Study Guide with Additional Support, p. 195
- Skillbuilder Practice: Summarizing, p. 206
- Section Quiz, p. 219
- Reteaching Activity, p. 221

eEdition with Audio DVD-ROM

ENGLISH LEARNERS

Pupil Edition in Spanish

eEdition with Audio DVD-ROM

Unit Resource Book
- RSG (Spanish), p. 199
- RSG with Additional Support (Spanish), p. 203

Multi-Language Glossary

Test Generator
- Section Quiz in Spanish

INCLUSION

Unit Resource Book
- RSG with Additional Support, p. 195
- Section Quiz, p. 219
- Reteaching Activity, p. 221

GIFTED & TALENTED

Unit Resource Book
- Interdisciplinary Projects, p. 214

PRE-AP

Unit Resource Book
- Connect Geography and Culture, p. 216

TECHNOLOGY

Unit Transparency Book
- 3-Minute Warm-Ups, TT27
- Essential Question, TT28
- Critical Thinking, TT30
- Map: Human Migration, TT31

Daily Test Practice Transparencies
- Section 2, TT13

Power Presentations

ClassZone.com

World Cultures and Geography Video on DVD

COMPARING

Storage Unit Innovations

Point out to students that the resources of the environment helped determine which types of storage units were created. Lifestyle was also a determining factor. Settled groups did not need storage units that were readily transportable.

- Which type of container would be the best for storing liquid foods such as olive oil or honey? *(A clay pot could contain liquids while the basket would let them run out. However, leather bags were used to store such liquids as water, wine, or milk.)*

CRITICAL THINKING ANSWERS

1. **Clay Pot** keeps out moisture, is not easily damaged
2. **Basket** Raw materials—grass or reeds—are easy to acquire and work with.
3. **Leather Bag** couldn't carry a lot, since they were relatively small for ease of transporting

Teacher-Tested Activities

Ann Christianson, John Muir Middle School, Wausau, Wisconsin

To help students understand that diffusion takes place even in their hometown, I have them graph restaurants. Using the phonebook of your community make a tally list of the culture represented by the food served. Next create a circle graph from the collected data. Then ask, What observations can be made about the cultures in your community? Ask, Would this information look different if it were collected in 1950? What about 1900? If possible try to get copies of the restaurant pages in old phonebooks for comparisons.

Innovation Something new that is introduced for the first time is called an **innovation**. New ideas, inventions, and patterns of behavior are types of innovations that change a culture. The computer is an example of an invention that changed cultures in the United States and around the world. Innovation may take place by accident, or it may be deliberate. The prehistoric control of fire was probably an accident, but it forever changed the way people lived. However, the use of existing resources and technology to solve an old or a new problem is a deliberate innovation. **Technology** refers to people's application of knowledge, tools, and inventions to meet their needs.

Some innovations can dramatically change the way people live. The domestication of wild plants thousands of years ago was such a change. **Domestication** means to raise or to tend plants or animals to be of use to humans. Dependence on agriculture resulted in more densely populated settlements.

Like most cultural changes, the agricultural revolution led to other changes. These settled societies needed to be organized differently from groups of **nomads**, who had no set home but moved from place to place to find food for their animals. This led to more innovations. For example, people needed to find ways to water crops in the field and to store the food once harvested. Look at the pictures below to see how different cultures used the resources and technology available to them to solve a storage problem. New or different tools were also needed to farm the land.

COMPARING Storage Unit Innovations

Clay Pot Where clay was plentiful, clay pots served as storage units. **1. What would be the advantage of a clay pot for storage?**

Basket Woven grass or reeds created light, portable storage units. **2. Why would this material be used for storage?**

Leather Bag Hides made into bags made storage units easy to transport. **3. What would be a disadvantage of this type of storage?**

98

DIFFERENTIATING INSTRUCTION

English Learners

Describe an Invention That Changed Lives

Ask students to review the information in the first paragraph on this page. Then discuss how innovations and technology can change people's lives. For example, how did the automobile change the way people traveled? How did electricity change people's daily lives? To conclude this activity, have students write a brief description of how one innovation (tool or invention) has changed their lives.

Struggling Readers

Chart Changes in Technology

After reviewing information on this page, ask students to prepare a chart that lists several changes in technology in various areas, such as transportation, medicine, communications, printing, heating and cooling homes, and cooking and storing food. Invite volunteers to share one item on their list and explain its cultural significance.

The Spread of Agriculture

The spread of agriculture is the most significant change in human history. Having available food year-round allowed people to settle and eventually to develop specialized labor and cities.

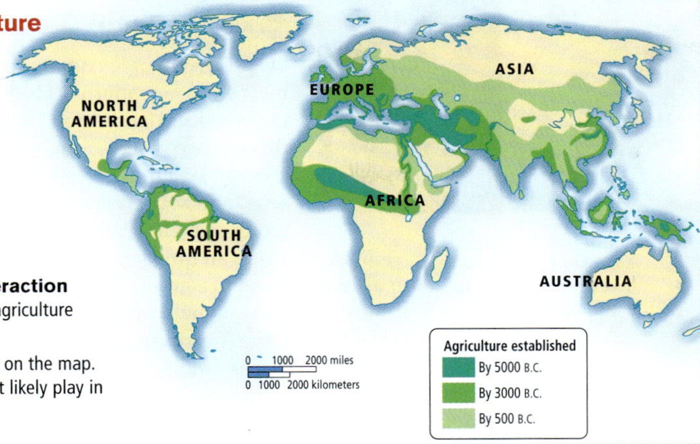

CRITICAL THINKING

1. **Human-Environment Interaction** Which continent did not have agriculture established by 500 B.C?
2. **Region** Look for river valleys on the map. What role did river valleys most likely play in the spread of agriculture?

Diffusion The spread of ideas, inventions, and patterns of behavior from one group to another is called **diffusion**. Whenever a group of people comes in contact with another group, diffusion is possible.

The spread of agriculture is an early example of diffusion. The spread of U.S. fast-food restaurants around the world is a modern example. In early times, traders often brought new ideas and inventions to other cultures. Written language, the use of coins, and religious beliefs all moved along ancient trade routes. Missionaries and invaders also carried cultural elements with them.

Geographers study diffusion to see patterns in the development of cultures across the Earth's surface. One pattern they have observed is diffusion from cultural hearths. A **cultural hearth** is an area with an advanced culture from which ideas or technology spread. The map above shows where early cultural hearths existed.

In the past, the spread of culture was usually slow because of geographic barriers. Large bodies of water, mountains, or deserts often made it difficult for people to interact with others. Sometimes political boundaries limited contact between peoples. In today's world, it's almost impossible to avoid some kind of interaction with other groups of people. Satellite television, the Internet, and other forms of mass communication speed new ideas, practices, and inventions around the globe.

FIND MAIN IDEAS Identify the ways that culture changes.
Culture changes as a result of innovation or diffusion.

DIFFERENTIATING INSTRUCTION: TIERED ACTIVITIES

OBJECTIVE Present a news broadcast on a cultural change.

Basic	On Level	Challenge
Brainstorm with students some important cultural changes in history, such as new forms of transportation or communication. Then ask groups of students to select one change and prepare a short newscast that reports on the change and how it changed people's lives.	Adapt the Basic activity, so that students are reporting on a cultural change that is taking place today.	Adapt the Basic activity, so that students are reporting on a cultural change that they predict will happen five years from today.

CONNECT Geography & Culture

The Spread of Agriculture

The spread of agriculture led to improved farming techniques, which in turn allowed people to pursue other types of activities. Today, less than 1.8% of people in the United States are employed in agriculture, compared to about 42% worldwide.

- Which continent had established agriculture in most of its land area by 3000 B.C.? *(Europe)*

CRITICAL THINKING ANSWERS

1. **Human-Environment Interaction** Australia
2. **Region** Some of the earliest agriculture was located in river valleys. Agriculture then spread from these areas to other areas.

Unit Resource Book: Introduction to World Geography
- Outline Map Activities: Early Cultural Centers, pp. 209–210
- Interdisciplinary Projects: The Domestication of Animals, pp. 214–215
- Connect Geography & Culture: The Spread of Buddhism, p. 216

More About . . .

Inuit and Technology

The lives of the isolated Inuit of northern Canada and Alaska have changed in recent years because of modern technology. Access to computers now makes communication among the Inuit themselves—and with outside cultures—easy and routine. The Inuit are using software that makes it possible for Web pages and other communications to be published in Inuktitut, the native language of the Inuit. Some Inuit, however, worry that the easy access to modern cultures will weaken the traditional culture.

Unit Resource Book: Introduction to World Geography
- Geography Project: Comparing Natural Environments, p. 241

Teach

Accepting Cultural Change

Talk About It

- **Analyze Cause and Effect** What are some effects of not having electricity in homes? *(no, TVs, computers, microwaves, or electric lights)*

CONNECT to the Essential Question

How does culture develop and how does it shape our lives?

Ask students what they have learned so far that can help them answer this question. Students might mention:

- Cultural innovations can improve people's lives.
- Cultural change can be rapid or slow and may or may not be accepted.

4 Assess & Reteach

Assess Have students complete the Section Assessment.

Unit Resource Book: Introduction
- Section Quiz, p. 219

Interactive Review @ClassZone.com
- Section Quiz

Power Presentations

Test Generator

Reteach Have students summarize the effects of innovation and cultural diffusion on culture.

Unit Resource Book: Introduction
- Reteaching Activity, p. 221

Amish Transportation Amish people choose not to drive automobiles, but use horse-drawn vehicles.

Accepting Cultural Change

KEY QUESTION Is cultural change always accepted?

Over time, people come in contact with different ideas, inventions, or patterns of behavior. If a cultural exchange takes place, the culture begins to change. Sometimes this change is slow, and people just become used to the change. When the effects of cultural change— such as a new food source or a tool—are positive, the lives of the group may improve.

But sometimes an innovation is unacceptable, such as use of lands or animals sacred to a group. Sometimes a group may need to decide if the change would help or harm their society. For example, in the United States, the Amish choose not to drive cars, not to have electricity in their homes, and to send their children to private, one-room schoolhouses. They do this because they reject the impact of modern life on their way of life.

Change may sometimes be forced on a group. This often happens when a region is invaded. For example, Spanish conquerors pressed their culture on the native peoples of the Americas in the 1500s.

Cultural change may be accepted or rejected, depending upon whether it has positive or negative effects. ▶

MAKE GENERALIZATIONS Explain why some cultural changes might be rejected.

Section 2 Assessment

ONLINE QUIZ For test practice, go to **Interactive Review @ClassZone.com**

TERMS & NAMES

1. **Explain the importance of**
   - innovation
   - technology
   - diffusion
   - cultural hearth

USE YOUR READING NOTES

2. **Analyze Cause and Effect** Use your completed web diagram to answer the following question:
   What role do cultural hearths play in changing cultures?

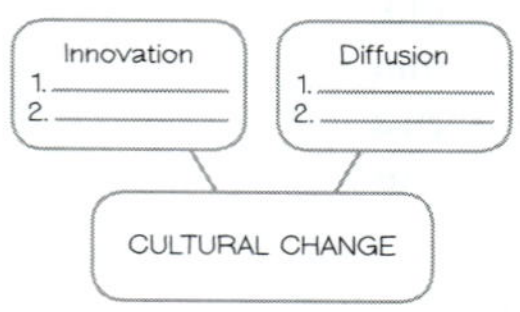

KEY IDEAS

3. What are two ways that bring about cultural change?
4. How does deliberate innovation take place?
5. What three groups helped to spread culture in earlier times?

CRITICAL THINKING

6. **Make Inferences** What are some reasons why a group may accept cultural change?
7. **Draw Conclusions** How has mass communication changed the way culture spreads?
8. **CONNECT to Today** How has the United States been affected by cultural exchange in recent times?
9. **TECHNOLOGY Make a Multimedia Presentation** Plan a power presentation slide show illustrating three inventions in transportation and three in communication that brought cultural change. Each slide should have a visual of the invention and a description of the change.

SECTION 2 ASSESSMENT ANSWERS

Terms & Names

1. innovation: something new that is introduced for the first time; technology: people's application of knowledge, tools, and inventions to meet their needs; diffusion: the spread of ideas, inventions, and patterns of behavior from one group to another; cultural hearth: an area where a culture originated and spread to other areas

Use Your Reading Notes

2. Cultural hearths are the locations where innovation or diffusion may take place.
   Diagram *Innovation: new ideas, inventions or patterns of behavior; Diffusion: spread of new ideas, inventions, or patterns of behavior*

Key Ideas

3. innovation and diffusion
4. People use the resources and technology that they have to solve new or old problems.
5. traders, missionaries, invaders

Critical Thinking

6. Possible answer: Cultural change may bring a better food, a new religion, or better technology.
7. Possible answer: Mass communication has overcome geographic barriers and moves rapidly to all parts of the globe.
8. Answers will vary but might give examples of foods and beliefs brought to the United States by recent immigrant groups and of elements of American culture that have spread worldwide.
9. Use the rubric to score multimedia presentations.

Presentation Rubric

	Text	Illustration
4	clear and informative on all six inventions	very helpful illustrations
3	five inventions included, adequate information	adequate illustration
2	3–5 inventions included, limited information	poor illustration
1	fewer than three inventions included, missing information	unclear or missing

Interactive Review

Click here to complete these and other activities online @ClassZone.com

CHAPTER SUMMARY

Key Idea 1
Human beings are members of social groups that have shared and unique behaviors and attitudes.

Key Idea 2
Cultures do not remain the same but change over time.

For Review and Study Notes, go to Interactive Review @ClassZone.com

NAME GAME

Use the Terms & Names list to complete each sentence on paper or online.

1. I am the spread of ideas, inventions, and patterns of behavior from one group to another. ___diffusion___
2. I am a scientist who studies culture. ________
3. I am a people that shares a language, customs, and a common heritage. ________
4. I am the methods, materials, or tools available to complete a task. ________
5. I am the total way of life held in common by a specific group of people. ________
6. I am something new that is introduced for the first time. ________
7. I am a group of languages that have a common origin. ________
8. I am an area where the transfer of elements of culture between two groups occurs. ________
9. I am human communication, either written, spoken, or signed. ________
10. I am an organized system of beliefs in a god or gods and a set of practices. ________

agricultural revolution
anthropologist
culture
cultural hearth
diffusion
domestication
ethnic group
innovation
language
language family
missionary
nomad
religion
technology

Activities

Flip Cards

Use the online flip cards to quiz yourself on terms and names introduced in this chapter.

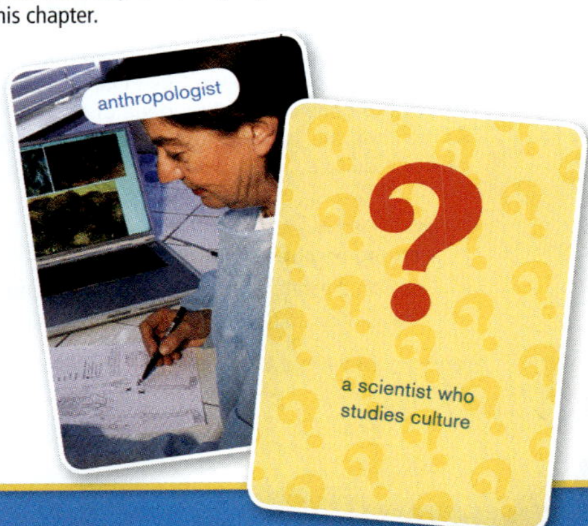

Crossword Puzzle

Complete an online crossword puzzle to test your knowledge of culture.

ACROSS
1. human communication, either written, spoken, or signed

101

ONLINE AND ON-SCREEN ACTIVITIES

Flip Cards

Vocabulary Flip Cards are virtual flash cards at **ClassZone.com.** Students can quiz each other or work independently to study the flip cards for the chapter, "People and Culture."

- They can quiz themselves on key terms and names by reviewing definitions.
- Demonstrate how to advance and shuffle the deck for another challenge.

Crossword Puzzle

Students can review important elements of culture and its change by completing this crossword puzzle online at **ClassZone.com.**

- As an extension of this activity, have students use the *Activity Maker* at **ClassZone.com** to make their own crossword puzzle for one section of the chapter.
- Then have students trade puzzles with classmates to solve each other's puzzles.

Chapter Summary

As a follow-up to the chapter, encourage students to pair up and go over the **Review and Study Notes** at **ClassZone.com.**

Alternatively, students pairs can review their Reading Study Guide notes, making sure that they can add important details to each section's "Key Idea."

Name Game

Teaching Tips

ON PAPER Have students work in pairs to complete this activity. As an extension, they can write questions for the terms that are not covered in the items shown here.

ONLINE Make sure that students understand how to drag and drop terms to fill in blanks. Ask them to print out their results and compare them with those of a classmate.

ANSWERS

1. diffusion
2. anthropologist
3. ethnic group
4. technology
5. culture
6. innovation
7. language family
8. cultural hearth
9. language
10. religion

Make It Fun

Teacher-Tested Activities

Ann Christianson, John Muir Middle School, Wausau, Wisconsin

To have students demonstrate their understanding of elements of culture, I have them create special postcards that include cultural traits and images. Either assign or have students select a country from which the postcard will come. Students will create a picture postcard representing some aspect of culture in the land. On the reverse side of the postcard students write a paragraph describing several cultural elements they encountered while staying in the country.

Vocabulary ANSWERS

1. Culture is the shared attitudes, a knowledge and behaviors of a group.
2. An ethnic groups is a people that share a language, customs, and a common heritage.
3. Religion is an organized system of beliefs and practices, often centered on one or more gods.
4. Language is human communication, either written, spoken, or signed.
5. Technology is the application of knowledge, tools, and inventions to meet needs.
6. Language and religion are two elements of culture.
7. Innovation and diffusion are ways that culture changes and they originate from cultural hearths.
8. They are two significant changes in human culture.

Key Ideas ANSWERS

9. because people in different parts of the world meet their basic needs in different ways
10. They observe their family, friends, and others, and they also learn from direct teaching.
11. religion and language
12. belief in one god, belief in more than one god, and belief in divine forces in nature
13. Language is important because it is the way people communicate ideas to one another.
14. Language changes when people interact or borrow words from each other.
15. by introducing something new, like an idea or invention, into a culture for the first time
16. by spreading ideas, inventions, or patterns of behavior from one culture to another
17. a place where elements of culture originate and then spread to other areas
18. There often were geographic barriers that prevented contact between cultures.
19. because the change might be something that is helpful
20. Usually forced cultural change comes when a people is conquered by another group.

CHAPTER 4 Assessment

VOCABULARY

Explain the significance of each of the following.

1. culture
2. ethnic group
3. religion
4. language
5. technology

Explain how the terms and names in each group are related.

6. culture, religion, and language
7. innovation, diffusion, and cultural hearth
8. domestication and agricultural revolution

KEY IDEAS

1 What Is Culture?

9. Why are there different cultures?
10. How do people learn about their culture?
11. What are two of the most important elements of culture?
12. What are the three types of religious beliefs?
13. Why is language important to a culture?
14. How does language change?

2 How Does Culture Change?

15. How does innovation change culture?
16. How does diffusion change culture?
17. What is a cultural hearth?
18. Why was culture change a slow process in earlier times?
19. Why might a people accept a cultural change?
20. In what way is cultural change forced upon a people?

102

CRITICAL THINKING

21. **Categorize** Create a table to list the factors that have helped or limited cultural change over the years.

HELP	LIMIT

22. **Make Generalizations** What makes each culture unique?
23. **Make Inferences** Why are families important in a culture?
24. **Connect to Economics** Why was the agricultural revolution such a dramatic cultural change?
25. **Five Themes: Movement** How have aspects of culture been spread throughout history?
26. **Connect Geography & Culture** How important is culture to a person's sense of identity?

Answer the ESSENTIAL QUESTION

How does culture develop, and how does it shape our lives?

Written Response Write a two- or three-paragraph response to the Essential Question. Be sure to consider the key ideas of each section and the fact that all human beings are a part of a culture.

Response Rubric
A strong response will:
- explain the nature of culture and its elements
- discuss how culture shapes the lives of individuals

ESSENTIAL QUESTION REVIEW

Before students begin their essays, have them review the key chapter elements that will help them answer the Essential Question. Direct them to significant features in the chapter, such as:

- Worldster: Culture and Your World, Section 1
- Comparing World Religions, Section 1
- World Language Families Today, Section 1
- Comparing Storage Unit Innovations, Section 2
- Connect Geography & Culture: The Spread of Agriculture, Section 2

Written Response Rubric

	Culture Elements	Shaping Effects
4	Many elements discussed	Clear identification of many effects
3	4–5 elements discussed	4–5 effects identified
2	2–3 elements discussed	2–3 effects ideinitfied
1	One element discussed	Little or no understanding of effects

STANDARDS-BASED ASSESSMENT

Test Practice
- Online Test Practice @ClassZone.com
- Test-Taking Strategies and Practice at the front of this book

THEMATIC MAP

Use the map and your knowledge of geography to answer questions 1 and 2 on your paper.

South Asian Religions

1. Which of the following countries has a large Buddhist population?

A. India
B. Sri Lanka
C. Bangladesh
D. Nepal

2. Which of the following countries has a large population of Muslims?

A. Bhutan
B. Sri Lanka
C. Bangladesh
D. India

CHART

Study the chart below. Use the information in the chart to answer questions 3 and 4.

Languages of Europe

Country	Languages Spoken
Austria	German, Slovene, Croatian, Hungarian
Belgium	Flemish, French, German
France	French
Germany	German
Netherlands	Dutch, Frisian
Switzerland	German, French, Italian, Romansch

Source: *Infoplease database 2006*

3. In which countries is a single language spoken?

4. If you only spoke German, in which countries would you need a translator?

GeoActivity

1. INTERDISCIPLINARY ACTIVITY–MATHEMATICS

With a small group, research world religions on the Internet. Use a computer program to create a database and a graph showing the three countries with the largest number of members for each of the major religions, and the total population of those countries.

2. WRITING FOR SOCIAL STUDIES

Reread the paragraphs on the elements of culture. Then write a paragraph describing the culture of your community. Discuss such elements as religion, language, government, and economic activities.

3. MENTAL MAPPING

Create a map of your town or neighborhood showing schools and places of worship. Be sure to include
- elementary, middle, and high schools, public and private
- any places of worship

DIFFERENTIATING CHAPTER ASSESSMENT

Unit Resource Book:
- **Chapter Test A,** pp. 222–226. *Use for Struggling Readers, Beginning-to-Intermediate-Level English Learners, and for Inclusion of students whose IEP suggests supported test-taking.*
- **Chapter Test B,** pp. 227–231. *Use for On-Level Learners and ESOL students with Advanced-Level English.*
- **Chapter Test C,** pp. 232–236. *Use for Gifted and Talented students or students assigned Pre-AP work.*

Test Generator
- Chapter Tests A, B, and C, in English and Spanish

McDougal Littell Assessment System

TEST **Test Generator** Use three levels of tests in English and Spanish tied to state standards. Plus, choose questions to customize tests for all populations.

SCORE Scan and Score plain paper student tests in minutes.

REPORT Create instant progress reports, sortable by student, by standard, or other criteria.

RETEACH Respond immediately to students' needs by using a wealth of reteaching resources.

Critical Thinking ANSWERS

21. Possible answer: Help: trade, immigration, mass communication; Limit: geographic barriers, political boundaries, rejection

22. Each culture is unique because it has common beliefs, values, and patterns of behavior that developed differently from other cultures.

23. Families are the basic unit of a culture's social structure and are the first place where children learn their culture.

24. because people not only altered the way that they obtained their food but also where and how they lived, which led to more innovations

25. In the past aspects of culture have been spread by traders, missionaries, and invaders. Today forms of mass communication are major movers of cultural diffusion.

26. A culture gives people their beliefs and values and tells them how they should act toward members of the group and to those who are outside the group.

Standards-Based Assessment ANSWERS

1. B
2. C
3. France and Germany
4. France and the Netherlands

GeoActivity

1. INTERDISCIPLINARY ACTIVITY RUBRIC

The database should:
- reflect research conducted
- include appropriate and accurate data
- be organized in a clear, readable fashion

2. WRITING FOR SOCIAL STUDIES RUBRIC

The paragraph should:
- clearly identify and describe elements of the student's community
- be executed in complete and correct English mechanics

3. MENTAL MAPPING RUBRIC

The map should:
- show an approximate neighborhood in the student's community
- correctly locate and label schools and places of worship

N
E
W
S

UNIT 2

The United States and Canada

Why It Matters:

The United States and Canada are two of the largest countries in the world. Because both nations have vast land areas, abundant resources, and well-educated populations, they have become two of the wealthiest countries in the world.

Statue of Liberty in New York

CHAPTER 5 The United States

Chateau Frontenac in Quebec

CHAPTER 6 Canada

Meet Kirima

In this unit, you will meet Kirima, a middle-school girl from Nunavut, Canada. Learn more about her on **Worldster** @ClassZone.com

Objectives

- Compare the differences in population density between the United States and Canada
- Explain the varied climates found in the United States and Canada
- Identify the major economic activities in the United States and Canada
- Name the 50 U.S. states and their capitals
- Name Canada's 10 provinces and 3 territories, and their capitals

Comparing Data

Although the landmass of Canada is about 20% more than that of the United States, Canada's population is only about 10% that of the United States. About 80% of the people in both countries live in urban areas. However, because of Canada's vast landmass and low total population, its average population density is only about nine people per square mile. In contrast, the overall population density of the United States is about 81 people per square mile.

ANSWERS

1. Canada is almost 700,000 square miles larger than the continental United States.
2. The U.S. population is almost ten times bigger that that of Canada's.
3. the Mississippi and Missouri rivers
4. Mt. McKinley is 796 feet higher than Mt. Logan.

UNIT 2 ATLAS

The United States and Canada

The United States and Canada occupy the central and northern four-fifths of the continent of North America. The two countries are bound together not only by physical geography and cultural heritage but also by strong economic and political ties. As you study the maps, notice geographic patterns and specific details about the region. Answer the questions on each map in your notebook.

As you study the graphs on this page, compare the landmass, population, rivers, and mountains of Canada with those of the United States (and in a few cases other regions). Then jot down the answers to the following questions in your notebook.

Comparing Data

1. How does Canada compare in size to the United States?
2. Approximately how many times larger is the population of the United States than that of Canada?
3. Which U.S. rivers are more than twice as long as the Mackenzie River in Canada?
4. What is the difference in height between the tallest peak in the United States and the tallest peak in Canada?

Comparing Data

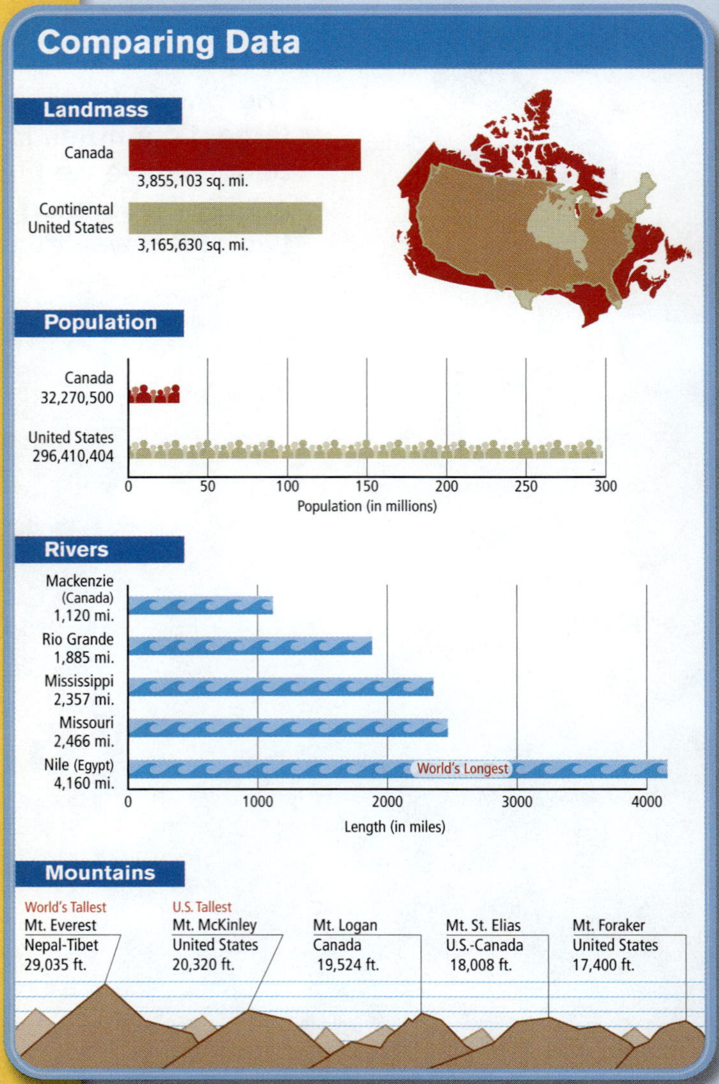

FIVE THEMES AND SIX ELEMENTS

Five Themes

Location

Most students may think of Canada as a "cold" country because of its northern location, but may not realize that the United States also has a portion of its territory located above the Arctic Circle—in Alaska. You may also want to point out that the United States and Canada share one of the world's longest borders. Both also border the Atlantic and Pacific oceans and the Great Lakes.

Six Elements

The World in Spatial Terms

Explain that the maps and data found in this Unit Atlas provide a lot of insight into each country. For example, by comparing the population density map and the climate map, one can see that much of each country's population lies in the humid continental climate zone that the United States and Canada share.

Population Density

THINK LIKE A GEOGRAPHER

1. **Human-Environment Interaction** Where are the areas of greatest population density?
2. **Region** Which color represents the most widespread population density level in the United States?

107

DIFFERENTIATING INSTRUCTION

English Learners

Comprehension: Self-Monitor

To help with comprehension of "persons per square miles" in the map legend, have students replace *per* with "for each" or "in each." Explain that *per* is a preposition—it helps explain the relationship between *persons* and *square mile.* Have students name other prepositions. *(at, by, in, to, from)* Then have them use the map to make a sentence with "persons per square mile" or another preposition.

Inclusion

Graphing Population Density

Students who are visually impaired may have difficulty with maps such as the one on this page. To help these students more easily compare the differences among states or regions, pair them with other students and have them create bar graphs or pie graphs from population data available in encyclopedias, almanacs, or government Web sites, such as www.census.gov.

Population Density

Point out to students that a population map can provide more information than just a look at persons per square mile. When combined with a knowledge of history, culture, and geography, it can help people better understand the settlement and expansion of a country or region. In this case, the population density map can help students connect the densely populated eastern United States with the early founding of the country by European colonists and subsequent waves of European immigration.

- True or False: Canada's most densely populated area is along Hudson Bay. *(False; along the U.S.–Canada border)*
- True or False: The most densely populated part of the United States lies west of the Mississippi River. *(False; east of the Mississippi River)*
- What factor might account for Los Angeles, New York, and Chicago being so densely populated? *(Possible Answer: All are port cities.)*

THINK LIKE A GEOGRAPHER ANSWERS

1. **Human Environment Interaction** In the United States, the greatest population density is in the eastern half of the country and urban areas in the western half. In Canada, the greatest population density is along the U.S.–Canada border.
2. **Region** pale yellow (1–24 persons per square mile)

Comparing Regions

Immigration in the United States and Canada

The history of both countries is a story of migration, from the earliest settlers to today. Recent immigration statistics from Canada show that many of its immigrants come from Asian countries such as India, the Philippines, and China. U.S. immigration statistics show that many immigrants come to the United States from Mexico, India, and China.

Of particular note is that 2.8% of U.S. immigrants come from Canada—and that 2.4% of Canadian immigrants come from the United States.

Climate

Point out the diversity of the climate zones in the United States. Invite students to discuss how this might be an economic advantage. *(Possible Answer: It might allow for more innovations, especially in technology, that could benefit the entire country.)*

- What types of climate does Alaska have? *(tundra, subarctic, and marine west coast)*
- What type of climate does the Great Lakes region have? *(humid continental and subarctic)*
- Why might Florida and Hawaii have similar climates? *(Both are close to the Tropic of Cancer and surrounded, or nearly surrounded in Florida's case, by the sea.)*

THINK LIKE A GEOGRAPHER ANSWERS

1. **Region** The humid continental climate is more suitable than the subarctic climate to the north.
2. **Place** subarctic

More About . . .

Subarctic Climates

This type of climate zone is unique in the following ways:

- It is found only in the Northern Hemisphere, between 50° to 70° latitude. This climate zone stretches across North America (from Newfoundland to Alaska) and across Europe and Asia (from Scandinavia to Siberia).
- It has the widest annual temperature range of any of Earth's climate zones, from -70°C (-94°F) to more than 40°C (104°F).
- Summers are short and cool (although the days are long). Winters are bitterly cold (with days that are generally very short and clear). Overall precipitation is low, especially in the winter.
- Frost can happen at any time of the year. On average, only about three months of the year are frost-free.

THINK LIKE A GEOGRAPHER

1. **Region** Why might most of the population of Canada be clustered along the border with the United States?
2. **Place** What climate is found throughout most of Canada?

FIVE THEMES AND SIX ELEMENTS

Five Themes

Regions

Point out that many population centers in the United States and Canada are located on or near large bodies of water, such as the Great Lakes and the Atlantic and Pacific oceans. Discuss how cities such as Toronto and Chicago are shaped by the Great Lakes; Vancouver and Los Angeles by the Pacific Ocean; Halifax and Miami by the Atlantic Ocean.

Six Elements

The Uses of Geography

Remind students that climate refers to long-term patterns in weather, or the typical weather of a place or region. Scientists use sophisticated technologies to collect data from the atmosphere, land, and oceans and powerful computers to make sense of the data. By studying this data over long periods of time, scientists hope to determine whether the Earth's climate is changing in a significant way—or whether the effects are short-term.

Economic Activity

THINK LIKE A GEOGRAPHER

1. **Human-Environment Interaction** What is the main land use in Canada?
2. **Location** Near what bodies of water might American or Canadian fishermen be likely to live?

109

DIFFERENTIATING INSTRUCTION

Struggling Readers

Compare Economies

To help students synthesize the information about the economic activity of the United States and Canada, review each of the land use terms with students. Have students point out the similarities and differences between the two countries. For example, ask them to explain why much of the land use in the United States is commercial agriculture while in Canada it's forestland. *(Canada's forestland reflects its colder, subarctic climate, which isn't suitable for agriculture.)*

Economic Activity

Point out to students the many oil resources in the United States. Yet the United States imports much of its oil. Discuss with students how the United States might make better use of its energy resources. *(Possible Answers: increase energy efficiency; drive more fuel-efficient vehicles; reduce energy consumption at home; make better use of renewable energy sources)*

- What resources are located along the Arctic Ocean coast of Alaska? *(natural gas and oil)*
- Where are most of the U.S timber resources located? *(in the Pacific Northwest)*
- What major energy resource is found in the eastern United States? *(coal)*

THINK LIKE A GEOGRAPHER ANSWERS

1. **Human-Environment Interaction** The main land use in Canada is forestland.
2. **Location** Both: Atlantic, Pacific, and Arctic oceans, Great Lakes; American only: Gulf of Mexico; Canadian only: Hudson Bay

Comparing Regions

U.S.–Canada Fishing Agreements

Fishers from the United States and Canada share many fishing grounds in the Atlantic and Pacific oceans. Although competing with each other, both countries also want to prevent over fishing—which would damage the oceans' ecosystems and hurt each country's economy. Because of these common goals, the United States and Canada have worked together to reach many agreements on fishing in both oceans.

One such recent agreement is the Pacific Hake/Whiting Agreement, ratified by the U.S. Senate in 2005. This agreement sets limits on how many Pacific whiting (a type of fish called Pacific hake in Canada) each country's fishers can catch.

Regional Overview

For much of their histories, both Canada and the United States have welcomed immigrants. Today, both countries are discussing their immigration policies, particularly how to better regulate immigration due to homeland security issues, such as terrorism.

- What are some reasons why the United States and Canada hold great appeal for immigrants? *(Possible Answers: richness of resources; diverse populations; personal freedoms; economic opportunities)*
- What type of economy does the United States and Canada both have? *(a market, or demand, economy that is also called capitalism)*
- What type of governments do both countries have? *(representative democracies and federal republics)* How is power divided in a federal republic? *(between the national government and various state, territorial, or provincial governments*)

More About . . .

Quebec City

Founded by Champlain as a trading post, Quebec City is now the capital of Quebec province—and one of North America's oldest cities settled by Europeans. Quebec City celebrates its 400th anniversary on July 3, 2008. At one point, Quebec City was the center of New France—France's North American empire, which stretched from Hudson Bay to Louisiana. Its historic neighborhood, Old Quebec, is a UNESCO World Heritage Site.

Regional Overview

The United States and Canada

The United States and Canada occupy the central and northern four-fifths of the continent of North America. The chapters in this unit provide information about the geography, history, culture, government, and economics of the United States and Canada.

GEOGRAPHY

The United States and Canada extend from the Atlantic to the Pacific oceans and north to the Arctic ocean. The United States also reaches to the Gulf of Mexico. In total area, each country ranks among the largest in the world—Canada ranks second and the United States is fourth. In addition to their huge landmass, both countries are rich in natural resources.

Samuel de Champlain
Champlain founded Quebec, the first permanent French settlement in North America.

HISTORY

Both the United States and Canada have been settled by immigrants from all over the world, beginning with the first settlers who probably migrated from Asia after the last Ice Age. The geographic richness of North America's resources has attracted immigrants from around the world and has enabled both the United States and Canada to develop into global economic powers. This continuing immigration is a recurring theme in the history of both countries; so is the constant migration of people within each country.

CULTURE

As you know, culture involves the following factors: food and shelter, religion, language, education, and political and social organization. In Canada, there are two official languages—English and French—because France played an important early role in the history and development of the country. In the United States, English is the predominant language. Both countries have highly diverse populations that have come from around the world to live within their borders.

FIVE THEMES AND SIX ELEMENTS

Five Themes

Movement

The United States and Canada are each other's biggest trading partners. Imports to Canada from the United States make up about 56% of all Canada's imports; exports from Canada to the United States make up about 84% of all Canada's exports. However, imports from Canada only make up about 17% of all imports into the United States; exports to Canada only make up about 23% of all U.S. exports.

Six Elements

Environment and Society

Just as Canada and the United States share ocean resources, so, too, do they share boundary waters, such as the Great Lakes. To ensure the protection of shared lakes and rivers, the United States and Canada signed the Boundary Waters Treaty in 1909.

GOVERNMENT

Both Canada and the United States are representative democracies. In such a system of government, citizens hold political power through elected representatives. Canada and the United States are also federal republics. In a federal republic, power is divided between the national government and various state, territorial, or provincial governments. The personal freedoms guaranteed by the Canadian and U.S. governments have drawn millions to their shores.

Mount Rushmore Massive likenesses of four American presidents were carved on Mount Rushmore in the Black Hills of South Dakota.

ECONOMICS

The United States and Canada both have market economies. This is an economy in which the production of goods and services is determined by the demand from consumers. This kind of economy is also called a demand economy, or capitalism. Migration across the continent, combined with industrialization and urbanization, stimulated economic development in the United States and Canada. The economies of these two nations have been transformed by technology and by the movement toward globalization.

Toronto People skate outdoors in Toronto, one of Canada's booming economic centers.

111

More About . . .

Mount Rushmore

Nearly three million people visit Mount Rushmore each year. The sculpture by Gutzon Borglum and hundreds of workers took about 14 years to complete, from 1927 to 1941. Students may be interested in trying to identify the presidents on the monument: George Washington, Thomas Jefferson, Theodore Roosevelt, and Abraham Lincoln.

Mount Rushmore played a prominent role in one of Alfred Hitchcock's most famous movies, *North by Northwest*. Because Hitchcock was refused permission to film on the monument, the giant presidents' heads were created in a Hollywood studio.

DIFFERENTIATING INSTRUCTION

Gifted & Talented

Create a Monument

Have students work in pairs to create a monument that would symbolize U.S.–Canadian history. Have them create a design or model for their monument and write an inscription for it.

Pre-AP

Analyze the Service Economy

The economies of the United States and Canada also depend on services such as banking and finances. Have pairs of students research service industries in one city in the United States and one in Canada. Then have them analyze the importance of service industries to each city's overall economy as well as compare and contrast service industries in each city.

Unit Writing Project
Teaching Tips

- As students prepare scripts for their news broadcast, encourage them to seek out a variety of reliable sources to use, including newspapers, magazines, and Web sites. You may also want to arrange for students to interview local experts at school.
- Provide students with an example of a script for a news broadcast, or have them research typical formats.
- Encourage students to create props and visuals to go with their news broadcasts. Remind them that maps, charts, diagrams, and other visuals will help to illustrate their key points as well as make their presentations more engaging.
- Encourage students to rehearse presenting their news broadcasts with a partner and to revise them based on his or her feedback.

Each map on the country cards illustrates the shape of a state, province, or country. Tell students that the maps do not reflect the comparative size of these landmasses within a particular region. To get an idea of the actual size of the state, province, or country, students should study the land area statistics provided on the card.

UNIT 2 ATLAS

Country Almanac

Click here to compare the most recent data on countries @ClassZone.com

Unit Writing Project

News Broadcast

As you read this unit, think of an issue to present in a television news broadcast. You might choose an issue like immigration, poverty, or outsourcing of jobs. Pretend you are a TV news reporter and gather information as you read each chapter.

Think About:

- how you will organize your script
- five things to be sure to tell the viewer: who, what, where, when, and why

United States

The United States is made up of 50 states, the District of Columbia, and 4 territories. Alaska, Hawaii, and the territories are separated from the 48 mainland states.

United States

GEOGRAPHY
Capital: Washington, D.C.
Total Area: 3,718,711 sq. mi.
Population: 296,410,404

ECONOMY
Imports: cars; oil; televisions; clothes
Exports: food; computers; machinery

CULTURE
Language: No official language
Religion: Christian 85%; Jewish 2%; Muslim 2%

District of Columbia

Total Area: 68 sq. mi.
Population: 550,521

Alabama

Capital: Montgomery
Total Area: 52,419 sq. mi.
Population: 4,557,808

Alaska

Capital: Juneau
Total Area: 663,267 sq. mi.
Population: 663,661

Arizona

Capital: Phoenix
Total Area: 113,998 sq. mi.
Population: 5,939,292

Arkansas

Capital: Little Rock
Total Area: 53,179 sq. mi.
Population: 2,779,154

California

Capital: Sacramento
Total Area: 163,696 sq. mi.
Population: 36,132,147

112 Unit 2

INTERDISCIPLINARY ACTIVITIES

CONNECT to Science

Research State Animals

Point out the bear on the California state flag and note that California's official state mammal is the grizzly bear. Have students choose a state and research its official state animals (i.e., bird, mammal, marine mammal, and so on). Have students report on the animals' natural habitats and any threats to these areas, as well as what steps are being taken to ensure the survival of these state animals.

Colorado
Capital: Denver
Total Area: 104,094 sq. mi.
Population: 4,665,177

Connecticut
Capital: Hartford
Total Area: 5,543 sq. mi.
Population: 3,510,297

Delaware
Capital: Dover
Total Area: 2,489 sq. mi.
Population: 843,524

Florida
Capital: Tallahassee
Total Area: 65,755 sq. mi.
Population: 17,789,864

Georgia
Capital: Atlanta
Total Area: 59,425 sq. mi.
Population: 9,072,576

Hawaii
Capital: Honolulu
Total Area: 10,931 sq. mi.
Population: 1,275,194

Idaho
Capital: Boise
Total Area: 83,570 sq. mi.
Population: 1,429,096

Illinois
Capital: Springfield
Total Area: 57,914 sq. mi.
Population: 12,763,371

Indiana
Capital: Indianapolis
Total Area: 36,418 sq. mi.
Population: 6,271,973

Corn Harvest A combine harvests corn in Indiana.

Country Almanac

The location for the capital of the United States, Washington, D.C., was chosen by George Washington, who had been authorized to do so by Congress in 1790. The North agreed to its location in the South because the the South agreed that the new nation would take on Revolutionary War debt.

The actual name was also a compromise: Washington, the city, is named after George Washington; the federal district, the District of Columbia, after Christopher Columbus.

- What is the smallest state in terms of landmass? *(Rhode Island)* The largest? *(Alaska)*
- Which state has the smallest population? *(Wyoming)* The largest? *(California)*
- What is the official language of the United States? *(There is no official language.)*

Comparing Regions

Corn in the United States

Native to North and South America, corn was first grown by Native Americans thousands of years ago. Although the Midwest is home to a region called the "Corn Belt" because of its high corn production, many states across the United States also grow corn. Some other "corn-growing" states include Delaware, Florida, Maine, Massachusetts, New Mexico, and Vermont.

Corn is used as a feed grain for animals and processed into many products, including sweeteners, corn oil—and ethanol for fuel. Although demand for corn could grow across the country—because of the expanding use of ethanol as an alternative fuel for cars and trucks—some experts warn that it could also drop, if other biomass is used as a source for ethanol.

Each map on the country cards illustrates the shape of a state, province, or country. Tell students that the maps do not reflect the comparative size of these landmasses within a particular region. To get an idea of the actual size of the state, province, or country, students should study the land area statistics provided on the card.

INTERDISCIPLINARY ACTIVITIES

CONNECT to Math

Graph State Populations

Have students create a bar graph to compare the populations of states within regions and between regions. For example, students may choose to graph the New England states and the Pacific Northwest states, then compare state populations.

CONNECT to Language Arts

Create Harvest Poems

The fall, or autumn, season is typically when many crops across the country are harvested. Brainstorm with students words and phrases that describe the fall harvest. Then have students work on their own to create poems about the fall harvest. Ask volunteers to read their poems, or create a classroom display.

More About . . .

The Gray Wolf

Once ranging across most of North America, the gray wolf was hunted to near extinction. When the United States passed the Endangered Species Act in 1973, only a few hundred wolves had survived in the continental United States, in northern Minnesota. Today, the gray wolf is still on the endangered species list, but its numbers are growing. The gray wolf's current range includes Alaska, Canada, Washington, Idaho, Montana, Minnesota, and Yellowstone National Park in Wyoming.

Each map on the country cards illustrates the shape of a state, province, or country. Tell students that the maps do not reflect the comparative size of these landmasses within a particular region. To get an idea of the actual size of the state, province, or country, students should study the land area statistics provided on the card.

UNIT 2 ATLAS

Iowa
Capital: Des Moines
Total Area: 56,272 sq. mi.
Population: 2,966,334

Kansas
Capital: Topeka
Total Area: 82,277 sq. mi.
Population: 2,744,687

Kentucky
Capital: Frankfort
Total Area: 40,409 sq. mi.
Population: 4,173,405

Louisiana
Capital: Baton Rouge
Total Area: 51,840 sq. mi.
Population: 4,523,628

Maine
Capital: Augusta
Total Area: 35,385 sq. mi.
Population: 1,321,505

Maryland
Capital: Annapolis
Total Area: 12,407 sq. mi.
Population: 5,600,388

Massachusetts
Capital: Boston
Total Area: 10,555 sq. mi.
Population: 6,398,743

Michigan
Capital: Lansing
Total Area: 96,716 sq. mi.
Population: 10,120,860

Minnesota
Capital: St. Paul
Total Area: 86,939 sq. mi.
Population: 5,132,799

Mississippi
Capital: Jackson
Total Area: 48,430 sq. mi.
Population: 2,921,088

Gray Wolf A gray wolf lopes through the snow in Minnesota.

INTERDISCIPLINARY ACTIVITIES

CONNECT to Art

Make Animal Posters

Have students choose one of the state animals they reported on in the previous Connect to Science activity and create a poster illustrating the animal and its habitat. Students should also include a slogan or title and other captions for their poster that help promote the survival of their chosen animal.

Lizard A mountain short-horned lizard near Santa Fe, New Mexico

Missouri

Capital: Jefferson City
Total Area: 69,704 sq. mi.
Population: 5,800,310

Montana

Capital: Helena
Total Area: 147,042 sq. mi.
Population: 935,670

Nebraska

Capital: Lincoln
Total Area: 77,354 sq. mi.
Population: 1,758,787

Nevada

Capital: Carson City
Total Area: 110,561 sq. mi.
Population: 2,414,807

New Hampshire

Capital: Concord
Total Area: 9,350 sq. mi.
Population: 1,309,940

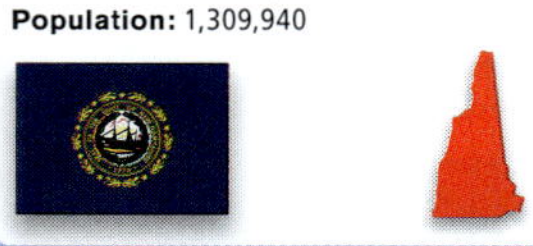

New Jersey

Capital: Trenton
Total Area: 8,721 sq. mi.
Population: 8,717,925

New Mexico

Capital: Santa Fe
Total Area: 121,590 sq. mi.
Population: 1,928,384

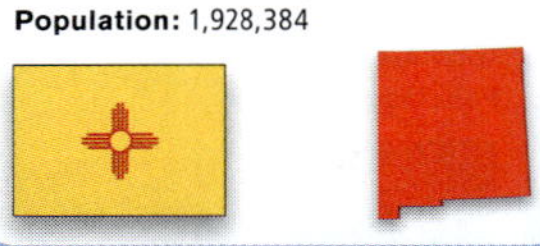

New York

Capital: Albany
Total Area: 54,556 sq. mi.
Population: 19,254,630

North Carolina

Capital: Raleigh
Total Area: 53,819 sq. mi.
Population: 8,683,242

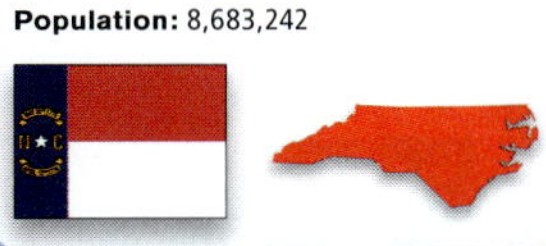

North Dakota

Capital: Bismarck
Total Area: 70,700 sq. mi.
Population: 636,677

Ohio

Capital: Columbus
Total Area: 44,825 sq. mi.
Population: 11,464,042

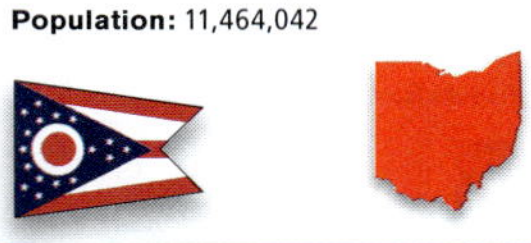

115

Comparing Regions

U.S. Deserts

Stretching from southern Oregon to northern Mexico, the North American Desert is a huge region of about 500,000 square miles (or over 700,000 square miles, depending on how the region is defined). Made up of the Great Basin Desert, the Sonoran Desert, the Mojave Desert, and the Chihuahuan Desert, the North American Desert is home to a wide variety of plant and animal life. Its plant life includes the Joshua tree, mesquite, and cacti. Its animal life includes reptiles, rodents, coyotes, and bighorn sheep (in the higher elevations).

Of note is that the Great Basin Desert is considered a "cold" desert because it is cooler while the Mojave, Sonoran, and Chihuahuan deserts are considered "hot" deserts (because of their more typical high temperatures and plants).

INTERDISCIPLINARY ACTIVITIES

CONNECT to Math

Compare State Sizes

To help students better grasp the large variations in the sizes of the states, have them use the land area data for each state to create paper rectangles that represent each state. For example, a one-inch-by-one-inch square could represent 10,000 square miles. Have each student choose ten states to represent and arrange them from biggest to smallest.

CONNECT to Civics

Hold a Flag Fundraiser

If your school or classroom does not have a state flag, suggest to students that they hold a fundraiser to purchase one as a class gift. (Follow all school guidelines for fundraising.) Hold a celebration—with volunteers explaining the history and symbolism of the state flag—when the flag is first raised or displayed.

Each map on the country cards illustrates the shape of a state, province, or country. Tell students that the maps do not reflect the comparative size of these landmasses within a particular region. To get an idea of the actual size of the state, province, or country, students should study the land area statistics provided on the card.

Comparing Regions

Populations of State Capitals

The state capital with the greatest population is Phoenix, Arizona, with over 1.3 million people. The state capital with the fewest people is Montpelier, Vermont, with about 8,000 people. Somewhat surprising, perhaps, is the fact that many state capitals are not their respective state's most populous city. In addition to Phoenix, the 16 other state capitals that are their state's most populated cities are Little Rock, Arkansas; Denver, Colorado; Atlanta, Georgia; Honolulu, Hawaii; Boise, Idaho; Indianapolis, Indiana; Des Moines, Iowa; Boston, Massachusetts; Jackson, Mississippi; Columbus, Ohio; Oklahoma City, Oklahoma; Providence, Rhode Island; Columbia, South Carolina; Salt Lake City, Utah; Charleston, West Virginia; and Cheyenne, Wyoming.

Each map on the country cards illustrates the shape of a state, province, or country. Tell students that the maps do not reflect the comparative size of these landmasses within a particular region. To get an idea of the actual size of the state, province, or country, students should study the land area statistics provided on the card.

UNIT 2 ATLAS

Heron A heron takes flight.

Oklahoma

Capital: Oklahoma City
Total Area: 69,898 sq. mi.
Population: 3,547,884

Oregon

Capital: Salem
Total Area: 98,381 sq. mi.
Population: 3,641,056

Pennsylvania

Capital: Harrisburg
Total Area: 46,055 sq. mi.
Population: 12,429,616

Rhode Island

Capital: Providence
Total Area: 1,545 sq. mi.
Population: 1,076,189

South Carolina

Capital: Columbia
Total Area: 32,020 sq. mi.
Population: 4,255,083

South Dakota

Capital: Pierre
Total Area: 77,117 sq. mi.
Population: 775,933

Tennessee

Capital: Nashville
Total Area: 42,143 sq. mi.
Population: 5,962,959

Texas

Capital: Austin
Total Area: 268,581 sq. mi.
Population: 22,859,968

Utah

Capital: Salt Lake City
Total Area: 84,899 sq. mi.
Population: 2,469,585

Vermont

Capital: Montpelier
Total Area: 9,614 sq. mi.
Population: 623,050

INTERDISCIPLINARY ACTIVITIES

CONNECT to Science

Conserve Water Resources

Discuss with students the importance of water resources—rivers, lakes, ponds, and so on—to humans and to other animals, such as herons (shown in the photograph on this page). Then have students form small groups and create a list of ways to protect and conserve local water resources. Encourage students to use these lists to create posters that can be displayed around the school or community.

Virginia

Capital: Richmond
Total Area: 42,774 sq. mi.
Population: 7,567,465

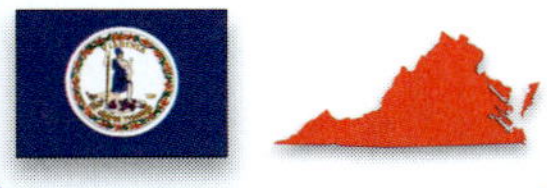

Washington

Capital: Olympia
Total Area: 71,300 sq. mi.
Population: 6,287,759

West Virginia

Capital: Charleston
Total Area: 24,230 sq. mi.
Population: 1,816,856

Wisconsin

Capital: Madison
Total Area: 65,498 sq. mi.
Population: 5,536,201

Wyoming

Capital: Cheyenne
Total Area: 97,814 sq. mi.
Population: 509,294

Rocky Mountain Bighorn Sheep Bighorn sheep are found in many Rocky Mountain states, including Wyoming.

U.S. Territories

American Samoa

Capital: Pago Pago
Total Area: 84 sq. mi.
Population: 62,700 (2004)

Guam

Capital: Agana
Total Area: 209 sq. mi.
Population: 165,000 (2004)

U.S. Virgin Islands

Capital: Charlotte Amalie
Total Area: 136 sq. mi.
Population: 109,000 (2004)

Puerto Rico

Capital: San Juan
Total Area: 3,515 sq. mi.
Population: 898,000 (2004)

More About . . .

U.S. Territories

What follows are some basic facts about the U.S. territories. Of particular note is the population density of these territories.

- American Samoa is comprised of seven islands in the south Pacific Ocean—five volcanic islands and two coral atolls. The United States acquired the American Samoan islands in 1904 and 1925. Population density: more than 750 people per square mile.
- Guam, in the north Pacific Ocean, is a volcanic island. The United States gained Guam after the Spanish-American War, then lost it to and regained it from the Japanese during World War II. In 1950, Guam became a U.S. territory. Population density: more than 775 people per square mile.
- Puerto Rico, in the Caribbean Sea, was ceded to the United States after the Spanish-American War. It became a commonwealth associated with the United States in 1952. Puerto Rican voters have voted against U.S. statehood three times. Population density: more than 1,100 people per square mile.
- U.S. Virgin Islands, in the Caribbean, were bought by the United States from Denmark in 1917 for $25 million. Population density: about 817 people per square mile.

INTERDISCIPLINARY ACTIVITIES

CONNECT to Math

Graph and Compare Population Densities

Have students create bar graphs of the population densities of Canada, the United states, its territories listed on this page, and five U.S. states. Then ask them to explain the differences in population density, using the Five Themes: Human Environment Interaction.

CONNECT to Language Arts

Describe Island Life

Point out to students that each of the U.S. territories listed on this page is also a tourist destination. Have students describe daily life, from the point of view of a native islander and from the point of view of a tourist. Ask volunteers to read one of their descriptions, then discuss as a class the differences in viewpoints.

Each map on the country cards illustrates the shape of a state, province, or country. Tell students that the maps do not reflect the comparative size of these landmasses within a particular region. To get an idea of the actual size of the state, province, or country, students should study the land area statistics provided on the card.

Canada

Point out that Canada has fewer people than California. However, Canada's most populated province, Ontario, would rank as the 6th-most populous "state" in the United States, after California, Texas, New York, Florida, and Illinois.

You might also want to tell students the significance of the maple leaf symbol on the Canadian flag. (The maple leaf represents the traditional emblem of Canada, which symbolized its land and people.) Red and white are Canada's official colors.

- What are the official languages of Canada? *(English and French)*
- What is the biggest territory or province in Canada, in terms of landmass? *(Nunavut)* Which has the smallest population? *(Nunavut)*

Comparing Regions

Alberta and British Columbia

Alberta, one of Canada's three prairie provinces (along with Saskatchewan and Manitoba), is partially separated from British Columbia by the Rocky Mountains. Canada's petroleum powerhouse, Alberta contains the country's richest deposits of oil and natural gas. Its energy resources fueled the growth of Calgary and Edmonton.

British Columbia, Canada's most western province, is known for its natural beauty, hydroelectric power, fisheries (especially Pacific salmon), and timber resources. Although relatively few people are employed in its mining industries, British Columbia earns several billion dollars in sales each year from its resources of coal, natural gas, copper, oil, gold, and zinc.

Along with Ontario, British Columbia and Alberta receive most of Canada's immigrants—most likely because of the many economic opportunities in these provinces.

Each map on the country cards illustrates the shape of a state, province, or country. Tell students that the maps do not reflect the comparative size of these landmasses within a particular region. To get an idea of the actual size of the state, province, or country, students should study the land area statistics provided on the card.

UNIT 2 ATLAS Country Almanac

Click here to compare the most recent data on countries @ClassZone.com

Puffins Atlantic puffins on Machias Seal Island in Canada

Canada

Canada is made up of 10 provinces and 3 territories. The 3 territories are sparsely populated but make up about 41 percent of the country's landmass.

Canada

GEOGRAPHY
Capital: Ottawa
Total Area: 3,855,103 sq. mi.
Population: 32,270,500

ECONOMY
Imports: machinery; chemicals; oil
Exports: food; natural gas; wood

CULTURE
Language: English; French
Religion: Christian 77%; Muslim 2%; Jewish 1%

Alberta

Capital: Edmonton
Total Area: 255,541 sq. mi.
Population: 3,256,800

British Columbia

Capital: Victoria
Total Area: 364,764 sq. mi.
Population: 4,254,500

Manitoba

Capital: Winnipeg
Total Area: 250,116 sq. mi.
Population: 1,177,600

New Brunswick

Capital: Fredericton
Total Area: 28,150 sq. mi.
Population: 752,000

Newfoundland and Labrador

Capital: St. John's
Total Area: 156,453 sq. mi.
Population: 516,000

Northwest Territories

Capital: Yellowknife
Total Area: 519,734 sq. mi.
Population: 43,000

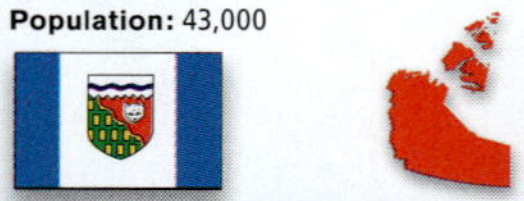

INTERDISCIPLINARY ACTIVITIES

CONNECT to Economics

Chart Canada's Natural Resources

Assign groups of students to one of Canada's provinces or territories. Have groups use classroom and library resources, such as atlases and encyclopedias, and reliable Internet Web sites to research the natural resources of their province/territory. Each group should then create a chart of the resources, including statistical data (employment, income from sales, and so on) and an assessment of potential damage to Canada's environment from each resource.

Nova Scotia

Capital: Halifax
Total Area: 21,345 sq. mi.
Population: 937,900

Nunavut

Capital: Iqaluit
Total Area: 808,185 sq. mi.
Population: 30,000

Ontario

Capital: Toronto
Total Area: 415,598 sq. mi.
Population: 12,541,400

Prince Edward Island

Capital: Charlottetown
Total Area: 2,185 sq. mi.
Population: 138,100

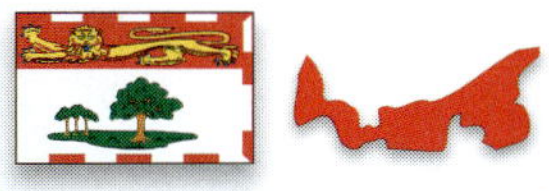

Quebec

Capital: Quebec City
Total Area: 595,391 sq. mi.
Population: 7,598,100

Saskatchewan

Capital: Regina
Total Area: 251,366 sq. mi.
Population: 994,100

Yukon Territory

Capital: Whitehorse
Total Area: 186,272 sq. mi.
Population: 31,000

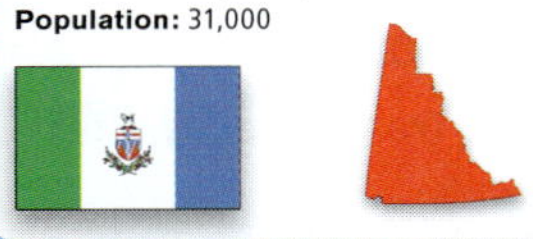

Polar Bear A polar bear in a field of tundra fireweed near Hudson Bay

More About . . .

Prince Edward Island (PEI)

Canada's smallest province or territory, PEI has been known for much of its history for its red soil and closeness to sea resources. Iron oxides in PEI's rock and soil account for the island's red color. Farming remains a mainstay of the island, with its chief crops being potatoes and dairy products. Fishing—for lobster and shellfish such as scallops and mussels—is another of PEI's key industries.

Not surprising, considering PEI's beauty, is the fact that tourism is also one of the islands main industries. One of the world's longest bridges—the Confederation Bridge—now directly links the island to Canada's mainland.

Although the vast majority of islanders are of British descent, PEI is also home to a significant number of islanders of French or Acadian descent.

INTERDISCIPLINARY ACTIVITIES

CONNECT to Civics

Debate U.S.–Canada Relationship

As a class, discuss the advantages and disadvantages of the relationship between the United States and Canada. Then ask students to take the position of either a Canadian citizen or a U.S. citizen, forming two debate teams. Have the teams prepare to debate this statement: The U.S.–Canada relationship benefits each country equally. Agree on the debate rules before holding the debate.

CONNECT to Language Arts

Explore Passages from *Anne of Green Gables*

One of Canada's best-know authors is Lucy Maud Montgomery, who wrote about an orphan girl on Prince Edward Island in her classic tale, *Anne of Green Gables*. Make available to students copies or passages from this book for them to read. For each reading, have students explain what the passage reveals about Canada's people, history, and culture.

Each map on the country cards illustrates the shape of a state, province, or country. Tell students that the maps do not reflect the comparative size of these landmasses within a particular region. To get an idea of the actual size of the state, province, or country, students should study the land area statistics provided on the card.

CHAPTER 5 The United States PLANNING GUIDE

Complete print resources all on one USB drive allow you to customize lesson plans for all students.

Section Titles & Key Ideas	Unit Resource Book The United States and Canada	Assessment Resources
1 From Coast to Coast **KEY IDEA** The physical geography of the United States varies greatly from coast to coast.	• **Reading Study Guide,** pp. 3–4 • **RSG with Additional Support,** pp. 11–12 • **Vocabulary Practice,** p. 35 • **Skillbuilder Practice,** p. 36 • **Outline Map Activities,** p. 39 • **Geography Project,** p. 147	**Unit Resource Book** • Section Quiz, p. 53 • Reteaching Activity, p. 57 **Test Generator CD-ROM** **Daily Test Practice Transparency,** TT14
2 Creating a Nation **KEY IDEA** Native Americans and peoples from other continents all came together to build a nation.	• **Reading Study Guide,** pp. 5–6 • **RSG with Additional Support,** pp. 13–14 • **Map and Graph Skills,** p. 37 • **History Makers,** p. 41 • **Primary Source,** p. 48 • **World Literature,** p. 143	**Unit Resource Book** • Section Quiz, p. 54 • Reteaching Activity, p. 58 **Test Generator CD-ROM** **Daily Test Practice Transparency,** TT15
3 A Diverse Culture **KEY IDEA** The United States is one of the most diverse nations in the world, and it has a rich culture blended from many world cultures.	• **Reading Study Guide,** pp. 7–8 • **RSG with Additional Support,** pp. 15–16 • **Skillbuilder Practice,** p. 36 • **History Makers,** p. 42 • **Connect to Math,** p. 47	**Unit Resource Book** • Section Quiz, p. 55 • Reteaching Activity, p. 59 **Test Generator CD-ROM** **Daily Test Practice Transparency,** TT16
4 Democracy and Free Enterprise **KEY IDEA** The United States purchases goods from and sells goods to many countries around the world.	• **Reading Study Guide,** pp. 9–10 • **RSG with Additional Support,** pp. 17–18 • **Skillbuilder Practice,** p. 36 • **Interdisciplinary Projects,** p. 44 • **Economics in the World,** p. 46	**Unit Resource Book** • Section Quiz, p. 56 • Reteaching Activity, p. 60 • Chapter Test (Forms A, B, & C), pp. 61–75 **Test Generator CD-ROM** **Daily Test Practice Transparency,** TT17

McDougal Littell Assessment System

TEST SCORE REPORT RETEACH

No Time? To focus students on important content in the chapter, use the **"Reading Study Guide with Additional Support"** in *Resources2GO.*

Teacher Presentation Options

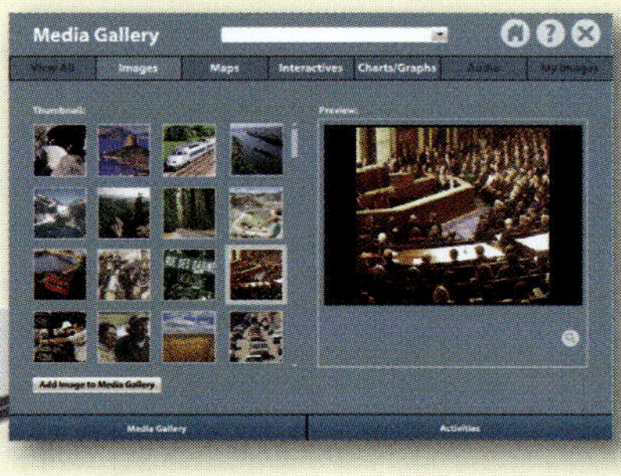

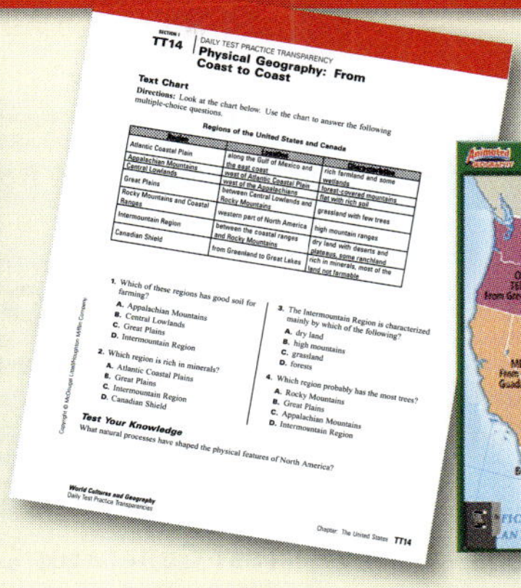

Electronic Books

eEdition DVD-ROM

eEdition Online

Presentation Toolkit

Video/DVD

- World Cultures and Geography

Power Presentations DVD-ROM

- Discussion Starters
- Media Gallery
- Review Game
- Interactive Review
- Animated Features

Unit Transparency Book

- 3-Minute Warm-Ups
- Essential Question
- Critical Thinking
- Maps
- Cultures

Daily Test Practice Transparencies

Animated Geography

- Present-Day United States Map
- Growth of the United States Map

OnlineActivities at ClassZone.com

Animated Geography

- Videos and Photos
- 3-D Animations
- Interactive Maps

Interactive Review

- Section Quizzes
- Review and Study Notes
- Vocabulary Flip Cards
- Graphic Organizers
- Name Game and Crossword Puzzles
- GeoGame

Online Test Practice

Map Center

- Atlas Maps
- Political Maps with Data
- Links to Mapping Web Sites

Activity Center

- Activity Maker—Review Games, Puzzles, Flip Cards, and Graphic Organizers
- Worldster

Research & Writing Center

- Web Sites
- Data Updates
- Primary Sources
- Writing Models

Current Events

Jim Easton
Roscoe Middle School
Roscoe, Illinois

Make It Fun — Teacher-Tested Activities

As you teach this chapter, look for my classroom-proven ideas for:

- making the lesson fun
- focusing students' attention and comprehension
- integrating technology with traditional teaching

Teaching Bell to Bell When you see this timing guide, have students discuss in pairs, in groups of 3–6, or as a class.

- **Think, Pair, Share** (pairs)
- **Reader, Recorder, Reporter** (3–6)
- **Talk About It** (any size)

Chapter 5 RESOURCES FOR DIFFERENTIATED INSTRUCTION

Struggling Readers

eEdition with Audio DVD-ROM

Lesson Plans
- Strategies for Inclusion

Unit Resource Book
- Reading Study Guide with Additional Support, pp. 11–18 B
- Reteaching Activity, pp. 57–60
- Chapter Test (Form A), pp. 61–65 A

Best Practices Toolkit C
- Interactive Reading Strategies, pp. 90–145
- Differentiation Activity Bank: Struggling Readers, pp. 146–149

Teacher's Edition Activities
- Clarify the Land Bridge Location, Sec. 2
- Create a Time Line of U.S. Acquisitions, Sec. 2
- Understanding Metaphors that Describe American Culture, Sec. 3
- Defining the Duties of Each Branch of Government, Sec. 4
- Tiered Activities, Sec. 1–4

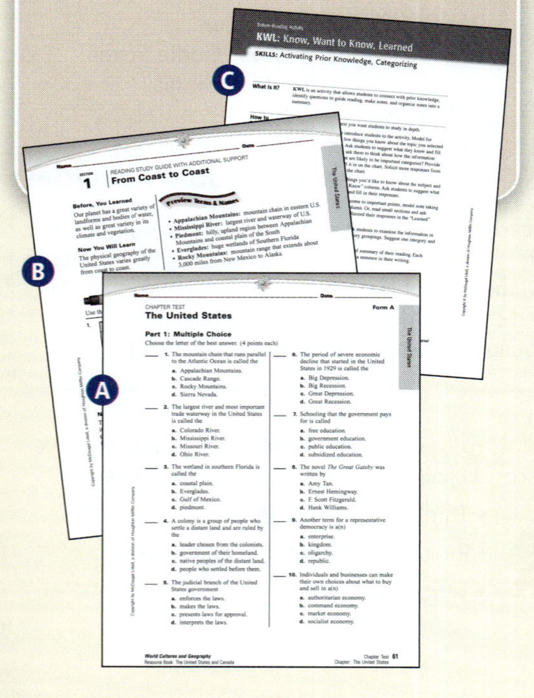

English Learners

Pupil's Edition in Spanish

eEdition with Audio DVD-ROM

Lesson Plans
- Strategies for English Learners

Unit Resource Book
- Reading Study Guide (Spanish), pp. 19–26
- RSG with Additional Support (Spanish), pp. 27–34 A B

Multi-Language Glossary C

Best Practices Toolkit
- Differentiation Activity Bank: English Learners, pp. 165–194

Test Generator CD-ROM
- Section Quizzes in Spanish
- Chapter Test in Spanish

Teacher's Edition Activities
- Pronounce and Preview, Sec. 1–4
- Comprehension, Sec. 1
- Vocabulary: Idioms, Sec. 2
- Vocbulary: Key Academic Vocabulary, Sec. 3
- Vocabulary: Suffixes, Sec. 4
- Tiered Activities, Sec. 1–4

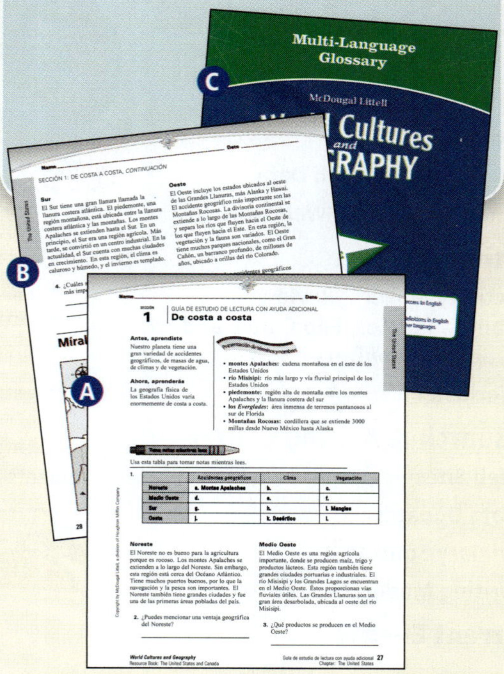

Inclusion

eEdition with Audio DVD-ROM

Lesson Plans
- Strategies for Inclusion

Unit Resource Book
- Reading Study Guide with Additional Support, pp. 11–18 A B
- Reteaching Activity, pp. 57–60
- Chapter Test (Form A), pp. 61–65

Best Practices Toolkit C
- Interactive Reading Strategies, pp. 90–145
- Differentiation Activity Bank: Struggling Readers, pp. 146–149

Teacher's Edition Activities
- Match Terms to Map, Sec. 1
- Match Terms and Definitions, Sec. 2
- Reading and Defining Terms Verbally , Sec. 3
- Appreciate Diffferent Types of American Music, Sec. 3
- Predicting ABCs, Sec. 4
- Tiered Activities, Sec. 1–4

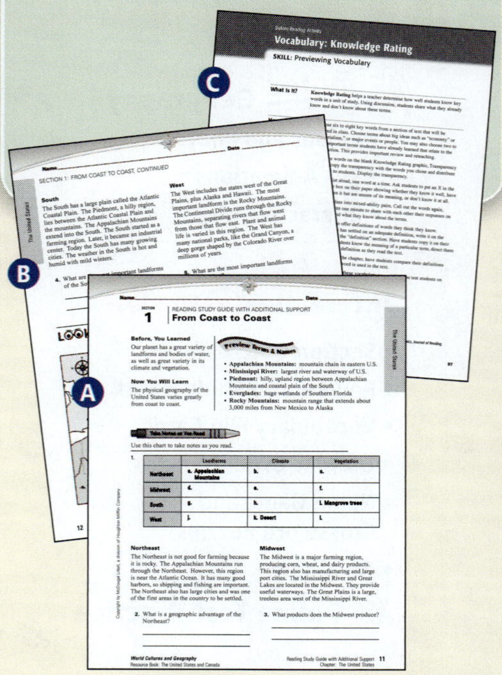

Gifted & Talented Students

Lesson Plans
- Core Lesson Plan

Unit Resource Book
- Outline Map Activities, p. 39 Ⓐ
- History Maker, pp. 41–42 Ⓑ Ⓒ
- Interdisciplinary Projects, pp. 43–45
- Economics in the World, p. 46
- Chapter Test (Form C), pp. 71–75

Best Practices Toolkit
- Student Products Rubrics, pp. 1–14
- Writing Templates, pp. 78–79

Teacher's Edition Activities
- Illustrate a Civil War Event, Sec. 2
- Perform Music Inspired by Other Cultures, Sec. 3
- Tiered Activities, Sec. 1–4

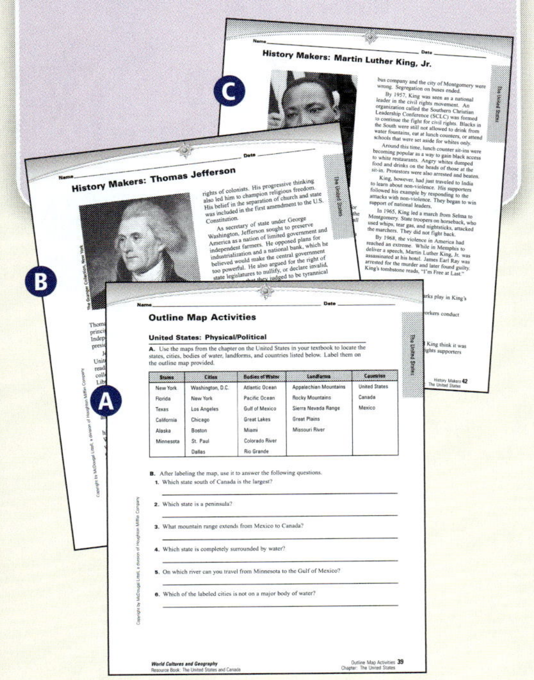

Pre-AP*

Lesson Plans
- Core Lesson Plan

Unit Resource Book
- Vocabulary Practice, p. 35
- Map and Graph Skills, p. 37
- Economics in the World, p. 46 Ⓐ
- Connect to Math, p. 47 Ⓑ
- Primary Source, p. 48 Ⓒ
- Chapter Test (Form C), pp. 71–75

Best Practices Toolkit
- Writing Templates, pp. 78–89
- Differentiation Activity Bank: Pre-AP Students, pp. 195–200

* Pre-AP is a registered trademark of the College Entrance Examination Board, which was not involved in the production of and does not endorse this product.

Teacher's Edition Activities
- Think Critically About the Atlantic Ocean, Sec. 1
- Clarify the South's Resistance to Ending Slavery, Sec. 2
- Compile a List of Books About Immigrant Life in America, Sec. 3
- Tiered Activities, Sec. 1–4

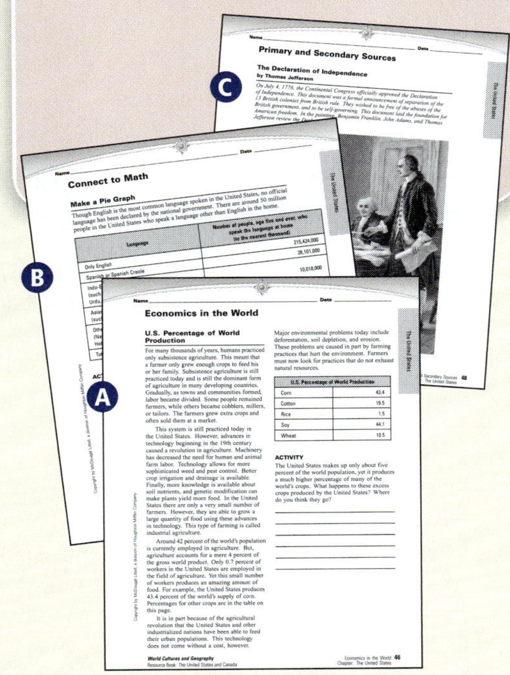

Interdisciplinary Activities

CONNECT to History

Review a Movie About U.S. History

CLASS TIME 30 minutes

TASK Prepare a review of a movie about an important event in U.S. history.

SUPPLIES
- list of age-appropriate films on U.S. history
- writing materials/computers

ACTIVITY Have students prepare a review of a movie about an important event in U.S. history. In their reviews, students should identify the setting (time and place) of the story, include a brief summary of the plot, point out possible historical inaccuracies, and explain what the movie teaches about the development and culture of the United States. Post the reviews and ask volunteers to read theirs out loud.

Good for Visual and Verbal Learners

CONNECT to Math

Pie Graph of Current U.S. Immigration Data

CLASS TIME 40 minutes

TASK Create a pie graph that shows the top five countries of origin of immigrants to the United States.

SUPPLIES
- recent data on U.S. immigration
- paper, colored markers or pencils

ACTIVITY Have students work in pairs to create a pie graph that shows the top five countries of origin of immigrants to the United States. Supply copies of immigration data for a recent year. Refer students to the instructions for making a pie graph in the Connect to Math feature in Section 3 of this chapter. The graph should include wedges for each of the five countries and a sixth wedge for "other." To calculate the number in the "other" category, students can add up the data on the top-five countries and subtract this number from the total number of immigrants. The remainder will be all "other" immigrants.

Good for Visual Learners

Introduce the ESSENTIAL QUESTION

- Identify the advantages and disadvantages of the United States' varied geography.
- Explain how the United States grew from a group of small colonies to a strong nation and world power.
- Point out ways in which America's diverse population has influenced its culture.
- Describe the government and the economy of the United States and explain the relationship between them.

Key Ideas

Section 1: The physical geography of the United States varies greatly from coast to coast.

Section 2: Native Americans and peoples from other countries all came together to build a nation.

Section 3: The United States is one of the most diverse nations in the world, and it has a rich culture blended from many world cultures.

Section 4: The United States purchases goods from and sells goods to many countries around the world.

CONNECT to the Essential Question

You may wish to revisit the Essential Question periodically with students as they work through the chapter.

Suggestions for reconnecting are provided at an appropriate point in the *Teacher's Edition* for each section.

Yosemite Falls in Yosemite National Park, California

CHAPTER 5 The United States

1 GEOGRAPHY From Coast to Coast

2 HISTORY Creating a Nation

3 CULTURE A Diverse Culture

4 GOVERNMENT & ECONOMICS Democracy and Free Enterprise

ESSENTIAL QUESTION

How did a developing nation grow into a world power?

CONNECT Geography & History

Use the map and the time line to answer the following questions.

1. Which of the Great Lakes is located entirely within the United States?
2. In what year did Hurricane Katrina occur and how did the location of New Orleans contribute to the damage caused by it?

1492

History
1565 Pedro Menéndes de Avilés founds the city of St. Augustine in Florida for Spain. ▶

Government
1787 The Constitution of the United States is written in Philadelphia.

History
1775 American Revolution begins with the battles of Lexington and Concord. ▶

TIME LINE DISCUSSION

Tell students that the development of what is now the United States began with the establishment of small colonies in the eastern part of the country, then expanded westward to the Pacific Ocean and beyond.

- Which event on the time line highlights the founding of an early settlement in the Americas? *(1565—the founding of St. Augustine)*
- In what way do you think the events highlighted for 1775 and 1787 are related? *(Possible Answer: The American Revolution, beginning in 1775, led to the establishment of the United States as an independent country and the writing of the U.S. Constitution in 1787.)*
- Which event on the time line highlights an example of how the United States grew geographically? *(1804—Lewis and Clark start to explore the Louisiana Purchase. Suggest students preview the Growth of United States 1783–1853 map in Section 2 to identify the Louisiana Purchase.)*

Animated GEOGRAPHY
Present-Day United States

Click here to explore the United States @ ClassZone.com

Culture

1965 Martin Luther King Jr., leads voting rights march in Selma, Alabama. ▶

Today

Geography

◀ **1804** Lewis and Clark begin expedition to explore Louisiana Purchase.

Geography

2005 Hurricane Katrina destroys much of New Orleans.

121

TEACHER MEDIA FAVORITES

Books for the Teacher

Gillon, Steven and Stephen Hoye. *10 Days That Unexpectedly Changed America.* Random House Audio, 2006. Audio book companion to the History Channel documentaries on ten pivotal moments in American history. (3 DVDs also available from A & E Home Video.)

Weinstein, Allen and David Rubel. *The Story of America.* DK Publishers, Inc., 2002. Chronicles the important events in the history of America from settlement to 21st-century superpower.

Video/DVD

Destination America. (DVD) PBS Home Video, 2005. Acclaimed TV documentary on the people and cultures that created and developed the United States. (Illustrated companion book by Chuck Wills also available.)

Lewis and Clark: Great Journey West. (DVD) National Geographic Video, 2002. Thrilling reenactment of Lewis and Clark's first expedition into the uncharted West. Narrated by Jeff Bridges.

Internet

Go to the *Research and Writing Center* at **ClassZone.com** to link to:

- data on the region
- a variety of chapter-specific content-reviewed sites
- "History Makers" biographies
- primary sources

Animated GEOGRAPHY
Present-Day United States
CLASSZONE.COM

Point out that this is a political map of the United States, showing the states, their capitals, and selected other cities. Explain why Alaska and Hawaii are displayed as insets and point out their locations on an expanded U.S. map or world map.

- What is our nation's capital? *(Washington, D.C.)* Where is it located? *(on the East Coast between Virginia and Maryland)* According to the time line, what historic event took place about 100 miles northeast of our nation's capital? *(the writing of the Constitution in Philadelphia, Pennsylvania)*
- Why do you think the Mississippi River was an important waterway in the development of the United States? *(Possible Answer: It was a transportation and shipping route through the South and Midwest for settlers, traders, and goods.)*
- Compare this map to the Growth of the United States 1783–1853 map in Section 2. What waterway might Lewis and Clark have traveled to explore the Louisiana Purchase? *(Missouri River)*

CONNECT GEOGRAPHY & HISTORY ANSWERS

1. Lake Michigan
2. Hurricane Katrina occurred in 2005. New Orleans's location—with many areas below or just above sea level—along the coast made it vulnerable to destructive flooding.

Presentation Options

Use the **Power Presentations** DVD-ROM to project the animated map.

- Prompt students to use the pop-up images and audio from the map to discover what they already know about the geography, culture, history, economics, or governments of the region.
- Challenge students to answer the map questions individually, in small groups, or as a class activity. Use all of the interactive material to help preview chapter content.

1 Plan & Prepare

Objectives

- Identify the physical features, climate, and life of the Northeast
- Identify the physical features, climate, and life of the Midwest
- Identify the physical features, climate, and life of the South
- Identify the physical features, climate, and life of the West

Read for the Essential Question

Help students read for a purpose by reminding them of the Essential Question: "How did a developing nation grow into a world power?" Point out that they can use text features such as Key Questions, special features, images, and maps to form answers to the question.

Vocabulary

Best Practices Toolkit

Use the ***Best Practices Toolkit*** to model strategies for vocabulary notetaking. Vary strategies throughout the year. Choose from: *Knowledge Rating, Predicting ABCs, Definition Mapping, Word Sort, Frayer Model (Word Squares), Magnet Words, Word Wheel,* and *Student VOC.*

Vocabulary Strategies, TT9–TT16

Reading Strategy

Best Practices Toolkit

Display the Three-Column Chart transparency.

- Explain to students that in categorizing they need to determine how the information in the section can be sorted into the categories shown in the chart.
- Model filling in categories on a chart using the Comparing U.S. Geographic Regions feature that immediately follows the section.
- Have students suggest entries using the categories in the chart.

Three-Column Chart, TT21

SECTION 1 Reading for Understanding

Key Ideas

BEFORE, YOU LEARNED

Our planet has a great variety of landforms and bodies of water, as well as great variety in its climate and vegetation.

NOW YOU WILL LEARN

The physical geography of the United States varies greatly from coast to coast.

Vocabulary

TERMS & NAMES

Appalachian (AP•uh•LAY•chee•uhn) **Mountains** a mountain chain in the eastern United States, running parallel to the Atlantic Ocean

Great Lakes five lakes forming the largest group of freshwater lakes in the world

Mississippi River the largest river and chief waterway of the United States

Great Plains a vast grassland in central North America

Gulf of Mexico an arm of the Atlantic Ocean that lies south of the United States

Piedmont (PEED•MAHNT) a hilly, upland region between the Appalachian Mountains and the coastal plain of the South

Everglades huge wetlands of southern Florida

Rocky Mountains a mountain range that extends about 3,000 miles from New Mexico to Alaska

Continental Divide a high ridge line in the Rocky Mountains that divides east-flowing from west-flowing rivers

Mt. McKinley (also called Denali) the tallest mountain of North America, located in Alaska

Grand Canyon a deep gorge cut through northern Arizona by the Colorado River

REVIEW

continent one of seven large landmasses on the surface of the earth

Reading Strategy

Re-create the chart that is shown at right. As you read and respond to the **KEY QUESTIONS**, use the chart to organize important details about the physical geography of the United States.

Skillbuilder Handbook, page R7

CATEGORIZE

	LANDFORMS	CLIMATE	VEGETATION
Northeast			
Midwest			
South			
West			

GRAPHIC ORGANIZERS
Go to Interactive Review @ClassZone.com

PRETEACHING VOCABULARY

English Learners

Pronounce and Preview

Pronounce each term for students. Review the meanings of words within the definitions, such as *upland*, *wetlands*, and *gorge*.

- To modify vocabulary learning, have students complete worksheets as they read, instead of afterward.

Unit Resource Book: The United States & Canada
- Vocabulary Practice, p. 35

Inclusion

Match Terms to Map

Read aloud terms and definitions. Then have students turn to the Physical Geography of the United States map. Have students find the term (or its location) on the map. To extend the activity, have students create a map (with pictures or symbols and labels for each term) or a portfolio (with images for each term drawn or cut out of magazines). They can add items for terms they learn in the other sections of the chapter.

SECTION 1 GEOGRAPHY

Oakland Bay Bridge, San Francisco

Moose in Denali National Park, Alaska

From Coast to Coast

Connecting to Your World

What outdoor activities do you enjoy? Sports such as surfing and scuba diving are popular in Hawaii. Many people go skiing and snowboarding in the snowy mountains of Colorado. The United States is a large country with nearly every type of landscape and climate. One thing that helps to define the major U.S. regions is the type of activities common to the area.

Surfing and Skiing These are two activities that can be enjoyed in the varied U.S. landscapes.

Northeast

KEY QUESTION What are the geographic advantages and disadvantages of the Northeast?

The United States stretches across the North American **continent**, or large landmass. The country can be divided into four regions: Northeast, Midwest, South, and West. The features that set the regions apart are their landforms, climates, plants, animals, and ways of life. The Northeast includes the New England states of Maine, New Hampshire, Vermont, Massachusetts, Rhode Island, and Connecticut, as well as New York, Pennsylvania, and New Jersey.

Appalachian Mountains These mountains extend for almost 2,000 miles in the eastern United States and Canada.

2 Focus & Motivate

3-Minute Warm-Up

Write on the board or display the transparency:

- Look at the images on this page. What types of landscapes are featured? *(coastal areas, lakes, forests, snow-covered terrain, mountains)* Describe the landscape in your region. *(Accept all reasonable answers.)*

Unit Transparency Book: The United States & Canada
- 3-Minute Warm-Ups, TT1

Connecting to Your World

More About . . .

Outdoor Activities in the United States

Some of the top outdoor activities that Americans enjoy are bicycling and mountain biking, walking and hiking, camping, visiting a nature center, nature trail, or zoo, bird-watching, and picnicking.

3 Teach

Northeast

Think, Pair, Share

- What are the two most important physical features of the Northeast? *(Atlantic Ocean, Appalachian Mountains)*
- Why is farming not very profitable in much of the Northeast? *(Much of the terrain is too hilly and rocky for productive farming.)*
- **Categorize** What are some important industries in the Northeast? *(fishing, shipping, manufacturing, mining, banking, finance, trade)*

SECTION 1 PROGRAM RESOURCES

Unit Resource Book: The United States and Canada

ON LEVEL

Unit Resource Book
- Reading Study Guide, p. 3
- Section Quiz, p. 53

STRUGGLING READERS

Unit Resource Book
- Reading Study Guide with Additional Support, p. 11
- Skillbuilder Practice: Sequencing Events, p. 36
- Section Quiz, p. 53
- Reteaching Activity, p. 57

eEdition with Audio DVD-ROM

ENGLISH LEARNERS

Pupil Edition in Spanish

eEdition with Audio DVD-ROM

Unit Resource Book
- RSG (Spanish), p. 19
- RSG with Additional Support (Spanish), p. 27

Multi-Language Glossary

Test Generator
- Section Quiz in Spanish

INCLUSION

Unit Resource Book
- RSG with Additional Support, p. 11
- Section Quiz, p. 53
- Reteaching Activity, p. 57

GIFTED & TALENTED

Unit Resource Book
- Outline Map Activities, p. 39
- Section Quiz, p. 53

PRE-AP
- Section Quiz, p. 53
- Geography Project, p. 147

TECHNOLOGY

Unit Transparency Book
- 3-Minute Warm-Ups, TT1
- Essential Question, TT2
- Critical Thinking, TT3
- Map: US & Canada: Population Density, TT7

Daily Test Practice Transparencies
- Section 1, TT14

Power Presentations

ClassZone.com

World Cultures and Geography Video on DVD

More About . . .

Appalachian Mountains

The Appalachians, the oldest mountain chain in North America, extend from Newfoundland, Quebec, and New Brunswick to Alabama, and are made up of many mountain ranges. Some of them include the White, Green, and Catskill mountains in the northern part of the chain and the Allegheny, Blue Ridge, Cumberland, and Great Smoky mountains farther south.

Make It Fun

Teacher-Tested Activities

Jim Easton, Roscoe Middle School, Roscoe, Illinois

My students enjoy playing the "Where Am I?" activity. Choose a building, landform, bridge, stadium, or statue, such as the Empire State Building in New York City.

- Make a copy of a photograph of the famous object that you have chosen.
- Display the picture on a wall or chalkboard in the classroom.
- Give hints to students. For example, for the Empire State Building, you might say: "Since September 11, 2001, this is the tallest structure in the city."
- After the object has been identified, have students learn more about it. For example, they might research when the Empire State Building was completed and find statistics about its construction.

Teach

Midwest

Talk About It

- What is the most important trade waterway in the United States? *(Mississippi River)*
- What geographic advantage do cities such as Minneapolis, Chicago, and Detroit have? *(They are located along waterways, which are valuable for shipping and transportation.)*
- **Categorize** What are the three most important physical features of the Midwest? *(Great Lakes, Mississippi River, Great Plains)*

Landforms, Climate, and Vegetation The two most important physical features of the Northeast are the Atlantic Ocean Ⓐ (shown on the map on the next page) and the **Appalachian** (AP•uh•LAY•chee•uhn) **Mountains** Ⓑ. The Atlantic coastline, which forms the region's eastern border, has deep-water harbors that support trade. The Appalachians are a mountain chain that runs parallel to the Atlantic Coast.

In much of the Northeast, especially in New England, the terrain is too hilly and rocky to make good farmland. Most of the region lacks mineral wealth, but Pennsylvania has coal and iron ore deposits.

Summers are warm and humid. Winters are cold and snowy, with the most snow falling in the northernmost states. Despite the effects of the logging industry, forests still stretch over much of the region.

Life in the Northeast The Northeast was one of the first areas of North America to be heavily settled by Europeans. Because of the area's rocky land and good harbors, fishing and shipping became more profitable than farming. Textile mills built in the Northeast in Lowell, Massachusetts, were among the first U.S. factories.

The Northeast is still the most densely populated and industrialized region of the United States. The coast is crowded with cities. Shipping and manufacturing are major parts of the economy, but service industries also play a large part. New York City, the largest city in the country, is a world center of finance, trade, and culture.

▲ **COMPARE AND CONTRAST** Describe the geographic advantages and disadvantages of the Northeast. Its main advantage is the sea and a long coastline with good harbors; its main disadvantage is little farmland.

Midwest

▼ **KEY QUESTION** What are the most important features of the Midwest's geography?

The central and northern part of North America has one of the largest areas of lowlands on the planet. Those plains make up most of the Midwest. The region stretches from Ohio west to the Dakotas and from Canada south to the Ohio River, Missouri, and Kansas.

because cities develop as ports and centers of trade along the river, enabling goods to flow into the heartland ▼

Mississippi River
Minneapolis, Minnesota is located on the Mississippi River. **Why might important cities be located along a major river?**

DIFFERENTIATING INSTRUCTION

English Learners

Comprehension: Connection

Students learning English may have difficulty understanding the sentence, "Because of the area's rocky land and good harbors, fishing and shipping became more profitable than farming." Point out the cause/effect relationships implied in the sentence. The land was not that productive for farming. However, the good harbors helped make fishing and shipping very profitable.

Pre-AP

Think Critically About the Atlantic Ocean

Have students categorize the advantages and disadvantages of living near or along the Atlantic Ocean. What benefits can the ocean provide? *(Possible Answer: source of food, jobs, recreation; means of transporting people and products)* What challenges can it present? *(Possible Answer: flooding and other damage from storms; coastal soil erosion; cleaning up polluted waters)*

Physical Geography of the United States

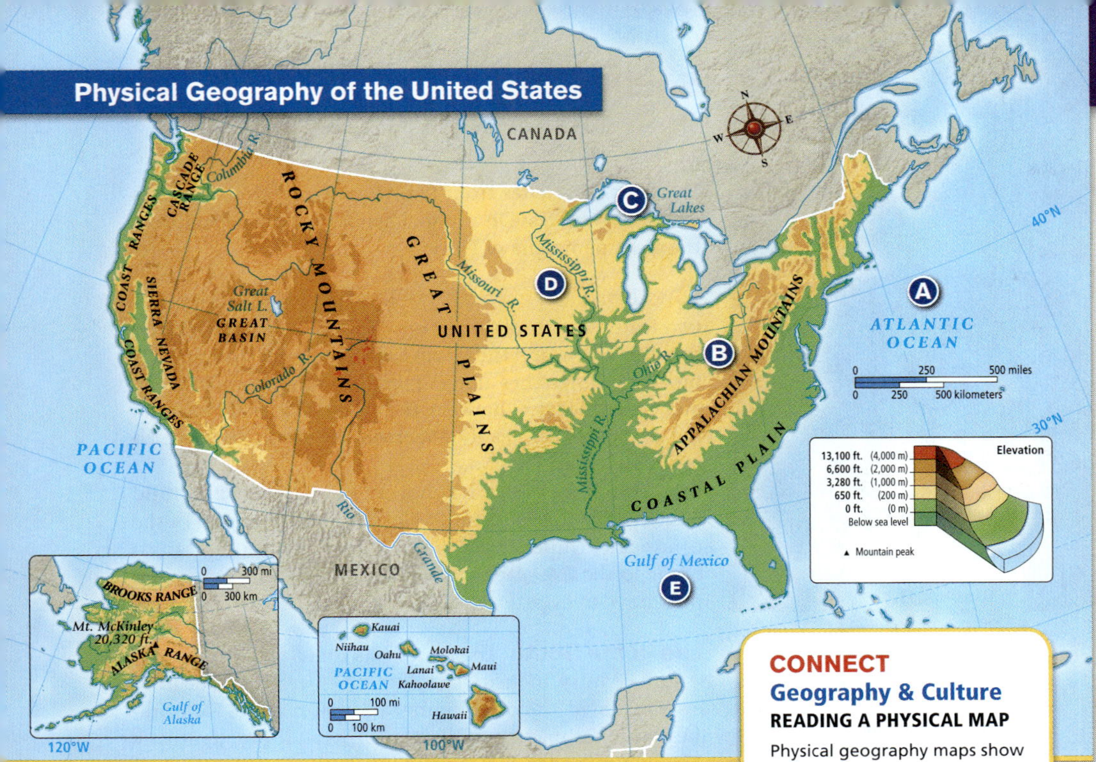

CONNECT Geography & Culture

READING A PHYSICAL MAP

Physical geography maps show landforms and bodies of water.

- Use the key to identify elevation.
- Use the scale to estimate distance.

1. **Movement** What landform might have restricted movement westward from the east coast?
2. **Location** Which rivers flow into the Pacific Ocean?

Landforms, Climate, and Vegetation The **Great Lakes** (C) make up the world's largest group of freshwater lakes. Four of the lakes—Ontario, Erie, Huron, and Superior—form part of the U.S.-Canada border. The **Mississippi River** (D) is the largest river and most important trade waterway in the United States. West of the Mississippi lies the vast, mostly treeless **Great Plains**. Its native plants are prairie grasses and wildflowers. Huge herds of bison once roamed the plains. The Midwest has a harsh climate, with hot summers and cold winters.

Life in the Midwest The Midwest is the major farming region of the United States, with corn, wheat, and dairy among its important products. The Midwest also contains major manufacturing areas, such as Detroit, Michigan, home of the automobile industry. The Midwest's largest city is Chicago, a port and transportation hub along Lake Michigan. The St. Lawrence Seaway connects Chicago to the Atlantic, and canals and rivers connect the city to the **Gulf of Mexico** (E).

EVALUATE Describe geographic features of the Midwest.
the Great Lakes, Mississippi River, and Great Plains

Physical Geography of the United States

CONNECT Geography & Culture

Reading a Physical Map Point out that this map highlights the physical features of the United States. Direct attention to the key. Make sure students understand the relationship between the colors in the map and the elevation levels shown in the key.

- Which mountain range has the highest elevation? *(Rocky Mountains)*
- What physical features besides land elevations and mountains are shown on the map? *(lakes, rivers, plains, basin)*

ANSWERS

1. **Movement** Appalachian Mountains
2. **Location** Columbia, Colorado

Unit Resource Book: The United States & Canada
- Outline Map Activities, p. 39

More About . . .

Bison

It is estimated that between 30 and 50 million bison roamed the Great Plains when the Europeans arrived. During that time, bison herds were aggressively hunted for the commercial fur trade, dramatically reducing their numbers. Population growth and the expansion of the railroad also contributed to their decline. By the end of the century, the total bison population was less than 1,000. Today, as a result of both government and private efforts, the bison population has increased to over 150,000. Modern herds live on private ranches and in national and state parks.

INTERDISCIPLINARY ACTIVITIES

CONNECT to Economics

Trace Water Routes for Shipping Goods

As the United States developed, businesses relied on inland waterways to receive and ship goods to shipping ports. Have students study the maps in this chapter, the section atlases, and other resources. Then have them list the inland water routes that might have been used to transport goods to and from America's ports. Help students fill in a chart like the one to the right for each section of the country.

Section of Country	Water Routes	Port(s)
Northeast		Boston
Midwest	Lake Michigan	
South		New Orleans
West	Columbia River	

Teach
South

Reader, Recorder, Reporter

- What factors contributed to the growth of industry in the South? *(mild climate, the invention of air conditioning, lower wages)*
- **Categorize** What are some significant physical features of the South? *(Atlantic Ocean coast, Everglades, Appalachians, Atlantic Coastal Plain, Piedmont region, Mississippi River, Gulf of Mexico coast)*

Comparing Regions

Florida Panthers and Snow Leopards

Both the Florida panther and the snow leopard are endangered cats. Once roaming over most of the southeastern United States, the Florida panther's habitat is now limited to southern Florida. The snow leopard has a much bigger habitat—across the cold, mountainous regions of Central Asia—but its numbers are thin for such a large area. Each animal has adapted to its environment. For example, the Florida panther has short fur, while the snow leopard has long fur.

Teach
West

Talk About It

- Would you characterize the physical features and life in the West as uniform or varied? Explain your answer. *(Possible Answer: They are varied. The West has a wide range of landforms, climate, and vegetation. It includes densely populated urban areas and large areas that are sparsely populated.)*
- Name some important industries in the West. *(ranching, mining, logging, agriculture, oil drilling, high-tech industries)*
- **Categorize** Categorize the various types of climate in the West. *(most of the West: arid or desert; Northwest coast: rainy; Hawaii: tropical; Alaska: cold)*

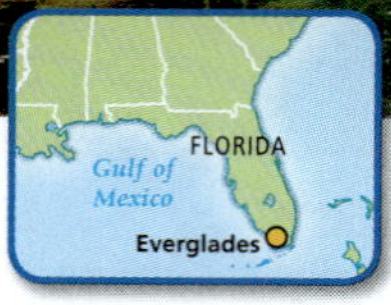

Everglades National Park A cluster of dwarf cypresses grows in this park in Florida. **What wildlife might thrive in a subtropical wetland such as the Everglades?**

alligators, snakes, birds, turtles, and so forth ▲

South

KEY QUESTION What is the climate of most of the South?

The South stretches from the Atlantic Ocean to Texas; it contains 16 states. On a map, its most noticeable feature is the giant peninsula of Florida.

Landforms, Climate, and Vegetation The South shares two features with other regions. The Appalachians extend into the South, and the Mississippi River runs through it. A broad plain stretches along the coasts of the Atlantic Ocean and Gulf of Mexico. Between the Atlantic Coastal Plain and the Appalachians lies the **Piedmont** (PEED•MAHNT), which is a hilly, upland region.

Most of the South has hot, rainy summers and mild winters. The southern tip of Florida is the only part of the continental United States to lie in the tropics. Florida is also home to the **Everglades**, a wetland filled with grasses and mangrove trees.

Life in the South The South was first a farming region. Then in the 20th century, industries moved there because of the mild climate, newly invented air conditioning, and lower wages for workers. Now the South has many booming cities. In addition, Kentucky and West Virginia mine for coal, and many of the Gulf States drill for oil. Deadly hurricanes sometimes hit places on the South's long coastline. In 2005, Hurricane Katrina struck the Gulf Coast, including New Orleans, Louisiana. Most of the city was flooded. Rebuilding New Orleans is a challenge that will take years to complete.

FIND MAIN IDEAS Describe the climate of the South. hot and humid

West

KEY QUESTION What are the climates found in the West?

A huge region of dramatic scenery, the West includes all the states from the Great Plains to the Pacific, plus Alaska and Hawaii.

Landforms, Climate, and Vegetation The West's most visible feature is the **Rocky Mountains**, which extend 3,000 miles from Alaska to New Mexico. The **Continental Divide**, a high ridge line that separates the east-flowing and west-flowing rivers of the continent, runs through the mountains. The tallest mountain in North America is **Mt. McKinley** (also called Denali) in Alaska.

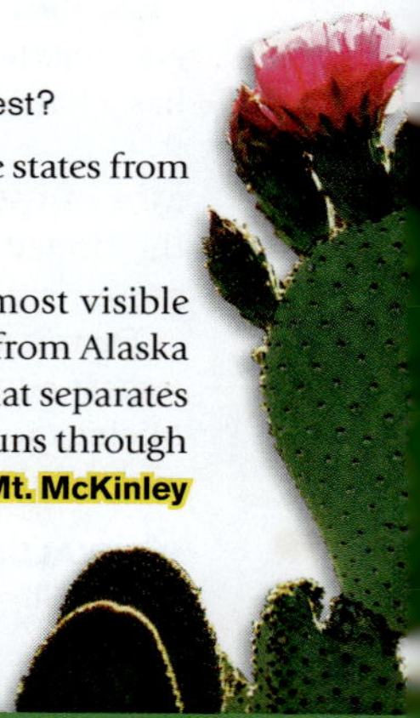

DIFFERENTIATING INSTRUCTION: TIERED ACTIVITIES

OBJECTIVE Prepare and present a visual display on the four regions of the United States.

Basic

Arrange the class into four groups. Assign each group one of the four regions of the country. Have each group create a poster with illustrations and labels for the landforms, climates, and vegetation of their region. Ask each group to present their posters to the class.

On Level

Arrange the class into four groups. Assign each group one of the four regions. Have each group create a map with illustrations and labels for landforms, climates, and vegetation, major bodies of water and waterways, and states and capital cities. Ask each group to present their illustrated maps to the class.

Challenge

Arrange the class into four groups. Assign each group one of the four regions. Have each group create a board game that tests knowledge of landforms, climates, and vegetation, major bodies of water and waterways, and states and capital cities. Have each group present its game to the class.

The West has varied climates, plants, and animals. Much of the region has an arid or desert climate, but the Northwest coast is rainy. Hawaii has a tropical climate; most of Alaska is cold. Cacti such as the many-armed saguaro (suh•GWAHR•oh) grow in the deserts, while the redwoods, tallest of all trees, grow along the Pacific coast. Wildlife includes snakes, lizards, wild cats, wolves, and the massive grizzly bear.

Life in the West Ranching, mining, and logging are important industries in the region. Parts of California and Hawaii are devoted to agriculture, while oil drilling is a major industry in Alaska. Recently, high-tech industries have clustered in California and Washington.

Some areas, such as southern California, are very urban, but most Western states are sparsely populated. The West has many national parks to preserve its landscapes. One is at the **Grand Canyon**, a deep, colorful gorge that the Colorado River cut through northern Arizona over many millions of years.

The Grand Canyon The Grand Canyon in Arizona was created by erosion caused by the Colorado River.

SUMMARIZE Describe the climates of the West.
The West has many climates: arid, rainy and mild, tropical, and very cold.

ONLINE QUIZ For test practice, go to **Interactive Review** @ClassZone.com

Section 1 Assessment

TERMS & NAMES

1. **Explain the importance of**
   - Great Lakes
   - Mississippi River
   - Great Plains
   - Rocky Mountains

USE YOUR READING NOTES

2. **Categorize** Use your completed chart to answer the following question:
   Which of the four regions have mountains? Name the mountains found in each region.

	LANDFORMS	CLIMATE	VEGETATION
Northeast			
Midwest			
South			
West			

KEY IDEAS

3. How did the geography of the Northeast affect its economic development?
4. In what way does the Mississippi River provide an economic link between the Midwest and South?
5. Why are there many national parks in the West?

CRITICAL THINKING

6. **Draw Conclusions** Why is the Northeast coast so crowded with cities?
7. **Analyze Causes** Why is the South in more danger from hurricanes than the Midwest or the Northeast?
8. **CONNECT to Today** In 2005, the U.S. government debated whether to allow companies to drill for oil in a part of Alaska that had been protected for wildlife. What arguments might be made for banning or allowing drilling?
9. **SCIENCE** **Create a Poster** Learn about the Grand Canyon in an encyclopedia or educational Web site. Then create a poster explaining how it was formed.

CONNECT to the Essential Question

How did a developing nation grow into a world power?

Ask students what they have learned so far that can help them answer this question. Students might mention:

- Physical features have aided the development of the United States.
- The diverse landforms, climates, and vegetation of the United States support a diverse economy.

4 Assess & Reteach

Assess Have students complete the Section Assessment.

Unit Resource Book: The United States & Canada
- Section Quiz, p. 53

Interactive Review @ClassZone.com
- Section Quiz

Power Presentation

Test Generator

Reteach Have pairs of students create one "True or False" statement about each of the four regions of the United States. Then have each pair exchange their statements with another pair and evaluate the statements. For each false statement, have students correct it, using their textbooks.

Unit Resource Book: The United States & Canada
- Reteaching Activity, p. 57

SECTION 1 ASSESSMENT ANSWERS

Terms & Names

1. Great Lakes: largest group of freshwater lakes in world; Mississippi River: largest U.S. river; Great Plains: grassland in North America; Rocky Mountains: range extending 3,000 miles

Use Your Reading Notes

2. The Appalachians are in the Northeast and the South, and the Rockies are in the West.

	LANDFORMS	CLIMATE	VEGETATION
Northeast	ocean, mountains	warm & humid, cold & snowy	rocky terrain, forests
Midwest	lakes, rivers, plains	hot summers, cold winters	plains, grasses, wildflowers
South	mountains, rivers	hot summers, mild winters	wetlands
West	mountains	arid, rain in Northwest	deserts, redwoods on Pacific coast

Key Ideas

3. The Northeast has rocky land but good harbors, so it turned to shipping and fishing.
4. The Mississippi is a major waterway on which goods are shipped.
5. People wanted to protect western landscapes.

Critical Thinking

6. Northeast had good harbors, so shipping is important; port cities that depend on shipping are usually located on a coast.
7. Coastlines put a region at risk from hurricanes. The South has a long coastline.
8. Some students will favor the economic growth offered by new oil resources; others will favor protecting wildlife over economic growth.
9. Use the rubric to score posters.

Poster Rubric

	Accuracy	Illustrations
4	thorough, accurate explanation	neat, clearly labeled, accurate
3	accurate	labeled, accurate
2	mostly accurate	some labels, mostly accurate
1	inaccurate	missing, inaccurate

1 Plan & Prepare

Objective

- Compare and contrast the landforms, bodies of water, climates, and economies of the four regions of the United States.

Springboard Tell students to close their texts. Ask volunteers to brainstorm words that would best describe each region. List all answers on the board or chart paper. After students have read the Comparing feature, ask what changes they would make to the lists for each region.

2 Teach

U.S. Geographic Regions

Talk About It

- Identify the scenes that illustrate each region. *(West: Grand Tetons in Wyoming; Midwest: Kansas farmland on the Great Plains; South: the Florida Everglades; Northeast: New England coast or Maine coast)*
- What other scenes could be used to illustrate each region? *(West: Rocky Mountains, Pacific coast; Midwest: Great Lakes, Mississippi River; South: Appalachian Mountains, Gulf of Mexico; Northeast: Appalachians, coastal plain)*
- **Compare and Contrast** In what ways are the regions similar economically? In what ways are they different? *(Possible Answers: trade in the Midwest and Northeast; mining in the West and coal mining and oil drilling in the South; manufacturing in the Midwest, South, and Northeast, but not in the West)*

More About . . .

Bald Eagles

Bald eagles are unique to North America. Large numbers of them live in Alaska and Canada, but they can also be found in most of the states in the continental United States. In 1782, the bald eagle was selected as the official national bird of the United States. By the mid-1900s, however, their numbers had declined, many killed by hunters and trappers. Others died as wilderness was converted to farmland, as urban development spread, and by pesticides. Today their numbers are increasing.

COMPARING U.S. Geographic Regions

The United States is divided into four main regions. They are distinguished from one another by their landforms, bodies of water, climates, and economies.

WEST

The West has a great variety of scenery (see Wyoming's Grand Tetons below), climate, and economic activities.

Landforms: Rocky Mountains
Bodies of Water: Pacific Ocean
Climate: varied—arid; rainy and mild; cold; tropical
Economy: ranching, mining, logging, farming, high-tech industries

Bears are found throughout the West.

Bald eagles thrive in the Alaskan wilderness

DIFFERENTIATING INSTRUCTION: TIERED ACTIVITIES

OBJECTIVE Create a PSA (public service announcement) about protecting natural areas in each U.S. region.

Basic

Assign one region to each of four groups. Explain what a PSA is (a TV or radio ad that provides information as a public service). Have each group create a one-minute PSA for its region on pollution. Have groups create a script, practice it, and then perform their PSAs for the class.

On Level

In addition to the Basic level instructions, have each group use available classroom resources to include information on how to help protect animal habitats common to its region in their PSAs. Encourage students to add music, sound effects, and visuals, as appropriate.

Challenge

In addition to the On Level instructions, have each group create a 15-second and 30-second PSA, in addition to their one-minute PSA. After each group has presented all their PSAs, discuss the pros and cons of PSAs of different time lengths.

MIDWEST

The Midwest has extremes in weather and ways of life, ranging from farming (see Kansas alfalfa farm below) to manufacturing.

Landforms: Great Plains
Bodies of Water: Great Lakes, Mississippi River
Climate: hot summers, cold winters
Economy: farming, manufacturing, transportation, trade

Prairie dogs are common on the Great Plains.

NORTHEAST

The Northeast was the first area of North America to be heavily settled by Europeans. It is still the most densely populated region.

Landforms: Appalachian Mountains and Coastal Plain
Bodies of Water: Atlantic Ocean (see Maine lighthouse on Atlantic coast above)
Climate: warm, humid summers; cold, snowy winters
Economy: trade, manufacturing, service industries

Seagulls are common along the Atlantic coast.

SOUTH

The South has attracted industries and people in part because of the warm climate.

Landforms: Appalachian Mountains, Everglades (see above)
Bodies of Water: Gulf of Mexico, Mississippi River, Atlantic Ocean
Climate: hot rainy summers and mild winters
Economy: farming, manufacturing, coal mining, oil drilling

Alligators thrive in the Everglades.

CRITICAL THINKING

1. **Compare and Contrast** Which region does not have mountains as one of its basic landforms?
2. **Compare and Contrast** In which region does farming not play a major role in the economy?

Comparing Regions

Farming in the United States

Here are how some states compare in terms of agriculture, according to the U.S. Census Bureau (2004):

- Alaska—1,000 farms; about $14 million per year in total farm earnings
- California—77,000 farms; about $8.5 billion per year
- Hawaii—6,000 farms; about $221 million per year
- Iowa—90,000 farms; about $1.2 billion per year
- New York—36,000 farms; about $800 million per year
- Texas—229,000 farms; about $3.6 billion per year

The total land area of farms in the United States is about 936 million acres, or about 443 acres per farm.

3 Wrap-Up

Encourage students to review their Reading Notes chart for Section 1 and add additional headings, such as *bodies of water* and *economy*, and details based on their reading of the Comparing feature. To extend this activity, ask for volunteers to share some of the additions they made to their charts.

CRITICAL THINKING ANSWERS

1. **Compare and Contrast** Midwest
2. **Compare and Contrast** Northeast

INTERDISCIPLINARY ACTIVITIES

CONNECT to Science

Learn About Animals

What animals are native or common to your region of the country? Have students brainstorm a list of wild animals where you live. Ask volunteers to do research on animals in your area to learn more about their characteristics. If time permits, schedule a field trip to a zoo. Solicit help from zoo staff members in identifying, viewing, and learning more about animals native to your area.

CONNECT to Science

Explain Weather Patterns

What weather patterns are typical for the area where you live? Have the class discuss year-round weather patterns. What factors do students think contribute to creating the weather patterns that are typical of your area? If time permits, have students do further research on this subject and/or invite a professional meteorologist to talk to your class.

1 Plan & Prepare

Objectives

- Explain how the United States was first settled
- Describe how the United States became an independent nation and expanded its territory
- Analyze the regional differences that helped lead to the Civil War and to its end
- Identify the challenges the United States faced in the 1900s and in the early 21st century

Read for the Essential Question

Help students read for a purpose by reminding them of the Essential Question: "How did a developing nation grow into a world power?" Point out that they can use text features, such as Key Questions, special features, images, and maps to form answers to the question.

Vocabulary

Best Practices Toolkit

Use the ***Best Practices Toolkit*** to model strategies for vocabulary notetaking. Vary strategies throughout the year. Choose from: *Knowledge Rating, Predicting ABCs, Definition Mapping, Word Sort, Frayer Model (Word Squares), Magnet Words, Word Wheel,* and *Student VOC.*

Vocabulary Strategies, TT9–TT16

Reading Strategy

Best Practices Toolkit

Display the Web Diagram transparency.

- Remind students that main ideas are important ideas in a block of text and that details support main ideas.
- Model filling in main ideas and details using the subheads and text in this section.
- Have students suggest entries. Remind students that main ideas in each block of text in the section will help answer a Key Question.

Web Diagram, TT28

SECTION 2 Reading for Understanding

Key Ideas

BEFORE, YOU LEARNED

The United States spans the North American continent from the Atlantic to the Pacific oceans.

NOW YOU WILL LEARN

Native Americans and peoples from other continents all came together to build a nation.

Vocabulary

TERMS & NAMES

land bridge dry land that appeared between Asia and Alaska during the Ice Age

immigrant a person who moves to one country from another

colony a group of people who settle a distant land but are ruled by their homeland

Constitution the document that is the basis for the U.S. government

Louisiana Purchase the action by which President Thomas Jefferson bought the Louisiana Territory from France

Lewis and Clark Expedition a journey made by Lewis and Clark to explore the Louisiana Terrritory

manifest destiny the idea that the United States should own all the land between, and even beyond, the Atlantic and Pacific oceans

Civil War (1861–1865) a conflict between North and South over the issues of states' rights and slavery

transcontinental railroad a railroad that crosses a continent

Great Depression a period of severe economic decline from 1929 into the early 1940s

Dust Bowl a region of the Great Plains that suffered drought and suffocating dust storms

Reading Strategy

Re-create the diagram shown at right. As you read and respond to the **KEY QUESTIONS**, use the outer ovals to note important details.

Skillbuilder Handbook, page R4

FIND MAIN IDEAS

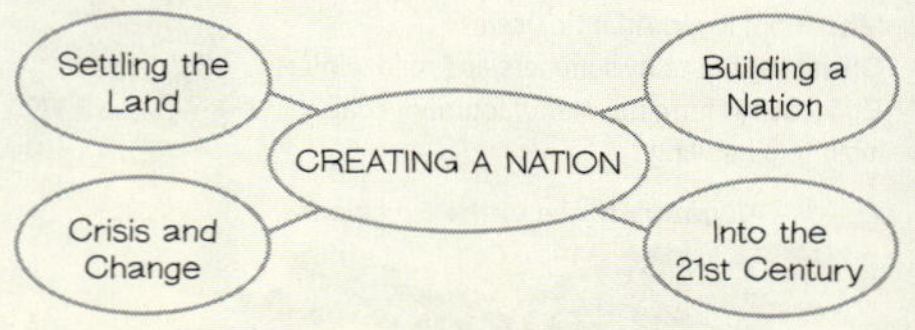

GRAPHIC ORGANIZERS Go to Interactive Review @ClassZone.com

PRETEACHING VOCABULARY

English Learners

Pronounce and Preview

Pronounce each term for students. Review the meanings of words within definitions, such as *document, decline,* and *drought.*

- To modify vocabulary learning, have students complete worksheets as they read, instead of afterward.

Unit Resource Book: The United States & Canada
- Vocabulary Practice, p. 35

Inclusion

Match Terms and Definitions

Have student pairs write the vocabulary words and their definitions on separate index cards. Then have them play a matching game. Ask one student to shuffle the cards and arrange them. After their partner has finished matching the terms and definitions, have the students work together to correct any errors. Students should then switch roles.

Oakland Bay Bridge, San Francisco

SECTION 2 HISTORY

Moose in Denali National Park, Alaska

Creating a Nation

Connecting to Your World

What do you think of when someone mentions U.S. history? Do you picture Native Americans hunting bison? Maybe you think of Pilgrims landing in New England or of the Liberty Bell. No matter what image comes to mind, it is probably connected in some way to geography. The ways that people chose to live on the land, the movement of people, and the interaction of cultures all have played a crucial role in shaping the past.

The Liberty Bell The Liberty Bell was rung to announce the first public reading of the Declaration of Independence in Philadelphia on July 8, 1776.

Settling the Land

KEY QUESTION Who were the first groups to settle the land that became the United States?

The history of the United States involves many different cultures, including Native Americans, Europeans, Asians, and Africans. Sometimes various groups such as Native Americans and European settlers fought, but more often, they learned from each other and helped to create a new, American culture. For example, the music of African slaves blended with that of Irish and Scottish settlers to form a new American music. The history of the United States is also the story of how a developing nation on the Atlantic coast grew to span a continent.

Bison Hunt Huge herds of bison once roamed the Great Plains.

2 Focus & Motivate

3-Minute Warm-Up

Write on the board or display the transparency:

- What challenges has America faced in the past and today? *(Possible Answers: disputes over land, war for independence, a civil war, world wars, terrorism)*

Unit Transparency Book
- 3-Minute Warm-Ups, TT1

Connecting to Your World

More About . . .

The Liberty Bell

The Liberty Bell is a symbol of freedom. It is inscribed with these words from the Bible: "Proclaim Liberty throughout all the land unto all the inhabitants thereof." It is on display in Philadelphia.

3 Teach

Settling the Land

Think, Pair, Share

- What groups inhabited North America before the arrival of European settlers? *(Inuit and Native Americans)* What ways of life did they pursue? *(They lived in close-knit clans and hunted, fished, and farmed.)*
- What groups of people settled colonies in North America in the 1500s? *(Spanish, English, French, African)*
- **Find Main Ideas** Why did Europeans first come to North America? *(for gold, land, and religious freedom)*

SECTION 2 PROGRAM RESOURCES

Unit Resource Book: The United States and Canada

Unit Resource Book
- Reading Study Guide, p. 5
- Map and Graph Skills, p. 37
- Section Quiz, p. 54

STRUGGLING READERS

Unit Resource Book
- Reading Study Guide with Additional Support, p. 13
- Skillbuilder Practice: Sequencing Events, p. 36
- Section Quiz, p. 54
- Reteaching Activity, p. 58

eEdition with Audio DVD-ROM

ENGLISH LEARNERS

Pupil Edition in Spanish

eEdition with Audio DVD-ROM

Unit Resource Book
- RSG (Spanish), p. 21
- RSG with Additional Support (Spanish), p. 29

Multi-Language Glossary

Test Generator
- Section Quiz in Spanish

INCLUSION

Unit Resource Book
- RSG with Additional Support, p. 13
- Section Quiz, p. 54
- Reteaching Activity, p. 58

GIFTED & TALENTED

Unit Resource Book
- History Makers, p. 41
- Interdisciplinary Projects, pp. 43, 45
- Section Quiz, p. 54

PRE-AP

Unit Resource Book
- Primary Source, p. 48
- World Literature, p. 143

TECHNOLOGY

Unit Transparency Book
- 3-Minute Warm-Ups, TT1
- Essential Question, TT2
- Critical Thinking, TT4
- Map: Selected Native Peoples of North America, c. 1600, TT8

Daily Test Practice Transparencies
- Section 2, TT15

Power Presentations

ClassZone.com

World Cultures and Geography Video on DVD

CONNECT Geography & History

A Land Bridge

Explain that the Ice Age caused the level of the sea to drop several hundred feet, exposing the land bridge connecting Siberia to Alaska. Scientists believe that the land bridge made it possible for many plants and animals to migrate to North America, in addition to humans.

- Why would people have migrated from Asia to North America? *(Possible Answers: to follow animal herds for hunting, to find new land for farming, to find new natural resources, such as timber)*

More About . . .

St. Augustine, Florida

Founded about 42 years before the English colony at Jamestown, St. Augustine today bills itself as the "nation's oldest city." The city's Spanish heritage can still be seen in its street pattern and architecture.

Teach

Building a Nation

Talk About It

- What did the colonists achieve by winning the Revolutionary War? *(They won their independence from Great Britain and the opportunity to build a new nation.)*
- What is the purpose of the U.S. Constitution? *(It provides a basis for the U.S. government.)*
- **Find Main Ideas** What is manifest destiny? *(the idea that the United States should own all land between and even beyond the two oceans)* How did the United States fulfill its manifest destiny? *(by acquiring territory piece by piece and expanding the country, such as by the Louisiana Purchase)*

CONNECT Geography & History

A Land Bridge

Today, Alaska and Siberia are separated by the Bering Strait, a strip of sea only 55 miles wide. During the last ice age, glaciers moved south from the North Pole, freezing up the waters of the Bering Strait and exposing more land. This formed the Beringia land bridge, over which the earliest Americans may have migrated from Asia.

Native Americans Scientists believe that the first inhabitants of North America came from Asia. According to one theory, the Ice Age glaciers held so much frozen water that ocean levels fell. The drop in sea level uncovered a **land bridge**, a stretch of dry land between Asia and Alaska. Many thousands of years ago, Asians hunting for food crossed into North America. Other people dispute the land bridge theory. Some scientists believe that people came from Asia on boats.

No matter how humans arrived, people eventually spread to all parts of the Americas. Those who settled North America are split into two main groups: the Inuit, who live in the far north, and Native Americans, who are divided into many smaller groups. By the year 1500, hundreds of cultural groups lived in North America. They spoke different languages. Some lived by hunting and fishing, while others farmed. Most lived in close-knit societies organized by clans.

The First Colonies In the 1500s, Europeans began to come to North America. This began a pattern of immigrants coming to the United States from around the world. An **immigrant** is a person who moves to one country from another. Some Europeans came looking for gold or land. Others came to find religious freedom. Many Africans were brought to the colonies by force to work as slaves.

Over time, Europeans started many American colonies. A **colony** is a group of people who settle a distant land but are ruled by their homeland. The Spanish founded St. Augustine, Florida, one of the oldest U.S. settlements. They also built missions in the Southwest. English settlements include Jamestown, Virginia, and Plymouth Colony in Massachusetts. The French colony, New France, included eastern Canada and lands near the Great Lakes and Mississippi River.

A SUMMARIZE Describe the first groups that settled the United States. Among the first groups of settlers were the Spanish, English, French, and Africans who encountered the Native Americans who were already here.

Building a Nation

KEY QUESTION How did the United States gain new territory?

In 1763, Great Britain defeated France in the French and Indian War and took over New France. To pay off the war debt, Britain's rulers imposed new taxes on the American colonists.

DIFFERENTIATING INSTRUCTION

English Learners

Vocabulary: Idioms

Students learning English may have difficulty with the expression *close-knit societies.* Explain that the word *knit* here means "firmly joined together" and that a close-knit society is a group of people united by common values and common interests.

Struggling Readers

Clarify the Land Bridge Location

Some students may be unaware how close parts of eastern Asia are to the western borders of the United States. Ask them to identify the locations of Siberia, the Bering Strait, and Alaska on a globe or world map to help them understand the relatively short distance separating eastern Asia from Alaska.

ANALYZING Primary Sources

Thomas Jefferson (1743–1826) was the third president of the United States. He was proud of having written the Declaration of Independence. Here is the most famous quotation from it.

> **We hold these truths to be self-evident [obvious], that all men are created equal, that they are endowed [provided] by their Creator with certain unalienable [not able to be taken away] Rights, that among these are Life, Liberty and the pursuit of Happiness.**
>
> Source: *The Declaration of Independence*, by Thomas Jefferson

DOCUMENT-BASED QUESTION

Why might it be more accurate to say that people have a right to pursue happiness instead of the right to be happy?

Benjamin Franklin reads Jefferson's Declaration. ▶

ONLINE PRIMARY SOURCE To read more of Thomas Jefferson's writing, go to the **Research & Writing Center** @ClassZone.com

Britain also passed laws to control the colonies. The colonists objected to the new laws because they had no representatives in the British lawmaking body to speak for their interests and concerns.

The Revolution and Constitution In 1775, the Revolutionary War began, and in 1776, the colonies proclaimed their freedom in the Declaration of Independence, written by Thomas Jefferson. By 1783, the United States had won the war against Great Britain and become a new nation.

Americans feared abuses of power, so they set up a weak national government. But leaders soon realized that the government was too weak to run the country well. In 1787, leaders from 12 states met to create a new government. They wrote the **Constitution** of the United States, the document that is the basis for the U.S. government.

The Constitution set up a federal government—a union of states with a central authority—divided into three equal branches.

1. The legislative branch, Congress, makes the laws. It consists of two houses—the Senate and the House of Representatives.
2. The executive branch is led by the president, the head of the government. George Washington, who won the 1788 presidential election, was the first president of the United States.
3. The judicial branch settles disputes about laws and includes all the courts in the country, including the Supreme Court.

ANALYZING Primary Sources

When the Second Continental Congress convened in May of 1776, some delegates hoped that the grievances against Britain could be reconciled. When it became clear that reconciliation was not possible, the Congress debated a proposal to declare the colonies free and independent states. It then appointed a committee to create the Declaration of Independence, and the committee chose Thomas Jefferson to write it.

- Why might July 4, 1776, be celebrated as the birthday of the United States? *(It honors the day that the American colonies adopted the Declaration, formally declaring their freedom and independence from Britain.)*

DOCUMENT-BASED QUESTION ANSWER

Possible answer: "To be happy" is a feeling that no government can ensure people will feel; "to pursue happiness," however, is something that the government can help ensure through fair laws and protection of personal liberties.

Unit Resource Book: The United States & Canada

- History Makers: Thomas Jefferson, p. 41
- Primary and Secondary Sources: The Declaration of Independence, p. 48

More About . . .

The Writing of the Constitution

The Articles of Confederation, the law of the nation until 1789, gave the states extensive powers. Believing that the federal government needed to be strengthened, delegates met in Philadelphia in late May of 1787 to create a new constitution. However, delegates disagreed on several major issues. The Great Compromise, which provided for equal representation in the Senate and representation based on population in the House of Representatives, resolved one of the most vexing disagreements—although it still left unresolved issues over slavery. Delegates signed the final document on September 17, 1787, and it was ratified in 1788. A promise to add a Bill of Rights as the first amendments to the new constitution helped ensure ratification.

INTERDISCIPLINARY ACTIVITY

CONNECT to Civics

Chart How Government Branches Work

Have students make charts, such as the one to the right, to help them better understand how the U.S. government works. Have them use their textbooks to complete their charts. As students finish each section of this chapter, have them update their charts. Encourage students to use past and present examples that they learn about from newspapers or television.

Branch	Functions	Examples
1. Legislative	1. Congress	1. makes laws
2. Executive	2. President	2. enforces laws
3. Judicial	3. Supreme Court	3. settles disputes

Animated GEOGRAPHY
Growth of the United States 1783–1853
CLASSZONE.COM

CONNECT Geography & Culture

Draw attention to the note in the bottom right corner of the map. Point out that the map shows contemporary state borders. The areas that are shaded in different colors identify territories the United States acquired *before* individual states were formed in those territories. Then have students study the dates and territories that the United States acquired.

- Using the dates territories were acquired, in which general direction did the United States expand? *(westward)*

ANSWERS

1. **Region** California, Nevada, Utah, Wyoming, Colorado, Arizona, New Mexico
2. **Region** Mississippi, Alabama

Unit Resource Book: The United States & Canada
- Map and Graph Skills, p. 37
- Outline Map Activities, p. 39

More About . . .

The Lewis and Clark Expedition

Lewis and Clark's exploration of the Louisiana and Oregon territories is considered one of the most exciting adventures in U.S. history. The expedition accomplished several purposes, including tracing the borders of the Louisiana territory, gathering scientific information, establishing contact with Native Americans, and initiating claims to the Oregon territory. Starting out from St. Louis, Lewis and Clark led a team of explorers up the Missouri River, across the Rocky Mountains on horseback, and down the Columbia River to the Pacific Ocean. Along the way, they survived mishaps, sickness, hunger, exhaustion, severe weather, Native American attacks, and encounters with dangerous animals. The expedition took more than two years to complete.

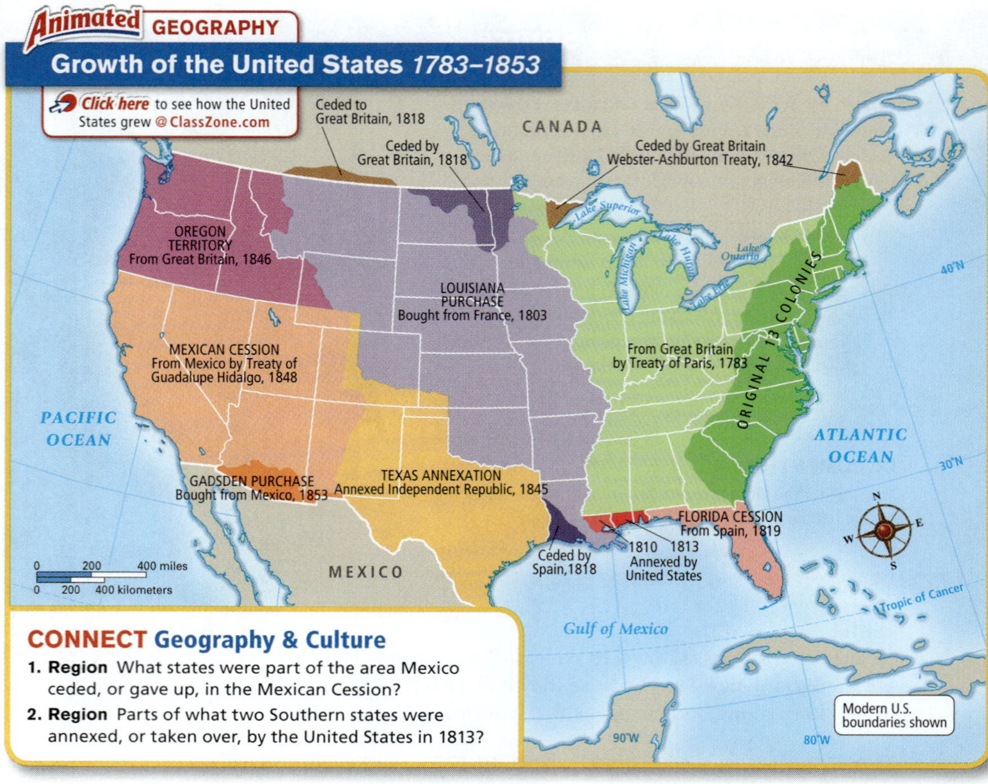

CONNECT Geography & Culture

1. **Region** What states were part of the area Mexico ceded, or gave up, in the Mexican Cession?
2. **Region** Parts of what two Southern states were annexed, or taken over, by the United States in 1813?

The Nation Grows Even before the war, Americans had begun to move west to new lands. After the war, the United States gained all of Britain's territory south of Canada and west to the Mississippi. In 1803, President Thomas Jefferson bought Louisiana, a large territory between the Mississippi River and the Rocky Mountains, from France. The **Louisiana Purchase** doubled the size of the United States. Jefferson sent Meriwether Lewis and William Clark to explore the region. The **Lewis and Clark Expedition** gained valuable knowledge of the West.

During the 1800s, many Americans believed in **manifest destiny**, the idea that the United States should own all the land between, and even some beyond, the oceans. The United States gained that territory piece by piece. Texas declared itself free from Mexico in 1836 and became a state in 1845. Then the United States and Mexico fought the Mexican War (1846–1848). The United States gained a huge region stretching from Texas west to the Pacific and north to Oregon.

FIND MAIN IDEAS Describe how the United States acquired new territory. through war and purchase

134 Chapter 5

DIFFERENTIATING INSTRUCTION

Struggling Readers

Create a Time Line of U.S. Acquisitions

Provide students with a sample time line, such as the one shown at right. Tell them to create a time line similar to it, highlighting the dates when the United States acquired territories using the map on this page. Students' time lines should include call outs that briefly explain each acquisition.

Unit Resource Book: The United States & Canada
- Skillbuilder Practice: Sequencing Events, p. 36

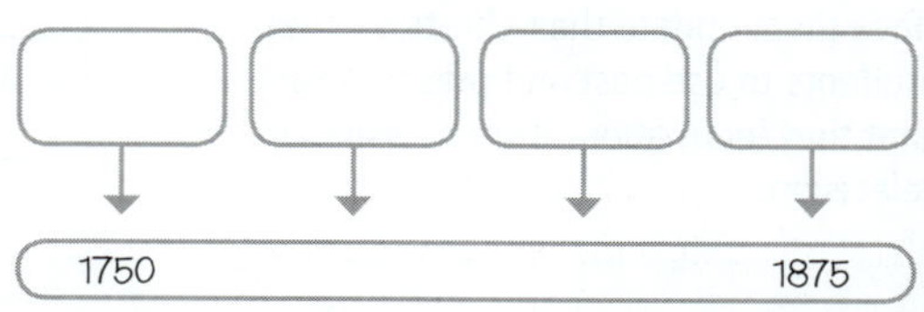

Crisis and Change

KEY QUESTION How did regional differences help to cause and end the Civil War?

The Northeast became the first manufacturing region. In contrast, large farms called plantations were common in the South. Plantations needed a lot of workers, and the South became a region that relied on African slave labor.

Growing Pains and Civil War From the nation's beginning, some Americans wanted to end slavery. In time, the country split into opposing sides: Northern free states and Southern slave states. The South came to believe that states had the right to reject federal laws and even to secede, or withdraw, from the Union.

In 1860, Abraham Lincoln, a Northerner who hated slavery, was elected president. In response, 11 Southern states seceded from the Union, forming the Confederate States of America. Fighting broke out over secession, and the **Civil War** began in 1861.

At first, the South won more battles. But the North had a larger population and factories to make the supplies the army needed, and so the North defeated the South in 1865. After the war, North and South were reunited.

Expansion In 1867, the U.S. government purchased Alaska from Russia. In 1898, the United States won the Spanish-American War and gained Puerto Rico, Guam, and the Philippines.

The economy also grew, especially in the Northeast. Immigrants poured into U.S. cities. A larger population needed more food, clothing, and housing. Factories replaced hand manufacturing with machines that made goods more quickly. In 1869, workers completed the first **transcontinental railroad** that crossed the continent from coast to coast. The United States continued to build railroads, making it easier to ship goods to market.

HISTORY MAKERS

Abraham Lincoln (1809–1865)

Born in a log cabin, Abraham Lincoln was a self-educated man. When he ran for president in 1860, his critics said he lacked experience. They mocked his rough appearance. Yet Lincoln became president and led the nation through its worst crisis.

During the Civil War, he helped plan the North's war strategy. His Gettysburg Address inspired the nation. Above all, he never forgot his main goal—to save the union. Lincoln achieved that goal but paid a terrible price. Days after the war ended, a Southerner killed him.

ONLINE BIOGRAPHY For more on the life of Abraham Lincoln, go to the Research & Writing Center @ClassZone.com

COMPARE AND CONTRAST Explain how regional differences helped to cause and end the Civil War.

The continuation of slavery in the South helped to cause the war; the development of industry in the North helped that side to win the war.

DIFFERENTIATING INSTRUCTION

Gifted & Talented

Illustrate a Civil War Event

Students skilled in drawing or painting might enjoy creating a picture to illustrate one or more events during or after the Civil War. Encourage them to use their imaginations to depict Lincoln at Gettysburg, the building of a section of railroad, or a collage containing multiple scenes. Display the artwork and invite students to explain why the scenes they depicted were important to U.S. history.

Pre-AP

Clarify the South's Resistance to Ending Slavery

Have students write a brief explanation of why so many people in the South opposed ending slavery. Explanations should focus on the cultural and economic reasons for this Southern opposition. Encourage students to use classroom and library resources to find support for their main ideas.

Teach

Crisis and Change

Reader, Recorder, Reporter

- What major crisis divided Americans in 1861? *(Bitter differences over slavery led to the Civil War.)*
- What factors helped the North defeat the South in the Civil War? *(The North had a larger population and more factories to make supplies for its army.)*
- **Find Main Ideas** What major events contributed to the growth of the United States in the last half of the 19th century? *(the purchase of Alaska; the acquisition of Puerto Rico, Guam, and the Philippines; the arrival of more immigrants; the introduction of machines in factories to make goods more quickly; the completion of the transcontinental railroad)*
- What were the main economic differences between the North and the South prior to the Civil War? *(The Northeast was the first manufacturing region while the South relied on plantations and slave labor.)*

RESEARCH & WRITING CENTER

HISTORY MAKERS

Abraham Lincoln

Lincoln is still regarded by many people in the United States and around the world as one of history's greatest heroes. Known as Honest Abe and the Great Emancipator, Lincoln was admired—even by his critics—for his common sense, fairness, and honesty. Some historians believe that his shrewd military strategies were the key to the Union victory in the Civil War. By preserving the Union, Lincoln helped ensure the continued unity and growth of the United States.

Unit Resource Book: The United States & Canada
- World Literature: The Gettysburg Address, p. 143

Teach

Into the 21st Century

Think, Pair, Share

- Identify four crises that Americans confronted in the first half of the 20th century. *(World War I, the Great Depression, the Dust Bowl, World War II)*
- Identify some social and technological advances that were achieved in the second half of the 20th century. *(Civil rights and women's liberation movements brought more equality. Space exploration spurred technological advances.)*
- **Draw Conclusions** What challenges and responsibilities has the United States faced as a world leader? *(Possible Answers: competing with the Soviet Union to help ensure freedom around the world; responding to the threats of terrorism; providing aid to people in need)*

Unit Resource Book: The United States & Canada

- Interdisciplinary Projects: Science: Competition for Space, pp. 43, 45

CONNECT Geography & History

The Dust Bowl

The major dust storms that created the Dust Bowl began in the early 1930s. Some of the hardest hit states included Kansas, Colorado, Texas, Oklahoma, and New Mexico. The winds were so strong and the dust clouds so thick that many people suffered severe lung damage. In addition to the loss of farmlands, farm machinery was also destroyed. As a result, thousands of families abandoned their farms and set out for California to seek jobs.

Fun Facts!

Seabiscuit

In 1938, Seabiscuit was the most featured news story in the country. It has been estimated that up to 40 million people listened to his races on the radio (about 1/3 of all Americans at the time). The film *Seabiscuit* (DVD Universal, 2003) tells his story and those of his owner, trainer, and jockey.

CONNECT Geography & History

The Dust Bowl

During World War I, farmers plowed up the Great Plains to plant crops. When drought hit in the 1930s, crops died. Wind storms blew away the soil, which no longer had the deep roots of prairie plants to hold it. Dust damaged farms across a 25,000-square-mile region called the Dust Bowl. Dust clouds are shown rolling in over Lamar, Colorado, in 1934.

Into the 21st Century

KEY QUESTION What changes did the United States experience during the 1900s?

During much of the 1700s and 1800s, Americans were more concerned with challenges that their own country faced than with events around the world. During the 1900s, however, the U.S. role in the world changed.

Turmoil and Change When World War I started in Europe in 1914, most Americans wanted to stay out of the fighting. Then Germany sank a number of ships, killing U.S. civilians. The United States joined the war in 1917 and help the Allies to win it.

After the war, the 1920s seemed to be an era of wealth and progress. For example, the affordable automobile was built and more people than ever owned stocks. But problems lurked behind the good times. In 1929, stock prices plunged, and many people lost their money. That began the **Great Depression**, a period of severe economic decline (1929–early 1940s) in the United States. Businesses failed, and millions were out of work. Life was hard in the **Dust Bowl**, a large area in the Great Plains that suffered drought and dust storms.

By the late 1930s, war was again raging in Europe and Asia, and again the United States wanted to avoid fighting. Then in 1941, Japan attacked Pearl Harbor, Hawaii, and the United States went to war. The fighting lasted for four more years until the war ended in 1945.

Fun Facts!

SEABISCUIT

During the dark days of the Depression, the story of Seabiscuit (at far right) lifted people's spirits. He was an unlikely looking race horse who managed to defeat some of the great racing champions of the day, such as War Admiral. Seabiscuit stood for the underdog overcoming adversity.

DIFFERENTIATING INSTRUCTION: TIERED ACTIVITIES

OBJECTIVE Create a chart listing the challenges and accomplishments of the United States in the 20th century.

Basic

On the board, list in mixed order U.S. challenges and accomplishments in the 20th century. Have students create a two-column chart, labeled *U.S. Challenges 1900s* and *U.S. Accomplishments 1900s.* Have students work in pairs to sort challenges from accomplishments.

On Level

Have students work individually to create a two-column chart, labeled *U.S. Challenges 1900s* and *U.S. Accomplishments 1900s.* Have students use their textbooks to complete their charts, listing each item in chronological order, under the appropriate heading.

Challenge

In addition to the On Level activity, have students write a summary of how the role of the United States in the world changed. To extend this activity, challenge students to explain what they think the role of the United States should be in the 21st century.

Nation and World U.S. life was transformed in the 20th century. The civil rights and women's liberation movements brought more equality. Rapid technological change was spurred by space exploration, including the moon flights and the space shuttle program.

After World War II, the United States became a world leader. For decades, it competed with the Communist Soviet Union to be the strongest political and military power. When the Soviet Union fell apart in 1991, the United States was the richest and most powerful nation in the world.

In the 21st century, the United States confronted the danger of terrorism, which is the use of violence for political purposes. On September 11, 2001, terrorists flew airplanes into the World Trade Center in New York City and into the Pentagon. Approximately 3,000 people died. The U.S. government responded by launching a war against terrorism.

The United States also tries to be a leader in helping people. For example, when a deadly tsunami struck South Asia in December 2004, Americans gave more than $1 billion to relief efforts. Americans also serve in agencies that improve life in developing countries.

Moon Landing Apollo 15 moon base, established between July 30 and August 7, 1971

FIND MAIN IDEAS Describe U.S. changes during the 1900s.
It became a world leader, it developed a more equal society, and it underwent rapid technological change.

Section 2 Assessment

ONLINE QUIZ For test practice, go to **Interactive Review** @ClassZone.com

TERMS & NAMES

1. **Explain the importance of**
   - land bridge
   - Louisiana Purchase
   - manifest destiny
   - Dust Bowl

USE YOUR READING NOTES

2. **Find Main Ideas** Use your completed diagram to answer the following question:
   How did the U.S. role in the world change during the 1900s?

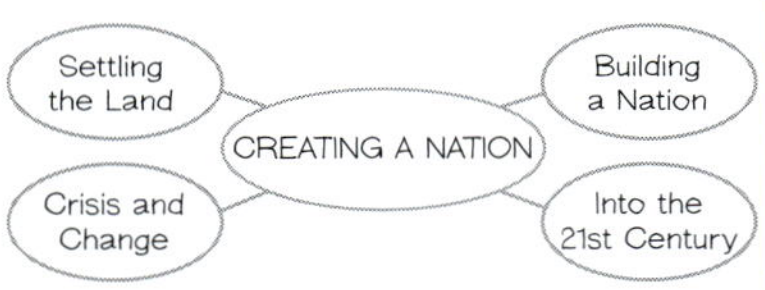

KEY IDEAS

3. Which three events added the largest amounts of land to the United States?
4. How did physical geography make human suffering during the Great Depression worse?
5. In what ways is the United States a world leader?

CRITICAL THINKING

6. **Compare** How was the U.S. response to the September 11 attacks similar to its response to the Pearl Harbor attack?
7. **Evaluate** Review the physical geography you studied in Section 1. What do you think was the biggest obstacle to building the transcontinental railroad?
8. **CONNECT to Today** Does technology make it more or less likely that the United States will return to its old policy of not taking part in world affairs? Explain.
9. **WRITING** **Write a Personal Response** Find one of Lincoln's Civil War speeches. Write a paragraph analyzing why that speech inspired people.

CONNECT to the Essential Question

How did a developing nation grow into a world power?

Ask students what they have learned so far that can help them answer this question. Students might mention:

- Native Americans and European settlers established settlements in North America.
- Colonists won their freedom from Britain, created a new nation, survived a bitter Civil War, and began expanding the nation's borders westward.
- As the nation grew, it overcame many challenges.

4 Assess & Reteach

Assess Have students complete the Section Assessment.

Unit Resource Book: The United States & Canada
- Section Quiz, p. 54

Interactive Review @ClassZone.com
- Section Quiz

Power Presentations

Test Generator

Reteach Arrange the class into four groups. Assign each group one of the subsections of Section 2. Have them list the important events and ideas covered in the subsection they have been assigned. Have each group present their lists to the class.

Unit Resource Book: The United States & Canada
- Reteaching Activity, p. 58

SECTION 2 ASSESSMENT ANSWERS

Terms & Names

1. land bridge: between Asia and Alaska during Ice Age; Louisiana Purchase: Jefferson bought from France; manifest destiny: United States should own all land between oceans; Dust Bowl: region of plains that suffered drought

Use Your Reading Notes

2. The United States went from a country that concentrated on its own affairs to a world leader.

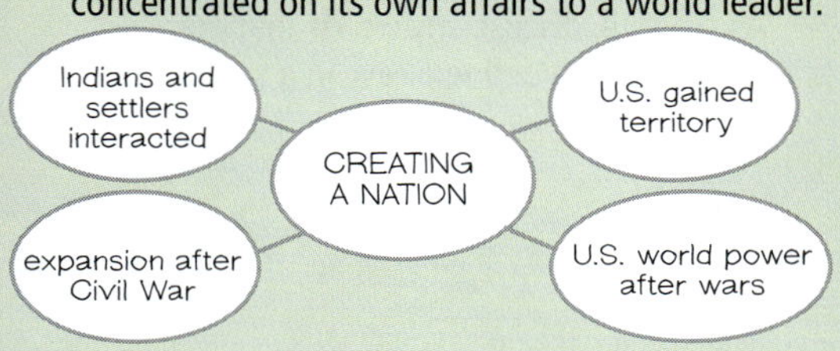

Key Ideas

3. Revolutionary War, Louisiana Purchase, war with Mexico
4. drought caused dust storms in Dust Bowl
5. strongest military and economic power; leads fight against terrorism

Critical Thinking

6. Possible answer: Both unprovoked attacks caused the United States to go to war.
7. Possible answer: the Rocky Mountains
8. Possible answer: It makes it less likely because technologies such as the Internet and wireless phones have made the world a much more interconnected place.
9. Use the rubric to score the paragraph.

Paragraph Rubric

	Structure	Content
4	logical and flowing	shows thorough understanding
3	clear	shows good understanding
2	somewhat confusing	shows basic understanding
1	confusing	shows incomplete understanding

1 Plan & Prepare

Objectives

- Identify the reasons for the diversity of the U.S. population, the richness that the diversity has provided, and the challenges facing the United States because of it
- Describe the arts and popular culture of the United States

Read for the Essential Question

Help students read for a purpose by reminding them of the Essential Question: "How did a developing nation grow into a world power?" Point out that they can use text features, such as Key Questions, special features, images, and maps to form answers to the question.

Vocabulary

Best Practices Toolkit

Use the ***Best Practices Toolkit*** to model strategies for vocabulary notetaking. Vary strategies throughout the year. Choose from: *Knowledge Rating, Predicting ABCs, Definition Mapping, Word Sort, Frayer Model (Word Squares), Magnet Words, Word Wheel,* and *Student VOC.*

Vocabulary Strategies, TT19–TT16

Reading Strategy

Best Practices Toolkit

Display the Three-Column Chart transparency.

- Point out to students that to categorize information in the section they must decide how it can be sorted into the categories shown in the chart.
- Model filling in entries from the Cultural Diversity subsection in the section.
- Have students suggest entries, using the categories in the chart.
- Display the completed chart for review.

Three-Column Chart, TT21

SECTION 3 Reading for Understanding

Key Ideas

BEFORE, YOU LEARNED

From the nation's beginnings until the present time, people from many countries have moved to the United States.

NOW YOU WILL LEARN

The United States is one of the most diverse nations in the world, and it has a rich culture blended from many world cultures.

Vocabulary

TERMS & NAMES

public education schooling that is paid for by the government

literacy the ability to read and write

cultural blending something new created from combining the elements of two or more cultures

jazz a type of music that developed from a blending of African rhythms, American band music, and the musical styles of African Americans and Europeans

blues a type of music with lyrics that express sorrow, usually about problems in love or the hardships of life

BACKGROUND VOCABULARY

official language a language that by law must be used for government, business, and education

REVIEW

culture the behavior patterns, arts, and beliefs of a group of people

Visual Vocabulary Jazz in New Orleans

Reading Strategy

Re-create the chart shown at right. As you read and respond to the **KEY QUESTIONS**, use the chart to organize important details about the diverse culture of the United States.

Skillbuilder Handbook, page R7

CATEGORIZE

RELIGION	LANGUAGE	ETHNICITY	ARTS & CULTURE
1.	1.	1.	1.
2.	2.	2.	2.
3.	3.	3.	3.

GRAPHIC ORGANIZERS
Go to **Interactive Review** @ClassZone.com

PRETEACHING VOCABULARY

English Learners

Pronounce and Preview

Pronounce each term for students. Review the meanings of words within definitions, such as *elements* and *styles.*

- To modify vocabulary learning, have students complete worksheets as they read, instead of afterward.

Unit Resource Book: The United States & Canada
- Vocabulary Practice, p. 35

Inclusion

Reading and Defining Terms Verbally

Have volunteers read one term and its definition aloud. Ask others to define the term in their own words verbally, and then use it in a sentence. Repeat for each term. As a group, discuss any words that are challenging to define or use in a sentence.

SECTION 3 CULTURE

Oakland Bay Bridge, San Francisco

Moose in Denali National Park, Alaska

A Diverse Culture

Connecting to Your World

Think about some of the most popular foods in the United States. Many of them are versions of food that came from another country. Many recreational activities, too, came from overseas. Martial arts developed in East Asia, yoga originated in India, skiing first began in Norway, and baseball grew out of an English game. The **cultures**—behavior, arts, beliefs—of countless immigrants have made American culture what it is today.

Martial Arts Students practice karate, a Japanese martial art.

Cultural Diversity

KEY QUESTION What are some ways that immigrants have made U.S. culture more diverse?

People have come to the United States from all over the world. As you learned in Section 2, the earliest colonists were Spanish, English, and French. In addition, hundreds of thousands of Africans were brought to America as slaves. In the mid 1800s, millions of Northern Europeans came, mainly from Ireland, Germany, and the United Kingdom. After 1900, immigrants came largely from Italy, as well as from Russia and other Eastern European countries. In recent years, most immigrants have come from Mexico, other Latin American nations, and Asia.

New York City A crowd strolls in the middle of Fifth Avenue in New York during a street festival.

139

2 Focus & Motivate

3-Minute Warm-Up

Write on the board or display the transparency:

- Look at the photograph of New York City. How does this street scene help you to understand the diverse culture of the United States? *(Possible Answer: shows people with different clothing styles, jewelry, and hairstyles, which are examples of diversity)*

Unit Transparency Book
- 3-Minute Warm-Ups, TT1

Connecting to Your World

More About . . .

Martial Arts

In addition to karate, popular forms in the United States include judo and aikido (Japanese), kung fu (Chinese), and tae kwon do (Korean).

3 Teach

Cultural Diversity

Reader, Recorder, Reporter

- What two beliefs do most immigrants to America have in common? *(belief in a happy, successful, and better life in the United States; belief in the value of education)*
- **Categorize** Name three categories of cultural blending in the United States and identify one example for each. *(categories: music, food, language; accept examples, such as jazz, salsa, and words such as café)*

SECTION 3 PROGRAM RESOURCES

Unit Resource Book: The United States and Canada

ON LEVEL

Unit Resource Book
- Reading Study Guide, p. 7
- Section Quiz, p. 55

STRUGGLING READERS

Unit Resource Book
- Reading Study Guide with Additional Support, p. 15
- Skillbuilder Practice: Sequencing Events, p. 36
- Section Quiz, p. 55
- Reteaching Activity, p. 59

eEdition with Audio DVD-ROM

ENGLISH LEARNERS

Pupil Edition in Spanish

eEdition with Audio DVD-ROM

Unit Resource Book
- RSG (Spanish), p. 23
- RSG with Additional Support (Spanish), p. 31

Multi-Language Glossary

Test Generator
- Section Quiz in Spanish

INCLUSION

Unit Resource Book
- RSG with Additional Support, p. 15
- Section Quiz, p. 55
- Reteaching Activity, p. 59

GIFTED & TALENTED

Unit Resource Book
- History Makers: Martin Luther King, Jr., p. 42
- Section Quiz, p. 55

PRE-AP

Unit Resource Book
- Connect to Math, p. 47
- Section Quiz, p. 55

TECHNOLOGY

Unit Transparency Book
- 3-Minute Warm-Ups, TT1
- Essential Question, TT2
- Critical Thinking, TT5
- Cultures, TT9, TT10

Daily Test Practice Transparencies
- Section 3, TT16

Power Presentations

ClassZone.com

World Cultures and Geography Video on DVD

Fun Facts!

Leetspeak

A variation of leetspeak involves using keystrokes to create pictures that convey emotions or that comment in other ways on the message being sent. A popular example is the familiar "smiley face," resting on its side. It is formed by a combination of the colon, dash, and close-parenthesis keys and looks like this: :-). To make a frowning face, the open-parenthesis key is used: :-(.

More About . . .

The Foundation of Religious Diversity

From colonial times onward, many immigrants came to the United States to escape religious persecution in their countries of origin. An individual's freedom to exercise his or her religious beliefs without interference from the government is guaranteed in the Bill of Rights, which were the first 10 amendments to the U.S. Constitution. The First Amendment to the U.S. Constitution states that "Congress shall make no law respecting an establishment of religion, or prohibiting the free exercise thereof. . . ." The other freedoms guaranteed by the First Amendment are freedom of speech, freedom of the press, freedom of assembly, and freedom to petition the government.

Diversity and Shared Beliefs Because the United States has accepted high numbers of immigrants for most of its history, U.S. society is quite diverse. Even so, most people who come to the United States have two things in common. First, they believe in the idea that anyone from any background can have a happy, successful life. Many immigrants come to the United States hoping to make life better for themselves and their children.

Second, most immigrants believe in the value of education. The United States offers free public education through high school. **Public education** is schooling for which the government pays. As a result, the United States has a high rate of **literacy**, or the ability to read and write. About 99 percent of all Americans are literate.

Fun Facts!

LEETSPEAK

Have you heard of "133t5p34k" or "leet-speak"? Leetspeak is a kind of made-up written dialect. It comes from the word *elite*, which means a special, privileged group.

Leet began as part of the underground culture centered around text messaging on the Internet. It involves making changes to standard written text. So, for example, the term "leet" itself is often written "1337" or "133t."

Language and Religion Although English is spoken by almost everyone, the United States does not have an **official language**—one that by law must be used for government, business, and education. Because of the large number of people from Latin America, Spanish is the second most common language.

The way immigrants learn English often follows a pattern. Some adult immigrants struggle with English. The children of those immigrants, however, usually speak two languages—English and the language of their parents. Finally, the children of the next generation use English as their first language.

Diversity is also seen in the variety of religions practiced in the United States. A majority of Americans identify themselves as Christian, with 52 percent being Protestant and 24 percent being Catholic. The remaining Americans are Jews, Mormons, Muslims, and people who practice other religions or no religion.

Cultural Blending Music, language, and food all provide examples of the coming together of the behavior patterns, arts, and beliefs of people from different cultures. In music, traditions from Africa and Europe have blended to create a new American music. The same blending has occurred with food. Different ingredients and recipes from Latin America, Africa, Asia, and Europe have come together to create American cuisine. When elements from two or more cultures combine into something new, it is called **cultural blending**.

American English is one of the best examples of cultural blending. Americans use many words that come from other languages. For example, the terms *kindergarten*, *frankfurter*, and *sauerbraten* come from German. Enslaved Africans brought many terms with them. Some of these evolved into words we use every day, such as *banana* and *cola*. *Rodeo* and *canyon* are examples of words that come from Spanish. Many place names are of Native American origin. For example, the name *Mississippi* means "big river."

DIFFERENTIATING INSTRUCTION

English Learners

Key Academic Vocabulary

Review with students some of the key academic vocabulary words that appear on this page. You may want to focus on *generation* ("Finally, the children of the next generation use English as their first language.") and *traditions* ("In music, traditions from Africa and Europe have blended to create a new American music.")

Gifted & Talented

Perform Music Inspired by Other Cultures

Students who possess musical talents might enjoy performing a work that reflects a blending of other cultural traditions in an American musical piece. Invite members of the class to prepare an example to present to the class. Suggest a variety of options, such as folk, country, reggae, or salsa styles. Tell performers to prepare a brief introduction for their performances.

COMPARING Languages, Religion, and Ethnicity

ETHNICITY

These maps show the distribution of certain ethnic minorities. African Americans, for example, make up about 12 percent of the total U.S. population but in the Southeast they are more heavily concentrated and make up a higher percentage of the population. Hispanic Americans now make up about 13 percent and Asian Americans about 4 percent of the total U.S. population.

0%–10% | 11%–25% | 26%–50% | 51%–87%

AFRICAN AMERICAN

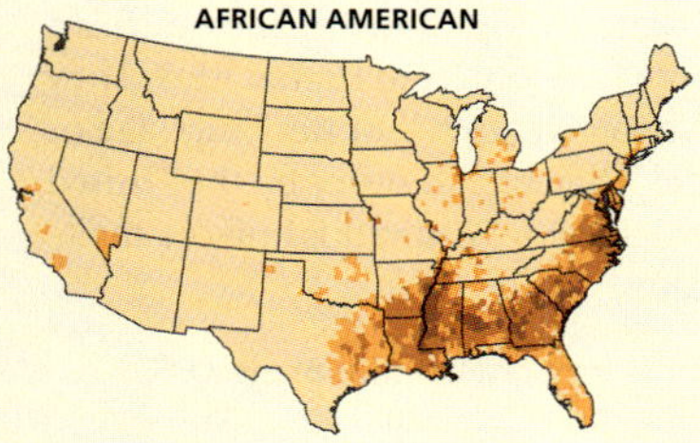

HISPANIC

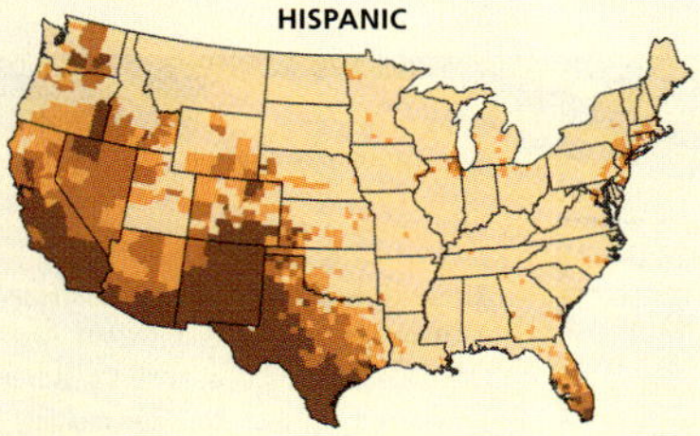

ASIAN

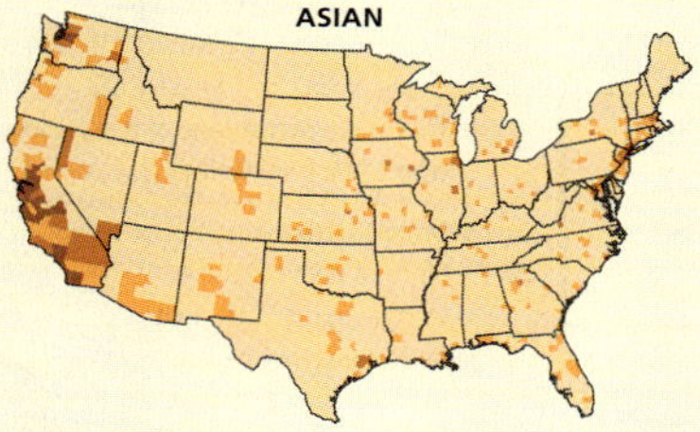

LANGUAGES

The chart below shows the most common languages spoken in U.S. households.

Rank	Language
1	English
2	Spanish
3	Chinese
4	French
5	German
6	Tagalog
7	Vietnamese
8	Italian
9	Korean
10	Russian

Source: U.S. Census Bureau (2000)

RELIGIONS

The pie graph below shows the religious preferences of Americans. Christians make up over three-quarters of the population.

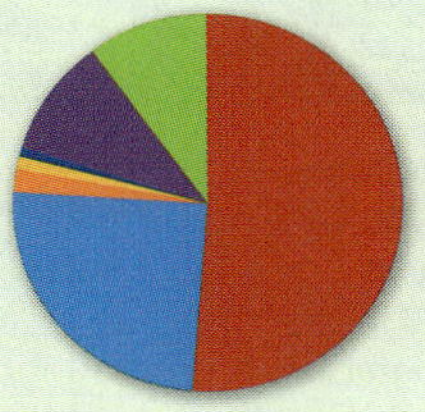

Source: *CIA World Factbook* (2002)

CRITICAL THINKING

1. **Categorize** What percentage of the population is Christian?
2. **Make Inferences** Why might Hispanics be so heavily concentrated in the West and Southwest?

INTERDISCIPLINARY ACTIVITIES

CONNECT to Math

Calculate World Religion Populations

About six and a half billion people live in the world today (6,500,000,000). Supply students with this number and the percentages given in the Comparing Regions side note to calculate the number of people who identify themselves as Christian, Muslim, Jewish, Other, or Nonreligious. Offer them the following suggestions:

- Convert each percentage in the Comparing Regions side note to a decimal (For example, 39% becomes .39).
- Multiple the total world religion population (6,500,000,000) by each of these decimals to obtain the number of people in each category.

To extend the activity, have students create a pie graph of world religions, following steps two and three in the Connect to Math feature activity on the next page.

COMPARING

Languages, Religion, Ethnicity

Have students examine the three maps. Ask them what reasons might explain the concentration of African Americans in the Southeast and Asians in the West. *(Possible answer: Large numbers of African Americans were originally brought to the South to work on plantations. The West would have been the nearest entry point for Asians arriving across the Pacific.)*

CRITICAL THINKING ANSWERS

1. **Categorize** about 76%
2. **Make Inferences** The West and Southwest are near countries, such as Mexico, from which Hispanics migrated.

Comparing Regions

World Religions

Have students compare U.S. statistics on religion with statistics for the world. About 33% of the world population identifies itself as Christian, 20% Muslim, about two tenths of 1% Jewish, 34% other (Hindus, Buddhists, and groups that practice local folk, tribal, or new-age religions), and 12% nonreligious.

More About . . .

Challenges of Diversity

One of the challenges of diversity in the United States is providing equal educational opportunities to all. In 1954, the Supreme Court ruled against segregation in public schools in its landmark case, *Brown* v. *Board of Education of Topeka.* However, equal education for all still remains one of the key goals of the today's civil rights movement.

Unit Resource Book: The United States & Canada
- History Makers: Martin Luther King, Jr., p. 42

More About . . .

Origins of Quilts and Mosaics

Although the origins of quilting were once thought by some to have been in America, researchers have now found that quilting was done by various peoples around the world throughout the past. With mosaics, some of the earliest evidence of its origins is from ancient Greece. However, other evidence shows that Assyrians, Persians, Egyptians, and other ancient civilizations also created mosaic art forms.

CONNECT to Math

Explain to students that Ellis Island was originally just three acres in size. Over the years, the addition of landfill expanded the island to about 27 acres to meet the growing numbers of immigrants.

The facilities on Ellis Island were first built in response to the massive influx of immigrants during the third great wave of immigration at the end of the 1800s. Its function as a processing center was reduced after 1924, when immigration rates began to decline. Ellis Island was closed permanently in 1954.

Today, the National Park Service manages the island. Its main building has been restored and houses a museum containing photographs, artifacts, and interactive exhibits honoring the millions of immigrants who entered the United States through Ellis Island.

Unit Resource Book: The United States & Canada
- Connect to Math: Make a Pie Graph, p. 47

Challenges of Diversity Immigrants have added much richness to American culture. But living in such a diverse country also has its difficulties. For example, immigrants tend to settle in cities where they can find other people from their homeland. As a result, some cities end up with high percentages of immigrants, many of whom don't speak English. These people face limited employment options and are more likely to live in poverty.

Diversity also affects education. Most immigrant children learn to speak and read English, but that learning doesn't happen overnight. In some places, such as California, teachers have classes in which students speak several different languages. Teaching such a group English—or any other subject—can be a tough job.

Finally, diversity sometimes leads to debate. Many people say that the United States is a melting pot. By that, they mean that over time, people who come to the United States adopt a uniform American culture. In recent times, some people have objected to that idea. They believe that a person does not have to give up his or her language or customs to be an American. Instead, they like to describe the United States as a quilt or a mosaic, in which many individual pieces make up the whole but still remain unique.

MAKE INFERENCES Explain how immigrants have made U.S. culture more diverse. They have introduced new words to English and introduced new foods.

CONNECT to Math

From 1892 to 1924, Ellis Island (shown below) in New York harbor was the main entry point for immigrants. An estimated 17 million immigrants passed through its facilities. After immigration services moved to New York City, Ellis Island served as a detention center for both aliens and deportees.

U.S. Immigration *1820–2004*

Country of Origin	Number of Immigrants (to the nearest thousand)
Germany	7,238,000
Mexico	6,849,000
Italy	5,446,000
United Kingdom	5,337,000
Ireland	4,788,000
Other	40,211,000
Total	69,869,000

Sources: Office of Immigration Statistics, Department of Homeland Security

Activity

Make a Pie Graph

1. Calculate the percentage of immigrants from each country on the chart. First divide the number of immigrants from a country by the total. Then multiply the answer by 100. For example:
   7,238,000 ÷ 69,869,000 = .104
   .104 × 100 = 10.4 percent
   That is the percentage of immigrants who came from Germany.
2. Draw a circle on a piece of paper. Draw a wedge on the circle that represents 10.4 percent of its total. Add wedges for each of the other countries and for Other.
3. Color each wedge a different color. Label the wedges with the country names. Add the numbers beneath the names.

DIFFERENTIATING INSTRUCTION

Struggling Readers

Understanding Metaphors That Describe American Culture

Explain that a metaphor is a figure of speech that is used to compare two unlike things. Both *melting pot* and *quilt* are metaphors that describe characteristics of our culture. Have students analyze these metaphors.

- In the phrase, "the United States is a melting pot," what does *pot* represent? *(the United States)* What does *melting* represent? *(the blending of many cultures)* What does the term *melting pot* mean? *(many cultures combining into one American culture)*
- In the phrase, "the United States as a quilt or mosaic," what does *quilt or mosaic* represent? *(American culture)* What do the individual pieces of the quilt or mosaic represent? *(individual ethnic groups or cultural influences)* What does the stitching or glue that binds the pieces together represent? *(shared values, customs, and beliefs that unite Americans)*

The Arts and Popular Culture

KEY QUESTION How has cultural blending affected U.S. arts?

Cultural diversity has had a huge impact on American arts. This impact can clearly be seen in literature and music.

Literature Many writers have produced novels about immigrants and their children. For example, Amy Tan wrote *The Joy Luck Club* about Chinese mothers and their Chinese-American daughters. Much American literature deals with American history or American experiences. For example, *The Great Gatsby* by F. Scott Fitzgerald portrays a man chasing the American dream during the 1920s.

Jazz, Blues, Country, and Rock American music has also been affected by cultural blending. One example of this is jazz. **Jazz** is a type of music that African Americans developed from a blending of African rhythms, American band music, and the musical styles of African Americans and Europeans. Another distinctive type of music related to jazz is the blues. **Blues** features lyrics that express sorrow, usually about problems in love or the hardships of life.

Two other types of American music developed from the blending of different styles. Country music grew out of folk music, the blues, and religious music. Great artists of country music include Hank Williams and Patsy Cline. Rock music evolved from elements of the blues, country music, and American popular songs. Chuck Berry and Elvis Presley were important names in the early history of rock music.

CONNECT Geography & Culture

The Blues and the Mississippi Delta
The blues developed in the South. One region closely associated with the music is the Delta, a broad, flat area in northwestern Mississippi. Many people there worked as sharecroppers, farming someone else's land for little reward. The blues were the perfect way for people in the Delta to express their troubles. Great blues artists include Robert Johnson (at right), Bessie Smith, and Muddy Waters.

DRAW CONCLUSIONS
What do you think some of the topics of Mississippi Delta blues might be?

DIFFERENTIATING INSTRUCTION

Pre-AP

Compile a List of Books About Immigrant Life in America

Have students compile a list of fictional or nonfictional books that they have read that involve immigrant experiences in the United States. Ask volunteers to describe the content of one book they have listed and what they learned from it. Compile the individual lists into a unified list, or class bibliography.

Inclusion

Appreciate Different Types of American Music

Students who have difficulty analyzing print materials might benefit from listening to the various types of music discussed on this page and the next page. Play examples of jazz, blues, country, rock, and hip-hop, and discuss the distinctive elements of each type. If time permits, invite students to play their own favorite recordings for the class.

Teach

The Arts and Popular Culture

Think, Pair, Share

- Jazz is a blend of what musical influences? *(African rhythms, American band music, and African-American and European musical styles)* Name two elements of hip-hop music. *(rapping and sampling)*
- In what way has the Internet made access to music easier? *(Online services provide a wide variety of music that can be easily downloaded onto compact hand-held players.)*
- **Categorize** List examples of the diversity that makes up the arts and popular culture that Americans enjoy today. *(Possible Answers: novels of immigrants and their children; jazz, blues, country, rock, and hip-hop music; shared leisure activities and entertainment, such as playing team sports and watching TV; hobbies, such as raising pets, gardening, and acquiring collectibles; volunteering)*

CONNECT Geography & Culture

The Blues and the Mississippi Delta

The term *the blues* refers to a state of sadness or depression. Downhome or country blues originated in the rural South and featured a solo male singer who accompanied himself on a guitar or other musical instrument.

- In the early 1900s, a form called classic blues achieved wide popularity. It incorporated elements of New Orleans jazz and featured female singers accompanied by bands.
- In the 1930s, Chicago became the center for a style of blues that combined elements of both country and classic blues.
- By the late 1940s, Chicago blues artists had begun to add electronic amplification to their performances. This style became immensely popular both nationally and internationally.

ANSWER

Draw Conclusions Possible Answers: family problems, interpersonal relationships, unemployment, poverty, social injustice

More About . . .

Recording Devices

Today's students may be unfamiliar with recording devices that preceded the invention of the compact disc (CD) and downloadable music. The first recording device, the phonograph, was invented by Thomas Edison in 1877. It made use of a tin-wrapped cylinder to record and play back sounds.

Flat disc records, played on a gramophone, were developed in the late 1890s. Long-playing records (LPs) were invented in the 1940s. Other recording formats included 8-track tapes and cassette tapes, which were introduced in the 1950s and 1960s. With the emergence of the CD in the early 1980s, phonograph records and tapes became less popular.

More About . . .

Televised Sports

The popularity of the Super Bowl can be illustrated by the intensity of the competition among television networks—and the huge sums of money they offer—for the rights to broadcast the game. Business sponsors also pay large sums to air commercials during the broadcast. Recent Super Bowl ads have cost over two-million dollars—for each 30-second ad. Many of these businesses use their Super Bowl ads to introduce new products or national advertising campaigns.

Today's American Music Scene Chances are that you and your friends listen to a wide range of music without even realizing it. Some of the most popular songs on the radio today echo the earlier blues and rock of your parents' generation, or even classical music. And there's a growing internationalism, with music from all parts of the world showing up in the CD racks. One of the most popular musical trends is hip-hop.

Hip-hop music developed in the South Bronx section of New York City in the 1970s. One technique involved in hip-hop is rapping, a form of chanted street poetry set to rhythmic music. Another technique is sampling, in which an existing song provides an underlying rhythm or melody and other sounds are superimposed over it—such as music made by scratching on a turntable or "beatboxing," creating beats and sound effects using the mouth. By the 1990s, hip-hop was a regular on the music charts and had begun to spread world-wide.

However, the biggest story in American music today has nothing to do with the kinds of music people are listening to—it's all about how we are listening. Access to music of all kinds and in huge numbers is higher than ever, thanks to the Internet. Online music services offer databases of inexpensive downloadable music from all genres with a click of the mouse. And all those songs can now be loaded right onto a player that easily fits into the palm of your hand. There's no predicting what the next change in music technology will be, but the evidence is clear: more people will be "wired" to a broad range of music that increases every day.

Sports, Entertainment, and Leisure Americans are avid sports fans. Many people take part in team sports and individual activities such as in-line skating or golf. In addition, millions follow team sports such as baseball, basketball, and football—both at stadiums and on television. In fact, the national football championship, the Super Bowl, is one of the most highly watched television programs each year.

Play Ball Fans cheer during a Seattle Mariners baseball game at Safeco Field in Seattle.

DIFFERENTIATING INSTRUCTION: TIERED ACTIVITIES

OBJECTIVE Design a series of storyboards to illustrate the diversity of life in the United States.

Basic

Have the class brainstorm examples of influences on American culture. Provide students with paper and markers. Have pairs of students create a four-panel storyboard on current influences on music, film, or literature. Have students present and display their storyboards.

On Level

Have pairs of students create eight-panel storyboards:

- The first four panels should show past cultural influences on one aspect of American culture.
- The second half should show current influences.

Have students present and display their storyboards.

Challenge

Have pairs of students create 12-panel storyboards:

- The first eight panels are as for the On Level activity.
- The last four panels should show what students predict will be future influences on American culture.

Have students present and display their storyboards.

Americans spend a lot of time on leisure activities and entertainment. Leisure is time that is free from work or other duties. One reason Americans have so much leisure time is that the United States is a prosperous nation. Perhaps the two most popular forms of entertainment are movies and television. A film or TV program can have an audience of millions. These two forms of entertainment have created experiences that the majority of Americans share. They help to unify the culture.

Other common pastimes in the United States include a wide variety of hobbies. Some of the most popular are playing computer games, camping, raising pets, gardening, and collecting things such as baseball cards, stamps, or autographs.

Finally, many Americans use part of their free time to improve their society. For example, some people cook meals at shelters for homeless people. Others volunteer to pick up litter along highways. Often, people find that helping others not only makes them feel better about themselves, but it makes their communities better too.

Skateboarding An 11-year-old boy skateboards in New York City.

ANALYZE CAUSES AND EFFECTS Explain how cultural blending has affected the arts in the United States. Cultural blending has produced great variety and diversity in the arts by combining elements from Europeans, Africans, Asians, and Native Americans in a rich and complex brew; jazz is a good example in music.

Section 3 Assessment

ONLINE QUIZ For test practice, go to Interactive Review @ ClassZone.com

TERMS & NAMES

1. **Explain the importance of**
   - public education
   - literacy
   - cultural blending
   - jazz

USE YOUR READING NOTES

2. **Categorize** Use your completed chart to answer the following question:

   What are some elements and examples of the diversity that makes up the culture of the United States?

RELIGION	LANGUAGE	ETHNICITY	ARTS & CULTURE
1.	1.	1.	1.
2.	2.	2.	2.
3.	3.	3.	3.

KEY IDEAS

3. Why do people come to the United States?
4. In what way is American English an example of cultural blending?
5. How has the physical geography of the United States affected its literature?

CRITICAL THINKING

6. **Draw Conclusions** How does public education help immigrants adapt to living in the United States?
7. **Analyze Effects** How did the development of the blues lead to additional cultural blending in music?
8. **CONNECT to Today** Do you think that the United States today is more like a melting pot or a mosaic? Use evidence to support your answer.
9. **TECHNOLOGY** **Plan a Web Page** Look for more examples of cultural blending in U.S. culture. Plan a Web page that would explain what cultural blending is and give examples from categories such as food, fashion, literature, music, and recreational activities.

CONNECT to the Essential Question

How did a developing nation grow into a world power?

Ask students what they have learned so far that can help them answer this question. Students might mention:

- Immigrants to the United States have enriched American life.
- The diversity of the United States can be illustrated by its many ethnicities, languages, and religions.

4 Assess & Reteach

Assess Have students complete the Section Assessment.

Unit Resource Book: The United States & Canada
- Section Quiz, p. 55

Interactive Review @ClassZone.com
- Section Quiz

Power Presentations

Test Generator

Reteach Have students prepare a slip of paper, containing one example of a cultural influence on life in America. Place the slips in an open container. Call on a volunteer to select a strip, read it out loud, and then briefly tell what he or she knows about the item. Repeat the process until all the slips have been discussed.

Unit Resource Book: The United States & Canada
- Reteaching Activity, p. 59

SECTION 3 ASSESSMENT ANSWERS

Terms & Names

1. public education: schooling paid for by government; literacy: ability to read, write; cultural blending: creating something new from elements of two cultures; jazz: music developed from African, European, and American elements

Use Your Reading Notes

2. Language, religion, and ethnicity are expressed in the literature, music, entertainment, and leisure of the various groups in the nation.

RELIGION	LANGUAGE	ETHNICITY	ARTS & CULTURE
1. Christian 2. Jewish 3. Muslim 4. Mormon	1. English 2. Spanish 3. Chinese 4. French	1. European American 2. African American 3. Hispanic 4. Asian	1. literature 2. music 3. sports 4. entertainment

Key Ideas

3. Most people want to improve life for themselves and their children.
4. American English has adopted words from many other languages.
5. The physical geography of the United States creates distinct regions, and each region has its own literature about life there.

Critical Thinking

6. Public education teaches immigrants to speak English and also helps them to learn about the United States.
7. Possible answer: The blues influenced later forms of music, such as country and rock.
8. Students should support their opinion with evidence drawn from this chapter, the news, their reading, or their lives.
9. Use the rubric to score Web pages.

Web Page Rubric

	Content	Organization
4	all elements	well organized, interesting
3	most elements	well organized
2	some elements	poorly organized
1	few elements	incomplete, disorganized

1 Plan & Prepare

Objectives

- Explain the functions of the three branches of the U.S. government
- Describe the basis of the U.S economy and the challenges and opportunities of globalization

Read for the Essential Question

Help students read for a purpose by reminding them of the Essential Question: "How did a developing nation grow into a world power?" Point out that they can use text features such as Key Questions, special features, images, and maps to form answers to the question.

Vocabulary

Best Practices Toolkit

Use the ***Best Practices Toolkit*** to model strategies for vocabulary notetaking. Vary strategies throughout the year. Choose from: *Knowledge Rating, Predicting ABCs, Definition Mapping, Word Sort, Frayer Model (Word Squares), Magnet Words, Word Wheel,* and *Student VOC.*

Vocabulary Strategies, TT9–TT16

Reading Strategy

Best Practices Toolkit

Display the T Chart/Two-Column Chart transparency.

- Remind students that facts and details can be organized into categories to help make sense of new information.
- Model filling in details about the U.S. government and economy.
- Have students suggest entries, based on what they know about the U.S. government and economy. Have students revise these entries as they read the section.

T Chart/Two-Column Chart , TT20

SECTION 4 Reading for Understanding

Key Ideas

BEFORE, YOU LEARNED
The United States is a country of great diversity that continues to draw immigrants from around the world.

NOW YOU WILL LEARN
The United States purchases goods from and sells goods to many countries around the world.

Vocabulary

TERMS & NAMES

amendment a written change to the U.S. Constitution that must go through an approval process

representative democracy a type of government in which citizens hold political power through elected representatives

republic another term for representative democracy

federal system a government system in which powers are divided between the national and state governments

free enterprise an economic system in which businesses are free to operate without much government involvement

market economy an economic system in which the production of goods and services is decided by supply and the demand of consumers

Interstate Highway System a network of more than 45,000 miles of roads that links every major U.S. city

global economy economy in which buying and selling occurs across national borders

multinational corporation a company that operates in more than one country

REVIEW

import to bring something from one country into another

export to send something from one country to another

Reading Strategy

Re-create the chart shown at right. As you read and respond to the **KEY QUESTIONS**, use the chart to note the main ideas about the government and economy of the United States.

Skillbuilder Handbook, page R7

CATEGORIZE

GOVERNMENT	ECONOMY
1.	1.
2.	2.
3.	3.

GRAPHIC ORGANIZERS
Go to **Interactive Review @ClassZone.com**

146 Chapter 5

PRETEACHING VOCABULARY

English Learners

Pronounce and Preview

Pronounce each term for students. Review the meanings of words within the definitions, such as *approval, operate,* and *network.*

- To modify vocabulary learning, have students complete worksheets as they read, instead of afterward.

Unit Resource Book: The United States & Canada
- Vocabulary Practice, p. 35

Inclusion

Predicting ABCs

Have students fill out the graphic organizer for terms/names in the section.
- Use the transparency to model the strategy.
- Have volunteers fill in the words for Sections, 1, 2, and 3.
- Then distribute copies of the organizer for students to work on with a partner.

Best Practices Toolkit
- Predicting ABCs, TT10

Oakland Bay Bridge, San Francisco

Moose in Denali National Park, Alaska

SECTION 4

GOVERNMENT & ECONOMICS

Democracy and Free Enterprise

Connecting to Your World

Have you ever had an election in your classroom? The candidate gives a speech and asks you and your classmates for their support and their votes. At national political conventions you may have seen on television, music is played, and balloons are dropped from the ceiling. People wave signs and cheer. Why do people become so excited? They hope that they have just chosen the person who will be elected the nation's next leader. In the United States, citizens play a role in selecting their own government. Not all countries recognize the people's right to choose their leaders.

The U.S. Government

KEY QUESTION What are the three branches of the U.S. government?

As Section 2 explained, the U.S. government is based on a written constitution. This document has changed since it was first written in 1787. Changes are added through **amendments**, written additions that must go through an approval process. There have been 27 amendments.

possibly the environment, war, and so forth ▾

Political Action A crowd waves American flags at a political convention. **What issues are most important to you and your classmates?**

147

2 Focus & Motivate

3-Minute Warm-Up

Write on the board or display the transparency:

- Look at the Agriculture and Industry of the United States map in this section. What does it tell you about the U.S. economy? *(Possible Answer: It's diverse with regional differences.)*

Unit Transparency Book
- 3-Minute Warm-Ups, TT1

Connecting to Your World

More About . . .

Political Action in the United States

Although those under 18 can't vote in U.S. elections, many young people still try to influence their outcomes by helping register voters and volunteering for campaigns.

3 Teach

The U.S. Government

Think, Pair, Share

- How would you define *federal system? (Possible Answer: a way to divide political power between the national and state governments)*
- Why do people say that our government is a system of checks and balances? *(The Constitution gives each branch of government the power to check, or limit, the other two.)*
- **Categorize** How would you categorize the type of government the United States has? *(representative democracy, or a republic)*

SECTION 4 PROGRAM RESOURCES

Unit Resource Book: The United States and Canada

ON LEVEL

Unit Resource Book
- Reading Study Guide, p. 9
- Section Quiz, p. 56

STRUGGLING READERS

Unit Resource Book
- Reading Study Guide with Additional Support, p. 17
- Skillbuilder Practice: Sequencing Events, p. 36
- Section Quiz, p. 56
- Reteaching Activity, p. 60

eEdition with Audio DVD-ROM

ENGLISH LEARNERS

Pupil Edition in Spanish

eEdition with Audio DVD-ROM

Unit Resource Book
- RSG (Spanish), p. 25
- RSG with Additional Support (Spanish), p. 33

Multi-Language Glossary

Test Generator
- Section Quiz in Spanish

INCLUSION

Unit Resource Book
- RSG with Additional Support, p. 17
- Section Quiz, p. 56
- Reteaching Activity, p. 60

GIFTED & TALENTED

Unit Resource Book
- Interdisciplinary Projects, pp. 44, 45
- Section Quiz, p. 56

PRE-AP

Unit Resource Book
- Economics in the World, p. 46
- Section Quiz, p. 56

TECHNOLOGY

Unit Transparency Book
- 3-Minute Warm-Ups, TT1
- Essential Question, TT2
- Critical Thinking, TT6

Daily Test Practice Transparencies
- Section 4, TT17

Power Presentations

ClassZone.com

World Cultures and Geography Video on DVD

COMPARING

Branches of the U.S. Government

Point out to students that the legislative branch of government was not always elected by a direct vote of the people. Before 1913 and the passage of the 17th Amendment to the Constitution, senators were selected by their state legislatures.

- How would you explain why senators were chosen by their state legislatures before 1913, rather than by a direct election? *(Possible Answer: because they were viewed as representing their state government and its interests, not necessarily those of the people)*

CRITICAL THINKING ANSWERS

Draw Conclusions by either signing laws or vetoing them

Comparing Regions

Republics Around the World

According to the CIA World Factbook, some countries that claim to be republics are not. For example, *Belarus* is a republic in name, but in reality is ruled by a dictator. *Libya* claims to be a "Jamahiriya" (a state of the masses), but is also ruled by a dictator. Other countries make no claims about being a republic or democracy. For example, *Burma* is ruled by a military junta, or group of military officers who seized power. *North Korea* is a communist state, ruled by a dictator.

COMPARING Branches of the U.S. Government

LEGISLATIVE	EXECUTIVE	JUDICIAL
• U.S. Congress creates, abolishes, and changes laws • approves appointments and treaties • imposes taxes • declares war • impeaches and tries officials for misconduct	• President either signs laws or vetoes [rejects] them • enforces laws • appoints judges and other officials • commands the military forces • directs foreign and national security affairs	• U.S. Supreme Court interprets laws • overturns laws that conflict with the Constitution • hears certain cases involving federal law or treaties • reviews cases on appeal

CRITICAL THINKING

Draw Conclusions What is the main way in which the president acts as a check on the Congress?

A Federal Republic The U.S. government is a **representative democracy**. A democracy is a government in which the people govern. The United States is much too large for every citizen to take part in making laws and policies, so U.S. citizens elect officials to represent them. Another term for a representative democracy is a **republic**.

The Constitution also created a **federal system** in which power is divided between the national and state governments. As you learned in Section 2, the national government is further divided into three branches: legislative, executive, and judicial.

Checks and Balances Recall that Americans fought for independence because they believed that Britain had ruled them unfairly. As a result, they tried to prevent future abuses of power. They wrote a Constitution that gives each branch of government the ability to act as a check, or limit, upon the other two. Those limits keep power in balance by spreading it across the branches. That's why people say our government is a system of checks and balances.

For example, Congress makes the laws, but the president must sign them before they go into effect. And the Supreme Court has the power to say that a law is not valid. These are just some of the ways that the branches of government act as checks on each other.

CATEGORIZE Describe the three branches of the U.S. government.
legislative, executive, judicial

DIFFERENTIATING INSTRUCTION

English Learners

Vocabulary: Suffixes

Point out that the words *representative, executive* and *legislative* each contain the suffix *-ive*. In general, this suffix means "performing an action." Thus, executive in this instance means "the branch of government that executes, or carries out, the laws." (It can also mean the person who is in charge of running a group or organization.) Have students form words with *-ive* and define them, such as *subject/subjective* and *elect/elective.*

Struggling Readers

Defining the Duties of Each Branch of Government

Have pairs of students choose one branch of government and define its duties in their own words. Encourage students to use a dictionary and other classroom resources for help in understanding terms in the Comparing feature on this page, such as *imposes, enforces,* and *appeal.* Ask for volunteers to share their definitions with the class and revise or add to them as a whole-group activity.

The U.S. Economy

KEY QUESTION Who controls the U.S. economy?

The U.S. economy is based on **free enterprise**—businesses are free to operate without much government involvement. Individuals and companies own most of the resources, equipment, and factories.

A Market Economy The United States has a **market economy**, in which individuals and businesses can make their own choices about what to buy and sell. The freedom people have to try to make money has enriched the country as a whole. The U.S. economy produces goods and services worth more than the output of any other country.

Service industries—businesses that provide services, not products—account for 78 percent of the money earned by the U.S. economy. Service industries include banks, insurance companies, hotels, restaurants, and repair shops. Manufacturing accounts for 21 percent and agriculture accounts for 1 percent.

Agriculture and Industry of the United States

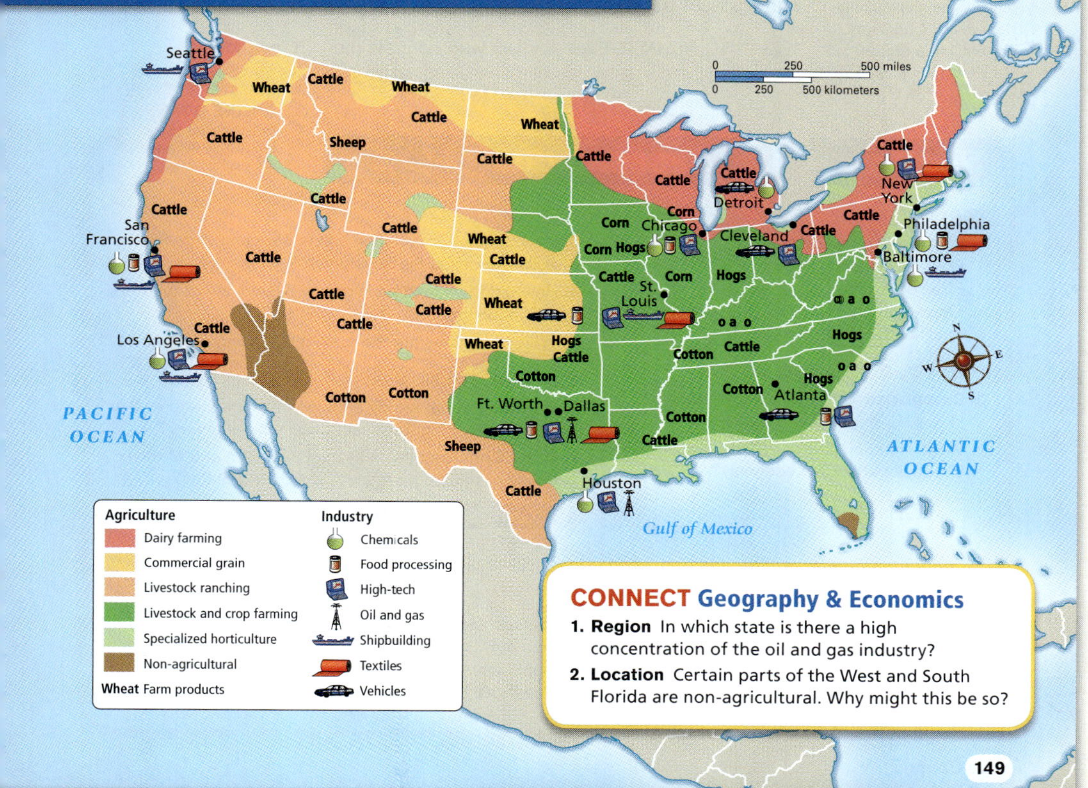

CONNECT Geography & Economics

1. **Region** In which state is there a high concentration of the oil and gas industry?
2. **Location** Certain parts of the West and South Florida are non-agricultural. Why might this be so?

DIFFERENTIATING INSTRUCTION: TIERED ACTIVITIES

OBJECTIVE Create a glossary of terms related to the U.S. economy.

Basic

Provide pairs of students with a list of vocabulary terms related to the U.S. economy. Tell them to use the list to prepare a glossary of economic terms and definitions. Glossaries should be in alphabetical order. They can be handwritten or computer generated and illustrated.

On Level

Have students work independently to complete the Basic level activity. In addition, have students organize their glossaries by categories that they create. Use the completed glossaries to review the features of the U.S. economy.

Challenge

In addition to the On Level activity, have students write a summary description of the U.S. economy focusing on one category that they created and using at least five terms from their glossaries. Use the completed glossaries and summaries to review the features of the U.S. economy.

Teach

The U.S. Economy

Talk About It

- Explain the relationship between free enterprise and a market economy. *(A market economy is an economic system in which individuals and businesses can make their own choices about what to buy and sell. To function effectively, a market economy depends on free enterprise, an economic system in which businesses are free to operate without much government involvement.)*
- What role does the U.S. government play in the economy? *(It promotes safety, fights unfair business practices, and builds structures to help economic growth.)* What specific examples can you think of to illustrate the government's role in the economy? *(Possible Answers: sets standards for product safety; bans harmful products from the marketplace; establishes agencies to regulate working conditions and promote equal opportunities; enforces laws to combat fraud; provides loans and other incentives to businesses)*
- **Categorize** What type of industries account for most of the money earned in the U.S. economy? *(service industries)* Identify some examples of this type of industry. *(banks, insurance companies, hotels, restaurants, and repair shops)*

Agriculture and Industry of the United States

CONNECT Geography & Economics

Encourage students to use their own knowledge and the information they have learned from the chapter to explain the relationship between types of agriculture/industry and their locations. *(Possible Answer: Agriculture appears to dominate the West and Midwest.)*

ANSWERS

1. **Region** Texas
2. **Location** Possible Answer: The geography (land features) of these areas might not be suitable for growing crops or raising livestock.

Unit Resource Book: The United States & Canada

- Economics in the World: U.S. Percentage of World Production, p. 46

CONNECT to the Essential Question

How did a developing nation grow into a world power?

Ask students what they have learned so far that can help them answer this question. Students might mention:

- The representative democracy of the United States protects people's freedoms and represents their interests.
- The U.S. economy has enriched the nation and made it a world leader.

Unit Resource Book: The United States & Canada
- Essential Question Graphic, TT2

4 Assess & Reteach

Assess Have students complete the Section Assessment.

Unit Resource Book: The United States & Canada
- Section Quiz, p. 56

Interactive Review @ClassZone.com
- Section Quiz

Power Presentations

Test Generator

Reteach Have students work in small groups and create one graphic organizer on the U.S. government and one on the U.S. economy. Have each group present their organizer to the class.

Unit Resource Book: The United States & Canada
- Reteaching Activity, p. 60

Highway System The Los Angeles freeways help to unify a widespread city. **How might a highway system link a region?**

by helping to transport goods and people ▲

The U.S. government does not run the economy, but it plays an important role. It promotes safety, fights unfair business actions, and builds structures to help economic growth. For example, the government paid to construct the **Interstate Highway System**, a network of more than 45,000 miles of roads that links major U.S. cities. The highways improved the transport of goods and people.

Globalization and a Changing Economy One of the biggest changes in recent times is the rise of a **global economy**, in which much buying and selling occurs across international borders. Technologies such as the Internet have linked businesses around the world.

The United States trades with many other countries. It **imports** goods such as automobiles, clothing, and petroleum. It **exports** goods such as airplanes, computers, and plastics. Canada, Mexico, Japan, and China are among the biggest U.S. trading partners.

The creation of **multinational corporations**, or companies that operate in more than one country, has added to the global economy. Such corporations often locate factories in countries where resources or labor are cheap. Many multinational corporations have their headquarters in the United States.

FIND MAIN IDEAS Explain who controls the U.S. economy.
mostly individuals and companies, with some government control

Section 4 Assessment

ONLINE QUIZ For test practice, go to **Interactive Review @ClassZone.com**

TERMS & NAMES

1. **Explain the importance of**
   - federal system
   - free enterprise
   - market economy
   - multinational corporation

USE YOUR READING NOTES

2. **Categorize** Use your completed chart to answer the following question:
   What are some duties of government that might affect the economy?

GOVERNMENT	ECONOMY
1.	1.
2.	2.
3.	3.

KEY IDEAS

3. How did the Constitution create a system of checks and balances in the government?
4. How did the Interstate Highway System affect the U.S. economy?
5. Which countries are the main U.S. trading partners?

CRITICAL THINKING

6. **Contrast** What are differences between the roles that individuals and government play in the economy?
7. **Draw Conclusions** Why might Canada and Mexico be major trading partners of the United States?
8. **CONNECT to Today** How do you think that the growth of multinational corporations might affect the government's role in the economy?
9. **MATH** **Create a Pie Graph** Look back through this section to learn what percentage of the U.S. economy is service industries, what percentage is manufacturing, and what percentage is agriculture. Use those percentages to create a pie graph.

SECTION 4 ASSESSMENT ANSWERS

Terms & Names

1. federal system: powers are divided between national and state governments; free enterprise: businesses operate without much government involvement; market economy: individuals and businesses make choices about what to buy and sell; multinational corporation: company that operates in more than one country

Use Your Reading Notes

2. creates laws; approves treaties; imposes taxes; declares war

GOVERNMENT	ECONOMY
1. Three branches	1. Market economy
2. Federal Republic	2. Global economy
3. Checks and balances	3. Multinational corporations

Key Ideas

3. Each branch puts checks, or limits, on the other two, which helps to spread or balance power.
4. It improved the transport of goods and people, which helped the economy to grow.
5. Canada, Mexico, Japan, China

Critical Thinking

6. Possible answers: Individuals own most resources, equipment, and factories and make choices to try to increase wealth; the government acts for the overall good of the economy.
7. Possible answer: They are located right next to the United States.
8. Answers will vary. Students might point out that it will be harder for the U.S. government to police businesses that have operations in foreign lands.
9. Use the rubric to score pie graphs.

Pie Graph Rubric

	Representation of Percentages	Style of Pie Graph
4	very precise	very clear
3	accurate	clear
2	fairly accurate	vague
1	inaccurate	unclear

Click here to complete these and other activities online @ClassZone.com

CHAPTER SUMMARY

Key Idea 1
The physical geography of the United States varies greatly from coast to coast.

Key Idea 2
Native Americans and peoples from other continents all came together to build a nation.

Key Idea 3
The United States is one of the most diverse nations in the world, and it has a rich culture blended from many world cultures.

Key Idea 4
The United States purchases goods from and sells goods to many countries around the world.

For **Review and Study Notes**, go to **Interactive Review** @ClassZone.com

NAME GAME

Use the Terms & Names list to complete each sentence on paper or online.

1. I am a swampy region located in southern Florida. ____**Everglades**____
2. I am vast grassland in north-central North America. ________
3. I am a mountain chain that runs from Alaska to New Mexico. ________
4. I am a mountain chain that runs parallel to the Atlantic coast from the Northeast to the South. ________
5. I cut through a plateau in Arizona. ________
6. Most of me lies on the U.S.–Canada border, although one part of me lies entirely inside the United States. ________
7. I run along the high ridge line of the Rocky Mountains. ________
8. I am located in Alaska. ________
9. I lie to the south of the United States. ________
10. I stretch across the United States and link every major city. ________

Appalachian Mountains
continental divide
Dust Bowl
Everglades
Grand Canyon
Great Lakes
Great Plains
Gulf of Mexico
Interstate Highway System
land bridge
Mississippi River
Mt. McKinley
Piedmont
Rocky Mountains

Activities

GeoGame

Use this online map to show what you know about the physical geography of the United States. Drag and drop each place name to its location on the map.

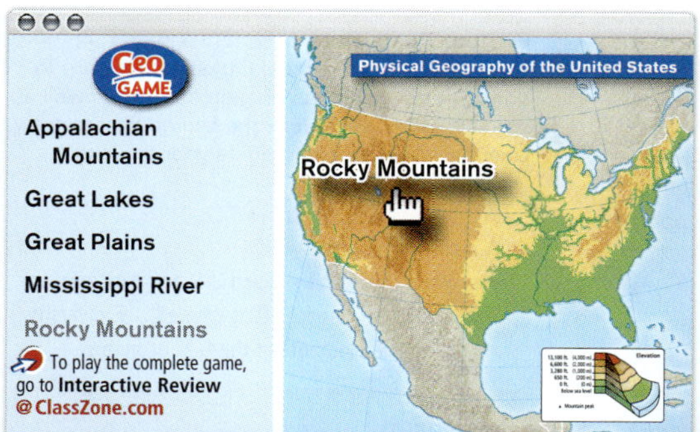

To play the complete game, go to **Interactive Review** @ClassZone.com

Crossword Puzzle

Complete the online crossword puzzle to test your knowledge of how the United States became a nation.

ACROSS
1. president during the Civil War

ONLINE AND ON-SCREEN ACTIVITIES

GeoGame

Use the interactive map on **Power Presentations** DVD-ROM or the **eEdition** to help students review the United States' geographic features.

- Demonstrate how to drag and drop the map labels listed onto the appropriate locations on the map.
- Advance to the scorecard to show how students have a chance to improve their scores.

Crossword Puzzle

Students can review important people and events in United States' history by completing this crossword puzzle online at **ClassZone.com**.

- As an extension of this activity, have students use the **Activity Maker** at **ClassZone.com** to make their own crossword puzzle for one section of the chapter.
- Then have students trade puzzles with classmates to solve answers for each other's puzzles.

Chapter Summary

As a follow-up to the chapter, encourage students to pair up and go over the **Review and Study Notes** at **ClassZone.com**.

Alternatively, students pairs can review their Reading Study Guide notes, making sure that they can add important details to each section's "Key Idea."

Name Game

Teaching Tips

ON PAPER Have students work in pairs to complete this activity. As an extension, they can write questions for the terms that are not covered in the items shown here.

ONLINE Make sure that students understand how to drag and drop terms to fill in blanks. Ask them to print out their results and compare them with those of a classmate.

ANSWERS

1. Everglades
2. Great Plains
3. Rocky Mountains
4. Appalachian Mountains
5. Grand Canyon
6. Great Lakes
7. continental divide
8. Mt. McKinley
9. Gulf of Mexico
10. Interstate Highway System

Make It Fun

Teacher-Tested Activities

Jim Easton, Roscoe Middle School, Roscoe, Illinois

My students often work with outline maps as a review activity. Several activities can be created using outline maps, depending on the interests and abilities of the students. Here is one such activity.

- Pass out a U.S. outline map.
- Give students a list of locations to label on the outline map.
- Once they have labeled map, assign them a task to complete. For example, they might map a flight from New York to Los Angeles through Chicago and then find the distance for separate legs of flight.

Vocabulary ANSWERS

1. Lewis and Clark Expedition explored the land acquired in the Louisiana Purchase.
2. The Great Lakes are five lakes that form the largest group of freshwater lakes in the world.
3. The Civil War (1861–1865) was a conflict between North and South over the issues of states' rights and slavery.
4. The Great Depression was a period of severe economic decline (1929–early 1940s).
5. Literacy is the ability to read and write.
6. Cultural blending means creating something new from the elements of two or more cultures.
7. A republic is a term for a representative democracy.
8. Free enterprise is an economic system in which businesses are free to operate without much government involvement.
9. The first immigrants came by way of the land bridge; immigrants started the first colonies.
10. The Constitution of the United States created a federal system in which power is shared between the national and state governments; the Constitution can be changed by amendment.

Key Ideas ANSWERS

11. the Appalachian Mountains, the Rocky Mountains, the Mississippi River
12. the Midwest; corn and wheat, dairy
13. the West
14. It left Britain with war debt that they tried to make the colonists pay, creating resentment that contributed to the Revolutionary War.
15. to explore the lands gained in the Louisiana Purchase
16. Mexico
17. Irish, Germans, Britons from United Kingdom
18. by writing literature and developing distinctive musical styles such as jazz and the blues
19. They want to keep their own culture, not adopt a uniform American culture.
20. The people rule by electing officials to represent them in government.
21. It is a government system that divides power between national and state governments.
22. the development of a more global economy

CHAPTER 5 Assessment

VOCABULARY

Explain the significance of each of the following.

1. Lewis and Clark Expedition
2. Great Lakes
3. Civil War
4. Great Depression
5. literacy
6. cultural blending
7. republic
8. free enterprise

Explain how the first term in each group relates to the terms that follow.

9. immigrant: land bridge, colony
10. Constitution: amendment, federal system

KEY IDEAS

1 From Coast to Coast

11. What are three major geographic features that run from north to south?
12. What is the nation's largest agricultural region, and what are its main crops?
13. Which region has the most mountainous terrain?

2 Creating a Nation

14. How did the French and Indian War help lead to the Revolutionary War?
15. What was the purpose of Lewis and Clark's journey?
16. Which neighboring country lost large amounts of land to the United States?

3 A Diverse Culture

17. Which immigrant groups came to the United States in large numbers during the mid-1800s?
18. How have African Americans contributed to culture?
19. Why might some immigrants object to the concept of the United States as a melting pot?

4 Democracy and Free Enterprise

20. Why is the U.S. government considered a representative democracy?
21. What is a federal system?
22. What economic change was spurred by technology?

CRITICAL THINKING

23. **Compare and Contrast** Create a table to compare and contrast the transcontinental railroad and the Interstate Highway System.

RAILROAD	HIGHWAY SYSTEM

24. **Distinguish Fact from Opinion** The Declaration of Independence lists several complaints about the British king, including "imposing Taxes on us without our Consent." Is that complaint fact or opinion? Explain.
25. **Draw Conclusions** When you consider the part farming plays in the overall economy, do you think most Americans live in urban or rural areas? Why?
26. **Five Themes: Location** In 1941, Japan bombed the U.S. naval base at Pearl Harbor, Hawaii. Why would the Japanese view Pearl Harbor as a big military threat?
27. **Connect Geography & Culture** The number of Spanish speakers is growing. How might U.S. culture change as a result?

Answer the ESSENTIAL QUESTION

How did a developing nation grow into a world power?

Written Response Write a two paragraph response to the Essential Question. Be sure to consider the key ideas of each section as well as specific ideas about how the United States grew. Use the rubric below to guide your thinking.

Response Rubric

A strong response will:

- explain the steps by which U.S. territory grew
- discuss the influences that created U.S. diversity
- analyze the elements of the U.S. economy

ESSENTIAL QUESTION REVIEW

Before students begin their essays, have them review the key chapter elements that will help them answer the Essential Question. Direct them to significant features in the chapter, such as:

- Comparing U.S. Geographic Regions, Section 1
- Animated Geography: Growth of the United States 1783–1853, Section 2
- Comparing Languages, Religion, and Ethnicity, Section 3
- Agriculture and Industry of the United States, Section 4

Written Response Rubric

	Coverage of Key Ideas	Explanation	Accuracy
4	all	clear, concise	no errors
3	most	good	few/minor errors
2	some	fair	several errors
1	two or fewer	poor	many errors

STANDARDS-BASED ASSESSMENT

Test Practice

- Online Test Practice @ClassZone.com
- Test-Taking Strategies and Practice at the front of this book

ELEVATION GRAPH

Use the elevation graph to answer questions 1 and 2 on your paper about the elevation of the western United States at 38° north latitude.

Elevation Profile of the Western United States

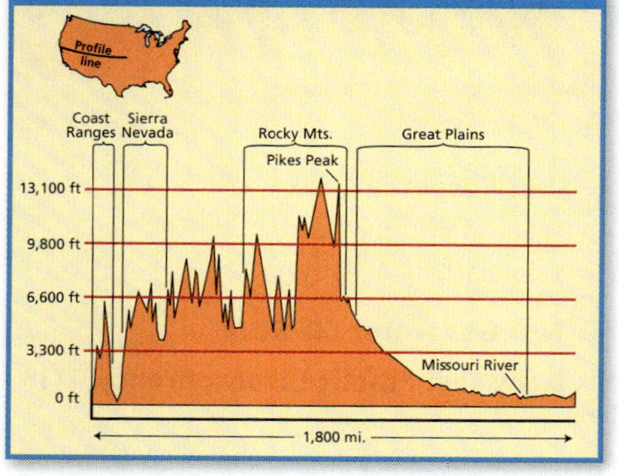

1. Which of the following has the highest general elevation?

- **A.** Coast Ranges
- **B.** Sierra Nevada
- **C.** Rocky Mountains
- **D.** Great Plains

2. Which has the lowest elevation?

- **A.** between Sierra Nevada and Rocky Mountains
- **B.** between Rocky Mountains and Great Plains
- **C.** between Coast Ranges and Sierra Nevada
- **D.** in the Coast Ranges

ANALYZING PRIMARY SOURCES

Examine the Great Seal of the United States shown below. Use the seal to answer questions 3 and 4 on your own paper.

3. Why might the motto *E Pluribus Unum* (one from many) be on the Great Seal of the United States?

4. Why might the eagle on the seal clutch an olive branch in one claw and arrows in the other claw?

GeoActivity

1. INTERDISCIPLINARY ACTIVITY–MUSIC

Working in a small group, find three or four examples of jazz that you like. Concentrate on instrumental music, not songs with lyrics. In a group discussion, compare and contrast the pieces. Then play the music and present your findings in class.

2. WRITING FOR SOCIAL STUDIES

Research to find out more about the American bison–including how large it is, how long it lives, what it eats, what its habits are, and where it still can be found. Create a booklet about the bison to teach younger students about this American animal.

3. MENTAL MAPPING

Create an outline map of the United States and label the following:

- Appalachian Mountains
- Rocky Mountains
- Atlantic Ocean
- Pacific Ocean
- Gulf of Mexico
- Mississippi River
- Great Lakes
- Great Plains

DIFFERENTIATING CHAPTER ASSESSMENT

Unit Resource Book

- **Chapter Test A,** pp. 61–65. *Use for Struggling Readers, Beginning-to-Intermediate-Level English Learners, and for Inclusion of students whose IEP suggests supported test-taking.*
- **Chapter Test B,** pp. 66–70. *Use for On-Level Learners and ESOL students with Advanced-Level English.*
- **Chapter Test C,** pp. 71–75. *Use for Gifted and Talented students or students assigned Pre-AP work.*

Test Generator

- Chapter Tests A, B, and C, in English and Spanish

McDougal Littell Assessment System

TEST **Test Generator** Use three levels of tests in English and Spanish tied to state standards. Plus, choose questions to customize tests for all populations.

SCORE Scan and Score plain paper student tests in minutes.

REPORT Create instant progress reports, sortable by student, by standard, or other criteria.

RETEACH Respond immediately to students' needs by using a wealth of reteaching resources.

Critical Thinking ANSWERS

23. transcontinental railroad: built in the 1800s, crossed the continent east to west, linking the two coasts; improved transport; spurred economic growth
Interstate Highway System: built in the 1900s; criss-crossed the whole country, linking major cities; improved transport; spurred economic growth

24. It is a fact; it can be proven that the colonists had no representatives in the British law-making body who could consent to taxes.

25. Because farming is such a small part of the economy, most Americans probably live in urban areas.

26. Possible answer: Hawaii is in the Pacific and therefore much closer to Japan than other U.S. bases.

27. Possible answers: More Spanish words might enter the language; foods from Latino countries might become more common; American music might be influenced by the music of Latino cultures.

Standards-Based Assessment ANSWERS

1. C

2. C

3. Out of many states one nation was formed; also, out of many people from many nations.

4. to symbolize a readiness for either peace or war if peace should fail

GeoActivity

1. INTERDISCIPLINARY ACTIVITY RUBRIC

The presentation should

- compare and contrast interesting examples of jazz
- give adequate excerpts from different kinds of jazz

2. WRITING FOR SOCIAL STUDIES RUBRIC

The booklet should

- show evidence of careful research
- present interesting information about bisons

3. MENTAL MAPPING RUBRIC

The outline map should

- represent the approximate shape of the United States
- include all geographic features listed in their correct location

CHAPTER 6

Canada PLANNING GUIDE

Complete print resources all on one USB drive allow you to customize lesson plans for all students.

Section Titles & Key Ideas	Unit Resource Book The United States and Canada	Assessment Resources
1 Mountains, Prairies, and Coastlands **KEY IDEA** The physical geography of Canada shares many landforms and bodies of water with the United States.	• **Reading Study Guide,** pp. 77–78 • **RSG with Additional Support,** pp. 85–86 • **Vocabulary Practice,** p. 109 • **Skillbuilder Practice,** p. 110 • **Outline Map Activities,** p. 113	**Unit Resource Book** • Section Quiz, p. 123 • Reteaching Activity, p. 127 **Test Generator CD-ROM** **Daily Test Practice Transparency,** TT18
2 First Peoples to a Modern Nation **KEY IDEA** Canada's diversity is the result of regional differences and events in Canada's history.	• **Reading Study Guide,** pp. 79–80 • **RSG with Additional Support,** pp. 87–88 • **Map and Graph Skills,** p. 111 • **History Makers,** pp. 115, 116 • **Connect to History,** p. 121 • **Primary Source,** p. 122	**Unit Resource Book** • Section Quiz, p. 124 • Reteaching Activity, p. 128 **Test Generator CD-ROM** **Daily Test Practice Transparency,** TT19
3 A Nation of Immigrants **KEY IDEA** Because Canada is a nation of immigrants, it has a diverse population and a variety of cultures shaped by many influences.	• **Reading Study Guide,** pp. 81–82 • **RSG with Additional Support,** pp. 89–90 • **Skillbuilder Practice,** p. 110 • **Interdisciplinary Projects,** pp. 117–119	**Unit Resource Book** • Section Quiz, p. 125 • Reteaching Activity, p. 129 **Test Generator CD-ROM** **Daily Test Practice Transparency,** TT20
4 Rich Resources in a Vast Land **KEY IDEA** Canada has a vast wealth of resources and an educated population to support its government and economy.	• **Reading Study Guide,** pp. 83–84 • **RSG with Additional Support,** pp. 91–92 • **Skillbuilder Practice,** p. 110 • **Economics in the World,** p. 120	**Unit Resource Book** • Section Quiz, p. 126 • Reteaching Activity, p. 130 • Chapter Test (Forms A, B, & C), pp. 131–142 **Test Generator CD-ROM** **Daily Test Practice Transparency,** TT21

McDougal Littell Assessment System

TEST SCORE REPORT RETEACH

INTEGRATED TECHNOLOGY

No Time? To focus students on important content in the chapter, use the **"Reading Study Guide with Additional Support"** in ***Resources2Go.***

Teacher Presentation Options

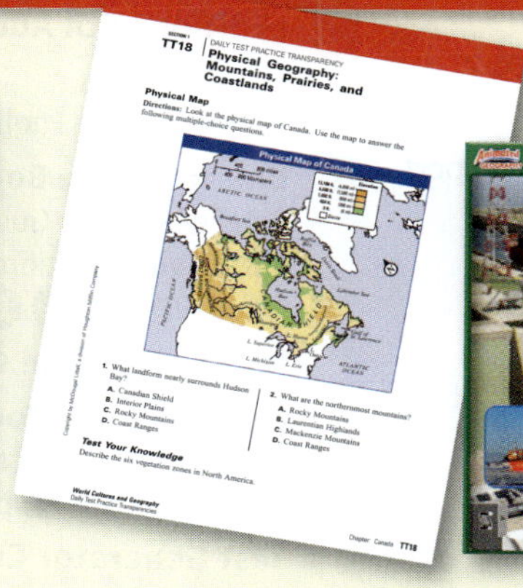

Electronic Books

eEdition DVD-ROM

eEdition Online

Presentation Toolkit

Video/DVD

- World Cultures and Geography

Power Presentations DVD-ROM

- Discussion Starters
- Media Gallery
- Review Game
- Interactive Review
- Animated Features

Unit Transparency Book

- 3-Minute Warm-Ups
- Essential Question
- Critical Thinking
- Maps
- Cultures

Daily Test Practice Transparencies

Animated Geography

- Present-Day Canada Map
- The St. Lawrence Seaway
- Worldster: Kirima's Web Page

OnlineActivities at ClassZone.com

Animated Geography

- Videos and Photos
- 3-D Animations
- Interactive Maps

Interactive Review

- Section Quizzes
- Review and Study Notes
- Vocabulary Flip Cards
- Graphic Organizers
- Name Game and Crossword Puzzles
- GeoGame

Online Test Practice

Map Center

- Atlas Maps
- Political Maps with Data
- Links to Mapping Web Sites

Activity Center

- Activity Maker—Review Games, Puzzles, Flip Cards, and Graphic Organizers
- Worldster

Research & Writing Center

- Web Sites
- Data Updates
- Primary Sources
- Writing Models

Current Events

Marci Smith Deal
Hurst-Euless-Bedford ISD
Bedford, Texas

Make It Fun — Teacher-Tested Activities

As you teach this chapter, look for my classroom-proven ideas for:

- making the lesson fun
- focusing students' attention and comprehension
- integrating technology with traditional teaching

Teaching Bell to Bell When you see this timing guide, have students discuss in pairs, in groups of 3–6, or as a class.

- **Think, Pair, Share** (pairs)
- **Reader, Recorder, Reporter** (3–6)
- **Talk About It** (any size)

CHAPTER 6 RESOURCES FOR DIFFERENTIATED INSTRUCTION

Struggling Readers

eEdition with Audio DVD-ROM

Lesson Plans
- Strategies for Inclusion

Unit Resource Book
- Reading Study Guide with Additional Support, pp. 85–92 (B)
- Reteaching Activity, pp. 127–130
- Chapter Test (Form A), pp. 131–134 (A)

Best Practices Toolkit (C)
- Interactive Reading Strategies, pp. 90–145
- Differentiation Activity Bank: Struggling Readers, pp. 146–149

Teacher's Edition Activities
- "Who Am I?," Sec. 2
- Mapping the Countries of Origin of Canadian Immigrants, Sec. 3
- Find Main Ideas and Supporting Details, Sec. 4
- Tiered Activities, Sec. 1–4

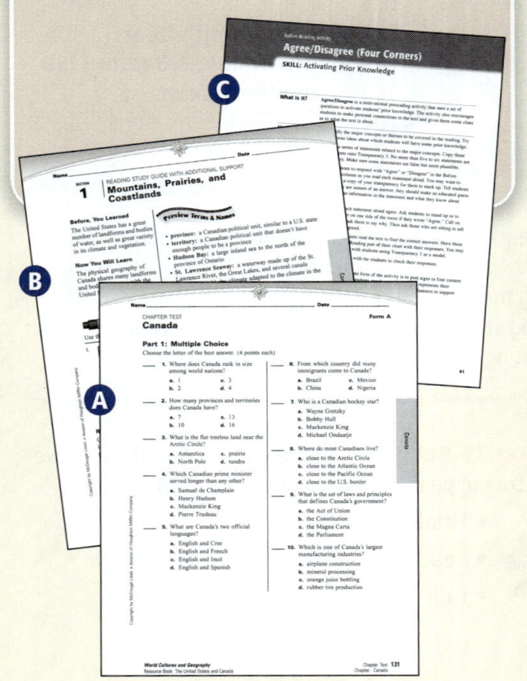

English Learners

Pupil's Edition in Spanish

eEdition with Audio DVD-ROM

Lesson Plans
- Strategies for English Learners

Unit Resource Book
- Reading Study Guide (Spanish), pp. 93–100
- RSG with Additional Support (Spanish), pp. 101–108 (A) (B)

Multi-Language Glossary (C)

Best Practices Toolkit
- Differentiation Activity Bank: English Learners, pp. 165–194

Test Generator CD-ROM
- Section Quizzes in Spanish
- Chapter Test in Spanish

Teacher's Edition Activities
- Pronounce and Preview, Sec. 1–4
- Vocabulary: Word Relationships, Sec. 1
- Comprehension: Self-Monitor, Sec. 2
- Vocabulary: Suffixes, Sec. 3
- Tiered Activities, Sec. 1–4

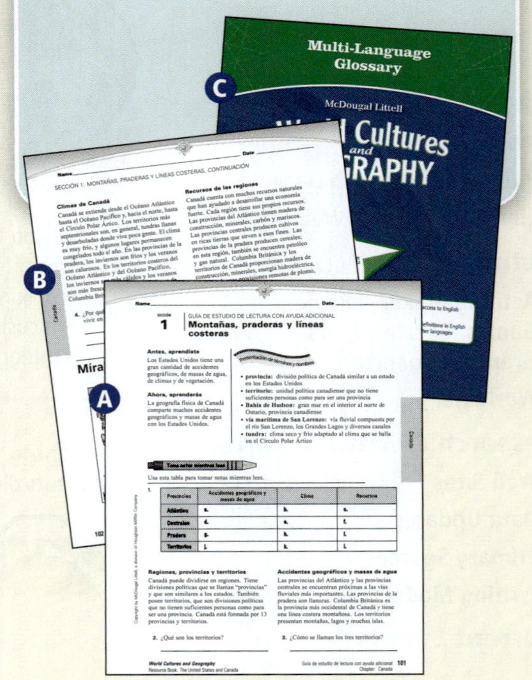

Inclusion

eEdition with Audio DVD-ROM

Lesson Plans
- Strategies for Inclusion

Unit Resource Book
- Reading Study Guide with Additional Support, pp. 85–92 (A) (B)
- Reteaching Activity, pp. 127–130
- Chapter Test (Form A), pp. 131–134

Best Practices Toolkit (C)
- Interactive Reading Strategies, pp. 90–145
- Differentiation Activity Bank: Struggling Readers, pp. 146–149

Teacher's Edition Activities
- Match Terms and Definitions, Sec. 1, 2
- Using Terms in Captions, Sec. 3
- Hear and Use Vocabulary, Sec. 4
- Tiered Activities, Sec. 1–4

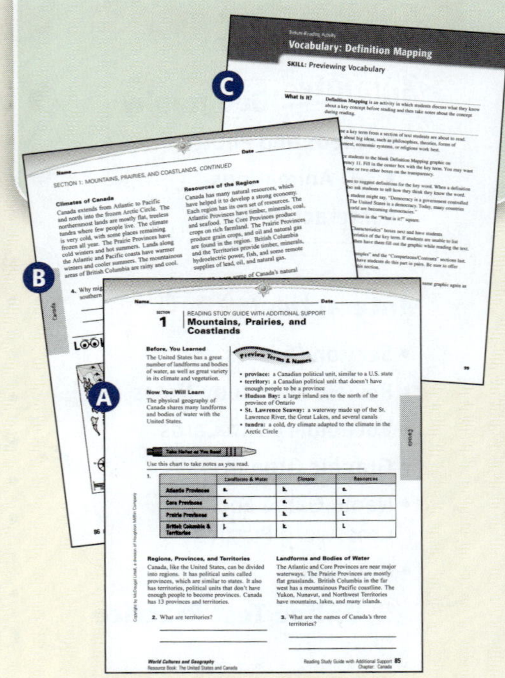

Gifted & Talented Students

Lesson Plans
- Core Lesson Plan

Unit Resource Book
- Outline Map Activities, p. 113
- History Maker, pp. 115–116 (A)
- Interdisciplinary Projects, pp. 117–119 (B)
- Economics in the World, p. 120
- Connect to History, p. 121 (C)
- Chapter Test (Form C), pp. 139–142

Best Practices Toolkit
- Student Products Rubrics, pp. 1–14
- Writing Templates, pp. 78–89

Teacher's Edition Activities
- Travel Journal of Canada, Sec. 1
- Tiered Activities, Sec. 1–4

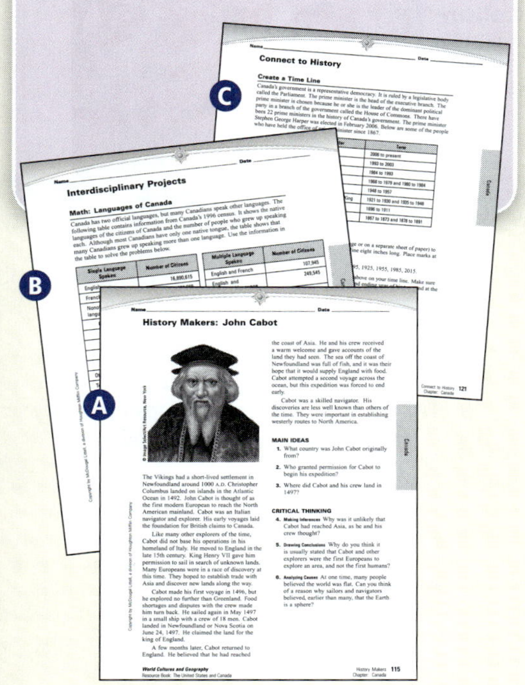

Pre-AP*

Lesson Plans
- Core Lesson Plan

Unit Resource Book
- Vocabulary Practice, p. 109
- Map and Graph Skills, pp. 111–112
- History Maker, pp. 115–116 (A)
- Economics in the World, p. 120 (B)
- Connect to History, p. 121
- Primary Source, p. 122 (C)
- Chapter Test (Form C), pp. 139–142

Best Practices Toolkit
- Differentiation Activity Bank: Pre-AP Students, pp. 195–200
- Writing Templates, pp. 78–89

* Pre-AP is a registered trademark of the College Entrance Examination Board, which was not involved in the production of and does not endorse this product.

Teacher's Edition Activities
- Tiered Activities, Sec. 1–4

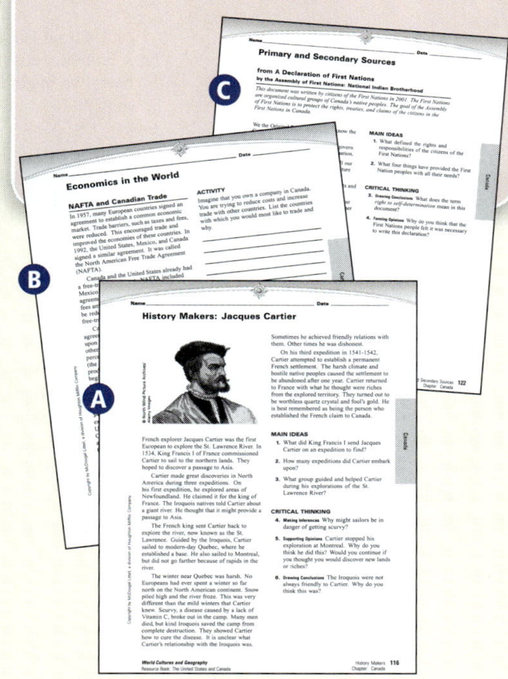

CONNECT to Language Arts

Describing Canada's Seasons

CLASS TIME 30 minutes

TASK Choose a province in Canada and describe the seasons it experiences.

SUPPLIES
- classroom resources on Canada's geography and climate
- writing materials/computers

ACTIVITY Have students use the "Physical Geography of Canada" map in Section 1 and select a province in Canada that interests them, such as Prince Edward Island or Newfoundland. Using information from Section 1 about the province and classroom resources, have students describe the province in terms of the seasons it experiences. Encourage students to use vivid details in their writing. Ask volunteers to read their descriptions out loud.

Good for Verbal Learners

CONNECT to Science

Winter Survival Pictorial Guides

CLASS TIME 30 minutes

TASK Create a guide for how to survive the winter in Canada's Arctic regions.

SUPPLIES
- resources on Canada's winter season
- paper, colored markers or pencils

ACTIVITY Have students work in pairs to create a pictorial guide that informs travelers to Canada's Arctic regions what they need to bring and explains why these are needed from a scientific perspective. You may want to have students focus on only one aspect of winter survival, such as clothing, food, shelter, or travel. Students' guides should include visuals (self-made or found), descriptions of each item, and a scientific explanation of why it's needed. Have students create a classroom display of their guides and/or present them to the class.

Good for Visual and Verbal Learners

Introduce the ESSENTIAL QUESTION

- Identify Canada's four geographic regions and each region's key features.
- Explain why Canada's diversity is the result of regional differences and historic events.
- Describe how immigration has influenced Canada's diversity and culture.
- Analyze how Canada's resources and educated population has shaped its government and economics.

Key Ideas

Section 1: The physical geography of Canada shares many landforms and bodies of water with the United States.

Section 2: Canada's diversity is the result of regional differences and events in Canada's history.

Section 3: Because Canada is a nation of immigrants, it has a diverse population and a variety of cultures shaped by many influences.

Section 4: Canada has a vast wealth of resources and an educated population to support its government and economy.

CONNECT to the Essential Question

You may wish to revisit the Essential Question periodically with students as they work through the chapter.

Suggestions for reconnecting are provided at an appropriate point in the *Teacher's Edition* for each section.

Banff National Park, Alberta, Canada

CHAPTER 6

Canada

1 GEOGRAPHY
Mountains, Prairies, and Coastlands

2 HISTORY
First Peoples to a Modern Nation

3 CULTURE
A Nation of Immigrants

4 GOVERNMENT & ECONOMICS
Rich Resources in a Vast Land

ESSENTIAL QUESTION

How is Canada managing its resources in the 21st century?

CONNECT Geography & History

Use the map and the time line to answer the following questions.

1. When was the city of Quebec founded?
2. On what body of water are both Quebec and Montreal located?

1600

Geography
1608 Samuel de Champlain founds Quebec City. ▶

Geography
1791 Great Britain creates Upper and Lower Canada.

Culture
1763 Great Britain takes over New France, causing Canada to have two major languages. ▶

TIME LINE DISCUSSION

Point out to students that the French and British fought over Canada, as they did for colonial America. Ask students what other European power influenced colonial America's early history, but not Canada's. *(Spain)*

- Which event on the time line highlights the event that caused Canada to have two main languages? *(1763—Great Britain takes over New France)*
- When did Canada become self-governing? *(1931)*
- Why do you think Canada agreed to join the North American Free Trade Agreement, or NAFTA? *(Possible Answer: Because of its many resources, free trade would help its economy expand.)*

History
1867 Canada becomes a dominion in the British Empire.

Government
1931 Canada becomes self-governing. (Parliament building in Ottawa) ►

Today

Culture
1908 Lucy Maud Montgomery publishes *Anne of Green Gables.*

Economics
1992 NAFTA (North American Free Trade Agreement) unites Canada, the United States, and Mexico in a free-trade zone.

155

TEACHER MEDIA FAVORITES

Books for Students

Carter, Anne Laurel. *Last Chance Bay.* Penguin Canada, 2004. Novel set during World War II about a girl in a Cape Breton mining village.

King, Thomas et al. *Our Story: Aboriginal Voices on Canada's Past.* Anchor Canada, 2005.

Moor, Christopher. *The Big Book of Canada.* Tundra Books, 2002. Richly illustrated, this book provides a wealth of information about Canada.

Video / DVD

Anne of Green Gables. (DVD) Sullivan, 2001. Acclaimed adaptation of Lucy Maud Montgomery's classic tale of an orphan girl growing up on Prince Edward Island in the late 1800s.

Pilot Guide to Canada. (DVD) Pilot Productions, 2005. Follows two modern-day travelers on separate journeys, exploring some of the wonderful sites, cultures, and traditions of the world's second largest country.

Internet

Go to the *Research and Writing Center* at **ClassZone.com** to link to:

- data on the region
- a variety of chapter-specific content-reviewed sites
- "History Makers" biographies
- primary sources

Animated GEOGRAPHY
Present-Day Canada
CLASSZONE.COM

Ask students what they notice about the location of most of Canada's provincial capitals and other cities. *(Most are located close to the U.S.-Canada border.)* Ask what they notice about the geographic location of most of Canada's cities on the map. *(Most are located on or near waterways.)*

- What is Canada's capital? *(Ottawa)* Where is it located? *(in the province of Ontario, near the St. Lawrence River)*
- What is Canada's northernmost provincial capital? *(Iqaluit, provincial capital of Nunavut)*
- Compare this map to the Agriculture and Industry of Canada map in Section 4. Why are there so few industries in northern Canada? *(Possible Answer: A cold climate and scarce population would make industry difficult and expensive.)*

CONNECT GEOGRAPHY & HISTORY ANSWERS

1. 1608

2. St. Lawrence River

Presentation Options

Use the **Power Presentations** DVD-ROM to project the animated map.

- Prompt students to use the pop-up images and audio from the map to discover what they already know about the geography, culture, history, economics, or governments of the region.
- Challenge students to answer the map questions individually, in small groups, or as a class activity. Use all of the interactive material to help preview chapter content.

1 Plan & Prepare

Objectives

- Identify the political units and geographic features of the four regions of Canada
- Analyze the influence of geographic features on Canada's climates
- List the main resources of Canada's four regions

Read for the Essential Question

Help students read for a purpose by reminding them of the Essential Question: "How is Canada managing its resources in the 21st century?" Point out that they can use text features such as Key Questions, special features, images, and maps to form answers to the question.

Vocabulary

Best Practices Toolkit

Use the ***Best Practices Toolkit*** to model strategies for vocabulary notetaking. Vary strategies throughout the year. Choose from: *Knowledge Rating, Predicting ABCs, Definition Mapping, Word Sort, Frayer Model (Word Squares), Magnet Words, Word Wheel,* and *Student VOC.*

Vocabulary Strategies, TT9–TT16

Reading Strategy

Best Practices Toolkit

Display the Three-Column Chart transparency.

- Point out to students that categorizing involves deciding how to sort information into categories, as shown on the chart.
- Model entering information on the chart using information about landforms and bodies of water in this section.
- Have students suggest entries as they read this section.
- Display the graphic organizer for students to review.

Three-Column Chart, TT21

SECTION 1 Reading for Understanding

Key Ideas

BEFORE, YOU LEARNED

The United States has a great number of landforms and bodies of water, as well as great variety in its climate and vegetation.

NOW YOU WILL LEARN

The physical geography of Canada shares many landforms and bodies of water with the United States.

Vocabulary

TERMS & NAMES

province a Canadian political unit, similar to a U.S. state

territory a Canadian political unit that doesn't have enough people to be a province

Canadian Shield a horseshoe-shaped, rocky plateau that covers much of eastern central Canada

Hudson Bay a large inland sea to the north of the province of Ontario

St. Lawrence Seaway a waterway made up of the St. Lawrence River, the Great Lakes, and several canals

hydroelectric having to do with electricity created by water-powered engines

Nunavut (NOO•nuh•VOOT) a territory created in 1999 from the eastern part of the Northwest Territories and home to many Inuit

REVIEW

Great Lakes five lakes forming the largest group of freshwater lakes in the world

tundra a cold, dry climate and vegetation adapted to the climate in the Arctic Circle

Visual Vocabulary tundra

Reading Strategy

Re-create the chart that is shown at right. As you read and respond to the **KEY QUESTIONS**, use the chart to organize important details about the physical geography of Canada.

Skillbuilder Handbook, page R7

CATEGORIZE

	LANDFORMS & WATER	CLIMATE	RESOURCES
Atlantic Provinces			
Core Provinces			
Prairie Provinces			
British Columbia & Territories			

GRAPHIC ORGANIZERS
Go to **Interactive Review** @ClassZone.com

156 Chapter 6

PRETEACHING VOCABULARY

English Learners

Pronounce and Preview

Pronounce each term for students. Review the meanings of words within the definitions, such as *political unit, horseshoe-shaped,* and *adapted.*

- To modify vocabulary learning, have students complete worksheets as they read, instead of afterward.

Unit Resource Book: The United States & Canada
- Vocabulary Practice, p. 109

Inclusion

Match Terms and Definitions

Read a definition out loud. Explain the skill of scanning, then write on the board one or two words to scan for (such as "U.S." in the definition of "province"). Have students locate the definition you just read, then read it and say the term or name it defines. You may want to model this again, then have students continue the activity, working in pairs.

SECTION 1 GEOGRAPHY

Totem poles in Stanley Park, Vancouver, British Columbia

Atlantic puffins on an island in Canada

Mountains, Prairies, and Coastlands

Connecting to Your World

The United States has long coastlines on its eastern and western borders. It also has vast plains in the center and towering mountains in the west. What you may not know is that Canada, located north of the United States, shares many of the same geographic features. In this section, you will learn more about the landscape of Canada, the second largest country in the world. You will also learn how its northern location affects its climate.

Landforms and Bodies of Water

KEY QUESTION What are the four regions of Canada?

You have learned that the United States can be divided into four regions, and that each region contains many states. Canada can also be divided into regions. It has political units called **provinces**, which are similar to states. It also has **territories**, political units that don't have enough people to become provinces. Canada's 13 provinces and territories are often divided into four geographic regions: the Atlantic Provinces, the Core Provinces, the Prairie Provinces, and British Columbia and Territories.

Niagara Falls Niagara Falls forms a natural boundary between Canada and the United States.

157

2 Focus & Motivate

3-Minute Warm-Up

Write on the board or display the transparency:

- Look at the inset map of Niagara Falls. How might being near Niagara Falls and the lakes affect people on both sides of the border? *(Possible Answers: creates tourism and shipping jobs; provides opportunities for transportation and recreation)*

Unit Transparency Book: The United States & Canada
- 3-Minute Warm-Ups, TT11

Connecting to Your World

More About . . .

Niagara Falls

Niagara Falls includes the American Falls and Horseshoe Falls (on the Canadian side). Both are over 150 feet high. The American Falls is over 1,000 feet wide, while Horseshoe Falls is over 2,000 feet wide.

3 Teach

Landforms and Bodies of Water

Think, Pair, Share

- Which region is better for seafaring than farming, and why? *(Atlantic Provinces, because of good harbors and lack of good farming soil)*
- **Categorize** List some types of water important to Canada and give examples of each. *(oceans—Atlantic, Pacific; bays—Hudson Bay; rivers—St. Lawrence, Mackenzie; lakes—Great Lakes, Great Bear, Great Slave)*

SECTION 1 PROGRAM RESOURCES

Unit Resource Book: The United States and Canada

ON LEVEL

Unit Resource Book
- Reading Study Guide, p. 77
- Outline Map Activities, p. 113
- Section Quiz, p. 123

STRUGGLING READERS

Unit Resource Book
- Reading Study Guide with Additional Support, p. 85
- Skillbuilder Practice: Analyzing Cause and Effect, p. 110
- Section Quiz, p. 123
- Reteaching Activity, p. 127

eEdition with Audio DVD-ROM

ENGLISH LEARNERS

Pupil Edition in Spanish

eEdition with Audio DVD-ROM

Unit Resource Book
- RSG (Spanish), p. 93
- RSG with Additional Support (Spanish), p. 101

Multi-Language Glossary

Test Generator
- Section Quiz in Spanish

INCLUSION

Unit Resource Book
- RSG with Additional Support, p. 85
- Section Quiz, p. 123
- Reteaching Activity, p. 127

GIFTED & TALENTED

Unit Resource Book
- Section Quiz, p. 123

PRE-AP

Unit Resource Book
- Section Quiz, p. 123

TECHNOLOGY

Unit Transparency Book
- 3-Minute Warm-Ups, TT11
- Essential Question, TT12
- Critical Thinking, TT13
- Map: US & Canada: Rivers, TT18, TT18a

Daily Test Practice Transparencies
- Section 1, TT18

Power Presentations

ClassZone.com

World Cultures and Geography Video on DVD

COMPARING

Regions of Canada

Have students examine the four images. Ask them how these images support the text's discussion about population of the Core Provinces. *(Image of Quebec shows buildings crowded together on land surrounding the St. Lawrence River, while the others show less populated landscapes.)*

CRITICAL THINKING ANSWER

Evaluate the Prairie Provinces, because they have flat, open, and fertile land

More About . . .

The Canadian Shield

The Canadian Shield, also called the Laurentian Plateau, is the largest area of exposed Precambrian rock on earth. The Precambrian era ranges from over 500 million to almost 4 billion years ago, when primitive bacteria were first formed. The Canadian Shield extends from the Great Lakes to the Arctic Ocean and includes most of Greenland. Part of it also extends into areas of the northern United States.

Comparing Regions

Fortified Colonial Cities: Quebec and San Juan

The cities of Quebec in Canada and San Juan in Puerto Rico are both built near water and have historic forts, protective walls, and other structures that are UNESCO World Heritage sites. Quebec's fortified areas were built soon after Samuel de Champlain settled on the banks of the St. Lawrence River in 1608. San Juan's defensive structures, on San Juan Bay in the Caribbean Sea, were started earlier, in the 1500s by Spanish troops.

Atlantic Provinces This region includes Newfoundland, Nova Scotia, New Brunswick, and Prince Edward Island. (See map opposite.) The Appalachian Mountains Ⓐ extend into the region. In most places, the soil is poor for farming. Because the Atlantic coast has good harbors, the Atlantic Provinces became a region of seafarers instead.

The western edge of the region lies on the **Canadian Shield** Ⓑ, a horseshoe-shaped plateau that covers much of east-central Canada. During the Ice Age, glaciers scraped the soil from the shield, leaving exposed rock. Much of the Canadian Shield is too barren to settle.

Core Provinces Bodies of water border the Core Provinces of Quebec and Ontario. **Hudson Bay** Ⓒ is a large inland sea to the north, and the St. Lawrence River Ⓓ and **Great Lakes** lie to the south. That river, those lakes, and several canals form the **St. Lawrence Seaway**, a waterway linking the Atlantic Ocean to the heart of North America.

Quebec and Ontario are the most densely populated provinces, even though most of their lands lie on the barren Canadian Shield. People have crowded into a narrow region of fertile land along the Great Lakes and St. Lawrence River. Many rivers flow from the Canadian Shield into this area. Canadians have built **hydroelectric** plants on the rivers, to create electricity from water-powered engines.

COMPARING Regions of Canada

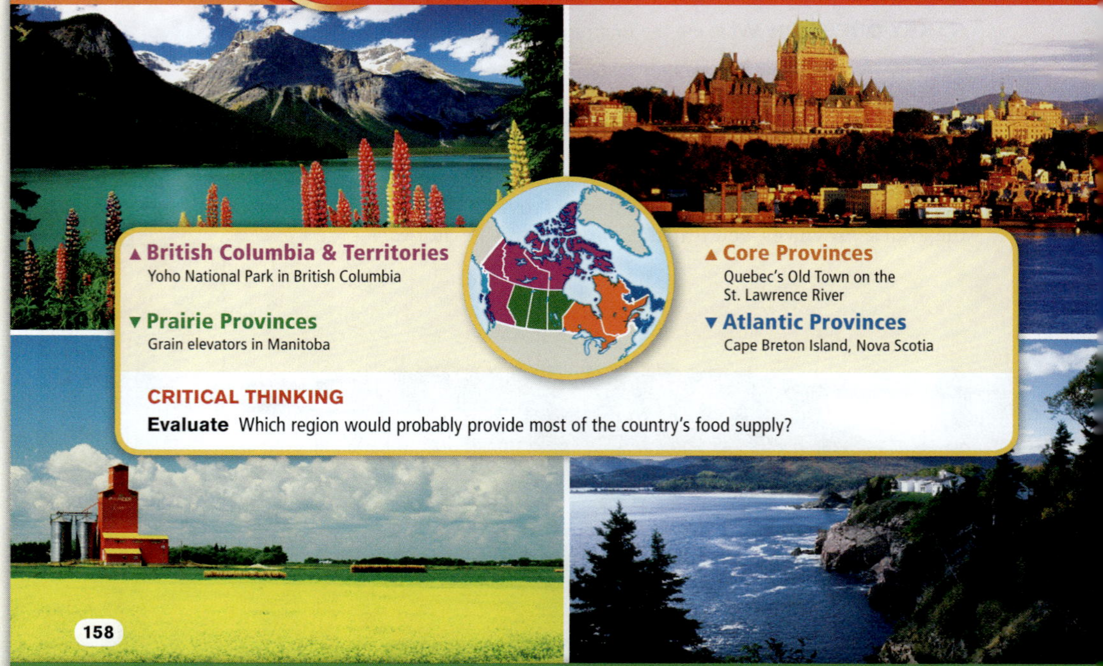

▲ **British Columbia & Territories** Yoho National Park in British Columbia

▼ **Prairie Provinces** Grain elevators in Manitoba

▲ **Core Provinces** Quebec's Old Town on the St. Lawrence River

▼ **Atlantic Provinces** Cape Breton Island, Nova Scotia

CRITICAL THINKING

Evaluate Which region would probably provide most of the country's food supply?

158

DIFFERENTIATING INSTRUCTION

English Learners

Vocabulary: Word Relationships

Have students find the following words each time they appear on this page: *soil, farming, barren, fertile.* Ask them to read the sentences containing the word(s) and then to define the identified word(s). Have them use the words to create two sentences about farming: one explaining the kind of land needed for farming, another describing land conditions that won't let farming take place.

Gifted & Talented

Travel Journal of Canada

Read aloud to students a book excerpt about a landscape (such as the opening to "The Encircled River" in John McPhee's *Coming into the Country),* and discuss how the author engages the senses. Then have students write a journal entry for each image on this page, as if they were traveling across Canada. In their last entry, have them compare and contrast previous scenes.

Physical Geography of Canada

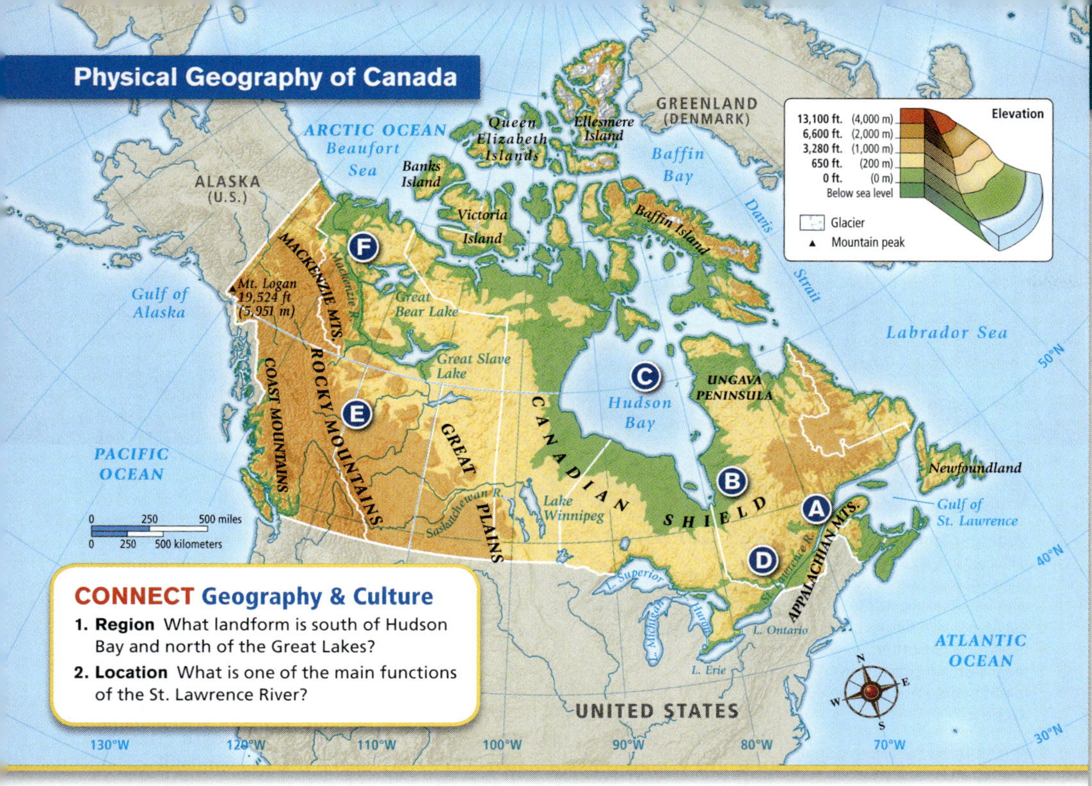

CONNECT Geography & Culture

1. **Region** What landform is south of Hudson Bay and north of the Great Lakes?
2. **Location** What is one of the main functions of the St. Lawrence River?

Prairie Provinces The Great Plains, which you studied in an earlier chapter, extend northward into the Prairie Provinces—Manitoba, Saskatchewan (sa•SKACH•uh•WAHN), and Alberta. To the west, this region is bordered by the Canadian Rockies (E). The Prairie Provinces are mostly flat grasslands and are one of Canada's major farming regions. Evergreen forests are found in the north.

British Columbia and Territories British Columbia is bounded by the Pacific Ocean to the west and the Canadian Rockies to the east. British Columbia has a scenic coast and jagged coastal mountains.

Canada's three territories are the Yukon, the Northwest Territories, and **Nunavut** (NOO•nuh•VOOT), created in 1999 from the eastern part of the Northwest Territories. Nunavut is home to many native peoples. It has a rocky mainland and thousands of Arctic Islands.

The Northwest Territories contain two of Canada's largest lakes, the Great Bear and Great Slave lakes. The Mackenzie River (F), Canada's longest, runs from the Great Slave Lake to the Arctic Ocean.

SUMMARIZE Name the four regions of Canada. Atlantic Provinces, Core Provinces, Prairie Provinces, and British Columbia and Territories

INTERDISCIPLINARY ACTIVITIES

CONNECT to Language Arts

Create "Where I Live" Dialogue

Have students pick a physical feature on the map from two different regions of Canada. Have them create a chart with the five themes of geography as column heads, and a row for each feature, and then fill in the chart using information from their textbooks and classroom resources. Then have them use their charts to create a dialogue between two people who "live" along these features, talking about what their lives are like.

CONNECT to Math

Graph Elevation West to East

Have students pick a certain latitude on the map to use for creating an elevations bar graph. Have them use a full sheet of paper to write the elevations from the map legend on the Y-axis. On the X-axis, have them start at the west and continue toward the east at their selected latitude, graphing each feature. Discuss how this is another method for helping to see elevation changes.

Physical Geography of Canada

CONNECT Geography & Culture

Have students create a chart with the elevations from the map legend heading each column. Have them identify features for each elevation and write these in the correct column. Have them write statements identifying similarities or differences in elevation between different features.

ANSWERS

1. **Region** Canadian Shield
2. **Location** As part of St. Lawrence Seaway, it links the Atlantic Ocean to the heart of North America.

More About . . .

Canadian Place Names

Places get their names for various reasons. For example: Lake Winnipeg gets its name from the Cree Indian words for "muddy waters." It is said that Greenland, an island mostly covered by a glacier today, was given that name by Eric the Red of Norway around the year 985, to make it seem more attractive for people to settle there.

Teacher-Tested Activities

Marci Smith Deal, Hurst-Euless-Bedford ISD, Bedford, Texas

My students enjoy looking over the kind of information provided in an atlas. Use the Unit Atlas to create a side by side chart of the United States and Canada on the chalkboard, showing size, population, population density, time zones, types of government, and GNP. Then ask the students the following questions:

- What similarities do you see between the United States and Canada?
- Looking at a map of North America, why do you think there is such a big difference in population between the United States and Canada?
- If you were the prime minister of Canada and using only the information on the board, what would be the focus of your term in office and why?

Teach

Climates of Canada

Reader, Recorder, Reporter

- About what percent of all Canadians live further than 200 miles from the U.S. border? *(about ten percent)*
- What are some reasons that most Canadians live within 200 miles of the U.S. Border? *(North is very cold with very short winter days and long periods of darkness; southern regions near oceans and lakes are milder and would have longer winter daylight.)*
- **Categorize** What specific causes affect temperatures in Canada? *(Arctic Circle—makes it colder; oceans and lakes—make winter warmer and summer cooler; prairie—contribute to cold winters and warm to hot summers)*

More About . . .

North American Sled Dog Races

Sled dog racing is a popular winter sport in parts of North America. Races cover huge distances in temperatures around or far below 0°F. The Canadian Challenge Race covers almost 300 miles in the longer of its two races. Teams of 12 dogs race from Prince Albert, Saskatchewan, north to LaRonge and back. This is a qualifying race for Alaska's Iditarod, that takes 10–17 days, covering about 1,100 miles. This race includes parts of the historic Iditarod Trail that was once a dog sled mail route for gold mining camps.

CONNECT to the Essential Question

How is Canada managing its resources in the 21st Century?

Ask students what they have learned so far that can help them answer this question. Students might mention:

- The Atlantic Provinces have made use of the coast's good harbors.
- Hydroelectric plants have been built in areas with rivers, such as British Columbia.
- The Core Provinces and Prairie Provinces farm on their fertile land.

Midnight Sun Under the midnight sun, snowmobiles pull sleds loaded with supplies back to camp across the Arctic ice in Nunavut, Canada. **What other means of transportation might people use on the Arctic ice?**

sleds pulled by dogs

Climates of Canada

KEY QUESTION What are some influences on Canada's climates?

Think of the coldest states that you know of within the United States. Canada is located even farther north than those states—so far north that it reaches into the Arctic Circle.

A Northern Land Canadians think of themselves as a people of the far north, but in reality the northernmost lands are thinly populated. Since the northern part of Canada lies much closer to the North Pole than to the equator, its climate is very cold. Some places remain frozen all year, and most of the region is **tundra**, flat treeless land near the Arctic Circle. Also, places that are far from the equator have a great variation in the hours of sunlight they receive during the year. Winter days are very short, and the long period of darkness increases the cold. Because life in the far north is so harsh, about 90 percent of all Canadians live within about 200 miles of the U.S. border.

Mild and Harsh Climates Bodies of water also influence Canada's climate. The Atlantic and Pacific oceans have a moderating effect on the coasts; winters are warmer and summers cooler there. Similarly, the Great Lakes make the climates of southern Ontario and Quebec milder. In contrast, the Prairie Provinces, located far from the oceans, have very cold winters and warm to hot summers.

The oceans and lakes make nearby lands humid, while the Prairie Provinces are drier. Mountains also play a role in causing this difference. Winds blowing in from the Pacific pass over the coastal mountains. As they rise, they drop moisture, causing high rainfalls in parts of British Columbia. The lands east of the mountains are arid.

FIND MAIN IDEAS Describe influences on Canada's climate.
its far northern location and the nearness of the ocean

DIFFERENTIATING INSTRUCTION: TIERED ACTIVITIES

OBJECTIVE Create a map showing climates of Canada.

Basic

Have students refer to the present-day Canada map to draw an outline map of Canada's provinces and territories. Have them label the major bodies of water. As a group, identify climate features, decide on symbols to show these features, and then label the map to show them.

On Level

Have students draw the same map as for the Basic activity. Tell them to create a legend for these climate features: very cold year round; moderated temperatures; very cold winters and warm to hot summers; humid; and arid. Have them label their map according to Canada's varied climates.

Challenge

Have students draw the same map as for the On Level activity. Have them identify climate features and create a legend. Have them label their map and add footnotes explaining the connection between a physical feature and climate (e.g., being east of the coastal mountains prevents rain from falling).

Resources of the Regions

KEY QUESTION What are the main resources of each region?

Canada has many natural resources, which have helped it to develop a strong economy. Each region has its own distinct set of resources.

- The Atlantic Provinces have an abundance of timber and mineral wealth, especially coal. The Gulf of St. Lawrence and coastal waters supply plentiful stocks of seafood for export.
- The Core Provinces have a flat fertile plain in the south, plus a mild, humid climate that is good for farming. The St. Lawrence Seaway is a resource that aids transportation and economic growth. Other rivers are used to create hydroelectric power.
- The fertile plains of the southern Prairie Provinces are used to raise grain. Oil and natural gas are found in the region, especially in Alberta.
- British Columbia has forests, minerals, hydroelectric power, ocean fishing, and some fertile lands and valleys. The territories also have resources, such as oil, natural gas, and lead, that have not been developed because they are located far from markets.

CATEGORIZE Describe the resources of each of Canada's regions.

tlantic: timber, ıinerals, fish; ore: fertile land, t. Lawrence eaway, hydroelec-ic power; Prairie: ertile land, oil, atural gas; British olumbia and erritories: timber, ıinerals, hydro-lectric power, il, natural gas ▸

Fun Facts!

BEARS ON ICE!

Churchill, Manitoba, is called the polar bear capital of the world. Polar bears live on sea ice and hunt seals. But in the southwestern Hudson Bay, the ice pack melts in summer, forcing the bears onto land. Every year in late fall, polar bears gather near Churchill waiting for the bay to freeze so they can return to the ice.

Section 1 Assessment

ONLINE QUIZ For test practice, go to Interactive Review @ClassZone.com

TERMS & NAMES

1. **Explain the importance of**
   - province
   - territory
   - Hudson Bay
   - Nunavut

USE YOUR READING NOTES

2. **Categorize** Use your completed chart to answer the following question:

   What mountain chain separates the Prairie Provinces and British Columbia?

	LANDFORMS & WATER	CLIMATE	RESOURCES
Atlantic Provinces			
Core Provinces			
Prairie Provinces			
British Columbia & Territories			

KEY IDEAS

3. Which two regions are the biggest farming areas of Canada?
4. Which two regions have ocean fishing as a major part of their economies?
5. Why haven't Canadians developed the mineral resources of the territories?

CRITICAL THINKING

6. **Analyze Effects** Give one economic advantage and one disadvantage of the Canadian Shield.
7. **Contrast** How is the climate of the center of the country different from the climates of the coasts?
8. **CONNECT to Today** What technologies might make it possible to use the mineral wealth of the territories?
9. **SCIENCE** **Sketch a Map** Learn about the Ice Age glaciers in an encyclopedia, a science book, or an educational Web site. Then sketch a map showing which parts of North America were covered by glaciers.

Teach

Resources of the Regions

Talk About It

- How important is ease of transportation to a region's use of its natural resources? Support your answer. *(Possible Answer: Territories may have many resources, but no close markets and difficult terrain, so their resources may remain largely untapped.)*
- **Categorize** What do you think are Canada's three most important resources? *(Possible Answer: St. Lawrence seaway; oceans; plains)*

4 Assess & Reteach

Assess Have students complete the Section Assessment.

Unit Resource Book: The United States & Canada
- Section Quiz, p. 123

Interactive Review @ClassZone.com
- Section Quiz

Power Presentations

Test Generator

Reteach Have students assume the role of Canada's Prime Minister. Have them write a speech, explaining how they plan to help strengthen each region.

Unit Resource Book: The United States & Canada
- Reteaching Activity, p. 127

SECTION 1 ASSESSMENT ANSWERS

Terms & Names

1. province: a Canadian political unit; territory: a political unit that doesn't have enough people to become a province; Hudson Bay: a large inland sea; Nunavut: a territory created in 1999

Use Your Reading Notes

2. the Rocky Mountains

   Diagram *Atlantic Provinces—Appalachians, Atlantic, Canadian Shield; milder near Atlantic; timber, coal. Core Provinces—Hudson Bay, Great Lakes, Canadian Shield; milder near Great Lakes; crops, hydroelectric power. Prairie Provinces—Great Plains; cold winters, hot summers; grain, oil gas. British Columbia and Territories—Pacific, Rockies; milder near Pacific; forests, minerals*

Key Ideas

3. Core Provinces, Prairie Provinces
4. Atlantic Provinces, British Columbia
5. resources located too far away from markets

Critical Thinking

6. Possible answer: Many rivers that flow from the shield are used to generate hydroelectric power. The Canadian Shield is not a fertile region, so few people can settle there.
7. The center of the country is arid and has more extreme temperatures; the coasts are more humid and have milder temperatures.
8. Students might suggest the need for technologies to make it easier to live in such a cold region as well as improved transportation.
9. Use the rubric to score maps.

Map Rubric

	Content on Map	Presentation of Map
4	all required elements	clear, neat, complete
3	most elements	neat, complete
2	some elements	somewhat sloppy or incomplete
1	little or no information	sloppy or mostly incomplete

1 Plan & Prepare

Objectives: Analyze how the technology of the St. Lawrence Seaway's locks and canals helps people accomplish the goal of increased shipping to inland areas.

Visual Vocabulary: Ask students to describe what they see in the illustration. Call out familiar items listed below, asking students to locate each one on the page. Then tell them the precise term, as listed in boldface. Add terms to a class vocabulary list, illustrated poster, or Word Wall.

- a view from above **(aerial)**
- a watertight chamber with gates where ships can be lifted or lowered **(lock)**
- a small channel for water to flow through **(sluice)**
- the front part of a ship **(bow)**
- a shaft or stack on a ship for ventilation or discharging smoke **(funnel)**
- a large chimney or upright pipe to discharge gases or smoke, often used in industry **(smokestack)**

2 Teach

Engage . . .

DISPLAY "The St. Lawrence Seaway" feature from the **Power Presentations** DVD-ROM or the *Animations Center* at **ClassZone.com.** As the animation loads on screen, ask students:

- What is happening with the ship?
- Why are there stairs?
- What else can you see on either side of the lock?
- What is the ship carrying?

Encourage all answers.

Explore . . .

ROLL OVER sections of the illustration or click on items in the menu. Ask students to choose which areas they would like to explore further.

- Then return to the discussion questions. Have any answers changed?
- Have students jot down a list of at least three new questions that arise from exploring the rollover features.

Animated GEOGRAPHY

The St. Lawrence Seaway

Click here to enter the St. Lawrence Seaway @ClassZone.com

SHIPS INTO THE HEARTLAND

The St. Lawrence Seaway connects the Great Lakes to the Atlantic Ocean by way of the St. Lawrence River and a system of canals and locks. The seaway enables ships to sail into the heart of the continent. It was completed in the 1950s as a joint project of the United States and Canada.

Click here to see how ships are raised and lowered 600 feet by a series of locks.

A A seaway lock is a watertight chamber with gates at either end that allows a ship to be lifted or lowered from one level to another. Here the lock gates open to allow a ship to enter.

B When all gates are shut, water is let into the lock through a sluice (small channel) to raise or lower the water level.

C Once the water in the chamber is level with the canal ahead, the second gate opens and the ship moves on.

Click here to see how millions of tons of goods are shipped on the seaway each year. Most of the freight, consisting of bulk cargo such as grains and minerals, travels from North America to Europe.

DIFFERENTIATING INSTRUCTION: TIERED ACTIVITIES

OBJECTIVE Create a "People and Technology: Locks and Canals" visual to illustrate how people use technology to accomplish a goal

Basic

Ask students to give examples of how a tool or machine—types of technology—helps them (e.g., using a hammer to build a doghouse). Have them review the images and text specific to the Seaway locks and canals. Have them identify one goal people wanted to accomplish with the Seaway and how a tool or machine helped (e.g., moving a ship between two bodies of water by building canals). Have them title their poster, write a brief caption for the goal and how the technology helped meet it, and illustrate their poster.

GeoActivity

Build a Lock Use a cardboard shoebox to make a lock. If you have two or more shoeboxes, you might show ships moving in two directions. Use small plastic boats and trucks to show how the boats move in the seaway. Cargo trucks and other vehicles might be positioned along the docks.

On Level

Discuss with students how the scene on these pages shows many ways in which people use a tool or machine—technology—to accomplish a goal. Have them review the entire scene and text, then identify at least three goals people wanted to accomplish and how technology helped. Have them create a chart with a title, a statement of each goal, and a brief explanation of how the technology helped meet each goal. Have students also include an illustration for each goal/technology.

Challenge

Have students follow the On Level activity instructions. In addition, have them create a schematic drawing—a detailed diagram that explains, sometimes step-by-step, how something works—for each goal/technology they list. (You may want to provide examples of schematic drawings for students to use as models or have students use library and online resources to search for their own examples.)

ANIMATED GEOGRAPHY

Explain . . .

CHOOSE topics to help students answer their questions, such as:

- **Lock and Canal Systems:** The St. Lawrence Seaway System has three major lock systems with a total of six canals (from 0.3 to 23.5 nautical miles long) and 19 locks.
- **Canadian and U.S. Operations:** The lock systems are in both countries. The Welland Canal is in Canada, and St. Mary's River is in the United States.
- **Ship Size:** Ships up to 740 feet long and 78 feet wide can pass through the locks, carrying cargoes of 56 million pounds.
- **Getting Through Locks:** It takes up to 10 minutes to fill a lock with about 24 million gallons of water, and about 45 minutes for a ship to pass through.

Unit Transparency Book
- Map: U.S. and Canada Rivers, TT18

3 Wrap Up

Elaborate . . .

ASSIGN roles for a role-play of the proposed Seaway locks and canals.

- Have several students act as a team of representatives for the project who are answering questions at a public meeting. Assign other students to be the audience members. Have audience members ask questions to find out the project's purpose and benefits, and details about it.

Evaluate . . .

Build a Lock
Activity Rubric

4 Excellent The student's construction shows many details of the lock operation, including vessel passage, and portrays surrounding context.

3 Strong The construction shows some details of the lock operation.

2 Fair The construction conveys a general idea of lock operation, but has few details.

1 Poor The construction lacks detail and fails to convey the lock operation.

1 Plan & Prepare

Objectives

- Identify Canada's earliest inhabitants and later European explorers and colonists
- Describe the steps that led to Canada's independence from British rule
- Analyze the importance of minority self-rule in Canada and Canada's relationship with the United States today

Read for the Essential Question

Help students read for a purpose by reminding them of the Essential Question: "How is Canada managing its resources in the 21st century?" Point out that they can use text features, such as Key Questions, special features, images, and maps to form answers to the question.

Vocabulary

Best Practices Toolkit

Use the ***Best Practices Toolkit*** to model strategies for vocabulary notetaking. Vary strategies throughout the year. Choose from: *Knowledge Rating, Predicting ABCs, Definition Mapping, Word Sort, Frayer Model (Word Squares), Magnet Words, Word Wheel,* and *Student VOC.*

Vocabulary Strategies, TT9–TT16

Reading Strategy

Best Practices Toolkit

Display the Web Diagram transparency.

- Remind students that main ideas are the most important points in a text.
- Have students find the main idea in the "Connecting to Your World" text at the start of this section.
- Model finding and entering main ideas on a chart using the "Connect to History" feature in this section.
- Have students suggest entries.

Web Diagram, TT28

SECTION 2 Reading for Understanding

Key Ideas

BEFORE, YOU LEARNED

Canada spans the continent, from the Atlantic to the Pacific oceans, and from the U.S. border to the Arctic Ocean.

NOW YOU WILL LEARN

Canada's diversity is the result of regional differences and events in Canada's history.

Vocabulary

TERMS & NAMES

bilingual using or able to use two languages equally

First Nations organized cultural groups of Canada's native peoples

Inuit (IHN•yoo•iht) the native peoples who inhabit the Arctic region of North America

Seven Years' War a conflict (1756–1763) between Britain and France that was fought in North America, Europe, and India

dominion in the British Empire, a nation that is allowed to govern its domestic affairs

William Lyon Mackenzie King Canadian prime minister who led his country to independence

Commonwealth of Nations an association including the United Kingdom and many of its former colonies

separatist a person who wants a region to break away from the nation it is a part of

BACKGROUND VOCABULARY

Loyalist a person who supported Great Britain during the American Revolutionary War

Visual Vocabulary bilingual

Reading Strategy

Re-create the web diagram shown at right. As you read and respond to the **KEY QUESTIONS**, use the outer ovals to note important details about Canada's history.

Skillbuilder Handbook, page R4

FIND MAIN IDEAS

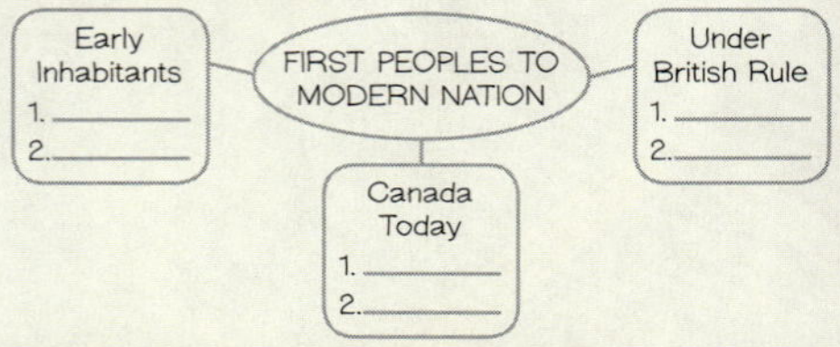

GRAPHIC ORGANIZERS
Go to **Interactive Review** @ClassZone.com

PRETEACHING VOCABULARY

English Learners

Pronounce and Preview

Pronounce each term for students. Review the meanings of words within the definitions, such as *equally, domestic,* and *association.*

- To modify vocabulary learning, have students complete worksheets as they read, instead of afterward.

Unit Resource Book: The United States & Canada

- Vocabulary Practice, p. 109

Inclusion

Match Terms and Definitions

Have pairs of students work together to create a set of index cards with each term written on a card. To review the terms, have pairs take turns reading them to each other. Then one partner should hold out the cards, fanned out, for the other to pick one and say the term. The partner holding the cards should then give the definition of the term in his or her own words or read it from the textbook. Have partners alternate roles.

SECTION 2 HISTORY

Totem poles in Stanley Park, Vancouver, British Columbia

Atlantic puffins on an island in Canada

First Peoples to a Modern Nation

Connecting to Your World

Marie is a student at a *polyvalente,* a secondary school in Quebec City. Most of her classes are taught in French, but she has one course in English conversation. Last year, Marie traveled as an exchange student to Manitoba. Unlike Quebec, most people in Manitoba conduct their daily business in English. Marie wants to become fluent in both languages to work in government. Canada is a **bilingual** country with two official languages—English and French.

Marie A young participant in Quebec City's *Fêtes de la Nouvelle France* (Festival of New France) parade

Early Inhabitants and First Explorers

KEY QUESTION Which European countries started colonies in Canada?

Earlier, you learned that many scientists believe that the first inhabitants of North America came from Asia by crossing over on a land bridge. A second theory suggests that some people journeyed to North America by boat. Over time, humans spread throughout the Americas. The people in different regions developed separate cultural groups whose ways of life were adapted to the places where they lived.

Quebec City Samuel de Champlain, an early explorer, founded Quebec City, shown below.

165

2 Focus & Motivate

3-Minute Warm-Up

Write on the board or display the transparency:

- How do you think Canada's history might have contributed to its diversity? *(Possible Answers: through its early settlement and later immigration; because of its long history of respecting other cultures)*

Unit Transparency Book
- 3-Minute Warm-Ups, TT11

Connecting to Your World

More About . . .

Totem Poles

Some totem poles depict family stories, such as the Beaver Crest Pole, carved by Nisga'a artist Norman Tait and his family members. It depicts how the family's Eagle clan adopted the beaver as their family emblem.

3 Teach

Early Inhabitants and First Explorers

Talk About It

- How were the lives of Canada's earliest cultural groups similar and different? *(similar: most hunted, land important; different: Iroquois farmed, differing climates meant different resources, such as hunting different animals)*
- **Find Main Ideas** How has human movement played a role in Canada's population? *(Possible Answer: Native peoples spread across the land; European exploration led to settlements.)*

SECTION 2 PROGRAM RESOURCES

Unit Resource Book: The United States and Canada

ON LEVEL

Unit Resource Book
- Reading Study Guide, p. 79
- Map and Graph Skills, p. 111
- Section Quiz, p. 124

STRUGGLING READERS

Unit Resource Book
- Reading Study Guide with Additional Support, p. 87
- Skillbuilder Practice: Analyzing Cause and Effect, p. 110
- Section Quiz, p. 124
- Reteaching Activity, p. 128

eEdition with Audio DVD-ROM

ENGLISH LEARNERS

Pupil Edition in Spanish
eEdition with Audio DVD-ROM
Unit Resource Book
- RSG (Spanish), p. 95
- RSG with Additional Support (Spanish), p. 103

Multi-Language Glossary
Test Generator
- Section Quiz in Spanish

INCLUSION

Unit Resource Book
- RSG with Additional Support, p. 87
- Section Quiz, p. 124
- Reteaching Activity, p. 128

GIFTED & TALENTED

Unit Resource Book
- History Makers, p. 115
- Connect to History, p. 121
- Section Quiz, p. 124

PRE-AP

Unit Resource Book
- History Makers, p. 116
- Primary Source, p. 122
- Section Quiz, p. 124

TECHNOLOGY

Unit Transparency Book
- 3-Minute Warm-Ups, TT11
- Essential Question, TT12
- Critical Thinking, TT14
- Map: Territorial Growth of Canada, TT17, TT17a, TT17b

Daily Test Practice Transparencies
- Section 2, TT19

Power Presentations
ClassZone.com
World Cultures and Geography Video on DVD

CONNECT to History

Later explorers continued to expand the areas in Canada that Europeans knew about. Henry Kelsey is said to be the first European to see Canada's prairies. He journeyed there in 1690–1692. Samuel Hearne became the first British citizen to reach the Arctic Ocean by land. He journeyed there in 1770–1772, led by Chipewyan leader Matonabbee. In both cases, knowledge of native people's languages and cultures helped the journey to succeed.

Unit Resource Book: The United States & Canada

- Map and Graph Skills, p. 111
- History Makers: John Cabot, p. 113
- History Makers: Jacques Cartier, p. 116
- Connect to History: Create a Time Line, p. 121

Comparing Regions

Traditional Hunting Methods in Canada's Regions

The native peoples of Canada hunted different animals using different methods, depending on the region.

- When Maritime peoples moved inland for increased warmth and shelter in the winter, they hunted moose using birch-bark moose calls and spears, and they ambushed caribou at the shores of frozen lakes.
- People in the Plains built bison pounds (pens) that had walls of wooden poles interwoven with brush. They built lanes of small rock and brush, sometimes more than half a mile long, to drive the bison into the pound.
- Inuit people in the Arctic hunted whales from the bow of a large skin boat (umiaq), using harpoons.

The Earliest Settlers In Canada, the organized cultural groups of native peoples are called **First Nations**. In the east lived the Algonquian (al•GAHNG•kwee•uhn) peoples, who hunted, and the Iroquois (IHR•uh•KWOY), who hunted and farmed. On the plains, the main groups were the Blackfoot, Cree, Ojibwa, and Sioux (soo). Most lived by hunting bison. The mild Pacific coast was heavily populated. Some groups were the Haida, Kwakiutl (KWAH•kee•OOT•uhl), and Salish. The frosty Arctic lands were inhabited by the **Inuit** (IHN•yoo•iht), who survived by hunting polar bears, whales, and seals.

First Colonies In 1497, the explorer John Cabot landed on the coast of Newfoundland and claimed it for England. In 1534, the French explorer Jacques Cartier (kahr•TYAY) arrived at the mouth of the St. Lawrence. He claimed the nearby lands for France.

Many of the French who came to Canada were single men who wanted to grow rich by trading for furs. Some British who came to Canada also wanted furs. The British Hudson's Bay Company set up many trading posts. But almost all the British who moved to North America came as colonists with families. Britain and France were bitter rivals for lands and wealth. The two nations fought several wars during the late 1600s and 1700s.

SUMMARIZE Describe which countries started colonies in Canada.
France and Great Britain

CONNECT to History

A number of different European explorers contributed to knowledge of the lands that later became Canada.

Activity

Create a Time Line

Materials
- paper
- pencil
- ruler

1. Draw a time line to visually express the dates at left. Draw a line that is 8 inches long. Place marks at each end and at two-inch intervals.
2. Put these dates on your time line: 1450, 1500, 1550, 1600, 1650.
3. Put the events at left on your time line. Place each one in the right segment at the correct distance from the other dates.

◀ *A replica of John Cabot's ship, the Matthew, in full sail during the making of a film.*

DIFFERENTIATING INSTRUCTION

English Learners

Comprehension: Self-Monitor

To help students visualize which native cultural groups lived in which areas, remind them to pay close attention to the terms denoting the regions *(east, plains, Pacific coast, Arctic)*. You may want to suggest that they write each region and the cultural groups that lived there on a large "sticky" note, and post it onto the Physical Geography of Canada map in the previous section.

Struggling Readers

"Who Am I?"

After students finish this page, model a "Who Am I?" game. Provide information about an explorer from the Connect to History feature, then ask, "Who am I?" (e.g., for de Champlain: "I founded Quebec City. Who am I?"). Then do the same with the text (e.g., "I hunt and farm in the east. Who am I?" *I am an Iroquois person.)* Then have students each pick a group or person, come up with their clue, and say it for the whole group to guess.

Under British Rule

KEY QUESTION What steps did Canada take to gain independence?

In 1754, fighting again broke out between the British and French in North America. The war, which spread to Europe and India, is called the **Seven Years' War** because it lasted from 1756 to 1763 in Europe. (In the United States, it is called the French and Indian War.) Britain won and took French lands, including the colony of Quebec.

Upper and Lower Canada Right after the war, French-speakers in Quebec outnumbered the English-speakers, but the population of English-speakers grew. After the American Revolution (1775–1783), thousands of **Loyalists**, people who had supported the British, moved to Canada. French-speakers and English-speakers mistrusted each other. So in 1791, the British divided Quebec into English-speaking Upper Canada (Ontario) and French-speaking Lower Canada (Quebec).

Canada Achieves Self-Government In the 1800s, Canada underwent many changes. Explorers pushed west looking for new places to find furs. Canadians increased their trade with the United States.

In 1867, Canada became a **dominion**, a nation in the British Empire that is self-governing. The British monarch remained Canada's head of state, and the legislature was modeled on the British Parliament.

During the late 1800s, Canada continued to grow. In 1885, a transcontinental railroad was finished. European immigrants moved to Canadian cities and the wide-open plains. The economy boomed.

Canadians began to question their role in the British Empire. After World War I, Canadian leaders such as Prime Minister **William Lyon Mackenzie King** sought independence. This was granted in 1931. Canada joined the **Commonwealth of Nations**, an association of the United Kingdom and many of its former colonies.

HISTORY MAKERS

William Lyon Mackenzie King (1874–1950)

Mackenzie King (top left) served as prime minister of Canada longer than anyone else. Perhaps his greatest accomplishment was leading Canada to independence. He also kept his country together during World War II by uniting French-speaking and English-speaking Canadians despite their opposing views. In the picture above, political leaders gather for the Quebec Conference in August, 1943. From the left: Mackenzie King, U.S. President Franklin D. Roosevelt, British Prime Minister Winston Churchill, and the Earl of Athlone (Governor General of Canada).

ONLINE BIOGRAPHY For more on the life of Mackenzie King, go to the Research & Writing Center @ClassZone.com

FIND MAIN IDEAS Explain how Canada gained independence.
First it became a dominion; then it became an independent nation and member of the Commonwealth.

INTERDISCIPLINARY ACTIVITIES

CONNECT to Civics

Make a Chart

Have students make a chart with columns labeled *colony, dominion, independent.* Have them label the first row: "What it means" and define each term. Have them create two more rows: *pros* and *cons.* You may want students to work in pairs or small groups to brainstorm pros and cons. Have them share what they came up with, and add any new ideas from others to their own charts.

CONNECT to Language Arts

Write a Letter to the Editor

Have students pick one of these scenarios: *from colony to dominion; from dominion to independence.* Have them write a letter to the editor supporting or opposing the change, from the perspective of a particular person (e.g., a French-speaking farmer in Quebec, or a descendant of a loyalist). They should give specific reasons for their view, using information from their pros and cons chart and from the text and images in this section.

Teach

Under British Rule

Reader, Recorder, Reporter

- How did the Seven Years' war affect Canada? *(Britain won and took French lands, including Quebec.)*
- Why was the colony of Quebec divided after the American Revolution? *(An increase in English-speaking loyalists from the United States caused tensions with French speakers.)*
- What economic changes took place in Canada in the 1800s? *(Explorers went west for furs; trade increased with the United States; the transcontinental railroad was finished; immigration grew.)*
- **Find Main Ideas** How did Canada's political status change from the 1700s to the 1930s? *(It went from colony to dominion, then to independence and membership in the Commonwealth of Nations.)*

RESEARCH & WRITING CENTER

HISTORY MAKERS

William Lyon Mackenzie King

Among Mackenzie King's other contributions, he helped improve living standards for Canadians in many ways. During his 22 years in office, an old age pension began in 1927, unemployment insurance (now called employment insurance) started in 1940, and the Family Allowance Act (1944) started a monthly payment to Canadian families with children, regardless of income.

Teach

Canada Today

Think, Pair, Share

- What caused Canada to create a law about services being offered in both French and English? *(Separatists wanted Quebec to break away in response to discrimination they felt.)*
- Why would the Inuit want self-rule? *(Possible Answers: to preserve their language and culture; to have more say in their affairs)*
- **Find Main Ideas** What two major political challenges has Canada faced since independence, and what are the major issues with each? *(minority groups: Quebec separation, Inuit self-rule; relations with the United States: views on terrorism, global warming, trade)*

Unit Resource Book: The United States & Canada
- Skillbuilder Practice: Analyze Causes and Effects, p. 100

RESEARCH & WRITING CENTER

ANALYZING Primary Sources

One result of the Canadian Multiculturalism Act was the Ethnic Diversity Survey. Fifteen years after the Act, the survey showed progress towards all Canadians having an equal chance to contribute to Canadian society, but also a need for improvement.

- What might a government do to help promote the goals of the Act? *(Possible Answers: fund programs to support people's heritage; pass and enforce discrimination law)*
- Why do you think the government decided to create this Act? *(Possible Answer: to preserve the heritage of all groups in Canada)*

DOCUMENT-BASED QUESTION ANSWER

Possible answer: It might apply to language, religion, and traditions such as hunting and fishing.

Unit Resource Book: The United States & Canada
- Primary Source: A Declaration of First Nations, p. 122

Canada Today

KEY QUESTION Which minority groups in Canada wanted to gain more self-rule?

One major political issue was resolved when Canada became independent, but the country has faced other ongoing concerns since then. Two of those issues have involved responding to minority groups within Canada and dealing with the United States.

Tensions over Quebec and Nunavut During the 1960s, some French Canadians campaigned against what they regarded as discrimination. A group of **separatists**—people who want a region to break away from a country it is a part of—wanted Quebec to break away from Canada. In response, the national government passed a law that required services to be offered in French and English in areas where 10 percent of the people speak French or English. The separatists continued to seek independence. The people of Quebec voted on the issue of separating in 1995, but the proposal was defeated.

The French-speakers were not the only Canadians to want more self-rule. The Inuit of the far north had a similar desire. But instead of seeking separation from Canada, the Inuit wanted self-rule within the country. They negotiated with the government to gain a separate Inuit territory as part of Canada, and Nunavut was created in 1999.

ONLINE PRIMARY SOURCE To learn more about the Canadian Multiculturalism Act of 1988, go to the **Research & Writing Center @ClassZone.com**

ANALYZING Primary Sources

The Canadian Multiculturalism Act (1988) set forth the government's intention to help all members of Canadian society preserve their unique cultures.

> It is hereby declared to be the policy of the Government of Canada to . . . recognize and promote the understanding that multiculturalism reflects the cultural and racial diversity of Canadian society and acknowledges the freedom of all members of Canadian society to preserve [save], enhance [develop], and share their cultural heritage.
>
> Source: *The Canadian Multiculturalism Act of 1988*

DOCUMENT-BASED QUESTION

What aspects of cultural heritage do you think this act might apply to?

Mural painted by children for Canada Day ▶

DIFFERENTIATING INSTRUCTION: TIERED ACTIVITIES

OBJECTIVE Create a fact sheet about Canada's diversity using journalists' questions.

Basic

Remind students of the six journalists' questions: *who, what, when, where, how,* and *why.* Use these words to brainstorm questions about Canada's diversity. Review the information on this page to find the answers. List them. Have students use the list to create a six-item question-and-answer fact sheet.

On Level

Remind students of the six journalists' questions. Have them use the words to brainstorm questions about Canada's diversity, and then find the answers in their textbooks. Every student should then create his or her own question-and-answer fact sheet with at least one item for each word.

Challenge

Remind students of the six journalists' questions. Have them work on their own to brainstorm as many questions about Canada's diversity as they can for each word. Every student should then create a question-and-answer fact sheet with one item for each word and a statement summarizing diversity.

Relations with the United States At times, tensions exist between Canada and the United States. For example, the two countries have some differences in their views of how best to fight terrorism. Global warming and trade are two other issues on which Canada and the United States occasionally have different opinions and follow divergent paths.

In spite of such disagreements, Canada and the United States have a good relationship. Their 5,525-mile border is the longest unprotected border in the world, although guards do check papers and inspect vehicles.

The U.S. and Canadian economies are closely linked. The two countries have signed agreements to promote trade with each other, such as the 1989 Free Trade Agreement and the 1994 North American Free Trade Agreement (NAFTA). From 1989 to 2004, U.S. exports to Canada more than doubled, and imports from Canada tripled.

CONNECT Geography & History

Environmental Partners

Canada and the United States work as partners to fight pollution. One area at risk is the heavily industrialized Great Lakes region. Air pollution was causing acid rain, which kills trees and fish. In 1991, the two nations signed an agreement to improve air quality. Since then, the gases that cause acid rain have been reduced by about a third.

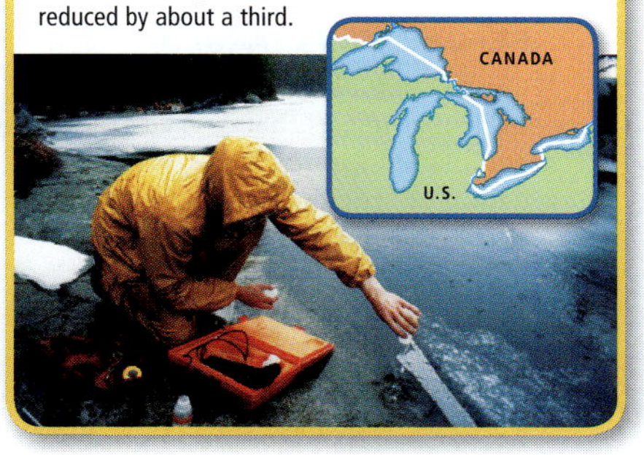

SUMMARIZE Explain which minorities in Canada wanted self-rule.
the French-speakers of Quebec and the Inuit in the far north

Section 2 Assessment

ONLINE QUIZ For test practice, go to **Interactive Review @ClassZone.com**

TERMS & NAMES

1. **Explain the importance of**
   - bilingual
   - Inuit
   - dominion
   - separatist

USE YOUR READING NOTES

2. **Find Main Ideas** Use your completed diagram to answer the following question:

   How has Canada accommodated the diversity of its population?

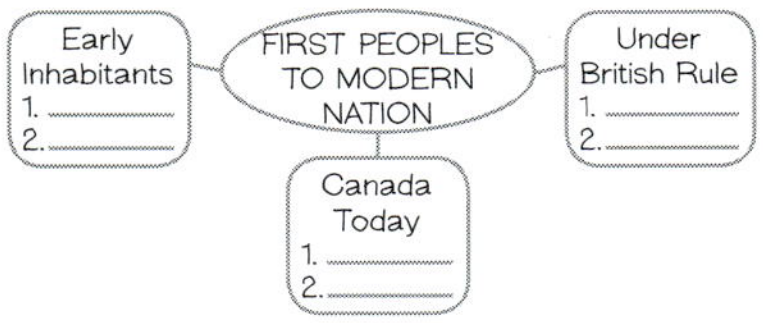

KEY IDEAS

3. How did the rivalry between Britain and France affect Canadian history?
4. What did dominion status mean for Canada?
5. What is unique about the U.S.-Canada border?

CRITICAL THINKING

6. **Analyze Causes** How did Canada's history cause it to become a bilingual country?
7. **Contrast** What was the main difference between the goals of the Quebec separatists and the Inuit?
8. **CONNECT to Today** How do you think the terrorist attacks of September 11, 2001, have affected the number of border guards and inspections on the U.S.-Canada border? Explain.
9. **WRITING** **Create a Venn Diagram** Make a Venn diagram showing how Canadian history and United States history are alike and different. Then write a paragraph summarizing the main ideas.

CONNECT to the Essential Question

How is Canada managing its resources in the 21st century?

Ask students what they have learned so far that can help them answer this question. Students might mention:

- Canada values human diversity as one of its greatest resources.
- Canada has tripled its exports to the United States since the signing of NAFTA.
- Canada and the United States are working together to reduce pollution.

4 Assess & Reteach

Assess Have students complete the Section Assessment.

Unit Resource Book: The United States & Canada
- Section Quiz, p. 124

Interactive Review @ClassZone.com
- Section Quiz

Power Presentations

Test Generator

Reteach Have pairs of students create one "True or False" statement about each of the subsections of the chapter. Then have each pair exchange and evaluate their statements with another pair. For each false statement, have students correct it, using their textbooks.

Unit Resource Book: The United States & Canada
- Reteaching Activity, p. 128

SECTION 2 ASSESSMENT ANSWERS

Terms & Names

1. bilingual: being able to use two languages; Inuit: native peoples who inhabit Arctic region of North America; dominion: nation allowed to govern its domestic affairs; separatist: person who wants region to break away from nation

Use Your Reading Notes

2. Canada has promoted multiculturalism to reflect the rich history of the country and the parts that make up the modern nation.

Key Ideas

3. It led to several wars, with the result that Great Britain took France's North American lands.
4. Canada self-governing within British Empire, with British monarch remaining head of state
5. It is the longest unprotected border in the world.

Critical Thinking

6. It was originally colonized by both Britain and France. Even after Britain took over French lands, large French-speaking population remained.
7. The separatists want to completely break away from Canada, while the Inuit simply wanted their own territory within Canada.
8. Possible Answer: numbers of guards and inspections have probably both increased in effort to catch terrorists trying to enter country.
9. Use the rubric to score diagrams.

Diagram Rubric

	Content on Diagram	Presentation of Diagram
4	all required elements	clear, neat, complete
3	most required elements	neat, complete
2	some required information	somewhat sloppy or incomplete
1	little or no information	sloppy or mostly incomplete

1 Plan & Prepare

Objectives

- Explain why Canada has such a diverse population
- Describe life in Canada today
- Identify the areas of arts and popular culture in Canada that have won worldwide fame

Read for the Essential Question

Help students read for a purpose by reminding them of the Essential Question: "How is Canada managing its resources in the 21st century?" Point out that they can use text features, such as Key Questions, special features, images, and maps to form answers to the question.

Vocabulary

Best Practices Toolkit

Use the ***Best Practices Toolkit*** to model strategies for vocabulary notetaking. Vary strategies throughout the year. Choose from: *Knowledge Rating, Predicting ABCs, Definition Mapping, Word Sort, Frayer Model (Word Squares), Magnet Words, Word Wheel,* and *Student VOC.*

Vocabulary Strategies, TT9–TT16

Reading Strategy

Best Practices Toolkit

Display the Three-Column Chart transparency.

- Point out to students that to categorize information in the section they must decide how it can be sorted into the categories shown in the chart.
- Model filling in entries from the "A Diverse Population" subsection in the section.
- Have students suggest entries, using the categories in the chart.
- Display the completed chart for review.

Three-Column Chart, TT21

SECTION 3 Reading for Understanding

Key Ideas

BEFORE, YOU LEARNED

Canada has a history that begins about the same time as that of the United States, and like the United States, Canada is a nation of immigrants.

NOW YOU WILL LEARN

Because Canada is a nation of immigrants, it has a diverse population and a variety of cultures shaped by many influences.

Vocabulary

TERMS & NAMES

Royal Canadian Mounted Police an organization of officers who were responsible for law enforcement in the Canadian West

Inuktitut the language of the Inuit

lacrosse a team sport invented by Native Americans

Visual Vocabulary lacrosse

BACKGROUND VOCABULARY

igloo a house made of snow

Visual Vocabulary igloo

Reading Strategy

Re-create the chart that is shown at right. As you read and respond to the **KEY QUESTIONS**, use the chart to organize information about the diverse culture of Canada.

Skillbuilder Handbook, page R7

CATEGORIZE

LANGUAGE	RELIGION	ETHNICITY	ARTS & CULTURE
1.	1.	1.	1.
2.	2.	2.	2.
3.	3.	3.	3.
4.	4.	4.	4.

GRAPHIC ORGANIZERS
Go to **Interactive Review** @ClassZone.com

PRETEACHING VOCABULARY

English Learners

Pronounce and Preview

Pronounce each term for students. Review the meanings of words within definitions, such as *enforcement* and *Inuit.*

- To modify vocabulary learning, have students complete worksheets as they read, instead of afterward.

Unit Resource Book: The United States & Canada
- Vocabulary Practice, p. 109

Inclusion

Using Terms in Captions

Have students choose one of the terms to illustrate. Then have them write a one- or two-sentence caption that describes their illustration and uses the term correctly. Have volunteers present their illustrations and captions or display them.

SECTION 3

CULTURE

Totem poles in Stanley Park, Vancouver, British Columbia

Atlantic puffins on an island in Canada

A Nation of Immigrants

Connecting to Your World

Have you ever got caught up in the gunfights and horse chases of a television western? Like the United States, Canada also has a rugged western region that was settled later than the East. However, the settling of Canada's West was more peaceful than that of the U.S. West. A group of officers known as the **Royal Canadian Mounted Police** (RCMP) provided law enforcement in the Canadian West. Famous for their bright red uniforms, the Mounties have become a symbol of Canada around the world.

Mounted Police A constable in the RCMP wears the ceremonial red uniform.

A Diverse Population

KEY QUESTION Where have immigrants to Canada come from?

One of the first jobs assigned to the Mounties was to keep peace between the First Nations and the settlers who moved into the Canadian West. Today, Canada is still seeking ways to prevent conflicts between cultural groups. In general, the government works to protect the diversity of Canadian society.

Nunavut An Inuit with his dog team stands outside his igloo in Nunavut near Hudson Bay.

2 Focus & Motivate

3-Minute Warm-Up

Write on the board or display the transparency:

- Look at the photograph of Nunavut. How does this scene help you to understand Canada's diversity? *(Possible Answer: By showing one of its native groups along with Canada's flag, this scene suggests that Canada and its people are proud of their diversity.)*

Unit Transparency Book: The United States & Canada
- 3-Minute Warm-Ups, TT11

Connecting to Your World

More About . . .

The Royal Canadian Mounted Police

Today, the RCMP serves as Canada's national police force. Some of its tasks include preventing organized crime and terrorism, working with young people, and assisting Canada's native communities.

3 Teach

A Diverse Population

Reader, Recorder, Reporter

- What challenges does Canada face in regard to its diversity? *(to prevent conflict between cultural groups and protect Canada's diversity)*
- **Categorize** Identify the first inhabitants of Canada. *(native groups such as Cree, Ojibwa, and Inuit)* Identify the first Europeans to arrive in Canada. *(Norse)* Identify three of Canada's religions. *(Catholic, Protestant, Muslim)*

SECTION 3 PROGRAM RESOURCES

Unit Resource Book: The United States and Canada

ON LEVEL

Unit Resource Book
- Reading Study Guide, p. 81
- Section Quiz, p. 125

STRUGGLING READERS

Unit Resource Book
- Reading Study Guide with Additional Support, p. 89
- Skillbuilder Practice: Analyzing Cause and Effect, p. 110
- Section Quiz, p. 125
- Reteaching Activity, p. 129

eEdition with Audio DVD-ROM

ENGLISH LEARNERS

Pupil Edition in Spanish

eEdition with Audio DVD-ROM

Unit Resource Book
- RSG (Spanish), p. 97
- RSG with Additional Support (Spanish), p. 105

Multi-Language Glossary

Test Generator
- Section Quiz in Spanish

INCLUSION

Unit Resource Book
- RSG with Additional Support, p. 89
- Section Quiz, p. 125
- Reteaching Activity, p. 129

GIFTED & TALENTED

Unit Resource Book
- Interdisciplinary Projects, pp. 117–119
- Section Quiz, p. 125

PRE-AP

Unit Resource Book
- Section Quiz, p. 125

TECHNOLOGY

Unit Transparency Book
- 3-Minute Warm-Ups, TT11
- Essential Question, TT12
- Critical Thinking, TT15
- Cultures, TT19, TT20

Daily Test Practice Transparencies
- Section 3, TT20

Power Presentations

ClassZone.com

World Cultures and Geography Video on DVD

Selected Ethnic Groups of Canada

Have students determine the percentage of Canada's total population of 30 million that each of the five ethnic groups shown on the bar graph represents. *(Chinese—about 3%; English—about 20%; French—about 15%; Scottish—about 13%; Ukrainian—about 3%)*

- **Make Inferences** Why are there more Canadians of English descent than French? *(Possible Answer: because Canada was ruled by Britain for many years and is still a member of the Commonwealth of Nations)*

Unit Resource Book: The United States & Canada

- Interdisciplinary Projects: Math: Languages of Canada, pp. 117, 119

More About . . .

The Pysanka Festival

Since 1973, this festival has been held each year during the first week of July. Conceived as part of the centennial celebrations of the Royal Canadian Mounted Police, the dedication message of the giant egg (to the left of the photograph) is written in four languages and states: "This Pysanka symbolizes the harmony, vitality and culture of the community and is dedicated as a tribute to the . . . Royal Canadian Mounted Police who brought peace and security to the largest multi-cultural settlement in all of Canada."

Teach

Life in Canada

Talk About It

- Why is life in Canada's rural areas changing? *(because technologies such as satellite dishes have been bringing modern life to these areas)*
- **Categorize** What is Canada's "third way of life"? *(Arctic life)* Why do you think it is called Canada's "third way of life"? *(Possible Answer: because it is a distinct way of life compared to Canadian life in urban or rural areas)*

Ukrainian Dancers Dancers perform at the Pysanka Festival in Vegreville, Alberta. A pysanka is a decorated Ukrainian Easter egg. **Why might people of British and French origins together make up more than one-third of the Canadian population?**

because Britain and France were the earliest colonial powers and had many settlers in Canada

Selected Ethnic Groups of Canada

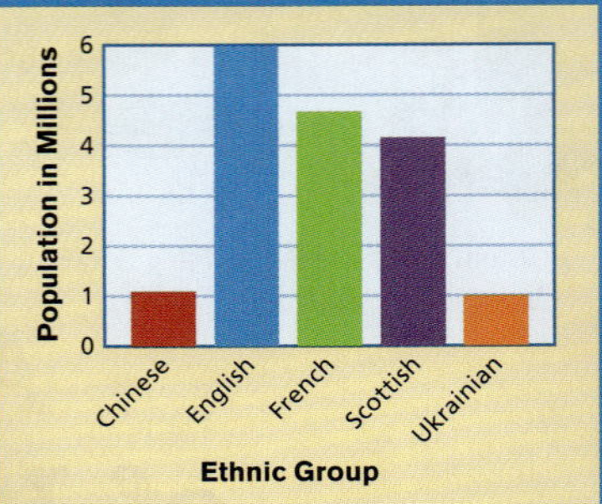

- The Canadian census provides information about 28 ethnic groups.
- The bar graph above shows the number (in millions) of five ethnic groups out of a total Canadian population of close to 30 million people.

Source: Statistics Canada

Canada has been diverse from its beginning. The first inhabitants of Canada were different native groups, such as the Cree, the Ojibwa, and the Inuit. The first Europeans to arrive were the Norse, and then the English and French. In the 1800s and 1900s, people from a number of other countries moved to Canada—including Scotland, Ireland, Germany, Italy, China, Ukraine, Norway, and Sweden.

Because of its history, Canada is a bilingual country with two official languages. About 59 percent of the people speak English as a first language, and about 23 percent speak French. About 18 percent, mostly immigrants, speak other languages. Some Canadians speak native languages such as **Inuktitut**, the language of the Inuit.

About 43 percent of Canadians are Roman Catholic, and about 23 percent are Protestant. Many French Canadians are Catholic, as are many Italian, Ukrainian, and Irish immigrants. Many people of English descent are Protestant. About 2 percent of Canadians are Muslim.

SUMMARIZE Describe where immigrants to Canada have come from. England, France, Scotland, Ireland, Germany, Italy, China, Ukraine, the Netherlands, Norway, and Sweden

Life in Canada

KEY QUESTION What are three distinct ways of life in Canada?

Ethnic groups aren't spread evenly across Canada. French Canadians are concentrated in the province of Quebec. Ukrainians tended to settle in the Prairie Provinces, where the land is similar to the plains of Ukraine. Many Chinese immigrants live on the Pacific coast.

DIFFERENTIATING INSTRUCTION

English Learners

Vocabulary: Suffixes

Point out one of the words on this textbook page that has the suffix *-ian,* such as *Italian.* Tell students that this suffix means "of, relating to, or resembling." Ask students to find other examples of the suffix *-ian* on this page. *(Canadians, Ukrainian)* Have students think of other terms that use this suffix. *(Possible Answers: Lithuanian, Georgian, Virginian, Columbian)*

Struggling Readers

Mapping the Countries of Origin of Canadian Immigrants

Have pairs of students list the groups who came to Canada in the 1800s and 1900s. On a world map or atlas, have them locate the countries where these groups immigrated from originally. Then have each pair create a sketch map or diagram that shows the flow of these immigrants from their countries of origins to Canada.

Where People Live As Section 1 explained, about 90 percent of all Canadians live near the southern border. One of the most densely populated regions of Canada is the area near the St. Lawrence River and the Great Lakes. Port cities are also densely populated.

More than 75 percent of Canadians live in cities. In general, cities are more modern and offer more jobs and cultural opportunities. Rural areas tend to have more traditional ways of life based on farming. But in recent times, technologies such as satellite dishes have been bringing modern life to the rural areas of Canada.

In Canada, a third way of life exists—Arctic life. For centuries, people in the Arctic region followed traditional ways. They lived in tents or **igloos**, traveled by dog sled, hunted and fished, and made clothes from animal skins. Many Inuit still hunt and fish, but they live in wooden houses and wear modern clothing designed for cold weather.

CONNECT Geography & Culture

Arctic Life and the Snowmobile

Traditionally, the Inuit traveled by dogsled. Now, most Inuit families use snowmobiles instead. This new method of transportation is one of the changes allowing them to find wage-paying jobs. Dogs can be fed with meat caught by the Inuit themselves. Snowmobiles need gasoline that must be bought.

A Distinct Identity Canada has more land than the United States, but it has only about 10 percent as many people. As a result, Canadians sometimes fear that they will be overwhelmed by U.S. culture and customs. In fact, during the 1950s, the government issued a report warning that U.S. film, radio, and magazines were invading Canadian culture. The government legislated percentages of television and radio programming that had to have Canadian content.

To fight U.S. cultural dominance, Canadians have worked to develop their national identity. Many are proud of being a northern people and of having such a beautiful country. And as you are about to read, Canada has made distinctive contributions to the arts.

COMPARE Describe three distinct ways of life in Canada.
city life, rural life, and Arctic life

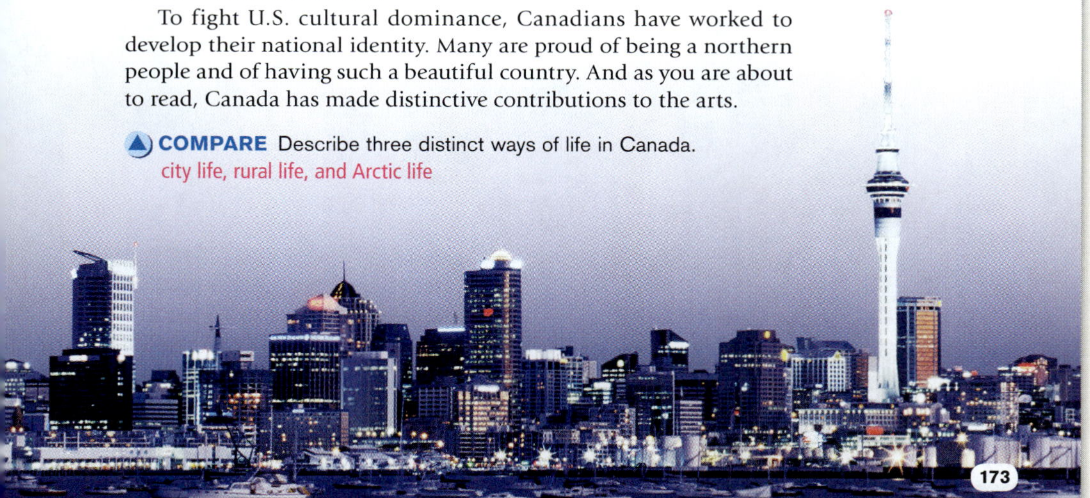

Toronto, Canada The Toronto skyline gleams at night.

173

INTERDISCIPLINARY ACTIVITIES

CONNECT to Science

Compare Traditional and Modern Winter Clothing

Provide students with the information from the More About above. Assign students a type of winter clothing, such as hats, boots, or coats. Have students use classroom and library resources to create a visual that depicts traditional and modern versions of their assigned item, along with information about the materials used for each and their advantages and disadvantages as winter gear. Have students display their visuals.

CONNECT to Civics

Debate U.S. Culture in Canada

Have students reread "A Distinct Identity" above. Have them form two teams and a group of judges. Have teams debate whether Canada should allow more access to Canadian airways for U.S. programs. Have judges vote on allowing more U.S. access, based on each team's arguments.

Unit Resource Book: The United States & Canada
- Skillbuilder Practice: Analyzing Cause and Effect, p. 110

CONNECT Geography & Culture

Arctic Life and the Snowmobile

Some of the first buyers of snowmobiles when they came on the market in the 1930s were doctors and veterinarians. In many areas of heavy snowfall, snowmobiles continue to be of great importance in providing emergency medical services. Today, Canada—even with its relatively small population—accounts for about 25% of all new snowmobiles sold in the world each year.

More About . . .

Traditional Snow Travel

Although snow can be a great barrier to movement, the Inuit people were able to travel widely in the Arctic region—long before the snowmobile was invented. Some of the key adaptations the Inuit made to allow for efficient snow travel were:
- light-weight, warm, waterproof clothing (made of animal skins)
- snow goggles, to prevent snow blindness (made of wood, leather, bone, or ivory)
- snowshoes (made of wood and rawhide)
- sleds (made of wood, with bone or antler sometimes used for the sled runners)—and pulled by hand or a team of huskies.

More About . . .

Toronto

The Town of York, on the north shore of Lake Ontario, was founded by the British in 1793. In 1834, this frontier village became the City of Toronto. Today, Toronto is the largest urban center in Canada—and is considered one of the world's most multicultural cities.

CONNECT to the Essential Question

How is Canada managing its resources in the 21st century?

Ask students what they have learned so far that can help them answer this question. Students might mention:
- Canada has a diverse population.
- Canadians are well-educated.

More About . . .

Caribou

With an estimated herd of over 3 million in North America, caribou continue to play a central role in the Inuit diet and culture. Caribou are still hunted for food and clothing. Caribou also provide shelter—their skin is used for tents during the summer.

Comparing Regions

Arctic Culture Groups: the Inuit and the Sámi

The Sámi are an Arctic culture group living in Finland, Sweden, Norway, and Russia. Similar to Nunavut, three northern areas in Finland were declared the Sámi Homeland. As Canada does, Finland also ensures the rights of the Sámi to maintain and develop their own culture and language. Like the Inuit, the Sámi also rely on fishing and hunting. The Sami also herd reindeer—domesticated Arctic deer—called caribou in North America.

Unit Resource Book: The United States & Canada

- Interdisciplinary Projects: Inuit Music, pp. 118, 119

Presentation Options

Go to the *Activity Center* at **ClassZone.com** to project Worldster.

- Click on the multimedia features of Kirima's Web page.
- Ask students which areas of Kirima's life they would like to explore. Listen to music, examine photographs, learn simple phrases in Inuktitut, and more.

CONNECT to Your Life

Journal Entry

Remind students that photographs often provide information not covered in the text. For example, the two children playing hockey are wearing rubber boots, not traditional boots made from animal skins. Tell students to look for other details in the photos before doing their journal entry.

Worldster

A Day in Kirima's Life

To learn more about Kirima and her world, go to the **Activity Center** @ClassZone.com

In Inuktitut, Kirima's name looks like this:

Hi! My name is Kirima. I am a 13-year-old girl, and I live in the community of Taloyoak in Nunavut. My school has 240 students. We have a modern school with two computer labs, a science lab, and a gymnasium. But we are also proud of our traditional ways. A display about Inuit life is the very first thing that visitors to our school see. Let me tell you about my day.

9:30 A.M. I have a computer literacy class in the morning. Computer skills are very important, because e-mail and the Internet help link my community to southern Canada.

12:00 P.M. One day a week, one class is invited to stay at school for lunch instead of going home. Today is my class's turn, and we eat caribou stew. I love caribou meat, even though it is stringy and gets caught in my teeth.

3:45 P.M. Some of the kids from my class are playing street hockey; they play right in the middle of the road because there are so few cars! My friends and I prefer to jump up and down on a trampoline made of a stretched animal skin.

7:30 P.M. After dinner, my grandmother gives me a sewing or knitting lesson. Inuit women are able to sew seams so tightly that our clothes are waterproof. Even though my family buys many of our clothes at the store, Grandmother says it is important to learn traditional skills.

CONNECT to Your Life

Journal Entry Think about the skills that you have learned. Which of them are modern and which are traditional? How is your life similar to Kirima's life? Record your ideas in your journal.

174 Chapter 6

DIFFERENTIATING INSTRUCTION: TIERED ACTIVITIES

OBJECTIVE Explore modern Inuit life by comparing and contrasting your daily life in winter with that of an Inuit adolescent.

Basic

Have students work in small groups to create a short skit about one activity during a typical winter school day. Each group should focus on one of the following activities: preparing and eating a special meal, taking part in an outdoor activity, or learning a new or traditional skill.

On Level

Adapt the Basic activity, with each group presenting a skit about a U.S. teen who has transferred to Kirima's school. The skit should convey his or her impressions and observations of his or her first school day attending a school in Nunavut, such as Kirima's.

Challenge

Have students work in small groups to present a TV newscast in which their news team reports on the activities of a typical winter day in your school and on those of a Nunavut school. The newscast should include interviews, live reports, and similar features of a daily newscast.

Arts and Popular Culture

KEY QUESTION How have Canadians won worldwide fame?

Education is an aspect of Canadian life that varies from region to region. Each province runs its own school system. Overall, Canada has well-educated people and a high literacy rate of 97 percent.

Canadian literature is written in both French and English. Several Canadian novelists, such as Michael Ondaatje and Margaret Atwood, are read around the world. A popular children's author is Lucy Maud Montgomery, who wrote *Anne of Green Gables*. The city of Stratford, Ontario, holds a yearly theater festival showcasing the plays of William Shakespeare.

Because of Canada's cold climate, winter sports are common. Ice hockey is one of Canada's most popular sports, and Canadian hockey teams have won many international competitions. Wayne Gretzky, possibly the greatest hockey player of all time, is Canadian. **Lacrosse**, a team sport invented by Native Americans, is widely played in Canada. In fact, Canada has made ice hockey its official winter sport and lacrosse its official summer sport.

Wayne Gretzky Canadian Wayne Gretzky (known as "the Great One") is shown at center competing in the 1998 Winter Olympics.

FIND MAIN IDEAS Describe some areas in which Canadians have won worldwide fame. literature, music, hockey

Section 3 Assessment

ONLINE QUIZ For test practice, go to **Interactive Review @ClassZone.com**

TERMS & NAMES

1. **Explain the importance of**
   - Royal Canadian Mounted Police
   - Inuktitut
   - lacrosse

USE YOUR READING NOTES

2. **Categorize** Use your completed chart to answer the following question:

   What are some elements and examples of the diversity that makes up the culture of Canada?

LANGUAGE	RELIGION	ETHNICITY	ARTS & CULTURE
1.	1.	1.	1.
2.	2.	2.	2.
3.	3.	3.	3.
4.	4.	4.	4.

KEY IDEAS

3. What languages are spoken in Canada?
4. What way of life is found in Canada but not in many other countries?
5. How has Canada's geography influenced its sports?

CRITICAL THINKING

6. **Make Inferences** Why do many Canadian performers pursue their careers in the United States?
7. **Summarize** What are some examples of Canada's contributions to literature?
8. **CONNECT to Today** You learned in Chapter 5 that many of today's immigrants to the United States come from Spanish-speaking countries. How would you explain the fact that Canada doesn't have a large number of Spanish-speaking immigrants?
9. **WORLD LANGUAGES** **Create a Travel Dictionary** Think of 20 words that travelers should know. Then use a French-English dictionary to create a French dictionary for travelers to Quebec.

Teach

Arts and Popular Culture

Think, Pair, Share

- Why are Canadians considered well-educated? *(high literacy rate of 97%)*
- **Categorize** What are Canada's official sports? *(ice hockey; lacrosse)*

4 Assess & Reteach

Assess Have students complete the Section Assessment.

Unit Resource Book: The United States & Canada
- Section Quiz, p. 125

Interactive Review @ClassZone.com
- Section Quiz

Power Presentations

Test Generator

Reteach Have students each prepare one question and answer about Canadian diversity.

Unit Resource Book: The United States & Canada
- Reteaching Activity, p. 129

SECTION 3 ASSESSMENT ANSWERS

Terms & Names

1. Royal Canadian Mounted Police: organization of officers responsible for law enforcement in Canadian West: Inuktitut: language of Inuit; lacrosse: sport invented by Native Americans

Use Your Reading Notes

2. The diversity of Canada results from language, religion, and ethnicity as expressed in the literature, music, entertainment, and leisure.

LANGUAGE	RELIGION	ETHNICITY	ARTS & CULTURE
1. French 2. English 3. Inuktitut 4. Other	1. Catholic 2. Protestant 3. Muslim 4. Other	1. First Peoples 2. English 3. French 4. Other	1. high literacy 2. literature in English and French 3. theater 4. winter sports

Key Ideas

3. French, English, and other languages, such as Italian and Chinese; native languages, such as Cree and Inuktitut
4. Arctic life
5. because of cold climate, winter sports popular

Critical Thinking

6. Possible answer: because the United States is a much bigger market
7. Michael Ondaatje, Margaret Atwood, Lucy Maud Montgomery, Shakespeare Festival in Ontario
8. Possible answer: because Spanish-speaking countries are south of United States, so United States is a closer destination for them than Canada
9. Use the rubric to score the dictionaries.

Dictionary Rubric

	Content of Dictionary	Presentation of Dictionary
4	16 to 20 words	accurately defines all words
3	11 to 15 words	defines most words
2	6 to 10 words	defines some words
1	0 to 5 words	defines few words

1 Plan & Prepare

Objectives

- Describe the structure of Canada's government
- Explain the key economic advantages of Canada

Read for the Essential Question

Help students read for a purpose by reminding them of the Essential Question: "How is Canada managing its resources in the 21st century?" Point out that they can use text features such as Key Questions, special features, images, and maps to form answers to the question.

Vocabulary

Best Practices Toolkit

Use the ***Best Practices Toolkit*** to model strategies for vocabulary notetaking. Vary strategies throughout the year. Choose from: *Knowledge Rating, Predicting ABCs, Definition Mapping, Word Sort, Frayer Model (Word Squares), Magnet Words, Word Wheel,* and *Student VOC.*

Vocabulary Strategies, TT9–TT16

Reading Strategy

Best Practices Toolkit

Display the Two-Column Chart transparency.

- Tell students that their answers to the Key Questions in this section will help them find main ideas.
- Tell students that identifying main ideas can help them understand a topic.
- Model filling in a portion of the chart using information from the beginning of the section.

Two-Column Chart, TT20

SECTION 4

Reading for Understanding

Key Ideas

BEFORE, YOU LEARNED

Although Canada is a land of great diversity, it has a small population for such a large geographical area.

NOW YOU WILL LEARN

Canada has a vast wealth of resources and an educated population to support its government and economy.

Vocabulary

TERMS & NAMES

constitution the set of laws and principles that defines the nature of a government

constitutional monarchy a government in which the powers of the king or queen are limited by the constitution

Parliament the legislature of Canada, named after the British Parliament

prime minister the head of the Canadian government

cabinet system a system of government that links the executive and legislative branches by choosing the prime minister and major department heads from the legislature

REVIEW

federal system a government system in which powers are divided between the national and state or provincial governments

free enterprise an economic system in which businesses are free to operate without much government interference

Reading Strategy

Re-create the chart that is shown at right. As you read and respond to the **KEY QUESTIONS**, use the chart to note the main ideas about the government and economy of Canada.

Skillbuilder Handbook, page R4

FIND MAIN IDEAS

GOVERNMENT	ECONOMY
1.	1.
2.	2.
3.	3.
4.	4.

GRAPHIC ORGANIZERS
Go to Interactive Review @ClassZone.com

176 Chapter 6

PRETEACHING VOCABULARY

English Learners

Pronounce and Preview

Pronounce each term for students. Review the meanings of words within definitions, such as *principles, executive, legislative,* and *provincial.*

- To modify vocabulary learning, have students complete worksheets as they read, instead of afterward.

Unit Resource Book: The United States & Canada
- Vocabulary Practice, p. 35

Inclusion

Hear and Use Vocabulary

Pair students to work on reading the terms and names and using them in sentences. Ask a student to read each word and definition to a visually impaired partner. Spelling out each word may help the students pronounce the words. Each student should then orally use the terms in sentences and discuss how the sentences reflect the definitions.

SECTION 4
GOVERNMENT & ECONOMICS

Totem poles in Stanley Park, Vancouver, British Columbia

Atlantic puffins on an island in Canada

Rich Resources in a Vast Land

Connecting to Your World

Have you ever seen the feature-length cartoons *Snow White* or *Alice in Wonderland?* *Snow White* features an evil, jealous queen. In *Alice*, the Queen of Hearts starts shouting "Off with her head" every time Alice upsets her. Many old stories portray kings and queens as having total power; they make the laws and punish anyone who disagrees with them. But some modern-day kings and queens have much less power than a president or other head of government. One of those is the monarch of Canada.

The Government of Canada

KEY QUESTION What is the structure of Canada's government?

Canada's monarch is the reigning king or queen of Great Britain. But under the law, the monarch really has no power. He or she is just a symbol of Canada's past ties to Britain. Because the **constitution** limits the powers of the ruler, Canada's government is called a **constitutional monarchy**.

Changing of the Guard This ceremony takes place on the lawn of Parliament Hill in Ottawa, Ontario, Canada.

2 Focus & Motivate

3-Minute Warm-Up

Write on the board or display the transparency:

- Look at the photograph and caption at the bottom of this page. What do the red uniforms and Parliament Hill suggest about Canada? *(its British heritage)*

Unit Transparency Book
- 3-Minute Warm-Ups, TT11

Connecting to Your World

More About . . .

The Queen and Canada

Queen Elizabeth II became queen in 1952 and celebrated her Golden Jubilee (50th anniversary) in 2002. In that time, she visited Canada 19 times.

3 Teach

The Government of Canada

Talk About It

- Why is Canada's government called a constitutional monarchy? *(because a constitution limits the power of the ruler—the reigning king or queen of Great Britain)*
- **Find Main Ideas** Explain the federal system of government. *(A federal system of government means that power is divided between the national government and the governments of the provinces and territories.)*

SECTION 4 PROGRAM RESOURCES

Unit Resource Book: The United States and Canada

ON LEVEL
Unit Resource Book
- Reading Study Guide, p. 83
- Section Quiz, p. 126

STRUGGLING READERS
Unit Resource Book
- Reading Study Guide with Additional Support, p. 91
- Skillbuilder Practice: Analyzing Cause and Effect, p. 110
- Section Quiz, p. 126
- Reteaching Activity, p. 130

eEdition with Audio DVD-ROM

ENGLISH LEARNERS
Pupil Edition in Spanish
eEdition with Audio DVD-ROM
Unit Resource Book
- RSG (Spanish), p. 99
- RSG with Additional Support (Spanish), p. 107

Multi-Language Glossary
Test Generator
- Section Quiz in Spanish

INCLUSION
Unit Resource Book
- RSG with Additional Support, p. 91
- Section Quiz, p. 126
- Reteaching Activity, p. 130

GIFTED & TALENTED
Unit Resource Book
- Section Quiz, p. 126

PRE-AP
Unit Resource Book
- Economics in the World, p. 120
- Section Quiz, p. 126

TECHNOLOGY
Unit Transparency Book
- 3-Minute Warm-Ups, TT11
- Essential Question, TT12
- Critical Thinking, TT16

Daily Test Practice Transparencies
- Section 4, TT21

Power Presentations
ClassZone.com
World Cultures and Geography Video on DVD

More About . . .

Senators and the Governor General

Senators are appointed to the Senate in the Canadian Parliament by the governor general of Canada. The governor general is appointed by the head of state—the queen or king currently reigning in Great Britain. Canada's prime minister advises the head of state as to which person might be best for governor general. The responsibilities of the governor general include appointing senators to the Parliament with the advice of the prime minister and representing the head of state in Canada. This means that the governor general does what the head of state would do if he or she were in Canada.

Teach

The Economy of Canada

Think, Pair, Share

- What is one advantage of a free enterprise system? *(Businesses operate without much interference from the government.)*
- In which part of the economy do most Canadians work? *(service industries)*
- **Find Main Ideas** Explain Canada's role in the world economy. *(Canada's economy relies on international trade with other countries by exporting natural resources and some manufactured goods such as automobiles and machinery.)*

Queen Elizabeth II The Queen accepts flowers while on tour in the province of Saskatchewan. **Why might the Queen be popular in Canada?**

She represents Canada's British heritage and as such is a symbol of Canadian history and identity as well as a symbol of Canadian unity.

Parliament and Government Canada's government is also a representative democracy, but its government is quite different from that of the United States. The legislative body of Canada is called the **Parliament**. Like the U.S. Congress, the Canadian Parliament has two houses—the Senate and the House of Commons. Canada's citizens vote for members of the House of Commons. Senators are appointed.

The head of the executive branch is the **prime minister**. The prime minister is chosen because he or she is the leader of the dominant political party in the House of Commons (the legislative branch). The heads of the government departments also come from Parliament. They form a group of official advisers to the prime minister called a cabinet; this is called a **cabinet system** of government.

Federalism The nation of Canada is divided into smaller political units called provinces and territories. As in the United States, power is divided between the national government and the governments of the provinces and territories. A government that divides power in this way is called a **federal system**.

Originally, Canada's national government had much more power than the governments of the provinces. The provincial governments have control over issues such as health care, education, and the use of resources. These issues have become vital to modern life, so the provincial governments have become more important.

FIND MAIN IDEAS Describe the structure of Canada's form of government. monarch with symbolic role, a two-house parliament, prime minister chosen from majority party in House of Commons, cabinet chosen from House of Commons

The Economy of Canada

KEY QUESTION What is Canada's biggest economic advantage?

Canada's economy is both similar to and different from the U.S. economy. One major similarity is that both economies are based on **free enterprise**, an economic system in which businesses are free to operate without much government interference. One difference is that the Canadian government plays a larger role in the economy (for example, in health care) than the U.S. government does.

What Canada Produces Canada has an abundance of natural resources such as forests, minerals, and land. This wealth of resources is Canada's biggest economic advantage. A second advantage that Canada has is a well-educated population. Some of the Canadian economy's main industries are listed on the next page.

DIFFERENTIATING INSTRUCTION

Struggling Readers

Find Main Ideas and Supporting Details

To help students understand federalism, model using the chart to analyze main ideas and details in the paragraphs about federalism on this page. After completing the chart, ask students to work in pairs to write a definition of federalism. Tell students that using this charting method can help them break down difficult concepts into manageable chunks.

Main Ideas	Details
Government power in Canada is divided.	1. called a federal system or federalism 2. divided between national and provinces and territories 3. provincial governments are important 4. power division is similar to United States

Agriculture and Industry of Canada

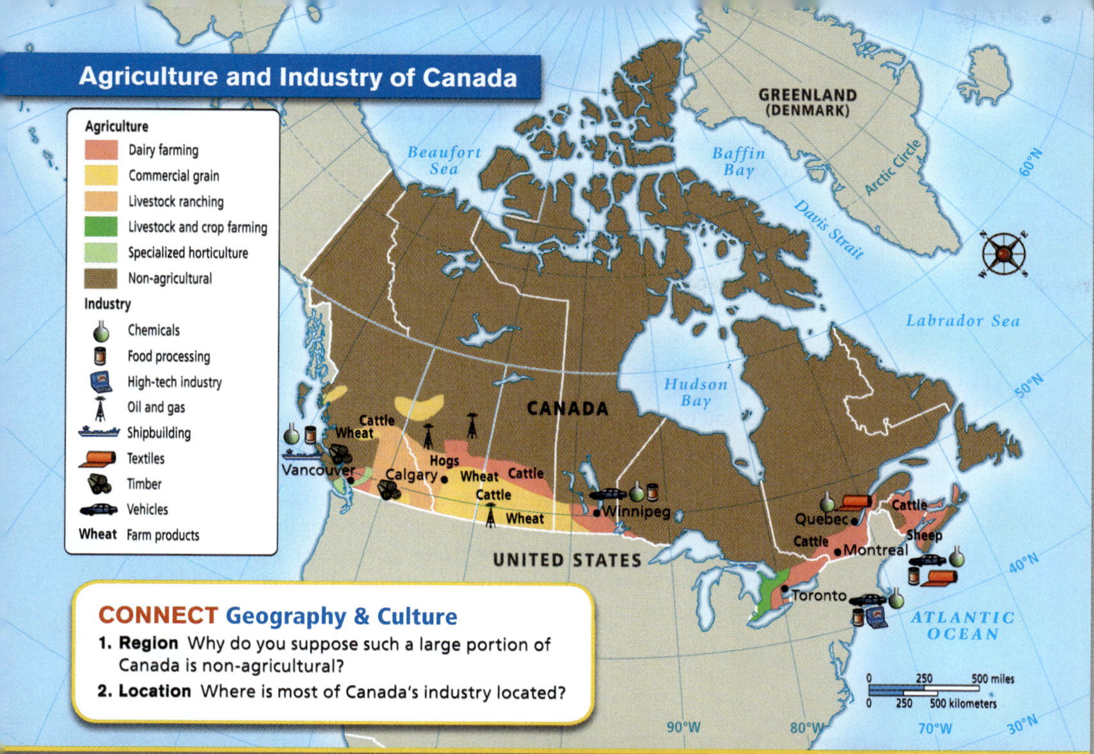

CONNECT Geography & Culture

1. **Region** Why do you suppose such a large portion of Canada is non-agricultural?
2. **Location** Where is most of Canada's industry located?

Agriculture, Forestry, and Fishing About two percent of Canadians work in agriculture, but Canada produces a great quantity of food. Wheat, beef, dairy products, and hogs are some of the main products. Forestry products include lumber, plywood, and paper. Fishing yields salmon, crab, and lobster.

Mining and Energy Canada is a major producer of uranium, zinc, nickel, copper, iron ore, coal, gold, and silver. It also produces oil and natural gas, as well as hydroelectricity.

Manufacturing About 14 percent of Canadians work in manufacturing, and five percent work in construction. One of the main industries is the processing of minerals and resources such as trees. Processing minerals involves removing useful metal from ore. Trees are processed into logs, boards, and wood pulp.

Service Industries About 75 percent of all Canadians work in service industries. These include government, education, health care, retail stores, hotels, restaurants, and banking.

Oil Pump Oil being pumped from the ground near Calgary, Alberta, Canada.

DIFFERENTIATING INSTRUCTION: TIERED ACTIVITIES

OBJECTIVE Draw a pie graph to represent information about Canada's agriculture and industries.

Basic

Instruct students to gather information from the page about the percentage of Canadians working in certain industries. Have students draw a pie graph with labels to represent percentages. Since the percentages don't equal 100, have students indicate the remaining percentage with an "other" segment.

On Level

Instruct students to gather information from the page about the percentage of Canadians working in certain industries. Have students draw a pie graph with labels to represent the percentages. Tell students to use colors to distinguish categories and remind them to use an "other" segment as necessary.

Challenge

Instruct students to gather information about the percentage of Canadians working in certain industries. Have students draw a pie graph with color-coded and detailed labels to represent percentages. Ask students to write a paragraph discussing how these percentages have shifted in the past century.

Agriculture and Industry of Canada

CONNECT Geography & Culture

Point out that this map shows significant agriculture and industry in Canada. However, this doesn't mean that every type of agriculture or industry is widely distributed. Some types of agriculture and industry are found only in certain regions. Ask students near which city specialized horticulture is centered. *(Vancouver)*

ANSWERS

1. **Region** Most of the colder regions of the north are either tundra or forest and are thus not suited to agriculture.
2. **Location** in cities near the border with the United States

More About . . .

Fishing

The map of Canada on this page shows that the country is bordered on three sides by water—the longest coastline of any country in the world. Canada also has a significant amount of fresh water. With these tremendous water resources, fishing is important to the economy. Canada exports about 85 percent of the fish and seafood caught in its waters. Approximately two thirds of the catch is exported to the United States. The federal government manages fishing in Canada. Its goals include regulating the Canadian fishery so that its resources are available in the future.

CONNECT to the Essential Question

How is Canada managing its resources in the 21st century?

Ask students what they have learned so far that can help them answer this question. Students might mention:

- Canada has vast natural resources that are an economic advantage.
- One of Canada's main industries is the processing of minerals and resources.

Comparing Regions

International Trade

Canada and the United States both have many trading partners. Canada has more than 100 trading partners, while the United States has nearly 200 trading partners. Their major trading partners include each other and the European Union, Mexico, Japan, and China. Canada imports the most merchandise from the United States.

Unit Resource Book: The United States & Canada
- Economics in the World, p. 120

4 Assess & Reteach

Assess Have students complete the Section Assessment.

Unit Resource Book: The United States & Canada
- Section Quiz, p. 126

Interactive Review @ClassZone.com
- Section Quiz

Power Presentations

Test Generator

Reteach Have students use their Find Main Ideas graphic organizers to discuss with a partner Canada's form of government and its role in the world economy.

Unit Resource Book: The United States & Canada
- Reteaching Activity, p. 130

Canada's Role in the World Economy Historically, the economy of Canada has always relied on trade. The fur trade between Canada's native peoples and European fishermen was just the start of what would become a key Canadian industry.

Today, Canada is a major participant in international trade. Because Canada has such a wealth of natural resources, it has always relied on exporting those materials to other nations. In recent decades, Canada has also begun to export manufactured goods such as automobiles and machinery.

Many Canadian industries, especially those that process minerals and wood products, are partially owned by people in other countries. Most of those foreign owners are from the United States.

As you might guess, Canada's biggest trading partner is the United States, which is its closest neighbor and which has a much larger population. The United States buys up to 85 percent of Canada's exports. Most of the products that Canada imports come from the United States. Trade between the two countries has boomed because of NAFTA (North American Free Trade Agreement), signed in 1992. Japan is Canada's second largest trading partner. However, Canada has developed a growing trade relationship with China in recent years.

FIND MAIN IDEA Explain Canada's biggest economic advantage.
its wealth of natural resources

Section 4 Assessment

ONLINE QUIZ For test practice, go to **Interactive Review @ClassZone.com**

TERMS & NAMES

1. **Explain the importance of**
   - constitutional monarchy
   - Parliament
   - prime minister
   - cabinet system

USE YOUR READING NOTES

2. **Find Main Ideas** Use your completed chart to answer the following question: What are some examples of Canada's government structure and economic advantages?

GOVERNMENT	ECONOMY
1.	1.
2.	2.
3.	3.
4.	4.

KEY IDEAS

3. What role does the British monarch play in the Canadian government?
4. How is the prime minister chosen?
5. How has the Canadian economy changed recently?

CRITICAL THINKING

6. **Draw Conclusions** Why do you suppose that Canadian industries process resources before exporting them instead of exporting raw materials?
7. **Evaluate** Is Canada's government a good example of a representative democracy? Explain.
8. **CONNECT to Today** In 2010, Vancouver, British Columbia, will host the winter Olympics. Which part of the economy will benefit the most from that?
9. **TECHNOLOGY** **Plan a Web Page** Imagine that you work for the Canadian government, and your job is to encourage foreign industries to move to Canada. Plan a Web page that shows Canada's advantages.

SECTION 4 ASSESSMENT ANSWERS

Terms & Names

1. Constitutional monarchy: powers of king or queen are limited by constitution; Parliament: legislature of Canada; prime minister: head of Canadian government; cabinet system: government that chooses prime minister and department heads from legislature

Use Your Reading Notes

2. Monarch, parliament, prime minister, cabinet; resources; well-educated population
   Diagram *Government—symbolic monarch, parliament, prime minister, cabinet, federal system; Economy—well-educated population, free enterprise, abundant resources, manufacturing, services*

Key Ideas

3. Monarch is reminder of Canada's ties to Britain.
4. Prime minister chosen as leader of dominant political party in House of Commons
5. Trade with United States has boomed, and export of manufactured goods has increased.

Critical Thinking

6. Possible answer: to increase the value of the resources so they sell for a higher price
7. Possible answers: Yes, members of House of Commons are elected, and prime minister and cabinet chosen from there; no, Senate and prime minister are not chosen by election.
8. Possible answer: Service industries, which include hotels and restaurants, will benefit most because tourists will come to see games.
9. Use the rubric to score Web pages.

Web Page Rubric

	Web Content	Organization
4	all elements	excellently organized
3	most elements	well organized
2	some elements	not well organized
1	little or no information	poorly organized

Interactive Review

Click here to complete these and other activities online @ ClassZone.com

CHAPTER SUMMARY

Key Idea 1
The physical geography of Canada varies greatly among its 13 provinces and territories, which has resulted in different ways of life.

Key Idea 2
The history of Canada reflects the diversity of its regions and its population.

Key Idea 3
Because Canada is a nation of immigrants, it has a diverse population and a culture that is shaped by many influences.

Key Idea 4
Canada has a vast wealth of resources and an educated population to support its government and economy.

For Review and Study Notes, go to Interactive Review @ ClassZone.com

NAME GAME

Use the Terms & Names list to complete each sentence on paper or online.

1. I am a team sport invented by Native Americans. lacrosse
2. I am an organized cultural group of Canada's native people. ________
3. I am a Canadian political unit, similar to a U.S. state. ________
4. I am a government in which the powers of the king or queen are limited by law. ________
5. I am a territory that is home to many Inuit people. ________
6. I am a type of government that chooses the prime minister and major department heads from the legislature. ________
7. I am a large inland sea in northern Canada. ________
8. I am a waterway linking the Atlantic Ocean to inland North America. ________
9. I am a nation in the British Empire that is allowed to govern its domestic affairs. ________
10. I am the legislature of Canada, named after a similar institution in Britain. ________

cabinet system
Commonwealth of Nations
constitutional monarchy
dominion
First Nations
Hudson Bay
Inuktitut
lacrosse
Nunavut
Parliament
province
Royal Canadian Mounted Police
separatist
Seven Years' War
St. Lawrence Seaway

Activities

Flip Cards

Use the online flip cards to quiz yourself on the terms and names introduced in this chapter.

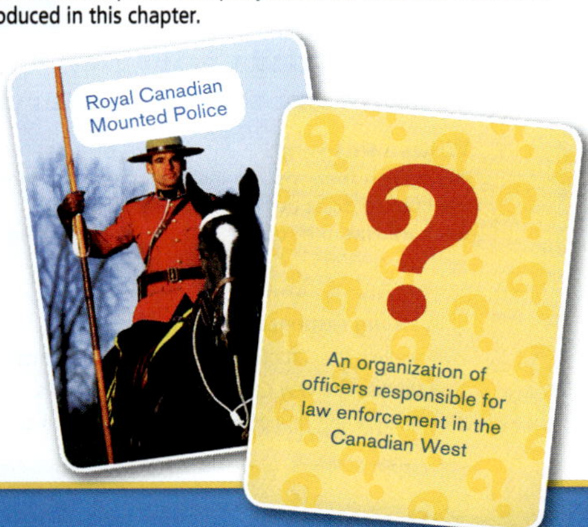

Crossword Puzzle

Complete an online puzzle to test your knowledge of Canada's path from its first peoples to a modern nation.

ACROSS

1. city founded by Samuel de Champlain in 1608

ONLINE AND ON-SCREEN ACTIVITIES

Flip Cards

Vocabulary Flip Cards are virtual flash cards at **ClassZone.com**

Students can quiz each other or work independently to study the flip cards for this chapter.

- They can quiz themselves on key terms and names by reviewing definitions.
- Demonstrate how to advance and shuffle the deck for another challenge.

Crossword Puzzle

Students can review important concepts of "The United States and Canada" by completing the online crossword puzzle at **ClassZone.com.**

- As an extension, or for more challenge, have students use the *Activity Maker* at **ClassZone.com** to make their own crossword puzzle for one section of the chapter.
- Then have them trade puzzles with classmates to solve answers for each other's sections.

Chapter Summary

As a follow-up to the chapter, encourage students to pair up and go over the **Review and Study Notes at ClassZone.com.**

Alternatively, students pairs can review their Reading Study Guide notes, making sure that they can add important details to each section's "Key Idea."

Name Game

Teaching Tips

ON PAPER Have students work in pairs to complete this activity. As an extension, they can write questions for the terms that are not covered in the items shown here.

ONLINE Make sure that students understand how to drag and drop terms to fill in blanks. Ask them to print out their results and compare them with those of a classmate.

ANSWERS

1. lacrosse
2. First Nations
3. province
4. constitutional monarchy
5. Nunavut
6. cabinet system
7. Hudson Bay
8. St. Lawrence Seaway
9. dominion
10. Parliament

Make It Fun

Teacher-Tested Activities

Marci Smith Deal, Hurst-Euless-Bedford ISD, Bedford, Texas

My students enjoy looking at photographs. Several activities can be created with pictures, depending on the interests and abilities of students. Here is one such review activity.

- Gather a number of photographs of Canada. Such photographs might include the Calgary Stampede, a dog-sled race, Parliament in Ottawa, an icy scene in Nunavut, Niagara Falls, and so forth.
- Write a list of locations on the chalkboard.
- Show students each picture and ask them to identify the location from the list.

Vocabulary ANSWERS

1. a Canadian political unit, similar to a U.S. state
2. a political unit that doesn't have enough people to become a province
3. a horseshoe-shaped, rocky plateau that covers much of eastern central Canada
4. a large inland sea to the north of the province of Ontario
5. a waterway made up of the St. Lawrence River, the Great Lakes, and several canals
6. using or being able to use two languages equally
7. an organized cultural group of Canada's native people
8. Canada went from being a colony to a dominion to an independent member of the Commonwealth of Nations whose government is a constitutional monarchy.
9. The Inuit speak Inuktitut and have a homeland named Nunavut.

Key Ideas ANSWERS

10. The Appalachian Mountains are in the east, the Rocky Mountains and coastal mountains are in the far west, and the plains are in the center of the country.
11. It connects the heart of North America to the Atlantic Ocean, allowing tons of goods to be shipped.
12. flat, treeless lands between the ice cap and the tree line of the Arctic regions
13. After the war, many Loyalists moved from the United States to Canada.
14. to divide it between English-speakers and French-speakers
15. He led Canada to independence from Great Britain.
16. It has grown more modern, with snowmobiles replacing dogsleds and houses replacing tents and igloos.
17. the Shakespeare Festival at Stratford
18. probably hockey; also possibly lacrosse
19. The prime minister and heads of the major departments are chosen from the House of Commons.
20. a wealth of resources and educated people
21. by producing oil, natural gas, and hydroelectric power

CHAPTER 6 Assessment

VOCABULARY

Explain the significance of each of the following.

1. province
2. territory
3. Canadian Shield
4. Hudson Bay
5. St. Lawrence Seaway
6. bilingual
7. First Nation

Explain how the terms in each group are related.

8. dominion, Commonwealth of Nations, constitutional monarchy
9. Nunavut, Inuit, Inuktitut

KEY IDEAS

1 Mountains, Prairies, and Coastlands

10. Where are the major mountains and plains of Canada located?
11. What is the economic significance of the St. Lawrence Seaway?
12. What is tundra?

2 First Peoples to a Modern Nation

13. How did the American Revolution affect Canada?
14. Why did Great Britain divide Quebec into Upper Canada and Lower Canada?
15. What was the most important accomplishment of William Lyon Mackenzie Smith?

3 A Nation of Immigrants

16. How has life in the Arctic changed recently?
17. What is an important annual event in Canadian theater?
18. What is the most popular sport in Canada?

4 Rich Resources in a Vast Land

19. How is the executive branch of the Canadian government linked to the legislative branch?
20. What are two economic advantages that Canada has?
21. How does Canada help the United States meet its energy needs?

CRITICAL THINKING

22. **Sequence Events** Create a chart like the one below on your own paper. Fill in the events that led to Canadian independence in the proper order.

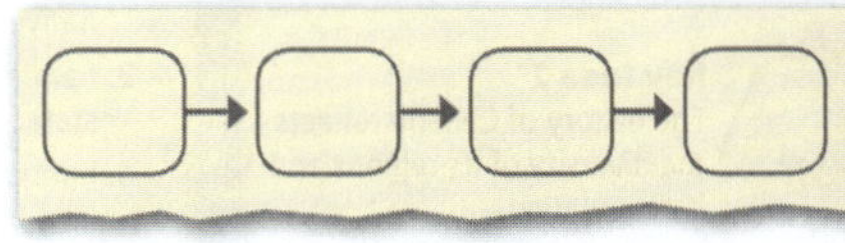

23. **Make Predictions** What problems do you think Quebec might face if it did become a separate nation from Canada?
24. **Analyze Effects** Do you think it will be easier or harder for the Inuit to preserve their traditional culture in the future? Explain.
25. **Five Themes: Region** Why do you think the United States is Canada's main trading partner?
26. **Connect Geography & History** How might Canada be different if France had won the Seven Years' War?
27. **Connect Geography & Culture** Canada has tried to build its population by encouraging immigration. What effect will this have on the cultural makeup of Canada and on Canada's multiculturalism?

Answer the ESSENTIAL QUESTION

How is Canada managing its resources in the 21st century?

Written Response Write a two- or three-paragraph response to the Essential Question. Be sure to consider the key ideas of each section. Use the rubric to guide your thinking.

Response Rubric

A strong response will:

- explain the structure of Canada's government
- discuss the way Canada uses its resources
- analyze the role of minority groups in Canada

ESSENTIAL QUESTION REVIEW

Before students begin their essays, have them review the key chapter elements hat will help them answer the Essential Question. Direct them to significant features in the chapter, such as

- Animated Geography: The St. Lawrence Seaway, Section 1
- Connect Geography & History, Section 2
- Connect Geography & Culture, Section 3
- Agriculture and Industry of Canada, Section 4

Written Response Rubric

	Coverage of Key Ideas	Explanation	Accuracy
4	all	clear, concise	no errors
3	most	good	few/minor errors
2	some	fair	several errors
1	two or fewer	poor	many errors

STANDARDS-BASED ASSESSMENT

Test Practice

- Online Test Practice @ClassZone.com
- Test-Taking Strategies and Practice at the front of this book

PIE GRAPHS

Use the pie graphs to answer question 1 on your paper.

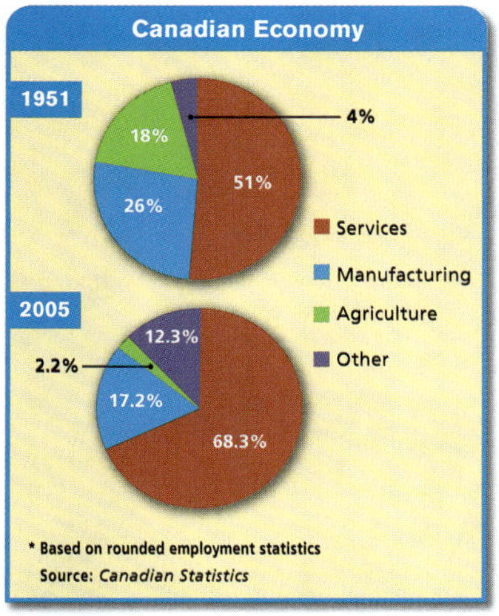

1. Which sector showed the greatest increase in growth from 1951 to 2005?

A. services
B. manufacturing
C. agriculture
D. other

CHART

Examine the chart below. Use the information and your knowledge of Canada to answer questions 2 and 3 on your paper.

Comparing the Canadian and U.S. Governments

Aspects of Government	CANADA	UNITED STATES
Type	Constitutional Monarchy (limited power)	Constitutional Republic (limited power)
Head of State	Monarch	President
Head of Government	Prime Minister	President
Legislature	Parliament	Congress
System	Federal (central and provinces)	Federal (central and states)

2. What is one difference in the federal systems of the United States and Canada?

3. Who is the head of state and the head of government in Canada?

GeoActivity

1. INTERDISCIPLINARY ACTIVITY–SCIENCE

With a partner, use library books, encyclopedias, or Internet sites to learn more about the tundra and what grows there. Find out more about the climate of Arctic lands. Create an illustrated poster showing what tundra looks like and what plants grow upon it. Include a bar graph of average monthly temperature.

2. WRITING FOR SOCIAL STUDIES

Unit Writing Project Revisit the television news broadcast that you developed to showcase an important issue. What other important issues would you like to present in a broadcast? Choose an issue and write a script that tells the viewer five things: who, what, where, when, and why.

3. MENTAL MAPPING

Create an outline map of Canada and label the following:

- Appalachian Mountains
- Canadian Rockies
- St. Lawrence River
- Mackenzie River
- Hudson Bay
- Great Lakes
- Great Slave Lake

Critical Thinking ANSWERS

22. Canadian colonies began to ask for local self-government. Canada became a dominion. Canadians began to question their role in the empire. Prime Minister King sought independence, which was granted in 1931.

23. It might be difficult for Quebec to defend itself and have a self-supporting economy.

24. Possible answers: Easier—Because Inuit have own territory, they will find it easier to preserve way of life; harder—Technologies like Internet will continue to change culture by bringing in outside influences.

25. U.S. is Canada's next-door neighbor, no huge obstacles between countries, share common language and heritage, and United States has more people and bigger economy.

26. Possible answer: French might be most common language, government would not be modeled after British government, Canada wouldn't belong to Commonwealth.

27. Possible answer: Canada will become more diverse, so government might do more to promote multiculturalism.

Standards-Based Assessment ANSWERS

1. A

2. The United States is made up of states; Canada is made up of provinces.

3. In Canada, the head of state is the monarch and the head of government is the prime minister; in the United States, both roles are filled by the president.

GeoActivity

1. INTERDISCIPLINARY ACTIVITY RUBRIC

The poster should
- reflect thorough research
- be neat, organized, and attractive

2. WRITING FOR SOCIAL STUDIES RUBRIC

The Unit Writing Project should
- present an interesting issue
- present a well-written script

3. MENTAL MAPPING RUBRIC

The outline map should
- represent the approximate shape of Canada
- include all geographic features listed in their correct location

DIFFERENTIATING CHAPTER ASSESSMENT

Unit Resource Book

- **Chapter Test A,** pp. 131–134. *Use for Struggling Readers, Beginning-to-Intermediate-Level English Learners, and for Inclusion of students whose IEP suggests supported test-taking.*
- **Chapter Test B,** pp. 135–138. *Use for On-Level Learners and ESOL students with Advanced-Level English.*
- **Chapter Test C,** pp. 139–142. *Use for Gifted and Talented students or students assigned Pre-AP work.*

Test Generator

- Chapter Tests A, B, and C, in English and Spanish

McDougal Littell Assessment System

TEST **Test Generator** Use three levels of tests in English and Spanish tied to state standards. Plus, choose questions to customize tests for all populations.

SCORE Scan and Score plain paper student tests in minutes.

REPORT Create instant progress reports, sortable by student, by standard, or other criteria.

RETEACH Respond immediately to students' needs by using a wealth of reteaching resources.

N
E
W
S

UNIT 3 Latin America

Why It Matters:

Latin America is a region of many countries. The region spans a large area north and south of the equator. The region faces political and economic challenges. How it meets these challenges will have an impact on other parts of the world.

The Andes Mountains, Chile

CHAPTER 7 **Latin America**
Physical Geography & History

Paseo de la Reforma, Mexico City

CHAPTER 8 **Mexico**

Marketplace in Guatemala

CHAPTER 9 **Middle America and Spanish-Speaking South America**

Amazon River

CHAPTER 10 **Brazil**

185

Objectives

- Identify the location and countries of Latin America
- Describe the geographic features of Latin America
- Explain the varied climates found in Latin America
- Identify the major economic activities in Latin America

Comparing Data

Latin America occupies an area that is only slightly larger than Canada and the United States combined, but it contains about 233 million more people. The name of the region comes from the Latin-based languages (Spanish, Portuguese, and French) spoken there. However, not all Latin American countries use these languages. English is the main language of Belize, Guyana, and several Caribbean island nations, and the official language of Suriname is Dutch.

ANSWERS

1. Latin America has 264,935,596 more people.
2. The Amazon River discharges about 5,720,000 cubic feet per second more than the Mississippi River discharges.

UNIT 3 ATLAS

The region of Latin America stretches from Mexico to the tip of South America. As you study the maps, notice geographic patterns and specific details about the region. Answer the GeoActivity questions on each map in your notebook.

As you study the graphs on this page, compare the landmass, population, rivers, and mountains of Latin America with those of the United States. Then jot down the answers to the following questions in your notebook.

Comparing Data

1. Compare the population of Latin America with that of the United States. How many more people live in Latin America than in the United States?
2. The discharge rate of a river measures how much water moves through a certain location in a certain amount of time. The rate in this graph is measured in cubic feet per second. How much greater is the discharge rate of the Amazon River than that of the Mississippi River?

Comparing Data

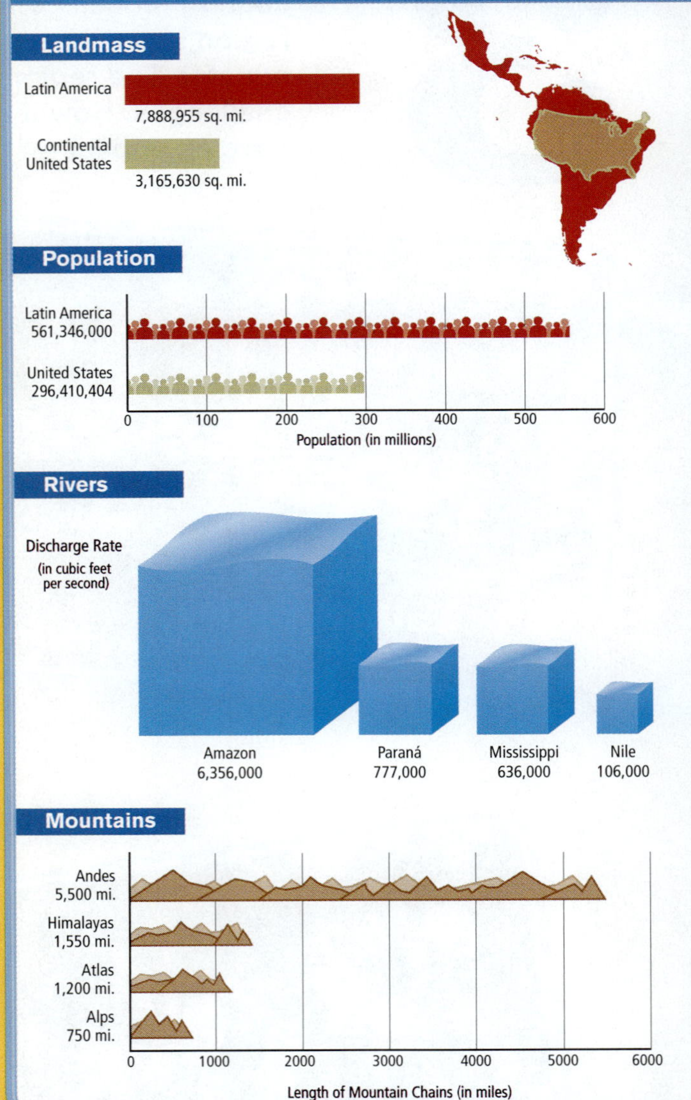

186 Unit 3

FIVE THEMES AND SIX ELEMENTS

Five Themes

Region

Explain that Latin America is spread across two continents: North and South America. The Caribbean islands and the countries from Mexico to Panama are all part of North America.

Six Elements

Places and Regions

Point out that the land between Mexico and Colombia is often referred to as Central America. This narrow region links North and South America.

Population Density

THINK LIKE A GEOGRAPHER

1. **Place** Where is the population of Mexico most dense?
2. **Region** Where do most people live in Latin America?

DIFFERENTIATING INSTRUCTION

Struggling Readers

Pronounce Place Names

Model the pronunciation of Latin American countries to help students who might encounter difficulties decoding them. In particular, focus on the varied sounds represented by the letter *u* in such names as Guatemala, Guyana, Cuba, Peru, and Uruguay.

Inclusion

Locate Population Centers

Review the categories of population density that appear in the map key. Then have students locate the equator. Ask them to find the area with the highest population density north of the equator *(near Mexico City and Puerto Rico)* and the area with the highest population density south of the equator *(near Rio de Janeiro and São Paulo).*

Population Density

Tell students that the Pacific coastline of this region is also part of the "Ring of Fire" that they'll learn more about in a later unit. For many people of Latin America, this means that they are at risk for earthquakes and volcanoes. Ask students if they think any good can come from a volcanic eruption. *(Possible answer: Volcanic ash and lava can help keep soil fertile.)*

- What color on the map legend indicates areas with zero persons per square mile? *(light yellow)*
- What is the most densely populated city in Argentina? *(Buenos Aires)*
- Why might Brazil have such a large area of its country with so few people? *(Possible answer: The Amazon rain forest is not suitable for a large population.)*

THINK LIKE A GEOGRAPHER ANSWERS

1. **Place** Mexico's most densely populated area is around Mexico City.
2. **Region** Most Latin Americans live in cities and along the coasts.

More About . . .

Mexico City

The capital of Mexico sits on the site of Tenochtitlán, an Aztec city that was first settled in the 1300s. Spanish conquistadors destroyed Tenochtitlán in 1521 and built Mexico City to serve as the capital of New Spain, which included a large area of present-day United States. Today Mexico City is the world's second-largest metropolitan area, with a population of more than 18 million.

Climate

Point out that although Latin America has almost twice as many people as the United States, much of the land is largely unsettled. Discuss some of the geographic features and climates that make much of this region hard to settle. *(Possible Answers: tall mountain ranges, such as the Andes; very hot and dry areas, such as the Atacama Desert; areas that are very hot and wet, such as the Amazon rain forest)*

- Compare the climates of Haiti and the Dominican Republic. *(Both have a tropical wet climate, but the Dominican Republic also has a tropical wet and dry climate zone.)*
- Compare the climates of Brazil and Argentina. *(Same: humid subtropical, marine west coast, and semiarid; Different: Argentina—desert and highland; Brazil—tropical wet and tropical wet and dry)*
- Why does much of Latin America have a tropical climate? *(because much of the region lies near the equator)*

THINK LIKE A GEOGRAPHER ANSWERS

1. **Place** The southernmost part of Latin America has a marine west coast climate.
2. **Region** Highland climates can be found near South America's western coast and in the central part of Mexico and Central America.

More About . . .

El Niño

Every three to seven years an unusual change in ocean temperatures off the western coast of tropical South America disrupts weather patterns all over the world. Trade winds that usually blow from the east weaken, and much warmer air and water flow in from the west. The rising ocean temperature increases convection in the atmosphere over the eastern Pacific, and this can bring changes to normal weather patterns everywhere.

In addition, many fish and marine animals die because the now warmer Pacific waters hold fewer nutrients. Peruvian fishers noticed this phenomenon and that it usually occurred near Christmas. They named the effect El Niño, or "the male child."

THINK LIKE A GEOGRAPHER

1. **Place** What is the climate of the southernmost tip of South America?
2. **Region** In what parts of Latin America can you find a highland climate?

188 Unit 3

FIVE THEMES AND SIX ELEMENTS

Five Themes

Region

Remind students that the term *tropics* refers to the area that lies between the Tropic of Cancer and the Tropic of Capricorn. Explain that this region contains all the places on Earth where the sun appears directly overhead at noon on at least one day each year.

Six Elements

Physical Systems

Have students use the Economic Activity map on the next page to locate areas of limited agriculture in western South America. Explain that this strip of land generally follows the Andes Mountains and has a highland climate, as shown on the Climate map. Ask volunteers to explain why it might be difficult to grow crops there.

Latin America
Economic Activity

THINK LIKE A GEOGRAPHER

1. **Place** What do the colors on the economic activity map represent?
2. **Human-Environment Interaction** What are the main economic activities of Mexico?

DIFFERENTIATING INSTRUCTION

English Learners

Comprehension: Connection

Model the use of the preposition *of* with cardinal and ordinal directions. Bolivia, for example, is *north of* Argentina and *southeast of* Peru. Encourage students to make similar statements about the relative locations of countries, cities, climate regions, and economic activities.

Gifted & Talented

Comparing Economic Activities

Have pairs of students use the other Unit Atlas maps to compare the economic activities of a Latin American country to that of the United States and one other country. Have students create a chart or other visual display to show these comparisons.

Economic Activity

Talk about how gaining access to much of Latin America's resources is difficult because of its geography and climate. Point out that although some countries—Brazil, for example—have robust economies, millions of people in this region are very poor, living without services that many people take for granted such as electricity, clean drinking water, and sanitation systems.

- What are the major resources of Venezuela? *(natural gas and oil)*
- Where is copper found? *(Chile)* Tin? *(Bolivia)*
- In what countries is silver a major resource? *(in Mexico and Peru)*

THINK LIKE A GEOGRAPHER ANSWERS

1. **Place** The colors show how the land is used.
2. **Human-Environment Interaction** Mexican economic activities include silver, oil, and natural gas production, commercial agriculture, and livestock raising.

Comparing Regions

Rain Forests

Latin America contains vast rain forests, large areas of tall trees in a climate that is warm all year and receives abundant rain. These tropical rain forests grow near the equator.

Rain forests also grow in temperate regions, in cooler coastal areas farther north or south of the equator. North America has the world's largest temperate rain forest, along the Pacific coast. However, development has reduced this temperate rain forest to about half of its original size.

Regional Overview

Remind students that while Latin America's economies have grown, many people still live below the poverty line, and there is also great income inequality. Discuss why many Latin Americans emigrate to the United States. *(Possible answers: to find better jobs and greater economic opportunities; to find more educational opportunities)*

- What geographic features in Latin America provide lands for grazing and farming? *(large plateaus and vast plains)*
- Which two European countries set up colonies in Latin America and controlled them for about 300 years? *(Spain and Portugal)*
- After gaining independence, why did many Latin American countries fall under the rule of wealthy landowners or military dictators? *(Under colonial rule, the people had little or no say in government. As a result, they had little experience in setting up stable governments.)*

More About . . .

The Maya

The Maya first established villages in Central America by 1800 B.C., and by A.D. 250 they had built a network of large cities throughout what is now southeastern Mexico, Belize, Guatemala, and Honduras. The Maya communicated with a written language, practiced astronomy, developed a 365-day calendar, and became one of the first civilizations to recognize the mathematical concept of zero. The Maya abandoned their cities around A.D. 900—for reasons that are not clear—and moved into the countryside. However, the ruins of many Mayan cities remain, and more than six million people still speak Mayan languages.

UNIT 3 ATLAS

Regional Overview

Latin America

Latin America is a far-reaching region of many countries. The region spans a great distance both north and south of the equator. Latin America is diverse in its land, climate, and people.

GEOGRAPHY

Mountain ranges and highlands make up a large part of Latin America. The Andes Mountains, along the western coast of South America, are the largest and the longest mountain range above sea level in the world. Volcanoes, some of them active, extend through parts of Mexico, Central America, and South America. Large plateaus and vast plains provide lands for grazing and farming.

Latin America's climate is as varied as its landscape. The climate ranges from dry deserts to hot, tropical regions to cold highlands. Vegetation in Latin America varies based on the climate.

HISTORY

The Maya, Aztec, and Inca developed advanced civilizations in Latin America before the arrival of the Spanish in the 15th century. They built complex cities and structures and changed their environment to meet their needs.

After the voyages of Columbus, Spanish *conquistadors* destroyed these civilizations. Spain and Portugal set up colonies in the region and controlled most of it for the next 300 years. By the early 1800s, most Latin American countries had gained their independence.

Machu Picchu
This ancient Inca city is located near present-day Cuzco, Peru.

FIVE THEMES AND SIX ELEMENTS

Five Themes

Place

Explain to students that, in addition to the five European languages spoken in Latin America, hundreds of indigenous languages are still spoken by about 40 million people. Some of the most common include Guaraní in Paraguay, Kekchi in Guatemala, Nahuatl in Mexico, and Quechua, an official language of Peru, spoken by about 8.5 million people.

Six Elements

Human Systems

Have students find the location of the Panama Canal on a map or globe. Ask them to explain why Panama presented a good location for the site of a canal, and describe ways that the canal might have transformed the economy and culture of Panama.

CULTURE

Latin American culture today is a blend of several cultures—Indian, European, African, Asian, and people of mixed ancestry. This blend is evident in the music, art, foods, and languages of the region. The majority of Latin Americans speak Spanish, but others speak Portuguese, French, English, and Dutch. Some Indian groups continue to speak their traditional languages.

View of Medellin, Colombia

GOVERNMENT

As colonies, Latin Americans had little or no say in how they were governed. This inexperience made it difficult to establish stable governments after gaining independence. As a result, the governments of many Latin American countries came under the rule of wealthy landowners or military dictators. By the 1980s, widespread protests against these governments occurred in several countries. Today more Latin American countries have democratic governments.

ECONOMICS

Since the 1800s, many Latin American countries have depended on one or two products for export, particularly agricultural products, and have imported most of their manufactured goods. Recently, however, many Latin American countries began to develop other industries. Some Latin American countries have become major exporters of manufactured goods. Latin American countries have also established organizations to promote trade among countries in the region.

Comparing Regions

Haiti and the United States

The United States is the oldest independent nation in the Western Hemisphere. Haiti is the second oldest, and the oldest in Latin America. Both countries were established after successful revolutions against colonial powers. Haiti is located on the western part of Hispaniola, the island on which Columbus first landed in 1492. Spain ceded the area to France in 1697. The French imported many enslaved people from Africa and forced them to work on sugar plantations. Haitian slaves began a revolution in 1791, and by 1804 they had defeated the French and established a republic. Since independence the United States has become one of the richest countries in the world today, while Haiti is the poorest in the Western Hemisphere.

DIFFERENTIATING INSTRUCTION

Pre-AP

Compare Government and Economics

Have students investigate the effects that foreign ownership of Latin American resources has had upon governments in the region, particularly during the 19th and 20th centuries. Ask them to find an example of a government that favored foreign corporate interests over the interests of its citizens.

Inclusion

Compare Cultures

Emphasize the cultural diversity of Latin America by having students listen to several of the region's distinct musical styles. Popular examples might include Brazilian samba, Argentine tango, Cuban son, Haitian konpa, and Jamaican reggae. After listening, have students create illustrations or poems to describe what they heard.

Unit Writing Project
Teaching Tips

- Help students begin the process of choosing a county or area by categorizing countries or areas according to their interests. For example, students interested in archaeology might select Guatemala or Peru, while those interested in marine life and coral reefs might select a nation in the Caribbean.
- Encourage students to consult family-friendly travel guides from the library or the Internet. This will help them increase their options for vacation activities and provide specific details to include in their journal entries.
- Remind students to consider the languages spoken in their destination countries. Explain that in non-English-speaking countries, many workers in the tourist industry can speak some English. Visitors to remote areas, however, should expect to communicate in the native language.

Each map on the country cards illustrates the shape of a state, province, or country. Tell students that the maps do not reflect the comparative size of these landmasses within a particular region. To get an idea of the actual size of the state, province, or country, students should study the land area statistics provided on the card.

UNIT 3 ATLAS Country Almanac

Click here to compare the most recent data on countries @ClassZone.com

Unit Writing Project

As you read this unit, choose a country or an area of Latin America that interests you. Imagine that you are on a week's vacation there. Write a daily journal entry describing what you've seen and learned about the area and your reactions to it.

Think About:

- an area that you find interesting
- what you learned about the area that makes it interesting to you

Latin America

The Almanac provides information about the geography, economy, and culture of the countries of Latin America.

Antigua and Barbuda

GEOGRAPHY
Capital: St. John's
Total Area: 171 sq. mi.
Population: 68,722

ECONOMY
Imports: machinery; agricultural products
Exports: petroleum products

CULTURE
Language: English
Religion: Anglican 32%; Moravian 12%; Catholic 11%

Argentina

GEOGRAPHY
Capital: Buenos Aires
Total Area: 1,068,302 sq. mi.
Population: 38,747,000

ECONOMY
Imports: chemicals; machinery
Exports: food and livestock

CULTURE
Language: Spanish
Religion: Catholic 80%; Protestant 5%

Bahamas

GEOGRAPHY
Capital: Nassau
Total Area: 5,382 sq. mi.
Population: 323,000

ECONOMY
Imports: machinery; food products
Exports: crustaceans and mollusks

CULTURE
Language: English
Religion: Baptist 18%; Catholic 17%; Anglican 11%

Barbados

GEOGRAPHY
Capital: Bridgetown
Total Area: 166 sq. mi.
Population: 270,000

ECONOMY
Imports: capital goods; food
Exports: food; sugar and molasses; rum

CULTURE
Language: English
Religion: Anglican 26%; Pentecostal 11%

Belize

GEOGRAPHY
Capital: Belmopan
Total Area: 8,867 sq. mi.
Population: 270,000

ECONOMY
Imports: machinery; mineral fuels; food
Exports: seafood; sugar; bananas

CULTURE
Language: English
Religion: Catholic 50%; Protestant 32%; nonreligious 9%

Bolivia

GEOGRAPHY
Capital: Sucre
Total Area: 424,164 sq. mi.
Population: 9,182,000

ECONOMY
Imports: machinery; chemicals; food
Exports: soybean products; natural gas

CULTURE
Language: Spanish; Aymara; Quechua
Religion: Catholic 89%; Protestant 9%

Brazil

GEOGRAPHY
Capital: Brasília
Total Area: 3,286,488 sq. mi.
Population: 186,405,000

ECONOMY
Imports: machinery; chemicals
Exports: meat; sugar; coffee; soybeans

CULTURE
Language: Portuguese
Religion: Catholic 72%; Protestant 23%

Chile

GEOGRAPHY
Capital: Santiago
Total Area: 292,260 sq. mi.
Population: 16,295,000

ECONOMY
Imports: machinery; metals; copper
Exports: copper; food products; fruit

CULTURE
Language: Spanish
Religion: Catholic 70%; Protestant 15%

INTERDISCIPLINARY ACTIVITIES

CONNECT to Language Arts

Persuasive Writing

Ask students to write a paragraph that persuades U.S. tourists to visit a certain Latin American city. Have them focus on the key features or attractions of the city that might appeal to U.S. visitors.

CONNECT to Art

Design an Advertisement

Have students create a one-page magazine advertisement that promotes tourism for a Latin American country, scenic area, or historic site. Ask them to combine headlines with drawn or clipped images in ways that will appeal to readers and capture their attention.

Colombia

GEOGRAPHY
Capital: Bogotá
Total Area: 439,736 sq. mi.
Population: 45,600,000

ECONOMY
Imports: capital goods; consumer goods
Exports: petroleum; chemicals; coal; food; machinery; coffee

CULTURE
Language: Spanish
Religion: Catholic 92%

Costa Rica

GEOGRAPHY
Capital: San José
Total Area: 19,730 sq. mi.
Population: 4,327,000

ECONOMY
Imports: general merchandise
Exports: bananas; coffee; tropical fruit

CULTURE
Language: Spanish
Religion: Catholic 86%; Protestant 9%; other Christian 2%

Cuba

GEOGRAPHY
Capital: Havana
Total Area: 42,803 sq. mi.
Population: 11,269,000

ECONOMY
Imports: food and livestock; cereals
Exports: raw sugar; nickel; seafood; medicines

CULTURE
Language: Spanish
Religion: Catholic 39%

Dominica

GEOGRAPHY
Capital: Roseau
Total Area: 291 sq. mi.
Population: 69,029

ECONOMY
Imports: food; machinery
Exports: coconut-based soaps; cosmetics

CULTURE
Language: English
Religion: Catholic 70%; six largest Protestant groups 17%

Dominican Republic

GEOGRAPHY
Capital: Santo Domingo
Total Area: 18,815 sq. mi.
Population: 8,895,000

ECONOMY
Imports: refined petroleum; food
Exports: ferronickel; ships' stores; raw sugar; cocoa

CULTURE
Language: Spanish
Religion: Catholic 82%; Protestant 6%

Ecuador

GEOGRAPHY
Capital: Quito
Total Area: 109,483 sq. mi.
Population: 13,228,000

ECONOMY
Imports: chemicals; food and live animals
Exports: fish and crustaceans; cut flowers

CULTURE
Language: Spanish; Quechua; Shuar
Religion: Catholic 94%; Protestant 2%

El Salvador

GEOGRAPHY
Capital: San Salvador
Total Area: 8,124 sq. mi.
Population: 6,881,000

ECONOMY
Imports: machinery; chemicals and chemical products; food; petroleum
Exports: clothing; coffee; paper; yarn

CULTURE
Language: Spanish
Religion: Catholic 78%; Protestant 17%

Grenada

GEOGRAPHY
Capital: St. George's
Total Area: 133 sq. mi.
Population: 89,502

ECONOMY
Imports: machinery; food; chemicals
Exports: electronic components; nutmeg; fish; paper products; cocoa beans

CULTURE
Language: English
Religion: Catholic 58%; Protestant 38%

Guatemala

GEOGRAPHY
Capital: Guatemala City
Total Area: 42,043 sq. mi.
Population: 12,599,000

ECONOMY
Imports: machinery; chemicals; crude and refined petroleum; road vehicles
Exports: coffee; sugar; bananas; spices

CULTURE
Language: Spanish
Religion: Catholic 76%; Protestant 22%

Rio de Janeiro Located on Brazil's southeast coast, Rio de Janeiro is Brazil's second-largest city.

Country Almanac

The countries of Latin America can also be defined as parts of other regions. Mexico, for example, is part of North America. Seven other North American countries—Belize, Guatemala, El Salvador, Honduras, Nicaragua, Costa Rica, and Panama—are often categorized as Central America. South America contains 12 independent nations and French Guiana, a French overseas department. Finally, the many island nations in the region are often spoken of in terms of the sea that surrounds them, the Caribbean.

- Which Latin American country has the largest land area? *(Brazil)* Which Latin American country has the smallest land area? *(St. Kitts and Nevis)*
- In which country is French a major language? *(Haiti)*
- Which country exports cut flowers? *(Ecuador)*
- Which countries export fossil fuel products, such as petroleum or natural gas? *(Antigua and Barbuda, Bolivia, Colombia, Suriname, and Venezuela)*

INTERDISCIPLINARY ACTIVITIES

CONNECT to Math

Make a Population Bar Graph

Ask students to compare the population of the countries of Latin America. Have pairs of students work together to construct a bar graph to compare these figures. Remind students to find the smallest and largest population in order to determine the divisions of the vertical axis. Have them include the names of the countries on the horizontal axis. As a follow-up activity, students might list the countries in order of population size.

Each map on the country cards illustrates the shape of a state, province, or country. Tell students that the maps do not reflect the comparative size of these landmasses within a particular region. To get an idea of the actual size of the state, province, or country, students should study the land area statistics provided on the card.

More About . . .

Rain Forest Wildlife

The incredible biodiversity of tropical rain forests makes them the world's most important animal habitats. Although these forests covers about seven percent of Earth's surface, they hold at least half of its wildlife species. However, many of these animals are becoming endangered as more and more of their environment is lost. Each year people destroy more than 50,000 square miles of rain forest. Especially at risk are the rain forests of South America.

Each map on the country cards illustrates the shape of a state, province, or country. Tell students that the maps do not reflect the comparative size of these landmasses within a particular region. To get an idea of the actual size of the state, province, or country, students should study the land area statistics provided on the card.

UNIT 3 ATLAS

Guyana

GEOGRAPHY
Capital: Georgetown
Total Area: 83,000 sq. mi.
Population: 751,000

ECONOMY
Imports: consumer goods; fuels
Exports: gold; sugar; shrimp; rice; timber

CULTURE
Language: English
Religion: Hindu 34%; Protestant 28%; Catholic 12%; Muslim 9%

Jamaica

GEOGRAPHY
Capital: Kingston
Total Area: 4,244 sq. mi.
Population: 2,651,000

ECONOMY
Imports: consumer goods; petroleum
Exports: alumina; bauxite; clothing; sugar; coffee; rum

CULTURE
Language: English
Religion: Protestant 61%; Catholic 3%

Panama

GEOGRAPHY
Capital: Panama City
Total Area: 30,193 sq. mi.
Population: 3,232,000

ECONOMY
Imports: mineral fuels; petroleum; machinery; transport equipment
Exports: bananas; seafoods

CULTURE
Language: Spanish
Religion: Catholic 82%; Christian 13%

Haiti

GEOGRAPHY
Capital: Port-au-Prince
Total Area: 10,714 sq. mi.
Population: 8,528,000

ECONOMY
Imports: food and livestock; machinery
Exports: clothing; mangoes; cacao; essential oils; leather goods

CULTURE
Language: Haitian Creole; French
Religion: Catholic 69%; Protestant 24%

Honduras

GEOGRAPHY
Capital: Tegucigalpa
Total Area: 43,278 sq. mi.
Population: 7,205,000

ECONOMY
Imports: food and livestock; machinery
Exports: shrimp; coffee; palm oil

CULTURE
Language: Spanish
Religion: Catholic 87%; Protestant 10%

Mexico

GEOGRAPHY
Capital: Mexico City
Total Area: 761,606 sq. mi.
Population: 107,029,000

ECONOMY
Imports: electronics; clothing; rubber and plastic products
Exports: road vehicles; machinery; textiles

CULTURE
Language: Spanish
Religion: Catholic 90%; Protestant 4%;

Nicaragua

GEOGRAPHY
Capital: Managua
Total Area: 49,998 sq. mi.
Population: 5,487,000

ECONOMY
Imports: consumer goods; fuels
Exports: food products; coffee

CULTURE
Language: Spanish
Religion: Catholic 85%; Protestant 12%

Paraguay

GEOGRAPHY
Capital: Asunción
Total Area: 157,047 sq. mi.
Population: 6,158,000

ECONOMY
Imports: machinery; chemicals; food
Exports: soybean products; meats

CULTURE
Language: Spanish; Guaraní
Religion: Catholic 90%; Protestant 5%

Green Crested Basilisk This lizard is one of many species of animals found in the rain forests of Latin America.

194 Unit 3

INTERDISCIPLINARY ACTIVITIES

CONNECT to Art

Create a Tropical Wildlife Poster

Have students find information about a tropical fish, bird, or other creature that interests them. Ask them to create posters (either by painting images or making collages) showing the animals in their natural habitats. Have them add titles and captions to complete their posters.

CONNECT to Civics

Latin American Governments

Ask students to choose a Latin American country and find out basic facts about its system of government. Have them identify the head of state and legislative bodies of their country and report their findings to the class.

Peru

GEOGRAPHY
Capital: Lima
Total Area: 496,226 sq. mi.
Population: 27,968,000

ECONOMY
Imports: machinery; chemicals and chemical products; petroleum; food
Exports: gold; animal feed; copper

CULTURE
Language: Spanish; Quechua; Aymara
Religion: Catholic 96%

St. Kitts and Nevis

GEOGRAPHY
Capital: Basseterre
Total Area: 101 sq. mi.
Population: 38,958

ECONOMY
Imports: machinery; food; metals
Exports: raw sugar; telecommunications equipment

CULTURE
Language: English
Religion: Protestant 85%; Catholic 7%

St. Lucia

GEOGRAPHY
Capital: Castries
Total Area: 238 sq. mi.
Population: 161,000

ECONOMY
Imports: food; machinery; manufactured goods
Exports: bananas; beer and ale; clothing

CULTURE
Language: English
Religion: Catholic 69%; Protestant 22%

St. Vincent and the Grenadines

GEOGRAPHY
Capital: Kingstown
Total Area: 150 sq. mi.
Population: 119,000

ECONOMY
Imports: food products; machinery
Exports: bananas; packaged flour; packaged rice; eddoes and dasheens

CULTURE
Language: English
Religion: Protestant 58%; Catholic 11%

Suriname

GEOGRAPHY
Capital: Paramaribo
Total Area: 63,039 sq. mi.
Population: 449,000

ECONOMY
Imports: machinery; food; road vehicles
Exports: alumina; gold; petroleum; rice

CULTURE
Language: Dutch
Religion: Christian 50%; Hindu 18%; Muslim 14%

Trinidad and Tobago

GEOGRAPHY
Capital: Port of Spain
Total Area: 1,980 sq. mi.
Population: 1,305,000

ECONOMY
Imports: petroleum; industrial machinery
Exports: floating docks; iron and steel

CULTURE
Language: English
Religion: six largest Protestant bodies 30%; Catholic 29%; Hindu 24%

Uruguay

GEOGRAPHY
Capital: Montevideo
Total Area: 68,039 sq. mi.
Population: 3,463,000

ECONOMY
Imports: chemicals; food and tobacco
Exports: leather goods; beef

CULTURE
Language: Spanish
Religion: Catholic 78%; atheist 6%

Venezuela

GEOGRAPHY
Capital: Caracas
Total Area: 352,144 sq. mi.
Population: 26,749,000

ECONOMY
Imports: machinery; chemicals; vehicles
Exports: petroleum products; aluminum

CULTURE
Language: Spanish; 31 indigenous Indian Languages
Religion: Catholic 90%; Protestant 2%

Caribbean Islands The islands of the Caribbean are among the world's most popular tourist destinations.

195

INTERDISCIPLINARY ACTIVITIES

CONNECT to Language Arts

Create a Postcard

Have students choose a Caribbean island as a travel destination. Have them create a postcard to "send" home. One side should contain an image or images that they create, the other a written message about the trip. To extend the activity, have students create a postage stamp for their postcards.

Comparing Regions

The Andes and the Himalayas

The Andes mountain range is the longest on Earth. The range stretches through seven countries, for over 5,000 miles, a distance longer than the combined length of the Rocky, Appalachian, and Sierra Nevada mountain ranges. One peak in the Andes, Aconcagua, is the highest in the Western Hemisphere. The Amazon River begins high in the Peruvian Andes and flows about 4,000 miles across South America to the Atlantic Ocean. The Amazon River is longer than the Mississippi and Missouri rivers combined.

The Himalayas are the world's highest mountains, but extend about 4,000 miles less than the Andes Mountains. Another of the world's most famous rivers, the Ganges, has its source in the western Himalayas. Although it is about 1,600 miles long—less than half the length of the Amazon River—the Ganges river provides water for more than 300 million people.

Each map on the country cards illustrates the shape of a state, province, or country. Tell students that the maps do not reflect the comparative size of these landmasses within a particular region. To get an idea of the actual size of the state, province, or country, students should study the land area statistics provided on the card.

Latin America Physical Geography and Early History

PLANNING GUIDE

Complete print resources all on one USB drive allow you to customize lesson plans for all students.

Section Titles & Key Ideas	Unit Resource Book Latin America	Assessment Resources
1 Physical Geography of Mexico, Central America, and the Caribbean **KEY IDEA** The geography of Mexico, Central America, and the Caribbean Islands, south of the United States, contains mountains, highlands, and plains.	• **Reading Study Guide,** pp. 3–4 • **RSG with Additional Support,** pp. 11–12 • **Vocabulary Practice,** p. 35 • **Skillbuilder Practice,** p. 36 • **Map and Graph Skills,** p. 37 • **Interdisciplinary Projects,** pp. 44–45	**Unit Resource Book** • Section Quiz, p. 49 • Reteaching Activity, p. 53 **Test Generator CD-ROM** **Daily Test Practice Transparency,** TT22
2 Physical Geography of South America **KEY IDEA** The physical geography of South America consists of a wide variety of landforms and climates.	• **Reading Study Guide,** pp. 5–6 • **RSG with Additional Support,** pp. 13–14 • **Outline Map Activities,** p. 39 • **Primary Source,** p. 48	**Unit Resource Book** • Section Quiz, p. 50 • Reteaching Activity, p. 54 **Test Generator CD-ROM** **Daily Test Practice Transparency,** TT23
3 Ancient Civilizations **KEY IDEA** People adapted to these challenging geographic and climatic conditions and developed great civilizations.	• **Reading Study Guide,** pp. 7–8 • **RSG with Additional Support,** pp. 15–16 • **Skillbuilder Practice,** p. 36 • **History Makers,** pp. 41–42 • **Interdisciplinary Projects,** pp. 43, 45	**Unit Resource Book** • Section Quiz, p. 51 • Reteaching Activity, p. 55 **Test Generator CD-ROM** **Daily Test Practice Transparency,** TT24
4 From Colonization to Independence **KEY IDEA** Life under the conquerors was hard, as the people of Latin America struggled to gain their independence.	• **Reading Study Guide,** pp. 9–10 • **RSG with Additional Support,** pp. 17–18 • **Economics in the World,** p. 46 • **Connect to Language Arts,** p. 47	**Unit Resource Book** • Section Quiz, p. 52 • Reteaching Activity, p. 56 • Chapter Test (Forms A, B, & C), pp. 57–68 **Test Generator CD-ROM** **Daily Test Practice Transparency,** TT25

McDougal Littell **Assessment System**

TEST SCORE REPORT RETEACH

No Time? To focus students on important content in the chapter, use the **"Reading Study Guide with Additional Support"** in ***Resources2Go.***

Teacher Presentation Options

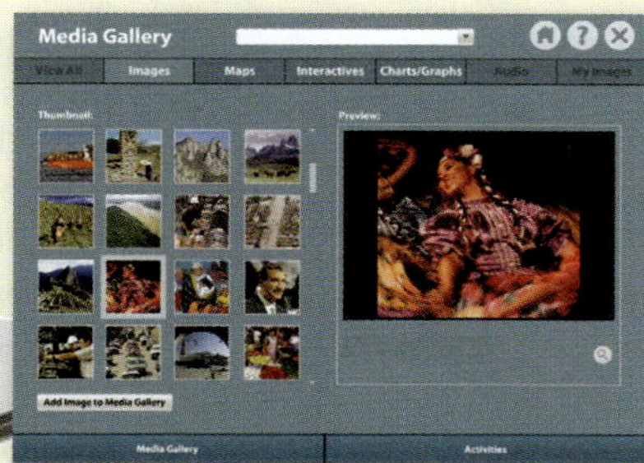

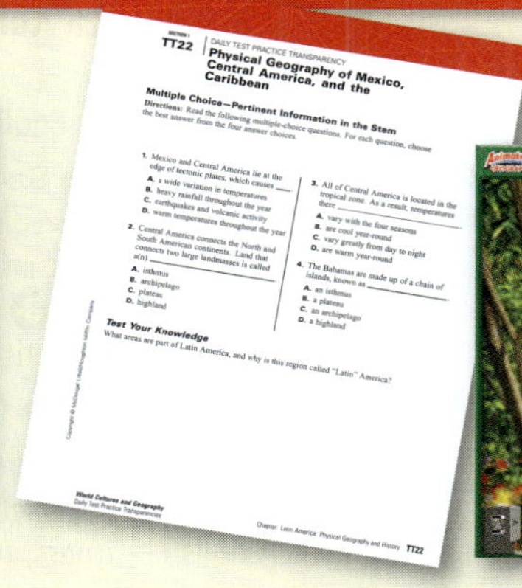

Electronic Books

eEdition DVD-ROM

eEdition Online

Presentation Toolkit

Video/DVD

- World Cultures and Geography

Power Presentations DVD-ROM

- Discussion Starters
- Media Gallery
- Review Game
- Interactive Review
- Animated Features

Unit Transparency Book

- 3-Minute Warm-Ups
- Essential Question
- Critical Thinking
- Maps
- Cultures

Daily Test Practice Transparencies

Animated Geography

- Present-Day Latin America Map
- Olmec, Maya, Aztec, and Inca Map
- *Chinampas*

OnlineActivities at ClassZone.com

Animated Geography

- Videos and Photos
- 3-D Animations
- Interactive Maps

Interactive Review

- Section Quizzes
- Review and Study Notes
- Vocabulary Flip Cards
- Graphic Organizers
- Name Game and Crossword Puzzles
- GeoGame

Online Test Practice

Map Center

- Atlas Maps
- Political Maps with Data
- Links to Mapping Web Sites

Activity Center

- Activity Maker—Review Games, Puzzles, Flip Cards, and Graphic Organizers
- Worldster

Research & Writing Center

- Web Sites
- Data Updates
- Primary Sources
- Writing Models

Current Events

Greta Frensley
Knox Doss Middle School
Gallatin, Tennessee

Make It Fun — Teacher-Tested Activities

As you teach this chapter, look for my classroom-proven ideas for:

- making the lesson fun
- focusing students' attention and comprehension
- integrating technology with traditional teaching

Teaching Bell to Bell When you see this timing guide, have students discuss in pairs, in groups of 3–6, or as a class.

- **Think, Pair, Share** (pairs)
- **Reader, Recorder, Reporter** (3–6)
- **Talk About It** (any size)

CHAPTER 7 RESOURCES FOR DIFFERENTIATED INSTRUCTION

Struggling Readers

eEdition with Audio DVD-ROM

Lesson Plans
- Strategies for Inclusion

Unit Resource Book
- Reading Study Guide with Additional Support, pp. 11–18 Ⓑ
- Reteaching Activity, pp. 53–56
- Chapter Test (Form A), pp. 57–60 Ⓐ

Best Practices Toolkit Ⓒ
- Interactive Reading Strategies, pp. 90–145
- Differentiation Activity Bank: Struggling Readers, pp. 146–149

Teacher's Edition Activities
- Identifying South America's Physical Features, Sec. 2
- Review Translations of Climate Zones, Sec. 2
- Chart Details About the Maya Civilization, Sec. 3
- Chart Details About Spanish Class System, Sec. 4
- Tiered Activities, Sec. 1–4

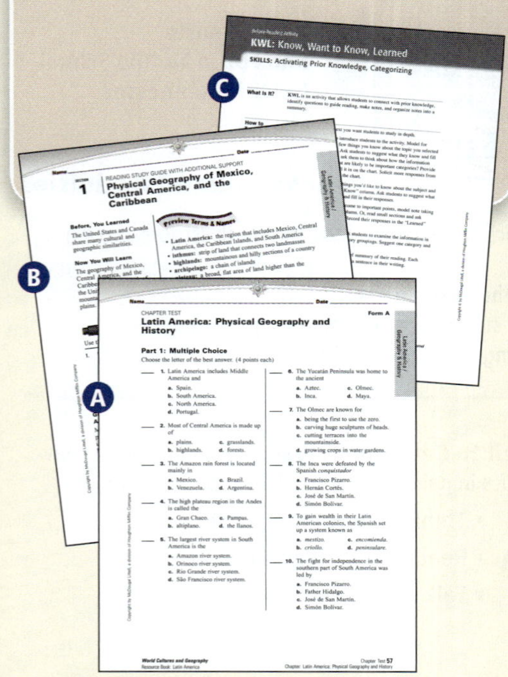

English Learners

Pupil's Edition in Spanish

eEdition with Audio DVD-ROM

Lesson Plans
- Strategies for English Learners

Unit Resource Book
- Reading Study Guide (Spanish), pp. 19–26
- RSG with Additional Support (Spanish), pp. 27–34 Ⓐ Ⓑ
- Vocabulary Practice, p. 35

Multi-Language Glossary Ⓒ

Best Practices Toolkit
- Differentiation Activity Bank: English Learners, pp. 165–194

Test Generator CD-ROM
- Section Quizzes in Spanish
- Chapter Test in Spanish

Teacher's Edition Activities
- Pronounce and Preview, Sec. 1–4
- Vocabulary: Roots, Sec. 1
- Vocabulary: Key Academic Vocabulary, Sec. 2, 3
- Tiered Activities, Sec. 1–4

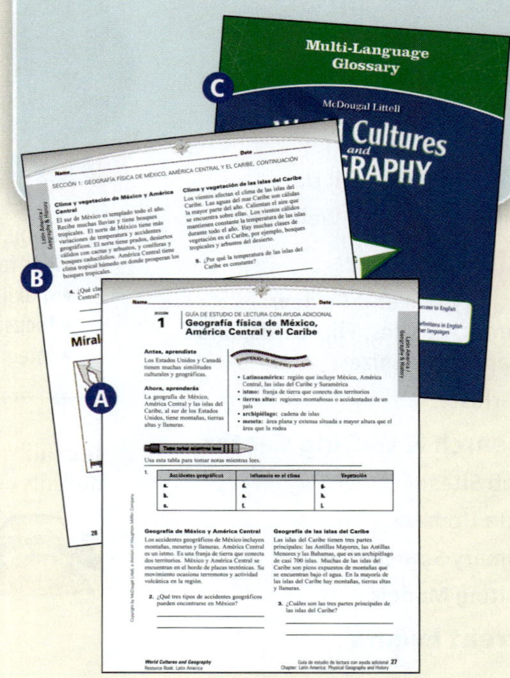

Inclusion

eEdition with Audio DVD-ROM

Lesson Plans
- Strategies for Inclusion

Unit Resource Book
- Reading Study Guide with Additional Support, pp. 11–18 Ⓐ Ⓑ
- Reteaching Activity, pp. 53–56
- Chapter Test (Form A), pp. 57–60

Best Practices Toolkit Ⓒ
- Interactive Reading Strategies, pp. 90–145
- Differentiation Activity Bank: Struggling Readers, pp. 146–149

Teacher's Edition Activities
- Create Visual Vocabulary, Sec. 1
- Vocabulary: Word Wheel, Sec. 2
- Match Terms and Definitions, Sec. 3
- Locate Places on World Map, Sec. 4
- Tiered Activities, Sec. 1–4

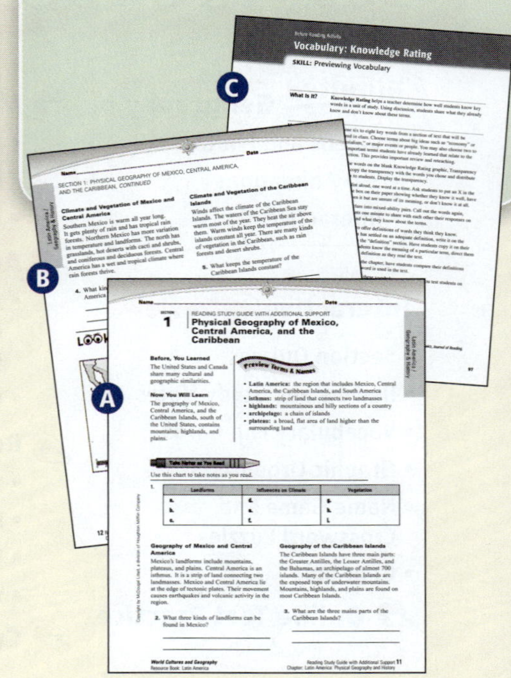

Interdisciplinary Activities

Gifted & Talented Students

Lesson Plans
- Core Lesson Plan

Unit Resource Book
- Outline Map Activities, p. 39
- History Makers, pp. 41–42 (A)
- Interdisciplinary Projects, pp. 44–45 (B)
- Connect to Language Arts, p. 47 (C)
- Chapter Test (Form C), pp. 65–68

Best Practices Toolkit
- Student Products Rubrics, pp. 1–14
- Writing Templates, pp. 78–79

Teacher's Edition Activities
- Write a Poem About Tierra Del Fuego, Sec. 2
- Illustrate Climate Zones, Sec. 2
- Describe Inca Challenges, Sec. 3
- Tiered Activities, Sec. 1–4

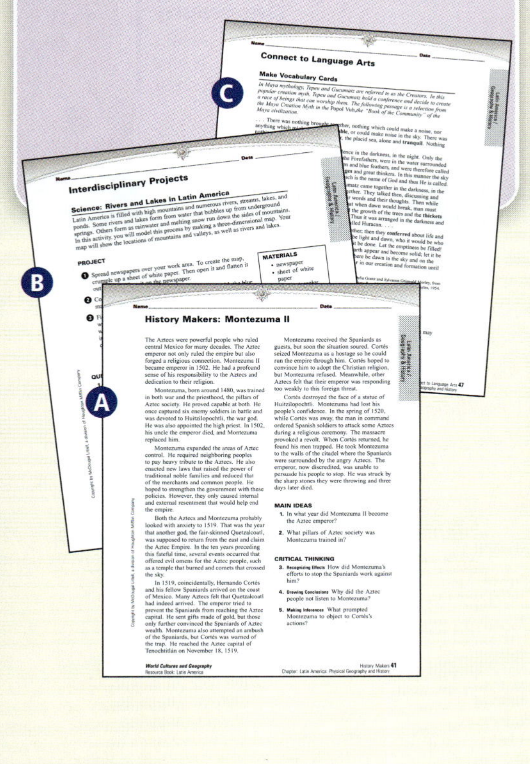

Pre-AP*

Lesson Plans
- Core Lesson Plan

Unit Resource Book
- Map and Graph Skills, p. 37
- Interdisciplinary Projects, pp. 43, 45 (A)
- Economics in the World, p. 46 (B)
- Primary Source, p. 48 (C)
- Chapter Test (Form C), pp. 65–68

Best Practices Toolkit
- Writing Templates, pp. 78–79
- Differentiation Activity Bank: Pre-AP Students, pp. 195–200

* Pre-AP is a registered trademark of the College Entrance Examination Board, which was not involved in the production of and does not endorse this product.

Teacher's Edition Activities
- Map Middle America, Sec. 1
- Estimate the Size of Lake Titicaca, Sec. 2
- Tiered Activities, Sec. 1–4

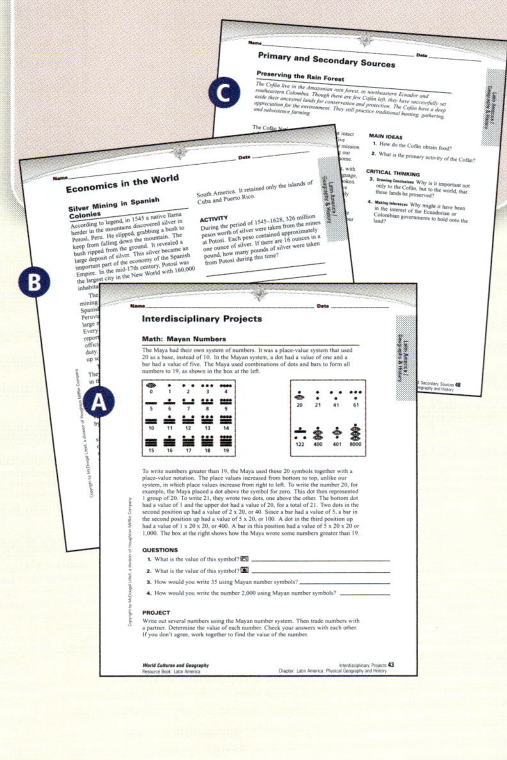

CONNECT to Art

Illustrate an Important Geographic Feature of Latin America

CLASS TIME 30 minutes

TASK Prepare an illustration of an important geographic feature of Latin America.

SUPPLIES
- drawing paper, colored markers or pencils

ACTIVITY Have students review the information in Sections 1 and 2 on the geographic features of Middle America and South America. Ask students to select an important geographic feature, such as the Amazon River, volcanoes in Central America, or the Andes Mountains, and prepare an illustration of it. Remind students to include a title and captions to explain the importance of the feature to Latin America.

Good for Visual Learners

CONNECT to Language Arts

Write a News Story

CLASS TIME 30 minutes

TASK Write a news story describing Machu Picchu when it was rediscovered in 1911.

SUPPLIES
- computer, pencils or pens, writing paper

ACTIVITY As students review the photograph of Machu Picchu in Section 3, explain that this Inca city was "lost" until 1911. Then have students assume the role of a newspaper reporter accompanying the explorer who rediscovered Machu Picchu and work individually to write a news story about the experience. Remind students to be sure their stories answer the "Who, What, Where, When, Why" questions of good reporting. Allow students to use library and other resources to gather background for their stories.

Good for Verbal Learners

Introduce the ESSENTIAL QUESTION

- Point out that physical geography can both promote and hinder the growth of civilizations.
- Predict some general climate characteristics in Latin America according to its location.
- Discuss reasons that Latin American civilizations remained isolated from other civilizations in Africa, Europe, and Asia.
- Explain why Latin America was attractive as an area for colonies for European countries and how the colonies gained their independence.

Key Ideas

Section 1: The geography of Mexico, Central America, and the Caribbean Islands, south of the United States, contains mountains, highlands, and plains.

Section 2: The physical geography of South America consists of a wide variety of landforms and climates.

Section 3: People adapted to these challenging geographic and climatic conditions and developed great civilizations.

Section 4: Life under the conquerors was hard, as the people of Latin America struggled to gain their independence.

CONNECT to the Essential Question

You may wish to revisit the Essential Question periodically with students as they work through the chapter.

Suggestions for reconnecting are provided at an appropriate point in the *Teacher's Edition* for each section.

Spanish-style church in Mexico

CHAPTER 7

Latin America

Physical Geography and History

ESSENTIAL QUESTION

How have Latin America's geography and resources helped shape its history?

CONNECT Geography & History

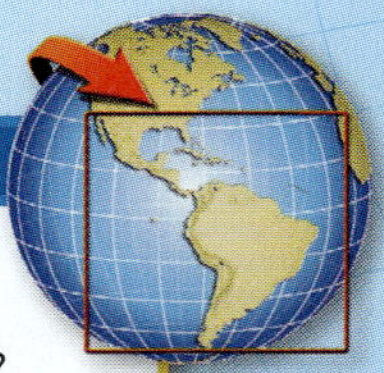

Use the map and the time line to answer the following questions.

1. What is the largest country in Latin America?
2. What country is named after Simón Bolívar, one of the leaders for South American independence?

1200 B.C.

Culture
◄ **1200 B.C.** Olmec build the first known civilization in southeastern Mexico.

History
A.D. 900 Classic period of Mayan civilization ends.

Geography
1200s Inca settle in Cuzco Valley.

Geography
1325 Aztecs establish their capital city, Tenochtitlán. ►

TIME LINE DISCUSSION

Note that the time line covers a span of about 3,000 years and that much of the early history of the region has only been discovered through recent archaeological discoveries.

- About how long were the Incas a powerful civilization in the Cuzco Valley? *(about 300 years until they were conquered by Pizarro in 1532)*
- When did the Aztecs fall to the Spanish? *(1521—when Hernán Cortés conquered them)*
- How are the events of the early 1500s related to the events of the early 1800s? *(Spanish armies defeated native people and began colonies; these colonies eventually fought for independence from Spain in early 1800s.)*

History
1532 Francisco Pizarro conquers the Inca.

History
1810–1825 Simón Bolívar leads South American countries in their fight for independence from Spain.

1850

History
1521 Hernán Cortés conquers the Aztecs.

History
1810 Father Hidalgo calls for Mexico's independence from Spain.

197

CHAPTER 7

Animated GEOGRAPHY
Present-Day Latin America
CLASSZONE.COM

Note that current political divisions between countries do not necessarily reflect areas of ancient civilizations. For example, the Inca civilization included areas of modern Ecuador, Peru, Chile, Bolivia, and Argentina.

- What physical feature is the most noticeable in Brazil? *(rivers draining into the Amazon River)*
- How would you describe the size of Mexico compared to the countries in Central America? *(Mexico is much larger.)*
- Which country on the map might be described as long and narrow? *(Chile)*

CONNECT GEOGRAPHY & HISTORY ANSWERS

1. Brazil
2. Bolivia

Presentation Options

Use the **Power Presentations** DVD-ROM to project the animated map.

- Prompt students to use the pop-up images and audio from the map to discover what they already know about the geography, culture, history, economics, or governments of the region.
- Challenge students to answer the map questions individually, in small groups, or as a class activity. Use all of the interactive material to help preview chapter content.

TEACHER MEDIA FAVORITES

Books for the Teacher

D'Altroy, Terence N. *The Incas* (The Peoples of America). Blackwell Publishing, new edition 2003. Includes both history and archaeology in charting the rise of this empire.

Mann, Charles C. *1491: New Revelations of the Americas Before Columbus.* Knopf, 2006. A new way of looking at the Americas before the European arrival.

Video / DVD

NOVA: Secrets of Lost Empires—Inca. (DVD) WGBH Boston, 2006 release. Scientists look at how Incas survived in the rugged Andes.

In Search of History—The Aztec Empire. (History Channel) (DVD) A&E Home Video, 2005 release. Includes the rise of the Aztecs, their rituals, and the loss to the Spanish.

Internet

Go to the *Research and Writing Center* at **ClassZone.com** to link to:

- data on the region
- a variety of chapter-specific, content-reviewed sites
- "History Makers" biographies
- primary sources

1 Plan & Prepare

Objectives

- Describe and compare the geography of Mexico, Central America, and the Caribbean Islands
- Identify the climates and vegetation of Mexico, Central America, and the Caribbean Islands

Read for the Essential Question

Help students read for a purpose by reminding them of the Essential Question: "How have Latin America's geography and resources helped shape its history?" Point out that they can use text features such as Key Questions, special features, images, and maps to form answers to the question.

Vocabulary

Best Practices Toolkit

Use the ***Best Practices Toolkit*** to model strategies for vocabulary notetaking. Vary strategies throughout the year. Choose from: *Knowledge Rating, Predicting ABCs, Definition Mapping, Word Sort, Word Wheel, Frayer Model (Word Squares), Magnet Words,* and *Student VOC.*

Vocabulary Strategies, TT9–TT16

Reading Strategy

Best Practices Toolkit

Display the Three-Column Chart transparency.

- Explain to students that in categorizing, they need to determine how the information in the section can be sorted into the categories shown on the chart.
- Model filling in categories on the chart, using the Landforms column.
- Have students suggest entries, using the other categories in the chart.
- Display the completed chart for review.

Three-Column Chart, TT21

Section 1 Reading for Understanding

Key Ideas

BEFORE, YOU LEARNED

The United States and Canada share many cultural and geographic similarities.

NOW YOU WILL LEARN

The geography of Mexico, Central America, and the Caribbean Islands, south of the United States, contains mountains, highlands, and plains.

Vocabulary

TERMS & NAMES

Latin America the region that includes Mexico, Central America, the Caribbean Islands, and South America

Sierra Madre Occidental mountain range that runs from north to south down the western part of Mexico

Sierra Madre Oriental mountain range that runs from north to south down the eastern part of Mexico

isthmus strip of land that connects two landmasses

highlands mountainous and hilly sections of a country

archipelago a chain of islands

Greater Antilles the northern largest Caribbean islands that include Cuba, Jamaica, Hispaniola (which includes Haiti and the Dominican Republic), and Puerto Rico

Lesser Antilles the smaller Caribbean islands southeast of the Greater Antilles

REVIEW

plateau a broad, flat area of land higher than the surrounding land

tectonic plate a large rigid section of the Earth's crust that is in constant motion

Reading Strategy

Re-create the chart shown at right. As you read and respond to the **KEY QUESTIONS**, use the chart to organize details about the geography of Mexico, Central America, and the Caribbean Islands.

See Skillbuilder Handbook, page R7

CATEGORIZE

LANDFORMS	INFLUENCES ON CLIMATE	VEGETATION
1.	1.	1.
2.	2.	2.
3.	3.	3.

GRAPHIC ORGANIZERS
Go to **Interactive Review** @ClassZone.com

198 Chapter 7

PRETEACHING VOCABULARY

English Learners

Pronounce and Preview

Pronounce each term for students. Review the meanings of words within definitions, such as *landmasses*, *surrounding*, and *rigid*.

- To modify vocabulary learning, have students complete worksheets as they read, instead of afterward.

Unit Resource Book: Latin America

- Vocabulary Practice, p. 35

Inclusion

Create Visual Vocabulary

Read the definition of each term aloud, and answer any questions students might have. Then assign each student to draw a simple illustration of each term, writing the word as the title and definition as the caption. Ask volunteers to share their illustrations when complete and to explain them as necessary. Post the illustrations to create a visual vocabulary display

SECTION 1 GEOGRAPHY

Great Bay Beach on the Caribbean Island of Saint Maarten

Waterfall in Rincon de la Vieja National Park, Costa Rica

Physical Geography of Mexico, Central America, and the Caribbean

Connecting to Your World

What would you do if the ground started to open up as you were walking? That is what happened to a Mexican farmer in February 1943. The farmer watched in amazement as the ground rumbled and split open while he worked. Within 24 hours, a small smoking cone had appeared. The cone was the beginning of the volcano Paricutín (pah•REE•koo•TEEN).

Volcanoes are just one feature of Latin America's geography. Let's take a look at some other geographic features.

Mountains and Islands

KEY QUESTION How is the geography of Mexico, Central America, and the Caribbean Islands alike?

The region of **Latin America** includes Mexico, Central America, the Caribbean Islands, which together make up Middle America, and South America. The Spanish and Portuguese colonized much of the region. Their languages derived from Latin, so they called the region *Latin America*. In this section, you will learn about the geography of Middle America.

Montserrat This Caribbean volcano lay dormant for 500 years, then erupted from 1995 through 1997.

199

2 Focus & Motivate

3-Minute Warm-Up

Write on the board or display the transparency:

- Look at the physical geography map in Section 1. What landform makes up much of Mexico? *(mountains)*

Unit Transparency Book: Latin America
- 3-Minute Warm-Ups, TT1

Connecting to Your World

More About . . .

Birth of a New Volcano

Since people have been keeping records, only two new volcanoes are known to have been born in North America and Middle America. Both are in central Mexico. Paricutin grew to almost 1,500 feet from its base and erupted for about 9 years.

3 Teach

Mountains and Islands

Talk About It

- How do tectonic plates affect Mexico and Central America? *(They cause earthquakes and volcanoes.)*
- What is the relationship between many Caribbean Islands and underwater mountains? *(The islands are actually the tops of the underwater mountains.)*
- **Categorize** What are the main geographic features in Central America? *(isthmus, highlands, volcanoes, and coastal plains)*

SECTION 1 PROGRAM RESOURCES

Unit Resource Book: Latin America

ON LEVEL

Unit Resource Book
- Reading Study Guide, p. 3
- Map and Graph Skills, p. 37
- Section Quiz, p. 49

STRUGGLING READERS

Unit Resource Book
- Reading Study Guide with Additional Support, p. 11
- Skillbuilder Practice: Finding Main Ideas, p. 36
- Section Quiz, p. 49
- Reteaching Activity, p. 53

eEdition with Audio DVD-ROM

ENGLISH LEARNERS

Pupil Edition in Spanish
eEdition with Audio DVD-ROM
Unit Resource Book
- RSG (Spanish), p. 19
- RSG with Additional Support (Spanish), p. 27

Multi-Language Glossary
Test Generator
- Section Quiz in Spanish

INCLUSION

Unit Resource Book
- RSG with Additional Support, p. 11
- Section Quiz, p. 49
- Reteaching Activity, p. 53

GIFTED & TALENTED

Unit Resource Book
- Interdisciplinary Projects, pp. 44–45
- Section Quiz, p. 49

PRE-AP

Unit Resource Book
- Section Quiz, p. 49

TECHNOLOGY

Unit Transparency Book
- 3-Minute Warm-Ups, TT1
- Essential Question, TT2
- Critical Thinking, TT3
- Map: Latin America: Physical, TT7

Daily Test Practice Transparencies
- Section 1, TT22

Power Presentations
ClassZone.com
World Cultures and Geography Video on DVD

More About . . .

Earthquakes in Mexico and Central America

Because of the physical geography, Mexico and Central America are especially prone to earthquakes. An earthquake devastated the central area of Managua, the capital of Nicaragua, in 1972, killing as many as 10,000 and leaving 300,000 homeless. A quake struck Mexico City in 1985, killing about 10,000 and leveling hundreds of buildings.

The response of the governments in both cases contributed to protests against them. The Somoza government in Nicaragua received millions in international aid, but much of it went into the pockets of government officials. These acts of corruption led to protests, which eventually felled the government. Mexican citizens protested against what they felt was an inadequate government response to the earthquake, helping to move their country toward greater democracy.

Comparing Regions

The Importance of an Isthmus

Because it connects two larger landmasses, an isthmus is important for both economic and military reasons.

- In the early history of the Mediterranean region, the Isthmus of Corinth connected the areas containing Athens and Sparta and included seaports.
- The Isthmus of Suez effectively connected Asia and Africa and included ports on the Mediterranean and the Red Sea, leading to the Indian Ocean. Although this isthmus allowed land transportation between the two continents, it also blocked water routes.

Canals through these isthmuses have been attempted since the earliest times, with the first work on a Suez canal beginning around 1850 B.C. The Romans considered building a canal through the Isthmus of Corinth around 600 B.C. The earliest attempts at a Panama canal did not begin until the 1880s.

Mexico Mexico's landforms include mountains, **plateaus**, and plains. You can see these on the map opposite. Two major mountain ranges are the **Sierra Madre Occidental** Ⓐ, located in western Mexico, and the **Sierra Madre Oriental** Ⓑ, located in eastern Mexico. Mexico sits on three large **tectonic plates**, and the movement of these plates causes earthquakes and volcanic activity.

Between the two mountain ranges lies the vast Mexican Plateau, making up about 40 percent of Mexico's land regions. Most of Mexico's people live in this area. The plateau rises about 4,000 feet above sea level in the northern part to about 8,000 feet in the southern part. Several volcanoes frame the southern edge of the plateau. East and west of the plateau, along the Gulf of Mexico and the Pacific Ocean, lie the coastal plains.

Mexico has few major rivers. Most of the rivers are not deep enough or wide enough for large boats to move through. Mexico's largest river is the Rio Grande, which is the northern border with Texas.

Sea Turtles These hawksbill sea turtles are swimming over reefs in the Caribbean Sea off the coast of the Cayman Islands.

Central America South of Mexico are the countries of Central America. Central America is an **isthmus**, a strip of land that connects two large landmasses. It connects the North and South American continents.

Highlands, the hilly sections of a country, make up most of Central America. Like Mexico, Central America lies at the edge of tectonic plates, so earthquakes occur often. A string of volcanoes, some of which are active, lines the Pacific Coast. Plains lie along both the Pacific and Caribbean coasts.

The Caribbean Islands East of Central America lie the Caribbean Islands. The islands are made up of three main parts. The Bahamas are an **archipelago**, a chain of nearly 700 islands. The **Greater Antilles** Ⓒ include the largest islands of Cuba, Jamaica, Hispaniola (which includes Haiti and the Dominican Republic), and Puerto Rico. The **Lesser Antilles** Ⓓ include the remaining smaller islands. Many of the Caribbean Islands are actually the exposed tops of underwater mountains. Mountains and highlands make up the interior of most Caribbean Islands, and plains circle the highlands.

A COMPARE AND CONTRAST Describe how the geography of Mexico, Central America, and the Caribbean Islands is alike.

Mountains and highlands make up much of Mexico, Central America, and the Caribbean Islands. All three have coastal plains. Location on the edge of tectonic plates makes earthquakes frequent occurrences. ◄

DIFFERENTIATING INSTRUCTION

English Learners

Vocabulary: Roots

Note the terms *Sierra Madre Occidental* and *Sierra Madre Oriental* in the second sentence on this page. Ask students to locate them on the physical map on the next page and to suggest possible meanings for "occidental" and "oriental" based on their locations. Then explain that Latin is the base language for both Spanish and English and that *occidens* means the direction of the setting sun and *oriens* the direction of the rising sun.

Pre-AP

Map Middle America

Invite students to prepare a map—from memory or to complete an outline map—of Middle America. Maps should include the correct locations of all countries, capitals, and geographic features. Then have students work with a partner to review and correct each other's map.

Physical Geography of Mexico, Central America, and the Caribbean Islands

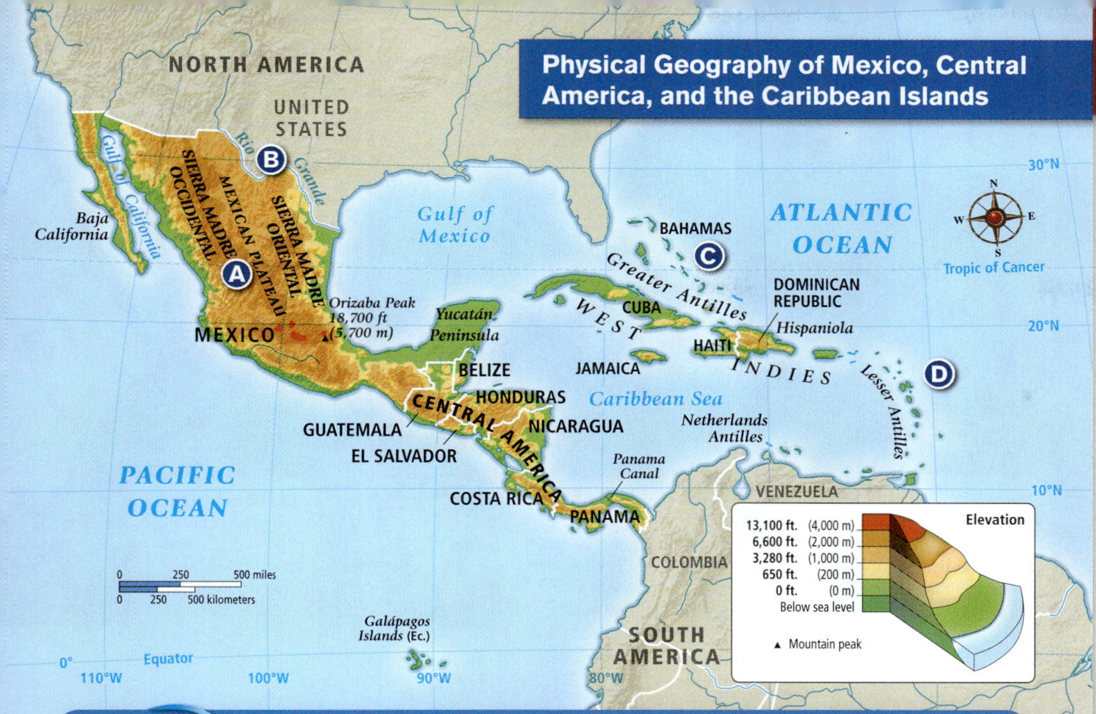

CONNECT Geography & Economics

Panama Canal The Panama Canal was built by cutting through the Central American land bridge, which connected the Atlantic and Pacific oceans. Before the canal was opened in 1914, a ship bringing copper from Peru to Florida had to sail south around South America and then north. Using the canal shortened the distance and time.

CRITICAL THINKING

1. **Movement** By how many miles did the canal shorten a ship's journey?
2. **Make Inferences** Why was Panama a good choice for building a canal?

201

INTERDISCIPLINARY ACTIVITIES

CONNECT to Civics

Deliver a Speech at the Opening of the Panama Canal

Ask students to write a speech that they might have given at a ceremony celebrating the opening of the Panama Canal in 1914. Point out that the canal was considered the greatest engineering feat in the world at the time. It was also expensive, costing about $350 million. Suggest that students write the speech from either the point of view of a U.S. citizen or a Panamanian, and consider the importance of the canal in terms of military security and economic wealth.

CONNECT to Language Arts

Journal Entry by a Ship's Captain After a First Trip Through the Panama Canal

Ask students to assume the role of a ship's captain who had often sailed around South America before the opening of the Panama Canal. Then have students write a journal entry of the captain's first passage through the Canal. Entries might include information about the time saved, the complexity of the construction, and the process of going through the canal. Invite students to present or display their entries.

Physical Geography of Mexico, Central America, and the Caribbean Islands

Have students use the elevation key to compare areas of high and low elevation in Middle America.

- Describe the elevation of the Yucatán Peninsula? *(at sea level)*
- Where is the highest elevation located? *(in Mexico)*

Unit Resource Book: Latin America

- Interdisciplinary Projects: Rivers and Lakes in Latin America, pp. 44–45

CONNECT Geography & Economics

Panama Canal

Today, Panama is considering expanding the canal to handle more and larger ships, partially because of China's strong international economy. The main destination for China's ships is the east coast of the United States. If the ships travel west from China instead of east, it takes them only one more day to arrive because they can use the Suez Canal. Expanding the Panama Canal would mean more ships could pass through it—with less wait time, which sometimes involves as much as a week on a first-come, first-served basis.

CRITICAL THINKING ANSWERS

1. **Movement** by about 10,000 miles
2. **Make Inferences** It is the narrowest part of the isthmus, making it most efficient for building a canal.

More About . . .

Ownership of the Panama Canal

In 1902, the United States wanted to build a canal across the isthmus, which was then part of Colombia. Colombia did not accept U.S. offers for the land. With backing from the United States, Panama declared its independence from Colombia. The U.S. military discouraged any response by Colombia. Panama then gave the United States a very favorable deal on the land, and construction began shortly after. But most people in Panama were never happy about the canal. Protests were common throughout the 20th century. In 1999, the United States finally turned over control of the canal to Panama.

COMPARING

Climates

Note that in general, Central America has two seasons: a rainy season, from about May through November, and a dry season, which begins in November and lasts into April.

- In which area that is pictured would you expect the most rainfall? *(the rain forest)*

ANSWERS

1. **Desert** They are able to store water.
2. **Rain Forest** hot and rainy
3. **Caribbean Islands** The warm water warms the air that blows across the islands, keeping temperatures warm-year round.

Teach

Climate and Vegetation

Think, Pair, Share

- What causes the variation in temperatures in the Mexican and Central American mountains and highlands? *(differences in elevation; higher elevations have cooler temperatures)*
- How does wind affect the climate of the Caribbean Islands? *(Warm winds keep temperatures warm and constant year-round; wind brings greater rain to areas that are not protected from the wind.)*
- **Categorize** Which areas lie south of the Tropic of Cancer? *(southern areas of Mexico, all of Central America, and the Caribbean Islands)*

More About . . .

Cloud Forests in Central America

One type of rain forest in Central America is called a "cloud forest." These forests, located on the sides of mountains at higher elevations, are often immersed in clouds. More moisture is transmitted directly to the trees and plants from the cloud and light mist than is transmitted through rain. Cloud forests are cooler than rain forests at lower elevations, and they support many mosses and ferns.

COMPARING Climates

Desert Barrel cacti grow in Mexico's deserts. **1. How are these plants able to thrive in this dry climate?**

Rain Forest La Fortuna Waterfall is located in Costa Rica. **2. What kind of climate would you find here?**

Caribbean Islands This volcanic peak is in St. Lucia. **3. How does the water affect climate in the islands?**

Climate and Vegetation

KEY QUESTION What is a major influence on the climate of the Caribbean Islands?

Latin America stretches across more than one-half of the Western Hemisphere. It has a wide variety of climates and vegetation.

Mexico About half of Mexico lies south of the tropic of Cancer. The temperatures in this part of Mexico are generally warm and constant year-round, with abundant rainfall. Tropical rain forests thrive here.

A greater variation in temperatures exists in places located north of the tropic of Cancer. The driest areas are in the deserts of northwestern Mexico, as well as in the northern part of the Mexican Plateau. Desert shrubs grow in this climate. In Mexico's mountains and highlands, average temperatures vary with increases in elevation. The vegetation here consists mostly of coniferous and deciduous forests.

Central America All of Central America is located in the tropical zone, so temperatures are generally warm year-round. The plains along the Caribbean coast receive the greatest amount of precipitation. The climate is perfect for the lush rain forests that grow there.

As in Mexico, places at higher elevations in Central America experience cooler average temperatures than those at lower elevations. Winds that blow inland from the warm Caribbean Sea bring heavy

DIFFERENTIATING INSTRUCTION: TIERED ACTIVITIES

OBJECTIVE Demonstrate an understanding of the climate and vegetation of Middle America.

Basic

Have students work in small groups to prepare a chart that compares the climate and vegetation of Mexico, Central America, and the Caribbean Islands.

On Level

Have students work in small groups to prepare a poster that illustrates the climate and vegetation of Mexico, Central America, and the Caribbean Islands. Direct students to add captions to accompany their illustrations. Display completed posters in the class.

Challenge

Have students work in small groups to prepare a crossword puzzle using information in the text about the climate and vegetation of Mexico, Central America, and the Caribbean Islands. Have groups exchange and complete their puzzles.

rains on the eastern side of the mountains. Places on the westward side and in the interior highlands receive the least amount of precipitation. Forests of oak and pine grow at higher elevations, while desert shrubs grow in the drier areas on the Pacific side.

The Caribbean Islands Winds play a big role in the climate of the Caribbean Islands. The waters in the Caribbean Sea stay warm most of the year and heat the air over them. Warm winds blow across the islands, keeping the temperatures warm and constant year-round. The Lesser Antilles are divided into islands that face the wind and those that are protected from the wind. Those protected from the wind receive less rain than those facing the wind.

As in Mexico and Central America, elevation influences climate in the islands' mountains. The islands' climates allow diverse vegetation, including rain forests, deciduous forests, and desert shrubs, to grow there.

SUMMARIZE Describe the major influence on the climate of the Caribbean Islands. the winds that blow from the Caribbean Sea across the islands

Tree Frog A red-eyed tree frog, a unique animal found in Costa Rica's rain forest, hops a ride on a caiman, an alligator-like reptile.

Section 1 Assessment

ONLINE QUIZ For test practice, go to **Interactive Review @ClassZone.com**

TERMS & NAMES

1. **Explain the importance of**
   - isthmus
   - archipelago
   - Greater Antilles
   - Lesser Antilles

USE YOUR READING NOTES

2. **Categorize** Use your completed chart to answer the following question:
   How does climate influence the location of rain forests in Central America?

LANDFORMS	INFLUENCES ON CLIMATE	VEGETATION
1. 2. 3.	1. 2. 3.	1. 2. 3.

KEY IDEAS

3. How did Latin America get its name?
4. Where do most people in Mexico live?
5. What affects climate in the mountains and highlands of Mexico and Central America?

CRITICAL THINKING

6. **Analyze Causes and Effects** How did the Panama Canal help trade in Latin America?
7. **Make Generalizations** What generalization can you make about climate in Mexico, Central America, and the Caribbean Islands?
8. **CONNECT to Today** In 1999, the United States gave control of the Panama Canal to Panama. How do you think the country of Panama will benefit from the canal?
9. **WRITING** **Create a Poster** Choose a country in Central America or one of the Caribbean Islands and prepare a captioned poster for a geography fair. Include information about the country's land, climate, and vegetation.

CONNECT to the Essential Question

How have Latin America's geography and resources helped shape its history?

Ask students what they have learned so far that can help them answer this question. Students might mention:

- Most of Mexico's people live on the plateau between the mountain ranges on the east and west.
- Earthquakes and volcanoes affect all of Mexico and Central America.
- Most of Middle America experiences a warm climate.

4 Assess & Reteach

Assess Have students complete the Section Assessment.

Unit Resource Book: Latin America
- Section Quiz, p. 49

Interactive Review @ClassZone.com
- Section Quiz

Power Presentations

Test Generator

Reteach Have pairs of students create one multiple-choice question about each of the two main headings, "Mountains and Islands" and "Climate and Vegetation." Then have pairs exchange questions and answer them. Have students use their textbooks to correct any incorrect answers.

Unit Resource Book: Latin America
- Reteaching Activity, p. 53

SECTION 1 ASSESSMENT ANSWERS

Terms & Names

1. isthmus: strip of land that connects two landmasses; archipelago: a chain of islands; Greater Antilles: Cuba, Jamaica, Hispaniola, and Puerto Rico; Lesser Antilles: the smaller Caribbean islands southeast of the Greater Antilles

Use Your Reading Notes

2. Rain forests in Central America are located in areas that have a tropical climate, where temperatures are warm year-round and rainfall is plentiful.

LANDFORMS	INFLUENCES ON CLIMATE	VEGETATION
1. mountains 2. plateaus 3. plains	1. location in tropical zone 2. elevation 3. winds	1. rain forests 2. deciduous forests 3. desert shrubs

Key Ideas

3. The European languages spoken by the colonizers were derived from Latin.
4. Most people live on the Mexican Plateau.
5. the elevation

Critical Thinking

6. Ships were able to make more trips in less time, thereby allowing for more trade and greater profits.
7. Possible answer: The climate is affected by the location in the tropical zone, elevation, and the winds.
8. Possible answer: Panama's economy would profit from the money charged for ships to pass through.
9. Use the rubric to score posters.

Poster Rubric

	Quality of Research	Presentation of Information
4	detailed and thorough	many and Informative
3	several good details	several
2	some accurate details	some
1	few details or inaccurate details	unclear

1 Plan & Prepare

Objectives

- Describe the main features of South America's geography
- Identify the climates and vegetation of South America

Read for the Essential Question

Help students read for a purpose by reminding them of the Essential Question: "How have Latin America's geography and resources helped shape its history?" Point out that they can use text features such as Key Questions, special features, images, and maps to form answers to the question.

Vocabulary

Best Practices Toolkit

Use the ***Best Practices Toolkit*** to model strategies for vocabulary notetaking. Vary strategies throughout the year. Choose from: *Knowledge Rating, Predicting ABCs, Definition Mapping, Word Sort, Word Wheel, Frayer Model (Word Squares), Magnet Words,* and *Student VOC.*

Vocabulary Strategies, TT9–TT16

Reading Strategy

Best Practices Toolkit

Display the Web Diagram transparency.

- Remind students that summarizing involves identifying main ideas and explaining them in short sentences or phrases.
- Model filling in summary phrases or sentences in the "Landforms, Rivers, and Lakes" summary area.
- Have students suggest summaries for the "Climate and Vegetation" summary area.
- Display the completed diagram for review.

Web Diagram, TT28

SECTION 2 Reading for Understanding

Key Ideas

BEFORE, YOU LEARNED

The physical geography of much of Mexico, Central America, and the Caribbean Islands consists of mountains and highlands.

NOW YOU WILL LEARN

The physical geography of South America consists of a wide variety of landforms and climates.

Vocabulary

TERMS & NAMES

Andes Mountains a mountain range located on South America's west coast and extending the full length of the continent

altiplano a high plateau

llanos (YAH•nohs) grasslands of western Venezuela and northeastern Colombia

Pampas grassy plains in south-central South America

Amazon River South America's longest river (about 4,000 miles, or 6,400 kilometers) and the second-longest river in the world

REVIEW

highlands mountainous or hilly sections of a country

rain forest a broadleaf tree region in a tropical climate

Visual Vocabulary *altiplano*

Reading Strategy

Re-create the diagram shown at right. As you read and respond to the **KEY QUESTIONS**, use the diagram to summarize each of the main parts of Section 2.

See Skillbuilder Handbook, page R5

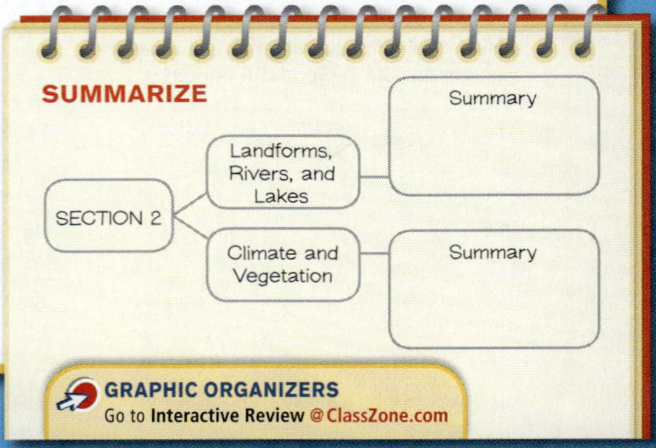

PRETEACHING VOCABULARY

English Learners

Pronounce and Preview

Pronounce each term for students. Review the meanings of words within definitions, such as *extending, full length,* and *broadleaf tree.*

- To modify vocabulary learning, have students complete worksheets as they read, instead of afterward.

Unit Resource Book: Latin America
- Vocabulary Practice, p. 35

Inclusion

Vocabulary: Word Wheel

Have students work in pairs to fill out the graphic organizer for the terms in the section.

- Use the transparency to model the strategy.
- Have volunteers suggest entries.
- Then distribute copies of the organizer for students to complete with a partner.

Best Practices Toolkit
- Vocabulary: Word Wheel, TT13

Great Bay Beach on the Caribbean Island of Saint Maarten

SECTION 2 GEOGRAPHY

Waterfall in Rincon de la Vieja National Park, Costa Rica

Physical Geography of South America

Connecting to Your World

Look carefully at a map of South America, and you will see many physical features similar to those that define North America, particularly the United States. Just as the Rocky Mountains rise in western North America, the Andes Mountains stretch the entire length of western South America. Great river systems drain both continents, and large plains are central to both North and South America. Of course, as you will soon learn, there are important differences as well.

Landforms, Rivers, and Lakes

KEY QUESTION What are the main features of South America's landscape?

South America has nearly every type of physical feature, including mountains, grasslands, plains, and **highlands**. The continent has few large lakes but is drained by five major river systems, including the Amazon River, one of the largest rivers in the world.

Patagonia Deserts, plateaus, and highlands make up this area.

205

2 Focus & Motivate

3-Minute Warm-Up

Write on the board or display the transparency:

- How do the Andes compare to other mountain ranges in the world? *(They are the world's longest and second highest range.)*

Unit Transparency Book
- 3-Minute Warm-Ups, TT1

Connecting to Your World

More About . . .

The Rockies and the Andes

Both mountain ranges run north and south. The tallest peak in the Rockies is Mount Elbert, at 14,433 feet. In the Andes it is Mount Aconcagua, at 22,834 feet.

3 Teach

Landforms, Rivers, and Lakes

Talk About It

- Which countries are drained by the Amazon River? *(most of Brazil and Peru; parts of Colombia, Ecuador, Bolivia, and Venezuela)*
- **Summarize** Describe each of the main areas of the Central Plains. *(The llanos are wide grasslands; the Amazon rain forest is a tropical rain forest; Gran Chaco has subtropical grasslands and low forests; the Pampas are a large grassy plain.)*

SECTION 2 PROGRAM RESOURCES

Unit Resource Book: Latin America

ON LEVEL

Unit Resource Book
- Reading Study Guide, p. 5
- Vocabulary Practice, p. 35
- Outline Map Activities, p. 39
- Connect to Science, p. 50

STRUGGLING READERS

Unit Resource Book
- Reading Study Guide with Additional Support, p. 13
- Skillbuilder Practice: Finding Main Ideas, p. 36
- Section Quiz, p. 50
- Reteaching Activity, p. 54

ENGLISH LEARNERS

Pupil Edition in Spanish

Pupil Edition in Audio DVD-ROM

Unit Resource Book
- RSG (Spanish), p. 21
- RSG with Additional Support (Spanish), p. 29

Multi-Language Glossary

Test Generator
- Section Quiz in Spanish

INCLUSION

Unit Resource Book
- RSG with Additional Support, p. 13
- Section Quiz, p. 50
- Reteaching Activity, p. 54

GIFTED & TALENTED

Unit Resource Book
- Section Quiz, p. 50

PRE-AP

Unit Resource Book
- Primary Source: Preserving the Rain Forest, p. 48
- Section Quiz, p. 50

TECHNOLOGY

Unit Transparency Book
- 3-Minute Warm-Ups, TT1
- Essential Question, TT2
- Critical Thinking, TT4
- Cultures: Machu Picchu, TT9
- Map: Latin America: Physical, TT7

Daily Test Practice Transparencies
- Section 2, TT23

Power Presentations

ClassZone.com

World Cultures and Geography Video on DVD

More About . . .

Altiplano

In Bolivia and Peru, the Andes separates into two ranges, the Cordillera Occidental and Cordillera Oriental. Between them lies the altiplano. This high plateau was formed millions of years ago, when sediments from the mountains collected in the valley between the mountain ranges. The altiplano includes land for llama grazing, mineral deposits, and salt flats. The salt flats, which are the dried remnants of ancient lakes, are perfectly flat and stretch for miles.

Fun Facts!

Angel Falls

Jimmy Angel was searching for gold when he saw Angel Falls. The falls drop off the edge of one of Venezuela's mesas, or flat-topped mountains. In 1937, Angel returned to the area with his wife and two passengers, to try to land on top of the mountain. He was able to land, but the plane nose-dived into the marshes. It took his group 11 days to hike back down the steep mountain. The plane remained on top of the mountain until 1970, when it was removed by helicopter and put on display at the airport of Ciudad Bolívar in Venezuela.

Make It Fun

Teacher-Tested Activities

Greta Frensley, Knox Doss Middle School, Gallatin, Tennessee

My classes play "Where in the World" to practice using latitude and longitude to find the location of places in the region they are studying. Choose several cities or other places, such as Kingston, Jamaica, or the Yucatán Peninsula and list their coordinate on the board. Have students use atlases and use the coordinates to locate the places. The first student to find it should write the name of the place on the board. When all places have been located, provide students with an outline map of the region. Have them label the lines of latitude and longitude and the places they have located on their outline maps.

The Andes As the map on the opposite page shows, the **Andes Mountains** (A) stretch along western South America for a distance of about 5,500 miles. Longer than any other mountain range above sea level, the Andes are the world's second-highest range. Only the Himalayas in Asia are higher. The tallest Andean peak, Mount Aconcagua, is the Western Hemisphere's tallest mountain. People living in the Andes grow crops on terraces cut into the mountains.

The Andes region also includes valleys and the ***altiplano***, or "high plateau," shared by Peru and Bolivia. Made up of a series of basins between mountains, the *altiplano* is one of the world's highest inhabited regions. Its southern half contains important deposits of copper, silver, tungsten, and tin.

Plains and Highlands The Central Plains extend eastward from the Andes and cover about three-fifths of South America. The *llanos,* Gran Chaco, Amazon rain forest, and Pampas make up the Central Plains. The ***llanos*** (YAH•nohs) are wide grasslands that stretch across northeastern Colombia and western Venezuela. There, cattle ranches and small farms are common.

A tropical **rain forest** in the Amazon Basin (B) covers about 40 percent of Brazil. The Gran Chaco is a largely uninhabited area consisting of subtropical grasslands and low forests. Nearly half of this region is located in Argentina. The remainder extends into Paraguay and Bolivia. Just south of the Gran Chaco lies a large grassy plain called the **Pampas**, Argentina's most populated area.

The Eastern Highlands are two separate regions. The Guiana Highlands (C) to the north of the Amazon Basin consist of tropical forests and grasslands. The Brazilian Highlands (D) to the south cover about one-fourth of South America and consist mostly of hills and plateaus.

Fun Facts!

ANGEL FALLS

Angel Falls in the Guiana Highlands of southeastern Venezuela, the world's tallest waterfall, drops 3,212 feet. The waterfall was named for Jimmy Angel, an American pilot who spotted it from his airplane in 1935.

INTERDISCIPLINARY ACTIVITIES

CONNECT to Math

Graph Mountain Heights

Have students construct a graph or illustration (similar to the one for the Fun Facts graphic on this page), which compares the tallest mountains on various continents. They can use the following data: Africa—Kilimanjaro, 19,340 feet; South America—Aconcagua, 22,834; North America—McKinley, 20,320; Asia—Everest, 29,035; Europe—Elbrus 18,510.

CONNECT to Language Arts

Journal Entry on First Seeing Angel Falls

Ask students to take on the role of Jimmy Angel the first time he spotted Angel Falls from his plane and write a journal entry describing the experience. Students may wish to include information on how remote it is, the difficulty of finding it again, the mesas, or flat-topped mountains in the area, and the sight of water seeming to "fall from the clouds." Ask volunteers to read their entries to the class.

Physical Geography of South America

CONNECT Geography & History

READING A PHYSICAL MAP

Use the key in this physical map to identify the elevations of Patagonia and the Amazon Basin.

1. **Place** What two oceans border South America?
2. **Region** Where are large areas of highlands found in South America?

207

DIFFERENTIATING INSTRUCTION

Struggling Readers

Identifying South America's Physical Features

Provide students with a blank outline map of South America. Have them cover the map on this page. Using only the information from their textbooks, direct students to work in pairs to label South America's main features as described in their textbooks. Then have students compare their maps to the map of South America on this page and make any necessary corrections.

Gifted & Talented

Write a Poem About Tierra del Fuego

Have students locate Tierra del Fuego on the map on this page. Provide students with an image of this area and/or historical background. (In 1520, the explorer Magellan spotted the native people carrying fire from place to place for warmth and named the area *Tierra del Fuego*, or "Land of Fire".) Then have students visualize what it might be like to live there, and create a poem about it. Invite students to present their poems.

Physical Geography of South America

CONNECT Geography & History

Reading a Physical Map Note that South America has many mountainous areas and highlands, as well as low areas. Review these low areas with students and ask them about the number of river valleys or basins they observe in these areas.

- Where would you expect to find the densest populations in relation to elevation? *(Possible Answer: along low-level areas, where there are river valleys and where rivers empty into the ocean)*

ANSWERS

1. **Place** the Atlantic and Pacific oceans
2. **Region** in the northern and eastern parts of South America

Unit Resource Book: Latin America
- Outline Map Activities: Latin America: Physical/Political, pp. 39–40

More About . . .

The Geography of Chile

Chile stretches north and south for a distance of over 2,600 miles, but its maximum width is only about 220 miles. Because of its length, it has three distinct climate zones. Even though it is extremely narrow, it has three geographic areas that cross its width: the high Andes to the east, lower coastal mountains to the west, and about a 25–40 mile wide valley in between, which is the country's most populated area and contains large amounts of valuable mineral deposits.

Teach

Climate and Vegetation

Talk About It

- What countries contain most of the grasslands in South America? *(Venezuela, Colombia, Argentina, Uruguay)*
- How is the rain forest climate of the Amazon affected by the humidity? *(The average temperature is only about 80 degrees, but the high humidity makes it seem much hotter.)*
- **Summarize** How do mountain climates and vegetation change as the elevation increases? *(Climates become colder as the elevation increases, and plant life consists of fewer but hardier species.)*

Unit Resource Book: Latin America
- Skillbuilder Practice: Finding Main Ideas, p. 36

More About . . .

The Amazon River

The Amazon is the second longest river in the world (after the Nile in Africa) and ranks first in the world in area drained, number of tributaries, and amount of water discharged.

The Amazon River may have gotten its name after a Spanish explorer fought a battle against Indians in the region in which women had fought alongside men. He thought the women resembled the Amazon women warriors in Greek myths. Others believe that the name came from the Indian word *amassona,* meaning "boat destroyer."

More About . . .

Duckweed in Lake Maracaibo

In 2004, a plant called duckweed began to cover increasing areas of Lake Maracaibo. No one was sure of its source or why it grew so quickly, sometimes doubling in size in just 48 hours. It doesn't harm fish, but tangles the propellers of small fishing boats. Some say that pollution of the lake—related to the large oil industry in the area—is one of the causes of the weeds.

Rivers and Lakes South America has five large river systems. The largest is the Amazon river system, which carries about one-fifth of the Earth's river water. The **Amazon River** flows from the Andes to the Atlantic Ocean and is about 4,000 miles long. The river and its tributaries drain much of Brazil and Peru, as well as parts of Colombia, Ecuador, Bolivia, and Venezuela. South America's other river systems are the Rio de la Plata, the Magdalena-Cauca, the Orinoco, and the São Francisco.

Lake Maracaibo A girl uses a pole to move her boat between houses on Lake Maracaibo. **How have people living on Lake Maracaibo adapted to their environment?**

They build their homes on the lake and use it for transportation.

South America has few large natural lakes. The largest, Lake Maracaibo in northwestern Venezuela, covers an area of more than 5,000 square miles. About two-thirds of Venezuela's total petroleum output comes from the Lake Maracaibo region. Lake Titicaca is the second largest lake in South America. Located in the altiplano at an elevation of 12,500 feet, Lake Titicaca is the world's highest navigable lake.

SUMMARIZE List the main geographic features of South America.

the Andes, the Central Plains, the Eastern Highlands, the Amazon river system and rain forest, Lake Maracaibo, and Lake Titicaca

Climate and Vegetation

KEY QUESTION What kinds of vegetation are found in the tropical and desert climates of South America?

South America has a wide range of climates—from the steamy rain forest of the Amazon Basin to the icy cold of the upper Andes. Because climate largely determines the vegetation of a region, South America has a great variety of plant life as well.

Mountain Climates The climate of the Andes Mountains changes with elevation. For example, near the equator, the climate at the lowest elevations is tropical. However, if you were to climb midway up the mountains, you would most likely find a more temperate, or mild, climate. At an elevation of more than 15,000 feet, you would encounter extremely low temperatures and icy winds. The only vegetation would be a sparse assortment of mosses, lichens, and dwarf shrubs. Climate zones at various elevations in the Andes are compared on the opposite page.

DIFFERENTIATING INSTRUCTION

English Learners

Key Academic Vocabulary

Note the word *output* in the third sentence of the second paragraph. Read the sentence out loud and ask students to suggest definitions. Then ask students to look up *output* in a dictionary. You may want to discuss other related words, such as *input* and *product*.

Pre-AP

Estimate the Size of Lake Titicaca

Have students locate Lake Titicaca on the map on the previous page. Using the map scale, have them estimate its size in square miles and kilometers *(about 3,200 square miles, or 8,300 sq. km)*. Remind students area is calculated by multiplying the length times the width.

COMPARING Mountain Climate Zones

Ⓐ Guanaco (a kind of llama) in the Chilean Andes

Ⓑ Potato fields in Peru

Ⓐ Tierra Helada (Frozen Land)
- Above 10,000 feet
- Livestock: llamas, sheep

Ⓑ Tierra Fría (Cold Land)
- 6,000–10,000 feet
- Crops: wheat, apples, potatoes, barley

Ⓒ Tierra Templada (Temperate Land)
- 3,000–6,000 feet
- Crops: corn, cotton, coffee, citrus fruits

Ⓓ Tierra Caliente (Hot Land)
- Sea level–3,000 feet
- Crops: bananas, cacao, sugar cane, rice

CRITICAL THINKING

Draw Conclusions What zones are most productive for growing crops?

Ⓒ Corn fields in Ecuador

Ⓓ Sugar cane fields in Colombia

209

COMPARING

Mountain Climate Zones

Have students examine the four climate zones. Point out that each zone corresponds to a certain elevation. Ask students why different crops are listed in different zones. *(The listed crops are ones that grow well in the particular zone.)*

CRITICAL THINKING ANSWER

Draw Conclusions zones B–D

Comparing Regions

Growing Coffee

Coffee tends to grow best at higher elevations, as in the Tierra Templada. Soil, sunlight, rain, and elevation all affect how the coffee grows and tastes. Much of the world's coffee comes from Central and South America, but countries such as Kenya, Ethiopia, India, and Indonesia are also major coffee producers.

DIFFERENTIATING INSTRUCTION

Struggling Readers

Review Translations of Climate Zones

Review the names of the climate zones with students. Then ask them for the translation of *tierra* (land) and the other words in each zone. *(helada—frozen; fria—cold; templada—temperate; caliente—hot)* Ask students what language they think the names of these climate zones are in and why. *(Spanish, because most people in South America speak Spanish)*

Gifted & Talented

Illustrate Climate Zones

Have students use the information in the feature on this page to create a model of mountain climate zones. Students should include the elevations for each of the four climate zones and the kinds of crops that are conducive to each zone. Encourage students to be creative in the way they design their model. Display completed models in the classroom.

More About . . .

Amazon Rain Forest

The Amazon rain forest is an important resource in the exchange of carbon dioxide and oxygen. In fact, experts estimate that the Amazon rain forest is responsible for about 20 percent of all the world's oxygen. Some also suggest that the Amazon rain forest may contain more than half of all the world's animal species, including over 500 mammals, 175 lizards, 300 reptiles, and one-third of the world's bird species—as well as about 30 million insect species.

Human use of the rain forest for economic reasons, however, put this key ecosystem at risk. As rain forest is cleared for cattle grazing and logging, some experts estimate that more than 130 species of plants and animals are being destroyed each day, totaling more than 50,000 each year.

Unit Resource Book: Latin America
- Primary and Secondary Sources: Preserving the Rain Forest, p. 48

More About . . .

Pampas Cats

The Pampas, sometimes called the "ocean of grass," is home to some unique animals. One—the pampas cat—is similar in size to a house cat. Pampas cats may have fur that's yellowish-white, brown, or silver gray. They live in the open grassland areas of the Pampas, although they sometimes take shelter in trees. They hunt at night for small mammals such as guinea pigs and ground birds. However, these wild cats face the loss of their habitat from development and activities such as cattle grazing.

Tropical Rain Forests Much of South America lies in the low latitudes, which are tropical. There, the Amazon rain forest covers more than 2 million square miles. Located primarily in Brazil, the rain forest is bounded to the north by the Guiana Highlands, to the south by Brazil's central plateau, to the west by the Andes, and to the east by the Atlantic Ocean. The rain forest climate is hot and wet. Although the average temperature remains a fairly constant 80°F, high humidity makes it seem hotter. Rain falls throughout the year, for a yearly average of between 50 to 175 inches.

The Amazon rain forest has the world's richest collection of life forms. In fact, many of its insects, birds, and plants have yet to be named. Wildlife includes jaguars, monkeys, and manatees. Plant life is abundant. The rain forest's many varieties of trees include rosewood, Brazil nut, rubber, mahogany, and cedar.

Grasslands Located in Venezuela and Colombia, the *llanos* cover an area of about 220,000 square miles. They extend north and west to the Andes and south to the Amazon Basin. The *llanos* have a warm climate, and their rainfall averages between 45 and 180 inches.

The Pampas cover an area of about 295,000 square miles in central Argentina, and they extend into Uruguay. The grasslands are good for raising cattle, and the rich soil produces a variety of crops, particularly soybeans and wheat. In addition to being an important agricultural area, the Pampas are also home to most of Argentina's cities.

The *Llanos* Cattle graze on the *llanos* in Venezuela. **Why is this area a good place to raise cattle?**

The grass provides food for the cattle.

DIFFERENTIATING INSTRUCTION: TIERED ACTIVITIES

OBJECTIVE Describe the major kinds of vegetation in South America.

Basic

Have pairs of students work together to create a web diagram that describes the vegetation of South America. Direct students to label the hub "Vegetation." Then have them extend ovals from the hub that identify and describe the three major kinds of vegetation.

On Level

Have pairs of students create a poster illustrating the vegetation in South America. Students may draw their own pictures or find pictures in magazines. Encourage students to add captions to their illustrations. Display completed posters.

Challenge

Have students choose a country in South America. Have them write an article to be included in a travel magazine that describes the vegetation in that country and gives reasons why travelers might be interested in visiting it.

Deserts Desert climates occur along much of Peru's coast, as well as in northern Chile's Atacama Desert and southern Argentina's Patagonian desert. Desert climates are dry, and temperatures may be hot or cold. The desert in southern Patagonia is an example of a cold desert, with temperatures averaging only about 44°F.

The Atacama Desert is warmer, with summer temperatures averaging in the mid-60s°F. Although an ocean current from the Antarctic mixes with warm air to produce clouds and fog in the Atacama Desert, there is little rain. In fact, in the center of the Atacama, no rainfall has ever been recorded. There, nothing grows. Yet in the desert's fog zones, cacti, ferns, and many other types of vegetation grow.

Jujuy Desert The Jujuy Desert is located in northwestern Argentina.

CATEGORIZE Identify the kinds of vegetation found in South America's tropical and desert climates. Tropical rain forests are found in the tropical climates and cacti and ferns are found in the desert climates.

Section 2 Assessment

ONLINE QUIZ For test practice, go to **Interactive Review @ClassZone.com**

TERMS & NAMES

1. **Explain the importance of**
   - *altiplano*
   - *llanos*
   - Pampas
   - Amazon River

USE YOUR READING NOTES

2. **Summarize** Use your completed chart to answer the following question:

   How is Lake Maracaibo important to South America?

SECTION 2 → Landforms, Rivers, and Lakes → Summary

SECTION 2 → Climate and Vegetation → Summary

KEY IDEAS

3. What are the main river systems of South America?
4. Where are the Guiana and Brazilian Highlands located?
5. What relationship exists between elevation and climates of the Andes?

CRITICAL THINKING

6. **Compare and Contrast** How are the four main regions of the Central Plains different from one another?
7. **Analyze Causes and Effects** Why is a variety of plant life able to grow in the Atacama Desert?
8. **CONNECT to Today** The Amazon rain forest is a popular destination for tourists today. What features do you think attract tourists to this area?
9. **ART** **Create a Mural** Draw a panoramic mural of the geographic features of South America. Include the continent's landforms and waterways.

CONNECT to the Essential Question

How have Latin America's geography and resources helped shape its history?

Ask students what they have learned so far that can help them answer this question. Students might mention:

- South America has a wide variety of geographic features, with many different resources and uses.
- Differing climates and vegetation support various crops and economic activities.

4 Assess & Reteach

Assess Have students complete the Section Assessment.

Unit Resource Book: Latin America
- Section Quiz, p. 50

Interactive Review @ClassZone.com
- Section Quiz

Power Presentations

Test Generator

Reteach Arrange the class into seven groups and assign each one of the subheads in the chapter (Andes, Plains and Highlands, and so on). Ask each group to prepare a written summary of their subhead section and present it to the class. Add or clarify ideas as needed.

Unit Resource Book: Latin America
- Reteaching Activity, p. 54

SECTION 2 ASSESSMENT ANSWERS

Terms & Names

1. *altiplano:* a high plateau; *llanos:* grasslands in parts of Venezuela and Colombia; Pampas: grassy plains in south-central South America; Amazon River: South America's longest river

Use Your Reading Notes

2. It provides about two-thirds of Venezuela's petroleum.

SECTION 2 → Landforms, Rivers, and Lakes → Summary: Andes Mountains, highlands, plains; few lakes, Lake Maracaibo largest and source of petroleum, Lake Titicaca highest navigable lake; five major river systems, the largest Amazon River

SECTION 2 → Climate and Vegetation → Summary: Mountain climate in Andes Mountains; tropical in low latitudes, rain forests; warm climate, grasslands; desert climate, cacti, ferns

Key Ideas

3. Amazon, Rio de la Plata, Magdalena-Cauca, Orinoco, São Francisco
4. Guiana Highlands are north of the Amazon Basin and the Brazilian Highlands are south of it.
5. The climate is colder with higher elevations.

Critical Thinking

6. The *llanos* are wide grasslands that support ranching and farming. The Gran Chaco is an area of subtropical grasslands and low forests, and is largely uninhabited. The Amazon rain forest is hot and wet, with many species of plants and animals. The Pampas is a large plain that is heavily populated.
7. Fog provides moisture for plant growth.
8. Tourists come to see unusual plants and animals.
9. Use the rubric to score murals.

Mural Rubric

	Presentation of Mural	Content of Mural
4	clear and easy to understand	excellent information
3	neat, fairly easy to understand	good information
2	sometimes hard to follow	some information
1	unclear, sloppy	little information

1 Plan & Prepare

Objectives

- Describe the Olmec and Mayan civilizations and their significant achievements
- Describe the Aztec and Inca civilizations and their significant achievements

Read for the Essential Question

Help students read for a purpose by reminding them of the Essential Question: "How have Latin America's geography and resources helped shape its history?" Point out that they can use text features such as Key Questions, special features, images, and maps to form answers to the question.

Vocabulary

Best Practices Toolkit

Use the ***Best Practices Toolkit*** to model strategies for vocabulary notetaking. Vary strategies throughout the year. Choose from: *Knowledge Rating, Predicting ABCs, Definition Mapping, Word Sort, Word Wheel, Frayer Model (Word Squares), Magnet Words, and Student VOC.*

Vocabulary Strategies, TT9–TT16

Reading Strategy

Best Practices Toolkit

Display the Venn Diagram transparency.

- Remind students that comparing means finding similarities and that contrasting means finding differences.
- Model filling in details from the text that are unique to the Aztec Empire.
- Have students suggest entries for the Inca and Both areas.
- Display the completed diagram for review.

Venn Diagram, TT22

Section 3 Reading for Understanding

Key Ideas

BEFORE, YOU LEARNED

Latin America contains a wide variety of landforms, climates, and vegetation.

NOW YOU WILL LEARN

People adapted to these challenging geographic and climatic conditions and developed great civilizations.

Vocabulary

TERMS & NAMES

Olmec an early civilization along the Gulf Coast of what is now southern Mexico

Maya an early civilization located in what is now the Yucatán Peninsula, Guatemala, and northern Belize

cultural hearth the heartland, or place of origin, of a major culture

glyph a carved or engraved symbol that stands for a syllable or a word

empire a political system in which people or lands are controlled by one ruler

Aztec an early civilization in the Valley of Mexico

chinampas artificial islands used for farming

Inca an early civilization in the Andes Mountains of Peru

REVIEW

culture the shared attitudes, knowledge, and behaviors of a group

BACKGROUND VOCABULARY

jaguar a large cat mainly found in Central and South America

Visual Vocabulary glyph

Reading Strategy

Re-create the Venn diagram shown at right. As you read and respond to the **KEY QUESTIONS**, use the diagram to show how the Aztec and Inca empires were alike and different.

Skillbuilder Handbook, page R9

COMPARE AND CONTRAST

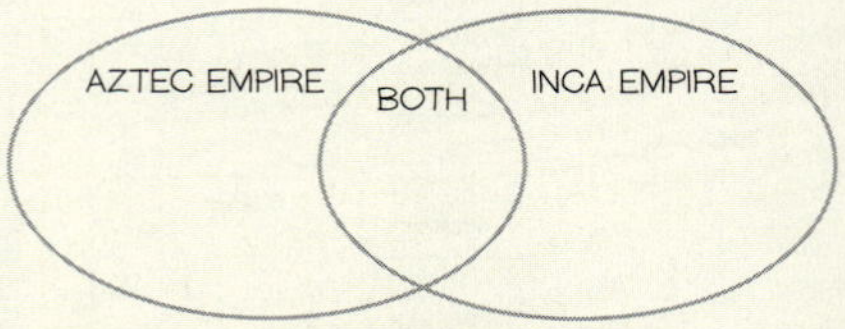

GRAPHIC ORGANIZERS
Go to **Interactive Review** @ClassZone.com

PRETEACHING VOCABULARY

English Learners

Pronounce and Preview

Pronounce each term for students. Review the meanings of words within definitions, such as *engraved*, *syllable*, and *artificial*.

- To modify vocabulary learning, have students complete worksheets as they read, instead of afterward.

Unit Resource Book: Latin America

- Vocabulary Practice, p. 35

Inclusion

Match Terms and Definitions

Have student pairs write the vocabulary words and their definitions on separate index cards. Then have them play a matching game. Ask one student to shuffle the cards and arrange them. After their partner has finished matching the terms and definitions, have the students work together to correct any errors. Students should then switch roles.

Great Bay Beach on the Caribbean Island of Saint Maarten

SECTION 3 HISTORY

Waterfall in Rincon de la Vieja National Park, Costa Rica

Ancient Civilizations

Connecting to Your World

How do you think the stone head pictured here was made? An ancient civilization in Latin America carved these heads about 3,000 years ago. What do you think of when you hear the word *pyramid*? Most likely, you think of the pyramids built about 4,500 years ago by the ancient Egyptians. Did you know that ancient civilizations in Latin America, such as the Maya, constructed these huge structures too? As you read this section, you will find out more about the Maya and other early civilizations that thrived in what is now Latin America.

Olmec Head The Olmec carved these large heads from a stone called basalt.

The Olmec and the Maya

KEY QUESTION How were the Olmec and Mayan civilizations alike?

The first known civilization to develop in Latin America were the **Olmec**. They lived along the Gulf Coast of what is now southern Mexico about 3,200 years ago. Another civilization, the **Maya**, developed in the highlands and flatlands of what is now the Yucatán Peninsula, Guatemala, and northern Belize. Archaeological evidence shows that these two cultures built well-laid-out cities and complex civilizations. They were farmers, artists, and architects.

The Tikal Pyramid The Maya built this pyramid in the rain forest of what is now Guatemala.

2 Focus & Motivate

3-Minute Warm-Up

Write on the board or display the transparency:

- Look at the map of the Olmec, Maya, Aztec, and Inca and at the photograph of the Inca road. What difficulties did the Inca face in building roads in their empire? *(Possible answer: They faced the difficulty of building roads through very rugged terrains.)*

Unit Transparency Book: Latin America
- 3-Minute Warm-Ups, TT1

Connecting to Your World

More About . . .

Tikal

The pyramid at left is one of five at Tikal, which is now part of Tikal National Park—and a UNESCO World Heritage site. The pyramids range in height, from 138 feet to 213 feet tall. Pyramid IV is thought to have been one of the tallest buildings in all of the ancient Western Hemisphere.

3 Teach

The Olmec and the Maya

Think, Pair, Share

- What were the main economic activities of the Olmec and Maya? *(farming and trade)*
- **Compare and Contrast** How were the end of the Olmec and Maya civilizations similar? *(Both abandoned region for unknown reasons.)*

SECTION 3 PROGRAM RESOURCES

Unit Resource Book: Latin America

ON LEVEL
Unit Resource Book
- Reading Study Guide, p. 7
- Section Quiz, p. 51

STRUGGLING READERS
Unit Resource Book
- Reading Study Guide with Additional Support, p. 15
- Skillbuilder Practice: Finding Main Ideas, p. 36
- Section Quiz, p. 51
- Reteaching Activity, p. 55

ENGLISH LEARNERS
Pupil Edition in Spanish
eEdition with Audio DVD-ROM
Unit Resource Book
- RSG (Spanish), p. 23
- RSG with Additional Support (Spanish), p. 31

Multi-Language Glossary
Test Generator
- Section Quiz in Spanish

INCLUSION
Unit Resource Book
- RSG with Additional Support, p. 15
- Section Quiz, p. 51
- Reteaching Activity, p. 55

GIFTED & TALENTED
Unit Resource Book
- History Makers: pp 41–42
- Section Quiz, p. 51

PRE-AP
Unit Resource Book
- Interdisciplinary Projects, pp. 43, 45
- Section Quiz, p. 51

TECHNOLOGY
Unit Transparency Book
- 3-Minute Warm-Ups, TT1
- Essential Question, TT2
- Critical Thinking, TT5
- Cultures, TT9

Daily Test Practice Transparencies
- Section 3, TT24

Power Presentations
ClassZone.com
World Cultures and Geography Video on DVD

CONNECT History to Today

A Recent Maya Discovery

The image of the woman was carved into a monument along with an inscription. However, the city was attacked by invaders, who hacked off the inscription. The monument was then buried near the city's temples. The work of archaeologists in Guatemala and elsewhere is endangered by looters who come to steal the valuable ancient artifacts. The archaeologists first noticed the monument after looters had dug a trench and exposed part of it.

Unit Resource Book: Latin America
- Skillbuilder Practice: Finding Main Ideas, p. 3

More About . . .

Olmec Stone Sculptures of Heads

There have been no discoveries that explain the purpose or origin of the stone heads of the Olmec. Some experts think that the heads are carvings of Olmec rulers. The Olmec heads were carved from single pieces of stone quarried 50 or more miles away—some weighing more than 20 tons—prompting speculation as to how the large stones were moved to their final sites.

More About . . .

Mayan Calendar and Chichén Itzá

An important Mayan ruin is the pyramid of Chichén Itzá, designed with 91 steps on each of its four sides. These steps correspond to days, and when combined with the top platform total 365 days in a year. The pyramid was placed so that the shadows cast at the equinoxes formed symbolic figures for the Mayans.

Unit Resource Book: Latin America
- Interdisciplinary Projects: Maya Numbers, pp. 43, 45

The Olmec, a Cultural Hearth The Olmec were traders and skilled farmers. As farming began to thrive, the Olmec could count on a steady food supply. Having enough food led to a larger population and allowed the Olmec to focus on tasks other than farming. An adequate food supply also led to the growth of cities.

Olmec cities included plazas, housing areas, and ceremonial centers. Their oldest known city is San Lorenzo. The Olmec are known for their huge stone sculptures of heads. Some sculptures show a half-human, half-animal **jaguar**, the Olmec's chief god.

The Olmec began to abandon their cities beginning around 600 B.C. for unknown reasons. However, historians consider the Olmec civilization a **cultural hearth**, the place of origin of a major **culture**. The Olmec culture shaped other cultures in the region, particularly the Maya.

CONNECT History to Today

A Recent Maya Discovery

In 2005, archaeologists working in Guatemala at the site of an ancient Mayan city made an exciting discovery—a stone monument with the carving of a woman's face. The carving dates back to around the A.D. 500s, making it the earliest known likeness of a woman that the Maya carved in stone. This is significant because it suggests that women had important leadership roles in Mayan society early on in Mayan history.

An archaeology student wraps the Mayan stone monument to take for further study.

Stone Carving
The drawing (right) helps to more clearly see what is carved in part of the stone monument.

The Maya, Masters of the City The Maya lived in villages in southern Mexico and northern Central America as early as 1500 B.C. At the height of their civilization, around A.D. 250, the Maya built impressive cities with stone temples, pyramids, plazas, palaces, and ball courts. They were farmers and traders. Corn, beans, and squash were important crops. The Maya also traded salt, chocolate, and cotton with other cultures.

The Maya were advanced in their knowledge of science and technology. They created a 365-day calendar by watching the stars. The calendar identified events throughout the year, such as planting times and holidays. The Maya used a mathematical system based on the number 20 and were the first people to use the zero. They also developed **glyph** writing, carved symbols that stood for a syllable or a word.

In about A.D. 900, the Maya started abandoning their cities, like the Olmec had done earlier. The reasons remain unclear. However, descendants of the Maya still live in the region today.

COMPARE AND CONTRAST Explain how the Olmec and Mayan civilizations were alike and different.

Both were traders and farmers; built plazas and temples; abandoned their cities. The Maya also built pyramids and made advances in science and mathematics.

214 Chapter 7

DIFFERENTIATING INSTRUCTION

Struggling Readers

Chart Details About the Mayan Civilization

Review with students the information about Mayan civilization on this page and discuss any questions they may have. Then help them to categorize details by completing a graphic organizer, such as the one at right, either as a large group on the board or individually on paper.

Mayan Civilization		
Economic Activities	Achievements	End of Civilization
• Farming corn, beans, squash • Trading	• Calendar • Math system	• Left cities A.D. 900 • Reasons unknown

Animated GEOGRAPHY

Olmec, Maya, Aztec, and Inca

Click here to explore these ancient civilizations @ ClassZone.com

CONNECT Geography & History

READING A THEMATIC MAP

Thematic maps focus on specific types of information. When reading this thematic map, use

- the title to find out what civilizations are featured in the map
- the key to identify what civilization the colors and symbols stand for

Place What civilization occupied the Yucatán Peninsula?

DIFFERENTIATING INSTRUCTION

English Learners

Key Academic Vocabulary

Note the word *thematic* in the skill title and ask students to try to identify its root word *(theme)*. Read the title out loud and ask students to suggest definitions for *theme*. Then ask students to look up the root word in a dictionary. You may want to discuss other common uses of the word such as *theme song* and *theme park*.

Gifted & Talented

Describe Inca Challenges

Have students locate the Inca civilization and its capital, Cuzco, on the map. Note that Cuzco was the center of government for the empire. Prompt students to describe the challenges faced by the Inca because of the size and geography of their land, and to explain how they might have adapted to them. Invite students to write a short essay, create a graphic organizer, or use an illustrated diagram to respond to the prompt.

Animated GEOGRAPHY

Olmec, Maya, Aztec, and Inca

CLASSZONE.COM

CONNECT Geography & History

Reading a Thematic Map Note that although there is a slight overlap between the territories of the Olmec, Maya, and Aztec, these civilizations did not exist during the same periods, so there were no conflicts among them over land.

- Considering the size of the Inca empire, the Spanish conquered it in a fairly short time. How might Inca roads have played a role in this conquest? *(Possible Answers: The roads enabled the Spanish to move their armies quickly and easily.)*

ANSWER

Place Maya

Presentation Options

Use the **Power Presentations** DVD-ROM to project the animated map.

- Click on each civilization in the key.
- Listen to the audio and examine the images to learn more about the Olmec, Maya, Aztec, and Inca.

More About . . .

Pyramid of the Sun

An early civilization existed before the Aztec settled in Tenochtitlán in the 1300s. These people built a great city northeast of Mexico City around 200 B.C. The city included the Pyramid of the Sun.

By about A.D. 600, the city was one of the largest in the ancient world. Around 750, the city was attacked, never to recover. The original name of the city has long been lost. Centuries after its fall, the Aztec named it "City of the Gods," or Teotihuacán, which is the name it is known as today.

More About . . .

Inca Roads

Some of the roads built by the Incas are intact today and still used as trails. In some cases, they are paved with stone and include steps to help the traveler get up and down the great heights of the Andes.

Teach

The Aztec and the Inca

Reader, Recorder, Reporter

- What actions made the Aztec a strong empire? *(warfare and the collection of tribute from conquered people)*
- What was the chief geographic challenge for the Inca? *(They lived in the harsh environment of the Andes.)*
- **Compare and Contrast** How were the capital cities of the Aztec and Inca similar? *(Both were the center of the civilization, complex cities, and built in difficult circumstances.)*

Unit Resource Book: Latin America
- History Makers: Montezuma II, p. 41
- History Makers: Pachacuti, p. 42

RESEARCH & WRITING CENTER

ANALYZING Primary Sources

The Aztec legend of the eagle on the prickly pear cactus is so powerful that is used on the flag of Mexico. The image of the eagle on the prickly pear is also seen in many types of ancient and contemporary art.

- How did the Aztecs know the eagle was a god? *(He bowed to them and spoke to them to confirm the legend.)*

DOCUMENT-BASED QUESTION ANSWER

The legend said that when they came to a cactus with an eagle sitting on top of it, they should build their city there.

The Aztec and the Inca

KEY QUESTION How did the Aztec and the Inca use their environments to develop their empires?

Two great civilizations, the Aztec and the Inca, developed vast **empires**, political systems in which people or lands are controlled by one ruler. Both civilizations were very powerful in their regions until the Spanish conquered them in the 1500s.

The Aztec, a Military Culture The **Aztec** moved from northern Mexico into the Valley of Mexico and what is now Mexico City in about the A.D. 1300s. They took control of the valley in about 1428 and ruled their empire until 1521. The Aztec built a strong military empire through warfare and by collecting tribute—money, goods, or crops—from the people they defeated.

The Aztec built Tenochtitlán (teh•NOHCH•tee•TLAHN), their capital city, on an island in Lake Texcoco. The city included houses and causeways, or roads made of earth. At its height, the population reached around 300,000. A great pyramid dedicated to Huitzilopochtli, the sun and war god, stood in the center of the city. The Aztec used the watery environment around Tenochtitlán to create **chinampas**, or island gardens, to grow food and flowers.

ONLINE PRIMARY SOURCE To read more legends of the Aztec and Inca, go to **Research & Writing Center @ ClassZone.com**

ANALYZING Primary Sources

The Eagle on the Prickly Pear Why would the Aztec build a city in the middle of a swampy lake? According to legend, one of their gods predicted that they would see a prickly pear cactus with an eagle sitting on top of it. The Aztec wandered until they saw this sign. There they built Tenochtitlán, which is now Mexico City.

> [A]s they passed through the reeds, there in front of them was the prickly pear with the eagle perched on top, . . . his claws punching holes in his prey. When he saw the Mexicans in the distance, he bowed to them. . . . The . . . spirit said, 'Mexicans, this is the place.' And with that they all wept. 'We are favored,' they said. 'We are blessed. We have seen where our city will be. Now let us go rest.'
>
> Source: *The Eagle on the Prickly Pear,* retold by John Bierhorst

DOCUMENT-BASED QUESTION

How did the Aztec know where to stop to build their capital city?

DIFFERENTIATING INSTRUCTION: TIERED ACTIVITIES

OBJECTIVE Identify significant events in the ancient Latin American civilizations.

Basic

Have students make a time line that shows important dates for the four main ancient Latin American civilizations. Suggest they use different colors or symbols to distinguish among the civilizations. Each date should be accompanied with a one-sentence caption. Display completed time lines.

On Level

In addition to the Basic activity, have students list significant achievements of each civilization using short, bulleted text. Encourage students to be creative in the way they illustrate the achievements. Display completed time lines.

Challenge

In addition to the On Level activity, have students write a three-paragraph essay, comparing and contrasting the civilizations. Display completed time lines, and invite volunteers to share their essays with the class.

Inca, Mountain Empire The **Inca** lived in the Andes Mountains in what is now Peru. They controlled a large empire from the early 1400s until 1533. The empire centered around Cuzco (KOOZ•koh), the Inca's mountain capital city. By 1500, the empire extended about 2,500 miles along the west coast of South America. The Inca built complex cities such as Machu Picchu. To farm the steep land, they cut terraces into the mountainsides, where they grew corn and potatoes. They built aqueducts, or canals, to irrigate the land. On higher areas, the Inca grazed llamas.

The Inca adapted to difficult conditions. They built 14,000 miles of roads on which runners carried messages. To keep records, they used *quipus* (KEE•poos), counting tools of knotted cords. Today's descendants of the Inca, the Quechua (KEHCH•wuh), make up about 45 percent of Peru's population. They still use terraces and aqueducts, raise corn and potatoes, and graze llamas.

SYNTHESIZE Describe how the Inca changed their environment to develop an empire. They built terraces to farm, aqueducts to irrigate the land, and roads.

Machu Picchu Flowers grow on the slopes of Machu Picchu today.

Section 3 Assessment

ONLINE QUIZ For test practice, go to **Interactive Review** @ClassZone.com

TERMS & NAMES

1. **Explain the importance of**
   - Olmec
   - Maya
   - Aztec
   - Inca

USE YOUR READING NOTES

2. **Compare and Contrast** Use your completed Venn diagram to answer the following question:

   How were the ways the Aztec and Inca modified their environments similar?

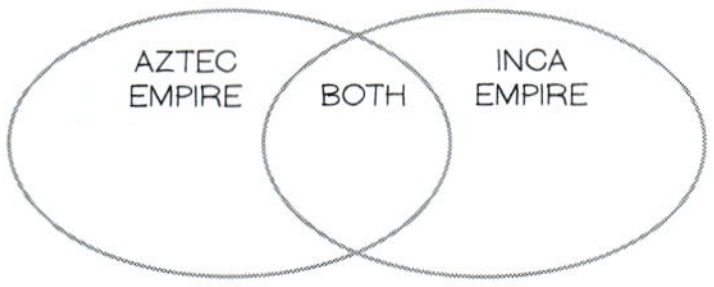

KEY IDEAS

3. What are the earliest known civilizations in what is now southern Mexico?
4. How did the Aztec build their large empire?
5. How were the Inca able to communicate throughout their large empire?

CRITICAL THINKING

6. **Compare and Contrast** How was the decline of the Mayan and Olmec civilizations different from the decline of the Aztec and Inca civilizations?
7. **Identify Problems and Solutions** What problems might an empire have in ruling millions of people?
8. **CONNECT to Today** Each year, around 300,000 people visit Machu Picchu. What problems do you think this presents for this archaeological site? Explain.
9. **WRITING** **Write a Newspaper Article** Imagine that you are a reporter visiting a city in one of the early Latin American civilizations. Write a short article that describes the city's architecture.

CONNECT to the Essential Question

How have Latin America's geography and resources helped shape its history?

Ask students what they have learned so far that can help them answer this question. Students might mention:

- The Olmec and Maya thrived by farming and trading with other cultures.
- The Aztec and Inca made the most of their available resources and geographic locations to build great empires.

4 Assess & Reteach

Assess Have students complete the Section Assessment.

Unit Resource Book: Latin America
- Section Quiz, p. 51

Interactive Review @ClassZone.com
- Section Quiz

Power Presentations

Test Generator

Reteach Have each student prepare one question and answer about the ancient civilizations of Latin America. They may create true or false, multiple choice, short answer, or essay questions. Ask volunteers to read their questions and others to answer them. Have students revise their questions and/or answers as necessary.

Unit Resource Book: Latin America
- Reteaching Activity, p. 55

SECTION 3 ASSESSMENT ANSWERS

Terms & Names

1. All are early civilizations. Location: Olmec: along the Gulf Coast of what is now southern Mexico; Maya: in what is now Yucatán, Guatemala, and northern Belize; Aztec: in the Valley of Mexico and present-day Mexico City; Inca: in the Andes Mountains of Peru

Use Your Reading Notes

2. They both modified the land in order to farm. **Diagram** *Aztec: in the Valley of Mexico; built through warfare; used* chinampas *to farm; Both: ruled large empire; planned capital city; conquered by the Spanish; Inca: in Andes Mountains; built roads; cut terraces into mountainsides; built aqueducts*

Key Ideas

3. Olmec and Maya
4. They built their empire through warfare and by collecting tribute.
5. They built an extensive system of roads and used runners to communicate messages.

Critical Thinking

6. The Olmec and Maya disappeared for unknown reasons. The Aztec and the Inca were defeated by the Spanish.
7. Possible answers: keeping in contact with all parts of the empire; having to fight wars to keep control; worrying about angry people wanting to defeat the empire; growing enough food to feed the people
8. People might destroy buildings, leave garbage, and be disrespectful of the ancient ruins. The high elevation might cause breathing problems.
9. Use the rubric to score newspaper articles.

Newspaper Article Rubric

	Facts About Architecture	Errors in Mechanics
4	very detailed	few or none
3	several details	some
2	few details	many
1	no details	major

1 Plan & Prepare

Objective: Explain how the Aztec built *chinampas*, how they lived on them, and how *chinampas* are used today.

Visual Vocabulary: Ask students to describe what they see in the illustration. Call out familiar items listed below, asking students to locate each one on the page. Then tell them the precise term, as listed in boldface. Add terms to a class vocabulary list, illustrated poster, or Word Wall.

- layer of earth under the topsoil **(subsoil)**
- Spanish name for stakes or tree trunks set in the subsoil **(huejotes)**
- Spanish name for corn **(maíz)**
- type of water plant **(reeds)**
- waterway between the *chinampas* **(canal)**

2 Teach

Engage . . .

DISPLAY the "*Chinampas*" feature from the **Power Presentations** DVD-ROM or the *Animations Center* at **ClassZone.com.** As the illustration loads on screen, have volunteers point to some of the vocabulary terms listed above. Then ask:

- How did the woven mats help build up the land?
- Why did flowers and certain crops grow so well on the *chinampas?*
- How did the Aztec live on and travel around the *chinampas?*
- How are the *chinampas* used today?

Encourage all answers.

Explore . . .

ROLL OVER sections of the illustration or click on items in the menu. Ask students to choose which areas they would like to explore further.

- Then return to the discussion questions. Have any answers changed?
- Have students jot down a list of at least three new questions that arise from watching the rollover features.

Animated GEOGRAPHY
Chinampas

Click here to enter the *chinampas* @ClassZone.com

ISLAND GARDENS

Much of the land around Tenochtitlán, where the Aztec settled, was swampy, posing a challenge for farming. The Aztec were resourceful at adapting to their environment. They built *chinampas,* human-made islands created for planting. The rich soil allowed the Aztec to grow crops and flowers.

Click here to see how the Aztec built the *chinampas.* Learn how the Aztec used woven mats and mud from the lake to build the chinampas.

Click here to see how the Aztec lived on the *chinampas.* Learn where they lived, what they wore, and what they ate.

Click here to see how *chinampas* are used today and the problems they face.

DIFFERENTIATING INSTRUCTION: TIERED ACTIVITIES

OBJECTIVE Explain how the Aztec built *chinampas*, how they lived on them, and how *chinampas* are used today.

Basic

Illustrate How the Aztec Lived and Farmed on *Chinampas*

Have students work in pairs to create a poster or other visual, illustrating how the Aztec lived and farmed on the *chinampas*. Posters or displays should have a title, along with at least three captions to explain key points. Ask volunteers to present their work, then create a display of all posters and other visuals.

GeoActivity

Make a Model Work with a partner to create a small model of a *chinampa*. Apply what you have learned about how the Aztec built *chinampas* to make a model using materials similar to those the Aztec used.

On Level

Diagram How *Chinampas* Were Built

Have pairs of students create a series of diagrams, illustrating step-by-step how *chinampas* were built. Each diagram should have a title, along with captions to explain key points. Ask volunteers to present their work, then create a display of all diagrams.

Challenge

Create a Magazine Article on *Chinampas* Today

Have pairs of students create a magazine article that explains how *chinampas* are used today and the challenges facing the people who live and work on them. Articles should include at least two visuals and one side bar. Ask volunteers to present their work, then create a display of all articles.

Explain . . .

CHOOSE topics to help students answer their questions, such as:

- **Building *Chinampas*:** Sometimes *chinampas* are called "floating gardens" but this is incorrect since they are anchored to the bottom of the lake. Trees also helped secure the islands, since their roots reached down through the soil and kept it in place.
- **Living on the *Chinampas*:** Fish were plentiful in the canals between the *chinampas*, and many birds were attracted to the flowering crops. The Aztec efficiently used the *chinampas* and relied on a plant nursery to ensure that only healthy plants were used.
- ***Chinampas* Today:** The water sources that originally kept the canals between the *chinampas* filled with water have been diverted to provide drinking water for the huge metropolis of Mexico City.

Elaborate . . .

ASSIGN students into small groups.

- Direct each group to write five true-false questions for the other groups to answer. Then have groups trade questions and answer them, using the animated features as needed. As students complete Tiered Activities, they may consult each other as experts.

Evaluate . . .

Make a Model

Activity Rubric

4 Excellent The student's model accurately illustrates a *chinampa* and includes several details.

3 Strong The student's model is fairly accurate and includes some detail.

2 Fair The student's model includes only a few details.

1 Poor The model lacks detail and fails to illustrate a *chinampa*.

1 Plan & Prepare

Objectives

- Summarize the history of Spanish rule in Latin America and its effects on the culture
- Describe the struggles for independence in Mexico and South America

Read for the Essential Question

Help students read for a purpose by reminding them of the Essential Question: "How have Latin America's geography and resources helped shape its history?" Point out that they can use text features such as Key Questions, special features, images, and maps to form answers to the question.

Vocabulary

Best Practices Toolkit

Use the *Best Practices Toolkit* to model strategies for vocabulary notetaking. Vary strategies throughout the year. Choose from: *Knowledge Rating, Predicting ABCs, Definition Mapping, Word Sort, Word Wheel, Frayer Model (Word Squares), Magnet Words, and Student VOC.*

Vocabulary Strategies, TT9–TT16

Reading Strategy

Best Practices Toolkit

Display the Time Line transparency.

- Explain to students that sequencing events, or arranging them in chronological order, will help them better understand the relationship among them, including cause-and-effect relationships.
- Model filling in events in the first few boxes on the time line.
- Have students suggest entries to complete the time line.
- Display the completed time line for review.

Time Line, TT33

Section 4 Reading for Understanding

Key Ideas

BEFORE, YOU LEARNED

Powerful civilizations arose in Latin America but eventually fell to conquerors.

NOW YOU WILL LEARN

Life under the conquerors was hard, as the people of Latin America struggled to gain their independence.

Vocabulary

TERMS & NAMES

Columbian Exchange the movement of plants and animals between Latin America and Europe after Columbus' voyage to the Americas in A.D. 1492

conquistador (kahn•KWIHS•tuh•DAWR) Spanish word for "conqueror"

colony overseas territory ruled by a nation

mestizo (mehs•TEE•zoh) person with mixed European and Indian ancestry

Father Hidalgo father of Mexican independence

Simón Bolívar leader for independence in northern South America

José de San Martín leader for independence in southern South America

REVIEW

empire a political system in which people or lands are controlled by one ruler

Visual Vocabulary *conquistador*

Reading Strategy

Re-create the time line shown at right. As you read and respond to the **KEY QUESTIONS**, use the time line to show the events that led to independence in Mexico and countries in South America.

See Skillbuilder Handbook, page R6

SEQUENCE EVENTS

A.D. 1521 — A.D. 1826

GRAPHIC ORGANIZERS
Go to Interactive Review @ClassZone.com

PRETEACHING VOCABULARY

English Learners

Pronounce and Preview

Pronounce each term for students. Review the meanings of words within definitions, such as *movement of plants and animals, conqueror,* and *ancestry*.

- To modify vocabulary learning, have students complete worksheets as they read, instead of afterward.

Unit Resource Book: Latin America

- Vocabulary Practice, p. 35

Inclusion

Locate Places on a World Map

Have volunteers take turns reading the vocabulary terms and names aloud. Instruct the rest of the class to follow along as they read, looking and listening for geographic places in the definitions. As geographic places are identified, have students take turns locating them on a world map.

SECTION 4 HISTORY

Great Bay Beach on the Caribbean Island of Saint Maarten

Waterfall in Rincon de la Vieja National Park, Costa Rica

From Colonization to Independence

Connecting to Your World

If you had lived in Europe before Christopher Columbus arrived in the Americas, you would have never enjoyed chocolate, corn, turkey, peppers, potatoes, or tomatoes. And if you had lived in the Americas, you would never have eaten oranges, bananas, beef, or pork. After the Spanish came to America, plants and animals were exchanged, or traded, between America and Europe. This came to be known as the **Columbian Exchange**.

Conquered Lands

KEY QUESTION How did Spanish rule affect life in Latin America?

After Columbus arrived in 1492, life for the Indians changed dramatically. In 1521, a Spanish ***conquistador*** (kahn•KWIHS•tuh•DAWR), or conqueror, Hernán Cortés, defeated the Aztec. And in 1533 Francisco Pizarro, another Spanish *conquistador*, defeated the Inca.

turnips and onions ▼

The Exchange
The graphic shows the goods and diseases transferred between the two hemispheres. **What vegetables did Europeans bring to the Americas?**

Columbian Exchange

221

2 Focus & Motivate

3-Minute Warm-Up

Write on the board or display the transparency:

- Look at the Columbian Exchange diagram. What do the arrows on the diagram represent? *(the direction of the exchange of products between North America and Europe)*

Unit Transparency Book: Latin America
- 3-Minute Warm-Ups, TT1

Connecting to Your World

More About . . .

The Columbian Exchange
Before Columbus, the Western Hemisphere was largely isolated from the rest of the world. As a result, its native peoples were especially vulnerable to diseases brought over by Europeans. In the first 100 years of contact with Europeans, disease killed more than 50 million native peoples.

3 Teach

Conquered Lands

Reader, Recorder, Reporter

- What was the basis of the Spanish class society? *(a person's ancestry and where he or she was born)* What system helped Spain grow wealthy? (encomienda, *in which Indians had to work for Spanish landlords)*
- **Sequence Events** When did the Aztec empire fall? *(1521)* The Inca empire? *(1532)*

SECTION 4 PROGRAM RESOURCES

Unit Resource Book: Latin America

ON LEVEL
Unit Resource Book
- Reading Study Guide, p. 9
- Section Quiz, p. 52

STRUGGLING READERS
Unit Resource Book
- Reading Study Guide with Additional Support, p. 17
- Skillbuilder Practice: Finding Main Ideas, p. 36
- Section Quiz, p. 52
- Reteaching Activity, p. 56

ENGLISH LEARNERS
Pupil Edition in Spanish
Pupil Edition Audio DVD-ROM
Unit Resource Book
- RSG (Spanish), p. 25
- RSG with Additional Support (Spanish), p. 33

Multi-Language Glossary
Test Generator
- Section Quiz in Spanish

INCLUSION
Unit Resource Book
- RSG with Additional Support, p. 17
- Section Quiz, p. 52
- Reteaching Activity, p. 56

GIFTED & TALENTED
Unit Resource Book
- Connect to Language Arts, p. 47
- Section Quiz, p. 52

PRE-AP
Unit Resource Book
- Economics in the World, p. 46
- Section Quiz, p. 52

TECHNOLOGY
Unit Transparency Book
- 3-Minute Warm-Ups, TT1
- Essential Question, TT2
- Critical Thinking, TT6
- Map: Latin American: Political, TT8
- Culture, TT10

Daily Test Practice Transparencies
- Section 4, TT25

Power Presentations
ClassZone.com
World Cultures and Geography Video on DVD

CONNECT to Language Arts

Tell students that another way that the Spanish helped control and manage the colonies was through the design of new towns or cities. King Philip II developed a set of rules in 1573 called the "Laws of the Indies," which specified how new settlements had to be laid out. Towns had to have a grid or checkerboard layout with a large central square and important government, church, and private buildings facing it. Santo Domingo, the capital of the Dominican Republic, is a very early example of this pattern.

- What other examples of Spanish influence might you expect to find in Latin America? *(Possible Answers: religion, clothing, food, arts and literature)*

Unit Resource Book: Latin America
- Connect to Language Arts: Make Vocabulary Cards, p. 47

More About . . .

Treasure from the Spanish Colonies

The Spanish established a system of sending two convoys of ships, called *flotas*, to the Latin American colonies each year. They would exchange European goods and return with treasure—gold and silver—from the New World. In some years, more than 100 ships traveled the route. The amount of treasure is not exactly known, but some ships might have carried 1,000 tons of silver.

The *flotas* sometimes suffered from storms, which sank multiple ships at a time, as well as from pirate attacks. One Spanish ship that sank carried 34 tons of silver bullion, 180,000 silver coins, and 125 gold bars and discs. Today, treasure hunters still search areas around Florida and the Caribbean for Spanish shipwrecks and their gold and silver treasures.

Unit Resource Book: Latin America
- Economics in the World: Silver Mining in Spanish Colonies, p. 47

Colonial Rule By the mid-1500s, the Spanish had set up **colonies**, or overseas territories ruled by a nation, in various parts of Latin America. Spain's goal was to take advantage of the resources and vast lands in its new **empire**.

To rule the empire, Spain set up a class society. At the top were the *peninsulares* (peh•neen•soo•LAHR•ehs), people born in Spain. They held the high government positions. Below them were the *criollos* (kree•OH•lohs), Spaniards born in Latin America. They were often wealthy, but could not hold high government offices. These two groups controlled land, wealth, and power in the colonies.

Below the *criollos* were the **mestizos** (mehs•TEE•zohs), people of Spanish and Indian ancestry. They had little power. Finally, African enslaved persons and Indians were at the bottom and had no power.

Colonial Economy One of Spain's main purposes in creating colonies was to make Spain wealthy. To do so, Spain set up a system known as *encomienda* (ehn•koh•mee•EHN•duh). Under this system, Indians mined, ranched, and farmed for Spanish landlords. The Indians lived in poverty and hardship, essentially enslaved.

Spain, however, grew wealthy. The Spanish established huge ranches to raise cattle and sheep and large plantations to grow sugar cane, coffee, and cacao in various parts of Latin America. Spain made huge profits from the gold and silver extracted from Mexican mines.

SYNTHESIZE Explain how life changed for people in Latin America under Spanish rule. After Spanish rule, society was divided into distinct social classes. Indians essentially became enslaved to the Spanish.

CONNECT to Language Arts

The time during which Spain ruled its colonies in Latin America is often referred to as the region's *colonial period*. If you visited Latin America today, you would see many examples of Spanish influence. In addition to hearing the Spanish language, you would see examples of Spanish architectural styles in buildings that the Spanish constructed throughout their colonies, such as the church and fort pictured below.

Activity

Make Vocabulary Cards

Materials
- index cards
- markers

1. Learn the new terms in this chapter by making vocabulary flip cards.

2. Write a term on one side of an index card.
3. Write the definition of the term on the other side.

4. If possible, draw a picture next to the definition to illustrate it.

DIFFERENTIATING INSTRUCTION

Struggling Readers

Chart Details About Spanish Class System

Help students review the information concerning the Spanish class system on this page and discuss any questions. Then guide them in completing a graphic organizer such as the one to the right, either on the board or individually, to help them categorize the information. (You may want to provide some examples to get students started.)

Spanish Class System		
Class	**Ancestry or Birth**	**Power or Wealth**
Peninsulare	*Spanish, born in Spain*	*High Government Positions*
Criollos		
Mestizos		
Africans and Indians		

Independence from the Spanish

KEY QUESTION What events led to independence in Mexico and South America?

In the 1800s, Spain began paying less attention to its Latin American colonies because it was fighting a war in Europe against Napoleon. Various groups in Latin America saw a chance to gain their freedom. The *criollos* and *mestizos* began to organize an independence movement. They called upon the Indians and enslaved persons to join their rebellion. By 1826, all of Latin America, except Cuba and Puerto Rico, became independent.

Mexico's Path to Independence In the late 1700s and early 1800s, freedom was in the air. The people of Mexico had heard about the American Revolution of 1776 and the French Revolution of 1789. The rights of all people were being talked about around the world. Many people in Mexico became excited by these ideas. They wanted their independence from Spain.

The first step toward independence happened in 1810 in a small village in north central Mexico. **Father Hidalgo**, a priest in the village, called on the people to rebel against Spain. The rebellion failed, however, and Father Hidalgo was captured and executed by the Spanish.

A new leader took Father Hidalgo's place. José María Morelos y Pavón also organized an army to fight the Spanish. He and other revolutionaries declared Mexico's independence in 1813. However, the Spanish defeated Morelos, and he was executed in 1815.

In 1821, the revolutionaries made a plan to win the support of all the groups in Mexico. The plan guaranteed independence, freedom of religion, and equality. Spain then declared Mexico independent with the Treaty of Cordoba in August 1821.

Mexico's struggle for independence lasted 11 years. But Mexico still had hard times ahead. You will learn more about Mexico's struggle toward democracy in the next chapter.

HISTORY MAKERS

Father Miguel Hidalgo
1753–1811

Father Hidalgo, a priest in the village of Dolores, sympathized with the Indians and *mestizos* and joined a secret society to work for Mexican independence. On September 16, 1810, Father Hidalgo rang the church bell in Dolores and urged his parishioners to fight for freedom. Thousands joined his army, but with clubs and farm tools as weapons, they were no match against the Spanish soldiers.

Today, Father Hidalgo is known as the father of Mexican independence. To honor him, Mexican Independence Day is celebrated on September 16.

ONLINE BIOGRAPHY
For more on the life of Father Hidalgo, go to the **Research & Writing Center @ClassZone.com**

Teach

Independence from the Spanish

Talk About It

- Why did the *criollos* and *mestizos* join the independence movement? *(They had little power or money under colonial rule and wanted their freedom.)*
- Who was Dom Pedro? *(He was the son of the Portuguese king, who declared Brazil independent and made himself emperor.)*
- **Sequence Events** How do you think the American and French Revolutions affected Simón Bolívar? *(Possible Answer: The success of these revolutions inspired him to fight for South American independence.)*

Unit Resource Book: Latin America
- Skillbuilder Practice: Finding Main Ideas, p. 36

RESEARCH & WRITING CENTER

HISTORY MAKERS

Father Miguel Hidalgo

Father Hidalgo was a *criollo*, born in Mexico (called New Spain at the time). When his group's plot was discovered, he called for a revolution with a battle cry, including the words, "*Viva Mexico!*"—translated as "Long live Mexico!" Today, the Mexican president rings Father Hidalgo's church bell (now in Mexico City) and repeats Hidalgo's cry to begin Mexican Independence Day.

CONNECT to the Essential Question

How have Latin America's geography and resources helped shape its history?

Ask students what they have learned so far that can help them answer this question. Students might mention:

- The resources of Latin America made the Spanish eager to conquer and rule the region.
- The unfairness of the Spanish class and economic systems spurred Latin Americans to fight for independence.

Unit Transparency Book: Latin America
- Essential Question, TT2

DIFFERENTIATING INSTRUCTION: TIERED ACTIVITIES

OBJECTIVE Explain why people in Latin America wanted independence from Spain.

Basic

Work with students to help them determine why many people in the Latin American colonies were eager for independence. List the reasons on the board. Then have students write a paragraph from the point of view of a Latin American colonist.

On Level

Have students work individually to list the reasons why many people in the Latin American colonies were eager for independence. Then have students select either Father Hidalgo or Simón Bolívar, and write a one-minute speech the person might have given in support of independence.

Challenge

Have students work in small groups to present a panel discussion regarding independence for Latin America from Spain. Students might present the views of Father Hidalgo or Simón Bolívar, as well as of people from the different classes of Latin American society.

CONNECT History & Economics

Coffee

Today, more than 100 countries export coffee. Developing countries earn more from coffee than any other export except oil. U.S. imports of coffee reached over $2 billion in 2004.

- How important do you think coffee is to Latin American economies? Why? *(Possible Answer: It's likely very important, since it produces two-thirds of all the coffee in the world.)*

4 Assess & Reteach

Assess Have students complete the Section Assessment.

Unit Resource Book: Latin America
- Section Quiz, p. 52

Interactive Review @ClassZone.com
- Section Quiz

Power Presentations

Test Generator

Reteach Have students work in pairs to write about two events discussed in this section, summarizing each one's importance to Latin America. Ask each pair to share one of their examples, then list the events in chronological order on the board or chart paper.

Unit Resource Book: Latin America
- Reteaching Activity, p. 56

Independence for South America Just as they did in Mexico, the American and French revolutions had inspired dreams of freedom among people throughout South America. Beginning in 1810, two leaders led the fight for independence from Spain.

Simón Bolívar, a Venezuelan general, led the fight in the northern part of South America. To honor his efforts, South Americans call him "the Liberator." **José de San Martín**, an Argentine general, led the fight for independence in the southern part of South America. By 1825, nearly all of South America was free from Spanish rule.

Brazil, a Portuguese colony, also gained independence at this time. When Brazilians demanded independence in 1822, Dom Pedro, the son of the Portuguese king, declared Brazil independent and made himself emperor. You will read more about Brazil in a later chapter.

CONNECT History & Economics

Coffee

Coffee was first grown in Africa. European colonists brought coffee trees, like those shown in the illustration below, to the Americas in the Columbian Exchange. Brazil and Colombia, along with the rest of Latin America, produce two-thirds of the world's coffee.

SUMMARIZE Discuss the events that led to independence in Mexico and South America.

People were upset with conditions under Spanish rule. Father Hidalgo urged Mexicans to fight for freedom. Bolívar and San Martín defeated the Spanish in South America.

Section 4 Assessment

ONLINE QUIZ For test practice, go to Interactive Review @ClassZone.com

TERMS & NAMES

1. **Explain the importance of**
   - Columbian Exchange
   - *conquistador*
   - colony
   - *mestizo*

USE YOUR READING NOTES

2. **Sequence Events** Use your time line to answer the following question:

   How long was much of Latin America under Spanish rule?

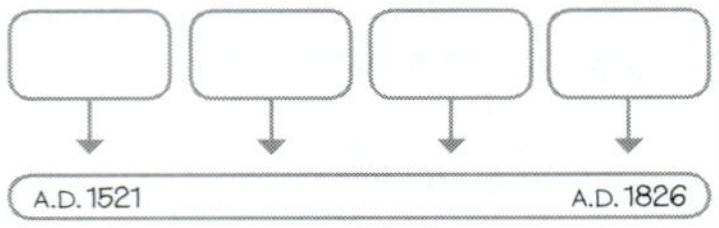

KEY IDEAS

3. Why were the *mestizos* important in Mexico's fight for independence?
4. Who was Father Hidalgo and why was he important?
5. How did most of South America gain its independence?

CRITICAL THINKING

6. **Analyze Causes and Effects** Why did Mexicans decide to fight for independence from Spain?
7. **Evaluate** What geographical challenges did South Americans have in their fight for independence?
8. **CONNECT to Today** What problems from the colonial days might still affect people in Latin America today?
9. **TECHNOLOGY** **Write a Biography** Use the Internet to find more information about one of the leaders for independence in Latin America discussed in this section. Then use a word processor to write a brief biography of that individual.

SECTION 4 ASSESSMENT ANSWERS

Terms & Names

1. Columbian Exchange: the movement of plants and animals between Latin America and Europe after Columbus's voyage; *conquistador*: Spanish word for "conqueror"; colony: overseas territory ruled by a nation; *mestizo*: person with mixed European and Indian ancestry

Use Your Reading Notes

2. The Spanish ruled for about 300 years. **Diagram** *Cortés defeats Aztec, A.D. 1521; Pizarro defeats Inca, 1532; Father Hidalgo calls for Mexico's independence, 1810; Mexico gains independence from Spain, 1821; Brazil gains independence from Portugal, 1822; Most of Latin America is independent, 1826.*

Key Ideas

3. They had little power. They were the main group to fight against Spain for Mexican independence.
4. He was a priest in a small town who gave the first speech to call for Mexican independence. He is the father of Mexican independence.
5. Two generals, Simón Bolívar and José de San Martín, led armies in the north and south to fight the Spanish forces.

Critical Thinking

6. They were tired of poverty and being under Spanish control. They also were inspired by independence movements in America and France.
7. The population was spread over a large area. The armies had to cross over the Andes Mountains.
8. Possible answers may include poverty and discrimination against Indians.
9. Use the rubric to score biographies.

Biography Rubric

	Use of Research	Content of Biography
4	many facts	excellent development
3	several facts	good development
2	few facts	some development
1	no relevant facts	little development

Interactive Review

Click here to complete these and other activities online @ClassZone.com

CHAPTER SUMMARY

Key Idea 1
The geography of Mexico, Central America, and the Caribbean Islands contains mountains, highlands, and plains.

Key Idea 2
The physical geography of South America consists of a wide variety of landforms and climates.

Key Idea 3
People adapted to these challenging geographic and climatic conditions and developed great civilizations.

Key Idea 4
Life under the conquerors was hard, as the people of Latin America struggled to gain their independence.

For **Review and Study Notes**, go to **Interactive Review** @ClassZone.com

NAME GAME

Use the Terms & Names list to complete each sentence on paper or online.

1. I am built to make farming on mountainsides easier. ___terrace___
2. I am the civilization that built a vast empire in Peru. ________
3. I am a large plain in Argentina. ________
4. I am the civilization that is considered the cultural hearth of southern Mexico. ________
5. I am the father of Mexican independence. ________
6. I am a large region that includes Mexico, Central and South America, and the Caribbean Islands. ________
7. I allowed plants, animals, and ideas to be traded between Europe and the Americas. ________
8. I am the large islands in the Caribbean Sea. ________
9. I am a landform on which the Panama Canal is located. ________
10. I am the place on which the Aztec grew food and flowers. ________

chinampas
Columbian Exchange
Father Hidalgo
Greater Antilles
Inca
isthmus
Latin America
Lesser Antilles
Olmec
Pampas
Simón Bolívar
terrace

Activities

GeoGame

Use this online map to show what you know about present-day Latin America. Click and drag each place name to its location on the map.

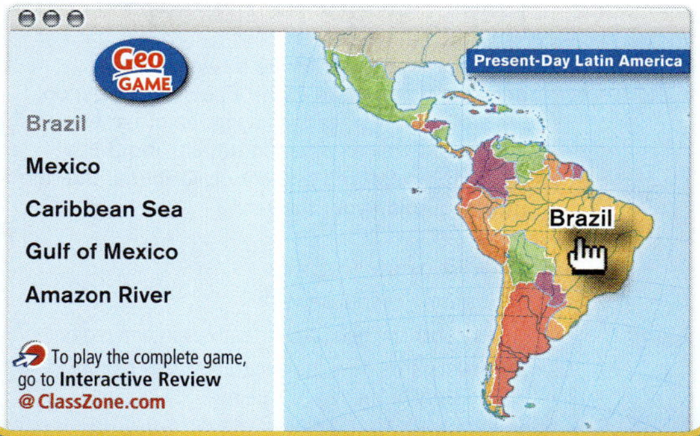

Crossword Puzzle

Complete an online crossword puzzle to test your knowledge of the geography and history of Latin America.

ACROSS
1. Mayan writing—a carved symbol that stands for a syllable or word

225

ONLINE AND ON-SCREEN ACTIVITIES

GeoGame

Use the interactive map on **Power Presentations** DVD-ROM or the **eEdition** to help students review the physical geography of Latin America.

- Demonstrate how to drag and drop the map labels listed onto the appropriate locations on the map.
- Advance to the scorecard to show how students have a chance to improve their scores.

Crossword Puzzle

Students can review important information about Latin America by completing this crossword puzzle online at **ClassZone.com.**

- As an extension of this activity, have students use the *Activity Maker* at **ClassZone.com** to make their own crossword puzzle for one section of the chapter.
- Then have students trade puzzles with classmates to solve each other's puzzles.

Chapter Summary

As a follow-up to the chapter, encourage students to pair up and go over the **Review and Study Notes** at **ClassZone.com.** Alternatively, student pairs can review their Reading Study Guide notes, making sure that they can add important details to each section's "Key Idea."

Name Game

Teaching Tips

ON PAPER Have students work in pairs to complete this activity. As an extension, they can write questions for the terms that are not covered in the items shown.

ONLINE Make sure that students understand how to drag and drop terms to fill in blanks. Ask them to print out their results and compare them with those of a classmate.

ANSWERS

1. terrace
2. Inca
3. Pampas
4. Olmec
5. Father Hidalgo
6. Latin America
7. Columbian Exchange
8. Greater Antilles
9. isthmus
10. *chinampas*

Make It Fun

Teacher-Tested Activities

Greta Frensley, Knox Doss Middle School, Gallatin, Tennessee

Try this activity, "Term Bingo," as a quick review of the terms covered in the chapter. On the board, write at least 20 terms or geographic features that have been covered. Then have students fold a sheet of paper into fourths, both vertically and horizontally. When the paper is unfolded, there should be 16 squares. Have each student choose 16 of the terms on the board, and write one term in each square on his or her "card." Then call out definitions or clues for each of the terms. The first person to match the definitions with a row of terms wins.

Vocabulary ANSWERS

1. the region that includes Mexico, Central america, the Caribbean Islands, and South America
2. strip of land that connects two landmasses, such as Central America connecting North and South America
3. mountain chain in South America
4. grassy plains in south-central South America
5. second-longest river in the world, flowing from northern Peru to the Atlantic Ocean
6. early civilization in the Valley of Mexico
7. early civilization in the Andes Mountains of Peru
8. Aztec island gardens
9. person with mixed European and Indian ancestry
10. leader for independence in northern South America
11. early complex civilizations in Mexico
12. geographic features in South America

Key Ideas ANSWERS

13. because the European languages spoken by the colonizers were derived from Latin
14. Sierra Madre Oriental and Sierra Madre Occidental
15. because it is located in the tropical zone
16. in the Amazon River basin
17. the Amazon
18. elevation
19. in what is now southern Mexico
20. The city had houses, causeways, a pyramid dedicated to the sun and war god, and *chinampas* around it.
21. They used quipus, counting tools of knotted cords.
22. the Spanish
23. persons with mixed European and Indian ancestry; they were instrumental in the independence movement
24. They followed their leaders—Simón Bolívar and José de San Martín—in their fight against the Spanish.

CHAPTER 7 Assessment

VOCABULARY

Explain the significance of each of the following.

1. Latin America
2. isthmus
3. Andes Mountains
4. Pampas
5. Amazon River
6. Aztec
7. Inca
8. *chinampas*
9. *mestizo*
10. Simón Bolívar

Explain how the terms and names in each group are related.

11. Olmec, Maya, and Aztec
12. Pampas, Andes Mountains, and Amazon River

KEY IDEAS

1 Physical Geography of Mexico, Central America, and the Caribbean

13. Why are Mexico, Central and South America, and the Caribbean Islands called Latin America?
14. What are two mountain ranges in Mexico?
15. Why is Central America's climate warm year-round?

2 Physical Geography of South America

16. Where is the world's largest tropical rain forest?
17. What is the largest river system in South America?
18. What affects climate in the Andes Mountains?

3 Ancient Civilizations

19. Where was the Mayan civilization located?
20. What features did Tenochtitlán have?
21. How did the Inca keep records?

4 From Colonization to Independence

22. Who defeated the Aztec and Inca empires?
23. Who were the *mestizos* and why are they important?
24. How did most South American countries become independent?

CRITICAL THINKING

25. **Compare and Contrast** Create a table to compare and contrast the landforms, farming methods, and building methods in the Aztec and Inca empires.

AZTEC	INCA

26. **Analyze Causes and Effects** How did the lives of Indians change after the Spanish conquered them?
27. **Identify Problems and Solutions** How might the Spanish have prevented rebellion?
28. **Connect to Economics** How did the *encomienda* system prevent the Mexican economy from being open to all people?
29. **Five Themes: Human-Environment Interaction** How did the Aztec and Inca use technology to change the environment?
30. **Make Inferences** What problems did Latin Americans likely face after independence?

Answer the ESSENTIAL QUESTION

How have Latin America's geography and resources helped shape its history?

Written response Write a two- or three-paragraph response to the Essential Question. Be sure to consider the key ideas of each section as well as specific details about how geography affected ancient civilizations. Use the rubric to guide your thinking.

Response Rubric
A strong response will:
- discuss the geographic features in one region
- summarize how the geography affected the history of the region's people

ESSENTIAL QUESTION REVIEW

Before students begin their essays, have them review the key chapter elements that will help them answer the Essential Question. Direct them to significant features in the chapter, such as

- Comparing Climates, Section 1
- Comparing Mountain Climate Zones, Section 2
- Animated Geography: *Chinampas*, Section 3
- History Makers: Father Miguel Hidalgo, Section 4

Written Response Rubric

	Coverage of Key Ideas	Explanation	Accuracy
4	all	clear, concise	few or none
3	most	good	some
2	some	fair	several errors
1	one or none	poor	many errors

STANDARDS-BASED ASSESSMENT

Test Practice
- Online Test Practice @ClassZone.com
- Test-Taking Strategies and Practice at the front of this book

SECONDARY SOURCE

Use context clues in the paragraph below to answer questions 1 and 2.

> The full knowledge of the Maya calendar must have been guarded . . . by the ruling elite, since it was undoubtedly a source of great power. . . . One might assume, however, that even the poorest farmer had some knowledge of the basic system, by which to guide his family's daily life.
>
> Source: The Ancient Maya, by Robert J. Sharer

1. What is the most likely reason that the ruling elite guarded knowledge of the calendar?

A. They did not want the farmers to be confused.
B. They wanted to stop other people from stealing the calendar.
C. They believed the knowledge gave them power over others.
D. They didn't want others to use the knowledge against them.

2. What statement does the paragraph support?

A. The Maya did not know about the calendar.
B. Most Maya had some knowledge of the calendar.
C. Most Maya did not care about the calendar.
D. The ruling elite wanted everyone to know about the calendar.

LINE GRAPH

Use the line graph below to answer questions 3 and 4 on your paper.

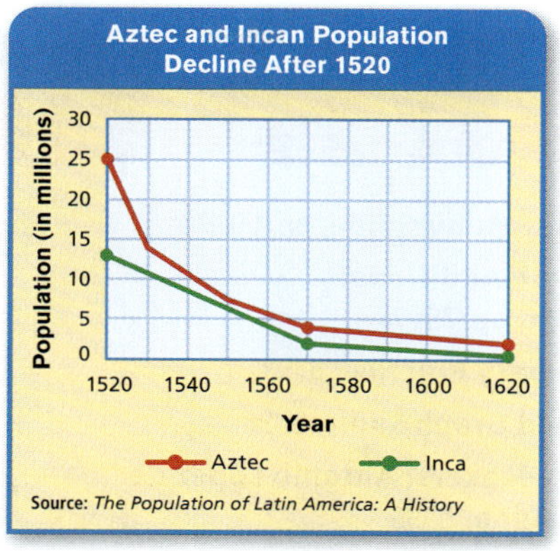

3. Between what years did the Aztec population show the sharpest decline?

4. What was the Inca population in 1520? In 1620?

GeoActivity

1. INTERDISCIPLINARY ACTIVITY–DRAMA

With a small group, find another legend from the ancient civilizations of Latin America, such as the Aztec legend about the volcano Popocatépetl. Create a skit of the legend to present to the class.

2. WRITING FOR SOCIAL STUDIES

Reread the part of Section 2 about Tenochtitlán. Imagine that you are visiting the city. Write a letter to a friend telling about your visit. Describe the features of the city and give your impression of it.

3. MENTAL MAPPING

Create an outline map of Latin America and label the following:
- Mexico
- Mexico City
- Central America
- Panama Canal
- Greater Antilles
- Lesser Antilles
- South America
- Amazon River

Critical Thinking ANSWERS

25. Table should show that the Aztec lived in the Valley of Mexico, and the Inca in the Andes Mountains; the Aztec used *chinampas*, and the Incas used terraces to farm; the Aztec built cities and causeways, and the Inca built cities, aqueducts, and roads.

26. They worked for the Spanish landlords and became essentially enslaved.

27. Possible answer: They could have given the *mestizos* a voice in the government. They could have treated the Indians more kindly and shared wealth with the people.

28. The system was set up to make Spain wealthy; the Indians did not benefit and were essentially enslaved.

29. The Inca developed ways to cut terraces into the land to farm on mountains. They built roads and aqueducts. The Aztec developed *chinampas* to raise food. They used plans to create a well-laid-out capital city.

30. Possible answer: They had to run the farms, ranches, and mines themselves. They had to learn how to rule themselves.

Standards-Based Assessment ANSWERS

1. C **3.** 1520–1540
2. B **4.** about 14 million; about 1 million

GeoActivity

1. INTERDISCIPLINARY ACTIVITY RUBRIC

The skit should
- present a legend from an ancient Latin American civilization
- use dialogue that realistically develops the legend
- be audible to the audience

2. WRITING FOR SOCIAL STUDIES RUBRIC

The letter should
- clearly describe the features of Tenochtitlán
- provide personal impressions of the city's features

3. MENTAL MAPPING RUBRIC

The outline map should
- represent the approximate shape of Latin America
- include all geographic features listed in their correct location

DIFFERENTIATING CHAPTER ASSESSMENT

Unit Resource Book
- **Chapter Test A**, pp. 57–60. *Use for Struggling Readers, Beginning-to-Intermediate-Level English Learners, and for Inclusion of students whose IEP suggests supported test-taking.*
- **Chapter Test B**, pp. 61–64. *Use for On-Level Learners and ESOL students with Advanced-Level English.*
- **Chapter Test C**, pp. 65–68. *Use for Gifted and Talented students or students assigned Pre-AP work.*

Test Generator
- Chapter Tests A, B, and C, in English and Spanish

McDougal Littell Assessment System

TEST **Test Generator** Use three levels of tests in English and Spanish tied to state standards. Plus, choose questions to customize tests for all populations.

SCORE Scan and Score plain paper student tests in minutes.

REPORT Create instant progress reports, sortable by student, by standard, or other criteria.

RETEACH Respond immediately to students' needs by using a wealth of reteaching resources.

CHAPTER 8

Mexico PLANNING GUIDE

Complete print resources all on one USB drive allow you to customize lesson plans for all students.

Section Titles & Key Ideas	Unit Resource Book Latin America	Assessment Resources
1 A Struggle Toward Democracy **KEY IDEA** Mexico had to overcome many obstacles as the country moved toward establishing a democracy.	• **Reading Study Guide,** pp. 69–70 • **RSG with Additional Support,** pp. 75–76 • **Vocabulary Practice,** p. 93 • **Map and Graph Skills,** p. 95 • **History Makers: Antonio López de Santa Anna,** p. 99	**Unit Resource Book** • Section Quiz, p. 107 • Reteaching Activity, p. 110 **Test Generator CD-ROM** **Daily Test Practice Transparency,** TT26
2 A Blend of Traditions **KEY IDEA** Ancient civilizations, colonial powers, and the modern world affect the culture and daily lives of the Mexican people.	• **Reading Study Guide,** pp. 71–72 • **RSG with Additional Support,** pp. 77–78 • **Skillbuilder Practice,** p. 94 • **Interdisciplinary Projects,** pp. 102–103 • **Primary Source,** p. 106	**Unit Resource Book** • Section Quiz, p. 108 • Reteaching Activity, p. 111 **Test Generator CD-ROM** **Daily Test Practice Transparency,** TT27
3 Creating a New Economy **KEY IDEA** Mexico faces new challenges as the country takes steps to modernize its economy.	• **Reading Study Guide,** pp. 73–74 • **RSG with Additional Support,** pp. 79–80 • **History Makers: Vicente Fox,** p. 100 • **Interdisciplinary Projects,** pp. 101, 103 • **Economics in the World,** p. 104 • **Connect to Science,** p. 105	**Unit Resource Book** • Section Quiz, p. 109 • Reteaching Activity, p. 112 • Chapter Test (Forms A, B, &C), pp. 113–124 **Test Generator CD-ROM** **Daily Test Practice Transparency,** TT28

McDougal Littell **Assessment System**

TEST SCORE REPORT RETEACH

No Time? To focus students on important content in the chapter, use the **"Reading Study Guide with Additional Support"** in *Resources2Go.*

Teacher Presentation Options

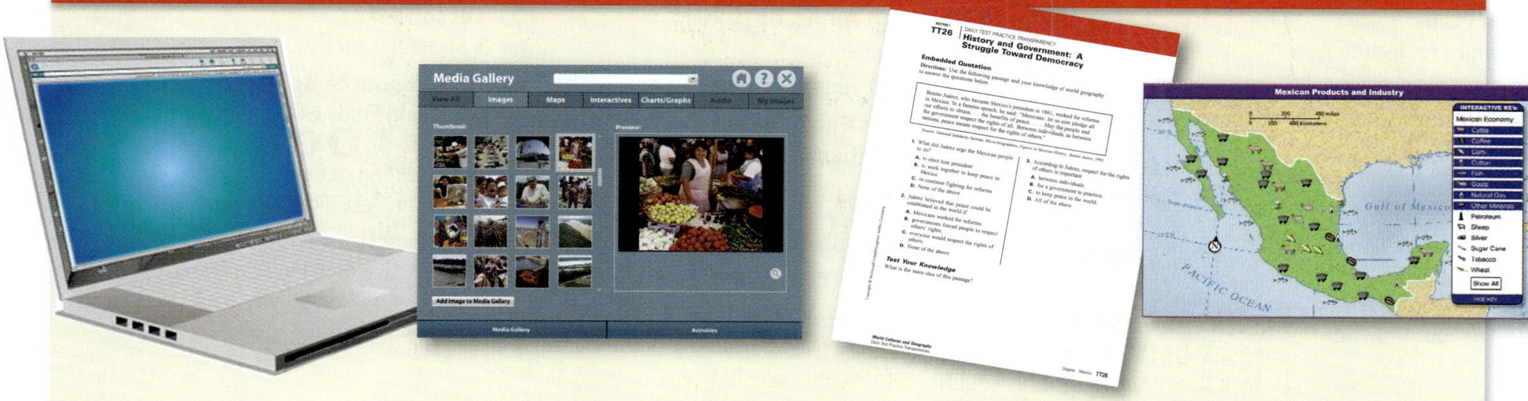

Electronic Books

eEdition DVD-ROM

eEdition Online

Presentation Toolkit

Video/DVD

- World Cultures and Geography

Power Presentations DVD-ROM

- Discussion Starters
- Media Gallery
- Review Game
- Interactive Review
- Animated Features

Unit Transparency Book

- 3-Minute Warm-Ups
- Essential Question
- Critical Thinking
- Maps
- Cultures

Daily Test Practice Transparencies

Animated Geography

- Present-Day Mexico Map
- Mexican Industry Map

OnlineActivities at ClassZone.com

Animated Geography

- Videos and Photos
- 3-D Animations
- Interactive Maps

Interactive Review

- Section Quizzes
- Review and Study Notes
- Vocabulary Flip Cards
- Graphic Organizers
- Name Game and Crossword Puzzles
- GeoGame

Online Test Practice

Map Center

- Atlas Maps
- Political Maps with Data
- Links to Mapping Web Sites

Activity Center

- Activity Maker—Review Games, Puzzles, Flip Cards, and Graphic Organizers
- Worldster

Research & Writing Center

- Web Sites
- Data Updates
- Primary Sources
- Writing Models

Current Events

Todd Harrison
Hardin County Middle School
Savannah, Tennessee

Make It Fun — Teacher-Tested Activities

As you teach this chapter, look for my classroom-proven ideas for:

- making the lesson fun
- focusing students' attention and comprehension
- integrating technology with traditional teaching

Teaching Bell to Bell When you see this timing guide, have students discuss in pairs, in groups of 3–6, or as a class.

- **Think, Pair, Share** (pairs)
- **Reader, Recorder, Reporter** (3–6)
- **Talk About It** (any size)

RESOURCES FOR DIFFERENTIATED INSTRUCTION

CHAPTER 8

Struggling Readers

eEdition with Audio DVD-ROM

Lesson Plans
- Strategies for Inclusion

Unit Resource Book
- Reading Study Guide with Additional Support, pp. 75–80 B
- Reteaching Activity, pp. 110–112
- Chapter Test (Form A), pp. 113–116 A

Best Practices Toolkit C
- Interactive Reading Strategies, pp. 90–145
- Differentiation Activity Bank: Struggling Readers, pp. 146–149

Teacher's Edition Activities
- Analyze Mexico's Loss, Sec. 1
- Tiered Activities, Sec. 1–3

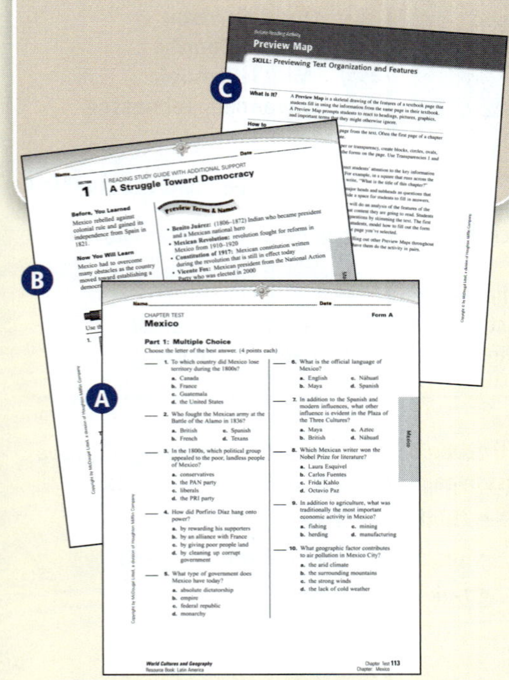

English Learners

Pupil's Edition in Spanish

eEdition with Audio DVD-ROM

Lesson Plans
- Strategies for English Learners

Unit Resource Book
- Reading Study Guide (Spanish), pp. 81–86
- RSG with Additional Support (Spanish), pp. 87–92 A B
- Vocabulary Practice, p. 93

Multi-Language Glossary C

Best Practices Toolkit
- Differentiation Activity Bank: English Learners, pp. 165–194

Test Generator CD-ROM
- Section Quizzes in Spanish
- Chapter Test in Spanish

Teacher's Edition Activities
- Pronounce and Preview, Sec. 1–3
- Key Academic Vocabulary, Sec. 1
- What Do I See?, Sec. 2
- Tiered Activities, Sec. 1–3

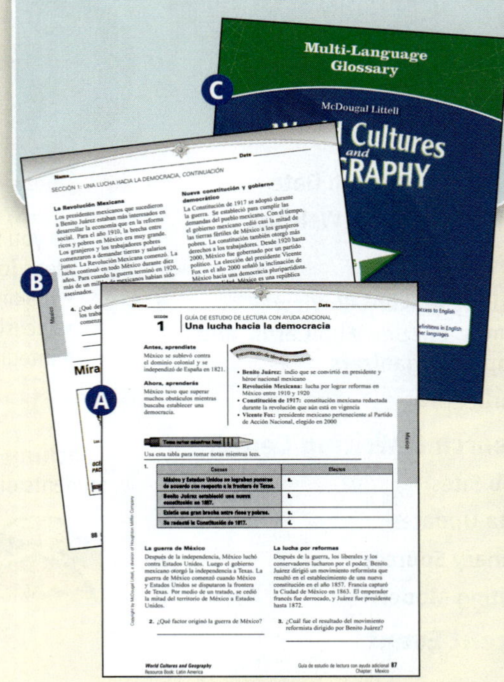

Inclusion

eEdition with Audio DVD-ROM

Lesson Plans
- Strategies for Inclusion

Unit Resource Book
- Reading Study Guide with Additional Support, pp. 75–80 A B
- Reteaching Activity, pp. 110–112
- Chapter Test (Form A), pp. 113–116

Best Practices Toolkit C
- Interactive Reading Strategies, pp. 90–145
- Differentiation Activity Bank: Struggling Readers, pp. 146–149

Teacher's Edition Activities
- Vocabulary: Definition Mapping, Sec. 1
- Guess the Terms and Definitions, Sec. 2
- Finding Terms in Context, Sec. 3
- Tiered Activities, Sec. 1–3

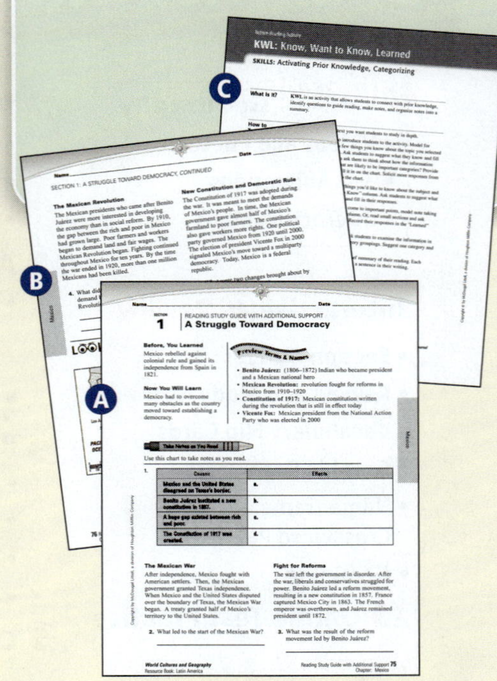

Interdisciplinary Activities

Gifted & Talented Students

Lesson Plans
- Core Lesson Plan

Unit Resource Book
- Outline Map Activities, p. 97
- History Makers, pp. 99–100 A B
- Interdisciplinary Projects, pp. 101, 103 C
- Chapter Test (Form C), pp. 121–124

Best Practices Toolkit
- Student Products Rubrics, pp. 1–14
- Writing Templates, pp. 78–79

Teacher's Edition Activities
- Plans for Improving Rural Life, Sec. 2
- Tiered Activities, Sec. 1–3

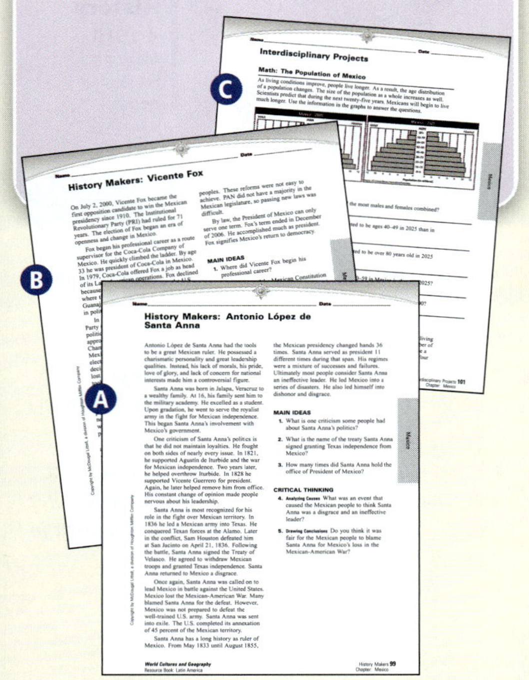

Pre-AP*

Lesson Plans
- Core Lesson Plan

Unit Resource Book
- Map and Graph Skills, pp. 95–96
- Interdisciplinary Projects, pp. 102–103 A
- Economics in the World, p. 104 B
- Connect to Science, p. 105
- Primary Source, p. 106 C
- Chapter Test (Form C), pp. 121–124

Best Practices Toolkit
- Writing Templates, pp. 78–79
- Differentiation Activity Bank: Pre-AP Students, pp. 195–200

* Pre-AP is a registered trademark of the College Entrance Examination Board, which was not involved in the production of and does not endorse this product.

Teacher's Edition Activities
- Tiered Activities, Sec. 1–3

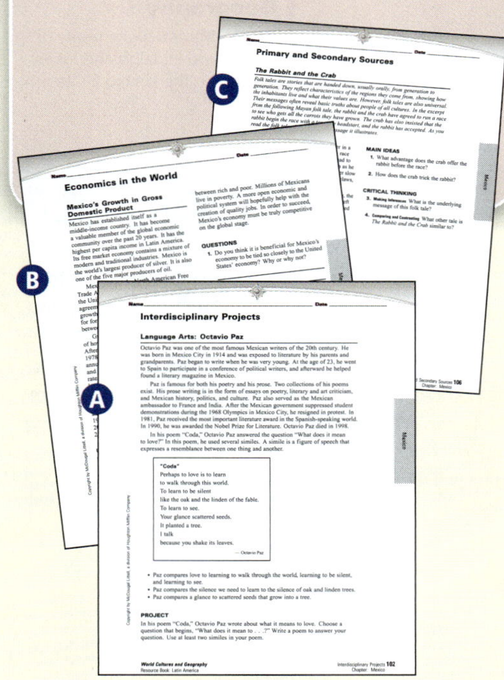

CONNECT to Language Arts

Poem About Mexico's Struggle Toward Democracy

CLASS TIME 30 minutes

TASK Create a narrative poem about Mexico's struggle for democracy.

SUPPLIES
- writing materials/computers

ACTIVITY Have students brainstorm a list of descriptive words that would help someone visualize the Mexican people's struggle for democracy. Then have students use their lists to create a narrative poem about the events and people in Mexico's history during its steps toward democracy in the 1800s and 1900s. Students may also choose to focus just on one event or person from this time period. Ask volunteers to read their poems aloud or create a classroom display.

Good for Visual and Verbal Learners

CONNECT to Economics

Planning a Trade Agreement

CLASS TIME 30 minutes

TASK Create a plan for a new trade agreement between Mexico and the United States.

SUPPLIES
- writing materials/computers

ACTIVITY Have students work in pairs to create a new trade agreement between Mexico and the United States. Topics for the agreement might be sharing oil fields in the Gulf of Mexico, requiring products to be made by workers who are treated the same on both sides of the borders, or exchanging technologies or resources. Agreements should be clearly and concisely written, and focus on the rules for trade, or the conditions each side must meet. Have all the students present their agreements, then discuss them as a class. Afterward, have students revise their agreements, based on the class discussion.

Good for Verbal and Logical Learners

Introduce the ESSENTIAL QUESTION

- Identify the challenges Mexico faced in creating a stable government.
- Explain that Mexico's culture is a blend of traditions from native peoples and Spanish colonists and modern practices.
- Discuss how Mexico is building a modern economy and addressing the challenges of job creation and pollution.

Key Ideas

Section 1: Mexico had to overcome many obstacles as the country moved toward establishing a democracy.

Section 2: Ancient civilizations, colonial powers, and the modern world affect the culture and daily lives of the Mexican people.

Section 3: Mexico faces new challenges as the country takes steps to modernize its economy.

CONNECT to the Essential Question

You may wish to revisit the Essential Question periodically with students as they work through the chapter.

Suggestions for reconnecting are provided at an appropriate point in the *Teacher's Edition* for each section.

Senita cactus in Baja California Sur, Mexico

CHAPTER 8

Mexico

1 HISTORY & GOVERNMENT
A Struggle Toward Democracy

2 CULTURE
A Blend of Traditions

3 ECONOMICS
Creating a New Economy

ESSENTIAL QUESTION

How does Mexico reflect both ancient traditions and the challenges of the modern world?

CONNECT Geography & History

Use the map and the time line to answer the following questions.

1. What Mexican states are located on a peninsula on the Pacific Coast?
2. What was one result of the Mexican Revolution?

1848

Geography
1848 Mexico signs treaty of Guadalupe Hidalgo and loses territory to the United States.

History
◄ **1864** Napoleon appoints Maximilian as Emperor of Mexico. (Maximilian)

History
◄ **1910** Mexican Revolution begins.

Economics
1917 New constitution redistributes land more equally among the people.

228 Chapter 8

TIME LINE DISCUSSION

Have students identify two countries the time line shows as having an interest in Mexico, and brainstorm why they might have been interested in Mexico. *(Possible answers: United States and France; for people to settle there; to profit from Mexico's resources; to have more power against other countries in North America)*

- What year was the treaty of Guadalupe Hidalgo signed and what did it do? *(1848—gave territory in Mexico to the United States)*
- How many years after the Mexican Revolution began did it take for the PRI to come to power? *(about 19 years—1910 to 1929)*
- What event in the time line points to a recent economic development in Mexico? *(the signing of the North American Free Trade Agreement in 1992)*

Animated GEOGRAPHY Present-Day Mexico

Click here to explore Mexico @ClassZone.com

Government
▲ **1929** The Institutional Revolutionary Party (Partido Revolucionario Institucional, or PRI) comes to power.

Culture
1990 Octavio Paz wins Nobel Prize for Literature.

Economics
1992 Mexico signs the North American Free Trade Agreement (NAFTA).

Today

Government
◀ **2000** Vicente Fox is elected president of Mexico.

229

Animated GEOGRAPHY Present-Day Mexico
CLASSZONE.COM

Discuss how water access may have encouraged other countries to have an interest in Mexico. *(Coastal access allows sea travel and rivers allow inland travel as well as water for irrigation.)* Have students identify the item on the map that indicates that much of Mexico's climate is tropical. *(Most of Mexico is along or south of the Tropic of Cancer.)*

- What is the name and location of Mexico's capital? *(Mexico City, Distrito Federal—federal district)*
- What Mexican states form Mexico's border with Guatemala? *(Chiapas, Tabasco, Campeche, Quintana Roo)*
- Locate the states of Durango and Nuevo León. What physical feature do they lack that most other states in Mexico have? *(access to a major body of water, such as the ocean or a river)*

CONNECT GEOGRAPHY & HISTORY ANSWERS

1. Baja California, Baja California Sur
2. Land was redistributed more equally among the people.

Presentation Options

Use the **Power Presentations** DVD-ROM to project the animated map.

- Prompt students to use the pop-up images and audio from the map to discover what they already know about the geography, culture, history, economics, or governments of the region.
- Challenge students to answer the map questions individually, in small groups, or as a class activity. Use all of the interactive material to help preview chapter content.

TEACHER MEDIA FAVORITES

Books for the Teacher

Ryan, Pam Munoz. *Esperanza Rising.* Blue Sky Press, 2002. Story set in 1930 of a Mexican girl's life after moving to California. (available on audio)

Saldaña, René Jr. *The Jumping Tree.* Laurel Leaf, 2002. Novel/vignettes about a young Mexican-American who keeps close ties with his Mexican relatives.

Hill, Laban Carrick. *Casa Azul: An Encounter with Frieda Kahlo.* Watson-Guptill, 2005. Story of a rural sister and brother befriended by Frida Kahlo in Mexico City.

Video / DVD

Globe Trekker—Mexico. (DVD) Pilot Productions, 2002. Follows two travelers to various areas in Mexico.

The Maya: Temples, Tombs And Time. (Video) Questar, 2002. History, archeology, and present life of the Maya.

The Sights & Sounds of Modern Mexico. (DVD) Fideo Knowledge, 2004. Sampling of daily life in Mexico with limited narration.

Internet

Go to the *Research and Writing Center* at **ClassZone.com** to link to:

- data on the region
- a variety of chapter-specific, content-reviewed sites
- "History Makers" biographies
- primary sources

1 Plan & Prepare

Objectives

- Explain how Mexico established a stable government after gaining independence
- Analyze the effects the revolution and constitutional change had on the Mexican people and their government

Read for the Essential Question

Help students read for a purpose by reminding them of the Essential Question: "How does Mexico reflect both ancient traditions and the challenges of the modern world?" Point out that they can use text features such as Key Questions, special features, images, and maps to form answers to the question.

Vocabulary

Best Practices Toolkit

Use the ***Best Practices Toolkit*** to model strategies for vocabulary notetaking. Vary strategies throughout the year. Choose from: *Knowledge Rating, Predicting ABCs, Definition Mapping, Word Sort, Word Wheel, Frayer Model (Word Squares), Magnet Words,* and *Student VOC.*

Vocabulary Strategies, TT9–TT16

Reading Strategy

Best Practices Toolkit

Display the T Chart/Two-Column Chart transparency.

- Point out to students that analyzing causes and effects involves identifying events or situations (causes) and what happened as a result of them (effects).
- Model entering information on the chart, using information from the "War and Reform" subsection.
- Have students suggest entries.
- Display the completed chart for students to review.

T Chart/Two-Column Chart, TT20

Section 1 Reading for Understanding

Key Ideas

BEFORE, YOU LEARNED

Mexico rebelled against colonial rule and gained its independence from Spain in 1821.

NOW YOU WILL LEARN

Mexico had to overcome many obstacles as the country moved toward establishing a democracy.

Vocabulary

TERMS & NAMES

Antonio López de Santa Anna (1794–1876) Mexican general, president, and leader of Mexican independence from Spain

Republic of Texas constitutional government of Texas after independence from Mexico

annex to add to an existing territory

cession surrendered territory

Benito Juárez (1806–1872) Indian who became president and a Mexican national hero

Mexican Revolution a fight for reforms in Mexico from 1910 to 1920

Constitution of 1917 Mexican constitution written during the revolution that is still in effect today

Vicente Fox Mexican president from the National Action Party who was elected in 2000

REVIEW

constitution a formal plan of government

revolution the overthrow of a ruler or government; a major change in ideas

Reading Strategy

Re-create the chart shown at right. As you read and respond to the **KEY QUESTIONS**, look for the effects of the causes that are listed.

See Skillbuilder Handbook, page R8

ANALYZE CAUSES AND EFFECTS

CAUSES	EFFECTS
Mexico and the United States could not agree on the border of Texas.	
Benito Juárez instituted a new constitution in 1857.	
Huge gap existed between rich and poor; most Mexicans did not own land.	
The Constitution of 1917 was created.	

GRAPHIC ORGANIZERS Go to **Interactive Review** @ClassZone.com

230 Chapter 8

PRETEACHING VOCABULARY

English Learners

Pronounce and Preview

Pronounce each term for students. Review the meanings of words within definitions, such as *constitutional, surrendered,* and *overthrow.*

- To modify vocabulary learning, have students complete worksheets as they read, instead of afterward.

Unit Resource Book: Latin America
- Vocabulary Practice, p. 93

Inclusion

Vocabulary: Definition Mapping

Have students work in pairs to fill out the graphic organizer for the terms in the section.

- Use the transparency to model the strategy.
- Have volunteers suggest entries.
- Then distribute copies of the organizer for students to complete with a partner.

Best Practices Toolkit
- Vocabulary: Definition Mapping, TT11

SECTION 1
HISTORY & GOVERNMENT

Pyramid of Kukulkan

Monarch butterflies

A Struggle Toward Democracy

Connecting to Your World

Think what an important part George Washington played in the U.S. fight for independence from Britain. In Mexican history, General **Antonio López de Santa Anna** played a key role in Mexico's fight for independence from Spain. He served as both a soldier and a president. As president, however, Santa Anna was not able to establish a secure government. Several factors made it difficult for the Mexican people to create a stable government.

Santa Anna

War and Reform

KEY QUESTION What challenges did Mexico have in establishing a stable government?

During Spanish rule, Mexicans had little control over their lives. Spain made many decisions for Mexico. As a result, after gaining independence in 1821, Mexico had trouble establishing a stable government. The Mexican people had no experience in governing themselves, and the nation had a weak economy. Invasion by foreign countries also prevented Mexico from establishing a strong government.

Reform Demands Protests, pictured in this mural by David Siquieros, led to the Mexican Revolution.

231

2 Focus & Motivate

3-Minute Warm-Up

Write on the board or display the transparency:

- Look at the mural image "Reform Demands" by David Siquieros and read the caption. What do you think the protestors wanted? *(Possible answer: more land, fairer work conditions and laws)*

Unit Transparency Book: Latin America
- 3-Minute Warm-Ups, TT11

Connecting to Your World

More About . . .

Pyramid of Kukulkan

The pyramid of Kukulkan, also known as El Castillo, is part of the ancient Mayan city of Chichén Itzá. The feathered serpent carved at its top was a common deity for the different cultures that lived in Chichén Itzá.

3 Teach

War and Reform

Reader, Recorder, Reporter

- How did Mexico's conservatives and liberals differ? *(conservatives: rich landowners and military leaders, opposed the constitution; liberals: poor and landless, supported the constitution)*
- **Analyze Causes and Effects** What was the result of the Treaty of Guadalupe Hidalgo? *(Mexico gave up the northern part of its territory to the United States.)*

SECTION 1 PROGRAM RESOURCES

Unit Resource Book: Latin America

ON LEVEL

Unit Resource Book
- Reading Study Guide, p. 69
- Map and Graph Skills, p. 95
- Section Quiz, p. 107

STRUGGLING READERS

Unit Resource Book
- Reading Study Guide with Additional Support, p. 75
- Skillbuilder Practice: Comparing and Contrasting, p. 94
- Section Quiz, p. 107
- Reteaching Activity, p. 110

eEdition with Audio DVD-ROM

ENGLISH LEARNERS

Pupil Edition in Spanish

eEdition with Audio DVD-ROM

Unit Resource Book
- RSG (Spanish), p. 81
- RSG with Additional Support (Spanish), p. 87

Multi-Language Glossary

Test Generator
- Section Quiz in Spanish

INCLUSION

Unit Resource Book
- RSG with Additional Support, p. 75
- Section Quiz, p. 107
- Reteaching Activity, p. 110

GIFTED & TALENTED

Unit Resource Book
- History Makers: Antonio López de Santa Anna, p. 99
- Section Quiz, p. 107

PRE-AP

Unit Resource Book
- Section Quiz, p. 107

TECHNOLOGY

Unit Transparency Book
- 3-Minute Warm-Ups, TT11
- Essential Question, TT12
- Critical Thinking, TT13

Daily Test Practice Transparencies
- Section 1, TT26

Power Presentations

ClassZone.com

World Cultures and Geography Video on DVD

Republic of Mexico *1824–1853*

CONNECT Geography & History

Have students examine the map and brainstorm how losing this land may have affected the Mexican people. *(Possible answer: They might have lost confidence in the Mexican government.)*

ANSWERS

1. **Location** the Red River
2. **Region** It greatly reduced the size of the Mexican territory.

More About . . .

The Gadsden Purchase

After the Treaty of Guadalupe Hidalgo was signed, Mexico and the United States still disagreed over an area called the Mesilla Valley. The United States wanted this land because it was the best route for a southern railroad to the Pacific coast. So James Gadsden, the U.S. minister to Mexico, negotiated to buy about 30,000 square miles of this land, which was eventually bought for $10 million. The agreement, the Treaty of La Mesilla, was signed by Gadsden and Mexican president Santa Anna.

Unit Resource Book: Latin America
- History Makers: Antonio López de Santa Anna, p. 99

Comparing Regions

Mexico and U.S. Land and Water Areas

Today, Mexico has about 742,000 square miles of land and about 19,000 square miles of water. The United States today is more than four times the size of Mexico, with a land area of about 3.5 million square miles and a water area of over 180,000 square miles.

Unit Resource Book: Latin America
- Map and Graph Skills: Reading a Climate Map, pp. 95–96

The Mexican War Until 1848, Mexican territory included what is now the southwestern part of the United States. In the early 1800s, few Mexicans lived there. To increase the population, Mexico encouraged settlers from the United States to move to Texas. To get land, the settlers had to follow Mexican law and pay a small fee.

Soon, the American settlers wanted more independence. When the Mexican government refused to grant independence, Texans revolted and broke away from Mexico. Santa Anna and his troops won early victories against the Texans, including the Battle of the Alamo in San Antonio, but eventually they were defeated at the Battle of San Jacinto in 1836. That same year, Santa Anna signed a treaty granting Texas independence, and the **Republic of Texas** was established.

Mexico faced another problem when the United States **annexed**, or added, Texas in 1845. The United States and Mexico could not agree on the southern boundary of Texas. By 1846, the dispute led to a war that lasted two years. Then in 1848, Mexico and the United States signed the Treaty of Guadalupe Hidalgo. As a result of this treaty, Mexico's surrendered territory, or **cession**, included the northern half of what was once Mexico.

Republic of Mexico *1824–1853*

CONNECT Geography & History

1. **Location** What river formed the border between Mexico and the United States before the Texas Annexation?
2. **Region** How did the Treaty of Guadalupe Hidalgo affect Mexican territory?

232 Chapter 8

DIFFERENTIATING INSTRUCTION

English Learners

Key Academic Vocabulary

Note the word *purchase* in the map legend (Gadsden Purchase). Explain that *purchase* can be used as a verb and as a noun. Ask students to suggest definitions or synonyms for purchase. *(verb: to buy, to pay for, to exchange money for; noun: goods that were bought)* Then ask students to look up the definitions and write two sentences, one using *purchase* as a verb, the other using it as a noun.

Struggling Readers

Analyzing Mexico's Losses

Have students work in pairs to analyze the effects on Mexico of its losses of territory to the United States. Suggest they focus on economic and political effects, including international relations. Each pair should decide how to format this information, such as in a chart or essay, then present it to the class. To extend the activity, have students hold a round table discussion on the effects of Mexico's losses.

Fight for Reforms The Mexican War drained Mexico's economy and left the government in disorder. After the war, two groups within the country, the liberals and conservatives, struggled for power.

Conservatives consisted of Mexico's rich landowners and military leaders. The liberals generally consisted of the nation's poor, landless people. A liberal leader, **Benito Juárez**, greatly influenced Mexican politics at this time. He led a reform movement that resulted in a new **constitution** in 1857. Among other things, the constitution guaranteed freedom of speech and called for a federal system of government. But it did not make Catholicism the official religion, as many church officials wanted. Juárez and other reformers fought against the opponents of the constitution. This struggle left Mexico weak and open to foreign invasion.

France, Spain, and Britain sent forces to Mexico in 1861. France captured Mexico City in 1863 and named Maximilian, a European nobleman, as emperor of Mexico. His rule ended when the Mexican people overthrew and executed him. Juárez, who had been elected president in 1861, returned to that office until his death in 1872.

ANALYZE CAUSES AND EFFECTS Explain why establishing a stable government was difficult in Mexico. Possible answer: Because of Spanish colonial rule, Mexicans had no experience in self-government or in running an economy. Then, Mexico had to fight the Texans and the U.S.

HISTORY MAKERS

Benito Juárez 1806–1872

Benito Juárez, an Indian, was born in Oaxaca (wuh•HAH•kuh), where he received a law degree in 1831. Politics became Juárez's life, and he served in many capacities, including governor of Oaxaca and president of Mexico. His work for fairness and equality made Juárez a hero in the eyes of many Mexican people.

ONLINE BIOGRAPHY For more on the life of Benito Juárez, go to the **Research & Writing Center @ ClassZone.com**

Revolution and Constitutional Change

KEY QUESTION Why was the election of President Vicente Fox significant?

After Juárez's death, the reform movement weakened. Juárez's successors were more interested in developing the economy than in reform. They believed in government controlled by a small group.

In 1876, General Porfirio Díaz became dictator. Unlike Juárez, he gave land, power, or favors to anyone who supported him. Those who did not support him were shut out of power. By 1910, the gap between the rich and poor had grown huge. Just one percent of landowners controlled more than 90 percent of the land. Most Mexicans owned no land at all. This gap set the stage for the Mexican Revolution.

RESEARCH & WRITING CENTER

HISTORY MAKERS

Benito Juárez

When the conservatives (along with Santa Anna) returned to power in the early 1850s, Benito Juárez and other liberals were exiled from Mexico. Juárez lived in New Orleans, where he continued his political activities. He returned to Mexico in 1855 when the liberals took power, and became Minister of Justice. One of his achievements as minister came to be known as the Juárez Law, which ended certain privileges, such as special courts, of the clergy and military.

Teach

Revolution and Constitutional Change

Think, Pair, Share

- Why is the Mexican Revolution also called a civil war? *(because armies within the country were fighting each other)*
- **Analyze Causes and Effects** How did the dictatorship of General Porfirio Díaz lead to the Mexican Revolution? *(Díaz only gave land, power, and favors to supporters; the gap between rich and poor grew huge.)*

Unit Resource Book: Latin America
- Skillbuilder Practice: Comparing and Contrasting, p. 94

Make It Fun

Teacher-Tested Activities

Todd Harrison, Hardin County Middle School, Savannah, Tennessee

My students create a "Round-the-Room Time Line." Using long pieces of chart paper, students make a time line of significant events of Mexico's history.

- Organize the class in small groups. Assign each group a time period, starting with 1821 and ending with the present.
- Students should use the events in the section and research other significant events to add to the time line. Illustrating the time line makes it more interesting.
- Discuss the significance of the events to Mexico.

INTERDISCIPLINARY ACTIVITIES

CONNECT to Language Arts

Benito Juárez Memorial Plaque

Ask small groups of students to assume the role of a committee, creating a memorial plaque for Benito Juárez. Groups should brainstorm possible information to include, write the text, and then revise it in a format appropriate for a memorial (such as with the name and dates at top, centered text, on a plaque-shaped paper). You may want to provide images of memorial plaques for review. Have groups present their work.

CONNECT to Civics

Annotated Time Line

Have students make a list of all the dates mentioned in the "War and Reform" subsection and note why they were significant to Mexico's development of a stable government. Then have students use this information to create a time line with a title and annotations for each entry. Post the time lines, then discuss what lessons can be learned today from Mexico's struggles.

CONNECT Geography & History

Pancho Villa: Knowing the Land

U.S. forces tried to find Villa for 11 months—without success. For this and other events, Villa is remembered in history and legend. Pancho Villa State Park in Columbus, New Mexico, contains historical exhibits depicting the raid. Villa is also the subject of many *corridos* (ballads) and films.

Reforms of the Mexican Constitution of 1917

Discuss how the Mexican and U.S. governments are similar and how they differ. *(Similarities: both are federal republics, with states and a federal district, have three branches of government, and elect a president; Differences: the Mexican president has more control.)*

- What were the key elements of land reform, and why were these so important to the Mexican people? *(Possible Answer: The Mexican Constitution broke up large estates, restricted foreign ownership of land, and gave the government control of resources; these reforms provided more economic opportunities for all Mexicans.)*

More About . . .

Farming in Mexico

About 14 percent of the land in Mexico is farmland, employing almost one in five workers. Fruits, vegetables, coffee, and cotton are among the country's major export products, mostly to the United States. In turn, most of Mexico's imports come from the United States, with one of Mexico's major import products being U.S. agricultural machinery.

The Mexican Revolution By 1910, Mexicans from various walks of life were protesting Díaz's rule and calling for reforms. Farmers wanted land, and workers wanted fair wages and better working conditions.

The **Mexican Revolution** began when Francisco Madero, a wealthy rancher, called for a **revolution** to defeat Díaz. Leaders arose in different parts of Mexico and gathered their own armies. Emiliano Zapata led an army in southern Mexico and fought for land ownership for poor farmers. Pancho Villa led forces in northern Mexico. He became popular for his policy of stealing from the rich to give to the poor. Madero became president, but was soon overthrown. The fighting among the various Mexican groups for control of the government continued, turning the Revolution into a civil war. By the time it was over in 1920, more than one million Mexicans had been killed.

CONNECT Geography & History

Pancho Villa: Knowing the Land

Pancho Villa's knowledge of northern Mexico's geography helped him to avoid being captured by the U.S. army. In 1916, Villa, angry over U.S. involvement in the Mexican Revolution, led an attack in Columbus, New Mexico, in which 18 U.S. citizens were killed. The United States sent soldiers to capture Villa, but Villa's knowledge of the land and his popularity helped him elude the U.S. army.

New Constitution The **Constitution of 1917** was adopted during the revolution to meet the demands of Mexico's various groups and regions. Land reform was the central issue. Eventually, the government redistributed nearly half of Mexico's farmland to poor people. Millions of acres of farmland were divided into *ejidos* (eh•HEE•thaws), community farms owned by villagers. The constitution also brought about changes regarding workers' rights and the relationship of the government and the Church. The Constitution of 1917 is still in effect in Mexico today.

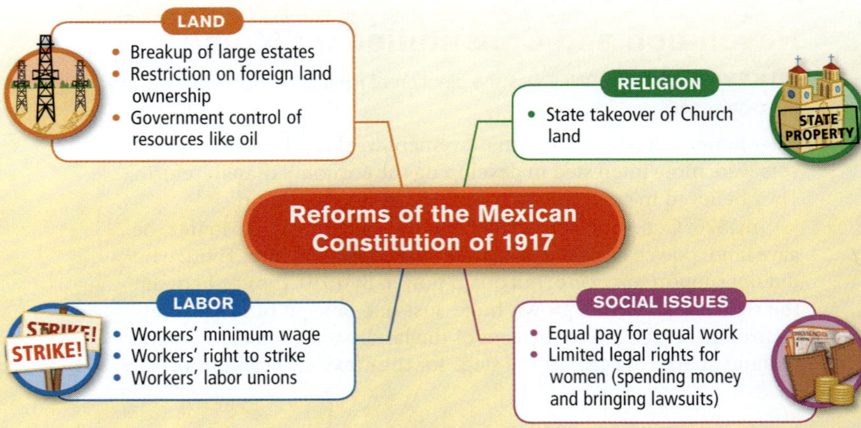

234

DIFFERENTIATING INSTRUCTION: TIERED ACTIVITIES

OBJECTIVE Create newspaper headlines about the Mexican Revolution.

Basic

Ask students to assume the role of a newspaper editor at the time of the Mexican Revolution. Have them review the information on the Revolution, identify three events, and create separate headlines about each one. Invite students to share one headline with the class.

On Level

In addition to the Basic activity, have students create two versions of the headline for each event. One headline should be from the perspective of a newspaper that supports the event; the other from one that does not. Invite students to share one set of headlines with the class.

Challenge

Have students create headlines as for the On Level activity. Have them also write the first sentence of the lead paragraph for each headline. Sentences should aim to make readers want to read on, and answer one or more of the six journalists' questions *(who, when, where, what, why, and how).*

Democratic Rule In the 1920s, one political party came to power by making peace among the various armies. Today it is called the Institutional Revolutionary Party (Partido Revolucionario Institucional, or PRI). The party controlled the government from 1929 until 2000. It helped introduce elements of democracy and stability to Mexico. But the party often blocked opposition to its policies and was accused of corruption.

In 2000, **Vicente Fox**, a member of the National Action Party (Partido Acción Nacional, or PAN), was elected president. His election signaled Mexico's move toward a multiparty democracy. In 2006, Felipe Calderon, also a member of PAN, won the presidential election. Today, besides PRI and PAN, there are at least five other political parties in Mexico.

Today, Mexico is a democracy organized much like the U.S. government. Mexico is a federal republic made up of 31 states and a federal district. Mexico also has three branches of government. But the Mexican president has more control than the U.S. president.

. was the rst time a resident was lected from party other han the PRI, narking the eginning of multiparty emocracy. ▶

MAKE INFERENCES Explain the significance of the election of Vicente Fox as Mexico's president.

Campaign Rally Supporters take part in an election rally for PAN in June 2000.

Section 1 Assessment

ONLINE QUIZ For test practice, go to **Interactive Review** @ClassZone.com

TERMS & NAMES

1. **Explain the importance of**
   - Antonio López de Santa Anna
   - Benito Juárez
   - Constitution of 1917
   - Vicente Fox

USE YOUR READING NOTES

2. **Analyze Causes and Effects** Use your completed chart to answer the following question:

   What effect did the Constitution of 1857 have on Mexico?

CAUSES	EFFECTS
Mexico and the United States could not agree on the border of Texas.	
Benito Juárez instituted a new constitution in 1857.	
Huge gap existed between rich and poor; most Mexicans did not own land.	
The Constitution of 1917 is created.	

KEY IDEAS

3. What dispute led to war between Mexico and the United States?
4. Why was the Mexican Revolution fought among different groups?
5. How did the Constitution of 1917 help farmers?

CRITICAL THINKING

6. **Analyze Points of View** Before 1910, how did different groups in Mexico view the need for reform?
7. **Make Inferences** Why do you think that Benito Juárez's background might have led him to support reform in Mexico?
8. **CONNECT to Today** How might having more than one political party benefit Mexico today?
9. **TECHNOLOGY** **Use the Internet** Find out more about a political leader mentioned in this lesson by using the Internet. Then prepare a bulleted list of the top five most interesting facts about the person.

CONNECT to the Essential Question

How does Mexico reflect both ancient traditions and the challenges of the modern world?

Ask students what they have learned so far that can help them answer this question. Students might mention:

- Mexico gained independence, ending Spain's long colonial rule.
- Mexico endured major conflicts, both internally and externally, in its struggle to become a democracy.
- Mexico's Constitution has lasted since 1917.

4 Assess & Reteach

Assess Have students complete the Section Assessment.

Unit Resource Book: Latin America
- Section Quiz, p. 107

Interactive Review @ClassZone.com
- Section Quiz

Power Presentations

Test Generator

Reteach Have the class form into small groups. Direct each group to create a five-question matching test on either Mexico before 1910 or after 1910, using chart paper. Then have groups take one another's test.

Unit Resource Book: Latin America
- Reteaching Activity, p. 110

SECTION 1 ASSESSMENT ANSWERS

Terms & Names

1. Santa Anna: Mexican general and president who defeated the forces at the Alamo; Benito Juárez: Indian who was a Mexican national hero and president; Constitution of 1917: Mexican constitution still in effect today; Vicente Fox: Mexican president elected in 2000

Use Your Reading Notes

2. It was opposed by several groups, which left Mexico weak and open to foreign invasion.
   Diagram *War between Mexico and the U.S.; Guaranteed freedom of speech, federal form of government; Mexican Revolution; Farmland redistributed, changes in workers' rights*

Key Ideas

3. It was a dispute over the border of Texas.
4. Leaders arose in different parts of Mexico and gathered their own armies. The armies had different goals and ended up fighting each other.
5. The government took the large estates of the rich landowners and gave parcels of land to small farmers and villages.

Critical Thinking

6. Each group had a different idea about equality, land ownership, and reforms needed in Mexico.
7. Juárez was an Indian. His family would have been among the landless people. His education enabled him to fight for reform.
8. Possible response: Having two or more parties can better represent the views of more people.
9. Use the rubric to score lists.

List Rubric

	Points of Interest	Use of Research
4	5 or more	many interesting facts
3	4	several facts
2	3	few facts
1	2	no relevant facts

1 Plan & Prepare

Objectives

- Describe the blend of people and lifestyles in Mexico today
- Identify common themes in works by Mexican artists and writers
- Summarize the popular celebrations and sports of Mexico

Read for the Essential Question

Help students read for a purpose by reminding them of the Essential Question: "How does Mexico reflect both ancient traditions and the challenges of the modern world?" Point out that they can use text features, such as Key Questions, special features, images, and maps to form answers to the question.

Vocabulary

Best Practices Toolkit

Use the ***Best Practices Toolkit*** to model strategies for vocabulary notetaking. Vary strategies throughout the year. Choose from: *Knowledge Rating, Predicting ABCs, Definition Mapping, Word Sort, Word Wheel, Frayer Model (Word Squares), Magnet Words,* and *Student VOC.*

Vocabulary Strategies, TT9–TT16

Reading Strategy

Best Practices Toolkit

Display the T Chart/Two-Column Chart transparency.

- Explain to students that a summary is a shorter version of a block of text, and that they can use the Key Questions to help them summarize each topic in this section.
- Model adding summary entries, using the information on "People and Lifestyle" in this section and the Key Question.
- Have students suggest entries.
- Display the completed chart for students to review.

T Chart/Two-Column Chart, TT20

SECTION 2 Reading for Understanding

Key Ideas

BEFORE, YOU LEARNED

Mexico's history reflects the impact of ancient civilizations, colonial powers, and the modern world.

NOW YOU WILL LEARN

These three influences affect the culture and daily lives of the Mexican people.

Vocabulary

TERMS & NAMES

Plaza of the Three Cultures plaza in Mexico City that shows parts of Aztec, Spanish, and modern influences in Mexico

colonia a neighborhood in Mexico

squatter person who settles on unoccupied land without having legal claim to it

mural a wall painting

Diego Rivera famous muralist who painted the history of Mexico on the walls of the National Palace

fiesta a holiday celebrated with parades, games, and food

Day of the Dead holiday to remember loved ones who have died

REVIEW

urban having to do with a city

rural having to do with the countryside

push factor a reason that causes people to leave an area

Visual Vocabulary fiesta

Reading Strategy

Re-create the chart shown at right. As you read and respond to the **KEY QUESTIONS**, use the chart to summarize each of the main sections of Section 2.

See Skillbuilder Handbook, page R5

SUMMARIZE

SECTION	SUMMARY
People and Lifestyle	
Mexico's Great Murals	
Celebrations and Sports	

GRAPHIC ORGANIZERS
Go to **Interactive Review** @ClassZone.com

PRETEACHING VOCABULARY

English Learners

Pronounce and Preview

Pronounce each term for students. Review the meanings of words within definitions, such as *influences, unoccupied,* and *legal claim.*

- To modify vocabulary learning, have students complete worksheets as they read, instead of afterward.

Unit Resource Book: Latin America

- Vocabulary Practice, p. 93

Inclusion

Guess the Terms and Definitions

Have students form small groups. Then read the terms and their definitions aloud. Have each group choose one term to paraphrase, dramatize, or illustrate while the other groups guess which one it is. Cycle through the terms more than once, challenging groups to present the terms in new ways.

Pyramid of Kukulkan

Monarch butterflies

A Blend of Traditions

Connecting to Your World

When you look around your community, do you see the influences of different cultures? Are there places that date back many years and places that are modern? Mexico today reflects a blend of different cultures, both traditional and modern. The **Plaza of the Three Cultures** in Mexico City displays this blend. The plaza contains the ruins of an Aztec city, a Spanish colonial church, and modern government buildings. Other parts of life in Mexico also reflect these influences.

People and Lifestyle

KEY QUESTION How do urban and rural life in Mexico differ?

Mexico today is a living blend of Indian, Spanish, and modern influences. The majority of Mexican people are *mestizos*. Almost all Mexicans speak Spanish, the nation's official language. Many Mexican Indians also speak their Indian languages, such as Maya and Náhuatl. Most Mexicans are Roman Catholics, but some belong to other religions.

The Plaza of the Three Cultures This plaza in Mexico City reflects the influences of the Aztec, Spanish, and modern cultures.

2 Focus & Motivate

3-Minute Warm-Up

Write on the board or display the transparency:

- How do the images in this section show a blend of traditional and modern? *(Possible answers: traditional and modern buildings and clothes; transportation by animal and car; ancient customs and modern sports)*

Unit Transparency Book: Latin America
- 3-Minute Warm-Ups, TT11

Connecting to Your World

More About . . .

Plazas of Mexico City

Many public places in Mexico have "plaza" in their name. Some of the ones in and around Mexico City are Plaza de la Constitución in the old city center, Plaza Mexico (the world's largest bullring), and Plaza Satélite (a shopping mall).

3 Teach

People and Lifestyle

Reader, Recorder, Reporter

- What are some ways to describe the people of Mexico today? *(Possible answers: Most are of Mexican Indian and/or Spanish backgrounds; speakers of Spanish; Catholic; urban dwellers; and value the family.)*
- **Summarize** Summarize the importance of family life in Mexico. *(Possible answer: Social life centers on the family; older members are honored; several generations may live together.)*

SECTION 2 PROGRAM RESOURCES

Unit Resource Book: Latin America

ON LEVEL

Unit Resource Book
- Reading Study Guide, p. 71
- Map and Graph Skills, p. 95
- Section Quiz, p. 108

STRUGGLING READERS

Unit Resource Book
- Reading Study Guide with Additional Support, p. 77
- Skillbuilder Practice: Comparing and Contrasting, p. 94
- Section Quiz, p. 1078
- Reteaching Activity, p. 111

eEdition with Audio DVD-ROM

ENGLISH LEARNERS

Pupil Edition in Spanish
eEdition with Audio DVD-ROM
Unit Resource Book
- RSG (Spanish), p. 83
- RSG with Additional Support (Spanish), p. 89

Multi-Language Glossary
Test Generator
- Section Quiz in Spanish

INCLUSION

Unit Resource Book
- RSG with Additional Support, p. 77
- Section Quiz, p. 108
- Reteaching Activity, p. 111

GIFTED & TALENTED

Unit Resource Book
- History Makers: Antonio López de Santa Anna, p. 99
- Section Quiz, p. 108

PRE-AP

Unit Resource Book
- Section Quiz, p. 108

TECHNOLOGY

Unit Transparency Book
- 3-Minute Warm-Ups, TT11
- Essential Question, TT12
- Critical Thinking, TT14

Daily Test Practice Transparencies
- Section 2, TT27

Power Presentations
ClassZone.com
World Cultures and Geography Video on DVD

COMPARING

Urban and Rural Life

Mexico has almost twice as many unpaved roads (over 144,000 miles) as paved roads (under 73, 000 miles). How do you think this contributes to rural and urban population numbers? *(Possible Answer: There are probably more paved roads in cities, which would attract more jobs and bring more people to live there.)*

CRITICAL THINKING ANSWER

Make Generalizations The population in a Mexican city is larger and more dense than in rural areas; rural life appears to be more traditional than urban life.

Unit Resource Book: Latin America
- Skillbuilder Practice: Comparing and Contrasting, p. 94

More About . . .

Use of Wood in Mexico

Deforestation is a world problem today. Mexico is one of the most threatened areas in this regard, and its government considers deforestation a national security issue. Researchers have identified a combination of causes of deforestation, legal and illegal.

Legal causes include:
- government sale of timber
- clearing land for agriculture and cattle raising

Illegal causes include:
- farming by poor people who have no jobs and no land of their own
- illegal clearing of land to sell timber or grow drug crops

City and Country Life Today, over three-fourths of Mexico's people live in **urban** areas. More than 22 million people live in and around Mexico City, making it one of the largest cities in the world. Mexican cities include high-rise office buildings and modern apartment buildings, as well as older houses built in the Spanish colonial style.

The cities' neighborhoods are called ***colonias***. Wealthy people generally live in prosperous neighborhoods, away from the city center. Poorer people live closer to the center of the city, sometimes in neighborhoods with unpaved streets and no running water. Some people work in factories or as street vendors, but many are unemployed.

In contrast to city life, people in Mexico's **rural** areas live in villages or on farms, near their fields. Homes are small, often having only one room and a dirt floor. Rural areas have few health-care services, roads, and schools. The Mexican government is working to improve the public services in rural areas.

The poor conditions in the rural areas are the **push factors** that cause people to move to cities. But once there, they become **squatters**, people who settle on unoccupied land without having legal claim to it. Over time, these areas develop into new *colonias*. Some rural Mexicans migrate to other countries, including the United States.

COMPARING Urban and Rural Life

URBAN	RURAL
• Densely populated; about 75% of the population	• Less densely populated; about 25% of the population
• Primary schools, secondary schools, and universities	• Primary schools; almost no secondary schools; no universities
• Major source of energy is electricity	• Major source of energy is firewood
• Clothes similar to those worn in the United States; bought in stores	• Clothes often traditional; sometimes homemade

CRITICAL THINKING

Make Generalizations What generalization could you make about rural life compared to life in a Mexican city?

DIFFERENTIATING INSTRUCTION

English Learners

"What do I see?"

Ask students to closely examine the urban and rural scenes on this page. Then have them list as many things (or nouns) in each picture as they can, using adjectives (for example, tall buildings, rough dirt road, and so on). Then read the "City and Country Life" text out loud, and ask volunteers to offer examples of how these scenes support the information in the text.

Gifted & Talented

Plans for Improving Rural Life

Have students form small groups and brainstorm some ideas for improving rural life in Mexico. Ask students to list the ideas on chart paper. Then have students choose one idea for improvement of rural conditions, and develop a written plan for carrying it out. Invite a spokesperson from each group to present its plan.

Family Life Family life is important in Mexico. In some families, several generations of family members may live together. Older members of the family are honored. Children are taught to respect adults.

The social life of many Mexicans centers on the family. Families gather together to celebrate birthdays and other kinds of holidays.

Mexican children are required to attend school for nine years—six years of primary school and three years of middle school. Some continue their education in a three-year high school and then a university. In some rural areas, students attend classes in one-room schools. They often travel to nearby towns to go to middle school or high school.

COMPARE AND CONTRAST Describe the differences between urban and rural life in Mexico. Urban: greater access to services and schools; rural: less congested

Mexico's Great Murals

KEY QUESTION What common subjects do Mexican artists and writers focus on in their works?

Creating art has a long history in Mexico. The Olmec carved sculptures. The Aztec wrote music and poetry. The Maya created paintings in their pyramids and the Spanish in their colonial churches. Mexican art today blends these influences.

After the Mexican Revolution, Mexican art experienced a great awakening. Just as ancient artists had created paintings on their pyramids, Mexican artists created **murals**, or wall paintings, that often depicted scenes from Mexico's history. Many important painters portrayed such scenes on murals in Mexico's public buildings. The most famous Mexican muralists are José Orozco, David Siquieros, and **Diego Rivera**. Frida Kahlo, another important Mexican painter, is known for paintings that show her personal feelings about events in her life.

Self-portrait with Monkey Frida Kahlo is known for her self-portraits. Monkeys and other animals roamed the gardens around her home and were often included in her artwork.

Mexican writers have often written about Mexico's social and political problems. Octavio Paz, Carlos Fuentes, and Laura Esquivel are three well-known Mexican writers. Paz won the 1990 Nobel Prize for Literature; he was the first Mexican to win this award. Fuentes writes novels about Mexican history. In one popular novel, Esquivel described life for rural Mexican women during the Mexican Revolution.

SUMMARIZE Describe the subjects that Mexican artists and writers focus on in their works. events from Mexican history, Mexico's social and political problems, and about life in Mexico at certain times

Comparing Regions

One-Room and Regional Schools in Mexico and the United States

As in Mexico, in some rural areas of the United States students still attend one-room schools and travel to other towns for high school. Gold Creek Elementary School in Montana is an example of one of about 400 such one-room schools throughout the United States. In western Massachusetts, Mohawk Trail Regional High School (grades 7 to 12) has students from nine different towns.

Teach

Mexico's Great Murals

Talk About It

- Why do you think Mexican art flourished after the Mexican Revolution? *(Possible answer: The revolution made people feel that everyone's lives and history were important, not just the lives of the rich and powerful.)*
- **Summarize** How can art help people feel more involved in their society? *(Possible answer: It can lead people to discuss topics such as history, social problems, and political struggles.)*

Unit Resource Book: Latin America
- Primary and Secondary Sources: "The Rabbit and the Crab," p. 106

More About . . .

Octavio Paz

In his 1990 Nobel Lecture, Octavio Paz talked about how Mexico's long history is still alive today: "In Mexico, the Spaniards encountered history as well as geography. That history is still alive: it is a present rather than a past. The temples and gods of pre-Columbian Mexico are a pile of ruins, but the spirit that breathed life into that world has not disappeared. . . . Being a Mexican writer means listening to the voice of that present, that presence. Listening to it, speaking with it, deciphering it: expressing it. . . ."

Unit Resource Book: Latin America
- Interdisciplinary Projects: Octavio Paz, pp. 102, 103

INTERDISCIPLINARY ACTIVITIES

CONNECT to Art

Design a Mexico Tourism Poster

Invite groups of students to design a Mexico tourism poster based on what they have read in this chapter so far. First, have each group brainstorm ideas of images and text to include, so the poster conveys Mexico's diverse history and culture. Second, have groups sketch their poster layout and share it with the class for constructive feedback. Finally, have groups complete their posters and present them to the class .

CONNECT to Language Arts

Writing Fiction About Local Life

Read to the class a brief fiction selection, such as the opening pages of the novel *The Jumping Tree* by René Saldaña, Jr. Ask students to identify details they heard, such as who the characters are, the setting and scene, and what is happening. Then invite each student to choose a historical or current event that took place in their community, and write the first paragraph of a fictional story about it.

Teach

Celebrations and Sports

Think, Pair, Share

- What is *Cinco de Mayo? (May 5 celebration of the Mexican army's defeat of the French in 1862)*
- **Summarize** What are some popular celebrations and sports in Mexico? *(Independence Day—Sept. 16;* Cinco de Mayo*—May 5; Day of the Dead—Nov. 1 and 2;* La Quinceañera*—girl's 15th birthday; soccer, bullfighting, jai alai)*

Fun Facts!

Piñatas

The piñatas of today have been adapted by people to reflect new times and places. They are sold over the Internet for parties in the United States, and not just for children's parties. There are piñatas in the shapes of many animals and insects, piñatas of TV and comic book characters, talking piñatas, and piñatas of wedding bells, wedding cakes, brides, and grooms.

CONNECT History & Culture

Day of the Dead

The Aztec version of Mexico's Day of the Dead was a nearly month-long celebration led by the goddess Mictecacihuatl ("Lady of the Dead"). After Catholicism came to Mexico, the date was changed to match the church's celebrations of All Souls' Day and All Saints' Day in early November.

CRITICAL THINKING ANSWER

Make Inferences The holiday is celebrated by families gathering together to remember family members who have died.

Fun Facts!

PIÑATAS

Piñatas are part of many Mexican fiestas. But where did piñatas come from? Some historians believe Marco Polo brought them from China to Italy in the 1200s. The Spanish brought them to Spain and then introduced them to Mexico.

Celebrations and Sports

KEY QUESTION What kinds of holidays do Mexicans celebrate?

Mexicans celebrate many holidays and events with a fiesta. A **fiesta** is a celebration with fireworks, parades, music, dancing, and foods. At some fiestas, children enjoy themselves by trying to break open a piñata. A piñata is a decorated container filled with candy and toys, usually hung from the ceiling. Children are blindfolded and given a stick to break open the piñata and gather its treats. Fiestas bring people together to have fun.

Celebrations Mexicans celebrate Mexico's Independence Day on September 16. Mexicans also celebrate *Cinco de Mayo* (May 5), which is the day in 1862 when the Mexican army defeated the French.

Mexicans celebrate religious holidays such as Easter with church services and processions. They celebrate the **Day of the Dead** on November 1 and 2 to honor family members who have died.

La Quinceañera is a celebration of a Mexican girl's 15th birthday. The girl dresses in a full-length gown. She first takes part in church services with family and friends. After church, everyone goes to a reception where a huge fiesta takes place.

CONNECT History & Culture

Day of the Dead This holiday has its roots in an ancient Aztec celebration, in which people remembered dead ancestors. Today, particularly in rural areas, relatives celebrate the Day of the Dead by gathering in cemeteries to decorate family graves with candles and flowers and to share stories about loved ones who have died.

CRITICAL THINKING

Make Inferences How does this holiday show the importance of family to Mexican people?

240 Chapter 8

DIFFERENTIATING INSTRUCTION: TIERED ACTIVITIES

OBJECTIVE Compare and contrast Mexican celebrations to your own.

Basic	On Level	Challenge
Have students work in pairs to create a poster that shows one of the holidays described on this page. Posters should indicate the day the holiday is celebrated and the reason for the celebration. Students' posters should also include illustrations of a similar holiday that they celebrate.	Have students work in pairs to present a "conversation" in which a Mexican student and an American student talk about the holidays they celebrate. The conversations should explain why each holiday is celebrated. It should also point out similarities and differences between the holidays.	Have students research other holidays celebrated in Mexico. Ask them to present their findings in an article about Mexican holidays and celebrations. Direct students to include comparisons of the Mexican holidays with ones celebrated in the United States.

Sports The number one sport in Mexico is *fútbol*, or soccer. Mexicans enjoy playing it and watching it. Fans cheer on their favorite teams in stadiums located in several Mexican cities. Aztec Stadium in Mexico City holds 114,000 fans! Mexicans also enjoy playing baseball and watching their professional baseball teams.

Many Mexicans enjoy bullfighting, a sport brought to Mexico by the Spanish. The number of bullfights has decreased in recent years because some people have led campaigns to ban them. However, people still attend bullfights in the bullrings found in many Mexican cities.

A sport growing in popularity in Mexico is jai alai, a fast-paced ball game played in a three-walled court. The equipment includes a hard rubber ball and wicker-basket scoops, which players use to catch and throw the ball.

Soccer in Aztec Stadium Fans watch as a player from the Mexican team (left) battles for the ball in a game with a player from Argentina.

FIND MAIN IDEAS Identify the kinds of holidays that Mexican people celebrate today.
historical holidays, religious holidays, and personal events

Section 2 Assessment

ONLINE QUIZ For test practice, go to **Interactive Review @ClassZone.com**

TERMS & NAMES

1. **Explain the importance of**
   - Plaza of the Three Cultures
   - *colonia*
   - mural
   - fiesta

USE YOUR READING NOTES

2. **Summarize** Use your completed chart to answer the following question:
   What are some ways Mexicans spend their leisure time?

SECTION	SUMMARY
People and Lifestyle	
Mexico's Great Murals	
Celebrations and Sports	

KEY IDEAS

3. What two cultures have influenced modern Mexico?
4. Where do most Mexican people live today?
5. What is the subject of many Mexican murals?

CRITICAL THINKING

6. **Draw Conclusions** Why do many people in Mexico's rural areas move to cities?
7. **Form and Support Opinions** What problems of urban and rural life do you think are most important for Mexico to address?
8. **CONNECT to Today** Some people have led campaigns to ban bullfighting. What kinds of campaigns does your area have on behalf of animals?
9. **ART** **Create a Mural Panel** Work with a group to research an important event in the history of your community. Decide what aspect of the event you want to illustrate. Then create a mural panel to depict the scene you have chosen.

CONNECT to the Essential Question

How does Mexico reflect both ancient traditions and the challenges of the modern world?

Ask students what they have learned so far that can help them answer this question. Students might mention:

- Old buildings and ruins exist in the midst of crowded, modern cities.
- Mexico is working to improve rural areas and poor urban areas.
- Mexico's arts, celebrations, and sports blend historical and modern elements.

4 Assess & Reteach

Assess Have students complete the Section Assessment.

Unit Resource Book: Latin America
- Section Quiz, p. 108

Interactive Review @ClassZone.com
- Section Quiz

Power Presentations

Test Generator

Reteach Have the class or small groups play a round-robin game using the information in this section. The first person starts a sentence, such as "Santa Anna was. . . ." The next person finishes it, and then starts a new sentence. Continue until all the section's key points have been covered.

Unit Resource Book: Latin America
- Reteaching Activity, p. 111

SECTION 2 ASSESSMENT ANSWERS

Terms & Names

1. Plaza of the Three Cultures: plaza shows the three cultures that blended to make modern Mexico; *colonia*: a Mexican neighborhood; mural: wall painting; fiesta: a holiday celebrated with parades, games, and food

Use Your Reading Notes

2. They spend leisure time participating in fiestas and in playing and watching sports.

SECTION	SUMMARY
People and Lifestyle	Most are mestizos; family is important
Mexico's Great Murals	Murals important art form; writers write about Mexico's problems
Celebrations and Sports	Historical and religious holidays; soccer most popular

Key Ideas

3. Spanish and native cultures, like the Aztec, have blended to form modern Mexico.
4. Most live in urban areas.
5. They depict scenes from Mexico's history.

Critical Thinking

6. They move to cities to find jobs and to take advantage of the services available there that are lacking in rural areas.
7. Answers will vary, but may include providing modern services in rural areas and dealing with the problem of unemployment in urban areas.
8. Possible response: Some people think the treatment of circus animals is cruel. Other groups work to save endangered animals or work against killing animals to make fur coats.
9. Use the rubric to score murals.

Mural Rubric

	Evidence of Research	Presentation of Information in Mural
4	thorough and detailed	excellent
3	many good details	good
2	some good details	adequate
1	incomplete or haphazard	weak

1 Focus & Motivate

Objective

Compare and contrast Mexico's Indian and Spanish cultures.

Springboard Have students consider the following statement as they review the Comparing feature: All groups of people are in some ways the same and in some ways different.

2 Teach

Comparing Mexico's Cultural Traditions

Talk About It

- Why was 1521 a significant date in Mexican history? *(the Spanish conquest, which brought new customs and traditions)*
- How might the site of the Mayan temple ruins have looked when the temples were still in use by the Maya? *(Possible answers: fewer trees; people in special clothing; no tourists or visitors from the cultures)*
- **Make Inferences** How important was religion for Mexico's Indian and Spanish cultures? *(Possible answer: very important, because both groups built impressive religious buildings in their cities)*

More About . . .

Native Languages in Mexico

The 2005 Mexican census showed that from 1950 to 2005, the percentage of people who spoke a native language decreased from slightly more than 11 percent to slightly less than seven percent. But because of overall population growth, the number of speakers of a native language more than doubled, from just under 2.5 million to over six million.

COMPARING Mexico's Cultural Traditions

Mexican culture is a blend of both Indian and Spanish traditions. The influences of both cultures are evident in Mexico today.

Indian Traditions

Before the Spanish conquest in 1521, Mexico's Indian groups had well-established, unique cultures. Indian groups created cities with huge structures. They held celebrations, such as festivals to remember those who died. Crops such as corn, chilies, and tomatoes contributed to their flavorful diet.

Food

Tortillas, a popular Mexican food made from corn, date back to the ancient Aztecs.

Clothing

These girls wear handmade, colorful dresses.

▲ *Ruins of Mayan temple at Palenque, Mexico*

Festivals

Mexican children of Indian ancestry dressed in traditional clothing participate in a local festival.

DIFFERENTIATING INSTRUCTION: TIERED ACTIVITIES

OBJECTIVE Compare and contrast Mexico's Indian and Spanish cultures.

Basic

Have students work individually or in pairs to create a Venn diagram that compares and contrasts Mexico's Indian and Spanish traditions. As a class or whole group activity, invite students to list one or two entries from their diagrams on a Venn diagram drawn on the board or chart paper.

On Level

Have students create a Venn diagram as for the Basic activity. Then have them write a paragraph, summarizing the information on their diagrams. Invite students to read their summaries to the class.

Challenge

Have students create a compare-and-contrast graphic organizer of their choice. Then have them write a short compare-and-contrast essay, based on the information in their organizers. Invite students to read their essays to the class or display completed essays in the class.

Food

Mexican sweet breads are made from wheat, which the Spaniards brought to Mexico.

Spanish Traditions

The Spanish conquest brought European customs and traditions to the region. The Spanish built their traditional square cities that surrounded plazas and included government buildings and churches. They honored their dead on All Saints' Day and All Souls' Day in November. They introduced foods such as oranges, wheat, beef, pork, sugar cane, coffee, and onions to the Mexican diet.

▲ *Spanish colonial church in Taxco, Mexico*

Clothing

These dancers are dressed in traditional Spanish clothing from the colonial period.

Festivals

Mexican children participate in a procession celebrating Easter, a Roman Catholic holiday.

CRITICAL THINKING

1. **Evaluate** Why are Mexican traditions today a blend of Native American and Spanish traditions?
2. **Compare and Contrast** Based on the large photos, how do Native American and Spanish architecture compare?

Comparing Regions

Church Architecture in Spain and Mexico

The Church of Santa Prisca in Taxco, with its many highly-ornamented altarpieces, is an example of the Churrigueresque-baroque architecture popular in Spain in the late 17th and early 18th centuries. Although other architects also used this elaborately decorated style, the name comes from the Churrigueras, a family of architects whose father was an altarpiece maker. One of Spain's most famous examples of this architecture is the sacristy—the room where a church's sacred vessels and vestments are kept—of the Cartuja in Granada.

3 Wrap Up

Remind students of the Essential Question: How does Mexico reflect both ancient traditions and the challenges of the modern world? Have students work in small groups to answer this question based on the information in Comparing Mexico's Cultural Traditions.

CRITICAL THINKING ANSWERS

1. **Evaluate** They are a blend of both traditions because both cultures have contributed to life in Mexico today.
2. **Compare and Contrast** Possible answer: Both buildings are places of worship; Spanish church made of brick, Mayan temple of stone; Spanish architecture appears more elaborate.

INTERDISCIPLINARY ACTIVITIES

CONNECT to Math

What shapes are found in buildings?

Challenge students to identify as many shapes as they can in the buildings shown on these two pages. Have them identify not just basic shapes (such as rectangles and circles), but also other kinds, such as trapezoids and pyramids. Have students share their findings.

CONNECT to Language Arts

Haiku About Mexico

Have students write a haiku, about the blending of traditions in Mexico in general or about one specific tradition. Remind students that a haiku is a poem consisting of three unrhymed lines of five, seven, and five syllables. You may also want to share examples of haikus. Have students share their poems and explain why they chose the theme they did. To extend this activity, have students create illustrated versions of their poems, using stylized writing.

1 Plan & Prepare

Objectives

- Explain how Mexico is building a strong economy
- Describe the challenges Mexico faces in strengthening its economy

Read for the Essential Question

Help students read for a purpose by reminding them of the Essential Question: "How does Mexico reflect both ancient traditions and the challenges of the modern world?" Point out that they can use text features such as Key Questions, special features, images, and maps to form answers to the question.

Vocabulary

Best Practices Toolkit

Use the ***Best Practices Toolkit*** to model strategies for vocabulary notetaking. Vary strategies throughout the year. Choose from: *Knowledge Rating, Predicting ABCs, Definition Mapping, Word Sort, Word Wheel, Frayer Model (Word Squares), Magnet Words,* and *Student VOC.*

Vocabulary Strategies, TT9–TT16

Reading Strategy

Best Practices Toolkit

Display the Web Diagram transparency.

- Remind students that main ideas are the key points about a topic discussed in a section of text.
- Model filling in entries from the "Building a Modern Economy" subsection in this section.
- Have students suggest entries.
- Display the completed diagram for review.

Web Diagram, TT28

SECTION 3 Reading for Understanding

Key Ideas

BEFORE, YOU LEARNED

The Mexican government has become more democratic with the elimination of the one-party system.

NOW YOU WILL LEARN

Mexico faces new challenges as the country takes steps to modernize its economy.

Vocabulary

TERMS & NAMES

maquiladora factory in which materials are imported and assembled into products for export

global economy economy in which buying and selling occurs across national borders

North American Free Trade Agreement (NAFTA) agreement that reduced trade barriers among Mexico, Canada, and the United States

REVIEW

immigration process of coming to another country to live

Visual Vocabulary X-ray table built in a *maquiladora*

Reading Strategy

Re-create the web diagram shown at right. As you read and respond to the **KEY QUESTIONS**, find supporting details for each main idea in Section 3.

See Skillbuilder Handbook, page R4

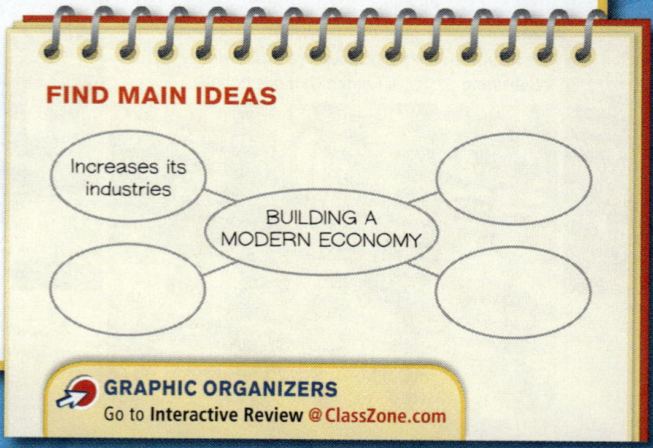

PRETEACHING VOCABULARY

English Learners

Pronounce and Preview

Pronounce each term for students. Review the meanings of words within definitions, such as *assembles, finished goods,* and *trade barriers.*

- To modify vocabulary learning, have students complete worksheets as they read, instead of afterward.

Unit Resource Book: Latin America

- Vocabulary Practice, p. 93

Inclusion

Finding Terms in Context

Have students work in pairs, with each partner scanning the text in this section for two of the terms. Each partner should read the information about one term out loud, and ask a question about it for the other partner to answer. Partners should agree that each answer is correct, reviewing the text together, if needed, before going on to the next term.

Pyramid of Kukulkan

SECTION 3

GOVERNMENT & ECONOMICS

Monarch butterflies

Creating a New Economy

Connecting to Your World

Do you add tomatoes to a salad or to your favorite kind of sandwich? Chances are that the tomatoes you eat were grown in Mexico. Tomatoes make up the largest percentage of vegetables that Mexico exports to the United States. Agriculture is an important part of Mexico's economy. In this section, you will learn how Mexico has expanded its economy.

Building a Modern Economy

KEY QUESTION How has Mexico created a strong modern economy?

Vicente Fox's election as president in 2000 changed Mexico from a one-party system to a multiparty democracy. With a stable government, Mexico is working to improve the quality of life for its people. It has been modernizing its industry and economy by cooperating with other North American countries.

Animated GEOGRAPHY

Mexican Industry Workers perform their jobs on a floating oil-rig platform off the Mexican coast in the Gulf of Mexico.

Click here for an interactive map of Mexican industry @ ClassZone.com

245

2 Focus & Motivate

3-Minute Warm-Up

Write on the board or display the transparency:

- Read this section's title. Why would a country want to create a new economy? *(Possible Answers: to create better jobs and improve the standard of living; to earn more money to pay for improvements in poor areas)*

Unit Transparency Book: Latin America
- 3-Minute Warm-Ups, TT11

Connecting to Your World

More About . . .

Mexican Industry

Both Mexico and the United States have oil fields in the Gulf of Mexico. These oil fields are expected to continue to help each country's economy, especially with the recent discoveries of new oil fields there.

3 Teach

Building a Modern Economy

Reader, Recorder, Reporter

- Why was Vicente Fox's election important to Mexico's economy? *(The change to a multiparty system made Mexico's government more stable, making it possible for the government to work on economic improvements.)*
- **Find Main Ideas** How has Mexico been building its economy? *(by industrialization—oil, factories; by taking part in the global economy—NAFTA)*

SECTION 3 PROGRAM RESOURCES

Unit Resource Book: Latin America

ON LEVEL

Unit Resource Book
- Reading Study Guide, p. 73
- Outline Map Activities, p. 97
- Section Quiz, p. 109

STRUGGLING READERS

Unit Resource Book
- Reading Study Guide with Additional Support, p. 79
- Skillbuilder Practice: Comparing and Contrasting, p. 94
- Section Quiz, p. 109
- Reteaching Activity, p. 112

eEdition with Audio DVD-ROM

ENGLISH LEARNERS

Pupil Edition in Spanish

eEdition with Audio DVD-ROM

Unit Resource Book
- RSG (Spanish), p. 85
- RSG with Additional Support (Spanish), p. 91

Multi-Language Glossary

Test Generator
- Section Quiz in Spanish

INCLUSION

Unit Resource Book
- RSG with Additional Support, p. 79
- Section Quiz, p. 109
- Reteaching Activity, p. 112

GIFTED & TALENTED

Unit Resource Book
- History Makers: Vicente Fox, p. 100
- Interdisciplinary Projects, pp. 101, 103
- Section Quiz, p. 109

PRE-AP

Unit Resource Book
- Economics in the World, p. 104
- Connect to Science, p. 105
- Section Quiz, p. 109

TECHNOLOGY

Unit Transparency Book
- 3-Minute Warm-Ups, TT11
- Essential Question, TT12
- Critical Thinking, TT15

Daily Test Practice Transparencies
- Section 3, TT28

Power Presentations

ClassZone.com

World Cultures and Geography Video on DVD

RESEARCH & WRITING CENTER

ANALYZING Primary Sources

Vicente Fox grew up on a ranch in the state of Guanajuato, in rural central Mexico. He saw many people living in poverty. Later, in his job as a route supervisor, he saw many remote regions of the country. These experiences helped shape his desire to improve Mexico's economy.

- What problems might Fox have faced in trying to make changes happen in a multiparty government? *(Possible answer: Members of the political party he defeated might not have supported him.)*

DOCUMENT-BASED QUESTION ANSWER

Possible answer: Fox thinks that the government cannot solve all social problems; he believes that education, jobs, and developing Mexico's regions can remove poverty.

Unit Resource Book: Latin America
- History Makers: Vicente Fox, p. 100

More About . . .

Maquiladoras

Currently, about one-fourth of the 43 million Mexican workers may be underemployed, doing only part-time or low-skilled work. To help boost employment, *maquiladoras* have become common in parts of Mexico (and in other countries as well). As the number of *maquiladoras* has grown, so has debate about their benefits.

While they boost a country's economy overall, some observers of *maquiladoras* claim that too little of the profit goes back to the workers, who may live and work in very harsh conditions.

Unit Resource Book: Latin America
- Economics in the World: Mexico's Growth in Gross Domestic Product, p. 104

ANALYZING Primary Sources

Vicente Fox promised as president to improve education, jobs, and opportunities in Mexico. For his official portrait, Fox posed with Mexican people from all walks of life to show, as the caption indicates, that "We are all Mexico." Read what Fox had to say in his inaugural speech.

> It is time we recognized that everything cannot be solved by the State [government]. . . . Quality education, employment and regional development are the levers to remove, once and for all, the signs of poverty.
>
> Source: Vicente Fox's inaugural speech

DOCUMENT-BASED QUESTION

What does Fox think is the best way to reduce poverty?

ONLINE PRIMARY SOURCE To read more works of Vicente Fox, go to the **Research & Writing Center @ClassZone.com**

Industrialization Traditionally, Mexico's economy depended upon agriculture and mining. But since the 1940s, Mexico has become more industrialized. Mexico continues to be one of the world's major producers of silver. Its most profitable industry is oil production. Today, Mexico is tied with China as the world's fifth largest producer of crude petroleum.

Other industries have also become important to Mexico's economy. In recent years, many factories have been located along Mexico's border with the United States. These factories are called *maquiladoras.* A ***maquiladora*** is a factory that assembles imported materials into finished goods that are exported. Jeans, appliances, car engines, and computers are some of the products manufactured in this way.

Global Economy In 1992, Mexico took a stronger role in the **global economy**, in which nations cooperate to trade goods and services. It signed the **North American Free Trade Agreement (NAFTA)** with the United States and Canada. NAFTA created rules about trade in North America. The agreement made it easier for the three countries to transport goods and services across their borders.

DRAW CONCLUSIONS Discuss the steps Mexico has taken to build its economy. Mexico has become more industrialized. Its most profitable industry is oil production. Mexico has increased production of manufactured goods and taken a greater role in the global economy.

INTERDISCIPLINARY ACTIVITIES

CONNECT to Civics

"Improving Our Community" Speech

Have the class discuss areas for improvement in their community and brainstorm possible solutions. List these on the board. Have them discuss the pros and cons of each solution. Then have students write and present a brief speech, as if speaking to fellow citizens, identifying the most important problem(s) for the community to address and the best solution(s). Encourage them to use language that is both engaging and convincing.

CONNECT to Math

Charting World Oil Production

Provide students with statistics for recent world oil production (from www.eia.doe.gov or another source). Have them chart the information for the top ten oil-producing countries, using a bar graph or pie chart. To extend this activity, have them graph the top-ten oil-using countries.

Unit Resource Book: Latin America
- Interdisciplinary Projects: The Population of Mexico, pp. 101, 103

Facing New Challenges

KEY QUESTION What are some problems Mexico must solve in order to continue to develop a stronger economy?

Mexico faces two major challenges in continuing to develop a strong economy. It has to reduce pollution, and it has to create more jobs in order to improve conditions for its people and to slow migration to other countries.

Pollution Like other large international cities, Mexico City has had to deal with increased air pollution, brought on in large part by industrialization. Mexico City sits in a valley, almost completely surrounded by mountains. The mountains help produce a layer of warm air above the city. This layer keeps gases from car exhaust and thousands of industries from blowing away. These gases react with sunlight to form smog, a thick brown haze, over the city. The smog causes health problems for many people.

Recently, the Mexican government has taken steps to deal with the pollution problem. It has urged automobile manufacturers to produce cars that use cleaner fuels and helped companies develop smog controls. It has also encouraged public transportation.

CONNECT to Science

Like many modern cities, Mexico City became industrialized in the last 50 years. While developing a modern economy helps a country, the increase in industries often causes pollution problems. Many other cities have experienced situations similar to Mexico City's.

Mexico City Smog blankets the city.

Activity

Create a Pollution Hot Spot Map

Materials

- newspapers and magazines
- large blank world map
- markers

1. Find information about pollution problems in other parts of the world.
2. Locate and label the places on the world map. Include Mexico City on the map.

3. Choose one of the locations and find out what is being done to reduce pollution there.
4. Present your map and findings to the class.

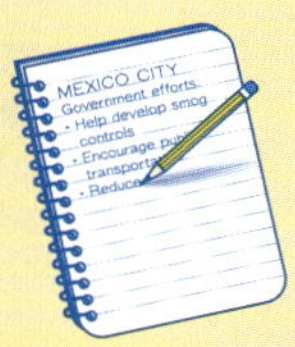

DIFFERENTIATING INSTRUCTION: TIERED ACTIVITIES

OBJECTIVE Diagram factors that contribute to pollution in Mexico City.

Basic

With students, identify the factors that add to Mexico City's pollution. List these on the board. Ask students how they could illustrate each factor in a labeled diagram to show its effects. As they suggest ideas, sketch them on the board, and then have each student create his or her own diagram based on one of the diagrams.

On Level

Have students complete the Basic activity, working in pairs. In addition, have students list at least one way each factor could be reduced or prevented from adding to Mexico City's pollution.

Challenge

Have students complete the On Level activity, working in pairs. In addition, have students research new technologies and inventions that could help reduce Mexico City's pollution and discuss their ideas with other pairs.

Teach

Facing New Challenges

Talk About It

- How does the pollution problem in Mexico City show a connection between physical features and human activity? *(Possible answer: The city is in a valley, nearly surrounded by mountains, which prevents industrial pollution from blowing away.)*
- **Find Main Ideas** How can creation of jobs in Mexico improve life for its people? *(Possible answers: may reduce overcrowding in cities; fewer people might leave to find work in the United States)*

Unit Resource Book: Latin America

- Skillbuilder Practice: Comparing and Contrasting, p. 94

Comparing Regions

Air Pollution in Mexico City, Los Angeles, and Santiago

Los Angeles, California, and Santiago, Chile—like Mexico City—are both cities with millions of people. Again like Mexico City, both have mountain barriers that trap pollution in their air.

CONNECT to Science

Air pollution consists of gases and tiny particles that get spread through the air. Some causes of air pollution, such as industrial emissions and vehicle exhaust, can be limited. Other causes—such as dust storms, some forest or brush fires, and volcanoes—usually cannot. Many countries are working to reduce pollution, for example, by making laws to lower industrial and vehicle emissions.

Unit Resource Book: Latin America

- Outline Map Activities: Mexico: Physical/Political, pp. 97–98
- Connect to Science: Create a Pollution Hot Spot Map, p. 105

CONNECT to the Essential Question

How does Mexico reflect both ancient traditions and the challenges of the modern world?

Ask students what they have learned so far that can help them answer this question. Students might mention:

- Mexico is trying to create more and better jobs through industry and global trade.
- Mexico is working to reduce pollution, especially in crowded Mexico City.

4 Assess & Reteach

Assess Have students complete the Section Assessment.

Unit Resource Book: Latin America
- Section Quiz, p. 109

Interactive Review @ClassZone.com
- Section Quiz

Power Presentations

Test Generator

Reteach Have students work in groups and create one graphic organizer on Mexico's steps to build a new economy and one on the challenges Mexico faces. Have each group present their graphic organizers to the class. Students should revise their graphic organizers as needed after the presentations.

Unit Resource Book: Latin America
- Reteaching Activity, p. 112

Providing Jobs These men are working on a large-scale construction project. **How might such projects help develop Mexico's economy?** They will help provide jobs.

Creation of Jobs The movement of people from rural to urban areas and across international borders presents a second challenge to Mexico's economy. Many Mexicans from rural areas move to cities for jobs. There, they often live in poor conditions, with no jobs or low-wage jobs.

Many Mexicans move to the United States for economic opportunities. Some risk the dangerous border crossing to the United States and enter the country illegally. Many earn money to help their families back home. Once in the United States, immigrants often work long hours at low-paying jobs, sometimes at more than one job.

As the Mexican economy creates more and better-paying jobs, the problem of illegal **immigration** may improve. In the meantime, Mexico and the United States continue to cooperate to strengthen security on their border and to find ways to resolve the immigration problems.

IDENTIFY PROBLEMS AND SOLUTIONS Identify the challenges Mexico faces in developing its economy. Mexico needs to control pollution and create jobs.

Section 3 Assessment

ONLINE QUIZ For test practice, go to **Interactive Review @ClassZone.com**

TERMS & NAMES

1. **Explain the importance of**
   - *maquiladora*
   - global economy
   - NAFTA
   - immigration

USE YOUR READING NOTES

2. **Find Main Ideas** Use your completed chart to answer the following question:

   What steps has Mexico taken to build a modern economy?

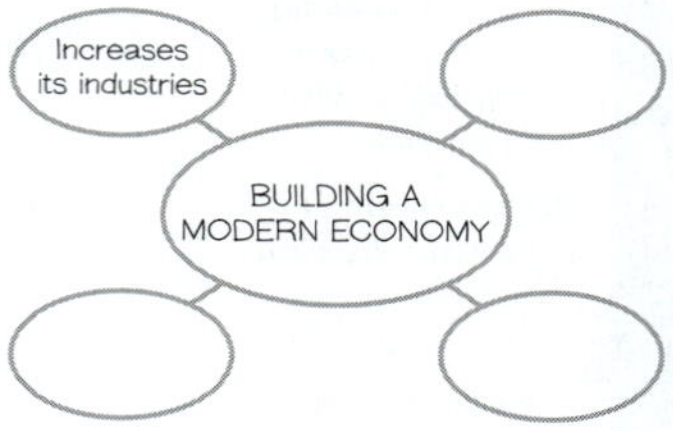

KEY IDEAS

3. What was Mexico's traditional economy based on until the 1940s?
4. What are Mexico's most important industries?
5. What are two challenges Mexico faces today?

CRITICAL THINKING

6. **Evaluate** What are the advantages and disadvantages of Mexico's membership in NAFTA?
7. **Analyze Causes and Effects** How did the growth of urban areas and industrialization contribute to Mexico's pollution problems?
8. **CONNECT to Today** Why is creating jobs important for Mexico today?
9. **WRITING** **Prepare a Report** Choose one of Mexico's major industries. Find out how many people are employed in the industry, its major locations, and how much money the industry brings to Mexico's economy. Prepare an illustrated report of your findings.

SECTION 3 ASSESSMENT ANSWERS

Terms & Names

1. *maquiladora*: factory that assembles imported materials into finished goods for export; global economy: system in which nations trade goods and services with other nations; NAFTA: agreement that reduced trade barriers among Mexico, Canada, and the U.S.; immigration: process of coming to another country to live

Use Your Reading Notes

2. developed oil production, increased manufacturing, role in global economy

Key Ideas

3. agriculture and mining
4. oil production, silver production, and manufacturing
5. pollution and creating jobs

Critical Thinking

6. Advantages: Mexico can sell products to a bigger area. It can more easily move its products easier across borders. Disadvantages: It might have conflicts with the other two countries.
7. Cities became crowded and exhaust from cars contributed to pollution. The burning of fuels to run Mexican industries also contributed to air pollution.
8. Creating jobs will help to improve living conditions. It will also help to reduce the number of people who migrate from rural areas to Mexican cities and to other countries in search of jobs.
9. Use the rubric to score reports.

Report Rubric

	Facts About Industries	Errors in Mechanics
4	very detailed	few or none
3	several details	some
2	few details	many
1	no details	major

Click here to complete these and other activities online @ClassZone.com

CHAPTER SUMMARY

Key Idea 1
Mexico overcame many obstacles as the country moved toward establishing a democracy.

Key Idea 2
Three cultures blended to create the heritage and daily life of the Mexican people.

Key Idea 3
Mexico continues to face challenges as it tries to modernize its economy.

For **Review and Study Notes**, go to **Interactive Review** @ClassZone.com

NAME GAME

Use the Terms & Names list to complete each sentence on paper or online.

1. I am the place that became independent from Mexico in 1837. **Republic of Texas**
2. I am the place where the cultural sources of modern Mexico can be seen. ______
3. I am a document that brought reforms to Mexico and is still in effect in Mexico today. ______
4. I am the president who broke Mexican one-party rule. ______
5. I am a neighborhood or suburb of a Mexican city. ______
6. I am an agreement made between North American countries. ______
7. I am a person who lives on land that is not my own. ______
8. I am a tax on imported goods. ______
9. I am a Mexican writer who won the Nobel Prize for Literature. ______
10. I am a famous Mexican mural painter. ______

colonia
Constitution of 1917
Vicente Fox
Benito Juárez
maquiladora
NAFTA
Octavio Paz
Plaza of the Three Cultures
Republic of Texas
Diego Rivera
squatter
tariff

Activities

Flip Cards

Use the online flip cards to quiz yourself on the terms and names introduced in this chapter.

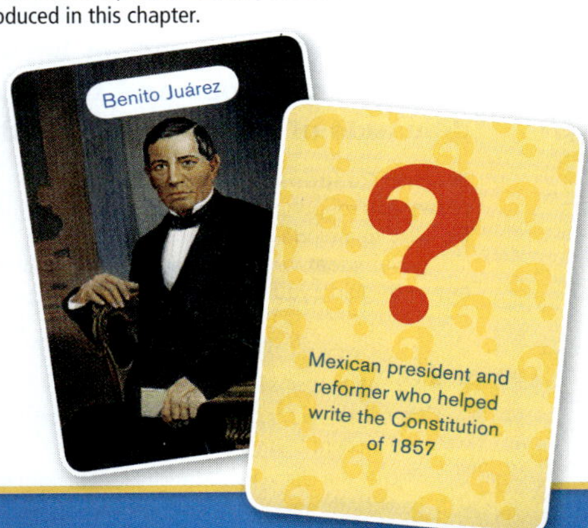

Crossword Puzzle

Complete an online crossword puzzle to test your knowledge of the history and culture of Mexico.

ACROSS
1. a holiday celebrated with parades, games, and food

Chapter Summary

As a follow-up to the chapter, encourage students to pair up and go over the **Review and Study Notes** at **ClassZone.com.**

Alternatively, students pairs can review their Reading Study Guide notes, making sure that they can add important details to each section's "Key Idea."

Name Game

Teaching Tips

ON PAPER Have students work in pairs to complete this activity. As an extension, they can write questions for the terms that are not covered in the items shown here.

ONLINE Make sure that students understand how to drag and drop terms to fill in blanks. Ask them to print out their results and compare them with those of a classmate.

ANSWERS

1. Republic of Texas
2. Plaza of the Three Cultures
3. Constitution of 1917
4. Vicente Fox
5. *colonia*
6. NAFTA
7. squatter
8. tariff
9. Octavio Paz
10. Diego Rivera

Make It Fun

Teacher-Tested Activities

Todd Harrison, Hardin County Middle School, Savannah, Tennessee

To learn the terms and names in the chapter, my class plays "What's the Question?"

- Write each term or name on a separate index card.
- Organize the class in two groups. Choose an index card, and say the term or name aloud.
- Have each group take turns giving a question to which the term you read aloud is the answer.
- The group with the most correct answers wins.

ONLINE AND ON-SCREEN ACTIVITIES

Flip Cards

Vocabulary Flip Cards are virtual flash cards at **ClassZone.com.**

Students can quiz each other or work independently to study the flip cards for the chapter, "Mexico."

- They can quiz themselves on key terms and names by reviewing definitions.
- Demonstrate how to advance and shuffle the deck for another challenge.

Crossword Puzzle

Students can review important people and events in Mexico's history by completing the online crossword puzzle at **ClassZone.com.**

- As an extension, or for more challenge, have students use the *Activity Maker* at **ClassZone.com** to make their own crossword puzzle for one section of the chapter.
- Then have them trade puzzles with classmates to solve answers for each other's sections.

Vocabulary ANSWERS

1. independent nation after winning independence from Spain
2. to add to an existing territory
3. plaza that illustrates Aztec, Spanish, and modern influences in Mexico
4. neighborhood in Mexico's cities
5. person who settles on unoccupied land without having legal claim to it
6. wall paintings, popular Mexican art
7. Mexico's famous muralist
8. holiday celebrated with parades, games, and food
9. holiday to remember loved ones who have died
10. the process of coming into another country to live
11. The Constitution of 1917 was the result of the reforms sought in the Mexican Revolution.
12. All served as president of Mexico.
13. important aspects of Mexico's modern economy

Key Ideas ANSWERS

14. Spain made all decisions, so Mexicans had no experience in governing themselves.
15. Juárez took part in the reform movement that created the Constitution of 1857, in which Mexicans were guaranteed freedom of speech, equality, and freedom from slavery.
16. People opposed the rule of Diaz. Madero called for revolution. Various groups called for reforms to deal with issues such as poverty, land ownership, and political power.
17. Vicente Fox was elected, and he was not a member of PRI, which had held power for more than 70 years. Mexico developed a multiparty system.
18. Indian, Spanish, and modern influences
19. lack of jobs, educational opportunities, and health services in rural areas
20. Ancient people created murals on pyramids and modern artists continued that tradition.
21. the day in 1862 when the Mexican army defeated the French
22. Oil is Mexico's most profitable industry.
23. Mexico supported its oil production and silver mines. It signed NAFTA and became trade partners with Canada and the United States. It set up *maquiladoras*, which increased the production of manufactured goods.
24. exhaust from cars and smoke from industries

CHAPTER 8 Assessment

VOCABULARY

Explain the significance of each of the following.

1. Republic of Texas
2. annex
3. Plaza of the Three Cultures
4. *colonia*
5. squatter
6. mural
7. Diego Rivera
8. fiesta
9. Day of the Dead
10. immigration

Explain how the terms and names in each group are related.

11. Mexican Revolution and Constitution of 1917
12. Antonio López de Santa Anna, Benito Juárez, and Vicente Fox
13. *maquiladoras*, global economy, and NAFTA

KEY IDEAS

1 A Struggle Toward Democracy

14. Why was it difficult for Mexicans to rule themselves after independence?
15. What contributions did Benito Juárez make to Mexico?
16. How and why did the Mexican Revolution happen?
17. Why was the 2000 presidential election in Mexico significant?

2 A Blend of Traditions

18. What are three influences on Mexican culture today?
19. What conditions led Mexican people to move from rural to urban areas?
20. Why is Mexican mural painting a continuation of ancient Mexican tradition?
21. What do Mexicans celebrate on *Cinco de Mayo*?

3 Creating a New Economy

22. Why is oil important to Mexico's economy?
23. What steps did Mexico take to modernize its economy?
24. What factors contribute to air pollution in Mexico City?
25. Why do many Mexicans migrate to the United States?

CRITICAL THINKING

26. **Analyze Causes and Effects** Complete a cause and effect diagram to explain the effects of Mexico's conflicts with Texas and the United States in the 1800s.

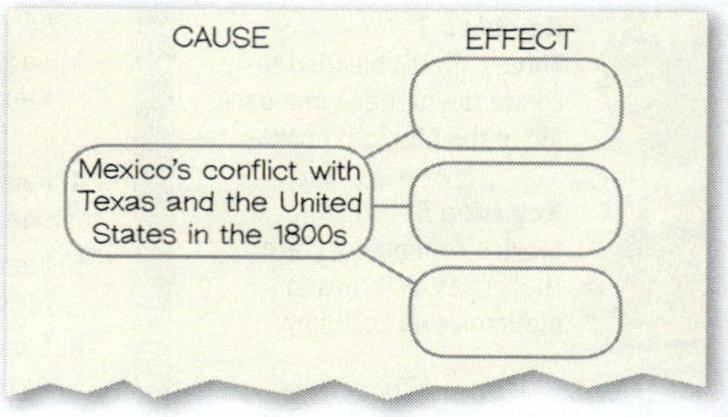

27. **Make Inferences** How do Mexico's holidays and arts show the influence of Mexico's history?
28. **Draw Conclusions** How did Mexico's frequent changes in leaders before the Mexican Revolution affect the country?
29. **Connect History & Art** Why did so many Mexican artists and writers paint and write about Mexico's past?
30. **Five Themes: Location** How does Mexico City's location contribute to the problem of smog in the city?

Answer the ESSENTIAL QUESTION

How does Mexico reflect both ancient traditions and the demands of the modern world?

Written Response Write a two- or three-paragraph response to the Essential Question. Be sure to include a discussion of Mexican life, government, and economy. Use the rubric below to guide your thinking.

Response Rubric

A strong response will:

- describe how Mexican life reflects both ancient and modern traditions
- discuss the challenges Mexico faced in creating a democratic government and a modern economy

ESSENTIAL QUESTION REVIEW

Before students begin their essays, have them review the key chapter elements that will help them answer the Essential Question. Direct students' attention to significant features in the chapter, such as:

- Republic of Mexico 1824–1853, Section 1
- Comparing Urban and Rural Life, Section 2
- Comparing Mexico's Cultural Traditions, Section 2
- Analyzing Primary Sources, Section 3

Written Response Rubric

	Coverage of Key Ideas	Explanation	Errors in Mechanics
4	all	clear and concise	few or none
3	most	good	some
2	some	fair	many
1	one or none	poor	many major

STANDARDS-BASED ASSESSMENT

Test Practice
- Online Test Practice @ClassZone.com
- Test-Taking Strategies and Practice at the front of this book

THEMATIC MAP

Use the map and your knowledge of Mexico to answer questions 1 and 2.

Selected Products of Mexico

1. In what region of Mexico is petroleum produced?

A. along the Pacific coast
B. along the U.S. border
C. along the coast of the Gulf of Mexico
D. along the interior of the country

2. In what part of Mexico is cotton grown?

A. in the southern part
B. in the northern part
C. only on the western coast
D. only on the eastern coast

CIRCLE GRAPH

Examine the graph below. Use the information in the graph to answer questions 3 and 4.

Mexico's Population Distribution

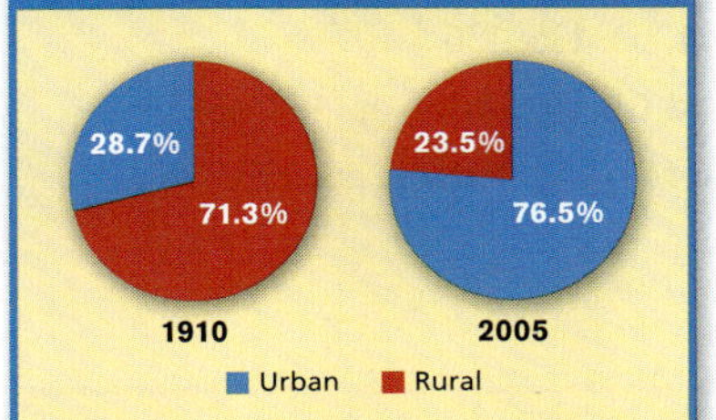

Source: National Institute of Statistics, Geography, and Informatics (Mexico)

3. What percentage of Mexico's population was urban in 1910? in 2005?

4. What does this tell you about changes that have taken place in Mexican society?

GeoActivity

1. INTERDISCIPLINARY ACTIVITY–SCIENCE

Find out about the habitat of the monarch butterfly in Mexico. Research the butterfly's migration pattern and life cycle. Find out what the Mexican government has done to protect these butterflies. Present your findings in an illustrated and captioned poster.

2. WRITING FOR SOCIAL STUDIES

Create a guide for visitors to Mexico City. Use the Internet to learn about places to visit, such as museums, and cultural institutions, such as the ballet. Write a description of each of these places, and include illustrations.

3. MENTAL MAPPING

Create an outline map of Mexico and label the following:
- Mexico City
- Rio Grande
- Sierra Madre Occidental
- Sierra Madre Oriental
- Gulf of Mexico
- Pacific Ocean

25. Mexico is working on creating a strong economy, but many people still do not have jobs. These people leave to make money to support their families.

Critical Thinking ANSWERS

26. Effects: Mexico left weakened; Mexico open to foreign invasion; Mexico lost large portion of its territory

27. Holidays such as Mexico's Independence Day and *Cinco de Mayo* celebrate historic events. The subject of many artists' and writers' work is Mexico's history.

28. It made it difficult for the country to establish a stable government.

29. The artists painted what was important to the Mexican people. Their work showed that Mexicans value both the Indian and Spanish influences.

30. Mexico City is almost completely surrounded by mountains, which help produce a layer of warm air above the city. This layer keeps gases from car exhausts and thousands of industries from blowing away. The gases react with sunlight to form smog.

Standards-Based Assessment ANSWERS

1. C
2. B
3. 28.7%; 76.5%
4. People have moved from rural areas to urban areas.

GeoActivity

1. INTERDISCIPLINARY ACTIVITY RUBRIC

The poster should
- show the butterfly's migration pattern and life cycle
- indicate government actions to protect the butterfly
- include explanatory captions

2. WRITING FOR SOCIAL STUDIES RUBRIC

The guide should
- clearly describe the places chosen
- include illustrations of the places

3. MENTAL MAPPING RUBRIC

The map should
- represent the approximate shape of Mexico
- include all geographic features listed in their correct location

DIFFERENTIATING CHAPTER ASSESSMENT

Unit Resource Book
- **Chapter Test A,** pp. 113–116. *Use for Struggling Readers, Beginning-to-Intermediate-Level English Learners, and for Inclusion of students whose IEP suggests supported test-taking.*
- **Chapter Test B,** pp. 117–120. *Use for On-Level Learners and ESOL students with Advanced-Level English.*
- **Chapter Test C,** pp. 121–124. *Use for Gifted and Talented students or students assigned Pre-AP work.*

Test Generator
- Chapter Tests A, B, and C, in English and Spanish

McDougal Littell Assessment System

TEST **Test Generator** Use three levels of tests in English and Spanish tied to state standards. Plus, choose questions to customize tests for all populations.

SCORE Scan and Score plain paper student tests in minutes.

REPORT Create instant progress reports, sortable by student, by standard, or other criteria.

RETEACH Respond immediately to students' needs by using a wealth of reteaching resources.

CHAPTER 9

Middle America and Spanish-Speaking South America
PLANNING GUIDE

Complete print resources all on one USB drive allow you to customize lesson plans for all students.

Section Titles & Key Ideas	Unit Resource Book Latin America	Assessment Resources
1 Central America **KEY IDEA** Central American countries face challenges as they develop democratic governments and improve their economies.	• **Reading Study Guide,** pp. 125–126 • **RSG with Additional Support,** pp. 135–136 • **Map and Graph Skills,** pp. 167–168 • **Outline Map Activities,** pp. 169–170	**Unit Resource Book** • Section Quiz, p. 179 • Reteaching Activity, p. 184 **Test Generator CD-ROM** **Daily Test Practice Transparency,** TT29
2 The Caribbean **KEY IDEA** The influences of various European countries and Africa are reflected in the Caribbean island cultures.	• **Reading Study Guide,** pp. 127–128 • **RSG with Additional Support,** pp. 137–138 • **History Makers: Fidel Castro,** p. 172 • **Interdisciplinary Projects,** pp. 173–175 • **Primary Source,** p. 178	**Unit Resource Book** • Section Quiz, p. 180 • Reteaching Activity, p. 185 **Test Generator CD-ROM** **Daily Test Practice Transparency,** TT30
3 Venezuela and Colombia **KEY IDEA** Venezuela's economy depends heavily on petroleum, while Colombia's economy depends more on agricultural products.	• **Reading Study Guide,** pp. 129–130 • **RSG with Additional Support,** pp. 139–140 • **Skillbuilder Practice,** p. 166	**Unit Resource Book** • Section Quiz, p. 181 • Reteaching Activity, p. 186 **Test Generator CD-ROM** **Daily Test Practice Transparency,** TT31
4 Peru, Bolivia, and Ecuador **KEY IDEA** Indians and *mestizos* make up a large part of the population of Peru, Bolivia, and Ecuador.	• **Reading Study Guide,** pp. 131–132 • **RSG with Additional Support,** pp. 141–142 • **Vocabulary Practice,** p. 165	**Unit Resource Book** • Section Quiz, p. 182 • Reteaching Activity, p. 187 **Test Generator CD-ROM** **Daily Test Practice Transparency,** TT32
5 The Southern Cone **KEY IDEA** Geography has influenced the economies of the nations in the Southern Cone.	• **Reading Study Guide,** pp. 133–134 • **RSG with Additional Support,** pp. 143–144 • **History Makers: Juan and Eva Peron,** p. 171 • **Economics in the World,** p. 176 • **Connect to Math,** p. 177	**Unit Resource Book** • Section Quiz, p. 183 • Reteaching Activity, p. 188 • Chapter Test (Forms A, B, & C), pp. 189–200 **Test Generator CD-ROM** **Daily Test Practice Transparency,** TT33

McDougal Littell **Assessment System**
TEST SCORE REPORT RETEACH

No Time? To focus students on important content in the chapter, use the **"Reading Study Guide with Additional Support"** in ***Resources2Go.***

Teacher Presentation Options

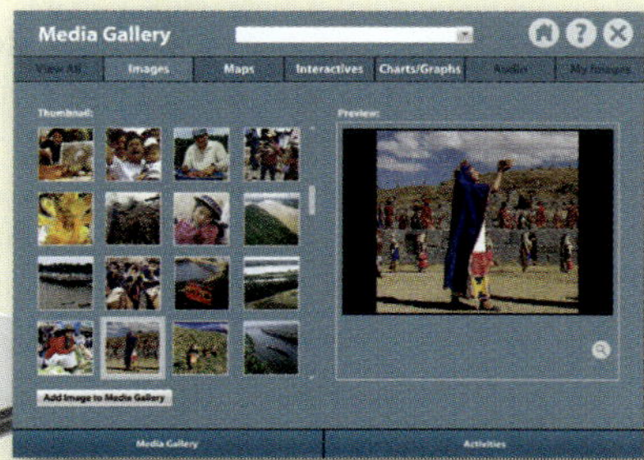

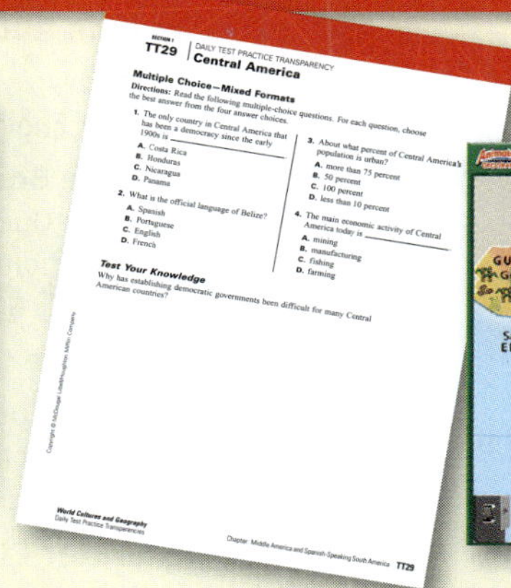

Electronic Books

eEdition DVD-ROM

eEdition Online

Presentation Toolkit

Video/DVD

- World Cultures and Geography

Power Presentations DVD-ROM

- Discussion Starters
- Media Gallery
- Review Game
- Interactive Review
- Animated Features

Unit Transparency Book

- 3-Minute Warm-Ups
- Essential Question
- Critical Thinking
- Maps
- Cultures

Daily Test Practice Transparencies

Animated Geography

- Present-Day Middle America and Spanish-Speaking South America Map
- Economy of Central America Map

OnlineActivities at ClassZone.com

Animated Geography

- Videos and Photos
- 3-D Animations
- Interactive Maps

Interactive Review

- Section Quizzes
- Review and Study Notes
- Vocabulary Flip Cards
- Graphic Organizers
- Name Game and Crossword Puzzles
- GeoGame

Online Test Practice

Map Center

- Atlas Maps
- Political Maps with Data
- Links to Mapping Web Sites

Activity Center

- Activity Maker—Review Games, Puzzles, Flip Cards, and Graphic Organizers
- Worldster

Research & Writing Center

- Web Sites
- Data Updates
- Primary Sources
- Writing Models

Current Events

Terry Rhodes
McNair Middle School
Fayetteville, Arkansas

Make It Fun — Teacher-Tested Activities

As you teach this chapter, look for my classroom-proven ideas for:

- making the lesson fun
- focusing students' attention and comprehension
- integrating technology with traditional teaching

Teaching Bell to Bell When you see this timing guide, have students discuss in pairs, in groups of 3–6, or as a class.

- **Think, Pair, Share** (pairs)
- **Reader, Recorder, Reporter** (3–6)
- **Talk About It** (any size)

CHAPTER 9 RESOURCES FOR DIFFERENTIATED INSTRUCTION

Struggling Readers

eEdition with Audio DVD-ROM

Lesson Plans
- Strategies for Inclusion

Unit Resource Book
- Reading Study Guide with Additional Support, pp. 135–144 (B)
- Reteaching Activity, pp. 184–188
- Chapter Test (Form A), pp. 189–192 (A)

Best Practices Toolkit (C)
- Interactive Reading Strategies, pp. 90–145
- Differentiation Activity Bank: Struggling Readers, pp. 146–149

Teacher's Edition Activities
- Use a Map, Sec. 4
- Tiered Activities, Sec. 1–5

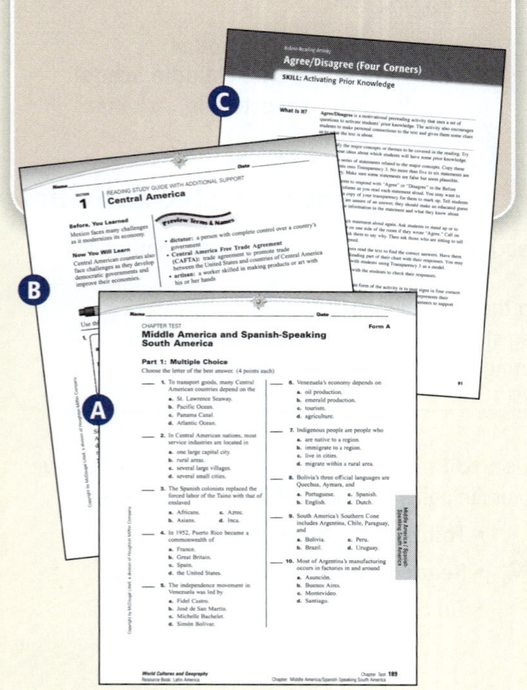

English Learners

Pupil's Edition in Spanish

eEdition with Audio DVD-ROM

Lesson Plans
- Strategies for English Learners

Unit Resource Book
- Reading Study Guide (Spanish), pp. 145–154
- RSG with Additional Support (Spanish), pp. 155–164 (A) (B)

Multi-Language Glossary (C)

Best Practices Toolkit
- Differentiation Activity Bank: English Learners, pp. 165–194

Test Generator CD-ROM
- Section Quizzes in Spanish
- Chapter Test in Spanish

Teacher's Edition Activities
- Pronounce and Preview, Sec. 1–5
- Language: Transitions, Sec. 1
- Comprehension: Connection, Sec.2
- Language: Pronoun Referents, Sec. 3
- Language: Verb Tense, Sec. 4
- Vocabulary: Word Relationships, Sec. 5
- Tiered Activities, Sec. 1–5

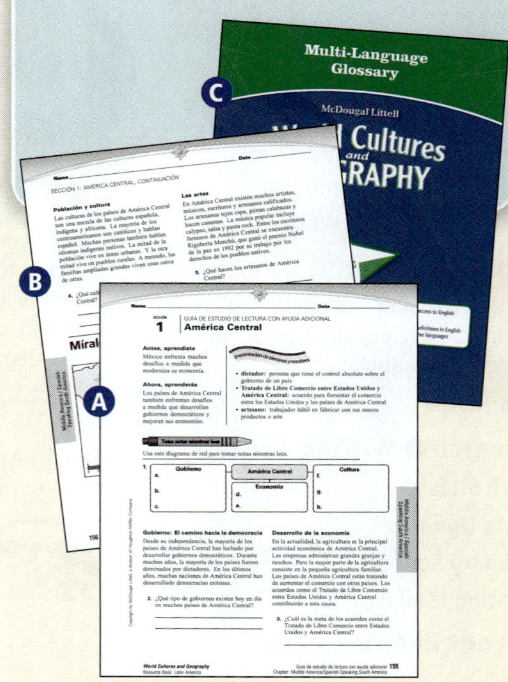

Inclusion

eEdition with Audio DVD-ROM

Lesson Plans
- Strategies for Inclusion

Unit Resource Book
- Reading Study Guide with Additional Support, pp. 135–144 (A) (B)
- Reteaching Activity, pp. 184–188
- Chapter Test (Form A), pp. 189–192

Best Practices Toolkit (C)
- Interactive Reading Strategies, pp. 90–145
- Differentiation Activity Bank: Struggling Readers, pp. 146–149

Teacher's Edition Activities
- Match Terms and Definitions, Sec. 1
- Identify Economic Features, Sec. 1
- Record Terms and Names, Sec. 2
- Sort Words into Categories, Sec. 3
- Use a Pronunciation Guide, Sec. 4
- Extend Visual Vocabulary, Sec. 5
- Tiered Activities, Sec. 1–5

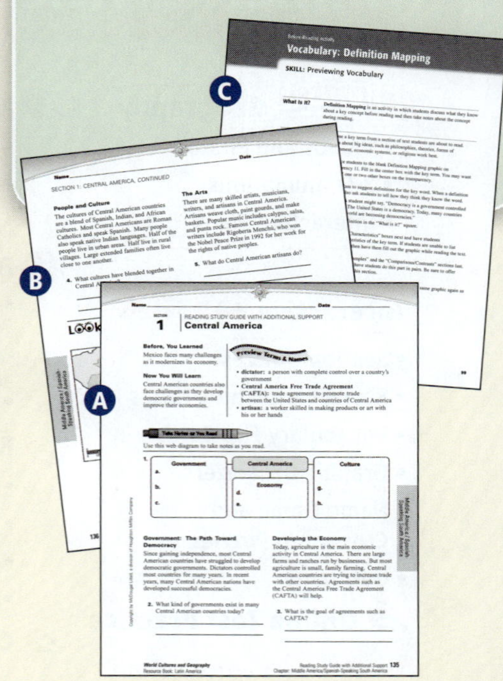

Interdisciplinary Activities

Gifted & Talented Students

Lesson Plans
- Core Lesson Plan

Unit Resource Book
- Map and Graph Skills, pp. 167–168 (A)
- Outline Map Activities, pp. 169–170
- History Makers, pp. 171–172 (B)
- Interdisciplinary Projects, pp. 174–175
- Economics in the World, p. 176
- Chapter Test (Form C), pp. 197–200

Best Practices Toolkit (C)
- Student Products Rubrics, pp. 1–14
- Writing Templates, pp. 78–79

Teacher's Edition Activities
- Write a Letter to the Editor, Sec. 3
- Tiered Activities, Sec. 1–5

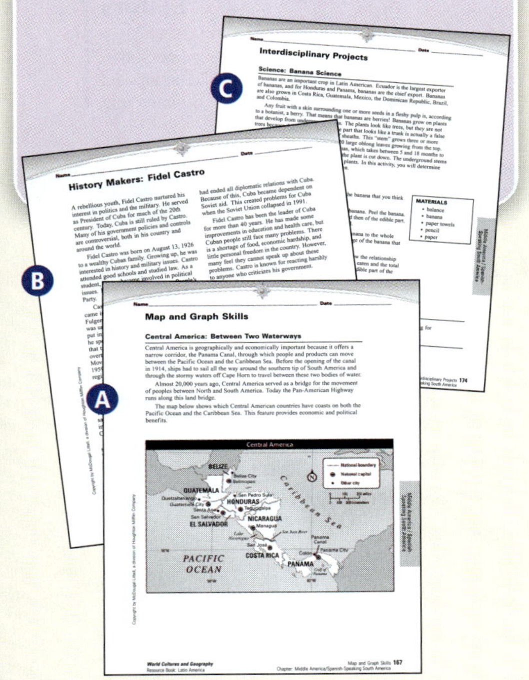

Pre-AP*

Lesson Plans
- Core Lesson Plan

Unit Resource Book
- Interdisciplinary Projects, pp. 173–175 (A)
- Connect to Math, p. 177 (B)
- Primary Source, pp. 178–179
- Chapter Test (Form C), pp. 197–200

Best Practices Toolkit (C)
- Writing Templates, pp. 78–79
- Differentiation Activity Bank: Pre-AP Students, pp. 195–200

* Pre-AP is a registered trademark of the College Entrance Examination Board, which was not involved in the production of and does not endorse this product.

Teacher's Edition Activities
- Relate GDP and Standard of Living, Sec. 5
- Tiered Activities, Sec. 1–5

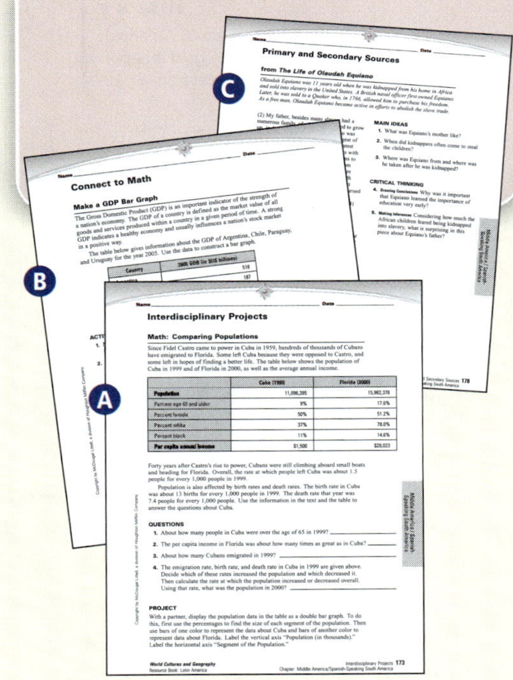

CONNECT to Language Arts

Put Yourself in the Picture

CLASS TIME 20 minutes

TASK Write a first-person narrative about a scene in Middle America or Spanish-speaking South America.

SUPPLIES
- writing materials/computers
- audiocassette recorder (optional)

ACTIVITY Have students choose a photograph shown on the opening spread of one of the sections. Ask them to put themselves in the scene and make mental images of what they are seeing, hearing, feeling, smelling, and possibly tasting. Then have them write a brief, first-person narrative describing their experience. Students may choose to write from the perspective of a person or an animal, a visitor or a resident. Students who have difficulty writing may record their stories.

Good for Visual and Verbal Learners

CONNECT to Math

Compare the Population of South American Cities and Countries

CLASS TIME 30 minutes

TASK Create a bar graph comparing the populations of capital cities to the population of South American countries.

SUPPLIES
- recent population data on the nine South American countries from Sections 3, 4, and 5, and their capitals (available from The World Factbook at www.cia.gov)
- paper, colored markers or pencils

ACTIVITY Distribute the population data for the nine South American countries and their capitals. Have students work in pairs to create a bar graph that compares the two populations for each country. Students could use two bars side by side, with one representing the city, the other the country, or could create one bar, with the population of the capital shown as a part of the total population of the country.

Good for Visual Learners

Introduce the ESSENTIAL QUESTION

- Explain that many countries in this region were once the colonies of European nations that controlled their governments and economies for years.
- Discuss how it often took many years after these countries gained independence before they were able to put democratic governments in place.
- Point out that geographic features, such as the oceans and the Andes Mountains, and natural resources, such as forests and oil, greatly influence the economies of this region.

Key Ideas

Section 1: Central American countries face challenges as they develop democratic governments and improve their economies.

Section 2: The influences of various European countries and Africa are reflected in the Caribbean island cultures.

Section 3: Venezuela's economy depends heavily on petroleum, while Colombia's economy depends more on agricultural products.

Section 4: Indians and *mestizos* make up a large part of the population of Peru, Bolivia, and Ecuador.

Section 5: Geography has also influenced the economies of the nations in the Southern Cone.

CONNECT to the Essential Question

You may wish to revisit the Essential Question periodically with students as they work through the chapter.

Suggestions for reconnecting are provided at an appropriate point in the *Teacher's Edition* for each section.

Scarlet macaws in Honduras

CHAPTER 9

Middle America and Spanish-Speaking South America

1 FOCUS ON Central America

2 FOCUS ON The Caribbean

3 FOCUS ON Venezuela and Colombia

4 FOCUS ON Peru, Bolivia, and Ecuador

5 FOCUS ON The Southern Cone

ESSENTIAL QUESTION

How are the countries of this region working to strengthen their governments and economies?

CONNECT Geography & History

Use the map and the time line to answer the following questions.

1. What country connects Central and South America?
2. Who led the Cuban independence movement?

1800

History
1804 Haiti is the first Caribbean Island to become independent. (Present-day Haitian flag) ►

History
1821 Central American countries gain independence from Spain.

History
◄ **1895** José Martí leads Cuban revolution against Spanish rule.

History
1902 Cuba gains independence.

TIME LINE DISCUSSION

Tell students that the turbulent history of this region since 1800 reflects the struggles of the people of this region to gain independence from foreign powers and establish democratic governments and diversified economies.

- About how long did it take many Central American countries to develop democracies after they became independent? *(about 170 years, from 1821 to the 1990s)*
- Which Caribbean island gained its independence first? *(Haiti, in 1804)*
- What important cultural site in Peru did an archaeologist discover in 1911? *(Machu Picchu)*

Culture
1911 Archaeologist Hiram Bingham locates the site of Machu Picchu.
(Incan watchtower) ▼

Government
1959 Fidel Castro overthrows government of Fulgencio Batista.
(Fulgencio Batista) ►

Culture
1992 Rigoberta Menchú wins the Nobel Peace Prize.

Today

History
1952 Puerto Rico becomes a U.S. commonwealth.

History
1990s Many Central American countries develop democracies.

TEACHER MEDIA FAVORITES

Books for the Teacher

Blouet, Brian W. and Olwyn M. Blouet, eds. *Latin America and the Caribbean.* 4th ed. Wiley, 2002. A thematic and regional survey of the area.

Skidmore, Thomas E. and Smith, Peter H. *Modern Latin America.* 6th ed. Oxford U P, 2005. Covers contemporary political, social, and economic issues.

Video / DVD

The Sights and Sounds of Central America. Video Knowledge, 2004. Explores the people and environment of the region.

The Sights and Sounds of South America. Video Knowledge, 2004. Includes images of the capitals and sounds of Spanish and Indian languages and music.

Internet

Go to the *Research and Writing Center* at **ClassZone.com** to link to:

- data on the region
- a variety of chapter-specific content-reviewed sites
- "History Makers" biographies
- primary sources

Animated GEOGRAPHY

Present-Day Middle America and Spanish-Speaking South America

CLASSZONE.COM

Explain that this is a political map of the region, showing the countries and their capital cities. Point out that most of the countries shown on the map are located in the tropical zone between the Tropic of Cancer and the Tropic of Capricorn.

- How does the map suggest why tourism is an important part of the economy in Central America and the Caribbean? *(Possible answer: The countries are all close to the ocean and tourists probably like to visit the warm-water beaches there.)*
- Which country on the map has two capitals and what are they? *(Bolivia: La Paz and Sucre)*
- What does the time line tell you about the relationship of Puerto Rico to the United States? *(It became a commonwealth of the United States in 1952.)*

CONNECT GEOGRAPHY & HISTORY ANSWERS

1. Panama
2. José Martí

Presentation Options

Use the **Power Presentations** DVD-ROM to project the animated map.

- Prompt students to use the pop-up images and audio from the map to discover what they already know about the geography, culture, history, economics, or governments of the region.
- Challenge students to answer the map questions individually, in small groups, or as a class activity. Use all of the interactive material to help preview chapter content.

1 Plan & Prepare

Objectives

- Explain the political and economic challenges faced by Central American countries
- Describe the people and culture of Central America

Read for the Essential Question

Help students read for a purpose by reminding them of the Essential Question: "How are the countries of this region working to strengthen their governments and economies?" Point out that they can use text features such as Key Questions, special features, images, and maps to form answers to the question.

Vocabulary

Best Practices Toolkit

Use the ***Best Practices Toolkit*** to model strategies for vocabulary notetaking. Vary strategies throughout the year. Choose from: *Knowledge Rating, Predicting ABCs, Definition Mapping, Word Sort, Word Wheel, Frayer Model (Word Squares), Magnet Words,* and *Student VOC.*

Vocabulary Strategies, TT9–TT16

Reading Strategy

Best Practices Toolkit

Display the Web Diagram transparency.

- Point out to students that to categorize information in the section they must decide how it can be sorted into the categories shown in the diagram.
- Model filling in entries for the Government category from the "Path Toward Democracy" subsection in this section.
- Have students suggest entries, using the categories in the diagram.
- Display the completed diagram for review.

Web Diagram, TT28

Section 1 Reading for Understanding

Key Ideas

BEFORE, YOU LEARNED

Mexico faces many challenges as it modernizes its economy.

NOW YOU WILL LEARN

Central American countries also face challenges as they develop democratic governments and improve their economies.

Vocabulary

TERMS & NAMES

ecotourism travel to a natural habitat in a way that does not damage the habitat

dictator a person with complete control over a country's government

subsistence farming a kind of farming in which farmers grow only enough to feed their families

Central America Free Trade Agreement (CAFTA) trade agreement to promote trade between the United States and countries of Central America

service industry an industry that provides services rather than objects

artisan a worker skilled in making products or art with his or her hands

REVIEW

maquiladora factory in which materials are imported and assembled into products for export

North American Free Trade Agreement (NAFTA) agreement that reduced trade barriers among Mexico, Canada, and the United States

mestizo person with mixed European and Indian ancestry

Reading Strategy

Re-create the web diagram shown at right. As you read and respond to the **KEY QUESTIONS**, use the diagram to outline the major aspects of Central America's government, economy, and culture.

See Skillbuilder Handbook, page R7

CATEGORIZE

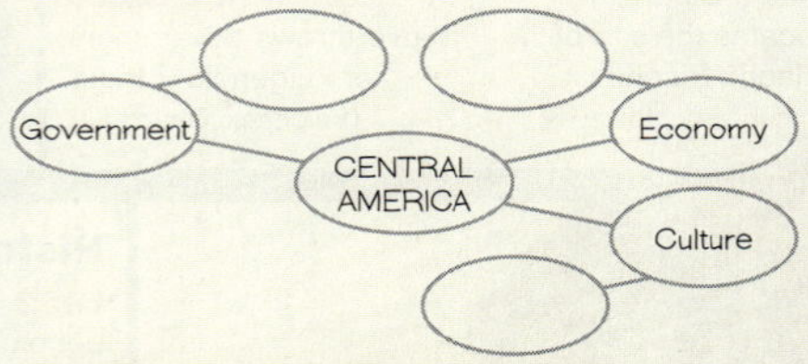

GRAPHIC ORGANIZERS
Go to **Interactive Review** @ClassZone.com

254 Chapter 9

PRETEACHING VOCABULARY

English Learners

Pronounce and Preview

Pronounce each term for students. Review the meanings of words within definitions, such as *habitat, complete control,* and *promote.*

- To modify vocabulary learning, have students complete worksheets as they read, instead of afterward.

Unit Resource Book: Latin America
- Vocabulary Practice, p. 165

Inclusion

Match Terms and Definitions

Have student pairs write the vocabulary words and definitions on index cards. Then have them cut out the words and definitions to create a matching game. Ask one student to shuffle the cards and arrange them. After their partner has finished matching the terms and definitions, have the students work together to correct any errors. Students should then switch roles.

Machu Picchu, Peru

Dolphin off the coast of Honduras

Central America

Connecting to Your World

What is the most exotic animal you have ever seen? If you were to visit the rain forest in Costa Rica's Corcovado National Park, you would see many exotic animals, such as macaws, coatis, and anteaters. People's interest in visiting this and other natural habitats has led some countries in Central America to promote **ecotourism**, or travel to natural habitats in a way that does not damage the habitat. Ecotourism is just one way that Central American countries are expanding their economies.

Keel-billed Toucan This bird makes its home in the treetops of the rain forests.

Government and Economy

KEY QUESTION What steps have Central American countries taken to improve their economies?

Central America today faces major challenges. In most countries, a wide gap exists between the small number of wealthy people and the large number of poor people. A large percentage of the population is unemployed. In many areas, people lack basic services, such as running water and electricity. In recent years, Central American countries have been working to improve their economies and to develop democratic governments.

Ecotourism The people shown here are standing on a bridge, looking down over a rain forest in Costa Rica.

255

2 Focus & Motivate

3-Minute Warm-Up

Write on the board or display the transparency:

- Look at the photograph of the people on the bridge. Why is the bridge a good way for tourists to see the rain forest? *(Possible answer: It allows people to see a wide area and it's less harmful to the environment.)*

Unit Transparency Book: Latin America
- 3-Minute Warm-Ups, TT20

Connecting to Your World

More About . . .

Ecotourism in Costa Rica

Ecotourism has helped Costa Rica diversify its economy. Tourism brought in $1.1 billion in revenue in 2000—second only to its manufacture of computer chips.

3 Teach

Government and Economy

Reader, Recorder, Reporter

- What problems do poor people face in Central America? *(unemployment, lack of basic services, little say in government)*
- **Categorize** Name two crops grown on plantations and two by subsistence farmers. *(plantations: bananas, sugar cane, coffee; subsistence farming: corn, vegetables)*

SECTION 1 PROGRAM RESOURCES

Unit Resource Book: Latin America

ON LEVEL
Unit Resource Book
- Reading Study Guide, p. 125
- Outline Map Activities, p. 169
- Section Quiz, p. 179

STRUGGLING READERS
Unit Resource Book
- Reading Study Guide with Additional Support, p. 135
- Skillbuilder Practice: Reading a Time Line, p. 166
- Section Quiz, p. 179
- Reteaching Activity, p. 184

eEdition with Audio DVD-ROM

ENGLISH LEARNERS
Pupil Edition in Spanish
eEdition with Audio DVD-ROM
Unit Resource Book
- RSG (Spanish), p. 145
- RSG with Additional Support (Spanish), p. 155

Multi-Language Glossary
Test Generator
- Section Quiz in Spanish

INCLUSION
Unit Resource Book
- RSG with Additional Support, p. 135
- Section Quiz, p. 179
- Reteaching Activity, p. 184

GIFTED & TALENTED
Unit Resource Book
- Map and Graph Skills, p. 167
- Section Quiz, p. 179

PRE-AP
Unit Resource Book
- Section Quiz, p. 179

TECHNOLOGY
Unit Transparency Book
- 3-Minute Warm-Ups, TT20
- Essential Question, TT21
- Critical Thinking, TT22

Daily Test Practice Transparencies
- Section 1, TT29

Power Presentations
ClassZone.com
World Cultures and Geography Video on DVD

More About . . .

Central American Governments

Problems of poverty, border disputes, and government corruption continue to plague many countries in Central America. Many people claim that reforms, including more equitable land distribution and access to jobs, have not been achieved with democratic governments. Some hope that the Central American Common Market might lead to a unified region similar to the European Union.

Unit Resource Book: Latin America
- Skillbuilder Practice: Reading a Time Line, p. 166

Animated GEOGRAPHY

Economy of Central America

CLASSZONE.COM

CONNECT Geography & Economics

Point out that this map highlights the economy in Central America. Direct attention to the key. Make sure students understand that the colors show different kinds of land use and that the symbols show different resources that provide income.

- What type of land use is related to timber? *(forests)*
- Where are sugar cane plantations located? *(in Guatemala, Belize, El Salvador, and Nicaragua)*

ANSWERS

1. **Location** Panama, Costa Rica, Belize
2. **Place** for crops and forests

Unit Resource Book: Latin America
- Outline Map Activities: Latin America: Political, pp. 169–170

Presentation Options

Use the **Power Presentations** DVD-ROM to project the animated map.

- Click on the key to activate the map.
- Discover how the resources of Central America relate to its economy.

Path Toward Democracy Since gaining independence from Spain in 1821, most Central American countries have struggled to develop democratic governments. As in Mexico, the wealthy in Central America controlled most aspects of government. Most of the population remained poor, with no say in how they were governed.

Costa Rica is the only country in the region that has been a democracy since the early 1900s. Nearly all the other Central American countries have been under the rule of **dictators**, or leaders with complete control over their governments. Starting in the 1950s, civil wars fought for equal rights led to years of suffering in countries such as Guatemala, El Salvador, and Nicaragua (NIHK•uh•RAH•gwuh). Since the 1990s, however, many Central American nations have developed democracies in which more people participate in government.

Developing the Economy During the colonial period, Spain set up large plantations that focused on growing one kind of crop. Today, agriculture is the main economic activity in Central America. Large plantations still produce crops, mainly bananas, sugar cane, and coffee, for export. Cattle are raised on large ranches in the drier western parts of Central America. But most agriculture consists of **subsistence farming**, in which poor farmers grow a variety of crops, such as corn and vegetables, on small plots to feed their own families.

Animated GEOGRAPHY

Economy of Central America

Click here to learn more about the resources of Central America @ClassZone.com

CONNECT Geography & Economics

1. **Location** In what countries is tourism a major industry?
2. **Place** How is the land used in Panama?

DIFFERENTIATING INSTRUCTION

English Learners

Language: Transitions

Help students identify the words that are used to signal a change between time periods in the "Path Toward Democracy" and "Developing the Economy" subsections. Explain that *since* is a synonym for *after*. Point out that another way of saying "since gaining independence" would be "after the colonial period." Ask students for another way to say "during the colonial period" *(before independence)*. Also point out that *Today* signals the present time.

Inclusion

Identify Economic Features

Have copies of the half-page map enlarged for learners with visual impairments. Allow the student to work with a sighted partner to identify and locate each of the major resources highlighted on the map. After partners have thoroughly studied the map, encourage them to make summary statements about the economy of each country and about the region as a whole.

COMPARING Human-Made Waterways

	PANAMA CANAL	ST. LAWRENCE SEAWAY
Location	Central America	North America
Length	50 miles	2,340 miles
Tonnage	279 million tons	48 million tons
Average time to cross	8 hours	8 days
Average ships per year	14,000	4,000

Cargo ships passing through the locks of the Panama Canal

CRITICAL THINKING

Compare and Contrast How do the two waterways compare in terms of the number of ships that pass through them every year?

As the map indicates, timber is an important economic activity in forested areas, such as in Costa Rica. Tourism, and in particular ecotourism, has become a major industry in Costa Rica and Guatemala. ***Maquiladoras*** in Guatemala and Honduras produce goods such as clothing for export. In recent years, Central American countries have begun to develop technology and telecommunications industries to help expand their economies.

Promoting International Trade As you have learned, the Panama Canal is important to world trade. It is a crossroads for goods that travel between the western and eastern hemispheres. The economies of many Central American countries depend on the canal for shipping and receiving goods.

Today, Central American nations are trying to increase trade with other countries of the world. The United States, Costa Rica, Guatemala, El Salvador, Honduras, and Nicaragua have signed the **Central America Free Trade Agreement (CAFTA)**. Like **NAFTA**, this trade agreement will lower tariffs and make trade easier among member nations. The five Central American nations hope the flow of money into Central America from CAFTA will help to develop more jobs and strengthen their economies.

EVALUATE Discuss the steps Central American countries have taken to strengthen their economies.

expanded some industries, such as tourism, manufacturing, and technology industries; joined CAFTA to increase trade with other countries and create jobs

COMPARING

Human-Made Waterways

Remind students that the St. Lawrence Seaway allows ships to travel between the Atlantic Ocean and the Great Lakes. Allow them to refer to the "Economy of Central America" map if needed to answer the questions.

- What two bodies of water are connected by the Panama Canal? *(Caribbean Sea and Pacific Ocean)*
- What common purpose do the two human-made waterways share? *(to make shipping easier and faster)*

CRITICAL THINKING ANSWER

Compare and Contrast Fewer ships pass through the St. Lawrence Seaway.

Unit Resource Book: Latin America

- Map and Graph Skills: Central America: Between Two Waterways, pp. 167–168

More About . . .

CAFTA

In addition to the five Central American countries, the Dominican Republic is now part of the free trade agreement known as DR-CAFTA. About 80 percent of goods from the region already entered the United States duty free. DR-CAFTA now opens up these markets to U.S. goods, services, and farm products.

The agreement offers phased-in tariff reductions on food products such as corn, rice, dairy, and poultry. Research by the World Bank indicates that most people in Nicaragua, Guatemala, and El Salvador will benefit from lower food prices because they buy more of these products than they produce.

INTERDISCIPLINARY ACTIVITIES

CONNECT to Math

Graph Waterway Statistics

Have students graph the information in the Comparing feature by using symbols. For example, they might use a clock to represent a day and show one-third of a clock for the Panama Canal and eight clocks for the St. Lawrence Seaway. Each ship symbol might represent 4,000 ships, and each ton symbol might represent 50 tons. Students can approximate amounts that are not exact by showing the symbol as a percentage that is less than full size, or 100 percent.

CONNECT to Language Arts

Business Writing About Trade

Invite students to choose one of the countries in Central America. Have them write a paragraph from the point of view of a businessperson in that country about how CAFTA and the Panama Canal affect the economy. Encourage students to think about markets for the country's products as well as other benefits such as lower import prices. Invite volunteers to share with the whole class.

Teach

People and Culture

Think, Pair, Share

- What are similarities and differences between rural and urban life in Central America? *(Differences: rural people live in small villages and farm, urban people live in a large city and work in factories and service industries; Similarities: both shop at open-air markets)*
- How do you know that family is an important part of the culture? *(Possible answer: Many family members live near one another and they get together often for celebrations.)*
- **Categorize** What languages are spoken in Central America? *(Spanish, English, Indian languages)*

Comparing Regions

Family Structures Around the World

The family living arrangements described in Central America are more common when people live in villages close to where they were born. Families also tend to be larger in such rural areas, resulting in more aunts, uncles, and cousins. As in Africa and the Middle East, traditional extended families are becoming less common as people move to cities to find work.

In the United States, where people may move great distances for college education or jobs, families have become more scattered. Families also tend to be smaller, resulting in fewer relatives. The 2000 Census found that only about 4 million U.S. households included more than two generations.

People and Culture

KEY QUESTION How is the present-day culture of Central America a blend of several cultures?

The cultures of Central American countries are a blend of Spanish, Indian, and African cultures. Two-thirds of Central Americans are ***mestizos***. Most Central Americans are Roman Catholics. Spanish is the official language of all the Central American countries except Belize, where English is the official language. Many people also speak Indian languages.

Rural and Urban Life Half of Central America's population is urban. Many people in rural areas live in small villages built on mountainsides. Because they plant small fields for food, they grow barely enough to feed their families.

Most Central American nations require children to attend school for nine years. However, rural schools are poorly equipped, and it is difficult to hire teachers to work in these areas. Children often do not attend school because they are needed to work on family farms.

Generally, Central American countries have one large capital city, where factories and service industries are located. **Service industries** provide services, such as banking, rather than objects. Open-air markets, where people buy and sell products, are found in large cities as well as in small villages throughout Central America. Market days are a time for people to shop and to socialize.

Daily Life The family is important in Central American life. A Central American family includes grandparents, parents, children, aunts, and uncles, who often live in houses built near one another. Family members frequently gather to celebrate family events and holidays.

Farmers are selling crops they've grown; people are shopping for food. ▼

Open-Air Market
Markets such as this one take place throughout Central America. **What kinds of activities are taking place in this market?**

DIFFERENTIATING INSTRUCTION: TIERED ACTIVITIES

OBJECTIVE Create a comparison of daily life in Central America and the United States.

Basic

Guide students in creating a two-column chart, with the headings *Central America* and *United States.* Then direct them to compare the regions on four different topics: school, shopping, family, and food. Allow students to share their charts with a partner and then revise their charts as needed.

On Level

Invite students to reflect on daily life in Central America and in the United States. Then have them write a brief essay that compares the two, using specific examples related to school, shopping, family, and food. Allow students to share their essays and then revise them as needed.

Challenge

Invite students to write a letter to a Central American exchange student, describing what to expect when he or she arrives at his or her school. Have them consider what might be familiar in the United States and what might seem unusual. Allow students to share their letters and then revise them as needed.

Corn is the main ingredient in many Central American foods, such as tortillas. Meals include beans and vegetables such as chili peppers and avocados, and fruits such as guavas and mangoes.

The Arts Central American countries are known for the work of their **artisans**, people skilled in making things with their hands. The works include Panamanian molas, Nicaraguan painted gourds, and Honduran baskets.

Popular music includes calypso, salsa, and punta rock, a type of music created by Central Americans. It combines traditional rhythms with modern instruments. A popular instrument is the marimba, a kind of xylophone played throughout the region.

Famous Central American writers include Rigoberta Menchú, an Indian Guatemalan woman. She won the Nobel Peace Prize in 1992 for her work for the rights of Indians in Central America.

Making a Mola This Panamanian woman is making a mola, a cloth panel used for decorating such things as clothing.

SUMMARIZE Explain how the culture of Central America is a blend of several cultures. People are a blend of Spanish, Indian, and African cultures. Most people are *mestizos* and Roman Catholic.

Section 1 Assessment

ONLINE QUIZ For test practice, go to **Interactive Review @ClassZone.com**

TERMS & NAMES

1. **Explain the importance of**
   - ecotourism
   - dictator
   - CAFTA
   - artisan

USE YOUR READING NOTES

2. **Categorize** Use your completed diagram to answer the following question: What kinds of music are popular in Central American countries today?

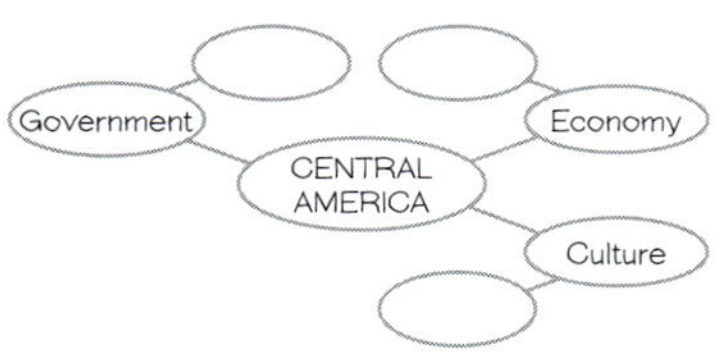

KEY IDEAS

3. What is the main economic activity in Central America?
4. What five Central American countries signed CAFTA?
5. What kinds of crafts do Central American artisans create?

CRITICAL THINKING

6. **Draw Conclusions** How did the civil wars in Central American countries affect attempts to establish strong governments?
7. **Identify Problems and Solutions** What problems do people in the rural areas of Central America face?
8. **CONNECT to Today** Many ecotourists visit the various natural habitats in Central America. Why do you think ecotourism has become so popular?
9. **TECHNOLOGY** **Make a Multimedia Presentation** Choose a Central American country and research its art, music, and other traditions. In a small group, prepare a multimedia presentation of your findings.

CONNECT to the Essential Question

How are the countries of this region working to strengthen their governments and economies?

Ask students what they have learned so far that can help them answer this question. Students might mention:

- Many Central American countries now have democratic governments that allow more people to participate.
- The region has a diverse economy based on agriculture, timber, manufacturing, tourism, and trade.
- Five countries and the United States signed CAFTA to promote trade, develop jobs, and strengthen the economy.

4 Assess & Reteach

Assess Have students complete the Section Assessment.

Unit Resource Book: Latin America
- Section Quiz, p. 179

Interactive Review @ClassZone.com
- Section Quiz

Power Presentations

Test Generator

Reteach For each subsection of the section, have a pair of students write three multiple-choice questions about the subsection, and then quiz other pairs.

Unit Resource Book: Latin America
- Reteaching Activity, p. 184

SECTION 1 ASSESSMENT ANSWERS

Terms & Names

1. ecotourism: travel to a natural habitat in a way that does not damage the habitat; dictator: person with complete control over a country's government; CAFTA: agreement that promotes trade between the U.S. and Central America; artisan: worker skilled in making products or art with his or her hands

Use Your Reading Notes

2. calypso, salsa, punta rock
   Diagram *Government: mostly democracies; Economy: mostly substinence farming, ecotourism important, CAFTA. Culture: population 1/2 urban, family important, known for works of artisans, music*

Key Ideas

3. Agriculture is the main economic activity.
4. Costa Rica, Guatemala, El Salvador, Honduras, and Nicaragua
5. molas, painted gourds, and baskets

Critical Thinking

6. The civil wars caused Central American countries to struggle to establish strong governments.
7. They grow barely enough food to feed their families. Rural schools are poorly equipped, it is difficult to hire teachers to work in these areas. Children often do not attend school because they are needed to work on family farms.
8. It helps the economy of the country, and gives an understanding of the people's lives.
9. Use the rubric to score presentations.

Presentation Rubric

	Use of Multimedia	Content of Text
4	excellent	well-organized; informative
3	good	organized; some good information
2	adequate	some organization; some facts
1	poorly chosen	little organization; few details

1 Plan & Prepare

Objectives

- Explain how Caribbean economies have changed over time
- Compare the governments of Cuba and Puerto Rico
- Analyze the influences on Caribbean culture

Read for the Essential Question

Help students read for a purpose by reminding them of the Essential Question: "How are the countries of this region working to strengthen their governments and economies?" Point out that they can use text features such as Key Questions, special features, images, and maps to form answers to the question.

Vocabulary

Best Practices Toolkit

Use the ***Best Practices Toolkit*** to model strategies for vocabulary notetaking. Vary strategies throughout the year. Choose from: *Knowledge Rating, Predicting ABCs, Definition Mapping, Word Sort, Word Wheel, Frayer Model (Word Squares), Magnet Words,* and *Student VOC.*

Vocabulary Strategies, TT9–TT16

Reading Strategy

Best Practices Toolkit

Display the Three-Column Chart transparency.

- Point out to students that to summarize information in the section they must identify the most important ideas about each topic in the chart and restate them in their own words.
- Model filling in entries from the "Economic Activities" subsection in the section.
- Have students suggest entries, using the topics in the chart.
- Display the completed chart for review.

Three-Column Chart, TT21

SECTION 2 Reading for Understanding

Key Ideas

BEFORE, YOU LEARNED

Mayan and Spanish influences are seen today in Central American cultures.

NOW YOU WILL LEARN

The influences of various European countries and Africa are reflected in the Caribbean island cultures.

Vocabulary

TERMS & NAMES

Taino Indian group in the Caribbean islands

dependency a place governed by a country that it is not officially a part of

one-crop economy an economy that depends on a single crop for income

tourism industry that provides services for travelers

Caribbean Community and Common Market (CARICOM) a trade organization of several Caribbean nations

commonwealth a self-governing political unit that is associated with another country

REVIEW

Central America Free Trade Agreement (CAFTA) trade agreement to promote trade between the United States and countries of Central America

dictator person with complete control over a country's government

communism a type of government in which the Communist Party holds all political power and controls the economy

Reading Strategy

Re-create the chart shown at right. As you read and respond to the **KEY QUESTIONS**, use the chart to summarize what you have learned about the economies, governments, and cultures of the Caribbean islands.

See Skillbuilder Handbook, page R5

SUMMARIZE

ECONOMY	GOVERNMENT	CULTURE
1.	1.	1.
2.	2.	2.
3.	3.	3.

GRAPHIC ORGANIZERS
Go to **Interactive Review** @ClassZone.com

PRETEACHING VOCABULARY

English Learners

Pronounce and Preview

Pronounce each term for students. Review the meanings of words within definitions, such as *officially, self-governing,* and *associated.*

- To modify vocabulary learning, have students complete worksheets as they read, instead of afterward.

Unit Resource Book: Latin America

- Vocabulary Practice, p. 165

Inclusion

Record Terms and Names

Have students read each term and its definition into an audiocassette recorder. As students listen to their recordings play back, have them follow along with the words in their textbooks. Then ask students to use each term and name in a sentence.

Machu Picchu, Peru

SECTION 2 FOCUS ON

Dolphin off the coast of Honduras

The Caribbean

Connecting to Your World

What images come to mind when you hear the words Caribbean islands? Many people think of beautiful sandy beaches and clear blue water. Thousands of people from all over the world travel to the Caribbean islands each year to enjoy the warm climate and outdoor activities. But the Caribbean islands are home for many thousands more. These people are shaping the economy and culture of the islands today.

History and Economy

KEY QUESTION How have Caribbean nations worked to develop their economies?

Although Spain ruled most of Central America, several other European nations claimed and settled the Caribbean islands. In addition, Africans were brought to the islands as enslaved persons. The influences of all these people contributed to the diverse cultures in the Caribbean islands today.

Beach in Varadero, Cuba

2 Focus & Motivate

3-Minute Warm-Up

Write on the board or display the transparency:

- Why do you think the Caribbean islands are so popular with tourists? *(Possible answer: These islands have nice beaches and a warm climate.)*

Unit Transparency Book: Latin America
- 3-Minute Warm-Ups, TT20

Connecting to Your World

More About . . .

Cuban Tourism

Since the 1990s, tourism has become more important than exports as part of the Cuban economy. Most tourists come from Canada, Europe, and Latin America. Varadero is one of four main tourist areas in the country featuring new resort facilities.

3 Teach

History and Economy

Talk About It

- How have the governments of Caribbean nations changed over time? *(Once ruled by other countries, most are independent today.)*
- How has tourism helped the economy of the Caribbean? *(It provides jobs; visitors bring money into the economy.)*
- **Summarize** Why did European nations bring Africans as slaves to the Caribbean islands? *(to work on sugar plantations, to replace Tainos who had died from disease and ill treatment)*

SECTION 2 PROGRAM RESOURCES

Unit Resource Book: Latin America

ON LEVEL

Unit Resource Book
- Reading Study Guide, p. 127
- Section Quiz, p. 180

STRUGGLING READERS

Unit Resource Book
- Reading Study Guide with Additional Support, p. 137
- Skillbuilder Practice: Reading a Time Line, p. 166
- Section Quiz, p. 180
- Reteaching Activity, p. 185

eEdition with Audio DVD-ROM

ENGLISH LEARNERS

Pupil Edition in Spanish
eEdition with Audio DVD-ROM
Unit Resource Book
- RSG (Spanish), p. 147
- RSG with Additional Support (Spanish), p. 157

Multi-Language Glossary
Test Generator
- Section Quiz in Spanish

INCLUSION

Unit Resource Book
- RSG with Additional Support, p. 137
- Section Quiz, p. 180
- Reteaching Activity, p. 185

GIFTED & TALENTED

Unit Resource Book
- History Makers, p. 172
- Section Quiz, p. 180

PRE-AP

Unit Resource Book
- Interdisciplinary Projects, pp. 173, 175
- Primary Source: from *The Life of Olaudah Equiano*, p. 178
- Section Quiz, p. 180

TECHNOLOGY

Unit Transparency Book
- 3-Minute Warm-Ups, TT20
- Essential Question, TT21
- Critical Thinking, TT23
- Map: The Caribbean Islands: Political, TT27

Daily Test Practice Transparencies
- Section 2, TT30

Power Presentations
ClassZone.com
World Cultures and Geography Video on DVD

Slave Trade from West Africa to Caribbean Islands

CONNECT Geography & History

Point out that this map highlights the movement of people from one place to another. Explain that the African slave trade lasted into the 19th century. Ask students the approximate number of enslaved persons brought to the United States directly from Africa *(500,000)*. How does this number compare to the numbers brought to other parts of the Americas? *(It is much smaller.)*

ANSWERS

1. **Region** from western Africa
2. **Movement** about 2 million

Unit Resource Book: Latin America

- Interdisciplinary Projects: Comparing Populations, pp. 173, 175
- Primary and Secondary Sources: from *The Life of Olaudah Equiano*, p. 178

Comparing Regions

One-Crop Economies

Starting in the early 1600s, Virginia and Maryland, English colonies in North America, relied on a one-crop economy, as did Caribbean nations. The crop was tobacco rather than sugar, and exporting tobacco became the region's primary source of income.

There were downsides to this one-crop economy, as well. Settlers in Jamestown were so driven by the desire to get rich from planting tobacco that they neglected to plant food crops and were unwilling to spend time on defense. Over time too much tobacco was grown, and the price fell. The Southern tobacco plantations also depended on the labor of enslaved Africans. In addition, the cultivation of one crop also depleted the soil and led to a need for more land. This desire for land conflicts with local Indians as growers moved west.

Slave Trade from West Africa to Caribbean Islands

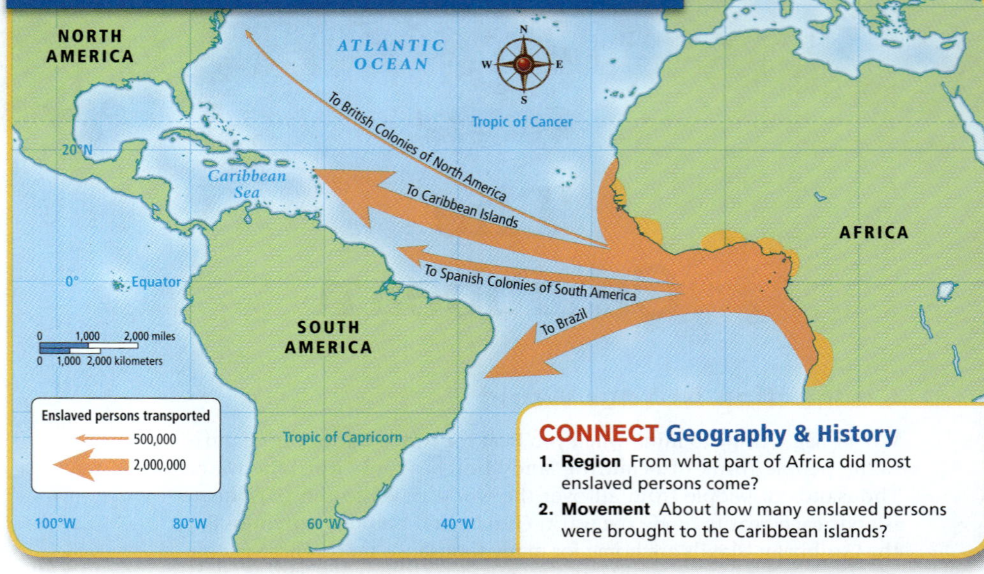

CONNECT Geography & History

1. **Region** From what part of Africa did most enslaved persons come?
2. **Movement** About how many enslaved persons were brought to the Caribbean islands?

From Colonization to Independence After Columbus arrived in the Caribbean islands in 1492, the Spanish settled some of the islands and set up sugar cane plantations. They used the native **Taino** people as forced labor. By the 19th century, the French, British, and Dutch also claimed islands to profit from the sugar trade.

After nearly all the Taino people died due to disease and cruel treatment, Europeans looked to enslaved Africans to replace the Taino. Forty percent of all enslaved persons brought to the Americas were brought to the Caribbean islands.

In 1804, Haiti was the first island to achieve independence. The independence movement there started as a slave revolt against the French, who forced Africans to work on the sugar plantations. In 1898, Cuba gained independence from Spain, while Puerto Rico became a dependency of the United States. A **dependency** is a place governed by or closely connected with a country that it is not officially part of. Today, only a few islands are still under the direction of European nations or the United States.

Economic Activities During colonial times, the Caribbean islands focused mainly on developing the sugar cane industry. Sugar was so profitable that European plantation owners raised few other crops. A country that depends on a single crop for most of its income has a **one-crop economy**. This kind of economy can be unstable. If a crop fails or prices for the crop are low, the entire economy suffers.

DIFFERENTIATING INSTRUCTION

English Learners

Comprehension: Connection

Help students understand the chain of events that began with Columbus's arrival in the Caribbean islands and ended with Haiti's independence in 1804. Read the subsection "From Colonization to Independence" aloud, pausing at appropriate places as students work in pairs to complete a cause-and-effect chain like the one shown below. Call on volunteers to present the information in their completed graphic organizers and clarify any misunderstandings as needed.

Columbus arrives in the Caribbean → [ ] → [ ] → Other Europeans claim Caribbean islands → [ ] → [ ] → Haiti gains indepen-dence

iversified
heir econo-
nies to stop
ependence
n a one-crop
conomy;
creased the
ourist indus-
y; signed
ARICOM ▸

In the late 1800s, the sugar industry in the Caribbean islands was in trouble. Places in other parts of the world that raised sugar cane charged lower prices for their sugar. As a result, the sugar trade in the Caribbean islands declined.

The Caribbean nations found they had to diversify their economies. They began to raise other crops, such as bananas, pineapples, and citrus fruits. They also developed industries such as fishing, mining, and chemical plants. **Tourism**, which provides services for travelers, has become a very important industry. Residents of the islands are able to find jobs in the hotels, restaurants, and resorts on the islands. They also work as guides on sailing trips, snorkeling, and other activities for tourists.

Several Caribbean nations today are members of the **Caribbean Community and Common Market (CARICOM)**, a trade organization similar to NAFTA and **CAFTA**. The purpose of the organization is to coordinate economic and trade relations among the member nations.

DRAW CONCLUSIONS Explain what Caribbean nations have done to develop their economies.

Two Different Governments

KEY QUESTION How are the Puerto Rican and Cuban governments different?

Most Caribbean nations have at some time been under the rule of **dictators**. Today, most nations have democratically elected governments. Two islands, however, Cuba and Puerto Rico, have developed two distinctly different forms of government.

Cuba Spain controlled Cuba until the United States defeated Spain in the Spanish-American War in 1898. The United States military occupied the island until Cuba became an independent country in 1902. However, U.S. influence continued until 1959, when rebel forces led by Fidel Castro overthrew an unpopular dictator. By 1961, Castro had established communism in Cuba, with close ties to the Soviet Union. **Communism** is a type of government in which the Communist Party holds all political power and controls the economy. Castro's government

HISTORY MAKERS

Fidel Castro born 1926

Castro was the son of a wealthy farmer. As a boy, he worked in his father's sugar cane fields. He attended private schools and received a law degree from the University of Havana in 1950. There he also developed an interest in politics. As a lawyer, Castro worked on behalf of poor people. In 1953, he unsuccessfully tried to overthrow Cuba's dictator, Batista. After succeeding in 1959, Castro himself has ruled as a dictator for more than 40 years.

ONLINE BIOGRAPHY
For more on the life of Fidel Castro, go to the **Research & Writing Center @ClassZone.com**

Teach

Two Different Governments

Reader, Recorder, Reporter

- Why did Cuba have strong ties to the Soviet Union? *(because both had Communist governments)*
- Give some examples of democracy in Puerto Rico. *(Possible answers: Voters approved the constitution; there are two political parties; the island is self-governing.)*
- **Summarize** What was the United States' involvement with Cuba's government? *(It helped the country gain independence from Spain and continued to influence Cuba's government until Castro took over in 1959.)*

RESEARCH & WRITING CENTER

HISTORY MAKERS

Fidel Castro

Castro was part of armed revolutionary groups from the time he was in college. In his campaign to overthrow Batista, his skill in spreading propaganda was instrumental in allowing his small group of 800 guerillas to defeat the Cuban army of 30,000 soldiers. Castro had gained the support of middle- and upper-class Cubans by promising a constitutional government that ensured rights such as a free press and private property. Once in power, however, he consolidated his power—and admitted he had deceived his supporters.

Unit Resource Book: Latin America
- History Makers: Fidel Castro, p. 172

DIFFERENTIATING INSTRUCTION

CONNECT to Language Arts

Write About a Job

Have students reread the "Economic Activities" subsection, focusing on the more diversified economies of the Caribbean nations. Ask them to write a paragraph about a job that they would like to have if they lived in a Caribbean country. They may choose a job specifically listed in the textbook or suggest others related to those industries. Direct them to include reasons why they would like the job and describe qualifications needed. Allow students to share their writing in small groups.

CONNECT to Art

Illustrate Changes in Cuba

Ask students to reread the "Cuba" subsection and create a collage or an illustrated time line that shows the change in Cuba's government from Spanish control to Castro's takeover. Encourage students to be creative in drawing symbols to show influences on Cuba's government. Display students' completed artwork.

Unit Resource Book: Latin America
- Skillbuilder Practice: Reading a Time Line, p. 166

Make It Fun

Teacher-Tested Activities

Terry Rhodes, McNair Middle School, Fayetteville, Arkansas

My students enjoy making a "Country Cube." Pairs of students choose a country and find six geographic facts about it. They make a cardboard cube and illustrate one fact on each of the six sides. They can be as creative as they want in how they illustrate each fact. The finished cubes can be used as a learning tool.

COMPARING

Puerto Rican and Cuban Governments

Point out that the head of Puerto Rico's government is a governor, similar to the head of an individual state government in the United States. Explain that Puerto Rico is divided into municipalities or cities, another similarity to U.S. states. Cuba's 14 provinces are also similar to the political division of the 50 U.S. states.

- Why is Cuba's head of government called president rather than governor? *(because Cuba is an independent country)*
- How is Puerto Rico's relationship to the U.S. government different from that of a U.S. state? *(Puerto Rico's commissioner is not able to vote in the House of Representatives, but state representatives are able to vote.)*

CRITICAL THINKING ANSWER

Evaluate Two parties can represent more than one point of view and give people a choice. With one party, there is no choice.

Teach

People and Culture

Think, Pair, Share

- What three cultures have influenced Caribbean culture? *(African, European, and Indian)*
- Why have different musical styles developed in different Caribbean nations? *(Possible answer: because the people of each nation have had different influences, depending on their early history and cultural groups)*
- What two art forms have made use of oil drums? *(music and folk art)*
- **Summarize** What are some types of instruments used in Caribbean music? *(steel drums and various kinds of guitars)*

COMPARING Puerto Rican and Cuban Governments

	PUERTO RICO	CUBA
Political Status	U.S. commonwealth	Independent country
Type of Government	Democracy	Communist state
Head of Government	Governor	President
Voting Age	18	16
Number of Political Parties	2	1
Political Divisions	Divided into 78 municipalities	Divided into 14 provinces
Relationship to U.S.	Non-voting commissioner in the U.S. House of Representatives	None

CRITICAL THINKING

Evaluate How does a two-party system help make a government more democratic?

improved health care and education for the Cuban people. However, Castro has ruled Cuba as a dictator and has denied Cubans many rights and freedoms.

Puerto Rico Like Cuba, Puerto Rico was a Spanish colony until 1898, when it became a U.S. territory. Puerto Rico, however, was never independent. In 1952, Puerto Rican voters approved a constitution that made Puerto Rico a commonwealth of the United States. As a **commonwealth**, Puerto Rico is self-governing but is still a part of the United States. Although Puerto Ricans are U.S. citizens, they cannot vote for the U.S. president.

COMPARE AND CONTRAST Describe the similarities and differences between the governments of Cuba and Puerto Rico. both once controlled by Spain; Cuba became an independent Communist nation. Puerto Rico became a commonwealth.

People and Culture

KEY QUESTION Why has African culture been a major influence on Caribbean life?

The cultures of the Caribbean islands reflect Indian, African, and European influences. Because the region was the center of the slave trade, African influences have left a mark on many aspects of

DIFFERENTIATING INSTRUCTION: TIERED ACTIVITIES

OBJECTIVE Create a classroom display on Puerto Rico's government as it relates to U.S. federal and state governments.

Basic

Give students fact sheets about their state government and comparable information for Puerto Rico, such as population and number of representatives in Congress. Have them use the material to create a visual comparing their state to Puerto Rico to include in the display.

On Level

Have students use resources such as encyclopedias to find information about their state government and Puerto Rico's government, such as names of the governors and how the political divisions are structured. Direct students to create visuals with captions to include in the display.

Challenge

Have students use library resources to find out how the government of Puerto Rico as a commonwealth is related to the U.S. government and how its executive and judicial branches are organized to handle self-government. Have them create captioned visuals to include in the display.

Caribbean life. People in the islands speak a variety of languages. Spanish is the official language in Cuba, French in Haiti, and English in Jamaica.

Music in the Caribbean also reflects a blend of cultures. Calypso music began in Trinidad. It combines styles from Africa, Spain, and the Caribbean. Steel drums and guitars accompany calypso songs. Reggae developed in Jamaica in the 1960s. It blends African music, Caribbean music, and U.S. music to make its own unique style. Caribbean music includes many guitar-like instruments that have been created in the region, such as the Puerto Rican *cuatro* and the Cuban *tres*.

Most people in the Caribbean live in urban areas, where they hope to find jobs in the tourist industry. People celebrate festivals, such as Carnival. Artisans create folk art, such as oil drum art and papier-mâché sculptures. Popular sports are football, known as soccer in the United States, and baseball.

Fun Facts!

STEEL DRUMS

The steel drum was first made in Trinidad from the end and part of the sides of a 55-gallon steel oil barrel. The surface was hammered in certain ways to produce various tones. The steel drum is one of the few acoustic musical instruments created in the 20th century.

EVALUATE Explain why African culture has been a major influence on life in the Caribbean. because the Caribbean was the center of the slave trade, which brought many Africans to the region

Section 2 Assessment

ONLINE QUIZ For test practice, go to **Interactive Review** @ClassZone.com

TERMS & NAMES

1. **Explain the importance of**
   - Taino
   - dependency
   - tourism
   - commonwealth

USE YOUR READING NOTES

2. **Summarize** Use your completed chart to answer the following question:
   Why did the Caribbean islands diversify their economies?

ECONOMY	GOVERNMENT	CULTURE
1. 2. 3.	1. 2. 3.	1. 2. 3.

KEY IDEAS

3. What was the main economic activity in the Caribbean islands during colonial times?
4. How did Fidel Castro come to power in Cuba?
5. Why do people in the Caribbean Islands speak a variety of languages?

CRITICAL THINKING

6. **Compare and Contrast** How is Cuba's government different from the U.S. government?
7. **Draw Conclusions** Why do most people in the Caribbean islands live in urban areas?
8. **CONNECT to Today** In recent years Puerto Ricans have debated the issue of becoming a state of the United States. Why might people support or oppose Puerto Rican statehood?
9. **WRITING** **Write a Country Profile** Choose one Caribbean island and prepare a country profile for its Web site. Include its location, the kind of government and economy it has, and tourist attractions.

CONNECT to the Essential Question

How are the countries of this region working to strengthen their governments and economies?

Ask students what they have learned so far that can help them answer this question. Students might mention:

- Governments of most Caribbean countries are now independent.
- Caribbean nations have diversified their economies.

Fun Facts!

Steel Drums

Caribbean bands play steel drums of different sizes to create sounds that range from high to low.

4 Assess & Reteach

Assess Have students complete the Section Assessment.

Unit Resource Book: Latin America
- Section Quiz, p. 180

Interactive Review @ClassZone.com
- Section Quiz

Power Presentations

Test Generator

Reteach Organize the class into three groups. Ask each group to prepare notes about one main heading in the section to teach the other groups.

Unit Resource Book: Latin America
- Reteaching Activity, p. 185

SECTION 2 ASSESSMENT ANSWERS

Terms & Names

1. Taino: an Indian group in the Caribbean; dependency: a place governed by or closely connected with a country that it is not officially part of; tourism: industry that provides services for travelers; commonwealth: a self-governing political unit that is associated with another country

Use Your Reading Notes

2. to not depend on a one-crop economy

ECONOMY	GOVERNMENT	CULTURE
Sugar cane important product, tourism important industry	once colonies, independent today	African culture a great influence

Key Ideas

3. the sugar cane industry
4. He overthrew the government of an unpopular dictator in 1959.
5. because the islands were settled by several European nations

Critical Thinking

6. Cuba: ruled by dictator, government owns property; U.S. has elected government, and property is privately owned.
7. to take advantage of jobs in the tourist industry
8. Answers will vary, but those opposed might want to keep independence; those in support might indicate that Puerto Rico would benefit economically.
9. Use the rubric to score country profiles.

Country Profile Rubric

	Evidence of Research	Presentation of Information
4	thorough and detailed	excellent
3	many good details	good
2	some good details	adequate
1	incomplete or haphazard	weak

1 Plan & Prepare

Objectives

- Describe the government, economy, and culture of Venezuela
- Describe the government, economy, and people of Colombia

Read for the Essential Question

Help students read for a purpose by reminding them of the Essential Question: "How are the countries of this region working to strengthen their governments and economies?" Point out that they can use text features such as Key Questions, special features, images, and maps to form answers to the question.

Vocabulary

Best Practices Toolkit

Use the ***Best Practices Toolkit*** to model strategies for vocabulary notetaking. Vary strategies throughout the year. Choose from: *Knowledge Rating, Predicting ABCs, Definition Mapping, Word Sort, Word Wheel, Frayer Model (Word Squares), Magnet Words,* and *Student VOC.*

Vocabulary Strategies, TT9–TT16

Reading Strategy

Best Practices Toolkit

Display the Web Diagram transparency.

- Remind students that main ideas are important ideas in a segment of text.
- Model filling in main ideas on a graphic organizer for Venezuela using the subheads throughout the section.
- Have students suggest entries. Remind students that main ideas in each block of text in the section will help them answer the Key Questions.
- Create a similar diagram for Colombia. Display the completed diagrams for review.

Web Diagram, TT28

SECTION 3 Reading for Understanding

Key Ideas

BEFORE, YOU LEARNED

Many Caribbean nations are working together to increase trade.

NOW YOU WILL LEARN

Venezuela's economy depends heavily on petroleum, while Colombia's economy depends more on agricultural products.

Vocabulary

TERMS & NAMES

federal republic form of government in which power is divided between a national government and state governments

Caracas capital city of Venezuela

joropo (huh•ROH•poh) Venezuelan national folk dance

Bogotá capital city of Colombia

Gabriel García Márquez Colombian author and Nobel Prize winner

Fernando Botero Colombian artist known for portraits of people with exaggerated forms

REVIEW

Simón Bolívar (boh•LEE•vahr) leader for independence in northern South America

dictator person with complete control over a country's government

llanos (YAH•nohs) grasslands of South America's Central Plains

BACKGROUND VOCABULARY

mosaic a picture made by placing small, colored pieces of tile or glass on a surface

Reading Strategy

Re-create the cluster diagram shown at right for Venezuela and Colombia. As you read and respond to the **KEY QUESTIONS**, use the diagram to record details about the two countries' government, economy, and people.

Skillbuilder Handbook, page R4

FIND MAIN IDEAS

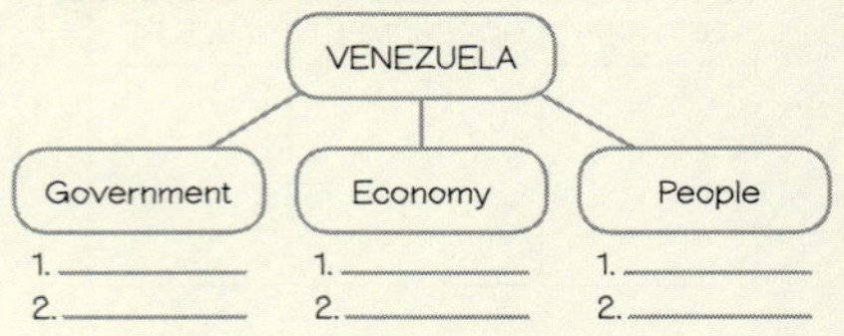

GRAPHIC ORGANIZERS Go to **Interactive Review @ClassZone.com**

PRETEACHING VOCABULARY

English Learners

Pronounce and Preview

Pronounce each term for students. Review the meanings of words within definitions, such as *national, folk dance,* and *exaggerated.*

- To modify vocabulary learning, have students complete worksheets as they read, instead of afterward.

Unit Resource Book: Latin America

- Vocabulary Practice, p. 165

Inclusion

Sort Words into Categories

Display a three-column chart transparency with the headings *Cities, Culture,* and *Government* or write the headings on the board. Call on volunteers to read the definition of each term and name and suggest which category is most appropriate for each. The term *llanos* does not easily fit these categories, but might best be related to culture.

Machu Picchu, Peru

Dolphin off the coast of Honduras

Venezuela and Colombia

Connecting to Your World

Many civilizations have created beautiful mosaic art. A **mosaic** involves placing small, colored pieces of stone, tile, or glass next to each other on a surface to make a picture or design. The countries of South America form a kind of mosaic of their own, a cultural mosaic. People from different cultural groups live near each other but keep their own cultural identities. *Mestizos,* Indians, and people of African ancestry form part of the cultural mosaic in Venezuela and Colombia.

Cultural Mosaic South Americans today reflect a blend of cultures.

Venezuela

KEY QUESTION What are the main economic activities of Venezuela?

Venezuela and Colombia are located in the northern and northwestern part of South America. Both countries border the Caribbean Sea, and Colombia also borders the Pacific Ocean. In both countries, most people live in urban areas. While Venezuela and Colombia share similar histories, their governments and economies have developed differently.

Caracas, Venezuela

2 Focus & Motivate

3-Minute Warm-Up

Write on the board or display the transparency:

- Look at the photos of Caracas, Venezuela, and Bogotá, Colombia. How are the two cities alike? How are they different? *(Possible answer: Alike—both modern-looking cities; different—Bogotá seems to be at a higher elevation than Caracas.)*

Unit Transparency Book: Latin America
- 3-Minute Warm-Ups, TT20

Connecting to Your World

More About . . .

Caracas

Caracas—Venezuela's largest city and capital—is almost completely surrounded by mountains. Caracas was established in 1567 and named after local Indians.

3 Teach

Venezuela

Talk About It

- How did government change in Venezuela during the 20th century? *(Years of civil war and dictators were followed by democratically-elected governments.)*
- **Find Main Ideas** What is life like in Venezuelan cities? *(Professionals live comfortably; poor people live as squatters in crowded areas on the outskirts of the cities.)*

SECTION 3 PROGRAM RESOURCES

Unit Resource Book: Latin America

ON LEVEL
Unit Resource Book
- Reading Study Guide, p. 129
- Section Quiz, p. 181

STRUGGLING READERS
Unit Resource Book
- Reading Study Guide with Additional Support, p. 139
- Skillbuilder Practice: Reading a Time Line, p. 166
- Section Quiz, p. 181
- Reteaching Activity, p. 186

eEdition with Audio DVD-ROM

ENGLISH LEARNERS
Pupil Edition in Spanish
eEdition with Audio DVD-ROM
Unit Resource Book
- RSG (Spanish), p. 149
- RSG with Additional Support (Spanish), p. 159

Multi-Language Glossary
Test Generator
- Section Quiz in Spanish

INCLUSION
Unit Resource Book
- RSG with Additional Support, p. 139
- Section Quiz, p. 181
- Reteaching Activity, p. 186

GIFTED & TALENTED
Unit Resource Book
- Section Quiz, p. 181

PRE-AP
Unit Resource Book
- Section Quiz, p. 181

TECHNOLOGY
Unit Transparency Book
- 3-Minute Warm-Ups, TT20
- Essential Question, TT21
- Critical Thinking, TT24

Daily Test Practice Transparencies
- Section 3, TT31

Power Presentations
ClassZone.com
World Cultures and Geography Video on DVD

ANALYZING Primary Sources

In his 20s, Simón Bolívar was influenced by studying Enlightenment thinkers and by a visit to the United States. He wrote the "Jamaica Letter" in 1815 while in exile after several attempts to liberate Venezuela had failed. In the letter he blamed the evils of the Spanish colonial system on South Americans' inability to establish republics. He favored a strong central government and a powerful executive.

- What does Bolívar mean by "institutions which are wholly representative"? *(Possible answer: ways for people to freely choose leaders to represent their interests)*
- Why might popular systems cause a country's downfall? *(Possible answer: if people elect unqualified or corrupt leaders)*

DOCUMENT-BASED QUESTION ANSWER

Possible answer: taking part in elections and knowing what the representatives being voted for stand for

Comparing Regions

Cities Built on Water

The Indians of Lake Maracaibo had houses, storage platforms, and islands where their animals were kept. Some modern houses on stilts still exist on the lake. Houses built on pilings over lakes and marshes are also found in many parts of the world. Archaeologists have discovered evidence of similar homes in parts of Europe dating to the 4th century B.C. Such dwellings also existed in parts of Africa, Southeast Asia, and Australia.

Perhaps the most famous city on water is Venice, Italy, which consists of more than 100 islands in a large lagoon, connected by canals and bridges. Fishers had established settlements there by the 5th century, and small townships had been established by the 7th century. Venice would become an important trading and cultural center and today is one of the most visited cities in Italy.

ANALYZING Primary Sources

Simón Bolívar (1783–1830) is known as the liberator of South America. But his dream of uniting South America into one nation failed. In a letter, he explained what kind of government he thought would be best for South America.

> Events . . . have proved that institutions which are wholly representative are not suited to our . . . customs, and present knowledge. . . . As long as our countrymen do not acquire the abilities and political virtues that distinguish our brothers of the north [the United States], . . . popular systems, far from working to our advantage, will, I greatly fear, bring about our downfall.
>
> Source: Jamaican Letter

Simón Bolívar

DOCUMENT–BASED QUESTION

What "abilities" would be needed to set up a representative system?

History and Government Venezuela became part of Spain's empire in the early 1500s. When Spanish explorers saw Indian villages built on stilts on Lake Maracaibo, they named the area "Venezuela" after Venice, an Italian city built on water. By the 1700s, Venezuela was a colony, and most residents had little control over their lives.

Led by the Venezuelan **Simón Bolívar**, Venezuela became the first colony to declare independence from Spain in 1811 and to win it in 1821. After independence, the country went through years of civil war and **dictators**. Although establishing a democracy was not easy, Venezuela's leaders have been democratically elected since 1958.

Today, Venezuela is a **federal republic**, a government in which power is divided between a national government and state governments. The national government consists of a president, a congress, and a supreme court. Venezuela has 22 states and a Federal District. Each state and the Federal District have a governor and a congress.

Economy Venezuela's economy today is dependent on oil production. Its oil fields are located in the Maracaibo Basin, the location of Lake Maracaibo (South America's largest lake), and on the eastern plains. Venezuela is one of the world's leading oil producers. About 75 percent of the nation's exports are oil exports, especially to the United States and Canada. But, like the Caribbean countries with one-crop economies, Venezuela's dependence on one product sometimes results in economic instability, as oil prices rise and fall.

INTERDISCIPLINARY ACTIVITIES

CONNECT to Civics

Compare Two Governments

Have students read about the federal republic in Venezuela. Direct them to create a two-column chart in which they compare the government of Venezuela to the government of the United States, which is also a federal republic. Invite volunteers to use their charts to summarize the similarities and differences between the two governments for the class.

CONNECT to Language Arts

Describe Life on Lake Maracaibo

Invite students to imagine what life might have been like in an Indian village built on stilts on Lake Maracaibo in the 1400s and 1500s. Then have students write a paragraph that describes "their life" in the village before and after the arrival of the Spanish explorers.

About ten percent of Venezuela's workers are farmers. More than half of them cultivate small farms, where they raise enough food and animals to support their families. Larger farms and ranches supply most of Venezuela's commercial products. Large farms grow bananas, coffee, corn, oranges, and rice. Large cattle ranches are an important part of the economy of Venezuela's **llanos** (YAH•nohs).

People and Culture Like the majority of Central Americans, most Venezuelans are *mestizos*. Other Venezuelans are Indians or people of European or African ancestry. Because Venezuela is a former Spanish colony, the official language is Spanish, and most Venezuelans are Roman Catholics.

Most Venezuelans live in cities, such as the capital city, **Caracas**. Many city dwellers live comfortably in houses and apartment buildings and work as professionals such as doctors, government workers, and lawyers. However, as in many parts of Latin America, the gap between the rich and the poor continues to be large. Many poor people from rural areas travel to cities in search of jobs. They live as squatters in crowded settlements outside the cities. To encourage people to stay in rural areas, Venezuela's government has paved roads, created health and education services, and provided electrical service in many rural areas.

Venezuelans enjoy sports, music, and dancing. Football (soccer) and baseball are among the most popular sports. Rodeos are popular, especially in cattle-raising regions in the llanos. The plains also inspired the ***joropo*** (huh•ROH•poh), Venezuela's national folk dance. While Venezuelans love to dance to Caribbean salsa, meringue, and calypso, they change each dance to make it truly Venezuelan.

CATEGORIZE Describe the economic activities of Venezuela.
The economic activities include oil production, farming, and ranching.

Baseball Fans Venezuelans cheer on their baseball team as it plays Italy in the 2006 World Baseball Classic.

More About . . .

Population of Venezuelan Cities

The population of Venezuela expanded rapidly during the oil boom of the 1940s and 1950s. Today, more than 80 percent of its population is urban, and about one-third lives in the four largest cities. The government built high-rise public housing to deal with the large number of rural migrants, but was unable to meet all housing needs. The areas where squatters live in Venezuela are called *ranchos.* Some of the *ranchos* have more substantial housing and neighborhood services than others, but many are still overcrowded and dangerous. Because the *ranchos* are usually built on hillsides, they are also vulnerable to floods and mudslides.

More About . . .

The *Joropo*

Popular throughout Venezuela and eastern Colombia, the *joropo* refers to a musical style as well as to a folk dance. The *joropo llanero* (from llanos) is the most popular regional style of the *joropo* in both countries. Traditionally the dance is similar to a waltz and is performed by couples. In communities with a more working-class population there tends to be more improvisation of steps, as compared to dances in a more elite setting. *Joropo* music is performed using a variety of stringed instruments and maracas, or rattles made from gourds.

DIFFERENTIATING INSTRUCTION

English Learners

Language: Pronoun Referents

Help students identify the referents for *them* and *they* used on this page. Remind students to look to the previous sentence to find the noun that is being replaced. For example, the pronouns in the second sentence refer back to farmers in the first. Point out that these two pronouns are plural so they should look for plural nouns. Explain that *they* is used for the subject of a sentence and *them* is used for objects of verbs or prepositions.

Gifted & Talented

Write a Letter to the Editor

Invite students to reflect on the descriptions of poor people who live in Venezuelan cities and the actions the government is taking to encourage people to stay in rural areas. Then have students write a letter to the editor of a Venezuelan newspaper about the problems of the poor and whether they agree or disagree with the government's actions. Students may write from the point of view of a city dweller or a rural person.

Teach
Colombia

Reader, Recorder, Reporter

- Why have the people of Colombia had a hard time deciding what form of government they want? *(because they cannot agree on whether the central government or regional governments should have more power)*
- How is the Colombian economy different from the Venezuelan economy? *(The Colombian economy is more diversified and includes agriculture, manufacturing, service industries, and mining.)*
- How are Colombian cities similar to those in Venezuela? *(They include professionals who work in industry and live comfortable lives along with poor people whose lack of skills makes it hard for them to find good jobs.)*
- **Find Main Ideas** What government and economic challenges do Colombians face? *(political violence, civil war, and ending the illegal cocaine trade)*

Unit Resource Book: Latin America
- Skillbuilder Practice: Reading a Time Line, p. 166

CONNECT Geography & Economics

Coffee

The "Colombian Mild" arabica coffee beans are grown primarily in Colombia, Kenya, and Tanzania. Most of the operations involved in getting Colombian coffee beans to a processing mill are still done by hand, as shown in the photograph.

The National Federation of Coffee Growers of Colombia (FNC) represents the interests of more than 500,000 coffee farmers. The FNC has long used the coffee-farmer character of Juan Valdez in its marketing and advertising campaigns to promote the superiority of Colombian coffee, which usually commands a premium price.

Colombia

KEY QUESTION What are the characteristics of Colombia's population?

Colombia was named after Christopher Columbus. It is second in population and fourth in size among South American countries. The Andes Mountains stretch across western Colombia. Hot lowlands are located along the coasts of the Caribbean Sea and the Pacific Ocean, and plains are found in the eastern part of the country.

History and Government Like Venezuela, Colombia was a Spanish colony until 1819, when Simón Bolívar gained Colombia's independence. Periods of violence and civil war followed. Colombians could not agree on what kind of government to establish. Some people wanted a strong central government, and others supported strong regional governments. Political violence and civil war continue to threaten the nation today.

Today, Colombia is a republic. The national government is made up of a legislative, executive, and judicial branch. A president, elected to a four-year term, heads the executive branch. Colombia is divided into 32 departments and the district of **Bogotá**, the nation's capital. Each department has an elected legislature and governor.

CONNECT Geography & Economics

Coffee

Coffee was brought to Colombia in 1808. Colombia today is the world's second-leading coffee producer. Its land and climate are perfect for growing a certain type of coffee—Arabica. This coffee grows best in rainy regions near the equator, at elevations between 3,600 and 6,300 feet. One coffee tree produces only enough beans to make one pound of coffee a year!

Economy Unlike Venezuela, Colombia's economy relies on agricultural products. Coffee is Colombia's leading legal export. Large plantations produce bananas, corn, cotton, and sugar. Colombian ranches raise cattle for meat and for leather goods. Stopping the illegal cocaine trade has been a major economic challenge for Colombia. Over the past 30 years, Colombia has supported the cut-flower industry as an alternative to growing cocaine-producing coca plants. It has become the second-largest exporter of cut flowers in the world.

Manufacturing and service industries have become increasingly important to Colombia's economy. Manufactured goods include clothing, chemicals, and processed foods. Service industries employ about 45 percent of Colombian workers.

Colombia produces large amounts of coal and petroleum. Emeralds from Colombian mines account for more than 90 percent of the world's supply.

DIFFERENTIATING INSTRUCTION: TIERED ACTIVITIES

OBJECTIVE Design a visual that promotes the Colombian economy.

Basic

Have students read the "Economy" subsection about Colombia. Then have them create posters that illustrate the Colombian economy. Students may summarize the whole economy or focus on a particular sector, such as agriculture, industry, or mining. Display student posters in the classroom.

On Level

Have students read the "Economy" subsection about Colombia. Then direct them to sketch out a design for a display at an international trade show promoting Colombian products that would be of interest to consumers and businesses in other countries. Have students share their designs.

Challenge

Invite students to plan a Web page that the department of economic development in Colombia might use to promote its products. Encourage students to include links from their main page to lead people to more information on different categories. Have students share their plans.

People and Culture Spanish is Colombia's official language, and most Colombians are Roman Catholics. Most people are *mestizos*. About 70 percent of Colombia's people live in the highland valley basins in western Colombia. Colombia's capital and largest city, Bogotá, is located in a basin in the Andes Mountains.

Most Colombians live in urban areas. People who work in the business, service, and government industries in the cities live quite comfortably. Since the mid-1900s, rural Colombians have moved to cities in search of a better life. However, their lack of education and skills make it difficult to find jobs. They sometimes end up living in poverty in squatter settlements that circle the cities.

Colombia has produced famous writers and artists. **Gabriel García Márquez**, a Nobel Prize winner, writes about Colombian life using a mixture of realism and fantasy. **Fernando Botero** is an artist known for portraits that show people in exaggerated forms.

Bogotá Colombia's capital city lies more than 8,000 feet above sea level and is home to nearly seven million people.

SUMMARIZE Describe the characteristics of Colombia's population. Most people live in urban areas in the highland valleys of the Andes mountains; most Colombians are *mestizos* and Roman Catholics.

Section 3 Assessment

ONLINE QUIZ For test practice, go to **Interactive Review @ClassZone.com**

TERMS & NAMES

1. **Explain the importance of**
   - federal republic
   - Caracas
   - *joropo*
   - Bogotá

USE YOUR READING NOTES

2. **Find Main Ideas** Use your completed diagram to answer the following question: What kind of government do Venezuela and Colombia have today?

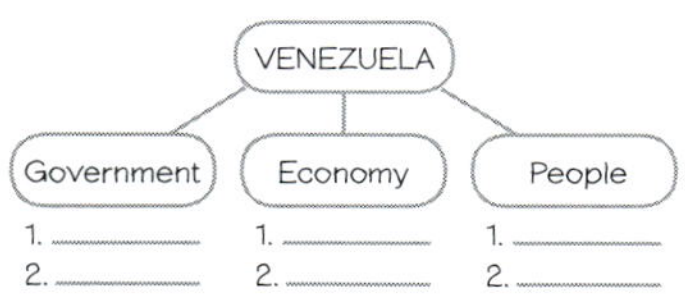

KEY IDEAS

3. What are Venezuela's main products and crops?
4. What three geographic regions are found in Colombia?
5. What industries are important to Colombia's economy?

CRITICAL THINKING

6. **Identify Problems and Solutions** How has the Venezuelan government attempted to curb the movement of people from rural to urban areas?
7. **Compare and Contrast** How are the histories of Venezuela and Colombia similar?
8. **CONNECT to Today** Much of Venezuela's income comes from oil production. How can that be both positive and negative for its economy?
9. **WRITING** **Write a Marketing Campaign Ad** Choose a product from Venezuela or Colombia, such as oil or cut flowers, and write an ad to convince other countries to buy this product.

CONNECT to the Essential Question

How are the countries of this region working to strengthen their governments and economies?

Ask students what they have learned so far that can help them answer this question. Students might mention:

- Venezuela and Colombia are both federal republics, trying to improve the lives of all their citizens.
- The Venezuelan economy relies on oil exports and some agricultural products.
- Colombia's diverse economy includes agriculture, manufacturing and service industries, and mining.

4 Assess & Reteach

Assess Have students complete the Section Assessment.

Unit Resource Book: Latin America
- Section Quiz, p. 181

Interactive Review @ClassZone.com
- Section Quiz

Power Presentations

Test Generator

Reteach Have students work in pairs to create Venn diagrams comparing Venezuela and Colombia. Call on volunteers to use their completed diagrams to summarize what they learned about the governments, economies, and people of the two countries.

Unit Resource Book: Latin America
- Reteaching Activity, p. 186

SECTION 3 ASSESSMENT ANSWERS

Terms & Names

1. federal republic: form of government in which power is divided between a national government and state governments; Caracas: capital of Venezuela; *joropo*: national folk dance of Venezuela; Bogotá: capital of Colombia

Use Your Reading Notes

2. They both are a republic.
   Diagram *Venezuela—Government: former Spanish colony, once ruled by dictators, federal republic today; Economy: dependent on oil, half of farmers work small farms, large farms supply commercial products; People: most* mestizos, *mostly urban. (similar answers for Colombia)*

Key Ideas

3. Venezuela's main product is oil. Its main crops are bananas, coffee, corn, oranges, and rice.
4. highlands, lowlands, and plains
5. manufacturing, service industries, mining, and the cut-flower industry

Critical Thinking

6. by paving roads, creating health and education services, and providing electrical service in many rural areas
7. both Spanish colonies, independent in the 1800s
8. Possible response: Positive because it is a good source of income; negative because a problem with oil production or the world's situation could negatively affect Venezuela's economy
9. Use the rubric to score ads.

Ad Rubric

	Quality of Ad	Effectiveness of Presentation
4	thorough and detailed	excellent
3	many good details	good
2	some good details	adequate
1	incomplete or haphazard	weak

1 Plan & Prepare

Objectives

- Describe the government, economy, and culture of Peru
- Compare the governments, economies, and peoples of Bolivia and Ecuador

Read for the Essential Question

Help students read for a purpose by reminding them of the Essential Question: "How are the countries of this region working to strengthen their governments and economies?" Point out that they can use text features such as Key Questions, special features, images, and maps to form answers to the question.

Vocabulary

Best Practices Toolkit

Use the ***Best Practices Toolkit*** to model strategies for vocabulary notetaking. Vary strategies throughout the year. Choose from: *Knowledge Rating, Predicting ABCs, Definition Mapping, Word Sort, Word Wheel, Frayer Model (Word Squares), Magnet Words,* and *Student VOC.*

Vocabulary Strategies, TT9–TT16

Reading Strategy

Best Practices Toolkit

Display the Three-Column Chart transparency.

- Point out to students that in categorizing they must decide how the information in the section can be sorted into the categories shown in the chart.
- Model filling in entries from the Peru subsection of the section.
- Have students suggest entries, using the categories in the chart.
- Display the completed chart for review.

Three-Column Chart, TT21

SECTION 4 Reading for Understanding

Key Ideas

BEFORE, YOU LEARNED

The cultures of Venezuela and Colombia have been greatly influenced by their Spanish colonial heritage.

NOW YOU WILL LEARN

Indians and *mestizos* make up a large part of the population of Peru, Bolivia, and Ecuador.

Vocabulary

TERMS & NAMES

llama a South American mammal related to the camel

alpaca a South American mammal related to the llama

selva Spanish name for the eastern Peruvian regions that contain rain forests

indigenous native to a region

landlocked surrounded by land with no access to a sea

quinoa a kind of weed from the Andean region that produces a small grain

REVIEW

altiplano the high plateau region of Bolivia

BACKGROUND VOCABULARY

edible fit for eating

Visual Vocabulary landlocked

Reading Strategy

Re-create the chart shown at right. As you read and respond to the **KEY QUESTIONS**, use the chart to categorize details about Peru, Bolivia, and Ecuador.

Skillbuilder Handbook, page R7

CATEGORIZE

	PERU	BOLIVIA	ECUADOR
Government			
Economy			
People			

GRAPHIC ORGANIZERS Go to Interactive Review @ClassZone.com

272 Chapter 9

PRETEACHING VOCABULARY

English Learners

Pronounce and Preview

Pronounce each term for students. Review the meanings of words within definitions, such as *mammal, access,* and *weed.*

- To modify vocabulary learning, have students complete worksheets as they read, instead of afterward.

Unit Resource Book: Latin America

- Vocabulary Practice, p. 165

Inclusion

Use a Pronunciation Guide

Encourage students with hearing impairments to use a dictionary to reinforce their understanding of how to pronounce the terms and names. Have students look up each term, and note how the word is divided into syllables and marked for pronunciation. Allow students to practice pronouncing the vocabulary words with a partner.

Machu Picchu, Peru

Dolphin off the coast of Honduras

SECTION 4

FOCUS ON

Peru, Bolivia, and Ecuador

Connecting to Your World

How would you adapt to living in a mountain region, where it was cold much of the time? For hundreds of years, Indians have lived in the Andes Mountains, where they developed innovative ways to deal with their harsh environment. They raised **llamas** and **alpacas**, small camel-like animals, on the mountainsides. The Inca used the freezing nighttime temperatures and the strong daytime sunlight to preserve and store potatoes and meat by freezing and drying them. Today, people in the Andes Mountains continue to adapt to their environment.

Peruvian Girl and Baby Alpaca

Peru

KEY QUESTION What products are made in Peru's coastal regions?

The Andes Mountains are located in Bolivia, Ecuador, and Peru. Peru is South America's third largest country. It has three land regions—the coast along the Pacific Ocean, the Andes Mountains, and the **selva**, an area of rain forests in eastern Peru.

Adapting to the Mountains Farmers in Bolivia plant crops on mountain terraces, such as these.

273

2 Focus & Motivate

3-Minute Warm-Up

Write on the board or display the transparency.

- What would be some advantages and disadvantages to living in the Andes Mountains environment? *(Advantages: good for coffee growing, mining, raising llamas and alpacas; Disadvantages: difficult for farming and transportation)*

Unit Transparency Book: Latin America
- 3-Minute Warm-Ups, TT20

Connecting to Your World

More About . . .

Llamas and Alpacas

The Indians of the Andes domesticated llamas and alpacas more than 1,000 years ago from related wild animal species, the guanaco and vicuña. These animals' blood systems are well-adapted to high altitudes.

3 Teach

Peru

Think, Pair, Share

- Who is Alejandro Toledo, and why is he important to Peru's modern history? *(He is the first Indian president elected to govern Peru.)*
- **Categorize** What are the three land regions of Peru? *(Pacific Coast, Andes Mountains, and selva, or rain forests)*

SECTION 4 PROGRAM RESOURCES

Unit Resource Book: Latin America

ON LEVEL

Unit Resource Book
- Reading Study Guide, p. 131
- Section Quiz, p. 182

STRUGGLING READERS

Unit Resource Book
- Reading Study Guide with Additional Support, p. 141
- Skillbuilder Practice: Reading a Time Line, p. 166
- Section Quiz, p. 182
- Reteaching Activity, p. 187

eEdition with Audio DVD-ROM

ENGLISH LEARNERS

Pupil Edition in Spanish

eEdition with Audio DVD-ROM

Unit Resource Book
- RSG (Spanish), p. 151
- RSG with Additional Support (Spanish), p. 161

Multi-Language Glossary

Test Generator
- Section Quiz in Spanish

INCLUSION

Unit Resource Book
- RSG with Additional Support, p. 141
- Section Quiz, p. 182
- Reteaching Activity, p. 187

GIFTED & TALENTED

Unit Resource Book
- Interdisciplinary Projects, pp. 174–175
- Section Quiz, p. 182

PRE-AP

Unit Resource Book
- Section Quiz, p. 182

TECHNOLOGY

Unit Transparency Book
- 3-Minute Warm-Ups, TT20
- Essential Question, TT21
- Critical Thinking, TT25
- Cultures, TT29

Daily Test Practice Transparencies
- Section 4, TT32

Power Presentations

ClassZone.com

World Cultures and Geography Video on DVD

More About . . .

Alejandro Toledo

Alejandro Toledo was born in 1946 and grew up in a large working-class family. He received scholarships, partially based on his soccer skills, that allowed him to attend college in California, where he eventually earned a Ph.D. in economics from Stanford University. Toledo worked as an economist for several international organizations before returning to Peru.

Toledo is of Quechuan ancestry, but it was his wife, an American anthropologist born in Belgium, who delivered campaign speeches in Quechua. Although the economy of Peru improved during Toledo's five-year term, he was not reelected in the 2006 election.

Comparing Regions

Soccer in South America and the United States

The term soccer is used in the United States to distinguish the game from gridiron football. British immigrants brought soccer to North America in the 1860s, but gridiron football became more popular in the early 20th century. Today, soccer is especially popular among U.S. high school and college students, both men and women. The United States hosted the World Cup finals in 1994, and the U.S. team won the Women's World Cup in 1999.

Europeans brought football (soccer) to South America in the late 19th and early 20th centuries, and it is now the most popular sport on the continent. A stadium in Brazil can hold more than 150,000 fans. Uruguay won the first World Cup in 1930, and since then South American teams have won 8 of the 17 world championships.

History and Government Between about 1200 and the early 1500s, Peru was home to the Inca civilization. The empire in the Andes Mountains extended from present-day Colombia to Argentina. In 1533, the Spanish conquered the Inca and ruled Peru for 300 years.

Since gaining independence in 1821, Peru has suffered military takeovers, dictatorships, and wars with neighboring countries and revolutionary groups. Today, Peru is a democratic republic. In 2001, Peruvians elected Alejandro Toledo, the first Indian president.

Economy Peru's rugged geography and scarce farmland have made it difficult for the country to create a strong economy. About one-third of the people are farmers. Coffee, potatoes, and grains are grown on highland terraces. Sugar cane, cotton, and asparagus are grown in coastal valleys. Peru is one of the world's leading producers of asparagus.

Peru is a leading producer of copper, lead, silver, and zinc, which are mined in the Peruvian mountains. It also has a profitable fishing industry. Factories that process metals, fish, and sugar cane have been built along Peru's coast. Peru also exports petroleum.

She might sell it to tourists. ▼

Peruvian Weaver This woman weaves colorful yarns with a hand loom. **What do you think she will do with the products she creates?**

People and Culture *Mestizos* and **indigenous** people, or people native to a region, make up most of Peru's population. About ten percent are of European, African, or Asian ancestry. Spanish and Quechua (KEHCH•wuh) are Peru's official languages. About 75 percent of Peruvians speak Spanish, and about 25 percent speak Quechua or another Indian language.

Most Peruvians live in cities. While the middle and upper classes live in houses and apartment buildings, the poor live in slums or squatter settlements. Peru's government has tried to develop these settlements by providing running water and sewer systems.

Peruvians create beautiful sculpture, pottery, and textiles. Music and dancing are popular throughout Peru. As elsewhere in Latin America, football (soccer) is the most popular sport.

CATEGORIZE Identify the kinds of products that are produced in Peru's coastal regions. Peru's coastal regions produce sugar cane, cotton, and asparagus, and factories process metals, fish, and sugar cane.

274 Chapter 9

DIFFERENTIATING INSTRUCTION

CONNECT to Art

Create a Brochure

Have students examine the photograph of the Peruvian weaver and create a brochure that could be used to promote her products to tourists and overseas buyers. Encourage students to include drawings of individual products with brief descriptions that may include ideas for how the products could be used, for example the long strips could be worn as belts or used as straps for guitars. Suggest that they emphasize that the products are handmade in traditional styles.

CONNECT to Math

Graph Economic and Cultural Data

Have students read the "Economy" and "People and Culture" subsections and notice where fractions or percentages are used to describe characteristics of Peru. Direct students to make three simple pie graphs to show the information about segments of the workforce, ethnic background, and languages. They should use the information in the text to place appropriate labels on each section of their graphs.

CONNECT Geography & Culture

The Lake Dwellers The Uros, descendants of an ancient people, live on floating islands on Lake Titicaca. Titicaca, the world's highest lake, is on the border of Peru and Bolivia. They use a kind of reed, or wetland grass, called *totora* (toh•TOH•rah) to make floating mats. They use the mats to create islands in the lake. The Uros also make reed boats and houses. They even have a floating soccer field!

CRITICAL THINKING

Make Inferences How might the Uros have contributed to making Lake Titicaca a popular tourist spot?

Bolivia and Ecuador

KEY QUESTION How are Bolivia's and Ecuador's geographies, histories, and economies alike?

Bolivia is a **landlocked** country, surrounded entirely by land with no access to a sea. The country has two capital cities—Sucre, which is the official capital, and La Paz, which is the administrative capital. Ecuador is one of South America's smallest countries. Large plains are located in northern and eastern Bolivia and on Ecuador's Pacific coastal plain. The Andes Mountains circle a high plateau in western Bolivia and run through the center of Ecuador.

History and Government Both Bolivia and Ecuador were colonized by the Spanish and achieved independence in the 1800s. Bolivia was named for Simón Bolívar. *Ecuador* is the Spanish word for "equator," which crosses the country.

Both Bolivia and Ecuador struggled to develop democracies. Each country has had several constitutions that provided for free elections. But dictators and military leaders often took control.

Today, both Bolivia and Ecuador are democratic republics. Both governments include a national legislature, a supreme court, and an elected president. In 2005, Bolivians elected Evo Morales as president, the first indigenous person to be elected to that office in Bolivia.

Evo Morales An Aymara, Morales was elected president of Bolivia by a large margin in December 2005. He took office in January 2006.

DIFFERENTIATING INSTRUCTION

Struggling Readers

Use a Map

Ask students to refer to maps of South America as they read about Bolivia and Ecuador. Explain that the maps will help them better distinguish the two countries. For example, they can identify the countries that surround Bolivia to better understand what it means to be landlocked. Have students locate the capitals and physical features of each country. Then ask volunteers to describe what they have read in their textbooks and seen on the maps in their own words.

English Learners

Language: Verb Tense

Point out that the first two paragraphs in the "History and Government" subsection are written in the past tense to describe incidents in the histories of Bolivia and Ecuador. Ask students to find the verb "has had" in the second paragraph. Explain that the verb tense here shifted to the present perfect to indicate something that happened over time and that continues today. Using the past tense would mean that the countries' current constitutions do not provide for free elections.

CONNECT Geography & Culture

The Lake Dwellers

Remind students that in Section 3 they read about the Indians of Lake Maracaibo in Venezuela, who built villages on stilts over the water. About 2,000 Uros occupy about 40 islands and make their living mostly from fishing, weaving, and tourism. Most visitors come to the Uros's floating islands by a two-hour boat ride from Puno, Peru.

CRITICAL THINKING ANSWER

Make Inferences Possible answer: People visit the area to see the Uros living on reed islands and making boats. They might like to ride in the boats on the highest lake in the world.

Teach

Bolivia and Ecuador

Talk About It

- How does Ecuador's geography affect its economy? *(Its coastal plains are good for farming and trade.)*
- Why are llamas and alpacas important to Bolivian farmers? *(Llamas allow farmers to transport goods. Alpacas provide food, clothing, and fertilizer.)*
- **Categorize** What natural resources of Bolivia and Ecuador could be developed to strengthen their economies? *(Bolivia: minerals, fertile soil; Ecuador: forests, farm land)*

Unit Resource Book: Latin America
- Skillbuilder Practice: Reading a Time Line, p. 166

More About . . .

Bolivia's Two Capitals

Sucre, founded by Spaniards in 1539, was established as Bolivia's capital in 1839. It is named for Antonio José de Sucre, Bolivia's first president. The Spanish founded La Paz in 1548. In 1898, civil war broke out when some Bolivians wanted La Paz to be the capital. It was then agreed that Sucre would remain the legal capital and home of the Supreme Court. La Paz became the administrative capital, with the legislative and executive branches.

COMPARING

Industries

Encourage students to read the feature as they would an illustrated chart. They should read down each "column" to gain an overview of each country's economy. Then they can read across each "row," to compare the three countries on each segment of the economy.

- What industries do Peru, Bolivia, and Ecuador have in common? *(petroleum)*
- How is Bolivia's major trading partner different from the other two countries? *(Bolivia's major trading partner is Brazil, not the United States.)*

CRITICAL THINKING ANSWER

Compare and Contrast All three produce coffee.

Unit Resource Book: Latin America
- Interdisciplinary Projects: Banana Science, pp. 174–175

More About . . .

Quinoa

Quinoa (pronounced KEEN-wa) is a Spanish word derived from the Quechua *kinua.* This grain is native to the Andes and was a staple of the Incas because it is high in protein, vitamins, and minerals, and grows well in the harsh mountain conditions. The seeds may be cooked and eaten whole like rice or may be ground into flour. The leaves of the plant may also be cooked or eaten in salads. There has been a resurgence of interest in quinoa and other ancient grains among health conscious consumers in North America and Europe, who have been encouraged to add more whole grains to their diets.

COMPARING Industries

PERU

- **Major industries:** mineral production, steel, petroleum refining, fishing (Above: fishing crew harvesting shellfish off the coast of Peru)
- **Major agricultural products:** coffee, cotton, sugar cane, asparagus
- **Major trading partner:** United States

BOLIVIA

- **Major industries:** mining, petroleum, food and beverages (Above: miner underground in Bolivian silver mine)
- **Major agricultural products:** soybeans, coffee, cotton, corn, potatoes
- **Major trading partner:** Brazil

ECUADOR

- **Major industries:** petroleum, food processing, textiles, wood products
- **Major agricultural products:** bananas, coffee, cacao, balsa wood, shrimp (Above: Ecuadorian workers in a banana-processing plant)
- **Major trading partner:** United States

CRITICAL THINKING

Compare and Contrast What agricultural products do Peru, Bolivia, and Ecuador have in common?

Economy Bolivia and Ecuador both have natural resources that have not been fully developed. Bolivia has many minerals and fertile soil. Ecuador's farm and timberland on the coast and in the eastern lowlands still need development. Both nations are working to develop their resources and help improve their economies.

One of Bolivia's leading exports is natural gas. Bolivia is also a leading producer of tin, but only a small percentage of the nation's workers are employed in this industry. About half of Bolivia's workers are farmers. Potatoes, wheat, and **quinoa**, a kind of weed that produces a small **edible** grain, are grown on the **altiplano**, the high plateau region of Bolivia, just east of the Andes Mountains. Bananas, beans, cacao, coffee, soybeans, and corn are also important agricultural products.

Ecuador's major export is petroleum. Gold is also an important mineral product. Most manufacturing in Ecuador takes place in Guayaquil (gwy•uh•KEEL), the nation's largest city, and in Quito, the capital. Ecuador's major manufacturing products include cement, processed foods, and textiles. Ecuador is the world's major supplier of balsa wood, used to make model airplanes. Like Bolivians, most Ecuadorians are farmers. They grow bananas, cacao, coffee, and sugar cane on the country's coastal plain.

DIFFERENTIATING INSTRUCTION: TIERED ACTIVITIES

OBJECTIVE Report on the economies of Peru, Bolivia, and Ecuador.

Basic

Organize students into three "expert" groups and assign each group one of the three countries. Have students take notes about the country's economy. Then have students form new groups of three with one expert on each country in each group. Experts should report about their country's economy.

On Level

Have students choose one of the three countries and review information about its economy. Direct students to prepare a brief speech about their country's economy. Have students form groups of three, with a representative from each country. Have students give their speeches to the members of their groups.

Challenge

Organize students into groups, each focused on one of the three countries. Have students prepare a news program about the economy of that country. Encourage students to prepare visuals to include in their program. Have groups present the program to the class.

People and Culture Spanish and the Indian languages Quechua and Aymara are Bolivia's three official languages. Most Bolivians practice subsistence farming. They have traditionally raised llamas to transport goods and alpacas to provide food, clothing, and fertilizer. In fact, Bolivia has more llamas than any other place in the world!

Most Ecuadorians are Indians or *mestizos*. Many Indians live in the Andes, speak Indian languages, and wear traditional clothes.

As they have for thousands of years, indigenous people in Bolivia today create jewelry, rugs, and shawls. Ecuadorian artists carve objects using the tagua plant. The tagua nut resembles ivory, which is a material that comes from elephant tusks. The tagua nut is used to make buttons and works of art. No animals are harmed to create tagua art.

Countries in the Andes Mountains are famous for panpipe music. Aymara and Quechua peoples play these flute-type instruments combined with drums and guitar-like instruments to create unique regional music.

COMPARE AND CONTRAST Compare the geographies, histories, and economies of Bolivia and Ecuador.

livia is land- ked, Ecuador not; both ve plains and ountains; both re Spanish onies and republics day; both ed to develop ources; most ople in both farmers. ►

Playing a Panpipe Peruvian musicians play panpipes, like the one this girl is playing.

Section 4 Assessment

ONLINE QUIZ For test practice, go to **Interactive Review** @ClassZone.com

TERMS & NAMES

1. **Explain the importance of**
   - llama
   - selva
   - landlocked
   - quinoa

USE YOUR READING NOTES

2. **Categorize** Use your chart to answer the following question:

   How are Bolivia and Ecuador working to improve their economies?

	PERU	BOLIVIA	ECUADOR
Government			
Economy			
People			

KEY IDEAS

3. What kind of government do Peru, Ecuador, and Bolivia have?
4. How are most people in Peru, Ecuador, and Bolivia employed?
5. What groups of people make up a large percentage of the population of all three countries?

CRITICAL THINKING

6. **Analyze Causes and Effects** How do you think being a landlocked nation affects Bolivia's economy?
7. **Make Inferences** Why do you think that so many Indian communities have been able to keep their traditional customs and ways of life?
8. **CONNECT to Today** Why is it important for Bolivia and Ecuador to fully develop their natural resources?
9. **WRITING** **Create a Picture Essay** Choose an indigenous group in Peru, Bolivia, or Ecuador. Research their customs and ways of life. Create a captioned picture essay to present the information.

CONNECT to the Essential Question

How are the countries of this region working to strengthen their governments and economies?

Ask students what they have learned so far that can help them answer this question. Students might mention:

- Peru, Bolivia, and Ecuador are all democratic republics.
- The Peruvian economy includes some agriculture, mining, and fishing.
- Bolivia and Ecuador are working to develop their natural resources to improve their economies.

4 Assess & Reteach

Assess Have students complete the Section Assessment.

Unit Resource Book: Latin America
- Section Quiz, p. 182

Interactive Review @ClassZone.com
- Section Quiz

Power Presentations

Test Generator

Reteach Have students work in pairs to create two true-false questions about each of the subsections of the section. Have partners read their questions to the rest of the class and call on volunteers to answer them. Ask students to supply the correct information that would make any false statements true.

Unit Resource Book: Latin America
- Reteaching Activity, p. 187

SECTION 4 ASSESSMENT ANSWERS

Terms & Names

1. llama: small camel-like animal raised in Andean regions; selva: eastern Peruvian regions that contain rain forests; landlocked: surrounded by land with no access to a sea; quinoa: a kind of weed that produces a small, edible grain

Use Your Reading Notes

2. by developing their underdeveloped resources
   Diagram *Peru—Government: once Spanish colony, democratic republic today; Economy: 1/3 of people are farmers, asparagus important export, profitable fishing industry; People: most* mestizos *and Indians, Spanish and Quechua official languages, most live in urban areas (similar answers for Bolivia and Ecuador)*

Key Ideas

3. democratic republics
4. The majority of people are farmers in Bolivia and Ecuador; one-third are farmers in Peru.
5. Indians and *mestizos*

Critical Thinking

6. It makes it difficult to transport products to other countries.
7. They have been able to keep their traditional practices most likely because they have lived together in the Andes, away from outside influences.
8. Developing their natural resources will help the two countries expand their economies.
9. Use the rubric to score picture essays.

Picture Essay Rubric

	Evidence of Research	Quality of Picture Essay
4	thorough and detailed	excellent
3	many good details	good
2	some good details	adequate
1	incomplete or haphazard	weak

1 Plan & Prepare

Objective

- Compare life in two different climate regions in Middle America and Spanish-speaking South America.

Springboard Ask students what they think the differences are in temperature and rainfall in the two climate regions based on the photographs on the two pages. *(Tropical regions are hot and rainy—the photos show scuba divers, a man wearing short sleeves, and lush green landscapes. Mountain regions are cold and some areas have less rainfall—the photos show snow and people wearing heavier clothing, and the mountains are more barren looking.)*

2 Teach

Climate Regions

Talk About It

- Which region practices subsistence agriculture? How do you know? *(mountain; because the people grow crops such as barley, potatoes, and wheat for families to live on)*
- What do the photos suggest about the temperature ranges in the mountain region? *(Possible Answer: There is a wide variation in the temperatures because some areas are cold and snowy while others are warm enough at times to grow some crops.)*
- What is similar about the sloth and the spectacled bear? *(Possible answer: Both are shown in a tree and have adapted to living in a particular climate.)*

COMPARING Climate Regions

Tropical and mountain climates are both found in Middle America and Spanish-speaking South America. Each climate supports its own unique vegetation and animal life.

Tropical

The climate in tropical regions is hot year-round, with abundant rainfall. This climate produces the thick, green rain forests and the many exotic plants and animals that live in them. Farmers plant crops, such as sugar cane and bananas, that thrive in the tropical conditions.

Outdoor Activities

Scuba divers explore the coral reefs off the coast of Bonaire, one of the Caribbean islands.

Agriculture

A farmer harvests sugar cane on the island nation of Barbados.

Nevis Peak and Botanical Gardens on the Caribbean island of Nevis ▼

Wildlife

Sloths, such as this one hanging upside down from a branch, live in the tropical forests of Central America and South America and spend most of their lives in trees.

DIFFERENTIATING INSTRUCTION: TIERED ACTIVITIES

OBJECTIVE Compare life in two different climate regions in Middle America and Spanish-speaking South America.

Basic

Ask students to divide a sheet of paper into *tropical* and *mountain* sections. Then have them study the images and write a summary statement for each area. Extend the activity by asking students to write a compare and contrast statement about each of the three topics shown in the inset visuals.

On Level

Have students study the images above and imagine what it would be like to live in each climate region. Ask them to write two letters between pen pals who live in each region. Encourage them to include details of daily life that they can infer from the information in the photos, captions, and introductory text.

Challenge

Invite students to study all the visuals and written material on these pages and think about what might appeal to tourists about each climate region. Have students create a brochure or similar product that promotes the variety of tourist attractions available in different areas and at different times of the year.

Outdoor Activities

Ice climbers make their way up Huayna Potosi, a mountain in Bolivia popular among mountain climbers.

Mountain

Unlike the fairly constant climate of the tropical areas, the highland climate of the Andes Mountains varies with changes in elevation. Temperatures range from warm in the lower elevations to freezing in the highest elevations. Farming is difficult in this climate, but several crops, such as potatoes, barley, and wheat, grow well there. Various species of plants and animals have adapted to the mountain conditions.

Agriculture

These Ecuadorian farmers are harvesting barley.

Sacsayhuamán, an ancient Inca fortress, overlooks the city of Cuzco, Peru. ▼

Wildlife

The spectacled bear makes its home in the Andes Mountains. It gets its name from the markings around its eyes.

CRITICAL THINKING

1. **Compare and Contrast** What differences can be seen between tropical and mountain climate regions?
2. **Form an Opinion** Which climate region is more appealing to you? Why?

INTERDISCIPLINARY ACTIVITIES

CONNECT to Language Arts

Write a Journal Entry

Have students think about a particular activity that they would enjoy while visiting one of the regions. Encourage them to review all the photos and captions before they start writing. They might focus on one of the outdoor activities or a visit to a specific site. Visiting with local people and seeing native plants and wildlife might also be part of their experience. Their journal entry should describe their experiences and their thoughts about it.

CONNECT to Science

Compare Crops

Have students choose one of the crops grown in the tropical region to one grown in the mountain region. Allow them to use classroom or library resources to find out about the growing conditions needed for each crop, including type of soil, rainfall, temperature, and length of time between planting and harvest. Ask students to create a fact sheet that summarizes what they've learned about the crops. Invite volunteers to share their information with the class.

More About . . .

Coral Reefs

Coral is a small sea animal similar to a sea anemone. Coral reefs are formed from the accumulation of the skeletons of these tiny animals over thousands of years. The reefs are sometimes likened to rain forests under the sea because of the diversity of the organisms that live in them. Coral reefs are found in the warm waters of the western Atlantic Ocean between Bermuda and Brazil and in the Indian and Pacific oceans region between Africa and Hawaii.

Comparing Regions

Rocky Mountains and the Andes

The Rocky Mountains extend about 3,000 miles throughout western Canada and the United States. The highest peak is Mt. Elbert in Colorado at 14,331 feet. Climate is coldest in Canada and Alaska and warmest in New Mexico, with variations linked to both latitude and altitude. Populations are concentrated in towns at the base of mountains or in river valleys.

The Andes Mountains extend for more than 4,000 miles from the Caribbean through western South America. Ten peaks are higher than 22,000 feet and many more reach 20,000 feet. Subsistence agriculture based on corn, wheat, and potatoes takes place up to 10,000 feet. Most of the population, including residents of the largest cities, are located in the Andes region.

3 Wrap Up

Encourage students to complete a two-column chart listing the main geographical features of each region based on the visuals and copy. They may want to list pros and cons for each region when considering their answer to Critical Thinking question 2.

CRITICAL THINKING ANSWERS

1. **Compare and Contrast** Possible answer: Tropical regions are lush and green, whereas mountain regions appear more barren.
2. **Form an Opinion** Answers will vary but students should give reasons for their opinion.

1 Plan & Prepare

Objectives

- Compare the governments, economies, and cultures of Argentina and Chile
- Describe the populations of Paraguay and Uruguay

Read for the Essential Question

Help students read for a purpose by reminding them of the Essential Question: "How are the countries of this region working to strengthen their governments and economies?" Point out that they can use text features such as Key Questions, special features, images, and maps to form answers to the question.

Vocabulary

Best Practices Toolkit

Use the ***Best Practices Toolkit*** to model strategies for vocabulary notetaking. Vary strategies throughout the year. Choose from: *Knowledge Rating, Predicting ABCs, Definition Mapping, Word Sort, Word Wheel, Frayer Model (Word Squares), Magnet Words,* and *Student VOC.*

Vocabulary Strategies, TT9–TT16

Reading Strategy

Best Practices Toolkit

Display the Web Diagram transparency.

- Remind students that main ideas are important ideas in a segment of text.
- Model filling in main ideas on a graphic organizer about Argentina, using the subheads in this section.
- Have students suggest entries. Remind students that main ideas in each block of text in the section will help them answer the Key Questions.
- Repeat for each country in the section. Display the completed diagrams for review.

Web Diagram, TT28

Section 5 Reading for Understanding

Key Ideas

BEFORE, YOU LEARNED

Geography has influenced the economies and people's ways of life in Peru, Bolivia, and Ecuador.

NOW YOU WILL LEARN

Geography has also influenced the economies of the nations in the Southern Cone.

Vocabulary

TERMS & NAMES

gaucho Argentinian cowboy

Southern Cone South American nations located in the cone-shaped southernmost part of South America

estancia (eh•STAHN•syah) large farm or ranch in Argentina

Mercosur association of several South American countries to promote trade among the countries

REVIEW

José de San Martín leader for independence in southern South America

gross domestic product (GDP) the total value of all the goods and services produced in a country in a year

Pampas grassy plains in south-central South America

landlocked surrounded by land with no access to sea

Visual Vocabulary
Southern Cone

Reading Strategy

Re-create the web diagram shown at right for each country in the section. As you read and respond to the **KEY QUESTIONS**, use the diagram to organize details about the nation's history and government, economy, and population.

Skillbuilder Handbook, page R4

FIND MAIN IDEAS

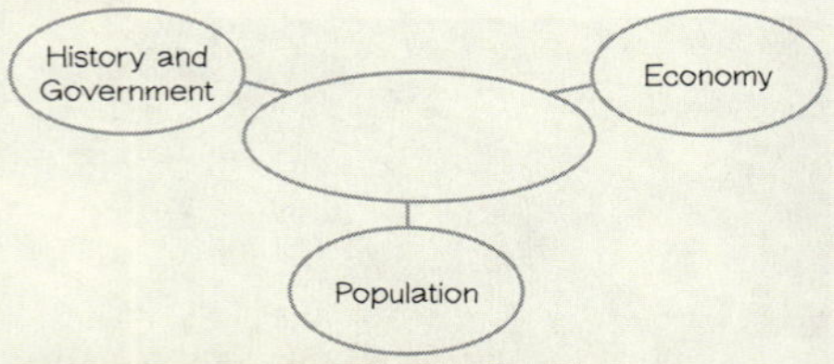

GRAPHIC ORGANIZERS
Go to **Interactive Review** @ClassZone.com

PRETEACHING VOCABULARY

English Learners

Pronounce and Preview

Pronounce each term for students. Review the meanings of words within definitions, such as *cone-shaped, southernmost,* and *promote.*

- To modify vocabulary learning, have students complete worksheets as they read, instead of afterward.

Unit Resource Book: Latin America

- Vocabulary Practice, p. 165

Inclusion

Extend Visual Vocabulary

Help visual and spatial learners more completely understand the name *Southern Cone* by making the connection between the visual vocabulary map and an actual paper cone. Give students paper that is cut, so it will form into a cone. Invite students to sketch out the map on the paper, and then shape the paper to form a cone. Point out that the tip of the cone represents the southernmost part of Chile.

SECTION 5

FOCUS ON

Machu Picchu, Peru

Dolphin off the coast of Honduras

The Southern Cone

Connecting to Your World

Cowboys played an important role in cattle ranching in the American West in the 1800s. In Argentina, **gauchos**, or cowboys, also play an important role in cattle ranching. As in the United States, the life and culture of gauchos is the subject of many Argentinian stories and movies. But the gaucho is just one aspect of the culture of Argentina and its neighbors.

Argentinian gaucho

Argentina and Chile

KEY QUESTION How do Argentina's plains and Chile's mountains affect their economies?

Together with Paraguay and Uruguay, Argentina and Chile form part of South America's **Southern Cone**, the cone-shaped southernmost area of South America. The largest of the four countries is Argentina, a country of plains, plateaus, mountains, forests, and a long coastline. The Andes Mountains make up much of Chile, the world's longest and narrowest country. Argentina is bordered by Paraguay on the north and Uruguay on the east.

Patagonia Sheep farmers round up their sheep. **How would you describe the land on which this sheep ranch is located?**

flat, not much vegetation

2 Focus & Motivate

3-Minute Warm-Up

Write on the board or display the transparency:

- Look at the photograph of sheep farmers in Patagonia. How are cowboys and sheep ranchers similar? *(Possible Answer: Both ride on horses to herd large numbers of animals.)*

Unit Transparency Book: Latin America
- 3-Minute Warm-Ups, TT20

Connecting to Your World

More About . . .

Gauchos

Gauchos were common in Argentina and Uruguay from the mid-1700s to the mid-1800s, and most were *mestizos*. Like American cowboys, gauchos came to symbolize the freedom of the open frontier. After the fencing of the grasslands, many gauchos became ranch hands.

3 Teach

Argentina and Chile

Reader, Recorder, Reporter

- What is similar about the economies of Argentina and Chile? *(Possible Answer: Both rely heavily on service industries, which make up most of their GDP and employ more than half of their people. Fishing is also important to both.)*
- **Find Main Ideas** Where do most people live in Argentina and Chile? *(in large cities)*

SECTION 5 PROGRAM RESOURCES

Unit Resource Book: Latin America

ON LEVEL

Unit Resource Book
- Reading Study Guide, p. 133
- Section Quiz, p. 183

STRUGGLING READERS

Unit Resource Book
- Reading Study Guide with Additional Support, p. 143
- Skillbuilder Practice: Reading a Time Line, p. 166
- Section Quiz, p. 183
- Reteaching Activity, p. 188

eEdition with Audio DVD-ROM

ENGLISH LEARNERS

Pupil Edition in Spanish

eEdition with Audio DVD-ROM

Unit Resource Book
- RSG (Spanish), p. 153
- RSG with Additional Support (Spanish), p. 163

Multi-Language Glossary

Test Generator
- Section Quiz in Spanish

INCLUSION

Unit Resource Book
- RSG with Additional Support, p. 143
- Section Quiz, p. 183
- Reteaching Activity, p. 188

GIFTED & TALENTED

Unit Resource Book
- History Makers, p. 171
- Section Quiz, p. 183

PRE-AP

Unit Resource Book
- Economics in the World, p. 176
- Connect to Math, p. 177
- Section Quiz, p. 183

TECHNOLOGY

Unit Transparency Book
- 3-Minute Warm-Ups, TT20
- Essential Question, TT21
- Critical Thinking, TT26
- Map: Argentina: Economic Activities, TT28
- Cultures, TT30

Daily Test Practice Transparencies
- Section 5, TT33

Power Presentations

ClassZone.com

World Cultures and Geography Video on DVD

Comparing Regions

Female Heads of State

In 1993, there were 15 women serving as presidents or prime ministers of their countries, the largest number at any one time. When Michelle Bachelet became Chile's president, there were only a total of 12 women leaders in the world. In addition to Bachelet, there were four female heads of state from Europe, four from Asia or the Pacific Islands, two from Africa, and one from the Caribbean, as of May 2006.

Unit Resource Book: Latin America

- History Makers: Juan and Eva Peron, p. 171

CONNECT to Math

Help students understand the meaning of billions of dollars by building up from 10 to 1 billion (1,000,000,000) by successively adding zeros. Point out that one billion is equal to one thousand million.

Explain that GDP reflects the production of all goods and services within a country (no matter who owns the resources that produced them) in a given period of time such as three months or a year. Economists look at changes in GDP over time to see whether a country's economy is growing or declining.

- When students have completed their graphs, ask them why they think Argentina's GDP is so much larger than that of the other countries. *(because its economy is stronger than the others)*

Unit Resource Book: Latin America

- Connect to Math: Make a GDP Bar Graph, p. 177

Michelle Bachelet
Michelle Bachelet won more than 53 percent of the votes. She promised to bring more jobs to Chile and to work for social justice.

History and Government The Spanish who first arrived in Argentina expected to find gold and silver. In fact, *Argentina* comes from the Latin word *argentum*, which means "silver." *Chile* likely comes from the Native American word meaning "where the land ends." In the 1500s, the Spanish came looking for gold and silver in both regions. Later, **José de San Martín** of Argentina helped both countries gain independence from Spain by 1818.

Like other South American nations, Argentina was controlled by dictators into the 20th century. Chile's government, on the other hand, was more stable. Presidents were often elected because they promised social reforms. In 1973, however, General Augusto Pinochet took control of Chile's government and ruled as a military dictator until 1990. Today, Argentina and Chile are republics. In 2006, Chile elected its first female president, Michelle Bachelet.

Economy In both Argentina and Chile, service industries make up most of the nations' **gross domestic product (GDP)** and employ more than half of the nations' workers. Most of Argentina's manufacturing occurs in factories in and around Buenos Aires. The **Pampas** and Patagonia are important to Argentina's agriculture. Beef cattle and grain and fruits, such as wheat and grapes, are raised on large ranches and farms, called **estancias** (eh•STAHN-syahs), on the Pampas. Thousands of sheep graze in Patagonia. Petroleum from Patagonia is Argentina's main mineral. Fishing along Argentina's long coastline is also an important part of Argentina's economy.

CONNECT to Math

The GDP is an important indicator of the strength of a nation's economy. A strong GDP indicates a healthy economy and usually influences a nation's stock market in a positive way.

Stock market in Buenos Aires, Argentina

Activity

Make a GDP Bar Graph

Materials

- Gross domestic product figures for Argentina, Chile, Paraguay, and Uruguay
- Graph paper

1. Research the latest GDP figures for Argentina, Chile, Paraguay, and Uruguay.

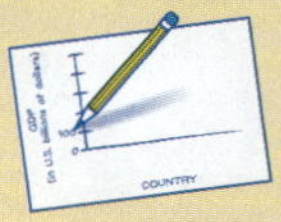

2. Make a graph. Label the horizontal axis "Country." Label the vertical axis "GDP in U.S. billions of dollars." Label the axis by 100s, starting from 0.

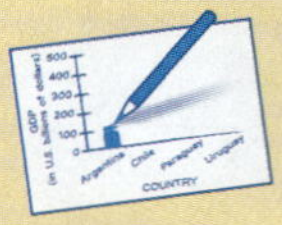

3. Draw a bar for each country to represent the GDP figure for that country.

DIFFERENTIATING INSTRUCTION

English Learners

Vocabulary: Word Relationships

Have students work in pairs to list all the words in the "Economy" subsection that relate to agriculture. Then suggest that they put them in two groups—crops or livestock. Clarify that the word *farm* is usually used for crops, and that *ranch* usually refers to places where livestock are raised. Ask students to identify the grain and fruit that are mentioned in the paragraph (wheat and grapes).

Pre-AP

Relate GDP and Standard of Living

Encourage students to research the population of each of the four countries in the Connect to Math feature. Have them calculate per capita GDP by dividing GDP by population. Then have students find information about other factors that indicate standard of living, such as life expectancy. Ask them to summarize their findings, and explain how per capita GDP is related to standard of living.

Because only about three percent of Chile's land can be farmed, Chile developed copper mining in its mountains and fishing industries along its coast. Today, Chile is the world's leading copper producer. Both Argentina and Chile are part of **Mercosur**, an association of several South American countries, to promote trade among the countries.

People and Culture Argentina and Chile have large urban populations, particularly in the major cities of Buenos Aires and Santiago. Like other South American cities, these cities are surrounded by squatter settlements.

About 85 percent of Argentina's population is of European ancestry. Although gauchos still work on Argentina's ranches today, they are also celebrated, much like American cowboys, in Argentinian poetry, literature, painting, and music. The tango, a dance that combines European and African influences, is Argentina's national dance.

Dancing the Tango The tango involves alternating long, slow steps with quick, short steps.

Chile's population, unlike Argentina's, is mostly *mestizo*. Many Chileans have left difficult conditions in rural areas for opportunities in the cities. Chile has produced many famous writers, such as poets Gabriela Mistral and Pablo Neruda. Chileans spend their leisure time going to the movies and playing and watching football (soccer), the most popular sport.

DRAW CONCLUSIONS Explain how geography affects parts of Argentina's and Chile's economies. In Argentina, the Pampas and Patagonia are conducive to farming and ranching. The mountains in Chile provide minerals, and the long coast provides fishing.

Paraguay and Uruguay

KEY QUESTION How is the population of Paraguay different from Uruguay's?

Paraguay is a small **landlocked** country of rivers, hills, and forests. Hilly grasslands cover much of Uruguay, located southeast of Paraguay. Both countries won independence from the Spanish, experienced political turmoil, and today are constitutional republics.

Paraguay Most people in Paraguay work in service industries and in agriculture. Cattle raising and most farming take place in eastern Paraguay, where the richest soil is found. Forests cover much of eastern Paraguay, and wood products are important industries. Paraguay is a founding member of Mercosur.

More About . . .

Mercosur

Mercosur is an acronym for the Spanish phrase Mercado Común del Sur, meaning Common Market of the South. In 1988, Argentina and Brazil, South America's two largest economies, first made a commitment to work toward a common market. Mercosur was then formed in 1991 when the two countries, along with Uruguay and Paraguay, signed the Treaty of Asunción. Venezuela was admitted as a full member in 2006; other South American countries are associate members.

The organization's goal is to improve economic development. It does this by promoting free trade among members and negotiating agreements with other trading blocs.

Unit Resource Book: Latin America
- Economics in the World: Latin American Economies, p. 176

Teach

Paraguay and Uruguay

Think, Pair, Share

- How are the economies of Paraguay and Uruguay similar? *(Both rely heavily on service industries and cattle raising.)*
- What are the Guaraní and how have they influenced the culture of Paraguay? *(They are the largest indigenous group and have influenced music and handicrafts, especially lace making.)*
- **Find Main Ideas** What is the most important influence on the culture of Uruguay? *(European traditions)*

DIFFERENTIATING INSTRUCTION: TIERED ACTIVITIES

OBJECTIVE Create a mural to portray the culture, history, and economics of Argentina and Chile.

Basic

Discuss what topics to cover in the mural, then organize students into groups assigned to different topics. For this activity, have students depict aspects of the countries' culture that they find most interesting, such as playing soccer. Then have students work together to complete their sections of the mural.

On Level

To complement the Basic activity, have students focus on depicting aspects of the economies of each country, based on the information in their textbooks and other classroom resources. Then have students work together to complete their sections of the mural.

Challenge

To complement the Basic and On Level activities, have students focus on depicting aspects of the histories of each country, based on their textbooks, classroom resources, and library resources. Then have students work together to complete their sections of the mural.

CONNECT to the Essential Question

How are the countries of this region working to strengthen their governments and economies?

Ask students what they have learned so far that can help them answer this question. Students might mention:

- All the Southern Cone nations are now republics.
- Service industries are important to the economies of Southern Cone countries.

Unit Transparency Book: Latin America
- Essential Question TT21

4 Assess & Reteach

Assess Have students complete the Section Assessment.

Unit Resource Book: Latin America
- Section Quiz, p. 183

Interactive Review @ClassZone.com
- Section Quiz

Power Presentations

Test Generator

Reteach Organize students into four groups and assign each group one of the four Southern Cone countries. Have students review how geography influences the economy of their country. Ask volunteers to summarize the group's findings.

Unit Resource Book: Latin America
- Reteaching Activity, p. 188

Montevideo The Plaza Independencia is located in the center of Montevideo in Uruguay. **What other culture makes use of plazas in its cities?**

the Mexican culture

Mestizos make up most of Paraguay's population, and the Guaraní (GWAH•ruh•NEE) are the nation's largest indigenous group. The Guaraní influence is reflected in Paraguay's arts, particularly in music and handicrafts. The Guaraní are especially known for nanduti lace, which incorporates a lace-making technique introduced by the Spanish.

Uruguay More than 60 percent of Uruguay's people work in service industries. Uruguay's plains and grasslands are perfect for raising cattle and sheep. Meat, hides, and wools are the nation's biggest exports. Like other Southern Cone nations, Uruguay is a member of Mercosur.

Most of Uruguay's people are of Spanish and Italian ancestry. Most people live in cities, half in Montevideo, the capital city. Unlike many other South American nations, Uruguay's culture is influenced more by European traditions than by native ones. Gaucho folklore has inspired Uruguay's music and art and Montevideo's Gaucho Museum.

COMPARE AND CONTRAST Compare the populations of Paraguay and Uruguay. *Mestizos* make up most of Paraguay's population; people of Spanish and Italian ancestry make up most of Uruguay's population.

ONLINE QUIZ For test practice, go to **Interactive Review @ClassZone.com**

Section 5 Assessment

TERMS & NAMES

1. **Explain the importance of**
   - gaucho
   - Southern Cone
   - *estancia*
   - Mercosur

USE YOUR READING NOTES

2. **Find Main Ideas** Use your completed diagram to answer the following question:
   What products account for most of Uruguay's exports?

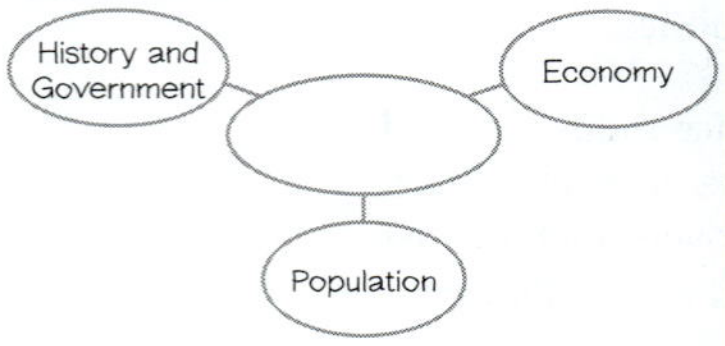

KEY IDEAS

3. How are the histories of the Southern Cone nations alike?
4. What industry employs a majority of the people in Argentina and Chile?
5. Why does most farming and cattle ranching occur in eastern Paraguay?

CRITICAL THINKING

6. **Make Inferences** Why have Southern Cone nations become members of Mercosur?
7. **Draw Conclusions** How might Argentina's geography and size contribute to its strong economy?
8. **CONNECT to Today** In 2006, Chileans elected their first female president. What do you think this says about women's role in Chilean politics?
9. **MATH Create a Pie Graph** Choose one of the Southern Cone nations and research its major exports. Make a pie graph to present the information.

SECTION 5 ASSESSMENT ANSWERS

Terms & Names

1. gaucho: Argentinian cowboy; Southern Cone: South American nations located in the cone-shaped part of southern South America; *estancia:* large farm or ranch in Argentina; Mercosur: association of several South American countries to promote trade among the countries

Use Your Reading Notes

2. meat, hides, and wools: one sample organizer

Key Ideas

3. They were founded as Spanish colonies and won independence from Spain in the 1800s.
4. service industries
5. This part of the country has the richest soil.

Critical Thinking

6. Possible response: They hope that trade within the region will strengthen the individual nations' economies and create profits for businesses.
7. There is more land to use to produce crops, to raise animals, and to expand industries.
8. Answers will vary. Students might indicate that women play an active role in politics in Chile.
9. Use the rubric to score pie graphs.

Pie Graph Rubric

	Representation of Percentages	Number of Errors
4	very precise	few or none
3	accurate	some
2	fairly accurate	many
1	inaccurate	many major

Interactive Review

Click here to complete these and other activities online @ ClassZone.com

CHAPTER SUMMARY

Key Idea 1
Central American countries face challenges as they improve their economies.

Key Idea 2
Europe and Africa have influenced Caribbean cultures.

Key Idea 3
Venezuela's economy depends heavily on petroleum, while Colombia's economy depends more on agricultural products.

Key Idea 4
Indians and *mestizos* make up a large part of Peru, Bolivia, and Ecuador.

Key Idea 5
Geography has influenced the economies of Southern Cone nations.

For **Review and Study Notes**, go to **Interactive Review** @ ClassZone.com

NAME GAME

Use the Terms & Names list to complete each sentence on paper or online.

1. I am a place that is ruled by or closely connected with another country. __dependency__
2. I describe the countries in the southern part of South America. ________
3. I describe the countries of Bolivia and Paraguay. ________
4. I am the national folk dance of Venezuela. ________
5. I am a measure of a country's economy. ________
6. I am a trade agreement made between Central American countries. ________
7. I am an important animal in the Andes Mountains. ________
8. I am a kind of grassy plain. ________
9. I draw tourists to a natural habitat without harming the environment. ________
10. I am a large farm in Argentina. ________

CAFTA
commonwealth
dependency
ecotourism
estancia
gaucho
gross domestic product, or GDP
joropo
landlocked
llama
llanos
Southern Cone

Activities

GeoGame

Use this online map to show what you know about the geography of Middle America and Spanish-Speaking South America. Drag and drop each place name to its location on the map.

To play the complete game, go to **Interactive Review** @ ClassZone.com

Crossword Puzzle

Complete an online crossword puzzle to test your knowledge of the region's history, culture, government, and economics.

ACROSS
1. a small camel-like animal, related to the llama

ONLINE AND ON-SCREEN ACTIVITIES

GeoGame

Use the interactive map on the **Power Presentations** DVD-ROM or at **ClassZone.com** to help students review the physical geography of Middle America and Spanish-speaking South America.

- Demonstrate how to drag and drop the map labels listed onto the appropriate locations on the map.
- Advance to the scorecard to show how students have a chance to improve their scores.

Crossword Puzzle

Students can review important information about Middle America and Spanish-speaking South America by completing this crossword puzzle online at **ClassZone.com**.

- As an extension of this activity, have students use the *Activity Maker* at **ClassZone.com** to make their own crossword puzzle for one section of the chapter.
- Then have students trade puzzles with classmates to solve answers for each other's puzzles.

Chapter Summary

As a follow-up to the chapter, encourage students to pair up and go over the **Review and Study Notes** at **ClassZone.com**.

Alternatively, students pairs can review their Reading Study Guide notes, making sure that they can add important details to each section's "Key Idea."

Name Game

Teaching Tips

ON PAPER Have students work in pairs to complete this activity. As an extension, they can write questions for the terms that are not covered in the items shown here.

ONLINE Make sure that students understand how to drag and drop terms to fill in blanks. Ask them to print out their results and compare them with those of a classmate.

ANSWERS

1. dependency
2. Southern Cone
3. landlocked
4. *joropo*
5. gross domestic product, or GDP
6. CAFTA
7. llama
8. llanos
9. ecotourism
10. *estancia*

Teacher-Tested Activities

Terry Rhodes, McNair Middle School, Fayetteville, Arkansas

My class plays "Find the Capital" to learn more about places in the region they are studying. They have to incorporate their knowledge of using coordinates to find places in Middle America and Spanish-speaking South America. Prepare a list of the capital cities of the countries in the region and their coordinates. Call out the coordinates of one of the cities, and have students find the country in which the coordinates are located. Ask students to then identify the capital of the country.

Vocabulary ANSWERS

1. travel to natural habitats in a way that does not damage the habitat; an important industry in some Central American countries
2. economy based on growing one crop, causing economic problems for some Central American nations
3. the Indian group in the Caribbean Islands when Columbus arrived; enslaved by the Spanish
4. form of government in Puerto Rico, in which it has self-rule but is associated with the United States
5. country that is surrounded by land with no access to the sea
6. trade agreements with Central and South American nations
7. terms that describe Cuba's government
8. terms that have to do with ranching and farming in Argentina

Key Ideas ANSWERS

9. agriculture
10. *mestizos*
11. to replace the Indians, who died from overwork and disease
12. Puerto Rico is a commonwealth of the United States and a democracy.
13. petroleum
14. coffee
15. mining, fishing, farming
16. Bolivia—natural gas and tin; Ecuador—petroleum and gold
17. service industries
18. on the Pampas and in Patagonia
19. They live in urban areas; half in Montevideo.

Critical Thinking ANSWERS

20. Answers:
- Venezuela—petroleum; bananas, coffee, corn, rice, oranges, cattle
- Peru—sugar cane, cotton, coffee, potatoes, grains, asparagus; copper, silver, lead, zinc
- Argentina—service industries; manufacturing; cattle, sheep, wheat, grapes; petroleum; fishing

CHAPTER 9 Assessment

VOCABULARY

Explain the significance of each of the following.

1. ecotourism
2. one-crop economy
3. Taino
4. commonwealth
5. landlocked

Explain how the terms in each group are related.

6. CAFTA and Mercosur
7. dictator and communism
8. Pampas, gaucho, and *estancia*

KEY IDEAS

1 Central America

9. What is Central America's main economic activity?
10. What group of people makes up most of Central America's population?

2 The Caribbean

11. Why did the Spanish bring enslaved Africans to the Caribbean islands?
12. What kind of government does Puerto Rico have?

3 Venezuela and Colombia

13. What product is the most important to the economy of Venezuela?
14. What is Colombia's leading export?

4 Peru, Bolivia, and Ecuador

15. What are three economic activities in Peru?
16. What are the major mineral exports of Bolivia and Ecuador?

5 Argentina and Chile

17. What economic activity provides most of Argentina's and Chile's gross domestic product?
18. In what parts of Argentina does farming and ranching take place?
19. Where do most people in Uruguay live?

CRITICAL THINKING

20. **Compare and Contrast** Create a chart to compare the industries and major products of Venezuela, Peru, and Argentina.

COUNTRY	MAJOR INDUSTRIES/PRODUCTS
Venezuela	
Peru	
Argentina	

21. **Evaluate** What challenges do Central American countries face today?
22. **Identify Problems and Solutions** What have Central American and Caribbean nations done to help solve the problem of one-crop economies?
23. **Five Themes: Movement** How has the settlement of Caribbean nations affected the culture of those nations?
24. **Connect to History** What conditions created during the Spanish conquest of South America made it difficult for nations there to form stable governments?

Answer the ESSENTIAL QUESTION

How are the countries of this region working to strengthen their governments and economies?

Written response Write a two- or three- paragraph response to the Essential Question. Be sure to consider the key ideas of each section. Use the rubric below to guide your thinking.

Response Rubric
A strong response will:
- discuss ways in which the countries have strengthened their governments
- explain how the nations have strengthened their economies

ESSENTIAL QUESTION REVIEW

Before students begin their essays, have them review the key chapter elements that will help them answer the Essential Question. Direct them to significant features in the chapter, such as:

- Animated Geography: Economy of Central America, Section 1
- Comparing Puerto Rican and Cuban Governments, Section 2
- Connect Geography and Economics, Section 3
- Comparing Industries, Section 4
- Comparing Climate regions, Section 4

Written Response Rubric

	Coverage of Key Ideas	Explanation	Errors in Mechanics
4	all	clear and concise	few or none
3	most	good	some
2	some	fair	many
1	one or none	poor	many major

STANDARDS-BASED ASSESSMENT

Test Practice
- Online Test Practice @ClassZone.com
- Test-Taking Strategies and Practice at the front of this book

THEMATIC MAP

Use the map to answer questions 1 and 2.

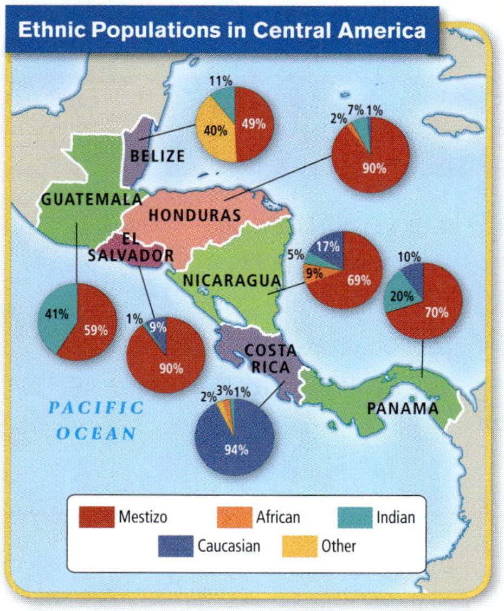

1. What percentage of Guatemalans are *mestizo*?

A. about 59 percent
B. about 33 percent
C. about 3 percent
D. about 1 percent

2. Which of the following countries has the largest percentage of *mestizos*?

A. Belize
B. Panama
C. El Salvador
D. Guatemala

TABLE

Use the table below to answer questions 3 and 4 on your paper.

Literacy and Life Expectancy in Selected Countries

	Literacy Rate	Life Expectancy
Haiti	53 percent	53 years
Costa Rica	96 percent	77 years
Ecuador	93 percent	76 years
Argentina	97 percent	76 years

Source: *The World Factbook*, 2006

3. Which country has the lowest life expectancy?

4. What conclusion can you draw about the relationship between literacy rates and life expectancy?

GeoActivity

1. INTERDISCIPLINARY ACTIVITY–SCIENCE

Use the library or visit a zoo to find out about an animal that makes its home in the region, such as the Andean condor, spectacled bear, tapir, or toucan. Find out how the animal adapts to its environment. Present the information in an illustrated, captioned poster.

2. WRITING FOR SOCIAL STUDIES

Write an ad that encourages people to visit one of the countries discussed in the chapter. Your ad should focus on the landforms, people, and cultural features that visitors to the country would find interesting.

3. MENTAL MAPPING

Create an outline map of South America and label the following:

- Southern Cone nations
- llanos
- altiplano
- Pampas
- Andes Mountains
- coastal plains
- equator

DIFFERENTIATING CHAPTER ASSESSMENT

Unit Resource Book

- **Chapter Test A,** pp. 189–192. *Use for Struggling Readers, Beginning-to-Intermediate-Level English Learners, and for Inclusion of students whose IEP suggests supported test-taking.*
- **Chapter Test B,** pp. 193–196. *Use for On-Level Learners and ESOL students with Advanced-Level English.*
- **Chapter Test C,** pp. 197–200. *Use for Gifted and Talented students or students assigned Pre-AP work.*

Test Generator

- Chapter Tests A, B, and C, in English and Spanish

McDougal Littell Assessment System

TEST **Test Generator** Use three levels of tests in English and Spanish tied to state standards. Plus, choose questions to customize tests for all populations.

SCORE Scan and Score plain paper student tests in minutes.

REPORT Create instant progress reports, sortable by student, by standard, or other criteria.

RETEACH Respond immediately to students' needs by using a wealth of reteaching resources.

21. They face the problems of a huge gap between rich and poor; unemployment; and lack of basic services in many places.

22. These nations have diversified their crops and industries.

23. People from many different countries settled the Caribbean nations; as a result, the culture of the Caribbean nations is a blend of the cultures of these nations.

24. The Spanish controlled the government and wealth. When South Americans fought for independence, they had little experience ruling themselves and no money to support an economy. They had to learn how to create both.

Standards-Based Assessment ANSWERS

1. A
2. C
3. Haiti
4. Places with higher literacy rates generally have higher life expectancies.

GeoActivity

1. INTERDISCIPLINARY ACTIVITY RUBRIC

The poster should

- provide details about the chosen animal
- indicate how the animal adapts to its environment
- include explanatory captions

2. WRITING FOR SOCIAL STUDIES RUBRIC

The ad should

- clearly describe the features of the chosen country
- provide reasons why visitors would enjoy visiting the chosen country

3. MENTAL MAPPING RUBRIC

The outline map should

- represent the approximate shape of South America
- include all geographic features listed in their correct location

Chapter 10 Brazil PLANNING GUIDE

Complete print resources all on one USB drive allow you to customize lesson plans for all students.

Section Titles & Key Ideas	Unit Resource Book Latin America	Assessment Resources
1 From Portuguese Colony to Modern Giant **KEY IDEA** Brazil's government is dealing with the problems resulting from urbanization.	• **Reading Study Guide,** pp. 201–202 • **RSG with Additional Support,** pp. 207–208 • **Outline Map Activities,** p. 229 • **Primary Source,** p. 238 • **World Literature,** pp. 260, 264 • **Geography Project,** p. 265	**Unit Resource Book** • Section Quiz, p. 239 • Reteaching Activity, p. 242 **Test Generator CD-ROM** **Daily Test Practice Transparency,** TT34
2 A Multicultural Society **KEY IDEA** Several cultures have influenced Brazil's unique culture.	• **Reading Study Guide,** pp. 203–204 • **RSG with Additional Support,** pp. 209–210 • **Vocabulary Practice,** p. 225 • **Map and Graph Skills,** p. 227 • **Interdisciplinary Projects,** pp. 234–235 • **Connect to Music,** p. 237	**Unit Resource Book** • Section Quiz, p. 240 • Reteaching Activity, p. 243 **Test Generator CD-ROM** **Daily Test Practice Transparency,** TT35
3 Developing an Abundant Land **KEY IDEA** An abundance of natural resources helps make Brazil a major industrial nation.	• **Reading Study Guide,** pp. 205–206 • **RSG with Additional Support,** pp. 211–212 • **Skillbuilder Practice,** p. 226 • **Interdisciplinary Projects,** pp. 233, 235 • **Economics in the World,** p. 236	**Unit Resource Book** • Section Quiz, p. 241 • Reteaching Activity, p. 244 • Chapter Test (Forms A, B, & C), pp. 245–259 **Test Generator CD-ROM** **Daily Test Practice Transparency,** TT36

McDougal Littell **Assessment System**

TEST | SCORE | REPORT | RETEACH

No Time? To focus students on important content in the chapter, use the **"Reading Study Guide with Additional Support"** in ***Resources2Go.***

Teacher Presentation Options

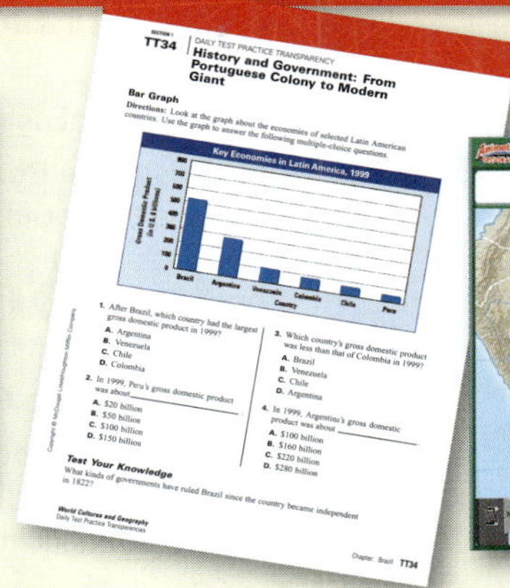

Electronic Books

eEdition DVD-ROM

eEdition Online

Presentation Toolkit

Video/DVD

- World Cultures and Geography

Power Presentations DVD-ROM

- Discussion Starters
- Media Gallery
- Review Game
- Interactive Review
- Animated Features

Unit Transparency Book

- 3-Minute Warm-Ups
- Essential Question
- Critical Thinking
- Maps
- Cultures

Daily Test Practice Transparencies

Animated Geography

- Present-Day Brazil Map
- Worldster: Pedro's Web Page

OnlineActivities at ClassZone.com

Animated Geography

- Videos and Photos
- 3-D Animations
- Interactive Maps

Interactive Review

- Section Quizzes
- Review and Study Notes
- Vocabulary Flip Cards
- Graphic Organizers
- Name Game and Crossword Puzzles
- GeoGame

Online Test Practice

Map Center

- Atlas Maps
- Political Maps with Data
- Links to Mapping Web Sites

Activity Center

- Activity Maker—Review Games, Puzzles, Flip Cards, and Graphic Organizers
- Worldster

Research & Writing Center

- Web Sites
- Data Updates
- Primary Sources
- Writing Models

Current Events

Esther Howse
Byron Center West Middle School
Byron Center, Michigan

Make It Fun — Teacher-Tested Activities

As you teach this chapter, look for my classroom-proven ideas for:

- making the lesson fun
- focusing students' attention and comprehension
- integrating technology with traditional teaching

Teaching Bell to Bell When you see this timing guide, have students discuss in pairs, in groups of 3–6, or as a class.

- **Think, Pair, Share** (pairs)
- **Reader, Recorder, Reporter** (3–6)
- **Talk About It** (any size)

10 RESOURCES FOR DIFFERENTIATED INSTRUCTION

Struggling Readers

eEdition with Audio DVD-ROM

Lesson Plans
- Strategies for Inclusion

Unit Resource Book
- Reading Study Guide with Additional Support, pp. 207–212 (B)
- Reteaching Activity, pp. 242–244
- Chapter Test (Form A), pp. 245–249 (A)

Best Practices Toolkit (C)
- Interactive Reading Strategies, pp. 90–145
- Differentiation Activity Bank: Struggling Readers, pp. 146–149

Teacher's Edition Activities
- Use Reciprocal Teaching, Sec. 1
- Tiered Activities, Sec. 1–3

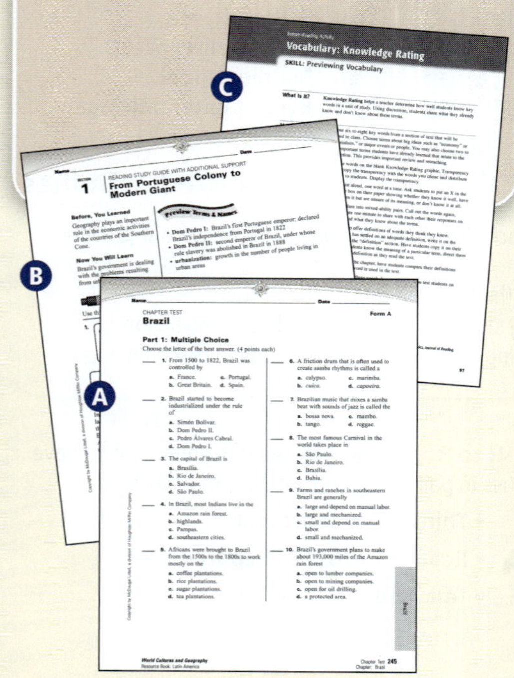

English Learners

Pupil's Edition in Spanish

eEdition with Audio DVD-ROM

Lesson Plans
- Strategies for English Learners

Unit Resource Book
- Reading Study Guide (Spanish), pp. 213–218
- RSG with Additional Support (Spanish), pp. 219–224 (A) (B)
- Vocabulary Practice, p. 225

Multi-Language Glossary (C)

Best Practices Toolkit
- Differentiation Activity Bank: English Learners, pp. 165–194

Test Generator CD-ROM
- Section Quizzes in Spanish
- Chapter Test in Spanish

Teacher's Edition Activities
- Pronounce and Preview, Sec. 1–3
- Comprehension: Connections, Sec. 1
- Vocabulary: Relationships, Sec. 2
- Vocabulary: Key Academic Vocabulary, Sec. 3
- Tiered Activities, Sec. 1–3

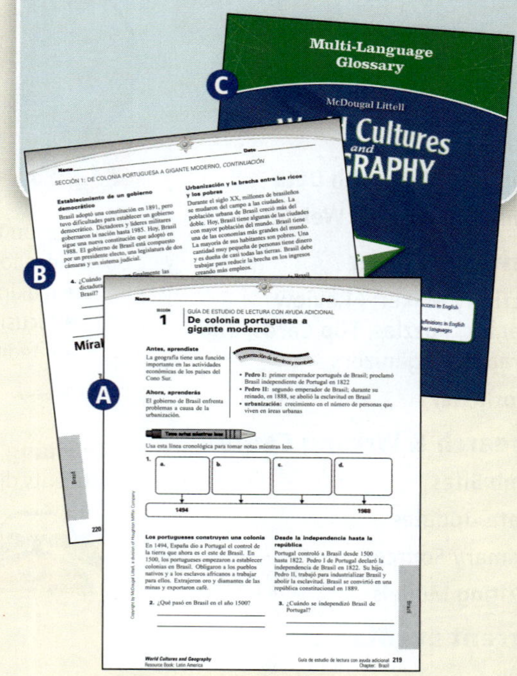

Inclusion

eEdition with Audio DVD-ROM

Lesson Plans
- Strategies for Inclusion

Unit Resource Book
- Reading Study Guide with Additional Support, pp. 207–212 (A) (B)
- Reteaching Activity, pp. 242–244
- Chapter Test (Form A), pp. 245–249

Best Practices Toolkit (C)
- Interactive Reading Strategies, pp. 90–145
- Differentiation Activity Bank: Struggling Readers, pp. 146–149

Teacher's Edition Activities
- Match Names to the Time Line and Map, Sec. 1
- Use Terms in Captions, Sec. 2
- Visualize and Compare Numbers, Sec. 2
- Vocabulary: Word Sort/Word Splash, Sec. 3
- Tiered Activities, Sec. 1–3

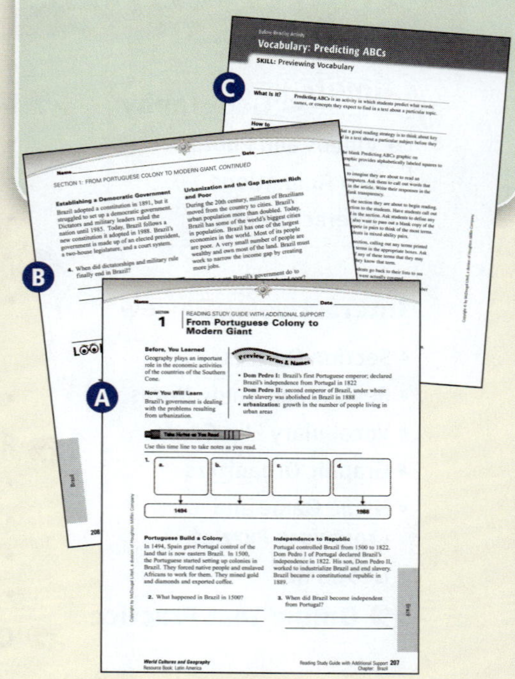

Interdisciplinary Activities

Gifted & Talented Students

Lesson Plans
- Core Lesson Plan

Unit Resource Book
- Outline Map Activities, pp. 229–230
- History Makers, pp. 231–232 (A)
- Interdisciplinary Projects, pp. 233–235 (B) (C)
- Economics in the World, p. 236
- Chapter Test (Form C), pp. 255–259

Best Practices Toolkit
- Student Products Rubrics, pp. 1–14
- Writing Templates, pp. 78–79

Teacher's Edition Activities
- Share Oral Traditions, Sec. 2
- Tiered Activities, Sec. 1–3

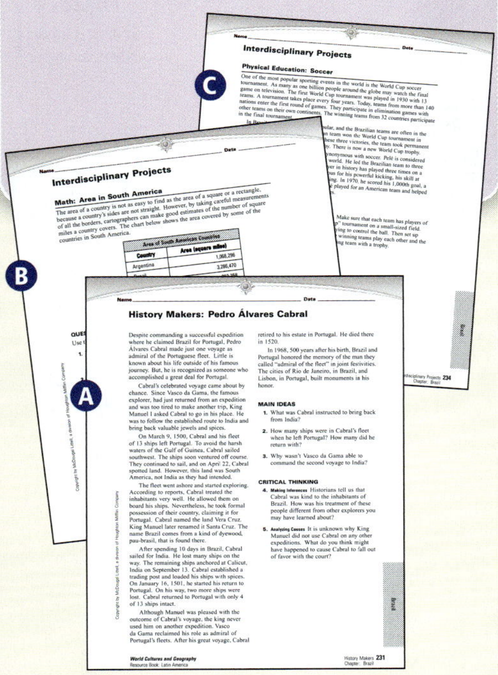

Pre-AP*

Lesson Plans
- Core Lesson Plan

Unit Resource Book
- Map and Graph Skills, pp. 227–228
- Connect to Music, p. 237 (A)
- Primary Source, p. 238 (B)
- Chapter Test (Form C), pp. 255–259
- World Literature, p. 261 (C)

Best Practices Toolkit
- Writing Templates, pp. 78–79
- Differentiation Activity Bank: Pre-AP Students, pp. 195–200

* Pre-AP is a registered trademark of the College Entrance Examination Board, which was not involved in the production of and does not endorse this product.

Teacher's Edition Activities
- Connect Economics to Population, Sec. 3
- Tiered Activities, Sec. 1–3

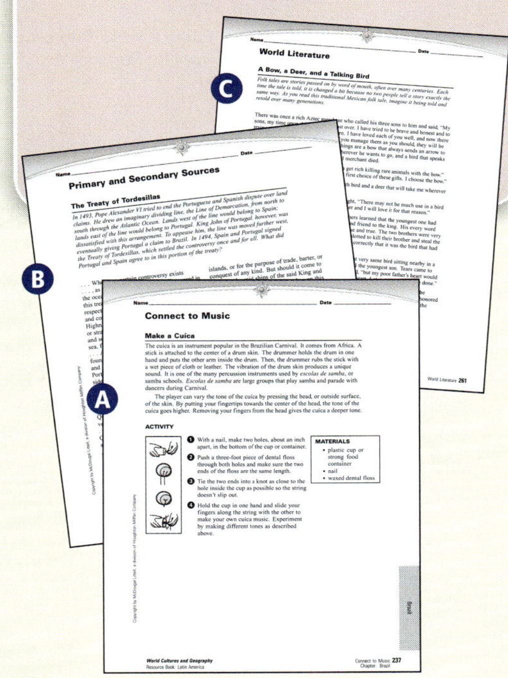

CONNECT to History

Follow the Leaders

CLASS TIME 20 minutes

TASK Trace the actions of European and Brazilian leaders in the 19th century on a world map.

SUPPLIES
- outline political map of the world
- colored pencils or markers

ACTIVITY Have students identify France, Portugal, and Brazil on their maps and label those countries. Then direct students to the subsection "Independence to Republic" in Section 1. Have them draw different colored lines to show Napoleon's invasion of Portugal and the movement between Portugal and Brazil of the different leaders described. Tell students to create a key to identify each line. Call on volunteers to summarize the information on their maps.

Good for Spatial and Visual Learners

CONNECT Math

Track Population Changes

CLASS TIME 20 minutes

TASK Create a line graph showing changes in Brazil's population between 1950 and 2000.

SUPPLIES
- graph paper
- colored markers or pencils

ACTIVITY Direct students to the "Comparing Urban and Rural Populations" feature in Section 1. Have students work in pairs to use the information in the pie graphs to create a line graph. First have students calculate the numbers of the urban and rural population for each of the three dates, rounded to the nearest 1,000. Then ask them to create three lines on their graphs to show changes in total population, urban population, and rural population. Suggest that they mark the vertical axis in increments of 20 million. Discuss what new information the line graph conveys.

Good for Visual and Mathematical Learners

Introduce the ESSENTIAL QUESTION

- Point out how Brazil's government is set up like the U.S. government, and is working to bridge the gap between rich and poor.
- Explain that Brazil's natural resources include soil suitable for agriculture and valuable minerals.
- Discuss how Brazil's diverse population, including Indians, Africans, and immigrants from Europe and Asia, contributes to its economy.
- Describe how Brazil's many rivers are also key resources, used to produce a great deal of electric power.

Key Ideas

Section 1: Brazil's government is dealing with the problems resulting from urbanization.

Section 2: Several cultures have influenced Brazil's unique culture.

Section 3: An abundance of natural resources helps make Brazil a major industrial nation.

CONNECT to the Essential Question

You may wish to revisit the Essential Question periodically with students as they work through the chapter.

Suggestions for reconnecting are provided at an appropriate point in the *Teacher's Edition* for each section.

Water lilies on a lake in the Pantanal Wetlands, Brazil

CHAPTER 10

Brazil

ESSENTIAL QUESTION

How have Brazil's people used the country's abundant natural resources to make Brazil an economic giant?

1 HISTORY & GOVERNMENT
From Portuguese Colony to Modern Giant

2 CULTURE
A Multicultural Society

3 ECONOMICS
Developing an Abundant Land

CONNECT Geography & History

Use the map and the time line to answer the following questions.

1. When was Brazil claimed by Portugal?
2. How far is Brasília from Rio de Janeiro?

1494

Geography
◄ **1494** Spain and Portugal sign the Treaty of Tordesillas.

History
1500 Pedro Álvares Cabral claims Brazil for Portugal. ►

History
1822 Brazil declares independence from Portugal.

History
1888 Brazil abolishes slavery.

TIME LINE DISCUSSION

Remind students that Spain colonized most countries in South America, but Brazil was a Portuguese colony.

- What two events on the time line explain why Portuguese, rather than Spanish, is the primary language in Brazil? *(Treaty of Tordesillas in 1494 and Cabral's claim of Brazil for Portugal in 1500)*
- How long did Portugal control Brazil? *(322 years, 1500–1822)*
- What year did Brazil outlaw slavery? *(1888)*
- When did Brazil begin to have democratically elected governments? *(after 1985, when military rule ended)*

Animated GEOGRAPHY
Present-Day Brazil

Click here to explore Brazil @ClassZone.com

Government
1960 Brasília becomes Brazil's new capital.

Government
1985 Military rule ends in Brazil.

Today

History
▲ **1889** Brazil becomes a republic.

Culture
◀ **2002** Brazil wins its fifth World Cup in soccer.

TEACHER MEDIA FAVORITES

Books for the Teacher

Barter, James. *The Amazon.* Lucent, 2003. In-depth coverage ranging from early civilizations to contemporary challenges.

Dicks, Brian. *Brazil.* Facts on File, 2003. Extensive information includes treatment of problems such as poverty.

Video / DVD

Beyond Our Borders: Brazil. New Dimension Media, 2005. Provides essential information on history, geography, and culture.

Brazil: Emerging Powers. Wall Street Journal Video, 1996. Focuses on Brazil's growing economy through a variety of success stories.

Internet

Go to the *Research and Writing Center* at **ClassZone.com** to link to:

- data on the region
- a variety of chapter-specific, content-reviewed sites
- "History Makers" biographies
- primary sources

Animated GEOGRAPHY
Present-Day Brazil
CLASSZONE.COM

Explain that this is a political map of Brazil that shows the states, the federal district (similar to the District of Columbus in the United States), and several important cities. Point out that it also shows that there are mountains in parts of Brazil, and names many major rivers.

- Where are most of Brazil's cities located? *(close to the coast)*
- Through what states does the Amazon River flow? *(through the states of Pará and Amazonas)*
- Where is the national capital of Brazil located? *(Distrito Federal, nearly in the center of Brazil)* According to the time line, when did it become the capital? *(1960)*

CONNECT GEOGRAPHY & HISTORY ANSWERS

1. 1500

2. about 600 miles

Presentation Options

Use the **Power Presentations** DVD-ROM to project the animated map.

- Prompt students to use the pop-up images and audio from the map to discover what they already know about the geography, culture, history, economics, or government of Brazil.
- Challenge students to answer the map questions individually, in small groups, or as a class activity. Use all of the interactive material to help preview chapter content.

1 Plan & Prepare

Objectives

- Outline how Brazil changed from a Portuguese colony to an independent nation
- Discuss the challenges faced by the Brazilian government today

Read for the Essential Question

Help students read for a purpose by reminding them of the Essential Question: "How have Brazil's people used the country's abundant natural resources to make Brazil an economic giant?" Point out that they can use text features such as Key Questions, special features, images, and maps to form answers to the question.

Vocabulary

Best Practices Toolkit

Use the ***Best Practices Toolkit*** to model strategies for vocabulary notetaking. Vary strategies throughout the year. Choose from: *Knowledge Rating, Predicting ABCs, Definition Mapping, Word Sort, Word Wheel, Frayer Model (Word Squares), Magnet Words,* and *Student VOC.*

Vocabulary Strategies, TT9–TT16

Reading Strategy

Best Practices Toolkit

Display the Time Line transparency.

- Point out to students that sequencing events involves placing them in the correct order as they occurred in time.
- Model filling in entries from the "Independence to Republic" subsection in the section.
- Have students suggest entries by locating the dates of important events described in the textbook.
- Display the completed time line for review.

Time Line, TT33

SECTION 1

Reading for Understanding

Key Ideas

BEFORE, YOU LEARNED

Geography plays an important role in the economic activities of the countries of the Southern Cone.

NOW YOU WILL LEARN

Brazil's government is dealing with the problems resulting from urbanization.

Vocabulary

TERMS & NAMES

Treaty of Tordesillas (TAWR•day•SEEL•yahs) 1494 treaty that gave Portugal control over land that is now part of Brazil

Pedro Álvares Cabral Portuguese explorer who in 1500 claimed land that is now part of Brazil for Portugal

Dom Pedro I Brazil's first Portuguese emperor; declared Brazil's independence from Portugal in 1822

Dom Pedro II second emperor of Brazil, under whose rule slavery was abolished in Brazil in 1888

favelas Brazilian name for the poor neighborhoods that surround the cities

Rio de Janeiro (REE•oh day zhuh•NAYR•OH) Brazil's capital city from 1763 to 1960

Brasília Brazil's current capital city

REVIEW

urbanization growth in the number of people living in urban areas

Visual Vocabulary
Dom Pedro I

Reading Strategy

Re-create the time line shown at right. As you read and respond to the **KEY QUESTIONS**, use the time line to show the major events in Brazil's history and the development of a democratic government.

See Skillbuilder Handbook, page R6

SEQUENCE EVENTS

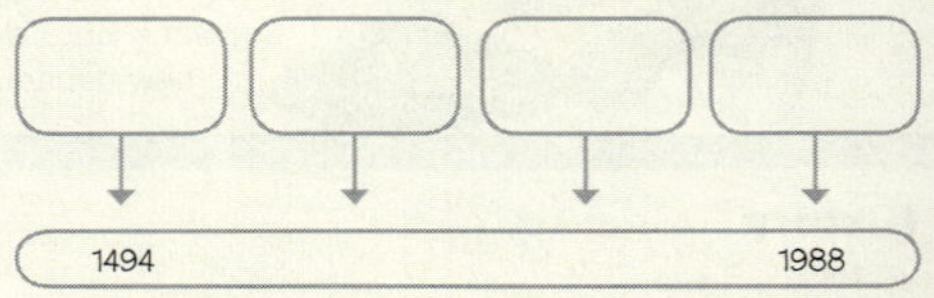

GRAPHIC ORGANIZERS
Go to Interactive Review @ClassZone.com

PRETEACHING VOCABULARY

English Learners

Pronounce and Preview

Pronounce each term for students. Review the meanings of words within definitions, such as *treaty, claimed,* and *abolished.*

- To modify vocabulary learning, have students complete worksheets as they read, instead of afterward.

Unit Resource Book: Latin America

- Vocabulary Practice, p. 225

Inclusion

Match Names to the Time Line and Map

Review the proper names and their definitions with students. Point out that *Dom* is a Portuguese title indicating a person of high rank. Have students identify where each name would be located on the time line on the chapter opener spread. Encourage them to locate Rio de Janeiro and Brasília on the map as well. Call on volunteers to explain how *favelas* are related to those cities.

SECTION 1

HISTORY & GOVERNMENT

Niemeyer Museum of Contemporary Arts, Rio de Janeiro

Collared anteater

From Portuguese Colony to Modern Giant

Connecting to Your World

What does the Statue of Liberty represent to you? For many people, it represents the United States, symbolizes freedom, and welcomes those who arrive at the nation's shores. In Brazil, a similar symbol is the statue of Christ the Redeemer, which welcomes people to Rio de Janeiro, Brazil's second largest city. Like the Statue of Liberty, it is one of the world's best-known and most-visited monuments.

Portuguese Build a Colony

KEY QUESTION What helped to make Brazil a profitable colony for Portugal?

After Columbus' expeditions, Portugal feared that if Columbus had found a route to Asia, Spain might claim lands that Portugal had already claimed. So in 1494, both countries signed the **Treaty of Tordesillas** (TAWR•day•SEEL•yahs), which drew an imaginary line from north to south around the world. Spain could claim all lands west of the line, and Portugal those east of the line. This gave Portugal control of the land in what is now eastern Brazil. The treaty line is illustrated on the map on the next page.

Rio de Janeiro The statue of Christ the Redeemer overlooks Rio de Janeiro and Guanabara Bay. Sugarloaf Mountain is seen in the distance.

2 Focus & Motivate

3-Minute Warm-Up

Write on the board or display the transparency:

- Look at the photograph of Rio de Janeiro. How does the city's geography affect the way it is laid out? *(Possible answer: The city is built between the ocean and the mountains. It doesn't have a lot of room to expand outward, so it has a lot of skyscrapers.)*

Unit Transparency Book: Latin America
- 3-Minute Warm-Ups, TT31

Connecting to Your World

More About . . .

Statue of Christ the Redeemer

Completed in 1931, the 98.5-foot tall statue is located on Mount Corcovado. A train carried the pieces of the statue to the summit—and it now transports thousands of visitors to the top each year.

3 Teach

Portuguese Build a Colony

Reader, Recorder, Reporter

- How did Dom Pedro II modernize Brazil? *(Possible answer: He encouraged industrialization and abolished slavery.)*
- **Sequence Events** How long after Brazil's independence from Portugal did it become a republic? *(67 years, 1822–1889)*

SECTION 1 PROGRAM RESOURCES

Unit Resource Book: Latin America

ON LEVEL
Unit Resource Book
- Reading Study Guide, p. 201
- Outline Map Activities, p. 229
- Section Quiz, p. 239

STRUGGLING READERS
Unit Resource Book
- Reading Study Guide with Additional Support, p. 207
- Skillbuilder Practice, p. 226
- Section Quiz, p. 239
- Reteaching Activity, p. 242

ENGLISH LEARNERS
Pupil Edition in Spanish
Pupil Edition Audio DVD-ROM
Unit Resource Book
- RSG (Spanish), p. 213
- RSG with Additional Support (Spanish), p. 219

Multi-Language Glossary
Test Generator
- Section Quiz in Spanish

INCLUSION
Unit Resource Book
- RSG with Additional Support, p. 207
- Section Quiz, p. 239
- Reteaching Activity, p. 242

GIFTED & TALENTED
Unit Resource Book
- Geography Project, p. 265
- Section Quiz, p. 239

PRE-AP
Unit Resource Book
- World Literature, pp. 260, 264
- Section Quiz, p. 239

TECHNOLOGY
Unit Transparency Book
- 3-Minute Warm-Ups, TT31
- Essential Question, TT32
- Critical Thinking, TT33
- Cultures, TT39

Daily Test Practice Transparencies
- Section 1, TT34

Power Presentations
ClassZone.com
World Cultures and Geography Video on DVD

Treaty of Tordesillas

CONNECT Geography & History

Have students look at a globe or world map that indicates longitude to see that the Treaty of Tordesillas line was located west of the dividing line (the prime meridian) between the eastern and western hemispheres.

- Why was Pedro Álvares Cabral able to claim what is now eastern Brazil for Portugal? *(because the land was east of the line drawn by the Treaty of Tordesillas)*
- Where else did Portugal have colonies at that time? *(in Africa)*

ANSWER

Region Spain

Unit Resource Book: Latin America

- History Makers: Pedro Álvares Cabral, p. 231
- History Makers: Dom Pedro II, p. 232
- Primary and Secondary Sources: The Treaty of Tordesillas, p. 238

Comparing Regions

Gold Rushes in Brazil and California

Gold was first discovered in Brazil in the state of Minas Gerais in 1695. With much of the mining work done by forced labor, a large part of the wealth went to the Portuguese rulers. Boom towns sprang up as the population moved to the mining region, and in 1763 the capital of Brazil was moved from Salvador to Rio de Janeiro.

The first modern gold rush began in northern California in 1848. Because of advances in communications and transportation—and the prospect of personal fortunes—the scope of the California gold rush was larger than any that had occurred before. Between 1848 and 1853, at least 250,000 workers from all over the world came in search of gold.

Treaty of Tordesillas

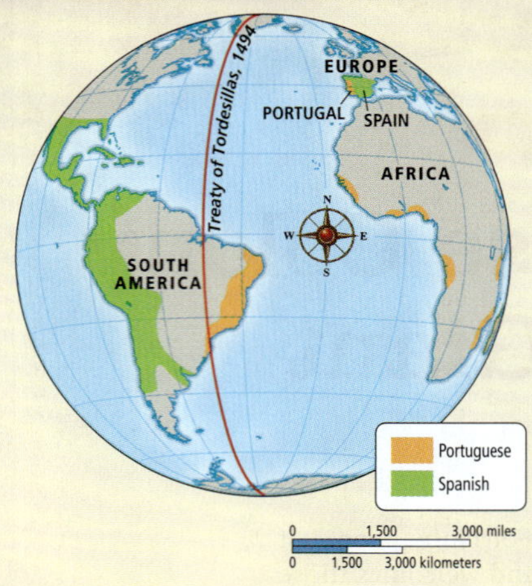

CONNECT Geography & History
Region In 1494, which country claimed most of the lands in North and South America?

The Colony Expands In 1500, **Pedro Álvares Cabral** landed on the coast of what is now eastern Brazil and claimed the land for Portugal. Like the Spanish in South America, the Portuguese came to find gold and silver. When they found neither, Portuguese colonists cleared out large areas of the land to establish sugar cane plantations. The huge demand for sugar made it an important export and a source of wealth for Portugal.

The Portuguese later developed tobacco and cotton plantations and cattle ranches for meat and hides. They forced the indigenous people to work on the plantations and ranches. Many died from disease and overwork, and others fled into the rain forest. African slaves were brought in to replace them.

The discovery of gold in the late 1600s and diamonds in the early 1700s west of present-day Rio de Janeiro attracted many people farther inland. Coffee plants were introduced in Brazil in 1727, and by the mid-1800s coffee had become Brazil's chief export.

Independence to Republic Portugal controlled Brazil from 1500 to 1822. In 1807, the French ruler, Napoleon, invaded Portugal. As a result, Prince John, the Portuguese ruler, fled to Brazil and established a monarchy there.

In 1821, Prince John returned to Portugal, leaving his son, Pedro, in charge of the colony. Pedro and other officials declared Brazil's independence from Portugal in 1822, and Pedro became **Dom Pedro I**, Brazil's first emperor. Unpopular with the people, he returned to Portugal in 1831, and his son became Emperor **Dom Pedro II**.

Under Dom Pedro II's rule, Brazil started to become industrialized. Railroads and telegraph lines improved transportation and communication. Pedro II worked to end slavery, which was abolished in 1888. This angered wealthy plantation owners, who forced Pedro II to give up the throne. Brazil became a constitutional republic in 1889.

Dom Pedro II

FIND MAIN IDEA Explain what made Brazil a profitable Portuguese colony. With the demand for sugar, sugar cane brought wealth to Portugal; tobacco, cotton, cattle, gold, diamonds, coffee, and rubber helped Brazil grow into a rich country.

DIFFERENTIATING INSTRUCTION

Struggling Readers

Use Reciprocal Teaching

Help students improve their comprehension of the history of Brazil described in the "The Colony Expands" and "Independence to Republic" subsections. Have them work in pairs and instruct them to take turns reading each paragraph aloud. Readers should pause at the end of the paragraph and ask their partners questions about what they just read. Encourage students to focus on the main ideas in each paragraph.

English Learners

Comprehension: Connections

Explain that the subsection "Independence to Republic" describes how different leaders came to power in Brazil. Point out that the last sentence in the first paragraph uses the phrase *as a result* to explicitly state an effect caused by Napoleon's invasion of Portugal. In the remaining paragraphs, ask students to infer causes and effects by looking at what came before and what happened after each event. Call on volunteers to identify what caused changes in Brazil's government.

COMPARING U.S. & Brazilian Governments

	UNITED STATES	BRAZIL
Type of Government	federal republic	federal republic
Branches of Government	legislative, executive, judicial	legislative, executive, judicial
Election of President	elected directly by the people and the electoral college	elected directly by the people; must win 50 percent plus one votes
Voting	voters must be at least 18 years old; voting is a choice	voters must be at least 16 years old; voting is mandatory for citizens ages 18 to 70

CRITICAL THINKING

1. **Compare** How are the Brazilian and U.S. governments alike?
2. **Contrast** How are elections different in the two countries?

Challenges of a Modern Nation

KEY QUESTION What challenges does the government of Brazil face today?

After adopting a constitution in 1891 that was based on the U.S. constitution, Brazil struggled to establish a democratic government. Dictators and military leaders ruled the nation until 1985. Today's government faces the problem of **urbanization** and the wide gap between the nation's rich and poor.

Establishing a Democratic Government After military rule ended, Brazilians voted for a president for the first time since 1960. Today, Brazil is ruled by the constitution established in 1988.

Brazil's federal government includes an executive branch led by a president, a two-house legislature, and a court system. The country is divided into 26 states and a federal district in Brasília. Since Brazil has many political parties with different viewpoints, elected officials have to work with all the parties in order to rule effectively.

Urbanization During the last half of the 20th century, millions of people from rural areas moved to Brazil's cities in search of jobs. Brazil's urban population more than doubled between 1950 and 2000. Several of Brazil's cities are among the largest cities in the world in population. All but two of Brazil's heavily populated cities are located near Brazil's eastern coast.

INTERDISCIPLINARY ACTIVITIES

CONNECT to Language Arts

Paraphrase an Historic Document

Provide students with a copy of the preamble to Brazil's 1988 constitution. (This document is available online.) Have students paraphrase the document in writing. Discuss with the class the similarities and differences with the preamble to the U.S. Constitution.

Unit Resource Book: Latin America
- World Literature: From *Journey Through a Tropical Jungle*, p. 260
- World Literature: from *Popul Vuh*, p. 264

CONNECT to Civics

Discuss Voting Practices

Organize the class into two groups: have one discuss voluntary voting (as done in the United States) and the other discuss mandatory voting (as in Brazil). Students in each group should list the pros and cons of the voting method used and draw conclusions about that method. Invite volunteers from each group to summarize their group's discussion for the class. Ask students to suggest ways to encourage more young adults to vote. To extend this activity, hold a class discussion on the U.S. voting age.

COMPARING

U.S. & Brazilian Governments

Explain that in the United States, someone might lose the popular vote, but still be elected president because of the electoral college. This result, although rare, has happened a few times in U.S. history, most recently in 2000. Many U.S. presidents may win only a plurality of the popular vote, but all are required to win a majority of the electoral college votes.

- Why do you think voting is not mandatory for 16- to 18-year olds in Brazil? *(Possible answer: because they are inexperienced voters)*

CRITICAL THINKING ANSWERS

1. **Compare** They are federal republics, with three branches, and the people elect the president.
2. **Contrast** Brazilians are younger by two years when they can vote. Also, Brazilians between 18 and 70 are required to vote.

Teach

Challenges of a Modern Nation

Think, Pair, Share

- How does the number of political parties in Brazil affect the government? *(Possible answer: Government leaders have to work with people from many different parties to get things done.)*
- Why did the government move the capital of Brazil from Rio de Janeiro to Brasília? *(Possible answer: to encourage settlement further inland; to make the capital more centrally located)*
- **Sequence Events** When did Brazil establish its current constitution? *(1988)*

Unit Resource Book: Latin America
- Skillbuilder Practice: Drawing Conclusions, p. 226
- Geography Project: Landscape in a Box, pp. 265–266

COMPARING

Urban and Rural Populations

- Which photo shows a densely populated area? *(São Paulo, because the photo shows buildings close together; the map also shows it as being in one of Brazil's most densely populated states)*
- How is the Distrito Federal different from the rest of the state of Goiás? *(It is a densely populated area while the state surrounding it has fewer than 60 persons per square mile.)*
- Which 25-year period saw the greatest shift in population? Why? *(1950–1975, because the changes in the percentages of rural and urban people were greater than in 1975–2000)*

Comparing Regions

U.S. and Brazilian Urbanization and Population Density

The United States was more urbanized earlier than Brazil was (64.2% in 1950), but was slightly less urbanized in 2000 (79.1%). The United Nations predicts these trends to continue with the United States becoming 83.7% urban by 2015 and Brazil 88.2%.

CRITICAL THINKING ANSWERS

1. **Draw Conclusions** between 0 and 60 people per square mile
2. **Evaluate** about 141 million

Unit Resource Book: Latin America

- Outline Map Activities: South America: Physical/Political, p. 229

COMPARING Urban and Rural Populations

São Paulo

Tresópolis

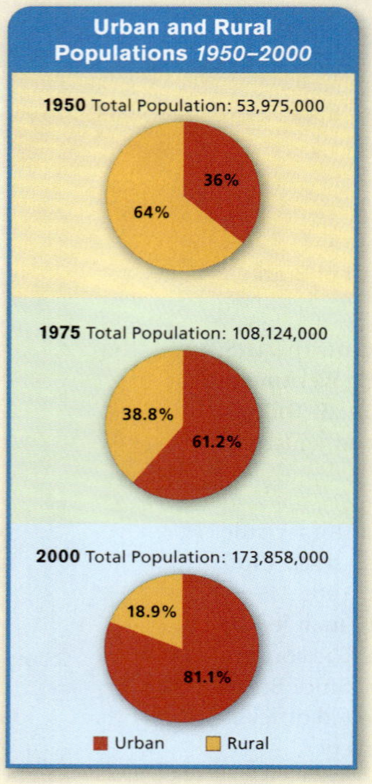

Source: Population Division of the Department of Economic and Social Affairs of the United Nations Secretariat, *World Population Prospects: The 2004 Revision* and *World Urbanization Prospects: The 2003 Revision*

CRITICAL THINKING

1. **Draw Conclusions** What is the population density in most of Brazil?
2. **Evaluate** About how many people lived in urban areas in Brazil in 2000?

DIFFERENTIATING INSTRUCTION: TIERED ACTIVITIES

OBJECTIVE Summarize the differences between urban and rural life in Brazil.

Basic

Have students write captions for each of the photos, the map, and the graphs. Their caption for the graphs should summarize the trend shown in all three graphs. Direct them to write a summary of the information about urban and rural population of Brazil. Have students share their summaries.

On Level

Have students write two paragraphs about what they think life would be like in São Paulo and Tresópolis based on the photographs, map, and graphs. Encourage them to think about the possible benefits and drawbacks of daily life in each place. Allow students to share their paragraphs with a partner.

Challenge

Have students write an essay on Brazil's population changes from 1950 to 2000, with an analysis of the problems that resulted from these changes. Ask them to consider issues such as housing and jobs. Invite them to predict what life might be like in 2015. Allow students to share their essays with a partner.

Many poor Brazilians who moved from rural areas work at low-income jobs or are unemployed. Most live in neighborhoods called ***favelas***, located on the outskirts of cities. In 1960 the Brazilian government moved the capital city from **Rio de Janeiro** (REE•oh day zhuh•NAYR•oh) to **Brasília**, about 600 miles inland. The Brazilian government has encouraged people to move inland, even offering land to people who are willing to move to the nation's interior.

Favela This *favela* is located outside of Rio de Janeiro. **Why might it be located there?**

It is less expensive to live on the outskirts of cities. People in *favelas* often work in cities.

'azil has to
eal with the
oblems that
sult from
banization,
ch as popula-
on increase;
e growing
umber of
velas; income
p between
h and poor.

Bridging the Gap Between Rich and Poor Today, Brazil has one of the largest economies in the world. However, one of the major challenges facing Brazil is how to bridge the large gap between the rich and poor. A small number of Brazilians live comfortably or in luxury, but most Brazilians are poor. Brazil has one of the most uneven distributions of land in the world. Two percent of landowners own most of the land, while more than half of Brazil's farmers work on less than three percent of the land. To help narrow the income gap, Brazil will have to create more jobs for the nation's poor.

SUMMARIZE Discuss some of the challenges facing Brazil today.

Section 1 Assessment

ONLINE QUIZ For test practice, go to **Interactive Review @ClassZone.com**

TERMS & NAMES

1. **Explain the importance of**
   - Pedro Álvares Cabral
   - *favela*
   - Rio de Janeiro
   - Brasília

USE YOUR READING NOTES

2. **Sequence Events** Use your time line to answer the following question:
   How long after becoming a republic did military rule end in Brazil?

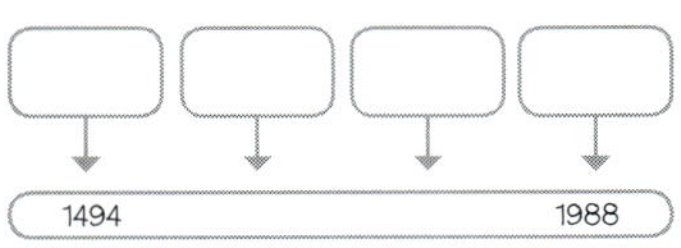

KEY IDEAS

3. What European nation developed Brazil as its colony?
4. Where do most people in Brazil live today?
5. Why did the Brazilian government offer land to people to move to the nation's interior?

CRITICAL THINKING

6. **Make Inferences** How has uneven land distribution contributed to poverty in Brazil?
7. **Analyze Causes and Effects** Why has urbanization resulted in the creation of *favelas* in Brazil's cities?
8. **CONNECT to Today** What is one action Brazil's government hopes to take to narrow the income gap between its rich and poor?
9. **WRITING** **Write a Newspaper Article** Research information about Brazil's government. Find out who is the current president, how the legislative branch is organized, and the term of office for both. Present your findings in the form of a newspaper article.

CONNECT to the Essential Question

How have Brazil's people used the country's abundant natural resources to make Brazil an economic giant?

Ask students what they have learned so far that can help them answer this question. Students might mention:

- The Portuguese gained wealth from sugar, tobacco, cotton, and cattle in Brazil.
- The discovery of gold and diamonds and the introduction of coffee plants diversified Brazil's economy.
- In the 1800s, Brazil industrialized and improved its communications and transportation systems.

4 Assess & Reteach

Assess Have students complete the Section Assessment.

Unit Resource Book: Latin America
- Section Quiz, p. 239

Interactive Review @ClassZone.com
- Section Quiz

Power Presentations

Test Generator

Reteach Call on volunteers to use the section photographs, maps, and graphic organizers to summarize Brazil's change from a colony to an economic giant. Record student information on the board. Invite other students to add further details.

Unit Resource Book: Latin America
- Reteaching Activity, p. 242

SECTION 1 ASSESSMENT ANSWERS

Terms & Names

1. Pedro Cabral: first Portuguese explorer to land in what became part of Brazil; *favela:* Brazilian neighborhood that surrounds the city; Brasília: Brazil's current capital city; Rio de Janeiro: Brazil's capital city before 1960

Use Your Reading Notes

2. 96 years
   Diagram *1494: Treaty of Tordesillas; 1500: Cabral claims Brazil for Portugal; 1822: Brazil declares independence; 1889: Brazil becomes republic; 1891: Brazil adopts constitution; 1985: Military rule ends; 1988: Constitution establishes present-day government*

Key Ideas

3. Portugal
4. Most live in urban areas.
5. to try to get people to move from crowded cities near the eastern coast to the interior of the country

Critical Thinking

6. Possible response: Because half the farmers work on less than three percent of the land, they do not make enough money, leaving many people in poverty.
7. Many poor people have moved from rural areas to the nation's cities. These people have low-income jobs or are unemployed and most live in slums on the outskirts of Brazil's cities.
8. create more jobs for the nation's poor
9. Use the rubric to score newspaper articles.

Newspaper Article Rubric

	Facts About Government	Errors in Mechanics
4	very detailed	few or none
3	several details	some
2	few details	many
1	no details	major

1 Plan & Prepare

Objectives

- Explain how people from many places have blended and influenced Brazil's culture
- Describe the arts and entertainment of Brazil

Read for the Essential Question

Help students read for a purpose by reminding them of the Essential Question: "How have Brazil's people used the country's abundant natural resources to make Brazil an economic giant?" Point out that they can use text features such as Key Questions, special features, images, and maps to form answers to the question.

Vocabulary

Best Practices Toolkit

Use the ***Best Practices Toolkit*** to model strategies for vocabulary notetaking. Vary strategies throughout the year. Choose from: *Knowledge Rating, Predicting ABCs, Definition Mapping, Word Sort, Word Wheel, Frayer Model (Word Squares), Magnet Words,* and *Student VOC.*

Vocabulary Strategies, TT9–TT16

Reading Strategy

Best Practices Toolkit

Display the Three-Column Chart transparency.

- Point out to students that to categorize information in the section they must decide how it can be sorted into the categories shown in the chart.
- Model filling in entries from the "Architecture and Art" subsection in the section.
- Have students suggest entries, using the categories in the chart.
- Display the completed chart for review.

Three-Column Chart, TT21

Section 2 Reading for Understanding

Key Ideas

BEFORE, YOU LEARNED

Brazil's government is dealing with problems resulting from urbanization.

NOW YOU WILL LEARN

Several cultures have influenced Brazil's unique culture.

Vocabulary

TERMS & NAMES

Candomblé (kahn•duhm•BLEH) African religious practices that are mixed with Roman Catholic beliefs to produce a unique belief

quilombos (kih•LOHM•buhs) communities created by escaped African slaves

samba (SAM•buh) music and dance, with roots in African rhythms; the most famous form of Brazilian music worldwide

cuíca (kwee•kuh) a friction drum used in the samba

bossa nova (BAHS•uh NOH•vuh) a jazz version of the samba

capoeira (KAP•oh•AY•ruh) a Brazilian dance combined with martial arts

REVIEW

immigrant a person who leaves one area to settle in another

mural a wall painting

Visual Vocabulary *capoeira*

Reading Strategy

Re-create the chart shown at right. As you read and respond to the **KEY QUESTIONS**, use the chart to organize important details about Brazil's people and culture.

See Skillbuilder Handbook, page R7

CATEGORIZE

ART AND ARCHITECTURE	MUSIC	ENTERTAINMENT
1.	1.	1.
2.	2.	2.
3.	3.	3.

GRAPHIC ORGANIZERS
Go to **Interactive Review** @ClassZone.com

296 Chapter 10

PRETEACHING VOCABULARY

English Learners

Pronounce and Preview

Pronounce each term for students. Review the meanings of words within definitions, such as *practices, roots,* and *friction.*

- To modify vocabulary learning, have students complete worksheets as they read, instead of afterward.

Unit Resource Book: Latin America

- Vocabulary Practice, p. 225

Inclusion

Use Terms in Captions

Have students choose one of the terms to illustrate. Then have them write a one- or two-sentence caption that describes their illustration and uses the term correctly. Invite volunteers to present their illustrations and captions or display them in the classroom.

Niemeyer Museum of Contemporary Arts, Rio de Janeiro

Collared anteater

SECTION 2 CULTURE

A Multicultural Society

Connecting to Your World

Think about some of the foods you eat. Think about the music that you and your family enjoy. Chances are that many of the foods you enjoy and the music you listen to have their roots in a variety of cultures. If you visited Brazil, you would see how a variety of cultures has also influenced the foods, music, and other aspects of Brazilian culture today.

A Blend of Many Cultures

KEY QUESTION What cultures have most influenced Brazil?

Brazilian culture includes European, African, Asian, and Indian influences. Portuguese is the official language. Most Brazilians are Roman Catholic. At the same time, Brazilians of African ancestry mix ***Candomblé*** (kahn•duhm•BLEH), West African religious practices, with Catholicism to create a unique Brazilian blend.

Brazilian Diversity These Brazilian girls are getting ready to participate in Carnival in Rio de Janeiro.

297

2 Focus & Motivate

3-Minute Warm-Up

Write on the board or display the transparency:

- What are the roots of Brazilian culture? Give an example of a cultural influence on Brazil. *(Roots: African, European, Asian, and Indian; Examples: African influences on religion, food, and the arts; Indian influences on the arts)*

Unit Transparency Book: Latin America
- 3-Minute Warm-Ups, TT31

Connecting to Your World

More About . . .

Candomblé

African practices used in *Candomblé* include outdoor worship, offerings of candles and flowers, and dance. Worshipers also give African names to Catholic saints.

3 Teach

A Blend of Many Cultures

Think, Pair, Share

- How did *quilombos* help Africans preserve their culture in Brazil? *(Escaped slaves set up communities called* quilombos *that were similar to their African homes.)*
- **Categorize** From what countries have European and Asian immigrants come to Brazil? *(Europe: Italy, Spain, Portugal, Germany; Asia: Japan, China, Korea)*

SECTION 2 PROGRAM RESOURCES

Unit Resource Book: Latin America

ON LEVEL
Unit Resource Book
- Reading Study Guide, p. 203
- Map and Graph Skills, p. 227
- Section Quiz, p. 240

STRUGGLING READERS
Unit Resource Book
- Reading Study Guide with Additional Support, p. 209
- Skillbuilder Practice: Drawing Conclusions, p. 226
- Section Quiz, p. 240
- Reteaching Activity, p. 243

eEdition with Audio DVD-ROM

ENGLISH LEARNERS
Pupil Edition in Spanish
eEdition with Audio DVD-ROM
Unit Resource Book
- RSG (Spanish), p. 215
- RSG with Additional Support (Spanish), p. 221

Multi-Language Glossary
Test Generator
- Section Quiz in Spanish

INCLUSION
Unit Resource Book
- RSG with Additional Support, p. 209
- Section Quiz, p. 240
- Reteaching Activity, p. 243

GIFTED & TALENTED
Unit Resource Book
- Interdisciplinary Projects, pp. 234–235
- Section Quiz, p. 240

PRE-AP
Unit Resource Book
- Connect to Music, p. 237
- World Literature, p. 261
- Section Quiz, p. 240

TECHNOLOGY

Unit Transparency Book
- 3-Minute Warm-Ups, TT31
- Essential Question, TT32
- Critical Thinking, TT34
- Cultures, TT38

Daily Test Practice Transparencies
- Section 2, TT35

Power Presentations
ClassZone.com
World Cultures and Geography Video on DVD

ANALYZING Primary Sources

Daniel Munduruku was born in the state of Pará in the Amazon rain forest. *Tales of the Amazon* includes information about his own experiences as an indigenous person coming to the city. His work is unusual because few people who are natives of the rain forest have written about life there.

- How do Daniel Munduruku's experiences as a child reflect an oral tradition? *(Possible answer: Stories were passed down from one generation to the next.)*
- Why might Daniel Munduruku have chosen to write books about a culture with a strong oral tradition? *(Possible answers: so that more people could learn about the culture; to help preserve the culture)*

CRITICAL THINKING ANSWER

Summarize The people's stories help keep the culture alive and give people an identity as a culture.

Unit Resource Book: Latin America

- World Literature: "A Bow, a Deer, and a Talking Bird," pp. 261–263

More About . . .

The Guarani

The Guarani in Brazil are located primarily in the southeastern part of the country. Many more Guarani live in Argentina, Uruguay, and especially in Paraguay, where Guarani is an official language along with Spanish. There may be as many as five million Guarani throughout South America.

When the Spanish arrived in South America, the Guarani were living as hunter-gatherers who also fished and grew crops such as corn and sweet potatoes. The Spanish came to their territory in search of gold. Intermarriage created a large *mestizo* population and a blended culture and language.

ANALYZING Primary Sources

Daniel Munduruku is a Munduruku Indian. He has written several books about his culture. Here he describes learning the ancient myths of his people.

> When I was a . . . boy, my grandfather would . . . tell the stories that explained the origins of our people and the vision of the universe. . . . Now . . . I realize that those ancient myths say what cannot be said. They are pure poetry, and through them I see how an identity can be created in the oral tradition.
>
> Source: *Tales of the Amazon, How the Munduruku Indians Live*

CRITICAL THINKING

Summarize How is storytelling important in the Munduruku culture?

Native Brazilians Up to 6 million Indians lived in Brazil before the Portuguese arrived. Tens of thousands died from diseases brought by the European colonists. Today, only about 700,000 Indians live in Brazil, mostly in the Amazon rain forest. They make up less than one percent of Brazil's population. The Tupi and Guarani make up the largest groups. Other tribes, such as the Yanomami and Bororo, still follow traditional ways of hunting and farming.

The Brazilian government has set up reservations to protect these cultures. But even isolated groups now have contact with outsiders, such as miners, loggers, and researchers, who move into the area to study native cultures. This increased contact threatens the groups' traditional ways of life, arts, crafts, and languages.

Yanomami Woman A Yanomami woman poses with her child. **How might being photographed threaten the Yanomami's traditional way of life?**

Being photographed exposes them to the outside world.

DIFFERENTIATING INSTRUCTION

Inclusion

Visualize and Compare Numbers

Help students to visualize the relationship between 100,000 and one million in order to better understand how the number of Indians in Brazil has declined. Remind them that one million is the same as ten hundred-thousands. Draw a stick figure on the board that represents 100,000 people. Ask students how many figures would be needed to represent six million and how many would represent the number of Indians in Brazil today. *(60; 7)*

Gifted & Talented

Share Oral Traditions

Encourage students to think about the role of oral traditions in their family, such as stories about earlier generations of their family or about family experiences that get repeated often. Have students work in pairs and share an example of such a story. Suggest that they include sound effects and gestures in telling their stories. Invite students to discuss how hearing a story is different from reading it. Have pairs form groups of four and share their thoughts about oral traditions.

Africans Today, most Brazilians with African ancestry live in the nation's northern and coastal regions and make up about 6 percent of the population. The Portuguese brought Africans to Brazil between the 1500s and 1800s to work on the sugar plantations. In fact, of all the slaves brought to North and South America, more than a third were brought to Brazil.

Some of these slaves escaped to freedom in Brazil's inland northern region, where they established communities that were similar to their African homes. They called the settlements **quilombos** (kih•LOHM•buhs), an African word for "housing." By the end of the 1600s, as many as 25,000 Africans may have lived in these communities. *Quilombos* can still be found in parts of Brazil today, particularly in the northern part, where people continue their traditional way of life. Today most Brazilians of African ancestry live in the northeastern state of Bahia. Its capital, Salvador, was the center of the Portuguese empire, sugar industry, and slave trade.

African influences are evident in various parts of Brazilian culture. African slaves created Brazil's national food, *feijoada* (fayh•zhoo•AH•duh), a bean stew. African rhythms play an important role in Brazilian music. Brazil's art and literature also reflect an African influence.

CONNECT History & Culture

African Influence on Brazilian Food

Since most slaves brought to Brazil arrived from West Africa, West African cooking has had a strong influence on Brazilian food. Palm oil, called *dendê*, and chili peppers, called *malagueta*, are basic West African cooking ingredients. Today, Bahian cooks make *moqueca*, a popular stew. It is made with seafood, coconut, garlic, onion, parsley, *malagueta*, and *dendê*.

Europeans and Asians In the 1700s, Europeans viewed Brazil as a place to acquire wealth. With the discovery of gold and diamonds, many immigrated to Brazil to make their fortunes. After Brazil declared its independence, many Europeans, including Italians, Spanish, Portuguese, and Germans, came to work in Brazil, mostly in the coffee industry. Today, people of European ancestry make up about half of Brazil's population. Most live in the southern part of the country.

Japanese **immigrants** first began arriving in Brazil in the early 1900s. Most came to work on coffee or tea farms. More Japanese arrived after World War II. Today Brazil has the largest Japanese population outside of Japan, with most making their homes in the state of São Paulo. In recent years, Asians from China and Korea have also immigrated to Brazil.

The cultures of indigenous people, Africans, Europeans, and Asians have influenced Brazilians' food, religion, music, art, literature, and economy. ◄

EVALUATE Explain how Brazil's culture is a blend of several cultures.

CONNECT History & Culture

African Influence on Brazilian Food

Enslaved African women were used as cooks on Brazilian sugarcane plantations. These African women used ingredients from Africa that the Portuguese brought to Brazil for trade. *Dendê* was originally used in Brazil for fuel, but eventually the African cooks used it in cooking because they had used it at home. In addition to *malagueta*, African cooks also used coconut milk, okra, and bananas as cooking ingredients. Dishes such as *moqueca* that originated in the Brazilian state of Bahia have remained regional, rather than spreading throughout the country as *feijoada* has. Certain ingredients were not available in all regions of Brazil and were often too expensive for most people to afford.

Unit Resource Book: Latin America

- Map and Graph Skills: Reading a Thematic Map, pp. 227–228

More About . . .

Quilombos

Escaped slaves from sugar plantations established the largest *quilombo*, Palmares, as a republic in what later became the state of Alagoas in northeastern Brazil. This republic was ruled by an elected chief and lasted from 1630 to 1697. The population of Palmares consisted of runaways and of slaves who had been captured from Portuguese plantations. Runaways became free, but those who had been captured remained slaves.

Palmares's population reached about 20,000 people in the 1690s. They successfully fought off several attempts by the Portuguese to take control of Palmares, but were finally defeated in 1694 and the republic ended in 1697.

INTERDISCIPLINARY ACTIVITIES

CONNECT to Math

Graph Ethnic Populations

Have students create pie graphs to reflect the ancestry of the people of Brazil. Students should use these percentages from the text for their graphs: Native Brazilians, one percent; Africans, six percent; Europeans, 50 percent. Have them add these percentages together to determine the size of the final segment of the pie graph. *(43 percent)* Ask them how this segment should be labeled. *("Other" or "Asians and others")*

CONNECT to Language Arts

Write a Restaurant Review

Invite students to assume the role of a food critic and "visit" a restaurant that specializes in Brazilian food such as *moqueca* and *feijoada.* Have them write a restaurant review describing their experience. They might include information about the taste of the food based on the descriptions and photograph in the textbook, as well as background information on how the food was influenced by African culture.

Teach

Arts and Entertainment

Talk About It

- How does Carnival reflect Brazil's cultural diversity? *(Possible answer: It combines a European Roman Catholic religious festival with aspects of African traditions, such as different kinds of music, dance, and drumming.)*
- What activities do tourists enjoy in Brazil? *(Possible answer: shopping for crafts made by Indian artists, attending Carnival, and swimming at beach areas such as Copacabana)*
- **Categorize** Name three examples of Brazilian music and dance that have been influenced by African culture. *(samba, samba-reggae, and capoeira)*

Unit Resource Book: Latin America
- Skillbuilder Practice: Drawing Conclusions, p. 226

RESEARCH & WRITING CENTER

History Makers

Oscar Niemeyer

Oscar Niemeyer graduated from the National School of Fine Arts in Rio de Janeiro in 1934. Around that time he began working with Lúcio Costa, the leading architect of the modern style in Brazil. The French architect Le Corbusier—who worked with Niemeyer and Costa on the Ministry of Education and Health building in Rio—also influenced his work.

Niemeyer's first major independent project was the design for a new suburb of Belo Horizonte in 1941. The mayor of Belo Horizonte—Juscelino Kubitschek de Oliveira—would later become president of Brazil and ask Niemeyer to be the major architect of Brasília.

HISTORY MAKERS

Oscar Niemeyer born 1907

Oscar Niemeyer is a famous architect. As Brasília's chief architect, he designed the National Congress building, pictured here, as well as the city's commercial buildings and Cathedral. He also helped design the UN headquarters in New York City. Niemeyer's building designs have been compared to Brazil's music as being graceful, curvy, and seeming to sway in the air.

ONLINE BIOGRAPHY
For more on the life of Oscar Niemeyer, go to the Research & Writing Center @ClassZone.com

Arts and Entertainment

KEY QUESTION What are some ways Brazilians enjoy holidays and leisure time?

Brazil's art, literature, music, and celebrations reflect the nation's blend of cultural traditions. Brazil's Carnival, in particular, highlights Brazil's cultural diversity.

Architecture and Art Brazil's earliest art included the crafts made by Brazil's Indian groups. Their work included such items as pottery, baskets, and jewelry. Native American groups today continue to make these handicrafts, many of which are sold to tourists in Brazilian markets.

António Francisco Lisboa, known as Aleijadinho, is a well known sculptor of the colonial period. He created religious sculptures for many churches in the mid- to late 1700s. Today, Mario Cravo Junior is an important sculptor who creates statues made with concrete.

Cândido Portinari is a famous Brazilian painter and muralist. His six largest **murals** are found in New York's United Nations building and in Washington's Library of Congress. Much of his art reflects rural life and social concerns.

Brazil is the home of Oscar Niemeyer, one of the world's most famous architects. He is best known for designing several buildings in Brasília, Brazil's capital. Niemeyer has also designed a variety of buildings in several countries throughout the world, including France, Ghana, Israel, and Lebanon. More recently, Niemeyer created a unique design for an art museum near the city of Rio de Janeiro.

Music and Dance Brazilian music is a blend of African, European, and Indian cultures. Brazilians enjoy dancing to the **samba** (SAM•buh), the most famous form of Brazilian music. The samba developed in the early 1900s and has its roots in African rhythms. The samba has many forms, but the most popular is the street samba danced during the celebrations of Carnival, a Brazilian holiday. The drum plays a major role in Brazilian music. Brazilian drums come in many shapes and sizes, each producing its own sound and rhythm. Drummers play the surdo, a bass drum, and the ***cuíca*** (kwee•kuh), a friction drum, to create samba rhythms.

300 Chapter 10

DIFFERENTIATING INSTRUCTION

English Learners

Vocabulary: Word Relationships

Help students understand the relationship between the creator involved in different types of artistic work and the art form or product that is created. Have students work with a partner to fill in a chart like the one shown. For the first three rows, both words are found in the text. For the remainder of the chart, only the missing word is found in the text. Encourage students to look for the common root word (such as *sculpt*) to find the missing words in the chart.

Type of Artist	Art Form
sculptor	
	mural
architect	
	drumming
	painting
dancer	
musician	
potter	
jeweler	

Worldster

A Day in Pedro's Life

To learn more about Pedro and his world, go to the **Activity Center @ClassZone.com**

This is how to say "Hello, my name is Pedro" in Portuguese:

"Oi, meu nome é Pedro."

Hi! My name is Pedro. I am 13 years old and live in São Paulo, the largest city in Brazil and the third largest city in the world in population. I attend a public school, a short walk from my apartment. Many of my friends live in the same apartment building, so there's always someone to do things with. Let me tell you about my day.

6 A.M. My family and I have breakfast before my parents go to work and my sister and I go to school. Today we're having cereal and papaya.

7 A.M.–noon My school day begins early. Because of overcrowded schools, many students go to school in two shifts. I like having the morning shift. My school year runs from March to December.

1 P.M. My family and I get together for the main meal of the day. Today I'm looking forward to *feijoada,* which is a dish made with black beans and beef.

6 P.M. After I finish my homework and have a small snack, I'm off to the local park to play soccer. Next to playing the game, my favorite thing to do is to watch our city's professional team play soccer.

CONNECT to Your Life

Journal Entry Think about your typical daily schedule. How are the meals you eat and the time you eat the same as and different from Pedro's? Record your ideas in your journal.

More About . . .

Feijoada

Feijoada—the stew created by African slaves that has become the national food of Brazil—may include a variety of meats. It is traditionally eaten with cooked greens, such as collards, kale, or cabbage, along with sliced oranges and *farofa,* which is toasted manioc flour. Manioc is a root vegetable native to South America that the Portuguese then introduced to Africa.

Comparing Regions

São Paulo and New York City

Pedro's home city of São Paulo covers 576 square miles, with a metropolitan area population of more than 18 million people. About one-half of the city's people are of European descent and one-third of African descent. There is a large Asian population as well. São Paulo is an important cultural, intellectual, and manufacturing center.

New York City, which covers about 300 square miles, has a metropolitan area population of more than 18 million. New York is often considered the cultural and financial capital of the United States. Its population is now about one-third people of European ancestry, one-third Hispanic, one-fourth African American, and one-tenth Asian.

Presentation Options

Go to the *Activity Center* at **ClassZone.com** to project Worldster.

- Click on the multimedia features of Pedro's Web page.
- Ask students which areas of Pedro's life they would like to explore. Listen to music, examine photographs, learn simple phrases in Portuguese, and more.

CONNECT to Your Life

Journal Entry

Students' entries will vary, but should compare their meals and meal times with Pedro's.

INTERDISCIPLINARY ACTIVITIES

CONNECT to Science

Compare Monthly Temperatures

Have students work in pairs, using available resources to find the monthly average temperatures in São Paulo and in their community. Have students create a chart comparing the temperatures. Discuss how this information relates to when Pedro has school vacation. *(Pedro lives in the southern hemisphere, and it is summer there in January and February.)*

CONNECT to Language Arts

Write a Letter

Encourage students to think about other aspects of their lives that are similar to or different from Pedro's. For example: Do they live in a city or a small town? Do they walk or ride to school? Have students write Pedro a letter about their day that includes some of these comparisons. Invite students to share their letters with a partner.

Unit Resource Book: Latin America
- Interdisciplinary Projects: Soccer: pp. 234, 235

CONNECT to Music

African slaves brought the *cuíca* to Brazil. Some people think the *cuíca* was developed in Africa for use in lion hunting because the sound can be similar to the roar of a female lioness.

Most *cuíca* vessels are made of metal, but they may also be made of wood or fiberglass. The stick is usually made from bamboo. By pressing a finger close to the center of the drumhead, the drummer can make the sound higher or lower.

Unit Resource Book: Latin America

- Connect to Music: Make a *Cuíca,* p. 237

Comparing Regions

Carnival and Mardi Gras

Carnival is celebrated in many countries in Europe and the Americas and has been celebrated in Rio since the 1840s. The largest celebration of Carnival in the United States—Mardi Gras—has been held in New Orleans, Louisiana, since 1857. Mardi Gras is a French phrase meaning "Fat Tuesday," the name given to the last day before Lent. Mardi Gras begins on the feast of Epiphany, January 6 (also known as Twelfth Night), with a parade of floats and marchers in costume. It ends on "Fat Tuesday."

Make It Fun

Teacher-Tested Activities

Esther Howse, Byron Center West Middle School, Byron Center, Michigan

My students enjoy creating a "Culture Fair." Organize the class into small groups. Have each group research an aspect of the nation's culture, such as art, music, or celebrations. Each group creates or brings an object that represents their assigned cultural aspect. For example, in studying Brazil, students might bring or draw an example of the artwork of one of the artists mentioned in the chapter. Students then display their creations for classroom viewing.

CONNECT to Music

The *cuíca* and other percussion instruments create the energetic sounds of the samba during Carnival celebrations (shown below). The *cuíca* is an African friction drum made by attaching a stick to the center of a drum skin. The drummer holds the drum in one hand and puts the other arm inside the drum. The sound is made by rubbing the stick with a wet piece of cloth or leather. When the stick is rubbed, the drum skin vibrates, producing a distinctive sound.

▲ The inside of a cuíca

Activity

Make a *Cuíca*

Materials

- plastic cup or strong food container
- nail
- waxed dental floss

1. With a nail, make two holes, about an inch apart, in the bottom of the cup or container.

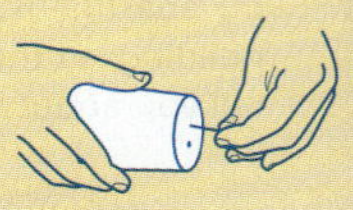

2. Push a three-foot piece of dental floss through both holes and make sure the two ends of the floss are the same length.

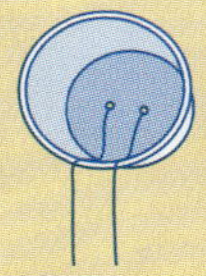

3. Tie the two ends into a knot as close to the hole inside the cup as possible so the string doesn't slip out.

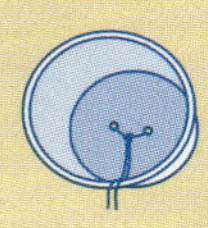

4. Hold the cup in one hand and slide your fingers along the string with the other to make your own *cuíca* music.

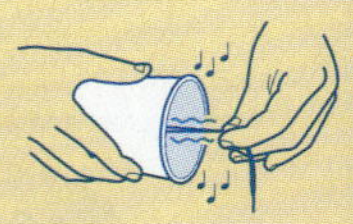

Another popular form of Brazilian music and dance is the **bossa nova** (BAHS•uh NOH•vuh). It mixes the samba beat with the sounds of jazz. Brazilian jazz musicians popularized the bossa nova in the rest of the world. Brazilian samba-reggae is a popular form of music based on African drum traditions. **Capoeira** (kap•oh•AY•ruh) combines dance and the martial arts. It developed in Brazil from African origins. During Portuguese rule, the Portuguese forbade the African slaves to fight. So the Africans created fight dances, which developed into *capoeira.*

Festivals Brazilians are known for Carnival, celebrated every year for four days before the beginning of the Christian season of Lent. The holiday combines a Roman Catholic festival with African celebrations and includes parades and street parties. The world's most famous Carnival takes place in Rio de Janeiro. Thousands of spectators line the streets to watch costumed people ride elaborately decorated floats. In clubs and other places, thousands of dancers, who have been practicing for months, participate in samba dance competitions.

DIFFERENTIATING INSTRUCTION: TIERED ACTIVITIES

OBJECTIVE Create a visual description of Carnival.

Basic

Have students review the text related to music, dance, and festivals. Encourage them to think about how Carnival reflects Brazil's diverse culture. Direct pairs of students to work together to create a poster that promotes Carnival in Rio de Janeiro to tourists. Have students display their posters.

On Level

Adapt the Basic activity so that pairs of students create an illustrated magazine article about Carnival in Rio de Janeiro. Direct students to include at least three visuals with captions along with a short essay about Carnival. Have students share their magazine articles in small groups or display them.

Challenge

Adapt the On Level activity, so that pairs of students create an illustrated magazine article that compares Carnival to a similar celebration in another part of the world. Allow time for additional research as needed. Have students share their articles in small groups or display them.

Beach in Recife, Brazil Children enjoy playing in the waters of Boa Viagem beach, a popular beach in Recife.

Recreation Brazil's thousands of miles of coastline provide extensive sandy beaches. On weekends, thousands of Brazilian families flock to the beaches to swim and enjoy picnics. Copacabana Beach in Rio de Janeiro attracts millions of tourists every year.

Brazil's most famous sport is football (soccer). It is played everywhere—in cities, beaches, and rural areas. People of all ages play it for fun. Professional football teams draw huge crowds of enthusiastic fans in football stadiums in many Brazilian cities. Maracana Stadium in Rio de Janeiro, one of the world's largest stadiums, holds more than 150,000 people. Brazil's teams have produced some of the world's top football players, such as Pelé and Ronaldo. Brazil's national football team has won 5 of the 17 World Cups ever awarded, the most for any football team. The World Cup is a trophy awarded to a nation's football team for winning the world championship.

SUMMARIZE Describe the ways Brazilians enjoy holidays and their leisure time. They enjoy music and dance, celebrate Carnival, enjoy recreational activities at beaches, and playing and watching soccer.

Section 2 Assessment

ONLINE QUIZ For test practice, go to **Interactive Review** @ClassZone.com

TERMS & NAMES
1. **Explain the importance of**
   - *Candomblé*
   - *quilombos*
   - *cuíca*
   - *capoeira*

USE YOUR READING NOTES
2. **Categorize** Use your completed chart to answer the following question.
   What contributions has Oscar Niemeyer made to Brazilian culture?

ART AND ARCHITECTURE	MUSIC	ENTERTAINMENT
1.	1.	1.
2.	2.	2.
3.	3.	3.

KEY IDEAS
3. What four cultures have contributed to modern Brazilian culture?
4. Where do most Brazilians of African ancestry live?
5. What type of music and dance is identified worldwide as Brazilian?

CRITICAL THINKING
6. **Make Inferences** Why are the traditional ways of life of some Indian groups in Brazil threatened?
7. **Draw Conclusions** How have Africans contributed to Brazilian culture?
8. **CONNECT to Today** Why do you think the Brazilian government has set up reservations to protect traditional Indian cultures?
9. **ART** **Make a Collage** Create a collage that illustrates ways people enjoy themselves in Brazil. Use the information in the section as well as in other sources. Then make a collage that features the various sources of entertainment in Brazil.

CONNECT to the Essential Question

How have Brazil's people used the country's abundant natural resources to make Brazil an economic giant?

Ask students what they have learned so far that can help them answer this question. Students might mention:

- Brazil's culture is a blend of European, African, Asian, and Indian influences.
- European and Asian immigrants came to Brazil to work in mines and in the coffee and tea industries.
- Tourists come to Brazil for Carnival, and to enjoy its beaches and diverse culture.

4 Assess & Reteach

Assess Have students complete the Section Assessment.

Unit Resource Book: Latin America
- Section Quiz, p. 240

Interactive Review @ClassZone.com
- Section Quiz

Power Presentations

Test Generator

Reteach Direct students to write two questions and answers that bring out important ideas about Brazil's culture. Have students take turns asking their questions.

Unit Resource Book: Latin America
- Reteaching Activity, p. 243

SECTION 2 ASSESSMENT ANSWERS

Terms & Names
1. *Candomblé*: religious practices brought by African slaves that are mixed with Roman Catholic beliefs; *quilombos*: communities created by escaped African slaves; *cuíca*: a friction drum used in the samba; *capoeira*: a Brazilian dance combined with martial arts

Use Your Reading Notes
2. He has designed several buildings in Brazil and around the world. He is Brasília's chief architect.

ART AND ARCHITECTURE	MUSIC	ENTERTAINMENT
1. crafts made by Brazil's Indian groups 2. important sculptors and muralists	1. blend of cultures 2. samba famous Brazilian music	1. Carnival 2. beaches

Key Ideas
3. European, African, Asian, and Indian cultures
4. They live in the northeast and coastal regions, especially in Bahia state.
5. samba

Critical Thinking
6. because these groups are exposed to outside groups and their cultures
7. Africans continued their religious beliefs by mixing them with Catholicism. They created the national food. They brought the roots of Brazil's music and dance.
8. Answers will vary, but may include that the government set up reservations to keep outside groups and ideas away from the traditional cultures in order for these cultures to preserve their traditional way of life.
9. Use the rubric to score collages.

Collage Rubric

	Design of Collage	Illustration and Effort
4	attractive	excellent
3	neat	good
2	somewhat neat	adequate
1	sloppy	weak

1 Plan & Prepare

Objectives

- Describe how natural resources have helped Brazil's economy
- Explain why it is important to preserve the rain forest in Brazil

Read for the Essential Question

Help students read for a purpose by reminding them of the Essential Question: "How have Brazil's people used the country's abundant natural resources to make Brazil an economic giant?" Point out that they can use text features such as Key Questions, special features, images, and maps to form answers to the question.

Vocabulary

Best Practices Toolkit

Use the ***Best Practices Toolkit*** to model strategies for vocabulary notetaking. Vary strategies throughout the year. Choose from: *Knowledge Rating, Predicting ABCs, Definition Mapping, Word Sort, Word Wheel, Frayer Model (Word Squares), Magnet Words,* and *Student VOC.*

Vocabulary Strategies, TT9–TT16

Reading Strategy

Best Practices Toolkit

Display the Web Diagram transparency.

- Remind students that main ideas are important ideas in a segment of text.
- Model filling in some main ideas on the graphic organizer using the subheads throughout this section.
- Have students suggest entries. Remind students that main ideas in each block of text in the section will help them answer the Key Questions.
- Display the completed diagram for review.

Web Diagram, TT28

SECTION 3 Reading for Understanding

Key Ideas

BEFORE, YOU LEARNED

The culture of Brazil today is a blend of many cultures.

NOW YOU WILL LEARN

An abundance of natural resources helps make Brazil a major industrial nation.

Vocabulary

TERMS & NAMES

hydroelectricity electric power generated by water

ethanol (EHTH•uh•NAWL) a liquid made by using a chemical process to convert sugar cane to a kind of alcohol that can be used for fuel

deforestation the cutting and clearing away of trees

boycott to stop buying and using products from certain sources as a way of protest

REVIEW

economy a system for producing and exchanging goods and services among a group of people

export a product or resource sold to another country

rain forest a broadleaf-tree region in a tropical climate

global warming an increase in the average temperature of the Earth's atmosphere

BACKGROUND VOCABULARY

mechanized equipped with machinery

self-sufficient able to provide for one's own needs without outside help

Reading Strategy

Re-create the web diagram shown at right. As you read and respond to the **KEY QUESTIONS**, use the diagram to organize main ideas and details about Brazil's economy.

See Skillbuilder Handbook, page R4

FIND MAIN IDEAS

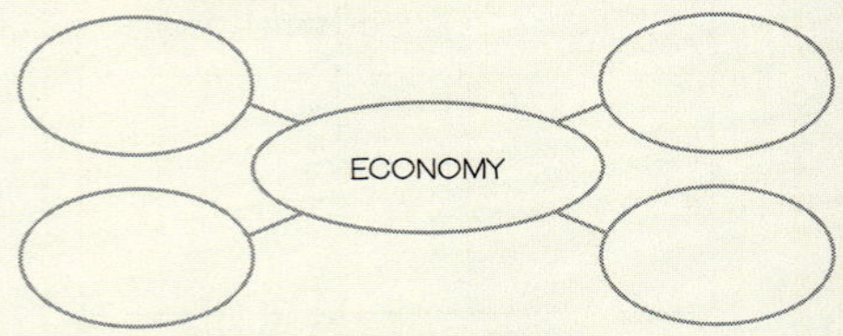

GRAPHIC ORGANIZERS
Go to Interactive Review @ClassZone.com

PRETEACHING VOCABULARY

English Learners

Pronounce and Preview

Pronounce each term for students. Review the meanings of words within definitions, such as *generated, convert,* and *protest.*

- To modify vocabulary learning, have students complete worksheets as they read, instead of afterward.

Unit Resource Book: Latin America

- Vocabulary Practice, p. 225

Inclusion

Vocabulary: Word Sort/ Word Splash

Have students fill out the graphic organizer for the terms in the section.

- Use the transparency to model the strategy.
- Have volunteers suggest entries.
- Then distribute copies of the organizer for students to complete with a partner.

Best Practices Toolkit

- Vocabulary: Word Sort/Word Splash, TT12

SECTION 3 ECONOMICS

Niemeyer Museum of Contemporary Arts, Rio de Janeiro

Collared anteater

Developing an Abundant Land

Connecting to Your World

In the early 1900s, Henry Ford invented a way to manufacture cars that would make them affordable for most people. His invention sparked the U.S. automobile industry and made the United States a major producer of automobiles. Today Brazil has become a leading maker of automobiles in the world and one of the most industrialized countries in South America.

Creating an Economic Giant

KEY QUESTION What factors have helped Brazil become an important industrial nation?

Brazil's **economy** is the largest in South America and one of the largest in the world. Brazil's climate makes it possible for the nation to grow a wide range of crops. The abundance of natural resources, used to develop a variety of industries and manufactured goods, has helped to make Brazil a growing economic power. **Hydroelectricity**, or electric power generated by water, is provided by power plants along the Amazon River and many other rivers flowing through Brazil. Hydroelectricity supplies most of the energy needed to run these industries.

Itaipu Dam The dam, built on the Paraná River, provides about 25 percent of Brazil's power.

305

2 Focus & Motivate

3-Minute Warm-Up

Write on the board or display the transparency:

- Look at the large photograph of the Itaipu Dam. What does it suggest about the importance of rivers in Brazil? *(Possible Answer: Rivers provide a major source of Brazil's electric power.)*

Unit Transparency Book: Latin America
- 3-Minute Warm-Ups, TT31

Connecting to Your World

More About . . .

Itaipu Dam

The Itaipu Dam, which is nearly five miles long, is a joint project of Brazil and Paraguay. It provides electricity for both countries. The Itaipu Dam is one of the largest hydroelectric projects in the world. Its energy output is expected to be surpassed by China's Three Gorges Dam in the near future.

3 Teach

Creating an Economic Giant

Reader, Recorder, Reporter

- Why is hydroelectricity important to the Brazilian economy? *(It supplies most of the energy used to run its industries.)*
- **Find Main Ideas** Which segment of Brazil's economy employs the largest percentage of workers? *(service industries—about half)*

SECTION 3 PROGRAM RESOURCES

Unit Resource Book: Latin America

ON LEVEL

Unit Resource Book
- Reading Study Guide, p. 205
- Section Quiz, p. 241

STRUGGLING READERS

Unit Resource Book
- Reading Study Guide with Additional Support, p. 211
- Skillbuilder Practice: Drawing Conclusions, p. 226
- Section Quiz, p. 241
- Reteaching Activity, p. 244

eEdition with Audio DVD-ROM

ENGLISH LEARNERS

Pupil Edition in Spanish

eEdition with Audio DVD-ROM

Unit Resource Book
- RSG (Spanish), p. 217
- RSG with Additional Support (Spanish), p. 223

Multi-Language Glossary

Test Generator
- Section Quiz in Spanish

INCLUSION

Unit Resource Book
- RSG with Additional Support, p. 21
- Section Quiz, p. 241
- Reteaching Activity, p. 244

GIFTED & TALENTED

Unit Resource Book
- Interdisciplinary Projects, pp. 233, 235
- Section Quiz, p. 241

PRE-AP

Unit Resource Book
- Economics in the World, p. 236
- Section Quiz, p. 241

TECHNOLOGY

Unit Transparency Book
- 3-Minute Warm-Ups, TT31
- Essential Question, TT32
- Critical Thinking, TT35
- Map: Rain Forests and Natural Resources of Latin America, TT37, TT37a

Daily Test Practice Transparencies
- Section 3, TT36

Power Presentations

ClassZone.com

World Cultures and Geography Video on DVD

Fun Facts!

Ethanol

It's easiest to make ethanol from sugar cane and sugar beets. However, ethanol can also be made from other crops that contain sugar or starch that can then be converted to sugar, such as corn, barley, wheat, rice, sunflowers, and potatoes. In the United States, most ethanol is made from corn. Both Brazil and the United States produce more than 4 billion gallons of ethanol per year. However, in the United States only a small percentage of cars currently run on a blend of 85 percent ethanol and 15 percent gasoline.

CONNECT Geography & Economics

Deforestation

Deforestation is not limited to tropical rain forests, but the clearing of trees in these forests has happened at a much more rapid pace than it has in temperate forests. For example, the forests of Europe and North America were cleared for settlement over several hundred years, but in the rain forests millions of acres are lost each year. The Tropical Forestry Action Plan was developed in 1985 to provide money to help reduce deforestation and replant damaged areas. Some developing countries may also have their foreign debts reduced in exchange for protecting their forests.

CRITICAL THINKING ANSWER

Make Inferences Possible answer: Farmers, loggers, and miners see opportunities to make money in their industries by using the rain forest. They clear away the trees in order to pursue their interests.

Unit Resource Book: Latin America

- Economics in the World: Major Coffee Producing Countries, p. 236

Fun Facts!

SUGAR-RUN CARS

Did you know that sugar can run cars? Many Brazilians use **ethanol** (EHTH•uh•NAWL), a liquid formed when sugar cane is converted to alcohol, as gasoline.

Brazil introduced a car engine that runs on either gasoline or ethanol. So today, almost all new cars in Brazil use ethanol.

Agriculture About one-fifth of Brazil's workers are employed in agriculture. Most farming and livestock grazing takes place in the southern and southeastern parts of Brazil, which have the nation's best soil and climate for agriculture. Farms and ranches in this part of Brazil are generally large and **mechanized**. Farms in the northern part of Brazil, however, are smaller and generally depend on manual labor to produce crops.

Brazil is one of the world's leading exporters of agricultural products, second only to the United States. Brazil is the world's largest producer of sugar cane and coffee. It is a major producer of soybeans (the nation's primary farm **export**), oranges, wheat, corn, and beef, as well as a world leader in the raising of cattle, hogs, and poultry.

Industries Brazil is rich in minerals, and a major producer of the world's gold and diamonds. Brazil has large deposits of iron, manganese, and bauxite, which are important raw materials used in manufacturing. Most large manufacturing plants are located in Brazil's southeastern section. In recent years, Brazil has invested in high-tech equipment to run its manufacturing industries, which employ about 10 percent of its workers. Brazil has one of the largest steel plants in Latin America and is a world leader in automobile production.

CONNECT Geography & Economics

Deforestation More than half of the Amazon rain forest is located in Brazil. As Brazil's economy developed, many people moved to the Amazon region to develop its resources. Several economic activities have contributed to deforestation, including:

- logging by the timber industry to export woods such as mahogany
- clearing the forests for raising crops and livestock and for housing
- mining minerals for export and as raw materials

CRITICAL THINKING

Make Inferences How might the opportunity to make money contribute to deforestation?

Slash-and-Burn Agriculture Trees in the Amazon rain forest are burned to clear the land for farming and ranching.

DIFFERENTIATING INSTRUCTION

English Learners

Key Academic Vocabulary

Have students work in pairs to locate key academic words (or their variations) that describe Brazil's agriculture. *(mechanized, manual, primary).* Invite volunteers to use these words to make their own statements about agriculture in Brazil. To extend this activity, have students use the terms to compare and contrast agriculture in northern and southern Brazil.

Pre-AP

Connect Economics to Population

Have students discuss the relationship between population changes in Brazil in the second half of the 20th century and the segments of Brazil's economy. Suggest that they consider how urbanization is related to the growth of manufacturing and service industries and how mechanization allows Brazil to be a major agricultural exporter even with a shrinking rural population.

Service industries employ about half of Brazil's workers. Many Brazilians work for government agencies, hotels and restaurants, and stores.

Brazil is leading the way in becoming energy **self-sufficient**. Hydroelectric power from Brazil's major rivers provide most of the nation's energy needs. In fact, Brazil expects to become energy-independent in the near future.

ANALYZE CAUSES AND EFFECTS Explain why Brazil has become an important industrial nation.

azil has undant natu- resources; w high-tech uipment for nufactur- ; new fuel urces; hydro- ctricity. ▶

Preserving the Rain Forest

KEY QUESTION Why is preserving the rain forest important?

The Amazon **rain forest**, the world's largest rain forest, covers one-third of South America. In the past, rain forests covered about 14 percent of the Earth's surface. Today, because of **deforestation**, the cutting down and clearing away of trees, rain forests cover only 6 percent of the Earth's surface.

Sloth The sloth makes its home in the treetops of the rain forest. **How might deforestation affect this animal?**

It might cause sloth populations to dwindle due to lack of habitat.

Logging in the Rain Forest Although today logging in the Amazon rain forest is controlled by the Brazilian government, illegal logging continues.

Endangering Ways of Life Deforestation threatens the homes and ways of life of some of the people who live in the rain forests.

DIFFERENTIATING INSTRUCTION: TIERED ACTIVITIES

OBJECTIVE Demonstrate an understanding of the relationship between deforestation and Brazil's economy.

Basic

Have students summarize the "Connect Geography and Economics" feature by adapting it for younger students. Give students copies of the visuals from the feature and have them work in groups to write a summary and captions. Ask each group to present its work.

On Level

Have students write a letter to the editor expressing a point of view about deforestation. They may write from any perspective, for example as a member of an environmental group, a logger, or a person who lives in the rain forest. Invite students to share their letters with a partner.

Challenge

Have pairs of students write a dialogue between a logger in the rain forest and someone who wants to protect the rain forest. Suggest that students consider how ecotourism or sustainable practices might allow for a win-win solution. Invite students to perform their dialogues for the class.

Comparing Regions

Hydroelectric Power Around the World

The use of hydroelectric power varies in different regions of the world. South America as a whole gets around 70 percent of its electric power from water, with Brazil and Paraguay getting more than 90 percent of their electricity from this source. Most of Brazil's hydroelectric plants are located in the south and southeast where its major cities and industries are located.

Although the United States produces more electricity from waterpower than Brazil does, it amounts to only about 10 percent of total electric power production in the United States.

Teach

Preserving the Rain Forest

Talk About It

- How has deforestation affected the size of the rain forest? *(It has decreased it from 14 percent of the Earth's surface to 6 percent.)*
- Why do some people organize boycotts of products that destroy the rain forest? *(Possible answer: to discourage producers of those products by hurting them economically)*
- **Find Main Ideas** How does government protection of parts of the rain forest help prevent deforestation? *(Possible answer: It keeps people from cutting down trees for logging, agriculture, or mining.)*

Unit Resource Book: Latin America

- Skillbuilder Practice: Drawing Conclusions, p. 226
- Interdisciplinary Projects: Area in South America, pp. 233, 235

CONNECT to the Essential Question

How have Brazil's people used the country's abundant natural resources to make Brazil an economic giant?

Ask students what they have learned so far that can help them answer this question. Students might mention:

- Brazil produces and exports a variety of agricultural products.
- Mineral deposits and hydroelectric power from rivers support the growth of Brazil's industries.
- Brazilians are working to save the rain forest and use its resources wisely.

Unit Transparency Book: Latin America
- Essential Question Graphic, TT32

4 Assess & Reteach

Assess Have students complete the Section Assessment.

Unit Resource Book: Latin America
- Section Quiz, p. 241

Interactive Review @ClassZone.com
- Section Quiz

Power Presentations

Test Generator

Reteach Have students make up a five-question quiz about Brazil's resources and economy. Have them exchange papers and complete each other's quizzes. Each student should correct his or her quiz by reviewing the text.

Unit Resource Book: Latin America
- Reteaching Activity, p. 244

Preserving the Brazilian Rain Forest These children are planting trees to reverse the effects of deforestation.

Why Save It? You read earlier in this book that the rain forest is an important resource. Rain forests help to regulate the world's climate by absorbing carbon dioxide and producing oxygen. As trees are cut down, less carbon dioxide is absorbed, which contributes to **global warming**. Rain forests are home to millions of species of plants and animals, more than half of the world's species. Many medicines are made from rain-forest plants. Destroying rain forests endangers the traditional ways of life of some of the people who live there.

What Is Being Done? Many governments, groups, and individuals are working to preserve the rain forests. Recently, Brazil's government presented plans to declare about 193,000 square miles of the Amazon rain forest a protected area. Some groups raise money to buy land to create large rain forest reserves. Some organize protests against plans that might result in deforestation. Others organize boycotts of products that destroy the forests. To **boycott** is to stop buying or using certain products as a form of protest. Still others work to educate people about the importance of rain forests to the world.

SUMMARIZE Explain why preserving rain forests is important.
Rain forests regulate climate; provide plant and animal habitats, medicines, and human habitats

ONLINE QUIZ For test practice, go to Interactive Review @ClassZone.com

Section 3 Assessment

TERMS & NAMES

1. **Explain the importance of**
   - hydroelectricity
   - ethanol
   - deforestation
   - boycott

USE YOUR READING NOTES

2. **Find Main Ideas** Use your web diagram to answer the following question:
   What industry employs half of Brazilian workers?

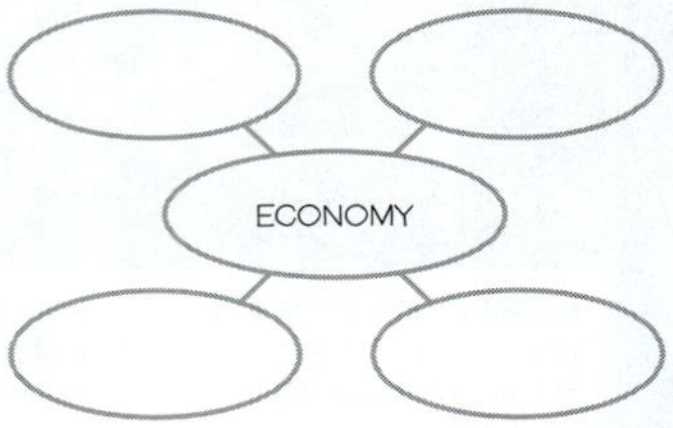

KEY IDEAS

3. What is Brazil's major export crop?
4. Where does most farming take place in Brazil?
5. How is Brazil becoming energy self-sufficient?

CRITICAL THINKING

6. **Analyze Causes and Effects** What effect has rain-forest deforestation had on the world's climate?
7. **Identify Problems and Solutions** What do you think is the best solution for curbing the deforestation of rain forests?
8. **CONNECT to Today** Brazil is becoming energy-independent. What effect might that have on Brazil's economy?
9. **LANGUAGE ARTS** **Write a Slogan** Imagine that you are working for an organization dedicated to preserving the rain forest. Write a slogan for the organization to use to educate people about the reasons that preserving rain forests is important.

SECTION 3 ASSESSMENT ANSWERS

Terms & Names

1. hydroelectricity: electric power generated by water; ethanol: a liquid formed when sugar cane is converted to alcohol to be used for fuel; deforestation: the cutting and clearing away of trees; boycott: to stop buying and using products from certain sources as a way of protest

Use Your Reading Notes

2. the service industry

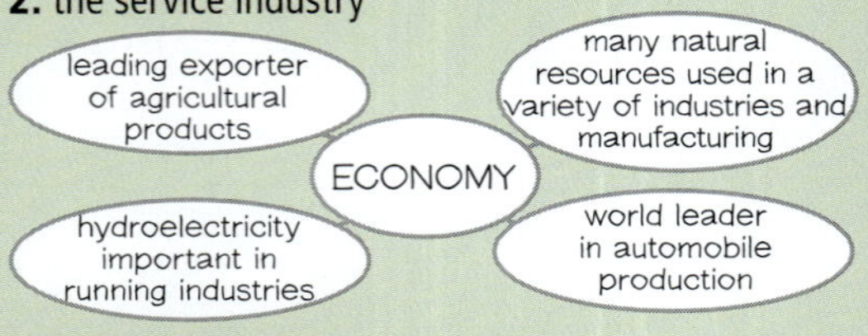

Key Ideas

3. Soybeans are Brazil's primary farm export.
4. in southern and southeastern parts of Brazil
5. Brazil is becoming energy-self-sufficient by making cars that run on ethanol and by producing its own hydroelectric power.

Critical Thinking

6. The lack of forests to absorb carbon dioxide has led to an increase in global warming.
7. Answers will vary. Students should indicate the solution they think will be most effective in curbing deforestation.
8. Answers will vary, but students should indicate that Brazil might not have to import oil from other countries to meet its energy needs.
9. Use the rubric to score slogans.

Slogan Rubric

	Message of Slogan	Effectiveness of Slogan
4	evidence of excellent understanding	effective
3	evidence of good understanding	good
2	evidence of some understanding	satisfactory
1	little evidence of understanding	ineffective

CHAPTER SUMMARY

Key Idea 1
Brazil's government is dealing with the problems resulting from urbanization.

Key Idea 2
Several cultures have influenced Brazil's unique culture.

Key Idea 3
An abundance of natural resources helps make Brazil a major industrial nation.

For **Review and Study Notes**, go to **Interactive Review** @ClassZone.com

NAME GAME

Use the Terms & Names list to complete each sentence on paper or online.

1. I am the capital of Brazil. Brasília
2. I am a community created by escaped African slaves. ________
3. I am the first emperor of Brazil. ________
4. I am a government official who helped abolish slavery in Brazil. ________
5. I am the city famous for Carnival. ________
6. I am the music of Brazil that is best known worldwide. ________
7. I am a neighborhood located in areas surrounding big cities. ________
8. I am a cleaner, renewable fuel for cars. ________
9. I am a religious practice brought from Africa to Brazil. ________
10. I destroy millions of acres of forest. ________

Brasília
Candomblé
capoeira
cuíca
deforestation
Dom Pedro I
Dom Pedro II
ethanol
favela
quilombo
Rio de Janeiro
samba

Activities

Flip Cards

Use the online flip cards to quiz yourself on the terms and names introduced in this chapter.

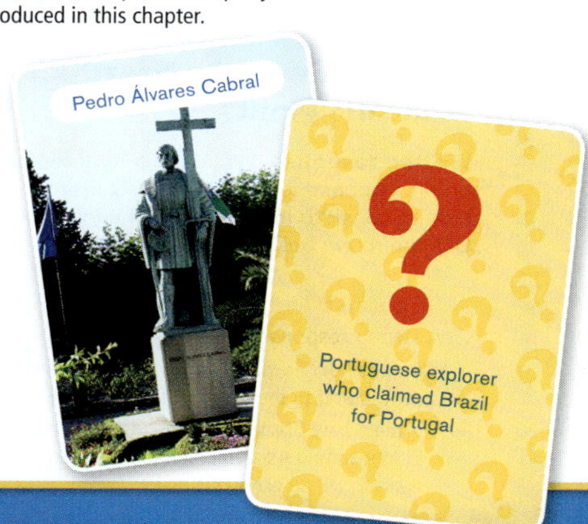

Crossword Puzzle

Complete an online puzzle to test your knowledge of Brazil's history, economy, and culture.

ACROSS
1. friction drum used in the samba

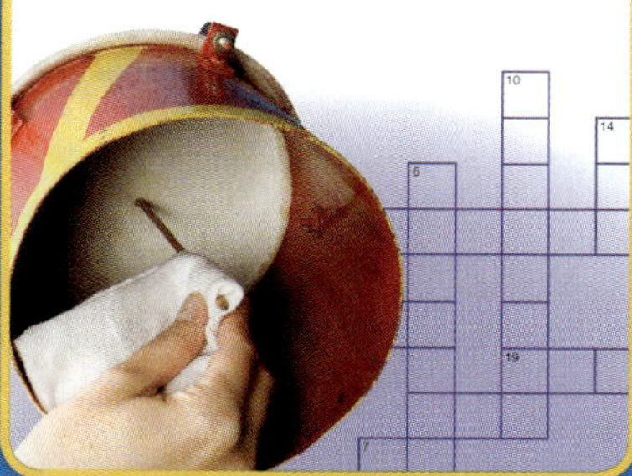

ONLINE AND ON-SCREEN ACTIVITIES

GeoGame

Vocabulary Flip Cards are virtual flash cards at **ClassZone.com.** Students can quiz each other or work independently to study the flip cards for the chapter, "Brazil."

- They can quiz themselves on key terms and names by reviewing definitions.
- Demonstrate how to advance and shuffle the deck for another challenge.

Crossword Puzzle

Students can review important information about Brazil's history, economy, and culture by completing this crossword puzzle online at **ClassZone.com.**

- As an extension of this activity, have students use the *Activity Maker* at **ClassZone.com** to make their own crossword puzzle for one section of the chapter.
- Then have them trade puzzles with classmates to solve each other's puzzles.

Chapter Summary

As a follow-up to the chapter, encourage students to pair up and go over the **Review and Study Notes** at **ClassZone.com.**

Alternatively, students pairs can review their Reading Study Guide notes, making sure that they can add important details to each section's "Key Idea."

Name Game

Teaching Tips

ON PAPER Have students work in pairs to complete this activity. As an extension, they can write questions for the terms that are not covered in the items shown here.

ONLINE Make sure that students understand how to drag and drop terms to fill in blanks. Ask them to print out their results and compare them with those of a classmate.

ANSWERS

1. Brasília
2. *quilombo*
3. Dom Pedro I
4. Dom Pedro II
5. Rio de Janeiro
6. samba
7. *favela*
8. ethanol
9. *Candomblé*
10. deforestation

Teacher-Tested Activities

Esther Howse, Byron Center West Middle School, Byron Center, Michigan

To review what they have learned about a particular country, my students make an "Acrostic Poem." For example, an acrostic about Brazil using ***RED*** might be ***R**io de Janeiro*, ***E**nergy self-sufficient*, ***D**ams provide electricity*. Have students make an acrostic using the word ***BRASÍLIA***. You might provide a word bank for students to use to make their acrostic, or they might complete the acrostic on their own. Have students share their acrostics with one another.

Vocabulary ANSWERS

1. treaty that gave Portugal control over land that is now eastern Brazil
2. Portuguese explorer who claimed Brazil for Portugal
3. first emperor of Brazil; declared Brazil independent from Portugal
4. emperor of Brazil; abolished slavery there
5. current capital city of Brazil
6. capital city until 1960
7. electric power generated by water; runs most of Brazil's industries
8. a liquid formed when sugar cane is converted to alcohol to be used for fuel
9. the cutting and clearing away of trees occurring in Amazon rain forest
10. to stop buying and using products from certain sources as a way of protest
11. All are related to Africans in Brazil.
12. Brazilian dances

Key Ideas ANSWERS

13. In 1500, Pedro Álvares Cabral claimed Brazil for Portugal.
14. Dom Pedro I declared Brazil independent from Portugal in 1822.
15. bridging the income gap between the rich and poor
16. Between the 1950s and 2000, the urban population more than doubled.
17. Brazilians speak Portuguese as the official language.
18. in São Paulo
19. African influences can be seen in Brazilian food, music, dances, instruments, and religion.
20. Carnival combines African and European music, dances, and instruments to celebrate before the Christian season of Lent.
21. The abundance of natural resources has helped to develop Brazilian industries, thereby making Brazil an economic power.
22. because this part of Brazil has the best soil and climate for agriculture and ranching
23. hydroelectricity
24. They have raised money to create rain-forest reserves, protest plans that might result in deforestation, organize boycotts of products that destroy the rain forest.

Critical Thinking ANSWERS

25. Possible causes: logging, clearing forests for crops, mining

CHAPTER 10 Assessment

VOCABULARY

Explain the significance of each of the following.

1. Treaty of Tordesillas
2. Pedro Álvares Cabral
3. Dom Pedro I
4. Dom Pedro II
5. Brasilia
6. Rio de Janeiro
7. hydroelectricity
8. ethanol
9. deforestation
10. boycott

Explain how the terms and names in each group are related.

11. *Candomblé*, *cuíca*, and *quilombos*
12. samba and *capoeira*

KEY IDEAS

1 From Portuguese Colony to Modern Giant

13. What explorer claimed Brazil for Portugal?
14. How did Brazil become independent?
15. What is Brazil's biggest economic challenge?
16. How has Brazil's urban population changed in the last half of the 20th century?

2 A Multicultural Society

17. What Portuguese influences are seen in Brazil today?
18. Where does the largest Japanese population live in Brazil today?
19. How have Africans influenced Brazilian culture?
20. Why is Carnival considered a blend of European and African traditions?

3 Developing an Abundant Land

21. How has having an abundance of natural resources benefited Brazil?
22. Why does most farming and ranching in Brazil take place in the southern and southeastern parts?
23. What provides most of Brazil's energy needs?
24. What actions are people taking today to help preserve the rain forest?

CRITICAL THINKING

25. **Analyze Causes and Effects** Create a diagram to explain the causes of deforestation of the Amazon rain forest.

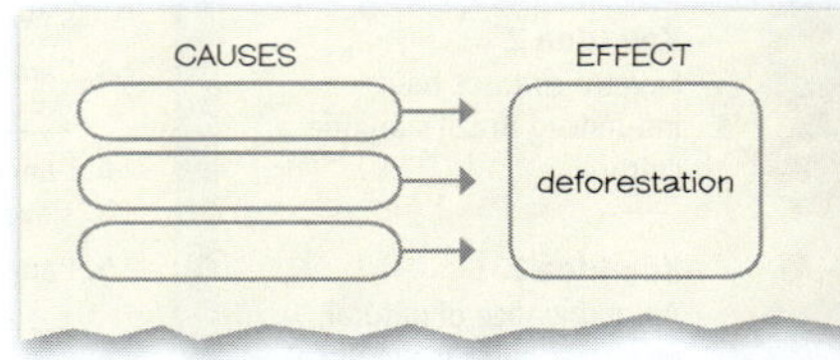

26. **Evaluate** How did the introduction of coffee plants in Brazil affect its economy?
27. **Summarize** What did Brazil's government do to try to relieve overcrowding in its cities?
28. **Five Themes: Human-Environment Interaction** How might deforestation affect the Indians of the Amazon rain forest?
29. **Connect to Economics** Why is narrowing the income gap important for Brazil's government to address?
30. **Connect Geography & History** What historical events have contributed to Brazil's cultural diversity?

Answer the ESSENTIAL QUESTION

How have Brazil's people used the country's abundant natural resources to make Brazil an economic giant?

Written response Write a two- or three-paragraph response to the Essential Question. Be sure to consider the key ideas of each section. Use the rubric below to guide your thinking.

Response Rubric
A strong response will:
- include how Brazil's natural resources have contributed to the growth of industries
- explain how Brazilian Indians, Africans, and European and Asian immigrants have helped make Brazil an economic power

ESSENTIAL QUESTION REVIEW

Before students begin their essays, have them review the key chapter elements that will help them answer the Essential Question. Direct them to significant features in the chapter, such as

- Comparing U.S. & Brazilian Governments, Section 1
- Comparing Urban and Rural Populations, Section 1
- Analyzing Primary Sources, Section 2
- Connect History & Culture: African Influence on Brazilian Food, Section 2
- Connect Geography and Economics, Section 3

Written Response Rubric

	Coverage of Key Ideas	Explanation	Errors in Mechanics
4	all	clear and concise	few or none
3	most	good	some
2	some	fair	many
1	one or none	poor	many major

STANDARDS-BASED ASSESSMENT

Test Practice
- Online Test Practice @ClassZone.com
- Test-Taking Strategies and Practice at the front of this book

PRIMARY SOURCE

Use the primary source below to answer questions 1 and 2.

> It is not just those who depend directly on the tropical forests who suffer from deforestation, but the entire population of tropical forest countries. The forests assist in the regulation of local climate patterns, protecting watersheds, preventing floods, guaranteeing and controlling huge flows of life-giving water. Strip away the forests and there is, first, too much water . . . and then too little.
>
> Source: Charles, Prince of Wales, from a speech given on February 6, 1990

1. What observation does the passage support?

- **A.** Deforestation hurts no one.
- **B.** Deforestation prevents flooding.
- **C.** Rain forests help to prevent floods.
- **D.** Rain forests affect only a few people.

2. What is the best title for this passage?

- **A.** Prince Charles Helps the Rain Forests
- **B.** Where Are the Rain Forests?
- **C.** Deforestation–Is It Serious?
- **D.** The Rain Forest and Water Supply

LINE GRAPH

Use the graph below to answer questions 3 and 4.

Population in São Paulo and Rio de Janeiro *1970–2000*

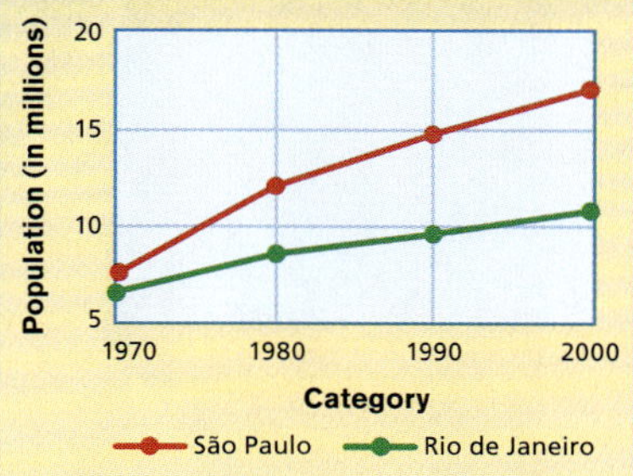

Source: United Nations Department of Economic and Social Affairs/Population Division, *World Urbanization Prospects: The 2003 Revision*

3. What was the population of Rio de Janeiro in 2000?

4. Between what years did São Paulo's population increase the most?

GeoActivity

1. INTERDISCIPLINARY ACTIVITY–SCIENCE

Research the four layers of a tropical rain forest. Find out the differences in the four layers, the kinds of plants that grow there, and the kinds of animals that live in each layer. Include a diagram to illustrate the four layers.

2. WRITING FOR SOCIAL STUDIES

Unit Writing Project Decide on the place from your journal that you found most interesting, and find more information about it. Then write an article for a travel magazine describing the place and giving reasons for people to visit it.

3. MENTAL MAPPING

Create an outline map of Brazil and label the following:

- Bahia
- Salvador
- São Paulo
- Brasília
- Rio de Janeiro
- The Amazon River

DIFFERENTIATING CHAPTER ASSESSMENT

Unit Resource Book

- **Chapter Test A,** pp. 245–249. *Use for Struggling Readers, Beginning-to-Intermediate-Level English Learners, and for Inclusion of students whose IEP suggests supported test-taking.*
- **Chapter Test B,** pp. 250–254. *Use for On-Level Learners and ESOL students with Advanced-Level English.*
- **Chapter Test C,** pp. 255–259. *Use for Gifted and Talented students or students assigned Pre-AP work.*

Test Generator

- Chapter Tests A, B, and C, in English and Spanish

McDougal Littell Assessment System

TEST **Test Generator** Use three levels of tests in English and Spanish tied to state standards. Plus, choose questions to customize tests for all populations.

SCORE Scan and Score plain paper student tests in minutes.

REPORT Create instant progress reports, sortable by student, by standard, or other criteria.

RETEACH Respond immediately to students' needs by using a wealth of reteaching resources.

26. After its introduction, coffee became Brazil's chief export; Brazil became a leading world coffee producer.

27. gave land to people who moved inland

28. Answers will vary, but might include that deforestation would reduce the resources the Indians use to maintain their traditional way of life, which would threaten their lifestyle.

29. Possible answer: It is important for the government to address the income gap because although Brazil has a strong economy, most Brazilians are poor and do not have jobs or their jobs are low-paying. This results in many Brazilians living in poverty.

30. Brazil was first colonized by the Portuguese, who established Brazil's large sugar plantations. At first the Portuguese forced the Indians to work the fields. Then they brought Africans to work as slaves. After independence, Europeans from Italy, Spain, Portugal, and Germany came to Brazil to work in the coffee industry. In the early 1900s Japanese immigrants also arrived to work on the coffee and tea farms. The descendants of all these people have contributed to Brazil's cultural diversity today.

Standards-Based Assessment ANSWERS

1. C

2. D

3. around 11 million

4. between 1970 and 1980

GeoActivity

1. INTERDISCIPLINARY ACTIVITY RUBRIC

The diagram should:

- indicate the differences in the four layers
- indicate the kinds of plants and animals in each layer
- illustrate the four layers

2. WRITING FOR SOCIAL STUDIES RUBRIC

The article should:

- clearly describe the features of the chosen place
- provide reasons why visitors would enjoy visiting it

3. MENTAL MAPPING RUBRIC

The outline map should:

- represent the approximate shape of Brazil
- include all geographic features listed in their correct location

N
E
W
S

UNIT 4

Europe

Why It Matters:

In the past, Europeans used the oceans and seas to make voyages for trade and to build empires. Their culture spread around the world. Today, Europeans still play a large part in world affairs.

The Alps

CHAPTER 11 Europe: Physical Geography and History

Notre Dame Cathedral

CHAPTER 12 Western Europe

London Street Scene

CHAPTER 13 United Kingdom

Prague, Czech Republic

CHAPTER 14 Eastern Europe

313

Objectives

- Identify the locations and countries of Europe
- Compare the differences in population density between Europe and the United States
- Describe the varied climates found in Europe
- Identify the major economic activities in Europe

Comparing Data

Although the landmass of Europe is smaller than that of the United States, the region has nearly twice the population. Remind students that until fairly recently, most immigrants to the United States came from Europe.

Throughout its history, Europe's rivers and mountains have served as natural boundaries as well as transportation routes and suppliers of resources. Europe's rivers, which are all shorter than the longest rivers in the United States, continue to play major roles in the continent's economic and cultural life today.

ANSWERS

1. Europe is smaller, by almost 900,000 square miles.
2. bigger; more densely populated
3. The Danube is 581 miles shorter than the Mississippi.
4. Mont Blanc; shorter

Unit 4 Atlas: Europe

Europe occupies the western portion of the Eurasian landmass. Many people view the Ural Mountains as the eastern border of Europe, which means that Europe includes part of Russia. However, for historic and cultural reasons, Russia and the Eurasian republics are not considered in this unit.

As you study the graphs on this page, compare the landmass, population, rivers, and mountains of Europe with those of the United States and the world. Then jot down the answers to the following questions in your notebook.

Comparing Data

1. How does Europe compare in size to the United States?
2. Is Europe's population bigger or smaller than that of the United States? Given what you know about Europe's size, do you think that makes Europe more or less densely populated than the United States?
3. How does the Danube River compare to the Mississippi River in the United States?
4. What is the tallest peak in Europe? Is it taller or shorter than Mt. McKinley, the tallest mountain in the United States?

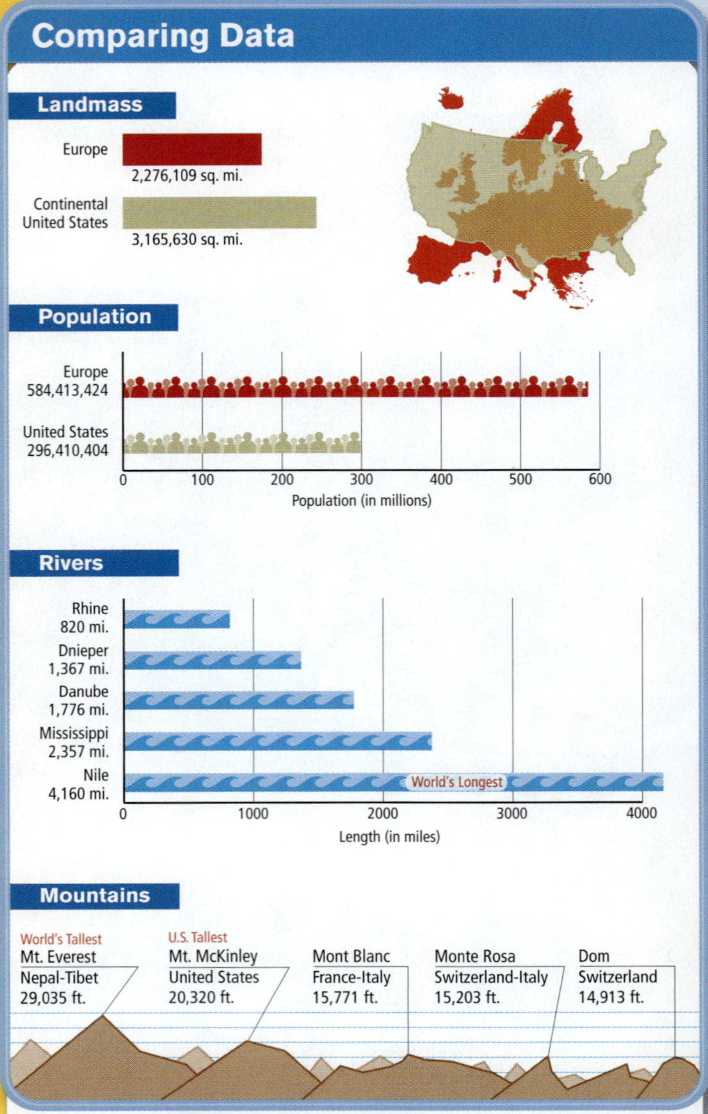

FIVE THEMES AND SIX ELEMENTS

Five Themes

Location

Have students consider the most densely populated parts of Europe. Ask volunteers to describe these locations in relative and absolute terms.

Six Elements

Places and Regions

Explain that some parts of Europe, including the United Kingdom and northern Europe, are separated from continental Europe by narrow channels of water. The island nation of Iceland, near the Arctic Circle, is considered part of far northern Europe.

Population Density

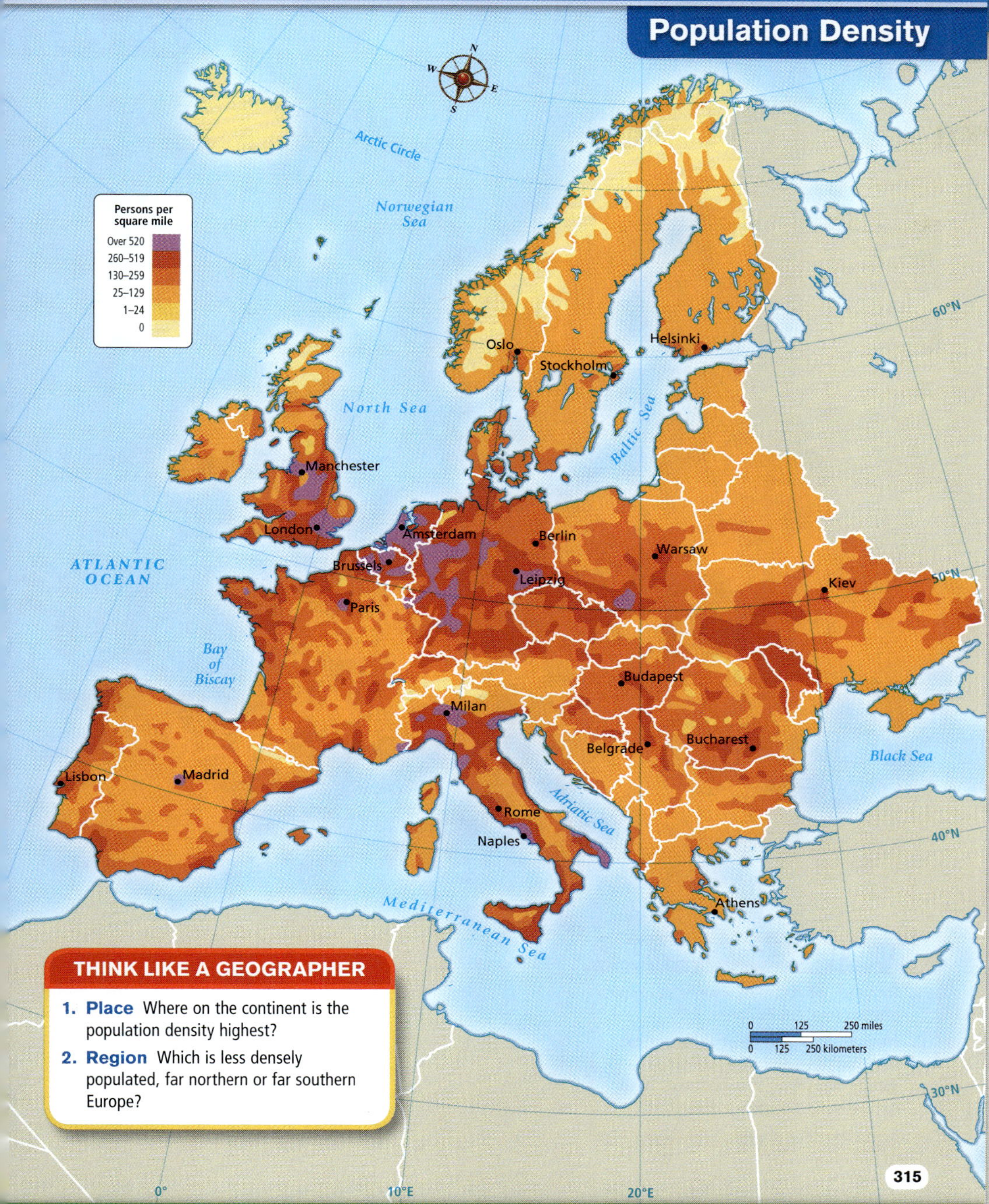

THINK LIKE A GEOGRAPHER

1. **Place** Where on the continent is the population density highest?
2. **Region** Which is less densely populated, far northern or far southern Europe?

Population Density

Much of Europe is densely populated. Invite students to make observations about Europe's population, based on that point. For example, they may note that the yellow area above Italy on the map is the Alps, which are generally not suited for human settlements.

- Which cities on the map are located in areas that have population densities of over 520 persons per square mile? *(Amsterdam, Brussels, Leipzig, London, Madrid, Manchester, Milan, Naples, Paris)*
- Why do you think northern Europe is less densely populated than the rest of Europe? *(Possible Answers: Its separation from continental Europe means it is geographically isolated; locations near the Arctic Circle have a harsher climate.)*

THINK LIKE A GEOGRAPHER ANSWERS

1. **Place** north-central Europe, mainly in Germany, Belgium, the Netherlands, and the United Kingdom
2. **Region** far northern Europe—Scandinavia and Iceland

More About . . .

London

London, the capital city of the United Kingdom, is one of the largest metropolitan areas in Europe. The population of the city itself is over 7 million people. When the suburban areas surrounding the city are included, the population increases to over 11 million. London consists of many distinct streets and neighborhoods and is ethnically and culturally diverse.

DIFFERENTIATING INSTRUCTION

English Learners

Comprehension: Self-Monitor

To help with comprehension of "persons per square miles" in the map legend, have students replace *per* with "for each" or "in each." Explain that *per* is a preposition—it helps explain the relationship between *persons* and *square mile.* Have students name other prepositions. *(at, by, in, to, from)* Then have them use the map to make a sentence with "persons per square mile" or another preposition.

Inclusion

Graphing Population Density

Students who are visually impaired may have difficulty interpreting maps such as the one on this page. To help them more easily compare the differences among European regions, pair them with other students. Have them work together to create bar or pie graphs from the map or from population data available in encyclopedias, almanacs, or government Web sites (such as www.cia.gov/cia/publications/factbook).

Climate

Europe has many different climate zones. The two largest climate zones are marine west coast in central and western Europe and humid continental in northern and eastern Europe. Note northern Europe's mix of climates, compared to southern Europe's largely Mediterranean climate.

- Which climate zones border the Black Sea? *(semiarid, Mediterranean, humid continental)*
- Note that part of northern Europe has a subarctic climate. Why do you think a large part of Iceland—located at about the same latitude—has a marine west coast climate? *(Possible Answer: Much of the area of northern Europe with a subarctic climate is farther away from the sea than the area of Iceland with a marine west coast climate.)*

THINK LIKE A GEOGRAPHER ANSWERS

1. Region Mediterranean

2. Human-Environment Interaction marine west coast

More About . . .

The North Sea

The North Sea is a northeastern extension of the Atlantic Ocean. The United Kingdom, Denmark, Norway, and Germany are some of the countries that border it. In 1970, a major oil field was discovered under the North Sea's floor. However, because the sea is more than 2,000 feet deep in places and often has severe weather, it was not until the 1980s and 1990s that the oil began to be extracted.

Climate

THINK LIKE A GEOGRAPHER

1. **Region** What climate zone occurs most often in southern Europe?
2. **Human-Environment Interaction** Compare this map to the population density map on the previous page. What climate is most densely populated?

FIVE THEMES AND SIX ELEMENTS

Five Themes

Region

Remind students that the term *Arctic Circle* refers to an imaginary line that lies south of the North Pole. It marks the northernmost point at which the sun can be seen during the winter solstice. During the summer solstice, the sun still can be seen overhead at the Arctic Circle at midnight.

Six Elements

The World in Spatial Terms

Explain that the maps and data found in this Unit Atlas can provide a lot of insight into each European country. For example, by comparing the population density map and the climate map, one can see that, in addition to the marine west coast climate zone, the Mediterranean climate zone is also one of Europe's most densely populated areas.

Europe

Economic Activity

THINK LIKE A GEOGRAPHER

1. **Human-Environment Interaction** What activity is the majority of the land in Europe used for?
2. **Region** Which sea is the source of many of Europe's energy resources? What countries are most likely to benefit from these resources?

DIFFERENTIATING INSTRUCTION

Struggling Readers

Compare Economies

To help students synthesize the information about the economic activity of Europe, review the land use terms with them. Then have students compare the similarities and differences in economic activities between two European countries. To extend this activity, ask them to explain differences in land use in terms of climate.

Pre-AP

Analyze the Service Economy

The economies of European countries also depend on service industries such as banking and tourism. Have pairs of students research service industries in a northern European country and a central or southern European country. Then have them analyze the importance of service industries to each country's overall economy, as well as compare and contrast service industries between the two countries.

Economic Activity

Point out that much of the land in central and southern Europe is rich in fertile farmland. Likewise, northern Europe is rich in forestland and fishing stocks. However, Europe's energy resources—coal, natural gas, and oil—are not evenly spread out across the continent.

- In addition to commercial agriculture, how is land used in Spain and Portugal? *(livestock raising, dairying)*
- What is the main economic resource in the United Kingdom? *(coal)*
- How does land use in Switzerland differ from land use in Iceland? *(Dairying is the main land use in Switzerland; livestock raising is the main use in Iceland.)*

THINK LIKE A GEOGRAPHER ANSWERS

1. **Human-Environment Interaction** commercial agriculture
2. **Region** North Sea; United Kingdom, Netherlands, Germany, Denmark, and Norway

Comparing Regions

Economic Activity in Finland and Greece

Both Finland, in northern Europe, and Greece, in southern Europe, used to have economies primarily based on agriculture. However, Finland modernized its economy during the 20th century to include manufacturing and trade. Greece attempted a similar change, but was less successful.

In 2005, agriculture accounted for about 8 percent of Greece's economy, but less than 2 percent of Finland's economy. Finland has become a major manufacturer of electronics, including cell phones. Greece manufactures metal products and construction materials, among other items. As coastal countries, both also have strong fishing industries.

Regional Overview

Despite its small land area, Europe has a large, diverse population and a thriving, modern economy. European art, politics, science, and other cultural influences have spread throughout the world. The European Union brings more than half of Europe's nations together, further helping to strengthen Europe's role in the world.

- How far north does Europe's geography extend? *(to the Arctic Ocean)*
- What are two types of government found in Europe? *(democracy, monarchy)*
- How might the European Union help European countries increase trade? *(Possible Answer: Having a common currency, as the European Union does, might make it easier for trade to occur within Europe and with other nations.)*

More About . . .

The Leaning Tower of Pisa

The Leaning Tower of Pisa is a famous marble bell tower in Pisa, Italy. The tower was intended to stand upright when construction began in the late 12th century. It soon began leaning to one side, most likely because of unstable soil underneath it. Many efforts have been made over the centuries to preserve the tower, which continues to sink and tilt. A recent attempt to remove soil from under the tower was successful, and the tower is now considered to be stable for the next 300 years.

UNIT 4 ATLAS Regional Overview

Europe

Europe is the world's second smallest continent in area, but one of the largest in population. The population is diverse, with many different cultures developing on the small landmass. The chapters in this unit provide more information about the geography, history, culture, government, and economics of Europe.

GEOGRAPHY

Europe extends from the Arctic Ocean in the north to the Mediterranean Sea in the south. Geographically, the European continent stretches from the Atlantic Ocean all the way to the Ural Mountains in Russia, but in this unit, we will mark Europe's western border where Russia begins.

HISTORY

For centuries, different groups of people settled, lived in, and fought over the lands of Europe. Though some regions are still troubled by conflict, Europe is more unified than ever before.

CULTURE

Since the time of the ancient Greeks and Romans, Europe's culture has had a global influence. European ideas about politics, science, art, philosophy, and religion have spread around the world.

GOVERNMENT

The governments of Europe come in all shapes and sizes. Some are democracies. Other nations have monarchs who govern alongside a parliament chosen by the people.

ECONOMICS

Europe has many natural resources and abundant farmland that strengthen its economies. Many European countries are also highly industrialized. The European Union has worked to continue Europe's role as an economic power.

Leaning Tower of Pisa Construction started on this Italian monument in 1173. It began to tilt when only three of its eight stories were completed.

FIVE THEMES AND SIX ELEMENTS

Five Themes

Movement

Explain to students that a railroad tunnel under the English Channel, known as the Channel Tunnel, links the United Kingdom and France. Before the Channel Tunnel, people and services could only cross the English Channel by air or boat. The Channel Tunnel opened in 1994.

Six Elements

Human Systems

Have students find the location of Athens on a map or globe. Ask them to explain why Athens seemed to ancient Greeks to be a good location for a city, and describe ways that the city might have transformed the culture of ancient Greece.

UNIT 4 ATLAS Country Almanac

Click here to compare the most recent data on countries @ ClassZone.com

Unit Writing Project

Imagine that your class is planning a trip to Europe. Choose a country from this unit that you would like to visit. Write a persuasive speech to convince your classmates to pick your country for their trip.

Think About:

- specific sites you would visit
- the people you would meet
- how it differs from your home

Europe

Europe is made up of 43 different countries.

Albania

GEOGRAPHY
Capital: Tiranë
Total Area: 11,100 sq. mi.
Population: 3,130,000

ECONOMY
Imports: food; machinery; minerals; clothing
Exports: clothing; metals

CULTURE
Language: Albanian
Religion: Muslim 39%; Catholic 17%

Andorra

GEOGRAPHY
Capital: Andorra la Vella
Total Area: 181 sq. mi.
Population: 70,549

ECONOMY
Imports: food; tobacco; machinery
Exports: motor vehicles; photo equipment

CULTURE
Language: Catalan
Religion: Catholic 89%; nonreligious 5%

Austria

GEOGRAPHY
Capital: Vienna
Total Area: 32,382 sq. mi.
Population: 8,189,000

ECONOMY
Imports: machinery; vehicles; chemicals
Exports: transportation equipment; steel

CULTURE
Language: German
Religion: Catholic 75%; nonreligious 9%; Protestant 5%

Belgium

GEOGRAPHY
Capital: Brussels
Total Area: 11,787 sq. mi.
Population: 10,419,000

ECONOMY
Imports: machinery; medicine; food
Exports: machinery; vehicles; medicine

CULTURE
Language: Dutch; French; German
Religion: Catholic 88%; Muslim 3%

Belarus

GEOGRAPHY
Capital: Minsk
Total Area: 80,155 sq. mi.
Population: 9,755,000

ECONOMY
Imports: petroleum; chemicals; food
Exports: food; petroleum; road vehicles

CULTURE
Language: Belarusian; Russian
Religion: Belarusian Orthodox 32%; Catholic 18%

Bosnia and Herzegovina

GEOGRAPHY
Capital: Sarajevo
Total Area: 19,741 sq. mi.
Population: 3,907,000

ECONOMY
Imports: machinery; chemicals; fuels
Exports: metals; clothing; wood products

CULTURE
Language: Bosnian
Religion: Sunni Muslim 43%; Serbian Orthodox 30%; Catholic 18%

Bulgaria

GEOGRAPHY
Capital: Sofia
Total Area: 42,823 sq. mi.
Population: 7,726,000

ECONOMY
Imports: textiles; crude petroleum; plastics
Exports: clothing; metals; mineral fuels

CULTURE
Language: Bulgarian
Religion: Bulgarian Orthodox 72%; Sunni Muslim 12%

Croatia

GEOGRAPHY
Capital: Zagreb
Total Area: 21,831 sq. mi.
Population: 4,551,000

ECONOMY
Imports: machinery; metals; petroleum
Exports: chemicals; clothing; petroleum

CULTURE
Language: Croatian
Religion: Catholic 89%; Eastern Orthodox 6%; Sunni Muslim 2%

Unit Writing Project

Teaching Tips

- As students prepare their speeches encourage them to consult family-friendly travel guides from the library or the Internet. This will help them come up with examples of fun and interesting places to see and things to do to persuade their classmates.
- Provide students with an example of a persuasive speech, or have them research typical formats.
- Encourage students to create props and visuals to go with their speech. Remind them that photographs and other visuals will help to illustrate their key points as well as make their speech more engaging.
- Encourage each student to work with a partner to practice presenting his or her speech and to revise the speech based on the feedback he or she receives.

INTERDISCIPLINARY ACTIVITIES

CONNECT to Art

Create a European Tower Visual

Have students research a European tower that interests them, such as the Eiffel Tower in Paris, the Leaning Tower of Pisa, or the Tower of London. Ask them to create a visual display, such as a diagram or collage, showing their tower and its location. Have them add titles and captions to complete their visual displays. Invite students to present or post their work.

CONNECT to Drama

Read a European Play

Provide students with copies of an accessible scene from an ancient Greek tragedy, such as *Antigone,* or a Shakespearean play, such as *Romeo and Juliet.* Have students read the play aloud, taking different roles. To extend the activity, have students rehearse and stage a performance of the scene.

Each map on the country cards illustrates the shape of a state, province, or country. Tell students that the maps do not reflect the comparative size of these landmasses within a particular region. To get an idea of the actual size of the state, province, or country, students should study the land area statistics provided on the card.

Country Almanac

Historically, Europe has been one of the world's most influential regions as a result of its global explorations and overseas colonization. Today, Europe continues to influence the world. It has a well-educated, healthy, productive population, and Europeans generally enjoy some of the highest standards of living in the world.

Europe, like all other regions, faces challenges. The European Union is relatively young and still evolving; its population is diverse, and countries struggle to incorporate new cultural influences into centuries-old traditions.

- Which European country has the smallest land area and population? *(Vatican City)*
- Which European country has the largest land area? *(Ukraine)*
- In which countries is Catholicism the main religion? *(Andorra, Austria, Belgium, Croatia, Czech Republic, France, Hungary, Ireland, Italy, Liechtenstein, Lithuania, Luxembourg, Malta, Monaco, Netherlands, Poland, Portugal, San Marino, Slovakia, Slovenia, Spain, Switzerland, and Vatican City)*

Each map on the country cards illustrates the shape of a state, province, or country. Tell students that the maps do not reflect the comparative size of these landmasses within a particular region. To get an idea of the actual size of the state, province, or country, students should study the land area statistics provided on the card.

UNIT 4 ATLAS

Czech Republic

GEOGRAPHY
Capital: Prague
Total Area: 30,450 sq. mi.
Population: 10,220,000

ECONOMY
Imports: machinery; chemicals; vehicles
Exports: computers; vehicles; metals

CULTURE
Language: Czech
Religion: Catholic 40%; nonreligious 32%; Protestant 3%

Denmark

GEOGRAPHY
Capital: Copenhagen
Total Area: 16,639 sq. mi.
Population: 5,431,000

ECONOMY
Imports: machinery; food; tobacco
Exports: agricultural products; swine

CULTURE
Language: Danish
Religion: Evangelical Lutheran 86%; Muslim 2%

Estonia

GEOGRAPHY
Capital: Tallinn
Total Area: 17,462 sq. mi.
Population: 1,330,000

ECONOMY
Imports: textiles; food
Exports: wood; textiles; paper

CULTURE
Language: Estonian
Religion: Orthodox 20%; Evangelical Lutheran 14%

Finland

GEOGRAPHY
Capital: Helsinki
Total Area: 130,559 sq. mi.
Population: 5,249,000

ECONOMY
Imports: machinery; mineral fuels; vehicles
Exports: paper products; wood products

CULTURE
Language: Finnish; Swedish
Religion: Evangelical Lutheran 85%; nonreligious 13%

France

GEOGRAPHY
Capital: Paris
Total Area: 211,209 sq. mi.
Population: 60,496,000

ECONOMY
Imports: machinery; chemicals; food
Exports: aircraft; perfumes; cosmetics

CULTURE
Language: French
Religion: Catholic 82%; Muslim 7%; Protestant 4%

Germany

GEOGRAPHY
Capital: Berlin
Total Area: 137,847 sq. mi.
Population: 82,689,000

ECONOMY
Imports: televisions; computers; food
Exports: televisions; medical instruments

CULTURE
Language: German
Religion: Protestant 36%; Catholic 34%; nonreligious 17%

Greece

GEOGRAPHY
Capital: Athens
Total Area: 50,942 sq. mi.
Population: 11,120,000

ECONOMY
Imports: chemicals; petroleum; ships
Exports: fruits and nuts; aluminum

CULTURE
Language: Greek
Religion: Eastern Orthodox 94%; Muslim 1%

Hungary

GEOGRAPHY
Capital: Budapest
Total Area: 35,919 sq. mi.
Population: 10,098,000

ECONOMY
Imports: machinery; vehicles
Exports: telecommunications equipment

CULTURE
Language: Hungarian
Religion: Catholic 58%; Reformed 18%; nonreligious 19%

Tour de France Each year, about 200 riders compete in this famous bicycle race.

320 Unit 4

INTERDISCIPLINARY ACTIVITIES

CONNECT to Language Arts

Describe Biking Through Europe

Point out to students that cycling is very popular in much of Europe, which is home to the world's most famous bike race, the Tour de France. Have students write a short essay that describes a scene they can visualize about a bike race or trip through the countryside of one European country.

Iceland

GEOGRAPHY
Capital: Reykjavik
Total Area: 39,769 sq. mi.
Population: 295,000

ECONOMY
Imports: machinery; food; clothing
Exports: fish; aluminum; medicines

CULTURE
Language: Icelandic
Religion: Evangelical Lutheran 87%; Catholic 2%

Ireland

GEOGRAPHY
Capital: Dublin
Total Area: 27,135 sq. mi.
Population: 4,148,000

ECONOMY
Imports: computers; electronics; food
Exports: computers; recording devices

CULTURE
Language: Irish; English
Religion: Catholic 88%; Church of Ireland (Anglican) 3%

Italy

GEOGRAPHY
Capital: Rome
Total Area: 116,306 sq. mi.
Population: 58,093,000

ECONOMY
Imports: machinery; chemicals; iron; steel
Exports: chemicals; textile yarn and fabrics; food

CULTURE
Language: Italian
Religion: Catholic 80%; nonreligious 13%

Latvia

GEOGRAPHY
Capital: Riga
Total Area: 24,938 sq. mi.
Population: 2,307,000

ECONOMY
Imports: machinery; chemicals; vehicles
Exports: wood; metals; textiles

CULTURE
Language: Latvian
Religion: Catholic 15%; Lutheran 15%; Orthodox 8%

Liechtenstein

GEOGRAPHY
Capital: Vaduz
Total Area: 62 sq. mi.
Population: 33,717

ECONOMY
Imports: machinery; glass and ceramics
Exports: precision tools; food products

CULTURE
Language: German
Religion: Catholic 80%; Protestant 8%; Muslim 3%

Lithuania

GEOGRAPHY
Capital: Vilnius
Total Area: 25,174 sq. mi.
Population: 3,431,000

ECONOMY
Imports: chemicals; clothing
Exports: mineral fuels; clothing; food

CULTURE
Language: Lithuanian
Religion: Catholic 79%; nonreligious 10%; Orthodox 5%

Luxembourg

GEOGRAPHY
Capital: Luxembourg
Total Area: 998 sq. mi.
Population: 465,000

ECONOMY
Imports: machinery; metals; chemicals
Exports: metals; transport equipment; food

CULTURE
Language: Luxemburgian; French
Religion: Catholic 91%; Protestant 2%

Macedonia

GEOGRAPHY
Capital: Skopje
Total Area: 9,781 sq. mi.
Population: 2,034,000

ECONOMY
Imports: mineral fuels; food; live animals
Exports: clothing; iron; tobacco; beverages

CULTURE
Language: Macedonian; Albanian
Religion: Orthodox 59%; Sunni Muslim 28%

Malta

GEOGRAPHY
Capital: Valletta
Total Area: 122 sq. mi.
Population: 402,000

ECONOMY
Imports: electronics; petroleum; food
Exports: clothing; children's toys and games

CULTURE
Language: Maltese; English
Religion: Catholic 95%

More About . . .

Italian Geography

The physical geography of Italy makes it one of the most scenic places in southern Europe. Much of Italy is a boot-shaped peninsula that lies between the Adriatic and Mediterranean Seas, and about 75 percent of Italy has mountains or hills. The Italian region of Tuscany is particularly known for its rolling hills and fertile soil, where olives and grapes are grown.

Through much of its history, Italy has been influenced by the Mediterranean Sea and has influenced other cultures along its shores. Italy's largest island is Sicily, once called "Trinacria" because of its triangular shape. Desired as much for its strategic location as for its agricultural value, Sicily has been invaded many times throughout history by Romans, Greeks, Phoenicians, and Allied forces during World War II.

INTERDISCIPLINARY ACTIVITIES

CONNECT to Math

Size Comparisons of European Countries

To help students better grasp the large variations in the sizes of European countries, have them use the land area data for each country to create paper rectangles that represent that country. For example, a one-inch-by-one-inch square could represent 10,000 square miles. Have each student choose ten countries to represent and arrange them from biggest to smallest.

Each map on the country cards illustrates the shape of a state, province, or country. Tell students that the maps do not reflect the comparative size of these landmasses within a particular region. To get an idea of the actual size of the state, province, or country, students should study the land area statistics provided on the card.

More About . . .

Monaco

Monaco is an exceptionally small Mediterranean country. Its land area is just one square mile and is divided into four districts. Its best-known district, Monte Carlo, is a major tourist destination, with luxurious hotels and a famous gambling casino that was built in the 1860s.

With a population of only about 32,000 people, tourism is Monaco's main industry. Its scenic views and warm beaches, along with events such as the famous auto race the Monaco Grand Prix, make it a popular tourist destination. French is the official language, but many people also speak English and Italian.

Each map on the country cards illustrates the shape of a state, province, or country. Tell students that the maps do not reflect the comparative size of these landmasses within a particular region. To get an idea of the actual size of the state, province, or country, students should study the land area statistics provided on the card.

UNIT 4 ATLAS

Republic of Moldova

GEOGRAPHY
Capital: Chişinău
Total Area: 13,067 sq. mi.
Population: 4,206,000

ECONOMY
Imports: minerals; chemicals; textiles
Exports: food; textiles

CULTURE
Language: Romanian
Religion: Orthodox 46%; Muslim 6%; Catholic 2%; Protestant 2%

Monaco

GEOGRAPHY
Capital: Monaco
Total Area: 1 square mile
Population: 32,409

ECONOMY
Imports: perfumes; clothing; publishing
Exports: plastic products; glass; paper

CULTURE
Language: French
Religion: Catholic 89%; Jewish 2%

Montenegro

GEOGRAPHY
Capital: Podgorica
Total Area: 5,415 sq. mi.
Population: 630,548

ECONOMY
Imports: no information available
Exports: no information available

CULTURE
Language: Serbian
Religion: Orthodox, Muslim, Catholic

Netherlands

GEOGRAPHY
Capital: Amsterdam
Total Area: 16,033 sq. mi.
Population: 16,299,000

ECONOMY
Imports: computers; chemicals; food
Exports: chemicals; food; mineral fuels

CULTURE
Language: Dutch
Religion: nonreligious 43%; Catholic 31%; Reformed 14%

Norway

GEOGRAPHY
Capital: Oslo
Total Area: 125,182 sq. mi.
Population: 4,620,000

ECONOMY
Imports: road vehicles; ships; metals
Exports: crude petroleum; metals; fish

CULTURE
Language: Norwegian
Religion: Church of Norway 86%; Muslim 2%, Catholic 1%

Poland

GEOGRAPHY
Capital: Warsaw
Total Area: 120,728 sq. mi.
Population: 38,530,000

ECONOMY
Imports: machinery; textiles
Exports: food; furniture; ships

CULTURE
Language: Polish
Religion: Catholic 91%; Polish Orthodox 1%

Portugal

GEOGRAPHY
Capital: Lisbon
Total Area: 35,672 sq. mi.
Population: 10,495,000

ECONOMY
Imports: telecommunications equipment
Exports: road vehicles; clothing; fabrics

CULTURE
Language: Portuguese
Religion: Catholic 87%; nonreligious 7%

Romania

GEOGRAPHY
Capital: Bucharest
Total Area: 91,699 sq. mi.
Population: 21,711,000

ECONOMY
Imports: fabrics; chemicals; petroleum
Exports: clothing; iron and steel

CULTURE
Language: Romanian
Religion: Romanian Orthodox 87%; Protestant 6%; Catholic 5%

San Marino

GEOGRAPHY
Capital: San Marino
Total Area: 24 sq. mi.
Population: 28,880

ECONOMY
Imports: electricity; gold
Exports: postage stamps; leather goods; ceramics; wine

CULTURE
Language: Italian
Religion: Catholic 89%

322 Unit 4

INTERDISCIPLINARY ACTIVITIES

CONNECT to Art

Design a Travel Advertisement

Have students create a one-page magazine advertisement promoting tourism to one of Europe's "lesser-known" destinations, such as San Marino. Ask them to combine headlines with drawn or clipped images in ways that will appeal to readers and capture their attention.

CONNECT to Economics

Charting Europe's Natural Resources

Have groups of students research the natural resources of one or more of Europe's countries. Each group should create a chart of the resources, what these resources might be used for, and an assessment of any potential damage to Europe's environment from the exploration, processing, or use of each resource.

Serbia

GEOGRAPHY
Capital: Belgrade
Total Area: 34,116 sq. mi.
Population: 9,396,400

ECONOMY
Imports: machinery; mineral fuels
Exports: food & live animals; machinery

CULTURE
Language: Serbian
Religion: Serbian Orthodox 63%; Muslim 19%; nonreligious 13%

Slovakia

GEOGRAPHY
Capital: Bratislava
Total Area: 18,859 sq. mi.
Population: 5,401,000

ECONOMY
Imports: machinery; fuels
Exports: road vehicles; machinery; metals

CULTURE
Language: Slovak
Religion: Catholic 69%; Slovak Evangelical 7%

Slovenia

GEOGRAPHY
Capital: Ljubljana
Total Area: 7,827 sq. mi.
Population: 1,967,000

ECONOMY
Imports: machinery; vehicles; fuels
Exports: medicines and pharmaceuticals; furniture

CULTURE
Language: Slovene
Religion: Catholic 84%; nonreligious 8%

Spain

GEOGRAPHY
Capital: Madrid
Total Area: 194,897 sq. mi.
Population: 43,064,000

ECONOMY
Imports: vehicles; chemicals; petroleum
Exports: fruits and vegetables; chemicals

CULTURE
Language: Castilian Spanish; Euskera (Basque); Catalan; Galician
Religion: Catholic 92%

Sweden

GEOGRAPHY
Capital: Stockholm
Total Area: 173,732 sq. mi.
Population: 9,041,000

ECONOMY
Imports: machinery; chemicals; vehicles
Exports: vehicles; electronics; medicines

CULTURE
Language: Swedish
Religion: Church of Sweden 87%; Muslim 2%; Catholic 2%

Switzerland

GEOGRAPHY
Capital: Bern
Total Area: 15,942 sq. mi.
Population: 7,252,000

ECONOMY
Imports: machinery; vehicles; food
Exports: precision instruments, watches

CULTURE
Language: French; German; Italian
Religion: Catholic 42%; Protestant 35%; Muslim 4%; Orthodox 2%

Ukraine

GEOGRAPHY
Capital: Kiev
Total Area: 233,090 sq. mi.
Population: 46,481,000

ECONOMY
Imports: natural gas; chemicals; food
Exports: metals; wood; food

CULTURE
Language: Ukrainian
Religion: Ukrainian Orthodox 29%; Catholic 7%; Protestant 4%

United Kingdom

GEOGRAPHY
Capital: London
Total Area: 94,526 sq. mi.
Population: 59,668,000

ECONOMY
Imports: radios; televisions; aircraft
Exports: computers; aircraft; petroleum

CULTURE
Language: English
Religion: Anglican 29%; nonreligious 16%; Catholic 11%

Vatican City

GEOGRAPHY
Capital: Vatican City
Total Area: 0.2 sq. mi.
Population: 921

ECONOMY
Imports: no information available
Exports: no information available

CULTURE
Language: Italian; Latin
Religion: Catholic

Majorca, Spain

323

Comparing Regions

Sweden and Switzerland

Sweden and Switzerland are two countries in Europe with populations under 10 million people. However, Sweden occupies more than ten times the land area of Switzerland. Citizens of both countries enjoy some of Europe's highest standards of living.

Sweden is a constitutional monarchy with a highly industrialized economy. In contrast, Switzerland is a federal republic with an economy largely based on international trade, especially banking. Sweden entered the European Union in 1995, while Switzerland remains outside the Union.

INTERDISCIPLINARY ACTIVITIES

CONNECT to Civics

Debating U.S.–British Relations

As a class, discuss the advantages and disadvantages of the relationship between the United States and the United Kingdom. Then ask students to take a position as either a British or U.S. government official, forming two debate teams. Have the teams prepare to debate this statement: The relationship between the United States and the United Kingdom benefits each country equally. Remind students to agree on the rules before holding the debate.

Each map on the country cards illustrates the shape of a state, province, or country. Tell students that the maps do not reflect the comparative size of these landmasses within a particular region. To get an idea of the actual size of the state, province, or country, students should study the land area statistics provided on the card.

CHAPTER 11 Europe Physical Geography and History
PLANNING GUIDE

Complete print resources all on one USB drive allow you to customize lesson plans for all students.

Section Titles & Key Ideas	Unit Resource Book Europe	Assessment Resources
1 Europe's Dramatic Landscape **KEY IDEA** Europe, too, has landforms ranging from mountains to plains. Its climate is influenced by its nearness to the ocean.	• **Reading Study Guide,** pp. 3–4 • **RSG with Additional Support,** pp. 11–12 • **Vocabulary Practice,** p. 35 • **Skillbuilder Practice,** p. 36 • **Interdisciplinary Projects,** pp. 43, 45	**Unit Resource Book** • Section Quiz, p. 49 • Reteaching Activity, p. 53 **Test Generator CD-ROM** **Daily Test Practice Transparency,** TT37
2 Classical Greece and Rome **KEY IDEA** Ancient Greek and Roman achievements in government, art and architecture, engineering, and law continue to influence Europe and the world today.	• **Reading Study Guide,** pp. 5–6 • **RSG with Additional Support,** pp. 13–14 • **History Makers: Julius Caesar,** p. 41 • **Interdisciplinary Projects,** pp. 44–45	**Unit Resource Book** • Section Quiz, p. 50 • Reteaching Activity, p. 54 **Test Generator CD-ROM** **Daily Test Practice Transparency,** TT38
3 The Middle Ages and Renaissance **KEY IDEA** Feudalism provided stability after the fall of the Roman Empire. The Renaissance marked a rebirth of creativity in Europe.	• **Reading Study Guide,** pp. 7–8 • **RSG with Additional Support,** pp. 15–16 • **Economics in the World,** p. 46 • **Connect to History,** p. 47 • **Primary Source,** p. 48	**Unit Resource Book** • Section Quiz, p. 51 • Reteaching Activity, p. 55 **Test Generator CD-ROM** **Daily Test Practice Transparency,** TT39
4 Modern European History **KEY IDEA** Revolutions in science, politics, and industry transformed Western Europe. After two world wars, European nations found new ways to cooperate.	• **Reading Study Guide,** pp. 9–10 • **RSG with Additional Support,** pp. 17–18 • **Map and Graph Skills,** p. 37 • **Outline Map Activities,** p. 39	**Unit Resource Book** • Section Quiz, p. 52 • Reteaching Activity, p. 56 • Chapter Test (Forms A, B, & C), pp. 57–68 **Test Generator CD-ROM** **Daily Test Practice Transparency,** TT40

INTEGRATED TECHNOLOGY

No Time? To focus students on important content in the chapter, use the **"Reading Study Guide with Additional Support"** in *Resources2Go.*

Teacher Presentation Options

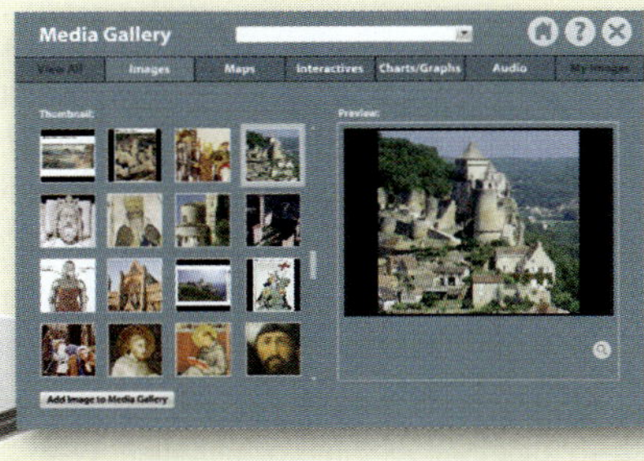

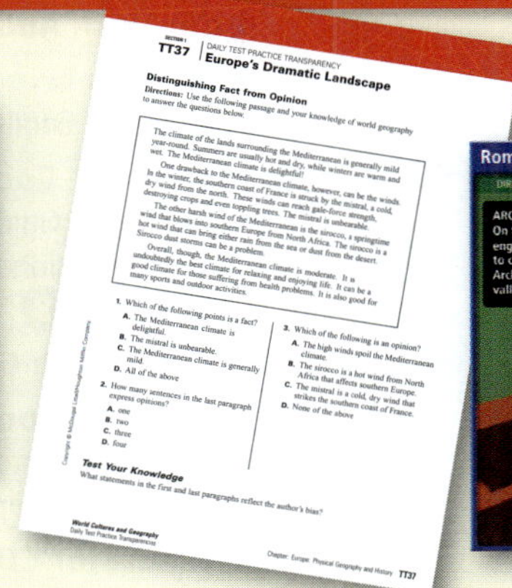

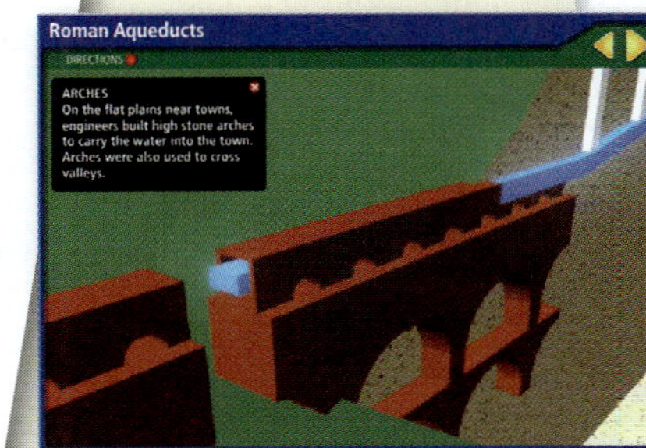

Electronic Books

eEdition DVD-ROM

eEdition Online

Presentation Toolkit

Video/DVD

- World Cultures and Geography

Power Presentations DVD-ROM

- Discussion Starters
- Media Gallery
- Review Game
- Interactive Review
- Animated Features

Unit Transparency Book

- 3-Minute Warm-Ups
- Essential Question
- Critical Thinking
- Maps
- Cultures

Daily Test Practice Transparencies

Animated Geography

- Present-Day Europe Map
- Roman Aqueducts
- European Feudalism
- Leonardo da Vinci and Renaissance Art

OnlineActivities at ClassZone.com

Animated Geography

- Videos and Photos
- 3-D Animations
- Interactive Maps

Interactive Review

- Section Quizzes
- Review and Study Notes
- Vocabulary Flip Cards
- Graphic Organizers
- Name Game and Crossword Puzzles
- GeoGame

Online Test Practice

Map Center

- Atlas Maps
- Political Maps with Data
- Links to Mapping Web Sites

Activity Center

- Activity Maker—Review Games, Puzzles, Flip Cards, and Graphic Organizers
- Worldster

Research & Writing Center

- Web Sites
- Data Updates
- Primary Sources
- Writing Models

Current Events

Judith K. Bock
Peter J. Palombi Middle School (retired)
Lake Villa, Illinois

Make It Fun — Teacher-Tested Activities

As you teach this chapter, look for my classroom-proven ideas for:

- making the lesson fun
- focusing students' attention and comprehension
- integrating technology with traditional teaching

Teaching Bell to Bell When you see this timing guide, have students discuss in pairs, in groups of 3–6, or as a class.

- **Think, Pair, Share** (pairs)
- **Reader, Recorder, Reporter** (3–6)
- **Talk About It** (any size)

CHAPTER 11 RESOURCES FOR DIFFERENTIATED INSTRUCTION

Struggling Readers

eEdition with Audio DVD-ROM

Lesson Plans
- Strategies for Inclusion

Unit Resource Book
- Reading Study Guide with Additional Support, pp. 11–18 **B**
- Reteaching Activity, pp. 53–56
- Chapter Test (Form A), pp. 57–60 **A**

Best Practices Toolkit **C**
- Interactive Reading Strategies, pp. 90–145
- Differentiation Activity Bank: Struggling Readers, pp. 146–149

Teacher's Edition Activities
- Understanding Physical Maps, Sec. 1
- Weather Forecaster Role-Play, Sec. 1
- Create a Visual Vocabulary, Sec. 3
- Relationships Between Text and Graphic Elements, Sec. 3
- Connecting Imperialism and Colonial Possessions, Sec. 4
- Tiered Activities, Sec. 1–4

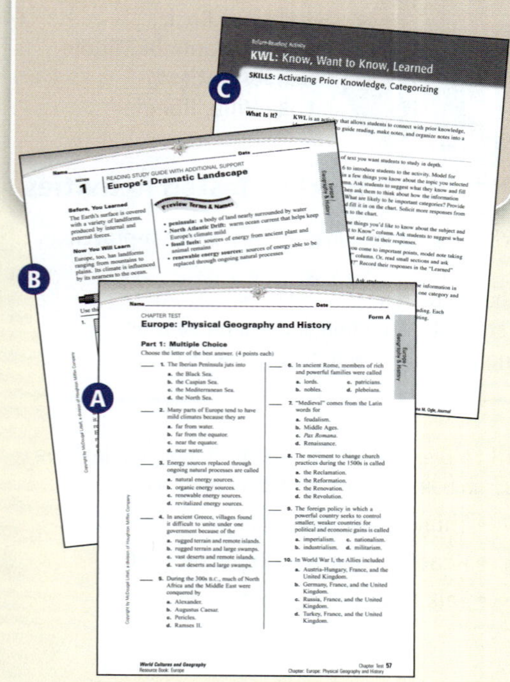

English Learners

Pupil's Edition in Spanish

eEdition with Audio DVD-ROM

Lesson Plans
- Strategies for English Learners

Unit Resource Book
- Reading Study Guide (Spanish), pp. 19–26
- RSG with Additional Support (Spanish), pp. 27–34 **A** **B**

Multi-Language Glossary **C**

Best Practices Toolkit
- Differentiation Activity Bank: English Learners, pp. 165–194

Test Generator CD-ROM
- Section Quizzes in Spanish
- Chapter Test in Spanish

Teacher's Edition Activities
- Pronounce and Preview, Sec. 1–4
- Vocabulary: Multiple-Meaning Words, Sec. 1
- Vocabulary: Cognates, Sec. 2
- Vocabulary: Suffixes, Sec. 3
- Language: Pronoun Referents, Sec. 3
- Vocabulary: Word Relationships, Sec. 4
- Tiered Activities, Sec. 1–4

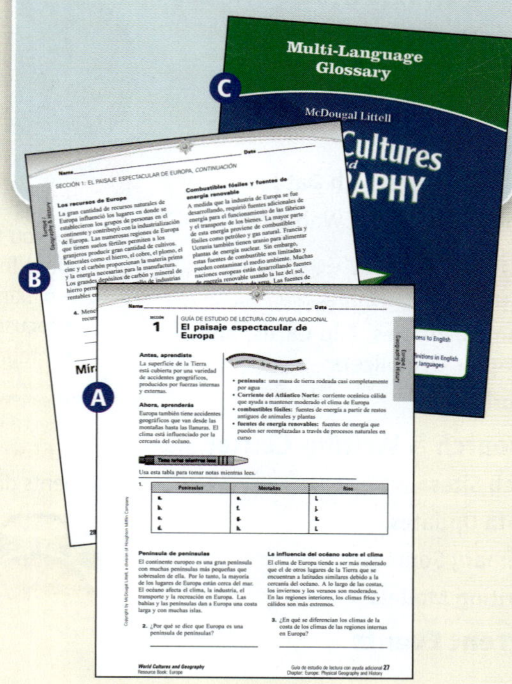

Inclusion

eEdition with Audio DVD-ROM

Lesson Plans
- Strategies for Inclusion

Unit Resource Book
- Reading Study Guide with Additional Support, pp. 11–18 **A** **B**
- Reteaching Activity, pp. 53–56
- Chapter Test (Form A), pp. 57–60

Best Practices Toolkit **C**
- Interactive Reading Strategies, pp. 90–145
- Differentiation Activity Bank: Struggling Readers, pp. 146–149

Teacher's Edition Activities
- Create Visual Vocabulary, Sec. 1
- Create Sentences Using Vocabulary, Sec. 2
- Match Terms and Definitions, Sec. 3
- Vocabulary in Context, Sec. 4
- Tiered Activities, Sec. 1–4

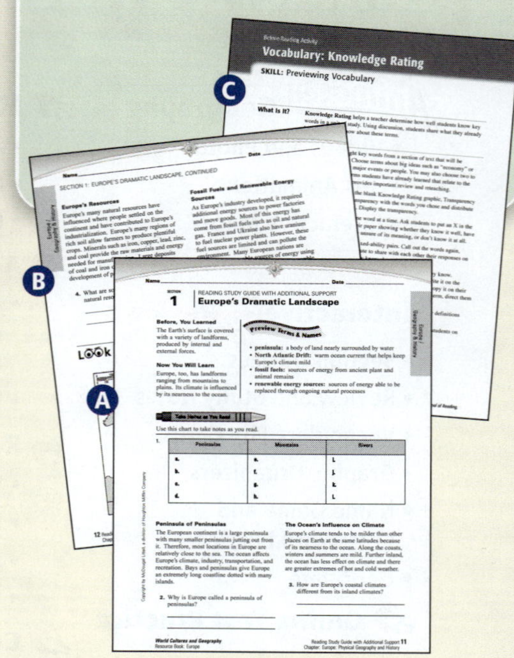

Interdisciplinary Activities

Gifted & Talented Students

Lesson Plans
- Core Lesson Plan

Unit Resource Book
- Outline Map Activities, p. 39 (A)
- History Maker, pp. 41, 42
- Interdisciplinary Projects, pp. 43–45 (B)
- Connect to History, p. 47 (C)
- Chapter Test (Form C), pp. 65–68

Best Practices Toolkit
- Student Products Rubrics, pp. 1–14
- Writing Templates, pp. 78–89

Teacher's Edition Activities
- Compare River Systems, Sec. 1
- Describe Climate, Sec. 1
- Create a Feudal Society Display, Sec. 3
- Tiered Activities, Sec. 1–4

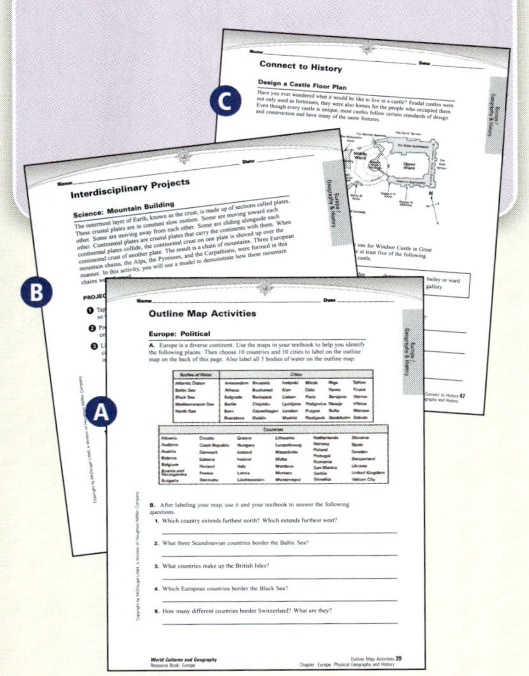

Pre-AP*

Lesson Plans
- Core Lesson Plan

Unit Resource Book
- Vocabulary Practice, p. 35
- Map and Graph Skills, pp. 37–38
- History Makers, pp. 41–42 (A) (B)
- Economics in the World, p. 46 (C)
- Chapter Test (Form C), pp. 65–68

Best Practices Toolkit
- Writing Templates, pp. 78–89
- Differentiation Activity Bank: Pre-AP Students, pp. 195–200

* Pre-AP is a registered trademark of the College Entrance Examination Board, which was not involved in the production of and does not endorse this product.

Teacher's Edition Activities
- Research Europe's Port Cities, Sec. 1
- Clarify Primary Source, Sec. 2
- Encyclopedia Entries on the Black Death, Sec. 3
- Tiered Activities, Sec. 1–4

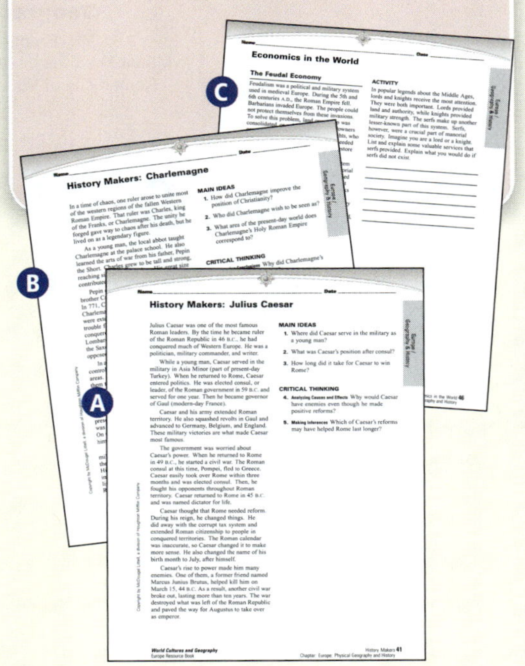

CONNECT to Math

Graphing Hot and Cold

CLASS TIME 30 minutes

TASK Create line graph displays showing the average January (or winter) and July (or summer) temperatures for three major cities in Europe and three in the United States.

SUPPLIES
- temperature data of three major European cities (Paris, London, Rome) and three major U.S. cities (New York, Chicago, Los Angeles)
- construction paper
- colored markers or pencils, rulers

ACTIVITY Draw students' attention to the graphs included with the Comparing Climate Regions of Europe feature in Section 1. Distribute the temperature data for European and U.S. cities. Have students work in pairs to construct line graph posters showing the data. They should label the axes, provide a title for the graph, and connect the data with a line between points. They may use different colors for each set of cities. Display the completed posters in the classroom.

Good for Visual and Mathematical Learners

CONNECT to History

What Happened to an Empire?

CLASS TIME 30 minutes

TASK Trace the outlines of the Roman Empire on a map of present-day Europe.

SUPPLIES
- outline map of present-day Europe
- colored markers or pencils

ACTIVITY Assist students in locating rivers on their maps and write the names of the rivers on the board. Tell them this will help them find Roman boundaries. Then have students study their map and the "Division of the Roman Empire" map in Section 2. Direct them to trace the outlines of the Roman Empire on their map using a different color for each half. Then ask them to identify several countries that once were parts of the Roman Empire.

Good for Spatial and Visual Learners

Introduce the ESSENTIAL QUESTION

- Identify the geographic features, climates, and resources of Europe.
- Describe how the classical cultures of Greece and Rome continue to influence the world today.
- Explain feudalism and the changes brought about by the Renaissance.
- Summarize the key ideas and events of modern European history.

Key Ideas

Section 1: Europe, too, has landforms ranging from mountains to plains. Its climate is influenced by its nearness to the ocean.

Section 2: Ancient Greek and Roman achievements in government, art and architecture, engineering, and law continue to influence Europe and the world today.

Section 3: Feudalism provided stability after the fall of the Roman Empire. The Renaissance marked a rebirth of creativity in Europe.

Section 4: Revolutions in science, politics, and industry transformed Western Europe. After two world wars, European nations found new ways to cooperate.

CONNECT to the Essential Question

You may wish to revisit the Essential Question periodically with students as they work through the chapter.

Suggestions for reconnecting are provided at an appropriate point in the *Teacher's Edition* for each section.

Geiranger Fjord in Norway

CHAPTER 11

Europe: Physical Geography and History

ESSENTIAL QUESTION

What changes have taken place in Europe since ancient times?

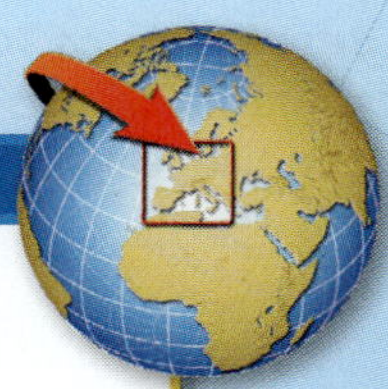

CONNECT Geography & History

Use the map and the time line to answer the following questions.

1. What countries share borders with France?
2. The European Union's headquarters are in the capital of Belgium. What city is that?

2000 B.C.

History
◄ **27 B.C.** Augustus forms the Roman Empire, which lasts until A.D. 476. (Coin showing Augustus)

Geography
A.D. 79 Eruption of Mount Vesuvius destroys Roman city of Pompeii.

History
400s Beginning of the Middle Ages ►

TIME LINE DISCUSSION

Point out to students that Europe is the world's second smallest continent in terms of landmass, but is very densely populated. Europeans are a diverse group, and enjoy a relatively mild climate and have abundant resources. Until recently, however, European history has included many conflicts, including two destructive world wars.

- What events changed Europe after the fall of the Roman Empire? *(400s—beginning of the Middle Ages; c. 1300—beginning of the Renaissance; 1799—Napoleon's seizure of power in France; 1992—treaty forming the European Union)*
- About how many years passed between the fall of the Roman Empire and the beginning of the Renaissance? *(824)*
- Which direction did Renaissance culture travel to reach Germany? *(north)*

Culture
c. 1300 The Renaissance begins in Italy.

Government
1799 Napoleon seizes power in France following the French Revolution. ►

Economics
▲ **1992** Treaty forming the European Union is signed.

Today

325

Animated GEOGRAPHY
Present-Day Europe
CLASSZONE.COM

Physical features such as numerous peninsulas and bays encouraged European countries to explore and build trade routes by sea. Fishing and international shipping helped develop European economies.

- Which country lies closest to Africa? *(Spain)* Which countries border the Black Sea? *(Ukraine, Romania, Bulgaria)*
- The Czech Republic is landlocked. What countries might it work with to gain access to a sea? *(Germany or Poland)*
- Find the closest points between Norway and Germany. Using the scale shown on the map, about how many miles apart are these countries? *(200)*

CONNECT GEOGRAPHY & HISTORY ANSWERS

1. Spain, Andorra, Monaco, Belgium, Luxembourg, Germany, Switzerland, and Italy
2. Brussels

Presentation Options

Use the **Power Presentations** DVD-ROM to project the animated map.

- Prompt students to use the pop-up images and audio from the map to discover what they already know about the geography, culture, history, economics, or governments of the region.
- Challenge students to answer the map questions individually, in small groups, or as a class activity. Use all of the interactive material to help preview chapter content.

TEACHER MEDIA FAVORITES

Books for the Teacher

Barnes, Ian and Robert Hudson. *The History Atlas of Europe*. MacMillan Publishing Company, 1998. Instructive atlas of European history.

Barzun, Jacques. *From Dawn to Decadence: 500 years of Western Cultural Life, 1500 to the Present*. Harper Perennial, 2001. A perceptive look at European civilization.

Video / DVD

The French Revolution. (History Channel) Doug Shultz. 2005. Sweeping examination of a seminal world event.

Modern Marvels: Athens Subway. (History Channel) Larayne Decoeur, Noah Morowitz. 2006. An encounter between modern needs and the priceless past.

Internet

Go to the *Research and Writing Center* at **ClassZone.com** to link to:

- data on the region
- a variety of chapter-specific content-reviewed sites
- "History Makers" biographies
- primary sources

1 Plan & Prepare

Objectives

- Explain why Europe is called the "peninsula of peninsulas"
- Describe the ocean's influence on Europe's climate
- Identify Europe's natural resources

Read for the Essential Question

Help students read for a purpose by reminding them of the Essential Question: "What changes have taken place in Europe since ancient times?" Point out that they can use text features such as Key Questions, special features, images, and maps to form answers to the question.

Vocabulary

Best Practices Toolkit

Use the ***Best Practices Toolkit*** to model strategies for vocabulary notetaking. Vary strategies throughout the year. Choose from: *Knowledge Rating, Predicting ABCs, Definition Mapping, Word Sort, Frayer Model (Word Squares), Magnet Words, Word Wheel,* and *Student VOC.*

Vocabulary Strategies, TT9–TT16

Reading Strategy

Best Practices Toolkit

Display the Three-Column Chart transparency.

- Point out to students that to categorize means they must decide how information in the section can be grouped or sorted by topic.
- Model filling in one or two details for each category on the chart.
- Have students suggest entries.
- Display the graphic organizer for students to review.

Three-Column Chart, TT21

Section 1 Reading for Understanding

Key Ideas

BEFORE, YOU LEARNED

The Earth's surface is covered with a variety of landforms, produced by internal and external forces.

NOW YOU WILL LEARN

Europe, too, has landforms ranging from mountains to plains. Its climate is influenced by its nearness to the ocean.

Vocabulary

TERMS & NAMES

peninsula a body of land nearly surrounded by water

Alps Europe's tallest mountain range, stretching across southern Europe

Northern European Plain vast area of flat or gently rolling land from France to Russia

North Atlantic Drift warm ocean current that helps keep Europe's climate mild

fossil fuels sources of energy from ancient plant and animal remains

renewable energy sources sources of energy able to be replaced through ongoing natural processes

hydroelectric power electricity made by water-powered engines

BACKGROUND VOCABULARY

seafaring using the sea for transportation

Visual Vocabulary The Alps

Reading Strategy

Re-create the chart shown at right. As you read and respond to the **KEY QUESTIONS,** use the chart to organize important details about the physical geography of Europe.

See Skillbuilder Handbook, page R7

CATEGORIZE

PENINSULAS	MOUNTAINS	RIVERS
1.	1.	1.
2.	2.	2.
3.	3.	3.

GRAPHIC ORGANIZERS
Go to **Interactive Review** @ClassZone.com

PRETEACHING VOCABULARY

English Learners

Pronounce and Preview

Pronounce each term for students. Review the meanings of words within the definitions, such as *fertile plain, ocean current,* and *natural processes.*

- To modify vocabulary learning, have students complete worksheets as they read, instead of afterward.

Unit Resource Book: Europe
- Vocabulary Practice, p. 35

Inclusion

Create Visual Vocabulary

Read aloud the definitions of each vocabulary term and name. Provide examples as needed and answer any questions. Then draw students' attention to the Visual Vocabulary image on this page. Ask students to create a Visual Vocabulary image for three other vocabulary terms.

SECTION 1 GEOGRAPHY

Neuschwanstein Castle, Bavaria, Germany

Swan in front of gardens at Giverny, France

Europe's Dramatic Landscape

Connecting to Your World

What comes to mind when you think of Europe? Snowcapped mountains on travel posters? Medieval castles next to winding rivers that you've seen in jigsaw puzzles? Or perhaps you've watched the Winter Olympics on television and seen skiers speeding down steep mountain slopes. Each of these images tells us something about Europe's physical geography. High mountain ranges, oceans, long broad rivers, and fertile plains are all a part of it. Let's see how they fit together.

Alcazar Castle Europe is famous for its castles, like this one in Segovia, Spain.

Peninsula of Peninsulas

KEY QUESTION Why is Europe called a peninsula of peninsulas?

A quick look at a map of Europe explains why it is called the "peninsula of peninsulas." A **peninsula** is a piece of land nearly surrounded by water. The entire continent of Europe is a peninsula with smaller peninsulas jutting out from it. Because of these peninsulas, most places in Europe are no more than 300 miles from an ocean or sea. Its nearness to these bodies of water has influenced Europe in many ways. The ocean modifies Europe's climate. Many Europeans use the ocean for both business and pleasure.

Opatija, Croatia This town is located on the Istria Peninsula in southern Europe.

2 Focus & Motivate

3-Minute Warm-Up

Write on the board or display the transparency:

- Look at the photographs on this page and the next. What do they suggest about the landscape of Europe? *(Possible Answers: Much of Europe lies along bodies of water; waterways are important in Europe; Europe has high mountains and dense forests.)*

Unit Transparency Book: Europe
- 3-Minute Warm-Ups, TT1

Connecting to Your World

More About . . .

Croatia and the Sea

The Adriatic Sea borders Croatia, and its coastline influences Croatia's economy. Along the Adriatic coast, ships are built at Pula, and Rijeka and Dubrovnik are key seaports. Tourism at Adriatic beaches also boosts the economy.

3 Teach

Peninsula of Peninsulas

Reader, Recorder, Reporter

- Which geographical feature influenced cultural differences in Europe? *(mountain ranges)*
- How did the peninsulas of Europe influence economic development? *(Long coastlines encouraged exploration, fishing, and trading.)*
- **Categorize** What are the main geographical features of Europe? *(peninsulas, mountains, rivers, plains)*

SECTION 1 PROGRAM RESOURCES

Unit Resource Book: Europe

ON LEVEL

Unit Resource Book
- Reading Study Guide, p. 3
- Section Quiz, p. 49

STRUGGLING READERS

Unit Resource Book
- Reading Study Guide with Additional Support, p. 11
- Skillbuilder Practice: Explaining Geographic Patterns, p. 36
- Section Quiz, p. 49
- Reteaching Activity, p. 53

eEdition with Audio DVD-ROM

ENGLISH LEARNERS

Pupil Edition in Spanish

eEdition with Audio DVD-ROM

Unit Resource Book
- RSG (Spanish), p. 19
- RSG with Additional Support (Spanish), p. 27

Multi-Language Glossary

Test Generator
- Section Quiz in Spanish

INCLUSION

Unit Resource Book
- RSG with Additional Support, p. 11
- Section Quiz, p. 49
- Reteaching Activity, p. 53

GIFTED & TALENTED

Unit Resource Book
- Interdisciplinary Projects, pp. 43, 45
- Section Quiz, p. 49

PRE-AP

Unit Resource Book
- Section Quiz, p. 49

TECHNOLOGY

Unit Transparency Book
- 3-Minute Warm-Ups, TT1
- Essential Question, TT2
- Critical Thinking, TT3
- Map: Europe: Physical, TT7

Daily Test Practice Transparencies
- Section 1, TT37

Power Presentations

ClassZone.com

World Cultures and Geography Video on DVD

More About . . .

The Alps

In addition to being the tallest mountain range in Europe, the Alps are also the largest. They cross south-central Europe in a large crescent shape about 750 miles long. Mont Blanc is the highest peak at 15,771 feet. The Alps range covers an area of about 80,000 square miles. Before roads and tunnels were built, the Alps formed natural barriers between European nations.

Unit Resource Book: Europe
- Interdisciplinary Projects: Mountain Building, pp. 43, 45

CONNECT Geography & Economics

Seafaring Economies

Waterways have long played an important role in European economies. For example, Bremen, the oldest port city in Germany, was founded about 1,200 years ago. It is located along the banks of the Weser River in northwest Germany and remains a center for shipping, trade, and industry.

Make It Fun

Teacher-Tested Activities

Judith K. Bock, Peter J. Palombi Middle School, Lake Villa, Illinois

I use an activity called "Bulletin Board" to introduce a chapter. Make sure you've got a map of Europe on a bulletin board, with string, scissors, and tacks available.

- Write key places, events, or people from the chapter on slips of paper and put them in a bowl. Have each student draw one.
- Have students briefly research their topics and find or draw a picture to illustrate them. They should mount their picture on construction paper and write a short caption describing their topic and what country it is associated with.
- Have students hang their pictures on the bulletin board, connecting them to the correct country on the map with string. Have each give a brief oral presentation.

Europe's Coastline Europe's peninsulas extend off the continent in all directions, as the map on the opposite page shows. Far to the north, the Scandinavian Peninsula **A** juts out into the North Sea. Spain and Portugal occupy the Iberian Peninsula **B**, which extends into the Mediterranean Sea in southern Europe. The Italian and Balkan Peninsulas also stretch into the Mediterranean Sea.

Europe's many bays and peninsulas give it a long, uneven coastline, dotted with islands and stretching almost 24,000 miles. Europe has more coastline than Africa, a much larger continent. Long ago, Europe's nearness to the sea and its many natural harbors encouraged **seafaring**, or the use of the sea for transportation. Europeans fished, traded, and in time set out to explore other parts of the world.

Mountains Mountains are another key landform in Europe. They affect climate, travel, and culture. The **Alps** **C**, Europe's tallest mountain range, stretch across eight countries in southern Europe. The Pyrenees **D** once formed a natural barrier between France and Spain. Europe also has smaller mountain ranges, such as the Apennines.

Europe's mountain ranges separated the groups of people who settled the land thousands of years ago. This is one reason different cultures developed in Europe. Mountain ranges also contain natural resources, such as timber, which influence where people settle.

Mountains surround the **Northern European Plain** **E**, a vast area of flat or gently rolling land from France to Russia. Parts of the plain have fertile soil, making them Europe's major farming areas. Other parts are rich with coal. Industrial centers developed near the coal deposits. These variations have made settlement across the plain uneven.

SUMMARIZE Explain why Europe is called a peninsula of peninsulas.
because it is a large peninsula with many smaller peninsulas jutting out from it

CONNECT Geography & Economics

Seafaring Economies

Europe's vast coastline boosted its economy by encouraging exploration, fishing, and trade. Europe provides over half of the world's international shipping. Countries with large fleets, such as Great Britain, ship goods around the world. To do so, they use ports on rivers like the Thames in London, shown here.

DIFFERENTIATING INSTRUCTION

English Learners

Vocabulary: Multiple-Meaning Words

Examine the term *mountain ranges* on this page. Explain that *range* or *ranges* has many meanings. Examples are: "open lands where cattle graze;" "a place for shooting at targets such as a shooting range;" and "to wander or roam through an area." In its usage here, *ranges* means "an extended group or series, particularly a row of mountains."

Pre-AP

Research Europe's Port Cities

Have students work in pairs to analyze the economy of a European port city. Students' analyses should be based on statistical data about the overall economy of the chosen city. Based on the data, have students explain the importance of shipping to the city's overall economy. Have students present their analyses or create a classroom display.

Physical Geography of Europe

CONNECT Geography & Economics

A Network of Rivers Europe's rivers are key trade and transportation networks. Goods are transported on rivers to and from coastal harbors. Cities sprang up along riverbanks to take advantage of these benefits.

- The **Rhine River** travels north from the Swiss Alps into the North Sea.
- The **Danube River** flows southeast from Germany into the Black Sea.
- The **Thames River** in the United Kingdom runs east across southern England to the North Sea.

CRITICAL THINKING

1. **Find Main Ideas** What advantages result from Europe's network of rivers?
2. **Draw Conclusions** Why are rivers important to landlocked countries in Europe?

329

Physical Geography of Europe

Point out that a physical map provides a geographical overview. A physical map can help in understanding the location of one place or region in relation to another, as well as the geographical features of that place or region. Draw students' attention to the map key. Ask:

- What does the map key help you understand about the geography of Europe? *(the elevations of different areas)*
- What is the elevation of northern Europe along the Baltic Sea? *(0 feet)*

CONNECT Geography & Economics

Point out to students how much area Europe's rivers cover when viewed as a network. Ask:

- How might goods be transported between rivers? *(Possible Answers: trucks, trains)*

CRITICAL THINKING ANSWERS

1. **Find Main Ideas** Rivers provide trade and transportation networks.
2. **Draw Conclusions** They offer landlocked countries outlets to the sea and links to other countries within Europe.

Unit Resource Book: Europe
- Skillbuilder Practice: Explaining Geographic Patterns, p. 36

More About . . .

The Rhine River

Many famous cities of Europe are situated along the Rhine River. These include Cologne and Mainz in Germany, Strasbourg in France, and Basel in Switzerland. The Rhine originates in two streams high in the Alps. Historically, the Rhine served as a political border of the Roman Empire. Since ancient times, the Rhine has been a major part of the inland waterway transportation system in Europe.

DIFFERENTIATING INSTRUCTION

Struggling Readers

Understanding Physical Maps

Make sure students understand what they can learn from the physical maps on this page. To help them understand what kind of information is included on a physical map, ask them to point out the different kinds of features that they see on the map at the top of the page. *(peninsulas, rivers, plains, mountains, islands, seas, oceans)* Name one example of each feature, and have them find two more.

Gifted & Talented

Compare River Systems

Have students choose one of the three major rivers identified in the Connect box. Have them use their textbooks and classroom resources to compare that river with the Mississippi River in the United States. Have students create a large illustrated chart or poster showing the comparisons. Information could include lock systems, types of ships, and uses of the waterway. Students should also explain the importance of each river.

Teach

The Ocean's Influence on Climate

Think, Pair, Share

- How does the North Atlantic Drift influence Europe's climate? *(milder temperatures than at the same latitudes in the United States)*
- What enables olive trees to grow in the dry Mediterranean climate? *(wax-coated leaves that retain water)*
- **Categorize** What types of plants survive in the Scandinavian Peninsula? *(in the spring, mosses and wildflowers)*

Comparing Regions

Tundra in the Scandinavian Peninsula and Rocky Mountains

Tundra can be found in the United States as well as in Finland and other parts of the Scandinavian Peninsula. Elevations and extreme cold in winter are similar in these regions. A large expanse of alpine tundra can be found at 12,000 feet of elevation in Rocky Mountain National Park in Colorado. Visitors can hike or drive to the high elevations to view the tundra. Scientists warn that tundra regions are at risk because of pollution and climate change.

COMPARING

Climate Regions of Europe

Remind students that much of Europe is close to the ocean, influencing all of its climate regions to some degree.

ANSWERS

1. Mediterranean Sea
2. different crops grow in different climates, can influence what a country can grow for export
3. greater variety of plants; some that are resources, are more likely to grow in a humid climate
4. snowfall covers the tundra, protects plants from bitter winter cold

The Ocean's Influence on Climate

KEY QUESTION How has nearness to the sea affected the climate and vegetation of Europe?

Usually, the farther a place is from the equator, the colder its climate is. Like the United States, the northern parts of Europe are colder than the southern parts. In Europe, however, the climate even in the far north is warmer than the same latitudes in the United States. Europe's climate tends to be milder because of its nearness to the ocean.

Westerly winds and a warm current in the Atlantic Ocean called the **North Atlantic Drift** influence Europe's climate. The warm ocean current heats the air above it. Winds blowing east across the Atlantic Ocean carry the heat and moisture to Europe. In the winter, these westerlies bring warm tropical air to parts of Europe. Although Norway is as far north as parts of Alaska and Canada, it gets little ice and snow on its coast during the winter months. During the summer, the winds shift and deliver cooler subarctic air, keeping temperatures from getting too hot. Farther inland, the ocean has less effect on climate. Winter temperatures in eastern Europe tend to be colder than those in western Europe.

COMPARING Climate Regions of Europe

Mediterranean Palm trees grow in this Spanish garden. **1. What feature is this climate zone named for?**

Marine West Coast Grapevines thrive in France's scenic Loire Valley. **2. How might climate affect a country's exports?**

Humid Continental Forests cover about one-fourth of Romania. **3. How could humidity affect plant life?**

Tundra Finland's tundra often blooms in the summer. **4. How does it change in winter?**

330

INTERDISCIPLINARY ACTIVITIES

CONNECT to Language Arts

Outlining Information

Have students work in pairs to outline the text on this page, including the Comparing graphic. They should organize the information in a standard outline format. Students should share their outlines with another pair and offer suggestions to each other for improvements.

CONNECT to Science

Create a Plant Fact Statement

Ask students to assume that they are botanists conducting research for a dictionary of plant life. Have them work in pairs to collect information about one of Europe's plants, such as grapes, palm trees, or tundra mosses. Students should present their findings in the form of a fact sheet with information about habitat needs and characteristic features of the plant. They should incorporate drawings or other illustrations into their fact sheet.

Europe's plant and animal life varies with its climate. Parts of the Scandinavian Peninsula in the north are covered with tundra. Few plants can survive the cold winters. There are no trees on the tundra, but in the spring, mosses and a few wildflowers appear. Thick forests of fir trees cover large areas of northern Europe. Such trees remain green year-round. Central and eastern Europe once had forests of maple, oak, and elm trees. Today, most have been cut down for lumber or fuel. Far to the south, along the Mediterranean Sea, the climate is too warm and dry for the trees seen farther north. Instead, the wax-coated leaves of cork and olive trees retain water, helping these plants grow here.

In the Arctic region, elk and reindeer still roam, but most animals that once inhabited Europe's dense northern forests are gone. European brown bears, deer, foxes, and wolves live in parts of Europe. Two goatlike animals, the chamois (SHAM•ee) and ibex, inhabit high mountain areas in southern and southwestern Europe.

ANALYZE EFFECTS Explain the effects of Europe's location on its climate and vegetation.

he nearness most of rope to the ean and the orth Atlantic ift reduce e effects location at gher, norally colder titudes. ▶

Fun Facts!

PALM TREES IN SCOTLAND?

Did you know that there are palm trees in Scotland? Even though Stranraer, Scotland is almost as far north as southern Alaska, palm trees thrive there. For this, Scots can thank the North Atlantic Drift. This ocean current brings mild weather to the region.

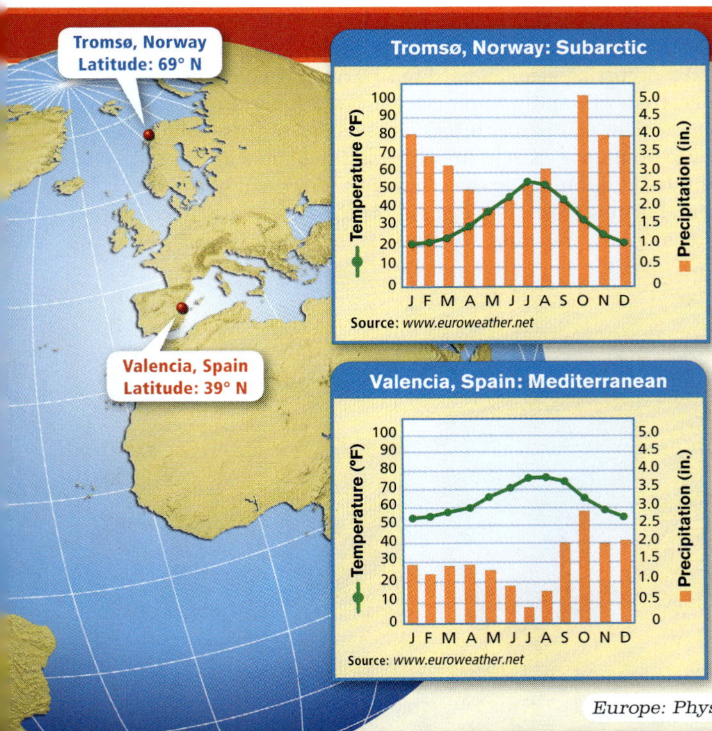

Europe's Many Climates

Much of Europe's climate is mild as a result of the ocean, but differences in latitude still affect climate. Tromsø, Norway, far to the north, experiences the cool temperatures of a subarctic climate. Valencia, Spain, has a warm Mediterranean climate.

CRITICAL THINKING

1. **Evaluate** In both locations, more rain falls in October than in any other month. How does Valencia's October precipitation compare to Tromsø's?
2. **Compare and Contrast** How do the ranges of temperatures that each city experiences during the year compare to each other?

Fun Facts!

Palm Trees in Scotland?

Like Stranraer, Scotland, other parts of the British Isles also have warm climates. The southernmost point of mainland England—called the Lizard Peninsula—has the warmest climate in England. It has been designated an Area of Outstanding Natural Beauty because of its variety of rare plants. Subtropical vegetation grows on the peninsula, and spring arrives very early.

COMPARING

Europe's Many Climates

Explain that climographs are composed of a bar graph and a line graph. Point out that the line graph represents the average temperatures for each month. The bar graph represents the average precipitation (rain, snow, and so on) for each month.

- In which month is precipitation lowest in Valencia? *(July)* In Tromsø? *(May)*
- In both locations, December and January are the coldest months. How does Tromsø's January temperature compare to Valencia's? *(Tromsø is about 30 degrees colder.)*

CRITICAL THINKING ANSWERS

1. **Evaluate** Tromsø receives over five inches of rain in October, while Valencia gets almost three inches.
2. **Compare and Contrast** Tromsø has about a 40-degree range while Valencia has about a 30-degree range.

DIFFERENTIATING INSTRUCTION

Gifted & Talented

Describe Climate

Have each student select a European city and research climate data for that city using classroom resources and the Web site listed in their textbooks (www.euroweather.net). Have them determine the best way to present their data, such as a chart or other visual display. Information should include annual precipitation, average temperature, latitude, and longitude.

Struggling Readers

Weather Forecaster Role-Play

Struggling readers may need additional help in understanding the graphs on this page. Guide them in creating statements about rainfall and temperature for each city. Use these statements to create a brief weather report. Then ask volunteers to role-play being a weather forecaster on a television news program and present the weather report.

Teach

Europe's Resources

Talk About It

- Which resource is more abundant in Europe than the world's average? *(soil for farming)*
- Which renewable energy sources are Europeans using? *(hydroelectric, wind, solar, geothermal)*
- **Categorize** Which energy sources used in Europe can harm the environment? *(uranium and fossil fuels, such as oil, gas, and coal)*

More About . . .

Geothermal Energy

Electricity that is generated by heat from Earth is called geothermal energy. Hot springs and geysers are sources of geothermal energy. Magma far below Earth's surface creates this type of energy by heating nearby rocks that in turn heat underground water.

CONNECT to the Essential Question

What changes have taken place in Europe since ancient times?

Ask students what they have learned so far that can help them answer this question. Students might mention:

- Europeans have used the extensive coastline to develop seafaring economies.
- Europeans have used rivers to develop inland transportation networks.
- Europeans today are turning to renewable energy sources.

Europe's Resources

KEY QUESTION Why are energy resources so important to Europe's growth?

Europe has a variety of natural resources, from fertile soils to mineral deposits. These resources have influenced where people settled on the continent. They have also contributed to Europe's industrialization.

Fertile Soil and Abundant Resources Soil is an important resource because it is used to grow food for people to eat. Europe's regions of rich soil allow farmers to produce plentiful crops. About one-third of Europe's land can be used for farming, three times more than the world average.

Europe's landscape also provides a wide variety of important mineral resources. Minerals like iron, copper, lead, zinc, and coal provide the raw materials and energy needed for manufacturing. For instance, both coal and iron ore are used to make steel. Large deposits of these minerals helped industry develop in the United Kingdom and the Ruhr Valley in Germany.

Fuel for Industry As Europe's industry developed, it required additional energy sources to power factories and move goods. Several countries in northern Europe tap into oil and natural gas deposits in the North Sea. Countries such as France and Ukraine have uranium, a rare, naturally occurring element used to fuel nuclear power plants.

However, coal, oil, and natural gas are **fossil fuels**, sources of energy from ancient plant and animal remains. All of these energy sources—even uranium—exist in limited quantities and will eventually run out. In addition, using these energy sources can harm the environment. Burning fossil fuels can pollute the air and is believed to contribute to global warming. Nuclear plants create toxic waste.

They provide clean, renewable sources of energy, and they make Italy less reliant on importing energy from other countries. ▼

Geothermal Power Plant This power plant in Larderello, Italy, produces enough energy each year to power half a million homes. **Why might Italy want to develop more geothermal plants?**

DIFFERENTIATING INSTRUCTION

OBJECTIVE Create a display on global warming and European responses to it.

Basic

Working in small groups using the information in their textbooks and classroom resources, have students create a collage, poster, or other graphic that shows the causes of global warming. Students should also include a summary that explains what global warming is and the harm it can cause.

On Level

Working in small groups, have students create a collage, poster, or other graphic that shows Europe's renewable energy sources. Students should also include a summary that explains the costs and benefits of renewable energy sources and how they can help reduce global warming.

Challenge

Working in small groups using classroom and library resources, have students create a collage, poster, or other graphic that shows what Europeans are doing to help reduce global warming. Students should also include a summary of other steps that could be taken to reduce global warming.

Renewable Energy As an alternative to fossil fuels, many European nations are trying to find renewable sources of energy. **Renewable energy sources** can be replaced through ongoing natural processes, such as sunshine, wind, and flowing water. These types of energy sources are also sustainable, meaning that they can be utilized without being used up.

Hydroelectric power is a renewable energy source that uses water-powered engines to make electricity. Austria generates energy by damming Alpine rivers. Coastal countries such as Ireland and Portugal are developing technology to harness wave energy. Germany, Spain, and Denmark lead the pack in producing electricity from wind energy. Other nations, including the Netherlands, generate heat and electricity by collecting solar energy from the sun. Italy and Iceland tap geothermal energy by drilling into naturally hot groundwater. These alternative energy sources have many benefits. They are sustainable. They reduce pollution. They mean that European nations can rely less on importing energy from other countries.

DRAW CONCLUSIONS Explain why energy resources are important to Europe's growth. They provide power for factories, businesses, and cities as well as fuel for farm machinery and vehicles.

European Renewable Energy Sources

In 2001, 15 European countries used renewable energy sources to supply 15.2 percent of the energy they consumed. This pie graph shows how much energy came from each source.

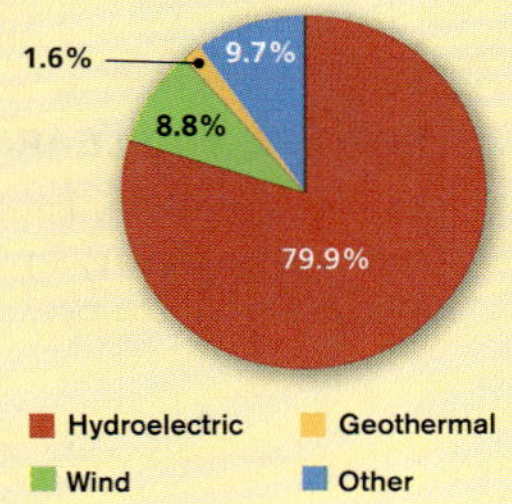

Source: European Commission, *Electricity from Renewable Energy Sources*, 2004

Section 1 Assessment

ONLINE QUIZ For test practice, go to **Interactive Review** @ClassZone.com

TERMS & NAMES

1. **Explain the importance of**
   - Alps
   - Northern European Plain
   - North Atlantic Drift
   - fossil fuels

USE YOUR READING NOTES

2. **Categorize** Use your completed chart to answer the following question:
   What are the major mountain ranges of Europe?

PENINSULAS	MOUNTAINS	RIVERS
1.	1.	1.
2.	2.	2.
3.	3.	3.

KEY IDEAS

3. Why is the Northern European Plain important to Europe's economy?
4. How does the North Atlantic Drift affect the climate of Europe?
5. What are Europe's most important mineral resources?

CRITICAL THINKING

6. **Draw Conclusions** Why might mountains be both an asset and a disadvantage?
7. **Make Inferences** Why are coal and iron important to industrializing nations?
8. **CONNECT to Today** Why is Europe's network of rivers still important to its economy?
9. **WRITING** **Write a Report** Choose one of Europe's major rivers to write a report about. List the countries the river runs through. Include some major cities you might see if you sailed along the river. Finally, explain one way the river is important to Europe's economy.

European Renewable Energy Sources

- Which renewable energy source is used most in Europe? *(hydroelectric power)*
- Which renewable energy source might be categorized as "other"? *(solar energy)*

4 Assess & Reteach

Assess Have students complete the Section Assessment.

Unit Resource Book: Europe
- Section Quiz, p. 49

Interactive Review @ClassZone.com
- Section Quiz

Power Presentations

Test Generator

Reteach Have students create index cards with one question and answer for each subsect or. Then have them pair up and quiz each other.

Unit Resource Book: Europe
- Reteaching Activity, p. 53

DIFFERENTIATING INSTRUCTION

Terms & Names

1. Alps: Europe's tallest mountain range; Northern European Plain: vast area of flat or gently rolling land from France to Russia; North Atlantic Drift: warm ocean current helps keep temperatures in most of Europe mild; fossil fuels: sources of energy from ancient plant and animal remains

Use Your Reading Notes

2. Alps, Pyrenees, and Apennines

PENINSULAS	MOUNTAINS	RIVERS
1. Scandinavian Peninsula 2. Iberian Peninsula 3. Italian Peninsula	1. The Alps 2. Pyrenees 3. Apennines	1. Rhine River 2. Danube River 3. Thames River

Key Ideas

3. because its flat land and rich soil make it Europe's major farming area
4. warming effect on climate, makes temperatures warmer than otherwise given their latitudes
5. iron ore, copper, lead, zinc, and coal

Critical Thinking

6. Assets: provide hydroelectric power, key tourist attractions for scenic beauty, hiking, skiing; Disadvantages: can be barriers to trade, travel, transportation and have limited use for farming
7. Coal and iron are used to make steel, important to industrialization.
8. provide transportation network to move goods to and from coastal seaports
9. Use the rubric to score reports.

Report Rubric

	Clarity	Geographic Details	Accuracy
4	clear, complete	many	no errors
3	clear	some	few/minor
2	some confusion	few	several
1	unclear	very few	many

1 Plan & Prepare

Objectives

- Explain the development and spread of ancient Greek culture
- Compare the governments of the Roman Republic and Roman Empire
- Identify ancient Greek and Roman cultural influences in modern life

Read for the Essential Question

Help students read for a purpose by reminding them of the Essential Question: "What changes have taken place in Europe since ancient times?" Point out that they can use text features such as Key Questions, special features, images, and maps to form answers to the question.

Vocabulary

Best Practices Toolkit

Use the ***Best Practices Toolkit*** to model strategies for vocabulary notetaking. Vary strategies throughout the year. Choose from: *Knowledge Rating, Predicting ABCs, Definition Mapping, Word Sort, Frayer Model (Word Squares), Magnet Words, Word Wheel,* and *Student VOC.*

Vocabulary Strategies, TT9–TT16

Reading Strategy

Best Practices Toolkit

Display the Three-Column Chart transparency.

- Remind students that comparing means looking for similarities (What is the same?), while contrasting means looking for differences (What is different?).
- Discuss the similarities and differences between Athens and Sparta. Then model filling in a few details of Greek democracy on the chart.
- Have students suggest entries.
- Display the graphic organizer for students to review.

Three-Column Chart, TT21

Section 2 Reading for Understanding

Key Ideas

BEFORE, YOU LEARNED

Europe has diverse landforms. Its climate and vegetation vary by latitude and distance from the ocean.

NOW YOU WILL LEARN

Ancient Greek and Roman achievements in government, art and architecture, engineering, and law continue to influence Europe and the world today.

Vocabulary

TERMS & NAMES

democracy a government in which the citizens make political decisions, either directly or through elected representatives

Peloponnesus (PEHL•uh•puh•NEE•suhs) the peninsula in southern Greece where Sparta was located

city-state a political unit made up of a city and its surrounding lands

tyrant someone who takes power illegally

oligarchy (AHL•ih•GAHR•kee) a government ruled by a few powerful individuals

republic a government in which citizens elect representatives to rule in their name

patrician (puh•TRIHSH•uhn) a wealthy landowner who held a high government position in ancient Rome

plebeian (plih•BEE•uhn) a commoner who was allowed to vote but not to hold government office in ancient Rome

BACKGROUND VOCABULARY

isolate to cut off or set apart from a group

REVIEW

monarchy a type of government in which a ruling family headed by a king or queen holds political power

Reading Strategy

Re-create the chart shown at right. As you read and respond to the **KEY QUESTIONS**, use the chart to help you compare and contrast the governments of ancient Greece, the Roman Republic, and the Roman Empire.

See Skillbuilder Handbook, page R9

COMPARE AND CONTRAST

GOVERNMENT	WHO RULED	HOW CHOSEN	WHO PARTICIPATED
Greek Democracy			
Roman Republic			
Roman Empire			

GRAPHIC ORGANIZERS Go to Interactive Review @ClassZone.com

PRETEACHING VOCABULARY

English Learners

Pronounce and Preview

Pronounce each term for students. Review the meanings of words within the definitions, such as *representatives, political unit,* and *commoner.*

- To modify vocabulary learning, have students complete worksheets as they read, instead of afterward.

Unit Resource Book: Europe

- Vocabulary Practice, p. 35

Inclusion

Create Sentences Using Vocabulary

Read each term and its definition aloud. Ask pairs of students to write a sentence that shows the correct usage of each term. Then have each member of the pair take turns and write a definition of the term in their own words, without looking at the sentence they wrote or their textbooks. Have the other member read the definition. Working together, have students make revisions to their definitions as necessary.

Neuschwanstein Castle, Bavaria, Germany

SECTION 2 HISTORY

Swan in front of gardens at Giverny, France

Classical Greece and Rome

Connecting to Your World

Lincoln Memorial Some buildings in Washington, D.C., reflect ancient Greek architecture.

Have you ever been to the nation's capital, Washington, D.C.? It is the center of the U.S. government. The United States is a **democracy**, a government in which citizens make political decisions. You are about to learn about ancient Greece, the place where modern democracy began. Ideas about what democracy should be have changed over time. Democratic government in Greece stands as a remarkable first step.

History of Ancient Greece

KEY QUESTION How did ancient Greek culture spread?

Between 5000 and 3000 B.C., groups of people began settling on the **Peloponnesus** (PEHL•uh•puh•NEE•suhs), a mountainous peninsula in southern Europe. Almost 2,000 small islands surrounded the peninsula. Villages were **isolated**, or cut off, from each other. The rugged terrain and remote islands made it difficult to unite the villages under one government. In time, people had settled throughout what is now Greece.

The Acropolis This part of Athens held important buildings constructed in the second half of the fifth century B.C.

335

2 Focus & Motivate

3-Minute Warm-Up

Write on the board or display the transparency:

- How can ancient civilizations still influence the modern world? *(Possible Answer: Their achievements may still have influence because they have helped people learn how to govern effectively, develop economies, and create new technologies.)*

Unit Transparency Book
- 3-Minute Warm-Ups, TT1

Connecting to Your World

More About . . .

Democracy

The word *democracy*, meaning rule by the people, comes from the Greek word *demokratia*, coined from *demos* ("people") and *kratos* ("rule") in the fifth century B.C.

3 Teach

History of Ancient Greece

Think, Pair, Share

- Where did ancient Greek civilization begin? *(on the Peloponnesus and the islands surrounding this mountainous peninsula)*
- In what way did Alexander spread Greek culture? *(by admiring Greek culture and taking it with him as he expanded his empire)*
- **Compare and Contrast** How were Greek city-states similar and different? *(They shared the same culture and language, but differed in their forms of government.)*

SECTION 2 PROGRAM RESOURCES

Unit Resource Book: Europe

ON LEVEL

Unit Resource Book
- Reading Study Guide, p. 5
- Section Quiz, p. 50

STRUGGLING READERS

Unit Resource Book
- Reading Study Guide with Additional Support, p. 13
- Skillbuilder Practice: Explaining Geographic Patterns, p. 36
- Section Quiz, p. 50
- Reteaching Activity, p. 54

eEdition with Audio DVD-ROM

ENGLISH LEARNERS

Pupil Edition in Spanish
eEdition with Audio DVD-ROM
Unit Resource Book
- RSG (Spanish), p. 21
- RSG with Additional Support (Spanish), p. 29

Multi-Language Glossary
Test Generator
- Section Quiz in Spanish

INCLUSION

Unit Resource Book
- RSG with Additional Support, p. 13
- Section Quiz, p. 50
- Reteaching Activity, p. 54

GIFTED & TALENTED

Unit Resource Book
- Interdisciplinary Projects, pp. 44–45
- Section Quiz, p. 50

PRE-AP

Unit Resource Book
- History Makers: Julius Caesar, p. 41
- Section Quiz, p. 50

TECHNOLOGY

Unit Transparency Book
- 3-Minute Warm-Ups, TT1
- Essential Question, TT2
- Critical Thinking, TT4
- Map: The Roman Empire, TT8
- Cultures, TT9

Daily Test Practice Transparencies
- Section 2, TT38

Power Presentations
ClassZone.com
World Cultures and Geography Video on DVD

More About . . .

Pericles

Athens achieved greatness in politics and culture under the leadership of Pericles. Among his democratic reforms, Pericles introduced a policy of paying salaries to citizens who served in public offices and provided jury services. The poor as well as the rich could enter state service. Literature and the arts flourished in the age of Pericles. The Parthenon was another cultural achievement under his leadership.

ANALYZING Primary Sources

The funeral oration of Pericles survives solely because Thucydides, a contemporary of Pericles, included speeches in his history of the Peloponnesian War. Thucydides is generally regarded as the first modern historian because he strove for accuracy and objectivity.

- What does Pericles say about someone who neglects the affairs of Athens? *(That person is useless.)*
- Why might Pericles praise democracy in a speech intended to honor fallen soldiers? *(Possible Answer: People who fight for the state should have a say in state affairs.)*

DOCUMENT-BASED QUESTION ANSWER

Possible Answer: In a democracy such as ancient Greece, which depended on its citizens to help make decisions about leadership and government, all citizens needed to be politically active in order for society to work well.

Rise of City-States: Athens and Sparta Eventually, Greek towns and cities became **city-states**, political units made up of a city and surrounding villages. Most people were farmers and herders, but where land was rocky and soil was poor, people made their living from the sea. They fished, sailed, and traded with other city-states.

Greek city-states shared a common culture and language, but each had unique features. They chose different forms of government. At first, many were monarchies, ruled by kings. Others were ruled by tyrants. In Greece, a **tyrant** was someone who took power illegally.

Athens and Sparta, the largest Greek city-states, had different governments. Sparta was an **oligarchy** (AHL•ih•GAHR•kee), a system ruled by a few powerful individuals. Two kings governed the state, making all major decisions with the help of a few officials. Sparta was a military state. Its large slave class farmed the land, freeing male citizens to serve in the army. Spartan boys began receiving military training at age seven. After decades of military service, they became citizens at 30. They faced losing their citizenship if they did not fight bravely.

At first, kings also governed Athens. Then, at the end of the sixth century B.C., Athens became a limited democracy. All citizens had the right to take part in the government and decide on laws. However, only free adult males were citizens. Women, slaves, and foreigners were not. Athens became a center of Greek culture. It attracted the finest scholars, artists, and philosophers from all over the Mediterranean.

ANALYZING Primary Sources

Pericles (495–429 B.C.) was an Athenian statesman and general. His "Funeral Oration" honors those who died in the Peloponnesian War and praises democracy. Pericles saw participation in government as a civic duty. Greek citizens often discussed politics in an agora, or marketplace, like the one shown at left.

> **An Athenian citizen does not neglect public affairs when attending to his private business. . . . We consider a man who takes no interest in the state not as harmless, but as useless.**
>
> Source: Thucydides, from *The Peloponnesian War*

Pericles ▲

DOCUMENT-BASED QUESTION

Why was an interest in politics considered a public duty?

336 Chapter 11

DIFFERENTIATING INSTRUCTION

English Learners

Vocabulary: Cognates

Explain that the English word *democracy* comes from the Greek word *demokratia*. Other languages share the root word in common with English. In Spanish, the word is *democracia*. In Italian, the word is *democrazia*. In French, the word is *démocratie*. In German, the word is *demokratie*.

Pre-AP

Clarify Primary Source

Point out that in the passage above, Pericles expressed his ideas negatively, using such words as "no" and "not." Have students rewrite the sentences without those words. Then ask them to use Pericles's ideas to write an answer to this question: "Are U.S. citizens who do not exercise their right to vote useless?"

Wars and Conquest In the early fifth century B.C., the rulers of Persia, the region in southwestern Asia that is now Iran, tried to conquer Greece. Led by Sparta and Athens, the Greeks resisted. Conflict between the Greeks and Persians lasted on and off for many years. The Greeks defeated the Persians, keeping Greek culture alive.

Decades later, Athens and Sparta fought each other in the Peloponnesian War. Sparta won the war, becoming the dominant power in Greece. Weakened by their internal conflict, the city-states paid little attention to neighboring Macedonia, a kingdom north of Greece that was preparing to attack.

In 338 B.C., King Philip II of Macedonia used his well-trained army to seize control of Greece. After Philip's death, his son Alexander took control. A brilliant military planner, Alexander conquered vast new territories and became known as Alexander the Great. His empire extended into North Africa, the Middle East, and Asia. As his empire expanded, Greek culture spread. When Alexander died, three of his generals divided his territory among themselves, ending one of the great empires of the ancient world.

EVALUATE Explain the differences between Athens and Sparta. Sparta: military state run by two kings; Athens: limited but direct democracy allowed all citizens to participate in government

History of Ancient Rome

KEY QUESTION How were the governments of the Roman Republic and the Roman Empire different?

While Athens was creating a democracy, the people of Rome, located west of Greece on the Italian Peninsula, were also making changes to their government. They overthrew the foreign kings that ruled them and set up a **monarchy** of their own. Then, in 509 B.C., the Romans rejected rule by kings and created a new form of government.

From Republic to Empire The Romans set up a **republic**, a form of government in which citizens elect representatives to rule in their name. It was not a democracy. All male citizens could vote, but only **patricians** (puh•TRISH•uhnz), members of rich and powerful families, could hold the highest government offices. Farmers, merchants, and craftspeople made up the class known as **plebeians** (plih•BEE•uhnz). Over time, the plebeians gained more political power.

HISTORY MAKERS

Alexander 356–323 B.C.

One important effect of Alexander's conquest was the spread of Greek culture to other parts of the world. When Alexander was young, the great Greek thinker Aristotle was his teacher. Alexander came to know and admire Greek culture. His many conquests enabled him to spread Greek language, ideas, and beliefs far beyond the Greek peninsula. Greek culture influenced life throughout his vast empire.

ONLINE BIOGRAPHY For more on the life of Alexander the Great, go to the **Research & Writing Center @ClassZone.com**

HISTORY MAKERS

Alexander the Great

Alexander was only 16 when his father, Philip II, named him regent, granting him the power to rule in Philip's absence. Alexander was just 18 when he led a cavalry charge that helped defeat the Greeks. When his father was assassinated, Alexander became the new ruler at age 20. He set out to conquer other lands. As he did so, Alexander blended Greek culture into the cultures of conquered regions.

Teach

History of Ancient Rome

Talk About It

- Why did the Romans accept the rule of one person in 27 B.C.? *(weary of civil wars)*
- What was the main purpose of dividing the empire into halves? *(to make it easier to govern)*
- **Compare and Contrast** What power in the Roman Republic did plebeians not share with patricians? *(Plebeians had no right to hold high government offices.)*

Unit Resource Book: Europe
- History Makers: Julius Caesar, p. 41

INTERDISCIPLINARY ACTIVITIES

CONNECT to Art

Images of Alexander the Great

Have students work in small groups to create a visual display that shows some of the major accomplishments of Alexander the Great. Encourage students to focus on one aspect of Alexander's achievements, such as his military campaigns or his spreading of Greek culture.

CONNECT to Civics

Compare Roman and U.S. Governments

Have students work in small groups to compare and contrast government and politics in the Roman Republic with that in the United States. Have students then present their findings as an oral report or as part of a classroom display.

More About . . .

Augustus

The Roman Senate gave Octavian the name Augustus, meaning "revered." Considered a wise ruler, Augustus restored buildings, constructed roads, and reformed coinage. He was also a patron of the arts. Under his rule, Rome experienced a period of peace and prosperity often called the Pax Romana.

Division of the Roman Empire *A.D. 395*

CONNECT Geography & History

After the fall of the Western Roman Empire, Constantinople became the center for Christianity and Greco-Roman culture.

- Using the gridlines on the map, describe the location of the Western Roman Empire. *(north-south, from above 50° N to 30° N; east-west, from 10° W to 20° E)*

ANSWER

Region Western: Rome; Eastern: Constantinople

Unit Resource Book: Europe
- Skillbuilder Practice: Explaining Geographic Patterns, p. 36

CONNECT to the Essential Question

What changes have taken place in Europe since ancient times?

Ask students what they have learned so far that can help them answer this question. Students might mention:

- Greek and Roman ideas influenced modern government.
- Ancient Greek and Roman cultural achievements still shape modern culture.

Over centuries the Roman Republic grew, until it controlled the entire Italian Peninsula. As they conquered, the Romans offered citizenship to many groups, a policy that strengthened the republic. After almost 500 years, however, the republic began to fall apart. By 27 B.C., military leaders were fighting civil wars. Worn down by the conflict and seeking order at the cost of liberty, the Romans allowed Octavian—later known as Augustus—to take over the government. The republic became an empire, united under a supreme leader.

Augustus and later emperors greatly expanded the empire, as the map below shows. Palestine, a Jewish kingdom on the eastern edge of the Mediterranean, came under Roman control. It was here that Jesus was born and Christianity began. The new religion quickly spread across the empire, becoming its official religion in A.D. 380.

The Empire Falls Apart By the third century A.D., the Roman Empire faced serious problems, which you can see in the chart below. To make the empire easier to govern, it was divided it into two halves. Constantinople became the capital of the eastern empire. Rome remained the capital of the western part. Germanic tribes who lived outside of Rome's borders began pushing into Roman territory. In A.D. 476, the Western Roman Empire fell. The Eastern Roman Empire lasted about 1,000 years more.

In the Roman Republic, citizens elected representatives to govern for them. Under the Roman Empire, the emperor took over the government and had all the power. ▶

COMPARE AND CONTRAST Explain how the governments of the Roman Republic and the Roman Empire were different.

Division of the Roman Empire *A.D. 395*

CONNECT Geography & History

Region What were the capital cities of the Western Roman Empire and the Eastern Roman Empire?

DIFFERENTIATING INSTRUCTION: TIERED ACTIVITIES

OBJECTIVE Form opinions about problems faced by the Roman Empire and possible solutions.

Basic

Refer students to the "Causes for Decline of the Roman Empire" chart included on the map. Discuss the list together. Then have students write a short statement of opinion about one problem and its possible solution from the point of view of an ordinary Roman citizen.

On Level

Refer students to the "Causes for Decline of the Roman Empire" chart included on the map. Then have them assume the role of an ordinary Roman citizen. Have them write a letter to the emperor explaining how their families are affected by these problems and making suggestions for solutions.

Challenge

Refer students to the "Causes for Decline of the Roman Empire" chart on the map. Then have pairs of students create a dialogue between a Roman citizen and the emperor about the problems and possible solutions. Have students practice and revise their dialogues before presenting them to the class.

Classical Culture

KEY QUESTION What accomplishments of ancient Greece and Rome still influence modern life?

The ancient Greeks and Romans left a powerful legacy. The Greeks excelled as artists, writers, and philosophers. Their ideas led to developments in theater, science, and government that still influence life today. Perhaps the greatest legacy of the Greeks is democracy. Greek society was one of the first to give people a voice in government. Democracy is a goal for many countries.

The Romans adopted many aspects of Greek culture, but they also had many practical skills of their own. Roman engineers designed and built roads, aqueducts, and public buildings such as the Colosseum. Their system of roads helped expand trade networks and spread culture, including Christianity. The Romans invented the idea of the republic and created a written code of law, the Law of the Twelve Tables. These ideas later shaped legal systems throughout Europe and the Americas.

eece: myths, ays, poetry, ilosophy, d democ- cy; Rome: gineering, chitecture, d law ▸

SUMMARIZE Describe the achievements of ancient Greece and Rome that continue to influence modern life.

Animated GEOGRAPHY

Roman Aqueduct Their advances in engineering allowed the Romans to build aqueducts, like this one in France, which carried fresh water from distant sources into cities and towns.

Click here to see how aqueducts work @ClassZone.com

Section 2 Assessment

ONLINE QUIZ For test practice, go to **Interactive Review** @ClassZone.com

TERMS & NAMES

1. **Explain the importance of**
   - democracy
   - oligarchy
   - patrician
   - plebeian

USE YOUR READING NOTES

2. **Compare and Contrast** Use your completed chart to answer the following question:
   Who participated in elections in the Roman Republic?

GOVERNMENT	WHO RULED	HOW CHOSEN	WHO PARTICIPATED
Greek Democracy			
Roman Republic			
Roman Empire			

KEY IDEAS

3. Why were the waters surrounding the Greek peninsulas an important resource for ancient Greece?
4. How did Alexander help spread Greek culture to foreign lands?
5. What role did patricians and plebeians play in the early Roman Republic?

CRITICAL THINKING

6. **Compare and Contrast** What are some differences between the achievements of the ancient Greek and Roman civilizations?
7. **Analyze Causes** Which of the causes of the fall of the Western Roman Empire do you think was most significant? Why?
8. **CONNECT to Today** How was the original Roman Republic similar to the U.S. government today?
9. **WRITING** **Rewrite a Myth** Pick a myth from Greek or Roman literature. Rewrite it as a poem or short story set in the present.

Teach

Classical Culture

Reader, Recorder, Reporter

- **Compare and Contrast** Compare Greek and Roman influences. *(Greek: democracy, arts and sciences; Roman: law and engineering)*

Unit Resource Book: Europe
- Interdisciplinary Projects: Greek Mythology, pp. 44–45

Animated GEOGRAPHY

Roman Aqueduct

CLASSZONE.COM

Roman aqueducts relied on gravity to supply and transport water. Although many people think of them as being above ground, much of the aqueduct system was actually underground.

4 Assess & Reteach

Assess Have students complete the Section Assessment.

Unit Resource Book: Europe
- Section Quiz, p. 50

Interactive Review @ClassZone.com
- Section Quiz

Power Presentations

Test Generator

Reteach Have pairs of students create "True or False" statements about the chapter. Have them exchange statements with another pair. Then have them correct each false statement.

Unit Resource Book: Europe
- Reteaching Activity, p. 54

SECTION 2 ASSESSMENT ANSWERS

Terms & Names

1. democracy: government in which citizens make decisions; oligarchy: government ruled by a few individuals; patricians: could hold government office in ancient Rome; plebeian: could vote but not hold office in ancient Rome

Use Your Reading Notes

2. Patricians and plebeians had the right to vote.

GOVERN-MENT	WHO RULED	HOW CHOSEN	WHO PARTICIPATED
Greek Democracy	citizens	elected	free adult male citizens
Roman Republic	patricians	elected	all male citizens
Roman Empire	an emperor	self-appointed	no one

Key Ideas

3. provided food; easiest routes for trade and travel
4. In the areas he conquered, Alexander encouraged the use of the Greek language and spread Greek ideas and beliefs.
5. patricians had most of the power, were the only ones allowed to hold high government offices; plebeians could vote but not hold office

Critical Thinking

6. Greeks were creative thinkers, made advances in theater, art, and philosophy; Romans were more practical, with advances in engineering and law.
7. accept all answers supported with reasons
8. In both, people elect representatives to voice their issues in government.
9. Use the rubric to score myths.

Myth Rubric

	Connection to Myth	Writing Style
4	very clear	lively and interesting
3	clear	interesting
2	some confusion	mainly factual in style
1	unclear	disorganized

1 Plan & Prepare

Objective

Gain an appreciation of the cultures of ancient Greece and Rome and understand their enduring values.

Springboard The ancient Greeks and Romans left many cultural influences. Did you know that the ancient Greeks started the Olympic games? Did you know that ancient Roman architects developed a way to build domes?

- Can you think of an example of a building in the United States that has a dome? *(Possible Answers: U.S. Capitol building; many places of worship; public buildings such as state government buildings)*

2 Teach

Classical Cultures

Talk About It

- What new art form did the Greeks create? *(the drama)* Can the influence of this art form be seen today? *(Possible Answer: Yes, in modern plays, TV shows, and films)*
- What kinds of places did ancient Greeks and Romans use for entertainment? *(The Greeks used amphitheaters, the Romans used arenas.)*
- What Greek ideas did the Romans use and transform? *(ideas about the gods and architecture)*

More About . . .

Greek Drama

The dramatic form developed in Greece from festivals held each year to celebrate the god Dionysus. Singing and dancing took place as part of this religious festival. The festivals evolved to include what we would now call drama.

Greek actors wore masks for performances, and men played all the parts, even women's parts. Historians believe that an Athenian called Thespis played the first character. He is considered the inventor of tragedy. Today, actors are sometimes called thespians in honor of Thespis.

COMPARING Classical Cultures

The cultures of ancient Greece and Rome shaped the people who came after them. The influence of both cultures can still be seen today.

Greece

The Greeks developed new ideas about architecture and the gods. They created a new art form—the Greek drama. Other cultures, including ancient Rome, picked up many of these ideas. We can still see ancient Greece in ruins like the Acropolis, shown below, and reflected in our own culture.

Religion

Poseidon is the Greek god of the sea. The Greeks often showed their gods as looking like human beings, but with divine powers.

Daily Life

Greek pottery provides a good record of daily life. This pot shows women collecting water from a public fountain.

Entertainment

Many Greek plays were performed during religious festivals, in large amphitheaters.

DIFFERENTIATING INSTRUCTION: TIERED ACTIVITIES

OBJECTIVE Gain a deeper understanding of classical culture and understand its enduring influence.

Basic

Have students study the photos of Greek and Roman architecture. Provide additional examples. Then have students write a paragraph comparing the two architectural styles. Ask them to include simple drawings of one architectural feature from each ancient culture.

On Level

Have students study the photos of Greek and Roman architecture. Provide additional examples. Then have students write a three-paragraph essay comparing the two architectural styles. Ask them to include drawings of one architectural feature from each ancient culture.

Challenge

Have students study the photos of Greek and Roman architecture. Provide other examples. Then have students write an essay comparing one of the architectural styles with a building in the United States. Ask them to include drawings of architectural features from each.

Religion

The Romans called the god of the sea Neptune. Roman depictions of their gods and goddesses frequently seem more realistic and three-dimensional than those of the Greeks.

Rome

Roman culture was based on values of strength and loyalty. The Romans picked up some Greek ideas about the gods and architecture. They transformed these ideas into styles that were uniquely Roman, as seen in the Forum, shown below.

Daily Life

Agriculture and trade formed the basis of the Roman economy. Shops occupied the ground story of many ancient buildings.

Entertainment

Chariot races, which took place in oval arenas called circuses, drew huge crowds.

CRITICAL THINKING

1. **Compare and Contrast** What do their forms of entertainment suggest about the differences between Greek and Roman culture?
2. **Make Inferences** Why might these cultures have saved their classical buildings?

More About . . .

Roman Construction

The Romans were innovative engineers. Some of their major innovations include

- Use of concrete—The Romans developed concrete by adding rocks or gravel to mortar. Along with arches, the development of concrete enabled the Romans to construct many public works projects such as aqueducts and sewer systems.
- Road networks—The Romans knew the importance of roads in maintaining their empire. These highways can be considered a model for today's highway system.
- Stadium architecture—The famous Roman arena, the Colosseum, was built using innovative methods of design and construction. It could hold about 50,000 spectators and still stands today.

Comparing Regions

Greek and Roman Architecture

The ancient Greeks developed architectural features such as certain types of columns. The Parthenon on the Athens Acropolis was built with numerous columns. The ruins of the Parthenon and its columns can be seen in Athens today. The Pantheon—built as a temple in Rome beginning in 27 B.C.—displays a Greek influence in a row of columns at the front of the building, but it was built with distinctly Roman features. Roman architects developed vaults and arches. The use of arches made it possible to build the Pantheon with a large dome.

3 Wrap Up

CRITICAL THINKING ANSWERS

1. **Compare and Contrast** Possible Answer: Romans preferred competitive sports for entertainment while the Greeks liked a more thoughtful entertainment, such as plays.
2. **Make Inferences** Possible Answer: The buildings serve as a reminder of past achievements and as the foundation upon which their present-day culture was built.

INTERDISCIPLINARY ACTIVITIES

CONNECT to Art

Design an Urn

Direct students' attention to the photograph of the Greek pottery, or urn. Discuss the features of the urn, such as its colors and designs. Then have students create a design for a vase. The design should reflect the influence of Greek or Roman styles. Assist students in developing ideas for a design by providing additional illustrations of classical art.

CONNECT to Language Arts

Discuss Classical Entertainment

Have students do additional research about entertainment in ancient Greece and Rome. Have them write a paragraph comparing Greek and Roman entertainments to each other, as well as to modern entertainment.

1 Plan & Prepare

Objectives

- Explain the importance of feudalism in the Middle Ages
- Analyze the effects of the Renaissance and how it spread
- Identify the causes and effects of the Reformation

Read for the Essential Question

Help students read for a purpose by reminding them of the Essential Question: "What changes have taken place in Europe since ancient times?" Point out that they can use text features such as Key Questions, special features, images, and maps to form answers to the question.

Vocabulary

Best Practices Toolkit

Use the ***Best Practices Toolkit*** to model strategies for vocabulary notetaking. Vary strategies throughout the year. Choose from: *Knowledge Rating, Predicting ABCs, Definition Mapping, Word Sort, Frayer Model (Word Squares), Magnet Words, Word Wheel,* and *Student VOC.*

Vocabulary Strategies, TT9–TT16

Reading Strategy

Best Practices Toolkit

Display the Cause-and-Effect Diagram (Multiple Causes) transparency.

- Have students brainstorm cause-and-effect signal words and phrases, such as *since, as a result, resulted in, led to, because, so,* and *caused.*
- Model filling in causes and effects using information under the subheading "Medieval Society."
- Have students suggest entries.
- Display the graphic organizer for students to review.

Cause-and-Effect Diagram (Multiple Causes), TT24

SECTION 3 Reading for Understanding

Key Ideas

BEFORE, YOU LEARNED

Ancient Greek and Roman achievements continue to influence our world today.

NOW YOU WILL LEARN

Feudalism provided stability after the fall of the Roman Empire. The Renaissance marked a rebirth of creativity in Europe.

Vocabulary

TERMS & NAMES

Middle Ages the period between the fall of the Roman Empire and the modern era, from about A.D. 476 to 1453

medieval from the Middle Ages

feudalism a political system in which lords gave land to vassals in exchange for services

lord a powerful landowner

vassal a less wealthy noble who paid taxes to and served a lord in exchange for land

knight a vassal trained in combat who fought on behalf of lords

serf a person who lived and worked on the manor of a lord or vassal

manor a noble's house and the villages on his land where the peasants lived

Renaissance (REHN•ih•SAHNS) a rebirth of creativity, literature, and learning in Europe from about 1300 to 1600

patron a wealthy or powerful person who provides money, support, and encouragement to an artist or a cause

secular worldly or not related to religion

perspective a technique used by artists to give the appearance of depth and distance

Reformation a movement in the 1500s to change practices in the Catholic Church

Protestant a member of a Christian Church founded on the principles of the Reformation

Reading Strategy

Re-create the cause-and-effect diagram shown at right. As you read and respond to the **KEY QUESTIONS**, use the diagram to help you find the effect of the events listed in the first ovals in each pair.

See Skillbuilder Handbook, page R8

ANALYZE CAUSE AND EFFECT

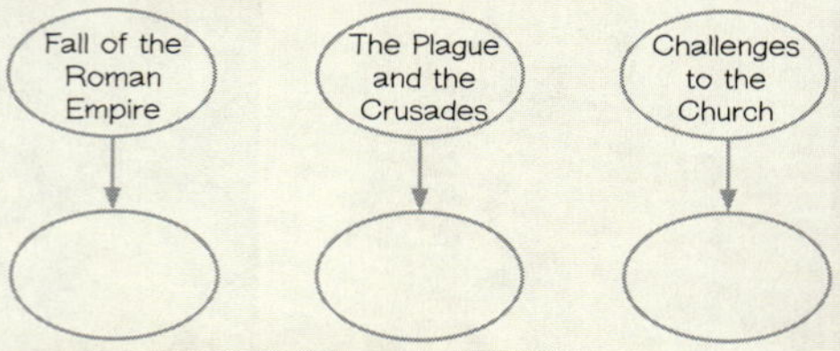

GRAPHIC ORGANIZERS
Go to **Interactive Review @ClassZone.com**

342 Chapter 11

PRETEACHING VOCABULARY

English Learners

Pronounce and Preview

Pronounce each term for students. Review the meanings of words within the definitions, such as *on behalf of, creativity,* and *technique.*

- To modify vocabulary learning, have students complete worksheets as they read, instead of afterward.

Unit Resource Book

- Vocabulary Practice, p. 35

Inclusion

Match Terms and Definitions

Have student pairs write each term or name and definition side-by-side on slips of paper. Then have them cut apart the words and definitions. Have one student take the definitions and the other student take the words. Direct students to take turns flashing a card, then matching it to the correct word or definition. Encourage the students to work together to correct any errors.

Neuschwanstein Castle, Bavaria, Germany

Swan in front of gardens at Giverny, France

SECTION 3 HISTORY

The Middle Ages and Renaissance

Connecting to Your World

Sometimes it can be difficult to imagine life without the ideas and technology that exist today. The pace of change seems to increase with every passing decade. Even 25 years ago, cell phones were just being invented. Now they are everywhere. At the beginning of the period in European history you are about to study, new ideas and inventions traveled much more slowly than they do today. Yet change did take place. In time, new ideas and discoveries challenged many Europeans' accepted beliefs about how the world worked and their place in it.

Cell phones Many people today use cell phones.

The Middle Ages

KEY QUESTION Why did feudalism develop in Europe?

Historians call the period of history between the fall of the Roman Empire and the beginning of the modern era the **Middle Ages**. It is also called the **medieval** period, from the Latin words for "Middle Ages." The collapse of the Western Roman Empire made many people fearful and uncertain. Europeans no longer had a strong central government, an army to protect them, or a common culture and belief system to unite them. Many advances of the ancient world were lost.

Bayeux Tapestry This textile shows feudal knights during the Norman Conquest of England from 1064 to 1066.

343

2 Focus & Motivate

3-Minute Warm-Up

Write on the board or display the transparency:

- Look at the image of the Bayeux Tapestry. What does this image suggest about life in Europe during the Middle Ages? *(Possible Answer: It was a time of conflict; knights were the key type of military force.)*

Unit Transparency Book: Europe
- 3-Minute Warm-Ups, TT1

Connecting to Your World

More About . . .

Bayeux Tapestry

The tapestry is a record in embroidery of the invasion of England by William, Duke of Normandy. It measures about 230 feet long by about 20 inches wide.

3 Teach

The Middle Ages

Talk About It

- Why was this period of time called the Middle Ages? *(because it came between the end of the Roman Empire and the start of the modern era)*
- What kind of organization was the Holy Roman Empire? *(a group of states loosely associated with the Roman Catholic Church)*
- **Analyze Cause and Effect** How did feudalism change the social structure of Europe? *(A king ruled over a small kingdom of people who were ranked in order from church officials and lords, to vassals and knights, to peasants.)*

SECTION 3 PROGRAM RESOURCES

Unit Resource Book: Europe

ON LEVEL

Unit Resource Book
- Reading Study Guide, p. 7
- Section Quiz, p. 51

STRUGGLING READERS

Unit Resource Book
- Reading Study Guide with Additional Support, p. 15
- Skillbuilder Practice: Explaining Geographic Patterns, p. 36
- Section Quiz, p. 51
- Reteaching Activity, p. 55

eEdition with Audio DVD-ROM

ENGLISH LEARNERS

Pupil Edition in Spanish

eEdition with Audio DVD-ROM

Unit Resource Book
- RSG (Spanish), p. 23
- RSG with Additional Support (Spanish), p. 31

Multi-Language Glossary

Test Generator
- Section Quiz in Spanish

INCLUSION

Unit Resource Book
- RSG with Additional Support, p. 15
- Section Quiz, p. 51

GIFTED & TALENTED

Unit Resource Book
- Connect to History, p. 47
- Primary Source: Elizabeth I, p. 48
- Section Quiz, p. 51

PRE-AP

Unit Resource Book
- History Makers: Charlemagne, p. 42
- Economics in the World, p. 46

TECHNOLOGY

Unit Transparency Book
- 3-Minute Warm-Ups, TT1
- Essential Question, TT2
- Critical Thinking, TT5
- Cultures, TT10

Daily Test Practice Transparencies
- Section 3, TT39

Power Presentations

ClassZone.com

World Cultures and Geography Video on DVD

More About . . .

Charlemagne

A court attendant described Charlemagne as at least six feet tall and strong, with piercing eyes and light hair. Charlemagne was alert, forceful, and concerned about the moral welfare of his empire, bringing Christianity to the peoples he conquered. He established a system of justice and made his court a center of learning and culture. Historians believe Charlemagne is a key figure in European history since he paved the way for later movements toward political, intellectual, and cultural unity.

Unit Resource Book: Europe
- History Makers: Charlemagne, p. 42

More About . . .

Knights

A boy would start training for knighthood at about age seven. He would leave home and apprentice himself to another noble. He was taught to use weapons and proper manners. As a teenager, the apprentice became valet, or manservant, to his mentor and rode into battle as his squire. Eventually the apprentice was dubbed a knight, usually in a ceremony that involved touching a sword blade to the shoulder. The usefulness of knights in battle ended as the introduction of artillery changed military strategy.

Advantages: protection from weapons; Disadvantages: heavy, restricts and slows movement, tires out the wearer ▼

European Knight Knights often wore a suit of armor to protect themselves in battle. These suits made it difficult to move quickly—the average suit weighed 65 pounds! **What might the advantages and disadvantages of armor be?**

Medieval Society The Germanic tribes in western Europe were quite different from the Romans. They had no tradition of central government. At first, many small kingdoms replaced the Roman Empire. Roads and water systems were not kept up. Trade declined. As their economies slowed, western European towns shrank. Residents abandoned them, heading to the countryside to become farmers. Literacy—and with it the educated middle class—all but disappeared.

During the early Middle Ages, invaders such as the Huns, Moors, and Vikings threatened Europe. Constant conflict and warfare plagued the region. In the 700s, a ruler named Charlemagne (SHAHR•luh•MAYN) brought much of France and Germany under his control. Charlemagne was a strong military leader. He worked with the pope, the leader of the Roman Catholic Church, to strengthen the church and his own empire. Pope Leo III crowned him emperor in 800. However, Charlemagne died in 814. In 843, his three grandsons divided his empire among themselves. Europe again became a disorderly group of small kingdoms.

Faced with such disorder, Europeans turned to a political system called **feudalism**, which would remain in place from about the 9th to the 14th century. Feudalism created a new social structure in Europe. The king ruled at the top of society. Nobles, church officials, knights, and peasants had their places below the king.

Feudalism depended on an agreement between two groups of nobles—lords and vassals.

- **Lords**, or powerful landowners, gave some of their land to less-wealthy nobles called vassals.
- **Vassals** pledged to serve their lords. They paid taxes to the lords in exchange for their plots of land, called fiefs (feefs).
- Some vassals were warriors known as **knights**, who began combat training as young boys. Knights fought on behalf of their lords.
- The feudal structure also included peasants. Most were **serfs**, who lived and worked on a noble's land. Serfs received housing and protection in return.

This political system kept Europe divided into many small kingdoms and estates, with little trade between them. As a result, the nobles' lands became the center of most economic activity. The main part of a noble's land was called a **manor**. Often the manor was a fortified house, surrounded by farmland worked by serfs. Manors supplied much of what their residents needed. Towns grew less important, as townspeople left to work on manors.

DIFFERENTIATING INSTRUCTION

Struggling Readers

Create a Visual Vocabulary of the Middle Ages

Have students create a Visual Vocabulary of one of the highlighted terms on this page. They should write the term at the top of a sheet of paper, draw or find an illustration, and write a short definition below. Have students create a classroom display of their Visual Vocabulary pages.

English Learners

Vocabulary: Suffixes

Point out that *feudalism* contains the suffix *-ism*. This suffix means "an action, process, or practice." Have students review the definition of *feudalism* on the Reading For Understanding page to be sure they understand it. Then have them think of other words that have the suffix *-ism* and define those terms in their own words. *(terrorism, racism, heroism)*

The Role of the Church The Roman Catholic Church was one institution that survived the fall of Rome. After the division of the Roman Empire, Christianity split into several different churches. The Roman Catholic Church developed in the Western Roman Empire. Many of the Germanic tribes that invaded Rome converted to Christianity. The religion spread slowly across Europe. The church became the main source of education during the Middle Ages. Church officials built universities where nobles could go and study.

Eventually, the territory once controlled by Charlemagne became the Holy Roman Empire. It was a loose confederation of states associated with the Catholic Church, rather than a unified empire with a strong central government. Even so, the Holy Roman Empire helped to bring Europe back together after the divisions of the Middle Ages.

ANALYZE CAUSES Explain how events in Europe contributed to the rise of feudalism. Feudalism developed to provide order, stability, and security at the local level, to fill the void left when the Western Roman Empire collapsed.

Animated GEOGRAPHY

European Feudalism

Click here to hear from each level of feudal society @ ClassZone.com

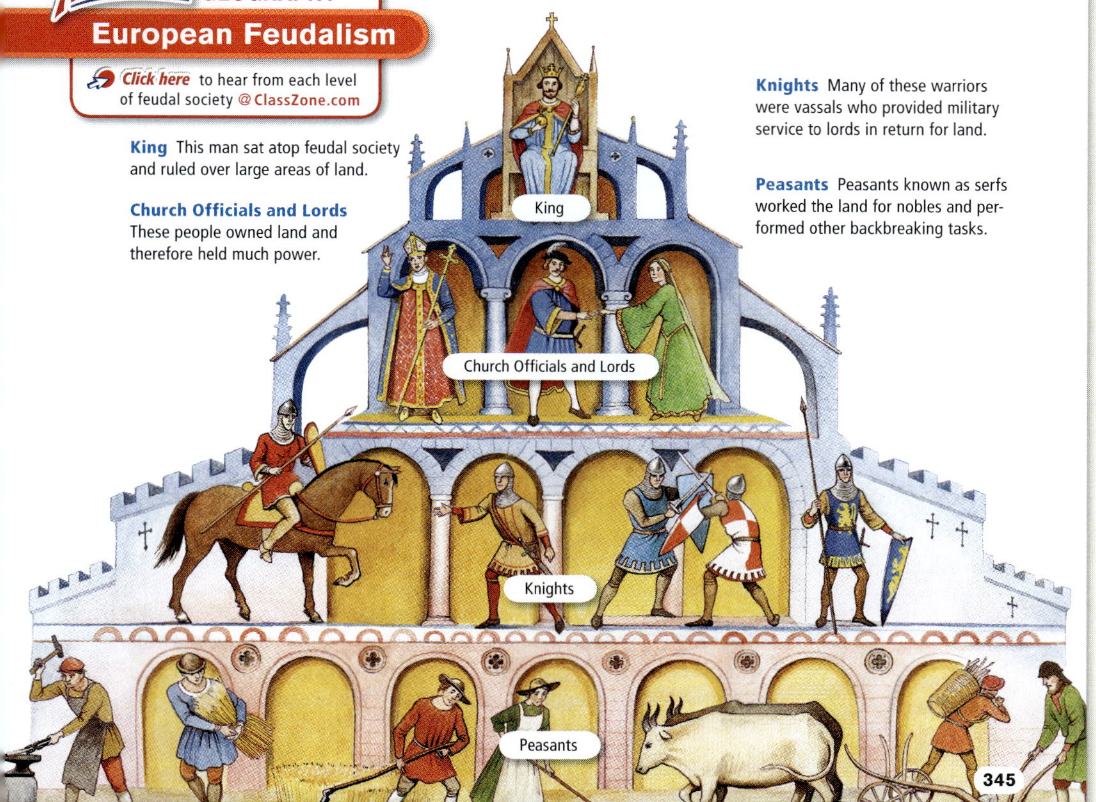

DIFFERENTIATING INSTRUCTION

Struggling Readers

Relationships Between Text and Graphic Elements

Help students learn the relationship between the text and graphic elements of the European Feudalism feature. Have students study the graphic. Then have volunteers read the text aloud. Ask questions to help them better establish the link between text and graphic, such as: Why are church officials ranked with lords? *(Possible Answers: both were of the same rank; held similar powers)*

Gifted & Talented

Create a Feudal Society Display

Brainstorm with students other ways to present the information shown in the European Feudalism feature. Would an illustrated encyclopedia entry work? A collage? Have students form small groups to discuss and create an alternative display appropriate for the presentation of the content.

More About . . .

Peasant Revolts

Peasants were often discontented with their lot. A major peasants' revolt known as Wat Tyler's Rebellion occurred in England during the reign of King Richard II. Peasants protested forced labor and payment of unfair taxes. Riots broke out in various places in England. Then, in 1381, the farm laborer Wat Tyler led thousands of peasants from the countryside to London. The king met with the peasant leaders and even agreed to their demands, but he later failed to fulfill his promises.

Animated GEOGRAPHY

European Feudalism

CLASSZONE.COM

Remind students that the political system of feudalism required a rigid social structure to be effective.

- In brief, what activities are performed at different levels of society? *(The king rules his kingdom, the priests pray, the lords rule their lands, the knights do battle, and the peasants perform the hard labor.)*
- What can you learn from the pyramid about which rank of people is greatest in number? *(The bottom and biggest level of the pyramid symbolizes that the peasants outnumber the higher ranks.)*
- How might people of one rank feel about people in another rank? *(Possible Answer: The people of higher rank might feel they are superior; people of lower rank might resent those above them.)*

Unit Resource Book: Europe
- Economics in the World, p. 46

Presentation Options

Use the **Power Presentations** DVD-ROM to project the animated graphic. Prompt students to use the pop-up images and audio to learn more about each level of feudal society.

CONNECT to History

We often think of castles as long-lasting fortresses built of stone. However, some castles, such as "motte and bailey" castles, were built of timber and only meant to be used as temporary defenses. These castles were composed of a wooden tower on a motte, or earthen mound. The bailey was the name for the grounds between the motte and the outer wall surrounding this "castle." William the Conqueror relied on these structures during the Norman Conquest of England.

Unit Resource Book: Europe
- Connect to History, p. 47

Teach

The Renaissance

Reader, Recorder, Reporter

- How did the Black Death affect the power of feudal lords? *(The high death rate led to a labor shortage on manors.)*
- What new interests were reflected in Renaissance art? *(secular, or worldly, and individual concerns)*
- **Analyze Cause and Effect** How did the Renaissance bring about demands for religious reform? *(New ideas led people to question existing institutions and practices.)*

Unit Resource Book: Europe
- Primary Source: A Conference with Elizabeth I, p. 48

More About . . .

The Black Death

In the 14th century, a plague known as the Black Death struck Europe. Historians believe it entered Europe from ships at ports along the Mediterranean Sea. The plague is an infectious disease that spreads quickly. The death toll from the plague in Europe has been estimated at 20 to 25 million people.

CONNECT to History

Feudal castles were designed for defense against enemy attacks. Many included the following defensive features:

- Moats Ⓐ, filled with water, prevented attackers from getting too close.
- Watchtowers Ⓑ allowed guards to fire on approaching enemies from a protected position.
- Thick stone walls Ⓒ kept enemies out of the castle's inner courtyard.

Bodiam Castle This castle, built by an English knight around 1385, has many defensive features.

Activity

Design a Castle Floor Plan

Materials
- paper
- pencil
- ruler

1. Be sure your castle has these features: walls, windows, and moat.
2. Research and add one other defensive feature to your castle.

The Renaissance

KEY QUESTION How did the Renaissance change Europe?

Peace and stability were returning to Europe. Merchants again felt safe traveling on the roads. Trade began again. Towns grew. Travel spread new ideas, which set the stage for change.

Forces of Change At the end of the 11th century, thousands of western European Christians took part in the Crusades, a series of military expeditions to take back the Holy Land, Palestine, from the Muslims. The Crusades led to centuries of mistrust between Christians and Muslims and imposed economic burdens on many Europeans. However, they resulted in economic growth, increasing trade between towns on the Mediterranean Sea and in the Middle East.

The importance of towns increased during the Crusades, when towns were needed to supply armies. As towns grew into cities and serfs left manors to find better work, feudal lords lost power. In the 1300s, the deadly plague, known as the Black Death, swept through Europe. About one-third of Europe's population died. The high death rate led to a labor shortage, which further weakened feudal ties.

All these forces helped bring about the Renaissance. The **Renaissance** (REHN•ih•SAHNS) was a 300-year period of renewed interest in learning and art from about 1300 to 1600. The rediscovery of ancient Greek and Roman knowledge influenced the Renaissance. Europeans developed new ideas about art, science, and humanity.

DIFFERENTIATING INSTRUCTION

English Learners

Language: Pronoun Referents

Point out the pronoun *they* in the sentence beginning, "However, they resulted in economic growth. . . ." Read aloud the previous sentence beginning, "The Crusades led to centuries of mistrust. . . ." Review the plural nouns that follow. Make sure students understand that *they* refers to the Crusades and not to other plural nouns in the previous sentence.

Pre-AP

Encyclopedia Entries on the Black Death

Have students research one aspect of the Black Death, such as its beginnings, how it spread, yearly estimates of deaths, or its consequences. Using their findings, have them create an encyclopedia entry. The entry should have a title. Have students combine their entries into one complete entry or create a classroom display.

The Rebirth of Europe The Renaissance began in the city-states of the Italian Peninsula. Increased trade between Italian towns and the Middle East after the Crusades had made many Italian merchants and bankers wealthy. They used their new wealth to build and furnish beautiful palaces. Some became **patrons** of the arts, supporting painters and writers. They showed their pride in their city by hiring architects to build churches, public fountains, and sculptures. City-states like Florence, Rome, and Venice competed to display the talents of Italy's finest artists, such as Michelangelo and Leonardo da Vinci.

Renaissance art reflected the beliefs of the period. Many scholars began to study humanity, prompting a new interest in the individual and in **secular**, or worldly, concerns. Many Renaissance paintings still had religious themes, but others depicted contemporary people instead of biblical figures. Painters also found new ways to create more lifelike portraits and realistic landscapes. The technique known as **perspective** gave objects in a painting the appearance of depth and distance. The Renaissance also produced notable writers. Many wrote in their national languages, rather than Latin, which was the practice before the Renaissance. For example, the poet Dante wrote his finest works in Italian.

Animated GEOGRAPHY

Mona Lisa Leonardo da Vinci's *Mona Lisa* is one of the most famous paintings of the Renaissance.

Click here to see more works by Renaissance artists @ClassZone.com

COMPARING Medieval and Renaissance Art

MEDIEVAL

Madonna Enthroned by Duccio di Buoninsegna

- Created art with religious themes, especially scenes from the Bible
- Created flat, two-dimensional art

RENAISSANCE

Peasant Wedding by Pieter Brueghel

- Created art about secular as well as religious themes, with more emphasis on the individual and daily life
- Created lifelike, realistic sculptures and paintings

CRITICAL THINKING

Form and Support Opinions Which artistic style seems more lifelike? Why?

INTERDISCIPLINARY ACTIVITIES

CONNECT to Art

Create a Da Vinci Poster

Point out that Leonardo da Vinci is one of the most famous figures of the European Renaissance. Have students work in pairs to create a poster about his life or work. Encourage students to try to imitate his style of drawing or painting.

CONNECT to Language Arts

Write an Art Review

Have students review the information in the subsection, "The Rebirth of Europe," and use it as a basis for an analysis of the three paintings on this page. Ask students to assume the role of reviewers for an art magazine. Have them write a short review comparing and contrasting the three paintings, incorporating ideas from the text in their reviews.

Animated GEOGRAPHY Leonardo da Vinci

CLASSZONE.COM

Leonardo da Vinci began painting the *Mona Lisa* in 1503, when he was in his early fifties. Although he is well-known today as a painter, he explored many other areas of interest. He filled notebooks with scientific observations and created drawings of ideas for inventions. He investigated architectural design, engineering, war machines, and possible aircraft. In 1516, da Vinci became "first painter, engineer, and architect" to King Francis I of France.

COMPARING

Medieval and Renaissance Art

Remind students that feudal society placed church officials and nobles above peasants. Have students discuss the similarities and differences between the two paintings. *(Similar: Both show many people; Different: The styles, settings, and themes differ.)*

- How does the medieval example reflect social roles in feudal society? *(Possible Answer: The church had great power, and peasants were not considered important enough to be subjects for art.)*
- How does the Renaissance example reflect a changed view of humanity? *(Possible Answer: Religious themes were becoming less of a focus, as secular interests started to take hold.)*

CRITICAL THINKING ANSWER

Form and Support Opinions Possible Answer: Renaissance; new Renaissance techniques such as perspective and an increased importance placed on the individual made capturing an accurate likeness more important.

CONNECT History & Culture

The Printing Press

Historians believe that Gutenberg invented his printing press about 1450. Before Gutenberg's press, Europeans had advanced to making type for printing by carving blocks of wood. Gutenberg improved upon this method. He adapted an existing kind of printing press to exert enough pressure to make the print sharper and more even.

CRITICAL THINKING ANSWER

Draw Conclusions Possible Answer: Ideas were made more accessible to more people, because books could be produced relatively quickly and cheaply.

Unit Resource Book: Europe
- Skillbuilder Practice: Explaining Geographic Patterns, p. 36

Comparing Regions

Printing Around the World

Here are some milestones in the advances of printing from around the world:

- c. 1045—Movable type was invented in China. A printer named Pi Sheng began making type from clay.
- c. 1450—Gutenberg invented his press.
- 1639—Stephen Daye set up the first printing press in North America.
- By 1814—Fredrich Koenig of Germany perfected a steam-powered press with revolving cylinders.
- 1880s—Ottmar Mergenthaler, a German resident of the United States, revolutionized printing by patenting the Linotype, a machine that eliminated the need to set type by hand.

CONNECT to the Essential Question

What changes have taken place in Europe since ancient times?

Ask students what they have learned so far that can help them answer this question. Students might mention:

- Feudalism brought order to Europe.
- The Renaissance rediscovered classical culture and led to new ways of thinking.
- Martin Luther started the Reformation, which led to the spread of Protestantism.

CONNECT History & Culture

The Printing Press The invention of the printing press had a huge impact on European society. Johann Gutenberg, a German, created a machine that pressed movable type against paper. Until then, Europeans copied books slowly by hand. The printing press allowed 500 times as many books to be printed in the same amount of time. The first book printed on Gutenberg's press was a Bible. Ideas spread as books became cheap enough for many to buy.

CRITICAL THINKING

Draw Conclusions How did the printing press help ideas spread during the Renaissance?

The Renaissance Spreads In the late 1400s, the Renaissance began to spread north from Italy to France, England, Germany, and Flanders, a region that today is part of Belgium. The Hundred Years' War, a series of battles between France and England, ended in 1453. With the conflict over, cities and trade routes expanded. A wealthy merchant class developed. Like the Italian merchants, they eagerly sponsored artists and writers. So did the monarchs of these countries, who viewed artistic achievements as a source of national pride.

Unlike the Italian artists, many northern European artists chose to paint scenes of everyday life. Pieter Brueghel the Elder, an artist from Flanders, painted peasants dancing and feasting. His paintings included many details of daily life. Authors such as William Shakespeare examined human nature. Many of Shakespeare's plays, including *Hamlet* and *Romeo and Juliet,* are still performed today. Scholars also made scientific advances, learning more about the human body and the minerals that make up the Earth's surface.

The scholars in northern Europe were also more interested in studying Christianity than ancient Greek and Roman art. Their studies led them to call for reforms to existing religious practices. They criticized the Catholic Church for caring more about wealth and power than about spiritual matters. Renaissance knowledge and ideas like these spread across Europe, aided by the printing press. People began to examine and question the institutions around them.

FIND MAIN IDEAS Explain how the ideas of the Renaissance changed Europe. New ideas and ways of thinking led Europeans to examine their beliefs and institutions around them. Wealthy patrons supported the arts, improving cities.

DIFFERENTIATING INSTRUCTION: TIERED ACTIVITIES

OBJECTIVE Create a presentation on historical figures and daily life of the Renaissance.

Basic

Guide pairs of students in identifying Renaissance artists and literary figures not discussed in this section. Have them choose one and use library or Internet resources to research their choice. Direct them to create a display about their artist or writer, including images and captions.

On Level

Have small groups of students research any historical person of the Renaissance period not discussed in this section. Direct groups to create an oral presentation about their person. Have students rehearse and revise their presentation before presenting it to the class.

Challenge

Have students research the daily lives of people from one European country during the Renaissance period. Direct students to create a multimedia presentation about daily life during the Renaissance. Have students present their work or set it up as a "self-guided" tour.

The Reformation

KEY QUESTION What concerns led Martin Luther and others to break with the Catholic Church?

During the 14th and 15th centuries, criticisms of the Catholic Church grew more intense. In the 1500s, these concerns set the stage for the **Reformation**, a movement to change church practices. Martin Luther, a German monk, led this movement. In 1517, Luther wrote his Ninety-Five Theses, a list of statements of belief. He attacked practices he saw as corrupt. Luther posted his theses on a church door in Wittenberg, Germany. His supporters sent copies throughout Europe.

Luther's ideas spread quickly. Many northern countries broke with the Catholic Church. Other reformers in Switzerland, Scotland, and England created their own **Protestant** sects, the name given to Christians who protested against the Catholic Church. The Catholic Church tried to slow the expansion of Protestantism. They began their own Catholic Reformation. Luther was excommunicated, or cast out of the Catholic Church. Church leaders sent missionaries overseas in an effort to spread Catholic ideas around the world.

SUMMARIZE Explain what concerns led Martin Luther and others to break with the Catholic Church. *Luther saw corruption in the Church and was critical of certain practices.*

Martin Luther Luther nails his Ninety-Five Theses to the church door in Wittenberg.

ONLINE QUIZ For test practice, go to **Interactive Review @ClassZone.com**

Section 3 Assessment

TERMS & NAMES

1. **Explain the importance of**
   - Middle Ages
   - feudalism
   - Renaissance
   - Reformation

USE YOUR READING NOTES

2. **Analyze Cause and Effect** Use your completed diagram to answer the following question:
   How did the plague help bring about the Renaissance?

Fall of the Roman Empire → ( )
The Plague and the Crusades → ( )
Challenges to the Church → ( )

KEY IDEAS

3. In what ways did the Crusades contribute to the start of the Renaissance?
4. Where did the Renaissance begin?
5. What was Martin Luther's role in the Reformation?

CRITICAL THINKING

6. **Compare and Contrast** What were the main differences between the Italian Renaissance and the Renaissance in northern Europe?
7. **Draw Conclusions** What did the leaders of the Catholic Church hope to achieve with the Catholic Reformation?
8. **CONNECT to Today** How is the Internet similar to the printing press? How are they different?
9. **WRITING** **Prepare an Art Lecture** Choose a painting by a Renaissance artist. Visit an art museum or library to do research. Prepare a short talk for the class on the artist and the significance of the painting. Show the class a picture of the artwork.

Teach

The Reformation

Think, Pair, Share

- What were Luther's Ninety-Five Theses? *(list of statements of belief, criticized the church)*
- **Cause and Effect** How did Luther's Ninety-Five Theses affect Europe? *(influenced other reformers to create their own Protestant sects)*

4 Assess & Reteach

Assess Have students complete the Section Assessment.

Unit Resource Book: Europe
- Section Quiz, p. 51

Interactive Review @ClassZone.com
- Section Quiz

Power Presentations

Test Generator

Reteach Assign small groups of students one part of the section. Have them identify causes and effects, then list them in a chart. Have students share and discuss their charts.

Unit Resource Book: Europe
- Reteaching Activity, p. 55

SECTION 3 ASSESSMENT ANSWERS

Terms & Names

1. Middle Ages: period between fall of the Roman Empire and the modern era; feudalism: system in which lords gave land and protection in return for services; Renaissance: rebirth of creativity and learning in Europe; Reformation: movement in the 1500s to change church practices

Use Your Reading Notes

2. caused instability in the feudal system

Fall of the Roman Empire → Middle Ages; feudalism
The Plague and the Crusades → The Renaissance
Challenges to the Church → The Reformation

Key Ideas

3. increased trade, made towns more important
4. Italian city-states
5. key leadership role; stated his concerns publicly and called on the church to make changes

Critical Thinking

6. Italian: focus on ancient Greece and Rome, painters created more realistic depictions of people; Northern: calls for church reform, painters focused on ordinary people, daily life
7. to prevent Catholics from leaving the church and joining Protestant sects; to gain new converts
8. Possible Answers: Internet makes more information available to more people; can cost more to get internet access than to buy a book
9. Use the rubric to score lectures.

Lecture Rubric

	Style of Presentation	Understanding of Concepts
4	well-organized	thorough, accurate
3	clear	accurate
2	disorganized	some inaccuracies
1	no organization	little or no understanding

1 Plan & Prepare

Objectives

- Describe how new ideas produced change throughout Europe
- Identify the reasons for Europe's expanding power around the world
- Explain European efforts to unite after war

Read for the Essential Question

Help students read for a purpose by reminding them of the Essential Question: "What changes have taken place in Europe since ancient times?" Point out that they can use text features such as Key Questions, special features, images, and maps to form answers to the question.

Vocabulary

Best Practices Toolkit

Use the ***Best Practices Toolkit*** to model strategies for vocabulary notetaking. Vary strategies throughout the year. Choose from: *Knowledge Rating, Predicting ABCs, Definition Mapping, Word Sort, Frayer Model (Word Squares), Magnet Words, Word Wheel,* and *Student VOC.*

Vocabulary Strategies, TT9–TT16

Reading Strategy

Best Practices Toolkit

Display the Time Line transparency.

- Point out that sequencing events, or arranging events in chronological order, will help students understand how the events are related.
- Model filling in a few events from the "New Ideas Produce Change" subsection.
- Have students suggest entries.
- Display the graphic organizer for students to review.

Time Line, TT33

SECTION 4 Reading for Understanding

Key Ideas

BEFORE, YOU LEARNED

Feudalism and religion provided stability during the Middle Ages. The Renaissance marked a rebirth of creativity in Europe.

NOW YOU WILL LEARN

Revolutions in science, politics, and industry transformed Western Europe. After two world wars, European nations found new ways to cooperate.

Vocabulary

TERMS & NAMES

Scientific Revolution a major change in European thinking in the mid-1500s that led to the questioning of old theories

Enlightenment a philosophical movement in the 1600s and 1700s that was characterized by the use of reason and scientific methods

French Revolution a conflict in France between 1789 and 1799 that ended the monarchy and led to changes in the way France was governed

nationalism pride in and loyalty to one's nation

Industrial Revolution the shift that began in Britain in the 1760s from making goods by hand to making them by machine

imperialism the practice of one country controlling the government and economy of another country or terrritory

Holocaust the systematic murder of Jews and other minorities by the Nazis during World War II

European Union (EU) an organization of European nations whose members cooperate on economic, social, and political issues

Reading Strategy

Re-create the time line shown at right. As you read and respond to the **KEY QUESTIONS**, use the time line to help you place the events of modern European history in order.

See Skillbuilder Handbook, page R6

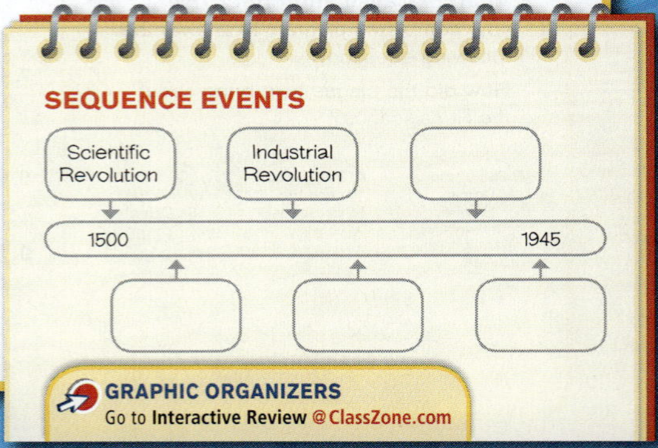

350 Chapter 11

PRETEACHING VOCABULARY

English Learners

Pronounce and Preview

Pronounce each term for students. Review the meanings of words within the definitions, such as *theories, philosophical movement,* and *systematic.*

- To modify vocabulary learning, have students complete worksheets as they read, instead of afterward.

Unit Resource Book: Europe
- Vocabulary Practice, p. 35

Inclusion

Vocabulary in Context

Have students work in pairs to locate the terms and names in the text. Ask one member of the pair to read the sentence aloud and the other to read the definition. Have pairs discuss the definition in context. Then have pairs write complete sentences using each term and name.

SECTION 4 HISTORY

Neuschwanstein Castle, Bavaria, Germany

Swan in front of gardens at Giverny, France

Modern European History

Connecting to Your World

How would you feel if you learned that what you knew about the world was wrong? Today, students aren't likely to disagree when a teacher describes how planets revolve around the sun. In the 1600s, when the Italian astronomer Galileo Galilei observed the sky through his telescope, he was looking for proof of this theory. Some things that seem obvious to us now weren't accepted as fact until a few hundred years ago. In this section, you will learn how new ideas changed Europeans' views of the world.

Telescope Galileo built his first telescope in 1609.

New Ideas Produce Change

KEY QUESTION How did Enlightenment ideas affect the struggle for independence in many European countries?

In the 1500s and 1600s, scientists such as Galileo examined accepted scientific ideas using reason and careful observation. Knowledge grew rapidly in astronomy, anatomy, and other fields. These discoveries were part of the **Scientific Revolution**, which caused scientists to re-examine old theories.

In the 1600s, European philosophers began to question traditional beliefs and accepted ideas. They argued that reason could be used to study both human behavior and the natural world. Because of the influence of these ideas, this era is known as the **Enlightenment**, or the Age of Reason.

Accademia del Cimento This scientific academy was founded by students of Galileo in 1657.

351

2 Focus & Motivate

3-Minute Warm-Up

Write on the board or display the transparency:

- Look at the painting of the Accademia del Cimento. What does it suggest about how change came about in modern Europe? *(Possible Answer: through inventions, observation, inquiry, and scientific discussion)*

Unit Transparency Book: Europe
- 3-Minute Warm-Ups, TT1

Connecting to Your World

More About . . .

Galileo

The telescope enabled Galileo to discover mountains on the moon and moons around Jupiter. Galileo also saw that Copernicus's theory that the Earth revolves around the sun was correct. The Catholic Church imprisoned Galileo for defending Copernicus.

3 Teach

New Ideas Produce Change

Think, Pair, Share

- How was France affected by the new ideas of the Enlightenment? *(French citizens revolted against monarchy and privilege.)*
- How did nationalism change Europe? *(Europeans began to see themselves as citizens of a nation, not subjects of monarchs, and pushed for democracy and self-government.)*
- **Sequence Events** What event helped spread nationalism across Europe? *(French Revolution)*

SECTION 4 PROGRAM RESOURCES

Unit Resource Book: Europe

ON LEVEL

Unit Resource Book
- Reading Study Guide, p. 9
- Map and Graph Skills, p. 37
- Section Quiz, p. 52

STRUGGLING READERS

Unit Resource Book
- Reading Study Guide with Additional Support, p. 17
- Skillbuilder Practice: Explaining Geographic Patterns, p. 36
- Section Quiz, p. 52
- Reteaching Activity, p. 56

eEdition with Audio DVD-ROM

ENGLISH LEARNERS

Pupil Edition in Spanish
eEdition with Audio DVD-ROM
Unit Resource Book
- RSG (Spanish), p. 25
- RSG with Additional Support (Spanish), p. 33

Multi-Language Glossary
Test Generator
- Section Quiz in Spanish

INCLUSION

Unit Resource Book
- RSG with Additional Support, p. 17
- Section Quiz, p. 52
- Reteaching Activity, p. 56

GIFTED & TALENTED

Unit Resource Book
- Outline Map Activities, p. 39
- Section Quiz, p. 52

PRE-AP

Unit Resource Book
- Section Quiz, p. 52

TECHNOLOGY

Unit Transparency Book
- 3-Minute Warm-Ups, TT1
- Essential Question, TT2
- Critical Thinking, TT6

Daily Test Practice Transparencies
- Section 4, TT40

Power Presentations
ClassZone.com
World Cultures and Geography Video on DVD

CONNECT History & Culture

The Guillotine

Point out that by the mid-1790s the guillotine was used repeatedly in France. The Revolution, begun in 1789, underwent stages during which legislatures formed, dissolved, and provoked power struggles. A new National Convention formed in 1792 ordered Louis XVI beheaded for treason. The Revolution then entered its bloodiest stage. During what is known as the Reign of Terror, hundreds of thousands of French citizens were arrested. Thousands of these people were executed by guillotine.

More About . . .

Napoleon Bonaparte

Napoleon rose swiftly to power. At the age of 16, he graduated from a military academy. In 1793, when Napoleon was 24, he received a promotion in the French army to brigadier general. In 1799, when he was just 30, he took control of France. Although Napoleon is perhaps best known for his military conquests and defeats, he revised the French judicial system and established a fairer set of laws called the Napoleonic Code. In general, historians view Napoleon as a military genius and as the ruler who laid the foundation for modern France.

CONNECT History & Culture

The Guillotine

Before the French Revolution, only nobles had the privilege of execution by beheading. Commoners faced hanging or more gruesome methods. In 1792, Dr. Joseph-Ignace Guillotin proposed a law requiring a machine to carry out executions. The guillotine, with its weighted blade that severed a victim's neck in one cut, was thought to be more humane and democratic than previous methods.

Enlightenment and Revolution The Enlightenment thinker John Locke argued that people had the rights to life, liberty, and property. The government's job was to protect these rights. When it failed to do so, people had the right to rebel. In 1789, Enlightenment ideas inspired French citizens to challenge the monarchy and the privileges of the wealthiest classes. Their protests led to the **French Revolution**. Radical revolutionary leaders took control of the government. They beheaded the king and abolished the monarchy.

Napoleon Seizes Power During the French Revolution, Napoleon Bonaparte distinguished himself as a brilliant leader of the French army. In 1799, Napoleon seized control of France. His goal was to create and rule a great empire. His army rapidly conquered most of Europe. However, his hopes of a long-lasting empire ended after a poorly planned attack on Russia killed many of his soldiers. In 1815, Napoleon faced his final defeat against allied European troops at the Battle of Waterloo.

Nationalism Sweeps Europe

The French Revolution helped spread **nationalism**, pride and loyalty to one's nation. Many Europeans began to see themselves as citizens of a nation, not subjects of a king. Alarmed by the French Revolution, Europe's leaders sought to stop the spread of democracy. They put kings back on their thrones, but ideas of democracy and nationalism were too powerful to fade quickly. Many European countries revolted against their rulers. By the 1870s, the smaller states that made up Italy and Germany had become unified nations. Much of western Europe had achieved self-government, inspired by the spirit of nationalism.

SUMMARIZE Explain how Enlightenment ideals contributed to European revolutions.
They moved people to question government and individual rights and inspired independence movements.

Napoleon's Coronation
Napoleon crowned himself emperor of France in an elaborate ceremony at Paris's Notre Dame Cathedral in 1804. **Why do you think Napoleon staged such an elaborate ceremony?**

to make his rule seem more legitimate to the French people

INTERDISCIPLINARY ACTIVITIES

CONNECT to Drama

Skit of the French Revolution

Have small groups of students prepare a brief skit about a protest by French citizens against the monarchy and privileged classes. Assign, or allow students to choose, roles such as king, noble, peasant, and merchant. Guide students by providing descriptions and illustrations of revolutionary protests. Then have each group prepare, rehearse, and revise its skit before presenting it to the class.

CONNECT to Language Arts

Comparing Revolutions

Have students write a compare-and-contrast essay of the French Revolution and another revolution of their choice, such as the American Revolution. Encourage students to create or find at least three visuals for their essay. Have student read their essays out loud. As a class, create a chart listing the similarities and differences among all the revolutions presented.

Europe's Expanding Power

KEY QUESTION How did the Industrial Revolution change Europe?

In the 1700s, new methods of making goods started a peaceful revolution. Industrialization led European nations to build empires.

Industrial Revolution Many of the inventions of the Scientific Revolution changed the way Europeans worked. New machines produced goods more quickly with fewer workers. The change became known as the **Industrial Revolution**. Factories were built near rivers, so that they could be powered by water. By the 1760s, steam powered the machines, and factories appeared in cities. People moved to cities from the countryside, in search of work. The Industrial Revolution began in Great Britain and spread to other European countries.

Imperialism The newly industrial nations of Europe needed raw materials and new markets for their products. Major European nations looked to Asia and Africa for valuable natural resources. Many nations made **imperialism** their foreign policy, seeking to control smaller, weaker countries politically and economically. European nations claimed overseas colonies without considering how their policies might affect the lives of the people living in these places.

New inventions led to a shift from producing goods by hand to producing goods in factories. Cities grew and workers moved from rural areas to cities. The Industrial Revolution also led to the rise of imperialism.

UNDERSTAND EFFECTS Explain how the Industrial Revolution changed Europe.

European Colonial Possessions *c. 1900*

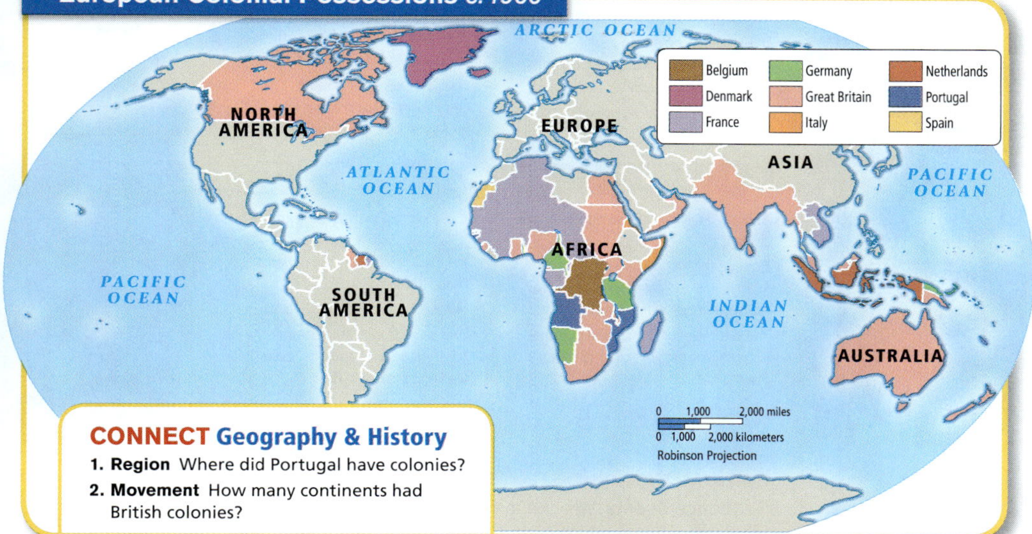

CONNECT Geography & History

1. **Region** Where did Portugal have colonies?
2. **Movement** How many continents had British colonies?

DIFFERENTIATING INSTRUCTION

English Learners

Vocabulary: Word Relationships

Point out that on the previous page students read about a violent revolution. They saw the words *revolution*, *revolt*, *rebel*, and *protest* used to tell about the overthrow of a government. Draw students' attention to the sentence on this page: "In the 1700s, new methods of making goods started a peaceful revolution." Explain that a revolution can also mean a sudden and important change for the better.

Struggling Readers

Connecting Imperialism and Colonial Possessions

The connection between imperialism and possessing colonies may be difficult for struggling readers. Review the phrases "seeking to control smaller, weaker countries politically and economically" and "European nations claimed overseas colonies." Invite students to try to explain these ideas and the connections between them in their own words.

3 Teach

Europe's Expanding Power

Reader, Recorder, Reporter

- How were goods produced by the late 1700s in Europe? *(by machine)*
- Describe how factories caused cities to grow. *(People moved from rural areas to cities to work in factories.)*
- **Sequence Events** What led European nations to establish overseas colonies? *(They needed natural resources for industrial production and markets for products.)*

European Colonial Possessions *c. 1900*

CONNECT Geography & History

Point out how the map shows the extent of European control over Africa by 1900. Tell students that the major European nations gathered for a conference at Berlin, Germany, in 1884 and 1885. There the European nations divided the continent of Africa among themselves without considering the wishes of the African people. The Berlin Conference is considered a defining event in the competition among European nations for African resources.

ANSWERS

1. **Region** Africa
2. **Movement** Five continents

Unit Resource Book: Europe
- Skillbuilder Practice: Explaining Geographic Patterns, p. 36
- Map and Graph Skills: Reading Historical Maps, p. 37

COMPARING

World War I & World War II

Both world wars divided Europe, yet also united nations in competing alliances. During World War II, nations of different foreign policies and political points of view came together to combat Nazi Germany. The Allied forces had an interest in democratic government with one exception: the Communist Soviet Union led by Joseph Stalin. Even so, Stalin cooperated with the Allies for the common good.

CRITICAL THINKING ANSWER

Sequence Events World War I: rise in nationalism and militarism; system of alliances; assassination of Archduke Ferdinand; World War II: global economic depression; Hitler taking power promising to expand German territory; Germany's invasion of Poland.

More About . . .

Trench Warfare

Much of the fighting of World War I occurred from defensive trenches dug into the ground. The invention of the machine gun and the sheer number of bullets flying through the air forced soldiers to abandon mobility in favor of the shelter and protection of trenches. A line of trenches extended from Belgium's coastline through northeastern France to Switzerland.

A typical World War I trench system included several parallel trenches stretching at least a mile back from the front lines. Communications trenches roughly perpendicular to the main trenches allowed troops, ammunition, and supplies to move between the main trenches. As the war progressed, both sides developed deeper and stronger trenches to prevent an enemy breach.

COMPARING World War I & World War II

Europe in Conflict During the 20th century, tensions between European nations led to two devastating wars, World War I (1914–1918) and World War II (1939–1945). These wars left much of Europe in ruins.

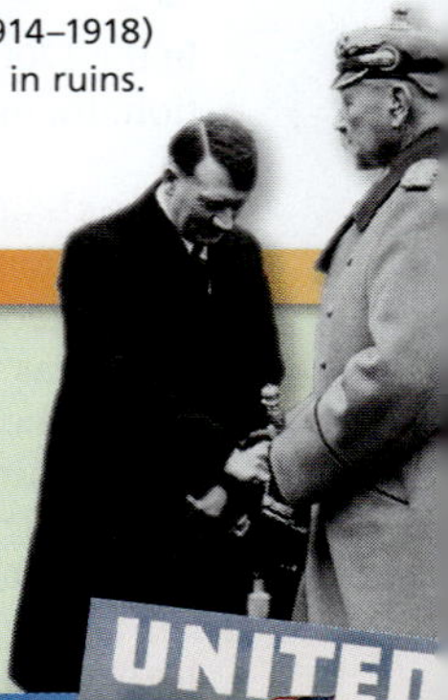

	WORLD WAR I 1914–1918	WORLD WAR II 1939–1945
Why	• A rise in nationalism, imperialism, and military buildup increases tensions. • European nations form mutual protection alliances. • The assassination of Archduke Franz Ferdinand, shown at left, forces allied nations into war.	• A global economic depression worsens conditions in Europe. • Adolf Hitler, leader of the Nazi Party, at right, gains control of Germany in 1933 and promises to expand German territory. • Under Hitler's leadership, Germany invades Poland in 1939.
Who	**Central Powers vs. Allies** • Central Powers: Germany, Austria-Hungary, and Turkey • Allies: Russia, France, the United Kingdom, and the United States (in 1917)	**Axis Powers vs. Allies** • Axis Powers: Germany, Italy, and Japan • Allies: The United Kingdom, France, the Soviet Union, and the United States (in 1941)
How	• Machine guns make it difficult for forces to advance. • Soldiers on both sides fight from defensive trenches. • Trench warfare leads to development of new weapons, including poison gas and tanks, shown at left.	• The German strategy of *blitzkrieg* uses fast-moving tanks and airplanes followed by ground troops to overwhelm enemies. • Airplanes like this one allow the war to be fought over great distances. • Nazi labor and death camps carry out the **Holocaust**, the mass murder of Jews and others.
Outcomes	• About 8.5 million soldiers and 13 million civilians die. • Fighting the war costs Europe over $330 billion and causes mass physical destruction, as seen in France at left. • The peace treaty blames and punishes Germany for the war, causing German resentment.	• Historians estimate total deaths between 35 and 60 million. The Holocaust claims the lives of 6 million Jews. (right: concentration camp survivors) • The war costs over $1 trillion. • Two superpowers emerge after the war, the United States and the Soviet Union, leading to the Cold War.

CRITICAL THINKING

Sequence Events What led to the start of Europe's two world wars?

DIFFERENTIATING INSTRUCTION: TIERED ACTIVITIES

OBJECTIVE Create a display on the consequences of World War II.

Basic

Have pairs of students use classroom resources to learn more about the outcomes of World War II. Then have them create a poster of the outcomes of World War II. Posters should include a visual and caption for each outcome listed. Have students present their posters or post them in the classroom.

On Level

Have pairs of students research the outcomes of World War II. Direct them to research the outcomes for one nation of Europe. Assign countries to ensure no overlap among groups. Then have them prepare in-depth fact sheets to present to the class or post in the classroom.

Challenge

Have five groups of students research the outcomes of World War II on North America, South America, Africa, Asia, and Oceania and Antarctica. Assign continents to ensure no overlap among groups. Then have them prepare a visual display of their choice to present to the class or post in the classroom.

Uniting After War

KEY QUESTION How did the two world wars encourage European nations to work together?

After World War II, Europe was devastated. European nations had been at war with each other for several years. Two goals emerged in the war's aftermath: to rebuild Europe's shattered nations and their economies and to work together to prevent future wars.

Creating a European Union In 1952, Belgium, France, Italy, Luxembourg, the Netherlands, and West Germany formed the European Coal and Steel Community (ECSC). Its six members agreed to combine their iron, steel, and coal industries. Their success with these commodities led them to drop trade barriers on others.

In 1967, they created the European Community (EC). Members worked to find ways to move goods, workers, and money more easily across their borders. Trade increased, and more countries wanted to join. In 1973, the EC began to admit other nations, paving the way for the European Union.

By 1992, 12 Western European nations belonged to the EC. That year, all of them signed the Maastricht (MAH-strikt) Treaty. The treaty formed the **European Union (EU)**, an organization of European nations whose members cooperate on economic, social, and political issues. By 2004, the EU had expanded to include Eastern European nations, bringing the total number of members to 25 countries. Candidate nations continue to apply.

How Does the EU Work? The goal of the European Union is to bring the people and countries of Europe closer together. Citizens of EU nations have European citizenship. They can travel freely throughout the EU. They can live and work anywhere in the union. They can vote in elections in the country they live in, even if they are citizens elsewhere.

The European Union member nations also work together on political and social matters such as immigration, law enforcement, and the environment. For instance, the EU sponsors many of Europe's efforts toward finding cleaner, sustainable sources of energy, which you read about earlier in the chapter. The EU also tries to protect the diverse cultures and traditions of its member nations. It funds cultural programs, including education in the languages of other EU nations.

Members of the European Union

Date	Country
By 1952 (ECSC)	Belgium
	France
	Italy
	Luxembourg
	Netherlands
	West Germany
By 1973 (EC)	Denmark
	Ireland
	United Kingdom
By 1995	Greece
	Portugal
	Spain
	East Germany*
	Austria
	Finland
	Sweden
By 2004	Cyprus
	Czech Republic
	Estonia
	Hungary
	Latvia
	Lithuania
	Malta
	Poland
	Slovakia
	Slovenia
By 2007 or 2008	Bulgaria
	Romania

*East Germany joined through German reunification and not as a separate state.

DIFFERENTIATING INSTRUCTION: TIERED ACTIVITIES

OBJECTIVE Lead a class discussion on recent European efforts toward union.

Basic

Have students review the content of this subsection. Then have them work in small groups to create a time line of dates and events that have taken place to unite Europe. Ask them to use their time line as a vehicle for a class discussion on steps Europe has taken to unite.

On Level

Have students use classroom and library resources to learn more about the advantages and disadvantages of European unity from the perspective of one country. After students have given their reports to the class, have volunteers lead a class discussion on European unity.

Challenge

Have students use classroom and library resources to research the European Union's proposed constitution. Lead them in a panel discussion about the similarities and differences between the EU and the U.S. constitutions. Have them comment on steps EU members might take to adopt a constitution.

Teach

Uniting After War

Talk About It

- Why did European nations begin to unite after World War II? *(to rebuild and prevent future wars)*
- What is the Maastricht Treaty? *(It formed the European Union.)*
- **Sequence Events** What events led up to the formation of the European Union? *(the devastation of war, creation of the European Coal and Steel Community in 1952, creation of the European Community in 1967)*

Unit Resource Book: Europe
- Outline Map Activities: Europe: Political, p. 39

Members of the European Union

Discuss with students why countries have wanted to join the European Union. *(Possible Answers: to improve their economies; to trade, work, and travel more freely)*

- What do most of the countries who joined after 1995 have in common? *(All but two are Eastern European countries.)*

CONNECT to the Essential Question

What changes have taken place in Europe since ancient times?

Ask students what they have learned so far that can help them answer this question. Students might mention:

- Enlightenment ideas sparked Europeans to challenge old traditions and replace governments.
- The Industrial Revolution changed how people worked and led to imperialism.
- European nations have worked to form closer ties after two devastating wars.

CONNECT History & Culture

The Euro

A European Economic and Monetary Union was set up to unify economic policies and currencies of EU member nations. The final step in this process was the creation and circulation of a common coinage, the euro. The European Central Bank, created in 1998, works with banks of EU members to manage monetary policies and the euro.

- How might the spread of a nation's image on coinage help European nations cooperate? *(Possible Answer: Citizens of other nations may become familiar with another nation's culture and values and learn to better appreciate them.)*

4 Assess & Reteach

Assess Have students complete the Section Assessment.

Unit Resource Book: Europe
- Section Quiz, p. 52

Interactive Review @ClassZone.com
- Section Quiz

Power Presentations

Test Generator

Reteach Have students work in pairs to create a list of changes in Europe in sequence for each subsection. Ask students to share and discuss each other's lists and revise them afterward.

Unit Resource Book: Europe
- Reteaching Activity, p. 56

CONNECT History & Culture

The Euro

The front sides of euro coins have the same image, showing the value of the coin. The backs of the coins have a national image chosen by the country that issued the coin. This is one small way that countries are able to belong to the European Union and share a common currency, but maintain their own cultural heritages.

Economic Unity One major area of EU cooperation is its economic policy. The EU is a single market, which means that its members can trade goods freely, without paying taxes on those goods. Many EU members also use the same currency, called the euro. In 1999, 11 EU nations adopted the euro. This decision made it easier for members to trade with each other.

As a single economic unit, the EU is powerful, more than any member nation on its own. However, the EU is weak in other ways. The countries that belong to the union remain independent. The EU does not yet have a constitution. Its leaders make decisions but have little authority to enforce them. So far, the member nations have cooperated on most issues, and the EU has been successful. However, keeping such a large organization unified could be challenging, especially as new countries join.

Seeing the effects of competition and conflict made them more willing to work together. ◄

SUMMARIZE Explain how two wars encouraged European nations to work together.

Section 4 Assessment

ONLINE QUIZ For test practice, go to Interactive Review @ClassZone.com

TERMS & NAMES

1. **Explain the importance of**
   - Enlightenment
   - nationalism
   - imperialism
   - European Union

USE YOUR READING NOTES

2. **Sequence Events** Use your completed time line to answer the following question:

   Which event occurred between the Scientific Revolution and the French Revolution?

   Scientific Revolution | Industrial Revolution | [ ]
   1500 — 1945
   [ ] | [ ] | [ ]

KEY IDEAS

3. How did Enlightenment ideas influence the French Revolution?
4. How did the Industrial Revolution affect imperialism?
5. What steps did the EU and its predecessors take to bring the nations of Europe closer together?

CRITICAL THINKING

6. **Analyze Cause and Effect** How did the Industrial Revolution change where people lived?
7. **Compare and Contrast** In what ways were World War I and World War II similar?
8. **CONNECT to Today** Why might Eastern European nations want to join the European Union?
9. **WRITING** **Write a Speech** Imagine that you have been asked to speak to EU members about why you think the nations of Europe should or should not form a "United States of Europe". Write a brief speech expressing your opinion.

SECTION 4 ASSESSMENT ANSWERS

Terms & Names

1. Enlightenment: movement in the 1600s and 1700s using reason; nationalism: pride in and loyalty to one's nation; imperialism: practice of controlling government and economy of another country; European Union: nations cooperate on economic, social, and political issues

Use Your Reading Notes

2. the Enlightenment

   Scientific Revolution | Industrial Revolution | World War I
   1500 — 1945
   The Enlightenment | French Revolution | World War II

Key Ideas

3. John Locke's theories on human rights inspired French revolutionaries to rebel against their rulers.
4. European nations built empires to provide raw materials and markets for manufactured goods.
5. lowered barriers to travel and trade; created a common currency, the euro, to facilitate trade

Critical Thinking

6. People began to move to cities from the countryside, seeking work in urban factories.
7. both fought in Europe, caused enormous loss of life and property damage for Europeans, and included many of the same major combatants
8. Benefits include increased trade and cooperation with neighbors and greater mobility for workers.
9. Use the rubric to score speeches.

Speech Rubric

	Clarity	Point of View
4	clear, complete	clearly identified
3	clear	clearly identified
2	some confusion	not clearly identified
1	unclear	not identified

Click here to complete these and other activities online @ ClassZone.com

CHAPTER SUMMARY

Key Idea 1
Europe has diverse landforms ranging from mountains to plains. Its climate is influenced by its nearness to the ocean.

Key Idea 2
The achievements of the ancient Greeks and Romans continue to influence our modern world.

Key Idea 3
Feudalism provided stability after the fall of the Roman Empire. The Renaissance marked a rebirth of creativity.

Key Idea 4
Revolutions in science, politics, and industry transformed Europe. After two world wars, European nations found new ways to cooperate.

For **Review and Study Notes**, go to **Interactive Review** @ ClassZone.com

NAME GAME

Use the Terms & Names list to complete each sentence on paper or online.

1. I am Europe's tallest mountain range. Alps
2. I am the vast area of flat or gently rolling land from France to Russia. ______
3. I make the climate in much of Europe warmer than it would be otherwise. ______
4. I am the form of government that originated in the Greek city-state of Athens. ______
5. I am the form of government that Rome had before it became an empire. ______
6. I am the medieval political system that gave nobles, peasants, and serfs protection in exchange for service. ______
7. I am a time of rebirth in creativity and the arts in Europe. ______
8. I am the religious protest and reform movement that split the Church in the 1500s. ______
9. I am a time of great interest in using reason to understand and improve society. ______
10. I am an organization made up of European nations that works together to solve common problems. ______

Alps
city-state
democracy
Enlightenment
European Union
feudalism
French Revolution
imperialism
Industrial Revolution
North Atlantic Drift
Northern European Plain
Peloponnesus
Reformation
Renaissance
republic

Activities

GeoGame

Use this online map to reinforce your understanding of Europe's physical geography. Drag and drop each place name in the list to its location on the map.

Crossword Puzzle

Complete an online crossword puzzle to test your knowledge of Europe's geography and history.

ACROSS
1. vassal trained in combat who fought on behalf of lords

ONLINE AND ON-SCREEN ACTIVITIES

GeoGame

Use the interactive map on **Power Presentations** DVD-ROM or the **eEdition** to help students review Europe's geographic features.

- Demonstrate how to drag and drop the map labels listed onto the appropriate locations on the map.
- Advance to the scorecard to show how students have a chance to improve their scores.

Crossword Puzzle

Students can review important people and events in European history by completing this crossword puzzle online at **ClassZone.com.**

- As an extension of this activity, have students use the *Activity Maker* at **ClassZone.com** to make their own crossword puzzle for one section of the chapter.
- Then have students trade puzzles with classmates to solve each other's puzzles.

Chapter Summary

As a follow-up to the chapter, encourage students to pair up and go over the **Review and Study Notes** at **ClassZone.com.**

Alternatively, students pairs can review their Reading Study Guide notes, making sure that they can add important details to each section's "Key Idea."

Name Game

Teaching Tips

ON PAPER Have students work in pairs to complete this activity. As an extension, they can write questions for the terms that are not covered in the items shown here.

ONLINE Make sure that students understand how to drag and drop terms to fill in blanks. Ask them to print out their results and compare them with those of a classmate.

ANSWERS

1. Alps
2. Northern European Plain
3. North Atlantic Drift
4. democracy
5. republic
6. feudalism
7. Renaissance
8. Reformation
9. Enlightenment
10. European Union

Make It Fun

Teacher-Tested Activities

Judith K. Bock, Peter J. Palombi Middle School, Lake Villa, Illinois

The game "I Spy" is a great way to review a chapter, using a map and some imagination.

- Have students look at a map of Europe.
- Start by making an "I spy" statement, such as, "I spy a country that rhymes with dance" *(France)* or "I spy the country where the Renaissance began" *(Italy)*.
- Call on a student to provide the answer. That student then gives a new "I spy" statement and calls on another student to answer. Continue around the classroom.
- It can be helpful to have students write a few "I spy" statements in advance.

Vocabulary ANSWERS

1. sources of energy from ancient plant and animal remains
2. Citizens elect representatives to rule in their name in a republic.
3. a noble's house and the villages on his land where the peasants lived
4. a rebirth of creativity in Europe from about A.D. 1300 to 1600
5. the systematic murder of Europe's Jews by the Nazis
6. Members of the European Union cooperate on economic, social, and political issues.
7. North Atlantic Drift
8. patrician
9. vassals
10. nationalism

Key Ideas ANSWERS

11. It has a moderating effect on climate, leading to mild weather throughout much of Europe.
12. They have promoted trade and travel and helped the economies of Europe thrive.
13. Parts of it have rich soil suited to farming, and other parts are rich with coal.
14. Although limited to male citizens, Athens had the first democratic government in the ancient world. All citizens had the right to participate directly in government, deciding what laws to make and voting on them.
15. They gave some conquered peoples full or partial citizenship.
16. Lords had workers to farm their lands and soldiers for their armies.
17. The Crusades increased trade and created economic opportunities for townspeople.
18. Italian merchants involved in overseas trade grew rich and had the money to support Renaissance artists and public building projects.
19. Goods could be made faster and more cheaply with fewer workers because they were made by machines in factories rather than by hand at home.
20. The removal and execution of the French king and the abolition of the monarchy threatened their power and made them afraid that their own subjects might do the same.

Chapter 11 Assessment

VOCABULARY

Explain the significance of each of the following.

1. fossil fuels
2. republic
3. manor
4. Renaissance
5. Holocaust
6. European Union

Choose the best answer from each pair.

7. This influences Europe's mild climate. (Northern European Plain/North Atlantic Drift)
8. This group of people could hold government office in ancient Rome. (patrician/plebeian)
9. This group of people pledged service to wealthy landowners in exchange for land under the feudal structure. (lords/vassals)
10. This is a feeling of pride for and loyalty to one's nation. (nationalism/imperialism)

KEY IDEAS

1 Europe's Dramatic Landscape

11. How does the North Atlantic Drift affect the climate of Europe?
12. How do Europe's waterways affect its economy?
13. Why is the Northern European Plain a valuable resource?

2 Classical Greece and Rome

14. Why is Athens considered the birthplace of modern democracy?
15. What did the Romans give conquered peoples?

3 The Middle Ages and Renaissance

16. How did feudalism benefit lords?
17. What contributed to the growth of towns before the Renaissance?
18. Why did the Renaissance begin in Italy?

4 Modern European History

19. How did the Industrial Revolution change the way goods were made?
20. Why were many European leaders upset by the French Revolution?

CRITICAL THINKING

21. **Form and Support Opinions** Create a graphic organizer like the one below. Include three factors that influence the mild climate of Europe.

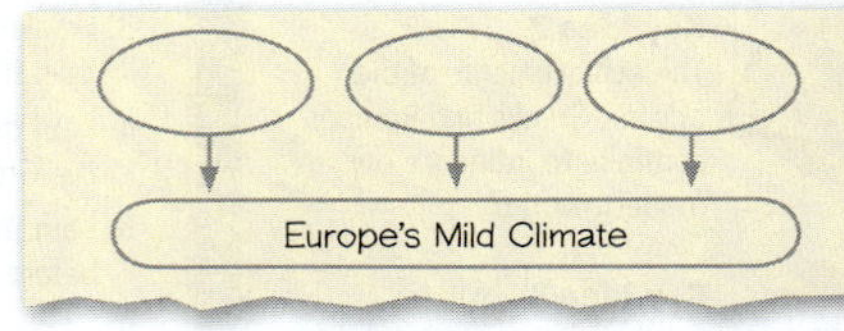

22. **Connect Geography & History** What effect have Europe's peninsulas had on its development?
23. **Five Themes: Place** How did Athens and Sparta differ in their views of citizens?
24. **Connect Geography & Culture** How did feudalism affect trade and daily life in medieval Europe?
25. **Analyze Causes and Effects** How are the Renaissance and the Reformation related to each other?
26. **Make Inferences** How did the World War I peace agreement contribute to the start of World War II?

Answer the ESSENTIAL QUESTION

What changes have taken place in Europe since ancient times?

Written Response Write a two- or three-paragraph response to the Essential Question. Consider the key ideas of each section as well as specific ideas about how Europe has changed. Use the rubric below to guide you.

Response Rubric

A strong response will:
- identify elements of ancient society and politics
- discuss three major historical events that occurred after the fall of the Roman Empire
- explain how each of those events changed European society and politics

ESSENTIAL QUESTION REVIEW

Before students begin their essays, have them review the key chapter elements that will help them answer the Essential Questions. Direct them to significant features in the chapter, such as:

- Connect Geography & Economics, Section 1
- Comparing Classical Cultures, Section 2
- Connect History & Culture, Section 3
- Comparing World War I & World War II, Section 4

Written Response Rubric

	Coverage of Key Ideas	Explanation	Accuracy
4	all	clear, concise	no errors
3	most	good	few/minor errors
2	some	fair	several errors
1	two or fewer	poor	many errors

STANDARDS-BASED ASSESSMENT

Test Practice
- Online Test Practice @ClassZone.com
- Test-Taking Strategies and Practice at the front of this book

PHYSICAL MAP

Use the map to provide short answers to questions 1 and 2 on your paper.

European Peninsulas

1. **Which of the peninsulas is located farthest north?**
2. **Which of the peninsulas has a coastline on the Black Sea?**

PRIMARY SOURCE

The following excerpt is from Pericles's "Funeral Oration." Use context clues within the quotation to answer questions 3 and 4 on your paper.

> Our constitution is named a democracy, because it is in the hands not of the few but of the many. But our laws secure equal justice for all in their private disputes, and our public opinion welcomes and honors talent in every branch of achievement, not as a matter of privilege but on grounds of excellence alone.
>
> Source: Thucydides, from *The Peloponnesian War*

3. **According to Pericles, what made Athens a democracy?**
   - **A.** It had laws that protect only the elite.
   - **B.** It granted privileges to the rich.
   - **C.** It was governed by a small group of leaders.
   - **D.** It was governed by the many.
4. **Whom did the laws of Athens protect?**
   - **A.** everyone
   - **B.** government officials
   - **C.** privileged citizens
   - **D.** peasants

GeoActivity

1. INTERDISCIPLINARY ACTIVITY–HISTORY

With a small group, learn more about the events of the French Revolution. Create a political cartoon about one of these events such as the rioting by the poor, or Napoleon's takeover of government. Your cartoon should show a clear point of view.

2. WRITING FOR SOCIAL STUDIES

Reread the part of Section 4 on the growth of nationalism in Western Europe. Imagine that you live in a European monarchy in the mid-1800s. Write a letter to a friend telling why you believe in independence for your country.

3. MENTAL MAPPING

Create an outline map of France and label the following:
- Rhone River
- Pyrenees
- Massif Central
- Seine River
- English Channel
- Atlantic Ocean

DIFFERENTIATING CHAPTER ASSESSMENT

Unit Resource Book
- **Chapter Test A,** pp. 57–60. *Use for Struggling Readers, Beginning-to-Intermediate-Level English Learners, and for Inclusion of students whose IEP suggests supported test-taking.*
- **Chapter Test B,** pp. 61–64. *Use for On-Level Learners and ESOL students with Advanced-Level English.*
- **Chapter Test C,** pp. 65–68. *Use for Gifted and Talented students or students assigned Pre-AP work.*

Test Generator
- Chapter Tests A, B, and C, in English and Spanish

McDougal Littell Assessment System

TEST **Test Generator** Use three levels of tests in English and Spanish tied to state standards. Plus, choose questions to customize tests for all populations.

SCORE Scan and Score plain paper student tests in minutes.

REPORT Create instant progress reports, sortable by student, by standard, or other criteria.

RETEACH Respond immediately to students' needs by using a wealth of reteaching resources.

CHAPTER 11 ASSESSMENT

Critical Thinking ANSWERS

21. Possible Answers: location near the ocean, westerly winds, North Atlantic Drift

22. Many peninsulas give Europe an extensive coastline, making it suited to trade, encouraging travel, and leading many Europeans to find occupations related to seafaring and shipbuilding.

23. Spartan citizens had to protect and support the aims of the state, and citizenship could be taken away if someone failed to do so. Athenians believed that citizens had a right and a responsibility to participate in government.

24. development of manors isolated people; many small estates with little trade between them; people left towns for manors

25. The Renaissance caused people to question their beliefs and the institutions around them, leading to the Reformation.

26. Many Germans resented the treaty, which might have made them more willing to follow Hitler.

Standards-Based Assessment ANSWERS

1. Scandinavian Peninsula
2. Balkan Peninsula
3. D
4. A

GeoActivity

1. INTERDISCIPLINARY ACTIVITY RUBRIC

The political cartoon should
- represent an event from the French Revolution
- show a clear point of view
- be nicely illustrated

2. WRITING FOR SOCIAL STUDIES RUBRIC

The letter should
- have a point of view about gaining independence
- make a clear argument

3. MENTAL MAPPING RUBRIC

The outline map should
- represent the approximate shape of France
- include all geographic features listed in their correct location

CHAPTER 12

Western Europe PLANNING GUIDE

Complete print resources all on one USB drive allow you to customize lesson plans for all students.

Section Titles & Key Ideas	Unit Resource Book Europe	Assessment Resources
1 Greece and Italy **KEY IDEA** For many centuries, Greece and Italy were collections of small states. Fueled by nationalism, each struggled to gain independence and unite as a nation.	• **Reading Study Guide,** pp. 69–70 • **RSG with Additional Support,** pp. 79–80 • **Vocabulary Practice,** p. 109 • **Outline Map Activities,** p. 113 • **Interdisciplinary Projects,** pp. 117, 119	**Unit Resource Book** • Section Quiz, p. 124 • Reteaching Activity, p. 129 **Test Generator CD-ROM** **Daily Test Practice Transparency,** TT41
2 Spain and Portugal **KEY IDEA** Like Greece and Italy, Spain and Portugal were ruled by foreigners. After gaining independence, both developed overseas empires to fuel their economies.	• **Reading Study Guide,** pp. 71–72 • **RSG with Additional Support,** pp. 81–82 • **Skillbuilder Practice,** p. 110 • **Primary Source,** p. 122	**Unit Resource Book** • Section Quiz, p. 125 • Reteaching Activity, p. 130 **Test Generator CD-ROM** **Daily Test Practice Transparency,** TT42
3 France and the Benelux Countries **KEY IDEA** Over the centuries, France's natural resources have helped it prosper. Through economic union, Belgium, Luxembourg, and the Netherlands have also prospered.	• **Reading Study Guide,** pp. 73–74 • **RSG with Additional Support,** pp. 83–84 • **Interdisciplinary Projects,** pp. 118–119 • **Connect to History,** p. 121	**Unit Resource Book** • Section Quiz, p. 126 • Reteaching Activity, p. 131 **Test Generator CD-ROM** **Daily Test Practice Transparency,** TT43
4 Germany and the Alpine Countries **KEY IDEA** Germany's central location helped it to dominate neighboring lands. Germany and the Alpine countries are linked in many ways.	• **Reading Study Guide,** pp. 75–76 • **RSG with Additional Support,** pp. 85–86 • **History Makers: Adolf Hitler,** p. 116 • **Economics in the World,** p. 120	**Unit Resource Book** • Section Quiz, p. 127 • Reteaching Activity, p. 132 **Test Generator CD-ROM** **Daily Test Practice Transparency,** TT44
5 The Nordic Countries **KEY IDEA** The Nordic countries have histories and cultures that are closely intertwined.	• **Reading Study Guide,** pp. 77–78 • **RSG with Additional Support,** pp. 87–88 • **Map and Graph Skills,** p. 111 • **History Makers: Leif Ericson,** p. 115	**Unit Resource Book** • Section Quiz, p. 128 • Reteaching Activity, p. 133 • Chapter Test (Forms A, B, & C), pp. 134–148 **Test Generator CD-ROM** **Daily Test Practice Transparency,** TT45

McDougal Littell **Assessment System**

TEST SCORE REPORT RETEACH

INTEGRATED TECHNOLOGY

No Time? To focus students on important content in the chapter, use the **"Reading Study Guide with Additional Support"** in ***Resources2GO.***

Teacher Presentation Options

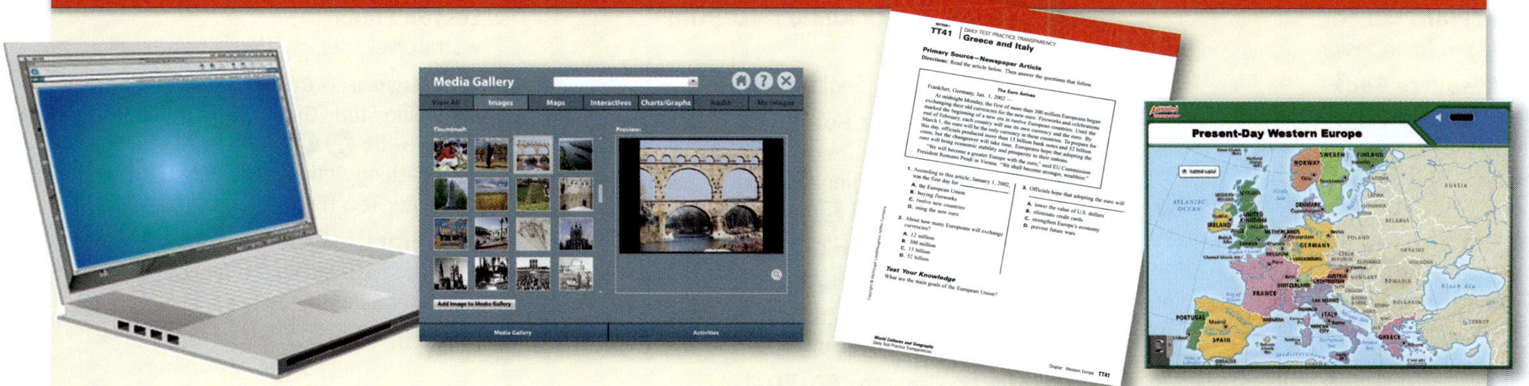

Electronic Books
eEdition DVD-ROM
eEdition Online

Presentation Toolkit
Video/DVD
- World Cultures and Geography

Power Presentations DVD-ROM
- Discussion Starters
- Media Gallery
- Review Game
- Interactive Review
- Animated Features

Unit Transparency Book
- 3-Minute Warm-Ups
- Essential Question
- Critical Thinking
- Maps
- Cultures

Daily Test Practice Transparencies

Animated Geography
- Present-Day Western Europe Map
- Worldster: Ulrika's Web Page

OnlineActivities at ClassZone.com

Animated Geography
- **Videos and Photos**
- **3-D Animations**
- **Interactive Maps**

Interactive Review
- **Section Quizzes**
- **Review and Study Notes**
- **Vocabulary Flip Cards**
- **Graphic Organizers**
- **Name Game and Crossword Puzzles**
- **GeoGame**

Online Test Practice

Map Center
- **Atlas Maps**
- **Political Maps with Data**
- **Links to Mapping Web Sites**

Activity Center
- **Activity Maker—Review Games, Puzzles, Flip Cards, and Graphic Organizers**
- **Worldster**

Research & Writing Center
- **Web Sites**
- **Data Updates**
- **Primary Sources**
- **Writing Models**

Current Events

Amber McVey
Antioch Middle School
Gladstone, Missouri

Make It Fun — Teacher-Tested Activities

As you teach this chapter, look for my classroom-proven ideas for:
- making the lesson fun
- focusing students' attention and comprehension
- integrating technology with traditional teaching

Teaching Bell to Bell When you see this timing guide, have students discuss in pairs, in groups of 3–6, or as a class.
- **Think, Pair, Share** (pairs)
- **Reader, Recorder, Reporter** (3–6)
- **Talk About It** (any size)

CHAPTER 12

RESOURCES FOR DIFFERENTIATED INSTRUCTION

Struggling Readers

eEdition with Audio DVD-ROM

Lesson Plans
- Strategies for Inclusion

Unit Resource Book
- Reading Study Guide with Additional Support, pp. 79–88 B
- Reteaching Activity, pp. 129–133
- Chapter Test (Form A), pp. 134–138 A

Best Practices Toolkit C
- Interactive Reading Strategies, pp. 90–145
- Differentiation Activity Bank: Struggling Readers, pp. 146–149

Teacher's Edition Activities
- Categorize Facts about Greece, Sec. 1
- Highlight Confusing Words and Phrases, Sec. 2
- Listen to France's Modern History, Sec. 3
- Flow Chart of Nordic History, Sec. 5
- Tiered Activities, Sec. 1–5

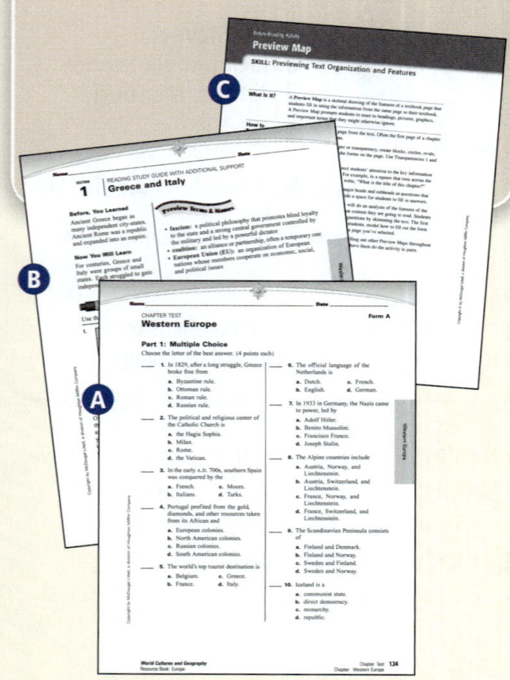

English Learners

Pupil's Edition in Spanish

eEdition with Audio DVD-ROM

Lesson Plans
- Strategies for English Learners

Unit Resource Book
- Reading Study Guide (Spanish), pp. 89–98
- RSG with Additional Support (Spanish), pp. 99–108 A B

Multi-Language Glossary C

Best Practices Toolkit
- Differentiation Activity Bank: English Learners, pp. 165–194

Test Generator CD-ROM
- Section Quizzes in Spanish
- Chapter Test in Spanish

Teacher's Edition Activities
- Pronounce and Preview, Sec. 1–5
- Vocabulary: Key Academic Vocabulary, Sec. 2
- Language: Verb Tenses, Sec. 5
- Tiered Activities, Sec. 1–5

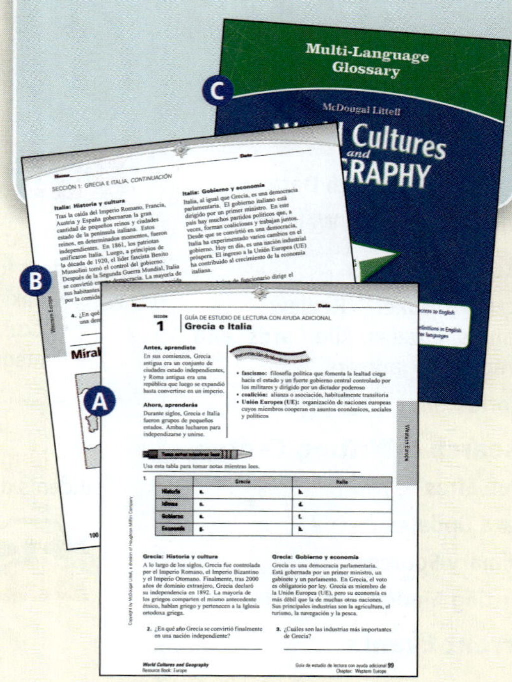

Inclusion

eEdition with Audio DVD-ROM

Lesson Plans
- Strategies for Inclusion

Unit Resource Book
- Reading Study Guide with Additional Support, pp. 79–88 A B
- Reteaching Activity, pp. 129–133
- Chapter Test (Form A), pp. 134–138

Best Practices Toolkit C
- Interactive Reading Strategies, pp. 90–145
- Differentiation Activity Bank: Struggling Readers, pp. 146–149

Teacher's Edition Activities
- Read Enlarged Vocabulary Definitions, Sec. 1
- Locate Places on a World Map, Sec. 2
- Break Down Word Meanings, Sec. 3
- Sculpt Terms & Names, Sec. 4
- Build a "Wall" in the Classroom, Sec. 4
- Act Out Silent Vocabulary Clues, Sec. 5
- Tiered Activities, Sec. 1–5

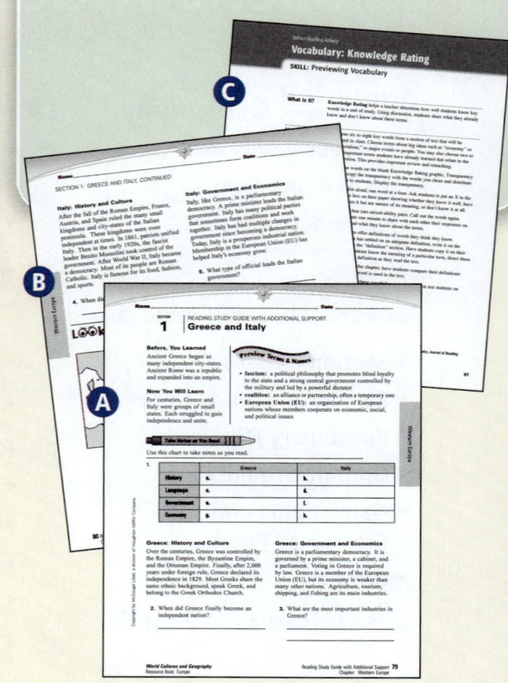

Gifted & Talented Students

Lesson Plans
- Core Lesson Plan

Unit Resource Book
- Outline Map Activities, p. 113
- History Makers, pp. 115, 116 A B
- Interdisciplinary Projects, pp. 117–119 C
- Chapter Test (Form C), pp. 144–148

Best Practices Toolkit
- Student Products Rubrics, pp. 1–14
- Writing Templates, pp. 78–89

Teacher's Edition Activities
- Board Game on Greek Culture, Sec. 1
- Write a "Life on the Other Side" Letter, Sec. 4
- Tiered Activities, Sec. 1–5

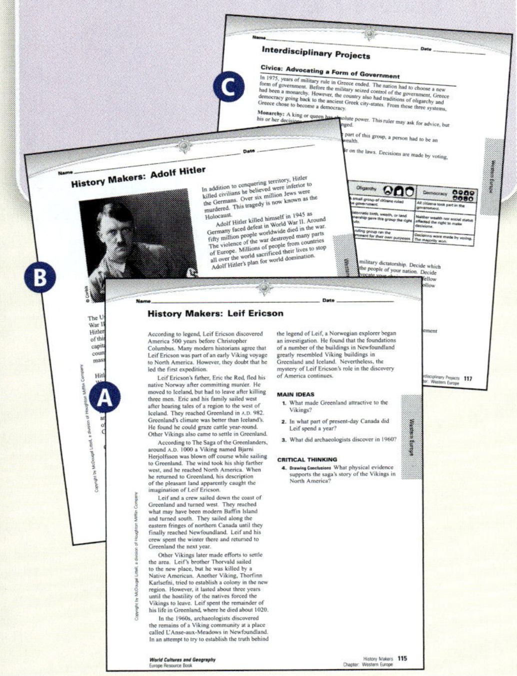

Pre-AP*

Lesson Plans
- Core Lesson Plan

Unit Resource Book
- Vocabulary Practice, p. 109
- Economics in the World, p. 120 A
- Connect to History, p. 121 B
- Primary Source, pp. 122–123
- Chapter Test (Form C), pp. 144–148

Best Practices Toolkit C
- Writing Templates, pp. 78–89
- Differentiation Activity Bank: Pre-AP Students, pp. 195–200

* Pre-AP is a registered trademark of the College Entrance Examination Board, which was not involved in the production of and does not endorse this product.

Teacher's Edition Activities
- Bastille Day Readers Theater, Sec. 3
- Tiered Activities, Sec. 1–5

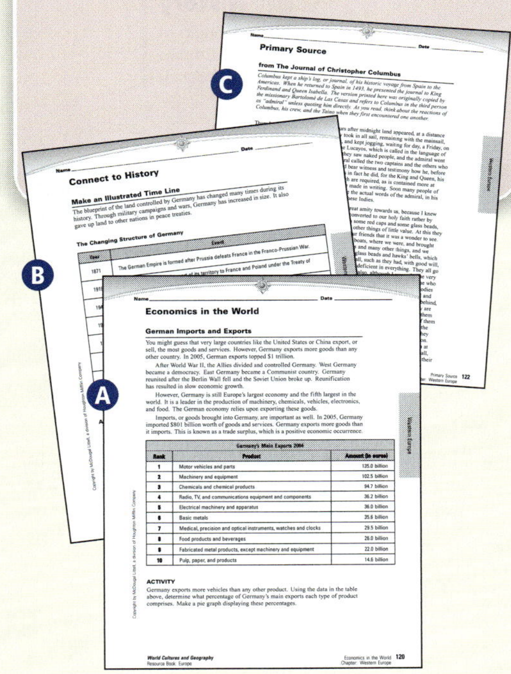

Interdisciplinary Activities

CONNECT to Language Arts

Plan an Itinerary for a Trip Through Western Europe

CLASS TIME 30 minutes

TASK Create a travel itinerary for a one-month trip to Western Europe.

SUPPLIES
- maps of Western Europe
- computer or pencil/paper

ACTIVITY Have students imagine they are taking a one-month trip to Western Europe. Have them create a day-by-day itinerary telling what country they'll visit that day and what activities they'll do there. Tell them they must visit all the Western European countries discussed in this chapter. As an extension, have students trace their proposed itineraries on photocopied political maps of Western Europe.

Good for Verbal/Spatial Learners

CONNECT to Music

Plan a Classical Music Concert

CLASS TIME 30 minutes (prep time)

TASK Organize a class concert featuring the classical music of Western European composers.

SUPPLIES
- a variety of classical music audio tapes or CDs containing the works of Beethoven, Bach, Brahms, Wagner, Pachelbel, Handel, Schubert, Strauss, Haydn, and Mozart among others
- audio tape or CD player(s)

ACTIVITY Have pairs of students sample various compositions and select one to include in a class concert. Pairs should then work together to write a short description about the composition and composer they selected. Prepare a program listing all the selections and composers. Before playing each piece, have the student pairs introduce it by reading their descriptions aloud. Discuss each selection afterward.

Good for Audio Learners

Introduce the ESSENTIAL QUESTION

- Identify the geographic characteristics that define the subregions of Western Europe.
- Identify the cultural characteristics that define the subregions of Western Europe.
- Summarize the similarities and differences among the countries of Western Europe.

Key Ideas

Section 1: For many centuries, Greece and Italy were collections of small states. Fueled by nationalism, each struggled to gain independence and unite as a nation.

Section 2: Like Greece and Italy, Spain and Portugal were ruled by foreigners. After gaining independence, both developed overseas empires to fuel their economies.

Section 3: Over the centuries, France's natural resources have helped it prosper. Through economic union, Belgium, Luxembourg, and the Netherlands have also prospered.

Section 4: Germany's central location helped it to dominate neighboring lands. Germany and the Alpine countries are linked in many ways.

Section 5: The Nordic countries have histories and cultures that are closely intertwined.

CONNECT to the Essential Question

You may wish to revisit the Essential Question periodically with students as they work through the chapter.

Suggestions for reconnecting are provided at an appropriate point in the *Teacher's Edition* for each section.

Nyhavn Harbor in Copenhagen, Denmark

CHAPTER 12

Western Europe

1 FOCUS ON Greece and Italy

2 FOCUS ON Spain and Portugal

3 FOCUS ON France and the Benelux Countries

4 FOCUS ON Germany and the Alpine Countries

5 FOCUS ON The Nordic Countries

ESSENTIAL QUESTION

What geographic and cultural characteristics define the subregions of Western Europe?

CONNECT Geography & History

Use the map and the time line to answer the following questions.

1. How many Western European nations have territory that extends east of 20° E longitude?
2. What country is Berlin the capital of, and when did the city reunite?

476

History

▲ **476** Beginning of the Byzantine Empire (Empress Theodora)

Geography

1418 Prince Henry of Portugal begins sponsoring expeditions of exploration.

Culture

1880 Bastille Day is set aside as a French national holiday. ▶

TIME LINE DISCUSSION

Point out to students that almost all of the countries of Western Europe have direct access to the sea. Ask students to identify the landlocked countries of Western Europe. *(Andorra, Austria, Liechtenstein, Luxembourg, Switzerland, Vatican City)*

- How many years separate the beginning of the Byzantine Empire and the first modern event covered on the time line? *(942 years)*
- What geographic advantage did Portugal and Spain have over Italy and Greece for exploring the Atlantic Ocean? *(Portugal and Spain are located on the Atlantic, so their explorers did not have to navigate the Mediterranean Sea or the Strait of Gibraltar to reach the ocean.)*
- What cultural characteristic on the time line relates to France? *(1880—Bastille Day)*

Animated GEOGRAPHY

Present-Day Western Europe

Click here to explore Western Europe @ ClassZone.com

Geography
1989 Berlin reunites after the opening of the Berlin Wall. ▶

Government
1920 The Nazi Party forms in Germany.

Government
2006 Finland's first female president, Tarja Halonen, is elected to a second term. ▶

Today

TEACHER MEDIA FAVORITES

Books for Students

Kotapish, Dawn. *Daily Life in Ancient and Modern Athens.* Lerner Publishing Group, 2000. Readable and richly illustrated.

Lowry, Lois. *Number the Stars.* Laurel Leaf, 1998. Story of the evacuation of Jews from Denmark during World War II.

Schmemann, Serge. *When the Wall Came Down.* New York Times, 2006. A journalist's first-hand report of the fall of the Berlin Wall.

Vincent, Mary, and R. A. Stradling. *Cultural Atlas of Spain and Portugal.* Checkmark Books, 1995. Good overview, with excellent maps.

Video/DVD

France and Benelux. Avalon Travel Publishing, 2005. Host Rick Steves highlights cultural and historical attractions on this 180-minute tour of France, the Netherlands, Belgium, and Luxembourg.

Visions of Italy. Acorn Media, 2005. Shot by a helicopter-mounted camera, this video offers narrated views of Italy.

Internet

Go to the *Research & Writing Center* at **ClassZone.com** to link to:

- data on the region
- a variety of chapter-specific content-reviewed sites
- "History Makers" biographies
- primary sources

Animated GEOGRAPHY

Present-Day Western Europe

CLASSZONE.COM

Remind students that Western Europe's mountain ranges, rivers, and nearness to the ocean have helped to shape the boundaries and cultures of its subregions.

- Which river forms part of the border between France and Germany? *(Rhine River)* In what country is the mouth of the Rhine located? *(Netherlands)*
- What four countries likely share a similar cool climate for most of the year? *(Iceland, Norway, Sweden, Finland)*
- Find Andorra and Liechtenstein. What geographic characteristics do these nations share? *(Both are very small, landlocked nations sandwiched between larger countries.)*

CONNECT GEOGRAPHY & HISTORY ANSWERS

1. four (Norway, Sweden, Finland, and Greece)
2. Germany; 1989

Presentation Options

Use the **Power Presentations** DVD-ROM to project the animated map.

- Prompt students to use the pop-up images and audio from the map to discover what they already know about the geography, culture, history, economics, or governments of the region.
- Challenge students to answer the map questions individually, in small groups, or as a class activity. Use all of the interactive material to help preview chapter content.

1 Plan & Prepare

Objectives

- Compare and contrast the governments of ancient and modern Greece
- Identify the geographic and cultural characteristics of modern Italy

Read for the Essential Question

Help students read for a purpose by reminding them of the Essential Question: "What geographic and cultural characteristics define the subregions of Western Europe?" Point out that they can use text features such as Key Questions, special features, images, and maps to form answers to the question.

Vocabulary

Best Practices Toolkit

Use the ***Best Practices Toolkit*** to model strategies for vocabulary notetaking. Vary strategies throughout the year. Choose from: *Knowledge Rating, Predicting ABCs, Definition Mapping, Word Sort, Frayer Model (Word Squares), Magnet Words, Word Wheel,* and *Student VOC.*

Vocabulary Strategies, TT9–TT16

Reading Strategy

Best Practices Toolkit

Display the T Chart/Two-Column Chart transparency.

- Remind students that to summarize means to focus on the essential information of their reading.
- Model filling in a few entries for Greece, using the textbook.
- Have students suggest entries.
- Display the graphic organizer for students to review.

T Chart/Two-Column Chart, TT20

SECTION 1 Reading for Understanding

Key Ideas

BEFORE, YOU LEARNED

Ancient Greece began as many independent city-states with different ways of life. Ancient Rome began as a republic and expanded into a vast empire.

NOW YOU WILL LEARN

For many centuries, Greece and Italy were collections of small states. Fueled by nationalism, each struggled to gain independence and unite as a nation.

Vocabulary

TERMS & NAMES

Byzantine Empire the eastern half of the Roman Empire that survived for a thousand years after the fall of Rome

fascism (FASH•IHZ•uhm) a political philosophy that promotes blind loyalty to the state and a strong central government controlled by a powerful dictator

Romance language any of the languages that developed from the Roman language, Latin, such as Spanish, Portuguese, French, Italian, and Romanian

Vatican the official residence of the pope in Vatican City, and the political and religious center of the Roman Catholic Church

coalition an alliance or partnership, often a temporary one

BACKGROUND VOCABULARY

compulsory required

REVIEW

nationalism pride in and loyalty to one's nation

European Union (EU) an organization of European nations whose members cooperate on economic, social, and political issues

Reading Strategy

Re-create the chart shown at right. As you read and respond to the **KEY QUESTIONS**, use the chart to help you summarize what you learn about Greece and Italy.

Skillbuilder Handbook, page R5

SUMMARIZE

	GREECE	ITALY
History		
Language		
Government		
Economy		

GRAPHIC ORGANIZERS Go to Interactive Review @ClassZone.com

362 Chapter 12

PRETEACHING VOCABULARY

English Learners

Pronounce and Preview

Pronounce each term for students. Review the meanings of words within the definitions, such as *philosophy, blind loyalty,* and *central government.*

- To modify vocabulary learning, have students complete worksheets as they read, instead of afterward.

Unit Resource Book: Europe
- Vocabulary Practice, p. 109

Inclusion

Read Enlarged Vocabulary Definitions

To help students who are visually impaired, make an enlarged photocopy of this page and distribute. Ask volunteers to take turns reading a word and its definition out loud. Discuss each term as a class and answer any questions students might have. Have students cut out the terms and names, and glue or tape them in a vocabulary notebook.

Chateau de Polignac, France

Cattle in the Alps

SECTION 1 FOCUS ON

Greece and Italy

Greek Pottery Tourists come to Greece to see artifacts like this fifth century B.C. pot.

Connecting to Your World

It's not surprising that Greece and Italy are popular tourist destinations. Stretching across southern peninsulas in the Mediterranean Sea, both countries have mild climates and warm, sunny weather much of the year. In both places, visitors can see the magnificent treasures of the distant past. While both countries were once part of the Roman Empire, after it divided, these nations developed different cultures. In the 1800s, both Greece and Italy fought for independence and tried to preserve their ancient heritage as they became modern nations.

Greece

KEY QUESTION How has Greece's government changed since the Byzantine Empire?

As you learned in Chapter 11, ancient Greece began as a series of independent city-states. Eventually, it became part of the Roman Empire. When the Roman Empire divided, Greece was in the eastern half, called the **Byzantine Empire**. Most of the Byzantine Empire's inhabitants spoke Greek, not Latin, and followed the eastern traditions of Christianity. By 1453, the Byzantines had been conquered by the Islamic Ottoman Turks. Before declaring their independence in 1829, Greeks had lived under foreign rule for over 2,000 years.

Byzantine Ruins in Greece

2 Focus & Motivate

3-Minute Warm-Up

Write on the board or display the transparency:

- What can you learn about the Byzantine Empire from the photograph and the small (inset) map? *(Possible Answer: The Byzantine Empire is gone, but its buildings still stand today; the Byzantine Empire covered much of southern Europe.)*

Unit Transparency Book: Europe
- 3-Minute Warm-Ups, TT11

Connecting to Your World

More About . . .

Greek Pottery

The designs on ancient Greek pottery were made without using paint. Instead, artisans used a watery clay mixture that, when fired, turned black. Areas not coated with the mixture turned reddish-brown.

3 Teach

Greece

Talk About It

- Do you think it is a good idea to make voting compulsory? *(Yes: because everyone should take part in the democratic process; No: because some people may be uninformed and may not vote for the best candidate.)*
- **Summarize** What cultural aspects of the Byzantine Empire still exist in Greece today? *(Most inhabitants speak Greek and practice a form of Christianity.)*

SECTION 1 PROGRAM RESOURCES

Unit Resource Book: Europe

ON LEVEL
Unit Resource Book
- Reading Study Guide, p. 69
- Map and Graph Skills, p. 111
- Section Quiz, p. 124

STRUGGLING READERS
Unit Resource Book
- Reading Study Guide with Additional Support, p. 79
- Skillbuilder Practice: Synthesizing, p. 110
- Section Quiz, p. 124
- Reteaching Activity, p. 129

eEdition with Audio DVD-ROM

ENGLISH LEARNERS
Pupil Edition in Spanish
eEdition with Audio DVD-ROM
Unit Resource Book
- RSG (Spanish), p. 89
- RSG with Additional Support (Spanish), p. 99

Multi-Language Glossary
Test Generator
- Section Quiz in Spanish

INCLUSION
Unit Resource Book
- RSG with Additional Support, p. 79
- Section Quiz, p. 124
- Reteaching Activity, p. 129

GIFTED & TALENTED
Unit Resource Book
- Interdisciplinary Projects, pp. 117, 119
- Section Quiz, p. 124

PRE-AP
Unit Resource Book
- Section Quiz, p. 124

TECHNOLOGY
Unit Transparency Book
- 3-Minute Warm-Ups, TT11
- Essential Question, TT12
- Critical Thinking, TT13
- Map: Europe's Elevation, TT18, TT18a

Daily Test Practice Transparencies
- Section 1, TT41

Power Presentations
ClassZone.com
World Cultures and Geography Video on DVD

More About . . .

Athens

Athens is famous for its Acropolis, a hill whose name in Greek means "high city." The Acropolis is home to the Parthenon, a temple built to honor the goddess Athena. The Greeks completed the Parthenon around 432 B.C.

Athens has long been known as a host of the Olympic games. A Greek tradition first held around 776 B.C. in the city of Olympia, the games entertained people throughout southern Europe. When emperor Theodosius discontinued the games in A.D. 393, they didn't reappear until 1896, when the first modern Olympic games were held in Athens. The city of Athens played host once again for the 2004 Olympic Games.

More About . . .

The European Union

The European Union is an organization that was created to promote peace, democracy, economic progress, and respect for human rights. More than half of Europe's 44 nations belong, and the EU continues to accept applications from "candidate countries." In order to be considered for the EU, a candidate country must have a stable form of democratic government, the ability to enforce its nation's laws and protect its people, and a working economy and competitive system of trade. Greece joined the European Union in 1981.

History When the Byzantine Empire fell, Greece became part of the Ottoman Empire for approximately 400 years. By the late 1700s, desire for independence was growing. **Nationalism** increased as more Greeks learned about their past. Interest and pride in Greece's history as a center of culture and democracy grew.

In 1829, after a long struggle, Greece broke free from Ottoman rule. Its leaders made the nation a monarchy. Then, in 1967, Greek military officers seized control of the government. When the military government collapsed, voters decided not to return to a monarchy. Since 1975, Greece has been a parliamentary democracy.

Culture Most Greeks share the same ethnic background, language, and religion. They speak Greek and belong to the Greek Orthodox Church, a form of Christianity. Every major town has a patron saint, and townspeople celebrate their saint with an annual festival. Easter is also an important religious holiday in Greece.

About two-thirds of Greeks live in cities. Many live in Athens, Greece's capital. Cities have both older sections with narrow streets and newer areas with high-rise apartments and shopping centers. Many cities have a coffee house, or *kafeneio*, where friends meet.

Government and Economics Greece is a parliamentary democracy. The president's job is mostly ceremonial. The country is governed by the prime minister, the cabinet, and the parliament. Voting in Greece is **compulsory**, or required. Failing to vote is against the law.

Greece has fewer high-tech industries and service jobs than many countries in Western Europe. Although it is a member of the **European Union (EU)**, its economy lags behind other nations. Agriculture makes up a greater percentage of its economy than wealthier nations. Because of Greece's position on the Mediterranean Sea, tourism, shipping, and fishing remain important.

SEQUENCE EVENTS Explain the changes to Greece's government since the Byzantine Empire.
became a monarchy in 1829, a dictatorship in 1967, and a parliamentary democracy in 1975

Greek Orthodox Church This church, with its typical whitewashed walls and blue domes, is located in Oia on the island of Santorini.

DIFFERENTIATING INSTRUCTION

Struggling Readers

Categorize Facts About Greece

Have each student review the page and write a fact about Greece on a slip of paper. Put completed slips in a hat. On the board, draw a four-column chart with the headings "History," "Culture," "Government," and "Economics." Ask students to draw a slip of paper from the hat, read it aloud, and decide in which category the fact belongs. Have them shorten the fact to a few key words and write it in the appropriate column.

Gifted & Talented

Board Game on Greek Culture

Have students work in small groups to create a board game that teaches facts about Greek culture. Students should:
- review the page to identify facts to be used in their game.
- plan and create the game.
- play their game, clarifying the directions and making adjustments as needed.
- play each other's games, testing their knowledge of Greek culture.

COMPARING Western European Economies

The percentage of a nation's labor force employed in agriculture is a good indication of that country's economic development. Take a look at how Greece compares to these other Western European countries.

	GDP (Per Capita*)	Agriculture	Industry	Services
		(Percentage Employed in Selected Sectors)		
Denmark	$34,600	3	21	76
France	$29,900	4	24	72
Spain	$25,500	5	30	65
Greece	$22,200	12	20	68
Portugal	$19,300	10	30	60

*GDP: Gross Domestic Product
Source: *CIA World Factbook, 2006*

CRITICAL THINKING

Compare and Contrast How do Greece's GDP per capita and agricultural employment compare to the other countries listed in the chart?

Italy

KEY QUESTION What changes has Italian society experienced in recent years?

After the Western Roman Empire fell in the late 400s, the Italian peninsula split into many small kingdoms and city-states. Sometimes they were independent. At other times, France, Austria, and Spain ruled over them.

History In 1796, France conquered much of the Italian Peninsula. French rule led to new laws, better roads, and a common currency. Italians saw the benefits of unity and began to work toward independence. In 1861, Italian patriots such as Giuseppe Garibaldi and Camillo Cavour succeeded in unifying Italy.

In the early 1920s, Benito Mussolini took control of the Italian government. He promoted **fascism** (FASH•IHZ•uhm), a political system based on fierce nationalism and a strong central government led by a dictator. Mussolini gained support by promising a return to a powerful Roman Empire. During World War II, Italy fought with Germany against the Allies. After Italian forces suffered major defeats, Mussolini was forced out and killed as he tried to escape Italy. In 1946, Italians replaced the monarchy with democracy.

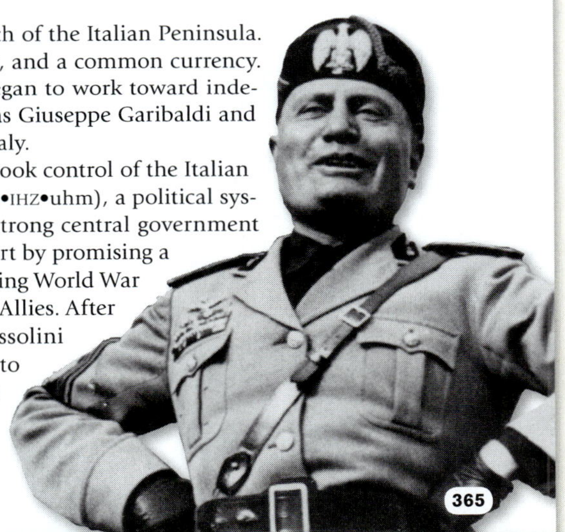

Benito Mussolini
Fascist dictator Mussolini took control of Italy in the early 1920s.

365

COMPARING

Western European Economies

Explain that GDP is often used as a measure of a nation's economic performance. At $42,000, the United States has one of the world's highest per capita GDPs.

CRITICAL THINKING ANSWER

Compare and Contrast Greece's GDP is below most other countries' except Portugal; Greece has the highest percentage of agricultural employees of the countries listed on the chart.

Unit Resource Book: Europe
- Skillbuilder Practice: Synthesizing, p. 110
- Map and Graph Skills: Europe's Annual Precipitation, p. 111

Teach

Italy

Reader, Recorder, Reporter

- What steps toward unity did Italy take in the 1800s? *(It used a common currency and grew closer from an improved transportation system under French rule; Italian patriots led Italy to independence in 1861.)*
- Do you think Benito Mussolini was a good leader? *(Yes: he continued to unite Italy and built up nationalism; No: he made false promises and allied with Germany in World War II.)*
- **Summarize** How does Italy continue to reflect the cultural traditions of the Roman Empire? *(Italian is the official language; most Italians are Roman Catholic; the country is a mix of rural-agricultural and modern-urban societies.)*

Unit Resource Book: Europe
- Interdisciplinary Projects: Advocating a Form of Government, pp. 117, 119

INTERDISCIPLINARY ACTIVITIES

CONNECT to Art

Create a Poster Promoting National Pride

Have students work in small groups to design and create posters that promote national pride for Italy. Posters should list characteristics about Italy that would make its citizens proud. Display the completed posters in the classroom. Then lead a discussion about the difference between nationalism and fascism.

CONNECT to Math

Pie Charts of Mediterranean Economies

Have students create a series of pie charts—one for each country listed in the graphic on this page. Pie charts should accurately reflect the percentages of employees working in agriculture, industry, and services.

More About . . .

Vatican City

The Italian government recognized Vatican City in 1929. It is ruled by the Pope, the head of the Roman Catholic Church. Most of its 932 citizens live within the boundaries of its 0.17 square-mile (0.44 km) area. However, the 3,000 workers who maintain the city-state live outside its walls.

Vatican City is partially funded through museum fees and the sale of postage stamps, coins, and tourist memorabilia. The Roman Catholic Church also supports the Vatican with an annual contribution. More modern forms of income, such as investments and real estate, provide additional money to boost the economy.

Languages of Western Europe

CONNECT Geography & Culture

Reading a Thematic Map Draw students' attention to the map key, and then read aloud the suggestions for reading a thematic map in the lower righthand corner of the page. As students study the map, ask them to list Celtic languages. *(Gaelic, Irish, Welsh, Breton)*

ANSWERS

1. **Place** Hellenic
2. **Region** Romance

Culture Like Greece, modern Italy has been influenced by traditions. The official language is Italian, a Romance language. **Romance languages** come from Latin and are spoken in places that used to be part of the Roman Empire. Take a look at the map below to see where other Romance languages are spoken.

Most Italians are Roman Catholics. For many centuries, Catholic Church leaders and the Italian government were closely allied. The political and religious center of the Church, the **Vatican**, is located in Italy. Although it lies within the city of Rome, it is an independent country, the smallest such country in the world.

In recent years, Italy has become more diverse as immigrants have come from nearby Morocco and Tunisia in North Africa and Albania on the Balkan Peninsula. Much of Italy has changed from a rural, agricultural nation to a modern, urban society. However, life differs between northern and southern Italy. Northern Italy is richer, with more people employed in manufacturing than in the south, where more people work in agriculture.

Italian food changes depending on the region of the country, from pasta with tomato sauce in the south to a dish made of rice called *risotto* in the north. Sports are also important to Italian culture. Many Italians enjoy watching race car driving, cycling, and soccer.

Languages of Western Europe

CONNECT Geography & Culture

READING A THEMATIC MAP

Language maps show what languages are spoken in a specific area.

- Use the labels to identify the language.
- Use the key to see what branch the language belongs to.

1. **Place** What branch does Greek belong to?
2. **Region** What branch is most common in southern Europe?

DIFFERENTIATING INSTRUCTION: TIERED ACTIVITIES

OBJECTIVE Write a travel journal, magazine article, or essay about Greece or Italy.

Basic

Have students choose either Greece or Italy and imagine they have just taken a trip there. They should write a series of journal entries in which they describe the sights and sounds of their trip. Remind students to base their entries on facts from the text and to use sensory details to make the passage come alive.

On Level

Have students choose either Greece or Italy and imagine they are journalists writing an article for a magazine. Articles should describe features that tourists can expect to experience. Remind them to base their articles on facts from the text and to use persuasive language to encourage travel.

Challenge

Have students choose either Greece or Italy and imagine they are applying for a scholarship to that country. Students should write an essay explaining why they want the scholarship. Remind them to use facts from the text, including geographic and cultural characteristics and main economic activities.

Government and Economics Italy is a parliamentary democracy. Voters elect three-fourths of parliament members. The rest are assigned by a complex system designed to make sure each of Italy's many political parties is represented in the government. A prime minister heads the government. Usually this leader comes from the party that wins the most votes in the election. Because Italy has so many parties, leaders often have to form coalition governments. A **coalition** is an alliance or partnership, which is often temporary. Members of several parties agree to work together. If parties withdraw from a coalition, leaders have to form a new one. As a result, Italy has had many changes in government.

Once a mainly agricultural country, Italy is now a prosperous industrial nation. Only five percent of the population has a farm-related job. European Union membership has helped Italy grow, opening new markets for its products, which include fashionable clothing, shoes, and cars. Milan, in northern Italy, is world-famous for its fashion shows. However, economic growth for the country has been uneven. The north has many of Italy's factories, while the south remains largely farmland. Italy continues to look for ways to bring greater prosperity to the southern region.

Runway Model A model walks the runway at a Milan fashion show.

SUMMARIZE Describe the changes that Italian society has experienced in recent years. has become more diverse as a result of immigration; has shifted from rural to urban

Section 1 Assessment

ONLINE QUIZ For test practice, go to **Interactive Review** @ClassZone.com

TERMS & NAMES

1. **Explain the importance of**
   - Byzantine Empire
   - fascism
   - Vatican
   - coalition

USE YOUR READING NOTES

2. **Summarize** Use your completed chart to answer the following question:

   How is Greece's economy different from Italy's economy?

	GREECE	ITALY
History		
Language		
Government		
Economy		

KEY IDEAS

3. In what ways did Greek culture develop differently from Italian culture after the division of the Roman Empire?
4. Why are so many Italians members of the Catholic Church?
5. How has Italy's economy changed over time?

CRITICAL THINKING

6. **Analyze Cause and Effect** How did the Byzantine Empire influence Greek culture?
7. **Make Inferences** What are the disadvantages for Italy of having had so many coalition governments?
8. **CONNECT to Today** Why would it be an advantage for Italy to make sure that southern Italy is as prosperous as northern Italy?
9. **MATH** **Create a Bar Graph** Look at the chart on Western European economies in this section. Create a bar graph showing the percentage of workers employed in agriculture in each country.

CONNECT to the Essential Question

What geographic and cultural characteristics define the subregions of Western Europe?

Ask students what they have learned so far that can help them answer this question. Students might mention:

- Greece and Italy share a common history as nations that once belonged to the Roman Empire.
- Greece and Italy both developed a sense of nationalism that led to independence.
- Agriculture remains an important part of the economy in Greece and southern Italy.

4 Assess & Reteach

Assess Have students complete the Section Assessment.

Unit Resource Book: Europe
- Section Quiz, p. 124

Interactive Review @ClassZone.com
- Section Quiz

Power Presentations

Test Generator

Reteach Name a cultural characteristic, such as language or religion, and have two volunteers—one representing Greece and one representing Italy—make a true statement about that characteristic for the rest of the class. Continue naming characteristics until all students have had a turn to represent both Greece and Italy.

Unit Resource Book: Europe
- Reteaching Activity, p. 129

SECTION 1 ASSESSMENT ANSWERS

Terms & Names

1. Byzantine Empire: eastern half of the Roman Empire; fascism: political philosophy promoting a strong, dictator-controlled government; Vatican: political and religious center of the Roman Catholic Church; coalition: an alliance or union

Use Your Reading Notes

2. Italy has been more successful in industrializing, giving it a stronger economic base than Greece.

	GREECE	ITALY
History	E. Roman Empire	W. Roman Empire
Language	Greek	Italian
Government	parliamentary democracy	parliamentary democracy
Economy	more agricultural; lagging	industrial, but prosperity uneven

Key Ideas

3. joined the Greek Orthodox Church and continued to speak the Greek language; Romance languages spoken and Roman Catholicism practiced in the W. Roman Empire
4. church and government closely allied; the Vatican, church headquarters, located in Rome
5. manufacturing increased prosperity; prosperity uneven, much of southern Italy still farmland

Critical Thinking

6. Most inhabitants spoke Greek instead of Latin and followed the Greek Orthodox Church.
7. can make accomplishing goals difficult
8. Possible answers: would increase prosperity; would reduce tensions between regions
9. Use the rubric to score graphs.

Graph Rubric

	Style of Bar Graph	Number of Errors
4	very clear	few or none
3	clear	some
2	vague	many
1	unclear	many major

❶ Plan & Prepare

Objective

Compare and contrast geographic and cultural characteristics of northern and southern Italy.

Springboard Only slightly larger than Arizona, Italy is visited by almost 40 million tourists each year. It is the fifth most visited country in the world behind France, the United States, Spain, and China.

❷ Teach

Comparing Regions of Italy

Talk About It

- How does climate influence the type of work people do in northern Italy? *(Because of northern Italy's cooler climate, people work more in factories and other industries rather than in agriculture.)*
- Why does southern Italy benefit more from its coastline than northern Italy? *(The climate is warmer in southern Italy, attracting many tourists to its sandy beaches along the Mediterranean and Adriatic seas.)*
- **Compare** What geographic features do both regions share? *(mountains, coastline)*

COMPARING Regions of Italy

Italy can be roughly divided into three regions: northern, central, and southern. While northern and southern Italy share aspects of Italian culture, they are also different.

Recreation
Like these hikers, many people enjoy the rugged trails winding through the Italian Alps.

Northern Italy

The craggy Alps of northern Italy are dotted with villages, such as St. Magdalena, shown below. Tourists often come to hike or ski in the mountains. The northern region is also highly industrialized, however, with major metropolitan areas such as Milan and Turin. Many of Italy's cities with populations of 100,000 or greater are in the north.

Work
Northern Italy has more factories than farms defining the region's major cities. Workers at this factory in Bologna manufacture Ducati motorcycles.

DIFFERENTIATING INSTRUCTION: TIERED ACTIVITIES

OBJECTIVE Create a class mural that compares and contrasts geographic and cultural characteristics of northern and southern Italy.

Basic

Have students work as a group to review the text and analyze a physical map of Italy to find information about the geographic characteristics of each region. Then have students create original artwork on a class mural depicting that information. Have students write a one-sentence caption for each illustration they create.

On Level

Have students work as a group to review the text and analyze a physical/political map of Italy to find information about cities and economic activities in each region. Then have students create original artwork on a class mural depicting that information. Have students write a one-sentence caption for each illustration they create.

Challenge

Have students work as a group to review the text and use an almanac or encyclopedia to find information about the people and cultural characteristics of each region. Then have students create original artwork on a class mural depicting that information. Have students write a one-sentence caption for each illustration they create.

Recreation

Sunbathers lounge on the beach in Sorrento, enjoying its warm Mediterranean climate.

Southern Italy

Southern Italy also has mountains, but it benefits more from the coastline along the Mediterranean and Adriatic seas than does northern Italy. Many of the tourists visiting southern Italy come to relax on sandy beaches in towns such as Minori, shown below.

The south is also much less industrialized than the north. Southern Italy has fewer big cities and more people are employed in agriculture, such as growing and harvesting grapes and olives.

Work

These men are gathering olives to make olive oil in an orchard in Bisceglie. It is located in the southern region of Puglia, where the arid climate is particularly suited to growing olives.

CRITICAL THINKING

1. **Compare and Contrast** What geographic features are characteristic of each region?
2. **Make Inferences** Which region do you think is wealthier and why?

INTERDISCIPLINARY ACTIVITIES

CONNECT to Language Arts

Write a Poem About Italy

Have students write a poem about Italy.

- Students may choose to write about northern Italy, southern Italy, or both.
- Remind students to use the images in the feature for inspiration. What words come to mind as they look at the mountains? What colors do they see as they look at the beach image?
- Remind them that poems can, but do not have to, rhyme. Ask volunteers to share their completed poems aloud.

CONNECT to Art

Create a Landscape of Italy

Have students create a landscape work of art using watercolors or pastels.

- Students should select one scene to recreate—the mountains of northern Italy or the seaside town of southern Italy—as depicted in the feature.
- Remind students that traditional landscape paintings usually did not include people.
- Have students use colored poster board to frame their pieces. Display them.

Comparing Regions

Italian Cities

The city of Venice in northeastern Italy is unusual in that it is built on top of 118 small islands. Its residents and visitors use a system of canals and bridges to move about the city. Boats traveling along the canals include slow-moving gondolas and modern river boats. Venice is a busy tourist and commercial center.

Naples, one of the biggest cities in southern Italy, is located along the west coast and is a major industrial center. Iron, steel, petroleum, textiles, and machinery are among the products manufactured in Naples. Naples also attracts many tourists due to its vibrant culture and beautiful location on the Bay of Naples.

More About . . .

Olives

Long an essential part of the Mediterranean diet, people have grown olives since prehistoric times. Most olive-producing nations, like Italy, grow the fruit for its oil, which can make up about one-fourth of an olive's flesh. Olive oil, used mainly for cooking, is produced by crushing and pressing ripe olives. Italy is one of the world's leading producers and exporters of olive oil.

3 Wrap Up

CRITICAL THINKING ANSWERS

1. **Compare and Contrast** Northern—mountainous, coastline not as attractive to tourists; Southern—not as mountainous, sandy coastline
2. **Make Inferences** Possible Answer: Northern Italy, because it has more factories and cities that create more income than the more agricultural southern region

1 Plan & Prepare

Objectives

- Identify regional differences that affect Spain's culture
- Explain how Portugal's location helped it become a leader in the Age of Exploration

Read for the Essential Question

Help students read for a purpose by reminding them of the Essential Question: "What geographic and cultural characteristics define the subregions of Western Europe?" Point out that they can use text features such as Key Questions, special features, images, and maps to form answers to the question.

Vocabulary

Best Practices Toolkit

Use the ***Best Practices Toolkit*** to model strategies for vocabulary notetaking. Vary strategies throughout the year. Choose from: *Knowledge Rating, Predicting ABCs, Definition Mapping, Word Sort, Frayer Model (Word Squares), Magnet Words, Word Wheel,* and *Student VOC.*

Vocabulary Strategies, TT9–TT16

Reading Strategy

Best Practices Toolkit

Display the Venn Diagram transparency.

- Remind students that comparing means looking for similarities (What is the same?), while contrasting means looking for differences (What is different?).
- Discuss the similarities and differences between Spain and Portugal. Then model filling in a few entries on the chart.
- Have students suggest entries.
- Display the graphic organizer for students to review.

Venn Diagram, TT22

SECTION 2 Reading for Understanding

Key Ideas

BEFORE, YOU LEARNED

After the division and fall of the Roman Empire, Greece and Italy developed into different nations.

NOW YOU WILL LEARN

Like Greece and Italy, Spain and Portugal were ruled by foreigners. After gaining independence, both developed overseas empires to fuel their economies.

Vocabulary

TERMS & NAMES

Moors the group of Muslims from North Africa who conquered Spain in the eighth century

Reconquista the successful effort by the Spanish to drive the Moors out of Spain

Christopher Columbus Italian navigator and explorer who sailed for Spain and explored the Caribbean and the coast of Central and South America

Basque (bask) an ethnic group living in the western Pyrenees and along the Bay of Biscay in Spain and France; also the name of their language

BACKGROUND VOCABULARY

separatist a person who wants a region to break away from the nation it is a part of

Visual Vocabulary Christopher Columbus reaches the New World.

Reading Strategy

Re-create the Venn diagram shown at right. As you read and respond to the **KEY QUESTIONS**, use the diagram to help you compare and contrast Spain and Portugal.

Skillbuilder Handbook, page R9

COMPARE AND CONTRAST

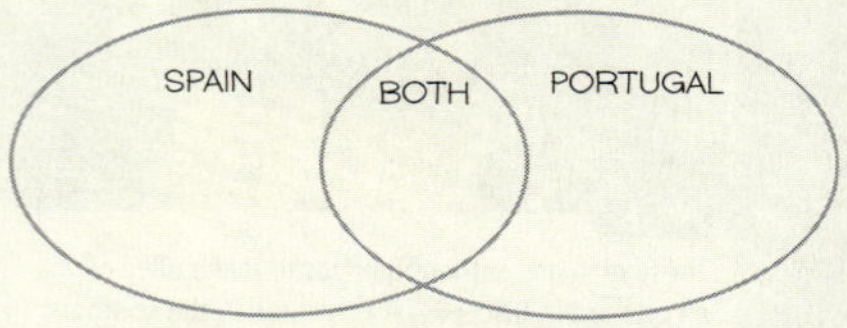

GRAPHIC ORGANIZERS Go to **Interactive Review** @ClassZone.com

PRETEACHING VOCABULARY

English Learners

Pronounce and Preview

Pronounce each term for students. Review the meanings of words within the definitions, such as *Muslims, ethnic group,* and *Pyrenees.*

- To modify vocabulary learning, have students complete worksheets as they read, instead of afterward.

Unit Resource Book: Europe
- Vocabulary Practice, p. 109

Inclusion

Locate Places on a World Map

Have volunteers take turns reading the vocabulary terms and names aloud. Instruct the rest of the class to follow along as they read, looking and listening for geographic places in the definitions. As geographic places are identified, have students take turns locating them on a world map.

SECTION 2 FOCUS ON

Chateau de Polignac, France

Cattle in the Alps

Spain and Portugal

Connecting to Your World

Without ever visiting Spain, most Americans know more about Spanish culture than they think. For over 300 years, Spain had an empire in what is today Mexico and the southwestern United States. In places where Spain once ruled, the impact of Spanish culture is easy to see. Many southwestern cities have Spanish names and buildings influenced by Spanish architecture. Throughout the United States, many people speak Spanish and eat Spanish food. Spanish culture itself was the product of influences from many other cultures.

Spanish Food Spanish food—including small appetizers called *tapas*, such as this squid dish—has become popular in the United States.

Spain

KEY QUESTION What regional differences affect Spain's culture?

For centuries, Spain was part of the Roman Empire. When the empire fell, a Germanic tribe conquered the peninsula and established a Christian kingdom. It lasted until the early A.D. 700s, when the **Moors**, Muslim peoples from North Africa, took control of southern Spain. Spain's Moorish rulers brought a more advanced culture to medieval Europe. Muslim scholars made new discoveries in medicine, mathematics, and other fields. They remained there for almost eight centuries, when groups of Christians still living in northern Spain succeeded in driving them out of the region.

The Alhambra Located in Granada, Spain, the Alhambra was a Moorish palace and fortress.

2 Focus & Motivate

3-Minute Warm-Up

Write on the board or display the transparency:

- Match the cultural characteristic with the country it represents.

1. bullfighting *(b)*	**a.** Portugal
2. flamenco *(b)*	**b.** Spain
3. Basque *(b)*	
4. mostly rural *(a)*	

Unit Transparency Book
- 3-Minute Warm-Ups, TT11

Connecting to Your World

More About . . .

Spanish in the United States

Some scholars believe as many as 10,000 words in American English have been borrowed from Spanish. A few examples include: *rodeo, salsa, cafeteria, canyon, patio, plaza, ranch, mosquito,* and *vanilla*.

3 Teach

Spain

Think, Pair, Share

- Why had much of Spain's power faded by the 1800s? *(All of its colonies had declared independence.)*
- **Compare and Contrast** What cultural differences separate people living in the Basque region from those living in the rest of Spain? *(The Basque have their own language and customs; they live in the Pyrenees, separate from the rest of the country.)*

SECTION 2 PROGRAM RESOURCES

Unit Resource Book: Europe

ON LEVEL

Unit Resource Book
- Reading Study Guide, p. 71
- Outline Map Activities, p. 113
- Section Quiz, p. 125

STRUGGLING READERS

Unit Resource Book
- Reading Study Guide with Additional Support, p. 81
- Skillbuilder Practice: Synthesizing, p. 110
- Section Quiz, p. 125
- Reteaching Activity, p. 130

eEdition with Audio DVD-ROM

ENGLISH LEARNERS

Pupil Edition in Spanish

eEdition with Audio DVD-ROM

Unit Resource Book
- RSG (Spanish), p. 91
- RSG with Additional Support (Spanish), p. 101

Multi-Language Glossary

Test Generator
- Section Quiz in Spanish

INCLUSION

Unit Resource Book
- RSG with Additional Support, p. 81
- Section Quiz, p. 125
- Reteaching Activity, p. 130

GIFTED & TALENTED

Unit Resource Book
- Section Quiz, p. 125

PRE-AP

Unit Resource Book
- Primary Source: from The Journal of Christopher Columbus, p. 122
- Section Quiz, p. 125

TECHNOLOGY

Unit Transparency Book
- 3-Minute Warm-Ups, TT11
- Essential Question, TT12
- Critical Thinking, TT14

Daily Test Practice Transparencies
- Section 2, TT42

Power Presentations

ClassZone.com

World Cultures and Geography Video on DVD

Iberian Peninsula Land Use *c. 1500*

CONNECT Geography & History

As students examine the map, encourage them to look for patterns or trends in the information provided.

- In what region did grain grow best? *(in the northern part of Spain and Portugal)*
- Was the majority of land used to grow crops or for other activities? *(more land for forests and grazing than for crops)*

ANSWER

Human-Environment Interaction the northern and coastal regions

Unit Resource Book: Europe

- Outline Map Activities: Western Europe: Physical/Political, p. 113

More About . . .

Christopher Columbus

Following his famous 1492 expedition to the Americas, Columbus led three more explorations in 1493, 1498, and 1502. His second and third voyages resulted in the discovery of Puerto Rico, Jamaica, Trinidad, and the Lesser Antilles (a chain of small islands southeast of Puerto Rico). While Columbus was impressed with the beauty and riches of the Caribbean, he never realized his dream of finding an all-water route to Asia.

Unit Resource Book: Europe

- Primary and Secondary Sources: from The Journal of Christopher Columbus, pp. 122–123

Iberian Peninsula Land Use *c. 1500*

CONNECT Geography & History

Human-Environment Interaction What parts of Spain could be used to grow crops?

History In the 1000s, Christians in northern Spain began the **Reconquista**, the effort to drive the Moors out of Spain. It lasted until the 1400s, when King Ferdinand and Queen Isabella conquered the last Muslim kingdom. Once unified under its Catholic monarchs, Spain began to look beyond its territory, which was poor in resources and farmland. Spain turned to the sea, and sponsored **Christopher Columbus's** first voyage to the Americas in 1492. By the 1500s, Spain's conquests in the Americas had made it rich. Its colonies provided Spain with resources it lacked. For a time, Spain was the world's greatest power. By the 1800s, Spain's power had faded, and its colonies declared independence.

Spain remained neutral during World War I. In the 1930s, the Spanish fought a civil war over whether the country should be a monarchy or a republic. During the struggle, Francisco Franco of Spain's fascist party won control of the government. He ruled as a dictator for almost 40 years.

Culture The Spanish share many cultural traits. Until a few decades ago, however, most people identified more with the region they lived in than the country. Spain has many regional languages, such as Catalan and Valencian, which emphasize cultural differences. The Roman Catholic Church was the biggest common tie. Almost all Spaniards are Catholic.

In the 1950s and 1960s, Spain's economy developed quickly and changed the way many people lived. Many people left farms to take manufacturing jobs in cities. Today, most of Spain's people live in cities or towns. Their homes are apartments rather than houses. Even in rural areas, better farming methods and labor-saving machinery have made life easier.

The steps, music, and costumes of flamenco all come from Spanish culture. Continuing performances keep them alive for new generations. ▼

Flamenco Flamenco, a traditional Spanish dance, consists of men and women in elaborate costumes performing intricate steps and gestures. **How do dances like the flamenco preserve Spanish culture?**

DIFFERENTIATING INSTRUCTION

English Learners

Key Academic Vocabulary

Challenge students to locate the following words on this page in their textbooks: *unified, cultural,* and *emphasize*. Have students work in pairs to write a definition for each word. As a class, come to a consensus about the definitions. Then have students work individually to write adapted forms of each word *(unify, culture, emphasis).* Discuss these as a class and answer any lingering questions.

Struggling Readers

Highlight Confusing Words and Phrases

Distribute a highlighter and a photocopy of this page to struggling readers. Instruct students to read the page silently, highlighting any unfamiliar words or confusing phrases as they read. Then have volunteers share a word or phrase they highlighted. Ask if anyone else highlighted the same word or phrase. Discuss these as a group, making sure all students understand the page's content.

CONNECT History & Culture

The Running of the Bulls Every July, tourists come to Pamplona in northern Spain for the Fiesta de San Fermín. The main attraction is the daily stampede of a half-dozen bulls through the city's narrow streets to the bullfighting ring. Runners sprint ahead, trying to avoid getting slashed by sharp horns. According to legend, the run may have started in the 1200s as a way to move the bulls through town to be sold at market.

CRITICAL THINKING

Make Inferences Why might some people object to the Running of the Bulls?

While the shift to an urban society threatened some Spanish traditions, many Spaniards still enjoy taking part in the old customs. *Paella* (pah•AY•yah), a flavorful dish of seafood, meat, and vegetables mixed with yellow rice, remains a popular choice. Spaniards like soccer, but bullfighting remains Spain's most famous traditional sport. Audiences also enjoy watching Spanish dances, such as the flamenco.

Government and Economics After Franco's death in 1975, Spain became a parliamentary monarchy. This change allowed the people to have a voice in their government. King Juan Carlos I has ruled the country alongside elected officials since the monarchy was formed.

Today, Spain's leaders face demands by some in the **Basque** (bask) ethnic group to create a separate Basque nation. The Basque people live in the Pyrenees Mountains. They have lived in Spain longer than any other group and have kept a distinct language and customs. Some Basque **separatist** groups have tried violence to further their cause.

For centuries, Spain's economy depended heavily on fishing and farming. It lagged behind most of Western Europe in industrial growth. After World War II, Spain's economy grew more rapidly. Factories began turning out cars and steel, and tourism increased. In the 1980s, Spain joined the EU. Membership boosted the economy by promoting trade and by making financial aid from the EU available.

EVALUATE Explain some of the regional differences that exist within Spanish culture. Spain has several regional languages such as Catalan and Valencian. The Basque ethnic group also maintains a separate culture and language from the rest of Spain, which has led to political tensions.

CONNECT History & Culture

The Running of the Bulls

The Running of the Bulls takes place each day of the Fiesta at 8:00 A.M. The length of the run is about a half mile and usually lasts about three minutes—unless one or more of the bulls get loose. The six bulls selected can weigh more than 1,300 pounds and are bred for the purpose of fighting. They are smart, fast, aggressive, and unpredictable. Since record-keeping began in 1924, 13 deaths and hundreds of serious injuries have resulted from the Running of the Bulls.

CRITICAL THINKING ANSWER

Make Inferences Possible Answers: They might object to the bulls being aggravated and teased; they might worry about injuries to runners or bulls.

Comparing Regions

Basque People in Spain and the United States

For much of their history in Spain, the Basque people have been noted shipbuilders, fishers, and sailors. Much of Columbus' crew was thought to have been made up of Basque sailors.

In the United States, many Basque immigrants moved to the West. They're considered to have played a major role in the development of the western range sheep-ranching business. Sheepherding was a common occupation in the Basque regions of the Pyrenees, so many immigrants found it a good way to make a living.

INTERDISCIPLINARY ACTIVITIES

CONNECT to Civics

Time Line of Spain's Government

Review the information on this page that describes Spain's current type of government—parliamentary monarchy. Discuss the difference between a parliamentary monarchy and a monarchy. (You may wish to have students research the difference.) Then have students use information from their textbooks to create an annotated time line of the history of Spain's government, beginning with the A.D. 700s and ending in 1975.

CONNECT to Language Arts

Running of the Bulls Limerick

Challenge students to write a limerick about the Running of the Bulls in Pamplona. Remind them that limericks must:

- have five lines, with a rhyming pattern of a-a-b-b-a.
- have three beats in lines 1, 2, and 5; have two beats in lines 3 and 4.
- contain hyperbole, onomatopoeia, puns, idioms, or other language devices.

If possible, locate a sample limerick to help students understand their form and pattern.

Teach

Portugal

Talk About It

- Why did Portugal lose its position as a world power by the 1800s? *(Larger European nations overpowered it.)*
- What caused Portugal to begin working harder to improve its economy in recent years? *(Portugal joined the EU, which helped increase trade and expanded its service industries.)*
- **Compare and Contrast** How does Portugal's culture compare to the rest of Western Europe? *(Possible Answer: Portugal is largely rural, with many small farming and fishing villages.)*

Unit Resource Book: Europe
- Skillbuilder Practice: Synthesizing, p. 110

RESEARCH & WRITING CENTER

HISTORY MAKERS

Prince Henry

The third son of King John I of Portugal, Prince Henry set up his own royal court at Sagres, a barren piece of land on Portugal's southwestern coast. Sagres became a school for oceanography and navigation, attracting sea captains, navigators, mapmakers, astronomers, shipbuilders, and instrument makers. Under Henry's guidance, cartography techniques and navigational instruments were improved.

Portugal

KEY QUESTION How did Portugal's location contribute to its role as a leader in the Age of Exploration?

Portugal shares the Iberian Peninsula with Spain, but occupies a much smaller area. It is located on Europe's western coastline along the Atlantic Ocean. Despite its small size, its location and keen interest in exploration helped it build an empire.

HISTORY MAKERS

Prince Henry 1349–1460

Historians call Prince Henry "the Navigator" because he made major contributions to maritime exploration. He wasn't a sailor, and he rarely traveled far from his homeland of Portugal. During his lifetime, he sponsored more than 50 voyages of exploration. His dedication to exploration eventually led Portugal to discover valuable trade routes to Asia. These voyages greatly advanced European knowledge of the world. Prince Henry's efforts helped make Portugal a major sea power and put the nation on the path to empire.

ONLINE BIOGRAPHY
For more on the life of Prince Henry, go to the **Research & Writing Center** @ClassZone.com

History For many centuries, the region that is now Portugal was, like Spain, under Muslim rule. In 1143, Portugal became an independent kingdom. Because of its coastal location, the Portuguese became skilled sailors, navigators, and shipbuilders. In the 1400s and 1500s, Portugal played a key role in Europe's Age of Exploration. Prince Henry of Portugal sponsored many expeditions. Portuguese explorers helped their country build an empire.

For a time, overseas colonies brought prosperity. Portugal profited from the spice trade with Asia and the gold, diamonds, and other resources taken from its African and South American colonies. By the 1800s, however, Portugal had lost its position as a world power to larger European nations. In 1822, Brazil, Portugal's richest colony, declared independence.

In 1908, revolutionaries overthrew Portugal's king and made the country a republic, but political unrest continued. In 1926, military leaders seized control. Dictators ruled until 1976, when Portugal became a parliamentary democracy in a peaceful revolution.

Culture The official language of Portugal is Portuguese. Like Spanish, Portuguese is a Romance language. Most Portuguese are Roman Catholics. Especially in rural areas, Catholic priests play a key role in political and social life. Traditional Catholic celebrations are important.

Unlike much of Western Europe, Portugal today is still a largely rural country with many small farming and fishing villages. Just under half of its citizens live in rural areas. Although Portugal remains a rural country, its cities are growing. Each year, many people move from the countryside in search of jobs in the cities.

DIFFERENTIATING INSTRUCTION: TIERED ACTIVITIES

OBJECTIVE Create a flow chart that depicts Portugal's history.

Basic	On Level	Challenge
Have students reread the text under "History" and list three events on their flow charts. Challenge students to come up with a symbol or simple picture that represents each event. Then have students write a label that identifies each image. Have students present their flow charts.	Have students reread the text under "History" and list at least five events on their flow charts. Challenge students to come up with a symbol or simple picture that represents each event. Then have students write a caption that summarizes the event. Have students present their flow charts.	Have students reread the text under "History" and list all events on their flow charts. Challenge students to come up with a symbol or image to represent each event. Then have students write a summary, analyzing causes and effects for each historical event. Have students present their flow charts.

One-third of the population lives in or near its two largest cities, Lisbon and Porto. Lisbon, in southern Portugal, is the nation's capital and largest city. It is a center of tourism and commerce. Porto is a major seaport and industrial center.

Government and Economics Since 1976, Portugal has been a parliamentary democracy. Citizens elect members of parliament and the president. Candidates from many political parties compete for seats in its parliament.

Portugal's colonies once made it the richest nation in Europe. Today, it is one of the poorest, with the lowest per capita GDP in Western Europe. For decades, leaders failed to strengthen its economy and neglected its roads, ports, and factories. Lack of energy resources and a poor educational system also slowed its growth. Since joining the EU in 1986, however, Portugal has worked hard to improve its economy. It has diversified and expanded the role of service industries. Membership in the EU has helped increase trade. Today, Portugal continues its efforts to develop its economy.

Albufeira, Portugal Two fishermen tend to their nets in this fishing village. Fishing remains an important part of the Portuguese economy. **What geographic trait makes Portugal a major fishing nation?**

Portugal's extensive coastline

ANALYZE EFFECTS Explain how Portugal's location contributed to its role as a leader in the Age of Exploration. Portugal's easy access to the sea helped its people become skilled sailors and shipbuilders.

Section 2 Assessment

ONLINE QUIZ For test practice, go to Interactive Review @ClassZone.com

TERMS & NAMES

1. **Explain the importance of**
   - Moors
   - Reconquista
   - Christopher Columbus
   - Basque

USE YOUR READING NOTES

2. **Compare and Contrast** Use your completed diagram to answer the following question:
   How is Spain's government different from Portugal's government?

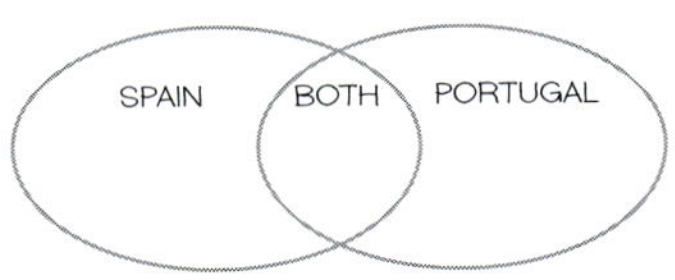

KEY IDEAS

3. What major event in Spain's history occurred in 1492?
4. What role did Portugal play in the Age of Exploration?
5. In what ways are the cultures and histories of Spain and Portugal similar?

CRITICAL THINKING

6. **Analyze Causes** What are some reasons for Portugal's slow economic growth?
7. **Summarize** How has EU membership changed Spain's economy?
8. **CONNECT to Today** How has increased urbanization affected the societies of Spain and Portugal?
9. **WRITING** **Plan an Itinerary** Plan a cultural and historical tour of a city in Spain. Choose two places that show its diverse religious and ethnic heritage. Tell why you have chosen each place and explain its importance to the history of the city.

CONNECT to the Essential Question

What geographic and cultural characteristics define the subregions of Western Europe?

Ask students what they have learned so far that can help them answer this question. Students might mention:

- Spain has retained its cultural traditions while improving its economy through industry and tourism.
- Portugal remains mostly rural and is working to improve its economy.
- Spain and Portugal share a similar history, geographic location, type of government, and some cultural aspects.

4 Assess & Reteach

Assess Have students complete the Section Assessment.

Unit Resource Book: Europe
- Section Quiz, p. 125

Interactive Review @ClassZone.com
- Section Quiz

Power Presentations

Test Generator

Reteach Divide students into two teams. Have one team write 10 questions and answers about Portugal. Have the other write 10 questions and answers about Spain. Take turns reading aloud a question. Allow the other team 30 seconds to answer. Award points for correct answers.

Unit Resource Book: Europe
- Reteaching Activity, p. 130

SECTION 2 ASSESSMENT ANSWERS

Terms & Names

1. Moors, Reconquista: Muslims from North Africa conquered Spain, were driven out by the Spanish; Christopher Columbus: Italian explorer for Spain; Basque: ethnic group that lives in the western Pyrenees and along the Bay of Biscay

Use Your Reading Notes

2. Spain has a parliamentary monarchy. Portugal has a parliamentary democracy.
   Diagram *Spain—speak Spanish; parliamentary monarchy; economy grew after World War II. Both—Moorish rule; had empires, then power faded; mostly Catholic; EU members. Portugal—speak Portuguese; parliamentary democracy; poor but developing economy*

Key Ideas

3. Columbus made his voyage for Spain and began the European exploration of the Americas.
4. Portuguese-sponsored expeditions advanced European knowledge of the world.
5. both part of the W. Roman Empire; under Muslim rule; part of the Age of Exploration with empires; mostly Catholic; speak Romance languages

Critical Thinking

6. failure to strengthen the economy by building infrastructure (roads, ports, factories), a poor educational system, and a lack of energy resources
7. made trading partners and financial aid available
8. life differs; people live in apartments instead of houses, old traditions are threatened
9. Use the rubric to score itineraries.

Itineraries Rubric

	Clarity	Geographic Details	Accuracy
4	clear, complete	many	no errors
3	clear	some	few/minor
2	less clear	few	several
1	unclear	very few	many

1 Plan & Prepare

Objectives

- Explain how France has benefited from its geography
- Describe the cultural characteristics that the Benelux countries share

Read for the Essential Question

Help students read for a purpose by reminding them of the Essential Question: "What geographic and cultural characteristics define the subregions of Western Europe?" Point out that they can use text features such as Key Questions, special features, images, and maps to form answers to the question.

Vocabulary

Best Practices Toolkit

Use the ***Best Practices Toolkit*** to model strategies for vocabulary notetaking. Vary strategies throughout the year. Choose from: *Knowledge Rating, Predicting ABCs, Definition Mapping, Word Sort, Frayer Model (Word Squares), Magnet Words, Word Wheel,* and *Student VOC.*

Vocabulary Strategies, TT9–TT16

Reading Strategy

Best Practices Toolkit

Display the Web Diagram transparency.

- Remind students that main ideas are the most important points in a text.
- Have students find the main idea in the "Connecting to Your World" text at the start of this section.
- Model finding and entering main ideas on a chart using the "Connect to History" feature in this section.
- Have students suggest entries.

Web Diagram, TT28

Section 3 Reading for Understanding

Key Ideas

BEFORE, YOU LEARNED

Spain and Portugal both created empires to supplement the natural resources of the Iberian Peninsula. Spain's economy is stronger than Portugal's economy.

NOW YOU WILL LEARN

Over the centuries, France's natural resources have helped it prosper. Through economic union, Belgium, Luxembourg, and the Netherlands have also prospered.

Vocabulary

TERMS & NAMES

Benelux (BEHN•uh•LUHKS) name for the economic union formed by Belgium, the Netherlands, and Luxembourg to ensure the fast and efficient movement of people, goods, and services within these nations

polders (POHL•durz) land reclaimed by draining it of water with the use of a dam and pump

duchy the territory ruled by a duke or duchess

multilingual able to speak many languages

BACKGROUND VOCABULARY

autonomy self-governance or independence

REVIEW

Northern European Plain vast area of flat or gently rolling land from France to Russia

Visual Vocabulary polders in the Netherlands

Reading Strategy

Re-create the web diagram shown at right. As you read and respond to the **KEY QUESTIONS**, use the diagram to help you find the main ideas about France and the Benelux countries.

Skillbuilder Handbook, page R4

FIND MAIN IDEAS

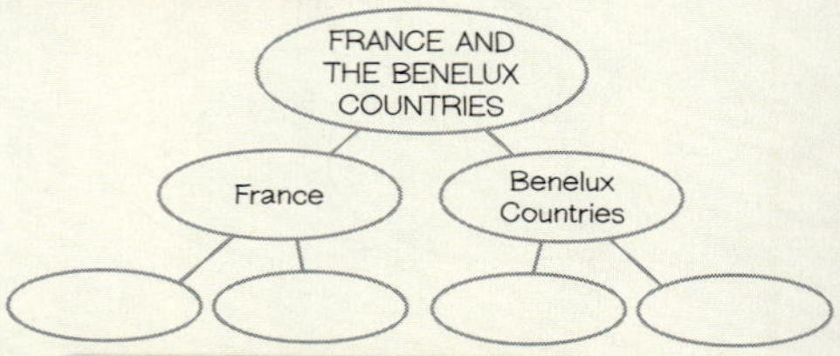

GRAPHIC ORGANIZERS
Go to **Interactive Review** @ClassZone.com

PRETEACHING VOCABULARY

English Learners

Pronounce and Preview

Pronounce each term for students. Review the meanings of words within the definitions, such as *efficient, reclaimed,* and *self-governance.*

- To modify vocabulary learning, have students complete worksheets as they read, instead of afterward.

Unit Resource Book: Europe
- Vocabulary Practice, p. 109

Inclusion

Break Down Word Meanings

To help students who have difficulty analyzing or synthesizing information, break down the vocabulary terms by syllable and identify their meanings separately. For example, have students write the word *multilingual* and draw a line through the word between *multi-* and *-lingual.* Explain that *multi-* means "many" and that *lingual* means "language". Repeat this process for *Benelux, duchy,* and *autonomy.*

Chateau de Polignac, France

SECTION 3 FOCUS ON

Cattle in the Alps

France and the Benelux Countries

Connecting to Your World

Can you imagine spending three weeks riding your bike from Chicago to Los Angeles? Every July, about 200 riders in the Tour de France bicycle race cover that distance and more in one of the world's most watched sporting events. The course of the race runs through much of France and some of the surrounding countries. Riders cover terrain from the mountainous roads of the Alps to the cobbled streets of Paris. The race is a great source of national pride for the French people.

Tour de France Cyclists race down a curvy mountain road in western France.

France

KEY QUESTION How has its colonial empire changed the population of France?

France is the largest country in Western Europe. Its landscape ranges from the peaks of the Alps and the beaches on the Mediterranean Sea to cities such as Paris and fishing villages along the Atlantic Ocean. For much of its history, kings ruled this territory. Then, in 1789, the French were inspired to fight for their independence. Since then, France has gone back and forth many times between monarchy and democracy.

Arc de Triomphe This monument in Paris celebrates military victories.

2 Focus & Motivate

3-Minute Warm-Up

Write on the board or display the transparency:

- Brainstorm cultural characteristics associated with France. *(Possible Answers: Eiffel Tower, Louvre, French language, French cooking, fashion, painting, Impressionism)*

Unit Transparency Book
- 3-Minute Warm-Ups, TT11

Connecting to Your World

More About . . .

The Tour de France

The race route changes every year, alternating between clockwise and counter-clockwise routes around France. It is made up of 20 or more stages. The cyclist with the fastest total time after each stage of the race wears the coveted yellow jersey—the *maillot jaune*—for the following day's ride.

3 Teach

France

Reader, Recorder, Reporter

- Summarize France's political history from 1792 to the present. *(The government has gone back and forth between monarchy and democracy several times.)*
- **Find Main Ideas** How has France used its natural resources to improve its economy? *(French farmers use the fertile soil in the Northern European Plain to grow produce for export; the landscape attracts many tourists.)*

SECTION 3 PROGRAM RESOURCES

Unit Resource Book: Europe

ON LEVEL

Unit Resource Book
- Reading Study Guide, p. 73
- Section Quiz, p. 126

STRUGGLING READERS

Unit Resource Book
- Reading Study Guide with Additional Support, p. 83
- Skillbuilder Practice: Synthesizing, p. 110
- Section Quiz, p. 126
- Reteaching Activity, p. 131

eEdition with Audio DVD-ROM

ENGLISH LEARNERS

Pupil Edition in Spanish
eEdition with Audio DVD-ROM
Unit Resource Book
- RSG (Spanish), p. 93
- RSG with Additional Support (Spanish), p. 103

Multi-Language Glossary
Test Generator
- Section Quiz in Spanish

INCLUSION

Unit Resource Book
- RSG with Additional Support, p. 83
- Section Quiz, p. 126
- Reteaching Activity, p. 131

GIFTED & TALENTED

Unit Resource Book
- Interdisciplinary Projects, pp. 118–119
- Section Quiz, p. 126

PRE-AP

Unit Resource Book
- Connect to History, p. 121
- Section Quiz, p. 126

TECHNOLOGY

Unit Transparency Book
- 3-Minute Warm-Ups, TT11
- Essential Question, TT12
- Critical Thinking, TT15
- Map: European Union, 2007, TT19
- Cultures, TT21

Daily Test Practice Transparencies
- Section 3, TT43

Power Presentations
ClassZone.com
World Cultures and Geography Video on DVD

More About . . .

Napoleon Bonaparte

A general in the French Revolution, Napoleon became a national hero in 1797. He was made First Consul (of three) when the French created a new government. With Napoleon at the helm, the new government made many reforms that improved the lives of the French people. Napoleon wielded so much power that he had effectively become the king of France, but without the title.

In 1804, Napoleon proposed a constitution that, if passed, would make him emperor of France. This title would allow Napoleon's family to take power after his death, which was not the case if he remained First Consul. The French people passed the constitution by a wide margin. Napoleon I became Emperor of France.

CONNECT to History

Tell students that French revolutionaries stormed the Bastille seeking the arms and munitions stored there. A symbol of the absolute power that the king had held over the people for so long, the Bastille was torn down completely in the months after the crowd broke in. Its rubble represented the old regime that the people had overthrown.

TIME LINE ANSWERS

July 14, 1789—a mob of French citizens storms the Bastille; August 4, 1789—nobles allow commoners greater role in government; 1791—constitution adopted; June 1791—King Louis XVI tries to flee France; August 1792—King Louis XVI is imprisoned; 1793—constitution adopted; January 1793—King Louis XVI is beheaded; 1795—constitution adopted; October 1799—Napoleon seizes power, ending the French Revolution

Unit Resource Book: Europe
- Connect to History: Make an Illustrated Time Line, p. 121

History In 1792, French revolutionaries decided to make France a democracy. Soon after, Napoleon Bonaparte seized power and ruled as emperor until his defeat in 1815. France then returned briefly to a monarchy. In 1848, a second, less violent revolt occurred that led to another democracy. Voters chose Louis Napoleon Bonaparte, a nephew of Napoleon, as president of the new government, but like his famous uncle, he declared himself emperor. After his armies suffered major defeats, he was forced to step down, and France once again became a republic. Political instability and frequent changes in government continued. Then, in 1958, French voters agreed to a new constitution that gave their president greater power.

Despite its many changes in government, France became a powerful nation. Like other major powers, it developed a colonial empire in Africa and Asia that boosted its economy. In the 1900s, however, many battles of World War I and World War II were fought on French soil. These conflicts devastated France. Millions died, and the destruction of factories, farms, and cities shattered the French economy. In the 1960s and 1970s, the French colonies overseas gained their independence, another blow to the French economy. It has since recovered, as a result of hard work and access to natural resources.

CONNECT to History

Bastille Day is celebrated each year on July 14. It marks the beginning of the French Revolution. On July 14, 1789, an angry mob took control of the Bastille, a French prison. On August 4, 1789, frightened French nobles allowed the common people a greater role in government. In June 1791, fearing for his life, King Louis XVI tried to escape from France. He was imprisoned in August 1792. In January 1793, revolutionaries beheaded the king. Three constitutions were adopted during the revolution, one in 1791, one in 1793, and the last in 1795. In October 1799, Napoleon Bonaparte seized power, ending the revolution.

Bastille Day parade in Paris

Activity

Make an Illustrated Time Line

Materials
- markers
- butcher paper

1. To create a French Revolution time line, pick beginning and end dates.
2. Draw a line and divide it into segments. Leave room to illustrate each event.
3. Choose three or four events to show on the time line.
4. Use markers to create drawings and write short captions describing each event.

DIFFERENTIATING INSTRUCTION

Struggling Readers

Listen to France's Modern History on CD

To support students in reading and understanding the text on this page, play the audio version on the **Pupil Edition Audio CD.** As they listen, suggest that students use a ruler to follow along with the text in their books. Remind students to move the ruler down as each line is read aloud to help prevent confusion and enhance the flow of the content.

Pre-AP

Bastille Day Readers Theater

Provide students with additional research about the events that took place in France on July 14, 1789. Have them read and highlight key information, then discuss it as a class. Working in three teams, have students write a readers theater. Assign teams various stages of the event, so that each team will create one "act." Allow time for students to write a script and practice, then have them perform their readers theater about Bastille Day.

Culture Most French citizens have a very strong cultural identity, and take pride in French culture, history, and language. In recent decades, France has become more diverse. People have emigrated from its former colonies in North Africa and Asia. Many Muslims now live in this primarily Roman Catholic nation.

France, like much of Western Europe, is highly urbanized and densely settled. Three out of four people live in cities. Because space is scarce, apartments are more common than houses. People often walk or ride public transportation. In fact, many cities limit car use and parking on city streets.

French culture has had a global influence. French painters such as Claude Monet and Pierre-Auguste Renoir influenced the popular Impressionist style of painting in the late 1800s. French chefs are well-known for creating culinary masterpieces. Chefs in restaurants around the world try to master French techniques. Each year on July 14, the French celebrate their nation and its culture on Bastille Day. Fireworks displays and parades commemorate French independence.

Government and Economics Today, France is a parliamentary democracy with power concentrated at the national level. Its government spreads power across executive, legislative, and judicial branches, like the United States. In France, however, the executive branch includes a president and a prime minister. France's president serves a five-year term and oversees foreign affairs. He or she appoints a prime minister to take care of the daily operation of the government.

After the devastation of World War II, France worked hard to modernize its economy. Now, its economy is one of the world's strongest. Although a small percentage of French workers are farmers, France is Europe's largest exporter of farm produce. This is because of the fertile soil of the **Northern European Plain** that covers much of France, as well as modern farming methods. France, an EU member, is a major producer of cars, high-speed trains, and airplanes.

Tourism is a key industry. France has been the world's top tourist destination for several years running. Many people travel to France to visit its museums, castles, and cathedrals. Visitors also wait in long lines to go to the top of the most famous landmark in Paris, the Eiffel Tower.

ANALYZE EFFECTS Explain how France's colonial empire has changed its population.

has increased cultural diversity, as people from former colonies have moved to France ▶

THE EIFFEL TOWER

Height: 1,063 feet

Weight: 10,100 tons

Date Completed: March 31, 1889

Elevators: Seven elevators travel over 64,000 miles each year, equal to more than two and a half times around the Earth.

Materials: The tower is made of iron, and is painted about every 5 years with 50 to 60 tons of paint to keep it from rusting.

Composition: It consists of 18,038 pieces, joined together by 2.5 million rivets.

INTERDISCIPLINARY ACTIVITIES

CONNECT to Math

Build a Scale Model of the Eiffel Tower

Have students review the facts and figures about the dimensions of the Eiffel Tower on this page. Refer students to an almanac or encyclopedia in order to gather additional information about its structure. As a class, discuss how to accurately calculate the length, width, and height of a scale model of the monument. Then, using toothpicks, clay, and poster board, have students build a scale model of the Eiffel Tower.

CONNECT to Art

Review an Impressionist Work of Art

Provide several art books or direct students to a reputable art Web site, such as the National Gallery, where students can view examples of French Impressionism. Discuss important characteristics of Impressionism, such as small, visible brush strokes; use of unmixed, primary colors; and the effect of reflected light. Then have each student choose one Impressionist painting and write an art review on that piece.

More About . . .

French Cooking

France's most famous school of culinary arts is the prestigious Le Cordon Bleu. Attended by students from more than 70 countries, Le Cordon Bleu's Master Chefs teach methods of preparing fine cuisine, creating delicate pastries, and baking bread. The school is internationally recognized as the world leader in culinary education. An aspiring chef can earn a diploma from Le Cordon Bleu in just nine months.

Unit Resource Book: Europe
- Interdisciplinary Projects: French Cooking, pp. 118–119

More About . . .

The Eiffel Tower

Until the Chrysler Building was built in New York City in 1930, the Eiffel Tower was the world's tallest structure.

Comparing Regions

French Geography

The Northern European Plain covers much of France, mainly the northern and western portions of the country. It is mostly flat with gently rolling hills. The climate varies from cool and moist to warm and dry, allowing a great variety of crops to be grown. Some of these include wheat, barley, corn, sugar beets, potatoes, apples, grapes, peaches, plums, apricots, and tomatoes.

The rest of France—especially the mountainous terrain of the Alps on the east and the Pyrenees to the south—is not as suitable for agriculture. The Massif Central region of France contains mountains, volcanoes, hills, valleys, thick forests, and rivers. This area is sparsely populated and ideal for nature lovers. Popular activities in the region include hiking, mountain biking, kayaking, downhill skiing, fishing, mountain climbing, and hang gliding.

Teach

The Benelux Countries

Talk About It

- In what ways are the Benelux countries alike? *(constitutional monarchies; highly urbanized with modern economies; belong to the EU; depend on trade with neighboring countries)*
- How have the Dutch adapted to the geography of the Netherlands? *(They have used dams, canals, and pumps to create polders.)*
- **Find Main Ideas** Why do you think Belgium, the Netherlands, and Luxembourg formed an economic union? *(Possible Answer: Separately, they are small and less powerful; united, they gain power and economic benefits.)*

Unit Resource Book: Europe
- Skillbuilder Practice: Synthesizing, p. 110

CONNECT Geography & Economics

Tulip Mania

The Netherlands currently grows over 9 billion tulip bulbs each year. Nearly 7 billion of those are exported, with the United States being the largest importer. The Netherlands earns $750 million a year from tulip exports.

Make It Fun

Teacher-Tested Activities

Amber McVey, Antioch Middle School, Gladstone, Missouri

As a review activity, my classes play "Red Light, Green Light." Here's how it works.

- First, put together a list of true/false questions about the section.
- Divide the class into teams. Give each team two pieces of paper, one that says *false* and one that says *true*. Have teams stand at the back of the classroom.
- Ask a question. Each team discusses and holds up one of the sheets of paper so that the other teams can't see.
- Each correct answer earns the team one giant step forward. The first team to reach the front of the classroom wins.

The Benelux Countries

KEY QUESTION How has belonging to an economic union helped the Benelux countries?

The Netherlands, Belgium, and Luxembourg belong to an economic union known as the **Benelux** (BEHN•uh•LUHKS) countries. The name was created by combining the first few letters of each country's name. The Benelux nations are alike in many ways. All are constitutional monarchies. All are highly urbanized with well-developed economies. All belong to the EU and depend on trade with larger neighbors.

The Netherlands Following centuries of foreign rule, the Dutch declared their independence from Spain in 1581. Spain finally recognized the Netherlands as an independent nation in 1648. The 1600s were an era of prosperity and achievement for the Dutch. They became a great sea power with colonies in Southeast Asia and the Americas. Dutch is the official language of the Netherlands. Many people also speak English, German, or French. Most Dutch are Christians, though many Muslim immigrants have come from former Dutch colonies in Indonesia and Suriname.

The Netherlands is very densely populated. Because nearly half of its land is below sea level, living space is scarce. The Dutch have used dikes, dams, canals, and pumping systems to reclaim land that is below sea level. The drained lands, called **polders** (POHL•durz), have rich soil and are often used for farming. The Dutch also build factories and towns on them. Forty percent of the land in the Netherlands is from polders.

The Netherlands is also highly urbanized. Because of the shortage of land, Dutch rural areas are close to cities. Bicycles are very common, both for recreation and as a means of transportation in crowded cities.

CONNECT Geography & Economics

Tulip Mania

In the early 1600s in the Netherlands, people went crazy over tulips imported from Turkey. Dutch families bought rare varieties of bulbs, not for their gardens but as an investment. Tulip prices soared. One man traded his mansion for just three bulbs. When prices dropped, many people lost everything. The craze ended, but tulips were in the Netherlands to stay. Today, the country is one of the world's biggest exporters of tulip bulbs.

Belgium and Luxembourg For many years, Belgium was part of the Netherlands. It gained its independence in 1830. Belgium's central location has been both an advantage and a disadvantage. It has enabled Belgium to prosper through trade with its neighbors. But it also made Belgium a battleground during the two world wars, which caused great destruction and loss of life.

DIFFERENTIATING INSTRUCTION: TIERED ACTIVITIES

OBJECTIVE Design a postcard from the Netherlands and write about Dutch culture on the back.

Basic	On Level	Challenge
Have students review the information about the Netherlands. Provide several images that represent Dutch culture. Have students select an image and draw it on the front of their postcards. On the back, have them write a sentence about how it represents the Netherlands.	Have students review the information about the Netherlands and make a list of all items that represent Dutch culture. Have students select an item from the list and draw it on the front of their postcards. On the back, have them write a paragraph about how their image represents the Netherlands.	Have students review the information about the Netherlands. Have students do additional research on Dutch culture. Ask them to select a cultural aspect not mentioned in their textbooks and draw it on the front of their postcards. On the back, have them write about how it represents the Netherlands.

At one time, cultural differences between Dutch-speaking Flemings in the north and French-speaking Walloons in the south caused serious tensions. In recent years, the government has eased tensions by giving both ethnic groups greater **autonomy**. Now, the Belgian constitution recognizes separate cultural communities based on language, as well as separate economic regions for the two groups.

Luxembourg is one of the smallest countries in Europe. In 1890, it broke away from the Netherlands to become a **duchy**, a state ruled by a duke or duchess. The country has three official languages: French, used in government; German, used in newspapers; and Luxembourgish, used for everyday matters. Many of its people are **multilingual**, or able to speak several languages. This has made Luxembourg attractive to foreign companies looking for new locations. At one time, steel was Luxembourg's most important product. Today, its economy is more diversified, with banking and other financial services making up a key segment of its economy.

Antwerp, Belgium
The cafés in this square in the heart of Antwerp's historic district are typical of Europe's outdoor cafés.

MAKE INFERENCES Explain how belonging to an economic union has helped the Benelux countries. It has lowered trade barriers and allowed them to combine their resources in trading with other countries.

Section 3 Assessment

ONLINE QUIZ For test practice, go to **Interactive Review** @ClassZone.com

TERMS & NAMES

1. **Explain the importance of**
   - Benelux
   - polders
   - duchy
   - multilingual

USE YOUR READING NOTES

2. **Find Main Ideas** Use your completed web diagram to answer the following question:
   How are France and the Benelux countries similar?

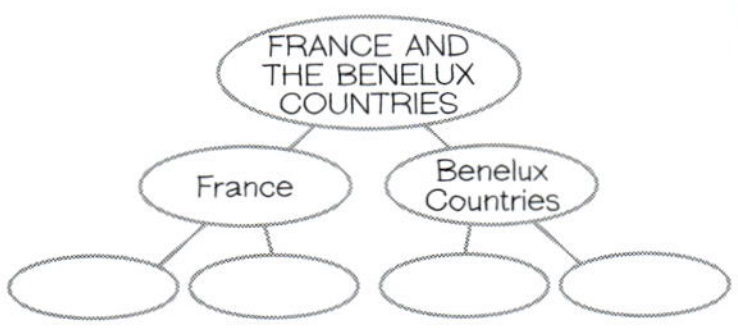

KEY IDEAS

3. What are France's leading industrial products and services?
4. How have the people of the Netherlands used technology to reshape their environment?
5. What cultural differences have caused problems within Belgium?

CRITICAL THINKING

6. **Make Inferences** How has France's location been beneficial for its economy?
7. **Analyze Effects** How do the Benelux countries benefit from their nearness to Germany and France?
8. **CONNECT to Today** Why might France object to trade barriers on farm produce?
9. **WRITING** **Create an Advertisement** You have been asked by a French travel agency to make a list of the top five reasons to visit France. Create an advertisement explaining why France is an interesting and fun place to visit.

CONNECT to the Essential Question

What geographic and cultural characteristics define the subregions of Western Europe?

Ask students what they have learned so far that can help them answer this question. Students might mention:

- France is growing more diverse and has one of the world's strongest economies.
- Belgium and the Netherlands have fertile land for farming, while Luxembourg's economy is more diversified.
- Belgium and Luxembourg were once part of the Netherlands; today, the three countries belong to an economic union.

4 Assess & Reteach

Assess Have students complete the Section Assessment.

Unit Resource Book: Europe
- Section Quiz, p. 126

Interactive Review @ClassZone.com
- Section Quiz

Power Presentations

Test Generator

Reteach Have students write five facts about France and the Benelux countries. Then have them take turns reading a fact out loud while another student determines which country the fact describes. Continue until everyone has had a turn in both roles.

Unit Resource Book: Europe
- Reteaching Activity, p. 131

SECTION 3 ASSESSMENT ANSWERS

Terms & Names

1. Benelux: economic union of Belgium, the Netherlands, and Luxembourg; polders: land reclaimed by draining water; duchy: territory ruled by a duke or duchess; multilingual: able to speak many languages

Use Your Reading Notes

2. All are EU members, and they are highly industrialized and largely urban.
   Diagram *France—strong cultural identity; urbanized and densely settled; parliamentary democracy; strong economy; EU member. Benelux Countries—small; highly urbanized; constitutional monarchies; well-developed economies; EU members*

Key Ideas

3. cars, high-speed trains, airplanes, and tourism
4. The Dutch have used technology to reclaim land that was underwater and create polders, providing more living space.
5. The Flemish ethnic group in the north speaks Dutch. The Walloons in the south speak French.

Critical Thinking

6. France has rich soil and a variety of natural resources.
7. France and Germany provide large markets for Benelux goods.
8. France is Europe's largest exporter of farm produce, so trade barriers on farm goods could hurt its economy.
9. Use the rubric to score advertisements.

Advertisement Rubric

	Level of Content	Language	Accuracy
4	detailed	vivid, effective	no errors
3	adequate	descriptive	few/minor
2	some details	somewhat descriptive	several
1	few details	general, imprecise	many

1 Plan & Prepare

Objectives

- Explain how Germany was divided after World War II, and then reunified
- Identify the cultural characteristics the Alpine countries share with Germany

Read for the Essential Question

Help students read for a purpose by reminding them of the Essential Question: "What geographic and cultural characteristics define the subregions of Western Europe?" Point out that they can use text features such as Key Questions, special features, images, and maps to form answers to the question.

Vocabulary

Best Practices Toolkit

Use the ***Best Practices Toolkit*** to model strategies for vocabulary notetaking. Vary strategies throughout the year. Choose from: *Knowledge Rating, Predicting ABCs, Definition Mapping, Word Sort, Frayer Model (Word Squares), Magnet Words, Word Wheel,* and *Student VOC.*

Vocabulary Strategies, TT9–TT16

Reading Strategy

Best Practices Toolkit

Display the Time Line transparency.

- Point out that sequencing events, or arranging events in chronological order, will help students understand how the events are related.
- Model filling in a few events from the "Modern History" subsection.
- Have students suggest entries.
- Display the graphic organizer for students to review.

Time Line, TT33

SECTION 4 Reading for Understanding

Key Ideas

BEFORE, YOU LEARNED

France prospered as a result of natural resources. The Benelux countries have benefited from their economic union.

NOW YOU WILL LEARN

Germany's central location helped it to dominate neighboring lands. Germany and the Alpine countries are linked in many ways.

Vocabulary

TERMS & NAMES

Prussia the most powerful German state in the Holy Roman Empire

Adolf Hitler German head of state from 1933 to 1945

Berlin Wall a wall built by East Germany's Communist government to close off East Berlin from West Berlin

Alpine having to do with the Alps mountain range

neutrality a policy of not taking part in war

BACKGROUND VOCABULARY

reunify to bring something that has been separated back together

REVIEW

Holocaust the systematic murder of Jews and other minorities by the Nazis during World War II

market economy an economic system in which the production of goods and services is decided by supply and the demand of consumers

Reading Strategy

Re-create the time line shown at right. As you read and respond to the **KEY QUESTIONS**, use the time line to help you sequence the events that have shaped Germany and the Alpine countries.

See Skillbuilder Handbook, page R6

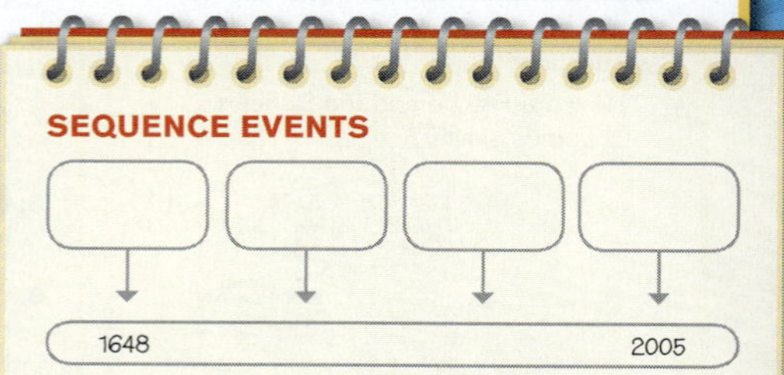

GRAPHIC ORGANIZERS Go to Interactive Review @ClassZone.com

PRETEACHING VOCABULARY

English Learners

Pronounce and Preview

Pronounce each term for students. Review the meanings of words within the definitions, such as *head of state, policy,* and *systematic.*

- To modify vocabulary learning, have students complete worksheets as they read, instead of afterward.

Unit Resource Book: Europe
- Vocabulary Practice, p. 109

Inclusion

Sculpt Terms and Names

Read aloud the terms and their definitions as a class. Provide students with a small amount of moldable clay. Challenge them to sculpt the following vocabulary terms: *Berlin Wall, Alpine,* and *reunify.* Suggest that students use toothpicks to add details to their sculptures. Have students suggest ways they could communicate the meanings of the terms *Adolf Hitler, Prussia,* and *neutrality* without using clay or writing the definitions down.

Chateau de Polignac, France

Cattle in the Alps

SECTION 4 FOCUS ON

Germany and the Alpine Countries

Connecting to Your World

Germany has had a big influence on the United States, and for good reason. In 1990, more Americans reported that their ancestors came from Germany than from any other country. It's no wonder that Americans enjoy eating hot dogs, or frankfurters—a sausage named after Frankfurt, Germany. American children go to kindergarten, a type of school borrowed from Germany. And in December, many Americans decorate trees for the holidays, a German tradition.

Germany

KEY QUESTION How was Germany reshaped after World War II?

German traditions are part of Germany's strong national identity. For a long time, military strength was another aspect of Germany's identity. Germany is a large country centrally located on the European continent. It is well-positioned to dominate other nations and has conquered surrounding lands several times.

Neuschwanstein Castle King Ludwig II began building this fairy-tale castle in 1869.

2 Focus & Motivate

3-Minute Warm-Up

Write on the board or display the transparency:

- Write true or false for each statement. If false, explain why.

1. Germany was once a collection of smaller states instead of a unified nation. *(True)*
2. Austria declared itself neutral just before World War II began. *(False; it became neutral after World War II ended.)*

Unit Transparency Book
- 3-Minute Warm-Ups, TT11

Connecting to Your World

More About . . .

German Culture in the United States

Germans popularized many forms of recreation in America, such as swimming pools, bowling alleys, and amusement parks.

3 Teach

Germany

Think, Pair, Share

- How is the German government similar to and different from that of the United States? *(It is a federal republic made up of a union of states like the U.S., but its head of government is a chancellor selected by a parliament.)*
- **Sequence Events** What event immediately preceded the fall of the Berlin Wall? *(East Germany's government lost power.)*

SECTION 4 PROGRAM RESOURCES

Unit Resource Book: Europe

ON LEVEL
Unit Resource Book
- Reading Study Guide, p. 75
- Section Quiz, p. 127

STRUGGLING READERS
Unit Resource Book
- Reading Study Guide with Additional Support, p. 85
- Skillbuilder Practice: Synthesizing, p. 110
- Section Quiz, p. 127
- Reteaching Activity, p. 132

eEdition with Audio DVD-ROM

ENGLISH LEARNERS
Pupil Edition in Spanish
eEdition with Audio DVD-ROM
Unit Resource Book
- RSG (Spanish), p. 95
- RSG with Additional Support (Spanish), p. 105

Multi-Language Glossary
Test Generator
- Section Quiz in Spanish

INCLUSION
Unit Resource Book
- RSG with Additional Support, p. 85
- Section Quiz, p. 127
- Reteaching Activity, p. 132

GIFTED & TALENTED
Unit Resource Book
- History Makers: Adolf Hitler, p. 116
- Section Quiz, p. 127

PRE-AP
Unit Resource Book
- Economics in the World, p. 120
- Section Quiz, p. 127

TECHNOLOGY
Unit Transparency Book
- 3-Minute Warm-Ups, TT11
- Essential Question, TT12
- Critical Thinking, TT16

Daily Test Practice Transparencies
- Section 4, TT44

Power Presentations
ClassZone.com
World Cultures and Geography Video on DVD

More About . . .

Adolf Hitler

Although Hitler controlled Germany as a dictator in World War II, he ran for President and won only about 30% of the popular vote in 1932. As part of a political deal, Hitler was named German chancellor instead. When the President passed away two years later, Hitler assumed control of both roles.

Unit Resource Book: Europe
- History Makers: Adolf Hitler, p. 116

RESEARCH & WRITING CENTER

ANALYZING Primary Sources

In his account of the fall of the Berlin Wall, Andreas Ramos describes his experience as he approached the East German border:

We finally reached the border just after midnight. The East German border was always a serious place. Armed guards kept you in your car, watching for attempts at escapes. Tonight was a different country. . . . We walked through the border. On both sides the guard towers were empty and the barbed wire was shoved aside in great piles. Large signs told us that we needed sets of car documents. The East German guard asked if we had documents. I handed him my Danish cat's vaccination documents, in Danish. He waved us through.

- What differences did Ramos notice when he first approached the East German border? *(The guard towers were empty; the barbed wire was pushed aside.)*

DOCUMENT-BASED QUESTION ANSWER

Possible Answer: He meant that Germany was no longer divided into separate countries, one Communist, the other a democracy, and that the two would once again become a united country.

History Like Italy, Germany was once a collection of small states instead of a united nation. **Prussia** was the most powerful German state. In 1871, Prussia unified the German states into one nation. Germany was a military power until its defeat in World War I, which weakened its economy and reduced its territory.

Many Germans, frustrated by the poor economy, began to support the fascist Nazi Party. The Nazis promised to make Germany the world power it had once been. In 1933, **Adolf Hitler**, the party's leader, became Germany's head of state. Hitler began a movement to gain more territory for Germany that eventually led to World War II. The war devastated Germany and much of Europe. The Nazis also carried out the **Holocaust**, the systematic murder of millions of Jews and other ethnic minorities.

The Allies defeated the Germans and divided Germany into four zones. The United States, France, and Britain merged their zones into the democratic West Germany. The Soviet Union's zone became East Germany, a Communist country. Berlin, located in East Germany, was divided into east and west sides. In 1961, the East German government built the **Berlin Wall** to formally divide the city in two. East Germany's government fell in 1989, and the Berlin Wall came down. In 1990, Germany decided to **reunify**, ending the east-west division.

ONLINE PRIMARY SOURCE To read more eyewitness accounts, go to the **Research & Writing Center @ClassZone.com**

ANALYZING Primary Sources

Andreas Ramos traveled from Denmark to Germany in 1989. He was one of the five million people in Berlin to witness the fall of the Berlin Wall. Afterwards, he wrote the following.

> Everything was out of control. . . . There were fireworks, kites, flags and flags and flags, dogs, children. The wall was finally breaking. . . . I saw an indescribable joy in people's faces. It was the end of the government telling people what not to do, it was the end of the Wall, the war, the East, the West.
>
> Source: Andreas Ramos, *A Personal Account of The Fall of the Berlin Wall*

DOCUMENT-BASED QUESTION

What do you think Ramos meant by "it was the end . . . of the East, the West"?

DIFFERENTIATING INSTRUCTION

Inclusion

Build a "Wall" in the Classroom

Read the primary source feature on this page as a class and discuss it. Then have students reenact the division of Germany by building a "wall" using empty cardboard boxes. The wall should extend down the middle of the classroom with students on both sides. Describe what life for Germans on each side of the wall may have been like. Then discuss the events leading to the fall of the Berlin Wall as students take down the wall.

Gifted & Talented

Write a "Life on the Other Side" Letter

Provide information about life in East and West Germany from several reputable sources. In small groups, have students read the information, then discuss it as a class. Ask students to imagine that they live in either East or West Germany. Have them write a letter to a friend on the other side of the wall, describing their daily lives and how they think their experiences are similar and different.

Culture Germany has the largest population in Western Europe. German is the official language, though different forms of the language are spoken. Catholics and Protestants each account for about one-third of Germany's population. About four percent are Muslims.

Today, approximately 90 percent of Germany's population lives in cities and their surrounding suburbs. Cities offer greater employment opportunities and more varied entertainment. German culture has a strong musical tradition. Some great classical composers, including Bach, Beethoven, and Brahms, were German. Richard Wagner wrote spectacular operas based on German myths. With teenagers, American music is popular. Many German pop stars sing in English.

Government and Economics Germany is a federal republic—a union of states similar to that of the United States. Germany's president acts as the country's formal chief of state, mainly performing ceremonial duties. The chancellor heads the government. The German people elect the president, but the chancellor is selected by parliament. In 2005, Angela Merkel became Germany's first female chancellor.

Germany has had to overcome some serious economic problems. At the time of reunification, East German industries were outdated. Almost half of East Germany's workers were unemployed. In 1990, East Germany adopted West Germany's currency and **market economy**. The German economy continues to adjust and is getting stronger.

Angela Merkel Germany's first female chancellor

EVALUATE Explain how World War II reshaped Germany.
Germany was divided into four occupation zones, which became democratic West Germany and Communist East Germany. Berlin, located in East Germany, was also split into two parts.

CONNECT Government & Culture

The Reichstag When East and West Germany reunited in 1990, they chose Berlin as their capital. In 1999, the German legislature began meeting in Berlin's renovated Reichstag building (shown at right). The building's prominent glass dome contains a ramp leading to an observation platform. It is a symbol of the people being able to raise themselves above their leaders, who meet in the chamber below.

CRITICAL THINKING

Make Inferences What else could the Reichstag building symbolize about Germany and its government?

More About . . .

The German Economy

Germany is one of the world's leaders in exports, earning over $1 trillion annually on the sale of its products. German car manufacturers, such as Audi, Daimler-Chrysler, Mercedes-Benz, BMW, Volkswagen, and Porsche, make a considerable contribution to the nation's economy. The auto industry employs about 1.26 million workers and can produce as many as 10 million vehicles each year.

Unit Resource Book: Europe
- Economics in the World: German Imports and Exports, p. 120

CONNECT Government & Culture

The Reichstag

The history of the Reichstag building is turbulent. Originally completed in 1894, part of the building was destroyed by an arsonist's fire in 1933. In 1945, the dome of the building was blown up. After World War II, the Reichstag was rebuilt using a simpler design, but it would not serve again as the meeting place of the German Parliament until 1999. After Germany's reunification in 1990, however, the building once again underwent renovations. With a modern design and a glass dome in place, the Reichstag now stands as an architectural achievement that represents Germany's past, present, and future.

CRITICAL THINKING ANSWER

Make Inferences Possible Answer: The building is very modern and progressive, reflecting the changes in Germany and its government.

INTERDISCIPLINARY ACTIVITIES

CONNECT to Music

Listen to German Music

Locate recordings of well-known works by Bach, Beethoven, Brahms, and Wagner. Have students draw a four-column chart on a sheet of paper. Columns should be labeled with the composers' names. Then play samples of each composer's music. As students listen, have them write comments about what they hear *(musical instruments, tempo, volume, how it makes them feel, what they think about as they listen, what they like or dislike about it, and so on).*

CONNECT to Math

Graph German Export Data

Have students use the 2005 data below to create pie graphs showing how much Germany exports to its trading partners.

Export Partners: France 10.2%, United States 8.8%, United Kingdom 7.9%, Italy 6.9%, Netherlands 6.1%, Belgium 5.6%, Austria 5.4%, Spain 5.1%

Then have student create three questions that can be answered from their graphs.

Teach

The Alpine Countries

Reader, Recorder, Reporter

- Why do you think Austria adopted a policy of neutrality after World War II? *(Possible Answer: Because Austria was so deeply involved in World War I and World War II , the people and government probably wanted peace in their country.)*
- Why does Switzerland have three official languages? *(Possible Answer: Because large German, French, and Italian groups settled there and retained their native languages.)*
- **Sequence Events** How many years separated Switzerland's and Liechtenstein's independence? *(218 years)*

Unit Resource Book: Europe
- Skillbuilder Practice: Synthesizing, p. 110

Comparing Regions

The Alps and the Rocky Mountains

The Alps run about 750 miles, starting in Southern France and ending in Albania. The highest peak in the Alps is Mont Blanc on the French–Italian border at 15,771 feet (4,807 meters). The Rocky Mountains extend about 3,000 miles, stretching from Alaska and Canada into New Mexico. The highest peak of the Rockies is Mt. Elbert in Colorado, at 14,433 feet (4,399 meters).

Both mountain ranges were formed over millions of years. Today, both are important water sources for major river systems—for example, the Rhine River starts in the Alps, the Colorado River in the Rockies.

The Alpine Countries

KEY QUESTION How has Austrian history been tied to war?

Austria, Switzerland, and Liechtenstein are **Alpine** countries, named for the Alps Mountains common to all three. The Alpine countries have historic and cultural ties to Germany. All three, like Germany, once were part of the Holy Roman Empire. All three share a common border with Germany. They also share the German language.

Austria Austria was once a powerful nation. In 1804, Austria became an empire. For a time, it was the leader of the German states. But in 1866, Austria lost a war with Prussia. The next year, it joined a dual monarchy with Hungary. The two countries cooperated in foreign affairs but had separate governments. Then, Austria fought on the losing side in World War I. Following its defeat, Austria separated from Hungary. After World War II, the nation adopted a policy of **neutrality**, meaning that it does not participate in military conflicts.

Almost all Austrians speak German, and the majority are Roman Catholics. Although most Austrians live in cities or towns, they still manage to enjoy the outdoors. Austria, about three-fourths of which is covered by mountains, is a nation of skiers. The Austrian Alps offer nearly perfect conditions for this winter sport. In the summer, the scenic beauty of the Alps attracts many hikers and backpackers.

After World War II, Austria became a federal republic with both a president and a chancellor. Austria belongs to the European Union. One of its major industries is iron and steel production. Tourism also plays a major role in Austria's economy. Many people visit the mountainous country to ski, hike, and admire the scenery.

Income from the industry is not consistent over the course of the year. People employed in the industry have less job stability and may have to find other jobs in the off-season. ▼

Skiing Like the other Alpine countries, many tourists visit Austria to ski its many slopes. Tourism is a billion-euro industry for Austria. **What are the economic disadvantages of a seasonal industry?**

DIFFERENTIATING INSTRUCTION: TIERED ACTIVITIES

OBJECTIVE Create a travel brochure for the Alpine countries.

Basic

Have students review the information about the Alpine countries. Then have them create a travel brochure promoting tourism to the region. Encourage students to make the brochure artistic and inviting. Suggest using short, bulleted text to provide information about the region.

On Level

Extend the Basic activity by having students conduct limited research to find three images for their brochures. For any images chosen, remind students to include the source and write a caption in the brochures. Encourage students to use a balance of text and images.

Challenge

Extend the On Level activity by having students conduct limited research to find additional information about the Alpine countries. Instruct students to focus on the country's geographic and cultural characteristics. Encourage students to use a balance of text and images.

Switzerland and Liechtenstein Switzerland and Liechtenstein are small countries that share a border. They were once part of the Holy Roman Empire. Switzerland became independent in 1648, and Liechtenstein in 1866. Both are officially neutral.

German, French, and Italian are Switzerland's official languages. Liechtenstein's official language is German. Switzerland has a mix of Catholics and Protestants, while most Liechtensteiners are Catholic. Switzerland is famous for the high-quality chocolate and watches it produces. Liechtenstein is known for its postage stamps, which are prized by collectors all over the world. Both countries are popular tourist destinations because of their beautiful Alpine landscapes. Both have reputations as centers of international finance. Tiny Liechtenstein uses Swiss money.

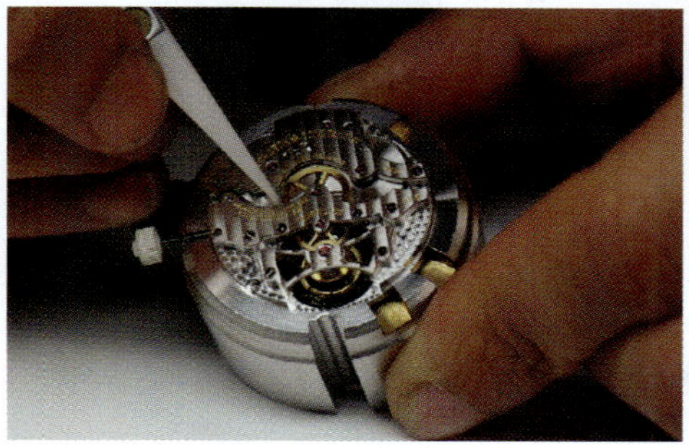

Swiss Watches The intricate works of many Swiss watches are still made and repaired by hand.

Switzerland is a federal republic, like Austria and Germany. The Swiss government is very democratic. The Swiss can vote to change laws passed by their legislature. Liechtenstein is a constitutional monarchy ruled by a prince. It also has a prime minister and parliament.

SUMMARIZE Explain how Austrian history has been tied to war.
After Austria lost a war with Prussia, it no longer led the German states. Austria was defeated in World War II, resulting in its adoption of neutrality.

Section 4 Assessment

ONLINE QUIZ For test practice, go to **Interactive Review @ClassZone.com**

TERMS & NAMES

1. **Explain the importance of**
   - Adolf Hitler
   - Berlin Wall
   - Alpine
   - neutrality

USE YOUR READING NOTES

2. **Sequence Events** Use your completed time line to answer the following question:
   What significant event occurred in Germany in 1961?

1648 — 2005

KEY IDEAS

3. What role has geographic location played in Germany's rise to power?
4. How did Prussia's actions strengthen the German states?
5. What do the Alpine countries have in common?

CRITICAL THINKING

6. **Make Inferences** What did the fall of the Berlin Wall symbolize for Germany?
7. **Draw Conclusions** How have the Alps helped shape the cultures and economies of the Alpine countries?
8. **CONNECT to Today** Why would it be important for the Alpine countries to maintain good relations with Germany?
9. **TECHNOLOGY Prepare an Oral Presentation** Choose one of Germany's many famous musicians–Bach, Beethoven, or Brahms–and do research to create a short biography of his life. Choose one of his songs to play during your presentation.

CONNECT to the Essential Question

What geographic and cultural characteristics define the subregions of Western Europe?

Ask students what they have learned so far that can help them answer this question. Students might mention:

- Germany has made significant changes since World War II to become a unified country with a strong economy.
- The Alpine countries share a similar geography, a policy of neutrality, and close cultural ties to Germany.

4 Assess & Reteach

Assess Have students complete the Section Assessment.

Unit Resource Book: Europe
- Section Quiz, p. 127

Interactive Review @ClassZone.com
- Section Quiz

Power Presentations

Test Generator

Reteach Have each student create a list of "fun facts" for each country discussed in this section. Lists should include facts about each country's:

- geographic features
- key historical events
- economic activities
- cultural characteristics

Unit Resource Book: Europe
- Reteaching Activity, p. 132

SECTION 4 ASSESSMENT ANSWERS

Terms & Names

1. Adolf Hitler: German head of state from 1933 to 1945; Berlin Wall: built to close off East Berlin from West Berlin; Alpine: having to do with the Alps mountain range; neutrality: a policy of not taking part in war

Use Your Reading Notes

2. The Berlin Wall was built.
   Diagram *1648, Switzerland independent; 1866, Liechtenstein independent; 1871, Prussia unifies German states; 1933, Adolf Hitler becomes German head of state; 1961, Berlin Wall built; 1989; Berlin Wall comes down; 1990, Germany reunifies; 2005, Angela Merkel becomes first female chancellor of Germany*

Key Ideas

3. Germany's location in Europe's center has facilitated its takeover of nearby countries.
4. Prussia unified the German states.
5. German language, central location in Europe, common border, official neutrality

Critical Thinking

6. the reunification of the city and the country and the end of East versus West
7. Skiing, mountaineering, and related types of recreation are part of the culture. The Alps also provide a strong basis for tourism.
8. They are small nations that could have problems defending themselves against Germany's military power, and they share a common border.
9. Use the rubric to score presentations.

Presentation Rubric

	Style of Presentation	Understanding of Concepts
4	well-organized	thorough, accurate understanding
3	clear	accurate understanding
2	disorganized	some inaccuracies in understanding
1	no organization	little or no understanding

1 Plan & Prepare

Objectives

- Explain why Norway and Sweden are so closely connected culturally
- Describe how Sweden influenced Finland's history and culture
- Identify the geographic and cultural traits shared by Denmark and Iceland

Read for the Essential Question

Help students read for a purpose by reminding them of the Essential Question: "What geographic and cultural characteristics define the subregions of Western Europe?" Point out that they can use text features such as Key Questions, special features, images, and maps to form answers to the question.

Vocabulary

Best Practices Toolkit

Use the ***Best Practices Toolkit*** to model strategies for vocabulary notetaking. Vary strategies throughout the year. Choose from: *Knowledge Rating, Predicting ABCs, Definition Mapping, Word Sort, Frayer Model (Word Squares), Magnet Words, Word Wheel,* and *Student VOC.*

Vocabulary Strategies, TT9–TT16

Reading Strategy

Best Practices Toolkit

Display the Three-Column Chart transparency.

- Point out to students that to categorize means to decide how information in the section can be grouped or sorted by topic.
- Model filling in one or two details for each category on the chart.
- Have students suggest entries.
- Display the graphic organizer for students to review.

Three-Column Chart, TT21

Section 5 Reading for Understanding

Key Ideas

BEFORE, YOU LEARNED

Germany and the Alpine countries share common borders and a common language. They are linked in many ways.

NOW YOU WILL LEARN

The Nordic countries have histories and cultures that are closely intertwined.

Vocabulary

TERMS & NAMES

Vikings a seafaring Scandinavian people who raided northern and western Europe from the 9th to the 11th century

Sami people of northern Scandinavia who traditionally herd reindeer; also the name of their language

ombudsman an official who investigates citizens' complaints against the government

welfare state a social system in which the government provides for many of its citizens' needs

BACKGROUND VOCABULARY

Nordic relating to Scandinavia

REVIEW

hydroelectric power electricity made by water-powered engines

Visual Vocabulary Sami with a reindeer

Reading Strategy

Re-create the chart shown at right. As you read and respond to the **KEY QUESTIONS**, use the chart to help you organize important details about the Nordic countries.

See Skillbuilder Handbook, page R7

CATEGORIZE

	Language	Government	Economy
Sweden			
Norway			
Finland			
Denmark			
Iceland			

GRAPHIC ORGANIZERS
Go to **Interactive Review** @ClassZone.com

388 Chapter 12

PRETEACHING VOCABULARY

English Learners

Pronounce and Preview

Pronounce each term for students. Review the meanings of words within the definitions, such as *seafaring, traditionally,* and *social system.*

- To modify vocabulary learning, have students complete worksheets as they read, instead of afterward.

Unit Resource Book: Europe
- Vocabulary Practice, p. 109

Inclusion

Act Out Silent Vocabulary Clues

Read and discuss each term and definition with students. Then have students take turns acting out silent clues for the vocabulary terms while the rest of the class guesses the correct answer. To challenge the class, suggest that students act out other important terms found within the rest of the section.

Chateau de Polignac, France

SECTION 5 FOCUS ON

Cattle in the Alps

The Nordic Countries

Leif Ericson

Connecting to Your World

Who were the first Europeans to reach North America? Many think that Columbus and his crew were the first to do so. In fact, Leif Ericson, a Viking, was probably the first. According to an ancient account from Greenland, Ericson hoped to find a forested land glimpsed by an Icelandic trader and sailor. Ericson and his men sailed west from Greenland. They landed and spent the winter on the North American continent, at a place Ericson called Vinland.

Sweden and Norway

KEY QUESTION Why do Sweden and Norway have so much in common?

Five nations in northernmost Europe make up the **Nordic**, or Scandinavian, countries: Sweden, Norway, Finland, Denmark, and Iceland. These lands were first settled thousands of years ago. Sweden and Norway share the same landmass, the Scandinavian Peninsula. Their histories are intertwined. Sweden and Norway were both influenced by the **Vikings**, seafaring Scandinavians who raided Europe from the 9th to the 11th century.

Viking Ship This Viking longship was recovered from a farm in Slagen, Norway, in 1904.

2 Focus & Motivate

3-Minute Warm-Up

Write on the board or display the transparency:

- What can you infer about Vikings by studying the photograph of the Viking ship? *(Possible Answers: They were skilled shipbuilders; they had to be strong to row the ships; their culture was fairly simple because there is not a lot of artistic detail on the ship.)*

Unit Transparency Book
- 3-Minute Warm-Ups, TT11

Connecting to Your World

More About . . .

Leif Ericson

Ericson's ship encountered a wrecked vessel on one of their voyages. He rescued the crew and was given the shipwreck's cargo as a reward. This act earned him the nickname "Leif the Lucky."

Unit Resource Book: Europe
- Vocabulary Practice, p. 115

3 Teach

Sweden and Norway

Reader, Recorder, Reporter

- Why are Swedish speakers and Norwegian speakers often able to understand each other? *(The two languages are similar.)*
- **Categorize** How would you categorize the early history of Sweden and Norway? *(Both were influenced by the Vikings; Sweden became a great power and ruled over Norway.)*

SECTION 5 PROGRAM RESOURCES

Unit Resource Book: Europe

ON LEVEL

Unit Resource Book
- Reading Study Guide, p. 77
- Section Quiz, p. 128

STRUGGLING READERS

Unit Resource Book
- Reading Study Guide with Additional Support, p. 87
- Skillbuilder Practice: Synthesizing, p. 110
- Section Quiz, p. 128
- Reteaching Activity, p. 133

eEdition with Audio DVD-ROM

ENGLISH LEARNERS

Pupil Edition in Spanish
eEdition with Audio DVD-ROM
Unit Resource Book
- RSG (Spanish), p. 97
- RSG with Additional Support (Spanish), p. 107

Multi-Language Glossary
Test Generator
- Section Quiz in Spanish

INCLUSION

Unit Resource Book
- RSG with Additional Support, p. 87
- Section Quiz, p. 128
- Reteaching Activity, p. 133

GIFTED & TALENTED

Unit Resource Book
- History Makers: Leif Ericson, p. 115
- Section Quiz, p. 128

PRE-AP

Unit Resource Book
- Section Quiz, p. 128

TECHNOLOGY

Unit Transparency Book
- 3-Minute Warm-Ups, TT11
- Essential Question, TT12
- Critical Thinking, TT17
- Cultures, TT20

Daily Test Practice Transparencies
- Section 5, TT45

Power Presentations
ClassZone.com
World Cultures and Geography Video on DVD

The Viking Invasions of Western Europe *820–941*

CONNECT Geography & History

As students study the map, point out that, except for open-ocean journeys to Iceland, the Vikings sailed mostly along the coastline of Europe.

- What cities did the Vikings reach by sailing along rivers? *(Paris, London, Lisbon)*
- Why do you think the Vikings didn't try to settle all the lands they explored? *(Possible Answers: They preferred to explore and trade, rather than settle; they may have been defeated by some people.)*

ANSWERS

1. **Region** Scandinavia
2. **Movement** Seville

More About . . .

The Sami

Reindeer herding, fishing, and hunting are some of the ways many Sami earn a living. However, these means of livelihood are being jeopardized by conflicts over other uses for forests and grazing lands, such as oil exploration mining, logging, tourism, and development. The governments of Sweden, Norway, and Finland are working to balance the rights of the Sami and of other users of these far northern lands, but some disputes have still ended up in court.

History The Vikings were some of the world's best sailors and shipbuilders. They also were pirates and fierce warriors. During the three centuries that the Vikings lived in Scandinavia, they invaded much of Europe, as you can see from the map here.

The Viking Invasions of Western Europe *820–941*

CONNECT Geography & History

1. **Region** Where was the Viking homeland?
2. **Movement** What was the southernmost city reached by the Vikings?

In the late 1000s, the Vikings converted to Christianity and settled down. Sweden and Norway became monarchies. Finland became part of Sweden in 1323. In 1397, Denmark's Queen Margaret formed a union with Sweden and Norway, which lasted until Sweden withdrew in 1523. Sweden had become a great power, often battling Denmark, Russia, and Poland for territory. In 1700, those three countries attacked Sweden. Sweden's king, Charles XII, fell in battle in 1718, bringing Sweden's domination to an end. In 1809, Sweden lost control of Finland to Russia.

Norway came under Sweden's control in 1814. In 1905, Sweden recognized Norway's independence. During the two world wars, Sweden and Norway remained neutral. However, during World War II, Norway was invaded and occupied by the Germans.

Culture Although Swedes and Norwegians speak different languages, Swedish and Norwegian are similar. As a result, speakers of the two languages often are able to communicate with each other. In both countries, Lutheranism is the official state religion. However, Swedes and Norwegians have the freedom to practice different religions.

In Sweden, Finns make up the largest immigrant population, while immigrants in Norway come from many nations. In the northern regions of Sweden and Norway, there is a large minority population of **Sami**, whose traditional way of life involves caring for herds of reindeer. Their native language is also called Sami.

Because of their climates, Sweden and Norway are less densely populated than other European countries. Most people live in the countries' southern and central regions. About 80 percent of Swedes and 75 percent of Norwegians live in urban areas. Many people in both countries own vacation homes. Swedes and Norwegians enjoy outdoor sports year round. Skiing and ice skating are both popular.

DIFFERENTIATING INSTRUCTION

English Learners

Language: Verb Tenses

Photocopy this page and distribute it to students. Explain that two different verb tenses are used. Under the subheading "History," the past tense is used because it describes events that already happened. Under the subheading "Culture," the present tense is used because it describes current characteristics or events. As students read, instruct them to highlight each past tense verb in one color and each present tense verb in another.

Struggling Readers

Flow Chart of Nordic History

To help students make sense of all the events on this page, have them make a flow chart. Start with a single box, for example: *1000s—Vikings settle down*. Arrows should point from that box to four boxes labeled *Sweden*, *Norway*, *Finland*, and *Denmark*. Smaller boxes under each country should contain other events that occurred there. Explain that one event could affect two countries, so arrows should be both horizontal and vertical.

Worldster
A Day in Ulrika's Life

To learn more about Ulrika and her world, go to the **Activity Center** @ClassZone.com

To introduce herself, Ulrika says:

"Hello, mitt namn är Ulrika."

Hi! My name is Ulrika. I am 14 years old and live in Uppsala, a city of 180,000 people. Uppsala is about 50 miles north of Stockholm, Sweden's capital. I like to read and play computer games. On the weekends, I leave the city with some of my friends to take horseback-riding lessons. Riding is really popular here. Of course, like you, I go to school. Here's what my day is usually like.

School At my school, all of our books and school supplies are free—even paper! We also get a free lunch every day. This year, one of my favorite classes is metal working. I'm making a modern sculpture for our yard.

Break Time We have a lot of breaks during the day. Usually, we go outside and play soccer, field hockey, or another game or sport. It's great to be able to go outdoors.

After School Today I'll be spending two hours at the hospital. No, I'm not sick! This is part of going to school here. In the eighth grade, we get to learn about different kinds of jobs. I want to be a doctor, so I work one afternoon a week at the hospital as an aide.

Dinner Time I'm meeting my family for dinner at my aunt and uncle's house. It's Thursday so I know what we're having—pea soup, with pancakes and lingonberries for dessert. That's a traditional Thursday meal in Sweden!

CONNECT to Your Life

Journal Entry Do you know what kind of job you want when you're older? Would volunteering like Ulrika does help you decide? Record your thoughts in your journal.

DIFFERENTIATING INSTRUCTION: TIERED ACTIVITIES

OBJECTIVE Develop a set of questions for a mock interview with Ulrika.

Basic

Have students review the information on this page and write three interview questions for Ulrika. Remind them that the answers to their questions should be found on the textbook page. Then have pairs of students role-play the interview. Switch roles so both students have a turn as Ulrika and the interviewer.

On Level

Have students use the information on this page and the section content about Sweden to write a set of five interview questions for Ulrika. Then have pairs role-play the interview. Switch roles so both students have a turn as Ulrika and the interviewer.

Challenge

Extend the On Level activity by having students add one question that can't be answered from their textbooks to their interview. Have them conduct outside research to find the answer. Have pairs role-play the interview. Switch roles so both students have a turn as Ulrika and the interviewer.

More About . . .

Lingonberries

A wild relative of blueberries and cranberries, lingonberries are native to Scandinavia, northern Germany, and northern areas of North America. Each year, about 20 million pounds of the tiny berries are harvested from the wild.

Comparing Regions

Compulsory Education Around the World

Children in Sweden are required by the Swedish government to go to school from ages 7 to 16. The school year must last at least 178 days and is divided into two terms. In the United States, individual states determine the ages at which children are required to attend school, ranging from ages 5–7 to ages 16–18. Although most states require students to attend school 180 days a year on average, some states require less, some more. One study of 43 countries, found that 33 had school years longer than 180 days, with some as high as 220 days.

Presentation Options

Go to the *Activity Center* at **ClassZone.com** to project Worldster.

- Click on the multimedia features of Ulrika's Web page.
- Ask students which areas of Ulrika's life they would like to explore. Listen to music, examine photographs, learn simple phrases in Swedish, and more.

CONNECT to Your Life

Journal Entry

Encourage students to examine the photos for details described in the text. For example, what sport is being depicted? *(field hockey)* What type of fruit eaten in the U.S. is similar to the lingonberries shown? *(blueberries, cranberries)* Tell students to look for clues in the photos before writing their journal entries.

More About . . .

Norway's North Sea Oil

When oil was discovered on the continental shelf of Norway in 1969, the country became one of the world's largest exporters of oil and natural gas. Although oil and gas exports have fueled the nation's economy, Norway is planning for when its North Sea oil and gas reserves run out. It is exploring the possibility of drilling in the Barents Sea in the far north Arctic Circle. Norway already relies on hydroelectric power for much of its energy needs.

3 Teach

Finland

Think, Pair, Share

- Why is Finland's culture so strongly influenced by Sweden? *(Finland was part of Sweden for about 700 years.)*
- Why might the people of Finland want a government with a president and a prime minister that share power? *(Possible Answer: The two leaders would serve as a system of checks and balances for each other.)*
- **Categorize** How would you categorize Finland's primary exports? *(natural resources: forest products; high-tech items: mobile phones)*

Unit Resource Book: Europe
- Skillbuilder Practice: Synthesizing, p. 110

Scandinavian Design
Scandinavian design, known for its use of simple forms, has become popular in the United States.

Government and Economics Sweden and Norway are constitutional monarchies. Although both have monarchs, their prime ministers and cabinets hold most of the executive power. Both Sweden and Norway appoint ombudsmen to protect their citizens. An **ombudsman** investigates complaints against the government and makes sure that governmental power is not abused.

Sweden and Norway are welfare states. **Welfare states** use taxes to provide a wide range of services to their citizens. Families with children under age 16 are given money to care for them. Workers receive good retirement and unemployment plans.

Sweden has a highly industrialized economy. It also has a reputation for modern design, which has spread to other countries through two major Swedish companies, household goods store IKEA and clothing store H&M. Norway took longer to develop manufacturing because it did not have enough energy resources to power factories. By 1900, Norway was able to use hydroelectric power from its rivers. Today, Norway also has access to North Sea oil. Sweden joined the European Union in 1995, but Norway's citizens voted not to join.

DRAW CONCLUSIONS Explain why Sweden and Norway share so many similarities. Both were influenced by their Viking heritage. Then, Sweden and Norway were united under Denmark. Later, Sweden ruled Norway for nearly 100 years.

Finland

KEY QUESTION How did Finland's neighbors affect its history?

Finland's location between Sweden and Russia has played a major role in Finland's history. Both countries have influenced its culture.

History and Culture In 1155, Finland became part of Sweden. It remained under Swedish control until 1809, when Russia conquered Finland. In 1917, Finland gained its independence from Russia. During World War II, Finland had to fight off two invasions by the Soviet Union.

Because of its history, Finland's culture is strongly influenced by Sweden. Finland's two official languages are Finnish and Swedish. Most people in Finland are Finnish. Swedes make up the largest minority group. Small groups of Sami also live in Finland.

About three out of five Finns live in urban areas. Still, many of them find opportunities to enjoy outdoor recreation. With snow on the ground for almost half the year, sports like cross-country skiing and ice hockey are especially popular. Northern Finland even has reindeer races!

INTERDISCIPLINARY ACTIVITIES

CONNECT to Science

Weather Report for Helsinki

Provide information about Finland's climate and discuss it. Remind students of the difference between *climate* and *weather.* As a class, determine what season it is in Finland. Then, have student pairs create a five-day weather forecast for Helsinki, Finland. They should include a range of high and low temperatures for each day as well as predict possible precipitation. When complete, have them compare their work with an actual-five day forecast.

CONNECT to Language Arts

Radio Spot on Reindeer Racing

Have students work in pairs to write a radio spot advertising the upcoming reindeer racing championships in Finland. They should use vivid language to describe the action of the races in order to grab the audience's attention and convince listeners to attend the event. After each pair has written their spot, have them record it into a tape recorder or other device. Then, as a class, listen to all the radio spots and have the class vote for the best ones.

Fun Facts!

REINDEER RACING

What do you do when you've got snow on the ground and a herd of reindeer? If you're in Finland, you race! Each year, the Sami in northern Finland hold a championship reindeer race. Drivers on skis wearing aerodynamic outfits and helmets are harnessed to reindeer. They compete in races against the clock and against other jockeys.

Government and Economics Finland is a democratic republic. A president and prime minister share executive power. Voters elect the president, who appoints the prime minister. In 2000, Finland elected its first female president, Tarja Halonen. Finland is a welfare state.

Finland's forests are its most plentiful natural resource. Forests cover almost two-thirds of the land, more than in any other European country. Paper and other forest products make up 30 percent of Finland's exports. Finland also produces mobile phones and other high-tech items. In 1995, Finland joined the European Union.

SUMMARIZE Explain how Finland's neighbors affected its history.
Finland was invaded by both Sweden and Russia. As a result, Finnish culture is strongly influenced by Sweden.

Denmark and Iceland

KEY QUESTION How are Denmark and Iceland similar?

Denmark occupies the Jutland Peninsula and over 400 small islands surrounding the peninsula. The nation also includes Greenland and the Faroe Islands. Denmark is a small country, but it has played a large role in Scandinavian history. Iceland is an island nation just south of the Arctic Circle. It was ruled by Denmark for over 500 years.

History Viking raids on other countries shaped Denmark's early history. Denmark's power expanded during the late 1100s and early 1200s. For a time, Denmark controlled Sweden, Norway, and Iceland. Sweden withdrew from the union in 1523. Denmark continued to rule Norway until 1814. Iceland gained its independence in 1944.

DIFFERENTIATING INSTRUCTION: TIERED ACTIVITIES

OBJECTIVE Design a book jacket for a nonfiction travel book about Denmark or Iceland.

Basic

Have students use the information in their textbooks to design a book cover for a travel book about Denmark or Iceland. Students should come up with a book title, draw an original cover image that represents a geographic or cultural aspect of the country, and write several "fun facts" for the back of the book cover.

On Level

Extend the Basic activity by having students create a book jacket with information about the country and the book's contents on the book flaps and back cover. You may wish to have students conduct limited outside research in order to gather more information to include on their book jackets.

Challenge

Extend the On Level activity to include a review of the book on the back cover. Students might evaluate the book's contents, explain how it could be useful to the reader, or write opinions about the country. Students should conduct limited research to gather more information to include on their book jackets.

Fun Facts!

Reindeer Racing

The reindeer championships began in the 1950s and have become a unique part of Finland's culture. Racing season usually lasts from February to April.

More About . . .

Tarja Halonen

Inaugurated to her first term in 2000, Halonen is Finland's 11th president. A member of Finland's Parliament since 1979 she was also the first woman in Finnish history to hold the position of Foreign Minister. Although she is considered one of the most powerful women and effective leaders in the world, some of her harshest critics have continued to focus on what she wears, not what she achieves.

3 Teach

Denmark and Iceland

Talk About It

- Why do you think such a large percentage of Iceland's population lives in cities? *(Possible Answers: Besides the fishing industry, most people probably work in cities; cities offer services that help people maintain a comfortable life in such a harsh environment.)*
- Do you think Iceland should join the European Union? Why or why not? (*Yes: The EU could improve Iceland's economy through trade, lowering the cost of living there; No: Iceland has a stable government and economy so it doesn't need to become a member of the EU.)*
- **Categorize** Would you categorize Denmark as an island nation? Why or why not? *(Yes: It includes more than 400 islands as well as Greenland and the Faroe Islands; No: much of it is located on a peninsula, and its islands are small, so it is not technically an island nation.)*

CONNECT to the Essential Question

What geographic and cultural characteristics define the subregions of Western Europe?

Ask students what they have learned so far that can help them answer this question. Students might mention:

- Sweden and Norway share the same climate, religion, and government type, and they have similar languages.
- Because Finland was under Swedish control for many years, Sweden has influenced Finland's language and history.
- Denmark and Iceland are similar geographically, share a common ancestry, and are both highly urbanized.

Unit Transparency Book: Europe
- Essential Question, TT12

4 Assess & Reteach

Assess Have students complete the Section Assessment.

Unit Resource Book: Europe
- Section Quiz, p. 128

Interactive Review @ClassZone.com
- Section Quiz

Power Presentations

Test Generator

Reteach Have students write a summary of the geography and culture of each Nordic country.

Unit Resource Book: Europe
- Reteaching Activity, p. 133

Nyhavn Harbor These colorful old sailors' quarters in Copenhagen, Denmark, have been converted to cafés and dance clubs.

Culture Most Danes have Danish ancestry, though some are descended from Germans. Icelanders are mainly of Norwegian or Celtic descent. Over 90 percent of Iceland's population lives in urban areas, as do most Danes. The Evangelical Lutheran Church is the official religion of both countries. Danish is spoken in Denmark, and Icelandic in Iceland.

Government and Economics Denmark is a constitutional monarchy. The Danish monarch has mainly ceremonial duties. The prime minister serves as the head of government. Like other Scandinavian nations, Denmark and Iceland have welfare systems that provide for many of their citizens' needs. Iceland is a republic. Its people elect a president, who serves a four-year term and has limited powers. The prime minister and cabinet perform most executive functions.

Despite lacking natural resources, Denmark has a strong, modern economy. The cost of living in Iceland is high because so many goods have to be imported to the small island nation. Denmark belongs to the EU but has not adopted the euro. Iceland is not an EU member.

both highly urbanized; most Danes share a common ancestry, as do most Icelanders; official religion of both is Evangelical Lutheran ▸

COMPARE AND CONTRAST Explain how Denmark and Iceland are similar to each other.

Section 5 Assessment

ONLINE QUIZ For test practice, go to **Interactive Review @ClassZone.com**

TERMS & NAMES

1. **Explain the importance of**
   - Vikings
   - Sami
   - ombudsman
   - welfare state

USE YOUR READING NOTES

2. **Categorize** Use your completed chart to answer the following question:
   Which of the Nordic countries are welfare states?

	Language	Government	Economy
Sweden			
Norway			
Finland			
Denmark			
Iceland			

KEY IDEAS

3. What cultural characteristics do Sweden and Norway share?
4. How has Finland's location affected its history?
5. Why is Iceland's cost of living so high?

CRITICAL THINKING

6. **Make Inferences** How has the climate of the Scandinavian countries helped define their cultures?
7. **Summarize** What benefits do welfare states offer their citizens?
8. **CONNECT to Today** How do you think having ombudsmen would change the U.S. government?
9. **WRITING** **Write a Speech** Use appropriate sources to research and prepare a speech explaining three advantages and disadvantages of welfare states like those in Scandinavia. Then explain why you would or would not like to live in a country with that kind of economy.

SECTION 5 ASSESSMENT ANSWERS

Terms & Names

1. Vikings: seafaring Scandinavians; Sami: northern Scandinavians who herd reindeer; ombudsman: official who investigates complaints against government; welfare state: system where government provides for citizens' needs

Use Your Reading Notes

2. all of them

	Language	Government	Economy
Sweden	Swedish	constitutional monarchy	EU member
Norway	Norwegian	constitutional monarchy	not part of EU
Finland	Finnish, Swedish	democratic republic	EU member
Denmark	Danish	constitutional monarchy	EU member
Iceland	Icelandic	republic	not part of EU

Key Ideas

3. same state religion (Lutheranism); love of outdoor sports, especially winter sports
4. located between Russia and Sweden, both have successfully invaded and controlled Finland
5. Most goods are imported, raising their cost.

Critical Thinking

6. Possible answers: Cold climate is the basis for the popularity of winter sports; long coastlines have made seafaring and shipbuilding traditions.
7. families get money to care for children under 16; excellent retirement and unemployment plans
8. Possible answers: People would be better represented, and they might feel more satisified with how the government works.
9. Use the rubric to score speeches.

Speech Rubric

	Understanding of Concepts	Style of Speech
4	clear, complete	well-organized
3	clear	organized
2	some confusion	disorganized
1	unclear	no organization

Interactive Review

Click here to complete these and other activities online @ClassZone.com

CHAPTER SUMMARY

Key Idea 1
Greece and Italy began as collections of small states that gained independence.

Key Idea 2
Both Spain and Portugal were ruled by foreigners and developed colonial empires.

Key Idea 3
France and the Benelux nations have prospered.

Key Idea 4
Germany and the Alpine countries are linked in many ways.

Key Idea 5
The Nordic countries have histories and cultures that are closely intertwined.

For **Review and Study Notes**, go to **Interactive Review** @ClassZone.com

NAME GAME

Use the Terms & Names list to complete each sentence on paper or online.

1. I was a barrier that divided a city in half. **Berlin Wall**
2. I investigate complaints against the government. ________
3. I was the eastern half of the Roman Empire after the fall of Rome. ________
4. I am a political system based on a strong central government led by a dictator. ________
5. I am one of a group of people living in Sweden's far north. ________
6. I am an ethnic group living in the northern Pyrenees. ________
7. I am the official home of the Pope. ________
8. I was Germany's leader during World War II. ________
9. I am an economic union formed by three western European nations. ________
10. I was a powerful German state in the Holy Roman Empire. ________

Adolf Hitler
Basque
Benelux
Berlin Wall
Byzantine Empire
Christopher Columbus
coalition
fascism
Moors
ombudsman
Prussia
Reconquista
Sami
Vatican
Vikings

Activities

GeoGame

Use this online map to reinforce your understanding of the countries and cities of Western Europe. Drag and drop each place name in the list to its location on the map.

To play the complete game, go to **Interactive Review** @ClassZone.com

Crossword Puzzle

Complete an online crossword puzzle to test your knowledge of Western Europe.

ACROSS
1. seafaring Scandinavians who raided Europe from the 9th to 11th century

395

ONLINE AND ON-SCREEN ACTIVITIES

GeoGame

Use the interactive map on **Power Presentations** DVD-ROM or the *eEdition* to help students review Western Europe's geographic features.

- Demonstrate how to drag and drop the map labels listed onto the appropriate locations on the map.
- Advance to the scorecard to show how students have a chance to improve their scores.

Crossword Puzzle

Students can review important people and events in Western Europe's history by completing this crossword puzzle online at **ClassZone.com.**

- As an extension of this activity, have students use the *Activity Maker* at **ClassZone**.com to make their own crossword puzzle for one section of the chapter.
- Then have them trade puzzles with classmates to solve each other's puzzles.

Chapter Summary

As a follow-up to the chapter, encourage students to pair up and go over the **Review and Study Notes** at **ClassZone.com.**

Alternatively, students pairs can review their Reading Study Guide notes, making sure that they can add important details to each section's "Key Idea."

Name Game

Teaching Tips

ON PAPER Have students work in pairs to complete this activity. As an extension, they can write questions for the terms that are not covered in the items shown here.

ONLINE Make sure that students understand how to drag and drop terms to fill in blanks. Ask them to print out their results and compare them with those of a classmate.

ANSWERS

1. Berlin Wall
2. ombudsman
3. Byzantine Empire
4. fascism
5. Sami
6. Basque
7. Vatican
8. Adolf Hitler
9. Benelux
10. Prussia

Make It Fun

Teacher-Tested Activities

Amber McVey, Antioch Middle School, Gladstone, Missouri

One great way to review a chapter is to have the students teach each other. We use "Ask an Expert" to answer the Key Ideas questions in the chapter review.

- Divide the class into teams, one team member for each section in the chapter.
- Assign one member of each team to be an "expert" on one section.
- Have the experts for each section work together to answer the questions for their section.
- Then have the students reassemble into their original teams. Each expert then shares his or her knowledge about the assigned section with the other team members.

Vocabulary ANSWERS

1. eastern half of the Roman Empire
2. comes from Latin, spoken in places that were part of the Roman Empire
3. an alliance or partnership, like those formed in the Italian government
4. an Italian explorer who sailed for Spain and journeyed to the Americas in 1492
5. Belgium, the Netherlands, and Luxembourg formed the Benelux economic union.
6. built by East Germany's Communist government to close off East Berlin
7. not participating in wars, such as Austria, Switzerland, and Liechtenstein
8. The Reconquista was the effort to drive the Moors out of Spain.
9. both ethnic groups; the Basque live in the Western Pyrenees and along the Bay of Biscay, the Sami in northern Sweden
10. Adolf Hitler was Germany's Nazi leader in World War II. Nazism led to the Holocaust, the systematic killing of Jews and others.
11. The Nordic nations are all welfare states that provide for many of their citizens' needs. Sweden and Norway also use ombudsmen as another way to care for citizens.

Key Ideas ANSWERS

12. Greek coastline facilitates trade, shipping, and fishing
13. They share historic ties. The Vatican, the center of the Church, is in Italy.
14. the successful effort by Catholic forces to drive the Muslims from Spain
15. Portugal diversified its economy and expanded the role of service industries. Membership in the EU increased trade.
16. Both have executive, legislative, and judicial branches. In France, however, an elected president oversees foreign affairs and a prime minister appointed by the president controls daily operation of the government.
17. All are constitutional monarchies; highly urbanized with well-developed economies; EU members; and dependent on trade with larger neighbors.
18. The war devastated Germany. After the split, East Germany was much less developed industrially than West Germany.
19. the Alps
20. Norway did not have enough energy resources to power factories.
21. seafaring Scandinavians who raided Europe's coastline during the 9th to 11th centuries

Chapter 12 Assessment

VOCABULARY

Explain the significance of each of the following.

1. Byzantine Empire
2. Romance language
3. coalition
4. Christopher Columbus
5. Benelux
6. Berlin Wall
7. neutrality

Explain how the terms and names in each group are related.

8. Moors, Reconquista
9. Basque, Sami
10. Adolf Hitler, Holocaust
11. ombudsman, welfare state

KEY IDEAS

1 Greece and Italy

12. How has Greece's geography defined some of its major economic activities?
13. Why has the Catholic Church had such a strong impact on Italian culture?

2 Spain and Portugal

14. What was the Reconquista?
15. How has Portugal improved its economy?

3 France and the Benelux Countries

16. How is the French government similar to and different from the U.S. government?
17. How are the three Benelux countries alike?

4 Germany and the Alpine Countries

18. How did World War II affect Germany's economy?
19. What geographical characteristic most clearly defines the Alpine countries?

5 The Nordic Countries

20. Why did Norway's manufacturing sector lag behind other Scandinavian countries?
21. Who were the Vikings?

CRITICAL THINKING

22. **Compare and Contrast** Create a table to compare and contrast the geography, work, and recreation of northern and southern Italy.

NORTHERN ITALY	SOUTHERN ITALY

23. **Analyze Causes and Effects** Why did Spain begin exploring the world by sea?
24. **Connect Geography & History** What geographical factor helped Portugal to establish an empire despite its small size?
25. **Compare** How do France's experiences with its eastern neighbor Germany compare to Finland's experiences with Russia?
26. **Connect Geography & Economics** How has the presence of offshore oil contributed to Norway's economy?
27. **Five Themes: Location** How has Scandinavia's northern location helped shape its culture?

Answer the ESSENTIAL QUESTION

What geographic and cultural characteristics define the subregions of Western Europe?

Written Response Write a two- or three-paragraph response to the Essential Question. Consider the key ideas of each section and the specific characteristics of each subregion. Use the rubric below to guide your thinking.

Response Rubric
A strong response will:
- identify each subregion
- explain how geographic characteristics define each subregion
- compare and contrast cultural characteristics within each subregion

ESSENTIAL QUESTION REVIEW

Before students begin their essays, have them review the key chapter elements that will help them answer the Essential Question. Direct them to significant features in the chapter such as

- Comparing Western European Economies, Section 1
- Languages of Western Europe, Section 1
- Connect to History, Section 3
- Analyzing Primary Sources, Section 4
- The Viking Invasions of Western Europe *820–941*, Section 5

Written Response Rubric

	Coverage of Key Ideas	Explanation	Accuracy
4	all	clear, concise	no errors
3	most	good	few/minor errors
2	some	fair	several errors
1	two or fewer	poor	many errors

STANDARDS-BASED ASSESSMENT

Test Practice
- Online Test Practice @ClassZone.com
- Test-Taking Strategies and Practice at the front of this book

THEMATIC MAP

Use the map and your knowledge of Europe to answer questions 1 and 2 on your paper.

Romance Languages in Europe Today

1. **The Romance languages are concentrated in which part of Europe?**
   - **A.** north
   - **B.** east
   - **C.** southwest
   - **D.** southeast

2. **Which statement best describes the pattern of Romance languages?**
   - **A.** They are not spoken along the Mediterranean Sea.
   - **B.** They are found in the former Roman Empire.
   - **C.** They are dying out.
   - **D.** They are spoken only in northern Europe.

BAR GRAPH

Use the information in the graph below to answer questions 3 and 4 on your paper.

Nuclear Energy Generation *2005*

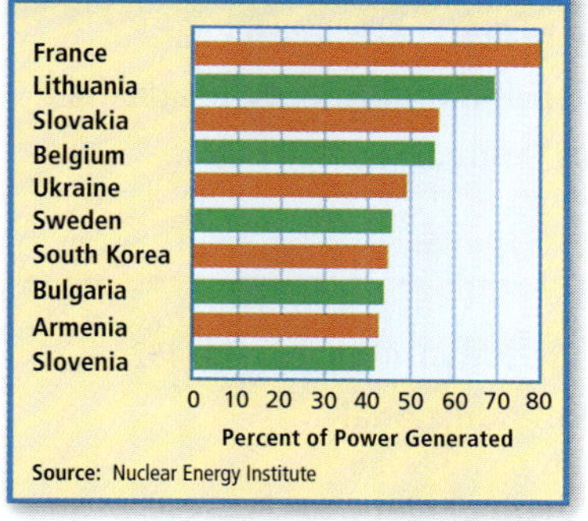

3. **How many countries get more of their energy from nuclear power than France?**
4. **How many of the top ten nuclear energy producers are in Western Europe?**

GeoActivity

1. INTERDISCIPLINARY ACTIVITY–HISTORY

With a small group, research Viking ships and create a poster about them. Be sure your poster shows how the ship was built, what materials were used, and when and where the Vikings traveled.

2. WRITING FOR SOCIAL STUDIES

Reread the part of Section 5 that describes Ulrika's day. Then write a letter to Ulrika about how you spend a typical day. Use interesting details to describe your activities and point out how your day differs from Ulrika's day.

3. MENTAL MAPPING

Create an outline map of Scandinavia and label the following:
- Denmark
- Finland
- Norway
- Iceland
- Sweden
- Greenland
- Faroe Islands

DIFFERENTIATING CHAPTER ASSESSMENT

Unit Resource Book
- **Chapter Test A,** pp. 134–138. *Use for Struggling Readers, Beginning-to-Intermediate-Level English Learners, and for Inclusion of students whose IEP suggests supported test-taking.*
- **Chapter Test B,** pp. 139–143. *Use for On-Level Learners and ESOL students with Advanced-Level English.*
- **Chapter Test C,** pp. 144–148. *Use for Gifted and Talented students or students assigned Pre-AP work.*

Test Generator
- Chapter Tests A, B, and C, in English and Spanish

McDougal Littell Assessment System

TEST **Test Generator** Use three levels of tests in English and Spanish tied to state standards. Plus, choose questions to customize tests for all populations.

SCORE Scan and Score plain paper student tests in minutes.

REPORT Create instant progress reports, sortable by student, by standard, or other criteria.

RETEACH Respond immediately to students' needs by using a wealth of reteaching resources.

Critical Thinking ANSWERS

22. Possible answers: Northern—mountainous; industrial; recreation appropriate to terrain such as hiking or skiing. Southern—mountainous, extensive coastline; agricultural; recreation appropriate to terrain, such as swimming and sunbathing

23. Spain's territory was poor in resources and farmland, so it began to look for additional territory to make up for those shortcomings.

24. Portugal's location on the Atlantic Ocean helped the Portuguese become skilled sailors, shipbuilders, and navigators capable of long sea voyages to establish colonies.

25. Both have been invaded by their eastern neighbors. France was invaded by Germany during World Wars I and II. Finland was under Russian rule for more than a century.

26. Oil has provided another energy source to power Norwegian factories.

27. Scandinavians enjoy many winter sports.

Standards-Based Assessment ANSWERS

1. C

2. B

3. None; France gets more energy from nuclear power than any other country in the world.

4. Three; France, Belgium, and Sweden

GeoActivity

1. INTERDISCIPLINARY ACTIVITY RUBRIC

The poster should
- show how the Viking ships were built
- show what materials were used
- explain when and where the Vikings traveled

2. WRITING FOR SOCIAL STUDIES RUBRIC

The letter should
- include several activities with interesting and detailed descriptions
- explain how these activities are different from Ulrika's day

3. MENTAL MAPPING RUBRIC

The outline map should
- represent the approximate shape of Scandinavia
- include all geographic features listed in their correct location

CHAPTER 13 United Kingdom PLANNING GUIDE

Complete print resources all on one USB drive allow you to customize lesson plans for all students.

Section Titles & Key Ideas	Unit Resource Book Europe	Assessment Resources
1 Building a British Empire **KEY IDEA** Many developments, including the Industrial Revolution and a colonial empire, helped the United Kingdom grow into a major world power.	• **Reading Study Guide,** pp. 149–150 • **RSG with Additional Support,** pp. 155–156 • **History Makers: Elizabeth I,** p. 179 • **Interdisciplinary Projects,** pp. 182–183 • **Primary Sources,** pp. 187–188	**Unit Resource Book** • Section Quiz, p. 189 • Reteaching Activity, p. 192 **Test Generator CD-ROM** **Daily Test Practice Transparency,** TT46
2 From Shakespeare to J. K. Rowling **KEY IDEA** Britain's history as an industrial and colonial power has shaped its culture. Britain's increasingly diverse population continues to enrich its cultural life.	• **Reading Study Guide,** pp. 151–152 • **RSG with Additional Support,** pp. 157–158 • **History Makers: Shakespeare,** p. 180 • **Interdisciplinary Projects,** pp. 181, 183 • **Connect to Language Arts,** p. 185	**Unit Resource Book** • Section Quiz, p. 190 • Reteaching Activity, p. 193 **Test Generator CD-ROM** **Daily Test Practice Transparency,** TT47
3 Parliament and Free Enterprise **KEY IDEA** The United Kingdom's government is a constitutional monarchy. Its economy is strong and adapts to global changes.	• **Reading Study Guide,** pp. 153–154 • **RSG with Additional Support,** pp. 159–160 • **Skillbuilder Practice,** p. 174 • **Map and Graph Skills,** p. 175 • **Economics in the World,** p. 184	**Unit Resource Book** • Section Quiz, p. 191 • Reteaching Activity, p. 194 • Chapter Test (Forms A, B, & C), pp. 195–206 **Test Generator CD-ROM** **Daily Test Practice Transparency,** TT48

McDougal Littell **Assessment System**

TEST SCORE REPORT RETEACH

No Time? To focus students on important content in the chapter, use the **"Reading Study Guide with Additional Support"** in ***Resources2GO.***

Teacher Presentation Options

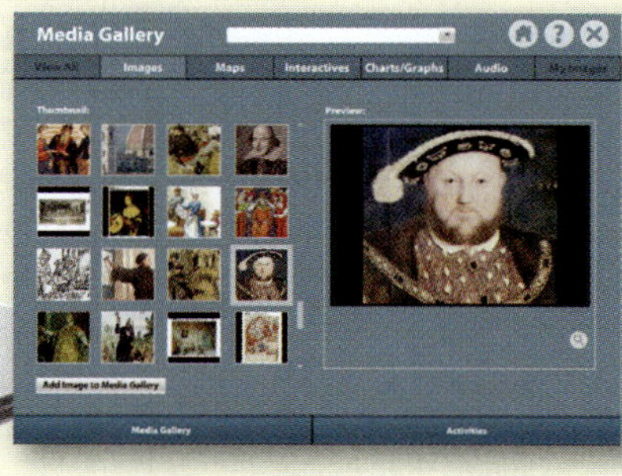

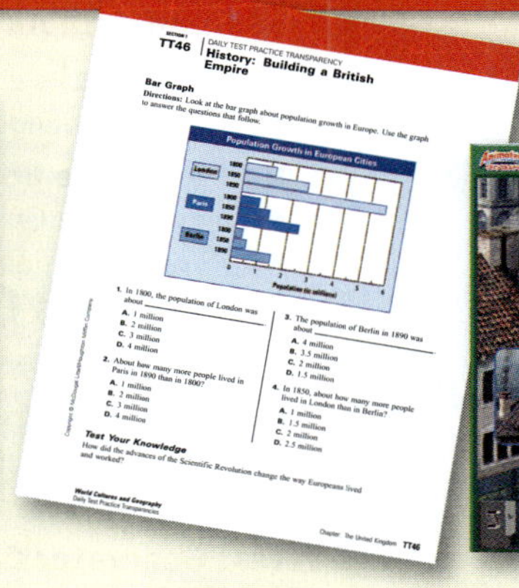

Electronic Books

eEdition DVD-ROM

eEdition Online

Presentation Toolkit

Video/DVD

- World Cultures and Geography

Power Presentations DVD-ROM

- Discussion Starters
- Media Gallery
- Review Game
- Interactive Review
- Animated Features

Unit Transparency Book

- 3-Minute Warm-Ups
- Essential Question
- Critical Thinking
- Maps
- Cultures

Daily Test Practice Transparencies

Animated Geography

- United Kingdom Map
- Victorian London

OnlineActivities at ClassZone.com

Animated Geography

- Videos and Photos
- 3-D Animations
- Interactive Maps

Interactive Review

- Section Quizzes
- Review and Study Notes
- Vocabulary Flip Cards
- Graphic Organizers
- Name Game and Crossword Puzzles
- GeoGame

Online Test Practice

Map Center

- Atlas Maps
- Political Maps with Data
- Links to Mapping Web Sites

Activity Center

- Activity Maker—Review Games, Puzzles, Flip Cards, and Graphic Organizers
- Worldster

Research & Writing Center

- Web Sites
- Data Updates
- Primary Sources
- Writing Models

Current Events

Ken Metz
Glacier Creek Middle School
Cross Plains, Wisconsin

Make It Fun — Teacher-Tested Activities

As you teach this chapter, look for my classroom-prover ideas for:

- making the lesson fun
- focusing students' attention and comprehension
- integrating technology with traditional teaching

Teaching Bell to Bell When you see this timing guide, have students discuss in pairs, in groups of 3–6, or as a class.

- **Think, Pair, Share** (pairs)
- **Reader, Recorder, Reporter** (3–6)
- **Talk About It** (any size)

CHAPTER 13 RESOURCES FOR DIFFERENTIATED INSTRUCTION

Struggling Readers

eEdition with Audio DVD-ROM

Lesson Plans
- Strategies for Inclusion

Unit Resource Book
- Reading Study Guide with Additional Support, pp. 155–160 (B)
- Reteaching Activity, pp. 192–194
- Chapter Test (Form A), pp. 195–198 (A)

Best Practices Toolkit (C)
- Interactive Reading Strategies, pp. 90–145
- Differentiation Activity Bank: Struggling Readers, pp. 146–149

Teacher's Edition Activities
- Focus on Effects, Sec. 2
- Respond to Focused Questions, Sec. 3
- Tiered Activities, Sec. 1–3

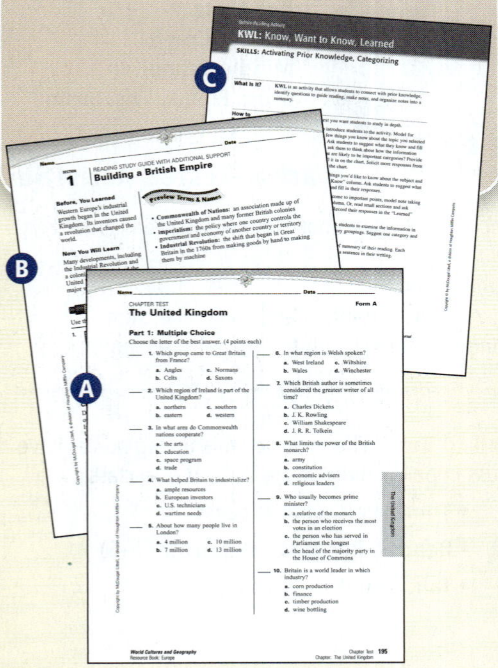

English Learners

Pupil's Edition in Spanish

eEdition with Audio DVD-ROM

Lesson Plans
- Strategies for English Learners

Unit Resource Book
- Reading Study Guide (Spanish), pp. 161–166
- RSG with Additional Support (Spanish), pp. 167–172 (A) (B)

Multi-Language Glossary (C)

Best Practices Toolkit
- Differentiation Activity Bank: English Learners, pp. 165–194

Test Generator CD-ROM
- Section Quizzes in Spanish
- Chapter Test in Spanish

Teacher's Edition Activities
- Pronounce and Preview, Sec. 1–3
- Vocabulary: Key Academic Vocabulary, Sec. 1
- Vocabulary: Prefixes, Sec. 2
- Tiered Activities, Sec. 1–3

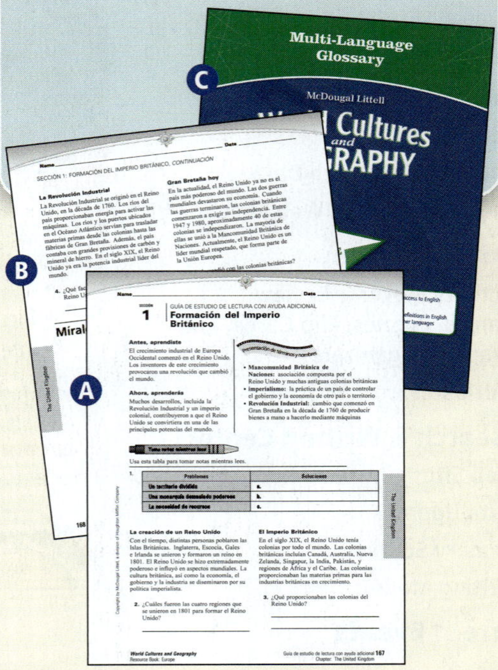

Inclusion

eEdition with Audio DVD-ROM

Lesson Plans
- Strategies for Inclusion

Unit Resource Book
- Reading Study Guide with Additional Support, pp. 155–160 (A) (B)
- Reteaching Activity, pp. 192–194
- Chapter Test (Form A), pp. 195–198

Best Practices Toolkit (C)
- Interactive Reading Strategies, pp. 90–145
- Differentiation Activity Bank: Struggling Readers, pp. 146–149

Teacher's Edition Activities
- Vocabulary: Predicting ABCs, Sec. 1
- Locate Terms and Names in the Text, Sec. 2
- Reading and Defining Terms Verbally, Sec. 3
- Tiered Activities, Sec. 1–3

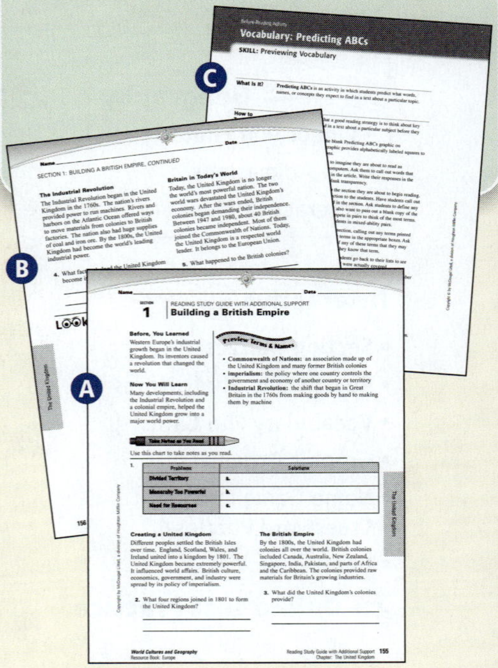

Gifted & Talented Students

Lesson Plans
- Core Lesson Plan

Unit Resource Book
- Outline Map Activities, p. 177
- History Makers, pp. 179–180 **A** **B**
- Interdisciplinary Projects, pp. 181–183 **C**
- Chapter Test (Form C), pp. 203–206

Best Practices Toolkit
- Student Products Rubrics, pp. 1–14
- Writing Templates, pp. 78–89

Teacher's Edition Activities
- Comparison Chart of U.S. and British Governments, Sec. 3
- Tiered Activities, Sec. 1–3

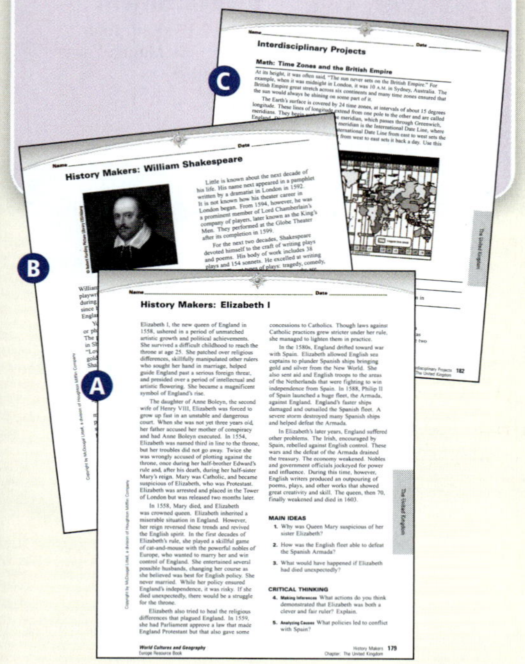

Pre-AP*

Lesson Plans
- Core Lesson Plan

Unit Resource Book
- Map and Graph Skills, pp. 175–176
- Economics in the World, p. 184 **A**
- Connect to Language Arts, p. 185 **B**
- Primary Sources, pp. 187–88 **C**
- Chapter Test (Form C), pp. 203–206

Best Practices Toolkit
- Writing Templates, pp. 78–89
- Differentiation Activity Bank: Pre-AP Students, pp. 195–200

* Pre-AP is a registered trademark of the College Entrance Examination Board, which was not involved in the production of and does not endorse this product.

Teacher's Edition Activities
- Analyze the Effects of Imperialism, Sec. 1
- Tiered Activities, Sec. 1–3

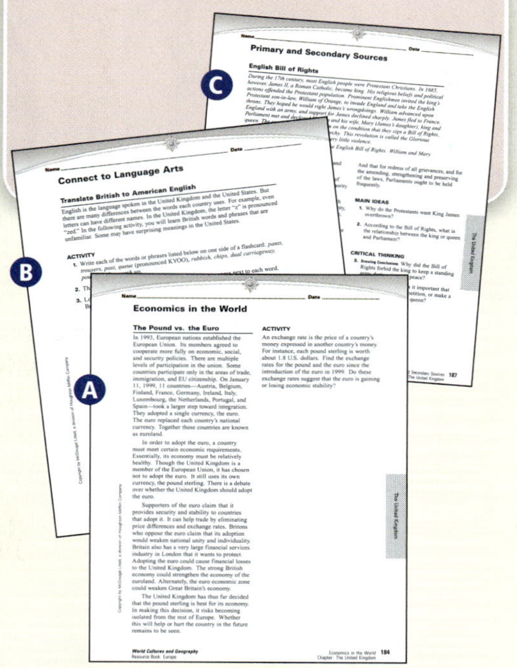

Interdisciplinary Activities

CONNECT to Language Arts

Review a British Book or Play

CLASS TIME 30 minutes

TASK Prepare a review of a book or play written by a British author.

SUPPLIES
- list of age-appropriate British literature
- writing materials/computers

ACTIVITY Have students prepare a review of a book or play written by a British author. In their reviews, students should identify the setting (time and place) of the story and include a brief summary of the plot. They should also comment on whether the book has any particularly British elements (setting, language, and so on) and explain any words or ideas that might be confusing to a non-British reader. Post the reviews or ask volunteers to read theirs out loud.

Good for Verbal Learners

CONNECT to History

You Be the Map Maker

CLASS TIME 30 minutes

TASK Prepare a historical map showing the extent of the British empire in 1930.

SUPPLIES
- historical atlases or Internet access
- blank world maps

ACTIVITY Have students prepare maps of the world showing which countries were part of the British empire in 1930. Have students use a key that allows them to indicate which countries became part of the Commonwealth of Nations.

Good for Visual and Spatial Learners

Introduce the ESSENTIAL QUESTION

- Describe the history of the United Kingdom, including how the British Empire affected the world.
- Summarize the United Kingdom's cultural heritage and the influences on its culture today.
- Discuss the government of the United Kingdom and the conflict in Northern Ireland.
- Explain how the United Kingdom's economy has changed.

Key Ideas

Section 1: Many developments, including the Industrial Revolution and a colonial empire, helped the United Kingdom grow into a major world power.

Section 2: Britain's history as an industrial and colonial power has shaped its culture. Britain's increasingly diverse population continues to enrich its cultural life.

Section 3: The United Kingdom's government is a constitutional monarchy. Its economy is strong and adapts to global changes.

CONNECT to the Essential Question

You may wish to revisit the Essential Question periodically with students as they work through the chapter.

Suggestions for reconnecting are provided at an appropriate point in the *Teacher's Edition* for each section.

Double-decker buses in London

CHAPTER 13

United Kingdom

1 HISTORY
Building a British Empire

2 CULTURE
From Shakespeare to J.K. Rowling

3 GOVERNMENT & ECONOMICS
Parliament and Free Enterprise

ESSENTIAL QUESTION

How did being an island nation influence the development of the United Kingdom?

CONNECT Geography & History

Use the map and the time line to answer the following questions.

1. What is the capital of the United Kingdom?
2. Under what body of water would a railroad tunnel between England and France run?

8000 B.C.

Geography
◂ **8000 B.C.** People begin settling the British Isles.
(rock formation in Cornwall)

Government
◂ **1215** King John signs the Magna Carta.

Government
1689 Parliament passes the English Bill of Rights.

TIME LINE DISCUSSION

Point out to students that the settlement of the British Isles began around 8000 B.C., but the Kingdom of Great Britain did not come into being until 1707.

- In what way are the events of 1215 and 1689 related? *(Possible Answer: Both are related to government and limited the king's power.)*
- Which event on the time line shows how Great Britain is connecting to other nations? *(1994—The Channel Tunnel is built between England and France.)*

Animated GEOGRAPHY

United Kingdom

Click here to explore the United Kingdom @ ClassZone.com

Economics
1994 The Channel Tunnel between England and France is completed.

History
▲ **1760s** Industrial Revolution begins in Great Britain

Geography
1930s The British Commonwealth of Nations is formed.

Today

TEACHER MEDIA FAVORITES

Books for the Teacher

Marshall, P. J. *The Cambridge Illustrated History of the British Empire.* Cambridge University Press, 2001. The empire after the loss of the American colonies.

Tonge, Jonathan and Gareth Schott. *Northern Ireland.* Polity Press, 2006. Examines the reasons for and the possible end to conflict in Northern Ireland.

Video / DVD

Britain: England, Scotland and Wales. (DVD) Analytical Software, 2004. Visual tour of Britain's culture, cities, and countryside.

A History of Britain. (DVD) A&E, 2002. Fifteen episodes of a History Channel/BBC production that chronicles British history from the beginnings of Britain through Winston Churchill.

Internet

Go to the *Research & Writing Center* at **ClassZone.com** to link to:

- data on the region
- a variety of chapter-specific content-reviewed sites
- "History Makers" biographies
- primary sources

Animated GEOGRAPHY

United Kingdom

CLASSZONE.COM

Point out that this is a political map of the United Kingdom, indicating its political divisions, national and regional capitals, and neighboring countries.

- Where is the capital of Great Britain located? *(in the southeastern part of England, along the Thames River)*
- What separates Northern Ireland from the rest of the United Kingdom? *(the Irish Sea)*
- What are the four divisions of Great Britain? *(Scotland, England, Wales, Northern Ireland)*
- What is the capital of Scotland? *(Edinburgh)*

CONNECT GEOGRAPHY & HISTORY ANSWERS

1. London

2. the English Channel

Presentation Options

Use the **Power Presentations** DVD-ROM to project the animated map.

- Prompt students to use the pop-up images and audio from the map to discover what they already know about the geography, culture, history, economics, or governments of the region.
- Challenge students to answer the map questions individually, in small groups, or as a class activity. Use all of the interactive material to help preview chapter content.

1 Plan & Prepare

Objectives

- Identify the aspects of Britain's culture that were influenced by its early settlers
- Analyze how British ideas and customs affected the rest of world
- Describe how Britain's role in the modern world has changed

Read for the Essential Question

Help students read for a purpose by reminding them of the Essential Question: "How did being an island nation influence the development of the United Kingdom?" Point out that they can use text features such as Key Questions, special features, images, and maps to form answers to the question.

Vocabulary

Best Practices Toolkit

Use the ***Best Practices Toolkit*** to model strategies for vocabulary notetaking. Vary strategies throughout the year. Choose from: *Knowledge Rating, Predicting ABCs, Definition Mapping, Word Sort, Frayer Model (Word Squares), Magnet Words, Word Wheel,* and *Student VOC.*

Vocabulary Strategies, TT9–TT16

Reading Strategy

Best Practices Toolkit

Display the T Chart/Two-Column Chart transparency.

- Explain that problems can have more than one solution. Discuss why each of the items listed in the chart is a problem.
- Model filling in solutions to the first problem, using the "Creating a United Kingdom" section that begins on the next page.
- Have students suggest entries.
- Display the completed chart for students to review.

T Chart/Two-Column Chart, TT20

SECTION 1

Reading for Understanding

Key Ideas

BEFORE, YOU LEARNED

Western Europe's industrial growth began in the United Kingdom. Its inventors caused a revolution that changed the world.

NOW YOU WILL LEARN

Many developments, including the Industrial Revolution and a colonial empire, helped the United Kingdom grow into a major world power.

Vocabulary

TERMS & NAMES

representative government a system of government with a legislature that is at least partly elected by the people

Magna Carta a charter, or document, signed by England's King John in 1215 that limited the power of the monarch and guaranteed nobles basic rights

Parliament the national legislature of the United Kingdom

Commonwealth of Nations an association made up of the United Kingdom and many former British colonies

BACKGROUND VOCABULARY

mainland the primary landmass of a continent or territory rather than its islands or peninsulas

REVIEW

imperialism the policy where one country controls the government and economy of another country or territory

Industrial Revolution the shift that began in Great Britain in the 1760s from making goods by hand to making them by machine

Reading Strategy

Re-create the chart shown at right. As you read and respond to the **KEY QUESTIONS**, use the chart to help you identify the solutions to problems faced by the United Kingdom.

Skillbuilder Handbook, page R10

IDENTIFY PROBLEMS AND SOLUTIONS

PROBLEMS	SOLUTIONS
Divided Territory	
Monarchy Too Powerful	
Need for Resources	

GRAPHIC ORGANIZERS
Go to Interactive Review @ClassZone.com

400 Chapter 13

PRETEACHING VOCABULARY

English Learners

Pronounce and Preview

Pronounce each term for students. Review the meanings of words within definitions, such as *legislature, charter,* and *nobles.*

- To modify vocabulary learning, have students complete worksheets as they read, instead of afterward.

Unit Resource Book: Europe

- Vocabulary Practice, p. 173

Inclusion

Vocabulary: Predicting ABCs

Have students work in pairs to fill out the graphic organizer for the terms in the section.

- Use the transparency to model the strategy.
- Have volunteers suggest entries.
- Then distribute copies of the organizer for students to complete with a partner.

Best Practices Toolkit

- Vocabulary: Predicting ABCs, TT10

SECTION 1 HISTORY

Bodiam Castle in East Sussex, England

Sheep in the Cotswolds

Building a British Empire

Connecting to Your World

Patriotic Teen An American teenager displays one of her country's names.

When someone asks you what country you live in, do you say America, the United States, or the U.S.? Our country has many names. So does the United Kingdom. Few use its official name, the United Kingdom of Great Britain and Northern Ireland. Some say the United Kingdom. Some call it Great Britain, the name of the island shared by England, Scotland, and Wales. Others just use Britain. Its different names reflect its history.

Creating a United Kingdom

KEY QUESTION What aspects of culture did Britain's settlers influence?

Different groups settled the British Isles over time. The Celts, the Romans, Germanic tribes called the Angles and the Saxons, and the Normans from France all inhabited the region. Each contributed to British culture. They shaped its language, government, and customs. By the late 1200s, English kings wanted to bring the British Isles under their control. They conquered Wales and much of Ireland. In 1707, England, Wales, and Scotland united as the Kingdom of Great Britain. Ireland became part of the union in 1801, creating the United Kingdom of Great Britain and Ireland. Ireland split into two parts in 1949, and Northern Ireland stayed in the union.

Stonehenge Early settlers of the British Isles built Stonehenge between about 3100 and 1500 B.C.

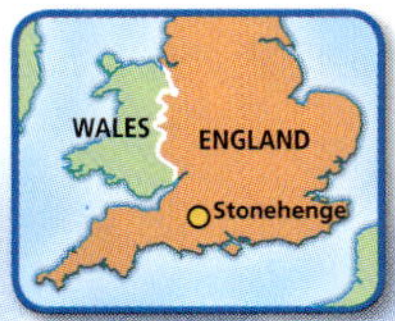

401

2 Focus & Motivate

3-Minute Warm-Up

Write on the board or display the transparency:

- True or false? Correct any false statements.

1. Different groups settled the British Isles. *(T)*
2. England, Scotland, and Northern Ireland make up the United Kingdom. *(F; also Wales)*

Unit Transparency Book: Europe
- 3-Minute Warm-Ups, TT22

Connecting to Your World

More About . . .

Northern Ireland

Whether Northern Ireland should remain part of the United Kingdom or join the Republic of Ireland has caused conflict for many years. Within the region, this conflict is commonly called "The Troubles."

3 Teach

Creating a United Kingdom

Think, Pair, Share

- What groups contributed to early British culture? *(Celts, Romans, Angles and Saxons, Normans)*
- What set apart the United Kingdom from mainland Europe? *(its location as an island group)*
- **Identify Problems and Solutions** Identify one problem that might arise from having many groups settle an area and a possible solution. *(Problems: People might have trouble communicating; there might be culture clashes. Solutions: People might blend languages; they might adopt aspects of other cultures.)*

SECTION 1 PROGRAM RESOURCES

Unit Resource Book: Europe

ON LEVEL

Unit Resource Book
- Reading Study Guide, p. 149
- Outline Map Activities, p. 177
- Section Quiz, p. 189

STRUGGLING READERS

Unit Resource Book
- Reading Study Guide with Additional Support, p. 155
- Skillbuilder Practice: Recognizing Effects, p. 174
- Section Quiz, p. 189
- Reteaching Activity, p. 192

eEdition with Audio DVD-ROM

ENGLISH LEARNERS

Pupil Edition in Spanish

eEdition with Audio DVD-ROM

Unit Resource Book
- RSG (Spanish), p. 161
- RSG with Additional Support (Spanish), p. 167

Multi-Language Glossary

Test Generator
- Section Quiz in Spanish

INCLUSION

Unit Resource Book
- RSG with Additional Support, p. 155
- Section Quiz, p. 189
- Reteaching Activity, p. 192

GIFTED & TALENTED

Unit Resource Book
- History Makers: Elizabeth I, p. 179
- Interdisciplinary Projects, pp. 182–183
- Section Quiz, p. 189

PRE-AP

Unit Resource Book
- Primary Sources, pp. 187–188
- Section Quiz, p. 189

TECHNOLOGY

Unit Transparency Book
- 3-Minute Warm-Ups, TT22
- Essential Question, TT23
- Critical Thinking, TT24
- Map: United Kingdom: Political, TT27

Daily Test Practice Transparencies
- Section 1, TT46

Power Presentations

ClassZone.com

World Cultures and Geography Video on DVD

Teach

Influencing the Modern Age

Reader, Recorder, Reporter

- Why was 1215 a significant year in British history? *(King John signed the Magna Carta, a step toward representative government.)*
- How did geography affect the Industrial Revolution in the United Kingdom? *(rivers provided power; rivers and harbors eased transportation; had needed natural resources)*
- **Identifying Problems and Solutions** Why did the United Kingdom create the Commonwealth of Nations? *(as a solution for British colonies wanting more control)*

Unit Resource Book: Europe

- History Makers: Elizabeth I, p. 179
- Primary and Secondary Sources: from the Magna Carta, p. 188

Make It Fun

Teacher-Tested Activities

Ken Metz, Glacier Creek Middle School, Cross Plains, Wisconsin

I find that "The Secret Spot" is a great way to keep kids listening as we work our way through a chapter.

- First, I pick a word that will be used often in class that day, a key term or important concept in the chapter.
- Without giving away the word, I explain that there is a secret spot in the room, and each time I am in that spot, I'll use the word.
- It keeps them listening, because they all want to be the first to figure it out. Once a student guesses correctly, you can ask them to explain why the word is important.

Although the United Kingdom is a small island group, it has had an influence on world affairs far greater than its size would suggest. At the height of its power, its colonies spanned the globe. As it became a world leader, the United Kingdom's location set it apart from nations on the European **mainland**, allowing it to create a unique identity.

SUMMARIZE Explain the aspects of culture that Britain's early settlers influenced. Early settlers shaped Britain's language, government, and customs.

Influencing the Modern Age

KEY QUESTION How did British governmental and economic ideas influence the rest of the world?

British ideas and customs have had a lasting influence on the way people around the world live. The British policy of **imperialism** helped spread new developments, such as representative government and the Industrial Revolution, which still affect political and economic life.

Representative Government In 1215, the kingdom took steps toward **representative government**, a system in which the legislature is at least partly elected by the people. British nobles forced King John to sign the **Magna Carta**. This document outlined nobles' rights and limited the king's powers. British kings gradually acknowledged that they needed the people's support to govern. A group of representatives called **Parliament** was established. Some call King Edward's 1295 Parliament a "model parliament" because it was more representative of British society than earlier versions. Today, several other European governments, such as Greece and Hungary, also have parliaments.

In 1689, the English Bill of Rights strengthened the rights of citizens. This document outlined the relationship between the monarchy and Parliament. In doing so, it limited the power of the monarch and guaranteed basic freedoms to English citizens. It became a model for other nations, including the United States. Many countries have been influenced by British advances in representative government.

impressive size, elaborate architecture ▼

Houses of Parliament This building in London, finished in 1860, contains the chambers where the United Kindom's Parliament meets. **How might the building's exterior show that an important group meets inside?**

DIFFERENTIATING INSTRUCTION

English Learners

Key Academic Vocabulary

Review with students some of the key academic vocabulary words that appear on this page. You may want to focus on *globe* ("At the height of its power, its colonies spanned the globe.") and *unique* ("As it became a world leader, the United Kingdom's location set it apart from nations on the European mainland, allowing it to create a unique identity.").

Pre-AP

Analyze the Effects of Imperialism

Have students analyze the likely effects of imperialism, considering them from the point of view of the country in control (the empire) and of the country being controlled by another (the colony). *(Empire: access to resources and markets; spreading of its culture; may face high costs in defending empire; Colony: gain new ideas, products and technologies; lose autonomy and traditions)*

ANALYZING Primary Sources

English Bill of Rights In 1689, Parliament passed the English Bill of Rights. They presented the document to King William III and Queen Mary, who agreed to uphold it.

> That the . . . suspending of laws, or the execution of laws, by regal authority, without consent of parliament, is illegal . . .
>
> That election of members of parliament ought to be free . . .
>
> And that for redress [remedy] of all grievances, and for the amending, strengthening, and preserving of the laws, parliaments ought to be held frequently.
>
> Source: English Bill of Rights

▲ King William III and Queen Mary with the English Bill of Rights

DOCUMENT-BASED QUESTION

How did the English Bill of Rights protect English citizens?

The British Empire In the 1500s, Britain joined other European nations in the race to claim overseas colonies in order to gain resources and expand trade networks. By the 1800s, Great Britain used its strong navy to control a prosperous empire. At its height, the empire covered one-fourth of the globe. British colonies included Canada, Australia, New Zealand, Singapore, India, and Pakistan, as well as parts of Africa and the Caribbean. They provided raw materials needed for industrialization and markets for British goods. As British control spread, so did its language and culture.

In the 1920s, several British colonies controlled their internal affairs, but also wanted to manage their foreign policy and defense. In response, the United Kingdom created the **Commonwealth of Nations**, an association of countries that had been part of the British empire. Original members included Australia, Canada, and South Africa. Members of the Commonwealth were independent, but agreed to cooperate in trade and political matters. The British empire still controlled many other territories.

Commonwealth of Nations This World War II poster shows soldiers from several Commonwealth nations, including India, South Africa, New Zealand, Canada, and Australia.

ANALYZING Primary Sources

The English Bill of Rights also gave subjects the right to petition the king without being prosecuted. It prohibited excessive bail, fines, and cruel and unusual punishment. It also gave Parliament the right to approve the levying of money and the raising of an army during peace time.

- What opinions do you think King William III and Queen Mary had about the English Bill of Rights? *(Possible Answer: They may not have liked it because it limited the power of the monarchy.)*

DOCUMENT-BASED QUESTION ANSWER

Possible answer: It placed limits on the power of the monarchy, strengthened Parliament, and guaranteed basic rights.

Unit Resource Book: Europe
- Primary and Secondary Sources: English Bill of Rights, p. 187

More About . . .

The British Empire

The British empire is sometimes divided into two time periods, with the first ending with the American Revolution. The "second" British empire reached its high point in the late 19th and early 20th centuries, reaching into all continents. It was so vast and widespread that people said the sun never set on the British empire.

Today, the remaining dependent territories of the British empire are Anguilla, Bermuda, British Antarctica Territory, British Indian Ocean Territory, British Virgin Islands, Cayman Islands, Falkland Islands, Gibraltar, Montserrat, Pitcairn Islands, South Georgia and the South Sandwich Islands, St. Helena, and Turks and Caicos Islands.

Unit Resource Book: Europe
- Interdisciplinary Projects: Time Zones and the British Empire, pp. 182–183

INTERDISCIPLINARY ACTIVITIES

CONNECT to Civics

Compare Bills of Rights

Have students list rights that they have as Americans. Ask them to compare these rights to the rights noted in the Analyzing Primary Sources feature on the English Bill of Rights on this page. You may also wish to share the additional information in the *Teacher's Edition* about the English Bill of Rights with them. Ask students to explain which set of Bill of Rights they feel better protects its citizens.

CONNECT to Art

Illustrate the Extent of the British Empire

Have student pairs create a visual display that shows the extent of the British empire at its peak. Students might include a map, graphs, and illustrations. Provide students with a list of the places included in the empire to supplement those listed in their text and in the *Teacher's Edition*.

Teach

Britain in Today's World

Think, Pair, Share

- Describe the effects of two world wars on Britain. *(They devastated the British economy and left Britain with large debts. Bombs destroyed large parts of many cities.)*
- **Identify Problems and Solutions** How did World War II affect British colonies? *(Great Britain was unable to maintain its colonies, which began demanding independence.)*

Unit Resource Book: Europe

- Skillbuilder Practice: Recognizing Effects, p. 174
- Outline Map Activities: United Kingdom: Political, pp. 177–178

RESEARCH & WRITING CENTER

HISTORY MAKERS

Winston Churchill

Winston Churchill became Prime Minister in 1940, after Great Britain was already involved in World War II. His determination and fighting spirit inspired an entire nation. In a speech before the House of Commons in 1940 he declared, "We shall fight on the beaches, we shall fight on the landing grounds, we shall fight in the fields and in the streets, we shall fight in the hills; we shall never surrender."

CONNECT to the Essential Question

How did being an island nation influence the development of the United Kingdom?

Ask students what they have learned so far that can help them answer this question. Students might mention:

- As an island nation, the United Kingdom developed a unique identity apart from mainland Europe.
- The United Kingdom became a naval power, which helped it build its empire.
- Britain maintains close ties with its former colonies and other European nations.

The Industrial Revolution The **Industrial Revolution** that you read about in Chapter 11 began in Great Britain in the 1760s. The country's geographic advantages helped make it the world's first industrial nation. Early on, rivers provided power to fuel machines. Rivers and harbors on the Atlantic Ocean offered ways to transport raw materials from British colonies to factories and finished goods to overseas markets. The nation also had ample supplies of coal and iron ore.

By the 1800s, the United Kingdom had become the world's leading industrial power. Although Britain tried to keep its new technology to itself, other nations sought to industrialize. British knowledge spread first to nearby Belgium and France, then to much of the world.

ANALYZE EFFECTS Explain how British governmental and economic ideas influenced the world. representative government influenced others; imperialism changed life in the colonies; Industrial Revolution improved manufacturing

HISTORY MAKERS

Winston Churchill 1874–1965

A brilliant speaker and enemy of Nazi Germany, Prime Minister Winston Churchill led the United Kingdom during World War II. His stirring speeches on the radio and in the British Parliament rallied the British people. As the British prepared for an expected invasion by Germany, Churchill urged them to show courage and make this "their finest hour." His confidence that democratic government would in the end win out over dictatorships inspired the United Kingdom and its allies to work hard for victory.

ONLINE BIOGRAPHY For more on the life of Winston Churchill, go to the **Research & Writing Center** @ClassZone.com

Britain in Today's World

KEY QUESTION How has Great Britain's position in world politics changed?

At the beginning of the 1900s, Great Britain was a major world power. Its leaders governed a prosperous nation and a vast empire. Today, the United Kingdom is no longer the world's richest or most powerful nation. However, it remains a respected world leader.

Fighting Two World Wars The United Kingdom played a major role in two world wars. As you learned in Chapter 11, it served as a leader of the Allies in both struggles. The British people showed great courage. During World War II, civilians faced massive German air raids on their cities. British and Commonwealth soldiers fought bravely in Europe, North Africa, and Asia. In stirring radio addresses, Prime Minister Winston Churchill inspired confidence as the British people faced the threat of a German invasion.

The two wars had a devastating effect on the United Kingdom's economy. Bombs had destroyed large parts of London and other cities. When the war ended, Britain had huge debts. All the industry that had been devoted to the war effort had to be converted back to manufacturing consumer goods. It took the British economy years to recover fully.

404 Chapter 13

DIFFERENTIATING INSTRUCTION: TIERED ACTIVITIES

OBJECTIVE Prepare a presentation on the courage of British citizens during World War II.

Basic

Review with students the information about Winston Churchill and Great Britain during World War II. Have student groups prepare an introduction for the presentation, giving a brief overview of Britain's role in World War II, what the British lived through, and Churchill's leadership.

On Level

Have students think about what it would have been like to face massive air raids, to live in damaged cities, or to be sent away from their city homes for their protection. Then have them write a paragraph about their thoughts. Have students include their writing as part of the presentation.

Challenge

Have students use the Winston Churchill quote in History Makers and additional research on Churchill to explain the importance of his leadership during this time. Students should prepare a brief report on Churchill's leadership and its influences on Britain and the world as part of the presentation.

SECTION 1 HISTORY

Bodiam Castle in East Sussex, England

Sheep in the Cotswolds

Building a British Empire

Connecting to Your World

When someone asks you what country you live in, do you say America, the United States, or the U.S.? Our country has many names. So does the United Kingdom. Few use its official name, the United Kingdom of Great Britain and Northern Ireland. Some say the United Kingdom. Some call it Great Britain, the name of the island shared by England, Scotland, and Wales. Others just use Britain. Its different names reflect its history.

Patriotic Teen An American teenager displays one of her country's names.

Creating a United Kingdom

KEY QUESTION What aspects of culture did Britain's settlers influence?

Different groups settled the British Isles over time. The Celts, the Romans, Germanic tribes called the Angles and the Saxons, and the Normans from France all inhabited the region. Each contributed to British culture. They shaped its language, government, and customs. By the late 1200s, English kings wanted to bring the British Isles under their control. They conquered Wales and much of Ireland. In 1707, England, Wales, and Scotland united as the Kingdom of Great Britain. Ireland became part of the union in 1801, creating the United Kingdom of Great Britain and Ireland. Ireland split into two parts in 1949, and Northern Ireland stayed in the union.

Stonehenge Early settlers of the British Isles built Stonehenge between about 3100 and 1500 B.C.

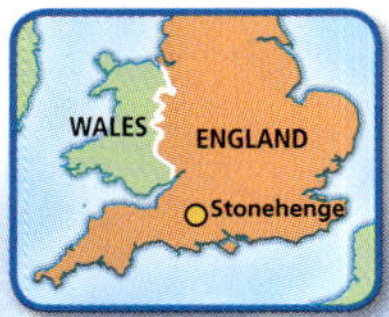

2 Focus & Motivate

3-Minute Warm-Up

Write on the board or display the transparency:

- True or false? Correct any false statements.

1. Different groups settled the British Isles. *(T)*
2. England, Scotland, and Northern Ireland make up the United Kingdom. *(F; also Wales)*

Unit Transparency Book: Europe
- 3-Minute Warm-Ups, TT22

Connecting to Your World

More About . . .

Northern Ireland

Whether Northern Ireland should remain part of the United Kingdom or join the Republic of Ireland has caused conflict for many years. Within the region, this conflict is commonly called "The Troubles."

3 Teach

Creating a United Kingdom

Think, Pair, Share

- What groups contributed to early British culture? *(Celts, Romans, Angles and Saxons, Normans)*
- What set apart the United Kingdom from mainland Europe? *(its location as an island group)*
- **Identify Problems and Solutions** Identify one problem that might arise from having many groups settle an area and a possible solution. *(Problems: People might have trouble communicating; there might be culture clashes. Solutions: People might blend languages; they might adopt aspects of other cultures.)*

SECTION 1 PROGRAM RESOURCES

Unit Resource Book: Europe

ON LEVEL

Unit Resource Book
- Reading Study Guide, p. 149
- Outline Map Activities, p. 177
- Section Quiz, p. 189

STRUGGLING READERS

Unit Resource Book
- Reading Study Guide with Additional Support, p. 155
- Skillbuilder Practice: Recognizing Effects, p. 174
- Section Quiz, p. 189
- Reteaching Activity, p. 192

eEdition with Audio DVD-ROM

ENGLISH LEARNERS

Pupil Edition in Spanish

eEdition with Audio DVD-ROM

Unit Resource Book
- RSG (Spanish), p. 161
- RSG with Additional Support (Spanish), p. 167

Multi-Language Glossary

Test Generator
- Section Quiz in Spanish

INCLUSION

Unit Resource Book
- RSG with Additional Support, p. 155
- Section Quiz, p. 189
- Reteaching Activity, p. 192

GIFTED & TALENTED

Unit Resource Book
- History Makers: Elizabeth I, p. 179
- Interdisciplinary Projects, pp. 182–183
- Section Quiz, p. 189

PRE-AP

Unit Resource Book
- Primary Sources, pp. 187–188
- Section Quiz, p. 189

TECHNOLOGY

Unit Transparency Book
- 3-Minute Warm-Ups, TT22
- Essential Question, TT23
- Critical Thinking, TT24
- Map: United Kingdom: Political, TT27

Daily Test Practice Transparencies
- Section 1, TT46

Power Presentations

ClassZone.com

World Cultures and Geography Video on DVD

Teach

Influencing the Modern Age

Reader, Recorder, Reporter

- Why was 1215 a significant year in British history? *(King John signed the Magna Carta, a step toward representative government.)*
- How did geography affect the Industrial Revolution in the United Kingdom? *(rivers provided power; rivers and harbors eased transportation; had needed natural resources)*
- **Identifying Problems and Solutions** Why did the United Kingdom create the Commonwealth of Nations? *(as a solution for British colonies wanting more control)*

Unit Resource Book: Europe

- History Makers: Elizabeth I, p. 179
- Primary and Secondary Sources: from the Magna Carta, p. 188

Make It Fun

Teacher-Tested Activities

Ken Metz, Glacier Creek Middle School, Cross Plains, Wisconsin

I find that "The Secret Spot" is a great way to keep kids listening as we work our way through a chapter.

- First, I pick a word that will be used often in class that day, a key term or important concept in the chapter.
- Without giving away the word, I explain that there is a secret spot in the room, and each time I am in that spot, I'll use the word.
- It keeps them listening, because they all want to be the first to figure it out. Once a student guesses correctly, you can ask them to explain why the word is important.

Although the United Kingdom is a small island group, it has had an influence on world affairs far greater than its size would suggest. At the height of its power, its colonies spanned the globe. As it became a world leader, the United Kingdom's location set it apart from nations on the European **mainland**, allowing it to create a unique identity.

SUMMARIZE Explain the aspects of culture that Britain's early settlers influenced. Early settlers shaped Britain's language, government, and customs.

Influencing the Modern Age

KEY QUESTION How did British governmental and economic ideas influence the rest of the world?

British ideas and customs have had a lasting influence on the way people around the world live. The British policy of **imperialism** helped spread new developments, such as representative government and the Industrial Revolution, which still affect political and economic life.

Representative Government In 1215, the kingdom took steps toward **representative government**, a system in which the legislature is at least partly elected by the people. British nobles forced King John to sign the **Magna Carta**. This document outlined nobles' rights and limited the king's powers. British kings gradually acknowledged that they needed the people's support to govern. A group of representatives called **Parliament** was established. Some call King Edward's 1295 Parliament a "model parliament" because it was more representative of British society than earlier versions. Today, several other European governments, such as Greece and Hungary, also have parliaments.

In 1689, the English Bill of Rights strengthened the rights of citizens. This document outlined the relationship between the monarchy and Parliament. In doing so, it limited the power of the monarch and guaranteed basic freedoms to English citizens. It became a model for other nations, including the United States. Many countries have been influenced by British advances in representative government.

impressive size, elaborate architecture ▼

Houses of Parliament This building in London, finished in 1860, contains the chambers where the United Kindom's Parliament meets. **How might the building's exterior show that an important group meets inside?**

402

DIFFERENTIATING INSTRUCTION

English Learners

Key Academic Vocabulary

Review with students some of the key academic vocabulary words that appear on this page. You may want to focus on *globe* ("At the height of its power, its colonies spanned the globe.") and *unique* ("As it became a world leader, the United Kingdom's location set it apart from nations on the European mainland, allowing it to create a unique identity.").

Pre-AP

Analyze the Effects of Imperialism

Have students analyze the likely effects of imperialism, considering them from the point of view of the country in control (the empire) and of the country being controlled by another (the colony). *(Empire: access to resources and markets; spreading of its culture; may face high costs in defending empire; Colony: gain new ideas, products and technologies; lose autonomy and traditions)*

The End of the Empire In the postwar period, many British colonies in Africa and Asia began demanding independence. Great Britain was struggling to recover from the war. Its leaders lacked the time or resources to maintain its colonies. Between 1947 and 1980, about 40 British colonies became independent. Almost all of them joined the Commonwealth of Nations.

Today, the United Kingdom works closely with its former colonies and with other European nations. Its ties to Commonwealth nations have strengthened its role in world affairs. The United Kingdom is also a member of the European Union. However, some British citizens think that their country should be less involved with the rest of Europe, which is a source of tension for the government.

EVALUATE Explain how Britain's role in world politics has changed.

no longer has the power or wealth it did as leader of British Empire, but still plays a key role in world affairs

The Struggle for Independence

Colonial Independence This time line shows when some of Great Britain's many colonies achieved their independence from the empire.

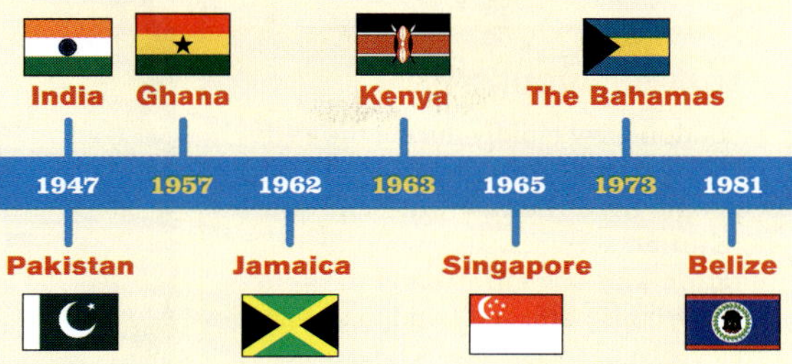

Section 1 Assessment

ONLINE QUIZ For test practice, go to **Interactive Review @ClassZone.com**

TERMS & NAMES

1. **Explain the importance of**
   - Magna Carta
   - Parliament
   - Commonwealth of Nations

USE YOUR READING NOTES

2. **Identify Problems and Solutions** Use your completed chart to answer the following question: How did the United Kingdom meet its needs for more natural resources?

PROBLEMS	SOLUTIONS
Divided Territory	
Monarchy Too Powerful	
Need for Resources	

KEY IDEAS

3. How did the creation of Parliament change the British monarchy?
4. What were the effects of Industrial Revolution on the way the British lived?
5. What benefits did the Commonwealth of Nations have for British colonies?

CRITICAL THINKING

6. **Analyze Causes and Effects** Why did the United Kingdom have to give up its colonial empire?
7. **Make Inferences** How did the British overcome the disadvantages of being an island nation?
8. **CONNECT to Today** Why might the United Kingdom want to maintain ties with former colonies?
9. **ART** **Design a Memorial** Research the experiences of British soldiers or civilians during World War II. Create a design for a war memorial honoring their part in the war.

The Struggle for Independence

Note that Hong Kong—one of Britain's smallest, but most lucrative, former colonies—was returned to China in 1997.

- Which countries achieved independence in the 1960s? *(Jamaica, Kenya, Singapore)*
- Why do you think British colonies wanted independence? *(Possible Answers: They wanted to rule themselves; they felt Britain was too far away or different to rule them.)*

4 Assess & Reteach

Assess Have students complete the Section Assessment.

Unit Resource Book: Europe
- Section Quiz, p. 189

Interactive Review @ClassZone.com
- Section Quiz

Power Presentations

Test Generator

Reteach Have pairs of students create one "True or False" statement about each of the three subsections in "Building a British Empire." Then have pairs exchange their statements with another pair and evaluate the statements. For each false statement, have students correct it, using their textbooks as a resource.

Unit Resource Book: Europe
- Reteaching Activity, p. 192

SECTION 1 ASSESSMENT ANSWERS

Terms & Names

1. Magna Carta: document limited the monarchy and gave nobles basic rights; Parliament: national legislature of the United Kingdom; Commonwealth of Nations: association of the United Kingdom and former British colonies

Use Your Reading Notes

2. It developed overseas colonies to provide resources.

PROBLEMS	SOLUTIONS
Divided Territory	British kings conquer and unite territory by 1801
Monarchy Too Powerful	the Magna Carta, creation of Parliament, and English Bill of Rights limit power of monarchy
Need for Resources	from the 1500s to the 1800s, Great Britain develops a colonial empire that spans the globe

Key Ideas

3. limited the monarch's power, who needed Parliament's support to govern
4. Many workers moved from rural areas to cities for factory jobs; cities became more crowded.
5. gained ability to control their foreign policy and defense; still benefited from trade with Great Britain

Critical Thinking

6. After World War II, British leaders lacked the resources to control the empire.
7. Possible answer: in earlier times, by using the advantages of the Industrial Revolution and its colonies; since World War II, by joining the EU, which has given it new trade opportunities
8. to strengthen its position in world affairs; to add possible trading partners
9. Use the rubric to score designs.

Design Rubric

	Understanding of Event	Design
4	clear, very complete	outstanding design
3	clear	acceptable design
2	some confusion	fair design
1	unclear	inappropriate design

1 Plan & Prepare

Objectives: Describe daily life in Victorian London.

Visual Vocabulary: Ask students to describe what they see in the illustration. Call out familiar items listed below, asking students to locate each one on the page. Then tell them the precise term, as listed in boldface. Add terms to a class vocabulary list, illustrated poster, or Word Wall.

- horse-drawn cart (**carriage, coach,** or **wagon)**
- lights used to illuminate public ways **(streetlights)**
- public safety official **(constable, bobby)**
- item used to protect women from the sun **(parasol)**

2 Teach

Engage . . .

DISPLAY the "Victorian London" feature from the **Power Presentations** DVD-ROM or the *Animations Center* at **ClassZone.com.** As the illustration loads on screen, have volunteers point to some of the vocabulary terms listed above. Then ask:

- What river flows through London?
- What public services did people need in Victorian London?
- What role did class play in Victorian London?
- What economic activities were common in Victorian London?

Explore . . .

ROLL OVER sections of the illustration or click on items in the menu. Ask students to choose which areas they would like to explore further.

- Then return to the discussion questions. Have any answers changed?
- Have students jot down a list of at least three new questions that arise from watching the rollover features.

Animated GEOGRAPHY Victorian London

Click here to enter Victorian London @ClassZone.com

LONDON AROUND 1890

London grew rapidly during the 19th century. By 1901, over 6.5 million people lived there. All these people required services, such as transportation and law enforcement. Rapid growth also affected London society as different social classes interacted.

Click here to experience what daily life in Victorian London would have been like. Learn more about the different social classes that shared the city, such as the upper-class residents shown above. See some of the public services needed to keep the city running, including electricity and the London Underground.

Click here to learn more about the economic activities that might have taken place in this important commercial center. Connected to the world by the Thames River, the city attracted all types of businesses, from street stalls to major factories.

DIFFERENTIATING INSTRUCTION: TIERED ACTIVITIES

OBJECTIVE Compare Victorian London to a modern city.

Basic

Make a Comparison Chart

Have students work in small groups to create and complete two-column charts, with columns labeled "Victorian London" and "The Modern City of [name]." Ask students to consider these aspects of city life, adding relevant details to their charts:

- transportation and other services
- law enforcement
- social classes and ethnic diversity
- economic activities
- overall atmosphere

Have groups discuss how city life has changed over time.

GeoActivity

A Victorian Talk Show
Break into groups of four. One person will pretend to be someone from this London street who is appearing on a talk show. The rest of the group will interview him or her. Choose the best question and answer from your group and share it with the class.

407

On Level

Compare and Contrast City Life—Then and Now

Have students work individually to create charts as described in the Basic activity. Then have students summarize their findings by writing compare and contrast paragraphs about life in Victorian London and life in a modern city.

Challenge

Create Different Perspectives on City Life

Have students work individually to create charts as described in the Basic activity. Then have them imagine what it would be like to live in Victorian London and in a modern city. Have them write descriptions of city life from the point of view of a person living in Victorian London and from the point of view of a person living in the modern city selected.

ANIMATED GEOGRAPHY

Explain . . .

CHOOSE topics to help students answer their questions, such as:

- **Necessary Public Services:** Inadequate sewage systems made Victorian London a smelly, unhealthy place. The term "bobbies" for a police officer comes from Sir Robert Peel who started London's Metropolitan Police in 1829. Railroads made travel faster and helped the city grow.
- **Role of Social Classes:** London grew as people from other parts of Britain and outside the country came to live there. Many of these new arrivals were poor. Londoners at the time generally could be divided into upper, middle, working, and under classes.
- **Economic Activities:** Unemployment and high food prices hurt many people. The Industrial Revolution changed the nature of the work most people did—from farming and craft work to machine operating. Children often worked grueling hours in factories or as chimney sweeps.

3 Wrap Up

Elaborate . . .

ASSIGN small groups of students a topic, such as transportation, industry, social classes, sanitation and health, or law and order. Have each group discuss their topic and summarize key points about it for the rest of the class.

Evaluate . . .

A Victorian Talk Show
Activity Rubric

4 Excellent The questions are relevant, and the answers are accurate and well developed. The presentation is clear and engaging.

3 Strong The questions are relevant and the answers are accurate. The presentation is clear.

2 Fair The questions are acceptable and the answers are somewhat accurate. The presentation is adequate.

1 Poor The questions are irrelevant. Answers are inaccurate. The presentation is poor.

1 Plan & Prepare

Objectives

- Describe life in the United Kingdom
- Identify examples of Great Britain's rich cultural heritage

Read for the Essential Question

Help students read for a purpose by reminding them of the Essential Question: "How did being an island nation influence the development of the United Kingdom?" Point out that they can use text features such as Key Questions, special features, images, and maps to form answers to the question.

Vocabulary

Best Practices Toolkit

Use the ***Best Practices Toolkit*** to model strategies for vocabulary notetaking. Vary strategies throughout the year. Choose from: *Knowledge Rating, Predicting ABCs, Definition Mapping, Word Sort, Frayer Model (Word Squares), Magnet Words, Word Wheel,* and *Student VOC.*

Vocabulary Strategies, TT9–TT16

Reading Strategy

Best Practices Toolkit

Display the Web Diagram transparency.

- Remind students that main ideas are important ideas in a block of text.
- Model filling in main ideas, using the subheads in this section.
- Have students suggest entries. Remind students that main ideas will help them answer Key Questions.
- Display the completed diagram for students to review.

Web Diagram, TT28

SECTION 2 Reading for Understanding

Key Ideas

BEFORE, YOU LEARNED

Great Britain became an industrial leader with a worldwide empire. Today, it no longer has a great empire, but it remains a respected European power.

NOW YOU WILL LEARN

Britain's history as an industrial and colonial power has shaped its culture. Britain's increasingly diverse population continues to enrich its cultural life.

Vocabulary

TERMS & NAMES

Briton a British person

Gaelic any of the Celtic family of languages spoken in Ireland or Scotland

multicultural relating to or including many different cultures

Church of England the official church of England headed by the Archbishop of Canterbury

William Shakespeare an English playwright and poet during the late 16th and early 17th centuries

REVIEW

immigrant a person who leaves one area to settle in another

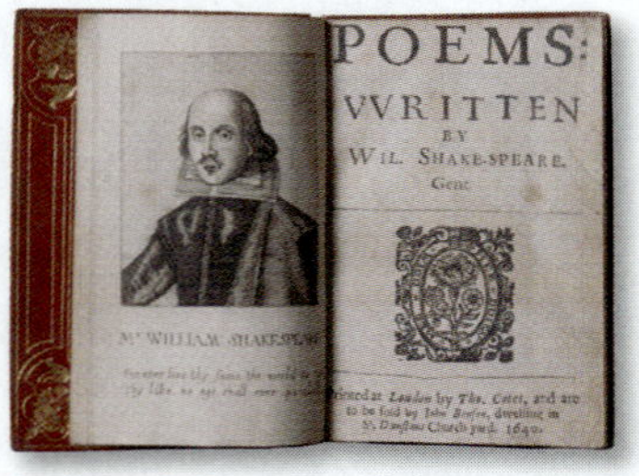

Visual Vocabulary William Shakespeare

Reading Strategy

Re-create the web diagram shown at right. As you read and respond to the **KEY QUESTIONS**, use the diagram to help you find main ideas about British culture.

Skillbuilder Handbook, page R4

FIND MAIN IDEAS

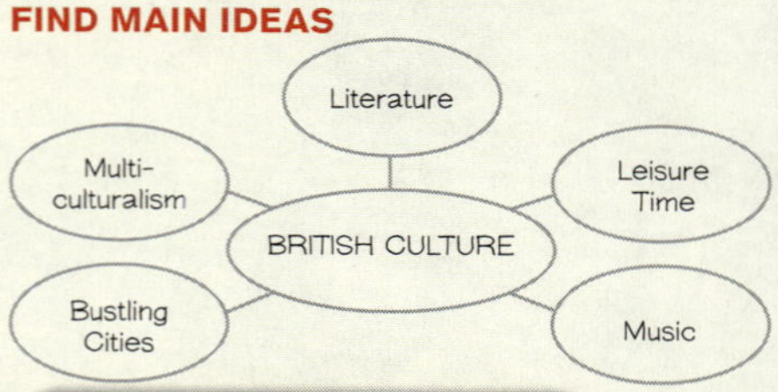

GRAPHIC ORGANIZERS Go to Interactive Review @ClassZone.com

PRETEACHING VOCABULARY

English Learners

Pronounce and Preview

Pronounce each term for students. Review the meanings of words within definitions, such as *family of languages, sect,* and *playwright.*

- To modify vocabulary learning, have students complete worksheets as they read, instead of afterward.

Unit Resource Book: Europe
- Vocabulary Practice, p. 173

Inclusion

Locate Terms and Names in the Text

Invite pairs of students work together to locate the terms and names in the section text. Have one member of the team read the sentence out loud, and then have the other read the definition aloud. Ask the pair to discuss any words or definitions that they have difficulty understanding. Encourage them to write the definitions in their own words.

SECTION 2 CULTURE

Bodiam Castle in East Sussex, England

Sheep in the Cotswolds

From Shakespeare to J. K. Rowling

Connecting to Your World

Have you ever read a Harry Potter book? One day in 1990, an idea for a book popped into the head of J. K. Rowling, a British author. She wanted to write about a young boy named Harry who could do magic. Today, millions of people around the world know about this boy. Rowling's Harry Potter books have made publishing history, selling out at bookstores hours after they are put on the shelves. Rowling is just one part of Great Britain's rich cultural heritage.

Harry Potter Fans Like these boys, many fans dress up as their favorite Harry Potter characters.

Life in the United Kingdom

KEY QUESTION What influences have made British culture more diverse in recent years?

Before World War II, most British people shared the same ethnicity and religious beliefs. Today, people of many different faiths and customs make their home in the United Kingdom. Many of them have immigrated from the former colonies of the British empire. The influence of these newcomers can be seen throughout Britain. Although some changes have caused tensions, the mix of old and new traditions is making many British cities lively places to live.

Tower Bridge The Tower Bridge, built in 1894, spans the Thames River near London's modern City Hall, the round building at right.

409

2 Focus & Motivate

3-Minute Warm-Up

Write on the board or display the transparency:

- What do Shakespeare and J. K. Rowling have in common? *(Possible Answers: Both are famous and part of Great Britain's literary heritage.)*

Unit Transparency Book: Europe
- 3-Minute Warm-Ups, TT22

Connecting to Your World

More About . . .

Harry Potter

Author J. K. Rowling got the idea for her first Harry Potter book while riding on a train from Manchester to London. She also worked on parts of the book in Portugal, before finishing it up in Edinburgh.

3 Teach

Life in the United Kingdom

Talk About It

- How has the United Kingdom's colonial history affected British culture? *(Possible Answer: Diverse people have immigrated from former colonies, bringing different faiths and customs.)*
- How did the Industrial Revolution affect life in Great Britain? *(Possible Answer: It led to the growth of cities; more people live in cities than in rural areas.)*
- **Find Main Ideas** How would you describe British culture today? *(Possible Answers: a multicultural society; one of the most diverse cultures in the world)*

SECTION 2 PROGRAM RESOURCES

Unit Resource Book: Europe

ON LEVEL

Unit Resource Book
- Reading Study Guide, p. 151
- Section Quiz, p. 190

STRUGGLING READERS

Unit Resource Book
- Reading Study Guide with Additional Support, p. 157
- Skillbuilder Practice: Recognizing Effects, p. 174
- Section Quiz, p. 190
- Reteaching Activity, p. 193

eEdition with Audio DVD-ROM

ENGLISH LEARNERS

Pupil Edition in Spanish

eEdition with Audio DVD-ROM

Unit Resource Book
- RSG (Spanish), p. 163
- RSG with Additional Support (Spanish), p. 169

Multi-Language Glossary

Test Generator
- Section Quiz in Spanish

INCLUSION

Unit Resource Book
- RSG with Additional Support, p. 157
- Section Quiz, p. 190
- Reteaching Activity, p. 193

GIFTED & TALENTED

Unit Resource Book
- History Makers, p. 180
- Interdisciplinary Projects, pp. 181, 183
- Section Quiz, p. 190

PRE-AP

Unit Resource Book
- Connect to Language Arts, p. 185
- Section Quiz, p. 190

TECHNOLOGY

Unit Transparency Book
- 3-Minute Warm-Ups, TT22
- Essential Question, TT23
- Critical Thinking, TT25
- Map: Population Density, TT28
- Cultures, TT29, TT30

Daily Test Practice Transparencies
- Section 2, TT47

Power Presentations

ClassZone.com

World Cultures and Geography Video on DVD

10 Biggest Cities in Great Britain

Point out that the capitals of England and Scotland—London and Edinburgh—are among the largest cities in Great Britain, while the capitals of Wales and Northern Ireland—Cardiff and Belfast—are not. According to a recent census, the population of Cardiff was 303,200 and the population of Belfast was 277,200.

- Compare the size of London to the second largest city in Great Britain. *(London is more than seven times the size of Birmingham.)*

Unit Resource Book: Europe

- Skillbuilder Practice: Recognizing Effects, p. 174

Comparing Regions

City Size: London and Tokyo

The United Kingdom and Japan are both island nations in the Northern Hemisphere. London is the largest city in the United Kingdom, with over 7 million people. Tokyo is Japan's largest city, home to about 8.5 million people. When the entire metropolitan area (which includes suburban areas outside the city proper) is taken into account, Tokyo's population jumps to about 35 million by some estimates. London's, however, only increases to about 11 million when its greater metropolitan area is taken into account.

Bustling Cities For many centuries, Great Britain was a largely rural country. The Industrial Revolution sparked urban growth. Today, nine out of ten **Britons**, or British people, live in cities. London, the capital of the United Kingdom, is by far the largest city, with over 7 million residents. While London is the biggest, the nation has several other major cities. The map below shows the location of Britain's largest cities by population, most of which fall within industrial areas.

By the 1800s, London was already a busy city. Today, it is a multilevel city with a subway system below ground and skyscrapers above. It is a global center of culture and commerce. It also has many tourist attractions that draw millions of visitors each year.

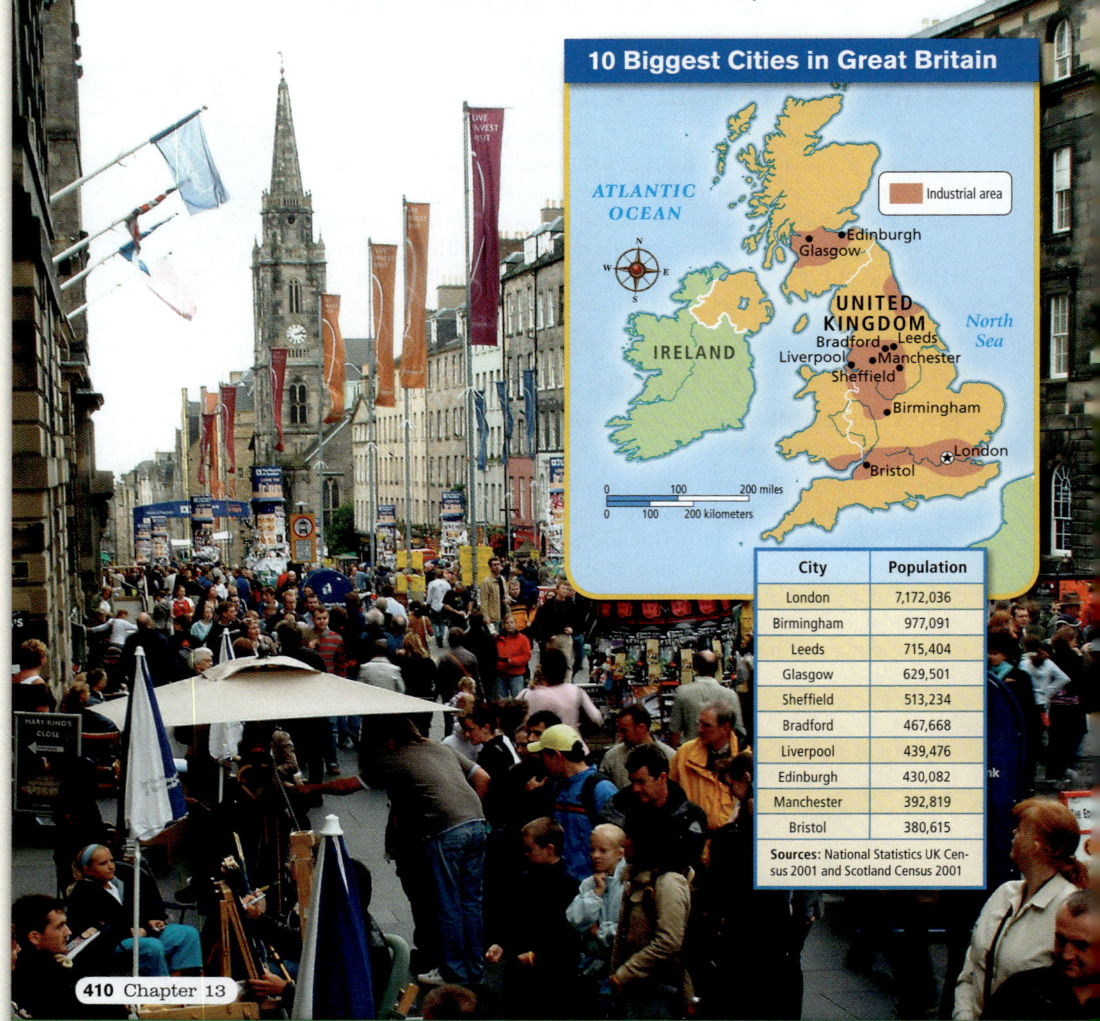

Edinburgh People stroll down a street in Edinburgh, Scotland, during a festival.

City	Population
London	7,172,036
Birmingham	977,091
Leeds	715,404
Glasgow	629,501
Sheffield	513,234
Bradford	467,668
Liverpool	439,476
Edinburgh	430,082
Manchester	392,819
Bristol	380,615

Sources: National Statistics UK Census 2001 and Scotland Census 2001

410 Chapter 13

INTERDISCIPLINARY ACTIVITIES

CONNECT to Math

Graph and Compare Populations

Have students make bar graphs to compare the populations of the ten cities in the chart above. Students should label the axes and use a logical scale. Then have students write five fill-in-the-blank statements comparing populations. For example: ________ has about 12,200 more people than Bristol" or "The population of Manchester is ________ less than that of Glasgow." (Manchester; 236,682) Have students switch statements with a partner and complete them.

CONNECT to Language Arts

Create a Print Advertisement for London

Have student groups create a print advertisement to attract tourists to London. Have them use what they have read about London and Great Britain in this section, along with any background knowledge. Have groups present their ads to the class.

CONNECT to Language Arts

Would you know what to do if someone told you to go find a bobby? If you were British, you would know to look for a police officer, like those shown below. Although British English and American English are similar, some of the words used by Britons are very different—and some can be difficult to translate.

Activity

Translate British to American English

Materials

- flipcards
- pencil
- dictionary

1. Write each of the words listed below on one side of a flipcard: *biscuit, bloke, boot, chemist, jumper, lift, lorry, mate, petrol,* and *zebra crossing.*
2. Think about what each word might mean. Write your guess next to each one.
3. Look up each word in the dictionary. Look for the definition identified as British. Write the definition on the other side of each flipcard.

Multiculturalism Each region in the United Kingdom—England, Scotland, Wales, and Northern Ireland—has its own customs. English is Great Britain's official language. In Scotland and Ireland, however, some speak **Gaelic**, the language brought to the British Isles by the Celtic people. In Wales, about one-fifth of the population speaks Welsh, another Celtic language.

Over the years, the United Kingdom has welcomed **immigrants** from around the world. After World War II, the nation saw an increase in immigration from former colonies in South Asia, Africa, and the Caribbean. Today, about one in ten people in Britain is an immigrant. These newcomers have added variety to British culture and influenced tastes in food and music. They have made the United Kingdom one of the most **multicultural** countries in the world, meaning that it includes many cultures. This has also caused tensions, however, as diverse customs and viewpoints occasionally clash.

Immigration has also had an impact on religion in the United Kingdom. Many Britons belong to the **Church of England**. As the nation's official church, it combines both Catholic and Protestant traditions. In Northern Ireland, about two-fifths of the population are Roman Catholics. However, Great Britain also has many religious minorities, such as Hindus, Muslims, and Sikhs. Their faiths are reflected in the temples and mosques found in many British cities.

EVALUATE Explain why British culture has become more diverse. Immigrants, especially from Asia, Africa, and the Caribbean, have brought different languages, religious beliefs, and customs to Britain.

Fun Facts!

CHICKEN TIKKA MASALA

At one time fish and chips was England's favorite dish, but no more. Now Britons prefer an Indian-influenced dish called chicken tikka masala. The dish is an Indian meal called chicken tikka, to which the Britons have added a masala sauce to satisfy their love of gravy. The new creation is a true combination of British and Indian culture.

CONNECT to Language Arts

In addition to using some different terms, the spelling of certain words, such as colour, centre, and centralise, differs between British and American English. The editors of the Harry Potter books caused a stir when they "Americanized" the language in the popular books for the U.S. editions.

ACTIVITY ANSWERS

Possible Answers: *biscuit*—cracker or cookie; *bloke*—man or guy; *boot*—trunk of a car; *chemist*—pharmacist; *jumper*—sweater; *lift*—elevator; *lorry*—truck; *mate*—friend or buddy; *petrol*—gasoline; *zebra crossing*—crosswalk (marked with white stripes)

Unit Resource Book: Europe

- Connect to Language Arts: Translate British to American English, pp. 185–186

Fun Facts!

Chicken Tikka Masala

Nobody is sure how chicken tikka masala came to be. One popular story claims that a restaurant diner demanded to know where the gravy was for his tandoori chicken. The chef opened a can of tomato soup and poured it over the chicken. Whether or not that is true, chicken tikka masala has gained such popularity that one British grocery chain, Sainsbury's, sells 1.6 million chicken tikka masala meals a week! The dish has even been exported back to India.

DIFFERENTIATING INSTRUCTION

Struggling Readers

Focus on Effects

Give students this question to help focus their reading of the subsection "Multiculturalism": What effects has the increase in immigration had on British culture? Have students work in pairs to find the effects. Suggest that students pause after reading a sentence or two, to discuss whether any effects of immigration were mentioned. *(Possible Answers: influenced tastes in food and music; made Britain one of the world's most multicultural countries; made religion more diverse)*

English Learners

Vocabulary: Prefixes

Explain that the prefix *multi-* means "many." Point out the vocabulary word *multicultural* in the second paragraph ("They have made the United Kingdom one of the most multicultural countries in the world, meaning that it includes many cultures."). Point out the context clue in the sentence that could also help students understand the meaning of the word. *(meaning)* Other *multi-* words include *multilingual* (speaking many languages) and *multifaceted* (having many sides).

Teach

A Rich Cultural Heritage

Reader, Recorder, Reporter

- Why is William Shakespeare significant in British cultural history? *(Possible Answer: many consider him the finest writer of all time)*
- What do the *Chronicles of Narnia,* the *Lord of the Rings* series, and the Harry Potter books have in common? *(All were written by British authors and include fantastic worlds.)*
- **Find Main Ideas** What is the connection among William Shakespeare, soccer, and the Rolling Stones? *(Possible Answer: Britain has a rich cultural heritage, including literature, sports, and music, that it has exported to many parts of the world.)*

Unit Resource Book: Europe
- History Makers: William Shakespeare, p. 180

COMPARING

Sports

Just as soccer is played in the United States, American football is also played in Europe. In fact, Great Britain has its own American football league. It's called the British American Football League, or BAFL, much like the NFL (National Football League) in the United States. Instead of the Super Bowl—the NFL's championship game—Britain's teams play in the BritBowl. Team names include the London Blitz and the Nottingham Caesars.

- Compare and contrast European soccer with American football. *(Similarities: 11 players on each team; both kick a ball for a goal or field goal; Differences: only the soccer goalie can use his or her hands while American football players can throw or run with the ball)*

ANSWERS

1. **European Football** Soccer balls are round; American footballs are oval.
2. **American Football** American football uses more padding, including helmets.

A Rich Cultural Heritage

KEY QUESTION What cultural achievements is Britain known for?

Many of the United Kingdom's most treasured exports don't come from factories. Literature, music, and popular culture make up a rich cultural heritage. With its history as an imperial power, the nation has been exporting its culture around the world for centuries.

Literature Many consider English playwright and poet **William Shakespeare** the finest writer of all time. His plays have entertained audiences for over 400 years. In 1997, a replica of the Globe Theatre—where Shakespeare's plays were originally performed—opened in London, introducing a new generation of Britons to his work.

Many talented authors followed Shakespeare. Jane Austen wrote about British life in the late 1700s and early 1800s in such books as *Pride and Prejudice.* In books such as *Oliver Twist* and *A Christmas Carol,* Charles Dickens explored Britain's social problems. Modern British authors have written many stories for young people. C. S. Lewis crafted fantastic tales in the *Chronicles of Narnia.* J. R. R. Tolkien created the magical world of the *Lord of the Rings* series. Most recently, J. K. Rowling's young wizard Harry Potter has charmed readers of all ages.

Leisure Time Many Britons are sports fans. Soccer, known as football in Europe, is the nation's favorite sport. The modern form of the game played worldwide is thought to have originated in England.

COMPARING Sports

European Football Ten players on each team try to kick a round ball into a goal defended by one player. **1. How do the balls used differ?**

American Football Teams of 11 heavily padded players carry or throw an oval ball across a goal line. **2. Which version uses more padding?**

DIFFERENTIATING INSTRUCTION: TIERED ACTIVITIES

OBJECTIVE Create a visual display showing Britain's contributions to world culture.

Basic

Remind students that the title of this section is "From Shakespeare to J. K. Rowling." Have students choose two examples of Britain's cultural heritage, using "From _____ to _____" as their title. They should illustrate the examples and include a caption for each. Invite students to present or display their work.

On Level

Have students make a visual display, with the headings *Literature, Sports, Movies,* and *Music.* Students should fill in their chart with examples from the section or additional research. One example from each category should be illustrated and captioned. Invite students to present or display their work.

Challenge

In addition to the On Level activity, have students write a summary that describes the contributions of Britain to world culture. Invite students to present or display their work. To extend this activity, have students discuss what they think is Britain's most significant contribution to world culture.

Two other games invented in Britain have become popular in its former colonies. Cricket is played with a bat and a ball. Teams from the United Kingdom and other Commonwealth nations such as India and Pakistan often compete in international games. Rugby, a sport like football, is popular in Australia and South Africa.

The United Kingdom has also produced some well-known films and television shows that have been exported around the world. One British reality television program, *Pop Idol*, allowed viewers to vote for their favorite singers each week. The show was an instant success. Versions of it spread to over 30 countries, including the United States.

Music In the 1960s, the United States faced an unusual kind of invasion, not by soldiers but by music. American teenagers went crazy for British rock 'n' roll. Mobs of screaming fans greeted "British Invasion" bands like the Beatles and the Rolling Stones wherever they performed. Since then, many other British bands, such as Coldplay, have gained popularity with Americans.

Playing Cricket Some Pakistani Muslim boys use a traffic cone to play cricket on a London street.

SUMMARIZE Describe British cultural achievements that have had a global impact. British achievements in literature, sports, and popular culture have spread around the world.

ONLINE QUIZ For test practice, go to **Interactive Review** @ClassZone.com

Section 2 Assessment

TERMS & NAMES

1. **Explain the importance of**
   - Briton
   - Gaelic
   - multicultural
   - Church of England

USE YOUR READING NOTES

2. **Find Main Ideas** Use your completed web diagram to answer the following question:

   What British authors have written books for young people?

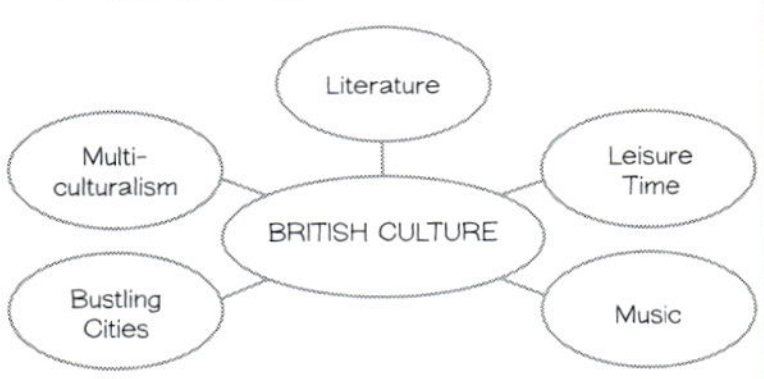

KEY IDEAS

3. In addition to English, what other languages are spoken in the United Kingdom?
4. How have immigrants changed British culture?
5. What early writer helped make the United Kingdom well known for its literary achievements?

CRITICAL THINKING

6. **Draw Conclusions** How did British sports become popular in countries like India and Australia?
7. **Form and Support Opinions** What are the advantages of having a diverse British population?
8. **CONNECT to Today** What challenges might the United Kingdom face if the diversity of its population continues to grow?
9. **WRITING** **Create a Brochure** In 2012, London will host the Summer Olympic Games. Use a word processor to create a brochure telling Americans why they should come to London for the 2012 Olympics. Find a picture to illustrate your brochure.

CONNECT to the Essential Question

How did being an island nation influence the development of the United Kingdom?

Ask students what they have learned so far that can help them answer this question. Students might mention:

- Until World War II, many Britons shared the same ethnicity and religious beliefs.
- Immigrants from former colonies have changed this island nation into a multicultural society.
- British culture continues to influence cultures around the world.

4 Assess & Reteach

Assess Have students complete the Section Assessment.

Unit Resource Book: Europe
- Section Quiz, p. 190

Interactive Review @ClassZone.com
- Section Quiz

Power Presentations

Test Generator

Reteach Have students write down as many main ideas and important details as they can remember from this section in two minutes. Then have pairs compare their lists and revise them as needed for completeness.

Unit Resource Book: Europe
- Reteaching Activity, p. 193

SECTION 2 ASSESSMENT ANSWERS

Terms & Names

1. Briton: a British person; Gaelic: Celtic family of languages spoken in Ireland or Scotland; multicultural: having many different cultures; Church of England: official church of England

Use Your Reading Notes

2. C. S. Lewis, J. R. R. Tolkien, and J. K. Rowling

   Diagram *Bustling Cities—90 percent live in cities; Multiculturalism—immigrants, many from former colonies, influence British culture; Literature—many talented authors; Leisure Time—British sports, TV shows, and films have become popular around the world; Music—British bands have become popular in the U.S. for decades*

Key Ideas

3. Gaelic and Welsh
4. brought new languages, food, music, religions
5. William Shakespeare

Critical Thinking

6. When colonies were part of the British empire, Britons living there played these sports. They continued to be played after the British left.
7. Possible answers: It might encourage creativity in the arts, fuel the economy as immigrants offer new products or services, or encourage closer trade and political ties with other nations.
8. Possible answers: It could be difficult to govern a nation with so many diverse viewpoints, or different groups could come into conflict.
9. Use the rubric to score brochures.

Brochure Rubric

	Level of Content	Type of Language
4	detailed	vivid and effective
3	adequate	descriptive
2	some details	somewhat descriptive
1	few or no details	general, imprecise

1 Plan & Prepare

Objectives

- Explain the government of the United Kingdom, including its constitutional monarchy and regional governments
- Identify changes that have taken place in the United Kingdom's economy

Read for the Essential Question

Help students read for a purpose by reminding them of the Essential Question: "How did being an island nation influence the development of the United Kingdom?" Point out that they can use text features such as Key Questions, special features, images, and maps to form answers to the question.

Vocabulary

Best Practices Toolkit

Use the ***Best Practices Toolkit*** to model strategies for vocabulary notetaking. Vary strategies throughout the year. Choose from: *Knowledge Rating, Predicting ABCs, Definition Mapping, Word Sort, Frayer Model (Word Squares), Magnet Words, Word Wheel,* and *Student VOC.*

Vocabulary Strategies, TT9–TT16

Reading Strategy

Best Practices Toolkit

Display the Web Diagram transparency.

- Point out to students that to summarize information, they should focus on the key ideas from their reading. Summaries should be shorter and less detailed than the original text.
- Model filling in key ideas for "Government" on the diagram.
- Have students suggest entries.
- Display the completed diagram for review.

Web Diagram, TT28

Section 3 Reading for Understanding

Key Ideas

BEFORE, YOU LEARNED

British culture is changing as the population becomes more diverse.

NOW YOU WILL LEARN

The United Kingdom's government is a constitutional monarchy. Its economy is strong and adapts to global changes.

Vocabulary

TERMS & NAMES

constitutional monarchy a government in which the powers of the king or queen are limited by a constitution

unwritten constitution a framework for government that is not a single written document but includes many different laws, court decisions, and political customs

devolution the process of shifting some power from national to regional government

Good Friday Agreement an agreement between Northern Ireland's unionists and nationalists that set up a new government

wind farm a power plant that uses windmills to generate electricity

Channel Tunnel the underwater railroad tunnel between France and England under the English Channel

Visual Vocabulary wind farm

Reading Strategy

Re-create the web diagram shown at right. As you read and respond to the **KEY QUESTIONS**, use the web diagram to help you summarize information about the government and economy of the United Kingdom.

Skillbuilder Handbook, page R5

SUMMARIZE

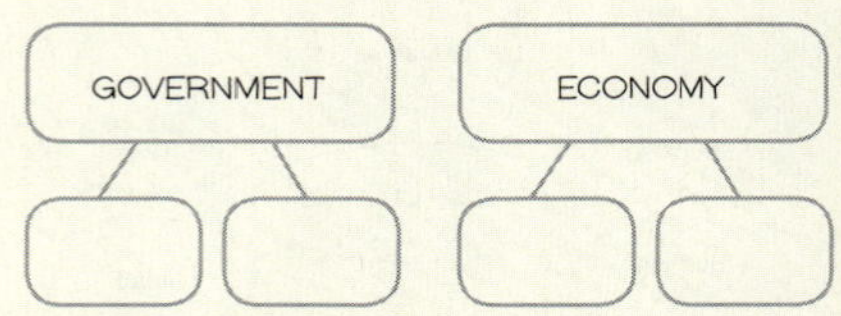

GRAPHIC ORGANIZERS
Go to **Interactive Review** @ClassZone.com

PRETEACHING VOCABULARY

English Learners

Pronounce and Preview

Pronounce each term for students. Review the meanings of words within definitions, such as *framework, unionists,* and *nationalists.*

- To modify vocabulary learning, have students complete worksheets as they read, instead of afterward.

Unit Resource Book: Europe
- Vocabulary Practice, p. 173

Inclusion

Reading and Defining Terms Verbally

Have volunteers read one term and its definition aloud. Ask others to define the term in their own words verbally and then use it in a sentence. Repeat for each term. As a group, discuss any words that are challenging to define or use in a sentence.

SECTION 3

GOVERNMENT & ECONOMICS

Bodiam Castle in East Sussex, England

Sheep in the Cotswolds

Parliament and Free Enterprise

Connecting to Your World

Most Americans know that the United Kingdom has a monarch, Queen Elizabeth II. You may have seen pictures of her, her son Prince Charles, or her grandsons Prince William and Prince Harry. While most Americans have heard of Queen Elizabeth II, many aren't sure exactly what role she plays in British political life. Does she help govern Britain? Does she control the nation's foreign policy or manage its economy? Keep reading to learn more about how the British government works and how monarchs and elected leaders fit into it.

The Royal Family Queen Elizabeth II (above) with her grandsons, William and Harry. The queen uses this gold coach (below) for special occasions, such as a parade celebrating her reign.

Queen and Parliament

KEY QUESTION How are the regions of the United Kingdom governed?

The British sovereign, or monarch, serves as an important symbol of the British nation, but not as the leader of the government. Elected leaders govern the country. The sovereign is not above the law, and he or she rules and acts with Parliament's approval. The king or queen performs mostly ceremonial duties, which can include attending formal dinners or appointing ambassadors to other countries.

415

2 Focus & Motivate

3-Minute Warm-Up

Write on the board or display the transparency:

- What ideas come to mind when you read about kings and queens? *(Possible Answers: They live in castles; they are non-democratic rulers; they appear in fairy tales; they reflect past times.)*

Unit Transparency Book: Europe
- 3-Minute Warm-Ups, TT1

Connecting to Your World

More About . . .

The Queen

In the United Kingdom, the Queen's official title is "Elizabeth the Second, by the Grace of God of the United Kingdom of Great Britain and Northern Ireland and of Her other Realms and Territories Queen, Head of the Commonwealth, Defender of the Faith."

3 Teach

Queen and Parliament

Reader, Recorder, Reporter

- Describe the United Kingdom's constitution. *(an unwritten constitution made up of laws, court decisions, and political customs)*
- What is the Good Friday Agreement? How effective was it? *(It set up a new government in Northern Ireland in 1998. The government was shut down in 2002 due to renewed violence.)*
- **Summarize** Summarize the role of the British monarch. *(The British monarch is a symbol of the nation who performs ceremonial duties.)*

SECTION 3 PROGRAM RESOURCES

Unit Resource Book: Europe

ON LEVEL

Unit Resource Book
- Reading Study Guide, p. 153
- Map and Graph Skills, p. 175
- Section Quiz, p. 191

STRUGGLING READERS

Unit Resource Book
- Reading Study Guide with Additional Support, p. 159
- Skillbuilder Practice: Recognizing Effects, p. 174
- Section Quiz, p. 191
- Reteaching Activity, p. 194

eEdition with Audio DVD-ROM

ENGLISH LEARNERS

Pupil Edition in Spanish

eEdition with Audio DVD-ROM

Unit Resource Book
- RSG (Spanish), p. 165
- RSG with Additional Support (Spanish), p. 171

Multi-Language Glossary

Test Generator
- Section Quiz in Spanish

INCLUSION

Unit Resource Book
- RSG with Additional Support, p. 159
- Section Quiz, p. 191
- Reteaching Activity, p. 194

GIFTED & TALENTED

Unit Resource Book
- Section Quiz, p. 191

PRE-AP

Unit Resource Book
- Economics in the World, p. 184
- Section Quiz, p. 191

TECHNOLOGY

Unit Transparency Book
- 3-Minute Warm-Ups, TT22
- Essential Question, TT23
- Critical Thinking, TT26

Daily Test Practice Transparencies
- Section 3, TT48

Power Presentations

ClassZone.com

World Cultures and Geography Video on DVD

Comparing Regions

The World's Constitutional Monarchies

In addition to the United Kingdom, many members of the Commonwealth of Nations, including Canada, Australia, the Bahamas, Jamaica, New Zealand, and Malaysia, are constitutional monarchies. Of these, many have the British monarch as their head of state. Others, such as Malaysia, have a separate monarch. Countries outside the Commonwealth with constitutional monarchies include Japan, Jordan, Nepal, the Netherlands, and Norway.

COMPARING

U.S. and British Governments

Remind students that like the British Parliament, the United States Congress is made up of two houses, the House of Representatives and the Senate. Unlike Parliament, both houses of Congress have real power.

- How is the British prime minister chosen? *(The leader of the majority party in the House of Commons usually becomes Prime Minister.)*
- What advantages and disadvantages might the British system of choosing a prime minister have? *(Advantages: The government might be able to do more with a prime minister and majority of the House of Commons from the same party; Disadvantages: There might be fewer checks on the prime minister's power.)*

CRITICAL THINKING ANSWER

Compare and Contrast The United States

Unit Resource Book: Europe

- Skillbuilder Practice: Recognizing Effects, p. 174

Constitutional Monarchy The United Kingdom is a **constitutional monarchy**, in which the monarch's power is limited by a constitution. Unlike most countries, it has an **unwritten constitution** made up of laws, court decisions, and political customs. Parliament, the United Kingdom's legislature, can change it as needed.

Parliament is made up of two houses and the sovereign. Members of the House of Lords are mostly appointed. This house has little real power. The House of Commons, the key lawmaking body, consists of elected officials. Usually, the leader of the party with the most seats in the House of Commons becomes prime minister. The prime minister heads the executive branch of the government. He or she governs with the support of the Cabinet. Tony Blair has served as prime minister since 1997.

Regional Governments All four regions of the United Kingdom—England, Wales, Scotland, and Northern Ireland—are represented in Parliament. In the late 1990s, Parliament shifted some power from the national level to regional governments, a process called **devolution**. Scotland and Wales set up parliaments. They took over their domestic affairs, while Parliament kept charge of foreign policy and defense.

COMPARING U.S. and British Governments

U.S. GOVERNMENT

U.S. citizens elect representatives and the President for the national government.

BRITISH GOVERNMENT

British citizens elect members of the House of Commons. The Prime Minister is the leader of the majority party.

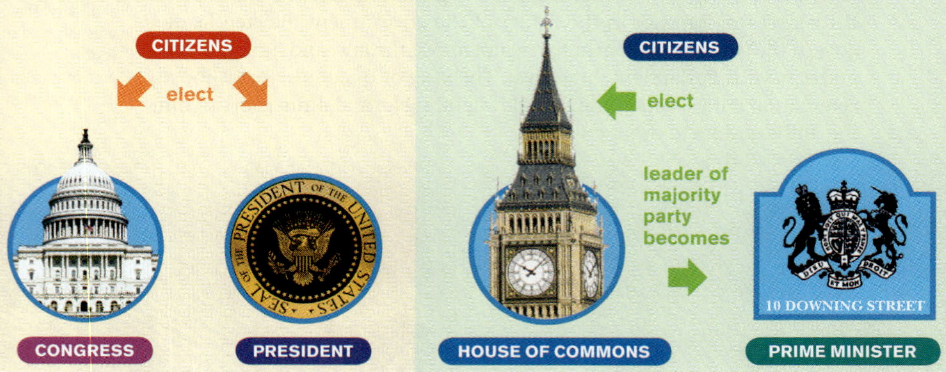

CRITICAL THINKING

Compare and Contrast Where do citizens more directly choose their leader?

DIFFERENTIATING INSTRUCTION

Struggling Readers

Respond to Focused Questions

Have pairs of students take turns reading aloud to one another, pausing to find answers to focused questions, such as:

- How is Britain's constitution different from most countries? *(It's unwritten.)*
- What are the two houses of Parliament? *(House of Commons, House of Lords)*
- How is the prime minister chosen? *(The leader of the political party with the most seats in the House of Commons usually is named prime minister.)*

Gifted & Talented

Comparison Chart of U.S. and British Governments

Have students create a Venn diagram showing similarities and differences between the U.S. and British governments. Students should draw upon the Comparing feature, information from this chapter and additional research. Ask students to consider each country's Bill of Rights, how leaders are chosen, constitutions, and other elements. Invite volunteers to present one entry from their diagrams.

Establishing a government in Northern Ireland has been more difficult. The English first expanded into Ireland in the 1100s. The Irish, who had their own culture, bitterly resented the English. In 1949, the southern part of Ireland became the Republic of Ireland. Since the 1920s, Northern Ireland has been divided by conflict between two groups. The nationalists, who are mostly Catholic, believe Northern Ireland should unite with the Republic of Ireland. The unionists, mainly Protestants, want to stay part of the United Kingdom. In 1998, both sides signed the **Good Friday Agreement**, which set up a new government. In 2002, however, renewed violence led to the government's shutdown. Recent efforts toward self-government have failed, but violence between the two sides has decreased and talks continue.

In the late 1990s, Scotland, Wales, and Northern Ireland established regional governments to take care of domestic policy. The government in Northern Ireland has been dissolved due to conflict. Parliament still controls the foreign policy and defense of all four regions. ◄

SUMMARIZE Explain how the United Kingdom governs its regions.

The British Economy

KEY QUESTION How has the United Kingdom's economy changed?

Today, the United Kingdom is a densely populated urban nation. Cities cover most areas that were once farmland. The country also lost its empire, which had boosted its economy. These changes have forced it to adapt.

A Powerful Economy For decades, manufacturing formed the basis of Great Britain's economy. It was fueled by natural resources such as iron ore, coal, and oil deposits in the North Sea. As these resources are depleted, British scientists have begun looking for sustainable energy sources, including wind energy. The map at right shows Britain's development of **wind farms**, power plants that use windmills to generate electricity.

At one time, Great Britain was known for its textile and steel industries. While it remains a leader in the production of textiles, chemicals, and motor vehicles, it has fewer manufacturing jobs than before. Over three-fourths of Britain's labor force now works in service industries. The nation is a world leader in insurance and financial services. London's banks and stock exchange have made it one of Europe's leading business centers.

Wind Farms in Britain

CONNECT Geography & Economics

1. **Human-Environment Interaction** Are more wind farms located inland or near the coast?
2. **Region** Which of the four regions has the most wind farms under construction?

Teach

The British Economy

Think, Pair, Share

- What two factors have forced the British economy to adapt? *(changing from a rural, farming nation to an urban, industrial nation; losing its empire)*
- What fueled British manufacturing for many years? *(natural resources, such as iron ore, coal, and oil)* Why are they seeking new forms of energy? *(These resources are being depleted.)*
- **Summarize** Summarize Great Britain's economy today. *(Possible Answer: Britain has a strong global economy and is strengthening ties with Europe.)*

Unit Resource Book: Europe
- Economics in the World: The Pound vs. the Euro, p. 184

Wind Farms in Britain

CONNECT Geography & Economics

Britain is using a variety of options to increase its energy sources and lower its greenhouse gas emissions. These include renewable energy sources, such as wind power, solar power, and tidal power. Not everyone is in favor of wind farms.

- Wind farms can be set up onshore or offshore. Offshore farms are more expensive to build and harder to maintain, but they may be more productive because of the greater wind power offshore.
- Some people claim that wind farms are noisy and ugly. Others find the wind farms interesting.
- Some people fear that birds are harmed by wind farms. In general, wind farms that are placed appropriately do not seem to endanger birds.

ANSWERS

1. **Human-Environment Interaction** near the coast
2. **Region** Scotland

Unit Resource Book: Europe
- Map and Graph Skills: Reading Bar Graphs, pp. 175–176

DIFFERENTIATING INSTRUCTION: TIERED ACTIVITIES

OBJECTIVE Demonstrate an understanding of the issues of wind power in Britain.

Basic

Have students create fact sheets about wind farms in Britain. The sheets should answer questions such as:
- Why is Britain using wind farms?
- Where are wind farms located in Britain?
- How and where are wind farms expanding?

On Level

Have students create a visual, such as a poster, promoting wind power in Britain. The visual should show where wind farms are currently located and being built, and describe the benefits of wind power. In addition, it should respond to concerns about wind power. Invite students to present their visuals.

Challenge

Have students do additional research about Britain's use of wind energy. Then have them write two editorials, one for wind farms and one opposed to them. Have students share their editorials in small groups. Ask students to assess which editorials are more effective and why.

CONNECT to the Essential Question

How did being an island nation influence the development of the United Kingdom?

Ask students what they have learned so far that can help them answer this question. Students might mention:

- The British government is shifting some power to regional governments.
- Northern Ireland, separate from Great Britain but part of the United Kingdom, has been a source of conflict.
- Britain is strengthening its ties to Europe and the global economy.

Unit Transparency Book: Europe
- Essential Question Graphic, TT23

4 Assess & Reteach

Assess Have students complete the Section Assessment.

Unit Resource Book: Europe
- Section Quiz, p. 191

Interactive Review @ClassZone.com
- Section Quiz

Power Presentations

Test Generator

Reteach Have student pairs create four main idea questions about Britain's government and economy. Collect the questions, read them aloud one at a time, and answer them as a class.

Unit Resource Book: Europe
- Reteaching Activity, p. 194

The Channel Tunnel Trains like this one, operated by a company called Eurotunnel, shuttle passengers and freight through the Channel Tunnel. In June 2006, the 100 millionth passenger traveled on the train.

Participating in the Global Economy Winston Churchill once said of Great Britain, "We are with Europe but not of it." In some ways the nation still stands apart. In 2002, most European Union members agreed to replace their own currency with the euro. Britain chose to continue using its national currency, the pound.

In most other ways, however, the United Kingdom is now closely tied to Europe through participation in the European Union and trade with industrial and developing nations. The nation has a trillion-dollar economy. Because of its prosperity, it can trade with other countries. It has goods and services that other nations want. Most of the United Kingdom's trade is with other EU members.

Technology has also drawn the United Kingdom closer to its neighbors in Europe. In 1994, British and French workers completed construction of the 31-mile undersea **Channel Tunnel**. The tunnel links Britain to France by train. Every day, an average of 45,000 people make the 35-minute journey to the European mainland.

ANALYZE EFFECTS Explain how the United Kingdom's economy has changed. It is no longer an agricultural nation. Manufacturing has declined, and much of the labor force now holds service jobs.

ONLINE QUIZ For test practice, go to **Interactive Review @ClassZone.com**

Section 3 Assessment

TERMS & NAMES

1. **Explain the importance of**
   - constitutional monarchy
   - unwritten constitution
   - Good Friday Agreement
   - Channel Tunnel

USE YOUR READING NOTES

2. **Summarize** Use your completed web diagram to answer the following question:
   What decision did the United Kingdom make that kept its economy separate from the rest of the European Union?

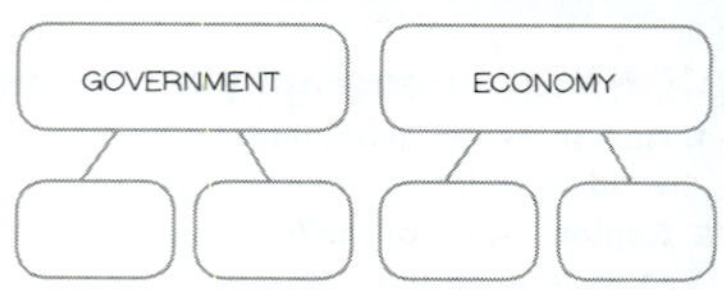

KEY IDEAS

3. Why is the United Kingdom said to have an unwritten constitution?
4. What are the roles of the monarch, prime minister, and Parliament within the British government?
5. What are the United Kingdom's most important service industries?

CRITICAL THINKING

6. **Compare and Contrast** How are the British Parliament and the U.S. Congress alike and different?
7. **Analyze Causes and Effects** Why might a long history of conflict make it hard for Northern Ireland to set up a workable government?
8. **CONNECT to Today** How might London's growing cultural diversity help it expand its global financial and business services?
9. **WRITING** **Write a Report** Choose a major British industry to research. Write a paragraph explaining why it is important to the United Kingdom's economy.

SECTION 3 ASSESSMENT ANSWERS

Terms & Names

1. constitutional monarchy: system in which a constitution limits the monarch's power; unwritten constitution: not a single document but includes many different laws and customs; Good Friday Agreement: between Northern Ireland's unionists and nationalists, set up a new government; Channel Tunnel: underwater railroad tunnel between France and England

Use Your Reading Notes

2. It did not adopt the euro.
   Diagram *Government—monarch's duties mostly ceremonial; Parliament has two houses; prime minister heads executive branch; regional parliaments. Economy—had to adapt; basis changed from manufacturing to services; trade, the EU, and the Channel Tunnel connect Britain to Europe.*

Key Ideas

3. not one written document, but a flexible collection of legal rulings and political customs
4. Parliament is legislature; prime minister is head of government, leads executive branch; monarch is head of state, duties mostly ceremonial
5. insurance, business, and finance

Critical Thinking

6. Similar: Both are the lawmaking branches of their governments. Different: Only one of the British houses is made up of elected officials.
7. People on both sides who have seen friends or family harmed and have made sacrifices might find it hard to cooperate and compromise.
8. many newcomers have connections to homelands; also have language skills needed to build trade
9. Use the rubric to score reports.

Writing Rubric

	Clarity of Text	Accuracy
4	clear, complete	no errors
3	clear	few/minor
2	some confusion	several
1	unclear	many

Click here to complete these and other activities online @ ClassZone.com

CHAPTER SUMMARY

Key Idea 1
Many developments, including the Industrial Revolution and a colonial empire, helped the United Kingdom grow into a major world power.

Key Idea 2
Britain's long history as an industrial and colonial power has shaped its culture. Britain's increasingly diverse population continues to enrich its cultural life.

Key Idea 3
The United Kingdom's government is a constitutional monarchy. Its economy is strong and adapts to global changes.

For **Review and Study Notes**, go to **Interactive Review** @ ClassZone.com

NAME GAME

Use the Terms & Names list to complete each sentence on paper or online.

1. I am a famous English playwright. **William Shakespeare**
2. I connect Britain with France by rail. ______
3. I am the United Kingdom's official church. ______
4. I am a British person. ______
5. I am a government in which the powers of the king and queen are limited by a constitution. ______
6. I am a document written in 1215 that limited the power of the English monarch. ______
7. I am the language spoken by some people in Scotland and Ireland along with English. ______
8. I am the process of shifting power from national to regional government. ______
9. I am the legislative branch of the British government. ______
10. I am the organization made up of former British colonies. ______

Briton
Channel Tunnel
Church of England
Commonwealth of Nations
constitutional monarchy
devolution
Gaelic
Good Friday Agreement
imperialism
Magna Carta
multicultural
Parliament
unwritten constitution
William Shakespeare

Activities

GeoGame

Use this online map to reinforce your understanding of the United Kingdom. Drag and drop each place name in the list to its location on the map.

To play the complete game, go to **Interactive Review** @ ClassZone.com

Crossword Puzzle

Complete an online crossword puzzle to test your knowledge of the United Kingdom.

ACROSS
1. a power plant that uses windmills to generate electricity

419

ONLINE AND ON-SCREEN ACTIVITIES

GeoGame

Use the interactive map on **Power Presentations** DVD-ROM or the **eEdition** to help students review the locations of important places in the United Kingdom.

- Demonstrate how to drag and drop the map labels listed onto the appropriate locations on the map.
- Advance to the scorecard to show how students have a chance to improve their scores.

Crossword Puzzle

Students can review important people and events in the United Kingdom's history by completing this crossword puzzle online at **ClassZone.com**.

- As an extension of this activity, have students use the *Activity Maker* at **ClassZone.com** to make their own crossword puzzle for one section of the chapter.
- Then have students trade puzzles with classmates to solve answers for each other's puzzles.

Chapter Summary

As a follow-up to the chapter, encourage students to pair up and go over the **Review and Study Notes at ClassZone.com.**

Alternatively, students pairs can review their Reading Study Guide notes, making sure that they can add important details to each section's "Key Idea."

Name Game

Teaching Tips

ON PAPER Have students work in pairs to complete this activity. As an extension, they can write questions for the terms that are not covered in the items shown here.

ONLINE Make sure that students understand how to drag and drop terms to fill in blanks. Ask them to print out their results and compare them with those of a classmate.

ANSWERS

1. William Shakespeare
2. Channel Tunnel
3. Church of England
4. Briton
5. constitutional monarchy
6. Magna Carta
7. Gaelic
8. devolution
9. Parliament
10. Commonwealth of Nations

Make It Fun

Teacher-Tested Activities

Ken Metz, Glacier Creek Middle School, Cross Plains, Wisconsin

For a different way to review a chapter, try the "Fish Bowl" method. Here's how.

- Break the students into small groups and have each group write a few chapter review questions on small slips of paper. You can add some you've written yourself to make sure key concepts are covered.
- Put all the questions in a fish bowl. Have a student draw a question, read it aloud, and call on someone to answer it.
- Once a student answers correctly, he or she should draw and ask the next question, until each student has gone.

Vocabulary ANSWERS

1. Parliament is the United Kingdom's national legislature.
2. When a country has many different cultures, it is called multicultural.
3. The Church of England is the official church of England.
4. William Shakespeare was a famous English playwright.
5. England is said to have an unwritten constitution that includes many different laws, court decisions, and political customs.
6. The two factions in Northern Ireland signed the Good Friday Agreement to establish a new government.
7. B
8. A
9. D
10. E
11. C

Key Ideas ANSWERS

12. England, Northern Ireland, Scotland, Wales
13. It provided raw materials for industry and a market for manufactured goods.
14. rivers for transportation and to power factories; good harbors; ample natural resources, such as coal and iron ore
15. Two world wars had left Britain with large debts and vast losses of life and property. After many colonies declared independence, Britain lacked the resources and power they had provided.
16. London
17. Welsh and Gaelic
18. He wrote plays that have been performed all over the world for centuries, spreading English language and culture.
19. music, television, and literature
20. It granted all three the right to form regional parliaments with control over their domestic affairs.
21. The service sector is expanding, but manufacturing is declining.
22. It participates through trade and involvement with the European Union.

CHAPTER 13 Assessment

VOCABULARY

Explain the significance of each of the following.

1. Parliament
2. multicultural
3. Church of England
4. William Shakespeare
5. unwritten constitution
6. Good Friday Agreement

Match the item on the left with its description on the right.

7. Magna Carta
8. Gaelic
9. Briton
10. devolution
11. Channel Tunnel

A. spoken in Scotland and Ireland
B. limited the power of the monarchy
C. connects Britain to France underwater
D. a British person
E. shifting power from national to regional government

KEY IDEAS

1 Building a British Empire

12. What four regions make up the United Kingdom?
13. How did Great Britain benefit from its empire?
14. What geographic factors led to the start of the Industrial Revolution in Great Britain?
15. Why did Great Britain's standing as a world power change after World War II?

2 From Shakespeare to Harry Potter

16. What is Great Britain's largest city?
17. What languages are spoken in Wales, Ireland, and Scotland besides English?
18. How did Shakespeare contribute to British culture?
19. In what areas has British culture influenced popular culture in the United States in recent decades?

3 Parliament and Free Enterprise

20. What has the United Kingdom done to give more self-government to the regions of Scotland, Wales, and Northern Ireland?
21. Which part of the United Kingdom's economy is expanding, and which has become smaller?
22. How does the United Kingdom participate in the global economy?

CRITICAL THINKING

23. **Draw Conclusions** Create a diagram and include two factors that led to British industrial growth.

24. **Five Themes: Movement** How did the Industrial Revolution contribute to the urbanization of the United Kingdom?
25. **Connect Geography & Culture** How do sports keep Great Britain tied to its former colonies?
26. **Compare and Contrast** How does the selection of the British prime minister differ from that of the U.S. president?
27. **Connect to Economics** What effect might a decline in North Sea oil have on the United Kingdom's economy?
28. **Draw Conclusions** How is Britain's role in the global economy different today from its role during the height of the British Empire?

Answer the ESSENTIAL QUESTION

How did being an island nation influence the development of the United Kingdom?

Written Response Write a two- or three-paragraph response to the Essential Question. Be sure to consider the key ideas of each section as well as specific ideas about the United Kingdom. Use the rubric below to guide your thinking.

Response Rubric

A strong response will:

- discuss how Britain's location, waterways, and natural resources influenced its development
- explain why Britain needed to look beyond its island home for resources and how it met this need through empire building and trade

ESSENTIAL QUESTION REVIEW

Before students begin their essays, have them review the key chapter elements that will help them answer the Essential Questions. Direct them to significant features in the chapter, such as:

- Analyzing Primary Sources, Section 1
- History Makers: Winston Churchill, Section 1
- Animated Geography, Section 1
- Connect to Language Arts, Section 2
- Comparing U.S. and British Governments, Section 3
- Wind Farms in Britain, Section 3

Written Response Rubric

	Coverage of Key Ideas	Explanation	Accuracy
4	all	clear, consise	no errors
3	most	good	few/minor errors
2	some	fair	several errors
1	two or fewer	poor	many errors

STANDARDS-BASED ASSESSMENT

Test Practice
- Online Test Practice @ClassZone.com
- Test-Taking Strategies and Practice at the front of this book

THEMATIC MAP

Use the map below to answer questions 1 and 2 on your paper.

North Sea Oil & Gas Fields

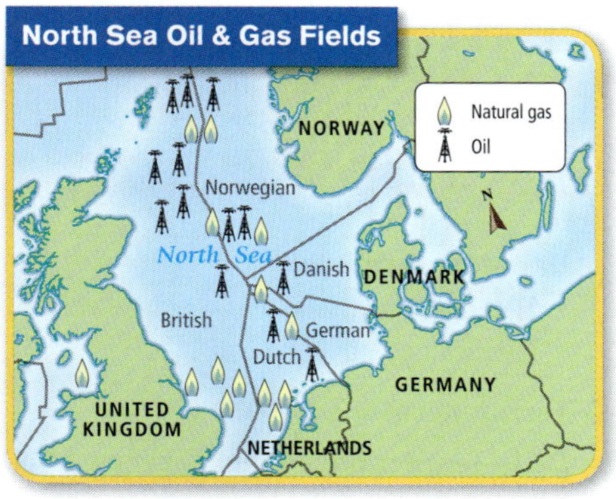

1. The map above shows sectors of natural gas and oil fields in the North Sea. Which sector has the fewest oil fields?

A. Norwegian Sector
B. German Sector
C. British Sector
D. Dutch Sector

2. How many oil fields are shown in the British Sector of the North Sea?

A. four
B. seven
C. nine
D. ten

LINE GRAPH

Examine the graph below. Use the graph to answer questions 3 and 4 on your paper.

Population Growth: Great Britain

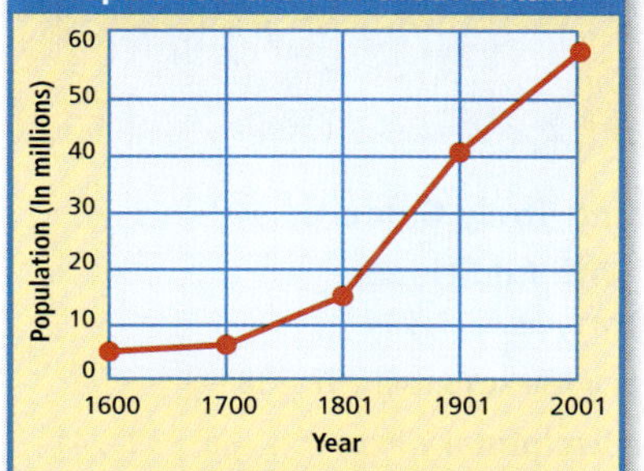

*The first census of the United Kingdom was taken in 1801. Population figures for 1600 and 1700 are estimates. Figures include Ireland, except for 2001.

Sources: GenDocs, "Population of Great Britain & Ireland 1570–1931" and National Statistics UK Census 2001

3. When did Great Britain experience the greatest population growth?

A. 1600s
B. 1700s
C. 1800s
D. 1900s

4. What was the British population in 1901?

A. 416,000
B. 4.8 million
C. 16.3 million
D. 41.6 million

GeoActivity

1. INTERDISCIPLINARY ACTIVITY–LANGUAGE ARTS

Review the section on British literature. Pick one writer in this section. Find a short excerpt from one of his or her works to read aloud to the class. Introduce your choice and provide some information about the writer.

2. WRITING FOR SOCIAL STUDIES

Reread the part of Section 2 on British culture. Pick three things described in this section and write a paragraph on each explaining how they are similar to and different from your own culture.

3. MENTAL MAPPING

Create an outline map of the United Kingdom and label the following:

- England
- Scotland
- Wales
- Northern Ireland
- London
- English Channel
- Atlantic Ocean

DIFFERENTIATING CHAPTER ASSESSMENT

Unit Resource Book

- **Chapter Test A,** pp. 195–198. *Use for Struggling Readers, Beginning-to-Intermediate-Level English Learners, and for Inclusion of students whose IEP suggests supported test-taking.*
- **Chapter Test B,** pp. 199–202. *Use for On-Level Learners and ESOL students with Advanced-Level English.*
- **Chapter Test C,** pp. 203–206. *Use for Gifted and Talented students or students assigned Pre-AP work.*

Test Generator

- Chapter Tests A, B, and C, in English and Spanish

McDougal Littell Assessment System

TEST **Test Generator** Use three levels of tests in English and Spanish tied to state standards. Plus, choose questions to customize tests for all populations.

SCORE Scan and Score plain paper student tests in minutes.

REPORT Create instant progress reports, sortable by student, by standard, or other criteria.

RETEACH Respond immediately to students' needs by using a wealth of reteaching resources.

Critical Thinking ANSWERS

23. Possible answers: Britain's geographic advantages; colonies providing raw materials and markets for manufactured goods

24. Workers moved to cities for factory jobs.

25. Sports such as soccer, cricket, and rugby are popular in former British colonies. Britain often competes against its former colonies.

26. Americans vote for candidates to become president. British voters elect members of the House of Commons. The leader of the majority party in the House of Commons usually becomes prime minister.

27. Possible answer: Britain will have to spend more of its money on energy and may have less to spend on other needs.

28. When ruling the British Empire, British leaders had a great deal of control over the economies of the colonies, as well as greater control over the global economy. Today, many factors determine decisions in the global marketplace. Britain has a much smaller role in making these decisions.

Standards-Based Assessment ANSWERS

1. B
2. B
3. C
4. D

GeoActivity

1. INTERDISCIPLINARY ACTIVITY RUBRIC

The oral presentation should

- provide an introduction to the selected author
- include an excerpt from the author's writings

2. WRITING FOR SOCIAL STUDIES RUBRIC

The paragraph should

- reflect three aspects of British culture
- compare and contrast them with American culture

3. MENTAL MAPPING RUBRIC

The outline map should

- represent the approximate shape of the United Kingdom
- include all geographic features listed in their correct location

CHAPTER 14

Eastern Europe PLANNING GUIDE

Complete print resources all on one USB drive allow you to customize lesson plans for all students.

Section Titles & Key Ideas	Unit Resource Book Europe	Assessment Resources
1 Poland, Ukraine, and the Baltic States **KEY IDEA** Poland, Ukraine, and the Baltic States faced many challenges after shaking off Communist rule.	• **Reading Study Guide,** pp. 207–208 • **RSG with Additional Support,** pp. 213–214 • **History Makers,** pp. 237, 238 • **Interdisciplinary Projects,** pp. 239–241 • **Connect to Science,** p. 243	**Unit Resource Book** • Section Quiz, p. 245 • Reteaching Activity, p. 248 **Test Generator CD-ROM** **Daily Test Practice Transparency,** TT49
2 Hungary and the Czech Republic **KEY IDEA** After decades of Communist rule, Hungary and the Czech Republic made economic and political reforms.	• **Reading Study Guide,** pp. 209–210 • **RSG with Additional Support,** pp. 215–216 • **Vocabulary Practice,** p. 231 • **Outline Map Activities,** pp. 235–236 • **Primary Source,** p. 244	**Unit Resource Book** • Section Quiz, p. 246 • Reteaching Activity, p. 249 **Test Generator CD-ROM** **Daily Test Practice Transparency,** TT50
3 The Balkans **KEY IDEA** In the 1990s, the breakup of Yugoslavia was violent. Ethnic divisions and economic and political issues led to war.	• **Reading Study Guide,** pp. 211–212 • **RSG with Additional Support,** pp. 217–218 • **Skillbuilder Practice,** p. 232 • **Map and Graph Skills,** pp. 233–234 • **Economics in the World,** p. 242	**Unit Resource Book** • Section Quiz, p. 247 • Reteaching Activity, p. 250 • Chapter Test (Forms A, B, & C), pp. 251–262 **Test Generator CD-ROM** **Daily Test Practice Transparency,** TT51

No Time? To focus students on important content in the chapter, use the **"Reading Study Guide with Additional Support"** in *Resources2Go.*

Teacher Presentation Options

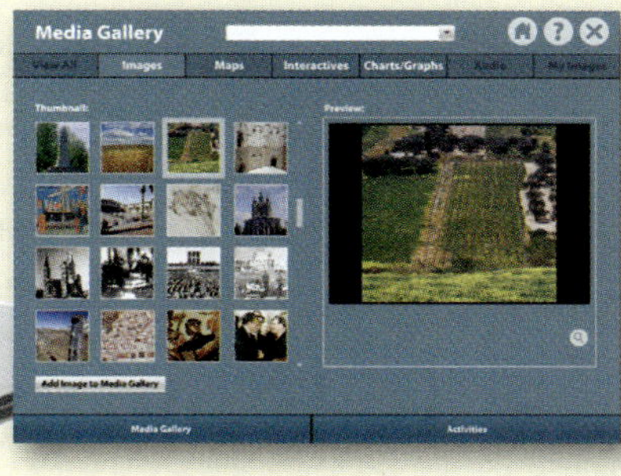

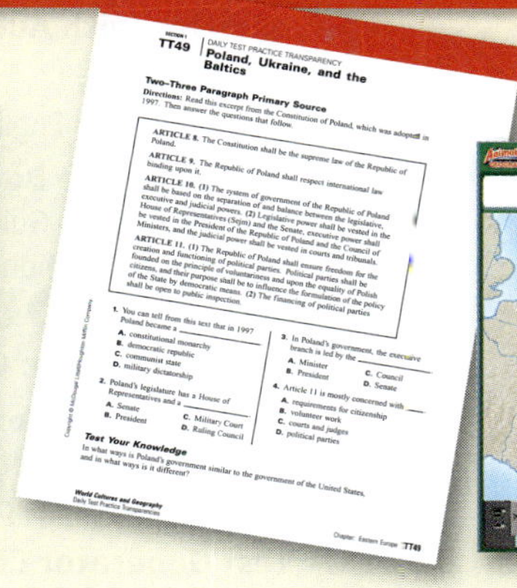

Electronic Books

eEdition DVD-ROM

eEdition Online

Presentation Toolkit

Video/DVD

- World Cultures and Geography

Power Presentations DVD-ROM

- Discussion Starters
- Media Gallery
- Review Game
- Interactive Review
- Animated Features

Unit Transparency Book

- 3-Minute Warm-Ups
- Essential Question
- Critical Thinking
- Maps
- Cultures

Daily Test Practice Transparencies

Animated Geography

- Present-Day Eastern Europe Map
- Worldster: Radek's Web Page

OnlineActivities at ClassZone.com

Animated Geography

- Videos and Photos
- 3-D Animations
- Interactive Maps

Interactive Review

- Section Quizzes
- Review and Study Notes
- Vocabulary Flip Cards
- Graphic Organizers
- Name Game and Crossword Puzzles
- GeoGame

Online Test Practice

Map Center

- Atlas Maps
- Political Maps with Data
- Links to Mapping Web Sites

Activity Center

- Activity Maker—Review Games, Puzzles, Flip Cards, and Graphic Organizers
- Worldster

Research & Writing Center

- Web Sites
- Data Updates
- Primary Sources
- Writing Models

Current Events

Suzanne Scheidler Moen
DeForest Middle School
DeForest, Wisconsin

Make It Fun — Teacher-Tested Activities

As you teach this chapter, look for my classroom-proven ideas for:

- making the lesson fun
- focusing students' attention and comprehension
- integrating technology with traditional teaching

Teaching Bell to Bell When you see this timing guide, have students discuss in pairs, in groups of 3–6, or as a class.

- **Think, Pair, Share** (pairs)
- **Reader, Recorder, Reporter** (3–6)
- **Talk About It** (any size)

RESOURCES FOR DIFFERENTIATED INSTRUCTION

CHAPTER 14

Struggling Readers

eEdition with Audio DVD-ROM

Lesson Plans
- Strategies for Inclusion

Unit Resource Book
- Reading Study Guide with Additional Support, pp. 213–218 (B)
- Reteaching Activity, pp. 248–250
- Chapter Test (Form A), pp. 251–254 (A)

Best Practices Toolkit (C)
- Interactive Reading Strategies, pp. 90–145
- Differentiation Activity Bank: Struggling Readers, pp. 146–149

Teacher's Edition Activities
- Sequencing Ukraine's History, Sec. 1
- Learning from Photographs, Sec. 2
- Tiered Activities, Sec. 1–3

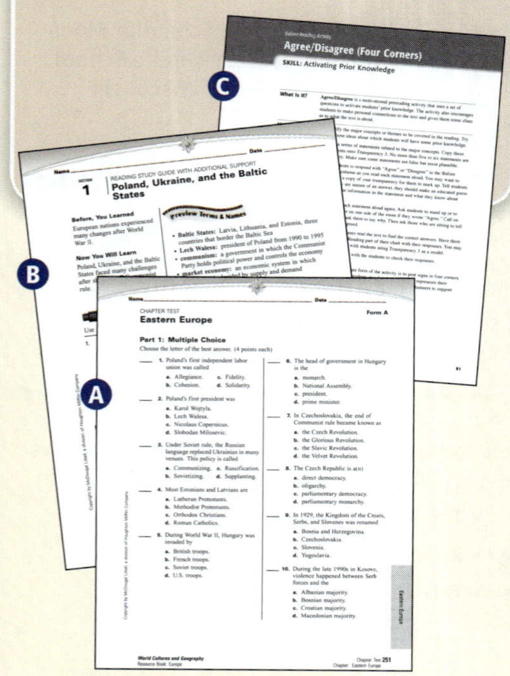
Agree/Disagree (Four Corners)
Poland, Ukraine, and the Baltic States
Eastern Europe

English Learners

Pupil's Edition in Spanish

eEdition with Audio DVD-ROM

Lesson Plans
- Strategies for English Learners

Unit Resource Book
- Reading Study Guide (Spanish), pp. 219–224
- RSG with Additional Support (Spanish), pp. 225–230 (A) (B)

Multi-Language Glossary (C)

Best Practices Toolkit
- Differentiation Activity Bank: English Learners, pp. 165–194

Test Generator CD-ROM
- Section Quizzes in Spanish
- Chapter Test in Spanish

Teacher's Edition Activities
- Pronounce and Preview, Sec. 1–3
- Culture: Clarify, Sec. 1
- Comprehension: Task Support, Sec. 1
- Comprehension: Connection, Sec. 2
- Vocabulary: Key Academic Vocabulary, Sec. 2
- Tiered Activities, Sec. 1–3

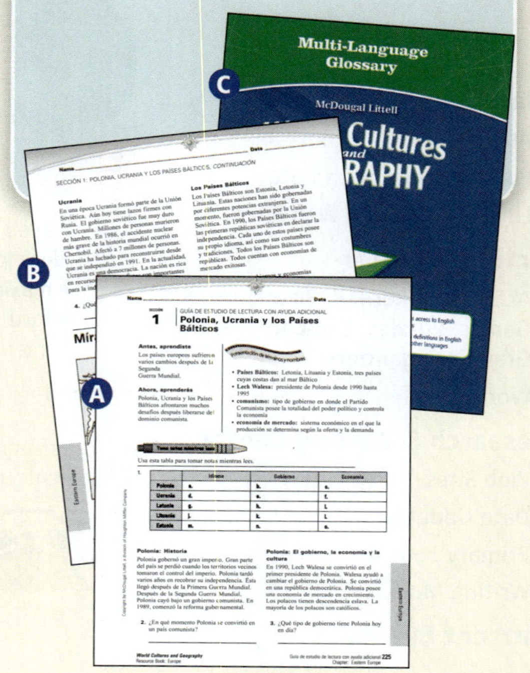
Multi-Language Glossary
Polonia, Ucrania y los Países Bálticos

Inclusion

eEdition with Audio DVD-ROM

Lesson Plans
- Strategies for Inclusion

Unit Resource Book
- Reading Study Guide with Additional Support, pp. 213–218 (A) (B)
- Reteaching Activity, pp. 248–250
- Chapter Test (Form A), pp. 251–254

Best Practices Toolkit (C)
- Interactive Reading Strategies, pp. 90–145
- Differentiation Activity Bank: Struggling Readers, pp. 146–149

Teacher's Edition Activities
- Fill in the Blanks, Sec. 1
- Define Terms in Your Own Words, Sec. 2
- Vocabulary: Student VOC Strategy, Sec. 3
- Tiered Activities, Sec. 1–3

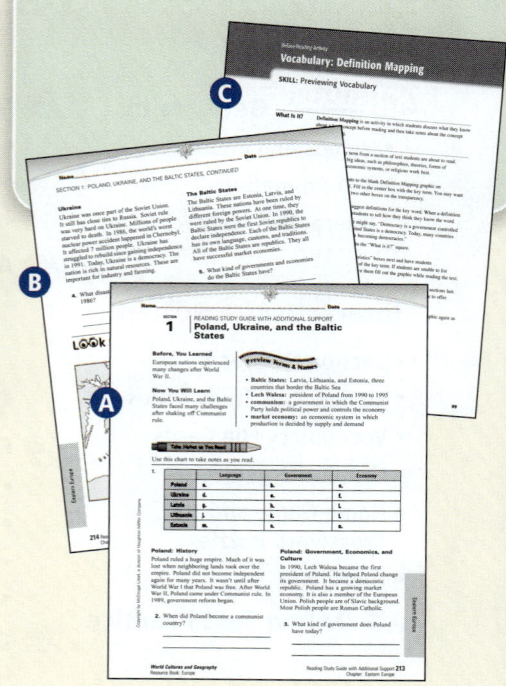
Vocabulary: Definition Mapping
Poland, Ukraine, and the Baltic States

Interdisciplinary Activities

Gifted & Talented Students

Lesson Plans
- Core Lesson Plan

Unit Resource Book
- Outline Map Activities, pp. 235–236
- History Makers, pp. 237, 238 A B
- Economics in the World, p. 242
- Connect to Science, p. 243 C
- Chapter Test (Form C), pp. 259–262

Best Practices Toolkit
- Student Products Rubrics, pp. 1–14
- Writing Templates, pp. 78–89

Teacher's Edition Activities
- Assess the Decision to Recognize Solidarity, Sec. 1
- Explore the Variety of Roma Music, Sec. 2
- Tiered Activities, Sec. 1–3

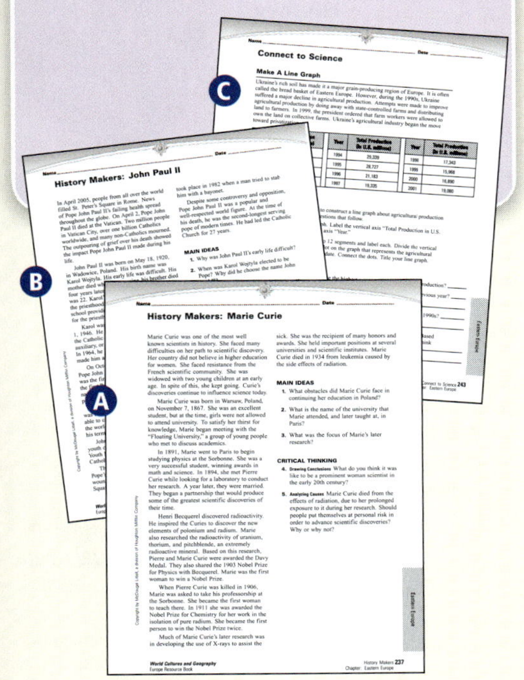
C
Connect to Science
Make A Line Graph
B
History Makers: John Paul II
A
History Makers: Marie Curie

Pre-AP*

Lesson Plans
- Core Lesson Plan

Unit Resource Book
- Vocabulary Practice, p. 231
- Map and Graph Skills, pp. 233–234
- Interdisciplinary Projects, pp. 239–241 A B
- Primary Source, p. 244 C
- Chapter Test (Form C), pp. 259–262

Best Practices Toolkit
- Writing Templates, pp. 78–89
- Differentiation Activity Bank: Pre-AP Students, pp. 195–200

* Pre-AP is a registered trademark of the College Entrance Examination Board, which was not involved in the production of and does not endorse this product.

Teacher's Edition Activities
- Debating International Cooperation, Sec. 1
- Analyzing River-Based Trade, Sec. 2
- Critique the Dancing House, Sec. 2
- Tiered Activities, Sec. 1–3

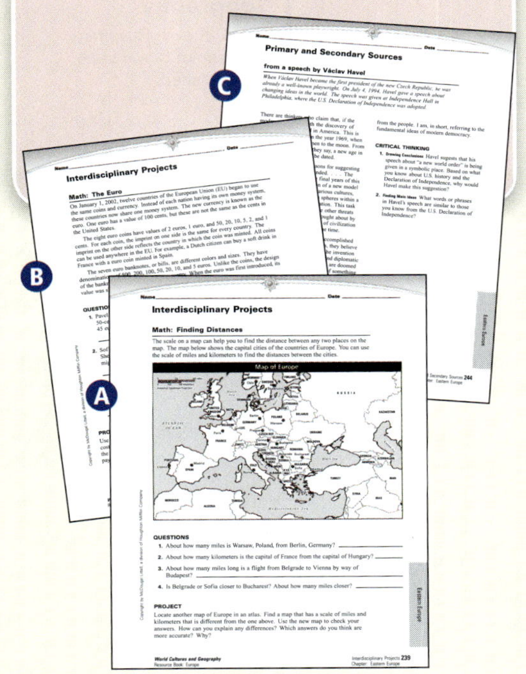
C
Primary and Secondary Sources
from a speech by Václav Havel
B
Interdisciplinary Projects
Math: The Euro
A
Interdisciplinary Projects
Math: Finding Distances

CONNECT to Language Arts

Create a Radio Profile

CLASS TIME 30 minutes

TASK Create a radio profile of Czech leader Václav Havel.

SUPPLIES
- information about Václav Havel
- tape recorder, or digital audio recorder and editing software

ACTIVITY Invite student groups to create a two-minute recording of a profile of Václav Havel. Encourage students to begin by gathering information about Havel, then writing a script. The script should include the voices of the reporters and of those assuming the roles of people who know Havel and Havel himself. Then have students record their scripts and edit the recording as needed. When finished, have each group "broadcast" their report.

Good for Audio and Verbal Learners

CONNECT to Math

Pie Graph of Eastern European Per Capita Income

CLASS TIME 30 minutes

TASK Create a bar graph that shows the per capita income for each of the countries of Eastern Europe.

SUPPLIES
- recent data on per capita income by nation (such as from The World Factbook at www.cia.gov)
- graph paper, colored markers or pencils

ACTIVITY Have student pairs create a bar graph that shows the annual per capita income for each of the nations of Eastern Europe. Before beginning, remind students to agree on a common scale of units, such as "one graph square equals $1,000." Completed bar graphs should include a title, such as "Per Capita Income for Eastern Europe." As an extension, students can create similar graphs for the nations of Western Europe and compare them to those for Eastern Europe.

Good for Visual and Mathematical/ Logical Learners

Introduce the ESSENTIAL QUESTION

- Point out that after World War II, Communist governments took control of the countries of Eastern Europe.
- Explain that the collapse of the Communist governments in the 1990s brought democracy to Eastern Europe.
- Discuss the challenges these newly democratic countries have faced.

Key Ideas

Section 1: Poland, Ukraine, and the Baltic States faced many challenges after shaking off Communist rule.

Section 2: After decades of Communist rule, Hungary and the Czech Republic made economic and political reforms.

Section 3: In the 1990s, the breakup of Yugoslavia was violent. Ethnic divisions and economic and political issues led to war.

CONNECT to the Essential Question

You may wish to revisit the Essential Question periodically with students as they work through the chapter.

Suggestions for reconnecting are provided at an appropriate point in the *Teacher's Edition* for each section.

Prague, Czech Republic

CHAPTER 14

Eastern Europe

1 FOCUS ON **Poland, Ukraine, and the Baltic States**

2 FOCUS ON **Hungary and the Czech Republic**

3 FOCUS ON **The Balkans**

ESSENTIAL QUESTION

In what ways has Eastern Europe changed since the end of World War II?

CONNECT Geography & History

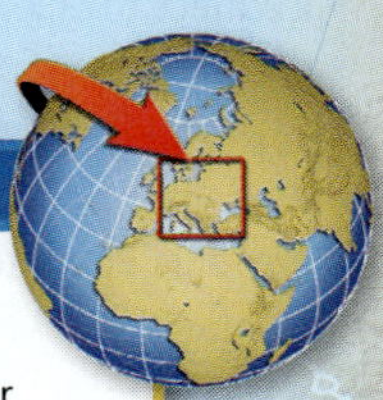

Use the map and the time line to answer the following questions.

1. What four Eastern European countries border the Baltic Sea?
2. What city in the Ukraine still carries the name of an early Slavic state?

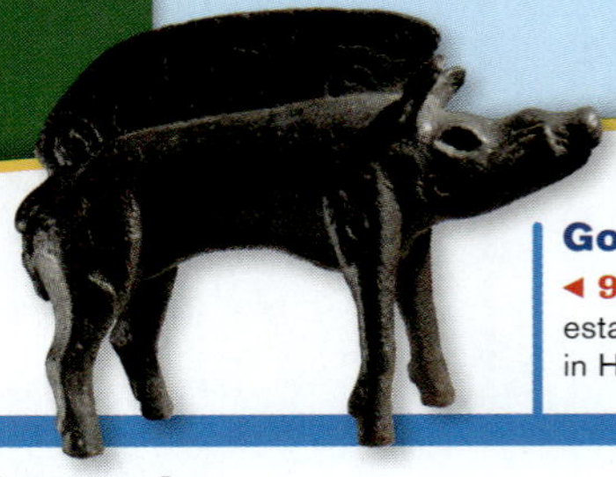

800

Geography
800s Kievan Rus, the first eastern Slavic state, is established.

Government
◀ **970** Magyar leaders establish a stable kingdom in Hungary. (Magyar artifact)

History
1918 Yugoslavia is formed at the end of World War I. (King Peter I) ▶

TIME LINE DISCUSSION

Point out to students that all of the Eastern European countries shown on the map were ruled by Communist governments from the 1940s until the 1990s. How many independent nations exist in Eastern Europe today? *(19)*

- What was the first eastern Slavic state and when was it formed? *(Kievan Rus, in the 800s)*
- How many years separate the establishment of a stable kingdom in Hungary and the creation of Yugoslavia? *(948 years)*
- Which event on the time line shows Eastern Europe starting to join with Western Europe? *(2004—The First Eastern European Nations—the Czech Republic, Hungary, Poland, Slovenia, and the Baltic States—join the EU.)*
- What geographic characteristic might have contributed to the Czech Republic, Hungary, Poland, and Slovenia being among the first Eastern European Nations to join the European Union? *(All of these countries share borders with Western Europe.)*

History

1940s Communists gain power in Eastern Europe. (Josip Broz Tito) ▼

Economics

2004 The first Eastern European nations–the Czech Republic, Hungary, Poland, Slovenia, and the Baltic states–join the EU.

Geography

1986 Chernobyl nuclear disaster in Ukraine ▶

Today

TEACHER MEDIA FAVORITES

Books for Students

Barber, Antonia and Shena Guild. *Hidden Tales from Eastern Europe.* Frances Lincoln, 2004. Folk tales from Croatia, Poland, Slovakia, Slovenia, Romania, Serbia, and Russia.

Cipkowski, Peter. *Revolution in Eastern Europe.* Wiley, 1991. A view of the collapse of communism in Poland, Hungary, East Germany, Czechoslovakia, Romania, and the Soviet Union.

Ingram, W. Scott. *The Chernobyl Nuclear Disaster.* Facts on File, 2005. A look at the 1986 disaster and its lasting consequences.

Video / DVD

Eastern Europe. Publishers Group West, 2005. A 180-minute tour through Poland, Czech Republic, Hungary, Slovenia, Croatia, and Turkey.

Jacques Cousteau River Explorations. Warner Home Video, 2006. Explores the Danube River through Eastern Europe, as well as the Mississippi and the Amazon.

Internet

Go to the *Research & Writing Center* at **ClassZone.com** to link to:

- data on the region
- a variety of chapter-specific, content-reviewed sites
- "History Makers" biographies
- primary sources

Animated GEOGRAPHY

Present-Day Eastern Europe

CLASSZONE.COM

Point out that the culture and the economy of Eastern Europe has been shaped by its geography. Rivers, mountains, large bodies of water, and location all have influenced the people in these diverse countries.

- Which is the largest nation of Eastern Europe? *(Ukraine)* Which is the smallest? *(Montenegro)*
- Find the nations of Estonia and Albania. Name one geographic characteristic that these nations share. *(Both have long stretches of coastline.)* Name one geographic characteristic that is different. *(Possible Answers: Estonia is located far to the north of Albania. Estonia includes two large islands; Albania has none.)*
- Odessa, a city located in Ukraine, is a center for trade in the region. How does its geography contribute to its importance? *(It is located on the Black Sea near the mouths of the Dneister and Dneiper rivers. Goods can be shipped by water to and from Odessa.)*

CONNECT GEOGRAPHY & HISTORY ANSWERS

1. Estonia, Latvia, Lithuania, and Poland

2. Kiev

Presentation Options

Use the **Power Presentations** DVD-ROM to project the animated map.

- Prompt students to use the pop-up images and audio from the map to discover what they already know about the geography, culture, history, economics, or governments of the region.
- Challenge students to answer the map questions individually, in small groups, or as a class activity. Use all of the interactive material to help preview chapter content.

1 Plan & Prepare

Objectives

- Describe the history, culture, government, and economy of Poland
- Describe the history, culture, government, and economy of Ukraine
- Compare and contrast the Baltic States

Read for the Essential Question

Help students read for a purpose by reminding them of the Essential Question: "In what ways has Eastern Europe changed since the end of World War II?" Point out that they can use text features such as Key Questions, special features, images, and maps to form answers to the question.

Vocabulary

Best Practices Toolkit

Use the ***Best Practices Toolkit*** to model strategies for vocabulary notetaking. Vary strategies throughout the year. Choose from: *Knowledge Rating, Predicting ABCs, Definition Mapping, Word Sort, Frayer Model (Word Squares), Magnet Words, Word Wheel,* and *Student VOC.*

Vocabulary Strategies, TT9–TT16

Reading Strategy

Best Practices Toolkit

Display the Three-Column Chart transparency.

- Explain to students that in categorizing, they need to determine how the information in the section can be sorted into the categories shown on the chart.
- Model filling in categories on the chart, using the "Language" column.
- Have students suggest entries, using the other categories in the chart.
- Display the completed chart for review.

Three-Column Chart, TT21

SECTION 1 Reading for Understanding

Key Ideas

BEFORE, YOU LEARNED

European nations experienced many changes after World War II.

NOW YOU WILL LEARN

Poland, Ukraine, and the Baltic States faced many challenges after shaking off Communist rule.

Vocabulary

TERMS & NAMES

Baltic States Latvia, Lithuania, and Estonia, three countries that border the Baltic Sea

Lech Walesa (lehk wah•LEHN•suh) Polish leader who cofounded Solidarity and served as president of Poland from 1990 to 1995

Solidarity Poland's first independent labor union, cofounded by Lech Walesa

Russification the effort to make countries occupied by the Soviet Union more Russian by replacing local languages and customs

bread basket an abundant grain-producing region

deport to expel from a country

BACKGROUND VOCABULARY

brain drain the loss of skilled workers who move in search of better opportunities

REVIEW

communism a type of government in which the Communist Party holds all political power and controls the economy

command economy an economic system in which the production of goods and services is decided by a central government

market economy an economic system in which the production of goods and services is decided by the supply of goods and the demand of consumers

Reading Strategy

Re-create the chart shown at right. As you read and respond to the **KEY QUESTIONS**, use the chart to help you categorize information about the Eastern European nations in this chapter.

Skillbuilder Handbook, page R7

CATEGORIZE

	LANGUAGE	GOVERNMENT	ECONOMY
Poland			
Ukraine			
Latvia			
Lithuania			
Estonia			

GRAPHIC ORGANIZERS
Go to Interactive Review @ClassZone.com

424 Chapter 14

PRETEACHING VOCABULARY

English Learners

Pronounce and Preview

Pronounce each term for students. Review the meanings of words within definitions, such as *union, occupied,* and *expel.*

- To modify vocabulary learning, have students complete worksheets as they read, instead of afterward.

Unit Resource Book: Europe
- Vocabulary Practice, p. 231

Inclusion

Fill in the Blanks

Invite students to work in pairs. One student finds a term in the text, then reads the sentence in which that term is used, substituting the word "blank" for the term. The other student must choose a term from the vocabulary list that correctly fills in the blank.

Prague, Czech Republic

SECTION 1 FOCUS ON

Horse-drawn wagon in Romania

Poland, Ukraine, and the Baltic States

Connecting to Your World

After the American Revolution, Americans quickly learned that building a strong independent nation was not easy. It took time to create a stable government and prosperous economy. In this section, you will read about Poland, Ukraine, and the three **Baltic States**, Latvia, Lithuania, and Estonia. Along with the rest of Eastern Europe, they gained independence from **communism** in the 1990s. Like the United States, they faced difficult choices. Many countries found the new freedom challenging.

Poland

KEY QUESTION What actions did government leaders take to prepare Poland for a market economy?

Poland's name comes from the Slavic word *Polanie*, which means plain or field. The Northern European Plain covers much of Poland. Its flat, fertile land began attracting Slavic settlers as early as 2000 B.C. During the Middle Ages, Poland ruled an empire that covered much of central and eastern Europe. By the late 18th century, however, Poland was conquered by its neighbors. It did not regain its independence until after World War I.

Polish Countryside Freshly plowed fields cover these gently rolling hills in the Roztocze region of southeast Poland.

2 Focus & Motivate

3-Minute Warm-Up

Write on the board or display the transparency:

- True or false? Correct any false statements.

1. Lech Walesa was elected Poland's first president. *(T)*
2. The Baltic States include Ukraine. *(F; The Baltic States are Estonia, Latvia, and Lithuania.)*

Unit Transparency Book: Europe
- 3-Minute Warm-Ups, TT31

Connecting to Your World

More About . . .

Farming in Poland

Most farms in Poland are small family farms. Still, Poland exports farm products. It is one of the world's largest producers of sugar beets.

3 Teach

Poland

Talk About It

- Why might the Communist government of Poland have been reluctant to recognize Solidarity? *(Possible Answer: because recognizing the union might mean giving up some power)*
- What economic challenges does Poland face? *(high unemployment and brain drain)*
- **Categorize** What groups have played a role in Poland's culture? *(Poles, Germans, Soviets, Ukrainians, Jews, and Catholics)*

SECTION 1 PROGRAM RESOURCES

Unit Resource Book: Europe

ON LEVEL
Unit Resource Book
- Reading Study Guide, p. 207
- Section Quiz, p. 245

STRUGGLING READERS
Unit Resource Book
- Reading Study Guide with Additional Support, p. 213
- Skillbuilder Practice: Creating a Multimedia Presentation, p. 232
- Section Quiz, p. 245
- Reteaching Activity, p. 248

eEdition with Audio DVD-ROM

ENGLISH LEARNERS
Pupil Edition in Spanish
eEdition with Audio DVD-ROM
Unit Resource Book
- RSG (Spanish), p. 219
- RSG with Additional Support (Spanish), p. 225

Multi-Language Glossary
Test Generator
- Section Quiz in Spanish

INCLUSION
Unit Resource Book
- RSG with Additional Support, p. 213
- Section Quiz, p. 245
- Reteaching Activity, p. 248

GIFTED & TALENTED
Unit Resource Book
- History Makers, pp. 237, 238
- Connect to Science, p. 243
- Section Quiz, p. 245

PRE-AP
Unit Resource Book
- Interdisciplinary Projects, pp. 239–241
- Section Quiz, p. 245

TECHNOLOGY
Unit Transparency Book
- 3-Minute Warm-Ups, TT31
- Essential Question, TT32
- Critical Thinking, TT33
- Map: Eastern Europe: Political, TT36
- Cultures, TT39

Daily Test Practice Transparencies
- Section 1, TT49

Power Presentations
ClassZone.com
World Cultures and Geography Video on DVD

More About . . .

Solidarity

Through much of the 1970s, Polish workers used strikes to protest poor economic conditions. Poland's Communist government responded by attacking or jailing strike leaders. However, the labor movement grew.

The Solidarity union was founded in September 1980, and soon 10 million Polish workers belonged to Solidarity. In late 1981, the Polish government declared Solidarity illegal and arrested its leaders. The Union continued to operate, mainly in secret. Workers kept up pressure on the government, and the union was allowed to resume its activities openly in 1989.

More About . . .

Pope John Paul II

Known as a political activist, Pope John Paul II traveled more extensively than all other popes before him combined to encourage popular political movements around the world. He visited nations such as Cuba, Brazil, the Philippines, and South Korea, and he called for the recognition of human rights. John Paul II died on April 2, 2005, after serving as Pope for 26 years—one of the longest tenures of any pope.

Unit Resource Book: Europe

- History Makers: John Paul II, p. 238

History During World War II, both Germany and the Soviet Union, a union of Communist republics led by Russia, invaded Poland. At the end of the war, Soviet troops drove the Germans out of Poland and supported Communists in gaining control of the government. By the 1950s, Communist rule was firmly established in Poland and throughout Eastern Europe. Communist leaders took over the economy and limited the rights of citizens.

During the 1970s, Poland faced food shortages and rising prices. As economic conditions worsened, people questioned Communist rule. In 1980, thousands of Polish workers went on strike, led by **Lech Walesa** (lehk wah•LEHN•suh). They demanded better pay, greater political freedom, and recognition of **Solidarity**, the nation's first independent labor union. Communist leaders saw public support for Solidarity growing. In 1989, they reached an agreement with Solidarity that led to government reform. Poland held its first free elections. The new government lifted many restrictions on Polish citizens.

Culture Before World War II, Poland was a diverse nation, with Germans, Jews, Ukrainians, and other ethnic groups. However, the war caused many groups to migrate and shifted Poland's borders. Almost all of the 3 million Jews living in Poland were killed during the Holocaust. Today, most people in Poland are Poles. Their ancestors belonged to the Slavic tribes that settled in Poland long ago. Polish, a Slavic language, is spoken throughout the country.

Catholic Mourners People gather in Pilsudski Square in Warsaw, Poland, to mourn the death of Pope John Paul II.

Most Poles are Roman Catholics, which unites the Polish people. During Communist rule, when religious practice was restricted, many continued to practice the Catholic faith. In 1978, Cardinal Karol Wojtyla (voy•TIH•wah) became the world's first Polish pope, Pope John Paul II. He made Catholic Poles very proud. In visits to Poland, he encouraged efforts to gain greater freedom.

Poland has a rich cultural tradition, with a strong emphasis on education. Many outstanding scientists, such as Nicolaus Copernicus and Marie Curie, were Polish. Like other nations, Poland has to work hard to preserve its folk culture in the face of mass media and urbanization.

DIFFERENTIATING INSTRUCTION

English Learners

Culture: Clarify

Students from non-Catholic cultures may not understand the position of the pope as a religious leader. Explain to them that the Pope is the leader of the Roman Catholic Church in all parts of the world. This was one of the reasons Poles were proud when Karol Wojtyla became pope.

Gifted & Talented

Assess the Decision to Recognize Solidarity

Have students assume the role of Communist leaders as they decide whether or not to end nationwide strikes and recognize the Solidarity union. You may wish to begin with a brief discussion of the purpose of unions—to organize workers to improve wages and working conditions. In their role, have students discuss the advantages and disadvantages of recognizing Solidarity, and vote on a final decision.

Government and Economics In 1990, Lech Walesa became Poland's first president. He helped Poland shift to democracy. In 1997, Poland adopted a new constitution to reflect its democratic government. Today, Poland is a republic. Voters elect the nation's president and legislature. The prime minister runs the government.

In the 1990s, Poland's leaders enacted a series of economic reforms to pave the way from a **command economy** to a **market economy**. Polish leaders sold government-owned industries to private companies. They shut down many old and inefficient factories and invited foreign companies to invest in Poland. Today, the country's major industries include production of coal, iron, steel, and machinery. Poland also has a large shipbuilding industry. Poland joined the European Union in 2004. Membership gave it new markets for such farm products as potatoes and rye and for manufactured goods.

Although Poland's transition to a market economy was more successful than that of many Eastern European nations, it still faces many economic challenges. High unemployment remains a serious problem. Factory closings have left many workers without jobs. Lack of good jobs for skilled workers has caused a serious **brain drain**, as many of Poland's bright young people leave the country in search of better job opportunities in Western Europe or the United States.

SUMMARIZE Explain how Poland became a market economy.
sold off government-owned industries to private firms, shut down old and inefficient factories, and sought foreign investment

ONLINE PRIMARY SOURCE To read the rest of Walesa's speech, go to the **Research & Writing Center** @ClassZone.com

ANALYZING Primary Sources

Lech Walesa (born 1943) led the struggle to form Solidarity. His peaceful resistance helped end Communist rule in Poland.

> **We respect the dignity and the rights of every man and every nation. The path to a brighter future of the world leads through honest reconciliation [resolution] of the conflicting interests and not through hatred and bloodshed.**
>
> Source: Lech Walesa, Nobel Peace Prize acceptance speech, 1983

DOCUMENT-BASED QUESTION

How does Walesa suggest conflicts among nations should be settled?

427

More About . . .

Brain Drain

While brain drain hurts the economies of Eastern Europe, it has often proved beneficial to the economies of Western Europe and the United States. Engineers, computer scientists, and other skilled workers bring their knowledge to their new homes, helping to develop new technologies and other business ideas and to create new jobs.

RESEARCH & WRITING CENTER

ANALYZING Primary Sources

For his work in helping the workers of Poland, Lech Walesa won the 1983 Nobel Peace Prize. His experience with Solidarity led Lech Walesa into politics. As Poland's first popularly elected president, Walesa helped Poland convert from communism to democracy. His presidential term ended in 1995, when he lost the presidential election to a challenger.

- Why might Lech Walesa have been awarded the Nobel Peace Prize? *(Possible Answer: His work with Solidarity helped create a peaceful solution to the conflict between workers and the Polish government.)*

DOCUMENT-BASED QUESTION ANSWER

He suggests that people should settle arguments by sorting out their differences peacefully.

INTERDISCIPLINARY ACTIVITIES

CONNECT to Art

Create an Ad to Stop Brain Drain

Discuss why Poland wants to avoid the loss of skilled workers. Then invite students to create a print ad to encourage workers to stay in Poland and help build their nation's economy, rather than take their skills elsewhere. Ads might appeal to patriotism, to contributing to a better economy for all Poles, or even to having the opportunity to become wealthy without leaving Poland. Have students present or post their ads.

CONNECT to Civics

Write a Campaign Speech

Remind students that Poland faces problems, but its economy has bright spots, too. Have students write campaign speeches in the voice of a candidate for the Polish presidency. In their speeches, candidates should outline the challenges facing Poland and share ideas for overcoming them. Have students present their speeches, then hold an election.

Teach

Ukraine

Think, Pair, Share

- What caused the state of Kievan Rus to collapse? *(It was invaded and conquered by nomads.)*
- Why might the people of Ukraine have protested Soviet rule from the 1960s to 1991? *(Soviet farm policies caused millions of farm families to starve. The Chernobyl disaster affected 7 million people in and around Ukraine.)*
- **Categorize** What are the two main ethnic groups in Ukraine? *(Ukrainians and Russians)*

Unit Resource Book: Europe

- Skillbuilder Practice: Creating a Multimedia Presentation, p. 232

CONNECT to Science

Point out to students that the damage from Chernobyl was widespread. Radioactivity from the explosion drifted across Ukraine, Russia, and Belarus, and as far west as France. More than 100,000 people were evacuated around Chernobyl. Over time, the radiation in many areas around Chernobyl has declined. But the area closest to the plant, called the Chernobyl Exclusion Zone, will be contaminated for decades to come.

Unit Resource Book: Europe

- Interdisciplinary Projects: Finding Distances, pp. 239, 241
- Connect to Science: Make a Line Graph, p. 243

More About . . .

The Soviet Union

The Union of Soviet Socialist Republics, also known as the Soviet Union, was formed in 1922, when Russia, Ukraine, Belarus, and Transcaucasia—which included Armenia, Azerbaijan, and Georgia—joined under a single Communist government. By 1940, 15 republics made up the Soviet Union. At its height, the Soviet Union covered 8,650,000 square miles, nearly one-sixth of Earth's surface.

Ukraine

KEY QUESTION How did Soviet control affect the economy and culture of Ukraine?

For Ukraine, once a part of the Soviet Union, independence has not been easy. Nearby Russia still plays a part in its political and economic life. Ukraine faces many challenges as it seeks ties with the West while maintaining good relations with Russia.

History In the 800s, the town of Kiev in Ukraine was the center of the first eastern Slavic state, Kievan Rus. After 400 years as a powerful state, invading nomads conquered the territory. Ukraine came under foreign rule for centuries. By the late 1700s, it was part of the Russian Empire. In 1922, Ukraine became one of the first Soviet republics.

Soviet policies caused great hardships for the nation. In the 1930s, the Soviets took land from Ukrainian farmers to form huge, government-run farms called collectives. They required almost all crops to be sent to cities to feed the workers there. Millions of farm families starved. On April 26, 1986, the world's worst nuclear power accident occurred in Ukraine at the Chernobyl power plant. Radiation from Chernobyl has affected an estimated 7 million people in and around Ukraine.

Ukrainians had been protesting Soviet rule since the 1960s. When the Soviet Union collapsed in 1991, Ukraine finally gained its independence and began to rebuild.

CONNECT to Science

The Chernobyl explosion put large amounts of radiation into the atmosphere. The chart below shows the radiation released by a form of radioactive iodine during the week after the accident. The petabecquerel, or PBq, is a measure of radioactivity. For reference, a household smoke detector, which is very safe, emits 0.00000000003 PBq of radiation.

Date of Release	Amount (in PBq)
April 26	704
April 27	204
April 28	150
April 29	102
April 30	69
May 1	62
May 2	102

Source: Nuclear Energy Agency, "Chernobyl: Assessment of Radiological and Health Impacts"

◄ *radiation inspectors*

Activity

Make a Line Graph

Materials
- Graph paper
- Markers

1. Draw two lines for your graph. Label the vertical axis "Level of Radiation" and the horizontal axis "Date of Radiation Release."
2. Divide the horizontal axis into ten segments and label each. Divide the vertical axis into units of 100. Put a dot on the graph that represents the radiation measured for each date. Connect the dots. Title your line graph.

DIFFERENTIATING INSTRUCTION

Struggling Readers

Sequencing Ukraine's History

To provide support for students who have difficulty with sequencing, work with them to create a time line of the history of Ukraine. Help them to select important events from the text and to link those events to dates on a time line. You might also encourage students to create drawings or symbols to illustrate particular time line events.

English Learners

Comprehension: Task Support

Encourage students to reread this page and make a list of any words or phrases they do not understand. Then ask volunteers to share their words and phrases as you compile a group list on the board. Once the list is compiled, read through it together, and discuss each word or phrase until students clearly understand its meaning.

Culture About three-fourths of Ukraine's population is ethnic Ukrainian. Russians are the largest ethnic minority, at about one-fifth of the population. Under Soviet rule, the government imposed a **Russification** policy on the country. The Russian language replaced Ukrainian in schools, newspapers, and the government. After decades of speaking Russian, many Ukrainians knew it better than their native language. In 1991, Ukrainian became the official language, but Russian is still widely spoken.

About two-thirds of Ukrainians live in Kiev and other cities. Western Ukraine is heavily rural. Most people in rural areas live in large villages. They are employed in farming or in making small handicrafts, such as the traditional decorated eggs known as *pysanky*.

CONNECT History & Culture

Pysanky

Beautifully decorated Easter eggs known as *pysanky* are among the most famous Ukrainian crafts. The word *pysanky* means "written egg." Artists create intricate and colorful designs by drawing patterns on the eggs with wax and then dipping them in different-colored dyes. The wax prevents parts of the egg from picking up color. Many of the images and colors on the egg have special meaning. Designs are often passed down within a family for generations. The eggs are given as gifts during the Easter holiday.

Government and Economics In 1990, Ukraine held its first free, multiparty election. Today, Ukraine is a democracy with a president, prime minister, and legislature. Ukrainians are committed to free elections. In 2004, tens of thousands of them took part in the Orange Revolution, a peaceful protest against fraud in the presidential election.

Ukrainians remain divided on plans for reform. Some favor closer ties to Russia. Others hope for EU membership and trade with Western Europe. Ukraine maintains good relations with Russia, whom it depends on for oil and natural gas. Ukraine belongs to the Commonwealth of Independent States (CIS), a group of former Soviet republics that cooperate on political and economic matters.

Ukraine is rich in natural resources and fertile soil. Most heavy industries are in eastern Ukraine near the nation's coal deposits. Ukraine's rich soil has made it the **bread basket**, or major grain-producing region, of Europe. However, Ukraine has lagged behind many former Communist countries in converting to private ownership. Many farms and factories also lack modern equipment.

posed a
ssification
icy requiring
at Russian
place the
al language
official busi-
ss and took
d from farm-
to create
te-owned
ms ▶

ANALYZE EFFECTS
Identify how Soviet control affected Ukraine's economy and culture.

Older, limited equipment reduces farm production. ▼

Ukrainian Farm
Farmers harvest wheat on a collective farm near Lvov, Ukraine. **How does equipment affect farm production?**

429

DIFFERENTIATING INSTRUCTION

Pre-AP

Debating International Cooperation

Invite students to explore the advantages and disadvantages of an Eastern European country joining the European Union or the Commonwealth of Independent States. Ask one group of students to list advantages. Ask another group to list disadvantages. Then have the students use their lists as the springboard for a debate over whether or not a nation should join such a group.

Joining with Other Countries	
Advantages	Disadvantages
economic cooperation	*loss of autonomy*

CONNECT History & Culture

Pysanky

The art of *pysanky* has spread to other regions of the world where Ukrainian emigrants now live. Although *pysanky* may be Ukraine's best-known art form, Ukraine is known for its folk arts, including woodcarving, pottery, and weaving. Ukrainian art can also be found in its churches. The Cathedral of St. Sophia in Kiev, built around the 12th century, has beautiful frescoes and mosaics that draw visitors from all over the world.

More About . . .

Ukraine's Economy

For many years, Ukraine's energy resources played an important role in supporting the economy of the Soviet Union. Large oil, natural gas, and coal deposits provided the energy that ran large Soviet factories, fueled vehicles, and heated homes. By about 1975, however, Soviet energy demands had caused a serious decline in Ukraine's oil and natural gas supplies, and production of these resources declined.

Today, coal remains a key component of Ukraine's economy. Ukrainians use it to convert the country's rich iron ore deposits into steel. Coal also fuels many of Ukraine's electrical power stations.

More About . . .

The Commonwealth of Independent States

The Commonwealth of Independent States was formed on December 8, 1991, to replace the crumbling Soviet Union. Twelve former Soviet states, including founding members Russia, Belarus, and Ukraine, comprise the organization. All the members are sovereign states who cooperate on political and economic issues, such as defense, immigration, and the environment.

Teach

The Baltic States

Reader, Recorder, Reporter

- Why might the Baltic States have been the first to declare independence from the former Soviet Union? *(Possible Answer: Because of their long history of foreign rule, they might have been more eager to finally gain freedom.)*
- Why have the people of the Baltic States had a hard time preserving their languages, customs, and traditions? *(They had a hard time because the Russians deported many Baltic people and brought in Russian workers to replace them. These Russians brought in their own traditions.)*
- **Categorize** What are the three major languages spoken in the Baltic States? *(Lithuanian, Latvian, and Estonian)*

Unit Resource Book: Europe
- Interdisciplinary Projects: The Euro, pp. 240, 241

More About . . .

The Baltic Sea

The Baltic Sea is a branch of the North Atlantic Ocean. It covers more than 160,000 square miles in area, and borders not only the Baltic States, but also Sweden, Norway, Denmark, Finland, Russia, Poland, and Germany. From the 13th to the 15th century, the Baltic Sea was the critical transportation link for the Hanseatic League, a powerful group of German merchants that traded fish, lumber, and other products throughout Europe. Today, although the Baltic Sea is no longer a major continent-wide trade link, it still serves as an important shipping route for the nations that surround it.

The Baltic States

KEY QUESTION How are the Baltic States alike and different from each other?

The Baltic States—Estonia, Latvia, and Lithuania—share a coastline on the Baltic Sea. They also share the challenge of becoming part of prosperous, modern Europe.

Striving for Independence The Baltic States have a long history of foreign rule. At different times, Danes, Swedes, Poles, and Germans have controlled them. By the late 1700s, they belonged to the vast Russian Empire. They remained a part of it until it ended in 1918. In 1940, the Soviet Union occupied the Baltic States and forced them to join it. In all three countries, the Soviets decided to **deport**, or expel, the thousands of people who resisted their rule. The Soviets brought in Russian workers to replace them. In 1990, the Baltic States became the first Soviet republics to declare independence. The Soviet Union formally recognized their claims in September 1991.

Culture Although they share a common history, each Baltic state has its own language and customs. In Lithuania, most people speak Lithuanian and are Roman Catholics. The people of Latvia speak Latvian, while Estonians speak Estonian, a language related to Finnish. Most Estonians and Latvians are Lutheran Protestants.

Throughout the Baltic States, the Soviet policy of Russification brought population changes. For instance, before Soviet rule, Latvians made up three-fourths of that country's population. Now they make up just over half. In each country, Russians are the largest minority. During the Soviet era, many people left farms and villages to work in urban factories. Today, two-thirds of the people in the Baltic States live in cities. The people of Latvia, Lithuania, and Estonia have had to work to preserve their own languages, customs, and traditions.

Song Festival These girls are dressed up to participate in the All-Estonian Song Festival, held every five years in the capital city of Tallinn.

DIFFERENTIATING INSTRUCTION: TIERED ACTIVITIES

OBJECTIVE Design a brochure to bring investments to Poland, Ukraine, or the Baltic States.

Basic

Arrange the class into one of three groups: Poland, Ukraine, or the Baltic States. Encourage the groups to scan the text for facts and images that would showcase opportunities to potential investors. Guide students in combining at least five facts and two visuals to create an investment brochure.

On Level

In addition to the Basic activity, have students include a chart or graphic organizer that organizes the facts they have chosen to include into logical categories.

Challenge

In addition to the On Level activity, have students include a map of their assigned country, highlighting the economic opportunities in different areas of the country.

Government and Economics The Baltic States are republics with an elected president and legislature. A prime minister heads the government and carries out government operations.

The Baltic States are now successful market economies. Most industries and farms are privately owned. In 2004 all became EU members. Germany has become a key trade partner, while trade with Russia is decreasing. Estonia's major industries are electronics, wood products, and textiles. Latvia's factories make vehicles, farm machinery, and fertilizers. Lithuania produces machine tools, appliances, and electric motors.

During decades of Communist rule, government leaders ignored many serious environmental problems. Old and faulty wastewater-treatment plants, leaking sewers, and fertilizer, oil, and chemical runoff from factories and farms have polluted the Baltic Sea. All Baltic countries are now involved in the cleanup.

ll share a altic Sea astline, are rmer Soviet publics, and ve become mocratic publics with ccessful arket economies. They ve different nguages, ligions, and stoms. ▶

COMPARE AND CONTRAST Explain how the Baltic States are similar to and different from each other.

Independence Rally
Many Eastern European nations held rallies to protest Communist rule, such as this one in Kaunas in southern Lithuania in 1989.

CONNECT to the Essential Question

In what ways has Eastern Europe changed since World War II?

Ask students what they have learned so far that can help them answer this question. Students might mention:

- Changes in government and political alliances resulted as Eastern Europe moved from communism to democracy after the Soviet Union fell.
- In the 1990s, Eastern European economies changed from command economies to market economies.

Section 1 Assessment

ONLINE QUIZ For test practice, go to **Interactive Review** @ClassZone.com

TERMS & NAMES

1. **Explain the importance of**
   - Baltic States
   - Solidarity
   - Russification
   - deport

USE YOUR READING NOTES

2. **Categorize** Use your completed chart to answer the following question:

   What kind of government does Poland have?

	LANGUAGE	GOVERNMENT	ECONOMY
Poland			
Ukraine			
Latvia			
Lithuania			
Estonia			

KEY IDEAS

3. What were the goals of Solidarity in Poland?
4. How did the Soviet Union carry out its Russification policy in the Ukraine?
5. What environmental issue do the Baltic States face today?

CRITICAL THINKING

6. **Form and Support Opinions** Why do many Eastern European countries want to join the European Union?
7. **Analyze Causes and Effects** Why are many people in the Baltic States concerned about each country's national identity?
8. **CONNECT to Today** Why might Polish leaders want to encourage young people to stay in Poland?
9. **ART Give a Presentation** Do research to learn more about one of the folk arts of the countries in this section. Give a short presentation on where and how this craft is made, how it is used, and any customs associated with it. Show pictures of the craft.

4 Assess & Reteach

Assess Have students complete the Section Assessment.

Unit Resource Book: Europe
- Section Quiz, p. 245

Interactive Review @ClassZone.com
- Section Quiz

Power Presentations

Test Generator

Reteach Have student groups play "fact tennis" about changes in Poland, Ukraine, and the Baltic States. To start, one group names a change in Poland since World War II. The other group must "volley" by naming another change. When one group fails to name a change, the other wins a point. When no group can return a volley, the game moves on to the next region.

Unit Resource Book: Europe
- Reteaching Activity, p. 248

SECTION 1 ASSESSMENT ANSWERS

Terms & Names

1. Baltic States: Latvia, Lithuania, and Estonia, the three small countries that border the Baltic Sea; Solidarity: Poland's first independent labor union; Russification: policy to make Soviet occupied countries more Russian by replacing local languages and cultures; deport: to expel from a country

Use Your Reading Notes

2. a republic

	LANGUAGE	GOVERNMENT	ECONOMY
Poland	Polish	republic	market
Ukraine	Ukrainian	democracy	market
Latvia	Latvian	republic	market
Lithuania	Lithuanian	republic	market
Estonia	Estonian	republic	market

Key Ideas

3. better pay, greater political freedom, and the right to form an independent labor union
4. replaced Ukrainian with Russian in schools, newspapers, and the government
5. pollution of the Baltic Sea resulting from aging wastewater-treatment plants, leaking sewers, and industrial and agricultural runoff

Critical Thinking

6. to increase trade links with Western Europe and find new markets for their exports
7. All have large Russian minorities and fear increased Russian influence on Baltic culture or decreased knowledge of local customs and languages.
8. economy benefits from keeping skilled young laborers in the country
9. Use the rubric to score presentations.

Presentation Rubric

	Style	Illustrations
4	well-organized and informative	very helpful
3	organized	adequate
2	disorganized and confusing	poor
1	no organization	unclear or missing

1 Plan & Prepare

Objectives

- Summarize the main ideas about the history, culture, government, and economy of Hungary
- Summarize the main ideas about the history, culture, government, and economy of the Czech Republic

Read for the Essential Question

Help students read for a purpose by reminding them of the Essential Question: "In what ways has Eastern Europe changed since the end of World War II?" Point out that they can use text features such as Key Questions, special features, images, and maps to form answers to the question.

Vocabulary

Best Practices Toolkit

Use the ***Best Practices Toolkit*** to model strategies for vocabulary notetaking. Vary strategies throughout the year. Choose from: *Knowledge Rating, Predicting ABCs, Definition Mapping, Word Sort, Frayer Model (Word Squares), Magnet Words, Word Wheel,* and *Student VOC.*

Vocabulary Strategies, TT9–TT16

Reading Strategy

Best Practices Toolkit

Display the Venn Diagram transparency.

- Remind students that comparing means finding similarities and that contrasting means finding differences.
- Model filling in details from the text that are unique to Hungary.
- Have students suggest entries for the Czech Republic and Both areas.
- Display the completed diagram for review.

Venn Diagram, TT22

SECTION 2 Reading for Understanding

Key Ideas

BEFORE, YOU LEARNED

Independence brought political and economic changes and new challenges for Poland, Ukraine, and the Baltic States.

NOW YOU WILL LEARN

After decades of Communist rule, Hungary and the Czech Republic made economic and political reforms.

Vocabulary

TERMS & NAMES

Magyar an ethnic Hungarian person or the Hungarian language

Czechoslovakia (CHEHK•uh•sluh•VAH•kee•uh) former country in Eastern Europe that existed from 1918 until 1993, when it split into the Czech Republic and Slovakia

Velvet Revolution the peaceful protest by the Czech people that led to the smooth end of communism in Czechoslovakia

BACKGROUND VOCABULARY

strategic relating to a plan of action designed to achieve a specific goal

standard of living an economic measure relating to the quality and amount of goods available to a group of people and how those goods are distributed across the group

eclectic made up of parts from a variety of sources

Visual Vocabulary Velvet Revolution

Reading Strategy

Re-create the Venn diagram shown at right. As you read and respond to the **KEY QUESTIONS**, use the diagram to help you compare and contrast Hungary and the Czech Republic.

Skillbuilder Handbook, page R9

COMPARE AND CONTRAST

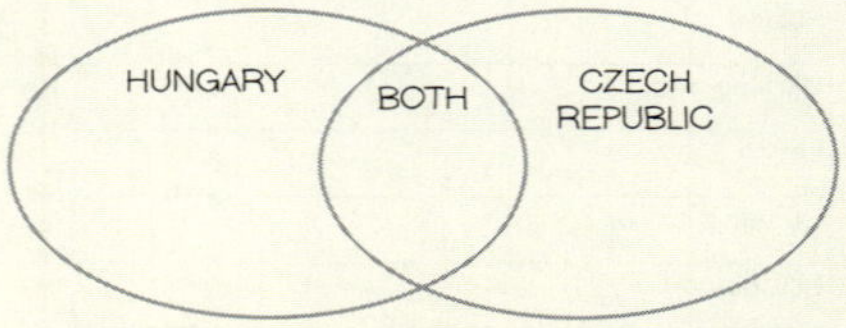

GRAPHIC ORGANIZERS
Go to Interactive Review @ClassZone.com

PRETEACHING VOCABULARY

English Learners

Pronounce and Preview

Pronounce each term for students. Review the meanings of words within definitions, such as *plan of action, specific goal,* and *economic measure.*

- To modify vocabulary learning, have students complete worksheets as they read, instead of afterward.

Unit Resource Book: Europe
- Vocabulary Practice, p. 231

Inclusion

Define Terms in Your Own Words

One at a time, read the Terms & Names aloud. After each term or name is read, have students write their own definitions for the term in their notebooks. Then invite students to read their definitions aloud. After each student has read his or her definition, the group can discuss its accuracy and how it might be improved.

Prague, Czech Republic

SECTION 2 FOCUS ON

Horse-drawn wagon in Romania

Hungary and the Czech Republic

Connecting to Your World

Have you ever played chess or checkers? In these games, the player who controls the middle of the board can have a **strategic** advantage. That player can move in many different directions. Hungary sits right between the eastern and western regions of Europe and shares borders with seven other countries. It can choose which neighbors to befriend. Unlike good chess players, however, Hungary did not always have a choice about which way it moved.

Playing Chess A boy learns to play chess from his father.

Hungary

KEY QUESTION How is Hungary's location an advantage in rebuilding its economy?

Hungary's central location attracted many groups. Hungary's earliest settlers were the **Magyar**, or ethnic Hungarian, people. In the late tenth century, Magyar leaders created a stable and prosperous kingdom. Then, in 1526, the Ottoman Turks and Austria conquered and divided Hungary's territory. Eventually, Austria drove out the Ottomans and took over. Hungary fought for independence from Austria's harsh rule.

Budapest The Széchenyi Chain Bridge was the first to cross the Danube River in Budapest.

2 Focus & Motivate

3-Minute Warm-Up

Write on the board or display the transparency:

- What forces would cause the political system of a nation to change? *(Possible Answers: war, political protest, revolution, election)*

Unit Transparency Book: Europe
- 3-Minute Warm-Ups, TT31

Connecting to Your World

More About . . .

The Danube River

The Danube flows for about 1,770 miles, from Germany through Eastern Europe to the Black Sea. It serves as an important trade route and provides running water that can be used to produce hydroelectricity.

3 Teach

Hungary

Talk About It

- What happened in 1956 in Hungary? *(People in Budapest protested the loss of freedoms; the Soviets crushed the protest, killing thousands.)*
- What religious and ethnic groups play a role in Hungary's culture? *(Religious: Roman Catholics and Protestants: Ethnic groups: Magyars, Croats, Roma, Romanians, and Slovaks)*
- **Compare and Contrast** What was the occupation of most Hungarians before World War II, and why did this change? *(Most were farmers; Communist rulers expanded industry, so many farmers moved to cities to work in factories.)*

SECTION 2 PROGRAM RESOURCES

Unit Resource Book: Europe

ON LEVEL

Unit Resource Book
- Reading Study Guide, p. 209
- Outline Map Activities, p. 235
- Section Quiz, p. 246

STRUGGLING READERS

Unit Resource Book
- Reading Study Guide with Additional Support, p. 215
- Skillbuilder Practice: Creating a Multimedia Presentation, p. 232
- Section Quiz, p. 246
- Reteaching Activity, p. 249

eEdition with Audio DVD-ROM

ENGLISH LEARNERS

Pupil Edition in Spanish

eEdition with Audio DVD-ROM

Unit Resource Book
- RSG (Spanish), p. 221
- RSG with Additional Support (Spanish), p. 227

Multi-Language Glossary

Test Generator
- Section Quiz in Spanish

INCLUSION

Unit Resource Book
- RSG with Additional Support, p. 215
- Section Quiz, p. 246
- Reteaching Activity, p. 249

GIFTED & TALENTED

Unit Resource Book
- Economics in the World, p. 242
- Section Quiz, p. 246

PRE-AP

Unit Resource Book
- Primary Source: Václav Havel, p. 244
- Section Quiz, p. 246

TECHNOLOGY

Unit Transparency Book
- 3-Minute Warm-Ups, TT31
- Essential Question, TT32
- Critical Thinking, TT34
- Map: Eastern Europe: Natural Resources, TT37
- Cultures, TT38

Daily Test Practice Transparencies
- Section 2, TT50

Power Presentations

ClassZone.com

World Cultures and Geography Video on DVD

Fun Facts!

Paprika and Vitamin C

Brought to Europe from the Americas, the chili pepper serves as the raw ingredient of paprika. To make paprika, the peppers are dried and ground into a powder. Paprika is the key ingredient in many traditional Hungarian recipes. Perhaps the best known is goulash, a spicy stew made with chunks of meat and onions.

More About . . .

The 1956 Revolution

The Revolution began on October 23, and the revolutionaries soon controlled all levels of the government. People who had been imprisoned for disagreeing with the government were freed, and farmers took back land that had been taken from them. The Soviet Union agreed to some reforms, until the new Hungarian government announced that Hungary would end its alliance with the Soviet Union. On November 4, the Soviets invaded Budapest to end the revolt. About 200,000 people fled Hungary to avoid Soviet persecution.

Make It Fun

Teacher-Tested Activities

Suzanne Scheidler Moen, DeForest Middle School, DeForest, Wisconsin

I often use the VIP (Very Important Points) strategy to help students focus their reading.

- Depending on the length of the reading, give each student the same number of sticky notes cut into thin strips. Usually three or four is an ample amount. Tell students they must use all of them.
- Have the students place the stickies next to Very Important Points as they read. These important ideas may be points of interest, things that are confusing, or a point where a connection has been made to prior learning.
- After they complete the reading, pair students to discuss the VIPs they marked.

History By 1867, Austria had been weakened by many costly wars, and it was forced to share power with Hungary. Together they ruled an empire known as Austria-Hungary. During World War I, Austria-Hungary fought on the losing side. After its defeat, its territory was divided up. Hungary was independent but lost much of its land.

During World War II, Soviet troops invaded Hungary. They were followed by Communist Party agents, who wanted to make sure the Communists gained control of the government. Many Hungarians opposed Communist rule. They resented the loss of their freedom. In 1956, thousands of protesters flooded the streets of Budapest, the nation's capital, demanding that Soviet troops leave. The Soviet Union responded by sending more troops and tanks to crush the uprising. Thousands of protesters died in the fighting.

In later decades, Communist Party officials gave Hungarians greater economic freedom, allowing some private ownership of business. In 1989, as the Soviet government began to fall apart, the Hungarian government agreed to allow other political parties. A year later, Hungarian reformers removed the Communists from power in the country's first free elections.

Fun Facts!

PAPRIKA AND VITAMIN C

Paprika is a very popular spice in Hungarian food. It comes from a pepper that flavors food and also fights the common cold. When a cold starts, many people drink orange juice for extra vitamin C. In 1932, Hungarian scientist Albert Szent-Gyorgyi found that a paprika pepper contains five to six times more vitamin C than an orange.

Culture Most Hungarians are Magyars, descendants of Hungary's early settlers. Hungarians speak Hungarian, also called Magyar, a language related to Finnish and Estonian. As you can see from the map at right, Hungarian is different from the languages spoken in surrounding countries. About two-thirds of Hungarians are Roman Catholics. One-fourth are Protestants. Other ethnic groups include Croats, Roma (also called Gypsies), Romanians, and Slovaks.

During the Communist era, many Hungarians moved from farms to cities. Everyday life changed as a result. In cities today, modern influences like blue jeans and pop music clash with traditional dress and folk songs. Budapest is a good example of Hungary's mix of old and new. It is actually two cities divided by the Danube River. Medieval Buda sits on the western bank and Pest, which underwent a surge of development in the 18th century, sits on the eastern bank.

Traditional and Modern (left) Teenagers perform at a Hungarian folk dance. (right) The crowd enjoys the show at the Sziget Music Festival in Budapest.

DIFFERENTIATING INSTRUCTION

Struggling Readers

Learning from Photographs

Invite students to carefully examine the photographs at the bottom of this page. Then ask the following questions:

- In which photograph does the young people's clothing look more modern? *(the one on the right)*
- In which photograph does the dress look more traditional? *(the one on the left)*

Point out that both photographs were taken in recent years, reflecting that Hungarian culture is a mix of old and new.

Gifted & Talented

Explore the Variety of Roma Music

Roma (gypsy) music is one of Eastern Europe's most popular musical exports. Ask students to gather recordings of different types of Roma music and play the recordings for the class. Before each selection, presenters should offer information about the piece, such as whether it is traditional or contemporary.

Languages of Eastern Europe

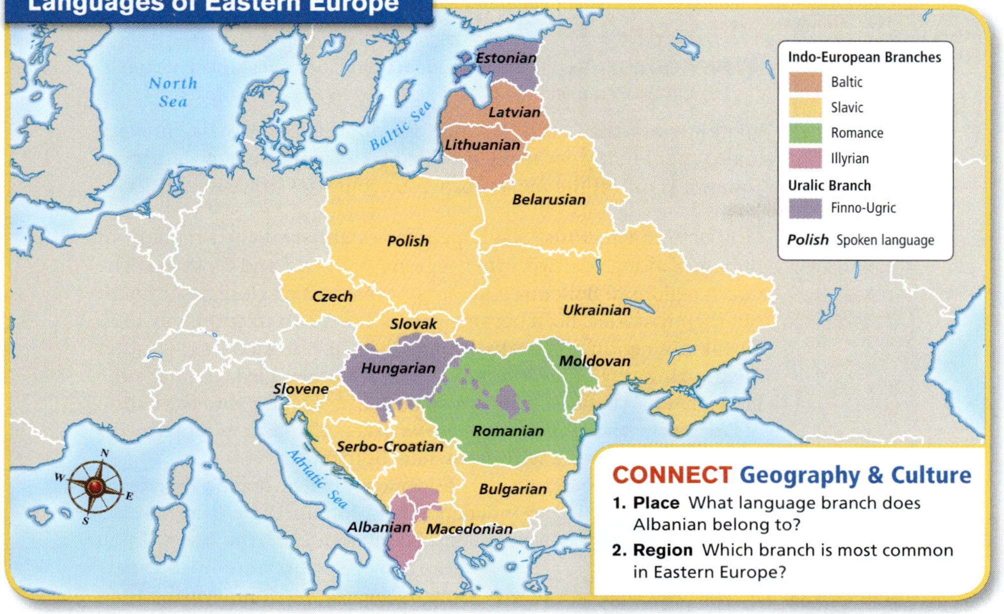

CONNECT Geography & Culture

1. **Place** What language branch does Albanian belong to?
2. **Region** Which branch is most common in Eastern Europe?

Government and Economics Hungary is a parliamentary democracy. Voters elect the members of parliament, called the National Assembly, to serve four-year terms. The National Assembly elects the president and the prime minister. The president—who serves a five-year term—is the commander in chief of the armed forces, but has little power. The prime minister is the head of government.

Before World War II, most Hungarians were farmers living in rural villages. Under Communist rule, officials worked hard to expand industry. As industry increased, many Hungarians moved to cities to work in government-run factories. By the 1990s, employment in agriculture decreased from over half to one-eighth of the population.

In the decades after the failed Hungarian revolt in 1956, the government began to allow some private businesses to operate. After the Communists lost the 1990 elections, Hungary's new leaders lifted many government controls on business and sold most remaining state-owned businesses and farms. The country moved rapidly toward a market economy. Today, it has one of the highest **standards of living** in Eastern Europe. Hungary joined the EU in 2004. The nation's central location and the Danube River help boost trade. EU members Germany, Austria, France, and Italy are its main trade partners.

FIND MAIN IDEAS Explain how Hungary's location is helpful in rebuilding its economy. central location and Danube River boost trade

Languages of Eastern Europe

CONNECT Geography & Culture

Invite students to look at the map and consider how language might effect political alliances and trade. Point out by way of example that the United States has a closer relationship with Britain than with France. One reason is that Great Britain and the United States share a common language.

- Do you think Romanians would find it easier to communicate with Belarussians or Moldovans? Why? *(Moldovans, because both speak Romance languages)*
- Who might Latvia be more likely to develop a closer relationship with—Lithuania or Estonia? Why? *(Lithuania, because their languages are from the same branch)*

ANSWERS

1. Place Illyrian

2. Region Slavic

Unit Resource Book: Europe

- Outline Map Activities: Eastern Europe, pp. 235–236
- World Literature: The Great Bear: The Story of Callisto and Arcus, pp. 263–264
- World Literature: from Piers Plowman, pp. 265–266

DIFFERENTIATING INSTRUCTION

English Learners

Comprehension: Connection

Help students navigate the text by using chronological cues. Guide students to locate such cues as "Before World War II," "Under Communist rule," "By the 1990s," "In the decades after the failed Hungarian revolt of 1956," and "After the Communists lost the 1990 elections." Make a chart on the board using these headers. Then under each heading, write down facts about the Hungarian economy during that time.

Pre-AP

Analyzing River-Based Trade

To help your class explore the role the Danube plays in Hungary's economic growth, invite students to give a presentation on the economic growth of an Eastern European river city in another country, such as Belgrade, Bratislava, or Warsaw. Encourage students to use maps to show trade routes and explain how river trade influenced the city.

Teach

The Czech Republic

Reader, Recorder, Reporter

- How was the nation of Czechoslovakia created? *(Bohemia, Moravia, and Slovakia were united by European leaders after World War I.)*
- **Compare and Contrast** What did Czechoslovakia's revolt of 1989 and its separation into Slovakia and the Czech Republic three years later have in common? *(Both happened smoothly and peacefully, so the term "velvet" is used to describe them.)*

Unit Resource Book: Europe

- Economics in the World: Tourism in Eastern Europe, p. 242

Comparing Regions

Slovaks and Czechs

Slovakia has an area about two-thirds the size of the Czech Republic and about half as many people. Due to their history, both nations have much in common. Roman Catholics are the largest religious group in both countries, though they account for about three-fifths of the Slovak population compared to less than two-fifths of Czechs. The languages in both countries, Slovak and Czech, are closely related, and speakers of one can understand the other.

RESEARCH & WRITING CENTER

HISTORY MAKERS

Václav Havel

Václav Havel's plays often used humor to point out problems with the Communist government. In his most famous play, the government creates an artificial language designed to improve communication. Instead of accomplishing its goal, the language makes communication impossible.

Unit Resource Book: Europe

- Primary and Secondary Sources: from a speech by Václav Havel, p. 244

The Czech Republic

KEY QUESTION How did freedom from Communist rule change the Czech Republic?

Like Hungary, the Czech Republic was once part of Austria-Hungary. It also spent many decades controlled by a Communist government. Today, it is one of the most prosperous countries in Eastern Europe.

History In the 900s, Slavic peoples established the kingdom of Bohemia alongside two other regions, Moravia and Slovakia. The Czech regions of Bohemia and Moravia maintained close ties throughout their histories. Both became part of the Austrian empire in 1526. Slovakia, a region of Slovaks, became part of Hungary. In 1918, after Austria-Hungary's defeat in World War I, Europe's leaders made the lands that had been Bohemia, Moravia, and Slovakia into the nation of **Czechoslovakia** (CHEHK•uh•sluh•VAH•kee•uh). Czechs and Slovaks were thrown together within the new borders.

Czechoslovakia lasted only a few decades as an independent nation. In 1948, Communists took over the country. In 1968, the Soviet Union crushed efforts by Czech Communist leaders to give citizens greater freedom. Fearing their own people might demand more freedom, the Soviets sent troops and tanks to end protests.

In 1989, thousands of protesters filled the streets of Prague calling for an end to Communist rule. Massive protests convinced Communist leaders to resign. The end of Communist rule took place so smoothly, that this protest became known as the **Velvet Revolution**. In 1990, voters elected protest leader Václav Havel president. Three years later, Czech and Slovak leaders agreed to divide the country into the Czech Republic and Slovakia. Some call this friendly split the "Velvet Divorce."

Culture Most people in the Czech Republic are Czechs. They speak Czech, a Slavic language. The country has many smaller ethnic groups including Slovaks, Germans, Roma (Gypsies), Hungarians, and Poles. Before World War II, Czechoslovakia had a large Jewish population. Almost all the nation's Jews were killed during the Holocaust. Roman Catholics are the largest religious group in the Czech Republic.

HISTORY MAKERS

Václav Havel (born 1936)

Václav Havel's (VAHT-slahv HAH-vuhl) admirers call him a hero for his role in leading the Velvet Revolution. During Czechoslovakia's years of Communist rule, the country's leaders called him a danger to society. A well-known playwright, he spent several years in prison for openly opposing Communist rule. After the end of communism, grateful citizens chose Havel as Czechoslovakia's first president. He later served as the first president of the new Czech Republic.

ONLINE BIOGRAPHY
For more on the life of Václav Havel, go to the **Research & Writing Center** @ClassZone.com

INTERDISCIPLINARY ACTIVITIES

CONNECT to Language Arts

Write a Satirical Skit About Government

Václav Havel wrote plays that used satire to criticize the government. Invite students to choose an aspect of the local, state, or federal government they disagree with, then write and present a satirical skit that presents their ideas. Point out that satire often uses humor, and that to be effective, satire clearly presents a situation and shows why it is "wrong." After the skits have been presented, discuss satire's role in shaping political opinions.

CONNECT to Civics

Create a Time Line of the Velvet Revolution

Point out how the Velvet Revolution, which took place in just six weeks, is a dramatic example of ordinary citizens changing the course of their government for the better. Have students do additional research and create a time line of the main events of the Velvet Revolution, starting with the first demonstrations on November 17, 1989, and ending with Vaclav Hável's election as president on December 29, 1989.

The Czech Republic has a tradition of fine art. Composer Antonin Dvorák (DVAWR•zhahk) wrote music inspired by the Czech landscape and folk music. Famous Czech writers include Franz Kafka and Milan Kundera, whose books explore the human condition. For centuries, Czech artists have crafted beautiful glassware. Glittering Czech crystal chandeliers hang in opera houses, palaces, and mansions around the world today.

Prague Residents and tourists alike use the Charles Bridge to cross the Vltava River and get around Prague.

The culture of the Czech Republic is a vibrant mix of old and new. Prague is the nation's capital and largest city. In the 1300s, it became a center of art and learning in Europe. The city has many beautiful Renaissance buildings. It is known as the "City of a Hundred Spires" because of the churches that dot its streets. Contemporary influences are making their marks on Prague. Young people dance to modern pop music in clubs that have opened in basements in the historic sections of town. Each year, millions of tourists add to the **eclectic** atmosphere of the Czech Republic.

CONNECT Geography & Culture

The Dancing House Prague is known for its architecture. A popular stop on any tour of Prague is the Rasin Building, nicknamed the Dancing House. Although it is made of concrete, steel, and glass, it seems to swirl and twist, like a couple dancing.

Not everyone is a fan of the building, built by American architect Frank Gehry and Czech architect Vlado Milunic in 1996. Some feel it clashes with the older historic buildings that surround it. However, supporters find it lively and exciting, a symbol of the new Czech Republic.

CRITICAL THINKING

Make Inferences Why might some Prague residents want the city to build more new buildings like the Dancing House?

More About . . .

Antonin Dvorák

The son of an innkeeper, Dvorák played the fiddle as a youth before going on to study music at the Prague Organ School. There, Dvorák began to write his own music, but rarely had his works performed. After years of struggle, his music became known around the world.

CONNECT Geography & Culture

The Dancing House

Many of Prague's most notable buildings are hundreds of years old. One such building is the Prague Castle. Its original parts were built in the 800s, but by the 1300s it needed major repairs. Remodeled a number of times over the centuries, it still stands today. If one looks closely, film fans might notice the castle in the film Mission Impossible. Ever since it was built, the castle has served as the seat of Prague's government.

CRITICAL THINKING ANSWERS

Make Inferences because it is new and different, a symbol of the modern city

Unit Resource Book: Europe

- Geography Project: The Compass in Europe, pp. 267–268

DIFFERENTIATING INSTRUCTION

English Learners

Key Academic Vocabulary

Point out to students the vocabulary word *eclectic*, and explain that an "eclectic atmosphere" is one in which a number of different styles have been picked out or put together. Have students think of other examples of places they could accurately describe as "eclectic." Then have students look up key academic vocabulary words such as *coherent* and compare their definitions to that of eclectic.

Pre-AP

Critique the Dancing House

Invite students to assume the role of architecture critics for a newspaper and write an architectural review on the Dancing House. In their reviews, students should comment on the materials chosen, the shape and look of the building, and whether the building's style is an appropriate complement to the traditional architectural styles of Prague. Have students present or post their reviews.

More About . . .

Brno

Evidence suggests that people have lived in the area of Brno since prehistoric times. The city of Brno was incorporated in 1243. Perhaps Brno's best known resident was the priest and scientist Gregor Mendel. His work on the theory of heredity began while he was living in a monastery in Brno in 1865.

Comparing Regions

Sports in the Czech Republic and the United States

In the Czech Republic, just as in the United States, football is the most popular spectator sport. But in Czech Republic, when people say "football," they mean the game Americans call "soccer." In the United States, basketball and baseball are other popular sports. This isn't the case in the Czech Republic, where hockey runs a close second to football. In fact, a number of Czech hockey players have gone on to enjoy professional careers in the United States' National Hockey League.

Presentation Options

Go the *Activity Center* at **ClassZone.com** to project Worldster.

- Click on the multimedia features of Radek's Web page.
- Ask students which areas of Radek's life they would like to explore. Listen to music, examine photographs, learn simple phrases in Czech, and more.

CONNECT to Your Life

Journal Entry

Before they write their entries, encourage students to look at the photos and think about the ways in which the special events Radek celebrates are similar to and different from theirs.

Worldster

A Day in Radek's Life

To learn more about Radek and his world, go to the **Activity Center** @ClassZone.com

To introduce himself, Radek says:

"Dobrý den. Jmenuju se Radek."

Hello! My name is Radek. I am 14 years old and live in Brno in the southeastern part of the Czech Republic. With almost 400,000 people, it is the second largest city after Prague. We live in a small apartment. On weekends, I like to read, listen to music, or watch television. Sometimes my family takes the train to Prague. Here's my day.

7 A.M. My sister and I eat breakfast with our parents. They both have to go to work and we go to school. I have rye bread or a roll with butter and jam and a glass of milk. Sometimes we have slices of cheese or salami, too.

8 A.M.–4 P.M. I have to work hard at school, especially in my English class. Everyone has to take courses in at least one foreign language. My school has many after-school activities. I joined the gymnastics club. I'm looking forward to our first competition!

6 P.M. In the evening, my family eats dinner. Sometimes, my grandmother comes over and helps my mother make roast duck with *knedlíky*, or potato dumplings, and cabbage. It's one of my favorite meals.

7 P.M. Today is my name day. In the Czech Republic, each day of the year has a different name assigned to it. It's traditional to celebrate your birthday and your name day. Some of my friends and I are going out for ice cream sundaes to celebrate.

CONNECT to Your Life

Journal Entry Think about your own birthday. How do you celebrate it? Do you have other special occasions, like Radek, that you celebrate with your family? Record your ideas in your journal.

DIFFERENTIATING INSTRUCTION: TIERED ACTIVITIES

OBJECTIVE Write a plan for filming a documentary about Hungary or the Czech Republic.

Basic

Invite students to list five scenes that when assembled into a documentary would tell the story of Hungary or the Czech Republic. Suggest students include scenes about culture, life under Communist rule, and contemporary life. Have students present their scene lists and explain their choices.

On Level

Invite students to list ten scenes that when assembled into a documentary would tell the story of Hungary or the Czech Republic. Suggest students include scenes about culture, life under Communist rule, and contemporary life. Have students present their scene lists and explain their choices.

Challenge

In addition to listing ten scenes from Hungary or the Czech Republic, have students include three people they would interview and a set of questions they would ask them. Have students present their scene lists and interview subjects, and explain their choices.

Government and Economics The Czech Republic is a parliamentary democracy. Voters elect members of parliament. The parliament elects the president, who serves as the head of state. The president appoints a prime minister to oversee the daily operations of government.

Before Communist rule, the Czech economy was diverse and balanced. Under communism, however, the government shifted the economy toward heavy industry. After Communist control ended, the country's new leaders had to regain the balance that it once had. This has led to some unemployment as factory jobs decline.

On the whole, however, the Czech Republic's transition from a command to a market economy has been a smooth one. Most farms and factories have been sold to private owners. The Czech Republic joined the EU in 2004. Foreign investment and exports of Czech products, especially to Germany, have helped the economy continue growing. It has been among the most successful economies of the former Communist nations in Eastern Europe.

eater political d economic edom, oices in litical parties, r elections, sing of vernment strictions on sinesses, vate citizens owed to own sinesses ▶

Škoda Auto This Czech company has been manufacturing cars for over 100 years. When Communist rule ended, the industry privatized and Volkswagen took over the company.

SUMMARIZE Explain how the Czech Republic changed after gaining freedom from Communist rule.

Section 2 Assessment

ONLINE QUIZ For test practice, go to **Interactive Review** @ClassZone.com

TERMS & NAMES

1. **Explain the importance of**
   - Magyar
   - Czechoslovakia
   - Velvet Revolution

USE YOUR READING NOTES

2. **Compare and Contrast** Use your completed chart to answer the following question:
   How did Hungary's fight for independence differ from the Czech Republic's and how were they similar?

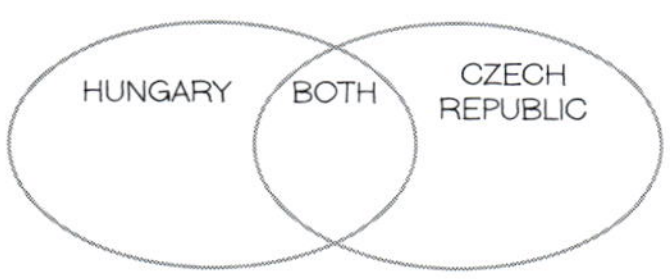

KEY IDEAS

3. What ethnic group do many Hungarians belong to?
4. How was Czechoslovakia formed?
5. How did the Velvet Revolution change Czechoslovakia?

CRITICAL THINKING

6. **Analyze Causes and Effects** How do you think the failure of the 1956 Hungarian revolt affected other Eastern European nations?
7. **Find Main Ideas** How did Václav Havel influence Czechoslovakia?
8. **CONNECT to Today** Why do you think the Czech Republic recently built new expressways and rail links to Germany?
9. **WRITING** **Express an Opinion** The young people who led the Hungarian revolt in 1956 were called freedom fighters by some and lawbreakers and traitors by others. Write a paragraph telling how you think they should be remembered.

CONNECT to the Essential Question

In what ways has Eastern Europe changed since World War II?

Ask students what they have learned so far that can help them answer this question. Students might mention:

- Hungary suffered under Communist rule, but now has a thriving economy and is part of the EU.
- After the Velvet Revolution, Czechoslovakia divided into Slovakia and the Czech Republic, now part of the EU.

4 Assess & Reteach

Assess Have students complete the Section Assessment.

Unit Resource Book: Europe
- Section Quiz, p. 246

Interactive Review @ClassZone.com
- Section Quiz

Power Presentations

Test Generator

Reteach Have students create almanac entries for Hungary and the Czech Republic. In their entries, they should include a map, key historical events and people, and facts about government, economy, and culture.

Unit Resource Book: Europe
- Reteaching Activity, p. 249

SECTION 2 ASSESSMENT ANSWERS

Terms & Names

1. Magyar: ethnic Hungarian; Czechoslovakia: Eastern European country from 1918 to 1993; Velvet Revolution: peaceful protest that ended communism in Czechoslovakia

Use Your Reading Notes

2. Both had revolutions put down by the Soviets; Czech transition from communism was smoother.
   Diagram *Hungary—speak Hungarian, 1956 uprising, rougher transition from communism; Czech Republic—speak Czech, 1968 revolution, smoother transition from communism; Both—parliamentary democracy, mix of old and new cultures, Roman Catholic majority, successful market economies.*

Key Ideas

3. Magyar
4. After World War I, European leaders joined Bohemia, Moravia, and Slovakia into one country, Czechoslovakia.
5. Protests forced Communist leaders to resign, ending Communist rule in Czechoslovakia.

Critical Thinking

6. Possible answers: stifled dissent; made people more cautious, afraid of protesting communism
7. He spoke out against communism and became the first president of Czechoslovakia after communism.
8. Germany is a major trading partner of the Czech Republic. Improved transportation makes both trade and tourism easier and probably less costly.
9. Use the rubric to score paragraphs.

Paragraph Rubric

	Understanding of Concepts	Accuracy
4	clear, complete	no errors
3	clear	few/minor
2	some confusion	several
1	unclear	many

1 Plan & Prepare

Objectives

- Describe how their history as part of Yugoslavia shaped the Balkans
- Explain the challenges the former Yugoslav republics face in forming stable governments and improving their economies

Read for the Essential Question

Help students read for a purpose by reminding them of the Essential Question: "In what ways has Eastern Europe changed since the end of World War II?" Point out that they can use text features such as Key Questions, special features, images, and maps to form answers to the question.

Vocabulary

Best Practices Toolkit

Use the ***Best Practices Toolkit*** to model strategies for vocabulary notetaking. Vary strategies throughout the year. Choose from: *Knowledge Rating, Predicting ABCs, Definition Mapping, Word Sort, Frayer Model (Word Squares), Magnet Words, Word Wheel,* and *Student VOC.*

Vocabulary Strategies, TT9–TT16

Reading Strategy

Best Practices Toolkit

Display the Web Diagram transparency.

- Remind students that main ideas are the most important points in a text.
- Model finding and entering main ideas, using the "Creation of Yugoslavia" subsection.
- Have students suggest entries.
- Display the completed diagram for review.

Web Diagram, TT28

Section 3 Reading for Understanding

Key Ideas

BEFORE, YOU LEARNED

The Czech Republic and Slovakia became independent countries without conflict or bloodshed.

NOW YOU WILL LEARN

In the 1990s the breakup of Yugoslavia was violent. Ethnic divisions and economic and political issues led to war.

Vocabulary

TERMS & NAMES

Yugoslavia a country on the Balkan Peninsula from 1918 to 1991

Josip Broz Tito the Communist leader of Yugoslavia from 1953 to 1980

ethnic cleansing removing an ethnic or religious group from an area by force or the mass killing of members of such a group

Kosovo a self-governing province within Serbia

Slobodan Milosevic president of Serbia from 1989 to 1997 and of Yugoslavia from 1997 to 2000; a key figure in the ethnic conflicts in the Balkans in the 1990s

BACKGROUND VOCABULARY

Bosniac an ethnic Muslim from Bosnia and Herzegovina

refugee a person who flees a place to find safety

Visual Vocabulary Josip Broz Tito

Reading Strategy

Re-create the web diagram shown at right. As you read and respond to the **KEY QUESTIONS**, use the diagram to help you find main ideas about the Balkans.

Skillbuilder Handbook, page R4

FIND MAIN IDEAS

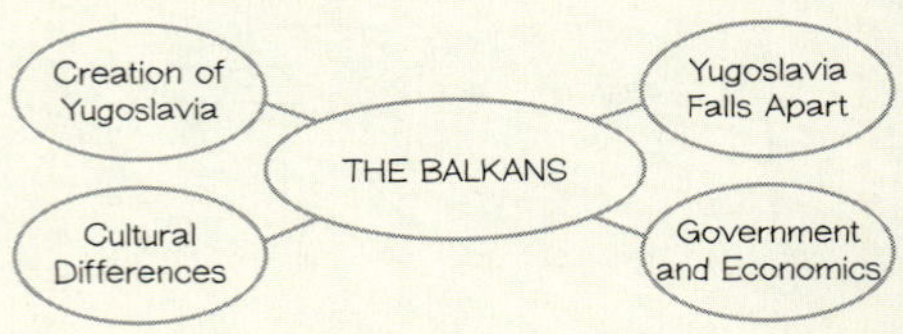

GRAPHIC ORGANIZERS
Go to Interactive Review @ ClassZone.com

PRETEACHING VOCABULARY

English Learners

Pronounce and Preview

Pronounce each term for students. Review the meanings of words within definitions, such as *self-governing, province,* and *flees.*

- To modify vocabulary learning, have students complete worksheets as they read, instead of afterward.

Unit Resource Book: Europe
- Vocabulary Practice, p. 231

Inclusion

Vocabulary: Student VOC Strategy

Have students fill out the graphic organizer for the terms and names in the section.

- Use the transparency to model the strategy.
- Have volunteers suggest entries.
- Then distribute copies of the organizer for students to work on with a partner.

Best Practices Toolkit
- Vocabulary: Student VOC Strategy, TT16

Prague, Czech Republic

SECTION 3 FOCUS ON

Horse-drawn wagon in Romania

The Balkans

Connecting to Your World

At the opening ceremonies of the Olympic Games, athletes from around the globe proudly display their national flags as they march around a packed stadium. The games demonstrate that people of different religions, languages, and beliefs can come together without conflict. In 1984, the Winter Olympic Games took place in Sarajevo, a city in what was then Yugoslavia on the Balkan Peninsula. In the 1990s, wars caused by ethnic conflict destroyed much of the city, including the Olympic facilities. Sarajevo has been rebuilt, but memories of the violence that destroyed it remain.

Sarajevo The flame is lit (above) at the 1984 Winter Olympics opening ceremonies. This Muslim cemetery (below) is in Sarajevo.

History and Culture

KEY QUESTION What caused the breakup of Yugoslavia?

In the 500s, various groups of Slavic peoples settled the Balkan Peninsula. Centuries of foreign rule created differences in the religion, language, and customs of these Slavic groups. When these groups joined together in one nation in the 20th century, their ethnic differences became a time bomb waiting to explode.

2 Focus & Motivate

3-Minute Warm-Up

Write on the board or display the transparency:

- What different methods can governments use to resolve disputes over territory? *(Possible Answers: negotiation and compromise, military threats, monetary payment for territory, war)*

Unit Transparency Book: Europe
- 3-Minute Warm-Ups, TT31

Connecting to Your World

More About . . .

The Olympic Games

Although the Olympic Games provide a peaceful way for nations to interact, they have been influenced by politics. In 1936, Adolph Hitler used the Berlin summer games to showcase the Nazi racist ideology. After the Soviet Union invaded Afghanistan, about 60 nations boycotted the 1980 Moscow summer games.

3 Teach

History and Culture

Talk About It

- Why might making a nation out of distinct groups be difficult? *(Possible Answer: because each group has its own traditions, and some of those traditions might clash)*
- **Find Main Ideas** What ethnic groups were involved in the violence in Kosovo in the 1990s? *(Serbs and Albanians)*

SECTION 3 PROGRAM RESOURCES

Unit Resource Book: Europe

ON LEVEL

Unit Resource Book
- Reading Study Guide, p. 211
- Map and Graph Skills, p. 233
- Section Quiz, p. 247

STRUGGLING READERS

Unit Resource Book
- Reading Study Guide with Additional Support, p. 217
- Skillbuilder Practice: Creating a Multimedia Presentation, p. 232
- Section Quiz, p. 247
- Reteaching Activity, p. 250

eEdition with Audio DVD-ROM

ENGLISH LEARNERS

Pupil Edition in Spanish
eEdition with Audio DVD-ROM
Unit Resource Book
- RSG (Spanish), p. 223
- RSG with Additional Support (Spanish), p. 229

Multi-Language Glossary
Test Generator
- Section Quiz in Spanish

INCLUSION

Unit Resource Book
- RSG with Additional Support, p. 217
- Section Quiz, p. 247
- Reteaching Activity, p. 250

GIFTED & TALENTED

Unit Resource Book
- Section Quiz, p. 247

PRE-AP

Unit Resource Book
- Section Quiz, p. 247

TECHNOLOGY

Unit Transparency Book
- 3-Minute Warm-Ups, TT31
- Essential Question, TT32
- Critical Thinking, TT35

Daily Test Practice Transparencies
- Section 3, TT51

Power Presentations
ClassZone.com
World Cultures and Geography Video on DVD

More About . . .

Josip Broz Tito

Tito controlled Yugoslavia by intimidating or imprisoning those who opposed his policies and by election fraud. Unlike most other Eastern European leaders, Tito clashed frequently with Joseph Stalin, the powerful leader of the Soviet Union. Stalin unsuccessfully attempted to reduce Tito's power. In response, Tito initiated closer relationships with Western Europe.

Comparing Regions

Ethnicity in the Balkans and the United States

In the United States, ethnic groups often mix within neighborhoods, communities, and states. Yet distinctly Asian, Irish, African American, Jewish, Latino and other ethnic neighborhoods also exist. Disputes between ethnic groups occur at times. In the Balkans, the divisions between ethnic groups are often more rigid. Centuries-long disputes between groups also fueled the violent upheavals of the 1990s.

Unit Resource Book: Europe
- Skillbuilder Practice: Creating a Multimedia Presentation, p. 232

More About . . .

Kosovo

During the Middle Ages, Serbs controlled the province of Kosovo. But after their defeat by the Turks in 1389, the ethnic makeup of the region changed. Eventually, more ethnic Albanians moved into Kosovo. By 1991, about four out of five people in Kosovo were Albanians. Only about one in ten were Serbs. Under Milosevic, Serbs once again attempted to take control of Kosovo. As of 2006, United Nations peacekeepers maintain a presence in Kosovo, but ethnic conflict is still part of life there.

Creation of Yugoslavia After World War I, Slavic groups formed a country on the Balkan Peninsula called the Kingdom of Croats, Serbs, and Slovenes. It was renamed **Yugoslavia** in 1929. When World War II ended, **Josip Broz Tito** took over Yugoslavia. It became a Communist state consisting of six republics: Bosnia-Herzegovina, Croatia, Macedonia, Montenegro, Serbia, and Slovenia. Within each republic, the majority of people belonged to the same ethnic group. Tito's tight control brought political stability and helped unite Yugoslavia.

Yugoslavia Falls Apart After Tito died in 1980, Yugoslavia faced many economic and political problems. The other republics felt Serbia was trying to take control. In 1991, Slovenia, Macedonia, Croatia, and Bosnia-Herzegovina each declared independence. War broke out in Bosnia, as Serbs clashed with **Bosniacs** and Croats. Serbian troops occupied parts of Bosnia, using force to drive all non-Serbs out. This tactic, known as **ethnic cleansing**, left thousands dead or homeless. A 1995 peace plan gave most of Bosnia to the Bosniacs and Croats and the rest to the Bosnian Serbs. In 1998, rebels in the Serbian province of **Kosovo** began fighting for independence. Violence between Serb forces and the Albanian majority in Kosovo lasted until late 1999. In 2002, Serbian leader **Slobodan Milosevic** was tried for war crimes related to ethnic cleansing in Bosnia and Kosovo. He died in prison before his trial ended. By 2003, Yugoslavia no longer existed.

Cultural Differences Five new nations resulted from the breakup of Yugoslavia: Bosnia and Herzegovina, Croatia, Macedonia, Serbia and Montenegro, and Slovenia. As the map on the next page suggests, most of the former Yugoslav republics are dominated by one ethnic group. Most Balkan peoples speak Slavic languages. The biggest difference among ethnic groups is in religion. These countries have had to find ways to accept their differences. They have also had to work to return their cities to the cultural centers they once were.

ANALYZE CAUSES Explain why Yugoslavia collapsed.

economic and political problems; conflict between Serbians and other ethnic groups in Bosnia and Herzegovina and Kosovo

Dubrovnik, Croatia
Dubrovnik is known for its terra cotta roofs, some of which were damaged in the ethnic conflict in the 1990s and needed to be repaired.

442

INTERDISCIPLINARY ACTIVITIES

CONNECT to Language Arts

Write a Poem Promoting Cooperation

Discuss how the violence in the Balkans was caused by the inability of different ethnic groups to solve problems peacefully. Invite students to write poems promoting cooperation among ethnic groups and peaceful problem-solving. Once students have finished their poems, invite them to share their work in a "Poems for Peace" poetry reading.

CONNECT to Art

Design a Peace Monument for the Balkans

Invite students to submit designs for a monument to peace that will be placed in cities across the Balkans. Student designs should both respectfully address the toll that war has taken on the Balkans and encourage people of all ethnicities to work toward a peaceful future. Have students submit their final designs on large sheets of drawing paper, then post the designs and let the class vote for a winning design.

COMPARING Ethnic Distribution in the Balkans

Similarities and Differences Yugoslavia brought many ethnic groups together in one nation. In the 1990s, differences in language, religion, and culture overwhelmed similarities, until the region erupted in conflict. Today, these groups must find ways to live together.

CRITICAL THINKING

1. **Make Inferences** Given the distribution of ethnic groups, which region do you think experienced the most conflict?
2. **Compare and Contrast** What are the main differences between Serbs, Croats, and Bosniacs?

443

DIFFERENTIATING INSTRUCTION: TIERED ACTIVITIES

OBJECTIVE Create a guide to the ethnic diversity of the Balkans.

Basic

Have student groups use the text and the map on this page to create a guidebook page for one of the former Yugoslav republics. Each page should include a map, along with text about the country's ethnic and religious groups and the languages spoken there. Have students present or post their work.

On Level

Have student groups use the text and the map on this page to create one guidebook page for each of two former Yugoslav republics. Each page should include a map, along with text about the ethnic and religious groups, languages spoken, economy, and government. Have students present or post their work.

Challenge

In addition to the On Level activity, have students write a brief compare-and-contrast essay about the two countries that their groups chose. Have students present or post their work.

COMPARING

Ethnic Distribution in the Balkans

The Balkan region is made up of six different nations and eight major ethnic groups. In 2006, it had a population of about 21 million. Yet at about 99,000 square miles in total area, it is slightly smaller than the state of Colorado, which has a population of about 4.7 million.

- How does ethnic distribution in Bosnia and Herzegovina compare to that in Serbia? *(Serbia contains a greater number of different ethnic groups, but ethnic distribution in Bosnia and Herzegovina is more fragmented.)*

CRITICAL THINKING ANSWERS

1. **Make Inferences** Bosnia and Herzegovina
2. **Compare and Contrast** They speak the same language, but use different alphabets, and all practice different religions.

Unit Resource Book: Europe
- Map and Graph Skills: Reading Thematic Maps, pp. 233–234

CONNECT to the Essential Question

In what ways has Eastern Europe changed since World War II?

Ask students what they have learned so far that can help them answer this question. Students might mention:

- After World War II, Communist leader Josip Broz Tito helped unite the different ethnic groups that made up Yugoslavia.
- During the 1990s, Communist rule collapsed and Yugoslavia broke up, leading to violence and ethnic cleansing.
- Today, the former Yugoslav republics are trying to form stable governments and rebuild their economies.

Unit Transparency Book: Europe
- Essential Question, TT32

Teach

Government and Economics

Talk About It

- How has Bosnia and Herzegovina tried to form a stable government? *(by electing the legislature and a three-member presidency to represent the country's major ethnic groups—Bosniac, Croat, and Serb)*
- **Compare and Contrast** Why is Slovenia the most prosperous Balkan nation? *(It has a stable government and had less involvement in the 1990s conflict.)*

4 Assess & Reteach

Assess Have students complete the Section Assessment.

Unit Resource Book: Europe
- Section Quiz, p. 247

Interactive Review @ClassZone.com
- Section Quiz

Power Presentations

Test Generator

Reteach Have students create one question and answer related to the Balkans about each of the following from this section: history, people/culture, government/economy. Take turns asking and answering questions, until all questions have been answered.

Unit Resource Book: Europe
- Reteaching Activity, p. 250

Ljubljana, Slovenia Visitors enjoy the sights on this street in Slovenia's capital. **Why is tourism important to the former Yugoslav republics' economies?**

It provides much-needed revenue for the recovering economies.

Government and Economics

KEY QUESTION What governmental and economic problems do many former Yugoslav republics face?

Today, the remaining ethnic tensions have made it difficult for the former Yugoslav republics to form stable governments. One concern is protecting the rights of all ethnic groups to prevent future conflicts. In Bosnia and Herzegovina, the people elect the legislature and a three-member presidency—one Bosniac, one Croat, and one Serb—to represent the country's ethnic majorities. Serbia and Montenegro are two republics that formed one country in 2003. In 2006, however, a majority of Montenegrins voted to become an independent nation.

Wartime destruction and the movement of **refugees** between nations have forced the Balkan countries to rebuild their economies. Market economies have been slow to develop. Macedonia and Bosnia and Herzegovina are the poorest. Like Serbia, they suffer from high unemployment. Slovenia is the most prosperous nation, as a result of stable government and less involvement in the 1990s conflict. It is the only EU member. Croatia was another wealthy republic. Today, it is slowly recovering from billions of dollars worth of war damage.

ANALYZE EFFECTS Explain the governmental and economic problems faced by the former Yugoslav republics.

recovery from wartime destruction, slow transition to market economies, high unemployment

ONLINE QUIZ For test practice, go to Interactive Review @ClassZone.com

Section 3 Assessment

TERMS & NAMES

1. **Explain the importance of**
   - Yugoslavia
   - Josip Broz Tito
   - ethnic cleansing
   - Kosovo

USE YOUR READING NOTES

2. **Find Main Ideas** Use your completed chart to answer the following question: How did Tito's death affect Yugoslavia?

KEY IDEAS

3. What kind of economic system did Yugoslavia have under Tito?
4. Why did much of the ethnic fighting occur in Bosnia and Kosovo?
5. What economic problem is shared by many of the former Yugoslav republics?

CRITICAL THINKING

6. **Analyze Causes and Effects** How did ethnic diversity make it harder to unite Yugoslavia?
7. **Identify Problems and Solutions** How might Balkan nations protect the rights of all ethnic groups equally in order to prevent future conflicts?
8. **CONNECT to Today** What might be the benefits of EU membership for Slovenia?
9. **GOVERNMENT Draft a Bill of Rights** Draft a Bill of Rights for the ethnic groups living in the Balkans. List ten rights these groups should have. Consider the rights of ethnic minorities in these countries.

SECTION 3 ASSESSMENT ANSWERS

Terms & Names

1. Yugoslavia: country on Balkan Peninsula from 1918 to 1991; Tito: Yugoslavia's Communist leader from 1953 to 1980; ethnic cleansing: forcing an ethnic or religious group from an area by violence; Kosovo: self-governing province in Serbia

Use Your Reading Notes

2. Tito was a uniting force, and when he died, the divisions between ethnic groups became clearer. **Diagram** *Creation of Yugoslavia—Slavic groups united six republics to form Yugoslavia after WWI; Cultural Differences—most speak Slavic languages, biggest difference is religion; Yugoslavia Falls Apart—After Tito's death, economic and political problems led to war as ethnic conflict escalated; Government and Economics—trying to form stable governments and rebuild economies.*

Key Ideas

3. command economy
4. When they declared independence, Serbia turned to violence to prevent them from gaining it.
5. high unemployment

Critical Thinking

6. Each group had a cultural identity with its own beliefs, customs, and language, making it hard for them to think of themselves as Yugoslavs.
7. Possible answers: education about importance of tolerance, rights of citizens, and dangers of prejudice; enforcement of laws that make discrimination based on ethnicity illegal
8. integration into the European economy leading to more trade and tourism
9. Use the rubric to score the bills of rights.

Bill of Rights Rubric

	Text	Accuracy
4	clear and informative	no errors
3	adequate to explain ideas	few/minor
2	confusing	several
1	unclear	many

Interactive Review

Click here to complete these and other activities online @ClassZone.com

CHAPTER SUMMARY

Key Idea 1
Poland, Ukraine, and the Baltic states faced challenges after shaking off Communist rule.

Key Idea 2
After decades of Communist rule, Hungary and the Czech Republic made economic and political reforms.

Key Idea 3
In the 1990s the breakup of Yugoslavia was violent. Ethnic divisions and economic and political issues led to war.

For **Review and Study Notes**, go to **Interactive Review** @ClassZone.com

NAME GAME

Use the Terms & Names list to complete each sentence on paper or online.

1. I am a self-governing province within Serbia and Montenegro. ___Kosovo___
2. I am the Communist leader who kept Yugoslavia from splitting apart for over 40 years. ________
3. I am a region where grain production thrives. ________
4. I am the policy practiced by the Soviet Union in an effort to make occupied countries more Russian. ________
5. I am another name for the countries of Latvia, Lithuania, and Estonia. ________
6. I am another term for ethnic Hungarians. ________
7. I mean to expel someone from a country. ________
8. I am the leader who helped start Poland's first independent trade union. ________
9. I am the Serbian leader who was charged with war crimes. ________
10. I am the Balkan country that split into six nations. ________

Baltic States
bread basket
crossroads
Czechoslovakia
deport
ethnic cleansing
Josip Broz Tito
Kosovo
Lech Walesa
Magyar
Russification
Slobodan Milosevic
Solidarity
Václav Havel
Yugoslavia

Activities

GeoGame

Use this online map to reinforce your understanding of Eastern Europe. Drag and drop each place name in the list to its location on the map.

Geo GAME

Poland
Kiev
Budapest
Czech Republic
Ukraine

To play the complete game, go to **Interactive Review** @ClassZone.com

Crossword Puzzle

Complete an online crossword puzzle to test your knowledge of Eastern Europe.

ACROSS
1. Communist leader of Yugoslavia from 1953 to 1980

445

Chapter Summary

As a follow-up to the chapter, encourage students to pair up and go over the **Review and Study Notes** at **ClassZone.com.**

Alternatively, students pairs can review their Reading Study Guide notes, making sure that they can add important details to each section's "Key Idea."

Name Game

Teaching Tips

ON PAPER Have students work in pairs to complete this activity. As an extension, they can write questions for the terms that are not covered in the items shown here.

ONLINE Make sure that students understand how to drag and drop terms to fill in blanks. Ask them to print out their results and compare them with those of a classmate.

ANSWERS

1. Kosovo	**6.** Magyar
2. Josip Broz Tito	**7.** deport
3. bread basket	**8.** Lech Walesa
4. Russification	**9.** Slobodan Milosevic
5. Baltic States	**10.** Yugoslavia

Make It Fun

Teacher-Tested Activities

Suzanne Scheidler Moen, DeForest Middle School, DeForest, Wisconsin

I find that playing "Hoops" is a great way to get students actively engaged in review.

- You'll need a soft ball, a trash can, and questions. Divide the class into two teams.
- Have a student from one team come to the front of the room and choose a two-point or three-point question. Three-point questions should be more difficult.
- A correctly answered question earns one or two points out of two or three possible. The student earns the final point by making a basket from a set distance.
- Continue, alternating teams, until each student has answered at least one question.

ONLINE AND ON-SCREEN ACTIVITIES

GeoGame

Use the interactive map on **Power Presentations** DVD-ROM or the **eEdition** to help students review important places in Eastern Europe.

- Demonstrate how to drag and drop the map labels listed onto the appropriate locations on the map.
- Advance to the scorecard to show how students have a chance to improve their scores.

Crossword Puzzle

Students can review important people and events in Eastern Europe's history by completing this crossword puzzle online at **ClassZone.com.**

- As an extension of this activity, have students use the *Activity Maker* at **ClassZone.com** to make their own crossword puzzle for one section of the chapter.
- Then have them trade puzzles with classmates to solve each other's puzzles.

Vocabulary ANSWERS

1. cofounder of Polish labor union Solidarity and Poland's president from 1990 to 1995
2. an abundant grain-producing region
3. an ethnic Hungarian person
4. the peaceful protest that ended communism in Czechoslovakia
5. using violence or murder to forcefully remove an ethnic or religious group from an area
6. a self-governing province within Serbia
7. When the Baltic States were Soviet republics, Soviet leaders instituted a policy of Russification to increase the influence of Russian culture on Baltic peoples.
8. Both were countries in Eastern Europe that were made up of independent regions and have divided into smaller nations.
9. Both were rulers of Yugoslavia.

Key Ideas ANSWERS

10. Ukraine, Estonia, Latvia, Lithuania (also, shown on map, Belarus and Moldova)
11. It increased resistance by becoming a rallying point for citizens who wanted to speak out against Communist rule.
12. Most farms and factories have shifted from state ownership to private ownership, though Ukraine has lagged behind in privatizing.
13. percentage of Russians living in these nations increased dramatically; native language and customs almost replaced by Russian language and traditions
14. The Soviets sent large numbers of troops and tanks to put down the revolt.
15. has greatly increased the number of people living in urban areas
16. Prague
17. The strong show of public support for the protesters convinced the Communist government that it could not maintain its hold on power.
18. It became a Communist country, the national identity of citizens increased, and ethnic rivalries decreased.
19. The ethnic Albanian majority in the Serbian province of Kosovo wanted to break away, and Serbia was determined to stop it from doing so.
20. Most are Slavic languages.
21. stable government and less involvement in the ethnic conflict of the 1990s

CHAPTER 14 Assessment

VOCABULARY

Explain the significance of each of the following.

1. Lech Walesa
2. bread basket
3. Magyar
4. Velvet Revolution
5. ethnic cleansing
6. Kosovo

Explain how the terms and names in each group are related.

7. Baltic States and Russification
8. Yugoslavia and Czechoslovakia
9. Josip Broz Tito and Slobodan Milosevic

KEY IDEAS

1 Poland, Ukraine, and the Baltic States

10. Which Eastern European countries were part of the Soviet Union before it collapsed?
11. What effect did Solidarity have on Polish resistance to Communist rule in the 1980s?
12. How did the ownership of farms and factories in Poland, Ukraine, and the Baltic States change after the collapse of communism?
13. How did the Soviet policy of Russification affect Ukraine and the Baltic States?

2 Hungary and the Czech Republic

14. Why did the 1956 Hungarian revolt fail?
15. How has industrial growth affected where people in Hungary and the Czech Republic live?
16. What is the Czech Republic's capital and largest city?
17. How was the Czech Republic able to move peacefully from Communist rule to democratic government?

3 The Balkans

18. How did Yugoslavia change during Tito's rule?
19. What led to the violence in Kosovo?
20. In what way are many of the languages spoken in the Balkans similar?
21. What helped Slovenia make a smooth transition to a market economy?

CRITICAL THINKING

22. **Compare and Contrast** How have Poland, Hungary, and the Czech Republic changed from command to market economies?

COMMAND ECONOMY	MARKET ECONOMY

23. **Form and Support Opinions** Why have many former Communist countries wanted closer ties with Western Europe?
24. **Compare and Contrast** How were the roles of Lech Walesa and Václav Havel in their nations' history similar?
25. **Connect Geography & Government** What effect did the collapse of Communist governments have on Eastern Europe?
26. **Connect Geography & History** How do the new governments of the former Yugoslav republics reflect their recent history?
27. **Five Themes: Place** How were the breakups of Czechoslovakia and Yugoslavia different?

Answer the ESSENTIAL QUESTION

In what ways has Eastern Europe changed since the end of World War II?

Written Response Write a two- or three-paragraph response to the Essential Question. Consider the key ideas of each section as well as specific ideas about Eastern Europe. Use the rubric below to guide your thinking.

Response Rubric

A strong response will:
- discuss changes in government and civil rights
- discuss changes in economies
- discuss the impact of Communist rule on language and national identity

ESSENTIAL QUESTION REVIEW

Before students begin their essays, have them review the key chapter elements that will help them answer the Essential Question. Direct them to significant features in the chapter such as

- Analyzing Primary Sources, Section 1
- Connect to Science, Section 1
- History Makers: Václav Havel, Section 2
- Connect Geography & Culture, Section 2
- Comparing Ethnic Distribution in the Balkans, Section 3

Written Response Rubric

	Coverage of Key Ideas	Explanation	Accuracy
4	all	clear, concise	no errors
3	most	good	few/minor errors
2	some	fair	several errors
1	two or fewer	poor	many errors

STANDARDS-BASED ASSESSMENT

Test Practice
- Online Test Practice @ClassZone.com
- Test-Taking Strategies and Practice at the front of this book

HISTORICAL MAP

Use the map to answer questions 1 and 2 on your paper.

Division of Poland *1795*

1. **What three countries divided Poland among them?**
   A. Russia, Prussia, and Austria
   B. Sweden, Prussia, and Hungary
   C. Sweden, Russia, and Prussia
   D. Russia, Hungary, and the Ottoman Empire

2. **Which country gained the majority of Polish territory?**
   A. Austria **C.** Prussia
   B. Russia **D.** Sweden

BAR GRAPH

Examine the graph below. Use the graph to answer questions 3 and 4 on your paper.

Selected Eastern European GDPs

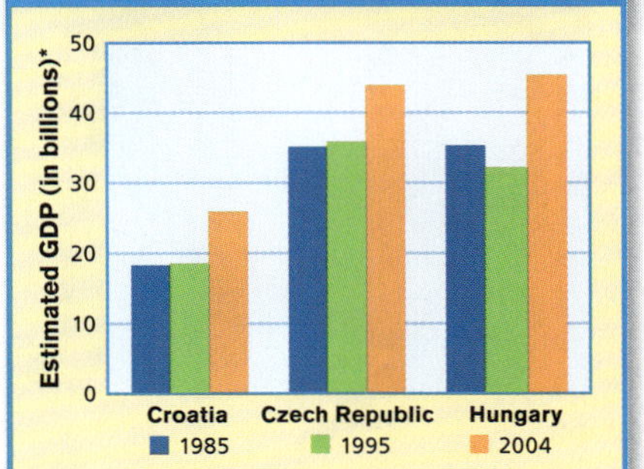

*Constant 1990 prices in U.S. dollars

Source: UN Statistics Division National Accounts Database

3. **Which country's GDP dropped between 1985 and 1995?**
4. **During which time period did all three nations see the biggest increase in GDP?**

GeoActivity

1. INTERDISCIPLINARY ACTIVITY–MATHEMATICS

With a small group, pick a country in this section. Find information on the religious makeup of the country, the percentage of people who live in urban and rural areas, and the percentage of people who can read. Use the information to create three pie graphs.

2. WRITING FOR SOCIAL STUDIES

Unit Writing Project Review your presentation about the European country you want to visit. Write a journal entry that describes how you spent your favorite day there, after your class decided to make the trip.

3. MENTAL MAPPING

Create an outline map of Eastern Europe and label the following:
- Hungary
- Poland
- Czech Republic
- Ukraine
- Latvia
- Lithuania
- Estonia
- Prague
- Chernobyl

Critical Thinking ANSWERS

22. Possible answers: Command—state-owned factories and farms; government controls all economic decisions; Market—factories and farms sold to private owners; supply and demand controls economic decisions

23. Western Europe can provide the Eastern European nations with a larger market for products as well as goods and technology needed to build their own economies.

24. Both were leaders of protest movements that toppled Communist governments.

25. They were able to become independent, create democratic governments, and establish market economies, though some transitions were smoother than others.

26. They were established with the goal of preserving the rights of all ethnic groups living in the country.

27. The breakup of Czechoslovakia occurred peacefully, while the breakup of Yugoslavia was violent with great loss of life and destruction of property.

Standards-Based Assessment ANSWERS

1. A
2. B
3. Hungary
4. 1995 to 2004

GeoActivity

1. INTERDISCIPLINARY ACTIVITY RUBRIC

The pie graphs should
- be correctly formatted as pie graphs
- contain correct information on each topic
- clearly represent the percentages

2. WRITING FOR SOCIAL STUDIES RUBRIC

The journal entry should
- include activities that relate to the country chosen
- describe the day in an interesting and detailed manner

3. MENTAL MAPPING RUBRIC

The outline map should:
- represent the approximate shape of Eastern Europe
- include all geographic features listed in their correct location

DIFFERENTIATING CHAPTER ASSESSMENT

Unit Resource Book

- **Chapter Test A,** pp. 251–254. *Use for Struggling Readers, Beginning-to-Intermediate-Level English Learners, and for Inclusion of students whose IEP suggests supported test-taking.*
- **Chapter Test B,** pp. 255–258. *Use for On-Level Learners and ESOL students with Advanced-Level English.*
- **Chapter Test C,** pp. 259–262. *Use for Gifted and Talented students or students assigned Pre-AP work.*

Test Generator

- Chapter Tests A, B, and C, in English and Spanish

McDougal Littell Assessment System

TEST **Test Generator** Use three levels of tests in English and Spanish tied to state standards. Plus, choose questions to customize tests for all populations.

SCORE Scan and Score plain paper student tests in minutes.

REPORT Create instant progress reports, sortable by student, by standard, or other criteria.

RETEACH Respond immediately to students' needs by using a wealth of reteaching resources.

N
E
W
S

UNIT 5

Russia and the Eurasian Republics

Why It Matters:

Since gaining their independence, Russia and the Eurasian republics have worked to stabilize their governments and economies. Achieving these goals will help ensure peace in the region and strengthen Russia's position as a world leader.

Moscow's Red Square

CHAPTER 15 **Russia**

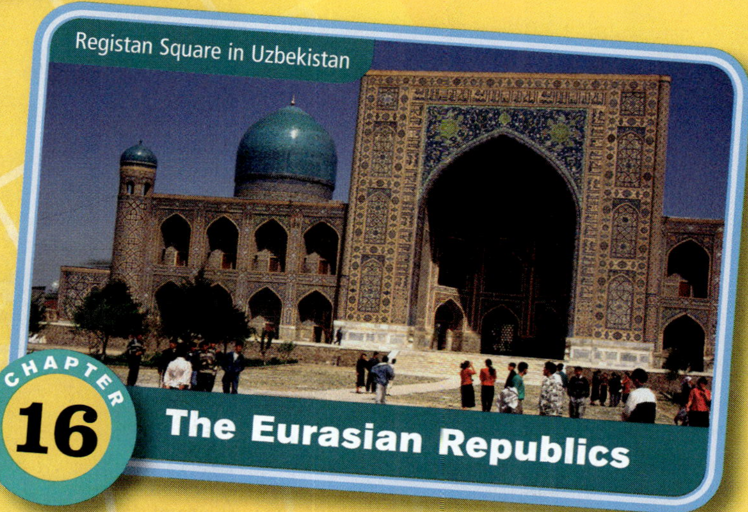

Registan Square in Uzbekistan

CHAPTER 16 **The Eurasian Republics**

Meet Zuhura

In this unit, you will meet Zuhura, a middle-school girl from Uzbekistan. Learn more about her on **Worldster @ClassZone.com**

Objectives

- Identify the physical and political features of Russia and the Eurasian republics
- Locate areas of high and low population density in Russia and the Eurasian republics
- Describe the climates found in Russia and the Eurasian republics
- Discuss the major economic activities in Russia and the Eurasian republics

Comparing Data

Although the landmass of Russia and the Eurasian republics is twice as large as that of the United States, the region has a smaller population. Russia itself—approximately 1.8 times the size of the United States—is the world's largest country in terms of area.

ANSWERS

1. United States; because it is smaller in size but has more people
2. Lake Baikal is 3,783 feet deeper than Crater Lake.

Russia and the Eurasian Republics

Russia and the Eurasian republics span two continents. The part of the region that lies to the west of the Ural Mountains is part of Europe. The part of the region that lies to the east of the Urals is part of Asia. As you study the maps, notice geographic patterns and specific details about the region. Answer the questions on each map in your notebook.

As you study the graphs on this page, compare the landmass, population, and lakes of Russia and the Eurasian republics with those of the United States. Then jot down the answers to the following questions in your notebook.

Comparing Data

1. Compare the landmass and population of Russia and the Eurasian republics with those of the United States. Based on these data, which region do you think has a higher population density? Why?
2. How much deeper is Lake Baikal than the deepest lake in the United States?

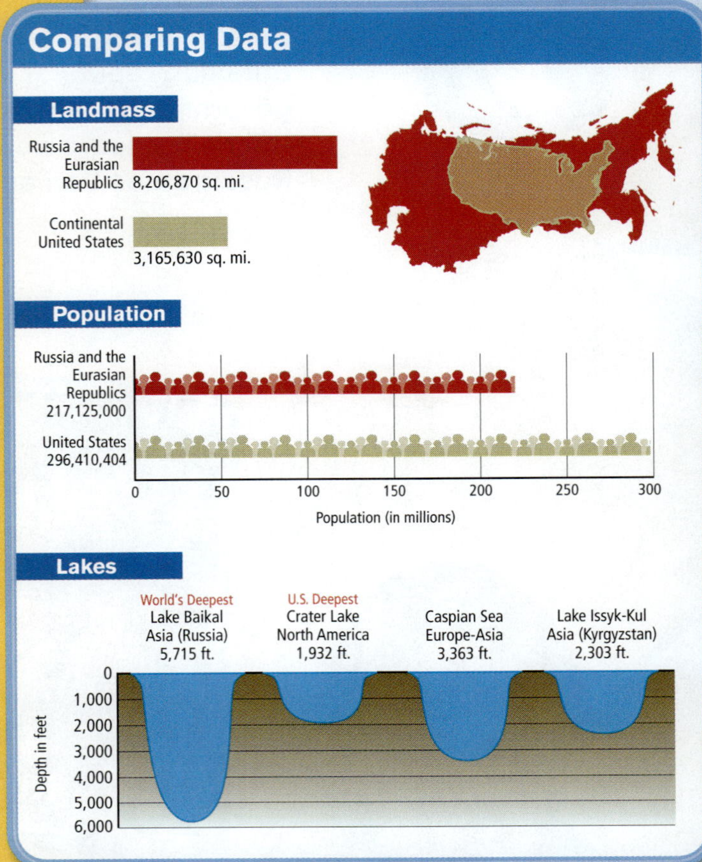

450 Unit 5

FIVE THEMES AND SIX ELEMENTS

Five Themes

Movement

Make sure that students comprehend the vast size of the region. Invite them to consider the challenges in transportation—of people and products—that such distances present.

Six Elements

Places and Regions

Explain that Russia and the Eurasian republics are spread across two continents. As a result, they interact closely with nations in Europe, the Middle East, and Asia.

Physical and Political

THINK LIKE A GEOGRAPHER

1. **Location** On which landform is Russia's capital located?
2. **Place** How would you describe much of the elevation in Tajikistan?

Physical and Political

Ask students to locate the Ural Mountains. Point out that this mountain range separates the Northern European Plain from the West Siberian Plain and has traditionally been considered to divide the continents of Europe and Asia.

- What is the highest mountain in this region? *(Mt. Elbrus at 18,510 feet, or 5,642 meters)*
- Describe the change in elevation as one moves from west to east. *(Plains and lower elevations give way to the Central Siberian Plateau and higher elevations, even along the eastern coast.)*

THINK LIKE A GEOGRAPHER ANSWERS

1. **Location** Northern European Plain
2. **Place** mostly very high, with much of it over 13,100 feet

More About . . .

The Union of Soviet Socialist Republics

Before 1991, the Russian Federation, the Eurasian republics, and the western republics were members of the Union of Soviet Socialist Republics (USSR). In the 1980s, political reforms introduced by Soviet leader Mikhail Gorbachev greatly relaxed the control of the central government. This created a growing sense of autonomy and nationalism within the 15 individual republics. The republics began to declare their independence, and by late December, the USSR officially ceased to exist.

DIFFERENTIATING INSTRUCTION

English Learners

Make Comparative Statements

Use information from the graphs on the opposite page to model the use of comparative and superlative adjectives. Explain that Lake Baikal is *deep.* It is *deeper than* Crater Lake. It is the *deepest* lake in the world. Encourage students to make comparative statements about the other lakes, populations, and landmasses.

Gifted & Talented

Track the Trans-Siberian Railroad

Explain that the world's longest continuous rail line connects Moscow and Vladivostok. Have students work in pairs to plan a trip on the Trans-Siberian Railroad. Ask them to plot the route, determine the amount of time their journey will require, and list some of the cities and sights they'll see along the way.

Population Density

Remind students that Russia is the world's largest country in terms of landmass. Yet much of its land is sparsely populated.

- Which cities in Russia are the most densely populated? *(Moscow and St. Petersburg)*
- Why do you think the population density is so low in eastern and northern Russia? *(Possible Answer: because of the cold, harsh climate)*
- How might St. Petersburg's location affect its population density? *(Possible Answer: Because it's a port city, it has access to sea resources and trade with other places, which help to support a larger population.)*

THINK LIKE A GEOGRAPHER ANSWERS

1. **Region** in the western part of the country
2. **Place** very sparsely populated

More About . . .

Moscow, the Capital of Russia

Located on the Moskova River, which runs between the Oks and Volga rivers, Moscow is Russia's political, industrial, and cultural center. About seven percent of Russia's total population—about ten million people—call Moscow their home. Moscow is also the largest city in Europe.

THINK LIKE A GEOGRAPHER

1. **Region** In what part of Russia is the population most dense?
2. **Place** How would you describe the population between the Caspian and Aral seas?

452 Unit 5

FIVE THEMES AND SIX ELEMENTS

Five Themes

Location

Point out that the most populated parts of the region are located in the same general area. Ask volunteers to describe these locations in relative and specific terms.

Six Elements

Human Systems

Point out that most of the densely populated areas in the region are located along rivers or coasts. Review the importance of waterways for transportation and emphasize the relationship between good transportation systems and population growth.

Climate

THINK LIKE A GEOGRAPHER

1. **Place** What is the climate of the area that is closest to the Arctic Ocean?
2. **Region** What area has a desert climate?

Climate

You may want to have students compare the climate map with the population density map. Ask them to note relationships between climate and population. Point out that the harsh conditions found in the tundra, subarctic, and desert climates probably make it difficult for these regions to support large human populations.

- What types of climates are found in Kyrgyzstan? *(highland and semiarid)*
- Where can you find a humid subtropical climate in this region? *(in Georgia, along the Black Sea and the country's southern coast)*
- What makes the Aral Sea significant? *(Possible answer: It's located entirely in a desert climate zone, which means it's most likely the major water source in that region.)*

THINK LIKE A GEOGRAPHER ANSWERS

1. **Place** tundra
2. **Region** the area between the Caspian Sea and Lake Balkhash

Comparing Regions

Western Russia and Eastern Russia

Although many of Russia's natural resources are found in eastern Russia, few of its people live there. The harsh climate and geography of much of eastern Russia has made accessing these resources—and developing population growth—difficult. In contrast, western Russia is part of a large, fertile plain with a climate that's more suitable for farming and for habitation.

DIFFERENTIATING INSTRUCTION

Inclusion

Compare Climates

Review the conditions that one would encounter in each climate region of Russia and the Eurasian republics. Explain that vast coastal areas in the far north remain frozen for most of the year. Have small groups find a coastal area that might be a good destination for a beach vacation, such as the humid subtropical region on the Black Sea or the Mediterranean region on the Caspian Sea.

Economic Activity

Russia has a vast amount of forestland. Discuss with students some of the advantages and disadvantages of harvesting these trees. *(Advantages: source of jobs and income; Disadvantages: could increase pollution and reduce animal habitats)*

- Where are most of Russia's gold resources located? *(in the eastern part of the country)*
- What are Uzbekistan's major resources? *(natural gas, gold, and other minerals)*

THINK LIKE A GEOGRAPHER ANSWERS

1. **Human-Environment Interaction** in western and central Russia
2. **Region** the Caspian Sea

More About . . .

The Caspian Consortium Pipeline

Kazakhstan, the largest Eurasian republic, contains large deposits of oil. For years, poor transportation systems prevented most of the country's oil from reaching foreign markets. Then in 2001, the Caspian Consortium pipeline opened. It carries oil from Tengiz in western Kazakhstan to the Russian port of Novorossiysk on the Black Sea. Three countries and ten private companies invested in this huge project. By 2015, the pipeline should deliver about 1.4 million barrels of oil per day.

UNIT 5 ATLAS

Economic Activity

THINK LIKE A GEOGRAPHER

1. **Human-Environment Interaction** Where in Russia is most of the land used for commercial agriculture?
2. **Region** Around what body of water is oil a major resource?

FIVE THEMES AND SIX ELEMENTS

Five Themes

Human-Environment Interaction

Point out the large areas of limited agriculture in the northeastern and southern parts of the region. Have students refer to the climate map on the previous page to help explain this lack of agricultural activity. Remind them that few types of crops can thrive in dry climates.

Six Elements

Environment and Society

Review the difference between renewable and nonrenewable resources. Explain that the economic activities indicated by the land-use areas on the map make use of renewable resources. The economic activities identified by symbols make use of nonrenewable resources. You may want to discuss how renewable resources still need to be managed properly—for example, poor farming practices can result in soil becoming depleted.

Regional Overview

Russia and the Eurasian Republics

Russia and the Eurasian republics span two continents: Europe and Asia. Both continents have had a significant influence on the regions' histories and cultures.

GEOGRAPHY

Both Russia and the Eurasian republics contain rugged areas, with towering mountains, wide deserts, and vast plains. The climates in these regions range from the frigid weather in Siberia to more mild temperatures in Georgia.

Frozen Water A group of boys plays hockey on Lake Baikal.

HISTORY

Many groups have influenced Russia and the Eurasian republics. Throughout their histories, the regions have often been divided between holding onto Eastern, or Asian, traditions and adopting more modern ideas from the West, or Europe.

CULTURE

The blend of East and West is reflected in the culture of both regions. More than 100 ethnic groups from both Asia and Europe live in Russia and the Eurasian republics.

GOVERNMENT

When the Soviet Union dissolved in 1991, Russia and 14 other republics formed. Since then, Russia and the Eurasian republics have struggled to establish democratic governments.

ECONOMICS

Russia's economy has boomed as a result of its abundant energy resources. The countries of the Eurasian republics also have vast supplies of oil and coal. However, these countries are only beginning to benefit economically from their natural resources.

Regional Overview

Remind students that other countries and regions have been influenced by other people and places. For example, the United States has been influenced by Asia, South America, Europe, and Africa. Talk about how this blending of cultures can have both positive and a negative results. *(Positive: enriches culture with new ideas and products; Negative: can lead to conflicts)*

- Why has Russia's economy boomed? *(because of its vast energy resources)*
- How many ethnic groups live in this region? *(more than 100, from both Asia and Europe)*
- What is the struggle that the governments of Russia and the Eurasian republics are facing? *(to establish democratic governments)*

Comparing Regions

Arctic Regions in North America and Siberia

Temperatures in arctic regions have risen significantly since the 1950s, resulting in an increased thawing of ice and permafrost. Northern North America and western Siberia have warmed more than other arctic regions—by 3.6 to 5.4 degrees Fahrenheit. Most scientists believe that global warming, as a result of human activities, is responsible for this arctic warming trend.

DIFFERENTIATING INSTRUCTION

Struggling Readers

Compare Land-Use Areas

Using the Economic Activity map, explain the meanings of the five land-use areas. Help students list specific examples of the economic activities that might be found in each area. Encourage them to explain the relationship between climate and land use.

Pre-AP

Investigate Climate Change

Explain that huge amounts of carbon and methane gas lie within the frozen soil and lakes of Siberia. As the permafrost melts, it releases these gases into the atmosphere. Have pairs of students research the thawing of permafrost in the region and report on its effects.

Unit Writing Project

Teaching Tips

- As students prepare their itineraries, encourage them to consult family-friendly travel guides from the library or the Internet. This will help them increase their sightseeing options and provide them with practical and logistical details.
- Provide students with an example or an outline of a chronologically organized travel itinerary. Explain that their itineraries should begin with their arrival in the country or region and end with their departure.
- Remind students to consider the travel times that their trips within the region might require. Vast distances and limited transportation options can make some journeys difficult and lengthy.

Each map on the country cards illustrates the shape of a state, province, or country. Tell students that the maps do not reflect the comparative size of these landmasses within a particular region. To get an idea of the actual size of the state, province, or country, students should study the land area statistics provided on the card.

UNIT 5 ATLAS

Country Almanac

Click here to compare the most recent data on countries @ClassZone.com

Unit Writing Project

Travel Itinerary

As you read this unit, pick a country or region that interests you. Then imagine that you are planning a trip to this place. Write an itinerary for a week-long stay.

Think About:

- the clothes you should bring
- what specific sites you would like to visit
- what foods you would like to eat
- the questions you would like to ask the people who live there

Russia and the Eurasian Republics

The country cards on these pages contain information about Russia and the Eurasian republics. As you study the land area of Russia, consider that the country is bigger than the continents of Europe and Antarctica combined.

Russian Federation

GEOGRAPHY
Capital: Moscow
Total Area: 6,592,772 sq. mi.
Population: 143,202,000

ECONOMY
Imports: food; chemicals; cars
Exports: fuels; metals; chemicals

CULTURE
Language: Russian
Religion: Christian 57%; Muslim 8%; nonreligious 27%; atheist 5%

Russian Bear The brown bear of Eurasia is sometimes considered the symbol of Russia.

INTERDISCIPLINARY ACTIVITIES

CONNECT to Language Arts

Write a Paragraph

Have students write a paragraph in the present tense that describes the photograph of the bear and its environment. After they have set the scene, ask them to imagine and describe what has captured the bear's attention.

Armenia

GEOGRAPHY
Capital: Yerevan
Total Area: 11,506 sq. mi.
Population: 3,016,000

ECONOMY
Imports: mineral fuels; rough diamonds
Exports: alcohol; cut diamonds

CULTURE
Language: Armenian
Religion: Armenian Apostolic 65%

Azerbaijan

GEOGRAPHY
Capital: Baku
Total Area: 33,436 sq. mi.
Population: 8,411,000

ECONOMY
Imports: natural gas; iron and steel
Exports: petroleum; food

CULTURE
Language: Azerbaijani
Religion: Muslim 93%; Russian Orthodox 1%; Armenian Apostolic 1%

Georgia

GEOGRAPHY
Capital: Tbilisi
Total Area: 26,911 sq. mi.
Population: 4,474,000

ECONOMY
Imports: food; mineral fuels
Exports: aircraft; food; mineral fuels

CULTURE
Language: Georgian
Religion: Christian 46%; Sunni Muslim 11%

Kazakhstan

GEOGRAPHY
Capital: Astana
Total Area: 1,049,155 sq. mi.
Population: 14,825,000

ECONOMY
Imports: mineral fuels; chemicals
Exports: petroleum; metals; food

CULTURE
Language: Kazakh
Religion: Muslim 47%; Russian Orthodox 8%; Protestant 2%

Kyrgyzstan

GEOGRAPHY
Capital: Bishkek
Total Area: 76,641 sq. mi.
Population: 5,264,000

ECONOMY
Imports: petroleum; food; chemicals
Exports: metals; electricity; tobacco

CULTURE
Language: Kyrgyz; Russian
Religion: Muslim 75%; Christian 7%

Tajikistan

GEOGRAPHY
Capital: Dushanbe
Total Area: 55,251 sq. mi.
Population: 6,507,000

ECONOMY
Imports: natural gas; grain and flour
Exports: aluminum; electricity

CULTURE
Language: Tajik
Religion: Sunni Muslim 80%; Shi'a Muslim 5%; Russian Orthodox 2%

Turkmenistan

GEOGRAPHY
Capital: Ashgabat
Total Area: 188,456 sq. mi.
Population: 4,833,000

ECONOMY
Imports: chemicals; food
Exports: natural gas; fabrics; raw cotton

CULTURE
Language: Turkmen
Religion: Muslim 87%; Russian Orthodox 2%

Uzbekistan

GEOGRAPHY
Capital: Tashkent
Total Area: 172,742 sq. mi.
Population: 26,593,000

ECONOMY
Imports: food; metalworking products
Exports: cotton fiber; food; gold; metals

CULTURE
Language: Uzbek
Religion: Muslim 76%; nonreligious 18%

Ural Mountains The Urals extend from the Arctic Ocean to the Caspian Sea and contain a wealth of mineral resources.

457

Country Almanac

Comprising a vast area that spans Europe and Asia, Russia and the Eurasian republics exhibit a fascinating mix of Eastern and Western influences. Traditionally, the Ural Mountains have served as the dividing line between the two continents. This 1,500-mile-long mountain range also provides an incredible source of minerals, including iron ore, coal, copper, gold, silver, and zinc.

- Which of the Eurasian republics has the largest population? *(Uzbekistan)*
- Which of the Eurasian republics has the largest land area? *(Kazakhstan)*
- In which countries do a majority of the people follow Islam? *(Azerbaijan, Kyrgyzstan, Tajikistan, Turkmenistan, and Uzbekistan)*

More About . . .

Armenian Literature

Armenian literature has its roots in the A.D. 400s with the development of its alphabet and the later translation of the Bible into Armenian. One of the major themes running throughout Armenian literature over the centuries is the struggle for freedom from oppression by outside rulers. Khachatur Abovian, considered the creator of the modern Armenian novel, focused on this theme in his masterpiece *The Wounds of Armenia*, published in the mid-1800s.

INTERDISCIPLINARY ACTIVITIES

CONNECT to Math

Graph Populations

Have students create a bar graph to compare the populations of the Eurasian republics. Tell them to number the vertical axis from zero to 30 in increments of five. Explain that the numbers represent millions of people. On the horizontal axis, have students list each country. They should then use the statistics on the cards to graph the countries' populations.

CONNECT to Science

Research Minerals

Have students choose one of the following mineral resources found in the Urals: amethyst, bauxite, coal, copper, gold, iron, lead, nickel, platinum, silver, topaz, zinc. Ask them to research the mineral they chose. Then invite volunteers to describe their mineral to the class, explain how it is extracted, and cite examples of how people use it.

Each map on the country cards illustrates the shape of a state, province, or country. Tell students that the maps do not reflect the comparative size of these landmasses within a particular region. To get an idea of the actual size of the state, province, or country, students should study the land area statistics provided on the card.

Chapter 15 Russia PLANNING GUIDE

Complete print resources all on one USB drive allow you to customize lesson plans for all students.

Section Titles & Key Ideas	Unit Resource Book Russia and the Eurasian Republics	Assessment Resources
1 Sweeping Across Eurasia **KEY IDEA** Russia is a vast country that contains a variety of landforms, climates, vegetation regions, and natural resources.	• **Reading Study Guide,** pp. 3–4 • **RSG with Additional Support,** pp. 11–12 • **Vocabulary Practice,** p. 35 • **Map and Graph Skills,** p. 37 • **Interdisciplinary Projects,** p. 43 • **Connect to Math,** p. 46	**Unit Resource Book** • Section Quiz, p. 49 • Reteaching Activity, p. 53 **Test Generator CD-ROM** **Daily Test Practice Transparency,** TT52
2 Governing a Vast Land **KEY IDEA** The sometimes conflicting influences of the East and West have helped shape Russian history.	• **Reading Study Guide,** pp. 5–6 • **RSG with Additional Support,** pp. 13–14 • **Outline Map Activities,** p. 39 • **History Makers,** p. 41	**Unit Resource Book** • Section Quiz, p. 50 • Reteaching Activity, p. 54 **Test Generator CD-ROM** **Daily Test Practice Transparency,** TT53
3 Blending Europe and Asia **KEY IDEA** The influences of Europe and Asia are reflected in the people and culture of Russia.	• **Reading Study Guide,** pp. 7–8 • **RSG with Additional Support,** pp. 15–16 • **Skillbuilder Practice,** p. 36	**Unit Resource Book** • Section Quiz, p. 51 • Reteaching Activity, p. 55 **Test Generator CD-ROM** **Daily Test Practice Transparency,** TT54
4 The Struggle for Reform **KEY IDEA** Russia is on its way to establishing a market economy but has struggled to change its government.	• **Reading Study Guide,** pp. 9–10 • **RSG with Additional Support,** pp. 17–18 • **History Makers,** p. 42 • **Economics in the World,** p. 45 • **Primary Sources,** pp. 47–48	**Unit Resource Book** • Section Quiz, p. 52 • Reteaching Activity, p. 56 • Chapter Test (Forms A, B, & C), pp. 57–71 **Test Generator CD-ROM** **Daily Test Practice Transparency,** TT55

McDougal Littell **Assessment System**
TEST SCORE REPORT RETEACH

No Time? To focus students on important content in the chapter, use the **"Reading Study Guide with Additional Support"** in ***Resources2Go.***

Teacher Presentation Options

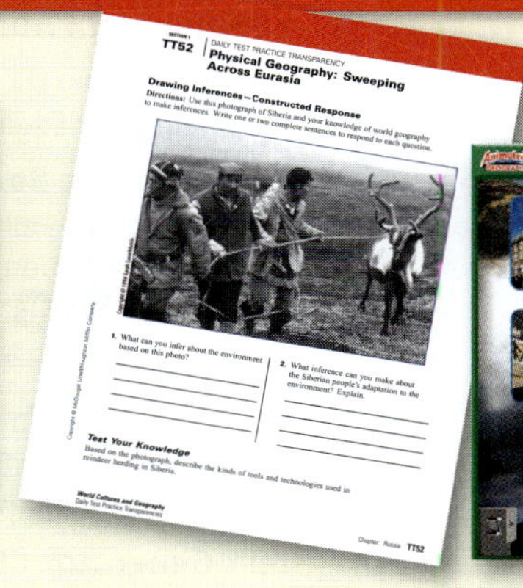

Electronic Books
eEdition DVD-ROM
eEdition Online

Presentation Toolkit
Video/DVD
- World Cultures and Geography

Power Presentations DVD-ROM
- Discussion Starters
- Media Gallery
- Review Game
- Interactive Review
- Animated Features

Unit Transparency Book
- 3-Minute Warm-Ups
- Essential Question
- Critical Thinking
- Maps
- Cultures

Daily Test Practice Transparencies

Animated Geography
- Present-Day Russia Map
- Russia's Natural Resources
- The Trans-Siberian Railroad

OnlineActivities at ClassZone.com

Animated Geography
- **Videos and Photos**
- **3-D Animations**
- **Interactive Maps**

Interactive Review
- **Section Quizzes**
- **Review and Study Notes**
- **Vocabulary Flip Cards**
- **Graphic Organizers**
- **Name Game and Crossword Puzzles**
- **GeoGame**

Online Test Practice

Map Center
- **Atlas Maps**
- **Political Maps with Data**
- **Links to Mapping Web Sites**

Activity Center
- **Activity Maker—Review Games, Puzzles, Flip Cards, and Graphic Organizers**
- **Worldster**

Research & Writing Center
- **Web Sites**
- **Data Updates**
- **Primary Sources**
- **Writing Models**

Current Events

Tim Mortensen
Patrick Marsh Middle School
Sun Prairie, Wisconsin

Make It Fun — Teacher-Tested Activities

As you teach this chapter, look for my classroom-proven ideas for:
- making the lesson fun
- focusing students' attention and comprehension
- integrating technology with traditional teaching

Teaching Bell to Bell When you see this timing guide, have students discuss in pairs, in groups of 3–6, or as a class.
- **Think, Pair, Share** (pairs)
- **Reader, Recorder, Reporter** (3–6)
- **Talk About It** (any size)

CHAPTER 15

RESOURCES FOR DIFFERENTIATED INSTRUCTION

Struggling Readers

eEdition with Audio DVD-ROM

Lesson Plans
- Strategies for Inclusion

Unit Resource Book
- Reading Study Guide with Additional Support, pp. 11–18 B
- Vocabulary Practice, p. 35
- Skillbuilder Practice, p. 36
- Reteaching Activity, pp. 53–56
- Chapter Test (Form A), pp. 57–61 A

Best Practices Toolkit C
- Interactive Reading Strategies, pp. 90–145
- Differentiation Activity Bank: Struggling Readers, pp. 146–149

Teacher's Edition Activities
- Compare and Contrast, Sec. 1
- Read a Map, Sec. 2
- Find Main Ideas, Sec. 3
- Tiered Activities, Sec. 1–4

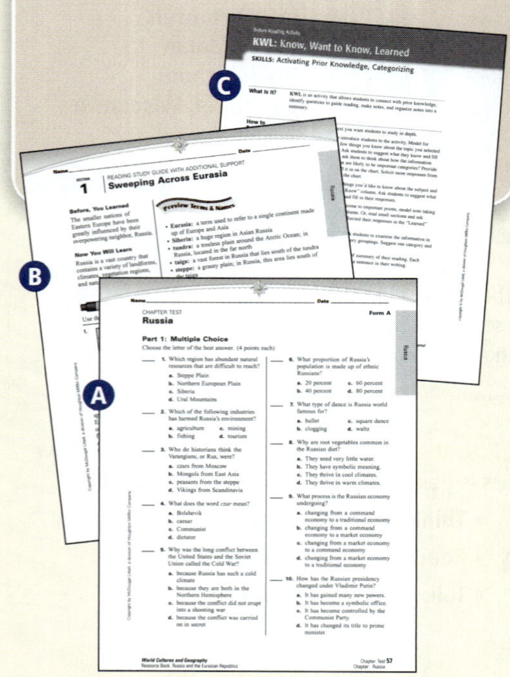

English Learners

Pupil Edition in Spanish

eEdition with Audio DVD-ROM

Lesson Plans
- Strategies for English Learners

Unit Resource Book
- Reading Study Guide (Spanish), pp. 19–26
- RSG with Additional Support (Spanish), pp. 27–34 A B

Multi-Language Glossary C

Best Practices Toolkit
- Differentiation Activity Bank: English Learners, pp. 165–194

Test Generator CD-ROM
- Section Quizzes in Spanish
- Chapter Test in Spanish

Teacher's Edition Activities
- Pronounce and Preview, Sec. 1–4
- Understand Vocabulary, Sec. 1, 4
- Build Vocabulary, Sec. 2
- Clarify Vocabulary, Sec. 3
- Tiered Activities, Sec. 1–4

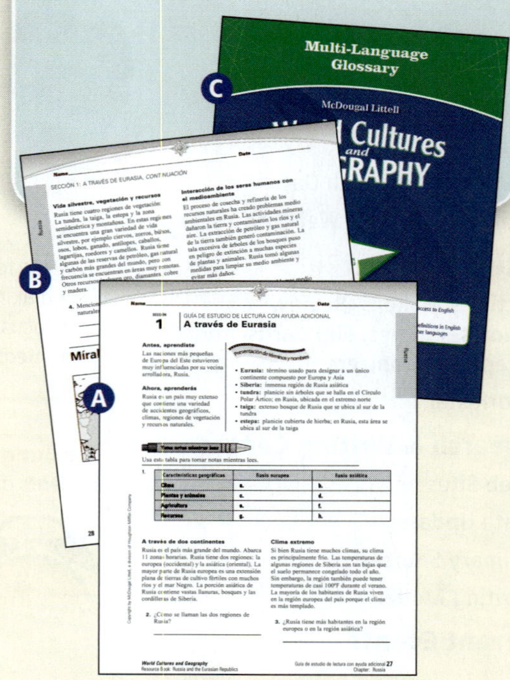

Inclusion

eEdition with Audio DVD-ROM

Lesson Plans
- Strategies for Inclusion

Unit Resource Book
- Reading Study Guide with Additional Support, pp. 11–18 A B
- Reteaching Activity, pp. 53–56
- Chapter Test (Form A), pp. 57–61

Best Practices Toolkit C
- Interactive Reading Strategies, pp. 90–145
- Differentiation Activity Bank: Struggling Readers, pp. 146–149

Teacher's Edition Activities
- Present the Terms, Sec. 1
- Connect Map and Text, Sec. 1
- Spell and Define Vocabulary, Sec. 2
- Create Vocabulary Posters, Sec. 3
- Locate the Terms and Names, Sec. 4
- Tiered Activities, Sec. 1–4

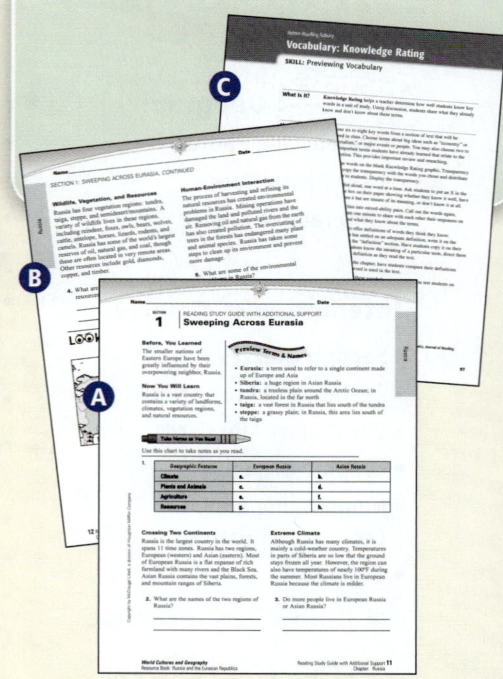

Interdisciplinary Activities

Gifted & Talented Students

Lesson Plans
- Core Lesson Plan

Unit Resource Book
- History Makers, pp. 41–42 (A) (B)
- Connect to Math, p. 46 (C)
- Chapter Test (Form C), pp. 67–71

Best Practices Toolkit
- Student Products Rubrics, pp. 1–14
- Writing Templates, pp. 78–89

Teacher's Edition Activities
- Create an Exhibit, Sec. 1
- Research Russian Culture, Sec. 3
- Tiered Activities, Sec. 1–4

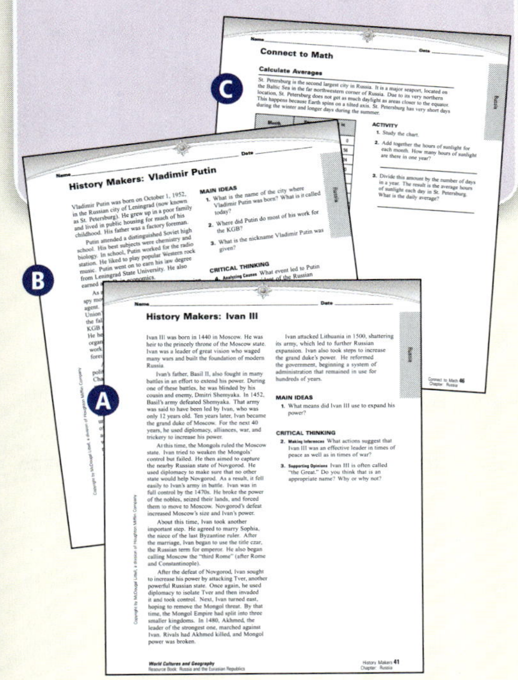
C Connect to Math
B History Makers: Vladimir Putin
A History Makers: Ivan III

Pre-AP*

Lesson Plans
- Core Lesson Plan

Unit Resource Book
- Interdisciplinary Projects, p. 43 (A)
- Economics in the World, p. 45 (B)
- Primary Sources, pp. 47–48 (C)
- Chapter Test (Form C), pp. 67–71

Best Practices Toolkit
- Writing Templates, pp. 78–89
- Differentiation Activity Bank: Pre-AP Students, pp. 195–200

* Pre-AP is a registered trademark of the College Entrance Examination Board, which was not involved in the production of and does not endorse this product.

Teacher's Edition Activities
- Compare Climates, Sec. 1
- Draw Conclusions, Sec. 4
- Tiered Activities, Sec. 1–4

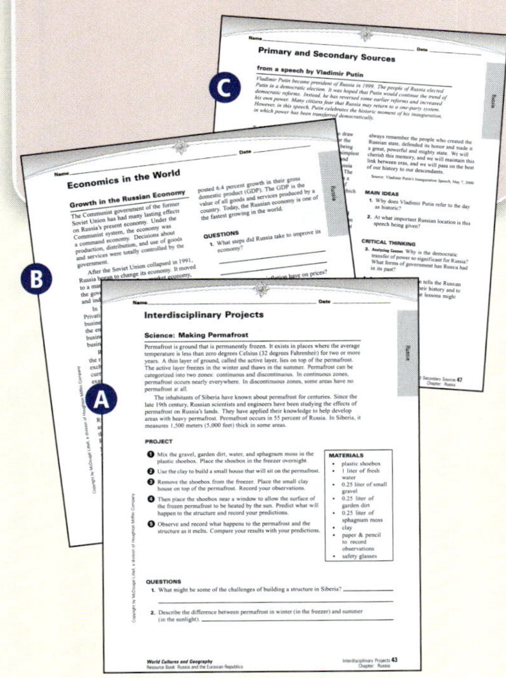
C Primary and Secondary Sources
B Economics in the World
A Interdisciplinary Projects

CONNECT to History

Review a Russian or Soviet Film

CLASS TIME 30 minutes

TASK Prepare a review of a Russian or Soviet film.

SUPPLIES
- list of age-appropriate Russian and Soviet films (start with the Library of Congress list of Russian films at www.loc.gov/rr/mopic/findaid/russian.html)
- writing materials/computers

ACTIVITY Have students prepare a review of a Russian/Soviet movie. In their reviews, students should identify the setting (time and place) of the story, include a brief summary of the plot, and explain what the movie teaches about Russia and/or the Soviet Union. Post the reviews. Ask volunteers to read theirs aloud.

Good for Visual and Verbal Learners

CONNECT to Economics

Graph Russian and U.S. Economic Growth

CLASS TIME 30 minutes

TASK Create a graph to compare economic growth in Russia from 1991 to 2005 with that of the United States for the same period.

SUPPLIES
- data on U.S. economic growth from 1991–2005 (search at www.cia.gov for data from the CIA World Factbook)
- paper
- colored markers or pencils

ACTIVITY Refer students to the Connect to History & Economics feature in Section 4 of this chapter for data on Russia; provide students with copies of data from the CIA World Factbook for data on the U.S. Then have students work in pairs to create bar graphs that show the years of Russia's economic growth in one color and the years of U.S. growth in another. Discuss the similarities and differences between the graphs, and have students offer explanations for these.

Good for Visual and Mathematical/Logical Learners

Introduce the ESSENTIAL QUESTION

- Explain that, according to some geographers, Russia lies on two continents.
- Analyze the influences of the East and West that have shaped Russian history.
- Identify the ways Russian life blends traditional and modern ways.
- Describe the recent changes in the Russian government and economy.

Key Ideas

Section 1: Russia is a vast country that contains a variety of landforms, climates, vegetation regions, and natural resources.

Section 2: The sometimes conflicting influences of the East and West have helped shape Russian history.

Section 3: The influences of Europe and Asia are reflected in the people and culture of Russia.

Section 4: Russia is on its way to establishing a market economy but has struggled to change its government.

CONNECT to the Essential Question

You may wish to revisit the Essential Question periodically with students as they work through the chapter.

Suggestions for reconnecting are provided at an appropriate point in the *Teacher's Edition* for each section.

Wooden onion-dome churches on Kizhi Island in northern Russia

CHAPTER 15 Russia

1 GEOGRAPHY **Sweeping Across Eurasia**

2 HISTORY **Governing a Vast Land**

3 CULTURE **Blending Europe and Asia**

4 GOVERNMENT & ECONOMICS **The Struggle for Reform**

ESSENTIAL QUESTION

How is Russia preserving its Eastern culture while adapting to Western influences?

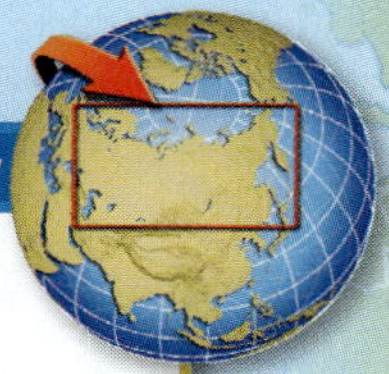

CONNECT Geography & History

Use the map and the time line to answer the following questions.

1. What is the capital of Russia today?
2. What was the capital in 1703?

950

Culture
◄ **988** Orthodox Christianity becomes the official religion of Russia.

History
1237 Mongols conquer Russian lands.

Geography
1703 St. Petersburg is founded and becomes Russia's capital. (Palace of Catherine the Great) ▼

TIME LINE DISCUSSION

Ask students to identify the time line entries that signified a change in Russia's government. *(takeover by Mongols in 1237; new capital in St. Petersburg in 1703; takeover by revolutionaries in 1917; formation of the Russian Federation in 1991)*

- Which early event shown on the time line probably had the greatest impact on Russian culture? *(the establishment of Orthodox Christianity as Russia's official religion)*
- Tell students that Vladimir Lenin brought communism to Russia. This system of government ended when the Russian Federation was formed. For how many years was Russia a Communist state? *(for 74 years)*

Government
1917 After revolutionaries overthrow the czar, Vladimir Lenin gains power. ►

Today

Culture
▲ **1869** Leo Tolstoy writes the novel *War and Peace.*

History
◄ **1991** The USSR collapses, and the Russian Federation forms.

459

TEACHER MEDIA FAVORITES

Books for the Teacher

Figes, Orlando. *Natasha's Dance: A Cultural History of Russia.* Picador USA, 2003. Examines Russian history through arts and culture.

Hosking, Geoffrey. *Russia: People and Empire, 1552–1917.* Harvard University Press, 1998. Explores the history of the Russian Empire.

Solzhenitsyn, Alexander. *One Day in the Life of Ivan Denisovich.* Blackstone Audio, Inc., 1992. Classic Russian novel about Soviet labor camps. (also available in paperback)

Video / DVD

Battleship Potemkin. Image Entertainment, 1925 (DVD release 1998). Eisenstein's classic film based on the mutiny aboard the *Potemkin.*

Journey Across Russia. Educational Video Network, 2004. Explores the diverse culture of Russia, including the traditions, arts, and architecture throughout this vast country.

Internet

Go to the *Research and Writing Center* at **ClassZone.com** to link to:

- data on the region
- a variety of chapter-specific, content-reviewed sites
- "History Makers" biographies
- primary sources

Animated GEOGRAPHY

Present-Day Russia

CLASSZONE.COM

The map shows Russia's location on two continents: Europe and Asia. Throughout its history, Russia has been greatly influenced by people and cultural traditions from both continents.

- In which continent are Moscow and St. Petersburg, Russia's present and former capitals, located? *(in Europe)* How might the location of these capitals have affected Russian government? *(The European, or Western, influence might have been stronger in the national government.)*
- In which continent is Mongolia located? *(in Asia)* What cultural influences did the Mongols probably bring to Russia in 1237? *(Eastern influences)*
- What body of water does Russia border to the north? *(the Arctic Ocean)* What conclusions can you draw about the climate in this part of Russia? *(It is probably very cold.)*

CONNECT GEOGRAPHY & HISTORY ANSWERS

1. Moscow

2. St. Petersburg

Presentation Options

Use the **Power Presentations** DVD-ROM to project the animated map.

- Prompt students to use the pop-up images and audio from the map to discover what they already know about the geography, culture, history, economics, or government of the region.
- Challenge students to answer the map questions individually, in small groups, or as a class activity. Use all of the interactive material to help preview chapter content.

1 Plan & Prepare

Objectives

- Compare and contrast western and eastern Russia
- Describe Russia's climate, vegetation regions, and natural resources
- Analyze the causes of environmental problems in Russia

Read for the Essential Question

Help students read for a purpose by reminding them of the Essential Question: "How is Russia preserving its Eastern culture while adapting to Western influences?" Point out that they can use items in the section, such as Key Questions, special features, images, and maps, to formulate answers to the question.

Vocabulary

Best Practices Toolkit

Use the ***Best Practices Toolkit*** to model strategies for vocabulary notetaking. Vary strategies throughout the year. Choose from: *Knowledge Rating, Predicting ABCs, Definition Mapping, Word Sort, Word Wheel, Frayer Model (Word Squares), Magnet Words*, and *Student VOC*.

Vocabulary Strategies, TT9–TT16

Reading Strategy

Best Practices Toolkit

Display the Three-Column Chart transparency.

- Remind students that when they categorize information, they must decide how the information in the section can be sorted into the categories shown in the chart.
- Model filling in categories on a chart using the "Crossing Two Continents" subsection.
- Have students suggest additional entries using the categories on the chart.

Three-Column Chart, TT21

SECTION 1 Reading for Understanding

Key Ideas

BEFORE, YOU LEARNED

The smaller nations of Eastern Europe have been greatly influenced by their overpowering neighbor, Russia.

NOW YOU WILL LEARN

Russia is a vast country that contains a variety of landforms, climates, vegetation regions, and natural resources.

Vocabulary

TERMS & NAMES

Eurasia a term used to refer to a single continent made up of Europe and Asia

Ural Mountains a north–south mountain range that forms the border between European and Asian Russia

Caspian Sea a large body of water bordering Russia to the south

Volga River the longest river in Europe, flowing through European Russia

Siberia a huge region in Asian Russia

tundra a treeless plain around the Arctic Ocean; in Russia, located in the far north

taiga (TY•guh) a vast forest in Russia that lies south of the tundra

steppe a grassy plain; in Russia, this area lies south of the taiga

BACKGROUND VOCABULARY

permafrost ground that is frozen throughout the year

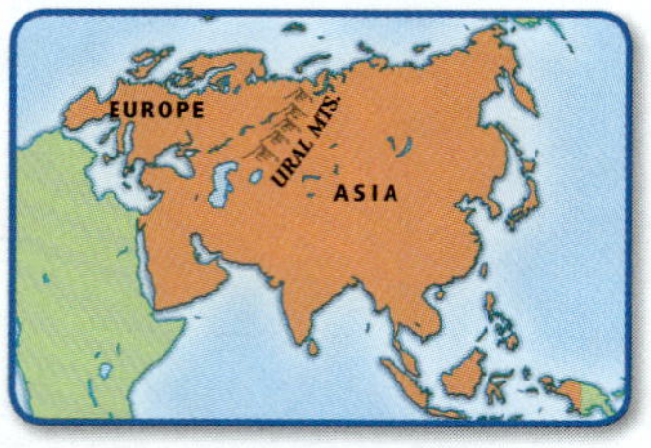

Visual Vocabulary Eurasia

Reading Strategy

Re-create the chart shown at right. As you read and respond to the **KEY QUESTIONS**, use the chart to organize important details about the physical geography of Russia.

Skillbuilder Handbook, page R7

CATEGORIZE

GEOGRAPHIC FEATURES	EUROPEAN RUSSIA	ASIAN RUSSIA
Climate		
Plants and Animals		
Agriculture		
Resources		

GRAPHIC ORGANIZERS Go to Interactive Review @ ClassZone.com

PRETEACHING VOCABULARY

English Learners

Pronounce and Preview

Pronounce each term for students. Review the meanings of words in the definitions, such as *continent, region*, and *plain.*

- To modify vocabulary learning, have students complete worksheets as they read the section.

Unit Resource Book: Russia & the Eurasian Republics

- Vocabulary Practice, p. 35

Inclusion

Present the Terms

Have students form small groups. Have each group choose one term to paraphrase, dramatize, or illustrate. The other groups should try to identify the term. Cycle through the terms more than once, challenging groups to present the terms in new ways.

SECTION 1 GEOGRAPHY

Orthodox Christian church in the Altai Shan in southern Siberia

Siberian pine trees, Karakol Lake, brown bear

Sweeping Across Eurasia

Connecting to Your World

Russia is huge. It is the largest country in the world and spans 11 time zones. However, many parts of the country have few people. In fact, even though Russia is nearly twice the size of the United States, it has less than half as many people. Like the United States, Russia contains a variety of landforms, including vast forests, snow-capped mountains, and wind-swept plains. But, because of its location, the climate is much colder. The size, geography, and climate of Russia have a major impact on its people.

Crossing Two Continents

KEY QUESTION How are western and eastern Russia similar and different?

Does Russia lie on one continent or on two? Some geographers recognize the Ural Mountains as a dividing line between Europe and Asia. Since Russia crosses the Urals, these geographers claim that the country lies in both Europe and Asia. Others view Europe and Asia as a single continent called **Eurasia**. These geographers divide Russia into western and eastern regions. This book shares that view. In this chapter, the two regions are referred to as European (or western) Russia and Asian (or eastern) Russia.

It's surrounded by mountains; the geography is rugged.

Lake Baikal This very deep lake in eastern Russia holds over 20 percent of Earth's fresh water. **What can you tell about the physical geography of the area from this picture?**

2 Focus & Motivate

3-Minute Warm-Up

Write on the board or display the transparency:

- Read the paragraph under Connecting to Your World. How many time zones does the continental United States have? *(three)* What differences in physical geography might there be in a country that spans 11 time zones? *(different landforms, climates, vegetation, and resources)*

Unit Transparency Book: Russia & the Eurasian Republics
- 3-Minute Warm-Ups, TT1

Connecting to Your World

More About . . .

Russia and the United States

The United States and Russia are separated at their closest points by the Bering Strait. Only about 55 miles separate Alaska's Cape Prince of Wales from Russia's Cape Dezhnev.

3 Teach

Crossing Two Continents

Reader, Recorder, Reporter

- What physical landform divides Europe from Asia? *(Ural Mountains)*
- Where is Siberia located? *(in Asian Russia)*
- **Categorize** Into which two regions do some geographers divide Russia? *(European, or western, Russia and Asian, or eastern Russia)*

SECTION 1 PROGRAM RESOURCES

Unit Resource Book: Russia and the Eurasian Republics

ON LEVEL

Unit Resource Book
- Reading Study Guide, p. 3
- Map and Graph Skills, p. 37
- Section Quiz, p. 49

STRUGGLING READERS

Unit Resource Book
- Reading Study Guide with Additional Support, p. 11
- Vocabulary Practice, p. 35
- Skillbuilder Practice, p. 36
- Section Quiz, p. 49
- Reteaching Activity, p. 53

eEdition with Audio DVD-ROM

ENGLISH LEARNERS

Pupil Edition in Spanish

eEdition with Audio DVD-ROM

Unit Resource Book
- RSG (Spanish), p. 19
- RSG with Additional Support (Spanish), p. 27

Multi-Language Glossary

Test Generator
- Section Quiz in Spanish

INCLUSION

Unit Resource Book
- RSG with Additional Support, p. 11
- Section Quiz, p. 49
- Reteaching Activity, p. 53

GIFTED & TALENTED

Unit Resource Book
- Interdisciplinary Projects, p. 43
- Connect to Math, p. 46
- Section Quiz, p. 49

PRE-AP

Unit Resource Book
- Section Quiz, p. 49

TECHNOLOGY

Unit Transparency Book
- 3-Minute Warm-Ups, TT1
- Essential Question, TT2
- Critical Thinking, TT3
- Map: Russia and the Eurasian Republics, TT7

Daily Test Practice Transparencies
- Section 1, TT52

Power Presentations

ClassZone.com

World Cultures and Geography Video on DVD

More About . . .

The Caspian Sea

Many geographers consider the Caspian a lake. Once part of the ancient world ocean, Tethys, the Caspian Sea became landlocked millions of years ago. It still retains a connection to its ocean past through the saltiness of its water—though ocean waters are about three times saltier than those of the Caspian.

Comparing Regions

Biodiversity of the Caspian Sea and Australia

It doesn't seem like the world's largest inland body of water and the world's smallest continent would have much in common, but they do. The Caspian Sea and Australia both have diverse species of rare plant and animal life, including kangaroos in Australia and beluga sturgeon in the Caspian. These rare species are sometimes called living fossils. In both places, isolation led to the unique life forms. Australia was separated from other landmasses by the ocean, and the Caspian Sea was separated from other oceans by land.

Fun Facts!

Freshwater Seals

Nerpas are yellow-green in color at birth. Within two weeks they turn white. Finally they turn to silvery gray. In winter, the seals scratch holes through the ice that covers Lake Baikal so that they can breathe.

European, or Western, Russia Study the map on the opposite page. Note that the borders of European Russia are formed by the Arctic Ocean and Baltic Sea to the north, the **Ural Mountains** Ⓐ to the east, and the Caucasus Mountains, the Black Sea, and the **Caspian Sea** Ⓑ to the south. To the west lie the European countries of Finland, Estonia, Latvia, Belarus, and Ukraine. Most of European Russia consists of a very flat expanse of land called the Northern European Plain Ⓒ. This plain contains rich farmland but does not provide any natural barriers to the west. As a result, the landform offers no protection from invaders.

Many rivers flow through European Russia, including the **Volga** Ⓓ, Europe's longest. Several canals and overland routes connect these rivers and form a transport system that carries two-thirds of Russia's waterway traffic. Another important body of water in western Russia is the Black Sea. It contains the country's busiest port.

Asian, or Eastern, Russia By far the largest part of Russia lies to the east of the Urals in Asian Russia. This region is bordered by the Urals to the west, the Arctic Ocean to the north, the Pacific Ocean to the east, and a series of mountain ranges to the south.

Most of Asian Russia is made up of a huge expanse of land called **Siberia** Ⓔ. Siberia contains the largest flat region in the world. Other parts of Siberia are not at all flat. Southeastern Siberia has mountain ranges with some of the world's highest peaks. Kamchatka Peninsula Ⓕ, in the far eastern part of Russia, has about 20 active volcanoes.

Asian Russia has several major rivers, including the Lena and the Yenisey rivers. Some of these rivers are frozen for much of the year and so limit transportation. Thousands of lakes are also found in Asian Russia. One of these, Lake Baikal Ⓖ, is the deepest in the world at 5,715 feet.

Possible answers: Similar—Both contain many rivers, mountains, and plains; Different—Western Russia has rich farmland and rivers that can be used for transport, while eastern Russia is larger and has rivers that are frozen for much of the year. ▲

▲ **COMPARE AND CONTRAST** Explain how western and eastern Russia are similar and different.

Fun Facts!

FRESHWATER SEALS

Lake Baikal is home to the world's only freshwater seal—the nerpa. Nerpas are very graceful and can remain underwater for nearly 70 minutes. Scientists don't know how the seals ended up in a lake in the middle of Asia. But in Lake Baikal, the seals have found an ideal habitat with plenty of food.

462

DIFFERENTIATING INSTRUCTION

Struggling Readers

Compare and Contrast

On a board or overhead, draw a chart similar to the one shown at right. Point out that the two subsections on this page are organized in the same way. Have students reread both subsections. Then ask student pairs to record details about each region. When they have completed the chart, have them compare and contrast the information.

European, or Western, Russia	Asian, or Eastern, Russia

Physical Geography of Russia

D Volga River, Western Russia

F Kamchatka, Eastern Russia

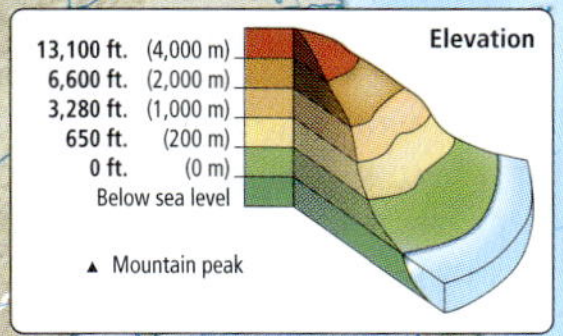

CONNECT Geography & Culture

READING A PHYSICAL MAP

Locate the Ural Mountains on the map above. This mountain range separates European Russia to the west from Asian Russia to the east. As you study these two regions,

- contrast their respective sizes
- compare their landforms and elevation
- consider how these features affect population

1. **Region** How do the landforms in Asian Russia differ from those in European Russia?
2. **Human-Environment Interaction** Which region do you think is more heavily populated? Why?

DIFFERENTIATING INSTRUCTION

Inclusion

Connect Map and Text

Help students see and recognize the significance of the annotated letters on the physical geography map of Russia. Tell students that thay can find the corresponding letters on the preceding page. Have students work in pairs to match the letters. One student should read a sentence containing a letter aloud, while the other locates the letter on the map. Then have the partners discuss the places on the map.

Gifted & Talented

Create an Exhibit

Have students use classroom and library resources to learn more about the Kamchatka Peninsula. Then have them work in small groups to create an exhibit. The exhibit should identify the region, explain why it is part of the Ring of Fire, and describe the geological processes that result on Kamchatka. Students should choose an appropriate medium to display what they have learned.

Physical Geography of Russia

CONNECT Geography & Culture

Reading a Physical Map Direct students' attention to the elevation key. Then ask them to identify other clues that can help them locate areas of high and low elevation. *(map labels such as Steppe, Central Siberian Plateau, and Caucasus Mountains)*

ANSWERS

1. **Region** The landforms in Asian Russia tend to have higher elevations.
2. **Human-Environment Interaction** European Russia is probably more heavily populated because the land is flatter and more hospitable.

Comparing Regions

Kamchatka and Micronesia: The Ring of Fire

What do Kamchatka, a northern peninsula, and the tropical islands of Micronesia have in common? Both regions are part of the Ring of Fire, an area prone to volcanic action and earthquakes. The Ring of Fire runs around the Pacific Ocean, from the Andes in South America to the Aleutian Islands in Alaska, and from Kamchatka to New Zealand. Active tectonic plates in this area explain the Ring of Fire's frequent earthquakes and active volcanoes.

Presentation Options

Use the **Power Presentations** DVD-ROM to project the animated map.

- Click on each place name in the menu.
- Listen to the audio and examine the photographs to learn how the geography of Russia affected its history, culture, and economics.

Teach

Climate, Vegetation, and Resources

Think, Pair, Share

- How would you describe the climate in parts of Russia? *(extreme in parts of Asian Russia and milder in European Russia)*
- Why do much of Russia's resources remain untouched? *(because they are located in remote areas of northern Siberia)*

CONNECT to Math

Tell students that, in the frigid Siberian winter, people's breath freezes into crystals that fall to the ground and make a noise they call "the whispering of the stars."

ACTIVITY ANSWERS

winter: –28.8°F; spring: 31°F; summer: 61.5°F; fall: 38.7°F

Unit Resource Book: Russia & the Eurasian Republics

- Skillbuilder Practice: Using a Database, p. 36
- Interdisciplinary Projects: Making Permafrost, p. 43
- Connect to Math: Calculate Averages, p. 46

Make It Fun

Teacher-Tested Activities

Tim Mortensen, Patrick Marsh Middle School, Sun Prairie, Wisconsin

To help students understand negative numbers, I draw a vertical line on the board and score 11 small horizontal lines through it. I label the middle line 0, the five lines to the right +1 to +5, and the five lines to the left –1 to –5. Then I use my number line to show students how to add positive and negative numbers. Finally, I test students' understanding by playing a math game. I call out simple addition problems and have volunteers show the class how to arrive at the answer using the number line.

Climate, Vegetation, and Resources

KEY QUESTION What types of natural resources are found in Russia?

Although Russia has many climates, it is mainly a cold-weather country. One Siberian town has recorded temperatures as low as –96°F. Such cold can burst tires and crack steel.

Extreme Climate All of Siberia has frigid weather during the winter. Temperatures remain so low in some areas that they are covered by a layer of permanently frozen ground called **permafrost**. However, the region can also have temperatures of nearly 100°F during the summer. These extreme temperature ranges are caused by two factors: Russia's northern latitude and its distance from any ocean. Much of the country lies many miles away from the oceans, which help moderate temperatures.

The climate is milder in European Russia. As a result, this region is more densely populated. However, even here the climate can be challenging. For example, the area around Moscow has snow on the ground for about five months of the year.

CONNECT to Math

Siberian winters are very cold. As you can see in the photo below, boiling water tossed into the region's wintry air can explode into a shower of ice crystals. Find out just how cold the winter is in Siberia. Follow the steps shown here to calculate the average temperature for each season in the Siberian town of Yakutsk.

Activity

Calculate Averages

1. Study the graph below.
2. Add the temperatures for the months in each season. For winter, you would add together: 18 + 40 + 45 + 33 + 8. Place a negative sign before the sum of these numbers.
3. Divide by the number of months in each season. For winter, you would divide by 5. Place the negative sign before the result. This is the average winter temperature in Yakutsk.

Average Temperatures in Yakutsk

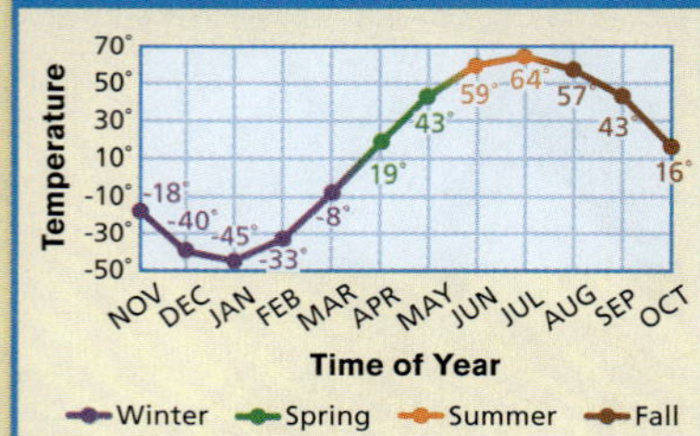

Source: *Gateway to Russia*

DIFFERENTIATING INSTRUCTION

English Learners

Understand Vocabulary

Read these sentences from the textbook page: "However, even here the climate can be challenging. For example, the area around Moscow has snow on the ground for about five months of the year." Point out that the word *challenging* in the sentence means "difficult" or "demanding." Explain that a challenge is something that is difficult to do or that tests a person's courage. Ask students to describe something they find challenging.

Pre-AP

Compare Climates

Have students research average monthly temperatures for two other cities in Russia and for a city near where they live and graph the results. Then ask them to determine the average seasonal temperatures for each city. Finally, have students write a paragraph, comparing the climates and explaining the differences in temperatures.

COMPARING Vegetation

A Reindeer in the tundra

B Autumn in the taiga

A Tundra The **tundra** in far northern Russia is a flat, treeless plain with long, harsh winters and short summers.

C Steppe The **steppe** is the grassland plain that stretches south of the taiga. Rich, fertile soil covers much of the steppe.

B Taiga The **taiga** (TY•guh) is a vast forest that lies south of the tundra. It is the largest forestland in the world.

D Semidesert & Mountain This area in the far southwest of Russia has dry lowlands around the Caspian Sea and rich vegetation near the Caucasus Mountains.

CRITICAL THINKING

Evaluate Which region probably provides the best farmland?

C Steppe in south-central Siberia

D Wildflowers in the central Caucasus Mountains

INTERDISCIPLINARY ACTIVITIES

CONNECT to Language Arts

Create a Brochure

Divide the class into four groups and assign each one a vegetation region shown on this page. Have each group use classroom resources to learn more about their region and create a brochure promoting it. Although some of these regions are not very populated and include extreme climates, students should look for unique ways to "sell" the region.

CONNECT to Science

Compare Vegetation

Have students use classroom and library resources to learn more about the main types of vegetation in each region shown. Have students write a paragraph about each region, describing its main types of vegetation and pointing out the similarities and differences among these types. Have students also include information explaining why certain types of vegetation are dominant in each region.

COMPARING

Vegetation

Remind students that vegetation is connected to the climate of an area. Along with climate, vegetation can affect the human and animal population in a region.

- What is the major vegetation of the taiga? *(trees or forests)* Which region of Russia has dry lowlands? *(the semidesert region in the far southwest of Russia around the Caspian Sea)*
- What do the tundra and steppe have in common? How do they differ? *(Both are large plains that are relatively flat. The tundra has a harsh climate with permafrost and limited plant life, while the steppe has rich, fertile soil.)*
- What factors contribute to the varied vegetation in Russia? *(Possible answers: Latitude, climate, altitude, and distance from large bodies of water can all affect the vegetation of a region.)*

CRITICAL THINKING ANSWER

Evaluate the steppe region

Unit Resource Book: Russia & the Eurasian Republics

- Map and Graph Skills: Reading Thematic Maps, p. 37

More About . . .

Taiga

Most of the trees in the taiga are conifers or evergreens. These trees are well-suited to their cold environment. Conifers generally have dark green, thin, needlelike leaves. Unlike deciduous trees, evergreens do not drop their leaves in the fall. Instead, they keep their leaves, so that they can start photosynthesizing, or using the sun to make food, as soon as possible in the spring. The dark green color also helps with photosynthesis. Thick, leathery leaves help the trees keep from drying out. The shape of the trees allows them to shed snow easily, helping to prevent broken branches from heavy snowfall.

More About . . .

Reindeer

Reindeer are strong swimmers, and young reindeer quickly learn to run. Reindeer are also well-adapted for extremely cold conditions. They have been domesticated by many people in Eurasia for use as work animals. They also provide food, clothing, and shelter. Unlike other deer, both males and females grow horns.

Animated GEOGRAPHY

Russia's Natural Resources

CLASSZONE.COM

Russia has a variety of natural resources, some of which are quite valuable, but location and climate make it difficult to access some of them.

- Sketch a map of Russia. Indicate on the map where each resource shown is located. *(Possible labels: coal, natural gas, and oil in Siberia; iron ore in the south and west; gold, diamonds, copper, and timber in the east and west)*
- How might technology and new materials help Russia use some of its remote resources? *(Possible answer: Technology and new materials might help people extract the resources.)*

CONNECT to the Essential Question

How is Russia preserving its Eastern culture while adapting to Western influences?

Ask students what they have learned so far that can help them answer this question. Students might mention the following:

- Russia is divided geographically into western and eastern regions.
- Although eastern Russia is much larger, most people live in western Russia because of its milder climate.
- Russia draws on natural resources from throughout the country.
- Russia is trying to balance its economic needs with environmental concerns.

Snowy Owl The snowy owl lives in the tundra for about half of the year. Although the birds are excellent fliers, they nest on the ground.

Varied Wildlife A wide variety of wildlife survives in Russia's four vegetation regions.

- Although few people live in the tundra, it is home to many animals, including reindeer, Arctic foxes, and snowy owls.
- The taiga also has a small human population. Bears, deer, and wolves thrive in the region because there are so few people.
- The animals in the steppe region are largely grazing animals. These include cattle, antelope, and wild horses.
- Only animals that can survive very dry conditions live in Russia's semidesert region. Lizards, snakes, rodents, and camels adapt well to this environment.

Abundant Natural Resources Russia has some of the largest reserves of natural gas, oil, and coal in the world. Despite this, the vast majority of these riches remains untouched. Many of Russia's resources are located in remote areas of northern Siberia. The geography and climate in these areas make it difficult to remove and transport the resources. However, Russia is able to take better advantage of other resources. These include iron ore in the south and west, as well as gold, diamonds, copper, and lumber in both the east and the west.

SUMMARIZE Describe Russia's natural resources. In eastern Russia, the natural resources consist of large reserves of natural gas, oil, and coal. In western Russia, the region has rich farmland.

Animated GEOGRAPHY

Russia's Natural Resources

Click here to see an interactive map of Russia's natural resources and vegetation @ClassZone.com

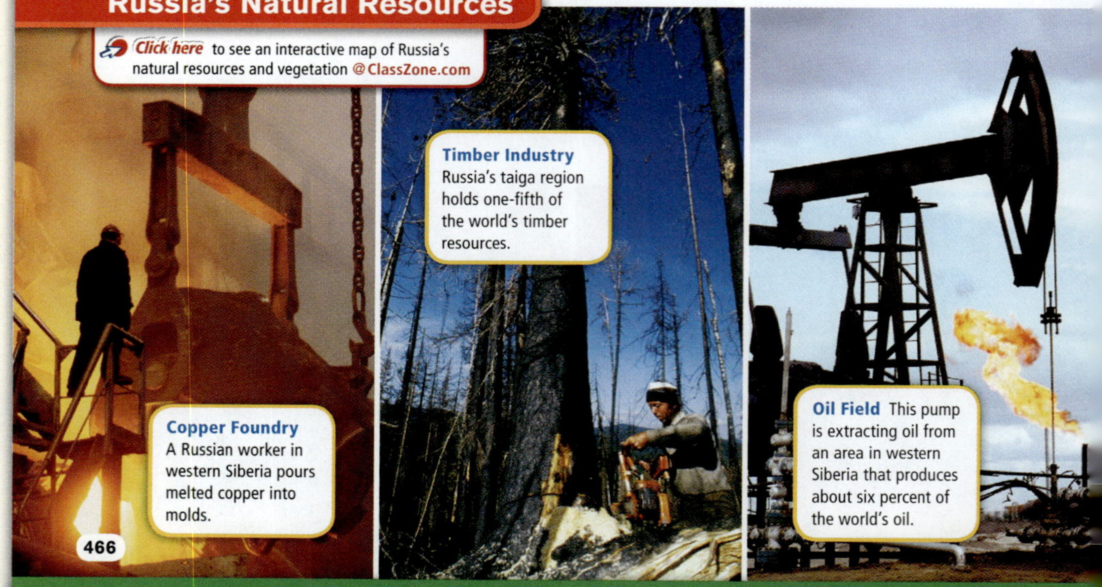

Copper Foundry A Russian worker in western Siberia pours melted copper into molds.

Timber Industry Russia's taiga region holds one-fifth of the world's timber resources.

Oil Field This pump is extracting oil from an area in western Siberia that produces about six percent of the world's oil.

DIFFERENTIATING INSTRUCTION: TIERED ACTIVITIES

OBJECTIVE Demonstrate an understanding of the environment in Russia's four vegetation zones.

Basic	On Level	Challenge
Have students work in groups to draw pictures illustrating the vegetation regions. Each group should focus on a region and illustrate its vegetation, wildlife, and resources.	Ask students to imagine that they live in one of Russia's vegetation regions and write a journal entry about life there. Students should describe the positive and negative impact of the region's vegetation, wildlife, and climate on daily life.	Have student groups select a region and research the impact that pollution and human activities have had on the region. Have each group create an annotated illustration of the condition of the region's environment and the effect on its vegetation and wildlife.

Human-Environment Interaction

KEY QUESTION What has caused some of the environmental problems in Russia?

As you have learned, Russia has some of the world's most abundant natural resources. Unfortunately, the process of extracting and developing some of these resources has created many environmental problems. Mining operations, in particular, have caused great damage. The mining sites have polluted the air in surrounding areas and contaminated nearby rivers. Industries that extract oil and gas have also added to the air and water pollution. In addition, overcutting trees in the taiga has caused problems. The loss of much of this forestland has endangered many plant and animal species.

Russia has taken steps to clean up its environment and prevent further damage. For example, the country has begun to enforce stricter rules on oil industries. The government has also discussed spending more money on solving environmental problems. Like other modern nations, Russia will have to balance its need for economic growth with its responsibility to protect the environment.

ANALYZE CAUSES Detail what has caused some of Russia's environmental problems. Mining and the oil and gas industries have polluted the air and contaminated rivers; overcutting trees has endangered many plant and animal species.

Section 1 Assessment

ONLINE QUIZ For test practice, go to **Interactive Review @ClassZone.com**

TERMS & NAMES

1. **Explain the importance of**
   - Ural Mountains
   - tundra
   - Siberia
   - steppe

USE YOUR READING NOTES

2. **Categorize** Use your completed chart to answer the following question:

   Why do more people live in European Russia than in Asian Russia?

GEOGRAPHIC FEATURES	EUROPEAN RUSSIA	ASIAN RUSSIA
Climate		
Plants and Animals		
Agriculture		
Resources		

KEY IDEAS

3. How do the rivers of European Russia provide the region with a transportation network?
4. Why does Siberia have extremely cold weather?
5. Why have many of Russia's natural resources remained untouched?

CRITICAL THINKING

6. **Draw Conclusions** Do you think many people live on the Kamchatka Peninsula? Why or why not?
7. **Analyze Causes and Effects** Why do you think that only plants with short roots grow in the tundra?
8. **CONNECT to Today** In 2006, Russia held a contest to develop a technology to turn coal-processing waste into usable energy. What impact might such a technology have on Russia's pollution problems?
9. **WRITING** **Make a Poster** Design and write a poster on the animal and plant life found in one of the four vegetation regions of Russia. Include pictures and captions.

Teach

Human-Environment Interaction

Talk About It

- What damage have mining operations in Russia caused? *(They have polluted the air and contaminated nearby rivers.)*
- What impact has overcutting trees in the taiga had on plant and animal life? *(It has endangered many plant and animal species.)*

4 Assess & Reteach

Assess Have students complete the Section Assessment.

Unit Resource Book: Russia & the Eurasian Republics
- Section Quiz, p. 49

Interactive Review @ClassZone.com
- Section Quiz

Power Presentations

Test Generator

Reteach Have students write down three facts that they have learned about Russia from this section. Then ask volunteers to share their lists. As a class, categorize each fact and place under an appropriate heading on the board.

Unit Resource Book: Russia & the Eurasian Republics
- Reteaching Activity, p. 53

SECTION 1 ASSESSMENT ANSWERS

Terms & Names

1. Ural Mountains: divide European and Asian Russia; Siberia: huge region in Asian Russia; tundra: treeless plain; steppe: grassy plain

Use Your Reading Notes

2. more moderate climate, suitable for agriculture, natural resources more easily available

GEOGRAPHIC FEATURES	EUROPEAN RUSSIA	ASIAN RUSSIA
Climate	milder than Asian Russia	harsh winters, short summers
Plants and Animals	rich vegetation; grazing animals	forests; reindeer
Agriculture	farming in the steppe	very little
Resources	iron ore, gold, lumber	natural gas, oil, coal

Key Ideas

3. The rivers have been connected by several canals and overland routes.
4. lies at a northern latitude and much of the land is far from the moderating influence of the oceans
5. Much of Russia's resources are located in remote areas with harsh climates.

Critical Thinking

6. Possible answer: No; the area's rugged terrain, cold weather, and many volcanoes probably prevent people from living there.
7. Possible answer: because the long winters prevent the growth of plants with deep roots
8. Possible answer: The technology might lessen the amounts of pollution in the air and in the rivers.
9. Use the rubric to score posters.

Poster Rubric

	Design	Captions	Errors
4	clear and inviting	shows thorough understanding	few or none
3	clear	shows good understanding	some
2	somewhat confusing	shows basic understanding	many
1	weak and unclear	shows little or no understanding	many major

1 Plan & Prepare

Objectives

- Explain how Russia expanded and became an empire
- Analyze why the Communist Soviet Union rose and fell

Read for the Essential Question

Help students read for a purpose by reminding them of the Essential Question: "How is Russia preserving its Eastern culture while adapting to Western influences?" Point out that they can use items in the section, such as Key Questions, special features, images, and maps to formulate answers to the question.

Vocabulary

Best Practices Toolkit

Use the ***Best Practices Toolkit*** to model strategies for vocabulary notetaking. Vary strategies throughout the year. Choose from: *Knowledge Rating, Predicting ABCs, Definition Mapping, Word Sort, Word Wheel, Frayer Model (Word Squares), Magnet Words,* and *Student VOC.*

Vocabulary Strategies, TT9–TT16

Reading Strategy

Best Practices Toolkit

Display the T Chart/Two-Column Chart transparency.

- Remind students that effects are things that happen as a result of an action or cause. Sometimes causes and effects are indicated by words or phrases such as *because, as a result,* and *due to.*
- Model the chart by filling in some causes and effects of not doing homework. Have students suggest entries.
- Remind students that they will be looking for the effect(s) of each cause listed in the chart. So, for example, they should note what happened as a result of the rise of the czars, not the reasons why the czars rose to power.

T Chart/Two-Column Chart, TT20

Section 2 Reading for Understanding

Key Ideas

BEFORE, YOU LEARNED

Russia sprawls across Eurasia, a landmass that includes both Europe and Asia.

NOW YOU WILL LEARN

The sometimes conflicting influences of the East and West have helped shape Russian history.

Vocabulary

TERMS & NAMES

czar (zahr) title for the rulers of Russia from the mid-1500s to the early 1900s

serf a peasant who is bound to the land

Peter the Great a czar who ruled Russia from 1682 to 1725 and tried to make Russia more European

Bolsheviks (BOHL•shuh•VIHKS) a group of revolutionaries who took control of the Russian government in 1917

communism a government and economic system in which nearly all political power and means of production are held by the government in the name of the people

Union of Soviet Socialist Republics (USSR), or **Soviet Union** a new nation with a Communist government created by the Bolsheviks in 1922; dissolved in 1991

Joseph Stalin a cruel 20th-century dictator who had millions of his political enemies executed

totalitarian (toh•TAL•ih•TAIR•ee•uhn) a government that controls public and private life

Cold War a conflict between the United States and the Soviet Union after World War II that never developed into open warfare

Mikhail Gorbachev (GAWR•buh•CHAWF) Soviet leader from 1985 to 1991 who increased freedoms for the Russian people

Reading Strategy

Re-create the chart shown at right. As you read and respond to the **KEY QUESTIONS**, use the chart to help you analyze causes and effects in Russia's history.

See Skillbuilder Handbook, page R8

ANALYZE CAUSES AND EFFECTS

CAUSES	EFFECTS
Rise of the czars	
Russian expansion	
Communist revolution	
Collapse of Soviet Union	

GRAPHIC ORGANIZERS
Go to Interactive Review @ ClassZone.com

468 Chapter 15

PRETEACHING VOCABULARY

English Learners

Pronounce and Preview

Pronounce each term for students. Review the meanings of words used in the definitions, such as *revolutionaries, means of production,* and *executed.*

- To modify vocabulary learning, have students complete worksheets as they read the section.

Unit Resource Book: Russia & the Eurasian Republics
- Vocabulary Practice, p. 35

Inclusion

Spell and Define Vocabulary

Have student pairs use letter tiles to spell all the vocabulary terms. One student should spell a term using the tiles, and then his or her partner should read its definition aloud. Students should trade roles, until all the terms and names have been spelled and defined.

SECTION 2 HISTORY

Orthodox Christian church in the Altai Shan in southern Siberia

Siberian pine trees, Karakol Lake, brown bear

Governing a Vast Land

Connecting to Your World

Think about a country that started out being settled by a few groups of people. Gradually the country expanded until it covered a vast area. As this happened, more and more ethnic groups influenced this country's culture. Sound familiar? If you guessed that the country is the United States, you're right. However, Russia has followed a similar pattern. Russia was settled mainly by the Slavs, people of eastern Europe. But as Russia expanded, it was strongly influenced by groups from both the East and the West.

The Rise of an Empire

KEY QUESTION How did the Russian Empire develop?

By the 800s, Slavic groups established many settlements in what is now called European Russia. In the 900s, the region came under the control of the Varangians. These people, also called Rus, were most likely Vikings from Scandinavia. The name "Russia" is taken from this group. Eventually, civil war weakened the Rus. In the 1200s, they fell to invaders from Mongolia, a region to the southeast.

Kremlin During the 1100s, the Rus began to develop a fortified town, or Kremlin. The walled fortress complex stands today in Moscow.

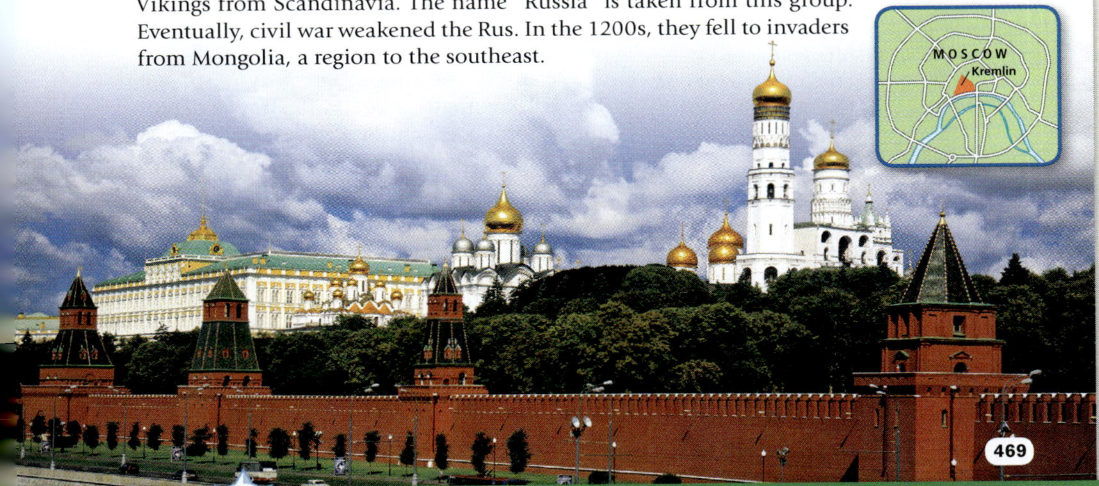

2 Focus & Motivate

3-Minute Warm-Up

Write on the board or display the transparency:

- Why might governing a vast land be difficult? *(People in different areas have different needs. It is hard to govern people far away. Resources need to be transported long distances.)*

Unit Transparency Book
- 3-Minute Warm-Ups, TT1

Connecting to Your World

More About . . .

The Settlement of Russia

Although the Slavs were the main group to settle Russia, Huns, Avars, Goths, and Magyars were also among the early settlers. The early settlers used the region's many rivers, such as the Volga and Don, to expand into Russia.

3 Teach

The Rise of an Empire

Think, Pair, Share

- How did Russia get its name? *(It's named after the Varangians, or Rus, who took control in the 900s.)*
- What changes did Peter the Great make in Russia? *(westernized Russia; reorganized the government; moved capital to St. Petersburg)*
- **Analyze Causes and Effects** How did Moscow gain power and break away from the Mongols? *(Moscow gained power through wealth from trade and taxes.)*

SECTION 2 PROGRAM RESOURCES

Unit Resource Book: Russia and the Eurasian Republics

ON LEVEL

Unit Resource Book
- Reading Study Guide, p. 5
- Outline Map Activities, p. 39
- Section Quiz, p. 50

STRUGGLING READERS

Unit Resource Book
- Reading Study Guide with Additional Support, p. 13
- Vocabulary Practice, p. 35
- Skillbuilder Practice, p. 36
- Section Quiz, p. 50
- Reteaching Activity, p. 54

eEdition with Audio DVD-ROM

ENGLISH LEARNERS

Pupil Edition in Spanish

eEdition with Audio DVD-ROM

Unit Resource Book
- RSG (Spanish), p. 21
- RSG with Additional Support (Spanish), p. 29

Multi-Language Glossary

Test Generator
- Section Quiz in Spanish

INCLUSION

Unit Resource Book
- RSG with Additional Support, p. 13
- Section Quiz, p. 50
- Reteaching Activity, p. 54

GIFTED & TALENTED

Unit Resource Book
- History Makers, p. 41
- Section Quiz, p. 50

PRE-AP

Unit Resource Book
- Section Quiz, p. 50

TECHNOLOGY

Unit Transparency Book
- 3-Minute Warm-Ups, TT1
- Essential Question, TT2
- Critical Thinking, TT4
- Cultures, TT9

Daily Test Practice Transparencies
- Section 2, TT53

Power Presentations

ClassZone.com

World Cultures and Geography Video on DVD

RESEARCH & WRITING CENTER

History Makers

Peter the Great

As part of his plan to modernize Russia, Peter traveled throughout Europe (often in disguise), sent other Russians to Europe, and encouraged European craftspeople, architects, merchants, and others to come to Russia. At about six and a half feet tall, Peter was a tall man and enjoyed physical work. Although he is said to have had a cruel streak—sometimes hitting government officials with a stick—Peter the Great's leadership helped Russia become a great empire.

Unit Resource Book: Russia & the Eurasian Republics
- History Makers: Ivan III, p. 41

More About . . .

St. Petersburg

St. Petersburg was built in 1703 to give Russia a port on the Baltic Sea. St. Petersburg became the Russian capital in 1712, although Sweden claimed the land it was built on until 1721. St. Petersburg also became a cultural center and is still home to one of the world's great art museums, the Hermitage. At different times in its history, St. Petersburg has been known as Petrograd and Leningrad.

Moscow's Princes Gain Power The Mongols allowed the grand prince of Moscow to collect taxes and keep a portion of them. This money, along with wealth gained through trade, helped Moscow to grow in power. At the same time, Mongol control weakened. Led by Ivan III, the Moscow state broke away from the Mongols in 1480.

Ivan IV became the grand prince of Moscow and, in 1547, the first ruler to be officially called **czar** (zahr). The word *czar* comes from "caesar," the title used for Roman emperors. Ivan IV came to be known as Ivan the Terrible because he was a harsh and suspicious ruler. He murdered hundreds of aristocrats and even killed his eldest son.

Ivan IV and the other czars who came after him passed several laws that bound peasants to the land. These peasants were called **serfs**. Serfdom formed the basis of the Russian economy until the late 1800s. Because of their poverty, harsh working conditions, and the strict controls on their lives, the serfs resented the czars' authority.

HISTORY MAKERS

Peter the Great 1672–1725

As a child, Peter had a lot of energy and curiosity. He would often organize his friends into an army and then play at war. As an adult, Peter still showed many of his childhood qualities. A dynamic ruler, he built up the navy and army. His enthusiasm for Western ways was extreme. He insisted that nobles shave off their beards, wear European fashions, and study the sciences. Those who opposed these changes were often punished.

ONLINE BIOGRAPHY
For more on the life of Peter the Great, go to the **Research & Writing Center @ClassZone.com**

Peter the Great After Ivan IV's death, Russia suffered under a series of weak rulers. Then in 1682, two half-brothers, Peter I (later known as **Peter the Great**) and Ivan, ruled together. Peter assumed full power when Ivan died in 1696.

The new czar was strongly influenced by the West and worked hard to modernize Russia. Peter believed that a more modern Russia would be better able to compete with Europe militarily and economically. As a result, he introduced European-style customs, factories, and schools in Russia. Peter even moved Russia's capital west from Moscow to an area on the Gulf of Finland. Architects modeled the new capital, called St. Petersburg, after European cities. Its construction cost huge amounts of money and the lives of about 30,000 workers.

Peter also reorganized Russia's government. He replaced inefficient administrative departments with a system of well-organized units. Each one had authority in a certain area, such as finances or justice.

In spite of Peter's efforts, many Russians remained suspicious of western Europeans. They considered western Europeans outsiders. Religious differences widened the gap. The Russians had adopted the Eastern Orthodox branch of Christianity, while western Europeans were mostly Catholics or Protestants.

DIFFERENTIATING INSTRUCTION

English Learners

Build Vocabulary

Read aloud this sentence from the History Makers feature: "A dynamic ruler, he built up the navy and army." Point out that the word *dynamic* in the sentence means "energetic and forceful." Explain that *dynamic* can also mean "marked by productive and frequent change." An antonym for the latter meaning is *static.*

Struggling Readers

Read a Map

Have students use the map at the beginning of the chapter to locate St. Petersburg. Ask students to describe the location. *(on the Black Sea; near Finland; in the northwestern part of the country)* Ask students to explain the advantages and disadvantages of this location. *(Advantages: port city; closer contact with the West; Disadvantages: isolated from most of Russia)*

Russia Expands After Peter's death in 1725, several rivals struggled to gain the throne. Eventually, in 1762, Empress Catherine II (known as Catherine the Great) came to power.

Catherine acquired the northern shore of the Black Sea and the Crimea through a series of wars with the Ottoman Empire. By gaining access to the Black Sea, Russia secured a warm-water port, one in which the water does not freeze in winter. As a result, the port could be used all year for trade. Also, through political deals, Russia obtained much of Poland and its territory.

Like Peter, Catherine admired European culture. She supported the arts, such as ballet and Italian opera. She also tried to pass legal reforms but did little to improve the life of the Russian serfs. In fact, when discontented serfs rebelled in 1773 and 1774, Catherine brutally crushed the revolt. She then tightened control over the serfs.

Catherine the Great died in 1796. Succeeding czars continued to expand Russia's territory. The empire also survived an invasion by the French emperor Napoleon I in 1812. By the late 1800s, Russia controlled land across Central Asia to the Pacific. To connect this vast expanse, Russia started to build the Trans-Siberian Railroad. Completed in 1916, it extended nearly 6,000 miles and greatly helped develop Siberia. You will learn more about this railroad in Animated Geography at the end of this section.

SUMMARIZE Explain how the Russian Empire developed. through war and political deals; by exploiting the serfs; by modernizing the region according to western influences

CONNECT History & Geography

Napoleon's 1812 March to Russia

In June 1812, Emperor Napoleon I of France and his army of about 600,000 men invaded Russia. Napoleon and his forces advanced on Moscow and occupied it for a time, but Napoleon knew the harsh Russian winter was quickly approaching. Soon he ordered his army to retreat. His soldiers fought against snowstorms, freezing temperatures, and attacking Russian soldiers. By the time Napoleon got out of Russia, only about 10,000 of his men had survived.

CRITICAL THINKING

Analyze Causes In what ways did Russia's geography bring about Napoleon's defeat?

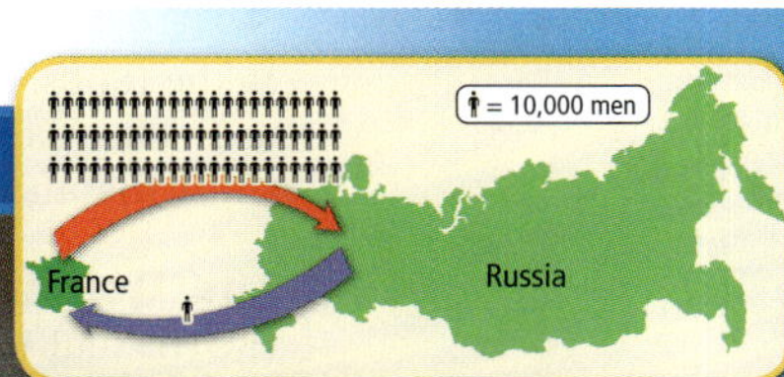

INTERDISCIPLINARY ACTIVITIES

CONNECT to Math

Represent Fractions and Calculate Percentages

Remind students that Napoleon started out with an army of about 600,000 and left with only 10,000. Ask students to write a fraction showing how many soldiers survived. *(1/60)* Then have them make a pie chart showing these results. Have students represent the number who died or deserted as a percentage. *(about 98%)*

CONNECT to Language Arts

Write a Journal Entry

Have students write a journal entry from the point of view of a soldier in Napoleon's army. Ask students to describe the Russian winter, their experiences as the French army retreated, and the results of their Russian campaign.

More About . . .

Catherine the Great

The marriage between Catherine and her husband Peter, the grandson of Peter the Great, was arranged. Both Catherine and Peter wanted to rule Russia and, as rivals, grew to despise one another. Peter was enthralled with all things from Prussia (Russia's neighbor), while Catherine held a deep love for all things Russian. Upon becoming emperor in 1761, Peter soon ended a war with Prussia and took other steps that made him very unpopular. Catherine, on the other hand, had the support of the Russian army and many of its people. In 1762, she deposed her husband and took over the throne of Russia. Shortly thereafter he was assassinated—most likely by Catherine's supporters—and Catherine was crowned empress in Moscow.

CONNECT History & Geography

Napoleon's 1812 March to Russia

When Napoleon invaded Russia, the Russians realized that they were vastly outnumbered. The French and Russian armies fought briefly at Borodino near Moscow, but the battle was inconclusive. The Russian army abandoned Moscow, and Napoleon's army took control of a deserted city short on supplies. On the day they took over Moscow, an extensive fire broke out, robbing the French army of shelter.

CRITICAL THINKING ANSWER

Analyze Causes Russia's harsh winter climate and rugged terrain helped defeat Napoleon.

Unit Resource Book: Russia & the Eurasian Republics
- Skillbuilder Practice: Using a Database, p. 36

Russian and Soviet Expansion

CONNECT Geography & History

In the mid-1400s, Russia consisted of only a small area around Moscow. Over roughly 300 years, Russia grew to encompass huge portions of Asia and extended farther west into Europe. Russia reached its greatest extent in 1947 under the Soviet Union. Since the breakup of the Soviet Union in 1991, Russia continues to wield influence over much of its former lands.

ANSWERS

1. **Location** Moscow
2. **Region** between 1599 and 1689

Unit Resource Book: Russia & the Eurasian Republics

- Outline Map Activities: Russia and the Eurasian Republics, p. 39

Teach

Revolution Brings Communism

Talk About It

- What type of government did the Bolsheviks establish when they took over Russia? *(a Communist form of government)*
- How did Stalin exert his control? *(by controlling artists, education, media, and religion)*
- **Analyze Causes and Effects** What unexpected result did Gorbachev's policy of openness have on the Soviet people? *(It allowed people to complain about economic problems and call for more freedom.)*

Russian and Soviet Expansion

CONNECT Geography & History

1. **Location** What city was contained in Russia's original territory?
2. **Region** When did the Russian Empire absorb most of Siberia?

Revolution Brings Communism

KEY QUESTION Why did the Soviet Union rise and fall?

In the 1800s, the czars continued their harsh rule. A few reforms, such as freeing the serfs, were carried out. Many people, though, did not think that these reforms went far enough.

The Bolshevik Revolution Then, in 1917, the Bolshevik (BOHL•shuh•VIHK) Party overthrew czar Nicholas II, and Vladimir Lenin came to power. The **Bolsheviks** believed that all political power and means of production should be held by the government in the name of the people. This type of system is called **communism**.

In 1922, the Bolsheviks created a new Communist nation called the **Union of the Soviet Socialist Republics (USSR)**, or **Soviet Union**. By 1929, **Joseph Stalin** had become dictator. Stalin was a cruel ruler who had millions of his political enemies executed. As dictator, Stalin also set up a **totalitarian** (toh•TAL•ih•TAIR•ee•uhn) government. This form of government controls every aspect of public and private life. Stalin controlled all newspapers, film, radio, and other sources of information. He also censored writers, composers, and other artists who did not glorify his achievements. Stalin's control even extended to education and religion. He built a police state to enforce his power.

DIFFERENTIATING INSTRUCTION: TIERED ACTIVITIES

OBJECTIVE Research what life was like under Stalin.

Basic

Provide student groups with articles on life under Stalin. Then have them discuss what they learned. Finally, have students summarize the key points of their discussion and share them with the rest of the class.

On Level

Have students read the first three sections of the "Requiem" by Anna Akhmatova. Invite groups of students to discuss the poem and then read it aloud to the class.

Challenge

Have students read a brief section from *The Gulag Archipelago* by Alexander Solzhenitsyn. Have them do research to learn more about the gulag, or forced labor system. Ask students to discuss their reading and then share what they learned.

Superpower and Collapse The Soviet Union joined the Allies during World War II and gained control of many countries in Eastern Europe during and after the war. Before long, deep distrust developed between the Soviet Union and the West. The United States and the Soviet Union each began to produce more and more powerful nuclear weapons. However, open warfare between the two never broke out. This conflict is called the **Cold War**. The buildup of military weapons during the Cold War drained money from the Soviet Union. As a result, food and other basic items were in short supply. Soviet citizens had to wait in line for hours to buy ordinary items.

In 1985, Soviet leader **Mikhail Gorbachev** (GAWR•buh•CHAWF) tried to revive the nation's economy by reducing government control. He also began allowing more freedom of expression and a free flow of information. However, Gorbachev's policy of openness had an unexpected result. It allowed Soviet citizens the right to complain about economic problems. They also began to call for more freedom.

The efforts to reform the Soviet Union led to its breakup. In 1991, the Soviet Union collapsed, and the Cold War ended. Russia continues to be a world leader. But, as you will learn in Section 4, Russia is still struggling to overcome many problems.

he Soviet nion rose response the czars' arsh rule nd inability reform; collapsed ecause its onomy was ruggling nd the eople were emanding ore freedom. ►

ANALYZE CAUSES AND EFFECTS Tell why the Soviet Union rose and fell.

Mikhail Gorbachev
Gorbachev brought about great change, but he did not want to end communism.

CONNECT to the Essential Question

How is Russia preserving its Eastern culture while adapting to Western influences?

Ask students what they have learned so far that can help them answer this question. Students might mention the following:

- Peter the Great and Catherine the Great were deeply influenced by Western culture and introduced European-style customs, factories, and schools in Russia.
- Distrust developed between the Soviet Union and the West during the Cold War, isolating Russia from the West.

Section 2 Assessment

ONLINE QUIZ For test practice, go to **Interactive Review @ClassZone.com**

TERMS & NAMES

1. **Explain the importance of**
   - czar
   - Bolsheviks
   - totalitarian
   - Cold War

USE YOUR READING NOTES

2. **Analyze Causes and Effects** Use your completed chart to answer the following question:

   What effect do you think the Cold War had on Western people and on the people of the Soviet Union?

CAUSES	EFFECTS
Rise of the czars	
Russian expansion	
Communist revolution	
Collapse of Soviet Union	

KEY IDEAS

3. Why was the Moscow state able to break away from Mongol rule?
4. What were the major accomplishments of Catherine the Great?
5. What caused the Cold War?

CRITICAL THINKING

6. **Make Inferences** Why do you think Peter the Great wanted to westernize Russia?
7. **Compare and Contrast** How were the collapses of czarist Russia and of the Soviet Union similar and different?
8. **CONNECT to Today** The Trans-Siberian Railroad continues to be a popular means of transportation today. Why do you think the railroad is still important?
9. **HISTORY Prepare an Oral Report** In your report, contrast western Europe and Eastern-influenced Russia during the time of Peter the Great. Include information on government, economy, and culture.

4 Assess & Reteach

Assess Have students complete the Section Assessment.

Unit Resource Book: Russia & the Eurasian Republics
- Section Quiz, p. 50

Interactive Review @ClassZone.com
- Section Quiz

Power Presentations

Test Generator

Reteach Divide the class into two groups. Have one group create a time line of events during czarist Russia. Have the other create a time line of events since the Bolshevik Revolution.

Unit Resource Book: Russia & the Eurasian Republics
- Reteaching Activity, p. 54

SECTION 2 ASSESSMENT ANSWERS

Terms & Names

1. czar: ruler of Russia; Bolsheviks: Communist revolutionaries; totalitarian: government that controls public and private life; Cold War: conflict between the United States and the Soviet Union

Use Your Reading Notes

2. made people fearful, distrustful

CAUSES	EFFECTS
Rise of the czars	serfs; westernized; expanded
Russian expansion	Black Sea port; Trans-Siberian Railroad
Communist revolution	ended czars; totalitarian government
Collapse of Soviet Union	Cold War ended

Key Ideas

3. Moscow gained in wealth and power through the collection of taxes and trade.
4. gained access to the Black Sea; continued to westernize Russia
5. the deep distrust between the Communist Soviet Union and the democratic countries of the West

Critical Thinking

6. to enable Russia to better compete politically, economically, and culturally with Europe
7. Both collapsed because people desired more freedom. Revolution overthrew czarist Russia, while the Soviet Union ended without warfare.
8. The railroad is most likely still the most efficient form of transportation into and out of Siberia.
9. Use the rubric to score oral reports.

Oral Reports Rubric

	Structure	Visuals
4	clear and informative	many helpful visuals
3	adequate to explain the ideas	several good visuals
2	confusing	few visuals
1	unclear	no visuals

1 Plan & Prepare

Objectives: Explain the importance of the Trans-Siberian Railroad and identify the challenges of building it.

Visual Vocabulary: Have students discuss the illustration. Then read the descriptions of items listed below and ask students to locate each one in the illustration. Then tell them the precise term shown in boldface. Add terms to a class vocabulary list, illustrated poster, or Word Wall.

- special boat used to cross water **(ferry)**
- tool used for cutting or chopping **(axe)**
- mixture of sand, gravel, and cement or mortar used for construction **(concrete)**
- framework used to support a structure, such as a bridge **(truss)**
- set of steel rails that a train runs on **(track)**
- distance between the set of rail tracks **(gauge)**

2 Teach

Engage . . .

DISPLAY "The Trans-Siberian Railroad" feature from the **Power Presentations** DVD-ROM or the *Animations Center* at **ClassZone.com.** As the animation loads on screen, ask students:

- What cities are located along the Trans-Siberian Railroad?
- Why was the Trans-Siberian Railroad built?
- What challenges did railroad workers face in building the railroad?
- What changes did the railroad bring to Russia?
- Why is the Trans-Siberian Railroad important to Russia today?

Explore . . .

ROLL OVER sections of the illustration or click on items in the menu. Ask students to choose which areas they would like to explore further.

- Then return to the discussion questions. Have any answers changed?
- Have students jot down a list of at least three new questions that arise from studying the rollover features.

Animated GEOGRAPHY

The Trans-Siberian Railroad

Click here to to explore the Trans-Siberian Railroad @ ClassZone.com

CONNECTING EAST AND WEST

In the 1800s, Siberia was Russia's frontier. The region had plentiful natural resources, but these remained largely untapped—until 1891. That's when construction on the Trans-Siberian Railroad began, and the process of linking East and West got under way.

Click here to see how the railroad helped small villages develop into large cities.

Click here to learn how the workers lived, laid tracks, and blasted through the ice.

474 Chapter 15

DIFFERENTIATING INSTRUCTION: TIERED ACTIVITIES

OBJECTIVE Explore the impact that the Trans-Siberian Railroad had on Russia's people and economy and consider its role in Russian and Soviet history.

Basic

Have students work in small groups to research the impact the railroad had on small towns and villages along its route. Ask students to find or draw before-and-after pictures that show how the small towns developed into large cities.

The Trans-Siberian Railroad Is Built

The first line of the Trans-Siberian Railroad was built between 1891 and 1916.

- Work began at both ends of the line: in Moscow and in Vladivostok.
- One of the greatest obstacles to completion was Lake Baikal, where for a time, a ferry carried the train across the water.
- When the line was completed, it became—and still is—the longest single track in the world.

GeoActivity

Interview Railroad Workers

With a partner, take turns role-playing a worker on the Trans-Siberian Railroad and a newspaper reporter. Ask the worker about the challenges and dangers of the job and life in the crowded barracks. Record your interviews and play them for the class.

On Level

Have student groups conduct the same research and include a brief written description of at least five cities or towns along the route. Ask students to focus on the impact the railroad had on each city and its people.

Challenge

Have students work together in groups to create a time line of important events in the life of the Trans-Siberian Railroad. Students should begin with the events that led to its construction and continue through the present. Make sure students include entries that describe the railroad's role during the Bolshevik Revolution, World War I, and World War II.

Explain . . .

CHOOSE topics to help students answer their questions, such as:

- **Method of Railroad Construction:** The Trans-Siberian railroad was not built from Moscow to the east. Instead, it was built in several directions at once: east from Moscow, west from Vladivostok, and in a separate segment in the middle.
- **How the Route Changed:** The route originally ran through Manchuria, but the Russians, fearing a Japanese takeover of the region, built a separate section that skirted Manchuria and went through Vladivostok. An alternate route was also built around Lake Baikal.
- **Length of the Railroad:** The Trans-Siberian Railroad is almost 6,000 miles long. The trip, traveling from one end to the other, takes about eight days.

3 Wrap-Up

Elaborate . . .

ASSIGN students partners. Have each pair create a word game with information about the Trans-Siberian Railroad. Students may create name games, crossword puzzles, fill-in-the-blank questions, or true-false questions. Have students swap with another pair and play each other's games.

Evaluate . . .

Interview Railroad Workers

Activity Rubric

4 Excellent The student's interview responses accurately describe working and living conditions and provide several details.

3 Strong The student's interview responses are fairly accurate and provide some detail.

2 Fair The student's interview responses provide only a few details.

1 Poor The interview responses lack detail and fail to describe the working and living conditions.

1 Plan & Prepare

Objectives

- Identify the elements that have shaped the daily lives of the Russian people
- Describe Russia's cultural heritage
- Explain how Soviet art was used as propaganda

Read for the Essential Question

Help students read for a purpose by reminding them of the Essential Question: "How is Russia preserving its Eastern culture while adapting to Western influences?" Point out that they can use items in the section, such as Key Questions, special features, images, and maps, to formulate answers to the question.

Vocabulary

Best Practices Toolkit

Use the ***Best Practices Toolkit*** to model strategies for vocabulary notetaking. Vary strategies throughout the year. Choose from: *Knowledge Rating, Predicting ABCs, Definition Mapping, Word Sort, Word Wheel, Frayer Model (Word Squares), Magnet Words,* and *Student VOC.*

Vocabulary Strategies, TT9–TT16

Reading Strategy

Best Practices Toolkit

Display the Web Diagram transparency.

- Remind students that main ideas are the most important ideas in a segment of text.
- Model filling in main ideas using the subheads throughout the section. Have students suggest entries.
- Remind students that main ideas in each block of text in the section will help them answer the Key Questions.
- Display the graphic organizer for students to review.

Web Diagram, TT28

SECTION 3 Reading for Understanding

Key Ideas

BEFORE, YOU LEARNED

Russia has struggled to unite European and Asian influences throughout its history.

NOW YOU WILL LEARN

The influences of Europe and Asia are reflected in the people and culture of Russia.

Vocabulary

TERMS & NAMES

Cyrillic (suh•RIHL•ihk) **alphabet** the official writing system of Russia

icon religious image used by Orthodox Christians in their worship

Leo Tolstoy (TOHL•stoy) a 19th-century writer who is one of Russia's greatest novelists

propaganda information that deliberately tries to influence opinion

BACKGROUND VOCABULARY

Orthodox Christianity the Eastern branch of Christianity that spread to Russia in the 900s and became the state religion

Islam a religion based on the teachings of the prophet Muhammad

Buddhism a religion that began in India and that is based on the teachings of Siddhartha Gautama

Judaism a religion based on the Hebrew Scriptures, whose followers are called Jews

REVIEW

ethnic group a group of people who share language, customs, and a common heritage

Visual Vocabulary dual-language sign depicting the Cyrillic alphabet

Reading Strategy

Re-create the diagram shown at right. As you read and respond to the **KEY QUESTIONS**, use the diagram to record main ideas about Russian culture. Use a new diagram for each aspect of Russian culture you learn about.

Skillbuilder Handbook, page R4

FIND MAIN IDEAS

GRAPHIC ORGANIZERS Go to **Interactive Review** @ClassZone.com

PRETEACHING VOCABULARY

English Learners

Pronounce and Preview

Pronounce each term for students. Review the meanings of words used in the definitions, such as *deliberately, state religion,* and *heritage.*

- To modify vocabulary learning, have students complete worksheets as they read the section.

Unit Resource Book: Russia & the Eurasian Republics
- Vocabulary Practice, p. 35

Inclusion

Create Vocabulary Posters

Write each vocabulary term in large print on poster board or chart paper. Read the definitions aloud and have students create an illustration on the poster that captures the meaning of the term. Display the posters in the room for reference as students read the section.

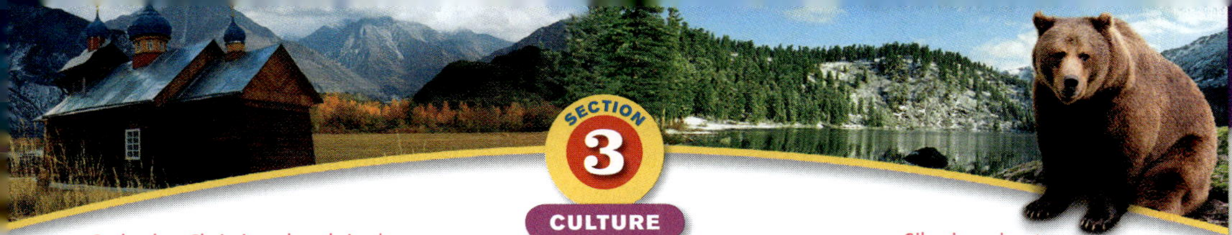

SECTION 3 CULTURE

Orthodox Christian church in the Altai Shan in southern Siberia

Siberian pine trees, Karakol Lake, brown bear

Blending Europe and Asia

Connecting to Your World

The United States is made up of **ethnic groups** from all over the world. Their influence has shaped our country. You can see this in many big cities. There, you will find signs and advertisements in many different languages, including Spanish, Korean, and Arabic. Another country that has been strongly influenced by ethnic groups is Russia. In fact, the immense territory of Russia contains more than a hundred different ethnic groups. Most of these groups come from Europe and Asia and have made many cultural contributions.

The Russian People

KEY QUESTION What elements shape the daily lives of the Russian people?

Most Russians live in western, or European, Russia. The rugged terrain and harsh climate in eastern Russia have kept populations low in that part of the country. Eighty percent of Russia's people are ethnic Russians, who are descended from the Slavs. Other ethnic groups include the Tatars, Ukrainians, Belarusians, and Chechens (CHECH•uhnz).

Ethnic Russians Russian women relax in front of Kazan Cathedral in St. Petersburg. Ethnic Russians form one of the largest ethnic groups in the world, with over 160 million people worldwide.

2 Focus & Motivate

3-Minute Warm-Up

Write on the board or display the transparency:

- What might motivate people of different ethnic groups to settle in an unfamiliar land? *(economic opportunities, greater freedoms, untenable conditions in homeland)*

Unit Transparency Book
- 3-Minute Warm-Ups, TT1

Connecting to Your World

More About . . .

Ethnic Groups

According to the *CIA World Factbook,* Russia's ethnic breakdown (as of 2002) looks like this: about 80% Russian, 4% Tatar, 2% Ukrainian, 1% Bashkir, 1% Chuvash, and 12% other.

3 Teach

The Russian People

Reader, Recorder, Reporter

- Why are populations low in the eastern, or Asian, part of Russia? *(because of the rugged terrain and harsh climate)*
- **Find Main Ideas** What religion is observed by most of the Russian people? *(Orthodox Christianity)*

SECTION 3 PROGRAM RESOURCES

Unit Resource Book: Russia and the Eurasian Republics

ON LEVEL
Unit Resource Book
- Reading Study Guide, p. 7
- Section Quiz, p. 51

STRUGGLING READERS
Unit Resource Book
- Reading Study Guide with Additional Support, p. 15
- Vocabulary Practice, p. 35
- Skillbuilder Practice, p. 36
- Section Quiz, p. 51
- Reteaching Activity, p. 55

eEdition with Audio DVD-ROM

ENGLISH LEARNERS
Pupil Edition in Spanish
eEdition with Audio DVD-ROM
Unit Resource Book
- RSG (Spanish), p. 23
- RSG with Additional Support (Spanish), p. 31

Multi-Language Glossary
Test Generator
- Section Quiz in Spanish

INCLUSION
Unit Resource Book
- RSG with Additional Support, p. 15
- Section Quiz, p. 51
- Reteaching Activity, p. 55

GIFTED & TALENTED
Unit Resource Book
- Section Quiz, p. 51

PRE-AP
Unit Resource Book
- Section Quiz, p. 51

TECHNOLOGY
Unit Transparency Book
- 3-Minute Warm-Ups, TT1
- Essential Question, TT2
- Critical Thinking, TT5
- Cultures, TT10

Daily Test Practice Transparencies
- Section 3, TT54

Power Presentations
ClassZone.com
World Cultures and Geography Video on DVD

Comparing Regions

Russia and the United States: Urban Areas

About 80 percent of the population of the United States lives in urban areas compared with about 75 percent of the Russian population. Russia's two largest cities are similar in size to the largest cities in the U.S. Moscow and St. Petersburg have populations of about 10.1 million and 4.7 million, respectively, while the largest cities in the U.S., New York and Los Angeles, have populations of about 8.1 million and 3.8 million, respectively.

Because Russia is about twice the size of the United States, the country is far less crowded. In fact, Russia's overall population density is only about 22 people per square mile—compared with the population density in the United States of about 80 people per square mile.

Unit Resource Book: Russia & the Eurasian Republics

- Skillbuilder Practice: Using a Database, p. 36

More About . . .

Orthodox Christianity

Eastern Orthodox Christianity developed over many years. It became a separate branch of Christianity in 1054, when the Great Schism occurred and Eastern Orthodox Christianity divided from Western Catholicism. The two churches differ in such areas as the use of icons and the dates of some Christian celebrations.

City and Country Life About three-fourths of Russia's people live in cities. The two largest cities are Moscow and St. Petersburg. Red Square in Moscow is the most famous city square in Russia. It was built in the late 1400s and served as a meeting place for the Russian people. The name of the square comes from an old Russian word for "beautiful," which today also means "red." You can see some of Red Square's most famous buildings on the opposite page.

During World War II, many of the buildings in Russia's urban areas were destroyed. After the war, Russia had housing shortages. The Soviet government tried to solve this shortage by building huge high-rises. Many people still live in these dwellings.

In the country, more people live in single-family homes. However, in remote areas, some homes lack plumbing, running water, gas, and electric power. The quality of health care and education is lower in rural areas than in cities. As a result, many people from rural areas have moved to the cities in search of jobs.

Religion By far the most common religion in Russia is **Orthodox Christianity**. During the 900s, missionaries brought this religion to Russia, and with it, the **Cyrillic** (suh•RIHL•ihk) **alphabet**. This alphabet was soon adopted and used to write the Russian language. Around 988, Vladimir I made Orthodox Christianity the state religion.

Before long, Russian Orthodox churches were constructed in the Byzantine style. These churches have distinctive onion-shaped domes. St. Basil's Cathedral in Moscow's Red Square, shown on the opposite page, is a famous example of Russia's Byzantine church style. The churches also usually contain religious paintings called **icons**. Icons are often painted according to strict rules set up by the church authorities and frequently depict God, Jesus, angels, or saints.

Orthodox Christmas Russian girls light candles inside a church near Moscow during a midnight Christmas service. Orthodox Christians celebrate Christmas on January 7.

DIFFERENTIATING INSTRUCTION

Struggling Readers

Find Main Ideas

To help students find main ideas about religion in Russia, have them fill in the web diagram shown. Students should fill in the ovals with terms that describe religion in Russia and brief explanations of each one.

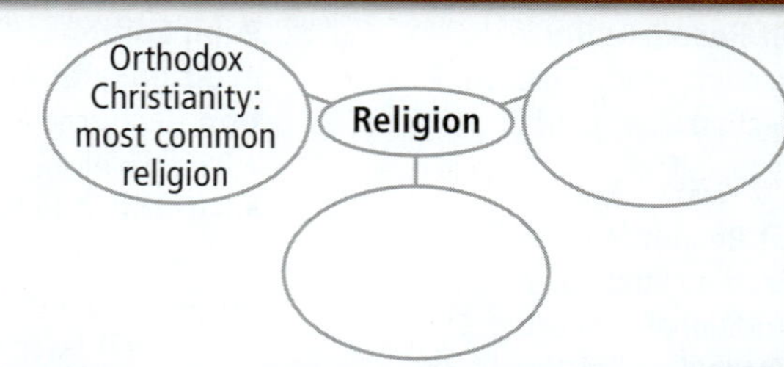

Moscow's Red Square

Red Square lies just outside the Kremlin's eastern wall. For over 500 years, the huge square (over 500,000 square feet) has served as a meeting place for the Russian people. Over the centuries, they have gathered there to celebrate festivals, hear announcements from the czar, and witness parades of Soviet military might. Today, Red Square is a popular spot for both Russians and tourists. The images here show some of the square's most legendary sites.

GUM Department Store Moscow's "State Department Store" is actually a huge mall that contains over 150 stores.

Lenin's Mausoleum Lenin's body—or at least a wax copy—lies in state in this red and black pyramid.

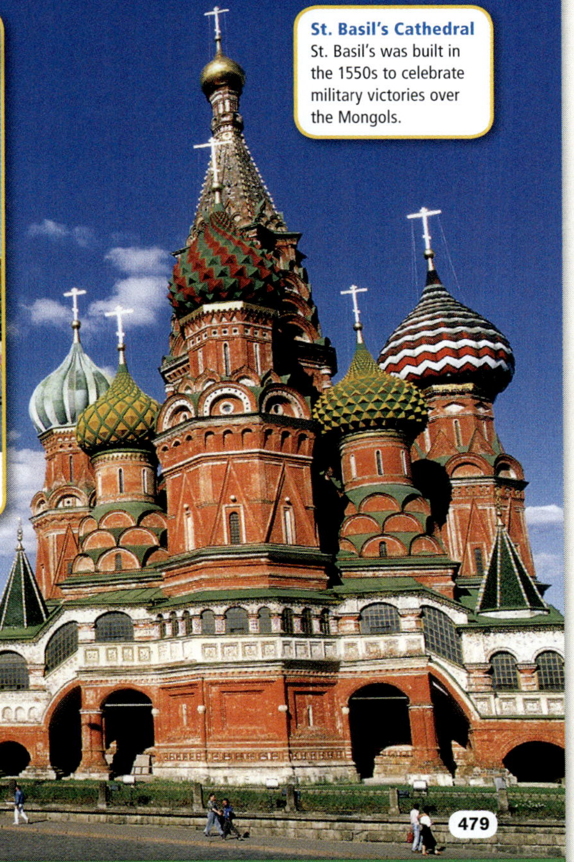

St. Basil's Cathedral St. Basil's was built in the 1550s to celebrate military victories over the Mongols.

DIFFERENTIATING INSTRUCTION: TIERED ACTIVITIES

OBJECTIVE Identify and describe some of the key sights in Moscow's Red Square.

Basic

Have students choose one of the images from Red Square. Ask them to imagine that the image is on the front of a postcard. Then have them write a note on the back of the card that describes the structure and tells the reader something about it.

On Level

Have groups of students choose one of the structures shown above and research to learn more about it. Encourage each group to share its findings in an oral report before the class.

Challenge

Have groups work together to create a virtual tour of Red Square. The tour should include visuals and information about the square and its key sights.

More About . . .

Moscow's Red Square

Moscow's Red Square, previously known as Trinity Square—then nicknamed "Fire Square" for its many fires—was given its current name in the 1600s. The proper name of St. Basil's Cathedral is Church of the Intercession. The cathedral's popular name honors a prophet named Basil, who predicted the Moscow fire of 1547.

Lenin's Mausoleum was built in 1924 to provide a resting place for the founder of the Communist state. During the Soviet era, Red Square became an important meeting place of the Communists. Members of the Communist leadership claimed that some of the buildings in Red Square weren't appropriate to the new regime and wanted to tear some of them down—including the GUM Department Store. Fortunately, the start of World War II diverted attention from this plan.

More About . . .

Lenin's Mausoleum

The red stones used in the construction of Lenin's Mausoleum represent communism, and the black stones represent mourning. For many years, soldiers guarded its entrance. During military parades, Soviet leaders would review the display of might from atop the mausoleum. And, after the defeat of Germany in World War II, hundreds of Nazi banners were thrown at its base. Today, however, the mausoleum is something of a curiosity. The long lines of people waiting to view Lenin's body have greatly declined. The dramatic changes that have swept through Russia have left the monument behind.

More About . . .

Russian Tea

Tea is thought to have been introduced in Russia by the 1600s. Russian-style tea is quite strong and often served with something sweet, such as honey, sugar, or jam. George Orwell wrote, "tea—unless one is drinking it in the Russian style—should be drunk without sugar." To make tea, Russians first prepare a tea concentrate called *zavarka.* Then hot water is added to a small amount of this concentrate in a glass or cup. Traditionally Russian tea is made in a samovar (Russian teapot) and served in glasses with silver holders.

CONNECT History & Culture

Russia and the Olympics

Politics played a major role in the 1980 and 1984 Olympics. After the 1979 Soviet invasion of Afghanistan, President Carter called for a U.S. boycott of the games in Moscow. Sixty other countries joined the boycott. As a result, the Soviet Union won 195 medals, 80 of them gold.

At the 1984 Olympics, held in Los Angeles, the Soviet Union and Eastern bloc countries held their own boycott. The United States won 174 medals, including a record 83 gold.

CRITICAL THINKING ANSWER

Make Inferences They are dedicated to their sports; they have a strong desire to win; they are hard working.

Islam, **Buddhism**, **Judaism**, and other forms of Christianity are also practiced. In fact, over 15 percent of the population practices Islam. However, many Jews left Russia for the United States and Israel because they were persecuted for practicing their religion.

Food Russians generally eat a hearty diet that uses a lot of root vegetables, such as beets, carrots, onions, and potatoes. These vegetables grow well in the cool Russian climate. They are also warm and filling foods for the cold Russian winters.

Traditional Russian dishes are eaten throughout the world. They include thin pancakes, served with smoked salmon and sour cream, called *blinis* (BLEE•neez), and beet soup called *borscht* (bawrsht). Tea is a traditional beverage, which Russians enjoy piping hot and strong.

Sports Not surprisingly, winter sports, such as hockey, ice skating, and skiing, are popular in Russia. However, the most popular sport is soccer. Tennis has also soared in popularity. In fact, in 2006, five of the top 15 women tennis players in the world were Russian, including tennis star Maria Sharapova. Russian athletes also compete seriously in the Olympics.

FIND MAIN IDEAS Identify some of the elements that shape the daily lives of the Russian people. religion, conditions in the city and country, food, sports

CONNECT History & Culture

Russia and the Olympics During the Soviet era, Russian Olympic athletes were supported by the government. The athletes excelled in ice skating, ice hockey, and gymnastics. Today, the old Soviet system of supporting athletes no longer exists. Nonetheless, between the time of the Soviet breakup and 2006, Russia has won four Olympic gold medals in men's figure skating and four in pairs skating. The Russians who won the 2006 Olympic gold medal in pairs figure skating are shown here.

CRITICAL THINKING

Make Inferences What can you infer about the dedication of the Russian athletes who compete in the Olympics?

Gold medalists Tatiana Totmianina and Maxim Marinin ▼

480 Chapter 15

DIFFERENTIATING INSTRUCTION

English Learners

Clarify Vocabulary

Read this sentence from the page: "Russians generally eat a hearty diet . . ." Point out that the word *diet* in the sentence means "the kinds of foods people usually eat." Explain that *diet* can be used to indicate the foods that a group typically eats or to indicate a special meal plan—usually one that is low in calories.

Gifted & Talented

Research Russian Culture

Have students work in groups to explore different aspects of Russian life. For example, one group could focus on food, another on sports, and another on religion. Groups should research the topic and give a presentation to the class. The presentation should include visual aids, such as images of athletes or Russian tea samovars.

Cultural Heritage

KEY QUESTION Who are some of the great Russian writers, artists, and composers of the 1800s and 1900s?

As you have learned, rulers such as Peter the Great and Catherine the Great encouraged the Russian people to adopt European customs. As a result, European culture gradually influenced Russian culture. Eventually, Russian literature and music that blended European and traditional styles began to emerge during the early 1800s. Soon Russia began to produce important works of literature, music, and art that are still highly regarded today.

Literature **Leo Tolstoy** (TOHL•stoy) is one of Russia's greatest novelists. In novels such as *War and Peace* (1869), Tolstoy showed deep concern for moral issues and for the welfare of the Russian people. In fact, the writer influenced Indian leader Mohandas Gandhi and his nonviolent independence movement. In turn, Gandhi inspired the nonviolent activism of civil rights leader Martin Luther King Jr.

Russia also boasted one of the best playwrights of the early 1900s. Anton Chekhov (CHEHK•awf) wrote insightful plays, including *The Three Sisters* (1901) and *The Cherry Orchard* (1904). Later in the century, Alexander Solzhenitsyn (SOHL•zhuh•NEET•sihn) described life in a Soviet labor camp under Stalin's rule in his short novel, *One Day in the Life of Ivan Denisovich* (1962). This novel led the Soviet Union to increase censorship.

ANALYZING Primary Sources

Alexander Solzhenitsyn (born 1918) is a Russian novelist who was arrested in 1945 for criticizing Stalin and spent eight years in prisons and labor camps like the one shown here. In this essay, Solzhenitsyn writes about what freedom means to him.

> This, I believe, is the single most precious freedom that prison takes away from us: the freedom to breathe freely, as I now can. No food on earth . . . is sweeter to me than this air steeped in the fragrance of flowers, of moisture and freshness.
>
> Source: "Freedom to Breathe," essay by Alexander Solzhenitsyn

DOCUMENT-BASED QUESTION

What do you think Solzhenitsyn means by "the freedom to breathe freely"?

Teach

Cultural Heritage

Think, Pair, Share

- From what sources did Russian literature and music evolve? *(European and traditional styles)*
- Describe the effect Tolstoy's writings had worldwide. *(Tolstoy's writings influenced Mohandas Gandhi in his nonviolent independence movement. Gandhi, in turn, inspired Martin Luther King Jr.)*
- What contribution did Kandinsky make to art in Russia? *(He was one of the first artists to paint in the abstract style. His work showed that visual arts were developing in Russia.)*
- **Find Main Ideas** How did literature, music, and art change in Russia after the rule of Peter the Great and Catherine the Great? *(The arts began to encompass both European and traditional influences. Russian works, especially literature and ballet, became popular and influential inside and outside of Russia.)*

ANALYZING Primary Sources

Alexander Solzhenitsyn was awarded the Nobel Prize in Literature in 1970. He did not travel to Sweden to receive the prize, however, because he feared that he would not be allowed to return to his homeland if he did so. Instead, he received his prize in Stockholm in 1974 after he had been exiled from the Soviet Union.

DOCUMENT-BASED QUESTION ANSWER

the freedom to express himself freely

INTERDISCIPLINARY ACTIVITIES

CONNECT to Civics

Discuss Nonviolent Protest

Have students work in groups to discuss the ideas of Tolstoy, Gandhi, and King regarding nonviolent movements. Have them identify key similarities and differences among the three figures and summarize their impact on world affairs today. Then initiate a class discussion about nonviolence as a method of protest.

CONNECT to Language Arts

Deliver a Speech

Have students prepare a speech on freedom. The speech should incorporate the quotation from Solzhenitsyn as well as students' own ideas about the meaning of freedom. Encourage students to practice their speeches before delivering them to the class.

More About . . .

Russian Music

Sergei Prokofiev is another major name in Russian music. Although he left Russia for a time, he returned to the Soviet Union when it was under Stalin's control. The ballet *Romeo and Juliet* and the children's symphony *Peter and the Wolf* are among his best-known works. Prokofiev visited Hollywood and studied music in film. His works include several film scores, including those for Sergei Eisenstein's *Alexander Nevsky* and *Ivan the Terrible*. Prokofiev died on the same day as Stalin in 1953.

CONNECT to the Essential Question

How is Russia preserving its Eastern culture while adapting to Western influences?

Ask students what they have learned so far that can help them answer this question. Students might mention the following:

- Russia has a diverse ethnic population, and many religions are represented.
- Russian arts draw upon both European and traditional styles.
- Russian literature, art, dance, and film have influenced the arts in other parts of the world.

Marc Chagall In this 1925 painting called *Peasant Life*, Chagall uses vivid colors to celebrate rural Jewish life.

Art Russian painting was slower to adopt European influences than was Russian literature. However, around 1911, Wassily Kandinsky (kan•DIHN•skee) became one of the first artists to paint in the abstract style. Abstract painting does not represent objects using realistic forms. Instead, Kandinsky used bold colors and shapes. Around the same time, Jewish artist Marc Chagall (shuh•GAHL) created imaginative paintings in which the figures often seem to float in the air.

Music and Dance Russian music flowered during the 1800s. Using traditional Russian melodies, Peter Ilich Tchaikovsky (chy•KAWF•skee) wrote many great works, including the ballet *Swan Lake* (1877). Igor Stravinsky introduced modern forms of musical expression with his bold compositions, including the ballet *Rite of Spring* (1913).

You may have noticed that both Tchaikovsky and Stravinsky wrote ballets. That is because dance, especially ballet, had become very popular in Russia. In fact, by the mid-1800s, Russia boasted two world-class ballet companies: the Bolshoi (BOHL•shoy) Ballet and a company that came to be known as the Kirov (KEE•RAWF).

SUMMARIZE Explain what Russian writers, artists, and composers of the 1800s and 1900s contributed to Russian culture.

Tolstoy stirred interest in moral issues; Solzhenitsyn pointed out the evils of censorship; composers introduced modern musical forms. ◄

Bolshoi Ballet *Bolshoi* is Russian for "grand." The company first performed *Swan Lake* in 1877. **Why do you think the Bolshoi is popular outside of Russia?**

because dance is universal; because the dancers are among the best in the world

DIFFERENTIATING INSTRUCTION: TIERED ACTIVITIES

OBJECTIVE Appreciate and respond to Russian music.

Basic	On Level	Challenge
Play part of a recording of Prokofiev's *Peter and the Wolf.* Ask students to listen to the music and identify the instruments and themes that represent Peter, the wolf, and other characters.	Play part of a recording of music or dance by a Russian composer, such as Tchaikovsky's *Swan Lake,* Stravinsky's *Rite of Spring,* or Prokofiev's *Peter and the Wolf.* Ask students to write a paragraph expressing their personal response to the music and its imagery.	Play part of a recording of music or dance by a Russian composer, such as Tchaikovsky's *Swan Lake,* Stravinsky's *Rite of Spring,* or Prokofiev's *Peter and the Wolf.* Ask students to write a poem evoked by the music.

Soviet Arts

KEY QUESTION What purpose did Soviet propaganda serve?

The establishment of the Soviet government in the early 1920s led to the censorship of literature in Russia. The Soviet era, though, did produce some strong works of art in other media. For instance, the films of Sergei Eisenstein, such as *Battleship Potemkim* (1925), influenced the art of filmmaking.

The Soviets also used the arts for propaganda. **Propaganda** is information that deliberately tries to influence opinion. Much of Soviet propaganda was conveyed through posters. Artists were required to create posters that supported and glorified the Communist Party and its programs. These posters were mass-produced and placed in very visible areas. They were constant reminders of Communist policy and guides for proper thought. Under Stalin's rule, the Communist Party outlawed artists who did not promote Communist ideals.

Soviet Poster This 1930s poster shows Soviet citizens waving a flag with Stalin's image and looking toward a bright future. **What feelings do you think posters like this one were meant to inspire in Soviet citizens?** Possible answers: They were meant to inspire pride, optimism, and a sense of superiority.

RECOGNIZE BIAS AND PROPAGANDA Discuss the purpose of propaganda and how the Soviets used it. to proclaim Communist policy and guide proper thought and behavior

ONLINE QUIZ For test practice, go to **Interactive Review @ClassZone.com**

Section 3 Assessment

TERMS & NAMES

1. **Explain the importance of**
   - Cyrillic alphabet
   - Leo Tolstoy
   - icon
   - ethnic group

USE YOUR READING NOTES

2. **Find Main Ideas** Use your completed diagram to answer the following question:
   How do Russian people blend Western and Eastern cultures?

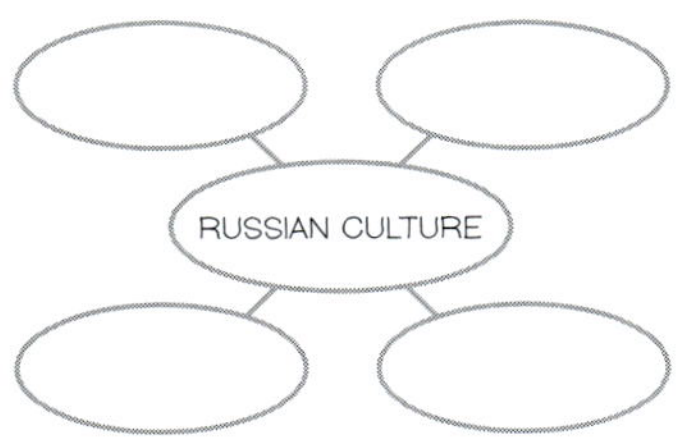

KEY IDEAS

3. What are some of the distinctive features of Russian Orthodox churches?
4. Why are winter sports so popular in Russia?
5. What impact did the Soviet era have on the arts?

CRITICAL THINKING

6. **Make Inferences** Why do you think Russia has developed so many great tennis players?
7. **Compare and Contrast** How does life in the country compare with city life in Russia?
8. **CONNECT to Today** Fast food and soft drinks are popular in Russia today. What does this fact suggest about Western influence in Russia?
9. **TECHNOLOGY** **Present a Slide Show** Use pictures to present a slide show about Russian culture. Listen to music by Tchaikovsky and Stravinsky and select your favorite piece. Play the piece as you present your slide show.

Teach

Soviet Arts

Talk About It

- How did the Communist government affect the arts in the Soviet Union? *(They censored literature. They made artists create Communist propaganda.)*
- **Find Main Ideas** What was the purpose of Soviet propaganda posters? *(to inspire people to work hard for the Soviet Union; to glorify the Communist party; to remind people who was in charge)*

4 Assess & Reteach

Assess Have students complete the Section Assessment.

Unit Resource Book: Russia & the Eurasian Republics
- Section Quiz, p. 51

Interactive Review @ClassZone.com
- Section Quiz

Power Presentations

Test Generator

Reteach Have students make a poster with four sections: people (ethnicity, religion), daily life (housing, food), sports, and arts. Have them add an appropriate illustration and caption (in the form of a main idea) to each section.

Unit Resource Book: Russia & the Eurasian Republics
- Reteaching Activity, p. 56

SECTION 3 ASSESSMENT ANSWERS

Terms & Names

1. Cyrillic alphabet: Russia's writing system; icon: Orthodox Christian religious image; Leo Tolstoy: 19th-century Russian writer; ethnic group: people sharing language, customs, and heritage

Use Your Reading Notes

2. Most Russians are Orthodox Christians and eat traditional Russian foods, but they have also adopted many foods and sports from the West.

Key Ideas

3. onion-shaped domes and icons
4. because the sports are well adapted to the Russian climate
5. During the Soviet era, literature was censored and creating propaganda was encouraged.

Critical Thinking

6. long tradition of training and supporting its athletes and inspiring them to compete
7. City dwellers live in big high-rise apartment buildings, while those in the country live in single-family houses; city dwellers have access to more amenities and services.
8. Western influence is still a powerful force in Russia.
9. Use the rubric to score slide shows.

Slide Show Rubric

	Content	Visuals
4	shows thorough understanding	many visuals
3	shows good understanding	several visuals
2	shows basic understanding	few visuals
1	shows little understanding	no appropriate visuals

1 Plan & Prepare

Objective

Compare traditional elements of Russian culture with modern, Western influences.

Springboard Ask students to discuss the cultural traditions that they observe in their families. Perhaps they wear special clothing for holidays or eat special foods. Then ask students why they think it's important to preserve such traditions.

2 Teach

Comparing Traditional and Modern Russia

Talk About It

- Which items are found on both pages? *(Matryoshka dolls)* What do you think might be nesting in the George W. Bush and Vladimir Putin dolls? *(Possible answer: earlier American presidents and Russian or Soviet leaders)*
- Which images represent things that could be found in the United States? *(Possible answers: the apartment building and the rock concert)*
- **Compare and Contrast** Based on these pictures, which Russia would you rather live in, traditional or modern Russia? *(Some students may find the simpler, bucolic peace of traditional Russia appealing, while others may feel more at home in modern Russia.)*

COMPARING Traditional and Modern Russia

Throughout its history, Russia has adopted cultural influences from the West. That cultural exchange continues today. But even as Russia changes, it maintains its character. As these pictures show, the country mixes Western styles with a definite Russian flavor.

Traditional Russia

For hundreds of years, Russians have retreated to their country houses, or dachas, to work in their vegetable gardens. They might also boil water for their tea in a samovar or listen to music played on a balalaika (BAL•uh•LY•kuh). Who knows? The house might even be decorated with nesting Matryoshka (MA•tree•OHSH•kuh) dolls.

Matryoshka Dolls
Traditional dolls nestle one inside the other and depict a round-faced peasant girl.

Beverages
Since the mid-1700s, beautifully decorated samovars have been used in Russia to prepare tea.

◀ A traditional dacha in western Russia

Music
This triangular folk guitar comes in many sizes, including the bass balalaika shown here.

DIFFERENTIATING INSTRUCTION: TIERED ACTIVITIES

OBJECTIVE Understand how Russian culture blends traditional and modern elements.

Basic

Have students draw a two-column chart and label one column Traditional and the other Modern. Then have them study the images above and write a summary statement for each one under the appropriate heading.

On Level

Have students study the two largest images on these pages. Ask them to write a paragraph, describing the similarities and differences in the two dwellings.

Challenge

Ask students to imagine that they are visiting Russia. What traditional aspects of life do they encounter? How does modern Russian life compare with their own? Have students write a travel diary about their experiences.

Modern Russia

Many people in Russian cities live in high-rise apartment buildings built during the Soviet era. Further Western influence can be seen in the vending machines and music clubs that dot the cities. Even some traditional Matryoshka dolls now have a Western twist.

Matryoshka Dolls
Today, Matryoshka dolls can depict anyone, including George W. Bush and Vladimir Putin.

Beverages
Russians enjoy American-style soft drinks, but the machines dispense the soda into real glasses.

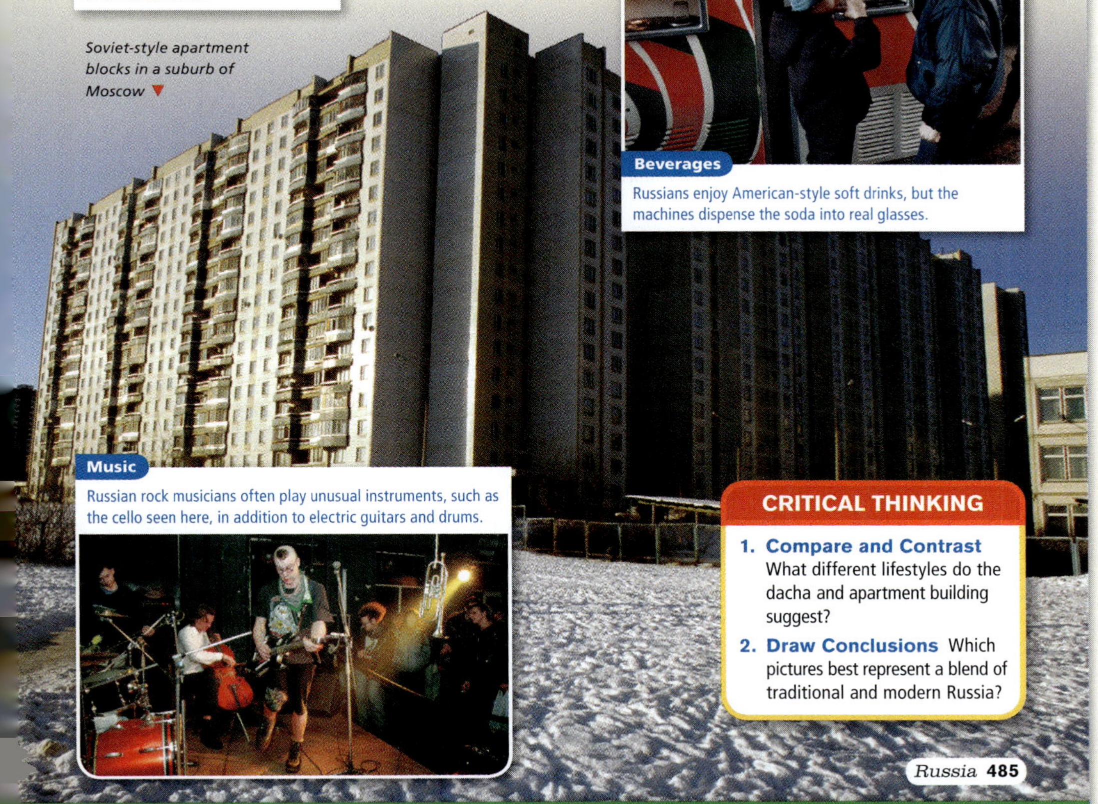

Soviet-style apartment blocks in a suburb of Moscow ▼

Music
Russian rock musicians often play unusual instruments, such as the cello seen here, in addition to electric guitars and drums.

CRITICAL THINKING

1. **Compare and Contrast** What different lifestyles do the dacha and apartment building suggest?
2. **Draw Conclusions** Which pictures best represent a blend of traditional and modern Russia?

INTERDISCIPLINARY ACTIVITIES

CONNECT to Language Arts

Write an Editorial

Think about these questions: How important are traditions and traditional practices? Why do traditional practices change or die out? What should be done to preserve traditional ways? Can traditional ways and modern influences co-exist? Write an editorial about American cultural influences on Russia and other countries.

CONNECT to Art

Compare Music

Find recordings of Russian music, both contemporary rock and traditional. Have students listen to both. Then have them compare and contrast the two types of music. Alternately, compare Russian folk music to American folk music or Russian contemporary music to American contemporary music.

More About . . .

Matryoshka Dolls

The largest Matryoshka doll had 72 pieces and was about three feet high! It was made in 1970 for a celebration of Lenin's birthday. Matryoshka dolls may actually have Japanese roots. During the Russo-Japanese War in the early 1900s, Russian prisoners learned to make a type of nesting doll from their Japanese captors. When they came home, they developed a doll that was uniquely Russian.

Comparing Regions

Russia and the United States: Influences

Russia has not only absorbed and adopted Western culture, it has also influenced it.

- Sergei Eisenstein's films, especially his use of montage, have influenced many American filmmakers.
- Russian choreographers Balanchine and Bolm came to the United States and merged Russian ballet with American ballet's speed, energy, and excitement.
- Russian actor Michael Chekhov taught the Stanislavsky System, or method acting, developed by Konstantin Stanislavsky to many American actors, including Gregory Peck, Clint Eastwood, Jack Nicholson, and Al Pacino.

3 Wrap-Up

Remind students of the Essential Question: How is Russia preserving its Eastern culture while adapting to Western influences? Have students work in small groups to answer this question based on the information in Comparing Traditional and Modern Russia.

CRITICAL THINKING ANSWERS

1. **Compare and Contrast** The dacha suggests an unhurried life that is tied to the land, while the apartment building suggests a more stressful, detached existence.
2. **Draw Conclusions** pictures of the Matryoshka dolls

1 Plan & Prepare

Objectives

- Compare the Russian government to the U.S. government and explain why Russia's reforms have slowed
- Describe the Russian economic boom and the economic challenges Russia now faces

Read for the Essential Question

Help students read for a purpose by reminding them of the Essential Question: "How is Russia preserving its Eastern culture while adapting to Western influences?" Point out that they can use items in the section, such as Key Questions, special features, images, and maps to formulate answers to the question.

Vocabulary

Best Practices Toolkit

Use the ***Best Practices Toolkit*** to model strategies for vocabulary notetaking. Vary strategies throughout the year. Choose from: *Knowledge Rating, Predicting ABCs, Definition Mapping, Word Sort, Word Wheel, Frayer Model (Word Squares), Magnet Words,* and *Student VOC.*

Vocabulary Strategies, TT9–TT16

Reading Strategy

Best Practices Toolkit

Display the T Chart/Two-Column Chart transparency.

- Remind students that there are often many solutions to a problem. In this case, they will be asked to identify the solutions chosen by the Russian government.
- Have students first brainstorm solutions to the problem of establishing a new government. Then use their ideas to model filling in the chart.
- Have students predict solutions for the problems listed in the chart. Have students revise their charts as they read the section.

T Chart/Two-Column Chart, TT20

Section 4 Reading for Understanding

Key Ideas

BEFORE, YOU LEARNED

The Communist government in Russia controlled the country's economy for most of the 20th century.

NOW YOU WILL LEARN

Russia is on its way to establishing a market economy but has struggled to change its government.

Vocabulary

TERMS & NAMES

Vladimir Putin (POOT•ihn) president of Russia who increased executive power in the 2000s

Chechnya (CHEHCH•nee•uh) a largely Islamic republic in southwestern Russia that continues to fight for its independence

command economy an economic system in which the production of goods and services is controlled by a central government

market economy an economic system in which the production of goods and services is decided by the supply and demand of consumers

privatization (pry•vuh•tih•ZAY•shuhn) the process of selling government-owned businesses to private individuals

Boris Yeltsin (YEHLT•sihn) president who abruptly transformed the Russian economy after the Soviet Union collapsed

shock therapy term applied to Yeltsin's economic plan, which called for an abrupt shift from a command economy to a market economy

Reading Strategy

Re-create the chart shown at right. As you read and respond to the **KEY QUESTIONS**, use the chart to identify the problems the Russian government faced after the collapse of the Soviet Union and the solutions they came up with.

Skillbuilder Handbook, page R10

IDENTIFY PROBLEMS AND SOLUTIONS

PROBLEMS	SOLUTIONS
Establishing a new government	
Responding to internal troubles	
Changing the economy	

GRAPHIC ORGANIZERS
Go to **Interactive Review** @ClassZone.com

PRETEACHING VOCABULARY

English Learners

Pronounce and Preview

Pronounce each term for students. Review the meanings of words used in the definitions, such as *executive power, central government,* and *abruptly.*

- To modify vocabulary learning, have students complete worksheets as they read the section.

Unit Resource Book: Russia & the Eurasian Republics
- Vocabulary Practice, p. 35

Inclusion

Locate the Terms and Names

Have pairs of students work together to locate the terms and names in the section text. Ask one student to read the sentence or passage that contains each term aloud. Then have students take turns restating the definition in their own words.

SECTION 4

GOVERNMENT & ECONOMICS

Orthodox Christian church in the Altai Shan in southern Siberia

Siberian pine trees, Karakol Lake, brown bear

The Struggle for Reform

Connecting to Your World

Freedom of speech. Freedom of the press. Freedom of religion. These are some of the freedoms we enjoy under a democracy. You probably take them for granted. But the Russian people are still trying to gain many of these freedoms. Since the breakup of the Soviet Union, Russia has worked to establish a new form of government. But internal conflicts and violence have threatened to tear the country apart. In addition, Russia has abruptly changed its economic system. In the midst of these problems and changes, democratic reform has proved difficult to achieve.

Vladimir Putin Putin worked for 15 years in the KGB, the Soviet Union's intelligence agency, before he entered politics.

A New Government

KEY QUESTION Why did Russian democratic reform slow at the beginning of the 21st century?

After the breakup of the Soviet Union in 1991, Russians introduced many democratic reforms to their government. They set up several political parties, elected a president, and reorganized the structure of the government. However, while Russia has taken some steps forward, the country has also taken some steps back. In the early part of the 21st century, the government reversed some earlier reforms and seized more control.

Pro-Putin Rally Young people demonstrate their support for Putin in a rally in 2002. The word on the banner means "Russia."

2 Focus & Motivate

3-Minute Warm-Up

Write on the board or display the transparency:

- What freedoms do you think all people should have? *(Possible answers: speech, press, religion, assembly)*

Unit Transparency Book
- 3-Minute Warm-Ups, TT1

Connecting to Your World

More About . . .

Freedom in Russia

Under President Putin, the government has taken steps to restrict freedoms, especially freedom of speech. Putin's government has prosecuted thousands of reporters for libel, shut down television stations, and increased control over many media outlets.

3 Teach

A New Government

Think, Pair, Share

- How does the Russian president compare with the U.S. president? *(The Russian president has much more power.)*
- **Identify Problems and Solutions** How did the Russian government deal with the problem of lawlessness that occurred after 1991? *(It restricted freedom.)*

SECTION 4 PROGRAM RESOURCES

Unit Resource Book: Russia and the Eurasian Republics

ON LEVEL

Unit Resource Book
- Reading Study Guide, p. 9
- Section Quiz, p. 52

STRUGGLING READERS

Unit Resource Book
- Reading Study Guide with Additional Support, p. 17
- Vocabulary Practice, p. 35
- Skillbuilder Practice, p. 36
- Section Quiz, p. 52
- Reteaching Activity, p. 56

eEdition with Audio DVD-ROM

ENGLISH LEARNERS

Pupil Edition in Spanish

eEdition with Audio DVD-ROM

Unit Resource Book
- RSG (Spanish), p. 25
- RSG with Additional Support (Spanish), p. 33

Multi-Language Glossary

Test Generator
- Section Quiz in Spanish

INCLUSION

Unit Resource Book
- RSG with Additional Support, p. 17
- Section Quiz, p. 52
- Reteaching Activity, p. 56

GIFTED & TALENTED

Unit Resource Book
- History Makers, p. 42
- Section Quiz, p. 52

PRE-AP

Unit Resource Book
- Economics in the World, p. 45
- Primary Sources, pp. 47–48
- Section Quiz, p. 52

TECHNOLOGY

Unit Transparency Book
- 3-Minute Warm-Ups, TT1
- Essential Question, TT2
- Critical Thinking, TT6
- Map: Russian and Soviet Expansion, TT8, TT8a, TT8b

Daily Test Practice Transparencies
- Section 4, TT55

Power Presentations

ClassZone.com

World Cultures and Geography Video on DVD

COMPARING

U.S. and Russian Governments

Have students tell what they know about the structure of the U.S. government before analyzing the graphic. Ask them to name the branches of the government *(executive, legislative, judicial)* and briefly discuss the role of each one.

CRITICAL THINKING ANSWER

Compare and Contrast Both are elected and submit or appoint candidates for the Supreme Court; Russia's president appoints the prime minister, can dismiss Parliament, and directs the military and security.

More About . . .

Vladimir Putin

Vladimir Putin worked for the Soviet Union's intelligence agency, the KGB, for many years. Then, under Russian president Boris Yeltsin, Putin headed up the country's new national security forces. Yeltsin made Putin prime minister in 1999, and Putin became acting president when Yeltsin left office. Putin was elected president in 2000 and again in 2004. According to the Russian constitution, he cannot serve a third consecutive term, though many have speculated that he might try.

Unit Resource Book: Russia & the Eurasian Republics

- History Makers: Vladimir Putin, p. 42
- Primary Sources: from a speech by Vladimir Putin; from a speech by Mikhail Gorbachev, pp. 47–48

COMPARING U.S. and Russian Governments

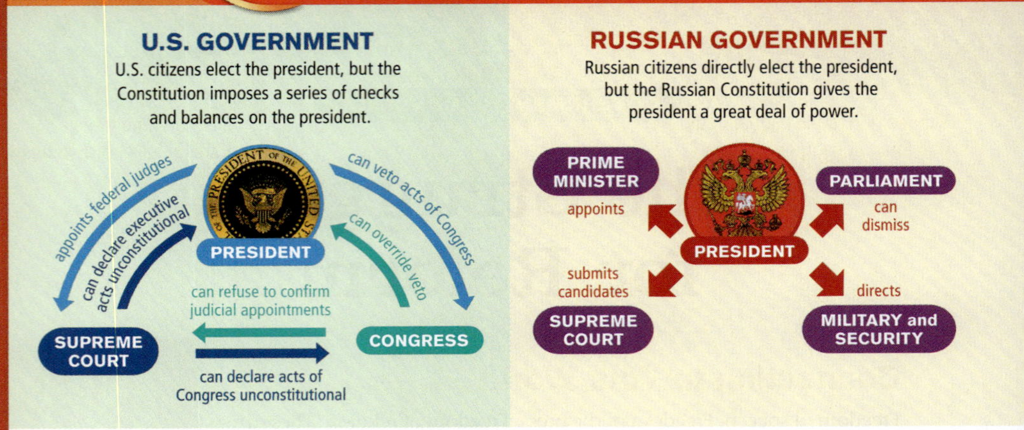

CRITICAL THINKING

Compare and Contrast What are some similarities and differences between the powers of the president of Russia and those of the president of the United States?

Presidential Power The structure of the Russian government has much in common with that of the U.S. government. For example, the legislative branch of each government consists of two sections. Also, both governments' legislatures propose laws that they then submit to the president. However, the powers of the presidents differ. As you can see from the graphic above, the Russian president has more power than the U.S. president. By the early 2000s, President **Vladimir Putin** (POOT•ihn) of Russia had steadily begun to increase those powers.

When he became president, Putin made many changes to Russia's political system. He decreased the power of rival political parties and greatly increased the power of his own. He also took powers away from elected representatives and restricted the freedom of Russia's newspapers and other media. Today, some fear that Russia may return to the one-party system that ruled the Soviet Union.

Internal Troubles The Russian government restricted freedom, in part, in response to the rise in lawlessness and organized crime that occurred after 1991. The sudden social, political, and economic changes that followed independence encouraged this lawlessness. To combat crime in their country, many Russians have been willing to sacrifice their own liberty for order and the rule of law.

DIFFERENTIATING INSTRUCTION

English Learners

Understand Vocabulary

Read the following sentence aloud: "[Putin] also took powers away from elected representatives and restricted the freedom of Russia's newspapers and other media." Explain that the word *media* means "the means of mass communications, such as television and radio stations." Point out that *media* is the plural form of *medium* and so takes the plural form of a verb.

Pre-AP

Draw Conclusions

Direct students to the statement: "Today, some fear that Russia may return to the one-party system that ruled the Soviet Union." Have students discuss why a one-party system might threaten democratic reforms and draw conclusions about the effects a one-party system could have on a nation and its people.

The Russian government also reversed some of its earlier reforms in reaction to terrorism and war with Chechnya. **Chechnya** (CHEHCH•nee•uh) is a largely Islamic region in southwestern Russia. When the people of Chechnya demanded their independence in 1991, Russia refused. Russia invaded the region in 1994 and again in 1999. Militants from Chechnya have also committed many acts of violence. The struggle between Russia and Chechnya continues today.

ANALYZE EFFECTS Explain why Russia's democratic reforms have slowed. because of a rise in lawlessness and organized crime and concerns over internal terrorism

Building a Market Economy

KEY QUESTION What led to Russia's economic boom?

Although Russia has struggled to establish a democratic government, the country has had greater success reforming its economy. Russia is changing from a **command economy**, which is controlled by the government, to a **market economy**. This is an economic system in which private individuals own most of the businesses and operate them with little government control. The process of selling government-owned businesses to private individuals is called **privatization**.

Making the Transition Privatization has not been easy, however. When the Soviet Union broke apart, the economy was in a state of confusion. Encouraged by Western economists, Russian president **Boris Yeltsin** (YEHLT•sihn) adopted a plan to establish a market economy called **shock therapy**. This plan called for abrupt, widespread privatization. Many workers, though, were unwilling to privatize industries and farms. Soon industrial production fell, prices rose, businesses closed, and thousands of people lost work. Some critics of Yeltsin's plan complained that Russia got a lot of "shock" but no "therapy." In time, however, the economy began to prosper.

It's very similar. ▼

Moscow Mall Russia's new prosperity can be seen in this Western-style shopping mall. **How does this mall compare with those in your town?**

DIFFERENTIATING INSTRUCTION: TIERED ACTIVITIES

OBJECTIVE Gain insight into the conflict between Russia and Chechnya.

Basic	On Level	Challenge
Have students use classroom resources to learn more about Chechnya. Then have them create a map and brief factual profile of the region. The profile should include type of government, major resources, and demographic information (such as ethnicity and religion).	Have students research to learn more about the major events of the Chechen conflict. Then have them create a time line of key events. Students should include a brief summary of major events in 1991, 1994, 1999, and 2003, and also include current events.	Have students research to learn about the causes of the Chechen conflict. Then have them write commentaries about the conflict. Possible topics include pleas for and against Chechen independence, discussion of human rights issues, and world reaction to the conflict.

Teach

Building a Market Economy

Reader, Recorder, Reporter

- What changes is Russia making in its economic system? *(Russia is changing from a command economy to a market economy through privatization.)*
- What was a major factor in Russia's economic boom? *(Russian exports of natural resources, especially oil)*
- What effects have improvements in the Russian economy had on the Russian people? *(Many people have seen higher salaries; more and more people have joined the middle class.)*
- **Identify Problems and Solutions** How did Yeltsin try to improve the Russian economy? How effective was it? *(Yeltsin tried a plan he called "shock therapy," which called for abrupt, widespread privatization. It was not very effective at first, but over time the Russian economy improved.)*

Unit Resource Book: Russia & the Eurasian Republics

- Skillbuilder Practice: Using a Database, p. 36
- Economics in the World: Growth in the Russian Economy, p. 45

CONNECT to the Essential Question

How is Russia preserving its Eastern culture while adapting to Western influences?

Ask students what they have learned so far that can help them answer this question. Students might mention the following:

- Russia has made democratic reforms but has also restricted freedom.
- Russia is reforming its economic system and building a market economy.

Unit Transparency Book

- Essential Question, TT2

CONNECT History & Economics

From Bust to Boom

Explain that most countries experience ups and downs in their economy. Political turmoil, worldwide events, and natural disasters are only some of the events that can affect a country's economy. For example, the U.S. economy had a major setback during the Great Depression.

4 Assess & Reteach

Assess Have students complete the Section Assessment.

Unit Resource Book: Russia & the Eurasian Republics
- Section Quiz, p. 52

Interactive Review @ClassZone.com
- Section Quiz

Power Presentations

Test Generator

Reteach Have students work in groups and create one graphic organizer on the Russian government and one on the Russian economy. Have each group present its graphic organizers to the class. Students should revise their graphic organizers as needed after the presentations.

Unit Resource Book: Russia & the Eurasian Republics
- Reteaching Activity, p. 56

CONNECT History & Economics

From Bust to Boom

The graph shown here measures Russia's economic growth. As you can see, the economy has experienced many ups and downs since the breakup of the Soviet Union. But, overall, Yeltsin's shock therapy plan and the country's move toward a market economy have been a success.

Russia's Economic Growth *1991–2005*

Source: *CIA World Factbook*

An Economic Boom Economic prosperity was not easy to achieve, though. In 1998, the country's money lost value, and the Russian stock market crashed. Economic hardships caused a steep rise in homelessness and a decline in life expectancy. But by 2001, Russia's economy showed steady growth and began to boom. The boom was largely due to Russian exports of oil.

As a result of the surging economy, salaries rose, unemployment decreased, and the Russian people could afford to buy goods. A growing middle class began to emerge. Unfortunately, not everyone has been able to take part in Russia's economic success. Many Russians are still unemployed or underpaid. But Russia has come a long way in a short amount of time.

SEQUENCE EVENTS Identify the series of events that led to Russia's economic boom. collapse of the Soviet Union, Yeltsin's shock therapy plan, initial decline of economy, economic rebound

Section 4 Assessment

ONLINE QUIZ For test practice, go to Interactive Review @ClassZone.com

TERMS & NAMES

1. **Explain the importance of**
   - shock therapy
   - market economy
   - command economy
   - privatization

USE YOUR READING NOTES

2. **Identify Problems and Solutions** Use your completed problem-solution chart to answer the following question: How successful was Boris Yeltsin's shock therapy plan?

PROBLEMS	SOLUTIONS
Establishing a new government	
Responding to internal troubles	
Changing the economy	

KEY IDEAS

3. How did President Putin increase the powers of the Russian president?
4. How did Russia begin the shift from a command economy to a market economy?
5. How have the Russian people benefited from their country's economic boom?

CRITICAL THINKING

6. **Make Inferences** Why might Russians be willing to accept a very powerful president?
7. **Evaluate** What connection can you draw between Russia's physical geography and its economic boom?
8. CONNECT to Today What appears to be the most positive development in today's Russia?
9. MATH **Create Graphs** Use the Internet and your library to look up data on the unemployment rates for several years in the Soviet Union and in Russia. Then create a graph for each. Write a summary that compares and contrasts the graphs.

SECTION 4 ASSESSMENT ANSWERS

Terms & Names

1. shock therapy: Yeltsin's economic plan; command economy: production controlled by government; market economy: businesses owned by people; privatization: sale of businesses to individuals

Use Your Reading Notes

2. caused many economic problems and crime but paved the way for Russia's economic boom

PROBLEMS	SOLUTIONS
Establishing a new government	reforms reversed; more presidential power
Responding to internal troubles	sacrifice liberty for rule of law
Changing the economy	shock therapy; economic boom

Key Ideas

3. decreased the powers of representatives and political parties, restricted freedom of the media
4. through shock therapy plan
5. unemployment decreased, salaries increased, growing middle class

Critical Thinking

6. They are used to being under a dominant central government with a powerful leader.
7. Russia's vast reserve of oil and other natural resources was, in large part, the reason for the country's economic boom.
8. large skilled labor force, many natural resources, a booming economy, and a government that is committed to a market economy
9. Use the rubric to score graphs.

Graph Rubric

	Graphs	Summary
4	all data accurate	draws correct conclusions
3	most data accurate	draws mostly correct conclusions
2	some data accurate	draws some correct conclusions
1	little or no accurate data	draws few or no correct conclusions

CHAPTER SUMMARY

Key Idea 1
Russia is a vast country that contains a variety of landforms, climates, vegetation regions, and natural resources.

Key Idea 2
The sometimes conflicting influences of the East and West have helped shape Russian history.

Key Idea 3
The influences of Europe and Asia are reflected in the people and culture of Russia.

Key Idea 4
Russia is on its way to establishing a market economy but has struggled to change its government.

For **Review and Study Notes**, go to **Interactive Review** @ ClassZone.com

NAME GAME

Use the Terms & Names list to complete each sentence on paper or online.

1. I separate European Russia from Asian Russia. Ural Mountains
2. I am a large body of water bordering Russia to the south. ______
3. I am the longest river in Europe. ______
4. I am a vast forest that lies south of the tundra in Russia. ______
5. I am a peasant who is bound to the land. ______
6. I am the title for the ruler of Russia from the mid-1500s to the early 1900s. ______
7. I am a type of government that controls every aspect of public and private life. ______
8. I am the official writing system of Russia. ______
9. I am an economic system in which the production of goods and services is controlled by a central government. ______
10. I am the process in which government-owned businesses are sold to private individuals. ______

Caspian Sea
command economy
Cyrillic alphabet
czar
Eurasia
icon
market economy
privatization
serf
Slav
steppe
taiga
totalitarian
Ural Mountains
Volga River

Activities

GeoGame

Use this online map to show what you know about Russia's location, geographic features, and important places. Drag and drop each place name to its location on the map.

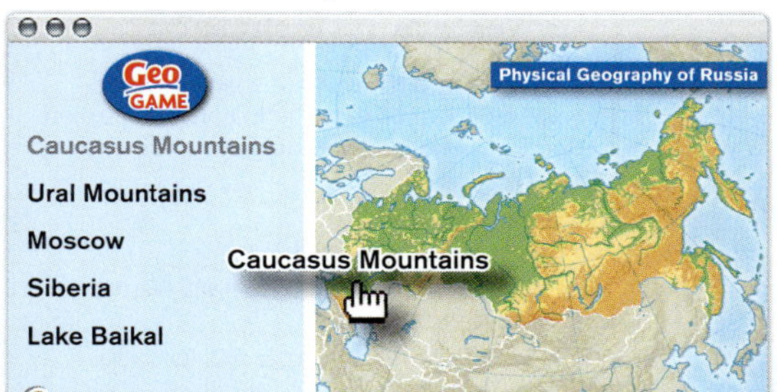

To play the complete game, go to **Interactive Review** @ ClassZone.com

Crossword Puzzle

Complete an online crossword puzzle to test your knowledge of Russia.

ACROSS
1. czar who tried to make Russia more European

ONLINE AND ON-SCREEN ACTIVITIES

GeoGame

Use the interactive map on **Power Presentations** DVD-ROM or at **ClassZone.com** to help students review Russia's geographic features.

- Demonstrate how to drag and drop the place names onto the map. Click on the Submit button to view results.
- Advance to the scorecard to show how students will be scored in their first and second attempts.

Crossword Puzzle

Students can review important people and events in Russian history by completing this crossword puzzle online at **ClassZone.com.**

- As an extension of this activity, have students use the *Activity Maker* at **ClassZone.com** to make their own crossword puzzle for one section of the chapter.
- Then have students trade puzzles with classmates and solve them.

Chapter Summary

As a follow-up to the chapter, encourage students to pair up and go over the **Review and Study Notes** at **ClassZone.com.**

Alternatively, student pairs can review their Reading Study Guide notes, making sure that they can add important details to each section's Key Ideas.

Name Game

Teaching Tips

ON PAPER Have students work in pairs to complete this activity. As an extension, they can write questions for the terms that are not covered in the items shown here.

ONLINE Make sure that students understand how to drag and drop terms to fill in blanks. Ask them to print out their results and compare them with those of a classmate.

ANSWERS

1. Ural Mountains
2. Caspian Sea
3. Volga River
4. taiga
5. serf
6. czar
7. totalitarian
8. Cyrillic alphabet
9. command economy
10. privatization

Teacher-Tested Activities

Tim Mortensen, Patrick Marsh Middle School, Sun Prairie, Wisconsin

My classes like to play an action version of the "Name Game." Here's how it works.

- First, I divide the class into ten small groups. Each group is assigned one of the "Name Game" items.
- Then each group acts out the clue. For instance, the students might form themselves into a mountain range to show the Ural Mountains. If the name is hard to enact, the students can use charade-type clues to help the class sound it out.
- The group that guesses the answer correctly, gets one point.

Vocabulary ANSWERS

1. single continent made up of Europe and Asia
2. mountain range that forms the border between European and Asian Russia
3. title for the ruler of Russia from the mid-1500s to the early 1900s
4. peasant who is bound to the land
5. group of revolutionaries who took control of the Russian government in 1917
6. cruel dictator who had millions of his political enemies executed
7. government that controls every aspect of public and private life
8. official writing system of Russia
9. religious image used by Orthodox Christians
10. system in which the production of goods and services is controlled by a central government
11. all vegetation regions in Russia
12. When Russia transformed to a market economy, formerly government-owned businesses were privatized.

Key Ideas ANSWERS

13. provides rich farmland but does not provide any natural barriers against invaders
14. They are frozen for much of the year.
15. It has harsh winters and short summers.
16. Ivan IV and the czars who came after him passed laws that bound peasants to the land.
17. He reorganized it, replacing badly organized administrative departments.
18. It was a conflict between the Soviet Union and the United States. Both sides produced powerful nuclear weapons, but open warfare never broke out between the two.
19. They are sacred images of Orthodox Christianity used in religious ceremonies.
20. hockey, ice skating, and skiing
21. Gandhi who, in turn, inspired Martin Luther King Jr.
22. appoints the prime minister, can dismiss the Parliament, directs the military and security
23. to convert the Russian economy from a command economy to a market economy
24. exports of Russia's vast natural resources, particularly oil

CHAPTER 15 Assessment

VOCABULARY

Explain the significance of each of the following.

1. Eurasia
2. Ural Mountains
3. czar
4. serf
5. Bolsheviks
6. Joseph Stalin
7. totalitarian
8. Cyrillic alphabet
9. icon
10. command economy

Explain how the terms and names in each group are related.

11. tundra, taiga, and steppe
12. market economy and privatization

KEY IDEAS

1 Sweeping Across Eurasia

13. Why has the Northern European Plain in Russia been both a blessing and a curse?
14. Why are many of the main rivers in Siberia hardly ever used for transportation?
15. Why do few people live on the tundra?

2 Governing a Vast Land

16. How did the serf system arise in Russia?
17. How did Peter the Great attempt to make Russia's government run efficiently?
18. What was the Cold War?

3 Blending Europe and Asia

19. What are icons and how are they used?
20. What winter sports are popular in Russia?
21. Who did Leo Tolstoy inspire?

4 The Struggle for Reform

22. What are some of the Russian president's powers?
23. What was the purpose of President Boris Yeltsin's shock therapy plan?
24. What helped bring about Russia's economic boom?

CRITICAL THINKING

25. **Compare and Contrast** Create a chart to compare and contrast the landforms, climate, and natural resources of European Russia and Asian Russia.

EUROPEAN RUSSIA	ASIAN RUSSIA

26. **Analyze Causes** What led to the overthrow of the czar in 1917?
27. **Draw Conclusions** How did the censorship of Solzhenitsyn reflect totalitarianism?
28. **Connect to Economics** How did privatization change Russia?
29. **Connect History & Culture** How do Peter the Great's efforts to westernize Russia affect the country's culture today?
30. **Five Themes: Movement** How did the Trans-Siberian Railroad help unite western and eastern Russia?

Answer the ESSENTIAL QUESTION

How is Russia preserving its Eastern culture while adapting to Western influences?

Written Response Write a two- or three-paragraph response to the Essential Question. Use the rubric below to guide your thinking.

Response Rubric

A strong response will:

- discuss the history of Eastern and Western influence in Russia
- identify Western characteristics Russia has adopted
- explain ways in which Russia maintains its Eastern flavor today

ESSENTIAL QUESTION REVIEW

Before students begin their essays, have them review the key chapter elements that will help them answer the Essential Question. Direct them to significant features in the chapter, such as:

- Comparing Vegetation, Section 1
- Russian and Soviet Expansion, Section 2
- Moscow's Red Square, Section 3
- Connect History and Culture: Russia and the Olympics, Section 3
- Comparing Traditional and Modern Russia, Section 3
- Comparing U.S. and Russian Governments, Section 4

Written Response Rubric

	Coverage of Key Ideas	Explanation	Errors in Mechanics
4	all	clear and concise	few or none
3	most	good	some
2	some	fair	many
1	one or none	poor	many major

STANDARDS-BASED ASSESSMENT

Test Practice

- Online Test Practice @ClassZone.com
- Test-Taking Strategies and Practice at the front of this book

THEMATIC MAP

Use the map to answer questions 1 and 2 on your paper.

1. Which vegetation region covers most of Russia?

A. tundra
B. taiga
C. steppe
D. semidesert

2. Which region is closest to the Arctic Ocean?

A. tundra
B. taiga
C. steppe
D. semidesert

LINE GRAPH

Examine the graph below. Use the information in the graph to answer questions 3 and 4.

Source: U.N. World Population Division

3. What was the life expectancy in 1990?

A. about 60
B. about 65
C. about 63
D. about 67

4. When did the life expectancy of Russian men decline the most?

A. between 1985 and 1990
B. between 1990 and 1995
C. between 1995 and 2000
D. between 2000 and 2005

GeoActivity

1. INTERDISCIPLINARY ACTIVITY–ECONOMICS

With a small group, prepare a short speech that President Boris Yeltsin might have given to the Russian people about his shock therapy plan. In the speech, explain the long-term benefits of the plan, but also warn the people about the short-term harm it may cause.

2. WRITING FOR SOCIAL STUDIES

Recall what you have learned about censorship under a totalitarian government. Do you think government should ever have the power to control what citizens say or create? Examine the conflicting viewpoints and then explain your position.

3. MENTAL MAPPING

Create a physical map of Russia and label the following:

- Volga River
- Ural Mountains
- Black Sea
- Caspian Sea
- Siberia
- Arctic Ocean
- Lake Baikal
- Kamchatka Peninsula

DIFFERENTIATING CHAPTER ASSESSMENT

Unit Resource Book

- **Chapter Test A,** pp. 57–61. *Use for Struggling Readers, Beginning-to-Intermediate-Level English Learners, and for Inclusion of students whose IEP suggests supported test-taking.*
- **Chapter Test B,** pp. 62–66. *Use for On-Level Learners and ESOL students with Advanced-Level English.*
- **Chapter Test C,** pp. 67–71. *Use for Gifted and Talented students or students assigned Pre-AP work.*

Test Generator

- Chapter Tests A, B, and C, in English and Spanish

McDougal Littell Assessment System

TEST **Test Generator** Use three levels of tests in English and Spanish tied to state standards. Plus, choose questions to customize tests for all populations.

SCORE Scan and Score plain paper student tests in minutes.

REPORT Create instant progress reports, sortable by student, by standard, or other criteria.

RETEACH Respond immediately to students' needs by using a wealth of reteaching resources.

Critical Thinking ANSWERS

25. European Russia: mountains, plains, rivers form transportation network; cold winters, hot summers; gold, diamonds, copper, lumber; Asian Russia: Siberia, Lake Baikal; long, cold winters, short hot summers; oil, natural gas, gold, diamonds, lumber

26. hundreds of years of oppression of the serfs, not enough reforms, huge difference between the rich and the poor

27. It reflects the power of a totalitarian government to control all aspects of public and private life.

28. It forced the government to sell industries and farmland to private individuals. Many workers were unwilling to privatize.

29. Western influence can be seen in Russia's current government, economy, and daily life.

30. It opened up Siberia, thereby increasing the population and the use of natural resources in this region.

Standards-Based Assessment ANSWERS

1. B
2. A
3. C
4. B

GeoActivity

1. INTERDISCIPLINARY ACTIVITY RUBRIC

The speech should

- reflect an understanding of Yeltsin's shock therapy plan
- describe its immediate consequences
- detail the benefits that eventually resulted

2. WRITING FOR SOCIAL STUDIES RUBRIC

The essay should

- describe censorship under a totalitarian government
- consider whether censorship is ever legitimate

3. MENTAL MAPPING RUBRIC

The outline map should

- represent the approximate shape of Russia
- include all geographic features listed in their correct location

CHAPTER 16 The Eurasian Republics PLANNING GUIDE

Complete print resources all on one USB drive allow you to customize lesson plans for all students.

Section Titles & Key Ideas	Unit Resource Book Russia and the Eurasian Republics	Assessment Resources
1 Center of a Landmass **KEY IDEA** The geography and climate of the Eurasian republics pose challenges, but the countries are rich in natural resources.	• **Reading Study Guide,** pp. 73–74 • **RSG with Additional Support,** pp. 79–80 • **Vocabulary Practice,** p. 97 • **Outline Map Activities,** p. 101 • **Connect to Science,** p. 108	**Unit Resource Book** • Section Quiz, p. 110 • Reteaching Activity, p. 113 **Test Generator CD-ROM** **Daily Test Practice Transparency,** TT56
2 Historic Crossroads **KEY IDEA** The geographic location of the Eurasian republics has attracted many different cultures, ideas, and conquerors.	• **Reading Study Guide,** pp. 75–76 • **RSG with Additional Support,** pp. 81–82 • **Skillbuilder Practice,** p. 98 • **History Makers,** p. 103 • **Interdisciplinary Projects,** p. 105	**Unit Resource Book** • Section Quiz, p. 111 • Reteaching Activity, p. 114 **Test Generator CD-ROM** **Daily Test Practice Transparency,** TT57
3 The Challenge of Independence **KEY IDEA** The republics are working to overcome internal problems and establish democratic governments and market economies.	• **Reading Study Guide,** pp. 77–78 • **RSG with Additional Support,** pp. 83–84 • **Economics in the World,** p. 107 • **Primary Sources,** p. 109 • **Geography Project,** p. 131	**Unit Resource Book** • Section Quiz, p. 112 • Reteaching Activity, p. 115 • Chapter Test (Forms A, B, & C), pp. 117–128 **Test Generator CD-ROM** **Daily Test Practice Transparency,** TT58

McDougal Littell Assessment System

TEST SCORE REPORT RETEACH

No Time? To focus students on important content in the chapter, use the **"Reading Study Guide with Additional Support"** in *Resources2Go.*

Teacher Presentation Options

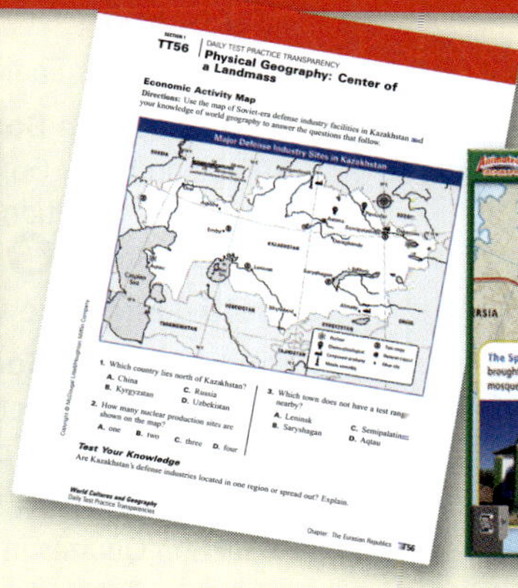
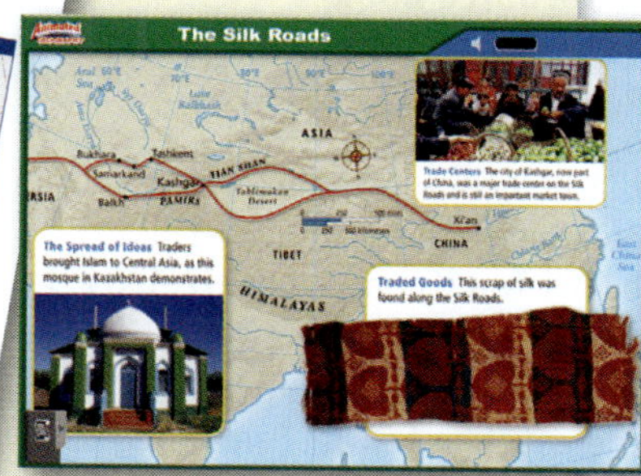

Electronic Books

eEdition DVD-ROM

eEdition Online

Presentation Toolkit

Video/DVD

- World Cultures and Geography

Power Presentations DVD-ROM

- Discussion Starters
- Media Gallery
- Review Game
- Interactive Review
- Animated Features

Unit Transparency Book

- 3-Minute Warm-Ups
- Essential Question
- Critical Thinking
- Maps
- Cultures

Daily Test Practice Transparencies

Animated Geography

- Present-Day Eurasian Republics Map
- The Silk Roads
- Worldster: Zuhura's Web Page

OnlineActivities at ClassZone.com

Animated Geography

- **Videos and Photos**
- **3-D Animations**
- **Interactive Maps**

Interactive Review

- **Section Quizzes**
- **Review and Study Notes**
- **Vocabulary Flip Cards**
- **Graphic Organizers**
- **Name Game and Crossword Puzzles**
- **GeoGame**

Online Test Practice

Map Center

- **Atlas Maps**
- **Political Maps with Data**
- **Links to Mapping Web Sites**

Activity Center

- **Activity Maker—Review Games, Puzzles, Flip Cards, and Graphic Organizers**
- **Worldster**

Research & Writing Center

- **Web Sites**
- **Data Updates**
- **Primary Sources**
- **Writing Models**

Current Events

Matt Parker
Eastgate Middle School
Kansas City, Missouri

Make It Fun Teacher-Tested Activities

As you teach this chapter, look for my classroom-proven ideas for:

- making the lesson fun
- focusing students' attention and comprehension
- integrating technology with traditional teaching

Teaching Bell to Bell When you see this timing guide, have students discuss in pairs, in groups of 3–6, or as a class.

- **Think, Pair, Share** (pairs)
- **Reader, Recorder, Reporter** (3–6)
- **Talk About It** (any size)

CHAPTER 16 RESOURCES FOR DIFFERENTIATED INSTRUCTION

Struggling Readers

eEdition with Audio DVD-ROM

Lesson Plans
- Strategies for Inclusion

Unit Resource Book
- Reading Study Guide with Additional Support, pp. 79–84 (B)
- Vocabulary Practice, p. 97
- Skillbuilder Practice, p. 98
- Reteaching Activity, pp. 113–115
- Chapter Test (Form A), pp. 117–120 (A)

Best Practices Toolkit (C)
- Interactive Reading Strategies, pp. 90–145
- Differentiation Activity Bank: Struggling Readers, pp. 146–149

Teacher's Edition Activities
- Locate Major Map Features, Sec. 1
- Trace the Silk Roads, Sec. 2
- Learn Terms, Sec. 3
- Tiered Activities, Sec. 1–3

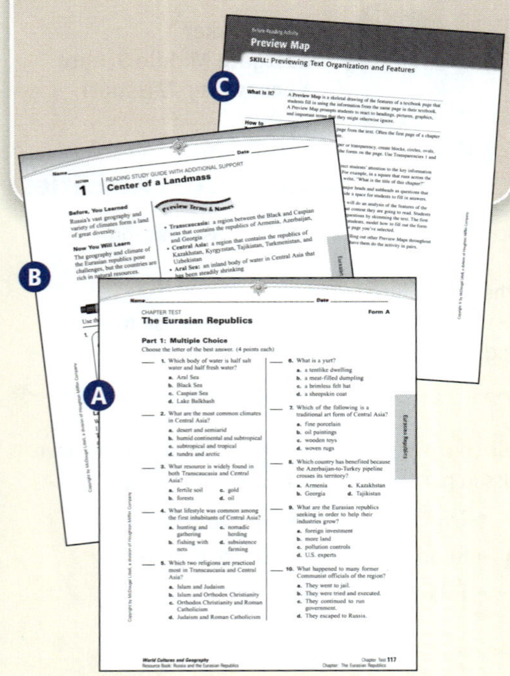

English Learners

Pupil Edition in Spanish

Lesson Plans
- Strategies for English Learners

Unit Resource Book
- Reading Study Guide (Spanish), pp. 85–90
- RSG with Additional Support (Spanish), pp. 91–96 (A) (B)

Multi-Language Glossary (C)

Best Practices Toolkit
- Differentiation Activity Bank: English Learners, pp. 165–194

Test Generator CD-ROM
- Section Quizzes in Spanish
- Chapter Test in Spanish

Teacher's Edition Activities
- Pronounce and Preview, Sec. 1–3
- Explore Vocabulary, Sec. 1
- Build Background, Sec. 2
- Use Graphic Organizers, Sec. 2
- Understand Prefixes, Sec. 3
- Tiered Activities, Sec. 1–3

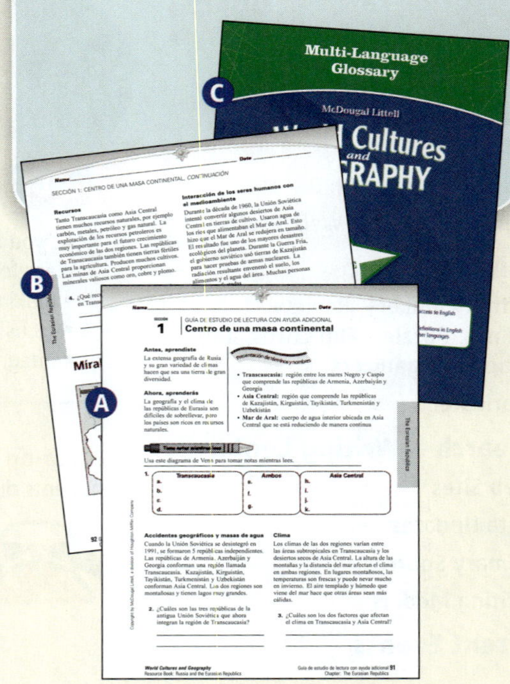

Inclusion

eEdition with Audio DVD-ROM

Lesson Plans
- Strategies for Inclusion

Unit Resource Book
- Reading Study Guide with Additional Support, pp. 79–84 (A) (B)
- Reteaching Activity, pp. 113–115
- Chapter Test (Form A), pp. 117–120

Best Practices Toolkit (C)
- Interactive Reading Strategies, pp. 90–145
- Differentiation Activity Bank: Struggling Readers, pp. 146–149

Teacher's Edition Activities
- Read Enlarged Vocabulary Definitions, Sec. 1
- Trace a Route on a Map, Sec. 1
- Visualize Terms, Sec. 2
- Use Predicting ABCs, Sec. 3
- Tiered Activities, Sec. 1–3

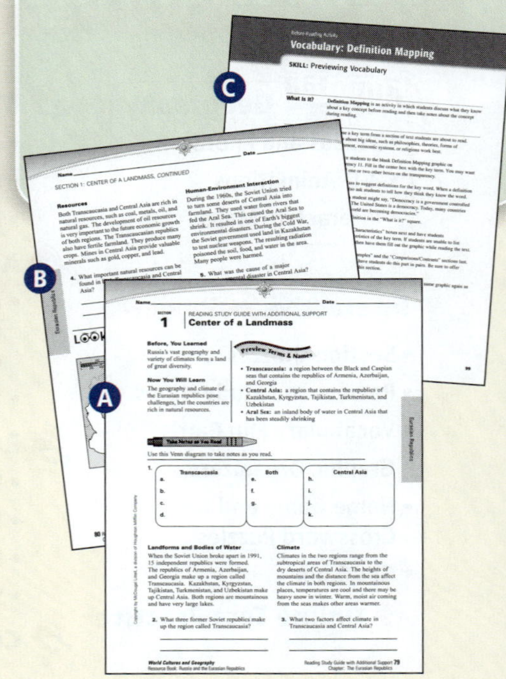

Gifted & Talented Students

Lesson Plans
- Core Lesson Plan

Unit Resource Book
- History Makers, pp. 103–104 (A)
- Interdisciplinary Projects, p. 105 (B)
- Economics in the World, p. 107
- Connect to Science, p. 108 (C)
- Chapter Test (Form C), pp. 125–128

Best Practices Toolkit
- Student Products Rubrics, pp. 1–14
- Writing Templates, pp. 78–89

Teacher's Edition Activities
- Draw a Physical Map, Sec. 1
- Tiered Activities, Sec. 1–3

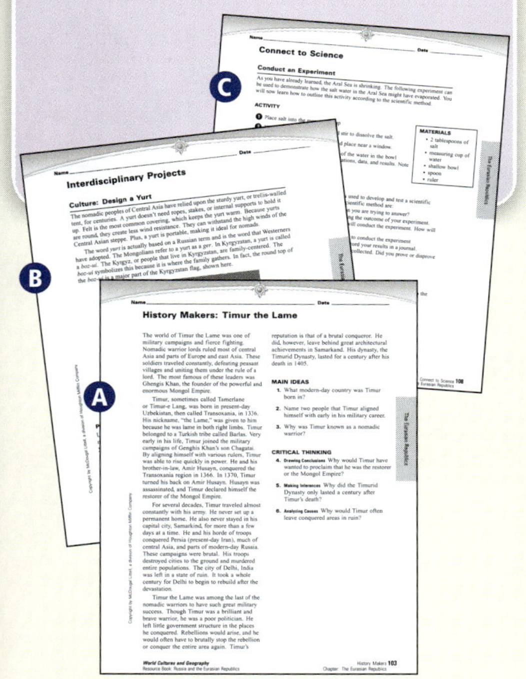

Pre-AP*

Lesson Plans
- Core Lesson Plan

Unit Resource Book
- Primary Sources, p. 109 (A)
- Chapter Test (Form C), pp. 125–128
- World Literature, p. 129 (B)
- Geography Project, p. 131 (C)

Best Practices Toolkit
- Writing Templates, pp. 78–89
- Differentiation Activity Bank: Pre-AP Students, pp. 195–200

* Pre-AP is a registered trademark of the College Entrance Examination Board, which was not involved in the production of and does not endorse this product.

Teacher's Edition Activities
- Trace a Shipping Route, Sec. 1
- Research the Dispute Over the Caspian, Sec. 1
- Make a Population Pie Graph, Sec. 2
- Tiered Activities, Sec. 1–3

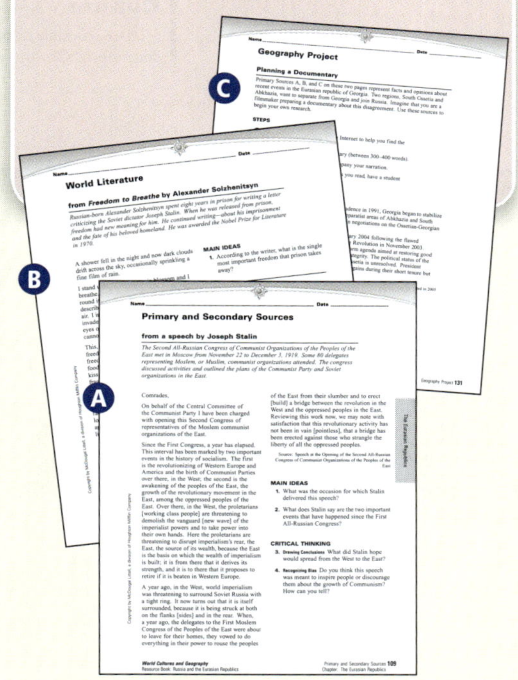

CONNECT to Art

Create Dioramas for Transcaucasia and Central Asia

CLASS TIME 30 minutes

TASK Prepare a diorama illustrating the features of either Transcaucasia or Central Asia.

SUPPLIES
- poster board, art paper, pens, markers, paint, and other art supplies

ACTIVITY Arrange students into two groups. Have one prepare a diorama for Transcaucasia. Have the other prepare one for Central Asia. Tell students to include scenic backgrounds, two- or three-dimensional features, and callouts describing the features, including landforms, bodies of water, and resources. Display the completed dioramas and invite other classes to view them.

Good for Visual and Verbal Learners

CONNECT to Math

Eurasian Republics Ethnic Distribution Bar Graphs

CLASS TIME 30 minutes

TASK Have students create bar graphs of the ethnic distribution of Transcaucasia and Central Aisa.

SUPPLIES
- textbooks, paper, pens, pencils, or markers

ACTIVITY Have students use the information in the Comparing Ethnic Distribution feature in Section 2 to create bar graphs for the ethnic distribution in Transcaucasia and in Central Asia.

Good for Visual and Mathematical/ Spatial Learners

Introduce the ESSENTIAL QUESTION

- Identify the geographical features, climates, and natural resources of the Eurasian republics.
- Describe the environmental problems in the Eurasian republics.
- Identify the ethnic groups and historical events that have influenced life in the Eurasian republics, and describe the cultures and traditions of the people living in this region today.
- Explain the problems that the Eurasian republics have faced in establishing democracies, and describe their efforts to develop market economies.

Key Ideas

Section 1: The geography and climate of the Eurasian republics pose challenges, but the countries are rich in natural resources.

Section 2: The geographic location of the Eurasian republics has attracted many different cultures, ideas, and conquerors.

Section 3: The republics are working to overcome internal problems and establish democratic governments and market economies.

CONNECT to the Essential Question

You may wish to revisit the Essential Question periodically with students as they work through the chapter.

Suggestions for reconnecting are provided at an appropriate point in the *Teacher's Edition* for each section.

Kalta Minor Minaret and Islamic learning center in Uzbekistan

CHAPTER 16

The Eurasian Republics

1 GEOGRAPHY
Center of a Landmass

2 HISTORY & CULTURE
Historic Crossroads

3 GOVERNMENT & ECONOMICS
The Challenge of Independence

ESSENTIAL QUESTION

How can the Eurasian republics meet the challenges of independence?

CONNECT Geography & History

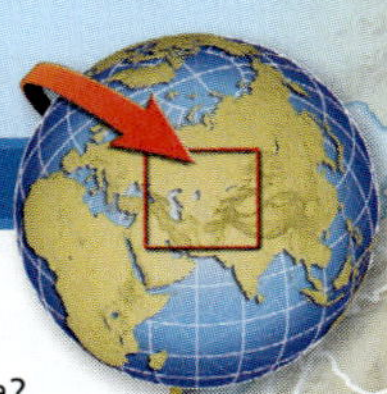

Use the map and the time line to answer the following questions.

1. When did the Mongols conquer Central Asia?
2. What body of water lies between Azerbaijan and Turkmenistan?

600

Culture
◄ **600s** Arabs spread Islam to Transcaucasia and Central Asia.
(Muslim prayer rug)

Culture
700s Turkish peoples migrate to Central Asia.

History
1200s Mongols conquer Central Asia and stabilize trade along Silk Roads.
(Mongol conqueror Timur the Lame) ►

TIME LINE DISCUSSION

Tell students that the diverse cultural makeup of the Eurasian republics reflects a rich history influenced by migration, conquest, and trade. Use a world map or globe to show why this region is considered a "crossroad."

- Which events on the time line highlight early migrations? *(600s: Arabs spread Islam; 700s: Turkish peoples migrate to Central Asia; 1200s: Mongols conquer Central Asia)*
- What event on the time line highlights an environmental problem in Central Asia? *(1960s–Today: shrinkage of the Aral Sea)*
- Which two countries were engaged in the "Great Game" for control of Central Asia? *(Russia and Great Britain)*
- Which event on the time line highlights a major political event affecting the future of the Eurasian republics? *(1991: The Soviet Union collapses, and the republics gain independence.)*

Geography
◄ 1960s–Today
Aral Sea steadily shrinks.

Today

History
1813–1907 Russia and Great Britain conduct the "Great Game," each trying to gain control of Central Asia.

Government
1991 Soviet Union collapses; Eurasian republics gain independence.

Animated GEOGRAPHY
Present-Day Eurasian Republics
CLASSZONE.COM

Point out that in this political map, the eight Eurasian republics are shown in contrasting colors. The surrounding countries are a uniform buff color.

- From what direction would Arab and Turkish cultures have come to Transcaucasia and Central Asia? *(from the west and south)*
- When did the Mongols conquer Central Asia? *(in the 1200s)*
- Which capital city is located on a body of water? *(Baku, Azerbaijan)*

CONNECT GEOGRAPHY & HISTORY ANSWERS

1. in the 1200s

2. Caspian Sea

Presentation Options

Use the **Power Presentations** DVD-ROM to project the animated map.

- Prompt students to use the pop-up images and audio from the map to discover what they already know about the geographies, cultures, histories, economics, or governments of the region.
- Challenge students to answer the map questions individually, in small groups, or as a class activity. Use all of the interactive material to help preview chapter content.

TEACHER MEDIA FAVORITES

Books for the Teacher

Libal, Joyce. *Uzbekistan.* Mason Crest Publishers, 2005. One volume in *The Growth and Influence of Islam in the Nations of Asia and Central Asia* series.

Major, John S., Belanus, Betty J., Ma, Yo-Yo. *Caravan to America: Living Arts of the Silk Road.* Cricket Books, 2002. This colorful book explores eight artists whose cultural roots come from Silk Road countries.

Sherbell, Shepard. *Soviets.* Yale University Press, 2001. Stark black and white photos (grouped by theme) depict the last days of the Soviet Union.

Video / DVD

Silk Road DVD Collection. (DVD) Central Park Media, 2002. Contains 12 episodes, exploring the history of and cultures along this ancient trade route.

Transcaucasus: Georgia, Armenia & Azerbaijan. (VHS) GlobeScope Expeditions, 2003. A 24-minute video tour of this unique region.

Internet

Go to the *Research and Writing Center* at **ClassZone.com** to link to:

- data on the region
- a variety of chapter-specific, content-reviewed sites
- "History Makers" biographies
- primary sources

1 Plan & Prepare

Objectives

- Identify the major landforms and bodies of water in Transcaucasia and Central Asia
- Describe the climates and resources of Transcaucasia and Central Asia
- Analyze the causes of some environmental problems in Central Asia

Read for the Essential Question

Help students read for a purpose by reminding them of the Essential Question: "How can the Eurasian republics meet the challenges of independence?" Point out that they can use items in the section, such as Key Questions, special features, images, and maps to formulate answers to the question.

Vocabulary

Best Practices Toolkit

Use the ***Best Practices Toolkit*** to model strategies for vocabulary notetaking. Vary strategies throughout the year. Choose from: *Knowledge Rating, Predicting ABCs, Definition Mapping, Word Sort, Word Wheel, Frayer Model (Word Squares), Magnet Words,* and *Student VOC.*

Vocabulary Strategies, TT9–TT16

Reading Strategy

Best Practices Toolkit

Display the Venn Diagram transparency.

- Remind students that comparing and contrasting involves recognizing similarities and differences.
- Model filling in the diagram by comparing and contrasting cats and dogs. Enter differences in the large outer areas and similarities in the area where the ovals overlap. Have students suggest entries.
- Point out that the similarities and differences they enter on Transcaucasia and Central Asia will help them answer the key questions in this section.

Venn Diagram, TT22

Section 1 Reading for Understanding

Key Ideas

BEFORE, YOU LEARNED

Russia's vast geography and variety of climates form a land of great diversity.

NOW YOU WILL LEARN

The geography and climate of the Eurasian republics pose challenges, but the countries are rich in natural resources.

Vocabulary

TERMS & NAMES

Transcaucasia (TRANS•kaw•KAY•zhuh) a region between the Black and Caspian seas that contains the republics of Armenia, Azerbaijan (AZ•uhr•by•JAHN), and Georgia

Central Asia a region that contains the republics of Kazakhstan (KAH•zahk•STAN), Kyrgyzstan (KEER•gee•STAN), Tajikistan (tah•JIHK•ih•STAN), Turkmenistan (TURK•mehn•ih•STAN), and Uzbekistan (uz•BEHK•ih•STAN)

Caucasus (KAW•kuh•suhs) **Mountains** a mountain range that runs between the Black Sea and the Caspian Sea

landlocked completely surrounded by land

Tian Shan (TYAHN SHAHN) a mountain range that runs across Kyrgyzstan and Tajikistan

Kara Kum a huge desert that covers most of Turkmenistan

Aral Sea an inland body of water in Central Asia that has been steadily shrinking

Lake Balkhash a lake in eastern Kazakhstan

REVIEW

Eurasia a term used to refer to a single continent that is made up of Europe and Asia

Caspian Sea a sea situated between Transcaucasia and Central Asia; the largest inland body of water in the world

Reading Strategy

Re-create the Venn diagram shown at right. As you read and respond to the **KEY QUESTIONS**, use the diagram to compare and contrast the landforms, climates, and resources of Transcaucasia and Central Asia.

See Skillbuilder Handbook, page R9

COMPARE AND CONTRAST

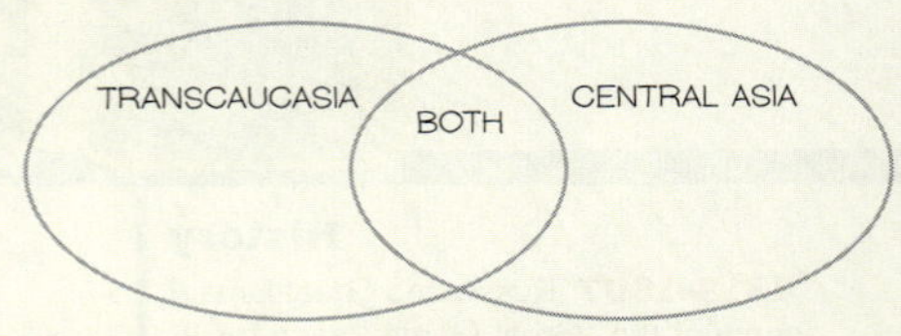

GRAPHIC ORGANIZERS Go to Interactive Review @ClassZone.com

496 Chapter 16

PRETEACHING VOCABULARY

English Learners

Pronounce and Preview

Pronounce each term for students. Review the meanings of words in the definitions, such as *inland, steadily,* and *situated.*

- To modify vocabulary learning, have students complete worksheets as they read the section.

Unit Resource Book: Russia & the Eurasian Republics

- Vocabulary Practice, p. 97

Inclusion

Read Enlarged Vocabulary Definitions

To help students who are visually impaired, make an enlarged photocopy of this page and distribute it. Ask volunteers to take turns reading a word and its definition aloud. Discuss each term as a class and answer any questions students might have. Have students cut out the enlarged terms and names and glue or tape them in a vocabulary notebook.

SECTION 1 GEOGRAPHY

Cemetery in Kyrgyzstan

Snow leopard in mountains of Central Asia

Center of a Landmass

Connecting to Your World

You have already learned that the Soviet Union was a Communist country that consisted of Russia and the surrounding republics. Remember that some geographers place this enormous nation in a region they call **Eurasia**, a landmass made up of both Europe and Asia. When the Soviet Union broke apart in 1991, 15 independent republics were formed. Three of these republics are located between the Black and Caspian seas in a region called Transcaucasia. Five more independent republics lie in Central Asia, a region east of the Caspian Sea.

Landforms and Bodies of Water

KEY QUESTION What are the major landforms and bodies of water in Transcaucasia and Central Asia?

Transcaucasia consists of the republics of Armenia, Azerbaijan, and Georgia. **Central Asia** contains the republics of Kazakhstan, Kyrgyzstan, Tajikistan, Turkmenistan, and Uzbekistan. Both regions are very mountainous and boast some of the largest lakes in the world.

Tian Shan The Tian Shan, which is Chinese for "Heavenly Mountains," stretch for nearly 1,500 miles.

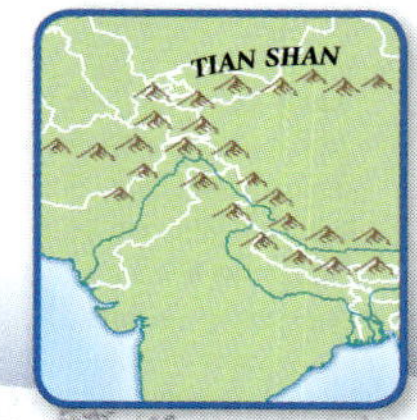

2 Focus & Motivate

3-Minute Warm-Up

Write on the board or display the transparency:

- What might daily life be like in mountainous or desert areas? *(might lack services; climate might be harsh; travel might be difficult)*

Unit Transparency Book: Russia & the Eurasian Republics
- 3-Minute Warm-Ups, TT11

Connecting to Your World

More About . . .

Tian Shan

The Tian Shan contains some of the tallest mountains in the world, including Pobeda Peak (over 24,000 feet). Mountain passes, such as the Terek Pass (over 12,000 feet), link Kyrgyzstan, Kazakhstan, and China.

3 Teach

Landforms and Bodies of Water

Think, Pair, Share

- Identify the republics in Transcaucasia and Central Asia. *(Transcaucasia: Armenia, Azerbaijan, Georgia; Central Asia: Kazakhstan, Kyrgyzstan, Tajikistan, Turkmenistan, Uzbekistan)*
- **Compare and Contrast** What is a major geographical difference between the two regions? *(Transcaucasia includes much fertile land; desert occupies much of Central Asia.)*

SECTION 1 PROGRAM RESOURCES

Unit Resource Book: Russia and the Eurasian Republics

ON LEVEL

Unit Resource Book
- Reading Study Guide, p. 73
- Outline Map Activities, p. 101
- Section Quiz, p. 110

STRUGGLING READERS

Unit Resource Book
- Reading Study Guide with Additional Support, p. 79
- Vocabulary Practice, p. 97
- Skillbuilder Practice, p. 98
- Section Quiz, p. 110
- Reteaching Activity, p. 113

eEdition with Audio DVD-ROM

ENGLISH LEARNERS

Pupil Edition in Spanish

eEdition with Audio DVD-ROM

Unit Resource Book
- RSG (Spanish), p. 85
- RSG with Additional Support (Spanish), p. 91

Multi-Language Glossary

Test Generator
- Section Quiz in Spanish

INCLUSION

Unit Resource Book
- RSG with Additional Support, p. 79
- Section Quiz, p. 110
- Reteaching Activity, p. 113

GIFTED & TALENTED

Unit Resource Book
- Connect to Science, p. 108
- Section Quiz, p. 110

PRE-AP

Unit Resource Book
- Section Quiz, p. 110

TECHNOLOGY

Unit Transparency Book
- 3-Minute Warm-Ups, TT11
- Essential Question, TT12
- Critical Thinking, TT13
- Map, TT16
- Map, TT17

Daily Test Practice Transparencies
- Section 1, TT56

Power Presentations

ClassZone.com

World Cultures and Geography Video on DVD

More About . . .

The Black Sea

The Black Sea is bordered by Georgia on the east, Russia and Ukraine on the north, Romania and Bulgaria on the west, and Turkey on the south. Its name reflects the stormy, inhospitable nature of the sea. Several bodies of water connect it to the Mediterranean Sea in the west. The Volga-Don canal in Russia connects it to the Caspian Sea in the east. The Black Sea is an important shipping hub, with major ports serving all of the countries it borders.

More About . . .

The Pamirs

Though situated primarily in Tajikistan, the Pamirs also extend north into southeastern Kyrgyzstan, south into northeastern Afghanistan, and east into western China. Several mountain ranges, including the Tian Shan, Kunlun, Himalayas, and Hindu Kush, meet at the Pamirs. Because this region includes a concentration of some of the tallest mountain peaks on Earth, it is sometimes referred to as "the roof of the world." The mountains in this region are mostly rocky and treeless, separated by deep canyons. The Amu Darya, a vital source of water for farmers in Central Asia, originates in the Pamirs.

Make It Fun

Teacher-Tested Activities

Matt Parker, Eastgate Middle School, Kansas City, Missouri

After my students have read a section in their textbook, we sometimes play a game that's like a cloze, or fill-in-the-blank, activity. I read aloud a paragraph from the section but leave out every fifth word. After every omission, I pause and invite students to call out the missing word. This activity not only encourages active listening, it also reinforces using context clues as a reading strategy and serves as a solid review of the material.

Transcaucasia As you can see on the map on the opposite page, Transcaucasia is bounded by Turkey and Iran to the south. The **Caucasus Mountains** (A) rise in the north and form the border between Russia and Transcaucasia. These mountains lie between two bodies of water. The Black Sea borders Georgia on the west. The **Caspian Sea** (B), which is actually a saltwater lake, borders Azerbaijan on the east. It is the largest inland body of water in the world.

Armenia lies between Georgia and Azerbaijan. In fact, Armenia is **landlocked**, or completely surrounded by land. The country contains fertile volcanic soil on a southern plateau. Fertile coastal plains lie in Georgia on the Black Sea and along the Caspian Sea in Azerbaijan.

Central Asia This region is bordered by Russia to the north, the Caspian Sea to the west, a series of mountain ranges to the south, and China to the east. Mountains dominate Kyrgyzstan and Tajikistan. Most of this rugged terrain is part of the Alay and **Tian Shan** (C) mountain ranges. The towering Tian Shan range contains some of the largest alpine glaciers in the world. An even higher mountain range, called the Pamirs, is located in southeastern Tajikistan.

Kazakhstan has a variety of landforms. The mountainous terrain in the east lowers to a large steppe in the north and a desert in the south. Then the land lowers even more toward the Caspian Sea, where it sinks to about 433 feet below sea level. The huge desert of **Kara Kum** (D) covers most of Turkmenistan. Another vast desert, the Kyzyl Kum, occupies much of Uzbekistan.

The two most important rivers of Central Asia, the Syr Darya and the Amu Darya, flow into the **Aral Sea** (E). *Darya* means "river." Much of the water from these rivers is used for irrigation. You will learn about the impact of this irrigation on the Aral Sea later in this section. **Lake Balkhash** (F) is situated in eastern Kazakhstan. It is unusual because the lake is composed of fresh water in its western half and salt water in its eastern half.

CATEGORIZE Describe the major landforms and bodies of water in Transcaucasia and Central Asia. Caucasus Mountains and the Black and Caspian seas in Transcaucasia; Tian Shan and Pamir mountains, Kara Kum and Kyzyl Kum deserts, Aral Sea, and Lake Balkhash in Central Asia

The desert is too barren, dry, and hot to support much human habitation. ▼

Kara Kum The few people who live in this desert are mainly sheep and camel herders. **Why do you think the Kara Kum is sparsely populated?**

498 Chapter 16

DIFFERENTIATING INSTRUCTION

Struggling Readers

Locate Major Map Features

Students reading below grade level may benefit from visually identifying the locations of major features in Transcaucasia and Central Asia. Have students use the physical map on the opposite page to compare the sizes and shapes of the Eurasian republics and identify the locations of their landforms and bodies of water.

English Learners

Explore Vocabulary

Explain that irrigation is the supply of water to land. Ask students why irrigation would be vital to a dry region. *(It's the main way farmers can water their fields since rainfall is so low.)* To help students understand irrigation, discuss common irrigation methods, such as digging ditches or building canals that direct the flow of water to farmland.

Physical Geography of the Eurasian Republics

Cotton farmers in Uzbekistan

CONNECT Geography & Culture

READING A PHYSICAL MAP

As you study the map,

- identify and locate the republics in Transcaucasia and Central Asia
- note which republics in each region are landlocked
- compare elevation across the two regions

1. **Region** Which two republics in Central Asia appear to be as mountainous as those in Transcaucasia?
2. **Human-Environment Interaction** What impact might this mountainous terrain have on the people there?

DIFFERENTIATING INSTRUCTION

Gifted & Talented

Draw a Physical Map

Students might enjoy creating their own physical maps of one of the republics. Refer them to the political map in this chapter or to an atlas map of each country to outline its shape. Have them use text information and the physical map on this page to place physical features. Encourage them to create their own legend and symbols for showing mountain regions, plains, deserts, and bodies of water.

Pre-AP

Trace a Shipping Route

Supply students with the information about the Black Sea in the More About on the previous page. Then ask this question:

- How might agricultural products and manufactured goods be transported from Central Asia to a country along the Mediterranean Sea?

Have students use a map of Europe and Asia or a world map to trace shipping routes that might be used over land and sea.

Physical Geography of the Eurasian Republics

CONNECT Geography & Culture

Reading a Physical Map Have students compare this map to the political map on the chapter opener to more clearly identify the borders of these countries. Ask volunteers to identify the landlocked countries *(Armenia, Uzbekistan, Kyrgyzstan, Tajikistan).* Then have them identify countries containing coastal areas along the Caspian Sea that are below sea level *(Azerbaijan, Turkmenistan, Kazakhstan)* and the vast area of Central Asia that is at, or not much above, sea level *(Turkmenistan, Uzbekistan, Kazakhstan)*.

ANSWERS

1. **Region** Tajikistan and Kyrgyzstan
2. **Human-Environment Interaction** People probably live difficult, isolated lives there.

Unit Resource Book: Russia & the Eurasian Republics

- Outline Map Activities: Transcaucasia: Physical/Political, p. 101

More About . . .

The Geography of Uzbekistan

The Kyzyl Kum in northern and central Uzbekistan is very sparsely populated. However, this area supports mining operations and some farming, including raising livestock. Cotton farming is heavily concentrated in the plains to the south and east and is a key part of Uzbekistan's economy. The country's urban population is mostly found in the east near the Tian Shan, such as in the cities of Tashkent and Samarkand.

COMPARING

Climates

Ask students to study the photographs.

- What type of climate is suggested in each photograph, and what comparison is being made? *(A mild climate in part of Transcaucasia is contrasted with a harsher climate in part of Central Asia.)*
- Have them list words that would describe each climate based on each image. *(Transcaucasia: hot, dry, sunny; Central Asia: cold, sunny, dry)* Then have students read about each region's climate and revise their lists as necessary.

ANSWERS

1. **Transcaucasia** People lying in the sun and wearing light clothing reflects Georgia's milder climate.
2. **Central Asia** Only plants that can survive with little rain can probably grow there.

Teach

Climate

Reader, Recorder, Reporter

- Which country in Transcaucasia has the mildest climate? *(Georgia)* What are some factors that influence its climate? *(The Caucasus Mountains protect the coastal areas from receiving too much rain; moist air from the Black Sea moderates its climate.)*
- What are some factors that contribute to the harsh climate of much of Central Asia? *(Much of Central Asia is landlocked. Mountain ranges block moist air, contributing to the semiarid and desert climates.)*
- **Compare and Contrast** What is the most noticeable difference between the climates of these two regions? *(In general, Central Asia has a much harsher climate than Transcaucasia, with extreme high and low temperatures.)*

COMPARING Climates

Transcaucasia Visitors enjoy a day on Georgia's Black Sea coast. **1. How does this image reflect the area's climate?**

Central Asia Kyrgyzstan averages less than 20 inches of rain a year. **2. How might this climate affect vegetation?**

Climate

KEY QUESTION How do the climates in Transcaucasia and Central Asia differ?

Climate in the two regions ranges from the subtropical areas of Transcaucasia to the semiarid and desert climates of Central Asia. The high elevations of the mountains and distance from the sea have a big impact on climate in both regions.

Transcaucasia In the lowlands of Azerbaijan, winters are mild and summers are hot. In the mountainous areas of Armenia, on the other hand, temperatures are colder, and snow in the winter can be heavy. Georgia has a milder climate than either Armenia or Azerbaijan. The Caucasus Mountains protect the coastal areas from too much rain. In addition, moist air from the Black Sea moderates Georgia's climate.

Central Asia Like Armenia, the nations of Central Asia are landlocked. They are many miles from the moderating influence of the world's oceans. As a result, Central Asia has a much harsher climate than Transcaucasia, with extreme high and low temperatures. Southern and southeastern mountain ranges block any moist air and contribute to the semiarid and desert climates of the region.

COMPARE AND CONTRAST Compare the climates of Transcaucasia and Central Asia. The climate of Central Asia is harsher and more extreme than that in Transcaucasia.

INTERDISCIPLINARY ACTIVITIES

CONNECT to Math

Calculate Averages

Provide students with temperature and precipitation statistics for each region (from atlases, encyclopedias, or Internet sources such as the *CIA World Factbook*). If the statistics are for individual countries or provide a range of temperatures or precipitation levels, have students calculate the average for each region, including annual average precipitation levels, winter temperature, and summer temperature. Tell them to enter the results of their calculations in a chart like the one shown.

	Transcaucasia	Central Asia
Annual Average Precipitation		
Average Winter Temperature		
Average Summer Temperature		

Resources

KEY QUESTION What resources are abundant in Transcaucasia and Central Asia?

Both Transcaucasia and Central Asia are rich in natural resources. The regions have huge reserves of coal and metals, but oil and natural gas are especially plentiful. In fact, the petroleum deposits around and under the Caspian Sea are among the world's largest. The development of the oil resources in both regions should help the young nations thrive.

Transcaucasia Most of the oil in Transcaucasia lies in Azerbaijan. In fact, the country's name means "land of flames." The name refers to fires that erupted from the rocks and the rivers leading to the Caspian Sea. The fires were the result of underground oil and gas deposits.

The Transcaucasian republics also have fertile farmland. In the rich soil along the coastal plain of Georgia, farmers grow grapes, citrus fruit, and tea. In the highlands of Armenia, farm products include peaches, apricots, walnuts, and wheat. Because of the lack of precipitation, many of the farms in Azerbaijan require irrigation.

Central Asia Much of the oil in Central Asia can be found in Kazakhstan and Turkmenistan. Nations all over the world are interested in buying this oil, which may eventually bring great wealth to the region. In addition, rich deposits of highly valued energy resources, such as coal and natural gas, are plentiful. Mines also provide valuable minerals, including gold, copper, and lead.

Unlike Transcaucasia, Central Asia has little fertile soil, and many areas do not receive much rain. For that reason, water often has to be brought in to irrigate crops. Cotton is an important crop for all of the Central Asian republics. Other crops include grain, fruits, and vegetables.

CONNECT Geography & Economics

Oil and the Caspian Sea

For years, the five countries that border the Caspian Sea have argued over rights to the oil-rich area. In 2003, Azerbaijan, Kazakhstan, and Russia agreed to a division of the sea and its resources. However, Iran and Turkmenistan did not agree to the plan. Nonetheless, progress has been made in extracting the oil. In 2005, a pipeline was built to pump oil from Azerbaijan to Turkey. In this photograph, a mosque in the Azerbaijan capital of Baku overlooks oil derricks in the Caspian.

SUMMARIZE Identify the resources that are abundant in Transcaucasia and Central Asia. oil, natural gas, coal, metals, and farmland in Transcaucasia; oil, coal, natural gas, minerals, and cotton in Central Asia

Teach

Resources

Think, Pair, Share

- What natural resources are common in both Transcaucasia and Central Asia? *(oil, gas, coal, and metals)*
- Which natural resource may eventually bring great wealth to both regions? *(oil)*
- **Compare and Contrast** In what ways does farming differ in the two regions? *(Much of Transcaucasia has fertile soil and receives sufficient rainfall to support its crops. Central Asia has little fertile soil, and many areas do not receive much rain. As a result, crops require irrigation.)*

CONNECT Geography & Economics

Oil and the Caspian Sea

Tell students that the Caspian Sea pipeline from Azerbaijan to Turkey is more than 1,000 miles long. It begins at Baku, goes across Azerbaijan to Tbilisi, Georgia, across Georgia into Turkey, and down Turkey to the port city of Ceyhan on the eastern edge of the Mediterranean. The pipeline is capable of delivering the equivalent of one million barrels of oil per day.

DIFFERENTIATING INSTRUCTION

Inclusion

Trace a Route on a Map

For visual learners, use the political map of the Eurasian republics to point out the countries that border the Caspian Sea. Outline each country's coastline in red to help students see the borders. Then use the information in the Connect Geography & Economics note above to roughly trace the route of the pipeline on the map. Use a world map or a globe to show the end of the route at Ceyhan.

Pre-AP

Research the Dispute over the Caspian

Tell students that the dispute over oil rights in the Caspian has much to do with whether the body of water is considered a sea or a lake. Have students research the question and share their findings with the rest of the class.

Teach

Human-Environment Interaction

Talk About It

- What caused the Aral Sea to shrink drastically? *(Using water from the Syr Darya and Amu Darya for an irrigation project reduced the amount of water that flows from these rivers into the Aral Sea.)*
- What caused a dramatic increase in leukemia, birth defects, and mental illness in Kazakhstan? *(radiation from Soviet nuclear-testing programs)* What other environmental problem has Kazakhstan faced? *(pollution from heavy industry)*
- **Compare and Contrast** Which of the environmental problems discussed in this section do you think might have the greatest long-term negative effects? *(Possible answer: radiation from nuclear testing because it is harder to remove or eliminate)*

Unit Resource Book: Russia & the Eurasian Republics
- Skillbuilder Practice: Sequencing Events, p. 98

CONNECT to Science

Explain to students that at one point, the damaging effects of the Aral Sea's shrinkage virtually wiped out what had been a thriving fishing industry and grounded most of the boats that sailed on its waters. Scientists feared that the Aral Sea might dry up completely by 2020.

Unit Resource Book: Russia & the Eurasian Republics
- Connect to Science: Conduct an Experiment, p. 108

Human-Environment Interaction

KEY QUESTION What has led to some of the environmental problems in Central Asia?

For decades the Soviet Union attempted to overcome its geographic limits by trying to increase its farmable land. The government also took advantage of some of its vast, uninhabited areas by carrying out experimental projects in them. Unfortunately, these projects sometimes had unintended negative results, particularly in Central Asia.

Aral Sea In the 1960s, the Soviet Union began using river water from the Syr Darya and Amu Darya for an irrigation project. This project was part of a plan to convert areas of Central Asian desert into thriving cotton farms. Indeed, the production of cotton soon soared. However, since the Aral Sea received most of its water from the two rivers, the irrigation caused the sea to drastically shrink. The sea even split into two, the small Aral and the large Aral. Recent efforts have helped raise the level of the small Aral. As a result, fish, jobs, and people have begun to return to the area. But the Soviet plan caused one of Earth's greatest environmental disasters.

CONNECT to Science

By the end of the 20th century, irrigation and evaporation had caused the Aral Sea to shrink about 75 percent. The retreating waters left behind dry beds filled with salt, fertilizers, and pesticides. These chemicals had washed into the sea from the cotton farms. Windstorms then blew the chemicals into the surrounding area. As a result, the rates of cancer and other diseases greatly increased in the nearby population.

Disappearing Sea A camel stands before an abandoned ship on what was once the floor of the Aral Sea.

Activity

Conduct an Evaporation Experiment

Materials
- 2 tablespoons of salt
- measuring cup of water
- spoon
- shallow bowl

1. Place salt into the measuring cup.
2. Fill the cup with warm water and stir to dissolve the salt.
3. Pour the mixture into the bowl and place near a window.
4. Check the bowl every day for a week. Note what evaporates and what remains behind.

DIFFERENTIATING INSTRUCTION: TIERED ACTIVITIES

OBJECTIVE Create posters illustrating the geography of the Eurasian republics.

Basic

Ask pairs of students to create a series of posters that illustrate the major geographic features of the Eurasian republics. Remind students to include the names of specific mountains, deserts, plains, and bodies of water.

On Level

In addition to the Basic activity, have student pairs present their posters to the class, and discuss the similarities and differences between the two regions.

Challenge

In addition to the On Level activity, have students include information about climate, resources, and environmental problems on their posters.

Nuclear Testing Other Soviet programs have caused problems in Central Asia. As you may recall, the Soviet Union and the United States engaged in a weapons race during the Cold War. Between 1949 and 1990, the Soviet government used part of the vast steppes in northern Kazakhstan to test their nuclear weapons. Unfortunately, radiation from this testing poisoned the soil, food, and water over about a 7,000-square-mile area. The radiation caused dramatic increases in the rates of leukemia, birth defects, and mental illness in the area.

Recently, the nations of Central Asia have taken steps to avoid more environmental problems. For example, Kazakhstan has started to develop more light industry to prevent an overdependence on heavy industries. This measure should help decrease the region's air and water pollution problems.

Nuclear Blast The Soviets conducted about 715 nuclear tests. In this picture, a test site is destroyed in Kazakhstan in 2000.

ANALYZE CAUSES Explain what led to some of the environmental problems in Central Asia.
A Soviet agricultural plan caused the Aral Sea to shrink; Soviet nuclear testing in Kazakhstan has poisoned the soil, food, and water in this area.

Section 1 Assessment

ONLINE QUIZ For test practice, go to **Interactive Review @ClassZone.com**

TERMS & NAMES

1. **Explain the importance of**
   - Transcaucasia
   - Central Asia
   - Kara Kum
   - Aral Sea

USE YOUR READING NOTES

2. **Compare and Contrast** Use your completed Venn diagram to answer the following question:

   What element of their physical geography offers great promise for the futures of both Transcaucasia and Central Asia?

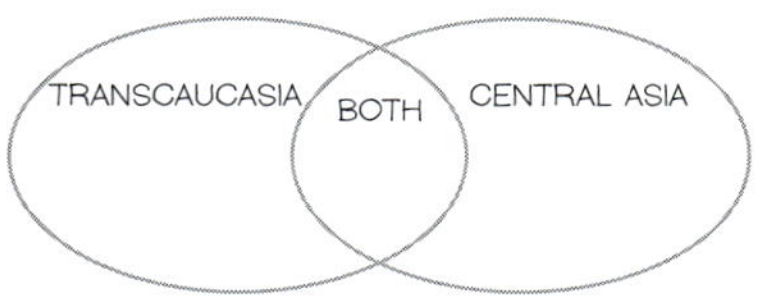

KEY IDEAS

3. What types of landforms are found in Kazakhstan?
4. Why does Central Asia have a harsh climate?
5. What areas of Transcaucasia have rich soil?

CRITICAL THINKING

6. **Make Inferences** Which region do you think is more densely populated, Transcaucasia or Central Asia? Explain your answer.
7. **Compare and Contrast** How are the landforms and bodies of water of Georgia and Armenia similar and different?
8. **CONNECT to Today** What lessons can be learned today from the Aral Sea environmental disaster?
9. **WRITING** **Prepare an Environmental Report** Do further research about the causes and effects of pollution in Central Asia. Also find out about possible solutions for these problems. Then write an environmental report that summarizes your research.

CONNECT to the Essential Question

How can the Eurasian republics meet the challenges of independence?

Ask students what they have learned so far that can help them answer this question. Students might mention the following:

- The terrain and climates of the republics make agriculture difficult.
- The republics are rich in natural resources.
- Central Asia has inherited many environmental problems.

4 Assess & Reteach

Assess Have students complete the Section Assessment.

Unit Resource Book: Russia & the Eurasian Republics
- Section Quiz, p. 110

Interactive Review @ClassZone.com
- Section Quiz

Power Presentations

Test Generator

Reteach Divide the class into four groups. Assign each group one of the subsections in Section 1. Have them list the important facts, events, and ideas covered in the subsection. Then have each group present its list to the class.

Unit Resource Book: Russia & the Eurasian Republics
- Reteaching Activity, p. 113

SECTION 1 ASSESSMENT ANSWERS

Terms & Names

1. Transcaucasia: Armenia, Azerbaijan, Georgia; Central Asia: Kazakhstan, Kyrgyzstan, Tajikistan, Turkmenistan, Uzbekistan; Kara Kum: huge desert in Turkmenistan; Aral Sea: an inland body of water in Central Asia

Use Your Reading Notes

2. Both regions have valuable resources.

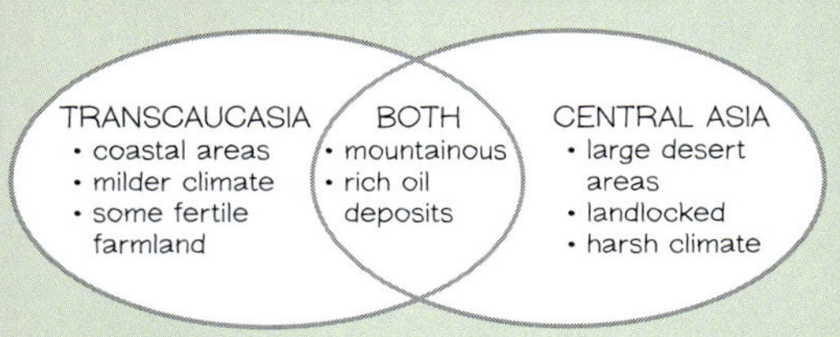

Key Ideas

3. mountainous land in the east; steppe in the north; desert in the south; lowlands by the Caspian Sea (in the west)
4. It is many miles from the seas.
5. in Georgia, the coastal plain that rims the Black Sea; the highlands in Armenia

Critical Thinking

6. Transcaucasia, because the weather isn't as harsh and the coastal areas are more hospitable
7. Both have mountains and areas with rich soil. In Georgia, fertile soil is on the coastal plain; the rich soil in Armenia is on the plateau. Armenia is landlocked, while Georgia is not.
8. Don't grow crops in areas where they would not naturally grow; be careful with water sources.
9. Use the rubric to score reports.

Report Rubric

	Content	Errors
4	many clear causes and effects	few or none
3	an adequate number of causes and effects	some
2	a few causes and effects	many
1	few or no causes and effects	many major

1 Plan & Prepare

Objectives

- Explain how Transcaucasia and Central Asia have been influenced by military conquest, migration, and trade
- Describe family life and traditions in Transcaucasia and Central Asia

Read for the Essential Question

Help students read for a purpose by reminding them of the Essential Question: "How can the Eurasian republics meet the challenges of independence?" Point out that they can use items in the text, such as Key Questions, special features, images, and maps to formulate answers to the question.

Vocabulary

Best Practices Toolkit

Use the ***Best Practices Toolkit*** to model strategies for vocabulary notetaking. Vary strategies throughout the year. Choose from: *Knowledge Rating, Predicting ABCs, Definition Mapping, Word Sort, Word Wheel, Frayer Model (Word Squares), Magnet Words,* and *Student VOC.*

Vocabulary Strategies, TT9–TT16

Reading Strategy

Best Practices Toolkit

Display the Web Diagram transparency.

- Remind students that main ideas are important ideas in a block of text and that details support the main ideas.
- Model filling in the web with the main idea and details drawn from the Connecting to Your World paragraph. Have students suggest entries.
- Remind students that main ideas in each block of text in the section will help them answer a Key Question.

Web Diagram, TT28

SECTION 2

Reading for Understanding

Key Ideas

BEFORE, YOU LEARNED

The Eurasian republics are largely mountainous and desert regions with harsh climates.

NOW YOU WILL LEARN

The geographic location of the Eurasian republics has attracted many different cultures, ideas, and conquerors.

Vocabulary

TERMS & NAMES

nomad member of a group who makes a living by herding animals and moving from place to place as the seasons change

Islam a religion that is based on the teachings of the prophet Muhammad

Muslim a follower of Islam

Silk Roads major trade routes that ran from China to the rest of East Asia, through Central Asia, and into Europe

Russification the process of making a culture more Russian

yurt a portable, tentlike structure used by the nomads of Central Asia

REVIEW

ethnic group a group of people who share language, customs, and a common heritage

Visual Vocabulary Russification

Reading Strategy

Re-create the web diagram shown at right. As you read and respond to the **KEY QUESTIONS**, use the diagram to record main ideas about the history and culture of the Eurasian republics. Add ovals as needed.

See Skillbuilder Handbook, page R4

FIND MAIN IDEAS

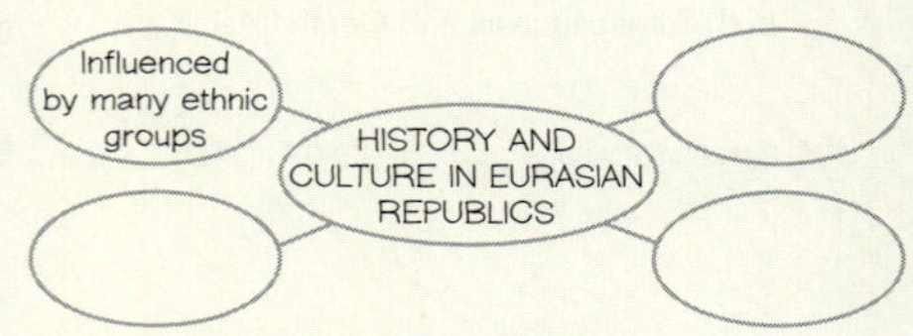

GRAPHIC ORGANIZERS
Go to **Interactive Review** @ClassZone.com

PRETEACHING VOCABULARY

English Learners

Pronounce and Preview

Pronounce each term for students. Review the meanings of words in the definitions, such as *process, portable,* and *heritage.*

- To modify vocabulary learning, have students complete worksheets as they read the section.

Unit Resource Book: Russia & the Eurasian Republics
- Vocabulary Practice, p. 97

Inclusion

Visualize Terms

Have a volunteer read each term and its definition aloud. Ask another volunteer to define the term in his or her own words and then use it in a sentence. With the class, discuss any words that are challenging to define or use in a sentence. Encourage students to use words and phrases that help them visualize the term.

Cemetery in Kyrgyzstan

SECTION 2

HISTORY & CULTURE

Snow leopard in mountains of Central Asia

Historic Crossroads

Connecting to Your World

Imagine that you had to move every few months. In the summer, you might live in New England. Then in the fall, you would move to the Midwest. Finally, in the winter, you would go to the Southwest. In the spring, you would start the cycle all over again. This way of life probably sounds very different from your own, but it is similar to the way thousands of nomads live. **Nomads** are people who make a living by herding animals and moving from place to place as the seasons change. The people who first lived in Transcaucasia and Central Asia led a nomadic way of life.

Nomadic Man This man's weather-beaten face reflects a life lived largely out of doors.

Regions of Exchange

KEY QUESTION How have various groups of people and trade influenced Transcaucasia and Central Asia?

For centuries, many groups of people moved through Transcaucasia and Central Asia. Some of them settled in these regions and had a major influence on culture. Three main reasons for this movement were military conquest, migration, and trade.

Nomadic Home Nomads live in yurts, portable tents that usually consist of several layers of felt stretched around a wooden frame.

2 Focus & Motivate

3-Minute Warm-Up

Write on the board or display the transparency:

- Look at the photograph of the nomad and yurt on this page. What do you imagine life would be like moving from place to place? *(lonely, exposed to the elements, primitive)*

Unit Transparency Book
- 3-Minute Warm-Ups, TT11

Connecting to Your World

More About . . .

Central Asian Nomadic Herders

Nomadic herders dominated rural Central Asia before the formation of the Soviet Union in 1922, after which many nomadic families were forced to settle in villages and become farmers. Today, only remnants of the nomadic culture exist in Central Asia.

3 Teach

Regions of Exchange

Think, Pair, Share

- What groups of people have had a major influence on Transcaucasia and Central Asia? *(Arabs, Turks, Mongols, Soviets)*
- **Find Main Ideas** How did the groups who moved through Transcaucasia and Central Asia influence the regions? *(ethnic makeup, languages spoken, religions practiced, new inventions and ideas)*

SECTION 2 PROGRAM RESOURCES

Unit Resource Book: Russia and the Eurasian Republics

ON LEVEL

Unit Resource Book
- Reading Study Guide, p. 75
- Section Quiz, p. 111

STRUGGLING READERS

Unit Resource Book
- Reading Study Guide with Additional Support, p. 81
- Vocabulary Practice, p. 97
- Skillbuilder Practice, p. 98
- Section Quiz, p. 111
- Reteaching Activity, p. 114

eEdition with Audio DVD-ROM

ENGLISH LEARNERS

Pupil Edition in Spanish

eEdition with Audio DVD-ROM

Unit Resource Book
- RSG (Spanish), p. 87
- RSG with Additional Support (Spanish), p. 93

Multi-Language Glossary

Test Generator
- Section Quiz in Spanish

INCLUSION

Unit Resource Book
- RSG with Additional Support, p. 81
- Section Quiz, p. 111
- Reteaching Activity, p. 114

GIFTED & TALENTED

Unit Resource Book
- History Makers, p. 103
- Interdisciplinary Projects, p. 105
- Section Quiz, p. 111

PRE-AP

Unit Resource Book
- Section Quiz, p. 111

TECHNOLOGY

Unit Transparency Book
- 3-Minute Warm-Ups, TT11
- Essential Question, TT12
- Critical Thinking, TT14
- Cultures, TT18
- Cultures, TT19

Daily Test Practice Transparencies
- Section 2, TT57

Power Presentations

ClassZone.com

World Cultures and Geography Video on DVD

Comparing Regions

Religious Practices in Transcaucasia and Central Asia

In Transcaucasia, most of the people in Georgia and Armenia are Christian and belong to the Georgian Orthodox Church and Armenian Orthodox Church, respectively. Islam is the dominant religion in Azerbaijan. A sizeable minority of Armenian and Russian Christians also live in Azerbaijan. In Central Asia, most of the people practice Islam, tracing their religious beliefs to their Arab and Turkish origins. Central Asia also has a small but significant Christian population, most of whom belong to the Russian Orthodox Church.

Animated GEOGRAPHY

The Silk Roads

CLASSZONE.COM

CONNECT Geography & History

Tell students that trade along these routes began about 100 B.C. and thrived until the late A.D. 700s. For a period of time after that, travel along the Silk Roads became very dangerous and trade declined until it revived during the Mongol occupation of Central Asia.

ANSWER

Human-Environment Interaction
Possible Answer: They were strong, determined, adaptable, and brave.

Unit Resource Book: Russia & the Eurasian Republics
- History Makers: Timur the Lame, p. 103

Presentation Options

Use the **Power Presentations** DVD-ROM to project the animated map.

- Click to animate the map.
- Listen to the audio and examine the photographs to learn how the Silk Roads affected Transcaucasia and Central Asia.

Migrations, Invasions, and Trade Three groups that had a major impact on the cultures of Transcaucasia and Central Asia were the Arabs, the Turks, and the Mongols.

In the 600s, Arabs moved into Transcaucasia in great numbers and introduced Islam. **Islam** is a religion based on the teachings of the prophet Muhammad. The followers of Islam are called **Muslims**. During the 1000s, tribes from Turkey moved into the region. Eventually, Transcaucasia became part of the Ottoman Empire, which ruled the area until World War I. In addition, Mongols raided Georgian lands from the 1200s to the early 1400s.

Central Asia followed a similar pattern of migration and invasion. Arabs spread Islam to parts of Central Asia. Turkish peoples began to migrate into Central Asia during the 700s. In the 1200s, the Mongols conquered the region.

The Mongol Empire brought stability to Central Asia. This stability encouraged trade in the region. A network of major trade routes, called the **Silk Roads**, flourished. They ran from China to the rest of East Asia, through Central Asia, and into Europe. Cities, such as Samarkand in present-day Uzbekistan and Merv in present-day Turkmenistan, sprang up along the Silk Roads. Central Asian nomads helped stimulate trade by traveling along the routes and exchanging livestock for food and manufactured products. Goods traded included jade, gold, horses, wool, and, of course, silk. Religions, including Islam, and inventions, such as the compass and gunpowder, also spread along these routes.

Animated GEOGRAPHY

The Silk Roads

Click here to take a journey on the Silk Roads @ClassZone.com

Travel along the demanding terrain of the Silk Roads was difficult and often dangerous. Traders avoided the harsh Taklimakan Desert as much as possible. Temperatures in the desert often reach 120°F in the summer. Even caravans that traveled along the edge of the desert often dealt with fierce sandstorms.

After the Taklimakan, caravans had to cross the steep Pamir Mountains. Caravans that traveled on southern routes had to cross over the Himalayas on narrow passes that dropped off into deep ravines.

CONNECT Geography & History

Human-Environment Interaction What do the difficulties along the Silk Roads suggest about the traders who traveled them?

Geographic Challenges A camel train winds through the Taklimakan Desert under the hot sun.

DIFFERENTIATING INSTRUCTION

Struggling Readers

Trace the Silk Roads

Use the map on this spread to help students better understand travel and trade along the Silk Roads. Trace the route, explaining that trade went in both directions. Have students locate cities along the routes, including Samarkand, Tashkent, and Kashgar, and note their relationship to nearby landforms.

English Learners

Build Background

English learners, as well as many other students, may not be aware that the Silk Roads were named in the 1870s for the large quantities of silk fabric that China once traded along the routes. Silk was highly valued all over the world, and until the A.D. 500s, China was the only country that made it.

Ethnic Groups The movement of the Arabs, Turks, and Mongols into Transcaucasia and Central Asia had a major impact on the ethnic makeup of these regions. For example, in Central Asia, many people are descended from Turkish and Mongol tribes. The Silk Roads also brought many different ethnic groups to the regions. Remember that an **ethnic group** is made up of people who share a common heritage.

As a result of this migration and trade, people belonging to many different ethnic groups live in the Eurasian republics today. The pie graphs at right show the ethnic distributions in Transcaucasia and Central Asia. Most of the people practice Islam or Orthodox Christianity.

In Transcaucasia, Armenian, Azerbaijani, and Georgian are the most common languages. Russian and various languages related to Turkish are commonly spoken in Central Asia.

Like people in other parts of the world, those in Transcaucasia and Central Asia do not always get along. Allegiance to their ethnic group is often of greater importance than loyalty to their country. For instance, in Georgia, the Ossetians continue to fight for their independence from the republic. In Central Asia, the Kyrgyz and the Uzbeks have had conflicts over territorial claims.

COMPARING Ethnic Distribution

These pie graphs show the major ethnic groups in Transcaucasia and Central Asia today. Compare the distribution within and between the two regions.

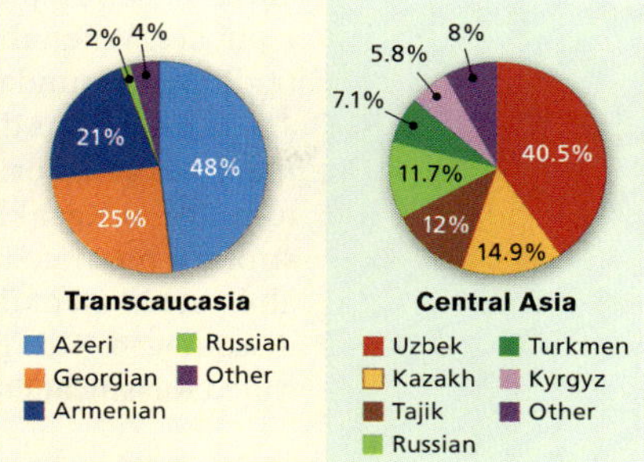

Source: *CIA World Factbook 2006*

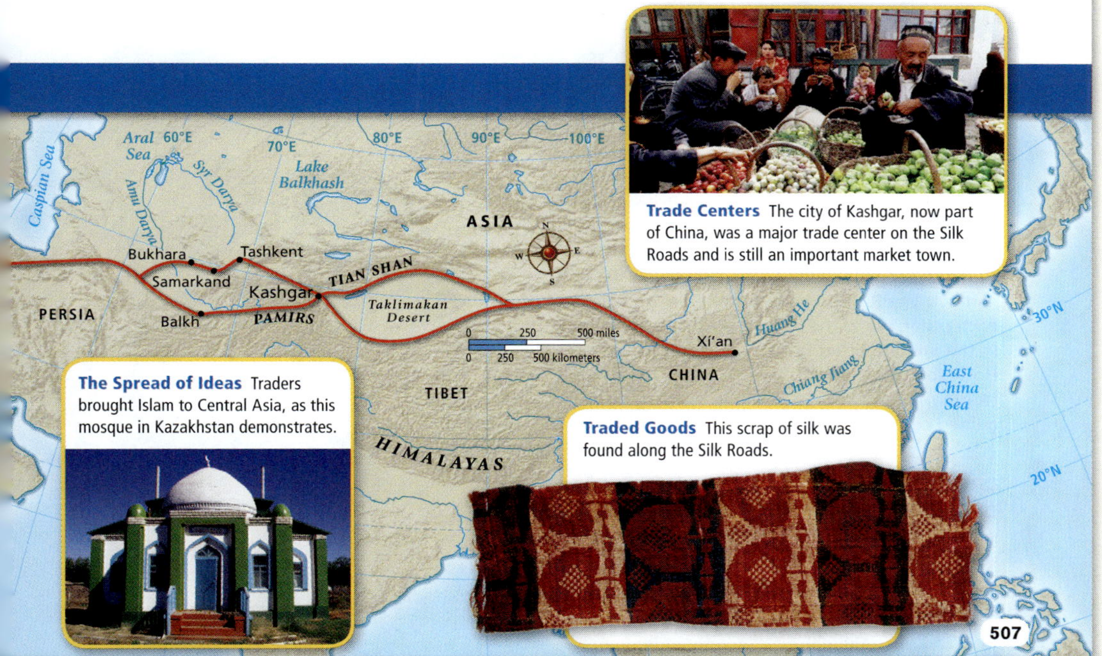

Trade Centers The city of Kashgar, now part of China, was a major trade center on the Silk Roads and is still an important market town.

The Spread of Ideas Traders brought Islam to Central Asia, as this mosque in Kazakhstan demonstrates.

Traded Goods This scrap of silk was found along the Silk Roads.

507

DIFFERENTIATING INSTRUCTION

Pre-AP

Make a Population Pie Graph

Have students research current population statistics for the Eurasian republics. Tell them to use the statistics to determine the total population of each region and then to calculate the percentage of the total that each country represents. Finally, have them create two population pie charts—one for Transcaucasia and another for Central Asia—modeling them on the ones shown in the Comparing feature. You may want to provide them with the following directions.

- Add the populations of the countries in a region to determine the total for the region.
- Divide the population of each country by the total. Multiply the answers by 100. The result is the percentage of the region's population that comes from that country.
- For Transcaucasia, draw a circle, then draw a wedge in the circle that represents the percent of the population from one country. Draw wedges representing the percents from the other two countries. Color and label the wedges. Repeat the process for Central Asia.

COMPARING

Ethnic Distribution

After students have had an opportunity to study the pie charts, ask the following questions:

- What is the largest ethnic group in Transcaucasia? *(Azeri)* in Central Asia? *(Uzbek)*
- What is the only ethnic group (besides "other") common to both regions? *(Russian)* Which region contains a larger percentage of this group? *(Central Asia)*

More About . . .

Travel on the Silk Roads

As the photo inset on the opposite page shows, camels were used to navigate the desert trails of the Silk Roads. Among the dangers of desert travel were gusts of wind that blew sudden dust storms across the terrain. Though they often lasted only a few moments, the storms could cause a person to suffocate. According to ancient Chinese sources, old camels, experienced in desert travel, could sense a dust storm before it arrived and would bury their mouths in sand to protect themselves. Seeing this, the men in the caravan would quickly cover their faces with cloth until the storm passed.

More About . . .

Kashgar

Today, the Sunday Market in Kashgar is one of the largest open-air bazaars in Central Asia. Tens of thousands of people from the surrounding areas descend on the city early in the morning to buy and sell horses, goats, sheep, camels, pigeons, chickens, timber, straw matting, roasted peppers, dumplings, spices, herbs, rice, breads, clothing, and many other goods.

More About . . .

Soviet Control of Kazakhstan

The Soviet Union considered the settlement of Kazakhstan a key factor in their strategy to control Central Asia. Beginning in 1927, their efforts to create settlements, promote farming, and transform the nomadic Kazakh's way of life was successful to a degree—but disastrous for the Kazakhs. From 1926 to 1939, more than a million Kazakhs died from starvation or disease. Hundreds of thousands of others abandoned the country, migrating to Uzbekistan, Turkmenistan, and China.

Then in the early 1950s, the Soviets launched a program to convert large areas of uncultivated land in Kazakhstan to wheat farming. More than a million Russian volunteers were sent to Kazakhstan. At about the same time, Kazakhstan also became a center for activities related to the Soviet space program and nuclear testing.

CONNECT Geography & History

The Great Game

Tell students that at the heart of the issue that led to the Great Game was the belief that control of Central Asia was a key to controlling the world. Great Britain feared that Russia was a threat to its imperial strength. It fought several punishing wars against Afghan rebels before abandoning its efforts in 1921. Russian and Soviet influence on Afghanistan lasted until 1989, when Muslim rebels drove out the Soviet troops.

CRITICAL THINKING ANSWER

Draw Conclusions After the Bolshevik Revolution, Central Asia became part of the Soviet Union.

Unit Resource Book: Russia & the Eurasian Republics
- Skillbuilder Practice: Sequencing Events, p. 98

Soviet Control In 1922, Russia joined with three other territories to form the Soviet Union. These territories came to include Transcaucasia, Central Asia, and the western republics—Belarus, Moldova, Ukraine, and the Baltic republics. The Soviets then divided their land into 15 republics. When they divided the land, however, they often ignored traditional boundaries. This policy created deep resentment.

As the Soviets developed their republics, they tried to make them more Russian. For example, they forced a large number of Russians to move to Kazakhstan to work on farms and in cities. They also strongly promoted the Russian language and the use of the Cyrillic alphabet in all of their republics. Traditional ethnic ways of life were discouraged. This process of making other cultures more Russian is called **Russification**.

Independence When the Soviet Union collapsed in 1991, the republics in Transcaucasia and Central Asia gained their independence. Since then, they have begun to move toward market economies. However, widespread poverty and ethnic disputes are still common. As a result, many people have migrated from the former Soviet republics into Russia, seeking better conditions.

EVALUATE Explain how trade and various groups of people have influenced Transcaucasia and Central Asia. Arabs brought Islam; Turks and Mongols affected ethnic makeup; Silk Roads stimulated trade; Soviet Union fueled ethnic resentment

CONNECT Geography & History

The Great Game From about 1813 to 1907, the Russian Empire and Great Britain competed for control of Central Asia. The Russian Empire wanted to use the region to invade India, a British colony. Battles between the two empires took place in the Kingdom of Afghanistan. Their struggle came to be called the "Great Game." The cartoon illustrates the conflict. The Russian bear and the British lion stare at each other, while the ruler of Afghanistan is caught in the middle. Great Britain managed to protect India. However, by the end of the 19th century, the Russian Empire had gained control of Central Asia.

CRITICAL THINKING

Draw Conclusions What lasting impact did the Great Game have on Central Asia in the 20th century?

The empires in 1900

INTERDISCIPLINARY ACTIVITIES

CONNECT to Economics

Describe Economic Effects of Changing Land Uses

Point out that the Soviet push to expand farmland in Central Asia is an example of changing land use. Tell students that another example of changing land use is turning farmland into housing developments. Then have students discuss the possible economic advantages and disadvantages of changing land use.

CONNECT to Language Arts

Write an Editorial

Have students discuss how people may be affected when the government prohibits them from observing their cultural traditions. Then have students work in pairs to write an editorial about the importance of retaining one's cultural heritage.

People and Traditions

KEY QUESTION What are some of the characteristics of life in Transcaucasia and Central Asia?

Many of the people of the Eurasian republics follow the same traditions as their ancestors. Most people in Central Asia live in rural areas, where they farm or herd livestock. In Transcaucasia, on the other hand, more than half the people live in urban areas. These people often work in offices or factories.

Family life is very important; multiple generations remain close in Central Asia. ▼

Family Life In both Transcaucasia and Central Asia, family life is very important. In many countries, extended families live together. These families include parents, grandparents, aunts, and uncles.

In the cities of Armenia and Azerbaijan, many families live in large apartment buildings. One- or two-story houses and apartment buildings are common in the cities of Georgia. Apartment buildings are also common in the cities of Central Asia.

In rural areas of Central Asia, however, nomads often live in a dwelling called a yurt. A **yurt** is a portable, tentlike structure covered with hides or textiles. Inside, yurts are lined with decorative rugs. When nomads search for good pasture for their livestock, they carry their yurts with them on horseback or in wagons.

In the Eurasian republics, families observe many ethnic and religious holidays. In Central Asia, the Muslim spring holiday of Noruz is widely celebrated. Orthodox Easter is a popular Christian holiday in Armenia and Georgia. Since gaining independence, the people of the Eurasian republics value the freedom to celebrate their ethnic and religious holidays. Under Soviet rule, these holidays were often banned.

Kazakh Picnic Many generations gather to enjoy a feast that includes meat dumplings, rice with roasted garlic, and bread. **What does the picture suggest about family life in Kazakhstan?**

509

DIFFERENTIATING INSTRUCTION

English Learners

Use Graphic Organizers

Students who are having difficulty identifying the main ideas in one or more subsections of this chapter may find it helpful to work in pairs creating outlines or graphic organizers, such as the one shown. Remind students that they can glean information from the section's maps, photographs, and special features, as well as from the text.

People and Traditions
↓
Central Asia
• Family Life . . .
• Traditions . . .
↓
Transcaucasia
• Family Life . . .
• Traditions . . .

Teach

People and Traditions

Think, Pair, Share

- In which region do more of the people live in rural areas? *(Central Asia)* In which region do more of the people live in urban areas? *(Transcaucasia)*
- What type of diet is part of the heritage of the nomadic life? *(meat and milk products)* What are some popular dishes in Transcaucasia? *(shish kabob, pilaf, grilled lamb and goat)*
- **Find Main Ideas** What characteristic of family life is common to both regions? *(In many of the countries, extended families live together and value large family gatherings to celebrate ethnic and religious holidays.)*

Unit Resource Book: Russia & the Eurasian Republics
- Interdisciplinary Projects: Design a Yurt, p. 105

More About . . .

The People of Central Asia

In all of the republics of Central Asia, the native populations are augmented by significant numbers of foreign-born residents, many of whom were sent to the republics to implement Russian farming, military, and scientific programs. The native population in these republics is mostly rural and Muslim. A large portion of the urban population of these republics, however, is non-native and Christian.

During Soviet rule of the republics, especially under Joseph Stalin, native artists and intellectuals were discriminated against, censored, and often imprisoned. Many died violent deaths. Since achieving independence, many of the Central Asian republics have experienced a resurgence of interest in their Muslim heritage and native traditions, customs, and art.

More About . . .

The Government of Uzbekistan

The head of Uzbekistan's government is Islam Karimov, who was also the leader of Soviet Uzbekistan. All media in the country is tightly controlled by Karimov's government—even though the country's constitution prohibits censorship and guarantees freedom of the press. As a result, some people in Uzbekistan get their news from foreign TV stations via satellite cable.

Presentation Options

Go to the *Activity Center* at **ClassZone.com** to project Worldster.

- Click on the multimedia features of Zuhura's Web page.
- Ask students which areas of Zuhura's life they would like to explore. Point out that they can listen to music, examine photographs, learn simple phrases in Uzbek, and more.

CONNECT to Your Life

Journal Entry To help students identify main ideas, tell them to focus on details of Zuhura's typical day—activities, chores, meals, school routine, and so on—before writing their journal entries.

Worldster

A Day in Zuhura's Life

To learn more about Zuhura and her world, go to the **Activity Center** @ClassZone.com

This is how to say "My name is Zuhura" in Uzbek:

"Mening ismim Zuhura."

Hi! My name is Zuhura. I am 14 years old and live in a rural town in Uzbekistan. School and doing chores are an important part of my life. But I also have fun with my friends. I enjoy watching Bollywood movies, which are films made in India. In fact, I have covered my school notebook with stickers of Bollywood movie stars. This is what my day is like.

6:30–7:30 A.M. I wake up and eat a breakfast of tea and bread with jam. Then I feed the chickens and milk the family cow. When I'm done, I head off to school with my brothers and sisters.

8 A.M.–2:30 P.M. School has seven 45-minute periods. Our school doesn't have heat. In the winter, it gets really cold. So we keep our coats on while we sit at our desks.

3 P.M. When I get home, I eat a snack of bread and tea and then help clean house and wash clothes by hand. Later, we all help my mother prepare dinner. Only then do I have time to spend with my friends.

8 P.M. We're having my favorite meal tonight: soup made of broth, carrots, potatoes, and meat. In the winter, we eat a lot of cabbage soup, which I don't like as well. After dinner, I do homework and watch TV.

CONNECT to Your Life

Journal Entry In your journal, write a schedule detailing your typical school day. Then compare your schedule with Zuhura's. How are the two schedules similar and how are they different?

DIFFERENTIATING INSTRUCTION: TIERED ACTIVITIES

OBJECTIVE Compare your life and culture with those of a middle-school girl from Uzbekistan.

Basic

Have students work in pairs to create a visual presentation about their lives similar to the one on this page. Tell them to illustrate at least three of the activities shown: breakfast, school day, dinner, time spent with friends.

On Level

Have pairs take turns role-playing two students—one from the United States and the other from Uzbekistan. Ask the students to talk about their school life, leisure activities, and cultural traditions.

Challenge

Have students write a series of journal entries in which Zuhura visits the United States and details her experiences and reactions.

Traditions People in the Eurasian republics eat a variety of foods. Many of the dishes in Central Asia and Transcaucasia use meat and milk products. This diet is part of the heritage of the nomadic life. Traditional foods of Kyrgyzstan include lamb, noodles with broth and mutton, and vegetable soup. In Kazakhstan, the people drink *kumiss,* which is made from horse's milk. In Transcaucasia, Georgians and Armenians enjoy shish kabob. In Azerbaijan, pilaf (a rice dish) and grilled goat and lamb are popular. Pilaf is also a popular dish in Central Asia.

Traditional arts thrive in Transcaucasia and Central Asia. For hundreds of years, Armenian artisans have constructed beautiful stone churches with many domes. The craftspeople of Turkmenistan, Uzbekistan, and Azerbaijan are known for weaving decorative rugs and tapestries.

Fun Facts!

HORSEBACK SPORTS

Many of Central Asia's sports involve horses. These include *kokpar,* a type of polo, and horseback wrestling. In Kazakhstan, riders still practice the ancient art of eagle hunting, as shown above. These traditional sports reflect the importance of horses and riding in the region's culture.

FIND MAIN IDEAS Identify some of the characteristics of life in Transcaucasia and Central Asia. In cities, many families live in large apartment buildings. In rural areas, nomad families live in yurts. Many people in both regions follow a traditional way of life.

ONLINE QUIZ For test practice, go to **Interactive Review @ClassZone.com**

Section 2 Assessment

TERMS & NAMES

1. **Explain the importance of**
   - nomad
   - Silk Roads
   - Russification
   - yurt

USE YOUR READING NOTES

2. **Find Main Ideas** Use your completed diagram to answer the following question: Why does traditional ethnic culture play such a large role in the lives of the people of the Eurasian republics?

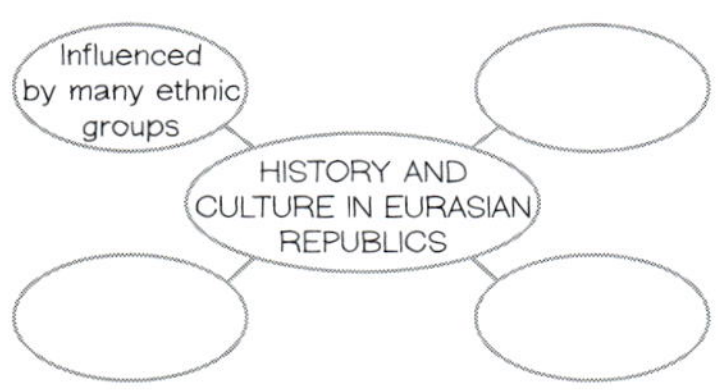

KEY IDEAS

3. What three groups had major impacts on the cultures of Transcaucasia and Central Asia?
4. How did the Soviet Union attempt to make its republics more Russian?
5. How do the dishes of the Eurasian republics reflect the region's nomadic heritage?

CRITICAL THINKING

6. **Make Inferences** Why did the Soviet Union ignore traditional boundaries when setting up its republics?
7. **Compare and Contrast** How do the dwellings of urban people and nomads in Central Asia compare?
8. **CONNECT to Today** After living under Soviet rule, why do you think the people in the Eurasian republics today value the freedom to celebrate their holidays?
9. **ART** **Design a Rug** Research rug making in Turkmenistan, Uzbekistan, and Azerbaijan. Draw a picture illustrating one of the rug designs. Then write a brief summary explaining what the design represents.

CONNECT to the Essential Question

How can the Eurasian republics meet the challenges of independence?

Ask students what they have learned so far that can help them answer this question. Students might mention the following:

- Migration, conquest, and trade have influenced life in the Eurasian republics.
- The ethnic diversity of the republics is reflected in people's daily life and traditions.

4 Assess & Reteach

Assess Have students complete the Section Assessment.

Unit Resource Book: Russia & the Eurasian Republics
- Section Quiz, p. 111

Interactive Review @ClassZone.com
- Section Quiz

Power Presentations

Test Generator

Reteach Have students write an example of a historical influence on life in the Eurasian republics on a slip of paper. Then have them jot down an example of a cultural tradition in the republics on another slip. Place the slips into a container. Have a volunteer select a strip and explain what he or she knows about the item on it. Repeat the process, until all the slips have been discussed.

Unit Resource Book: Russia & the Eurasian Republics
- Reteaching Activity, p. 114

SECTION 2 ASSESSMENT ANSWERS

Terms & Names

1. nomad: person who lives by herding and moving from place to place; Silk Roads: major trade routes; Russification: process of making a culture more Russian; yurt: a portable, tentlike structure

Use Your Reading Notes

2. because the Eurasian republics have been home to many different ethnic groups

Key Ideas

3. Arabs, Turks, and Mongols
4. forced a large number of Russians to move to Kazakhstan, promoted the Russian language, discouraged traditional ethnic ways of life
5. The dishes feature meat and milk, products related to herding.

Critical Thinking

6. The Soviets wanted to make all the cultures in their republics the same as the Russian culture.
7. In cities in both regions, people often live in apartment buildings. Nomads live in yurts.
8. Under Soviet rule, they were often prohibited from celebrating their ethnic and religious holidays.
9. Use the rubric to score rug designs.

Design Rubric

	Rug Design	Summary
4	excellent attempt at representing authentic design	very concise
3	good attempt	fairly clear
2	fair attempt	confusing
1	poor attempt	unclear

1 Plan & Prepare

Objectives

- Identify the obstacles that have prevented the Eurasian republics from developing true democracies and the efforts made to overcome those obstacles
- Describe how the Eurasian republics are changing from a command economy to a market economy

Read for the Essential Question

Help students read for a purpose by reminding them of the Essential Question: "How can the Eurasian republics meet the challenges of independence?" Point out that they can use items in the section, such as Key Questions, special features, images, and maps, to formulate answers to the question.

Vocabulary

Best Practices Toolkit

Use the ***Best Practices Toolkit*** to model strategies for vocabulary notetaking. Vary strategies throughout the year. Choose from: *Knowledge Rating, Predicting ABCs, Definition Mapping, Word Sort, Word Wheel, Frayer Model (Word Squares), Magnet Words,* and *Student VOC.*

Vocabulary Strategies, TT9–TT16

Reading Strategy

Best Practices Toolkit

Display the T Chart/Two-Column Chart transparency.

- Remind students that identifying problems and solutions means identifying challenges and discussing how people are trying to overcome those challenges.
- Model this strategy by filling in the chart with solutions to the following problem: finding time for homework. Have students suggest entries.
- Remind students that the information in their completed charts will help them answer the Key Questions in the section.

T Chart/Two-Column Chart, TT20

SECTION 3

Reading for Understanding

Key Ideas

BEFORE, YOU LEARNED

The Eurasian republics gained their independence after many years under the control of the Soviet Union.

NOW YOU WILL LEARN

The republics are working to overcome internal problems and establish democratic governments and market economies.

Vocabulary

TERMS & NAMES

Rose Revolution a peaceful uprising in Georgia that helped force a corrupt president to resign

Nagorno-Karabakh (nuh•GAWR•noh KAHR•uh•BAHK) a province in Azerbaijan that Armenians believe should be a part of their country

foreign investment money put into a business by people from another country

one-crop economy an economy that relies on one crop for much of its earnings

REVIEW

Joseph Stalin Soviet Union dictator who was born in Georgia

command economy an economic system in which the production of goods and services is controlled by a central government

market economy an economic system in which the production of goods and services is decided by the supply and demand of consumers

privatize to sell government-owned businesses to private individuals

Reading Strategy

Re-create the problem-solution chart shown at right. As you read and respond to the **KEY QUESTIONS**, use the chart to identify some of the solutions the Eurasian republics have come up with to settle their problems.

See Skillbuilder Handbook, page R10

IDENTIFY PROBLEMS AND SOLUTIONS

PROBLEMS	SOLUTIONS
Establishing democracies	
Changing their economies	
Developing their resources	

GRAPHIC ORGANIZERS
Go to Interactive Review @ClassZone.com

PRETEACHING VOCABULARY

English Learners

Pronounce and Preview

Pronounce each term for students. Review the meanings of words used in the definitions, such as *uprising, corrupt,* and *earnings.*

- To modify vocabulary learning, have students complete worksheets as they read the section.

Unit Resource Book: Russia & the Eurasian Republics
- Vocabulary Practice, p. 97

Inclusion

Use Predicting ABCs

Have students work in pairs to fill out the graphic organizer on the transparency for the terms and names in the section.

- Use the transparency to model the strategy.
- Have volunteers fill in the words for Sections 1, 2, and 3.
- Then distribute copies of the organizer for students to work on with a partner.

Best Practices Toolkit
- Predicting ABCs, TT10

SECTION 3

GOVERNMENT & ECONOMICS

Cemetery in Kyrgyzstan

Snow leopard in mountains of Central Asia

The Challenge of Independence

Connecting to Your World

Imagine that you are one of the founders of a new country. What challenges will you face? Well, to begin with, you will have to form a new government and economic system. Suppose you want to establish a democracy. What will you do if your people have no previous experience with democratic rule? How will you make democracy work? This is one of the challenges faced by the people of the Eurasian republics today.

Mixed Success with Democracy

KEY QUESTION What obstacles have the Eurasian republics faced in their efforts to establish democracies?

For centuries, the peoples of Transcaucasia and Central Asia have sought independence. In the 800s, for example, Armenia established an independent kingdom that lasted for more than one hundred years. In the early 1900s, many Eurasian groups fought for independence—first from czarist Russia and then from the Soviet Union. The republics did not gain their independence until 1991, when the Soviet Union collapsed.

Elected Presidents Many of the presidents of former Soviet republics pose before their countries' flags in 2004.

2 Focus & Motivate

3-Minute Warm-Up

Write on the board or display the transparency:

- Imagine that your country has just gained its independence, but you still have little political power. What would you do about it? *(write protest letters; vote officials out of office; organize a peaceful protest)*

Unit Transparency Book
- 3-Minute Warm-Ups, TT11

Connecting to Your World

More About . . .

The Collapse of the Soviet Union

Reforms granted to the Soviet republics in the late 1980s made the republics hungry for more. When the Soviet republics declared their independence in 1991, the Soviet Union fell just four months later.

3 Teach

Mixed Success with Democracy

Think, Pair, Share

- What obstacles to democracy have Eurasian republics faced since gaining independence? *(corrupt officials, ethnic conflicts, civil war)*
- **Identify Problems and Solutions** How did the people of Georgia help to deal with political corruption? *(They forced the corrupt president of Georgia to resign.)*

SECTION 3 PROGRAM RESOURCES

Unit Resource Book: Russia and the Eurasian Republics

ON LEVEL

Unit Resource Book
- Reading Study Guide, p. 77
- Map and Graph Skills, p. 99
- Section Quiz, p. 112

STRUGGLING READERS

Unit Resource Book
- Reading Study Guide with Additional Support, p. 83
- Vocabulary Practice, p. 97
- Skillbuilder Practice, p. 98
- Section Quiz, p. 112
- Reteaching Activity, p. 115

eEdition with Audio DVD-ROM

ENGLISH LEARNERS

Pupil Edition in Spanish

eEdition with Audio DVD-ROM

Unit Resource Book
- RSG (Spanish), p. 89
- RSG with Additional Support (Spanish), p. 95

Multi-Language Glossary

Test Generator
- Section Quiz in Spanish

INCLUSION

Unit Resource Book
- RSG with Additional Support, p. 83
- Section Quiz, p. 112
- Reteaching Activity, p. 115

GIFTED & TALENTED

Unit Resource Book
- History Makers, p. 104
- Economics in the World, p. 107

PRE-AP

Unit Resource Book
- Primary Sources, p. 109
- World Literature, p. 129
- Geography Project, p. 131

TECHNOLOGY

Unit Transparency Book
- 3-Minute Warm-Ups, TT11
- Essential Question, TT12
- Critical Thinking, TT15

Daily Test Practice Transparencies
- Section 3, TT58

Power Presentations

ClassZone.com

World Cultures and Geography Video on DVD

RESEARCH & WRITING CENTER

HISTORY MAKERS

Joseph Stalin

During his years as secretary-general of the Communist Party's Central Committee, Stalin became one of the most powerful political leaders of all time and is credited with making the Soviet Union a world power. He industrialized the republics and organized millions of private farmers into a system of state-run collectives. His tactics were ruthless. He eliminated officials whom he perceived as potential rivals to power, destroyed individual freedoms, and instituted brutal programs of terror to squash dissent. It is believed that he was responsible for the deaths of millions of ordinary citizens in addition to perceived political opponents.

Unit Resource Book: Russia & the Eurasian Republics

- History Makers: Joseph Stalin, p. 104
- Primary Sources: from a speech by Joseph Stalin, p. 109
- World Literature: from "Freedom to Breathe" by Alexander Solzhenitsyn, p. 129

More About . . .

The Rose Revolution

The movement to oust Georgia's president was spurred by reports from international observers, including the United States, that the results of the November 2003 parliamentary elections, which gave then President Shevardnaze's party the victory, had been falsified. On November 22, supporters of the opposition, led by Saakashvili, staged a large but peaceful demonstration as Shevardnaze addressed the opening session of the legislature. Shevardnaze declared a state of emergency and resigned the next day. In January 2004, Saakashvili was elected president of Georgia—capturing more than 96 percent of the votes.

Unit Resource Book: Russia & the Eurasian Republics

- Skillbuilder Practice: Sequencing Events, p. 98

HISTORY MAKERS

Joseph Stalin 1879–1953

Joseph Stalin was born in a small town near Tbilisi, Georgia. He spent most of his childhood in this region. However, after he became dictator of the Soviet Union in 1929, he did not show a preference for Georgia. Indeed, he believed that it would be better off under Soviet rule. Stalin even enforced the same economic policy with Georgia that he put into effect in other Soviet republics. This policy required Georgia to change from a farm-based to an industry-based society.

ONLINE BIOGRAPHY For more on the life of Joseph Stalin, go to the Research & Writing Center @ClassZone.com

Overcoming the Soviet Past The Eurasian republics had little experience with democracy. Under Soviet dictators such as **Joseph Stalin**, the Communist Party had controlled all aspects of the government. As a result, most of the politicians in the Eurasian republics were Communists. After independence, these officials often remained in office and ran the governments. In addition, Russia has continued to try to exert its influence over the Eurasian republics.

Many times, the Eurasian republics have taken on the appearance of democracies. Kyrgyzstan, for instance, adopted a parliament and allowed a free press. But in reality, the people had little voice in the government. Throughout the Eurasian republics, in fact, most people have very little political power.

Recently, however, some progress has been made. For example, in 2003, Mikhail Saakashvili (SAH•kahsh•VEE•lee) led a campaign against government corruption under Georgia's president and forced him to resign. Saakashvili himself became president. This peaceful uprising is known as the **Rose Revolution**.

Dealing with Conflicts As you know, the Soviet Union paid little attention to traditional boundaries when setting up the republics. For example, the Soviet government created a province named **Nagorno-Karabakh**, in which 76 percent of the people were Armenians, but placed it within the borders of Azerbaijan. After Armenia and Azerbaijan gained their independence, Nagorno-Karabakh remained in Azerbaijan. Fighting between Armenia and Azerbaijan over this territory soon began. A cease-fire was declared in 1994, but the dispute remains unresolved.

In Georgia, two ethnic groups—the Ossetians and the Abkhazians (ab•KAY•zhuhnz)—believe they should each have independent states. Both of these groups have fought with the Georgians over this issue, which also remains unresolved.

During the 1990s in Tajikistan, former Communist officials fought anti-Communists and Islamic groups for control of the country. This civil war killed thousands of people. Eventually, a peace agreement was reached.

IDENTIFY PROBLEMS Identify the obstacles the Eurasian republics have faced in their efforts to establish democracies.
few politicians experienced with democracy; ethnic conflicts; civil war

514 Chapter 16

DIFFERENTIATING INSTRUCTION

English Learners

Understand Prefixes

Point out the word *unresolved* in the sentence "A cease-fire was declared in 1994, but the dispute remains unresolved." Explain that the prefix *-un* is a prefix meaning "not." Ask students to define *unresolved. (not resolved, or not agreed upon)* Then have students brainstorm other examples of words that begin with *-un* and use them in sentences.

Struggling Readers

Learn Terms

Point out the term *traditional boundaries* in the sentence "As you know, the Soviet Union paid little attention to traditional boundaries when setting up the republics." You may want to use other examples from history—such as Western settlers moving onto Native American hunting grounds in the United States—to help students grasp the concept.

ANALYZING Primary Sources

Heydar Aliyev (1923?–2003) served as president of Azerbaijan for the last ten years of his life. His son became president when he died. The image here shows an Azeri man casting his ballot in a parliamentary election under a portrait of Heydar Aliyev. In the following speech, Aliyev talks about the difficulties of establishing a democracy.

> Some people think we should be able to establish democracy in a short time, but that's impossible. Azerbaijan is a young nation and democracy is a new concept. . . . Democracy is not an apple you buy at the market and bring back home.
>
> Source: Heydar Aliyev, Speech at Georgetown University, July 30, 1997

DOCUMENT-BASED QUESTION

What does Aliyev suggest about the process of establishing a democracy?

Changing to a Market Economy

KEY QUESTION What steps have the Eurasian republics taken toward developing their economies?

The Eurasian republics have many highly desired resources, including oil, natural gas, and coal. To take greater advantage of these resources, the Eurasian republics are attempting to develop market economies.

Controlling the Economy How does a country change from a command economy to a market economy? As you may recall, a **command economy** is one in which the government controls all businesses. In a **market economy**, private individuals own most of the businesses. To develop this type of economy, many Eurasian republics have started to **privatize** their businesses, or sell government-owned businesses to private individuals.

In addition, most of them have aggressively tried to obtain **foreign investments**. That means that they have encouraged people from other countries to put money into their businesses. Unfortunately, some republics are held back by being one-crop economies. A country with a **one-crop economy** relies on one crop for much of its earnings. The economy of Uzbekistan, for example, relies heavily on the country's production of cotton.

DIFFERENTIATING INSTRUCTION: TIERED ACTIVITIES

OBJECTIVE Understand the economic challenges facing the Eurasian republics.

Basic

Have student pairs create flip cards for the five key terms on this page. Encourage students to take turns quizzing each other on the terms until they thoroughly understand each one.

On Level

Have student pairs get together to discuss the difficulties that might arise when a country changes from a command economy to a market economy.

Challenge

Have groups of students research to find out how Uzbekistan's reliance on cotton has affected the country's economy. Invite students to present their findings to the class.

ANALYZING Primary Sources

Over a period of more than 30 years, Heydar Aliyev became very powerful in Azerbaijan. He first served as deputy chairman and then chairman of the regional KGB, or intelligence agency. In 1969, he was promoted to secretary of the Communist Party of Azerbaijan. From 1993 until 2003, he served as president of the country. During his time in office, he was credited with pursuing extended peace talks with Armenia regarding the dispute over Nagorno-Karabakh and with attracting foreign investments.

DOCUMENT-BASED QUESTION ANSWER

It takes time, care, patience, and experience.

Unit Resource Book: Russia & the Eurasian Republics
- Geography Project: Planning a Documentary, p. 131

Teach

Changing to a Market Economy

Talk About It

- Explain the difference between a command economy and a market economy. *(command. government owns/controls businesses; market: private individuals own most of the businesses)*
- **Identify Problems and Solutions** Explain the advantages and disadvantages in developing natural resources in the Eurasian republics. *(Advantage: can help these countries develop effective market economies; Disadvantage: may lead to pollution, as under the Soviets)*

Unit Resource Book: Russia & the Eurasian Republics
- Map and Graph Skills: Reading Line and Pie Graphs, p. 99
- Economics in the World: Central Asian Republics Economic Indicators, p. 107

CONNECT to the Essential Question

How can the Eurasian republics meet the challenges of independence?

Ask students what they have learned so far that can help them answer this question. Students might mention the following:

- Some Eurasian republics have taken steps to strengthen their fragile democracies.
- The Eurasian republics have taken initiatives to reform their economies.

Unit Transparency Book
- Essential Question, TT12

4 Assess & Reteach

Assess Have students complete the Section Assessment.

Unit Resource Book: Russia & the Eurasian Republics
- Section Quiz, p. 112

Interactive Review @ClassZone.com
- Section Quiz

Power Presentations

Test Generator

Reteach Have student pairs create one question and answer about the governments and economies of the Eurasian republics. Have students take turns asking and answering the questions.

Unit Resource Book: Russia & the Eurasian Republics
- Reteaching Activity, p. 115

Azerbaijan Pipeline The president of Turkey has called the 1,100-mile oil pipeline "the Silk Road of the 21st century." **How will the pipeline encourage trade in the region?** It will bring more money to the region.

Energy Resources Drive the Future The people of the Eurasian republics believe that developing their resources will greatly benefit their economies. As a result, Azerbaijan has worked for several years with many countries to build a $4 billion pipeline from Baku to Turkey's Mediterranean coast. In 2006, one million barrels of oil a day began to pump through it. Georgia also benefits from the pipeline because it extends through that republic. The pipeline has brought Georgia many job opportunities and much foreign investment. In addition, Kazakhstan reached an agreement with China in 2005 that involves building an oil pipeline from the Caspian Sea eastward into China.

Tajikistan is also attempting to develop its natural resources. The republic has a great potential for water power. As a result, it is planning to build several huge dams to increase its production of electricity. If completed, these dams would be the largest in the world.

EVALUATE Describe the steps that the Eurasian republics have taken toward developing their economies. privatized businesses, pursued foreign investments, marketed resources

ONLINE QUIZ For test practice, go to **Interactive Review @ClassZone.com**

Section 3 Assessment

TERMS & NAMES

1. **Explain the importance of**
   - Rose Revolution
   - Nagorno-Karabakh
   - foreign investment
   - one-crop economy

USE YOUR READING NOTES

2. **Identify Problems and Solutions** Use your completed chart to answer the following question:

   Why was the Rose Revolution a good step toward establishing a democracy in Georgia?

PROBLEMS	SOLUTIONS
Establishing democracies	
Changing their economies	
Developing their resources	

KEY IDEAS

3. How did the Nagorno-Karabakh conflict develop?
4. Why have many of the Eurasian republics pursued foreign investment?
5. How has the Azerbaijan pipeline benefited Georgia?

CRITICAL THINKING

6. **Make Inferences** Why might Uzbekistan's reliance on cotton production hold back its economy?
7. **Form and Support Opinions** Do you think there are any drawbacks to obtaining foreign investments? Explain your answer.
8. **CONNECT to Today** In 2006, Azerbaijan began to pump one million barrels of oil a day through its pipeline. What effect do you think this pipeline will have on Azerbaijan's economy?
9. **TECHNOLOGY** **Create a Multimedia Presentation** Show how an oil pipeline works. Include a diagram showing where it starts and where it ends, and illustrations of the machinery used.

SECTION 3 ASSESSMENT ANSWERS

Terms & Names

1. Rose Revolution: uprising in Georgia; Nagorno-Karabakh: Armenian province in Azerbaijan; foreign investment: foreign money in a domestic business; one-crop economy: reliance on one crop

Use Your Reading Notes

2. people helped root out corruption

PROBLEMS	SOLUTIONS
Establishing Democracies	Mixed success, with people gaining little political power
Changing their Economies	Changed from command to market economies
Developing their Resources	Oil pipelines in Azerbaijan and Kazakhstan; dams in Tajikistan

Key Ideas

3. The Soviet government placed this largely Armenian province within Azerbaijan. It remained in Azerbaijan after independence.
4. to increase the inflow of cash
5. brought job opportunities and foreign investment

Critical Thinking

6. If the crop fails, it could have a devastating effect on Uzbekistan's economy.
7. Yes; a country that seeks foreign investment may have to give up some control of its business.
8. It will enable Azerbaijan to sell much more oil and thereby greatly increase its cash inflow. This money can be used to improve the country's infrastructure, industry, and schools.
9. Use the rubric to score multimedia presentations.

Presentation Rubric

	Representation of Pipeline	Presentation of Material
4	very precise	very complete
3	accurate	fairly complete
2	fairly accurate	some items missing
1	inaccurate	few or no materials presented

Interactive Review

Click here to complete these and other activities online @ClassZone.com

CHAPTER SUMMARY

Key Idea 1
The geography and climate of the Eurasian republics pose challenges, but the countries are rich in natural resources.

Key Idea 2
The geographic location of the Eurasian republics has attracted many different cultures, ideas, and conquerors.

Key Idea 3
The republics are working to overcome internal problems and establish democratic governments and market economies.

For **Review and Study Notes**, go to **Interactive Review** @ClassZone.com

NAME GAME

Use the Terms & Names list to complete each sentence on paper or online.

1. I describe a country that is completely surrounded by land. landlocked
2. I contain some of the largest alpine glaciers in the world. ______
3. I have been steadily shrinking as a result of irrigation. ______
4. I make my living by moving from place to place. ______
5. I am a huge desert that covers much of Turkmenistan. ______
6. I am a peaceful uprising that took place in Georgia. ______
7. I am the process of making a culture more Russian. ______
8. I am a portable, tentlike dwelling. ______
9. I am an area that Azerbaijan and Armenia have fought over. ______
10. I am money put into businesses in other countries. ______

Aral Sea
Caucasus Mountains
foreign investment
Islam
Kara Kum
Lake Balkhash
landlocked
Muslim
Nagorno-Karabakh
nomad
Rose Revolution
Russification
Silk Roads
Tian Shan
yurt

Activities

Flip Cards

Use the online flip cards to quiz yourself on the terms and names introduced in this chapter.

Crossword Puzzle

Complete an online crossword puzzle to test your knowledge of the Eurasian republics.

ACROSS
1. major trading routes in Central Asia

517

Chapter Summary

As a follow-up to the chapter, encourage students to pair up and go over the **Review and Study Notes** at **ClassZone.com.**

Alternatively, student pairs can review their Reading Study Guide notes, making sure that they can add important details to each section's Key Ideas.

Name Game

Teaching Tips

ON PAPER Have students work in pairs to complete this activity. As an extension, they can write questions for the terms that are not covered in the items shown here.

ONLINE Make sure that students understand how to drag and drop terms to fill in blanks. Ask them to print out their results and compare them with those of a classmate.

ANSWERS

1. landlocked
2. Tian Shan
3. Aral Sea
4. nomad
5. Kara Kum
6. Rose Revolution
7. Russification
8. yurt
9. Nagorno-Karabakh
10. foreign investment

Make It Fun

Teacher-Tested Activities

Matt Parker, Eastgate Middle School, Kansas City, Missouri

I have my students play a variation on standard flip cards using causes and effects.

- First, I create a set of cards by writing a cause (for example, "irrigation project in Central Asia") on one side and its effect ("Aral Sea shrinkage") on the other.
- Then I divide and distribute the cards to student pairs who take turns quizzing each other on the causes and effects.
- You can also use cards to test sequencing events and finding main ideas.

ONLINE AND ON-SCREEN ACTIVITIES

Flip Cards

Vocabulary Flip Cards are virtual flash cards at **ClassZone.com.** Students can quiz each other or work independently to study the flip cards for the Eurasian republics.

- They can quiz themselves on key terms and names by reviewing definitions.
- Demonstrate how to advance and shuffle the deck for another challenge.

Crossword Puzzle

Students can review important people and events in the history of the Eurasian republics by completing this crossword puzzle online at **ClassZone.com.**

- As an extension, have students use the *Activity Maker* at **ClassZone.com** to make their own crossword puzzle for one section of the chapter.
- Then have students trade puzzles with classmates and solve each other's puzzles.

Vocabulary ANSWERS

1. region that contains the republics of Armenia, Azerbaijan, and Georgia
2. region that contains the republics of Kazakhstan, Kyrgyzstan, Tajikistan, Turmenistan, and Uzbekistan
3. mountain range that runs between the Black Sea and the Caspian Sea
4. an inland body of water in Central Asia that has been steadily shrinking
5. member of a group who makes a living by moving from place to place
6. religion that is based on the teachings of the prophet Muhammad
7. major trade routes that ran from China, through Central Asia, and into Europe
8. process of making a culture more Russian
9. peaceful uprising in Georgia that helped force a corrupt president to resign
10. a foreign business or person putting money into a domestic business to make a profit
11. The Soviet Union imposed Russian culture on the people living in the regions of present-day Transcaucasia and Central Asia.
12. The Rose Revolution refers to the peaceful uprising in Georgia; Nagorno-Karabakh refers to a bitter dispute over a predominantly Armenian province located in Azerbaijan.

Key Ideas ANSWERS

13. the Black Sea and the Caspian Sea
14. They prevent coastal areas from receiving too much rain.
15. Kara Kum and Kyzyl Kum
16. increased cotton production; caused the Aral Sea to shrink; polluted water and food; harmed the soil; caused disease
17. provided stability, which encouraged trade along the Silk Roads
18. silk, jade, gold, horses, wool; Islam and other religions; compass, gunpowder
19. ignored traditional boundaries; pursued a policy of Russification
20. decorative woven rugs and tapestries
21. They remained in office and ran the government.
22. the Rose Revolution and the dispute with the Ossetians and the Abkhazians
23. privatized businesses, pursued foreign investments, worked with foreign countries to expand their economies
24. Azerbaijan, Kazakhstan, Tajikistan

CHAPTER 16 Assessment

VOCABULARY

Explain the significance of each of the following.

1. Transcaucasia
2. Central Asia
3. Caucasus Mountains
4. Aral Sea
5. nomad
6. Islam
7. Silk Roads
8. Russification
9. Rose Revolution
10. foreign investment

Explain how the terms and names in each group are related.

11. Transcaucasia, Central Asia, and Russification
12. Rose Revolution and Nagorno-Karabakh

KEY IDEAS

1 Center of a Landmass

13. What bodies of water border Transcaucasia?
14. How do mountains affect Georgia's precipitation?
15. What deserts lie in Turkmenistan and Uzbekistan?
16. What effect did Soviet policies have on Central Asia?

2 Historic Crossroads

17. How did the Mongol Empire affect trade in Central Asia?
18. What was exchanged on the Silk Roads?
19. How did the Soviets create resentment among ethnic groups in the Eurasian republics?
20. What are the craftspeople of Turkmenistan, Uzbekistan, and Azerbaijan known for?

3 The Challenge of Independence

21. Why did many Communists continue to run the governments in the Eurasian republics after 1991?
22. What conflicts have arisen in Georgia?
23. How have the Eurasian republics attempted to change from command economies to market economies?
24. Which countries are working to market their natural resources?

CRITICAL THINKING

25. **Analyze Causes and Effects** Create a cause-and-effect chart to identify some of the policies of the Soviet Union and their impact on the Eurasian republics.

CAUSES	EFFECTS

26. **Identify Problems** What do you think are the biggest problems facing the Eurasian republics today?
27. **Make Inferences** Why do you think the Soviet Union promoted Russification?
28. **Evaluate** What aspects of Central Asian culture reflect its nomadic heritage?
29. **Connect Geography & Economics** How have geographic factors helped and hindered the republics' economies?
30. **Five Themes: Location** How did the location of Central Asia promote trade in the region?

Answer the ESSENTIAL QUESTION

How can the Eurasian republics meet the challenges of independence?

Written Response Write a two- or three-paragraph response to the Essential Question. Be sure to consider specific ideas about the challenges the Eurasian republics face. Use the rubric below to guide your thinking.

Response Rubric
A strong response will:
- identify the resources of the Eurasian republics
- discuss the problems confronting the republics
- explain the steps the republics have taken to modernize their governments and economies

ESSENTIAL QUESTION REVIEW

Before students begin their essays, have them review the key chapter elements that will help them answer the Essential Question. Direct them to significant features in the chapter, such as:

- Physical Geography of the Eurasian Republics, Section 1
- Connect Geography & Economics, Section 1
- Animated Geography: The Silk Roads, Section 2
- Comparing Ethnic Distribution, Section 2
- History Makers: Joseph Stalin, Section 3
- Analyzing Primary Sources, Section 3

Written Response Rubric

	Coverage of Key Ideas	Explanation	Number of Errors
4	all	clear and concise	few or none
3	most	good	some
2	some	fair	many
1	one or none	poor	many major

STANDARDS-BASED ASSESSMENT

Test Practice
- Online Test Practice @ClassZone.com
- Test-Taking Strategies and Practice at the front of this book

THEMATIC MAP

Use the map to answer questions 1 and 2 on your paper.

Industries in the Eurasian Republics

1. Which republic contains the only coal industry in the two regions?

A. Kazakhstan
B. Azerbaijan
C. Turkmenistan
D. Uzbekistan

2. Where is most of the oil and gas produced?

A. in Kazakhstan
B. in Turkmenistan
C. around the Caspian Sea
D. around the Aral Sea

PHOTOGRAPH

Examine the photograph below showing the inside of a yurt. Use the photograph to write brief answers for questions 3 and 4 on your paper.

3. What might the carpets in this yurt be used for?

4. Why is it important that the yurt and all of its furnishings be lightweight?

GeoActivity

1. INTERDISCIPLINARY ACTIVITY–SCIENCE

With a small group, review the information you read on the Aral Sea in Section 1. Then research further to learn about the evaporation of the sea. Present your findings on a poster. Include pictures and information on the effects of the evaporation.

2. WRITING FOR SOCIAL STUDIES

Unit Writing Project Review the itinerary you created for a week-long stay in a country in this unit. Add a hike along the old Silk Roads to your travel plans. Write a journal entry about your experiences and the people you meet there.

3. MENTAL MAPPING

Create an outline map of Transcaucasia and Central Asia and label the following:

- Tian Shan
- Kara Kum
- Kyzyl Kum
- Caspian Sea
- Aral Sea
- Lake Balkhash
- Syr Darya
- Amu Darya

Critical Thinking ANSWERS

25. Possible answers: Cause: irrigation of crops in desert areas/Effect: shrinkage of Aral Sea; Cause: Russification/Effect: suppression of traditional ethnic ways of life; Cause: arbitrary division of Soviet republic boundaries/Effect: ethnic conflicts today

26. Possible answers: establishing truly democratic governments; resolving disputes among ethnic groups; developing economies

27. Possible answer: The Soviet Union wanted to unify their nation, so they tried to make everyone Russian.

28. Possible answers: foods, sports, rug-weaving

29. Possible answers: Natural resources have the potential to enhance the republics' economies; arid land has led to policies that have caused more harm than good.

30. Central Asia lies between China and Europe, so the trade routes connecting the two had to go through Central Asia.

Standards-Based Assessment ANSWERS

1. A

2. C

3. Possible answers: for sleeping, floor coverings, wall linings, insulation, decoration

4. so that the dwelling can be easily packed up, moved, and set up again

GeoActivity

1. INTERDISCIPLINARY ACTIVITY RUBRIC

The poster should
- explore causes and effects of the evaporation
- use maps, charts, and pictures to convey the impact of the environmental disaster

2. WRITING FOR SOCIAL STUDIES RUBRIC

The journal entry should
- describe the terrain along the route
- include conversations with people
- relate personal impressions and experiences

3. MENTAL MAPPING RUBRIC

The outline map should
- represent the approximate shape of Transcaucasia and Central Asia
- include all geographic features listed in their correct location

DIFFERENTIATING CHAPTER ASSESSMENT

Unit Resource Book

- **Chapter Test A,** pp. 117–120. *Use for Struggling Readers, Beginning-to-Intermediate-Level English Learners, and for Inclusion of students whose IEP suggests supported test-taking.*
- **Chapter Test B,** pp. 121–124. *Use for On-Level Learners and ESOL students with Advanced-Level English.*
- **Chapter Test C,** pp. 125–128. *Use for Gifted and Talented students or students assigned Pre-AP work.*

Test Generator

- Chapter Tests A, B, and C, in English and Spanish

McDougal Littell Assessment System

TEST **Test Generator** Use three levels of tests in English and Spanish tied to state standards. Plus, choose questions to customize tests for all populations.

SCORE Scan and Score plain paper student tests in minutes.

REPORT Create instant progress reports, sortable by student, by standard, or other criteria.

RETEACH Respond immediately to students' needs by using a wealth of reteaching resources.

UNIT 6 Africa

Why It Matters:

Human life began millions of years ago in Africa. It is the second-largest continent in the world, and its resources could lead to great growth throughout the continent.

Mt. Kilimanjaro

CHAPTER 17 Africa: Physical Geography and Early History

Great Sphinx, Egypt

CHAPTER 18 Egypt and North Africa

Marketplace on the Niger River

CHAPTER 19 West Africa

Village in the Democratic Republic of Congo

CHAPTER 20 East, Central, and Southern Africa

521

Objectives

- Identify the location and countries of Africa
- Describe the population density of Africa
- Identify the major economic activities in Africa
- Explain the varied life expectancies found in Africa

Comparing Data

Africa occupies an area that is significantly larger than the United States and contains about 600 million more people. The Mississippi River is similar in length to the longest rivers in Africa except for the Nile, which is the world's longest river. However, the largest desert in the United States, the Mojave, is not even one-hundredth the size of the largest desert in Africa, the Sahara, which is also the world's largest.

ANSWERS

1. Africa is nearly four times the size of the continental United States.
2. The Mississippi and Zambezi rivers are closest in length.

UNIT 6 ATLAS

Africa

Africa is the world's second-largest continent in terms of both landmass and population. The equator runs across the middle of the continent. The vast Sahara Desert dominates Africa's northern half. As you study the maps, notice geographic patterns and details about the region. Answer the questions on each map in your notebook.

As you study the graphs on this page, compare the landmass, population, rivers, and deserts of Africa with those of the United States. Then jot down the answers to the following questions in your notebook.

Comparing Data

1. Compare Africa's landmass to that of the United States. About how much larger is Africa than the United States?
2. Of the rivers listed, which two are closest in length?

Comparing Data

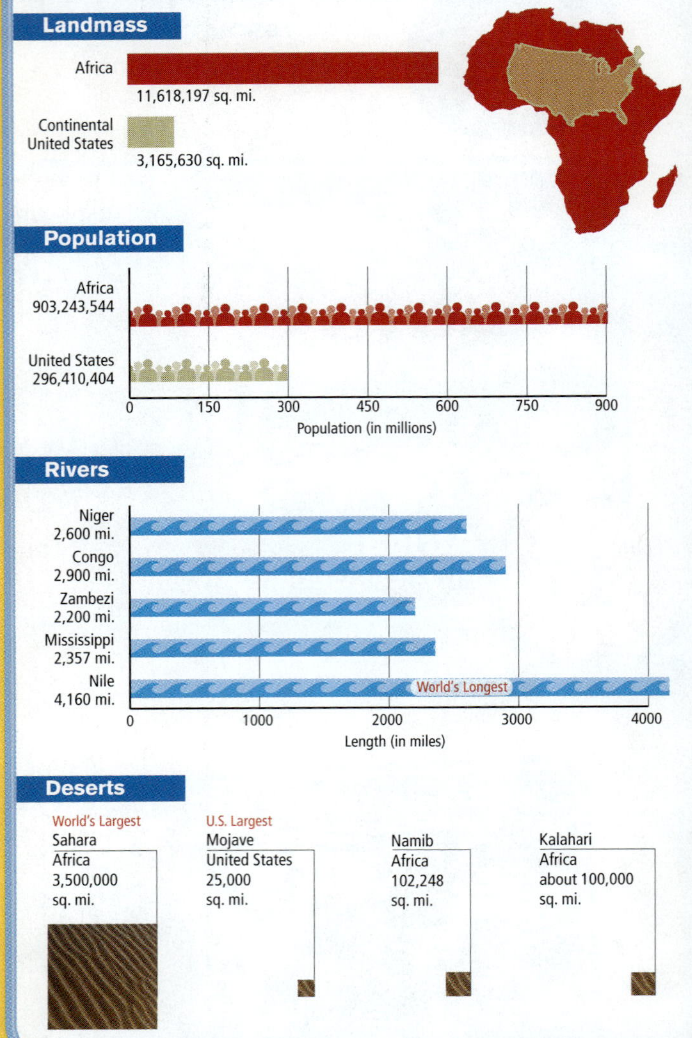

FIVE THEMES AND SIX ELEMENTS

Five Themes

Movement

Make sure that students comprehend that Africa is made up of 53 different countries. Invite them to consider the challenges in transportation—of people and products—that so many national borders present.

Six Elements

The World in Spatial Terms

Explain that the maps and data found in this Unit Atlas provide a lot of insight into each country. For example, by comparing the political map and the population density map, one can see that northern African countries such as Mali, Niger, and Libya have very low population densities. This area is also where the Sahara Desert is located.

THINK LIKE A GEOGRAPHER

1. **Place** What African country has three capitals?
2. **Region** Including both the Blue and White Nile, how many countries does the Nile run through?

Political

The 53 countries of Africa include several island nations as well. The largest of these African island countries is Madagascar in the Indian Ocean. Other African island countries include Mauritius in the Indian Ocean, and the Cape Verde islands in the Atlantic Ocean. Other islands or island chains off the coasts of Africa are territories of other countries. These include the Canary Islands (Spain), Madeira Islands (Portugal), St. Helena (the United Kingdom), and Réunion island (France).

- Which countries on the African continent are closest to Madagascar? *(Mozambique, Tanzania, and South Africa)*
- Find the Cape Verde islands. The islands were part of Portugal until 1975, when Cape Verde became an independent nation. Why do think Portugal colonized Cape Verde? *(Possible Answer: because of its location on trade routes)*
- What capital city is located on Lake Victoria? *(Kampala, Uganda)*

THINK LIKE A GEOGRAPHER ANSWERS

1. **Place** South Africa
2. **Region** four (Ethiopia, Sudan, Uganda, Egypt)

More About . . .

Mauritius

This island was occupied by the Dutch during the 1600s, who named it after Prince Maurice of Nassau. However, when the French claimed the island in 1715, they changed its name to Île de France. In 1810, the island was taken over again, this time by the British, who brought back its Dutch name. Today, Mauritius is an independent nation with a growing economy that includes manufacturing and tourism.

DIFFERENTIATING INSTRUCTION

Struggling Readers

Pronounce Place Names

Model the pronunciation of African countries to help students who might encounter difficulties decoding them. In particular, focus on the varied sounds represented by the letter *a* in such names as Mali, Angola, Madagascar, and Chad.

Inclusion

Locate National Capitals

Review the categories of cities that appear in the map key. Then have students locate the Equator, the Nile River and the Zambezi River. *(Nile: north of the Equator; Zambezi: south of the Equator)* Ask students to name the national capitals that each river runs through or near. *(Nile: Cairo, Khartoum, Addis Ababa, Kampala; Zambezi: Lilongwe, Harare, Lusaka)*

Population Density

Note that most of Africa's population is found in central and southern Africa, as well as coastal areas. Northern Africa, which includes the Sahara Desert, is far less densely populated than the rest of Africa.

- Which African cities on the map have population densities of over 520 persons per square mile? *(Cairo, Lagos, Johannesburg)*
- Why do you think areas of coastal northern Africa are more densely populated than the interior of northern Africa? *(Possible Answer: because sea access allows for easier travel and trade than the Sahara)*
- Compare Egypt and Sudan on this map with the political map on the previous page. What does this comparison reveal about these two countries? *(Possible Answer: Their most densely-populated areas are along the Nile River.)*

THINK LIKE A GEOGRAPHER ANSWERS

1. **Place** access to water sources such as coasts, rivers, and lakes
2. **Movement** the Sahara

More About . . .

Cairo, Egypt

Cairo is the capital city of Egypt and is located near the end of the Nile River. Cairo is a modern city whose streets are filled with famed mosques, palaces, and universities. The city also includes two islands in the Nile—Zamalik Island and Roda Island—that are connected to the rest of the city by bridges. Roda Island contains the Nilometer, an ancient column that is still used to measure the river's water level.

UNIT 6 ATLAS

Population Density

THINK LIKE A GEOGRAPHER

1. **Place** What geographical features exist in areas with high population densities?
2. **Movement** What would account for the low population in large areas of northern Africa?

FIVE THEMES AND SIX ELEMENTS

Five Themes

Human-Environment Interaction

Point out the large areas of limited population in northern Africa where the Sahara is located. Have students refer to the political map on the previous page to see which countries are located in the Sahara. Remind students that resources needed to support large populations—especially water and fertile soil—are generally scarce in desert regions.

Six Elements

Human Systems

Point out that most of the densely populated areas in Africa are located along rivers or coasts. Review the importance of waterways for transportation and emphasize the link between good transportation systems and economic and population growth.

Economic Activity

THINK LIKE A GEOGRAPHER

1. **Place** Name the three countries on the map with uranium as a major resource.
2. **Human-Environment Interaction** What kind of agriculture is most widely practiced in Africa?

525

DIFFERENTIATING INSTRUCTION

English Learners

Comprehension: Connection

Model the use of the preposition *of* with location and direction. Angola, for example, is *south of* Cameroon and *northwest of* Zimbabwe. Encourage students to make similar statements about the relative locations of African countries, cities, and economic activities.

Gifted & Talented

Comparing Economic Activities

Have pairs of students use the other Unit Atlas economic activity maps to compare the economic activities of an African country to that of the United States and one other country. Have students create a chart or other visual display to show these comparisons.

Economic Activity

Point out that Africa's economic wealth is based on several different major resources including coal, diamonds, and oil. Commercial agriculture is very limited in much of northern Africa because of the Sahara. Subsistence agriculture is common in central and southern Africa—and most of these countries are struggling economically.

- List the major economic resources in South Africa. *(fishing, diamonds, gold, uranium, coal, other minerals)*
- What is the main form of land use in Sudan, Chad, and Somalia? *(nomadic herding)*
- What advantage do Cape Town, Lagos, and Casablanca offer as manufacturing centers? *(All three cities are on the Atlantic Ocean, giving them good access to shipping routes.)*

THINK LIKE A GEOGRAPHER ANSWERS

1. **Place** Niger, Namibia, South Africa
2. **Human-Environment Interaction** subsistence agriculture

Comparing Regions

Major Industries in North Africa and Southern Africa

While North Africa and Southern Africa share the same continent, oil and natural gas drilling are the major industries in North African countries such as Algeria, Libya, and Tunisia. In contrast, mining for diamonds or gold are key industries in the southern African countries of South Africa, Namibia, and Botswana.

Until the late 1800s, the economy of the country of South Africa was based on agriculture. However, the discovery of large amounts of diamonds, gold, and other minerals led to rapid economic growth. By the 1990s, agriculture accounted for about five percent of South Africa's economy.

Life Expectancy

Explain that life expectancy in Africa varies a great deal according to region. Northern African countries generally have the highest life expectancies, while southern African nations generally have the lowest life expectancies. Many diseases that have been largely eliminated in the United States—such as malaria—are still common in Africa.

- What is the life expectancy of people in Angola? *(39 years and below)* In Uganda? *(50–59 years)*
- Which countries on the map have life expectancies of 70 and above? *(Morocco, Algeria, Tunisia, Libya, Egypt)*
- Why do you think childhood and infant diseases are the third leading cause of death in Africa? *(Possible Answer: because children do not receive proper nutrition or medical care)*

THINK LIKE A GEOGRAPHER ANSWERS

1. **Region** Morocco, Algeria, Tunisia, Libya, Egypt
2. **Human-Environment Interaction** 9%

Comparing Regions

Life Expectancy in Morocco vs. Life Expectancy in Angola

The North African country of Morocco has nearly double the life expectancy of the South African country of Angola. In Morocco, a person's life expectancy averages 70.9 years, while in Angola life expectancy averages only 38.6 years. Morocco, despite high unemployment, is generally a stable country with a young population that lives primarily in cities. Angola, one of the poorest countries in Africa, is slowly rebuilding itself following a 27-year-long civil war that ended in 2002. The government estimates that 4.7 million people were internally displaced by the war. Since 2002, more than 300,000 of Angola's original 450,000 refugees have returned home.

Life Expectancy

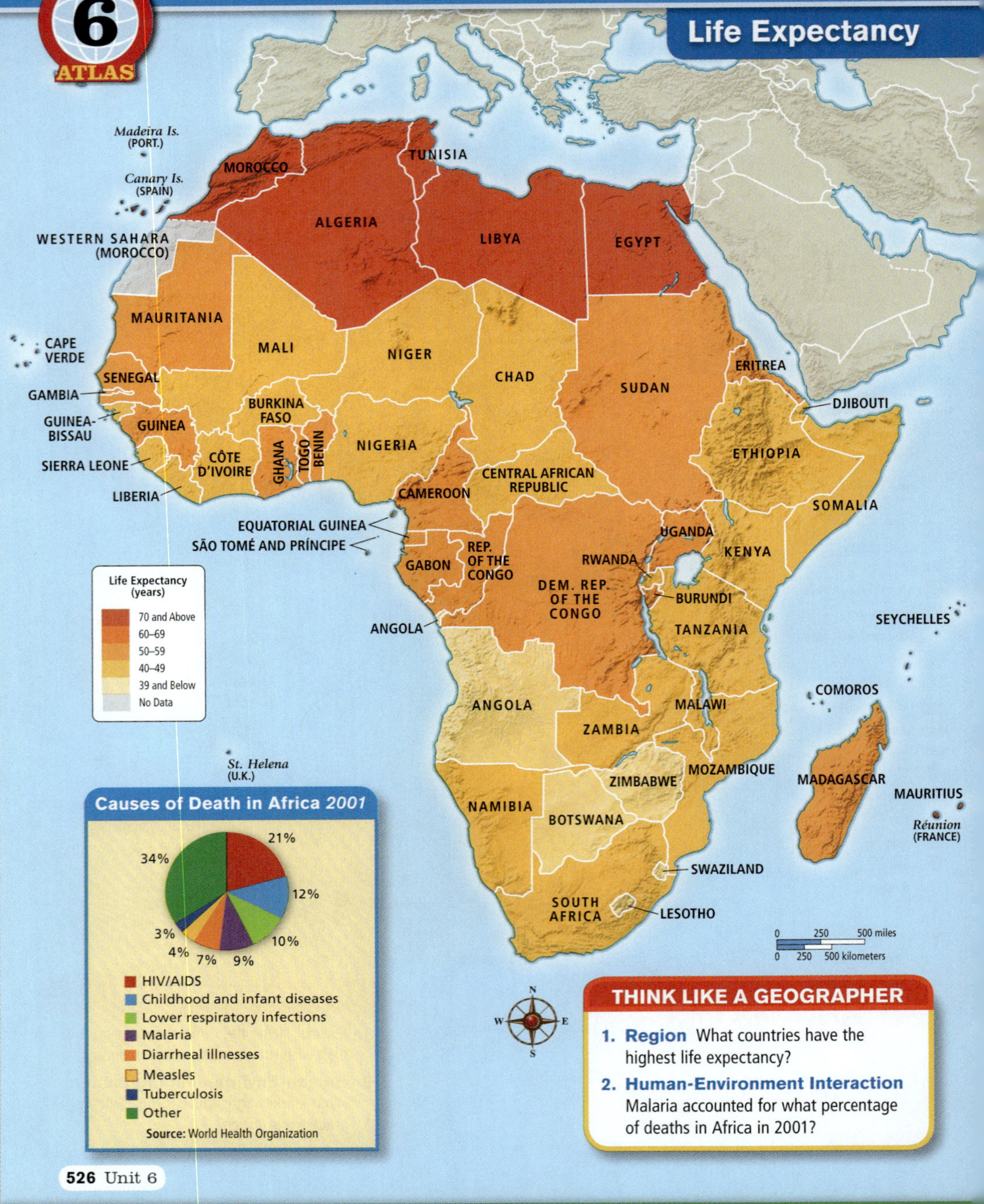

Causes of Death in Africa 2001

- HIV/AIDS 21%
- Childhood and infant diseases 12%
- Lower respiratory infections 10%
- Malaria 9%
- Diarrheal illnesses 7%
- Measles 4%
- Tuberculosis 3%
- Other 34%

Source: World Health Organization

THINK LIKE A GEOGRAPHER

1. **Region** What countries have the highest life expectancy?
2. **Human-Environment Interaction** Malaria accounted for what percentage of deaths in Africa in 2001?

FIVE THEMES AND SIX ELEMENTS

Five Themes

Region

Tell students that malaria kills almost one in ten Africans. This disease is carried by mosquitoes, which thrive in wet climates, such as those found in tropical regions. Point out that much of Africa lies in the tropics—between the Tropic of Cancer and the Tropic of Capricorn.

Six Elements

Place

Point out that there is no data shown on the map for Western Sahara. This territory is claimed by Morocco, but an internal group called the Polisario Front has vied for control of the area since 1976.

Regional Overview

Africa

The four chapters in this unit provide information on the geography, history, culture, government, and economics of selected African countries.

GEOGRAPHY

Savanna covers much of Africa, with rain forests located on the equator. The vast Sahara dominates northern Africa. The world's longest river, the Nile, flows through the northeast.

HISTORY

Great empires arose in Africa, including Egypt, Aksum, and Songhai. In the 18th and 19th centuries, European nations divided up most of Africa, creating colonies to enrich their own countries. African colonies began winning independence after World War II, but colonial history still affects Africa.

Nelson Mandela Nelson Mandela won South Africa's first multiracial election.

CULTURE

Africa is home to hundreds of ethnic groups and more than 1,000 languages. Major religions in Africa include Islam, Christianity, and traditional local faiths. Soccer is popular throughout the continent. Western and African cultures have affected each other, particularly in art and music.

GOVERNMENT

When African nations won independence, many kept the European colonial borders, leading some traditional ethnic enemies to clash. Some nations with young democracies are learning to share power among different groups.

ECONOMICS

Africa has potential for great wealth. But years of instability have made it hard for some countries to improve their economies. Other nations are working to help African economies, lending aid and trying to erase debt.

DIFFERENTIATING INSTRUCTION

Inclusion

Compare African Cultures Through Music

Emphasize the cultural diversity of Africa by having students listen to several distinct African musical styles. Popular examples might include Juju from Nigeria, Mbalax from Senegal, Kwela from South Africa, Makassi from Cameroon, and Morna from Cape Verde. After listening to the recordings, have students create illustrations or poems to describe what they heard.

Pre-AP

Compare Colonialism and African Colonies

Have students investigate the effects that European colonialism had on African colonies, particularly during the 19th and early 20th centuries. Ask them to find an example of a colonial practice or influence that still affects an African country today.

Regional Overview

Much of modern African history and culture has been shaped by western and colonial influences. As a legacy of the colonial era, disputed borders, corrupt government leaders, and struggles over ownership of property and businesses have resulted. In turn, these factors have contributed to civil wars and weak economies in many areas of Africa. Some African countries that have overcome these challenges—such as South Africa and Botswana—have healthy economies.

- What African river is the longest in the world? *(the Nile)* What African desert is the largest in the world? *(the Sahara)*
- How many different languages are spoken in Africa? *(more than 1,000)*
- Why do you think ethnic conflicts occur in Africa? *(Possible Answer: With so many different ethnic groups, spread across countries with weak economies, conflicts are more likely to result.)*

More About . . .

Nelson Mandela

Nelson Mandela, the first black African President of South Africa, was imprisoned from 1962 to 1990 for opposing the South African government's policy of apartheid. During his years in prison, he became the most important black leader in South Africa. He encouraged people to resist the government's policies and refused offers to alter his beliefs in exchange for freedom. In 1993, Mandela won the Nobel Peace Prize.

Unit Writing Project
Teaching Tips

- As students prepare a CD or song review, encourage them to seek out a variety of reliable sources to use, including newspapers, magazines, and Web sites. You may also want to arrange for students to interview local musical experts, such as from a community cultural organization or local university, at school.
- Provide students with an example of a song or CD review, or have them research typical formats.
- Encourage students to create props and visuals to go with their reviews. Remind them that CD covers, instrument photos, and other visuals will help to illustrate their key points as well as make their presentations more engaging.
- Encourage students to rehearse presenting their reviews with a partner and to revise them based on his or her feedback.

Each map on the country cards illustrates the shape of a state, province, or country. Tell students that the maps do not reflect the comparative size of these landmasses within a particular region. To get an idea of the actual size of the state, province, or country, students should study the land area statistics provided on the card.

UNIT 6 ATLAS Country Almanac

Click here to compare the most recent data on countries @ClassZone.com

Unit Writing Project

Music Review

Research and listen to some of the music from a region mentioned in this unit. Write a review of a CD or song. Learn what you can about the artist, and styles or instruments unique to the region.

Think About:

- how African musical styles or instruments influence music you regularly listen to
- how American or European music influenced African music

Africa

The country cards on these pages present information about the 53 nations in Africa.

Algeria

GEOGRAPHY
Capital: Algiers
Total Area: 919,595 sq. mi.
Population: 32,854,000

ECONOMY
Imports: industrial equipment; food; consumer goods
Exports: petroleum; natural gas

CULTURE
Language: Arabic; Tamazight
Religion: Sunni Muslim 99%

Angola

GEOGRAPHY
Capital: Luanda
Total Area: 481,354 sq. mi.
Population: 15,941,000

ECONOMY
Imports: consumer and capital goods
Exports: petroleum; diamonds; coffee

CULTURE
Language: Portuguese
Religion: Catholic 62%; Protestant 15%; traditional beliefs 5%

Benin

GEOGRAPHY
Capital: Porto-Novo
Total Area: 43,483 sq. mi.
Population: 8,439,000

ECONOMY
Imports: food; petroleum
Exports: cotton yarn

CULTURE
Language: French
Religion: traditional beliefs 35%; Catholic 26%; Muslim 21%

Botswana

GEOGRAPHY
Capital: Gaborone
Total Area: 231,804 sq. mi.
Population: 1,765,000

ECONOMY
Imports: machinery; food; tobacco; wood
Exports: diamonds; textiles; meat

CULTURE
Language: English; Tswana
Religion: traditional beliefs 39%; African Christian 31%; Protestant 11%

Burkina Faso

GEOGRAPHY
Capital: Ouagadougou
Total Area: 105,869 sq. mi.
Population: 13,228,000

ECONOMY
Imports: petroleum; food
Exports: cotton; shea nuts; gold

CULTURE
Language: French
Religion: Muslim 49%; traditional beliefs 34%; Christian 17%

Burundi

GEOGRAPHY
Capital: Bujumbura
Total Area: 10,745 sq. mi.
Population: 7,548,000

ECONOMY
Imports: consumer and capital goods
Exports: coffee; tea

CULTURE
Language: Rundi; French
Religion: Catholic 57%; Protestant 20%; traditional beliefs 7%

Cameroon

GEOGRAPHY
Capital: Yaoundé
Total Area: 183,568 sq. mi.
Population: 16,322,000

ECONOMY
Imports: minerals; industrial equipment
Exports: petroleum; lumber; cocoa beans

CULTURE
Language: French; English
Religion: Christian 47%; traditional beliefs 24%; Muslim 21%

Cape Verde

GEOGRAPHY
Capital: Praia
Total Area: 1,557 sq. mi.
Population: 507,000

ECONOMY
Imports: food; machinery; metals
Exports: shoes; clothing; fish

CULTURE
Language: Portuguese
Religion: Catholic 91%; Muslim 3%

INTERDISCIPLINARY ACTIVITIES

CONNECT to Language Arts

Persuasive Travel Writing

Ask students to write a paragraph that would help persuade U.S. tourists to visit a certain African country. Have them focus on the key features or attractions of the country that might appeal to U.S. visitors.

CONNECT to Art

Design a Travel Magazine Advertisement

Have students create a one-page magazine advertisement that promotes tourism to an African country, city, or historic site. First, have students study examples of travel ads that they find (or you supply). Then, ask them to create their own travel ads, combining headlines with drawn or photocopied images in ways that will appeal to readers and capture their attention.

Yellow-billed hornbill
found in southern Africa

Central African Republic

GEOGRAPHY
Capital: Bangui
Total Area: 240,535 sq. mi.
Population: 4,038,000

ECONOMY
Imports: road vehicles; raw cotton; food
Exports: wood; diamonds; cotton; coffee

CULTURE
Language: French; Sango
Religion: Catholic 18%; Muslim 16%; Protestant 14%; other Christian 35%

Chad

GEOGRAPHY
Capital: N'Djamena
Total Area: 495,755 sq. mi.
Population: 9,749,000

ECONOMY
Imports: petroleum
Exports: cattle, sheep, and goats; cotton

CULTURE
Language: Arabic; French
Religion: Muslim 54%; Catholic 20%; Protestant 14%

Comoros

GEOGRAPHY
Capital: Moroni
Total Area: 838 sq. mi.
Population: 798,000

ECONOMY
Imports: rice; meat and fish; vehicles
Exports: vanilla; cloves; ylang-ylang

CULTURE
Language: Comorian; Arabic; French
Religion: Sunni Muslim 98%

Democratic Republic of Congo

GEOGRAPHY
Capital: Kinshasa
Total Area: 905,568 sq. mi.
Population: 57,549,000

ECONOMY
Imports: petroleum products
Exports: diamonds; cobalt; copper; gold

CULTURE
Language: French; English
Religion: Catholic 41%; Protestant 32%

Republic of Congo

GEOGRAPHY
Capital: Brazzaville
Total Area: 132,047 sq. mi.
Population: 3,999,000

ECONOMY
Imports: petroleum products
Exports: wood; sugar

CULTURE
Language: French; Lingala; Monokutuba
Religion: Catholic 49%; Protestant 17%; African Christian 13%

Côte d'Ivoire

GEOGRAPHY
Capital: Yamoussoukro
Total Area: 124,503 sq. mi.
Population: 18,154,000

ECONOMY
Imports: petroleum; food products
Exports: cocoa beans; wood; coffee beans

CULTURE
Language: French
Religion: Muslim 39%; Christian 30%; nonreligious 17%; animist 12%

Djibouti

GEOGRAPHY
Capital: Djibouti
Total Area: 8,880 sq. mi.
Population: 793,000

ECONOMY
Imports: food; transport equipment
Exports: hides and skins; leather

CULTURE
Language: Arabic; French
Religion: Sunni Muslim 97%; Christian 3%

Egypt

GEOGRAPHY
Capital: Cairo
Total Area: 386,662 sq. mi.
Population: 74,033,000

ECONOMY
Imports: machinery; food; iron and steel
Exports: petroleum; food; raw cotton

CULTURE
Language: Arabic
Religion: Muslim 84%; Christian 15%

Equatorial Guinea

GEOGRAPHY
Capital: Malabo
Total Area: 10,831 sq. mi.
Population: 504,000

ECONOMY
Imports: petroleum products; machinery
Exports: methanol; wood; cocoa beans

CULTURE
Language: Spanish; French
Religion: Catholic 80%; Muslim 4%; African Christian 4%

529

INTERDISCIPLINARY ACTIVITIES

CONNECT to Civics

Compare African Governments

Ask students to choose an African country and find out basic facts about its system of government. Have them identify the head of state and legislative bodies of their chosen country and list their findings on the board or chart paper. Then, as a class activity, classify each government or compare and contrast them together.

Country Almanac

The second largest continent, Africa is made up of 53 independent nations including several island nations. The Northern African nation of Egypt is often categorized as being part of the Middle East, although it contains the largest city in Africa—Cairo. Africa enjoys a wealth of natural resources, including diamonds, oil, and uranium, which offers hope for improving Africa's economic future. Unfortunately, life expectancy varies greatly throughout the continent—with HIV/AIDS, other diseases, and wars and violent conflicts being significant factors. Finally, Africa enjoys one of the world's most diverse mix of cultures, religions, and languages.

- Which African country has the largest population? *(Nigeria)*
- Which African country has the smallest land area? *(Seychelles)*
- In which countries is the primary language English? *(Botswana, The Gambia, Ghana, Liberia, Mauritius, Namibia, Nigeria, Sierra Leone, Uganda, Zambia, and Zimbabwe)*
- Which countries export petroleum? *(Algeria, Angola, Cameroon, Egypt, Libya, Nigeria, Senegal, Seychelles, and Sudan)* Diamonds? *(Angola, Botswana, Central African Republic, Democratic Republic of Congo, Guinea, Namibia, Sierra Leone, South Africa, and Tanzania)*

Each map on the country cards illustrates the shape of a state, province, or country. Tell students that the maps do not reflect the comparative size of these landmasses within a particular region. To get an idea of the actual size of the state, province, or country, students should study the land area statistics provided on the card.

More About . . .

Kenyan Coffee

Much of Kenya's economy is based on subsistence farming. Two of the country's largest cash crops are tea and coffee, which Kenya exports throughout the world. However, varying prices on these and other crops often hurts Kenyan farmers. In 2006, Kenya became eligible for Fair Trade certification that helps protect Kenyan farmers against world price fluctuations, guaranteeing them a reasonable price for their coffee and tea.

Each map on the country cards illustrates the shape of a state, province, or country. Tell students that the maps do not reflect the comparative size of these landmasses within a particular region. To get an idea of the actual size of the state, province, or country, students should study the land area statistics provided on the card.

UNIT 6 ATLAS

Eritrea

GEOGRAPHY
Capital: Asmara
Total Area: 46,842 sq. mi.
Population: 4,401,000

ECONOMY
Imports: live animals; cereals; iron
Exports: raw sugar; vegetables and fruits; fish; sesame seeds

CULTURE
Language: Afar; Arabic; Tigre; Kunama
Religion: Christian 51%; Muslim 45%

Ethiopia

GEOGRAPHY
Capital: Addis Ababa
Total Area: 435,186 sq. mi.
Population: 77,431,000

ECONOMY
Imports: machinery; road vehicles; iron
Exports: coffee; leather; sesame seeds

CULTURE
Language: Amharic
Religion: Ethiopian Orthodox 50%; Muslim 33%; Protestant 10%

Gabon

GEOGRAPHY
Capital: Libreville
Total Area: 103,347 sq. mi.
Population: 1,384,000

ECONOMY
Imports: petroleum products
Exports: wood; manganese

CULTURE
Language: French
Religion: Catholic 57%; Protestant 18%; Muslim 3%

The Gambia

GEOGRAPHY
Capital: Banjul
Total Area: 4,363 sq. mi.
Population: 1,517,000

ECONOMY
Imports: food; machinery; petroleum
Exports: peanuts; fruits and vegetables

CULTURE
Language: English
Religion: Muslim 95%; Christian 4%

Ghana

GEOGRAPHY
Capital: Accra
Total Area: 92,456 sq. mi.
Population: 22,113,000

ECONOMY
Imports: petroleum; machinery; food
Exports: gold; cocoa beans; aluminum

CULTURE
Language: English
Religion: traditional beliefs 24%; Muslim 20%; Protestant 17%; African Christian 14%

Guinea

GEOGRAPHY
Capital: Conakry
Total Area: 94,926 sq. mi.
Population: 9,402,000

ECONOMY
Imports: refined petroleum; food
Exports: bauxite; gold; diamonds; fish

CULTURE
Language: French
Religion: Muslim 85%; Christian 10%

Guinea-Bissau

GEOGRAPHY
Capital: Bissau
Total Area: 13,946 sq. mi.
Population: 1,586,000

ECONOMY
Imports: food; transport equipment
Exports: cashews; cotton; logs

CULTURE
Language: Portuguese
Religion: traditional beliefs 45%; Muslim 40%; Christian 13%

Kenya

GEOGRAPHY
Capital: Nairobi
Total Area: 224,962 sq. mi.
Population: 34,256,000

ECONOMY
Imports: petroleum; machinery; chemicals
Exports: horticultural products; coffee

CULTURE
Language: Swahili; English
Religion: Catholic 22%; African Christian 21%; Protestant 20%

Lesotho

GEOGRAPHY
Capital: Maseru
Total Area: 11,720 sq. mi.
Population: 1,795,000

ECONOMY
Imports: food products; commodities
Exports: clothing; machinery

CULTURE
Language: Sotho; English
Religion: Catholic 38%; Protestant 13%; African Christian 12%

530 Unit 6

INTERDISCIPLINARY ACTIVITIES

CONNECT to Math

Graph Populations in a Region

Have each student select a region of Africa using the political map. Then have them create a pie graph comparing the populations of the countries in the region. If individuals require assistance, help them convert population figures into percentages of the region's total population. Have students present their graphs to the class.

CONNECT to Economics

Charting Natural Resources

Assign groups of students to one of the countries listed on this page. Have groups use available resources to learn more about the natural resources of their country. Each group should then create a chart of the resources, including statistical data (employment, income from sales, and so on) and an assessment of any potential damage to the country's environment from the exploration, processing, or use of each resource.

Liberia

GEOGRAPHY
Capital: Monrovia
Total Area: 43,000 sq. mi.
Population: 3,283,000

ECONOMY
Imports: food and live animals; petroleum
Exports: logs and timber; rubber

CULTURE
Language: English
Religion: traditional beliefs 63%; Muslim 16%; Protestant 14%

Libya

GEOGRAPHY
Capital: Tripoli
Total Area: 679,362 sq. mi.
Population: 5,853,000

ECONOMY
Imports: machinery; food; road vehicles
Exports: crude and refined petroleum; natural gas

CULTURE
Language: Arabic
Religion: Sunni Muslim 96%

Madagascar

GEOGRAPHY
Capital: Antananarivo
Total Area: 226,657 sq. mi.
Population: 18,606,000

ECONOMY
Imports: petroleum; machinery
Exports: textiles; vanilla; cloves; shellfish

CULTURE
Language: Malagasy; French
Religion: traditional beliefs 48%; Protestant 23%; Catholic 20%

Malawi

GEOGRAPHY
Capital: Lilongwe
Total Area: 45,745 sq. mi.
Population: 12,884,000

ECONOMY
Imports: food; machinery; chemicals
Exports: tobacco; sugar; tea; clothing

CULTURE
Language: Chichewa
Religion: Protestant 39%; Catholic 25%; Muslim 15%

Mali

GEOGRAPHY
Capital: Bamako
Total Area: 478,767 sq. mi.
Population: 13,518,000

ECONOMY
Imports: machinery; petroleum products
Exports: gold; raw cotton; live animals

CULTURE
Language: French
Religion: Muslim 82%; traditional beliefs 16%

Mauritania

GEOGRAPHY
Capital: Nouakchott
Total Area: 397,955 sq. mi.
Population: 3,069,000

ECONOMY
Imports: food products; vehicles
Exports: iron ore; fish

CULTURE
Language: Arabic; Fulani; Soninke; Wolof
Religion: Sunni Muslim 99%

Mauritius

GEOGRAPHY
Capital: Port Louis
Total Area: 788 sq. mi.
Population: 1,245,000

ECONOMY
Imports: fabrics; yarn; food; live animals
Exports: sugar; fabric; ships' stores

CULTURE
Language: English
Religion: Hindu 50%; Christian 32%; Muslim 17%

Morocco

GEOGRAPHY
Capital: Rabat
Total Area: 172,414 sq. mi.
Population: 31,478,000

ECONOMY
Imports: food; tobacco; cotton fabric
Exports: knitwear; phosphoric acid

CULTURE
Language: Arabic
Religion: Muslim 98%

Mozambique

GEOGRAPHY
Capital: Maputo
Total Area: 309,496 sq. mi.
Population: 19,792,000

ECONOMY
Imports: machinery; refined petroleum
Exports: aluminum; electricity; cotton

CULTURE
Language: Portuguese
Religion: traditional beliefs 50%; Catholic 16%; Muslim 11%

Meerkats Meerkats, like this family in South Africa, live in groups of between 3 and 25.

More About . . .

Liberian Culture

Liberia's population includes members of more than 16 different ethnic groups. English is the official language, but many African languages are also spoken. Freed American slaves settled in Liberia in the 19th century. Most people live in towns as opposed to cities and work in agriculture, growing coffee, cocoa, rice, and other crops.

Liberia has been struggling to overcome corruption and civil war. In 2004, peace was established and later that year Ellen Johnson Sirleaf was elected president, becoming Africa's first woman president.

INTERDISCIPLINARY ACTIVITIES

CONNECT to Language Arts

Descriptive Writing on Meerkats

Have students write a paragraph in the present tense that describes the photograph of the meerkats and their environment. After they have set the scene, ask them to also describe what they think has captured the meerkats' attention.

Each map on the country cards illustrates the shape of a state, province, or country. Tell students that the maps do not reflect the comparative size of these landmasses within a particular region. To get an idea of the actual size of the state, province, or country, students should study the land area statistics provided on the card.

Comparing Regions

Namibia and Rwanda

Namibia and Rwanda are both in Southern Africa, but have many differences. Namibia is a largely rural country with a small population. Rwanda has approximately 1/30th the land area of Namibia, but more than four times its population. Much of Namibia is covered by desert, while Rwanda is mostly mountainous. Namibia's population includes several ethnic groups, with English as the official language. Most of Rwanda's population is made up two specific groups, Hutu and Tutsi, both of whom speak Rwandan (Kinyarwanda). Namibia is generally stable with a government similar to the United States' system of government, while Rwanda has dealt with ethnic violence and fighting for more than 40 years.

Each map on the country cards illustrates the shape of a state, province, or country. Tell students that the maps do not reflect the comparative size of these landmasses within a particular region. To get an idea of the actual size of the state, province, or country, students should study the land area statistics provided on the card.

UNIT 6 ATLAS

Namibia

GEOGRAPHY
Capital: Windhoek
Total Area: 318,696 sq. mi.
Population: 2,031,000

ECONOMY
Imports: food; machinery; metals
Exports: diamonds; metals; fish; meat

CULTURE
Language: English
Religion: Protestant 48%; Catholic 18%; African Christian 11%

Niger

GEOGRAPHY
Capital: Niamey
Total Area: 489,191 sq. mi.
Population: 13,957,000

ECONOMY
Imports: food; capital goods
Exports: uranium; cattle; onions; cowpeas

CULTURE
Language: French
Religion: Sunni Muslim 91%

Nigeria

GEOGRAPHY
Capital: Abuja
Total Area: 356,669 sq. mi.
Population: 131,530,000

ECONOMY
Imports: machinery; chemicals; food
Exports: crude petroleum

CULTURE
Language: English
Religion: Muslim 44%; independent Christian 15%; African indigenous 10%

Rwanda

GEOGRAPHY
Capital: Kigali
Total Area: 10,169 sq. mi.
Population: 9,038,000

ECONOMY
Imports: capital goods; food
Exports: niobium and tantalum; tea; coffee

CULTURE
Language: Rwanda; French; English
Religion: Catholic 51%; Protestant 29%; traditional beliefs 9%

São Tomé and Príncipe

GEOGRAPHY
Capital: São Tomé
Total Area: 386 sq. mi.
Population: 157,000

ECONOMY
Imports: investment goods; agricultural products; petroleum
Exports: cocoa beans; copra; coffee

CULTURE
Language: Portuguese
Religion: Catholic 90%

Senegal

GEOGRAPHY
Capital: Dakar
Total Area: 75,749 sq. mi.
Population: 11,658,000

ECONOMY
Imports: food and live animals
Exports: petroleum; fresh crustaceans

CULTURE
Language: French
Religion: Muslim 88%; traditional beliefs 6%; Christian 6%

Seychelles

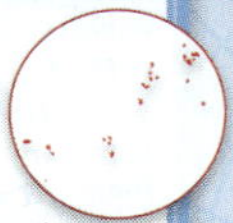

GEOGRAPHY
Capital: Victoria
Total Area: 176 sq. mi.
Population: 81,188

ECONOMY
Imports: food; mineral fuels; machinery
Exports: canned tuna; fresh and frozen fish; petroleum

CULTURE
Language: Creole; English; French
Religion: Catholic 90%; Anglican 7%

Sierra Leone

GEOGRAPHY
Capital: Freetown
Total Area: 27,699 sq. mi.
Population: 5,525,000

ECONOMY
Imports: food and live animals; fuels
Exports: diamonds; cacao; rutile

CULTURE
Language: English
Religion: Sunni Muslim 46%; traditional beliefs 40%; Christian 11%

Somalia

GEOGRAPHY
Capital: Mogadishu
Total Area: 246,201 sq. mi.
Population: 8,228,000

ECONOMY
Imports: raw sugar; cereals
Exports: goats; sheep; bovines; camels

CULTURE
Language: Somali; Arabic
Religion: Sunni Muslim 99%

532 Unit 6

INTERDISCIPLINARY ACTIVITIES

CONNECT to Math

Size Comparisons of African Countries

To help students better grasp the large variations in the sizes of African countries, have them use the total area data for each country to create paper rectangles that represent that country. For example, a one-inch-by-one-inch square could represent 10,000 square miles. Have each student choose five to ten countries to represent and order them from biggest to smallest.

South Africa

GEOGRAPHY
Capital: Pretoria/Cape Town/Bloemfontein
Total Area: 471,011 sq. mi.
Population: 47,432,000

ECONOMY
Imports: machinery; petroleum; chemicals
Exports: diamonds; gold; iron and steel

CULTURE
Language: Afrikaans; English; Xhosa; Zulu
Religion: independent churches 39%; Protestant 32%; traditional beliefs 8%

Sudan

GEOGRAPHY
Capital: Khartoum
Total Area: 967,499 sq. mi.
Population: 36,233,000

ECONOMY
Imports: machinery; food; chemicals
Exports: petroleum; sesame seeds; gold

CULTURE
Language: Arabic; English
Religion: Sunni Muslim 70%; traditional beliefs 12%; Catholic 8%

Swaziland

GEOGRAPHY
Capital: Mbabane
Total Area: 6,704 sq. mi.
Population: 1,032,000

ECONOMY
Imports: food; machinery; chemicals
Exports: soft drink concentrates; sugar

CULTURE
Language: Swati; English
Religion: African indigenous Christian 46%; Protestant 15%

Tanzania

GEOGRAPHY
Capital: Dodoma
Total Area: 364,900 sq. mi.
Population: 36,766,356

ECONOMY
Imports: food; machinery
Exports: gold; diamonds; gemstones; cashews; coffee; tea

CULTURE
Language: Swahili; English
Religion: Christian 47%; Muslim 32%

Togo

GEOGRAPHY
Capital: Lomé
Total Area: 21,925 sq. mi.
Population: 6,145,000

ECONOMY
Imports: food; petroleum; cement
Exports: cement; phosphates; cotton

CULTURE
Language: French
Religion: traditional beliefs 38%; Catholic 24%; Muslim 19%

Tunisia

GEOGRAPHY
Capital: Tunis
Total Area: 63,170 sq. mi.
Population: 10,102,000

ECONOMY
Imports: fabric; food; machinery
Exports: clothing; knitwear; phosphates

CULTURE
Language: Arabic
Religion: Sunni Muslim 99%

Uganda

GEOGRAPHY
Capital: Kampala
Total Area: 91,136 sq. mi.
Population: 28,816,000

ECONOMY
Imports: machinery; food
Exports: fish products; tea; cereal; cotton

CULTURE
Language: English
Religion: Catholic 33%; Protestant 33%; traditional beliefs 18%; Muslim 16%

Zambia

GEOGRAPHY
Capital: Lusaka
Total Area: 290,586 sq. mi.
Population: 11,668,000

ECONOMY
Imports: machinery; printed matter; cereals
Exports: refined copper; base metals; food

CULTURE
Language: English
Religion: traditional beliefs 27%; Protestant 23%; Catholic 17%

Zimbabwe

GEOGRAPHY
Capital: Harare
Total Area: 150,804 sq. mi.
Population: 13,010,000

ECONOMY
Imports: machinery; chemicals; food
Exports: tobacco; ferroalloys; cotton lint

CULTURE
Language: English
Religion: animist 41%; Protestant 24%; African indigenous Christian 14%

Hippopotamuses Once common across the continent, hippos are now found mostly in East and Central Africa.

Comparing Regions

South Africa and Sudan

Sudan, with more than twice the land area of South Africa, has a smaller population. South Africa has many different ethnic groups and 11 official languages. Sudan has only one official language, Arabic, which is used by two of its three main ethnic groups. Both countries have struggled with racial and ethnic divisions.

South Africa's government enforced a policy of apartheid for over 40 years to separate whites and non-whites. The policy led to significant violence and civil strife before it ended in 1994. Since then, South Africa has enjoyed a stable government and fairly healthy economy.

In Sudan, a civil war that began in the 1980s destroyed much of the country and displaced more than 4 million people. A separate violent struggle that began in the Darfur region of western Sudan continues today. Sudan's economy—much of it controlled by the government—is struggling, with most people engaged in subsistence farming.

INTERDISCIPLINARY ACTIVITIES

CONNECT to Science

Research Africa's Minerals

Have students choose one of the following mineral resources found in Africa: bauxite, coal, diamonds, gold, or iron ore. Ask them to conduct research on the mineral they chose. Have volunteers describe their mineral to the class, explain how it is extracted, and cite examples of how people use it.

CONNECT to Language Arts

Varying Descriptions of Daily Life in Africa

Have students write about daily life in an African country of their choice, from the perspective of a student, farmer, miner, or other worker. Ask volunteers to read one of their descriptions, then discuss as a class the differences in viewpoints.

Each map on the country cards illustrates the shape of a state, province, or country Tell students that the maps do not reflect the comparative size of these landmasses within a particular region. To get an idea of the actual size of the state, province, or country, students should study the land area statistics provided on the card.

Chapter 17 Africa: Physical Geography and Early History

PLANNING GUIDE

Complete print resources all on one USB drive allow you to customize lesson plans for all students.

Section Titles & Key Ideas	Unit Resource Book Africa	Assessment Resources
1 Africa's Geographic Extremes **KEY IDEA** Africa, the world's second largest continent, lies almost entirely in the tropics.	• **Reading Study Guide,** pp. 3–4 • **RSG with Additional Support,** pp. 7–8 • **Vocabulary Practice,** p. 19 • **Skillbuilder Practice,** p. 20 • **Map and Graph Skills,** p. 21 • **Outline Map Activities,** pp. 23–24 • **Interdisciplinary Activities,** pp. 26–27	**Unit Resource Book** • Section Quiz, p. 31 • Reteaching Activity, p. 33 **Test Generator CD-ROM** **Daily Test Practice Transparency,** TT59
2 Human Beginnings in Africa **KEY IDEA** Human life first emerged on this warm continent.	• **Reading Study Guide,** pp. 5–6 • **RSG with Additional Support,** pp. 9–10 • **History Makers,** p. 25 • **Connect to History,** p. 28 • **Primary and Secondary Sources,** p. 29	**Unit Resource Book** • Section Quiz, p. 32 • Reteaching Activity, p. 34 • Chapter Test (Forms A, B, & C), pp. 35–46 **Test Generator CD-ROM** **Daily Test Practice Transparency,** TT60

INTEGRATED TECHNOLOGY

No Time? To focus students on important content in the chapter, use the **"Reading Study Guide with Additional Support"** in ***Resources2Go.***

Teacher Presentation Options

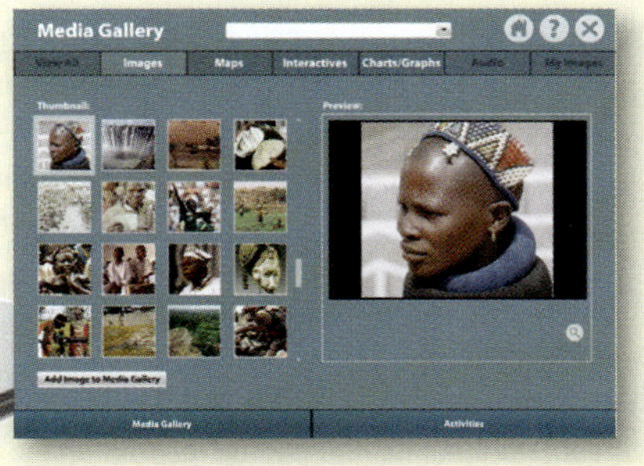

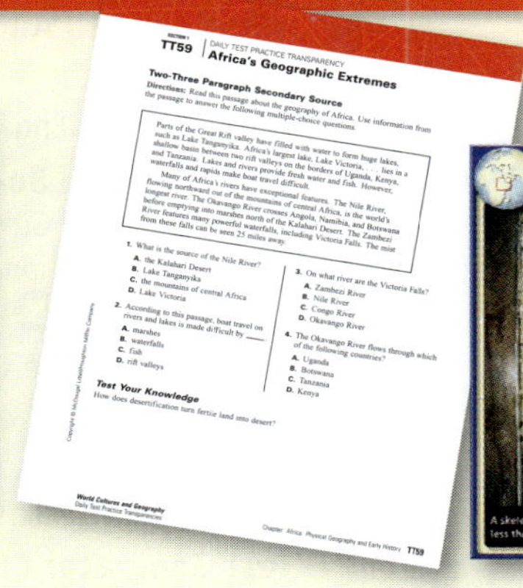

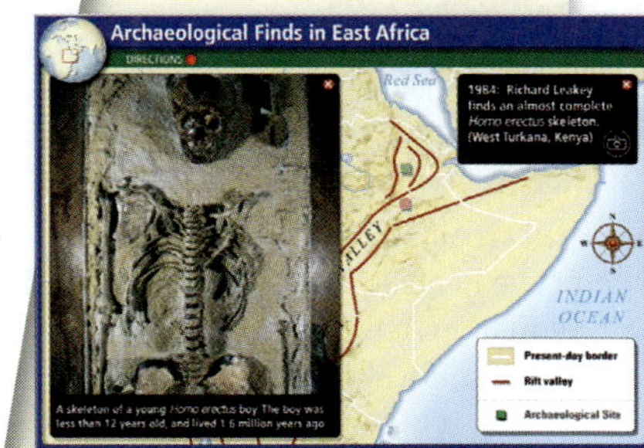

Electronic Books

eEdition DVD-ROM

eEdition Online

Presentation Toolkit

Video/DVD

- World Cultures and Geography

Power Presentations DVD-ROM

- Discussion Starters
- Media Gallery
- Review Game
- Interactive Review
- Animated Features

Unit Transparency Book

- 3-Minute Warm-Ups
- Essential Question
- Critical Thinking
- Maps
- Cultures

Daily Test Practice Transparencies

Animated Geography

- Physical Geography of Africa Map
- Vegetation Regions of Africa map
- Archaeological Finds in East Africa

OnlineActivities at ClassZone.com

Animated Geography

- Videos and Photos
- 3-D Animations
- Interactive Maps

Interactive Review

- Section Quizzes
- Review and Study Notes
- Vocabulary Flip Cards
- Graphic Organizers
- Name Game and Crossword Puzzles
- GeoGame

Online Test Practice

Map Center

- Atlas Maps
- Political Maps with Data
- Links to Mapping Web Sites

Activity Center

- Activity Maker—Review Games, Puzzles, Flip Cards, and Graphic Organizers
- Worldster

Research & Writing Center

- Web Sites
- Data Updates
- Primary Sources
- Writing Models

Current Events

Terry Rhodes
McNair Middle School
Fayetteville, Arkansas

Make It Fun **Teacher-Tested Activities**

As you teach this chapter, look for my classroom-proven ideas for:

- making the lesson fun
- focusing students' attention and comprehension
- integrating technology with traditional teaching

Teaching Bell to Bell When you see this timing guide, have students discuss in pairs, in groups of 3–6, or as a class.

- **Think, Pair, Share** (pairs)
- **Reader, Recorder, Reporter** (3–6)
- **Talk About It** (any size)

CHAPTER 17

RESOURCES FOR DIFFERENTIATED INSTRUCTION

Struggling Readers

eEdition with Audio DVD-ROM

Lesson Plans
- Strategies for Inclusion

Unit Resource Book
- Reading Study Guide with Additional Support, pp. 7–10 B
- Reteaching Activity, pp. 33–34
- Chapter Test (Form A), pp. 35–38 A

Best Practices Toolkit C
- Interactive Reading Strategies, pp. 90–145
- Differentiation Activity Bank: Struggling Readers, pp. 146–149

Teacher's Edition Activities
- Use a Map, Sec. 1
- Sequencing the Development of Early Humans, Sec. 2
- Tiered Activities, Sec. 1–2

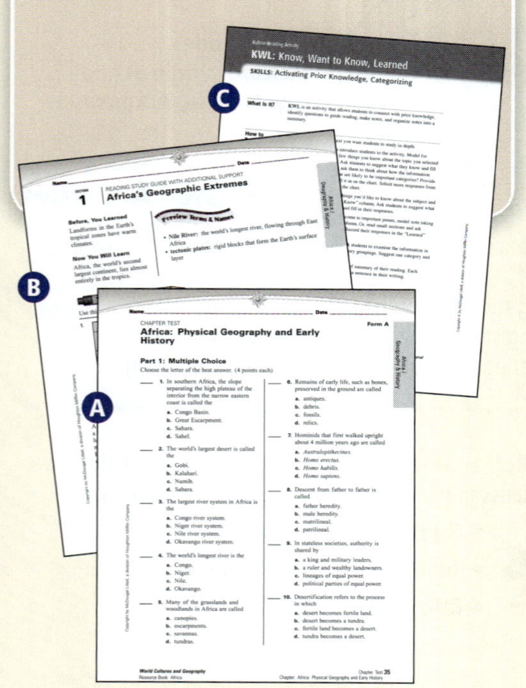

English Learners

Pupil's Edition in Spanish

eEdition with Audio DVD-ROM

Lesson Plans
- Strategies for English Learners

Unit Resource Book
- Reading Study Guide (Spanish), pp. 11–14
- RSG with Additional Support (Spanish), pp. 15–18 A B

Multi-Language Glossary C

Best Practices Toolkit
- Differentiation Activity Bank: English Learners, pp. 165–194

Test Generator CD-ROM
- Section Quizzes in Spanish
- Chapter Test in Spanish

Teacher's Edition Activities
- Pronounce and Preview, Sec. 1–2
- Key Academic Vocabulary, Sec. 1
- Tiered Activities, Sec. 1–2

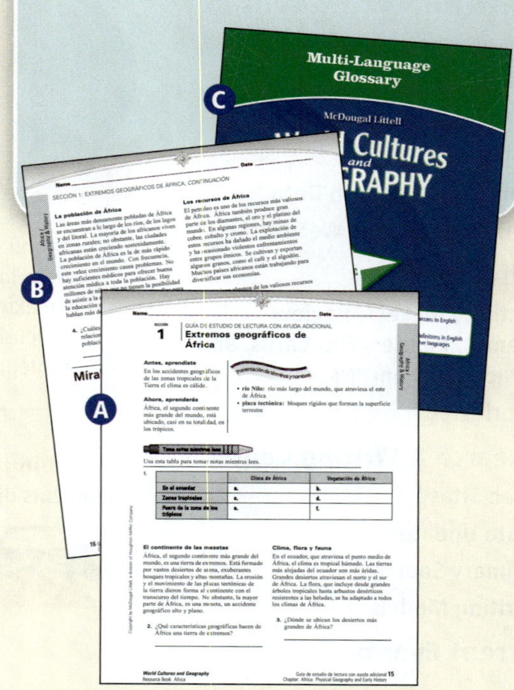

Inclusion

eEdition with Audio DVD-ROM

Lesson Plans
- Strategies for Inclusion

Unit Resource Book
- Reading Study Guide with Additional Support, pp. 7–10 A B
- Reteaching Activity, pp. 33–34
- Chapter Test (Form A), pp. 35–38

Best Practices Toolkit C
- Interactive Reading Strategies, pp. 90–145
- Differentiation Activity Bank: Struggling Readers, pp. 146–149

Teacher's Edition Activities
- Match Terms and Definitions, Sec. 1
- Records Terms and Names, Sec. 2
- Tiered Activities, Sec. 1–2

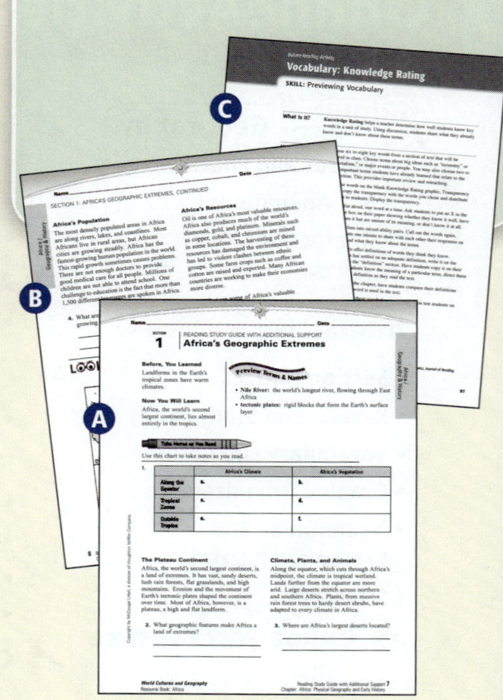

Interdisciplinary Activities

Gifted & Talented Students

Lesson Plans
- Core Lesson Plan

Unit Resource Book
- Outline Map Activities, pp. 23–24 (A)
- History Makers, p. 25 (B)
- Chapter Test (Form C), pp. 43–46

Best Practices Toolkit (C)
- Student Products Rubrics, pp. 1–14
- Writing Templates, pp. 78–89

Teacher's Edition Activities
- Compare Climate Disasters, Sec. 1
- Tiered Activities, Sec. 1–2

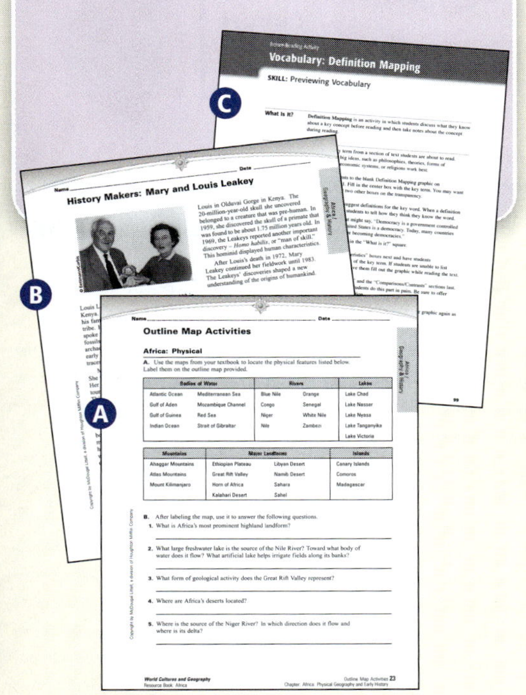

Pre-AP*

Lesson Plans
- Core Lesson Plan

Unit Resource Book
- Vocabulary Practice, p. 19
- Map and Graph Skills, pp. 21–22
- Interdisciplinary Projects, pp. 26–27 (A)
- Connect to History, p. 28 (B)
- Primary and Secondary Sources, p. 29 (C)
- Chapter Test (Form C), pp. 43–46

Best Practices Toolkit
- Writing Templates, pp. 78–89
- Differentiation Activity Bank: Pre-AP Students, pp. 195–200

* Pre-AP is a registered trademark of the College Entrance Examination Board, which was not involved in the production of and does not endorse this product.

Teacher's Edition Activities
- Report on Desertification, Sec. 1
- Tiered Activities, Sec. 1–2

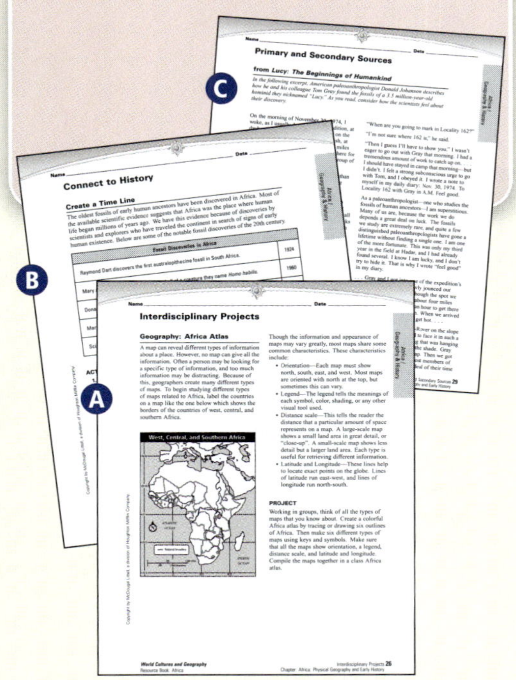

CONNECT to Language Arts

Poem About the Serengeti Plain

CLASS TIME 30 minutes

TASK Create a poem about Africa's Serengeti Plain.

SUPPLIES
- visuals depicting the Serengeti, such as the photograph in Section 1 of this chapter
- writing materials/computers

ACTIVITY Have students analyze images of the Serengeti and brainstorm a list of descriptive words that would help someone who has never seen the Serengeti visualize it Then have students use the images and their lists to create a descriptive poem about the Serengeti. Students may want to focus on one aspect of the Serengeti, such as a particular animal, or describe the plain as a whole. Ask volunteers to read their poems aloud or create a classroom display.

Good for Visual and Verbal Learners

CONNECT to Science

Planning an African Fossil Dig

CLASS TIME 30 minutes

TASK Create a plan for a fossil-digging expedition in Africa.

SUPPLIES
- resources on fossil digging in Africa
- writing materials/computers

ACTIVITY Have students work in pairs to create a plan for going on a two-week fossil dig in Africa. Plans should include a list of needed supplies, a route map to the digging site, and an intinerary, or schedule, of each day's events. Have all the students present their plans, then discuss them as a class. Afterward, have students revise their plans, based on the class discussion.

Good for Visual/Verbal and Logical/ Mathematical Learners

Introduce the ESSENTIAL QUESTION

- List the needs human beings must meet to survive in any environment.
- Describe the challenges human beings face in interacting with the extreme African environment.
- Explain how humans have adapted to the extreme African environment.

Key Ideas

Section 1: Africa, the world's second largest continent, lies almost entirely in the tropics.

Section 2: Human life first emerged on this warm continent.

CONNECT to the Essential Question

You may wish to revisit the Essential Question periodically with students as they work through the chapter.

Suggestions for reconnecting are provided at an appropriate point in the *Teacher's Edition* for each section.

More About . . .

Homo sapiens

Because we're so used to the plural-s construction in English, a common mistake is to try to construct a singular term *Homo sapien*. There is no such term. Although ending in "s," *Homo sapiens* is the singular, and can also be plural. *Homo*, meaning "man," is the noun in the Latin construction, while *sapiens*, meaning "wise," is the adjective.

Mount Kilimanjaro

CHAPTER 17

Africa

Physical Geography and Early History

1 GEOGRAPHY
Africa's Geographic Extremes

2 HISTORY
Human Beginnings in Africa

ESSENTIAL QUESTION

How do humans interact with the extreme African environment?

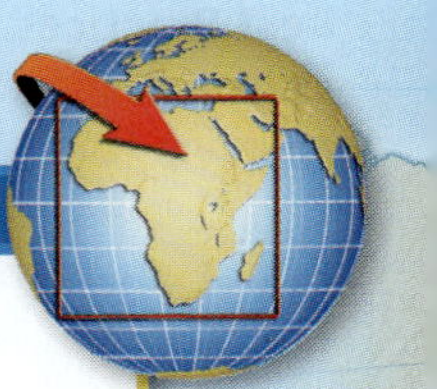

CONNECT Geography & History

Use the map and the time line to answer the following questions.

1. The highest mountain in Africa is closest to what coast?
2. From what locations might it have been easiest for *Homo erectus* to migrate out of Africa?

8,000,000 B.C.

History
6–7 Million Years Ago
First hominids appear.

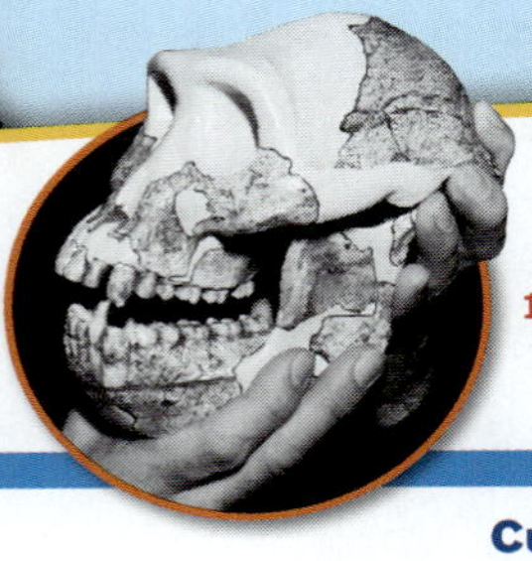

History
1.8 Million Years Ago
Homo erectus begins to migrate out of Africa.

Culture
2.5 Million Years Ago
Hominids develop first tools, beginning the Stone Age.

TIME LINE DISCUSSION

To help students appreciate the time span of early humans, have students compare how long the United States has been an independent country to when the first hominids appeared. *(about 230 years ago to about 7,000,000 years ago)*

- Ask students to name the different terms used for human ancestors as they developed and the time periods that correspond to these terms. *(hominids—6 to 7 million years ago; homo erectus—1.8 million years ago; homo sapiens—195,000 years ago)*
- When do human ancestors begin to migrate out of Africa? What is the likeliest place to which they migrated? Why? *(1.8 million years ago; Asia; because Asia is connected by land to Africa)*
- How did the development of farming along the Nile River in Egypt make it possible for an advanced civilization to develop there? *(Possible Answer: Farming ensured a stable society that could thrive and organize itself into a more complex society.)*

Animated GEOGRAPHY

Physical Geography of Africa

Click here to explore Africa @ClassZone.com

History

195,000 Years Ago
Earliest known *Homo sapiens* appears in Ethiopia.

Geography

3100 B.C. Upper Egypt and Lower Egypt are united. ▸

2000 B.C.

Culture

◂ **4000 B.C.** Early sickle blades are used for farming in Egypt.

535

TEACHER MEDIA FAVORITES

Books for the Teacher

Iliffe, John. *Africans: The History of a Continent*. Cambridge University Press, 1995. Engaging survey from a leading African historian.

Reader, John. *Africa: A Biography of the Continent*. Vintage, 1999. Narrative by a top photojournalist.

Video/DVD

National Geographic—Africa. National Geographic Video, 2001 (DVD). Eight-hour series for public television that took two years to film, presenting Africa as seen by its people.

Wonders of the African World. PBS, 2003. Exploration of the cultures, traditions, and history of Africa with African-American scholar Henry Louis Gates, Jr.

Internet

Go to the *Research and Writing Center* at **ClassZone.com** to link to:

- data on the region
- a variety of chapter-specific content-reviewed sites
- primary sources

Animated GEOGRAPHY

Physical Geography of Africa

CLASSZONE.COM

Africa is the continent on which the human species originated. From there, humans migrated to Asia, which is connected to Africa at its northeast region. Africa's physical features, such as vast deserts, have presented challenges to its human population throughout human history.

- What geographical features to the east and west of the Nile River protected the ancient Egyptian civilization from invaders? *(deserts)*
- Where and when do the first *homo sapiens* appear? Is this region in a high or low elevation? *(195,000 years ago, in Ethiopia; high)*
- What conclusions might you draw about the civilizations in Upper Egypt and Lower Egypt before they united in 3100 B.C.? *(Possible Answer: that they had been developing for thousands of years before being ready to unite)*

CONNECT GEOGRAPHY & HISTORY ANSWERS

1. the east coast, or Indian Ocean coast
2. the northeast, by the land connection to Asia

Presentation Options

Use the **Power Presentations** DVD-ROM to project the animated map.

- Prompt students to use the pop-up images and audio from the map to discover what they already know about the geography, culture, history, economics, or governments of the region.
- Challenge students to answer the map questions individually, in small groups, or as a class activity. Use all of the interactive material to help preview chapter content.

1 Plan & Prepare

Objectives

- Identify Africa's major features
- Describe Africa's climates and the kinds of plant and animal life they support
- Summarize the challenges faced by Africa's population
- Explain the importance of natural resources to the people of Africa

Read for the Essential Question

Help students read for a purpose by reminding them of the Essential Question: "How do humans interact with the extreme African environment?" Point out that they can use text features, such as Key Questions, special features, images, and maps to form answers to the question.

Vocabulary

Best Practices Toolkit

Use the ***Best Practices Toolkit*** to model strategies for vocabulary notetaking. Vary strategies throughout the year. Choose from: *Knowledge Rating, Predicting ABCs, Definition Mapping, Word Sort, Frayer Model (Word Squares), Magnet Words, Word Wheel,* and *Student VOC.*

Vocabulary Strategies, TT9–TT16

Reading Strategy

Best Practices Toolkit

Display the Three-Column Chart transparency.

- Explain to students that to summarize information means to reinforce key points and main ideas.
- Model filling in a few summary statements on the chart using the Climate, Plants, and Animals subsection.
- Have students suggest entries to the chart. Encourage students to expand their charts to include summaries of other subsections within the chapter.

Three-Column Chart, TT21

Section 1 Reading for Understanding

Key Ideas

BEFORE, YOU LEARNED

Landforms in the Earth's tropical zones have warm climates.

NOW YOU WILL LEARN

Africa, the world's second largest continent, lies almost entirely in the tropics.

Vocabulary

TERMS & NAMES

basin a region drained by a river system

escarpment a steep slope separating two flat areas of different heights

Mount Kilimanjaro the highest mountain in Africa, in northeast Tanzania (TAN•zuh•NEE•uh)

Sahara the world's largest desert

Victoria Falls a 355-foot waterfall on the Zambezi River in South Central Africa

Nile River the longest river in the world, which empties into the Mediterranean Sea in northeast Egypt

canopy the highest layer in a forest, formed by the treetops

Sahel (suh•HAYL) a semiarid region just south of the Sahara

desertification the change of fertile land to desert

REVIEW

tectonic plates a large rigid section of the Earth's crust that is in constant motion

savanna a grassland in the tropics or subtropics

Visual Vocabulary Sahara

Reading Strategy

Re-create the chart shown at right. As you read and respond to the **KEY QUESTIONS**, use the chart to summarize information about the physical geography of Africa.

Skillbuilder Handbook, page R5

SUMMARIZE

	AFRICA'S CLIMATE	AFRICA'S VEGETATION
Along Equator		
Tropical Zones		
Outside Tropics		

GRAPHIC ORGANIZERS
Go to Interactive Review @ClassZone.com

PRETEACHING VOCABULARY

English Learners

Pronounce and Preview

Pronounce each term for students. Review the meanings of words within the definitions, such as *drained*, *steep*, and *semiarid*.

- To modify vocabulary learning, have students complete worksheets as they read, instead of afterward.

Unit Resource Book: Africa
- Vocabulary Practice, p. 19

Inclusion

Match Terms and Definitions

Have pairs of students write each term on an index card and its definition on another index card. Then have one student shuffle the cards and arrange them. After the partner finishes matching the terms and definitions, have the students work together to create complete sentences with the terms.

Victoria Falls and Zambezi River Gorge, Zimbabwe

Lions, East Africa

Africa's Geographic Extremes

Connecting to Your World

Africa, the world's second largest continent, is a land of geographic extremes. Parts of the vast Sahara may not see a drop of rain for years. In contrast, areas in Africa's rain forest region may receive more than five feet of rain a year. On Africa's savanna, herds of zebras, gazelles, and elephants graze, hunted by lions and leopards. They are just a few of Africa's thousands of animal and plant species.

Lion Most African lions, like this one in Kenya, now live in protected wildlife reserves.

The Plateau Continent

KEY QUESTION What are some of Africa's geographic features?

Most of Africa is a plateau, a high and flat landform. The African plateau overlooks a thin strip of coastline. Several places on the plateau contain bowl-shaped areas called **basins**. The Niger River flows through the Niger Basin, the Sudan River flows through the Sudan Basin, and the Congo River flows through the Congo Basin. In the Chad Basin in West Central Africa, water collected to form Africa's fourth largest lake, Lake Chad.

Africa's Serengeti Wildebeests and zebras graze on the Serengeti Plain in Tanzania.

2 Focus & Motivate

3-Minute Warm-Up

Write on the board or display the transparency:

- Look at the photograph of Africa's Serengeti in Section 1. What might life be like for people who live there? *(Possible Answer: They might rely on hunting and live nomadic lives.)*

Unit Transparency Book: Africa
- 3-Minute Warm-Ups, TT1

Connecting to Your World

More About . . .

The Sahara

Although much of the Sahara is now dry, it was not always so. Between about 8,000 and 3,000 B.C. most of the Sahara received enough rain to support a savanna environment.

3 Teach

The Plateau Continent

Talk About It

- What caused Africa's Great Rift to form? *(Sections of the African tectonic plate are slowly pulling apart.)*
- **Summarize** What are Africa's major geographic features? *(plateaus, river basins, Victoria Falls, the Great Escarpment, the Sahara, Mount Kilimanjaro, the Great Rift, and the Nile)*

SECTION 1 PROGRAM RESOURCES

Unit Resource Book: Africa

ON LEVEL

Unit Resource Book
- Reading Study Guide, p. 3
- Map and Graph Skills, p. 21
- Section Quiz, p. 31

STRUGGLING READERS

Unit Resource Book
- Reading Study Guide with Additional Support, p. 7
- Skillbuilder Practice: Drawing Conclusions, p. 20
- Section Quiz, p. 31
- Reteaching Activity, p. 33

eEdition with Audio DVD-ROM

ENGLISH LEARNERS

Pupil Edition in Spanish

eEdition with Audio DVD-ROM

Unit Resource Book
- RSG (Spanish), p. 11
- RSG with Additional Support (Spanish), p. 15

Multi-Language Glossary

Test Generator
- Section Quiz in Spanish

INCLUSION

Unit Resource Book
- RSG with Additional Support, p. 7
- Section Quiz, p. 31
- Reteaching Activity, p. 33

GIFTED & TALENTED

Unit Resource Book
- Outline Map Activities, p. 23
- Section Quiz, p. 31

PRE-AP

Unit Resource Book
- Interdisciplinary Projects, pp. 26, 27
- Section Quiz, p. 31

TECHNOLOGY

Unit Transparency Book
- 3-Minute Warm-Ups, TT1
- Essential Question, TT2
- Critical Thinking, TT3
- Map: Africa: Physical, TT5

Daily Test Practice Transparencies
- Section 1, TT59

Power Presentations

ClassZone.com

World Cultures and Geography Video on DVD

More About . . .

The Great Rift

The Afar Triangle is an area near the Horn of Africa, along the Great Rift in Ethiopia just south of the Red Sea. In 2005, geologists studying the area were amazed to see the earth cracking and trembling beneath their feet. Recently, hundreds of crevices have opened up in the area, and in some places the earth has sunk about 328 feet (about 100 meters). Fumes as hot as 752°F (400°C) are being released, along with magma. These geologists believe this area will continue to split apart and sink until it is eventually flooded by the Red Sea—cutting off the Horn of Africa from the rest of the continent.

Comparing Regions

Volcanoes of Eastern Africa and the Pacific Northwest

In eastern Africa a tectonic plate is being pulled apart, allowing magma to build volcanoes such as Mount Kilimanjaro (now dormant). In the Pacific Northwest, a tectonic plate off the coast of Washington and Oregon is pushing eastward under the edge of the continent, causing it to fold. One of these "folds" is the Cascade Mountains, which contain active volcanoes such as Mount Saint Helens and Mount Rainier.

Congo River from Space

Point out that the area shown in this satellite image is part of the Congo Basin.

- Why would this area be a good location for the two capital cities shown? *(Possible Answer: The river provides a source of water and helps keep land fertile, both factors valuable for farming.)*

Distinctive Landforms Slow, dramatic forces shaped Africa. Erosion gradually molded southern Africa's **Great Escarpment**, a slope separating the high plateau of the interior from the narrow eastern coast.

Earlier you learned that **tectonic plates** move inside the earth. In eastern Africa, sections of the African Plate are slowly pulling apart. This change, called divergence, opened the Great Rift, a series of valleys running 4,000 miles down most of the continent. Long, deep lakes formed in the Great Rift, including Lake Nyasa and Lake Tanganyika. However, Lake Victoria, Africa's largest lake, lies in a shallow basin between two rift valleys. The Earth's plates pushing together produced Africa's volcanic mountains. These include Mount Kenya, Mount Cameroon, and Africa's highest mountain, **Mount Kilimanjaro**.

The **Sahara**, the world's largest desert, stretches across North Africa. Sahara is the Arabic word for "desert." Its total area is roughly equal to that of the continental United States. Other African deserts include the Kalahari and Namib, both located in southern Africa.

forests ▼

Congo River This satellite photo shows Kinshasa and Brazzaville, two capital cities on opposite banks of the Congo River. The city areas are the lighter areas in the photo. **What might the dark green areas represent?**

Rivers of Africa Most of Africa's rivers have rapids and waterfalls, making them hard to navigate. The Congo River winds through Central Africa. Many tributaries feed into the Congo, making the Congo river system the largest in Africa. While much of the upper Congo is navigable, 32 high waterfalls dot the river near its mouth. People who live near **Victoria Falls**, a 355-foot waterfall on southern Africa's Zambezi River, call it Mosi-oa-Tunya (MOH•see•oh•ah•TOON•yah), "the smoke that thunders," because of its roaring mist.

Congo River from Space

DIFFERENTIATING INSTRUCTION

English Learners

Key Academic Vocabulary

Point out the words *series* (second paragraph, third line) and *area* (third paragraph, second line) from the Academic Word List. Discuss the meaning of these words and have pairs of students create new sentences for each word.

Struggling Readers

Use a Map

Read aloud the subsection called "Distinctive Landforms," pausing to have students find the locations mentioned on the map in the chapter opener ("Physical Geography of Africa"). Discuss the forces (erosion, divergence, pushing together) that molded the various locations.

COMPARING Landforms

Great Rift Tectonic plates pulling apart created Africa's Great Rift. **1. What is this tectonic process called?**

Mount Kilimanjaro Tectonic shifts also created volcanic mountains, such as Tanzania's Kilimanjaro. **2. How can plate movement create volcanic mountains?**

The world's longest river, the **Nile**, flows north through northeast Africa for more than 4,000 miles. For thousands of years, people have depended on the Nile's annual floods to enrich their soil for farming. But for most of that time, no one knew where the river began. The ancient Greek geographer Ptolemy guessed the Nile's source was in a snowy area he called the "Mountains of the Moon." In the 1870s explorers found that Lake Victoria was the main source of the Nile.

SUMMARIZE Describe the key geographic features in Africa. wide plateaus, deep rift valleys, and the world's largest desert and longest river

Climate, Plants, and Animals

KEY QUESTION How does Africa's location shape its climate?

The Equator divides the Earth into two hemispheres, and also runs right through Africa's midpoint. Turn the page to see climate and vegetation maps of Africa.

A Warm Continent Along the Equator, the climate is tropical wetland. Bands of tropical wet and dry climate border these wetlands. Next come strips of semiarid climate, and then deserts. In the southeast part of the continent lies a small area of humid subtropical and marine west coast climates. Areas of higher elevation have a highland climate, with rainy winters, dry summers, and typically cooler temperatures than tropical and semiarid climates.

INTERDISCIPLINARY ACTIVITIES

CONNECT to Science

Create Visuals of the Nile

Have students work in pairs to research the effects of the flooding of the Nile in ancient times and today. Students should prepare a visual, such as a poster or chart, showing their findings. Students' visuals should include maps and other images, including ones of artifacts of ancient Egyptian civilizations that flourished on the Nile and of life on the Nile today.

CONNECT to Language Arts

Write a Poem

Have students write a poem on one of the distinctive landforms or major rivers of Africa. Poems should take as their focal point some fact about the landform or river, such as how it was created, its location, or its size. Have volunteers read their poems or create a classroom display.

Teach

Climate, Plants, and Animals

Think, Pair, Share

- What do plants that live in the desert need to survive? *(long root systems that can reach underground water sources)*
- How does the climate at higher elevations differ from areas with a tropical climate? *(Higher elevations have a highland climate, with rainy winters, dry summers, and cooler temperatures than tropical climates.)*
- **Summarize** What are some of the things people have done that have contributed to desertification of the Sahel region? *(farming, overgrazing of livestock, and collecting wood for fuel)*

COMPARING

Landforms

The Great Rift escarpments, shown on the left, are in some places as high as 9,000 feet (about 2,700 meters) above the valley floor. The Great Rift valley will likely continue to get deeper and wider, until it is some day flooded with sea water. On the right, Mount Kilimanjaro, now dormant, was created when magma, always close to the surface under the Great Rift valley, broke through the surface about 750,000 years ago.

ANSWERS

1. **Great Rift** divergence
2. **Mount Kilimanjaro** Tectonic plates can push together to form mountains, or pull apart. Either movement can form volcanoes by bringing magma to the surface.

Fun Facts!

Extreme Insects

The African Goliath Beetle may be a heavyweight, but the longest insect in the world is the Stick Insect. There are about 3,000 stick insect species in the world. Many of them can fly. The Giant African Stick Insect is the longest in Africa, measuring as much as 7.25 inches long (185 millimeters). The longest in the world is found on the island of Borneo, reaching a length of about 13 inches (about 328 millimeters).

CONNECT Geography & Science

Desertification

Remind students that human activity can contribute to desertification, but that humans can also take action to reverse the process. Ask students to think of things people can do to reclaim land that has turned into desert. *(Possible Answers: planting trees, irrigating the land, planting crops that are good at holding soil and moisture, letting land renew its nutrients)*

CRITICAL THINKING ANSWER

Draw Conclusions Possible Answer: People may have to abandon raising livestock and farming, and eventually may have to relocate to support themselves.

Make It Fun

Teacher-Tested Activities

Terry Rhodes, McNair Middle School, Fayetteville, Arkansas

Section 1 talks about the different climate and vegetation regions in Africa. Have students select a region in Africa, and research its features and climate. Then have them make a postcard of what they think a day in the region might be like. Have them draw a picture on the front and write to a classmate on the back.

Fun Facts!

EXTREME INSECTS

The Goliath Beetle, found in Africa's rain forest, is one of the heaviest insects in the world. It is about five inches long and can weigh up to 100 grams, or 3.5 ounces. Goliath beetles are harmless to humans, feeding on dead plant matter, other insects, and animal waste.

Africa's vegetation mirrors the climate, with rain forests **A** (on the map, opposite) along the Equator. Areas of grasslands and woodlands, called **savannas** **B**, are in the wet, dry, and semiarid regions. Scattered shrubs and some grasses grow in the desert **C**. The plants most able to survive in the desert have long root systems that can reach underground water sources. Mediterranean vegetation grows along Africa's northern coast and southern tip. In those areas, native plants include shrubs, small trees, and grasses.

Animal and Plant Life Ferns and mosses make up the rain forest's ground floor, below a middle story of trees and palms. The top layer, towering to about 150 feet, is called the **canopy**. There, leaf-filled branches sprout dazzling orchids. The canopy is also home to fruit-eating parrots and pigeons, seed-eating beetles, and gorillas and chimpanzees.

Grazing animals feed on the grasses and trees of the savanna. Bordering both the savanna and the Sahara Desert is a semiarid region called the **Sahel** (suh•HAYL), from the Arabic word for "shore." But **desertification**, a process in which fertile land becomes desert, is turning the Sahel into the Sahara. The illustrations below show how human activity is speeding desertification.

DRAW CONCLUSIONS Describe how Africa's location affects its vegetation. In Africa's desert, scattered shrubs grow. The rain forests are home to towering trees, including fruit trees; ferns, mosses, and orchids.

CONNECT Geography & Science

Desertification Climate change expanded the Sahara in ancient times. Today, human activity is speeding up the process.

1. The Sahel receives very little rainfall. The vegetation lives in a fragile state.
2. Farming, overgrazing by livestock, and the collection of wood for fuel can erode and expose soil.
3. Vegetation gradually disappears. The remaining plants do not cover the soil as well as the grass did.
4. With less vegetation covering the fertile topsoil, the soil eventually dries up and blows away.

CRITICAL THINKING

Draw Conclusions What impact might desertification have on people living in the affected areas?

DIFFERENTIATING INSTRUCTION

Gifted & Talented

Compare Climate Disasters

Have students create a visual, showing a comparison of conditions causing desertification to those that caused the dustbowl in the United States during the Great Depression. The visual should show comparisons of causes, impact on humans, and solutions for making the soil fertile again.

Pre-AP

Report on Desertification

Have students work in pairs or small groups to research information about desertification in a specific country of Africa. Information should include maps, photos, and notes on the problems that desertification causes and the measures being taken to solve them. One student from each pair or group should present the research findings to the class.

Animated GEOGRAPHY
Vegetation Regions of Africa

Click here to see regions of Africa @ClassZone.com

Climates of Africa

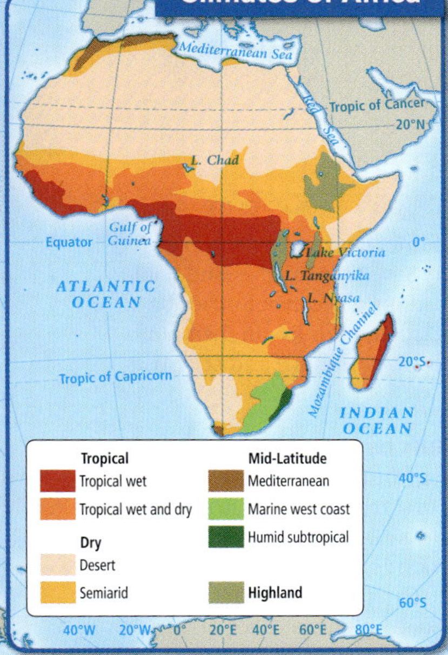

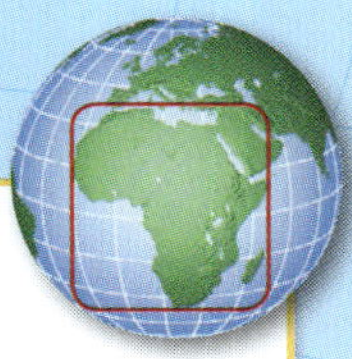

CONNECT Geography & Culture

READING A VEGETATION MAP

Africa's vegetation regions closely mirror the continent's climate zones. As you study Africa's vegetation,

- compare the two maps shown
- note how the climate and vegetation bands are similar
- think about how these vegetation zones affect human activity

1. **Location** What type of vegetation zone lies between Africa's northern coast and the rest of the continent?
2. **Human-Environment Interaction** Which vegetation zones would best support livestock?

541

Animated GEOGRAPHY
Vegetation Regions of Africa
CLASSZONE.COM

CONNECT Geography & Culture

Reading a Vegetation Map Ask students to use the map key to help them identify the largest and smallest vegetation regions. *(Largest: tropical grassland; smallest: Mediterranean shrub)*

ANSWERS

1. **Location** desert and dry shrub
2. **Human-Environment Interaction** tropical grassland and temperate grassland

Unit Resource Book: Africa
- Outline Map Activities, p. 23

Presentation Options

Use the **Power Presentations** DVD-ROM to project the animated map.

- Click to animate the map.
- Examine the photographs to learn about the vegetation regions of Africa.

Climates of Africa

Point out to students how varied the climates of Africa are, and how different regions have the same climate. For example, the northwestern tip of Africa, nearest Europe, has a Mediterranean climate, as does the southwestern tip.

- What is the largest climate zone in Africa? *(desert)*
- What is the climate of eastern Madagascar? *(tropical wet)*

Unit Resource Book: Africa
- Interdisciplinary Projects: Africa Atlas, pp. 26–27

DIFFERENTIATING INSTRUCTION: TIERED ACTIVITIES

OBJECTIVE Compare the vegetation regions map of Africa with the climates map.

Basic

Have students work in pairs to compare the two maps shown on this page. Ask them to prepare complete sentences about the climate and vegetation around these geographical features: Nile River, Mediterranean Sea, Congo River.

On Level

Have small groups of students choose two climate zones and research one country in each of them. Have them prepare reports on the climate and vegetation of each country, noting the impact they have on the country's economy and population.

Challenge

Have pairs of students research the economies of the three richest and three poorest African countries and prepare a report describing their climates and vegetation. Have students present their reports to the class, focusing on the relationship between each country's prosperity and its climate and vegetation.

Teach

Africa's Population

Reader, Recorder, Reporter

- Imagine that there were 200 different languages spoken in your country. How might that affect daily life? *(Possible Answers: Newspapers and books would be printed in many different languages; you would only be able to understand some television and radio stations; people might have difficulty traveling away from their home region.)*
- Why do you think Africa is most densely populated along rivers, lakes, and coastlines? *(Possible Answer: to be close to water for drinking, irrigation, and transportation)*
- **Summarize** What are some challenges of Africa's fast-growing population? *(There are few doctors to care for the sick; many children cannot go to school; not enough teachers.)*

Unit Resource Book: Africa
- Skillbuilder Practice: Drawing Conclusions, p. 20

More About . . .

Second Languages in Africa

Non-African languages such as French, English, and Arabic are official languages in some African countries, used in government and for business. Some African languages are used for the same purposes. While English is the official language of Zambia, about one third of Zambians say Bemba is their first language. Swahili is a language that is used widely in East Africa for trade.

Teach

Africa's Resources

Think, Pair, Share

- Why have African countries established national parks? *(to protect wildlife from poachers and attract tourists)*
- **Summarize** Why does the United Nations want to keep "conflict diamonds" off the world market? *(because they are mined with slave labor and are used to fund civil wars)*

Unit Resource Book: Africa
- Map and Graph Skills: African Land Use, p. 21

Africa's Population

KEY QUESTION What challenges does Africa's growing population face?

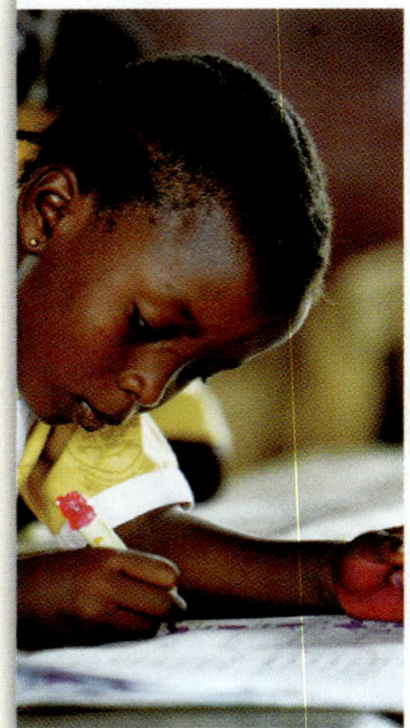

Schoolgirl in Ghana
Although school is free for children under 16 in Ghana, nearly 25 percent of children in the country have never attended school.

Population distribution in Africa varies dramatically. Large areas of the Sahara, for example, have no people at all. But more than 95 percent of Egypt's population live along the Nile Valley, making it one of the most densely populated places on Earth. The most densely populated areas in Africa are along rivers, lakes, and coastlines. Although most Africans live in rural areas, African cities are growing steadily.

Africa's population is the second-largest, but fastest-growing, in the world. This rapid growth sometimes causes problems. Most countries have few doctors to care for the sick. In addition, the United Nations estimates that about 40 million children south of the Sahara do not attend primary school. African nations and the UN are working to improve education by training more teachers and making school more affordable.

More than 1,500 languages are spoken in Africa. The Democratic Republic of the Congo, in Central Africa, has more than 200 languages and dialects just within its own borders. Many Africans also use a second language to ease communication. For example, French is a language commonly spoken in some West and Central African countries. The language remains in Africa from the years when France held colonies in the region.

FIND MAIN IDEAS Describe some of the challenges for Africa's growing population. scarce health care, poor education, overcrowding in densely-populated areas

Africa's Resources

KEY QUESTION What are some of Africa's natural resources?

Africa's savannas once teemed with wildlife. But hunters and poachers killed animals for game, and farmers have turned grassland habitats into farmland. To halt this decline, many African countries have established national parks. There, protected animals live in their natural habitats, tourists can take pictures of the animals, and scientists can study their behavior.

Energy and Other Mineral Resources Oil is one of Africa's most valuable resources. Oil-producing countries include Algeria, Kenya, Libya, Egypt, Angola, and Nigeria. Africa also produces about 40 percent of the world's diamonds and more than half of its gold. South Africa has abundant gold, diamonds, and platinum. Other diamond-producing countries include Ghana and Namibia. Zambia produces copper and cobalt, while Zimbabwe mines chromium.

DIFFERENTIATING INSTRUCTION: TIERED ACTIVITIES

OBJECTIVE Present information on the natural resources of African countries.

Basic	On Level	Challenge
Have pairs of students use the information in their textbooks and classroom resources to create a chart of 10 African countries of their choice and their resources.	Have small groups of students do additional research on the countries in Africa mentioned in their textbooks and create a chart, ranking them in order of greatest producers of oil, diamonds, gold, platinum, copper, cobalt, and chromium.	Have students work in pairs to create a report on African countries, such as Nigeria, that are engaged in conflict over wealth from natural resources. Have them present their reports to the class.

Are Resources Enough? Even the richest mineral resources do not guarantee economic success. For example, the oil that enriches Nigeria has also led to conflicts between ethnic groups over who profits from the resources. Oil production also brings natural gas deposits to the surface, which are often burned off. This "gas flaring" wastes an energy source and harms the environment. Oil companies are now looking for ways to better use natural gas resources. In Angola, rebel groups have mined diamonds, possibly using slave labor, and have sold them to fund civil wars. The United Nations and other groups are working to keep these "conflict diamonds" off the world market.

African countries that export farm crops usually concentrate on one or two crops, a practice put in place during Africa's colonial era. For example, Ethiopia and Uganda export coffee, while Chad exports cotton. But, as with oil wealth, relying on any single crop puts a country at risk when prices fall or resources run out. Many countries are working to make their economies more diverse.

Drying Food These Nigerian women are using the heat from gas flares to dry food on round trays. However, the flaring releases pollutants into the air.

SUMMARIZE Identify some of Africa's key resources. oil, gold, diamonds, copper, chromium, natural gas

Section 1 Assessment

ONLINE QUIZ For test practice, go to **Interactive Review** @ClassZone.com

TERMS & NAMES

1. **Explain the importance of**
   - basin
   - savanna
   - Sahel
   - desertification

USE YOUR READING NOTES

2. **Summarize** Use your completed chart to answer the following question:

   How does Africa's vegetation reflect the climate it grows in?

	AFRICA'S CLIMATE	AFRICA'S VEGETATION
Along Equator		
Tropical Zones		
Outside Tropics		

KEY IDEAS

3. The Nile is the world's longest river, but the Congo river system is Africa's longest. How is this difference possible?
4. What process formed the Great Escarpment?
5. What have been some of the problems associated with Africa's oil resources?

CRITICAL THINKING

6. **Analyze Causes and Effects** How have plate tectonics shaped Africa's landscape?
7. **Identify Problems and Solutions** Why is it risky for a country to depend on a single resource?
8. **CONNECT to Today** How might increasing oil consumption in the United States affect African oil-exporting countries?
9. **ART Tourism Brochure** Create a brochure for an African national park. Include information on the animals and habitats people will see, and explain how the park will protect those animals and habitats.

CONNECT to the Essential Question

How do humans interact with the extreme African environment?

Ask students what they have learned so far that can help them answer this question. Students might mention:

- People have tended to settle near sources of water, such as rivers, lakes, and coasts.
- Countries have established national parks to protect habitats, but desertification of the Sahel has resulted from farming, overgrazing, and collecting wood.

4 Assess & Reteach

Assess Have students complete the Section Assessment.

Unit Resource Book: Africa
- Section Quiz, p. 31

Interactive Review @ClassZone.com
- Section Quiz

Power Presentations

Test Generator

Reteach Assign students to ten groups, one for each of the nine subsections and one for the Animated Geography map. Have them make flash cards for all of the vocabulary terms that fall within their section, plus any other term or facts that they feel are important to know. Have them pair off with other groups and quiz each other with the flash cards they have created.

Unit Resource Book: Africa
- Reteaching Activity, p. 33

SECTION 1 ASSESSMENT ANSWERS

Terms & Names

1. basin: region drained by a river system; savanna: grassland in the tropics or subtropics; Sahel: semiarid region just south of the Sahara; desertification: change of fertile land to desert

Use Your Reading Notes

2. Climate and vegetation mirror each other closely.

	AFRICA'S CLIMATE	AFRICA'S VEGETATION
Along Equator	tropical wet	rain forest, grassland
Tropical Zones	tropical wet/dry	tropical grassland
Outside Tropics	desert, semiarid	desert, mediterranean

Key Ideas

3. Smaller tributaries feed into the Congo, creating the river system. The Nile is a single river.
4. erosion
5. Oil wealth has led to conflicts between ethnic groups, and pollution, such as gas flaring, which is a byproduct of oil production.

Critical Thinking

6. Earthquakes formed Africa's Great Rift and volcanoes formed many of Africa's mountains.
7. Fluctuations in the market price for that resource could put a country's economy in jeopardy.
8. It would bring them increased wealth, but it might also increase pollution.
9. Use the rubric to score brochures.

Brochure Rubric

	Points Addressed	Content	Accuracy
4	5 or more	clear, concise	no errors
3	4	good	few/minor
2	3	fair	several
1	2 or fewer	poor	many

1 Plan & Prepare

Objectives

- Identify the stages of the development of human ancestors in Africa
- Explain the structure of society in traditional African cultures

Read for the Essential Question

Help students read for a purpose by reminding them of the Essential Question: "How do humans interact with the extreme African environment?" Point out that they can use text features, such as Key Questions, special features, images, and maps, to form answers to the question.

Vocabulary

Best Practices Toolkit

Use the ***Best Practices Toolkit*** to model strategies for vocabulary notetaking. Vary strategies throughout the year. Choose from: *Knowledge Rating, Predicting ABCs, Definition Mapping, Word Sort, Frayer Model (Word Squares), Magnet Words, Word Wheel,* and *Student VOC.*

Vocabulary Strategies, TT9–TT16

Reading Strategy

Best Practices Toolkit

Display the Web Diagram transparency.

- Remind students that main ideas are important ideas in a block of text and that details support main ideas.
- Model filling in main ideas and details using the section subheads and text.
- Have students suggest entries. Remind students that the main ideas in each block of text in the section will help answer a Key Question.

Web Diagram, TT28

SECTION 2 Reading for Understanding

Key Ideas

BEFORE, YOU LEARNED

Africa is home to thousands of plant and animal species, and has a wealth of natural resources.

NOW YOU WILL LEARN

Human life first emerged on this warm continent.

Vocabulary

TERMS & NAMES

hominid a human primate; an early human

bipedal walking on two feet

Stone Age the earliest period of human culture, in which people used stone tools

animism the belief that spirits exist in animals, plants, natural forces, and ancestors

griot (gree•OH) a West African storyteller

stateless societies groups that have no single authority, but make decisions based on compromise

lineage descent from an ancestor

REVIEW

fossil remains of early life preserved in the ground

culture the shared attitudes, knowledge, and behaviors of a group

BACKGROUND VOCABULARY

ancestor an early relative

domesticate to adapt or train plants or animals for human use

patrilineal descent from father to father

Reading Strategy

Re-create the web diagram shown at right. As you read and respond to the **KEY QUESTIONS**, use the diagram to record the main ideas about African societies.

Skillbuilder Handbook, page R4

FIND MAIN IDEAS

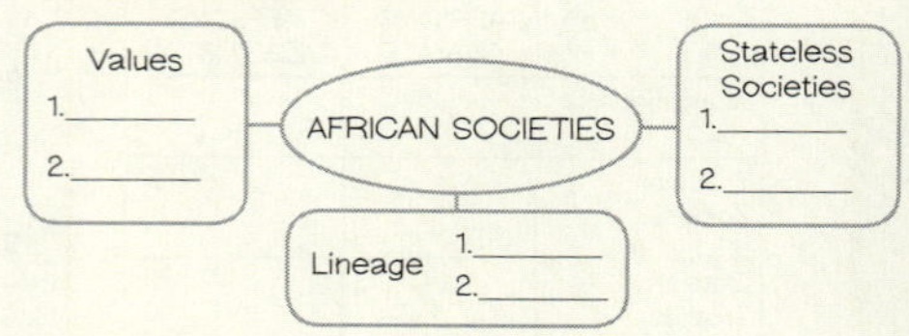

GRAPHIC ORGANIZERS
Go to **Interactive Review** @ClassZone.com

544 Chapter 17

PRETEACHING VOCABULARY

English Learners

Pronounce and Preview

Pronounce each term for students. Review the words within definitions, such as *primate, compromise,* and *descent.*

- To modify vocabulary learning, have students complete worksheets as they read, instead of afterward.

Unit Resource Book: Africa
- Vocabulary Practice, p. 19

Inclusion

Record Terms and Names

Have students read each term and its definition into an audio cassette recorder. As students play back their recordings, have them follow along with the words in their textbooks. Then have students use each term in a complete sentence.

Victoria Falls and Zambezi River Gorge, Zimbabwe

Lions, East Africa

SECTION 2 HISTORY

Human Beginnings in Africa

Connecting to Your World

Have you ever made an exciting discovery? Perhaps you found a treasure on the street or learned something that made you decide to become a drummer, dancer, hockey player, or scientist.

If you're really lucky, you may make a find that changes the way people think. Anthropologist Donald Johanson had that kind of luck in Ethiopia in 1974 when he and his assistants took a new route back to their vehicle. They stumbled on the bones of a **fossil**, which they named Lucy after a popular song. Lucy was an early human, called a **hominid**. She lived about 3.2 million years ago.

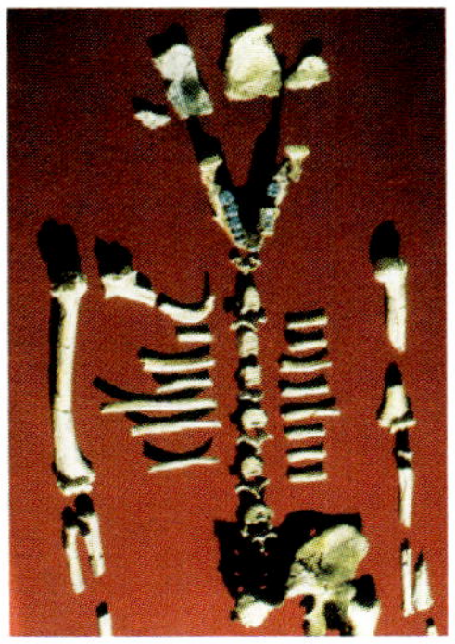

Lucy This early human is the most-studied fossil to date.

Human Ancestors

KEY QUESTION What were some of the important hominid finds in Africa?

Lucy, an adult female, probably stood about three and a half feet tall and weighed a little over 60 pounds. She and other hominids were **bipedal**, meaning they walked on two feet. If early hominids used tools, the tools have since disappeared. The first tools that scientists have found—stones to cut meat—date back a mere 2.5 million years.

Animated GEOGRAPHY

Archaeological Dig in East Africa

Scientists continue to look to Africa for clues about early human development.

Click here to see fossils found in East Africa @ClassZone.com

545

2 Focus & Motivate

3-Minute Warm-Up

Write on the board or display the transparency:

- What were some of the important discoveries made about early humans? *(Possible Answers: language, making and using tools, controlling fire, walking upright)*

Unit Transparency Book: Africa
- 3-Minute Warm-Ups, TT1

Connecting to Your World

More About . . .

Hunting for Fossils in Africa

Although Africa may be better known for its human fossil finds, scientists are also digging for other—and older—fossils that date from about 200 million to 300 million years ago, before the supercontinent known as Pangea broke apart.

3 Teach

Human Ancestors

Talk About It

- Is the use of stone tools enough to define a species as human? *(Yes: Only humans have the intelligence to use tools, so any species using them must be a direct ancestor; No: Species not in our direct line, such as chimpanzees, have also been observed using basic tools.)*
- **Find Main Ideas** What are the traits that scientists believe indicate that a fossil is an ancestor to modern humans? *(walking on two feet, using tools, using language)*

SECTION 2 PROGRAM RESOURCES

Unit Resource Book: Africa

ON LEVEL

Unit Resource Book
- Reading Study Guide, p. 5
- Section Quiz, p. 32

STRUGGLING READERS

Unit Resource Book
- Reading Study Guide with Additional Support, p. 9
- Skillbuilder Practice: Drawing Conclusions, p. 20
- Section Quiz, p. 32
- Reteaching Activity, p. 34

eEdition with Audio DVD-ROM

ENGLISH LEARNERS

Pupil Edition in Spanish

eEdition with Audio DVD-ROM

Unit Resource Book
- RSG (Spanish), p. 13
- RSG with Additional Support (Spanish), p. 17

Multi-Language Glossary

Test Generator
- Section Quiz in Spanish

INCLUSION

Unit Resource Book
- RSG with Additional Support, p. 9
- Section Quiz, p. 32
- Reteaching Activity, p. 34

GIFTED & TALENTED

Unit Resource Book
- History Makers, p. 25
- Section Quiz, p. 32

PRE-AP

Unit Resource Book
- Section Quiz, p. 32

TECHNOLOGY

Unit Transparency Book
- 3-Minute Warm-Ups, TT1
- Essential Question, TT2
- Critical Thinking, TT4
- Map: Origin of Early Humans, TT6
- Cultures, TT7

Daily Test Practice Transparencies
- Section 2, TT60

Power Presentations

ClassZone.com

World Cultures and Geography Video on DVD

More About . . .

The Search for New Species

Scientists continually review previous findings, updating other scientists' work and, where appropriate, correcting it. One such example occurred in 2006, when scientists disputed fossil findings introduced two years earlier.

In 2004, a team announced what appeared to be a new human species on the island of Flores in Indonesia. Called *Homo floresiensis*, the skeleton was only about 39 inches tall, and had a brain about one-third the size of a modern human brain. In 2006, another team of researchers determined that the fossil was not a new species, but rather a modern human with a genetic condition causing a smaller head.

RESEARCH & WRITING CENTER

ANALYZING Primary Sources

Mary Leakey and her husband Louis spent much of their adult lives hunting human fossils, so it is not surprising that fossil hunting became a family tradition. Their son Richard and his wife Meave have also made important discoveries, and took their daughter Louise on her first field expedition when she was two weeks old. Louise, not surprisingly, has also become an expert fossil hunter.

- Why was Mary Leakey so pleased that the footprint trail was so long? *(Possible Answer: A long trail, with many footsteps, would make it easier to take important measurements, such as the size of the feet and stride length.)*
- What does the fact that there was a dual trail of footprints tell us? *(Two people were walking together, which might provide information about early social structures.)*

DOCUMENT-BASED QUESTION ANSWER

The footprints showed that our hominid ancestors walked upright on two feet—3.5 million years ago.

Unit Resource Book: Africa
- Primary and Secondary Sources: *from Lucy: The Beginnings of Humankind*, p. 29

Chad Skull A reconstruction of the 6.5-million-year-old skull found in Chad in 2001

Important Discoveries Older hominids have been found since Lucy's discovery. Scientists unearthed a six- to seven-million-year-old skull in Chad in 2001. It resembles both a human skull and an ape skull. Scientists wonder whether this and other finds could be early human **ancestors**. Other important discoveries include:

Australopithecines (aw•STRAY•loh•PIHTH•uh•seenz) Like Lucy, these hominids first walked upright in eastern, southern, and central Africa about 4 million years ago.

Homo habilis (HOH•moh–HAB•uh•luhs) Tool-using hominids appeared between 1.5 and 2 million years ago in Africa. In 1960, Mary and Louis Leakey found the skull of a creature they named *Homo habilis*, or "handy man," near chipped pebble tools.

Homo erectus (HOH•moh–ih•REHK•tus) In 1984, scientists found the skeleton of "Turkana Boy" in Kenya. This pre-teen lived about 1.6 million years ago. *Homo erectus* fossils have also been discovered in Asia. They were the first hominids to leave Africa.

Homo sapiens (HOH•moh–SAY•pee•EHNZ) People identical to modern humans first appeared about 150,000 years ago.

Stone Age The **Stone Age**, during which hominids first began using stone tools, began 2.5 million years ago. During this important period, people also learned how to control fire and use language.

ONLINE PRIMARY SOURCE To read accounts of important discoveries, go to the Research & Writing Center @ClassZone.com

SUMMARIZE Describe the key hominid finds in Africa. *Australopithecines,* which first walked upright 4 million years ago; *Homo habilis,* which used stone tools; and *Homo erectus,* the first hominids to leave Africa

ANALYZING Primary Sources

Oldest human footprints Mary Leakey and the footprints she discovered in 1978

Mary Leakey (1913–1996), as a young girl, was determined to study the Stone Age. Her most famous discovery was a set of hominid footprints, preserved in dried lava in Tanzania. In a 1978 letter, she wrote:

> We have found magnificent hominid footprints this year, in a dual trail 23 metres [75 feet] long! (confidential) It is a most important find and demonstrates that 3½ m.y. [million years] ago the fully bipedal, striding gait had been completely developed.
>
> Mary Leakey, letter to paleontologist Dick Hooijer

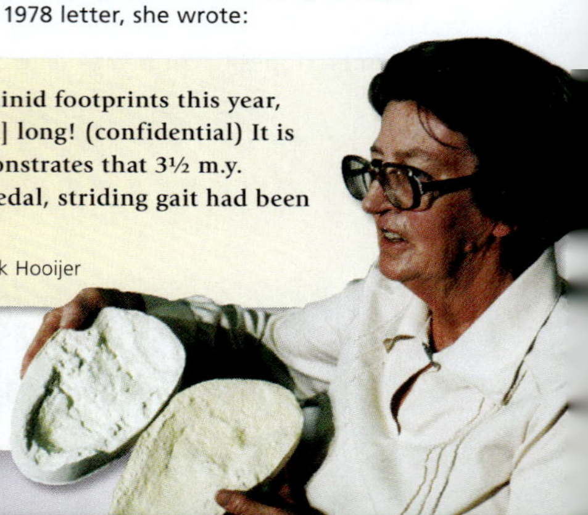

DOCUMENT-BASED QUESTION:
Why were these footprints so important?

546 Chapter 17

DIFFERENTIATING INSTRUCTION

Struggling Readers

Sequencing the Development of Early Humans

Help students understand the sequence of early human development by having them complete a graphic organizer, such as the one shown at right. Then have them read sentences from the text that give additional details on each stage of development.

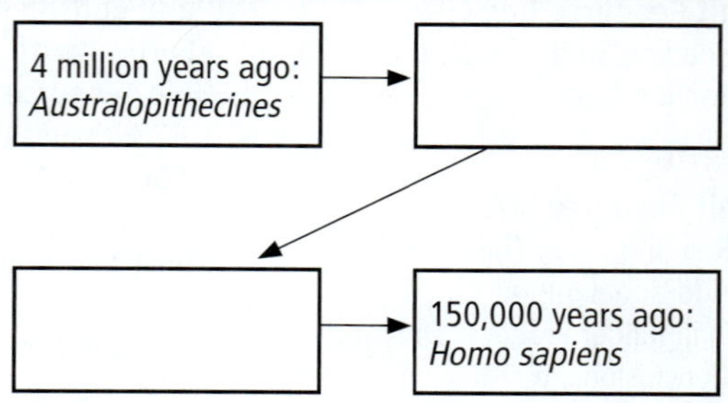

CONNECT to History

The Stone Age in Africa is divided into two periods. The Paleolithic period lasted from 2.5 million years ago to 8000 B.C. During this period, hominids hunted and gathered food. Early humans also chopped meat with sharpened stone tools. Later, they tamed fire, used harpoons and needles, and developed language. The Neolithic period lasted from 8000 B.C. to 3500 B.C. People living at this time polished stone tools and made pottery. They began to grow crops, **domesticate** animals, and live in villages. Around 3500 B.C. they developed writing.

Hand axes These stone tools from the Paleolithic period were found in Tanzania.

Activity

Create a Time Line

Materials
- paper
- pencil
- ruler

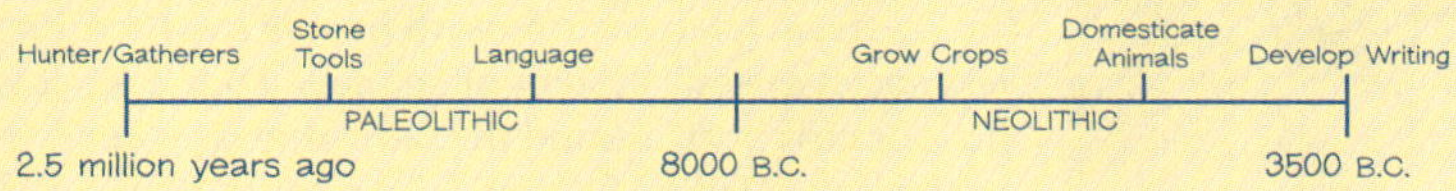

1. Reread this chapter and add the dates for the sequence of hominid discoveries in Africa.
2. Look for specific dates in the text. Make sure to match the correct events and dates.

Characteristics of African Societies

KEY QUESTION What is the nature of an extended family in African society?

In the previous section, you learned that hundreds of separate languages may be spoken in a single African country. Still, many African **cultures** share certain traits.

Beliefs and Values A typical African family probably wouldn't fit around your dinner table. The extended family of Africa consists of as many as five generations of relatives and spouses who all obey a single leader. This large family works together, protects its members from outsiders, and cares for its weakest members.

Today many Africans are Muslims or Christians. But Africans may also practice traditional religions. Traditional African religions maintain that an all-powerful God created everything. They also believe that spirits exist in animals, plants, natural forces, and ancestors, an idea called **animism**.

One generation of Africans passes information to the next by telling stories. West Africans call their storytellers **griots** (gree•OHZ). You need a good memory to be a griot because some of the best stories about ancestors and other heroes go on for hours or even days. Griots usually accompany their stories with music. Today many popular griots perform on television and record their stories on CDs.

DIFFERENTIATING INSTRUCTION: TIERED ACTIVITIES

OBJECTIVE Explore West African griot storytelling.

Basic

Have students work alone to create a brief story that they can memorize as a griot would and tell in one minute. Have students write down their stories first, commit them to memory, then present them with a partner.

On Level

In addition to the Basic activity instructions, have students work in pairs to link their stories. Have students memorize this new, combined story, and rehearse presenting it with their partner. Encourage students to include visuals and/or music when presenting to the class.

Challenge

Have students research a traditional folktale from West Africa. Have them adapt or revise it, or use it as the basis for a modern tale that they create. Then have students memorize and present their tales. Encourage students to include visuals and/or music when presenting to the class.

CONNECT to History

Remind students that people did not start living in cities until after they learned to grow crops. Ask students why being able to grow crops would help cities to develop. *(Possible Answer: If one part of the population could provide food for everyone, then other people could devote their energy to other activities, such as making new tools or building roads.)*

TIME LINE ANSWERS

stone tools—2.5 million years ago; language—2.5 million years ago; grow crops—8000 B.C. to 3500 B.C.; domesticate animals—8000 B.C. to 3500 B.C.; 8000 B.C. to 3500 B.C.—made polished stone tools and pottery, lived in villages

Unit Resource Book: Africa
- Connect to History: Create a Time Line, p. 28

Teach

Characteristics of African Societies

Reader, Recorder, Reporter

- How are African societies similar and different to our society? *(Similar: We have similar religious beliefs, and also enjoy stories accompanied by music; we go to school with members of our age set; Different: Families are more extended, African stateless societies are organized around smaller groups; authority is based on lineage, not election to office.)*
- When telling a story that takes several days to complete, what might a griot do to help keep the audience engaged? *(Possible Answer: The beginning of each session might include a recap of what happened before, the story might be told in rhyme and with a meter or rhythm, and many lines might repeat, until the audience also knows them by heart and can recite along with the griot.)*
- **Find Main Ideas** What are some of the advantages of having an extended family? *(A large family can work together, protect its members from outsiders, and care for its weakest members.)*

CONNECT to the Essential Question

How do humans interact with the extreme African environment?

Ask students what they have learned so far that can help them answer this question. Students might mention:

- Discoveries of early human fossils show that humans hunted and gathered food, made tools, grew crops, domesticated animals, and built villages.
- The large family organization and stateless societies of traditional African cultures helped them work together and protect and care for each other.

4 Assess & Reteach

Assess Have students complete the Section Assessment.

Unit Resource Book: Africa
- Section Quiz, p. 32

Interactive Review @ClassZone.com
- Section Quiz

Power Presentations

Test Generator

Reteach Have students work in groups to create a time line covering 7 million years, with one inch representing 100,000 years. Have them tape paper together to accommodate the nearly six-foot long time line that will result. Have them then add the discoveries about early humans mentioned in the chapter, including the events of the Neolithic period, which fall within the last one-tenth of an inch or so of the time line.

Unit Resource Book: Africa
- Reteaching Activity, p. 34

Elders in African Societies Elders often fill teaching roles, like this Masai elder in Kenya.

Stateless Societies Traditional African cultures didn't organize themselves into states, with one person at the top. Instead they formed stateless societies. In a **stateless society**, two or more equal groups would cooperate.

The extended families that made up these groups were based on lineage. **Lineage** means descent from one parent back several generations to a single ancestor. For example, your own **patrilineal** descent goes back from your father, to his father, and so on.

In traditional African stateless societies, lineages with equal power would share authority. If members clashed, elders from the warring lineage groups would meet to make peace.

To create relationships with people in other lineages, a stateless society uses a system of age sets. Your age set would be the boys or girls born around the same time as you. If you were a boy in the Nandi tribe of Kenya, for example, you would go through military training with your age set. The warrior set, responsible for fighting and governing the tribe, would stay in power for about 15 years. Then it would become a set of elders, and the next younger age set would move up.

FIND MAIN IDEAS Describe the structure of an African extended family. up to five generations living together, obeying a single leader, cooperating with other extended families in stateless societies

Section 2 Assessment

ONLINE QUIZ For test practice, go to **Interactive Review @ClassZone.com**

TERMS & NAMES

1. **Explain the importance of**
   - hominid
   - animism
   - stateless societies
   - lineage

USE YOUR READING NOTES

2. **Find Main Ideas** Use your completed diagram to answer the following question: What are the roles of elders in traditional African societies?

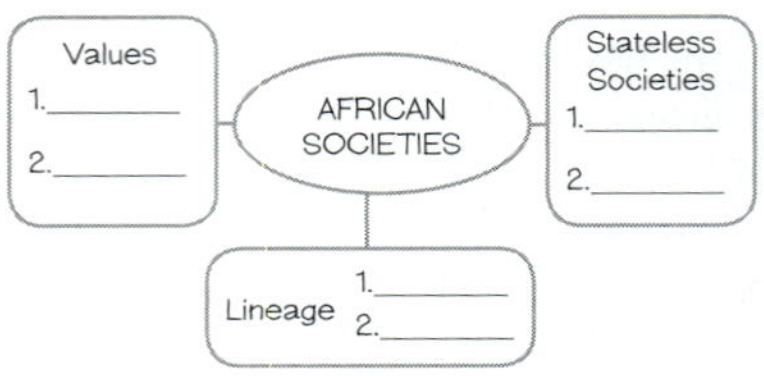

KEY IDEAS

3. Who is Lucy, and how old is she?
4. Which was the first human to leave Africa?
5. How would warring lineage groups make peace?

CRITICAL THINKING

6. **Summarize** What major human advances took place during the Stone Age?
7. **Compare and Contrast** What similarities do African cultures share?
8. **CONNECT to Today** What systems of age sets exist in the United States?
9. **MATH Draw a Scale Time Line** Find a way to demonstrate how long ago Lucy lived. You might start by making a scale with 100 years equal to 1 inch. How many inches, feet, or yards would it take to show 1,000 or 1 million years? How about 3.5 million years?

SECTION 2 ASSESSMENT ANSWERS

Terms & Names

1. hominid: early human; animism: belief of spirits in living things, nature, and ancestors; stateless societies: groups lacking a single authority; lineage: descent back to a single ancestor

Use Your Reading Notes

2. They share decision-making authority.

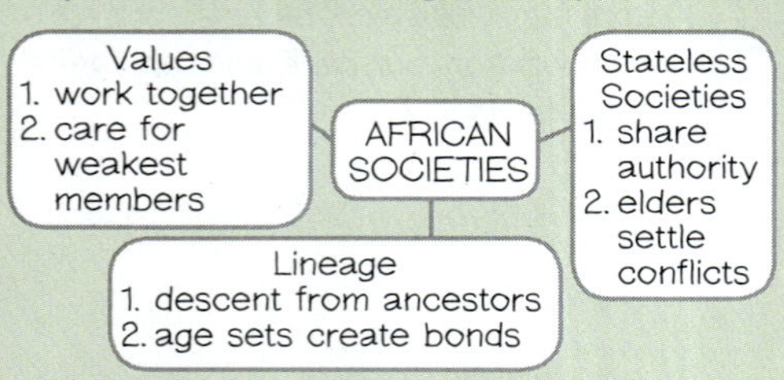

Key Ideas

3. Lucy is an early hominid discovery, and is about 3.5 million years old.
4. *Homo erectus*
5. Elders from clashing groups would meet to talk over differences.

Critical Thinking

6. mastery of fire, language, stone tool making
7. Many Africans live in extended families. They may also follow traditional animist religions, and pass down their history orally.
8. school grades, the military, workers, retirees
9. Use the rubric to score time lines.

Time Line Rubric

	Scale	Presentation
4	accurate math	excellent
3	few errors	good
2	some errors	fair
1	many errors	poor

Interactive Review

Click here to complete these and other activities online @ClassZone.com

CHAPTER SUMMARY

Key Idea 1
Africa, the world's second largest continent, lies almost entirely in the tropics.

Key Idea 2
Human life first emerged on this warm continent.

For **Review and Study Notes**, go to **Interactive Review** @ClassZone.com

NAME GAME

Use the Terms & Names list to complete each sentence on paper or online.

1. I am the longest river in the world. ________ Nile River
2. I am a process that changes grasslands and semi-arid regions into deserts. ________
3. I am a landform some Africans call Mosi-oa-Tunya, "the smoke that thunders." ________
4. I am the top layer of the rain forest. ________
5. I am Africa's grasslands, where antelopes, giraffes, and elephants graze. ________
6. We are the slowly shifting blocks that make up the Earth's surface layer. ________
7. I can tell stories that last for many hours or even days. ________
8. I am the belief that trees, animals, and ancestors have living spirits. ________
9. We do not have a single leader, but resolve problems by having our elders meet and talk until they agree. ________
10. I am the earliest period of human history. ________

animism
canopy
desertification
griot
hominid
Mount Kilimanjaro
Nile River
Sahel
savanna
stateless societies
Stone Age
tectonic plates
Victoria Falls

Activities

Flip Cards

Use the online flip cards to quiz yourself on the terms and names introduced in this chapter.

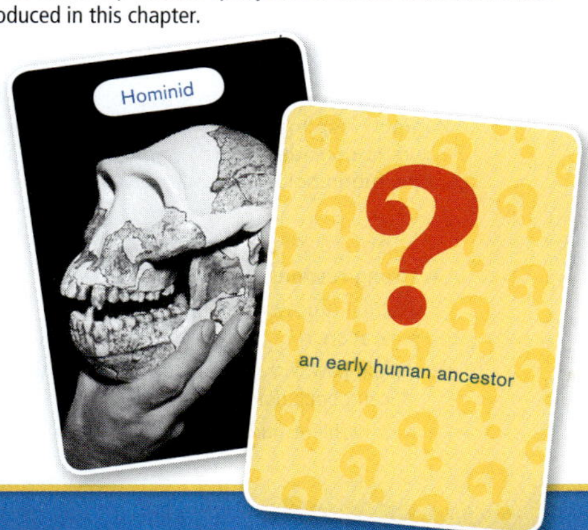

Crossword Puzzle

Complete an online crossword puzzle to test your knowledge of Africa's geography and early history.

ACROSS
1. the world's largest desert

ONLINE AND ON-SCREEN ACTIVITIES

Flip Cards

Vocabulary Flip Cards are virtual flash cards at **ClassZone.com.**

Students can quiz each other or work independently to study the flip cards for this chapter.

- They can quiz themselves on key terms and names by reviewing definitions.
- Demonstrate how to advance and shuffle the deck for another challenge.

Crossword Puzzle

Students can review important people and events in Africa's early history by completing this crossword puzzle online at **ClassZone.com.**

- As an extension of this activity, have students use the *Activity Maker* at **ClassZone.com** to make their own crossword puzzle for one section of the chapter.
- Then have students trade puzzles with classmates to solve answers for each other's puzzles.

Chapter Summary

As a follow-up to the chapter, encourage students to pair up and go over the **Review and Study Notes** at **ClassZone.com.**

Alternatively, students pairs can review their Reading Study Guide notes, making sure that they can add important details to each section's "Key Idea."

Name Game

Teaching Tips

ON PAPER Have students work in pairs to complete this activity. As an extension, they can write questions for the terms that are not covered in the items shown here.

ONLINE Make sure that students understand how to drag and drop terms to fill in blanks. Ask them to print out their results and compare them with those of a classmate.

ANSWERS

1. Nile River
2. desertification
3. Victoria Falls
4. canopy
5. savanna
6. tectonic plates
7. griot
8. animism
9. stateless societies
10. Stone Age

Make It Fun

Teacher-Tested Activities

Terry Rhodes, McNair Middle School, Fayetteville, Arkansas

Wind-Up Activity — As students learned about Africa's physical geography in this chapter, they also learned that the words *Sahara* and *Sahel* came from Arabic. In the chapters to follow, they will learn that many places and features in Africa use names given by non-Africans during exploration or colonization. Using the political map in the unit atlas, list some African places and features, mixing European and African names (examples: Côte d'Ivoire, Zimbabwe, Cape Verde, Lake Tanganyika), and have students guess or research the origins of their names.

Vocabulary ANSWERS

1. the highest mountain in Africa, located in Tanzania
2. the world's largest river
3. grassland in the tropics or subtropics
4. a semiarid region just south of the Sahara
5. the change of fertile land to desert
6. an early human ancestor
7. walking on two feet
8. the belief that spirits exist in plants, animals, natural forces, and ancestors
9. societies with no single form of authority
10. descent from a parent back to a single ancestor
11. The Sahel is the semiarid region that borders the Sahara and the savanna. Desertification gradually turns the Sahel region into desert.
12. In stateless societies, culture is based on the cooperative relations between lineage groups.

Key Ideas ANSWERS

13. divergence, or tectonic plates slowly pulling apart
14. rapids and waterfalls
15. They use a common language to communicate. One example is French.
16. gold, diamonds, and platinum
17. Lucy's bones showed that she was bipedal, providing early evidence of this human trait.
18. A griot memorizes and tells stories to pass on information about a culture's ancestors.
19. Extended families are structured around lineages, or descent from ancestors.
20. The societies create age sets, grouping boys and girls of roughly the same ages together.

CHAPTER 17 Assessment

VOCABULARY

Explain the significance of each of the following.

1. Mount Kilimanjaro
2. Nile River
3. savanna
4. Sahel
5. desertification
6. hominid
7. bipedal
8. animism
9. stateless societies
10. lineage

Explain how the terms and names in each group are related.

11. Sahel, desertification, savanna
12. stateless societies, culture, lineage

KEY IDEAS

1 Africa's Geographic Extremes

13. What geographic process formed Africa's Great Rift Valley?
14. What makes many African rivers hard to navigate?
15. How do many Africans communicate with people who don't speak their language?
16. What minerals make South Africa a wealthy country?

2 Human Beginnings in Africa

17. Why was Lucy such an important discovery?
18. What does a griot do?
19. How are African extended families structured?
20. How do stateless societies create relationships between people of different lineages?

Goliath Beetle

CRITICAL THINKING

21. **Analyze Causes and Effects** Create a table to show some of the effects of human settlement on rain forests and grasslands.

CAUSES	EFFECTS
cut down rain forests for farmland	
cut trees for firewood and use grass for grazing	

22. **Summarize** What events led to the formation of game parks in Africa?
23. **Compare and Contrast** How are traditional African religions similar to and different from Christianity and Islam?
24. **Five Themes: Movement** How might family members moving to a city affect a traditional extended family structure?
25. **Connect Geography & Science** Why can't scientists say for sure whether *Australopithecines* like Lucy used tools?

Answer the ESSENTIAL QUESTION

How do humans interact with the extreme African environment?

Written Response Write a two- or three-paragraph response to the Essential Question. Be sure to consider the key ideas of each section. Use the rubric below to guide your thinking.

Response Rubric

A strong response will:

- describe Africa's landforms, climate, and vegetation
- explain how climate and terrain have influenced African population density
- identify the resources available in Africa

ESSENTIAL QUESTION REVIEW

Before students begin their essays, have them review the key chapter elements that will help them answer the Essential Questions. Direct them to significant features in the chapter, such as:

- Connect Geography and Science, Section 1
- Animated Geography: Vegetation Regions of Africa, Climates of Africa, Section 1
- Analyzing Primary Sources, Section 2
- Connect to History, Section 2

Written Response Rubric

	Coverage of Key Ideas	Explanation	Errors in Mechanics
4	all	clear and concise	few or none
3	most	good	some
2	some	fair	many
1	one or none	poor	many major

STANDARDS-BASED ASSESSMENT

Test Practice
- Online Test Practice @ClassZone.com
- Test-Taking Strategies and Practice at the front of this book

LINE GRAPH

Use the line graph and your knowledge of Africa to answer questions 1 and 2.

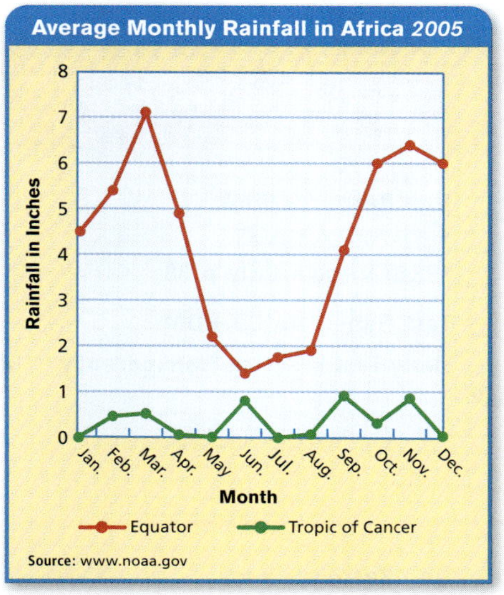

1. When is the "dry season" along the Equator?

A. September through December
B. January through March
C. March through May
D. June through August

2. In what month did the most rain fall along the Tropic of Cancer?

A. March
B. June
C. September
D. November

ELEVATION PROFILE

Use the elevation profile at the Equator to answer questions 3 and 4 on your own paper.

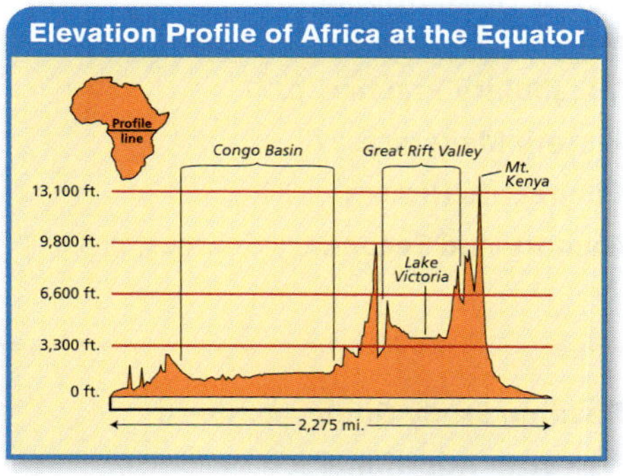

3. At what elevation is Mount Kenya?

4. Based on this elevation profile, why might Africa be called the "plateau continent"?

GeoActivity

1. INTERDISCIPLINARY ACTIVITY–SCIENCE

With a small group, choose one of Africa's native animals to research. Find out where the animal lives, what it eats, whether it has predators, and if its habitat has been threatened by hunting or farming. Present your findings to the class.

2. WRITING FOR SOCIAL STUDIES

Reread the description of griots in Section 2. Write a story, poem, or rap about one of your ancestors or an American folk hero. Memorize your story and present it to your class without using notes.

3. MENTAL MAPPING

Create an outline map of Africa and label the following:
- Lake Victoria
- Lake Chad
- Nile River
- Great Rift
- Great Escarpment
- Sahara Desert
- Sahel

Critical Thinking ANSWERS

21. Cause and effect chart could show:

Cause: clearing rain forest
Effects: plant and animal habitats destroyed; farmers must move on to other land because rain forest soil is poor

Cause: clearing grasslands
Effects: loss of protective plants turns grassland into desert.

22. Hunting killed animals, while farming destroyed animal habitats. National parks restored habitats and animal populations.

23. Similar: belief in one creator. Different: belief in spirits

24. It may be difficult for large extended families to work and live together in cities.

25. Simple tools made of wood would not have survived.

Standards-Based Assessment ANSWERS

1. D
2. C
3. more than 13,100 feet
4. Much of Africa has a high, flat elevation.

GeoActivity

1. INTERDISCIPLINARY ACTIVITY RUBRIC

The report should
- contain all requested information
- follow a clear structure

2. WRITING FOR SOCIAL STUDIES RUBRIC

The piece should
- be performed without notes
- cover the appropriate subject matter

3. MENTAL MAPPING RUBRIC

The outline map should
- represent the approximate shape of Africa
- include all geographic features listed in their correct location

DIFFERENTIATING CHAPTER ASSESSMENT

Unit Resource Book
- **Chapter Test A,** pp. 35–38.
  Use for Struggling Readers, Beginning-to-Intermediate-Level English Learners, and for Inclusion of students whose IEP suggests supported test-taking.
- **Chapter Test B,** pp. 39–42.
  Use for On-Level Learners and ESOL students with Advanced-Level English.
- **Chapter Test C,** pp. 43–46.
  Use for Gifted and Talented students or students assigned Pre-AP work.

Test Generator
- Chapter Tests A, B, and C, in English and Spanish

McDougal Littell Assessment System

TEST **Test Generator** Use three levels of tests in English and Spanish tied to state standards. Plus, choose questions to customize tests for all populations.

SCORE Scan and Score plain paper student tests in minutes.

REPORT Create instant progress reports, sortable by student, by standard, or other criteria.

RETEACH Respond immediately to students' needs by using a wealth of reteaching resources.

Chapter 18 Egypt and North Africa PLANNING GUIDE

Complete print resources all on one USB drive allow you to customize lesson plans for all students.

Section Titles & Key Ideas	Unit Resource Book Africa	Assessment Resources
1 Ancient Egypt: Pyramids and Pharaohs **KEY IDEA** The Nile River made the great civilizations of ancient Egypt possible.	• **Reading Study Guide,** pp. 47–48 • **RSG with Additional Support,** pp. 53–54 • **Vocabulary Practice,** p. 71 • **History Makers,** p. 77 • **Connect to History,** p. 83 • **Primary and Secondary Sources,** p. 84	**Unit Resource Book** • Section Quiz, p. 85 • Reteaching Activity, p. 88 **Test Generator CD-ROM** **Daily Test Practice Transparency,** TT61
2 Modern Egypt and Sudan **KEY IDEA** Egypt has a leadership role in North Africa, as well as in the larger Arab world.	• **Reading Study Guide,** pp. 49–50 • **RSG with Additional Support,** pp. 55–56 • **Skillbuilder Practice,** p. 72 • **Map and Graph Skills,** pp. 73–74 • **History Makers,** p. 78	**Unit Resource Book** • Section Quiz, p. 86 • Reteaching Activity, p. 89 **Test Generator CD-ROM** **Daily Test Practice Transparency,** TT62
3 Other Nations of North Africa **KEY IDEA** The Ottomans and Islam also influenced other North African nations, as did European and colonizing powers.	• **Reading Study Guide,** pp. 51–52 • **RSG with Additional Support,** pp. 57–58 • **Skillbuilder Practice,** p. 72 • **Outline Map Activities,** p. 75 • **Interdisciplinary Projects,** pp. 79–81	**Unit Resource Book** • Section Quiz, p. 87 • Reteaching Activity, p. 90 • Chapter Test (Forms A, B, & C), pp. 91–105 **Test Generator CD-ROM** **Daily Test Practice Transparency,** TT63

McDougal Littell Assessment System
TEST SCORE REPORT RETEACH

No Time? To focus students on important content in the chapter, use the **"Reading Study Guide with Additional Support"** in ***Resources2Go.***

Teacher Presentation Options

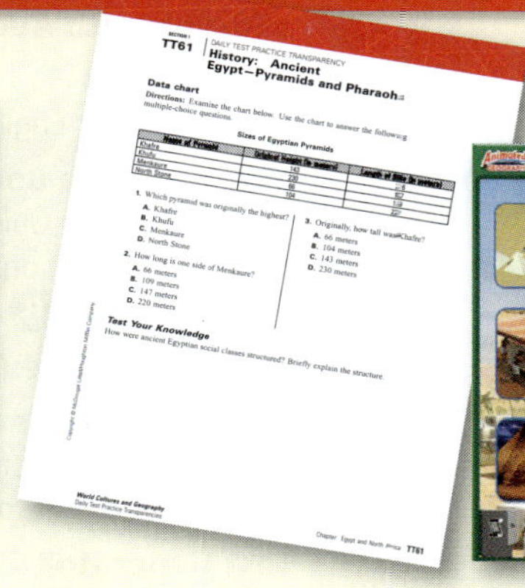

Electronic Books

eEdition DVD-ROM

eEdition Online

Presentation Toolkit

Video/DVD

- World Cultures and Geography

Power Presentations DVD-ROM

- Discussion Starters
- Media Gallery
- Review Game
- Interactive Review
- Animated Features

Unit Transparency Book

- 3-Minute Warm-Ups
- Essential Question
- Critical Thinking
- Maps
- Cultures

Daily Test Practice Transparencies

Animated Geography

- Water Sources of North Africa Map
- Egyptian Social Roles
- Egyptian and Kush Kingdom Map
- The Nile Valley
- The Great Pyramid

OnlineActivities at ClassZone.com

Animated Geography

- **Videos and Photos**
- **3-D Animations**
- **Interactive Maps**

Interactive Review

- **Section Quizzes**
- **Review and Study Notes**
- **Vocabulary Flip Cards**
- **Graphic Organizers**
- **Name Game and Crossword Puzzles**
- **GeoGame**

Online Test Practice

Map Center

- **Atlas Maps**
- **Political Maps with Data**
- **Links to Mapping Web Sites**

Activity Center

- **Activity Maker—Review Games, Puzzles, Flip Cards, and Graphic Organizers**
- **Worldster**

Research & Writing Center

- **Web Sites**
- **Data Updates**
- **Primary Sources**
- **Writing Models**

Current Events

Matthew J. Scheidler
Wayzata East Middle School
Plymouth, Minnesota

Make It Fun — Teacher-Tested Activities

As you teach this chapter, look for my classroom-proven ideas for:

- making the lesson fun
- focusing students' attention and comprehension
- integrating technology with traditional teaching

Teaching Bell to Bell When you see this timing guide, have students discuss in pairs, in groups of 3–6, or as a class.

- **Think, Pair, Share** (pairs)
- **Reader, Recorder, Reporter** (3–6)
- **Talk About It** (any size)

CHAPTER 18 RESOURCES FOR DIFFERENTIATED INSTRUCTION

Struggling Readers

eEdition with Audio DVD-ROM

Lesson Plans
- Strategies for Inclusion

Unit Resource Book
- Reading Study Guide with Additional Support, pp. 53–58 (B)
- Reteaching Activity, pp. 88–90
- Chapter Test (Form A), pp. 91–95 (A)

Best Practices Toolkit (C)
- Interactive Reading Strategies, pp. 90–145
- Differentiation Activity Bank: Struggling Readers, pp. 146–149

Teacher's Edition Activities
- Clarifying Egypt's Situation, Sec. 2
- Create a Time Line for North Africa, Sec. 3
- Tiered Activities, Sec. 1–3

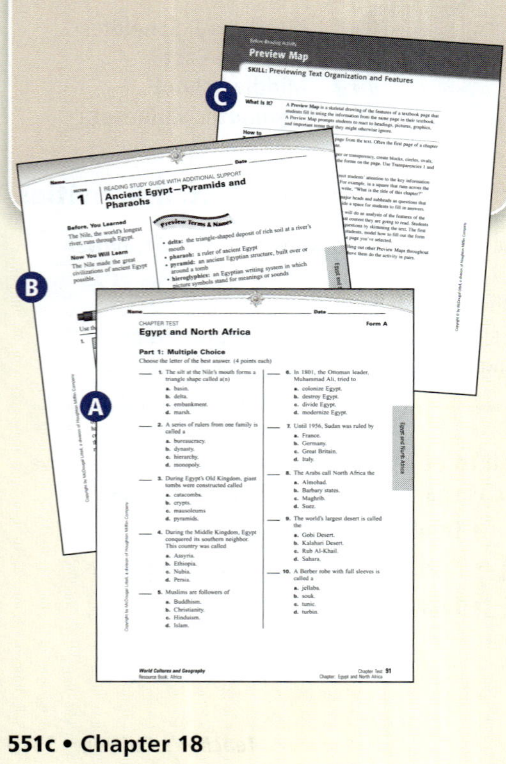

English Learners

Pupil's Edition in Spanish

eEdition with Audio DVD-ROM

Lesson Plans
- Strategies for English Learners

Unit Resource Book
- Reading Study Guide (Spanish), pp. 59–64
- RSG with Additional Support (Spanish), pp. 65–70 (A) (B)

Multi-Language Glossary (C)

Best Practices Toolkit
- Differentiation Activity Bank: English Learners, pp. 165–194

Test Generator CD-ROM
- Section Quizzes in Spanish
- Chapter Test in Spanish

Teacher's Edition Activities
- Pronounce and Preview, Sec. 1–3
- Key Academic Vocabulary, Sec. 1
- Word Relationships, Sec. 2
- Prefixes, Sec. 3
- Tiered Activities, Sec. 1–3

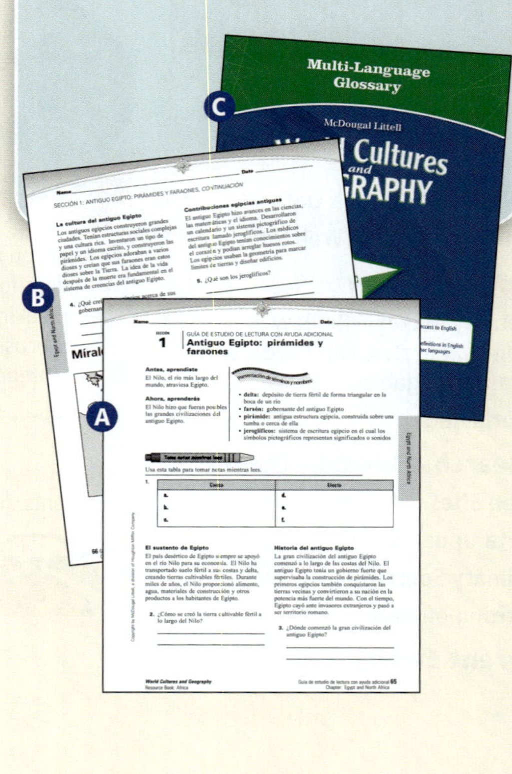

Inclusion

eEdition with Audio DVD-ROM

Lesson Plans
- Strategies for Inclusion

Unit Resource Book
- Reading Study Guide with Additional Support, pp. 53–58 (A) (B)
- Reteaching Activity, pp. 88–90
- Chapter Test (Form A), pp. 91–95

Best Practices Toolkit (C)
- Interactive Reading Strategies, pp. 90–145
- Differentiation Activity Bank: Struggling Readers, pp. 146–149

Teacher's Edition Activities
- Vocabulary Classification, Sec. 1, 3
- Knowledge Rating, Sec. 2
- Vocabulary Categories, Sec. 3
- Tiered Activities, Sec. 1–3

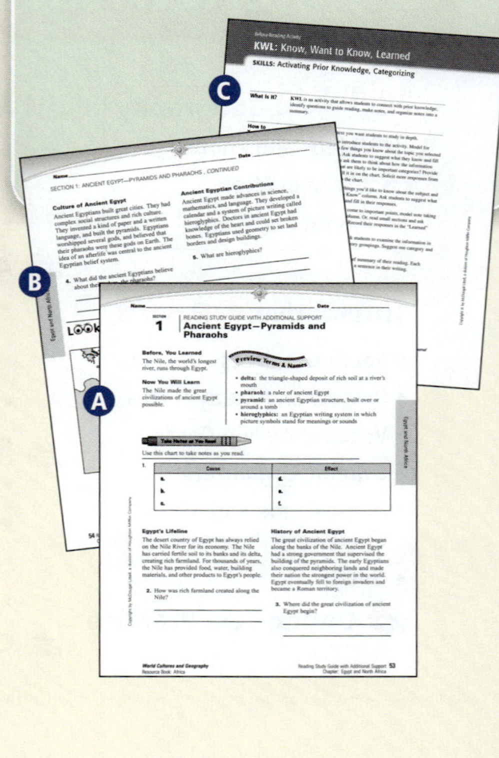

Gifted & Talented Students

Lesson Plans
- Core Lesson Plan

Unit Resource Book
- Outline Map Activities, pp. 75–76
- History Makers, pp. 77–78 (A) (B)
- Interdisciplinary Projects, pp. 79–81 (C)
- Chapter Test (Form C), pp. 101–105

Best Practices Toolkit
- Student Products Rubrics, pp. 1–14
- Writing Templates, pp. 78–89

Teacher's Edition Activities
- Describing the Immigration Experience, Sec. 3
- Tiered Activities, Sec. 1–3

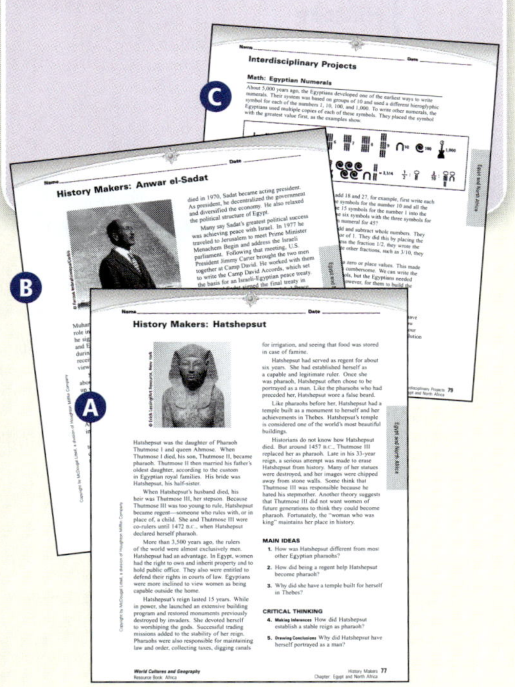

Pre-AP*

Lesson Plans
- Core Lesson Plan

Unit Resource Book
- Vocabulary Practice, p. 71
- Map and Graph Skills, pp. 73–74
- Economics in the World, p. 82 (A)
- Connect to History, p. 83 (B)
- Primary and Secondary Sources, p. 84 (C)
- Chapter Test (Form C), pp. 101–105

Best Practices Toolkit
- Writing Templates, pp. 78–89
- Differentiation Activity Bank: Pre-AP Students, pp. 195–200

* Pre-AP is a registered trademark of the College Entrance Examination Board, which was not involved in the production of and does not endorse this product.

Teacher's Edition Activities
- Critical Discussion About Dams, Sec. 1
- Tiered Activities, Sec. 1–3

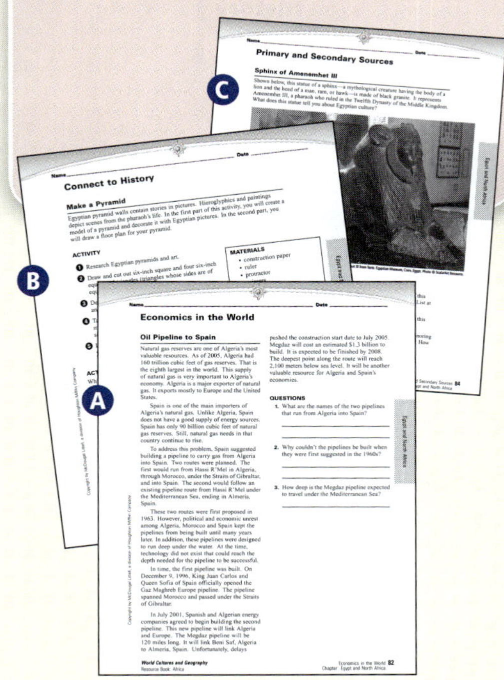

Interdisciplinary Activities

CONNECT to Math

How Do African Economies Compare?

CLASS TIME 30 minutes

TASK Create pie graph displays showing economic data for Egypt and two countries in other regions of Africa.

SUPPLIES
- economic data for Egypt and two countries of Africa (in percentages), such as from The World Factbook available at www.cia.gov
- construction paper, colored markers or pencils, rulers

ACTIVITY Draw students' attention to the pie graphs included in the Economic Activity 2000–2001 feature in Section 3. Have students work in pairs to construct pie graph posters using the supplied data. Ask students to prepare brief statements comparing their pie graphs with those in the feature. Display the completed graphs.

Good for Visual and Mathematical Learners

CONNECT to Science

Explore Desert Wildlife in a Mural

CLASS TIME 30 minutes

TASK Create a display of Sahara desert plants and animals.

SUPPLIES
- information and illustrations from printed or online sources about wildlife in the Sahara
- small boxes for a three-dimensional display, construction paper, colored markers or pencils, rulers

ACTIVITY Have students work in small groups to create a display that shows Saharan wildlife. The display should include a combination of at least five plants and animals. Students should also be able to explain how their wildlife adapts to the arid desert climate. Have students present their completed projects.

Good for Visual and Verbal Learners

Introduce the ESSENTIAL QUESTION

- Explain that Egyptian civilization was tied to the Nile and that the cultural advances of ancient Egypt are still appreciated today.
- Describe Egypt as a developing nation and its neighbor Sudan as a nation mired in conflict.
- Identify the shared cultural influences in North Africa and the different modern national governments and economies.

Key Ideas

Section 1: The Nile made the great civilizations of ancient Egypt possible.

Section 2: Egypt has a leadership role in North Africa, as well as in the larger Arab world.

Section 3: The Ottomans and Islam also influenced other North African nations, as did European and colonizing powers.

CONNECT to the Essential Question

You may wish to revisit the Essential Question periodically with students as they work through the chapter.

Suggestions for reconnecting are provided at an appropriate point in the *Teacher's Edition* for each section.

Great Sphinx

CHAPTER 18

Egypt and North Africa

ESSENTIAL QUESTION

How have people in North Africa adapted to the region's arid climate?

CONNECT Geography & History

Use the map and time line to answer the following questions.

1. Which country receives the most rainfall?
2. About how many years separate the building of the Great Pyramid and Suez Canal?

3100 B.C.

History
3100 B.C. Upper and Lower Egypt united

Culture
◄ 2600 B.C.–2500 B.C. Great Pyramid built

History
A.D. 642 Muslim invaders spread Islam to North Africa.

History
1516 Selim I rules the Ottoman Empire at the time of Egypt's conquest. ►

TIME LINE DISCUSSION

Point out that Egypt and North Africa include vast areas of desert, but that many millions of people live along the Nile River. Geography is key to understanding the history and modern life of this region. Its geographical location made Egypt an entry point for people of different cultures into and out of Africa centuries ago. Cultural differences have kept Sudan embroiled in regional conflicts since independence in 1956.

- Which event highlights a great advance in trade and transport between the Mediterranean Sea and the Red Sea? *(1869: Suez Canal completed)*
- Which years signal the end of European domination over North African nations? *(1951–1962: Libya, Morocco, Tunisia, and Algeria become independent from European nations.)*
- In what way do you think the events headed "Culture" on the time line might be related? *(Possible Answer: The Rosetta Stone is a cultural artifact from ancient Egypt. It enabled scholars to decipher ancient Egyptian hieroglyphics.)*

Animated GEOGRAPHY
Water Sources of North Africa

Click here to explore Egypt and North Africa @ ClassZone.com

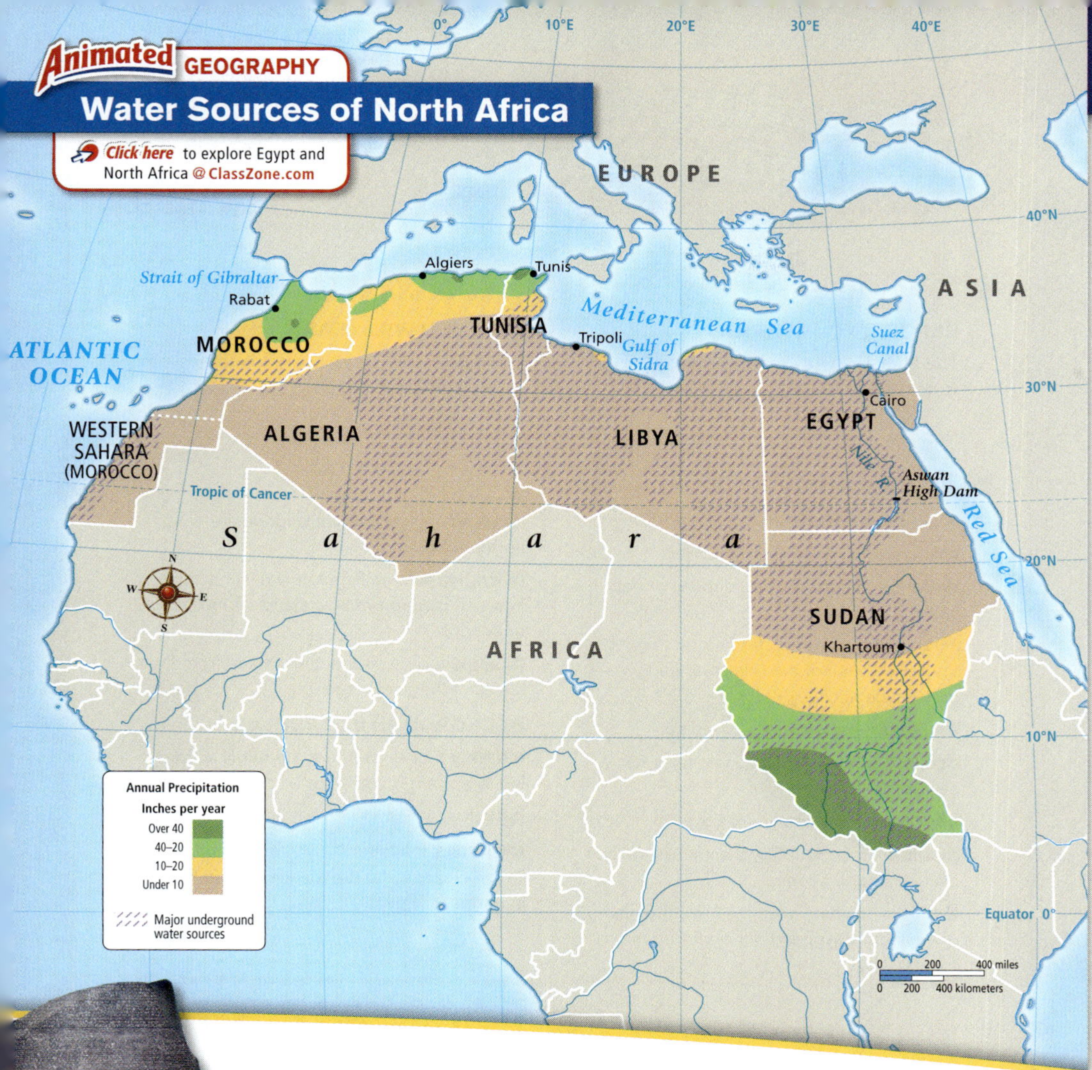

Culture
◄ **1799** Rosetta Stone discovered in Egypt

Government
1951–1962 Libya, Morocco, Tunisia, and Algeria become independent from European nations.

Today

Economics
1869 Suez Canal completed ►

History
2003 Darfur crisis erupts in Sudan.

TEACHER MEDIA FAVORITES

Books for the Student

Winget, Mary. *Cooking the North African Way*. Lerner Publishing Group, 2003. In addition to recipes for young readers, includes information about the food and culture of this region.

Mead, Alice. *Year of No Rain*. Recorded Books, Audio, 2003. A novel about a Sudanese boy who flees from rebel soldiers to a refugee camp.

Harris, Geraldine. *Ancient Egypt*. Cultural Atlas for Young People Series. Facts on File, Inc., 1990. Explores Egypt's history and culture; includes illustrations.

Video / DVD

Modern Marvels—Aswan Dam. (DVD) A&E Home Video, 2000. Documents the construction of the dam and its effects on Egyptian society and culture.

Mysteries of Egypt. (videocassette) National Geographic Video, 1999. Tales of pyramids and treasures of ancient Egypt.

Internet

Go to the *Research and Writing Center* at **ClassZone.com** to link to:

- data on the region
- a variety of chapter-specific content-reviewed sites
- "History Makers" biographies
- primary sources

Animated GEOGRAPHY
Water Sources of North Africa

CLASSZONE.COM

Point out that this is a thematic map because it focuses on the specific topic of water sources of North Africa. Draw students' attention to annual precipitation ranges in the legend.

- Identify the North African desert. *(Sahara)* What is the annual precipitation in most of North Africa? *(under ten inches)* Which areas receive the most precipitation? *(southern Sudan, northern Morocco, northern Algeria, and northern Tunisia)*
- Locate the Aswan High Dam on the map. Which nation controls the dam? *(Egypt)*
- In which nation is Khartoum located? *(Sudan)* What do you notice about the Nile River at Khartoum? *(It divides into two branches south of Khartoum.)*

CONNECT GEOGRAPHY & HISTORY ANSWERS

1. Sudan
2. about 4,359 years (2500 + 1859)

Presentation Options

Use the **Power Presentations** DVD-ROM to project the animated map.

- Prompt students to use the pop-up images and audio from the map to discover what they already know about the geography, culture, history, economics, or governments of the region.
- Challenge students to answer the map questions individually, in small groups, or as a class activity. Use all of the interactive material to help preview chapter content.

1 Plan & Prepare

Objectives

- Explain how the Nile is Egypt's lifeline
- Summarize the history of ancient Egypt
- Describe the culture of ancient Egypt

Read for the Essential Question

Help students read for a purpose by reminding them of the Essential Question: "How have people in North Africa adapted to the region's arid climate?" Point out that they can use text features such as Key Questions, special features, images, and maps to form answers to the question.

Vocabulary

Best Practices Toolkit

Use the ***Best Practices Toolkit*** to model strategies for vocabulary notetaking. Vary strategies throughout the year. Choose from: *Knowledge Rating, Predicting ABCs, Definition Mapping, Word Sort, Frayer Model (Word Squares), Magnet Words, Word Wheel,* and *Student VOC.*

Vocabulary Strategies, TT9–TT16

Reading Strategy

Best Practices Toolkit

Display the T Chart/Two-Column Chart transparency.

- Remind students that causes and effects are often signaled by prepositions such as *due to, as a result, since, after,* and verbs such as *made* and *caused.*
- Model filling in causes and effects using the subsection "Egypt's Lifeline."
- Have students suggest entries.
- Display the graphic organizer for students to review.

T Chart/Two-Column Chart , TT20

SECTION 1 Reading for Understanding

Key Ideas

BEFORE, YOU LEARNED

The Nile, the world's longest river, runs through Egypt.

NOW YOU WILL LEARN

The Nile made the great civilizations of ancient Egypt possible.

Vocabulary

TERMS & NAMES

silt fertile soil deposited by a river

delta the triangle-shaped deposit of rich soil at a river's mouth

Aswan High Dam a structure to control the floodwaters of the Nile River and provide electricity

cataract a high waterfall or rapids

dynasty a family or group that rules for several generations

pharaoh a ruler of ancient Egypt

pyramid an ancient Egyptian structure, built over or around a tomb

Nubia in ancient times, the Nile River valley of southern Egypt and northern Sudan

Kush an ancient Nubian kingdom in northern Sudan

hieroglyphics an Egyptian writing system in which picture symbols stand for meanings or sounds

papyrus a paperlike material made from the stems of the papyrus reed

BACKGROUND VOCABULARY

mummy a body prepared for burial according to ancient Egyptian practice

REVIEW

Nile River the longest river in the world, which empties into the Mediterranean Sea in northeast Egypt

Reading Strategy

Re-create the chart shown at right. As you read and respond to the **KEY QUESTIONS**, use the chart to analyze causes and effects related to the Nile River's annual floods.

See Skillbuilder Handbook, page R8

ANALYZE CAUSE AND EFFECT

CAUSE	EFFECT
1.	1.
2.	2.
3.	3.

GRAPHIC ORGANIZERS
Go to Interactive Review @ ClassZone.com

554 Chapter 18

PRETEACHING VOCABULARY

English Learners

Pronounce and Preview

Pronounce each term for students. Review the meanings of words within definitions, such as *deposit, structure,* and *generations.*

- To modify vocabulary learning, have students complete worksheets as they read, instead of afterward.

Unit Resource Book: Africa
- Vocabulary Practice, p. 71

Inclusion

Vocabulary Classification

Read aloud the terms and definitions. Have pairs of students find the terms and names in the Section 1 text and illustrations. Then help them classify the terms into one of two categories: whether it refers to the Nile River or to the lives of the ancient people who lived along the river. Discuss the decisions with each pair.

Philae Temple, Egypt

Nile crocodile

Ancient Egypt: Pyramids and Pharaohs

Connecting to Your World

In 1993, the Mississippi River overflowed after heavy rains. Levees broke, submerging farms, towns, and roads under the muddy water. About 50 people died, and thousands more had to leave their homes. The Mississippi River flood of 1993 was a disaster. But in ancient Egypt, annual river floods were seen as a good thing.

Egypt's Lifeline

KEY QUESTION Why is the Nile called Egypt's lifeline?

Egypt, in Africa's northeast corner, receives little rainfall. As a result, the country has relied on the **Nile River** for its economy. Each year, the Nile flooded due to upstream rain, depositing rich black soil called **silt** along its banks. The silt at the river's mouth formed a triangular shape, called a **delta**. Most of Egypt's population clustered in the Nile's river valley and delta. The Nile's yearly floods made these regions so fertile that ancient Egyptians called their country Kemet, or the Black Land.

Animated GEOGRAPHY

The Nile Valley The world's longest river creates a ribbon of fertile soil, cutting through the desert.

Click here to take a journey down the ancient Nile @ClassZone.com

2 Focus & Motivate

3-Minute Warm-Up

Write on the board or display the transparency:

- Look at the image of the Nile River in Section 1. If you lived in ancient Egypt, how would you use the Nile to help your economy? *(Possible Answers: to water crops, catch fish, carry goods for trade on boats, to explore for other resources)*

Unit Transparency Book: Africa
- 3-Minute Warm-Ups, TT8

Connecting to Your World

More About . . .

The Longest River in the World

The Nile flows south to north for about 4,160 miles. In the south the river is called the upper Nile because it has a higher elevation than the lower Nile in the north. Branches south of Khartoum are called the White Nile and the Blue Nile.

3 Teach

Egypt's Lifeline

Think, Pair, Share

- How did Nile flooding help farmers in ancient Egypt? *(Floods deposited silt on the riverbanks.)*
- How did ancient Egyptians use the Nile's water for farming? *(They dug canals for irrigation.)*
- **Analyze Cause and Effect** How has the Aswan High Dam changed Egypt's agricultural production? *(It helps control flooding and provides year-round irrigation. As a result, agricultural production has doubled.)*

SECTION 1 PROGRAM RESOURCES

Unit Resource Book: Africa

ON LEVEL

Unit Resource Book
- Reading Study Guide, p. 47
- Map and Graph Skills, p. 73
- Section Quiz, p. 85

STRUGGLING READERS

Unit Resource Book
- Reading Study Guide with Additional Support, p. 53
- Skillbuilder Practice: Comparing Maps, p. 72
- Section Quiz, p. 85
- Reteaching Activity, p. 88

eEdition with Audio DVD-ROM

ENGLISH LEARNERS

Pupil Edition in Spanish

Pupil Edition Audio DVD

Unit Resource Book
- RSG (Spanish), p. 59
- RSG with Additional Support (Spanish), p. 65

Multi-Language Glossary

Test Generator
- Section Quiz in Spanish

INCLUSION

Unit Resource Book
- RSG with Additional Support, p. 53
- Section Quiz, p. 85
- Reteaching Activity, p. 88

GIFTED & TALENTED

Unit Resource Book
- History Makers, p. 77
- Interdisciplinary Projects, pp. 79–81
- Section Quiz, p. 85

PRE-AP

Unit Resource Book
- Connect to History, p. 83
- Primary Source, p. 84
- Section Quiz, p. 85

TECHNOLOGY

Unit Transparency Book
- 3-Minute Warm-Ups, TT8
- Essential Question, TT9
- Critical Thinking, TT10
- Map: New Kingdom of Egypt, TT13
- Cultures, TT15

Daily Test Practice Transparencies
- Section 1, TT61

Power Presentations

ClassZone.com

World Cultures and Geography Video on DVD

CONNECT Geography & History

Aswan High Dam

The Aswan High Dam was built at a cost of about $1 billion. Its height is 364 feet. In comparison, Hoover Dam on the Colorado River in the United States measures 726 feet high. The Aswan High Dam reservoir is named for Gamal Abdel Nasser, Egypt's president during the years of dam construction. Lake Nasser stretches about 300 miles south from the dam into Sudan, where it is called Lake Nubia.

When rising waters created Lake Nasser, 90,000 to 100,000 people of the Nubian region were forced to relocate. Also, important archaeological sites had to be moved or would be lost. For example, at Abu Simbel, an international team of scientists and engineers dismantled and moved two temples and four gigantic statues carved into a cliff. These carvings had been made to honor the pharaoh Ramses II.

- Why is the Nile delta decreasing in size? *(The dam limits how much silt is reaching the Nile delta.)*

Teach

History of Ancient Egypt

Reader, Recorder, Reporter

- According to legend, what did King Menes accomplish for Egypt? *(Tradition says that he united Upper and Lower Egypt and founded the first Egyptian dynasty.)*
- During which kingdom did ancient Egypt become most powerful? *(the New Kingdom)* Which pharaohs ruled during that kingdom? *(Queen Hatshepsut and Ramses II, among others)*
- **Analyze Cause and Effect** How did geography isolate the earliest Egyptians? *(Nile River cataracts in the south disrupted navigation, deserts lay to the east and west, and the Mediterranean Sea bordered Egypt on the north.)*

Unit Resource Book: Africa
- Map and Graph Skills: Egypt and the Nile Delta, pp. 73–74

CONNECT Geography & History

Aswan High Dam

The Aswan High Dam contains the Nile floodwaters in a reservoir called Lake Nasser. In addition to increasing irrigated land, the dam provides electricity for factories and farm villages. But with little silt reaching the Nile's mouth and flood plain, farmers must now use costly chemical fertilizers, and the Nile delta is actually decreasing in size.

The Nile Then and Now In ancient Egypt, the Nile Valley provided various grasses and reeds used for huts, sandals, baskets, simple boats, and other products. Ancient Egyptians also dug canals to irrigate their fields and channel the Nile's water. Each June, floods would leave behind silt in a narrow strip along the flood plain and at the delta. Today the **Aswan High Dam** controls the Nile's flooding and provides year-round irrigation. Since the Aswan High Dam opened in 1970, Egypt has doubled its agricultural production.

FIND MAIN IDEAS Describe how the Nile has acted as Egypt's lifeline. The Nile's floods enriched the soil and irrigated crops.

History of Ancient Egypt

KEY QUESTION What other civilizations competed with ancient Egypt for power?

Ancient Egypt, one of the world's first great civilizations, arose along the banks of the Nile. For centuries, geography kept Egypt isolated. Six **cataracts**, or high waterfalls and rapids, make the Nile hard to navigate in the south. Deserts surround the Nile in the west and east, and the Mediterranean Sea borders Egypt to the north. People created farming villages along the Nile by about 4000 B.C. Eventually these villages became two kingdoms, Upper Egypt in the south and Lower Egypt at the Nile delta.

Uniting Upper and Lower Egypt Tradition says that around 3100 B.C., King Menes of Upper Egypt conquered Lower Egypt, uniting the two kingdoms. Modern historians believe King Menes may have been a legend, and that it probably took several rulers to join the kingdoms. King Menes is said to have founded the first Egyptian **dynasty**, or series of rulers from one family. These rulers came to be known as **pharaohs**, from the Egyptian words for "great house."

Old, Middle, and New Kingdoms Historians divide ancient Egyptian history into three periods. During the Old Kingdom, a strong central government supervised the construction of gigantic tombs called **pyramids**. Farmers and laborers built these structures when the Nile's floodwaters covered their fields. During the Middle Kingdom, Egypt conquered its southern neighbor, **Nubia**. Nubia provided Egypt with slaves, soldiers, and gold. Instead of pyramids, Middle Kingdom

DIFFERENTIATING INSTRUCTION

English Learners

Vocabulary: Key Academic Vocabulary

Read this sentence: "But with little silt reaching the Nile's mouth and flood plain, farmers must now use costly chemical fertilizers, and the Nile delta is actually decreasing in size." The key academic word in this sentence, *chemical*, signals that the delta soil must now be fertilized with these products, which are created by scientists, or chemists.

Pre-AP

Critical Discussion About Dams

Ask students to consider the effects of constructing dams. Hold a discussion about the following questions:

- What are the benefits? *(Possible Answers: flood control, better irrigation, "clean" hydroelectric power for homes and industry, a lake for recreation)*
- What are the disadvantages? *(Possible Answers: displacement of people, destruction of historic sites, changes in landscape, disappearance of silt, pollution)*

pharaohs built tombs in the desert sands. During the New Kingdom, Egypt became the strongest power in the world. A powerful pharaoh named Queen Hatshepsut led armies, sent off trade expeditions, and built splendid temples. Egypt's power peaked under the 66-year reign of Ramses II. He regained Asian territories that earlier pharaohs had lost. Statues of Ramses II still stand all over Egypt.

Conquest by Foreigners After Ramses' death in 1213 B.C., Egypt began to decline. Eventually, it fell to foreign invaders, including the **Kush** kingdom of Nubia. After an Assyrian invasion ended Kushite rule of Egypt, the Kush civilization withdrew south. In 332 B.C., Alexander the Great of Macedonia conquered Egypt. After Alexander's death, one of his generals, Ptolemy (TAHL•uh•mee), became king of Egypt, starting a dynasty that lasted nearly 300 years. One of the last Ptolemies, Queen Cleopatra, married the co-ruler of Rome, Mark Antony. Antony and Cleopatra's attempt to seize power from Antony's co-ruler failed in 31 B.C. The next year, Egypt became a Roman territory.

Animated GEOGRAPHY
Egyptian and Kush Kingdoms

Click here to learn more about Queen Hatshepsut and the New Kingdom @ ClassZone.com

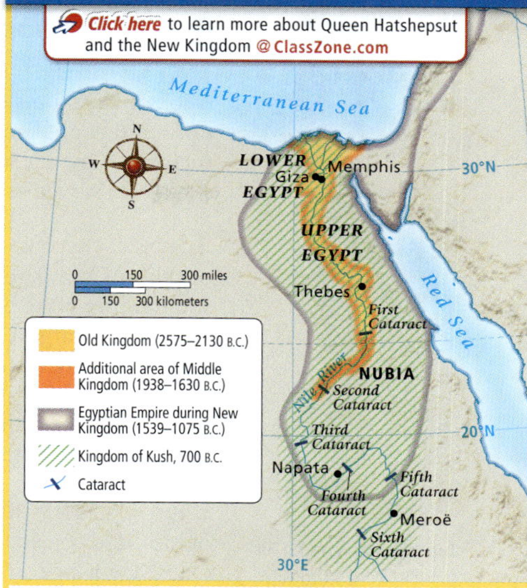

CONNECT Geography & History

Movement Which kingdom extended farthest north?

SUMMARIZE List the civilizations that interacted with ancient Egypt. Nubia, Kush, Macedonia, Rome

CONNECT to History

Egyptians decorated pyramid walls with elaborate paintings. They honored pharaohs by giving them "perfect" features, no matter how they really looked. Scenes might show pharaohs' accomplishments in life, or show them enjoying the afterlife.

Activity
Make a Pyramid

Materials
- construction paper
- ruler
- scissors
- tape
- markers

1. Research Egyptian pyramids and art.
2. Draw and cut out a six-inch square and four six-inch equilateral triangles (triangles whose sides are of equal length).
3. Decorate one side of each triangle with hieroglyphics and Egyptian paintings.
4. Tape the four sides together so that their top points meet. Then tape three of the bottom sides to the square base.
5. Leave one of the triangles unattached to the square so you can see inside.

Animated GEOGRAPHY
Egyptian and Kush Kingdoms
CLASSZONE.COM

CONNECT Geography & History

Point out that this map highlights the political borders of the ancient kingdoms in what is now Egypt and Sudan. Draw attention to the key and to the word Nubia on the map. Make sure students understand that Nubia is the name of a region, and that Kush is the name of the kingdom that occupied the region. Napata first, and later Meroë, were Kushite centers.

- Which kingdom controlled Egypt around 700 B.C.? *(Kush)*
- During which years did Egypt control the most land? *(1539–1075 B.C.)*

ANSWER

Movement the New Kingdom

Unit Resource Book: Africa
- Skillbuilder Practice: Comparing Maps, p. 72
- History Makers: Hatshepsut, p. 77

Presentation Options

Use the **Power Presentations** DVD-ROM to project the animated map.

- Listen to the audio and examine the images to learn more about the ancient Egyptian and Kush kingdoms.
- Click to animate the map.

CONNECT to History

Although the pyramid seems synonymous with Egypt, pyramid-shaped structures were built in other regions. The Kushite people of Nubia (later Sudan) built pyramids long after the Egyptians stopped using them. In the Americas, the ancient Olmec and Mayan peoples of Mexico and Central America built pyramids, as did the Moche people of Peru.

Unit Resource Book: Africa
- Connect to History: Make a Pyramid, p. 83

INTERDISCIPLINARY ACTIVITIES

CONNECT to Art
Design a Poster Fit for a Queen

Have students reread the paragraph about the New Kingdom. Divide them into small groups and ask them to design a poster honoring the accomplishments of Queen Hatshepsut. Posters should include drawings or other illustrations with captions.

CONNECT to Math
Egyptian Geometric Shapes

Have students work in pairs. They should study the instructions for making a pyramid. Then have pairs prepare a brief statement describing how the Egyptians used mathematics, or geometry, in construction and art.

Teach

Culture of Ancient Egypt

Talk About It

- Which social groups made up the middle class in ancient Egypt? *(merchants, manufacturers, and artists)*
- How did ancient Egyptians contribute to language? *(They invented a form of picture writing called hieroglyphics.)* What did they achieve in mathematics? *(They used geometry to determine land boundaries and to design buildings.)*
- **Analyze Cause and Effect** How did the belief in an earthly kind of afterlife affect the way Egyptians dealt with deaths and burials? *(Bodies of the wealthy and elite were preserved by making them into mummies, and burial chambers were filled with clothes, food, and personal items for use in the afterlife.)*

Unit Resource Book: Africa
- Primary and Secondary Sources: Sphinx of Amenemhet III, p. 84

Animated GEOGRAPHY Egyptian Social Roles

CLASSZONE.COM

Explain that nobles were typically among the ruling class because they owned land.

- What does the level of priests tell you about the role of religion in ancient Egypt? *(Possible Answer: Priests had a high position because religion was very important in Egyptian life.)*
- Why do you think Egyptians placed farmers above laborers and slaves? *(Possible Answer: Farmers had to know how to grow crops for food while laborers and slaves needed less knowledge.)*

CRITICAL THINKING ANSWER

probably "Priests and Nobles" and "Scribes and Government Officials"; priests and scribes would need to be able to read and write

Unit Resource Book: Africa
- Interdisciplinary Projects: Astronomy in Ancient Egypt, pp. 80, 81

Tutankhamen "King Tut" was nine years old when he became pharaoh, and he ruled with the help of advisors.

Culture of Ancient Egypt

KEY QUESTION What were some of the accomplishments of ancient Egyptian culture?

Ancient Egyptians built great cities, where architects, doctors, artisans, and engineers worked. They also invented a form of paper. But their most lasting achievement is probably the pyramids. These giant structures served as both monuments and tombs for the pharaohs.

Social Systems Ancient Egyptian social classes formed a kind of pyramid, as shown below. At the top was the pharoah, then a small upper class. Skilled or educated workers made up the middle classes. The largest groups, farmers, laborers and slaves, were at the pyramid's base.

Most boys followed their fathers' occupations, and girls usually became wives and mothers. Women could own, buy, sell, and inherit property, and divorce their husbands.

Egypt's Gods and the Afterlife Egyptians worshiped several deities, and considered their pharaohs to be earthly versions of them. One New Kingdom pharaoh, Akhenaton, promoted the worship of only one deity, the sun god Aton. When Akhenaton died, the new king, Tutankhamen, brought back worship of the old gods.

Animated GEOGRAPHY Egyptian Social Roles

Click here to hear from each level of Egyptian society @ClassZone.com

Social Systems Egyptian society was arranged with the largest group at the bottom and a small ruling class on top.

CRITICAL THINKING
What levels of the pyramid would require the most education?

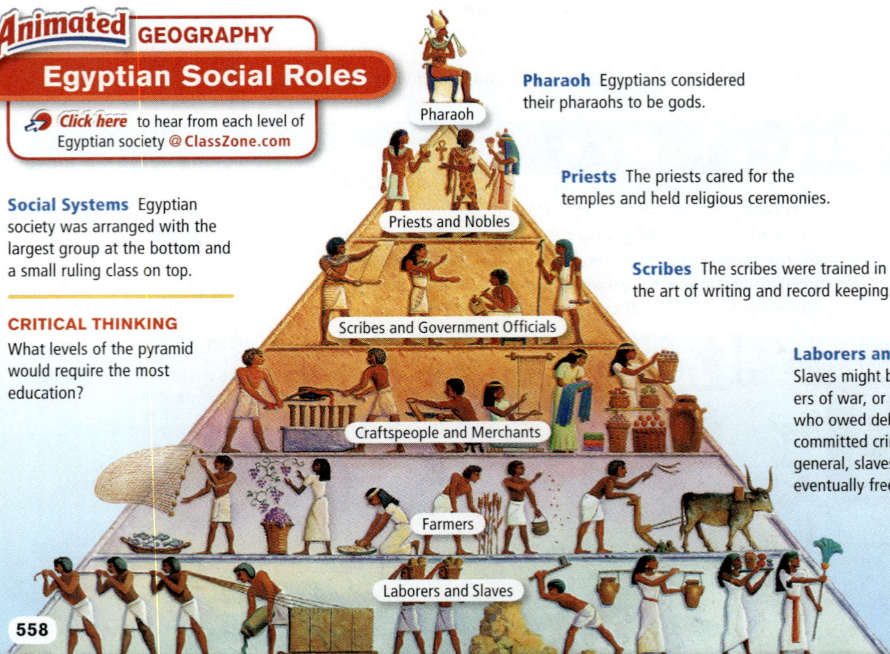

Pharaoh Egyptians considered their pharaohs to be gods.

Priests The priests cared for the temples and held religious ceremonies.

Scribes The scribes were trained in the art of writing and record keeping.

Laborers and Slaves Slaves might be prisoners of war, or people who owed debts or had committed crimes. In general, slaves were eventually freed.

558

DIFFERENTIATING INSTRUCTION: TIERED ACTIVITIES

OBJECTIVE Describe ancient Egyptian social roles and social systems.

Basic

Use the Social Roles feature to brainstorm answers to these questions with students:
- What qualifications were needed for each role?
- What would daily life be like for each class?

Have students work in pairs to summarize the answers. Have pairs exchange and discuss their summaries.

On Level

Adapt the Basic activity, so that students work independently to summarize the answers. In addition, have them write a paragraph describing Egyptian society. Have students exchange their summaries and paragraphs and discuss them.

Challenge

Adapt the On Level activity, so that students develop an essay of several paragraphs based upon describing Egyptian society. Essays should expand on issues of political power and social prestige. Ask volunteers to present their summaries and paragraphs and discuss them as a class.

Egyptians believed in an afterlife that resembled life on Earth. To preserve a person's body for the next life, they embalmed and dried it, and wrapped it in linen bandages, making a **mummy**. Egyptians also mummified pets, such as cats and monkeys. They packed their family members' tombs with items they thought they would need in the afterlife: clothes, food, makeup, and jewelry. Egyptian rulers had the fanciest tombs of all. The pharaohs' burial chambers were full of items such as sparkling gold treasures and statues of servants to care for them in the afterlife.

Hieroglyphics The word hieroglyphics comes from the Greek word for "sacred carving." **Why might the Greeks have called hieroglyphics "sacred"?**

because they were used in temples and monuments ▲

Ancient Egyptian Contributions Ancient Egypt made advances in language, science, and mathematics. Egyptians invented a form of picture writing called **hieroglyphics** which was used in temples, tombs, and on monuments. A different script was used on scrolls of **papyrus**, a paperlike material made from papyrus grass. The Egyptian calendar was based on the Nile farming cycle of flooding, planting, and harvesting. Doctors in ancient Egypt understood that a person's pulse was connected to the heart, and knew how to set broken bones. Egyptians used geometry to set land borders and design buildings.

SUMMARIZE Name some of the advancements of ancient Egypt.
Ancient Egyptians erected great cities, built pyramids, and gave women many modern rights.

Section 1 Assessment

ONLINE QUIZ For test practice, go to **Interactive Review** @ClassZone.com

TERMS & NAMES

1. **Explain the importance of**
   - Nile River
   - delta
   - Kush
   - hieroglyphics

USE YOUR READING NOTES

2. **Analyze Cause and Effect** Use your completed chart to answer the following question:

   What were the results of the Nile's annual floods?

CAUSE	EFFECT
1.	1.
2.	2.
3.	3.

KEY IDEAS

3. Why did ancient Egyptians cluster around the valley and delta of the Nile River?
4. What three periods make up ancient Egyptian history?
5. What was the largest social class in ancient Egypt?

CRITICAL THINKING

6. **Summarize** What are some positive and negative effects of the Aswan High Dam?
7. **Evaluate** What rights did women have in ancient Egypt?
8. **CONNECT to Today** What are the risks and benefits of living near a great river?
9. **SCIENCE** **Create a Multimedia Presentation** Using the library or Internet, research mummification and create a slide show to demonstrate the process. Show all the steps from removal of organs and embalming to drying and wrapping.

CONNECT to the Essential Question

How have people in North Africa adapted to the region's arid climate?

Ask students what they have learned so far that can help them answer this question. Students might mention:

- Ancient Egyptians dug canals to use the Nile for irrigation.
- Modern Egyptians built the Aswan High Dam for flood control and year-round irrigation.
- Ancient Egyptians built cities and pyramids, developed hieroglyphics, and made advances in science and mathematics.

4 Assess & Reteach

Assess Have students complete the Section Assessment.

Unit Resource Book: Africa
- Section Quiz, p. 85

Interactive Review @ClassZone.com
- Section Quiz

Power Presentations

Test Generator

Reteach Assign small groups of students one or more subheadings in Section 1. Ask groups to create two true/false questions from their assigned text. Then have a member of each group quiz other groups.

Unit Resource Book: Africa
- Reteaching Activity, p. 88

SECTION 1 ASSESSMENT ANSWERS

Terms & Names

1. Nile River: the world's longest river; delta: the deposit of rich soil at a river's mouth; Kush: ancient Nubian kingdom in northern Sudan; hieroglyphics: ancient Egyptian picture writing

Use Your Reading Notes

2. The annual floods brought water for Egypt's soil and left behind rich deposits of silt.

CAUSE	EFFECT
1. Ancient Nile flooded yearly	1. Silt deposited in river valley
2. Aswan High Dam built	2. Nile no longer floods; farm production doubled
3. Six cataracts along Southern Nile	3. Nile River difficult to navigate

Key Ideas

3. The land is fertile near the river.
4. the Old Kingdom, the Middle Kingdom, and the New Kingdom
5. laborers, farmers, and slaves

Critical Thinking

6. Positives: year-round irrigation, increased productivity, electricity, and flood relief; negatives: need for costly chemical fertilizers, shrinkage of Nile delta
7. They took traditional roles, but they could also own, buy, and sell property, and get divorced. Hatshepsut was a powerful female ruler.
8. Risks: flooding; benefits: rich farmland and transportation
9. Use the rubric to score presentations.

Presentation Rubric

	Slide Content
4	all steps illustrated
3	most steps illustrated
2	most steps missing
1	only a few steps illustrated

1 Plan & Prepare

Objectives: Describe the Great Pyramid and explain how it was built, including resources used; and its purpose in ancient Egyptian society.

Visual Vocabulary: Ask students to describe what they see in the illustration. Call out familiar items listed below, asking students to locate each one on the page. Then tell them the precise term, as listed in boldface. Add terms to a vocabulary list, illustrated poster, or Word Wall.

- sloping roadway **(ramp)**
- structure shaped like a triangle **(pyramid)**
- kind of rock used for building **(limestone blocks)**
- inside area **(interior)**
- room used as a tomb **(burial chamber)**

2 Teach

Engage . . .

DISPLAY "The Great Pyramid" feature from the **Power Presentations** DVD-ROM or the *Animations Center* at **ClassZone.com.** Have volunteers point to some of the vocabulary terms listed above. Then ask:

- What is the landscape like around the Great Pyramid?
- Who is the person with the red head covering and the worker with the scrolls?
- What equipment and methods were used for construction?
- What is the Great Pyramid like inside?

Encourage all answers.

Explore . . .

ROLL OVER sections of the illustration or click on items in the menu. Ask students to choose which areas they would like to explore further.

- Return to the discussion questions. Have any answers changed?
- Have students jot down a list of at least three new questions that arise from watching the rollover features.

Animated GEOGRAPHY
The Great Pyramid

Click here to enter the Great Pyramid @ClassZone.com

PYRAMIDS

The Great Pyramid of Khufu is Egypt's largest pyramid, originally standing 481 feet high. For more than 4,000 years it stood as the world's tallest building.

Click here to visit the pyramids of Egypt today. Learn about the rulers of ancient Egypt and their belief in the afterlife.

Click here to see how a pyramid was built. Learn about the materials and methods and the people who built them.

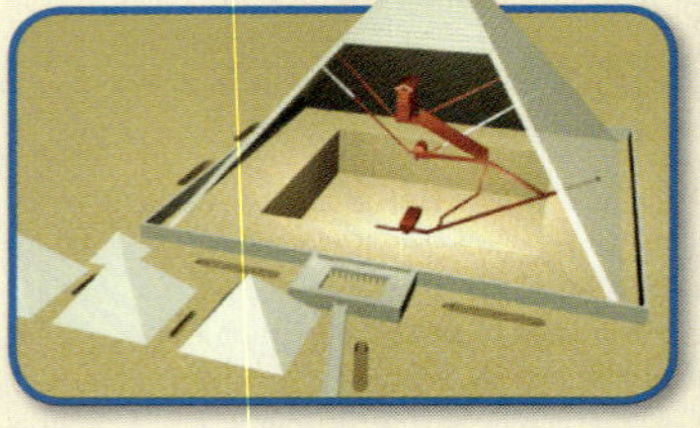

Click here to explore the passageways and chambers of the Great Pyramid's interior.

DIFFERENTIATING INSTRUCTION: TIERED ACTIVITIES

OBJECTIVE Describe the Great Pyramid and explain how it was built, including resources used, and its purpose in ancient Egyptian society.

Basic

Illustrating Construction Methods

Brainstorm with students the materials and methods used for modern building construction. Ask them to study the Animated Geography feature about the Great Pyramid. Then have pairs of students create one drawing about the Great Pyramid's construction and another drawing about modern building construction. Students should write a title for each drawing and include at least one caption that highlights a similarity and one that highlights a difference between construction methods and/or materials.

GeoActivity

Plan a Pyramid Split into groups of three and plan a pyramid for a newly crowned pharaoh. Consider how to keep the burial chamber safe from thieves and what sort of facilities the workers will need.

On Level

Comparing Architects and Construction Workers

Brainstorm with students the work life of modern architects and construction workers. Ask them to study the Animated Geography feature about the Great Pyramid. First, have pairs of students list the challenges faced by architects and workers of the Great Pyramid and of today. Next, have students list the role of technology then and now. Last, have students create a chart or visual display that organizes their findings.

Challenge

Describing the Importance of Great Construction Projects

Have students work in pairs to create a descriptive report about the role of the Great Pyramid in ancient Egyptian society and of a modern building, such as the Empire State building. The report should include details such as: each building's purpose, available construction materials and methods, and the challenges the architects and workers faced and how they overcame those challenges.

Engage . . .

CHOOSE topics to help students answer their questions, such as:

- **The Pharaoh:** Khufu is also known as Cheops. He lived about 2600 B.C., during the Fourth Dynasty. His pyramid was built with several passages and rooms inside.
- **The Giza Site:** The Great Pyramid stands on the west side of the Nile River outside Cairo. Beside it are pyramids built for the kings Khafre and Menkaure. Nearby is a colossal statue called the Great Sphinx.
- **Architecture:** Imhotep is the architect credited with developing the pyramid by using limestone blocks instead of mud bricks. He lived during the Third Dynasty.
- **Construction Workers:** Scholars estimate that 100,000 farm workers built the pyramids, working in shifts of three to four months. Farmers worked on the pyramids when the Nile flooded their fields.

3 Wrap-Up

Elaborate . . .

ASSIGN students into several groups to develop a skit about a pharaoh inspecting his pyramid construction site.

- Groups should assign each member a role, such as pharaoh, servant, architect, worker, and water carrier.
- Have groups perform their skits.

Evaluate . . .

Plan a Pyramid
Activity Rubric

4 Excellent The group's plan accurately describes steps in pyramid construction and includes many details about safety features and workers' facilities.

3 Strong The plan is fairly accurate and includes several details.

2 Fair The plan is somewhat accurate and includes a few details.

1 Poor The plan is inaccurate and lacks details.

1 Plan & Prepare

Objectives

- Identify foreign powers in Egypt since ancient times and the steps to Egyptian independence
- Describe life in Egypt today
- Explain how cultural differences cause conflicts in Sudan

Read for the Essential Question

Help students read for a purpose by reminding them of the Essential Question: "How have people in North Africa adapted to the region's arid climate?" Point out that they can use text features such as Key Questions, special features, images, and maps to form answers to the question.

Vocabulary

Best Practices Toolkit

Use the ***Best Practices Toolkit*** to model strategies for vocabulary notetaking. Vary strategies throughout the year. Choose from: *Knowledge Rating, Predicting ABCs, Definition Mapping, Word Sort, Frayer Model (Word Squares), Magnet Words, Word Wheel,* and *Student VOC.*

Vocabulary Strategies, TT9–TT16

Reading Strategy

Best Practices Toolkit

Display the Time Line transparency.

- Remind students that dates in the text, headings, and subheadings can help them sequence events, or place events in chronological order.
- Display the graphic organizer.
- Model filling in a few events in sequence, using the section subheads and text.
- Have students suggest entries.
- Display the graphic organizer for students to review.

Time Line, TT33

SECTION 2 Reading for Understanding

Key Ideas

BEFORE, YOU LEARNED

Ancient Egyptians built a great civilization, but were conquered by foreign powers.

NOW YOU WILL LEARN

A blending of cultures has made Egypt a leader for North Africa, as well as for Southwest Asia.

Vocabulary

TERMS & NAMES

Rosetta Stone an ancient Egyptian stone that provided a key to decipher hieroglyphics

Islam a religion believing in one god, called Allah, and his prophet on Earth, Muhammad

Muslim a follower of Islam

Suez Canal a waterway linking the Red Sea with the Mediterranean Sea

protectorate a weaker country or area controlled by a stronger country

Anwar el-Sadat president of Egypt from 1970 to 1981

Camp David Accords peace treaty between Egypt and Israel, signed in 1979

Sudan a country of northeast Africa, located south of Egypt

BACKGROUND VOCABULARY

modernize to accept modern ways or ideas

Visual Vocabulary Suez Canal

Reading Strategy

Re-create the time line shown at right. As you read and respond to the **KEY QUESTIONS**, use the time line to order important events in Egypt and Sudan.

See Skillbuilder Handbook, page R6

SEQUENCE EVENTS

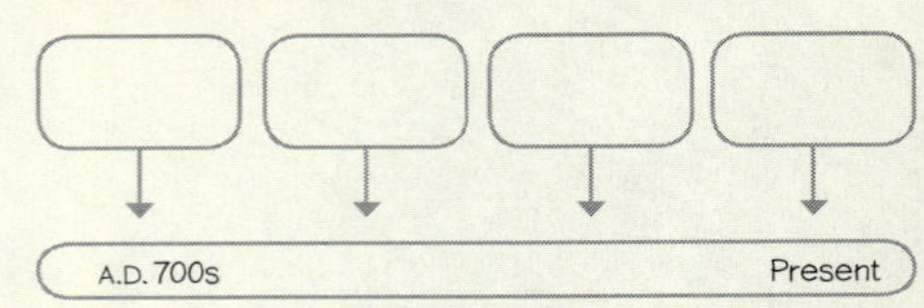

GRAPHIC ORGANIZERS Go to **Interactive Review** @ClassZone.com

PRETEACHING VOCABULARY

English Learners

Pronounce and Preview

Pronounce each term for students. Review the meanings of words within definitions, such as *decipher, prophet,* and *treaty.*

- To modify vocabulary learning, have students complete worksheets as they read, instead of afterward.

Unit Resource Book: Africa
- Vocabulary Practice, p. 71

Inclusion

Vocabulary: Knowledge Rating

Have students work in pairs to fill out the graphic organizer for the terms in the section.

- Use the transparency to model the strategy.
- Have volunteers suggest entries.
- Then distribute copies of the organizer for students to complete with a partner.

Best Practices Toolkit
- Vocabulary: Knowledge Rating, TT9

Philae Temple, Egypt

SECTION 2 FOCUS ON

Nile crocodile

Modern Egypt and Sudan

Connecting to Your World

Have you ever created a secret code? Typically, one set of letters stands for letters you already know. Hieroglyphics were like a mysterious code until the 1799 discovery of a carved stone called the **Rosetta Stone**. It contained the same text in ancient Greek, Egyptian script, and hieroglyphics. Since scholars knew the first two scripts, they were able to crack the code of hieroglyphics. But no one in the world today knows how ancient Egyptian was spoken.

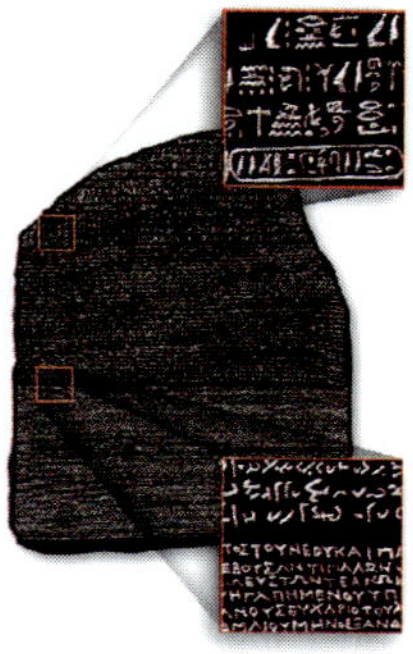

Rosetta Stone This discovery was the key to decoding Egyptian hieroglyphics.

From Ancient to Modern Times

KEY QUESTION What foreign powers ruled Egypt?

The Arabic language became the dominant language in Egypt and North Africa because of a culture shift in the 600s and 700s. Around that time, Arabs from Southwest Asia introduced a new religion called Islam.

Temple of Luxor A white mosque stands amid the ruins of this ancient temple.

563

2 Focus & Motivate

3-Minute Warm-Up

Write on the board or display the transparency:

- Egypt and Sudan are considered developing countries today. What challenges do developing countries face? *(Possible Answers: slums, overcrowding, unemployment, widespread poverty)*

Unit Transparency Book: Africa
- 3-Minute Warm-Ups, TT8

Connecting to Your World

More About . . .

The Temple of Luxor

Luxor is a modern Egyptian city, built along the Nile at the site of ancient Thebes. Amenhotep III started the Temple of Luxor during the 18th dynasty.

3 Teach

From Ancient to Modern Times

Think, Pair, Share

- Which religion was introduced into Egypt and North Africa in the 600s and 700s? *(Islam)* What is the name for the followers of this religion? *(Muslims)*
- Which foreign powers controlled Egypt after 1798? *(France, the Ottomans, and Britain)*
- **Sequence Events** What action by Egyptian president Anwar el-Sadat helped Israel agree to a peace treaty? *(Anwar el-Sadat visited Israel.)*

SECTION 2 PROGRAM RESOURCES

Unit Resource Book: Africa

ON LEVEL

Unit Resource Book
- Reading Study Guide, p. 49
- Section Quiz, p. 86

STRUGGLING READERS

Unit Resource Book
- Reading Study Guide with Additional Support, p. 55
- Skillbuilder Practice: Comparing Maps, p. 72
- Section Quiz, p. 86
- Reteaching Activity, p. 89

ENGLISH LEARNERS

Pupil Edition in Spanish
Pupil Edition Audio DVD
Unit Resource Book
- RSG (Spanish), p. 61
- RSG with Additional Support (Spanish), p. 67

Multi-Language Glossary
Test Generator
- Section Quiz in Spanish

INCLUSION

Unit Resource Book
- RSG with Additional Support, p. 55
- Section Quiz, p. 86
- Reteaching Activity, p. 89

GIFTED & TALENTED

Unit Resource Book
- History Makers, p. 78
- Section Quiz, p. 86

PRE-AP

Unit Resource Book
- Economics in the World, p. 82
- Section Quiz, p. 86

TECHNOLOGY

Unit Transparency Book
- 3-Minute Warm-Ups, TT8
- Essential Question, TT9
- Critical Thinking, TT11
- Map: Northeastern Africa: Physical, TT14

Daily Test Practice Transparencies
- Section 2, TT62

Power Presentations
ClassZone.com
World Cultures and Geography Video on DVD

RESEARCH & WRITING CENTER

HISTORY MAKERS

Muhammad Ali Pasha

The sultan of the Ottoman Empire gave Muhammad Ali the right to rule Egypt. Muhammad Ali was born in the Ottoman Empire at Kavala, which is now in Greece. As Egypt's ruler, he reorganized the military and took control of the economy. He placed agricultural lands under state administration and introduced crops such as cotton to export for a high return. In 1841 the Ottoman Empire granted Muhammad Ali the hereditary title of *khedive*. Muhammad Ali's fourth son, Said, took power in 1854.

Unit Resource Book: Africa
- History Makers: Anwar el-Sadat, p. 78

More About . . .

The Suez Canal

An effort had been made in ancient times to connect the Red Sea to the Mediterranean Sea. A canal existed for a period of time across the Isthmus of Suez, but was filled in during the 7th century. Muhammad Ali's son Said gave Ferdinand de Lesseps of France permission to plan construction of a new Suez Canal. A Suez Canal Company was soon formed to operate the project. The Suez Canal took ten years to construct and opened November 17, 1869. It is about 100 miles long and saves thousands of miles of travel for ships that would otherwise have to sail around the continent of Africa.

More About . . .

The Camp David Accords

The agreement is called the Camp David Accords because Egyptian president Anwar el-Sadat met prime minister Menachem Begin of Israel at Camp David, Maryland. President Jimmy Carter of the United States hosted the meeting, held in September 1978. Sadat and Begin shared the Nobel Peace Prize for their efforts to forge the treaty, which was signed in 1979. The treaty proved very unpopular in Egypt. Although Sadat attempted to squelch his opposition, he was assassinated in 1981.

Influence of Islam **Islam** is the belief in one god, called Allah, and his prophet on Earth, Muhammad. Followers of Islam are called **Muslims**. (You will learn more about the spread of Islam in the next unit.) Arab Muslims invaded Egypt in A.D. 639. Afterward, Egyptians began to speak Arabic, and many adopted the Muslim faith. Islamic empires ruled Egypt for more than 1,000 years. The Ottoman Empire took control in 1517, but other groups contended for power.

During Napoleon Bonaparte's rule, France briefly occupied Egypt, beginning in 1798. The Ottomans regained control in 1801 under a military leader, Muhammad Ali. He tried to **modernize** Egypt. Though Muhammad Ali's efforts failed, his son Said (sa•EED) made a more lasting change. Under Said's reign, in 1859, workers began to dig the **Suez Canal**, a waterway linking the Red Sea with the Mediterranean. The canal continues to be a key trade route for Egypt, Europe, and Southwest Asia.

HISTORY MAKERS

Muhammad Ali Pasha 1769–1849

As an officer in the Ottoman army, Muhammad Ali helped rid Egypt of French invaders. In 1805, he became Egypt's governor, or Pasha. During his 44-year reign, he attempted to transform Egypt into a modern country. Muhammad Ali brought European advisors and teachers to Egypt and sent Egyptians to study in Europe. He also tried to bring industry to this agricultural land. Muhammad Ali's descendants ruled under British control, but his family held power in Egypt until 1953.

ONLINE BIOGRAPHY
For more on the life of Muhammad Ali Pasha, go to the Research & Writing Center @ClassZone.com

Becoming Independent The valuable trade route created by the Suez Canal led more foreign powers to try to control Egypt. Britain invaded Egypt in 1882. During World War I, Britain declared Egypt a **protectorate**, or country controlled by a stronger nation.

After the war, Egyptians began to call for independence. Britain granted independence in 1922, but kept troops in Egypt to guard the Suez Canal. During World War II, Egypt and other Arab nations formed the Arab League, which opposed the formation of Israel in 1948. Egypt lost four wars with Israel between 1948 and 1973. In 1977, Egyptian President **Anwar el-Sadat** traveled to Israel. In 1979, Egypt and Israel signed the **Camp David Accords**, the first peace treaty between Israel and an Arab state. The Arab League expelled Egypt for signing the treaty, but readmitted Egypt in 1989.

Egypt remains a leader in North Africa and the Arab world. Its capital, Cairo, is Africa's biggest city. In Africa, only Nigeria has more people. Although Egypt is not a big oil producer, Saudi Arabia is the only Middle Eastern country with a stronger economy.

SEQUENCE EVENTS Identify the foreign powers that ruled Egypt.
Arab Muslims, the Ottoman Empire, France, Great Britain

DIFFERENTIATING INSTRUCTION

Struggling Readers

Clarifying Egypt's Situation

Guide students by pointing to the Suez Canal on a map or globe. Have them answer questions such as the following:
- Why was the Suez Canal important? *(Possible Answer: It served as a major shipping passage for trade and military might.)*
- Why would the canal cause foreign powers to try to control Egypt? *(Possible Answer: By controlling Egypt, they could control the Suez Canal.)*

English Learners

Vocabulary: Word Relationships

To help students understand the terms about ruling on this page, review related words and phrases such as *took control, contended for power, rule,* and *granted independence*. Focus on the sentence in the History Makers feature, "Muhammad Ali's descendants ruled under British control, but his family held power in Egypt until 1953." Explain that family members governed Egypt, but they also had to obey the British.

Life in Egypt Today

KEY QUESTION How do most Egyptians live?

Egyptian city life is hectic and noisy. Huge slums adjoin wealthy residential areas, and the streets teem with traffic and overcrowded streetcars. People might wear Western clothes or traditional Egyptian clothing: a shirt and pants for men and a long, colorful flowing gown for women. Culturally, most Egyptians consider themselves Arabs and practice Islam.

More than half of Egyptians live in the countryside. Most villagers tend small farms on land often owned by wealthier landowners. Electric power is common in rural areas. Poor city dwellers and villagers alike eat a diet of bread, beans, and vegetable stew. Sweetened coffee or tea are popular drinks.

National Government Egypt's president serves a six-year term and may be re-elected indefinitely. The powerful president also leads the armed forces, appoints a prime minister and cabinet, and even appoints some members of the parliament. In 2005, Egypt allowed multiple candidates to run for president. Previously, parliament had selected single candidates for citizens to vote on.

Economic Structure and Activities Despite its role as a regional leader, Egypt is still a developing nation, struggling with unemployment and with how to feed its growing population. Agriculture employs more than one-third of Egyptian workers, but accounts for only one-sixth of the economy.

upstream, because the Nile flows from south to north ▼

Cairo The Nile winds through Egypt's capital city. **If you are traveling south along the Nile, are you headed upstream or downstream?**

INTERDISCIPLINARY ACTIVITIES

CONNECT to Civics

Analyze Egyptian Government

Have students work in small groups to discuss the advantages and disadvantages of the Egyptian system of government and its election change in 2005. Ask them to make comparisons between the U.S. government and Egypt's. Then form a panel of one or two members from each group to present the comparisons as part of a panel discussion on Egypt's government.

CONNECT to Language Arts

Compare and Contrast City Life

Ask them to compare city and country life in Egypt with that in the United States in a four-paragraph essay as follows:

- Introduction
- Comparison of country life
- Comparison of city life
- Conclusion

Have students trade their essays with a partner and edit each other's work. Allow students time to revise their essays, then ask volunteers to read them to the class.

Teach

Life in Egypt Today

Reader, Recorder, Reporter

- How much of Egypt's economy do service industries account for? *(more than half)* How much of the economy does agriculture account for? *(one-sixth, but one-third of Egyptian workers)*
- What change occurred recently in the election of Egypt's president? *(Multiple candidates can run instead of single candidates chosen by parliament.)*
- **Sequence Events** What has Egypt done to boost its economy? *(The Egyptian government tried to attract foreign investment in Egypt's energy and tourism industries.)*

Unit Resource Book: Africa
- Economics in the World: Oil Pipeline to Spain, p. 82

More About . . .

Cairo

Cairo has long been regarded as more than "just" Egypt's capital. It has been the cultural, industrial, and commercial center for a much wider region—and remains an educational center for the Arab world. Cairo University is Egypt's largest university; Al-Azhar University, founded in 970, is the oldest. Cairo also has many mosques, or Islamic places of worship, notable for their religious and architectural importance.

Cairo was founded in 969 by the Fatimid Islamic group, near the ancient site of Memphis and the pyramids at Giza. Today, Cairo is a sprawling metropolis with the problems of traffic congestion, poverty, and pollution that typically plague major cities. However, major changes have occurred in recent decades. Cairo has an international airport, an improved system of highways and overpasses, and an extensive underground rail network.

Teach

Sudan: Egypt's Neighbor

Talk About It

- What cultural differences exist between north and south in Sudan? *(Most northerners are Muslims who speak Arabic while most southerners are Africans who observe traditional religions or in a few cases Christianity.)*
- What activity is most important in Sudan's economy? *(farming)* What is Sudan's leading export? *(oil)*
- **Sequence Events** When did the Darfur conflict erupt? *(2003)* How has Sudan been affected by the conflict? *(Tens of thousands of Sudanese civilians have died, about two million have been displaced, and disruption continues.)*

Unit Resource Book: Africa
- Skillbuilder Practice: Comparing Maps, p. 72

CONNECT History & Culture

Lost Boys of Sudan

The lost or orphaned refugees were primarily boys because they were in the fields herding cattle when militias attacked their villages. The boys walked about 1,000 miles, sometimes for days with no water or proper food. Many saw their companions die from starvation, sickness, lion attacks, or drownings at river crossings. Some Lost Boys spent years in refugee camps.

More About . . .

Refugees and the Darfur Crisis

An estimated two million Sudanese refugees live in overcrowded camps such as the Kalma Refugee Camp. Although the refugees have attracted international attention, they still lack sufficient food, water, and health services. African Union forces have policed the Darfur region, but Arab militias called "janjaweed" continue to attack civilians. The U.S. Congress declared the Darfur situation an attempted genocide.

Service industries, such as banking, education, and trade, employ half of Egypt's workers and account for more than half of the economy. Thanks to Egypt's high-grade cotton, textile production provides one-third of the economy. In this socialist country, the government owns most major industries. But change is under way. To stimulate Egypt's economy, the government tries to attract foreign investment in Egypt's energy and tourism industries. Tourists, lured by ancient temples and pyramids, provide the largest share of Egypt's foreign funds.

SUMMARIZE Describe daily life in Egypt. Most people live in the countryside, but cities are crowded. Unemployment is a problem. Clothing is a mix of Western and traditional dress.

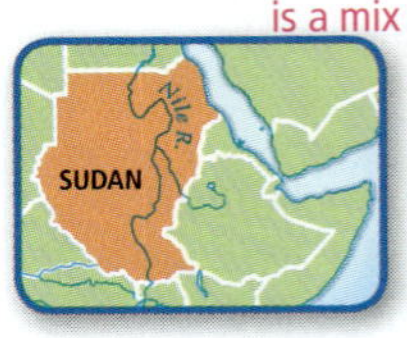

Sudan: Egypt's Neighbor

KEY QUESTION What cultural differences are behind the conflicts in Sudan?

You already learned that ancient Egypt controlled Nubia, or northern Sudan. After ancient Egypt lost control, Nubia's Kush kingdom endured until its collapse around A.D. 350. Modern **Sudan**, Egypt's southern neighbor, is Africa's largest nation in area. Part of Sudan extends into central Africa, with its savannas and rain forests. Northern Sudan is covered by desert.

Culture: Regional Differences Most people who live in the northern two-thirds of Sudan speak Arabic, the country's official language, and follow Islam. In the southern third of Sudan, various African groups practice traditional animist religions. A few are Christians.

Most Sudanese people live on subsistence farms or herd animals in the desert. Three out of ten Sudanese live in cities and towns. Sudan's main urban center—the area surrounding the capital, Khartoum—is home to more than six million people. Near the capital, two million more Sudanese live in refugee camps, displaced by civil wars.

CONNECT History & Culture

Lost Boys of Sudan

The Lost Boys were orphans of Sudan's civil war. These children and young adults banded together to make a harrowing journey from Sudan to Ethiopia, back to Sudan, and then to refugee camps in Kenya. In 2001, the United States took in 3,600 Lost Boys across the country, primarily in Texas, Michigan, and Arizona. Now mostly adults, many Lost Boys work one or two jobs while attending high school or college classes.

Modern History of Sudan In the 1800s, Sudan again came under Egyptian control. But in 1881, a Sudanese Muslim teacher led a four-year revolt, overturning Egyptian rule. In 1898, Britain and Egypt joined forces to conquer Sudan. Britain ruled Sudan until 1956.

Between independence and 1998, military governments ruled Sudan. Conflicts between Sudan's Arab Muslim north and non-Muslim,

566 Chapter 18

DIFFERENTIATING INSTRUCTION: TIERED ACTIVITIES

OBJECTIVE Describe the Sudanese refugee experience.

Basic

Review the text headed "Sudan: Egypt's Neighbor" and information about the Lost Boys. Have students work in pairs and think about what life was like for the Lost Boys before coming to the United States. Have students create short stories about the experiences of the Lost Boys.

On Level

Adapt the Basic activity, so that students work in pairs to create a skit or Readers Theater about the experiences of the Lost Boys and present it to the class.

Challenge

Have students work alone to create an interview with a Lost Boy. Then have them pair up, rehearse and revise their interviews, and present both to the class. (Students should play one role—interviewer or interviewee—in the first interview presentation, then switch roles for the second.)

non-Arab south led to two civil wars during the 20th century. Millions of people lost their lives. A separate conflict in a western region of Sudan called Darfur erupted in 2003. This war pitted Arab herders, aided by government militias, against non-Muslim ethnic groups. Tens of thousands of civilians died, and roughly two million farmers were forced from their homes.

Herding Sudan's government donated these goats to people in refugee camps.

Government and Economy Drought and famine periodically plague Sudan, which remains mired in the Darfur conflict. Under Sudan's 1998 constitution, the president serves a five-year term and appoints a cabinet. Voters elect most members of the legislature, who serve terms of four years.

This developing economy depends heavily on agriculture. Sudanese factories, which employ about four percent of the population, manufacture cement, food products, food, and textiles. In 1999, Sudan began producing oil, currently the country's leading export. But most of Sudan's roads remain unpaved, and four out of ten people live in poverty.

FIND MAIN IDEAS Describe the cultural differences leading to Sudan's civil wars and conflicts. ethnic (Arab vs. non-Arab Africans) and religious (Muslim vs. non-Muslim)

Section 2 Assessment

ONLINE QUIZ For test practice, go to **Interactive Review** @ClassZone.com

TERMS & NAMES
1. **Explain the importance of**
   - Rosetta Stone
   - Suez Canal
   - protectorate
   - modernize

USE YOUR READING NOTES
2. **Sequence Events** Use your completed time line to answer the following question: What was the last foreign power to control Egypt?

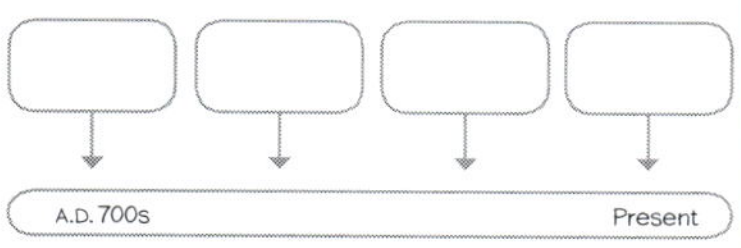

KEY IDEAS
3. Why were the Camp David Accords important?
4. With which culture and religion do most Egyptians identify?
5. What natural and political problems have plagued Sudan?

CRITICAL THINKING
6. **Draw Conclusions** Why do you think the Suez Canal has been so important to Egypt?
7. **Make Inferences** Why has the Sudanese government's attempt to enforce Islamic law caused problems in this mostly Muslim country?
8. **CONNECT to Today** Terrorist bombings at popular Egyptian tourist sites took place in 2004 and 2005. Why do such attacks threaten Egypt's economy?
9. **MATH** **Make a Graph** Make a bar graph showing Egypt's major industries. Your graph should clearly show which industries employ most people and which industries are most profitable.

CONNECT to the Essential Question

How have people in North Africa adapted to the region's arid climate?

Ask students what they have learned so far that can help them answer this question. Students might mention:

- Egypt became a regional leader, modernized, improved trade with the opening of the Suez Canal, and achieved independence.
- Egypt developed service and textile industries and is trying to attract tourists and foreign investment.
- Sudan overcame foreign rule and began exporting oil, but continues to face regional conflicts.

4 Assess & Reteach

Assess Have students complete the Section Assessment.

Unit Resource Book: Africa
- Section Quiz, p. 86

Interactive Review @ClassZone.com
- Section Quiz

Power Presentations

Test Generator

Reteach Have students compare the histories and current situations of Egypt and Sudan. Ask volunteers to record information on the board in a Venn diagram. Then have students use this information and their textbooks to create time lines for each country.

Unit Resource Book: Africa
- Reteaching Activity, p. 89

SECTION 2 ASSESSMENT ANSWERS

Terms & Names
1. Rosetta Stone: stone that allowed scholars to decode hieroglyphics; Suez Canal: important trade waterway linking the Red Sea and Mediterranean Sea; protectorate: a weaker region or country controlled by a stronger country; modernize: to accept or adopt modern ways

Use Your Reading Notes
2. Britain, which granted independence in 1922

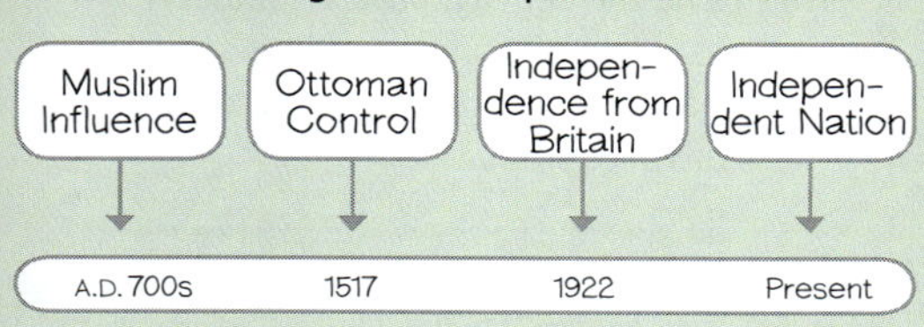

Key Ideas
3. After four wars, the Camp David Accords were the first peace treaty between an Arab state and Israel.
4. Most consider themselves Arabs and Muslims.
5. drought, famine, civil wars, and government-sponsored conflicts

Critical Thinking
6. It links the Red Sea to the Mediterranean, making it a strategic route for regional shipping.
7. Most people in southern Sudan aren't Muslims and resent government interference with their own religious customs and beliefs.
8. Tourism is Egypt's main industry; if tourists become fearful and stay away, Egypt's economy will suffer.
9. Use the rubric to score graphs.

Graph Rubric

	Industries Covered	Presentation
4	5 or more	attractive, accurate
3	4	neat
2	3	adequate, some errors
1	2 or fewer	sloppy, many errors

1 Plan & Prepare

Objectives

- Describe the cultural influences in North Africa
- Compare the governments and economies in North Africa

Read for the Essential Question

Help students read for a purpose by reminding them of the Essential Question: "How have people in North Africa adapted to the region's arid climate?" Point out that they can use text features such as Key Questions, special features, images, and maps to answer the question.

Vocabulary

Best Practices Toolkit

Use the ***Best Practices Toolkit*** to model strategies for vocabulary notetaking. Vary strategies throughout the year. Choose from: *Knowledge Rating, Predicting ABCs, Definition Mapping, Word Sort, Frayer Model (Word Squares), Magnet Words, Word Wheel,* and *Student VOC.*

Vocabulary Strategies, TT9–TT16

Reading Strategy

Best Practices Toolkit

Display the Web Diagram transparency.

- Remind students that comparing and contrasting means looking for examples of similarities and differences.
- Model filling in a few entries from the subsection "Government and Economics."
- Have students suggest entries. Remind them that the information in their completed charts will help them answer the Key Questions in the section.

Web Diagram, TT28

Section 3 Reading for Understanding

Key Ideas

BEFORE, YOU LEARNED

Egypt's modern culture was shaped by many outside forces, including the spread of Islam and conquest by the Ottoman Empire.

NOW YOU WILL LEARN

The Ottomans and Islam also influenced other North African nations, as did European and colonizing powers.

Vocabulary

TERMS & NAMES

Berbers (BUR•buhrz) the original inhabitants of North Africa and the Sahara

Carthage (KAR•thihj) an ancient North African city-state

Punic Wars (PYOO•nik) wars fought between Rome and Carthage during the 3rd and 2nd centuries B.C.

Maghrib (MUHG•ruhb) the Arab name for North Africa, which means "the West"

Atlas Mountains series of mountain ranges stretching from Morocco to Tunisia

souk (sook) open-air market

jellaba a traditional Berber garment; a robe with full sleeves

dictatorship a state or government ruled by a leader with absolute power

Visual Vocabulary souk

Reading Strategy

Re-create the web diagram shown at right. As you read and respond to the **KEY QUESTIONS**, use the diagram to compare and contrast important information about Morocco, Algeria, and Libya.

See Skillbuilder Handbook, page R9

COMPARE AND CONTRAST

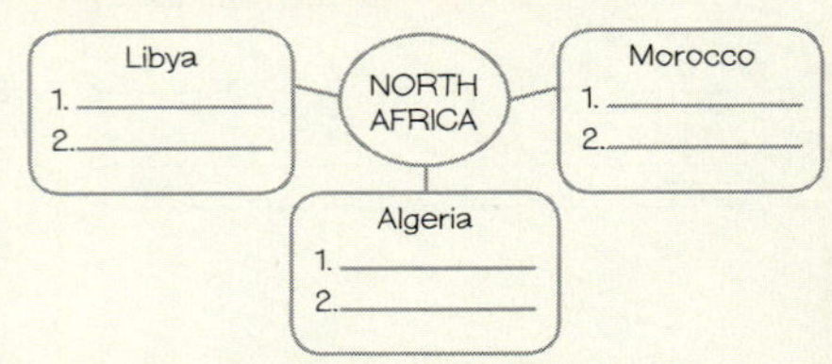

GRAPHIC ORGANIZERS
Go to **Interactive Review** @ClassZone.com

568 Chapter 18

PRETEACHING VOCABULARY

English Learners

Pronounce and Preview

Pronounce each term for students. Review the meanings of words within definitions, such as *inhabitants, traditional,* and *absolute power.*

- To modify vocabulary learning, have students complete worksheets as they read, instead of afterward.

Unit Resource Book: Africa
- Vocabulary Practice, p. 71

Inclusion

Vocabulary Categories

Read one term and its definition aloud. Discuss the definitions as a group. Then have a volunteer categorize one term, according to whether it refers to people, places, or events. Repeat for each term. To extend the activity, have volunteers call out a category and ask others to supply a vocabulary term or another word from this section that fits.

Philae Temple, Egypt

Nile crocodile

Other Nations of North Africa

Connecting to Your World

What happens in your home town on July 4? More than 200 years after declaring independence from Britain in 1776, U.S. citizens still celebrate the 4th of July with picnics, parades, concerts, and fireworks. By contrast, Libya celebrated its first Independence Day in 1951. Algeria won its freedom in 1962. Like many African nations, these North African countries have enjoyed less than 100 years of independence.

A Muslim Region

KEY QUESTION What cultures have most influenced North Africa?

Hazardous desert trade routes connected ancient North Africa to the sub-Saharan bulk of the continent. In contrast, just eight miles separate Europe from North Africa at the Strait of Gibraltar. As a result, ancient North Africans looked to Europe for trade and cultural contacts.

Ramadan Ramadan is a holy month of fasting in the Muslim faith. Here, men pray in the streets of Algiers.

2 Focus & Motivate

3-Minute Warm-Up

Write on the board or display the transparency:

- Before you read further, what have you already learned about North Africa? *(Possible Answers [within context of Egypt]: most North Africans are Muslims; religion is very important to North Africans; men and women may have traditional roles.)*

Unit Transparency Book: Africa
- 3-Minute Warm-Ups, TT8

Connecting to Your World

More About . . .

Gibraltar

The Rock of Gibraltar at the tip of Spain rises 1,398 feet and faces Morocco. Gibraltar is a British colony.

3 Teach

A Muslim Region

Reader, Recorder, Reporter

- How did geography help Carthage become a trading empire? *(Carthage was on a peninsula and had harbors on the Mediterranean Sea.)*
- Identify two Arab cultural practices adopted centuries ago by North African Berbers. *(the Muslim religion and the Arabic language)*
- **Compare and Contrast** How do North African countries today compare in improving opportunities for women? *(Tunisia leads in improving women's opportunities for schooling, jobs, and divorce reform.)*

SECTION 3 PROGRAM RESOURCES

Unit Resource Book: Africa

ON LEVEL
Unit Resource Book
- Reading Study Guide, p. 51
- Outline Map Activities, p. 75
- Section Quiz, p. 87

STRUGGLING READERS
Unit Resource Book
- Reading Study Guide with Additional Support, p. 57
- Skillbuilder Practice: Comparing Maps, p. 72
- Section Quiz, p. 87
- Reteaching Activity, p. 90

ENGLISH LEARNERS
Pupil Edition in Spanish
Pupil Edition Audio DVD-ROM
Unit Resource Book
- RSG (Spanish), p. 63
- RSG with Additional Support (Spanish), p. 69

Multi-Language Glossary
Test Generator
- Section Quiz in Spanish

INCLUSION
Unit Resource Book
- RSG with Additional Support, p. 57
- Section Quiz, p. 87
- Reteaching Activity, p. 90

GIFTED & TALENTED
Unit Resource Book
- Section Quiz, p. 87

PRE-AP
Unit Resource Book
- Section Quiz, p. 87

TECHNOLOGY
Unit Transparency Book
- 3-Minute Warm-Ups, TT8
- Essential Question, TT9
- Critical Thinking, TT12
- Cultures, TT16

Daily Test Practice Transparencies
- Section 3, TT63

Power Presentations
ClassZone.com
World Cultures and Geography Video on DVD

COMPARING

Climates

All deserts are not alike. Some deserts are found in cold regions. Freezing temperatures keep these deserts covered with snow and ice. Deserts in warm regions can have extreme variations in temperature. The Sahara becomes cool at night and hot during the day. The highest temperature ever recorded was 136°F in Libya in 1922, on the northern edge of the Sahara. The highest temperature recorded in the United States was 134°F at Death Valley in southeastern California.

ANSWERS

1. **Mediterranean** They moderate temperatures, keeping summer temperatures cooler.
2. **Desert** Cold air originating from higher elevations might keep temperatures lower; melting mountain snow could be a water source.

More About . . .

Carthage and Tunis

After the third defeat of Carthage, Romans burned the city and spread salt in the ruins to prevent rebuilding. During the Middle Ages, the Berber Hafsid Dynasty developed the area as a royal city and a center for trade between Africa and Europe.

Today, the Tunisian capital city of Tunis stands a short distance from the site of ancient Carthage. Tunis is a financial and industrial center. Food processing, textile making, and tourism are important parts of the economy. Among the tourist attractions at Tunis are the ruins of a Roman aqueduct.

COMPARING Climates

Mediterranean Morocco's coast has a rainy winter season and a dry summer. Breezes blow in from the Atlantic. **1. What effect would ocean breezes have on the temperature?**

Desert This Moroccan village lies between the Atlas Mountains and the Sahara. **2. Why would it be a good idea to settle a village at the foot of mountains?**

Precolonial North Africa **Berbers**, the first people to settle North Africa and the Sahara, have farmed and herded in these lands since before 2000 B.C. Around 750 B.C., traders from Phoenicia, now Lebanon, began to create a major city-state called **Carthage** near what is now Tunis, Tunisia. Carthage's location on a peninsula with two harbors helped it build a great trading empire on both sides of the Mediterranean. Soon Rome and Carthage, two mighty empires, came into conflict. Carthage fell to Rome after three wars, called the **Punic Wars**, fought from 264 B.C. to 146 B.C. Rome ruled Carthage until about A.D. 430. In A.D. 698, invaders from the Arabian peninsula swept into Carthage.

By A.D. 750, the Arab newcomers had spread their Arabic language and Muslim religion to most of the Berbers of North Africa. Europeans called the area the "Barbary states" after the Berbers. The Arabs called North Africa the **Maghrib** (MUHG•ruhb), which means "west." For centuries, Muslim Berbers or Arabs controlled the Maghrib. Two empires, the Almoravids and then the Almohads, ruled from about 1050 to 1269. Afterward, the Maghrib was divided into three Muslim states. These states fought several wars with each other and with Christian invaders from Europe. Eventually, Berbers and Arabs intermarried. Today, Berber customs and language are found mostly in remote mountain and desert areas. Beginning in the 1500s, the powerful Muslim Ottoman Empire controlled much of North Africa. But European powers such as Spain and Portugal made their own claims too.

DIFFERENTIATING INSTRUCTION

Struggling Readers

Create a Time Line for North Africa

To help students organize the facts and dates about "Precolonial North Africa," assist them in creating a time line. On the board or a transparency, copy a time line, such as the one to the right. Model filling in an entry, using the textbook. Have students complete the rest of the time line, as a class or small groups.

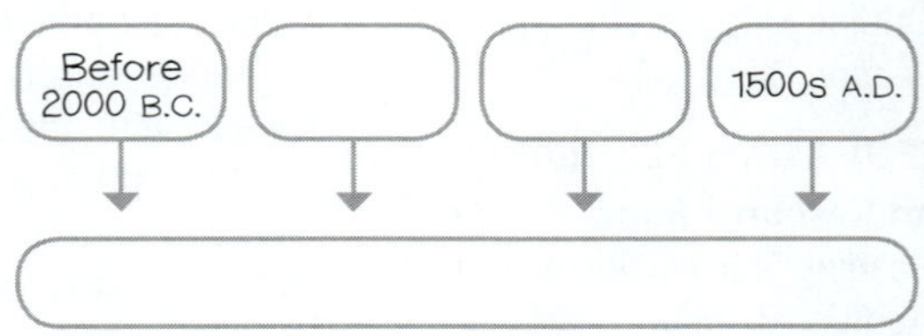

Modern North Africa As the Ottoman Empire weakened in the 1800s and 1900s, European countries seized land in North Africa. France invaded Algeria in 1830, and France and Spain each had zones of control in Morocco. Until the end of World War II, Italy controlled Libya. You'll learn more about European control of Africa in the next two chapters.

After World War II, independence movements strengthened in North Africa. Morocco won independence from France in 1956, and Spain also gave up most claims on Moroccan territory. Algeria won independence in 1962, after a bloody eight-year struggle with France. Since independence, many Algerians and Moroccans have immigrated to Europe for work. Today five million Muslims, mostly from Morocco and Algeria, live in France.

Maghrib Geography North Africa's main features are

- the Sahara, the world's largest desert
- the **Atlas Mountains**, a chain of several mountain ranges stretching from Morocco to Tunisia
- the coastal area, where most North Africans live

The Atlas Mountains dominate Morocco, and make up much of Algeria's coastline. The coastal and mountain regions of the Maghrib have rainy winters and hot, dry summers. Mountain rainfall forms rivers flowing to the coast and desert, but many desert rivers dry out during the hot summer months.

The Sahara is not all sand. While about a quarter of the desert is sandy, rocks and gravel cover much of the land. Some Saharan areas receive less than two inches of rain a year, but underground water sources also exist. About 2.5 million people live in the Sahara, traveling as nomadic herders or gathering at oasis towns.

Daily Life and Culture Centuries of intermarriage between Arabs and Berbers have made the two groups look indistinguishable. Almost all North Africans are Muslim. While most North Africans speak a dialect of Arabic, many people also speak French, Spanish, English, or Berber. North African writers often publish in French or Arabic.

Fun Facts!

BARBARY PIRATES

For hundreds of years, pirates operated out of North African ports. They would seize ships' cargo and enslave the crews or hold them for ransom. The pirates often had the support of North African rulers, who would demand tribute from countries in exchange for protection. In the early 1800s, the United States waged several wars against Barbary pirates.

Algiers Algeria's modern-day capital was once a base for the Barbary pirates.

571

DIFFERENTIATING INSTRUCTION

English Learners

Vocabulary: Prefixes

Review the first sentence under "Daily Life and Culture." Explain that *inter-* means "between" or "among." Here, *inter-* reinforces the idea of marriage "between Arabs and Berbers." Explain that *in-* before *-distinguish* means "not." The word *indistinguishable* means "not able to distinguish." Rephrase the sentence: Arabs and Berbers seem no different because the groups have married among each other for so long.

Gifted & Talented

Describing the Immigration Experience

Point out the last two sentences under "Modern North Africa." Discuss emigrating to a new country: What would it be like to leave everyone and everything you know behind to encounter a different way of life? Have students write a short story or narrative poem, describing the experience of a North African immigrant moving to or living in a European country.

Fun Facts!

Barbary Pirates

After the United States won its independence, it paid a tribute of $80,000 in 1784 to the Barbary states (Tripoli, Tunis, Morocco, and Algiers) to protect its shipping. In 1795, the United States paid a ransom of about $1 million to free 115 sailors from Algiers. In the early 1800s, under president Thomas Jefferson, U.S. forces tried to end Barbary piracy and the paying of ransoms and tribute. However, it wasn't until 1815 that U.S. forces under William Bainbridge and Stephen Decatur finally succeeded. European nations didn't fare as well—they continued paying annual tribute into the 1830s.

More About . . .

Casablanca, Morocco

The largest city and major port in Morocco is Casablanca, but most people think of it as the name of a classic movie. *Casablanca* was released during World War II. At the time, Casablanca served as a landing site for Allied forces entering North Africa.

More About . . .

Marrakech, Morocco

The city of Marrakech has a history that goes back to the Almoravid Empire. An Almoravid Berber leader named Yusuf ibn Tashufin founded Marrakech in 1062. It then became the capital city of the Empire, but the Almohads took control of it in 1147. The Koutoubia mosque, a famed landmark today, was completed during that century. A French quarter was developed after 1912 when France controlled the city.

Today Marrakech is a center for tourists. Marrakech is famous for its mosques, palace, extensive gardens, city square filled with entertainment—and world-famous *souk* or bazaar packed with stalls and goods of all kinds, especially fine leather goods.

Make It Fun

Teacher-Tested Activities

Matthew J. Scheidler, Wayzata East Middle School, Plymouth, Minnesota

Be the "________" (fill in the blank with a geographic feature, a person, etc.). This might sound a little unorthodox, but asking students to take on the role of a pyramid, or the Sahara, might help imprint the significance and aspects of the term in students' minds.

- Ask for a student volunteer (they may stand up for their performance).
- Ask the volunteer to "be the Sahara." Ask the student what the Sahara sounds like. (The student may make a wind sound.) You may ask the student what the Sahara "sees" each day. (Answers could include oases, nomads, camels.) You may ask the student what the Sahara "feels." (Students may comment on the effects of the scorching sun, without much shade to protect itself.)
- This idea could also be used for people, cities, and so forth.

Souk Colorful goods line the streets in this market in Marrakech, Morocco.

Between 1960 and 1990, the percentage of city dwellers in North Africa jumped from less than one-third to about one-half of the population. Urban living quarters include small attached houses, large apartment buildings, and slum towns near larger cities. In the countryside, large extended families may still share one all-purpose room. At an open-air market, called a **souk** (sook), people buy and sell produce or handicrafts, and visit with friends. They may also share a plate of couscous, a wheat dish, covered with a tagine, or stew.

In the countryside, Moroccan men and women wear traditional Berber clothing, such as a **jellaba**, or robe with full sleeves, or a hooded cloak called a burnoose. Men also wear a turban, a skullcap, or a red Moroccan hat called a fez. In cities, people often combine traditional and Western clothing. Mosques are popular North African gathering places. So are movie theaters and soccer stadiums.

Most North African countries offer free elementary and at least some secondary education. Girls are more likely to stay at home instead of working or going to school, especially in rural areas. Tunisia leads the region in improving opportunities for women, including better schooling and jobs, and divorce reform. In Libya, only about one in ten women work outside the home.

SUMMARIZE Identify the cultures that have most influenced North Africa. Berbers, Arab Muslims, the Ottoman Empire, Europeans

INTERDISCIPLINARY ACTIVITIES

CONNECT to Language Arts

Compare Opportunities

Have students write a paragraph based on information on this page about schooling and about opportunities for women in North African countries. They should compare and contrast North African educational opportunities and opportunities for women with those in the United States.

CONNECT to Art

Create a Marketplace Visual

Have small groups of students create marketplace displays. Have groups study the photograph of the Marrakech *souk* and plan what they would sell if they were in an open-air market. Remind them that a marketplace can include crafts, foods, and other products. Have students create displays by one of two methods: posterboard, illustrated with cutout images and drawings; or a small box in which to develop a three-dimensional scene.

Government and Economics

KEY QUESTION What product is most important to North Africa?

Because of occupation and colonization, Europe and Southwest Asia influenced culture across North Africa. But despite a shared cultural history, each North African country has developed its own government and economy. Algeria, Libya, and Morocco are good examples.

Algeria Algeria has an elected president, a prime minister the president appoints, and a Council of Ministers the prime minister selects. Voters elect one house of the legislature, with the other house partly elected from local assemblies and partly appointed by the president. Algeria became a multi-party state in 1989.

Algeria's government controls its major industries, including natural gas and petroleum production. To export its natural gas, Algeria built two pipelines running under the Mediterranean Sea to Spain and Sicily. Other industries include construction materials, steel, and textiles. These industries and small-scale farming, which occupies about one out of five Algerians, don't provide enough jobs. Hundreds of thousands of Algerians have emigrated to find work in developed countries such as France.

Libya In 1969, Colonel Muammar Gaddafi (guh•DAH•fee) led a military revolt that overthrew the country's monarchy. Gaddafi still heads Libya, though he holds no official title. Opposition groups exist, but only one political party is legal. In this **dictatorship**, or rule by an unelected, absolute leader, the government also controls television and the press. Local groups elect members of a General People's Congress.

Most Libyans live in overcrowded cities, lured by the discovery of petroleum in 1959. The government controls most economic activity. Although almost one in five Libyans are farmers, just about one percent of the land is farmable. Libya exports its oil, but the oil industry employs few workers.

Libya's strained relationship with the United States broke after Libyan terrorists blew up a U.S. airplane in 1988. In 2003, these relations improved when Libya compensated the victims' families and gave up its pursuit of nuclear weapons. The United States and United Nations agreed to end economic sanctions against Libya.

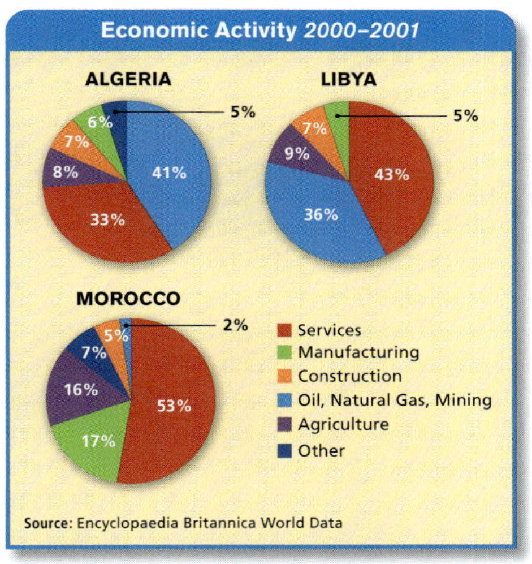

DIFFERENTIATING INSTRUCTION: TIERED ACTIVITIES

OBJECTIVE Develop graphic organizers to clarify economic information about Algeria and Libya.

Basic

Have pairs of students create two simple web diagrams of economic facts, one for each country. Students should use the text and pie graphs to provide at least five facts for each country. Have them compare their webs with other pairs.

On Level

Adapt the Basic activity by having students work independently to create spider maps for each country. Students should include two or three facts for each category. Have students display their spider maps and compare them to other students' spider maps.

Challenge

Adapt the On Level activity by having students choose the kind of graphic organizer that best displays the facts they choose. They should write a summary paragraph to include as a caption on each organizer. Have students display their organizers and compare them to other students' organizers.

Teach

Government and Economics

Talk About It

- How important are farming and fishing in providing jobs for Moroccans? *(These activities employ about four out of ten Moroccan workers.)* How does tourism help Morocco? *(It boosts the economy by bringing in nearly two million visitors annually.)*
- Why do many thousands of Algerians work in other countries? *(Algerian industry and farming don't provide enough jobs.)*
- **Compare and Contrast** How do heads of state differ among Libya, Morocco, and Algeria? *(Libya has a dictator, Morocco has an elected president, and Algeria has a constitutional monarch.)*

Unit Resource Book: Africa
- Skillbuilder Practice: Comparing Maps, p. 72
- Outline Map Activities: North Africa: Physical/Political, pp. 75–76

Economic Activity *2000–2001*

Interpreting Pie Graphs

Remind students that graphs make data easier to interpret. Here, the economies of three African countries can be compared using the pie graphs.

- In which country is oil, natural gas, and mining the smallest part of the economy? *(Morocco, 2%)*
- When percentages for all three countries are combined in each category, which economic activity is most significant? *(services)* Which activity besides "other" s least significant? *(construction)*

CONNECT to the Essential Question

How have people in North Africa adapted to the region's arid climate?

Ask students what they have learned so far that can help them answer this question. Students might mention:

- Ancient Egyptians built their civilization near an important water source, the Nile. Most North Africans live along the Nile or along the Mediterranean coast.
- Modern North Africans have harnessed the Nile through irrigation and damming.
- North African industries, such as oil production and tourism, take advantage of the region's history and resources.

Unit Transparency Book: Africa
- Essential Question, TT9

4 Assess & Reteach

Assess Have students complete the Section Assessment.

Unit Resource Book: Africa
- Section Quiz, p. 87

Interactive Review@ClassZone.com
- Section Quiz

Power Presentation

Test Generator

Reteach Assign small groups of students one of the following topics: culture, history, geography, government, or economics. Have groups list three key facts from their textbooks to present to the class.

Unit Resource Book: Africa
- Reteaching Activity, p. 90

Berbers Morocco has the largest Berber population. Here, they sell goods in an open market.

Morocco A king heads Morocco's constitutional monarchy. His extensive powers include commanding the armed forces, issuing laws, and controlling government agencies. He also appoints a prime minister and cabinet of other ministers. The largest parties support the king, though smaller opposition parties also exist. Voters elect one house of the legislature. Local and professional groups choose the other house.

Morocco's mining, communications, and some manufacturing industries are publicly owned. But most of its businesses and farms have private owners. This country's developing economy mostly depends on mining and agriculture. Farming and fishing employ about four out of ten Moroccan workers. Morocco exports most of the phosphate rock used for fertilizer. The country also produces wheat, barley, wine, citrus fruits, olive oil, sheep, and goats, while its factories turn out textiles, leather goods, fertilizers, and petroleum products. Tourism, one of Morocco's main industries, brings nearly two million visitors each year, mostly from Europe.

FIND MAIN IDEAS Describe the important product or products in North African economies. oil in Algeria and Libya, tourism and agriculture in Morocco

Section 3 Assessment

ONLINE QUIZ For test practice, go to **Interactive Review @ClassZone.com**

TERMS & NAMES

1. **Explain the importance of**
   - Berbers
   - Carthage
   - Punic Wars
   - dictatorship

USE YOUR READING NOTES

2. **Compare** Use your completed chart to answer the following question.
   What are some of the differences between Algeria, Morocco, and Libya?

NORTH AFRICA
Libya: 1. ______ 2. ______
Morocco: 1. ______ 2. ______
Algeria: 1. ______ 2. ______

KEY IDEAS

3. Which people first settled in North Africa?
4. What percentage of North Africans live in cities?
5. How does Arab Muslim culture affect life in North Africa?

CRITICAL THINKING

6. **Find Main Ideas** What ties do North African countries still have with Europe?
7. **Compare and Contrast** How does Libya's government compare with Algeria's?
8. **CONNECT to Today** In 2006, France had Europe's largest Muslim community, about five million North Africans. What problems might Muslim North Africans face in Europe?
9. **HISTORY Research a Report** Research a report on the Barbary Pirates. Include information on European and American attempts to stop piracy out of North Africa.

SECTION 3 ASSESSMENT ANSWERS

Terms & Names

1. Berbers: original inhabitants of North Africa and Sahara; Carthage: ancient North African city-state; Punic Wars: series of wars fought between Carthage and Rome during 2nd and 3rd centuries B.C.; dictatorship: state or government ruled by a leader with absolute power

Use Your Reading Notes

2. Answers may vary.

NORTH AFRICA
Libya: 1. mostly desert climate 2. strong oil industry
Morocco: 1. Mediterranean climate 2. mining, agriculture
Algeria: 1. mostly desert climate 2. strong oil industry

Key Ideas

3. Berbers
4. about 50 percent
5. It affects almost all aspects of life, including religion, language, customs, laws, and holidays.

Critical Thinking

6. Many North Africans speak and write European languages.
7. Libya is a dictatorship with complete government control over communication, and only one political party. Algeria is freer, with a popularly elected president and multiple political parties.
8. prejudice based on religion or ethnicity; lack of jobs; difficulty adjusting to a new environment; cultural conflicts over dress, social relations, etc.
9. Use the rubric to score reports.

Reports Rubric

	Information	Explanation	Number of Errors
4	very complete	clear and concise	few or none
3	complete	clear	some
2	barely accurate	clear but limited	many
1	little	unclear	many major

Click here to complete these and other activities online @ClassZone.com

CHAPTER SUMMARY

Key Idea 1
The Nile made possible the great civilizations of ancient Egypt.

Key Idea 2
A blending of cultures has made Egypt a leader for North Africa as well as for Southwest Asia.

Key Idea 3
The Ottomans and Islam also influenced North African nations, as did Europeans and colonizing powers.

For **Review and Study Notes**, go to **Interactive Review** @ClassZone.com

NAME GAME

Use the Terms & Names list to complete each sentence on paper or online.

1. I am the rich land at the mouth of a river. delta
2. I am the tomb of a pharaoh. ________
3. I am one of the six waterfalls along the Nile. ________
4. Ancient Egypt conquered my empire; later I conquered Egypt. ________
5. I helped scholars translate the hieroglyphics of ancient Egypt. ________
6. I signed the Camp David Accords, a peace treaty with Israel. ________
7. I am Africa's largest country. ________
8. I am the Arab name for a region that means "west." ________
9. I was an ancient city-state in North Africa. ________
10. I am a system of absolute rule by one unelected person. ________

Berber
Carthage
cataract
delta
dictatorship
dynasty
Kush
Maghrib
Nubia
papyrus
pyramid
Rosetta Stone
Anwar el-Sadat
Sudan
Suez Canal

Activities

GeoGame

Use this online map to reinforce your understanding of North African geography. Drag and drop each place name in the list to its location on the map.

To play the complete game, go to **Interactive Review** @ClassZone.com

Crossword Puzzle

Complete an online crossword puzzle to test your knowledge of North Africa.

ACROSS

1. a traditional Berber garment; a robe with full sleeves

ONLINE AND ON-SCREEN ACTIVITIES

GeoGame

Use the interactive map on **Power Presentations** DVD-ROM or the **eEdition** to help students review Egypt's and North Africa's geographic features.

- Demonstrate how to drag and drop the map labels listed onto the appropriate locations on the map.
- Advance to the scorecard to show how students have a chance to improve their scores.

Crossword Puzzle

Students can review important people and events in the history of Egypt and North Africa by completing this crossword puzzle online at **ClassZone.com**.

- As an extension of this activity, have students use the *Activity Maker* at **ClassZone.com** to make their own crossword puzzle for one section of the chapter.
- Then have students trade puzzles with classmates to solve answers for each other's puzzles.

Chapter Summary

As a follow-up to the chapter, encourage students to pair up and go over the **Review and Study Notes** at **ClassZone.com.**

Alternatively, student pairs can review their Reading Study Guide notes, making sure that they can add important details to each section's "Key Idea."

Name Game

Teaching Tips

ON PAPER Have students work in pairs to complete this activity. As an extension, they can write questions for the terms that are not covered in the items shown here.

ONLINE Make sure that students understand how to drag and drop terms to fill in blanks. Ask them to print out their results and compare them with those of a classmate.

ANSWERS

1. delta
2. pyramid
3. cataract
4. Kush
5. Rosetta Stone
6. Anwar el-Sadat
7. Sudan
8. Maghrib
9. Carthage
10. dictatorship

Make It Fun

Teacher-Tested Activities

Matthew J. Scheidler, Wayzata East Middle School, Plymouth, Minnesota

Experts' Ball—Everyone's Invited: If a lecture or discussion is planned for the next day, this activity can minimize the homework load while giving students the freedom to investigate a term, person, or geographic feature.

- Give each student an "admission ticket" before they leave class. On the ticket will be the student's term they are responsible to research for the Experts' Ball.
- Encourage students to investigate the history, significance, current situation, the future of their topic. For instance, if a topic was King Tut, they might discuss King Tut's young age, when he perished, the fact that King Tut's tomb is a global attraction, and new DNA findings.

Vocabulary ANSWERS

1. a family or group that rules for several generations
2. a structure built to control the floodwaters of the Nile
3. Egypt's president from 1970 to 1981, who signed the first peace agreement between Israel and an Arab state
4. Arab name for North Africa, meaning "the west"
5. the original inhabitants of North Africa and the Sahara
6. a religion believing in one god, called Allah, and his prophet on Earth, Muhammad
7. a series of mountain ranges stretching from Morocco to Tunisia
8. Several pharaohs from one family made up a dynasty; a pharaoh who died might be entombed in a pyramid.
9. The ancient region of Nubia, with its powerful Kush kingdom, was located in what is now Northern Sudan.

Key Ideas ANSWERS

10. They produced water for irrigation and left behind silt, which fertilized the crops.
11. Kush, Assyria, Alexander the Great (Greece/Macedonia), and Rome
12. They considered them gods and erected giant pyramids and statues (by order of the pharaohs) to honor them after death.
13. Answers will vary, but may include: building the pyramids, inventing writing, great cities, a long-lasting civilization, giving advanced rights to women
14. the Ottoman Empire and Great Britain
15. Muhammad Ali Pasha
16. Egypt boasts the second largest population in Africa, the continent's largest city, and the second most powerful economy in the Arab world.
17. Arab herders and government militia against non-Muslims
18. in the 1950s and 1960s
19. wheat, barley, wine, citrus fruits, olive oil, sheep, goats, textiles, leather goods, fertilizers, petroleum products, and phosphates
20. Oil is a major import, but the oil industry employs few workers.
21. Libya took responsibility for the 1988 airline bombing and abandoned its pursuit of nuclear weapons.

CHAPTER 18 Assessment

VOCABULARY

Explain the significance of each of the following.

1. dynasty
2. Aswan High Dam
3. Anwar el-Sadat
4. Maghrib
5. Berbers
6. Islam
7. Atlas Mountains

Explain how the terms and names in each group are related.

8. dynasty, pharaoh, and pyramid
9. Sudan, Nubia, and Kush

KEY IDEAS

1 Ancient Egypt: Pyramids and Pharaohs

10. Why were the yearly floods so important to Egyptian farmers?
11. What powers invaded and subdued ancient Egypt?
12. How did ancient Egyptians honor their pharaohs?
13. What do you think was ancient Egypt's greatest accomplishment?

2 Modern Egypt and Sudan

14. Which powerful empires clashed over control of Egypt between 1517 and 1922?
15. Who founded the dynasty that ruled modern Egypt?
16. Why is modern Egypt considered a leader in the region?
17. Which groups have been fighting in the Darfur region of Sudan?

3 Other Nations of North Africa

18. When did most North African countries win their independence from European colonizing powers?
19. What are some of Morocco's major products?
20. Why hasn't Libya's oil wealth brought its people out of poverty?
21. What hopeful developments improved the United States' relationship with Libya in 2003?

CRITICAL THINKING

22. **Analyze Cause and Effect** Create a table to assess the causes, intended effects, and unintended effects of the move of North Africans to cities.

CAUSES	INTENDED EFFECTS	UNINTENDED EFFECTS
move of North Africans to cities		

23. **Summarize** Summarize the history of Egypt's connection to Britain.
24. **Evaluate** What push-pull factor has led North Africans to immigrate to Europe?
25. **Five Themes: Place** What body of water was created by the Aswan High Dam?
26. **Connect Geography & History** Why did North Africa's history differ from that of the rest of Africa?

Answer the ESSENTIAL QUESTION

How have people in North Africa adapted to the region's arid climate?

Written Response Write a two- or three-paragraph response to the Essential Question. Be sure to consider the key ideas of each section. Use the rubric below to guide your thinking.

Response Rubric
A strong response will:
- describe North Africa as primarily desert
- discuss the yearly flooding of the Nile
- explain how the Aswan High Dam and Suez Canal contributed to the economy of Egypt
- mention the trans-Mediterranean trade between North Africa and Europe

ESSENTIAL QUESTION REVIEW

Before students begin their essays, have them review the key chapter elements that will help them answer the Essential Question. Direct them to significant features in the chapter, such as:

- Connect Geography & History: Aswan High Dam, Section 1
- Animated Geography: Egyptian & Kush Kingdoms, Section 1
- Animated Geography: The Great Pyramid, Section 1
- History Makers: Muhammad Ali Pasha, Section 2
- Connect History & Culture: Lost Boys of Sudan, Section 2
- Comparing Climate: Mediterranean & Desert, Section 3
- Economic Activity 2000–2001, Section 3

Written Response Rubric

	Coverage of Key Ideas	Explanation	Errors in Mechanics
4	all	clear and concise	few or none
3	most	good	some
2	some	fair	many
1	one or none	poor	many major

STANDARDS-BASED ASSESSMENT

Test Practice
- Online Test Practice @ClassZone.com
- Test-Taking Strategies and Practice at the front of this book

POPULATION MAP

Use the map and your knowledge of Africa to answer questions 1 and 2 on your paper.

North African Population Density

1. **What geographic feature explains the winding band of heavy population in the east?**
2. **Why are there fewer people scattered throughout northern Sudan than in the south?**

BAR GRAPH

Examine the graph below. Use the information in the graph to answer questions 3 and 4 on your paper.

North African Agriculture *2003*

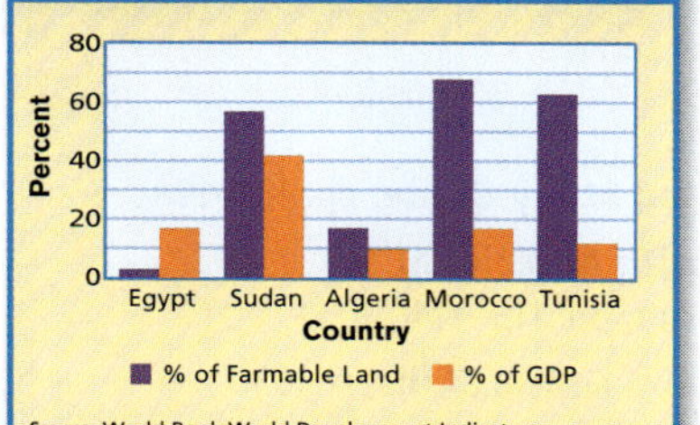

Source: World Bank World Development Indicators

3. **Agriculture is most important to which country's economy?**
   A. Egypt
   B. Sudan
   C. Morocco
   D. Tunisia
4. **Which country uses the smallest percentage of its land for agriculture?**
   A. Algeria
   B. Sudan
   C. Morocco
   D. Egypt

GeoActivity

1. INTERDISCIPLINARY ACTIVITY–SCIENCE

In addition to weakening the soil and increasing erosion, the Aswan High Dam has also caused the increase of a disease caused by tiny worms that breed in snails discharged into the Nile and its canals. With a small group, research other environmental effects of the Aswan High Dam.

2. WRITING FOR SOCIAL STUDIES

Reread the description of the packing of an Egyptian tomb in this section. Compile a list of the things you would put in such a tomb for a famous person, a favorite pet, or someone you know. Explain why you would include each of these things.

3. MENTAL MAPPING

Create an outline map of North Africa and label the following:

- Aswan High Dam
- Mediterranean Sea
- Sahara
- Egypt
- Sudan
- Tunisia
- Morocco
- Algeria
- Libya
- Maghrib

Critical Thinking ANSWERS

22. Intended Effects:
get better jobs
Unintended Effects:
overcrowding
pollution
break-up of extended families

23. Britain invaded Egypt in 1882. During World War I, Britain declared Egypt, part of the Ottoman Empire, a protectorate. In 1922, Egypt won independence from Britain, though British soldiers remained in charge of the Suez Canal.

24. The major factor is the lack of employment in North African countries.

25. Lake Nasser

26. North Africa's geographic isolation from the rest of Africa by the Sahara Desert and its accessibility to the Arab invaders as well as European colonizers contributed to its separate history.

Standards-Based Assessment ANSWERS

1. the Nile River valley
2. because the northern half of Sudan lies in the desert
3. B
4. D

GeoActivity

1. INTERDISCIPLINARY ACTIVITY RUBRIC

The report should
- provide adequate research into the Dam's effects
- avoid bias and explain trade-offs between beneficial and harmful effects

2. WRITING FOR SOCIAL STUDIES RUBRIC

The list should
- explain the ancient Egyptian reasons for leaving worldly objects in burial chambers
- contain appropriate items, and reasons for their inclusion

3. MENTAL MAPPING RUBRIC

The outline map should
- represent the region's approximate shape
- include all geographic features listed in their correct locations

DIFFERENTIATING CHAPTER ASSESSMENT

Unit Resource Book

- **Chapter Test A,** pp. 91–95.
  Use for Struggling Readers, Beginning-to-Intermediate-Level English Learners, and for Inclusion of students whose IEP suggests supported test-taking.
- **Chapter Test B,** pp. 96–100.
  Use for On-Level Learners and ESOL students with Advanced-Level English.
- **Chapter Test C,** pp. 101–106.
  Use for Gifted and Talented students or students assigned Pre-AP work.

Test Generator

- Chapter Tests A, B, and C, in English and Spanish

McDougal Littell Assessment System

TEST **Test Generator** Use three levels of tests in English and Spanish tied to state standards. Plus, choose questions to customize tests for all populations.

SCORE Scan and Score plain paper student tests in minutes.

REPORT Create instant progress reports, sortable by student, by standard, or other criteria.

RETEACH Respond immediately to students' needs by using a wealth of reteaching resources.

CHAPTER 19 West Africa PLANNING GUIDE

Complete print resources all on one USB drive allow you to customize lesson plans for all students.

Section Titles & Key Ideas	Unit Resource Book Africa	Assessment Resources
1 Three Trading Empires **KEY IDEA** Colonizers traveled to West Africa, dividing up land once held by great trading empires.	• **Reading Study Guide,** pp. 107–108 • **RSG with Additional Support,** pp. 113–114 • **Map and Graph Skills,** p. 133 • **History Makers,** pp. 137–138 • **Primary Source,** pp. 145–146	**Unit Resource Book** • Section Quiz, p. 147 • Reteaching Activity, p. 150 **Test Generator CD-ROM** **Daily Test Practice Transparency,** TT64
2 Nigeria and Ghana **KEY IDEA** Nigeria and Ghana have struggled to maintain stable governments and productive economies.	• **Reading Study Guide,** pp. 109–110 • **RSG with Additional Support,** pp. 115–116 • **Skillbuilder Practice,** p. 132 • **Economics in the World,** p. 142 • **Connect to Art,** p. 143	**Unit Resource Book** • Section Quiz, p. 148 • Reteaching Activity, p. 151 **Test Generator CD-ROM** **Daily Test Practice Transparency,** TT65
3 Mali and Côte d'Ivoire **KEY IDEA** Mali and Côte d'Ivoire are developing countries with lively cultural traditions.	• **Reading Study Guide,** pp. 111–112 • **RSG with Additional Support,** pp. 117–118 • **Skillbuilder Practice,** p. 132 • **Interdisciplinary Projects,** pp. 139–141	**Unit Resource Book** • Section Quiz, p. 149 • Reteaching Activity, p. 152 • Chapter Test (Forms A, B, & C), pp. 153–164 **Test Generator CD-ROM** **Daily Test Practice Transparency,** TT66

McDougal Littell Assessment System
TEST SCORE REPORT RETEACH

No Time? To focus students on important content in the chapter, use the **"Reading Study Guide with Additional Support"** in *Resources2Go.*

Teacher Presentation Options

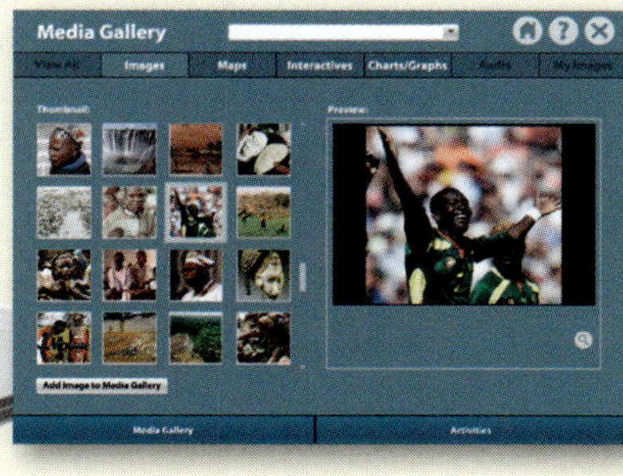

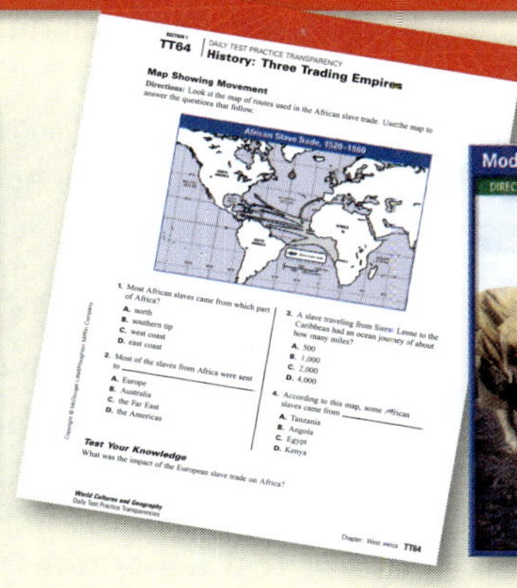

Electronic Books
eEdition DVD-ROM
eEdition Online

Presentation Toolkit
Video/DVD
- World Cultures and Geography

Power Presentations DVD-ROM
- Discussion Starters
- Media Gallery
- Review Game
- Interactive Review
- Animated Features

Unit Transparency Book
- 3-Minute Warm-Ups
- Essential Question
- Critical Thinking
- Maps
- Cultures

Daily Test Practice Transparencies

Animated Geography
- Colonies in Africa, 1913 Map
- West African Trading Empires Map
- Modern Griots
- Worldster: Ali's Web Page

OnlineActivities at ClassZone.com

Animated Geography
- **Videos and Photos**
- **3-D Animations**
- **Interactive Maps**

Interactive Review
- **Section Quizzes**
- **Review and Study Notes**
- **Vocabulary Flip Cards**
- **Graphic Organizers**
- **Name Game and Crossword Puzzles**
- **GeoGame**

Online Test Practice

Map Center
- **Atlas Maps**
- **Political Maps with Data**
- **Links to Mapping Web Sites**

Activity Center
- **Activity Maker—Review Games, Puzzles, Flip Cards, and Graphic Organizers**
- **Worldster**

Research & Writing Center
- **Web Sites**
- **Data Updates**
- **Primary Sources**
- **Writing Models**

Current Events

Don Stringfellow
Mobile County Public School System
Mobile, Alabama

Make It Fun — Teacher-Tested Activities

As you teach this chapter, look for my classroom-proven ideas for:
- making the lesson fun
- focusing students' attention and comprehension
- integrating technology with traditional teaching

Teaching Bell to Bell When you see this timing guide, have students discuss in pairs, in groups of 3–6, or as a class.
- **Think, Pair, Share** (pairs)
- **Reader, Recorder, Reporter** (3–6)
- **Talk About It** (any size)

CHAPTER 19

RESOURCES FOR DIFFERENTIATED INSTRUCTION

Struggling Readers

eEdition with Audio DVD-ROM

Lesson Plans
- Strategies for Inclusion

Unit Resource Book
- Reading Study Guide with Additional Support, pp. 113–118 (B)
- Reteaching Activity, pp. 150–152
- Chapter Test (Form A), pp. 153–156 (A)

Best Practices Toolkit (C)
- Interactive Reading Strategies, pp. 90–145
- Differentiation Activity Bank: Struggling Readers, pp. 146–149

Teacher's Edition Activities
- Clarifying the Division of Africa, Sec. 1
- Clarifying Past and Present, Sec. 3
- Tiered Activities, Sec. 1–3

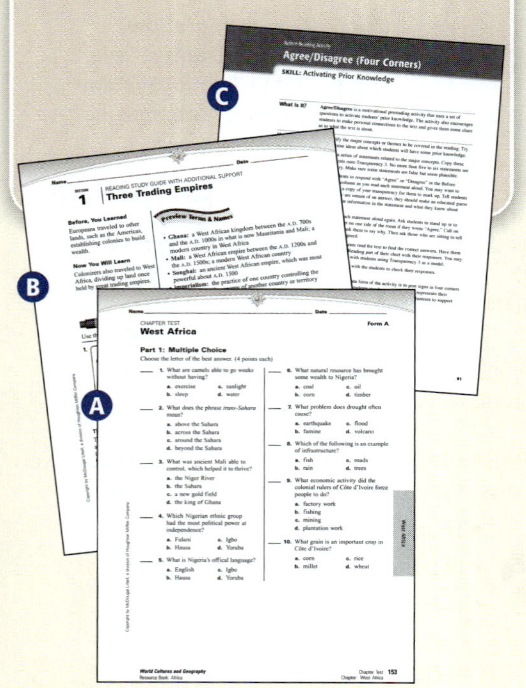

English Learners

Pupil's Edition in Spanish

eEdition with Audio DVD-ROM

Lesson Plans
- Strategies for English Learners

Unit Resource Book
- Reading Study Guide (Spanish), pp. 119–124
- RSG with Additional Support (Spanish), pp. 125–130 (A) (B)

Multi-Language Glossary (C)

Best Practices Toolkit
- Differentiation Activity Bank: English Learners, pp. 165–194

Test Generator CD-ROM
- Section Quizzes in Spanish
- Chapter Test in Spanish

Teacher's Edition Activities
- Pronounce and Preview, Sec. 1–3
- Key Academic Vocabulary, Sec. 1
- Vocabulary: Cognates, Sec. 2–3
- Tiered Activities, Sec. 1–3

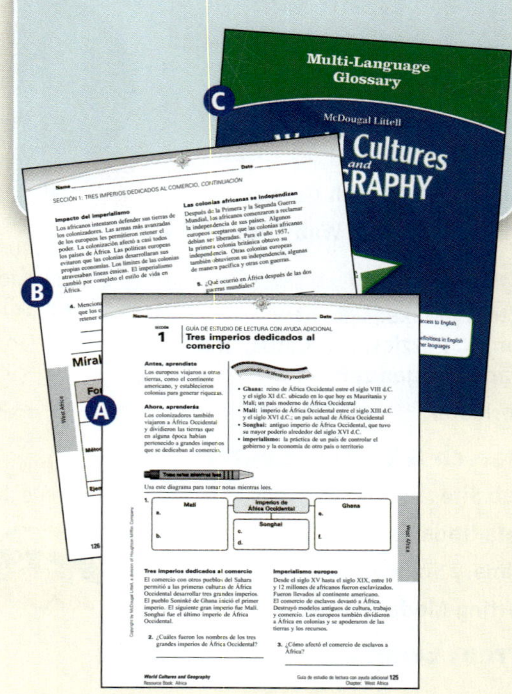

Inclusion

eEdition with Audio DVD-ROM

Lesson Plans
- Strategies for Inclusion

Unit Resource Book
- Reading Study Guide with Additional Support, pp. 113–118 (A) (B)
- Reteaching Activity, pp. 150–152
- Chapter Test (Form A), pp. 153–156

Best Practices Toolkit (C)
- Interactive Reading Strategies, pp. 90–145
- Differentiation Activity Bank: Struggling Readers, pp. 146–149

Teacher's Edition Activities
- Vocabulary Quiz, Sec. 1
- Play a Vocabulary Card Game, Sec. 2
- Vocabulary: Word Wheel, Sec. 3
- Tiered Activities, Sec. 1–3

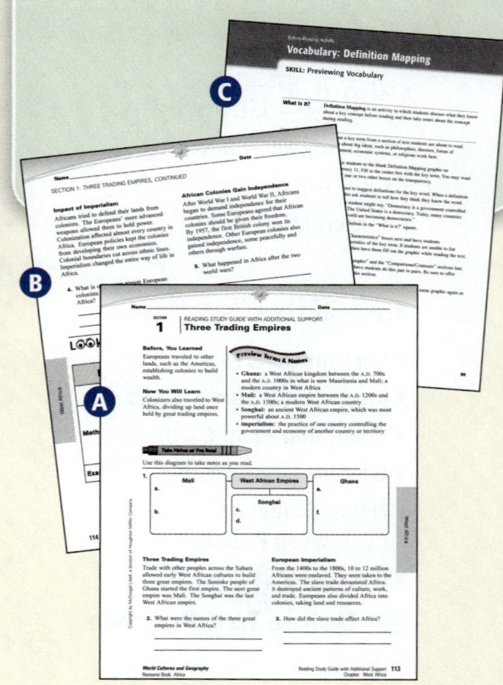

Interdisciplinary Activities

Gifted & Talented Students

Lesson Plans
- Core Lesson Plan

Unit Resource Book
- Outline Map Activities, pp. 135–136
- History Makers, p. 137 A
- Interdisciplinary Projects, pp. 139–141 B
- Economics in the World, p. 142 C
- Chapter Test (Form C), pp. 161–164

Best Practices Toolkit
- Student Products Rubrics, pp. 1–14
- Writing Templates, pp. 78–89

Teacher's Edition Activities
- Write a Citizen's Op-Ed Piece, Sec. 2
- Tiered Activities, Sec. 1–3

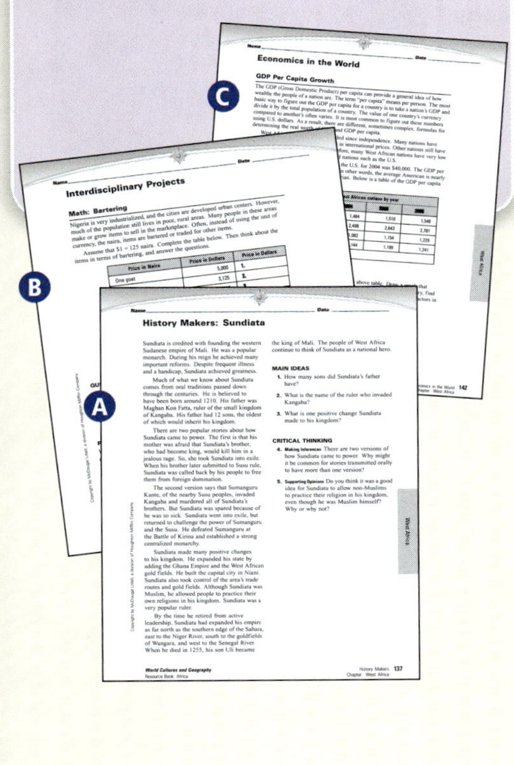

Pre-AP*

Lesson Plans
- Core Lesson Plan

Unit Resource Book
- Vocabulary Practice, p. 131
- Map and Graph Skills, pp. 133–134
- History Makers, p. 138 A
- Connect to Art, p. 143 B
- Primary Source, pp. 145–146
- Chapter Test (Form C), pp. 161–164

Best Practices Toolkit C
- Writing Templates, pp. 78–89
- Differentiation Activity Bank: Pre-AP Students, pp. 195–200

* Pre-AP is a registered trademark of the College Entrance Examination Board, which was not involved in the production of and does not endorse this product.

Teacher's Edition Activities
- Tiered Activities, Sec. 1–3

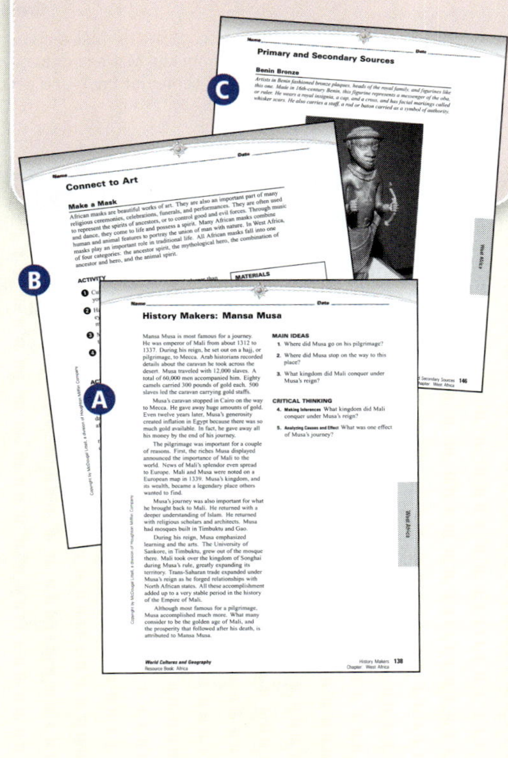

CONNECT to Science

Raise a Millet Crop on Paper

CLASS TIME 30 minutes

TASK Create a chart to explain raising and preparing millet.

SUPPLIES
- information and images about growing and preparing millet
- construction paper, colored markers or pencils

ACTIVITY Draw students' attention to the Visual Vocabulary and text about millet crops in Section 3. Have students work in pairs to create a chart about the botanical requirements of millet. They should supply details about climate, planting, cultivating, and preparing millet for food. Students should also try to explain how regional droughts in Mali might affect millet crops. Have students display their charts for class comment.

Good for Visual and Verbal Learners

CONNECT to Art

Map the Culture of Nigeria and Ghana

CLASS TIME 30 minutes

TASK Create a display map of art and music in Nigeria and Ghana.

SUPPLIES
- images of Nigerian musical instruments and Nigerian and Ghanian arts
- outline maps of Nigeria and Ghana
- colored markers or pencils

ACTIVITY Draw students' attention to the text about Nigerian music/art and the text and image of Ghana's arts in Section 2. Have students work in small groups to create a visual cultural display for each country. They should choose appropriate images and arrange them attractively upon the outline maps. Have groups present their maps accompanied by a brief summary of each country's cultural achievements.

Good for Visual and Spatial Learners

Introduce the ESSENTIAL QUESTION

- Describe West African trading empires and analyze the effects of European imperialism.
- Explain ethnic conflict in Nigeria and summarize the cultures, governments, and economies of Nigeria and Ghana since independence.
- Describe cultural life in Mali and Côte d'Ivoire, and identify the civil and economic challenges faced by these countries since independence.

Key Ideas

Section 1: Colonizers also traveled to West Africa, dividing up land once held by great trading empires.

Section 2: Nigeria and Ghana have struggled to maintain stable governments and productive economies.

Section 3: Mali and Côte d'Ivoire are developing countries with lively cultural traditions.

CONNECT to the Essential Question

You may wish to revisit the Essential Question periodically with students as they work through the chapter.

Suggestions for reconnecting are provided at an appropriate point in the *Teacher's Edition* for each section.

The Great Mosque at Djenné, Mali

CHAPTER 19

West Africa

1 HISTORY **Three Trading Empires**

2 FOCUS ON **Nigeria and Ghana**

3 FOCUS ON **Mali and Côte d'Ivoire**

ESSENTIAL QUESTION

How have centuries of colonialism affected West Africa?

CONNECT Geography & History

Use the map and the time line to answer the following questions.

1. What two European powers held the most land in Africa?
2. How did colonial boundaries appear to compare to traditional African ethnic boundaries?

A.D. 700

History
c. 700 Ghana empire founded by Soninke people.

History
c. 1300 Songhai empire begins to rise.

History
c. 1500 The Songhai empire reaches its peak.

Geography
1300s Mali empire at its peak; Timbuktu, Mali's largest city, becomes center for trade and scholarship. ▶

TIME LINE DISCUSSION

Point out that West Africa became a powerful trading center centuries ago because of its natural resources and location along trans-Saharan trade routes. In the 1500s, West Africa had become subject to foreign interests. Since then, West Africa has faced the challenges of European domination, struggles for independence, and civil conflict.

- Which city was the Mali empire's largest? *(Timbuktu)*
- Which event on the time line highlights the end of empire building in West Africa? *(1591—Moroccan invaders destroy Songhai.)*
- Which event on the time line explains the European presence in Africa in 1913? *(1884–1885—The Berlin Conference divides Africa among European nations.)*
- Which event on the time line highlights the beginning of African movements for independence? *(1957—Modern Ghana wins independence from Great Britain.)*

History
1591 Moroccan invaders destroy Songhai.

Government
▲1884–1885 The Berlin Conference divides Africa among European nations.

Government
1957 Modern Ghana wins independence from Great Britain, beginning a continent-wide independence movement.

Today

History
◄ 2005 Signed agreement ends fighting in Côte d'Ivoire.

579

TEACHER MEDIA FAVORITES

Books for the Teacher

Achebe, Chinua. *Things Fall Apart*. Heinemann, 1958. Award-winning novel about traditional Nigerian life and colonialism's impact.

Meredith, Martin. *The Fate of Africa*, Public Affairs, 2005. Acclaimed analysis of Africa's history since independence.

Mendonsa, Eugene L. *West Africa*. Carolina Academic Pr., 2002. Examines West African problems and traditional beliefs.

Video / DVD

National Geographic Africa (1996), National Geographic Video, 2001. Four-disc documentary about diverse African regions.

Families of Ghana, Master Communications, Inc., 2006. Part of an award-winning series showing rural and urban family life in different parts of the world.

Internet

Go to the *Research and Writing Center* at **ClassZone.com** to link to:

- data on the region
- a variety of chapter-specific, content-reviewed sites
- "History Makers" biographies
- primary sources

Animated GEOGRAPHY
Colonies in Africa *1913*
CLASSZONE.COM

Point out that this is a historical map of Africa. Draw attention to the color key and inset map. Display a current political map of Africa for comparison.

- How many European nations had African colonies in 1913? *(seven)*
- Which African colonies did Portugal control? *(Portuguese Guinea, Angola, and Mozambique)*
- Which European nation controlled the smallest area of West Africa? *(Portugal)*
- Which West African nation remained free of European control? *(Liberia)*

CONNECT GEOGRAPHY & HISTORY ANSWERS

1. France and Britain
2. Many different ethnic groups became part of single colonies.

Presentation Options

Use the **Power Presentations** DVD-ROM to project the animated map.

- Prompt students to use the pop-up images and audio from the map to discover what they already know about the geography, culture, history, economics, or governments of West Africa.
- Challenge students to answer the map questions individually, in small groups, or as a class activity. Use all of the interactive material to help preview chapter content.

1 Plan & Prepare

Objectives

- Explain the rise of the three West African trading empires
- Describe European imperialism in Africa and its impact
- Summarize how African colonies gained independence

Read for the Essential Question

Help students read for a purpose by reminding them of the Essential Question: "How have centuries of colonialism affected West Africa?" Point out that they can use text features such as Key Questions, special features, images, and maps to form answers to the question.

Vocabulary

Best Practices Toolkit

Use the ***Best Practices Toolkit*** to model strategies for vocabulary notetaking. Vary strategies throughout the year. Choose from: *Knowledge Rating, Predicting ABCs, Definition Mapping, Word Sort, Frayer Model (Word Squares), Magnet Words, Word Wheel,* and *Student VOC.*

Vocabulary Strategies, TT9–TT16

Reading Strategy

Best Practices Toolkit

Display the Web Diagram transparency.

- Remind students that main ideas are the most important points in a segment of text.
- Model filling in a main idea for each of the headings—Mali, Songhai, and Ghana—shown on the graphic organizer.
- Have students suggest entries.
- Display the completed diagram for students to review.

Web Diagram, TT28

SECTION 1

Reading for Understanding

Key Ideas

BEFORE, YOU LEARNED

Europeans traveled to other lands, such as the Americas, establishing colonies to build wealth.

NOW YOU WILL LEARN

Colonizers also traveled to West Africa, dividing up land once held by great trading empires.

Vocabulary

TERMS & NAMES

Ghana a West African kingdom between the A.D. 700s and the A.D. 1000s in what is now Mauritania and Mali; a modern country in West Africa

trans-Saharan across the Sahara

Mali a West African empire between the A.D. 1200s and the A.D. 1500s; a modern West African country

Mansa Musa (MAHN•sah moo•SAH) leader of the Mali empire from 1312 to about 1337

Songhai (sawng•hy) an ancient West African empire, which was most powerful about A.D. 1500

imperialism the practice of one country controlling the government and economy of another country or territory

Pan-Africanism a movement calling for the unity of all people of African descent

BACKGROUND VOCABULARY

camel a humped mammal used in the desert as a pack animal and provider of meat, milk, and wool

self-rule independent government

Reading Strategy

Re-create the diagram shown at right. As you read and respond to the **KEY QUESTIONS**, use the diagram to organize important details about West Africa.

Skillbuilder Handbook, page R4

FIND MAIN IDEAS

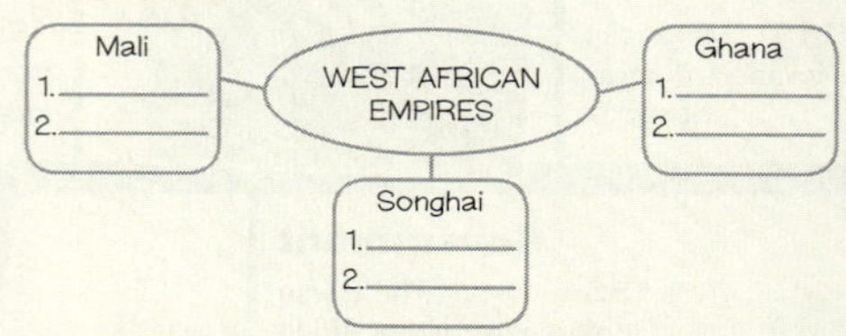

GRAPHIC ORGANIZERS
Go to **Interactive Review @ClassZone.com**

PRETEACHING VOCABULARY

English Learners

Pronounce and Preview

Pronounce each term for students. Review the meanings of words within definitions, such as *empire*, *unity*, and *descent*.

- To modify vocabulary learning, have students complete worksheets as they read, instead of afterward.

Unit Resource Book: Africa
- Vocabulary Practice, p. 131

Inclusion

Vocabulary Quiz

Have volunteers read aloud each term or name and definition. Guide students in correct pronunciation and discuss lengthy or difficult definitions. Then ask pairs of students to quiz each other. One member of the pair should read a vocabulary item and the partner should supply the definition. Students should then switch roles.

Maize crop, Benin

Mandrill

SECTION 1 HISTORY

Three Trading Empires

Connecting to Your World

What makes a country strong? Is it the size of a nation or population? The United States' strength results partly from its size, population, and military power, but also from trade. Trade helps a nation build wealth, which makes it strong. Trade with other peoples allowed early West African cultures to build empires in Africa.

Three Trading Empires

KEY QUESTION How did West African trading empires become wealthy?

Crossing the Sahara was always difficult, but it became easier around 2,000 years ago. That's when southwest Asian traders introduced camels to the region. **Camels** are perfectly adapted to the desert. Their feet can navigate soft sand, and they can survive without food or water for weeks. For centuries, camel caravans crossed the Sahara along established trade routes. These routes gave rise to three great empires.

Camel Caravan Nomadic Tuareg people lead their camels across the Sahara.

2 Focus & Motivate

3-Minute Warm-Up

Write on the board or display the transparency:

- How does the image of the camel caravan help explain how people adapted to this region? *(Possible Answers: People protected themselves from the hot sun and sand by wearing clothing that covered as much of the body as possible, and used camels to help them cross the desert.)*

Unit Transparency Book: Africa
- 3-Minute Warm-Ups, TT17

Connecting to Your World

More About . . .

Camels in West Africa

West Africans today still rely on the camel as a pack animal and as a source of milk, hides, and wool—and for its manure or dung as a fuel source.

3 Teach

Three Trading Empires

Think, Pair, Share

- How did gold help build West African trading empires? *(Possible Answer: Ghana taxed the gold trade, Mali controlled a large gold field, Songhai controlled trans-Saharan trade routes.)*
- Why was Timbuktu an important city? *(It was the capital of Mali's empire and a center of Muslim learning.)*
- **Find Main Ideas** How did Islam spread in West Africa? *(Possible Answer: Arab caravans introduced Islam to Ghana; Mali and Songhai leaders promoted Islam in their empires.)*

SECTION 1 PROGRAM RESOURCES

Unit Resource Book: Africa

ON LEVEL

Unit Resource Book
- Reading Study Guide, p. 107
- Map and Graph Skills, p. 133
- Section Quiz, p. 147

STRUGGLING READERS

Unit Resource Book
- Reading Study Guide with Additional Support, p. 113
- Skillbuilder Practice: Reading Charts, p. 132
- Section Quiz, p. 147
- Reteaching Activity, p. 150

eEdition with Audio DVD-ROM

ENGLISH LEARNERS

Pupil Edition in Spanish

eEdition with Audio DVD-ROM

Unit Resource Book
- RSG (Spanish), p. 119
- RSG with Additional Support (Spanish), p. 125

Multi-Language Glossary

Test Generator
- Section Quiz in Spanish

INCLUSION

Unit Resource Book
- RSG with Additional Support, p. 113
- Section Quiz, p. 147
- Reteaching Activity, p. 150

GIFTED & TALENTED

Unit Resource Book
- Map and Graph Skills, p. 133
- Section Quiz, p. 147

PRE-AP

Unit Resource Book
- History Makers, p. 138
- Primary Source, p. 145
- Section Quiz, p. 147

TECHNOLOGY

Unit Transparency Book
- 3-Minute Warm-Ups, TT20
- Essential Question, TT21
- Critical Thinking, TT22

Daily Test Practice Transparencies
- Section 1, TT29

Power Presentations

ClassZone.com

World Cultures and Geography Video on DVD

Animated GEOGRAPHY

West African Trading Empires

CLASSZONE.COM

CONNECT Geography & History

Explain that this map shows the areas held by each of the three ancient trading empires in West Africa. Direct attention to the key. Point out that the boundaries of the trading empires are superimposed upon a current political map of Africa.

- Which ancient empire covered the least area? *(Ghana)* Is ancient Ghana part of Ghana today? *(no)*
- Which ancient empire extended farthest east? *(Songhai)*

ANSWERS

1. **Movement** Mali, because of access to the sea, rivers, and trans-Saharan routes
2. **Location** Mali

Unit Resource Book: Africa
- Map and Graph Skills: Annual Rainfall in Africa, pp. 133–134

Presentation Options

Use the **Power Presentations** DVD-ROM to project the animated map.

- Click on each empire in the key.
- Listen to the audio and study the map to learn about each empire and ancient trade routes.

More About . . .

Mansa Musa and Timbuktu

Also known as Kankan Musa, Mansa—or Emperor—Musa became famous in his time for a splendid pilgrimage to Mecca in 1324. On his journey, Musa visited Cairo, where he spent so much gold that its value dropped and remained low for years. Musa brought back learned and professional people to his empire. He also commissioned the building of mosques, including the Djinguereber mosque that still stands in Timbuktu.

Unit Resource Book: Africa
- History Makers: Sundiata, p. 137
- History Makers: Mansa Musa, p. 138

Animated GEOGRAPHY

West African Trading Empires

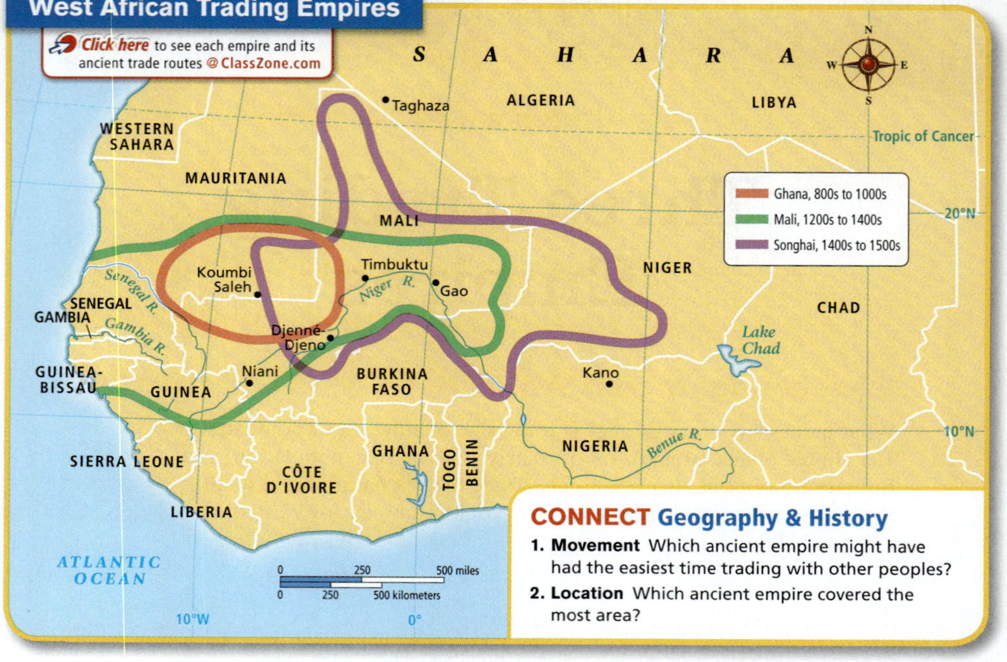

CONNECT Geography & History

1. **Movement** Which ancient empire might have had the easiest time trading with other peoples?
2. **Location** Which ancient empire covered the most area?

Ghana The Soninke people of **Ghana** founded the first West African empire around A.D. 700. Ancient Ghana began as a trading center for grains, cattle, and metals. Camel caravans from the Arabic world arrived at Ghana's markets, introducing Islam to the region and establishing a **trans-Saharan** trade. *Trans-Sahara* means "across the Sahara." Caravans brought books, salt, cloth, and tools from Arabia to Ghana, and carried back slaves, ivory, and gold. Taxing the gold trade enriched Ghana's rulers. In 1076, Moroccan Berbers conquered Ghana's major city. More defeats followed, and Ghana collapsed around 1200.

Mali As Ghana declined, several small states competed to control the empire. Between 1235 and 1240, a king named Sundiata conquered the other states to found a new empire east of Ghana. This empire, **Mali**, thrived because it controlled a large new gold field at the center of the trans-Saharan trade.

Sundiata's great-nephew **Mansa Musa** (MAHN•sah moo•SAH) ruled Mali from 1312 to about 1337, spreading Islam throughout the empire. Mosques, courts, and schools in Timbuktu, the capital of Mali's trading empire, made the city a major center of Muslim learning. After 1400, Mali gave way to a stronger empire.

DIFFERENTIATING INSTRUCTION

CONNECT to Art

Illustrating the West African Empires

Have students use the text on this page to retell the history of the Ghana and Mali trading empires in pictures. Students should choose two or three main ideas or events to depict. They may draw their own pictures or use images from other sources. They should identify the empire related to each of their pictures by writing brief captions. *(before independence).* Also point out that *Today* signals the present time.

CONNECT to Language Arts

A Trader's Diary

Have students write a diary entry from the point of view of a trader in ancient West Africa who is either (a) crossing the desert in a caravan, or (b) visiting Timbuktu for the first time. Suggest they consider questions such as the following: What are the sights and sounds of the desert? Of Timbuktu? Have volunteers read their entries, then discuss the likely accuracy of the content.

Songhai The **Songhai** empire developed from large towns near the present-day border between Mali and Niger. Songhai's strength came from its control of the trans-Saharan trade. Sunni Ali, a great military leader, ruled Songhai from 1464 to 1492. In 1493, Askia Muhammad became king, eventually enlarging Songhai to its greatest extent. Askia, known as Askia the Great, promoted the spread of both Islam and trade. The last of West Africa's three great empires declined after Moroccan fighters defeated Songhai in 1591.

SUMMARIZE Describe how the three empires of West Africa gained wealth. Each of the three empires in turn grew wealthy by controlling and/or taxing the trans-Saharan trade.

European Imperialism

KEY QUESTION Why did Europeans come to Africa?

In the 1500s, Portuguese traders made their way to Africa. Soon Holland and other European powers also vied for the rich trade in slaves, gold, and ivory.

The Slave Trade Shortly after the Portuguese arrived in West Africa, they started enslaving Africans and shipping them to Europe. Before this time, African powers had sent some slaves to Asia and Europe. But the Atlantic slave trade exceeded anything that had come earlier. Scholars say it was history's single largest forced movement of people. African merchants and rulers sold enslaved Africans to Europeans, who sent the slaves to the Americas. The slave trade devastated Africa. Between the late 1400s and the middle 1800s, 10 to 12 million enslaved persons came to the Americas. Slave raiding and warfare emptied towns and villages, destroying ancient patterns of culture, work, and trade.

Europeans Carve Up Africa Even after the slave trade ended, Europeans continued to raid Africa for cheap labor, natural resources, and markets. European **imperialism**, or control of weaker nations, grew with the European industrialization of the late 19th century. Railroads, steamships, and the repeating rifle helped Europeans control much of Africa. In the 1880s, European powers carved Africa into a colonial map for European convenience. The new colonial boundaries ignored traditional ethnic boundaries. In many cases, this grouped long-standing enemies together in colonies.

HISTORY MAKERS

Askia Muhammad 1441?–1538

Askia (AS•kee•uh) Muhammad, shown here in an artist's rendering, ruled the West African Songhai empire at its most powerful. He became king by overthrowing the son of Sunni Ali, an earlier ruler. Askia enlarged Songhai by seizing big chunks of Mali. He also conquered several states ruled by the Hausa people, and colonized Berber towns. In 1528, Askia Muhammad's son took power.

ONLINE BIOGRAPHY
For more on the rulers of West Africa, go to the **Research & Writing Center** @ClassZone.com

INTERDISCIPLINARY ACTIVITIES

Struggling Readers

Clarifying the Division of Africa

Some students may need clarification of the subsection "Europeans Carve Up Africa." To help them grasp the main ideas, have students offer suggestions for a summary of the paragraph. *(Possible Answer: European industrialization created the need for cheap labor, natural resources, and markets. In the 1880s, European powers divided Africa into colonies. This division of Africa placed enemy ethnic groups within the same colony.)*

English Learners

Key Academic Vocabulary

Read this sentence from the subsection "The Slave Trade": "But the Atlantic slave trade exceeded anything that had come earlier." Point out that "exceeded" means "was greater than," especially in terms of number. Explain that "anything that had come earlier" refers to the preceding statement. Ask students to rephrase the sentence in their own words. *(Possible Answer: During the Atlantic slave trade, more Africans were shipped overseas than ever before.)*

RESEARCH & WRITING CENTER

HISTORY MAKERS

Askia Muhammad

After Askia Muhammad took power in 1493, he went on a pilgrimage to Mecca. Then he set out to transform his government. He established directorships in various administrative areas such as agriculture, justice, and finance. Askia Muhammad ruled until 1528, when his son banished him to an island. He died in 1538 in the Songhai trading city of Gao, which is now part of Mali. His tomb still exists and attracts many visitors.

Teach

European Imperialism

Talk About It

- How long did the Atlantic slave trade last? *(almost four centuries, from the late 1400s to the middle 1800s)* How was Africa affected by the slave trade? *(Raids and warfare on towns and villages destroyed ancient patterns of culture, work, and trade.)*
- Why did Europeans continue to raid Africa after the slave trade ended? *(to have cheap labor, resources, and markets)*
- **Find Main Ideas** How has European imperialism affected the traditional ethnic life of Africa? *(Possible Answer: Colonial boundaries remained after independence and still cross ethnic lines; colonizers discouraged economic development.)*

Unit Resource Book: Africa
- Primary and Secondary Sources: Benin Bronze, p 146

Teach

African Colonies Gain Independence

Think, Pair, Share

- How did serving with Allied forces in World Wars I and II influence African soldiers? *(Africans began demanding independence for their own countries.)*
- **Find Main Ideas** Why did British colonies establish self-rule more easily than other European colonies in Africa? *(Britain's indirect form of colonial government helped to prepare African leaders in its colonies for self-rule.)*

Unit Resource Book: Africa
- Primary and Secondary Sources: Benin Bronze, p. 145

COMPARING

Forms of Colonial Rule
Point out that Africans remained under overall European control in both systems. Also, well into the 20th century, European countries forced Africans to pay taxes for the costs of colonial government.

CRITICAL THINKING ANSWER

Draw Conclusions Possible Answer: If the end goal was to prepare colonies for self-rule, then indirect rule would be the best method.

Unit Resource Book: Africa
- Skillbuilder Practice, p. 132

Comparing Regions

Belgium's Congo Colony
King Leopold II of Belgium claimed a large part of Central Africa as his personal possession, naming it the Congo Free State. Leopold's agents used extremely harsh tactics, such as amputating a laborer's hand if a quota wasn't met. By 1908, international protests forced Leopold to turn over the Congo to his nation, Belgium.

Impact of Imperialism By 1914, only Liberia, founded by former U.S. slaves, and Ethiopia were free of European control. Africans resisted colonization, but superior weapons helped Europeans hold power. Different colonial powers had different styles of control, as shown in the chart below. The British ruled indirectly, letting Africans take over some parts of government. France refused to share any power, considering its colonies part of a global French nation.

The scars of imperialism still mark Africa. Colonizers discouraged industrial development, keeping colonies dependent. Wars of conquest and resistance cost many lives. So did the brutal demands of plantation work. Colonial boundaries remained after independence and still cut across ethnic lines.

ANALYZE POINTS OF VIEW Describe why Europeans wanted to colonize Africa. They saw imperialism as an opportunity to take natural resources and to gain wealth through trade of goods.

African Colonies Gain Independence

KEY QUESTION How did African colonies win independence?

In the early 1900s, some colonies sent African students overseas to universities. These well-educated Africans began to dream of freedom for their own people. They joined with former slaves and descendents of slaves in the Americas in calling for freedom and unity for all people of African descent. Their movement became known as **Pan-Africanism**. African soldiers fought alongside Allied forces in World War I and World War II. After the wars, Africans began to demand independence for their own countries. Some Europeans also

COMPARING Forms of Colonial Rule

	INDIRECT RULE of African Colonies	DIRECT RULE of African Colonies
Methods	• Local governments used • Limited self-rule • Goal: to develop future leaders • Government based on European styles, but may have local influences	• Foreign officials brought in to rule • No self-rule • Goal: to get colonies to adopt European culture • Government based only on European styles
Examples	British colonies such as Nigeria	French colonies such as French West Africa

CRITICAL THINKING

Draw Conclusions Which form of rule do you think would be the best long-term method, and why?

DIFFERENTIATING INSTRUCTION: TIERED ACTIVITIES

OBJECTIVE Explore issues faced by a newly independent African nation.

Basic
Brainstorm with students the effects of imperialism in Africa, covering topics such as European rule, brutal demands of plantation work, and low pay. Then help student pairs prepare a list of problems that a newly independent African nation would have needed to solve. Have pairs exchange and compare lists.

On Level
In addition to the Basic brainstorming, have students work alone to prepare written statements about one key issue that they think needed to have been addressed first by the leaders of a newly independent African nation. Students should include problem-solving ideas in their statements.

Challenge
In addition to the On Level activity, have students form small groups and role-play a discussion of their ideas as members of a new national assembly. Invite students to present summaries of their discussions to the class.

began to question whether it was right to maintain colonies. Britain's indirect control of its colonies helped prepare African leaders for eventual **self-rule**. More tightly controlled colonial governments left newly independent nations to struggle on their own.

Methods of Attaining Independence The first West African colony to win independence was the Gold Coast. In 1957, it peacefully separated from Great Britain and reclaimed the ancient name of Ghana. Nigeria won independence in 1960. Independence for some other West African countries came only after years of guerrilla warfare. For example, war erupted in the Portuguese colony of Guinea-Bissau in the early 1960s. The conflict dragged on until 1974. By the end of 1975, Portugal had given up the last of its African claims.

COMPARE Describe the paths different colonies took toward independence. Some countries won independence peacefully after preparing for self-rule. Others fought wars for independence.

Independence Guinea-Bissau citizens celebrate the end of Portuguese colonization in 1974.

Section 1 Assessment

ONLINE QUIZ For test practice, go to **Interactive Review** @ClassZone.com

TERMS & NAMES

1. **Explain the importance of**
   - Ghana
   - Mansa Musa
   - Songhai
   - Pan-Africanism

USE YOUR READING NOTES

2. **Find Main Ideas** Use your completed diagram to answer the following question:

   What did the ancient West African empires have in common?

   WEST AFRICAN EMPIRES
   - Mali: 1. ______ 2. ______
   - Ghana: 1. ______ 2. ______
   - Songhai: 1. ______ 2. ______

KEY IDEAS

3. What were the names of West Africa's great empires?
4. How did industrialization promote European imperialism in Africa?
5. How did West African countries win independence from their colonizers?

CRITICAL THINKING

6. **Evaluate** How did World Wars I and II lead some Europeans to question colonialism?
7. **Make Inferences** Why do you think the modern nation of Ghana chose the name of an ancient empire in a different location?
8. **CONNECT to Today** In 1994 in Rwanda, 800,000 people were killed in ethnic fighting. How has Africa's colonial past promoted such conflicts?
9. **TECHNOLOGY** **Create a Multimedia Presentation** Create a slide show about historical Timbuktu. Focus on Timbuktu's history as a prominent center of trade and Islamic learning.

CONNECT to the Essential Question

How have centuries of colonialism affected West Africa?

Ask students what they have learned so far that can help them answer this question. Students might mention:

- Great trading empires enriched West Africa, but European raids for slaves and goods ruined ancient patterns of work, trade, and culture.
- European nations fueled ethnic conflicts by ignoring traditional boundaries when they carved up Africa.
- Colonial control caused some African nations to wage violent struggles to achieve self-rule.

4 Assess & Reteach

Assess Have students complete the Section Assessment.

Unit Resource Book: Africa
- Section Quiz, p. 147

Interactive Review @ClassZone.com
- Section Quiz

Power Presentations

Test Generator

Reteach Have students work in small groups to create a time line for "Three Trading Empires." Groups should choose events they consider most important for their time line. Then have groups compare time lines and explain their choices.

Unit Resource Book: Africa
- Reteaching Activity, p. 150

SECTION 1 ASSESSMENT ANSWERS

Terms & Names

1. Ghana: ancient trading kingdom; Mansa Musa: Mali's leader in the 1300s; Songhai: empire most powerful about 1500; Pan-Africanism: movement for freedom for people of African descent

Use Your Reading Notes

2. wealth built on trade, and expansion of Islam

   WEST AFRICAN EMPIRES
   - Mali: 1. Islamic 2. wealth from new gold field 3. created major center of learning
   - Ghana: 1. Islamic 2. wealth based on taxing trade 3. conquered by Berbers
   - Songhai: 1. Islamic 2. controlled trans-Saharan trade 3. conquered Mali

Key Ideas

3. Ghana, Mali, Songhai
4. It permitted Europeans to build railroads, sail steamships up African rivers, and defeat African insurgencies with superior weapons.
5. Some countries, such as Ghana, won independence peacefully. Others, such as Guinea-Bissau, had to fight lengthy wars.

Critical Thinking

6. Africans fought alongside Europeans in the wars. The wars were viewed as fights for freedom from conquest, so some Europeans came to agree that colonies deserved their freedom as well.
7. Answers may vary; e.g., Ghanaians may have wanted to reclaim Africa's pre-colonial glory.
8. African colonial boundaries, organized for European convenience, cut across ethnic lines, uniting (or in this case creating) enemies.
9. Use the rubric to score presentatons.

Presentation Rubric

	Content and presentation
4	excellently researched
3	well researched
2	adequately researched
1	poorly researched or incomplete

1 Plan & Prepare

Objectives

- Describe Nigeria's culture, government, and economics, and the challenges it faces
- Describe Ghana's government, culture, and economics and the challenges it faces

Read for the Essential Question

Help students read for a purpose by reminding them of the Essential Question: "How have centuries of colonialism affected West Africa?" Point out that they can use text features such as Key Questions, special features, images, and maps to form answers to the question.

Vocabulary

Best Practices Toolkit

Use the ***Best Practices Toolkit*** to model strategies for vocabulary notetaking. Vary strategies throughout the year. Choose from: *Knowledge Rating, Predicting ABCs, Definition Mapping, Word Sort, Frayer Model (Word Squares), Magnet Words, Word Wheel,* and *Student VOC.*

Vocabulary Strategies, TT9–TT16

Reading Strategy

Best Practices Toolkit

Display the T Chart/Two-Column Chart transparency.

- Remind students that summarizing means restating a paragraph, a passage, or a section in fewer sentences or words.
- Model summarizing the subsections about the histories of Nigeria and Ghana since independence.
- Have students suggest entries. Remind students that summarizing will help them answer Key Questions.
- Display the completed chart for students to review.

T Chart/Two-Column Chart, TT20

SECTION 2

Reading for Understanding

Key Ideas

BEFORE, YOU LEARNED

Colonizers traveled to West Africa, dividing up land once held by great trading empires.

NOW YOU WILL LEARN

Nigeria and Ghana have struggled to maintain stable governments and productive economies.

Vocabulary

TERMS & NAMES

Hausa (HOW•suh) a people of northern Nigeria and southern Niger

Yoruba (YAWR•uh•buh) a people of southwestern Nigeria, parts of Benin, and Togo

Igbo (IHG•bo) a people of southeastern Nigeria

coup (koo) military takeover

Biafra (bee•AF•ruh) a largely Igbo region of Nigeria that declared its independence in 1967, but surrendered after losing a civil war in 1970

Kwame Nkrumah (KWAHM•eh uhn•KROO•muh) modern Ghana's first president

Kente cloth a brightly colored, woven cloth made by the Ashanti and Ewe peoples

BACKGROUND VOCABULARY

stability durability; resistance to change

subsistence farming farming that produces just enough to feed the farmer's family

Ashanti (uh•SHAN•tee) a people of Ghana

Visual Vocabulary kente cloth

Reading Strategy

Re-create the chart shown at right. As you read and respond to the **KEY QUESTIONS**, use the chart to summarize information about Nigeria and Ghana.

Skillbuilder Handbook, page R5

SUMMARIZE

NIGERIA	GHANA
1.	1.
2.	2.
3.	3.
4.	4.

GRAPHIC ORGANIZERS
Go to Interactive Review @ClassZone.com

586 Chapter 19

PRETEACHING VOCABULARY

English Learners

Pronounce and Preview

Pronounce each term for students. Review the meanings of words within definitions, such as *takeover*, *surrendered*, and *durability*.

- To modify vocabulary learning, have students complete worksheets as they read, instead of afterward.

Unit Resource Book: Africa
- Vocabulary Practice, p. 131

Inclusion

Play a Vocabulary Card Game

Have student pairs write vocabulary items and definitions on separate index cards, shuffle them, then take turns drawing a card until all are drawn. Students should match any cards in hand, and set them aside after both partners check for accuracy. Then partners should take turns showing their unmatched cards, one at a time. The other partner should produce the correct match. Points may be tallied for all correct matches.

Maize crop, Benin

SECTION 2 FOCUS ON

Mandrill

Nigeria and Ghana

Connecting to Your World

How does your family make important decisions, such as where to take a vacation or whether to move to a new apartment? In some families, a parent or both parents make the decisions. Some families seek out the advice of an older relative. However your family makes decisions, you're probably used to your own system. Like your family, West African societies had their own familiar ways of ruling themselves and solving conflicts. But when a familiar system disappears, things get confusing.

Nigeria

KEY QUESTION How have Nigeria's main ethnic groups gotten along since independence?

When European powers colonized Africa, they didn't recognize the systems of governing the Africans used. Instead, they imposed European-style centralized governments to replace African stateless systems. Most ruling officials were Europeans. When African nations gained freedom in the mid-1900s, few Africans knew how to run a European-style government. As a result, many African governments lacked **stability**.

because it runs into the ocean ▼

River Port The Niger River remains an important West African transportation route. **Why might the Niger River be important to international trade?**

2 Focus & Motivate

3-Minute Warm-Up

Write on the board or display the transparency:

- Why do you think the Niger River is still an important West African transportation route? *(Possible Answer: It may be less expensive and more convenient.)*

Unit Transparency Book: Africa
- 3-Minute Warm-Ups, TT17

Connecting to Your World

More About . . .

The Niger River

The Niger is also used for hydroelectric power, irrigation, and fishing. The Niger Delta region is a source of oil, although pollution from its production threatens the fishing industry.

3 Teach

Nigeria

Reader, Recorder, Reporter

- Identify the three largest ethnic groups in Nigeria. *(Hausa, Yoruba, Igbo)* Why did other groups resent the Hausa? *(The Hausa held more political power.)*
- Why was Biafra created, and what was the result? *(Possible Answer: The Igbo wanted their own state. It led to a civil war.)*
- **Summarize** How has oil production helped and hurt Nigeria? *(Possible Answer: Oil has brought wealth, but it has also slowed growth of other industries and created oil spills.)*

SECTION 2 PROGRAM RESOURCES

Unit Resource Book: Africa

ON LEVEL

Unit Resource Book
- Reading Study Guide, p. 109
- Outline Map Activities, p. 135
- Section Quiz, p. 148

STRUGGLING READERS

Unit Resource Book
- Reading Study Guide with Additional Support, p. 115
- Skillbuilder Practice: Creating a Multimedia Presentation, p. 132
- Section Quiz, p. 148
- Reteaching Activity, p. 151

ENGLISH LEARNERS

Pupil Edition in Spanish
Pupil Edition Audio DVD-ROM
Unit Resource Book
- RSG (Spanish), p. 121
- RSG with Additional Support (Spanish), p. 127

Multi-Language Glossary
Test Generator
- Section Quiz in Spanish

INCLUSION

Unit Resource Book
- RSG with Additional Support, p. 115
- Section Quiz, p. 148
- Reteaching Activity, p. 151

GIFTED & TALENTED

Unit Resource Book
- Economics in the World, p. 142
- Section Quiz, p. 148

PRE-AP

Unit Resource Book
- Connect to Art, p. 143
- Primary Source, p. 145

Unit Transparency Book
- 3-Minute Warm-Ups, TT17
- Essential Question, TT18
- Critical Thinking, TT20

TECHNOLOGY

Unit Transparency Book
- Map: Population Density, TT23

Daily Test Practice Transparencies
- Section 2, TT65

Power Presentations
ClassZone.com
World Cultures and Geography Video on DVD

Ethnic Groups of Nigeria

CONNECT Geography & History

Tell students that this is a thematic map. It shows regions dominated by major Nigerian ethnic groups. Point out Abuja. This new capital city replaced Lagos as the capital in 1991, built because of congestion in the former capital. Also draw students' attention to the key.

- Have volunteers identify locations for each group named in the accompanying text. *(Hausa, north; Yoruba, southwest; Igbo, south and east)*
- What can you learn from the map about oil resources in Nigeria? *(Possible Answers: Nigeria's oil fields are along the Atlantic coast and Niger River.)*

ANSWER

Place Possible Answer: Abuja is easy to reach from all parts of Nigeria. It appears to be situated in a neutral area, one not occupied by a major ethnic group. Other factors may be a favorable climate and room to expand.

Unit Resource Book: Africa

- Skillbuilder Practice: Reading Charts, p. 132
- Outline Map Activities: West Africa: Physical/Political, pp. 135–136

More About . . .

Chinua Achebe and Wole Soyinka

The works of Achebe and Soyinka explore issues of colonialism and independence. Achebe was born into an Igbo family. His acclaimed novel *Things Fall Apart*, published in 1958, portrays tensions between traditional Igbo society and European influences.

Soyinka writes from the perspective of Yoruba culture and his love of human liberty, often criticizing colonialism and condemning the repression and corruption in African politics. Best known for his plays, in 1986 Soyinka became the first African to win the Nobel Prize for Literature.

History Since Independence In Section 1, you learned that colonial powers drew Africa's borders without considering the culture or history of African peoples. Nigeria is just one of the African countries where conflict has grown from cultural divisions.

The three largest groups in Nigeria are the **Hausa**, who live in the north; the **Yoruba**, who live in the southwest; and the **Igbo** (IHG•bo), who live in the east. After independence in 1960, Nigeria's major ethnic groups clashed. The Igbo and other southern groups resented the Hausa's greater political power.

Ethnic Groups of Nigeria

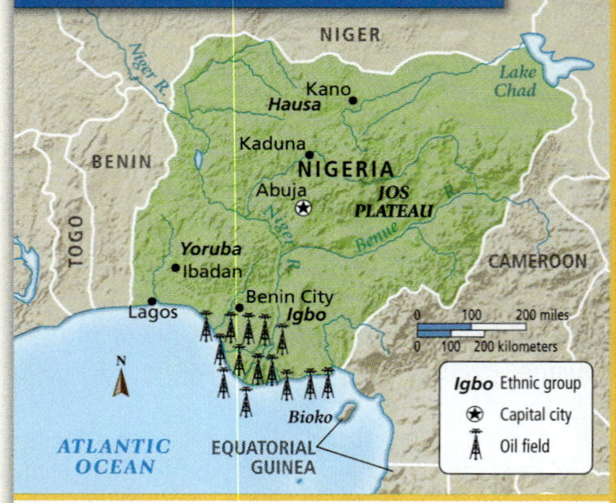

CONNECT Geography & History

Place Abuja, Nigeria's capital, was a planned city. Why might Nigeria want its capital in the center of the country?

Ethnic conflict between the groups fueled an Igbo-led **coup**, or military takeover, in 1966. In 1967, the Igbo broke away from Nigeria to create their own state, called **Biafra**. A three-year civil war followed, causing many deaths from fighting and starvation. Biafra surrendered and rejoined Nigeria in 1970. Military leaders ruled Nigeria from 1966 to 1979, and again from 1983 until 1999. A new constitution and democratic elections restored a civilian government in 1999. But ethnic and religious tensions still simmer.

Nigerian Culture and Daily Life More than 250 ethnic groups, who speak many different languages, live in Nigeria, whose official language is English. Nigeria's 132 million people make it Africa's most populous nation, and one of the world's largest. About half of Nigerians are Muslim, while 40 percent are Christian. Many people also practice traditional animist religions or mix them with Islam or Christianity.

At least two-thirds of Nigerians survive on less than one dollar a day. Most are poor farmers who live in grass, mud, or wood homes in clusters called compounds. In rural areas, people wear long, loose robes. Wealthy city dwellers live in modern houses or apartments and often wear Western clothes. Poor Nigerians, who have flooded the cities since the 1960s, often crowd into slums.

Nigerian authors Chinua Achebe (CHIHN•wah ah•CHAY•bay) and Wole Soyinka (WOH•lay shaw•YING•kuh) wrote their most famous works in English. In 1986, Soyinka became the first African to win the Nobel Prize for Literature. Nigerian music pulses with drums, xylophones, and string and wind instruments. The Yoruba are famous for their wood carvings. Nigerian art, especially sculptures, have inspired Western artists.

DIFFERENTIATING INSTRUCTION

English Learners

Vocabulary: Cognates

Point out the phrase on this page: "Nigeria's 132 million people make it Africa's most populous nation." Say the word *populous*. The Latin translation is *populosus*, derived from *populus*, meaning "people." In French the word is *populeux* or *populeuse*, and in Spanish *populoso* or *populosa*. Similar words related to *people* also derive from the Latin *populus*. For example, the English word *popular* is *popular* in Spanish; *populaire* in French.

Gifted & Talented

Write a Citizen's Op-Ed Piece

Have students use the text headed "History Since Independence" and the map of Nigeria's ethnic groups to write an op-ed piece that answers the following question: "What are the advantages and disadvantages of a unified Nigeria?" Have students present their op-ed pieces, and then hold a group discussion.

ANALYZING Primary Sources

Ken Saro-Wiwa was a Nigerian activist fighting for the rights of his Ogoni people. He and eight others were arrested and hanged in 1995 by Nigeria's military government. Many around the world believed Saro-Wiwa was convicted on false charges. Shortly before his death, Saro-Wiwa smuggled writings out of prison.

> It is also very important that we have chosen the path of non-violent struggle. Our opponents are given to violence and we cannot meet them on their turf, even if we wanted to. Non-violent struggle offers weak people the strength which they otherwise would not have. The spirit becomes important, and no gun can silence that.
>
> Source: Ken Saro-Wiwa, A Month and a Day: A Detention Diary

DOCUMENT-BASED QUESTION

What does "we cannot meet them on their turf" mean?

ONLINE PRIMARY SOURCE To read more of Saro-Wiwa's works, go to the **Research & Writing Center** @ClassZone.com

Government and Economics An elected president and vice president head Nigeria's government. They're assisted by an appointed cabinet and a two-house National Assembly.

Nigerians depend largely on agriculture for their livelihood. Two out of three Nigerian workers are farmers, who produce much of the world's cocoa, peanuts, and rubber. But most of Nigeria's farmers practice **subsistence farming**. Their food crops include maize, beans, and other vegetables and fruits. While manufacturing industries are developing, mining and petroleum produce the majority of the nation's wealth. In addition to producing coal, gold, iron ore, and natural gas, Nigeria is also one of the world's biggest oil exporters. Oil has contributed to Nigeria's wealth—and to its problems. In the 1970s, oil profits funded government programs to improve the standard of living. But relying too heavily on oil exports slowed development of Nigeria's other industries. Oil spills have also polluted certain areas, ruining farmland and poisoning fish.

Elections A new Nigerian constitution and free elections in 1999 ended nearly 30 years of military rule.

SUMMARIZE Summarize Nigeria's ethnic conflicts since independence. A civil war broke out in 1967, and ethnic tensions continue to simmer.

INTERDISCIPLINARY ACTIVITIES

CONNECT to Drama

Roleplay for Human Rights

Have students work in groups to develop and roleplay for the class a scene based on the primary source about the Ogoni people and Ken Saro-Wiwa. Possible roles could include a human rights activist, members of an oppressed group, and a military leader. Students should include in their presentations a clear statement of the rights being claimed.

CONNECT to Language Arts

Election Diary Entries

Draw students' attention to the photograph on this page, captioned "Elections." Ask students to imagine they are voting in a free election for the first time. Have them write a diary entry describing the election and their feelings before and after voting. Then have volunteers read their entries aloud.

RESEARCH & WRITING CENTER

ANALYZING Primary Sources

To many Ogoni people, Saro-Wiwa was a hero in the struggle against the pollution of their homeland in the Niger Delta. A writer by trade, Saro-Wiwa was a leader of the Movement for the Survival of the Ogoni People, which had been formed in 1990. In 1993, some 300,000 Ogoni people—about 60 percent of the entire group—gathered peacefully to protest against oil drilling and oil spills in their homeland.

- What is meant by, "The spirit becomes important, and no gun can silence that"? *(Possible Answer: A gun can kill people, but an idea—the spirit or cause that moves the people—lives on in others.)*

DOCUMENT-BASED QUESTION ANSWER

Possible Answer: The government is too strong for the Ogoni to oppose with violence.

Unit Resource Book: Africa

- Primary and Secondary Sources: from *The Travels of Ibn Battuta*, p. 146

More About...

Democratic Elections in Nigeria

In 1999 and in 2003 Nigerians elected Olusegun Obasanjo of the People's Democratic Party to the presidency. Obasanjo was a well-known military commander, having led Nigerian forces during the war with Biafra. He became leader of Nigeria in 1976, but voluntarily yielded power to a democratically-elected president in 1979.

Nigeria is divided into 36 states with an area reserved for the federal capital. A Nigerian president serves a four-year term and can serve one additional term if re-elected. The National Assembly consists of a Senate and a House of Representatives, but—unlike in the United States—Nigerian members of both houses are elected to four-year terms.

Teach Ghana

Think, Pair, Share

- What was the effect of Ghana's constitution of 1992? *(Ghana became a multiparty democracy.)*
- How large is Ghana's Ashanti group? *(Members outnumber Ghana's 100 other groups.)* What Ashanti crafts might you find at a marketplace in Ghana? *(clothing made of Kente cloth, carvings, and stools)*
- **Summarize** How do the economies of Ghana and Nigeria compare? *(Subsistence farming is a major activity in both countries, but Nigeria relies heavily on exporting oil while Ghana exports cocoa, minerals, and timber.)*

Unit Resource Book: Africa
- Interdisciplinary Projects: Bartering, p. 139

More About . . .

Kwame Nkrumah

Nkrumah was also a strong advocate of Pan-Africanism. In 1963 he helped to found the Organization of African Unity.

CONNECT to Art

People have made masks for various uses since ancient times. The Egyptian pharaoh Tutankhamen was buried with a gold mask. Ancient Greek actors used masks to convey facial expressions. A North American Iroquois group called the False Face Society wore masks for healing work.

In Africa, many ethnic groups are known for mask making. In West Africa, the Dogon people use masks in rituals to honor their ancestors. In the early 1900s, African masks influenced artists in Europe. In 1907 Pablo Picasso of Spain finished a painting in which the face of a seated woman resembles an African mask.

Unit Resource Book: Africa
- Connect to Art: Make a Mask, p. 143

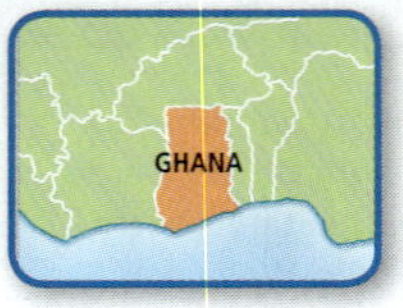

Ghana

KEY QUESTION What are the key points of Ghana's art and culture?

When Portuguese sailors landed on the west coast of Africa in 1471, they named this gold-rich coastal territory the Gold Coast. In 1957, the Gold Coast was the first African state to win independence. The new republic's leaders chose the name Ghana to recall the greatness of the ancient empire, even though no part of ancient Ghana was located within the modern country.

History Since Independence During the late 1940s and early 1950s, the British who controlled Ghana began to encourage educated Africans to participate in governing. **Kwame Nkrumah** (KWAHM•eh uhn•KROO•muh), a Ghanaian who had served as prime minister under British rule, became Ghana's first president in 1960. Nkrumah tried to boost living conditions, build factories, and expand education. Corruption and a failing economy triggered a coup in 1966.

A series of military leaders revolved through Ghana's capital, Accra. One of these soldiers, Jerry Rawlings, headed Ghana's government between 1979 and 2001. Ghana's new constitution, accepted in 1992, turned Ghana into a multiparty democracy.

Ghana's Culture and Daily Life Like Nigerians, most Ghanaians speak African languages and study Ghana's official language, English, in school. **Ashanti**, or Akan, members outnumber Ghana's 100 other groups, such as the Ewe, Ga, and Moshi-Dagomba. Many Ghanaians

CONNECT to Art

Many African cultures use masks like these for religious ceremonies, celebrations, funerals, and theatrical performances. They often depict spirits and ancestors.

Activity

Make a Mask

Materials
- cardboard
- scissors
- ruler and glue
- paint, colored paper, and other decorations

1. Cut the cardboard into an oval slightly larger than your face.
2. Have a classmate measure the distance between your eyes. Use the measurement to create eye holes on your mask.
3. Make holes for your nose and mouth, and decorate the mask in any way you like.
4. Punch holes on each side of the mask and attach elastic.

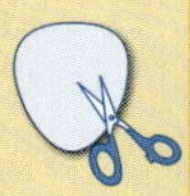

DIFFERENTIATING INSTRUCTION: TIERED ACTIVITIES

OBJECTIVE Increase understanding of the importance of culture in African life.

Basic

Brainstorm with students the kinds of costumes, masks, and music that might be used for an African celebration. Then have students work in pairs to plan an annual celebration of Ghana's independence day. Ask pairs to present their ideas in a list illustrated with two or three simple drawings.

On Level

Adapt the Basic activity by having students work individually. Students should draw upon prior knowledge as well as their textbook to complete a plan for the celebration. They should present their plan to the class.

Challenge

Adapt the On Level activity by having students complete a plan that supplies details about kinds of costumes, masks, and music. The plan should include a schedule of events and a paragraph promoting the event. Have students display their written work after presenting their plans to the class.

wear Western-style clothing. For special social and religious occasions, Ashanti and Ewe craftspeople weave **Kente cloth**, a brightly colored ceremonial cloth made into wraps, skirts, and blouses. More than six out of ten Ghanaians are Christian, just over 20 percent follow traditional African religions, and 16 percent are Muslim.

Marketplaces are central meeting places in Ghana. At a market, you might find carvings and copies of Ashanti stools. In Ashanti culture, the head of each lineage group has a stool to symbolize his power.

Government and Economics Ghana's democratic government includes a president, cabinet, and an elected parliament. Compared to Nigeria, whose economy rises or falls with the price of oil, Ghana boasts several exports. In addition to its biggest export, cocoa, Ghana also supplies the world with diamonds, gold, manganese, and timber. But as in Nigeria, most Ghanaians work as subsistence farmers.

FIND MAIN IDEAS Describe major points of Ghana's art. textiles, carvings, and Ashanti stools

Ghana's Arts Wearing Kente cloth, a Ghanaian boy holds an Ashanti stool.

Section 2 Assessment

ONLINE QUIZ For test practice, go to **Interactive Review** @ClassZone.com

TERMS & NAMES

1. **Explain the importance of**
   - Igbo
   - subsistence
   - Kwame Nkrumah
   - Kente cloth

USE YOUR READING NOTES

2. **Summarize** Use your completed chart to answer the following question: What are some of Nigeria and Ghana's main products?

NIGERIA	GHANA
1.	1.
2.	2.
3.	3.
4.	4.

KEY IDEAS

3. How do most Nigerians and Ghanaians support themselves?
4. What religions predominate in Nigeria and Ghana?
5. What does an Ashanti stool symbolize?

CRITICAL THINKING

6. **Categorize** What kind of governments do Nigeria and Ghana have today?
7. **Draw Conclusions** Why do you think Nigeria has struggled for so long to keep a stable government?
8. **CONNECT to Today** Nigeria is the fifth-largest supplier of oil to the United States. What problems could an unstable Nigerian government pose to the United States?
9. **SCIENCE** **Draw a Flow Chart** Research cocoa production in Ghana. Construct a flow chart to show how cocoa beans are harvested, processed, and turned into chocolate for cocoa and candy.

CONNECT to the Essential Question

How have centuries of colonialism affected West Africa?

Ask students what they have learned so far that can help them answer this question. Students might mention:

- Since independence in 1960 Nigeria has been challenged by ethnic conflict.
- After years of military rule, Nigeria and Ghana each had a constitution and democratic government by the 1990s.
- Nigeria exports oil and Ghana exports cocoa, minerals, and timber, although both rely mostly on farming.

4 Assess & Reteach

Assess Have students complete the Section Assessment.

Unit Resource Book: Africa
- Section Quiz, p. 148

Interactive Review @ClassZone.com
- Section Quiz

Power Presentations

Test Generator

Reteach Assign groups one paragraph from the subsection on Nigeria and one from the subsection on Ghana. Ask them to prepare a summary of each. Have volunteers from each group read their summaries aloud.

Unit Resource Book: Africa
- Reteaching Activity, p. 151

SECTION 2 ASSESSMENT ANSWERS

Terms & Names

1. Igbo: people of southeastern Nigeria; subsistence farming: growing just enough to exist; Kwame Nkrumah: modern Ghana's first president; Kente cloth: brightly colored cloth made by Ashanti and Ewe peoples

Use Your Reading Notes

2. Main products—Nigeria: oil; Ghana: cocoa

NIGERIA	GHANA
1. independence from Britain in 1960	1. independence from Britain in 1957
2. civil war in 1960s, military leaders for years	2. Kwame Nkrumah first president, coup in 1966
3. 250+ ethnic groups	3. new constitution in 1992
4. oil, coal, gold, iron, cocoa	4. cocoa, diamonds, gold, manganese, timber

Key Ideas

3. as subsistence farmers
4. Islam, Christianity, and traditional African religions
5. the power of the head of a lineage group

Critical Thinking

6. constitutional democracies
7. The country was cobbled together out of three major conflicting ethnic groups.
8. A military takeover by forces hostile to the United States could cut off a major portion of our oil supply, leading to terrorism or war.
9. Use the rubric to score the flow charts.

Chart Rubric

	Content	Organization
4	well researched	well organized
3	adequately researched	adequate
2	missing several steps	somewhat unclear
1	most steps missing	poorly organized, confusing, incomplete

1 Focus & Motivate

Objective

Compare and contrast desert life with coastal life in West Africa.

Springboard Many people in West Africa still live and work in traditional ways. Look closely at the photographs on these pages. Can you find electric poles and wires? Do you see any paved roads? Describe what your daily life would be like if you lived in one of these villages in West Africa.

2 Teach

West African Regions

Talk About It

- How is it possible for a settlement to develop in the desert? *(A water source is nearby, mud-brick houses insulate from desert heat, and camels provide transportation for trade.)*
- Why are some West African countries banning foreign fishing boats from their coasts? *(Commercial fishing is depleting stocks of fish, and people in coastal villages rely upon fishing for food.)*
- **Compare and Contrast** In what ways are the regions similar in opportunities for work? *(Possible Answers: Both economies are largely undeveloped; residents rely upon traditional ways of earning a living—herding and fishing.)*

Comparing Regions

Herding Cattle

Nomadic peoples such as the Tuareg of West Africa are traditional herders. The Maasai of East Africa are herders as well. African herders move with their cattle when new grazing land and a fresh water supply are needed. The animals are usually raised for personal use. They provide their owners with milk, hides, and, less often, meat.

In the western United States, ranchers raise livestock for different purposes. Most breed beef cattle to sell for the nation's food supply.

COMPARING West African Regions

While most West African countries have some sort of coastline, the Sahara still dominates countries such as Mauritania, Mali, and Niger.

Desert Life

Desert settlements, like this one in Mali, form near water sources. Mud-brick structures can provide insulation from the desert heat. Still, many African desert-dwellers live nomadically, moving from place to place.

Transportation
For centuries, camels have been the traditional means of crossing the Sahara.

Work
Cattle herders move their livestock from one place to another.

Desert village near Mopti, Mali ▼

Traditional mud-brick houses ▶

DIFFERENTIATING INSTRUCTION: TIERED ACTIVITIES

OBJECTIVE Compare and contrast desert life with coastal life in West Africa.

Basic

Have students study the images labeled "Work" and "Transportation" on each page. What differences and similarities do they see? Form pairs and ask them to complete a graphic organizer, such as a Venn diagram, for each category, based on their observations.

On Level

Adapt the Basic activity by having students work individually and include the category of housing. Ask students to also write a brief summary comparing the two ways of life.

Challenge

Adapt the On Level activity, to have students write an encyclopedia entry titled "Desert and Coastal Life in West Africa." Ask students to make inferences about the ways of life they describe. Have students read their entries aloud.

Work

A fisherman casts a net in Ghana.

Coastal Life

West African coastal villages, like this one in Côte d'Ivoire, rely heavily on fishing for food. Commercial fishing further out to sea can jeopardize fish stocks. Countries such as Mauritania are banning some foreign fishing boats from fishing off their coasts.

Fishing village near Fresco, Côte d'Ivoire ▼

Transportation

Fishing boat in Ghana

◄ *Coastal housing*

CRITICAL THINKING

1. **Compare and Contrast** How do people in these regions adapt their living arrangements to their surroundings?
2. **Make Inferences** Why would cattle herders need to move from place to place?

INTERDISCIPLINARY ACTIVITIES

CONNECT to Language Arts

Describe Adaptations to Local Geography

Brainstorm with students the geographical influences on life in their region, such as nearby mountains and rivers, and your region's climate. Then have them write a paragraph about how people in their region have adapted to its geography.

CONNECT to Art

Create a Display of Local Daily Life

Have students work in small groups to create a visual display about their community. They should pattern their display on this feature, incorporating the topics of transportation, work, and housing. Ask students to present their displays to the class, and then discuss how their community is similar to and different from communities in West Africa.

More About . . .

Transportation in West Africa

Waterways are essential for transportation in West Africa. In most West African countries, boats are used to transport people and goods on major rivers and along the seacoast. However, a landlocked country, such as Mali, is at a disadvantage. Railroad networks exist, but they are not well developed. Except for major cities, roads are few and in need of repair generally throughout all of Africa.

In West Africa's desert areas, the traditional method of desert travel—by camel caravan—is still used. However, this method can be dangerous and slow. West Africans also use four-wheel-drive vehicles. Although faster than camels, this method can be expensive. It usually requires traveling in convoys and the carrying of food, water, fuel, and spare tires for emergencies. The fastest—but most expensive—form of desert travel is by air.

Comparing Regions

West African Housing

Rectangular houses are typical of many traditional villages in West Africa. The houses are built with walls of mud or concrete, and tin or thatched roofs. The houses in Malinke villages in Senegal, however, are cylindrical. The roofs are conical and constructed of thatched straw. The Tuareg people make tents for dwellings using goatskins or woven mats of leaves.

3 Wrap Up

CRITICAL THINKING ANSWERS

1. **Compare and Contrast** They both build living quarters out of what's available. In the desert, mud-brick homes are common. On the coast, structures use more wooden and metal materials.
2. **Make Inferences** Grazing depletes the grasses that grow sparsely in desert and semi-desert regions, so new grazing areas must be found.

1 Plan & Prepare

Objectives

- Describe Mali's history since independence, economy, and culture
- Describe Cote d'Ivoire's history since independence, economy, and culture

Read for the Essential Question

Help students read for a purpose by reminding them of the Essential Question: "How have centuries of colonialism affected West Africa?" Point out that they can use text features such as Key Questions, special features, images, and maps to form answers to the question.

Vocabulary

Best Practices Toolkit

Use the ***Best Practices Toolkit*** to model strategies for vocabulary notetaking. Vary strategies throughout the year. Choose from: *Knowledge Rating, Predicting ABCs, Definition Mapping, Word Sort, Frayer Model (Word Squares), Magnet Words, Word Wheel,* and *Student VOC.*

Vocabulary Strategies, TT9–TT16

Reading Strategy

Best Practices Toolkit

Display the Venn Diagram transparency.

- Remind students that when they compare and contrast, they find both similarities and differences.
- Model filling in entries for Mali, using information from the subsection, "Mali's Daily Life and Culture."
- Have students suggest entries for the other two categories in the diagram.
- Display the completed diagram for students to review.

Venn Diagram, TT22

SECTION 3 Reading for Understanding

Key Ideas

BEFORE, YOU LEARNED

Nigeria and Ghana have struggled to maintain stable governments and productive economies.

NOW YOU WILL LEARN

Mali and Côte d'Ivoire are developing countries with lively cultural traditions.

Vocabulary

TERMS & NAMES

federation colonies, states, or countries that have joined together

inflation a price increase or a fall in the purchasing power of money

infrastructure basic services a community needs, such as roads, railroads, water and power lines, schools, and post offices

millet a grain raised in West Africa, often made into porridge

BACKGROUND VOCABULARY

rig to influence or fix the results of an election

REVIEW

drought a long period of little or no rainfall

famine a severe food shortage

Visual Vocabulary millet

Reading Strategy

Re-create the Venn diagram shown at right. As you read and respond to the **KEY QUESTIONS**, use the diagram to compare and contrast information about Mali and Côte d'Ivoire.

Skillbuilder Handbook, page R9

COMPARE AND CONTRAST

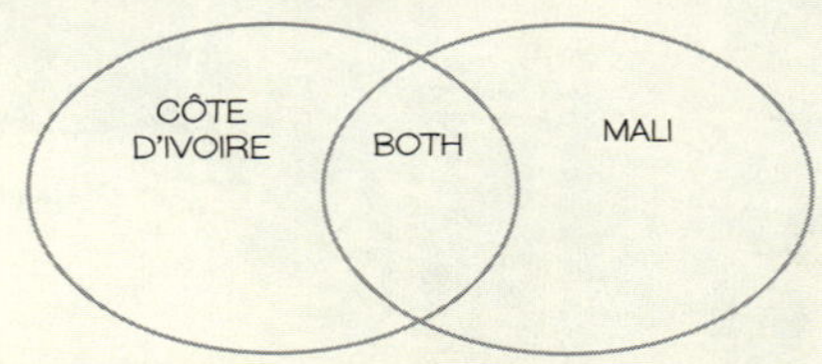

GRAPHIC ORGANIZERS
Go to **Interactive Review** @ClassZone.com

PRETEACHING VOCABULARY

English Learners

Pronounce and Preview

Pronounce each term for students. Review the meanings of words within definitions, such as *purchasing power*, *grain*, and *porridge*.

- To modify vocabulary learning, have students complete worksheets as they read, instead of afterward.

Unit Resource Book: Africa
- Vocabulary Practice, p. 131

Inclusion

Vocabulary: Word Wheel

Have students work in pairs to fill out the graphic organizer for the terms in the section.

- Use the transparency to model the strategy.
- Have volunteers suggest entries.
- Then distribute copies of the organizer for students to complete with a partner.

Best Practices Toolkit
- Vocabulary: Word Wheel, TT13

Maize crop, Benin

SECTION 3 FOCUS ON

Mandrill

Mali and Côte d'Ivoire

Connecting to Your World

Have you ever worked hard on something, only to feel like you got none of the benefit or credit? Mali produces more gold than any other African country except South Africa and Ghana. But despite Mali's natural wealth, local mining communities keep little of the wealth that their mines produce.

Panning for Gold Gold helped build the West African trading empires and is a key resource in modern Mali.

Mali

KEY QUESTION What does Mali's economy rely on?

Mali, located where three powerful African empires once reigned, remains one of the world's poorest countries. Nearly two-thirds of Mali's people live in poverty. Residents of Mali cluster in the basin of the Niger River because most of the country is located within the Sahara.

After overcoming fierce resistance from Africans, France seized a huge territory in West Africa in 1895. This territory included present-day Mali, then called French Sudan. French Sudan joined the Federation of West Africa in 1904. A **federation** is a group of colonies or states with a single government. In 1958, French Sudan won the right to govern itself within the larger French community. In 1960, the Republic of Mali gained full independence from France.

Mud-Brick Mosque Rebuilt in 1907, the mud-brick Great Mosque dominates the ancient trading town of Djenné.

2 Focus & Motivate

3-Minute Warm-Up

Write on the board or display the transparency:

- Find Mali on a map and think about West Africa's history. How might location affect trading in present-day Mali? *(Possible Answer: Mali has no seaport, so trade goods must be shipped by air, or through other countries.)*

Unit Transparency Book: Africa
- 3-Minute Warm-Ups, TT17

Connecting to Your World

More About . . .

Djenné

The importance of Djenné as a West African trading center has diminished, but people from small villages around Djenné bring their crops to trade or sell at local markets.

3 Teach

Mali

Talk About It

- Identify three challenges to Mali's economy. *(Possible Answers: having no seacoast; lack of fertile farmland; periodic droughts; insufficient mining of iron, bauxite, and uranium)*
- **Compare and Contrast** Compare the government of the French Sudan in 1904 to the government of Mali today. *(French Sudan was part of the Federation of West Africa, not independent, while Mali today is independent with a democratically-elected government.)*

SECTION 3 PROGRAM RESOURCES

Unit Resource Book: Africa

ON LEVEL

Unit Resource Book
- Reading Study Guide, p. 111
- Section Quiz, p. 149

STRUGGLING READERS

Unit Resource Book
- Reading Study Guide with Additional Support, p. 117
- Skillbuilder Practice: Reading a Time Line, p. 132
- Section Quiz, p. 149
- Reteaching Activity, p. 152

ENGLISH LEARNERS

Pupil Edition in Spanish
eEdition with Audio DVD-ROM
Unit Resource Book
- RSG (Spanish), p. 123
- RSG with Additional Support (Spanish), p. 129

Multi-Language Glossary
Test Generator
- Section Quiz in Spanish

INCLUSION

Unit Resource Book
- RSG with Additional Support, p. 117
- Section Quiz, p. 149
- Reteaching Activity, p. 152

GIFTED & TALENTED

Unit Resource Book
- Interdisciplinary Projects, pp. 140, 141
- Section Quiz, p. 149

PRE-AP

Unit Resource Book
- Section Quiz, p. 149

TECHNOLOGY

Unit Transparency Book
- 3-Minute Warm-Ups, TT17
- Essential Question, TT18
- Critical Thinking, TT21
- Cultures, TT24, TT25

Daily Test Practice Transparencies
- Section 3, TT66

Power Presentations
ClassZone.com
World Cultures and Geography Video on DVD

Animated GEOGRAPHY

More About . . .

Modern Griots

A West African griot, or jeli, performs at births, harvest festivals, celebrations of various kinds, and other important events. A griot may play music while retelling history or praising the event's sponsor.

The instruments and vocal styles of the griot influence popular music in West Africa today. Internationally known performers Youssou N'Dour of Senegal and Salif Keita of Mali draw upon the griot tradition. They use Western electronics and guitars along with traditional West African instruments, such as xylophones, flutes, and koras, which are a kind of harp-lute.

More About . . .

Salif Keita

Often called the golden voice of Africa, Salif Keita blends traditional Malian music with Western pop and rock. Keita was born in 1949 in Djoliba, Mali. In his late teens he began singing in the clubs of Mali's capital city, Bamako. In the 1970s, as a member of the band Les Ambassadeurs, Keita helped to fuse Malian styles with Afro-Cuban music. He first attracted international interest after moving to Paris in the 1980s.

To achieve musical success, Keita had to overcome many prejudices. He faced discrimination because of his albino skin color. As a descendant of Sundiata, Keita was also not expected to pursue a career traditionally reserved for people of a lower class.

History Since Independence This young country has faced many difficulties. During the 1960s, Mali was left in debt as the result of worldwide inflation. **Inflation** means that costs rise and the value of money falls. In the 1970s and 1980s, rainfall shortages caused regional **droughts**. Thousands starved or died of malnutrition from the resulting **famines**. Mali lived under one-party rule or military dictatorship until 1992. That year, Mali adopted a constitution allowing a multiparty system and a democratically elected government.

Animated GEOGRAPHY

Modern Griots Part musician, part storyteller, part historian, griots play a key role in West African culture.

Click here to learn more about and see griots today @ClassZone.com

Mali's Daily Life and Culture Most Malians are Muslims. The others follow traditional African religions or Christianity. French, the official language, is taught in schools, but most people speak African languages. Only a small percentage of Mali's school-age children attend schools, so fewer than half of the country's adults can read and write. Mali also has few doctors to treat many cases of malaria and waterborne illnesses. Buildings in Mali range from modern structures to buildings made of sun-dried mud bricks. The Great Mosque of Djenné, shown on the previous page, is an example of mud-brick architecture.

Despite its poverty, Mali's blend of Muslim cultures in the north and black African cultures in the south produces a rich cultural stew. Griots, or Jelis, as they are called in Mali's Bambara language, are important parts of Mali's culture. A famous musician is Salif Keita, a descendant of ancient Mali's founder, Sundiata. Mali's Malinka and Songhai peoples specialize in music and dancing, while the Bambara and Voltaic groups make fine masks, statues, and stools.

Economy Mali faces many challenges. Since it has no coast, it must rely on roads and ports leading to neighboring countries. Most Malians are livestock herders, fishers, or subsistence farmers, but less than four percent of Mali's land is fertile. The inland delta of the Niger River is an important source of fish. Farmers grow millet, rice, and sorghum for their families, or export cotton, peanuts, or sugar cane. However, the country suffers periodic droughts.

Mali has few factories, and the country's **infrastructure**—basic services such as roads and schools—remains undeveloped. Only about ten percent of Mali's roads are paved. Mali has important mineral resources, including gold, salt, and limestone. Gold is its most important mineral export. There are also large deposits of iron, bauxite, and uranium, but they are not significantly mined yet.

CATEGORIZE Describe Mali's economy. Mali's economy relies heavily on agriculture and gold mining.

INTERDISCIPLINARY ACTIVITIES

CONNECT to Civics

Become a Civic-Minded Citizen

Ask students to imagine that they are involved in a civic group working to improve the quality of life with Mali's residents. Tell them that group members are trying to define problems in order to begin finding solutions. Then have group members complete a graphic organizer, such as the one shown at right. Ask students to present their charts and discuss them as a class.

Mali's Challenges

	Challenges	Solutions
Economy		
Health Care		
Infrastructure		

Worldster
A Day in Ali's Life

To learn more about Ali and his world, go to the **Activity Center** @ClassZone.com

To introduce himself, Ali might say:

"Salut! Je m'appelle Ali."

Hi! My name is Ali. I'm 12 years old and live in Mali's capital city of Bamako. My city has almost 1.5 million people. My big brother and I share one room, and my parents share our second room with my younger sister. Five other families live in my compound.

7 A.M. It's still dark when I hear my mom cooking outside. For breakfast, we usually eat porridge. Sometimes, we also have beans and coffee. We eat together right outside our rooms.

8 A.M.–12 noon My seventh-grade class has 85 students—lots more boys than girls. I share a desk and a reading book with two other students. We also learn science, math, and French.

2 P.M.–6 P.M. I go to work. I'm an apprentice auto mechanic. I spend about four hours there every afternoon, doing whatever small jobs I'm given. I get paid only a little money, but I'm learning things I can use to make money later.

6 P.M.–8 P.M. Before dinner, I ride bikes or play soccer with my friends. We eat around eight o'clock. Our usual dinner is rice served with fish or meat sauce. Chocolate for dessert is a special treat.

CONNECT to Your Life

Journal Entry Think about your school and outside activities. What sort of things are you learning? How are they similar to what Ali is learning? Record your ideas in your journal.

INTERDISCIPLINARY ACTIVITIES

CONNECT to Art

Make a Daily Life Poster

Ask students to create a poster depicting a day in their own life based on the content of this Worldster. They may illustrate the poster with drawings or photographs. Students should include a suitable caption for each visual. Have students display their posters.

Unit Resource Book: Africa
- Interdisciplinary Projects: Cuisine of West Africa, pp. 140, 141

CONNECT to Language Arts

An Interview-Based Essay

Have students interview a family member, other trusted adult, or friend not in the same class at school. Students should write questions for the interview, using the content on this page as a starting point. Direct them to take notes during the interview, and then write them up in an essay format. Ask students to submit the notes from their interview along with their essay.

Comparing Regions

Education in Mali and Côte d'Ivoire

Because of Mali's poverty, many children work instead of go to school. Between 54% and 80% of Mali's people are considered illiterate; only about 40% of school-age children in Mali actually attend school. Côte d'Ivoire greatly improved its literacy rate with the implementation in the 1970s of a vast program for televised education. Today, about half of Côte d'Ivoire's adults are literate.

Comparing Regions

Soccer in Mali and Côte d'Ivoire

Soccer in Mali and Côte d'Ivoire is more than an afterschool pastime. Both countries have world-ranked national men's teams: the Eagles of Mali and the Elephants of Côte d'Ivoire. The Eagles and Elephants compete every two years in the Africa Cup of Nations tournament—akin to the World Cup—for bragging rights as the top team in Africa. Côte d'Ivoire won this tournament in 1992, while Mali has never won it. Egypt, with five victories since the tournament's inception in 1957, has won the most African Cups.

CONNECT to Your Life

Journal Entry

Remind students to study the photographs carefully since they may contain other clues about the life of the student portrayed. For example, the photograph of apprentice auto mechanics shows that Western-style clothing is worn by some people in Bamako.

Tell students to look for other details in the photographs before writing their journal entry.

Teach

Côte d'Ivoire

Think, Pair, Share

- What are the main exports of Côte d'Ivoire? *(coffee, cocoa, and palm oil)* Why is the export of these few products sometimes a disadvantage to Côte d'Ivoire? *(Shifting worldwide prices may cause its economy to weaken.)*
- How have political instability and civil warfare affected the economy and society of Côte d'Ivoire? *(Profits have been threatened, many Europeans have left, and crime has risen.)*
- **Compare and Contrast** How does national government compare to village government in Côte d'Ivoire? *(Côte d'Ivoire has a president elected by popular vote, an appointed prime minister, and a one-house National Assembly; some villages are governed by councils of elders, some by single chiefs.)*

Unit Resource Book: Africa
- Skillbuilder Practice: Reading Charts, p. 132

Fun Facts!

Cocoa

Next to cocoa, coffee is a major export of Côte d'Ivoire. The coffee tree has white flowers that give way to red berries or fruits. These berries usually hold two beans. A typical coffee tree produces enough beans each year to make 1.5 to 2 pounds of coffee. After oil, coffee is the most valuable commodity to developing countries.

More About . . .

Yamoussoukro

Our Lady of Peace was built in Yamoussoukro during the presidency of Félix Houphouët-Boigny. It is an example—another being an impressive presidential palace—of the many construction projects that he encouraged in Yamoussoukro. In 1983, the National Assembly designated Yamoussoukro as the capital city because it was the birthplace and family home of Houphouët-Boigny.

Fun Facts!

COCOA

Cocoa is one of Côte d'Ivoire's biggest exports. Cocoa beans are the seeds of the cacao tree. Large pods grow on the trees, ranging from yellow to purple in color. These pods contain 20 to 60 seeds inside a sticky white pulp. The seeds are gathered and processed as cocoa beans.

Côte d'Ivoire

KEY QUESTION How has Côte d'Ivoire's economy suffered in recent years?

Côte d'Ivoire (KOHT dee•vwahr) lies west of Ghana and east of Liberia on Africa's west coast. Like Mali, its northern neighbor, Côte d'Ivoire was once part of French West Africa. In the early 1700s, the French established coastal trading posts in present-day Côte d'Ivoire, exchanging European goods for ivory and slaves. Later, French colonizers forced West Africans to build roads and railroads, and to toil on cocoa, coffee, and logging plantations. These thriving plantations made Côte d'Ivoire West Africa's wealthiest French colony. The colony won independence from France in 1960, 14 years after the French ended forced labor.

History Since Independence Félix Houphouët-Boigny (OOF•WAY BWAH•nyuh), an African doctor, led the independence movement of French West African colonies. He became Côte d'Ivoire's first president in 1960, and held office until he died in 1993. After Houphouët-Boigny's death, political instability hurt Côte d'Ivoire's economy. The military took over the government in 1999. In 2000, a **rigged**, or fixed, election caused mass protest, leading to a three-year civil war. In 2003, a power-sharing government came to office, but occasional violence continued. In 2005, South Africa hosted talks between the Ivoirian government and rebel leaders. Although the war formally ended, rebels still control the northern half of the country.

Our Lady of Peace This cathedral in Côte d'Ivoire's capital is the world's largest Christian church.

DIFFERENTIATING INSTRUCTION

Struggling Readers

Clarifying Past and Present

To help students understand the paragraph on this page about Côte d'Ivoire before independence, read this sentence aloud: "In the early 1700s, the French established coastal trading posts in present-day Côte d'Ivoire, exchanging European goods for ivory and slaves." Explain to students that the paragraph records events during the French control of West Africa. The adjective "present-day" can be explained as "what is now" Côte d'Ivoire.

English Learners

Vocabulary: Cognates

Students learning English may need help with the name *Côte d'Ivoire*. Explain that the name is French, and it means "Ivory Coast" in English. The English word *coast* comes from the Latin root *costa*, meaning "side." The French and Spanish words for *coast* have the same Latin root. In French, the word is *côte* and in Spanish, *costa*. Point out the similarity between the English word *ivory* and the French word that means the same, *ivoire*.

Government President Houphouët-Boigny spent millions transforming his hometown, Yamoussoukro (YAH•moo•SOO•kroh), into Côte d'Ivoire's capital. But most business still takes place in Abidjan, the nation's largest city. Under Côte d'Ivoire's constitution, the president is elected by popular vote for a five-year term. He appoints a prime minister, who governs with the aid of a one-house National Assembly. Until 1990, Côte d'Ivoire had a single-party government.

Economics Two out of three Ivoirians are farmers, and the country is one of the world's biggest exporters of coffee, cocoa, and palm oil. But Côte d'Ivoire's dependence on a few products means the economy rises or falls with shifting worldwide prices. Continuing warfare threatens profits too. Côte d'Ivoire has offshore oil reserves, which may feed this developing nation's growth. The country's infrastructure is well-developed, with more than 4,000 miles of paved roads, cell phone access, and some Internet availability. Abidjan's seaport is one of the most modern in West Africa.

World Cup Côte d'Ivoire's national soccer team earned its first World Cup trip in 2006.

Ivoirian Culture and Daily Life Our Lady of Peace, the world's largest Christian church, dominates the skyline of Côte d'Ivoire's capital. But Christians in the southern regions of the country are outnumbered by Muslims and people following traditional African religions, especially in rural regions and the north. Most Ivoirians belong to the Akan, Kru, Mandes, or Voltaic peoples. Soccer is popular in Côte d'Ivoire. The Elephants, Côte d'Ivoire's national team, surprised the world by qualifying for their first-ever World Cup in 2006. About 25 percent of the population are immigrants, mostly from poorer neighboring countries. French is the country's official language, but most people speak ethnic languages or a common tongue, Dioula. Côte d'Ivoire's literacy rate—over 50 percent—is among the highest in West Africa.

Nearly half of Ivoirians live in cities, and many urban areas have French and Southwest Asian communities. However, the recent civil war led many Europeans to leave the country. There is a large divide in the cities between the poor and wealthier people. Crime has risen because of recent fighting. Côte d'Ivoire had more than 215,000 homeless children in 2004, including 50,000 in Abidjan.

DIFFERENTIATING INSTRUCTION: TIERED ACTIVITIES

OBJECTIVE Create materials to promote Côte d'Ivoire, using its soccer success.

Basic

Brainstorm with the class about the value of sports in promoting good will toward a nation. Then have pairs of students work together to create a one-page brochure designed to promote travel to Côte d'Ivoire. Students should include information about the World Cup team. Have students display their brochures.

On Level

Adapt the Basic activity, so that pairs of students create illustrated two-page brochures. They may wish to incorporate borders, drawings, or other features with their text. Have students present and display their brochures.

Challenge

Adapt the On Level activity, so that pairs of students create well-designed, illustrated brochures, incorporating information about the World Cup team, to promote trade with and travel to Côte d'Ivoire. Have students present and display their finished brochures.

More About . . .

Félix Houphouët-Boigny

Born the son of a wealthy Baule chief, Houphouët-Boigny was both a physician and a planter. He entered politics by organizing the African Agricultural Syndicate, to defend the interests of local planters against European colonial policies. Later, he formed the Democratic Party of Côte d'Ivoire. Houghouët-Boigny won worldwide recognition as a skilled administrator whose presidency brought years of relative peace and prosperity to Côte d'Ivoire.

More About . . .

Abidjan, Côte d'Ivoire

Before Yamoussoukro, Abidjan was the capital city of Côte d'Ivoire. Félix Houghouet-Boigny served as its mayor before he became the nation's president. Abidjan is still an important site for government business—and serves as the country's cultural, financial, and industrial center. It is also a major West African seaport, linked by a lagoon and canal to the Gulf of Guinea on the Atlantic Ocean. Abidjan has an international airport and is a center of wealth, but it also has many people living in poverty, in crowded slums.

More About . . .

World Cup Soccer and Côte d'Ivoire

The Elephants—the nation's national men's team—were defeated in the 2006 World Cup competition, but team members have said that their participation had a positive impact upon everyone in Côte d'Ivoire. Although the country has been politically divided between north and south since 2002, the 2006 Elephants came from both regions. People all over Côte d'Ivoire cheered for the Elephants, which has now come to represent national unity.

CONNECT to the Essential Question

How have centuries of colonialism affected West Africa?

Ask students what they have learned so far that can help them answer this question. Students might mention:

- Mali and Côte d'Ivoire gained independence from France in 1960.
- Mali faces many challenges, but it has valuable mineral resources and a rich, diverse culture.
- Côte d'Ivoire is a major exporter of cocoa and coffee, but recent civil conflict threatens its economy and stability.

4 Assess & Reteach

Assess Have students complete the Section Assessment.

Unit Resource Book: Africa
- Section Quiz, p. 149

Interactive Review @ClassZone.com
- Section Quiz

Power Presentations

Test Generator

Reteach Assign groups of students either Mali or Côte d'Ivoire. Ask groups to construct a true-false quiz of five questions for their nation. Then have sets of groups take turns quizzing each other orally, while other groups referee.

Unit Transparency Book: Africa
- Reteaching Activity, p. 152

Mask Festival Yacouba people celebrate the Festival of Ignames, or yams.

Villagers live with extended families in compounds. The compounds are usually dried clay structures with thatched or metal roofs. Ethnic groups govern their villages in different ways; some with councils of elders, some with single chiefs. Trading, herding, and farming are the main activities. Important crops in the region are yams, root vegetables similar to sweet potatoes; and **millet**, a grain often made into porridge.

Ivoirian arts include music, storytelling by griots, and mask-making. The Dan people use masks to symbolize a chief's authority. The popular Festival of Masks is held in Côte d'Ivoire's forest region each November. Other mask ceremonies celebrate harvests. Alpha Blondy, a world-famous Ivoirian musician, blends African and Caribbean sounds in a style called "Afro-reggae." Zouglou is a musical style originating in Abidjan. Sung in French slang and local languages, it is similar to American hip-hop.

SEQUENCE EVENTS Explain what led to Côte d'Ivoire's economic struggles. Côte d'Ivoire was successful under its first president, who kept the country stable. After he died, people competed for power, leading to civil war.

Section 3 Assessment

ONLINE QUIZ For test practice, go to **Interactive Review @ClassZone.com**

TERMS & NAMES

1. **Explain the importance of**
   - inflation
   - infrastructure
   - drought
   - famine

USE YOUR READING NOTES

2. **Draw Conclusions** Use your completed diagram to answer the following question:
   Why is Côte d'Ivoire's economy more successful than Mali's?

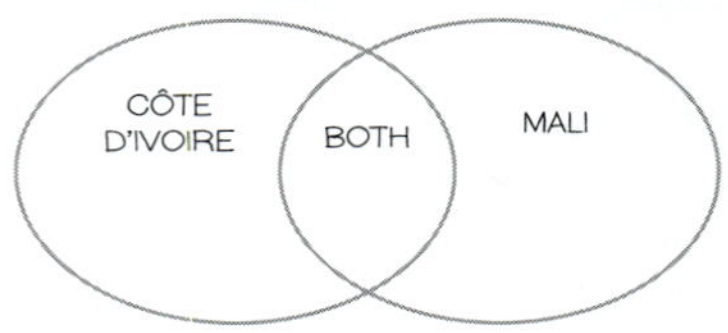

KEY IDEAS

3. In what year did both Mali and Côte d'Ivoire win their independence from France?
4. What are Mali's main exports?
5. What natural resource may improve Côte d'Ivoire's economy?

CRITICAL THINKING

6. **Compare and Contrast** How did Mali's taking the name of an ancient empire differ from Ghana's?
7. **Analyze Cause and Effect** What happened to spark Côte d'Ivoire's civil war?
8. **CONNECT to Today** Tensions remain high between Côte d'Ivoire's government and opposition leaders. What do you think might happen if French and West African peacekeeping troops leave the country?
9. **ART** **Draw a Mask** Research Bambara masks on the Internet. Copy a picture of your favorite mask and label it to describe its key elements and their symbolic meanings.

SECTION 3 ASSESSMENT ANSWERS

Terms & Names

1. inflation: drop in money's value; infrastructure: a community's basic services; drought: period of little or no rainfall; famine: severe food shortage

Use Your Reading Notes

2. Mali is landlocked, and has a small percentage of fertile land. Côte d'Ivoire has better agricultural land, and sea access.

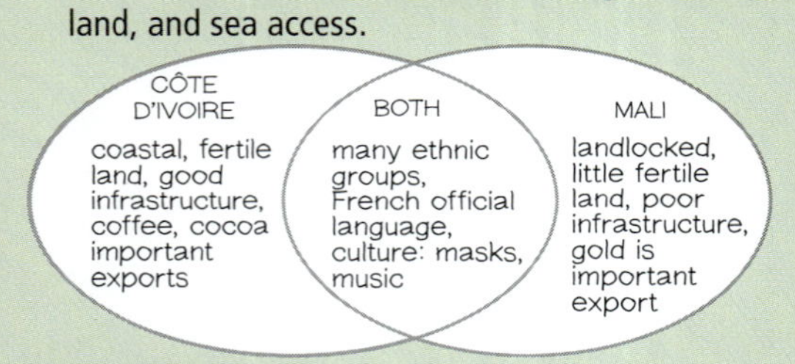

Key Ideas

3. 1960
4. gold, cotton, peanuts, and sugar cane
5. oil

Critical Thinking

6. Mali is located on the actual site of three ancient empires, while Ghana is not.
7. the rigged election of 2000
8. Possible response: Côte d'Ivoire might sink back into civil war.
9. Use the rubric to score the masks.

Mask Rubric

	Content	Presentation
4	describes 5 or more elements	neat, attractive
3	describes 4 elements	neat
2	describes 3 elements	somewhat neat
1	describes 2 or fewer elements	sloppy

Interactive Review

Click here to complete these and other activities online @ ClassZone.com

CHAPTER SUMMARY

Key Idea 1
Colonizers traveled to West Africa, dividing up land once held by great trading empires.

Key Idea 2
Nigeria and Ghana have struggled to maintain stable governments and productive economies.

Key Idea 3
Mali and Côte d'Ivoire have a wealth of natural resources, but are two of Africa's poorest countries.

For **Review and Study Notes**, go to **Interactive Review** @ ClassZone.com

NAME GAME

Use the Terms & Names list to complete each sentence on paper or online.

1. I am an ancient empire as well as a modern country that used to be known as the Gold Coast. ___Ghana___
2. I led the Mali Empire in the 14th century. ______
3. I am perfectly equipped for desert travel. ______
4. I am an ancient empire, whose main city was Timbuktu, as well as a modern West African country. ______
5. I am a region of Nigeria that broke away in a civil war in 1967. ______
6. I was modern Ghana's first president. ______
7. I am used by the Ashanti people to make clothing for religious ceremonies. ______
8. I am a nation's roads, bridges, schools, and post offices. ______
9. I am rising prices or a fall in the value of money.
10. I am a collection of states, colonies, or countries. ______

Biafra
camel
federation
Ghana
Igbo
imperialism
inflation
infrastructure
Kente cloth
Mali
Mansa Musa
Kwame Nkrumah
Songhai
subsistence
trans-Saharan
Yoruba

Activities

GeoGame

Use this map to reinforce your knowledge of West African geography. Drag and drop each name in the list to its location on the map.

Crossword Puzzle

Complete an online crossword puzzle to test your knowledge of the region.

ACROSS
1. a woven ceremonial fabric from Ghana

ONLINE AND ON-SCREEN ACTIVITIES

GeoGame

Use the interactive map on **Power Presentations** DVD-ROM or the **eEdition** to help students review the geography of West Africa.

- Demonstrate how to drag and drop the map labels listed onto the appropriate locations on the map.
- Advance to the scorecard to show how students have a chance to improve their scores.

Crossword Puzzle

Students can review important information about West Africa's history, economy, and culture by completing this crossword puzzle online at **ClassZone.com.**

- As an extension of this activity, have students use the *Activity Maker* at **ClassZone.com** to make their own crossword puzzle for one section of the chapter.
- Then have them trade puzzles with classmates to solve each other's puzzles.

Chapter Summary

As a follow-up to the chapter, encourage students to pair up and go over the **Review** and **Study Notes** at **ClassZone.com.**

Alternatively, students pairs can review their Reading Study Guide notes, making sure that they can add important details to each section's "Key Idea."

Name Game

Teaching Tips

ON PAPER Have students work in pairs to complete this activity. As an extension, they can write questions for the terms that are not covered in the items shown here.

ONLINE Make sure that students understand how to drag and drop terms to fill in blanks. Ask them to print out their results and compare them with those of a classmate.

ANSWERS

1. Ghana	**6.** Kwame Nkrumah
2. Mansa Musa	**7.** Kente cloth
3. camel	**8.** infrastructure
4. Mali	**9.** inflation
5. Biafra	**10.** federation

Make It Fun

Teacher-Tested Activities

Don Stringfellow, Mobile County Public School System, Mobile, Alabama

News Conference—Divide the class into different groups representing Nigeria, Ghana, and Mali. Have each student group arrange a mock news conference featuring an important representative from their selected country. Students will research important facts, location, critical issues, and important people that pertain to their country. Have each group act as student reporters and interview the other country's representative at a news conference in order to gather facts about that country.

Vocabulary ANSWERS

1. West African empire, most powerful around 1500
2. colonies, states, or countries that have joined together
3. basic services a community needs, such as roads and water
4. control of a weaker nation through seizing its territory
5. ancient West African trading empire, also a modern African country
6. brightly colored, woven cloth made by Ghana's Ashanti or Ewe people
7. military takeover
8. modern Ghana's first president
9. Ghana and Mali were two great African empires that participated in the trans-Saharan trade, supplying gold and other goods to trading partners across the Sahara.
10. The Igbo people of Nigeria led a military takeover, or coup, of the Nigerian government in 1966, and in 1967 broke away to form their own state, Biafra.

Key Ideas ANSWERS

11. He enlarged Songhai to its greatest extent; in addition, he promoted the spread of Islam and trade.
12. Portugal and Holland, or the Netherlands
13. Liberia and Ethiopia
14. Soldiers who had fought with Allied forces wanted independence for their own countries.
15. English
16. Nigeria
17. sculptures
18. to honor the great West African empire
19. debt caused by worldwide inflation, drought, and famine
20. 1992
21. Forced African labor on cocoa, coffee, and logging plantations brought great wealth to French colonizers.
22. coffee, cocoa, and palm oil

CHAPTER 19 Assessment

VOCABULARY

Explain the significance of each of the following.

1. Songhai
2. federation
3. infrastructure
4. imperialism
5. Ghana
6. Kente cloth
7. coup
8. Kwame Nkrumah

Explain how the terms and names in each group are related.

9. trans-Saharan, Ghana, and Mali
10. coup, Igbo, and Biafra

KEY IDEAS

1 Three Trading Empires

11. Why is Askia Muhammad known as Askia the Great?
12. Which European countries first explored Africa?
13. Which West African country remained free in the era of European imperialism?
14. What effect did World War II have on the African struggle for independence?

2 Nigeria and Ghana

15. What is the official language of Nigeria and Ghana?
16. Which African nation is the continent's most populous country?
17. What Nigerian art form inspired Western painters?
18. Why did the Gold Coast choose the ancient name of Ghana?

3 Mali and Côte d'Ivoire

19. What difficulties has Mali faced since independence?
20. In what year did Mali get its first democratic government?
21. What made Côte d'Ivoire the wealthiest colony in French West Africa?
22. What are Côte d'Ivoire's main exports?

CRITICAL THINKING

23. **Compare and Contrast** Create a chart to compare and contrast Ghana and Mali. Show their dates of independence, major products or industries, and chief challenges.

	GHANA	MALI
challenges		

24. **Make Predictions** What factors could affect Côte d'Ivoire's economy in the future?
25. **Connect to Economics** Which West African country would you predict will have the greatest economic success in the coming years? Why?
26. **Five Themes: Region** In Nigeria and Côte d'Ivoire, why does Islam dominate in northern regions, while Christianity and African religions prevail in the south or along the coast?
27. **Connect Geography & Culture** What makes Mali a cultural crossroads?

Answer the ESSENTIAL QUESTION

How have centuries of colonialism affected West Africa?

Written Response Write a two- or three-paragraph response to the Essential Question. Be sure to consider the key ideas of each section. Use the rubric below to guide your thinking.

Response Rubric
A strong response will:
- summarize colonialism in West Africa
- note colonialism's lingering effects on industry, ethnic conflict, and government corruption
- explain how concentration on a single crop or resource benefited colonizers, not Africans

ESSENTIAL QUESTION REVIEW

Before students begin their essays, have them review the key chapter elements that will help them answer the Essential Question. Direct them to significant features in the chapter, such as

- Animated Geography: West African Trading Empires, Section 1
- Comparing Forms of Colonial Rule, Section 1
- Analyzing Primary Sources, Section 2
- Comparing West African Regions, Section 2
- Animated Geography: Modern Griots, Section 3

Written Response Rubric

	Coverage of Key Ideas	Explanation	Errors in Mechanics
4	all	clear and concise	few or none
3	most	good	some
2	some	fair	many
1	one or none	poor	many major

STANDARDS-BASED ASSESSMENT

Test Practice

- **Online Test Practice @ClassZone.com**
- **Test-Taking Strategies and Practice** at the front of this book

BAR GRAPH

Use the bar graph and your knowledge of geography to answer questions 1 and 2.

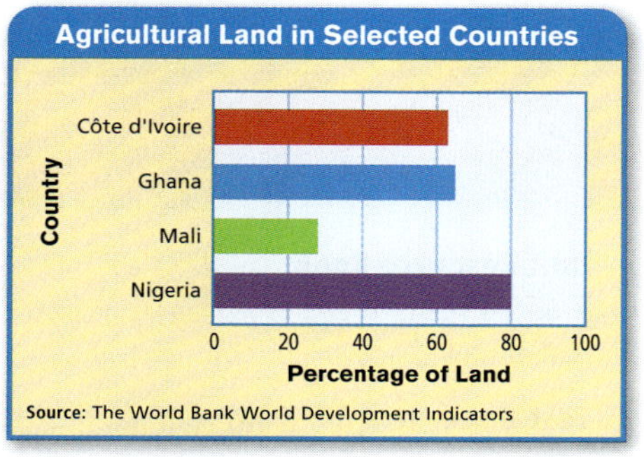

1. **Which nation has the highest percentage of agricultural land?**
   - **A.** Côte d'Ivoire
   - **B.** Ghana
   - **C.** Mali
   - **D.** Nigeria
2. **Why might Mali have such a low percentage of agricultural land?**
   - **A.** because Mali is landlocked
   - **B.** because Mali's economy is more focused on industry
   - **C.** because the northern part of Mali is in the Sahara
   - **D.** because Mali has no rivers for irrigation

PRIMARY SOURCE

Nigerian author Chinua Achebe gained fame for *Things Fall Apart*, his novel about Nigeria under colonial rule. In this passage, he discusses why his book is read worldwide. Use the passage to answer questions 3 and 4.

> I knew I had a story, but how it fit into the story of the world—I really had no sense of that. Its meaning for my Igbo people was clear to me, but I didn't know how other people elsewhere would respond to it. Did it have any meaning or resonance for them? I realized it did when . . . the whole class of a girls' college in South Korea wrote to me and each one expressed an opinion about the book. And then I learned something, which was that they had a history that was similar to the story of *Things Fall Apart*—the history of colonization.
>
> Source: Chinua Achebe, interview in *Atlantic Monthly*, August 2, 2000

3. **Why might people from other countries relate to Achebe's novel?**
4. **What can you infer about Achebe's reaction to the book's success?**

GeoActivity

1. INTERDISCIPLINARY ACTIVITY–SCIENCE

With a small group, research malaria and the waterborne illnesses that plague much of West Africa. Make a brief oral presentation to your class, describing preventive measures.

2. WRITING FOR SOCIAL STUDIES

Research the beginnings of Pan-Africanism. Write a short essay on how the movement for freedom in Africa was related to the struggle for African-American civil rights.

3. MENTAL MAPPING

Create an outline map of West Africa and label the following:

- Atlantic Ocean
- Ghana
- Accra
- Nigeria
- Abuja
- Lagos
- Côte d'Ivoire
- Yamoussoukro
- Mali
- Bamako

Critical Thinking ANSWERS

23. Ghana's problems and challenges include a relatively new constitution. Mali's problems/challenges include poor infrastructure, droughts, little fertile land, and no coastal access.

24. Answers will vary, but might include weather, worldwide prices, and civil warfare.

25. Answers will vary. Some students may say Nigeria because of its large population and vast oil reserves; others may say Côte d'Ivoire because of its potential oil wealth, decent infrastructure, and higher literacy rates.

26. Islam spread from North Africa to the south. Students may infer that Christianity was spread from the coastline northward by European colonizers.

27. Its location at the meeting point of the Muslim north and non-Muslim south gives it this status.

Standards-Based Assessment ANSWERS

1. D
2. C
3. because many areas around the world were colonized
4. Achebe was surprised at how people around the world reacted to his story.

GeoActivity

1. INTERDISCIPLINARY ACTIVITY RUBRIC

The report should

- describe how malaria and similar diseases are transmitted
- accurately explain preventive measures

2. WRITING FOR SOCIAL STUDIES RUBRIC

The essay should

- explain African struggle for freedom from imperialism
- explain African-American struggles in context of slavery (across Americas) and post-slavery civil rights
- describe how causes became connected

3. MENTAL MAPPING RUBRIC

The outline map should

- represent the approximate shape of the region
- include all geographic features listed in their correct locations

DIFFERENTIATING CHAPTER ASSESSMENT

Unit Resource Book

- **Chapter Test A,** pp. 153–156. *Use for Struggling Readers, Beginning-to-Intermediate-Level English Learners, and for Inclusion of students whose IEP suggests supported test-taking.*
- **Chapter Test B,** pp. 157–160. *Use for On-Level Learners and ESOL students with Advanced-Level English.*
- **Chapter Test C,** pp. 161–164. *Use for Gifted and Talented students or students assigned Pre-AP work.*

Test Generator

- Chapter Tests A, B, and C, in English and Spanish

McDougal Littell Assessment System

TEST **Test Generator** Use three levels of tests in English and Spanish tied to state standards. Plus, choose questions to customize tests for all populations.

SCORE Scan and Score plain paper student tests in minutes.

REPORT Create instant progress reports, sortable by student, by standard, or other criteria.

RETEACH Respond immediately to students' needs by using a wealth of reteaching resources.

CHAPTER 20 East, Central, and Southern Africa PLANNING GUIDE

Complete print resources all on one USB drive allow you to customize lesson plans for all students.

Section Titles & Key Ideas	Unit Resource Book Africa	Assessment Resources
1 Ethiopia **KEY IDEA** Ethiopia remained an independent nation throughout most of its history.	• **Reading Study Guide,** pp. 165–166 • **RSG with Additional Support,** pp. 175–176 • **Vocabulary Practice,** p. 205 • **Outline Map Activities,** pp. 209–210	**Unit Resource Book** • Section Quiz, p. 219 • Reteaching Activity, p. 224 **Test Generator CD-ROM** **Daily Test Practice Transparency,** TT67
2 Kenya **KEY IDEA** Trade was important in East Africa, particularly in what is now Kenya.	• **Reading Study Guide,** pp. 167–168 • **RSG with Additional Support,** pp. 177–178 • **History Makers,** p. 211 • **Connect to Culture,** p. 216 • **Primary and Secondary Sources,** pp. 217–218	**Unit Resource Book** • Section Quiz, p. 220 • Reteaching Activity, p. 225 **Test Generator CD-ROM** **Daily Test Practice Transparency,** TT68
3 Democratic Republic of the Congo **KEY IDEA** The Democratic Republic of the Congo is still struggling with setting up a true democracy.	• **Reading Study Guide,** pp. 169–170 • **RSG with Additional Support,** pp. 179–180 • **Skillbuilder Practice,** p. 206 • **Interdisciplinary Projects,** pp. 212–214	**Unit Resource Book** • Section Quiz, p. 221 • Reteaching Activity, p. 226 **Test Generator CD-ROM** **Daily Test Practice Transparency,** TT69
4 South Africa **KEY IDEA** South Africa made a successful change from a minority-rule nation to a majority-rule government, but still faces many challenges.	• **Reading Study Guide,** pp. 171–172 • **RSG with Additional Support,** pp. 181–182 • **Map and Graph Skills,** pp. 207–208 • **Economics in the World,** p. 215	**Unit Resource Book** • Section Quiz, p. 222 • Reteaching Activity, p. 227 **Test Generator CD-ROM** **Daily Test Practice Transparency,** TT70
5 Zimbabwe, Botswana, and Angola **KEY IDEA** Zimbabwe, Botswana, and Angola have all taken different paths since the end of colonialism.	• **Reading Study Guide,** pp. 173–174 • **RSG with Additional Support,** pp. 183–184	**Unit Resource Book** • Section Quiz, p. 223 • Reteaching Activity, p. 228 • Chapter Test (Forms A, B, & C), pp. 229–240 **Test Generator CD-ROM** **Daily Test Practice Transparency,** TT71

No Time? To focus students on important content in the chapter, use the **"Reading Study Guide with Additional Support"** in ***Resources2Go.***

Teacher Presentation Options

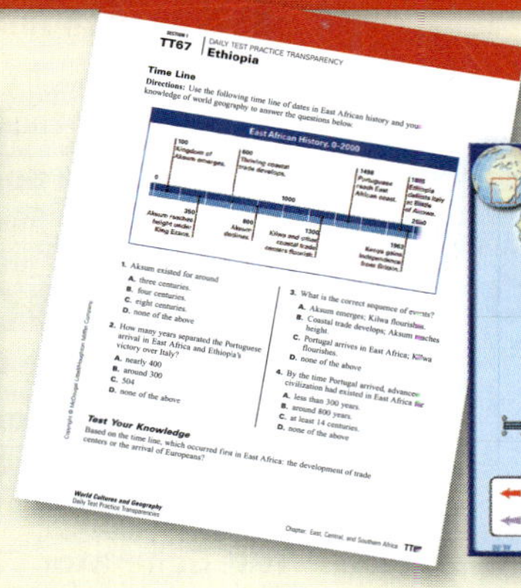

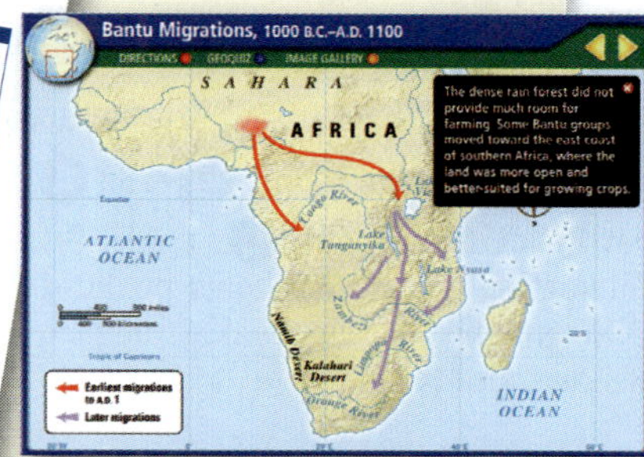

Electronic Books
eEdition DVD-ROM
eEdition Online

Presentation Toolkit
Video/DVD
- World Cultures and Geography

Power Presentations DVD-ROM
- Discussion Starters
- Media Gallery
- Review Game
- Interactive Review
- Animated Features

Unit Transparency Book
- 3-Minute Warm-Ups
- Essential Question
- Critical Thinking
- Maps
- Cultures

Daily Test Practice Transparencies

Animated Geography
- East, Central, and Southern Africa Map
- Bantu Migrations 1000 B.C.–A.D. 1100 Map

OnlineActivities at ClassZone.com

Animated Geography
- Videos and Photos
- 3-D Animations
- Interactive Maps

Interactive Review
- Section Quizzes
- Review and Study Notes
- Vocabulary Flip Cards
- Graphic Organizers
- Name Game and Crossword Puzzles
- GeoGame

Online Test Practice

Map Center
- Atlas Maps
- Political Maps with Data
- Links to Mapping Web Sites

Activity Center
- Activity Maker—Review Games, Puzzles, Flip Cards, and Graphic Organizers
- Worldster

Research & Writing Center
- Web Sites
- Data Updates
- Primary Sources
- Writing Models

Current Events

Erik Branch
James C. Wright Middle School
Madison, Wisconsin

Make It Fun — Teacher-Tested Activities

As you teach this chapter, look for my classroom-proven ideas for:
- making the lesson fun
- focusing students' attention and comprehension
- integrating technology with traditional teaching

Teaching Bell to Bell When you see this timing guide, have students discuss in pairs, in groups of 3–6, or as a class.
- **Think, Pair, Share** (pairs)
- **Reader, Recorder, Reporter** (3–6)
- **Talk About It** (any size)

CHAPTER 20 RESOURCES FOR DIFFERENTIATED INSTRUCTION

Struggling Readers

eEdition with Audio DVD-ROM

Lesson Plans
- Strategies for Inclusion

Unit Resource Book
- Reading Study Guide with Additional Support, pp. 175–184 B
- Reteaching Activity, pp. 224–228
- Chapter Test (Form A), pp. 229–232 A

Best Practices Toolkit C
- Interactive Reading Strategies, pp. 90–145
- Differentiation Activity Bank: Struggling Readers, pp. 146–149

Teacher's Edition Activities
- Chart Ethiopia's Government, Economy, and Challenges, Sec. 1
- Clarify Political Terminology, Sec. 2
- Clarify Cultural Terminology, Sec. 4
- Make Inferences about Botswana's Culture, Sec. 5

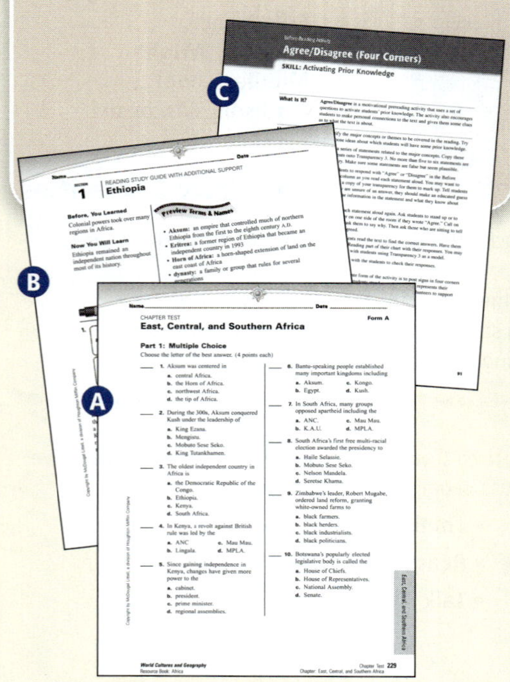

English Learners

Pupil's Edition in Spanish

eEdition with Audio DVD-ROM

Lesson Plans
- Strategies for English Learners

Unit Resource Book
- Reading Study Guide (Spanish), pp. 185–194
- RSG with Additional Support (Spanish), pp. 195–204 A B

Multi-Language Glossary C

Best Practices Toolkit
- Differentiation Activity Bank: English Learners, pp. 165–194

Test Generator CD-ROM
- Section Quizzes in Spanish
- Chapter Test in Spanish

Teacher's Edition Activities
- Pronounce and Preview, Sec. 1–5
- Vocabulary: Idioms, Sec. 1, Sec. 5
- Comprehension: Connection, Sec. 2
- Vocabulary: Latin Derivatives, Sec. 3
- Vocabulary: Key Academic Vocabulary, Sec. 4

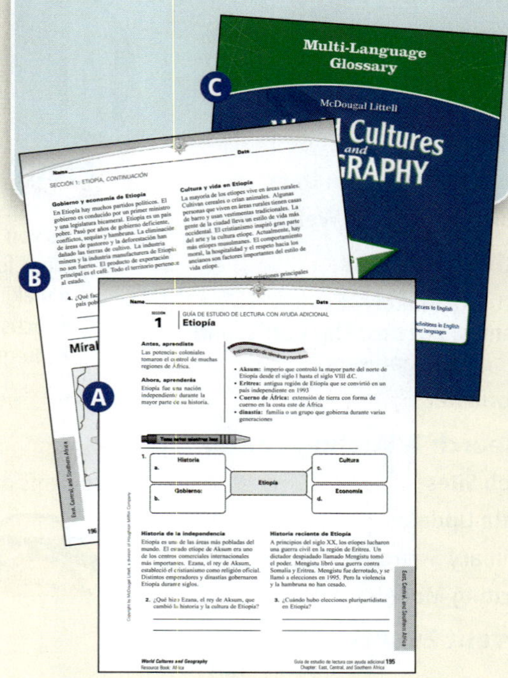

Inclusion

eEdition with Audio DVD-ROM

Lesson Plans
- Strategies for Inclusion

Unit Resource Book
- Reading Study Guide with Additional Support, pp. 175–184 A B
- Reteaching Activity, pp. 224–228
- Chapter Test (Form A), pp. 229–232

Best Practices Toolkit C
- Interactive Reading Strategies, pp. 90–145
- Differentiation Activity Bank: Struggling Readers, pp. 146–149

Teacher's Edition Activities
- Record Terms and Definitions, Sec. 1
- Vocabulary: Definition Mapping, Sec. 2
- Match Terms and Definitions, Sec. 3
- Terms in Context, Sec. 4
- Classify Vocabulary, Sec. 5
- Tiered Activities, Sec. 2–5

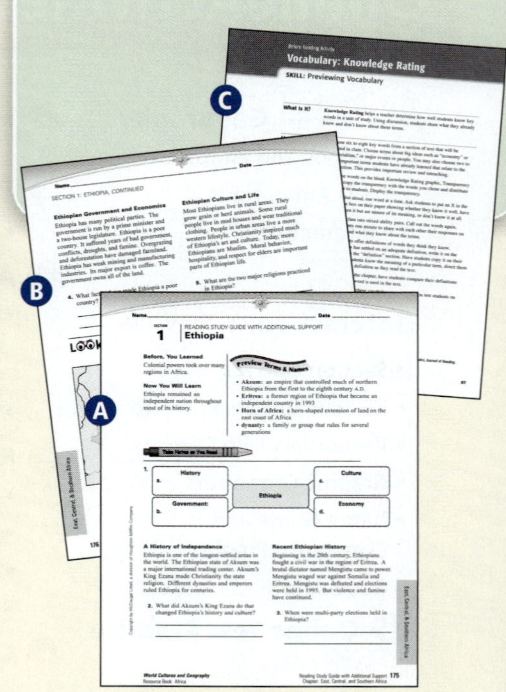

Interdisciplinary Activities

Gifted & Talented Students

Lesson Plans
- Core Lesson Plan

Unit Resource Book
- Outline Map Activities, pp. 209–210
- History Makers, p. 211 (A)
- Interdisciplinary Projects, pp. 212–214 (B)
- Chapter Test (Form C), pp. 237–240
- World Literature, p. 241 (C)

Best Practices Toolkit
- Student Products Rubrics, pp. 1–14
- Writing Templates, pp. 78–89

Teacher's Edition Activities
- Map African and Italian Trade Routes, Sec. 2
- Illustrate the Congo's History, Sec. 3
- Map Great Zimbabwe, Sec. 5
- Tiered Activities, Sec. 2–5

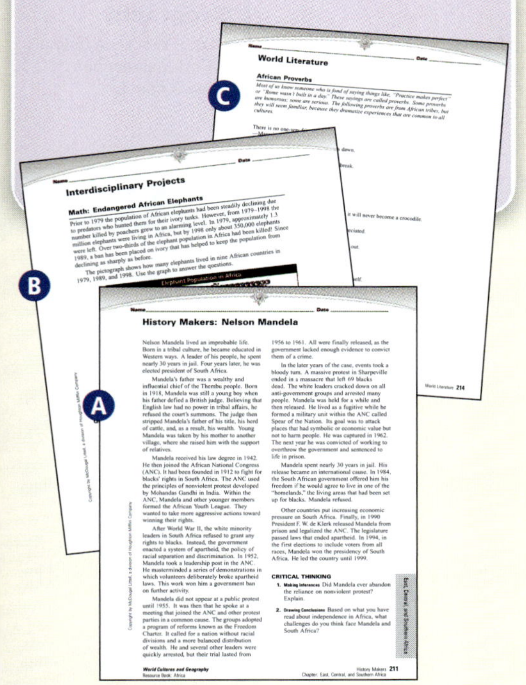

Pre-AP*

Lesson Plans
- Core Lesson Plan

Unit Resource Book
- Vocabulary Practice, p. 205
- Map and Graph Skills, pp. 207–208
- Economics in the World, p. 215 (A)
- Connect to Culture, p. 216 (B)
- Primary Source, pp. 217–218 (C)
- Chapter Test (Form C), pp. 237–240

Best Practices Toolkit
- Writing Templates, pp. 78–89
- Differentiation Activity Bank: Pre-AP Students, pp. 195–200

* Pre-AP is a registered trademark of the College Entrance Examination Board, which was not involved in the production of and does not endorse this product.

Teacher's Edition Activities
- Analyze Ethiopia's Geographic Location, Sec. 1
- Research Contemporary Kenyan Issues, Sec. 2
- Compare Forms of Government, Sec. 5
- Tiered Activities, Sec. 2–5

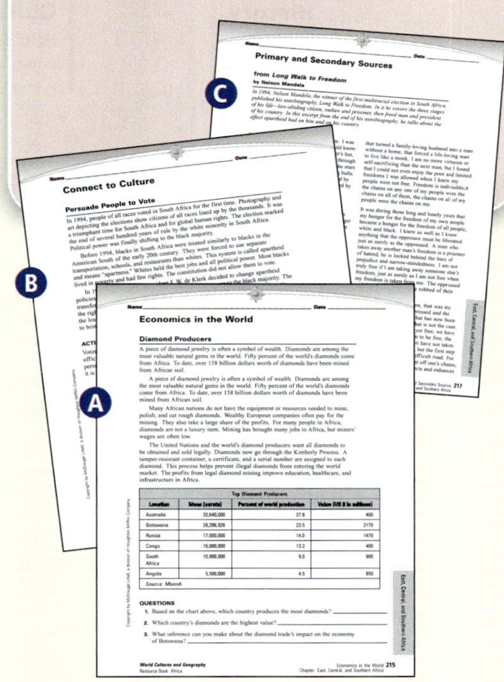

CONNECT to Math

Graph African Disease Rates

CLASS TIME 30 minutes

TASK Create a graphic display showing current disease rates for selected African countries.

SUPPLIES
- current African disease data (available from WHO—World Health Organization: www.who.int/hiv)
- paper, colored pencils or markers

ACTIVITY Supply students with current data on disease rates for selected African countries from East, Central, and Southern Africa. Ask students to prepare a graphic display (bar graphs or pie graphs) illustrating the statistics. To extend this activity, students can research statistics for one country over multiple years and prepare a line graph to illustrate them.

Good for Visual and Mathematical Learners

CONNECT to Language Arts

Write a Descriptive Essay on One East, Central, or Southern African Country

CLASS TIME 30 minutes

TASK Prepare an essay describing the history, culture, government, and economy of a country from one of these regions.

SUPPLIES
- writing materials/computers

ACTIVITY Students' essays should include an opening paragraph introducing the subject, separate paragraphs for each category, and a concluding paragraph offering a personal assessment of the country's challenges and possible solutions. Offer students the choice of reading their essays aloud or posting them for silent reading.

Good for Verbal Learners

Introduce the ESSENTIAL QUESTION

- Describe the history, government, economy, and culture of Ethiopia.
- Describe the history, culture, government, and economy of Kenya.
- Describe the history, culture, government, and economy of the Democratic Republic of the Congo.
- Describe the history, culture, government, and economy of South Africa.
- Compare the histories, cultures, governments, and economies of Zimbabwe, Botswana, and Angola.

Key Ideas

Section 1: Ethiopia remained an independent nation throughout most of its history.

Section 2: Trade was also important in East Africa, particularly in what is now Kenya.

Section 3: The Democratic Republic of the Congo is still struggling with setting up a true democracy.

Section 4: South Africa made a successful change from a minority-rule nation to a majority-rule government, but still faces many challenges.

Section 5: Zimbabwe, Botswana, and Angola have all taken different paths since the end of colonialism.

CONNECT to the Essential Question

You may wish to revisit the Essential Question periodically with students as they work through the chapter.

Suggestions for reconnecting are provided at an appropriate point in the *Teacher's Edition* for each section.

Victoria Falls

CHAPTER 20

East, Central, and Southern Africa

1 FOCUS ON **Ethiopia**

2 FOCUS ON **Kenya**

3 FOCUS ON **Democratic Republic of the Congo**

4 FOCUS ON **South Africa**

5 FOCUS ON **Zimbabwe, Botswana, and Angola**

ESSENTIAL QUESTION

How do boundaries set during colonial times affect stability in these regions today?

CONNECT Geography & History

Use the map and the time line to answer the following questions.

1. What nation is surrounded completely by the country of South Africa?
2. What sea route would Portuguese explorers have had to take to reach Kenya's coast in 1498?

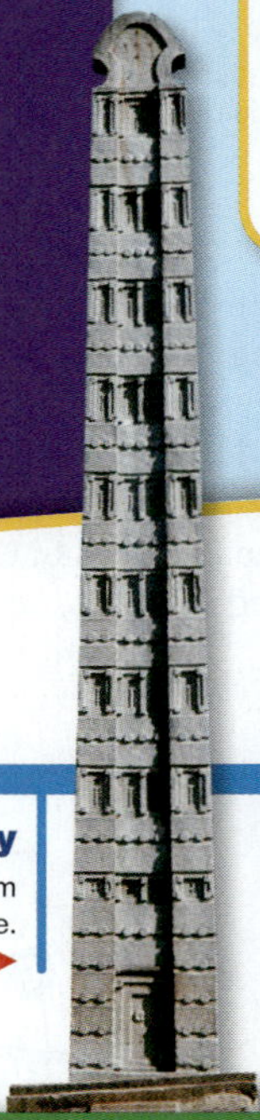

A.D. 50

History
50 Aksum begins to emerge. (Obelisk of Aksum) ▶

Culture
330 King Ezana establishes Christianity as Aksum's state religion.

Geography
1000 Great Zimbabwe is built. ▶

Geography
1498 Vasco da Gama of Portugal reaches Kenya's coast.

TIME LINE DISCUSSION

Point out that, with few exceptions, colonialism in east, central, and southern Africa has adversely affected their attempts to achieve political and economic stability.

- Which event on the time line highlights an early influence of Christianity in the region? *(A.D. 330—King Ezana establishes Christianity as Aksum's state religion.)*
- Which events on the time line describe foreign influences in these regions? *(1498—Vasco da Gama of Portugal reaches Kenya's coast; 1908—Belgium takes control of Congo Free State.)*
- Which events on the time line highlight attempts by countries to exercise political freedoms? *(1963—Kenya wins independence; 1994—South Africa holds its first free, democratic election; 2006—Angola and Congo hold long-awaited national elections.)*

Animated GEOGRAPHY
East, Central, and Southern Africa

Click here to explore East, Central, and Southern Africa @ClassZone.com

History
1908 Belgium takes control of Congo Free State.

Government
1994 South Africa holds its first free, democratic election. ▶

Today

History
1963 Kenya wins independence. (Jomo Kenyatta) ▶

Government
2006 Angola and Congo (Kinshasa) hold long-awaited national elections.

605

TEACHER MEDIA FAVORITES

Books for the Student

Kurtz, Jane. Saba: *Under the Hyena's Foot.* Pleasant Company Publications, 2003. In 1840s Ethiopia, 12-year-old Saba is kidnapped from her rural home and thrust in the center of intrigue at the royal palace.

Mwangi, Meja. *The Mzungu Boy.* Groundwood Books, 2005. A tale of friendship between a Kenyan boy and the grandson of a British farmer set in the 1950s during the Mau Mau rebellion.

Video / DVD

Wonders of the African World. (DVD) PBS Home Video, 2000. Documentary series explores Africa's history, including trips to Kenya's coast, the ruins of Aksum and Great Zimbabwe, and South Africa. Narrated by Henry Louis Gates, Jr.

Ethiopia: A Portrait. (DVD) Worlds Together, 2004. A comprehensive tour of the country, covering both its culture and history.

Internet

Go to the *Research and Writing Center* at **ClassZone.com** to link to:

- data on the region
- a variety of chapter-specific, content-reviewed sites
- "History Makers" biographies
- primary sources

Animated GEOGRAPHY
East, Central, and Southern Africa
CLASSZONE.COM

Make sure students understand that in this chapter they will be learning about some, but not all, of the countries of east, central, and southern Africa. Point out the countries to be covered—Ethiopia, Kenya, the Democratic Republic of the Congo, South Africa, Zimbabwe, Botswana, and Angola—and then ask students to identify some of the other countries on the map that are part of these regions of Africa. *(Possible Answers: Cameroon, Swaziland, and Somalia)*

- Which countries that are covered in this chapter are part of east Africa? *(Ethiopia, Kenya)* Of southern Africa? *(Angola, Zimbabwe, Botswana, South Africa)* Which country that is covered in this chapter is part of central Africa? *(Democratic Republic of the Congo)*
- Which countries on the map appear to be completely landlocked? *(Central African Republic, Ethiopia, Uganda, Rwanda, Burundi, Malawi, Zambia, Zimbabwe, Botswana, Swaziland, and Lesotho)*

CONNECT GEOGRAPHY & HISTORY ANSWERS

1. Lesotho
2. They would have had to sail down the west coast of Africa, around the southern tip, and back north along the eastern coast.

Presentation Options

Use the **Power Presentations** DVD-ROM to project the animated map.

- Prompt students to use the pop-up images and audio from the map to discover what they already know about the geography, culture, history, or economics of east, central, and southern Africa.
- Challenge students to answer the map questions individually, in small groups, or as a class activity. Use all of the interactive material to help preview chapter content.

1 Plan & Prepare

Objectives

- Summarize the important events in Ethiopia's history of independence
- Identify the challenges of Ethiopia's government and economy
- Describe the culture of contemporary Ethiopia

Read for the Essential Question

Help students read for a purpose by reminding them of the Essential Question: "How do boundaries set during colonial times affect stability in these regions today?" Point out that they can use text features such as Key Questions, special features, images, and maps to form answers to the question.

Vocabulary

Best Practices Toolkit

Use the ***Best Practices Toolkit*** to model strategies for vocabulary notetaking. Vary strategies throughout the year. Choose from: *Knowledge Rating, Predicting ABCs, Definition Mapping, Word Sort, Frayer Model (Word Squares), Magnet Words, Word Wheel,* and *Student VOC.*

Vocabulary Strategies, TT9–TT16

Reading Strategy

Best Practices Toolkit

Display the Web Diagram transparency.

- Remind students that main ideas are important ideas in a block of text and that details support main ideas.
- Model filling in main ideas and details using the "A History of Independence" subsection.
- Have students suggest entries. Remind students that main ideas in each block of text in the section will help answer a Key Question.
- Display the completed diagram for students to review.

Web Diagram, TT28

SECTION 1

Reading for Understanding

Key Ideas

BEFORE, YOU LEARNED

Colonial powers took over many regions in Africa.

NOW YOU WILL LEARN

Ethiopia remained an independent nation throughout most of its history.

Vocabulary

TERMS & NAMES

Aksum an empire that controlled much of northern Ethiopia from the first to the eighth century A.D.

obelisk a pillar-shaped stone monument

Haile Selassie (HY•lee suh•LAS•ee) emperor of Ethiopia from 1930 to 1974

Eritrea a former region of Ethiopia that became an independent country in 1993

BACKGROUND VOCABULARY

Horn of Africa a horn-shaped extension of land on the east coast of Africa

modernize to accept new ways or ideas

Red Sea a narrow sea that divides northeast Africa from the Arabian Peninsula

REVIEW

Kush an ancient Nubian kingdom in northern Sudan

dynasty a family or group that rules for several generations

Visual Vocabulary Red Sea

Reading Strategy

Re-create the web diagram shown at right. As you read and respond to the **KEY QUESTIONS**, use the diagram to organize important ideas about Ethiopia.

Skillbuilder Handbook, page R4

FIND MAIN IDEAS

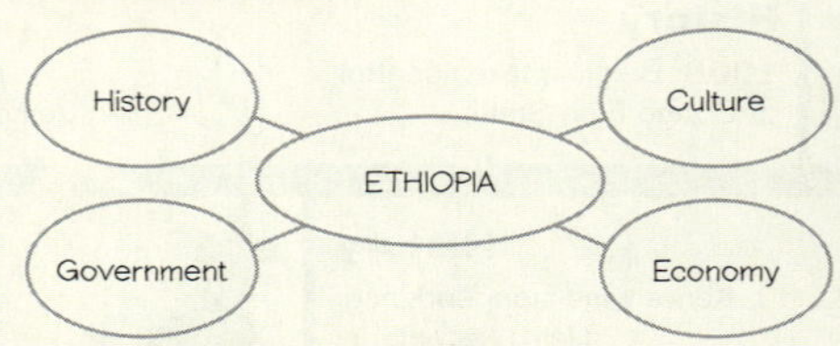

GRAPHIC ORGANIZERS
Go to **Interactive Review** @ClassZone.com

PRETEACHING VOCABULARY

English Learners

Pronounce and Preview

Pronounce each term for students. Review the meanings of words used in definitions, such as *extension, peninsula,* and *accept.*

- To modify vocabulary learning, have students complete worksheets as they read, instead of afterward.

Unit Resource Book: Africa
- Vocabulary Practice, p. 205

Inclusion

Record Terms and Definitions

Have students read each term and its definition into an audiocassette recorder. As students listen to their recordings play back, have them follow along with the words in their textbooks. Then have students use each term in a sentence.

Northern Ethiopia

African elephants, Tanzania

SECTION 1 FOCUS ON

Ethiopia

Connecting to Your World

Hundreds of workers took more than six years to create 60-foot carvings of four presidents at Mount Rushmore, South Dakota. Ethiopia's mountains contain an even more amazing monument. Eight hundred years ago, using nothing but hand tools, workers carved 11 Christian churches out of solid rock.

Rock Church One of 11 churches carved out of solid rock in Lalibela, Ethiopia.

A History of Independence

KEY QUESTION How did ancient Aksum's location make it a powerful trading empire?

Ethiopia is one of the longest-settled regions in the world. **Aksum**, the first major state to develop in Ethiopia, became important about A.D. 100. This kingdom achieved its greatest strength between the 300s and 600s.

Ancient Aksum Aksum was located near the **Horn of Africa** and expanded into parts of southwest Arabia. This major trade center exported spices, ivory, ebony, animal skins, and tortoiseshell. Trading partners included Arabia, India, and perhaps China. Aksumites built more than 100 carved stone pillars, called **obelisks**, in their capital. Some stood as monuments over rulers' tombs.

A Rugged Country Ethiopia's landforms include rift valleys and highlands.

607

2 Focus & Motivate

3-Minute Warm-Up

Write on the board or display the transparency:

- View the photograph of the Rock Church and the Ethiopian rift valley. What do they tell you about Ethiopia? *(Possible Answer: It has rugged landforms, and people adapted to them, such as by carving churches out of solid rock.)*

Unit Transparency Book: Africa
- 3-Minute Warm-Ups, TT26

Connecting to Your World

More About . . .

The Rock Churches at Lalibela

The rock churches were carved out of solid granite blocks buried under the Earth's surface. To get to the granite, workers dug trenches around an area of rock and worked from the top to the bottom. The churches are connected by underground passages.

3 Teach

A History of Independence

Think, Pair, Share

- What was the first major state to develop in Ethiopia? *(Aksum)*
- Who ruled Ethiopia from 1930 to 1974? *(Haile Selassie)*
- **Find Main Ideas** What events have set back Ethiopia's recent history? *(the brutal rule of dictator Mengistu, conflict with Eritrea, severe droughts in the 1990s and early 2000s, and election violence in 2005)*

SECTION 1 PROGRAM RESOURCES

Unit Resource Book: Africa

ON LEVEL

Unit Resource Book
- Reading Study Guide, p. 165
- Section Quiz, p. 219

STRUGGLING READERS

Unit Resource Book
- Reading Study Guide with Additional Support, p. 175
- Skillbuilder Practice: Identifying Problems and Solutions, p. 206
- Section Quiz, p. 219
- Reteaching Activity, p. 224

eEdition with Audio DVD-ROM

ENGLISH LEARNERS

Pupil Edition in Spanish

eEdition with Audio DVD-ROM

Unit Resource Book
- RSG (Spanish), p. 185
- RSG with Additional Support (Spanish), p. 195

Multi-Language Glossary

Test Generator
- Section Quiz in Spanish

INCLUSION

Unit Resource Book
- RSG with Additional Support, p. 175
- Section Quiz, p. 219
- Reteaching Activity, p. 224

GIFTED & TALENTED

Unit Resource Book
- Section Quiz, p. 219

PRE-AP

Unit Resource Book
- Section Quiz, p. 219

TECHNOLOGY

Unit Transparency Book
- 3-Minute Warm-Ups, TT26
- Essential Question, TT27
- Critical Thinking, TT28
- Map: Africa South of the Sahara: Political, TT33

Daily Test Practice Transparencies
- Section 1, TT67

Power Presentations

ClassZone.com

World Cultures and Geography Video on DVD

Ancient Aksum A.D. 300–700

CONNECT Geography & History

Make sure students understand that the area encompassed by Aksum is the shaded portion roughly in the center of the map along the southwestern coast of the Red Sea. Mention that Aksum's location made it a major trade center, and have students point out details on the map to help explain why this would be true.

ANSWERS

1. **Movement** The Arabian peninsula; students may also say Europe (via the Mediterranean) and India.
2. **Location** Nubia (Kush)

More About . . .

Haile Selassie and Mengistu

During his rule of Ethiopia, Haile Selassie worked to improve the country's educational system, strengthen law enforcement, and abolish oppressive feudal tax practices. The coup that removed him from power was led by a group of disgruntled military officers—including Mengistu—who took advantage of economic hard times and political unrest.

Haile Selassie was placed under house arrest in his palace. He died a year later, reportedly of natural causes, but there is some evidence to suggest that he had been murdered. Mengistu is also thought to have arranged the assassinations of rivals within the government, allowing him to become head of state. He then launched a repressive regime that lasted until 1991, when the Soviet Union withdrew its support.

Ancient Aksum A.D. 300–700

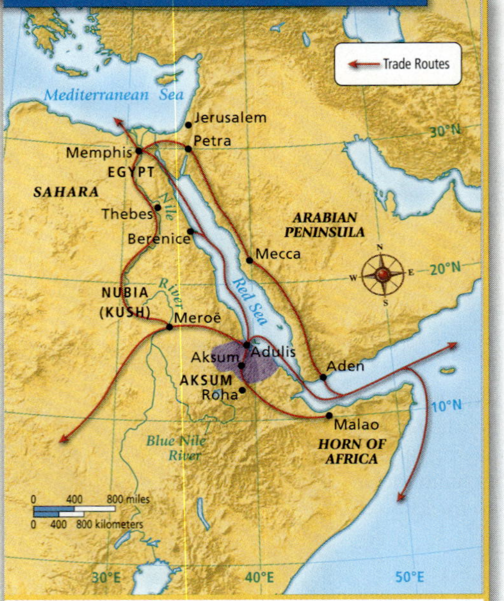

CONNECT Geography & History

1. **Movement** Aksum had trade access to what areas outside of Africa?
2. **Location** Traveling by land, what was Aksum's closest trading partner?

During the 300s, Aksum's King Ezana conquered **Kush** and made Christianity Aksum's state religion. But Muslim conquests in North Africa in the 600s hurt Aksum's trading power by taking over sea routes. The kingdom shifted to the south and moved its capital to Roha, the city that would become Lalibela. As Aksum declined, a new **dynasty**, the Zagwe, rose beginning in 1137. During the Zagwe reign, workers created 11 Christian churches from the rock at Lalibela.

Remaining Independent The Zagwe dynasty lasted less than 100 years, replaced by a line of emperors said to descend from the biblical King Solomon. These emperors had the backing of the Ethiopian Christian church and built a national identity. Menelik II became emperor of Ethiopia in 1889 and defeated Italy's 1896 invasion attempt.

Ethiopia's next important leader, **Haile Selassie**, took power in 1930. Haile Selassie's attempts to **modernize** Ethiopia included ending legal slavery and giving the country its first constitution. Ethiopia briefly fell to Italy during World War II but regained independence afterwards. In the later years of Haile Selassie's rule, Ethiopia fought a civil war in a region called **Eritrea**, along the **Red Sea**. In 1974, a coup removed Haile Selassie from power.

Recent Ethiopian History Ethiopia has suffered many setbacks in recent years. The coup that removed Haile Selassie brought a brutal military dictator named Mengistu to power. The state took control of land, industry, and banking. Mengistu waged war with neighboring Somalia and continued the conflict with Eritrea. The Mengistu government also badly mishandled a famine in 1984. A rebel group defeated Mengistu in 1991 and agreed to Eritrea's independence. But fighting along this tense border continued during 2005. More droughts in the 1990s and early 2000s resulted in millions of deaths. Ethiopia adopted a new constitution in 1994 and held multiparty elections the following year. But in November 2005, election violence led to scores of shootings.

DRAW CONCLUSIONS Describe how location helped Aksum dominate trade. Aksum's location on the Horn of Africa made it a crossroads between the Mediterranean powers and Egypt to the north and west, and Arabia and India to the east.

DIFFERENTIATING INSTRUCTION

English Learners

Vocabulary: Idioms

Explain that in the sentence, "Ethiopia briefly *fell to* Italy during World War II . . . ," the expression "fell to" is an idiom that means "was defeated by" or "was conquered by." To clarify the political situation addressed in this sentence, point out that during the war, Italy invaded Ethiopia and took control of it. Haile Selassie was forced to live in exile in the Sudan. After the war, Ethiopia was freed, and Haile Selassie returned to resume his rule.

Pre-AP

Analyze Ethiopia's Geographic Location

Italy attempted more than once to seize control of Ethiopia. Challenge students to explain why Italy was interested in controlling Ethiopia. Discuss critical-thinking questions such as the following:

- What resources might Italy gain from Ethiopia?
- What economic, political, or military advantages might Ethiopia's location offer Italy?

Government and Economics

KEY QUESTION What problems have hurt Ethiopia's economy?

Ethiopia is the oldest independent country in Africa, and one of the oldest in the world. Yet its first multiparty elections weren't held until 1995. The party that won in 1995 repeated its victories in 2000 and 2005. But some Ethiopians believed the 2005 election was dishonest, and they disputed the results in violent protests.

Structure of the Government and Economy Ethiopia has many political parties. The party with the most members in the legislature chooses the prime minister, who runs the executive branch and chooses the cabinet ministers. Ethiopia's legislature has two houses. The members of one house are elected directly by the voters, while representatives of the other house of the legislature are selected by state councils.

Despite its fertile soil and plentiful highland rainfall, Ethiopia remains poor. It has suffered from bad government; conflicts, such as the wars with Eritrea; natural disasters, including droughts; and limited use of its resources. The government owns all the land. A few farmers grow cash crops, but most Ethiopians are subsistence farmers. Overgrazing and deforestation have eroded the soil, destroying productive land while leaving fertile ground uncultivated. This landlocked country has no commercial fishing, and little mining and industry. Political instability has discouraged foreign investment in Ethiopia's economy. Most of the roads are unpaved, and only one railroad operates.

Like many other developing countries, Ethiopia has inadequate educational and health systems. Since schooling isn't required, the country's literacy rate hovers at just about 40 percent. Malaria, typhoid, and other diseases pose serious health risks.

What the Economy Produces Ethiopia's factories produce textiles, cement, food products, and shoes for domestic use. The country's top export is coffee. Other exports include animal skins and hides, oilseed, and sugarcane.

Like coffee, the cereal product teff is native to Ethiopia. Teff is the world's smallest grain. Its name comes from the Amharic word for "lost." Teff is well adapted to Ethiopia's highland soil, high in nutrients, and easily digested by people who can't eat wheat and other grains. Sorghum, another major grain product, is favored for its ability to grow in very dry conditions.

ANALYZE CAUSES Identify causes of Ethiopia's economic troubles. a combination of bad government, war, natural disasters, and limited use of its plentiful land and rainfall

Fun Facts!

RETURNING A STOLEN OBELISK

In 2004, Italy promised to return a 1,700-year-old obelisk removed from Ethiopia in 1937 during the Italian occupation. But the monument couldn't be returned by boat, as Ethiopia no longer had a port. Workers cut the obelisk into three sections, weighing up to 70–80 tons. Then Italy paid to upgrade the airport near Aksum to handle cargo planes big enough to carry the heavy chunks.

609

DIFFERENTIATING INSTRUCTION

Struggling Readers

Chart Ethiopia's Government, Economy, and Challenges

Help students summarize the information on this page. As you discuss the details of Ethiopia's government and economics, model filling in the appropriate columns in a chart, such as the one at right. Students can complete the chart and use it to review important facts about Ethiopia.

Ethiopia		
Government	Economy	Challenges
many parties	*subsistence farming*	*political and military conflicts*

Teach

Government and Economics

Reader, Recorder, Reporter

- Describe Ethiopia's form of government. *(Possible Answer: It has many political parties; the legislature has two houses; the party with the most members in the legislature chooses the prime minister, who heads the executive branch.)*
- Name two grain products grown in Ethiopia. *(teff and sorghum)*
- **Find Main Ideas** What are some of the economic problems that Ethiopian society is facing today? *(Possible Answers: unstable government, political and military conflicts, droughts, destruction of productive land, inadequate educational and health systems, and spread of disease)*

Unit Resource Book: Africa
- Skillbuilder Practice: Identifying Problems and Solutions, p. 206

Fun Facts!

Returning a Stolen Obelisk

Most of the obelisks (also called stelae) at Aksum are only slightly taller than an adult human. A few, however, are remarkably towering monuments. Beneath them are the tombs of Aksum's rulers. Archeologists believe that the obelisks may have been intended to represent stairways for the rulers to ascend to heaven. The tallest one—more than 100 feet high—is considered to be the largest carved-stone monument in the world. It lies in ruin in the royal cemetery.

The obelisk featured here is the second tallest one (more than 75 feet high). During World War II, Italy's ruler, Benito Mussolini, had it shipped to Rome, where it was installed in front of the Ministry of Italian Africa. The obelisk finally returned to Ethiopia in 2005, and work began in 2006 to restore its original site.

Teach

Ethiopian Culture and Life

Think, Pair, Share

- Where do most Ethiopians live, and what do most of them do for a living? *(Most Ethiopians live in rural areas and pursue farming or herding.)*
- Contrast the life styles of rural and urban Ethiopians. *(City dwellers live in Western-style houses, work in tall office buildings, and wear western clothing. Rural dwellers live in round, mud or wood houses, or rectangular stone houses, farm or herd, and sometimes wear shammas over their clothing.)*
- **Find Main Ideas** What are some of the moral values that Ethiopians of all faiths emphasize? *(hospitality, politeness, respect for elders, and the pursuit of wisdom)*

COMPARING

African Crops

Tell students that teff is used to make *injera*, a cooked bread. Remind students that a key advantage of teff is its high nutritional value—of great importance to people who are struggling to make ends meet.

- Which grain would you choose to grow if you were a farmer? Explain why. *(Answers may vary. Sorghum is less expensive to grow, but harder for animals to digest.)*
- Which grain might be easier to grow as a cash crop? Explain why. *(sorghum, because it's likely cheaper to grow)*

CRITICAL THINKING ANSWERS

1. **Evaluate** teff, because sorghum is difficult to digest
2. **Make Inferences** sorghum, because it can be used to produce ethanol

Ethiopian Culture and Life

KEY QUESTION What effects has religion had on Ethiopian culture?

When ancient Aksum moved south, its Christian population was separated from other Christian states. This led Ethiopian Orthodox Christianity to develop apart from other types of Christianity. For one thing, Ethiopian Orthodox Christians observe the Jewish laws of diet. They also pray in their ancient language of Ge'ez (gee•EHZ).

Daily Life Most Ethiopians live in rural areas, where they farm or herd cattle, goats, or sheep. Using wooden plows and oxen, subsistence farmers depend on rain to grow their cereal crops, such as teff, corn, wheat, or sorghum. Herders, like the Borana in the arid south, often must migrate to find water for their animals.

Ethiopian city dwellers live in Western-style houses, work in tall office buildings, and wear Western clothing. In rural regions, some people live in round mud and wood houses with thatched or tin roofs, while others live in rectangular stone houses. Many adults wear a white cotton cloth called a *shamma* over their shirts or dresses.

COMPARING African Crops

	TEFF common cereal crop in Ethiopia and Eritrea	SORGHUM common cereal crop in dry areas of Africa
Advantages	high in protein, fiber, and iron, as well as calcium, potassium, and other minerals	thrives in dry growing conditions
Disadvantages	expensive to produce grain; low yields	difficult for animals to digest, but still used for animal feed
Other Uses	stalks reinforce thatched roofs and mud bricks; animal feed	source of ethanol for gasoline; stalks used for fences; waste products make plastic

CRITICAL THINKING

1. **Evaluate** Which grain would you grow for animal feed, and why?
2. **Make Inferences** Which grain might be useful in reducing dependence on oil, and why?

DIFFERENTIATING INSTRUCTION: TIERED ACTIVITIES

OBJECTIVE Prepare a brochure highlighting Ethiopia's history, government, economy, and culture.

Basic

With students, list some facts related to Ethiopia's history, government, economy, and culture on the board or chart paper. Then have pairs prepare a four-page brochure on Ethiopia, using one page for each category. Each page should include one visual and a brief summary. Display the brochures.

On Level

Adapt the Basic activity so pairs include, as part of each summary, some of the problems and possible solutions related to that topic. Display the brochures. As a class, discuss Ethiopia's problems and possible solutions to them.

Challenge

Adapt the Basic activity so students work individually and include, as part of each summary, some of the problems and possible solutions related to that topic. Display the brochures. As a class, discuss Ethiopia's problems and possible solutions to them.

Ethiopians eat spicy meat or vegetable stews called *wat*, scooped up by hand with *injera* (ihn•JEER•uh), a flat sour bread made from fermented teff. Soccer and volleyball are popular sports. A game called *genna*, which resembles field hockey, is traditionally played at Christmas.

Cultural Traits Until 1974, Christianity was the official religion of the country's rulers, and it inspired much of Ethiopia's art and literature. In today's Ethiopia, however, Muslims slightly outnumber Christians. About nine out of ten Ethiopians are Muslim or Christian, while the rest follow traditional African religions.

Ethiopians of all faiths place a high value on moral behavior, hospitality, politeness, respect for elders, and the pursuit of wisdom. Some of Ethiopia's hundred languages are related to Arabic and Hebrew. Others are African.

he region's arly hristianity as influenced s ancient nd current rt, and cur- ent music. thiopian nguages re related to rabic and ebrew. ▶

The Arts Traditional Ethiopian arts include obelisks, the rock churches of Lalibela, church walls covered with paintings called frescoes, and lavishly illustrated Bibles. Modern artists and musicians continue to draw on their Ethiopian Orthodox faith for artistic inspiration. Today's Ethiopian writers write in modern Ethiopian languages, including Amharic.

SUMMARIZE Explain how religion has influenced Ethiopian culture.

Injera A woman pours injera batter onto a hot griddle. **How does injera appear to differ from sandwich bread?**

It's cooked on a griddle instead of baked in an oven.

Section 1 Assessment

ONLINE QUIZ For test practice, go to **Interactive Review @ClassZone.com**

TERMS & NAMES

1. **Explain the importance of**
   - Aksum
   - Horn of Africa
   - Haile Selassie
   - Red Sea

USE YOUR READING NOTES

2. **Find Main Ideas** Use your completed diagram to answer the following question:
   What is Ethiopia's main export?

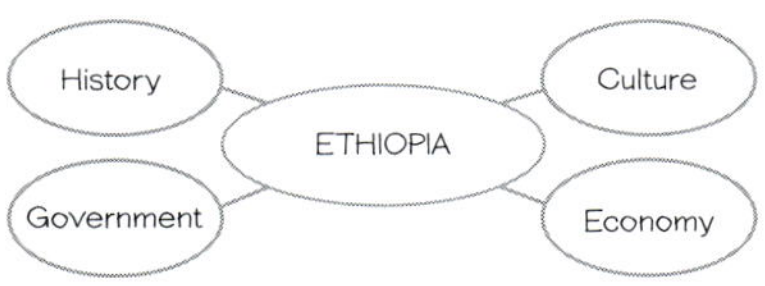

KEY IDEAS

3. What major accomplishment occurred during the Zagwe dynasty?
4. What features differentiate Ethiopian Orthodox Christianity from other Christian sects?
5. Why is there little commercial fishing in Ethiopia?

CRITICAL THINKING

6. **Make Inferences** Why do you think Menelik II is a national hero in Ethiopia?
7. **Identify Solutions** What changes could improve conditions in Ethiopia? Explain your answer.
8. **CONNECT to Today** In 2006, Ethiopia announced that its economy had grown due to a shift away from its concentration on agriculture. How could diversifying Ethiopia's products improve its economy?
9. **WRITING** **Create a Menu** Design a menu for an Ethiopian restaurant, including simple illustrations. Research the Ethiopian dishes *injera* and *wat*, so that you can describe their ingredients.

CONNECT to the Essential Question

How do boundaries set during colonial times affect stability in these regions today?

Ask students what they have learned so far that can help them answer this question. Students might mention:

- Ethiopia has remained an independent country for most of its history.
- In recent years, Ethiopians have faced political and military conflicts, droughts, and other challenges.
- Ethiopia's population is mostly rural, Muslim or Christian, and dependent on agriculture as a way of life.

4 Assess & Reteach

Assess Have students complete the Section Assessment.

Unit Resource Book: Africa
- Section Quiz, p. 219

Interactive Review @ClassZone.com
- Section Quiz

Power Presentations

Test Generator

Reteach Write the words *history*, *government*, *economy*, and *culture* on the board or chart paper. Call on a volunteer to supply one main idea about Ethiopia's history and list it under *history*. Continue this process, alternating categories, until all main ideas have been listed. Suggest that students copy the lists for review.

Unit Resource Book: Africa
- Reteaching Activity, p. 224

SECTION 1 ASSESSMENT ANSWERS

Terms & Names

1. Aksum: ancient empire controlling northern Ethiopia; Horn of Africa: horn-shaped extension of land on the east coast of Africa; Haile Selassie: emperor of Ethiopia, 1930–1974; Red Sea: sea dividing northeast Africa from Arabian peninsula

Use Your Reading Notes

2. coffee

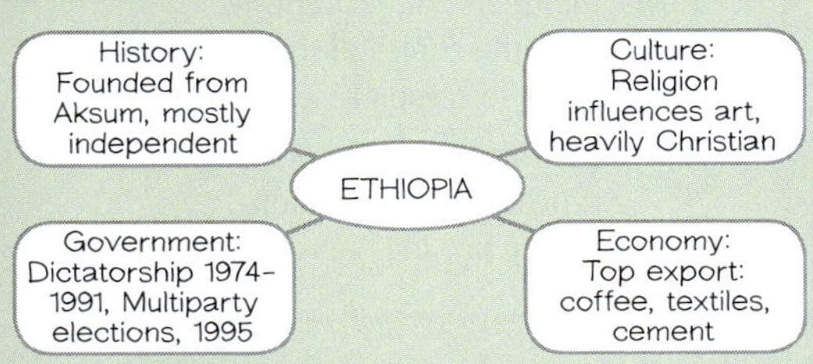

Key Ideas

3. Eleven churches were carved out of stone, using only hand tools.
4. These Christians follow Jewish dietary laws and pray in the ancient language of Ge'ez.
5. The country is landlocked.

Critical Thinking

6. His defeat of the Italian army prevented the colonization that afflicted almost every other African nation.
7. Answers will vary, but might include: universal education, which would improve literacy and perhaps health conditions; cultivate more arable land, which would ease the effects of droughts; put land in private hands, which would spur business investment
8. Diversification would allow profits even during the bad weather that can destroy farm crops.
9. Use the rubric to score menus.

Menu Rubric

	Illustrations	Descriptions
4	accurate, attractive	excellent
3	adequate	adequate
2	fair	fair
1	poor	poor

1 Plan & Prepare

Objectives

- Describe how Kenya was controlled by foreign powers and how it eventually achieved independence
- Summarize key aspects of Kenyan society today
- Explain how Kenya's government and economy work

Read for the Essential Question

Help students read for a purpose by reminding them of the Essential Question: "How do boundaries set during colonial times affect stability in these regions today?" Point out that they can use text features such as Key Questions, special features, images, and maps to form answers to the question.

Vocabulary

Best Practices Toolkit

Use the ***Best Practices Toolkit*** to model strategies for vocabulary notetaking. Vary strategies throughout the year. Choose from: *Knowledge Rating, Predicting ABCs, Definition Mapping, Word Sort, Frayer Model (Word Squares), Magnet Words, Word Wheel,* and *Student VOC.*

Vocabulary Strategies, TT9–TT16

Reading Strategy

Best Practices Toolkit

Display the T-Chart/Two-Column Chart transparency.

- Remind students that making an inference involves taking facts (and using reason and prior knowledge) to make an educated guess, or prediction, based on those facts.
- Model filling in entries, listing facts from the "History of a Trading Region" subsection and inferences that can be drawn from them.
- Have students suggest additional entries.
- Display the completed chart for students to review.

T-Chart/Two-Column Chart, TT20

SECTION 2

Reading for Understanding

Key Ideas

BEFORE, YOU LEARNED

West African kingdoms built vast trading empires.

NOW YOU WILL LEARN

Trade was also important in East Africa, particularly in what is now Kenya.

Vocabulary

TERMS & NAMES

safari an overland expedition for exploring, hunting, or photography

monsoon a seasonal wind bringing heavy rainfall

Swahili an African language, commonly spoken in East Africa

Mau Mau a Kenyan independence movement that began in the 1940s

Jomo Kenyatta independent Kenya's first president

BACKGROUND VOCABULARY

porridge a soft food made by boiling cereal grain in milk or water

Visual Vocabulary safari

Reading Strategy

Re-create the chart shown at right. As you read and respond to the **KEY QUESTIONS**, use the chart to make inferences about Kenya's government, history, and culture.

Skillbuilder Handbook, page R11

MAKE INFERENCES

FACTS	INFERENCES
1.	1.
2.	2.
3.	3.
4.	4.

GRAPHIC ORGANIZERS
Go to **Interactive Review** @ClassZone.com

612 Chapter 20

PRETEACHING VOCABULARY

English Learners

Pronounce and Preview

Pronounce each term for students. Review the meanings of words used in definitions, such as *expedition, seasonal,* and *commonly.*

- To modify vocabulary learning, have students complete worksheets as they read, instead of afterward.

Unit Resource Book: Africa
- Vocabulary Practice, p. 205

Inclusion

Vocabulary: Definition Mapping

Have students pair up and fill out the graphic organizer for this section's terms.
- Use the transparency to model the strategy.
- Have volunteers suggest entries.
- Then distribute copies of the organizer for students to complete with a partner.

Best Practices Toolkit
- Vocabulary: Definition Mapping, TT11

Northern Ethiopia

African elephants, Tanzania

SECTION 2 FOCUS ON

Kenya

Connecting to Your World

One hundred years ago, Europeans and Americans traveled to Africa on hunting or exploring expeditions called **safaris**. This year, thousands of people will go on safari to Kenya's Masai Mara, famous for lions, or Amboseli National Reserve, noted for elephants. But the only shooting tourists will do is with their cameras. Safaris bring Kenya millions of tourist dollars each year.

Elephants Elephants graze at Amboseli National Reserve.

History of a Trading Region

KEY QUESTION How did the Swahili language develop?

The ancient kingdom of Aksum controlled trade to Arabia. After the 600s, that trade fell into the hands of Muslim invaders. Now you'll learn how Muslim trading cities spread down the East African coast, affecting inland African populations.

Inland Villages, Coastal Cities Earlier you learned that the Bantu-speaking people of Africa migrated across the southern part of the continent between 1000 B.C. and A.D. 1100. Bantu speakers, seeking land for their farms and herds, eventually reached Africa's east coast. Their iron-working skills helped them produce farming tools.

Mombasa Arches representing elephant tusks welcome visitors to Kenya's main port city.

2 Focus & Motivate

3-Minute Warm-Up

Write on the board or display the transparency:

- For most of its history, foreign powers controlled the area that is now Kenya. What do you think life was like under those conditions? *(Possible Answers: Life would have been difficult; political and economic opportunities limited; resentments fueled.)*

Unit Transparency Book: Africa
- 3-Minute Warm-Ups, TT26

Connecting to Your World

More About . . .

Mombasa

In 1901, the British built a railroad from Mombasa to Lake Victoria, to encourage European settlement in the interior. As a result, Mombasa became a port of entry for many European immigrants to Africa.

3 Teach

History of a Trading Region

Talk About It

- What foreign powers controlled parts of Kenya during its history? *(Arabs, Portuguese, British)* How did the British expand the settlement of Kenya? *(It encouraged Europeans to establish large farms.)*
- **Make Inferences** What reasons do you think east Africans had for opposing British rule? *(Possible Answer: Europeans controlled the land and resources, as well as the profits.)*

SECTION 2 PROGRAM RESOURCES

Unit Resource Book: Africa

ON LEVEL
Unit Resource Book
- Reading Study Guide, p. 167
- Map and Graph Skills, p. 207
- Section Quiz, p. 220

STRUGGLING READERS
Unit Resource Book
- Reading Study Guide with Additional Support, p. 177
- Skillbuilder Practice: Identifying Problems and Solutions, p. 206
- Section Quiz, p. 220
- Reteaching Activity, p. 225

eEdition with Audio DVD-ROM

ENGLISH LEARNERS
Pupil Edition in Spanish
eEdition with Audio DVD-ROM
Unit Resource Book
- RSG (Spanish), p. 187
- RSG with Additional Support (Spanish), p. 197

Multi-Language Glossary
Test Generator
- Section Quiz in Spanish

INCLUSION
Unit Resource Book
- RSG with Additional Support, p. 177
- Section Quiz, p. 220
- Reteaching Activity, p. 225

GIFTED & TALENTED
Unit Resource Book
- Section Quiz, p. 220
- World Literature, p. 241

PRE-AP
Unit Resource Book
- Section Quiz, p. 220
- Geography Project, p. 243

TECHNOLOGY
Unit Transparency Book
- 3-Minute Warm-Ups, TT26
- Essential Question, TT27
- Critical Thinking, TT29
- Map: Languages, TT34
- Cultures, TT35

Daily Test Practice Transparencies
- Section 2, TT68

Power Presentations
ClassZone.com
World Cultures and Geography Video on DVD

COMPARING

Trade Across Regions *1400–1500*

Remind students that the East African trading cities were situated along the coast. Point out that although none of the goods traded were the same in these cities, silk, precious stones, and certain spices would have been considered luxury goods in Italy.

- In what way did East African trading routes differ from Italian trading routes? *(East African routes made use of just the sea—the Indian Ocean—to reach India. Italian routes included both overland and sea routes.)*
- **Make Inferences** Who were the slaves that were exported to India? *(likely native East Africans)*

CRITICAL THINKING ANSWERS

1. **Analyze Effects** They did not have the overland routes that European traders did.
2. **Make Inferences** Traders sold goods for more money than they paid for them. By carrying goods farther away, merchants were able to charge better prices.

Unit Resource Book: Africa
- Interdisciplinary Projects: Speak Swahili, p. 213

More About . . .

The Mau Mau Rebellion

During British occupation, native East African groups led by the Kikuyu established the Kenya African Union to protest British rule. Within this organization, some Kikuyu members—referred to as the Mau Mau—secretly pursued more aggressive actions to force the British to enact reforms.

In response, the British launched a military campaign to crush this freedom movement. The fighting resulted in the deaths of more than 11,000 Kenyans and about 2,000 other Africans and those who supported the British.

COMPARING **Trade Across Regions** *1400–1500*

	EAST AFRICAN TRADING CITIES	TRADING CITIES OF RENAISSANCE ITALY
Goods Traded	Exported gold, iron, ivory, beads, tortoiseshell, and slaves. Imported cotton and luxury goods, such as porcelain	Exported salt, wool, olive oil, grain. Imported silk, precious stones, pepper and other spices
Trading Partners	Persian Gulf states, India, China	northern and western Europe, Persia, India, and China
Trade winds/ patterns enabling trade	seasonal winds of the western half of the Indian Ocean that blow between West India and East Africa	overland trade with Europe, sailing routes across the Mediterranean, and overland to the Far East along the Silk Roads

CRITICAL THINKING

1. **Analyze Effects** Why did East African trading cities depend on trade winds?
2. **Make Inferences** How did merchants become wealthy?

Along this same coast, traders from Arabia had established several dozen thriving trading cities by the 1400s, including Mombasa and Malindi in today's Kenya. **Monsoons**, seasonal rain-bearing winds, propelled Arab trade ships to India. These ships carried gold, leopard skins, beads, copper, and ivory, brought to the coast by inland African traders. Other than trade, the inland villagers—farmers and herders who practiced African religions—had little to do with the Muslim city dwellers. But their trade contacts produced one important cultural effect. By the 1400s, most townspeople spoke a blend of Bantu and Arabic, an ancestor of today's African language **Swahili**.

European Domination By the early 1500s, Portugal controlled the wealthy East African cities. The Portuguese traded small numbers of slaves. In the late 1800s, Europe began the scramble for African colonies you learned about in the last chapter. After Germany and Britain divided East Africa, the territory that would become Kenya was called British East Africa. The British encouraged Europeans to settle in British East Africa and establish huge farms.

Independence and Afterward During the 1940s, East Africans began uniting to oppose British rule. Britain's refusal to consider their demands triggered a revolution in 1952. A resistance group called the **Mau Mau** led this revolt. The British imprisoned thousands of rebels, including **Jomo Kenyatta**, a Kenyan political leader. Kenyatta spent seven years in prison, despite denying Mau Mau involvement.

DIFFERENTIATING INSTRUCTION

English Learners

Comprehension: Connection

Explain that there are many ethnic groups in Kenya, most of whom have their own language. Discuss, or have students respond in writing to, the following questions:

- How would the development of Swahili have aided trade?
- How might Swahili and English help unite Kenyans today?

Gifted & Talented

Map African and Italian Trade Routes

Supply students with a globe, or map that shows Europe, Africa, and Asia. Have them use it to create their own maps, marking likely trade routes between Italy and Asia, and between Africa and India. Students should include labels for countries, bodies of water, and routes. Display the completed maps.

By the late 1950s, Britain agreed to national elections. Kenya won independence in 1963, and the following year Jomo Kenyatta became the nation's first president. Kenyatta's attempts to modernize Kenya included expanding educational opportunities and limiting property ownership to Kenyans. Kenyatta died in 1978 and was succeeded by his vice president, Daniel Moi, who led until 2002. For most of its recent history, Kenya has been a one-party state, despite a 1991 amendment establishing a multiparty system. In 2002, the head of a multi-ethnic party won a peaceful democratic presidential election.

SUMMARIZE Describe how East African trade created Swahili. Muslim traders began speaking an Arabic-Bantu blend, which became an early form of Swahili.

Life in Kenya

KEY QUESTION What interests do many Kenyans share?

Since independence, the Kenyan government has tried to promote national pride. That's difficult in a country where some 40 ethnic groups struggle to get along and no single culture dominates.

Daily Life and Activities Kenya's largest ethnic group is the Kikuyu, followed by the Luhya (LOO•yah), Kalenjin (KAL•uhn•jihn), Luo, and Kamba. Most groups have their own language, but many Kenyans also know Swahili and English, the country's official languages. At least two-thirds of Kenyans are Christian. About 10 percent follow traditional African religions. Another 10 percent are Muslim, living mostly along the coast.

One of Kenya's most famous writers, Ngugi wa Thiong'o, wrote novels in English and in his native Kikuyu. Most Kenyans enjoy dancing and dance performances and follow their favorite soccer teams. Several Kenyan runners have won Olympic medals for track and field. Kenyan runners also regularly win the yearly Boston Marathon.

About three out of four Kenyans live in rural areas. A farming family typically occupies a small house with a thatched roof, mud walls, and dirt floors. A few Kenyans, like the Masai, are nomadic herders. Every day, most Kenyans eat maize, or corn, which they grind and cook as **porridge**, then mix with other vegetables to make a stew.

SUMMARIZE Identify the common interests of many Kenyans. a love for dance and soccer, and pride in their country's world-class runners

the Kenyan flag ▼

Kenyan Runners Catherine Ndereba celebrates her marathon victory at the 2003 world track and field championships. **What is Ndereba holding behind herself?**

DIFFERENTIATING INSTRUCTION

Struggling Readers

Clarify Political Terminology

Some students may have difficulty understanding the expression *one-party state*, to describe a government with a multiparty system of democracy. Explain that the inference to be drawn is that in recent years political power in Kenya has been monopolized by one party, even though Kenya's political system now allows many political parties to take part in elections.

Pre-AP

Research Contemporary Kenyan Issues

Challenge students to learn more about recent political events that are influencing life in Kenya today. Invite students to research and report on the issues that are dividing Kenyans today and the efforts the Kenyan government and others are making to resolve those issues.

More About . . .

Kenya's President, Mwai Kibaki

In recent years, political and economic problems in Kenya continued to grow. In 2004, despite President Mwai Kibaki's attempts to improve the government, the United Nations issued a report contending that Kenya was one of the most corrupt countries in Africa. In 2005, ethnic clashes, organized protests, increased charges of corruption, the withdrawal of some foreign aid, a severe drought, and flash floods magnified the problems facing Kibaki.

Teach

Life in Kenya

Reader, Recorder, Reporter

- Which people represent the largest ethnic group in Kenya? *(the Kikuyu)*
- Where do most Kenyans live and what occupations do most of them pursue? *(The majority of Kenyans live in rural areas. Many of them pursue farming. A few are nomadic herders.)*
- **Make Inferences** What reason might explain why the majority of Kenya's population is Christian? *(Possible Answer: Christianity may have been brought to Kenya by European missionaries as early as the 1500s.)*

Unit Resource Book: Africa
- World Literature: African Proverbs, p. 241

More About . . .

Kenyan Track and Field Achievements

As of the end of 2006, Kenyan runners held the world track and field outdoor records for the 1,000-meter, 3,000-meter, and marathon races in the men's division, and the 20,000- and 30,000-meter races in the women's division. Kenyan men finished first in the Boston Marathon from 1991–2000, 2002–2004, and 2006; Kenyan women won in Boston from 2000–2002 and from 2004–2006.

RESEARCH & WRITING CENTER

ANALYZING Primary Sources

As early as 1930, Jomo Kenyatta predicted the possibility of violent protests directed at Britain. In a letter to *The Times of London,* he presented a list of government reforms sought by the Kikuyu people, and he warned that ignoring them "must inevitably result in a dangerous explosion—the one thing all sane men wish to avoid."

DOCUMENT–BASED QUESTION ANSWER

Africans' right to elect their own representatives

Unit Resource Book: Africa
- Skillbuilder Practice: Identifying Problems and Solutions, p. 206

Teach

Government and Economy

Talk About It

- What are Kenya's main cash crops? *(coffee, tea)* What other industry is important to Kenya's economy? *(tourism)*
- **Make Inferences** Do you think that many Kenyan citizens are economically secure? Why or why not? *(Possible Answer: The information in this subsection suggests that a large number of Kenyans are not economically secure.)*

Unit Resource Book: Africa
- Geography Project: Water Lifting Machine, pp. 243–244

CONNECT to the Essential Question

How do boundaries set during colonial times affect stability in these regions today?

Ask students what they have learned so far that can help them answer this question. Students might mention:

- For much of its history, Kenya was controlled by foreign interests.
- Kenya's government is working to promote national pride and improve the economy.
- Kenya is composed of many ethnic groups, most of whom are struggling to get out of poverty.

ANALYZING Primary Sources

Jomo Kenyatta (1894–1978) headed the Kenya African Union (K.A.U.), a pro-liberty party, before Kenya's independence. The British said that the K.A.U. was responsible for the Mau Mau rebellion, and they arrested Kenyatta after this speech:

> K.A.U. is you and you are the K.A.U. If we unite now, each and every one of us, and each tribe to another, we will cause the implementation in this country of that which the European calls democracy. True democracy has no colour distinction. It does not choose between black and white. We are here in this tremendous gathering under the K.A.U. flag to find which road leads us from darkness into democracy. In order to find it we Africans must first achieve the right to elect our own representatives.
>
> Source: Jomo Kenyatta, speech at the Kenya African Union Meeting, July 26, 1952

DOCUMENT–BASED QUESTION

What right did Kenyatta say was the first step to democracy?

Jomo Kenyatta used a traditional flywhisk, or flyswatter, as a presidential symbol. ▶

ONLINE PRIMARY SOURCE To see more of Jomo Kenyatta's works, go to the **Research & Writing Center** @ClassZone.com

Government and Economy

KEY QUESTION How has Kenya's government changed since independence?

Kenya's national motto is "Harambee," the Swahili word for cooperation. But pulling together this ethnically diverse developing country continues to pose a challenge. Both Kenya's government and its economy have been plagued by corruption.

Regional assemblies had many powers after independence. Changes since then have given much more power to the president. For example, the president heads the military. In addition, the president's cabinet, chosen from the National Assembly, helps pass laws the president favors. The president and the National Assembly are popularly elected for five-year terms. The Kenyan constitution guarantees many freedoms, but the president can jail people who threaten public security.

Private citizens own most of Kenya's land and businesses, but the government keeps partial ownership of some businesses and regulates others. In the past, government regulators have taken bribes.

DIFFERENTIATING INSTRUCTION: TIERED ACTIVITIES

OBJECTIVE Prepare a time line of important events in Kenya's history.

Basic

Have students work in pairs to make a time line that lists at least five important events in Kenya's history, ranging from pre-colonial times to the present. Have students present their time lines and discuss why they choose the events to list that they did.

On Level

Adapt the Basic activity so that student pairs make a time line that lists at least ten important events in Kenya's history, ranging from pre-colonial times to the present. Have students present their time lines and discuss why they chose the events to list that they did.

Challenge

Adapt the On Level activity so that students include a summary of how Kenya's past history has shaped its recent history when they present their time lines.

Kenya's new government is trying to root out corruption to promote the country's economic growth.

Kenya's economy is based on agriculture. Four out of ten Kenyans were unemployed in 2000, and half its people live in poor conditions. Kenya's subsistence farmers use traditional tools to grow enough maize to feed their families. But some large farms turn out Kenya's main cash crops, coffee and tea, and food processing is a growing industry.

Coffee and tea are Kenya's top moneymakers, but tourism is a key industry. Half a million tourists visit Kenya's national parks and game reserves each year. Tourism pumps $200 million into the economy annually. Some Kenyans resist tourism, saying this industry exploits resources to benefit relatively few Kenyans and recalls Kenya's colonial past. But other Kenyans feel that tourism is a good replacement for subsistence farming in a dry country with unpredictable rain.

SUMMARIZE Describe how Kenya's government has changed since independence. Most powers have shifted from regional assemblies to the president.

Economic Activities of Kenya *2004*

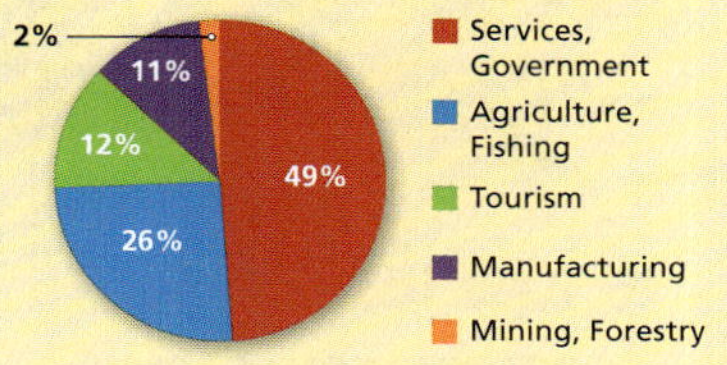

Source: Central Bank of Kenya Statistical Bulletin

"Say cheetah" A photographer gets a close-up view of a cheetah at Masai Mara National Reserve.

Section 2 Assessment

ONLINE QUIZ For test practice, go to **Interactive Review** @ClassZone.com

TERMS & NAMES

1. **Explain the importance of**
   - Safari
   - Swahili
   - Mau Mau
   - Jomo Kenyatta

USE YOUR READING NOTES

2. **Make Inferences** Use your completed diagram to answer the following question: Why did Kenya's one-party system offer only limited democracy?

FACTS	INFERENCES
1.	1.
2.	2.
3.	3.
4.	4.

KEY IDEAS

3. What European country colonized Kenya in the late 1800s?
4. How did President Kenyatta try to modernize Kenya?
5. What are Kenya's key industries?

CRITICAL THINKING

6. **Make Predictions** Could a multi-ethnic party end Kenya's history of corrupt government? Explain.
7. **Evaluate** Do you think tourism is a good idea or a bad idea for Kenya's economy? Explain.
8. **CONNECT to Today** Kenyan Nobel Prize winner Wangari Maathai said, "Protecting the global environment is directly related to securing peace." What do you think she meant by this statement?
9. **TECHNOLOGY** **Make a Mini-Dictionary** Use the Internet to research 20 Swahili words and phrases for a mini-dictionary of useful terms for a safari. List the word or phrase, its English meaning, and its origin.

Economic Activities of Kenya *2004*

- What is the second largest source of revenue for Kenya's economy? *(agriculture, fishing)* What important exports are included in this category? *(coffee, tea)*
- What is the third largest source of revenue? *(tourism)* Why do some Kenyans consider this source of revenue controversial? *(They believe that it exploits resources, benefits few Kenyans, and recalls Kenya's colonial past.)*

Unit Resource Book: Africa
- Map and Graph Skills: African Population Distributions, pp. 207–208

4 Assess & Reteach

Assess Have students complete the Section Assessment.

Unit Resource Book: Africa
- Section Quiz, p. 220

Interactive Review @ClassZone.com
- Section Quiz

Power Presentations

Test Generator

Reteach Have students work in pairs and create one graphic organizer on Kenya's culture and one on its economy. Have each pair present one of their graphic organizers to the class. Students should revise their graphic organizers as needed after the presentations.

Unit Resource Book: Africa
- Reteaching Activity, p. 225

SECTION 2 ASSESSMENT ANSWERS

Terms & Names

1. safari: expedition for exploring; Swahili: African language commonly spoken in East Africa; Mau Mau: a Kenyan independence movement that began in the 1940s; Jomo Kenyatta: independent Kenya's first president

Use Your Reading Notes

2. Lack of opposition may lead to government abuse.

FACTS	INFERENCES
1. Single-party government until 1991	1. political opposition had no power
2. Government regulators have taken bribes	2. government grew corrupt with no opposition
3. Forty ethnic groups with no single dominant culture	3. Swahili and English are common tongues

Key Ideas

3. Great Britain
4. He expanded education and limited property ownership to Kenyans.
5. farming and tourism

Critical Thinking

6. Students should predict that it might, because a multi-ethnic alliance should put a check on favoritism to any one group.
7. Answers will vary. Some students may say that Kenya should develop its manufacturing or encourage more farmers to grow cash crops. Others may say tourism is a good way to enrich the economy and encourage subsistence farmers to take up more profitable, dependable work.
8. Answers will vary, but should suggest that protecting the environment means there will be more resources to go around, so people will have less to fight about.
9. Use the rubric to score dictionaries.

Dictionary Rubric

	Dictionary Entries	Quality of Research
4	all entries present	excellent
3	one entry missing	good
2	2–3 entries missing	fair
1	4 or more entries missing	poor

1 Plan & Prepare

Objectives

- Summarize the history of the Congo region leading to the establishment of the Democratic Republic of the Congo
- Describe the culture and life of Congo
- Identify Congo's chief economic resources and the make-up of its government

Read for the Essential Question

Help students read for a purpose by reminding them of the Essential Question: "How do boundaries set during colonial times affect stability in these regions today?" Point out that they can use text features such as Key Questions, special features, images, and maps to form answers to the question.

Vocabulary

Best Practices Toolkit

Use the ***Best Practices Toolkit*** to model strategies for vocabulary notetaking. Vary strategies throughout the year. Choose from: *Knowledge Rating, Predicting ABCs, Definition Mapping, Word Sort, Frayer Model (Word Squares), Magnet Words, Word Wheel,* and *Student VOC.*

Vocabulary Strategies, TT9–TT16

Reading Strategy

Best Practices Toolkit

Display the Cluster Diagram transparency.

- Remind students that identifying solutions involves recognizing the factors that contribute to the problems and ways to eliminate or overcome those factors.
- Model filling in one problem, the factors contributing to it, and a solution from the "Independence and Afterward" subsection.
- Have students suggest entries. Point out that their diagrams will help them answer some of the key questions in this section.
- Display the completed diagram for students to review.

Cluster Diagram, TT28

SECTION 3 Reading for Understanding

Key Ideas

BEFORE, YOU LEARNED

As some African nations won political freedom from colonial powers, they fell under dictatorships or military rule.

NOW YOU WILL LEARN

The Democratic Republic of the Congo is still struggling with setting up a true democracy.

Vocabulary

TERMS & NAMES

Congo River Africa's second-longest river

Pygmy a member of a people with an average height of less than five feet

hunter-gatherer a person who hunts animals and gathers plants for food

Kongo a member of a people living along the lower Congo River; a Bantu-speaking kingdom of the Congo River region arising in the 1300s

Leopold II King of Belgium, who was forced to give up his private ownership of the Congo Free State

Mobutu Sese Seko dictator of the Democratic Republic of the Congo from 1965 to 1997

Zaire name of the Democratic Republic of the Congo between 1971 and 1997

BACKGROUND VOCABULARY

quota a target amount for production

Reading Strategy

Re-create the diagram shown at right. As you read and respond to the **KEY QUESTIONS**, use the diagram to help you make generalizations about the Democratic Republic of the Congo.

Skillbuilder Handbook, page R12

MAKE GENERALIZATIONS

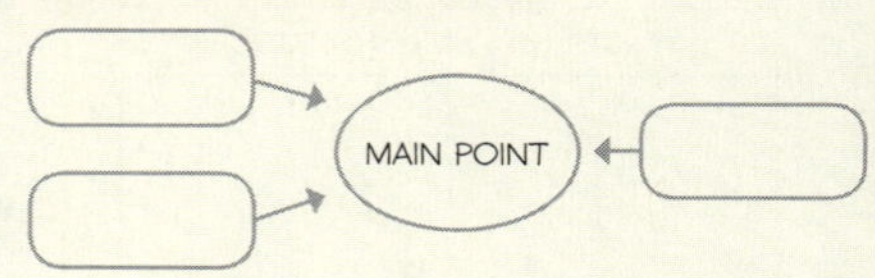

GRAPHIC ORGANIZERS
Go to Interactive Review @ClassZone.com

PRETEACHING VOCABULARY

English Learners

Pronounce and Preview

Pronounce each term for students. Review the meanings of words used in definitions, such as *gathers, dictator,* and *production.*

- To modify vocabulary learning, have students complete worksheets as they read, instead of afterward.

Unit Resource Book: Africa
- Vocabulary Practice, p. 205

Inclusion

Match Terms and Definitions

Have student pairs write the vocabulary words and definitions on index cards. Then have them cut out the words and definitions to create a matching game. Ask one student to shuffle the cards and arrange them. After the partner has finished matching the terms and definitions, have the students work together to correct any errors. Students should then switch roles.

Northern Ethiopia

SECTION 3 FOCUS ON

African elephants, Tanzania

Democratic Republic of the Congo

Connecting to Your World

Have you ever known two classmates with the same first name? To avoid confusion, did people add last names or initials when talking about them? Two African countries, both called Congo, lie along the Congo River. The capital of the Democratic Republic of the Congo is Kinshasa; the Republic of the Congo's capital is Brazzaville. In this section you'll learn about the larger of the two, the Democratic Republic of the Congo, also called Congo (Kinshasa).

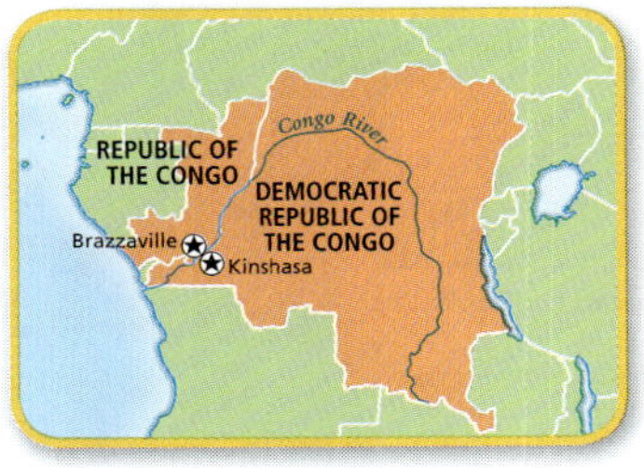

A River Culture

KEY QUESTION What groups have controlled what is now Congo?

The **Congo River**, the continent's second-longest, rises near the country's southeast corner and flows in a giant counterclockwise circle before draining into the Atlantic Ocean. The river's name comes from the Kongo kingdom that arose here in the 1300s. This mighty river flows through Congo (Kinshasa), and along the border with Congo (Brazzaville). Four deep lakes, including Lake Tanganyika, line Congo's eastern border.

Congo River The Congo and its tributaries form Africa's largest river system.

619

2 Focus & Motivate

3-Minute Warm-Up

Write on the board or display the transparency:

- Compare and contrast the Congo River to the Nile River. *(Similar: Both are important waterways and sources of water; Different: The Congo flows east to west, the Nile south to north.)*

Unit Transparency Book: Africa
- 3-Minute Warm-Ups, TT26

Connecting to Your World

More About . . .

The Congo River

The Congo—2,900 miles long—is the fifth longest river in the world. Though rapids and waterfalls make parts of it unnavigable, it is a major transportation route for Congo (Kinshasa), especially in areas of the country that lack serviceable roads.

3 Teach

A River Culture

Reader, Recorder, Reporter

- Who inhabited the Congo region before the arrival of Bantu-speaking people? *(Pygmies)*
- What foreign powers enslaved the people of the Congo region? *(Portugal, then Belgium's King Leopold)*
- **Identify Problems and Solutions** What problems arose after Congo achieved independence? *(Possible Answer: Europeans left behind an inexperienced government and an undisciplined army.)*

SECTION 3 PROGRAM RESOURCES

Unit Resource Book: Africa

ON LEVEL

Unit Resource Book
- Reading Study Guide, p. 169
- Section Quiz, p. 221

STRUGGLING READERS

Unit Resource Book
- Reading Study Guide with Additional Support, p. 179
- Skillbuilder Practice: Identifying Problems and Solutions, p. 206
- Section Quiz, p. 221
- Reteaching Activity, p. 226

eEdition with Audio DVD-ROM

ENGLISH LEARNERS

Pupil Edition in Spanish

eEdition with Audio DVD-ROM

Unit Resource Book
- RSG (Spanish), p. 189
- RSG with Additional Support (Spanish), p. 199

Multi-Language Glossary

Test Generator
- Section Quiz in Spanish

INCLUSION

Unit Resource Book
- RSG with Additional Support, p. 179
- Section Quiz, p. 221
- Reteaching Activity, p. 226

GIFTED & TALENTED

Unit Resource Book
- Interdisciplinary Projects, pp. 212, 214
- Section Quiz, p. 221

PRE-AP

Unit Resource Book
- Section Quiz, p. 221

TECHNOLOGY

Unit Transparency Book
- 3-Minute Warm-Ups, TT26
- Essential Question, TT27
- Critical Thinking, TT30

Daily Test Practice Transparencies
- Section 3, TT69

Power Presentations

ClassZone.com

World Cultures and Geography Video on DVD

Animated GEOGRAPHY

Bantu Migrations
1000 B.C.–A.D. 1100

CLASSZONE.COM

CONNECT Geography & History

Scholars believe that the Bantu migration—considered one of the greatest migrations in history—first began about 2,000 years ago. Today, Bantu-speaking people occupy an area extending from Central Africa all the way down to the southern tip of the continent. They represent hundreds of separate, diverse, cultural groups, including the Zulu, Swahili, Kikuyu, and Kongo.

- What geographic feature likely limited the Bantu migration from moving farther south? *(Kalahari Desert)*

ANSWER

Movement because the mostly uninhabitable Sahara lies to the north

Presentation Options

Use the **Power Presentations** DVD-ROM to project the animated map.

- Click to animate the map.
- Listen to the audio and examine the photographs to learn more about the ancient Bantu migrations and Bantu-speaking people today.

More About . . .

Mobutu Sese Seko

Born in what was then the Belgian Congo, Mobutu was educated by missionaries. He joined the military in 1949 and was eventually promoted to the highest officer rank open to an African in the Belgian Congolese army. After Congo achieved independence, Mobutu was named secretary of state for national defense.

During the turbulent years that followed, Mobutu assumed control of the government twice, removing Premier Lumumba from power and, later, President Joseph Kasavubu. He may have been involved in the death of Lumumba, who many believe was assassinated by military troops. Mobutu died in exile in 1997, only months after he was overthrown.

Animated GEOGRAPHY

Bantu Migrations *1000 B.C.–A.D. 1100*

Click here to learn about Bantu migrations and see modern Bantu speakers today @ ClassZone.com

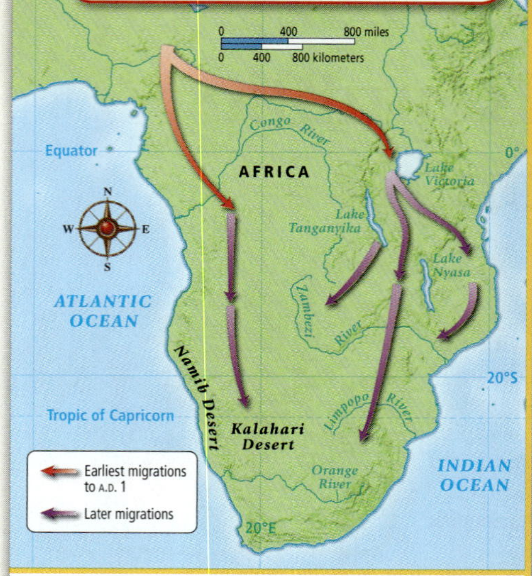

CONNECT Geography & History

Movement Based on what you have learned about Africa, why didn't Bantu speakers migrate northward?

Early History Congo's earliest inhabitants were **Pygmies**, very short people who still live as hunter-gatherers. **Hunter-gatherers** hunt animals and collect wild food instead of practicing agriculture and herding. The Bantu migrations brought Bantu-speaking peoples into the region between the 900s and 1300s. They established several important kingdoms, including Luba, Lunda, and **Kongo**. The Kongo kingdom ruled the coast. Kongo traded with Portugal, and members of the Kongo royal family became Catholic. But Portugal came to view Kongo as a source for slaves.

Belgian Control European enslavement of Kongo's people continued through the 1800s. In 1885, Belgium's King **Leopold II** convinced other European leaders to let him rule his own personal colony, to be called the Congo Free State. Instead of telling the truth—that he wanted to get richer—Leopold promised to bring Christianity and progress to Africa and end slavery. After forcing Africans to build a railroad, the king's agents cruelly punished those who failed to produce their **quota** of rubber. Protests from the United States and Britain led the Belgian government to take control of the Congo Free State in 1908, slightly improving living conditions.

Independence and Afterward When Belgian Congo suddenly won independence in 1960, its leaders were unprepared to govern. During the 1960s, Congolese troops rebelled against their Belgian officers, leaving behind a dangerous, undisciplined army. The inexperienced government couldn't unify the country, and several of Congo's provinces left or threatened to secede. A coup in 1965 gave Joseph Mobutu the presidency. President Mobutu tried to reduce ethnic divisions and instill pride in the country's heritage. He changed his own name to **Mobutu Sese Seko** and the country's name to **Zaire**. But Mobutu was a corrupt dictator who made himself rich at his country's expense. He was overthrown in 1997. The new leader, Laurent Kabila, renamed the country the Democratic Republic of the Congo. Kabila's 2001 assassination left his son Joseph in power.

SEQUENCE EVENTS Name the powers who controlled the Congo region from its earliest history to 1960. Pygmies, the Kongo people, King Leopold, Belgium as a colony, then an independent country

DIFFERENTIATING INSTRUCTION

English Learners

Vocabulary: Latin Derivatives

Review the definition of *quota,* "a target amount for production," and explain that it refers to any number or amount that is needed or asked for to achieve a designated goal. Its Latin root, *quot,* means "how many." Other words with the same Latin root are *quote, quotation,* and *quotient.* Provide students with examples of different types of quotas—for example: number of signatures needed on a candidate's petition to run for office.

Gifted & Talented

Illustrate the Congo's History

Challenge students to use art paper and drawing supplies to create scenes illustrating specific times or events in the Congo region's history. Encourage them to use their imaginations to depict hunter-gatherers, the Kongo royal family, Portuguese trading, King Leopold II, or other people and situations. Students can use illustrations from reference sources for their scenes if appropriate sources are available.

Culture and Life in Congo

KEY QUESTION What are Congo's four national languages?

As in many other African nations created by colonial powers, ethnic conflicts have plagued Congo (Kinshasa). Central Africa's largest country is home to more than 200 different languages.

A Land of Diversity Most of today's Congolese people descend from Bantu speakers who migrated to Congo centuries ago. Of the country's many local languages, almost all belong to the Bantu group and are closely related. French, the country's official language, is used in government and taught in school. Many Congolese also speak Swahili or another of the country's national languages, Kikongo, Lingala, or Tshiluba.

About four out of five Congolese are Christians. Most of these are Roman Catholic. Ten percent of Congolese are Muslims, and a similar number follow African religions.

Congolese Music Boys from the Bodjaba tribe play a Mokoto drum.

Daily Life More than a dozen cities in Congo have at least 100,000 people. Kinshasa's population has grown to more than six million since independence. Still, most of Congo's 62.5 million people live in villages containing a few dozen to a few hundred people. These villagers live in mud brick or dried mud-and-stick houses with thatched roofs. A wealthy family might install a metal roof.

Following independence, many Congolese businessmen traded Western-style dress for a national outfit of pants and a collarless jacket worn without shirt or tie. Women typically wear long, one-piece cotton dresses or a long skirt and blouse. Most Congolese meals feature porridge made of corn, rice, or cassava meal, topped with a spicy sauce. People who can afford meat or fish add it to the porridge. But many poor Congolese suffer malnutrition from a lack of protein.

Congolese Arts Centuries ago, enslaved Africans brought their musical styles to the Caribbean. Modern Congolese music takes those Afro-Caribbean rhythms and mixes them with African sounds. *Soukous,* a popular guitar-based musical style, developed in Kinshasa. The name comes from the French word *secouer,* or "to shake."

INTERDISCIPLINARY ACTIVITIES

CONNECT to Music

Listen to Congolese Music

Play examples of contemporary Congolese music and traditional African music for students to compare and discuss, based on questions such as the following:

- What are the distinctive qualities of each type of music?
- What influences of traditional music can students detect on contemporary forms?
- What influence, if any, does contemporary Congolese music have on popular music in the United States?

CONNECT to Civics

Create a Congolese Reform Plan

Ask students to assume the roles of members of a political group in the Congo dedicated to bringing reform to the country. Suggest students focus on steps that would restore confidence in the government and improve the economy. In small groups, discuss ideas and organize a plan of action.

Teach

Culture and Life in Congo

Think, Pair, Share

- What is Congo's official language? *(French)* What national languages are spoken in Congo? *(Swahili, Kikongo, Lingala, Tshiluba)*
- Where do most of Congo's people live? *(in small villages)* Describe their dwellings. *(mud-brick or dried mud-and-stick houses with thatched or tin roofs)*
- **Identify Problems and Solutions** Why do many poor Congolese suffer malnutrition? *(They lack protein in their diet.)* How could they solve this problem? *(Possible Answer: If there were better job opportunities, they could afford to buy meat or fish.)*

More About . . .

Kinshasa

The Democratic Republic of the Congo's capital and largest city, Kinshasa is situated on the south bank of the Congo River. This area was first settled around the 1400s and became a center of trade in the 18th and 19th centuries. In the late 19th century, the area was named Leopoldville in honor of King Leopold II. It was renamed Kinshasa in 1966, after Congo achieved independence.

Today, Kinshasa's government buildings, industry, commercial centers, and expensive housing lie closest to the river. The majority of the city's population lives further inland in modest housing.

Teacher-Tested Activities

Erik Branch, James C. Wright Middle School, Madison, Wisconsin

The textiles feature on the next page demonstrates how art and woven textiles convey messages. Have students look at what they are wearing and analyze what messages their clothing might communicate. What does a necktie communicate about the wearer? A skirt? A uniform? A sports logo?

CONNECT to Art

African Textiles

Explain that authentic, hand-woven kente garments are worn by the Ashanti for ceremonial occasions today. Hand-woven kente is made in strips only a few inches wide. The strips are sewn together to form the cloth. Early kente weavers made the cloth from raffia fibers.

Today, silk, cotton, and rayon are used, and some kente fabrics include metallic threads. Designs woven in the cloth range from geometric shapes and letters to likenesses of animals and plants. Factory-made fabrics, similar to kente cloth, feature designs that have been printed on the fabric instead of woven into it.

CRITICAL THINKING ANSWER

Compare and Contrast Patterns and designs are stamped onto Adinkra cloth, while they are woven or sewn into Kuba and Kente cloth.

Teach

Economics and Government

Talk About It

- What natural resources in Congo could help create a strong economy? *(copper, industrial diamonds, and petroleum)*
- Describe the composition of Congo's current government. *(a president, four vice presidents, and a 300-member assembly appointed by the president)*
- **Identify Problems and Solutions** What problems have adversely affected the economy and the lives of most Congolese? *(a dictatorial government, wars, ongoing violence, and government neglect of the country's infrastructure)* Based on what you have learned in this section, what recommendations would you make to improve Congo's government and economy? *(Possible Answers: establishing a true democracy; eliminating corruption; attracting foreign investors; building and maintaining transportation systems; developing natural resources)*

Unit Resource Book: Africa
- Skillbuilder Practice: Identifying Problems and Solutions, p. 206

CONNECT Culture & the Arts

African Textiles Woven textiles play important roles in many African cultures. A design may symbolize a culture's myths, or say something about the social status of the person wearing it.

Adinkra cloth originated with West Africa's Ashanti and Gyaman peoples. Patterns and symbols are stamped onto the cloth using dyes.

Kuba cloth comes from the Congo region. Weavers make the basic cloth from a palm fiber called raffia, then designs are sewn into it.

Kente cloth, also made by the Ashanti people, uses colored threads in the weaving process. Only royalty could wear certain kente patterns.

CRITICAL THINKING

Compare and Contrast What are some of the differences in the way these textiles are created?

Congolese art styles differ from region to region. In the southwest, the Kongo people make statues studded with stones and nails. The Kuba people craft statues of their royal leaders. Luba artists make small statues of mothers. Mangbetu sculptors of the north specialize in long, stylized heads.

SUMMARIZE Explain how the Congolese communicate when more than 200 languages are spoken in the country. Many Congolese use one of the official or national languages: French, Swahili, Kikongo, Lingala, or Tshiluba.

Economics and Government

KEY QUESTION What social effects have resulted from Congo's long dictatorship and frequent wars?

Congo's rich deposits of copper, industrial diamonds, and petroleum make it a potentially strong economy. But decades of dictatorship and wars have hurt the country. Ongoing violence has scared off investors, reduced production, driven up the debt, and killed millions.

Despite Congo's resources, most Congolese live as subsistence farmers. Their chief food crops include bananas, cassava, corn, rice, and peanuts, which Africans call "groundnuts." A few farmers also grow cash crops of cocoa, coffee, cotton, and tea. But years of war and government neglect have damaged the country's few paved roads and

DIFFERENTIATING INSTRUCTION: TIERED ACTIVITIES

OBJECTIVE Report on Congo's history, culture, or government and economy.

Basic	On Level	Challenge
Assign groups of students one of these topics: Congo's history, culture, or government and economy. Have each group prepare a brief oral report and present it to the class. Encourage groups to find or draw pictures to illustrate their reports. Assign roles for the presentation, or have students choose them.	Adapt the Basic activity so that students focus on identifying problems and solutions for their assigned topics. Have each group prepare a brief oral report and present it to the class. Encourage groups to find or draw pictures to illustrate their reports. Assign roles for the presentation, or have students choose them.	Adapt the On Level activity so that students include an interview they create and roleplay as part of their presentation. Have students assign their own roles.

made some railroads unusable. As a result, Congolese farmers continue to transport their goods on the Congo River. Fertile soil along the riverbanks remains largely unused for large-scale agriculture.

The country's chief economic activity is diamond mining. Congo also has petroleum reserves off the country's small coastline, as well as cadmium, cobalt, gold, manganese, iron ore, and copper. There's more potential wealth in Congo's forests, which cover more than half the country. Its rushing rivers teem with fish and could provide half of Africa's electric power needs.

New Constitution A woman in Kinshasa celebrates the Democratic Republic of the Congo's new constitution in 2006.

Congo's current government consists of a president, four vice presidents, and a 300-member assembly appointed by the president. In 2006 Congolese voters held the first general elections under a new constitution. The future will show whether Congo's new government will be strong enough to overcome continuing regional unrest.

UNDERSTAND EFFECTS Describe how dictatorship and civil war have hurt Congo. They have led to widespread poverty and debt, a lack of investors, and the death of millions of people.

Section 3 Assessment

ONLINE QUIZ For test practice, go to **Interactive Review** @ClassZone.com

TERMS & NAMES

1. **Explain the importance of**
   - Kongo
   - Leopold II
   - Mobutu Sese Seko
   - Zaire

USE YOUR READING NOTES

2. **Make Generalizations** Use your completed graphic organizer to answer the following question:

   How could Congo's resources help its economy?

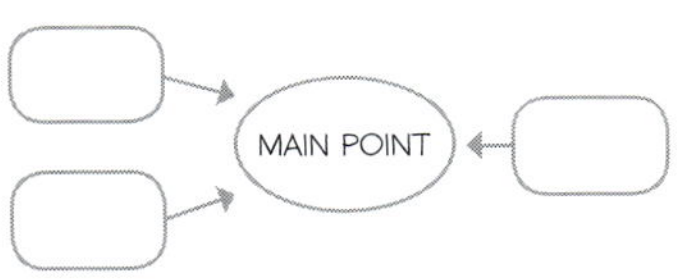

KEY IDEAS

3. Who were Kongo's earliest inhabitants?
4. Where do most Congolese live?
5. What is Congo's main product?

CRITICAL THINKING

6. **Make Inferences** Why might Congolese business people have rejected Western dress shortly after independence?
7. **Summarize** Why is the Congo River so important to the Congolese people?
8. **CONNECT to Today** In 2006, the Congolese elected a new government. What steps might the new leadership take to improve the Democratic Republic of the Congo's economy?
9. **GEOGRAPHY** **Map Congo's Resources** Sketch a map of Congo (Kinshasa). Label your map with the country's major regional resources, using a symbol for each resource that you explain in a key.

CONNECT to the Essential Question

How do boundaries set during colonial times affect stability in these regions today?

Ask students what they have learned so far that can help them answer this question. Students might mention:

- The Congo's history has been marked by foreign control, then an independence marred by internal strife.
- The Congo's population is composed of diverse ethnic groups, who mostly share Bantu origins.
- The Congo, hurt by corruption and governmental neglect, is working to improve its economy.

4 Assess & Reteach

Assess Have students complete the Section Assessment.

Unit Resource Book: Africa
- Section Quiz, p. 221

Interactive Review @ClassZone.com
- Section Quiz

Power Presentations

Test Generator

Reteach Have pairs of students create five "True or False" statements about the Congo. Then have each pair exchange their statements with another pair and evaluate them. For each false statement, have students correct it, using their textbooks.

Unit Resource Book: Africa
- Reteaching Activity, p. 226

SECTION 3 ASSESSMENT ANSWERS

Terms & Names

1. Kongo: Bantu-speaking kingdom in the Congo River region in the 1300s; Leopold II: Belgian king who formed the Congo Free State; Mobutu Sese Seko: dictator of the Democratic Republic of the Congo, 1965–1997; Zaire: the Democratic Republic of the Congo's name from 1971–1997

Use Your Reading Notes

2. develop hydroelectricity and large-scale agriculture along the rivers; develop forests for logging

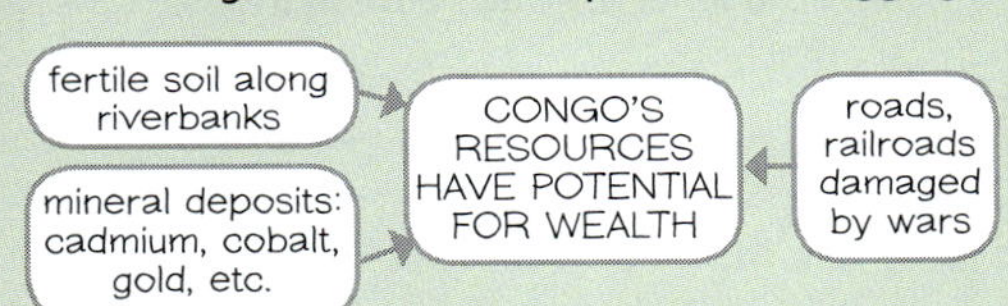

Key Ideas

3. Pygmy hunter-gatherers
4. in villages
5. diamonds

Critical Thinking

6. The change in dress may have been related to Mobutu's attempts to instill national pride and reduce ethnic divisions.
7. The river has long served as a means of transportation and a source of fish and fertile land. In the future, it could provide hydroelectric power.
8. Answers may vary; one step would be to repair roads and rail lines.
9. Use the rubric to score maps.

Map Rubric

	Map Presentation	Map Content
4	neat	5 or more resources labled
3	adequate	4 resources labeled
2	fair	2 or 3 resources labeled
1	sloppy	1 or no resources labeled

1 Plan & Prepare

Objectives

- Explain how Boer and British settlers affected the lives of black South Africans
- Describe contemporary South African society
- Evaluate the effectiveness of South Africa's government and economy today

Read for the Essential Question

Help students read for a purpose by reminding them of the Essential Question: "How do boundaries set during colonial times affect stability in these regions today?" Point out that they can use text features such as Key Questions, special features, images, and maps to form answers to the question.

Vocabulary

Best Practices Toolkit

Use the ***Best Practices Toolkit*** to model strategies for vocabulary notetaking. Vary strategies throughout the year. Choose from: *Knowledge Rating, Predicting ABCs, Definition Mapping, Word Sort, Frayer Model (Word Squares), Magnet Words, Word Wheel,* and *Student VOC.*

Vocabulary Strategies, TT9–TT16

Reading Strategy

Best Practices Toolkit

Display the Time Line transparency.

- Remind students that sequencing events involves placing them in the order in which they happen.
- Model filling in the first event: "Dutch East India Company workers arrive."
- Have students suggest entries.
- Display the completed diagram for students to review.

Time Line, TT33

SECTION 4 Reading for Understanding

Key Ideas

BEFORE, YOU LEARNED

When power in a nation changes hands, there are often violent conflicts.

NOW YOU WILL LEARN

South Africa made a successful change from a minority-rule nation to a majority-rule government but still faces many challenges.

Vocabulary

TERMS & NAMES

Boer a Dutch colonist or descendant in South Africa

Cape Colony South African land settled by Dutch colonists called Boers in the 1600s and 1700s

Afrikaner a South African of European ancestry who speaks a language called Afrikaans

apartheid (uh•PAHRT•hyt) the official policy of racial segregation practiced in South Africa

African National Congress (ANC) South African political party opposing apartheid

sanction a penalty or pressure applied to a country for not obeying an order or law

Nelson Mandela South Africa's first African president

BACKGROUND VOCABULARY

Commonwealth of Nations an association of British colonies

segregation the policy of separating people of different races

Reading Strategy

Re-create the time line shown at right. As you read and respond to the **KEY QUESTIONS**, use the time line to order events related to South Africa.

Skillbuilder Handbook, page R6

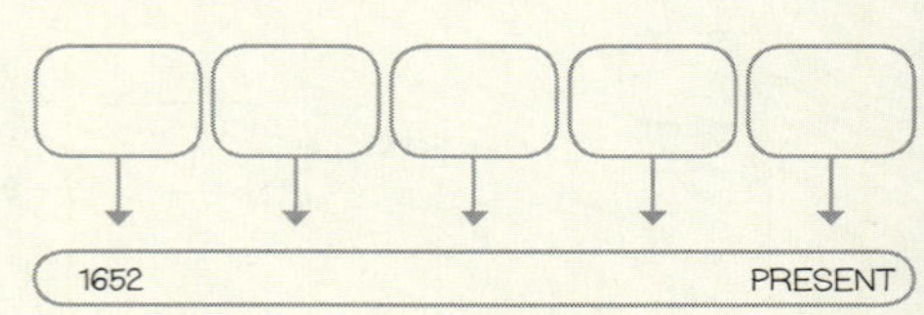

GRAPHIC ORGANIZERS Go to Interactive Review @ClassZone.com

PRETEACHING VOCABULARY

English Learners

Pronounce and Preview

Pronounce each term for students. Review the meanings of words used in definitions, such as *descendant, ancestry,* and *segregation.*

- To modify vocabulary learning, have students complete worksheets as they read, instead of afterward.

Unit Resource Book: Africa
- Vocabulary Practice, p. 205

Inclusion

Terms in Context

Read each term and its definition aloud. Ask students to look through the section and find illustrations or passages that contain the terms. Discuss the illustrations and passages, so that students better understand the terms in context. Then have students use each term in a sentence.

Northern Ethiopia

African elephants, Tanzania

SECTION 4 FOCUS ON

South Africa

Connecting to Your World

In the presidential election of 2004, just over half the eligible adults in the United States cast ballots. In contrast, in the South African elections of 1994, almost 90 percent of eligible voters came to the polls. Millions of people stood in long lines, patiently awaiting their turn to vote. Some had to wait for a day or more, but they didn't mind. The 1994 election in South Africa was the first time most South Africans had ever been allowed to vote.

Voting Lines South Africans wait to vote in the 1994 election.

Competition for the Land

KEY QUESTION What forms of external and internal protest helped bring an end to South Africa's minority rule?

The San, hunter-gatherers who lived in small bands, were the earliest modern human beings to populate the area. Around A.D. 100, a related group called the Khoikhoi (KOY•koy) migrated from the north. Today these groups are called the Khoisan. The great Bantu migration later brought Bantu speakers from the north. These peoples included the Zulu, the Xhosa (KOH•sah), the Sotho, and the Tswana. They herded cattle, raised grain, made iron tools and weapons, and traded with other peoples.

Johannesburg The Hillbrow Tower dominates the skyline of South Africa's largest metropolitan area.

2 Focus & Motivate

3-Minute Warm-Up

Write on the board or display the transparency:

- View the images and captions that accompany them in this section. What do they suggest about South Africa today? *(Possible Answer: South Africa has modern cities, politically-involved citizens, dynamic leaders, popular cultural traditions, and a thriving economy.)*

Unit Transparency Book
- 3-Minute Warm-Ups, TT26

Connecting to Your World

More About . . .

South African Elections

In the 1994 elections, candidates representing the African National Congress (ANC) won more than 60 percent of the seats to the National Assembly. The ANC repeated its success in the 1999 and 2004 elections, winning a majority of the seats both times.

3 Teach

Competition for the Land

Think, Pair, Share

- What effect did the arrival of Boer settlers have on the native South African population? *(Many died of diseases; others became slaves or servants of the Boers.)*
- **Sequence Events** When did South Africa adopt a new constitution? *(1996)* How many years was this after Europeans first landed in South Africa? *(344 years)*

SECTION 4 PROGRAM RESOURCES

Unit Resource Book: Africa

ON LEVEL

Unit Resource Book
- Reading Study Guide, p. 171
- Outline Map Activities, p. 209
- Section Quiz, p. 222

STRUGGLING READERS

Unit Resource Book
- Reading Study Guide with Additional Support, p. 181
- Skillbuilder Practice: Identifying Problems and Solutions, p. 206
- Section Quiz, p. 222
- Reteaching Activity, p. 227

eEdition with Audio DVD-ROM

ENGLISH LEARNERS

Pupil Edition in Spanish

eEdition with Audio DVD-ROM

Unit Resource Book
- RSG (Spanish), p. 191
- RSG with Additional Support (Spanish), p. 201

Multi-Language Glossary

Test Generator
- Section Quiz in Spanish

INCLUSION

Unit Resource Book
- RSG with Additional Support, p. 181
- Section Quiz, p. 222
- Reteaching Activity, p. 227

GIFTED & TALENTED

Unit Resource Book
- History Makers, p. 211
- Section Quiz, p. 222

PRE-AP

Unit Resource Book
- Connect to Culture, p. 216
- Primary Source, p. 217
- Section Quiz, p. 222

TECHNOLOGY

Unit Transparency Book
- 3-Minute Warm-Ups, TT26
- Essential Question, TT27
- Critical Thinking, TT31

Daily Test Practice Transparencies
- Section 4, TT70

Power Presentations

ClassZone.com

World Cultures and Geography Video on DVD

RESEARCH & WRITING CENTER

HISTORY MAKERS

Nelson Mandela

When Mandela was arrested, he was sentenced to life in prison. His long imprisonment attracted worldwide attention, and he became a symbol of the black African struggle to achieve racial equality. In 1990, in response to mounting international criticism, economic sanctions, and the threat of civil war, South Africa's president, F. W. de Klerk, released Mandela.

In the years that followed, leading up to the 1994 election, Mandela and the ANC met with de Klerk and members of his National Party to discuss political reforms. When Mandela was awarded the presidency in 1994, he formed a government dedicated to national unity that included members of the ANC, de Klerk's National Party, and other parties.

Unit Resource Book: Africa

- History Makers: Nelson Mandela, p. 211
- Primary and Secondary Sources: from *Long Walk to Freedom* by Nelson Mandela, pp. 217–218

More About . . .

The African National Congress

A group of black South Africans dedicated to seeking political equality founded the ANC in 1912 under the name South African Native National Congress (shortened in 1923 to African National Congress).

Inspired by the efforts of Mohandas Gandhi to improve the rights of Indians in South Africa, the ANC initially adopted his philosophy of nonviolent resistance, engaging in demonstrations, boycotts, and other forms of peaceful protest. Later, as tensions between the government and the ANC grew, the confrontations became increasingly violent.

Colonizers Clash European settlers working for the Dutch East India Company landed in South Africa in 1652. In 1657 the company let some workers start their own farms. These farmers became known as **Boers**. By the end of the 1700s, the Boer population had expanded into a large section of land called the **Cape Colony**.

This European intrusion shattered the Khoikhoi and San. Many died of diseases, such as smallpox. The survivors fought the colonists or became their slaves and servants. Gradually the Boers' Dutch language changed, blending with San, Khoikhoi, and other languages into a new language called Afrikaans. By 1770, white Afrikaans speakers, called **Afrikaners**, began to spread beyond the Cape Colony.

British settlers began arriving in 1820, after Britain won control of the Cape Colony. The British clashed with the Boers, defeating them in 1902. Five years earlier, Britain had defeated the last independent Zulu kingdom in South Africa. The Union of South Africa, which was established in 1910, joined the British **Commonwealth of Nations** in 1931.

HISTORY MAKERS

Nelson Mandela (born 1918)

Nelson Mandela, head of the African National Congress, spent almost 30 years in prison for resisting apartheid. In 1993, three years after his release from prison, Mandela accepted the Nobel Peace Prize on behalf of all South Africans who had sacrificed so much for freedom. In 1994, Mandela won South Africa's first free democratic election, and he served as the country's president until 1998.

ONLINE BIOGRAPHY For more on the life of Nelson Mandela, go to the Research & Writing Center @ClassZone.com

Apartheid and Its Overthrow Afrikaners and English-speaking whites controlled South African politics, excluding the black majority from power. Some Afrikaners wanted to make Afrikaans a national language and to enforce stricter **segregation**, or separation, between blacks and whites. After World War II, the Afrikaner National Party put into place a harsh program of **apartheid** (uh•PART•hyt), which made segregation a law. The new laws classified South Africans by race and separated the groups into their own schools and neighborhoods.

In South Africa, many groups opposed apartheid, including the **African National Congress (ANC)**, a black political party. Internal protests took the form of boycotts, strikes, marches, and occasional violence. The United States and other countries applied **sanctions**, or economic pressure, on South Africa, refusing to trade with the country. South Africa left the Commonwealth of Nations. It was banned from competing in the Olympics for nearly 30 years. In 1990 and 1991, the South African government repealed all its apartheid laws. The country's first free multiracial election in 1994 awarded the presidency to ANC leader **Nelson Mandela**. In 1996, South Africa adopted a new constitution guaranteeing basic freedoms to all its citizens.

A COMPARE Compare the internal and external protests that helped end apartheid. External protest took the form of sanctions, while internal protests included boycotts, strikes, and protest marches.

DIFFERENTIATING INSTRUCTION

Struggling Readers

Clarify Cultural Terminology

The terms used to identify the groups that controlled South Africa may be confusing to some students. Explain that the terms *Boers, Afrikaners,* and *white Afrikaans speakers* are synonymous. Point out that most of the Boer settlers were Dutch and became known as Afrikaners, after developing a new language (Afrikaans) in their new home.

English Learners

Key Academic Vocabulary

Draw attention to the sentence "Internal protests took the form of . . . ." Point out that the word *internal* refers to protests by groups within South Africa. The sanctions mentioned in the next sentence represent external forms of protest—those made by groups outside the country. Have students use *internal* and *external* in sentences.

South African Life and Culture

KEY QUESTION How are South Africans adjusting to life after apartheid?

The effects of apartheid linger in today's South Africa. Whites still have higher-paying jobs and enjoy a better standard of living than other groups. The government wants to correct these inequalities, but it takes time.

Cultural Characteristics Black Africans make up about 79 percent of South Africa's population. Ten percent of the population is white, about 9 percent are mixed race, and 2 percent are Asian. Four out of five South Africans are Christians. Many black South Africans belong to African independent churches that mix Christian and African beliefs. Before apartheid ended, English and Afrikaans were the country's only official languages. Today nine African languages also have official status, including Zulu, Xhosa, Sepedi, and Setswana. More than half of black South Africans and about 90 percent of whites live in urban areas.

Ladysmith Black Mambazo This popular ten-member group takes its name from a Zulu village called Ladysmith. "Mambazo" is a Zulu word for "axe."

Varied Ways of Life Many white South Africans live in roomy single-family suburban homes. English and Afrikaans speakers have recently begun to mix, moving into each other's neighborhoods. While there is a growing black urban middle class, many blacks still live in poor, segregated areas far from city centers, or in shelters. In rural areas, some blacks work as subsistence farmers on small farms, or on larger white-owned farms that raise cash crops.

South African artists excel in dance, painting, music, and writing. This country's wealth of languages has produced a rich literature in English, Zulu, Sotho, and Afrikaans. The vocal group Ladysmith Black Mambazo introduced the world to South African harmony singing, performed in an African humming style. Another popular South African group, the African Jazz Pioneers, performed for President Mandela.

IDENTIFY CAUSES Describe how life has changed in South Africa since the end of apartheid. The government has increased the number of official languages. English and Afrikaans speakers have begun to mix, and a black middle class is growing.

Teach

South African Life and Culture

Talk About It

- What groups make up South Africa's population, and which group is the largest? *(Black Africans represent the majority of the population. Whites, Asians, and people of mixed races make up the rest.)*
- In what ways are the effects of apartheid still apparent in today's South Africa? *(Whites still have the best jobs and enjoy a better standard of living.)*
- **Sequence Events** At what point in South Africa's history were English and Afrikaans its only official languages? *(during apartheid)* When were nine African languages granted official status? *(after apartheid ended)*

Unit Resource Book: Africa
- Outline Map Activities: Eastern and Southern Africa, pp. 209–210

More About . . .

Ladysmith Black Mambazo

The music performed by the Durban-based group Ladysmith Black Mambazo is an example of a popular form of *a capella* singing called *isicathamiya*, developed by Zulu migrant workers.

In the cities of Durban and Johannesburg, *isicathamiya* choirs participate in all-night competitions. Performances include intricately choreographed footwork and consist of three parts: an entrance (lasting until all members of the choir have reached the stage), an extended song, and a recessional.

Ladysmith Black Mambazo became internationally known in the mid-1980s, after appearing with U.S. singer Paul Simon on his *Graceland* concert tour and on the subsequent video and record album.

INTERDISCIPLINARY ACTIVITIES

CONNECT to Math

Calculate Population Groups

Provide students with the actual or estimated population of South Africa for the most recent year available. Direct them to use it and the statistics on this page to make up their own questions, or to answer questions such as:

- How many black Africans, whites, mixed race people, and Asians live in South Africa? *(Have students multiply South Africa's total population by the percentage given for each group.)*
- How many South Africans practice Christianity? *(Multiply total by 80%.)* How many practice other religions or no religion? *(Multiply total by 20%.)*
- About how many whites live in urban areas? In rural areas? *(Multiply total number of whites by 90% and 10%, respectively.)*
- About how many black South Africans would you estimate live in urban areas? In rural areas? *(Multiply by 50%.)*

Teach

Government and Economics

Reader, Recorder, Reporter

- In what ways is South Africa's government similar to the U.S. government? *(It has a constitution modeled after the U.S. Constitution, including a Bill of Rights; and three branches—executive, judicial, and a two-house legislature.)*
- Does South Africa have a strong or weak economy? *(Possible Answer: It has a strong economy, accounting for 25% of the goods and services produced by the entire continent.)*
- **Make Inferences** Do citizens choose South Africa's president? *(No; citizens elect members to the lower house of parliament, which then selects the president.)*

Unit Resource Book: Africa
- Skillbuilder Practice: Identifying Problems and Solutions, p. 206

CONNECT to Economics

For the pie graph activity, consider assigning specific countries to students. Ask students to determine which countries compare most favorably with South Africa and which ones compare least favorably, and to explain why.

Unit Resource Book: Africa
- Connect to Culture: Persuade People to Vote, p. 216

CONNECT to the Essential Question

How do boundaries set during colonial times affect stability in these regions today?

Ask students what they have learned so far that can help them answer this question. Students might mention:

- For over three centuries, the Boers and the British ruled South Africa, excluding blacks from participation.
- South Africa today has a stable, democratic government and Africa's most advanced economy.

Government and Economics

KEY QUESTION How is South Africa's national government like and unlike that of the United States?

While creating its new constitution, South Africa gave its citizens a large role in the decision-making process. Local leaders met with the public and gathered ideas on how the new government should work. South Africa's constitution, adopted in 1996, includes a Bill of Rights which guarantees rights of religious freedom, security, health care, property, education, and self-expression.

Like the United States, South Africa has three branches of government: executive, judicial, and a two-house legislative branch. However, the legislative branch operates in a British parliamentary style. The lower house of parliament, the National Assembly, chooses the president from among its members. The National Council of Provinces, or upper house, protects regional interests such as ethnic traditions. Also, South Africa has a separate capital for each branch of government. The parliament meets in Cape Town. The executive branch is based in Pretoria, while South Africa's highest court, the Constitutional Court, meets in Bloemfontein.

South Africa enjoys Africa's most advanced economy and turns out one-fourth of the goods and services produced by the entire continent. Having many different economic activities keeps the country's economy strong and protects against a fall in demand for any one product.

CONNECT to Economics

A country's economic activities can tell you about its level of development. Instead of a GDP based heavily on agriculture, developed economies have more industry and service jobs. In South Africa's developed economy, less than 3 percent of GDP comes from agriculture, 30 percent from industries such as manufacturing, and over 67 percent from service industries.

Activity

Make a Pie Graph

Materials
- paper
- colored pens
- protractor
- compass
- scratch paper or calculator

1. Research the economic activities of an African country. Look for data given as a percentage of the country's GDP.
2. Convert the percentages into decimals, then multiply each decimal by 360. This gives you the angles of each "slice" of your pie chart.
3. Use a compass and protractor to draw a large circle and angles within the circle.
4. Label each slice of the pie chart, then give your chart a title.

DIFFERENTIATING INSTRUCTION: TIERED ACTIVITIES

OBJECTIVE Create a visual display to show what South Africa is like today.

Basic	On Level	Challenge
Have students work individually to create a poster about South Africa's history, culture, government, or economy. Posters should include a title and captions. Invite students to present or display their posters.	Have students work individually to create an illustrated time line or chart about South Africa's history, culture, government, or economy. Time lines and charts should include a title and captions. Invite students to present or post their work.	Have students work independently to create an illustrated map about South Africa's history, government, culture, or economy. Maps should include a title, legend, labels, and captions. Have students present their maps.

South Africa's strong industries include mining, agriculture, fishing, and manufacturing. South Africa supplies one-third of the world's gold and produces coal, copper, diamonds, iron, and other minerals. South Africa's farms grow most of the country's food, and export fruits, corn, potatoes, sugar, wheat, and wool. Rich coastal waters produce fish and shellfish for export, while South African factories turn out chemicals, clothing and textiles, steel, machinery, motor vehicles, and processed foods. South Africa's trade is helped by the continent's best-developed infrastructure.

Unfortunately, black poverty and unemployment remain as effects of apartheid. Blacks are moving into better-paying jobs, but whites still occupy the most powerful roles.

Mineral Resources of South Africa

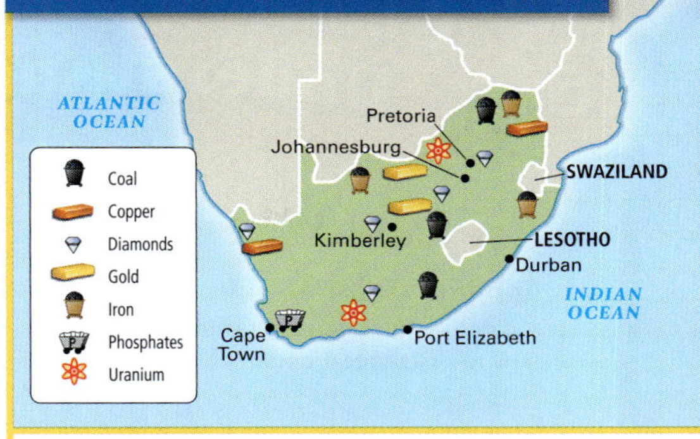

CONNECT Geography & History

Location Which resource would require the most transportation infrastructure, and why?

COMPARE AND CONTRAST Describe the similarities between the United States government and South Africa's. Both governments have three branches and a Bill of Rights.

Section 4 Assessment

ONLINE QUIZ For test practice, go to **Interactive Review @ClassZone.com**

TERMS & NAMES

1. **Explain the importance of**
   - Boer
   - Afrikaner
   - apartheid
   - Nelson Mandela

USE YOUR READING NOTES

2. **Sequence Events** Use your completed graphic organizer to answer the following question:

   For how long were there European colonies in what is now South Africa?

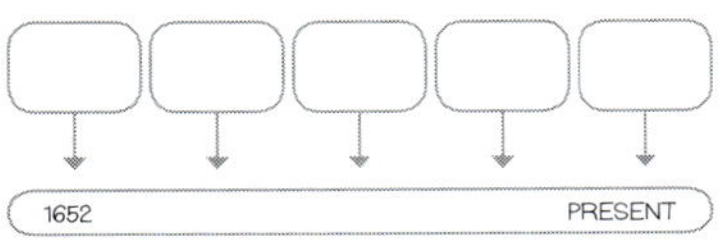

KEY IDEAS

3. Which two European countries colonized South Africa?
4. What are three of South Africa's official languages?
5. What are some of South Africa's leading industries?

CRITICAL THINKING

6. **Make Inferences** What makes Afrikaans an African language?
7. **Draw Conclusions** How do you think trade sanctions pressured South Africa to end apartheid?
8. **CONNECT to Today** South Africa's coastal waters produce tons of fish and seafood each year. But overfishing has severely reduced fish stocks. What can South Africa do to protect the fishing industry in the future?
9. **LANGUAGE ARTS** **Give a Speech** Research the writings and speeches of Nelson Mandela. Memorize a paragraph from a speech that you especially like and deliver it to your class. Explain the circumstances of the speech in a brief introduction.

Resources of South Africa

CONNECT Geography & History

Point out that a key economic advantage for South Africa is its many port cities.

- Where are most of South Africa's resources located? *(in the northeast)*

Based on the number of symbols for each item shown on the map, which resource would be South Africa's most plentiful? Least plentiful? *(diamonds; phosphates)*

ANSWER

Location Which resource would require the most transportation infrastructure, and why? *(gold, because none of it is located near the coast)*

4 Assess & Reteach

Assess Have students complete the Section Assessment.

Unit Resource Book: Africa
- Section Quiz, p. 222

Interactive Review @ClassZone.com
- Section Quiz

Power Presentations

Test Generator

Reteach Assign student groups one of these terms: *Boer settlers, British settlers, African National Congress, Nelson Mandela.* Then ask each group: "What influence did [*blank*] have on South Africa?" Allow groups time to prepare an answer that they then present to the class. Repeat the process with *people, ways of life, government, economy,* asking: "Describe South Africa's [*blank*]."

Unit Resource Book: Africa
- Reteaching Activity, p. 227

SECTION 4 ASSESSMENT ANSWERS

Terms & Names

1. Boer: Dutch colonist or descendant in South Africa; Afrikaner: South African of European descent who speaks Afrikaans; apartheid: the former policy of racial segregation practiced in South Africa; Nelson Mandela: South Africa's first African president

Use Your Reading Notes

2. about 258 years, from 1652 to 1910

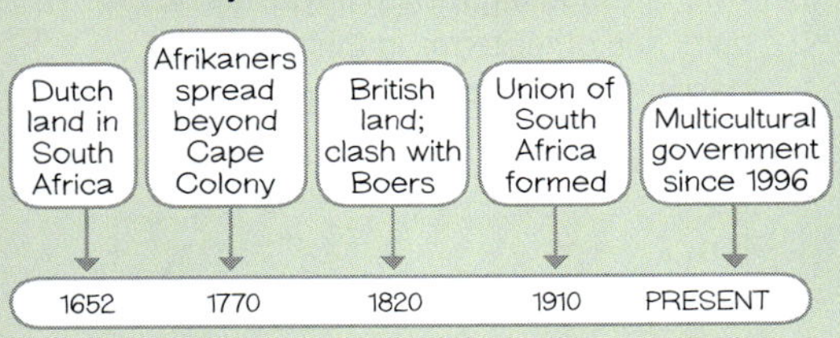

Key Ideas

3. the Netherlands and Great Britain
4. any three of the following: English, Afrikaans, Zulu, Xhosa, Sepedi, and Setswana
5. mining, agriculture, fishing, and manufacturing

Critical Thinking

6. While Afrikaans is based on Dutch, it contains African words and developed in Africa.
7. Answers will vary, but may include: Countries under trade sanctions suffer when they cannot trade their own goods or acquire the goods they do not produce themselves.
8. The government can regulate how much and what kind of fish can be caught, making sure that the fishing industry observes limits and doesn't destroy the fish on which it depends.
9. Use the rubric to score speeches.

Speech Rubric

	Speech Delivery	Circumstances of Speech
4	fully memorized	well explained
3	relies a little on notes	adequately explained
2	relies mostly on notes	fairly explained
1	relies wholly on notes	poorly explained or not understood

1 Plan & Prepare

Objectives

- Describe Zimbabwe's history, culture, government, and economy
- Describe Botswana's history, culture, government, and economy
- Describe Angola's history, culture, government, and economy

Read for the Essential Question

Help students read for a purpose by reminding them of the Essential Question: "How do boundaries set during colonial times affect stability in these regions today?" Point out that they can use text features such as Key Questions, special features, images, and maps to form answers to the question.

Vocabulary

Best Practices Toolkit

Use the ***Best Practices Toolkit*** to model strategies for vocabulary notetaking. Vary strategies throughout the year. Choose from: *Knowledge Rating, Predicting ABCs, Definition Mapping, Word Sort, Frayer Model (Word Squares), Magnet Words, Word Wheel,* and *Student VOC.*

Vocabulary Strategies, TT9–TT16

Reading Strategy

Best Practices Toolkit

Display the Cluster Diagram transparency.

- Remind students that comparing and contrasting involves recognizing similarities and differences.
- Model filling in the center oval with one similarity among Angola, Botswana, and Zimbabwe. Then fill in one difference in the boxes for each country.
- Have students suggest entries.
- Display the completed diagram for students to review.

Cluster Diagram, TT28

SECTION 5 Reading for Understanding

Key Ideas

BEFORE, YOU LEARNED

A common problem after colonialism is how to distribute a nation's wealth.

NOW YOU WILL LEARN

Zimbabwe, Botswana, and Angola have all taken different paths since the end of colonialism.

Vocabulary

TERMS & NAMES

Great Zimbabwe an empire built by the Shona people

Rhodesia the colonial name for the countries that are today Zambia and Zimbabwe

land reform dividing up large plots of land to distribute land more evenly

Kalahari a desert in southern Africa

Bechuanaland colonial name for Botswana

Marxist a person who supports the philosophy behind communism, including government ownership of the land and the means of production

BACKGROUND VOCABULARY

guerrilla a member of an irregular army that operates in small bands

Visual Vocabulary Great Zimbabwe

Reading Strategy

Re-create the web diagram shown at right. As you read and respond to the **KEY QUESTIONS**, use the diagram to compare and contrast Zimbabwe, Botswana, and Angola's exports

Skillbuilder Handbook, page R9

COMPARE AND CONTRAST

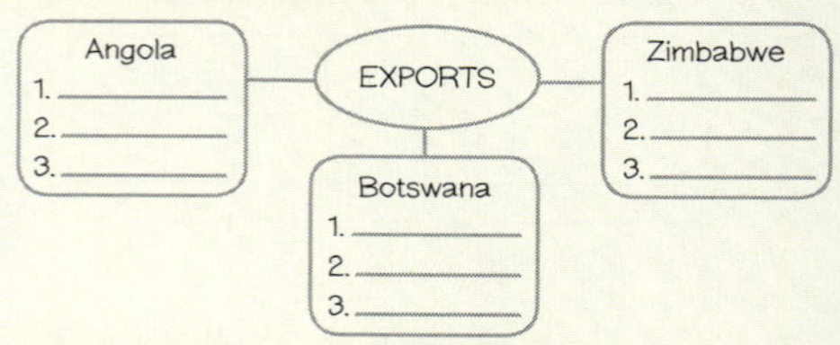

GRAPHIC ORGANIZERS Go to **Interactive Review** @ClassZone.com

PRETEACHING VOCABULARY

English Learners

Pronounce and Preview

Pronounce each term for students. Review the meanings of words used in definitions, such as *distribute, philosophy,* and *means of production.*

- To modify vocabulary learning, have students complete worksheets as they read, instead of afterward.

Unit Resource Book: Africa
- Vocabulary Practice, p. 205

Inclusion

Classify Vocabulary

Distribute a three-column chart to help students classify vocabulary according to whether the terms refer to people, places, or events/ideas. Read each term and its definition aloud, and help students decide where it should go. Then have student pairs find other terms in this section to classify and add to their charts.

Northern Ethiopia

African elephants, Tanzania

Zimbabwe, Botswana, and Angola

Connecting to Your World

Do you know someone who gets a big allowance but always runs short of money? Sometimes people who have less money realize that they need to plan more carefully. Indeed, careful planning can help someone with less money end up with more. You'll learn from studying about Zimbabwe, Botswana, and Angola—three countries in Southern Africa—that having abundant resources doesn't always lead to wealth.

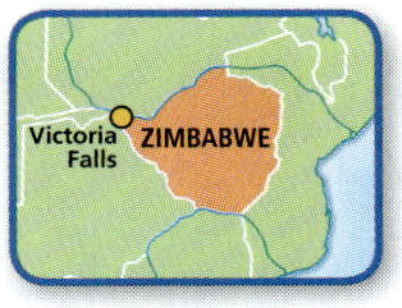

Zimbabwe

KEY QUESTION How has Robert Mugabe stayed in power?

Zimbabwe, a nation to the north of South Africa, is situated on a high plateau. Although it lies in the tropics, Zimbabwe's elevation gives it a temperate climate. Almost all of Zimbabwe's people are black Africans, and most belong to the Shona people. Since the late 1800s, Zimbabwe's political history has been troubled and violent.

Victoria Falls Africa's widest waterfall is twice as deep and wide as Niagara Falls on the U.S.–Canadian border.

2 Focus & Motivate

3-Minute Warm-Up

Write on the board or display the transparency:

- Preview the Reading for Understanding terms for this section. Predict what you will learn about Zimbabwe, Botswana, and Angola. *(Possible Answers: Some or all were first settled by black Africans; they were later colonized before gaining independence.)*

Unit Transparency Book: Africa
- 3-Minute Warm-Ups, TT26

Connecting to Your World

More About . . .

Victoria Falls

Victoria Falls—regarded as one of Earth's most spectacular natural sights—was named by British explorer David Livingstone to honor Queen Victoria.

3 Teach

Zimbabwe

Reader, Recorder, Reporter

- What great walled city was built in Zimbabwe around A.D. 1000? *(Great Zimbabwe)*
- Who colonized Zimbabwe and what did they name it? *(The British colonized it and named it Rhodesia.)*
- **Compare and Contrast** Compare Zimbabwe's agriculture before and after its recent land reforms. *(Before: large, productive farms; After: chaotic, less productive, many farmers out of work)*

SECTION 5 PROGRAM RESOURCES

Unit Resource Book: Africa

ON LEVEL

Unit Resource Book
- Reading Study Guide, p. 173
- Section Quiz, p. 223

STRUGGLING READERS

Unit Resource Book
- Reading Study Guide with Additional Support, p. 183
- Skillbuilder Practice: Identifying Problems and Solutions, p. 206
- Section Quiz, p. 223
- Reteaching Activity, p. 228

eEdition with Audio DVD-ROM

ENGLISH LEARNERS

Pupil Edition in Spanish

eEdition with Audio DVD-ROM

Unit Resource Book
- RSG (Spanish), p. 193
- RSG with Additional Support (Spanish), p. 203

Multi-Language Glossary

Test Generator
- Section Quiz in Spanish

INCLUSION

Unit Resource Book
- RSG with Additional Support, p. 183
- Section Quiz, p. 223
- Reteaching Activity, p. 228

GIFTED & TALENTED

Unit Resource Book
- Interdisciplinary Projects, pp. 212–213
- Section Quiz, p. 223

PRE-AP

Unit Resource Book
- Economics in the World, p. 215
- Section Quiz, p. 223

TECHNOLOGY

Unit Transparency Book
- 3-Minute Warm-Ups, TT26
- Essential Question, TT27
- Critical Thinking, TT32
- Cultures, TT36

Daily Test Practice Transparencies
- Section 5, TT71

Power Presentations

ClassZone.com

World Cultures and Geography Video on DVD

More About . . .

Guerrilla Warfare

Guerrilla warfare is often used in conflicts where one side is significantly inferior to the other in terms of troops, weaponry, and other resources. Guerrilla fighters tend to engage in small-scale maneuvers, such as ambushes, surprise attacks, and acts of sabotage. They use land features, such as thick forests, rivers, mountainous regions, and rugged terrain, to hide from the enemy.

In the 20th and 21st centuries, guerrilla warfare has been used in many conflicts around the world. Two notable examples are Vietnam and Afghanistan. In Rhodesia, guerrilla attacks against the white government were coordinated by two separate groups: the Zimbabwe African People's Union (ZAPU) and the Zimbabwe African National Union (ZANU).

ZAPU guerillas operated out of neighboring Zambia and ZANU guerrillas out of Mozambique. During the almost ten years of sporadic fighting in Rhodesia, thousands of people were killed—and hundreds of thousands displaced.

CONNECT to History

Great Zimbabwe

The oldest ruins at Great Zimbabwe date to the 700s, but people lived in the area for hundreds of years prior to that time. Between 1000 and 1500, the site was a center of a Karangan empire (a branch of the Shona people). This empire engaged in the mining of gold and copper for trade on Africa's east coast. Today, Great Zimbabwe is a national monument and a designated World Heritage site.

CRITICAL THINKING ANSWER

Draw Conclusions Possible Answers: The walls are not high enough; there do not appear to be any places for people to stand guard; the lack of mud or mortar binding between the stones could make the wall easy for an enemy to topple.

Precolonial History By about A.D. 1000, the Shona people had built **Great Zimbabwe**, which means "house of stone." This great city's ruins include an 800-foot-long wall and a 30-foot-high tower. The precisely cut stones nestle together without mud or mortar. Great Zimbabwe declined around 1500. The large population around it had probably used up the area's fertile land and clean water. Smaller Shona trading empires succeeded Great Zimbabwe.

British Rule The British South Africa Company, owned by Cecil Rhodes, built a railroad in what is now Zimbabwe in the 1890s. The new railroad encouraged British settlement. Settlers named the territory **Rhodesia** and defeated Africans who resisted colonization. In 1964, the northern part of Rhodesia broke off to form the independent nations of Malawi and Zambia. White Rhodesians defiantly refused to share power with blacks, despite worldwide sanctions. Through the 1970s, small groups of Africans engaged in **guerrilla** warfare, launching surprise attacks. A peace settlement in 1979 established Rhodesia's independence. The country was renamed Zimbabwe the following year. Since independence, Robert Mugabe has ruled the country as a dictator. He has kept power by jailing opponents and rigging elections. In the early 2000s, Mugabe ordered **land reform**, granting white-owned farms to black farmers. In the resulting chaos, food production dipped, and many whites fled the country.

CONNECT to History

Great Zimbabwe The largest remaining section of Great Zimbabwe is called the Great Enclosure. Its maximum diameter is about the length of a football field, and its walls are about 36 feet high.

1. Scholars believe kings and queens lived in the Great Enclosure, while farmers and workers lived outside.
2. A passage between the outer and inner walls leads to a 30-foot cone-shaped tower. Historians speculate that it may have had religious purposes.

CRITICAL THINKING

Draw Conclusions Scholars think the Great Enclosure's walls were not built for defense. Why do you think this is the case?

DIFFERENTIATING INSTRUCTION

English Learners

Vocabulary: Idioms

Students familiar with the literal meaning of the word *break* (to damage or injure) may be confused by its use in the sentence "In 1964, the northern part of Rhodesia broke off. . . ." Point out that the phrase *broke off* is used here as an idiomatic expression meaning "separated from." Ask students to rephrase the sentence. *(Northern Rhodesia separated from Rhodesia.)*

Gifted & Talented

Map Great Zimbabwe

Tell students that the walled city pictured in the Connect to History feature is located in a valley. The entire Zimbabwe ruin also includes the remains of a large fort, with many rooms and passages, set on a hill overlooking the valley. Have students use this information, along with the their textbooks, to draw a map that shows what they visualize Great Zimbabwe looked like.

Life and Culture English is the official language in Zimbabwe. Most people also speak Chishona or another tribal language called isiNdebele (ee•see•UHN•duh•BEHL•ay). Most of Zimbabwe's people are subsistence farmers. Most people follow traditional African religions or combine them with Christianity.

Government and Economics In Zimbabwe, the president is elected to a six-year term. Opponents and observers accused President Mugabe of manipulating the 2002 elections. The legislature consists of a 150-member House of Assembly and a 66-member Senate.

Most of Zimbabwe's income comes from large farms, which export beef, cotton, sugar, and tobacco. But Mugabe's land reform schemes threw many farmers out of work. In addition, the country's inflation rate has skyrocketed. Before Zimbabwe fixed its exchange rate in 2006, one U.S. dollar was worth 250,000 Zimbabwe dollars.

SUMMARIZE Describe Robert Mugabe's rule since independence.
He jailed opponents and rigged elections.

Botswana

KEY QUESTION How has Botswana been a success?

Botswana, located to the west and south of Zimbabwe and north of South Africa, is also landlocked. Its main river is the Okavango, which forms a swampy inland delta in the northeast. Much of Botswana lies in the **Kalahari** Desert, and surrounding regions are dry. Despite these disadvantages, Botswana is a success. For the past 40 years, this poor nation has maintained peace, economic growth, and a democratic government.

History Like South Africa and Zimbabwe, Botswana was first settled by the San. The Tswana, a Bantu-speaking people, migrated into this region between A.D. 1 and A.D. 1000. By the 1700s, several stone-walled villages and large towns had developed. During the 1800s, the Tswana sought British aid in their struggles with rival African groups. Britain brought the area under its protection in the late 1880s, naming it **Bechuanaland**. Britain governed Bechuanaland until the 1960s, refusing South Africa's attempts to take over. But Britain did little to develop the colony. In fact, Bechuanaland's capital was actually in Mafeking, South Africa, for nearly 70 years.

because of plant growth and changing water levels ▼

Okavango Delta Most of the swampy Okavango cannot be navigated. **Why might a swampy river be hard to navigate?**

More About . . .

Mugabe's Land Reforms

In the 1980s, after achieving independence, the Zimbabwe government began a program to settle black farmers on lands formerly owned by whites, who had begun to leave the country. Early in 2000, with Robert Mugabe's support, blacks began to forcibly take over the lands of white farmers who had chosen to remain. Within months hundreds of white-owned farms had been seized. In 2002, Mugabe's land reform demands had displaced thousands of white farmers and devastated Zimbabwe's agricultural economy.

Teach

Botswana

Think, Pair, Share

- Describe Botswana's physical features. *(large areas of swampy delta, desert, and dry surrounding regions)*
- What major health crisis do both Zimbabwe and Botswana share? *(a high rate of HIV/AIDS infection among their citizens)*
- **Compare and Contrast** In what ways does Botswana differ from Zimbabwe? *(Possible Answer: Botswana achieved independence peacefully; it has overcome geographical disadvantages to strengthen its economy; it has an effective democratic government.)*

Unit Resource Book: Africa
- Skillbuilder Practice: Identify Problems and Solutions, p. 206

INTERDISCIPLINARY ACTIVITIES

CONNECT to Health

Analyze Africa's High Rates of HIV/AIDS

Ask students to discuss the following questions:

- What reasons might explain why many countries in Africa have suffered high rates of HIV/AIDS infection among their people? *(Possible Answers: lack of public awareness regarding ways to prevent infection; inability of people who are poor to afford treatment; inadequate public funding for prevention/treatment programs)*
- Why do you think Botswana has been able to develop a model program to combat the HIV/AIDS epidemic? *(Possible Answer: It has a democratic government dedicated to serving its people, a growing economy to support funding of programs, and cooperates with world health organizations.)*

RESEARCH & WRITING CENTER

HISTORY MAKERS

Seretse Khama

Seretse Khama became chief of the Ngwato people when he was four years old, inheriting the title from his father. Khama's grandfather—Khama III, the Good—was the Ngwato chief who secured Britain's protection in the late 1800s when the territory became known as Bechuanaland.

Seretse Khama received his early education in South Africa and later attended the University of Oxford, where he studied law. During his tenure as president of Botswana, Khama championed multiracial democracy, greatly expanded educational opportunities, and promoted economic growth.

More About . . .

The Economy of Botswana

Botswana's two main industries are a study in contrasts. More than one-third of Botswana's gross domestic product (GDP) comes from mining, but mining provides employment for only a very small part of the country's working population. Agriculture—dominated by cattle raising—provides employment for a far greater number of people, but it represents only a small part of Botswana's GDP.

In recent years Botswana's economic growth has been disrupted as a result of the HIV epidemic. Beginning in the early 1990s, the spread of the disease placed a severe financial burden on government-funded health services, slowing the economic progress that the country had enjoyed from the late 1960s onward.

The country won independence peacefully in 1966, taking the name Botswana, a modern spelling of Bechuana. Seretse Khama, a former chief of the Ngwato people, was elected president. The new nation had a difficult road ahead of it. Bechuanaland had been poor; most people lived by raising cattle or by migrating to South Africa for work. Botswana relied on Britain for financial support for its first five years. The discovery of diamonds soon afterward gave Botswana the funds to improve its infrastructure and build other industries.

HISTORY MAKERS

Seretse Khama 1921–1980

While studying overseas, Seretse Khama married Ruth Williams, a white British woman, in 1948. The idea of an interracial couple ruling in Bechuanaland upset South Africa, which had recently set up its apartheid laws. Britain exiled the Khamas from the colony in 1950.

The Khamas returned to Bechuanaland in 1956, after Seretse gave up his title of chief. Seretse Khama became active in national politics in 1961 and was elected Botswana's first president five years later.

ONLINE BIOGRAPHY For more on the life of Seretse Khama, go to the **Research & Writing Center @ClassZone.com**

Life and Culture English is the official language in Botswana. Most people also speak Setswana. Since the 1980s, novels have been published in this African language. More than half of Botswana's population lives in cities. People in Botswana adopted Christianity and Western dress in the late 1800s.

Zimbabwe, Botswana, and South Africa share a common problem with many other African countries: a serious disease called HIV or AIDS. One in four adults in Zimbabwe and almost two in five adults in Botswana are affected. HIV/AIDS has been a medical and social tragedy because it strikes young adults, strains medical budgets in poor countries, and orphans many children. Botswana's model program to deal with this disease includes providing HIV/AIDS therapy to all its sick citizens.

Government and Economics Seretse Khama held office until his death in 1980, when he was replaced by his vice-president. Although Botswana has a multiparty system, Khama's Botswana Democratic Party (BDP) has won every national election since independence.

Botswana's republic has a popularly elected legislative body called the National Assembly, which selects the president. A House of Chiefs, made up of leaders from eight major tribes, advises the government on ethnic matters. Between the late 1960s and the 1990s, when the AIDS epidemic hit, Botswana enjoyed rapid economic growth. Mining and raising livestock are this country's main industries. In addition to beef, Botswana exports diamonds, copper, nickel, hides and skins, and textiles. It must import food.

DRAW CONCLUSIONS Describe how Botswana has succeeded since independence. The government has remained stable, and the country's economy has grown steadily.

634 Chapter 20

DIFFERENTIATING INSTRUCTION

Struggling Readers

Make Inferences About Botswana's Culture

Ask this question: Why do you think people in Botswana adopted Christianity and western dress in the late 1800s? *(It seems likely that they were influenced by their association with the British, which began around this time.)* If students have difficulty answering the question, provide clues to help them, such as "What important event took place in Botswana in the late 1800s?"

Pre-AP

Compare Forms of Government

Have students write a brief comparison essay on Botswana's legislative system and South Africa's. *(Possible Answer: Both countries have a National Assembly, elected by the people, which elects the president. Both countries have a government body whose purpose is to protect regional and ethnic groups. The two bodies differ: South Africa's National Council of Provinces is part of its two-house legislature; Botswana's House of Chiefs is a separate advisory board).*

Angola

KEY QUESTION Why has Angola had only one election since 1992?

Angola offers a stark contrast to Botswana. Angola has greater natural wealth, and it enjoyed prosperity in colonial times. But a quarter-century of conflict has destroyed much of its economy.

History Portugal colonized Angola in the 1500s, sending enslaved Africans to work on its colony in Brazil. Angolans began to demand independence in the 1950s. After three rebel groups finally won Angola's independence in 1975, civil war erupted among them. One group received aid from Cuba and the Soviet Union. These communist countries hoped that Angola would follow **Marxist** ideas, such as government ownership of the land and central economic planning. The winning rebel group, the MPLA, formed a Marxist government in Angola, but the other groups joined forces to continue fighting. Angola's government gradually cut its Marxist ties and allowed multiparty elections in 1992. Still, fighting between the MPLA and the main rebel group, UNITA, continued despite the signing of several peace agreements. The death of UNITA's leader in 2002 finally ended 27 years of civil war.

COMPARING African Governments

ZIMBABWE
- One president since independence; accused of rigging elections and jailing opposition to keep power
- Two-house legislature
- Multiparty system
- 2005 election allowed ruling party to change constitution

BOTSWANA
- Open, multiparty system, but one party dominates
- People elect legislature, which selects president
- House of Chiefs advises on ethnic matters
- Constitution limits president to 10 years in office

ANGOLA
- Government emerging from 27-year civil war
- One president since 1979
- Has not had national elections since 1992
- One-house legislature
- May write new constitution after elections scheduled for 2006

CRITICAL THINKING

Make Inferences Each country has had one party or person dominate its government since independence. How has Botswana remained more stable than Zimbabwe or Angola?

Teach

Angola

Talk About It

- Who colonized Angola in the 1500s? *(Portugal)* What has troubled political life in Angola since the country achieved independence? *(internal strife among rival political factions)*
- Describe Angola's culture. *(Many native languages are spoken; Portuguese is the official language; most Angolans are Christian; warfare and censorship have taken a toll on the country's heritage.)*
- **Compare and Contrast** In what ways are Angola and Botswana different? *(Angola was colonized against its will, and its people were exploited. Botswana voluntarily asked for and received the help and protection of Britain. Botswana has made the best of its limited resources to improve its economy. Angola has more natural wealth, but internal conflict has hurt its economy.)*

Unit Resource Book: Africa
- Economics in the World: Diamond Producers, p. 215

COMPARING

African Governments

Point out that Angola went without elections from 1992 to 2006.

- Which country limits its president to 10 years in office? *(Botswana)*
- Which country has had one president since independence? *(Zimbabwe)*

CRITICAL THINKING ANSWER

Make Inferences Unlike Angola, Botswana has been able to hold regular elections. Unlike Zimbabwe, Botswana's elections have been fair.

DIFFERENTIATING INSTRUCTION: TIERED ACTIVITIES

OBJECTIVE Illustrate the cultures of Zimbabwe, Botswana, and Angola.

Basic

Have students work individually to make a collage of various cultural aspects of one country of their choice. Ask students to include a title and labels or captions. Invite students to present their collages.

On Level

Adapt the Basic activity so that students' collages include a written summary of the culture of their chosen country. Invite students to present their collages and summaries.

Challenge

Adapt the On Level activity so that students' collages show one cultural aspect of their choice about all three countries, such as music or literature. Invite students to present their collages and summaries.

CONNECT to the Essential Question

How do boundaries set during colonial times affect stability in these regions today?

Ask students what they have learned so far that can help them answer this question. Students might mention:

- Zimbabwe, since achieving independence, has been ruled by a corrupt dictator.
- Botswana coexisted peacefully under British rule; it currently enjoys a democratic form of government.
- Angola's independence has been marred by violent internal strife.

4 Assess & Reteach

Assess Have students complete the Section Assessment.

Unit Resource Book: Africa
- Section Quiz, p. 223

Interactive Review @ClassZone.com
- Section Quiz

Power Presentations

Test Generator

Reteach Have students use their graphic organizers to discuss with a partner the similarities and differences among Zimbabwe, Botswana, and Angola.

Unit Resource Book: Africa
- Reteaching Activity, p. 228

Angolan Art Called *O Pensador*, or "The Thinker," this image has become a national symbol in Angola.

Angolan Culture Forty native languages are spoken in Angola, with Portuguese serving as an official language. After the government dropped its anti-religious Marxist doctrines, Christianity gained popularity. Now about nine out of ten Angolans are Christians. Like other Bantu-speaking Africans, Angolans share a rich culture in music and art. A Chokwe sculpture of a person apparently deep in thought has been widely reproduced. But decades of warfare have ruined many fine collections of traditional art, and MPLA censorship during the civil war years kept a literary tradition from growing.

Government and Economics If Angola's next national election for president and National Assembly proceeds as planned in 2006, it will be the country's first election since 1992. Angola's lengthy war destroyed the diversified, prosperous economy it enjoyed during colonial times. Currently, one in two Angolans practice subsistence farming, and the country imports half its food. Oil production currently accounts for 90 percent of its exports. Only Nigeria produces more oil south of the Sahara. Angola's other rich natural resources include gold, diamonds, forests, and fish. But benefiting from these resources will require a long period of peace and good government.

SUMMARIZE Describe what has caused Angola's instability since independence. Angola was in a civil war for 27 years.

ONLINE QUIZ For test practice, go to **Interactive Review @ClassZone.com**

Section 5 Assessment

TERMS & NAMES

1. **Explain the importance of**
   - Great Zimbabwe
   - Rhodesia
   - land reform
   - Kalahari

USE YOUR READING NOTES

2. **Compare and Contrast** Use your completed graphic organizer to answer the following question:
   What export is important to Angola, but less so to Zimbabwe and Botswana?

EXPORTS
- Angola: 1. ____ 2. ____ 3. ____
- Zimbabwe: 1. ____ 2. ____ 3. ____
- Botswana: 1. ____ 2. ____ 3. ____

KEY IDEAS

3. Who built Great Zimbabwe?
4. What was Botswana's colonial name?
5. Who fought in Angola's lengthy civil war?

CRITICAL THINKING

6. **Draw Conclusions** What were some negative effects of Mugabe's land reform?
7. **Compare and Contrast** What geographic advantage does Angola have that Botswana and Zimbabwe lack?
8. **CONNECT to Today** In Botswana, soldiers prevent poachers from killing elephants on game preserves. Today the elephant population in Botswana exceeds 120,000. What are some benefits and risks of a growing elephant population?
9. **MATH** **Make a Graph** Research Angola's oil imports over the past several decades. Graph the recent rise in production, and include predicted future production.

SECTION 5 ASSESSMENT ANSWERS

Terms & Names

1. Great Zimbabwe: an empire built from the 11th to the 16th century A.D.; Rhodesia: the colonial name for the countries that are today Zambia and Zimbabwe; land reform: dividing up plots of land to distribute land more evenly; Kalahari: desert in southern Africa

Use Your Reading Notes

2. oil

EXPORTS
- Angola: 1. oil-90% of exports 2. gold, diamonds
- Zimbabwe: 1. beef 2. sugar 3. tobacco
- Botswana: 1. diamonds 2. livestock

Key Ideas

3. Shona people living in Zimbabwe
4. Bechuanaland
5. the Marxist government and two rebel groups, who later banded together

Critical Thinking

6. It drove out white settlers, disrupted commercial farming, and threw black farmers out of work.
7. a coastline, allowing it to have fisheries
8. Benefits might be saving the animals' natural environment and the extra dollars ecotourism could pump into the economy; risks might be dangers to people and damage elephants do to trees.
9. Use the rubric to score graphs.

Graph Rubric

	Graph Content	Graph Presentation
4	well researched	neat
3	adequately researched	adequate
2	fairly researched	fair
1	poorly researched	poor

Interactive Review

Click here to complete these and other activities online @ClassZone.com

CHAPTER SUMMARY

Key Idea 1
Ethiopia remained independent throughout most of its history.

Key Idea 2
Trade was important in East Africa, particularly in what is now Kenya.

Key Idea 3
The Democratic Republic of the Congo is still struggling with setting up a true democracy.

Key Idea 4
South Africa made a successful change to a majority-rule government.

Key Idea 5
Zimbabwe, Botswana, and Angola have all taken different paths since the end of colonialism.

For Review and Study Notes, go to **Interactive Review** @ClassZone.com

NAME GAME

Use the Terms & Names list to complete each sentence on paper or online.

1. I was the legal separation of races in South Africa. **apartheid**
2. I am an extension of land on Africa's east coast, across the Gulf of Aden from the Arabian peninsula. ________
3. I am a language spoken by many East Africans. ________
4. I was independent Kenya's first president. ________
5. I was the name of the Democratic Republic of the Congo from 1971 to 1997. ________
6. I get my food by stalking animals and collecting roots, berries, and nuts. ________
7. I am a Dutch farmer who settled in South Africa. ________
8. I was South Africa's first democratically elected president. ________
9. I am a desert in southern Africa. ________
10. I am a member of an irregular army that fights in small bands. ________

Aksum
apartheid
Boer
guerrilla
Eritrea
Haile Selassie
Horn of Africa
hunter-gatherer
Jomo Kenyatta
Kalahari
Mau Mau
Nelson Mandela
Swahili
Zaire

Activities

GeoGame

Use this online map to show what you know about East, Central, and Southern Africa. Drag and drop each place name to its location on the map.

Geo GAME
Angola
Botswana
Ethiopia
Kenya
Eritrea

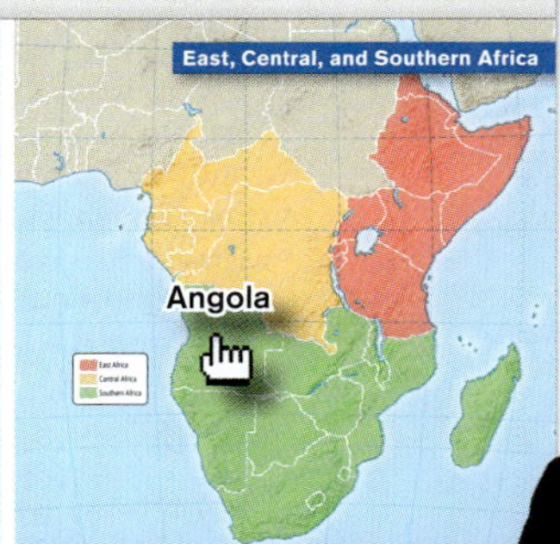

To play the complete game, go to **Interactive Review** @ClassZone.com

Crossword Puzzle

Complete an online crossword puzzle to test your knowledge of African regions.

ACROSS
1. ruler of Congo (Kinshasa) from 1965 to 1997

Chapter Summary

As a follow-up to the chapter, encourage students to pair up and go over the **Review and Study Notes** at **ClassZone.com.**

Alternatively, students pairs can review their Reading Study Guide notes, making sure that they can add important details to each section's "Key Idea."

Name Game
Teaching Tips

ON PAPER Have students work in pairs to complete this activity. As an extension, they can write questions for the terms that are not covered in the items shown here.

ONLINE Make sure that students understand how to drag and drop terms to fill in blanks. Ask them to print out their results and compare them with those of a classmate.

ANSWERS

1. apartheid
2. Horn of Africa
3. Swahili
4. Jomo Kenyatta
5. Zaire
6. hunter-gatherer
7. Boer
8. Nelson Mandela
9. Kalahari
10. guerrilla

Make It Fun
Teacher-Tested Activities

Erik Branch, James C. Wright Middle School, Madison, Wisconsin

Cultural Sanctions and the Olympics— Section 4 mentions the worldwide sanctions that pressured South Africa to end apartheid, including a ban on Olympic participation. Have students research examples of nations being banned from the Olympics, and of countries boycotting the games to protest actions by other countries. For example, Afghanistan was banned from the 2000 summer games because of its Taliban government, and 25 African nations boycotted the 1976 summer games after New Zealand's rugby team toured South Africa, violating the cultural boycott.

ONLINE AND ON-SCREEN ACTIVITIES

GeoGame

Use the interactive map on the **Power Presentations** DVD-ROM or at **ClassZone.com** to help students review locations and geographic features in east, central, and southern Africa.

- Demonstrate how to drag and drop the map labels listed onto the appropriate locations on the map.
- Advance to the scorecard to show how students have a chance to improve their scores.

Crossword Puzzle

Students can review important information about east, central, and southern Africa by completing this crossword puzzle online at **ClassZone.com.**

- As an extension of this activity, have students use the *Activity Maker* at **ClassZone.com** to make their own crossword puzzle for one section of the chapter.
- Have them trade puzzles with classmates to solve each other's puzzles.

Vocabulary ANSWERS

1. Haile Selassie was emperor of Ethiopia from 1930 to 1974.
2. Swahili is a language widely spoken in east and central Africa.
3. Leopold II was the Belgian king who made his own central African colony, called the Congo Free State.
4. Afrikaners are white, Afrikaans-speaking South Africans.
5. Apartheid was South Africa's system of legal racial segregation.
6. Great Zimbabwe was an empire built by the Shona people.
7. Bechuanaland was the former name of Botswana while it was a British colony.
8. Ezana brought Christianity to Aksum, an empire located along the Horn of Africa.
9. Long before the Kongo kingdom arose on the Congo River, Pygmies lived as hunter-gatherers in Congo's forests.

Key Ideas ANSWERS

10. They value moral behavior, hospitality, politeness, respect for elders, and the pursuit of wisdom.
11. fertile soil and plentiful rain in the highlands
12. Swahili and English
13. coffee and tea
14. first the Portuguese, later the Belgians
15. music and sculpture
16. the San people
17. They had better educational opportunities during apartheid and did not suffer racial discrimination.
18. The large population that lived there probably exhausted the area's fertile land and drinkable water.
19. 1992

Critical Thinking ANSWERS

20. Colonial languages: English, Portuguese, French, Dutch
African languages combining colonial and African tongues: Swahili, Afrikaans
Original African languages: Amharic, Kikuyu, Kikongo, Zulu, Setswana
21. Corruption has meant that money for infrastructure improvements has gone to bribes, so roads remain unpaved and bridges unbuilt. Violence has killed workers, scared off investors, reduced production, and increased debt.

CHAPTER 20 Assessment

VOCABULARY

Explain the significance of each of the following.

1. Haile Selassie
2. Swahili
3. Leopold II
4. Afrikaner
5. apartheid
6. Great Zimbabwe
7. Bechuanaland

Explain how the terms and names in each group are related.

8. Ezana, Aksum, and Horn of Africa
9. Pygmy, hunter-gatherer, and Kongo

KEY IDEAS

1 Ethiopia

10. What values do most Ethiopians share?
11. What natural advantages does Ethiopia have?

2 Kenya

12. What languages do Kenyans from different ethnic groups speak to each other?
13. What cash crops are most important to Kenya's economy?

3 The Democratic Republic of the Congo

14. Who enslaved the people of Congo?
15. What art forms do the Congolese specialize in?

4 South Africa

16. Who were South Africa's earliest human inhabitants?
17. Why do whites still hold most of the highest positions in business in South Africa?

5 Zimbabwe, Botswana, and Angola

18. What factors probably caused the decline of Great Zimbabwe?
19. When was Angola's last election before its planned election in 2006?

CRITICAL THINKING

20. **Categorize** Fill in this chart, categorizing these languages of East, Central, and Southern Africa: Portuguese, Swahili, English, Amharic, Dutch, Kikuyu, Kikongo, Zulu, French, Setswana, Afrikaans.

COLONIAL LANGUAGES	CONTEMPORARY AFRICAN LANGUAGES	ORIGINAL AFRICAN LANGUAGES

21. **Five Themes: Region** How have corruption and violence interfered with economic growth in countries like Kenya, Zimbabwe, Angola, and Congo (Kinshasa)?
22. **Evaluate** Why might it be difficult for South Africa to guarantee the rights in its Bill of Rights?
23. **Connect to Economics** Which southern African country would probably be most affected by a drop in world oil prices? Explain.

Answer the ESSENTIAL QUESTION

How do boundaries set during colonial times affect stability in these regions today?

Written Response Write a two- or three-paragraph response to the Essential Question. Be sure to consider the key ideas of each section. Use the rubric below to guide your thinking.

Response Rubric

A strong response will:

- summarize European colonialism in East, Central, and Southern Africa
- note negative effects of colonialism
- discuss transitions to independent rule that led to years of violence
- include a description of South Africa's apartheid government and its overthrow

ESSENTIAL QUESTION REVIEW

Before students begin their essays, have them review the key chapter elements that will help them answer the Essential Question. Direct them to significant features in the chapter, such as

- Ancient Aksum, *A.D. 300–700*, Section 1
- Comparing Trade Across Regions *1400–1500*, Section 2
- Animated Geography, Bantu Migrations *1000 B.C.–A.D. 1100*, Section 3
- History Makers: Nelson Mandela, Section 4
- Connect to Economics, Section 4
- Comparing African Governments, Section 5

Written Response Rubric

	Coverage of Key Ideas	Explanation	Errors in Mechanics
4	all	clear and concise	few or none
3	most	good	some
2	some	fair	many
1	one or none	poor	many major

STANDARDS-BASED ASSESSMENT

Test Practice
- Online Test Practice @ClassZone.com
- Test-Taking Strategies and Practice at the front of this book

THEMATIC MAP

Use the map to answer questions 1 and 2.

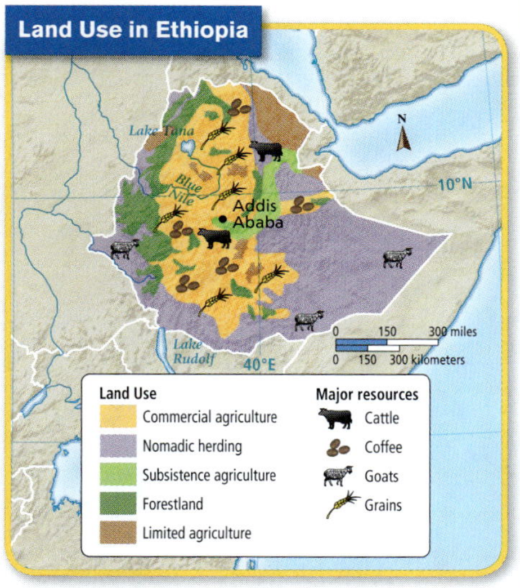

1. Where are most goats raised?

A. in commercial agriculture areas
B. in nomadic herding areas
C. in subsistence agriculture areas
D. in forest areas

2. Which major resource is most common in commercial agricultural areas?

A. cattle
B. coffee
C. goats
D. grains

PRIMARY SOURCE

Use the editorial cartoon and your knowledge of Africa to answer questions 3 and 4.

Source: Patrick Chappatte, *International Herald-Tribune*

3. What country does the flag on the polling station represent?

4. What does the cartoon imply about the man on the poster, Robert Mugabe?

GeoActivity

1. INTERDISCIPLINARY ACTIVITY–SCIENCE

The introduction of modern farming techniques in developing Asian and Latin American countries has been called the "Green Revolution" and has helped feed millions of people. Research the difficulties involved and the progress made in bringing the Green Revolution to Africa.

2. WRITING FOR SOCIAL STUDIES

Unit Writing Project Find a protest song in the African music you studied to complete the Unit Writing Project. Research the issue the song addressed and how the song dealt with the issue.

3. MENTAL MAPPING

Create an outline map of East, Central, and Southern Africa and label the following:

- Ethiopia
- Eritrea
- Kenya
- Lake Victoria
- Zimbabwe
- Congo (Kinshasa)
- Angola
- Botswana
- Kalahari Desert
- South Africa

22. It guarantees many rights, such as education and health care, that are difficult and expensive to deliver.

23. Angola; after Nigeria, it exports more oil than any other sub-Saharan African country.

Standards-Based Assessment ANSWERS

1. B
2. D
3. Zimbabwe
4. Mugabe is using false promises of food to gain votes during elections.

GeoActivity

1. INTERDISCIPLINARY ACTIVITY RUBRIC

The report should
- outline the advances and pros and cons involved in the Green Revolution
- describe obstacles to agricultural improvements in Africa, and progress made toward removing them
- avoid bias and explain tradeoffs between beneficial and possibly harmful unintended effects

2. WRITING FOR SOCIAL STUDIES RUBRIC

The review should
- summarize the issue protested against in the song
- explain the song or artist's response to the issue
- describe any reaction from the people or government criticized in the song

3. MENTAL MAPPING RUBRIC

The outline map should
- represent the approximate shape of the region
- include all geographic features listed in their correct locations

DIFFERENTIATING CHAPTER ASSESSMENT

Unit Resource Book
- **Chapter Test A,** pp. 229–232. *Use for Struggling Readers, Beginning-to-Intermediate-Level English Learners, and for Inclusion of students whose IEP suggests supported test-taking.*
- **Chapter Test B,** pp. 233–236. *Use for On-Level Learners and ESOL students with Advanced-Level English.*
- **Chapter Test C,** pp. 237–240. *Use for Gifted and Talented students or students assigned Pre-AP work.*

Test Generator
- Chapter Tests A, B, and C, in English and Spanish

McDougal Littell Assessment System

TEST **Test Generator** Use three levels of tests in English and Spanish tied to state standards. Plus, choose questions to customize tests for all populations.

SCORE Scan and Score plain paper student tests in minutes.

REPORT Create instant progress reports, sortable by student, by standard, or other criteria.

RETEACH Respond immediately to students' needs by using a wealth of reteaching resources.

N
E
W
S

UNIT 7

Southwest Asia and South Asia

Why It Matters:

Southwest Asia is the home of oil rich lands, vast deserts, and difficult political problems. South Asia is a region of ancient cultures, spectacular landforms, and rapidly growing populations.

Arabian Desert

CHAPTER 21 **Southwest Asia and South Asia**

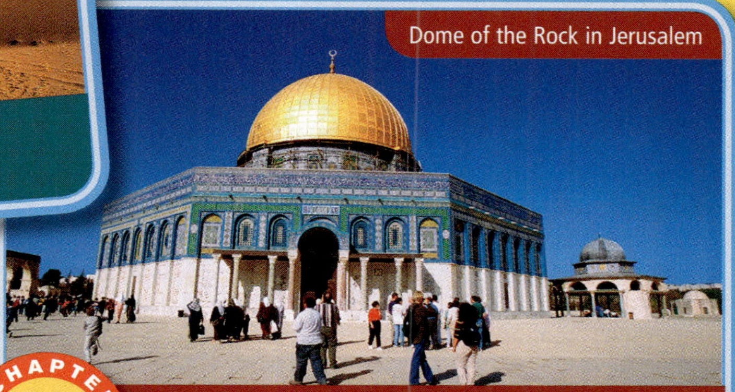

Dome of the Rock in Jerusalem

CHAPTER 22 **Southwest Asia**

Taj Mahal in Agra

CHAPTER 23 **India**

Marketplace in Bangladesh

CHAPTER 24 **South Asian Neighbors**

641

Objectives

- Compare the differences in population density between Southwest Asia and South Asia and the United States
- Describe the varied climates found in Southwest Asia and South Asia
- Identify the major economic activities in Southwest Asia and South Asia
- Summarize the geography, history, culture, government, and economics of Southwest Asia and South Asia
- Locate Southwest Asia's and South Asia's 23 countries on a map and identify their capitals

Comparing Data

Although the landmass of Southwest Asia and South Asia is only about 25% (one quarter) bigger in size than the continental United States, its population is significantly greater than that of the United States. The Mississippi River in the United States is nearly 25% longer than the longest rivers in Southwest Asia and South Asia, the Indus and the Brahmaputra. The largest desert in the world, the Sahara, is about 14 times greater than the largest desert in Southwest Asia and South Asia, the Rub al-Khali.

ANSWERS

1. Southwest Asia and South Asia
2. approximately six times larger
3. Nile is 2,300 miles longer.
4. Rub al-Khali is 225,000 square miles larger, or 10 times larger.

UNIT 7 ATLAS

Southwest Asia and South Asia

Southwest Asia and South Asia span a large area of an enormous continent. As you study the maps, notice geographic patterns and specific details about the region. Complete the GeoActivity with each map.

As you study the graphics and data on this page, compare the landmass, population, rivers, and deserts of Southwest and South Asia with those of the United States and, in a few cases, other regions. Then jot down answers to the following questions in your notebook.

Comparing Data

1. Which has the larger landmass—the continental United States or Southwest and South Asia?
2. Approximately how many times larger is the combined population of Southwest and South Asia than that of the United States?
3. How much longer is the Nile than the longest rivers in South Asia?
4. How much larger than the Mojave Desert is the Rub al-Khali?

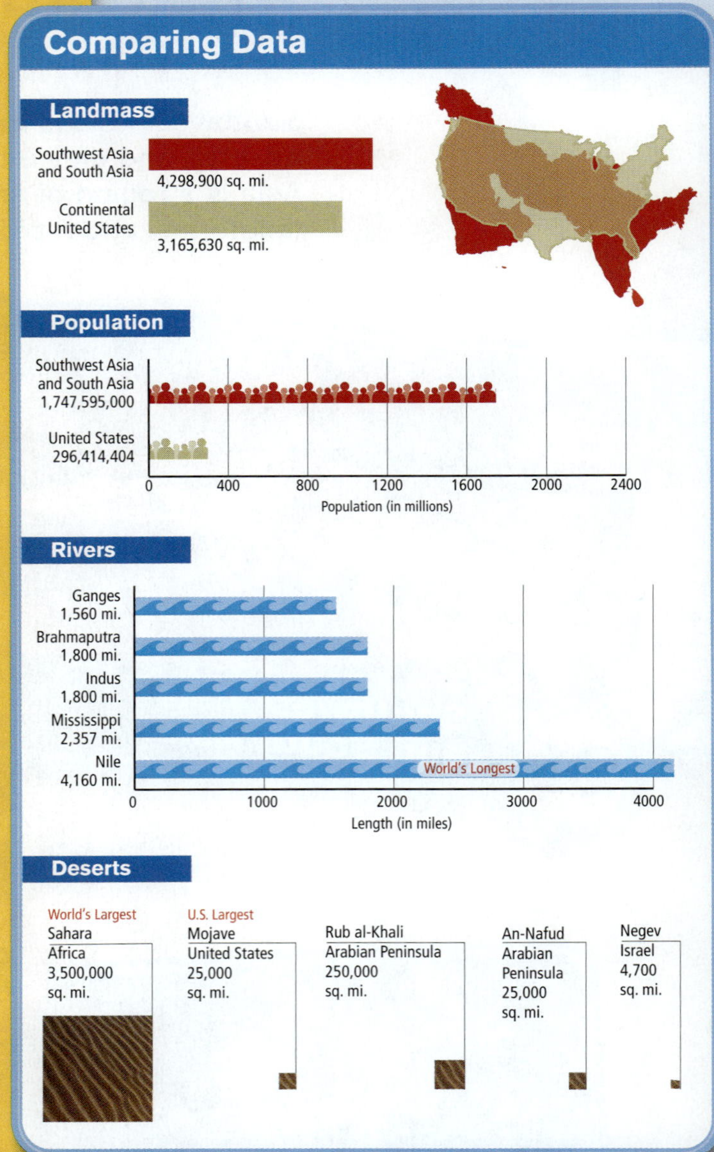

642 Unit 7

FIVE THEMES AND SIX ELEMENTS

Five Themes

Movement

Make sure that students comprehend that Southwest Asia and South Asia are made up of 23 separate countries, spread out across a huge area. Invite them to consider the challenges in transportation—of people and goods—that this region's geography and national borders present.

Six Elements

Places and Regions

Point out that the largest country in Southwest Asia is Saudi Arabia, while the largest country in South Asia is India. However, the population density of Saudi Arabia is very low, while the population density of India is very high.

Population Density

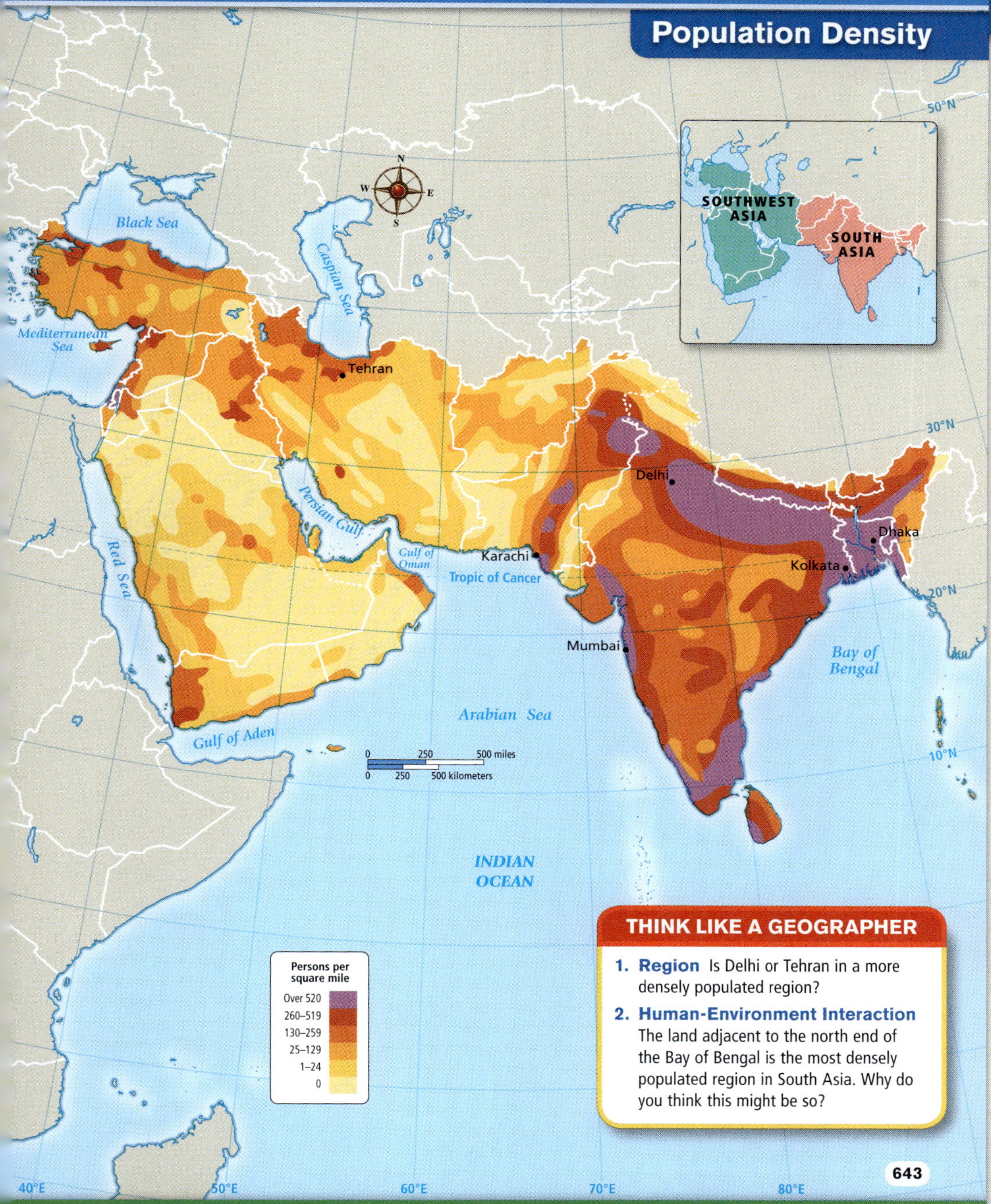

THINK LIKE A GEOGRAPHER

1. **Region** Is Delhi or Tehran in a more densely populated region?
2. **Human-Environment Interaction** The land adjacent to the north end of the Bay of Bengal is the most densely populated region in South Asia. Why do you think this might be so?

DIFFERENTIATING INSTRUCTION

English Learners

Comprehension: Self-Monitor

To help with the comprehension of "persons per square miles" in the map legend, have students replace *per* with "for each" or "in each." Explain that *per* is a preposition—it helps explain the relationship between *persons* and *square mile.* Have students name other prepositions. *(at, by, in, to, from)* Then have them use the map to make a sentence with "persons per square mile" or another preposition.

Inclusion

Graph Population Density

To help students who are visually-impaired more easily compare the differences in population density among countries or regions, pair them with another student. Have them work together to create bar graphs or pie graphs from the map, or from population data available in encyclopedias, almanacs, or government Web sites, such as http://www.cia.gov/cia/publications/factbook.

Population Density

Point out that India and Bangladesh are two of the world's most densely populated countries. India is the world's most populous democracy and has one of the world's fastest-growing economies. In contrast, Bangladesh remains one of the world's poorest countries, with a government that is generally considered corrupt and inefficient.

- Is Saudi Arabia or Iraq more densely populated? *(Iraq)*
- Compare Southwest Asia to South Asia in terms of population density. *(South Asia is much more densely populated than Southwest Asia. Many areas of Southwest Asia have 0 or 1–24 persons per square mile.)*
- Why would so much of Southwest Asia have so few people? *(Much of the region is desert and has limited water resources.)*

THINK LIKE A GEOGRAPHER ANSWERS

1. **Region** Delhi
2. **Human-Environment Interaction** It is close to the water resources of rivers, especially the mouths of the Ganges, and the bay.

More About . . .

Delhi

The city of Delhi—often called Old Delhi to distinguish it from New Delhi, India's capital—was surrounded by tall stone walls that were built in 1638, when it was the capital of the Mughal Empire. Delhi was invaded by several armies throughout the 1700s and 1800s, and served as India's capital from about 1912–1931.

Today, Delhi is also the name of a territory in northern India, the National Capital Territory of Delhi. The population of the Territory of Delhi is approximately 13.8 million, with most people residing in old Delhi, which has a population of about 7.2 million. Most people in the Territory of Delhi speak Hindi and/or Urdu. The region around old Delhi is especially hot and dry, with temperatures reaching over 110°F during the summer.

Climate

Tell students that the monsoon rains of the region are both vital to farming and yet a source of destructive flooding. In Bangladesh, for example, about one-third of the land floods each year during the rainy monsoon season.

- What are the climates of Nepal and Bhutan? *(humid subtropical and highland)*
- Compare the climates of Pakistan and Afghanistan. *(Both have semiarid and highland climates, but Pakistan also has desert and humid subtropical climates.)*
- Which island country—Cyprus or Sri Lanka—would you predict receives more rain? Why? *(Sri Lanka, because it has a tropical climate)*

THINK LIKE A GEOGRAPHER ANSWERS

1. **Region** Southwest Asia has more arid and desert areas; South Asia has more humid and tropical areas.
2. **Human-Environment Interaction** the Mediterranean and semiarid zones, because they are more suitable for habitation than the deserts of the region

More About . . .

Yemen

Yemen is a small republic in Southwest Asia, bordering Oman and Saudi Arabia. It has a narrow coastline along the Arabian Sea, the Gulf of Aden, and the Red Sea. It also has highlands, with fertile soil, even though most of Yemen is hot and dry. Seasonal monsoons bring rain to these mountain areas in the summer; however, there is little rain along the coasts. Yemen also has no permanent rivers or streams, forcing many people to rely on oases and underground springs for water.

UNIT 7 ATLAS

Climate

THINK LIKE A GEOGRAPHER

1. **Region** Compare the climate zones in Southwest and South Asia. Which region has more arid and desert areas and which has more humid and tropical areas?
2. **Human-Environment Interaction** In which climate zone of Southwest Asia might you expect to find most of the region's population?

644 Unit 7

FIVE THEMES AND SIX ELEMENTS

Five Themes

Region

Point out that many population centers in Southwest Asia and South Asia are located on or near rivers or the sea. You may also want to note how India's most densely populated areas match up to its wet or humid climate zones. Students may enjoy comparing climates and populations of different countries, such as India and Saudi Arabia, and Pakistan and Turkey, using the two Atlas maps.

Six Elements

Physical Systems

Remind students that climate refers to long-term patterns in weather, or the typical weather of a place or region. Droughts, or long periods without enough rain, are less likely to occur in tropical wet or tropical wet and dry climates, but are still possible. In India, a region with both climate zones, many of its people rely heavily on seasonal monsoons for water for farming. As a result, droughts have caused crop failures and many deaths in India.

Economic Activity

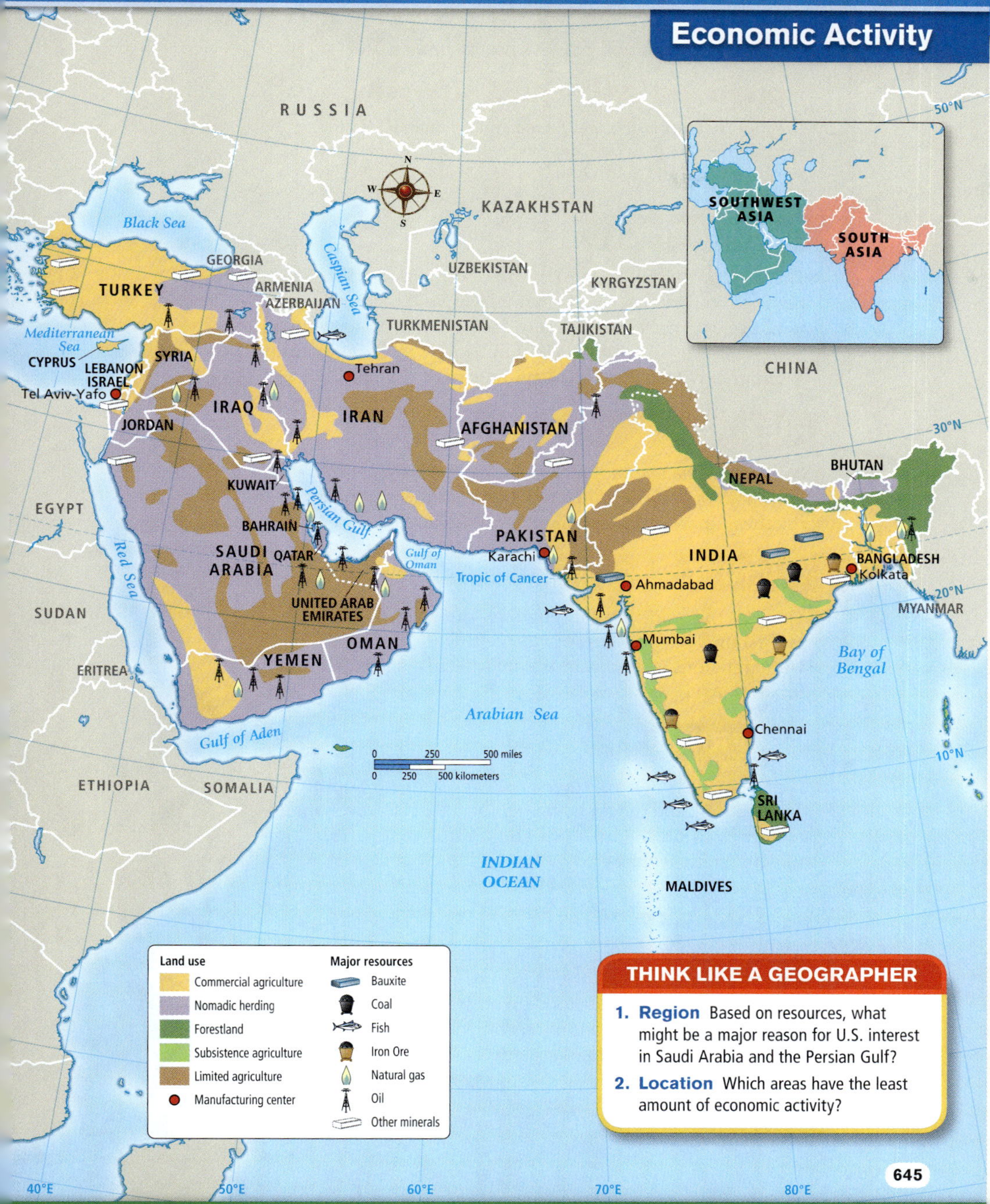

THINK LIKE A GEOGRAPHER

1. **Region** Based on resources, what might be a major reason for U.S. interest in Saudi Arabia and the Persian Gulf?
2. **Location** Which areas have the least amount of economic activity?

645

DIFFERENTIATING INSTRUCTION

Struggling Readers

Pronounce Place Names

Model the pronunciation of Southwest Asian countries to help students who might encounter difficulties decoding them. In particular, focus on the varied sounds represented by the letter *a* in such names as Kuwait, Bahrain, Qatar, Saudi Arabia, and Oman.

Gifted & Talented

Compare Economic Activities

Have pairs of students use the other Unit Atlas maps to compare the economic activities of a Southwest Asian or South Asian country to that of the United States and one other country. Have students create a chart or other visual display to show these comparisons.

Economic Activity

While Southwest Asia and South Asia share the same continent, there are significant differences in their major economic activities. Oil and natural gas drilling are the major industries in Southwest Asia, while agriculture and manufacturing are major economic activities in South Asia.

- What is the primary land use in India and Bangladesh? *(agriculture)*
- Which country has most of the two regions' coal resources? *(India)*
- Besides oil and natural gas, what are the major resources of Iran? *(other minerals and fishing)*

THINK LIKE A GEOGRAPHER ANSWERS

1. **Region** The U.S. economy relies on the oil and natural gas resources in Saudi Arabia and the Persian Gulf region.
2. **Location** Large areas of Southwest Asia and western South Asia have only nomadic herding and limited agricultural activity because of their dry climates.

Comparing Regions

Saudi Arabia and India

Saudi Arabia, the largest country in Southwest Asia, is the world's leading exporter of oil, with 14 major oil fields. However, its economy is dependent on its oil resources.

India, the largest country in South Asia, relies on farming in rural areas and factories in large cities. India's factories produce steel, tools, and computer software, among other items. India is also home to many talented English-speaking workers. As a result, it has become one of the world's leading outsourcing centers. Companies from around the world have moved their call centers, accounting services, and other services to India.

Regional Overview

Another flower—the poppy—is grown in South Asia, primarily in Afghanistan. Poppies can be converted into heroin. In fact, poppies grown in Afghanistan account for most of the world's heroin trade—and are thought to be funding terrorist activities in the region.

- Which religions have their roots in Southwest Asia? *(Judaism, Christianity, Islam)* In South Asia? *(Hinduism and Buddhism)*
- How much of the world's oil reserves are found in Southwest Asia? *(about half)*
- Which countries in these two regions have strong, independent democracies? *(Israel, Turkey, and India)*

More About . . .

Hinduism

Hinduism, one of the oldest religions in the world, had no single founder. It developed over 4,000 years in response to other religious movements in South Asia. Hinduism is based on the Veda, the oldest text in Hinduism and the main authority on religious truth. It is still accepted by most Hindus.

In 1949, the Indian constitution outlawed the caste system, which ranked people according to their birth and largely determined their job and social standing. However, its effects are still felt throughout India. For example, about 160 million people in India are still considered "untouchables"—as the outcasts of society. Although India's government is trying to improve life for these "untouchables," they are often treated harshly and face discrimination. Some are trying to organize themselves into political groups—and call themselves *Dalits,* which means "oppressed," to highlight their unfair treatment.

UNIT 7 ATLAS

Regional Overview

Southwest Asia and South Asia

The four chapters in this unit provide information about the geography, history, culture, government, and economics of Southwest Asia and South Asia.

GEOGRAPHY

The region of Southwest Asia and South Asia includes the highest mountain range, the Himalayas, and highest mountain, Everest, in the world. It also includes one of the world's largest deserts, the Rub al-Khali in Saudi Arabia. The monsoons greatly affect the weather and climate of South Asia.

HISTORY

Southwest Asia is sometimes called the cradle of civilization. A number of ancient civilizations arose in the region, including the Sumerian, Babylonian, Assyrian, and Hebrew.

CULTURE

Three of the world's great religions—Judaism, Christianity, and Islam—have their roots in Southwest Asia. Two other world religions—Hinduism and Buddhism—have their roots in South Asia.

GOVERNMENT

Israel and Turkey in Southwest Asia and India in South Asia are all strong and independent democracies. However, many countries in both Southwest and South Asia are ruled by undemocratic regimes.

ECONOMICS

About one-half of the world's oil reserves are found in Southwest Asia. In South Asia, people rely heavily on farming and fishing to provide food. In both Southwest and South Asia there is a growing emphasis on education as the key to compete in the global marketplace.

Flower Market
An Indian farmer sells marigolds for use at Hindu temples to honor deities.

FIVE THEMES AND SIX ELEMENTS

Five Themes

Movement

Saudi Arabia's major trading partners are the United States, Japan, Great Britain, and other European Union countries. Following the discovery of oil in Saudi Arabia in 1936, major oil refineries were built on the Red Sea and the Persian Gulf. Major railroads and ports were also constructed throughout the country. Today, oil is also shipped to the island country of Bahrain for refining, then carried by tanker to trading partners throughout the world.

Six Elements

Human Systems

Point out that many countries in Southwest Asia and South Asia are located along or near large bodies of water such as the Gulf of Aden, the Arabian Sea, and the Persian Gulf. Access to the sea allows these countries to harness the sea's resources as well as build shipping centers for trade. Review the importance of waterways for transportation and emphasize the relationship between good transportation systems and economic growth.

Country Almanac

Click here to compare the most recent data on countries @ ClassZone.com

Unit Writing Project

Lobbying Plan

As you read this unit, choose an important cause that you would like to lobby for—that is, convince local, state, or federal government officials to support. Pretend you are a lobbyist and create a written plan for your cause.

Think About:

- which leaders to contact
- a detailed plan of action that you would like them to support

Southwest Asia and South Asia

The Almanac provides information about the geography, economy, and culture of the countries of Southwest Asia and South Asia.

Afghanistan

GEOGRAPHY
Capital: Kabul
Total Area: 250,001 sq. mi.
Population: 29,863,000

ECONOMY
Imports: machinery; medicine; clothing
Exports: rugs, dried fruit

CULTURE
Language: Dari; Pashto
Religion: Sunni Muslim 89%; Shiite Muslim 9%

Bahrain

GEOGRAPHY
Capital: Manama
Total Area: 257 sq. mi.
Population: 727,000

ECONOMY
Imports: machinery; food; petroleum
Exports: aluminum; clothing; petroleum

CULTURE
Language: Arabic
Religion: Sunni Muslim 41%; Shiite Muslim 41%; Christian 11%; Hindu 6%

Bangladesh

GEOGRAPHY
Capital: Dhaka
Total Area: 55,599 sq. mi.
Population: 141,822,000

ECONOMY
Imports: textiles; cotton; rice and wheat
Exports: clothing; frozen fish; knitwear

CULTURE
Language: Bengali
Religion: Muslim 86%; Hindu 12%

Bhutan

GEOGRAPHY
Capital: Thimphu
Total Area: 18,147 sq. mi.
Population: 2,163,000

ECONOMY
Imports: computers; cars; food; petroleum
Exports: electricity; cement

CULTURE
Language: Dzongkha
Religion: Buddhist 74%; Hindu 21%

Cyprus

GEOGRAPHY
Capital: Nicosia
Total Area: 3,571 sq. mi.
Population: 835,000

ECONOMY
Imports: mineral fuels; consumer goods
Exports: pharmaceuticals; clothing

CULTURE
Language: Greek; Turkish
Religion: Greek Orthodox 95%; Catholic 2%; Muslim 1%

India

GEOGRAPHY
Capital: New Delhi
Total Area: 1,269,346 sq. mi.
Population: 1,103,371,000

ECONOMY
Imports: petroleum; gold and silver
Exports: jewelry; chemicals; food; cotton

CULTURE
Language: Hindi, English
Religion: Hindu 74%; Sunni Muslim 9%; Christian 6%; Shiite Muslim 3%

Iran

GEOGRAPHY
Capital: Tehran
Total Area: 636,296 sq. mi.
Population: 69,515,000

ECONOMY
Imports: machinery; cars; chemicals; food
Exports: petroleum; carpets; nuts

CULTURE
Language: Farsi
Religion: Shiite Muslim 90%; Sunni Muslim 6%; Zoroastrian 3%

INTERDISCIPLINARY ACTIVITIES

CONNECT to Math

Graph Populations

Have students create a bar graph to compare the populations of countries within Southwest Asia and South Asia. For example, students may choose to graph Southwest Asian countries, such as Bahrain and Iran, or South Asian countries, such as Bhutan and Bangladesh, then compare their populations. To extend this activity, have students write a brief explanation of possible reasons for the differences in populations.

Unit Writing Project

Teaching Tips

- As students prepare a written plan lobbying for an important cause, encourage them to seek out reliable sources to use, including newspapers, magazines, and Web sites. You may also want to arrange for students to interview local experts at school.
- Provide students with an example of a written plan that lobbies for a cause, or have them research various formats, such as a Power Presentation.
- Encourage students to create visuals to go with their written plans. Remind them that maps, charts, graphs, and other visuals will help to illustrate their key points as well as make their lobbying plan more effective.
- Encourage students to work with partners to review and help revise their plans.

Each map on the country cards illustrates the shape of a state, province, or country. Tell students that the maps do not reflect the comparative size of these landmasses within a particular region. To get an idea of the actual size of the state, province, or country, students should study the land area statistics provided on the card.

Comparing Regions

The Himalayan Kingdoms of Nepal and Bhutan

Isolated from much of the world for centuries, both countries are now struggling with modernization efforts. Technically constitutional monarchies, both countries have faced recent threats from violence: Bhutan because of foreign terrorists operating on its soil; Nepal from rebels seeking to overthrow its government.

Although both countries offer spectacular views and mountain-climbing, Nepal earns 14 times as much as Bhutan from tourism. (Bhutan limits visitors to help protect its cultural traditions.)

Each map on the country cards illustrates the shape of a state, province, or country. Tell students that the maps do not reflect the comparative size of these landmasses within a particular region. To get an idea of the actual size of the state, province, or country, students should study the land area statistics provided on the card.

UNIT 7 ATLAS

Iraq

GEOGRAPHY
Capital: Baghdad
Total Area: 168,754 sq. mi.
Population: 28,807,000

ECONOMY
Imports: consumer goods
Exports: petroleum; food

CULTURE
Language: Arabic; Kurdish
Religion: Shiite Muslim 62%; Sunni Muslim 34%; Christian 3%

Israel

GEOGRAPHY
Capital: Jerusalem
Total Area: 8,019 sq. mi.
Population: 6,725,000

ECONOMY
Imports: machinery; diamonds; chemicals
Exports: diamonds; chemicals; electronics

CULTURE
Language: Hebrew; Arabic
Religion: Jewish 76%; Muslim 16%; Christian 2%

Jordan

GEOGRAPHY
Capital: Amman
Total Area: 35,637 sq. mi.
Population: 5,703,000

ECONOMY
Imports: food; machinery; petroleum
Exports: clothing; chemicals; potash

CULTURE
Language: Arabic
Religion: Sunni Muslim 94%; Christian 4%

Kuwait

GEOGRAPHY
Capital: Kuwait
Total Area: 6,880 sq. mi.
Population: 2,687,000

ECONOMY
Imports: machinery; food; chemicals
Exports: petroleum; ethylene

CULTURE
Language: Arabic
Religion: Muslim 85%

Lebanon

GEOGRAPHY
Capital: Beirut
Total Area: 4,015 sq. mi.
Population: 3,577,000

ECONOMY
Imports: machinery; food; chemicals
Exports: jewelry; machinery; chemicals

CULTURE
Language: Arabic
Religion: Christian 38%; Shiite Muslim 34%; Sunni Muslim 21%; Druze 7%

Maldives

GEOGRAPHY
Capital: Male
Total Area: 116 sq. mi.
Population: 329,000

ECONOMY
Imports: food; oil; constuction goods
Exports: clothing; tropical fish; jet fuel

CULTURE
Language: Divehi
Religion: Sunni Muslim 100%

Nepal

GEOGRAPHY
Capital: Kathmandu
Total Area: 54,363 sq. mi.
Population: 27,133,000

ECONOMY
Imports: machinery; fabrics; fuels
Exports: clothing; carpets; shawls

CULTURE
Language: Nepali
Religion: Hindu 81%; Buddhist 11%; Muslim 4%; Kirat 4%

Oman

GEOGRAPHY
Capital: Muscat
Total Area: 82,031 sq. mi.
Population: 2,567,000

ECONOMY
Imports: machinery; cars; food; chemicals
Exports: petroleum; natural gas

CULTURE
Language: Arabic
Religion: Ibadiyah Muslim 75%; Hindu 6%; Christian 5%; Buddhist 1%

Pakistan

GEOGRAPHY
Capital: Islamabad
Total Area: 310,403 sq. mi.
Population: 157,935,000

ECONOMY
Imports: machinery; petroleum; chemicals
Exports: textiles; leather; rice; carpets

CULTURE
Language: Urdu
Religion: Sunni Muslim 79%; Shiite Muslim 17%; Christian 3%; Hindu 1%

INTERDISCIPLINARY ACTIVITIES

CONNECT to Language Arts

Write a Paragraph

Ask students to write a paragraph that convinces a U.S. business to open an office in one Southwest Asian or South Asian country. Have them focus on key characteristics, such as a country's resources, climate, and location, that might appeal to U.S. businesses. Invite volunteers to present or post their work.

CONNECT to Civics

Compare Governments

Ask students to choose one Southwest Asian country and one South Asian country, and find out basic facts about each one's system of government. Have them compare the head of state and legislative bodies of each country, and other features of government, and report their findings to the class.

Qatar

GEOGRAPHY
Capital: Doha
Total Area: 4,416 sq. mi.
Population: 813,000

ECONOMY
Imports: machinery; cars; food; chemicals
Exports: natural gas; petroleum; iron; steel

CULTURE
Language: Arabic
Religion: Muslim 83%; Christian 10%; Hindu 3%

Saudi Arabia

GEOGRAPHY
Capital: Riyadh
Total Area: 756,985 sq. mi.
Population: 24,573,000

ECONOMY
Imports: cars; machinery; food; chemicals
Exports: petroleum; chemicals

CULTURE
Language: Arabic
Religion: Sunni Muslim 84%; Shiite Muslim 10%; Christian 4%

Sri Lanka

GEOGRAPHY
Capital: Colombo
Total Area: 25,332 sq. mi.
Population: 20,743,000

ECONOMY
Imports: textiles; petroleum; food
Exports: clothing; tea; precious stones

CULTURE
Language: Sinhala, Tamil
Religion: Buddhist 77%; Muslim 9%; Hindu 8%; Christian 7%

Syria

GEOGRAPHY
Capital: Damascus
Total Area: 71,498 sq. mi.
Population: 19,043,000

ECONOMY
Imports: food; chemicals; machinery; iron
Exports: petroleum; cotton; vegetables

CULTURE
Language: Arabic
Religion: Sunni Muslim 74%; Shiite Muslim 12%; Christian 6%; Druze 3%

Turkey

GEOGRAPHY
Capital: Ankara
Total Area: 301,384 sq. mi.
Population: 73,193,000

ECONOMY
Imports: chemicals; machinery; petroleum
Exports: textiles; cars; machinery; iron

CULTURE
Language: Turkish
Religion: Sunni Muslim 67%; Shiite Muslim 30%; Christian 1%

United Arab Emirates

GEOGRAPHY
Capital: Abu Dhabi
Total Area: 32,000 sq. mi.
Population: 4,496,000

ECONOMY
Imports: machinery; food; textiles
Exports: petroleum; natural gas; gold

CULTURE
Language: Arabic
Religion: Sunni Muslim 80%; Shiite Muslim 16%

Yemen

GEOGRAPHY
Capital: San'a
Total Area: 203,850 sq. mi.
Population: 20,975,000

ECONOMY
Imports: food; machinery; petroleum
Exports: petroleum; fish; fruits

CULTURE
Language: Arabic
Religion: Sunni Muslim 60%; Shiite Muslim 40%

United Arab Emirates the Burj Al Arab Hotel in Dubai in the United Arab Emirates

Southwest Asia and South Asia

Water resources—especially rivers—have long been a vital resource for each region. Today, conflicts over the two regions' rivers are simmering as demand for water increases: between Israel and Jordan over the Jordan River system; among Turkey, Syria, and Iraq over the Tigris-Euphrates system; between India and Bangladesh over the Ganges system; and between India and Pakistan over the Indus system.

- Which continental South Asian country has the smallest population? *(Bhutan)* Which of the Southwest Asian countries has the smallest land area? *(Bahrain)*
- In which countries do people speak Arabic? *(Bahrain, Iraq, Israel, Jordan, Kuwait, Lebanon, Oman, Qatar, Saudi Arabia, Syria, United Arab Emirates, and Yemen)*
- Which countries export cotton? *(India and Syria)* Which country exports tropical fish? *(Maldives)*

More About . . .

United Arab Emirates

The United Arab Emirates is a federation of seven small kingdoms run by emirs, or sheiks. The UAE, as it is known, was formed in 1971 from a group of independent states that had been aligned with Great Britain. These states also included Bahrain and Qatar.

At one time, the kingdoms of the UAE were considered poor and isolated. Today, the UAE is a modern state and has a high standard of living, centered on oil exports. The crown prince of Dubai, Sheik Mohammed bin Rashid al-Makhtoum, is one of the world's top horse breeders. His vision for Dubai is to make it into a new Singapore, or economic powerhouse, including plans to build the world's tallest building—the Burj Dubai skyscraper.

Each map on the country cards illustrates the shape of a state, province, or country. Tell students that the maps do not reflect the comparative size of these landmasses within a particular region. To get an idea of the actual size of the state, province, or country, students should study the land area statistics provided on the card.

INTERDISCIPLINARY ACTIVITIES

CONNECT to Language Arts

Describe Daily Life

Discuss with students the likely similarities and differences in daily life among the countries of Southwest Asia and South Asia. Then have students choose a country, such as Turkey, Sri Lanka, or United Arab Emirates, and write a brief description of daily life from the point of view of a resident and from the point of view of a tourist. Ask volunteers to read their descriptions, then discuss the differences in viewpoints.

CONNECT to Economics

Chart Natural Resources

Assign small groups of students one of the countries listed on this page. Then guide them in using classroom and library resources, such as atlases and encyclopedias, to research more about the natural resources of their assigned country. Each group should create a chart of the resources, including statistical data (such as employment) and an assessment of any potential damage to the country's environment from the exploration, processing, or use of each resource.

CHAPTER 21 Southwest Asia and South Asia

PLANNING GUIDE

Complete print resources all on one USB drive allow you to customize lesson plans for all students.

Section Titles & Key Ideas	Unit Resource Book Southwest Asia and South Asia	Assessment Resources
1 Physical Geography of Southwest Asia **KEY IDEA** Southwest Asia has vast deserts and extensive mountain ranges that are similar to those in certain parts of Africa.	• **Reading Study Guide,** pp. 3–4 • **RSG with Additional Support,** pp. 9–10 • **Vocabulary Practice,** p. 27 • **Outline Map Activities,** p. 31 • **Interdisciplinary Projects,** p. 35 • **Connect to Math,** p. 38	**Unit Resource Book** • Section Quiz, p. 41 • Reteaching Activity, p. 44 **Test Generator CD-ROM** **Daily Test Practice Transparency,** TT72
2 Physical Geography of South Asia **KEY IDEA** The physical geography of South Asia is made up of mountains, plateaus, and plains, and there are wet and dry seasons.	• **Reading Study Guide,** pp. 5–6 • **RSG with Additional Support,** pp. 11–12 • **Interdisciplinary Projects: Understanding Monsoons,** p. 36	**Unit Resource Book** • Section Quiz, p. 42 • Reteaching Activity, p. 45 **Test Generator CD-ROM** **Daily Test Practice Transparency,** TT73
3 River Valley Civilizations **KEY IDEA** In both regions, civilizations developed first in river valleys where there was an abundant supply of water.	• **Reading Study Guide,** pp. 7–8 • **RSG with Additional Support,** pp. 13–14 • **Skillbuilder Practice,** p. 28 • **History Makers,** pp. 33–34 • **Primary Sources,** pp. 39–40	**Unit Resource Book** • Section Quiz, p. 43 • Reteaching Activity, p. 46 • Chapter Test (Forms A, B, & C), pp. 47–58 **Test Generator CD-ROM** **Daily Test Practice Transparency,** TT74

McDougal Littell **Assessment System**

TEST SCORE REPORT RETEACH

INTEGRATED TECHNOLOGY

No Time? To focus students on important content in the chapter, use the **"Reading Study Guide with Additional Support"** in ***Resources2Go.***

Teacher Presentation Options

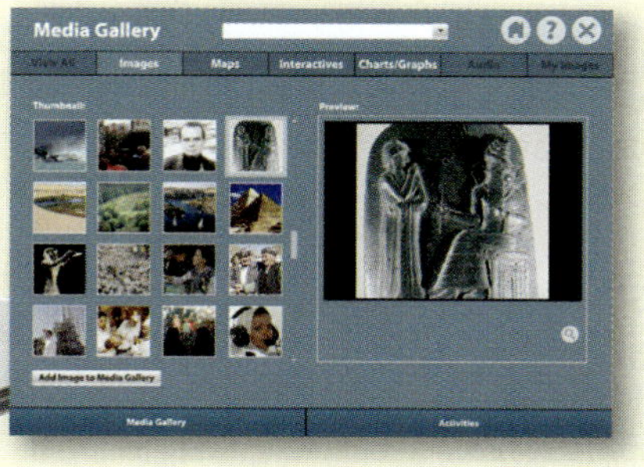

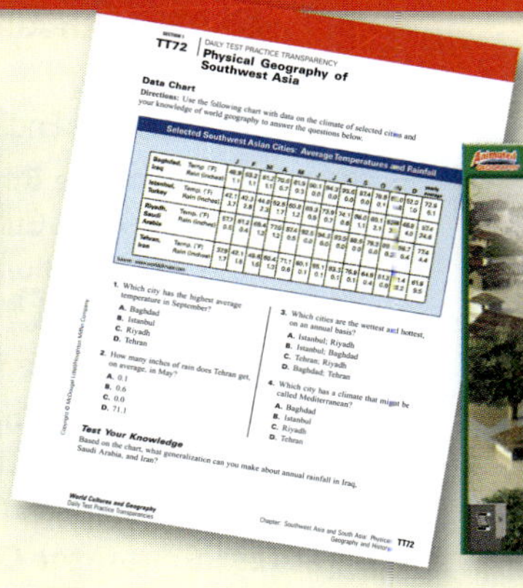

Electronic Books

eEdition DVD-ROM

eEdition Online

Presentation Toolkit

Video/DVD

- World Cultures and Geography

Power Presentations DVD-ROM

- Discussion Starters
- Media Gallery
- Review Game
- Interactive Review
- Animated Features

Unit Transparency Book

- 3-Minute Warm-Ups
- Essential Question
- Critical Thinking
- Maps
- Cultures

Daily Test Practice Transparencies

Animated Geography

- Present-Day Southwest Asia and South Asia Map
- Qanats: An Ancient Irrigation Technology
- Monsoons
- Mesopotamia

OnlineActivities at ClassZone.com

Animated Geography

- Videos and Photos
- 3-D Animations
- Interactive Maps

Interactive Review

- Section Quizzes
- Review and Study Notes
- Vocabulary Flip Cards
- Graphic Organizers
- Name Game and Crossword Puzzles
- GeoGame

Online Test Practice

Map Center

- Atlas Maps
- Political Maps with Data
- Links to Mapping Web Sites

Activity Center

- Activity Maker—Review Games, Puzzles, Flip Cards, and Graphic Organizers
- Worldster

Research & Writing Center

- Web Sites
- Data Updates
- Primary Sources
- Writing Models

Current Events

Amber L. McVey
Antioch Middle School
Gladstone, Missouri

Make It Fun — Teacher-Tested Activities

As you teach this chapter, look for my classroom-proven ideas for:

- making the lesson fun
- focusing students' attention and comprehension
- integrating technology with traditional teaching

Teaching Bell to Bell When you see this timing guide, have students discuss in pairs, in groups of 3–6, or as a class.

- **Think, Pair, Share** (pairs)
- **Reader, Recorder, Reporter** (3–6)
- **Talk About It** (any size)

RESOURCES FOR DIFFERENTIATED INSTRUCTION

CHAPTER 21

Struggling Readers

eEdition with Audio DVD-ROM

Lesson Plans
- Strategies for Inclusion

Unit Resource Book
- Reading Study Guide with Additional Support, pp. 9–14 (B)
- Reteaching Activity, pp. 44–46
- Chapter Test (Form A), pp. 47–50 (A)

Best Practices Toolkit (C)
- Interactive Reading Strategies, pp. 90–145
- Differentiation Activity Bank: Struggling Readers, pp. 146–149

Teacher's Edition Activities
- Main Ideas and Details, Sec. 1
- Define and Illustrate Terms, Sec. 2
- Discuss Traits of Civilization, Sec. 3
- Tiered Activities, Sec. 1–3

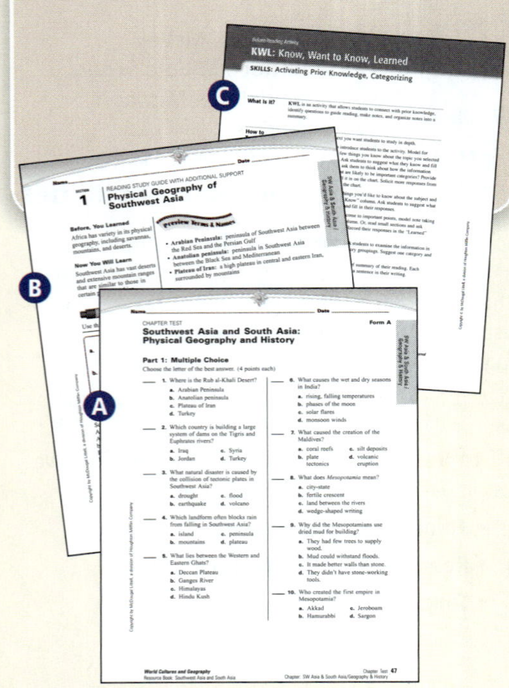

English Learners

Pupil's Edition in Spanish

eEdition with Audio DVD-ROM

Lesson Plans
- Strategies for English Learners

Unit Resource Book
- Reading Study Guide (Spanish), pp. 15–20
- RSG with Additional Support (Spanish), pp. 21–26 (A) (B)

Multi-Language Glossary (C)

Best Practices Toolkit
- Differentiation Activity Bank: English Learners, pp. 165–194

Test Generator CD-ROM
- Section Quizzes in Spanish
- Chapter Test in Spanish

Teacher's Edition Activities
- Pronounce and Preview, Sec. 1–3
- Vocabulary: Prefixes, Sec. 2
- Tiered Activities, Sec. 1–3

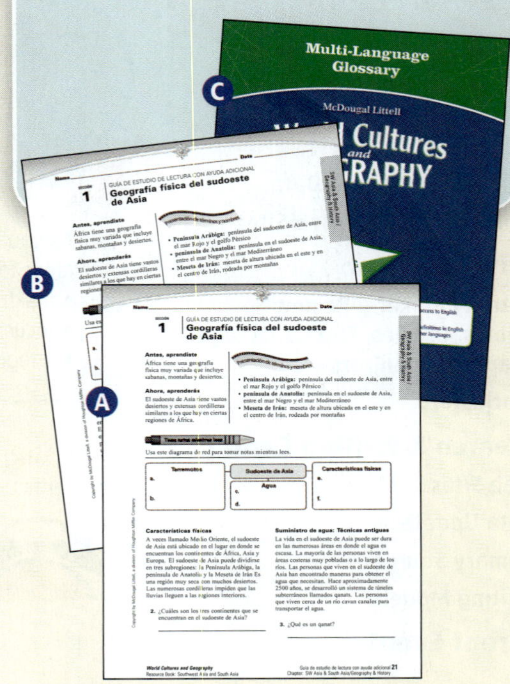

Inclusion

eEdition with Audio DVD-ROM

Lesson Plans
- Strategies for Inclusion

Unit Resource Book
- Reading Study Guide with Additional Support, pp. 9–14 (A) (B)
- Reteaching Activity, pp. 44–46
- Chapter Test (Form A), pp. 47–50

Best Practices Toolkit (C)
- Interactive Reading Strategies, pp. 90–145
- Differentiation Activity Bank: Struggling Readers, pp. 146–149

Teacher's Edition Activities
- Match Terms and Definitions, Sec. 1
- Locate Vocabulary Terms on a Map, Sec. 2
- Create Audio or Visual Vocabulary, Sec. 3
- Tiered Activities, Sec. 1–3

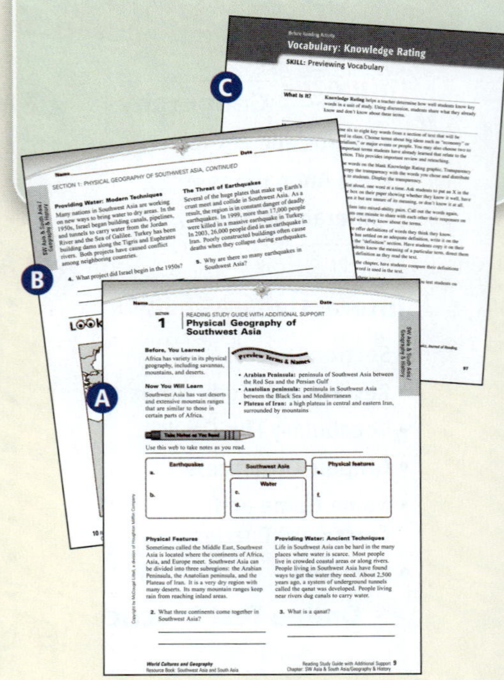

Interdisciplinary Activities

Gifted & Talented Students

Lesson Plans
- Core Lesson Plan

Unit Resource Book
- Outline Map Activities, p. 31
- History Makers, p. 34 (A)
- Connect to Math, p. 38 (B)
- Primary Source, p. 39 (C)
- Chapter Test (Form C), pp. 55–58

Best Practices Toolkit
- Student Products Rubrics, pp. 1–14
- Writing Templates, pp. 78–89

Teacher's Edition Activities
- Create a Cuneiform System, Sec. 3
- Tiered Activities, Sec. 1–3

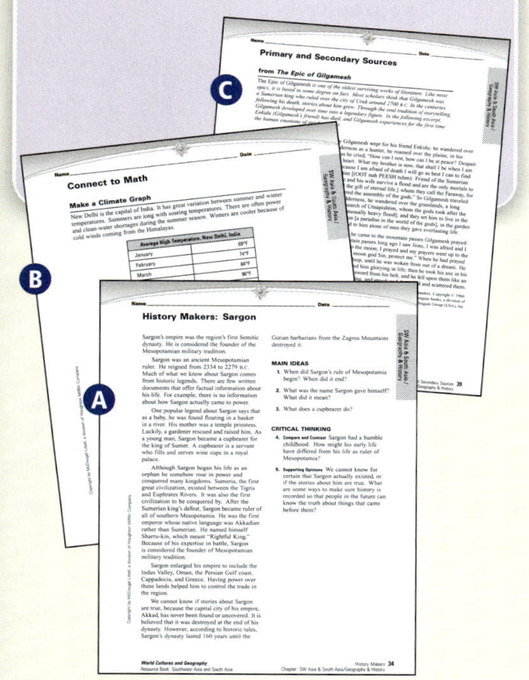

Pre-AP*

Lesson Plans
- Core Lesson Plan

Unit Resource Book
- Vocabulary Practice, p. 27
- Map and Graph Skills, pp. 29–30
- History Makers, p. 33 (A)
- Interdisciplinary Projects, pp. 35–36 (B)
- Primary Source, p. 40 (C)
- Chapter Test (Form C), pp. 55–58

Best Practices Toolkit
- Writing Templates, pp. 78–89
- Differentiation Activity Bank: Pre-AP Students, pp. 195–200

* Pre-AP is a registered trademark of the College Entrance Examination Board, which was not involved in the production of and does not endorse this product.

Teacher's Edition Activities
- Conserving Water in the Middle East, Sec. 1
- Tiered Activities, Sec. 1–3

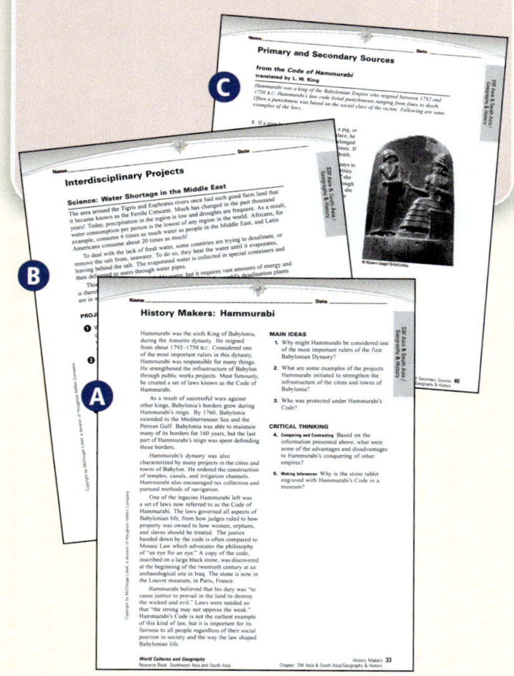

CONNECT to Language Arts

Poem About the Ancient Fertile Crescent or Indus River Valley

CLASS TIME 30 minutes

TASK Create a poem about the ancient Fertile Crescent or Indus River Valley.

SUPPLIES
- visuals depicting the Fertile Crescent and Indus River Valley, such as the photographs in Section 3 of this chapter
- writing materials/computers

ACTIVITY Have students analyze images of the ancient Fertile Crescent or Indus River Valley and brainstorm a list of descriptive words to describe its landscapes, people, and history. Then have students use the images and their lists to create a descriptive poem about one of these ancient regions. Ask volunteers to read their poems aloud or create a classroom display.

Good for Visual and Verbal Learners

CONNECT to Science

Planning a Water Project

CLASS TIME 30 minutes

TASK Create a plan for a water project for a country in either Southwest Asia or South Asia.

SUPPLIES
- resources on water projects around the world
- writing materials/computers

ACTIVITY Have students work in pairs to create a plan for a water project that would benefit the people of a country in Southwest Asia or South Asia. Plans should include reasons why the project is needed, how it will work and how people will benefit from it, estimated costs and who will pay for those costs, and the steps being taken to avoid environmental harm. Have all the students present their plans, then discuss them as a class. Afterward, have students revise their plans, based on the class discussion.

Good for Visual/Verbal and Logical/Mathematical Learners

Introduce the ESSENTIAL QUESTION

- Explain that the physical features and climate of Southwest and South Asia have strongly influenced the people who live there.
- Point out that Southwest Asia is located where the continents of Africa, Asia, and Europe meet—and it's been a geographic and cultural crossroads throughout history.
- Explain that the geography of Southwest and South Asia contributes to the region's extreme weather, climates, and natural disasters.

Key Ideas

Section 1: Southwest Asia has vast deserts and extensive mountain ranges that are similar to those in certain parts of Africa.

Section 2: The physical geography of South Asia is made up of mountains, plateaus, and plains, and there are wet and dry seasons.

Section 3: In both regions, civilizations developed first in river valleys where there was an abundant supply of water.

CONNECT to the Essential Question

You may wish to revisit the Essential Question periodically with students as they work through the chapter.

Suggestions for reconnecting are provided at an appropriate point in the *Teacher's Edition* for each section.

Indus River at Stakna Ldakh in India

CHAPTER 21

Southwest Asia and South Asia

Physical Geography and History

1 GEOGRAPHY **Physical Geography of Southwest Asia**

2 GEOGRAPHY **Physical Geography of South Asia**

3 HISTORY **River Valley Civilizations**

ESSENTIAL QUESTION

How have physical geography and history interacted in Southwest and South Asia?

CONNECT Geography & History

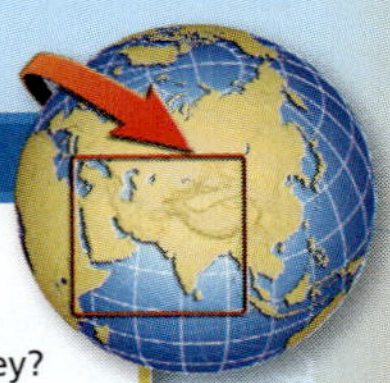

Use the map and the time line to answer the following questions.

1. When did cities appear in the Indus River valley?
2. In ancient times, people of the Indus Valley traded with people on the Tigris and Euphrates rivers. What water route would they take between the regions?

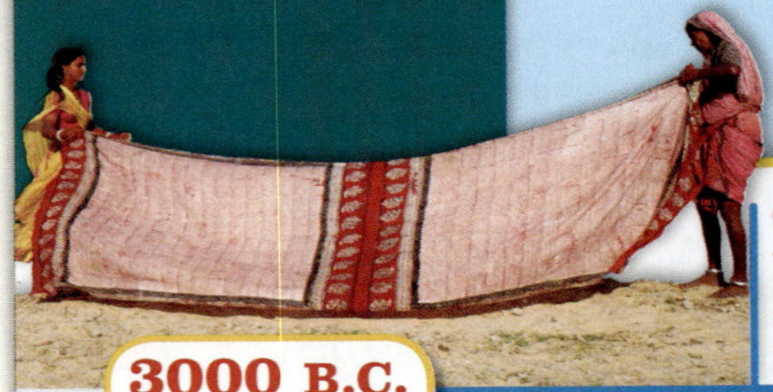

3000 B.C.

Culture
◄ **c. 3000 B.C.** Cotton cloth is woven in South Asia (and is still being woven today, as the photo shows).

Culture
c. 2400 B.C. Sumerians use the cuneiform writing system. ▼

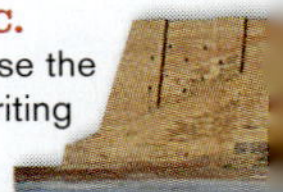

Geography
c. 2500 B.C. Cities develop in the Indus River valley of South Asia.

TIME LINE DISCUSSION

Point out to students that some of the oldest civilizations on earth began in the river valleys of Southwest Asia and South Asia. Discuss how centuries-old traditions may still influence the people of this region. *(Possible Answers: importance of family and religious beliefs; holidays celebrated; common foods and clothing styles)*

- What are some skills the ancient people of Southwest Asia and South Asia used that we still use today? *(c. 3000 B.C.—woven cloth; c. 2400 B.C.—writing)*
- Where were ziggurats, a special type of temple, first built? *(in Sumerian cities, c. 2200 B.C.)*
- About how many years ago did Hammurabi create a law code? *(about 3,760 years ago)*

Geography
c. 1900 B.C. The Sarasvati River of South Asia disappears, perhaps because of earthquakes.

History
▲ **c. 2200 B.C.**
Ziggurats are first built in Sumerian cities.

1500 B.C.

Government
◀ **c. 1750 B.C.** Hammurabi of Babylon compiles a law code.

TEACHER MEDIA FAVORITES

Books for the Teacher

Mansfield, Peter. *A History of the Middle East.* Penguin, 2004. History of the region (available on audio).

McIntosh, Jane. *A Peaceful Realm: The Rise and Fall of the Indus Civilization.* Westview Press, 2002. A vivid account, with maps and photographs.

Mines, Diane P. *Everyday Life in South Asia.* Indiana University Press, 2002. Anthology focusing on the daily lives of people.

Video / DVD

Globe Trekker—Middle East. Pilot Productions (double DVD), 2004. A unique tour of historical sites that have shaped the region and its people.

Lawrence of Arabia: Battle for the Arab World. PBS, 2003. A biography of T.E. Lawrence (teacher resources available).

Internet

Go to the *Research and Writing Center* at **ClassZone.com** to link to:

- data on the region
- a variety of chapter-specific, content-reviewed sites
- "History Makers" biographies
- primary sources

Animated GEOGRAPHY

Present-Day Southwest Asia and South Asia

CLASSZONE.COM

Southwest Asia includes countries such as Iraq, Iran, Syria, and Israel and is often called the Middle East. South Asia generally includes Afghanistan, Pakistan, India, Sri Lanka, the Maldives, Nepal, Bhutan, and Bangladesh.

- Rivers have played a major role in the culture, history, and economy of Southwest Asia and South Asia. What two major rivers provide water to Turkey, Syria, and Iraq? *(Tigris, Euphrates)*
- What national capitals are located on the Persian Gulf? *(Kuwait, Kuwait; Manama, Bahrain; Doha, Qatar; Abu Dhabi, United Arab Emirates; Muscat, Oman)* Why do you think so many capitals are located along the Persian Gulf? *(Possible Answer: allows for easy access to the Arabian Sea and Indian Ocean; port cities are good for trade)*

CONNECT GEOGRAPHY & HISTORY ANSWERS

1. about 2500 B.C.
2. the Arabian Sea to the Persian Gulf

Presentation Options

Use the **Power Presentations** DVD-ROM to project the animated map.

- Prompt students to use the pop-up images and audio from the map to discover what they already know about the geography, culture, history, economics, or governments of the region.
- Challenge students to answer the map questions individually, in small groups, or as a class activity. Use all of the interactive material to help preview chapter content.

1 Plan & Prepare

Objectives

- Describe the physical features of Southwest Asia
- Compare ancient and modern techniques of providing water to the people of the region
- Identify reasons why Southwest Asia is prone to earthquakes

Read for the Essential Question

Help students read for a purpose by reminding them of the Essential Question: "How have physical geography and history interacted in Southwest and South Asia?" Point out that they can use text features such as Key Questions, special features, images, and maps to form answers to the question.

Vocabulary

Best Practices Toolkit

Use the ***Best Practices Toolkit*** to model strategies for vocabulary notetaking. Vary strategies throughout the year. Choose from: *Knowledge Rating, Predicting ABCs, Definition Mapping, Word Sort, Frayer Model (Word Squares), Magnet Words, Word Wheel,* and *Student VOC.*

Vocabulary Strategies, TT9–TT16

Reading Strategy

Best Practices Toolkit

Display the Web Diagram transparency.

- Remind students that main ideas are the most important ideas in a segment of text.
- Model filling in main ideas on the graphic organizer, using the subheads throughout this section.
- Have students suggest entries. Remind students that main ideas in each block of text in the section will also help them answer the Key Questions.
- Display the graphic organizer for students to review.

Web Diagram, TT28

SECTION 1

Reading for Understanding

Key Ideas

BEFORE, YOU LEARNED

Africa has variety in its physical geography, including savannas, mountains, and deserts.

NOW YOU WILL LEARN

Southwest Asia has vast deserts and extensive mountain ranges that are similar to those in certain parts of Africa.

Vocabulary

TERMS & NAMES

Arabian Peninsula peninsula of Southwest Asia between the Red Sea and Persian Gulf

Anatolian peninsula in Southwest Asia between the Black Sea and Mediterranean

Plateau of Iran a high plateau in central and eastern Iran, surrounded by mountains

Tigris River a river of Southwest Asia that flows from a plateau on the Anatolian peninsula to the Persian Gulf

Euphrates (yoo•FRAY•teez) **River** a river of Southwest Asia that flows from a plateau on the Anatolian peninsula to the Persian Gulf

oasis fertile or green spot in desert created by underground water coming to the surface

REVIEW

plateau a high area of flat land

Visual Vocabulary Women carry water from an oasis near Khimsar, Rajasthan, India.

Reading Strategy

Re-create the diagram shown at right. As you read and respond to the **KEY QUESTIONS**, use the outer ovals to note important details about the region's physical features, water shortage, and earthquake threats.

See Skillbuilder Handbook, page R4

FIND MAIN IDEAS

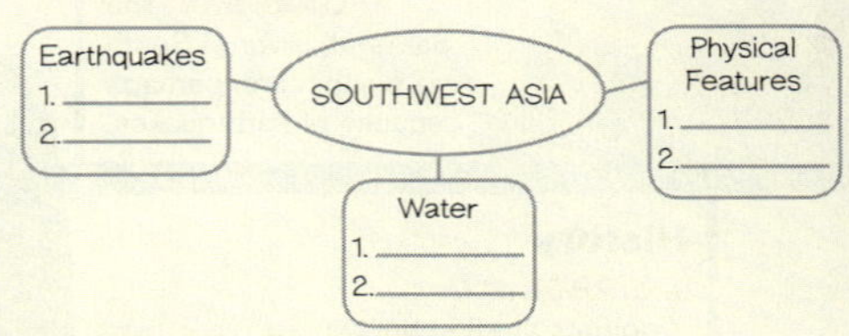

GRAPHIC ORGANIZERS
Go to Interactive Review @ClassZone.com

PRETEACHING VOCABULARY

English Learners

Pronounce and Preview

Pronounce each term for students. Review the meanings of words within the definitions, such as *peninsula, plateau,* and *fertile.*

- To modify vocabulary learning, have students complete worksheets as they read, instead of afterward.

Unit Resource Book: Southwest Asia & South Asia
- Vocabulary Practice, p. 27

Inclusion

Match Terms and Definitions

Have students write the vocabulary words on a blank sheet of paper and cut out each one. Have them do the same with the definitions. Instruct students to shuffle the terms and definitions. Then have them work with a partner to match the terms with the correct definitions.

SECTION 1

GEOGRAPHY

Rural mountainous landscape in Southwest Asia

Camels and camel cart at fair in Rajasthan, India

Physical Geography of Southwest Asia

Connecting to Your World

Have you ever seen pictures of the moon's surface? Many of NASA's moon photographs show plains of powdery grey dust with mountains looming in the background. The moon is a very dry place with no water or plants. Although Earth has water, did you know that some regions on Earth look almost as desolate? Southwest Asia has harsh deserts and tall mountains. Yet Southwest Asia also has livable areas, small watered regions where millions of people live.

Mt. Damavand Snow-capped Mount Damavand is located in Iran.

Physical Features

KEY QUESTION What are the three subregions of Southwest Asia?

Perhaps you have heard the name *Middle East* on the news. That term is often used for the region of Southwest Asia. If you look at the map in this section, you will see that Southwest Asia is located where three continents—Africa, Asia, and Europe—meet. As a result, it has historically been a crossroads, a place where roads meet and a place through which many travelers journey.

Rub al-Khali Deserts such as the Rub al-Khali cover most of the Arabian Peninsula.

653

2 Focus & Motivate

3-Minute Warm-Up

Write on the board or display the transparency:

- Look at the images on this page. What are two geographical formations present in Southwest Asia? *(Possible Answers: deserts, mountains)* What can you learn about this region from these images? *(Possible Answer: It's a harsh environment, with extremes in temperatures.)*

Unit Transparency Book: Southwest Asia & South Asia
- 3-Minute Warm-Ups, TT1

Connecting to Your World

More About . . .

Mount Damavand

Located in the Elburz Range of northern Iran, Mount Damavand is the highest point in the country at 18,934 feet. It is actually a dormant volcano, but there have been no recorded eruptions.

3 Teach

Physical Features

Think, Pair, Share

- What landform covers most of the Arabian Peninsula? *(desert)*
- What is one reason that Southwest Asia has a dry climate? *(Mountains prevent rain from reaching inland areas.)*
- **Find Main Ideas** What are the three most important freshwater sources in Southwest Asia? *(Tigris, Euphrates, and Jordan rivers)*

SECTION 1 PROGRAM RESOURCES

Unit Resource Book: Southwest Asia and South Asia

ON LEVEL

Unit Resource Book
- Reading Study Guide, p. 3
- Vocabulary Practice, p. 27
- Map and Graph Skills: Reading an Elevation Map, pp. 29–30
- Section Quiz, p. 41

STRUGGLING READERS

Unit Resource Book
- Reading Study Guide with Additional Support, p. 9
- Outline Map Activities: Southwest Asia, pp. 31–32
- Section Quiz, p. 41
- Reteaching Activity, p. 44

eEdition with Audio DVD-ROM

ENGLISH LEARNERS

Pupil Edition in Spanish

eEdition with Audio DVD-ROM

Unit Resource Book
- RSG (Spanish), p. 15
- RSG with Additional Support (Spanish), p. 21

Multi-Language Glossary

Test Generator
- Section Quiz in Spanish

INCLUSION

Unit Resource Book
- RSG with Additional Support, p. 9
- Section Quiz, p. 41
- Reteaching Activity, p. 44

GIFTED & TALENTED

Unit Resource Book
- Connect to Math, p. 38
- Section Quiz, p. 41

PRE-AP

Unit Resource Book
- Interdisciplinary Projects: Water Shortage in the Middle East, p. 35
- Section Quiz, p. 41

TECHNOLOGY

Unit Transparency Book
- 3-Minute Warm-Ups, TT1
- Essential Question, TT2
- Critical Thinking, TT3
- Map: Southwest Asia: Physical, TT6

Daily Test Practice Transparencies
- Section 1, TT72

Power Presentations

ClassZone.com

World Cultures and Geography Video on DVD

More About . . .

The Rub al-Khali Desert

Covering about 250,000 square miles on the southern Arabian peninsula, the Rub al-Khali is one of the world's largest deserts. In fact, it takes up over one-quarter of Saudi Arabia. Winds have blown the sand into huge dunes, some over 660 feet high. It is extremely dry in this desert. Bedouin nomads still travel through the Rub al-Khali, but very few people live there.

Fun Facts!

The Dead Sea

Surrounded by desert, the Dead Sea lies between Jordan and Israel. At 1,300 feet below sea level, it is the lowest body of water on Earth. Most of the fresh water entering the Dead Sea comes from the Jordan River, but the sea has no outlet. In the desert heat, the fresh water evaporates quickly, leaving behind a concentrated solution of sodium chloride (salt), magnesium, calcium, and other minerals.

Teach

Providing Water

Talk About It

- Why do most people in Southwest Asia live near the coast, in river valleys, or near oases? *(to have access to water)*
- Why do some of the modern water projects cause tension or conflict in this region? *(Because water is scarce, countries fear that other countries in the region may divert and use too much water.)*
- **Find Main Ideas** In what two general ways have the people of Southwest Asia gained access to the water they need? *(by diverting water from freshwater sources, such as mountains and river valleys using ancient and modern technology; by living near water sources, such as rivers or oases)*

Unit Resource Book: Southwest Asia & South Asia

- Interdisciplinary Projects: Water Shortage in the Middle East, p. 35

Fun Facts!

THE DEAD SEA

The Dead Sea is a land-locked salt lake, so salty that almost nothing can live in the water. It is so salty that you cannot sink. Salt concentration in the Dead Sea is around 31.5 percent, about nine times higher than in the world's oceans. The evaporation rate of the water is more than 55 inches per year. This keeps the water very salty.

Peninsulas and Plateaus Southwest Asia can be divided into three subregions, which are listed below. (See map on opposite page.)

1. The **Arabian Peninsula** Ⓐ is bordered by the Red Sea, Arabian Sea, and Persian Gulf. A peninsula is a piece of land linked to a larger landmass but nearly surrounded by water. Deserts cover most of the Arabian Peninsula. The Rub al-Khali is the largest.
2. The **Anatolian peninsula** Ⓑ, or Anatolia (also called Asia Minor), is home to most of Turkey. This peninsula is bounded by the Black Sea to the north and the Mediterranean to the south.
3. Extending east from Anatolia is the **Plateau of Iran** Ⓒ. A **plateau** is a high area of uplifted, then eroded land. The Persian Gulf lies to the southwest and the Caspian Sea to the north.

Deserts, Mountains, and Rivers Southwest Asia is an arid region with many deserts. The dry climate is partially caused by the terrain. For example, mountains lining the Mediterranean and Black Sea coasts of Anatolia prevent rain from reaching inland areas.

Mountains also affect the climate of the Plateau of Iran. The Elburz Mountains stop moist winds blowing from the Caspian Sea, and the Zagros Mountains block winds from the Persian Gulf.

Rivers are the major freshwater sources in the region. The **Tigris River** and the **Euphrates River** both flow from the Anatolian peninsula to the Persian Gulf. The land between these rivers is a farming region. To the southwest, the Jordan River borders parts of Syria, Jordan, the West Bank, and Israel. The river flows into a lake that is called the Dead Sea because it is so salty that almost nothing can live in it.

SUMMARIZE Identify the three subregions of Southwest Asia.
Arabian Peninsula, Anatolian peninsula, Plateau of Iran

Providing Water

KEY QUESTION What methods have people used to provide water to arid Southwest Asia?

Living in an arid region can be challenging because people need water for drinking, bathing, and farming. Most people in Southwest Asia are crowded into coastal areas that receive rain or places with sources of water, such as river valleys and oases. An **oasis** is a fertile or green spot in the desert where underground water comes to the surface.

Ancient Techniques Residents of Southwest Asia have found ways to get the water they need to live in dryer areas. About 2,500 years ago, the people living in what is now Iran developed a technology called the qanat (KA•naht). A qanat Ⓓ, a type of underground water supply system, is illustrated on the next page.

DIFFERENTIATING INSTRUCTION

Struggling Readers

Main Ideas and Details

To help students differentiate between the two peninsula subregions of Southwest Asia, have them follow the outlines of the Arabian and Anatolian peninsulas on the map on the next page. Have students point to and pronounce the names of the bodies of water surrounding each peninsula. *(Arabian Peninsula: Red Sea, Arabian Sea, Persian Gulf; Anatolian Peninsula: Mediterranean and Black Seas)*

Pre-AP

Conserving Water in the Middle East

Have students list the advantages and disadvantages of the main methods people have obtained water in this region. Then have them identify ways people could conserve water and use water more efficiently. To extend this activity, have students research new methods and technologies that could benefit this region.

Physical Geography of Southwest Asia

Animated GEOGRAPHY

Qanats: An Ancient Irrigation Technology

Click here to see how a qanat works @ ClassZone.com

The ancient Iranians dug 30- to 100-foot shafts at the foot of mountains to tap into the water table. They built underground tunnels that followed the slope of the land. These qanats collect water that seeps into the ground from melting snow and rivers and streams. They can carry water to villages as much as 50 miles away for drinking and irrigating fields. They can supply most of a village's water.

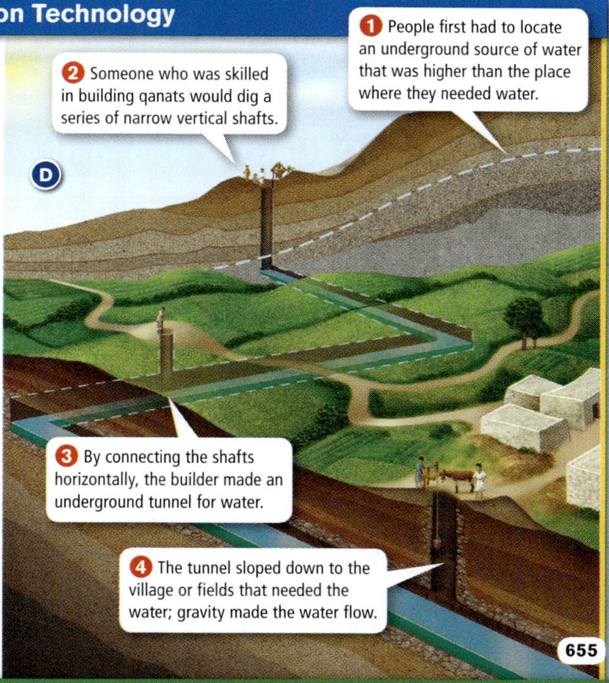

CRITICAL THINKING

1. **Make Inferences** What basic fact about the physical geography of the region has created a need for qanats?
2. **Draw Conclusions** What advantages do qanats have over canals that are located on the surface?

655

INTERDISCIPLINARY ACTIVITIES

CONNECT to Drama

Whose Water Is It?

Have students work in small groups to create a five-minute play about building a qanat, an irrigation canal, or a dam. Encourage students to focus on the tensions or conflicts that might arise, and how they could be resolved. Have each group perform their play for the rest of the class.

CONNECT to Language Arts

Write About It

Have students think about where the water they use every day comes from and how it gets to them. Then have them write a paragraph on how their water sources are both similar and different from water sources in Southwest Asia. Students can read their paragraphs aloud.

Physical Geography of Southwest Asia

Point out that this map highlights the physical features of Southwest Asia. Direct attention to the key. Make sure students understand the relationship between the colors in the map and the elevation levels shown in the key.

- At what elevation is the area where the Tigris and Euphrates flow into the Persian Gulf? *(0 feet, or sea level)*
- What body of water lies north of the Elburz Mountains? *(Caspian Sea)*
- What geographic feature important to life is not found in Saudi Arabia? *(large rivers)*

Unit Resource Book: Southwest Asia & South Asia

- Map and Graph Skills: Reading an Elevation Map, pp. 29–30
- Outline Map Activities: Southwest Asia, pp. 31–32

Animated GEOGRAPHY

Qanats

CLASSZONE.COM

In ancient times, qanats provided most of Iran's drinking water and irrigated the majority of Iran's crops. What might have been some challenges to building qanats? *(Possible Answers: It's difficult and dangerous to build underground tunnels; it requires skilled workers; people have to be able to locate an underground water source.)*

CRITICAL THINKING ANSWERS

1. **Make Inferences** Southwest Asia is an arid region with few water sources.
2. **Draw Conclusions** Qanats use underground sources of water and slope downhill so water can flow many miles to where it's needed, lessening water loss from evaporation.

Presentation Options

Use the **Power Presentations** DVD-ROM to project the animated map.

- Click the forward arrow to animate the creation of a qanat.
- Listen to the audio to learn more about how qanats were built and worked.

CONNECT to Math

Despite its desert location, Damascus is home to over 2,228,000 people, and some believe it to be the world's oldest inhabited settlement. Ask students to look at their graphs and say what months generally are the rainiest and what months are the driest. *(Damascus' mild winter season runs from December to February and has the most monthly rainfall. Summer runs from June to September, and has the least amount of rainfall.)*

Unit Resource Book: Southwest Asia & South Asia

- Connect to Math: Make a Climate Graph, p. 38

CONNECT to the Essential Question

How have physical geography and history interacted in Southwest and South Asia?

Ask students what they have learned so far that can help them answer this question. Students might mention:

- The region lies where Africa, Europe, and Asia meet, and has historically been a crossroads for many people.
- People settled between the Tigris and Euphrates because the soil was good for farming and there was a plentiful water supply.
- The arid climate has challenged people to find methods to provide water to drier areas, but also causes conflict over water sources.

In ancient times, the knowledge of how to make qanats spread from Iran to many other places. People living along rivers such as the Tigris and the Euphrates relied on surface water to grow their crops. They irrigated by building large canals to carry water from the river to their villages. Smaller canals carried water to the fields. In some arid regions such as Iran and Afghanistan, people still use qanats.

Modern Techniques Many nations in Southwest Asia have taken on huge projects to bring water to dry areas. Such water projects can make it possible to live on and farm previously unusable land. But sometimes the projects cause tension and even conflict.

In the 1950s, Israel began to build a giant system of canals, pipelines, and tunnels. Its purpose was to carry water from the Jordan River and Sea of Galilee to the Negev Desert. The project caused tensions with Syria and Jordan, who share those two water sources and worried that Israel would use too much water. Water and oil are two resources in the region that cause competition over scarce supplies.

Turkey has been working on its own large water project. Both the Tigris and Euphrates rivers begin in Turkey on the Anatolian peninsula. Turkey has been building a system of dams to create hydroelectric power and to divert water from the rivers to some of its farmlands. Syria and Iraq are located farther down the Tigris and Euphrates rivers. Both countries fear that if Turkey uses too much water for irrigation, they won't receive enough to meet their needs.

FIND MAIN IDEAS Describe the methods people have used to provide water in Southwest Asia. building qanats, irrigation canals, dams

CONNECT to Math

One definition of a desert is a region that receives less than 10 inches of precipitation a year. By that definition, many cities of Southwest Asia are built in deserts. One such city is Damascus, Syria, which receives approximately seven inches of rain a year (see chart). One way to learn more details about the climate of a place is to create a graph in which the bars show average monthly rainfall.

Average Monthly Rainfall
(Damascus, Syria)

Month	Rainfall (inches)
January	1.5
February	1.3
March	0.9
April	0.5
May	0.2
June	0.0
July	0.0
August	0.0
September	0.0
October	0.4
November	1.0
December	1.7

Source: www.worldclimate.com

Activity

Make a Rainfall Graph

Materials

- graph paper
- ruler
- pencil
- colored pencils

1. On a piece of paper, draw the horizontal axis and the vertical axis. On the vertical axis, place 20 marks, evenly spaced. Mark them 0.1, 0.2, 0.3, 0.4, etc., on up to 2.0.
2. Below the horizontal axis, write the names of all 12 months. You may abbreviate them. Space them evenly.
3. For each month, draw a bar whose height indicates the average amount of rainfall. Color all of the bars the same color.

DIFFERENTIATING INSTRUCTION: TIERED ACTIVITIES

OBJECTIVE Prepare and present a visual display on the three subregions of Southwest Asia.

Basic

Arrange the class into three groups. Assign each group one of the three subregions of Southwest Asia. Have each group create a poster with illustrations and labels for the geographical features, countries, climates, and waterways of its subregion. Ask each group to present its posters to the class.

On Level

Arrange the class into three groups. Assign each group one of the three subregions. Have each group create a map with illustrations and labels for the features, countries, climates, and important waterways of its subregion, and a depiction of how water is supplied. Have each group present its illustrated maps.

Challenge

Arrange the class into three groups. Assign each group one of the three subregions. Have each group create a card game with illustrations that tests knowledge of geographical features, countries, climates, important waterways, and methods of providing water. Ask each group to present its game.

The Threat of Earthquakes

KEY QUESTION Why does Southwest Asia have many earthquakes?

Southwest Asia has long been a crossroads where people meet, but it is also a place where several of the great plates that make up Earth's crust meet in a kind of continental collision zone. As a result, Southwest Asia is prone to deadly earthquakes. In August 1999, a major earthquake hit Izmit, Turkey. More than 17,000 people were killed, many of them crushed when buildings collapsed on them. Thousands more lost their homes. In December 2003, the city of Bam, Iran, experienced an earthquake that killed about 26,000 people. Sometimes, faulty building construction, rather than the strength of the earthquakes, causes many of these deaths.

SUMMARIZE Explain why Southwest Asia has many earthquakes. Continental plates meet in the region in a collision zone.

Earthquake Damage People walk past rubble from collapsed buildings in western Turkey after the 1999 earthquake.

Section 1 Assessment

ONLINE QUIZ For test practice, go to **Interactive Review @ClassZone.com**

TERMS & NAMES

1. **Explain the importance of**
   - Arabian Peninsula
   - Plateau of Iran
   - Euphrates River
   - oasis

USE YOUR READING NOTES

2. **Find Main Ideas** Use your completed chart to answer the following question: What are some possible solutions to the problems caused by living in such an arid region?

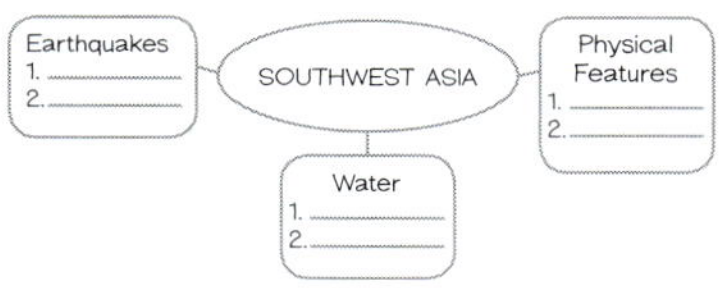

KEY IDEAS

3. How do the mountains of Southwest Asia affect its climate?
4. How did people in ancient times move water from the Tigris and Euphrates rivers to their crops?
5. What might be done to limit earthquake damage in the region?

CRITICAL THINKING

6. **Make Inferences** When two countries rely on the same river, what issues might arise?
7. **Compare and Contrast** How is a qanat different from a surface irrigation system?
8. **CONNECT to Today** What might be some problems with building dams in a region prone to earthquakes?
9. **SCIENCE Create an Informative Poster** Working with a partner, research in the library or on the Internet to learn about the Richter scale, which is used to measure earthquakes. Then create a poster listing how much damage happens at each level on the scale.

Teach

The Threat of Earthquakes

Think, Pair, Share

- Why is Southwest Asia prone to earthquakes? *(because several plates of the Earth's crust collide there)*
- What other factor may make earthquakes in Southwest Asia particularly deadly? *(poor building construction)*

4 Assess & Reteach

Assess Have students complete the Section Assessment.

Unit Resource Book: Southwest Asia & South Asia
- Section Quiz, p. 41

Interactive Review @ClassZone.com
- Section Quiz

Power Presentations

Test Generator

Reteach Have pairs of students reread one subsection of Section 1. Then have each pair write down the main ideas from that subsection in their own words. Have students share their main ideas and revise them as needed.

Unit Resource Book: Southwest Asia & South Asia
- Reteaching Activity, p. 44

SECTION 1 ASSESSMENT ANSWERS

Terms & Names

1. Arabian Peninsula: lies in Southwest Asia between the Red Sea and the Persian Gulf; Plateau of Iran: lies in central and eastern Iran and is surrounded by mountains; Euphrates River: flows from a plateau on the Anatolian peninsula to the Persian Gulf; oasis: a fertile or green spot in the desert created by underground water coming to the surface

Use Your Reading Notes

2. qanats, irrigation canals, dams, pipelines, tunnels, and huge projects to bring water to dry areas
   Diagram *Earthquakes: collision zone, improved building construction; Water: arid, oases, qanats, water projects; Physical Features: peninsulas, plateaus, deserts, mountains*

Key Ideas

3. They block moisture-bearing winds, making the region more arid.
4. by irrigation canals; qanats
5. Faulty building construction might be improved.

Critical Thinking

6. Possible answers: There could be conflict if the country that lies upstream pollutes the river. There could also be disputes about transportation rights. A country might fear that it won't receive enough water.
7. Possible answer: A qanat uses an underground tunnel.
8. Possible answer: Students may suggest that an earthquake will break the dam and cause a destructive flood.
9. Use the rubric to score posters.

Poster Rubric

	Elements of Poster	Effectiveness of Design
4	reflects thorough research	neat, organized, attractive
3	mostly complete research	neat and organized
2	some research	somewhat organized
1	little or no research	unorganized, messy

1 Plan & Prepare

Objectives

- Identify the physical features of the Indian subcontinent
- Describe the extreme weather and natural disasters that are common in South Asia

Read for the Essential Question

Help students read for a purpose by reminding them of the Essential Question: "How have physical geography and history interacted in Southwest and South Asia?" Point out that they can use text features, such as Key Questions, special features, images, and maps to form answers to the question.

Vocabulary

Best Practices Toolkit

Use the ***Best Practices Toolkit*** to model strategies for vocabulary notetaking. Vary strategies throughout the year. Choose from: *Knowledge Rating, Predicting ABCs, Definition Mapping, Word Sort, Frayer Model (Word Squares), Magnet Words, Word Wheel,* and *Student VOC.*

Vocabulary Strategies, TT9–TT16

Reading Strategy

Best Practices Toolkit

Display the Three-Column Chart transparency.

- Remind students that when they compare and contrast, they analyze how two or more items are both similar and different.
- Model filling in a few entries for physical features of Southwest Asia.
- Have students suggest entries for the rest of the chart.
- Display the graphic organizer for students to review.

Three-Column Chart, TT21

SECTION 2 Reading for Understanding

Key Ideas

BEFORE, YOU LEARNED

The physical geography of Southwest Asia is made up of deserts and mountains, and the region is mostly arid.

NOW YOU WILL LEARN

The physical geography of South Asia is made up of mountains, plateaus, and plains, and there are wet and dry seasons.

Vocabulary

TERMS & NAMES

subcontinent a large landmass that is part of a continent but is considered a separate region

Himalayas (HIHM•uh•LAY•uhz) the highest mountains in the world, which stretch along northern India, separating it from the rest of Asia

Mount Everest the world's tallest mountain, located in the Himalayas

Deccan (DEHK•uhn) **Plateau** a high area of land at the center of the Indian subcontinent

Indus River a river in South Asia that flows from the Himalayas to the Arabian Sea

Ganges (GAN•JEEZ) **River** a river of South Asia that flows southeast from the Himalayas to the Bay of Bengal

monsoon a seasonal wind system that produces a wet or dry period in a region

cyclone a violent storm with fierce winds that rotate in a circular pattern like a hurricane

tsunami (tsu•NAH•mee) a series of giant, destructive ocean waves caused by underwater earthquakes, volcanoes, or landslides

BACKGROUND VOCABULARY

famine a severe shortage of food that causes widespread hunger

Reading Strategy

Re-create the chart shown at right. As you read and respond to the **KEY QUESTIONS**, use the chart to compare and contrast the physical geography of South Asia with that of Southwest Asia.

See Skillbuilder Handbook, page R9

COMPARE AND CONTRAST

	SOUTH ASIA	SOUTHWEST ASIA
Physical Features		
Climate		

GRAPHIC ORGANIZERS
Go to Interactive Review @ ClassZone.com

658 Chapter 21

PRETEACHING VOCABULARY

English Learners

Pronounce and Preview

Pronounce each term for students. Review the meanings of words within definitions, such as *seasonal* and *shortage.*

- To modify vocabulary learning, have students complete worksheets as they read, instead of afterward.

Unit Resource Book: Southwest & South Asia
- Vocabulary Practice, p. 27

Inclusion

Locate Vocabulary Terms on a Map

Provide support for students as they take turns reading the definitions of the terms out loud. Use a large wall map of South Asia or the Physical Geography of South Asia map in this section, and challenge students to locate the geographic features and places mentioned in the vocabulary terms.

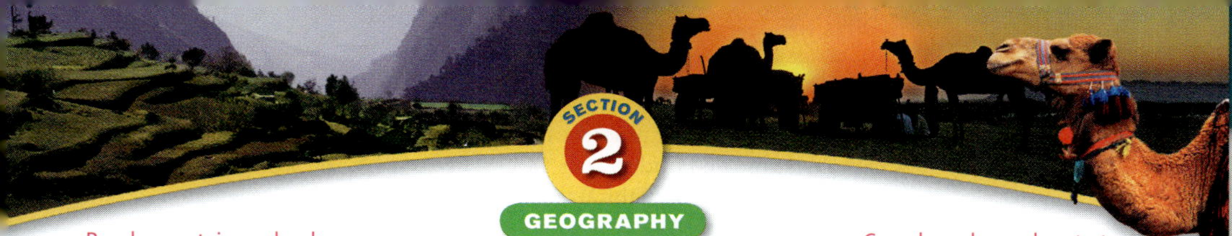

SECTION 2 GEOGRAPHY

Rural mountainous landscape in Southwest Asia

Camels and camel cart at fair in Rajasthan, India

Physical Geography of South Asia

Connecting to Your World

In many places, just five inches of rain can cause floods. In Mumbai (also known as Bombay), India, 37 inches of rain fell on July 26, 2005. Streets flooded and landslides buried neighborhoods. India has a climate with separate dry and wet seasons, but even during the wet season, 37 inches of rain in one day is rare.

Mumbai, India Man buying vegetables from a street vendor during the 2005 monsoon

Physical Features of the Peninsula

KEY QUESTION What are the three main rivers of the Indian subcontinent?

Millions of years ago, a huge supercontinent, Pangaea, broke apart into separate continents. Scientists believe that when that happened, India became a separate landmass. Over time, it inched north until it hit Asia. The collision pushed up high mountains where the two landmasses met. Those mountains form the northern boundary of the kite-shaped Indian subcontinent. A **subcontinent** is a large landmass that is part of a continent, yet is considered a separate region or entity.

Indian Subcontinent Scientists predict that as the two tectonic plates that formed the Himalayas continue to push against each other, India will become one huge mountain range.

Formation of the Himalayas

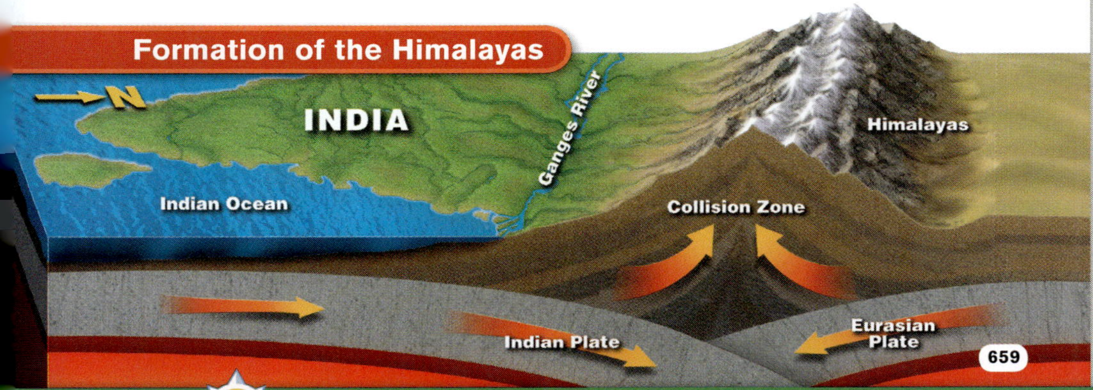

659

2 Focus & Motivate

3-Minute Warm-Up

Write on the board or display the transparency:

- Look at the illustration on this page. What two plates on the Earth's crust collided to form the Himalayas? *(the Indian plate and the Eurasian plate)*

Unit Transparency Book: Southwest Asia & South Asia
- 3-Minute Warm-Ups, TT1

Connecting to Your World

More About . . .

Rainfall in India

The months of heaviest rainfall in Mumbai are June thru September. July has the heaviest average rainfall—about 25 inches.

3 Teach

Physical Features of the Peninsula

Think, Pair, Share

- What geographical feature lies between the Western and Eastern Ghats? *(the Deccan Plateau)*
- What large bodies of water surround India? *(Arabian Sea, Bay of Bengal, Indian Ocean)*
- **Compare and Contrast** What is one major difference between the physical geography of Southwest Asia and South Asia? *(South Asia is considered a subcontinent.)*

SECTION 2 PROGRAM RESOURCES

Unit Resource Book: Southwest Asia and South Asia

ON LEVEL

Unit Resource Book
- Reading Study Guide, p. 5
- Vocabulary Practice, p. 27
- Section Quiz, p. 42

STRUGGLING READERS

Unit Resource Book
- Reading Study Guide with Additional Support, p. 11
- Section Quiz, p. 42
- Reteaching Activity, p. 45

eEdition with Audio DVD-ROM

ENGLISH LEARNERS

Pupil Edition in Spanish

eEdition with Audio DVD-ROM

Unit Resource Book
- RSG (Spanish), p. 17
- RSG with Additional Support (Spanish), p. 23

Multi-Language Glossary

Test Generator
- Section Quiz in Spanish

INCLUSION

Unit Resource Book
- RSG with Additional Support, p. 11
- Section Quiz, p. 42
- Reteaching Activity, p. 45

GIFTED & TALENTED

Unit Resource Book
- Section Quiz, p. 42

PRE-AP

Unit Resource Book
- Interdisciplinary Projects: Understanding Monsoons, p. 36
- Section Quiz, p. 42

TECHNOLOGY

Unit Transparency Book
- 3-Minute Warm-Ups, TT1
- Essential Question, TT2
- Critical Thinking, TT4
- Map: South Asia: Physical, TT7

Daily Test Practice Transparencies
- Section 2, TT73

Power Presentations

ClassZone.com

World Cultures and Geography Video on DVD

More About . . .

Mount Everest

Mount Everest is located on the border of Nepal and Tibet. Its elevation is 29,035 feet, making it the highest point on earth. Weather on the summit is extreme with temperatures dipping as low as -76°F. During July, the warmest month, daytime temperatures average just -2°F. Because of geological forces, Mount Everest is moving a few inches annually to the northeast. It's getting taller, too—by less than an inch every year.

COMPARING

River Systems

Remind students that river valleys have played a crucial role in the development of civilization. The Indus River Valley was home to some of the world's most ancient civilizations. It remains a vital source of water for the surrounding arid regions. The Ganges River flows through one of the most fertile areas of the world. Ancient civilizations prospered here, too. Hindus consider the Ganges holy, and millions make annual pilgrimages to the Ganges to bathe in its sacred waters.

CRITICAL THINKING ANSWER

Compare Possible Answers: Both rivers begin in the Himalayas; they both flow mostly southward; they both flow for over 1,500 miles.

Make It Fun

Teacher-Tested Activities

Amber L. McVey, Antioch Middle School, Gladstone, Missouri

I like to start each class with some sort of a daily journal or warm-up. It's normally a question regarding something the students learned the day before or a preview of what they are going to learn that day. I also use it as a skill builder. For example:

- Pop up a map on the overhead and ask students to find the latitude or longitude of a city.
- Present a passage from a primary source and have the students respond to it.

Mountains and Plateaus The mountains of South Asia include the Hindu Kush in the northwest. The **Himalayas** (HIHM•uh•LAY•uhz) extend east of the Hindu Kush. **Mount Everest** Ⓐ, the world's tallest peak, is located in the Himalayas. (See map on opposite page.)

Mountains called the Western Ghats and Eastern Ghats run along India's two coastlines, which form a "V" in the south. Between them stretches an area of high land called the **Deccan Plateau** Ⓑ. It is a largely arid region. One of the world's largest deserts, the Thar Desert, lies northwest of the Deccan.

Bodies of Water and Islands The Arabian Sea borders India's west coast and the Bay of Bengal borders the east. Both are part of the Indian Ocean, which lies to the south. Several rivers water the subcontinent. In the west, the **Indus River** Ⓒ runs from the Himalayas to the Arabian Sea. The **Ganges** (GAN•JEEZ) **River** Ⓓ—one of the largest, most important rivers of India—flows from the Himalayas to the Bay of Bengal. The Brahmaputra (BRAH•muh•POO•truh) River Ⓔ also starts in the Himalayas, but it joins the Ganges before reaching the bay. Together, the rivers have deposited silt and created an immense fertile delta in what is now Bangladesh (BAHN•gluh•DEHSH).

COMPARING River Systems

Indus River, Stakna Ldakh, India

Ganges River, Varanasi, India

	INDUS RIVER	GANGES RIVER
Source	the Himalayas	the Himalayas
Route	mostly south through Pakistan	mostly southeast through India
Length	1,800 miles	1,560 miles
Destination	Arabian Sea	Bay of Bengal

CRITICAL THINKING

Compare What do the two rivers have in common?

DIFFERENTIATING INSTRUCTION

Struggling Readers

Define and Illustrate Terms

Have students work in small groups to review the "Physical Features of the Peninsula" subsection and features, and define these terms in their own words: *tectonic plates, collision zone, delta, silt.* Have students create simple drawings or symbols to help them remember each term. Then have them share their definitions and drawings/symbols with the rest of the class.

English Learners

Vocabulary: Prefixes

Explain that the prefix *super-* means "extreme" or "more than." Point out the word *supercontinent* on the previous page (second paragraph, first line). Ask students to reread the sentence to see if they can figure out what it means. Suggest that they pay attention to the words *huge* and *broke apart*, to help them figure out the meaning. Ask students to add the prefix super- to other words, such as *hero*, *power*, *highway*, and *human*.

Physical Geography of South Asia

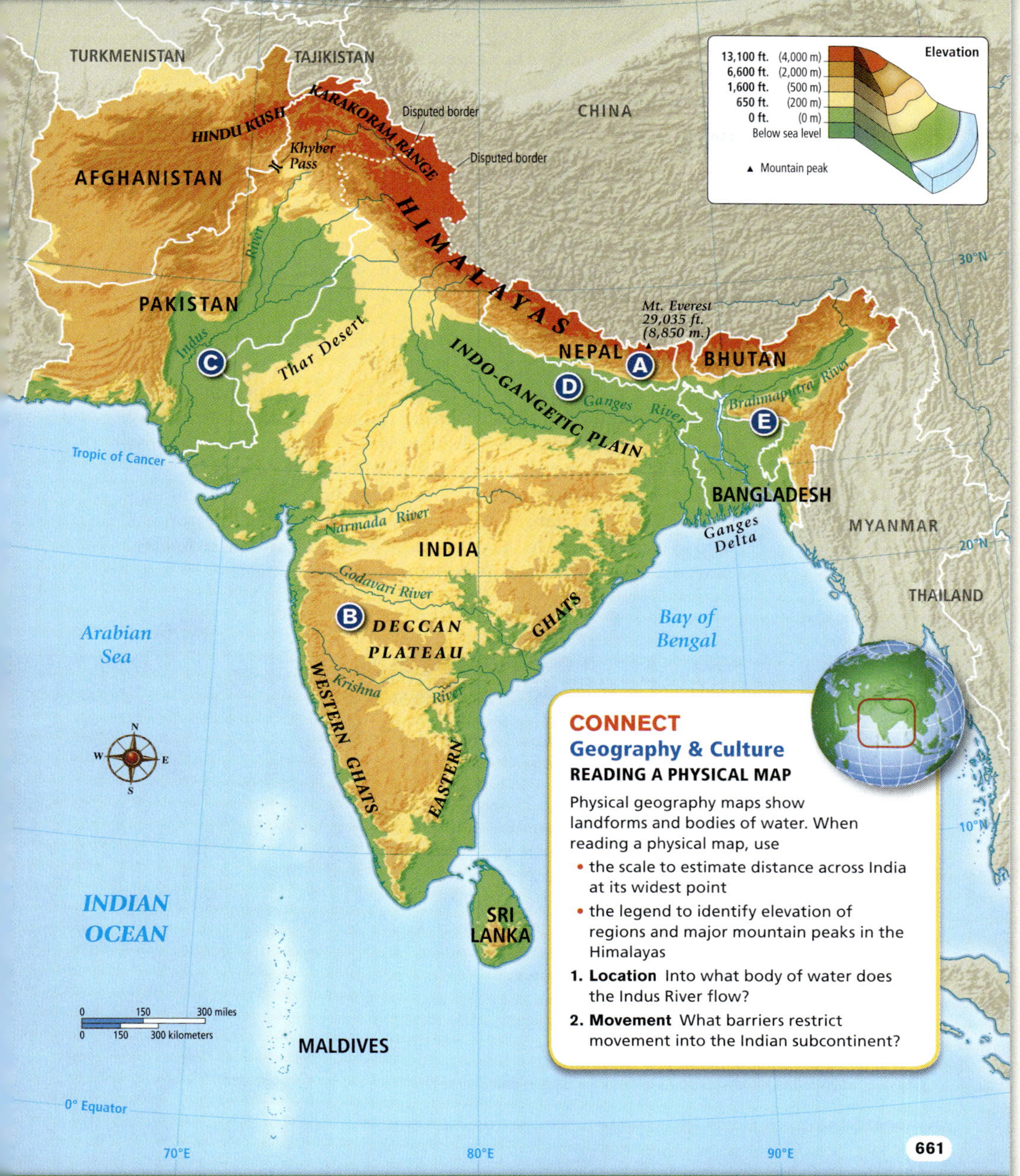

CONNECT
Geography & Culture
READING A PHYSICAL MAP

Physical geography maps show landforms and bodies of water. When reading a physical map, use

- the scale to estimate distance across India at its widest point
- the legend to identify elevation of regions and major mountain peaks in the Himalayas

1. **Location** Into what body of water does the Indus River flow?
2. **Movement** What barriers restrict movement into the Indian subcontinent?

661

INTERDISCIPLINARY ACTIVITIES

CONNECT to Art

South Asian Islands Posters

Have students work in pairs to create a poster about either Sri Lanka or the Maldives. Students can use library or Internet resources to find out more about the islands. The posters should illustrate the geography, physical features, and climates of these islands. Have students provide captions for each illustration and write a title for their posters.

CONNECT to Math

Estimating Distance

Refer students to the mileage scale on the map on this page. First, have pairs of students take turns estimating distances on the map without measuring. Students can choose to measure distances between geographical features, across the widest points of countries, between islands, and so on. Then have students use the scale and measure these distances, comparing them to their estimates. Finally, discuss strategies for making better estimates.

Physical Geography of South Asia

CONNECT Geography & Culture

Reading a Physical Map Ask students to identify several physical features on the map that extend across more than one country. *(Possible Answers: Khyber Pass connects Afghanistan and Pakistan; Himalayas extend across Pakistan, India, Nepal and Bhutan; Ganges Delta in Bangladesh and India)* Then talk about how many physical features of this region serve as barriers to the movement of people, goods, and ideas, while others help connect people to each other.

- What are some features that separate people? *(mountains, such as the Himalayas; deserts, such as the Thar Desert)*
- Which geographic features can aid movement? *(rivers, such as the Ganges; mountain passes, such as the Khyber Pass)*

ANSWERS

1. **Location** Arabian Sea
2. **Movement** Himalayan Mountains

More About . . .

The Thar Desert

Although the Thar Desert is one of the driest regions on earth—some parts receiving an average of less than four inches of rain per year—it is home to many species of animals. Its many small rodents serve as prey for desert foxes, caracals (a type of wild cat), and falcons. Perhaps most surprisingly, the Thar Desert is home to many people who graze sheep and goats. In fact, out of all the world's deserts, the Thar is the most densely populated.

Teach

Extreme Weather and Natural Disasters

Reader, Recorder, Reporter

- Why do summer monsoons bring rain, while winter monsoons do not? *(Winter winds blow from the northeast, across vast stretches of dry land, while summer monsoons blow from the southwest and pick up moisture from the Indian Ocean.)*
- **Compare and Contrast** What are some differences in the climate of South Asia as compared to Southwest Asia? *(Possible Answers: South Asia has monsoons, while Southwest Asia does not; South Asia has dry and wet seasons, while Southwest Asia is mostly dry.)*

ANALYZING Primary Sources

The Associated Press—or AP, for short—is the largest news agency in the world. Founded in 1848, it is also the oldest. With 3,700 employees and 242 bureaus, or offices, around the world, it provides global, around-the-clock news to about one billion people every day via newspapers, television stations, and radio stations.

- Why would a news agency such as AP be important during a natural disaster? *(Possible Answer: Its reports could generate donations and prompt volunteers to help.)*
- What countries of South Asia were affected by the 2002 flooding from monsoon rains? *(India, Bangladesh, and Nepal)*
- How many people in South Asia were displaced or trapped by the monsoon flooding? *(about 25 million)*

DOCUMENT-BASED QUESTION ANSWER

Possible Answer: The Brahmaputra River starts in the Himalayas. In addition, the Ganges, which joins the Brahmaputra, also starts in the Himalayas.

Unit Resource Book: Southwest Asia & South Asia
- Interdisciplinary Projects: Understanding Monsoons, p. 36

Two island nations lie off India's coast. To the southeast is Sri Lanka, a tropical island with some mountains. To the southwest is a group of islands called the Maldives, flung over 500 miles of ocean. The Maldives are made of reefs, ridges of coral formed on the slopes of undersea volcanoes. They barely rise above the water's surface.

SUMMARIZE List the three main rivers of the Indian subcontinent. the Indus, the Ganges, the Brahmaputra

Extreme Weather and Natural Disasters

KEY QUESTION What are some natural disasters in South Asia?

Floods are common in South Asia. Even though the climate has wet and dry seasons, no one can predict yearly rainfall. The variable rainfall sometimes causes problems.

Monsoons India's climate is affected by monsoons. A **monsoon** is a seasonal wind system that produces wet and dry seasons. In India, winter winds blow from the northeast and cross vast stretches of land and high mountains before reaching India. As a result, the winter monsoon is dry. Summer winds blow from the southwest and cross the Indian Ocean, so they pick up moisture and bring rain.

If the summer monsoon brings too much rain, floods occur. But that is not the only danger. A drought can result if the monsoon has failed to bring normal levels of moisture. Drought can lead to **famine**, a severe shortage of food that causes widespread hunger.

ANALYZING Primary Sources

Associated Press Newspapers cover disasters as they are happening. This story tells of monsoon flooding in India.

> *August 16, 2002* Monsoon rains have sent India's Brahmaputra River surging from its channel, swallowing up villages, drowning hundreds of people and leaving millions homeless. . . . Annual monsoon flooding has wreaked havoc across South Asia, killing more than 900 people in India, Bangladesh, and Nepal since June and displacing or trapping about 25 million more.
>
> Source: *The Associated Press*

DOCUMENT-BASED QUESTION

Why might rain in the Himalayas contribute to flooding in Bangladesh?

DIFFERENTIATING INSTRUCTION: TIERED ACTIVITIES

OBJECTIVE Create a chart listing recent natural disasters in South Asia.

Basic

Have students work independently to make a two-column chart labeled *Recent Natural Disasters of South Asia.* In the first column students should list natural disasters that have occurred recently in South Asia. In the second column, have students list which countries or regions were affected by that event.

On Level

Have each student create a four-column chart labeled *Recent Natural Disasters of South Asia.* In addition to the first two columns for the Basic activity, have students add a column to record the date of the event, and a fourth column to record the number of people affected. Each event should be listed in chronological order.

Challenge

In addition to the On Level activity, have students use reliable Internet Web sites to find out how well South Asia has recovered from the one of these disasters, such as the tsunami of December 2004. Students should summarize what has improved and what help is still needed.

Another kind of weather that causes hardship and destruction is a **cyclone**, a violent storm with fierce, rotating winds, similar to a hurricane. Cyclones can form in the ocean and then blow ashore. In 1970, a severe cyclone hit Bangladesh, and about 300,000 people were killed.

Natural Disasters India became part of Asia when one great landmass collided with another. Those two plates of Earth's crust are still pushing against each other, which makes South Asia prone to earthquakes. In October 2005, an earthquake that measured 7.6 on the Richter scale rocked Kashmir. Many buildings collapsed or were buried by landslides. About 75,000 people died.

2004 Tsunami Aerial photo shows damage caused by 2004 tsunami to seaside town in Sri Lanka.

Earthquakes in the ocean floor can lift part of Earth's crust, shifting a huge amount of seawater. The result is a **tsunami**, a series of giant ocean waves that wash over coastal areas. In December 2004, a powerful undersea earthquake occurred in the Indian Ocean near Indonesia. This triggered a tsunami that raced west across the ocean and caused damage in 12 nations, including a number in South Asia. The waves left destroyed villages, ruined fields, and dead bodies. An estimated 300,000 people died.

CATEGORIZE List some natural disasters in South Asia.
cyclones, earthquakes, tsunamis

Section 2 Assessment

ONLINE QUIZ For test practice, go to **Interactive Review @ClassZone.com**

TERMS & NAMES

1. **Explain the importance of**
   - Himalayas
   - Ganges River
   - cyclone
   - tsunami

USE YOUR READING NOTES

2. **Compare and Contrast** Use your completed chart to answer the following question:
   What are some of the differences between South and Southwest Asia in terms of physical features and climate?

	SOUTH ASIA	SOUTHWEST ASIA
Physical Features		
Climate		

KEY IDEAS

3. Why is India considered a subcontinent?
4. Why is the winter monsoon dry and the summer monsoon wet?
5. Why is South Asia so prone to earthquakes?

CRITICAL THINKING

6. **Make Inferences** How might farmers in a monsoon climate plan their work around the seasons?
7. **Evaluate** What might be some advantages and disadvantages of living in Bangladesh?
8. **CONNECT to Today** Many scientists claim that the global climate is getting warmer and polar icecaps might melt, raising the level of the oceans. Predict how that might affect the Maldives and Bangladesh.
9. **WRITING** **Write an Appeal** Review the descriptions of the disasters in this section. Choose one disaster, and imagine that you work for a relief agency such as UNICEF. Write an appeal or advertisement asking for aid for the victims.

CONNECT to the Essential Question

How have physical geography and history interacted in Southwest and South Asia?

Ask students what they have learned so far that can help them answer this question. Students might mention:

- The Indus, Ganges, and Brahmaputra rivers have created fertile regions.
- High mountains and the Indian Ocean influence the region's climate, creating monsoon seasons.

4 Assess & Reteach

Assess Have students complete the Section Assessment.

Unit Resource Book: Southwest Asia & South Asia
- Section Quiz, p. 42

Interactive Review @ClassZone.com
- Section Quiz

Power Presentations

Test Generator

Reteach Have each student write on a slip of paper one "True or False" statement about South Asia's physical geography or climate. Then collect the statements in a box. Have each student select a slip of paper, read a statement, and say whether it is true or false. For each false statement, have the student correct it.

Unit Resource Book: Southwest Asia & South Asia
- Reteaching Activity, p. 45

SECTION 2 ASSESSMENT ANSWERS

Terms & Names

1. Himalayas: the highest mountains in the world, stretch along northern India, separating it from the rest of Asia; Ganges River: flows southeast from the Himalayas to the Bay of Bengal; cyclone: a violent storm with fierce winds that rotate in a circular pattern; tsunami: a series of giant, destructive ocean waves caused by an underwater earthquake, landslide, or volcanic eruption.

Use Your Reading Notes

2. South Asia has higher mountains and is subject to monsoons and cyclones.

Key Ideas

3. part of Asia but separate landform because mountains separate it from rest of Asia
4. Winter monsoon crosses land and mountains before reaching India; summer monsoon crosses ocean.
5. Several plates of Earth's crust collide there.

Critical Thinking

6. The farmers will probably plant at the beginning of the rainy season or right before it.
7. Fertile soil of delta is an advantage, but occurrence of floods and cyclones is disadvantage.
8. Since the islands of the Maldives barely rise above sea level, higher ocean levels might cause them to disappear. Bangladesh is also low lying so that rising sea levels will cause severe flooding.
9. Use the rubric to score appeals.

Appeal Rubric

	Accuracy of Description	Errors in Mechanics
4	thoughtful and detailed	few or none
3	detailed	some
2	shows some knowledge	many
1	unclear	many major

1 Plan & Prepare

Objectives: Identify why monsoons are a necessary, but sometimes destructive, part of life in South Asia, and how the region's geography affects monsoons.

Visual Vocabulary: Ask students to describe what they see in the illustration. Call out familiar items listed below, asking students to locate each one on the page. Then tell them the precise term, as listed in boldface. Add terms to a class vocabulary list, illustrated poster, or Word Wall.

- gathered crops **(harvest)**
- material deposited by water **(sediment)**
- large mountain range in the north of India **(Himalayas)**
- something harmful that happens suddenly **(disaster)**

2 Teach

Engage . . .

DISPLAY the feature "Monsoons" from **Power Presentations** or **ClassZone.com.** As the illustration loads on screen, have volunteers point to some of the vocabulary terms listed above. Then ask

- What are the people planting? Why are they planting this crop?
- Why are the people in the bottom photograph waiting in line?
- What are the challenges farmers and others face from monsoon rains?
- What can people do to prevent flooding, or limit its damage?

Encourage all answers.

Explore . . .

ROLL OVER sections of the illustration or click on items in the menu. Ask students to choose which areas they would like to explore further.

- Then return to the discussion questions. Have any answers changed?
- Have students jot down a list of at least three new questions that arise from watching the rollover features.

Animated GEOGRAPHY
Monsoons

Click here to explore the effects of monsoons @ClassZone.com

LIVING WITH EXTREME WEATHER

Monsoons shape the lives of people in South Asia. If the summer monsoons bring the right amount of rain, then crops grow and there is a good harvest. If monsoons bring too much rain, then crops are flooded and the harvest is spoiled. The illustration at right shows both positive (near right) and negative (far right) effects of summer monsoons.

Click here to see how Asia benefits from summer monsoons. Learn how monsoons provide relief from months of scorching heat. Rain helps crops to grow (as shown above in a rice field), nourishes the rain forest, and produces floodwaters that deposit sediment to replenish the soil. (A)

Click here to witness the harmful effects of summer monsoons. See how too much monsoon rain causes floods that ruin crops and lives. (B)

DIFFERENTIATING INSTRUCTION: TIERED ACTIVITIES

OBJECTIVE Identify why monsoons are a necessary, but sometimes destructive, part of life in South Asia, and how the region's geography affects monsoons.

Basic

Have students take a sheet of paper and draw a line down the center. They should label one side *Benefits of Monsoons* and the other side *Harmful Effects of Monsoons.* Using their textbooks and other classroom resources, have students complete their charts. To extend this activity, have students illustrate one benefit and one harmful effect, and write a caption for each.

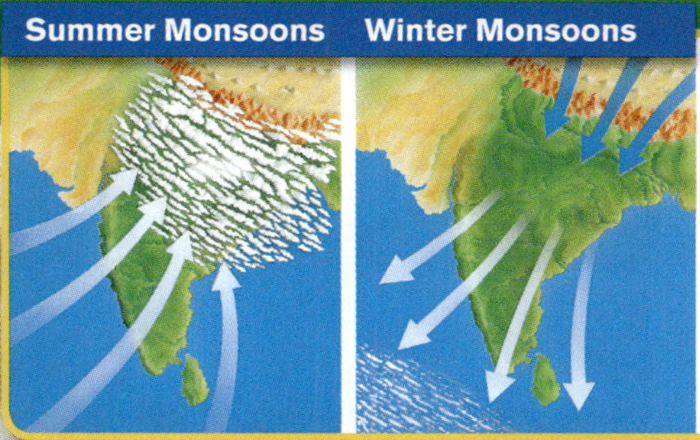

The summer and winter monsoons of South Asia blow from different directions. The summer monsoons blow from the southwest and the winter monsoons from the northeast.

- The summer monsoons are wet winds.
- The summer monsoons pick up water from the ocean.
- The winter monsoons are dry winds.
- The winter monsoons blow dry and cool because they cross the Himalayas.

GeoActivity

Make a Poster Working with a partner, use the Internet to research summer monsoons. Create a poster showing ways of coping with flooding from monsoon rains, such as disaster shelters. Add a map showing the extent of monsoons in South Asia. Write a paragraph explaining how summer monsoons affect the people of the region.

On Level

Show students what a flow chart looks like by drawing one on the board or overhead. Have students work individually to create their own flow charts of what happens when monsoons bring too much rain, just the right amount of rain, or not enough rain. Charts should be clearly labeled.

Challenge

Have students use library or Internet sources to find out what atmospheric, oceanic, and geographic conditions create monsoons. Have students draw a schematic diagram—a detailed diagram that explains, sometimes step-by-step, how something works—to show how these influences interact. (You may want to provide examples of schematic drawings for students to use as models or have students search for their own examples.)

Explain . . .

CHOOSE topics to help students answer their questions, such as:

- **Rainfall:** Monsoons provide anywhere from 50%–90% of the annual rainfall in India, depending on the region. The highest annual rainfall anywhere in the world was reported in 1860–1861 when over 1,000 inches of rain fell in the foothills of the Himalayas.
- **Floods:** Farmers depend on the monsoons. If monsoons bring less rain than expected, drought destroys the crops. Floods can even be beneficial, to some extent. During the summer monsoons, for example, rivers in Bangladesh overflow and flood about 20% of the countryside, bringing water and nourishing silt deposits to the crops.

3 Wrap Up

Elaborate . . .

ASSIGN students to groups of 3 or 4 to report on the topic, "International Aid and Monsoons."

- Have pairs of student use the Internet to visit one Web site of a reputable international aid organization (such as the Red Cross, Doctors Without Borders, UNICEF, or Oxfam) and research how it has helped victims of monsoon flooding in South Asia. Students can summarize their findings and share them with the rest of the class.

Evaluate . . .

Make a Poster

Activity Rubric

4 Excellent The student's poster has numerous details of how people cope with monsoon flooding; the paragraph includes all key points.

3 Strong The poster has some details; the paragraph includes most key points.

2 Fair The poster has few details; the paragraph includes some key points.

1 Poor The poster lacks detail; the paragraph fails to include key points.

1 Plan & Prepare

Objectives

- Describe how civilization arose in Mesopotamia and the Fertile Crescent
- Compare the accomplishments of the rulers Sargon and Hammurabi
- Identify features of the Indus River Valley Civilization

Read for the Essential Question

Help students read for a purpose by reminding them of the Essential Question: "How have physical geography and history interacted in Southwest and South Asia?" Point out that they can use text features, such as Key Questions, special features, images, and maps to form answers to the question.

Vocabulary

Best Practices Toolkit

Use the ***Best Practices Toolkit*** to model strategies for vocabulary notetaking. Vary strategies throughout the year. Choose from: *Knowledge Rating, Predicting ABCs, Definition Mapping, Word Sort, Frayer Model (Word Squares), Magnet Words, Word Wheel,* and *Student VOC.*

Vocabulary Strategies, TT9–TT16

Reading Strategy

Best Practices Toolkit

Display the T Chart/Two-Column Chart transparency

- Remind students that when they summarize, they restate a paragraph, a passage, or a section in fewer words.
- Model summarizing, using the five traits of civilization as an example. Then model filling in entries on the chart.
- Have students suggest entries.
- Display the graphic organizer for students to review.

T Chart/Two-Column Chart, TT20

SECTION 3 Reading for Understanding

Key Ideas

BEFORE, YOU LEARNED

The physical geography of both Southwest Asia and South Asia is rich and varied.

NOW YOU WILL LEARN

In both regions, civilizations developed first in river valleys where there was an abundant supply of water.

Vocabulary

TERMS & NAMES

Mesopotamia the land between the Tigris and Euphrates rivers

silt the fine soil carried by rivers and deposited on nearby lands

civilization an advanced form of culture that developed in cities

Sumer a region of city-states in Mesopotamia that was home to the first civilization

city-state a political unit made up of a city and its surrounding lands

ziggurat a temple built atop a series of increasingly smaller platforms

polytheistic believing in many gods

cuneiform the first-known writing system, which used wedge-shaped symbols and was developed in Sumer

Fertile Crescent a region stretching from the Persian Gulf northwest up the Tigris and Euphrates rivers and west over to the Mediterranean Sea

Hammurabi's Code one of the world's first law codes, compiled by the ruler Hammurabi

Harappan civilization an ancient civilization that developed along the Indus River

BACKGROUND VOCABULARY

grid a network of horizontal and vertical lines that create squares or rectangles

Reading Strategy

Re-create the chart shown at right. As you read and respond to the **KEY QUESTIONS**, use the chart to note the main achievements of the civilizations discussed in this section.

See Skillbuilder Handbook, page R5

SUMMARIZE

SUMERIAN	HARAPPAN

GRAPHIC ORGANIZERS
Go to Interactive Review @ClassZone.com

PRETEACHING VOCABULARY

English Learners

Pronounce and Preview

Pronounce each term for students. Review the meanings of words within definitions, such as *deposited, series, horizontal,* and *vertical.*

- To modify vocabulary learning, have students complete worksheets as they read, instead of afterward.

Unit Resource Book: Southwest Asia & South Asia
- Vocabulary Practice, p. 27

Inclusion

Create Audio or Visual Vocabulary

Have students work in pairs. Students with strong visual/reading skills can read the vocabulary terms and definitions aloud to their partners. Students should then work together to create either a visual review of the vocabulary terms, such as an illustrated glossary, or an audio review, such as a jingle or rhyming phrase about each term.

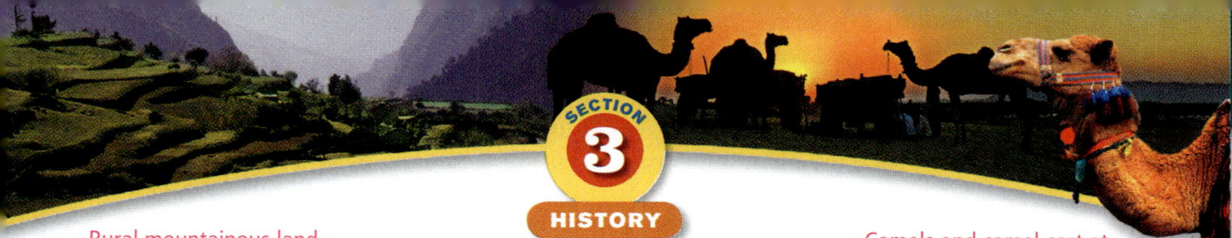

Rural mountainous landscape in Southwest Asia

Camels and camel cart at fair in Rajasthan, India

SECTION 3 HISTORY

River Valley Civilizations

Connecting to Your World

How would you get to sporting events or to the mall if the wheel had never been invented? How would you keep track of important information if you had no way to write it down? Think about what life would be like if your home didn't have indoor toilets, or if you couldn't wear cotton T-shirts or jeans. The people of ancient river valley civilizations invented many things that we rely on every day, including the wheel, writing, indoor plumbing, and cotton cloth.

Wheel The two-wheeled chariot was invented by the Sumerians in Mesopotamia around 3000 B.C.

Mesopotamia and the Fertile Crescent

KEY QUESTION What are the main characteristics of civilization?

In Section 1, you learned that the Tigris and the Euphrates rivers flow from Anatolia to the Persian Gulf. What you didn't learn is that the land between the two has its own name—**Mesopotamia**, which means "land between the rivers." In the spring, rain and melting snow in the Anatolian highlands fed the Tigris and Euphrates. As the swollen rivers rushed downhill, they picked up soil. When the rivers reached the flat region of Mesopotamia, they overflowed and spread over a wide area. The fine soil in the water, called **silt**, settled on the land. Silt is very fertile and makes good soil for farming.

Animated GEOGRAPHY

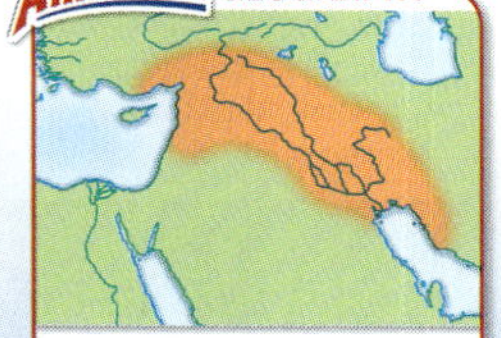

Mesopotamia This floating village is located in the marshes of southern Iraq near the eastern end of Mesopotamia.

Click here to learn more about Mesopotamia @ClassZone.com

667

2 Focus & Motivate

3-Minute Warm-Up

Write on the board or display the transparency:

- What caused the Tigris and Euphrates to overflow in the spring? *(rain and melting snow in the Anatolian Highlands)* What was so important about this overflow? *(It spread silt over a wide area, creating fertile soil for farming.)*

Unit Transparency Book
- 3-Minute Warm-Ups, TT1

Connecting to Your World

More About . . .

The Fertile Crescent

Primitive farming communities as far back as 8000 B.C. inhabited the Fertile Crescent, then an extensive marshland. Scientists today warn that these marshlands are disappearing from dams and drainage projects on the Tigris and Euphrates rivers.

3 Teach

Mesopotamia and the Fertile Crescent

Talk About It

- Why is agriculture one of the most important developments in human history? *(Farming allowed people to settle in one place. These settlements grew into complex civilizations.)*
- **Summarize** What are some ancient Sumerian inventions that are still in use today? *(wheel, plow, a writing system)*

SECTION 3 PROGRAM RESOURCES

Unit Resource Book: Southwest Asia and South Asia

ON LEVEL

Unit Resource Book
- Reading Study Guide, p. 7
- Vocabulary Practice, p. 27
- Skillbuilder Practice: Synthesizing, p. 28
- Section Quiz, p. 43

STRUGGLING READERS

Unit Resource Book
- Reading Study Guide with Additional Support, p. 13
- Section Quiz, p. 43
- Reteaching Activity, p. 46

eEdition with Audio DVD-ROM

ENGLISH LEARNERS

Pupil Edition in Spanish
eEdition with Audio DVD-ROM
Unit Resource Book
- RSG (Spanish), p. 19
- RSG with Additional Support (Spanish), p. 25

Multi-Language Glossary
Test Generator
- Section Quiz in Spanish

INCLUSION

Unit Resource Book
- RSG with Additional Support, p. 13
- Section Quiz, p. 43
- Reteaching Activity, p. 46

GIFTED & TALENTED

Unit Resource Book
- History Makers: Sargon, p. 34
- Primary Sources, p. 39
- Section Quiz, p. 43

PRE-AP

Unit Resource Book
- History Makers: Hammurabi, p. 33
- Primary Sources, p. 40
- Section Quiz, p. 43

TECHNOLOGY

Unit Transparency Book
- 3-Minute Warm-Ups, TT1
- Essential Question, TT2
- Critical Thinking, TT5
- Cultures: Cuneiform Artifact, TT8
- Cultures: Harappan Artifact, TT9

Daily Test Practice Transparencies
- Section 3, TT74

Power Presentations
ClassZone.com
World Cultures and Geography Video on DVD

More About . . .

Ziggurats

Ziggurats were a common feature of Mesopotamian cities from about 2200 B.C. until about 500 B.C. Sumerians may have built ziggurats to be homes of the gods, but nobody knows for sure how the Sumerians actually used them.

A solid mud brick core made up the interior of a ziggurat, so there were no interior chambers. Some ziggurats had exterior steps or a ramp winding to the top. But many ziggurats that are still standing have no obvious path to the top. On some ziggurats, Sumerians may have planted trees and shrubs to create "hanging gardens."

When religious beliefs changed, it's thought that the ziggurats were left to deteriorate—and were pillaged for their bricks. The largest ziggurat still in existence is at Chogha Zanbil in southwestern Iran. It stands about 80 feet tall, but its original height was more than twice as high.

CONNECT Geography & Culture

Cuneiform

Explain to students that some cuneiform systems had over 600 symbols that represented words or parts of words. In the 1600s, European explorers found ancient cuneiform writing in Iranian ruins, but it would be about 200 years before scholars were able to decipher some of the symbols. Translating these and other more recently discovered cuneiform writings has contributed immensely to knowledge about the ancient world.

Unit Resource Book: Southwest Asia & South Asia

- Primary and Secondary Sources: from *The Epic of Gilgamesh*, p. 39

Civilization Arises The earliest humans lived by hunting animals and by gathering nuts, fruits, and seeds. Over time, they learned to save and plant some seeds to ensure they would have a food supply. That was the start of agriculture. When people began to irrigate their fields, their ability to grow crops improved and they had extra food.

Farming changed human life. Instead of wandering to find food, people settled in one place. In time, villages grew into cities. City dwellers had to figure out how to defend their city and how to do jobs such as building irrigation canals. To solve such problems, societies developed a more complex way of life. The advanced form of culture that arose in cities is called **civilization**. Civilization has five traits:

Advanced Cities Cities arose as places where farmers could store and trade their extra food.

Specialized Workers When a society has extra food, not everyone has to farm. Some can build houses or make pottery.

Complex Institutions A civilization has institutions such as an army, schools, a government, and a temple for worship.

Record Keeping People in cities needed to record information such as how much food they had stored. So civilizations developed record keeping, which involved writing.

Advanced Technology Early civilizations developed new technologies such as irrigation and bronze tools.

CONNECT Geography & Culture

Cuneiform

For writing supplies, Sumerians used a material they had a lot of—tablets made of clay from the riverbeds. At first, Sumerians drew pictures that stood for words. Later, they used a sharpened reed to make wedge-shaped marks in soft clay. The marks were combined into symbols that became the writing system called cuneiform. When the clay dried, it hardened, making a permanent record.

Sumerian Civilization The first civilization of Mesopotamia arose in **Sumer**, which was a land of many city-states. A **city-state** was made up of a city and nearby farmlands.

Walls surrounded the cities of Sumer. Buildings were made of dried mud because few trees grew in Mesopotamia. In many Sumerian cities, one building towered over all the others. It was the **ziggurat**, a temple built on a series of increasingly smaller platforms that made the ziggurat peak like a mountain. The Sumerian people were **polytheistic**, which means they believed in many gods.

The Sumerians invented many things we still use today, including the wheel and the plow. The Sumerians also developed the first writing system, which is called **cuneiform**.

A FIND MAIN IDEAS Give the main characteristics of civilization. advanced cities, specialized workers, complex institutions, record keeping, advanced technology

DIFFERENTIATING INSTRUCTION

Struggling Readers

Discuss Traits of Civilization

Help students relate the five traits of civilization to the present day world. Have students reread the five traits of civilization. Working in pairs, each student can pick one of the traits. Then students can take turns giving examples of that trait in both ancient civilization (using their textbooks as a reference if necessary), and in modern civilizations of today.

Gifted & Talented

Create a Cuneiform System

Students can use symbols to create their own simple cuneiform writing system. Each student should pick a sentence from the text. Then they create their own symbols for the words or syllables in that sentence. On a large sheet of paper, have students write each symbol they create, and next to it, the meaning of the symbol. Then students can write the sentence using the cuneiform symbols. Have students present or post their cuneiform sentences, and see if others can "read" them.

City-States and Empires

KEY QUESTION Who were two rulers of Mesopotamian empires?

The first rulers of city-states were priests, because people believed priests communicated with the gods. However, cities often came under attack because they stored extra food and wealth that outsiders wanted. In wartime, military leaders gained power. Eventually, those leaders became permanent rulers called kings, chosen with the approval of the priests.

The First Empire In 2334 B.C., a ruler named Sargon came to power. He conquered many cities and became the first king to rule all of Mesopotamia. His empire stretched from the Persian Gulf up the two great rivers and over to the Mediterranean Sea, a region called the **Fertile Crescent**. Sargon built a wealthy capital city called Akkad. During his grandson's rule, invaders broke up the empire.

Hammurabi About 500 years after Sargon, another great empire rose in Mesopotamia. A king of Babylon named Hammurabi conquered other city-states and built a vast and powerful empire. Because of his conquests, Hammurabi was the ruler of many peoples with different customs. He wanted a consistent way to govern his subjects, so he compiled a set of laws that applied to everyone under his rule. **Hammurabi's Code** was one of the earliest law codes in the world. His concept of justice—that punishment should fit the crime—has continued to evolve.

SUMMARIZE Identify two rulers of Mesopotamian empires. Sargon and Hammurabi

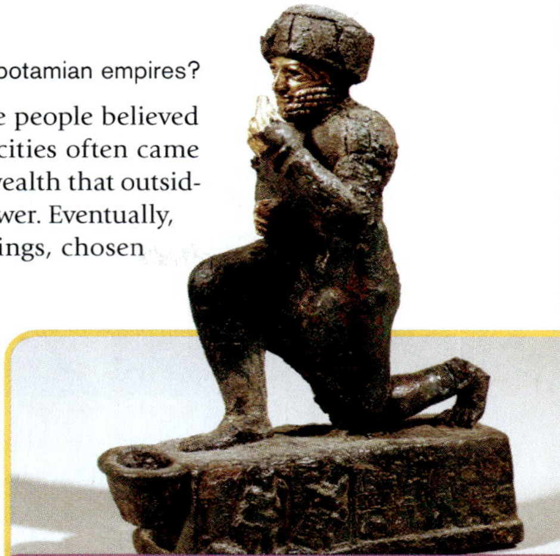

HISTORY MAKERS

Hammurabi (died about 1750 B.C.)

When Hammurabi became king, Babylon was a small state. Hammurabi led Babylon in conquering surrounding lands. One tactic he used was to dam main waterways to deprive his enemies of water.

During his reign, Hammurabi oversaw the creation of one of the world's oldest set of laws. He made Babylon so magnificent that for centuries after his death, it remained one of the most important cities of the ancient world.

ONLINE BIOGRAPHY
For more on the life of Hammurabi, go to the Research & Writing Center @ ClassZone.com

Indus River Valley Civilization

KEY QUESTION What were some of the accomplishments of the Harappan civilization?

An ancient civilization also developed on the Indus River in what is now Pakistan. As you know, monsoons cause South Asia to have a rainy season and a dry season. During the rainy season, the Indus River often flooded and deposited fertile silt on the land. Farming villages sprang up, and some gradually grew into cities.

DIFFERENTIATING INSTRUCTION: TIERED ACTIVITIES

OBJECTIVE Create a glossary of geographical and biographical names related to river valley civilizations

Basic

As a class, list the names of people and places in this section. Have pairs of students use the list to create a glossary of the names with a definition for each one. The glossaries should be in alphabetical order. Students can use the glossaries to review the relationships between places and people among the river valley civilizations.

On Level

Adapt the Basic activity, so that students work independently to complete the glossaries. In addition, have them organize their glossaries into categories, such as *Regions, Cities, Rulers,* or *Civilizations*. Students can use the glossaries to review the relationships between places and people among the river valley civilizations.

Challenge

Have students complete the On Level activity. In addition, ask them to find five new names to add to their glossaries, using classroom resources. Ask them to include how the names they added connect with what they have learned in this section. Students can use the glossaries to review the relationships between places and people among civilizations.

Teach

City-States and Empires

Reader, Recorder, Reporter

- Over what region did Sargon's empire stretch? *(the Fertile Crescent, or from the Persian Gulf, up the Tigris and Euphrates, and to the Mediterranean Sea)*
- What were the concepts behind Hammurabi's Code? *(Laws should be consistent and apply to everyone; punishment should fit the crime.)*
- **Summarize** Why was Hammurabi considered a great ruler? *(He built a vast empire that included Babylon; he conquered other city-states; he created Hammurabi's Code.)*

Unit Resource Book: Southwest Asia & South Asia
- History Makers: Sargon, p. 34

RESEARCH & WRITING CENTER

HISTORY MAKERS

Hammurabi

Hammurabi was known as a shrewd military leader and diplomat. In addition, he oversaw the building and repair of irrigation canals, temples, walls, and other structures. Discovered in 1901, Hammurabi's Code is inscribed on a black stone slab that is 8 feet tall. It is now at the Louvre in Paris.

Unit Resource Book: Southwest Asia & South Asia
- History Makers: Hammurabi, p. 33
- Primary and Secondary Sources: from the Code of Hammurabi, p. 40

Teach

Indus River Valley Civilization

Think, Pair, Share

- What are the two greatest cities of this region? *(Mohenjo-Daro and Harappa)*
- **Summarize** What may have caused the Harappan civilization to disappear? *(Possible Answers: earthquakes; invaders; a change in the river's course; an environment that could no longer support the population)*

Unit Resource Book: Southwest Asia & South Asia
- Skillbuilder Practice: Synthesizing, p. 28

CONNECT to the Essential Question

How have physical geography and history interacted in Southwest and South Asia?

Ask students what they have learned so far that can help them answer this question. Students might mention:

- The first civilizations developed in the Tigris, Euphrates, and Indus river valleys because of their fertile soil.
- The Harappan people created a great civilization, then declined, possibly because of changes in the course of the Indus River or from earthquakes.

4 Assess & Reteach

Assess Have students complete the Section Assessment.

Unit Resource Book: Southwest Asia & South Asia
- Section Quiz, p. 43

Interactive Review @ClassZone.com
- Section Quiz

Power Presentations

Test Generator

Reteach Have students work in groups to review subsections and list dates of events mentioned. Have each group use these dates to construct a time line of River Valley Civilizations. Encourage students to include illustrations and labels, then share their time lines.

Unit Resource Book: Southwest Asia & South Asia
- Reteaching Activity, p.46

Mohenjo-Daro Ruins of the ancient city of Mohenjo-Daro, located in present-day Pakistan

Ruins of more than 100 settlements have been found spread across a vast area of South Asia. The two greatest cities of this civilization are called Mohenjo-Daro and Harappa, and the culture is called the **Harappan civilization**. It is also known as the Indus Valley civilization. The ruins of those cities show that the Harappan people had advanced technology and a system of measurement. The cities were laid out on a **grid** with straight streets that crossed at right angles. Houses had indoor plumbing, and there was a sewer system. The Harappan civilization used a system of weights and bricks that were a standard size. The Harappan people engaged in trade with Afghanistan, Persia, and Mesopotamia. The Harappans developed their own writing system to keep records. Around 2000 to 1500 B.C., the Harappan civilization declined, and people left the cities. No one knows why, but possible causes include earthquakes, a shift in the river's course, and attacking invaders. The Harappan people might simply have exceeded the environment's ability to support the population.

SUMMARIZE Describe accomplishments of Harappan civilization.
cities built on a grid with a sewer system, indoor plumbing, accurate weights, a writing system, long-distance trade

ONLINE QUIZ For test practice, go to Interactive Review @ClassZone.com

Section 3 Assessment

TERMS & NAMES

1. **Explain the importance of**
   - Mesopotamia
   - civilization
   - Sumer
   - Harappan civilization

USE YOUR READING NOTES

2. **Summarize** Use your completed chart to answer the following question:
   What were some achievements of the civilizations mentioned in this section?

SUMERIAN	HARAPPAN

KEY IDEAS

3. What were examples of complex institutions in Sumer?
4. What was Sargon's important achievement?
5. What are some possible causes for the decline of the Harappan civilization?

CRITICAL THINKING

6. **Make Inferences** Which developed first, specialized workers or complex institutions? Explain.
7. **Compare** How were the Sumerian and Harappan civilizations similar?
8. **CONNECT to Today** In Southwest Asia, farmers dig up objects from ancient ruins and sell them. Do you think historic objects should be kept in museums or owned by individuals? Explain.
9. **TECHNOLOGY** **Plan a Multimedia Presentation** Choose either the Sumerian or Harappan civilization and use books, the Internet, or a field trip to a museum to learn more about it. Plan a multimedia presentation showing its art, geography, and technology.

SECTION 3 ASSESSMENT ANSWERS

Terms & Names

1. Mesopotamia: the land between the Tigris and Euphrates rivers; civilization: an advanced form of culture that developed in cities; Sumer: a land of city-states in Mesopotamia that was home to the first civilization; Harappan civilization: an ancient civilization that developed along the Indus River

Use Your Reading Notes

2. Sumerian: wheel, plow, cuneiform, ziggurats, law code. Harappan: sewer system and indoor plumbing, writing system, trade, weights and measures

Key Ideas

3. army, schools, governments, temples
4. He was the first king to unite all of Mesopotamia.
5. earthquake, shifting river, invaders, inability to support population

Critical Thinking

6. Possible answer: Specialized workers developed first because complex institutions can't exist if everyone is working as a farmer.
7. Possible answer: Both arose near rivers, both had cities, and both had a writing system.
8. Possible answer: Some students will say the objects should be in museums so everyone can learn about them; other students will say that if someone can afford to buy an ancient object, he or she has the right to own it.
9. Use the rubric to score multimedia presentations

Multimedia Presentation Rubric

	Organization of Text	Variety of Graphics
4	well organized, includes three information points	variety in links, maps, and graphics; easy to follow
3	adequately organized; includes two information points	variety in links, maps, and graphics; fairly clear
2	vague; includes one point	little variety; hard to follow
1	badly organized	no variety

CHAPTER SUMMARY

Key Idea 1
Southwest Asia has vast deserts and extensive mountain ranges that are similar to those in certain parts of Africa.

Key Idea 2
The physical geography of South Asia is made up of mountains, plateaus, and fertile plains, and the climate has wet and dry seasons.

Key Idea 3
In both regions, civilizations developed first in river valleys where there was an abundant supply of water.

For **Review and Study Notes**, go to **Interactive Review @ ClassZone.com**

NAME GAME

Use the Terms & Names list to complete each sentence on paper or online.

1. I am a mountain range running between the Indian subcontinent and Asia. Himalayas
2. I am the region that lies between the Tigris and Euphrates rivers. ________
3. I am a landform covered mostly by desert. ________
4. I flow from the Himalayas to the Bay of Bengal. ________
5. I lie in the center of the Indian subcontinent. ________
6. I am home to most of Turkey in Asia. ________
7. I am the tallest of my kind in the world. ________
8. I flow from the Himalayas to the Arabian Sea. ________
9. I flow from Anatolia to the Persian Gulf not far from the Euphrates. ________
10. I stretch from the Persian Gulf up two rivers and over to the Mediterranean Sea. ________

Anatolian peninsula
Arabian Peninsula
Deccan Plateau
Euphrates River
Fertile Crescent
Ganges River
Himalayas
Indus River
Plateau of Iran
Mesopotamia
Mount Everest
Sumer
Tigris River

Activities

Flip Cards

Use the online flip cards to quiz yourself on the terms and names introduced in this chapter.

Crossword Puzzle

Complete an online crossword puzzle to test your knowledge of Southwest Asia and South Asia.

ACROSS
1. Sumerian writing system that used clay tablets

671

ONLINE AND ON-SCREEN ACTIVITIES

Flip Cards

Vocabulary Flip Cards are virtual flash cards at **ClassZone.com.** Students can quiz each other or work independently to study the flip cards for the chapter, "Southwest Asia and South Asia."

- They can quiz themselves on key terms and names by reviewing definitions.
- Demonstrate how to advance and shuffle the deck for another challenge.

Crossword Puzzle

Students can review important people and events in Southwest Asia and South Asia's history by completing this crossword puzzle online at **ClassZone.com.**

- As an extension of this activity, have students use the *Activity Maker* at **ClassZone.com** to make their own crossword puzzle for one section of the chapter.
- Then have students trade puzzles with classmates to solve each other's puzzles.

Chapter Summary

As a follow-up to the chapter, encourage students to go over the **Review and Study Notes** at **ClassZone.com.**

Alternatively, student pairs can review their Reading Study Guide notes, making sure that they can add important details to each section's "Key Idea."

Name Game

Teaching Tips

ON PAPER Have students work in pairs to complete this activity. As an extension, they can write questions for the terms that are not covered in the items shown here.

ONLINE Make sure that students understand how to drag and drop terms to fill in blanks. Ask them to print out their results and compare them with those of a classmate.

ANSWERS

1. Himalayas
2. Mesopotamia
3. Arabian Peninsula
4. Ganges River
5. Deccan Plateau
6. Anatolian peninsula
7. Mount Everest
8. Indus River
9. Tigris River
10. Fertile Crescent

Make It Fun

Teacher-Tested Activities

Amber L. McVey, Antioch Middle School, Gladstone, Missouri

When you are looking for someone to answer a question, give an opinion, or provide feedback, you can use the popsicle stick activity.

- Write the name of each student on a popsicle stick and draw sticks randomly.
- If the student is unable to provide the answer, put the stick back in the container so it can be drawn again later.
- This method allows you to keep track of who has participated. It also keeps the students on their toes because they never know when they are going to be called upon.

Vocabulary ANSWERS

1. The Arabian Peninsula lies between the Red Sea and the Persian Gulf.
2. The Anatolian peninsula lies between the Black Sea and the Mediterranean.
3. Plateau of Iran lies in central eastern Iran.
4. The Tigris River flows from a plateau on the Anatolian peninsula to the Persian Gulf.
5. A subcontinent is a large landmass that is part of a continent but is a separate region.
6. Indus River flows from Himalayas to Arabian Sea.
7. A monsoon is a wind system that produces a wet or dry period in a region.
8. Mesopotamia is the land between the Tigris and Euphrates rivers.
9. A ziggurat is a temple atop a series of platforms that was built in Sumerian cities.
10. The Fertile Crescent stretches from the Persian Gulf up the Tigris and Euphrates rivers and over to the Mediterranean.
11. Civilization began in cities, which grew to city-states; two characteristics of civilization are record keeping, such as cuneiform, and complex buildings, such as the ziggurat.
12. Himalayas separate India from rest of Asia; Deccan Plateau is in center of subcontinent.

Key Ideas ANSWERS

13. Arabian Peninsula in center, Anatolian peninsula northwest, Plateau of Iran east
14. A qanat is an underground tunnel that goes from a water source to a village or field; it runs downhill, so gravity moves the water.
15. a desert climate
16. They are afraid that Turkey will divert so much water from the rivers that they won't have enough to meet their needs.
17. When the Indian landmass collided with Asia, the force pushed up mountains, some of which are the Himalayas.
18. Coral reefs built up on the slopes of undersea volcanoes, forming the Maldives.
19. Too much rain causes floods, while too little rain causes drought and famine.
20. fierce winds, heavy rains, storms
21. They deposit fertile silt on the land.
22. They learned to save gathered seeds and plant them.
23. Military leaders gained power and became permanent rulers.
24. They were well ordered, with streets laid out on grids and with sewers.

CHAPTER 21 Assessment

VOCABULARY

Explain the significance of each of the following.

1. Arabian Peninsula
2. Anatolian peninsula
3. Plateau of Iran
4. Tigris River
5. subcontinent
6. Indus River
7. monsoon
8. Mesopotamia
9. ziggurat
10. Fertile Crescent

Explain how the first term in each group relates to the terms that follow.

11. civilization: city-state, cuneiform, ziggurat
12. subcontinent: Himalayas, Deccan Plateau

KEY IDEAS

1 Physical Geography of Southwest Asia

13. Where are the three subregions of Southwest Asia located in relationship to each other?
14. What is a qanat, and how does it work?
15. What is the main climate on the Arabian Peninsula?
16. Why does Turkey's water project worry its neighbors?

2 Physical Geography of South Asia

17. How were the Himalayas formed?
18. How were the Maldives formed?
19. What happens when the summer monsoon brings either too much or too little rain?
20. What are the dangers associated with cyclones?

3 River Valley Civilizations

21. What is one positive result of having rivers that overflow their banks during spring floods?
22. How did humans change from hunting and gathering to farming?
23. How did the leadership of city-states in Sumer change from priests to kings?
24. What were Harappan cities like?

CRITICAL THINKING

25. **Compare and Contrast** Create a table to compare and contrast the physical geography of Southwest Asia and South Asia.

SOUTHWEST ASIA	SOUTH ASIA

26. **Draw Conclusions** How did Sumerian priests retain some power under kings?
27. **Compare** What do the Indus, Ganges, and Brahmaputra rivers all have in common?
28. **Connect to Language Arts** An ancient Sumerian proverb says, "Whoever has walked with truth generates life." What do you think that means?
29. **Five Themes: Human-Environment Interaction** Cities with the gridlike design of Harappan culture were found across South Asia. How might that have happened?

Answer the ESSENTIAL QUESTION

How have physical geography and history interacted in Southwest and South Asia?

Written Response Write a two- or three-paragraph response to the Essential Question. Be sure to consider the key ideas of each section about the rise of civilization in the region. Use the rubric below to guide your thinking.

Response Rubric
A written response will
- discuss the physical geography of each region
- relate physical geography to Sumerian civilization
- relate the physical geography of South Asia to the Harappan civilization

ESSENTIAL QUESTION REVIEW

Before students begin their essays, have them review the key chapter elements that will help them answer the Essential Questions. Direct them to significant features in the chapter, such as:

- Physical Geography of Southwest Asia Map, Section 1
- Physical Geography of South Asia Map, Section 2
- Animated Geography: Monsoons, Section 2
- Animated Geography: Fertile Crescent, Section 3
- History Makers: Hammurabi, Section 3

Written Response Rubric

	Coverage of Key Ideas	Explanation	Accuracy
4	all	clear, concise	no errors
3	most	good	few/minor errors
2	some	fair	several errors
1	two or fewer	poor	many errors

STANDARDS-BASED ASSESSMENT

Test Practice
- Online Test Practice @ClassZone.com
- Test-Taking Strategies and Practice at the front of this book

BAR GRAPH

Use the bar graph and your knowledge of Southwest Asia to answer questions 1 and 2 on your paper.

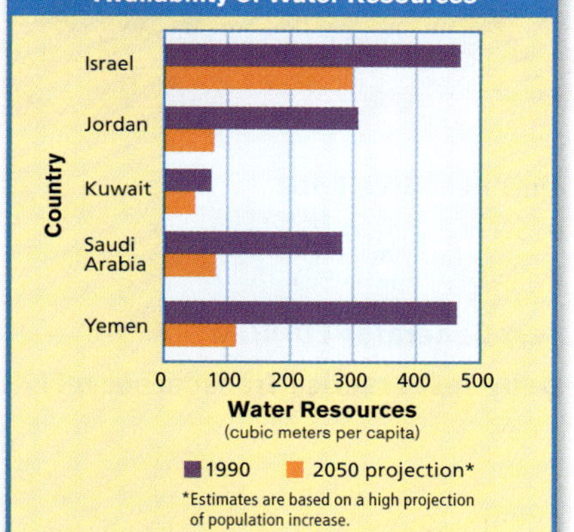

Source: Adapted from Robert Engelman and Pamela LeRoy, *Sustaining Water: An Update*, Population Action International, Washington D.C. 1995

1. Which country is projected to have the greatest supply of water by 2050?

A. Yemen
B. Israel
C. Kuwait
D. Jordan

2. Which country is projected to have the smallest supply of water by 2050?

A. Kuwait
B. Jordan
C. Saudi Arabia
D. Yemen

INFOGRAPHIC

Use this Indian farming calendar to answer questions 3 and 4 on your paper.

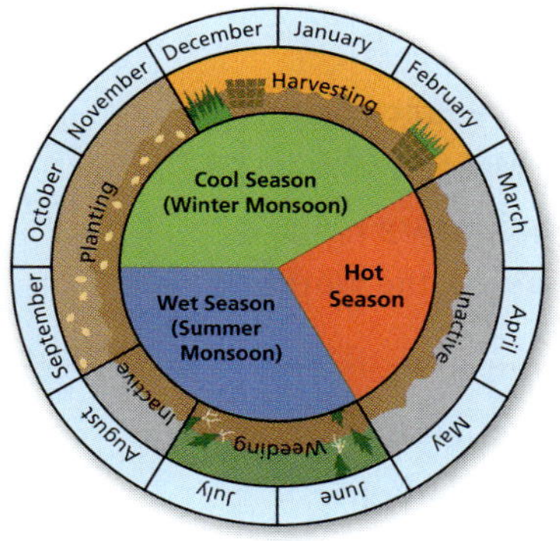

Source: India Country Studies

3. Which season is the most productive?

4. Which season is the least productive?

GeoActivity

1. INTERDISCIPLINARY ACTIVITY–ART

Working in a small group, find photographs of the statues and jewelry created by the ancient Sumerian and Harappan civilizations. Choose three from each region, and plan a multimedia presentation that will show and compare the art of the two regions.

2. WRITING FOR SOCIAL STUDIES

Research to learn what some of the laws in Hammurabi's code were. Write a review of those laws, quoting two or three examples and then explaining what you think their effect on society was and whether they were fair.

3. MENTAL MAPPING

Create an outline map of South Asia and label the following:

- Himalayas
- Deccan Plateau
- Indus River
- Ganges River
- Brahmaputra River
- Arabian Sea
- Bay of Bengal
- Indian Ocean
- Sri Lanka

Critical Thinking ANSWERS

25. Southwest Asia: three subregions (Arabian Peninsula, Anatolian peninsula, Plateau of Iran); main rivers are Tigris, Euphrates, Jordan; mostly a dry climate; prone to earthquakes
South Asia: subcontinent bordered by mountains; main rivers are Indus, Ganges, Brahmaputra; a monsoon climate with wet and dry seasons; prone to earthquakes, cyclones, tsunamis

26. Possible answer: They had to give their approval before a king could rule.

27. Possible answers: They all originate in the Himalayas and flow through South Asia.

28. Possible answers: Sumerians thought that truth was very important; truth was life-enhancing and affirming, while lies undermine the trust that is essential to a civilized life.

29. Possible answers: Harappan culture was widespread; the gridlike design worked well in that environment so people copied it; people moved from the original cities and carried the design with them.

Standards-Based Assessment ANSWERS

1. B
2. A
3. cool season
4. hot season

GeoActivity

1. INTERDISCIPLINARY ACTIVITY RUBRIC

The presentation should
- include all three information points
- have variety in links, maps, and graphics

2. WRITING FOR SOCIAL STUDIES RUBRIC

The review should
- give many details and examples from Hammurabi's code
- contain no, or very few, mechanical errors

3. MENTAL MAPPING RUBRIC

The outline map should
- represent the approximate shape of South Asia
- include all geographic features listed in their correct location

DIFFERENTIATING CHAPTER ASSESSMENT

Unit Resource Book

- **Chapter Test A,** pp. 45–50.
  Use for Struggling Readers, Beginning-to-Intermediate-Level English Learners, and for Inclusion of students whose IEP suggests supported test-taking.
- **Chapter Test B,** pp. 51–54.
  Use for On-Level Learners and ESOL students with Advanced-Level English.
- **Chapter Test C,** pp. 55–58.
  Use for Gifted and Talented students or students assigned Pre-AP work.

Test Generator

- Chapter Tests A, B, C, in English/Spanish

McDougal Littell Assessment System

TEST **Test Generator** Use three levels of tests in English and Spanish tied to state standards. Plus, choose questions to customize tests for all populations.

SCORE Scan and Score plain paper student tests in minutes.

REPORT Create instant progress reports, sortable by student, by standard, or other criteria.

RETEACH Respond immediately to students' needs by using a wealth of reteaching resources.

CHAPTER 22

Southwest Asia PLANNING GUIDE

Complete print resources all on one USB drive allow you to customize lesson plans for all students.

Section Titles & Key Ideas	Unit Resource Book Southwest Asia and South Asia	Assessment Resources
1 Saudi Arabia, Iraq, and Iran **KEY IDEA** The physical geography of the region has influenced the history and culture of many countries.	• **Reading Study Guide,** pp. 59–60 • **RSG with Additional Support,** pp. 65–66 • **Vocabulary Practice,** p. 83 • **History Makers: Muhammad,** p. 89 • **Interdisciplinary Projects,** p. 91 • **Economics in the World,** p. 94	**Unit Resource Book** • Section Quiz, p. 99 • Reteaching Activity, p. 102 **Test Generator CD-ROM** **Daily Test Practice Transparency,** TT75
2 Israel and the Palestinian Territories **KEY IDEA** Israel is a thriving democracy, and the Palestinian territories are trying to develop a representative government.	• **Reading Study Guide,** pp. 61–62 • **RSG with Additional Support,** pp. 67–68 • **Outline Map Activities,** p. 87 • **Primary Sources,** pp. 96–97	**Unit Resource Book** • Section Quiz, p. 100 • Reteaching Activity, p. 103 **Test Generator CD-ROM** **Daily Test Practice Transparency,** TT76
3 Turkey, Syria, and Lebanon **KEY IDEA** Turkey, Syria, and Lebanon are all struggling with similar issues of how to reconcile representative government with their traditions.	• **Reading Study Guide,** pp. 63–64 • **RSG with Additional Support,** pp. 69–70 • **History Makers: Mustafa Kemal,** p. 90 • **Interdisciplinary Projects,** p. 92 • **Connect to History,** p. 95	**Unit Resource Book** • Section Quiz, p. 101 • Reteaching Activity, p. 104 • Chapter Test (Forms A, B, & C), pp. 105–116 **Test Generator CD-ROM** **Daily Test Practice Transparency,** TT77

McDougal Littell **Assessment System**

TEST SCORE REPORT RETEACH

No Time? To focus students on important content in the chapter, use the **"Reading Study Guide with Additional Support"** in *Resources2GO.*

Teacher Presentation Options

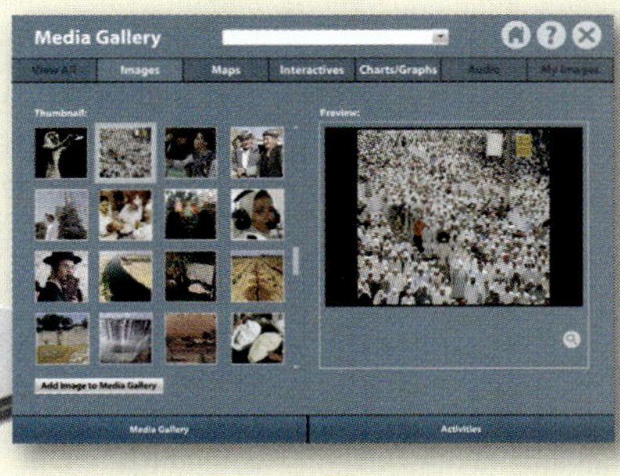

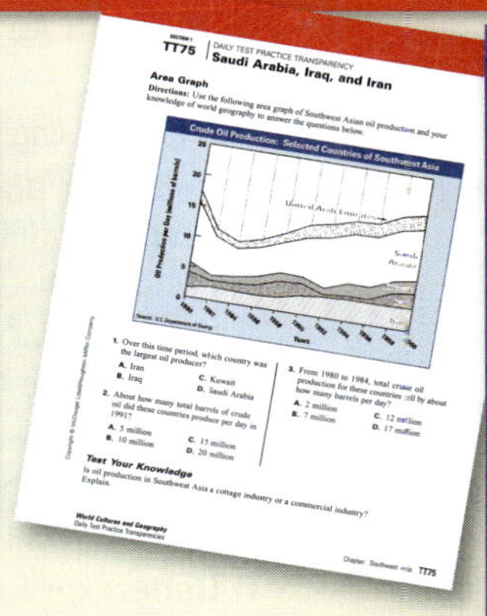

Electronic Books
eEdition DVD-ROM
eEdition Online

Presentation Toolkit
Video/DVD
- World Cultures and Geography

Power Presentations DVD-ROM
- Discussion Starters
- Media Gallery
- Review Game
- Interactive Review
- Animated Features

Unit Transparency Book
- 3-Minute Warm-Ups
- Essential Question
- Critical Thinking
- Maps
- Cultures

Daily Test Practice Transparencies

Animated Geography
- Economic Activity of Southwest Asia Map
- Holy Sites of Jerusalem

OnlineActivities at ClassZone.com

Animated Geography
- Videos and Photos
- 3-D Animations
- Interactive Maps

Interactive Review
- Section Quizzes
- Review and Study Notes
- Vocabulary Flip Cards
- Graphic Organizers
- Name Game and Crossword Puzzles
- GeoGame

Online Test Practice

Map Center
- Atlas Maps
- Political Maps with Data
- Links to Mapping Web Sites

Activity Center
- Activity Maker—Review Games, Puzzles, Flip Cards, and Graphic Organizers
- Worldster

Research & Writing Center
- Web Sites
- Data Updates
- Primary Sources
- Writing Models

Current Events

Don Stringfellow
Mobile City Public School System
Mobile, Alabama

Make It Fun — Teacher-Tested Activities

As you teach this chapter, look for my classroom-proven ideas for:
- making the lesson fun
- focusing students' attention and comprehension
- integrating technology with traditional teaching

Teaching Bell to Bell When you see this timing guide, have students discuss in pairs, in groups of 3–6, or as a class.
- **Think, Pair, Share** (pairs)
- **Reader, Recorder, Reporter** (3–6)
- **Talk About It** (any size)

RESOURCES FOR DIFFERENTIATED INSTRUCTION

CHAPTER 22

Struggling Readers

eEdition with Audio DVD-ROM

Lesson Plans
- Strategies for Inclusion

Unit Resource Book
- Reading Study Guide with Additional Support, pp. 65–70 (B)
- Reteaching Activity, pp. 102–104
- Chapter Test (Form A), pp. 105–116 (A)

Best Practices Toolkit (C)
- Interactive Reading Strategies, pp. 90–145
- Differentiation Activity Bank: Struggling Readers, pp. 146–149

Teacher's Edition Activities
- Define Economic Terms, Sec. 1
- Tiered Activities, Sec. 1–3

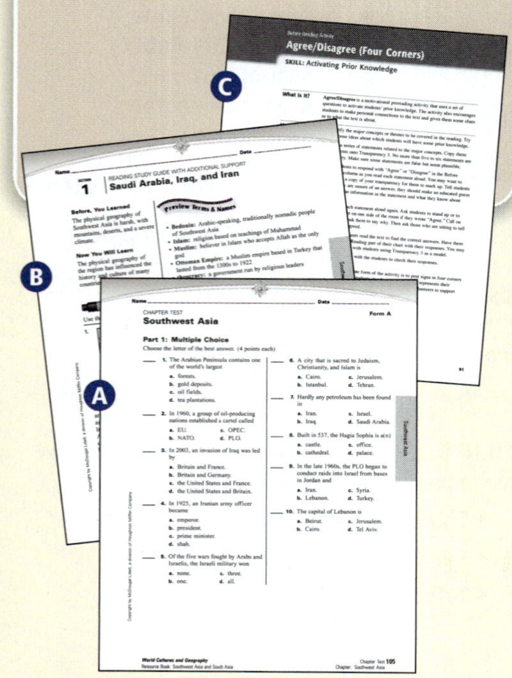

English Learners

Pupil's Edition in Spanish

eEdition with Audio DVD-ROM

Lesson Plans
- Strategies for English Learners

Unit Resource Book
- Reading Study Guide (Spanish), pp. 71–76
- RSG with Additional Support (Spanish), pp. 77–82 (A) (B)

Multi-Language Glossary (C)

Best Practices Toolkit
- Differentiation Activity Bank: English Learners, pp. 165–194

Test Generator CD-ROM
- Section Quizzes in Spanish
- Chapter Test in Spanish

Teacher's Edition Activities
- Pronounce and Preview, Sec. 1–3
- Language: Punctuation and Print Cues, Sec. 1
- Vocabulary: Idioms, Sec. 2
- Culture: Clarify, Sec. 3
- Tiered Activities, Sec. 1–3

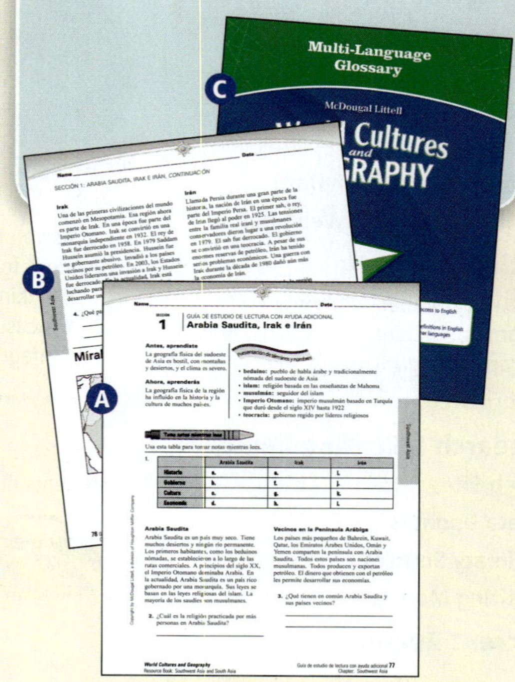

Inclusion

eEdition with Audio DVD-ROM

Lesson Plans
- Strategies for Inclusion

Unit Resource Book
- Reading Study Guide with Additional Support, pp. 65–70 (A) (B)
- Reteaching Activity, pp. 102–104
- Chapter Test (Form A), pp. 105–116

Best Practices Toolkit (C)
- Interactive Reading Strategies, pp. 90–145
- Differentiation Activity Bank: Struggling Readers, pp. 146–149

Teacher's Edition Activities
- Create Visual Vocabulary, Sec. 1
- Vocabulary: Frayer Model (Word Squares), Sec. 2
- Using Letter Tiles, Sec. 3
- Tiered Activities, Sec. 1–3

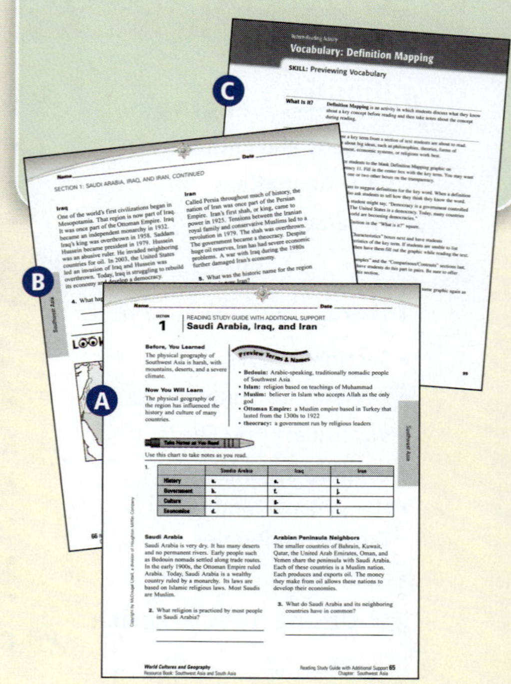

Interdisciplinary Activities

Gifted & Talented Students

Lesson Plans
- Core Lesson Plan

Unit Resource Book
- Map and Graph Skills, pp. 85–86 **A**
- Outline Map Activities, p. 87 **B**
- Interdisciplinary Projects, pp. 91–93 **C**
- Chapter Test (Form C), pp. 113–116

Best Practices Toolkit
- Student Products Rubrics, pp. 1–14
- Writing Templates, pp. 78–89

Teacher's Edition Activities
- Negotiate Regional Conflicts, Sec. 2
- Tiered Activities, Sec. 1–3

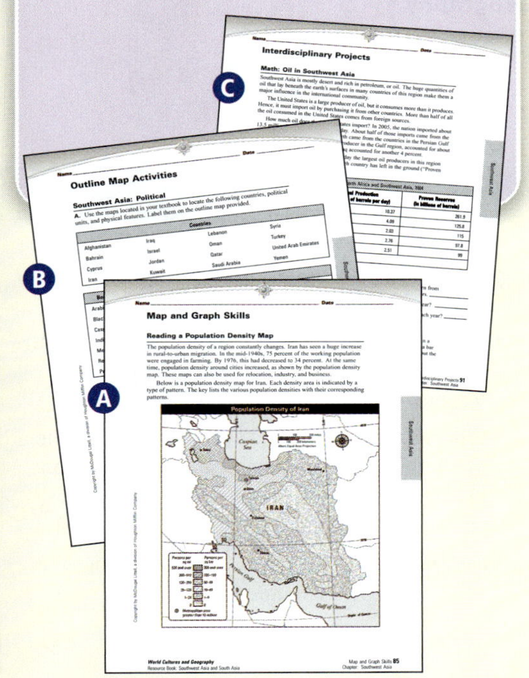
C — Interdisciplinary Projects: Math: Oil in Southwest Asia

B — Outline Map Activities: Southwest Asia: Political

A — Map and Graph Skills: Reading a Population Density Map

Pre-AP*

Lesson Plans
- Core Lesson Plan

Unit Resource Book
- Vocabulary Practice, p. 83
- History Makers, pp. 89–90 **A**
- Economics in the World, p. 94 **B**
- Connect to History, p. 95
- Primary Source, pp. 96–97 **C**
- Chapter Test (Form C), pp. 113–116

Best Practices Toolkit
- Writing Templates, pp. 78–89
- Differentiation Activity Bank: Pre-AP Students, pp. 195–200

* Pre-AP is a registered trademark of the College Entrance Examination Board, which was not involved in the production of and does not endorse this product.

Teacher's Edition Activities
- Summarize Istanbul's Geographic Importance, Sec. 3
- Tiered Activities, Sec. 1–3

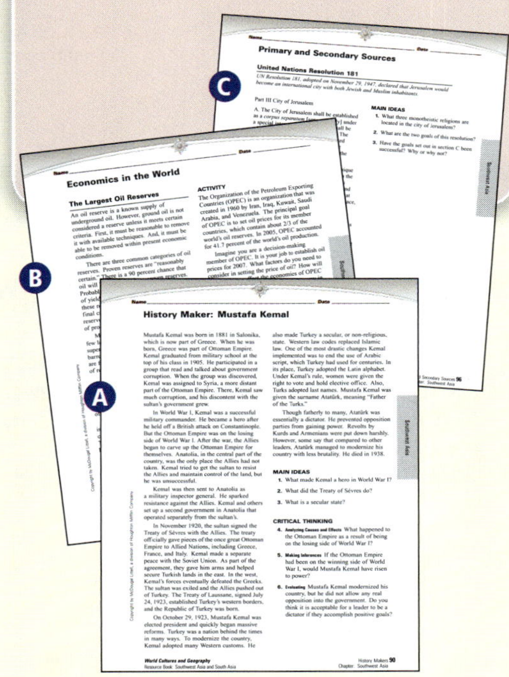
C — Primary and Secondary Sources: United Nations Resolution 181

B — Economics in the World: The Largest Oil Reserves

A — History Maker: Mustafa Kemal

CONNECT to Music

Plan a Southwest Asia Music Concert

CLASS TIME 30 minutes

TASK Organize a class concert featuring the music of Southwest Asian countries.

SUPPLIES
- a variety of music audiotapes or CDs of music from Turkey, Iraq, Iran, Syria, Lebanon, Israel, and the Palestinian territories
- audio tape or CD player(s)

ACTIVITY Have pairs of students listen to various songs and select one to include in a class concert. Pairs should then work together to write a short description about the song, the performer(s), and an overview of music from the country they selected. Have students create a program of all their selections. Before playing each song at the class concert, have student pairs introduce it by reading their descriptions aloud.

Good for Audio Learners

CONNECT to Language Arts

Plan a Trip to Visit Southwest Asia

CLASS TIME 30 minutes

TASK Create an itinerary to visit Southwest Asia during a 2-week trip.

SUPPLIES
- maps of Southwest Asia
- information about cultural and historic sites in Southwest Asia
- computers or pencil/paper

ACTIVITY Have students assume the role of a tour guide, planning a two-week trip to Southwest Asia. Have them create a day-by-day itinerary, listing what sites they'll visit that day and information about each place, such as its historical significance. To extend this activity, have students include visuals of the sites they plan to visit, using library or Internet resources.

Good for Verbal and Spatial Learners

Introduce the ESSENTIAL QUESTION

- Explain that Southwest Asia is the birthplace of three major world religions and home to many different ethnic groups.
- Point out that the global demand for oil has played a major role in the economies of some nations in the region.
- Explain that as the region has moved toward economic, political, and cultural modernization, conflict has ensued among religious and ethnic groups.

Key Ideas

Section 1: The physical geography of the region has influenced the history and culture of many countries.

Section 2: Israel is a thriving democracy, and the Palestinian territories are trying to develop a representative government.

Section 3: Turkey, Syria, and Lebanon are all struggling with similar issues of how to reconcile representative government with their traditions.

CONNECT to the Essential Question

You may wish to revisit the Essential Question periodically with students as they work through the chapter.

Suggestions for reconnecting are provided at an appropriate point in the *Teacher's Edition* for each section.

Dome of the Rock Mosque, Jerusalem

CHAPTER 22

Southwest Asia

1 FOCUS ON Saudi Arabia, Iraq, and Iran

2 FOCUS ON Israel and the Palestinian Territories

3 FOCUS ON Turkey, Syria, and Lebanon

ESSENTIAL QUESTION

How might changes in political culture affect the stability of the region?

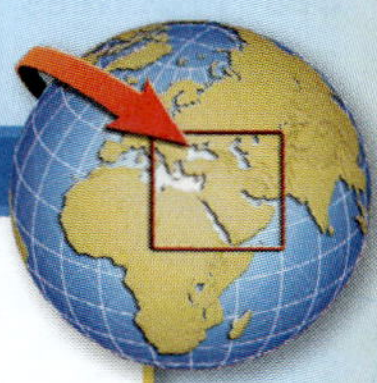

CONNECT Geography & History

Use the map and the time line to answer the following questions.

1. Which country in the region has the world's largest oil reserves?
2. What body of water is surrounded by oil-producing nations?

1000 B.C.

History
◄ **c. 960s B.C.** Solomon, king of Israel, builds a temple.

Culture
about 4 B.C. Jesus of Nazareth is born.

Geography
c. 330 B.C. Alexander the Great (at right) conquers Palestine. ►

674 Chapter 22

TIME LINE DISCUSSION

Tell students that Southwest Asia's political, economic, and religious history spans thousands of years. Describe how this region is one of the major crossroads of the world, as three continents meet here (Europe, Asia, and Africa). Point that three of the world's major religions—Judaism, Christianity, and Islam—emerged from this region.

- How many years after the birth of Jesus was Muhammad born? *(about 574 years)*
- What is most likely Saudi Arabia's chief export, and why? *(oil; It has the world's largest oil reserves.)*
- Which event on the time line is an early example of Southwest Asia being conquered by foreign invaders? *(c. 330 B.C.—Alexander the Great conquers Palestine.)*
- Define *sultan* in your own words. *(Possible Answer: an Ottoman ruler)*

Animated GEOGRAPHY

Economic Activity of Southwest Asia

Click here to explore Southwest Asia @ClassZone.com

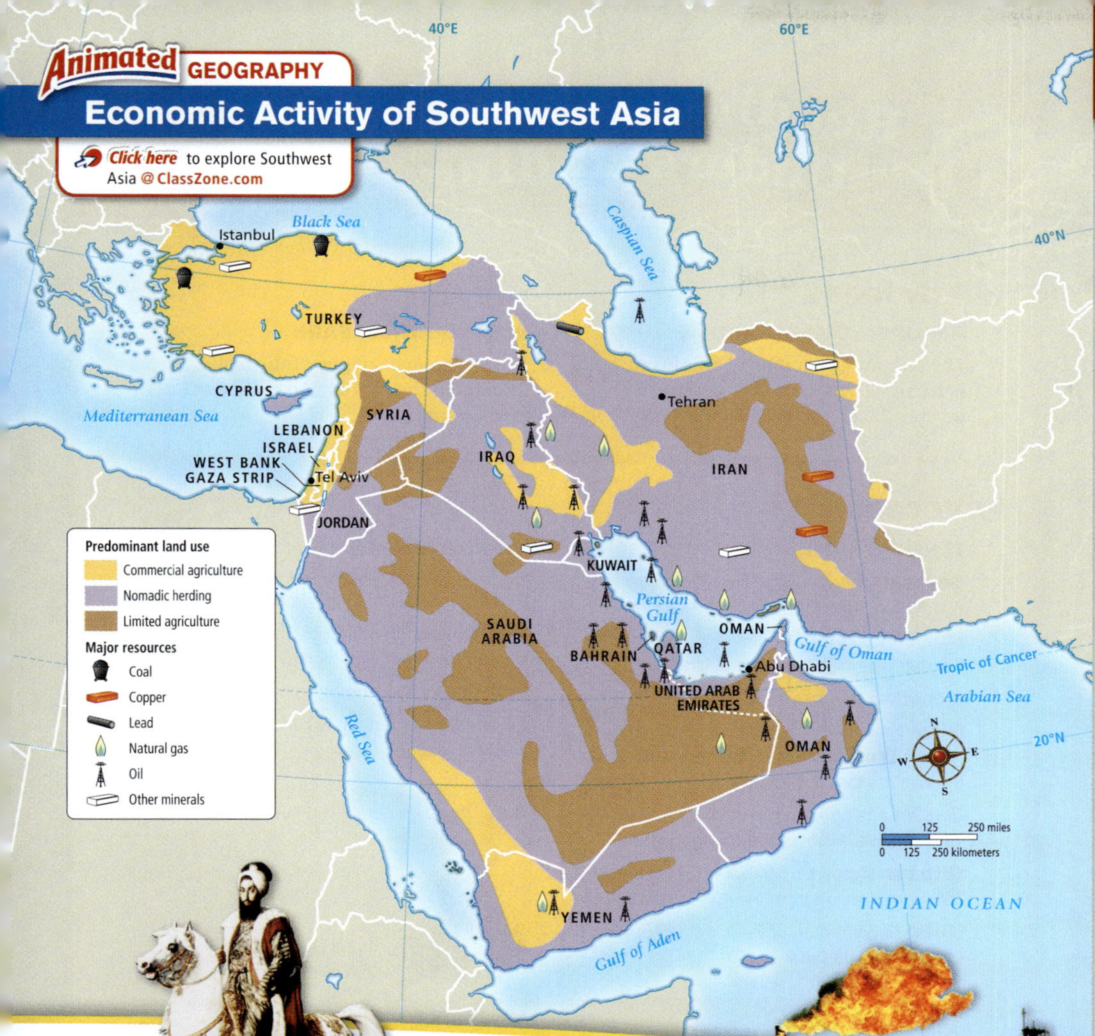

History
◄ **1520** Suleyman I becomes sultan of the Ottoman Empire.

Culture
about A.D. 570 The prophet Muhammad is born.

Economics
▲ **2006** Saudi Arabia has the world's largest oil reserves.

TEACHER MEDIA FAVORITES

Books for the Teacher

Nye, Naomi Shihab, ed. *The Space Between Our Footsteps.* Simon & Schuster Children's Publishing, 1998. Anthology of poems and paintings by writers from Middle Eastern countries.

Garcia, Laura Gallego. *Legend of the Wandering King*. Arthur A. Levine Books, 2005. In pre-Islamic Arabia, a prince journeys through Bedouin lands and distant cities. (available on audio)

Video / DVD

Islam: Empire of Faith. (DVD) PBS Home Video, 2003. Epic TV documentary charting the history of Islam.

Jerusalem: Within These Walls. National Geographic Video, 1997. Documentary examining the lives of Jewish, Muslim, Christian, and Armenian families living in Jerusalem.

Internet

Go to the *Research and Writing Center* at **ClassZone.com** to link to:

- data on the region
- a variety of chapter-specific, content-reviewed sites
- "History Makers" biographies
- primary sources

Animated GEOGRAPHY Economic Activity of Southwest Asia

CLASSZONE.COM

Point out that this is a map of some of the economic activity of Southwest Asia, showing the region's land use and where its major resources are located. Some economic activities, such as banking and trade, are not shown.

- In addition to oil, what other energy source is produced by some of the countries that border the Persian Gulf? *(natural gas)* What energy source does Turkey produce? *(coal)*
- What is the predominant land use in Turkey? *(commercial agriculture)* In Iran? *(nomadic herding)*
- Why is so much of Saudi Arabia's land used for nomadic herding or limited agriculture? *(Possible Answers: because of its large desert area and arid climate; lack of water or limited irrigation; lack of grasslands to graze livestock in many areas)*

CONNECT GEOGRAPHY & HISTORY ANSWERS

1. Saudi Arabia
2. Persian Gulf

Presentation Options

Use the **Power Presentations** DVD-ROM to project the animated map.

- Prompt students to use the pop-up images and audio from the map to discover what they already know about the geography, culture, history, economics, or governments of the region.
- Challenge students to answer the map questions individually, in small groups, or as a class activity. Use all of the interactive material to help preview chapter content.

1 Plan & Prepare

Objectives

- Describe the history, government, culture, and economics of Saudi Arabia
- Describe the history, government, culture, and economics of Iraq
- Describe the history, government, culture, and economics of Iran

Read for the Essential Question

Help students read for a purpose by reminding them of the Essential Question: "How might changes in political culture affect the stability of the region?" Point out that they can use text features such as Key Questions, special features, images, and maps to form answers to the question.

Vocabulary

Best Practices Toolkit

Use the *Best Practices Toolkit* to model strategies for vocabulary notetaking. Vary strategies throughout the year. Choose from: *Knowledge Rating, Predicting ABCs, Definition Mapping, Word Sort, Frayer Model (Word Squares), Magnet Words, Word Wheel,* and *Student VOC*.

Vocabulary Strategies, TT9–TT16

Reading Strategy

Best Practices Toolkit

Display the Three-Column Chart transparency.

- Remind students that main ideas are important ideas in a segment of text.
- Model filling in entries for Saudi Arabia.
- Have students suggest entries. Remind students that main ideas in each block of text in the section will help them answer the Key Questions.
- Display the completed chart for students to review.

Three-Column Chart, TT21

SECTION 1 Reading for Understanding

Key Ideas

BEFORE, YOU LEARNED

The physical geography of Southwest Asia is harsh, with mountains, deserts, and a severe climate.

NOW YOU WILL LEARN

The physical geography of the region has influenced the history and culture of many countries.

Vocabulary

TERMS & NAMES

Bedouin an Arabic-speaking, traditionally nomadic people of Southwest Asia

Muhammad (also spelled Mohammed) the founder and major prophet of Islam

monotheism a belief in one god

Islam a monotheistic religion based on the teachings of Muhammad and the writings of the Qur'an, the Muslim holy book

Muslim a believer in the religion of Islam who accepts Allah as the only god

Ottoman Empire a Muslim empire based in Turkey that lasted from the 1300s to 1922

OPEC Organization of Petroleum Exporting Countries, founded in 1960

Kurd a member of an ethnic group that does not have its own country but whose homeland lies in parts of Turkey, Iraq, and Iran

Persia a historic region of Southwest Asia located mostly in what is now Iran

shah title for the king of Persia or Iran

theocracy a government run by religious leaders

Visual Vocabulary Shah of Persia (Iran)

Reading Strategy

Re-create the chart shown at right. As you read and respond to the **KEY QUESTIONS**, use the chart to note the main ideas about the region's history, government, culture, and economics.

Skillbuilder Handbook, page R4

FIND MAIN IDEAS

	SAUDI ARABIA	IRAQ	IRAN
History			
Government			
Culture			
Economics			

GRAPHIC ORGANIZERS Go to Interactive Review @ClassZone.com

PRETEACHING VOCABULARY

English Learners

Pronounce and Preview

Pronounce each term for students. Review the meanings of words within definitions, such as *nomadic*, *prophet*, and *homeland*.

- To modify vocabulary learning, have students complete worksheets as they read, instead of afterward.

Unit Resource Book: Southwest Asia & South Asia

- Vocabulary Practice, p. 83

Inclusion

Create Visual Vocabulary

Read the definition of each term aloud, and answer any questions students might have. Then assign each student to draw a simple illustration of each term, writing the word as the title and definition as the caption. Ask volunteers to share their illustrations when complete and to explain them as necessary. Post the illustrations to create a Visual Vocabulary display.

Skyline of Yazd, Iran

SECTION 1 FOCUS ON

A herd of addax (antelopes) in Bahrain

Saudi Arabia, Iraq, and Iran

Connecting to Your World

Have you ever heard people say that they can't wake up until they've had their morning cup of coffee? Drinking coffee is just one of many things that our culture gained from Southwest Asia. The people of Southwest Asia (along with people in Egypt) invented algebra, recorded many observations about the stars, and discovered medicines that we still use today. Southwest Asia was also the birthplace of three major religions—Judaism, Christianity, and Islam. You will learn about Islam in this section.

Saudi Arabia

KEY QUESTION How did the Arabian Peninsula become a Muslim region?

Saudi Arabia occupies more than three-fourths of the Arabian Peninsula. As you learned earlier, deserts such as the Rub al-Khali cover most of the peninsula. Hardly anyone lives in the deserts; in fact, the name *Rub al-Khali* means "empty quarter." Lack of water is one of Saudi Arabia's biggest problems. No permanent rivers exist, and the country is using up its supplies of underground water.

People would probably cluster around water sources. ▼

Saudi Arabia Sand dunes are common in the deserts of the Arabian Peninsula, with an occasional water source such as the springs shown here. **How might scarcity of water affect settlement in the Arabian Peninsula?**

677

2 Focus & Motivate

3-Minute Warm-Up

Write on the board or display the transparency:

- Assume the role of a leader of an oil-rich country. What would you do to diversify your country's economy? *(Possible Answer: develop new industries, such as tourism)*

Unit Transparency Book: Southwest Asia & South Asia
- 3-Minute Warm-Ups, TT10

Connecting to Your World

More About . . .

The Lost Desert City of Ubar

According to Bedouin folklore, the magnificent city of Ubar disappeared beneath the sands of the Rub al-Khali centuries ago. Archeologists think Uber may have sunk into the sand when an underground limestone cavern collapsed.

3 Teach

Saudi Arabia

Think, Pair, Share

- What is one of Saudi Arabia's biggest problems? *(lack of water)*

  What are the goals of OPEC? *(to coordinate policies on selling petroleum and to influence worldwide oil supply and prices)*
- **Find Main Ideas** Why is oil so important to the Arabian Peninsula nations? *(Possible Answer: Oil makes up a large share of their GDP.)*

SECTION 1 PROGRAM RESOURCES

Unit Resource Book: Southwest Asia and South Asia

ON LEVEL

Unit Resource Book
- Reading Study Guide, p. 59
- Vocabulary Practice, p. 83
- History Makers: Muhammad, p. 89
- Section Quiz, p. 99

STRUGGLING READERS

Unit Resource Book
- Reading Study Guide with Additional Support, p. 65
- Section Quiz, p. 99
- Reteaching Activity, p. 102

eEdition with Audio DVD-ROM

ENGLISH LEARNERS

Pupil Edition in Spanish

eEdition with Audio DVD-ROM

Unit Resource Book
- RSG (Spanish), p. 71
- RSG with Additional Support (Spanish), p. 77

Multi-Language Glossary

Test Generator
- Section Quiz in Spanish

INCLUSION

Unit Resource Book
- RSG with Additional Support, p. 65
- Section Quiz, p. 99
- Reteaching Activity, p. 102

GIFTED & TALENTED

Unit Resource Book
- Map and Graph Skills, pp. 85–86
- Interdisciplinary Projects: Oil in Southwest Asia, pp. 91, 93
- Section Quiz, p. 99

PRE-AP

Unit Resource Book
- Economics in the World: The Largest Oil Reserves, p. 94

TECHNOLOGY

Unit Transparency Book
- 3-Minute Warm-Ups, TT10
- Essential Question, TT11
- Critical Thinking, TT12
- Map:Oil in Southwest Asia, TT15

Daily Test Practice Transparencies
- Section 1, TT75

Power Presentations

ClassZone.com

World Cultures and Geography Video on DVD

More About . . .

Islam

About 1.2 billion people practice Islam, but the majority of Muslims live outside the Arab world. Islamic religion includes practicing the Five Pillars of Islam—belief in God and Muhammad, prayer, charity, pilgrimage, and fasting. Practicing Muslims pray five times a day and attend mosques—Muslim places of worship—on Fridays. Believers are also required to make a pilgrimage to the holy city of Mecca at least once during their lifetime. There are different branches of Islam; the Sunni and Shiite branches are the most significant.

Unit Resource Book: Southwest Asia & South Asia

- History Makers: Muhammad, p. 89

Fun Facts!

The Bedouin

Traditional Bedouins live on meat and dairy products from raising herds of camels, sheep, or goats and sleep in tents made of cloth woven from goat or camel hair. Bedouins are renowned in the Arab world for their oral tradition of reciting poetry. In recent times, national governments have pressured Bedouins to give up their nomadic lifestyle. Currently, only about one in 10 Bedouins are nomadic year-round.

Make It Fun

Teacher-Tested Activities

Don Stringfellow, Mobile County Public Schools, Mobile, Alabama

Provide a higher-order thinking question on the board at the beginning of the period. Students must then provide an answer to the question as either a ticket out of class that day or into class the next day.

- The question should be answerable from information they learned that day in class.
- Challenge students to write their own higher-order thinking questions and collect their questions at the end of the class period.

Fun Facts!

THE BEDOUIN

Since about 1950, many Bedouin have given up their nomadic life. They have taken up farming or moved to cities in search of health care, education, and jobs such as police officers and oil workers.

To maintain their identity, some Bedouin return to the desert each year to live in tents and renew their sense of roots.

History, Government, and Culture Among the early inhabitants of Saudi Arabia were **Bedouin** nomads who moved around to find grazing for their herds. However, many trade routes crossed Saudi Arabia, and over time towns grew up along those routes. At first, most people in the towns believed in many gods. In the 600s, the prophet **Muhammad** began to preach a belief in one god, called **monotheism**. Muhammad's teachings and the writings of the Qur'an (the Muslim holy book) became the religion of **Islam**. Most people in Arabia became **Muslims**, or followers of Islam. Mecca, Muhammad's birthplace, became the holiest city of Islam. Various Muslim empires rose and fell over the next thousand years.

In the early 1900s, the **Ottoman Empire**, a Muslim empire founded in Turkey in the 1300s, ruled Arabia. A young Arab in the Saudi clan, Ibn Sa'ud, led a fight for independence, and after he succeeded in 1932, he became the king of Saudi Arabia. His family still rules the country. The king controls the government. The laws are based on Islamic religious law.

Economics and Oil Saudi Arabia has more petroleum deposits than any other country, and oil production creates about half the nation's wealth. Service industries, manufacturing, construction, and utilities account for much of the rest.

Beginning in the late 1800s, oil companies searched all over the world for oil resources. In 1938, these companies found large oil fields in the Arabian Peninsula. In 1948, portions of what would become one of the world's largest oil fields were discovered at al-Ghawar, just on the eastern edge of the Rub al-Khali. This field contains approximately one-quarter of Saudi Arabia's oil reserves.

Arabian Peninsula Neighbors A number of very small countries share the peninsula with Saudi Arabia. To the east are several states on the Persian Gulf—Bahrain, Kuwait, Qatar, and United Arab Emirates. To the south, Oman and Yemen face the Indian Ocean. Each of these countries is a Muslim nation that relies on oil production.

The region grew in importance as the global demand for oil increased. Arabian Peninsula nations make almost all of their export money and a large share of GDP (Gross Domestic Product) from oil. Large increases in oil prices allow oil-producing nations to funnel money into the development of other parts of their economies.

In 1960, a group of oil-rich nations, including Saudi Arabia and Kuwait, established a cartel, or organization of independent groups, to coordinate policies on selling petroleum. The group is the Organization of Petroleum Exporting Countries, also known as **OPEC**. The purpose of OPEC is to influence worldwide oil supply and prices.

DIFFERENTIATING INSTRUCTION

Struggling Readers

Define Economic Terms

Have students work in pairs to define these terms from the text in their own words: *service industries, oil reserves, global demand, export, gross domestic product*, and *cartel*. Encourage them to use the context of the sentences in which the terms appear to help them figure out the definitions. Then have students use a dictionary or other reference book to compare definitions. Ask each group to write a definition in their own words and a sentence for each term.

English Learners

Language: Punctuation and Print Cues

Tell students that parentheses are used when adding information that doesn't otherwise fit into the sentence. Ask students to locate and read the two sentences on this page that include parentheses. For each sentence, ask students to say what information is in the parentheses and how it adds to the text.

COMPARING Ethnic & Language Groups of Southwest Asia

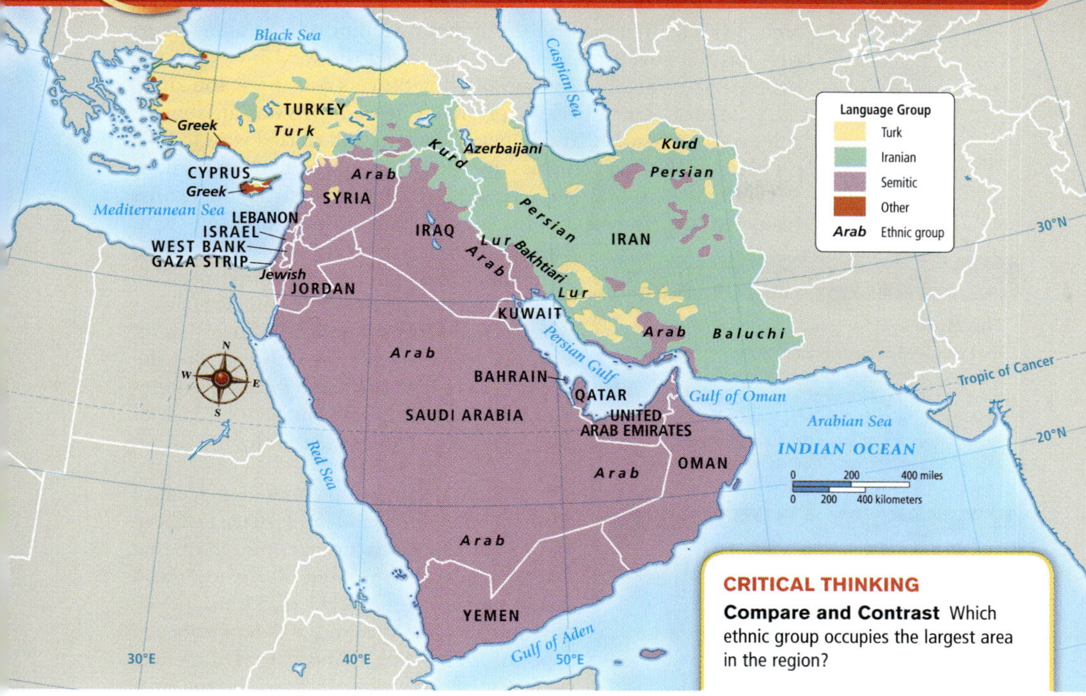

CRITICAL THINKING

Compare and Contrast Which ethnic group occupies the largest area in the region?

They do this by adjusting production quotas, or supply of oil. OPEC is a powerful force in international trade. Other members include Qatar, the United Arab Emirates, Iran, Iraq, and countries outside the region such as Venezuela, Indonesia, Libya, and Algeria.

SUMMARIZE Explain how the Arabian Peninsula became Muslim. At first, many people worshiped multiple gods. A prophet named Muhammad taught that there was only one god, and most people in the region adopted his faith.

Iraq

KEY QUESTION How effective a ruler was Saddam Hussein?

The region in Mesopotamia between the Tigris and Euphrates rivers is now part of the nation of Iraq, another Muslim country whose people speak Arabic. Islam spread to Mesopotamia during the 600s. After the rise and fall of a number of Muslim empires, the Ottoman Empire ruled the region at the start of the 1900s. But the Ottomans were on the losing side in World War I, so their empire ended. Great Britain took over Iraq temporarily.

INTERDISCIPLINARY ACTIVITIES

CONNECT to Language Arts

Summarize Information About Southwest Asia

Have students study the Ethnic & Language Groups of Southwest Asia map on this page. Then have students write a short paragraph that summarizes the different ethnic populations of each country. To the extent possible, students should try to identify the geographical locations of the groups. *(For example, the Kurds occupy northern Iraq, southeastern Turkey, and northern Iran.)*

CONNECT to Math

Graph Oil Exports and Imports

Tell students that oil makes up the following percentages of total exports of these countries: Saudi Arabia—89%, Iran—84%, Iraq—95%. Have students create bar graphs to compare figures.

Unit Resource Book: Southwest Asia & South Asia
- Interdisciplinary Projects: Oil in Southwest Asia, pp. 91, 93
- Economics in the World: The Largest Oil Reserves, p. 94

COMPARING

Ethnic & Language Groups of Southwest Asia

Tell students that Southwest Asia is home to many ethnic groups, including Arabs, Persians, Kurds, Turks, and Jews. The concentrations of ethnic and language groups vary among the countries of Southwest Asia. For example, Saudi Arabia is 90% Arab, with the majority speaking Arabic. Other countries in the region are more diverse. In Iran, for example, 51% of the population is of Persian descent, 24% are Azeri, and the remaining 25% are various other ethnic groups.

Ask students what they think might be some issues that countries might face when several different ethnic populations or language groups live close together. *(Possible Answers: conflicts due to cultural misunderstandings; discrimination or prejudice, such as in the workplace or school)*

CRITICAL THINKING ANSWER

Compare and Contrast Arab ethnic group

Unit Resource Book: Southwest Asia & South Asia
- Map and Graph Skills: Reading a Population Density Map, pp. 85–86

Teach

Iraq

Reader, Recorder, Reporter

- What ancient region is now part of the nation of Iraq? *(Mesopotamia)*
- What has crippled Iraq's economy in recent years? *(damage to Iraq's pipelines and refineries, and lack of trade)*
- **Find Main Ideas** What two branches of Islam are practiced in Iraq? *(Sunni Islam and Shi'a Islam)* What two ethnic groups make up the majority of the Iraqi population? *(Arabs and Kurds)*

Unit Resource Book: Southwest Asia & South Asia
- Skillbuilder Practice: Analyzing Causes and Effects, p. 84

CONNECT Geography & History

The Kurds

There are over 20 million Kurds living in Kurdistan, an area that includes parts of Turkey, Iraq, Iran, Syria, and Armenia. Most Kurds are Sunni Muslim. The dissolution of the Ottoman Empire and subsequent creation of nation-states left the Kurds without lands on which to pursue their nomadic way of life. Since then, they have attempted to establish their own statehood, but have been severely oppressed by the Turkish, Iraqi, and Iranian governments.

In 1988, Saddam Hussein began a resettlement program that drove over 1.5 million Kurds from their homes. In 1991, one million Iraqi Kurds were forced to flee to Turkey and Iran. Since the end of Saddam Hussein's regime in 2003, some Kurds have tried to reclaim homes taken from them in Iraq, leading to increased tension between them and other ethnic groups.

Teach

Iran

Talk About It

- For how many years did foreign powers intermittently rule Persia? *(about 2,000 years, from the 300s B.C. to A.D. 1700s)*
- Why was the shah of Iran overthrown in 1979? *(Conservative Muslims thought the shah's reforms violated Islamic law.)*
- **Find Main Ideas** What problems has Iran faced since the revolution of 1979? *(Possible Answer: It has faced economic problems because skilled workers have fled the country, the war with Iraq damaged oil refineries, and the United States cut off trade because of Iran's possible support of terrorists.)*

Government and Economics Iraq became an independent monarchy in 1932. Iraq made much money by selling oil, but many Iraqis thought the government should have used the wealth to help the people more. In 1958, a group of army officers overthrew the king.

Saddam Hussein became president in 1979. He was an abusive ruler who imprisoned and killed opponents and started wars to take oil fields from other countries. By 2003, the United States and Britain feared that Hussein might use weapons of mass destruction against his neighbors. They led an invasion of Iraq and overthrew Hussein. They found no weapons of mass destruction, but Hussein was put on trial for past crimes against his own people. The United States is helping Iraq form a democratic government. Iraq has great petroleum reserves. But the recent war in Iraq has damaged its refineries and pipelines. The war damage and lack of trade have crippled Iraq's economy.

CONNECT Geography & History

The Kurds

The Kurds lived as nomads for centuries, so they never founded a nation-state. Their home region includes lands that are legally part of Iraq, Iran, Turkey, and other nations that don't want to give up land to form a new nation of Kurdistan. The Kurdish refugees below fled Saddam Hussein's Iraq for safety in a refugee camp in Turkey.

Culture Arabs make up about 75 percent of the Iraqi population, and Kurds make up about 20 percent. The **Kurds** are an ethnic group that does not have a country but whose homeland lies in parts of Turkey, Iraq, and Iran.

Most Iraqis are Muslim, but they practice two types of Islam—Sunni and Shi'a. Differences between the two, originally based on the issue of choosing leaders, have caused conflicts.

SUMMARIZE Describe Saddam Hussein. He was a brutal ruler who imprisoned and killed opponents and started wars to gain oil fields.

Iran

KEY QUESTION What international conflicts has Iran been involved in since the 1979 revolution?

Iranians are a non-Arab people. Their ancestors were Aryans, who settled the Plateau of Iran in ancient times. That region had been called **Persia** throughout much of history. Today the country is called Iran, which means "land of the Aryans."

History, Government, and Economics The Persian Empire controlled Southwest Asia and Egypt during the 500s B.C. and was a great rival of ancient Greece. From the 300s B.C. onward, a series of foreign powers ruled Persia off and on, but Iranians regained control in the A.D. 1700s. In 1925, an army officer took over the government and became **shah**, or king. He and later his son ruled for 54 years.

DIFFERENTIATING INSTRUCTION: TIERED ACTIVITIES

OBJECTIVE Describe how Islam has contributed to Southwest Asia's history, politics, and culture.

Basic	On Level	Challenge
Ask students to divide a sheet of paper into three columns. Label each column Saudi Arabia, Iraq, and Iran. Have students write one sentence about how the Islamic religion has influenced politics, history, or culture in each country. Encourage students to use the text as a reference.	Ask students to pick one of the three countries discussed in this section. Have students write a one-paragraph summary on how Islam has influenced the politics, history, or culture of that country. Invite volunteers to read or post their summaries.	Ask students to write a short essay describing the beginnings of Islam, its influence on the region's people, culture, and governments, and its current role in region today. Invite students to read or post their essays.

They promoted reforms. Conservative Muslims believed the reforms violated Islamic law, so revolutionaries overthrew the shah in 1979. A religious leader became the ruler of Iran and established a **theocracy**, a government run by clergy according to religious law. U.S. support for the shah led the revolutionaries to hold a group of U.S. citizens hostage for nearly 15 months.

Like many nations of Southwest Asia, Iran has petroleum reserves and an economy that depends on oil. Yet it has severe economic problems. The revolution of 1979 caused many trained workers to flee Iran, creating a shortage of skilled labor. In the 1980s, Iraq and Iran fought an eight-year war that was in part about oil. Bombing during the war damaged refineries. In 1995, the United States accused Iran of helping terrorists and cut off trade.

Persian Rug Weavers
Iranian girls work on a Persian rug.

Culture People around the world prize Persian rugs, which are woven in highly decorative and colorful patterns, often featuring flowers. Mosques, Muslim places of worship, show Persian architecture at its best. Onion-shaped domes and slender towers called minarets give the buildings a graceful beauty.

SUMMARIZE Describe conflicts Iran has had since 1979.
U.S. hostage crisis, war with Iraq, U.S. trade boycott

Section 1 Assessment

ONLINE QUIZ For test practice, go to **Interactive Review** @ClassZone.com

TERMS & NAMES

1. **Explain the importance of**
   - Muhammad
   - Islam
   - Kurd
   - Persia

USE YOUR READING NOTES

2. **Find Main Ideas** Use your completed chart to answer the following question:
   What valuable resource do all three countries discussed in this section possess, and why is it so valuable?

	SAUDI ARABIA	IRAQ	IRAN
History			
Government			
Culture			
Economics			

KEY IDEAS

3. Why does Saudi Arabia face a water problem?
4. Why did the United States and Britain lead an invasion of Iraq in 2003?
5. What type of government was set up in Iran in 1979?

CRITICAL THINKING

6. **Analyze Causes** Why did towns grow in the deserts of the Arabian Peninsula?
7. **Compare and Contrast** How is Iran different from Saudi Arabia and Iraq?
8. **CONNECT to Today** How might newspapers and schools help Iraq become a more democratic country? Explain.
9. **LANGUAGE ARTS** **Give a News Broadcast** Working with a partner, research in the library or on the Internet to learn more about the U.S.-Iran hostage crisis of 1979 to 1981. Use a word-processing program to create, format, and present a news broadcast on the topic.

CONNECT to the Essential Question

How might changes in political culture affect the stability of the region?

Ask students what they have learned so far that can help them answer this question. Students might mention:

- Oil plays a key role in the economies of Saudi Arabia, Iraq, and Iran, yet it has often led to instability.
- Conflicts among different ethnic groups in the region have often led to war.
- Islam is a major influence on the region's cultures, politics and governments.

4 Assess & Reteach

Assess Have students complete the Section Assessment.

Unit Resource Book: Southwest Asia & South Asia
- Section Quiz, p. 99

Interactive Review @ClassZone.com
- Section Quiz

Power Presentations

Test Generator

Reteach Have pairs of students create five "True or False" statements about Saudi Arabia, Iraq, and Iran. Then have each pair exchange their statements with another pair and evaluate the statements. For each false statement, have students correct it, using their textbooks.

Unit Resource Book: Southwest Asia & South Asia
- Reteaching Activity, p. 102

SECTION 1 ASSESSMENT ANSWERS

Terms & Names

1. Muhammad: founder of Islam; Islam: religion based on teachings of Muhammad; Kurd: ethnic group whose homeland lies in parts of Turkey, Iraq, and Iran; Persia: historic region in Iran

Use Your Reading Notes

2. oil; Modern industrial nations depend upon oil.

	SAUDI ARABIA	IRAQ	IRAN
History	Bedouins; Saudi clan	site of Mesopotamian civilization	Persian empire
Government	monarch	Saddam Hussein overthrow	Theocracy
Culture	Islam	Muslim and Arabic	Persian and Muslim
Economics	huge oil reserves	large oil reserves	large oil reserves

Key Ideas

3. It has no permanent rivers and is using up its underground water.
4. They believed Iraq had weapons of mass destruction.
5. a theocracy, run by members of the clergy

Critical Thinking

6. Trade routes crossed Arabia, and towns sprang up on the routes to benefit from that trade.
7. Possible answer: A non-Arab people settled Iran, and it was never ruled by the Ottoman Empire.
8. Better schools will help to educate the people and open their eyes to a broader world; better newspapers will help the circulation of ideas and opinions to counter religious orthodoxy.
9. Use the rubric to score broadcasts.

Broadcast Rubric

	Details	Writing Style
4	realistic, compelling	clear, thorough
3	realistic	reasonable presentations
2	unrealistic; uses inaccurate facts	barely adequate
1	unrelated to actual events	incomplete, major errors

1 Plan & Prepare

Objectives

- Summarize the history of Israel and the Palestinian territories
- Explain why Jerusalem is considered the birthplace of three religions
- Describe how Israel is building its economy

Read for the Essential Question

Help students read for a purpose by reminding them of the Essential Question: "How might changes in political culture affect the stability of the region?" Point out that they can use text features such as Key Questions, special features, images, and maps to form answers to the question.

Vocabulary

Best Practices Toolkit

Use the ***Best Practices Toolkit*** to model strategies for vocabulary notetaking. Vary strategies throughout the year. Choose from: *Knowledge Rating, Predicting ABCs, Definition Mapping, Word Sort, Frayer Model (Word Squares), Magnet Words, Word Wheel,* and *Student VOC*.

Vocabulary Strategies, TT9–TT16

Reading Strategy

Best Practices Toolkit

Display the Web Diagram transparency.

- Remind students that identifying problems and solutions means finding and understanding the challenges faced by a group of people during a certain time and how they tried to overcome those challenges.
- Model filling in entries for a problem and one or more solutions from the section.
- Have students suggest entries.
- Display the completed diagram for students to review.

Web Diagram, TT28

SECTION 2 Reading for Understanding

Key Ideas

BEFORE, YOU LEARNED

Saudi Arabia, Iraq, and Iran are Muslim countries rich in oil resources with a history of nondemocratic governments.

NOW YOU WILL LEARN

Israel is a thriving democracy, and the Palestinian territories are trying to develop a representative government.

Vocabulary

TERMS & NAMES

Holocaust the systematic killing of millions of Jews and other groups by the Nazis during World War II

Gaza Strip a narrow territory along the Mediterranean Sea

West Bank a region west of the Jordan River

Palestinian territories the West Bank and Gaza Strip combined

Palestine Liberation Organization (PLO) a political group that claims to represent all Palestinians and to be working toward gaining an independent Palestinian nation

Jerusalem (juh•ROO•suh•luhm) the current capital of Israel and an ancient city that is holy to Judaism, Christianity, and Islam

Judaism (JOO•day•IHZ•uhm) the monotheistic religion of the Jews, based on the writings of the Hebrew Bible

Christianity a monotheistic religion based on the life and teachings of Jesus of Nazareth and on the writings of the Christian Bible

kibbutz (kih•BUTS) a type of settlement in Israel in which the community shares all wealth and property

Reading Strategy

Re-create the chart shown at right. As you read and respond to the **KEY QUESTIONS**, use the chart to record possible solutions to the problems that are listed.

Skillbuilder Handbook, page R10

IDENTIFY PROBLEMS AND SOLUTIONS

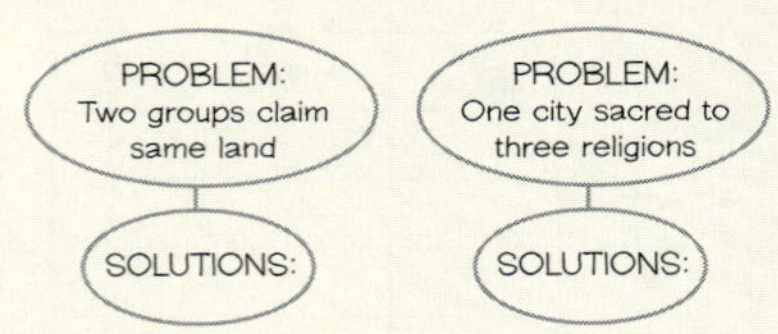

GRAPHIC ORGANIZERS Go to Interactive Review @ClassZone.com

PRETEACHING VOCABULARY

English Learners

Pronounce and Preview

Pronounce each term for students. Review the meanings of words within definitions, such as *territory*, *monotheistic*, and *settlement*.

- To modify vocabulary learning, have students complete worksheets as they read, instead of afterward.

Unit Resource Book: Southwest Asia & South Asia

- Vocabulary Practice, p. 83

Inclusion

Vocabulary: Frayer Model (Word Squares)

Have students work in pairs to fill out the graphic organizer for terms in the section.

- Use transparency to model the strategy.
- Have volunteers suggest entries.
- Then distribute copies of the organizer for students to complete with a partner.

Best Practices Toolkit

- Vocabulary: Frayer Model (Word Squares), TT14

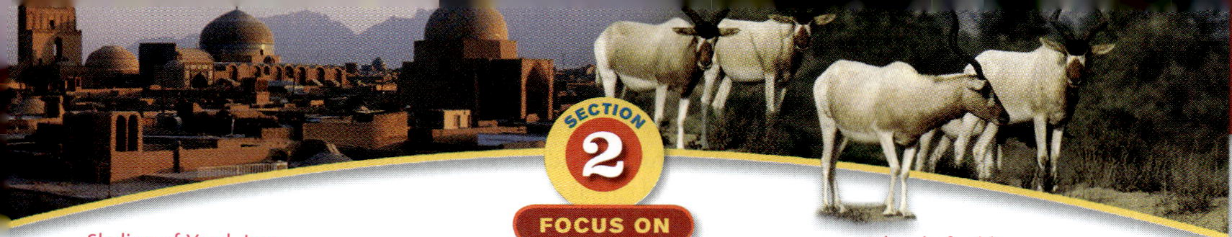

Skyline of Yazd, Iran

SECTION 2 FOCUS ON

A herd of addax (antelopes) in Bahrain

Israel and the Palestinian Territories

Connecting to Your World

Have you ever seen a TV show in which two people who share a room have a fight and then draw a line down the middle of the floor? Usually, they argue about the fairest way to divide the room. Now, think about what happens when two different cultural groups want to establish a nation on the same land. That situation exists between the Israelis and the Palestinians.

Israeli Security Barrier The Israeli security barrier separates parts of Israel from the West Bank.

History and Government

KEY QUESTION What foreign rulers have controlled Palestine?

Today, the land between the Mediterranean Sea and Jordan River is divided between Israel and the Palestinian territories. But for much of the last 3,000 years, that land was considered a single region. During ancient times, it was called Canaan. A people called the Hebrews moved there from Mesopotamia and eventually became known as Israelites. About 1000 B.C., they set up a kingdom, which soon afterward split into two kingdoms, Israel and Judah.

Animated GEOGRAPHY

Dome of the Rock This Jerusalem mosque covers the site where Muslims believe that Muhammad rose into heaven. Jews believe it is the site where Abraham prepared to sacrifice his son Isaac.

Click here for more about the holy sites of Jerusalem @ClassZone.com

2 Focus & Motivate

3-Minute Warm-Up

Write on the board or display the transparency:

- In your opinion, what is the main issue facing the Israelis and the Palestinians? Explain. *(Possible Answer: control over Jerusalem, because Israelis and Palestinians both want to control it)*

Unit Transparency Book: Southwest Asia & South Asia
- 3-Minute Warm-Ups, TT10

Connecting to Your World

Animated GEOGRAPHY

Dome of the Rock

Muslims built this octagonal shrine sometime between a.d. 685 and 691. It is richly adorned—inside and out—with marble, mosaic tiles, painted wood, carved stone, and carpets. The dome, made of wood, is about 60 feet in diameter.

3 Teach

History and Government

Think, Pair, Share

- What two regions make up the Palestinian territories? *(Gaza Strip and West Bank)*
- **Identify Problems and Solutions** Why was Palestine divided into two nations? *(Because of the Holocaust, many people felt that the Jews should have their own country. So after WWII, the UN divided Palestine into Arab and Jewish states.)*

SECTION 3 PROGRAM RESOURCES

Unit Resource Book: Southwest Asia and South Asia

ON LEVEL

Unit Resource Book
- Reading Study Guide, p. 61
- Vocabulary Practice, p. 83
- Section Quiz, p. 100

STRUGGLING READERS

Unit Resource Book
- Reading Study Guide with Additional Support, p. 67
- Section Quiz, p. 100
- Reteaching Activity, p. 103

eEdition with Audio DVD-ROM

ENGLISH LEARNERS

Pupil Edition in Spanish

eEdition with Audio DVD-ROM

Unit Resource Book
- RSG (Spanish), p. 73
- RSG with Additional Support (Spanish), p. 79

Multi-Language Glossary

Test Generator
- Section Quiz in Spanish

INCLUSION

Unit Resource Book
- RSG with Additional Support, p. 67
- Section Quiz, p. 100
- Reteaching Activity, p. 103

GIFTED & TALENTED

Unit Resource Book
- Outline Map Activities: Southwest Asia, pp. 87–88
- Section Quiz, p. 100

PRE-AP

Unit Resource Book
- Primary and Secondary Sources, pp. 96–97
- Section Quiz, p. 100

TECHNOLOGY

Unit Transparency Book
- 3-Minute Warm-Ups, TT10
- Essential Question, TT11
- Critical Thinking, TT13
- Map: Creation of Israel, TT16

Daily Test Practice Transparencies
- Section 2, TT76

Power Presentations

ClassZone.com

World Cultures and Geography Video on DVD

More About . . .

The Holocaust and the Founding of Israel

Soon after Adolf Hitler became chancellor of Germany in 1933, he implemented a policy of systematic persecution of Jews in Nazi Germany. He believed the Jews were responsible for all the problems facing Germany. During World War II, Hitler and other Nazi leaders ordered Jews confined to ghettos, and then sent them to concentration camps where most were executed, or died from starvation or disease.

By the time World War II came to an end, six million Jews had been killed, making it the worst genocide in history. After the war, about 250,000 Jewish refugees sought to emigrate to Palestine, which was then under British rule. Eventually, the United Nations took control of the situation, leading to the creation of the state of Israel.

Unit Resource Book: Southwest Asia & South Asia

- Primary and Secondary Sources: The Balfour Declaration, p. 97
- Primary and Secondary Sources: United Nations Resolution 181, p. 96

3 Teach

Culture and Religion

Talk About It

- What two major religions began in Israel? *(Judaism and Christianity)* On what teachings are they based? *(Judaism is based on the Hebrew Bible. Christianity is based on the teachings of Jesus.)*
- What is one reason for conflict between Jews and Muslims over Jerusalem? *(A mosque stands where an ancient Jewish temple once stood.)*
- **Identify Problems and Solutions** How did conductor Daniel Barenboim and author Edward Said work to help solve the Israeli-Palestinian conflict? *(They have started an orchestra made up of both Israelis and Palestinians.)*

Unit Resource Book: Southwest Asia & South Asia

- Skillbuilder Practice: Analyzing Causes and Effects, p. 84

Foreign Rule Beginning in the 700s B.C., a series of ancient empires ruled Canaan, including Alexander the Great in the 300s B.C. The last of these was Rome, which invaded in 63 B.C. The Romans renamed the region Palestine and called the people there Jews (after the name *Judah*). When the Jews staged uprisings, the Romans defeated them and drove them from Jerusalem. During the A.D. 600s, Muslim Arabs moved into the region. In 1099, Christian armies from Europe conquered Jerusalem. That began a series of wars known as the Crusades between Christians and Muslims to control the region. In the early 1500s, the Ottoman Empire conquered Palestine. After World War I, the League of Nations authorized Great Britain to govern Palestine.

A New Israel During World War II, Nazi Germany killed six million Jews in Europe in the **Holocaust**. Afterward, many people felt that the Jews should have their own country in Southwest Asia, site of their ancestral roots. The United Nations proposed dividing Palestine into two nations—Arab and Jewish—and the Jewish state of Israel was born in 1948. Its official language is Hebrew. The Arabs opposed a Jewish state in the region, so war broke out. Israel won the war, and many Palestinians fled to Arab countries as refugees.

The Palestinian Territories Arab-Israeli wars were also fought in 1956, 1967, and 1973, and each time Israel won. In the 1967 war, Israel took the **Gaza Strip**, a narrow territory on the Mediterranean coast, and the **West Bank**, a region west of the Jordan River. The two regions together make up the **Palestinian territories**. A political group called the **Palestine Liberation Organization (PLO)** began to fight to have a Palestinian state on those lands. The PLO and Israelis began to hold peace talks in the 1990s. Israel withdrew its troops from Gaza and much of the West Bank, and the Palestinians elected a government. Even so, the situation remains unstable, and issues need to be resolved. For example, the two sides disagree over who should control **Jerusalem**, which is sacred to Judaism, Christianity, and Islam.

CATEGORIZE Describe the rulers who have controlled Palestine. ancient empires, Alexander, Rome, the Ottoman Empire, Great Britain

West-East Orchestra Daniel Barenboim rehearses an orchestra made up of Israelis and Palestinians.

Culture and Religion

KEY QUESTION Why do three religions regard Jerusalem as holy?

Some people are working to help the two main groups in the region, Arabs and Jews, better understand each other. For example, Israeli conductor Daniel Barenboim and Palestinian-American author Edward Said started an orchestra for Israeli and Palestinian musicians.

684

DIFFERENTIATING INSTRUCTION

English Learners

Vocabulary: Idioms

Point out the sentences "The Arabs opposed a Jewish state in the region, so war broke out" and "The two regions together make up the Palestinian territories." Then have students to identify the idiom in the above sentences. *(broke out; make up)* Ask volunteers to replace the idioms with words that do not alter the meaning of the sentences. *(started; form)* Encourage students to write new sentences using each idiom.

Gifted & Talented

Negotiate Regional Conflicts

Have the class form groups of four students each. Have two students in each group represent Israel and two the Palestinian territories. First, have each group list the problems causing conflicts between the Israelis and Palestinians. Second, have each group brainstorm possible solutions to each problem, including the pros and cons of each. Last, have students agree on one solution for each problem and be prepared to explain why they choose that solution.

Religious Groups of Southwest Asia

Islamic Mosque Dome of the Rock, Jerusalem

Jewish Synagogue Old Synagogue in Mazkeret Batyz, Israel

Christian Church Getsemani Church, Jerusalem

CONNECT Geography & Culture

READING A THEMATIC MAP

Thematic maps show specific types of information, such as population density or economic activity. This map shows the diversity and distribution of religious groups in Southwest Asia. When reading a thematic map, remember that

- qualitative maps use colors (as well as dots and lines) to help you see patterns related to a specific idea
- an inset map makes a small or congested area on a thematic map easier to see

1. **Place** Which places have holy sites for three major religions?
2. **Location** Why might the location of these places be a problem?

INTERDISCIPLINARY ACTIVITIES

CONNECT to Art

Architecture of Holy Places

Have students choose one holy place in Southwest Asia, such as a well-known mosque, synagogue, or church. Using library and Internet resources, have students research its architectural construction and details. Then have students create an illustration or diagram of the site. Encourage students to include a title and captions for significant details about its architecture. Have students present or display their work.

CONNECT to Math

Calculate Population Density

Give students these statistics: **Israel**: area—8,367 square miles, population—6,352,000; **West Bank:** area—2,270 square miles, population—2,460,000; **Gaza Strip**: area—140 square miles, population—1,428,000. Then ask students to calculate the population densities for each. Have students create a chart or bar graph, showing their results.

Religious Groups of Southwest Asia

CONNECT Geography & Culture

Reading a Thematic Map Point out to students that the majority of the population of Southwest Asia is Muslim. Explain to students that the Druze religion is a small, monotheistic religious sect. Ask them what information the map is giving them about religious groups in this region. *(the locations of religious groups and holy places)*

- Which countries are home to Shiite Muslims? *(Iran, Iraq, Lebanon, Yemen, Syria, and Turkey)*
- Identify the countries that are home to Jews *(Israel)*, Christians *(Lebanon, Cyprus)*, and Druze *(Lebanon, Syria).*

ANSWERS

1. **Place** West Bank, Israel
2. **Location** Possible Answer: The religious groups may not get along.

Unit Resource Book: Southwest Asia & South Asia

- Outline Map Activities: Southwest Asia: Political, pp. 87–88

Comparing Regions

Religious Groups in the United States

Unlike Israel and the Palestinian territories, the predominant religion in the United States is Christianity. Approximately 84% of the American people identify themselves as Christian. In contrast, only an estimated 2% of the United States population is Jewish and 1.6% are Muslim.

RESEARCH & WRITING CENTER

ANALYZING Primary Sources

Kenneth Woodward joined *Newsweek* in 1964 and won numerous journalism awards for his articles on religion. His article, "A City That Echoes Eternity," appeared in the July 24, 2000, issue of *Newsweek*. In it, Woodward stressed that politics alone would not solve the tensions that exist over the city of Jerusalem.

Today, the conflict about control over Jerusalem and its holy sites continues. Plans by Muslims to add a fifth minaret at the Temple Mount site are being met with resistance, as are Jewish plans to add a synagogue there.

- Why does Jerusalem have meaning to people who may never even see it? *(Possible Answer: because significant events in the development of Judaism, Christianity, and Islam took place there)*

DOCUMENT-BASED QUESTION ANSWER

Possible Answer: because of its importance to Muslims, Jews, and Christians around the world

CONNECT to the Essential Question

How might changes in political culture affect the stability of the region?

Ask students what they have learned so far that can help them answer this question. Students might mention:

- For centuries, numerous foreign rulers have controlled the area that is now Israel and the Palestinian territories.
- The division of Palestine and establishment of Israel after World War II resulted in many conflicts among Jews and Arabs.
- Because three religious groups hold strong ties to this region, its stability remains fragile.

ANALYZING Primary Sources

Kenneth L. Woodward is a *Newsweek* editor who wrote about who should control Jerusalem. He mentioned Jerusalem's importance to Judaism, Christianity, and Islam.

> For billions of believers who may never see it, Jerusalem remains a city central to their sacred geography. This is why the future of the city is not just another Middle Eastern conflict between Arabs and Jews. . . . There will be no enduring solution to the question of Jerusalem that does not respect the attachments to the city formed by each faith.
>
> Source: "A City That Echoes Eternity," by Kenneth L. Woodward

DOCUMENT-BASED QUESTIONS

According to Woodward, why is the future of Jerusalem an especially important issue?

Market in Muslim Quarter of Jerusalem ▶

ONLINE PRIMARY SOURCE For more writings on Jerusalem today, go to the **Research & Writing Center @ClassZone.com**

Birthplace of Religions People from three religions have strong ties to the land that Israel occupies. It was the birthplace of two monotheistic religions, Judaism and Christianity. **Judaism**, the religion of the Jews, is based on the writings of the Hebrew Bible. This Bible includes Jewish law and historical and prophetic writings. It teaches that the Jews are descended from Abraham, who obeyed God's command to move from Mesopotamia to Canaan (later Palestine).

Christianity is based on the teachings of Jesus, who lived in Palestine under Roman rule. As a Jew, Jesus taught Jewish ideas. But he also taught ideas that some people found dangerous, so the Romans killed him. Christians believe he rose from the dead and ascended to heaven. Israel is special to Christians because Jesus taught there.

Jerusalem Jews and Muslims have their own reasons for regarding Jerusalem as holy. Jews hold it sacred because the ancient Jewish temple stood there. The temple was destroyed by Babylonian conquerors, rebuilt by Jews, and destroyed again by the Romans. Only one wall remains. One reason for conflict over Jerusalem is that a mosque now stands where the Jewish temple used to be. Muslims chose the site for a house of worship because they believe the prophet Muhammad was taken into heaven from that very spot.

FIND MAIN IDEAS Explain why three religions regard Jerusalem as holy. Jews revere it as the site of their ancient temple, Christians revere it as the place where Jesus lived and died, and Muslims view it as the place where Muhammad was taken into heaven.

DIFFERENTIATING INSTRUCTION: TIERED ACTIVITIES

OBJECTIVE Report on events leading to the current conflicts in Israel and the Palestinian territories.

Basic

Have pairs of students review the section, listing important events in the region from 1948 to today. Then ask students to use their lists to create a time line. Display the time lines.

On Level

Have small groups of students review the section, listing important events that have led up to the current conflicts. Then have each group prepare a five-minute news broadcast, explaining these events and their significance. Students should write a script, assign roles, and rehearse their broadcast before presenting it.

Challenge

In addition to the On Level activity, have students research current events to include in their broadcasts. Then have each group prepare a five-minute news broadcast, explaining these events and their significance. Students should write a script, assign roles, and rehearse their broadcast before presenting it.

Israel Builds Its Economy

KEY QUESTION What are the main economic activities in Israel?

Israel has few water resources and, unlike its Arab neighbors, hardly any petroleum. But from the start, it did have a highly trained workforce. These workers helped Israel to develop the most advanced economy in Southwest Asia.

First, Israel developed agriculture, using irrigation in dry areas. Today, many farms use drip irrigation, in which computers control the flow of water. Many Israelis moved onto a **kibbutz**, a settlement whose members share wealth and property. They receive housing and food in exchange for labor. The first kibbutzim were farms, although some have switched to manufacturing.

Farming is now a small part of the economy, which relies more on service industries and manufacturing. Israel's most important industries include the production of electronics and chemicals and the cutting of imported diamonds. Tourism also brings a lot of money to Israel, as people from many faiths visit the holy sites there.

Electronics A computer technician inspects hardware in Israel.

SUMMARIZE Describe the main economic activities in Israel.
production of electronics and chemicals, diamond cutting, tourism

Section 2 Assessment

ONLINE QUIZ For test practice, go to Interactive Review @ClassZone.com

TERMS & NAMES

1. **Explain the importance of**
   - Palestinian territories
   - Jerusalem
   - Judaism
   - Christianity

USE YOUR READING NOTES

2. **Identify Problems and Solutions** Use your completed chart to answer the following question:
   How might three religions share Jerusalem peacefully?

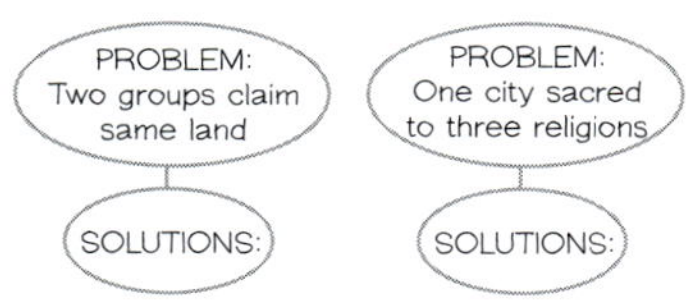

KEY IDEAS

3. How did Palestine become an Arab region?
4. Why do some Jews regard Jerusalem as holy?
5. How has life on a kibbutz changed?

CRITICAL THINKING

6. **Make Inferences** Why do you think the Holocaust convinced many people that there needed to be a Jewish state?
7. **Evaluate** What do you think is the most serious challenge to Arab-Israeli peace?
8. **CONNECT to Today** As part of the peace process, in 2005 Israel forced Jewish settlers to withdraw from the Gaza Strip and parts of the West Bank. The Israeli economy improved. Why might that have happened?
9. **TECHNOLOGY** **Plan a Virtual Tour** With a partner, research Jerusalem and make a list of its holy places from three religions. Plan a Web page that offers a tour of the city. Decide on links to other Web sites, design your Web page, and write captions.

Teach

Israel Builds Its Economy

Reader, Recorder, Reporter

- What are Israel's most important industries? *(production of electronics and chemicals, imported diamond-cutting, tourism)*
- What natural resources does Israel lack? *(water, petroleum)*
- **Identify Problems and Solutions** What has been a crucial factor in Israel developing the most advanced economy in Southwest Asia? *(Israel has a highly-trained workforce.)*

4 Assess & Reteach

Assess Have students complete the Section Assessment.

Unit Resource Book: Southwest Asia & South Asia
- Section Quiz, p. 100

Interactive Review @ClassZone.com
- Section Quiz

Power Presentations

Test Generator

Reteach Have students work in pairs to write three questions and answers about Israel and the Palestinian territories. Have pairs trade questions, answer each other's questions, and then compare answers. Encourage students to use their textbooks to revise any answers as needed.

Unit Resource Book: Southwest Asia & South Asia
- Reteaching Activity, p. 103

SECTION 1 ASSESSMENT ANSWERS

Terms & Names

1. Palestinian territories: West Bank and Gaza Strip; Jerusalem: capital of Israel and city holy to Jews, Christians, Muslims; Judaism: religion of Jews, based on Hebrew Bible; Christianity: religion based on life and teachings of Jesus

Use Your Reading Notes

2. **Diagram** *Problem: Two groups claim same land. Solutions: International conference to settle disputes. Two groups reach agreement. Problem: One city sacred to three religions. Solutions: International organization such as the UN might administer the city. Or UN, European Union (EU), and the United States might guarantee the safety of the city and access of all religions to its sacred sites.*

Key Ideas

3. The Romans drove out most of the Jews, and in the 600s, Muslim Arabs moved into the region.
4. The ancient Jewish temple stood there.
5. Some have switched to manufacturing.

Critical Thinking

6. Possible answer: It would be easier for Jews to protect themselves in the future if they had their own land, government, and army.
7. Many students will mention the challenge of having holy sites for both Judaism and Islam located on the same spot. Other students might mention that the Palestinians recently elected an anti-Israel party.
8. Possible answers: Business people probably assumed there would be less fighting, creating a better climate for trade. Also, tourists are more likely to visit during peaceful times.
9. Use the rubric to score virtual tours.

Tour Rubric

	Content Included on Web Page	Organization of Web Page
4	all elements	well organized, interesting
3	most elements	well organized
2	some elements	poorly organized
1	few elements	incomplete, disorganized

1 Plan & Prepare

Objectives

- Summarize Turkey's history, government, culture, and economy
- Compare and contrast the history, culture, government, and economy of Syria with those of Lebanon

Read for the Essential Question

Help students read for a purpose by reminding them of the Essential Question: "How might changes in political culture affect the stability of the region?" Point out that they can use text features such as Key Questions, special features, images, and maps to form answers to the question.

Vocabulary

Best Practices Toolkit

Use the ***Best Practices Toolkit*** to model strategies for vocabulary notetaking. Vary strategies throughout the year. Choose from: *Knowledge Rating, Predicting ABCs, Definition Mapping, Word Sort, Frayer Model (Word Squares), Magnet Words, Word Wheel,* and *Student VOC.*

Vocabulary Strategies, TT9–TT16

Reading Strategy

Best Practices Toolkit

Display the Three-Column Chart transparency.

- Point out to students that to categorize information in the section they must decide how it can be sorted into the categories shown in the chart.
- Model filling in entries from one of the subsections in the section.
- Have students suggest entries, using the categories in the chart.
- Display the completed chart for review.

Three-Column Chart, TT21

SECTION 3 Reading for Understanding

Key Ideas

BEFORE, YOU LEARNED

Various countries in the region are struggling with how to preserve past legacies as they organize their societies to move into the future.

NOW YOU WILL LEARN

Turkey, Syria, and Lebanon are all struggling with similar issues of how to reconcile representative government with their traditions.

Vocabulary

TERMS & NAMES

Istanbul the largest city of Turkey, formerly called Constantinople (capital of the Byzantine, or Eastern Roman, Empire) and originally called Byzantium in ancient times

European Union an organization of European nations whose members cooperate on economic, social, and political issues

Golan Heights a region that was formerly part of southwest Syria that Israel has occupied since the 1967 war

peacekeeping referring to military forces whose purpose is to prevent or end war

extended family the family unit that contains more relatives than just parents and children, including grandparents and other close family

Beirut capital of Lebanon and a center of banking and finance

REVIEW

Ottoman Empire a Muslim empire based in Turkey that lasted from the 1300s to 1922

Palestine Liberation Organization (PLO) a political group that claims to represent all Palestinians and to be working toward gaining an independent Palestinian nation

Reading Strategy

Re-create the chart shown at right. As you read and respond to the **KEY QUESTIONS**, use the chart to organize important facts about Turkey, Syria, and Lebanon.

Skillbuilder Handbook, page R7

CATEGORIZE

	TURKEY	SYRIA	LEBANON
History			
Culture			
Government			
Economics			

GRAPHIC ORGANIZERS
Go to **Interactive Review** @ClassZone.com

PRETEACHING VOCABULARY

English Learners

Pronounce and Preview

Pronounce each term for students. Review the meanings of words within definitions, such as *occupied, family unit,* and *finance.*

- To modify vocabulary learning, have students complete worksheets as they read, instead of afterward.

Unit Resource Book: Southwest Asia & South Asia
- Vocabulary Practice, p. 83

Inclusion

Use Letter Tiles

Have student pairs use letter tiles to spell out all the vocabulary words. When a person spells out a word, his or her partner should repeat the word, and then read the definition out loud. Students should trade roles, repeating the steps for each vocabulary word.

Skyline of Yazd, Iran

A herd of addax (antelopes) in Bahrain

Turkey, Syria, and Lebanon

Connecting to Your World

Think about the other students who were in your class last year. Even if you come from very different backgrounds or families, you have something in common. You know what that teacher was like and what the classroom rules were. And you share some of the same memories. The countries of Turkey, Syria, and Lebanon have a similar link. Their people have very different ethnic heritages, but they were all part of the Ottoman Empire that you read about in Section 1.

Turkey

KEY QUESTION What steps did Mustafa Kemal take to modernize Turkey?

The **Ottoman Empire** began in the 1300s when a group of Turkish people took over most of Anatolia and parts of the Balkan Peninsula in Europe. During the next two centuries, the empire expanded to include more of Europe and most of Southwest Asia and North Africa. But over time, the empire grew weak because of losses in wars and revolts within the empire. After being defeated in World War I, the empire came to an end. In its place, the much-smaller nation of Turkey was founded in 1923 on the Anatolian Peninsula.

It spans Europe and Asia and is connected to a number of important waterways. ▼

Istanbul Shown is the Galata Bridge in Istanbul, Turkey. The western part of the city is in Europe and the eastern part is in Asia. **Why might Istanbul be a good location for trade?**

689

2 Focus & Motivate

3-Minute Warm-Up

Write on the board or display the transparency:

- Match the pairs that go together.

1. Ottoman capital	**a.** Mustafa Kemal
2. Turkish president	**b.** Beirut
3. Lebanon capital	**c.** Istanbul

(1. c; 2. a; 3. b)

Unit Transparency Book: Southwest Asia & South Asia
- 3-Minute Warm-Ups, TT10

Connecting to Your World

More About . . .

The Galata Bridge

The Galata Bridge spans the Golden Horn, a narrow channel of the Bosporus strait. Every day before dawn, the bridge swings open to allow large ships to pass. Ferries and water buses leave from under the bridge, shuttling people back and forth between Europe and Asia.

3 Teach

Turkey

Think, Pair, Share

- What are some well-known buildings in Istanbul? *(the Hagia Sophia, Topkapi Palace, and Suleiman Mosque)*
- **Categorize** According to the European Union, what are some ways in which Turkey differs from Europe? *(history, culture, religion)*

SECTION 4 PROGRAM RESOURCES

Unit Resource Book: Southwest Asia and South Asia

ON LEVEL

Unit Resource Book
- Reading Study Guide, p. 63
- Vocabulary Practice, p. 83
- Section Quiz, p. 101

STRUGGLING READERS

Unit Resource Book
- Reading Study Guide with Additional Support, p. 69
- Section Quiz, p. 101
- Reteaching Activity, p. 104

eEdition with Audio DVD-ROM

ENGLISH LEARNERS

Pupil Edition in Spanish

Pupil Edition Audio DVD-ROM
- RSG (Spanish), p. 75
- RSG with Additional Support (Spanish), p. 81

Multi-Language Glossary

Test Generator
- Section Quiz in Spanish

INCLUSION

Unit Resource Book
- RSG with Additional Support, p. 69
- Section Quiz, p. 101
- Reteaching Activity, p. 104

GIFTED & TALENTED

Unit Resource Book
- Interdisciplinary Projects: Designing a Turkish Kilim, pp. 92–93
- Section Quiz, p. 101

PRE-AP

Unit Resource Book
- History Maker: Mustafa Kemal, p. 90
- Connect to History, p. 95
- Section Quiz, p. 101

TECHNOLOGY

Unit Transparency Book
- 3-Minute Warm-Ups, TT10
- Essential Question, TT11
- Critical Thinking, TT14
- Culture: Hagia Sophia, TT17
- Culture: Ceramic Tiles, TT18

Daily Test Practice Transparencies
- Section 3, TT77

Power Presentations
ClassZone.com
World Cultures and Geography Video on DVD

More About . . .

Modern Turkey's Government

When Mustafa Kemal came to power in Turkey in 1923, he established an authoritarian government ruled by only one political party, the Republican People's Party. In 1946, however, a newly-founded Democratic party emerged and won national elections. Tensions between the two parties resulted in the Turkish military taking over the government in the early 1960s.

During the 1960s to 1980s, the military intervened in Turkey's politics several times, but power always returned to a civilian government. A national constitution has been in effect since 1982. Today, Turkey is governed by a 550-member legislature—elected by popular vote—a prime minister, and a president. The government has banned extremist and religious political parties.

Unit Resource Book: Southwest Asia & South Asia

- History Makers: Mustafa Kemal, p. 90
- Interdisciplinary Projects: Designing a Turkish Kilim, pp. 92–93

CONNECT to History

Tell students that Istanbul is the largest city in Turkey—and the country's most important commercial center and seaport. Explain that because of its strategic location, Istanbul has been sought-after by foreign conquerers for over 2,500 years, and has been ruled by some of the world's most powerful empires. As the city grew—and to improve its defenses—its walls were rebuilt and expanded in the A.D. 400s. Many of these walls and turrets still stand today. *(Some dates and events students might add to the time line: 657 B.C.–Byzantium founded; A.D. 330–Constantine made Byzantium his capital; 537–Hagia Sophia built; 1453–Ottomans conquered Constantinople)*

Unit Resource Book: Southwest Asia & South Asia

- Connect to History: Make a Time Line, p. 95

History, Government, and Culture The first president of the Turkish republic was Mustafa Kemal (known as Ataturk), who wanted to modernize Turkey. His government made reforms such as updating the Turkish alphabet, encouraging economic development, creating modern courts and schools, and giving women the right to vote.

The Ottoman capital was **Istanbul**, which had earlier been the capital of the Byzantine Empire. It has many notable religious and government buildings:

- The Hagia Sophia is a cathedral built in 537, which is famous for its large, beautiful dome. The building is now a museum.
- The Topkapi Palace is a complex of several buildings and gardens. It was the home of the Ottoman ruler, or sultan.
- The Suleiman Mosque is similar in design to the Hagia Sophia; the Blue Mosque has six minarets.

Economics After World War II, several European nations formed what is now called the **European Union (EU)**, a political and economic organization to promote trade and peace. Since 1959, Turkey has tried to join the EU, but it has not been accepted. One reason is that Turkey is very different from Europe in terms of history, culture, and religion. Because of economic and trade ties, discussion continues.

SUMMARIZE Explain how Mustafa Kemal modernized Turkey. His government encouraged economic development, created modern courts and schools, and gave women the right to vote.

CONNECT to History

Built astride the waterway that links the Mediterranean and Black seas, Istanbul stands in Europe and Asia. It was founded about 657 B.C. as Byzantium. In A.D. 330, the Roman emperor Constantine made the city his capital, and it was renamed Constantinople. When the Ottomans conquered the city in 1453, they began to call it Istanbul. In 1923, Turkey moved its capital from Istanbul to Ankara.

The Blue Mosque in Istanbul was one of many mosques built by Muslim architects after the Ottomans conquered the city in 1453.

Activity

Make a Time Line

Materials
- paper
- pencil and eraser
- ruler

1. Draw a time line, starting with the date 1000 B.C. Add these dates: 500 B.C., A.D. 1, 500, 1500, 2000.
2. Find dates and events in this section to add to the time line.

690 Chapter 22

DIFFERENTIATING INSTRUCTION

English Learners

Culture: Clarify

Define the word *minaret*, by explaining that it is a tower attached to a mosque. Have students look at the picture of the Blue Mosque and identify the minarets. Explain that around the Muslim world, people call worshippers to prayer five times a day from the minarets. Tell students *minaret* is derived from the Arabic word *manara*, which means "lighthouse." Ask them to explain how a minaret is similar to a lighthouse.

Pre-AP

Summarize Istanbul's Geographic Importance

Have students write a summary of how Istanbul's geography has influenced its history, that of the Ottoman Empire, and that of Turkey. To extend this activity, have students create an annotated historical map, with captions and labels for key events in Istanbul's history, waterways, geographical landforms, and neighboring cities and countries.

Syria and Lebanon

KEY QUESTION What is the major cultural difference between Syria and Lebanon?

Like much of Southwest Asia, Syria and Lebanon were ruled by a series of empires ending with the Ottomans. After World War I, the Ottoman Empire was forced to give up most of its territory. The League of Nations issued mandates (commissions) to Britain and France to administer former Ottoman lands. France controlled Syria and Lebanon until they gained independence—Lebanon in 1943 and Syria in 1946. Even though Lebanon and Syria share a border and a history, Syria is almost entirely Muslim, while Lebanon's population is about 60 percent Muslim and 40 percent Christian.

History and Government At independence, Syria became a republic headed by a president. It opposed having a Jewish state in Southwest Asia and took part in three of the Arab-Israeli wars. In 1967, Syria lost a region called the **Golan Heights** to Israel. Syria has demanded the return of the region and the creation of a Palestinian state.

Lebanon, which shares borders with Syria and Israel, is the only nation in the region to have a large number of Christians. When the country gained independence, Muslims and Christians agreed to share power. In the late 1960s, the **Palestine Liberation Organization (PLO)** began to conduct raids into Israel from bases in Jordan and later Lebanon. Lebanese Muslims supported those actions, but Lebanese Christians did not. Muslims demanded a larger role in government because their numbers had grown.

In 1975, civil war broke out between Muslims and Christians. Syria sent troops into Lebanon, while Israel invaded Lebanon to drive out the PLO. In 1978, the UN sent **peacekeeping** forces to Lebanon, but the war dragged on. During the 1990s, steps were taken to disarm various groups of fighters, and elections were held in 1998. In 2005, a Lebanese politician who opposed Syria's influence was killed. Many people blamed Syria, so Syria withdrew from Lebanon.

Ottoman Empire and Post WWI Mandates *1926*

CONNECT Geography & History

1. **Location** In 1926, with what countries did Syria share borders?
2. **Region** Which countries in the region were controlled by France and which by Britain?

Teach
Syria and Lebanon

Talk About It

- What are some of the conflicts between Syria and Israel? *(Possible Answers: Syria opposed the creation of a Jewish state in the region; Syria lost the Golan Heights to Israel in 1967 and has demanded its return; Syria and Israel sent troops into Lebanon in the 1975 civil war)*
- Who fought in the civil war that started in Lebanon in 1975? *(Lebanese Muslims and Lebanese Christians; Syria; Israel; PLO)*
- Why is Lebanon's economy struggling? *(Possible Answer: Warfare has damaged businesses and much of Beirut, Lebanon's capital and financial center.)*
- **Categorize** List the three main products that make up Syria's economy. *(petroleum, cotton, wheat)*

Unit Resource Book: Southwest Asia & South Asia
- Skillbuilder Practice: Analyzing Causes and Effects, p. 84

Ottoman Empire and Post WWI Mandates *1926*

CONNECT Geography & History

Explain to students that after World War I, the League of Nations—an early version of the United Nations—had to decide what to do with the territories of the defeated Ottoman Empire. The League mandated that these territories would be provisionally governed by Allied member nations—mainly France and Great Britain—until the territories were capable of independence. Former Ottoman territories under this mandate included Iraq, Syria, Lebanon, and Palestine. By 1949, all former Ottoman territories had been granted independence.

ANSWERS

1. **Location** Turkey, Iraq, Transjordan, Palestine, Lebanon
2. **Region** France: Syria and Lebanon; Britain: Iraq, Transjordan, Palestine, Cyprus

DIFFERENTIATING INSTRUCTION: TIERED ACTIVITIES

OBJECTIVE Create time lines of key events in the conflicts among Syria, Lebanon, and Israel.

Basic

Have students list the following events in chronological order: Syria withdraws from Lebanon. *(2005)* A civil war breaks out in Lebanon. *(1975)* Israel invades Lebanon to fight Hezbollah terrorists. *(2006)* Israel invades Lebanon to drive out the PLO. *(1975)* Syria loses the Golan Heights to Israel. *(1967)*

On Level

In addition to the Basic activity, have students include the following information on their time lines:
- a brief explanation of why each event occurred
- what the results of each event have been for the countries involved

Challenge

In addition to the On Level activity, have students work in pairs to discuss Lebanon's economic and political problems and solutions, and include these on time lines. To extend this activity, hold a round-table discussion in which each pair presents one problem and solution for the group to discuss.

CONNECT to the Essential Question

How might changes in political culture affect the stability of the region?

Ask students what they have learned so far that can help them answer this question. Students might mention:

- The end of the Ottoman Empire resulted in the modernization of Turkey.
- Tensions between Syria and Israel have led to conflicts.
- Christians and Muslims share power in Lebanon, but have had many conflicts.

Unit Transparency Book: Southwest Asia & South Asia
- Essential Question, TT11

4 Assess & Reteach

Assess Have students complete the Section Assessment.

Unit Resource Book: Southwest Asia & South Asia
- Section Quiz, p. 101

Interactive Review @ClassZone.com
- Section Quiz

Power Presentations

Test Generator

Reteach Assign pairs of students to write two True/False statements about either Turkey, Syria, or Lebanon. Then have pairs trade their statements and solve each other's statements. For false statements, students should explain why they are false.

Unit Resource Book: Southwest Asia & South Asia
- Reteaching Activity, p. 104

Beirut This view of downtown Beirut shows the Mediterranean Sea.

Culture About half of all Syrians live in cities and half live in small villages. Many Syrians live in **extended families** composed of older parents and their sons' families. People in villages build houses of stone or sun-dried brick. People in cities live in modern housing and wear Western clothes.

More than 80 percent of Lebanon's people live in cities. During the 1975 civil war, bombs destroyed much of **Beirut**, the capital and financial center. The city was rebuilt, but warfare between Israel and the terrorist group Hezbollah that began in July 2006 inflicted heavy damage on the city.

Economics Both Syria and Lebanon have developing economies. Syria has petroleum deposits, but not as much as Saudi Arabia or Iraq. It has some fertile land, so agriculture is important; the main crops are cotton and wheat. Syria has little manufacturing. Until 1975, Lebanon's chief industries were the service industries of finance and banking. Warfare has greatly interfered with business, and these once-thriving industries are now struggling to recover.

COMPARE AND CONTRAST Describe the main cultural difference between Syria and Lebanon. Lebanon's population is 40 percent Christian, while Syria is almost entirely Muslim.

Section 3 Assessment

ONLINE QUIZ For test practice, go to **Interactive Review** @ClassZone.com

TERMS & NAMES

1. **Explain the importance of**
   - Istanbul
   - Golan Heights
   - peacekeeping
   - Beirut

USE YOUR READING NOTES

2. **Categorize** Use your completed chart to answer the following question:

   How were Turkey, Syria, and Lebanon affected by World War I?

	TURKEY	SYRIA	LEBANON
History			
Culture			
Government			
Economics			

KEY IDEAS

3. What did Mustafa Kemal do to modernize Turkey?
4. How is Lebanon different from every other country in Southwest Asia?
5. What role did Syria play in the Lebanese civil war?

CRITICAL THINKING

6. **Compare and Contrast** How was the culture of the Ottoman rulers similar to and different from the cultures of the people they ruled in Southwest Asia?
7. **Analyze Causes** Why does tension continue between Syria and Israel?
8. **CONNECT to Today** What are some of the reasons that Turkey has not yet been accepted into the EU?
9. **WRITING** **Create a Tour Guide** Using the information in this section, write a paragraph describing Istanbul for tourists.

SECTION 3 ASSESSMENT ANSWERS

Terms & Names

1. Istanbul: largest city of Turkey; Golan Heights: part of Syria but Israel has occupied since 1967 war; peacekeeping: military forces whose purpose is to prevent or end war; Beirut: capital of Lebanon, banking and finance center

Use Your Reading Notes

2. *Turkey was what remained of the Ottoman Empire after World War I; Syria and Lebanon were administered by France after World War I.* **Diagram** *Turkey has Ottoman culture and is a republic. Syria is Muslim in culture, a republic and a developing economy. Lebanon is Muslim and Christian; it has experienced turmoil and civil war; it has a developing economy.*

Key Ideas

3. encouraged economic development, created modern courts and schools, updated the alphabet used to write the Turkish language, and gave women the right to vote
4. It has a large number of Christians.
5. It sent troops into Lebanon to back the Muslims.

Critical Thinking

6. Possible answer: The Ottomans were Turks. They ruled people of Arabic, Kurdish, and Iranian culture. They all spoke different languages. The one thing they had in common was Islam.
7. Syria wants Israel to return the Golan Heights and to allow the Palestinians to have a nation. Syria never wanted a Jewish state in the region.
8. Differences in history, culture, and religion have made it difficult to fit Turkey into a union of European nations.
9. Use the rubric to score guides.

Tour Guide Rubric

	Accuracy of Description	Errors in Mechanics
4	thoughtful and detailed	few or none
3	detailed	some
2	shows some knowledge	many
1	unclear	many major

Interactive Review

Click here to complete these and other activities online @ClassZone.com

CHAPTER SUMMARY

Key Idea 1
The physical geography of Southwest Asia has influenced the history and culture of many countries.

Key Idea 2
Israel is a thriving democracy, and the Palestinian territories are trying to develop a representative government.

Key Idea 3
Turkey, Syria, and Lebanon are all struggling with similar issues of how to reconcile representative government with their traditions.

For **Review and Study Notes**, go to **Interactive Review** @ClassZone.com

NAME GAME

Use the Terms & Names list to complete each sentence on paper or online.

1. I am an ancient city that is holy to three major religions. **Jerusalem**
2. I am the religion based on the teachings of Muhammad. ______
3. I am the religion that is based on the life and teachings of Jesus of Nazareth. ______
4. I am the religion that is based on the writings of the Hebrew Bible. ______
5. I am a member of an Arabic-speaking nomadic people. ______
6. I am a member of an ethnic group that lives in Southwest Asia but does not have its own country. ______
7. I am the group of oil-producing countries that tries to coordinate its policies. ______
8. I am a settlement in Israel in which the community shares all wealth and property. ______
9. I am the largest city of Turkey and its former capital. ______
10. I used to be the title for the king of Iran. ______

Bedouin
Christianity
Islam
Istanbul
Gaza Strip
Jerusalem
Judaism
kibbutz
Kurd
monotheism
OPEC
Ottoman Empire
Palestinian territories
shah
West Bank

Activities

GeoGame

Use this online map to show what you know about the geography of Southwest Asia. Drag and drop each place name to its location on the map.

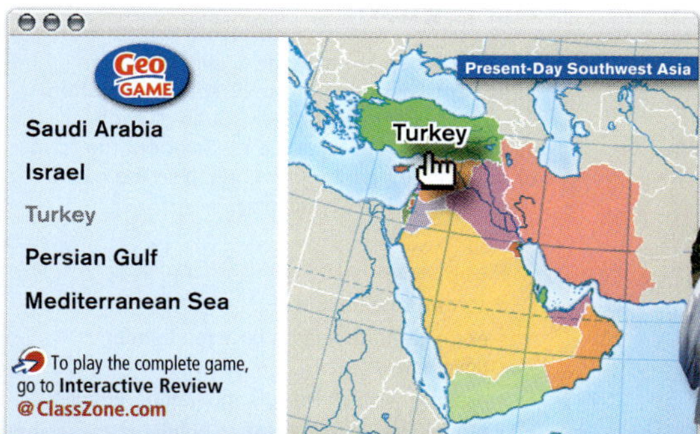

Crossword Puzzle

Complete an online puzzle to test your knowledge of Southwest Asia's history, culture, government, and economics.

ACROSS
1. member of Arabic-speaking nomadic people

693

ONLINE AND ON-SCREEN ACTIVITIES

GeoGame

Use the interactive map on **Power Presentations** DVD-ROM or at **ClassZone.com** to help students review Southwest Asia's geographic features.

- Demonstrate how to drag and drop the map labels listed onto the appropriate locations on the map.
- Advance to the scorecard to show how students can improve their scores.

Crossword Puzzle

Students can review important people and events in Southwest Asia's history by completing this crossword puzzle online at **ClassZone.com**.

- As an extension of this activity, have students use the *Activity Maker* at **ClassZone.com** to make their own crossword puzzle for one section of the chapter.
- Then have students trade puzzles with classmates to solve answers for each other's puzzles.

Chapter Summary

As a follow-up to the chapter, encourage students to pair up and go over the **Review and Study Notes** at **ClassZone.com**.

Alternatively, student pairs can review their Reading Study Guide notes, making sure that they can add important details to each section's "Key Idea."

Name Game

Teaching Tips

ON PAPER Have students work in pairs to complete this activity. As an extension, they can write questions for the terms that are not covered in the items shown.

ONLINE Make sure that students understand how to drag and drop terms to fill in blanks. Ask them to print out their results and compare them with those of a classmate.

ANSWERS

1. Jerusalem
2. Islam
3. Christianity
4. Judaism
5. Bedouin
6. Kurd
7. OPEC
8. kibbutz
9. Istanbul
10. shah

Teacher-Tested Activities

Don Stringfellow, Mobile County Public School System, Mobile, Alabama

You can make use of local newspapers or national news magazines to find articles and pictures about Southwest Asia.

- Students should find out where the stories take place and locate them on a map.
- Students should summarize the main idea of each article in a sentence.
- Students should check for common themes in the articles.

Vocabulary ANSWERS

1. Islam is a monotheistic religion based on the teachings of Muhammad.
2. Istanbul is the largest city in Turkey (formerly named Byzantium and Constantinople).
3. Jerusalem is the current capital of Israel and an ancient city holy to Jews, Christians, and Muslims.
4. A Kurd is a member of an ethnic group that does not have its own country but whose homeland is in parts of Turkey, Iraq, and Iran.
5. Monotheism means a belief in one god.
6. OPEC, the Organization of Petroleum Exporting Countries, was founded in 1960.
7. The Ottoman Empire was a Muslim empire based in Turkey.
8. Persia is a historic region of Southwest Asia located mostly in what is now Iran.
9. These two regions combined make up the Palestinian territories.
10. Both words are related to government in Iran; it used to be ruled by a shah and now is ruled by a theocracy.

Key Ideas ANSWERS

11. It is located where continents meet, so its many trade routes helped cities to grow.
12. damage caused by several wars
13. Iran and Iraq fought over disputed oil fields.
14. It is a tense or hostile relationship, as shown by the hostage crisis and the recent trade embargo by the United States.
15. The Holocaust led to support for a Jewish state because the Nazis murdered millions of Jews.
16. It was the land where Jesus taught.
17. Members of a kibbutz share all wealth and property.
18. questions about the control of Jerusalem and the recent election of a new Palestinian government
19. He did a lot to modernize the country.
20. The 1975 civil war destroyed much of Beirut, the capital and financial center of Lebanon.
21. When a Lebanese politician who opposed Syrian influence was assassinated, people protested until Syria withdrew its troops.
22. oil production and agriculture

CHAPTER 22 Assessment

VOCABULARY

Explain the significance of each of the following.

1. Islam
2. Istanbul
3. Jerusalem
4. Kurd
5. monotheism
6. OPEC
7. Ottoman Empire
8. Persia

Explain what the terms in each pair have in common.

9. Gaza Strip and West Bank
10. shah and theocracy

KEY IDEAS

1 Saudi Arabia, Iraq, and Iran

11. How did Arabia's geographic location lead to the growth of ancient cities?
12. What has prevented the Iraqi economy from benefiting from its oil fields?
13. Which two of the countries in this section fought a lengthy war, and why?
14. What type of relationship exists between Iran and the United States? Explain.

2 Israel and the Palestinian Territories

15. What historic event of the 20th century caused many people to support the idea of a Jewish state? Why?
16. Why is Palestine considered the birthplace of Christianity?
17. How is a kibbutz different from most other communities?
18. What obstacles remain to the Israeli-Palestinian peace process?

3 Turkey, Syria, and Lebanon

19. Why might the Turkish people honor Mustafa Kemal?
20. What damaged Lebanon's status as a financial center?
21. Why did Syria finally withdraw its troops from Lebanon?
22. What are the main economic activities in Syria?

CRITICAL THINKING

23. **Sequence Events** Create a time line showing the order in which nations of Southwest Asia gained independence.

24. **Draw Conclusions** Which site in Jerusalem is most likely to cause conflict?
25. **Five Themes: Region** The economies of many Southwest Asian nations depend on oil, a resource that may run out. What might they do to prepare for the future?
26. **Connect to Science** Companies in Israel are developing technology to take salt out of seawater. Why would they want to do this?
27. **Connect Geography & History** How did the 2003 invasion of Iraq affect Iraq's most valuable natural resource and its economy?
28. **Connect Geography & Culture** What do the religions of Judaism, Christianity, and Islam have in common?

Answer the ESSENTIAL QUESTION

How might changes in political culture affect the stability of the region?

Written Response Write a two- or three-paragraph response to the Essential Question. Be sure to consider the key ideas of each section as well as specific ideas about the forms of government each nation has. Use the rubric below to guide your thinking.

Response Rubric

A strong response will:

- examine which nations have representative governments
- examine which have authoritarian governments
- discuss the impact of war on politics in the region

ESSENTIAL QUESTION REVIEW

Before students begin their essays, have them review the key chapter elements that will help them answer the Essential Questions. Direct them to significant features in the chapter, such as:

- Comparing Ethnic & Language Groups of Southwest Asia, Section 1
- Religious Groups of Southwest Asia, Section 2
- Ottoman Empire and Post WWI Mandates *1926*, Section 3

Written Response Rubric

	Coverage of Key Ideas	Explanation	Accuracy
4	all	clear, concise	no errors
3	most	good	few/minor errors
2	some	fair	several errors
1	two or fewer	poor	many errors

STANDARDS-BASED ASSESSMENT

Test Practice
- Online Test Practice @ClassZone.com
- Test-Taking Strategies and Practice at the front of this book

PIE GRAPH

Use this pie graph to answer questions 1 and 2 on your paper.

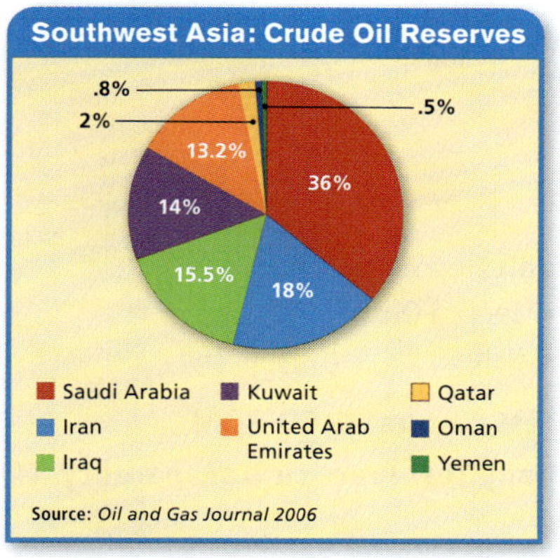

1. Which two countries have the second and third largest oil reserves in the region?

A. Iraq and Iran
B. Kuwait and Iran
C. Yemen and Oman
D. United Arab Emirates and Yemen

2. Which country has the largest oil reserves?

A. Iran
B. Iraq
C. Saudi Arabia
D. Kuwait

BAR GRAPH

Examine the bar graph below. Use the information in the graph to answer questions 3 and 4 on your paper.

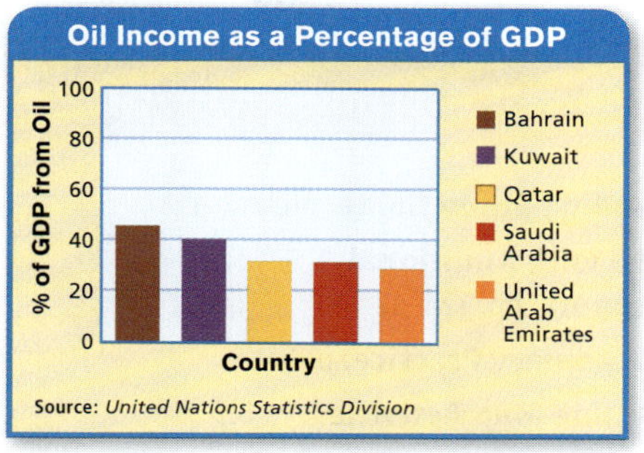

3. In which country is oil income the largest percentage of gross domestic product (GDP)?

A. Bahrain
B. Kuwait
C. Qatar
D. Saudi Arabia

4. In which countries is oil income the smallest percentage of GDP?

A. Bahrain and Kuwait
B. Kuwait and Saudi Arabia
C. Qatar and Bahrain
D. Saudi Arabia and United Arab Emirates

GeoActivity

1. INTERDISCIPLINARY ACTIVITY–ART

Using books and the Internet, look at photographs of Persian rugs. Research to learn about traditional patterns and colors used in the rugs. Then using crayons, colored pencils, or felt-tip markers, design your own rug.

2. WRITING FOR SOCIAL STUDIES

Research to learn about Suleyman I, the greatest ruler of the Ottoman Empire. Find out about his military campaigns and government, and about Ottoman culture under his rule. Then write a "History Maker" about him.

3. MENTAL MAPPING

Create an outline map of Southwest Asia and label the following:

- Iran
- Iraq
- Israel
- Palestinian territories
- Lebanon
- Saudi Arabia
- Syria
- Turkey

DIFFERENTIATING CHAPTER ASSESSMENT

Unit Resource Book

- **Chapter Test A,** pp. 105–108.
  Use for Struggling Readers, Beginning-to-Intermediate-Level English Learners, and for Inclusion of students whose IEP suggests supported test-taking.
- **Chapter Test B,** pp. 109–112.
  Use for On-Level Learners and ESOL students with Advanced-Level English.
- **Chapter Test C,** pp. 113–116.
  Use for gifted and talented students or students assigned Pre-AP work.

Test Generator

- Chapter Tests A, B, C, in English and Spanish

McDougal Littell Assessment System

TEST **Test Generator** Use three levels of tests in English and Spanish tied to state standards. Plus, choose questions to customize tests for all populations.

SCORE Scan and Score plain-paper student tests in minutes.

REPORT Create instant progress reports, sortable by student, by standard, or other criteria.

RETEACH Respond immediately to students' needs by using a wealth of reteaching resources.

Critical Thinking Answers

23. 1923—Turkey; 1932—Saudi Arabia; 1932—Iraq; 1943—Lebanon; 1946—Syria 1948—Israel

24. Possible answer: the site of the former Jewish temple, which now has a mosque standing on it

25. Possible answers: Students may note that they should prepare to have a more broadly based economy by developing other industries, by educating their citizens, or by focusing on advanced technologies.

26. Possible answers: Southwest Asia has a shortage of water. The ability to make seawater usable will help solve that problem.

27. Possible answer: The invasion and war damaged the oil refineries, pipelines, and economy.

28. Possible answers: All are monotheistic; all originated in Southwest Asia; all view Jerusalem as holy.

Standards-Based Assessment Answers

1. A
2. C
3. A
4. D

GeoActivity

1. INTERDISCIPLINARY ACTIVITY RUBRIC

The rug design should

- show an interesting and traditional pattern
- be neat and attractive

2. WRITING FOR SOCIAL STUDIES RUBRIC

The History Maker should

- be detailed and colorful
- be free of major mechanical errors

3. MENTAL MAPPING RUBRIC

The outline map should

- represent the approximate shape of Southwest Asia
- include all geographic features listed in their correct location

Chapter 23 India PLANNING GUIDE

Complete print resources all on one USB drive allow you to customize lesson plans for all students.

Section Titles & Key Ideas	Unit Resource Book Southwest Asia and South Asia	Assessment Resources
1 Empires and Independence **KEY IDEA** India is the dominant country of South Asia, and it has a rich and complicated history.	• **Reading Study Guide,** pp. 117–118 • **RSG with Additional Support,** pp. 123–124 • **Vocabulary Practice,** p. 141 • **Outline Map Activities,** p. 145 • **History Makers,** pp. 147–148	**Unit Resource Book** • Section Quiz, p. 157 • Reteaching Activity, p. 160 **Test Generator CD-ROM** **Daily Test Practice Transparency,** TT78
2 A Rich and Ancient Legacy **KEY IDEA** The culture of India is a blend of all of the many influences that have shaped its history.	• **Reading Study Guide,** pp. 119–120 • **RSG with Additional Support,** pp. 125–126 • **Skillbuilder Practice,** p. 142 • **Map and Graph Skills,** pp. 143–144 • **Primary Sources,** pp. 154–155	**Unit Resource Book** • Section Quiz, p. 158 • Reteaching Activity, p. 161 **Test Generator CD-ROM** **Daily Test Practice Transparency,** TT79
3 The World's Largest Democracy **KEY IDEA** Out of India's diversity, a robust democracy has developed.	• **Reading Study Guide,** pp. 121–122 • **RSG with Additional Support,** pp. 127–128 • **Interdisciplinary Projects,** pp. 149–151 • **Economics in the World,** p. 152 • **Primary Source,** p. 156	**Unit Resource Book** • Section Quiz, p. 159 • Reteaching Activity, p. 162 • Chapter Test (Forms A, B, & C), pp. 163–177 **Test Generator CD-ROM** **Daily Test Practice Transparency,** TT80

INTEGRATED TECHNOLOGY

No Time? To focus students on important content in the chapter, use the **"Reading Study Guide with Additional Support"** in ***Resources2Go.***

Teacher Presentation Options

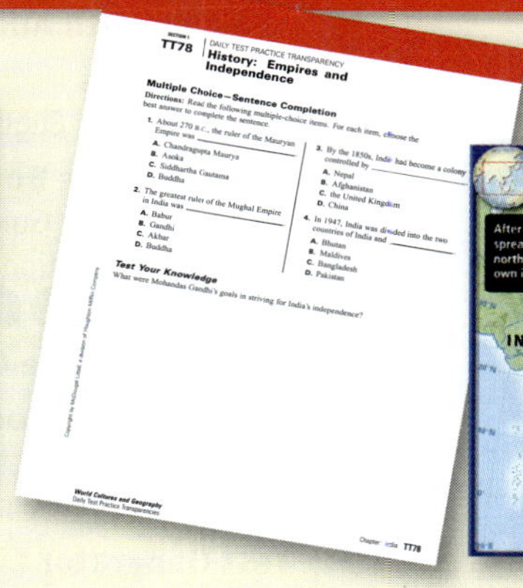

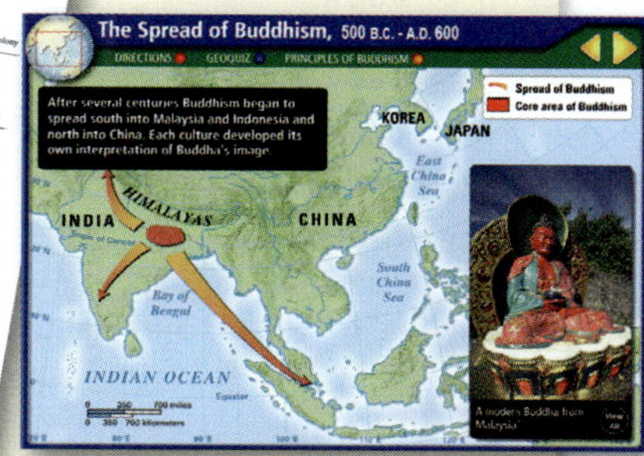

Electronic Books
eEdition DVD-ROM
eEdition Online

Presentation Toolkit
Video/DVD
- World Cultures and Geography

Power Presentations DVD-ROM
- Discussion Starters
- Media Gallery
- Review Game
- Interactive Review
- Animated Features

Unit Transparency Book
- 3-Minute Warm-Ups
- Essential Question
- Critical Thinking
- Maps
- Cultures

Daily Test Practice Transparencies

Animated Geography
- Population Density of India Map
- Spread of Hinduism and Buddhism Map
- Worldster: Savita's Web Page
- Agriculture & Industry Map

OnlineActivities at ClassZone.com

Animated Geography
- **Videos and Photos**
- **3-D Animations**
- **Interactive Maps**

Interactive Review
- **Section Quizzes**
- **Review and Study Notes**
- **Vocabulary Flip Cards**
- **Graphic Organizers**
- **Name Game and Crossword Puzzles**
- **GeoGame**

Online Test Practice

Map Center
- **Atlas Maps**
- **Political Maps with Data**
- **Links to Mapping Web Sites**

Activity Center
- **Activity Maker—Review Games, Puzzles, Flip Cards, and Graphic Organizers**
- **Worldster**

Research & Writing Center
- **Web Sites**
- **Data Updates**
- **Primary Sources**
- **Writing Models**

Current Events

Ann Christianson
John Muir Middle School
Wausau, Wisconsin

Make It Fun — Teacher-Tested Activities

As you teach this chapter, look for my classroom-proven ideas for:
- making the lesson fun
- focusing students' attention and comprehension
- integrating technology with traditional teaching

Teaching Bell to Bell When you see this timing guide, have students discuss in pairs, in groups of 3–6, or as a class.

- **Think, Pair, Share** (pairs)
- **Reader, Recorder, Reporter** (3–6)
- **Talk About It** (any size)

CHAPTER 23 RESOURCES FOR DIFFERENTIATED INSTRUCTION

Struggling Readers

eEdition with Audio DVD-ROM

Lesson Plans
- Strategies for Inclusion

Unit Resource Book
- Reading Study Guide with Additional Support, pp. 123–128 B
- Reteaching Activity, pp. 160–162
- Chapter Test (Form A), pp. 163–177 A

Best Practices Toolkit C
- Interactive Reading Strategies, pp. 90–145
- Differentiation Activity Bank: Struggling Readers, pp. 146–149

Teacher's Edition Activities
- Create a Graphic Organizer About Social Structures, Sec. 1
- Create a Time Line of Indian Independence, Sec. 1
- Pro and Con Chart About City Life, Sec. 2
- Use Transitions, Sec. 3
- Tiered Activities, Sec. 1–3

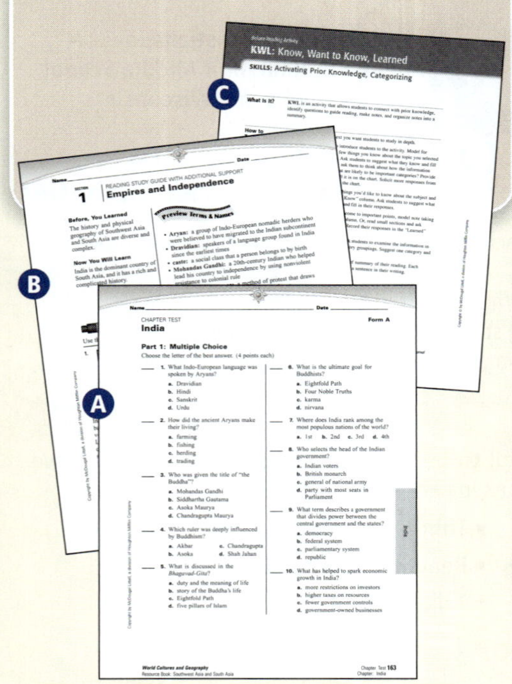

English Learners

Pupil's Edition in Spanish

eEdition with Audio DVD-ROM

Lesson Plans
- Strategies for English Learners

Unit Resource Book
- Reading Study Guide (Spanish), pp. 129–134
- RSG with Additional Support (Spanish), pp. 135–140 A B

Multi-Language Glossary C

Best Practices Toolkit
- Differentiation Activity Bank: English Learners, pp. 165–194

Test Generator CD-ROM
- Section Quizzes in Spanish
- Chapter Test in Spanish

Teacher's Edition Activities
- Pronounce and Preview, Sec. 1–3
- Vocabulary: Multiple-Meaning Words, Sec. 1
- Vocabulary: Roots, Sec. 1
- Vocabulary: Key Academic Vocabulary, Sec. 2, Sec. 3
- Tiered Activities, Sec. 1–3

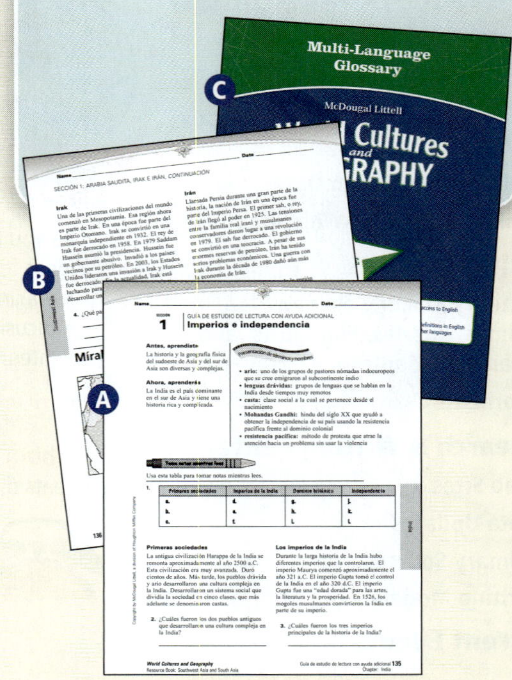

Inclusion

eEdition with Audio DVD-ROM

Lesson Plans
- Strategies for Inclusion

Unit Resource Book
- Reading Study Guide with Additional Support, pp. 123–128 A B
- Reteaching Activity, pp. 160–162
- Chapter Test (Form A), pp. 163–177

Best Practices Toolkit C
- Interactive Reading Strategies, pp. 90–145
- Differentiation Activity Bank: Struggling Readers, pp. 146–149

Teacher's Edition Activities
- Terms in Context, Sec. 1
- Vocabulary: Word Sort/Word Splash, Sec. 2
- Reading and Defining Terms Verbally, Sec. 3
- Tiered Activities, Sec. 1–3

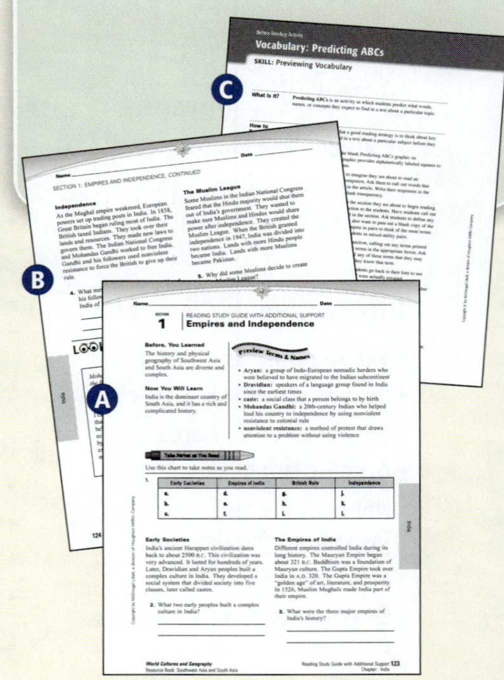

Gifted & Talented Students

Lesson Plans
- Core Lesson Plan

Unit Resource Book
- Outline Map Activities, p. 145
- History Makers, p. 147 (A)
- Interdisciplinary Projects, p. 149 (B)
- Primary Source, p. 154 (C)
- Chapter Test (Form C), pp. 173–177

Best Practices Toolkit
- Student Products Rubrics, pp. 1–14
- Writing Templates, pp. 78–89

Teacher's Edition Activities
- Argue For or Against Saving the Taj Mahal, Sec. 1
- Compare Hinduism and Buddhism, Sec. 2
- Letter from a Kashmiri, Sec. 3
- Tiered Activities, Sec. 1–3

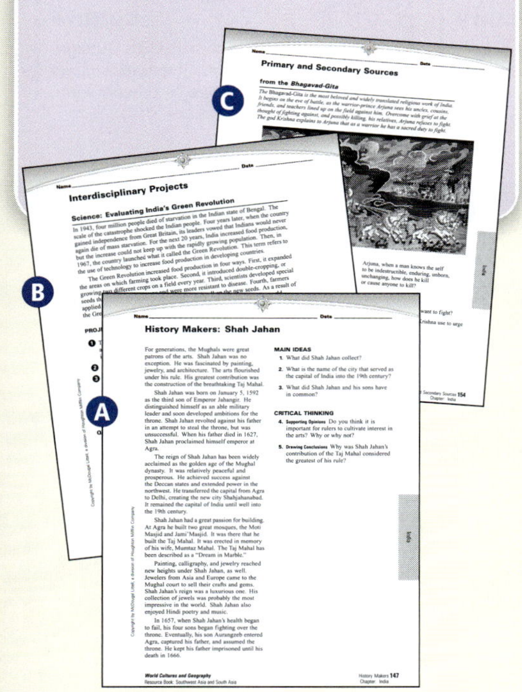
Primary and Secondary Sources
Interdisciplinary Projects
History Makers: Shah Jahan

Pre-AP*

Lesson Plans
- Core Lesson Plan

Unit Resource Book
- Vocabulary Practice, p. 141
- Map and Graph Skills, pp. 143–144
- History Makers, p. 148 (A)
- Economics in the World, p. 152 (B)
- Connect to Math, p. 153
- Primary Source, pp. 154–156 (C)
- Chapter Test (Form C), pp. 173–177

Best Practices Toolkit
- Writing Templates, pp. 78–89
- Differentiation Activity Bank: Pre-AP Students, pp. 195–200

* Pre-AP is a registered trademark of the College Entrance Examination Board, which was not involved in the production of and does not endorse this product.

Teacher's Edition Activities
- Nominate Gandhi for a Peace Prize, Sec. 1
- Write a Compare and Contrast Essay, Sec. 2
- Comparison Essay, Sec. 3
- Tiered Activities, Sec. 1–3

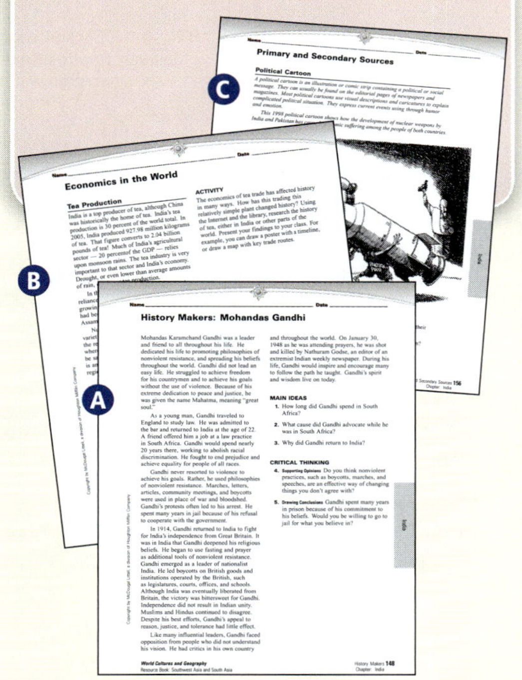
Primary and Secondary Sources
Economics in the World
History Makers: Mohandas Gandhi

Interdisciplinary Activities

CONNECT to Arts

Review a Bollywood Movie

CLASS TIME 30 minutes

TASK Prepare a review of a Bollywood movie.

SUPPLIES
- age-appropriate Bollywood film
- writing materials/computers

ACTIVITY Play part of an age-appropriate Bollywood film for students. Give students a synopsis of the whole plot. Then have them write a review of the film, comparing it to a Hollywood film with a similar plot or theme. Encourage students to try to persuade an American audience that may be unfamiliar with Bollywood to view the film.

Good for Visual and Verbal Learners

CONNECT to Language Arts

Brochure for an Educational Trip to India

CLASS TIME 40 minutes

TASK Create a brochure for a trip to India that highlights the history, geography, and culture of the country.

SUPPLIES
- paper, colored markers or pencils, or computer access

ACTIVITY Have students work in pairs to create their brochures. Brochures should feature a logical organization; clear, brief, persuasive text; and appealing images, including a map and other visuals. If time permits, students may want to research specific cities or sites within India. Have students display their brochures for the class.

Good for Visual and Verbal Learners

Introduce the ESSENTIAL QUESTION

- Identify some of the key aspects of traditional Indian life.
- Discuss how India's modern culture reflects India's past.
- Point out some of the differences between urban and rural life in India today.
- Describe the government and the economy of India, and explain the challenges the government faces.

Key Ideas

Section 1: India is the dominant country of South Asia, and it has a rich and complicated history.

Section 2: The culture of India is a blend of all of the many influences that have shaped its history.

Section 3: Out of India's diversity, a robust democracy has developed.

CONNECT to the Essential Question

You may wish to revisit the Essential Question periodically with students as they work through the chapter.

Suggestions for reconnecting are provided at an appropriate point in the *Teacher's Edition* for each section.

Tomb of Humayun in Delhi, India

CHAPTER 23 India

1 HISTORY Empires and Independence

2 CULTURE A Rich and Ancient Legacy

3 GOVERNMENT & ECONOMICS The World's Largest Democracy

ESSENTIAL QUESTION

How is India's traditional way of life changing in response to modern forces?

CONNECT Geography & History

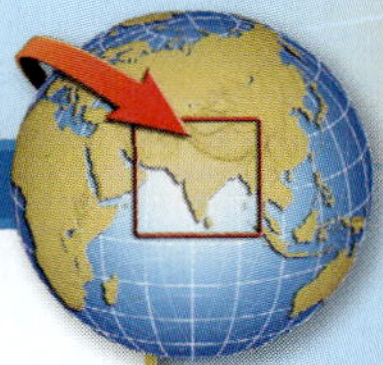

Use the map and the time line to answer the following questions.

1. Which Indian cities have the greatest populations?
2. The Taj Mahal was built in Agra, halfway between Delhi and Kanpur. In what year was it built? What is the population density in that area?

1500 B.C.

Geography
c. 1500 B.C. According to theory, the Aryans migrate to India where they encounter the Dravidians.

History
273 B.C. Asoka becomes king of Mauryan empire and builds pillars inscribed with his laws. ▶

Culture
▲ **c. A.D. 400s** Ancient Sanskrit epic the *Mahabharata* reaches its present form.

TIME LINE DISCUSSION

Tell students that an extensive urban civilization had evolved in India around 2500 B.C. and that many groups have contributed to and ruled the country over time. Remind students that India is also the home of two major world religions: Hinduism and Buddhism.

- What year was the Taj Mahal built? *(c. 1648)*
- How many years on the time line are there between Asoka becoming king and the Mughal empire being established? *(1,799 years)*
- What event on the time line shows the early contributions of India to world literature? *(c. A.D. 400s—Ancient Sanskrit epic the* Mahabharata *reaches its present form.)*
- Identify two empires that ruled India. *(Mauryan empire and Mughal empire)*

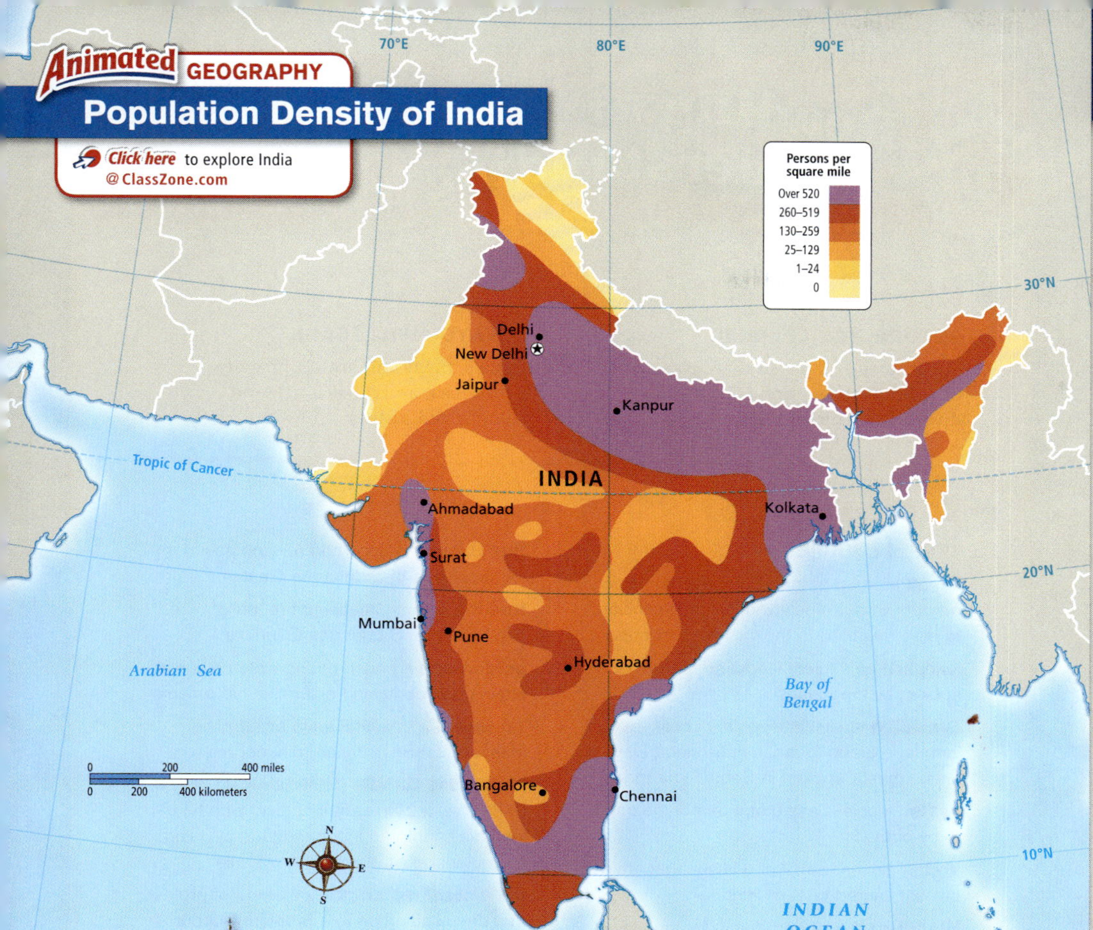

Culture
◄ **c. 1648** The Taj Mahal is built in Agra, India.

History
1526 The Mughal empire is established.

Today

Government
◄ **1947** Led by Gandhi, India gains independence from Great Britain.

TEACHER MEDIA FAVORITES

Books for the Teacher

Gandhi, Mohandas K. *My Experiments with Truth.* Highbridge Audio, 2001. Abridged audio book of the autobiography of Gandhi.

Kishore, Prem, and Anuradha Kishore Ganpati. *India: An Illustrated History.* Hippocrene Books, 2003. Overview of events that have shaped India.

R.K. Narayan. *The Mahabharata.* University of Chicago Press, 2000. Modern prose version.

Video / DVD

The Jewel in the Crown. (DVD) A&E Home Video, 2002. Acclaimed miniseries, based on Paul Scott's Raj Quartet novels, that examines the last years before India gains independence in 1947.

Mughal E Azam. (DVD) Shemraoo, 2003. A love story and a Bollywood classic.

Internet

Go to the *Research and Writing Center* at **ClassZone.com** to link to:

- data on the region
- a variety of chapter-specific content-reviewed sites
- "History Makers" biographies
- primary sources

Animated GEOGRAPHY

Population Density of India

CLASSZONE.COM

Point out that this map also identifies certain cities for points of reference. Note that some cities in India have changed names over time (for example, Bombay is now called Mumbai).

- Which city shown is in an area with the lowest population density? *(Bangalore)*
- What level of population density do most of the areas around the cities shown have? *(over 520 persons per square mile)*
- What is the population density of the area around Hyderabad? *(260–519 persons per square mile)* What other areas around the listed cities have this population density? *(Pune and Jaipur)*

CONNECT GEOGRAPHY & HISTORY ANSWERS

1. Delhi, New Delhi, Kanpur, Kolkata, Chennai, Mumbai, Surat, Ahmadabad
2. c. 1648; highest population density

Presentation Options

Use the **Power Presentations** DVD-ROM to project the animated map.

- Prompt students to use the pop-up images and audio from the map to discover what they already know about the geography, culture, history, economics, or governments of India.
- Challenge students to answer the map questions individually, in small groups, or as a class activity. Use all of the interactive material to help preview chapter content.

1 Plan & Prepare

Objectives

- Describe the two early societies that contributed to India's cultural development
- Identify the three main empires that ruled India
- Explain how India achieved independence and what led to partition

Read for the Essential Question

Help students read for a purpose by reminding them of the Essential Question: "How is India's traditional way of life changing in response to modern forces?" Point out that they can use text features such as Key Questions, special features, images, and maps to form answers to the question.

Vocabulary

Best Practices Toolkit

Use the ***Best Practices Toolkit*** to model strategies for vocabulary notetaking. Vary strategies throughout the year. Choose from: *Knowledge Rating, Predicting ABCs, Definition Mapping, Word Sort, Frayer Model (Word Squares), Magnet Words, Word Wheel,* and *Student VOC.*

Vocabulary Strategies, TT9–TT16

Reading Strategy

Best Practices Toolkit

Display the Three-Column Chart transparency.

- Remind students that main ideas are important ideas in a block of text and that details support main ideas.
- Model filling in main ideas using the subheads and text in this section.
- Have students suggest entries. Remind students that main ideas in each block of text in the section will help answer a Key Question.
- Display the completed chart for students to review.

Three-Column Chart, TT21

SECTION 1

Reading for Understanding

Key Ideas

BEFORE, YOU LEARNED

The history and physical geography of Southwest Asia and South Asia are diverse and complex.

NOW YOU WILL LEARN

India is the dominant country of South Asia, and it has a rich and complicated history.

Vocabulary

TERMS & NAMES

Aryan a group of Indo-European nomadic herders who are believed by many scholars to have migrated to the Indian subcontinent

Sanskrit an ancient language of India, first spoken by the Aryans

Dravidian (druh•VIHD•ee•uhn) speakers of a language group found in India since earliest times; mostly spoken in South India today

caste a social class that a person belongs to by birth

Vedas ancient Sanskrit writings that are the earliest sacred texts of Hinduism

Hinduism the modern name for the major religion of India

Buddhism a major world religion based on the teachings of Siddhartha Gautama

Buddha a name, meaning the "enlightened one," used for Siddhartha Gautama

golden age a period during which a society attains prosperity and cultural achievements

Taj Mahal a beautiful tomb built by the Mughal ruler Shah Jahan to honor his wife

Mohandas Gandhi (MOH•hehn•DAHS GAHN•dee) a 20th-century Indian who helped lead his country to independence by using nonviolent resistance to colonial rule

nonviolent resistance a method of protest that draws attention to a problem without using violence

Reading Strategy

Re-create the chart that is shown at right. As you read and respond to the **KEY QUESTIONS**, use the chart to note the main ideas about India's long history.

Skillbuilder Handbook, page R4

FIND MAIN IDEAS

EARLY SOCIETY	ANCIENT EMPIRES	BRITISH RULE	INDEPENDENCE
1.	1.	1.	1.
2.	2.	2.	2.
3.	3.	3.	3.

GRAPHIC ORGANIZERS
Go to Interactive Review @ClassZone.com

PRETEACHING VOCABULARY

English Learners

Pronounce and Preview

Pronounce each term for students. Review the meanings of words within definitions, such as *nomadic, migrated,* and *subcontinent.*

- To modify vocabulary learning, have students complete worksheets as they read, instead of afterward.

Unit Resource Book: Southwest Asia & South Asia
- Vocabulary Practice, p. 141

Inclusion

Terms in Context

Read "Sanskrit" and its definition aloud. Direct students to the image of a Sanskrit scroll in Section 1. Then read the other terms and definitions aloud. Ask students to look through the section and find images of each term or passages that contain the terms. Discuss the images and passages so that students understand the terms in context. Finally have students use each term in a sentence.

SECTION 1 HISTORY

Taj Mahal in Agra, India

Tigers are found in India.

Empires and Independence

Connecting to Your World

When you learned about U.S. history in the past, where did you start? Perhaps you began in 1492 with Columbus. The history of India begins further back in time. In fact, the Harappan civilization dates back to about 2500 B.C. So when you study Indian history, you are studying a time period of 4,500 years. During those years, many changes have occurred on the Indian subcontinent.

Lions Asoka's laws were carved on pillars, some of which were decorated with lions—a symbol of India.

Early Societies

KEY QUESTION Which two groups contributed to the development of a complex culture in India?

As you read in an earlier chapter, the Harappan civilization had many advanced cities with streets and sewers. After several centuries, the civilization went into decline and people left the cities. Scholars still don't know why. But many believe that another important event occurred shortly after the cities were abandoned. According to this theory, a group of people from the north or northwest came through a pass in the Hindu Kush Mountains onto the Indian subcontinent by about 1500 B.C.

Hindu Kush This mountain range in eastern Afghanistan and northern Pakistan borders India.

2 Focus & Motivate

3-Minute Warm-Up

Write on the board or display the transparency:

- Write T for true or F for false.
  1. The Hindu Kush is the tallest mountain range in India. *(F; the Himalayas)*
  2. Lions are a symbol of India. *(T)*
  3. Taj Mahal was a famous ruler of India. *(F; Shah Jahan)*

Unit Transparency Book: Southwest Asia & South Asia
- 3-Minute Warm-Ups, TT19

Connecting to Your World

More About . . .

The Harappan Civilization

The Harappan civilization—also known as the Indus Valley civilization—included the cities of Harappa, from which the civilization gets its name, and Mohenjo-Daro.

3 Teach

Early Societies

Think, Pair, Share

- What group probably came over the Hindu Kush into India about 1500 B.C.? *(the Aryans)*
- **Find Main Ideas** What is the main idea of the subsection "Early Religion"? *(Possible Answer: The Brahmanism of the Aryans evolved into Hinduism, the major religion in India.)*

SECTION 1 PROGRAM RESOURCES

Unit Resource Book: Southwest Asia and South Asia

ON LEVEL

Unit Resource Book
- Reading Study Guide, p. 117
- Vocabulary Practice, p. 141
- Outline Map Activities: India, pp. 145–146
- Section Quiz, p. 157

STRUGGLING READERS

Unit Resource Book
- Reading Study Guide with Additional Support, p. 123
- Section Quiz, p. 157
- Reteaching Activity, p. 160

eEdition with Audio DVD-ROM

ENGLISH LEARNERS

Pupil Edition in Spanish

eEdition with Audio DVD-ROM

Unit Resource Book
- RSG (Spanish), p. 129
- RSG with Additional Support (Spanish), p. 135

Multi-Language Glossary

Test Generator
- Section Quiz in Spanish

INCLUSION

Unit Resource Book
- RSG with Additional Support, p. 123
- Section Quiz, p. 157
- Reteaching Activity, p. 160

GIFTED & TALENTED

Unit Resource Book
- History Maker: Shah Jahan, p. 147
- Section Quiz, p. 157

PRE-AP

Unit Resource Book
- History Maker: Mohandas Gandhi, p. 148
- Section Quiz, p. 157

TECHNOLOGY

Unit Transparency Book
- 3-Minute Warm-Ups, TT19
- Essential Question, TT20
- Critical Thinking, TT21
- Map: Ancient Empires of South Asia, TT24, TT24a, TT24b

Daily Test Practice Transparencies
- Section 1, TT78

Power Presentations

ClassZone.com

World Cultures and Geography Video on DVD

More About . . .

Aryans

Aryan is a Sanskrit word meaning "noble." Aryans settled not only in India, but also in Iran. In the 19th century, the word Aryan came to mean the same as *Indo-European.* Some people even came to believe that Aryans originally came from northern Europe. Thus the Aryan ideal became linked with Germanic or Nordic people.

Comparing Regions

Social Structure in India, Kenya, and Japan

Although caste is often associated with Hinduism in India, many societies have divided people into classes based on birth and occupation. Although the name "Untouchables" was banned in India in 1950, the people, now known as Dalits, still face discrimination. Like the Untouchables, who did jobs nobody else wanted, the Burakumin of Japan also did "unclean" jobs, such as working in slaughterhouses or morgues. The Watta are considered the lowest caste by other Borana people in northeastern Kenya.

Make It Fun

Teacher-Tested Activities

Ann Christianson, John Muir Middle School, Wausau, Wisconsin

Students can write an acrostic to review some of the main concepts found in the chapter. An acrostic is a poem or series of lines in which the first letter of each line forms a name when read in sequence. The first letter of each line might, for example, spell out the name India, as shown below.

- **I** Indian Ocean surrounds the peninsula India occupies.
- **N** Nirvana is the state of wisdom that breaks the cycle of incarnation.
- **D** Dravidian is a language group of India.
- **I** Indo-European languages were spoken by the Aryans.
- **A** Aryans are believed to have migrated to the Indian subcontinent.

First Inhabitants The group of people who are believed to have migrated to India was the **Aryans**. They spoke an Indo-European language called **Sanskrit**, and their way of life was very different from that of the **Dravidians** (druh•VIHD•ee•uhn), the people living in India when the Aryans arrived. The term Dravidian refers to the family of languages they spoke. The Aryans were nomadic herders. The Dravidians might have been descended from the urban Harappan civilization, although until the Harappan script is deciphered, the language of the Indus Valley remains a mystery. The Aryans taught the Dravidians their language and religion and the Dravidians taught the Aryans about city life. A new and complex culture arose in India.

The idea of the Aryan migration is still a theory because no one has found physical evidence of such a move. Historians formed the theory because of clues in ancient religious stories and also because they noticed patterns in the languages that people speak.

Social Structure In earliest times, Aryan society was organized into three classes: warriors, priests, and commoners. In time, a system developed that divided society into five groups based broadly on work or occupation. The first group was the Brahmans, which included priests and teachers. The second was the Ksatriya (kuh•SHAT•ree•uh), which included rulers, nobles, and warriors. The third group, the Vaisya (VY•shuh), included bankers, merchants, and farmers. The fourth group, the Sudra, was made of artisans and laborers. The fifth group, which came into being much later, was the Untouchables. They did the jobs that no one else wanted to do. When Europeans first came to India in the 1500s, they called the social classes of India **castes**, classes to which people belonged by birth.

Early Religion Today, people call the early religion of the Aryans Brahmanism, after the priests, or Brahmans. Most of what we know about this early religion comes from ancient Sanskrit sacred texts called the **Vedas**, which contain hymns and teachings. The Aryans worshiped nature deities, and the priests made sacrifices to them.

Over time, Brahmanism gradually became **Hinduism**, which is the major religion of India. One big change was that many people came to believe that only one spirit or supreme being governs the universe. That supreme being, God, has different powers, so Hindus worship God in different forms.

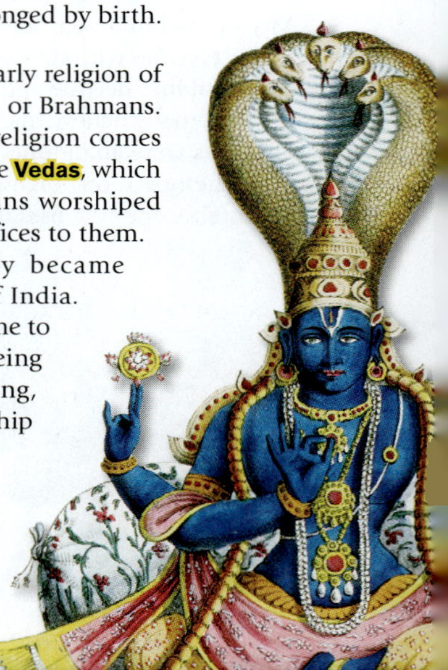

Vishnu Vishnu the preserver is an important Hindu deity. He is blue, the color of infinity for Hindus. ►

CATEGORIZE Describe the two groups that contributed to a complex culture in India. the Aryans and the Dravidians

DIFFERENTIATING INSTRUCTION

English Learners

Vocabulary: Multiple-Meaning Words

In the sentence "When Europeans first came to India in the 1500s, they called the social classes of India castes. . . ," class refers to a group of people who share social or economic status. Write *caste* and *cast* on the board. Define each. The word *caste* means a class to which somebody belongs by birth. The noun *cast* means a group of actors, or a mold or a stiff device used to set a broken bone.

Struggling Readers

Create a Graphic Organizer About Social Structures

Have students reread the subsection "Social Structure," and create two graphic organizers: one that shows the three classes of Aryan society and one that shows the five castes of later Aryan society. For the latter, have students include information about who was included in each caste and organize castes by order or rank.

CONNECT History & Culture

Sanskrit Sanskrit has been used by the educated classes in India for thousands of years. It is important in the history of India because it is the language of the sacred texts. Sanskrit is part of a major group of languages called the Indo-European family. Languages from the same family often have similar words for basic things.

English	father	mother	two	three	mouse
Sanskrit	pitar	matar	dva	trayas	mooshak
Greek	patros	matros	duo	treis	mūs
Latin	pater	mater	duo	tres	mūs
Spanish	padre	madre	dos	tres	raton
German	vater	mutter	zwei	drei	maus

CRITICAL THINKING

Compare Which word do you think is most alike in all of these Indo-European languages?

Scroll of Sanskrit text with illustrations of Hindu stories ▶

The Empires of India

KEY QUESTION What were the three main empires to rule India?

Different powers controlled India during its long history. Of the various empires that ruled India, some were Indian and some were not.

The Mauryan Empire About 321 B.C., an Indian ruler named Chandragupta Maurya (CHUNH•druh•GUHP•tuh MAH•ur•yuh) built an empire that would control most of India. About 50 years later, his grandson Asoka became the ruler of the Mauryan empire. Asoka adopted a religion called **Buddhism**, based on the teachings of Siddhartha Gautama, also called the **Buddha**, or "enlightened one." Asoka gave up warfare and sent missionaries to spread Buddhism, but he also allowed non-Buddhists to practice their faiths.

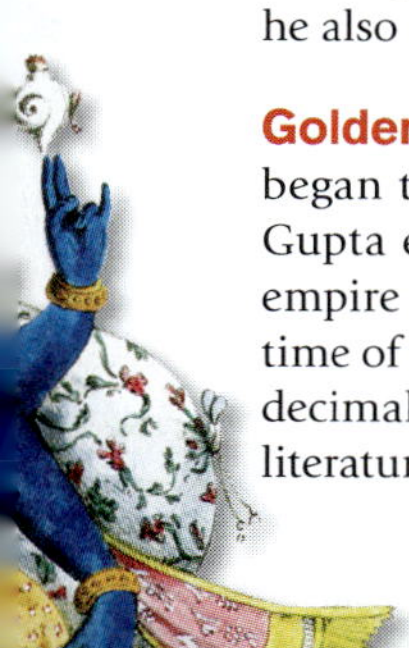

Golden Age of the Guptas After Asoka died, the Mauryan empire began to decline. The last Mauryan emperor died in 185 B.C. The Gupta empire took over India in A.D. 320. Trade made the Gupta empire wealthy. Historians call the Gupta empire a **golden age**—a time of prosperity and cultural achievements. Indians developed the decimal system and zero. Kalidasa wrote plays that are great works of literature. Beautiful temples and cave paintings were created.

CONNECT History & Culture

Sanskrit

- Sanskrit is one of 18 official languages of India, and many schools require students to learn it as a written language.
- Sanskrit is not considered a spoken language, but some organizations have been trying to bring spoken Sanskrit back into use.
- Sanskrit has been written using the Devanagari alphabet for many years, though it can be written using other alphabets or transliteration systems. The use of the Devanagari has made it difficult for people to put Sanskrit texts and instruction on the Internet, though new software is making this easier.
- Some Sanskrit words that have made their way into English are *guru*, *mantra*, *nirvana*, and *yoga*.

CRITICAL THINKING ANSWER

Compare Possible Answer: *Mother*—all begin with the same letter and have similar *t* or *d* sounds in the middle with *r* sounds near the end. Unlike *two*, which has *zwei*, and *mouse*, which has *raton*, as dissimilar words, all the words for *mother* are similar.

Teach

The Empires of India

Reader, Recorder, Reporter

- What significant change did Asoka bring to India? *(He spread Buddhism throughout India.)*
- What are some of the accomplishments of the golden age of the Guptas? *(Indians developed the decimal system and zero, wrote great plays, and created beautiful temples and cave paintings.)*
- **Find Main Ideas** What accomplishment did the Mughal empire have in common with Asoka? *(Both spread a new religion in India.)*

Unit Resource Book: Southwest Asia & South Asia

- Skillbuilder Practice: Making Generalizations, p. 142

INTERDISCIPLINARY ACTIVITIES

CONNECT to Art

Illustrate the Golden Age of the Guptas

Have students create illustrations for a history of India that focus on the Golden Age of the Guptas. Students should focus on specific accomplishments of the Guptas and identify the cause(s) of prosperity. Direct students to add captions to their illustrations. Encourage students to look at the images in this chapter to get a sense of the styles of Indian art, which they may wish to incorporate in their illustration style.

CONNECT to World Languages

Discuss Learning a New Language

Have students examine the chart of words from Indo-European languages. Ask them to think about what they need to know to read and speak a language. If they have learned or are learning a second language, have them consider their experiences. Then have students discuss in small groups, or write individually about, what effects language families might have on learning a new language.

Ancient Empires of India

Point out that none of the three major empires that ruled India expanded all the way to the southern tip of the subcontinent or to what is now Sri Lanka.

- Which empire covered the smallest area? *(Gupta Empire)* Which empire expanded farthest south? *(Mughal Empire)* Which empire spread farthest west? *(Mauryan Empire)*
- **Make Inferences** Which natural feature likely protected all three empires from China? *(the Himalayas)*

Unit Resource Book: Southwest Asia & South Asia
- Outline Map Activities: India: Physical/Political, pp. 145–146

CONNECT Geography & History

The Mughal Empire at Its Height

You may wish to share the following facts with your students:

- Shah Jahan's wife, Mumtaz Mahal, whose name means "Chosen One of the Palace," died after giving birth to their 14th child.
- The Taj Mahal—which includes several buildings and covers 42 acres—took over 22 years to build. More than 20,000 workers from India, Persia, the Ottoman Empire, and Europe worked on it.
- The Taj Mahal is decorated with calligraphy and an inlay of semiprecious stones in geometric designs and flower patterns.
- About ten years after the Taj Mahal was completed, Shah Jahan's son deposed him—and imprisoned him about a mile west of the Taj Mahal, on the opposite side of the river.

CRITICAL THINKING ANSWER

Make Inferences Possible Answer: Because of its architectural distinction and beautiful craftsmanship, the Taj Mahal stands today as a monument to the lasting accomplishments of the Mughals.

Unit Resource Book: Southwest Asia & South Asia
- History Makers: Shah Jahan, p. 147

Ancient Empires of India

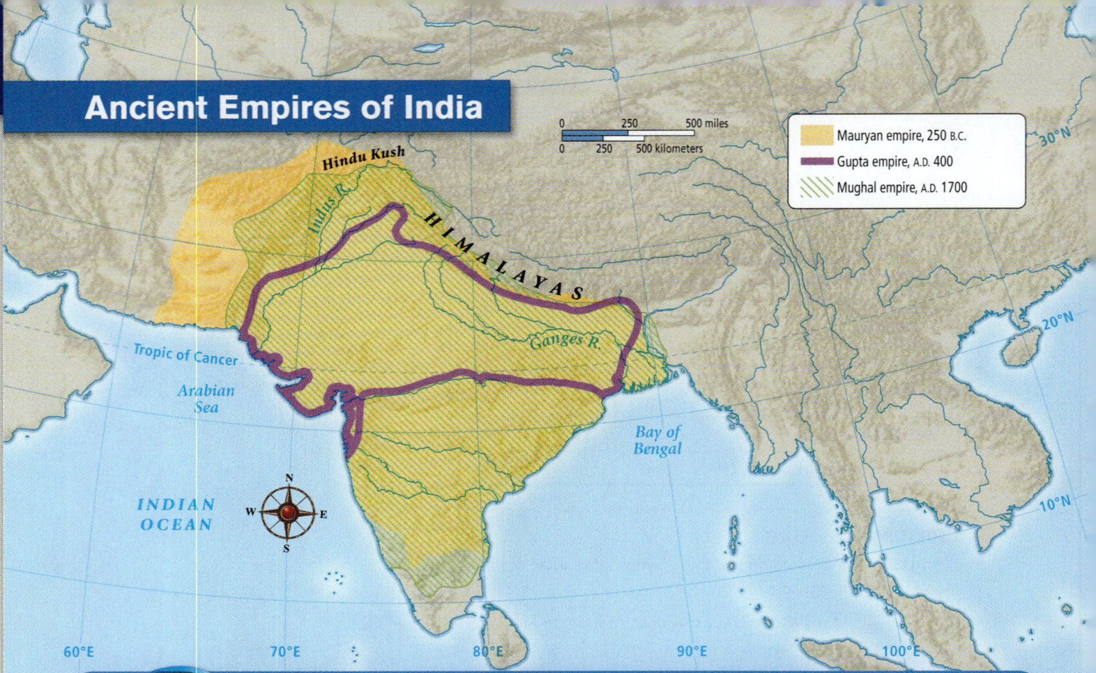

CONNECT Geography & History

The Mughal Empire at Its Height

The Taj Mahal Shah Jahan was a descendant of Akbar. Shah Jahan's wife died in 1631, and he ordered that a tomb be built "as beautiful as she was beautiful." This memorial, the Taj Mahal, is considered one of the most beautiful buildings in the world.

CRITICAL THINKING

Make Inferences Why might the Taj Mahal be considered a monument to the Mughal empire as well as to Shah Jahan's wife?

Taj Mahal Fine white marble and fabulous jewels were gathered from many parts of the Mughal empire to build the Taj Mahal. Its towering marble dome and slender minaret towers look like lace and seem to change color as the sun moves across the sky.

702 Chapter 23

DIFFERENTIATING INSTRUCTION

English Learners

Vocabulary: Roots

Point out the word *memorial* in the sentence "This memorial, the Taj Mahal, is considered one of the most beautiful buildings in the world." Explain that the root of *memorial* is *memor-*, which means "mindful." Other words with this root include *memory, commemorate,* and *memoir.* Have students look up these words in the dictionary and write their definitions in their notebooks.

Gifted & Talented

Argue for or Against Saving the Taj Mahal

Inform students that the Taj Mahal has been damaged by air pollution. Some steps—including closing foundries and banning vehicles near the monument—have been taken. Ask students to write a persuasive piece—an editorial, an ad, or a speech—for or against additional measures to save the Taj Mahal from any more damage from pollution.

Muslims and Mughals As you learned earlier, the religion of Islam began in Arabia during the 600s. Starting in the 700s, different groups of Muslims invaded India. The Muslims who came to India brought their beliefs with them, and some Indians converted to Islam. Over time, Muslims became a large minority group in India.

In 1526, a Muslim group from Central Asia called the Mughals made India part of their empire. The greatest ruler of this empire was Akbar, who ruled India fairly and gave many Hindus jobs in his government. A later Mughal ruler, Shah Jahan, is famous for having built the **Taj Mahal**, a beautiful memorial tomb for his wife. It is one of the most famous and most frequently visited buildings in the world.

SUMMARIZE Describe the ancient empires that ruled India. Mauryan, Gupta, Mughal

Independence

KEY QUESTION What steps led to Indian independence?

During the 1600s, the Mughal emperor fought wars that weakened his empire. Meanwhile, traders from Portugal, the Netherlands, Great Britain, and France had set up trading posts on India's coasts. As the Mughal empire weakened, Europeans seized more power. Of these, the British took over the most Indian territory.

British Rule In 1858, Britain established direct rule over most of India. Britain allowed local princes to retain their thrones as long as they swore allegiance to the British throne. Britain imposed taxes, took over Indian lands, and passed laws that gave the British more rights than Indians. In addition, Britain used Indian crops and products to enrich the British Empire. Indian resentment against British rule grew.

Indian National Congress In 1885, the Indian National Congress was formed to gain equal status for Indians. Its members included people from all over India. The Congress became a nationalistic organization and adopted the goal of independence for India. By 1920, an Indian lawyer named **Mohandas Gandhi** became one of the Indian National Congress's best-known leaders for independence from Britain.

HISTORY MAKERS

Mohandas Gandhi (1869–1948)

Gandhi (pictured above with his granddaughters) was born in India and studied law in England. His beliefs were influenced by Hinduism and his studies of Christianity and Islam. A few years after his return to India in 1915, Gandhi convinced the Indian National Congress to use nonviolent methods in the struggle for independence. Gandhi himself, the father of modern India, died violently when an assassin shot him. His methods and leadership influenced others such as Martin Luther King, Jr.

ONLINE BIOGRAPHY
For more on the life of Mohandas Gandhi, go to the Research & Writing Center @ClassZone.com

DIFFERENTIATING INSTRUCTION

Struggling Readers

Create a Time Line of Indian Independence

Have pairs of students create a time line of events that were part of the India's quest for independence. Students should include key events and dates from the subsection "Independence and Partition" and from the History Makers feature. Encourage them to include events, in proper chronological order, even if they don't have an exact date for the event.

Pre-AP

Nominate Gandhi for a Peace Prize

Students may be interested to know that Gandhi was nominated for the Nobel Peace Prize five times, but was never awarded the prize. The committee did not award a prize in 1948, the year Gandhi was killed, saying there was not a worthy living candidate. Have students write a speech or letter that explains why Gandhi should be awarded a posthumous peace prize.

RESEARCH & WRITING CENTER

History Makers

Mohandas Gandhi

Mohandas Gandhi, sometimes known as Mahatma, or "Great-Soul," grew up in a Jainist family, which believed in nonviolence. Gandhi worked for a time in South Africa. There he faced humiliation, assault, and other acts of prejudice. He refused to accept this injustice, and he rallied other Indians there to do the same.

Years later, Gandhi returned to India and began working for India's independence. He soon became a popular figure—and is considered by many to be the father of modern India.

- What methods did Gandhi favor in seeking independence for India? *(nonviolent methods)*
- In what way did Gandhi influence others? *(Others adopted his nonviolent methods and leadership style.)*

Unit Resource Book: Southwest Asia & South Asia
- History Makers: Mohandas Gandhi, p. 148

Teach

Independence

Talk about It

- Why did Indians grow resentful of the British? *(Possible Answers: Britain forced Indian princes to swear allegiance to Britain; imposed taxes, took over land, and passed laws giving the British more rights than Indians; used Indian crops and products to enrich the British Empire.)*
- How did Gandhi propose to win independence for India? *(through nonviolent resistance: using methods such as marches, refusing to buy British products, and refusing to pay taxes)*
- Why was the Muslim League formed? *(Some Muslims feared they would be discriminated against and shut out of India's new government by Hindus if India gained independence.)*
- **Find Main Ideas** Why was India partitioned? *(Possible Answer: India divided into two countries because the conflict between Muslims and Hindus remained unresolved.)*

RESEARCH & WRITING CENTER

ANALYZING Primary Sources

Mohandas Gandhi was arrested many times for his actions, both in India and in South Africa.

- What effect does Gandhi believe the British had on India? *(The British made India more helpless politically and economically than it had been before.)*
- Why do you think Gandhi came to his conclusion reluctantly? *(Possible Answer: He tried to think the best of the British or to think of the situation in the best light, but found he could no longer form a positive opinion of it.)*

DOCUMENT-BASED QUESTION ANSWER

Possible Answer: exploited them politically and economically

Comparing Regions

Nonviolent Resistance in India and the United States

Gandhi famously used nonviolent resistance to fight for India's independence from Great Britain. Martin Luther King, Jr. used similar tactics to fight against racial segregation in the United States. King said that Gandhi's philosophy was the "only morally and practically sound method open to oppressed people in their struggle for freedom."

Like Gandhi, King also used protest marches and boycotts—perhaps most notably the Montgomery bus boycott—in his efforts. Tragically, as also had happened to Gandhi, King was jailed for his activities and later assassinated.

ANALYZING Primary Sources

Mohandas Gandhi (1869–1948) was arrested by the British for his protests against colonial rule. In 1922, he was brought to trial for organizing opposition to the government. This excerpt comes from his testimony.

> I came reluctantly [unwillingly] to the conclusion that that British connection had made India more helpless than she ever was before, politically and economically. . . . The government established by law in British India is carried on for this exploitation [unfair treatment for gain] of the masses.
>
> Source: *India Emerges* by Steven Warshaw

DOCUMENT-BASED QUESTION

According to Gandhi, what did the British government do to the majority of Indians?

Gandhi leads a protest march ▶

ONLINE PRIMARY SOURCE To read more of Gandhi's writing, go to the **Research & Writing Center** @ClassZone.com

Gandhi wanted the British to leave India, but he did not believe in fighting to make that happen. He urged Congress to practice **nonviolent resistance**, which meant that only peaceful methods would be used to force the British to give up their rule. Such methods included holding marches, refusing to buy British products, and refusing to pay taxes. Gandhi led many marches himself. His emphasis on nonviolence also included the teaching that Indians of all religions could live peacefully together.

Muslim League Many Muslim members of the Indian National Congress shared Gandhi's dream of a free India where Muslims and Hindus could live in harmony. But other Muslims within Congress feared that the Hindu majority would shut them out of government after India gained independence. These Indians created a second nationalistic group called the Muslim League. An important member of the Muslim League was Muhammad Ali Jinnah.

At first, the Muslim League's goal was to give Muslims a role in Indian government. Over time, however, they began to fear that Muslims would suffer discrimination under majority rule by the Hindus. The Muslim League began to demand that India be partitioned, or split, into two independent countries—one Hindu and the other Muslim.

DIFFERENTIATING INSTRUCTION: TIERED ACTIVITIES

OBJECTIVE Make an argument for and against boycotts as a form of nonviolent resistance.

Basic

Arrange the class into small groups. Have each group create a pro and con chart. Within their groups, students should discuss reasons for and against supporting a boycott as a means of gaining freedom. When groups have completed their charts, have them compare ideas.

On Level

Have students write letters to the editor of an Indian newspaper supporting Gandhi's use of boycotts. Then have students swap letters with a partner and write a rebuttal to the arguments their partner presented.

Challenge

Arrange students into two groups. Assign one group a pro-boycott stance and the other an anti-boycott stance. Have each group prepare arguments and counter-arguments to support their position. Then have students debate the issue in front of the rest of the class.

Independence and Partition Talks between British and Indian leaders went on for decades. Finally, the British government granted India independence in 1947. Because of the disagreements between the Indian National Congress and the Muslim League, India was partitioned into two nations. The lands that had a Hindu majority became India. The lands that had a Muslim majority became Pakistan—which was made of two regions separated by over a thousand miles of Indian territory. West Pakistan lay on the northwest border of India, while East Pakistan (now Bangladesh) lay in the east.

Although there was a clear religious majority in most areas, both Hindus and Muslims lived in every part of India and Pakistan. After the partition, millions of Indian Muslims decided to move to Pakistan, and millions of Pakistani Hindus decided to move to India. This sudden migration of ten million people reawakened old angers, and about one million people were killed. Today, although conflicts continue, the two countries are learning to live and work together.

Partition Muslim refugees cram into coaches and climb onto the roof of a train leaving New Delhi, India, for Pakistan.

FIND MAIN IDEAS Explain the steps that led to Indian independence. formation of Indian National Congress; nonviolent protests; formation of Muslim League

Section 1 Assessment

ONLINE QUIZ For test practice, go to **Interactive Review @ClassZone.com**

TERMS & NAMES

1. **Explain the importance of**
   - Sanskrit
   - Hinduism
   - Buddhism
   - nonviolent resistance

USE YOUR READING NOTES

2. **Find Main Ideas** Use your completed chart to answer the following question: What are some of the elements that contributed to India's rich history?

EARLY SOCIETY	ANCIENT EMPIRES	BRITISH RULE	INDEPENDENCE
1.	1.	1.	1.
2.	2.	2.	2.
3.	3.	3.	3.

KEY IDEAS

3. What were the five major social classes that developed in ancient Indian society?
4. What cultural achievements occurred under the Guptas?
5. What two nationalistic groups worked for Indian independence, and what did they achieve?

CRITICAL THINKING

6. **Contrast** What is one major difference between Brahmanism and Hinduism?
7. **Analyze Effects** Which of Gandhi's goals were achieved, and which were not?
8. **CONNECT to Today** Recently, some cities changed their names to sound more Indian. For instance, the city that Europeans called Bombay was renamed Mumbai. Why do you think Indian cities did this?
9. **WRITING** **Write an Editorial** Choose a view held by members of the Indian National Congress, such as equal rights for Indians. Write an editorial expressing that viewpoint.

CONNECT to the Essential Question

How is India's traditional way of life changing in response to modern forces?

Ask students what they have learned so far that can help them answer this question. Students might mention:

- India has undergone many changes, as part of several empires, before gaining its independence.
- India was partitioned into two independent countries in an effort to resolve a conflict between India's Muslims and Hindus.

4 Assess & Reteach

Assess Have students complete the Section Assessment.

Unit Resource Book: Southwest Asia & South Asia
- Section Quiz, p. 157

Interactive Review @ClassZone.com
- Section Quiz

Power Presentations

Test Generator

Reteach Form students into three groups. Assign each group one of the subsections in "Empires and Independence." Have each group make a time line of key events in their subsection. Have each group present their time line to the class.

Unit Resource Book: Southwest Asia & South Asia
- Reteaching Activity, p. 160

SECTION 1 ASSESSMENT ANSWERS

Terms & Names

1. Sanskrit: ancient language of India; Hinduism: major religion of India; Buddhism: major religion based on teachings of Siddhartha Gautama; nonviolent resistance: method of protest

Use Your Reading Notes

2. Early inhabitants (Dravidian-speaking peoples), Aryans, ancient empires (Mauryan, Gupta, and Mughal), and British presence all contributed to the rich diversity of Indian history.
   Diagram *Early Society: caste, Sanskrit, religion; Ancient Empires: Mauryan, Gupta, Mughal; British Rule: British control, Indian National Congress, Muslim League; Independence: British-Indian talks, partition, conflict*

Key Ideas

3. Brahmans, Ksatriya, Vaisya, Sudra, Untouchables
4. the development of the concepts of zero and the decimal system, the writing of Kalidasa, and the creation of beautiful temples and cave paintings
5. Indian National Congress and Muslim League; independence for India and partition of India and Pakistan

Critical Thinking

6. Possible answer: Brahmanism believed in many gods; Hinduism believes in one supreme force but worships it in many forms.
7. Peaceful independence was achieved, but peace between Hindus and Muslims was not.
8. Possible answers: to express national pride, to get rid of European influence
9. Use the rubric to score editorials.

Editorial Rubric

	Level of Persuasiveness	Errors in Mechanics
4	very persuasive; gives many details	zero or one
3	persuasive; gives some details	two or three
2	somewhat persuasive; few details	four or five
1	not persuasive; little detail	more than five

1 Plan & Prepare

Objectives

- Identify the religions that originated in India
- Compare village and urban life in India today
- Describe popular culture in India today

Read for the Essential Question

Help students read for a purpose by reminding them of the Essential Question: "How is India's traditional way of life changing in response to modern forces?" Point out that they can use text features such as Key Questions, special features, images, and maps to form answers to the question.

Vocabulary

Best Practices Toolkit

Use the ***Best Practices Toolkit*** to model strategies for vocabulary notetaking. Vary strategies throughout the year. Choose from: *Knowledge Rating, Predicting ABCs, Definition Mapping, Word Sort, Frayer Model (Word Squares), Magnet Words, Word Wheel*, and *Student VOC*.

Vocabulary Strategies, TT9–TT16

Reading Strategy

Best Practices Toolkit

Display the Web Diagram transparency.

- Explain to students that in categorizing they need to determine how the information in the section can be sorted into the categories shown in the chart.
- Model filling in entries, using the "Birthplace of Many Religions" subsection.
- Have students suggest entries.
- Display the completed diagram for students to review.

Web Diagram, TT28

SECTION 2 Reading for Understanding

Key Ideas

BEFORE, YOU LEARNED

India has a diverse history to which many different peoples have contributed over thousands of years.

NOW YOU WILL LEARN

The culture of India is a blend of all of the many influences that have shaped its history.

Vocabulary

TERMS & NAMES

Bhagavad-Gita one of the sacred writings of Hinduism

reincarnation the rebirth of a soul in another body

karma in Hinduism, the consequences of a person's actions in this life, which determines his or her fate in the next life

nirvana in Buddhism, a state of wisdom that breaks the cycle of reincarnation

REVIEW

Hinduism the modern name for the major religion of India, which developed from Brahmanism

Buddhism a major world religion based on the teachings of Siddhartha Gautama

Buddha a name, meaning the "enlightened one," used for Siddhartha Gautama

Visual Vocabulary Buddha statue from 10th-century Kashmir

Reading Strategy

Re-create the web diagram shown at right. As you read and respond to the **KEY QUESTIONS**, use the diagram to organize important details about the culture of India.

Skillbuilder Handbook, page R7

CATEGORIZE

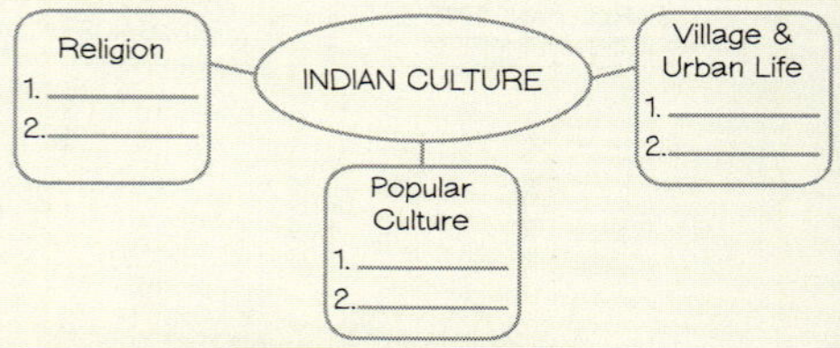

GRAPHIC ORGANIZERS
Go to **Interactive Review** @ClassZone.com

PRETEACHING VOCABULARY

English Learners

Pronounce and Preview

Pronounce each term for students. Review the meanings of words within definitions, such as *sacred, consequences,* and *fate.*

- To modify vocabulary learning, have students complete worksheets as they read, instead of afterward.

Unit Resource Book: Southwest Asia & South Asia
- Vocabulary Practice, p. 141

Inclusion

Vocabulary: Word Sort/ Word Splash

Have students work in pairs to fill out the graphic organizer for terms in the section.

- Use transparency to model the strategy.
- Have volunteers suggest entries.
- Then distribute copies of the organizer for students to complete with a partner.

Best Practices Toolkit
- Vocabulary: Word Sort/Word Splash, TT12

Taj Mahal in Agra, India

SECTION 2 CULTURE

Tigers are found in India.

A Rich and Ancient Legacy

Connecting to Your World

How many languages are spoken in your town or city? How many different ethnic groups live there? Because of the many groups that have migrated to India throughout its history, it has a larger diversity of languages than any other country in the world. More than 1,000 languages are spoken there. Many ethnic groups live in India. These groups have enriched Indian culture and have given it great variety. But even with all their differences, most Indians are linked by the strong ties of religion and daily life.

Shiva Shiva the destroyer is an important Hindu deity.

Birthplace of Many Religions

KEY QUESTION Which religions originated in India?

The religions of Hinduism and Buddhism, which you first read about in Section 1, both originated in India. Today, about 80 percent of all Indians practice Hinduism. Buddhism has practically disappeared from the land of its birth, but it is widely practiced in other parts of Asia. Other religions, including Sikhism and Jainism, have also played an important role in the history and culture of India.

Ganges Bathers descend steps called ghats to reach the Ganges River at Varanasi. To Hindus, it is the holiest river in India.

2 Focus & Motivate

3-Minute Warm-Up

Write on the board or display the transparency:

- Why is water often an important symbol in the world's religions? *(Possible Answers: It's a symbol of life; it's needed for survival.)*

Unit Transparency Book: Southwest Asia & South Asia
- 3-Minute Warm-Ups, TT19

Connecting to Your World

More About . . .

Languages in India

India recognizes 18 official languages in its constitution. Hindi, the most commonly spoken language, is considered the national language. English is used for most official communication. Some other languages include Assamese, Bengali, Gujarati, Kannada, Kashmiri, Konkani, Malayalam, Marathi, Nepali, Oriya, Punjabi, Sanskrit, Sindhi, Tamil, Telugu, and Urdu.

3 Teach

Birthplace of Many Religions

Think, Pair, Share

- Which religion do most Indians practice? *(Hinduism)* How significant is Buddhism in India? *(It is significant in that it originated in India, but few people in India practice it now.)*
- **Categorize** Which religions practiced in India originated in India? *(Hinduism, Buddhism, Sikhism, Jainism)*

SECTION 2 PROGRAM RESOURCES

Unit Resource Book: Southwest Asia and South Asia

ON LEVEL

Unit Resource Book
- Reading Study Guide, p. 119
- Vocabulary Practice, p. 141
- Skillbuilder Practice: Making Generalizations, p. 142
- Map and Graph Skills, pp. 143–144
- Section Quiz, p. 158

STRUGGLING READERS

Unit Resource Book
- Reading Study Guide with Additional Support, p. 125
- Section Quiz, p. 158
- Reteaching Activity, p. 161

eEdition with Audio DVD-ROM

ENGLISH LEARNERS

Pupil Edition in Spanish

eEdition with Audio DVD-ROM

Unit Resource Book
- RSG (Spanish), p. 131
- RSG with Additional Support (Spanish), p. 137

Multi-Language Glossary

Test Generator
- Section Quiz in Spanish

INCLUSION

Unit Resource Book
- RSG with Additional Support, p. 125
- Section Quiz, p. 158
- Reteaching Activity, p. 161

GIFTED & TALENTED

Unit Resource Book
- Primary and Secondary Sources: from the *Bhagavad-Gita*, p. 154
- Section Quiz, p. 158

PRE-AP

Unit Resource Book
- Primary and Secondary Sources, p. 155
- Section Quiz, p. 158

TECHNOLOGY

Unit Transparency Book
- 3-Minute Warm-Ups, TT19
- Essential Question, TT20
- Critical Thinking, TT22
- Cultures: Wearing Saris, TT26
- Cultures: Brass Work, TT27

Daily Test Practice Transparencies
- Section 2, TT79

Power Presentations

ClassZone.com

World Cultures and Geography Video on DVD

More About . . .

The *Bhagavad-Gita*

The *Bhagavad-Gita* is part of book VI of the great Indian epic, the *Mahabharata*. The *Bhagavad-Gita*, whose title means "song of God," is mainly a dialogue between Prince Arjuna and Krishna, an incarnation of the god Vishnu. In addition to the *Bhagavad-Gita*, the *Brahma Sutras* and the *Upanishads* are major Hindu texts.

Unit Resource Book: Southwest Asia & South Asia
- Primary and Secondary Sources: from the *Bhagavad-Gita*, p. 154
- Primary and Secondary Sources: from the *Rig Veda*, p. 155

Animated GEOGRAPHY

Spread of Hinduism and Buddhism *500 B.C.–A.D. 600*

CLASSZONE.COM

CONNECT Geography & Culture

Have students read the key to understand how the areas of Hinduism and Buddhism are being represented on this map.

- Describe the spread of Hinduism. *(Hinduism spread mainly within India.)*
- Describe the spread of Buddhism. *(Buddhism spread throughout Asia.)*

ANSWERS

1. **Region** northeast
2. **Movement** Buddhism

Presentation Options

Use the **Power Presentations** DVD-ROM to project the animated map.

- Use the forward arrow to animate the map.
- Listen to the audio and examine the photographs to learn more about the spread of Buddhism.

More About . . .

Siddhartha Gautama

Much of what is known about Siddhartha Gautama comes from Buddhist texts. Historians believe he lived sometime between the sixth and fourth centuries B.C., in what was then a kingdom of India and is now Nepal.

Hinduism As you learned, **Hinduism** developed from Brahmanism. The Vedas are the oldest of Hinduism's holy writings, but Hindus also have other sacred texts, including the ***Bhagavad-Gita***, which discusses duty, the meaning of life, and devotion to God. Hindus believe that one supreme spirit is the only reality in the universe. All living things have a soul, which comes from that one spirit. Hinduism does not focus on physical life, but teaches that souls should seek union with God. Most Hindus believe in **reincarnation**, the idea that the soul is reborn into a new body. **Karma**, the consequences of a person's actions in this life, decides what will happen to the soul in the next life.

Buddhism The religion of **Buddhism** is based on the teachings of Siddhartha Gautama, who gave up being a prince to seek the meaning of life. He taught the Four Noble Truths. First, life is filled with suffering. Second, people suffer because they want worldly things. Third, people can stop suffering if they stop wanting. And fourth, they can stop wanting by following the Eightfold Path, which teaches right living. When people give up desire, they achieve **nirvana**, a state of wisdom that breaks the cycle of reincarnation. Many people adopted the teachings of Siddhartha Gautama, whom they called the **Buddha**, or "enlightened one." Buddhism spread throughout Asia, in part because of missionaries whom the Mauryan ruler Asoka sent out.

Animated GEOGRAPHY

Spread of Hinduism and Buddhism *500 B.C.–A.D. 600*

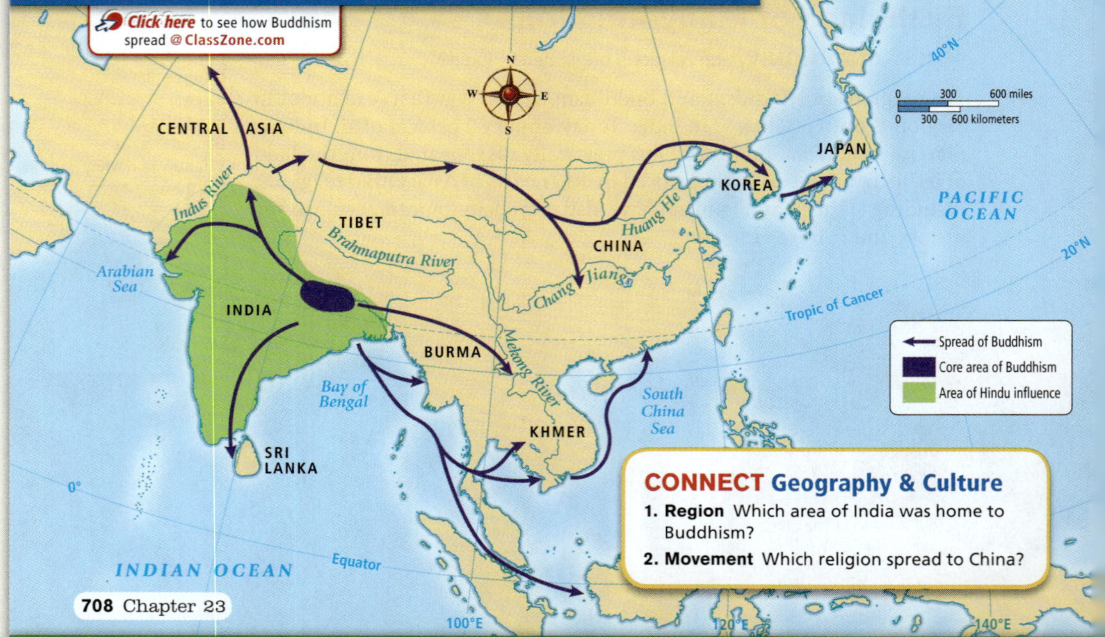

CONNECT Geography & Culture

1. **Region** Which area of India was home to Buddhism?
2. **Movement** Which religion spread to China?

708 Chapter 23

DIFFERENTIATING INSTRUCTION

English Learners

Key Academic Vocabulary

Read the item from the map key in the textbook: "Core area of Buddhism." Point out that the word *core* here means "central or main." Students should understand the purple area is the main area of Buddhism—where it started and then spread to other parts of the world.

Gifted & Talented

Compare Hinduism and Buddhism

Have students use the text and map on this page to compare Hinduism and Buddhism. Ask them to consider the beliefs or tenets of the two religions, their origins, and their areas of influence. Students may want to compare one or both of these religions to another religion in India or another part of the world.

Sikhism and Other Religions As you know, Islam arrived in India during the 700s. Today, about 13 percent of Indians are Muslims, and about 2 percent are Christians. Another 2 percent practice Sikhism, which originated in northern India. Sikhs believe in one god, in reincarnation, and in leading good, simple lives. Another religion that began in India is Jainism, which emphasizes the importance of doing no harm to other living beings.

CATEGORIZE Describe which religions originated in India.
Hinduism, Buddhism, Sikhism, Jainism

Village and Urban Life

KEY QUESTION How are city life and village life different?

With more than one billion people, India is the second most populous country in the world after China. Six of the world's 100 largest cities are in India. Even so, more than 70 percent of Indians live in rural areas where life is largely unchanged.

Family Life Most Indians live in extended families, which include several generations together. When a woman marries, she moves in with her husband's family—his parents, his brothers and their wives, and any unmarried sisters. Even today, parents often choose their children's spouses, because marriage is viewed as a union of two families.

Village Life Village life varies a great deal across India. In some regions, people live in small one-story structures made of mud. In more prosperous villages, the houses might be made of brick. In rice-growing regions, houses are built on bamboo stilts above the flooded paddies. Many villages across India still have no electricity or running water.

The tradition of sons who learn their fathers' occupations remains strong in many villages. This keeps villages stable by making sure that all the necessary jobs continue to be done. But in recent decades, many young villagers have decided to move to cities, where they can earn more money and choose different ways of life.

Traditional Clothing: The Sari

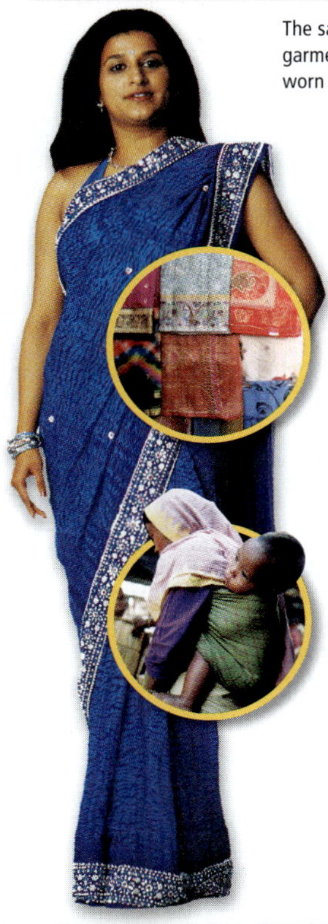

The sari is a traditional Indian garment that is still commonly worn today.

Folding A sari is a single six- to nine-yard piece of cloth. A young Indian woman must learn to wrap, pleat, and drape the sari.

Fabric Saris are made from a variety of fabrics, including natural fibers such as cotton and silk, as well as synthetic fibers. Brides traditionally wear a red silk sari.

Use The sari can be used to cradle babies, lift hot cooking utensils, protect from the sun, preserve modesty, hold shopping purchases, wipe sweat, dust a table, and protect from smog.

INTERDISCIPLINARY ACTIVITIES

CONNECT to Math

Graph Percentages of Religions in India

Have students fill in a chart, such as the one at right, that lists the percentages of people in India who practice each religion, using information from their textbooks. Tell them to use 3% for the "other" category. Once students have completed their charts, have them create a pie graph showing the percentage of each religion in India. Graphs should be clearly labeled and include a key.

Religion	Percentage of Indians
Hinduism	
Islam	
Christianity	
Sikhism	
Other (includes Buddhism and Jainism)	

Teach

Village and Urban Life

Reader, Recorder, Reporter

- Why are marriages often arranged by parents for their children in India? *(Marriage is seen as a union of two families.)*
- How might the decision of many young people to move to the cities affect village life? *(Possible Answer: It may destabilize villages as some jobs, which are often handed down from father to son, go undone.)*
- What problems have occurred as a result of rapid city growth? *(Possible Answers: There is a high poverty rate and a shortage of housing, transportation, and services. Many people have to live in slums, which lack electricity and running water and have open sewage spreading disease.)*
- **Categorize** Identify some of the positive aspects of city life. *(Possible Answers: Cities are centers of creativity and technological innovation; there is a greater variety of jobs.)*

Unit Resource Book: Southwest Asia & South Asia
- Skillbuilder Practice: Making Generalizations, p. 142

Traditional Clothing: The Sari

Saris are often worn with a tight-fitting top, called a choli, and a petticoat. Saris can be wrapped and draped many ways. The style tends to vary by region or background of the wearer.

- Explain how saris are a functional garment. *(Saris can be used to carry children, lift hot cooking utensils, hold purchases, dust, wipe sweat, preserve modesty, and protect from smog.)*

Comparing Regions

Traditional Clothing: Saris and Kimonos

Saris are traditional in India, and kimonos are traditional in Japan. Both are simple garments that are adjusted for the wearer: saris, by the way they are draped, and kimonos, by the way they are belted. Both take practice to put on properly. Indian women still wear saris, but in Japan kimonos are usually worn for special occasions.

More About . . .

Changing City Names

In the 1990s, Bombay changed its name to Mumbai. Other Indian cities made changes, too, including Kolkata, once Calcutta, and Chennai, previously Madras. Generally the changes were a way to restore or replace names that had been anglicized during the British reign with names in local languages. In 2005, Bangalore announced that it too was changing its name. Its new name, Bengaluru, literally means "City of Boiled Beans."

COMPARING

Rural and Urban India

Remind students that despite the large size of India's cities, about 70 percent of India's people still live in rural areas or villages.

- What are some of the similarities in village and city life? *(Possible Answer: People live in houses in both; both have street markets and shops; hand tools are used in production in both places.)*
- Based on the photographs, what are some other differences you can infer between rural and urban life? *(Possible Answers: Urban life is crowded, has lots of traffic, and is noisy; rural life is quieter and less crowded.)*

CRITICAL THINKING ANSWERS

1. **Compare** Possible Answers: the population, the kinds of jobs, the importance of the weather in villages and rural areas to farmers
2. **Draw Conclusions** Possible Answer: There are more people in a city to benefit from electricity. People are closer together in a city and may be nearer sources of production, making it more cost-effective to provide electricity.

Unit Resource Book: Southwest Asia & South Asia
- Map and Graph Skills: Reading a Weather Map, pp. 143–144

Urban Life Because of that migration, India's cities have grown. With 18 million residents, Mumbai (Bombay) is the world's third largest city. Kolkata (Calcutta) has more than 14 million residents. In addition, Hyderabad, Chennai (Madras), and Bangalore all have more than five million people.

Rapid urban growth causes shortages of housing, transportation, and services. In addition, because there are not enough jobs for everyone in the growing cities, there is a high rate of poverty. Many new arrivals can find no place to live except slums, where people construct shacks of flimsy materials. Slums have no electricity or running water, and sewage runs in the open, spreading disease.

In villages, young people are more likely to learn the same occupation as their fathers; in cities, they have opportunities to do different work. ▶

However, the government is dealing with population growth. (See Section 3.) Despite their problems, the cities are also centers of creativity and technological innovation. That is one of the reasons that young people in particular have been drawn to the cities.

COMPARE AND CONTRAST Describe how city and village life differ.

COMPARING Rural and Urban India

Itanagar, Arunachal Pradesh

Mumbai

	VILLAGES	CITIES
Population	5,000 or less	up to 18 million
Housing	mostly small houses	houses, apartments, shacks
Shopping	street markets and shops	street markets, shops, and stores
Jobs	mostly farming	manufacturing, services, high-tech
Production	hand tools, oxen-pulled plows	machines and technology, some hand tools

CRITICAL THINKING

1. **Compare** What aspect of life is most different in villages and cities?
2. **Draw Conclusions** Why do you think it is more common to have electricity in a city than in a village?

DIFFERENTIATING INSTRUCTION

Struggling Readers

Pro and Con Chart About City Life

Have students work in pairs to make a pro and con chart about city life in India, first setting up a two-column chart and labeling one column "Pro" and the other "Con." Have them begin by reviewing the chart in the Comparing Rural and Urban India feature. Then have them reread the text of "Village Life "and "Urban Life," pausing frequently to add items to their charts.

Pre-AP

Write a Compare and Contrast Essay

Have students write a compare and contrast essay, exploring the similarities and differences between rural and urban life in India. Alternately, invite students to write a compare and contrast essay exploring the similarities and differences in rural or urban life in India with those in the United States.

Popular Culture

KEY QUESTION What are the major religious festivals of India?

Many groups have shaped India's culture, including the Dravidians and Aryans, the invading Muslims, and the British colonists. As a result, some of India's culture is very old and some is modern. For example, the *Bhagavad-Gita* and the epic poem the *Ramayana* were originally written in the ancient language of Sanskrit. Millions of people still read those poems today, but in modern translations.

Bollywood Movies are a very popular form of entertainment in India, which has the largest motion picture industry in the world. India's movie industry is often called Bollywood, a word that combines the names *Hollywood* and *Bombay* (the former name of Mumbai). About 5 million people work in Indian movies. India's cultural diversity is reflected in its films, which are produced in 52 different languages. Indian movies are often based on ancient myths and folk tales. Frequently, they feature love stories, crime dramas, and stories about social issues. One thing that sets Indian movies apart is that many films feature song-and-dance numbers.

Sports One positive way that British rule influenced Indian culture is in sports, especially cricket. The sport of cricket is played by teams of 11 on a field with wickets, which are made of three upright sticks and two horizontal ones. The game involves defending the wickets by batting away balls that are pitched at them. Players score runs by running between two wickets while the opponent chases down a batted ball. Cricket is the most popular sport in India. Athletes become celebrities, and people follow the matches. Other popular sports include soccer and field hockey.

Fun Facts!

BOLLYWOOD VS. HOLLYWOOD

The Indian film industry is the largest in the world.

Films Released in 2004:

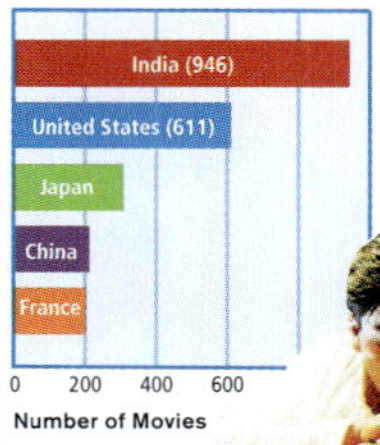

Source: *Screen Digest*, June 2005

Film Stars Shahrukh Khan and Aishwarya Rai star in the Indian film *Devdas*.

Cricket Indian boys play cricket in the southern Indian city of Nagapattinam.

Teach

Popular Culture

Think, Pair, Share

- Explain the origins of the name Bollywood for India's movie industry. *(It is a combination of Hollywood, the United States movie capital, with Bombay, the former name of Mumbai.)* How do Bollywood movies tend to differ from Hollywood movies? *(Bollywood movies tend to feature more song-and-dance numbers.)*
- What sports are popular in India? *(cricket, field hockey, and soccer)* How was cricket introduced to India? *(It was introduced during the British colonial period.)*
- **Categorize** Which two festivals can be categorized as celebrations of the triumph of good over evil? *(Dussehra, or Durga Puja, and Holi)*

Fun Facts!

Bollywood vs. Hollywood

Bollywood movies are becoming more popular in the United States. What started among the Indian population in the United States has begun to spread across the country, though Bollywood films are still generally considered a niche market.

The popularity of Bollywood films in the United States paved the way for a Bollywood-Hollywood collaboration: *Bride and Prejudice.* The film is a version of the British novelist Jane Austen's *Pride and Prejudice*. It features Bollywood star Aishwarya Rai and Hollywood actor Martin Henderson.

INTERDISCIPLINARY ACTIVITIES

CONNECT to Art

Create a Movie Poster

Using what they have learned about Bollywood movies, have students create movie posters for a Bollywood film festival for American audiences. They may wish to use actual film titles in their posters, such as *Kabhi Khushi Kabhi Gham (Sometimes Happiness, Sometimes Sadness)* or *K3G*, *Chocolate*, or *Bride and Prejudice.*

CONNECT to Language Arts

Write a Postcard

Have students imagine they are visiting an Indian family and experiencing Indian pop culture. Have students write a postcard to a friend at home, explaining what they have seen and done on their trip. You may wish to have students read Worldster: A Day in Savita's Life on the next page before they complete this activity.

More About . . .

Panna National Park

Panna National Park is in the state of Madhya Pradesh. This park is home to more than 200 bird species—and is best known for its large cats, including tigers and leopards. Other large animals within the park include antelopes, wolves, and hyenas.

Comparing Regions

Rice Dishes

Biryani is an Indian rice dish, usually flavored with the spices saffron or turmeric. It may include meat, fish, and/or vegetables. Paella is a Spanish rice dish that is also flavored with spices, including saffron. Also like biryani, paella may be made with meat, fish, and/or vegetables.

Presentation Options

Go to the *Activity Center* at **ClassZone.com** to project Worldster.

- Click on the multimedia features of Savita's Web page.
- Ask students which areas of Savita's life they would like to explore. Listen to music, examine photographs, learn simple phrases in Hindi, and more.

CONNECT to Your Life

Journal Entry

Ask students to first consider whether things about Savita's life appeal to them because they are similar to, or different from, their own lives. Then have students complete their journal entries.

Worldster: A Day in Savita's Life

To learn more about Savita and her world, go to the **Activity Center** @ClassZone.com

In Hindi, Savita's name looks like this:

सविता

Hi! My name is Savita. I am 12 years old, and I live in the town of Panna in north-central India. I have an 18-year-old sister and a 13-year-old brother. Let me tell you about my day.

9:30 A.M. I have a math test today, and I'm worried that I didn't study enough last night. My other classes are Hindi, English, science, geography, history, and art.

12:30 P.M. I go home for lunch. My mother cooks a vegetarian biryani, which is a rice dish. After I eat, I carry lunch to my father, who owns a snack shop near the bus station.

3:30 P.M. After school, my brother goes to the shop to give my father a break. I hurry home to see my older sister, who is bringing her baby for a visit. Then I go out to play soccer.

6:00 P.M. After my sister leaves, I do my homework. When I finish, I do household chores such as washing dishes and helping my mother cook. We eat dinner when my father comes home.

CONNECT to Your Life

Journal Entry What appeals to you about Savita's life? In your journal, describe those activities and ways of life.

712

DIFFERENTIATING INSTRUCTION: TIERED ACTIVITIES

OBJECTIVE Explore modern Indian culture by comparing and contrasting daily life.

Basic

Have students list what they would be doing during the following time periods or activities:

- 9:30 A.M. (at school)
- 12:30 P.M. (lunch)
- 3:30 P.M. (after school)
- 6:00 P.M. (dinner)

Then have students discuss how their day is similar to and different from Savita's.

On Level

For each item in the Worldster feature, have students describe in writing what they would do during a similar time period or activity. Have them write their responses in a "Like Savita" or "Unlike Savita" format. Have students share their responses in small groups.

Challenge

Have students write a letter to Savita, comparing their typical day to her day. Students should respond to the specific activities that Savita talks about in the Worldster feature.

Celebrations Many of the celebrations that take place throughout the year are religious festivals. Some of the most popular include:

- Dussehra, which celebrates the triumph of good over evil, is celebrated in different ways across India. In the north, people create likenesses of the evil king Ravana, stuff them with firecrackers, and explode them. In the east, the festival is called Durga Puja and involves worship of a deity named Durga.
- Diwali occurs only 20 days after Dussehra. Known as a festival of lights, it is celebrated by lighting oil lamps and setting off fireworks. During the festival, people worship Lakshmi, the deity of wealth. Because of its association with prosperity, the festival is very important to business people.
- Holi, which celebrates the end of winter, is a joyful festival. To express that joy, people throw colored water and powder on each other. They also exchange sweets. Originally a harvest festival, Holi also celebrates the triumph of good over evil.

Holi Hindus in India celebrate the joyous festival of Holi. People dance in the streets and shower each other with colored powder and dyed water.

CATEGORIZE Describe the major religious festivals of India.
Dussehra, Diwali, Holi

Section 2 Assessment

ONLINE QUIZ For test practice, go to **Interactive Review** @ClassZone.com

TERMS & NAMES

1. **Explain the importance of**
   - *Bhagavad-Gita*
   - reincarnation
   - karma
   - nirvana

USE YOUR READING NOTES

2. **Categorize** Use your completed chart to answer the following question:

   How do Indian films blend both ancient and modern culture?

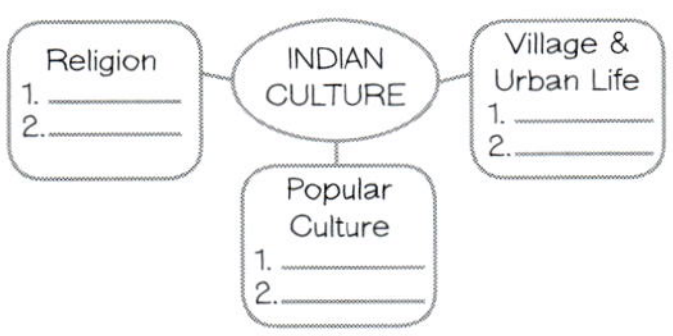

KEY IDEAS

3. Which religion was introduced to India from another region?
4. Why do many young villagers move to cities?
5. What does the festival of Dussehra celebrate?

CRITICAL THINKING

6. **Draw Conclusions** Which of Buddhism's teachings might have most appealed to people and helped the religion to spread throughout Asia?
7. **Make Inferences** Why do you think that Indian society regards marriage as the union of two families rather than two individuals?
8. **CONNECT to Today** If you were a government official in an Indian city, what would be the first service you would try to provide to the slum areas—electricity, running water, sewers, or something else? Explain.
9. **TECHNOLOGY** **Plan a Web Site** With a partner, learn more about Bollywood films. Design a Web site that would appeal to fans of those movies.

CONNECT to the Essential Question

How is India's traditional way of life changing in response to modern forces?

Ask students what they have learned so far that can help them answer this question. Students might mention:

- More young people are moving to cities, where there are more economic opportunities.
- People are reading ancient Indian texts in modern translations.
- Although British colonialism came to an end, cricket, a game introduced by the British, remains very popular.

4 Assess & Reteach

Assess Have students complete the Section Assessment.

Unit Resource Book: Southwest Asia & South Asia
- Section Quiz, p. 158

Interactive Review @ClassZone.com
- Section Quiz

Power Presentations

Test Generator

Reteach Have each student list three facts that they learned about Indian culture. Go around the room and have students share their lists. As a class, categorize each fact on the board or chart paper. Encourage students to identify appropriate categories.

Unit Resource Book: Southwest Asia & South Asia
- Reteaching Activity, p. 161

SECTION 2 ASSESSMENT ANSWERS

Terms & Names

1. *Bhagavad-Gita:* one of the sacred writings of Hinduism; reincarnation: the rebirth of a soul in another body; karma: consequences of person's actions in life, which determines fate in next life; nirvana: state of wisdom that breaks cycle of reincarnation

Use Your Reading Notes

2. They are often based on ancient myths and folk tales as well as love stories, crime dramas, and stories about social issues; many feature musical numbers.

   Diagram *Religion: Hinduism, Buddhism; Popular Culture: Bollywood, religious festivals; Village & Urban Life: migration, population growth*

Key Ideas

3. Islam
4. to find work
5. the triumph of good over evil

Critical Thinking

6. Possible answer: It explained how to put an end to the suffering that most people experience.
7. Possible answer: Most Indians live in extended families, and young couples do not move out on their own. So marriage does bring a lot of family interaction.
8. Many students will choose sewers to try to prevent disease. Others might choose running water because it is necessary for life.
9. Use the rubric to score Web sites.

Web Site Rubric

	Content Included on Web Page	Organization of Web Page
4	all required elements	well organized, interesting
3	most required elements	well organized
2	some required elements	poorly organized
1	little or no required information	incomplete, disorganized

1 Plan & Prepare

Objective

Compare and contrast traditional and modern Indian life.

Springboard Tell students to close their texts. Ask volunteers to brainstorm one thing they associate with traditional Indian life. Then have them brainstorm what they associate with modern Indian life. List responses on the board or chart paper. Then have students open their text and analyze the Comparing feature. What changes or additions would they make to the lists?

2 Teach

Traditional and Modern India

Talk About It

- List some aspects of traditional Indian life. *(saris, rickshaws, the Taj Mahal and Hindu temples, hand-thrown pottery)*
- List some aspects of modern Indian life. *(contemporary architecture, stock exchange, high-tech jobs, jeans, jets, traffic jams)*
- What is one way tradition plays a part in cities? *(Possible Answer: Rickshaws can be seen in busy city streets.)*

More About . . .

Varanasi

Varanasi is an ancient Indian city. During medieval times, the city was known as Banaras, but an official government order changed the name back to Varanasi in 1956. Located on the Ganges, the holy river for Hindus, Varanasi is a place of pilgrimage. Sarnath, where the Buddha preached his first sermon, is nearby to Varanasi, making this area special to both Hindus and Buddhists.

COMPARING Traditional and Modern India

India is one of the oldest cultures in the world. It continues to observe many of its traditions in terms of beliefs, clothing, diet, and work. At the same time, it is rushing in the 21st century to embrace the modern world and improve the standard of living of its people.

Traditional India

Traditional India is a land of the Taj Mahal, Hindu temples, beautiful saris, and rickshaws careening down narrow village streets. Although much of the developing middle class is modernizing, millions continue to follow the traditional ways in clothing, food, work, and daily life.

Architecture

Carvings cover this Hindu temple in Chennai.

Transportation

Even in cities, tradition still plays a part, as shown by bicycle rickshaws in Varanasi.

The village of Gulmarg, a former hill station to which the British retreated to escape the heat of the plains, sits on a hillside beneath the Himalayas. ▼

Jobs

A potter throws pots on wheel in the village of Gunupur.

DIFFERENTIATING INSTRUCTION: TIERED ACTIVITIES

OBJECTIVE Create an itinerary for a trip to India that explores both traditional and modern life.

Basic

Have students work in groups of four. Have two students from each group put together a list of traditional cultural aspects of India, while the other pair puts together a list of modern cultural aspects of India. Have students then work together to make a plan for a six-day trip, identifying one traditional and one modern aspect to visit each day.

Modern India

Modern India is racing to become a center of high-tech jobs and widespread prosperity. India's growing middle class is increasingly willing to abandon the sari for jeans, the railway for jet flights across the country, and the farming village for high-tech centers such as Bangalore.

Architecture

The Mumbai Stock Exchange is an example of contemporary architecture in the city of Mumbai.

Transportation

Traffic jams the street in front of the Capitol movie theater in Mumbai.

▲ *Aerial view of Mumbai, India*

Jobs

Call-center workers in Bangalore serve customers in the United States and Britain.

CRITICAL THINKING

1. **Compare and Contrast** What differences can be seen in the kinds of jobs that are available in the city and in the village?
2. **Make Inferences** How would a person's life change in moving from village to city?

Comparing Regions

Jobs from the United States in India

Many United States companies—as well as companies from other countries—have been outsourcing work to India for years. Outsourced work includes IT (information-technology), call service centers, and back-office work, such as accounting. Many companies outsource work to India because they can get it done at a lower cost than in their own country.

India also has an advantage for outsourcing compared to other countries in that it has a large, well-educated, English-speaking workforce. India has gained many jobs this way, but is facing a need to improve its infrastructure to continue to attract outsourced jobs.

3 Wrap Up

Ask students to refer to the Essential Question for the chapter. Ask them to identify at least three responses, based on the information they gathered from this Comparing feature. Have volunteers share their responses with the class.

CRITICAL THINKING ANSWERS

1. **Compare and Contrast** Possible Answer: Village jobs are traditional trades or crafts, usually for the local economy; city jobs are often high-tech or service-oriented, and part of the global economy.
2. **Make Inferences** Possible Answer: They would likely have a different kind of job. Their surroundings might be busier and noisier. They might have electricity for the first time. Their standard of living might change.

On Level

In addition to the Basic level activity, have each group write a brief description of each item they include on their itineraries. Encourage students to draw upon knowledge they have gained from the chapter so far and any prior knowledge or research, not just from the Comparing feature.

Challenge

In addition to the On Level instructions, have each group use an atlas to determine where places are in relation to each other. Have students create an outline map of India showing the sites they would visit to include with their itineraries.

1 Plan & Prepare

Objectives

- Describe the current challenges faced by India's parliamentary democracy
- Explain how India's resources are shaping the Indian economy

Read for the Essential Question

Help students read for a purpose by reminding them of the Essential Question: "How is India's traditional way of life changing in response to modern forces?" Point out that they can use text features such as Key Questions, special features, images, and maps to form answers to the question.

Vocabulary

Best Practices Toolkit

Use the ***Best Practices Toolkit*** to model strategies for vocabulary notetaking. Vary strategies throughout the year. Choose from: *Knowledge Rating, Predicting ABCs, Definition Mapping, Word Sort, Frayer Model (Word Squares), Magnet Words, Word Wheel,* and *Student VOC.*

Vocabulary Strategies, TT9–TT16

Reading Strategy

Best Practices Toolkit

Display the Three-Column Chart transparency.

- Remind students that main ideas are important ideas in a block of text.
- Model filling in main ideas about government using the "Government Structure" subsection.
- Have students suggest additional entries. Remind students that main ideas in each block of text in the section will help answer a Key Question.
- Display the completed chart for review.

Three-Column Chart, TT21

SECTION 3 Reading for Understanding

Key Ideas

BEFORE, YOU LEARNED

India's rich history and culture have contributed to its present diversity.

NOW YOU WILL LEARN

Out of India's diversity, a robust democracy has developed.

Vocabulary

TERMS & NAMES

parliamentary relating to a government with the parliament form of legislature

prime minister the head of the cabinet and chief executive of government in a parliamentary system

Kashmir a region in the northwestern part of the Indian subcontinent; India and Pakistan dispute control of Kashmir

Green Revolution the use of special seeds, irrigation, fertilizers, and pesticides to produce high crop yields and food production

cottage industry a small business in which the workers manufacture items in their homes

Bangalore Indian city that is home to many high-tech industries

BACKGROUND VOCABULARY

legislature the law-making body of a government

monarch a king or queen who rules a country or territory

Visual Vocabulary Kashmir: boats on Dal Lake

Reading Strategy

Re-create the chart shown at right. As you read and respond to the **KEY QUESTIONS**, use the chart to note the main ideas about India's government and economy.

Skillbuilder Handbook, page R4

FIND MAIN IDEAS

GOVERNMENT	ECONOMY	ISSUES AND TRENDS
1.	1.	1.
2.	2.	2.
3.	3.	3.
4.	4.	4.

GRAPHIC ORGANIZERS
Go to **Interactive Review** @ClassZone.com

PRETEACHING VOCABULARY

English Learners

Pronounce and Preview

Pronounce each term for students. Review the meanings of words within definitions, such as *cabinet, dispute,* and *pesticides.*

- To modify vocabulary learning, have students complete worksheets as they read, instead of afterward.

Unit Resource Book: Southwest Asia & South Asia
- Vocabulary Practice, p. 141

Inclusion

Reading and Defining Terms Verbally

Have volunteers read one term and its definition aloud. Ask others to define the term in their own words verbally, and then use it in a sentence. Repeat for each term. As a group, discuss any words that are challenging to define or use in a sentence.

Taj Mahal in Agra, India

Tigers are found in India.

SECTION 3

GOVERNMENT & ECONOMICS

The World's Largest Democracy

Connecting to Your World

Imagine what the United States might be like if it had 20 political parties. It might be difficult for any party to win control of government. Countries with a large number of political parties sometimes use a **parliamentary** system. This is a system in which the party with the most seats in the **legislature**, or Parliament, selects the head of government. India has a parliamentary system with many political parties.

Indian Elections Government official shows voters how to use a voting machine.

A Giant Republic

KEY QUESTION What are some of the challenges faced by the Indian government?

After India gained independence, its leaders decided that it should be a representative democracy. In fact, because India is the second most populous nation, it is the world's largest democracy. The structure of the national government is similar to that of Great Britain, except that India does not have a **monarch**. Like the United States, India has a federal system of government, which divides power between the central government and the states. India is divided into 28 states and 7 union territories.

New Delhi The Viceroy's House in New Delhi (now the Presidential Palace and Parliament), designed by Sir Edwin Lutyens

717

2 Focus & Motivate

3-Minute Warm-Up

Write on the board or display the transparency:

- What challenges might a multiparty political system face? How might a multiparty system be beneficial? *(Challenges: It is difficult for one party to win control, which may make getting things done difficult. Benefits: No one party controls things. Parties might work better together to get necessary votes.)*

Unit Transparency Book: Southwest Asia & South Asia
- 3-Minute Warm-Ups, TT19

Connecting to Your World

More About . . .

The Viceroy's House

Built as part of India's new capital in New Delhi, which was started in 1911, the Viceroy's House was completed in 1930.

3 Teach

A Giant Republic

Talk About It

- Explain the organization of India's parliament. *(India's parliament is divided into the Lok Sabha, elected by the people, and Rajya Sabha, chosen by the state and territorial legislatures.)*
- **Find Main Ideas** Why are other nations worried about the conflict between India and Pakistan? *(People worry that India and Pakistan may start a nuclear war over Kashmir.)*

SECTION 3 PROGRAM RESOURCES

Unit Resource Book: Southwest Asia and South Asia

ON LEVEL

Unit Resource Book
- Reading Study Guide, p. 121
- Vocabulary Practice, p. 141
- Section Quiz, p. 159

STRUGGLING READERS

Unit Resource Book
- Reading Study Guide with Additional Support, p. 127
- Section Quiz, p. 159
- Reteaching Activity, p. 162

eEdition with Audio DVD-ROM

ENGLISH LEARNERS

Pupil Edition in Spanish

eEdition with Audio DVD-ROM

Unit Resource Book
- RSG (Spanish), p. 133
- RSG with Additional Support (Spanish), p. 139

Multi-Language Glossary

Test Generator
- Section Quiz in Spanish

INCLUSION

Unit Resource Book
- RSG with Additional Support, p. 127
- Section Quiz, p. 159
- Reteaching Activity, p. 162

GIFTED & TALENTED

Unit Resource Book
- Connect to Math, p. 153
- Interdisciplinary Projects, pp. 150–151
- Section Quiz, p. 159

PRE-AP

Unit Resource Book
- Economics in the World, p. 152
- Primary and Secondary Sources: Political Cartoon, p. 156
- Section Quiz, p. 159

TECHNOLOGY

Unit Transparency Book
- 3-Minute Warm-Ups, TT19
- Essential Question, TT20
- Critical Thinking, TT23
- Map: Natural Resources of South Asia, TT25

Daily Test Practice Transparencies
- Section 3, TT80

Power Presentations

ClassZone.com

World Cultures and Geography Video on DVD

More About . . .

The Lok Sabha

The Lok Sabha, or House of the People, can have up to 552 members. Of these, 530 members represent states, 20 represent union territories, and two represent the Anglo-Indian community. The number of seats per state depends on the population of the state, with a goal of having roughly the same ratio of seats-to-population for all states. The number of seats per state varies from 1 to 80.

Members of the Lok Sabha must be at least 25 years old. The Lok Sabha term lasts five years from the date of its first meeting. It may be dissolved earlier, or extended if a Proclamation of Emergency is issued.

COMPARING

U.S. and Indian Governments

Remind students that both India and the United States have democratic governments, but the form of their governments differs.

- Compare the role of president in the United States with the roles of prime minister and president in India. *(Possible Answer: The president of the United States is both head of state and head of the executive branch. In India, the president is head of state, but it is a ceremonial position. The prime minister is the head of government.)*
- Identify one similarity and one difference between the legislatures of India and the United States. *(Similarity: Both are two-house congresses. Difference: In the United States, both houses are popularly elected, while in India only one house is.)*

CRITICAL THINKING ANSWER

Make Inferences Possible Answer: India has many parties, so it might be difficult for any one party to have a clear majority. Also, India was once a British colony. Though now independent, it may be more familiar with the British system.

Unit Resource Book: Southwest Asia & South Asia
- Primary and Secondary Sources: Political Cartoon, p. 156

Government Structure The Parliament of India consists of two houses. The first is the Lok Sabha (LAWK SUHB•hah)—House of the People. The second is the Rajya Sabha (RAHJ•yuh SUHB•hah)—Council of the States. The people of India vote for the members of the Lok Sabha, who serve five-year terms. The state and territorial legislatures choose almost all the members of the Rajya Sabha.

Parliament is led by a president, who is chosen by the elected members of Parliament and the state and territorial legislatures. In India, the president is the head of state, which is a largely ceremonial position of limited power. The president's duties are to give his assent to all bills before they become law, to name the prime minister, and to appoint 12 members of the Rajya Sabha.

The head of government and most powerful person in India is the **prime minister**. Usually, the prime minister is the leader of the party with the most seats in Parliament. The prime minister heads a Council of Ministers, which oversees the daily work of government. The prime minister and council usually serve for five years. But if the prime minister loses the legislature's support, the president may dissolve the Lok Sabha and call for new elections. In that case, a new parliament is formed and a new prime minister chosen.

Indian Flag In the center of the flag of modern India is a Buddhist symbol—the wheel of dharma.

Government Challenges As you read in Section 1, when the British colony of India gained independence, it was divided into India and Pakistan. The partition triggered mass migrations of refugees and outbreaks of violence between Muslims and Hindus. Since then, India and Pakistan have worked to resolve problems such as the conflict over Kashmir and the expanding populations in both countries.

COMPARING U.S. and Indian Governments

	UNITED STATES	INDIA
Type	representative democracy	parliamentary democracy
Head of State	president	president (symbolic position)
Head of Government	president	prime minister
Legislature	two-house congress (both popularly elected)	two-house congress (only one popularly elected)
System	federal	federal

CRITICAL THINKING

Make Inferences How would you explain India's preference for a parliamentary system of government rather than a presidential system?

DIFFERENTIATING INSTRUCTION

English Learners

Key Academic Vocabulary

Review with students the term *migration* on this page, and compare it to the key academic vocabulary word *migrate.* Point out the term *migrations* in the text ("The partition triggered mass migrations of refugees. . ."). Define *migrate* as "to move from one place to another" and *migration* as "the movement from one place to another." Point out that the suffix *-tion* signifies a noun.

Pre-AP

Comparison Essay

After students have read the subsection "Government Structure" and the Comparing U.S. and Indian Governments feature, have them write an essay comparing the structure of the U.S. government with that of India. Key points could include the difference between type of government (parliamentary vs. representative), the position of head of state and head of government, and the election of the legislature.

Kashmir An ongoing source of tension between the two nations is the region of **Kashmir** (KASH•meer) in the northwestern part of the subcontinent. The region is important because through it flows the Indus River, an important water source for both nations.

Kashmir was one of the princely states, which Britain had allowed to be ruled by a local prince. As such, it had the right to decide whether to join India or Pakistan. The ruler of Kashmir was Hindu, but most of the people were Muslims who wanted to join Pakistan. Some of those Muslims started an uprising, and Pakistani forces invaded Kashmir. The prince then announced Kashmir would join India, causing war to erupt between India and Pakistan. Eventually, the two countries divided Kashmir, but since then, fighting over the region has broken out several times.

The simmering tension between the two countries is a source of concern. During the late 1990s, India and Pakistan each tested nuclear weapons. Other nations worry that if a new war starts between India and Pakistan, they might resort to using those weapons.

Kashmir

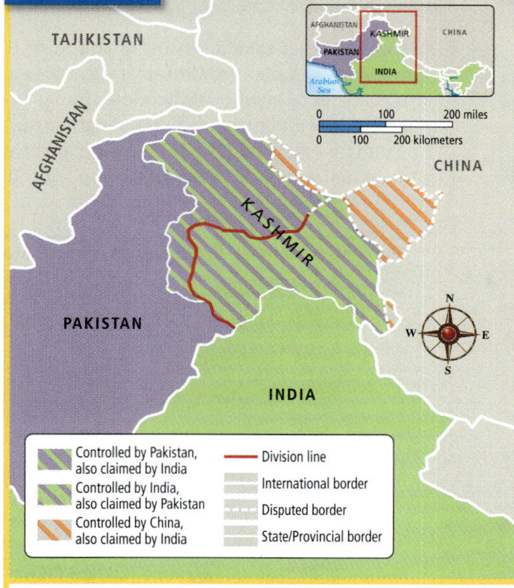

CONNECT Geography & History

1. **Region** Which countries does Kashmir border?
2. **Location** Where was the division line drawn?

Population Explosion In 2005, India had 1.1 billion people. Experts estimate that by 2050, India will have 1.6 billion people, the world's largest population. Providing for all those people will be a challenge. They will need housing, food, water, jobs, health care, and government services.

The Indian government is taking several steps to deal with this challenge. First, it has run publicity campaigns to persuade people to have fewer children. Second, it has tried to improve women's lives by offering them more education and job opportunities. Finally, many experts believe that India must improve its education system. Many children drop out of school to work, so only about 60 percent of Indians can read and write. Education helps people to get better jobs, which enable them to afford better health care and housing. As families gain more wealth, they tend to have fewer children and their children tend to stay in school longer.

FIND MAIN IDEAS Describe challenges faced by the Indian government. Possible answers: ending the conflicts with Pakistan over Kashmir, slowing population growth, and improving education

Kashmir

CONNECT Geography & History

Explain that Pakistan and India both have their own versions of maps showing Kashmir. This map shows the territory under dispute. Note that in addition to Pakistan, India, and China claiming all or part of Kashmir, some Kashmiris want independence and self-rule.

- What country in addition to Pakistan and India controls part of Kashmir? *(China)*
- Why might Pakistan and India create different maps of the region? *(Possible Answer: to support their claim to Kashmir)*

ANSWERS

1. **Region** Afghanistan, China, India, Pakistan
2. **Location** Possible Answer: roughly down the middle, or center

Unit Resource Book: Southwest Asia & South Asia

- Interdisciplinary Projects: Understanding Conflicting Points of View, pp. 150–151

Comparing Regions

Population in India and China

According to mid-2006 estimates, China is the world leader in population at 1,313,973,713, with India following at 1,095,351,995. According to population projections, they'll switch positions by 2050. Both China and India, however, have plans to curb population growth in the coming years.

China has had a long-standing one-child policy, designed to help curb population growth. Such policies are controversial as they can be considered to infringe upon the rights of people to bear children. At the same time, efforts to limit population generally have helped China deal with limited resources.

DIFFERENTIATING INSTRUCTION

Struggling Readers

Use Transitions

Have students focus on the transitions in the subsection "Population Explosion." Words such as *first, second,* and *finally* will help them identify a list of related items. Have them review the topic sentence of the second paragraph of this subsection to understand the connection between these items.

Gifted & Talented

Letter from a Kashmiri

Have students assume the role of a person living in Kashmir, where India, Pakistan, and China all claim some control. Have students consider options such as the division line in Kashmir becoming a permanent border and Kashmir becoming independent. Ask students to write a letter explaining how they think the situation should be handled and why. Have students compare their letters in small groups.

CONNECT to Math

Explain to students that population estimates generally draw on a number of factors, including birth rates, life expectancy, infant mortality rates, and population ages.

- Compare the changes in population from 1950 to 2010 in the United States and in India. *(Possible Answer: The U.S. population is predicted to double, while India's population will be more than three times greater.)*
- Which country will see the most growth over the 100 year period? *(India)*

Unit Resource Book: Southwest Asia & South Asia

- Connect to Math: Make a Line Graph, p. 153

Teach

A Booming Economy

Think, Pair, Share

- Describe agriculture in India. *(Possible Answer: Farms are generally small; many use plows pulled by oxen; crops include rice, wheat, oilseed, cotton, tea, sugar cane, and jute.)*
- Why hasn't land reform made much progress? *(Possible Answer: Land reform would distribute land more equitably, and large landowners, who have political influence, don't like that idea.)*
- How do India's natural resources affect its economy? *(Possible Answer: Numerous natural resources are helping India's economy grow.)*
- What are some key industries in India? *(Possible Answers: cotton textiles, steel manufacturing, chemicals, processed foods, transportation equipment, cement, mining, and cottage industries, like hand-crafted goods)*
- **Find Main Ideas** What is the main idea of the subsection "High Tech"? *(Possible Answer: High-tech industries are growing quickly in India and may be the future of India's economy.)*

Unit Resource Book: Southwest Asia & South Asia

- Skillbuilder Practice: Making Generalizations, p. 142
- Interdisciplinary Projects: Evaluating India's Green Revolution, pp. 149, 151

CONNECT to Math

Most countries take a census, or count, of their population. For example, the United States takes a census every 10 years. People who study population also use mathematical formulas to predict what the future population of a country might be. The table below gives statistics about the actual and predicted population for India, China, and the United States from 1950 to 2050.

Population of India, China, and the United States (in millions)

Country	1950	1970	1990	2010*	2030*	2050*
India	357	554	849	1,183	1,449	1,592
China	554	830	1,155	1,354	1,446	1,392
United States	157	210	255	312	360	394

Sources: World Population Prospects * Estimated Figures

Activity

Make a Line Graph

1. Draw the horizontal axis of your graph. Along that axis, make six marks at even intervals. Label them 1950, 1970, 1990, 2010, 2030, and 2050.
2. Draw the vertical axis of your graph. Along that axis, make 17 marks at even intervals. Label them 100, 200, and so on up to 1,700.
3. Choose a different colored pencil for each country. Using the color for India, place dots on the graph to mark the population for each year listed on the chart. Using your ruler, connect the dots with lines. Label the line "India." Then repeat for China and the United States.

A Booming Economy

KEY QUESTION What resources does India have to help its economy?

As you just read, creating jobs helps a country to slow population growth and reduce poverty. In recent years, India has succeeded in promoting economic growth, which in turn led to the creation of new jobs. One reason for the growth is that the government lessened its control of the economy and encouraged private ownership of companies. This decision caused foreign investment to increase, and the inflow of money helped business to boom.

Food for Peace Norman Borlaug won the Nobel Peace Prize for his work as an agricultural scientist in developing new seeds to feed the world's people. Here he stands with Pradeep Singa, a progressive Indian farmer, in a wheat field.

Agriculture In spite of the growth of business, most people in India make their living from farming. About half of all the land in India is used for agriculture. Farms are generally small. The average size is five acres, with more than half of all farms less than three acres. Many farmers cannot afford machinery, so they use plows that are pulled by oxen. Some of the main crops of India are rice, wheat, oilseed, cotton, tea, sugar cane, and a fiber called jute. Most farm families struggle to survive on what they grow for themselves.

One solution being considered for this problem is land reform—a more balanced distribution of land among farmers than now exists. Around the year 2000, 5 percent of India's farm families owned nearly 25 percent of India's farmland. Because the large landowners have great political influence, land-reform proposals have never made much progress.

INTERDISCIPLINARY ACTIVITIES

CONNECT to Math

Compare Population Growth Rates

Assign small groups of students one of these countries: India, China, or the United States. Have students determine the population growth rate from generation to generation (i.e., every 20 years) by dividing the later year by the earlier year. Have students note trends within their country of focus. Then have them compare their results with those of the other groups in the class.

CONNECT to Civics

Create a Poster About Land Reform

Have students create a poster to help promote or prevent land reform in India. The poster should have vivid visuals and a clear message. Wording should be limited, but the poster should provide information about land reform (positive or negative) and be persuasive. Display the posters for review, then discuss the issues as a class.

After a series of famines in the 1960s, scientists introduced new farming techniques. India underwent a **Green Revolution**, which was the introduction of seeds that yielded more crops per acre. This increased food production, especially of wheat and rice. In addition, better methods of farming, fertilizers, pesticides, and irrigation have helped farmers to grow more. Today, India produces enough crops to feed all its people.

Industry Although agriculture is the main economic activity in India, industry is also an important element. Cotton textiles are a major product of India. In the early 20th century, other industries began to develop.

India has a number of natural resources that can help its economy grow. It has iron ore and coal to make steel, and other minerals such as manganese, bauxite, and chromite. Besides steel, India manufactures cloth, chemicals, processed foods, transportation equipment, and cement. The map on this page shows where some of India's industries are located. The main industrial regions are centered around Kolkata in the east, Mumbai and Ahmadabad in the west, Chennai in the south, and Delhi in the north.

In addition, India has a very large number of **cottage industries**, businesses in which workers produce goods in their homes. Millions of craftspeople create textiles, brassware, jewelry, leather goods, matches, and incense in these industries.

In addition to manufacturing, India benefits from the mining industry. India exports some of its iron ore to other countries. Indian mines also produce precious metals and gemstones, such as silver, diamonds, and emeralds. India exports many cut diamonds to various parts of the world.

High Tech Perhaps the fastest growing industry in India is the information technology industry. During the period from 1991 to 2000, the computer software industry increased its earnings from $150 million to $5.7 billion. That is an increase of 38 times!

Animated GEOGRAPHY

Agriculture and Industry

Click here to compare the ancient and modern resources of India @ClassZone.com

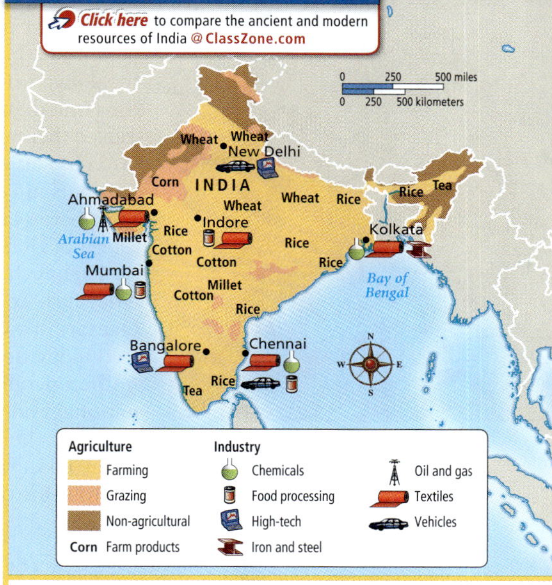

CONNECT Geography & Economics

Location Why might most of India's industrial cities be on the coasts?

DIFFERENTIATING INSTRUCTION: TIERED ACTIVITIES

OBJECTIVE Design storyboards that illustrate the diversity of India's industries.

Basic

Assign student groups agriculture, industry, or high tech. Have groups work together to create a four-panel storyboard on their assigned aspect of India's economy. Panels may focus on facts about the industry, its challenges, and its successes. Have students present and display their storyboards.

On Level

Adapt the Basic activity so that groups of students create six-panel storyboards:

- The first three panels should show the basic facts about the industry.
- The second three panels should show the industry's challenges and successes.

Have students present and display their storyboards.

Challenge

Adapt the On Level activity so that groups of students create eight-panel storyboards:

- The first six panels are as for the On Level activity.
- The last two panels should show what students predict will be future of the industry in India.

Have students present and display their storyboards.

Animated GEOGRAPHY

Agriculture and Industry

CLASSZONE.COM

CONNECT Geography & Economics

Have students use the key to identify different types of agriculture and industry represented on the map.

- Which crop is predominant in the east? *(rice)* Is more of India farmland or grazing land? *(farm)*
- Where are the main centers of high-tech? *(Bangalore and Delhi)* What industries are found in Chennai? *(chemicals, food processing, automobiles, and textiles)*

ANSWER

Location Possible Answer: Sea access makes it easier to ship exports and receive imports.

Unit Resource Book: Southwest Asia & South Asia

- Economics in the World: Tea Production, p. 152

Presentation Options

Use the **Power Presentations** DVD-ROM to project the animated map.

- Click on each item in the key.
- Explore the map to compare the ancient and modern resources of India.

Comparing Regions

Diamonds

Most people assume that South Africa is the main source of diamonds, yet diamonds are mined all over the world, except in Antarctica and Europe. The first record of diamonds comes from India, which was the only source of diamonds for about 1,000 years, beginning in the fourth century B.C. Now diamonds are found only in India's Panna region. India still remains one of the world's major diamond polishers.

With the discovery of major diamond mines in South Africa in the 1870s, the diamond industry changed. South Africa mined more diamonds in about 15 years than India had mined in about 2,000.

CONNECT to the Essential Question

How is India's traditional way of life changing in response to modern forces?

Ask students what they have learned so far that can help them answer this question. Students might mention:

- India's government faces many challenges, including resolving differences with Pakistan and a rapidly growing population.
- New methods of farming and types of seeds are improving Indian agriculture.
- New jobs in India's high tech industry are increasing.

Unit Transparency Book: Southwest Asia & South Asia
- Essential Question, TT20

4 Assess & Reteach

Assess Have students complete the Section Assessment.

Unit Resource Book: Southwest Asia & South Asia
- Section Quiz, p. 159

Interactive Review @ClassZone.com
- Section Quiz

Power Presentations

Test Generator

Reteach Have students list three main ideas each about India's government and economy. Invite students to share one main idea at a time. Ask students to scan the text for any main ideas they may have overlooked or misstated.

Unit Resource Book: Southwest Asia & South Asia
- Reteaching Activity, p. 162

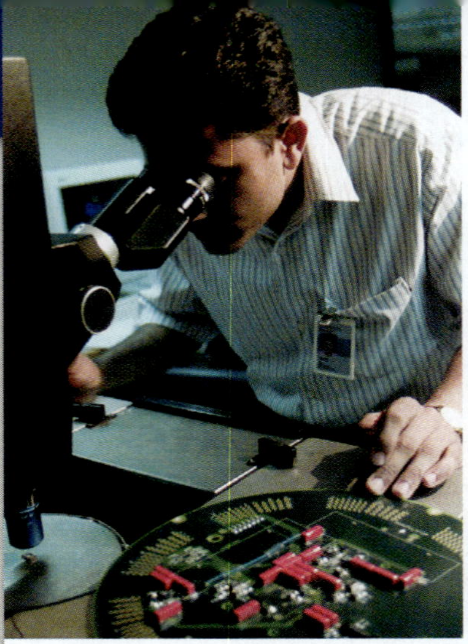

Software A chip designer looks through a microscope in Bangalore.

One reason for the rapid growth of high-tech industries is that a large percentage of Indians speak English, which has become the international language of business. India has the second-largest number of English-speaking, technically skilled workers in the world (after the United States). In fact, some U.S. companies have moved their technical support departments to India and hired Indian workers to help customers over the phone. This outsourcing of American jobs to India, as well as to other places in the world, is an increasingly controversial aspect of global economic interdependence.

High-tech industries are spread across India in many cities such as Hyderabad, Mumbai, Kolkata, and Delhi. One city in particular, **Bangalore**, is home to many high-tech industries. Bangalore is a large city of more than 6 million people located in southern India. Many international and Indian high-tech businesses have offices there. It is home to over a thousand computer software companies that are taking advantage of India's low wages and highly skilled workers. To some observers, Bangalore represents the future of the Indian economy.

SUMMARIZE Describe India's resources. farmland, iron ore, coal, other minerals, educated workers

ONLINE QUIZ For test practice, go to Interactive Review @ClassZone.com

Section 3 Assessment

TERMS & NAMES

1. **Explain the importance of**
   - parliamentary
   - Kashmir
   - Green Revolution
   - Bangalore

USE YOUR READING NOTES

2. **Find Main Ideas** Use your completed chart to answer the following question:
   What challenges does India's expanding population present?

GOVERNMENT	ECONOMY	ISSUES AND TRENDS
1.	1.	1.
2.	2.	2.
3.	3.	3.
4.	4.	4.

KEY IDEAS

3. What happens if the Indian prime minister loses the support of the legislature?
4. Why is Pakistan so interested in controlling Kashmir?
5. How did the Green Revolution help Indian agriculture?

CRITICAL THINKING

6. **Compare and Contrast** In India, how do the positions of president and prime minister differ?
7. **Analyze Causes** Why do other countries consider the Kashmir conflict a global problem rather than just a South Asian problem?
8. **CONNECT to Today** What modern technologies have made it possible for U.S. high-tech companies to hire Indian workers to do their customer support?
9. **SCIENCE** **Prepare an Educational Poster** With a partner, research the Green Revolution. Prepare a poster that shows what crops scientists developed high-yielding seeds for and how much more they produced than regular seeds.

SECTION 3 ASSESSMENT ANSWERS

Terms & Names

1. parliamentary: government with parliament as legislature; Kashmir: region disputed by India and Pakistan; Green Revolution: introduction of seeds that produced high yields; Bangalore: Indian city home to high-tech industries

Use Your Reading Notes

2. Population will require housing, food, water, jobs, health care, and government services.
   Diagram *Government: Parliament, prime minister, Council of Ministers, federal system; Economy: Green Revolution, industry, cottage industry, high tech; Issues and Trends: Kashmir, population growth, migration to cities, jobs*

Key Ideas

3. President dissolves Lok Sabha, elections are held, and a new prime minister is chosen.
4. because the majority of people there are Muslims and because the Indus River runs through Kashmir
5. It enabled farmers to produce more crops per acre.

Critical Thinking

6. President is chosen by members of Parliament and state and territorial legislatures and has limited powers. Prime minister is chosen as leader of party with most seats in Parliament.
7. If another war erupts over Kashmir, India and Pakistan might use nuclear weapons, which could cause damage to other countries.
8. satellite phones, the Internet, and other technologies that foster global communication
9. Use the rubric to score posters.

Poster Rubric

	Elements of Poster	Effectiveness of Design
4	reflects thorough research	neat, organized, attractive
3	mostly complete research	neat and organized
2	some research	somewhat organized
1	little or no research	unorganized, messy

Click here to complete these and other activities online @ ClassZone.com

CHAPTER SUMMARY

Key Idea 1
India is the dominant country of South Asia, and it has a rich and complicated history.

Key Idea 2
The culture of India is a blend of all of the many influences that have shaped its history.

Key Idea 3
Out of India's diversity, a robust democracy has developed.

For **Review and Study Notes**, go to **Interactive Review** @ ClassZone.com

NAME GAME

Use the Terms & Names list to complete each sentence on paper or online.

1. I am a region that India and Pakistan have fought over. Kashmir
2. I am the state of wisdom that breaks the cycle of reincarnation. ________
3. I am the chief executive of government in a parliamentary system. ________
4. I am an ancient language of India. ________
5. I am one of the people living in India before the arrival of the Aryans. ________
6. I am a city in southern India that is home to many high-tech industries. ________
7. I am the sum of a person's actions, which determines his or her fate in the next life. ________
8. I am a small business in which the workers manufacture items in their homes. ________
9. I am a method of protest that draws attention to a problem without using violence. ________
10. I am the rebirth of the soul in another body. ________

Aryan
Bangalore
Buddhism
cottage industry
Dravidian
Hinduism
karma
Kashmir
nirvana
nonviolent resistance
parliamentary
prime minister
reincarnation
Sanskrit
Vedas

Activities

Flip Cards

Use the online flip cards to quiz yourself on the terms and names introduced in this chapter.

Crossword Puzzle

Complete an online puzzle to test your knowledge of India's path from early societies to a modern nation.

ACROSS
1. leader of Indian independence movement

723

ONLINE AND ON-SCREEN ACTIVITIES

Flip Cards

Vocabulary Flip cards are virtual flash cards at **ClassZone.com.**

Students can quiz each other or work independently to study the flip cards for this chapter.

- They can quiz themselves on key terms and names by reviewing definitions.
- Demonstrate how to advance and shuffle the deck for another challenge.

Crossword Puzzle

Students can review important people and events in India's history by completing this crossword puzzle online at **ClassZone.com**.

- As an extension of this activity, have students use the *Activity Maker* at **ClassZone.com** to make their own crossword puzzle for one section of the chapter.
- Then have students trade puzzles with classmates to solve answers for each other's puzzles.

Chapter Summary

As a follow-up to the chapter, encourage students to pair up and go over the **Review and Study Notes** at **ClassZone.com.**

Alternatively, students pairs can review their Reading Study Guide notes, making sure that they can add important details to each section's "Key Idea."

Name Game

Teaching Tips

ON PAPER Have students work in pairs to complete this activity. As an extension, they can write questions for the terms that are not covered in the items shown here.

ONLINE Make sure that students understand how to drag and drop terms to fill in blanks. Ask them to print out their results and compare them with those of a classmate.

ANSWERS

1. Kashmir
2. nirvana
3. prime minister
4. Sanskrit
5. Dravidian
6. Bangalore
7. karma
8. cottage industry
9. nonviolent resistance
10. reincarnation

Make It Fun

Teacher-Tested Activities

Ann Christianson, John Muir Middle School, Wausau, Wisconsin

Drawing pictures can help students grasp the key terms and names in the chapter.

- Students write the term and its definition.
- Students draw a picture to represent the term. For "Green Revolution," for example, they might draw a picture of a field of rice before and after the revolution.
- Finally, students write the term in a sentence that shows they understand the term.

Vocabulary ANSWERS

1. Buddhism is a major world religion based on the teachings of Siddhartha Gautama.
2. A cottage industry is a small business in which the workers manufacture items in their homes.
3. A golden age is a period during which a society attains great prosperity and cultural achievements.
4. The Green Revolution means the introduction of seeds that produced very high yields, increasing food production.
5. Hinduism is modern name for major religion of India, which developed from Brahmanism.
6. Kashmir is region in northwestern Indian subcontinent disputed by India and Pakistan.
7. Mohandas Gandhi was a 20th-century Indian who helped lead his country to independence by using nonviolent resistance to colonial rule.
8. Nonviolent resistance is a method of protest that draws attention to a problem without using violence.
9. Parliamentary refers to a government with the parliament form of legislature.
10. Prime minister is head of cabinet and chief executive in a parliamentary system.
11. Reincarnation is the rebirth of the soul; karma determines what sort of rebirth one will have; attaining nirvana releases one from that cycle.
12. Sanskrit was the language of the Aryans; the *Bhagavad-Gita* was written in Sanskrit.

Key Ideas ANSWERS

13. Aryans were nomadic herders; Dravidians were used to settled, urban life.
14. He tried to give up warfare, improve his subjects' lives, and spread Buddhism.
15. Akbar is known as the greatest Mughal ruler for his fair government; Shah Jahan is famous for building the Taj Mahal.
16. Ten million people moved between India and Pakistan, and the hardships that resulted caused old angers to flare up between Hindus and Muslims.
17. Hinduism, Islam, Sikhism, Christianity, Jainism
18. A person can stop suffering by not wanting, and one can stop wanting by following the Eightfold Path.
19. In recent decades, many young people are leaving villages for cities where they can earn more money.
20. They almost always have musical numbers.
21. Power is divided between the national and state governments.

Chapter 23 Assessment

VOCABULARY

Explain the significance of each of the following.

1. Buddhism
2. cottage industry
3. golden age
4. Green Revolution
5. Hinduism
6. Kashmir
7. Mohandas Gandhi
8. nonviolent resistance
9. parliamentary
10. prime minister

Explain how the terms in each group are related.

11. karma, reincarnation, and nirvana
12. Aryan, Sanskrit, and *Bhagavad-Gita*

KEY IDEAS

1 Empires and Independence

13. How were the Aryans and Dravidians different?
14. How did converting to Buddhism affect Asoka?
15. Why are Akbar and Shah Jahan famous Mughal rulers?
16. Why did partition lead to violence?

2 A Rich and Ancient Legacy

17. What religions are practiced in India?
18. According to the Buddha, how does a person stop suffering?
19. How is village life changing?
20. How are Indian movies different from most American movies?

3 The World's Largest Democracy

21. In what way does India have a federal system?
22. How are the members of the parliament chosen?
23. How is the Indian government trying to deal with the population explosion?
24. What are India's main crops?

CRITICAL THINKING

25. **Sequence Events** Create a time line showing the various empires and major powers that ruled India.

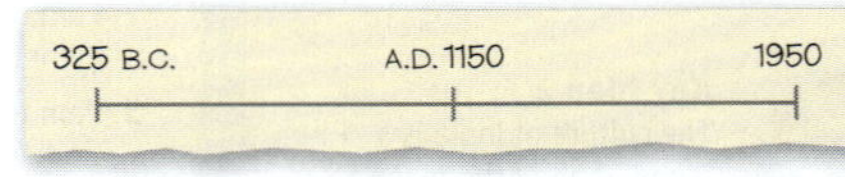

26. **Analyze Causes** Why do you think Indians objected to British rule?
27. **Compare and Contrast** How is Indian government similar to and different from British government?
28. **Connect to Economics** Having many English speakers gives India an economic advantage. Why do you think so many Indians speak English?
29. **Connect Geography & History** What geographic difficulty did the new nation of Pakistan face?
30. **Five Themes: Movement** More Dravidian-speaking people live in southern India than in the north. What might explain that pattern?

Answer the ESSENTIAL QUESTION

How is India's traditional way of life changing in response to modern forces?

Written Response Write a two- or three-paragraph response to the Essential Question. Be sure to consider the key ideas of each section, as well as specific ideas about traditional ways of life and modern changes. Use the rubric below to guide your thinking.

Response Rubric

A strong response will:

- discuss traditional ways of life
- examine the modern forces for change
- give examples of how Indian culture blends the two

ESSENTIAL QUESTION REVIEW

Before students begin their essays, have them review elements that will help answer the Essential Question. Direct them to features such as:

- Connect History and Culture: Sanskrit, Section 1
- Analyzing Primary Sources: Mohandas Gandhi, Section 1
- Animated Geography: Spread of Hinduism and Buddhism, Section 2
- Comparing Rural and Urban India, Section 2
- Comparing Traditional and Modern India, Section 2
- Animated Geography: Agriculture and Industry, Section 3

Written Response Rubric

	Coverage of Key Ideas	Explanation	Accuracy
4	all	clear, concise	no errors
3	most	good	few/minor errors
2	some	fair	several errors
1	two or fewer	poor	many errors

STANDARDS-BASED ASSESSMENT

Test Practice
- Online Test Practice @ClassZone.com
- Test-Taking Strategies and Practice at the front of this book

THEMATIC MAP

Use the map and your knowledge of India to answer questions 1 and 2 on your paper.

Languages of India

1. In what states are the Dravidian languages mainly spoken?

2. What language would you hear spoken in the state of Rajasthan?

3. What language would you hear spoken in the state of Kerala?

PRIMARY SOURCE

Use the following quote from Gandhi to answer question 4 on your paper.

> I do not want my house to be walled in on all sides and my windows to be stuffed. I want the cultures of all lands to be blown about my house as freely as possible. But I refuse to be blown off my feet by any.
>
> Source: Mohandas Gandhi

4. Which of the following choices best restates Gandhi's opinion?

A. He believed that people of all nations should integrate, or mix, cultures.

B. He believed that it is dangerous for people to try to mix cultures.

C. He wanted to appreciate other cultures, yet he didn't want people of a foreign culture to rule him.

D. He wanted to keep elements of his culture in the privacy of his home, and not to display it in public.

GeoActivity

1. INTERDISCIPLINARY ACTIVITY–MATH

Using textbooks, encyclopedias, or the Internet, look up information about the decimal system—how it represents fractions and how numbers are written. Create a poster that illustrates this by converting fractions into decimals.

2. WRITING FOR SOCIAL STUDIES

Research to learn more about the life of Mohandas Gandhi. Focus especially on the methods he used to protest British rule. Then write a longer "History Maker" about him.

3. MENTAL MAPPING

Create an outline map of India and label the following:
- Mumbai
- Kolkata
- Hyderabad
- Chennai
- Bangalore
- New Delhi

22. People vote for the members of the Lok Sabha, and the state and territorial legislatures choose the members of the Rajya Sabha.

23. It runs publicity campaigns to encourage smaller families, it is trying to improve the opportunities for women, and it is trying to improve its education system.

24. rice, wheat, oilseed, cotton, jute, tea, sugar cane

Critical Thinking ANSWERS

25. About 321 B.C.—Mauryan Empire, A.D. 320—Gupta Empire, 1526—Mughal Empire, 1858—British rule, 1947—independence

26. British were foreigners, they passed laws that discriminated against Indians, and they enriched Britain, not India.

27. It is similar in being a parliamentary system, but differs in not having a monarch.

28. During British rule, schools taught in English.

29. It was made up of two separate regions that were separated by India, a hostile nation.

30. If theory of Aryan migration is correct, then Aryans may have pushed Dravidians south.

Standards-Based Assessment ANSWERS

1. Andhra Pradesh, Karnataka, Kerala, Tamil Nadu

2. Hindi

3. Malayalam

4. C

GeoActivity

1. INTERDISCIPLINARY ACTIVITY RUBRIC

The poster should
- reflect thorough research
- be neat, organized, and attractive

2. WRITING FOR SOCIAL STUDIES RUBRIC

The History Maker should
- include interesting and colorful details
- be relatively free of mechanical errors

3. MENTAL MAPPING RUBRIC

The outline map should
- represent the approximate shape of India
- include all geographic features listed in their correct location

DIFFERENTIATING CHAPTER ASSESSMENT

Unit Resource Book
- **Chapter Test A,** pp. 163–167. *Use for Struggling Readers, Beginning-to-Intermediate-Level English Learners, and for Inclusion of students whose IEP suggests supported test-taking.*
- **Chapter Test B,** pp. 168–172. *Use for On-Level Learners and ESOL students with Advanced-Level English.*
- **Chapter Test C,** pp. 173–177. *Use for Gifted and Talented students or students assigned Pre-AP work.*

Test Generator
- Chapter Tests A, B, and C, in English and Spanish

McDougal Littell Assessment System

TEST **Test Generator** Use three levels of tests in English and Spanish tied to state standards. Plus, choose questions to customize tests for all populations.

SCORE Scan and Score plain paper student tests in minutes.

REPORT Create instant progress reports, sortable by student, by standard, or other criteria.

RETEACH Respond immediately to students' needs by using a wealth of reteaching resources.

CHAPTER 24

South Asian Neighbors PLANNING GUIDE

Complete print resources all on one USB drive allow you to customize lesson plans for all students.

Section Titles & Key Ideas	Unit Resource Book Southwest Asia and South Asia	Assessment Resources
1 Afghanistan, Pakistan, and Bangladesh **KEY IDEA** Other nations and cultures in the region besides India have managed to develop their own distinct identities.	• **Reading Study Guide,** pp. 179–180 • **RSG with Additional Support,** pp. 183–184 • **Vocabulary Practice,** p. 195 • **Skillbuilder Practice,** p. 196 • **Map and Graph Skills,** pp. 197–198 • **History Makers,** p. 201	**Unit Resource Book** • Section Quiz, p. 209 • Reteaching Activity, p. 211 **Test Generator CD-ROM** **Daily Test Practice Transparency,** TT81
2 Mountain Kingdoms and Island Nations **KEY IDEA** Some of India's smaller neighbors have also developed in distinctive and unique ways.	• **Reading Study Guide,** pp. 181–182 • **RSG with Additional Support,** pp. 185–186 • **Outline Map Activities,** pp. 199–200 • **Interdisciplinary Projects,** p. 204 • **Connect to Math,** p. 207	**Unit Resource Book** • Section Quiz, p. 210 • Reteaching Activity, p. 212 • Chapter Test (Forms A, B, & C), pp. 213–224 **Test Generator CD-ROM** **Daily Test Practice Transparency,** TT82

McDougal Littell **Assessment System**

TEST | SCORE | REPORT | RETEACH

No Time? To focus students on important content in the chapter, use the **"Reading Study Guide with Additional Support"** in ***Resources2Go.***

Teacher Presentation Options

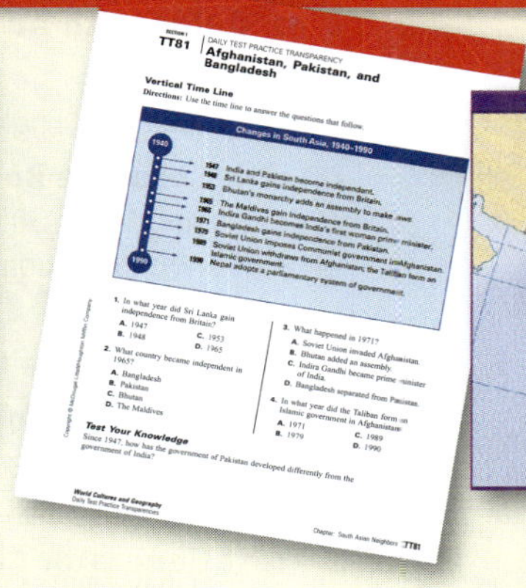

Electronic Books

eEdition DVD-ROM

eEdition Online

Presentation Toolkit

Video/DVD

- World Cultures and Geography

Power Presentations DVD-ROM

- Discussion Starters
- Media Gallery
- Review Game
- Interactive Review
- Animated Features

Unit Transparency Book

- 3-Minute Warm-Ups
- Essential Question
- Critical Thinking
- Maps
- Cultures

Daily Test Practice Transparencies

Animated Geography

- Natural Resources of South Asia Map
- Land Use in South Asia Map

OnlineActivities at ClassZone.com

Animated Geography

- Videos and Photos
- 3-D Animations
- Interactive Maps

Interactive Review

- Section Quizzes
- Review and Study Notes
- Vocabulary Flip Cards
- Graphic Organizers
- Name Game and Crossword Puzzles
- GeoGame

Online Test Practice

Map Center

- Atlas Maps
- Political Maps with Data
- Links to Mapping Web Sites

Activity Center

- Activity Maker—Review Games, Puzzles, Flip Cards, and Graphic Organizers
- Worldster

Research & Writing Center

- Web Sites
- Data Updates
- Primary Sources
- Writing Models

Current Events

Tim Mortensen
Patrick Marsh Middle School
Sun Prairie, Wisconsin

Make It Fun — Teacher-Tested Activities

As you teach this chapter, look for my classroom-proven ideas for:
- making the lesson fun
- focusing students' attention and comprehension
- integrating technology with traditional teaching

Teaching Bell to Bell When you see this timing guide, have students discuss in pairs, in groups of 3–6, or as a class.

- **Think, Pair, Share** (pairs)
- **Reader, Recorder, Reporter** (3–6)
- **Talk About It** (any size)

CHAPTER 24 RESOURCES FOR DIFFERENTIATED INSTRUCTION

Struggling Readers

eEdition with Audio DVD-ROM

Lesson Plans
- Strategies for Inclusion

Unit Resource Book
- Reading Study Guide with Additional Support, pp. 183–186 B
- Reteaching Activity, pp. 211–212
- Chapter Test (Form A), pp. 213–216 A

Best Practices Toolkit C
- Interactive Reading Strategies, pp. 90–145
- Differentiation Activity Bank: Struggling Readers, pp. 146–149

Teacher's Edition Activities
- Discuss Hydroelectricity and Exporting Energy, Sec. 2
- Tiered Activities, Sec. 1–2

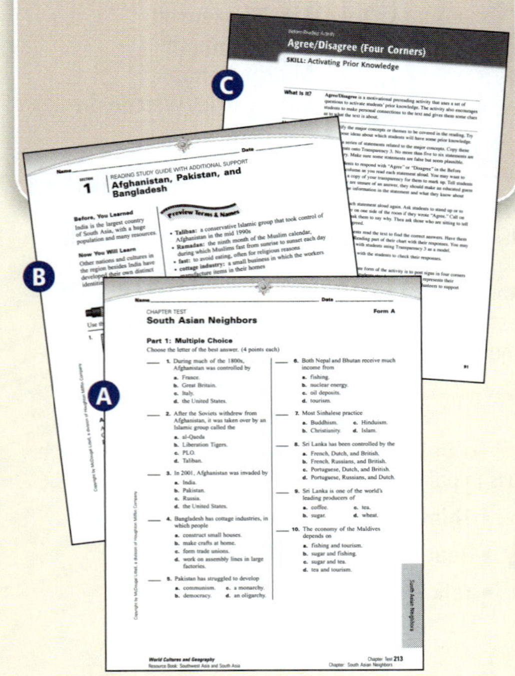

English Learners

Pupil's Edition in Spanish

eEdition with Audio DVD-ROM

Lesson Plans
- Strategies for English Learners

Unit Resource Book
- Reading Study Guide (Spanish), pp. 187–190
- RSG with Additional Support (Spanish), pp. 191–194 A B

Multi-Language Glossary C

Best Practices Toolkit
- Differentiation Activity Bank: English Learners, pp. 165–194

Test Generator CD-ROM
- Section Quizzes in Spanish
- Chapter Test in Spanish

Teacher's Edition Activities
- Pronounce and Preview, Sec. 1–2
- Vocabulary: Key Academic Vocabulary, Sec. 1
- Tiered Activities, Sec. 1–2

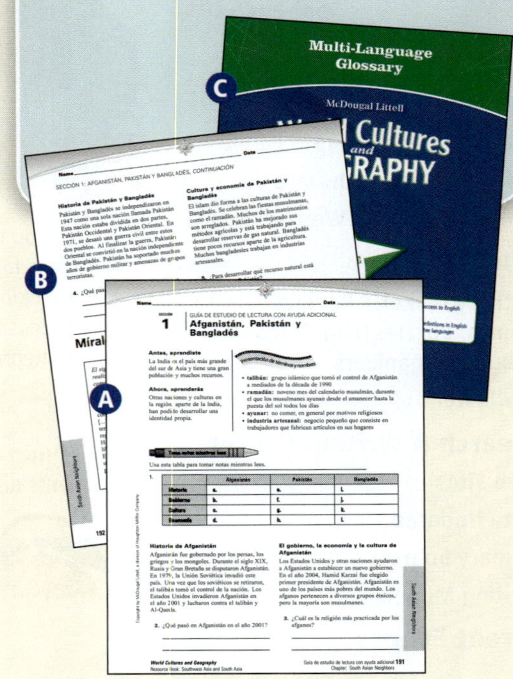

Inclusion

eEdition with Audio DVD-ROM

Lesson Plans
- Strategies for Inclusion

Unit Resource Book
- Reading Study Guide with Additional Support, pp. 183–186 A B
- Reteaching Activity, pp. 211–212
- Chapter Test (Form A), pp. 213–216

Best Practices Toolkit C
- Interactive Reading Strategies, pp. 90–145
- Differentiation Activity Bank: Struggling Readers, pp. 146–149

Teacher's Edition Activities
- Vocabulary: Predicting ABCs, Sec. 1
- Reading and Defining Terms Verbally, Sec. 2
- Tiered Activities, Sec. 1–2

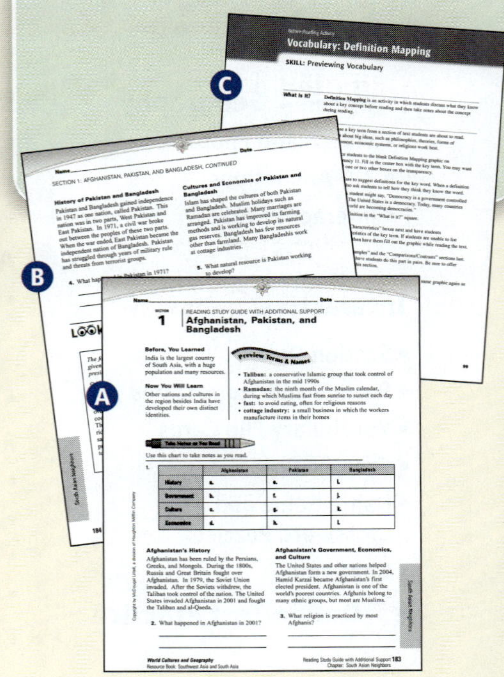

Gifted & Talented Students

Lesson Plans
- Core Lesson Plan

Unit Resource Book
- Outline Map Activities, p. 199
- History Makers, p. 201 (A)
- Interdisciplinary Projects, p. 204 (B)
- Primary Source, p. 208 (C)
- Chapter Test (Form C), pp. 221–224

Best Practices Toolkit
- Student Products Rubrics, pp. 1–14
- Writing Templates, pp. 78–89

Teacher's Edition Activities
- Describe the Importance of Literacy, Sec. 1
- Tiered Activities, Sec. 1–2

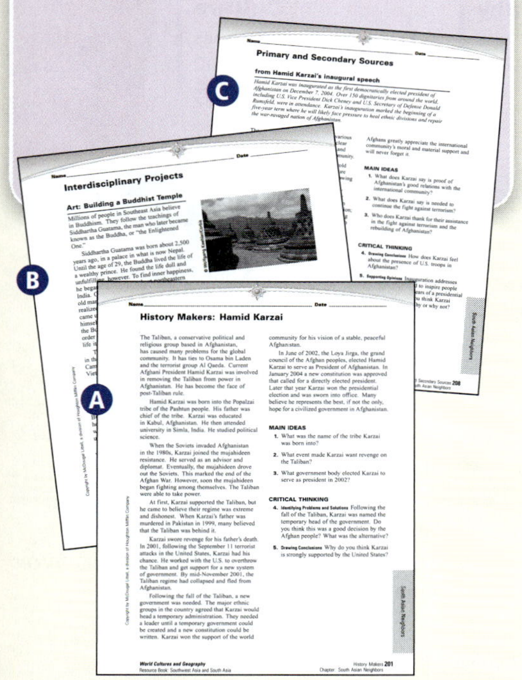

Pre-AP*

Lesson Plans
- Core Lesson Plan

Unit Resource Book
- Vocabulary Practice, p. 195
- Map and Graph Skills, pp. 197–198
- Interdisciplinary Projects, p. 203 (A)
- Economics in the World, p. 206 (B)
- Connect to Math, p. 207 (C)
- Chapter Test (Form C), pp. 221–224

Best Practices Toolkit
- Writing Templates, pp. 78–89
- Differentiation Activity Bank: Pre-AP Students, pp. 195–200

* Pre-AP is a registered trademark of the College Entrance Examination Board, which was not involved in the production of and does not endorse this product.

Teacher's Edition Activities
- Map Economic Activities, Sec. 2
- Tiered Activities, Sec. 1–2

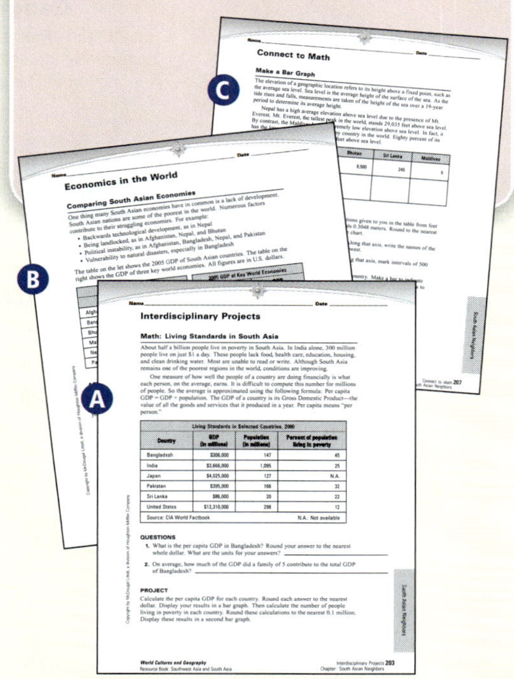

Interdisciplinary Activities

CONNECT to Math

Compare Economies Using Graphs

CLASS TIME 30 minutes

TASK Prepare a graph to compare the economies of the countries from this chapter as well as those of India and the United States.

SUPPLIES
- graph paper, colored pencils or markers

ACTIVITY Have students graph the Gross Domestic Products (per capita) of the countries in the chapter, as well as those of India and the United States. Provide students with the following figures: Afghanistan $800; Pakistan $2,400; Bangladesh $2,100; Nepal $1,400; Bhutan $1,400; Sri Lanka $4,300; Maldives $3,900; India $3,400; U.S. $41,600. After graphing, have students show how the countries compare, such as by making comparison statements or listing countries in order of GDP, from high to low.

Good for Mathematical Learners

CONNECT to Language Arts

News from the Neighbors

CLASS TIME 30 minutes

TASK Create a short news story on a historic event from the chapter.

SUPPLIES
- writing supplies (computer, or paper and pencils or pens)

ACTIVITY Have students select a historic event from the time line or the section text, and then write a short newspaper article about it. They should include as much background as possible, taken from their textbooks or other sources, and be sure to cover the six questions for a news story (who, what, when, where, why, how). Suggest they also add some analysis about the significance of the event for the future.

Good for Verbal Learners

Introduce the ESSENTIAL QUESTION

- Remind students how cultures may affect their neighbors through cultural diffusion.
- Discuss the importance of religion in India's history and predict how religion might also affect India's South Asian neighbors.
- Suggest some ways that India's South Asian neighbors might affect India, the rest of the region, and the world.

Key Ideas

Section 1: Other nations and cultures in the region besides India have managed to develop their own distinct identities.

Section 2: Some of India's smaller neighbors have also developed in distinctive and unique ways.

CONNECT to the Essential Question

You may wish to revisit the Essential Question periodically with students as they work through the chapter.

Suggestions for reconnecting are provided at an appropriate point in the *Teacher's Edition* for each section.

Coral atoll near island in the Maldives

CHAPTER 24

South Asian Neighbors

1 FOCUS ON **Afghanistan, Pakistan, and Bangladesh**

2 FOCUS ON **Mountain Kingdoms and Island Nations**

ESSENTIAL QUESTION

How do the history and cultures of India's neighbors affect the world?

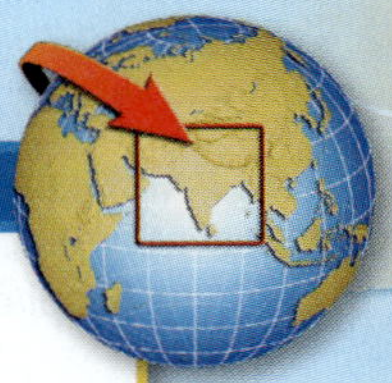

CONNECT Geography & History

Use the map and the time line to answer the following questions.

1. What countries separated East Pakistan (now Bangladesh) from Pakistan?
2. Which resource is common to Afghanistan, Pakistan, India, and Bangladesh?

1945

Geography
1947 The colony of India is partitioned, and Pakistan becomes an independent nation.

History
◄ **1979** The Soviet Union invades Afghanistan.

Geography
1971 East Pakistan wins independence as the new nation of Bangladesh.

Culture
1983 Ethnic tension causes a Tamil rebel group to start civil war in Sri Lanka. ►

726 Chapter 24

TIME LINE DISCUSSION

Note that the time line entries begin after World War II, when some countries in the region gained their independence. Point out how the events on the time line note conflicts and disasters, which have generally slowed the development of India's neighbors.

- Which event on the time line indicates interference in the region by a foreign country? *(1979—The Soviet Union invades Afghanistan.)*
- Which event on the time line indicates the importance of religion in Afghanistan? *(1996—The Taliban, a conservative Muslim group, takes control of Afghanistan.)*
- How do you think the 2004 tsunami affected the region? *(Possible Answer: It hurt the economy, forced people to relocate and rebuild, and forced countries to seek foreign aid.)*

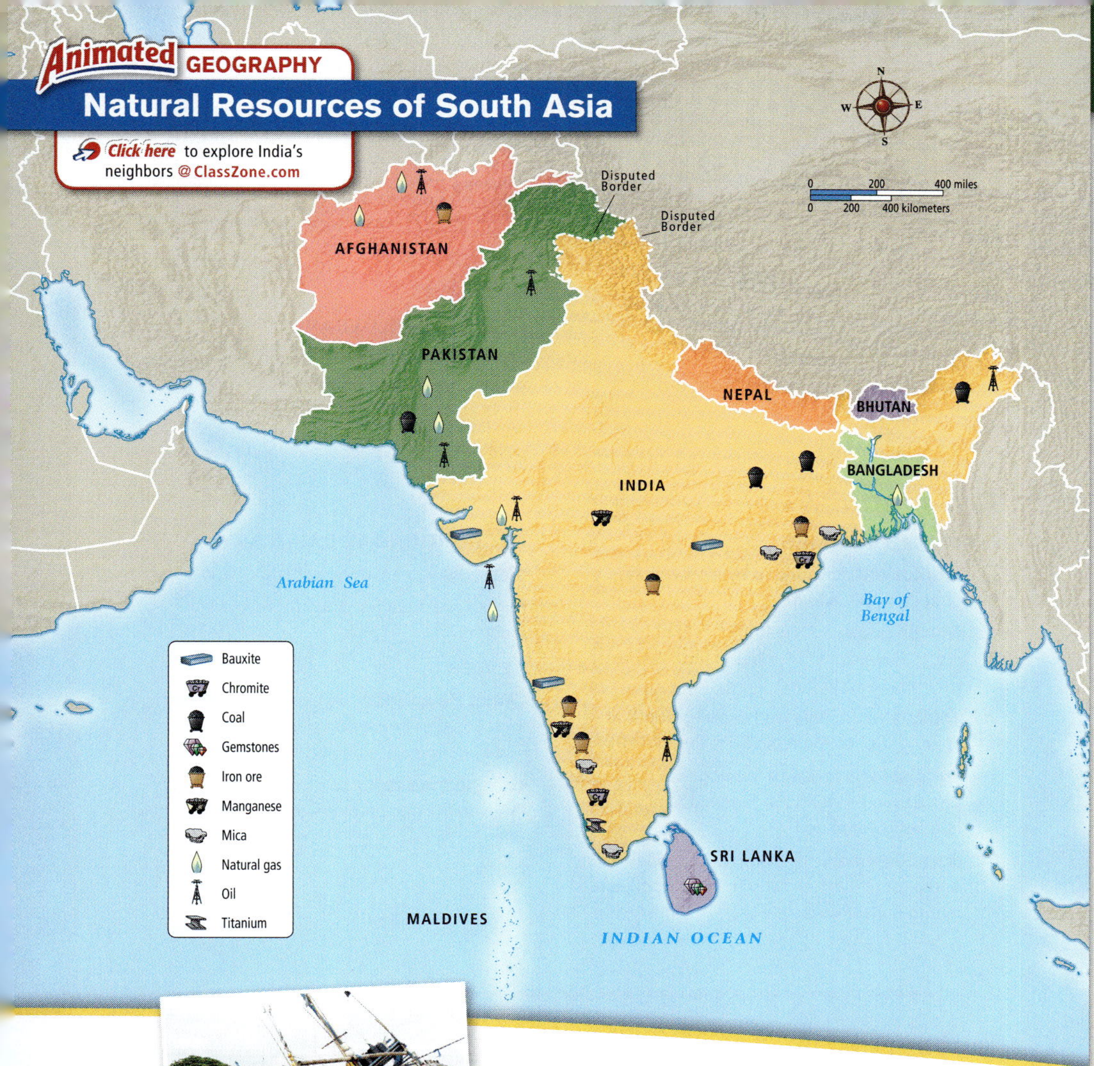

Geography

◂ **2004** A devastating tsunami hits much of Southeast and South Asia.

Today

Government

1996 The Taliban, a conservative Muslim group, takes control of Afghanistan.

Government

◂ **2006** Protests force Nepal's king to restore parliamentary government.

727

TEACHER MEDIA FAVORITES

Books for the Teacher

Mines, Diane and Sarah Lamb, eds. *Everyday Life in South Asia*. Indiana University Press, 2002. An anthology that focuses on how people of the region manage everyday affairs such as arranging marriages.

Stewart, Rory. *The Places in Between*. Harvest Books, 2006. An account of the author's walk across much of Southwest and South Asia, including Afghanistan, only a few months after the Taliban were defeated.

Video / DVD

Daughters of Afghanistan. (DVD) Choices, Inc., 2004. Explores the status of women through the ongoing struggles of five women and their children since the Taliban were ousted.

Bhutan—Gross National Happiness. (DVD) Going Dutch Distribution, 2006. Focuses on the unique, stated goal of the government of Bhutan to concentrate its development on the happiness of its people rather than economic measures.

Internet

Go to the *Research and Writing Center* at **ClassZone.com** to link to:

- data on human geography
- a variety of chapter-specific, content-reviewed sites
- "History Makers" biographies
- primary sources

Animated GEOGRAPHY

Natural Resources of South Asia

CLASSZONE.COM

Note the variety and amount of natural resources concentrated in India. Discuss how this might help India to dominate the region. Also note that the other countries in the region have few or no natural resources as listed in the legend.

- What natural resource is shown for Sri Lanka? *(gemstones)*
- How are Nepal and Bhutan similar on this map? *(Neither has any resources shown on map key.)*
- Why might the area of India that is to the north and east of Bangladesh be a potential source of conflict between the two countries? *(Possible Answer: because it has natural resources that both may want)*

CONNECT GEOGRAPHY & HISTORY ANSWERS

1. India, also possibly Nepal

2. natural gas

Presentation Options

Use the **Power Presentations** DVD-ROM to project the animated map.

- Prompt students to use the pop-up images and audio from the map to discover what they already know about the geography, culture, history, economics, or governments of India's South Asian neighbors.
- Challenge students to answer the map questions individually, in small groups, or as a class activity. Use all of the interactive material to help preview chapter content.

1 Plan & Prepare

Objectives

- Summarize the history, government, culture, and economics of Afghanistan
- Summarize the challenges facing Pakistan and Bangladesh since their independence and compare their cultures and economies

Read for the Essential Question

Help students read for a purpose by reminding them of the Essential Question: "How do the history and cultures of India's neighbors affect the world?" Point out that they can use text features such as Key Questions, special features, images, and maps to form answers to the question.

Vocabulary

Best Practices Toolkit

Use the ***Best Practices Toolkit*** to model strategies for vocabulary notetaking. Vary strategies throughout the year. Choose from: *Knowledge Rating, Predicting ABCs, Definition Mapping, Word Sort, Frayer Model (Word Squares), Magnet Words, Word Wheel,* and *Student VOC.*

Vocabulary Strategies, TT9–TT16

Reading Strategy

Best Practices Toolkit

Display the Three-Column Chart transparency.

- Explain to students that in categorizing they need to determine how the information in the section can be sorted into the categories shown in the chart.
- Model filling in categories on a chart using the Afghanistan subsection.
- Have students suggest entries using the other categories in the chart.
- Display the completed chart for students to review.

Three-Column chart, TT21

SECTION 1 Reading for Understanding

Key Ideas

BEFORE, YOU LEARNED

India is the largest country of South Asia, with a huge population and many resources.

NOW YOU WILL LEARN

Other nations and cultures in the region besides India have managed to develop their own distinct identities.

Vocabulary

TERMS & NAMES

landlocked surrounded by land with no access to the ocean

Soviet Union a Communist nation, consisting of Russia and 14 other states, that existed from 1922 to 1991

Taliban a conservative Islamic group that took control of Afghanistan in the mid-1990s

subsistence farming farming that produces just enough to feed the farmer's family with little left over to sell

Ramadan the ninth month of the Muslim calendar, during which Muslims fast from sunrise to sunset each day

BACKGROUND VOCABULARY

strategic helpful in the fighting of a war

fast to avoid eating, often for religious reasons

REVIEW

Green Revolution the use of special seeds, irrigation, fertilizers, and pesticides to produce high crop yields and food production

cottage industry a small business in which the workers manufacture items in their homes

Reading Strategy

Re-create the chart shown at right. As you read and respond to the **KEY QUESTIONS**, use the chart to organize important details about the history, government, culture, and economies of three nations.

Skillbuilder Handbook, page R7

CATEGORIZE

	AFGHANISTAN	PAKISTAN	BANGLADESH
History			
Government			
Culture			
Economics			

GRAPHIC ORGANIZERS Go to Interactive Review @ClassZone.com

PRETEACHING VOCABULARY

English Learners

Pronounce and Preview

Pronounce each term for students. Review the meanings of words within definitions, such as *access*, *conservative*, and *fast*.

- To modify vocabulary learning, have students complete worksheets as they read, instead of afterward.

Unit Resource Book: Southwest Asia & South Asia
- Vocabulary Practice, p. 195

Inclusion

Vocabulary: Predicting ABCs

Have students work in pairs to fill out the graphic organizer for the terms in the section.

- Use the transparency to model the strategy.
- Have volunteers suggest entries.
- Then distribute copies of the organizer for students to complete with a partner.

Best Practices Toolkit
- Vocabulary: Predicting ABCs, TT10

SECTION 1 FOCUS ON

Annapurna I in Nepal, one of the highest peaks in the world

Asian elephants in Sri Lanka

Afghanistan, Pakistan, and Bangladesh

Connecting to Your World

India's South Asian neighbors have been in the news often. In some cases, they make headlines because of natural disasters, such as the 2005 earthquake in Pakistan. Political events have also put a spotlight on the region. In the fall of 2001, for example, the United States went to war in Afghanistan to fight the terrorists who planned the attacks of September 11, 2001.

Afghanistan

KEY QUESTION Why have empires throughout history wanted to rule Afghanistan?

Afghanistan is a **landlocked** country with no access to the ocean. However, it has a **strategic** location where Southwest Asia, Central Asia, and South Asia meet. The mountain pass that allows people to cross the Hindu Kush lies on the Afghanistan-Pakistan border. Because of this, any empire that ruled India usually also tried to control Afghanistan to prevent invasions. Afghanistan's foreign rulers have included Persians, Greeks, and Mongols.

because there is a supply of water from snow melt in mountains ▼

Afghanistan Farmland among the mountain valleys in Jabul Saraj, Afghanistan. **Why might mountain valleys be suitable for farming?**

729

2 Focus & Motivate

3-Minute Warm-Up

Write on the board or display the transparency:

- Why do you think India and Pakistan (East and West) formed two separate countries in 1947? (Possible Answer: because India was mostly Hindu while Pakistan was mostly Muslim)

Unit Transparency Book: Southwest Asia & South Asia
- 3-Minute Warm-Ups, TT28

Connecting to Your World

More About . . .

Afghanistan in the News Before 9/11/01

In March, 2001, Afghanistan's Taliban government destroyed the world's two largest figures of Buddha, which had been carved in cliffs about 1,400 years ago, well before the area had become predominantly Muslim. Most Muslim countries in the world tried to stop this destruction, as did the UN, but all were unsuccessful.

3 Teach

Afghanistan

Talk About It

- Why is Afghanistan's location considered strategic? *(It is where South Asia, Central Asia, and Southwest Asia meet.)*
- **Categorize** What aspect of Afghan culture unites various ethnic groups? *(All share the Muslim faith.)*

SECTION 1 PROGRAM RESOURCES

Unit Resource Book: Southwest Asia and South Asia

ON LEVEL
Unit Resource Book
- Reading Study Guide, p. 179
- Vocabulary Practice, p. 195
- Skillbuilder Practice: Comparing, p. 196
- Map and Graph Skills, pp. 197–198
- Section Quiz, p. 209

STRUGGLING READERS
Unit Resource Book
- Reading Study Guide with Additional Support, p. 183
- Section Quiz, p. 209
- Reteaching Activity, p. 211

eEdition with Audio DVD-ROM

ENGLISH LEARNERS
Pupil Edition in Spanish
eEdition with Audio DVD-ROM
Unit Resource Book
- RSG (Spanish), p. 187
- RSG with Additional Support (Spanish), p. 191

Multi-Language Glossary
Test Generator
- Section Quiz in Spanish

INCLUSION
Unit Resource Book
- RSG with Additional Support, p. 183
- Section Quiz, p. 209
- Reteaching Activity, p. 211

GIFTED & TALENTED
Unit Resource Book
- History Makers: Hamid Karzai, p. 201
- Primary and Secondary Sources, p. 208
- Section Quiz, p. 209

PRE-AP
Unit Resource Book
- Interdisciplinary Projects, pp. 203, 205
- Economics in the World, p. 206
- Section Quiz, p. 209

TECHNOLOGY
Unit Transparency Book
- 3-Minute Warm-Ups, TT28
- Essential Question, TT29
- Critical Thinking, TT30
- Map, TT32
- Cultures: Playing the Drums in Pakistan, TT34

Daily Test Practice Transparencies
- Section 1, TT81

Power Presentations
ClassZone.com
World Cultures and Geography Video on DVD

More About . . .

Results of the Soviet Invasion of Afghanistan

In response to the invasion of Afghanistan by the Soviet Union, U.S. president Jimmy Carter proposed and enforced a U.S. boycott of the 1980 Olympics, which were held in Moscow. When the Soviet Union finally withdrew troops after a ten-year war, its government was severely weakened from the economic drain of the war—and from internal opposition to the war by some of the Soviet people.

Some experts claim that the great loss of lives and money in Afghanistan was a major factor in the fall of the Soviet Union in 1991, just two years after its troops were withdrawn.

RESEARCH & WRITING CENTER

ANALYZING Primary Sources

There was some concern about the Afghan election in October 2004 because of the threat of violence. However, Afghanistan's elections were generally peaceful, with 18 candidates competing. Although there were some problems with ensuring that people only voted once, international observers affirmed the election as fair.

- What was the result of the fight against terrorism in Afghanistan at the time of Hamid Karzai's election? *(Terrorism had been reduced to a lesser threat in Afghanistan.)*
- How did Karzai feel about the need for international cooperation? *(He felt it was necessary in fighting terrorism and in rebuilding Afghanistan.)*

DOCUMENT–BASED QUESTION ANSWER

He probably feels grateful because international cooperation was needed to rid Afghanistan of terrorist rule and is needed for greater peace, stability, and security.

Unit Resource Book: Southwest Asia & South Asia

- History Makers: Hamid Karzai, p. 201
- Primary and Secondary Sources: from Hamid Karzai's inaugural speech, p. 208

History and Government During the 1800s, Russia and Great Britain fought over Afghanistan. Britain wanted to protect its colony in India; Russia wanted a warm-water port and was trying to extend its empire to the Indian Ocean. Britain won the conflict and took partial control of Afghanistan. The country did not win its independence from Britain until 1919.

In 1979, the **Soviet Union**, a Communist nation that included Russia, invaded Afghanistan. Many Afghans rejected Communist beliefs and fought the invaders with aid from the United States. By 1989, the Soviets had withdrawn. Local groups fought for power until the **Taliban**, a conservative Islamic group, took over Afghanistan.

The Taliban allowed al-Qaeda (ahl•KY•dah)—the group that carried out the terrorist attacks of September 11, 2001—to train in Afghanistan. In late 2001, the United States invaded the country, suppressed the Taliban, and bombed al-Qaeda's bases. The United States and other countries helped the Afghans form a government. In 2004, Hamid Karzai (hah•MEED KAHR•zy) was elected president.

ONLINE PRIMARY SOURCE To read the entire speech of Hamid Karzai, go to the **Research & Writing Center** @ClassZone.com

ANALYZING Primary Sources

Hamid Karzai (born 1957) became the first elected president in Afghanistan's history in 2004. This quotation comes from the speech he gave when he was sworn into office.

> Our fight against terrorism is not yet over, even though we have succeeded to reduce this common enemy of humanity to a lesser threat in this country. . . . A decisive victory over terrorism requires serious and continued cooperation at regional and international levels.
>
> Three years ago, . . . international cooperation rid Afghanistan from the rule of terrorism. The same cooperation has led to the rebuilding of the Afghan state, and significant progress in restoring peace, stability, and security to our country.
>
> Source: Inaugural Speech by President Hamid Karzai

DOCUMENT-BASED QUESTION

How do you think Karzai feels about the continued presence of the U.S. troops in his country?

INTERDISCIPLINARY ACTIVITIES

CONNECT to Art

Illustrate the Strategic Location of Afghanistan

Ask students to review the importance of Afghanistan's location throughout history. Then provide students with an outline map of the area. Ask them to label the Hindu-Kush pass, show how invaders might have used the pass, and explain why countries fought over Afghanistan in the 1800s. Have students review each other's maps.

CONNECT to Language Arts

Journal Entry for the Soviet Invasion of Afghanistan

Ask students to visualize life in Afghanistan when the Soviets invaded in 1979. Then have students write a journal entry about the invasion. Suggest that some students write about the invasion from the point of view of those who support it while others write from the point of view of those who oppose it. Invite students to present or post their entries.

Animated GEOGRAPHY

Land Use in South Asia

Click here to learn more about land use and resources in South Asia @ ClassZone.com

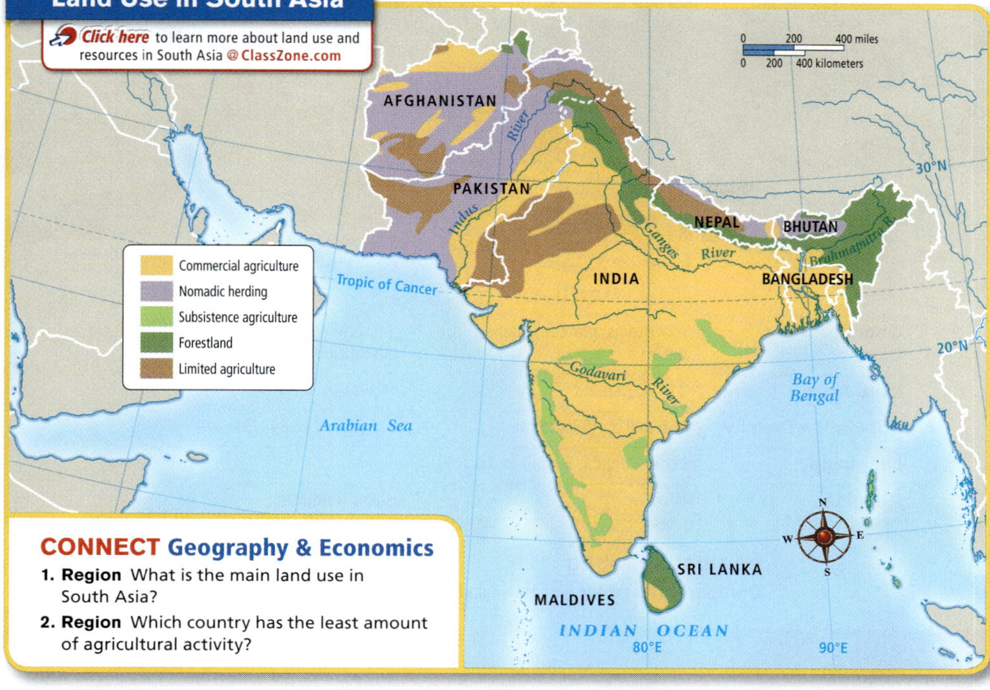

CONNECT Geography & Economics

1. **Region** What is the main land use in South Asia?
2. **Region** Which country has the least amount of agricultural activity?

Economics and Culture Afghanistan is one of the world's poorest countries. About 85 percent of Afghans live by **subsistence farming**, which means they raise only enough food to feed their families. The long years of war severely damaged the economy and prevented the development of industry. Since the fall of the Taliban, foreign nations have sent aid to Afghanistan, and the economy is starting to grow.

Both Islam and history strongly influence Afghan culture. The Afghan people belong to many ethnic groups and speak several languages, but almost all share the Muslim faith. Many rural people live in homes made of mud bricks and dress in traditional clothing. Few people can read, so folk tales, folk songs, and dances are important art forms. A popular sport is competitive kite flying, in which contestants try to cut down each other's kites by using strings coated with ground glass. Another sport is *buzkashi*, which dates from Mongol times. In it, horsemen compete to grab an animal carcass and drag it across a goal line.

ANALYZE CAUSES Explain why empires have wanted to rule Afghanistan. It has a strategic location where three regions meet and also has the mountain pass that provides access to India.

DIFFERENTIATING INSTRUCTION

English Learners

Key Academic Vocabulary

Note the word *traditional* in the third sentence of the second paragraph, to describe Afghan clothing. Read the sentence out loud and ask students to identify the root word *(tradition)* and suggest definitions. Then ask students to look up *tradition* and *traditional* in a dictionary. You may want to discuss how the words might also be used to describe modern American culture.

Gifted & Talented

Describe the Importance of Literacy

Ask students to review the paragraph describing Afghan culture and to note the importance of folk songs and tales. Discuss the importance of literacy and how written language allows people to share cultural traditions. Then have students write a short paragraph, comparing the importance of folk songs and tales in Afghan culture with their importance in American culture.

Animated GEOGRAPHY

Land Use in South Asia

CLASSZONE.COM

CONNECT Geography & Economics

Remind students that agricultural activity is greatly dependent on physical geography and climate. Ask students to mentally review what they know of the physical geography and climate of the region as they analyze this map.

- What geographic feature corresponds to the narrow band of limited agricultural activity along the northern edge of the map? *(the Himalaya Mountains)*
- Which country on the map appears to have the highest percentage of commercial agriculture? *(India or Bangladesh)*

ANSWERS

1. **Region** agriculture
2. **Region** Nepal

Unit Resource Book: Southwest Asia & South Asia

- Map and Graph Skills: Reading a Population Pyramid, pp. 197–198
- Interdisciplinary Projects: Living Standards in South Asia, pp. 203, 205

Presentation Options

Use the **Power Presentations** DVD-ROM to project the animated map.

- Click on the interactive key to activate the resources of South Asia.
- Study the pop-ups to learn more about land use in South Asia.

More About . . .

Kite Flying in Afghanistan

Two people fly a kite in the competitions. It takes many hours to make the kite itself, the coated wire, and the reel. The Taliban banned kite-flying, but kites were sold in secret throughout the country. Once the string of a kite is cut, it is carried away by the wind—and can be claimed by whoever finds it.

COMPARING

Pakistan and Bangladesh

Point out that Pakistan and Bangladesh are different in nearly every aspect on the chart. Note that although their total population is relatively similar (166 and 147 million, respectively), their population density is very different. In fact, Bangladesh is one of the world's most densely-populated countries.

- How do the photographs highlight the differences in geography and climate? *(Pakistan shows a high mountain climate, which is dry and cold, while Bangladesh shows a hot, wet lowland climate.)*

CRITICAL THINKING ANSWER

Compare Both practice Islam.

Unit Resource Book: Southwest Asia & South Asia
- Skillbuilder Practice: Comparing, p. 196

Teach

Pakistan and Bangladesh

Reader, Recorder, Reporter

- Why did the people of East Pakistan fight to become the independent country of Bangladesh? *(They resented being ruled by people so far away and who were so different.)*
- How are the cultures of Pakistan and Bangladesh similar? *(Both are dominated by Islamic practices and ideas.)*
- **Categorize** How does the amount of rainfall affect the agriculture of Pakistan and Bangladesh? *(Both rely on farming, but Pakistan is drier and grows wheat while Bangladesh, which often gets too much rain, grows rice.)*

Unit Resource Book: Southwest Asia & South Asia
- Economics in the World: Comparing South Asian Economies, p. 206

COMPARING Pakistan and Bangladesh

Baltoro Glacier

Karnaphuli River

	PAKISTAN	BANGLADESH
Geography & Climate	mountains, plateau, plains, desert	flat, low-lying plain, humid climate, monsoons
Land Area	310,403 sq. mi.	55,600 sq. mi.
Population Density	166 million (535/sq. mi.)	147 million (2,644/sq. mi.)
Language	Urdu, Punjabi, Sindhi, Pashto	Bengali
Religion	Islam	Islam, Hinduism

CRITICAL THINKING

Compare What is the biggest similarity between Pakistan and Bangladesh?

Pakistan and Bangladesh

KEY QUESTION Why did Bangladesh break away from Pakistan?

Pakistan and Bangladesh both have large populations. In 1947, they gained independence as one nation—Pakistan—that had two different regions separated by 1,000 miles of Indian territory.

Independence and Challenges Although they had the religion of Islam in common, West Pakistan and East Pakistan were very different in terms of culture, politics, and economics.

The capital of the country was located in West Pakistan, which controlled the government, army, and economy. The people of East Pakistan resented being ruled by people who were so far away and so different. In 1970, a cyclone and storm surge hit East Pakistan, killing more than 300,000 people. The survivors believed that West Pakistan delayed sending help. In 1971, a civil war broke out. East Pakistan won and became the parliamentary democracy of Bangladesh.

DIFFERENTIATING INSTRUCTION: TIERED ACTIVITIES

OBJECTIVE Chart important events and current challenges of Afghanistan, Pakistan, or Bangladesh.

Basic

Have student groups create a chart for Afghanistan that lists important:
- historical events
- recent events
- challenges facing the country today

Have students present their completed charts.

On Level

Adapt the Basic activity so that students' charts are on Pakistan. In addition, have students include historical figures from Pakistan's history on their charts. Have students present or display their completed charts.

Challenge

Adapt the Basic activity so that students' charts are on Bangladesh. In addition, have students include historical figures from Bangladesh's history on their charts and at least three solutions to meet the country's challenges. Have students present or display their completed charts.

Pakistan has had several periods of military rule and has struggled to develop democracy. One problem facing the Pakistani government today is the threat of terrorist groups. Like Afghanistan, much of Pakistan is mountainous. Pakistani officials are trying to find members of al-Qaeda and the Taliban hiding in those mountains.

Culture and Developing Economies Islam has shaped the cultures of both Pakistan and Bangladesh. Men and women have little contact with each other outside their families, and many marriages are arranged. A major holiday is **Ramadan**, a month when Muslims **fast**, or avoid eating, each day from sunrise to sunset.

Because Pakistan has a dry climate, its farmers rely on irrigation from its rivers. Pakistan underwent a **Green Revolution** that improved yields. Wheat is the main farm product. Pakistan is trying to develop its natural gas reserves. Bangladesh's greatest resource is fertile soil from silt deposited by rivers. Bangladesh often has too much water, which floods fields. Rice is the main crop. Bangladesh has **cottage industries**, in which people weave cloth or make crafts at home.

he people of ast Pakistan sented being led by a peo- e who were 000 miles vay and very fferent cul- rally. ▸

Cottage Industry A young girl makes matchboxes in Khulna, Bangladesh.

SUMMARIZE Explain why Bangladesh broke away from Pakistan.

Section 1 Assessment

ONLINE QUIZ For test practice, go to **Interactive Review** @ClassZone.com

TERMS & NAMES

1. **Explain the importance of**
   - Taliban
   - Ramadan

USE YOUR READING NOTES

2. **Categorize** Use your completed chart to answer the following question:
   Although the countries of Afghanistan, Pakistan, and Bangladesh differ in many ways, what binds them together?

	AFGHANISTAN	PAKISTAN	BANGLADESH
History			
Government			
Culture			
Economics			

KEY IDEAS

3. Why did many Afghans rebel against Soviet rule?
4. How do Muslims observe the holiday of Ramadan?
5. How have Bangladesh's rivers both helped and harmed the country?

CRITICAL THINKING

6. **Make Inferences** Why is the United States trying to help build a democratic government in Afghanistan?
7. **Identify Solutions** How might West Pakistan have prevented civil war with East Pakistan?
8. **CONNECT to Today** Why do you think the Pakistani government might be willing to help the United States capture terrorists who are hiding in the mountains of Pakistan?
9. **WRITING** **Write an Analysis** Write a paragraph explaining the relationship between natural disasters and economic growth in Bangladesh.

CONNECT to the Essential Question

How do the history and cultures of India's neighbors affect the world?

Ask students what they have learned so far that can help them answer this question. Students might mention:

- Afghanistan's strategic location has made it the site of many armed conflicts.
- Afghanistan, Pakistan, and Bangladesh share an Islamic culture, and all face challenges in developing modern economies.

4 Assess & Reteach

Assess Have students complete the Section Assessment.

Unit Resource Book: Southwest Asia & South Asia
- Section Quiz, p. 209

Interactive Review @ClassZone.com
- Section Quiz

Power Presentations

Test Generator

Reteach Have student pairs create one multiple-choice question for Afghanistan, for Pakistan, and for Bangladesh. Then have pairs trade questions and answer them. For any incorrect answers, have students find the correct answers, using their textbooks.

Unit Resource Book: Southwest Asia & South Asia
- Reteaching Activity, p. 211

SECTION 1 ASSESSMENT ANSWERS

Terms & Names

1. Taliban: conservative Muslim group that took control of Afghanistan in mid 1990s; Ramadan: month during which Muslims fast

Use Your Reading Notes

2. They differ in ethnicity, government, resources, climate, but are linked by Islam.
   Diagram *Afghanistan: fought over by Russia and Britain; new government after Taliban; Muslim; subsistence farming. Pakistan: gained independence in 1947 from India; military rule, struggling democracy; Muslim; wheat, natural gas. Bangledesh: in 1971 split from Pakistan; democracy; Muslim; rice, cottage industries.*

Key Ideas

3. They rejected the Soviet belief in communism.
4. They fast from sunrise to sunset for a month.
5. They deposit silt, creating fertile soil, but they also flood and ruin fields.

Critical Thinking

6. The United States is a democracy and believes that is the best form of government; also, U.S. officials may believe that a democratic government will not support terrorist bases.
7. West Pakistan might have done more to share power with East Pakistan, and it might have sent disaster relief more quickly in 1970.
8. The Pakistani government may fear that the terrorists will try to take over Pakistan.
9. Use the rubric to score analyses.

Analysis Rubric

	Structure	Content
4	logical, flowing	shows thorough understanding
3	clear	shows good understanding
2	not very clear	shows basic understanding
1	weak, unclear	little or no understanding

1 Plan & Prepare

Objective

Compare and contrast the causes, effects, and responses to natural disasters in Pakistan and Bangladesh.

Springboard Ask students to identify the population of their town or city. Then ask them to compare the deaths of 75,000 and 300,000 as listed in the text. (For example, 75,000 deaths is equal to five times an entire town with a population of 15,000, or 300,000 deaths is equal to half of an entire city of 600,000.)

2 Teach

Disaster Responses

Talk About It

- How can destroyed infrastructure endanger people? *(makes transportation dangerous; prevents supplies from reaching area; pollutes water supplies)*
- How do the images of aid for each disaster illustrate the responses of local people? *(Possible Answer: The job of providing aid to people falls overwhelmingly on local people, more so than on outside help.)*
- **Summarize** What is the basic reason why Pakistan and Afghanistan are prone to earthquakes? *(Two landmasses collided many years ago and their plates are still pushing against each other.)*

More About . . .

A History of Natural Disasters in Bangladesh

Bangladesh has been hit with many natural disasters. Besides the winds of cyclones and the floods from monsoon rains, Bangladesh has been struck by severe drought and epidemics as well. A drought in 1943 was so widespread that almost two million people died as a result. An epidemic in the early part of the 20th century killed nearly 400,000 more.

COMPARING Disaster Responses

Different parts of South Asia are vulnerable to different natural disasters. Pakistan and Afghanistan are prone to earthquakes. India and Bangladesh are prone to floods.

Earthquakes

The Indian subcontinent became part of Asia when two great landmasses collided with each other. The Indian-Australian Plate and the Eurasian Plate are still pushing against one another, which makes South Asia prone to earthquakes. In October 2005, an earthquake devastated Pakistan, killing approximately 75,000 people.

Aid
Pakistani paramilitary work at a collapsed building in Islamabad in 2005.

Temporary Housing
Earthquake survivors live in a tent camp.

Aerial view shows collapsed buildings in the village of Bagh, Pakistan, after the October 2005 earthquake. ▼

Destroyed Infrastructure
Earthquake survivors in 2005 cross a damaged bridge over the Neelum River.

DIFFERENTIATING INSTRUCTION: TIERED ACTIVITIES

OBJECTIVE Write a letter describing one of the region's natural disasters.

Basic

Ask each student to review the images in the feature. Then have them write a one-paragraph letter to a friend, describing the scene in one of the photographs in the Comparing feature. Invite students to present or post their letters.

On Level

Ask each student to review the images in the feature. Then have them write a letter to a friend, as follows:

- the first paragraph describes the scene
- the second paragraph explains the help needed in response

Invite students to present or post their letters.

Challenge

Adapt the On Level activity so that students write a three-page letter. Their third paragraph should offer suggestions for how to limit or reduce damage from a similar natural disaster in the future. Ask students to present their letters.

Floods

There are two main causes of flooding in South Asia: monsoon rains and cyclones. During the monsoons, heavy rains can cause severe flooding. But the most extreme weather pattern of South Asia is the cyclone, a violent storm with fierce winds and heavy rain. Cyclones are most destructive in low-lying Bangladesh, where high waves can swamp large parts of the country. In 1970, a cyclone struck Bangladesh. The winds, rains, and floods claimed at least 300,000 lives in the 20th century's worst cyclone.

Aid

Bangladeshi policewoman pours drinking water for flood victims in August, 2004.

Temporary Housing

Bangladeshi flood victims take shelter along a highway.

▲ *Aerial view of town of Brahmanbaria, Bangladesh, shows damage from 2004 floods.*

Destroyed Infrastructure

Bangladeshis cross a railroad track damaged by floodwaters.

CRITICAL THINKING

1. **Make Inferences** What probably causes the most deaths in earthquakes and floods?
2. **Identify Problems and Solutions** What different methods might be used to deal with earthquakes and floods?

INTERDISCIPLINARY ACTIVITIES

CONNECT to Science

Analyze Earthquakes

Earthquakes actually happen all over the United States, although not usually with the severity of the quakes that have caused great damage in California in recent years. Discuss the underlying causes of earthquakes (the continual shifting of tectonic plates) and how these same forces contribute to the creation of the earth's landscape, through mountain ranges and hills.

CONNECT to Science

Chart Local Weather Patterns

Have the class discuss the weather patterns of your area. Do these patterns cause large-scale damage, threaten crops, business, and so on? What can people do to help prevent damage? After the discussion, have students create a chart that reflects the key points of your discussion. To extend this activity, have students do further research, and/or invite a local emergency-preparedness professional to speak with your class.

Comparing Regions

The Effects of Disasters

The United States is a very large country compared to Bangladesh and Pakistan. The U.S. population is currently about double that of Pakistan and of Bangladesh as well. However, since 1900, the effects of natural disasters have been much worse in terms of deaths for the Asian countries.

For example, the United States has suffered about 40,000 deaths from natural disasters since 1900, while Bangladesh has suffered almost 3 million deaths, and Pakistan about 160,000.

One reason why Bangladesh is affected so severely by natural disasters is that it is very densely populated. Developed countries also have greater resources, so they are better able to predict problems before they happen, take steps to avoid injuries and deaths, and assist people afterward.

3 Wrap Up

Ask students to reread the Essential Question for the chapter. Discuss how natural disasters can occur anywhere on Earth and how weather patterns and geological events such as earthquakes are hard to predict—even with sophisticated technology.

CRITICAL THINKING ANSWERS

1. **Make Inferences** Possible Answer: from drowning in a flood or cyclone; from collapsing buildings in an earthquake; from lack of clean water and medical care after the disaster in both.
2. **Identify Problems and Solutions** Possible Answers: Strict building codes and not building in vulnerable areas, such as directly over a fault, might help prevent deaths from earthquakes; sturdy shelters built on stilts, levees, and dams might help prevent damage from floods.

1 Plan & Prepare

Objectives

- Compare the government, economy, culture, and religion of Nepal and of Bhutan
- Summarize the government, history, culture, government, and economy of Sri Lanka and of the Maldives

Read for the Essential Question

Help students read for a purpose by reminding them of the Essential Question: "How do the history and cultures of India's neighbors affect the world?" Point out that they can use text features such as Key Questions, special features, images, and maps to form answers to the question.

Vocabulary

Best Practices Toolkit

Use the ***Best Practices Toolkit*** to model strategies for vocabulary notetaking. Vary strategies throughout the year. Choose from: *Knowledge Rating, Predicting ABCs, Definition Mapping, Word Sort, Frayer Model (Word Squares), Magnet Words, Word Wheel*, and *Student VOC*.

Vocabulary Strategies, TT9–TT16

Reading Strategy

Best Practices Toolkit

Display the Web Diagram transparency.

- Explain to students that in categorizing they need to determine how the information in the section can be sorted into the categories shown in the diagram.
- Model filling in categories on a chart using the Nepal heading.
- Have students suggest entries, using the other categories on the diagram.
- Display the completed diagram for review.

Web Diagram TT28

Section 2 Reading for Understanding

Key Ideas

BEFORE, YOU LEARNED

India's larger neighbors have developed their own distinct cultures and identities.

NOW YOU WILL LEARN

Some of India's smaller neighbors have also developed in distinctive and unique ways.

Vocabulary

TERMS & NAMES

constitutional monarchy a government in which the powers of the king or queen are limited by the constitution

ecotourism travel to unique environments by people who take care to preserve them in their natural state

plantation agriculture the use of large farms to raise cash crops

REVIEW

subsistence farming agriculture that produces just enough to feed the farmer's family with little left over to sell

archipelago (AHR•kuh•PEHL•uh•GOH) group of islands

tsunami (tsu•NAH•mee) very large, destructive ocean waves caused by an underwater earthquake or volcanic eruption

Visual Vocabulary archipelago

Reading Strategy

Re-create the diagram shown at right. As you read and respond to the **KEY QUESTIONS**, use the outer ovals to organize important details about the economies of the mountain kingdoms and island nations.

Skillbuilder Handbook, page R7

CATEGORIZE

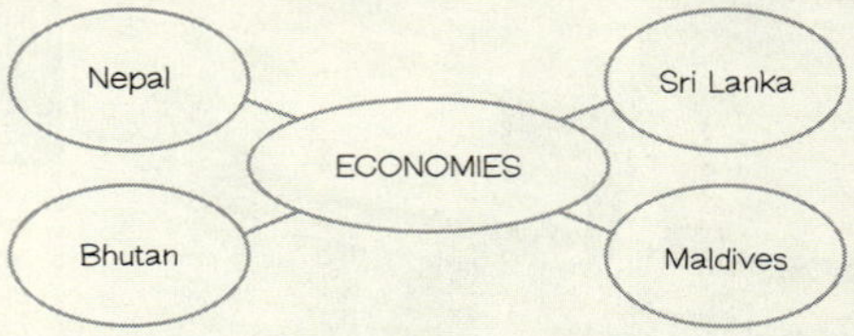

GRAPHIC ORGANIZERS Go to Interactive Review @ClassZone.com

PRETEACHING VOCABULARY

English Learners

Pronounce and Preview

Pronounce each term for students. Review the meanings of words within definitions, such as *constitution*, *unique environments*, and *destructive*.

- To modify vocabulary learning, have students complete worksheets as they read, instead of afterward.

Unit Resource Book: Southwest Asia & South Asia

- Vocabulary Practice, p. 195

Inclusion

Reading and Defining Terms Verbally

Have volunteers read one term and its definition aloud. Ask others to define the term in their own words verbally, and then use it in a sentence. Repeat for each term. As a group, discuss any words that are challenging to define or use in a sentence.

SECTION 2

FOCUS ON

Annapurna I in Nepal, one of the highest peaks in the world

Asian elephants in Sri Lanka

Mountain Kingdoms and Island Nations

Connecting to Your World

Imagine seeing a travel poster that shows a village of small stone houses built in a high valley among jagged, snowy mountains. Now, imagine a poster of a low-lying island with palm trees, sandy beaches, and lagoons. These two images will help you to understand the different types of countries you will learn about in this section.

Island Nation A holiday resort in the Maldives

Nepal and Bhutan

KEY QUESTION What are some of the main economic activities of the two countries?

Nepal and Bhutan are two landlocked kingdoms in the Himalayas. Because travel in the mountains is difficult, both countries are isolated. Few links to the outside world existed before the 1960s. However, Nepal and Bhutan have always been influenced by India and China.

Mountain Kingdom Village of Jharkot in Nepal

2 Focus & Motivate

3-Minute Warm-Up

Write on the board or display the transparency:

- What would you do if your island nation—such as the Maldives—was in danger of disappearing under rising ocean levels? *(Possible Answer: work to prevent global warming)*

Unit Transparency Book: Southwest Asia & South Asia
- 3-Minute Warm-Ups, TT28

Connecting to Your World

More About . . .

Geographic Differences—Elevations

The highest point in the Maldives is only about six feet. In Nepal, the highest peak in the Himalayas towers over 29,000 feet. The Kali Gandaki River gorge is generally considered the world's deepest canyon, with mountains rising over 16,400 feet above the water.

3 Teach

Nepal and Bhutan

Talk About It

- How do the rivers in Nepal and Bhutan help their economies? *(They generate hydroelectricity, which is exported.)*
- **Categorize** How would you categorize the governments of Nepal and Bhutan? *(They are both constitutional monarchies.)*

SECTION 2 PROGRAM RESOURCES

Unit Resource Book: Southwest Asia and South Asia

ON LEVEL

Unit Resource Book
- Reading Study Guide, p. 181
- Vocabulary Practice, p. 195
- Outline Map Activities, pp. 199–200
- Section Quiz, p. 210

STRUGGLING READERS

Unit Resource Book
- Reading Study Guide with Additional Support, p. 185
- Section Quiz, p. 210
- Reteaching Activity, p. 212

eEdition with Audio DVD-ROM

ENGLISH LEARNERS

Pupil Edition in Spanish
eEdition with Audio DVD-ROM
Unit Resource Book
- RSG (Spanish), p. 189
- RSG with Additional Support (Spanish), p. 193

Multi-Language Glossary
Test Generator
- Section Quiz in Spanish

INCLUSION

Unit Resource Book
- RSG with Additional Support, p. 185
- Section Quiz, p. 210
- Reteaching Activity, p. 212

GIFTED & TALENTED

Unit Resource Book
- Interdisciplinary Project, p. 204
- Section Quiz, p. 210

PRE-AP

Unit Resource Book
- Connect to Math, p. 207
- Section Quiz, p. 210

TECHNOLOGY

Unit Transparency Book
- 3-Minute Warm-Ups, TT28
- Essential Question, TT29
- Critical Thinking, TT31
- Map, TT33
- Cultures: Swayambbhunath, Nepal TT35

Daily Test Practice Transparencies
- Section 2, TT82

Power Presentations
ClassZone.com
World Cultures and Geography Video on DVD

CONNECT to Math

Note that the Himalayas are the highest mountain range in the world and contain a number of the world's tallest peaks. To extend this activity, have students also include a bar for the following peaks, which are the highest on their continents: Africa—Kilimanjaro, 19,341 feet; South America—Aconcagua, 22,834; North America—McKinley, 20,320; Europe—Elbrus 18,510.

Unit Resource Book: Southwest Asia & South Asia

- Outline Map Activities: Nepal and Bhutan: Physical/Political, pp. 199–200
- Connect to Math: Make a Bar Graph, p. 207

Fun Facts!

Thunder Dragon

Because of geographic barriers, little was known about Bhutan for much of its history. Foreigners considered it mysterious and referred to it by a variety of names before it became a formal, independent country: Hidden Holy Land, Southern Valley of Medicinal Herbs, and Lotus Garden of the Gods. It wasn't until a coronation ceremony in 1974 that members of the media and tourists were allowed to enter the country.

More About . . .

Ecotourism

Some regions with poor economies also contain unique physical or biological environments. However, large-scale tourism can harm the very environments that draw people in the first place—through increased vehicle traffic, building construction, and so on.

In Africa, Australia, and Central and South America, careful and limited ecotourism can generally avoid doing any lasting harm to rare natural environments. For example, people travel in the rain forest and view volcanoes in Costa Rica, help scientists track cheetahs in Namibia, and camp in the Australian Outback.

CONNECT to Math

A number of the highest mountains in the world are found in the Himalayas within the borders of Nepal. At left, climbers trek through the snow on a path near Mount Everest in Nepal.

Highest Mountain Peaks in the Himalayas	
Cho-Yo	26,906 feet
Everest	29,029 feet
Kanchenjunga	28,169 feet
Lhotse	27,940 feet
Makalu	27,766 feet

Source: www.welcomenepal.com

Activity: Make a Bar Graph

1. On a piece of notebook paper, rearrange the list of five mountains in order of height, with the tallest first.
2. Draw the horizontal axis of your graph. Along that axis, write the names of the five mountains from tallest to shortest.
3. Draw the vertical axis of your graph. Along that axis, mark intervals of 1,000 feet up to 30,000.
4. Choose a different colored pencil for each mountain. Make a bar to indicate height. Use the vertical axis to determine the correct height of the bar.

Government Nepal is a **constitutional monarchy**, in which a constitution limits the ruler's power. In recent years, Communist rebels have led uprisings against Nepal's government. In fighting the rebels, the king dissolved the government and began to rule alone. But early in 2006, protests forced him to restore democratic government.

Bhutan's monarchy has traditionally been a powerful one, with the monarch choosing the members of the legislature. However, in 2001, the king asked a commission to write a constitution, which was finished in 2005.

Fun Facts!

THUNDER DRAGON
The flag of Bhutan features a dragon, and in the Bhutanese language, the country is called "Land of the Thunder Dragon." Traditionally, the Bhutanese have believed the sound of thunder echoing through the mountains is the voice of dragons roaring.

Developing Economies Both Nepal and Bhutan are poor countries that rely on agriculture. In each country, the southern region near the Indian border has fertile lowlands that are mostly used for **subsistence farming**. The main crops are rice and wheat. Nepal also produces herbs and spices, while Bhutan produces rice, corn, cattle, and yaks. Both countries export hydroelectricity, generated by their fast-flowing mountain rivers. India is their main trading partner. In addition, both of the kingdoms benefit from tourism. Every year, climbers go to Nepal to try to reach the top of Mount Everest and other tall peaks. Bhutan promotes **ecotourism**, which occurs when people travel to unique environments but take care to preserve them.

Culture and Religion Each of the mountain kingdoms has a mixed population. The official language of Nepal is Nepali, which half the people speak. But Nepal is also home to ethnic groups that speak other languages. The two largest groups of Bhutan are the Bhote and the Nepalese. The official language is Dzongkha (DZAHNG•kah).

DIFFERENTIATING INSTRUCTION

Struggling Readers

Discuss Hydroelectricity and Exporting Energy

With students, review the sentence "Both countries export hydroelectricity . . ." in the "Developing Economies" subsection. Ask students to describe hydroelectricity and how it is produced. Discuss how mountains can hinder economic activity and how they can help it. Then have students explain the likely process involved in exporting electricity.

Pre-AP

Map Economic Activities

Ask students to review the paragraph on "Developing Economies." Then have each student use this information to prepare an economic-activity map for the two countries, with a legend and icons to illustrate different activities. Invite volunteers to display their work for the class.

India has influenced the cultures of both countries, especially their religions. Nepal is officially a Hindu country, but it also has a Buddhist minority of about 10 percent. Siddhartha Gautama, founder of Buddhism, was born in Nepal, and many Nepalese Hindus combine Buddhist practices with their Hindu faith. Bhutan, on the other hand, is a mostly Buddhist country with a Hindu minority of 25 percent. Nepal has many Hindu temples and Buddhist shrines, while Bhutan has Buddhist monasteries.

MAKE GENERALIZATIONS Describe the main economic activities of Nepal and Bhutan. *farming, tourism, the production of hydroelectricity*

Sri Lanka and the Maldives

KEY QUESTION What do Sri Lanka and the Maldives have in common?

Two of India's neighbors are island nations that are physically quite different from each other. Sri Lanka (formerly Ceylon) has one main island just 20 miles off the southeast coast of India. The Maldives is an **archipelago** of about 1,200 small islands that stretch over a distance of about 500 miles.

Religions of South Asia

CONNECT Geography & Culture

1. **Region** What other country besides India is mostly Hindu?
2. **Region** What is the predominant religion in Bhutan?

History, Culture, and Government People from India settled Sri Lanka more than 2,000 years ago. The Sinhalese, whose ancestors came from northern India, make up 74 percent of the population. Most Sinhalese are Buddhist. The next largest group is the Tamils, who originally came from southern India and who are mostly Hindu.

Starting in the 1500s, Sri Lanka was controlled by Europeans—first the Portuguese, then the Dutch, then the British. Sri Lanka gained independence in 1948 and set up a parliamentary government. Ethnic conflict is a serious problem there. Because the Sinhalese run the government, the Tamils believe that they are denied opportunities. A group called the Liberation Tigers of Tamil Eelam is fighting for a separate Tamil state in northern and eastern Sri Lanka. Civil war began in 1983 and continued until a cease-fire was signed in 2002. Since then, violence has resumed.

DIFFERENTIATING INSTRUCTION: TIERED ACTIVITIES

OBJECTIVE Organize information about mountain kingdoms and island nations.

Basic	On Level	Challenge
Review with students the usefulness, process, and format for making an outline. Then have pairs of students create an outline that organizes the key facts about the four countries in this section. Have pairs trade outlines and offer positive feedback and three suggestions for improvement.	In addition to the Basic activity, have students use their outlines to write a summary of this section. Have pairs trade outlines and summaries, and offer positive feedback and three suggestions for improvement.	Adapt the On Level activity so that students work individually to prepare their outlines. In place of a summary, have students use their outlines to create an essay. Have students trade outlines and essays, and offer positive feedback and three suggestions for improvement.

Religions of South Asia

CONNECT Geography & Culture

This map clearly shows the religious differences that led to the separation of India, Pakistan, and Bangladesh. It also shows the Buddhist areas in Bhutan and Nepal.

- Why do you think that Sri Lanka became predominantly Buddhist rather than Hindu like its nearest neighbor, India? *(Possible Answer: Most people in Sri Lanka are descendants of those who immigrated years ago from northern India, which was Buddhist at the time.)*

ANSWERS

1. **Region** Nepal
2. **Region** Buddhism

Unit Resource Book: Southwest Asia & South Asia
- Interdisciplinary Projects: Building a Buddhist Temple, pp. 204–205

Teach

Sri Lanka and the Maldives

Think, Pair, Share

- What European countries controlled Sri Lanka in the past? *(Portugal, the Netherlands, Britain)*
- Why does global warming threaten the Maldives? *(Possible Answer: The islands are low; if the ocean levels rise, the islands may be covered by water.)*
- **Categorize** What is the source of civil war and violence in Sri Lanka? *(The Sinhalese run the government and the Tamils believe they are denied equal opportunities.)*

Unit Resource Book: Southwest Asia & South Asia
- Skillbuilder Practice: Comparing, p. 196
- Geography Project: Understanding Destructive Storms, pp. 227–228

CONNECT to the Essential Question

How do the history and cultures of India's neighbors affect the world?

Ask students what they have learned so far that can help them answer this question. Students might mention:

- Nepal and Bhutan offer spectacular natural beauty and are working to improve their economies.
- Disasters such as the 2004 tsunami have had a huge economic impact on Sri Lanka and the Maldives.

Unit Resource Book: Southwest Asia & South Asia
- Section Quiz, p. 210

4 Assess & Reteach

Assess Have students complete the Section Assessment.

Unit Resource Book: Southwest Asia & South Asia
- Section Quiz, p. 210

Interactive Review @ClassZone.com
- Section Quiz

Power Presentations

Test Generator

Reteach Have student pairs create four "True or False" statements, one for each country in this section. Then have pairs trade statements and evaluate them. For each false statement, have students correct it, using their textbooks.

Unit Resource Book: Southwest Asia & South Asia
- Reteaching Activity, p. 212

Gone Fishing Stilt fishermen fish for sardines at high tide off a beach north of Weligama, Sri Lanka.

Most Maldivians are related culturally to the Sinhalese of Sri Lanka. However, the Maldivians are Muslim because traders brought Islam to the islands. In 1968, the country became an independent republic.

Economics of Plantations and Tourism The economy of Sri Lanka is based mainly on farming. The coastal areas have fertile plains that support **plantation agriculture**, or the growing of cash crops on estates. Sri Lanka is one of the world's leading producers of tea.

The geography of the Maldives affects its economy. Most of the islands are very small and rise no more than six feet above the ocean. The economy depends on fishing and tourism, as people visit for the lagoons, beaches, and warm climate.

Both nations suffered damage from the 2004 **tsunami**. Sri Lanka lost two-thirds of its fishing fleet and 99,000 homes. The Maldives were swamped, and property was destroyed. Tourism declined. Some geographers think that the Maldives may be in danger from global warming, a rise in world temperatures caused by pollution. If the polar ice caps melt, the oceans might rise and cover the Maldives.

Both are island nations; both were ruled by Europeans for about 400 years before independence ; both have a large Sinhalese population; both were damaged by the 2004 tsunami. ▶

COMPARE AND CONTRAST Explain what Sri Lanka and the Maldives have in common.

Section 2 Assessment

ONLINE QUIZ For test practice, go to Interactive Review @ClassZone.com

TERMS & NAMES

1. **Explain the importance of**
   - ecotourism
   - archipelago
   - plantation agriculture

USE YOUR READING NOTES

2. **Categorize** Use your completed chart to answer the following question:
   What are the main economic activities of Nepal, Bhutan, Sri Lanka, and the Maldives?

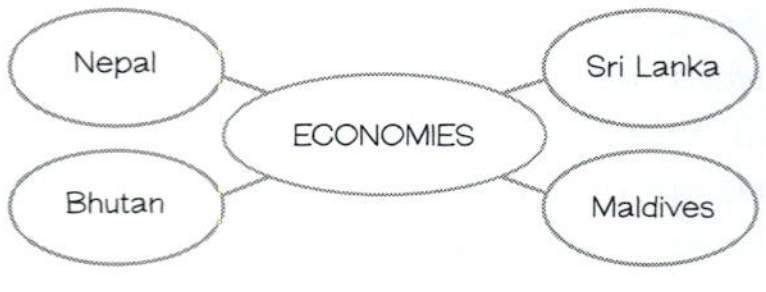

KEY IDEAS

3. Why did the king of Nepal disband the government and then later restore it?
4. What caused the civil war in Sri Lanka?
5. Why did the Maldives suffer so much damage from the 2004 tsunami?

CRITICAL THINKING

6. **Analyze Effects** How do you think the tsunami affected Sri Lanka's economy? Explain.
7. **Compare** How are the political problems in Nepal and Sri Lanka similar?
8. **CONNECT to Today** What are some rules that you think visitors might have to follow in ecotourism?
9. **TECHNOLOGY** **Plan a Multimedia Presentation** Plan a presentation on the tsunami damage to Sri Lanka and the Maldives. Be sure to include photographs, graphs or charts, and quotations or interviews.

SECTION 2 ASSESSMENT ANSWERS

Terms & Names

1. ecotourism: travel to unique environments by people who take care to preserve them in their natural state; archipelago: a group of islands; plantation agriculture: the use of large farms to raise cash crops

Use Your Reading Notes

2. Nepal and Bhutan: farming, hydroelectricity, tourism; Sri Lanka: plantation agriculture; Maldives: fishing and tourism
   Diagram *Economies: Nepal: farming, tourism; Bhutan: farming, tourism; Sri Lanka: agriculture; Maldives: fishing, tourism.*

Key Ideas

3. He took over the government to have more power to deal with the rebels but had to restore democratic government because of protests.
4. The Tamils believe the Sinhalese deny them opportunities, so a Tamil group is fighting to gain their own country.
5. The islands are only six feet above the ocean's surface, so they were swamped.

Critical Thinking

6. It caused an economic slowdown because much of the fishing fleet was destroyed and because the economy depends on farming the coastal plains, which were flooded.
7. Both governments are fighting rebel groups.
8. They are probably told not to leave trash, not to pick rare plants, not to harm animals, and not to use machinery that can cause pollution.
9. Use the rubric to score multimedia presentations.

Presentation Rubric

	Elements in Presentation	Organization of Presentation
4	includes all elements	well-organized, interesting
3	most elements	well-organized
2	missing elements	not well-organized
1	few elements	disorganized

Interactive Review

Click here to complete these and other activities online @ ClassZone.com

CHAPTER SUMMARY

Key Idea 1
Other nations and cultures in the region besides India have managed to develop their own distinct identities.

Key Idea 2
Some of India's smaller neighbors have also developed in distinctive and unique ways.

For **Review and Study Notes**, go to **Interactive Review** @ ClassZone.com

NAME GAME

Use the Terms & Names list to complete each sentence on paper or online.

1. I am a conservative Muslim group that ruled Afghanistan from the mid-1990s until 2002. Taliban
2. I am a small business in which the workers manufacture items in their homes. ______
3. I am a Muslim holiday in which people fast for a whole month. ______
4. I am a system of growing cash crops on large estates. ______
5. I am a Communist nation that existed from 1922 to 1991. ______
6. I am a series of ocean waves caused by an undersea earthquake or eruption. ______
7. I am the practice of visiting a region while taking care to preserve its environment. ______
8. I am a group of islands. ______
9. I produce just enough to feed a family with little left over to sell. ______
10. I refer to a region that has no access to the sea. ______

archipelago
constitutional monarchy
cottage industry
ecotourism
Green Revolution
landlocked
plantation agriculture
Ramadan
Soviet Union
subsistence farming
Taliban
tsunami

Activities

GeoGame

Use this online map to show what you know about South Asia. Click and drag each place name to its location on the map.

Crossword Puzzle

Complete the online crossword puzzle to test your knowledge of the history and cultures of South Asia.

ACROSS
1. first elected president of Afghanistan

741

ONLINE AND ON-SCREEN ACTIVITIES

GeoGame

Use the interactive map on the **Power Presentations** DVD-ROM or at **ClassZone.com** to help students review South Asia's geographic features and important places.

- Demonstrate how to drag and drop the map labels listed onto the appropriate locations on the map.
- Advance to the scorecard to show how students have a chance to improve their scores.

Crossword Puzzle

Students can review South Asia's history and culture by completing this crossword puzzle online at **ClassZone.com.**

- As an extension of this activity, have students use the *Activity Maker* at **ClassZone.com** to make their own crossword puzzle for one section of the chapter.
- Then have them trade puzzles with classmates to solve each other's puzzles.

Chapter Summary

As a follow-up to the chapter, encourage students to pair up and go over the **Review and Study Notes** at **ClassZone.com**.

Alternatively, student pairs can review their Reading Study Guide notes, making sure that they can add important details to each section's "Key Idea."

Name Game

Teaching Tips

ON PAPER Have students work in pairs to compete this activity. As an extension, they can write questions for terms that are not covered in the items shown here.

ONLINE Make sure that students understand how to drag and drop terms to fill in blanks. Ask them to print out their results and compare them with those of a classmate.

ANSWERS

1. Taliban
2. cottage industry
3. Ramadan
4. plantation agriculture
5. Soviet Union
6. tsunami
7. ecotourism
8. archipelago
9. subsistence farming
10. landlocked

Make It Fun

Teacher-Tested Activities

Tim Mortensen, Patrick Marsh Middle School, Sun Prairie, Wisconsin

Try a class Venn diagram activity to compare Sri Lanka and the Maldives.

- Divide students into groups.
- Give student groups five minutes to complete a Venn diagram that compares and contrasts the two countries.
- Groups with the most items will be considered the winner.

Vocabulary ANSWERS

1. Subsistence farming means farming that produces just enough to feed the farmer's family with nothing left over to sell.
2. An archipelago is a group of islands.
3. Ecotourism refers to travel to unique environments by people who take care to preserve them in their natural state.
4. Constitutional monarchy refers to a government in which the powers of the king or queen are limited by the constitution.
5. Plantation agriculture means the use of large farms to raise cash crops.
6. Ramadan is the ninth month of the Muslim calendar, during which Muslims fast from sunrise to sunset each day.
7. The Taliban is a conservative Muslim group that took control of Afghanistan in the mid 1990s.
8. Plantation agriculture means the use of large farms to raise cash crops for sale; subsistence farming means agriculture that produces just enough food to feed the farmer's family with nothing left to sell.
9. Archipelago means a group of islands surrounded by water; landlocked means surrounded by land with no access to the ocean.

Key Ideas ANSWERS

10. Islam
11. It is a meeting place of three regions and provides access to India, so many empires sought to control it.
12. The United States invaded Afghanistan, overthrew government, and helped set up new one; Pakistanis with U.S. support are working to find terrorists in their country.
13. earthquakes, cyclones
14. Green Revolution improved yields.
15. Communist rebels
16. by having a commission write a constitution
17. India has influenced the religions of both.
18. Sri Lanka relies on agriculture, and the Maldives on fishing and tourism.
19. to gain a Tamil nation on the island

Critical Thinking ANSWERS

20. Nepal and Bhutan Venn Diagram:
    - Nepal and Bhutan: both kingdoms, both in the Himalayas, both influenced by Indian religion, both rely on agriculture and tourism
    - Nepal only: mostly Hindu, king tried to increase power, fighting with Communist rebels, visited by mountain climbers

CHAPTER 24 Assessment

VOCABULARY

Explain the significance of each of the following.

1. subsistence farming
2. archipelago
3. ecotourism
4. constitutional monarchy
5. plantation agriculture
6. Ramadan
7. Taliban

Explain how the terms in each group differ.

8. plantation agriculture and subsistence farming
9. archipelago and landlocked

KEY IDEAS

1 Afghanistan, Pakistan, and Bangladesh

10. What cultural influence do Afghanistan, Pakistan, and Bangladesh share?
11. How has Afghanistan's location affected its history?
12. How has the U.S. war on terrorism affected Afghanistan and Pakistan?
13. What natural disasters are common in the region?
14. How does Pakistan meet its agricultural needs?

2 Mountain Kingdoms and Island Nations

15. Who are the rebels fighting the Nepalese government?
16. How has the king of Bhutan recently tried to change government?
17. What major influence has India had on Nepal and Bhutan?
18. What is the biggest economic difference between Sri Lanka and the Maldives?
19. What is the goal of the Liberation Tigers of Tamil Eelam?

CRITICAL THINKING

20. **Compare and Contrast** Create two Venn diagrams. Use one to compare and contrast Nepal and Bhutan. Use the other to compare and contrast Sri Lanka and the Maldives.

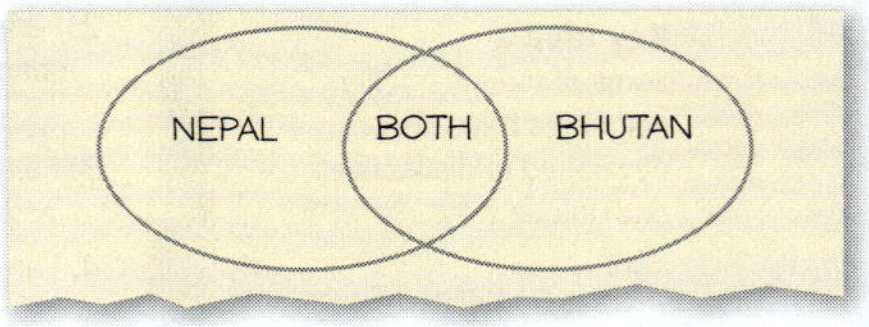

21. **Make Inferences** How was East Pakistan able to win independence from West Pakistan when the latter controlled the army?
22. **Compare and Contrast** How have the recent actions of the kings of Nepal and Bhutan differed?
23. **Connect to Economics** Would subsistence farming or plantation agriculture do more to develop a nation's economy?
24. **Five Themes: Region** What geographic reasons explain why India has been a strong influence on many countries?
25. **Connect Geography & Culture** What is the origin of the cultural differences between the Sinhalese and Tamil peoples?

Answer the ESSENTIAL QUESTION

How do the history and cultures of India's neighbors affect the world?

Written Response Write a three-paragraph response to the Essential Question. Be sure to give at least three examples of how countries in this region have affected recent world events. Use the rubric to guide your thinking.

Response Rubric
A strong response will:
- discuss natural disasters and their impacts
- examine political changes in the region
- analyze the region's role in war against terrorism

ESSENTIAL QUESTION REVIEW

Before students begin their essays, have them review the key chapter elements that will help them answer the Essential Question. Direct them to significant features in the chapter such as

- Animated Geography: Land Use in South Asia, Section 1
- Comparing Pakistan and Bangladesh, Section 1
- Comparing Disaster Responses, Section 1
- Religions of South Asia, Section 2

Written Response Rubric

	Coverage of Key Ideas	Explanation	Accuracy
4	all	clear, concise	no errors
3	most	good	few/minor errors
2	some	fair	several errors
1	two or fewer	poor	many errors

STANDARDS-BASED ASSESSMENT

Test Practice
- Online Test Practice @ClassZone.com
- Test-Taking Strategies and Practice at the front of this book

PHYSICAL MAP

Use the map to answer questions 1 and 2 on your paper.

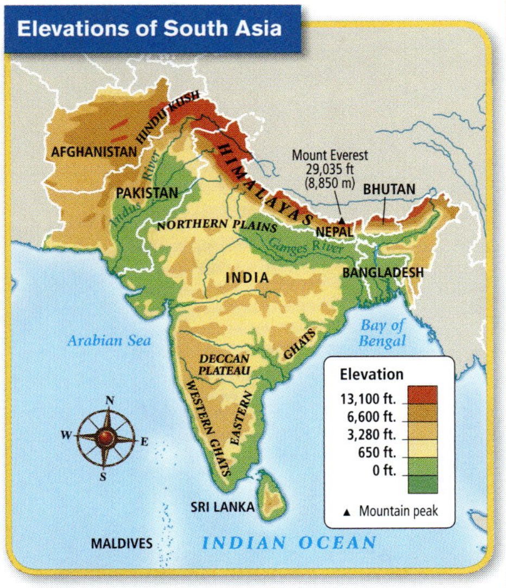

1. Which mountain range includes Mount Everest?

A. Hindu Kush
B. Western Ghats
C. Himalayas
D. Eastern Ghats

2. Which mainland country in South Asia has overall the lowest elevations?

A. Bangladesh
B. Afghanistan
C. Nepal
D. Bhutan

PIE GRAPHS

Examine the pie graphs below. Use the information in the graphs to answer questions 3 and 4 on your paper.

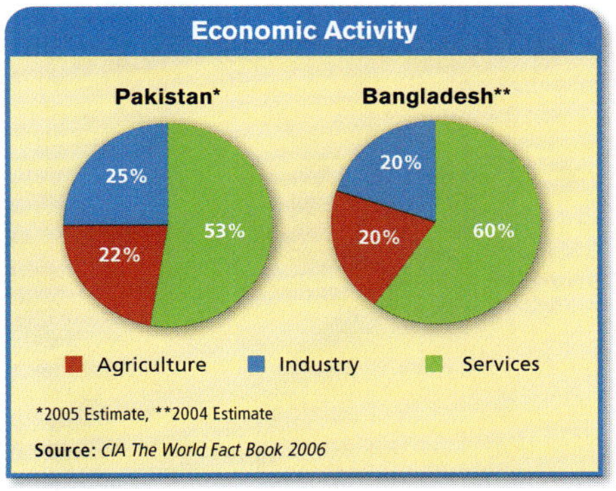

3. Which country is less industrialized?

4. Which economic sector in both countries employs the most people?

GeoActivity

1. INTERDISCIPLINARY ACTIVITY–SCIENCE

Using textbooks, encyclopedias, or the Internet, look up information about the possible causes of global warming and what might be done to prevent it. Create a poster that illustrates what individuals can do to deal with global warming.

2. WRITING FOR SOCIAL STUDIES

Unit Writing Project Revisit the plan of action that you developed as a lobbyist to support a cause. What other causes would you like to support? Develop a written plan of action for one or two of your favorite causes.

3. MENTAL MAPPING

Create an outline map of South Asia and label the following:

- Afghanistan
- Bangladesh
- Bhutan
- India
- the Maldives
- Nepal
- Pakistan
- Sri Lanka

DIFFERENTIATING CHAPTER ASSESSMENT

Unit Resource Book

- **Chapter Test A,** pp. 213–216. *Use for Struggling Readers, Beginning-to-Intermediate-Level English Learners, and for Inclusion of students whose IEP suggests supported test-taking.*
- **Chapter Test B,** pp. 217–220. *Use for On-Level Learners and ESOL students with Advanced-Level English.*
- **Chapter Test C,** pp. 221–224. *Use for Gifted and Talented students or students assigned Pre-AP work.*

Test Generator

- Chapter Tests A, B, and C, in English and Spanish

McDougal Littell Assessment System

TEST **Test Generator** Use three levels of tests in English and Spanish tied to state standards. Plus, choose questions to customize tests for all populations.

SCORE Scan and Score plain paper student tests in minutes.

REPORT Create instant progress reports, sortable by student, by standard, or other criteria.

RETEACH Respond immediately to students' needs by using a wealth of reteaching resources.

- Bhutan only: mostly Buddhist, king is trying to give up power to constitutional government, promotes ecotourism

Sri Lanka and the Maldives Venn Diagram:

- Sri Lanka and the Maldives: both island nations, both have large Sinhalese population, both devastated by tsunami
- Sri Lanka only: large Tamil minority, Sinhalese are Buddhist and Tamils are Hindu, long civil war, plantation agriculture
- Maldives only: Muslim, in danger from global warming

21. The two were separated by 1,000 miles, which made it difficult for West Pakistan to maintain control.

22. Possible answer: The king of Nepal tried to seize more power, while the king of Bhutan is promoting constitutional government, which will likely limit his power.

23. Possible answer: Plantation agriculture develops an economy more because it provides a living for more than just an individual farmer's family, and it also has a better chance of producing extra income that can be used to grow the economy.

24. Possible answer: It is a much larger, more populous country, and it is centrally located among Pakistan, Nepal, Bhutan, Bangladesh, Sri Lanka, and the Maldives.

25. Possible answer: Sinhalese originated in northern India, and Tamils in southern India.

Standards-Based Assessment ANSWERS

1. C
2. A
3. Bangladesh
4. services

GeoActivity

1. INTERDISCIPLINARY ACTIVITY RUBRIC
The poster should
- reflect thorough research
- be neat, organized, and attractive

2. WRITING FOR SOCIAL STUDIES RUBRIC
The Unit Writing Project should
- develop a plausible plan of action
- be well organized and practical

3. MENTAL MAPPING RUBRIC
The outline map should
- represent the shape of South Asia
- include all geographic features listed in their correct location

N
E
S
W

UNIT 8
East Asia and Southeast Asia

Why It Matters:

East Asia is home to several countries that are important members of the world economy. The nations of Southeast Asia are working hard to strengthen their economies. In the coming decades, both regions are expected to have a strong voice in world affairs.

Forbidden City in Beijing

CHAPTER 25 **China**

Blend of old and new in Seoul, South Korea

CHAPTER 26 **The Korean Peninsula**

Floating Shinto gate near Hiroshima

CHAPTER 27 **Japan**

Mosque and Petronas Towers in Kuala Lumpur, Malaysia

CHAPTER 28 **Southeast Asia**

745

Objectives

- Identify the political and physical features of East Asia and Southeast Asia
- Describe the climates found in East Asia and Southeast Asia
- Discuss the major economic activities in East Asia and Southeast Asia
- Locate areas of high and low population density in East Asia and Southeast Asia

Comparing Data

East Asia and Southeast Asia are about twice the size of the continental United States in landmass, but they have more than six times the number of people. The mainland of Asia has more than 50 mountains that rise over 20,000 feet. Many of these are part of the Himalayas, the tallest mountain range in the world. Mt. McKinley in Alaska is the only mountain in the United States over 20,000 feet.

ANSWERS

1. East Asia and Southeast Asia, because these regions have many more people—about six times as many as the United States—in an area that is only about twice as large as the United States
2. Chang Jiang; Mount Kongur

UNIT 8 ATLAS

East Asia and Southeast Asia

Use the Unit Atlas to learn about the geography of East Asia and Southeast Asia. As you study the maps in the atlas, notice geographic patterns and specific details about the two regions. Answer the questions on each map in your notebook.

As you study the graphs on this page, compare the landmass, population, rivers, and mountains of East Asia and Southeast Asia with those of the United States. Then jot down the answers to the following questions in your notebook.

Comparing Data

1. Compare the landmass and population of East Asia and Southeast Asia with those of the United States. Based on these data, which region do you think has a higher population density? Why?
2. What is the second longest river shown here? Which East Asian mountain is taller than Mount McKinley?

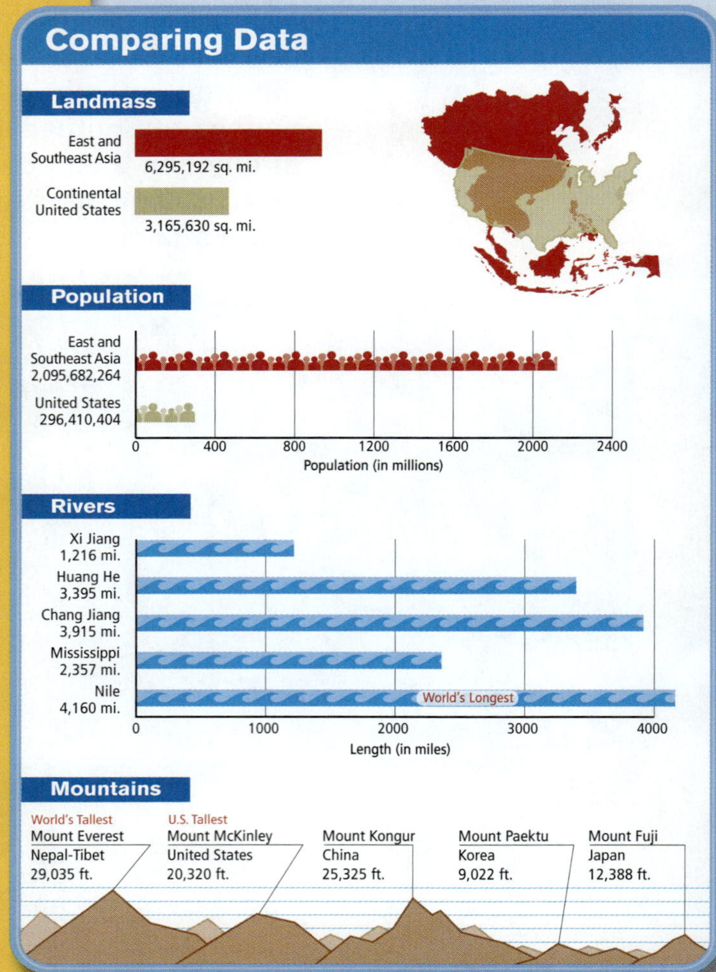

FIVE THEMES AND SIX ELEMENTS

Five Themes

Location

The Chang Jiang is China's most important river. Have students locate it on the political and physical map. Have them trace its origin in the mountains of western China and its path to the East China Sea. Point out that it is a major shipping route. Help students see that many of China's largest cities, such as Shanghai, are located on or near it.

Six Elements

The World in Spatial Terms

China was an important influence on the ancient cultures of many other East Asian and Southeast Asian countries. Have students study the physical relationship of China to its East Asian and Southeast Asian neighbors. Discuss the land and water routes that would have been used to travel from China to these countries.

Political and Physical

THINK LIKE A GEOGRAPHER

1. **Region** What is the biggest country in East Asia?
2. **Location** Which countries in East Asia and in Southeast Asia border China?

747

DIFFERENTIATING INSTRUCTION

English Learners

Identify Map Characteristics

Review the characteristics of a physical map (elevation, landforms, bodies of water) and a political map (political boundaries, national capitals, and other cities). Then have students find examples of both on this map. Invite volunteers to identify countries of their family or cultural heritage.

Struggling Readers

Locate the Countries of the Two Regions

Help struggling readers distinguish the countries of East Asia from the countries of Southeast Asia. Focus attention on the inset map, which shows the division between the two regions. Then point out the countries of East Asia and Southeast Asia on the larger map. Draw a circle around each region and have students locate and identify its countries.

Political and Physical

China's location helps explain in part the cultural and historical influence that it has had on other countries in the region. For example, China conquered the northern half of Korea in 108 B.C. Though driven out by native Korean tribes in the A.D. 300s, China continued to assert its influence until 1910. Over the centuries, the Chinese introduced Confucianism and Buddhism to Korea and had a major impact on its art and language.

- What river forms part of the border between North Korea and China? *(Yalu River)*
- What river in Southeast Asia flows along three major cities? What are those major cities? *(Mekong River; Vientiane, Phnom Penh, Ho Chi Minh City)*
- Describe the change in elevation as one moves west to east across China. *(Mountains and plateaus give way to lower elevations, such as plains.)*

THINK LIKE A GEOGRAPHER ANSWERS

1. **Region** China
2. **Location** North Korea, Vietnam, Laos, Myanmar

More About . . .

China's Influence on Japan

Japan came into contact with Chinese civilization as early as the 100s B.C. During a period extending to about A.D. 800, the Japanese incorporated a variety of Chinese influences into their culture, including agricultural techniques and management, forms of government, and religion. Today, Chinese influence can still be seen in Japan's art and architecture. Buddhism, introduced in the A.D. 500s, is one of Japan's two major religions.

Climate

Point out that East Asia is characterized by a variety of climates that reflect the diverse nature of its geographic features. For example, the deserts in western and northern China are fiercely hot in the summer but bitterly cold in the winter, while southeastern China has humid summers and mild winters. In contrast to East Asia, the climate of Southeast Asia is relatively uniform, with most countries in tropical climate zones.

- What is the climate of Cambodia? *(tropical wet and dry)*
- How does South Korea's climate differ from that of North Korea? *(South Korea has both humid subtropical and humid continental climates, while North Korea has a humid continental climate.)*
- How many climate zones does China have and what are they? *(seven; subarctic, humid continental, semiarid, desert, highland, humid subtropical, tropical wet)*

THINK LIKE A GEOGRAPHER ANSWERS

1. **Place** tropical wet and tropical wet and dry
2. **Region** The inland climate is generally drier.

Climate

THINK LIKE A GEOGRAPHER

1. **Place** What are the two main climates in Southeast Asia?
2. **Region** How does East Asia's inland climate compare with that along its coasts?

FIVE THEMES AND SIX ELEMENTS

Five Themes

Place

Review the characteristics of a highland climate. Then have students identify on the map the places in East Asia and Southeast Asia that experience this type of climate and discuss the reasons why.

Six Elements

Environment and Society

Have students discuss how the climate in different regions might affect the people who live there. For instance, you might contrast the lifestyles and challenges faced by people who live in semiarid areas of China with those of people who live in tropical wet Singapore.

Economic Activity

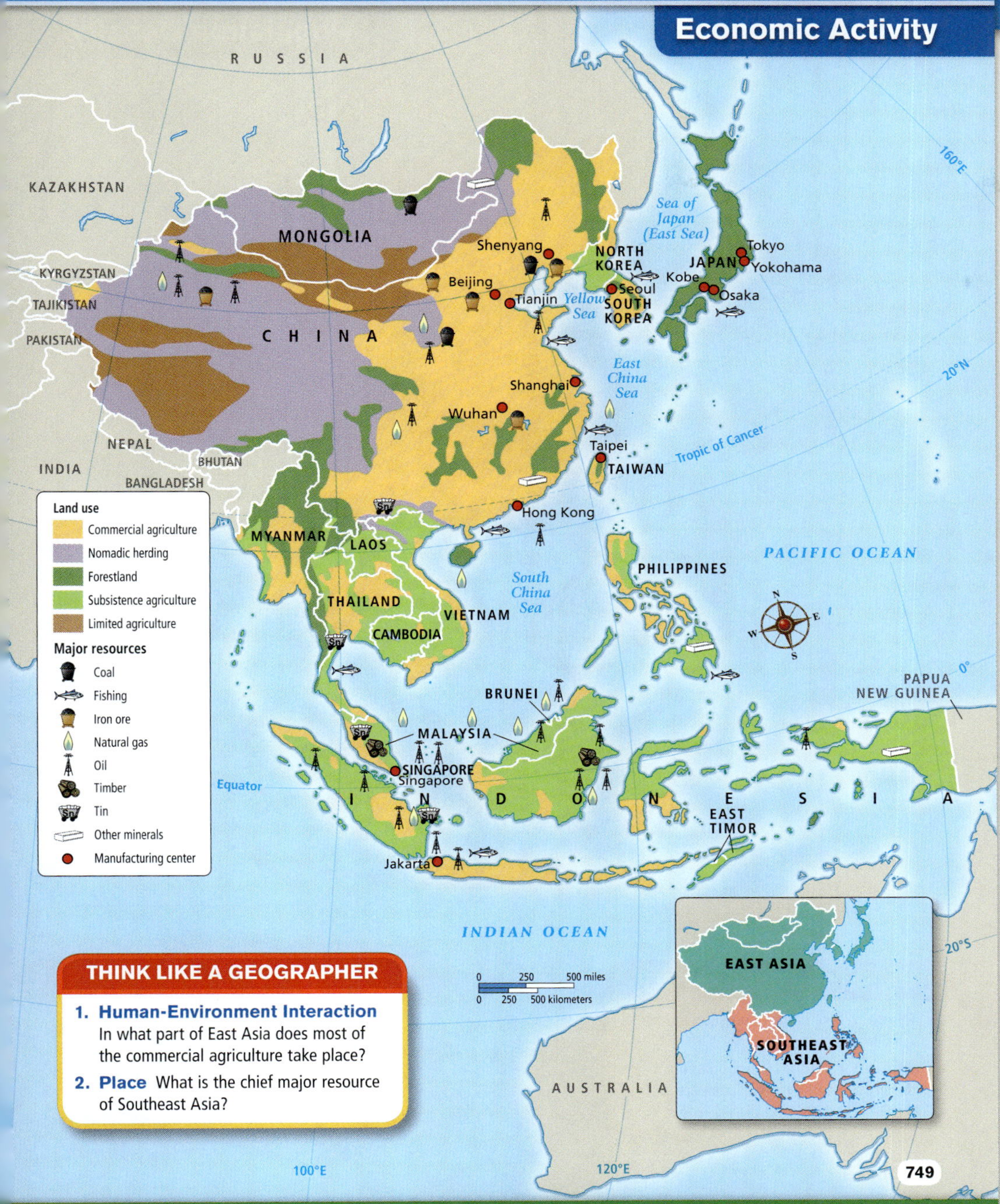

THINK LIKE A GEOGRAPHER

1. **Human-Environment Interaction** In what part of East Asia does most of the commercial agriculture take place?
2. **Place** What is the chief major resource of Southeast Asia?

749

DIFFERENTIATING INSTRUCTION

English Learners

Review Vocabulary

For students who have difficulty differentiating among the classifications of agriculture identified on the map, review the definitions of *commercial, subsistence,* and *limited.* Explain that commercial farmers produce crops they will sell, and subsistence farmers produce crops primarily for family consumption. Tell students that *limited agriculture* refers to areas in which the geography of the land makes farming difficult.

Pre-AP

Research Japan's Economy

Point out that Japan has one of the world's strongest economies, even though it has few natural resources. Have students investigate Japan's economy and explain how Japan acquires and pays for the raw materials it needs to produce its goods.

Economic Activity

Explain that China's gross domestic product (GDP) is one of the largest in the world, but its per capita GDP (GDP per person) is low. Traditionally, the government has maintained strict control over industry and banking. In recent years, however, China has established economic reforms that have led to an increase in privately owned businesses. It has also launched a massive public works program to support economic growth.

- What are the major resources of the Philippines? *(agriculture, fishing, and minerals)*
- What is the major land use of Japan? *(forestland)*
- Which country has the most manufacturing cities? *(China with six, followed by Japan with four)*
- Where would you expect to find the most densely populated areas of China, and why? *(in the east, because western China has limited resources, little commercial farming, and no manufacturing centers)*

THINK LIKE A GEOGRAPHER ANSWERS

1. **Human-Environment Interaction** in the eastern part
2. **Place** oil

More About . . .

China's Economy and the Three Gorges Dam

Scheduled to begin operation in 2009, the Three Gorges Dam on the Chang Jiang will be the largest dam in the world but is the subject of much controversy. Supporters boast that the dam will supply needed electric power, control flooding, and help to develop the area along the river. Critics claim that the cost of the dam will devastate China's economy and that its construction will have serious environmental consequences. The creation of the dam's huge reservoir (370 miles long) will wipe out hundreds of towns and many historical sites—and force the relocation of over 1 million people.

Population Density

Point out that although China is still the world's most populous country, India is predicted to surpass it in the coming years. By contrast, tell students that Japan has a very low birth rate that may hurt its economy in the future.

- Which part of East Asia appears to have the lowest population density? *(the north, central, and western parts)*
- Which parts of Southeast Asia have the highest population density? *(around Jakarta and Manila)*

THINK LIKE A GEOGRAPHER ANSWERS

1. **Place** in the eastern part, around major cities
2. **Region** There are fewer areas of very dense population in Southeast Asia, but there are also fewer areas with 0 persons per square mile.

More About . . .

Java

The most densely populated island in Indonesia is Java (where Jakarta is located). Though it covers only about seven percent of the total land area of Indonesia, it contains almost 50 percent of Indonesia's population. Java is also one of the most densely populated areas in all of East Asia and Southeast Asia.

Indonesia's three largest cities are in Java, including the capital, Jakarta, whose more than 11 million residents make it one of the largest cities in the world. To ease the crowded conditions on the island, the government launched a program in the 1960s to encourage residents to relocate to other, less-crowded islands. In spite of the government's efforts, Java's population is still increasing.

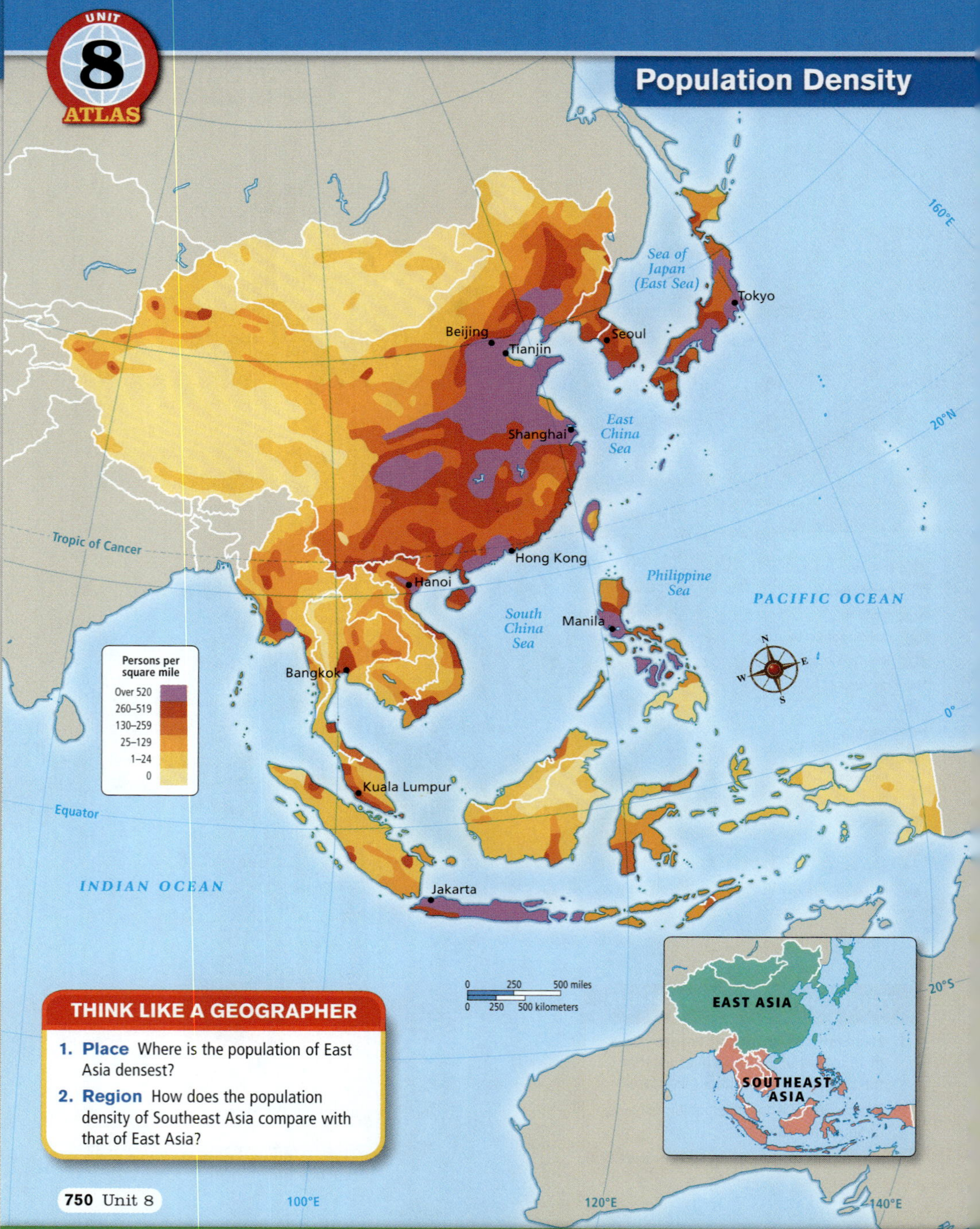

THINK LIKE A GEOGRAPHER

1. **Place** Where is the population of East Asia densest?
2. **Region** How does the population density of Southeast Asia compare with that of East Asia?

750 Unit 8

FIVE THEMES AND SIX ELEMENTS

Five Themes

Region

Tell students that, relative to its land area, tiny Singapore is by far the most densely populated country in all of East Asia and Southeast Asia. It averages more than 16,000 people per square mile. Have students discuss the impact such a high population density might have on everyday life, including housing, public transportation, and traffic management.

Six Elements

Human Systems

Explain that about 90 percent of Japan's population lives along the narrow plains that border the coasts. Point out that many of the cities on the map on this page are located in coastal areas or beside bodies of water. Then ask students why people often settle along rivers or coasts.

UNIT 8 ATLAS Regional Overview

East Asia and Southeast Asia

The countries of East Asia and Southeast Asia are home to about one-third of all the people on Earth. Many of the countries share cultural traditions and a common historical past.

GEOGRAPHY

Both East Asia and Southeast Asia consist of a mainland and a number of important islands off their coasts. The regions are rugged, with many high mountain ranges. Volcanic eruptions and earthquakes frequently occur in these areas.

HISTORY

China has played a large role in both regions. In East Asia, China influenced the ancient civilizations of Japan and Korea. Many cultures had an impact on ancient Southeast Asia, including China and India. The region also changed when European colonists began to take control in the 1500s.

CULTURE

In spite of these historical influences, the countries of East Asia and Southeast Asia have retained their own character. In East Asia, countries such as China and Japan have distinct cultural traditions. Countries such as Indonesia in Southeast Asia are more culturally diverse.

GOVERNMENT

The two regions contain several different forms of government, including democracy, communism, and constitutional monarchy. Many of the countries have strong executive branches that enforce strict control over their people.

ECONOMICS

Most of the countries of East Asia have strong economies. The economies of Southeast Asia are not quite as powerful. Some are still developing, while others—including those in Singapore and Malaysia—are thriving.

Great Buddha The Great Buddha of Kamakura, Japan, is about 40 feet high and was cast in 1252. The statue was originally located inside a temple, which was washed away in a tsunami in the 1400s.

Regional Overview

The Southeast Asian countries located entirely on the mainland of Asia—Cambodia, Laos, Myanmar, Thailand, and Vietnam—are remarkably similar in a number of ways: they share similar climates, a majority of the population is rural, most of the people pursue farming as a way of life, and the dominant religion is Buddhism.

- Which cultures have influenced Southeast Asia? *(Chinese, Indian, European)*
- What are some of the different forms of government in these two regions? *(democracy, communism, constitutional monarchy)*
- What are some of the natural disasters that occur in these two regions? *(volcanic eruptions and earthquakes)*

Comparing Regions

Singapore and Malaysia

Although its predominant religion is also Buddhism, Singapore—barely separated from Malaysia by the narrow Johore Strait—differs significantly from the mainland Southeast Asian countries. Singapore is urban, a major international port, and a world center of commerce, banking, and manufacturing.

Malaysia has part of its country on mainland Southeast Asia, while the other part is separated from the mainland by the South China Sea. It has a more diverse economy and a mostly urban population like Singapore. Malaysia differs from Singapore and mainland Southeast Asia in that a majority of its people are Muslims.

DIFFERENTIATING INSTRUCTION

Inclusion

Summarize Map Information

Pair visually impaired students (and/or students who may need help with analytical skills) with classmates who can provide coherent, verbal summaries of the key information contained in the maps in this atlas. Pairs can tape-record the summaries and replay them to review important facts.

Gifted & Talented

Compare and Contrast Countries

Ask students to use information from the unit atlas maps to compare and contrast one country from East Asia with one country from Southeast Asia. Have them design a poster-sized chart or other visual display to show the similarities and differences between the two countries. To extend this activity, have students choose a third country—from a different region of the world—to compare.

Unit Writing Project
Teaching Tips

- As students prepare their slide shows, suggest that they research information in encyclopedias, travel brochures, and country- or city-sponsored Web sites.
- Provide students with examples of appropriate formats for their slide shows, and encourage students who are proficient in photography and/or computerized design programs to help classmates create slides for their presentations.
- Remind students to include specific details, such as interesting foods, historic sites, dramatic landscapes, and seasonal festivals, to engage viewers' interest in the country.
- Encourage students to work with a partner to review and help revise each other's slide show.

Each map on the country cards illustrates the shape of a state, province, or country. Tell students that the maps do not reflect the comparative size of these landmasses within a particular region. To get an idea of the actual size of the state, province, or country, students should study the land area statistics provided on the card.

UNIT 8 ATLAS Country Almanac

Click here to compare the most recent data on countries @ClassZone.com

Unit Writing Project

Slide Show

Imagine that your family is considering a trip to a country in East Asia or Southeast Asia. Choose the country from this unit that you would most like to visit and prepare a slide show that will convince your family to go there.

Think About:

- how to convey the country's geography and climate
- what cultural and historical information to include
- what makes this place special

East Asia and Southeast Asia

As you study these cards, note that Southeast Asia contains one of the smallest countries in the word—Singapore—and East Asia contains one of the largest—China.

Brunei

GEOGRAPHY
Capital: Bandar Seri Begawan
Total Area: 2,228 sq. mi.
Population: 374,000

ECONOMY
Imports: machinery; food; chemicals
Exports: petroleum; natural gas

CULTURE
Language: Malay
Religion: Muslim 64%; traditional beliefs 11%; Buddhist 9%; Christian 8%

Cambodia

GEOGRAPHY
Capital: Phnom Penh
Total Area: 69,900 sq. mi.
Population: 14,071,000

ECONOMY
Imports: petroleum products; gold
Exports: clothing; rubber; timber

CULTURE
Language: Khmer
Religion: Buddhist 85%; Chinese folk-religionist 5%; traditional beliefs 4%

China

GEOGRAPHY
Capital: Beijing
Total Area: 3,705,407 sq. mi.
Population: 1,315,844,000

ECONOMY
Imports: machinery; petroleum; plastics
Exports: machinery; clothing; textiles

CULTURE
Language: Mandarin Chinese
Religion: nonreligious 42%; Chinese folk-religionist 29%; Buddhist 8%; atheist 8%

East Timor

GEOGRAPHY
Capital: Dili
Total Area: 5,794 sq. mi.
Population: 1,040,880

ECONOMY
Imports: food; construction materials
Exports: coffee; livestock; food; clothing

CULTURE
Language: Tetum; Portuguese
Religion: Catholic 87%; Protestant 5%; Muslim 3%; traditional beliefs 3%

Indonesia

GEOGRAPHY
Capital: Jakarta
Total Area: 741,100 sq. mi.
Population: 222,781,000

ECONOMY
Imports: machinery; petroleum; food
Exports: natural gas; petroleum

CULTURE
Language: Indonesian
Religion: Muslim 77%; Christian 13%; Hindu 3%; traditional beliefs 3%

Japan

GEOGRAPHY
Capital: Tokyo
Total Area: 145,883 sq. mi.
Population: 128,085,000

ECONOMY
Imports: machinery; petroleum; food
Exports: machinery; cars; metals

CULTURE
Language: Japanese
Religion: Shinto 93%; Buddhism 70% (many practice both); Christian 1%

Laos

GEOGRAPHY
Capital: Vientiane
Total Area: 91,429 sq. mi.
Population: 5,924,000

ECONOMY
Imports: mineral fuels; textiles
Exports: electricity; clothing; wood

CULTURE
Language: Lao
Religion: Buddhist 49%; traditional beliefs 42%; nonreligious 4%; Christian 2%

Malaysia

GEOGRAPHY
Capital: Kuala Lumpur
Total Area: 127,317 sq. mi.
Population: 25,347,000

ECONOMY
Imports: microcircuits; computers
Exports: transistors; computers

CULTURE
Language: Malay
Religion: Muslim 60%; Buddhist 19%; Christian 9%; Hindu 6%

INTERDISCIPLINARY ACTIVITIES

CONNECT to Math

Create Population Bar Graphs

Have students use the statistics on these two pages to create one bar graph showing the populations of countries in East Asia and another graph showing the populations of Southeast Asian countries. Invite students to compare the populations of countries within and across regions.

Mongolia

GEOGRAPHY
Capital: Ulaanbaatar
Total Area: 603,909 sq. mi.
Population: 2,646,000

ECONOMY
Imports: machinery; food; fuels; textiles
Exports: copper concentrate; gold

CULTURE
Language: Khalkha Mongolian
Religion: Tantric Buddhist 96%; Muslim 4%

Myanmar

GEOGRAPHY
Capital: Yangon
Total Area: 261,970 sq. mi.
Population: 50,519,000

ECONOMY
Imports: machinery; chemicals; fuels; food
Exports: food; fuels; wood

CULTURE
Language: Burmese
Religion: Buddhist 73%; traditional beliefs 13%; Christian 8%; Muslim 2%; Hindu 2%

North Korea

GEOGRAPHY
Capital: Pyongyang
Total Area: 46,541 sq. mi.
Population: 22,488,000

ECONOMY
Imports: food; machinery; fuels; textiles
Exports: animals; textiles; machinery

CULTURE
Language: Korean
Religion: nonreligious 56%; atheist 16%; Ch'ondogyo 13%; traditional beliefs 12%

Philippines

GEOGRAPHY
Capital: Manila
Total Area: 115,831 sq. mi.
Population: 83,054,000

ECONOMY
Imports: electronics; petroleum; chemicals
Exports: electronics; clothing; food

CULTURE
Language: Filipino; English
Religion: Catholic 81%; Protestant 7%; Muslim 5%; indigenous Christian 4%

Singapore

GEOGRAPHY
Capital: Singapore
Total Area: 267 sq. mi.
Population: 4,326,000

ECONOMY
Imports: electronics; petroleum; computers
Exports: electronics; computers; chemicals

CULTURE
Language: Chinese; Malay; Tamil; English
Religion: Buddhist 43%; Muslim 15%; Christian 15%; nonreligious 15%; Taoist 9%

South Korea

GEOGRAPHY
Capital: Seoul
Total Area: 38,023 sq. mi.
Population: 47,817,000

ECONOMY
Imports: electronics; petroleum
Exports: electronics; machinery; chemicals

CULTURE
Language: Korean
Religion: nonreligious 49%; Buddhist 23%; Protestant 20%; Catholic 7%

Taiwan

GEOGRAPHY
Capital: Taipei
Total Area: 13,892 sq. mi.
Population: 22,894,384

ECONOMY
Imports: machinery; minerals; chemicals
Exports: machinery; textiles; plastics

CULTURE
Language: Mandarin Chinese
Religion: Buddhist 22%; Taoism 21%; I-kuan Tao 4%; Protestant 2%

Thailand

GEOGRAPHY
Capital: Bangkok
Total Area: 198,457 sq. mi.
Population: 64,233,000

ECONOMY
Imports: machinery; chemicals; petroleum
Exports: food; electronics; chemicals

CULTURE
Language: Thai
Religion: Buddhist 94%; Muslim 5%; Christian 1%

Vietnam

GEOGRAPHY
Capital: Hanoi
Total Area: 127,244 sq. mi.
Population: 84,238,000

ECONOMY
Imports: machinery; petroleum; fertilizers
Exports: petroleum; clothing; fish

CULTURE
Language: Vietnamese
Religion: Buddhist 67%; Christian 9%; Cao Dai 4%; Hoa Hao 2%

Laos Towering limestone cliffs line the Nam Song River in Laos.

753

Country Almanac

Rivers are vital to the people and economies of these two regions. Like the Three Gorges Dam in China, dams are being planned for many rivers in Southeast Asia to boost the region's economies. As in China, these projects are also controversial. For example, one project in Vietnam, the Son La Dam, would force about 100,000 people to relocate.

- Which country in these two regions of Asia has the largest population and largest total land area? *(China)*
- Which country has the smallest population and the second smallest total land area? *(Brunei)* Which country has the smallest total land area? *(Singapore)*
- Which country and its capital have the same name? *(Singapore)*
- In which countries is some form of Christianity the dominant religion? *(East Timor and the Philippines)*
- Which country exports gold? *(Mongolia)* Imports gold? *(Cambodia)*

More About . . .

Skyscrapers in East Asia and Southeast Asia

One visible symbol of economic growth is the modern skyscraper. These tall buildings identify the business centers of most of the major cities around the world. More than half of the 50 tallest buildings in the world are located in East Asia and Southeast Asia. The tallest building in the world is Taipei 101 in Taipei, Taiwan. It is 1,670 feet high. The second tallest buildings are the Petronas Towers I and II (1,483 ft.) in Kuala Lumpur, Malaysia.

INTERDISCIPLINARY ACTIVITIES

CONNECT to Government

Learn About Flags

Briefly discuss the history and design of the American flag: how it came into being, how it has evolved, and the significance of its elements. Then have students research the flag of one East Asian or Southeast Asian country and report what they learn about its history and design.

CONNECT to Language Arts

Deliver a Speech

Ask students this question: *If I were asked to choose an East Asian or Southeast Asian country to live in as an exchange student, which one would I choose?* Allow students time to review the information in this atlas. Then ask several volunteers to deliver a short speech in which they explain what country they would choose and why.

Each map on the country cards illustrates the shape of a state, province, or country. Tell students that the maps do not reflect the comparative size of these landmasses within a particular region. To get an idea of the actual size of the state, province, or country, students should study the land area statistics provided on the card.

China PLANNING GUIDE

Complete print resources all on one USB drive allow you to customize lesson plans for all students.

Section Titles & Key Ideas	Unit Resource Book East Asia and Southeast Asia	Assessment Resources
1 Land of Three Rivers **KEY IDEA** China has varied landforms, ranging from high mountains to river valleys. It has diverse climates, plants, and animals.	• **Reading Study Guide,** pp. 3–4 • **RSG with Additional Support,** pp. 11–12 • **Vocabulary Practice,** p. 35 • **Outline Map Activities,** p. 39	**Unit Resource Book** • Section Quiz, p. 51 • Reteaching Activity, p. 55 **Test Generator CD-ROM** **Daily Test Practice Transparency,** TT83
2 Middle Kingdom to Communist Power **KEY IDEA** Throughout history, China's rulers have struggled with the problem of feeding so many people.	• **Reading Study Guide,** pp. 5–6 • **RSG with Additional Support,** pp. 13–14 • **History Makers,** pp. 41–42 • **Connect to Science,** p. 47 • **Primary Sources,** p. 48	**Unit Resource Book** • Section Quiz, p. 52 • Reteaching Activity, p. 56 **Test Generator CD-ROM** **Daily Test Practice Transparency,** TT84
3 Changing Chinese Culture **KEY IDEA** Some aspects of traditional life survive in China, while new ways of life have been introduced.	• **Reading Study Guide,** pp. 7–8 • **RSG with Additional Support,** pp. 15–16 • **Skillbuilder Practice,** p. 36 • **Interdisciplinary Projects,** pp. 43–44 • **Primary Sources,** p. 49	**Unit Resource Book** • Section Quiz, p. 53 • Reteaching Activity, p. 57 **Test Generator CD-ROM** **Daily Test Practice Transparency,** TT85
4 A Rising Power **KEY IDEA** Small changes in the Chinese government and big changes in the Chinese economy are taking place.	• **Reading Study Guide,** pp. 9–10 • **RSG with Additional Support,** pp. 17–18 • **Map and Graph Skills,** p. 37 • **Economics in the World,** p. 46	**Unit Resource Book** • Section Quiz, p. 54 • Reteaching Activity, p. 58 • Chapter Test (Forms A, B, & C), pp. 59–73 **Test Generator CD-ROM** **Daily Test Practice Transparency,** TT86

McDougal Littell Assessment System

TEST SCORE REPORT RETEACH

INTEGRATED TECHNOLOGY

No Time? To focus students on important content in the chapter, use the **"Reading Study Guide with Additional Support"** in ***Resources2Go.***

Teacher Presentation Options

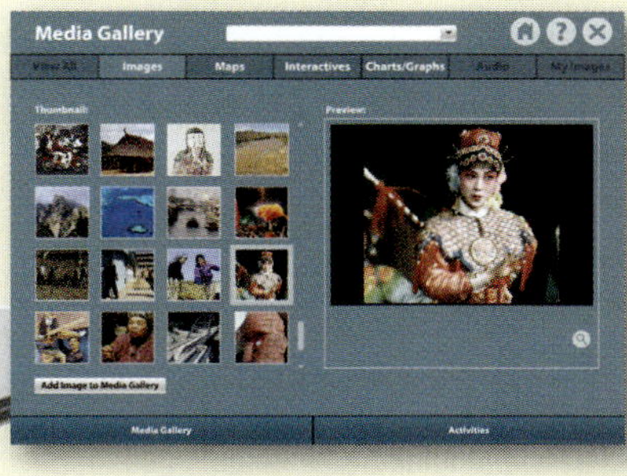

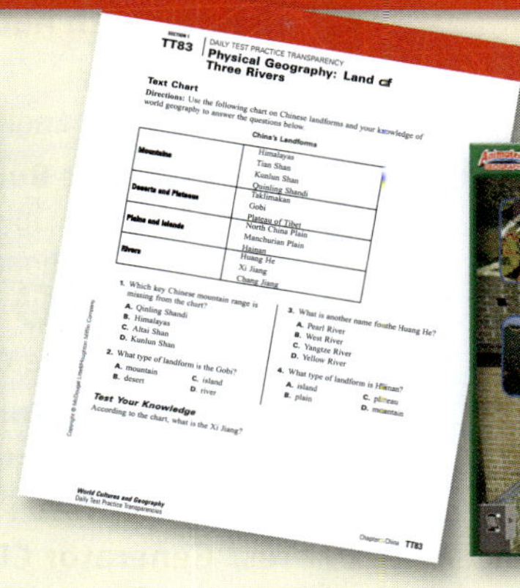

Electronic Books

eEdition DVD-ROM

eEdition Online

Presentation Toolkit

Video/DVD

- World Cultures and Geography

Power Presentations DVD-ROM

- Discussion Starters
- Media Gallery
- Review Game
- Interactive Review
- Animated Features

Unit Transparency Book

- 3-Minute Warm-Ups
- Essential Question
- Critical Thinking
- Maps
- Cultures

Daily Test Practice Transparencies

Animated Geography

- Present-Day China Map
- Physical Geography of China Map
- Three Gorges Dam
- The Great Wall of China
- Worldster: Cheng's Web Page

OnlineActivities at ClassZone.com

Animated Geography

- Videos and Photos
- 3-D Animations
- Interactive Maps

Interactive Review

- Section Quizzes
- Review and Study Notes
- Vocabulary Flip Cards
- Graphic Organizers
- Name Game and Crossword Puzzles
- GeoGame

Online Test Practice

Map Center

- Atlas Maps
- Political Maps with Data
- Links to Mapping Web Sites

Activity Center

- Activity Maker—Review Games, Puzzles, Flip Cards, and Graphic Organizers
- Worldster

Research & Writing Center

- Web Sites
- Data Updates
- Primary Sources
- Writing Models

Current Events

Matthew J. Scheidler
Wayzata East Middle School
Plymouth, Minnesota

Make It Fun — Teacher-Tested Activities

As you teach this chapter, look for my classroom-proven ideas for:

- making the lesson fun
- focusing students' attention and comprehension
- integrating technology with traditional teaching

Teaching Bell to Bell When you see this timing guide, have students discuss in pairs, in groups of 3–6, or as a class.

- **Think, Pair, Share** (pairs)
- **Reader, Recorder, Reporter** (3–6)
- **Talk About It** (any size)

CHAPTER 25 RESOURCES FOR DIFFERENTIATED INSTRUCTION

Struggling Readers

eEdition with Audio DVD-ROM

Lesson Plans
- Strategies for Inclusion

Unit Resource Book
- Reading Study Guide with Additional Support, pp. 11–18 (B)
- Vocabulary Practice, p. 35
- Reteaching Activity, pp. 55–58
- Chapter Test (Form A), pp. 59–63 (A)

Best Practices Toolkit (C)
- Interactive Reading Strategies, pp. 90–145
- Differentiation Activity Bank: Struggling Readers, pp. 146–149

Teacher's Edition Activities
- Categorize, Sec. 1
- Draw an Illustration, Sec. 2
- Read a Bar Graph, Sec. 4
- Tiered Activities, Sec. 1–4

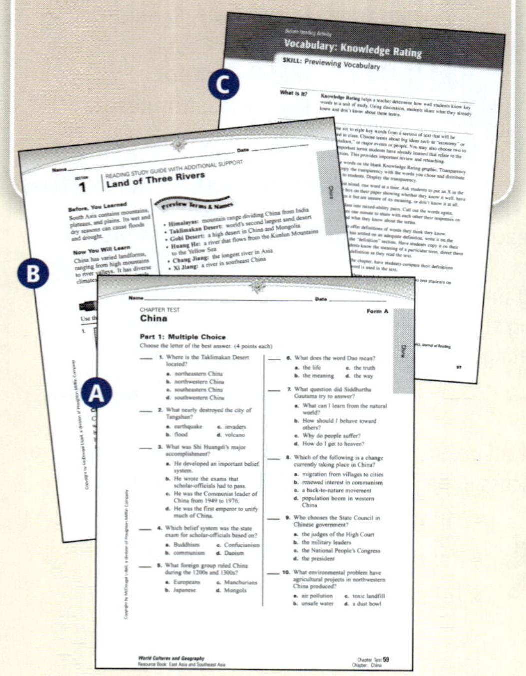

English Learners

Pupil Edition in Spanish

eEdition with Audio DVD-ROM

Lesson Plans
- Strategies for English Learners

Unit Resource Book
- Reading Study Guide (Spanish), pp. 19–26
- RSG with Additional Support (Spanish), pp. 27–34 (A) (B)

Multi-Language Glossary (C)

Best Practices Toolkit
- Differentiation Activity Bank: English Learners, pp. 165–194

Test Generator CD-ROM
- Section Quizzes in Spanish
- Chapter Test in Spanish

Teacher's Edition Activities
- Pronounce and Preview, Sec. 1–4
- Study Vocabulary, Sec. 1
- Understand Scientific Terms, Sec. 2
- Define Terms, Sec. 3
- Knowledge Rating Vocabulary, Sec. 4
- Tiered Activities, Sec. 1–4

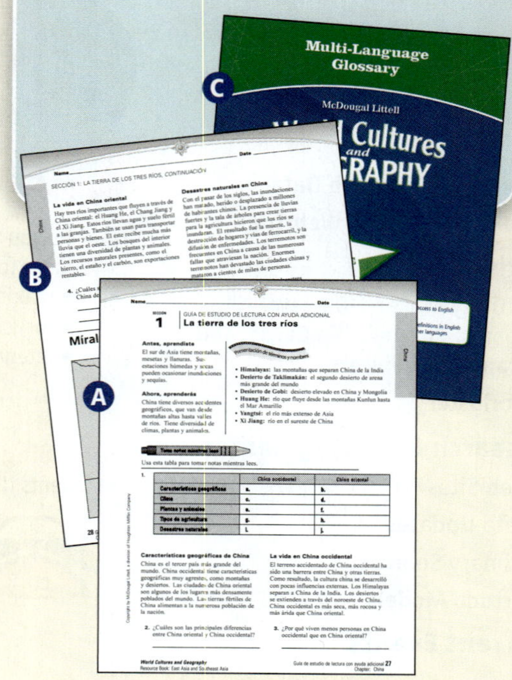

Inclusion

eEdition with Audio DVD-ROM

Lesson Plans
- Strategies for Inclusion

Unit Resource Book
- Reading Study Guide with Additional Support, pp. 11–18 (A) (B)
- Reteaching Activity, pp. 55–58
- Chapter Test (Form A), pp. 59–63

Best Practices Toolkit (C)
- Interactive Reading Strategies, pp. 90–145
- Differentiation Activity Bank: Struggling Readers, pp. 146–149

Teacher's Edition Activities
- Record Terms & Names, Sec. 1
- Locate Geographic Features, Sec. 1
- Predict ABCs, Sec. 2
- Match Terms and Definitions, Sec. 3, 4
- Make a Belief Systems Chart, Sec. 3
- Create a Checkers Bar Graph, Sec. 4
- Tiered Activities, Sec. 1–4

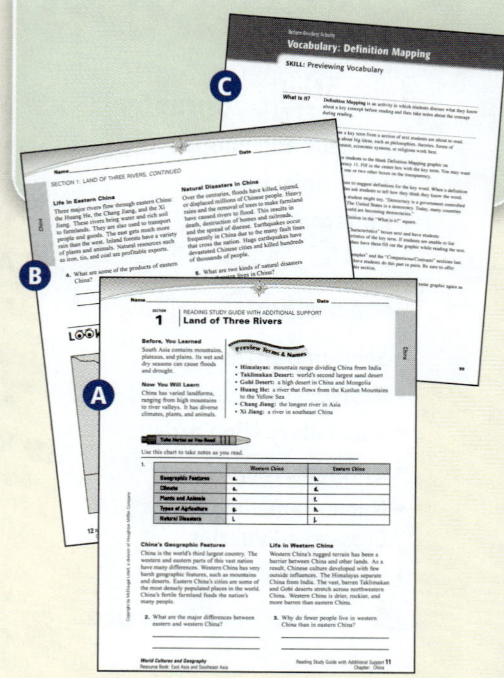

Gifted & Talented Students

Lesson Plans
- Core Lesson Plan

Unit Resource Book
- Outline Map Activities, p. 39 (A)
- History Makers, pp. 41–42 (B)
- Interdisciplinary Projects, pp. 43–44
- Economics in the World, p. 46 (C)
- Chapter Test (Form C), pp. 69–73

Best Practices Toolkit
- Student Products Rubrics, pp. 1–14
- Writing Templates, pp. 78–89

Teacher's Edition Activities
- Compare Canals, Sec. 1
- Create a Board Game, Sec. 2
- Research the Arts, Sec. 3
- Promote a Western Business, Sec. 4
- Tiered Activities, Sec. 1–4

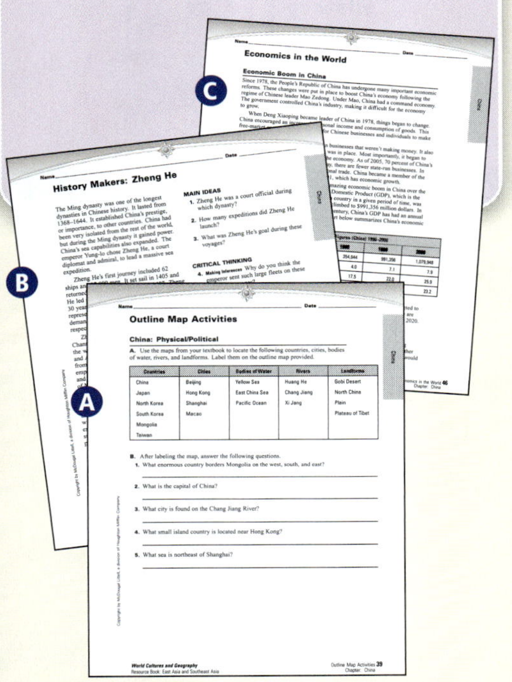

Pre-AP*

Lesson Plans
- Core Lesson Plan

Unit Resource Book
- Vocabulary Practice, p. 35 (A)
- Map and Graph Skills, pp. 37–38
- Connect to Science, p. 47
- Primary Sources, pp. 48–49 (B) (C)
- Chapter Test (Form C), pp. 69–73

Best Practices Toolkit
- Writing Templates, pp. 78–89
- Differentiation Activity Bank: Pre-AP Students, pp. 195–200

* Pre-AP is a registered trademark of the College Entrance Examination Board, which was not involved in the production of and does not endorse this product.

Teacher's Edition Activities
- Write a Dialogue, Sec. 1
- Write About Confucianism, Sec. 2
- Report on Zheng He, Sec. 2
- Debate the Farm Policy, Sec. 3
- Tiered Activities, Sec. 1–4

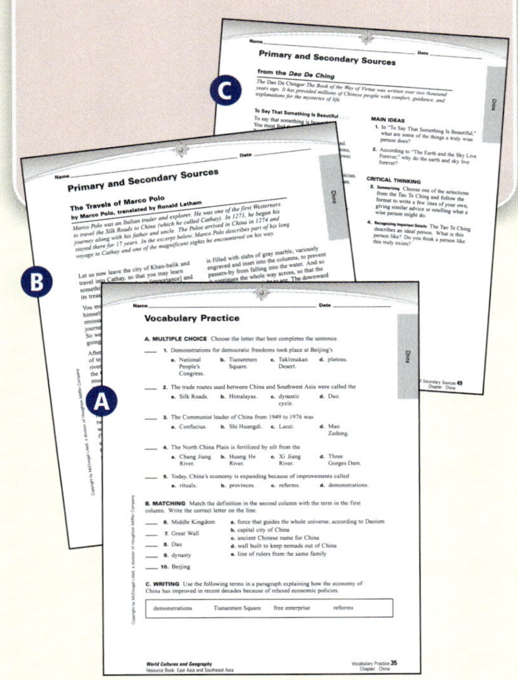

Interdisciplinary Activities

CONNECT to History

You Be the Mapmaker

CLASS TIME 20 minutes

TASK Trace the path of the Great Wall during the Ming Dynasty on a map of China.

SUPPLIES
- blank physical map of China
- colored pencils or markers

ACTIVITY Tell students they will trace the path of the Great Wall onto a blank physical map. Direct them to the map in Section 1 of the text. Have the class identify the major physical features of China and write those features on the board. Then have students write the names on the correct locations on their maps. When they are done, have them turn to the Animated Geography feature and study the map of the Great Wall. They should focus on the Ming Dynasty wall (the red lines). Using a red pencil or marker, have students add the Great Wall path to their maps. Then ask them to identify the physical features that lie inside the wall and those outside of the wall.

Good for Spatial Learners

CONNECT to Math

Graphing for Giants

CLASS TIME 20 minutes

TASK Create large bar graph displays showing the population of the top ten cities in China and the United States.

SUPPLIES
- data on the population of the top ten Chinese cities and the top five U.S. cities
- posterboard and construction paper
- glue and rulers

ACTIVITY Discuss the idea of China's enormous population compared to that of the United States. Distribute population data for Chinese and U.S. cities. Have students work in small groups to construct bar graph posters showing the data. Remind them to use the ruler to create the correct scale for the graphs. Suggest that they use one color of paper for the Chinese cities and a different color for the U.S. cities. Display the completed bar graph posters in the classroom.

Good for Visual Learners

Introduce the ESSENTIAL QUESTION

- Explain that Chinese culture is based on thousands of years of tradition.
- Point out that modern global trends, such as advanced information technology and an emerging worldwide economic system, are changing China's culture.
- Identify specific areas where Chinese life is changing.

Key Ideas

Section 1: China has varied landforms, ranging from high mountains to river valleys. It has diverse climates, plants, and animals.

Section 2: Throughout history, China's rulers have struggled with the problem of feeding so many people.

Section 3: Some aspects of traditional life survive in China, while new ways of life have been developed.

Section 4: Small changes in the Chinese government and big changes in the Chinese economy are taking place.

CONNECT to the Essential Question

You may wish to revisit the Essential Question periodically with students as they work through the chapter.

Suggestions for reconnecting are provided at an appropriate point in the *Teacher's Edition* for each section.

Chinese junk sails in Hong Kong's harbor.

CHAPTER 25

China

1 GEOGRAPHY — Land of Three Rivers

2 HISTORY — Middle Kingdom to Communist Power

3 CULTURE — Changing Chinese Culture

4 GOVERNMENT & ECONOMICS — A Rising Power

ESSENTIAL QUESTION

How is China's traditional way of life changing in response to global trends?

CONNECT Geography & History

Use the map and the time line to answer the following questions.

1. What is the capital of China?
2. On what river is the Three Gorges Dam?

500 B.C.

Culture
479 B.C. Confucius, China's greatest teacher, dies. ►

History
221 B.C. Shi Huangdi unifies China into an empire. ►

Culture
A.D. 65 Buddhism begins to travel to China from India.

TIME LINE DISCUSSION

Invite students to identify the items on the time line that helped shape China's traditional way of life. *(teachings of Confucius, unification of the Chinese empire, introduction of Buddhism into Chinese culture)*

- How many years are covered by the time line? *(2,428)* Point out that the items on the time line are only a few of the many significant events in China's extremely long history.
- In what direction did Buddhism travel to reach China from India? *(east)*

Animated GEOGRAPHY
Present-Day China

Click here to explore China @ClassZone.com

Geography
◄ **1421** Beijing becomes the capital of China. (Forbidden City)

Geography
609 The Grand Canal links the Huang He and Chang Jiang rivers.

Today

Government
◄ **1949** The Communists win the civil war; Mao Zedong becomes head of government.

Animated GEOGRAPHY
Present-Day China
CLASSZONE.COM

China's location and physical features, such as mountains and deserts, helped to limit outside influences. In today's globally connected world, physical barriers no longer effectively limit cultural exchange.

- Why do you think most of China's largest cities are located in the eastern part of the country? *(The western part must have physical features that discouraged settlement.)*
- From which country did a cultural influence enter China and what was that influence? *(India; Buddhism)* Where is that country in relation to China? *(in the southwest)*
- Find the island nation of Taiwan. It was a part of China until 1949, when an event occurred that caused the people on the island to break away. According to the time line, what event occurred then that created this new nation? *(The Communists won the civil war.)*

CONNECT GEOGRAPHY & HISTORY ANSWERS

1. Beijing
2. Chang Jiang

Presentation Options

Use the **Power Presentations** DVD-ROM to project the animated map.

- Prompt students to use the pop-up images and audio from the map to discover what they already know about the geography, culture, history, economics, or government of the region.
- Challenge students to answer the map questions individually, in small groups, or as a class activity. Use all of the interactive material to help preview chapter content.

TEACHER MEDIA FAVORITES

Books for the Teacher

Gray, Jack. *Rebellions and Revolutions.* Oxford U.P., 2003. Focuses on China's development problems.

Spence, Jonathan D. *The Search for Modern China,* 2nd ed. Norton, 1999. Brings China's modern history to life.

Video / DVD

China: A Century of Revolution. PBS, 2002. Three-part series on China's history.

Frontline: China in the Red, 2003. Follows 10 Chinese citizens caught up in China's transformation.

Internet

Go to the *Research and Writing Center* at **ClassZone.com** to link to:

- data on the region
- a variety of chapter-specific, content-reviewed sites
- "History Makers" biographies
- primary sources

1 Plan & Prepare

Objectives

- Compare the geographic features of western and eastern China
- Describe China's climate and resources
- Analyze the effects of China's natural disasters

Read for the Essential Question

Help students read for a purpose by reminding them of the Essential Question: "How is China's traditional way of life changing in response to global trends?" Point out that they can use items in the section, such as Key Questions, special features, images, and maps, to formulate answers to the question.

Vocabulary

Best Practices Toolkit

Use the ***Best Practices Toolkit*** to model strategies for vocabulary notetaking. Vary strategies throughout the year. Choose from: *Knowledge Rating, Predicting ABCs, Definition Mapping, Word Sort, Word Wheel, Frayer Model (Word Squares), Magnet Words,* and *Student VOC.*

Vocabulary Strategies, TT9–TT16

Reading Strategy

Best Practices Toolkit

Display the Three-Column Chart transparency.

- Point out to students that in categorizing they must decide how the information in the section can be sorted into the categories shown in the chart.
- Model filling in categories on a chart using the Comparing Western and Eastern China feature in the section.
- Have students suggest entries using the categories in the chart.

Three-Column Chart, TT21

Section 1 Reading for Understanding

Key Ideas

BEFORE, YOU LEARNED

South Asia contains mountains, plateaus, and plains. Its wet and dry seasons can cause floods and drought.

NOW YOU WILL LEARN

China has varied landforms, ranging from high mountains to river valleys. It has diverse climates, plants, and animals.

Vocabulary

TERMS & NAMES

Three Gorges Dam a barrier built on the Chang Jiang to control floods

hydroelectric power electricity made by water-powered engines

Himalayas (HIHM•uh•LAY•uhz) a mountain range that divides China from India

Taklimakan (TAH•kluh•muh•KAHN) **Desert** the second largest sand desert in the world

Gobi (GOH•bee) **Desert** a high desert in China and Mongolia

Huang He (hwahng huh) a river that flows from the Kunlun Mountains to the Yellow Sea

Chang Jiang (chahng jyahng) the longest river in Asia, flowing through eastern China

Xi (she) **Jiang** a river in southeast China

seismic (SYZ•mihk) having to do with earthquakes

REVIEW

plateau a high area of flat land

Visual Vocabulary plateau

Reading Strategy

Re-create the chart shown at right. As you read and respond to the **KEY QUESTIONS**, use the chart to organize important details about the physical geography of China.

See Skillbuilder Handbook, page R7

CATEGORIZE

	WESTERN CHINA	EASTERN CHINA
Geographic Features		
Climate		
Plants and Animals		
Types of Agriculture		
Natural Disasters		

GRAPHIC ORGANIZERS
Go to Interactive Review @ ClassZone.com

PRETEACHING VOCABULARY

English Learners

Pronounce and Preview

Pronounce each term for students. Review the meanings of words used in the definitions, such as *barrier, water-powered,* and *mountain range.*

- To modify vocabulary learning, have students complete worksheets as they read the section.

Unit Resource Book: East Asia & Southeast Asia

- Vocabulary Practice, p. 35

Inclusion

Record Terms & Names

Have students read each term and its definition into an audiocassette recorder. As students listen to their recordings, have them follow along in their textbooks. Then have students use each term and name in a sentence.

The Five-Phoenix Temple in south-central China's Black Dragon Pool Park

SECTION 1 GEOGRAPHY

China's giant pandas

Land of Three Rivers

Connecting to Your World

Much like the Mississippi in the United States, the Chang Jiang (chahng jyahng) River in eastern China has been both China's best friend and worst enemy. The fertile farmland and the transportation that the river provides help China feed its enormous population. But seasonal floods take a terrible toll on the million or so people who live near the river's banks. The **Three Gorges Dam** should prevent the floods. The dam will also generate **hydroelectric power**, electricity from water-powered engines. The dam was completed in 2006 but is not projected to begin operating until 2009.

China's Geographic Features

KEY QUESTION What are the main geographic features of western China and eastern China?

Everything about the People's Republic of China is on a big scale. China is the third largest country in area and has a population of 1.3 billion—the largest in the world. Western China contains some of the world's harshest geographic features, while eastern China is one of the world's most densely populated places.

may cause flooding upstream; has displaced nearby population ▼

Animated GEOGRAPHY

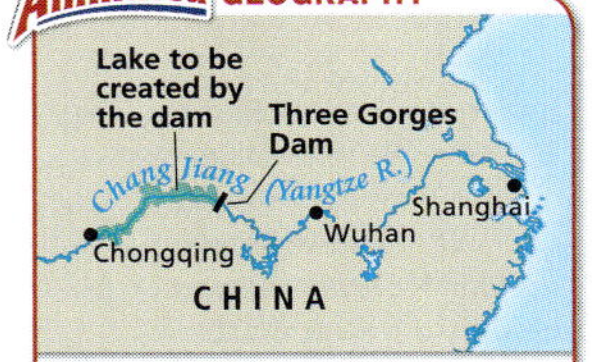

Three Gorges Dam Officials say the dam will relieve flooding on the Chang Jiang and produce more energy. Some people disagree. **What problems do you think the dam could cause?**

Click here to see more of the Three Gorges Dam project @ ClassZone.com

757

2 Focus & Motivate

3-Minute Warm-Up

Write on the board or display the transparency:

- Look at the small map on this page. On what river is the Three Gorges Dam located? *(Chang Jiang)* Why might China want to dam this river? *(to prevent flooding or to use for hydroelectric power)*

Unit Transparency Book
- 3-Minute Warm-Ups, TT1

Connecting to Your World

More About . . .

Three Gorges Dam

The dam is a mile and a half wide, and its reservoir will be 370 miles long. It took about 40,000 workers, $25 billion, and 13 years to complete. The dam's generating capacity will equal 18 nuclear reactors.

3 Teach

China's Geographic Features

Talk About It

- What geographic feature separates China from India? *(Himalayas)*
- Why do most Chinese farmers live in eastern China? *(China's major rivers, which deposit fertile soil along the riverbanks, are located there.)*
- **Categorize** What are the main landforms found in China? *(mountains, plateaus, deserts, and river valleys)*

SECTION 1 PROGRAM RESOURCES

Unit Resource Book: East Asia and Southeast Asia

ON LEVEL

Unit Resource Book
- Reading Study Guide, p. 3
- Outline Map Activities, p. 39
- Section Quiz, p. 51

STRUGGLING READERS

Unit Resource Book
- Reading Study Guide with Additional Support, p. 11
- Vocabulary Practice, p. 35
- Section Quiz, p. 51
- Reteaching Activity, p. 55

eEdition with Audio DVD-ROM

ENGLISH LEARNERS

Pupil Edition in Spanish

eEdition with Audio DVD-ROM

Unit Resource Book
- RSG (Spanish), p. 19
- RSG with Additional Support (Spanish), p. 27

Multi-Language Glossary

Test Generator
- Section Quiz in Spanish

INCLUSION

Unit Resource Book
- RSG with Additional Support, p. 11
- Section Quiz, p. 51
- Reteaching Activity, p. 55

GIFTED & TALENTED

Unit Resource Book
- Section Quiz, p. 51

PRE-AP

Unit Resource Book
- Section Quiz, p. 51

TECHNOLOGY

Unit Transparency Book
- 3-Minute Warm-Ups, TT1
- Essential Question, TT2
- Critical Thinking, TT3
- Map: East Asia Physical, TT7
- Map: The Silk Roads, TT8

Daily Test Practice Transparencies
- Section 1, TT83

Power Presentations

ClassZone.com

World Cultures and Geography Video on DVD

More About . . .

The Himalayas

The Himalayas were formed millions of years ago when India—driven by tectonic plate movement—collided with the Asian Plate. The collision caused rock to be thrust upward, and the Himalayas were born. The Himalayas, the world's tallest mountains, are continuing to rise. Due to the movement of tectonic plates, the Himalayas rise more than one centimeter (0.39 inch) every year. At that rate, the Himalayas will be 6.2 miles taller in a million years! However, scientists believe that sinking caused by gravity may offset this growth.

Fun Facts!

Dinosaur Cemetery

In 1995, scientists discovered the first known nesting dinosaur in the Gobi Desert. The creature's bones, which were similar to those of an ostrich, were found in a nest with about 20 eggs. Scientists at first believed that the dinosaur fed on eggs, so they named it oviraptor, or "egg snatcher." Later studies revealed that one of the eggs contained the embryo of an oviraptor, which shed new light on the animal's behavior. Instead of planning to eat the eggs, the dinosaur was protecting its future young.

Comparing Regions

The Nile and the Chang Jiang

The Nile is the world's longest river; its length surpasses that of the Chang Jiang by about 217 miles. Both rivers have created valleys rich in silt that provide prime farmland. The Nile's huge river basin is about one-tenth the size of Africa. Like the Chang Jiang, the Nile has traditionally served to transport goods and people. In 1970, the completion of the Aswan Dam in Egypt resulted in the Nile's use as a source of hydroelectric power.

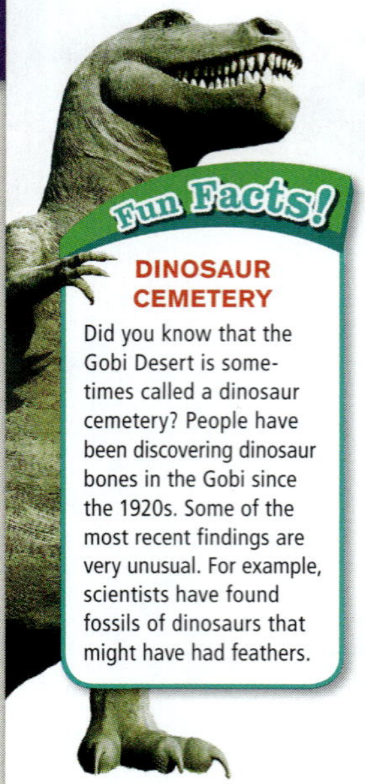

Fun Facts!

DINOSAUR CEMETERY

Did you know that the Gobi Desert is sometimes called a dinosaur cemetery? People have been discovering dinosaur bones in the Gobi since the 1920s. Some of the most recent findings are very unusual. For example, scientists have found fossils of dinosaurs that might have had feathers.

Western China One important fact about western China is that its rugged terrain acted as a barrier between China and other lands. As a result, Chinese culture developed with limited outside influences.

Western China is dotted with dramatic mountain ranges. The **Himalayas** (HIHM•uh•LAY•uhz), which separate China from India, contain the world's highest mountains. The Plateau of Tibet lies between the Himalayas and the Kunlun mountains. This **plateau**, or high area of flat land, is more than 13,000 feet above sea level and is called the "Roof of the World."

Two vast deserts stretch across northwestern China. The **Taklimakan** (TAH•kluh•muh•KAHN) **Desert** is the world's second largest sand desert. And the **Gobi** (GOH•bee) **Desert**, a high desert of China and Mongolia, is about five times the size of Colorado. With so much harsh landscape, western China has far fewer people than eastern China.

Eastern China An important feature of eastern China is its rivers. Three major rivers start in the highlands of western China and flow through eastern China. You can see all three on the map opposite:

1. The **Huang He** (hwahng huh) Ⓐ, or Yellow River, flows through northeast China. The river's floods deposit silt on the North China Plain, making it one of China's richest farming areas.
2. The **Chang Jiang** Ⓑ, or Yangtze River, is 3,915 miles long. It is the longest river in Asia and is a very busy waterway.
3. The **Xi** (she) **Jiang** Ⓒ, or West River, flows through southeast China. It is a major transportation route in southern China.

These rivers are critical to productive farming in eastern China. Spring floods deposit silt, a fertile soil good for growing crops, along the riverbanks. Because of the rich, flat river basins that result, most Chinese farmers live in eastern China's river valleys. The rivers also transport people and goods, connecting inland China with its coast.

A SUMMARIZE Describe the main geographic features of western China and eastern China. Western China is very mountainous; it also contains a high plateau and two vast deserts. Eastern China has many rivers and fertile river valleys.

Huang He (Yellow River) The Huang He carries fine dirt that gives it a yellowish color. **Based on the geographic features shown here, is the river most likely flowing through eastern or western China?** eastern China

758 Chapter 25

DIFFERENTIATING INSTRUCTION

Struggling Readers

Categorize

To help students differentiate among China's geographic features, have them complete a chart like this one. After correctly identifying the features, have students point them out on a classroom map.

Rivers	Highlands	Deserts
1. Huang He	1.	1.
2. Chang Jiang	2.	2.
3. Xi Jiang	3.	3.

CONNECT Geography & Culture

READING A PHYSICAL MAP

Locate China's three main rivers on the map. As you study the map,

- note the direction the rivers flow
- think about how the direction of flow might affect travel in China
- consider how areas of high and low elevation might affect local populations

1. **Location** Which of the three main rivers labeled on this map does not start in the Plateau of Tibet?
2. **Movement** Would it be easier for boats to travel from east to west along the rivers of China or from west to east? Explain.

China 759

Animated GEOGRAPHY

Physical Geography of China

CLASSZONE.COM

CONNECT Geography & Culture

Reading A Physical Map Ask students to use the legend to help them identify China's highest and lowest areas of elevation. *(highest: mountains and Plateau of Tibet; lowest: North China and Manchurian plains, coastline)*

ANSWERS

1. **Location** Xi Jiang
2. **Movement** west to east; Because China's rivers originate in the west and flow east, boats traveling east would be aided by the current.

Unit Resource Book: East Asia & Southeast Asia

- Outline Map Activities, p. 39

Presentation Options

Use the **Power Presentations** DVD-ROM to project the animated map.

- Listen to the audio and examine the photographs to learn how the geography of China affected its history, culture, and economics.
- Click to animate the map.

Make It Fun

Teacher-Tested Activities

Matthew J. Scheidler, Wayzata East Middle School, Plymouth, Minnesota

My students enjoy creating a "sculpture freeze" to depict a geographic concept. I assign groups an object or event to depict in a physical way. The group has 10 minutes to plan, then "freezes" their sculpture for 30 seconds. One group did the Three Gorges Dam. They had a student slither across the floor to represent the Chang Jiang until he reached the rest of the students who knelt in a tight row to represent the dam. This activity helps students remember important facts and terms related to China. It's more fun if students are not allowed to use props because it forces them to think creatively.

DIFFERENTIATING INSTRUCTION

Inclusion

Locate Geographic Features

Have copies of the full page map enlarged for learners with visual impairments. Have each student work with a sighted partner to find the features highlighted in the text. When they locate each feature, the pairs should circle it on the enlarged map. When finished, the pairs can review the locations of all the features.

Gifted & Talented

Compare Canals

Tell students that the oldest canal in the world—the Grand Canal—runs across the North China Plain. Have students research the canal and prepare an illustrated poster comparing that canal with the Suez and Panama canals. Information should include comparisons of length and other features such as locks, types of ships using the canals today, and the purpose or value of the canals.

Teach

China's Varied Climate and Resources

Think, Pair, Share

- In which area of China is herding a common activity? *(western highlands)*
- Why is bamboo important to Chinese culture? *(used to make chopsticks; the only food source of the panda)*
- Why is southeast China especially suited for growing rice? *(It has rich soil and a mild, rainy climate.)*

CONNECT Geography & History

The Silk Roads

In use as early as the second century B.C., the Silk Roads were traveled not only by traders but by many others as well. Some Chinese were sent as diplomats to forge alliances with neighboring peoples. Chinese Buddhist monks traveled the roads to obtain sacred texts from India.

During the Middle Ages, Marco Polo was one of many Westerners who traveled east in search of trade and adventure. In the 19th century, archaeologists and geographers used the roads in their exploration of Asian culture.

Unit Transparency Book
- Map: The Silk Roads, TT8

China's Varied Climate and Resources

KEY QUESTION How are the climate and resources of western and eastern China different?

Western and eastern China have very different climates and natural resources. Because of its climate differences, China also has a variety of plants and animals.

CONNECT Geography & History

The Silk Roads

Trade moved along routes on the edges of the Taklimakan and Gobi deserts. Traders traveled the routes by camel caravans. Among other items, they carried silver, jade, medicines, foods, and animals. But they carried so much silk that the routes came to be called the Silk Roads.

Life in Western China You've learned that western China is thinly populated because it is mostly mountains and deserts. The highest mountains are cold year-round. At lower elevations, summers are cool but sunny. Animals such as wild sheep and mountain antelope roam the highlands. Forests grow at elevations of around 12,000 feet and provide a home for wildcats.

As you might guess, almost nothing survives in the Gobi and Taklimakan deserts. They get little rain and have very sparse vegetation. Temperatures dip below 0°F in winter and rise to higher than 120°F in summer.

Many highlanders are nomadic herders of horses and of grazing animals, such as yaks, sheep, and goats. Because vegetation is scarce—mostly mosses and short shrubs—grazing herds must keep moving to find new sources of food.

Western China's main resources are petroleum, natural gas, and lead. Pipelines carry both oil and natural gas to the east.

Life in Eastern China Eastern China has a milder climate than western China. Northeastern China has a climate like the Northeast and Midwest of the United States, with cold winters, warm summers, and enough rain for farming. Mixed and broadleaf forests provide homes for leopards, deer, and more than 265 kinds of birds. In contrast, southeastern China has mild winters and hot, rainy summers. The region has mixed forests of wild tea and citrus trees, oaks, cypresses, and pines. Alligators, giant salamanders, monkeys, and pangolins—scaly animals that look like anteaters—live inland.

Wherever rain falls in China, bamboo grows. Bamboo is important to Chinese culture. Its long, woody stems are used to make chopsticks, among other things. Bamboo is also a vital food for the giant panda, China's most treasured animal. The panda, an endangered species, spends about 12 hours a day eating—and it eats mainly bamboo.

INTERDISCIPLINARY ACTIVITIES

CONNECT to Science

Create Animal Posters

Have students work in pairs to research one of the following animals found in China: yaks, pangolins, or pandas. Students should include the region where the animal is found, what it eats, and any distinguishing characteristics and habits. Students should create a poster showing their findings and include either a drawing of the animal or a collage of photos.

CONNECT to Language Arts

Write About the Silk Roads

Have students write a paragraph describing what the people in the photograph on this page might be experiencing as they travel the Silk Roads. Tell them to base their writing on information from the photo and the map as well as from the text.

Most of China's farmland is in eastern China. Many people think that all Chinese farmers grow rice, but in northern China they grow wheat, soybeans, and millet. These crops need less water than rice. Rice is grown in the rainy southeast, where the climate is so mild and the soil is so rich that farmers grow two or three crops a year.

Chinese farmers use traditional methods that are centuries old. They don't have the huge tractors or other large farm equipment that U.S. farmers use. The use of farm machinery, such as small hand-held tractors, has been on the rise since the 1990s, but China still relies on long hours of human labor in the fields.

Eastern China has abundant natural resources, such as tin and iron. In 2003, China produced about 35 percent of the world's coal, but at a deadly cost; 80 percent of all deaths in coal mines worldwide occur in China. The government is using clean coal technologies and modernizing equipment to improve safety.

COMPARE AND CONTRAST Compare the climate and resources of western and eastern China.

Eastern China has a milder, rainier, and less extreme climate than western China. Western China has petroleum, natural gas, and lead, while eastern China has coal and other resources and much farmland. ◄

COMPARING Western and Eastern China

	WESTERN	EASTERN
Population	thinly populated	heavily populated
Geographic Features	an arid climate and little usable farmland	important rivers, plentiful rain, rich farmland
Land Use	used mostly for grazing animals	used for agriculture
Resources	supplies of petroleum, natural gas, and lead	rich with coal and other resources

CRITICAL THINKING

1. **Analyze Effects** How is land use in each part of China affected by its geographic features?
2. **Make Inferences** Which part of China would most likely contribute more to the country's overall economy?

More About . . .

Rice

Rice has been grown in China for at least 4,000 years. In fact, the Chinese words for *agriculture* and *rice culture* are the same. However, China's economic development has changed the country's agricultural practices. Millions of farmers have switched to crops that yield higher profits. Polluted groundwater and unclear land rights have contributed to a decline in rice production. In 2004, rice production fell short by more than 4 million tons. As a result, China has begun relying on imported rice, which many families cannot afford.

COMPARING

Western and Eastern China

Remind students that people are attracted to areas where they can support themselves. Ask students to identify aspects of geography in both eastern and western China that would attract settlement. *(eastern: rich farmland, abundant resources; western: good grazing land, some resources)*

CRITICAL THINKING ANSWERS

1. **Analyze Effects** Mountains and deserts in western China make farming difficult, so most inhabitants are nomadic herders. River valleys in eastern China provide rich soil for productive farming.
2. **Make Inferences** eastern China

DIFFERENTIATING INSTRUCTION

English Learners

Study Vocabulary

To help students learning English as a second language understand words used in the comparing feature above, pair an English speaker with a non-English speaker. Have the English speaker explain the following terms: *thinly populated, arid climate, grazing animals,* and *plentiful rain.*

Pre-AP

Write a Dialogue

Have pairs of students write a dialogue between farmers discussing whether to use newer, more modern farming methods or to retain the more traditional methods. Then have the students read their dialogues to the class.

Teach

Natural Disasters in China

Reader, Recorder, Reporter

- What does the government hope to achieve by building the Three Gorges Dam? *(to prevent the Chang Jiang from flooding)*
- Why is there often greater loss of life from earthquakes in eastern China than in western China? *(Eastern China is more densely populated.)*

Comparing Regions

Disasters in Southeast Asia and the U.S. Gulf Coast

China isn't the only country to experience large-scale disasters. Remind students that in December 2004, a massive earthquake struck in the Indian Ocean off the coast of Indonesia. The resulting tsunamis, or huge ocean waves, flooded land from Africa to Thailand and Indonesia and killed more than 200,000 people.

On August 29, 2005, Hurricane Katrina tore into the central Gulf Coast of the southeastern United States. With sustained winds of 140 miles per hour, the hurricane devastated a 90,000-square-mile area. An accompanying storm surge, with tsunami-like waves, flooded 80 percent of New Orleans. Thousands of people were killed or injured and more than half a million had to leave their homes.

Flooded City Floods continue to plague China. In 2004, heavy rains caused flash floods along the Chang Jiang. The raging waters swamped cities and destroyed homes, roads, and power lines. **What special problems might occur when large cities experience floods?**

people made homeless; people stranded; utilities downed; water not safe to drink; businesses halted

Natural Disasters in China

KEY QUESTION How have floods and earthquakes affected China?

China has had its share of natural disasters. Severe floods and **seismic** (SYZ•mihk), or earthquake, activity often threaten Chinese lives.

Floods China's damaging floods have killed, injured, or displaced millions of people over the centuries. In fact, the Huang He has flooded so often that the Chinese call it "China's Sorrow."

One of the worst floods of recent times took place along the Chang Jiang in 1998. Heavy rains were one cause. Another cause was the cutting down of forests for farmland and urban growth. Without trees to hold the soil in place, the rainwater rushed into the river, causing it to rise. More than 3,000 people died in the resulting flood. Fourteen million people in four provinces lost homes, and railways and roads were destroyed. Deadly diseases such as cholera spread.

The Chinese government hopes the Three Gorges Dam will prevent future floods on the Chang Jiang. However, critics of the project raise tough issues:

- The dam will force more than 1 million people to leave their farms and homes. Hundreds of cities, towns, and villages will be under water.
- The lake will cover more than 1,000 important historical sites.
- The dam may create pollution and increase health risks because the river will no longer wash sewage away.

Will the benefits of the dam outweigh the risks? We won't know until the dam begins operating in 2009. Meanwhile, the Chinese people make the best of the ebb and flow of their powerful rivers.

DIFFERENTIATING INSTRUCTION: TIERED ACTIVITIES

OBJECTIVE Demonstrate an understanding of the impact of natural disasters on China.

Basic

Have students work in small groups to create a poster summarizing the devastation caused by the flooding of the Chang Jiang or the 1976 earthquake in Tangshan. Display the posters on the bulletin board.

On Level

Have students write a letter from the point of view of a Chinese citizen describing the effects of the building of the Three Gorges Dam or the earthquake in Tangshan.

Challenge

Provide student groups with a data sheet about the Three Gorges Dam. Students should decide whether they think the dam will improve life in China or create more problems than it will solve. One student from each group should report to the class.

Earthquakes Eastern and western China both experience earthquakes because many fault lines, or cracks in the earth's crust, cross the country. But earthquakes in eastern China carry a much higher death toll because the region is so heavily populated.

One example is a devastating earthquake that hit Tangshan (tahng•shahn), about 90 miles from the capital city of Beijing (bay•jing), in July 1976. The quake destroyed roads, canals, sewers, hospitals, factories, and homes. The quake hit at about 4:00 A.M., when people were sleeping. About 250,000 people died, and more than 500,000 were injured. A strong aftershock killed many people already trapped in the rubble.

More than 90 percent of the city's housing was lost. Survivors lived in tents with no electricity or clean water. Despite the hardships, the government began to rebuild the city right away. Today the Chinese call Tangshan the "Brave City of China."

Earthquake Detector In this ancient Chinese invention, an earthquake would cause a ball to fall from a dragon's mouth into a frog's mouth. The frog's position revealed the direction of the quake.

ANALYZE EFFECTS Explain how floods and earthquakes have affected China. Floods have killed, injured, and displaced millions of people over the centuries. Earthquakes have also caused a great deal of destruction.

Section 1 Assessment

ONLINE QUIZ For test practice, go to **Interactive Review** @ClassZone.com

TERMS & NAMES

1. **Explain the importance of**
   - Three Gorges Dam
   - Taklimakan Desert
   - Huang He
   - Chang Jiang

USE YOUR READING NOTES

2. **Categorize** Use your completed chart to answer the following question:

   Which part of the country is more in danger of floods, the western part or the eastern part? Why?

	WESTERN CHINA	EASTERN CHINA
Geographic Features		
Climate		
Plants and Animals		
Types of Agriculture		
Natural Disasters		

KEY IDEAS

3. Why does the majority of Chinese people live in the eastern third of the country?
4. How did the 1998 flood affect people?
5. Why is it difficult for humans to live in the Taklimakan and Gobi deserts?

CRITICAL THINKING

6. **Analyze Effects** How does the silt carried by rivers both help farmers and create problems for them?
7. **Make Inferences** Why do you think Chinese civilization developed near the rivers?
8. **CONNECT to Today** In 2002, China imported nearly half the oil it used. How do you think China's lack of oil affects its use of farm machinery?
9. **WRITING** **Prepare a Travel Brochure** Choose a geographical area of China and prepare a brochure for tourists. Include pictures and describe the best places to see.

CONNECT to the Essential Question

How is China's traditional way of life changing in response to global trends?

Ask students what they have learned so far that can help them answer this question. Students might mention the following:

- Use of farm machinery has increased.
- More than one million people will be resettled due to the construction of the Three Gorges Dam.

4 Assess & Reteach

Assess Have students complete the Section Assessment.

Unit Resource Book: East Asia & Southeast Asia
- Section Quiz, p. 51

Interactive Review @ClassZone.com
- Section Quiz

Power Presentations

Test Generator

Reteach Divide students into six teams and provide markers, paper, and scissors. Assign one part of the section to each team and tell them to compose cause-and-effect statements. Have teams cut their statements apart, separating causes from effects. Redistribute the effects slips to teams. Then read each cause aloud and have students identify the corresponding effect.

Unit Resource Book: East Asia & Southeast Asia
- Reteaching Activity, p. 55

SECTION 1 ASSESSMENT ANSWERS

Terms & Names

1. Three Gorges Dam: barrier built on Chang Jiang; Taklimakan Desert: world's second largest sand desert; Huang He: river in northeast China; Chang Jiang: longest river in Asia

Use Your Reading Notes

2. the eastern part because it has three large rivers

	WESTERN CHINA	EASTERN CHINA
Geographic Features	mountains, deserts	river valleys
Climate	dry	mild winters; hot rainy summers
Plants and Animals	short shrubs, grazing animals	bamboo, forests, monkeys
Types of Agriculture	used for grazing animals	growing rice, millet, wheat
Natural Disasters	earthquakes	floods

Key Ideas

3. Its climate is milder. The rich soil of the river valleys is good for farming.
4. More than 3,000 people died and 14 million lost homes. Cholera and other diseases spread.
5. The climate is severe and the landscape harsh.

Critical Thinking

6. The silt creates rich soil but can also inundate villages during floods.
7. Rivers supplied water for drinking, bathing, and the raising of crops. In time, surpluses of food allowed the economy to expand, cities to emerge, and people to do nonagricultural jobs.
8. Possible answers: makes the use of the machinery more expensive; discourages use
9. Use the rubric to score brochures.

Brochure Rubric

	Points of Interest	Description	Errors
4	5 or more	clear, concise	few or none
3	4	good	some
2	3	fair	many
1	2	poor	many major

1 Plan & Prepare

Objectives

- Explain the dynastic cycle
- Analyze the influence of scholars and inventors in China
- Identify groups who fought to control China in the early 20th century

Read for the Essential Question

Help students read for a purpose by reminding them of the Essential Question: "How is China's traditional way of life changing in response to global trends?" Point out that they can use items in the section, such as Key Questions, special features, images, and maps, to formulate answers to the question.

Vocabulary

Best Practices Toolkit

Use the ***Best Practices Toolkit*** to model strategies for vocabulary notetaking. Vary strategies throughout the year. Choose from: *Knowledge Rating, Predicting ABCs, Definition Mapping, Word Sort, Word Wheel, Frayer Model (Word Squares), Magnet Words*, and *Student VOC.*

Vocabulary Strategies, TT9–TT16

Reading Strategy

Best Practices Toolkit

Display the Two-Column Chart transparency.

- Remind students that identifying problems means finding and understanding the difficulties faced by a group of people during a certain time.
- Tell students that solving problems means understanding how people tried to remedy those problems.
- Model this strategy by filling in a problem from the section and identifying the solution.

Two-Column Chart, TT20

SECTION 2 Reading for Understanding

Key Ideas

BEFORE, YOU LEARNED

China is one of the largest countries in the world, but much of its farmland lies only in the eastern part.

NOW YOU WILL LEARN

Throughout history, China's rulers have struggled with the problem of feeding so many people.

Vocabulary

TERMS & NAMES

Shi Huangdi (shee hwahng•dee) the first emperor to unify much of China

emperor a ruler of an empire, which is a group of different territories or cultures led by a single all-powerful authority

dynasty (DY•nuh•stee) a line of rulers from the same family

North China Plain the fertile region between the Huang He and Chang Jiang rivers

Middle Kingdom an ancient Chinese name for China

dynastic (dy•NAS•tihk) **cycle** the pattern of the rise and fall of dynasties

Great Wall a huge and long wall built to keep nomads out of China

Confucianism (kuhn•FYOO•shuh•nihz•uhm) a belief system based on the teachings of Confucius, a Chinese scholar

scholar-official an educated person who worked in China's government

communism a government and economic system in which the government holds nearly all political power and the means of production in the name of the people

Mao Zedong (MOW dzuh•dahng) Communist leader of China from 1949 to 1976

Reading Strategy

Re-create the problem-solution chart shown at right. As you read and respond to the **KEY QUESTIONS**, use the chart to explain how Chinese leaders tried to solve some of the problems they faced.

See Skillbuilder Handbook, page R10

IDENTIFY PROBLEMS AND SOLUTIONS

PROBLEMS	SOLUTIONS
Gaining Power	
Keeping Order	
Feeding a Large Population	

GRAPHIC ORGANIZERS Go to **Interactive Review** @ClassZone.com

PRETEACHING VOCABULARY

English Learners

Pronounce and Preview

Pronounce each term for students. Review the meanings of words used in the definitions, such as *unify, fertile,* and *belief system.*

- To modify vocabulary learning, have students complete worksheets as they read the section.

Unit Resource Book: East Asia & Southeast Asia
- Vocabulary Practice, p. 35

Inclusion

Predict ABCs

Have students work in pairs to fill out the chart for the terms in the section.

- Use the transparency to model the strategy.
- Have volunteers fill in the words for Section 2.
- Then distribute copies of the organizer for students to work with a partner.

Best Practices Toolkit
- Predicting ABCs, TT10

SECTION 2 HISTORY

The Five-Phoenix Temple in south-central China's Black Dragon Pool Park

China's giant pandas

Middle Kingdom to Communist Power

Connecting to Your World

In democracies like the United States, people elect leaders, but that isn't the only way that rulers come to power. **Shi Huangdi** (shee hwahng•dee) became king of the state of Qin (chihn) when he was only 13 because he was the son of a king. Older people actually ran the government until 238 B.C., when Shi Huangdi became the sole ruler at the age of 21. He employed smart generals and advisers, who helped him conquer other Chinese kingdoms. By 221 B.C., Shi Huangdi became China's first **emperor**, or ruler of many states.

Shi Huangdi Shi Huangdi ruled as emperor from 221 to 210 B.C.

Emperors and Dynasties

KEY QUESTION What was the dynastic cycle?

Shi Huangdi ruled harshly. To keep control, he killed hundreds of people who criticized him. He took away the nobles' power and forced peasants to build roads, forts, and canals. Many died from overwork.

Historians divide Chinese history into dynasties. A **dynasty** (DY•nuh•stee) is a line of rulers from one family. Shi Huangdi began the Qin Dynasty, but Chinese history began long before Qin rule.

Terra Cotta Army Shi Huangdi had an army of 7,500 life-size terra cotta (clay) soldiers created to guard his tomb. **What do you think he wanted this army to convey about him?** Possible answer: He wanted to convey a sense of his power.

765

2 Focus & Motivate

3-Minute Warm-Up

Write on the board or display the transparency:

- In the United States, how does a person become president? *(elected)* What might be some other ways a person might come to power in a country? *(born into it, seized through military means)*

Unit Transparency Book
- 3-Minute Warm-Ups, TT1

Connecting to Your World

More About . . .

Shi Huangdi

The terra cotta army was found buried in large pits near Emperor Shi Huangdi's tomb. One pit contained 6,000 soldiers. In a second pit, terra cotta charioteers, archers, cavalrymen, and foot soldiers were arranged in battle formation.

3 Teach

Emperors and Dynasties

Think, Pair, Share

- Where did a major Chinese civilization begin? *(on the North China Plain)*
- What was Shi Huangdi's major contribution to the history of China? *(He united rival kingdoms into one empire.)*
- **Identify Problems and Solutions** What problems were brought about by Shi Huangdi's method of rule? *(forced labor, cruel treatment of his subjects)*

SECTION 2 PROGRAM RESOURCES

Unit Resource Book: East Asia and Southeast Asia

ON LEVEL

Unit Resource Book
- Reading Study Guide, p. 5
- Section Quiz, p. 52

STRUGGLING READERS

Unit Resource Book
- Reading Study Guide with Additional Support, p. 13
- Vocabulary Practice, p. 35
- Section Quiz, p. 52
- Reteaching Activity, p. 56

eEdition with Audio DVD-ROM

ENGLISH LEARNERS

Pupil Edition in Spanish
eEdition with Audio DVD-ROM
Unit Resource Book
- RSG (Spanish), p. 21
- RSG with Additional Support (Spanish), p. 29

Multi-Language Glossary
Test Generator
- Section Quiz in Spanish

INCLUSION

Unit Resource Book
- RSG with Additional Support, p. 13
- Section Quiz, p. 52
- Reteaching Activity, p. 56

GIFTED & TALENTED

Unit Resource Book
- History Makers, pp. 41–42
- Section Quiz, p. 52

PRE-AP

Unit Resource Book
- Connect to Science, p. 47
- Primary Sources, p. 48
- Section Quiz, p. 52

TECHNOLOGY

Unit Transparency Book
- 3-Minute Warm-Ups, TT1
- Essential Question, TT2
- Critical Thinking, TT4
- Cultures: The Great Wall, TT9

Daily Test Practice Transparencies
- Section 2, TT84

Power Presentations
ClassZone.com
World Cultures and Geography Video on DVD

More About . . .

The Middle Kingdom

The ancient Mandarin Chinese word *Zhongguo* means "Middle Kingdom." It is also the Chinese word for *China.* The term dates back to 1000 B.C. and the time of China's early civilization, located on the North China Plain. People there had no knowledge of other advanced civilizations. Surrounded by nomads whom they considered to be barbarians, the Chinese had good reason to believe that they were, at the very least, the cultural center, or middle, of the Earth—and were therefore the "Middle Kingdom."

Dynastic Cycle

CRITICAL THINKING ANSWER

Make Inferences Possible answer: because they are very powerful, commanding creatures who inspire fear

Comparing Regions

Rule by Divine Right in Europe and China

In medieval Europe, rulers adopted the doctrine of the Divine Right of Kings. This doctrine stated that the will of God determined a monarch's right to rule. As in China, the right to rule was passed on through the family. However, unlike the Mandate of Heaven, the concept of the Divine Right of Kings did not include the idea that a poor ruler could lose divine approval and therefore be overthrown.

Early Culture A major Chinese civilization developed on the **North China Plain**, which lies between the Huang He and Chang Jiang. As you have learned, the rivers deposit silt, so the land is fertile. Chinese farming started there, and around 2000 B.C., cities began to grow.

Some traits of that early culture have been part of Chinese culture ever since. For example, the Chinese had invented a writing system by 1000 B.C. Those characters evolved into the ones that are still used today. Also, the ancient Chinese placed great value on the family. Respect for the family is still central to Chinese life.

Uniting and Ruling China Other dynasties ruled northern China before the Qin, but Shi Huangdi was the first ruler to unite the kingdoms into one empire. Wars between the separate kingdoms ended. In the peace that followed, farming and trade grew.

A united China could better protect itself from attacks. Nomads from the north often invaded China. The Chinese saw the nomads as uncivilized, and in contrast, thought their own culture was very advanced. In fact, the Chinese called their country the **Middle Kingdom** because they thought they were the center of the world.

The Chinese believed that their rulers needed the Mandate of Heaven, or the gods' approval. A dynasty had the mandate as long as it ruled well. When a dynasty ruled badly, it lost the mandate. The people then rebelled, and a new dynasty took power. This pattern, called the **dynastic** (dy•NAS•tihk) **cycle**, was repeated many times.

For example, after Shi Huangdi died, his son ruled badly. War broke out, and the Han Dynasty took power in 202 B.C. At first, the Han ruled well, and people believed they had the mandate. After 400 years, however, rebellions and other problems brought Han rule to an end.

Dynastic Cycle

Chinese Dragon The Chinese consider dragons to be powerful but helpful creatures that bring good luck. This dragon decorates the wall of a palace.

CRITICAL THINKING

Make Inferences Why do you suppose dragons came to symbolize the emperor?

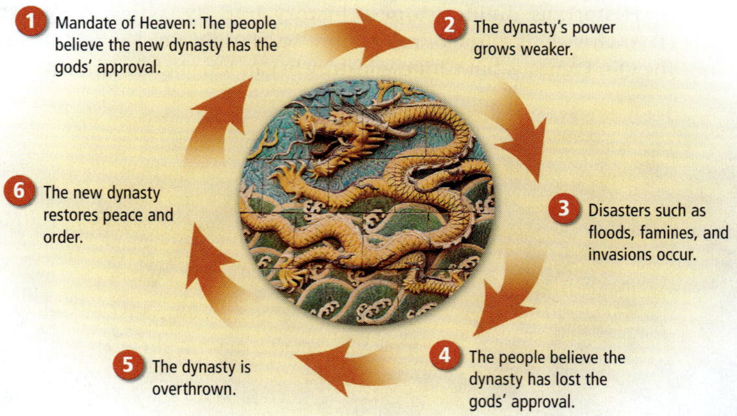

DIFFERENTIATING INSTRUCTION

Struggling Readers

Draw an Illustration

Have students use the diagram on this page to draw a six-panel illustration showing the stages of the dynastic cycle. Each panel should have a short caption explaining the stage in the cycle.

Gifted & Talented

Create a Board Game

Have students create a board game using the principles of the dynastic cycle. Participants in the game will attempt to remain in power while facing various problems within the empire. The students should design a board, write the rules, and demonstrate play.

The Great Wall China's history caused the Chinese to erect a unique structure. In ancient times when small kingdoms often fought each other, many of them built defensive walls. After Shi Huangdi united China, he decided to connect existing walls to stop invading nomads. Hundreds of thousands of peasants were forced to work on this **Great Wall**. Some later dynasties added to and rebuilt the wall. You will learn more about the Great Wall at the end of this section.

FIND MAIN IDEAS Describe the dynastic cycle. The dynastic cycle was the pattern that brought new emperors to rule. The cycle was said to be determined by the Mandate of Heaven.

Scholars and Inventors

KEY QUESTION How did scholars and inventions influence life in China?

China's rulers faced two difficult problems. First, they had to keep order in a huge country. Second, China had a large population to feed. Scholars and inventions helped Chinese emperors solve these two problems.

Confucius To help produce order, a teacher named Confucius (kuhn•FYOO•shuhs) created a code of behavior that was based on traditional values. Confucius taught that people must do what is right. However, he stressed different virtues for people in different social roles. For instance, rulers should be kind and set a good example. People who work in the government should be polite, honest, and fair. All subjects should be obedient and loyal to the ruler. Confucius also urged all men to get an education.

Starting with the Han Dynasty, China's rulers adopted Confucius' ideas. The rulers wanted well-trained people to help them run the country. They also wanted people who were virtuous and loyal. A state exam based on **Confucianism** (kuhn•FYOO•shuh•nihz•uhm), or the teachings of Confucius, was used to test men for government work. The exam was very difficult. It tested people's knowledge of history, law, and poetry as well as Confucianism. Most people failed the test. Those who passed became **scholar-officials**, or educated people who worked in the government. This orderly approach to choosing officials helped the Chinese emperors rule their large country.

HISTORY MAKERS

Confucius 551–479 B.C.

Confucius was the most important teacher in Chinese history. As a boy, he worked hard to learn all subjects. In his 30s, Confucius became a popular teacher. Later, he became a minor court official and tried to put his ideas about duty and social order into practice. Other officials resented Confucius' talk of virtue, so he did not keep his job for long. As a result, he felt like a failure. Little did he know that his ideas would guide rulers for centuries after his death.

ONLINE BIOGRAPHY
For more on the life of Confucius, go to the Research & Writing Center @ClassZone.com

Teach

Scholars and Inventors

Talk About It

- What role did the ideas of Confucius play in the early government of China? *(People who wanted government jobs had to pass a difficult examination based on the teachings of Confucius.)*
- Which early inventions of the Chinese made farming more productive? *(iron plow, horse collar, wheelbarrow)*
- What change did China's Mongol rulers make in China's foreign policy? *(Mongol rulers encouraged the trade of ideas.)*
- **Identify Problems and Solutions** What were some of the problems that were solved by Chinese inventors? *(plow loosened the soil, horse collar harnessed animal power, water mills ground grain, paper could be used to keep records, printing could preserve records)*

RESEARCH & WRITING CENTER

HISTORY MAKERS

Confucius

Confucius was not only a philosopher but a musician as well. Historians believe that Confucius sang folk songs and odes from classic poetry while accompanying himself on a type of zither, a stringed instrument.

DIFFERENTIATING INSTRUCTION

Pre-AP

Write about Confucianism

Have students choose one of the sayings by Confucius listed here. Then have each student write a paragraph in which he or she explains what the saying means. The student should also indicate whether the saying might apply to a teacher, a government official, or both.

Sayings:

- They must often change, who would be constant in happiness or wisdom.
- It is not possible for one to teach others who cannot teach his own family.
- He who merely knows right principles is not equal to him who loves them.
- If you enjoy what you do, you'll never work another day in your life.

RESEARCH & WRITING CENTER

Analyzing Primary Sources

Marco Polo was about 17 when he first traveled to Asia with his father and uncle. The Mongol emperor Kublai Khan, who ruled China at that time, took a liking to Polo and appointed him to serve in several government positions. Upon returning to Europe, Polo wrote *The Travels of Marco Polo,* a book that strongly influenced geographic exploration.

- What material did the Chinese first use to make "paper" money? *(tree bark)*
- Why did the Chinese make paper money in different sizes? *(to indicate different values)*

DOCUMENT-BASED QUESTION ANSWER

Other cultures adopted paper money because it is lighter than coins and because it established a common currency and value.

More About . . .

Silk

For thousands of years the Chinese carefully guarded the secret of the production of silk. Silk is produced when silkworms hatch from a silk moth's eggs. The worms eat only mulberry leaves until they are ready to spin a cocoon of silk. The larvae is fumigated and the cocoons are soaked to loosen the fibers. After the cocoon is wound onto a spool, the fine silk filaments are spun together to make silk thread, which is then dyed and woven into cloth. China is the world's leading producer of silk.

ANALYZING Primary Sources

Marco Polo (1254–1324) was an Italian merchant who lived in China for 17 years during the 1200s. Here is a description he wrote about one Chinese invention.

> [The emperor] has money made . . . out of the bark of trees. . . . [The money is] cut up into rectangles of various sizes, longer than they are broad [and worth different amounts]. . . . With this currency [money] he orders all payments to be made throughout every province and kingdom and region of his empire.
>
> Source: *The Travels of Marco Polo,* translated by Ronald Latham

Marco Polo

DOCUMENT-BASED QUESTION

Why do you think paper money was adopted by other cultures?

ONLINE PRIMARY SOURCE To read more of Marco Polo's writing, go to the **Research & Writing Center** @ClassZone.com

Chinese Inventions The Chinese people invented many things that solved problems or created prosperity. Some inventions helped the Chinese to feed their large population. The iron plow loosened soil to prepare it for planting crops. The horse collar enabled farmers to use horses to pull plows and carts. The wheelbarrow made it easier to move heavy loads. Water mills used the power of rivers to grind grain.

Some inventions aided government and education. For example, the invention of paper helped officials to keep records. More books became available after the invention of block printing—printing from large wooden blocks on which a whole page of information has been carved. Later, printing became easier with the invention of movable type—small blocks carved with a single character. The blocks can be rearranged to print new pages. The inventions of paper and printing allowed the Chinese to make records that exist to this day.

China and the World Other cultures adopted some of China's inventions such as paper money, the compass, and gunpowder. For centuries, however, the Chinese kept the secret of how to make silk, a luxurious fabric, and porcelain, a hard white ceramic also called china. Silk and porcelain were rare and beautiful, so the Chinese could trade them in foreign lands to gain wealth.

INTERDISCIPLINARY ACTIVITIES

CONNECT to Science

Create an Illustrated Chart

Have student pairs create an illustrated two-column chart with the headings "Problem" and "Solution." Students should list the inventions discussed on this page under "Solutions" and then identify the problem each one solved. Have them draw illustrations for each invention.

CONNECT to Language Arts

Role-Play Marco Polo's Stories

Have students research some of the stories Marco Polo is reputed to have told about the wonders of the emperor's court. Then ask them to prepare and act out a scene from one of the stories.

Unit Resource Book: East Asia & Southeast Asia

- Primary Sources: from *The Travels of Marco Polo*, p. 48

In contrast, the Chinese had little interest in obtaining foreign goods or technology because they thought their culture was more advanced. For centuries, the Chinese engaged in trade but adopted few foreign ideas.

During the 1200s and 1300s, a foreign people called the Mongols ruled China and encouraged the trade of ideas. They also allowed merchants, such as Marco Polo, to come to China. In the early 1400s, the Ming Dynasty ruled China and sponsored great voyages of exploration to Arabia and Africa. These were led by an admiral named Zheng He (juhng huh).

After the 1430s, scholar-officials urged the emperor to cut off contact with foreigners. As a result, China turned inward and was not interested in developments in other lands. By the 1800s, the West had developed more advanced technology and weapons than China had.

ANALYZE EFFECTS Discuss how scholars and inventions influenced life in China. Confucius developed a code of behavior that helped produce order in Chinese society. Inventions helped the Chinese feed their large population and aided government and education.

CONNECT to Science

Some time in the 1100s, sailors in China discovered how to make a magnetic compass. Earth is like a gigantic magnet. Lodestone, a magnetic rock, will point north if it is placed on a stick floating in water. After identifying where north is, sailors can find any other direction. So the invention of the compass greatly helped sailors and other travelers to plan their routes.

People in other countries also began to use the magnetic compass. This new technology helped to bring about an increase in trade. It also helped European explorers to go on voyages around the world.

Magnetic Compass
The Chinese characters around the rim of this compass give directions.

Activity

Make a Compass

Materials

- paper clip
- magnet
- piece of cork
- marker
- water
- bowl

1. Unfold a paper clip and rub one end on magnet.

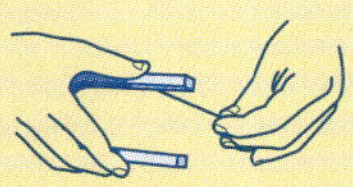

2. Stick the paper clip through a piece of cork or foam and mark magnetized end with marker.

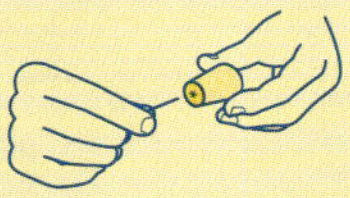

3. Pour water in bowl and float cork with paper clip on top.

4. Magnetized end of paper clip will point north.

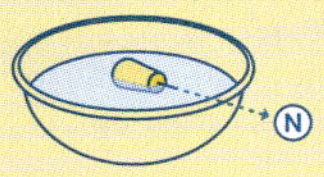

DIFFERENTIATING INSTRUCTION

English Learners

Understand Scientific Terms

Non-native English speakers may not be familiar with the term *magnetic compass*, nor with other words used in the activity on this page, such as *magnet, lodestone,* and *magnetized.* Have students search for these words in their own languages before completing the Connect to Science activity. Pair these students with native English speakers to make a compass.

Pre-AP

Report on Zheng He

Have students work in pairs or small groups to research information about the travels of Zheng He. Information should include a map of the routes traveled, notes on cultural exchanges between China and the countries visited, and highlights of the journeys. One student from each pair or group should present the research findings to the class.

More About . . .

Mongol Rule

The Mongols were the first foreign peoples to rule China. They established the Yuan Dynasty in 1271. However, the Mongols had little success in ruling China, perhaps because they assigned all important government positions to Mongolians.

Although the Mongolian rulers encouraged trade with other lands and generally welcomed foreigners, they also made it difficult for Chinese traders to travel outside of China.

The Yuan Dynasty—and with it Mongol rule—lasted less than 100 years. In 1368, after popular uprisings by both peasants and nobles, the last of the Mongol emperors was expelled.

CONNECT to Science

Tell students that during the earlier stages of airplane development, aviators relied on the magnetic compass to find their way. Because the magnetic compass does not require an external source of power, it continues to be used on today's airplanes as a backup or emergency instrument.

Unit Resource Book: East Asia & Southeast Asia

- History Makers: Zheng He, p. 41
- Connect to Science: Make a Compass, p. 47

Teach

The Rise of Communism

Reader, Recorder, Reporter

- Name the two sides that fought in China's civil war and explain why they were fighting. *(Nationalists and Communists; they fought for control of China.)*
- What was Chairman Mao's plan for producing food? *(Mao sent peasants to work on government-owned farms.)*
- **Identify Problems and Solutions** What problems prompted the rise of the Chinese Communist Party? *(widespread hunger, presence of foreigners, lack of farmland)*

The Long March *1934–1935*

CONNECT Geography & History

Point out that the Long March was a military maneuver—a retreat from Nationalist forces. Ask students what factors may have determined the route taken by the Communists. *(The Communists would have tried to avoid areas held by the Nationalists. They may have chosen to march through difficult terrain to make pursuit more difficult.)*

ANSWERS

1. **Movement** west, north, northeast
2. **Place** A mountain base is difficult to reach, and its defenders generally can see invaders approaching from a distance. Because invaders must climb to a mountain base, they are usually easier to repel.

More About . . .

Mao Zedong

Although Mao came from peasant stock, he was well educated. He received a modern education as well as training in the Chinese classics. Mao was one of the first to join the Chinese Communist Party. He helped to build the Red Army, which he led on the Long March. Although thousands died along the way, Mao emerged as the most important of the Communist leaders.

Unit Resource Book: East Asia & Southeast Asia
- History Makers: Mao Zedong, p. 42

The Rise of Communism

KEY QUESTION What impact did communism have on China?

A new dynasty called the Qing (chihng) took over China in 1644. During the 1800s, European nations wanted to trade with China. When the Qing rulers refused, the Europeans defeated China in a series of wars. The Europeans forced the Qing to open China and let them trade in several port cities. The United States and Japan were also given the right to trade in China.

China's population had grown rapidly, but food production did not increase at the same rate. People became angry about the widespread hunger and also about China's defeat at the hands of foreigners. As a result, many rebellions against the Qing took place.

In 1912, a group called the Nationalist Party defeated the Qing and set up a republic. The Nationalists then had to fight many warlords. By 1928, they had united China.

The Long March 1934–1935

CONNECT Geography & History

1. **Movement** What direction or directions did the Communists travel during their Long March?
2. **Place** What advantage did they gain by moving to a new base in the mountains? Explain how this would affect their ability to defend themselves.

Communist Revolution In 1921, a group of students and others formed the Chinese Communist Party. **Communism** is a system in which the government holds nearly all political power and the means of production in the name of the people. The Communists wanted to take over China. They set up bases in the countryside and gave land to peasants, many of whom joined them.

By 1930, civil war had broken out between the Nationalists and Communists. In 1934, the Nationalist army surrounded the Communists. To escape, the Communists fled on the Long March, walking for 6,000 miles.

Japan invaded China in 1937. The Nationalists and Communists stopped fighting and worked together to fight Japan. After Japan was defeated at the end of World War II, the civil war began again. In 1949, the Communists won and formed the People's Republic of China. The Nationalists fled to Taiwan (TY•wahn), an island off of China's southeast coast.

DIFFERENTIATING INSTRUCTION: TIERED ACTIVITIES

OBJECTIVE Create a time line of events leading to the Communist takeover of China.

Basic

Have students list these events on a time line:
- Japanese invasion, 1937
- Formation of the Chinese Communist Party, 1921
- Beginning of civil war, 1930
- Formation of the People's Republic of China, 1949
- Long March begins, 1934

On Level

Have students list the same events in chronological order on a time line. Entries should include the date, the event, and what occurred as a result of the event.

Challenge

Have students create a time line of at least five events leading to the Communist takeover of China. Entries should include the date, the event, what occurred as a result of the event, and a sentence explaining why that event was relevant in China's history.

Chairman Mao In 1949, **Mao Zedong** (MOW dzuh•dahng) became chairman, or head, of China's Communist government. He seized farmland from large landholders and divided it among the peasants. The government also took over all industry and business. It even controlled the personal lives of the Chinese people.

In 1958, Mao started an economic plan called the Great Leap Forward. He believed in using the hard work of peasants instead of modern technology to increase food production. Thousands of peasants were sent to work on huge, government-owned farms. Mao's program failed to increase food production, so in 1961 it was stopped. The Great Leap Forward resulted in over 20 million deaths from starvation. Some leaders began to talk about reversing Mao's policies.

In 1966, Mao called for a Cultural Revolution to stop those plans. He urged millions of high school and college students to form groups called Red Guards. They attacked educated people and criticized leaders who wanted change. The Red Guards closed schools and factories. They caused so much chaos that Mao finally had to send the army to stop them. After Mao died in 1976, the leaders who wanted change took over China.

Mao Zedong For 27 years, Mao Zedong was the most powerful man in China. This portrait still hangs outside the old palace of the emperors. **Why do you think Mao's portrait hangs outside a palace?** Possible answer: because Mao was considered an emperor of sorts

DRAW CONCLUSIONS Explain the impact of communism on China. Communism transformed China's government, economy, and way of life.

Section 2 Assessment

ONLINE QUIZ For test practice, go to **Interactive Review @ClassZone.com**

TERMS & NAMES

1. **Explain the importance of**
   - Shi Huangdi
   - dynastic cycle
   - Confucianism
   - Mao Zedong

USE YOUR READING NOTES

2. **Identify Problems and Solutions** Use your completed problem-solution chart to answer the following question:
   How successful was Mao Zedong at solving the two problems of keeping order and feeding people? Explain.

PROBLEMS	SOLUTIONS
Gaining Power	
Keeping Order	
Feeding a Large Population	

KEY IDEAS

3. Why did China's emperors think it was a good idea to employ scholar-officials who had studied Confucianism?
4. How did the iron plow and the horse collar help to feed China's large population?
5. How did the inventions of paper and block printing make record-keeping easier?

CRITICAL THINKING

6. **Draw Conclusions** How was the Mandate of Heaven connected to the dynastic cycle?
7. **Analyze Effects** In what ways did the Communists promote change, and in what ways did they oppose it?
8. **CONNECT to Today** Why do you think the Chinese people still honor Mao?
9. **TECHNOLOGY** **Create a Multimedia Presentation** Create a slide show about a group of inventions that changed farming, government, education, or trade. Show how the inventions work and explain why they made a difference.

CONNECT to the Essential Question

How is China's traditional way of life changing in response to global trends?

Ask students what they have learned so far that can help them answer this question. Students might mention the following:

- China is no longer under dynastic rule.
- Following Mao's death, China's new leadership wanted change.

4 Assess & Reteach

Assess Have students complete the Section Assessment.

Unit Resource Book: East Asia & Southeast Asia
- Section Quiz, p. 52

Interactive Review @ClassZone.com
- Section Quiz

Power Presentations

Test Generator

Reteach Divide students into small groups to create a section review. One group might present a panel discussion about emperors and dynasties. Another could present a panel on Chinese scholars and inventors, while a third could present a panel on the rise of communism in China. Panelists should be prepared to answer questions from the class.

Unit Resource Book: East Asia & Southeast Asia
- Reteaching Activity, p. 56

SECTION 2 ASSESSMENT ANSWERS

Terms & Names

1. Shi Huangdi: first emperor to unify much of China; dynastic cycle: pattern of the rise and fall of dynasties; Confucianism: belief system based on the teachings of Confucius; Mao Zedong: Communist leader of China

Use Your Reading Notes

2. Mao had mixed success at keeping order and was not successful at feeding the Chinese people.

PROBLEMS	SOLUTIONS
Gaining Power	Dynasties: Mandate of Heaven
Keeping Order	Scholar-officials: code of behavior
Feeding a Large Population	Farm tool inventions

Key Ideas

3. Scholar-officials understood the principles of Confucianism.
4. iron plow helped to prepare soil; horse collar allowed horses to do farm work
5. Paper offered a medium on which to keep accounts. Block printing speeded up printing.

Critical Thinking

6. When a dynasty ruled badly, the people believed that the dynasty had lost the mandate, and so they overthrew the dynasty. A new dynasty—believed to have the mandate—replaced the old.
7. Promoted change: took over privately owned lands, increased food production; Opposed change: refused to use modern technology
8. Possible answer: because Mao was the first head of the People's Republic of China
9. Use the rubric to score presentations.

Presentation Rubric

	Images	Description
4	engaging and informative	clear, concise
3	reasonably informative	reasonably clear
2	somewhat informative	fair
1	uninformative and uninteresting	poor

1 Plan & Prepare

Objectives: Identify the architectural and defensive features of the Great Wall and summarize its evolution.

Visual Vocabulary: Have students discuss the illustration. Then read the descriptions of items listed below and ask students to locate each one in the illustration. Then tell them the precise term shown in boldface. Add terms to a class vocabulary list, illustrated poster, or Word Wall.

- elevated lookout point **(watchtower)**
- a notch in a wall **(crenellation)**
- a low wall at the edge of a roof **(parapet)**
- an attack warning sign **(signal fire)**
- surrounding physical geography **(terrain)**
- a military machine for throwing rocks **(catapult)**

2 Teach

Engage . . .

DISPLAY the "The Great Wall of China" feature from the **Power Presentations** DVD-ROM or the *Animations Center* at **ClassZone.com**. As the animation loads on screen, ask students:

- In what part of China is the wall located?
- What is the geography of the land like where the wall is built?
- What materials were used to build the wall?
- How are the enemy soldiers attacking?
- How are the Chinese defenders keeping the attackers out?

Encourage all answers.

Explore . . .

ROLL OVER sections of the illustration or click on items in the menu. Ask students to choose which areas they would like to explore further.

- Then return to the discussion questions. Have any answers changed?
- Have students jot down a list of at least three new questions that arise from studying the rollover features.

Animated GEOGRAPHY
The Great Wall of China

Click here to enter the Great Wall of China @ClassZone.com

DEFENDING CHINA

Many barriers protect China: mountains to the west and south, an ocean to the east. But in the north, nomads were able to invade China, so the Chinese built a barrier—the Great Wall—to try to stop them.

Click here to see the Great Wall as it snakes across mountains, plains, and deserts. Learn about the challenges of building and maintaining about 4,000 miles of walls.

Click here to see the weapons used in a Great Wall battle. Learn about watchtowers, lookout soldiers, and the signals used to warn of an attack.

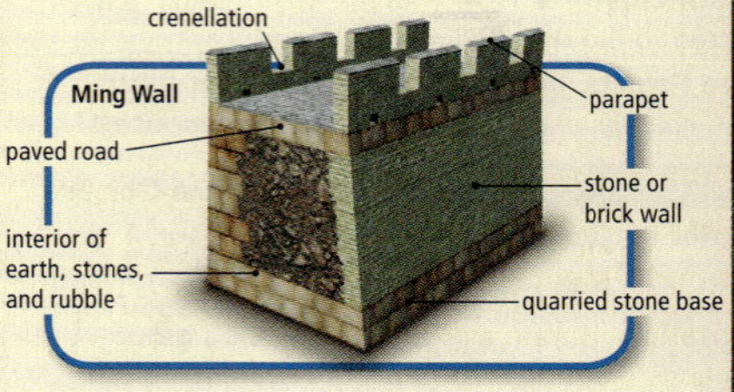

Click here to see inside the Great Wall. The construction methods were so sound that the Wall has lasted for centuries.

DIFFERENTIATING INSTRUCTION: TIERED ACTIVITIES

OBJECTIVE Identify dimensions of the Great Wall and its key defensive features, and consider how the purpose of the Great Wall evolved within the context of history.

Basic

Have students work in small groups to create a diagram or series of diagrams showing the dimensions of the Great Wall. First students should study the illustration and then research dimensions of the wall. Finally, have each group create diagrams illustrating various dimensions of the Great Wall, such as its width, height, the distance between watchtowers, and height of the towers.

The Great Wall of China Over Time

The Great Wall of China was built over many years.
- During the Warring States period (403 to 221 B.C.), rival kingdoms built separate walls.
- After Shi Huangdi united China (221 B.C.), a more connected wall was built.
- Ming Dynasty (A.D. 1368 to 1644) rulers repaired and added new sections to the wall.

GeoActivity

Sketch a Battle Plan You have received word that a group of invading nomads are bearing down on the part of the Great Wall that you defend. Sketch a battle plan to defeat the nomads, and use the surrounding terrain in your strategy.

On Level

Have students work in pairs. Tell them to study the illustration to identify aspects of the Great Wall that gave its defenders an advantage in battle and provided special challenges to attackers. Have them list their observations and note the advantage of each feature. Then student pairs should exchange lists with another pair and compare their findings.

Challenge

Have students work in pairs to write a report on the evolution of the Great Wall. The report should cover the intent of the first builders and explain the different objectives of later builders. The report should also explain how advances in weapon technology caused changes in construction methods. Have students present their reports orally.

Explain . . .

CHOOSE topics to help students answer their questions, such as:

- **Wall construction:** In its early stages, the interior part of the walls was made of packed earth and gravel. The exterior was constructed of either stone or brick.
- **Defense:** Signal fires were lit on the towers to announce the arrival of enemy forces. To prevent the enemy from scaling the wall, soldiers shot arrows using crossbows.
- **Life on the Wall:** Soldiers lived in the watchtowers. They bought food from the local villagers and often kept gardens near the wall to provide additional food.
- **The Wall:** Today only about 30 percent of the Great Wall remains in good condition. The Great Wall is China's number one tourist attraction.

Unit Transparency Book
- Cultures: The Great Wall, TT9

3 Wrap Up

Elaborate . . .

DIVIDE students into four teams to compete in a geography bee. Have each team write a five-question quiz on the Great Wall. Then have each team exchange their quiz with another team. Team members should read the questions and work together to determine the answers. The team with the most correct answers wins the geography bee.

Evaluate . . .

Sketch a Battle Plan

Activity Rubric

4 Excellent The student's sketch accurately accounts for the terrain and includes several details.

3 Strong The student's sketch is fairly accurate and includes some detail.

2 Fair The student's sketch includes only a few details.

1 Poor The sketch lacks detail and fails to address the challenges of the terrain.

1 Plan & Prepare

Objectives

- Identify values that were important in traditional Chinese society
- Analyze changes in Chinese life following the Communist takeover
- Compare life in rural and urban China

Read for the Essential Question

Help students read for a purpose by reminding them of the Essential Question: "How is China's traditional way of life changing in response to global trends?" Point out that they can use items in the section, such as Key Questions, special features, images, and maps, to formulate answers to the question.

Vocabulary

Best Practices Toolkit

Use the ***Best Practices Toolkit*** to model strategies for vocabulary notetaking. Vary strategies throughout the year. Choose from: *Knowledge Rating, Predicting ABCs, Definition Mapping, Word Sort, Word Wheel, Frayer Model (Word Squares), Magnet Words,* and *Student VOC.*

Vocabulary Strategies, TT9–TT16

Reading Strategy

Best Practices Toolkit

Display the Web Diagram transparency.

- Remind students that main ideas are important ideas in a segment of text.
- Model filling in main ideas on a graphic organizer using the subheads throughout this section.
- Have students suggest entries. Tell students that the main ideas in each block of text in the section will help them answer the Key Questions.

Web Diagram, TT28

SECTION 3 Reading for Understanding

Key Ideas

BEFORE, YOU LEARNED

China has a very long history. The people lived a traditional life during much of that history—until the Communist Revolution.

NOW YOU WILL LEARN

Some aspects of traditional life survive in China, while new ways of life have been developed.

Vocabulary

TERMS & NAMES

Daoism (DOW•IHZ•uhm) a Chinese belief system based on the idea of natural order in the world

Dao (DOW) according to the belief system of Daoism, a force that guides the whole universe

Silk Roads trade routes used between China and Southwest Asia

Buddhism a religion that is based on the teaching of Siddhartha Gautama

Siddhartha Gautama (sihd•DAHR•tuh GAW•tuh•muh) the founder of Buddhism also known as the Buddha, or "enlightened one"

Buddha the founder of Buddhism who taught that people could end their suffering by stopping their desire for wordly goods

BACKGROUND VOCABULARY

ritual a set of regularly followed ceremonies

communal shared ownership of property

REVIEW

Confucianism a Chinese belief system based on the teachings of Confucius

Reading Strategy

Re-create the web diagram shown at right. As you read and respond to the **KEY QUESTIONS**, use the outer ovals to record main ideas about changes in Chinese culture. Add ovals as needed.

See Skillbuilder Handbook, page R4

FIND MAIN IDEAS

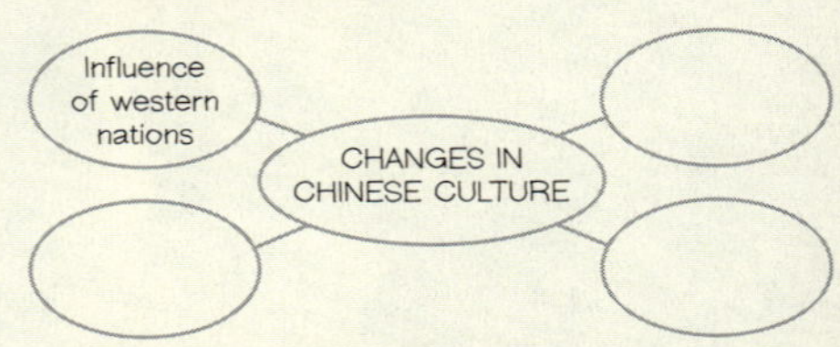

GRAPHIC ORGANIZERS
Go to **Interactive Review** @ClassZone.com

PRETEACHING VOCABULARY

English Learners

Pronounce and Preview

Pronounce each term for students. Review the meanings of words used in the definitions, such as *universe, trade routes,* and *ceremonies*.

- To modify vocabulary learning, have students complete worksheets as they read the section.

Unit Resource Book: East Asia & Southeast Asia
- Vocabulary Practice, p. 35

Inclusion

Match Terms and Definitions

Have student pairs write the vocabulary terms on one set of index cards and the definitions on another set. Then have pairs play a matching game. One student should shuffle the cards and place them face up on a desk. The other student should then match each term with its correct definition. Encourage partners to work together to correct any errors. Students should then switch roles.

SECTION 3 CULTURE

The Five-Phoenix Temple in south-central China's Black Dragon Pool Park

China's giant pandas

Changing Chinese Culture

Connecting to Your World

Have you ever asked your grandparents what life was like when they were your age? Cultural change can happen at any place and time. For example, Mei (may) Ling's grandmother, who lives in the country, talks of the village school she attended. Mei Ling's parents moved to Shanghai in search of better jobs. As a result, Mei Ling attends a big urban school instead of a village school. The move from the farm to the city is just one type of change in China. In this section, you'll learn how traditional Chinese culture is affected by these changes.

Chinese Family Like Mei Ling, this Chinese girl has a strong relationship with her grandmother.

Traditional Chinese Culture

KEY QUESTION What are some aspects of traditional Chinese culture?

Chinese culture is thousands of years old. Though life has changed in modern China, traditional values—such as close family ties and respect for one's parents and ancestors—remain strong.

Martial Arts Chinese students practice traditional martial arts.

775

2 Focus & Motivate

3-Minute Warm-Up

Write on the board or display the transparency:

- Look at the photographs of the people on this page and the next. How does their clothing help you understand the changes in China? *(People on this page are wearing Western-style clothing; the people on the next page are wearing traditional clothing but only because they are dressed for a celebration.)*

Unit Transparency Book
- 3-Minute Warm-Ups, TT1

Connecting to Your World

More About . . .

Martial Arts and Cultural Change

Martial arts began as a way to prepare soldiers for war. All martial arts training includes a series of movements to stretch, build stamina, and strengthen muscles, bones, and tendons.

3 Teach

Traditional Chinese Culture

Talk About It

- What are the defining characteristics of traditional Chinese culture? *(close family ties and respect for parents and ancestors)*
- **Find Main Ideas** How did the three main belief systems influence Chinese life? *(Each helped shape the values of the Chinese people.)*

SECTION 3 PROGRAM RESOURCES

Unit Resource Book: East Asia and Southeast Asia

ON LEVEL

Unit Resource Book
- Reading Study Guide, p. 7
- Section Quiz, p. 53

STRUGGLING READERS

Unit Resource Book
- Reading Study Guide with Additional Support, p. 15
- Vocabulary Practice, p. 35
- Skillbuilder Practice, p. 36
- Section Quiz, p. 53
- Reteaching Activity, p. 57

eEdition with Audio DVD-ROM

ENGLISH LEARNERS

Pupil Edition in Spanish

eEdition with Audio DVD-ROM

Unit Resource Book
- RSG (Spanish), p. 23
- RSG with Additional Support (Spanish), p. 31

Multi-Language Glossary

Test Generator
- Section Quiz in Spanish

INCLUSION

Unit Resource Book
- RSG with Additional Support, p. 15
- Section Quiz, p. 53
- Reteaching Activity, p. 57

GIFTED & TALENTED

Unit Resource Book
- Interdisciplinary Projects, pp. 43–44
- Section Quiz, p. 53

PRE-AP

Unit Resource Book
- Primary Sources, p. 49
- Section Quiz, p. 53

TECHNOLOGY

Unit Transparency Book
- 3-Minute Warm-Ups, TT1
- Essential Question, TT2
- Critical Thinking, TT5
- Cultures: Landscape Painting, TT10

Daily Test Practice Transparencies
- Section 3, TT85

Power Presentations

ClassZone.com

World Cultures and Geography Video on DVD

More About . . .

Honoring Ancestors

One way in which the Chinese have traditionally honored their ancestors is by treasuring their ancestors' portraits. These images are central to ceremonies in which ancestors are honored. The ceremonies are based, in part, on the belief that ancestors have the power to affect the health and happiness of their descendants. Portraits of ancestors are part of a family's legacy and, as such, are never sold and rarely shown to those outside the family. The Chinese believe that owning a portrait of someone else's ancestor would result in bad luck. Ancestor portraits of some Qing Dynasty members are 11 feet tall and 4 feet wide.

Unit Resource Book: East Asia & Southeast Asia

- Interdisciplinary Projects: Ancient Chinese Medicine, p. 43

More About . . .

Confucius' Birthday Celebration

Confucius' birthday is traditionally celebrated on September 28. Because Confucius was a teacher and believed that education was an important part of a fulfilled life, China honors its teachers around his birthday. The dancers in the picture on this page are wearing red, a traditional color for a Chinese holiday. The bird on the front of the costume is a white crane, which is traditionally used to identify a person of the first rank in society.

Unit Resource Book: East Asia & Southeast Asia

- Interdisciplinary Projects: from the *Analects of Confucius*, p. 44

Family and Village Life In traditional Chinese culture, family was very important. Older men made all the major decisions for the family. Men were taught to perform **rituals** honoring their ancestors. Women were taught that it was their duty to marry, have children, and care for their husbands' parents. Children were expected to respect their parents and care for them in old age.

For thousands of years, most Chinese people were farmers. Each day the villagers would go out to the fields to work. Nearly all Chinese lived in small rural villages of about 200 people. Houses in the village were crowded closely together. Most had a courtyard where chickens and pigs wandered in search of food. Small vegetable gardens provided food for the family meals. The villagers shared a well and bought some goods at local markets.

Beliefs Three different belief systems—Confucianism, Daoism, and Buddhism—have influenced Chinese life. **Confucianism** is the belief system based on the teachings of Confucius, whom you learned about in Section 2. In Confucianism, the person who is older or has more power has a responsibility to be kind to others. The younger or less powerful person is expected to be respectful and obedient. Confucianism also teaches that people can help to better themselves through education.

The second Chinese belief system, **Daoism** (DOW•IHZ•uhm), is based on the ideas of a teacher named Laozi. He taught that the world has a natural order and that a force called the **Dao** (DOW), meaning "the Way," guides all things. Laozi taught that people should observe nature and use those observations to find a way to peace and harmony.

Confucius' Birthday Celebration Dancers in traditional costumes celebrate the birth of Confucius in 551 B.C. **Why do the Chinese still honor Confucius?**
because his ideas are still very influential in China

INTERDISCIPLINARY ACTIVITIES

CONNECT to Language Arts

Describe Family Traditions

Point out to students that honoring one's ancestors is traditional in China. Allow students time to discuss the importance of this tradition, as well as ways in which Americans honor their ancestors. *(family photos, tending graves, telling stories that keep family history alive)* Then have students write a paragraph describing how they honor their own ancestors.

CONNECT to Art

Make a Nature Collage

Have students work in pairs to collect or draw images of nature that would show natural order or lead to feelings of peace and harmony. Students should complete the project by adding a title, a caption, or both. Display the collages in the classroom.

Unit Resource Book: East Asia & Southeast Asia

- Primary Sources: from *Dao De Ching*, p. 49

COMPARING Belief Systems

CONFUCIANISM

- Social order and good government should be based on strong family relationships.
- Respect for parents and elders helps to create order in society.
- Education helps to improve both individuals and society.

DAOISM

- The natural order—or the relationship among all living things in the universe—matters more than the social order.
- A universal force called the Dao (meaning "the Way") guides all things.
- Humans should learn to live simply and in harmony with nature.

BUDDHISM

- People suffer because of emotional attachments and their desire for worldly goods.
- People can stop suffering and achieve peace (or nirvana) by not wanting.
- To achieve nirvana, people should follow the Eightfold Path: right views, goals, speech, actions, job, effort, concentration, and meditation.

CRITICAL THINKING

Compare and Contrast How are the three belief systems similar? How are they different?

Missionaries who traveled the **Silk Roads**, the trade routes used between China and Southwest Asia, brought **Buddhism** to China. Buddhism is a religion based on the teachings of **Siddhartha Gautama** (sihd•DAHR•tuh GAW•tuh•muh), also known as the **Buddha**, or "enlightened one." Buddha taught that people suffer because their minds are not at ease. This suffering comes from wanting life to be different. The Buddha said people could stop the suffering by accepting life as it is and following the Eightfold Path, detailed above.

Written Language The development of a written Chinese language allowed the government to communicate with people in all parts of the empire, no matter what language they spoke. Chinese characters do not represent sounds, as Western letters do. They represent ideas or words. People who spoke differently could still read each other's letters because they used the same characters to write.

Chinese Character This character stands for the Dao, which means "the way."

The Arts Some of traditional China's greatest achievements in the arts were in bronze casting, jade carving, and dance. Bronze objects such as bells and containers featured geometric designs and were decorated with tigers, snakes, birds, and oxen. Jade was used to make jewelry, containers, and statues. There were two kinds of traditional Chinese dance: one based on proper behavior as taught by Confucius, and one based on Chinese warrior exercises, which we call martial arts.

SUMMARIZE Discuss some of the aspects of traditional Chinese culture. close family ties; respect for elders; belief systems of Confucianism, Daoism, and Buddhism; bronze casting; jade carving

DIFFERENTIATING INSTRUCTION

Inclusion

Make a Belief Systems Chart

To help visually impaired students, record the textbook narrative on the three belief systems from this section. After students have had a chance to listen to the recording, have a sighted student help the impaired student construct a large print chart of the three belief systems.

Gifted & Talented

Research the Arts

Have students work in small groups and research landscape painting or traditional Chinese dance and make a presentation to the class. Reports should include information on the history of the art form, training of artisans or performers, and illustrations of the art.

Unit Transparency Book

- Cultures: Landscape Painting, TT10

COMPARING

Belief Systems

Tell students that when Buddhism was first introduced in China from India, it was not readily accepted by all members of society. The nobles in particular did not understand how reaching a state of inner peace, or Nirvana, would be of benefit to society. Eventually, Buddhism became more widely practiced in China than in India. Today, many Buddhist temples can be found in China's cities and rural areas.

Some of Daoism's ideas were absorbed into Buddhism. Daoism is practiced in China and in other parts of the world today. Millions also follow the ethical teachings of Confucianism.

CRITICAL THINKING ANSWER

Compare and Contrast All three belief systems seek to help people find peace and order in their lives. However, Confucianism emphasizes social order, Daoism stresses the importance of all living things, and Buddhism stresses order on a personal level.

More About . . .

Chinese Writing

It is believed that Chinese writing was invented between 2500 and 2000 B.C. Although Chinese characters began as pictures, they gradually became stylized and no longer resembled what they represent. Although there is no limit to the number of characters used in Chinese writing, one needs to know about 3,000 characters to read a magazine or a newspaper, and 6,000 characters to read Chinese literature or technical material. Twelve basic strokes are used to write Chinese characters, with each character requiring between 1 and 64 strokes. However, all characters, regardless of the number of strokes they require, must be the same size on a printed page.

Population Density of China

CONNECT Geography & Culture

Point out that there are few cities in western China. Ask students to consider how this affects population density. *(Cities are more densely populated than rural areas. An area with few cities is unlikely to be densely populated.)*

ANSWERS

1. **Place** purple
2. **Region** eastern China

Teach

Changes in Chinese Life

Think, Pair, Share

- How did the establishment of industries affect the people who lived in rural areas? *(People were allowed to move from farms and fill industry jobs in towns and cities.)*
- How did education change following the economic reforms of 1978? *(School attendance through ninth grade became compulsory; very high standards were set for education.)*
- **Find Main Ideas** In what ways did life in China change as a result of the Communist takeover? *(Family farms ceased to exist; reforms brought industry and many people to the cities; educational opportunities improved.)*

Comparing Regions

The U.S. Great Migration

Like the population shift currently occurring in China, the Great Migration of African Americans in the United States was a population shift from rural areas to urban ones. In the early 20th century, millions of African Americans began moving from the rural south to the industrial north in search of employment.

By 1970, the African-American population of the United States was no longer predominantly southern and rural. Instead, migration had transformed it into a population that was largely northern and urban.

Population Density of China

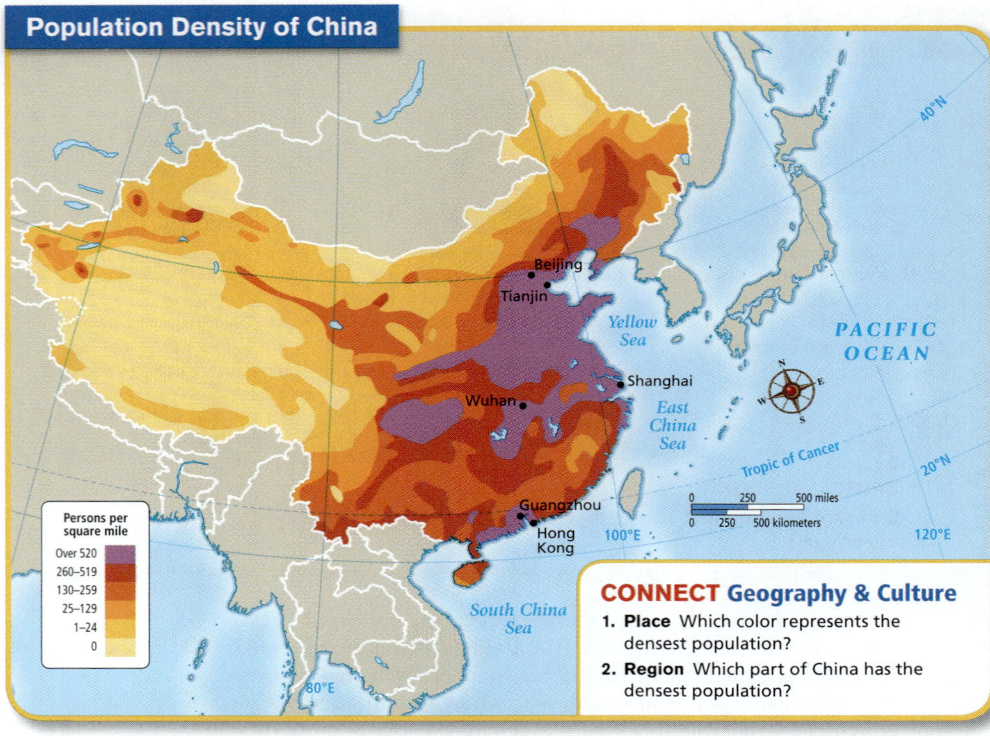

CONNECT Geography & Culture

1. **Place** Which color represents the densest population?
2. **Region** Which part of China has the densest population?

Changes in Chinese Life

KEY QUESTION How did life change in China after the Communists took over?

The traditional Chinese way of life changed dramatically with the rise of communism. Family farms ceased to exist, and farms were reorganized into **communal** farms where ownership was shared. The farms were strictly controlled by the government. Every family had a registration booklet listing its members. Families needed the booklet to get food, medical services, housing, schooling, and jobs.

Moving to Cities As a part of the new way of life under the Communists, people had to get government permission to move. But little movement was allowed until after 1978, when economic reforms were made. These reforms brought new industry to the cities. As a result, workers were allowed to move from farms into towns or cities to fill jobs there. Workers also moved into cities on the east coast where the booming economy required many construction workers. Industry even began to emerge in rural areas.

DIFFERENTIATING INSTRUCTION

English Learners

Define Terms

To help students understand the terms *family farms*, *communal farms*, *registration booklet*, *economic reforms*, and *construction workers*, have students work in pairs. Tell them to find each term in the text and to read the sentence in which it appears. Then have students write each term on the front of a card and provide a brief definition on the reverse side.

Pre-AP

Debate the Farm Policy

Divide students into two groups to debate the following question: Did the creation of communal farms improve the lives of the Chinese people? Students should research their position and debate the issue before the class.

Beijing China's capital is a crowded, rapidly modernizing city. Today, China has about 1.3 billion people, compared with about 298 million in the United States.

The shift in population to the cities was a major change in Chinese society. Today, about 60 percent of Chinese people live in rural areas, compared to almost 88 percent in 1950. Part of the reason for this spectacular urban growth is due to the expanded industrial sites in smaller towns, especially in eastern China. China has 34 cities with over a million people. In comparison, the United States has 9 cities of that size. China's two largest cities, Beijing and Shanghai, are huge. At the end of 2004, Beijing had 14.9 million residents and Shanghai, China's largest city, had 17.4 million. The population of these two cities is about equal to that of the state of California.

Population in China

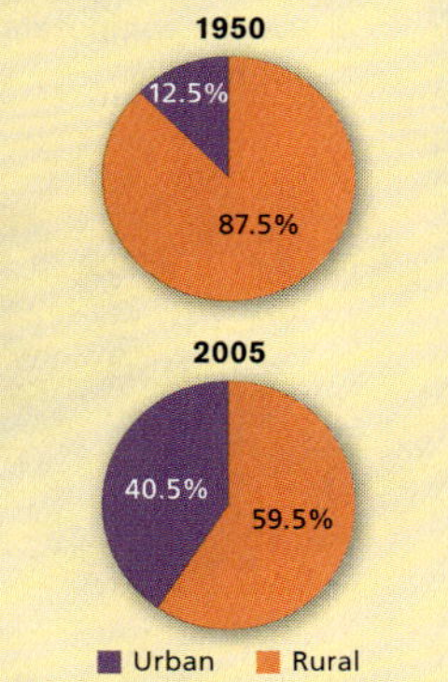

Source: *China Statistical Yearbook 2004* and *Globalis*

CRITICAL THINKING

Compare and Contrast
By how much did the urban population increase between 1950 and 2005?

Jobs and Education After the economic reforms of 1978, China's new leaders set very high standards for education. Today, students are expected to study many hours; they have little time for extracurricular activities. The law calls for school attendance through ninth grade. More rural children now complete nine years of schooling. Students must then do well on national exams to be able to go on to trade school, high school, or college. However, education is not free. Parents must pay fees and buy textbooks.

Cities have more educational opportunities than rural areas. Many city teenagers go on to high school. However, in the past, only a very small percentage of students went to college. Today, China's goal is to have 20 percent of its students in college by 2010.

EVALUATE Explain how life changed in China after the Communists took over. Farms became communal; the population shifted to the cities after economic reforms were instituted; education became more widely available.

INTERDISCIPLINARY ACTIVITIES

CONNECT to Language Arts

Compare Educational Systems

Have students reread the paragraphs on education in China. Then ask them to write a couple of paragraphs comparing Chinese education with American education.

CONNECT to Math

Make a Population Bar Graph

Have students work in pairs to make a bar graph comparing the 2004 populations of Beijing and Shanghai—China's two largest cities—with that of New York City—the largest city in the United States. Tell them that the 2004 population of New York City was 8.1 million. Remind them to include a title, a key, and any needed labels.

Population in China

Have students agree on the percentage of change in rural to urban population in the 55 years shown in the pie graphs. *(28%)* Ask them to think about how an increase this large would impact life in city areas. *(need for housing and basic services, such as water, sewer, roads, police protection)*

CRITICAL THINKING ANSWER

Compare and Contrast by 28 percent

Unit Resource Book: East Asia & Southeast Asia
- Skillbuilder Practice: Reading a Population Density Map, p. 36

More About . . .

Education in China

In an attempt to improve China's literacy rate, the government instituted a program of language reform. Because written Chinese is difficult to master, an official committee was formed to simplify the language. In 1964, the Committee for Reforming the Chinese Written Language released a list of more than 2,000 written characters, with each one simplified by eliminating some of the strokes originally required to write it.

CONNECT to the Essential Question

How is China's traditional way of life changing in response to global trends?

Ask students what they have learned so far that can help them answer this question. Students might mention the following:

- Some of the population has shifted from rural areas to urban centers.
- High educational standards have been set.

More About . . .

Basketball in China

Basketball has been a favorite pastime of city kids in China since the 1920s. YMCA missionaries introduced the game to China in the late 19th century, and it quickly became popular. It also had an unexpected cultural impact. Male players who wore their hair in a traditional braid found that the braid got in the way of play. As a result, many adopted shorter hairstyles.

Comparing Regions

Protecting Wildlife

Cheng lives in Jiangxi Province, where wild South China tigers have recently been sighted. Prior to this sighting, South China tigers were thought to be extinct. Today the tigers live in nature preserves and are protected by the government.

By 1967, the American red wolf was an endangered species. Then, the U.S. Fish and Wildlife Service rounded up the few remaining survivors and placed them in captivity. After a successful breeding program, they were reintroduced into their natural habitat. Today, about 100 red wolves roam North Carolina.

Presentations Options

Go to the **Activity Center** at **ClassZone.com** to project Worldster.

- Click on the multimedia features of Cheng's Web page.
- Ask students which areas of Cheng's life they would like to explore. Point out that they can listen to the music, examine photographs, learn simple phrases in Mandarin, and more.

CONNECT to Your Life

Journal Entry Remind students that photographs often provide information not covered in the text. For example, the photo of Cheng's family at dinner shows them using chopsticks, a traditional Chinese utensil. Tell students to look for other details in the photos before writing their journal entry.

Worldster

A Day in Cheng's Life

To learn more about Cheng and his world, go to the **Activity Center** @ClassZone.com

In Chinese, Cheng's name looks like this:

成

Hi! My name is Cheng. I am 14 years old and live in Yihuang, a rural town of about 20,000 in eastern China. Studying is very important to me. But I don't study all the time! On weekends, my friends and I play table tennis, basketball, or use the computers at the community center. Let me tell you about my day.

7 A.M. I have breakfast with my parents before they go to work and I leave for school. Today we're having eggs, peanuts, and steamed buns for breakfast.

8 A.M.–5:30 P.M. I have a long school day. We begin by saluting the flag and exercising. During the day, I have classes in Chinese, math, science, and history.

6 P.M. I'm meeting my parents and other family members at my grandparents' house for dinner. Tonight we're having one of my favorites, Four Happiness Pork along with rice and vegetables.

7 P.M. I have a basketball game tonight. I play on my school team. We're pretty good! After-school activities for other kids in my class are sports like gymnastics or music lessons. When I get home, I'll have homework to do.

CONNECT to Your Life

Journal Entry Think about your own daily schedule. What do you spend most of your day doing? How is your life different from Cheng's life? Record your ideas in your journal.

DIFFERENTIATING INSTRUCTION: TIERED ACTIVITIES

OBJECTIVE Explore modern Chinese life by comparing and contrasting social and cultural elements of your life with those of a Chinese adolescent.

Basic

Have students work in pairs to create a visual presentation about their lives similar to the one on this page. Tell them to illustrate at least three of the activities shown: breakfast, school day, dinner, or extracurricular activities.

On Level

Have student groups put on a talk show in which two students, one American and one Chinese, are interviewed about their leisure activities, their interests at school, and their dreams for the future.

Challenge

Have students write an interview in which Cheng asks his grandfather what life was like when he was a boy. Remind students that Cheng's grandfather probably lived on a farm and led a much more traditional life than Cheng.

Clashes Over Popular Culture

KEY QUESTION How has the government tried to control popular culture?

Life in modern China is a blend of old and new ways. Many people still live in rural areas and lead a more traditional life. But even that way of life is changing now that they have more access to television, computers, and cell phones. The Chinese government attempts to control the content available to people through technology.

Recently, the Communists began tighter control of foreign television programs. The government is taking this action because it is worried about the spread of democratic ideas. Television programs are forbidden to use the English language or show violence.

People in cities and even in many small towns have access to e-mail and the Internet. However, they cannot visit Web sites the government does not approve of or send messages that criticize their leaders.

In the past, the government demanded that art spread Communist ideas. Now theaters and art galleries show plays and works about a variety of subjects in the artists' own styles. Writers are beginning to discuss social issues but are not allowed to criticize the government.

DRAW CONCLUSIONS Draw conclusions about how the Chinese government controls popular culture.

The government controls foreign television programs, limits access to Web sites, and does not allow messages critical of leaders. The Communists also do not allow artists to criticize the government.

ONLINE QUIZ For test practice, go to **Interactive Review @ClassZone.com**

Section 3 Assessment

TERMS & NAMES

1. **Explain the importance of**
   - Daoism
   - Silk Roads
   - Buddhism

USE YOUR READING NOTES

2. **Find Main Ideas** Use your completed web to answer the following question:
   How have Chinese society and culture changed since dynastic times?

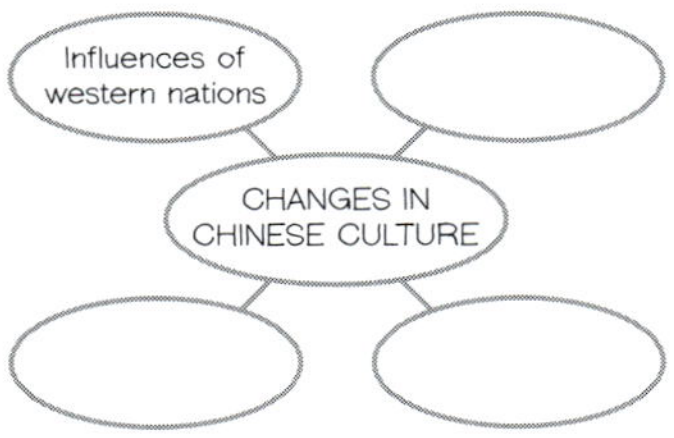

KEY IDEAS

3. What are the most important traditional Chinese values?
4. Which three belief systems influenced Chinese society?
5. How has the distribution of population in China changed since the 1950s?

CRITICAL THINKING

6. **Draw Conclusions** Why do you think the Chinese government is encouraging more students to go to college?
7. **Summarize** In what ways has life in China changed since the new leadership took over after 1978?
8. **CONNECT to Today** Which Confucian idea continues to influence Chinese society?
9. **ART** **Sketch a Design** Use the Internet or art books to find examples of Chinese bronze vessels. Create and sketch a design that could be used to create a bronze vessel.

Teach

Clashes Over Popular Culture

Talk About It

- What aspects of technology does the government try to control? *(television, Internet)*
- **Find Main Ideas** Why does the Chinese government believe it must control technology? *(because so much information is available on television and the Internet)*

4 Assess & Reteach

Assess Have students complete the Section Assessment.

Unit Resource Book: East Asia & Southeast Asia
- Section Quiz, p. 53

Interactive Review @ClassZone.com
- Section Quiz

Power Presentations

Test Generator

Reteach Have pairs of students write down the headings found in the text. Tell each pair to list at least one or two pieces of information for each heading. When they are finished, they can use the information to create quiz questions. Then the pair can exchange their questions with another pair and answer the questions.

Unit Resource Book: East Asia & Southeast Asia
- Reteaching Activity, p. 57

SECTION 3 ASSESSMENT ANSWERS

Terms & Names

1. Daoism: belief system based on the idea of natural order in the world; Silk Roads: routes used by traders moving goods from China to Southwest Asia; Buddhism: religion based on the teachings of Siddhartha Gautama

Use Your Reading Notes

2. many more people in cities; high educational standards; industry expanded; greater access to television, computers, and cell phones

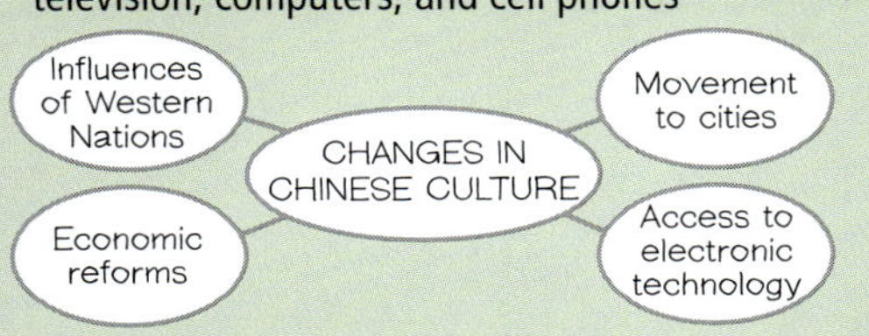

Key Ideas

3. respect for one's parents and ancestors
4. Confucianism, Buddhism, and Daoism
5. Many more people live in cities.

Critical Thinking

6. Possible answer: A well-educated population will help China compete in the global marketplace.
7. workers allowed to move into towns and cities; population shift to the cities; more educational opportunities
8. respect for one's elders
9. Use the rubric to score designs.

Design Rubric

	Presentation of Sketch	Style of Sketch
4	very cleanly drawn	shows thorough understanding
3	neatly drawn	shows good understanding
2	poorly drawn	shows basic understanding
1	sloppily drawn	shows little or no understanding

1 Plan & Prepare

Objective

- Compare and contrast the way of life in the Chinese countryside with that in China's urban areas.

Springboard Ask students to discuss what they know about life in rural and urban areas in the United States. What kinds of jobs do people typically hold in both areas? What different means of transportation are available? What are the advantages and disadvantages of living in the countryside or the city? At the end of the discussion, have students read the feature on China. How does life in rural and urban China compare with their image of life in the United States?

2 Teach

Rural and Urban China

Talk About It

- How is the farmer in rural China transporting his geese and chickens to market? *(by bicycle)*
- Why is construction booming in urban China? *(because more and more people are moving to the cities to live and work)*
- **Compare and Contrast** Based on the images, do you think life is harder in rural areas of China or in urban areas? *(Possible answer: Life is harder in rural areas because much of the work is done by hand or with the help of animals.)*

COMPARING Rural and Urban China

Until the 1970s, many more Chinese lived in the countryside than in cities. Then people began moving into the cities in record numbers. Today more than 40 percent of all Chinese live in the cities.

Traditional Countryside

Market day in small towns and villages is an important day for rural Chinese to do their shopping or to sell goods. Most of the goods are produced in the area near the marketplace. Market day is also a time to see friends and relatives who live in the area.

Jobs
A young boy feeds his water buffalo. Water buffalo still provide power for farm work.

Production
Peasants gather tomatoes for sale at the local market.

Outdoor vegetable market in eastern China ▼

Transportation
A farmer transports geese and chickens to market.

DIFFERENTIATING INSTRUCTION: TIERED ACTIVITIES

OBJECTIVE Gain a deeper understanding of how China's traditional way of life is changing.

Basic

Ask students to draw two columns on a sheet of paper and label one "Rural" and the other "City." Then have them study the images on these pages and write a summary statement for each area.

On Level

Have students study the two largest images. Ask them to identify similarities and differences in the way of life shown. Then have them write a brief description of the two scenes, pointing out differences and similarities.

Challenge

Ask students to imagine they are from rural China visiting a large city for the first time. What aspects of traditional life might they find? What might they see that is different? Have students record their impressions in a journal entry.

Jobs

China's economic boom has triggered many construction projects, especially in cities.

Booming Cities

Life in bustling, noisy cities like Shanghai and Beijing is very different from the less hectic life of the rural areas. Many people work in factories, as street vendors, at office jobs, or at their own businesses. Shopping takes place in Western-style malls or in stores in city neighborhoods.

Production

High-tech assembly plants like this digital camera plant have popped up in many Chinese cities.

▲ *Crowded, busy street in Shanghai*

Transportation

With a top speed of 267 miles per hour, Shanghai's maglev (magnetic levitation) train is the fastest in the world.

CRITICAL THINKING

1. **Compare and Contrast** What differences can be seen in the way goods are sold in the city and in the country?
2. **Make Inferences** In what ways would a person's life change if he or she moved from the country to the city?

INTERDISCIPLINARY ACTIVITIES

CONNECT to Technology

Create a Multimedia Presentation

Have students work in small groups to create a multimedia presentation about rural life and city life in China. Images and accompanying narrative should include the following subjects: jobs, production, transportation, leisure activities, family life, and school.

CONNECT to Art

Create a Persuasive Poster

Have groups of students create a poster designed to persuade people to move to a rural or urban area of China. The posters should focus on the positive aspects of life in the area. Encourage students to include pictures and slogans that will capture the viewers' imaginations. Display the finished posters in the classroom.

More About . . .

Rural China

It is not hard to imagine why young people in China would choose to leave the rural areas and go to the cities. Life in the countryside of China is dramatically different from life in one of the country's megacities. Country life is slower, has less technology, and generally is poorer. Few if any villages have automobiles, and much of the labor on farms is either done by people or animals. However, life in the villages is changing as electricity and modern tools are introduced.

Comparing Regions

Farmers' Markets

In the United States, outdoor markets have become increasingly popular. These markets, known as "farmers' markets," are similar to markets in China's rural areas. Typically open in the growing season, farmers' markets sell produce in every state in the union and can be found in both cities and rural areas. By selling directly to the consumer, farmers often receive a better price for their produce.

In the United States, these markets are a novelty and not the way most Americans purchase their food. However, in rural China, outdoor markets are the way most people buy food regularly.

3 Wrap Up

Ask students to come up with descriptive words and phrases for each picture in the feature. List the words on the board. Write the words that describe the rural pictures in one column and those that describe the urban pictures in another. Then ask volunteers to come up with a sentence that summarizes life in each area, using as many of the descriptive words as possible.

CRITICAL THINKING ANSWERS

1. **Compare and Contrast** country: goods sold outside by the producers; city: goods sold in stores
2. **Make Inferences** Life would be noisier, more crowded, and pollution might be a bigger problem. Traditional values and customs might become neglected. Farmers would no longer raise food but would probably work in industry.

1 Plan & Prepare

Objectives

- Explain the role the Chinese Communist Party plays in the Chinese government
- Describe changes in the Chinese economy since reforms began in 1978
- Identify the challenges China faces in the 21st century

Read for the Essential Question

Help students read for a purpose by reminding them of the Essential Question: "How is China's traditional way of life changing in response to global trends?" Point out that they can use items in the section, such as Key Questions, special features, images, and maps, to formulate answers to the question.

Vocabulary

Best Practices Toolkit

Use the ***Best Practices Toolkit*** to model strategies for vocabulary notetaking. Vary strategies throughout the year. Choose from: *Knowledge Rating, Predicting ABCs, Definition Mapping, Word Sort, Word Wheel, Frayer Model (Word Squares), Magnet Words,* and *Student VOC.*

Vocabulary Strategies, TT9–TT16

Reading Strategy

Best Practices Toolkit

Display the Three-Column Chart transparency.

- Remind students that when they summarize, they restate a paragraph, a passage, or a section in fewer words.
- Model summarizing China's government. Have students practice summarizing using the subheads within the section.
- Display the graphic organizer for students to review.

Three-Column Chart, TT21

SECTION 4 Reading for Understanding

Key Ideas

BEFORE, YOU LEARNED

Although some traditional ways of life are still practiced, life has changed in China. More people are living in cities.

NOW YOU WILL LEARN

Small changes in the Chinese government and big changes in the Chinese economy are taking place.

Vocabulary

TERMS & NAMES

province a governmental division like a state

National People's Congress (NPC) China's national law-making body

demonstration a public gathering to protest a policy or action

Tiananmen (tyahn•ahn•mehn) **Square** the location of a huge demonstration for democratic rights that occurred in 1989

command economy an economy in which the government owns all land, factories, and businesses, and decides what will be produced

free enterprise an economy in which business is privately owned

reform action to improve social or economic conditions

Visual Vocabulary Tiananmen Square

Reading Strategy

Re-create the chart shown at right. As you read and respond to the **KEY QUESTIONS**, use the chart to summarize information about China's government and economy.

See Skillbuilder Handbook, page R5

SUMMARIZE

GOVERNMENT	ECONOMY	ISSUES AND TRENDS

GRAPHIC ORGANIZERS Go to **Interactive Review** @ClassZone.com

PRETEACHING VOCABULARY

English Learners

Pronounce and Preview

Pronounce each term for students. Review the meanings of words used in the definitions, such as *governmental, democratic rights,* and *privately.*

- To modify vocabulary learning, have students complete worksheets as they read the section.

Unit Resource Book: East Asia & Southeast Asia
- Vocabulary Practice, p. 35

Inclusion

Match Terms and Definitions

Have students read each term and its definition into an audiocassette recorder. As students listen to the recordings, have them follow along in their textbooks. Then have students use each term and name in a sentence.

The Five-Phoenix Temple in south-central China's Black Dragon Pool Park

China's giant pandas

SECTION 4

GOVERNMENT & ECONOMICS

A Rising Power

Connecting to Your World

Take a look at the labels in your clothing or shoes, or on a small electronic device such as an inexpensive calculator. Many goods that Americans use every day are made in China. This is a big change from the days of Mao when most products produced in China stayed there. Chinese people had little contact with foreigners or their businesses. Today, China, a rising power in Asia and the world, welcomes foreigners and foreign companies.

2008 Summer Olympics When China hosts the Summer Olympics in 2008, thousands of foreigners will have an opportunity to see how much more modern China has become.

A Strong Central Government

KEY QUESTION Who controls China's central government?

For centuries, China has had a strong national government. It controlled **provinces**, state-like governmental units, and local areas. Today, the national government makes decisions that the provinces and local governments must carry out. The Chinese national law-making body is called the **National People's Congress**, or **NPC**. Local and provincial People's Congresses elect members of this body. However, in reality, the Communist Party sets the country's policies by selecting candidates for elections and controlling the NPC.

Shanghai Harbor Shanghai's harbor is one of the largest ports in the world.

785

2 Focus & Motivate

3-Minute Warm-Up

Write on the board or display the transparency:

- Look at the photographs in this section. What do these photographs suggest about modern China? *(Modern China has up-to-date cities, shopping malls, and a space program.)*

Unit Transparency Book
- 3-Minute Warm-Ups, TT1

Connecting to Your World

More About . . .

Beijing Olympics

To prepare for the Beijing Olympics, China expanded the subway system, built hundreds of hotels, and constructed new athletic facilities. Reducing the air pollution in the heavily polluted city was also a goal. People in Beijing were encouraged to join in a learn-English campaign.

3 Teach

A Strong Central Government

Reader, Recorder, Reporter

- How is China's government different from that of the United States? *(China has a strong central government that controls smaller units. The U.S. government shares power with the smaller units.)*
- **Summarize** What is the role of the Communist Party in China's government? *(The Party sets China's policies.)*

SECTION 4 PROGRAM RESOURCES

Unit Resource Book: East Asia and Southeast Asia

ON LEVEL

Unit Resource Book
- Reading Study Guide, p. 9
- Section Quiz, p. 54

STRUGGLING READERS

Unit Resource Book
- Reading Study Guide with Additional Support, p. 17
- Vocabulary Practice, p. 35
- Section Quiz, p. 54
- Reteaching Activity, p. 58

eEdition with Audio DVD-ROM

ENGLISH LEARNERS

Pupil Edition in Spanish

eEdition with Audio DVD-ROM

Unit Resource Book
- RSG (Spanish), p. 25
- RSG with Additional Support (Spanish), p. 33

Multi-Language Glossary

Test Generator
- Section Quiz in Spanish

INCLUSION

Unit Resource Book
- RSG with Additional Support, p. 17
- Section Quiz, p. 54
- Reteaching Activity, p. 58

GIFTED & TALENTED

Unit Resource Book
- Economics in the World, p. 46
- Section Quiz, p. 54

PRE-AP

Unit Resource Book
- Map and Graph Skills, p. 37
- Section Quiz, p. 54

TECHNOLOGY

Unit Transparency Book
- 3-Minute Warm-Ups, TT1
- Essential Question, TT2
- Critical Thinking, TT6

Daily Test Practice Transparencies
- Section 4, TT86

Power Presentations

ClassZone.com

World Cultures and Geography Video on DVD

More About . . .

Tiananmen Square

In 1989, Chinese university students were joined by other citizens in Beijing's Tiananmen Square to demand more democracy in the government and to end government corruption. The military brutally crushed the demonstrations and killed hundreds of protestors. Later, other known leaders of the pro-democracy movement were arrested, and some were executed. Millions around the world viewed the demonstration and military reaction on television.

COMPARING

Chinese and U.S. Governments

Point out that the chart presents a clear contrast between democracy and communism. Have students study the difference between how the president of each country is selected. Ask students which form of government guarantees voters more direct control over their leadership. *(United States)* Then ask why it is important for people to be able to choose their own leaders. *(People should be able to choose leaders that they think will do the best job in leading the country.)*

CRITICAL THINKING ANSWERS

1. **Compare and Contrast** In the United States, the voters elect the president; in China, the president is selected by the National People's Congress.
2. **Evaluate** Power in the United States is divided among the three branches, while power in China is concentrated in the NPC and the Communist Party.

Government and the Communist Party As you can see in the diagram below, the NPC chooses the president and the State Council, or cabinet. The president chooses a prime minister to run the government. The NPC also selects judges for the courts. However, every government office is watched over by Communist Party officials. They make sure the government follows the Party's plans.

Political Trends You learned that the government keeps tight control of people's lives. Beginning in April 1989, **demonstrations** against corruption and for more democratic freedoms took place at Beijing's **Tiananmen** (tyahn•ahn•mehn) **Square**. In early June, the government brutally cracked down on the demonstrators. But the event reflected increasing citizen pressure for reforms in China.

Today, people are a little freer to protest against government policies. But the government still limits many basic rights. Demonstrations are often broken up by force. Protestors are beaten or jailed if they continue to speak out against the government.

A FIND MAIN IDEAS Tell who controls China's central government.
the Communist Party through the NPC

COMPARING Chinese and U.S. Governments

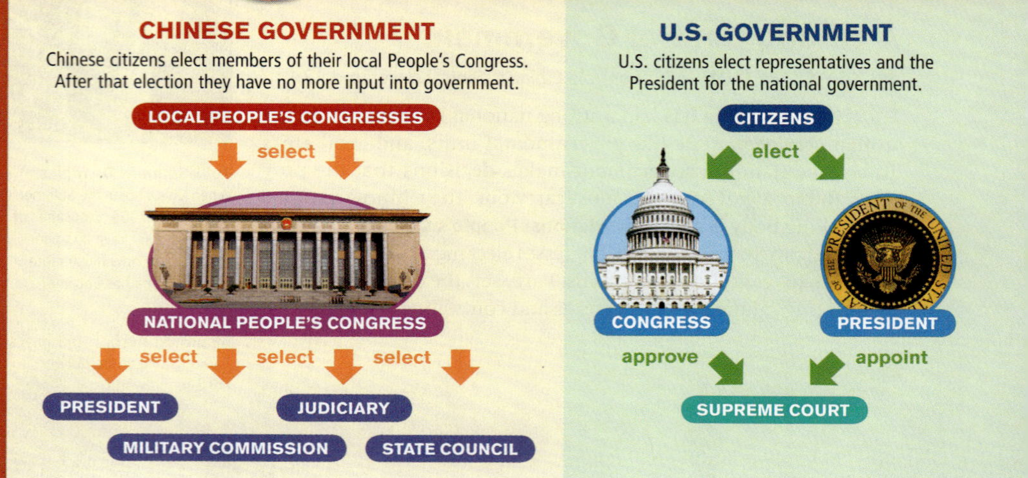

CRITICAL THINKING

1. **Compare and Contrast** How is the selection of a president in China different from the selection of a president in the United States?
2. **Evaluate** How does division of power in the U.S. government differ from China's?

786 Chapter 25

INTERDISCIPLINARY ACTIVITIES

CONNECT to Government

Discuss Government Systems

Have pairs of students study the comparing governments chart. Ask them to count the steps between the voter and the president in China. *(two: local People's Congress, National Congress)* Then have them count the steps between the voters and president in the United States. *(none)* Have the class discuss which set of voters has more control over the choice of president.

CONNECT to Civics

Compare Political Parties

Ask students, "What role do political parties play in the selection of local or national leaders?" Have students research the composition and role of political parties in the United States and in China. They should create a comparison chart or poster showing the differences in the parties.

Economic Reforms

KEY QUESTION How did economic reforms bring Western influences to China?

Under Mao, China had a **command economy**. That means the central government owned all the land, materials, and factories. It decided what goods would be produced. However, in 1978, leader Deng Xiaoping (duhng show•pihng) began to allow elements of **free enterprise**. In this type of economy, people can own a business and sell their products for a profit. The government invited foreign companies to invest in big projects and build factories in China.

International Trade The **reforms**, or improvements, described above dramatically changed China's economy. Today, China makes more low-cost goods than any other country and is a big player in international trade. Most of China's trade is with the United States, Taiwan, and Japan. As you can see in the graph below, in 15 years China's exports and imports skyrocketed.

Industries of China

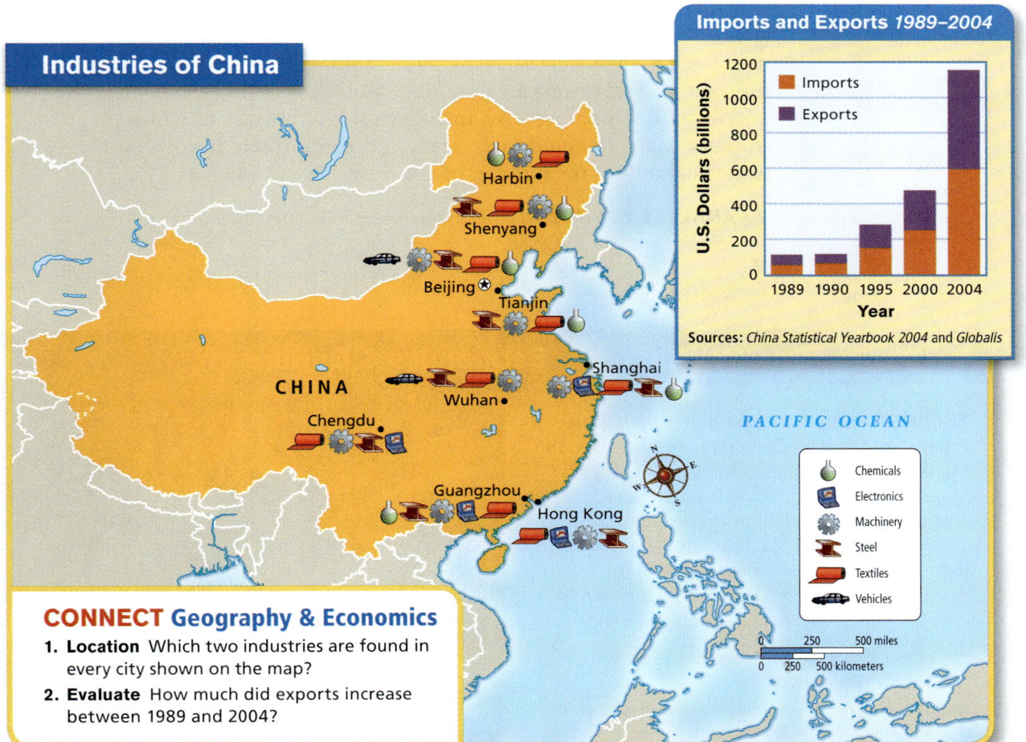

Sources: *China Statistical Yearbook 2004* and *Globalis*

CONNECT Geography & Economics

1. **Location** Which two industries are found in every city shown on the map?
2. **Evaluate** How much did exports increase between 1989 and 2004?

DIFFERENTIATING INSTRUCTION

Struggling Readers

Read a Bar Graph

Tell students that the bar graph on this page contains two sets of information, one about exports and the other about imports. Ask the students which color represents imports. *(orange)* Remind them what an import is. Ask what other information is shown on the graph. *(value of the goods in U.S. dollars)* Then have students estimate the value of the imports for each of the years shown. Repeat the process for exports.

Inclusion

Create a Checkers Bar Graph

Have students represent the information in the bar graph using checkers. Red checkers should represent imports, and black checkers should represent exports. Tell students that each checker has a value of 50 billion dollars. Then have them use the information in the graph to stack five columns of checkers representing China's imports and exports.

Teach

Economic Reforms

Talk About It

- What does it mean to make economic reforms? *(to change aspects of the economy in order to improve it)*
- What changes have been made in the Chinese economy? *(more privately owned businesses, more foreign businesses allowed into the country, more Western-style ideas and customs)*
- **Summarize** Summarize China's economic reforms. *(moving to free enterprise, more private ownership, greater international trade)*

Industries of China

CONNECT Geography & Economics

Have students study the graph first. Ask what the information in the graph suggests about the changes in China's economy. *(Both exports and imports have increased greatly.)* Remind them that exports are the products sold to buyers outside the country. Next have students look at the map. Ask them where most of the industries are located. *(in eastern China)* Then ask students where they think the products shown are exported to. *(Answers will vary but should include the United States.)*

ANSWERS

1. **Location** machinery and textiles
2. **Evaluate** by about 12 times

Unit Resource Book: East Asia & Southeast Asia

- Map and Graph Skills: Reading a Thematic Map, p. 37
- Economics in the World: Economic Boom in China, p. 46

More About . . .

Western Fast Food

Western fast food started as a novelty in China and has become a thriving industry. In addition to American foods, the Chinese people enjoy French, German, and Italian fast food. The two most popular fast food companies are McDonald's and Kentucky Fried Chicken. Recently, Pizza Hut and Rainforest Café have also expanded into China. Some of the foods offered are the same as those found in American restaurants, and others are adapted to suit local tastes. For instance, McDonald's sells a red bean pie, and KFC created a special soup for Chinese consumption.

CONNECT Economics & Culture

Remind students that at the end of the last section, they saw an image of a rural market. Tell them that, for many years, most Chinese people shopped for food and other products in such markets. At first, many Chinese people resisted Western-style malls as a way to shop, but malls have become increasingly popular.

CRITICAL THINKING ANSWER

Make Inferences Possible answers: the many levels of stores, the signs and bright colors, the modern layout of the mall

CONNECT to the Essential Question

How is China's traditional way of life changing in response to global trends?

Ask students what they have learned so far that can help them answer this question. Students might mention the following:

- More economic opportunities have led to an increase in Chinese companies that are privately owned.
- Western influences have had a strong impact on China's culture and economy.

More Economic Opportunities The Chinese government still owns part of most large Chinese firms, but now there are also many privately owned Chinese companies. In a move to attract foreign businesses, Chinese cities give companies tax breaks. They have also built airports, hotels, malls, and public transportation. Many foreign companies build factories in China. Some sell their products in China, and others buy products made by Chinese companies. As a result, many new jobs have been created.

Now the government is starting to develop western China. It is building roads and railways to the area and drilling there for oil and natural gas. Foreign companies are opening offices, factories, and stores in western Chinese cities.

Western Influences Economic changes have brought more Chinese people into contact with Europeans and Americans. Some of these people are students, scientists, and business people who travel to the West. Others work with Westerners in China. As a result, these groups have become more comfortable with Western ideas.

American and European chains opened stores in China—some in large, new malls like the one shown below. The Chinese now eat at many well-known Western fast-food restaurants. Water parks, miniature golf, and ski resorts are now all part of Chinese life.

Western companies also influence the treatment of Chinese workers. Many demand safe, clean, and comfortable factories for workers and limit the number of hours people can work.

ANALYZE EFFECTS Detail the Western influences that economic reforms brought to China.

more privately owned businesses; cities became more modern; malls, fast-food restaurants, and Western recreational facilities opened; treatment of workers has improved ▸

CONNECT Economics & Culture

Chinese Malls Shopping in malls is a new experience for most Chinese. In recent years, the Chinese have constructed some of the world's largest malls. In fact, China is projected to have 7 of the 10 largest malls in the world by 2010.

The South China Mall, currently the world's largest, is about three times the size of the Mall of America in the United States. Luchu Commercial City, the mall shown here, is located in Gangzhou.

CRITICAL THINKING

Make Inferences How do Chinese malls reflect Western influences?

788 Chapter 25

DIFFERENTIATING INSTRUCTION

English Learners

Knowledge Rating Vocabulary

Place the following words on a Knowledge Rating model: *privately owned, foreign businesses, tax breaks, public transportation.* Pair students and have them define the words. Students should compare their definitions and create a definition that all can agree on.

Best Practices Toolkit
- Knowledge Rating, TT9

Gifted & Talented

Promote a Western Business

Have students select a Western business that might expand to China. Then a team should plan and create a multimedia presentation showing how that business could gain popularity and prosper in China. The presentation should be shown to the class.

Current Issues and Trends

KEY QUESTION What challenges does China face in the 21st century?

With a population of over a billion, China needs to make sure its economy continues to grow. That growth is needed to supply enough jobs for its workers. The transportation and communication systems need to be expanded, and the country also needs to generate more electricity. As you learned in Section 1, the Three Gorges Dam will help produce more electricity. China has plans for 46 additional dams. However, this expansion does not come without problems. Huge projects like the dams or the creation of immense farms in western China often have negative effects on the environment.

Environment China faces serious air and water pollution problems. The country has 16 of the 20 most polluted cities in the world. China ranks behind only the United States in the creation of greenhouse gases. Because most industries use coal for fuel, China also creates large amounts of sulphur dioxide, a gas that contributes to acid rain. Agricultural projects in the northwestern regions of China have turned some farms into dust bowls. Each year, dust blowing out of China sweeps into Korea, Japan, and the United States.

Many rivers and lakes are polluted with wastes from manufacturing processes and raw sewage. Overall, about 700 million people only have contaminated water to use for drinking, cooking, and bathing. The Chinese government has begun to actively enforce environmental protection laws.

Air Pollution Getting to work in Chinese cities often requires masks to filter out harmful pollution. Lung-related diseases are on the rise in polluted cities. **Why has pollution increased so much in Chinese cities?** as a result of the growth of industries and of the cities

Expanding Freedoms In part, the Chinese government began to work on environmental issues as a result of pressure from groups within the country. Changes in China's government have made some environmental protests possible.

Though Chinese society is changing, concerns about basic personal freedoms remain. For example, many workers are forced to work long hours in unhealthy conditions. Some religious groups are not allowed to worship as they wish. As a result, newly formed civic organizations are looking for ways to gain democratic freedoms for the Chinese people.

DIFFERENTIATING INSTRUCTION: TIERED ACTIVITIES

OBJECTIVE Research the types and levels of pollution in Chinese cities.

Basic

Provide articles about air pollution in China. Then have students work in groups to study the problem and propose solutions to reduce emissions. Have students construct a table listing the types and levels of pollution in the most polluted cities in China.

On Level

Provide students with articles about air pollution solutions in China. Then have students create an oral presentation on China's attempts to reduce vehicle emissions. Encourage students to include a visual aid in their presentation.

Challenge

Have students debate the following issue: "Developing countries should not have to meet the emission standards required by developed countries." Provide the students with articles both pro and con on this issue. Students should present their debate to the class.

Teach

Current Issues and Trends

Think, Pair, Share

- Why is pollution a major concern in China? *(Extensive air and water pollution affects millions of people in China.)*
- Why does China want a successful space program? *(Such a program would suggest that China is a modern and powerful nation.)*
- **Summarize** Look at the subheads under the main heading. What general categories best summarize the content of the subheads? *(environment, government, technology)*

Comparing Regions

Air Pollution

Both Mexico and China face environmental challenges. The largest city in each country has extremely high rates of air pollution. Mexico City is the most polluted city in the world, according to the World Heath Organization. Beijing ranks in the top ten. Other large cities in each country, such as Shanghai (China) and Guadalajara (Mexico), are also among the most polluted in the world.

The major cause of air pollution in both countries is automobile emissions. As more and more individuals own cars, the levels of pollution have risen. Both governments have made efforts to reduce air pollution caused by vehicles by tightening standards for emissions and encouraging the conversion of trucks and cars to cleaner burning fuel, such as natural gas or liquefied petroleum. Since both countries are continuing to develop economically, it is likely that the air pollution issue will continue for some time in each country.

4 Assess & Reteach

Assess Have students complete the Section Assessment.

Unit Resource Book: East Asia & Southeast Asia
- Section Quiz, p. 54

Interactive Review @ClassZone.com
- Section Quiz

Power Presentations

Test Generator

Reteach Divide the class into groups and assign each group one segment of the section (using the red subheads). Have students reread their assigned segment and collectively write a main idea statement for each paragraph in the segment. Students should then combine the statements into one paragraph and write a main idea statement for that paragraph. Assist groups in making photocopies of their work to distribute to the entire class for use as a study aid.

Unit Resource Book: East Asia & Southeast Asia
- Reteaching Activity, p. 58

Chinese Space Program Two years after its first manned space flight, China put two more astronauts in space. They spent five days performing experiments and testing equipment. **What does having a space program tell you about the nation?**

The country has advanced scientifically and technologically, and its economy is doing well.

Space Program As China develops, it has tried to show the world that it is a modern nation. Its space program, which China began in the 1950s, is one of those efforts. By 1970, it had sent a satellite into orbit. China's spacecraft are launched and returned to China's space center located in Inner Mongolia.

Only three nations in the world—China, the United States, and Russia—have a manned space program. In 2003, the spacecraft Shenzhou V (shen•joh five) carried a Chinese taikonaut, or astronaut, into space. The name of the spacecraft means "Divine Vessel." In 2005, millions of Chinese watched on television as two men were sent into space in Shenzhou VI. China's space program has a goal of reaching the moon with an unmanned probe by the year 2010. Eventually, China will launch a space station. The Chinese government hopes their space program will build Chinese pride and show the rest of the world how far China has advanced in science and technology.

A CATEGORIZE Discuss the challenges that China faces in the future. environmental problems, expanding freedoms, further development of the space program

ONLINE QUIZ For test practice, go to **Interactive Review @ClassZone.com**

Section 4 Assessment

TERMS & NAMES

1. **Explain the importance of**
   - NPC
   - Tiananmen Square
   - command economy
   - free enterprise

USE YOUR READING NOTES

2. **Summarize** Use your completed chart to answer the following question:
   What economic challenges does China face?

GOVERNMENT	ECONOMY	ISSUES AND TRENDS

KEY IDEAS

3. What role does the Chinese Communist Party play in the government of China?
4. How has China's economy changed since the death of Mao?
5. Why are some Chinese concerned about basic personal freedoms?

CRITICAL THINKING

6. **Synthesize** In what way is the Communist Party more powerful than the Chinese government?
7. **Draw Conclusions** How might the presence of Western companies in China change working conditions?
8. **CONNECT to Today** How could environmental problems hold back the growth of China's economy?
9. **MATHEMATICS Create a Graph** Review the chart on exports and imports. Create a line graph with one line showing China's exports and one line showing its imports.

SECTION 4 ASSESSMENT ANSWERS

Terms & Names

1. NPC: China's national law-making body; Tiananmen Square: site of a 1989 protest for democratic rights; command economy: one in which government owns all land, factories, and businesses; free enterprise: economy in which business is privately owned

Use Your Reading Notes

2. huge population; needs more public services

GOVERNMENT	ECONOMY	ISSUES AND TRENDS
Communist control of NCP, which sets policy	Some free enterprise and expanded international trade	Environment; expanding personal freedoms

Key Ideas

3. makes all policy decisions; ensures that the NPC and the president carry out the policies
4. The country moved away from a command economy and closer to free enterprise.
5. because many workers are forced to work long hours in unhealthy conditions, and some religious groups are not allowed to worship as they wish

Critical Thinking

6. The Party, not the government, makes the policy decisions for the country.
7. The treatment of employees in Western-owned companies might help improve conditions for employees in Chinese companies.
8. Continued pollution may discourage industries from expanding or building new plants in China, thereby slowing its economic growth.
9. Use the rubric to score graphs.

Graph Rubric

	Style of Line Graph	Number of Errors
4	very clear	few or none
3	clear	some
2	vague	many
1	unclear	many major

CHAPTER SUMMARY

Key Idea 1
China has varied landforms, ranging from high mountains to river valleys. It has diverse climates, plants, and animals.

Key Idea 2
Throughout history, China's rulers have struggled with the problem of feeding so many people.

Key Idea 3
Some aspects of traditional life survive in China, while new ways of life have been introduced.

Key Idea 4
Small changes in the Chinese government and big changes in the Chinese economy are taking place.

For **Review and Study Notes**, go to **Interactive Review** @ ClassZone.com

NAME GAME

Use the Terms & Names list to complete each sentence on paper or online.

1. I am also known as the Yellow River or "China's Sorrow." **Huang He**
2. I am China's longest river and the Three Gorges Dam was built to control my floods. ______
3. I am the highest mountain range in the world and divide China from India. ______
4. I am the emperor who united China. ______
5. I am the second-largest sand desert in the world. ______
6. I am a high desert that divides Mongolia and China. ______
7. I will provide hydroelectric power to parts of China. ______
8. I am the area where China's earliest civilization began. ______
9. I am the routes that brought traders and Buddhism to China. ______
10. I am the location of an important political demonstration for democratic rights. ______

Buddhism
Chang Jiang
Confucianism
Gobi Desert
Great Wall
Himalayas
Huang He
Mao Zedong
Middle Kingdom
North China Plain
Shi Huangdi
Silk Roads
Taklimakan Desert
Three Gorges Dam
Tiananmen Square

Activities

GeoGame

Use this online map to show what you know about China's location, geographic features, and important places. Drag and drop each place name to its location on the map.

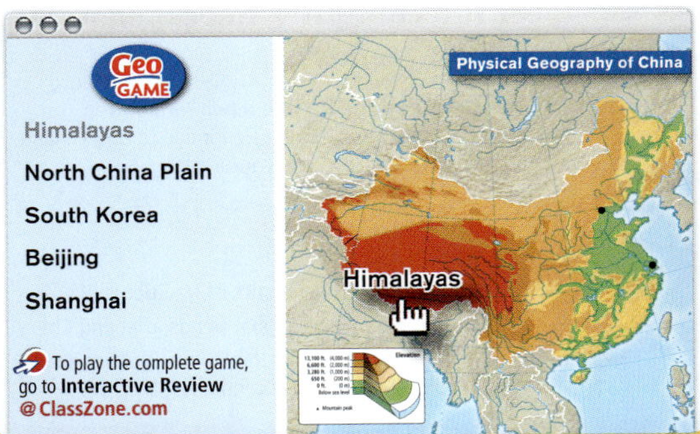

Crossword Puzzle

Complete an online crossword puzzle to test your knowledge of China's road to the present.

ACROSS
1. the first emperor to rule China

ONLINE AND ON-SCREEN ACTIVITIES

GeoGame

Use the interactive map on **Power Presentations** DVD-ROM or at **ClassZone.com** to help students review China's geographic features.

- Demonstrate how to drag and drop the place names onto the map. Click on the Submit button to view results.
- Advance to the scorecard to show how students will be scored in their first and second attempts.

Crossword Puzzle

Students can review important people and events in China's history by completing this crossword puzzle online at **ClassZone.com**.

- As an extension, have students use the *Activity Maker* at **ClassZone.com** to make their own crossword puzzle for one section of the chapter.
- Then have students trade puzzles and try solving them.

Chapter Summary

As a follow-up to the chapter, encourage students to pair up and go over the **Review and Study Notes** at **ClassZone.com.**

Alternatively, student pairs can review their Reading Study Guide notes, making sure that they can add important details to each section's Key Ideas.

Name Game

Teaching Tips

ON PAPER Have students work in pairs to complete this activity. As an extension, they can write questions for the terms that are not covered in the items shown here.

ONLINE Make sure that students understand how to drag and drop terms to fill in blanks. Ask them to print out their results and compare them with those of a classmate.

ANSWERS

1. Huang He
2. Chang Jiang
3. Himalayas
4. Shi Huangdi
5. Taklimakan Desert
6. Gobi Desert
7. Three Gorges Dam
8. North China Plain
9. Silk Roads
10. Tiananmen Square

Make It Fun

Teacher-Tested Activities

Matthew J. Scheidler, Wayzata East Middle School, Plymouth, Minnesota

My classes play "Pronto Geo" as a review activity. Here's how the game works.

- First, go to **ClassZone.com** and project the crossword puzzle on a whiteboard.
- Then divide the class in two. Make two sets of numbers and give each student a number. There should be a #1, a #2, and so forth, on each team.
- Make sure the path to the computer or whiteboard is clear.
- Call out a question and number such as, "Which Chinese scholar is known as China's greatest teacher? Number 12." The two students with number 12 race to the whiteboard. Whichever one legibly writes "Confucius" first wins.

Vocabulary ANSWERS

1. barrier built on the Chang Jiang to control floods and provide power for China
2. the second largest sand desert in the world
3. the longest river in Asia, flowing through eastern China
4. the first emperor to unify much of China
5. a belief system based on the teachings of Confucius
6. Communist leader of China from 1949 to 1976
7. Chinese belief system based on the idea of natural order in the world
8. routes used by traders moving goods from China to Southwest Asia
9. religion based on the teaching of Siddhartha Gautama
10. China's national law-making body
11. All are belief systems that have had a major influence on Chinese culture.
12. Both were leaders in China: Shi Huangdi united China, and Mao Zedong moved China to communism.

Key Ideas ANSWERS

13. It depends on them for water, for silt, and for transportation.
14. wheat, soybeans, millet, and rice
15. floods and earthquakes
16. The mandate was the approval of the gods for a dynasty to rule.
17. His code of behavior stressed different virtues for people in different social roles.
18. Mao wanted the government to control the economy and halt modernization. To do this, he started the Great Leap Forward and the Cultural Revolution.
19. Family has been an important aspect of Chinese life for centuries.
20. Confucianism, because it promotes respect, family values, and education
21. in eastern China; in rural areas
22. The NPC is the national law-making body but is watched over by the Communist Party.
23. moved toward a free enterprise system; developed international trade; developed western China; expanded services
24. China has serious air and water pollution problems.

CHAPTER 25 Assessment

VOCABULARY

Explain the significance of each of the following.

1. Three Gorges Dam
2. Taklimakan Desert
3. Chang Jiang
4. Shi Huangdi
5. Confucianism
6. Mao Zedong
7. Daoism
8. Silk Roads
9. Buddhism
10. National People's Congress

Explain how the terms and names in each group are related.

11. Confucianism, Daoism, and Buddhism
12. Shi Huangdi and Mao Zedong

KEY IDEAS

1 Land of Three Rivers

13. How does China depend on its rivers?
14. What crops are grown in eastern China?
15. What natural disasters are most common in China?

2 Middle Kingdom to Communist Power

16. What was the Mandate of Heaven?
17. How did Confucius' teaching help produce order?
18. What was Mao Zedong's goal for China, and how did he try to accomplish it?

3 Changing Chinese Culture

19. What is the role of family in Chinese life?
20. Which of China's three belief systems had the greatest influence on Chinese society and government? Why?
21. Where do the majority of Chinese live?

4 A Rising Power

22. What is the role of the NPC in China's government?
23. What has China done to improve its economy?
24. Why are the Chinese concerned about their environment?

CRITICAL THINKING

25. **Compare and Contrast** Create a chart to compare and contrast the landforms and water, climate, and population density of eastern China and western China.

EASTERN CHINA	WESTERN CHINA

26. **Analyze Causes and Effects** How did China change after Shi Huangdi unified it?
27. **Draw Conclusions** What parts of Chinese culture have changed little since dynastic times?
28. **Connect to Economics** How has the right to own private businesses changed the economy in China?
29. **Five Themes: Human-Environment Interaction** Explain which natural barriers helped to shape Chinese civilization.
30. **Connect Geography & Economics** How will the presence of oil and natural gas in western China influence its economy?

Answer the ESSENTIAL QUESTION

How is China's traditional way of life changing in response to global trends?

Written Response Write a two- or three-paragraph response to the Essential Question. Use the rubric below to guide your thinking.

Response Rubric

A strong response will:

- identify the characteristics of traditional life
- discuss the modern global trends affecting China
- explain ways in which the Chinese culture has adapted to the global trends

ESSENTIAL QUESTION REVIEW

Before students begin their essays, have them review the key chapter elements that will help them answer the Essential Question. Direct them to significant features in the chapter, such as

- Comparing Western and Eastern China, Section 1
- History Makers: Confucius, Section 2
- Animated Geography: The Great Wall of China, Section 2
- Comparing Rural and Urban China, Section 3

Written Response Rubric

	Coverage of Key Ideas	Explanation	Errors
4	all	clear, concise	few or none
3	most	good	some
2	some	fair	many
1	one or none	poor	many major

STANDARDS-BASED ASSESSMENT

Test Practice

- **Online Test Practice @ClassZone.com**
- **Test-Taking Strategies and Practice** at the front of this book

THEMATIC MAP

Use the map to answer questions 1 and 2 on your paper.

Energy and Mineral Resources of China

1. Which source of energy is found throughout China?

A. coal
B. hydroelectric power
C. natural gas
D. oil

2. Where is most of China's hydroelectric power found?

A. in the far west
B. in the southeast
C. in central China
D. in the far north

LINE GRAPH

Examine the graph below. Use the information in the graph to write short answers for questions 3 and 4.

China's Iron and Steel Trade *1999–2003*

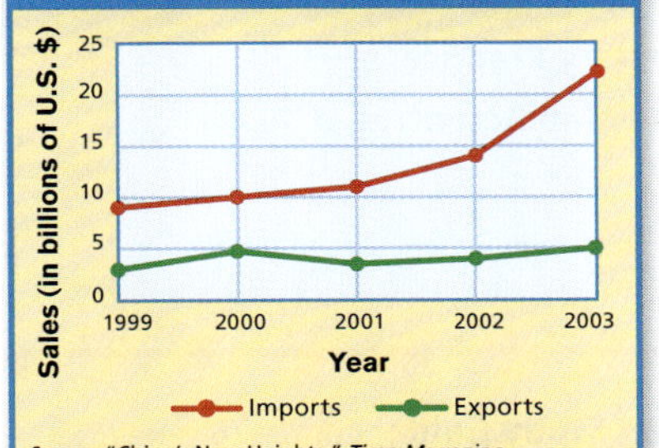

Source: "China's New Heights," *Time Magazine*, June 27, 2005

3. What was the approximate value of China's exports in 2001?

4. Based on the graph, how would you describe China's iron and steel trade between 1999 and 2003?

GeoActivity

1. INTERDISCIPLINARY ACTIVITY–SCIENCE

With a small group, review the Chinese inventions discussed in Section 2. Choose one invention to research further and create a museum-style display. Be sure your display shows how the invention works, and explain how it changed life in China or elsewhere.

2. WRITING FOR SOCIAL STUDIES

Reread the part of Section 3 on Chen's day. Imagine that Chen is coming to visit your town or city. Write a tour schedule describing the places you would take Chen so that he would experience life the way you do.

3. MENTAL MAPPING

Create an outline map of China and label the following:

- Himalayas
- Huang He
- Chang Jiang
- Gobi Desert
- Taklimakan Desert
- East China Sea
- Yellow Sea
- Beijing

Critical Thinking ANSWERS

25. Chart should show that eastern China is lower, wetter, and warmer than western China and has fertile soil and good farmland. It should also show that many more people live in the east than in the west.

26. Unification brought peace and prosperity.

27. emphasis on family, respect for ancestors, still largely a rural society, history of strong leaders, importance of group over the individual

28. The economy is booming. More businesses have started up, increasing the number of jobs available.

29. The Himalayas have protected China from invaders; the deserts in western China have kept the population in that region low; the rivers on the North China Plain helped the development of Chinese civilization.

30. Energy sources are in great demand, which will make this area more valuable.

Standards-Based Assessment ANSWERS

1. D

2. B

3. about $5 billion

4. Imports have gone up, while exports have stayed about the same.

GeoActivity

1. INTERDISCIPLINARY ACTIVITY RUBRIC

The display should

- show an example of an invention
- clearly demonstrate how the invention works
- accurately describe the invention's impact

2. WRITING FOR SOCIAL STUDIES RUBRIC

The tour schedule should

- clearly describe the places chosen
- explain how each place fits into the student's life

3. MENTAL MAPPING RUBRIC

The outline map should

- represent the approximate shape of China
- include all geographic features listed in their correct location

DIFFERENTIATING CHAPTER ASSESSMENT

Unit Resource Book

- **Chapter Test A,** pp. 59–63. *Use for Struggling Readers, Beginning-to-Intermediate-Level English Learners, and for Inclusion of students whose IEP suggests supported test-taking.*
- **Chapter Test B,** pp. 64–68. *Use for On-Level Learners and ESOL students with Advanced-Level English.*
- **Chapter Test C,** pp. 69–73. *Use for Gifted and Talented students or students assigned Pre-AP work.*

Test Generator

- Chapter Tests A, B, and C, in English and Spanish

McDougal Littell Assessment System

TEST **Test Generator** Use three levels of tests in English and Spanish tied to state standards. Plus, choose questions to customize tests for all populations.

SCORE Scan and Score plain paper student tests in minutes.

REPORT Create instant progress reports, sortable by student, by standard, or other criteria.

RETEACH Respond immediately to students' needs by using a wealth of reteaching resources.

CHAPTER 26 The Korean Peninsula PLANNING GUIDE

Complete print resources all on one USB drive allow you to customize lesson plans for all students.

Section Titles & Key Ideas	Unit Resource Book East Asia and Southeast Asia	Assessment Resources
1 A Rugged Peninsula **KEY IDEA** Korea lies on a peninsula, which provides a moderate climate and plentiful resources.	• **Reading Study Guide,** pp. 75–76 • **RSG with Additional Support,** pp. 81–82 • **Vocabulary Practice,** p. 99 • **Outline Map Activities,** p. 103	**Unit Resource Book** • Section Quiz, p. 114 • Reteaching Activity, p. 117 **Test Generator CD-ROM** **Daily Test Practice Transparency,** TT87
2 Strong Traditions, Modern Innovations **KEY IDEA** The Korean Peninsula has been invaded by China, Russia, and Japan but has still maintained its cultural identity.	• **Reading Study Guide,** pp. 77–78 • **RSG with Additional Support,** pp. 83–84 • **History Makers,** pp. 105–106 • **Interdisciplinary Projects,** pp. 107–108 • **Primary Sources,** pp. 112–113	**Unit Resource Book** • Section Quiz, p. 115 • Reteaching Activity, p. 118 **Test Generator CD-ROM** **Daily Test Practice Transparency,** TT88
3 Contrasting Countries **KEY IDEA** North Korea and South Korea have very different government systems and economic structures.	• **Reading Study Guide,** pp. 79–80 • **RSG with Additional Support,** pp. 85–86 • **Skillbuilder Practice,** p. 100 • **Map and Graph Skills,** p. 101 • **Economics in the World,** p. 110 • **Connect to Math,** p. 111	**Unit Resource Book** • Section Quiz, p. 116 • Reteaching Activity, p. 119 • Chapter Test (Forms A, B, & C), pp. 121–132 **Test Generator CD-ROM** **Daily Test Practice Transparency,** TT89

McDougal Littell Assessment System

TEST SCORE REPORT RETEACH

INTEGRATED TECHNOLOGY

No Time? To focus students on important content in the chapter, use the **"Reading Study Guide with Additional Support"** in ***Resources2Go.***

Teacher Presentation Options

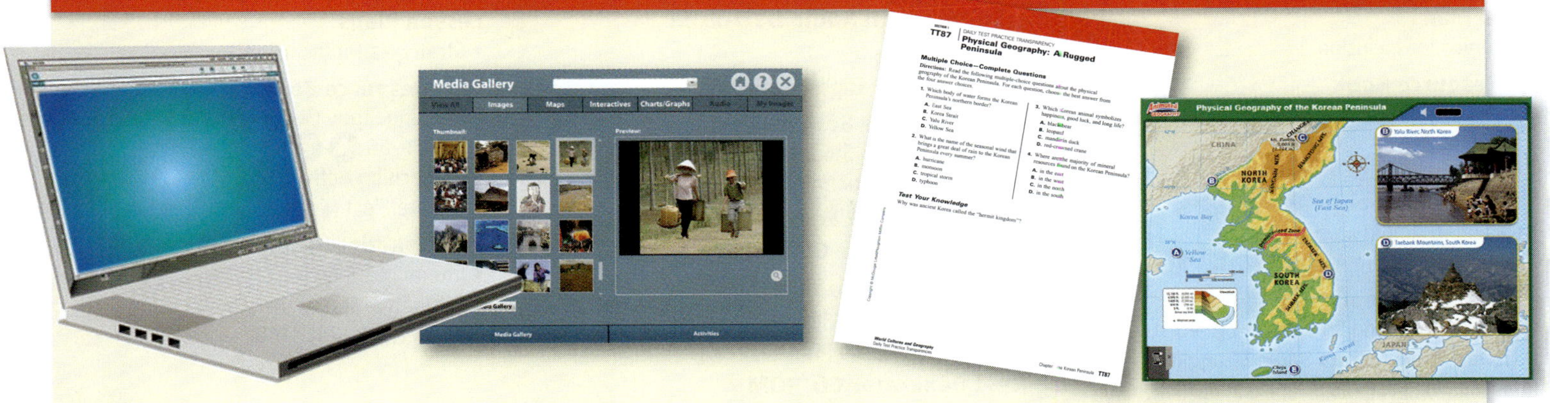

Electronic Books

eEdition DVD-ROM

eEdition Online

Presentation Toolkit

Video/DVD
- World Cultures and Geography

Power Presentations DVD-ROM
- Discussion Starters
- Media Gallery
- Review Game
- Interactive Review
- Animated Features

Unit Transparency Book
- 3-Minute Warm-Ups
- Essential Question
- Critical Thinking
- Maps
- Cultures

Daily Test Practice Transparencies

Animated Geography
- Present-Day Korean Peninsula Map
- Physical Geography of the Korean Peninsula Map
- Worldster: Yumi's Web page

OnlineActivities at ClassZone.com

Animated Geography
- Videos and Photos
- 3-D Animations
- Interactive Maps

Interactive Review
- Section Quizzes
- Review and Study Notes
- Vocabulary Flip Cards
- Graphic Organizers
- Name Game and Crossword Puzzles
- GeoGame

Online Test Practice

Map Center
- Atlas Maps
- Political Maps with Data
- Links to Mapping Web Sites

Activity Center
- Activity Maker—Review Games, Puzzles, Flip Cards, and Graphic Organizers
- Worldster

Research & Writing Center
- Web Sites
- Data Updates
- Primary Sources
- Writing Models

Current Events

Todd Harrison
Hardin County Middle School
Savannah, Tennessee

Make It Fun

Teacher-Tested Activities

As you teach this chapter, look for my classroom-proven ideas for:
- making the lesson fun
- focusing students' attention and comprehension
- integrating technology with traditional teaching

Teaching Bell to Bell When you see this timing guide, have students discuss in pairs, in groups of 3–6, or as a class.
- **Think, Pair, Share** (pairs)
- **Reader, Recorder, Reporter** (3–6)
- **Talk About It** (any size)

CHAPTER 26 RESOURCES FOR DIFFERENTIATED INSTRUCTION

Struggling Readers

eEdition with Audio DVD-ROM

Lesson Plans
- Strategies for Inclusion

Unit Resource Book
- Reading Study Guide with Additional Support, pp. 81–86 (B)
- Vocabulary Practice, p. 99
- Skillbuilder Practice, p. 100
- Reteaching Activity, pp. 117–119
- Chapter Test (Form A), pp. 121–124 (A)

Best Practices Toolkit (C)
- Interactive Reading Strategies, pp. 90–145
- Differentiation Activity Bank: Struggling Readers, pp. 146–149

Teacher's Edition Activities
- Build Vocabulary, Sec. 1
- Understand Comparisons, Sec. 1
- Sequence Korean History, Sec. 2
- Understand Graphics, Sec. 3
- Tiered Activities, Sec. 1–3

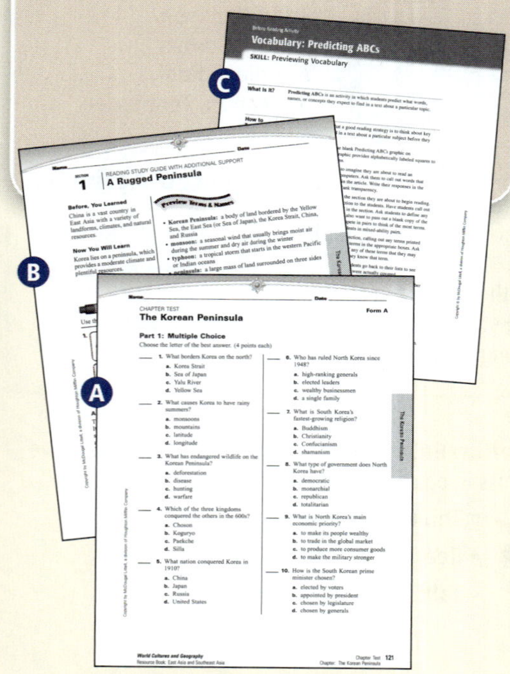

English Learners

Pupil Edition in Spanish

eEdition with Audio DVD-ROM

Lesson Plans
- Strategies for English Learners

Unit Resource Book
- Reading Study Guide (Spanish), pp. 87–92
- RSG with Additional Support (Spanish), pp. 93–98 (A) (B)

Multi-Language Glossary (C)

Best Practices Toolkit
- Differentiation Activity Bank: English Learners, pp. 165–194

Test Generator CD-ROM
- Section Quizzes in Spanish
- Chapter Test in Spanish

Teacher's Edition Activities
- Pronounce and Preview, Sec. 1–3
- Monitor Comprehension, Sec. 2
- Understand Multiple-Meaning Words, Sec. 3
- Re-Create Graphics, Sec. 3
- Tiered Activities, Sec. 1–3

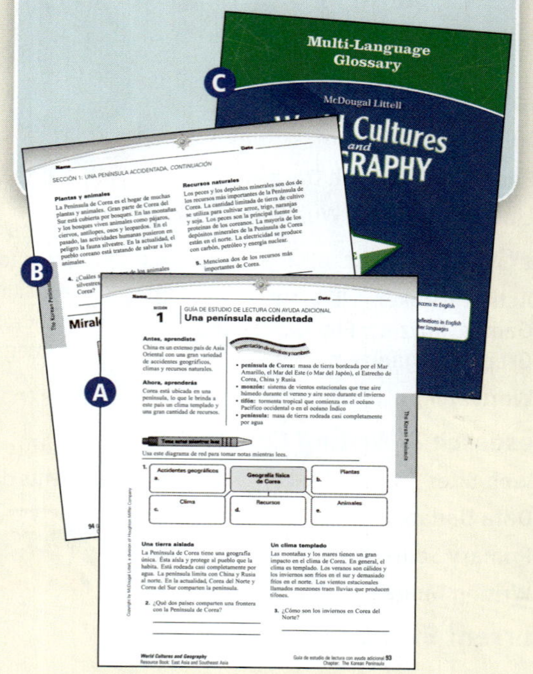

Inclusion

eEdition with Audio DVD-ROM

Lesson Plans
- Strategies for Inclusion

Unit Resource Book
- Reading Study Guide with Additional Support, pp. 81–86 (A) (B)
- Reteaching Activity, pp. 117–119
- Chapter Test (Form A), pp. 121–124

Best Practices Toolkit (C)
- Interactive Reading Strategies, pp. 90–145
- Differentiation Activity Bank: Struggling Readers, pp. 146–149

Teacher's Edition Activities
- Locate Terms, Sec. 1
- Sculpt the Korean Peninsula, Sec. 1
- Use Magnet Words, Sec. 2
- Act Out Terms, Sec. 3
- Tiered Activities, Sec. 1–3

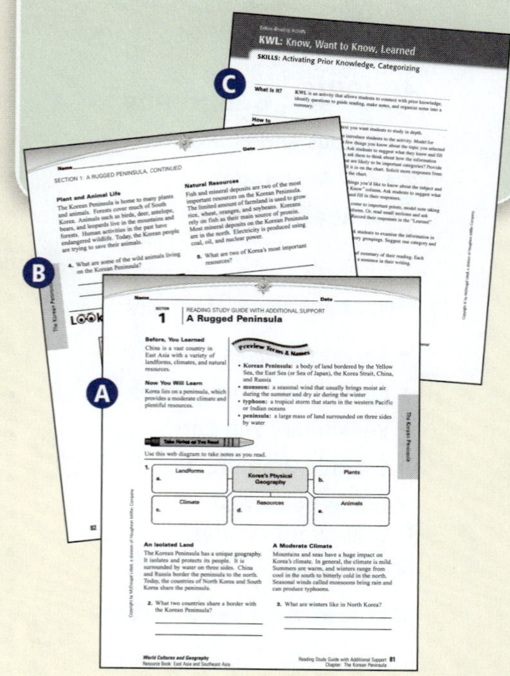

Gifted & Talented Students

Lesson Plans
- Core Lesson Plan

Unit Resource Book
- History Makers, pp. 105–106 (A)
- Interdisciplinary Projects, pp. 107–108 (B)
- Connect to Math, p. 111 (C)
- Chapter Test (Form C), pp. 129–132

Best Practices Toolkit
- Student Products Rubrics, pp. 1–14
- Writing Templates, pp. 78–89

Teacher's Edition Activities
- Create a Board Game, Sec. 1
- Write a Speech, Sec. 2
- Tiered Activities, Sec. 1–3

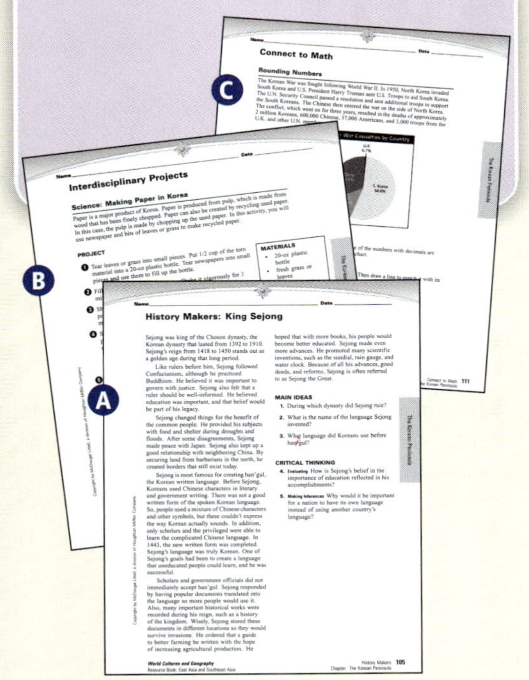
Connect to Math

Interdisciplinary Projects

History Makers: King Sejong

Pre-AP*

Lesson Plans
- Core Lesson Plan

Unit Resource Book
- Economics in the World, p. 110 (A)
- Primary Sources, pp. 112–113 (B) (C)
- Chapter Test (Form C), pp. 129–132

Best Practices Toolkit
- Writing Templates, pp. 78–89
- Differentiation Activity Bank: Pre-AP Students, pp. 195–200

* Pre-AP is a registered trademark of the College Entrance Examination Board, which was not involved in the production of and does not endorse this product.

Teacher's Edition Activities
- Make Inferences, Sec. 1
- Analyze Reunification of the Two Koreas, Sec. 2
- Analyze North Korea's Nuclear Weapons Program, Sec. 3
- Tiered Activities, Sec. 1–3

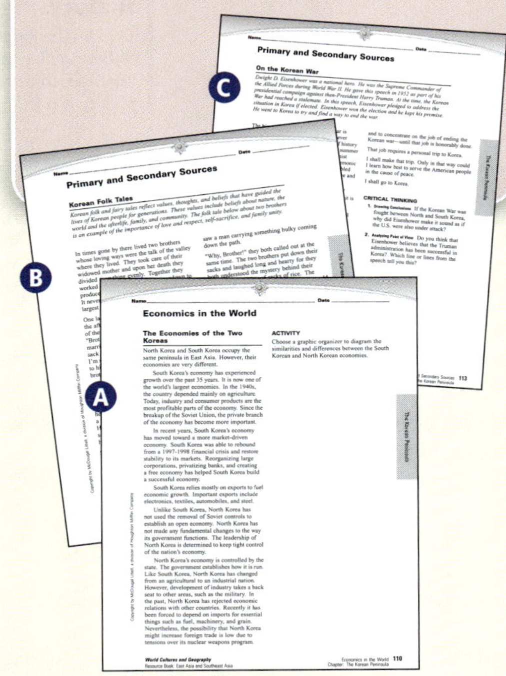
Primary and Secondary Sources

Primary and Secondary Sources

Economics in the World

Interdisciplinary Activities

CONNECT to Language Arts

Create a Travel Brochure on the Korean Peninsula

CLASS TIME 30 minutes

TASK Create a travel brochure on the culture and history of the Korean Peninsula.

SUPPLIES
- map of the Korean Peninsula
- computer or paper and pens/pencils

ACTIVITY Have groups of students work together to create a six-panel travel brochure on the Korean Peninsula. Their brochures should tell readers why the Korean Peninsula is a good place to visit and highlight interesting destinations, such as Cheju Island, Kyongbok Palace, and Haein-sa Temple. Brochures should include at least one visual and block of text on each panel. Ask student groups to present or post their brochures.

Good for Visual and Verbal Learners

CONNECT to Arts

Plan a Korean Art Show

CLASS TIME 30 minutes

TASK Organize an exhibition of Korean art in your classroom.

SUPPLIES
- library books and magazines with color images of various types of Korean arts: tomb mural paintings, Buddha sculptures, celadon pottery, and contemporary arts and crafts

ACTIVITY Have students view the art books and choose their favorite piece of Korean art. Then have students make a color photocopy or their own illustration of the work of art, including a label that explains its significance. Hang these images around the room and hold a "gallery-opening reception" in which students have time to view all the works and discuss them.

Good for Visual and Spatial Learners

Introduce the ESSENTIAL QUESTION

- Point out that the geography of the Korean Peninsula both isolated and protected its people.
- Explain that, until 1945, Korea had been a united country for over 1,000 years.
- Identify the cultural characteristics that North Korea and South Korea share.
- Summarize the reasons for conflict between North Korea and South Korea.

Key Ideas

Section 1: Korea lies on a peninsula, which provides a moderate climate and plentiful resources.

Section 2: The Korean Peninsula has been invaded by China, Russia, and Japan but has still maintained its cultural identity.

Section 3: North Korea and South Korea have very different government systems and economic structures.

CONNECT to the Essential Question

You may wish to revisit the Essential Question periodically with students as they work through the chapter.

Suggestions for reconnecting are provided at an appropriate point in the *Teacher's Edition* for each section.

Pulkuk Temple in Kyongju, South Korea

CHAPTER 26

The Korean Peninsula

1 GEOGRAPHY
A Rugged Peninsula

2 HISTORY & CULTURE
Strong Traditions, Modern Innovations

3 GOVERNMENT & ECONOMICS
Contrasting Countries

ESSENTIAL QUESTION

How can North Korea and South Korea overcome their differences and live in peace?

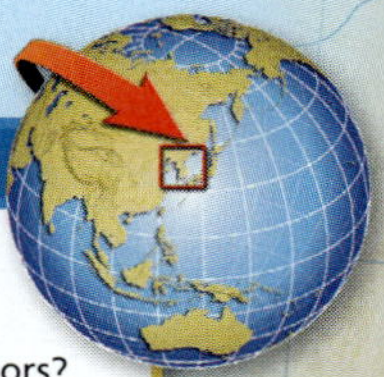

CONNECT Geography & History

Use the map and the time line to answer the following questions.

1. Which countries are the Koreas' closest neighbors?
2. What major religion spread to the Korean Peninsula from China?

A.D. 300

Culture
◀ **300s** Buddhism is introduced from China.

History
660s The kingdom of Silla unifies the Korean Peninsula.
(Silla crown) ▶

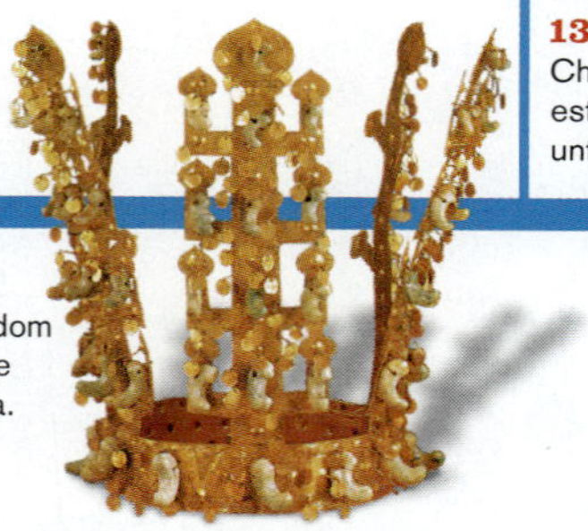

Government
1392 The Choson Dynasty is established and rules until 1910.

TIME LINE DISCUSSION

Invite students to identify the items on the time line that show how contact with other countries influenced Korea's history and culture. *(300s—Buddhism is introduced from China; 1910–1945—Japan has complete control of Korea; 1960s–1980s—South Korea experiences an economic boom.)*

- How many years separate the beginning of the kingdom of Silla from the establishment of the Choson Dynasty on the time line? *(about 732 years)*
- What event ended the Choson Dynasty? *(Japan's occupation of Korea in 1910)*
- In which direction did Buddhism travel to reach Korea? *(south)*
- How might South Korea's location have helped the country's economic boom? *(Possible answer: Direct access to Japan across the Korea Strait may have helped South Korea import and export goods.)*

Animated GEOGRAPHY

Present-Day Korean Peninsula

Click here to explore the Korean Peninsula @ClassZone.com

History
1910–1945 Japan has complete control of Korea.

Economics
◄ 1960s–1980s South Korea experiences an economic boom. (Ship carrying exports)

Today

History
◄ 1953 The Korean War ends with a truce, but North Korea and South Korea remain divided.

795

TEACHER MEDIA FAVORITES

Books for the Teacher

Granfield, Linda and Russell Freedman. *I Remember Korea: Veterans Tell Their Stories of the Korean War,* 1950–1953. Clarion Books, 2003. A collection of oral histories and poems on the Korean War, with photographs.

Stickler, John. *Land of the Morning Calm: Korean Culture Then and Now.* Shen's Books, 2003. An illustrated introduction to many aspects of Korean culture.

Video / DVD

Korean War Stories. PBS Video, 2001. Walter Cronkite and others present interviews and film footage from the Korean War in this documentary.

Families of Korea. Master Communication, 2006. Follows the daily life of two Korean children and their families, giving insight into Korean culture and daily life.

Internet

Go to the *Research and Writing Center* at **ClassZone.com** to link to:

- data on the region
- a variety of chapter-specific, content-reviewed sites
- "History Makers" biographies
- primary sources

Animated GEOGRAPHY

Present-Day Korean Peninsula

CLASSZONE.COM

Explain that this is a political map of the Korean Peninsula, showing country boundaries, capitals and other major cities, and other political features. Point out how much of the Korean Peninsula is mountainous.

- What bodies of water surround the Korean Peninsula? *(Sea of Japan, or East Sea; Korea Strait; Yellow Sea; Korea Bay)*
- What political feature divides North Korea and South Korea? *(the demilitarized zone)* Explain that this is a 2.5-mile-wide buffer zone set up to separate the two countries.
- What is North Korea's capital? *(Pyongyang)* What is South Korea's capital? *(Seoul)* What geographic features do these cities share? *(Both are located along rivers and in their countries' interior)*

CONNECT GEOGRAPHY & HISTORY ANSWERS

1. China, Russia, and Japan

2. Buddhism

Presentation Options

Use the **Power Presentations** DVD-ROM to project the animated map.

- Prompt students to use the pop-up images and audio from the map to discover what they already know about the geography, cultures, histories, economics, or governments of the Korean Peninsula.
- Challenge students to answer the map questions individually, in small groups, or as a class activity. Use all of the interactive material to help preview chapter content.

1 Plan & Prepare

Objectives

- Describe the geographic characteristics of the Korean Peninsula
- Identify the climate, plants, and animals found on the Korean Peninsula
- Summarize the natural resources found in North Korea and South Korea

Read for the Essential Question

Help students read for a purpose by reminding them of the Essential Question: "How can North Korea and South Korea overcome their differences and live in peace?" Point out that they can use items in the section, such as Key Questions, special features, images, and maps to formulate answers to the question.

Vocabulary

Best Practices Toolkit

Use the ***Best Practices Toolkit*** to model strategies for vocabulary notetaking. Vary strategies throughout the year. Choose from: *Knowledge Rating, Predicting ABCs, Definition Mapping, Word Sort, Word Wheel, Frayer Model (Word Squares), Magnet Words,* and *Student VOC.*

Vocabulary Strategies, TT9–TT16

Reading Strategy

Best Practices Toolkit

Display the Web Diagram transparency.

- Remind students that main ideas are important ideas in a block of text and that details support the main ideas.
- Model filling in a main idea and details using the subheads and text in this section.
- Have students suggest entries. Remind students that main ideas in each block of text in the section will help answer a Key Question.
- Display the completed diagram for review.

Web Diagram, TT28

SECTION 1 Reading for Understanding

Key Ideas

BEFORE, YOU LEARNED

China is a vast country in East Asia with a variety of landforms, climates, and natural resources.

NOW YOU WILL LEARN

Korea lies on a peninsula, which provides a moderate climate and plentiful resources.

Vocabulary

TERMS & NAMES

Korean Peninsula a body of land bordered by the Yellow Sea to the west, the East Sea (or Sea of Japan) to the east, the Korea Strait to the south, and China and Russia to the north

Yalu River the longest river on the Korean Peninsula; forms part of the border between China and North Korea

Mount Paektu (PAHK•too) the highest mountain (9,003 feet) on the Korean Peninsula

Cheju (CHEH•JOO) **Island** largest island off the southern coast of the Korean Peninsula

monsoon a seasonal wind that usually brings moist air during the summer and dry air during the winter

typhoon a tropical storm that starts in the western Pacific or Indian oceans

BACKGROUND VOCABULARY

peninsula a large mass of land surrounded on three sides by water

Visual Vocabulary satellite image of a typhoon

Reading Strategy

Re-create the web diagram shown at right. As you read and respond to the **KEY QUESTIONS**, use the diagram to record main ideas about the physical geography of North Korea and South Korea. Add ovals or start a new diagram as needed.

Skillbuilder Handbook, page R4

FIND MAIN IDEAS

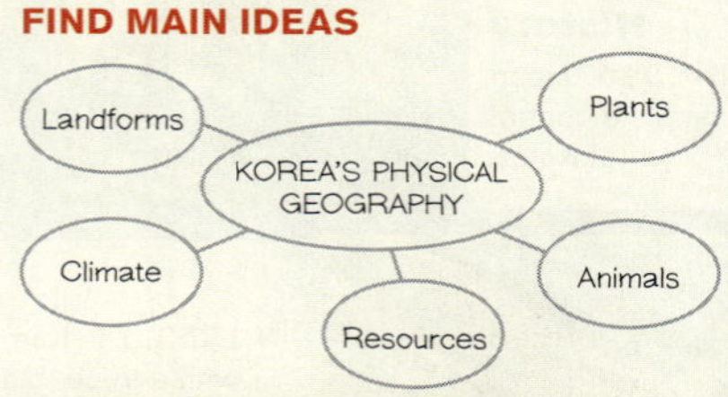

GRAPHIC ORGANIZERS
Go to **Interactive Review** @ **ClassZone.com**

PRETEACHING VOCABULARY

English Learners

Pronounce and Preview

Pronounce each term for students. Review the meanings of words in the definitions, such as *bordered, seasonal,* and *tropical.*

- To modify vocabulary learning, have students complete worksheets as they read the section.

Unit Resource Book: East Asia & Southeast Asia

- Vocabulary Practice, p. 99

Inclusion

Locate Terms

Have students take turns reading each term and its definition aloud. For each term that refers to a geographic location, have students find it on the physical geography map of the Korean Peninsula in this section. Then have them use each term in a sentence.

Kyongbok Palace in Seoul, South Korea

SECTION 1 GEOGRAPHY

Mandarin ducks in forested mountain area

A Rugged Peninsula

Connecting to Your World

Have you ever seen a photograph of a sunrise over a mountain range, such as the Rockies or Appalachians in the United States? These pictures often convey a sense of calm and peace. Korea is dominated by mountainous terrain. The peaceful sunrises over its mountains have inspired people to call Korea the "Land of the Morning Calm." The peninsula's rocky landscape has played a large part in shaping the history and culture of the Korean people.

Cheju Waterfall According to legend, heavenly creatures bathed under this fall, which is surrounded by lush vegetation.

An Isolated Land

KEY QUESTION What are some of the geographic features of the Korean Peninsula?

Korea is a **peninsula**, which means that it is surrounded on three sides by water: the East Sea (or Sea of Japan) to the east, the Yellow Sea to the west, and the Korea Strait to the south. China and Russia border Korea to the north. The **Korean Peninsula** has both isolated and protected its people. In fact, ancient Korea was called the "hermit kingdom." Today, the Korean Peninsula contains two nations—North Korea and South Korea. The tradition of isolation continues in North Korea, which has little contact with other nations.

Mount Daedunsan This mountain in South Korea is famous for its rocky cliffs.

797

2 Focus & Motivate

3-Minute Warm-Up

Write on the board or display the transparency:

- Look at the photograph of Mount Daedunsan. What can you learn about South Korea's geography from this photograph? *(South Korea is mountainous with dramatic cliffs.)*

Unit Transparency Book
- 3-Minute Warm-Ups, TT11

Connecting to Your World

More About . . .

Mount Daedunsan

Considered one of the Korean Peninsula's scenic wonders, Mount Daedunsan is also known as "Little Geumgang." Mount Geumgang in the north of the peninsula is regarded as the region's most beautiful mountain.

3 Teach

An Isolated Land

Talk About It

- What is the longest river on the Korean Peninsula? *(Yalu River)* What is the largest island? *(Cheju Island)*
- Which outside culture has had the strongest influence on the Korean people? *(Chinese)*
- **Find Main Ideas** What have been the advantages and disadvantages of Korea's isolation? *(preserved their culture and protected the country; limited what Koreans could learn from other cultures)*

SECTION 1 PROGRAM RESOURCES

Unit Resource Book: East Asia and Southeast Asia

ON LEVEL
Unit Resource Book
- Reading Study Guide, p. 75
- Outline Map Activities, p. 103
- Section Quiz, p. 114

STRUGGLING READERS
Unit Resource Book
- Reading Study Guide with Additional Support, p. 81
- Vocabulary Practice, p. 99
- Skillbuilder Practice, p. 100
- Section Quiz, p. 114
- Reteaching Activity, p. 117

eEdition with Audio DVD-ROM

ENGLISH LEARNERS
Pupil Edition in Spanish
eEdition with Audio DVD-ROM
Unit Resource Book
- RSG (Spanish), p. 87
- RSG with Additional Support (Spanish), p. 93

Multi-Language Glossary
Test Generator
- Section Quiz in Spanish

INCLUSION
Unit Resource Book
- RSG with Additional Support, p. 81
- Section Quiz, p. 114
- Reteaching Activity, p. 117

GIFTED AND TALENTED
Unit Resource Book
- Section Quiz, p. 114

PRE-AP
Unit Resource Book
- Section Quiz, p. 114

TECHNOLOGY
Unit Transparency Book
- 3-Minute Warm-Ups, TT11
- Essential Question, TT12
- Critical Thinking, TT13
- Map: Korea: Physical/Political, TT16
- Map: Climate and Agricultural Products of East Asia, TT17, TT17a

Daily Test Practice Transparencies
- Section 1, TT87

Power Presentations
ClassZone.com
World Cultures and Geography Video on DVD

More About . . .

The Yalu River

In addition to being a political boundary between North Korea and China, the Yalu River is an important source of hydroelectric power. One of the largest hydroelectric plants in the region is located along the river, supplying energy for a large part of North Korea as well as for a small region of southern China. Because of its geographic location, the Yalu River became a crucial battleground during the Korean War in the 1950s.

CONNECT Geography & Culture

The Hangul Alphabet

The Chinese writing system includes thousands of characters, making it very difficult to learn, while the hangul alphabet consists of only 24 letters: 14 consonants and 10 vowels. Nevertheless, because of the strong influence of Chinese culture in Korea, hangul was not used by educated and upper-class Koreans until after 1945.

- Why would a king want to help his people become more educated? *(Possible answer: to help them become more skilled workers, which would help improve the economy)*

The Korean Peninsula The map on the opposite page shows the bodies of water and landforms on the Korean Peninsula. As you have already learned, the Yellow Sea Ⓐ is situated to the west of the peninsula. The Korea Strait lies between the peninsula's southern coast and the west coast of Japan. The eastern side of the Korean Peninsula borders the East Sea (also known as the Sea of Japan). The **Yalu River** Ⓑ—known as the Amnok River in Korean—is the longest river on the peninsula. The Yalu and the Tumen rivers form the border between China and North Korea.

The rugged terrain of the Korean Peninsula consists of several mountain ranges. The Hamgyong (HAHM•GYUHNG) range extends along the peninsula's northern border and contains **Mount Paektu** (PAHK•too) Ⓒ. At 9,003 feet, this peak is the highest on the peninsula. The Nangnim Mountains run from north to south through the center of North Korea. Farther south, the Taebaek (TAY•BACK) Mountains Ⓓ run north to south along the eastern coast. Branching off from this range are the Sobaek Mountains, which extend across the center of South Korea.

More than 3,000 islands lie off the coasts of the Korean Peninsula. Many of them are small and uninhabited. But the largest, **Cheju** (CHEH•JOO) **Island** Ⓔ, has a population of about 530,000. Cheju lies near the southern part of the peninsula and plays an active role in the culture and economy of South Korea.

CONNECT Geography & Culture

The Hangul Alphabet

For over a thousand years, educated Koreans used a complicated writing system based on Chinese characters. Then in the 1440s, King Sejong, shown below, created a simple Korean writing script. The result was the hangul alphabet, which allowed uneducated Koreans to learn to read and write. The characters below express the word *hello*.

안녕하세요

Impact of Geography on Korea The isolation and protection provided by Korea's geography have allowed the Koreans to develop their own distinct culture. Of course, Chinese culture has had a strong influence. But Koreans have always adapted this culture to fit their own. Koreans even adapted the Chinese language to create their own, as you'll learn at left.

In general, Koreans have been content to remain in their land and follow their own way of life. However, maintaining a peaceful Korean state has not always been easy. Korea's aggressive neighbors, China, Japan, and Russia, have each attempted to control the peninsula. You will learn more about how these superpowers affected Korea's history in Section 2.

SUMMARIZE Describe some of the geographic features of the Korean Peninsula. four major mountain ranges; bordered by the Yellow Sea, East Sea, and Yalu and Tumen rivers

DIFFERENTIATING INSTRUCTION

Struggling Readers

Build Vocabulary

Have each student write any unfamiliar terms on index cards. Put the cards in a box and have a student pick one. He or she should then look up the definition of the term in the dictionary, write the definition on the other side of the index card, and present the term and its definition to the class. The collection of index cards can be added to at any time and used as flashcards for additional vocabulary practice.

Gifted & Talented

Create a Board Game

Have students work in pairs or small groups to create a board game that involves traveling to and from different places on the Korean Peninsula. The board game should include:

- a "mission" for each player (For example, one mission could be: "Travel from Pyongyang to Pusan.")
- cards that identify obstacles along the way, based on the geographic features of the region

Animated GEOGRAPHY: Physical Geography of the Korean Peninsula

Click here to see geographic features of the Korean Peninsula @ClassZone.com

B Yalu River, North Korea

D Taebaek Mountains, South Korea

CONNECT Geography & Culture

READING A PHYSICAL MAP

As you study the map,

- note the Demilitarized Zone, which marks the boundary between North and South Korea
- think about the impact on Korea of its neighbors
- note locations of high and low elevation in Korea

1. **Region** In what ways is the physical geography of North and South Korea similar?
2. **Movement** What impact might this physical geography have on both nations?

DIFFERENTIATING INSTRUCTION

Inclusion

Sculpt the Korean Peninsula

For visual learners, give students some modeling clay of different colors and a tray or sheet of poster board to work on. Have them work in pairs to sculpt a three-dimensional representation of the Korean Peninsula's elevations.

Pre-AP

Make Inferences

Encourage students to think critically about the demilitarized zone. Ask the following questions: Why might the border between North Korea and South Korea be called the demilitarized zone? *(because the two countries were at war)* What can you infer from the fact that this border is still called the demilitarized zone? *(The countries are still at war; a lasting peace hasn't been finalized yet.)*

Animated GEOGRAPHY: Physical Geography of the Korean Peninsula

CLASSZONE.COM

CONNECT Geography & Culture

Reading a Physical Map Point out that this map highlights the physical features of the Korean Peninsula. Direct students' attention to the key. Make sure they understand the relationship between the colors in the map and the elevation levels shown in the key.

- Which country is more mountainous? *(North Korea)*
- Where is the highest point of elevation on the map? *(Mount Paektu in North Korea)*
- Where is most of the low-elevation land found? *(along the west and south coasts)*

ANSWERS

1. **Region** Both are mountainous; both are surrounded by water on three sides; both are about the same size.
2. **Movement** The mountains would make travel on the ground difficult.

Unit Resource Book: East Asia & Southeast Asia

- Outline Map Activities: The Koreas: Physical/Political, p. 103

Presentation Options

Use the **Power Presentations** DVD-ROM to project the animated map.

- Click to animate the map.
- Listen to the audio and examine the photographs to learn about the geography of the Korean Peninsula.

Teach

Climate, Plants, and Animals

Think, Pair, Share

- In what way do the seasons in North Korea and South Korea differ? *(Winters are milder in South Korea than in North Korea.)*
- Why is wildlife on the Korean Peninsula under threat? *(Deforestation has destroyed many of the animals' homes.)*
- **Find Main Ideas** What is the result of efforts to reforest the land in the Korean Peninsula? *(Some endangered animals, such as the black bear, are making a comeback.)*

COMPARING

Climates

Ask students to think about areas in the United States that they have learned about that have climates similar to those described in the captions. To help students answer the questions in this feature, ask them how differences in climate affect human and plant life in the United States.

ANSWERS

1. **North Korea** People must dress warmly; their activities may be restricted; travel is more difficult; agriculture would halt.
2. **South Korea** It produces lush vegetation.

Unit Resource Book: East Asia & Southeast Asia

- Skillbuilder Practice: Comparing and Contrasting, p. 100

Climate, Plants, and Animals

KEY QUESTION What types of climates, plants, and animals are found on the Korean Peninsula?

The mountains of the Korean Peninsula and the seas that surround it both have a major impact on Korea's climate. This climate, in turn, has played a large role in determining the type of plant and animal life on the peninsula.

A Moderate Climate In general, the climate on the peninsula is moderate. Summer weather differs little throughout Korea, with average July temperatures ranging from 70°F to 80°F. But in the winter, the difference in temperature between North Korea and South Korea can be extreme. For example, the average January temperature in southeast Korea is around 35°F. During the same month, the average temperature in the northern mountains is only about –8°F.

Most of the Korean Peninsula receives around 40 inches of precipitation each year. The majority of this rain arrives during the summer monsoons. A **monsoon** is a seasonal wind that usually brings moist air during the summer and dry air during the winter. At times, summer monsoons help produce **typhoons**, fierce tropical storms that start in the western Pacific or Indian oceans. The Korean Peninsula is usually hit with one or two typhoons a year. Typhoons are similar to hurricanes, which you have probably heard about. Hurricanes are tropical storms that start in the northern Atlantic Ocean, the Caribbean Sea, or the eastern Pacific Ocean.

COMPARING Climates

North Korea Winter in North Korea can be long, cold, and dry. **1. How do you think the winters affect the North Korean people?**

South Korea Spring temperatures average about 50°F in South Korea. **2. What impact does this climate appear to have on the country's vegetation?**

DIFFERENTIATING INSTRUCTION

Struggling Readers

Understand Comparisons

To help struggling readers comprehend the comparisons being drawn on this page, help them break down the information using a Venn diagram, such as the one shown at right. Then ask them to write two sentences summarizing the similarities and differences in climate between North Korea and South Korea.

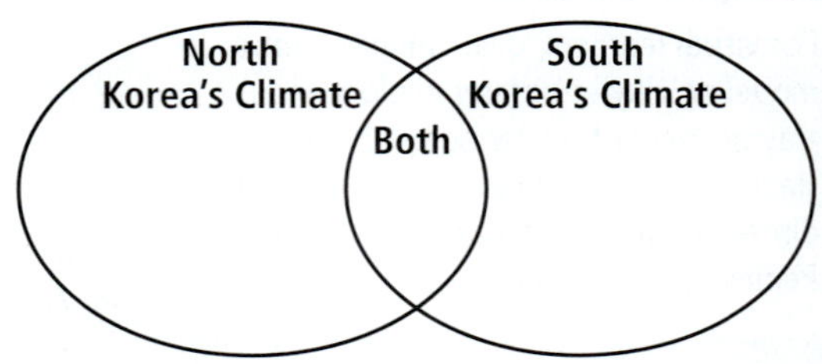

Plant and Animal Life The Korean Peninsula contains much diverse plant life, but the region's natural vegetation is largely forest. Today, forests cover about two-thirds of the land area of South Korea. Fir, spruce, pine, and elm trees grow in the northern and central parts of the country. Bamboo and evergreens grow along the southern coast.

The Korean Peninsula also has a wide variety of wildlife. Deer, mountain antelope, bears, and leopards make their home in mountainous areas and in the forests. Many types of birdlife also live on the peninsula, including herons, cranes, and wild pigeons. In addition, the peninsula is a popular stopover for ducks, wild geese, and swans migrating to Japan, Southeast Asia, and Australia.

As in much of the world, human activities have threatened the survival of wildlife on the Korean Peninsula, particularly in forested areas. People have cut down many of the trees for construction, heating, and cooking. This deforestation destroyed many of the animals' homes. Since the 1970s, however, strong efforts have been made to reforest the land. As a result, some animals that had been considered endangered are making a comeback. The black bear, for example, has returned to some remote mountain areas.

Some of the animals have been rescued by chance. In Section 2, you will learn about a wildlife preserve that accidentally developed along the border between North Korea and South Korea. This accidental preserve has become the Korean Peninsula's most important wildlife refuge.

Red-Crowned Crane In Korean culture, the red-crowned crane symbolizes happiness, good luck, and long life, and often appears in traditional paintings. **Why might a bird represent such positive attributes?**

because birds are beautiful in flight; because their ability to fly inspires thoughts of happiness, freedom, and soaring beyond boundaries

CATEGORIZE Tell what types of climates, plants, and animals are found on the Korean Peninsula. moderate climate; winters much colder in the north; vegetation is largely forest; animals include deer, bears, and birds

Natural Resources

KEY QUESTION How are the natural resources of North Korea and South Korea similar and different?

The rainfall and warm temperatures during the summer make Korea's climate ideal for growing rice. But the rocky terrain of the peninsula makes farmland scarce. Koreans have been able to squeeze tiny fields along hillsides by using terraced farming. Even so, the overall lack of good farmland has led Koreans to rely on other resources, including the abundant fish along the coastlines and mineral deposits.

Comparing Regions

Conservation Efforts in the Korean Peninsula and the United States

The Korean Peninsula isn't the only part of the world that has lost wildlife through deforestation. Deforestation threatens wildlife throughout the world, and the United States is no exception. Less than 5% of ancient forests remain intact in the continental United States. As on the Korean Peninsula, much of these U.S. forests were lost to logging for the timber industry and to development.

3 Teach

Natural Resources

Reader, Recorder, Reporter

- What challenges do Korean farmers face? *(Farmland is scarce, and overuse has made existing farmland infertile, requiring extensive irrigation and fertilizers.)*
- Why was South Korea hit especially hard by the oil crisis of the 1970s? *(because most of the country's electricity was produced in petroleum-burning plants, and the country had to import all of the oil to fuel those plants)*
- **Find Main Ideas** Why do Koreans rely on fishing to meet their nutritional needs? *(Because of limited farmland, they can't raise enough livestock.)*

INTERDISCIPLINARY ACTIVITIES

CONNECT to Language Arts

Design a Reforestation Poster

Have students work in small groups to design and create posters that promote reforestation efforts on the Korean Peninsula. Posters should list the harmful effects of deforestation and the benefits of reforestation. Display the completed posters in the classroom, and then lead a discussion about deforestation and reforestation.

CONNECT to Science

Create a Field Guide

As a class, create a field guide to the plants and wildlife of the Korean Peninsula. Divide the species mentioned on this page among small groups. Students can then look up their assigned species in an encyclopedia or nature guide and create an entry for the guide. Each entry should include visuals of each species (drawn, cut out, or photocopied), and several key facts. Compile the entries into a book to share with the class.

More About . . .

Cheju Island

Cheju Island, also known as Jeju Island or Quelpart Island, was once the independent kingdom of Tamra-guk. During its rule under the Choson Dynasty, political exiles were often sent to the island from the mainland. Today, the island is a popular resort destination, boasting a mild climate and scenic beaches, waterfalls, and volcanic rock formations.

CONNECT Geography & Culture

Women Divers of Cheju Island

The seafood the women of Cheju Island dive for doesn't just provide nourishment—it provides their livelihood. Exports of seafood has made diving a lucrative job for the women of Cheju Island.

Until the 19th century, men were the primary divers, but heavy taxes made the job unprofitable for them. Women took over the job because they did not have to pay taxes. Women's role as breadwinners on Cheju was unusual in Korea's traditional Confucian culture, which had often treated women as subservient to men.

CRITICAL THINKING ANSWER

Make Inferences They are strong; they can hold their breath for long periods of time; they command respect; they are proud of what they do.

Farmland and Agriculture The largest portion of farmland on the Korean Peninsula is found on the western and southern coastal plains. Because of overuse, however, the soil is not very fertile. To remedy this, farmers have extensively used irrigation and fertilizers.

Since most of the best farmland is located in South Korea, this nation has a much higher crop yield than North Korea. Rice is the most important crop in both North Korea and South Korea. These nations also grow wheat, potatoes, and barley. South Koreans, in addition, grow Chinese cabbage, melons, onions, and soybeans. Cheju Island is famous for its mandarin oranges. In fact, Cheju oranges constitute South Korea's major cash fruit crop.

Fishing The lack of usable land has affected Korea's fishing industry. Since the Korean Peninsula does not have enough land to raise much livestock, Koreans rely on fish as an important source of protein. As a result, the fishing industries of the Koreas have risen to keep up with the nutritional needs of the rising population. Both North Korea and South Korea have taken full advantage of their extensive coastal areas. In fact, South Korea has become one of the major deep-sea fishing countries of the world.

CONNECT Geography & Culture

Women Divers of Cheju Island

For more than 1,500 years, the women divers, or *haenyo*, of Cheju Island have caught shellfish, squid, and sea urchins, such as the one shown here. This seafood has provided the main source of nourishment for themselves and their families. Their husbands often stay at home and watch the children.

The women divers work until they are well into their 60s—and beyond. They can plunge down to depths of 60 feet and stay underwater for three to five minutes. And all of this is done without the aid of any breathing equipment. This tradition of diving has been handed down from mother to daughter for generations.

CRITICAL THINKING

Make Inferences What inferences can you make about the women divers?

DIFFERENTIATING INSTRUCTION: TIERED ACTIVITIES

OBJECTIVE Create a visual display of Korea's natural resources.

Basic

Have students create a visual display that shows Korea's natural resources in order of importance. Have them include photographs (cut out or copied) or illustrations of each resource and a label. Have students present or post their displays.

On Level

In addition to the Basic activity, have students include a one- or two-sentence caption under each resource that explains its importance. Have students present or post their displays.

Challenge

Adapt the Basic activity and have students create an illustrated map of the Korean Peninsula, showing where resources can be found. Encourage students to use symbols to indicate resources. They should also include captions for all illustrations and a map key. Have students present or post their maps.

Mineral Resources The vast majority of mineral resources on the Korean Peninsula are in the north. In fact, North Korea has the largest magnesite deposit in the world. Other minerals include iron ore, coal, gold, lead, and zinc. North Koreans have taken advantage of these rich deposits by mining them extensively. Mineral resources in South Korea are few. However, the nation does have some large graphite and tungsten deposits.

Most of North Korea's electric power is produced by coal-burning plants. In contrast, the majority of electricity in South Korea is provided by petroleum-burning plants. However, this nation has to import all of its oil. As a result, the oil crisis of the 1970s hit South Korea hard. To reduce South Korea's dependency on oil, the government had built 20 nuclear power plants by 2006. By 2015, the government hopes to have about 30 such plants.

COMPARE AND CONTRAST Explain how the natural resources of North Korea and South Korea are similar and different. South Korea has more farmland; both have extensive fishing resources; North Korea has many more minerals.

Minerals in the Koreas

Minerals	Uses
Graphite	• pencils • applications that involve high heat
Magnesite	• flooring materials • fertilizers • bricks
Tungsten	• filament in light bulbs • heating element in electric furnaces
Zinc	• coating on iron and steel to protect from corrosion • batteries

Section 1 Assessment

ONLINE QUIZ For test practice, go to **Interactive Review @ClassZone.com**

TERMS & NAMES

1. **Explain the importance of**
   - Korean Peninsula
   - Yalu River
   - Mount Paektu
   - Cheju Island

USE YOUR READING NOTES

2. **Find Main Ideas** Use your completed web to answer the following question:

   What impact have people had on the vegetation on the Korean Peninsula?

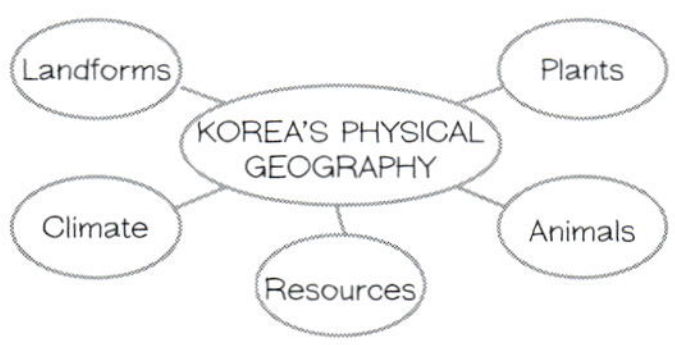

KEY IDEAS

3. What are some of the major mountain ranges of the Korean Peninsula?
4. Why did deforestation occur on the Korean Peninsula?
5. What have Korean farmers done to try to make their soil more fertile?

CRITICAL THINKING

6. **Draw Conclusions** How has the geography of the Korean Peninsula influenced Korea's culture?
7. **Compare and Contrast** How do the climates of North Korea and South Korea differ?
8. **CONNECT to Today** How has South Korea recently reduced its dependency on oil?
9. **WRITING Compose an Essay** Research and write an essay about the deep-sea fishing industry in South Korea. Include information about the types of ships used, the methods used to catch various fish, and the effect of this fishing on the environment.

CONNECT to the Essential Question

How can North Korea and South Korea overcome their differences and live in peace?

Ask students what they have learned so far that can help them answer this question. Students might mention the following:

- The Korean Peninsula has a tradition of isolation, which North Korea still maintains.
- Korea's aggressive neighbors have all attempted to take control of the peninsula in the past, making peace difficult.

4 Assess & Reteach

Assess Have students complete the Section Assessment.

Unit Resource Book: East Asia & Southeast Asia
- Section Quiz, p. 114

Interactive Review @ClassZone.com
- Section Quiz

Power Presentations

Test Generator

Reteach Divide the class into teams. Write "Korean Peninsula: Physical Features" on the board. Have students take turns calling out facts about its physical features. List these on the board. Do the same for "Climate, Plants, and Animals" and "Natural Resources."

Unit Resource Book: East Asia & Southeast Asia
- Reteaching Activity, p. 117

SECTION 1 ASSESSMENT ANSWERS

Terms and Names

1. Korean Peninsula: body of land bordered by the Yellow Sea to the west, the East Sea to the east, the Korea Strait to the south, and China to the north; Yalu River: longest river on the Korean Peninsula; Mount Paektu: highest mountain on the Korean Peninsula; Cheju Island: largest island off the southern coast of the Korean Peninsula

Use Your Reading Notes

2. People have destroyed much of the forestland.

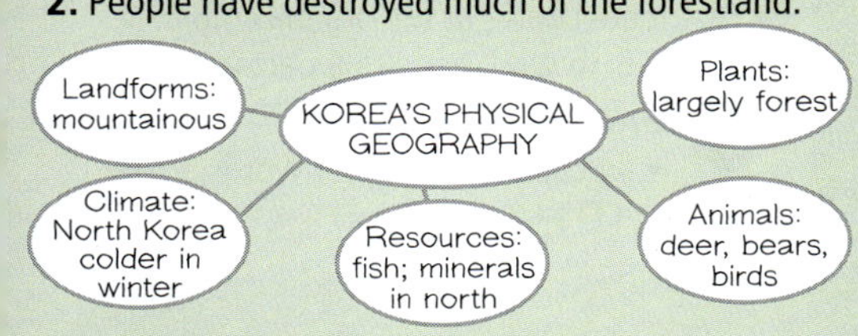

Key Ideas

3. the Hamgyong, Nangnim, Taebaek, and Sobaek ranges
4. People cut down trees for construction, heating, and cooking.
5. extensive use of irrigation and fertilizers

Critical Thinking

6. The mountains and surrounding seas have discouraged invasion and isolated and protected Korea's people and distinct culture.
7. North Korea tends to have much colder winters.
8. By building many new nuclear power plants, South Korea will not have to rely as much on petroleum-burning plants. As a result, the South Koreans will not have to import as much oil.
9. Use the rubric to score essays.

Essay Rubric

	Structure	Content	Errors
4	logical and flowing	shows thorough understanding	few or none
3	clear	shows good understanding	some
2	somewhat confusing	shows basic understanding	many
1	weak and unclear	shows little or no understanding	many major

1 Plan & Prepare

Objectives

- Identify the major events in Korean history
- Describe how Korean culture has been preserved

Read for the Essential Question

Help students read for a purpose by reminding them of the Essential Question: "How can North Korea and South Korea overcome their differences and live in peace?" Point out that they can use items in the text, such as Key Questions, special features, images, and maps to formulate answers to the question.

Vocabulary

Best Practices Toolkit

Use the ***Best Practices Toolkit*** to model strategies for vocabulary notetaking. Vary strategies throughout the year. Choose from: *Knowledge Rating, Predicting ABCs, Definition Mapping, Word Sort, Word Wheel, Frayer Model (Word Squares), Magnet Words*, and *Student VOC*.

Vocabulary Strategies, TT9–TT16

Reading Strategy

Best Practices Toolkit

Display the T Chart/Two-Column Chart transparency.

- Point out to students that when they summarize information, they should focus on the key ideas from their reading. Summaries should be shorter and less detailed than the original text.
- Model filling in key ideas for the paragraph under "Korean History" on the chart. Have students suggest entries.
- Display the completed chart for review.

T Chart/Two-Column Chart, TT20

Section 2 Reading for Understanding

Key Ideas

BEFORE, YOU LEARNED

Korea is surrounded by powerful neighbors, including China, Russia, and Japan.

NOW YOU WILL LEARN

The Korean Peninsula has been invaded by China, Russia, and Japan but has still maintained its cultural identity.

Vocabulary

TERMS & NAMES

Three Kingdoms the three kingdoms that had formed on the Korean Peninsula by A.D. 300

Korean War (1950–1953) a conflict between North Korea and the Soviet Union on one side and South Korea, the United States, and the UN on the other

demilitarized zone a buffer zone between North Korea and South Korea

shamanism (SHAH•muh•NIHZ•uhm) a belief system in which a person called a shaman is believed to be able to communicate with spirits and heal the sick

celadon (SEHL•uh•DAHN) Korean ceramic pottery with a thin blue or green glaze

hanbok a traditional Korean costume

kimchi (KIHM•chee) a Korean dish made of pickled vegetables

REVIEW

Confucianism a belief system based on the teachings of Confucius

Buddhism a religion that began in India and is based on the teachings of Siddhartha Gautama

Visual Vocabulary *hanbok*

Reading Strategy

Re-create the chart shown at right. As you read and respond to the **KEY QUESTIONS**, use the chart to summarize what you have learned about the history and culture of the Korean Peninsula.

Skillbuilder Handbook, page R5

SUMMARIZE

KOREAN HISTORY	KOREAN CULTURE
The first state. . .	

GRAPHIC ORGANIZERS Go to **Interactive Review** @ClassZone.com

PRETEACHING VOCABULARY

English Learners

Pronounce and Preview

Pronounce each term for students. Review the meanings of words in the definitions, such as *conflict, shaman*, and *glaze.*

- To modify vocabulary learning, have students complete worksheets as they read the section.

Unit Resource Book: East Asia & Southeast Asia

- Vocabulary Practice, p. 99

Inclusion

Use Magnet Words

Have students work in pairs to fill out the graphic organizer for the terms and names in the section.

- Use the transparency to model the strategy.
- Have volunteers suggest entries.
- Distribute copies of the organizer for students to work on with a partner.

Best Practices Toolkit

- Vocabulary: Magnet Words, TT15

SECTION 2 HISTORY & CULTURE

Kyongbok Palace in Seoul, South Korea

Mandarin ducks in forested mountain area

Strong Traditions, Modern Innovations

Connecting to Your World

You have probably learned that the United States was torn apart by a civil war between the North and the South during the 1860s. After four years of fighting, the two sides reunited to form a stronger union. In the 1950s, Korea became a battleground for a war waged by the United States and the Soviet Union. The war pitted North Korea against South Korea. But this conflict had a different ending: North and South Korea remained divided. As you will see, however, the Korean Peninsula has been united for much of its history.

Korean History

KEY QUESTION What are the major events in Korean history?

The first state on the Korean Peninsula was established around 2000 B.C. It took the name *Choson*. Around 108 B.C., China conquered the northern part of the peninsula. However, Korean tribes gradually won back the territory. In the northern area of the Korean Peninsula, several tribes united during the A.D. 100s and formed the state of Koguryo (KOH•gur•YOO). By about 300, two other states had been created on the peninsula. Historians refer to these states as the **Three Kingdoms**.

Koguryo Painting A Korean noblewoman is waited on by her attendants in this fourth-century tomb mural painting.

805

2 Focus & Motivate

3-Minute Warm-Up

Write on the board or display the transparency:

- How might the division of Korea affect the Korean people? *(Contact may be difficult; governments may clash.)*

Unit Transparency Book
- 3-Minute Warm-Ups, TT11

Connecting to Your World

More About . . .

Three Kingdoms Art

The earliest known paintings in Korea were tomb murals. These paintings were done in fresco—a painting technique that uses watercolors on plaster. Buddhism strongly influenced Korean art. Experts believe that the first Buddhist sculptures originated in the Koguryo Kingdom in the fifth century

3 Teach

Korean History

Think, Pair, Share

- Why are many Koreans still bitter about the Japanese occupation of Korea? *(Japan ruled Korea very harshly and tried to diminish the influence of Korean culture.)*
- **Summarize** Summarize the main differences between North Korea and South Korea. *(North Korea is ruled under a strict Communist dictatorship; South Korea is a democracy with a thriving economy.)*

SECTION 2 PROGRAM RESOURCES

Unit Resource Book: East Asia and Southeast Asia

ON LEVEL

Unit Resource Book
- Reading Study Guide, p. 77
- Section Quiz, p. 115

STRUGGLING READERS

Unit Resource Book
- Reading Study Guide with Additional Support, p. 83
- Vocabulary Practice, p. 99
- Skillbuilder Practice, p. 100
- Section Quiz, p. 115
- Reteaching Activity, p. 118

eEdition with Audio DVD-ROM

ENGLISH LEARNERS

Pupil Edition in Spanish

eEdition with Audio DVD-ROM

Unit Resource Book
- RSG (Spanish), p. 89
- RSG with Additional Support (Spanish), p. 95

Multi-Language Glossary

Test Generator
- Section Quiz in Spanish

INCLUSION

Unit Resource Book
- RSG with Additional Support, p. 83
- Section Quiz, p. 115
- Reteaching Activity, p. 118

GIFTED & TALENTED

Unit Resource Book
- History Makers, pp. 105–106
- Interdisciplinary Projects, pp. 107–108
- Section Quiz, p. 115

PRE-AP

Unit Resource Book
- Primary Sources, pp. 112–113
- Section Quiz, p. 115

TECHNOLOGY

Unit Transparency Book
- 3-Minute Warm-Ups, TT11
- Essential Question, TT12
- Critical Thinking, TT14
- Cultures, TT18, TT19

Daily Test Practice Transparencies
- Section 2, TT88

Power Presentations

ClassZone.com

World Cultures and Geography Video on DVD

More About . . .

Korea Under Yi Song-gye

Yi Song-gye made many changes as ruler of Korea. He replaced Buddhism with Confucianism as the official religion. He made farming the center of Korea's economy and had close ties with China's Ming dynasty. He also established a land redistribution policy that assigned land ownership based on a person's status, or rank. This policy helped Yi and his supporters gain economic and political power.

Unit Resource Book: East Asia & Southeast Asia

- History Makers: King Sejong, p. 105
- History Makers: Queen Sondok, p. 106

ANALYZING Primary Sources

The Proclamation of Korean Independence was signed by 33 prominent Koreans and began what is known as the March 1st Movement, which called for independence from Japanese rule. Several massive demonstrations took place as part of this movement. However, the Koreans' push for independence was brutally suppressed by the Japanese government.

- How would you describe the tone of this document? *(Possible answers: angry, defiant, demanding, assertive)*
- What is the primary message of this passage? *(Possible answer: Koreans must look toward a future of independence and not dwell on past wrongdoings.)*

DOCUMENT-BASED QUESTION ANSWER

They wanted to express their unhappiness with Japanese occupation and make a plea for self-government.

Three Kingdoms Period The Three Kingdoms were Koguryo in the northeast, Paekche (PAHK•CHAY) in the southwest, and Silla (SIHL•uh) in the southeast. The kingdoms had much in common. For example, they all had a strong military system, which they used to help them expand their territory. In addition, each had powerful kings and aristocracies made up of tribal chiefs.

During the 660s, Silla conquered Paekche and Koguryo and took control of the entire peninsula. Unified Silla was broken apart by rebels in the 800s. But the state reunited around 935 and called itself *Koryo. Korea* comes from *Koryo.* In 1392, a general named Yi Song-gye became the ruler of Korea. He gave the country back its ancient name of Choson. The Choson Dynasty lasted until 1910, more than 500 years.

Japanese Conquest Starting in the 1600s, the rulers of the Choson Dynasty closed Korea to all foreigners. In 1876, however, Japan forced Korea to open a few ports to trade. Soon, the United States, Russia, and several European nations signed trade agreements with Korea. In 1904 and 1905, Japan fought Russia for control over Manchuria. Japan won the Russo-Japanese War and then attacked Korea. By 1910, the Japanese had gained complete control of Korea.

ANALYZING Primary Sources

The Proclamation of Korean Independence was published on March 1, 1919, during the Japanese occupation of Korea. It was inspired by the Fourteen Points of U.S. President Woodrow Wilson, which supported returning self-government to some occupied nations following World War I. The painting shown here depicts Japanese soldiers marching into Seoul.

> We have no desire to accuse Japan of breaking many solemn treaties since 1836, nor to single out specially the teachers in the schools or government officials who treat the heritage of our ancestors as a colony of their own, . . . finding delight only in beating us down and bringing us under their heel. . . . Our urgent need today is the settling up of this house of ours and not a discussion of who has broken it down, or what has caused its ruin.
>
> Source: Proclamation of Korean Independence

DOCUMENT-BASED QUESTION

What did the writers of this document probably hope to achieve?

DIFFERENTIATING INSTRUCTION

Struggling Readers

Sequence Korean History

Ask students who have trouble sequencing events to work in pairs to create a time line listing the major events in Korean history. Using the time line in the beginning of the chapter as a starting point, have them fill in other major events from the "Korean History" section. As a whole-group activity, have students share and discuss their time lines.

Gifted & Talented

Write a Speech

After students have read the excerpt from the Proclamation of Korean Independence, have them try to visualize the demonstrations that took place during the March 1st Movement. Then have students work in small groups to write a short speech aimed at inspiring support for Korean independence from Japan. Have students present and discuss the speeches.

During the Japanese occupation, Japan turned Korea into a colony to benefit its own interests. The Japanese government managed Korean farms and businesses and reduced the Koreans to low-paid workers in their own country. As the Japanese prepared for war during the 1930s, they tried to stamp out Korean culture. They even ordered Koreans to speak only Japanese and to take Japanese names. As World War II raged, Koreans were forced to help Japan's war effort and to endure many hardships. Japan's harsh rule is still a source of bitterness for many Koreans.

The Korean War Korea remained under Japanese rule until Japan was defeated in World War II in 1945. The Koreans wanted independence. But U.S. forces occupied southern Korea, and Soviet troops occupied the northern half. Soon, North Korea established a Communist government, and South Korea set up a republic.

In 1950, North Korean troops invaded South Korea, starting the **Korean War**. The Soviet Union had supplied North Korea with tanks, airplanes, and money in an attempt to take over the peninsula. South Korea fought back with Allied forces consisting of U.S. and UN troops. In time, China entered the war on the side of North Korea. You can follow the course of the war by studying the maps on this page. A cease-fire was declared in 1953, but no peace treaty was ever agreed on. In time, a 2.5-mile-wide buffer zone, called the **demilitarized zone** (DMZ), was set up. It separates North Korea and South Korea to this day.

The Allies lost about 500,000 troops, and the Communists lost around 1,500,000 troops. Civilian deaths were also very high. A total of about two million civilians died in North Korea and South Korea.

The Korean War *1950–1953*

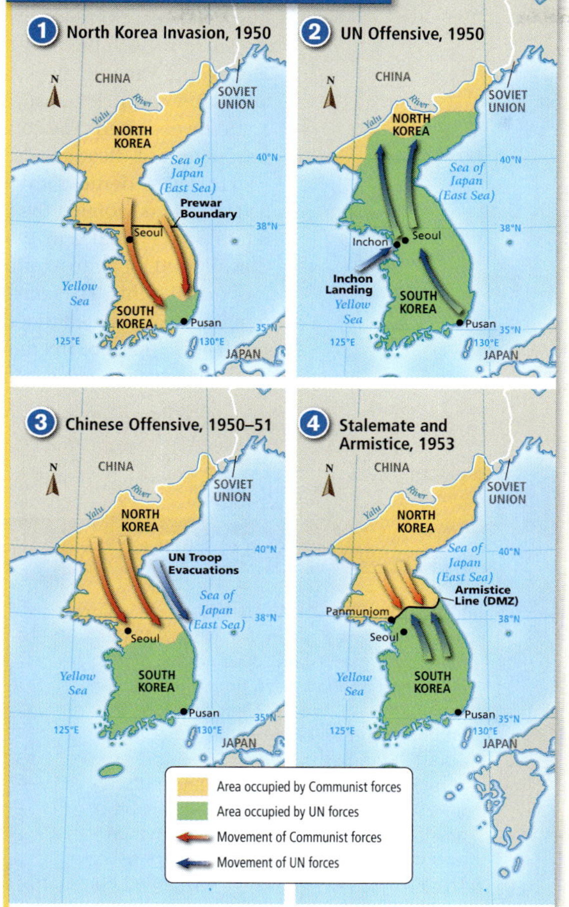

CONNECT Geography & History

1. **Movement** Which forces moved south almost to Pusan?
2. **Region** Compare maps 1 and 4. Which side, if any, gained more territory?

INTERDISCIPLINARY ACTIVITIES

CONNECT to Math

Chart Casualties of the Korean War

Have students create either bar or pie charts that compare the number of Allied, Communist, and civilian deaths in the Korean war, using the information on this page. To extend this activity, have students research and graph the casualties for each year of the war.

CONNECT to Language Arts

Report on the Korean War

Have students convert "The Korean War *1950–1953*" into a series of news reports on the war. For each small map shown, students should write a short news bulletin reporting on the progress of the war. Have students present their reports to the class.

The Korean War *1950-1953*

CONNECT Geography & History

Point out that these maps show the movement of troops during stages of the Korean War. Direct students' attention to the key. Ask students how far the UN forces advanced during their offensive. *(almost all the way to the Yalu River)*

ANSWERS

1. **Movement** Communist forces
2. **Region** neither side; The border remained near the 38°N parallel.

Unit Resource Book: East Asia & Southeast Asia
- Primary Sources: On the Korean War, p. 113

More About . . .

The Korean War Cease-Fire

Negotiations for a cease-fire began in 1951 and continued on and off for two years. The final sticking point in the negotiations was the UN principle that prisoners of war should not be returned to their original country against their will. The Communists did not want to agree to this policy. Eventually, it was agreed that prisoners of war would be placed in neutral custody for 90 days while their country of origin tried to persuade them to return.

Make It Fun

Teacher-Tested Activities

Todd Harrison, Hardin County Middle School, Savannah, Tennessee

I like to broaden a history lesson to include other parts of the world.

- First, I draw a time line on the board and label the beginning and end dates. For the lesson on these pages, I would label the dates 1910 and 1953.
- Then I invite volunteers to enter the dates and events they have just learned about.
- Finally, we discuss the worldwide events that occurred during the period shown on the time line. I use prompts such as "What began in 1914 and ended in 1918?"

CONNECT History & Geography

Wildlife Preserve in the DMZ

The DMZ has been the site of military tension and conflict since the end of the Korean War. Each country maintains troops along its border, and North Korea has been accused of digging underground tunnels through the zone. However, in recent years, the two countries have discussed building a bridge over the DMZ and have agreed to restore roads and railways connecting the two countries.

CRITICAL THINKING ANSWER

Make Predictions The preserve might be destroyed.

Unit Resource Book: East Asia & Southeast Asia

- Skillbuilder Practice: Comparing and Contrasting, p. 100

More About . . .

The Asiatic Black Bear

The Asiatic black bear can be found in the Himalayas, Southeast Asia, and parts of East Asia. These bears are omnivores but mostly eat meat. Asiatic black bears are worshiped as gods by the Ainu people of Japan. However, they are also hunted throughout Asia for their gall bladders and bile, which are considered very valuable in traditional Asian medicine.

Two Koreas In 1948, Kim Il Sung became North Korea's leader and continued to rule after the Korean War. Kim strove to make North Korea as self-sufficient as possible by emphasizing the development of heavy industry. He established a strict dictatorship. Kim died in 1994 and was succeeded by his son, Kim Jong Il. In the 2000s, North Korea started a nuclear weapons program, despite opposition from the United States, South Korea, and many other countries. North Korea has also been charged with numerous human rights violations.

Since the 1960s, South Korea has enjoyed an economic boom. The boom was due, in part, to financial aid from the United States and other countries. Also, South Korea began to produce goods that could be exported. The country used the resulting income to increase production of these goods.

However, democracy in South Korea has faced a rockier road. Several rulers violated the 1948 constitution and used the military to retain power. Finally, in 1987, South Korea adopted a new constitution that allowed for almost unlimited political freedom and granted greater individual rights. The constitution declared South Korea a democratic republic. It also stated that the country would seek reunification based on the principles of freedom and democracy. Some efforts have been made to reunify the two Koreas, as you will learn in Section 3.

A SEQUENCE EVENTS Discuss the major events in Korean history. Three Kingdoms period, Japanese conquest, World War II, the Korean War

CONNECT History & Geography

Wildlife Preserve in the DMZ Since the end of the Korean War, the DMZ has been a no-man's-land. As a result, plants, birds, and animals have flourished in this region. In fact, the DMZ has developed into an important wildlife refuge. Extending about 150 miles from coast to coast, the zone has become home to many endangered animals, including the Asiatic black bears shown here. Some people have proposed turning the DMZ into a peace park. The park would honor those who died during the war and save the wildlife preserve.

CRITICAL THINKING

Make Predictions What might happen to the preserve if the two Koreas reunite?

DIFFERENTIATING INSTRUCTION

English Learners

Monitor Comprehension

Non-native English speakers may have trouble following some of the more complex sentences on this page. Remind students to stop after each sentence and ask themselves if they understand the ideas expressed. They can check their own understanding by rewriting each sentence in their own words or by listing the main ideas in each paragraph.

Pre-AP

Analyze Reunification of the Two Koreas

Point out that, in contrast with South Koreans, North Koreans have little or no experience with democracy. Have students work in pairs to analyze how these different experiences might affect the possible reunification of the Koreas. Each pair should then present their analysis in writing or as an oral report.

Worldster

A Day in Yumi's Life

To learn more about Yumi and her world, go to the **Activity Center** @ClassZone.com

In Korean, Yumi's name looks like this:

유미

Hi! My name is Yumi. I am 14 years old and live in Seoul, the capital and largest city in South Korea. Seoul is a combination of old and new. For example, I live in a modern apartment complex, but nearby is Kyongbok, a medieval palace. My daily life also combines the old and new. Let me show you by describing a typical day.

6 A.M. Today, I am preparing for exams, so I get up early to study. Then I eat a traditional breakfast of rice, soup, kimchi, and fish. After breakfast, I ride my bike to school.

8 A.M.–3 P.M. At school, I study about new discoveries in science class and about my country's past in history class. Other courses include math and English. I eat a light lunch, which is provided by the school.

5 P.M. Sometimes, I spend a couple of hours after school with my friends. We go to shopping malls and play video games. But usually I have after-school classes in violin and English conversation.

7 P.M. I eat a traditional dinner with my family that includes rice and many *banchan* (side dishes). Then I study some more. When I can, I make time to watch my favorite TV show.

CONNECT to Your Life

Journal Entry Make a list of the recreational activities that you do on a regular basis. Then compare your activities with Yumi's. How are they different? How are they the same?

DIFFERENTIATING INSTRUCTION: TIERED ACTIVITIES

OBJECTIVE Compare and contrast Yumi's daily life with that of teens in other countries.

Basic

Have students create a two-column chart in which they compare their daily life to Yumi's. They should compare the following categories: wake-up times, meals, school schedule, and family routines.

On Level

Have students write a pen-pal letter to Yumi in which they describe their daily life. In the letter, they should point out similarities and differences between their daily routines and Yumi's, and tell Yumi about anything that surprised them about her life. They should also ask at least one question about Yumi's daily life.

Challenge

Have students assume the role of Yumi, and write letters in response to those written for the On-Level activity. Encourage students to think about how Yumi might view American daily life, what might surprise her, and what questions she might have for her pen pal.

More About . . .

Kyongbok Palace

This ancient palace was built in 1394 during the Choson Dynasty. It is a 40-acre complex with several buildings and pavilions. Today, the palace is a major tourist attraction in Seoul.

Comparing Regions

Education in South Korea and the United States

While the United States requires children to attend school until at least age 16, children in South Korea are only required to go to school for six years. As in the United States, elementary school and middle school are free and open to all children in South Korea. South Korea offers two types of high schools—general and vocational—as well as small schools specializing in the arts or sciences. Unlike American students, Korean students must apply for entrance into high school and pay tuition.

Presentation Options

Go to the *Activity Center* at **ClassZone.com** to project Worldster.

- Click on the multimedia features of Yumi's Web page.
- Ask students which areas of Yumi's life they would like to explore. Point out that they can listen to music, examine photographs, learn simple phrases in Korean, and more.

CONNECT to Your Life

Journal Entry

Encourage students to examine the photos for details described by Yumi in the text before writing their entry. For example, ask students what Yumi is doing in the third photo. *(laughing with a friend at a mall)* How does Yumi's dinner compare with a typical dinner in your home? *(Students may say her dinner has many more dishes than theirs.)*

Unit Resource Book: East Asia & Southeast Asia
- Interdisciplinary Projects: Korean Food, p. 107

Teach

Preserving Korean Culture

Talk About It

- What is the role of religion in South Korea? *(Confucianism still influences South Korea; many South Koreans practice Buddhism; Christianity is its fastest-growing religion; it has strong religious tolerance.)* In North Korea? *(Religion is discouraged.)*
- What is the focus of life in rural villages? *(farming, especially rice-growing)*
- **Summarize** How do Koreans integrate modern and traditional aspects of their culture into daily life? *(They wear modern clothing and use modern conveniences, but may wear traditional clothes for special occasions, eat traditional foods, and live in houses heated with ondols.)*

Unit Resource Book: East Asia & Southeast Asia

- Interdisciplinary Projects: Making Paper in Korea, p. 108
- Primary Sources: Korean Folk Tales, p. 112

CONNECT to the Essential Question

How can North Korea and South Korea overcome their differences and live in peace?

Ask students what they have learned so far that can help them answer this question. Students might mention the following:

- The Korean Peninsula has a long tradition of being united.
- South Korea's constitution declares that it will seek reunification with North Korea under democratic principles.
- All Koreans are united by a common culture, including ethnicity and language.

Haein-sa Temple This 8th-century temple houses the Tripitaka Koreana: woodblocks containing the most complete collection of Buddhist texts. **What do the woodblocks suggest about Buddhism in Korean culture?**

Buddhism is very important in Korean history and culture.

Preserving Korean Culture

KEY QUESTION How have Koreans tried to preserve their distinct culture?

Over the past century, Korea has faced much adversity. But the Korean people are united by a common ethnicity and language—and the determination to preserve their culture.

Religious Beliefs The original belief system practiced on the Korean Peninsula is **shamanism**. A shaman is a person who is believed to be able to communicate with the spirit world and heal the sick. Another belief system, **Confucianism**, entered the peninsula around 100 B.C. from China. Confucianism stresses social order and good government and influences Korean society today.

Buddhism, an Indian religion that seeks to put an end to suffering, arrived during the A.D. 300s. It was the dominant religion in Korea for several hundred years. Today, many South Koreans still practice Buddhism. Christianity was introduced around 1784 and is South Korea's fastest-growing religion.

Many Koreans practice aspects of several religions. Because of this, South Korea has strong religious tolerance. North Korea, on the other hand, discourages any religious practice.

Korean Arts Buddhist influence can be seen in the many temples found throughout South Korea, such as the one above. These temples often contain stone towers called pagodas. They also include stone sculptures of Buddha.

Korean pottery has been strongly influenced by the Chinese. During the Koryo Dynasty, Korean artisans used Chinese techniques to develop a unique type of ceramic pottery called **celadon**. This pottery has a thin glaze that usually ranges in color from blue to green.

Celadon The ceramic pottery originated in China, but Korean potters made it their own. Even Chinese visitors remarked on the beauty of Korean celadon. ►

DIFFERENTIATING INSTRUCTION: TIERED ACTIVITIES

OBJECTIVE Research and present an overview of shamanism, Confucianism, and Buddhism on the Korean Peninsula.

Basic

Have students choose shamanism, Confucianism, or Buddhism to research. Provide basic resources for them to use. Students should then create a chart or graphic organizer, listing at least five key facts about the belief system/religion they chose.

On Level

In addition to the Basic activity, have students summarize the influence the belief system/religion they chose has had on Korean society.

Challenge

Adapt the On-Level activity and have students complete a chart or graphic organizer for two of the belief systems/religions. Then have students use the chart to write a brief compare-and-contrast essay on the two.

Daily Life Because North Korea is a closed country, not much is known about the daily life of its people. Much of the following, therefore, describes life in South Korea.

Most Koreans wear modern-style clothing. However, on special occasions, South Koreans often wear traditional clothes, such as the **hanbok**. The woman's *hanbok* is made up of a short jacket and a long skirt. For men, a *hanbok* consists of roomy pants tied at the ankles and a jacket. The basic food for most Koreans is rice. Meals also often include soup and **kimchi**, a side dish made of pickled vegetables. Sports, including the martial art of *tae kwon do,* are popular pastimes.

Life in rural villages centers on farming, especially rice-growing. Many people live in houses with heated floors called *ondols.* These floors are traditionally heated by channels carrying hot air underneath them. Because the floors are so warm, Koreans often prefer to sit and even sleep on them. Many rural houses have modern conveniences, including satellite dishes. In cities, millions of people live in high-rise apartment buildings.

A FIND MAIN IDEAS Discuss how Koreans have tried to preserve their distinct culture. by practicing aspects of several religions, by still wearing *hanboks*, by eating kimchi, by still using heated floors called *ondols*, by developing unique pottery called celadon

Fun Facts!

KIMCHI

Kimchi is eaten at almost every meal. There are more than 200 varieties of the pickled vegetable side dish. Traditionally, Koreans get together in the fall to prepare huge batches of kimchi. They then store the kimchi in large earthenware jars, which they bury in their yards. Kimchi is so popular that it is celebrated in its own museum.

Section 2 Assessment

ONLINE QUIZ
For test practice, go to **Interactive Review** @ClassZone.com

TERMS & NAMES

1. **Explain the importance of**
   - Three Kingdoms
   - demilitarized zone
   - celadon
   - *hanbok*

USE YOUR READING NOTES

2. **Summarize** Use your completed chart to answer the following question:

   Why did the Koreans deeply resent the Japanese occupation?

KOREAN HISTORY	KOREAN CULTURE
The first state. . .	

KEY IDEAS

3. What are some of the traits that the Three Kingdoms had in common?
4. How did Japan gain control over Korea?
5. What does life in rural villages center on?

CRITICAL THINKING

6. **Make Inferences** Do you think the Korean people benefited from the Korean War? Explain why or why not.
7. **Sequence Events** In what chronological order did major belief systems enter the Korean Peninsula?
8. **CONNECT to Today** Why do you think North Korea has recently started a nuclear weapons program?
9. **TECHNOLOGY** **Create a Slide Show** Using the Internet and your library, research the various artworks found in Buddhist temples, such as pagodas and sculptures. Then do a slide presentation that shows these works, and provide a brief history about each slide.

Comparing Regions

Traditional Clothing in Korea and India

Unlike the Koreans, many people in India wear traditional clothes every day. Traditional Indian clothing varies by religion. The traditional dress for Hindu women in India is a *sari,* a five-to-seven-yard-long piece of brightly colored fabric that's worn wrapped around the body over a shirt and petticoat. Hindu men in India often wear short coats, called *angarkha.* Indian Muslims—men and women—traditionally wear a long white cotton shirt, called a *kurtah,* and pants called *pa'ijama.* Indian Muslim women also may wear a veil, called a *burkah,* that covers their faces and bodies.

4 Assess & Reteach

Assess Have students complete the Section Assessment.

Unit Resource Book: East Asia & Southeast Asia
- Section Quiz, p. 115

Interactive Review @ ClassZone.com
- Section Quiz

Power Presentations

Test Generator

Reteach Have students write a summary of the history and culture of North Korea and South Korea. Then have students trade summaries and offer one positive comment and one suggestion for an improvement of the summaries.

Unit Resource Book: East Asia & Southeast Asia
- Reteaching Activity, p. 118

SECTION 2 ASSESSMENT ANSWERS

Terms & Names

1. Three Kingdoms: the three kingdoms that had formed on the Korean Peninsula by A.D. 300; demilitarized zone: buffer zone between North Korea and South Korea; celadon: Korean ceramic pottery with thin blue to green glaze; *hanbok*: traditional Korean costume

Use Your Reading Notes

2. because the Japanese tried to eradicate Korean culture and treated the Koreans with cruelty

KOREAN HISTORY	KOREAN CULTURE
The first state, called Choson; Three Kingdoms; Silla; Koryo; Choson Dynasty; Japan conquered Korea; peninsula split into North and South Korea	shamanism, Buddhism, Confucianism, Christianity; pagodas, Buddha stone sculptures, celadon pottery; *hanbok*, kimchi, *ondols*

Key Ideas

3. strong military system, powerful kings, aristocracies made up of tribal chiefs
4. Japan fought Russia for control of Manchuria and then attacked Korea.
5. farming, especially rice-growing

Critical Thinking

6. Possible answer: no, because the Korean people suffered a great deal during the war, and the two countries continue to have problems today
7. shamanism; Confucianism around 100 B.C.; Buddhism during early A.D. 300s; Christianity around 1784
8. Possible answers: North Korea feels threatened; it stresses a strong military.
9. Use the rubric to score slide shows.

Slide Show Rubric

	Presentation of Artwork	Explanation of Artwork
4	clear and interesting	excellent and concise
3	clear	very good
2	not very clear	fair
1	unclear	confusing

1 Plan & Prepare

Objectives

- Describe the government and economy of North Korea
- Compare the government and economy of South Korea with those of North Korea
- Summarize the steps North Korea and South Korea have taken toward reunification

Read for the Essential Question

Help students read for a purpose by reminding them of the Essential Question: "How can North Korea and South Korea overcome their differences and live in peace?" Point out that they can use items in the text, such as Key Questions, special features, images, and maps to formulate answers to the question.

Vocabulary

Best Practices Toolkit

Use the ***Best Practices Toolkit*** to model strategies for vocabulary notetaking. Vary strategies throughout the year. Choose from: *Knowledge Rating, Predicting ABCs, Definition Mapping, Word Sort, Word Wheel, Frayer Model (Word Squares), Magnet Words,* and *Student VOC.*

Vocabulary Strategies, TT9–TT16

Reading Strategy

Best Practices Toolkit

Display the Venn Diagram transparency.

- Remind students that comparing and contrasting means looking for similarities (What is the same?), while contrasting means looking for differences (What is different?).
- Discuss the similarities and differences between North Korea and South Korea that students have already learned. Then model filling in a few entries on the chart. Have students suggest entries.
- Display the graphic organizer for students to review.

Venn Diagram, TT22

SECTION 3 Reading for Understanding

Key Ideas

BEFORE, YOU LEARNED

After the Korean War, the peninsula was divided into two countries: North Korea and South Korea.

NOW YOU WILL LEARN

North Korea and South Korea have very different government systems and economic structures.

Vocabulary

TERMS & NAMES

totalitarian a type of government that controls every aspect of public and private life

Korean Workers' Party (KWP) a Communist political group that controls the government of North Korea

dictator a ruler who exercises complete political power

Kim Jong Il the leader of North Korea

chaebol (JEH•buhl) a family-owned conglomerate made up of related businesses

BACKGROUND VOCABULARY

conglomerate an organization made up of several companies in different businesses

Visual Vocabulary Samsung is a major *chaebol* in South Korea.

Reading Strategy

Re-create the Venn diagram shown at right. As you read and respond to the **KEY QUESTIONS**, use the diagram to compare and contrast North Korea and South Korea.

Skillbuilder Handbook, page R9

COMPARE AND CONTRAST

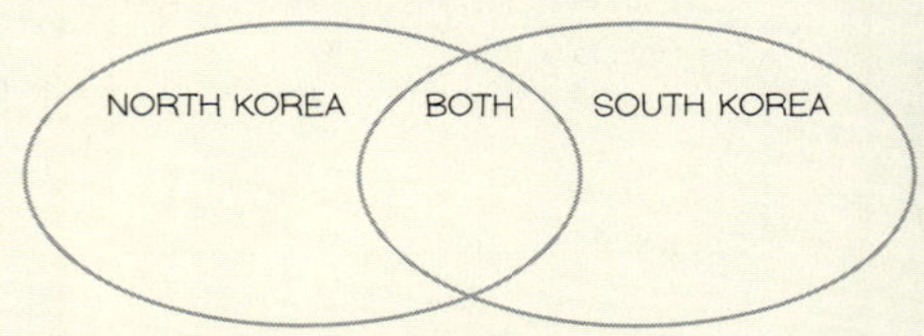

GRAPHIC ORGANIZERS Go to **Interactive Review** @ClassZone.com

PRETEACHING VOCABULARY

English Learners

Pronounce and Preview

Pronounce each term for students. Review the meanings of words in the definitions, such as *aspect, private life,* and *exercises.*

- To modify vocabulary learning, have students complete worksheets as they read the section.

Unit Resource Book: East Asia & Southeast Asia

- Vocabulary Practice, p. 99

Inclusion

Act Out Terms

After you have reviewed the definitions of each term, have students write them down on slips of paper. Then have them use the papers to play a game of charades. You could have students take turns acting out a term for the class, or you might have groups of students form charades teams.

SECTION 3

GOVERNMENT & ECONOMICS

Kyongbok Palace in Seoul, South Korea

Mandarin ducks in forested mountain area

Contrasting Countries

Connecting to Your World

Have you ever visited relatives in a nearby state? In the United States, making such a trip is pretty easy. However, similar visits have not been allowed between North Korea and South Korea for decades. When Korea split into two nations, many families were separated. As a result, Koreans could not visit relatives across the border at all. Recently, some movement between the two nations has been permitted, but only on a limited basis.

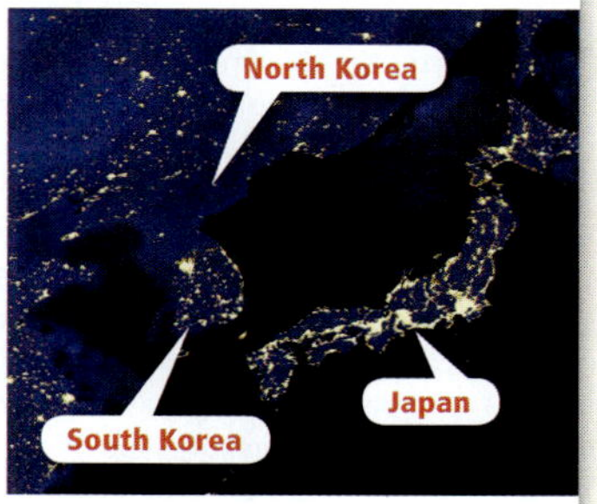

Satellite Image This nighttime image contrasts well-lit South Korea with the darkness of North Korea.

North Korea

KEY QUESTION How are the government and economy of North Korea structured?

As you have learned, a Communist government was set up in North Korea that isolated its people from the rest of the world. As a result, most North Koreans have no contact with ideas that contradict their government. Also, North Korea became economically isolated, especially from the West. Instead of encouraging interaction with non-Communist economies, the government promotes a policy of *juche* (joo•cheh), or self-reliance.

North Korea May Day People hold colored placards to form a picture of Communist strength and victory in a show of national unity.

2 Focus & Motivate

3-Minute Warm-Up

Write on the board or display the transparency:

- What would your life be like if you lived in an isolated country? *(fewer products to choose from; limited access to television, news sources, and Internet Web sites)*

Unit Transparency Book
- 3-Minute Warm-Ups, TT11

Connecting to Your World

More About . . .

May Day

In 1889, the International Socialist Congress declared May 1 as a day to support workers. After World War II, Communist governments began marking the day with parades, showing their political and military strength. Today, May Day is still observed as a workers' holiday in many countries.

3 Teach

North Korea

Reader, Recorder, Reporter

- What is the cause of much of North Korea's economic problems? *(The government spends more on military development than on producing goods for the people.)*
- **Compare and Contrast** Compare and contrast North Korea's government on paper with the way it is truly run. *(A constitution gives power to the people, but in reality, the country is run by a totalitarian government.)*

SECTION 3 PROGRAM RESOURCES

Unit Resource Book: East Asia and Southeast Asia

ON LEVEL

Unit Resource Book
- Reading Study Guide, p. 79
- Map and Graph Skills, p. 101
- Section Quiz, p. 116

STRUGGLING READERS

Unit Resource Book
- Reading Study Guide with Additional Support, p. 85
- Vocabulary Practice, p. 99
- Skillbuilder Practice, p. 100
- Section Quiz, p. 116
- Reteaching Activity, p. 119

eEdition with Audio DVD-ROM

ENGLISH LEARNERS

Pupil Edition in Spanish

eEdition with Audio DVD-ROM

Unit Resource Book
- RSG (Spanish), p. 91
- RSG with Additional Support (Spanish), p. 97

Multi-Language Glossary

Test Generator
- Section Quiz in Spanish

INCLUSION

Unit Resource Book
- RSG with Additional Support p. 85
- Section Quiz, p. 116
- Reteaching Activity, p. 119

GIFTED & TALENTED

Unit Resource Book
- Connect to Math, p. 111
- Section Quiz, p. 116

PRE-AP

Unit Resource Book
- Economics in the World, p. 110
- Section Quiz, p. 116

TECHNOLOGY

Unit Transparency Book
- 3-Minute Warm-Ups, TT11
- Essential Question, TT12
- Critical Thinking, TT15

Daily Test Practice Transparencies
- Section 3, TT89

Power Presentations

ClassZone.com

World Cultures and Geography Video on DVD

RESEARCH & WRITING CENTER

HISTORY MAKERS

Kim Jong Il

Kim Jong Il was born in Siberia, while his father was living as an exile in the former Soviet Union. But the official North Korean account declares that he was born at his father's guerilla base on Mount Paektu in 1942. A North Korean legend has it that a double rainbow and a bright star appeared in the sky when Kim Jong Il was born.

More About . . .

North Korea's Leadership Positions

Kim Il Sung's title was President of North Korea. After he died, the Supreme People's Assembly named him "eternal president," eliminating the position of president from the government. Kim Jong Il's position as Chairman of the National Defense Commission was made the highest post in North Korean government

HISTORY MAKERS

Kim Jong Il born 1941

After Kim Il Sung (above left) died in 1994, his son, Kim Jong Il (above right), became the dictator of North Korea. Soon a cult developed within North Korea around the new leader. Official government accounts praised him as a godlike hero and created legends about his birth. To the outside world, however, Kim Jong Il is a secretive and dangerous figure who starves his people into submission. Above all, many regard Kim Jong Il's determination to develop nuclear weapons as a real threat to the United States and its allies.

ONLINE BIOGRAPHY
For more on the life of Kim Jong Il, go to the **Research & Writing Center** @ClassZone.com

Government North Korea is ruled by a highly centralized, totalitarian form of government. A **totalitarian** government controls every aspect of public and private life. The country does have a constitution, which, on paper, gives political power to the people. However, in reality, the people have almost no freedom. All of the power lies with a Communist group called the **Korean Workers' Party (KWP)**. This party controls elections and chooses who may run for office. The KWP itself is controlled by the chairman of the National Defense Commission. In effect, the chairman is a **dictator**, or a ruler who has complete political power. Since 1994, **Kim Jong Il** has been the chairman of the National Defense Commission and the dictator of North Korea. The country does have a law-making body called the Supreme People's Assembly. This legislature, though, has little power.

Since the early 2000s, the North Korean government has been dealing with an international dispute over its program to develop nuclear weapons. Many nations object to the program, but North Korea has refused to halt it. North Korea claims that it needs nuclear weapons to defend its people against possible aggressors.

Economy Like the government, all means of production in North Korea are controlled by the state. In addition, the economy stresses the good of the state over the welfare of the people. For example, the government spends huge amounts on military development. As a result, it has little money to invest in producing goods for its people.

The North Korean economy is also very isolated. True to its policy of self-reliance, the country has generally discouraged foreign investment. However, North Korea did accept economic aid from the former Soviet Union and from China.

North Korea's economic plans have not been a success. Since the late 1990s, the North Korean people have had one of the lowest standards of living in Asia. But hopeful signs have begun to emerge. During the last few years, South Korean businesses have started to move into North Korea. The money that these businesses make may help the North Korean people. The economic partnership may also help improve relations between the two countries.

DIFFERENTIATING INSTRUCTION

English Learners

Understand Multiple-Meaning Words

Students learning English may have difficulty with the expression *the good of the state over the welfare of the people.* Point out that, in this sentence, the word *over* means "more than," not "above." Then have students look up *welfare* in the dictionary, write down the definition as it is used in their textbook, and use the word in a sentence.

Pre-AP

Analyze North Korea's Nuclear Weapons Program

Have students research current news articles about the international dispute over North Korea's nuclear weapons program. Students should then write a brief explanation of the conflict. Explanations should focus on why North Korea is developing nuclear weapons, what the international objections are, and how the conflict could be resolved.

Industries and Agriculture Throughout its history, North Korea has promoted heavy industry, especially for military production. Because of this, North Korea produces much military equipment, general machinery, steel, cement, textiles, and chemicals. The nation also heavily mines its many mineral deposits.

However, in the 1990s, the development of heavy industry in North Korea began to rapidly decline. One reason for this lies with the state's emphasis on military spending. North Korea spends so much on developing weapons that it does not have enough money to invest in its industrial equipment and infrastructure.

North Korea produces a variety of crops, including rice, barley, corn, wheat, and potatoes. Recently, though, crop production has fallen on hard times. During the mid-1990s, floods and droughts destroyed crops, causing thousands of people to starve to death. North Korean agriculture has never recovered. The lack of good farmland, the failures of state-run farms, and shortages of fertilizer and fuel have all added to the country's agricultural difficulties. In the early 2000s, North Korea has had to rely on aid from around the world to prevent its people from starving.

Fabric The production of textiles is an important industry in North Korea. The sales clerk shown here displays her colorful fabric in a Pyongyang department store.

EVALUATE Discuss the structure of North Korea's government and economy. All power lies with the chairman of the National Defense Commission; the economy is run by the state.

CONNECT to Math

In the image below, a North Korean woman makes textiles for the trade market. Study the chart of North Korea's major trading partners, and then use the decimal numbers in it to play a rounding numbers matching game. Remember to round down any number that contains a decimal of 0.4 or less. Round up any number that contains a decimal of 0.5 or more.

Activity

Rounding Numbers

1. Study the numbers in the box below. The numbers with decimals are taken from the trading partners pie chart.
2. Copy the Matching Game box on your paper.
3. Circle each number that contains a decimal. Then draw a line to match it with its rounded number.

Matching Game

8	15.2	9	16
3	19.6	39.0	4
7	6	10	19
39	3.8	40	20
9.3	15	7.1	6.0

Comparing Regions

Economic Development in China and North Korea

China is another Communist country that, until fairly recently, was economically isolated from the rest of the world. However, economic reforms have helped make China a major player in international trade since the 1970s. China has also developed a business climate that is friendly to foreign companies.

Nonetheless, China faces some of the same economic challenges as North Korea. In particular, China's infrastructure—including energy supplies, transportation systems, and communication—is not considered strong enough to support China's full potential in economic growth.

CONNECT to Math

Point out that the United States is not one of North Korea's major trading partners. Explain that the United States imposed an economic embargo on North Korea in 1950. Sanctions were eased greatly between 1989 and 2000, but trade between North Korea and the United States is still very limited. One reason is that goods manufactured in North Korea are subject to a high tariff rate when imported to the United States.

- Which country is North Korea's biggest trading partner? *(China)*
- What is the combined percentage of trade with North Korea for Japan and South Korea? *(19.6 plus 7.1, or 26.7)*

Unit Resource Book: East Asia & Southeast Asia

- Map and Graph Skills: Reading Pie Graphs, p. 101
- Connect to Math: Rounding Numbers, p. 111

INTERDISCIPLINARY ACTIVITIES

CONNECT to Economics

Write a Newspaper Editorial

Invite students to write an editorial about North Korea's approach to economic development. You may want to bring in a few examples of newspaper editorials for students to use as models.

CONNECT to Art

Create a Textile Pattern

Have students study the photograph of the fabric on this page. If possible, provide a few additional photographs or pictures of North Korean textiles. Then invite students to design a fabric pattern that might be sold in a North Korean market. They can use a variety of materials to create their designs—for floral designs, for example, they could either paint flowers or paste photos of flowers on paper in a geometric pattern.

Teach

South Korea

Think, Pair, Share

- How is the South Korean government similar to the U.S. government? *(South Korea has a democratic government, divided into three branches, with a democratically elected president.)*
- What role do *chaebols* play in South Korea's economy? *(They are major employers and control much of the economy. When they make bad decisions, it affects the entire economy.)*
- **Compare and Contrast** Compare the role of agriculture with that of industry in South Korea's economy. *(Possible answer: It plays a lesser role because there is not much farmland in South Korea and because the country has focused more on developing heavy and light industry.)*

COMPARING

Korean Governments

Remind students that although citizens elect the Supreme People's Assembly in North Korea, that body has very little real power. Point out the direction of the arrows in both diagrams and be sure students understand what they indicate.

- Which term appears with the arrows in the North Korea diagram but does not appear in the South Korea diagram? *(controls)* Why is this term significant? *(It highlights the fact that South Korea is a democracy, which is not controlled by its leaders, whereas North Korea is under such control.)*

CRITICAL THINKING ANSWERS

1. **Compare and Contrast** The chairman in North Korea is not elected by the people but controls the entire government. The president in South Korea is elected by the people and does not control the entire government.
2. **Draw Conclusions** virtually none

Unit Resource Book: East Asia & Southeast Asia

- Skillbuilder Practice: Comparing and Contrasting, p. 100

Voting South Koreans cast their ballots during the 2002 presidential election. **Does a similar election take place in North Korea?**

no, because North Koreans do not elect their leader

South Korea

KEY QUESTION How do the government and economy of South Korea compare with those of North Korea?

In many ways, the government and economy of South Korea are the opposite of North Korea's government and economy. You can compare the two countries' governments in the feature below.

Government Like the United States, the South Korean government is divided into three branches—the executive, the legislative, and the judicial. The president heads the executive branch and serves a five-year term. He or she is chosen by popular election and cannot be reelected. The president appoints a prime minister, who handles the running of the government. South Korea's legislature has a single house called the National Assembly. Citizens elect the members of the assembly to four-year terms.

During the 1980s and 1990s, South Korea began to allow more political freedom. It developed a multiparty system. The country also allowed fairer elections and greater freedom for its people.

COMPARING Korean Governments

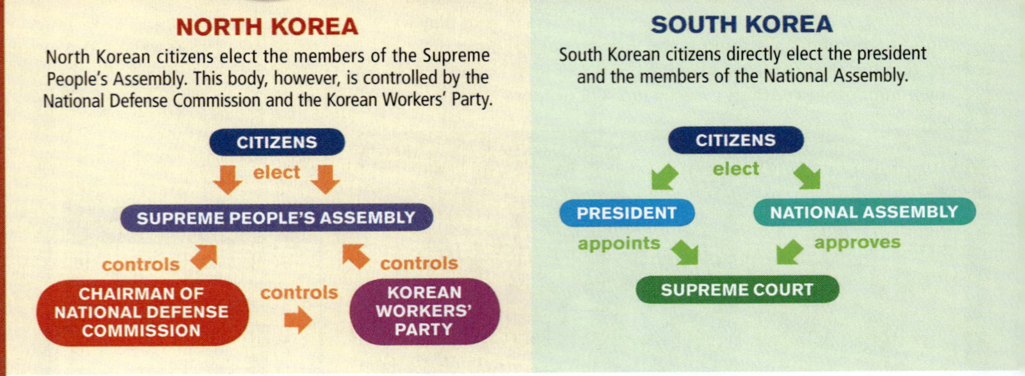

CRITICAL THINKING

1. **Compare and Contrast** How do the roles of chairman in North Korea and president in South Korea differ?
2. **Draw Conclusions** How much power do the people of North Korea have?

DIFFERENTIATING INSTRUCTION

Struggling Readers

Understand Graphics

Ask students who are having trouble understanding the diagram on this page to restate each relationship shown in their own words. For example, they might say: "In North Korea, citizens elect the Supreme People's Assembly. The Supreme People's Assembly is controlled by the Chairman of the National Defense and by the Korean Workers' Party."

English Learners

Re-Create Graphics

Have student pairs write all the labels from the Comparing Korean Governments feature on separate index cards. They should also make arrows using index cards. Then have the pairs work together to arrange their cards and arrows and re-create the graphic. After they have done it once successfully, encourage them to arrange the cards without looking at their textbooks.

Economy After the Korean War, South Korea developed its economy with a series of five-year plans. The goal of these plans was to export goods and use the resulting income to increase production. The government expressed this approach through the slogan "Production, Exports, Construction!"

Soon South Korean businesses began to thrive. In fact, some companies expanded into family-owned conglomerates called **chaebols** (JEH•buhlz). A **conglomerate** is made up of several companies in different businesses. The businesses within a *chaebol* are related. For example, Samsung is one of South Korea's largest *chaebols*. It is made up of companies that produce electronic equipment such as televisions, microwave ovens, and computers.

Eventually, *chaebols* controlled much of South Korea's economy and greatly contributed to its growth. These organizations, though, had drawbacks. For instance, *chaebols* sometimes underpaid their workers, causing many to demand higher wages. Also, some *chaebols* made bad economic decisions that led to an economic downturn in 1997. After the government received foreign loans and closed poorly run businesses, the economy began to rebound. *Chaebols* continue to play a vital role in South Korea's economy today.

Industries and Agriculture South Korea has developed both heavy and light industries. Products include chemicals, steel, shipbuilding, automobiles, electronics, and telecommunications. You can learn about the locations of South Korea's industries by studying the map on this page.

South Korea's economy has shifted away from agriculture. And, as you have learned, South Korea's mountainous terrain does not provide much land for farming. Nevertheless, South Korean farms produce a variety of crops, including rice, root crops, barley, vegetables, and fruit.

COMPARE AND CONTRAST Compare the government and economy of South Korea with those of North Korea. The South Korean government is more democratic and free; the economy of South Korea is doing much better than the North Korean economy.

Industries of North Korea and South Korea

CONNECT Geography & Economics

1. **Location** Where are most of the industries in North Korea and South Korea located?
2. **Region** What are North Korea's major industries?

Industries of North Korea and South Korea

CONNECT Geography & Economics

Review the key with students. Discuss what other symbols could have been used. *(Possible answers: color-coded dots, instead of icons; different icons, such as a test tube to indicate chemicals)*

- What are South Korea's major industries? *(vehicles, chemicals, steel, electronics, machinery, shipbuilding)*
- Why do you think South Korea's economy is more diverse than North Korea's? *(Possible answer: South Korea is focused on economic development and exporting goods, while North Korea is focused on developing its military strength.)*

ANSWERS

1. **Location** around each country's major cities
2. **Region** chemicals, machine building, mining, steel

Unit Resource Book: East Asia & Southeast Asia
- Economics in the World: The Economies of the Two Koreas, p. 110

CONNECT to the Essential Question

How can North Korea and South Korea overcome their differences and live in peace?

Ask students what they have learned so far that can help them answer this question. Students might mention the following:

- People in both countries have a desire to reunite.
- The economies of each country could benefit from reunification.

Unit Transparency Book
- Essential Question, TT12

DIFFERENTIATING INSTRUCTION: TIERED ACTIVITIES

OBJECTIVE Understand South Korea's economic development.

Basic

Have students illustrate an advertisement with the slogan "Production, Exports, Construction!" Tell students that their audience will be South Korean business owners and foreign investors.

On Level

Adapt the Basic activity but have students come up with a slogan and illustrations that will appeal to South Korean *chaebols*.

Challenge

Have groups of students write a television or radio ad that describes and promotes the diversity of South Korea's economy.

Teach

Efforts Toward Reunification

Reader, Recorder, Reporter

- What obstacles do North and South Korea face before they can be reunited? *(They have different cultures, North Korea sees South Korea's government as unlawful, and they have very different economic systems.)*
- **Compare and Contrast** What signs point to future reunification? *(Both countries desire it and have taken some steps to reunite, such as allowing family visits.)*

4 Assess & Reteach

Assess Have students complete the Section Assessment.

Unit Resource Book: East Asia & Southeast Asia
- Section Quiz, p. 116

Interactive Review @ClassZone.com
- Section Quiz

Power Presentations

Test Generator

Reteach Assign student pairs one of the following categories: government, economy, and Industry. Have the pairs write at least three compare-and-contrast statements about North Korea and South Korea for their assigned category. Then have each pair share their statements with the rest of the class.

Unit Resource Book: East Asia & Southeast Asia
- Reteaching Activity, p. 119

Unified Korea Flag Korean athletes carry a flag showing a map of the entire Korean Peninsula in the 2006 Olympics.

Efforts Toward Reunification

KEY QUESTION What steps have been taken toward the reunification of North Korea and South Korea?

Many obstacles remain before North Korea and South Korea can be reunited. For one thing, North Korea considers the South Korean government unlawful. Also, over the past decades, North Korea and South Korea have developed very different cultures.

But gradually, contacts between the two Koreas have increased. For example, more visits between family members have been allowed. Also, as you have already learned, a few South Korean businesses have moved into North Korea. The desire for reunification has even been reflected at the Olympic games. During the 2000, 2004, and 2006 games, North and South Korean athletes marched under the same flag during the opening ceremony. The two Koreas have agreed to compete as a single team in the 2008 Olympics.

FIND MAIN IDEAS Discuss the steps that have been taken toward the reunification of North and South Korea. more visits of family members between nations; some South Korean businesses now in North Korea; single team in the 2008 Olympic Games

ONLINE QUIZ For test practice, go to Interactive Review @ClassZone.com

Section 3 Assessment

TERMS & NAMES

1. **Explain the importance of**
   - totalitarian
   - Korean Workers' Party
   - dictator
   - *chaebol*

USE YOUR READING NOTES

2. **Compare and Contrast** Use your completed Venn diagram to answer the following question:

   How do North Korea and South Korea differ in their economic relationships with other countries?

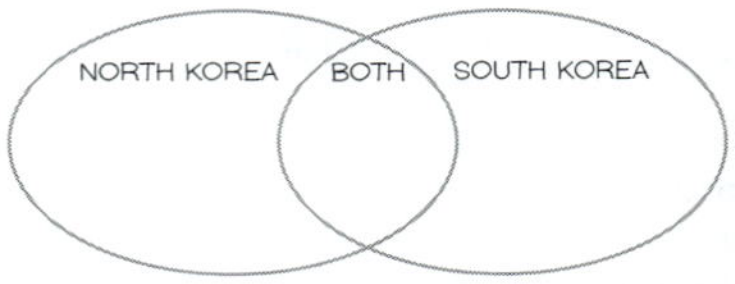

KEY IDEAS

3. What is *juche*?
4. How much power does the chairman of the National Defense Commission have in North Korea?
5. What have been some of the benefits and drawbacks of the *chaebols*?

CRITICAL THINKING

6. **Recognize Bias and Propaganda** Reread the History Makers on Kim Jong Il. How do official accounts of the leader reflect bias and propaganda?
7. **Evaluate** What might result from allowing the Korean people to travel freely within North and South Korea?
8. **CONNECT to Today** What message would North Korea and South Korea convey by competing in the Olympics as a single team?
9. **ECONOMICS** **Research a *Chaebol*** Choose a *chaebol* and research the type of products it makes and how well they sell in the United States. Summarize your findings in a one-paragraph report.

SECTION 3 ASSESSMENT ANSWERS

Terms & Names

1. totalitarian: government that controls public and private life; Korean Workers' Party: Communist political group controlled by the chairman; dictator: ruler with complete power; *chaebol:* family-owned conglomerate made up of related businesses

Use Your Reading Notes

2. South Korea has many more dealings with other countries in economic affairs.

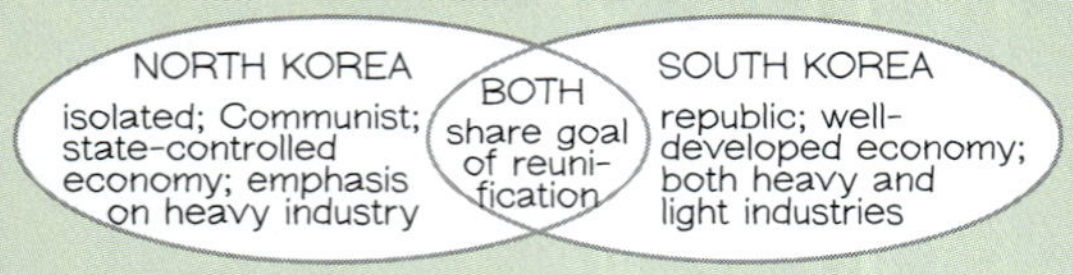

Key Ideas

3. North Korea's policy of self-reliance
4. absolute power
5. stimulated and carried South Korea's economy; underpaid workers and made bad decisions

Critical Thinking

6. details in the official accounts, such as "praised him as a godlike hero" and "created legends about his birth"
7. Travel might encourage greater understanding between the people and cultures of the two countries.
8. It would show that the Koreans have a strong desire to settle their differences and reunify.
9. Use the rubric to score reports.

Report Rubric

	Structure	Content	Errors
4	logical and flowing	shows thorough understanding	few or none
3	clear	shows good understanding	some
2	somewhat confusing	shows basic understanding	many
1	weak and unclear	shows little or no understanding	many major

Interactive Review

Click here to complete these and other activities online @ClassZone.com

CHAPTER SUMMARY

Key Idea 1
Korea lies on a peninsula, which provides a moderate climate and plentiful resources.

Key Idea 2
The Korean Peninsula has been invaded by China, Russia, and Japan but has still maintained its cultural identity.

Key Idea 3
North Korea and South Korea have very different government systems and economic structures.

For **Review and Study Notes**, go to **Interactive Review** @ClassZone.com

NAME GAME

Use the Terms & Names list to complete each sentence on paper or online.

1. I am the highest mountain on the Korean Peninsula. Mount Paektu
2. I am the largest island off the coast of the Korean Peninsula. ______
3. I am the dictator of North Korea. ______
4. I am made up of the states of Koguryo, Paekche, and Silla. ______
5. I am the buffer zone between North Korea and South Korea. ______
6. I am the original belief system practiced on the Korean Peninsula. ______
7. I am a traditional Korean costume. ______
8. I am a Korean side dish. ______
9. I am the political group that controls the North Korean government. ______
10. I am a family-owned conglomerate of businesses. ______

Buddhism
chaebol
Cheju
conglomerate
demilitarized zone
hanbok
kimchi
Kim Jong Il
Korean Workers' Party
Mount Paektu
shamanism
Three Kingdoms

Activities

GeoGame

Use this online map to show what you know about the location, geographic features, and important places of the Korean Peninsula. Drag and drop each place name to its location on the map.

Geo GAME
North Korea
South Korea
demilitarized zone
Yalu River
Cheju Island

To play the complete game, go to **Interactive Review** @ClassZone.com

Crossword Puzzle

Complete an online crossword puzzle to test your knowledge of Korean history and culture.

ACROSS
1. traditional Korean clothing

ONLINE AND ON-SCREEN ACTIVITIES

GeoGame

Use the interactive map on **Power Presentations** DVD-ROM or the **eEdition** to help students review the Korean Peninsula's geographic features.

- Demonstrate how to drag and drop the map labels listed onto the map. Click on the Submit button to view results.
- Advance to the scorecard to show how students will be scored in their first and second attempts.

Crossword Puzzle

Students can review important people and events in the history of the Korean Peninsula by completing this crossword puzzle online at **ClassZone.com.**

- As an extension of this activity, have students use the *Activity Maker* at **ClassZone.com** to make their own crossword puzzle for one section of the chapter.
- Then have students trade puzzles with classmates and try solving them.

Chapter Summary

As a follow-up to the chapter, encourage students to pair up and go over the **Review and Study Notes** at **ClassZone.com**.

Alternatively, student pairs can review their Reading Study Guide notes, making sure that they can add important details to each section's Key Ideas.

Name Game

Teaching Tips

ON PAPER Have students work in pairs to complete this activity. As an extension, they can write questions for the terms that are not covered in the items shown here.

ONLINE Make sure that students understand how to drag and drop terms to fill in blanks. Ask them to print out their results and compare them with those of a classmate.

ANSWERS

1. Mount Paektu
2. Cheju
3. Kim Jong Il
4. Three Kingdoms
5. demilitarized zone
6. shamanism
7. *hanbok*
8. kimchi
9. Korean Workers' Party
10. *chaebol*

Make It Fun

Teacher-Tested Activities

Todd Harrison, Hardin County Middle School, Savannah, Tennessee

Have your class play the following variation of the GeoGame.

- Display or project a large world map.
- Ask volunteers to find their town and a city in North Korea or South Korea on the map. Have them mark the sites.
- Then challenge students to use the map's scale to figure out the distance between the two places.
- Finally, have students study the map to determine the means of transportation they might use to travel from their hometown to the Korean Peninsula.

Vocabulary ANSWERS

1. a body of land bordered by the Yellow Sea to the west, the East Sea to the east, the Korea Strait to the south, and China to the north
2. highest mountain on the Korean Peninsula
3. largest island off the southern coast of the Korean Peninsula
4. conflict between North Korea, Soviet Union, and China on one side and South Korea, the United States, and the UN on the other
5. buffer zone between North and South Korea
6. belief system in which a person called a shaman is believed to be able to communicate with spirits and heal the sick
7. traditional Korean costume
8. Korean ceramic pottery with a thin blue or green glaze
9. three kingdoms that had formed on the Korean Peninsula by A.D. 300
10. a family-owned conglomerate made up of related businesses
11. Shamanism, Confucianism, and Buddhism are three of the major belief systems that influenced the Korean people.
12. Kim Jong Il is a dictator who controls the Korean Workers' Party and North Korea.

Key Ideas ANSWERS

13. Yellow Sea, East Sea, Korea Strait
14. the border between China and North Korea
15. usually bring large amounts of rain
16. petroleum-burning plants
17. by closing Korea to all foreigners
18. North Korean troops invaded South Korea.
19. Buddhism, Confucianism, and Christianity influence South Korea. Religious practice is discouraged in North Korea.
20. Buddhism has inspired Korean temples and statues; China inspired Korean pottery.
21. The KWP controls elections and chooses who may run for office. However, the KWP is totally controlled by the chairman.
22. Both are divided into the executive, legislative, and judicial branches; both have a multiparty system; both grant their people great political freedom.
23. to export goods and use the resulting income to increase production
24. During the 2000, 2004, and 2006 Olympics, North and South Korean athletes marched in the opening ceremony under the same flag. For the 2008 Olympics, the two Koreas have agreed to compete as a single team.

CHAPTER 26 Assessment

VOCABULARY

Explain the significance of each of the following.

1. Korean Peninsula
2. Mount Paektu
3. Cheju Island
4. Korean War
5. demilitarized zone
6. shamanism
7. *hanbok*
8. celadon
9. Three Kingdoms
10. *chaebol*

Explain how the terms and names in each group are related.

11. shamanism, Confucianism, and Buddhism
12. Korean Workers' Party, Kim Jong Il, and dictator

KEY IDEAS

1 A Rugged Peninsula

13. What bodies of water surround the Korean Peninsula?
14. What do the Yalu and Tumen rivers form?
15. How do summer monsoons affect the Korean Peninsula's climate?
16. How is the majority of the electricity produced in South Korea?

2 Strong Traditions, Modern Innovations

17. How did the Choson Dynasty isolate Korea?
18. How did the Korean War start?
19. What belief systems influence Korean society today?
20. What influences have inspired Korean arts?

3 Contrasting Countries

21. What is the role of the Korean Workers' Party in North Korea?
22. How is South Korea's government similar to the U.S. government?
23. What was the goal of South Korea's five-year economic plans?
24. How has the desire for the reunification of North and South Korea been reflected in the Olympics?

CRITICAL THINKING

25. **Compare and Contrast** Create a chart to compare and contrast the natural resources of North Korea and South Korea.

NORTH KOREA'S RESOURCES	SOUTH KOREA'S RESOURCES

26. **Evaluate** What steps has South Korea taken to overcome some of its problems?
27. **Make Inferences** Why do you think many South Koreans practice aspects of several religions?
28. **Identify Problems** What are some of the problems faced by North Korea?
29. **Connect Geography & Economics** How has North Korea's farmland affected its economy?
30. **Five Themes: Human-Environment Interaction** How have the geographic features of the Korean Peninsula affected this region's history?

Answer the ESSENTIAL QUESTION

How can North Korea and South Korea overcome their differences and live in peace?

Written Response Write a two- or three-paragraph response to the Essential Question. Use the rubric below to guide your thinking.

Response Rubric
A strong response will:
- identify the countries' differences and conflicts
- discuss the steps the two Koreas have taken toward overcoming these differences
- offer solutions about what more the two countries could do

ESSENTIAL QUESTION REVIEW

Before students begin their essays, have them review the key chapter elements that will help them answer the Essential Question. Direct them to significant features in the chapter, such as:

- Physical Geography of the Korean Peninsula, Section 1
- Comparing Climates, Section 1
- Analyzing Primary Sources, Section 2
- The Korean War *1950–1953*, Section 2
- Comparing Korean Governments, Section 3
- Industries of North Korea and South Korea, Section 3

Written Response Rubric

	Coverage of Key Ideas	Explanation	Errors in Mechanics
4	all	clear and concise	few or none
3	most	good	some
2	some	fair	many
1	one or none	poor	many major

STANDARDS-BASED ASSESSMENT

Test Practice

- Online Test Practice @ClassZone.com
- Test-Taking Strategies and Practice at the front of this book

LINE GRAPH

Use the graph to answer questions 1 and 2 on your paper.

Stock Market in South Korea

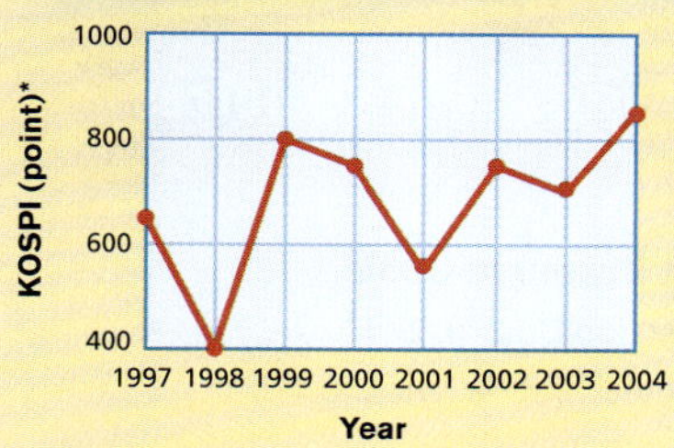

*KOSPI—Korea's Stock Price Index

Source: Republic of Korea, Ministry of Finance and Economy

1. In what year did the stock market in South Korea reach its lowest level?

A. 1997
B. 1998
C. 2001
D. 2003

2. When did the stock market reach 850 points?

A. 1999
B. 2000
C. 2002
D. 2004

CHART

Examine the chart below. Use the information in the chart to answer questions 3 and 4 on your paper.

Spread of Chinese Influence
A.D. 600–1400

Korea	Japan	Southeast Asia
Buddhism	Buddhism	Buddhism
Writing system	Writing system	Ideas
Civil service	Printing	
Printing		
Porcelain		

3. Which aspect of Chinese culture was most widely spread?

A. writing system
B. printing
C. Buddhism
D. porcelain

4. Which region was most influenced by China?

A. Korea
B. Japan
C. Southeast Asia
D. They were all equally influenced by China.

GeoActivity

1. INTERDISCIPLINARY ACTIVITY–SCIENCE

With a small group, use appropriate resources to research the causes and effects of typhoons on the Korean Peninsula. Display your findings on a poster. Include pictures, charts, and captions that explain the causes and effects.

2. WRITING FOR SOCIAL STUDIES

Do you think North Korea should be allowed to develop a nuclear weapons program? Examine the conflicting viewpoints of North Korea and of the United States and others who oppose it. Write your opinion in a brief essay.

3. MENTAL MAPPING

Create an outline map of the Korean Peninsula and label the following:

- Yellow Sea
- East Sea
- Korea Strait
- Yalu River
- Cheju Island
- Mount Paektu
- Nangnim Mountains
- Taebaek Mountains

DIFFERENTIATING CHAPTER ASSESSMENT

Unit Resource Book

- **Chapter Test A,** pp. 121–124. *Use for Struggling Readers, Beginning-to-Intermediate-Level English Learners, and for Inclusion of students whose IEP suggests supported test-taking.*
- **Chapter Test B,** pp. 125–128. *Use for On-Level Learners and ESOL students with Advanced-Level English.*
- **Chapter Test C,** pp. 129–132. *Use for Gifted and Talented students or students assigned Pre-AP work.*

Test Generator

- Chapter Tests A, B, and C, in English and Spanish

McDougal Littell Assessment System

TEST **Test Generator** Use three levels of tests in English and Spanish tied to state standards. Plus, choose questions to customize tests for all populations.

SCORE Scan and Score plain paper student tests in minutes.

REPORT Create instant progress reports, sortable by student, by standard, or other criteria.

RETEACH Respond immediately to students' needs by using a wealth of reteaching resources.

Critical Thinking ANSWERS

25. North Korea: large mineral deposits, little farmland, not much livestock, excellent fishing; South Korea: few mineral deposits, more farmland than in North Korea, not much livestock, excellent fishing

26. South Korea has accepted help from other nations, allowed its people more freedom, adjusted the economy to stimulate growth, and reached out to North Korea.

27. Some South Koreans might use the practices of different religions for different purposes and so want to practice several.

28. their policy of self-reliance, dictatorial leadership, emphasis on military

29. has had to rely on mining its large mineral deposits and using them in heavy industry; food shortages have resulted in reliance on aid from other countries

30. Korea's rugged terrain and surrounding seas have protected and isolated the Korean people from much of the world. However, the surrounding superpowers of Japan, China, Russia, and the former Soviet Union have all tried to control the Korean Peninsula.

Standards-Based Assessment ANSWERS

1. B
2. D
3. C
4. A

GeoActivity

1. INTERDISCIPLINARY ACTIVITY RUBRIC

The poster should

- explore the causes and effects of typhoons
- use maps, charts, and pictures to convey the impact of typhoons

2. WRITING FOR SOCIAL STUDIES RUBRIC

The essay should

- examine why North Korea wants to develop nuclear weapons
- explain why others oppose the plan
- support an opinion on the issue

3. MENTAL MAPPING RUBRIC

The outline map should

- represent the approximate shape of the Korean Peninsula
- include all geographic features listed in their correct location

CHAPTER 27 Japan PLANNING GUIDE

Complete print resources all on one USB drive allow you to customize lesson plans for all students.

Section Titles & Key Ideas	Unit Resource Book East Asia and Southeast Asia	Assessment Resources
1 Land of the Rising Sun **KEY IDEA** The mountainous islands that make up Japan provide the country's people with scant living space, a varied climate, and limited resources.	• **Reading Study Guide,** pp. 133–134 • **RSG with Additional Support,** pp. 139–140 • **Vocabulary Practice,** p. 157 • **Map and Graph Skills,** p. 159 • **Outline Map Activities,** p. 161	**Unit Resource Book** • Section Quiz, p. 173 • Reteaching Activity, p. 176 **Test Generator CD-ROM** **Daily Test Practice Transparency,** TT90
2 Samurai to Animé **KEY IDEA** Over the centuries, Japan has adapted ideas and technologies from other lands and created a unique culture.	• **Reading Study Guide,** pp. 135–136 • **RSG with Additional Support,** pp. 141–142 • **History Makers,** pp. 163–164 • **Connect to Culture,** p. 169 • **Primary Sources,** pp. 171–172	**Unit Resource Book** • Section Quiz, p. 174 • Reteaching Activity, p. 177 **Test Generator CD-ROM** **Daily Test Practice Transparency,** TT91
3 Economic Giant **KEY IDEA** The Japanese have drawn on their culture to create a modern government and become an economic giant.	• **Reading Study Guide,** pp. 137–138 • **RSG with Additional Support,** pp. 143–144 • **Skillbuilder Practice,** p. 158 • **Interdisciplinary Projects,** p. 165 • **Economics in the World,** p. 168	**Unit Resource Book** • Section Quiz, p. 175 • Reteaching Activity, p. 178 • Chapter Test (Forms A, B, & C), pp. 179–190 **Test Generator CD-ROM** **Daily Test Practice Transparency,** TT92

McDougal Littell Assessment System

TEST SCORE REPORT RETEACH

INTEGRATED TECHNOLOGY

No Time? To focus students on important content in the chapter, use the **"Reading Study Guide with Additional Support"** in ***Resources2Go.***

Teacher Presentation Options

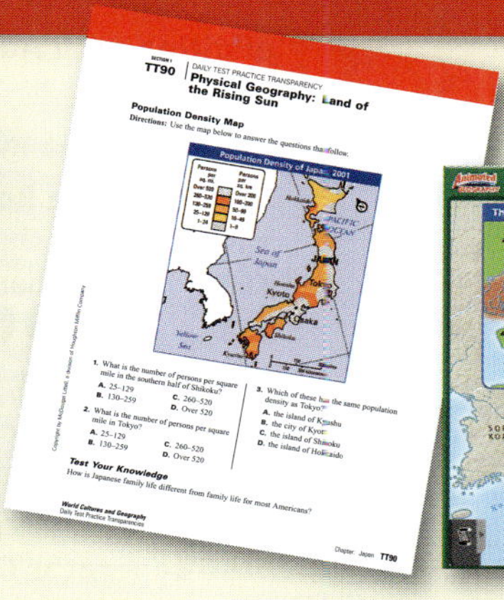

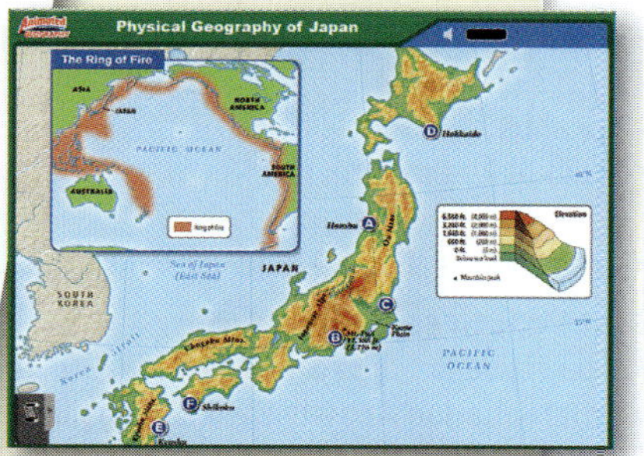

Electronic Books

eEdition DVD-ROM

eEdition Online

Presentation Toolkit

Video/DVD

- World Cultures and Geography

Power Presentations DVD-ROM

- Discussion Starters
- Media Gallery
- Review Game
- Interactive Review
- Animated Features

Unit Transparency Book

- 3-Minute Warm-Ups
- Essential Question
- Critical Thinking
- Maps
- Cultures

Daily Test Practice Transparencies

Animated Geography

- Present-Day Japan Map
- Physical Geography of Japan Map

OnlineActivities at ClassZone.com

Animated Geography

- Videos and Photos
- 3-D Animations
- Interactive Maps

Interactive Review

- Section Quizzes
- Review and Study Notes
- Vocabulary Flip Cards
- Graphic Organizers
- Name Game and Crossword Puzzles
- GeoGame

Online Test Practice

Map Center

- Atlas Maps
- Political Maps with Data
- Links to Mapping Web Sites

Activity Center

- Activity Maker—Review Games, Puzzles, Flip Cards, and Graphic Organizers
- Worldster

Research & Writing Center

- Web Sites
- Data Updates
- Primary Sources
- Writing Models

Current Events

Greta Frensley
Knox Doss Middle School
Gallatin, Tennessee

Make It Fun — Teacher-Tested Activities

As you teach this chapter, look for my classroom-proven ideas for:

- making the lesson fun
- focusing students' attention and comprehension
- integrating technology with traditional teaching

Teaching Bell to Bell When you see this timing guide, have students discuss in pairs, in groups of 3–6, or as a class.

- **Think, Pair, Share** (pairs)
- **Reader, Recorder, Reporter** (3–6)
- **Talk About It** (any size)

RESOURCES FOR DIFFERENTIATED INSTRUCTION

CHAPTER 27

Struggling Readers

eEdition with Audio DVD-ROM

Lesson Plans
- Strategies for Inclusion

Unit Resource Book
- Reading Study Guide with Additional Support, pp. 139–144 (B)
- Vocabulary Practice, p. 157
- Skillbuilder Practice, p. 158
- Reteaching Activity, pp. 176–178
- Chapter Test (Form A), pp. 179–182 (A)

Best Practices Toolkit (C)
- Interactive Reading Strategies, pp. 90–145
- Differentiation Activity Bank: Struggling Readers, pp. 146–149

Teacher's Edition Activities
- Create a Chart of Island Facts, Sec. 1
- Use Visuals to Compare, Sec. 2, 3
- Summarize Main Ideas, Sec. 2
- Tiered Activities, Sec. 1–3

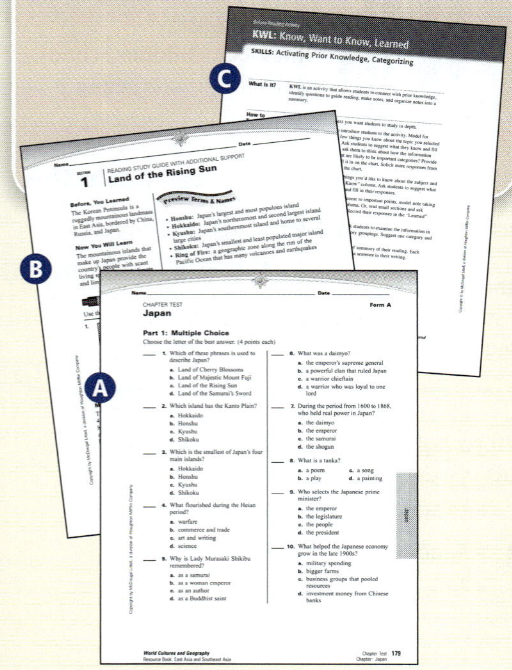

English Learners

Pupil Edition in Spanish

eEdition with Audio DVD-ROM

Lesson Plans
- Strategies for English Learners

Unit Resource Book
- Reading Study Guide (Spanish), pp. 145–150
- RSG with Additional Support (Spanish), pp. 151–156 (A) (B)

Multi-Language Glossary (C)

Best Practices Toolkit
- Differentiation Activity Bank: English Learners, pp. 165–194

Test Generator CD-ROM
- Section Quizzes in Spanish
- Chapter Test in Spanish

Teacher's Edition Activities
- Pronounce and Preview, Sec. 1–3
- Learn Key Vocabulary, Sec. 2
- Create an Economics Time Line, Sec. 3
- Tiered Activities, Sec. 1–3

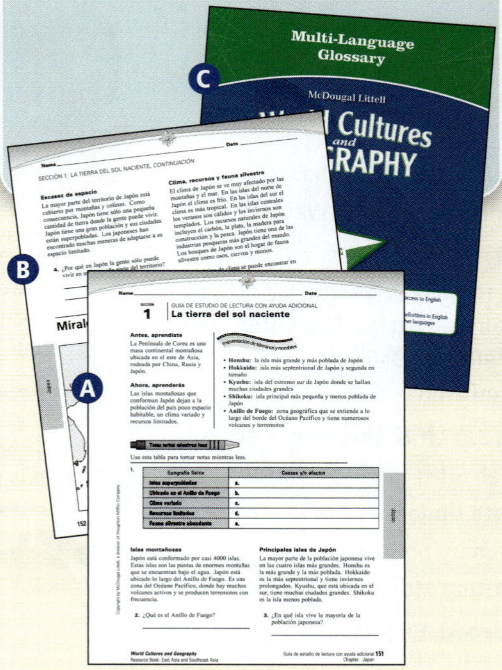

Inclusion

eEdition with Audio DVD-ROM

Lesson Plans
- Strategies for Inclusion

Unit Resource Book
- Reading Study Guide with Additional Support, pp. 139–144 (A) (B)
- Reteaching Activity, pp. 176–178
- Chapter Test (Form A), pp. 179–182

Best Practices Toolkit (C)
- Interactive Reading Strategies, pp. 90–145
- Differentiation Activity Bank: Struggling Readers, pp. 146–149

Teacher's Edition Activities
- Locate Vocabulary Words on Maps, Sec. 1
- Use Knowledge Rating, Sec. 2
- Read Enlarged Vocabulary Definitions, Sec. 3
- Tiered Activities, Sec. 1–3

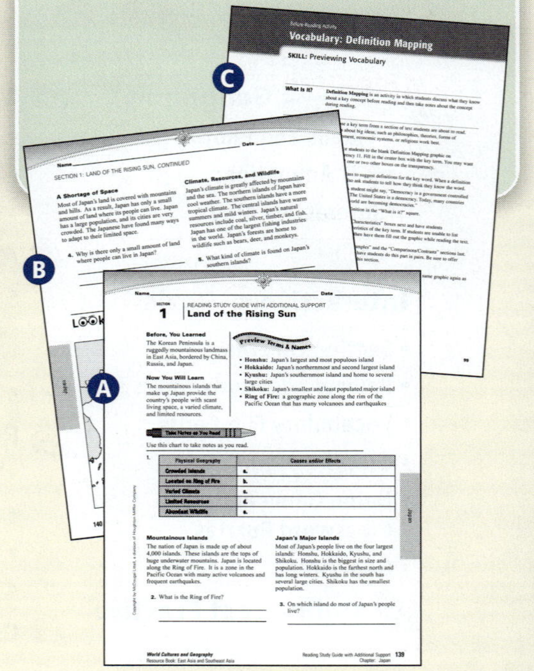

Gifted & Talented Students

Lesson Plans
- Core Lesson Plan

Unit Resource Book
- Outline Map Activities, p. 161 (A)
- History Makers, pp. 163–164 (B)
- Interdisciplinary Projects, pp. 165–166 (C)
- Chapter Test (Form C), pp. 187–190

Best Practices Toolkit
- Student Products Rubrics, pp. 1–14
- Writing Templates, pp. 78–89

Teacher's Edition Activities
- Build a Model of Japan's Islands, Sec. 1
- Write a Diary Entry, Sec. 2
- Create Origami Art, Sec. 2
- Analyze Militarism in Japan, Sec. 3
- Tiered Activities, Sec. 1–3

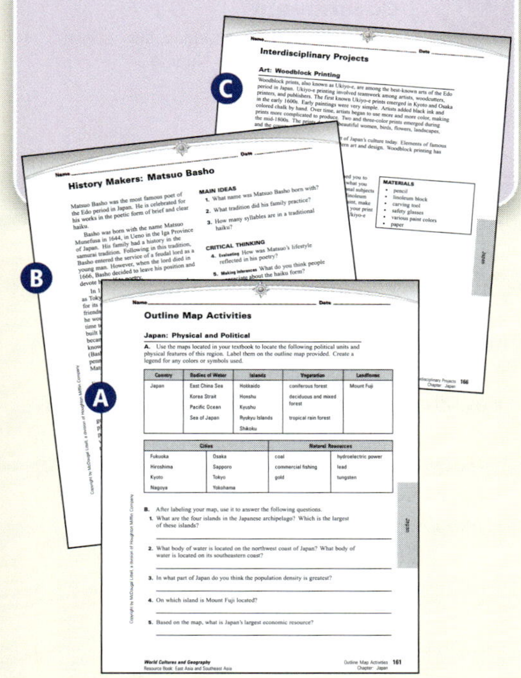

Pre-AP*

Lesson Plans
- Core Lesson Plan

Unit Resource Book
- Economics in the World, p. 168 (A)
- Connect to Culture, p. 169 (B)
- Primary Sources, pp. 171–172 (C)
- Chapter Test (Form C), pp. 187–190
- World Literature, p. 247

Best Practices Toolkit
- Writing Templates, pp. 78–89
- Differentiation Activity Bank: Pre-AP Students, pp. 195–200

* Pre-AP is a registered trademark of the College Entrance Examination Board, which was not involved in the production of and does not endorse this product.

Teacher's Edition Activities
- Write Newspaper Articles, Sec. 2
- Write a Radio Script, Sec. 3
- Tiered Activities, Sec. 1–3

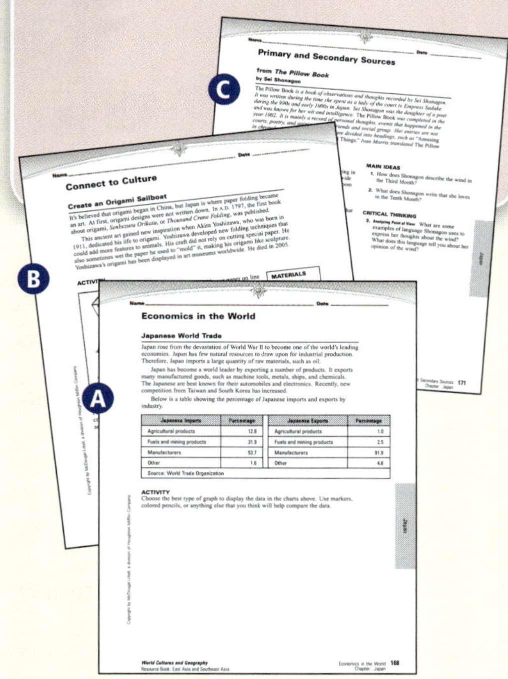

Interdisciplinary Activities

CONNECT to Language Arts

Create a Tour Itinerary

CLASS TIME 30 minutes

TASK Create a detailed itinerary for a one-week tour of Japan.

SUPPLIES
- information about historic and contemporary cultural sites in Japan
- audiocassette recorder or digital audiorecorder and editing software

ACTIVITY Have students create a seven-day itinerary for a tour of Japan. Encourage students to consider sites that tourists might be interested in, such as Mount Fuji, Kyoto, and the Hiroshima Peace Park. Remind students that they are responsible for choosing a series of tour stops that give tourists a close look at Japan's geography, culture, and economy. Students' finished itineraries should include a list of stops, an image for each, and a brief explanation of what tourists can learn by visiting the stop.

Good for Visual and Spatial Learners

CONNECT to Art

Creating Japanese Artwork

CLASS TIME 30 minutes

TASK Create an exhibition of Japanese-style artwork.

SUPPLIES
- images of historic and contemporary Japanese nature art
- art paper, watercolor paints and brushes, colored markers and pencils

ACTIVITY Have students show the close relationship between people and nature by analyzing traditional and contemporary Japanese art. Then have them choose nature-related subjects and create artwork of their own, imitating the styles used by Japanese artists. When finished, students should display their work and explain how their art demonstrates a connection between people and nature.

Good for Visual and Kinesthetic Learners

Introduce the ESSENTIAL QUESTION

- Describe Japan's ancient traditions and analyze the influence of China.
- Discuss the impact that Japan's periods of isolation has had on the country's culture.
- Explain that, today, Japan is a democratic country with a very strong global economy.

Key Ideas

Section 1: The mountainous islands that make up Japan provide the country's people with scant living space, a varied climate, and limited resources.

Section 2: Over the centuries, Japan has adapted ideas and technologies from other lands and created a unique culture.

Section 3: The Japanese have drawn on their culture to create a modern government and become an economic giant.

CONNECT to the Essential Question

You may wish to revisit the Essential Question periodically with students as they work through the chapter.

Suggestions for reconnecting are provided at an appropriate point in the *Teacher's Edition* for each section.

Kiyomizu Temple and cherry blossoms in Kyoto

CHAPTER 27 Japan

1 GEOGRAPHY **Land of the Rising Sun**

2 HISTORY & CULTURE **Samurai to Animé**

3 GOVERNMENT & ECONOMICS **Economic Giant**

ESSENTIAL QUESTION

How has Japan balanced its ancient traditions with its modern economy?

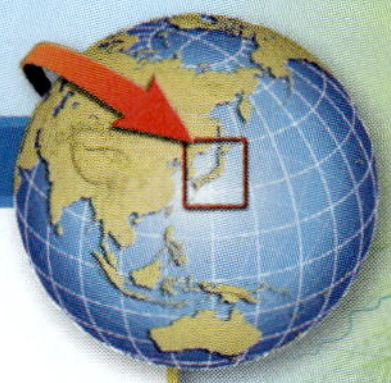

CONNECT Geography & History

Use the map and the time line to answer the following questions.

1. When did Heian become Japan's capital?
2. What is the capital of Japan today?

A.D. 300

History
◄ **300s** The Yamato clan gains control of Japan. (Seventh-century ruler Shotoku Taishi)

History
1100s Shogun rule is established. (Japan's first shogun, Minamoto no Yoritomo) ►

Government
◄ **794** Heian period begins, and the capital moves from Nara to Heian, modern Kyoto. (Phoenix Hall)

TIME LINE DISCUSSION

Point out to students that for many years, Japan developed with little direct contact with other cultures. Japan's isolation, during which time its government and economy were largely stable, had been preceded by several centuries of violent internal conflict.

- About how many years after the Yamato clan gained control was Shogun rule established? *(about 800 years)*
- How many years after the attack on Pearl Harbor did Japan become the largest producer of automobiles worldwide? *(39 years)*
- What neighboring nations likely had an influence on Japanese culture, history, and economics? *(China, North Korea, South Korea, and Russia)*
- What geographic characteristics might have contributed to Japan's many years of isolation? *(Possible answers: Japan is an island nation; the Pacific Ocean served as a barrier between Japan and Western cultures.)*

Animated GEOGRAPHY

Present-Day Japan

Click here to explore Japan @ClassZone.com

History

1941 Japan attacks Pearl Harbor, forcing the United States into World War II. (Japan's World War II naval battle flag) ►

Today

Culture

◄ **1600s** Poetry called haiku develops. (Haiku poet Matsuo Basho)

Economics

1980 Japan becomes the largest producer of automobiles worldwide.

823

Animated GEOGRAPHY

Present-Day Japan

CLASSZONE.COM

Refer students to the physical map of Japan. Point out that the culture, history, and economy of Japan have been greatly influenced by its geography as an island nation.

- What bodies of water surround Japan? *(Pacific Ocean, Sea of Okhotsk, Sea of Japan (East Sea), Korea Strait, East China Sea)*
- What are the four main islands of Japan? *(Kyushu, Shikoku, Honshu, and Hokkaido)*
- Which major cities listed on the map are not port cities? *(Sapporo and Kyoto)*

CONNECT GEOGRAPHY & HISTORY ANSWERS

1. 794

2. Tokyo

Presentation Options

Use the **Power Presentations** DVD-ROM to project the animated map.

- Prompt students to use the pop-up images and audio from the map to discover what they already know about the geography, culture, history, economics, or government of Japan.
- Challenge students to answer the map questions individually, in small groups, or as a class activity. Use all of the interactive material to help preview chapter content.

TEACHER MEDIA FAVORITES

Books for the Teacher

Gordon, Andrew. *A Modern History of Japan: From Tokugawa Times to the Present*. Oxford University Press, 2003. An overview of Japan's history, from the 1800s to the present.

Morton, W. Scott; Olenik, J. Kenneth; Lewis, Charlton. *Japan: Its History and Culture*. McGraw-Hill, 2005. A concise history, with emphasis on Japanese culture.

Video / DVD

Japan: Samurai, Geisha, and Technology. (DVD) Analytical Software, 2003. A look at the cities, historical sites, natural areas, and culture of contemporary Japan.

Japan: Memoirs of a Secret Empire. (DVD) PBS-Paramount, 2004. One of the PBS Empires series, this 170-minute video examines the history of Japan from the 16th to the 19th centuries.

Internet

Go to the *Research and Writing Center* at **ClassZone.com** to link to:

- data on the region
- a variety of chapter-specific, content-reviewed sites
- "History Makers" biographies
- primary sources

1 Plan & Prepare

Objectives

- Describe Japan's physical features and scant living space
- Summarize the climate and resources of Japan

Read for the Essential Question

Help students read for a purpose by reminding them of the Essential Question: "How has Japan balanced its ancient traditions with its modern economy?" Point out that they can use items in the section, such as Key Questions, special features, images, and maps to formulate answers to the question.

Vocabulary

Best Practices Toolkit

Use the ***Best Practices Toolkit*** to model strategies for vocabulary notetaking. Vary strategies throughout the year. Choose from: *Knowledge Rating, Predicting ABCs, Definition Mapping, Word Sort, Word Wheel, Frayer Model (Word Squares), Magnet Words,* and *Student VOC.*

Vocabulary Strategies, TT9–TT16

Reading Strategy

Best Practices Toolkit

Display the T-Chart/Two-Column Chart transparency.

- Remind students that causes and effects are often signaled by surrounding words and phrases. Have students brainstorm some of these signal words and phrases, such as *due to, as a result, since, because, so, which,* and *cause.*
- Model filling in causes and effects from the paragraph under "Mountainous Islands." Have students suggest entries.
- Display the completed chart for review.

T Chart/Two-Column Chart, TT20

Section 1 Reading for Understanding

Key Ideas

BEFORE, YOU LEARNED

The Korean Peninsula is a ruggedly mountainous landmass in East Asia, bordered by China, Russia, and Japan.

NOW YOU WILL LEARN

The mountainous islands that make up Japan provide the country's people with scant living space, a varied climate, and limited resources.

Vocabulary

TERMS & NAMES

Honshu Japan's largest and most populous island

Mount Fuji Japan's highest mountain peak

Kanto Plain the largest lowland of Japan, which extends from the Japanese Alps east to the Pacific

Hokkaido (hah•KY•doh) Japan's northernmost and second largest island

Kyushu (kee•OO•shoo) Japan's southernmost island and home to several large cities

Shikoku (shee•KAW•koo) Japan's smallest and least populated major island

Ring of Fire a geographic zone that extends along the rim of the Pacific Ocean and has numerous volcanoes and earthquakes

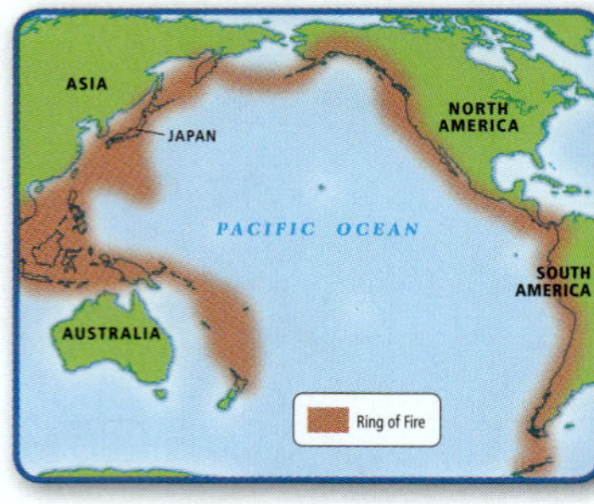

Visual Vocabulary Ring of Fire

Reading Strategy

Re-create the chart shown at right. As you read and respond to the **KEY QUESTIONS**, use the chart to analyze the causes and effects of Japan's physical geography.

Skillbuilder Handbook, page R8

ANALYZE CAUSES AND EFFECTS

PHYSICAL GEOGRAPHY	CAUSES AND/OR EFFECTS
Crowded Islands	
Located on Ring of Fire	
Varied Climate	
Limited Resources	
Abundant Wildlife	

GRAPHIC ORGANIZERS Go to **Interactive Review** @ClassZone.com

824 Chapter 27

PRETEACHING VOCABULARY

English Learners

Pronounce and Preview

Pronounce each term for students. Review the meanings of words in the definitions, such as *populous, lowland, northernmost,* and *geographic zone.*

- To modify vocabulary learning, have students complete worksheets as they read the section.

Unit Resource Book: East Asia & Southeast Asia
- Vocabulary Practice, p. 157

Inclusion

Locate Vocabulary Words on Maps

Invite students to familiarize themselves with vocabulary words and their definitions by locating the terms on the maps of Japan in this section. Ask students to point to each term as the vocabulary word is read, and describe the location of each place in their own words.

SECTION 1 GEOGRAPHY

Himeji Castle, also known as White Heron Castle, in Himeji City

Crane eating fish beside Kamo River in Kyoto

Land of the Rising Sun

Connecting to Your World

Think about the most beautiful scenery you've ever seen. You might recall majestic mountains, spectacular sunsets or sunrises, thundering waterfalls, or rapidly flowing rivers cutting through canyons. Japan possesses all of this natural beauty. Perhaps as a result, many of the Japanese people have a strong emotional bond with nature. In fact, the Japanese name for their country, *Nippon,* refers to nature. *Nippon* means "source of the sun" and is often translated as "Land of the Rising Sun."

Cherry Blossoms Cherry trees from Japan blossom in Washington, D.C.

Mountainous Islands

KEY QUESTION What are the main geographic features of Japan's four major islands?

Japan is made up of about 4,000 islands. It lies to the east of the Korean Peninsula, China, and Russia. The island nation is surrounded by the East Sea, or Sea of Japan, to the west and the Pacific Ocean to the south and east. The islands of Japan are actually the tops of a huge, mostly underwater, mountain range.

Mount Fuji A high-speed train runs past Mount Fuji.

2 Focus & Motivate

3-Minute Warm-Up

Write on the board or display the transparency:

- Study the photo of Mount Fuji. What contrast does the picture convey about Japan? *(the contrast between its natural beauty and its modern, hi-tech economy)*

Unit Transparency Book: East Asia & Southeast Asia
- 3-Minute Warm-Ups, TT20

Connecting to Your World

More About . . .

Mount Fuji

Mount Fuji is an active volcano, yet it hasn't erupted since 1707. The Japanese people consider it a spiritual site, and many temples are located on the mountain. Thousands of people climb its peak every summer.

3 Teach

Mountainous Islands

Talk About It

- How might the sea influence life in Japan? *(The Japanese probably depend on the sea for shipping, travel, and fishing.)*
- **Analyze Causes and Effects** What is a result of the limited living space in Japan? *(Many Japanese live in small apartments; cities are crowded.)*

SECTION 1 PROGRAM RESOURCES

Unit Resource Book: East Asia and Southeast Asia

ON LEVEL

Unit Resource Book
- Reading Study Guide, p. 133
- Map and Graph Skills, p. 159
- Outline Map Activities, p. 161
- Section Quiz, p. 173

STRUGGLING READERS

Unit Resource Book
- Reading Study Guide with Additional Support, p. 139
- Vocabulary Practice, p. 157
- Section Quiz, p. 173
- Reteaching Activity, p. 176

eEdition with Audio DVD-ROM

ENGLISH LEARNERS

Pupil Edition in Spanish
eEdition with Audio DVD-ROM
Unit Resource Book
- RSG (Spanish), p. 145
- RSG with Additional Support (Spanish), p. 151

Multi-Language Glossary
Test Generator
- Section Quiz in Spanish

INCLUSION

Unit Resource Book
- RSG with Additional Support, p. 139
- Section Quiz, p. 173
- Reteaching Activity, p. 176

GIFTED & TALENTED

Unit Resource Book
- Section Quiz, p. 173

PRE-AP

Unit Resource Book
- Section Quiz, p. 173

TECHNOLOGY

Unit Transparency Book
- 3-Minute Warm-Ups, TT20
- Essential Question, TT21
- Critical Thinking, TT22
- Map: Japan: Physical/Political, TT25
- Map: The Ring of Fire, TT26

Daily Test Practice Transparencies
- Section 1, TT90

Power Presentations
ClassZone.com
World Cultures and Geography Video on DVD

More About . . .

Winter Olympics in Japan

Japan's snow-peaked mountains have attracted people from around the world who love such winter sports as skiing and sledding. In 1972, the city of Sapporo on Hokkaido Island hosted the Winter Olympic Games. In 1998, less than 30 years later, the Winter Olympic Games returned to Japan. This time, they were held at the city of Nagano on Honshu, Japan's largest island.

More About . . .

Shikoku

The island of Shikoku may be small, but the northern area of the island is home to a significant agricultural region. Among the crops grown on Shikoku are rice, barley, wheat, and oranges. Because much of the land is mountainous, farming is often done on terraced hillsides.

CONNECT Geography & History

Japan and the Ring of Fire

Architects, builders, and scientists throughout the world are working to lessen earthquake damage by creating more earthquake-resistant buildings. These buildings are meant to stand up to the strong shaking that occurs during an earthquake, yet be flexible enough to avoid shattering when an earthquake bends their walls. The designers hope that their work will both reduce property damage and save lives when a powerful earthquake strikes.

CRITICAL THINKING ANSWER

Summarize buildings damaged and collapsed

Major Islands Although Japan has thousands of islands, most of its people live on the four largest ones: Honshu, Hokkaido, Kyushu, and Shikoku. Locate the islands on the map on the opposite page.

- **Honshu** Ⓐ is Japan's largest island and is home to about 80 percent of its people. Honshu is a land of extreme elevations. The towering Japanese Alps run down the center of the island. Honshu also contains **Mount Fuji** Ⓑ, Japan's highest peak. The largest lowland of Japan, the **Kanto Plain** Ⓒ, extends from the Japanese Alps east to the Pacific.
- **Hokkaido** (hah•KY•doh) Ⓓ lies to the north and is the second largest of Japan's four major islands. It has many mountains and thick forests. This island is sparsely populated, but its long winters make it ideal for winter sports.
- **Kyushu** (kee•OO•shoo) Ⓔ is the southernmost of the major islands. A steep chain of mountains runs down the center of this island. The heavily populated northwestern plains are home to several large cities.
- **Shikoku** (shee•KAW•koo) Ⓕ is the smallest of Japan's main islands. Because almost all of Shikoku is covered with mountains, only about 3 percent of Japan's population lives there.

CONNECT Geography & History

Japan and the Ring of Fire As you can see on the inset map on the opposite page, Japan is located along the **Ring of Fire**. This zone, which extends along the rim of the Pacific Ocean, has many active volcanoes and is subject to many earthquakes. Japan has about 80 active volcanoes and experiences about 1,500 earthquakes each year. Most of these earthquakes are minor—but not all. In 1923, the Great Kanto earthquake struck the Tokyo area, resulting in the deaths of about 140,000 people. The Kobe earthquake of 1995 killed about 6,000 people in that city. Some of the damage caused by the Kobe earthquake is shown here.

CRITICAL THINKING

Summarize What damage is apparent in this photograph of the Kobe earthquake?

826 Chapter 27

DIFFERENTIATING INSTRUCTION

Struggling Readers

Create a Chart of Island Facts

Help students differentiate among the four main islands of Japan by creating a four-column chart on the board. Invite students to review the text to identify facts about the islands' geography and human population. Then ask volunteers to take turns filling in facts about each island on the four-column chart.

Gifted & Talented

Build a Model of Japan's Islands

Have students work in small groups to create clay models of Honshu, Hokkaido, Kyushu, and Shikoku. Students can take measurements from the chapter maps and use information from the text to design their models. Groups should work together to ensure that the models they create share a common scale. When groups are finished, have them assemble their models into one island group.

CONNECT Geography & History

READING A THEMATIC MAP

The Ring of Fire map shows the location of active volcanoes in the Pacific Rim. As you study the map and key,

- use the color symbol to find the Ring of Fire
- note Japan's position along the Ring of Fire
- think about the impact the Ring of Fire might have on Japan's landforms

1. **Region** How much of Japan is on the Ring of Fire?
2. **Place** Compare the two maps. What can you conclude about Mount Fuji?

Japan **827**

Animated GEOGRAPHY
Physical Geography of Japan
CLASSZONE.COM

CONNECT Geography & History

Reading a Thematic Map Have students use the legend to trace the Ring of Fire on the inset map and to identify the area it covers on the large map. Then ask students to name the body of water that lies in the middle of the Ring of Fire. *(the Pacific Ocean)*

ANSWERS

1. **Region** all of it
2. **Place** Mount Fuji is a volcano.

Unit Resource Book: East Asia & Southeast Asia

- Map and Graph Skills: Creating a Relief Map, p. 159
- Outline Map Activities: Japan: Physical/Political, p. 161

Presentation Options

Use the **Power Presentations** DVD-ROM to project the animated map.

- Click to animate the map.
- Listen to the audio and examine the photographs to learn how the geography of Japan affected its history, culture, and economics.

INTERDISCIPLINARY ACTIVITIES

CONNECT to Science

Create Diagrams of Volcanic Island Formation

Invite students to work in small groups to create diagrams showing how volcanoes have formed the main islands of Japan along the Ring of Fire. Student diagrams should show several volcanic processes in sequence, including the initial eruption on the sea floor, the rising of the volcano above the ocean's surface, and the building of a habitable island.

CONNECT to Language Arts

Write a Poem About Mount Fuji

Some Japanese feel a deep spiritual connection to Mount Fuji. Others are simply struck by its profound natural beauty. Invite students to compose poems that reflect on the majesty of Mount Fuji and on their personal connection to the natural world.

Fun Facts!

Capsule Hotels

Capsule hotels are primarily used by businessmen who have missed the last train home and need a place to stay for one night. The men change their clothes in a locker room and often bathe in large—and sometimes luxurious—community baths. Each capsule contains a console of buttons the control the lights, television, radio, and alarm clock.

Teach

Climate and Resources

Think, Pair, Share

- How does the climate vary from northern Japan to southern Japan? *(The climate in northern Japan is generally colder than the more tropical climate in southern Japan.)*
- Where can you find large wildlife populations in Japan? *(in its thickly forested regions)*
- From what source does Japan get much of its food? *(the sea)*
- **Analyze Causes and Effects** Why does Japan have to import many of the natural resources it needs for industries? *(because Japan's resources are of poor quality)*

More About . . .

Japan's Fishing Industry

Japan depends on the sea for food, but its fishing industry has long been in decline. The demand for seafood has caused a significant decline in fish stocks off Japan's coast, forcing fishing boats to sail to more distant waters to fish. However, the establishment of a 200-mile offshore limit by neighboring countries has reduced the waters in which Japan can fish. As a result, Japan has begun to rely more and more on imported seafood to meet demand.

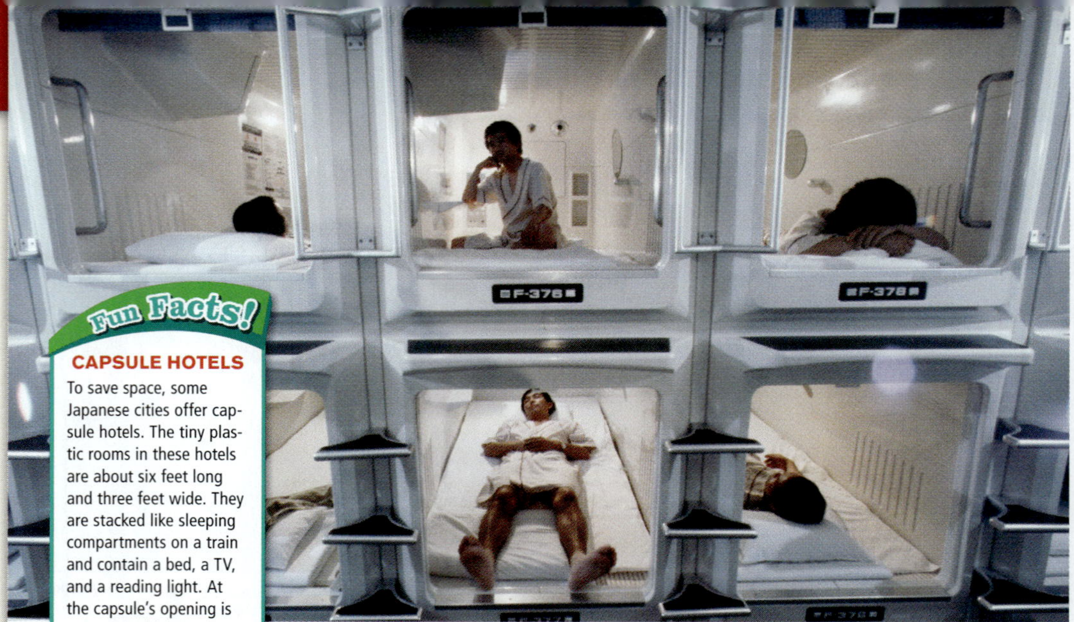

Fun Facts!

CAPSULE HOTELS
To save space, some Japanese cities offer capsule hotels. The tiny plastic rooms in these hotels are about six feet long and three feet wide. They are stacked like sleeping compartments on a train and contain a bed, a TV, and a reading light. At the capsule's opening is a screen, which can be pulled down for privacy.

A Shortage of Space About 70 percent of Japan is covered with mountains and hills. As a result, Japan has only a small amount of land where its people can live. The nation also has a high population density. To give you an idea, Japan is just a little smaller than California, but it has about four times as many people. As you can imagine, Japanese cities are very crowded.

The Japanese have learned to adapt to their limited space. For one thing, Japanese homes tend to be smaller than those in the United States. Also, many people live in apartments. It is not uncommon for a family of four to live in a one-bedroom apartment. Since these dwellings are often noisy and crowded, students usually study at their schools or in public libraries.

SUMMARIZE Describe the main geographic features of Japan's four major islands. Japanese Alps and Kanto Plain in Honshu; mountains and thick forests in Hokkaido; steep chain of mountains and plains in Kyushu; Shikoku almost entirely covered in mountains

Climate and Resources

KEY QUESTION What impact do mountains have on Japan's climate, resources, and wildlife?

Since Japan consists of many islands, the sea strongly affects its climate. The mountainous terrain of these islands also influences Japan's climate and its resources.

DIFFERENTIATING INSTRUCTION: TIERED ACTIVITIES

OBJECTIVE Create encyclopedia entries for Japan's islands.

Basic	On Level	Challenge
Have student pairs create an encyclopedia entry for one of Japan's main islands. Each atlas entry should include a map of the island, identifying major cities and geographic features, and a paragraph that summarizes the island's climate and resources.	Build on the Basic activity, and, in addition, have student pairs include at least two visuals and captions for each island, along with their map and paragraph.	Build on the On-Level activity, and have student pairs create entries for two islands and write a brief compare-and-contrast essay on them. Invite all students to present or display their entries.

A Varied Climate The climates on the major islands are influenced by their latitudes and by the currents of the Pacific Ocean. The northern islands of Japan have generally cool weather, while the southern ones have a more tropical climate. The central islands have a moderate climate, with warm summers and mild winters. The mountains throughout the islands have colder temperatures because of their higher elevations. Most of Japan gets at least 40 inches of rain each year.

Resources and Wildlife Japan has deposits of coal, lead, and silver, but many of these are of poor quality. Forests are abundant, but they are located in remote mountain areas and are difficult to harvest. Not surprisingly, given its vast coastal waters, Japan looks to the sea for food. In fact, Japan has one of the largest fishing industries in the world.

Despite its dense human population, Japan has a great deal of wildlife, especially in regions that are thickly forested. Bears, wild boars, raccoon dogs, and deer can be found in these areas. Japan's wilderness also contains a large, shaggy-haired monkey called the Japanese macaque (muh•KAK).

Japanese Macaques
Also called snow monkeys, Japanese macaques live farther north than any other monkeys in the world.

ANALYZE EFFECTS Discuss the impact of mountains on Japan's climate, resources, and wildlife. mountainous areas tend to be cooler; forestland and much wildlife in remote mountain areas

ONLINE QUIZ For test practice, go to **Interactive Review** @ClassZone.com

Section 1 Assessment

TERMS & NAMES

1. **Explain the importance of**
   - Honshu
   - Kanto Plain
   - Mount Fuji
   - Ring of Fire

USE YOUR READING NOTES

2. **Analyze Causes and Effects** Use your completed chart to answer the following question:
   Why does Japan have such limited living space?

PHYSICAL GEOGRAPHY	CAUSES AND/ OR EFFECTS
Crowded Islands	
Located on Ring of Fire	
Varied Climate	
Limited Resources	
Abundant Wildlife	

KEY IDEAS

3. On which two islands do most of the Japanese people live?
4. Why does Japan have so many earthquakes?
5. How does latitude affect Japan's climate?

CRITICAL THINKING

6. **Evaluate** What characteristics do the Japanese reveal in their ability to live on their crowded island nation?
7. **Draw Conclusions** What do you think is Japan's most important natural resource? Explain.
8. **CONNECT to Today** Subway employees push passengers into Tokyo subways and trains during rush hour. What does this tell you about the city?
9. **WRITING** **Pen a Journal Entry** Imagine that you are on a hike in the Japanese Alps. Write a journal entry about your experiences in these mountains. Record the sights you might see, the type of weather you might deal with, and the vegetation and wildlife you might encounter.

CONNECT to the Essential Question

How has Japan balanced its ancient traditions with its modern economy?

Ask students what they have learned so far that can help them answer this question. Students might mention the following:

- Many Japanese share a strong emotional bond with their natural landscape.
- People in Japan have adapted to living on an island, where usable land is scarce.

4 Assess & Reteach

Assess Have students complete the Section Assessment.

Unit Resource Book: East Asia & Southeast Asia
- Section Quiz, p. 173

Interactive Review @ClassZone.com
- Section Quiz

Power Presentations

Test Generator

Reteach Have pairs of students create three "true or false" statements about Japan's geographic features, climate, and resources. Then have each pair exchange their statements with another pair and evaluate them. Encourage students to correct any false statements.

Unit Resource Book: East Asia & Southeast Asia
- Reteaching Activity, p. 176

SECTION 1 ASSESSMENT ANSWERS

Terms & Names

1. Honshu: Japan's largest island; Mount Fuji: Japan's highest mountain; Kanto Plain: Japan's largest lowland; Ring of Fire: zone along Pacific rim with many volcanoes and earthquakes

Use Your Reading Notes

2. because about 70 percent covered by mountains

PHYSICAL GEOGRAPHY	CAUSES AND/ OR EFFECTS
Crowded Islands	largely mountains
Located on Ring of Fire	subject to volcanic eruptions and earthquakes
Varied Climate	colder in mountains
Limited Resources	resources in remote areas
Abundant Wildlife	forests protect animals

Key Ideas

3. Honshu and Kyushu
4. It is located in an unstable geographic zone called the Ring of Fire.
5. The northern islands tend to be cooler than the southern ones.

Critical Thinking

6. Possible answers: adaptability, patience, ingenuity, intelligence
7. Possible answers: fish, because Japan is surrounded by water; timber, because Japan has a great deal of forestland (although most of it is difficult to harvest)
8. Possible answers: The city is very crowded; the people are used to being compressed into small spaces.
9. Use the rubric to score journal entries.

Journal Entry Rubric

	Content	Errors
4	engaging and filled with details	few or none
3	interesting and detailed	some
2	somewhat detailed	many
1	few or no details	many major

1 Plan & Prepare

Objectives

- Explain Japan's rise from an isolated nation to a world power
- Identify aspects of traditional and modern Japanese culture

Read for the Essential Question

Help students read for a purpose by reminding them of the Essential Question: "How has Japan balanced its ancient traditions with its modern economy?" Point out that they can use items in the section, such as Key Questions, special features, images, and maps to formulate answers to the question.

Vocabulary

Best Practices Toolkit

Use the ***Best Practices Toolkit*** to model strategies for vocabulary notetaking. Vary strategies throughout the year. Choose from: *Knowledge Rating, Predicting ABCs, Definition Mapping, Word Sort, Word Wheel, Frayer Model (Word Squares), Magnet Words,* and *Student VOC.*

Vocabulary Strategies, TT9–TT16

Reading Strategy

Best Practices Toolkit

Display the Time Line transparency.

- Point out that sequencing events, or arranging events in chronological order, will help them understand how the events are related.
- Model filling in a few events from the paragraph under "Isolated Nation to World Power." Have students suggest some entries.
- Display the graphic organizer for students to review.

Time Line, TT33

SECTION 2

Reading for Understanding

Key Ideas

BEFORE, YOU LEARNED

Japan is a chain of islands that lies off the coast of China and the Korean Peninsula.

NOW YOU WILL LEARN

Over the centuries, Japan has adapted ideas and technologies from other lands and created a unique culture.

Vocabulary

TERMS & NAMES

Heian (hay•ahn) **period** the era in Japanese history from A.D. 794 to 1185; arts and writing flourished during this time

samurai (SAM•uh•RY) professional soldiers of Japan

shogun the leader of the military government of Japan beginning in 1192

daimyo (DY•mee•OH) powerful samurai who became warrior-chieftains

Tokugawa Ieyasu (TOH•koo•GAH•wah EE•yeh•YAH•soo) shogun who unified Japan

Shinto Japan's original religion; involves worshiping gods believed to be found in nature

BACKGROUND VOCABULARY

clan a group of families related through a common ancestor

REVIEW

emperor a man who is the ruler of an empire

Visual Vocabulary statue of daimyo draped with war flag

Reading Strategy

Re-create the time line shown at right. As you read and respond to the **KEY QUESTIONS**, use the time line to keep track of some of Japan's important historical and cultural events.

See Skillbuilder Handbook, page R6

SEQUENCE EVENTS

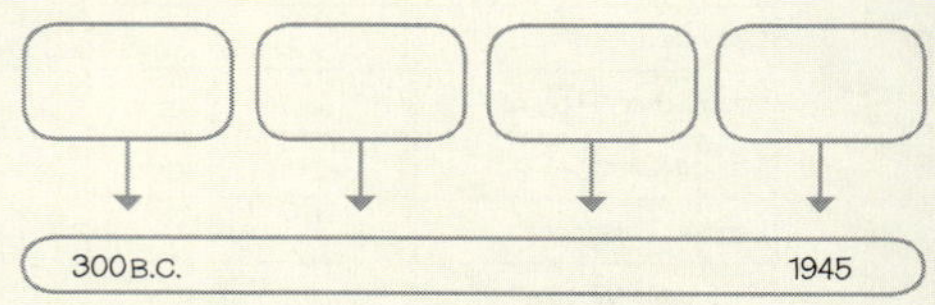

GRAPHIC ORGANIZERS Go to **Interactive Review @ ClassZone.com**

PRETEACHING VOCABULARY

English Learners

Pronounce and Preview

Pronounce each term for students. Review the meanings of words in the definitions, such as *flourished*, *warrior-chieftains*, and *unified*.

- To modify vocabulary learning, have students complete worksheets as they read the section.

Unit Resource Book: East Asia & Southeast Asia

- Vocabulary Practice, p. 157

Inclusion

Use Knowledge Rating

Have students work in pairs to fill out the graphic organizer for the terms in the section. First, model the strategy using the transparency. Have volunteers suggest entries. Then distribute copies of the organizer for students to complete with a partner.

Best Practices Toolkit

- Vocabulary: Knowledge Rating, TT9

Himeji Castle, also known as White Heron Castle, in Himeji City

SECTION 2

HISTORY & CULTURE

Crane eating fish beside Kamo River in Kyoto

Samurai to Animé

Connecting to Your World

Have you ever played a video game or watched an animated film? If so, you have probably used a product from Japan. Japanese culture has strongly influenced culture in the United States and throughout the world. But the reverse has also been true. Hundreds of years ago, several cultures played an important role in shaping Japanese culture. In this section, you will learn what these cultures were and how they affected Japan.

Video Games Only the United States sells more video games than Japan.

Isolated Nation to World Power

KEY QUESTION How did Japan change from an isolated nation to a world power?

The first influences on Japanese culture probably arrived from China and Korea around 300 B.C. At this time, Japanese families were organized into **clans**, who fought over land. Eventually, the Yamato family became the leading clan and established a member of its family as ruler, or **emperor**. The Yamato line of emperors continues to this day.

Influence from China Buddhism arrived in Japan from China. This statue of Buddha in Kyoto reflects China's influence.

2 Focus & Motivate

3-Minute Warm-Up

Write on the board or display the transparency:

- What aspects of Japanese culture are you most familiar with? *(Possible answers: animé, origami, haiku, sushi)*

Unit Transparency Book
- 3-Minute Warm-Ups, TT20

Connecting to Your World

More About . . .

Korean Influences

One custom that likely came to Japan from the Korean Peninsula was the building of *kofun*, or large funeral mounds. Inside these tombs, which began to appear in Japan in the 3rd century A.D., bodies were placed in wooden coffins and surrounded by tools, weapons, and personal items.

3 Teach

Isolated Nation to World Power

Reader, Recorder, Reporter

- From which family have all of Japan's emperors descended? *(the Yamato clan)*
- **Sequence Events** What happened in Japan in 1868? *(The emperor was restored as head of the government.)*

SECTION 2 PROGRAM RESOURCES

Unit Resource Book: East Asia and Southeast Asia

ON LEVEL

Unit Resource Book
- Reading Study Guide, p. 135
- Section Quiz, p. 174

STRUGGLING READERS

Unit Resource Book
- Reading Study Guide with Additional Support, p. 141
- Vocabulary Practice, p. 157
- Skillbuilder Practice, p. 158
- Section Quiz, p. 174
- Reteaching Activity, p. 177

eEdition with Audio DVD-ROM

ENGLISH LEARNERS

Pupil Edition in Spanish

eEdition with Audio DVD-ROM

Unit Resource Book
- RSG (Spanish), p. 147
- RSG with Additional Support (Spanish), p. 153

Multi-Language Glossary

Test Generator
- Section Quiz in Spanish

INCLUSION

Unit Resource Book
- RSG with Additional Support, p. 141
- Section Quiz, p. 174
- Reteaching Activity, p. 177

GIFTED & TALENTED

Unit Resource Book
- History Makers, pp. 163–164
- Interdisciplinary Projects, p. 166
- Section Quiz, p. 174

PRE-AP

Unit Resource Book
- Connect to Culture, p. 169
- Primary Sources: pp. 171–172
- Section Quiz, p. 174
- World Literature, p. 247

TECHNOLOGY

Unit Transparency Book
- 3-Minute Warm-Ups, TT20
- Essential Question, TT21
- Critical Thinking, TT23
- Cultures, TT27, TT28

Daily Test Practice Transparencies
- Section 2, TT91

Power Presentations

ClassZone.com

World Cultures and Geography Video on DVD

Make It Fun

Teacher-Tested Activities

Greta Frensley, Knox Doss Middle School, Gallatin, Tennessee

My students like to play a Jeopardy-style game show to test their knowledge of history.

- First, I prepare a stack of index cards with the answers on one side and the correct questions on the other.
- Then three "contestants" come to the front of the class to play the game. I give each one a small bell, which serves as a buzzer.
- Finally, I choose a card and read an answer, such as "This shogun unified Japan." The first student to ring in and give the correct question (in this case, "Who is Tokugawa Ieyasu?") is awarded five points.
- The first student to reach 25 points wins the game.

COMPARING

Samurai & Knights

Knights generally came from upper-class families. They began their training at a very young age, often while they worked as servants for their fathers. Samurai came from families of all social classes. They, too, were often trained in fighting by older men in their families. Ask students what kind of training a samurai or knight might have needed. *(Possible answers: training in fighting, horseback riding, weapon making)*

CRITICAL THINKING ANSWER

Compare and Contrast Both valued bravery and loyalty and fought wearing armor. However, samurai expected women to be courageous as well, while knights considered women weak and in need of protection.

Heian Period In the late 700s, the emperor moved his court from the capital of Nara to Heian (hay•ahn), which is modern Kyoto (kee•OH•toh). Heian remained the center of Japan's government for about 400 years. This era in Japanese history, which lasted from 794 to 1185, is called the **Heian period**.

Art and writing flourished during this period. Men and women of the nobility filled their days by painting and writing poetry and prose. In fact, one of the best accounts of Heian society was written by Lady Murasaki Shikibu (MOO•rah•SAH•kee SHEE•kee•boo). Her masterpiece, *The Tale of Genji*, is an account of the life of a prince in the emperor's court. The work is considered the world's first novel.

Samurai and Shoguns During the Heian period, the central government was relatively strong. However, by the mid-1000s, this power had begun to decrease. Wealthy clans bought large areas of land and set up private armies. The countryside became lawless and dangerous. As a result, smaller landowners sold parts of their land to strong warlords in exchange for protection. With more land, the lords gained more power. Soon, wars between rival lords became commonplace. To defend himself, each lord surrounded himself with a bodyguard of loyal warriors called **samurai** (SAM•uh•RY).

COMPARING Samurai & Knights

Samurai were similar to knights, a class of warriors that formed in medieval Europe. Here is how the two groups compare.

JAPANESE SAMURAI	EUROPEAN KNIGHTS
Live by code of honor and value bravery and loyalty to their lord above all else	Value bravery and loyalty to heavenly God, earthly lord, and chosen lady
Prefer death to defeat	Show humility
Expect women to live up to same values of honor and courage; women can fight as samurai	Regard women as weak creatures to be idolized and defended
Enter into battle with iron and leather armor, swords, and bows and arrows	Fight battles with chain mail or plate armor, broadsword, and lance

CRITICAL THINKING

Compare and Contrast What are some similarities and differences between Japanese samurai and European knights?

832 Chapter 27

DIFFERENTIATING INSTRUCTION

Struggling Readers

Use Visuals to Compare

Encourage students to first compare samurai and knights by analyzing the images of each in their textbooks. Invite students to name all the similarities and differences they can see in these images. Use this information as the springboard for a discussion comparing the lives of samurai and knights.

Gifted & Talented

Write a Diary Entry

Challenge students to create several diary entries that detail the end of the high culture of the Heian period and the transition to the period of danger and lawlessness that followed. Entries should be written from the point of view of an artist or government official and include details about daily life and the changes taking place.

During the 1100s, two powerful clans fought for control—the Taira (TY•rah) and the Minamoto. Each clan had a large samurai army. After about 30 years of war, the Minamoto gained control. This clan set up a military government in Kamakura. In 1192, the leader of this government was given the title of **shogun**, which means "supreme general of the emperor's army." Although the emperor still reigned, the shogun held real power. This pattern of government, in which shoguns ruled through puppet emperors, lasted until 1867.

Unified Japan Between 1192 and 1600, Japan was torn by internal warfare. Peasants had to fight to defend their villages. Sometimes, they hired samurai to protect them. Very powerful samurai arose and became warrior-chieftains called **daimyo** (DY•mee•OH), who became lords in their own right. The daimyo formed armies and fought for military supremacy.

Finally, in 1600, a daimyo named **Tokugawa Ieyasu** (TOH•koo•GAW•wah EE•yeh•YAH•soo) restored order. Three years later, he became shogun and unified Japan. He then moved the capital to his base at Edo, a small fishing village that would later become the city of Tokyo.

The Tokugawa shogunate held power until 1868. During this time, Japan enjoyed a period of stability—and isolation. For more than 200 years, the country was closed to Westerners.

HISTORY MAKERS

Tokugawa Ieyasu 1543–1616

Tokugawa Ieyasu could be merciless in the defense of his clan. In the late 1500s, his wife and eldest son were accused of conspiring against the family. As a result, Ieyasu ordered his son to commit suicide and had his wife executed. He also never forgot a grudge. As an adult, he executed a prisoner who had insulted him in childhood. On the other hand, Ieyasu could also be kind and generous. He rewarded those who were loyal to him and even showed compassion to his enemies.

ONLINE BIOGRAPHY
For more on the life of Tokugawa Ieyasu, go to the **Research & Writing Center** @ClassZone.com

Himeji Castle Tokugawa Ieyasu's son-in-law built the castle after he helped win the battle that brought Ieyasu to power.

833

RESEARCH & WRITING CENTER

HISTORY MAKERS

Tokugawa Ieyasu

It is little wonder that Tokugawa Ieyasu became one of Japan's great daimyos. From the time of his birth, war played an important role in his life. Ieyasu's father was a warrior, and often left the family to go to battle. When Ieyasu was just a boy, he was sent as a hostage to a powerful rival warrior clan.

As a daimyo, Ieyasu learned that it took more than military might to gain real power. He helped keep control of lands he conquered by reorganizing local governments and by giving some land and power to men who supported him.

More About . . .

Japan's Years of Isolation

In the early 1600s, the Tokugawa shogunate was open to trade and did business with the Dutch, English, Portuguese, and Spanish. The shogunate also allowed Christian missionaries to visit and preach in Japan.

However, by about 1610, Ieyasu Tokugawa became convinced that the missionaries threatened his power by introducing new ideas to Japan. He began to expel Christian missionaries and prohibited all Christian activity among the Japanese people. Eventually, Japan would also severely restrict trade with other nations, beginning a long period of isolation.

Unit Resource Book: East Asia & Southeast Asia
- Skillbuilder Practice: Identifying Problems and Solutions, p. 158

INTERDISCIPLINARY ACTIVITIES

CONNECT to Language Arts

Write News Features for Samurai

Have groups of students work together to research and write features for Japanese samurai living in 1185. Some students could write an article about the latest in samurai armor. Others could write about the importance of the samurai's sword. Still others could write about the importance of Zen Buddhism to the samurai. Assemble the features into a news bulletin.

CONNECT to Government

Stage a Debate

Invite students to assume the role of the Tokugawa shogun as he and his advisors debate whether to isolate Japan and close it to Westerners. Before the debate, students should research some of the reasons for the isolation policy. In their debate, students should discuss both the advantages and disadvantages of becoming a closed society.

Japanese Empire *1942*

CONNECT Geography & History

The geography of Japan played a critical role in how World War II was fought in Asia and the Pacific. While the war in Europe was fought mostly on land, much of the war in Asia was fought using ships and planes launched from them. Ask students to estimate the total size of the area (land and water) that Japan controlled. *(about 17,000,000 square miles)*

ANSWERS

1. **Location** Manchuria or eastern China
2. **Region** the Pacific Ocean

Teach

Traditional and Modern Japanese Culture

Talk About It

- How does Japan's written language reflect a Chinese influence? *(Chinese characters are used to write nouns in Japanese.)*
- What are the two main religions of Japan? *(Shinto and Buddhism)*
- **Sequence Events** When did the haiku develop? *(in the 1600s)*

More About . . .

The Ainu

The word *Ainu* means "human." Long ago, the Ainu made their living by hunting, fishing, and small-scale farming. At one time, the Ainu lived on all four of Japan's main islands, but they were driven northward following battles with ethnic Japanese. For many years, the Japanese government discouraged the practice of Ainu customs and even passed laws that discriminated against the Ainu. Today, discrimination remains a problem for the 25,000 Ainu who still live in Japan.

Rise of Modern Japan In 1853, the United States forced Japan to end its isolation. The government sent Commodore Matthew Perry to Japan with several warships to frighten the Tokugawa shogun. He soon agreed to open several Japanese ports to Western powers.

The Japanese people were angry that the shogun had given in to the foreigners' demands. As a result, the emperor was restored as head of government in 1868. The emperor chose the name *Meiji* (MAY•jee), which means "enlightened rule," for his reign. In an effort to counter Western influence, the emperor began a policy of modernization. Before long, Japan became a powerful nation.

As Japan's power grew, the nation sought to obtain an empire. In the early 1900s, Japan gained control of Korea. By 1937, Japan had invaded China and formed an alliance with Nazi Germany and Fascist Italy. Between 1939 and 1945, Japan joined and fought with these powers during World War II—and lost.

After the United States and the Allies won the war, they occupied Japan for several years and set up a democracy. In time, Japan developed into an economic superpower.

SUMMARIZE Explain how Japan changed from an isolated nation to a world power.

in mid-1800s, United States forced Japan to open its ports to trade; Japan began modernizing; eventually began pursuing imperialistic ambitions

Japanese Empire *1942*

CONNECT Geography & History

1. **Location** What part of mainland China did Japan control?
2. **Region** What body of water borders all of the conquered territories?

Traditional and Modern Japanese Culture

KEY QUESTION How does Japanese culture blend traditional and modern influences?

As you have learned, Japan has been isolated for long periods of time throughout its history. Because of this, 99 percent of the people who live there are of Japanese descent. Other groups include Chinese, Koreans, and the Ainu (EYE•noo). Some historians believe that the Ainu were Japan's original people. Most Ainu live on Hokkaido.

Japanese is the country's only official language. The written form reflects the influence of China on Japan. Chinese characters, called *kanji,* are used for nouns. Japanese characters, known as *kana,* can be used for changing nouns into adjectives and verbs.

DIFFERENTIATING INSTRUCTION

English Learners

Learn Key Vocabulary

Review this sentence from the textbook with students: "In an effort to counter Western influence, the emperor began a policy of modernization." Have students look up the definition of *policy* in a dictionary and write it down in their notebooks. Then have them write a sentence using the word.

Pre-AP

Write Newspaper Articles

Have students assume the role of reporters traveling to Japan just after it has been opened to trade with the United States. Encourage students to write newspaper articles that inform U.S. readers about Japanese life, including customs, architecture, government, and daily life. Invite students to read aloud or post their articles.

Religious Traditions Japan's religious traditions also demonstrate China's influence—and Japan's ability to take in new ideas and make them its own. The two major religions of Japan are Shinto and Buddhism. **Shinto** is Japan's original religion and is rarely practiced anywhere else in the world. Followers of this belief worship many gods, called *kami*. These gods are believed to be found in aspects of nature, such as rivers, trees, and rocks.

Buddhism, which originated in India, spread from China to Japan during the 500s. A form of this religion, called Zen Buddhism, became important in Japan. Zen stresses strict discipline of the body and mind as the path to wisdom. Many Japanese use rituals from both religions in their daily lives.

Torii Gate Like all torii gates, this one on Miyajima Island near Hiroshima serves as an entry to a Shinto shrine. **How does the gate reflect the importance of nature in Shinto?**

The gate is placed in an outdoor setting where nature can be worshiped.

Japanese Arts The arts in Japan blend traditional and modern influences. Kabuki is a traditional form of Japanese drama that features colorful costumes and an exaggerated acting style. Other traditional Japanese arts include flower arranging and the tea ceremony. Both reflect the importance of traditional Japanese virtues such as simplicity, discipline, and love of nature.

Modern arts are also popular in Japan. These include animated films called *animé*, which often feature characters with large, expressive eyes, and comic books called *manga*. Japan has many manga cafes, where people can read manga over a cup of tea or coffee.

CONNECT to Culture

Origami is an art form that involves folding paper into decorative shapes. Although origami started in China, it has thrived in Japan. One of the most popular origami shapes is the crane, shown here. The crane is considered a sacred bird in Japan. A finished sailboat is also shown.

Activity

Create an Origami Sailboat

Materials
- colored paper

1. With colored side down, fold your paper on line A–C. Open and fold point B to the diagonal crease. Then fold D to the crease.
2. Turn your paper around so that A is nearest to you. Fold A so that it meets points D and B. Then fold G–H up and crease on line E–F.
3. Finally, fold E–F up and crease on line G–H.

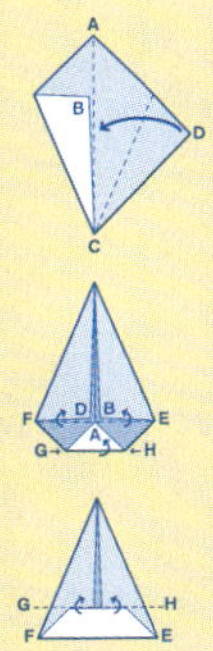

More About . . .

Shinto

The most important *kami* in the Shinto religion is Amaterasu Omikami, the goddess of the sun. According to Shinto beliefs, her descendents, including Jimmu, the first emperor, unified the Japanese people. All Japanese emperors are said to be descended from Amaterasu. The Grand Shrine of Ise, built to honor Amaterasu, is considered the most important Shinto shrine in the world.

CONNECT to Culture

Tell students that, thanks to one Japanese girl, origami cranes have become an international symbol of peace. Explain that Sadako Sasaki developed leukemia due to exposure to radiation from the U.S. atomic bomb dropped on Hiroshima in 1945 at the end of World War II. In an attempt to get well, she folded more than 1,000 paper cranes because, according to Japanese legend, folding 1,000 would grant her one wish. Sasaki died at the age of 12, but her legacy lives on. Every year, people send thousands of paper cranes to Hiroshima Peace Park, where a statue of Sasaki stands, to promote world peace.

Unit Resource Book: East Asia & Southeast Asia
- Connect to Culture: Create an Origami Sailboat, p. 169
- Interdisciplinary Projects: Woodblock Printing, p. 166

DIFFERENTIATING INSTRUCTION

Struggling Readers

Summarize Main Ideas

Help struggling readers understand traditional and modern Japanese culture by having them prepare and read aloud a brief summary of the main ideas from the "Traditional and Modern Japanese Culture" subsection. To check understanding, ask students questions about Japanese culture or invite them to define terms such as Shinto, Buddhism, kabuki, or animé.

Gifted & Talented

Create Origami Art

Encourage students to use directions found on the Internet or in books to create a work of origami art. As a challenge, invite students to create original origami shapes of their own design. Once students have become familiar with an origami pattern, invite them to teach it to the rest of the class.

More About . . .

Matsuo Basho

Basho's real name was Matsuo Munefusa. The name *Basho* comes from the Japanese name for banana or plantain plant. The plant grew around the poet's hut. As a result, he became known as "Master Basho," and the nickname stuck.

In addition to haiku, Basho wrote a type of linked verse called renga. He was considered a master of this genre as well. Basho's poetry is renowned for its ability to contain both natural beauty and deep meaning in just a few lines.

Unit Resource Book: East Asia & Southeast Asia

- History Makers: Matsuo Basho, p. 163
- History Makers: Japanese Woodblock Artists, p. 164
- World Literature: Haiku from Buson and Issa, p. 247

More About . . .

Lady Murasaki Shikibu

Historians do not know the real name of the author known as Lady Murasaki Shikibu. The name *Shikibu* comes from a title once held by her father. *Murasaki* is the name of one of the characters in *The Tale of Genji*. Details of the author's life are revealed in diaries she kept between 1007 and 1010, when she served in the court of Empress Akiko. In addition to writing what many consider the world's first novel, Murasaki Shikibu also wrote many poems.

Unit Resource Book: East Asia & Southeast Asia

- Primary Sources: from *The Tale of Genji* by Murasaki Shikibu, p. 171
- Primary Sources: from *The Pillow Book* by Sei Shonagon, p. 172

Woodblock Print Haiku convey images of nature that are often illustrated with landscape prints.

Japanese Literature The earliest known Japanese poetry, called tanka, was the dominant verse form from the 700s until the 1500s. Tanka means "short songs" and consists of 31 syllables divided among five lines. During the Heian period, aristocrats routinely composed and exchanged tanka with their loved ones. In the 1600s, another type of poetry called haiku (HY•koo) developed. This verse has 17 syllables arranged in three lines. The following haiku is by Matsuo Basho, who is considered a master of the form.

Cool crescent moon
shining faintly high above
Feather Black Mountain

In addition to poetry, Japanese writers have made many contributions to fiction and drama. As you have learned, Lady Murasaki Shikibu is credited with writing the first novel. Zeami Motokiyo, who wrote in the late 1300s, is one of Japan's greatest playwrights.

Life in Japan Because of Japan's shortage of living space, Japanese cities are very crowded. About three-fourths of the Japanese people live in cities. These urban areas are filled with dazzling, high-rise buildings, elegant department stores, and trendy shops. Even though the people live in small houses and apartments, overcrowding remains a constant problem. During rush hour, for instance, white-gloved employees push as many people into subways and trains as possible. Passengers have suffered broken bones from holding their arms in awkward positions in these cramped conditions.

Tokyo at Night Cars and people crowd the Ginza district in Tokyo, a popular area for entertainment and shopping.

DIFFERENTIATING INSTRUCTION: TIERED ACTIVITIES

OBJECTIVE Create an exhibit on Japanese history using a time line.

Basic

Have student pairs begin to create an exhibit on Japanese history by drawing a time line and filling in four or five key dates and events.

On Level

Have student pairs build on the exhibit by filling in several more dates and events on the time line and creating or finding images that illustrate some of the events.

Challenge

Have student pairs complete the time line by adding more events and images to it. Students should also write paragraphs summarizing important events in Japanese history. Once the time line is finished, display it in the classroom.

wear Western clothing and kimonos; practice tea ceremonies and enjoy animé and manga; eat Western foods and sushi; watch sumo wrestling and baseball ▶

In Japan's cities, most Japanese wear clothes that are similar to those worn in the United States. However, for special occasions, Japanese men and women might wear a traditional dress called a kimono. These garments are tied around the waist with a sash and are worn with sandals.

Although Western fast foods are popular in Japan, many Japanese still enjoy traditional fare, including rice with almost every meal. Other traditional foods include tofu (soybean curd cake) and sushi—vinegared rice usually topped with raw fish. Soups made with a soybean paste called *miso* are also popular.

You may be surprised to learn that the number one spectator sport in Japan is baseball. However, traditional sumo wrestling also has a wide following. Other popular traditional sports include *kendo* (fencing with bamboo sticks) and martial arts.

Japanese Baseball Japan won the first World Baseball Classic in 2006. Two of the players on the Japanese team played in the U.S. major leagues during the regular season.

FIND MAIN IDEAS Describe how Japanese culture blends traditional and modern influences.

Section 2 Assessment

ONLINE QUIZ For test practice, go to **Interactive Review @ClassZone.com**

TERMS & NAMES

1. **Explain the importance of**
   - samurai
   - shogun
   - daimyo
   - Shinto

USE YOUR READING NOTES

2. **Sequence Events** Use your completed time line to answer the following question:

   During which period was the haiku developed?

   [ ] [ ] [ ] [ ]
   300 B.C. ——— 1945

KEY IDEAS

3. How did the nobility fill their time during the Heian period?
4. Why did Japan begin to modernize during the mid- to late 1800s?
5. What is daily life like in Japanese cities?

CRITICAL THINKING

6. **Make Inferences** Why do you think the Japanese wanted to live in isolation from the West for so long?
7. **Synthesize** Reread the haiku in this section. How does the poem reflect the traditional Japanese values of simplicity, discipline, and love of nature?
8. **CONNECT to Today** What aspects of Japanese culture are popular in the United States today?
9. **TECHNOLOGY** **Give a Kabuki Slide Show** Research to learn about kabuki theater. Find pictures of costumes, make-up, and the set designs used. Then give a slide show about the drama form, using authentic kabuki music in the background.

CONNECT to the Essential Question

How has Japan balanced its ancient traditions with its modern economy?

Ask students what they have learned so far that can help them answer this question. Students might mention the following:

- Japan's isolation helped it develop a strong, unified culture.
- When Japan's isolation ended, it began to modernize and become a world power.
- Japanese culture today is a blend of traditional and modern influences.

4 Assess & Reteach

Assess Have students complete the Section Assessment.

Unit Resource Book: East Asia & Southeast Asia
- Section Quiz, p. 174

Interactive Review @ClassZone.com
- Section Quiz

Power Presentations

Test Generator

Reteach Create a two-column chart. Label one column "Japan before 1853," the other, "Japan after 1853." Invite students to take turns describing one aspect of Japanese history or culture and writing it under the appropriate column. Note that some aspects might be placed in both columns.

Unit Resource Book: East Asia & Southeast Asia
- Reteaching Activity, p. 177

SECTION 2 ASSESSMENT ANSWERS

Terms & Names

1. samurai: professional soldiers of Japan; shogun: leader of the military government of Japan beginning in 1192; daimyo: powerful samurai who became warrior-chieftains; Shinto: Japan's original religion

Use Your Reading Notes

2. during the Tokugawa shogunate

China and Korea influence Japan	Heian period begins	Japan unifies	Japan defeated in World War II
300 B.C.	A.D. 794	1603	1945

Key Ideas

3. by painting and writing poetry and prose
4. to counter and compete with Western powers
5. crowded, busy, modern, blended with traditional pursuits

Critical Thinking

6. Possible answers: because the Japanese didn't want any outside interference; because they wanted to live in peace; because they distrusted the West
7. Possible answers: It expresses its message in simple words; it requires discipline to write a poem containing only 17 syllables and three lines; it conveys its message using images of nature.
8. Possible answers: animé, manga, sushi
9. Use the rubric to score slide shows.

Slide Show Rubric

	Presentation of Slide Show	Images Used
4	logical and clear	very engaging
3	clear	interesting
2	somewhat confusing	somewhat interesting
1	weak and unclear	uninteresting

1 Plan & Prepare

Objective

Compare traditional elements of Japanese culture with modern styles and influences.

Springboard Ask students to close their texts and brainstorm things they associate with traditional Japanese life. Then have them brainstorm what they associate with modern Japanese life. List responses on the board. Finally, have students open their text and read the Comparing feature. What changes or additions would they make to their lists?

2 Teach

Comparing Traditional and Modern Japan

Talk About It

- What are the main features of a kabuki drama? *(colorful costumes, revolving stages, and an all-male cast)*
- What type of clothing was worn in traditional Japan? *(kimonos)*
- **Compare and Contrast** How is modern Japan different from traditional Japan? *(People wear more modern clothing, make music with electronics, watch animé for entertainment, and live a more busy, hectic life.)*

Comparing Regions

Traditional Theater in Japan and Vietnam

Kabuki plays were originally designed to show off the singing and dancing skills of the actors. Later, the plays took on serious subjects, from historical dramas to tales of doomed romance, but the actors' skills remained the centerpiece. Kabuki plays usually contain a moral lesson in which good behavior is rewarded and bad behavior is punished.

In contrast, traditional Vietnamese *hat cheo* plays are often humorous. The plays are frequently satires that mock government officials or other authority figures. Like kabuki, *hat cheo* plays incorporate music and dancing.

Japan is a land of ancient traditions. But it is also an ultra-modern country, where new styles are set. Today, as you can see from these pictures, the past and present live side by side in Japan.

Traditional Japan

Since the 1600s, the Japanese people have enjoyed kabuki drama, with its colorful costumes and revolving stages. The dramas recall a time when the Japanese wore kimonos and musicians played a three-stringed instrument called the *shamisen*. In this traditional Japan, people have time to sit and reflect in the peaceful calm of a Zen garden.

Dress
Many different types of kimonos can be worn by men, women, and children.

Entertainment
All of the roles in a kabuki drama are played by men.

Zen garden in Kyoto ▼

Music
The sound of the *shamisen* is similar to that of a banjo.

DIFFERENTIATING INSTRUCTION: TIERED ACTIVITIES

OBJECTIVE Understand the contrast between traditional and modern Japan.

Basic

Have groups of students study the two large images on these pages and write down words in a two-column chart that describe the different moods conveyed. Point out that the words they choose for each image will probably have opposite meanings.

On Level

Have students compare the actors shown in the kabuki drama with the characters in the animé film. Ask them to write a paragraph comparing their dress, gestures, facial expressions, and possible relationship.

Challenge

Ask students to imagine that they have gone back in time to visit Japan in the 1600s. What would they tell the people about Japanese culture in the 21st century? Have students write their ideas in the form of a dialogue.

Modern Japan

The worldwide popularity of animé attests to Japan's ability to set cultural trends. Other exports include karaoke and stylish, modern fashions. As you can see, modern life in Japan is also characterized by busy rush hours.

Dress

Today, Japanese fashion designers are copied throughout the world.

Entertainment

Many animé films tell sophisticated stories about love, growing up, and female empowerment.

Crowded Tokyo subway ▼

Music

In Japan, people can sing along with a karaoke player in public bars or in private shops.

CRITICAL THINKING

1. **Compare and Contrast** Select two corresponding images from traditional and modern Japan. How do they differ? What do they have in common?
2. **Analyze Causes** Why do you think animé and karaoke have become so popular in other countries?

More About . . .

Animé

Thanks in part to American movie fans, animé has become one of Japan's most important exports. A number of animé feature films have come to the United States on DVD/video, and some have even been shown in movie theaters nationwide. In 2003, the animé film *Spirited Away*, directed by Hayao Miyazaki, won the Academy Award for Best Animated Feature.

3 Wrap Up

Remind students of the Essential Question: How has Japan balanced its ancient traditions with its modern economy? Have students work in small groups to identify at least three responses, based on the information in Comparing Traditional and Modern Japan.

CRITICAL THINKING ANSWERS

1. **Compare and Contrast** Possible answer: The kimono and modern dress are both tasteful and elegant; however, unlike the dress, the kimono does not allow the wearer much freedom of movement.
2. **Analyze Causes** Possible answer: because film and music have universal appeal

INTERDISCIPLINARY ACTIVITIES

CONNECT to Language Arts

Write a Haiku About Traditional or Modern Japan

Have students choose an aspect of either traditional or modern Japanese culture and write a haiku describing it. You might wish to encourage students to try writing several haikus. When students have finished their haikus, invite them to share their work in a poetry reading session.

CONNECT to Art

Create an Animé Storyboard

Have students sketch out a series of images that show a short scene from an animé movie. Remind students that each image should be in sequence and show a different action. When students have finished their sketches, they should add captions for each image. Invite students to present or display their storyboards.

1 Plan & Prepare

Objectives

- Describe the modernization of Japan's government
- Explain how Japan created an economic powerhouse
- Describe the products and services the Japanese economy produces

Read for the Essential Question

Help students read for a purpose by reminding them of the Essential Question: "How has Japan balanced its ancient traditions with its modern economy?" Point out that they can use items in the text, such as Key Questions, special features, images, and maps to formulate answers to the question.

Vocabulary

Best Practices Toolkit

Use the ***Best Practices Toolkit*** to model strategies for vocabulary notetaking. Vary strategies throughout the year. Choose from: *Knowledge Rating, Predicting ABCs, Definition Mapping, Word Sort, Word Wheel, Frayer Model (Word Squares), Magnet Words,* and *Student VOC.*

Vocabulary Strategies, TT9–TT16

Reading Strategy

Best Practices Toolkit

Display the T-Chart/Two-Column Chart transparency.

- Remind students that main ideas are the most important points in a text.
- Model finding and entering main ideas on a chart using the paragraph under "Modernizing Japan's Government." Have students suggest entries.
- Display the completed chart for review.

Chart/Two-Column Chart , TT20

SECTION 3 Reading for Understanding

Key Ideas

BEFORE, YOU LEARNED
Japanese culture blends traditional and modern elements.

NOW YOU WILL LEARN
The Japanese have drawn on their culture to create a modern government and become an economic giant.

Vocabulary

TERMS & NAMES

militarism a government's aggressive use of its armed forces

Diet (DY•iht) Japan's law-making body, which consists of a House of Representatives and a House of Councillors

zaibatsu (ZY•baht•SOO) the large family-controlled banking and industrial groups that owned many companies in Japan before World War II

Junichiro Koizumi (JU•NEE•chee•roh KOH•EE•zoo•mee) Japan's prime minister from 2001 to 2006; Shinzo Abe became prime minister in 2006.

Visual Vocabulary The bombing of Pearl Harbor demonstrated Japanese militarism.

Reading Strategy

Re-create the chart shown at right. As you read and respond to the **KEY QUESTIONS**, use the chart to jot down main ideas about Japan's government and economy.

See Skillbuilder Handbook, page R4

FIND MAIN IDEAS

JAPAN'S GOVERNMENT	JAPAN'S ECONOMY
1.	1.
2.	2.
3.	3.

GRAPHIC ORGANIZERS
Go to **Interactive Review** @ClassZone.com

PRETEACHING VOCABULARY

English Learners

Pronounce and Preview

Pronounce each term for students. Review the meanings of words in the definitions, such as *aggressive*, *industrial groups*, and *prime minister*.

- To modify vocabulary learning, have students complete worksheets as they read the section.

Unit Resource Book: East Asia & Southeast Asia
- Vocabulary Practice, p. 157

Inclusion

Read Enlarged Vocabulary Definitions

To help students who are visually impaired, create enlarged photocopies of the Reading for Understanding page. Have volunteers take turns reading each vocabulary word. Ask the rest of the class to do a pronunciation check. Then have them read the word's definition aloud. Suggest that students keep the photocopies in a notebook for reference.

Himeji Castle, also known as White Heron Castle, in Himeji City

SECTION 3

GOVERNMENT & ECONOMICS

Crane eating fish beside Kamo River in Kyoto

Economic Giant

Connecting to Your World

You have probably seen stories on the news about people attempting to rebuild their homes in the wake of a disaster. In 2005, for example, the people of New Orleans faced this challenge after being hit by Hurricane Katrina. After World War II, Japan also faced horrible devastation. U.S. atomic bombs had reduced two cities, Hiroshima and Nagasaki, to rubble. Other cities had been hit hard by firebombing raids. But out of this destruction, Japan developed a modern government and a thriving economy.

Modernizing Japan's Government

KEY QUESTION How did the new constitution of Japan make the nation more democratic?

For centuries, many Japanese had regarded their emperor as a divine figure. However, after Japan's defeat in World War II, the emperor denied any claim to divine origins. He also gave up his authority to rule and became a symbolic head of state. This change was just one of many in the process of making the Japanese government more democratic.

Legislature Japan's legislature meets in the National Diet Building in Tokyo. The emperor and empress of Japan are pictured in the oval portrait.

2 Focus & Motivate

3-Minute Warm-Up

Write on the board or display the transparency:

- What Japanese products are you familiar with? *(cars, audio and video equipment, computers)*

Unit Transparency Book
- 3-Minute Warm-Ups, TT20

Connecting to Your World

More About . . .

Japan After World War II

From 1945 to 1952, Japan was occupied and under the control of Allied troops—most from the United States under the command of General Douglas MacArthur. War crimes trials were held, and about 900 Japanese were executed, including former prime minister Hideki Tojo.

3 Teach

Modernizing Japan's Government

Think, Pair, Share

- What did Japan's emperor do after Japan lost World War II? *(He denied claim to divine origins and became a symbolic head of state.)*
- **Find Main Ideas** What rights are guaranteed under Japan's constitution? *(freedom of speech, religion, press)*

SECTION 3 PROGRAM RESOURCES

Unit Resource Book: East Asia and Southeast Asia

ON LEVEL
Unit Resource Book
- Reading Study Guide, p. 137
- Section Quiz, p. 175

STRUGGLING READERS
Unit Resource Book
- Reading Study Guide with Additional Support, p. 143
- Vocabulary Practice, p. 157
- Skillbuilder Practice, p. 158
- Section Quiz, p. 175
- Reteaching Activity, p. 178

eEdition with Audio DVD-ROM

ENGLISH LEARNERS
Pupil Edition in Spanish
eEdition with Audio DVD-ROM
Unit Resource Book
- RSG (Spanish), p. 149
- RSG with Additional Support (Spanish), p. 155

Multi-Language Glossary
Test Generator
- Section Quiz in Spanish

INCLUSION
Unit Resource Book
- RSG with Additional Support, p. 143
- Section Quiz, p. 175
- Reteaching Activity, p. 178

GIFTED & TALENTED
Unit Resource Book
- Interdisciplinary Projects, p. 165
- Section Quiz, p. 175

PRE-AP
Unit Resource Book
- Economics in the World, p. 168
- Section Quiz, p. 175

TECHNOLOGY

Unit Transparency Book
- 3-Minute Warm-Ups, TT20
- Essential Question, TT21
- Critical Thinking, TT24

Daily Test Practice Transparencies
- Section 3, TT92

Power Presentations
ClassZone.com
World Cultures and Geography Video on DVD

More About . . .

Japan and Militarism

Although Japan is prohibited from having a military that can wage war, it does have a military force known collectively as the Self Defense Forces (SDF). This force usually operates within Japan's borders.

In addition, the 1960 Treaty of Mutual Cooperation and Security between the United States and Japan allows the United States to station troops in Japan. In return, the United States is obligated to help defend Japan in time of war. Since 1992, the SDF has been allowed to act outside of Japan's borders as part of the United Nations peacekeeping operations.

COMPARING

U.S. & Japanese Governments

Point out that Japan's legislative body, the Diet, consists of two houses. Ask students to name the two houses of the U.S. Congress *(House of Representatives and Senate)* and discuss what they know about the size and function of each one.

CRITICAL THINKING ANSWER

Compare and Contrast Both are divided into three branches—executive, legislative, and judicial—but in the United States, the president is elected by the people, not chosen by the legislature.

A New Constitution In 1947, Japan adopted a new constitution, which set up a parliamentary government. It also guaranteed the Japanese people many rights, including freedom of speech and religion and freedom of the press.

The new constitution prohibited Japan from having a military that can wage war. This policy was an intentional rejection of the militarism that led Japan into World War II. **Militarism** refers to a government's aggressive use of its armed forces. In 2004, however, this policy began to change when Japanese troops were sent to Iraq for humanitarian and construction work. It was the first time Japanese soldiers had entered an active foreign war zone since World War II.

Government Structure Japan's government is divided into three branches—executive, legislative, and judicial. The legislature consists of a law-making body called the **Diet** (DY•iht). Members of the Diet are elected by the Japanese people. The prime minister heads the executive branch and is chosen by members of the Diet. The judicial branch of Japan is headed by the Supreme Court.

ANALYZE EFFECTS Tell how Japan's new constitution made the nation more democratic. It set up a parliamentary government. It also guaranteed the Japanese people many rights, including freedom of speech, religion, and the press.

COMPARING U.S. & Japanese Governments

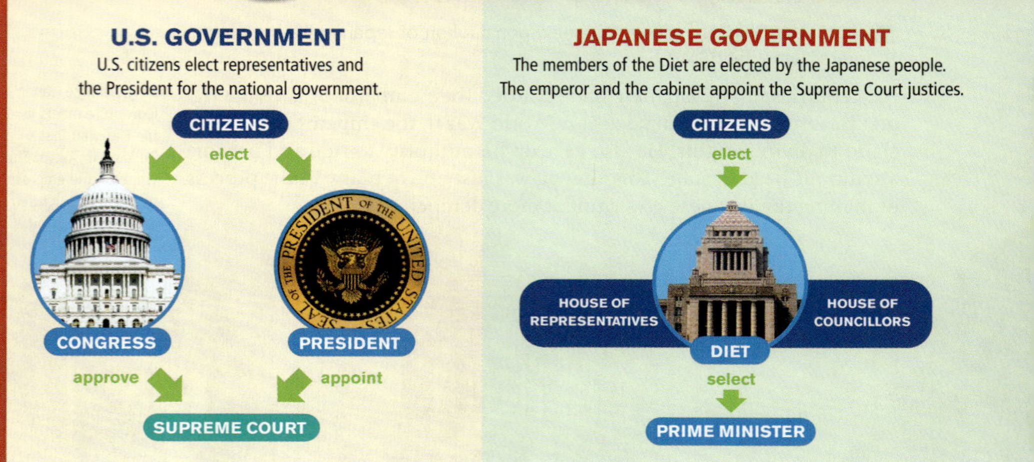

CRITICAL THINKING

Compare and Contrast How does the structure of Japan's government compare with that of the U.S. government?

DIFFERENTIATING INSTRUCTION

Struggling Readers

Use Visuals to Compare

Help students understand the structure of Japan's government by having them complete a chart like the one below.

Japan's Government	
Branch	Responsibilities

Gifted & Talented

Analyze Militarism in Japan

Some people feel that Japan should have a military force capable of waging war; others do not. Invite students to choose one side of this debate and write a persuasive essay supporting their position. Remind students that in order for their arguments to be effective, they should use facts and sound reasoning.

Creating an Economic Powerhouse

KEY QUESTION How did Japan create an economic powerhouse?

Like the government, Japan's economy was in chaos after World War II. Bombs had destroyed factories, and many people were unemployed. Despite all these problems, Japan managed to become an economic giant that is second only to the United States.

Japanese Companies and Workers Before the war, Japan's economy was run by large business groups called *zaibatsu* (ZY•baht•SOO). ***Zaibatsu*** were large, family-controlled banking and industrial groups. The Allies dissolved the *zaibatsu* after the war. However, groups composed of various businesses soon formed that were similar to the old *zaibatsu*. These groups pooled their resources to make investments in developing industries. Their cooperation greatly helped speed up Japan's economic growth.

CONNECT Economics & Culture

Recipe for Economic Success Much of Japan's success has to do with the country's emphasis on education and its willingness to reward hard work. The ability of the Japanese to adapt technologies and create new ones has also been an important ingredient.

Dedication Workers often put in 12-hour days. Many place the needs of the company before their own.

Adaptation The Japanese have always adapted to the influences of other cultures. Today, they often improve on imported technologies.

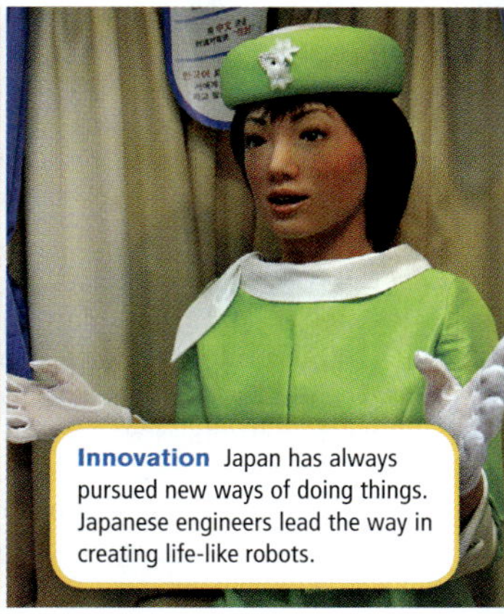

Innovation Japan has always pursued new ways of doing things. Japanese engineers lead the way in creating life-like robots.

CRITICAL THINKING

Evaluate How has the Japanese worker contributed to Japan's economic success?

Teach

Creating an Economic Powerhouse

Talk About It

- Why was Japan's economy in chaos after World War II? *(Bombs had destroyed factories, and many people were unemployed.)*
- Why did Japan's economy skyrocket between the 1950s and 1970s? *(greater investment cooperation, increased trade, and the switch to producing goods such as electronics, which required fewer raw materials)*
- Why did Japan's economy decline in the 1990s? *(because of the high price of Japanese goods and competition from low-cost businesses in developing nations)*
- **Find Main Ideas** What social changes were caused by Japan's economic growth? *(the movement of rural people to cities, the decline of farming, an increase in incomes, and decline of pressure to conform to traditional ways)*

Unit Resource Book: East Asia & Southeast Asia
- Skillbuilder Practice: Identifying Problems and Solutions, p. 158
- Interdisciplinary Projects: Currency Conversion in Japan, p. 165

CONNECT Economics & Culture

Recipe for Economic Success

Have students study and discuss the images. Then ask students what words they would use to describe each one. Do they think dedication, adaptation, and innovation are responsible for America's economic success as well?

CRITICAL THINKING ANSWER

Evaluate Japanese workers produce high-quality goods. Many place the group before the individual.

INTERDISCIPLINARY ACTIVITIES

CONNECT to Science

Design a New Electronics Product

Invite students to display their capacity for innovation by designing a new electronics product. Have students start by surveying products they are familiar with and listing what they like and dislike about each one. They can use their lists as a springboard for their innovative product. Have students create a drawing of their product, along with a written description of why it is needed and how it would work.

CONNECT to Language Arts

Create a Job Recruiting Poster

Have students design a job recruiting poster that would encourage qualified workers to apply for work at a Japanese company. Remind students that their poster should emphasize traits valued by Japanese companies.

Comparing Regions

Business Groups in Asia

In South Korea, modern business groups called *chaebols* have played a major role in its economy. Modeled on the *zaibatsu* of Japan, these Korean business groups are also family-owned. Hyundai and Samsung are among the most important *chaebols* in Korea today.

One reason for the *chaebols*' success is their support by the South Korean government, which helps arrange large foreign loans for the companies. This government support has also caused problems because the money loaned to the companies has not always been used wisely, and the companies have sometimes taken on more debt than they can handle.

ANALYZING Primary Sources

Junichiro Koizumi was born into a political family—both his father and his grandfather served in the Japanese Diet. Koizumi used the media very effectively to promote his popularity and policies. As a result, he was considered much more accessible and personable than traditional Japanese politicians.

DOCUMENT-BASED QUESTION ANSWER

Possible answer: embrace difficulties, free themselves from past limitations, and take on the spirit of "no fear, no hesitation, and no constraint"

Economic Miracle and Setbacks From the early 1950s to the early 1970s, Japan's economy skyrocketed. One reason for this economic boom was trade. Japan imported raw materials at low costs and exported manufactured goods. The nation also switched from producing products such as textiles to products such as electronics, which require fewer raw materials.

The success of the economy brought many social changes to Japan. Many people moved from rural areas to cities to work in factories and service industries. As a result, fewer people earned a living by farming. Some families saw their income doubling or tripling. In cities, the pressure to conform to traditional ways lessened. Because of this, young people often felt freer to develop their individuality.

By the 1990s, though, Japan's economy started to suffer some setbacks. The high price of Japanese products made them difficult to sell overseas. Also, Japanese industry faced stiff competition from low-cost businesses in developing nations. Soon, unemployment rose, and consumer spending declined. Many economists refer to the 1990s as Japan's "lost decade." In 2001, Prime Minister **Junichiro Koizumi** (JU•NEE•chee•roh KOH•EE•zoo•mee) made the economy his top priority. By 2006, many companies were again reporting record profits.

FIND MAIN IDEAS Explain how Japan developed its economy. It imported raw materials for a low cost and exported highly priced manufactured goods.

ANALYZING Primary Sources

Junichiro Koizumi (born 1942) served as prime minister of Japan from 2001 to 2006. At one point, his popularity reached almost rock-star status. In this speech, he tried to inspire confidence in Japan's economic future.

> The top priority that I must address is to rebuild our economy and reconstruct our society into ones full of pride and confidence. . . . We must embrace difficulties ahead . . . and free ourselves of past limitations as we create an economic social structure befitting the 21st century in the spirit of "No fear, no hesitation, and no constraint."
>
> Source: Speech to the Diet, May 7, 2001, Junichiro Koizumi

DOCUMENT-BASED QUESTION

What does Koizumi think the Japanese must do to achieve economic reforms?

DIFFERENTIATING INSTRUCTION

English Learners

Create an Economics Time Line

Help students get a clearer picture of modern Japan's economic development by encouraging them to create a time line. The time line should include entries for the early 1950s to the early 1970s, the mid 1970s to the 1990s, the 1990s to 2001, and 2001 to 2006. Each entry should describe Japan's economy during that time.

Pre-AP

Write a Radio Script

Ask students to write a radio script explaining how Japan rose from economic devastation following World War II to an economic powerhouse. In their scripts, student should stress the ingredients that enabled Japan to succeed, such as hard work, innovation, an educated workforce, and the pooling of resources by business groups. Once students have practiced and revised their scripts, have students record and play them for the class.

What the Economy Produces

KEY QUESTION What are the major manufacturing, agricultural, and service industries of Japan?

As you have learned, Japan's economic strategy called for producing many high-quality products that could be made with materials imported for a low cost. As a result, the Japanese focused on making electronic equipment and automobiles.

Manufacturing During the 1950s, the Japanese got a foothold in the electronics industry by producing transistor radios. These small radios became very popular with young people, especially in the United States and Europe. Soon, the Japanese began to manufacture televisions, tape recorders, VCRs, and computers. Innovations in the manufacturing process spurred the development of electronics.

Innovations also helped the growth of Japan's auto industry. For example, as early as 1970, the Japanese used digital technology with assembly lines. The rise of the auto industry in Japan is incredible. In 1962, Japan was sixth in the world in automobile production. By 1967, it had overtaken West Germany to rank second. And then in 1980, Japan surpassed the United States to become the largest producer of automobiles in the world.

Agriculture As you may recall, Japan has little farmland. However, Japanese farmers have been able to make this land as productive as possible. First of all, it has helped that Japan's climate is ideal for growing certain crops, such as rice. In addition, farmers extensively use irrigation and fertilizers.

The primary crop by far is rice. Other products include eggs, potatoes, fruits, and vegetables. Japan, though, does need to import some agricultural products, such as animal feed, wheat, and beans.

Major Industries of Japan

CONNECT Geography & Economics

1. **Region** On which island are most of the industries located?
2. **Human-Environment Interaction** Why do you think shipbuilding is a major industry in Japan?

Teach

What the Economy Produces

Reader, Recorder, Reporter

- How did Japan gain a foothold in the electronics industry? *(In the 1950s, they made small transistor radios, which were very popular with young people.)*
- How has Japan been able to make its limited farmland productive? *(taking advantage of its favorable climate and use of irrigation and fertilizer)*
- **Find Main Ideas** What was the key to Japan's manufacturing success? *(Possible answer: innovations, such as using digital technology with assembly lines)*

Unit Resource Book: East Asia & Southeast Asia
- Economics in the World: Japanese World Trade, p. 168

Major Industries of Japan

CONNECT Geography & Economics

Have students use the map and key to identify the major industries in each city shown. Which industries are located in many of the cities on the map? *(chemicals, high-tech, machinery, vehicles)* Which of the four main islands do not have major industries? *(Kyushu and Shikoku)*

ANSWERS

1. **Region** Honshu
2. **Human-Environment Interaction** because it is an island nation

DIFFERENTIATING INSTRUCTION: TIERED ACTIVITIES

OBJECTIVE Learn about Japan's economy.

Basic

Have students use the information on this page to create a three-column chart showing what the Japanese economy produces. Students should label one column Manufactured Goods, the second one Agricultural Goods, and the third one Service Industries.

On Level

Have student groups work together to find out which industries (manufacturing, agricultural, or service) produce the most revenue for Japan. Ask students to identify the top five industries in a list or chart.

Challenge

Have groups of students research to learn about Japan's major imports and exports and trading partners. Ask the groups to organize their findings in charts and pie graphs and share what they have learned with the rest of the class.

CONNECT to the Essential Question

How has Japan balanced its ancient traditions with its modern economy?

Ask students what they have learned so far that can help them answer this question. Students might mention the following:

- After World War II, the Japanese constitution set up a new democratic government.
- Japan began to manufacture high-quality goods, such as electronics products and automobiles, and its economy skyrocketed.

4 Assess & Reteach

Assess Have students complete the Section Assessment.

Unit Resource Book: East Asia & Southeast Asia
- Section Quiz, p. 175

Interactive Review @ClassZone.com
- Section Quiz

Power Presentations

Test Generator

Reteach Have students list events related to Japan's government and economy on small slips of paper. Place the slips of paper in a bag. Then invite students to draw the papers from the bag and use them to create a time line on the board.

Unit Resource Book: East Asia & Southeast Asia
- Reteaching Activity, p. 178

Tokyo Disneyland Tokyo Disneyland was the first Disney park to be built outside of the United States. It is one of the most visited theme parks in the world, attracting tourists from all over the globe.

Service Industries As manufacturing industries thrived in Japan, so too did the industries that served them. Banks, for example, have a lot of power in Japan. In fact, the Bank of Japan strongly influences the government's economic policies. Also, some of Japan's banks have a major impact on international banking.

In addition, Japan has many service industries that deal with entertainment, such as nightclubs, movie theaters, and restaurants. Although Western fast-food restaurants have become popular, traditional Japanese restaurants continue to attract many customers. Japan has a great variety of traditional eating places, including sushi restaurants. It also has a healthy tourist industry. Tourism was an important part of Junichiro Koizumi's policies. His efforts paid off. In 2005, Japan hosted more than 6 million visitors from overseas.

SUMMARIZE Identify Japan's major manufacturing, agricultural, and service industries. manufacturing: automobiles and electronics; agriculture: rice; service: banks, nightclubs, movie theaters, restaurants, tourism

Section 3 Assessment

ONLINE QUIZ For test practice, go to **Interactive Review @ClassZone.com**

TERMS & NAMES

1. **Explain the importance of**
   - militarism
   - Diet
   - *zaibatsu*
   - Junichiro Koizumi

USE YOUR READING NOTES

2. **Find Main Ideas** Use your completed chart to answer the following question: What economic obstacles did Japan have to overcome after World War II?

JAPAN'S GOVERNMENT	JAPAN'S ECONOMY
1.	1.
2.	2.
3.	3.

KEY IDEAS

3. What does Japan's constitution prohibit?
4. How did the *zaibatsu* influence Japan's businesses after World War II?
5. What are Japan's major manufacturing industries?

CRITICAL THINKING

6. **Form and Support Opinions** Do you think Japan's constitution should be amended to allow the Japanese to wage war? Explain why or why not.
7. **Analyze Effects** How do you think families are affected by the long workdays put in by many Japanese employees?
8. **CONNECT to Today** What impact do Japanese products have on the everyday lives of Americans?
9. **ECONOMICS** **Compare Cars** Research a car produced by a Japanese company. Then research a comparable car made by a U.S. company. Compare the two cars and write a report including information on style, price, features, and consumer ratings.

SECTION 3 ASSESSMENT ANSWERS

Terms & Names

1. militarism: aggressive use of armed forces; Diet: Japan's law-making body; *zaibatsu*: large family-controlled banking and industrial groups; Junichiro Koizumi: Japan's former prime minister

Use Your Reading Notes

2. The country was economically devastated.

JAPAN'S GOVERNMENT	JAPAN'S ECONOMY
1. new constitution after World War II	1. *zaibatsu* help economic growth
2. prime minister heads executive branch	2. brings social changes
3. Diet comprises legislative branch	3. focuses on electronics and automobiles

Key Ideas

3. It prohibits Japan from having a military that can wage war.
4. Groups similar to the old *zaibatsu* pooled their resources to invest in developing industries.
5. automobiles and electronics

Critical Thinking

6. Possible answers: No; after aggressively pursuing conflict in World War II, Japan has lost its right to wage war. Yes; any free country should have the option to wage war.
7. Possible answer: Families are probably neglected by the workers.
8. Possible answer: Americans often use Japanese cars and electronics and eat Japanese foods.
9. Use the rubric to score reports.

Report Rubric

	Content	Errors
4	thoroughly compares two similar cars	few or none
3	makes a good comparison	some
2	leaves out some information	many
1	leaves out much information	many major

CHAPTER SUMMARY

Key Idea 1
The mountainous islands that make up Japan provide the country's people with scant living space, a varied climate, and limited resources.

Key Idea 2
Over the centuries, Japan has adapted ideas and technologies from other lands and created a unique culture.

Key Idea 3
The Japanese have drawn on their culture to create a modern government and become an economic giant.

For **Review and Study Notes**, go to **Interactive Review** @ClassZone.com

NAME GAME

Use the Terms & Names list to complete each sentence on paper or online.

1. I am Japan's highest mountain. Mount Fuji
2. I am the largest lowland of Japan. ________
3. I am a powerful samurai who became a warrior-chieftain. ________
4. I am the shogun who unified Japan and moved the capital to Edo. ________
5. I am Japan's original religion. ________
6. I am the aggressive use of a government's armed forces. ________
7. I am Japan's law-making body, which consists of two houses. ________
8. I am a large family-controlled business group that owned many companies in Japan before World War II. ________
9. I am Japan's largest and most populous island. ________
10. I am a period in Japanese history in which arts and writing flourished. ________

daimyo
Diet
Heian
Honshu
Kanto Plain
Junichiro Koizumi
militarism
Mount Fuji
Shinto
shogun
Tokugawa Ieyasu
zaibatsu

Activities

GeoGame

Use this online map to show what you know about Japan's location, geographic features, and important places. Drag and drop each place name to its location on the map.

Crossword Puzzle

Complete an online crossword puzzle to test your knowledge of Japan.

ACROSS
1. professional soldier of Japan

ONLINE AND ON-SCREEN ACTIVITIES

GeoGame

Use the interactive map on **Power Presentations** DVD-ROM or at **ClassZone.com** to help students review Japan's geographic features.

- Demonstrate how to drag and drop the map labels listed onto the map. Click on the Submit button to view results.
- Advance to the scorecard to show how students will be scored in their first and second attempts.

Crossword Puzzle

Students can review important information about Japan by completing this crossword puzzle online at **ClassZone.com.**

- As an extension of this activity, have students use the *Activity Maker* at **ClassZone.com** to make their own crossword puzzle for one section of the chapter.
- Then have students trade puzzles and try solving them.

Chapter Summary

As a follow-up to the chapter, encourage students to pair up and go over the **Review and Study Notes** at **ClassZone.com.**

Alternatively, student pairs can review their Reading Study Guide notes, making sure that they can add important details to each section's Key Ideas.

Name Game

Teaching Tips

ON PAPER Have students work in pairs to complete this activity. As an extension, they can write questions for the terms that are not covered in the items shown here.

ONLINE Make sure that students understand how to drag and drop terms to fill in blanks. Ask them to print out their results and compare them with those of a classmate.

ANSWERS

1. Mount Fuji
2. Kanto Plain
3. daimyo
4. Tokugawa Ieyasu
5. Shinto
6. militarism
7. Diet
8. *zaibatsu*
9. Honshu
10. Heian

Make It Fun

Teacher-Tested Activities

Greta Frensley, Knox Doss Middle School, Gallatin, Tennessee

When my students play the GeoGame, I like to add the following twist:

- First, I divide the class into two teams.
- Then I display a large map of the country or region we have been studying.
- Finally, I name a place or feature, and the teams take turns locating it on the map. As an added challenge, I ask the teams to provide one fact about the place.
- The team that correctly locates the most places and comes up with appropriate facts wins the GeoGame.

Vocabulary ANSWERS

1. Japan's highest mountain peak
2. largest lowland of Japan
3. era in Japanese history from 794 to 1185
4. professional soldiers of Japan
5. the leader of the military government of Japan beginning in 1192
6. powerful samurai who became warrior-chieftains
7. Japan's original religion
8. a government's aggressive use of its armed forces
9. Japan's law-making body
10. Japan's prime minister between 2001 and 2006
11. three important political and military positions for hundreds of years in Japan
12. the four major islands of Japan

Key Ideas ANSWERS

13. the tops of a huge, mostly underwater mountain range
14. a zone that extends along the rim of the Pacific Ocean and has many volcanoes and earthquakes
15. The mountains throughout the islands tend to have cooler temperatures.
16. from the sea; from its large fishing industry
17. Yamato clan
18. They pursued a policy of isolation from European powers.
19. Shinto and Buddhism
20. Tanka is a 31-syllable poem divided among five lines that was the dominant verse form from the 700s to the 1500s.
21. The emperor is a symbolic head of state with no authority to rule.
22. family-controlled banking and industrial groups that ran Japan's economy before World War II
23. Many people moved from rural areas to the cities; some families' incomes doubled or tripled; traditional ways lapsed in urban areas, and young people often felt freer to develop their individuality.
24. Because Japan did not have many raw materials, they produced high-quality products that could be made with materials imported for a low cost.

CHAPTER 27 Assessment

VOCABULARY

Explain the significance of each of the following.

1. Mount Fuji
2. Kanto Plain
3. Heian period
4. samurai
5. shogun
6. daimyo
7. Shinto
8. militarism
9. Diet
10. Junichiro Koizumi

Explain how the terms and names in each group are related.

11. samurai, shogun, and daimyo
12. Honshu, Hokkaido, Kyushu, and Shikoku

KEY IDEAS

1 Land of the Rising Sun

13. What geographic feature forms the islands of Japan?
14. What is the Ring of Fire?
15. How does the climate of Japan's mountains differ from other regions of Japan?
16. Where does Japan get much of its food?

2 Samurai to Animé

17. To which clan have Japanese emperors belonged throughout history?
18. How did the Tokugawa shogunate affect Japan's foreign relations?
19. What are the two major religions of Japan?
20. What is tanka poetry?

3 Economic Giant

21. What is the role of the Japanese emperor today?
22. What were the *zaibatsu*?
23. What social changes did Japan's booming economy bring to the country between the 1950s and the 1970s?
24. Why did Japan focus on producing electronic equipment and automobiles?

CRITICAL THINKING

25. **Compare and Contrast** Use a Venn diagram to compare Japan in the early 1900s before World War II with the country after the war.

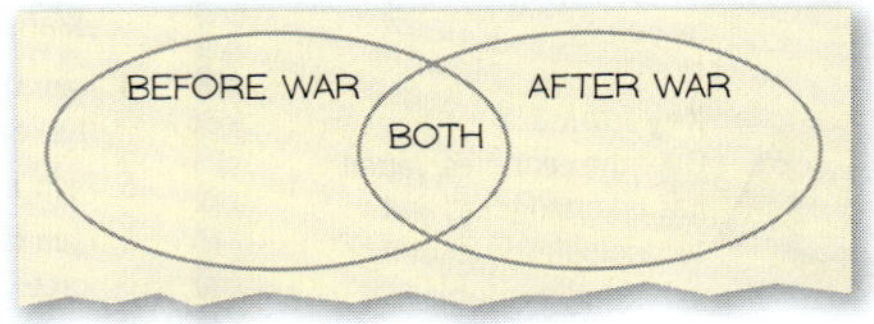

26. **Evaluate** How is the Japanese love of nature reflected in the country's culture?
27. **Form Opinions** Do you think the Tokugawa shogunate did more to help or to hurt Japanese society? Explain your answer.
28. **Make Inferences** Why do you think the Allies dissolved the *zaibatsu* after World War II?
29. **Connect Geography & Economics** How have Japan's natural resources affected its economy?
30. **Five Themes: Human-Environment Interaction** How has Japan's geography affected the way the Japanese live in cities?

Answer the ESSENTIAL QUESTION

How has Japan balanced its ancient traditions with its modern economy?

Written Response Write a two- or three-paragraph response to the Essential Question. Use the rubric below to guide your thinking.

Response Rubric

A strong response will:

- discuss the importance of Japan's cultural traditions
- explain how Japan has built up its strong economy
- emphasize Japan's ability to adapt to new ideas and changing times

ESSENTIAL QUESTION REVIEW

Before students begin their essays, have them review the key chapter elements that will help them answer the Essential Question. Direct them to significant features in the chapter such as

- Comparing Samurai & Knights, Section 2
- Connect to Culture: Origami, Section 2
- Comparing Traditional and Modern Japan, Section 2
- Comparing U.S and Japanese Governments, Section 3
- Connect Economics & Culture: Recipe for Economic Success, Section 3

Written Response Rubric

	Coverage of Key Ideas	Explanation	Errors in Mechanics
4	all	clear and concise	few or none
3	most	good	some
2	some	fair	many
1	one or none	poor	many major

STANDARDS-BASED ASSESSMENT

Test Practice
- Online Test Practice @ClassZone.com
- Test-Taking Strategies and Practice at the front of this book

THEMATIC MAP

Use the earthquake map to answer questions 1 and 2 on your paper.

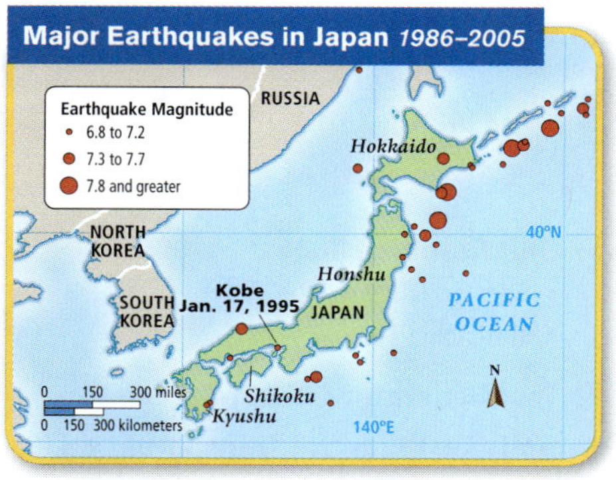

1. What was the magnitude of the 1995 earthquake in Kobe?

A. less than 6.8
B. 6.8 to 7.2
C. 7.3 to 7.7
D. 7.8 and greater

2. Near which island have most of the greatest magnitude earthquakes occurred?

A. Hokkaido
B. Honshu
C. Kyushu
D. Shikoku

PIE GRAPH

Use the pie graph on Japan's exports to answer questions 3 and 4 on your paper.

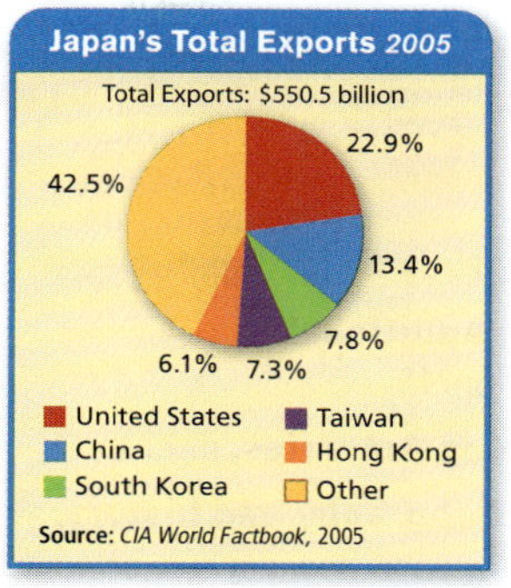

3. Which country among those listed below is Japan's biggest export partner?

A. South Korea
B. China
C. United States
D. Taiwan

4. What percentage of Japan's total exports were sent to China?

A. 6.1 percent
B. 7.3 percent
C. 13.4 percent
D. 42.5 percent

GeoActivity

1. INTERDISCIPLINARY ACTIVITY–ART

With a small group, design and create a miniature Japanese Zen garden. Use your library and the Internet to research this art form. Find out about patterns used in gravel and what rocks can represent. Present your Japanese Zen garden to your class.

2. WRITING FOR SOCIAL STUDIES

Imagine that you are a tourist in a Japanese city. Write a diary entry describing a typical day. Discuss the crowds, the sights, the food, and the blend of old and new you witness.

3. MENTAL MAPPING

Create an outline map of Japan and label the following:

- Pacific Ocean
- East Sea
- Honshu
- Hokkaido
- Kyushu
- Shikoku
- Japanese Alps
- Kanto Plain

DIFFERENTIATING CHAPTER ASSESSMENT

Unit Resource Book

- **Chapter Test A,** pp. 179–182. *Use for Struggling Readers, Beginning-to-Intermediate-Level English Learners, and for Inclusion of students whose IEP suggests supported test-taking.*
- **Chapter Test B,** pp. 183–186. *Use for On-Level Learners and ESOL students with Advanced-Level English.*
- **Chapter Test C,** pp. 187–190. *Use for Gifted and Talented students or students assigned Pre-AP work.*

Test Generator

- Chapter Tests A, B, and C, in English and Spanish

McDougal Littell Assessment System

TEST **Test Generator** Use three levels of tests in English and Spanish tied to state standards. Plus, choose questions to customize tests for all populations.

SCORE Scan and Score plain paper student tests in minutes.

REPORT Create instant progress reports, sortable by student, by standard, or other criteria.

RETEACH Respond immediately to students' needs by using a wealth of reteaching resources.

Critical Thinking ANSWERS

25. Possible answers: Before War—militaristic, *zaibatsu*, rural population; After War—more democratic government, *zaibatsu* dissolve, urban population, industries; Both—Shinto and Buddhism, traditional culture

26. Possible answers: in Shinto, in tanka and haiku, in the art of flower arranging

27. Possible answer: By isolating Japan, the Tokugawa shogun severely hurt foreign relations. This antagonism toward foreigners, in part, led Japan into World War II.

28. Possible answer: The *zaibatsu* had a lot of industrial power in pre-World War II Japan. Because of this, they may have been largely responsible for preparing Japan for war.

29. Possible answer: A lack of natural resources has led Japan to import low-cost materials and be innovative in using them to produce high-quality goods.

30. Possible answer: Cities in Japan are very crowded because the country does not have much terrain for cities to develop.

Standards-Based Assessment ANSWERS

1. B
2. A
3. C
4. C

GeoActivity

1. INTERDISCIPLINARY ACTIVITY RUBRIC

The Zen garden should

- be made of gravel and rocks
- contain traditional patterns
- reflect the serenity of a large garden

2. WRITING FOR SOCIAL STUDIES RUBRIC

The diary entry should

- describe the city's crowded conditions
- point out the blend of traditional and modern features
- provide details about specific sites and experiences

3. MENTAL MAPPING RUBRIC

The outline map should

- represent the approximate shape of Japan
- include all geographic features listed in their correct location

CHAPTER 28

Southeast Asia
PLANNING GUIDE

Complete print resources all on one USB drive allow you to customize lesson plans for all students.

Section Titles & Key Ideas	Unit Resource Book East Asia and Southeast Asia	Assessment Resources
1 Physical Geography of Southeast Asia **KEY IDEA** Southeast Asia is a mountainous region with a tropical climate and much volcanic activity on its many islands.	• **Reading Study Guide,** pp. 191–192 • **RSG with Additional Support,** pp. 197–198 • **Outline Map Activities,** p. 219 • **Connect Geography and History,** p. 227 • **Geography Project,** p. 249	**Unit Resource Book** • Section Quiz, p. 229 • Reteaching Activity, p. 232 **Test Generator CD-ROM** **Daily Test Practice Transparency,** TT93
2 The Mainland Countries **KEY IDEA** The nations of mainland Southeast Asia have closely connected histories and diverse cultures.	• **Reading Study Guide,** pp. 193–194 • **RSG with Additional Support,** pp. 199–200 • **Skillbuilder Practice,** p. 216 • **Map and Graph Skills,** p. 217 • **Economics in the World,** p. 226 • **Primary Sources,** p. 228	**Unit Resource Book** • Section Quiz, p. 230 • Reteaching Activity, p. 233 **Test Generator CD-ROM** **Daily Test Practice Transparency,** TT94
3 Indonesia and the Philippines **KEY IDEA** Indonesia and the Philippines both have long histories as centers of trade and rich cultural traditions.	• **Reading Study Guide,** pp. 195–196 • **RSG with Additional Support,** pp. 201–202 • **Vocabulary Practice,** p. 215 • **History Makers,** p. 222 • **Interdisciplinary Projects,** p. 224	**Unit Resource Book** • Section Quiz, p. 231 • Reteaching Activity, p. 234 • Chapter Test (Forms A, B, & C), pp. 235–246 **Test Generator CD-ROM** **Daily Test Practice Transparency,** TT95

McDougal Littell **Assessment System**

TEST | SCORE | REPORT | RETEACH

INTEGRATED TECHNOLOGY

No Time? To focus students on important content in the chapter, use the **"Reading Study Guide with Additional Support"** in ***Resources2Go.***

Teacher Presentation Options

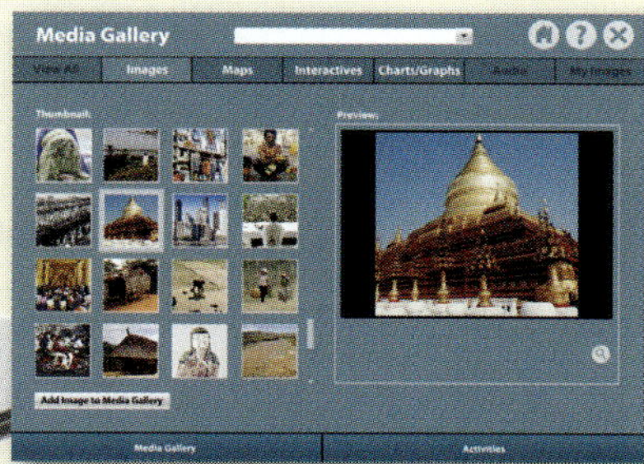

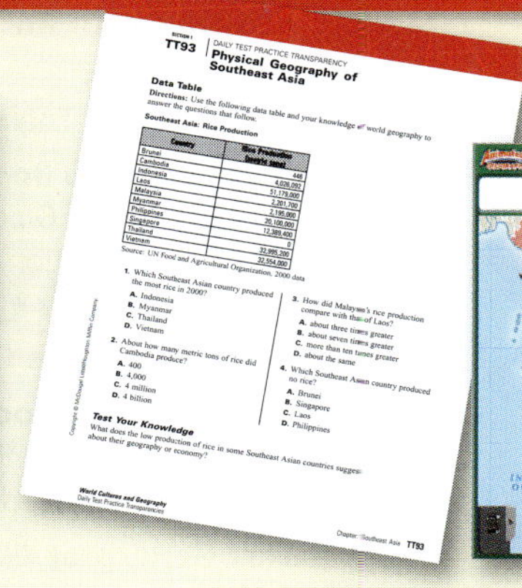

Electronic Books

eEdition DVD-ROM
eEdition Online

Presentation Toolkit

Video/DVD

- World Cultures and Geography

Power Presentations DVD-ROM

- Discussion Starters
- Media Gallery
- Review Game
- Interactive Review
- Animated Features

Unit Transparency Book

- 3-Minute Warm-Ups
- Essential Question
- Critical Thinking
- Maps
- Cultures

Daily Test Practice Transparencies

Animated Geography

- Present-Day Southeast Asia Map

OnlineActivities at ClassZone.com

Animated Geography

- Videos and Photos
- 3-D Animations
- Interactive Maps

Interactive Review

- Section Quizzes
- Review and Study Notes
- Vocabulary Flip Cards
- Graphic Organizers
- Name Game and Crossword Puzzles
- GeoGame

Online Test Practice

Map Center

- Atlas Maps
- Political Maps with Data
- Links to Mapping Web Sites

Activity Center

- Activity Maker—Review Games, Puzzles, Flip Cards, and Graphic Organizers
- Worldster

Research & Writing Center

- Web Sites
- Data Updates
- Primary Sources
- Writing Models

Current Events

Matt Parker
Eastgate Middle School
Kansas City, Missouri

Make It Fun — Teacher-Tested Activities

As you teach this chapter, look for my classroom-proven ideas for:

- making the lesson fun
- focusing students' attention and comprehension
- integrating technology with traditional teaching

Teaching Bell to Bell When you see this timing guide, have students discuss in pairs, in groups of 3–6, or as a class.

- **Think, Pair, Share** (pairs)
- **Reader, Recorder, Reporter** (3–6)
- **Talk About It** (any size)

CHAPTER 28

RESOURCES FOR DIFFERENTIATED INSTRUCTION

Struggling Readers

eEdition with Audio DVD-ROM

Lesson Plans
- Strategies for Inclusion

Unit Resource Book
- Reading Study Guide with Additional Support, pp. 197–202 (B)
- Vocabulary Practice, p. 215
- Skillbuilder Practice, p. 216
- Reteaching Activity, pp. 232–234
- Chapter Test (Form A), pp. 235–238 (A)

Best Practices Toolkit (C)
- Interactive Reading Strategies, pp. 90–145
- Differentiation Activity Bank: Struggling Readers, pp. 146–149

Teacher's Edition Activities
- Outline Southeast Asia's Geography, Sec. 1
- Tiered Activities, Sec. 1–3

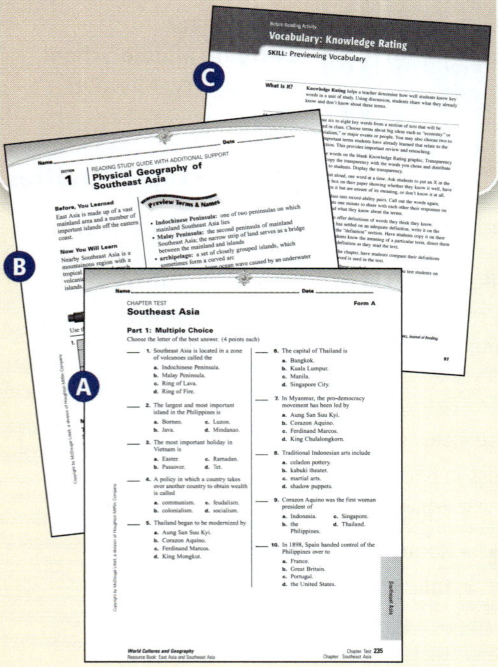

English Learners

Pupil Edition in Spanish

eEdition with Audio DVD-ROM

Lesson Plans
- Strategies for English Learners

Unit Resource Book
- Reading Study Guide (Spanish), pp. 203–208
- RSG with Additional Support (Spanish), pp. 209–214 (A) (B)

Multi-Language Glossary (C)

Best Practices Toolkit
- Differentiation Activity Bank: English Learners, pp. 165–194

Test Generator CD-ROM
- Section Quizzes in Spanish
- Chapter Test in Spanish

Teacher's Edition Activities
- Pronounce and Preview, Sec. 1–3
- Review Unfamiliar Words, Sec. 2
- Tiered Activities, Sec. 1–3

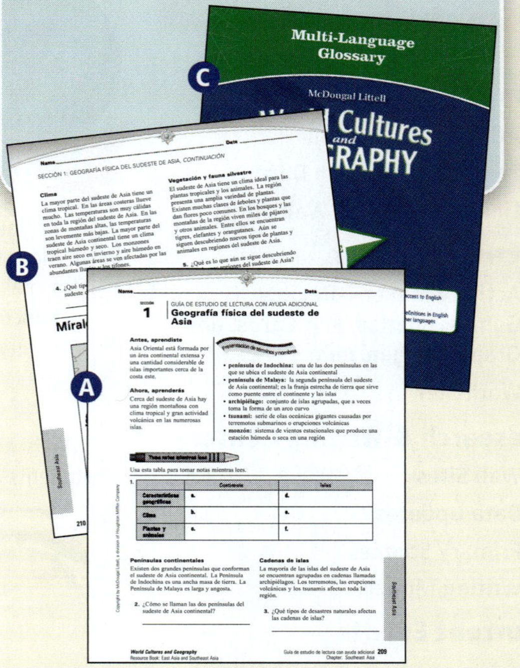

Inclusion

eEdition with Audio DVD-ROM

Lesson Plans
- Strategies for Inclusion

Unit Resource Book
- Reading Study Guide with Additional Support, pp. 197–202 (A) (B)
- Reteaching Activity, pp. 232–234
- Chapter Test (Form A), pp. 235–238

Best Practices Toolkit (C)
- Interactive Reading Strategies, pp. 90–145
- Differentiation Activity Bank: Struggling Readers, pp. 146–149

Teacher's Edition Activities
- Use Visual Vocabulary, Sec. 1
- Fill in the Blanks, Sec. 2
- Use Student VOC Strategy, Sec. 3
- Tiered Activities, Sec. 1–3

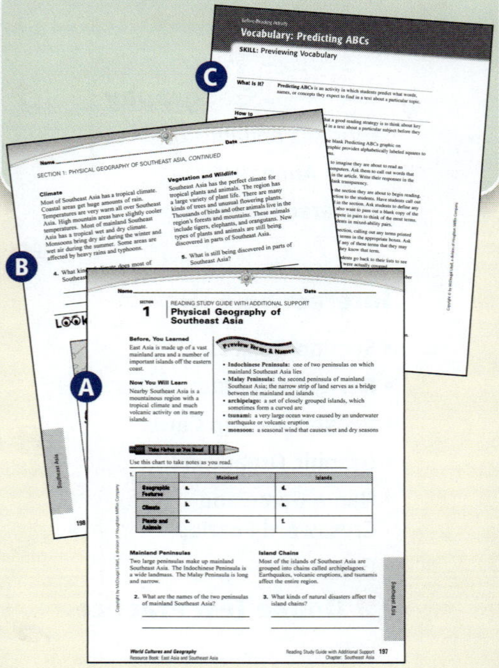

Gifted & Talented Students

Lesson Plans
- Core Lesson Plan

Unit Resource Book
- Map and Graph Skills, p. 217
- History Makers, pp. 221–222 A B
- Interdisciplinary Projects, p. 223
- Connect Geography and History, p. 227 C
- Chapter Test (Form C), pp. 243–246

Best Practices Toolkit
- Student Products Rubrics, pp. 1–14
- Writing Templates, pp. 78–89

Teacher's Edition Activities
- Present a Shadow Puppet Show, Sec. 3
- Tiered Activities, Sec. 1–3

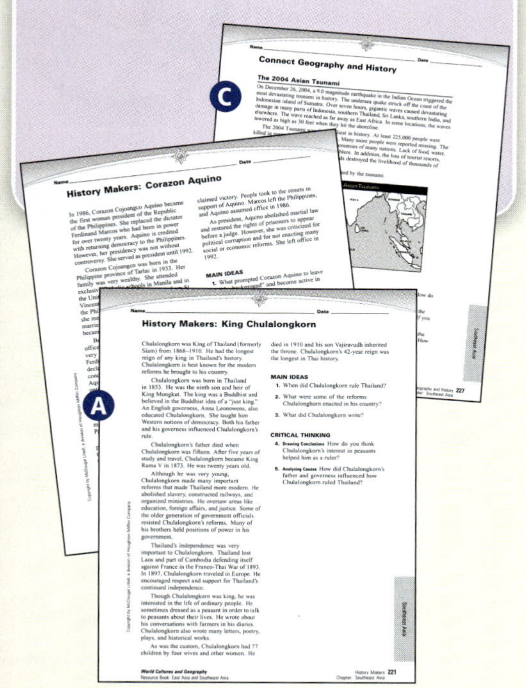

Pre-AP*

Lesson Plans
- Core Lesson Plan

Unit Resource Book
- Interdisciplinary Projects, p. 224
- Economics in the World, p. 226 A
- Primary Sources, p. 228 B
- Geography Project, p. 249 C
- Chapter Test (Form C), pp. 243–246

Best Practices Toolkit
- Writing Templates, pp. 78–89
- Differentiation Activity Bank: Pre-AP Students, pp. 195–200

* Pre-AP is a registered trademark of the College Entrance Examination Board, which was not involved in the production of and does not endorse this product.

Teacher's Edition Activities
- Compare and Contrast, Sec. 2
- Write an Argument, Sec. 3
- Tiered Activities, Sec. 1–3

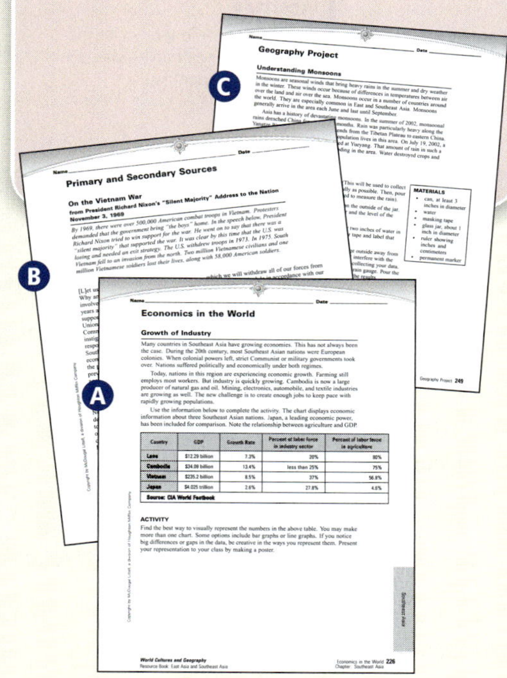

Interdisciplinary Activities

CONNECT to Language Arts

Review a Book About Southeast Asia

CLASS TIME 30 minutes

TASK Prepare a review of a book or movie about Southeast Asia.

SUPPLIES
- list of age-appropriate books or movies that take place in Southeast Asia or have a theme related to the region
- writing materials/computers

ACTIVITY Have students prepare a review of a book or movie about Southeast Asia. In their reviews, students should identify the setting of the story and include a brief summary of the plot. They should also express their opinions and explain why they did or did not like the book or movie. Post the reviews or ask volunteers to read theirs aloud.

Good for Verbal Learners

CONNECT to History

You Be the Map Maker

CLASS TIME 30 minutes

TASK Create one map of Southeast Asia as it appeared during the colonial period of the early 1900s and another map of Southeast Asia as it appears today.

SUPPLIES
- historical and modern atlases or Internet access
- outline maps of Southeast Asia
- colored pencils or markers

ACTIVITY Have students prepare a map showing which empires controlled the countries of Southeast Asia in the early 1900s and another map showing what the region looks like today. Be sure students include a title and key with their maps. Discuss the impact colonialism had—and continues to have—on the region.

Good for Visual and Spatial Learners

Introduce the ESSENTIAL QUESTION

- Explain how the geography of Southeast Asia influences the region's economy.
- Describe the history, government, and cultural traditions of Southeast Asia.
- Identify the major industries in each Southeast Asian country.
- Talk about the key political leaders in Southeast Asia and how they influence government.

Key Ideas

Section 1: Southeast Asia is a mountainous region with a tropical climate and much volcanic activity on its many islands.

Section 2: The nations of mainland Southeast Asia have closely connected histories and diverse cultures.

Section 3: Indonesia and the Philippines both have long histories as centers of trade and rich cultural traditions.

CONNECT to the Essential Question

You may wish to revisit the Essential Question periodically with students as they work through the chapter.

Suggestions for reconnecting are provided at an appropriate point in the *Teacher's Edition* for each section.

Buddhist sculptures near Bayon Temple, Cambodia

CHAPTER 28

Southeast Asia

1 GEOGRAPHY **Physical Geography of Southeast Asia**

2 FOCUS ON **The Mainland Countries**

3 FOCUS ON **Indonesia and the Philippines**

ESSENTIAL QUESTION

How can Southeast Asian nations establish stable governments and economies?

CONNECT Geography & History

Use the map and the time line to answer the following questions.

1. What two island chains are part of Southeast Asia?
2. Which Southeast Asian country was once known as Siam?

Culture
◀ **1300s** Muslim traders bring Islam to Indonesia.
(Mosque in Jakarta)

500

History
500s A people known as the Khmers establish an empire on the mainland peninsula.

History
1688 Siam, or present-day Thailand, begins to limit relations with Europeans.
(King Chulalongkorn of Siam) ▶

TIME LINE DISCUSSION

Tell students that Southeast Asia's history includes periods of conquest, colonization, and war. Its location and geography have affected its development.

- How did trade affect culture in Southeast Asia? *(In the 1300s, traders brought Islam to Indonesia.)*
- What might have caused the events of 1688? *(Possible answers: fear of being conquered or colonized; desire to maintain a traditional culture; the ruler's interest in isolationism)*
- What type of government ruled Vietnam after the Vietnam War? *(Communist)*
- How does the map reflect the event listed for 1975? *(Vietnam is shown on the map as a united country.)*
- What natural disasters have affected the region in the past few decades? *(In 1991, Mount Pinatubo erupted in the Philippines, and in 2004, a tsunami hit Southeast Asia.)*

Animated GEOGRAPHY
Present-Day Southeast Asia

Click here to explore Southeast Asia @ClassZone.com

Government
▲ **1975** Vietnam War ends; in time, north and south unite under Communist rule.

Geography
◄ **1991** Mount Pinatubo in the Philippines erupts.

Geography
2004 Devastating tsunamis hit Southeast Asia.

Today

TEACHER MEDIA FAVORITES

Books for the Teacher

Sexton, Colleen. *Philippines in Pictures.* Twenty-First Century Books, 2006. Contains vivid images of the Philippines.

Stewart, Whitney. *Aung San Suu Kyi: Fearless Voice of Burma.* Lerner Publications, 1997. Biography of the 1991 Nobel Peace Prize winner.

Video / DVD

The History Channel—Tsunami 2004: Waves of Death. (DVD) A&E Home Video, 2005. Documentary blends survivor experiences with scientific explanations.

Vietnam's Unseen War. (DVD) National Geographic Video, 1985. Focuses on photographs from North Vietnamese photographers and related personal stories to provide a unique view of the Vietnam War.

Internet

Go to the *Research and Writing Center* at **ClassZone.com** to link to:

- data on the region
- a variety of chapter-specific, content-reviewed sites
- "History Makers" biographies
- primary sources

Animated GEOGRAPHY Present-Day Southeast Asia
CLASSZONE.COM

Point out that this is a political map of Southeast Asia, showing the region's countries, their capitals, and selected other cities. The map also shows the various bodies of water that surround and run through this region. Point out the longitude and latitude lines, especially the equator.

- What is the capital of Vietnam? *(Hanoi)* What other city is identified in this country? *(Ho Chi Minh City)*
- What nation is both part of the mainland and part of an island? *(Malaysia)*
- Based on the latitude of Indonesia, predict what the climate is like. *(Possible answer: Indonesia lies on or close to the equator, so the climate is probably tropical.)*

CONNECT GEOGRAPHY & HISTORY ANSWERS

1. Indonesia and the Philippines
2. Thailand

Presentation Options

Use the **Power Presentations** DVD-ROM to project the animated map.

- Prompt students to use the pop-up images and audio from the map to discover what they already know about the geography, cultures, histories, economics, or governments of Southeast Asia.
- Challenge students to answer the map questions individually, in small groups, or as a class activity. Use all of the interactive material to help preview chapter content.

1 Plan & Prepare

Objectives

- Identify the physical features and common natural disasters of Southeast Asia's mainland and islands
- Describe the climate, vegetation, and wildlife of Southeast Asia

Read for the Essential Question

Help students read for a purpose by reminding them of the Essential Question: "How can Southeast Asian nations establish stable governments and economies?" Point out that they can use items in the section, such as Key Questions, special features, images, and maps to formulate answers to the question.

Vocabulary

Best Practices Toolkit

Use the ***Best Practices Toolkit*** to model strategies for vocabulary notetaking. Vary strategies throughout the year. Choose from: *Knowledge Rating, Predicting ABCs, Definition Mapping, Word Sort, Word Wheel, Frayer Model (Word Squares), Magnet Words,* and *Student VOC.*

Vocabulary Strategies, TT9–TT16

Reading Strategy

Best Practices Toolkit

Display the Three-Column Chart transparency.

- Remind students that main ideas are important ideas in a block of text and that details support main ideas. Text headings can often help them focus on the main idea of a section.
- Model filling in main ideas on the chart using the "Mainland and Islands" subsection.
- Have students suggest additional entries. Remind students that main ideas in each block of text in the section will help answer a Key Question.
- Display the completed chart for review.

Three-Column Chart, TT21

Section 1 Reading for Understanding

Key Ideas

BEFORE, YOU LEARNED

East Asia is made up of a vast mainland area and a number of important islands off the eastern coast.

NOW YOU WILL LEARN

Nearby Southeast Asia is a mountainous region with a tropical climate and much volcanic activity on its many islands.

Vocabulary

TERMS & NAMES

Indochinese Peninsula one of two peninsulas on which mainland Southeast Asia lies; the peninsula is located south of China

Malay Peninsula the second peninsula of mainland Southeast Asia; the narrow strip of land serves as a bridge between the mainland and islands

Mekong River a major river that runs south from southern China through Laos, Cambodia, and Vietnam

archipelago (AHR•kuh•PEHL•uh•GOH) a set of closely grouped islands, which sometimes form a curved arc

tsunami (tsu•NAH•mee) one or a series of large ocean waves caused by an underwater earthquake or volcanic eruption

REVIEW

monsoon a seasonal wind that causes wet and dry seasons

Visual Vocabulary Malay Archipelago

Reading Strategy

Re-create the chart shown at right. As you read and respond to the **KEY QUESTIONS**, use the chart to jot down main ideas about Southeast Asia's geographic features, climate, plants, and animals.

Skillbuilder Handbook, page R4

FIND MAIN IDEAS

	MAINLAND	ISLANDS
Geographic Features		
Climate		
Plants and Animals		

GRAPHIC ORGANIZERS
Go to **Interactive Review @ClassZone.com**

PRETEACHING VOCABULARY

English Learners

Pronounce and Preview

Pronounce each term for students. Review the meanings of words in the definitions, such as *peninsula*, *grouped*, and *arc*.

- To modify vocabulary learning, have students complete worksheets as they read the section.

Unit Resource Book: East Asia & Southeast Asia

- Vocabulary Practice, p. 215

Inclusion

Use Visual Vocabulary

Have students turn to the Physical Geography of Southeast Asia map. Read aloud the terms and definitions on the opposite page. As you read a definition, have students find the term on the map. (Note: This will work for all terms except *tsunami*.) To extend the activity, have students create visual aids, such as an illustration, for each term. They can add to their visual vocabulary as they read the other sections on Southeast Asia.

SECTION 1 GEOGRAPHY

Terraced rice fields in Bali, Indonesia

Adult and young orangutan, which means "man of the forest" in the Malay language

Physical Geography of Southeast Asia

Connecting to Your World

You may have heard of Mount St. Helens, the volcano that erupted in 1980 in the state of Washington. The blast killed 57 people and sent ash thousands of feet into the air. In 1991, a volcanic eruption that was ten times larger occurred on Mount Pinatubo in the Philippines. More than 300 people died in this eruption, and many thousands more lost their homes. In this section, you will learn why volcanic eruptions and other natural disasters occur frequently in Southeast Asia.

Mainland and Islands

KEY QUESTION What are some of the advantages and disadvantages of Southeast Asia's physical geography?

Southeast Asia has two distinct subregions: the southeastern corner of the Asian mainland and a great number of islands. Both regions have many mountains. Most of the mountains on the islands were formed from volcanoes. Southeast Asia is part of the Ring of Fire, a zone of volcanoes around the rim of the Pacific Ocean. As a result, volcanic eruptions and earthquakes are common.

Island Nation A temple built in the 17th century to honor a Hindu water goddess floats on the water in Bali, one of the islands that makes up Indonesia.

853

2 Focus & Motivate

3-Minute Warm-Up

Write on the board or display the transparency:

- Study the image at the bottom of this page. What might be the advantages and disadvantages of living in an island nation? *(Advantages: fishing, trade, beauty; Disadvantages: isolation, difficulty traveling)*

Unit Transparency Book: East Asia & Southeast Asia
- 3-Minute Warm-Ups, TT29

Connecting to Your World

More About . . .

Mount Pinatubo

When Mount Pinatubo erupted in 1991, it affected weather all over the world. It was the second-largest volcanic eruption in the 20th century. The largest was the 1912 eruption of Mount Novarupta in Alaska.

3 Teach

Mainland and Islands

Think, Pair, Share

- Which two physical features form the mainland and islands of Southeast Asia? *(peninsulas and archipelagoes)*
- **Find Main Ideas** What natural disasters are common in the region? *(volcanic eruptions and earthquakes)*

SECTION 1 PROGRAM RESOURCES

Unit Resource Book: East Asia and Southeast Asia

ON LEVEL

Unit Resource Book
- Reading Study Guide, p. 191
- Outline Map Activities, p. 219
- Section Quiz, p. 229

STRUGGLING READERS

Unit Resource Book
- Reading Study Guide with Additional Support, p. 197
- Vocabulary Practice, p. 215
- Skillbuilder Practice, p. 216
- Section Quiz, p. 229
- Reteaching Activity, p. 232

eEdition with Audio DVD-ROM

ENGLISH LEARNERS

Pupil Edition in Spanish

eEdition with Audio DVD-ROM

Unit Resource Book
- RSG (Spanish), p. 203
- RSG with Additional Support (Spanish), p. 209

Multi-Language Glossary

Test Generator
- Section Quiz in Spanish

INCLUSION

Unit Resource Book
- RSG with Additional Support, p. 197
- Section Quiz, p. 229
- Reteaching Activity, p. 232

GIFTED & TALENTED

Unit Resource Book
- Connect Geography and History, p. 227
- Section Quiz, p. 229

PRE-AP

Unit Resource Book
- Section Quiz, p. 229
- Geography Project, p. 249

TECHNOLOGY

Unit Transparency Book
- 3-Minute Warm-Ups, TT29
- Essential Question, TT30
- Critical Thinking, TT31
- Map, TT34
- Map, TT35

Daily Test Practice Transparencies
- Section 1, TT93

Power Presentations

ClassZone.com

World Cultures and Geography Video on DVD

Comparing Regions

The Aleutian Islands and the Malay Archipelago

Like the Malay Archipelago, Alaska's Aleutian Islands form an archipelago. The Aleutian chain, which is made up of 14 large islands, 55 small islands, and many islets, is much smaller than the Malay Archipelago, which includes about 20,000 islands.

The Aleutian Islands extend in an arc for about 1,100 miles between the tip of the Alaska Peninsula and Russia. The Malay Archipelago, on the other hand, extends about 3,800 miles. The Malay Archipelago lies along or very close to the equator, while the Aleutians lie close to the Arctic Circle.

Still, the archipelagos have something in common. Both island groups are found along the Ring of Fire, the volcanic zone around the rim of the Pacific Ocean. As a result, both are prone to volcanic activity. In fact, most of the islands in the Aleutian archipelago have volcanic origins.

Make It Fun

Teacher-Tested Activities

Matt Parker, Eastgate Middle School, Kansas City, Missouri

The visual learners in my class enjoy acting out terms and names. Here are some ideas for the terms on this page.

- For the Mekong River, one student might snake past three others holding labels for Laos, Cambodia, and Vietnam.
- To illustrate an archipelago, a group of students could form a curved arc.
- To demonstrate a tsunami, one group of students could act out an underwater earthquake while another rises in a great wave.

Mekong River The Mekong River is the heart and soul of mainland Southeast Asia. Millions of people rely on the river for farming, fishing, water, and transport. **What does this Vietnamese farmer seem to be using the Mekong for?**

to transport farm produce

Peninsulas As shown on the map on the opposite page, mainland Southeast Asia lies on two peninsulas. The **Indochinese Peninsula** Ⓐ is located south of China. The **Malay Peninsula** Ⓑ is a narrow strip of land about 700 miles long. It serves as a bridge between the mainland and the islands.

The Indochinese Peninsula has a series of north-south mountain ranges separated by river valleys and lowlands. To the west of these mountains flows one of the world's great rivers—the **Mekong** Ⓒ. It runs south for about 2,600 miles from southern China through Laos, Cambodia, and Vietnam. The river opens onto a wide fertile delta on Vietnam's coast.

Island Chains Most of the islands of Southeast Asia are grouped into archipelagoes. An **archipelago** (AHR•kuh•PEHL•uh•GOH) is a set of closely grouped islands. The largest of these island groups is the Malay Archipelago, which includes Indonesia Ⓓ and the Philippines Ⓔ. Indonesia is made up of about 17,500 islands. Java, Sumatra, and Borneo are the most important islands. The Philippines lies to the north of Indonesia. This nation consists of about 7,100 islands. The largest and most important of these islands is Luzon.

Natural Disasters You have already learned about the volcanic eruption of Mount Pinatubo in 1991. An even greater disaster occurred on December 26, 2004, when a series of tsunamis hit many parts of Southeast Asia. A **tsunami** (tsu•NAH•mee) is a very large ocean wave caused by an underwater earthquake or volcanic eruption. The Asian tsunami, as the disaster was called, was one of the deadliest in modern history. The tsunamis were caused by an earthquake that released energy equal to about 23,000 atomic bombs.

Advantage: The Mekong River allows much farming and fishing; Disadvantage: Southeast Asia's position along the Ring of Fire makes the region susceptible to natural disasters. ▸

ANALYZE EFFECTS Describe some of the advantages and disadvantages of Southeast Asia's physical geography.

DIFFERENTIATING INSTRUCTION

Struggling Readers

Outline Southeast Asia's Geography

To help students focus on main ideas and details, have them outline the information about Southeast Asia's mainland and islands. Students can use the text headings to help them identify main ideas. Show them how to set up an outline, using the example at right. Model how to fill in an outline. Remind students to add additional numbers or letters as needed.

Geography of Southeast Asia Outline

I. Peninsula
  A. Indochinese Peninsula
    1. North-south mountain ranges
    2.
  B.
II. Island chains
III. Natural disasters

Physical Geography of Southeast Asia

CONNECT Geography & History

The 2004 Asian Tsunami On December 26, 2004, a 9.0-magnitude earthquake in the Indian Ocean triggered tsunamis up to 50 feet high. The gigantic waves caused devastating damage in many parts of Indonesia, southern Thailand, Sri Lanka, southern India, and elsewhere. The inset map shows the worst-affected areas. Besides taking more than 200,000 lives, the tsunamis also destroyed thousands of miles of coastline. The image here shows the devastation in Banda Aceh, a city in Sumatra.

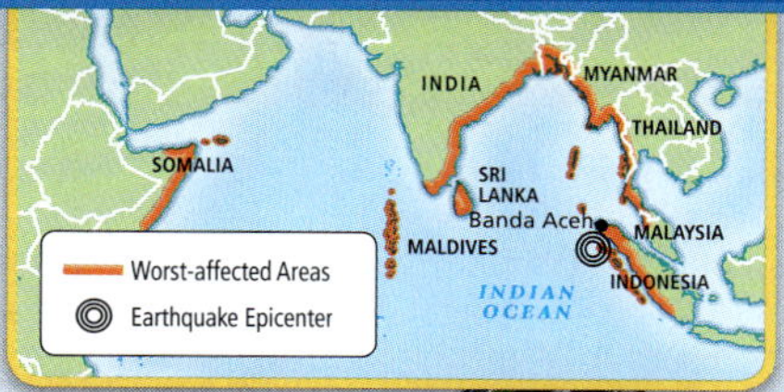

CRITICAL THINKING

Draw Conclusions Study the large map above. Why do you think Sumatra and southern Thailand were especially affected by the tsunamis?

Physical Geography of Southeast Asia

Have students use the elevation key to compare areas of high and low elevation in Southeast Asia.

- Describe the elevation of most coastal areas in Southeast Asia. *(at or below sea level)*
- Which areas in Southeast Asia have the highest elevation? *(central New Guinea and northern Myanmar)*

Unit Resource Book: East Asia & Southeast Asia

- Outline Map Activities: Southeast Asia, p. 219

CONNECT Geography & History

The 2004 Asian Tsunami

Point out the small, inset map and explain that it shows areas affected by the 2004 Asian tsunami.

CRITICAL THINKING ANSWER

Draw Conclusions They lie beside the Indian Ocean, unprotected by other land; they were close to the epicenter.

Unit Resource Book: East Asia & Southeast Asia

- Connect Geography and History: 2004 Asian Tsunami, p. 227

More About . . .

The 2004 Asian Tsunami

There was no official warning system in place when the tsunami hit, but natural signs alerted people in some places. For example, some animals fled for high ground shortly before the tsunami arrived. In Thailand, the sea retreated from the beach, stranding sea life and boats. Unfortunately, many people didn't recognize the sign and ran out onto the exposed sea floor.

INTERDISCIPLINARY ACTIVITIES

CONNECT to Language Arts

Write a Memorial Speech

Have students write a speech or poem that could be read at a memorial ceremony for victims of the 2004 Asian tsunami. In their speech or poem, students should express what they know about the tsunami and its effects. You may wish to share with students the additional information about the tsunami provided in the notes on this page before they begin writing.

CONNECT to Civics

Plan a Disaster Relief Effort

The 2004 Asian tsunami was a major catastrophe, killing more than 200,000 people. Ask students to work in groups to come up with ideas for disaster relief for a future tsunami or other natural disaster. Have students think about what help would be needed (services, supplies, and so on) and what organizations might assist with their plan. Invite students to present their plans.

Teach

Climate, Vegetation, and Wildlife

Reader, Recorder, Reporter

- What are the two main climates in Southeast Asia? *(tropical wet and tropical wet and dry)*
- **Find Main Ideas** How does Southeast Asia's climate affect plant and animal life in the region? *(The hot, humid climate supports a wide variety of vegetation and animal life.)*

Unit Resource Book: East Asia & Southeast Asia
- Skillbuilder Practice: Making Inferences, p. 216

COMPARING

Regional Rainfall

Review reading a bar graph with students. Point out the title, which tells the subject or purpose of the graph, and the labeled axes, which identify what information will be studied.

CRITICAL THINKING ANSWER

Compare and Contrast Manila, Hanoi, and Bangkok

Comparing Regions

Monsoons in Southeast Asia and North America

While Southeast Asia and India are the parts of the globe most frequently associated with monsoons, these seasonal winds occur all over the world. The North American monsoon affects northwestern Mexico and the southwestern part of the United States. Scientists are studying the North American monsoon to help them predict the extreme weather that can occur during the summer monsoon season, such as hail, flooding, and severe lightning.

Unit Resource Book: East Asia & Southeast Asia
- Geography Project: Understanding Monsoons, p. 249

Rafflesia The Rafflesia, which is native to Indonesia, is the world's largest flower. It can grow to be about three feet across.

Climate, Vegetation, and Wildlife

KEY QUESTION How does Southeast Asia's climate affect its vegetation and wildlife?

Most of Southeast Asia has a tropical climate characterized by heavy rainfall and generally warm temperatures. Tropical climates fall into two categories: a tropical wet climate and a tropical wet and dry climate.

Rain and Monsoons The coastal areas of Myanmar, Thailand, and Vietnam and most of Malaysia, Indonesia, and the Philippines have a tropical wet climate. These areas receive abundant rainfall. Parts of Malaysia get an average of about 100 inches of rain per year. In addition, temperatures are very warm, with an average annual temperature of 80°F. However, areas with higher elevation have cooler temperatures.

Much of mainland Southeast Asia has a tropical wet and dry climate that is shaped by monsoons. Remember that **monsoons** are seasonal winds. Monsoons bring dry air during the winter and moist air during the summer. Rainfall brought by the wet monsoon winds is often very heavy. The winds can also cause typhoons. Vietnam experiences about five typhoons each year.

COMPARING Regional Rainfall

In the image shown here, Malaysian schoolgirls walk in the rain. As you can see in the graph, Malaysia receives more rain than any other country in Southeast Asia. Compare the amounts of rainfall received by these Southeast Asian cities with that received by Seattle, one of the rainiest cities in the United States.

CRITICAL THINKING

Compare and Contrast Which three cities receive about the same amount of rainfall?

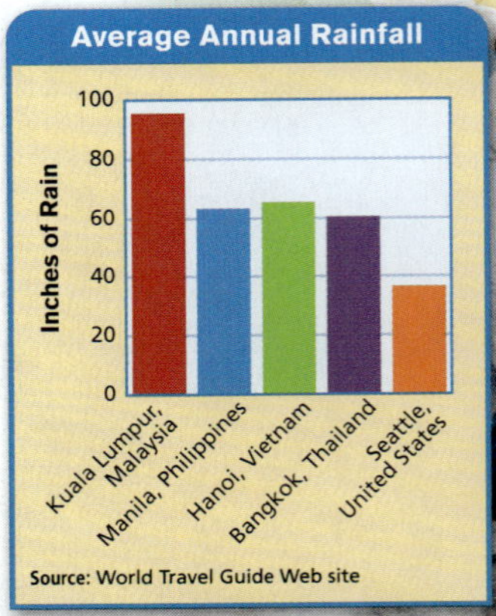

DIFFERENTIATING INSTRUCTION: TIERED ACTIVITIES

OBJECTIVE Prepare a group presentation on monsoons.

Basic

Have groups of students begin a presentation on monsoons. Ask them to explain what a monsoon is and use maps to identify areas affected by monsoons.

On Level

Have other student groups prepare a chart showing the effects of winter and summer monsoons in Southeast Asia. The charts should indicate rainfall amounts in both seasons.

Challenge

Have other student groups create a visual display that shows the factors and processes that cause monsoons. Students should include brief caption explanations. Combine all of the students' work into a single presentation.

Vegetation and Wildlife Southeast Asia's hot, humid climate supports tropical vegetation. In fact, Southeast Asia has one of the greatest varieties of vegetation on Earth. For example, it has a remarkable number of tree species. Tropical evergreen forests are situated in areas with much rainfall near the equator. Deciduous forests are common in the wet and dry climate zone. Teak, a wood used in shipbuilding and furniture, comes from these forests. Some unusual plants thrive in Southeast Asia. Many of these plants have strong-smelling flowers and fruits.

Southeast Asia also has thousands of varieties of birds and animals, and new species are still being discovered. In 2005, scientists found dozens of new plant and animal species in an isolated forest area of Indonesia. Not even people from nearby villages had ever visited the area. In addition, Southeast Asia is home to such large animals as the rhinoceros, tiger, and elephant. The orangutan is native to Indonesia and Malaysia.

Sumatran Rhinoceros
This small species of rhino has two horns and a coat of coarse hair. Most of these rhinos live in remote mountain areas of Indonesia.

warm temperatures and abundant rainfall produce variety of tropical plants; many animals thrive in tropical climate ▸

EVALUATE Explain how Southeast Asia's climate affects its vegetation and wildlife.

Section 1 Assessment

ONLINE QUIZ
For test practice, go to
Interactive Review
@ClassZone.com

TERMS & NAMES

1. **Explain the importance of**
   - Indochinese Peninsula
   - Malay Peninsula
   - Mekong River
   - archipelago

USE YOUR READING NOTES

2. **Find Main Ideas** Use your completed chart to answer the following question:
   Which Southeast Asian subregion receives more rainfall, the mainland or the islands?

	MAINLAND	ISLANDS
Geographic Features		
Climate		
Plants and Animals		

KEY IDEAS

3. Why are earthquakes and volcanic eruptions common in Southeast Asia?
4. Why is the Mekong an important river?
5. How do monsoons affect the climate of Southeast Asia?

CRITICAL THINKING

6. **Make Inferences** Based on the geographic features of Indonesia, what might be some of the major economic activities of this nation?
7. **Draw Conclusions** Why do you think new species of animals are still being discovered in Southeast Asia?
8. **CONNECT to Today** What do you think people in Southeast Asia could do to prevent a huge loss of life in the event of another tsunami?
9. **TECHNOLOGY** **Create a Multimedia Presentation** Create a multimedia presentation that shows how a volcano in the Ring of Fire forms and erupts. Use library and Internet sources. Include descriptions with your visuals.

CONNECT to the Essential Question

How can Southeast Asian nations establish stable governments and economies?

Ask students what they have learned so far that can help them answer this question. Students might mention the following:

- Physical features such as rivers, coastlines, and oceans have both aided and slowed the development of Southeast Asia.
- Natural disasters are common, and often devastating, in Southeast Asia.
- The region has diverse vegetation and wildlife, some of which benefit its economy.

4 Assess & Reteach

Assess Have students complete the Section Assessment.

Unit Resource Book: East Asia & Southeast Asia
- Section Quiz, p. 229

Interactive Review @ClassZone.com
- Section Quiz

Power Presentations

Test Generator

Reteach Have students create illustrations and captions of Southeast Asia's peninsulas, island chains, natural disasters, monsoons, vegetation, and wildlife. Have student pairs review and discuss their illustrations and captions.

Unit Resource Book: East Asia & Southeast Asia
- Reteaching Activity, p. 232

SECTION 1 ASSESSMENT ANSWERS

Terms & Names

1. Indochinese Peninsula: peninsula on which mainland Southeast Asia lies; Malay Peninsula: bridge between the mainland and islands; Mekong River: major river in mainland Southeast Asia; archipelago: closely grouped islands

Use Your Reading Notes

2. the islands

MAINLAND	ISLANDS
lies on two peninsulas, mountainous	archipelagoes
tropical wet and dry	tropical wet
tropical vegetation; great variety of animals	tropical vegetation; thousands of varieties of animals

Key Ideas

3. because the region is on the Ring of Fire
4. because millions of people rely on the river for farming and fishing
5. They bring dry air during the winter and moist air during the summer.

Critical Thinking

6. Possible answers: fishing; forestry; agriculture; mineral deposits; tourism
7. Possible answer: Much of Southeast Asia probably has thick forests, which are difficult to penetrate. As a result, some species in these forests are well concealed.
8. Possible answers: develop some sort of warning system; develop an evacuation plan
9. Use the rubric to score multimedia presentations.

Presentation Rubric

	Style	Descriptions
4	presents excellent visuals	shows thorough understanding
3	presents good visuals	shows good understanding
2	presents fair visuals	shows basic understanding
1	presents poor or no visuals	shows little or no understanding

1 Plan & Prepare

Objectives

- Describe the histories, governments, cultures, and economies of Cambodia, Laos, and Vietnam and explain how they are connected
- Describe the history, government, culture, and economy of Thailand
- Compare the histories, governments, cultures, and economies of Myanmar and Malaysia
- Identify significant details about Singapore's economy and government

Read for the Essential Question

Help students read for a purpose by reminding them of the Essential Question: "How can Southeast Asian nations establish stable governments and economies?" Point out that they can use items in the section, such as Key Questions, special features, images, and maps to formulate answers to the question.

Vocabulary

Best Practices Toolkit

Use the ***Best Practices Toolkit*** to model strategies for vocabulary notetaking. Vary strategies throughout the year. Choose from: *Knowledge Rating, Predicting ABCs, Definition Mapping, Word Sort, Word Wheel, Frayer Model (Word Squares), Magnet Words,* and *Student VOC.*

Vocabulary Strategies, TT9–TT16

Reading Strategy

Best Practices Toolkit

Display the Web Diagram transparency.

- Point out to students that when they summarize information, they present only key ideas. A summary should be shorter and less detailed than the original.
- Model summarizing key ideas from the "Cambodia, Laos, and Vietnam" subsection.
- Have students suggest additional entries.
- Display the completed diagram for review.

Web Diagram, TT28

SECTION 2 Reading for Understanding

Key Ideas

BEFORE, YOU LEARNED

Southeast Asia is made up of mainland countries and island chains.

NOW YOU WILL LEARN

The nations of mainland Southeast Asia have closely connected histories and diverse cultures.

Vocabulary

TERMS & NAMES

Indochina the name for a former French colony in Southeast Asia made up of Cambodia, Laos, and Vietnam

Khmer (kmair) **Empire** an empire that began in the 500s and had gained control of much of mainland Southeast Asia by the 800s

colonialism the control by one power over a dependent area or people

Tet the Vietnamese New Year and most important holiday in Vietnam

Siam the former name of Thailand

constitutional monarchy a government in which the powers of the king or queen are limited by a constitution

Aung San Suu Kyi (awng sahn soo chee) leader of the democracy movement in Myanmar

microstate an independent country that is very small in area and population

Visual Vocabulary celebration of Tet

Reading Strategy

Re-create the web diagram shown at right. As you read and respond to the **KEY QUESTIONS**, use a different copy of the diagram to summarize the history, government, economy, and culture of each mainland country.

See Skillbuilder Handbook, page R5

SUMMARIZE

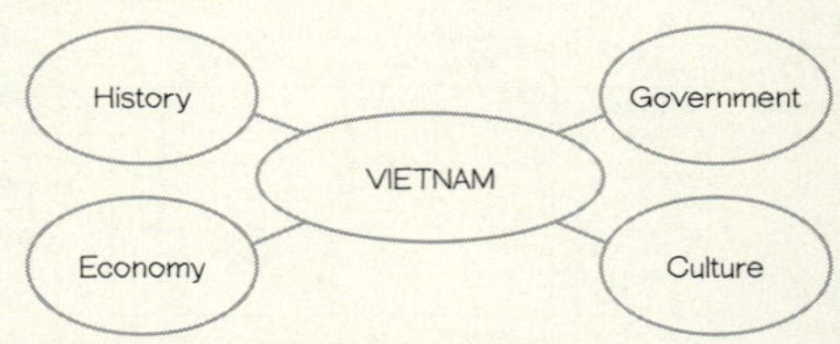

GRAPHIC ORGANIZERS
Go to Interactive Review @ ClassZone.com

PRETEACHING VOCABULARY

English Learners

Pronounce and Preview

Pronounce each term for students. Review the meanings of words in the definitions, such as *former*, *dependent*, and *limited*.

- To modify vocabulary learning, have students complete worksheets as they read the section.

Unit Resource Book: East Asia & Southeast Asia

- Vocabulary Practice, p. 215

Inclusion

Fill in the Blanks

Have student pairs take turns doing the following activity. Have one student find a term in the text and read the sentence in which it is used, substituting the word "blank" for the term. The other student should choose a term from the vocabulary list that correctly fills in the blank.

Terraced rice fields in Bali, Indonesia

SECTION 2 FOCUS ON

Adult and young orangutan, which means "man of the forest" in the Malay language

The Mainland Countries

Connecting to Your World

What have you already learned about the Vietnam War? Maybe you know that the war was fought between Communist and anti-Communist forces in North and South Vietnam. You might also know that the United States got involved in the conflict and that this involvement deeply divided the country. Vietnam is one of the mainland countries of Southeast Asia. In this section, you'll learn about the history and politics of mainland Southeast Asia and some of the causes that led to the Vietnam War.

Cambodia, Laos, and Vietnam

KEY QUESTION How are the histories, cultures, and economies of Indochina connected?

Indochina is the name given to a former French colony made up of Cambodia, Laos, and Vietnam. Indochina got its name because of its location east of India and south of China. India and China have also had a strong influence on Indochinese culture. The Mekong River flows through Cambodia, Laos, and Vietnam, forming a highway that connects them. For centuries, this river has been used to transport goods and ideas. Not surprisingly, the histories of these nations are also closely connected.

India and China ▼

Angkor Wat This temple in Cambodia was built in the 1100s as a Hindu temple but became a Buddhist temple about 100 years later. **Which two countries probably influenced the temple's religious uses?**

859

2 Focus & Motivate

3-Minute Warm-Up

Write on the board or display the transparency:

- Which two Vietnamese nations fought the Vietnam War? *(North and South Vietnam)* On which side did the United States fight? *(on the side of South Vietnam)*

Unit Transparency Book: East Asia & Southeast Asia
- 3-Minute Warm-Ups, TT29

Connecting to Your World

More About . . .

The Vietnam War

The Vietnam War is sometimes called the Second Indochina War. The U.S. government believed that if Vietnam fell to communism, so would the rest of Southeast Asia, an idea called the "domino theory."

3 Teach

Cambodia, Laos, and Vietnam

Think, Pair, Share

- Which river connects Cambodia, Laos, and Vietnam? *(the Mekong River)*
- **Summarize** Describe the economies of Laos, Cambodia, and Vietnam. *(Laos and Cambodia have developing economies based on small industries, while Vietnam has a rapidly growing economy based on oil and coffee exports.)*

SECTION 2 PROGRAM RESOURCES

Unit Resource Book: East Asia and Southeast Asia

ON LEVEL

Unit Resource Book
- Reading Study Guide, p. 193
- Section Quiz, p. 230

STRUGGLING READERS

Unit Resource Book
- Reading Study Guide with Additional Support, p. 199
- Vocabulary Practice, p. 215
- Skillbuilder Practice, p. 216
- Section Quiz, p. 230
- Reteaching Activity, p. 233

eEdition with Audio DVD-ROM

ENGLISH LEARNERS

Pupil Edition in Spanish
eEdition with Audio DVD-ROM
Unit Resource Book
- RSG (Spanish), p. 205
- RSG with Additional Support (Spanish), p. 211

Multi-Language Glossary
Test Generator
- Section Quiz in Spanish

INCLUSION

Unit Resource Book
- RSG with Additional Support, p. 199
- Section Quiz, p. 230
- Reteaching Activity, p. 233

GIFTED & TALENTED

Unit Resource Book
- Map and Graph Skills, p. 217
- History Makers, p. 221
- Interdisciplinary Projects, p. 223
- Section Quiz, p. 230

PRE-AP

Unit Resource Book
- Economics in the World, p. 226
- Primary Sources, p. 228
- Section Quiz, p. 230

TECHNOLOGY

Unit Transparency Book
- 3-Minute Warm-Ups, TT29
- Essential Question, TT30
- Critical Thinking, TT31
- Map, TT34
- Map, TT35

Daily Test Practice Transparencies
- Section 2, TT94

Power Presentations
ClassZone.com
World Cultures and Geography Video on DVD

Colonies in Southeast Asia *1895*

CONNECT Geography & History

Explain that European colonization of Southeast Asia was at its height in the 19th and early 20th centuries. Only Thailand was able to maintain its independence. In the 1930s, many anti-colonial rebellions took place, though they were largely unsuccessful. Most Southeast Asian countries did not begin to gain their independence until after World War II.

ANSWERS

1. **Location** French Indochina
2. **Region** the Dutch (Netherlands)

Unit Resource Book: East Asia & Southeast Asia

- Primary Sources: On the Vietnam War, p. 228

More About . . .

The Dictator of Cambodia

From 1975 to 1979, Cambodia was ruled by a ruthless dictator called Pol Pot. Born Saloth Sar, he took the name Pol Pot as he was building up the Communist Party in Cambodia in the early 1960s. Upon gaining power, Pol Pot and his Khmer Rouge government tried to create a completely agricultural society. Pol Pot forced people out of cities; abolished money, private property and religion; and had anyone suspected of being an intellectual killed. During his rule, about one to two million Cambodians died.

Pol Pot's government fell in 1979 when Vietnam invaded Cambodia. The dictator retreated to the Thai-Cambodian border and continued to fight against the Vietnamese. In July 1997, Pol Pot was charged with treason and sentenced to life imprisonment. He died in April 1998.

Colonies in Southeast Asia *1895*

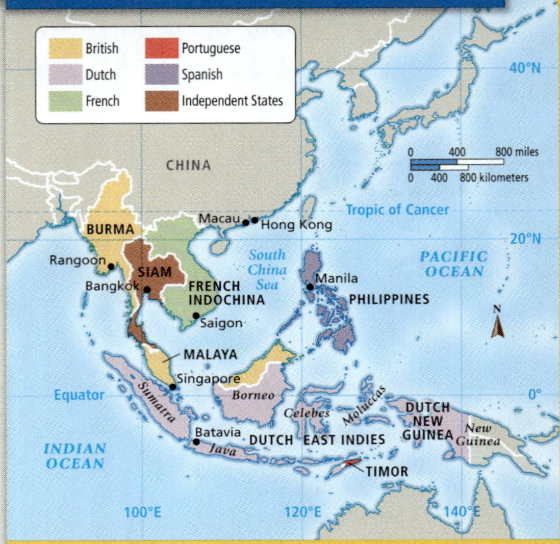

CONNECT Geography & History

1. **Location** What colony was controlled by France?
2. **Region** Which European country could access both the Indian and the Pacific oceans from its colony?

History and Government By the 800s, the **Khmer** (kmair) **Empire** had gained control of much of mainland Southeast Asia. The empire reached its peak around 1200. At different times, the Chinese also exerted political influence over parts of the mainland.

European colonists had been arriving in Southeast Asia since 1509. In the 1850s, France began to take control of Vietnam. The French also seized Laos and Cambodia. They took over the countries' political, economic, and social systems in order to obtain wealth, a policy called **colonialism**. By the 20th century, Europeans had colonized all of Southeast Asia except Siam (now Thailand).

In 1953, Laos and Cambodia gained their independence from France. The following year, a peace agreement ended fighting between Vietnam and France, dividing the country into a Communist north and a non-Communist south. But in 1957, a war between North Vietnam and South Vietnam broke out. In time, the fighting involved the United States and also spilled over into Cambodia and Laos. Finally, in 1975, South Vietnam was taken over by North Vietnam. Both Vietnam and Laos established Communist governments. Cambodia was ruled by a brutal Communist dictator until it became a democracy in 1993.

Culture and Economy More than 80 percent of the people in the three nations of Indochina live in rural areas. The majority of Indochinese practice Buddhism. They celebrate religious holidays as well as other traditional holidays. In Vietnam, the most important holiday is **Tet**, the Vietnamese New Year.

Laos and Cambodia have developing economies made up of small industries, such as clothing and textiles. Most of the people work as rice and peanut farmers. Vietnam, on the other hand, has one of the fastest-growing economies in the world. It is developing its oil industry and exports more coffee than any other country except Brazil.

All have been influenced by China, India, and the Khmer Empire; all were colonized; all three have developing economies, although the economy of Vietnam is growing the fastest. ▸

EVALUATE Explain how the histories, cultures, and economies of Indochina are connected.

DIFFERENTIATING INSTRUCTION

English Learners

Review Unfamiliar Words

Review with students some of the words that appear on this page and that may be unfamiliar. You may want to focus on *establish* ("Both Vietnam and Laos established Communist governments") and *export* ("It is developing its oil industry and exports more coffee than any other country except Brazil"). To check for understanding, ask students to use these words in sentences.

Pre-AP

Compare and Contrast

Have students write an essay comparing and contrasting Vietnam with Cambodia and/or Laos. Their essays should focus on the history and current culture and economy of each nation. Invite students to read their essays aloud.

Thailand

KEY QUESTION What events in Thailand's past have helped the country prosper?

Thailand is one of the most prosperous countries in Southeast Asia. For most of its history, it was known as **Siam**. Some scholars believe that the word *Siam* means "golden." The nation changed its name to Thailand in 1949.

History and Government The first people to live in the region of Thailand called themselves the Thai, which means "free." The Thai remained free, even after the Europeans began to arrive in Southeast Asia in the 1500s.

In the mid-1800s, Thai leader King Mongkut began to modernize his country. He and his son, King Chulalongkorn, started schools, built railroads, and abolished slavery in Siam. Through their leaders' guidance, the Siamese people escaped the harsh treatment that many other Southeast Asians experienced under foreign control.

Thailand has been a **constitutional monarchy** since 1932, a government in which the powers of the king or queen are limited by a constitution. Thailand came under the control of Japan in 1941. After World War II, a series of military dictators ruled the country. In the mid-1970s, Thailand began to develop a more democratic government. This process was halted in 2006, however, when the military overthrew the prime minister.

Culture and Economy Most of Thailand's people live in rural areas. In the early 1960s, though, large numbers of Thai began to move to cities in search of work. By 2006, the capital city of Bangkok had over 9 million people.

Thai food, including a noodle dish called pad thai, has become very popular in the United States and other countries. In fact, Thai cuisine ranks as the fourth most popular in the world.

Nearly 50 percent of the workers in Thailand earn a living by farming or fishing. Rice, corn, and sugar cane are some of the crops grown. Thailand's industries produce automobiles, electronic equipment, clothing, and plastics. Tourism is also a booming industry.

DRAW CONCLUSIONS Identify the events in Thailand's history that have helped the country prosper. remaining independent, avoiding colonization, developing its economy

CONNECT Geography & History

Neutral Zone

Siam lay between British-controlled Burma (now Myanmar) and French Indochina. Siamese kings skillfully promoted their country as a neutral zone between the two powers. However, King Chulalongkorn was influenced by European architecture. He had the palace shown here decorated with marble imported from Italy.

Teach

Thailand

Reader, Recorder, Reporter

- What does Thai mean? *(free)*
- What changes did King Mongkut and his son make? *(They modernized the country, started schools, built railroads, and abolished slavery.)*
- **Summarize** How do most people in Thailand live today? *(Most people live in rural areas, and many make their living by farming or fishing.)*

CONNECT Geography & History

Neutral Zone

King Chulalongkorn was influenced by more than Western architecture. His education blended Thai and Western ideas. As a result, many of his reforms were based on Western models. He studied how European colonies were administered and used Western practices as a model for the government he established.

King Chulalongkorn believed reforms were right for his people and, perhaps more important, thought they would help prevent European colonization. By consolidating power and showing a "civilized" face to the world, he strengthened Siam's independence.

Unit Resource Book: East Asia & Southeast Asia

- History Makers: King Chulalongkorn, p. 221

INTERDISCIPLINARY ACTIVITIES

CONNECT to Math

Graph City Populations

Have students create bar graphs showing the populations of these capitals: Phnom Penh, Cambodia—1,157,000; Vientiane, Laos—194,200; Yangon, Myanmar—4,455,500; Singapore, Singapore—4,253,000; Bangkok, Thailand—6,320,174; Hanoi, Vietnam—2,931,400.

Unit Resource Book: East Asia & Southeast Asia

- Map and Graph Skills: Reading Bar Graphs, p. 217

CONNECT to Art

Make an Illustrated Time Line

Have students make illustrated time lines of Thai history based on the information in their textbooks. Students should include cultural and economic events on their time lines. If time permits, students can use classroom resources to find additional information and images for their time lines.

Teach

Myanmar and Malaysia

Think, Pair, Share

- What were the early names of Myanmar and Malaysia? *(Burma and Melaka)*
- What types of governments do Myanmar and Malaysia have? *(military dictatorship and constitutional monarchy)*
- **Summarize** Describe the economies of Myanmar and Malaysia. *(Malaysia has a thriving market economy while Myanmar is poor with an agricultural-based economy.)*

Unit Resource Book: East Asia & Southeast Asia
- Skillbuilder Practice: Making Inferences, p. 216
- Economics in the World: Growth of Industry, p. 226

RESEARCH & WRITING CENTER

HISTORY MAKERS

Aung San Suu Kyi

Influenced by Mohandas Gandhi and Martin Luther King Jr., Aung San Suu Kyi led nonviolent rallies and protests against the Myanmar government. Although Suu Kyi was released from house arrest in 1995 and 2002, she was placed back under house arrest in 2003, where—despite worldwide protest—she remains today.

CONNECT to the Essential Question

How can Southeast Asian nations establish stable governments and economies?

Ask students what they have learned so far that can help them answer this question. Students might mention the following:

- Laos, Cambodia, and Myanmar have developing economies; Vietnam, Thailand, Malaysia, and Singapore have more modern, market-oriented ones.
- Although no longer colonies, some mainland nations have oppressive governments.

Myanmar and Malaysia

KEY QUESTION How do Myanmar and Malaysia compare?

Thailand's neighbors, Myanmar and Malaysia, present contrasting pictures. Malaysia has a thriving market economy and freely elected officials. However, Myanmar has widespread poverty and suffers under a military dictatorship.

History and Government In 1044, the Myanmar region united into a kingdom. Its people adopted Buddhism, which arrived from India. Since the main ethnic group of the kingdom was Burman, the region came to be called Burma. During the 1800s, Great Britain colonized Burma. Although Burma won its independence in 1948, it soon came under the control of a military dictator. In 1989, a new military government changed the country's name to Myanmar. A pro-democracy movement led by **Aung San Suu Kyi** (awng sahn soo chee) has struggled against this government ever since.

For hundreds of years, much of Malaysia was known as Melaka. This country was founded around 1400 and was strongly influenced by Arab Muslim traders. In the 1500s, Melaka came under the control of European colonizers and other rulers. Finally, in 1963, the states of Malay, Sabah, Sarawak, and Singapore united to form the Federation of Malaysia. Singapore left this group in 1965. Today, Malaysia is governed by a constitutional monarchy.

HISTORY MAKERS

Aung San Suu Kyi born 1945

Suu Kyi and other peace activists traveled throughout Myanmar, calling for democratic reforms and free elections. In 1989, after thousands of peaceful demonstrators had been shot and killed, Suu Kyi herself came face to face with an army unit that had been ordered to shoot her. At the last minute, however, an army major stepped in and withdrew the order. In 1991, while under house arrest, Suu Kyi's efforts to establish democracy in her country earned her the Nobel Peace Prize.

ONLINE BIOGRAPHY
For more on the life of Aung San Suu Kyi, go to the Research & Writing Center @ ClassZone.com

Culture and Economy Most of the people of Myanmar live in rural areas, while more than half the population of Malaysia lives in cities. About two-thirds of Myanmar's population are Burmese who practice Buddhism. Malaysia is ethnically much more diverse. About 60 percent of the people are Muslim.

Myanmar's economy is based on agriculture. Rice and teak are important crops. Malaysia, on the other hand, has an industrialized economy. The country produces a wide range of goods, including electronics and automobiles, which it trades on the international market.

COMPARE AND CONTRAST Compare and contrast Myanmar and Malaysia. Malaysia: democracy, thriving economy, ethnically diverse, about 60 percent Muslim; Myanmar: military government, agricultural economy, mostly Burmese and Buddhist

DIFFERENTIATING INSTRUCTION: TIERED ACTIVITIES

OBJECTIVE Learn more about the life, struggles, and achievements of Aung San Suu Kyi.

Basic

Have student pairs review the information in their textbooks on Suu Kyi. Then have them create a time line of her life. Invite students to research online to find additional information.

On Level

Have students write brief biographies of Suu Kyi, using print and online resources. Encourage students to add illustrations or other visuals to their biographies.

Challenge

Have students write a poem, song, or other creative piece about Suu Kyi's life, achievements, and struggles. Encourage students to include illustrations or other visuals with their work. Display students' work at all levels in the classroom.

Singapore

KEY QUESTION What type of economy and government does Singapore have?

Singapore is a densely populated island nation located at the southern tip of the Malay Peninsula. It is sometimes called a **microstate** because of its very small area—about 225 square miles. Singapore is a wealthy and powerful country. Its industrialized economy depends heavily on its exported goods, including electronics, chemicals, and clothing. In addition, the nation is a major financial center and has many banks and a stock exchange.

Officially, Singapore is a republic, but the government is dominated by one political party. This party limits freedom of speech and controls economic activities. Because Singapore favors order over liberty, the government enforces strict rules on its people. As a result, the country has a clean, corruption-free image throughout the world.

ndustrial-
zed economy
hat depends
on exports;
epublic, but
government
dominated by
one party ▸

SUMMARIZE Describe Singapore's economy and government.

Singapore The Central Area district is the hub of economic activity in Singapore.

Section 2 Assessment

ONLINE QUIZ For test practice, go to **Interactive Review @ClassZone.com**

TERMS & NAMES

1. **Explain the importance of**
   - Indochina
   - Khmer Empire
   - Siam
   - Aung San Suu Kyi

USE YOUR READING NOTES

2. **Summarize** Use your completed web diagrams to answer the following question:

   Which mainland countries appear to have the strongest economies?

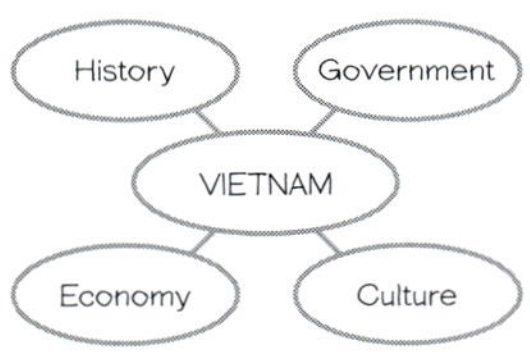

KEY IDEAS

3. How did the war between North Vietnam and South Vietnam affect Indochina?
4. How did Siam avoid colonization?
5. What is the Federation of Malaysia?

CRITICAL THINKING

6. **Identify Problems** In your opinion, what are the most difficult problems facing the mainland countries of Southeast Asia?
7. **Make Inferences** What influences might account for the fact that about 60 percent of the population of Malaysia is Muslim?
8. **CONNECT to Today** How might colonialism have affected the ability of some Southeast Asian countries to establish stable governments and economies?
9. **WRITING** **Write a Feature Article** Imagine that you have been assigned to write a feature on the Vietnamese holiday of Tet. Research to find images and information on the holiday's history and traditions.

Teach

Singapore

Reader, Recorder, Reporter

- **Summarize** Describe Singapore's government. *(Singapore is a republic that values order and restricts its people's freedom.)*

Unit Resource Book: East Asia & Southeast Asia
- Interdisciplinary Projects: Economic Growth in Singapore, p. 223

4 Assess & Reteach

Assess Have students complete the Section Assessment.

Unit Resource Book: East Asia & Southeast Asia
- Section Quiz, p. 230

Interactive Review @ClassZone.com
- Section Quiz

Power Presentations

Test Generator

Reteach Arrange the class into four groups. Assign each group one of the subsections of "The Mainland Countries." Have them list key ideas covered in the subsection and share their lists with the class.

Unit Resource Book: East Asia & Southeast Asia
- Reteaching Activity, p. 233

SECTION 2 ASSESSMENT ANSWERS

Terms & Names

1. Indochina: former French colony of Cambodia, Laos, and Vietnam; Khmer Empire: empire that once dominated the Southeast Asian mainland; Siam: former name of Thailand; Aung San Suu Kyi: pro-democracy leader in Myanmar

Use Your Reading Notes

2. Singapore, Malaysia, Vietnam

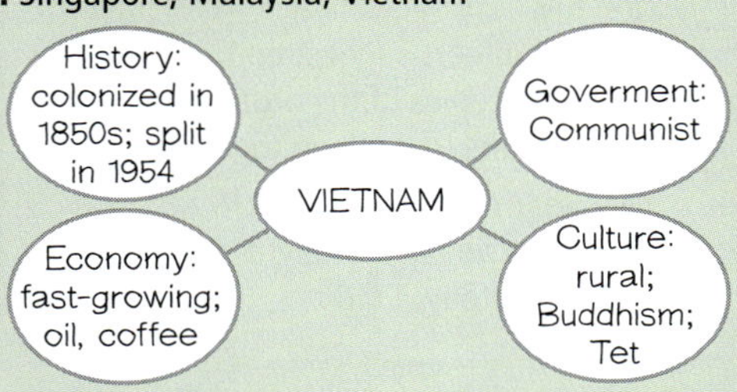

Key Ideas

3. The fighting began in Vietnam but spilled over into Cambodia and Laos.
4. It acted as a neutral zone between British-controlled Burma and French Indochina.
5. the united states of Malay, Sabah, Sarawak, and Singapore that formed in 1963

Critical Thinking

6. Possible answers: developing their economies, establishing more democratic governments
7. the strong influences of Muslim Arab traders
8. Possible answer: As a result of colonialism, they had little or no experience in managing their economies and governments.
9. Use the rubric to score articles.

Article Rubric

	Content	Errors
4	thorough understanding	few or none
3	good understanding	some
2	basic understanding	many
1	little or no understanding	many major

1 Plan & Prepare

Objectives

- Describe the history, government, culture, and economy of Indonesia
- Describe the history, government, culture, and economy of the Philippines

Read for the Essential Question

Help students read for a purpose by reminding them of the Essential Question: "How can Southeast Asian nations establish stable governments and economies?" Point out that they can use items in the section, such as Key Questions, special features, images, and maps to formulate answers to the question.

Vocabulary

Best Practices Toolkit

Use the ***Best Practices Toolkit*** to model strategies for vocabulary notetaking. Vary strategies throughout the year. Choose from: *Knowledge Rating, Predicting ABCs, Definition Mapping, Word Sort, Word Wheel, Frayer Model (Word Squares), Magnet Words,* and *Student VOC.*

Vocabulary Strategies, TT9–TT16

Reading Strategy

Best Practices Toolkit

Display the Three-Column Chart transparency.

- Tell students that when they categorize information, they organize similar material together.
- Model filling in the chart using the first paragraph under the "Indonesia" subsection.
- Have students suggest entries.
- Display the completed chart for review.

Three-Column Chart, TT21

Section 3 Reading for Understanding

Key Ideas

BEFORE, YOU LEARNED

The mainland countries of Southeast Asia are a mix of rural and urban cultures and wealthy and poor economies.

NOW YOU WILL LEARN

Indonesia and the Philippines both have rich cultural traditions and long histories as centers of trade.

Vocabulary

TERMS & NAMES

Bahasa Indonesia (bah•HAH•suh in•duh•NEE•zhuh) the language of Indonesia

shadow puppet traditional Indonesian art, in which puppets are handled from behind a screen lit by a bright light

Filipino an inhabitant of the Philippines

Ferdinand Marcos dictatorial president of the Philippines between 1965 and 1986

Corazon Aquino (uh•KEE•noh) politician who opposed Marcos and was president of the Philippines between 1986 and 1992

REVIEW

archipelago a set of closely grouped islands, which sometimes form a curved arc

Visual Vocabulary Ferdinand Marcos

Reading Strategy

Re-create the chart shown at right. As you read and respond to the **KEY QUESTIONS**, use the chart to categorize facts about the history, government, culture, and economy of Indonesia and the Philippines.

See Skillbuilder Handbook, page R7

CATEGORIZE

	INDONESIA	THE PHILIPPINES
History		
Government		
Culture		
Economy		

GRAPHIC ORGANIZERS
Go to **Interactive Review** @ClassZone.com

864 Chapter 28

PRETEACHING VOCABULARY

English Learners

Pronounce and Preview

Pronounce each term for students. Review the meanings of words used in the definitions, such as *traditional*, *dictatorial*, and *opposed*.

- To modify vocabulary learning, have students complete worksheets as they read the section.

Unit Resource Book: East Asia & Southeast Asia
- Vocabulary Practice, p. 215

Inclusion

Use Student VOC Strategy

Have students work in pairs to fill out the graphic organizer on the transparency for the terms and names in the section.

- Use the transparency to model the strategy.
- Have volunteers suggest entries.
- Distribute copies of the organizer for student pairs to work on together.

Best Practices Toolkit
- Vocabulary: Student VOC Strategy, TT16

Terraced rice fields in Bali, Indonesia

SECTION 3 FOCUS ON

Adult and young orangutan, which means "man of the forest" in the Malay language

Indonesia and the Philippines

Connecting to Your World

If you wanted to travel to a friend's house, would you try to get there by boat? Probably not. Most Americans journey from one place to another by car, bus, or train. However, on the islands of Southeast Asia, people often travel by boat. As a result, the surrounding seas serve as a sort of highway connecting the islands in this region. The seas, though, have also made it difficult to unify the island nations of Southeast Asia.

Indonesia

KEY QUESTION What unites and divides the people of Indonesia?

Indonesia is an **archipelago** that stretches for about 3,200 miles and consists of more than 17,500 islands. As you have already learned, the main islands are Java, Sumatra, and Borneo. Some of the smaller islands are uninhabited, but others are densely populated. Java, for instance, contains about 60 percent of Indonesia's population. Many cities in Indonesia are huge. For example, Jakarta, the country's capital, has more than 8 million people. Indonesia is the fourth most populous country in the world.

Jakarta Jakarta has broad boulevards, but traffic is a problem. To reduce congestion, some roads are only open to vehicles carrying three or more passengers.

865

2 Focus & Motivate

3-Minute Warm-Up

Write on the board or display the transparency:

- Look at the image of Jakarta on this page. How does Jakarta compare to a U.S. city? *(Possible answer: Both are crowded, with tall buildings and lots of traffic.)*

Unit Transparency Book: East Asia & Southeast Asia
- 3-Minute Warm-Ups, TT29

Connecting to Your World

More About . . .

Traveling by Boat

Pelni is the national Indonesian shipping company. Its modern ships are equipped with dorm-style sleeping areas and private cabins, canteens and dining rooms, mosques, and movie theaters.

3 Teach

Indonesia

Reader, Recorder, Reporter

- Which religions were introduced in Indonesia from distant lands? *(Buddhism, Hinduism, and Islam)*
- **Categorize** What are some of Indonesia's major industries? *(natural gas, oil, manufacturing, fishing, farming)*

SECTION 3 PROGRAM RESOURCES

Unit Resource Book: East Asia and Southeast Asia

ON LEVEL

Unit Resource Book
- Reading Study Guide, p. 195
- Section Quiz, p. 231

STRUGGLING READERS

Unit Resource Book
- Reading Study Guide with Additional Support, p. 201
- Vocabulary Practice, p. 215
- Section Quiz, p. 231
- Reteaching Activity, p. 234

eEdition with Audio DVD-ROM

ENGLISH LEARNERS

Pupil Edition in Spanish
eEdition with Audio DVD-ROM
Unit Resource Book
- RSG (Spanish), p. 207
- RSG with Additional Support (Spanish), p. 213

Multi-Language Glossary
Test Generator
- Section Quiz in Spanish

INCLUSION

Unit Resource Book
- RSG with Additional Support, p. 201
- Section Quiz, p. 231
- Reteaching Activity, p. 234

GIFTED & TALENTED

Unit Resource Book
- History Makers, p. 222
- Section Quiz, p. 231

PRE-AP

Unit Resource Book
- Interdisciplinary Projects, p. 224
- Section Quiz, p. 231

TECHNOLOGY

Unit Transparency Book
- 3-Minute Warm-Ups, TT29
- Essential Question, TT30
- Critical Thinking, TT31
- Map, TT34
- Map, TT35

Daily Test Practice Transparencies
- Section 3, TT95

Power Presentations
ClassZone.com
World Cultures and Geography Video on DVD

Comparing Regions

Shadow Puppets Around the World

Indonesia is not the only country that claims shadow puppetry as an art form. Shadow puppets also have a long tradition in India, Turkey, and China—and they have spread to Europe, too.

In China, the term *pi ying*, meaning "leather shadows," is used for the puppets, which are traditionally made of donkey skin decorated with intricate cut-outs. The puppets tend to be quite thin. Unlike in most other shadow puppet traditions, the color often shows on the screen.

In China, one person works the puppets, one person sings all the parts, and three others play instruments. Chinese shadow puppet shows often involve military subjects or recount folk tales or religious stories.

Fun Facts!

Shadow Puppets

One type of Javanese shadow puppet is called *Wayang Kulit*. *Wayang* means "shadow" or "ghost" and *Kulit* means "leather." The word for leather refers to the material used to make the puppets. Shadow puppets are made of colored animal hides, usually buffalo or goat, and have highly exaggerated features.

The *dalang*, or puppet master, moves the puppets and provides all the voices. An orchestra known as the *gamelan* accompanies the puppet show. Favorite Indonesian shadow puppet shows tell stories from the Indian epics the *Ramayana* and the *Mahabharata*.

Unit Resource Book: East Asia & Southeast Asia

- Interdisciplinary Projects: Making Batik, p. 224

History and Government As early as 1000 B.C., the region of Indonesia had become a crossroads for trade. Over the years, many goods and ideas arrived from distant lands. Indian traders brought Hinduism, and traders from both India and China carried Buddhism. By the 1300s, Muslim traders from Arabia had introduced Islam.

During the 1500s, Portugal gained control of present-day Indonesia and set up trade. The Dutch took over in 1641. They established the Dutch East India Company, a group that dominated trade on most of the islands. Indonesia gained independence from the Dutch in 1949 but was controlled by a military regime for many years. Today, Indonesia is a republic. In 2004, it held its first direct presidential elections. The new president pledged to promote democratic reforms.

Culture and Economy Most of the people of Indonesia live in rural areas. The population of the country is one of the most ethnically diverse in the world. About 300 ethnic groups make this island nation their home and speak about 250 different languages. For years, Indonesia lacked a common language. Then in the late 1920s, the nation created a language called **Bahasa Indonesia** that helped unify the people. Cultural traditions also unite Indonesians. Traditional arts include **shadow puppets**, in which puppets, such as those below, are worked from behind a screen lit by a bright light Ⓐ. On the other side of the screen, the audience sees the puppets as shadows Ⓑ.

From 1987 to 1997, Indonesia's economy grew rapidly. Then the nation suffered a financial slump when its money became devalued and its stock market fell. Indonesia's economy is still recovering. Today, oil and natural gas produce the greatest wealth in Indonesia. In fact, it is the world's largest exporter of liquefied natural gas. Manufacturing is also a major activity. Nonetheless, fishing and farming are a huge part of the economy, employing about half of all Indonesians.

DRAW CONCLUSIONS Identify some of the factors that unite and divide Indonesians. Bahasa Indonesia and cultural traditions unite; ethnic and religious diversity divide

Fun Facts!

SHADOW PUPPETS

The ancient art of shadow puppets is over 1,000 years old and probably originated in India and Egypt. Tiny holes are punched in the flat puppets to cast interesting shadow patterns. In the first image below, a handler works the puppets behind a screen under a bright light. The second image shows what the audience sees on the other side of the screen.

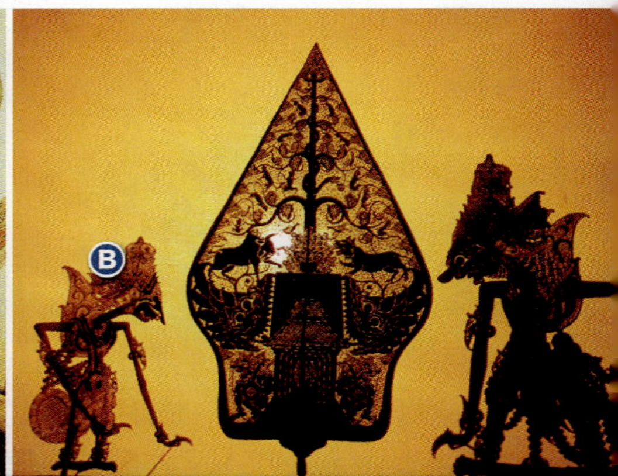

866

DIFFERENTIATING INSTRUCTION

Gifted & Talented

Present a Shadow Puppet Show

Invite students to work in groups to present a shadow puppet show. They can create simple shadow puppets using poster board and present them against a screen lit from behind. Have students create a short story to tell with their puppets and then present it to the class.

Pre-AP

Write an Argument

Have students work in small groups to discuss the advantages and disadvantages of a nation without a unifying language. Then have students work individually to write a brief argument in response to the following: Should countries have an official national language? Why or why not?

The Philippines

KEY QUESTION How has the Philippines improved its government and economy?

The Philippines is made up of around 7,100 islands, but only about 1,000 of them are inhabited. The nation is named for King Philip II of Spain, who sent explorers to the region during the 1500s. The people who live in the Philippines are called **Filipinos**.

History and Government Since ancient times, traders from China and the Malay region have had a strong influence on the culture of the Philippines. By the late 1500s, the Spanish Empire had colonized the Philippines. The Spanish controlled the area until 1898. After being defeated in the Spanish-American War, Spain handed control of the Philippines over to the United States. U.S. rule of the region lasted until 1946, when the Philippines gained its independence.

After independence, however, the country faced many economic problems. In 1965, **Ferdinand Marcos** became president and improved the economy. But in 1973, Marcos seized complete control of the government. **Corazon Aquino** (uh•KEE•noh) opposed Marcos during the 1986 election. Although the government declared that Marcos had won, many people suspected fraud. After much protest, Marcos left the country and Aquino took over as president. Today, the Philippines has a democratic government headed by a president.

ONLINE PRIMARY SOURCE To read more of Aquino's speeches, go to the **Research & Writing Center** @ClassZone.com

ANALYZING Primary Sources

Corazon Aquino (born 1933) was married to Benigno Aquino, Jr., a political rival of Ferdinand Marcos. When her husband was assassinated in 1983, Aquino suspected that Marcos was behind the death. She ran against Marcos in 1986 and became the first woman president of the Philippines. In the following, Aquino describes her political philosophy.

> I don't have any formula for ousting [overthrowing] a dictator or building democracy. . . . All I can suggest is to forget about yourself and just think of your people. It's always the people who make things happen.
>
> Source: *Time Asia*, August 1999, article by Sandra Burton

DOCUMENT-BASED QUESTION

According to Corazon Aquino, who builds a democracy?

DIFFERENTIATING INSTRUCTION: TIERED ACTIVITIES

OBJECTIVE Design and create a magazine feature on Corazon Aquino.

Basic

Have student pairs create a feature story (with a headline and a visual and caption) on Corazon Aquino's background, using information from their textbooks and online sources.

On Level

Have students work in groups to write a feature story (with a headline and a visual and caption) on Aquino's election. Encourage students to use print and online resources.

Challenge

Have students divide these tasks among their group: write a commentary about Aquino's term; create a political cartoon; write an account of Marcos fleeing the country. Then have all students work together to put the magazine feature together.

Teach

The Philippines

Think, Pair, Share

- What early traders strongly influenced the Philippines? *(traders from China and the Malay region)* What nations controlled the Philippines from the late 1500s to 1946? *(Spain and the United States)*
- **Categorize** List examples of some key characteristics of Filipino culture. *(Possible answers: includes Malay, Chinese, Spanish, and American influences; official languages include Filipino and English; much of the population is Roman Catholic; less than 50 percent of the people live in cities)*

RESEARCH & WRITING CENTER

ANALYZING Primary Sources

As president of the Philippines, Corazon Aquino restored her country's constitution and congress. She had less success with economic reforms, however. There were also several coup attempts during her time in office.

DOCUMENT-BASED QUESTION ANSWER

the people

Unit Resource Book: East Asia & Southeast Asia

- History Makers: Corazon Aquino, p. 222

CONNECT to the Essential Question

How can Southeast Asian nations establish stable governments and economies?

Ask students what they have learned so far that can help them answer this question. Students might mention the following:

- After gaining independence, Indonesia became a republic, with a strong president, while the Philippines formed a democratic government.
- Indonesia has rich oil and gas resources; the Philippines relies on farming but is developing its manufacturing and mining.

Unit Transparency Book
- Essential Question, TT30

4 Assess & Reteach

Assess Have students complete the Section Assessment.

Unit Resource Book: East Asia & Southeast Asia
- Section Quiz, p. 231

Interactive Review @ClassZone.com
- Section Quiz

Power Presentations

Test Generator

Reteach Have students write a summary of the information on Indonesia and the Philippines and share it with a partner.

Unit Resource Book: East Asia & Southeast Asia
- Reteaching Activity, p. 234

Rice Terraces in Luzon
Farmers first carved rice terraces into the mountains of Luzon about 2,000 years ago. The levels were reinforced with stone walls.

Culture and Economy Filipino culture blends Malay, Chinese, Spanish, and American influences. These influences can be seen in the country's languages and religions. The official languages of the country are Filipino and English. More than 80 percent of the people are Roman Catholic. In fact, the Philippines has more Christians than any other nation in Asia. Christianity was introduced when the Philippines became a Spanish colony. The country also has a large Muslim population. Forty-seven percent of the people live in cities. The largest city is the capital, Manila.

Traditionally, the economy of the Philippines has depended on farming. Over the last few decades, however, manufacturing and mining have developed significantly. Factories produce clothing, electronic goods, and wood products for export. The fertile volcanic soil of the islands helps farmers grow many crops, such as rice, sweet potatoes, bananas, and sugar cane. Kapok trees, which produce a fiber used for insulation and mattresses, are plentiful.

IDENTIFY SOLUTIONS Explain how the Philippines has improved its government and economy.

forced Marcos to step down and became more democratic; developed manufacturing and mining ◄

ONLINE QUIZ For test practice, go to **Interactive Review @ClassZone.com**

Section 3 Assessment

TERMS & NAMES

1. **Explain the importance of**
   - Bahasa Indonesia
   - Filipino
   - Ferdinand Marcos
   - Corazon Aquino

USE YOUR READING NOTES

2. **Categorize** Use your completed chart to answer the following question:

   What probably accounts for the great ethnic diversity in Indonesia?

	INDONESIA	THE PHILIPPINES
History		
Government		
Culture		
Economy		

KEY IDEAS

3. Why did Indonesia create a new language?
4. What problems did the Philippines face after gaining independence?
5. How did Corazon Aquino come to power?

CRITICAL THINKING

6. **Make Inferences** How did Corazon Aquino further the development of democracy in the Philippines?
7. **Compare and Contrast** How are the economies of Indonesia and the Philippines similar and different?
8. **CONNECT to Today** What do you think the people of Indonesia could do today to become a more united nation?
9. **ART** **Draw a Shadow Puppet** Using the Internet and library sources, find pictures of Indonesian shadow puppets. Then draw a shadow puppet of your own. Explain the role the puppet would play in a traditional performance.

SECTION 3 ASSESSMENT ANSWERS

Terms & Names

1. Bahasa Indonesia: language of Indonesia; Filipino: an official language of the Philippines; Ferdinand Marcos: dictatorial Filipino president; Corazon Aquino: president of the Philippines in 1986

Use Your Reading Notes

2. crossroad of trade; colonial past; islands developed distinct ethnic cultures

INDONESIA	THE PHILIPPINES
colonized in 1500s	colonized in 1500s
republic	democracy
ethnically diverse, rural	Malay, Chinese, Spanish, American influences
natural gas and oil	farming, manufacturing, mining

Key Ideas

3. to attempt to unite the diverse ethnic groups in Indonesia
4. economic problems and an unstable government
5. Aquino stepped in as president after voters protested Marcos's claim of victory.

Critical Thinking

6. by challenging and defeating the dictator Marcos
7. Possible answers: Both are based mostly on agriculture; however, the Philippines is trying to develop an economy that is based more on manufacturing.
8. Possible answers: improve communications and transportation; continue developing a government that represents all the people
9. Use the rubric to score drawings.

Drawing Rubric

	Drawing	Explanation
4	excellent attempt	shows thorough understanding
3	good attempt	shows good understanding
2	fair attempt	shows basic understanding
1	poor attempt	shows little or no understanding

Click here to complete these and other activities online @ClassZone.com

CHAPTER SUMMARY

Key Idea 1
Southeast Asia is a mountainous region with a tropical climate and much volcanic activity on its many islands.

Key Idea 2
The nations of mainland Southeast Asia have closely connected histories and diverse cultures.

Key Idea 3
Indonesia and the Philippines both have rich cultural traditions and long histories as centers of trade.

For **Review and Study Notes**, go to **Interactive Review** @ClassZone.com

NAME GAME

Use the Terms & Names list to complete each sentence on paper or online.

1. I am a narrow strip of land that serves as a bridge between the mainland and islands. **Malay Peninsula**
2. I was the dictatorial president of the Philippines between 1965 and 1986. ______
3. I am a very large ocean wave caused by an underwater earthquake or volcanic eruption. ______
4. I am a region that contains Cambodia, Laos, and Vietnam. ______
5. I had gained control of much of mainland Southeast Asia by the 800s. ______
6. I am the former name of Thailand. ______
7. I am a term for an independent country that is very small in area. ______
8. I am the leader of the pro-democracy movement in Myanmar. ______
9. I am the common language of Indonesia. ______
10. I am a set of closely grouped islands. ______

archipelago
Bahasa Indonesia
Indochina
Indochinese Peninsula
Khmer Empire
Malay Peninsula
Ferdinand Marcos
microstate
Siam
Aung San Suu Kyi
tsunami

Activities

Flip Cards

Use the online flip cards to quiz yourself on the terms and names introduced in this chapter.

Crossword Puzzle

Complete an online crossword puzzle to test your knowledge of the history and culture of Southeast Asia.

ACROSS
1. the most important holiday in Vietnam

Chapter Summary

As a follow-up to the chapter, encourage students to pair up and go over the **Review and Study Notes** at **ClassZone.com**.

Alternatively, student pairs can review their Reading Study Guide notes, making sure that they can add important details to each section's Key Ideas.

Name Game
Teaching Tips

ON PAPER Have students work in pairs to complete this activity. As an extension, they can write questions for terms that are not covered in the items shown here.

ONLINE Make sure that students understand how to drag and drop terms to fill in blanks. Ask them to print out their results and compare them with those of a classmate.

ANSWERS

1. Malay Peninsula
2. Ferdinand Marcos
3. tsunami
4. Indochina
5. Khmer Empire
6. Siam
7. microstate
8. Aung San Suu Kyi
9. Bahasa Indonesia
10. archipelago

ONLINE AND ON-SCREEN ACTIVITIES

Flip Cards

Vocabulary Flip Cards are virtual flash cards at **ClassZone.com.** Students can quiz each other or work independently to study the flip cards on the terms and names for Southeast Asia.

- They can quiz themselves on key terms and names by reviewing definitions.
- Demonstrate how to advance and shuffle the deck for another challenge.

Crossword Puzzle

Students can review the history and culture of Southeast Asia by completing this crossword puzzle online at **ClassZone.com.**

- As an extension of this activity, have students use the *Activity Maker* at **ClassZone.com** to make their own crossword puzzle for one section of the chapter.
- Then have students trade puzzles and try solving them.

Make It Fun

Teacher-Tested Activities

Matt Parker, Eastgate Middle School, Kansas City, Missouri

When we finish a chapter that covers several countries, I like to have my students make posters that summarize what they have learned.

- First, I divide the class into groups and assign each one a country.
- Then I have the groups decide how they will represent their countries. They can use text, charts and graphs, images with captions, and time lines.
- Once the posters are completed, I display them in the classroom.

Vocabulary ANSWERS

1. one of two peninsulas on which mainland Southeast Asia lies
2. the second peninsula, which serves as a bridge between the mainland and islands
3. major river that runs south from China through Laos, Cambodia, and Vietnam
4. former French colony made up of Laos, Cambodia, and Vietnam
5. empire that had gained control of much of the Southeast Asian mainland by the 800s
6. former name of Thailand
7. pro-democracy leader in Myanmar
8. independent country that is very small in area and population
9. the common language of Indonesia
10. an inhabitant of the Philippines and the island's language
11. Unlike Indochina, Siam avoided being taken over by colonialism.
12. Corazon Aquino became president of the Philippines after Marcos left the country.

Key Ideas ANSWERS

13. Indonesia and the Philippines
14. an earthquake or volcanic eruption on the ocean floor
15. a tropical wet climate in the coastal areas of Myanmar, Thailand, and Vietnam and in most of Malaysia, Indonesia, and the Philippines; a tropical wet and dry climate in much of mainland Southeast Asia
16. The tropical climate fosters tropical vegetation.
17. colonialism and the Vietnam War
18. started schools, built railroads, and abolished slavery
19. a wide range of products, including electronics and automobiles
20. It limits freedom, enforces strict rules on its people, and controls economic activities.
21. Indian traders brought Hinduism; traders from both India and China brought Buddhism; Muslim traders from Arabia brought Islam.
22. in rural areas
23. Many Filipinos suspected that the 1986 election was dishonest; they protested and drove Marcos out of the Philippines.
24. Chinese, Malay, Spanish, American

CHAPTER 28 Assessment

VOCABULARY

Explain the significance of each of the following.

1. Indochinese Peninsula
2. Malay Peninsula
3. Mekong River
4. Indochina
5. Khmer Empire
6. Siam
7. Aung San Suu Kyi
8. microstate
9. Bahasa Indonesia
10. Filipino

Explain how the terms and names in each group are related.

11. Indochina, Siam, and colonialism
12. Ferdinand Marcos and Corazon Aquino

KEY IDEAS

1 Physical Geography of Southeast Asia

13. Which island nations are located on the Malay Archipelago?
14. What causes a tsunami?
15. Which two types of climate predominate in Southeast Asia?
16. How does the climate in Southeast Asia affect its vegetation?

2 The Mainland Countries

17. What historical events have affected all of Indochina?
18. How did King Chulalongkorn modernize Siam?
19. What does Malaysia trade on the international market?
20. In what ways is the government of Singapore powerful?

3 Indonesia and the Philippines

21. What religions did traders bring to Indonesia?
22. Where do most of the people of Indonesia live?
23. Why did Ferdinand Marcos leave the Philippines?
24. What cultures have strongly influenced the Philippines?

CRITICAL THINKING

25. **Analyze Causes and Effects** Create a chart to identify the impact in Southeast Asia of colonization, the war between North and South Vietnam, and the Asian tsunami.

CAUSES	EFFECTS
Colonization	
War between North and South Vietnam	
Asian tsunami	

26. **Form Opinions** Why do you think Southeast Asia attracted so many colonists?
27. **Make Inferences** Why do you think Aung San Suu Kyi is considered a hero throughout much of the world?
28. **Draw Conclusions** How are Spanish and American influences reflected in Filipino culture?
29. **Five Themes: Location** How has Singapore's location benefited its economy?
30. **Connect Geography & Culture** Why do you think the tourism industry is booming in Thailand?

Answer the ESSENTIAL QUESTION

How can Southeast Asian nations establish stable governments and economies?

Written Response Write a two- or three-paragraph response to the Essential Question. Use the rubric below to guide your thinking.

Response Rubric

A strong response will:

- describe the governments and economies of the Southeast Asian nations today
- discuss the problems the nations have faced in the past
- detail the steps the nations are taking to improve their governments and economies

ESSENTIAL QUESTION REVIEW

Before students begin their essays, have them review the key chapter elements that will help them answer the Essential Question. Direct them to significant features in the chapter such as:

- Connect Geography & History: The 2004 Asian Tsunami, Section 1
- Comparing Regional Rainfall, Section 1
- Connect Geography & History: Colonies in Southeast Asia 1895, Section 2
- History Makers: Aung San Suu Kyi, Section 2
- Analyzing Primary Sources: Corazon Aquino, Section 3

Essential Question Rubric

	Coverage of Key Ideas	Explanation	Errors in Mechanics
4	all	clear and concise	few or none
3	most	good	some
2	some	fair	many
1	one or none	poor	many major

STANDARDS-BASED ASSESSMENT

Test Practice
- **Online Test Practice @ClassZone.com**
- **Test-Taking Strategies and Practice** at the front of this book

THEMATIC MAP

Use the map to answer questions 1 and 2 on your paper.

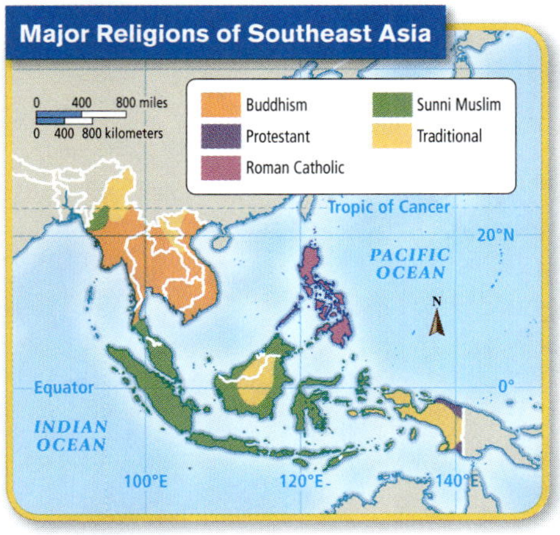

1. What is the predominant religion of mainland Southeast Asia?

A. Roman Catholic **C.** Traditional
B. Buddhism **D.** Sunni Muslim

2. Which country has the largest population of Sunni Muslims?

A. the Philippines **C.** Indonesia
B. Cambodia **D.** Thailand

BAR GRAPH

Examine the bar graph below. Use the graph to answer questions 3 and 4 on your paper.

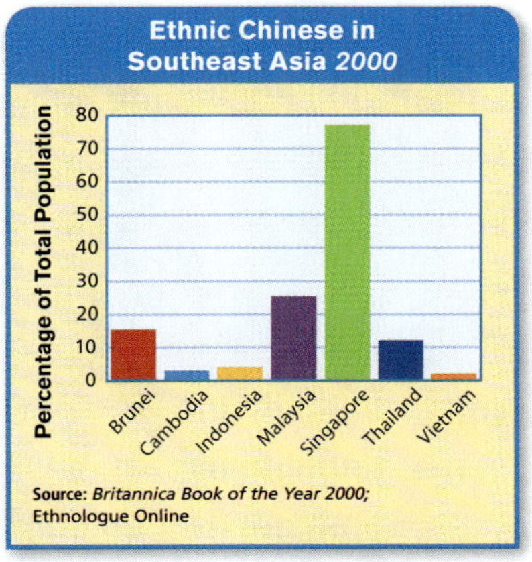

3. Which Southeast Asian nation has the highest percentage of ethnic Chinese?

A. Singapore **C.** Indonesia
B. Malaysia **D.** Thailand

4. What percentage of Vietnam's population is made up of ethnic Chinese?

A. 77.0 **C.** 12.1
B. 15.4 **D.** 1.5

GeoActivity

1. INTERDISCIPLINARY ACTIVITY–SCIENCE

With a small group, prepare a poster on tsunamis. Research to find information on the size and speed of the waves, how they are formed, and possible warning systems. Use pictures and diagrams to illustrate the information.

2. WRITING FOR SOCIAL STUDIES

Unit Writing Project Revisit your country slide show. To further convince your family to go to the country, write a detailed description of a place that best represents it. For Cambodia, for instance, you might write about Angkor Wat.

3. MENTAL MAPPING

Create an outline map of Southeast Asia and label the following:
- Pacific Ocean
- Mekong River
- Indochinese Peninsula
- Indochina
- Malay Peninsula
- Malay Archipelago
- Indonesia
- the Philippines

Critical Thinking ANSWERS

25. Possible answers: Colonization: control of economies was taken away; War between North and South Vietnam: brought communism to Vietnam, Laos, and Cambodia; Asian tsunami: devastated much of Indonesia and parts of Thailand

26. Possible answer: They were attracted by the region's resources and proximity to East Asia.

27. because she continues to lead a peaceful pro-democracy movement in Myanmar

28. Spanish: large percentage of Roman Catholics; American: English is an official language.

29. Singapore is located near the middle of Southeast Asia and has easy access to both the Indian and Pacific oceans.

30. warm climate, unique culture, Thai food is popular

Standards-Based Assessment ANSWERS

1. B
2. C
3. A
4. D

GeoActivity

1. INTERDISCIPLINARY ACTIVITY RUBRIC

The poster should
- detail the causes of tsunamis
- use visuals to explain the formation
- describe possible warning systems

2. WRITING FOR SOCIAL STUDIES RUBRIC

The description should
- create vivid word pictures of the place
- help persuade the student's family to visit the subject of the slide shows

3. MENTAL MAPPING RUBRIC

The outline map should
- represent the approximate shape of Southeast Asia
- include all geographic features listed in their correct location

DIFFERENTIATING CHAPTER ASSESSMENT

Unit Resource Book
- **Chapter Test A,** pp. 235–238. *Use for Struggling Readers, Beginning-to-Intermediate-Level English Learners, and for Inclusion of students whose IEP suggests supported test-taking.*
- **Chapter Test B,** pp. 239–242. *Use for On-Level Learners and ESOL students with Advanced-Level English.*
- **Chapter Test C,** pp. 243–246. *Use for Gifted and Talented students or students assigned Pre-AP work.*

Test Generator
- Chapter Tests A, B, and C, in English and Spanish

McDougal Littell Assessment System

TEST **Test Generator** Use three levels of tests in English and Spanish tied to state standards. Plus, choose questions to customize tests for all populations.

SCORE Scan and Score plain paper student tests in minutes.

REPORT Create instant progress reports, sortable by student, by standard, or other criteria.

RETEACH Respond immediately to students' needs by using a wealth of reteaching resources.

N
E
W
S

UNIT

9

Oceania and Antarctica

Why It Matters:

As modern technology shrinks our world, the people of the lands of Oceania have become much closer neighbors. Their lands include resources important to world trade. Their unique cultures add to the diversity of our world.

Two friends in Western Australia

CHAPTER 29 Australia and New Zealand

A War Canoe in the Solomon Islands

CHAPTER 30 Oceania and Antarctica

Meet Mick

In this unit, you will meet Mick, a middle-school boy from the Outback in Australia. Learn more about him on **Worldster** @ClassZone.com

873

Objectives

- Locate Antarctica and the 14 countries of Oceania on a map
- Compare the differences in population density among the countries of Oceania
- Explain the varied climates found in Oceania and Antarctica
- Identify the major economic activities in Oceania
- Summarize the geography, history, culture, government, and economics in Oceania

Comparing Data

Remind students that both Australia and Antarctica are considered continents. Although the landmass of Oceania and Antarctica is larger than the landmass of the United States, the region has only about 1/10th the population of the United States.

Next to Australia, the largest island nation in Oceania, Papua New Guinea, is about 85 times larger than Hawaii, the largest island in the United States.

The combined number of volcanoes in Oceania and Antarctica is larger than the number of volcanoes in the United States.

Note that no rivers are listed for these regions, as their rivers are tiny compared to the world's major rivers.

ANSWERS

1. The landmass is about the same.
2. about ten times as large
3. Greenland is larger by about 781,560 square miles, or about 14 times as large.
4. The region has about 180 active volcanoes compared to 156 in the United States.

UNIT 9 ATLAS Oceania and Antarctica

Use the Unit Atlas to learn about the geography of Oceania and Antarctica. As you study the maps in the atlas, notice geographic patterns and specific details about the two regions. Answer the questions on each map in your notebook.

As you study the graphs on this page, compare the landmass, population, islands, and volcanoes of Oceania and Antarctica with those of the United States. Then jot down the answers to the following questions in your notebook.

Comparing Data

1. How does the landmass of Oceania compare with that of the United States?
2. How much larger is the population of the United States than that of Oceania?
3. How does the South Island of New Zealand compare with the world's largest island?
4. How does the number of active volcanoes for this entire region compare to the United States?

Comparing Data

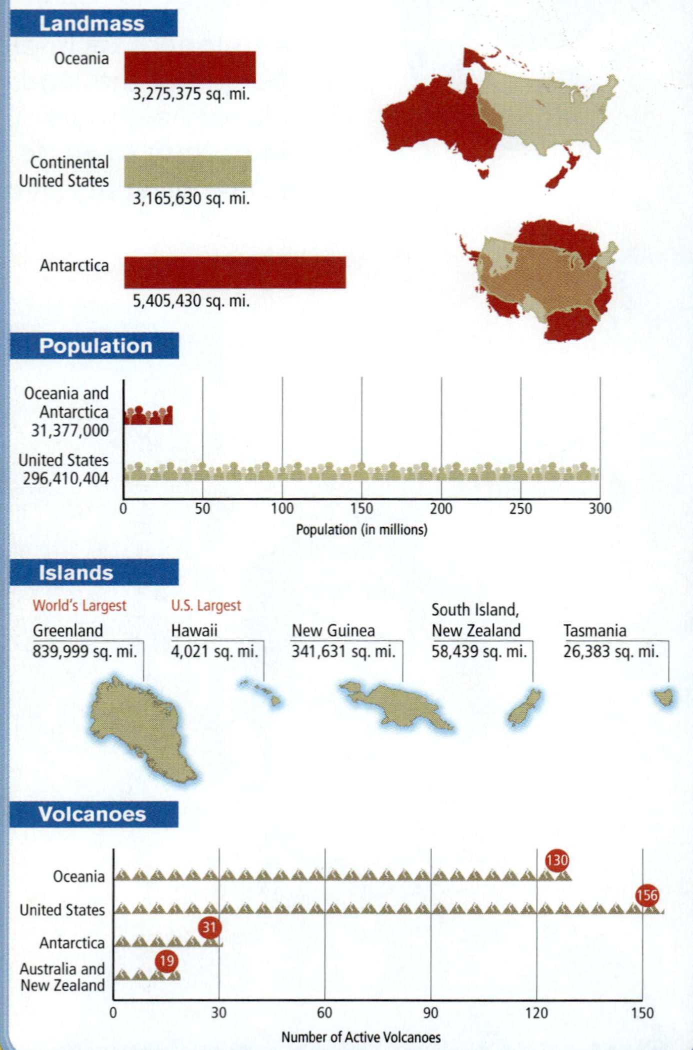

FIVE THEMES AND SIX ELEMENTS

Five Themes

Movement

Make sure that students comprehend that Oceania is made up of the continent of Australia and many island nations spread across hundreds of thousands of square miles of ocean. Invite them to consider the challenges of transporting people and products across such vast distances.

Six Elements

Places and Regions

Explain that not all of the islands in the region are independent nations. Some remain territories of other nations, including Pitcairn Island *(United Kingdom)*, New Caledonia, Wallis and Futuna, and French Polynesia *(France)*, and Northern Mariana Island and American Samoa *(United States)*.

Population Density

THINK LIKE A GEOGRAPHER

1. **Place** Which island group is most densely populated?
2. **Human-Environment Interaction** How would you describe the population density pattern of Australia?

875

Population Density

Much of Oceania is sparsely populated. The Marshall Islands, a chain of many small islands, is one of the more densely populated nations. Australia and New Zealand, two nations with larger land areas, are both lightly populated.

- Which countries on the map have areas with population densities of 1–24 persons per square mile? *(Australia, New Zealand, and Papua New Guinea)*
- Why do you think large areas of central Australia are less densely populated than the rest of Australia? *(Possible Answer: The hot, dry climate and physical geography make it difficult for people to live there.)*

NOTE: Areas of color on small islands have been exaggerated to help students read the population densities more easily.

THINK LIKE A GEOGRAPHER ANSWERS

1. the Marshall Islands
2. Australia is very lightly populated, except for a few cities on its coast.

More About . . .

Sydney, Australia

Sydney is the largest city in Australia, with a population of about four million people. A major port, as well as an industrial and cultural center, it has experienced tremendous growth and redevelopment since the end of World War II. Today Sydney has a variety of manufacturing businesses, universities, and museums. It's also known for its architectural landmarks such as the Sydney Opera House, a dramatic structure that resembles a sailing ship. In 2000, Sydney was the site for the Summer Olympic Games.

DIFFERENTIATING INSTRUCTION

Struggling Readers

Pronounce Place Names

Model the pronunciation of Oceanic countries to help students who might encounter difficulties decoding them. In particular, focus on the varied sounds represented by the letter *u* in such country names as Palau, Nauru, Papua New Guinea, and Australia.

Inclusion

Locate Low Population Densities

Review the categories of population density that appear in the map key. Then have students locate the equator. Ask them to find the area with the lowest population density north of the Equator *(Palau)* and the area with the lowest population density south of the Equator *(Australia)*.

Climate

Oceania and Antarctica have extreme differences in climate zones. Much of Australia's climate is desert and semiarid, while New Zealand has a marine west coast climate. Many island nations in the Pacific have a tropical wet or tropical wet and dry climate. Antarctica, on the other hand, is covered almost completely by an icecap climate, with a small area of tundra.

- Which climate zone does French Polynesia have? *(tropical wet and dry)*
- Look at the map. Why do you think Australia has so many different climate zones? *(Possible Answer: because of its size and location along the Tropic of Capricorn)*

THINK LIKE A GEOGRAPHER ANSWERS

1. Marine west coast
2. Tropical Wet
3. Desert

More About . . .

The Southern Ocean

The Southern Ocean is a world-encircling stretch of water that for a long time was just considered an extension of the Atlantic, Indian, and Pacific oceans. Have students look at a globe from the South Pole. It will be obvious that this is an enormous stretch of water. However, its limits are not bounded by continental coasts as with other oceans. Geographers set the boundaries based on a marine transition where the cold, dense waters of the Southern Ocean meet the warmer waters of the other three oceans. There is some disagreement about the location of the northern boundary, but between 50 and 60 degrees south latitude is commonly accepted.

Climate

THINK LIKE A GEOGRAPHER

1. **Region** Which climate do Australia and New Zealand have in common?
2. **Region** What is Australia's smallest climate region?
3. **Region** Which climate region occupies the greatest area of Australia?

FIVE THEMES AND SIX ELEMENTS

Five Themes

Location

Remind students that the term *Tropic of Capricorn* refers to a line of latitude south of the equator that marks the southern border of the tropics. During the summer solstice in the Southern Hemisphere (around December 22), the sun can be seen directly overhead at noon. The name Capricorn originally came from the arrival of the sun in the constellation Capricorn.

Six Elements

Places and Regions

Point out to students that many scientific researchers are currently working in Antarctica because its icecap climate is useful for studying global warming and other issues related to climate change. It was not until 1997 that the first satellite images of the entire continent were taken, which allowed scientists to better study the Antarctic icecap as well as features below its surface to gain more reliable data for their studies.

Economic Activity

THINK LIKE A GEOGRAPHER

1. **Human-Environment Interaction** Which of the economic activities occupies the most land in Australia and New Zealand?
2. **Place** In which nations is commercial fishing important?
3. **Place** What resources are found in the nation of Papua New Guinea?

877

Economic Activity

The economic activities of Oceania's countries are based in large part on their major resources, including coal, gold, minerals, timber and forest products, and fish. Many island nations rely on subsistence agriculture and forestland for economic resources.

- In addition to timber, what is the major resource in Samoa? *(fish)*
- Where is the major manufacturing center in New Zealand located? *(Auckland)*
- What are three types of economic resources found in Australia? *(Possible Answers: bauxite, coal, gold, iron ore, natural gas, other minerals, silver, tin, copper, and diamonds)*
- How does land use in French Polynesia differ from land use in Papua New Guinea? *(Subsistence agriculture is the main land use in French Polynesia; forestland is the main use in New Guinea.)*

THINK LIKE A GEOGRAPHER ANSWERS

1. livestock raising
2. Northern Mariana Islands, Solomon Islands, Vanuatu, American Samoa
3. gold, oil, natural gas, and fish

Comparing Regions

Economic Activity in Australia and Papua New Guinea

Both Australia and Papua New Guinea have significant mineral resources. Oil, copper, and gold account for a large part of Papua New Guinea's exports. Minerals and agricultural products make up a large portion of Australia's exports, but it also exports manufactured goods that Papua New Guinea does not.

Obtaining available minerals in Papua New Guinea is difficult because the terrain is rugged and the country lacks a solid infrastructure to transport or to manufacture goods from the raw materials.

Australia has the capability to manufacture and to ship both finished goods or raw materials to customers, particularly in countries on the Pacific Rim. Both nations count Japan and China as significant trading partners.

DIFFERENTIATING INSTRUCTION

English Learners

Comprehension: Connection

Model the use of the preposition *of* with direction. New Caledonia, for example, is *south of* the Marshall Islands and *north of* New Zealand. Encourage students to make similar statements about the relative locations of Oceania's countries, cities, and economic activities.

Gifted & Talented

Comparing Economic Activities

Have pairs of students use the Unit Atlas maps of the United States and one other country to compare the economic activities with a country in Oceania. Have students create a chart or other visual display to show these comparisons.

Regional Overview

Point out that the region of Oceania and Antarctica is a vast area made up of diverse geographical features including deserts, volcanoes, and a polar icecap. Although some people in Oceania have had little contact with the outside world until recently, others, such as Australians and New Zealanders, have long had close international relationships with other people around the world through trade and most recently tourism.

- Why does this region have unique plants and animals? *(The region was isolated from other parts of the world for a long period of time.)*
- How did people migrating from eastern Asia settle Oceania? *(They moved across land bridges or the ocean.)*
- How would you describe Antarctica? *(It's a frozen desert.)*
- How do advances in communications technology help scientists working in Antarctica? *(Possible Answer: Communicating through the Internet allows scientists in Antarctica to more easily send data back and forth to other scientists.)*

More About . . .

Roald Amundsen

In the early 1900s, Norwegian explorer Roald Amundsen was focused on trying to reach the North Pole. However, when he heard that another explorer had reached the North Pole ahead of him, Amundsen headed south. He reached the South Pole on December 14, 1911—just 34 days ahead of another explorer, Britain's Robert Scott. Amundsen and his team completed the last leg of their journey by dog team and on skis.

Unit 9 Atlas: Regional Overview

Oceania and Antarctica

This region contains two continents and thousands of islands spread across hundreds of thousands of square miles of the Pacific Ocean. Until recently the area was considered distant and remote. But with electronic devices and fast transportation, even remote lands have been integrated into the world at large.

GEOGRAPHY

This region contains a wide range of physical geography. Australia is a vast dry land, while New Zealand has active volcanoes and earthquakes. Antarctica is a frozen desert. Thousands of islands in the Pacific are atolls or parts of volcanic cones.

The isolation from the rest of the world led to the development of unique plants and animals in Australia and New Zealand.

Roald Amundsen
This Norwegian explorer was the first to reach the South Pole.

HISTORY

People migrating from eastern Asia populated the lands of this region. They moved across land bridges or the ocean and settled on the islands. Australia and New Zealand and the islands of Oceania were later colonized mostly by European nations.

CULTURE

The cultures of Australia and New Zealand reflect some of the customs of the United Kingdom, which colonized them. However, their cultures also have many aspects unique to their way of life. The people of the islands of Oceania had little contact with the outside world so they developed their own ways. Only recently have these island cultures begun to connect with the outside world.

FIVE THEMES AND SIX ELEMENTS

Five Themes

Location

Point out that the most densely populated areas of the region are island nations or territories. Ask volunteers to describe these islands in relative and specific terms. Then discuss the advantages and disadvantages of being an island nation in Oceania.

Six Elements

Environment and Society

Have students use the Climate map in this section to identify the types of climate found in Australia. Next have them use the Economic Activity map on the previous page to locate areas of limited agriculture in western Australia. Ask volunteers to explain why it might be difficult to grow crops there. Discuss the possible advantages and disadvantages of Australia's physical geography and climates in terms of its economic activities.

GOVERNMENT

Many of the countries of this region were colonies of European nations or the United States. Many of them gained their independence after World War II. The governments of the two largest nations, Australia and New Zealand, are linked to the United Kingdom. The governments of most of the nations are democratic.

ECONOMICS

Australia and New Zealand have well developed economies that focus primarily on producing raw materials such as wool, meat, and minerals. These products are traded in the international marketplace. Both Australia and New Zealand, along with the island nations of Oceania, have begun to expand their tourism industries.

Emperor Penguins
Colonies of many types of penguins are found in the Antarctic region.

879

More About . . .

Emperor Penguins

Emperor penguins huddle together on the open ice of Antarctica to protect themselves against cold temperatures and wind. They are the only creatures that do not leave Antarctica during its brutal winter, which begins in March.

The emperor penguins' breeding season occurs in winter, unlike most birds. Their unusual breeding ritual involves male penguins guarding eggs while females travel up to 50 miles across the ice in search of food. While the females search for food, the males stay standing up—and have nothing to eat—for about two months. Once the females return, the males leave for the sea to find food for themselves. The females then take over the care of the newly-hatched chicks.

DIFFERENTIATING INSTRUCTION

Struggling Readers

Comparing Land-Use Areas

Using the Economic Activity map, explain the meanings of the six land-use areas. Help students list specific examples of the economic activities that might be found in each area. Encourage them to explain the likely relationships between climate and land use.

English Learners

Comprehension: Task Support

To help students better understand the relationship between climate and economics, review the conditions that one would encounter in each climate zone shown on the map. Have small groups find areas that the travel industry might consider good destinations for beach vacations, such as the humid subtropical region in Australia.

Unit Writing Project

Teaching Tips

- As students prepare their illustrated time lines, encourage them to seek out a variety of reliable sources to use, including newspapers, magazines, and Web sites. You may also want to arrange for students to interview local experts on Oceania at school.
- Provide students with an example of an illustrated time line, or have them research typical formats.
- Remind students that use of color and interesting visuals will help to illustrate key points on their time lines as well as make their presentations more engaging.
- Encourage students to rehearse presenting their time lines with a partner and to revise them based on his or her feedback.

Each map on the country cards illustrates the shape of a state, province, or country. Tell students that the maps do not reflect the comparative size of these landmasses within a particular region. To get an idea of the actual size of the state, province, or country, students should study the land area statistics provided on the card.

UNIT 9 ATLAS Country Almanac

Click here to compare the most recent data on countries @ClassZone.com

Unit Project

Illustrated Time Line

As you read this unit, look for a description of the migration of people across the region. Do research to find information about people settling the lands. Create an illustrated time line showing the arrival of different groups in the subregions and countries of Oceania.

Think About:

- the first people to arrive in the region
- European colonizers

Oceania and Antarctica

Many of the nations in this region are tiny. The smallest nation is Nauru, which covers only 8.2 square miles with about 13,000 people. The region of Oceania, on the other hand, is gigantic—3,275,375 square miles and 32 million people.

Water Lily Water Lily on an island in Fiji

Australia

GEOGRAPHY
Capital: Canberra
Total Area: 2,967,909 sq. mi.
Population: 20,155,000

ECONOMY
Imports: machinery; chemicals; fuels; food
Exports: fuels; metals; textiles; food

CULTURE
Language: English
Religion: Catholic 27%; Protestant 37%; Orthodox 3%; no religion 16%

Fiji

GEOGRAPHY
Capital: Suva
Total Area: 7,054 sq. mi.
Population: 848,000

ECONOMY
Imports: fuels; machinery; textiles; animals
Exports: clothing; sugar; fish; gold

CULTURE
Language: English; Fijian; Hindustani
Religion: Christian 57%; Hindu 33%; Muslim 7%; nonreligious 1%

Kiribati

GEOGRAPHY
Capital: Tarawa
Total Area: 313 sq. mi.
Population: 103,092

ECONOMY
Imports: food; machinery; fuels
Exports: seaweed; fish; copra

CULTURE
Language: English
Religion: Catholic 55%; Protestant 37%; Mormon 3%; Baha'i 2%

Marshall Islands

GEOGRAPHY
Capital: Majuro
Total Area: 70 sq. mi.
Population: 59,071

ECONOMY
Imports: fuels; machinery; food
Exports: copra; coconut oil; aquarium fish

CULTURE
Language: Marshallese; English
Religion: Protestant 63%; Catholic 7%; Mormon 3%

Federated States of Micronesia

GEOGRAPHY
Capital: Palikir
Total Area: 271 sq. mi.
Population: 110,000

ECONOMY
Imports: food; fuels; machinery; tobacco
Exports: fish; bananas

CULTURE
Language: English; Trukese
Religion: Catholic 53%; Protestant 42%; Mormon 1%

Nauru

GEOGRAPHY
Capital: Yaren District
Total Area: 8 sq. mi.
Population: 13,048

ECONOMY
Imports: agricultural products
Exports: phosphate

CULTURE
Language: Nauruan; English
Religion: Protestant 54%; Catholic 28%

INTERDISCIPLINARY ACTIVITIES

CONNECT to Math

Calculating Population Percentages

Review with students the method for calculating percentages of a whole number. Point out that the entire region of Oceania has 32 million people. Then have students calculate the percentage of that 32 million made up by Australia, New Zealand, Papua New Guinea, and five other nations in the region. Students may also create a pie chart using the percentages.

CONNECT to Science

Create a Map of Antarctic Wildlife Migrations

Have students find information about a fish, bird, or other creature that migrates to or from Antarctica or Oceania. Have them draw an outline map of the region and put in the migration paths. Encourage students to include visuals as part of their maps or as a supplement (such as a poster or collage). Have students display their maps.

New Zealand

GEOGRAPHY
Capital: Wellington
Total Area: 103,738 sq. mi.
Population: 4,028,000

ECONOMY
Imports: machinery; petroleum; cars
Exports: beef; dairy products; wood

CULTURE
Language: English; Maori
Religion: Protestant 27%; Catholic 13%; no religion 27%

Palau

GEOGRAPHY
Capital: Koror
Total Area: 177 sq. mi.
Population: 20,303

ECONOMY
Imports: machinery; food; fuels; chemicals
Exports: high-grade tuna; clothing

CULTURE
Language: Palauan; English
Religion: Catholic 42%; Protestant & other Christian 30%; Modekngei 9%

Papua New Guinea

GEOGRAPHY
Capital: Port Moresby
Total Area: 178,704 sq. mi.
Population: 5,887,000

ECONOMY
Imports: petroleum; food; machinery
Exports: gold; petroleum; copper; logs

CULTURE
Language: English; Motu; Tok Pisin
Religion: Protestant 63%; Catholic 30%; traditional beliefs 4%; Baha'i 1%

Samoa

GEOGRAPHY
Capital: Apia
Total Area: 1,137 sq. mi.
Population: 185,000

ECONOMY
Imports: petroleum; machinery
Exports: fish; clothing; coconut products

CULTURE
Language: Samoan; English
Religion: Protestant 57%; Mormon 26%; Catholic 21%

Solomon Islands

GEOGRAPHY
Capital: Honiara
Total Area: 10,985 sq. mi.
Population: 478,000

ECONOMY
Imports: food; petroleum; machinery
Exports: timber; fish; cacao beans

CULTURE
Language: English
Religion: Protestant 74%; Catholic 11%; traditional beliefs 3%

Tonga

GEOGRAPHY
Capital: Nuku'alofa
Total Area: 289 sq. mi.
Population: 102,000

ECONOMY
Imports: food; fuels; machinery
Exports: squash; fish; vanilla beans

CULTURE
Language: Tongan; English
Religion: Free Wesleyan 41%; Catholic 16%; Mormon 14%

Tuvalu

GEOGRAPHY
Capital: Funafuti
Total Area: 10 sq. mi.
Population: 11,636

ECONOMY
Imports: food; fuels; machinery; metals
Exports: copra; stamps; handicrafts

CULTURE
Language: Tuvalan; English; Samoan
Religion: Church of Tuvalu 85%; Seventh-day Adventist 4%; Catholic 1%

Vanuatu

GEOGRAPHY
Capital: Port-Vila
Total Area: 4,710 sq. mi.
Population: 211,000

ECONOMY
Imports: machinery; food; chemicals; fuels
Exports: coconut oil; timber; beef; kava

CULTURE
Language: Bislama; French; English
Religion: Protestant 54%; Catholic 16%; traditional beliefs 4%; Baha'i 3%

Outrigger canoe The lands of Oceania were settled by people who came there in oceangoing outrigger canoes.

881

Country Almanac

Oceania and Antarctica are vast regions including three oceans and two continents—Australia, the smallest continent, and Antarctica, the only continent with no permanent residents. Antarctica is also the world's largest desert region, even though it is largely covered by ice year round. (Generally a desert region gets less than 10 inches of rain each year, as Antarctica does. Unlike most deserts, Antarctica's precipitation generally forms ice rather than evaporating.)

- Which Oceanic country has the second smallest land area? *(Tuvalu)*
- Which Oceanic country has the second largest population? *(Papua New Guinea)*
- Which countries export copra (dried coconut meat)? *(Kiribati, Marshall Islands, and Tuvalu)*
- In which countries is English the main language? *(Australia, Fiji, Kiribati, Federated States of Micronesia, New Zealand, Papua New Guinea, and Solomon Islands)*

INTERDISCIPLINARY ACTIVITIES

CONNECT to Language Arts

Describe Ocean Travel in a Small Boat

Have students write a paragraph in the present tense that describes the photograph of the settlers in the outrigger canoe. After they have set the scene, ask them to imagine and describe what an ocean voyage might be like in such a small boat.

Each map on the country cards illustrates the shape of a state, province, or country. Tell students that the maps do not reflect the comparative size of these landmasses within a particular region. To get an idea of the actual size of the state, province, or country, students should study the land area statistics provided on the card.

CHAPTER 29

Australia and New Zealand PLANNING GUIDE

Complete print resources all on one USB drive allow you to customize lesson plans for all students.

Section Titles & Key Ideas	Unit Resource Book Oceania and Antarctica	Assessment Resources
1 Unique Lands **KEY IDEA** Australia and New Zealand have plant and animal life found nowhere else on Earth.	• **Reading Study Guide,** pp. 3–4 • **RSG with Additional Support,** pp. 9–10 • **Vocabulary Practice,** p. 27 • **Outline Map Activities,** p. 31	**Unit Resource Book** • Section Quiz, p. 43 • Reteaching Activity, p. 46 **Test Generator CD-ROM** **Daily Test Practice Transparency,** TT96
2 Australia **KEY IDEA** Both aborigines and European immigrants have shaped Australia's history, culture, and economic life.	• **Reading Study Guide,** pp. 5–6 • **RSG with Additional Support,** pp. 11–12 • **History Makers: Eddie Mabo,** p. 33 • **Connect to Science,** p. 39 • **Primary Source,** p. 40	**Unit Resource Book** • Section Quiz, p. 44 • Reteaching Activity, p. 47 **Test Generator CD-ROM** **Daily Test Practice Transparency,** TT97
3 New Zealand **KEY IDEA** The Maori and European immigrants have influenced New Zealand's history, culture, and economy.	• **Reading Study Guide,** pp. 7–8 • **RSG with Additional Support,** pp. 13–14 • **Skillbuilder Practice,** p. 28 • **Interdisciplinary Projects,** pp. 35–37 • **Primary Source,** p. 41	**Unit Resource Book** • Section Quiz, p. 45 • Reteaching Activity, p. 48 • Chapter Test (Forms A, B, & C), pp. 49–63 **Test Generator CD-ROM** **Daily Test Practice Transparency,** TT98

No Time? To focus students on important content in the chapter, use the **"Reading Study Guide with Additional Support"** in ***Resources2Go.***

Teacher Presentation Options

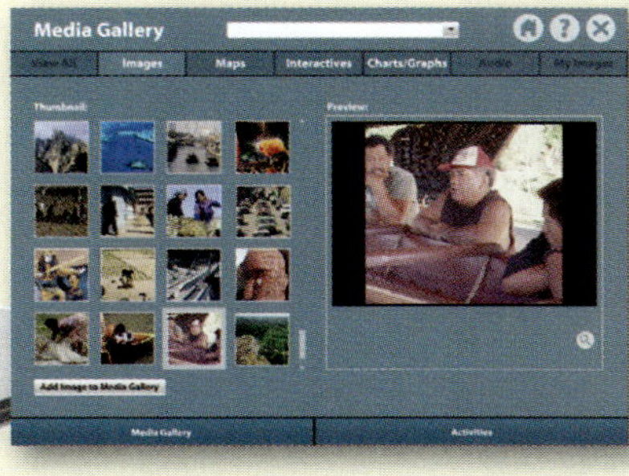

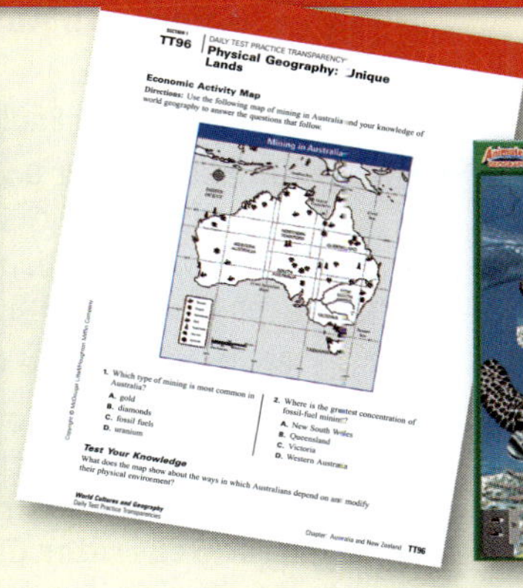

Electronic Books

eEdition DVD-ROM

eEdition Online

Presentation Toolkit

Video/DVD

- World Cultures and Geography

Power Presentations DVD-ROM

- Discussion Starters
- Media Gallery
- Review Game
- Interactive Review
- Animated Features

Unit Transparency Book

- 3-Minute Warm-Ups
- Essential Question
- Critical Thinking
- Maps
- Cultures

Daily Test Practice Transparencies

Animated Geography

- Present-Day Australia and New Zealand Map
- The Great Barrier Reef
- Worldster: Mick's Web Page

OnlineActivities at ClassZone.com

Animated Geography

- Videos and Photos
- 3-D Animations
- Interactive Maps

Interactive Review

- Section Quizzes
- Review and Study Notes
- Vocabulary Flip Cards
- Graphic Organizers
- Name Game and Crossword Puzzles
- GeoGame

Online Test Practice

Map Center

- Atlas Maps
- Political Maps with Data
- Links to Mapping Web Sites

Activity Center

- Activity Maker—Review Games, Puzzles, Flip Cards, and Graphic Organizers
- Worldster

Research & Writing Center

- Web Sites
- Data Updates
- Primary Sources
- Writing Models

Current Events

Jim Easton
Roscoe Middle School
Roscoe, Illinois

Make It Fun — Teacher-Tested Activities

As you teach this chapter, look for my classroom-proven ideas for:

- making the lesson fun
- focusing students' attention and comprehension
- integrating technology with traditional teaching

Teaching Bell to Bell When you see this timing guide, have students discuss in pairs, in groups of 3–6, or as a class.

- **Think, Pair, Share** (pairs)
- **Reader, Recorder, Reporter** (3–6)
- **Talk About It** (any size)

CHAPTER 29

RESOURCES FOR DIFFERENTIATED INSTRUCTION

Struggling Readers

eEdition with Audio DVD-ROM

Lesson Plans
- Strategies for Inclusion

Unit Resource Book
- Reading Study Guide with Additional Support, pp. 9–14 B
- Reteaching Activity, pp. 46–48
- Chapter Test (Form A), pp. 49–53 A

Best Practices Toolkit C
- Interactive Reading Strategies, pp. 90–145
- Differentiation Activity Bank: Struggling Readers, pp. 146–149

Teacher's Edition Activities
- Create Vocabulary Questions, Sec. 1
- Create an "Aussie" Glossary, Sec. 2
- Create a Time Line, Sec. 3
- Create Question & Answer Sheets, Sec. 3
- Tiered Activities, Sec. 1–3

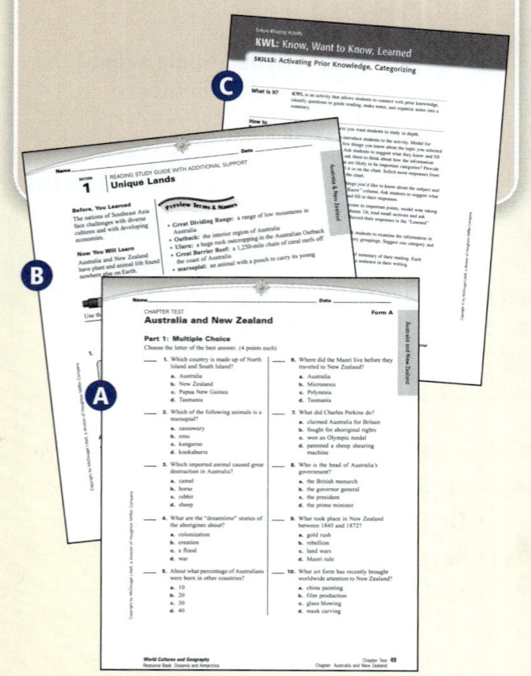

English Learners

Pupil's Edition in Spanish

eEdition with Audio DVD-ROM

Lesson Plans
- Strategies for English Learners

Unit Resource Book
- Reading Study Guide (Spanish), pp. 15–20
- RSG with Additional Support (Spanish), pp. 21–26 A B

Multi-Language Glossary C

Best Practices Toolkit
- Differentiation Activity Bank: English Learners, pp. 165–194

Test Generator CD-ROM
- Section Quizzes in Spanish
- Chapter Test in Spanish

Teacher's Edition Activities
- Prounounce and Preview, Sec. 1–3
- Key Academic Vocabulary, Sec. 1
- Word Relationships, Sec. 2
- Conversational Patterns: Dialogue, Sec. 2
- Comprehension: Connection, Sec. 3
- Tiered Activities, Sec. 1–3

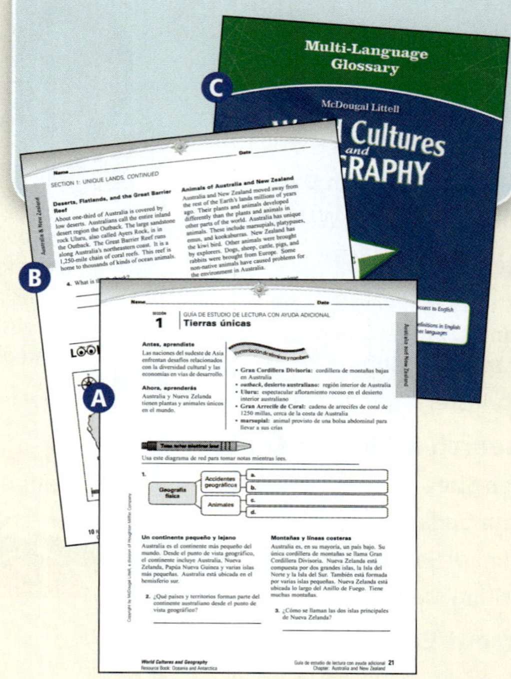

Inclusion

eEdition with Audio DVD-ROM

Lesson Plans
- Strategies for Inclusion

Unit Resource Book
- Reading Study Guide with Additional Support, pp. 9–14 A B
- Reteaching Activity, pp. 46–48
- Chapter Test (Form A), pp. 49–53

Best Practices Toolkit C
- Interactive Reading Strategies, pp. 90–145
- Differentiation Activity Bank: Struggling Readers, pp. 146–149

Teacher's Edition Activities
- Locate Images that Illustrate Vocabulary, Sec. 1
- Vocabulary: Frayer Model, Sec. 2
- Record and Listen to Terms and Definitions, Sec. 3
- Tiered Activities, Sec. 1–3

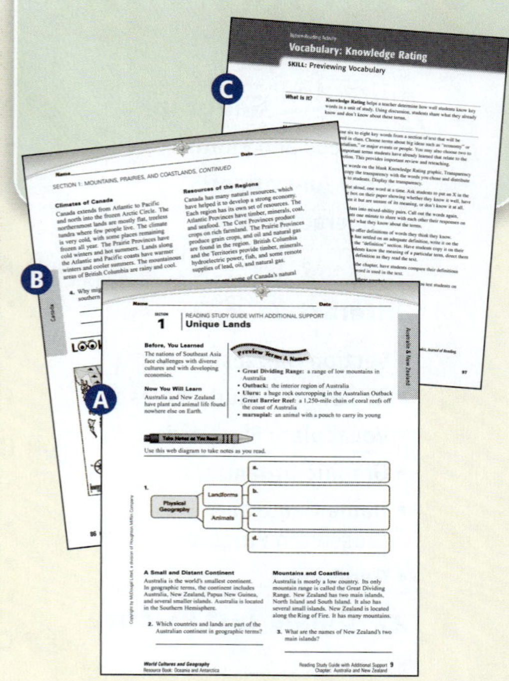

Gifted & Talented Students

Lesson Plans
- Core Lesson Plan

Unit Resource Book
- Outline Map Activities, p. 31 (A)
- History Makers, pp. 33–34 (B)
- Interdisciplinary Projects, pp. 35–37
- Connect to Science, p. 39 (C)
- Chapter Test (Form C), pp. 59–63

Best Practices Toolkit
- Student Products Rubrics, pp. 1–14
- Writing Templates, pp. 78–89

Teacher's Edition Activities
- Write from Different Points of View, Sec. 3
- Tiered Activities, Sec. 1–3

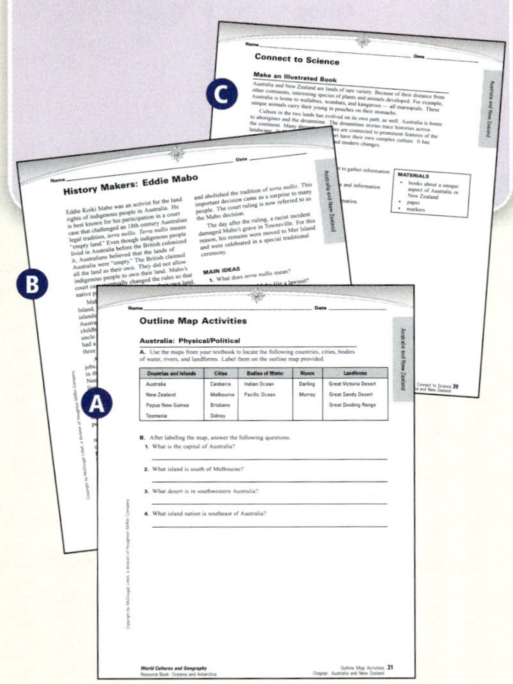

Pre-AP*

Lesson Plans
- Core Lesson Plan

Unit Resource Book
- Map and Graph Skills, pp. 29–30
- Economics in the World, p. 38 (A)
- Primary Source, pp. 40–41 (B) (C)
- Chapter Test (Form C), pp. 59–63

Best Practices Toolkit
- Writing Templates, pp. 78–89
- Differentiation Activity Bank: Pre-AP Students, pp. 195–200

* Pre-AP is a registered trademark of the College Entrance Examination Board, which was not involved in the production of and does not endorse this product.

Teacher's Edition Activities
- Create an Aboriginal Lands Plan, Sec. 2
- Tiered Activities, Sec. 1–3

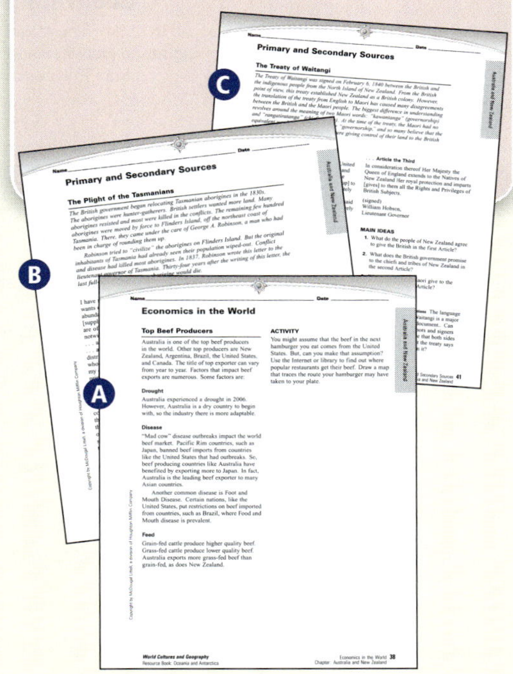

Interdisciplinary Activities

CONNECT to Math

Graphing Comparative Data

CLASS TIME 30 minutes

TASK Create graphs comparing population or economic data for Australia and New Zealand.

SUPPLIES
- graph paper
- colored pencils or markers
- data resources

ACTIVITY Provide student pairs with data on Australia's and New Zealand's population and economy (from The World Factbook at www.cia.gov or another reliable source). Have students review the data and decide which set to graph (for example, literacy rates and infant mortality or export partners and GDP). Each pair should also decide on the type of graph to use, and then create graphs to compare data for the two countries. Invite students to present or display their graphs.

Good for Mathematical/Logical and Visual Learners

CONNECT to Art

Movie Script Storyboard

CLASS TIME 30 minutes

TASK Create a storyboard for a movie based in Australia or New Zealand.

SUPPLIES
- large sheets of paper
- colored pencils or markers

ACTIVITY Have students work in groups to create a concept for a movie that takes place throughout Australia or New Zealand. They may want to choose a theme, such as a travel documentary, or a different story line that has scenes throughout the country. Have them decide on several scenes to include, and then create a storyboard that explains each scene using a simple visual (self-made or found) and summary text.

Good for Visual and Verbal Learners

Introduce the ESSENTIAL QUESTION

- Point out that because of its location, the continent of Australia (which includes New Zealand and other islands) has some features not found anywhere else.
- Explain that Australia's and New Zealand's cultures are based on more than a thousand years of aboriginal settlement and more recent European settlement.
- Identify specific areas where each country's culture reflects its aboriginal and European heritage.
- Describe how Australia and New Zealand are using their natural resources.

Key Ideas

Section 1: Australia and New Zealand have plant and animal life found nowhere else on Earth.

Section 2: Australia's history, culture, and economic life have been shaped by both aborigines and European immigrants.

Section 3: New Zealand's history, culture, and economy have been influenced by the Maori and European immigrants.

CONNECT to the Essential Question

You may wish to revisit the Essential Question periodically with students as they work through the chapter.

Suggestions for reconnecting are provided at an appropriate point in the *Teacher's Edition* for each section.

Sydney Opera House

CHAPTER 29 Australia and New Zealand

1 GEOGRAPHY **Unique Lands**

2 FOCUS ON **Australia**

3 FOCUS ON **New Zealand**

ESSENTIAL QUESTION

How has the human geography of Australia and New Zealand changed since European settlement?

CONNECT Geography & History

Use the map and the time line to answer the following questions.

1. Which body of water did Captain Cook cross to get from Australia to New Zealand?
2. What is the approximate location in latitude and longitude of the 2000 Olympic Games?

40,000 B.C.

Culture
◀ **40,000 B.C.** Australia is gradually settled by aboriginal people. (Aboriginal art–Iguana painted on tree bark)

Geography
1000 A.D. The Maori migrate to New Zealand from Polynesia.

Geography
1769–1770 Captain James Cook explores New Zealand and Australia for Great Britain. ▶

TIME LINE DISCUSSION

Have students review the time line and locate Australia and New Zealand on the map. Why would Great Britain have had an interest in these places? *(Possible Answers: both are remote, so good for a penal colony; climate would have allowed for longer agriculture season than in Great Britain)*

- What skill would the Maori have needed in order to migrate to New Zealand? *(Possible Answer: water navigation, since New Zealand is an island)*
- How many years after Cook explored New Zealand and Australia did it take for Britain to start a penal colony in Australia? *(about 18, 1770 to 1788)*
- In terms of compass points, in what part of Australia were the 2000 Olympic Games held? *(southeast)*
- Who settled in the region first—aboriginal people in Australia or the Maori in New Zealand? *(Aboriginal people settled Australia in 40,000 B.C. while the Maori settled in A.D. 1000.)*

Animated GEOGRAPHY

Present-Day Australia and New Zealand

Click here to explore Australia & New Zealand @ClassZone.com

History
1788 Britain starts a penal colony in Australia.

History
1840 Treaty of Waitangi gives Britain control over New Zealand.

Culture
◄ **2000** The Olympic Games are held in Sydney, Australia. (Cathy Freeman, aboriginal Australian, gold medalist)

Today

TEACHER MEDIA FAVORITES

Books for the Teacher

Chatwin, Bruce. *The Songlines.* Penguin Books, 1987. Explores the invisible aboriginal ancestral paths. (also available on audio)

Welsh, Frank. *Australia.* Overlook Hardcover, 2006. Comprehensive, up-to-date single-volume history.

Chambers, John H. *A Traveller's History of New Zealand and the South Pacific Islands.* Interlink Books, 2003. Covers the islands' history and present.

Video / DVD

Whale Rider. (DVD/VHS) Sony Pictures, 2003. A young girl in a coastal New Zealand village fights tradition to fulfill her destiny.

Australia's Great Barrier Reef. (VHS) National Geographic, 1998. Underwater video with music and narration.

Australia/New Zealand. (DVD) Educational Video Network, Inc., 2004. Explores the two countries' cultures, traditions, and history.

Internet

Go to the *Research and Writing Center* at **ClassZone.com** to link to:

- data on the region
- a variety of chapter-specific content-reviewed sites
- "History Makers" biographies
- primary sources

Animated GEOGRAPHY

Present-Day Australia and New Zealand

CLASSZONE.COM

Have students review the map legend and explain what type of map this is. *(political—shows national capitals and other cities)* Point out the lack of major rivers in most of Australia and the rest of the region. Ask students if they can identify what the symbols along Australia's northeast coast represent. *(the Great Barrier Reef)*

- What are some ways to describe the relative locations of Australia and New Zealand? *(Possible Answers: southern hemisphere, south of the equator, southern Pacific Ocean, surrounding Tasman Sea)*
- Where are most of Australia's major cities located? Why do you think this is? *(in the southeast; because they are near major rivers and ocean)*
- What New Zealand waterway might have a lot of traffic? Why? *(Cook Strait; because it's between the islands, and the capital is located along its coast)*

CONNECT GEOGRAPHY & HISTORY ANSWERS

1. Tasman Sea
2. approximately 151° E, 42° S, Sydney, Australia

Presentation Options

Use the **Power Presentations** DVD-ROM to project the animated map.

- Prompt students to use the pop-up images and audio from the map to discover what they already know about the geography, culture, history, economics, or governments of Australia and New Zealand.
- Challenge students to answer the map questions individually, in small groups or as a class activity. Use all of the interactive material to help preview chapter content.

1 Plan & Prepare

Objectives

- Identify the main features of the landscapes of Australia and New Zealand
- Categorize Australia's and New Zealand's native and imported animals

Read for the Essential Question

Help students read for a purpose by reminding them of the Essential Question: "How has the human geography of Australia and New Zealand changed since European settlement?" Point out that they can use text features such as Key Questions, special features, images, and maps to form answers to the question.

Vocabulary

Best Practices Toolkit

Use the ***Best Practices Toolkit*** to model strategies for vocabulary notetaking. Vary strategies throughout the year. Choose from: *Knowledge Rating, Predicting ABCs, Definition Mapping, Word Sort, Frayer Model (Word Squares), Magnet Words, Word Wheel,* and *Student VOC.*

Vocabulary Strategies, TT9–TT16

Reading Strategy

Best Practices Toolkit

Display the Web Diagram transparency.

- Remind students that to categorize they will need to sort the information in this section into the categories shown on the diagram.
- Model entering information on the diagram, using the text, images, and maps in this section, "Unique Lands."
- Have students suggest entries.
- Display the completed diagram for students to review.

Web Diagram, TT28

Section 1 Reading for Understanding

Key Ideas

BEFORE, YOU LEARNED

The nations of Southeast Asia face challenges with diverse cultures and with developing economies.

NOW YOU WILL LEARN

Australia and New Zealand have plant and animal life found nowhere else on Earth.

Vocabulary

TERMS & NAMES

Great Dividing Range a range of low mountains in Australia

Southern Alps a mountain chain in New Zealand

geyser (GY•zuhr) a hot spring that shoots sprays of steam and boiling water into the air

Outback the interior region of Australia

Uluru (oo•LOO•roo) a spectacular rock outcropping in the Australian Outback

Great Barrier Reef a 1,250-mile chain of coral reefs off the coast of Australia

marsupial (mahr•SOO•pee•uhl) an animal with an abdominal pouch to carry its young

BACKGROUND VOCABULARY

outcropping bedrock that rises above the soil

draft animal an animal used to pull a heavy load

Visual Vocabulary geyser

Reading Strategy

Re-create the web diagram shown at right. As you read and respond to the **KEY QUESTIONS**, use the diagram to categorize important details about the physical geography of Australia and New Zealand.

See Skillbuilder Handbook, page R7

CATEGORIZE

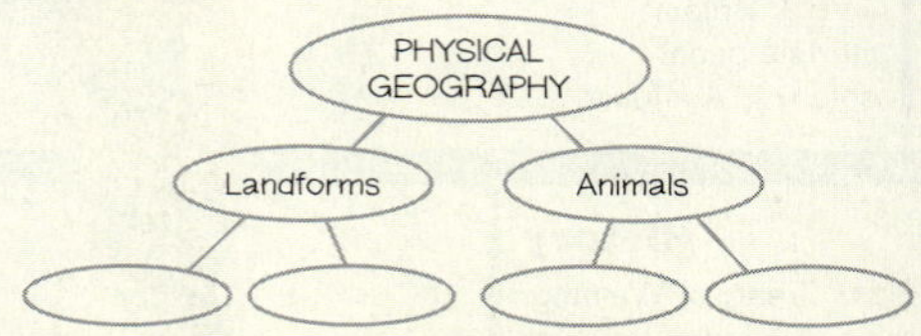

GRAPHIC ORGANIZERS
Go to Interactive Review @ClassZone.com

884 Chapter 29

PRETEACHING VOCABULARY

English Learners

Pronounce and Preview

Pronounce each term for students. Review the meanings of words within definitions, such as *interior, abdominal,* and *bedrock.*

- To modify vocabulary learning, have students complete worksheets as they read, instead of afterward.

Unit Resource Book: Oceania & Antarctica

- Vocabulary Practice, p. 27

Inclusion

Locate Images that Illustrate Vocabulary

Have students work in pairs or small groups. Have them take turns reading a term and definition out loud, and then work together to find an example of the term in the images and map in this section. Have them continue until all terms have been used.

Sydney Opera House

SECTION 1 GEOGRAPHY

Koala with eucalyptus

Unique Lands

Connecting to Your World

In some ways, the first explorations of Australia and New Zealand by Europeans were like the first explorations of the moon. The explorers and astronauts had to cross enormous distances to their destinations and knew little about actual conditions in the unknown lands. The early European explorers in Australia found a land very different from their native countries of England and the Netherlands. Australia was a vast, dry region inhabited by many exotic creatures. Think about how strange it would be to encounter kangaroos, if you had never seen one before!

A girl and a wallaby

Landforms

KEY QUESTION What are the landscapes of Australia and New Zealand like?

Australia is the world's smallest continent. Geographically, the continent consists of the island of Australia, the islands of New Zealand, Papua New Guinea, and several nearby islands. Australia is located about 7,000 miles southwest of North America. The continent of Australia is completely in the Southern Hemisphere. In fact, the continent is so far away from the other inhabited continents that it has developed some characteristics not found anywhere else in the world.

Uluru This enormous sandstone rock in the middle of the Australian desert is 5 miles around, and more than 1,100 feet high.

2 Focus & Motivate

3-Minute Warm-Up

Write on the board or display the transparency:

- View the images in this section. What features of Australia and New Zealand might have seemed unique to European explorers? *(Possible Answers: vast desert and rock formations, marsupials, geysers, the Great Barrier Reef, different sea life)*

Unit Transparency Book: Oceania & Antarctica
- 3-Minute Warm-Ups, TT1

Connecting to Your World

More About . . .

Sydney Harbor

The city of Sydney is built around a huge, deep harbor. One of the world's longest steel-arch bridges spans the harbor.

3 Teach

Landforms

Reader, Recorder, Reporter

- Where does Australia rank in size among world continents? *(It's the smallest.)*
- Why are geysers common in New Zealand? *(It's part of the volcanic Ring of Fire.)*
- **Categorize** How are Australia's and New Zealand's landscapes similar and different? *(Similar: surrounded by the ocean and have mountains; Different: New Zealand has higher mountains and geysers while Australia has deserts and reefs.)*

SECTION 1 PROGRAM RESOURCES

Unit Resource Book: Oceania and Antarctica

ON LEVEL

Unit Resource Book
- Reading Study Guide, p. 3
- Map and Graph Skills, p. 29
- Section Quiz, p. 42

STRUGGLING READERS

Unit Resource Book
- Reading Study Guide with Additional Support, p. 9
- Skillbuilder Practice: Analyzing Primary Sources, p. 28
- Section Quiz, p. 42
- Reteaching Activity, p. 45

eEdition with Audio DVD-ROM

ENGLISH LEARNERS

Pupil Edition in Spanish

eEdition with Audio DVD-ROM

Unit Resource Book
- RSG (Spanish), p. 15
- RSG with Additional Support (Spanish), p. 21

Multi-Language Glossary

Test Generator
- Section Quiz in Spanish

INCLUSION

Unit Resource Book
- RSG with Additional Support, p. 9
- Section Quiz, p. 42
- Reteaching Activity, p. 45

GIFTED & TALENTED

Unit Resource Book
- Outline Map Activities, p. 31
- Section Quiz, p. 42

PRE-AP

Unit Resource Book
- Section Quiz, p. 42

TECHNOLOGY

Unit Transparency Book
- 3-Minute Warm-Ups, TT1
- Essential Question, TT2
- Critical Thinking, TT3
- Maps: Australia and New Zealand: TT6, TT7

Daily Test Practice Transparencies
- Section 1, TT96

Power Presentations

ClassZone.com

World Cultures and Geography Video on DVD

More About . . .

Ring of Fire

The Pacific Ring of Fire is a major zone of earthquakes as well as volcanoes. The reason lies deep under the ocean. The Ring of Fire is at the margin of the Pacific Plate, one of Earth's major tectonic plates (massive, moveable segments made up mainly of Earth's crust). In some areas, this plate has been meeting and sinking under the adjacent plate over time, forming:

- island arcs (curved stretches of volcanoes, which created some island chains)
- long curving trenches in the ocean's deepest parts—where undersea earthquakes can happen, if one plate moves suddenly under the other

Comparing Regions

Sacred Spaces in Other Lands

As with Uluru (now a UNESCO World Heritage site), many natural sites around the world have long been held sacred by their native peoples. These sites are often awe-inspiring spots of great beauty. For example:

- Nachi Waterfall in Japan is a sacred space for Japan's Shinto religion.
- Hawaii's Mauna Kea volcano, especially its summit, is a sacred place in Hawaiian culture and religion.

Today, these sites and many others are trying to balance preservation needs with tourism and other uses.

A Geyser Erupts New Zealand's location on the Ring of Fire makes geysers a part of the landscape. **What other physical features are found on the Ring of Fire?**

volcanoes and hot springs

Mountains and Coastlands Australia has few highland areas and only one range of low mountains called the **Great Dividing Range** Ⓐ. This range lies on the eastern side of the continent. The area east of the range is quite hilly. See the map on the opposite page.

New Zealand consists of two main islands—North Island and South Island—and several smaller islands. Located on the volcanic Ring of Fire, much of the country is covered by mountains and hills. A mountain chain, the **Southern Alps** Ⓑ, runs down the middle of South Island. Seventeen peaks in this chain rise higher than 10,000 feet. Hundreds of rivers flow into narrow, deep-sided inlets created by melting glaciers. North Island has dense forests, active volcanoes, and many hot springs. Some hot springs called **geysers** (GY•zuhrz) shoot sprays of steam and boiling water into the air.

Deserts and Flatlands One-third of Australia is a low-lying desert region. The Great Dividing Range blocks moisture-bearing winds, making the interior dry. Australia has four major deserts, which range from sand to gravel and rock. Australians call the entire isolated region the **Outback**. Within the Outback is an immense rock **outcropping** called **Uluru** (oo•LOO•roo) Ⓒ, also called Ayers Rock. This enormous piece of red sandstone is 1.5 miles wide, 2.2 miles long, and rises more that 1,100 feet above the desert floor. Uluru is sacred to Australian native peoples.

The Great Barrier Reef One of Australia's most spectacular physical features is the **Great Barrier Reef** Ⓓ. The reef is located off the northeast coast of Australia and is a 1,250-mile chain of about 2,900 coral reefs formed by about 400 species of coral. Thousands of species of fish, mollusks, and sponges, as well as endangered green and loggerhead turtles, inhabit the reefs. Humpback whales also come to the reef's warm waters to bear their young. You can learn more about the reef in Animated Geography at the end of this section.

A COMPARE AND CONTRAST Describe the landscapes of Australia and New Zealand.

Australia is mainly flat and has much desert. New Zealand has mountain ranges, forests, volcanoes, and many rivers. ◄

A Silvertip Shark Patrols the Great Barrier Reef Among many thousands of species of fish and animals found on or around the reef, the shark is one of the largest.

886

DIFFERENTIATING INSTRUCTION

Struggling Readers

Create Vocabulary Questions

Have students work in pairs to create questions about the vocabulary terms. Model this using *Great Barrier Reef,* asking a question such as, "How long is the Great Barrier Reef"? Have students locate the term in this section and read the text to answer your question. Have pairs choose another term, locate it in the text, and create a question. Have them ask their question to the class.

English Learners

Key Academic Vocabulary

Write *located* and *location* on the board. Ask students to find these terms on this page and discuss their meaning to confirm understanding. Point out the preposition of place—*on* or *off*—that follows the term in each instance. Have students brainstorm other place prepositions *(by, near)* and then create sentences using *located* or *location* with one of these prepositions to describe something in the classroom.

Physical Geography of Australia & New Zealand

CONNECT Geography & Culture

READING A PHYSICAL MAP

Study the elevation legend on the map. As you study the map,

- find the three desert areas in Australia and look at their elevation
- compare the desert elevations with that of the Great Dividing Range

Place How would you describe the elevation of Australia's deserts?

Australia and New Zealand **887**

INTERDISCIPLINARY ACTIVITIES

CONNECT to Language Arts

Design a Nature Travel Tour

Have students work in small groups. Ask them to imagine they run a nature tour company and design a tour of Australia and New Zealand that explores each one's variety of physical features. Suggest they use the map and the information in this section. Have them design the tour from a starting point to an ending point, with stops on the way. They should briefly describe each place and say what activities the group will do there.

CONNECT to Math

Calculate Travel Tour Mileage

Have students use the map's scale to add mileages to their travel tour. They can include miles covered between stops and also along the course of any one destination (such as a tour up a river in Australia, or a drive up a mountain in New Zealand). They may also want to use their calculations to determine information such as the planned length of time to spend at each spot and whether to fly to certain places to save time.

Physical Geography of Australia & New Zealand

CONNECT Geography & Culture

Reading a Physical Map Ask students to examine Australia and New Zealand against the map's elevation legend. Have them write a sentence or brief paragraph comparing and contrasting the elevations of the two countries. *(Similar: Both have low coastal areas and higher interiors; Different: Australia's interior also includes some low elevations while New Zealand does not.)*

ANSWER

Place Compared to the mountain range, the deserts are not very high—ranging between 0 and 3,280 feet.

Unit Resource Book: Oceania & Antarctica

- Map and Graph Skills: Reading a Physical Map, pp. 29–30
- Outline Map Activities: Australia: Physical/Political, pp. 31–32

Make It Fun

Teacher-Tested Activities

Jim Easton, Roscoe Middle School, Roscoe, Illinois

To give students a chance to practice mapping skills, I have them use an outline map I provide or do a sketch map from scratch.

- Create a list of physical and political features to place on the map. This could include cities, countries, rivers, mountains, and so on. Place the list on the board.
- Hand out the outline maps or blank sheets of paper. Tell students to use the maps in this section as reference maps.
- Ask students to add coloring to match elevation and sea depths.
- Remind the students to create a legend for their map.
- Have students add the map to their personal atlas booklet or post the maps in the classroom.

CONNECT to Science

Tasmanian Devil

The Tasmanian devil, about the size of a small dog, is found only on the Australian island of Tasmania, though at one time it also lived on mainland Australia. It's a carnivorous marsupial—basically eating whatever it can find—from birds and small mammals to wallabies (alive or dead), to the carcasses of cattle and sheep. With its strong jaws and teeth, it eats its whole prey or find, including bones.

Early European settlers gave it the name "devil" in part from the fierce noises it makes, including snarls, coughs, and high-pitched screeches. Protected by law for years, the Tasmanian devil population is now threatened by a disease that is killing it in large numbers.

Unit Resource Book: Oceania & Antarctica

- Connect to Science: Make an Illustrated Book, p. 39

Teach

Australia's Animals

Think, Pair, Share

- What is a marsupial? *(an animal with an abdominal pouch to carry its young)*
- How is New Zealand's kiwi bird unique? *(It's the only bird in the world with nostrils on the tip of its bill.)*
- Why were non-native animals brought to Australia and New Zealand? *(for farming, ranching, hunting, travel, hauling)*
- **Categorize** What are some native and non-native animals in Australia and New Zealand? *(Native: Australia—marsupials, platypus, echidna, emu, cassowary, kookaburra; New Zealand—two species of bats, kiwi, and other birds; Non-native: Australia—dingo, rabbits, camels; both—sheep, goats, cattle, pigs)*

CONNECT to Science

Australia is home to marsupials, a unique class of animals that carry their young in pouches. There are about 125 species of these animals. They range in size from a mouse, to an average-sized dog, to kangaroos that can reach a height of five feet.

Activity

Make an Illustrated Book

Materials
- animal books
- paper
- markers

1. Study the animal books or use the Internet to gather information about marsupials.
2. Plan your booklet to include pictures and information about 3 or 4 animals.
3. Draw pictures and write out the information.
4. Bind the book together.

Australia's Animals

KEY QUESTION What types of animals are found only in Australia and New Zealand?

Australia and New Zealand separated from the rest of the Earth's landmasses millions of years ago. Their isolation from other continents allowed plants and animals there to develop differently. For example, Australia has **marsupials** (mahr•SOO•pee•uhlz), animals with abdominal pouches to carry their young. Many of these marsupials, such as the kangaroo, are not found elsewhere.

Native Animals Australia's marsupials are not its only interesting native animals. Two species of mammals that hatch their young from eggs, the platypus (PLAT•ih•puhs) and the echidna (ih•KIHD•nuh), are found there. Two flightless birds—the emu and the cassowary—are also native to Australia. The most famous Australian bird is the kookaburra. An Australian folk song celebrating the kookaburra is known around the world.

Except for two species of bats, New Zealand has no native mammals. Almost all other species were brought there. New Zealand's many native birds include the kiwi, the only bird in the world with nostrils on the tip of its bill. In fact, New Zealanders' slang term for themselves is "kiwi."

DIFFERENTIATING INSTRUCTION: TIERED ACTIVITIES

OBJECTIVE Create a categories guessing game.

Basic	On Level	Challenge
As a class, brainstorm categories of information found on this page (animals, marsupials, and so on). Model creating a clue ("I am Australia's most famous bird . . .") for the class to categorize. Then have each student create a card with a clue. Have the class give as many correct categories for each clue as possible.	Have students work in groups of three to brainstorm categories. Have the groups share their ideas to add any new categories. Then have each group write a clue card for each category and play the game. In each round, one student asks, one gives the answer(s), and one checks the answer(s).	Have student pairs brainstorm and then share ideas with the class. Next, have pairs create clue cards, and on the back of each card write as many correct answers as they can. Have them put the cards in a stack, clue side up. Partners take turns answering each clue, trying to get all the correct answers.

Non-Native Animals Over the years, many animals were brought to Australia and New Zealand. Geographers believe that the dingo, a doglike, meat-eating animal, probably came to Australia with the earliest settlers at least 3,500 years ago. More recent imports to both Australia and New Zealand are sheep, goats, cattle, and pigs. They were brought to establish farming and ranching operations.

Perhaps the most dramatic story of non-native animals concerns rabbits in Australia. In 1859, Thomas Austin released 24 rabbits into Australia so that he could hunt them. The rabbit population exploded because rabbits had few natural enemies. By 1900, Australia had more than a billion rabbits! The Australian government built a 2,000-mile fence to keep the rabbits from spreading. Rabbits destroyed crops and competed with grazing animals for grasslands. They are still a problem today, in spite of efforts to keep them under control.

CONNECT Geography & Culture

Australia's Camels

Camels were introduced into Australia by explorers who used them to travel its deserts. Later, camels became farm animals. They also were used in construction projects as **draft animals**, animals used to pull heavy loads. When machines replaced camels, they were released into Australia's semiarid lands. Today, there are free-range camel herds in the Outback.

SUMMARIZE Identify unique animals of Australia and New Zealand. Australia has marsupials and both have birds that are found nowhere else in the world.

ONLINE QUIZ For test practice, go to Interactive Review @ClassZone.com

Section 1 Assessment

TERMS & NAMES

1. **Explain the importance of**
   - Great Dividing Range
   - Southern Alps
   - Outback
   - Great Barrier Reef

USE YOUR READING NOTES

2. **Categorize** Use your completed web diagram to answer the following question:

   What physical features are found in New Zealand?

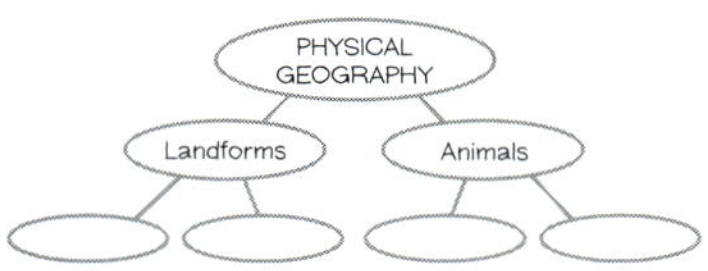

KEY IDEAS

3. What is the Australian Outback?
4. What are the two main islands of New Zealand?
5. Why did unique animal species develop in Australia and New Zealand?

CRITICAL THINKING

6. **Find Main Ideas** What is one of the many differences between the physical landscapes of Australia and New Zealand?
7. **Analyze Causes and Effects** What geographical factor accounted for the early isolation of Australia and New Zealand?
8. **CONNECT to Today** Why might governments want to limit the importation of non-native animals?
9. **SCIENCE** **Draw a Species Map** Research the plants and animals native to your state. Create an illustrated map showing five or six of these native species.

CONNECT to the Essential Question

How has the human geography of Australia and New Zealand changed since European settlement?

Ask students what they have learned so far that can help them answer this question. Students might mention:

- Europeans founded cities along the coasts.
- Non-native animals were introduced for work, travel, and sport.

CONNECT Geography & Culture

Australia's Camels

Most of Australia's camels are one-humped, rather than the two-humped type. Camels are used for tourism, camel racing, or meat.

4 Assess & Reteach

Assess Have students complete the Section Assessment.

Unit Resource Book: Oceania & Antarctica
- Section Quiz, p. 42

Interactive Review @ClassZone.com
- Section Quiz

Power Presentations

Test Generator

Reteach Have students write three facts about Australia and New Zealand, then list them on the board or chart paper and have the class sort them into categories.

Unit Resource Book: Oceania & Antarctica
- Reteaching Activity, p. 45

SECTION 1 ASSESSMENT ANSWERS

Terms & Names

1. Great Dividing Range: a range of low mountains in Australia; Southern Alps: a mountain chain in New Zealand; Outback: the interior region of Australia; Great Barrier Reef: a 1,250 mile chain of coral reefs off the coast of Australia

Use Your Reading Notes

2. mountains, hills, glaciers, and geysers
   Diagram *Physical Geography; Landforms: mountains, deserts, flatlands, coastlands. Animals: marsupials, imported animals*

Key Ideas

3. the interior region of Australia
4. North Island and South Island
5. because both Australia and New Zealand developed in isolation for so many years

Critical Thinking

6. Answers may vary but might mention that Australia is much larger than New Zealand, that Australia is much flatter than New Zealand, that Australia has much desert land while New Zealand is mountainous.
7. Possible answer: Both Australia and New Zealand were separated from the rest of the Earth's landmasses thousands of years ago.
8. Governments may want to prevent a situation like that which happened in Australia when a non-native species flourished and endangered other species.
9. Use the rubric to score illustrated maps.

Map Rubric

	Information	Accuracy
4	includes 5 animals and 5 plants	very precise
3	includes 3–4 animals and 3–4 plants	correct placement
2	includes 2 animals and 2 plants	incorrect placement of one element
1	includes 1 plant and 1 animal or none	incorrect placement of several elements

1 Plan & Prepare

Objectives: Analyze the diverse sea life found in and around the Great Barrier Reef.

Visual Vocabulary: Ask students to describe what they see in the illustration. Call out familiar items listed below, asking students to locate each one on the page. Then tell them the precise term, as listed in boldface. Add terms to a class vocabulary list, illustrated poster, or Word Wall.

- sea animal with a snout wider than its face **(dugong)**
- sea animal with a dome-like shell on its back **(sea turtle)**
- coral formation with many antler-like branches **(staghorn coral)**
- orange fish with three white stripes **(clown anemone fish)**
- air tank on a diver's back **(scuba—self-contained underwater breathing apparatus—tank)**

2 Teach

Engage . . .

DISPLAY "The Great Barrier Reef" feature from the **Power Presentations** DVD-ROM, or the *Animations Center* at **ClassZone.com.** As the illustration loads on screen, have volunteers point to some of the vocabulary terms listed above. Then ask:

- What different features make up this undersea environment?
- Why do you think the smaller fish tend to be closer to the reef?
- What positive and negative impacts could scuba diving and snorkeling have on the Great Barrier Reef?

Encourage all answers.

Explore . . .

ROLL OVER sections of the illustrations or click on items in the menu. Ask students to choose which parts of the scene to explore further.

- Then return to the discussion questions. Have any answers changed?
- Have students jot down a list of at least three new questions that arise from exploring the rollover features.

Animated GEOGRAPHY
The Great Barrier Reef

Click here to dive into the Great Barrier Reef @ClassZone.com

THE GREAT BARRIER REEF

The Great Barrier Reef is the grandest of all the coral reefs found in the ocean. Chains of coral reefs stretch 1,250 miles off the coast of Australia. The reef is home to thousands of species of plants and animals.

Click here to learn about sea life that has been around for 400 million years.

Click here Learn how the skeletons of millions of tiny animals and plants build a reef.

Click here to learn why the Great Barrier Reef is a World Heritage Area.

DIFFERENTIATING INSTRUCTION: TIERED ACTIVITIES

OBJECTIVE Collect and present data on sea life found in and around the Great Barrier Reef.

Basic

List and Count Types of Sea Life

Have students assume the role of the diver in the illustration. Their job is to count the animals (fish and coral) on this area of the reef, to help a local organization with its reef research. Have them identify as many types as they can. They may need to make up names for some, if they don't know them, or they can just describe features (as divers often do, until they can consult a resource to identify it). Have students list each type and how many of each they find.

GeoActivity

Design a Conservation Poster Research the environmental threats to the Great Barrier Reef. Design a poster that encourages people to take care of the reef and avoid harmful practices.

891

ANIMATED GEOGRAPHY

Explain . . .

CHOOSE topics to help students answer their questions, such as:

- **Dugongs:** Dugongs are a sea mammal. They are related to manatees found in the Atlantic Ocean. Dugongs live in warm, shallow waters and mainly eat seagrass. They surface to breathe air.
- **Sea Turtles:** The Great Barrier Reef has six species of sea turtles. Sea turtles lay their eggs in sand on the coast and then return to the water. Sand temperature determines gender of the hatchling (cooler for males, warmer for females).
- **Coral:** Corals are animals with no backbone. Each tiny coral is called a "polyp." Hundreds of polyps often join to form structures, such as brain coral, staghorn coral, sea fans, and table coral.
- **Anemone fishes:** These fish live in the tentacles of sea anemones. Sea anemones sting their prey to eat it. It's thought that something in the coating on anemone fish keeps them from being stung.

Elaborate . . .

ASSIGN students to work in small groups to write the script for a scene for an underwater video of the Great Barrier Reef.

- Have each group pick a theme (fish, coral, ecosystem, conservation, and so on).
- Have them write a script to enhance what the viewer is seeing (by pointing out features or noting interesting facts, for example).
- Have groups share their scripts with the class.

Evaluate . . .

Design a Conservation Poster

Activity Rubric

4 Excellent Attractively illustrated with effective messages that encourage care for the reef.

3 Good The illustration is adequate and most messages are clear.

2 Fair The illustration and messages are somewhat adequate.

1 Poor The illustration and messages are incomplete, messy, or missing.

On Level

Create a Sea Life Bar Graph

Have students identify and count sea life as for the Basic activity. Then have each student transfer the data into bar graphs.

Challenge

Create Sea Life Pie Charts

Have students identify and count sea life as for the Basic activity. Then have each student take the data and create two pie charts: one for coral and one for fish (with the number counted for each type next to its name).

1 Plan & Prepare

Objectives

- Discuss how European settlement changed Australia
- Describe the aboriginal and European influences that make up Australian culture today
- Summarize Australia's economic resources and activities

Read for the Essential Question

Help students read for a purpose by reminding them of the Essential Question: "How has the human geography of Australia and New Zealand changed since European settlement?" Point out that they can use text features such as Key Questions, special features, images, and maps to form answers to the question.

Vocabulary

Best Practices Toolkit

Use the ***Best Practices Toolkit*** to model strategies for vocabulary notetaking. Vary strategies throughout the year. Choose from: *Knowledge Rating, Predicting ABCs, Definition Mapping, Word Sort, Frayer Model (Word Squares), Magnet Words, Word Wheel,* and *Student VOC*.

Vocabulary Strategies, TT9–TT16

Reading Strategy

Best Practices Toolkit

Display the Web Diagram transparency.

- Remind students that main ideas are the most important points in a body of text.
- Model filling in entries, using the "History and Government" subsection.
- Have students suggest entries.
- Display the completed diagram for students to review.

Web Diagram, TT28

SECTION 2 Reading for Understanding

Key Ideas

BEFORE, YOU LEARNED

Australia is home to unique animals and physical features.

NOW YOU WILL LEARN

Australia's history, culture, and economic life have been shaped by both aborigines and European immigrants.

Vocabulary

TERMS & NAMES

James Cook a British explorer who claimed land in Australia

aborigine (ab•uh•RIHJ•uh•nee) an original inhabitant of Australia

penal colony a settlement that serves as a prison

Charles Perkins an aboriginal rights leader

BACKGROUND VOCABULARY

mechanized equipped with machines

REVIEW

hunter-gatherer a person who hunts animals and gathers plants for food

constitutional monarchy a government in which the powers of the king or queen are limited by a constitution

Visual Vocabulary Charles Perkins and aborigines

Reading Strategy

Re-create the web diagram shown at right. As you read and respond to the **KEY QUESTIONS**, use the diagram to record main ideas and details about the history, culture, and economy of Australia.

Skillbuilder Handbook, page R4

FIND MAIN IDEAS

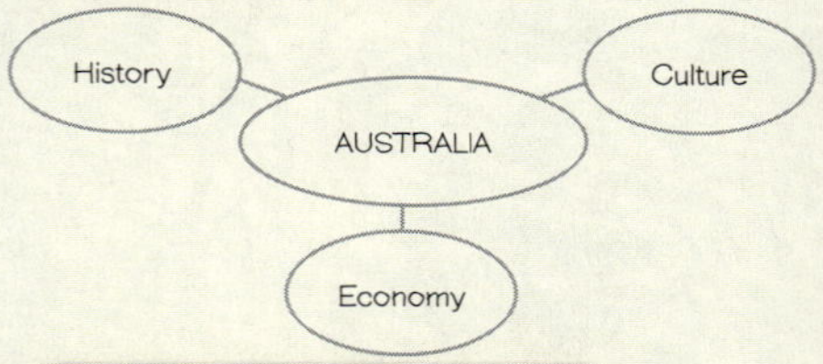

GRAPHIC ORGANIZERS
Go to **Interactive Review** @ClassZone.com

PRETEACHING VOCABULARY

English Learners

Pronounce and Preview

Pronounce each term for students. Review the meanings of words within definitions, such as *inhabitant, aboriginal,* and *equipped.*

- To modify vocabulary learning, have students complete worksheets as they read, instead of afterward.

Unit Resource Book: Oceania & Antarctica
- Vocabulary Practice, p. 27

Inclusion

Vocabulary: Frayer Model (Word Squares)

Have student pairs to fill out the graphic organizer for the terms in the section.

- Use the transparency to model the strategy.
- Have volunteers suggest entries.
- Then distribute copies of the organizer for students to complete with a partner.

Best Practices Toolkit
- Vocabulary: Frayer Model (Word Squares), TT14

Sydney Opera House

SECTION 2 FOCUS ON

Koala with eucalyptus

Australia

Connecting to Your World

Did you know that Australia is called "The Land Down Under"? It lies completely in the Southern Hemisphere. Its seasons are reversed from those in the Northern Hemisphere. So, people there go to the beach from December through March and often celebrate New Year's Day with a backyard barbecue. Now you will learn how similar and how different Australia is from North America.

History and Government

KEY QUESTION How did the arrival of British settlers change life in Australia?

Early European explorers believed there was a "great southern land" somewhere in the vast area of the Pacific and Indian Oceans. In the 1600s, explorers investigated the west, southwest, and northwest coasts of Australia but found it a dry desert, worth little. Then in 1770, British sea captain **James Cook** claimed the fertile east coast for Britain. The European explorers were not the first people to arrive in this southern land. Peoples from Southeast Asia had migrated there across water and by a land bridge at least 40,000 years before.

Aboriginal Art This modern wall mural is based on traditional aboriginal styles of painting.

2 Focus & Motivate

3-Minute Warm-Up

Write on the board or display the transparency:

- View the image of aboriginal art in this section. How do you think natural resources play a part in Australia's cultural life? *(Possible Answer: Animals and water may be combined with traditional aboriginal styles of painting in Australian art today.)*

Unit Transparency Book: Oceania & Antarctica
- 3-Minute Warm-Ups, TT1

Connecting to Your World

More About . . .

Cook's Pacific Sailing Voyages

Cook's voyages were also useful for what they *didn't* find—a rumored vast southern continent thought to balance northern landmasses, and a hoped-for northwest passage around Canada and Alaska.

3 Teach

History and Government

Reader, Recorder, Reporter

- How long have people been living in Australia? *(over 40,000 years)*
- What type of government does Australia have? *(a constitutional monarchy)*
- **Find Main Ideas** What events caused Australia's population to change over time? *(Possible Answers: Cook's arrival; Britain's penal colony and land grants; the discovery of gold)*

SECTION 2 PROGRAM RESOURCES

Unit Resource Book: Oceania and Antarctica

ON LEVEL

Unit Resource Book
- Reading Study Guide, p. 5
- Section Quiz, p. 43

STRUGGLING READERS

Unit Resource Book
- Reading Study Guide with Additional Support, p. 11
- Skillbuilder Practice: Analyzing Primary Sources, p. 28
- Section Quiz, p. 43
- Reteaching Activity, p. 46

eEdition with Audio DVD-ROM

ENGLISH LEARNERS

Pupil Edition in Spanish

eEdition with Audio DVD-ROM

Unit Resource Book
- RSG (Spanish), p. 17
- RSG with Additional Support (Spanish), p. 23

Multi-Language Glossary

Test Generator
- Section Quiz in Spanish

INCLUSION

Unit Resource Book
- RSG with Additional Support, p. 11
- Section Quiz, p. 43
- Reteaching Activity, p. 46

GIFTED & TALENTED

Unit Resource Book
- History Makers: Eddie Mabc, p. 33
- Interdisciplinary Projects, pp. 35–37
- Connect to Science, p. 39
- Section Quiz, p. 43

PRE-AP

Unit Resource Book
- Economics in the World, p. 38
- Primary Source, p. 40

TECHNOLOGY

Unit Transparency Book
- 3-Minute Warm-Ups, TT1
- Essential Question, TT2
- Critical Thinking, TT4
- Cultures, TT8, TT9

Daily Test Practice Transparencies
- Section 2, TT97

Power Presentations

ClassZone.com

World Cultures and Geography Video on DVD

RESEARCH & WRITING CENTER

HISTORY MAKERS

Charles Perkins

While at the University of Sydney in the mid-1960s, Charles Perkins and several other student activists organized a freedom ride through rural New South Wales. Inspired by freedom bus rides taking place in the United States at that time, their purpose was similar to U.S. civil rights leaders: to expose racism, prejudice, and discrimination—in Australia's case, against its aboriginal people.

Unit Resource Book: Oceania & Antarctica

- History Makers: Eddie Mabo, p. 33
- Primary and Secondary Sources: The Plight of Tasmanians, p. 40

More About . . .

A Flag for the New Commonwealth

Australia's flag is based on the winning entry from a 1901 competition. Its symbols show Australia's history and geography:

- the United Kingdom's Union Jack flag in the top left
- the seven-pointed "Commonwealth Star" in the bottom left, representing Australia's six original states and all its territories
- five stars on the right in the shape of the Southern Cross—a constellation visible mainly in the southern hemisphere that was a navigation aid for explorers

Today, many Australians want a more distinctly Australian flag, but there has been no agreement on a design.

HISTORY MAKERS

Charles Perkins 1936–2000

Charles Perkins, an important aboriginal leader, has been compared to U.S. civil rights leader Dr. Martin Luther King, Jr. Both of them worked for the rights of their people, and both led protests against racial discrimination.

Perkins was born on a reservation in the Outback and was sent away to school at age ten. There, he saw firsthand the unfair treatment of the aborigines. He was one of the first of his ethnic group to graduate from college. In 1984, Perkins became head of the Department of Aboriginal Affairs, the first aborigine to head a government department.

ONLINE BIOGRAPHY For more on the life of Charles Perkins, go to the **Research & Writing Center** @ClassZone.com

Aborigines: The First Settlers The British called the first settlers of Australia **aborigines** (ab•uh•RIHJ•uh•neez), meaning "original inhabitants." They were **hunter-gatherers**, who spread across the vast island continent to hunt animals and gather plants for food. They lived in distinct groups that had their own languages. When European colonists arrived, over 300,000 aborigines lived on the continent.

British Colony In 1786, Britain established a **penal colony**, or prison settlement, in this remote land. Men and women who had been imprisoned in England were transported by ship to Australia. The first group arrived in 1788, complete with guards and their families. Over the years, a total of about 160,000 came.

In 1790, the British government began to offer land grants to settlers, army officers, and even freed convicts to help populate the land. Four colonies were set up along the coasts and one on the island of Tasmania. In 1851, gold was discovered. Thousands of people from around the world flocked to Australia, and the population tripled. Clashes broke out between aborigines and settlers over the ownership of land. Many aborigines were killed or driven into the Outback. In the later part of the 20th century, **Charles Perkins**, an aborigine himself, led the fight for aboriginal rights. Aborigines continue to struggle to regain lands taken from them by later immigrants.

Independent Nation In 1901, Australia gained independence from Britain. The six colonies merged into a single government. Australia is a **constitutional monarchy**, a government in which the authority of the monarch is limited by a constitution. The monarch of the United Kingdom is Australia's monarch, but the monarch has little to do with running the government. An appointed governor general represents the monarch. The prime minister heads Australia's parliamentary-style government. The political party that wins the most seats in the two-house federal parliament selects the prime minister. All Australians 18 years of age and older are required to vote.

SUMMARIZE Explain how the arrival of Europeans changed Australia.
European colonists settled on lands claimed by aborigines, and merged colonies into a government that is a constitutional monarchy.

894 Chapter 29

DIFFERENTIATING INSTRUCTION

English Learners

Word Relationships

Have students find these phrases on this page in their textbooks: "*freed* convicts" and "*heads* Australia's . . . government." Explain that each phrase contains a word (in italics here) used as one part of speech (adjective and verb, respectively), but one that could also be used as another part of speech (verb and noun, respectively). Talk about each word's meaning and use here, then have students create sentences for each word.

Pre-AP

Creating an Aboriginal Lands Plan

Have students work in small groups to brainstorm possible ways to restore land to aboriginal peoples and/or compensate them for loss of lands. For each proposed solution, have them list possible advantages and disadvantages. Have groups share their ideas with the class. To extend the activity, have students research what actions Australians are currently taking on this issue.

COMPARING Governments

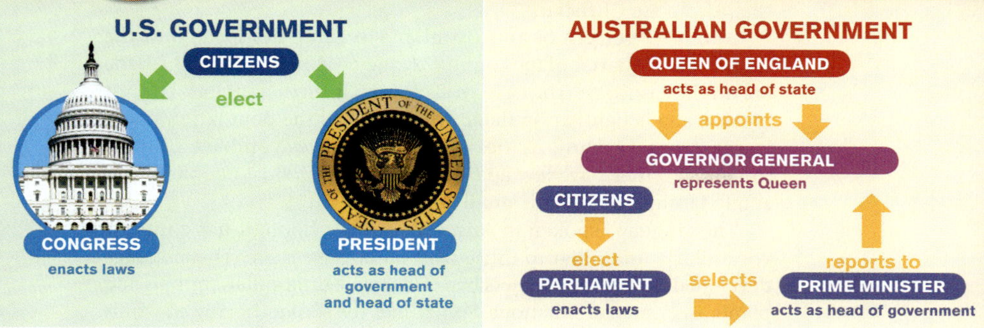

CRITICAL THINKING

Compare and Contrast How does the Prime Minister's role differ from that of the U.S. president?

Aussie Culture

KEY QUESTION What makes Australian culture unique?

Australia's distinctive culture has mainly been influenced by its earliest inhabitants, the aborigines, and its British settlers. Australia's isolation and its unusual plants and animals have also helped to develop a distinctive "Aussie" culture.

Aboriginal Heritage Each aborigine group had traditional lands that were defined by mountains, rivers, or lakes. Aborigines believe there is a close connection between people and the land, plants, and animals. This special relationship with the land forms the basis of their spiritual life. It helps them to understand and care for their environment.

Aboriginal people describe the creation of their land in a series of traditional stories known as "dreamtime" stories. The stories explain the lives of their ancestors and also the aborigines' relationship to the land. Often, the stories are linked to sacred sites, such as Uluru. Today, about 2 percent of the population is aborigine.

Modern Australia Australia is home to immigrants from many parts of the world. Most of these immigrant groups have only come since the end of World War II. About 20 percent of all Australians were born in other countries. However, Australia's early settlement was by immigrants who came from British Isles. Many elements of Australian culture are similar to the cultures of those countries. For

INTERDISCIPLINARY ACTIVITIES

CONNECT to Civics

Debate Use of Money for Roads

Pair or group students into eight groups and assign each one an Australian state or territory. Explain that they will take on the role of members of Parliament from that state or territory. Their objective is to create an argument for why money for roads should go to their state or territory. Then debate the issues as a class. Encourage lively, but respectful, debate and compromises, as needed.

CONNECT to Art

Design a New Flag for Australia

Show students an image of Australia's flag and discuss its symbolism (see the More About on the previous page). Invite them to create a new flag design, and then share their designs and explain them.

Unit Resource Book: Oceania & Antarctica

- Interdisciplinary Projects: Myths, pp. 36, 37

COMPARING

Governments

Explain that when a parliament chooses the prime minister, the person selected is typically the leader of the majority party. What are some of the pros and cons this system could have? *(Possible Answer: It could allow for getting more things done, but the majority may not always fully consider the views of minority parties.)*

CRITICAL THINKING ANSWER

Compare and Contrast Both act as head of government, but the U.S. president also serves as head of state while the Queen of England has this role in Australia.

Teach

Aussie Culture

Think, Pair, Share

- What cultural groups have traditionally had the most influence on Australian culture? *(aborigines, European settlers)*
- Where do most Australians live? *(in cities, located in the southern and eastern areas of the country)* What is the lightly-populated interior of Australia called? *(the Outback, or the bush)*
- **Find Main Ideas** What are some distinct features of Australian culture? *(Possible Answers: aboriginal heritage, English language, unique words, sports and outdoor recreation, the "Aussie" film industry, and life in the Outback)*

Unit Resource Book: Oceania & Antarctica

- Skillbuilder Practice: Analyzing Primary Sources, p. 28

Comparing Regions

Sand Art Around the World

One traditional form of Australian aboriginal art is ceremonial drawings in the desert sand. Sand art is also important in other cultures. Tibetan monks make *mandalas* or spiritual diagrams out of sand. Navajo people in the American southwest use sand paintings as part of healing ceremonies. Commercial versions of sand art—sold to tourists—are now an important source of income for some cultural groups.

Fun Facts!

The Boomerang

The nonreturning type of boomerang is less curved and heavier than the better-known returning type (which is shown in the photograph at right). Because of its shape, a non-returning boomerang can be aimed to fly in a straight line to hit a target, making it very useful in hunting or warfare. Non-returning throwing sticks were also used by many ancient peoples, including those in Egypt, India, and the American southwest.

More About . . .

Australian-rules Football

The playing field in Australian rules football is an oval with the following dimensions:

- a length between 135 and 185 meters
- a width between 110 and 155 meters
- a center square, 50 meters on each side
- two circles in the middle of the center square—an inner circle 3 meters in diameter and an outer circle 10 meters in diameter
- two goal posts, 6.4 meters apart, at each end of the oval
- two "behind posts"—one post 6.4 meters to the side of each goal post
- a goal square (actually a rectangle shape) extending 9 meters from each goal post.

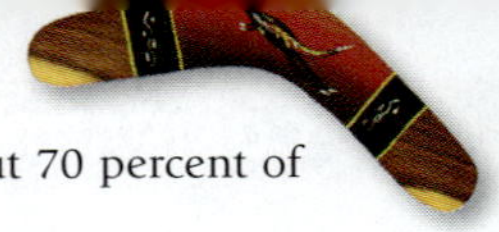

example, English is the official language, and about 70 percent of the people are Christians.

About 85 percent of all Australians live in cities located in the southeastern quarter of the country. Sydney, Melbourne, and Brisbane are on the east coast. The capital, Canberra, is further inland. The rest of Australia is lightly populated, with most of the population living in coastal areas. Unlike cities in other countries you have studied, Australian cities have few apartment buildings. Most people live in single-family homes with small gardens.

The language spoken in Australia is English, but it has a unique vocabulary unfamiliar to those who do not live there. The language developed because the settlers had to describe unfamiliar animals and plants and unusual situations. Sometimes the settlers borrowed words from the aborigines, such as *kangaroo* and *koala*. Other words, such as *brumbies* for wild horses and *outback*, were invented by the settlers.

Sports and Entertainment Because of their country's generally sunny, warm climate, Australians enjoy outdoor sports and recreation. Australians like team games such as rugby and cricket, which were introduced by British settlers. A popular spectator game is Australian-rules football, a combination of soccer and rugby played on a cricket field. Australians also enjoy sun and surf activities. They are very proud of their world-class athletes in tennis, swimming, and golf. Australians have also won many Olympic medals in track and field events, swimming, and diving.

The film industry in Australia is more than 100 years old. In the early years, movies tried to create a separate Aussie identity by using themes from the country's history. In recent years, a number of Australian actors and directors have gained international fame for their acting and productions.

Fun Facts!

THE BOOMERANG

Aborigines of Australia created the boomerang. They used the curved, flat wooden stick to hunt animals, as a tool, and as a weapon in warfare. "Returning" boomerangs can be thrown into the wind, go hundreds of feet, and then return to the thrower. Today, the sports-loving Australians hold boomerang competitions.

Living in the Bush Australians call the lightly populated interior of their country the Outback, or the bush. People in the bush frequently live on huge ranches called "stations." Neighbors may be many miles away. People living on these stations often have a small plane that they use the way most people use a car.

Children in the bush may live 200 miles from a school. These children are educated by correspondence schools and "schools of the air." They listen to instruction broadcast over two-way radio several times a week. At least once a week, most have an on-air private lesson with a teacher. Students receive their assignments by mail, complete them, and return them to the teacher. Today, they use e-mail and the Internet for their lessons as well.

FIND MAIN IDEAS Explain why Australian culture is unique.
Australian culture blends both aboriginal and European influences.

896 Chapter 29

INTERDISCIPLINARY ACTIVITIES

CONNECT to Math

Diagram an Australian Football Field

Provide students with the information in the side margin at left about the Australian rules football field dimensions. Then have student pairs convert the figures into feet/yards, and create a scaled diagram of the field.

Unit Resource Book: Oceania & Antarctica

- Interdisciplinary Projects: Australia's Population Growth, pp. 35, 37

CONNECT to Science

Illustrate Boomerang Flight Principles

Explain that boomerangs fly because they're more curved on top and flatter on bottom (like airplane wings). This causes greater air pressure below, and therefore a lift. Using this information and additional research, have students create a poster that shows how boomerangs fly. To extend the activity, have them also show why a boomerang "returns."

Worldster

A Day in Mick's Life

To learn more about Mick and his world, go to the **Activity Center** @ClassZone.com

To tell you about his day Mick might say:

"G'day, mates! Here's a Captain Cook (a look) at my day."

G'day! My name is Mick. I am 13 years old and live with my family on a sheep station (you probably know it as a ranch) in the Outback of Australia. We Aussies call it "living in the bush." Sometimes I miss not being with other kids. But I play games and sports with my two sisters and my brothers. Let me tell you about my day.

8 A.M. I get up, get dressed, and go have brekkie (breakfast). We usually eat eggs and snags (sausages) and put Vegemite (a yeast and vegetable spread) on our toast.

9 A.M.–3 P.M. Time for school. One room in our home is a school room for me and my brothers and sisters. We use computers and two-way radios to be in contact with our teacher, who is hundreds of miles away. Our mum helps us with our lessons.

4 P.M. I am going with my dad, who is flying our family's small plane over the station to check on our sheep. He lets me sit in the copilot seat and, sometimes, even take the controls.

6 P.M. My mum is grilling on the barbie (barbeque grill) tonight. After dinner, my family and I are going to watch our country's favorite sport, Australian-rules football, on the telly (television).

CONNECT to Your Life

Journal Entry Think about your school day. How is your school experience different from Mick's? What else is different about your life? Record your ideas in your journal.

DIFFERENTIATING INSTRUCTION

Struggling Readers

Create an "Aussie" Glossary

Explain the term *glossary* to students. Ask them to identify Australian English terms on this page and the equivalent American English term. Then have students create their own glossary of these terms. To extend this activity, have them make up sentences using the terms for others to translate.

English Learners

Conversational Patterns: Dialogue

Discuss as a group the activities that are included in Mick's day. Encourage various ways to describe them. *(after you get up, breakfast, after breakfast, school day, at school, after school, before dinner)* Then have students ask a partner a question that uses one such phrase, with the partner responding, using a different, but related, phrase. Alternate roles. Encourage students to use these phrases in their journal entries.

More About . . .

Vegemite

In 1922, Dr. Cyril P. Callister developed "vegemite" from brewer's yeast. Brewer's yeast is a rich source of B vitamins, such as thiamin (B1), riboflavin (B2), niacin (B3), and folate (B9—also called folic acid).

B vitamins help provide energy, make red blood cells, maintain the nervous system, heart, and keep body tissue in areas such as the skin, mouth, and eyes healthy.

The name "vegemite" wasn't used until the next year when the Fred Walker Cheese Company, where Callister was a chemist, ran a competition to name the spread.

More About . . .

Sheep Farming in Australia

Australia has over 100 million sheep and lambs (for wool and meat). Sheep and cattle grazing take up most of Australia's rangeland, with most farms between 100 hectares (just under 250 acres) and 499 hectares (over 1,200 acres—just under 2 square miles). Some grazing farms are among the country's largest farms at over 2,499 hectares (about 6,175 acres—almost 10 square miles). As in the United States, smaller family farms are finding it harder to make a profit, so some farmers have to take other jobs for extra income.

CONNECT to Your Life

Journal Entry

To help students decide what to write, suggest that they first create a chart with three columns: one that includes time blocks similar to how Mick's day is broken out, one labeled "Activities," and one "Notes." Encourage them to list their typical activities for each time block, and then in the "Notes" column to write about how their activities differ from Mick's.

Teach

A Well-Developed Economy

Talk About It

- What are some ways Australia's raw materials are used? *(for manufacturing, as energy sources, for exporting)*
- What factors have contributed to Australia's wealth? *(abundant mineral resources, productive agricultural lands, well-educated workforce)*
- **Find Main Ideas** What are some important economic activities in Australia? *(mining, farming, manufacturing local goods, tourism)*

COMPARING

Economic Activities

Australia has a labor force of about 10.4 million people. Of its workers:

- about 4% work in agriculture.
- about 21% work in industry.
- about 75% work in services

Have students create a pie chart or other graph to illustrate these figures, noting the numbers each represents.

Unit Resource Book: Oceania & Antarctica

- Economics in the World: Top Beef Producers, p. 38

More About . . .

One of Australia's National Parks: Kakadu

Kakadu National Park—in the Northern Territory—is one of six Commonwealth National Parks. It's managed jointly by Australia's government and by traditional aboriginal owners. Among the park's notable natural features are its huge termite mounds, which rise up from the ground, often to a height taller than a person.

The park, a UNESCO World Heritage Site, is important also for its cultural features. Among these are thousands of examples of ancient rock art that still hold significance for aboriginal people in Australia today.

A Well-Developed Economy

KEY QUESTION What are Australia's economic activities?

Australia has abundant mineral resources and productive agricultural lands. It also has a well-educated workforce. As a result, it is one of the world's wealthier nations. But unlike many of the developed nations, much of Australia's wealth comes from the processing of raw materials rather than the manufacturing of finished goods.

Opals The stones are valued for their fiery appearance. This photo shows a raw piece at right and a finished piece at left.

Economic Resources Raw materials needed in many different manufacturing processes and as energy sources are easily found in Australia. Coal, natural gas, and oil provide energy for the country. Natural gas is also exported. Minerals, such as copper, nickel, lead, and bauxite (used for producing aluminum), are in great supply. Gold, diamonds, and opals (a gemstone) are also found here.

In addition to its mineral resources, about 65 percent of Australia's land is suitable for agriculture. Most of the agricultural land is dry grazing land that supports large sheep and cattle ranches. Less than 10 percent of the agricultural land is cropland.

Economic Activities Mining and farming are important to Australia's economy. Large quantities of raw materials, such as coal,

COMPARING Economic Activities

Ranching Raising cattle and sheep is a major economic activity.

Opal Mining Australia's mines produce diamonds, lead, zinc, and opals for the world market.

Tourism Millions of people travel to Australia to visit unique parks.

DIFFERENTIATING INSTRUCTION: TIERED ACTIVITIES

OBJECTIVE Describe human-environment interaction in Australia.

Basic

As a class, brainstorm adjectives to describe the natural and human features for each image in the Comparing Economic Activities graphic on this textbook page. *(flat land, enclosed pens)* List these on the board, and then have students choose one image and describe it.

On Level

Adapt the Basic activity, so that students work in pairs to brainstorm adjectives that describe the natural and human features for each image in the Comparing Economic Activities graphic on this textbook page. Then have students choose two images and compare them.

Challenge

Adapt the On Level activity, so that students work in pairs to brainstorm adjectives that describe the natural and human features for each image in the Comparing Economic Activities graphic on this textbook page. Then have students compare and contrast all three images.

petroleum, wheat, wool, meat, minerals, and gemstones, are exported. Australian exports are sold on the world market, but especially to Japan, China, and South Korea.

Because Australia was so far from other industrialized countries, it had to either make products locally or import them at great cost. So, Australia produces many of the goods that it needs. Such goods include automobiles, food products, iron and steel, textiles, and household appliances. It imports the machines used to make these products and also imports construction equipment and mining machines. Australia's farms are highly **mechanized**, or equipped with machines, so the demand for farm workers is small. In addition to wool and cattle hides, Australians also produce dairy products, sugar cane, and fruits such as bananas, pineapples, and apples.

Many Australians work in the tourism industry. The Great Barrier Reef and Australia's many beaches attract about 5 million visitors a year. The largest numbers of tourists come from Japan, New Zealand, and nations in Southeast Asia.

Australia's Top Export Markets

Country	Exports from Australia
China	wool, iron ore, copper, coal
Japan	coal, iron ore
New Zealand	motor vehicles, chemicals
South Korea	wheat, zinc, wool, beef, sugar
United States	beef, crude petroleum

Source: *Australian Government Bureau of Statistics 2005*

SUMMARIZE Identify Australia's major economic activities.
mining, farming, manufacturing, and tourism

Section 2 Assessment

ONLINE QUIZ For test practice, go to **Interactive Review @ClassZone.com**

TERMS & NAMES

1. **Explain the importance of**
   - James Cook
   - aborigine
   - penal colony
   - Charles Perkins

USE YOUR READING NOTES

2. **Find Main Ideas** Use your completed web diagram to answer the following question:
   How did immigration change the culture of Australia?

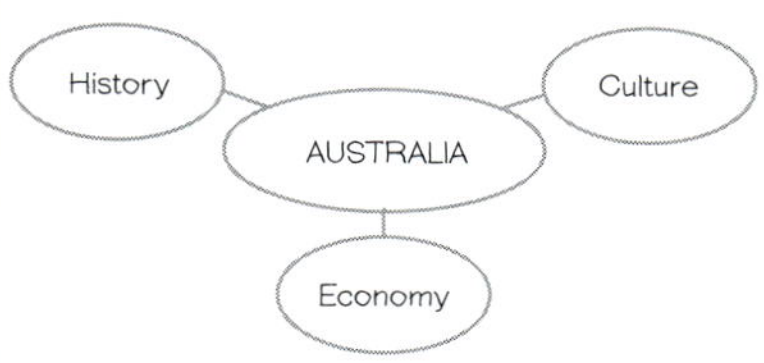

KEY IDEAS

3. What was the original reason for British colonization of Australia?
4. What are the two main cultural influences on Australia?
5. How do people in Australia make their living?

CRITICAL THINKING

6. **Make Inferences** How did European settlement influence Australia's development?
7. **Analyze Cause and Effect** What effect did Australia's location far from other inhabited areas have on its economic development?
8. **CONNECT to Today** How does the situation of the aborigines today compare to that of Native Americans?
9. **WRITING** **Create an Australian-English Dictionary** Use the Internet and library books to research Australian English. List ten unique Australian words and the word in American English that has the same meaning.

CONNECT to the Essential Question

How has the human geography of Australia and New Zealand changed since European settlement?

Ask students what they have learned so far that can help them answer this question. Students might mention:

- Australia's culture reflects European and Aboriginal influences.
- Australia's economy is based on plentiful mineral resources, productive agricultural lands, and a well-educated workforce.
- Tourism, a major industry in Australia, attracts millions of visitors.

4 Assess & Reteach

Assess Have students complete the Section Assessment.

Unit Resource Book: Oceania & Antarctica
- Section Quiz, p. 43

Interactive Review @ClassZone.com
- Section Quiz

Power Presentations

Test Generator

Reteach Ask student pairs to create a "Know Before You Go" fact sheet that will give tourists basic information about Australia's history, culture, and economy. Invite pairs to share their fact sheets.

Unit Resource Book: Oceania & Antarctica
- Reteaching Activity, p. 46

SECTION 2 ASSESSMENT ANSWERS

Terms & Names

1. James Cook: British explorer who claimed land in Australia; aborigine: the original settlers of Australia; penal colony: a settlement that serves as a prison; Charles Perkins: aboriginal rights leader

Use Your Reading Notes

2. Europeans brought their culture, took over the land from the aborigines, and began farming it.

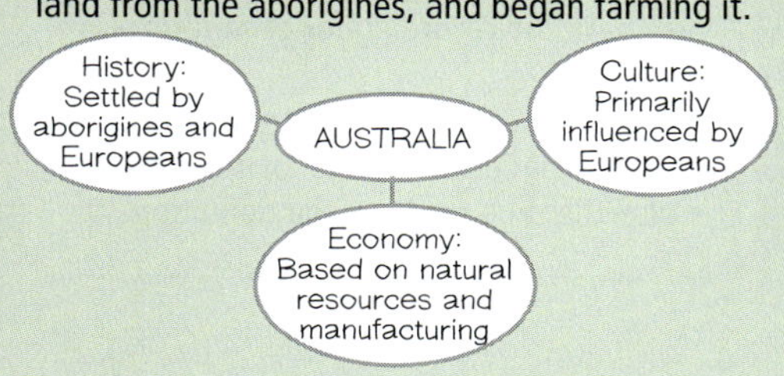

Key Ideas

3. to set up a penal colony
4. influenced by aborigines and early British settlers
5. They work in mining, manufacturing, agriculture, and service industries.

Critical Thinking

6. European settlers brought their language, religion, type of government, and other ways of living to Australia.
7. It produced most of the products it needed because it was so far from other countries.
8. Answers will vary. In both original inhabitants are a small percentage of the population; have land issues with their governments; and have full citizenship rights.
9. Use the rubric to score dictionaries.

Dictionary Rubric

	List and Define Australian Word	Identification of American English Word
4	clear and precise	correct
3	covers adequately	correct
2	barely explains	incorrect
1	unclear or missing	missing

1 Plan & Prepare

Objectives

- Summarize New Zealand's history and government
- Identify the major influences on New Zealand's culture
- Explain how natural resources influence New Zealand's economic activities

Read for the Essential Question

Help students read for a purpose by reminding them of the Essential Question: "How has the human geography of Australia and New Zealand changed since European settlement?" Point out that they can use text features such as Key Questions, special features, images, and maps to form answers to the question.

Vocabulary

Best Practices Toolkit

Use the ***Best Practices Toolkit*** to model strategies for vocabulary notetaking. Vary strategies throughout the year. Choose from: *Knowledge Rating, Predicting ABCs, Definition Mapping, Word Sort, Frayer Model (Word Squares), Magnet Words, Word Wheel,* and *Student VOC.*

Vocabulary Strategies, TT9–TT16

Reading Strategy

Best Practices Toolkit

Display the Web Diagram transparency.

- Remind students that they can use headings and subheadings to help them determine main ideas in a section of text.
- Model filling in entries, using the "Culture" subsection.
- Have students suggest entries. Remind students that main ideas in each block of text in the section will help them answer the Key Questions.
- Display the completed diagram for review.

Web Diagram, TT28

SECTION 3

Reading for Understanding

Key Ideas

BEFORE, YOU LEARNED

Life in Australia has been shaped by both aboriginal and European influences.

NOW YOU WILL LEARN

New Zealand's history, culture, and economy have been influenced by the Maori and European immigrants.

Vocabulary

TERMS & NAMES

Maori (MOW•ree) a Polynesian people who were the first inhabitants of New Zealand

Polynesia islands of the central and southern Pacific

Treaty of Waitangi (WY•TAHNG•gee) a treaty between the British and the Maori signed in 1840

Pakeha (PAH•kee•HAH) a Maori word that means "white people"

hydroelectric power electricity produced by moving water

geothermal energy power produced by the internal heat of the Earth

REVIEW

geyser an erupting hot spring

Visual Vocabulary Maori in traditional costume

Reading Strategy

Re-create the web diagram shown at right. As you read and respond to the **KEY QUESTIONS**, use the diagram to record important main ideas and details about the history, culture, and economy of New Zealand.

Skillbuilder Handbook, page R4

FIND MAIN IDEAS

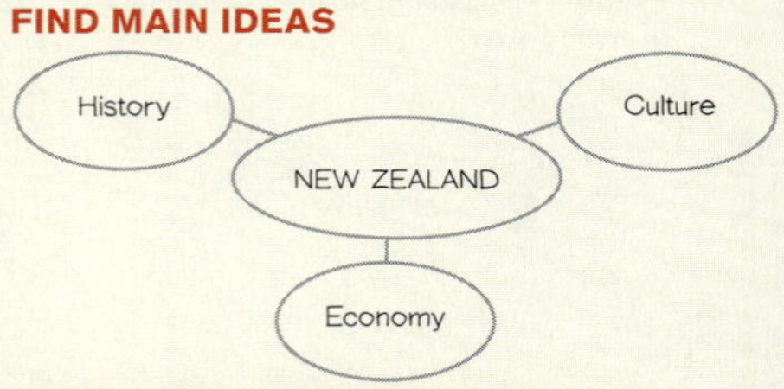

GRAPHIC ORGANIZERS
Go to **Interactive Review** @ClassZone.com

PRETEACHING VOCABULARY

English Learners

Pronounce and Preview

Pronounce each term for students. Review the meanings of words within definitions, such as *inhabitants, internal,* and *erupting.*

- To modify vocabulary learning, have students complete worksheets as they read, instead of afterward.

Unit Resource Book: Oceania & Antarctica
- Vocabulary Practice, p. 27

Inclusion

Record and Listen to Terms and Definitions

Have volunteers take turns reading the terms and definitions into an audiocassette or digital recorder. As they play back the recording for each term, have them:

- pause the tape and repeat the term.
- say the definition in their own words.
- play the tape to check the definition.

Sydney Opera House

Koala with eucalyptus

SECTION 3 FOCUS ON

New Zealand

Connecting to Your World

Imagine a place so far away that it would take you 17 hours to fly there from New York or 12 hours from Los Angeles. That place is New Zealand, about a thousand miles southeast of Australia and about 30 times smaller than Australia. Now imagine trying to find that small bit of land by ship in a vast ocean without good maps. That is what the earliest explorers had to do in their search for the southern islands. Their voyages took years to complete. But their discoveries expanded the knowledge of the planet and its people.

Waiting for a Flight Flight times to New Zealand from the United States are more than 12 hours.

History and Government

KEY QUESTION How did the Maori lose control of New Zealand?

Earlier you read that two large islands—North Island and South Island—and several smaller outlying islands make up the country of New Zealand. The islands were inhabited mostly by birds until the **Maori** (MOW•ree) people migrated there some time between A.D. 950 and 1150. The Maori people came by canoes from other islands of **Polynesia**, which are in the central and southern Pacific Ocean.

Marlborough Sounds South Island, New Zealand

2 Focus & Motivate

3-Minute Warm-Up

Write on the board or display the transparency:

- How would you describe life in New Zealand today? *(Possible Answers: can enjoy nature; Maori cultural traditions; a major movie industry; raising sheep; energy from hydroelectricity)*

Unit Transparency Book: Oceania & Antarctica
- 3-Minute Warm-Ups, TT1

Connecting to Your World

More About . . .

New Zealand and the Date Line

Travelers coming to New Zealand from the east cross the International Date Line and move ahead a full day. This imaginary line curves around New Zealand, so that it's the same date on every island.

3 Teach

History and Government

Think, Pair, Share

- Who were the first people to inhabit New Zealand? *(the Maori)*
- Which country first guaranteed women the right to vote: New Zealand or the United States? *(New Zealand)*
- **Find Main Ideas** What events led to Britain's control of New Zealand? *(Possible Answers: British settlers arrive; conflicts lead to the Treaty of Waitangi; land wars take place; the Maori suffer great losses as Britain gains control.)*

SECTION 3 PROGRAM RESOURCES

Unit Resource Book: Oceania and Antarctica

ON LEVEL

Unit Resource Book
- Reading Study Guide, p. 7
- Section Quiz, p. 44

STRUGGLING READERS

Unit Resource Book
- Reading Study Guide with Additional Support, p. 13
- Skillbuilder Practice: Analyzing Primary Sources, p. 28
- Section Quiz, p. 44
- Reteaching Activity, p. 47

eEdition with Audio DVD-ROM

ENGLISH LEARNERS

Pupil Edition in Spanish
eEdition with Audio DVD-ROM
Unit Resource Book
- RSG (Spanish), p. 19
- RSG with Additional Support (Spanish), p. 25

Multi-Language Glossary
Test Generator
- Section Quiz in Spanish

INCLUSION

Unit Resource Book
- RSG with Additional Support, p. 13
- Section Quiz, p. 44
- Reteaching Activity, p. 47

GIFTED & TALENTED

Unit Resource Book
- History Makers, p. 34

PRE-AP

Unit Resource Book
- Primary Source: The Treaty of Waitangi, p. 41

TECHNOLOGY

Unit Transparency Book
- 3-Minute Warm-Ups, TT1
- Essential Question, TT2
- Critical Thinking, TT5

Daily Test Practice Transparencies
- Section 3, TT98

Power Presentations
ClassZone.com
World Cultures and Geography Video on DVD

More About . . .

The Moa

At one point, there were many species of this wingless bird. They ranged from the size of a turkey, to more than 400 pounds and up to 10 feet tall with their neck extended. Along with hunting the moa for food, the Maori used the bones for tools and the huge eggs to carry water. Living relatives of the moa include the ostrich, Australia's emu and cassowary, and New Zealand's kiwi.

Studies of moa skeletal remains have allowed researchers to learn more about the moa's anatomy and habits. The birds had no trace of wings (including no sign of any wing bones). They ate berries, leaves, and grasses—and like many birds, they swallowed stones to grind up what they ate.

Comparing Regions

Land and Coast in the United States and New Zealand

In land area, New Zealand is about the size of the state of Colorado—that is, less than 1/30th as big as the total land area of the United States. But because of all of its islands, New Zealand has about three-quarters as much coastline—more than 9,000 miles, compared to the 12,000-plus miles of U.S. coast.

Access to the sea is important to the economy of both countries, especially fishing. Both also claim a 200 nautical mile *exclusive economic zone,* as allowed by the United Nations 1982 Convention on the Law of the Sea, which took effect in November 1994.

Maori Wood Sculpture
Wood sculptures tell stories to the viewer. The surface carvings have specific meaning to those who can read them. The carvings tell the history of the Maori people.

Maori Settlement The Maori called the land Aotearoa (ow•tay•ah•ROH•ah). It means "land of the long white cloud" and refers to the thick clouds and snow-capped mountains. The Maori, like the aborigines of Australia, were hunter-gatherers and fishers. They settled mainly on North Island and hunted a large flightless bird called the moa. By the 1500s, they had killed almost all of these birds. Over the years, the Maori learned to be farmers. They lived in separate tribes of about 500 members. Their culture was rich and included skilled woodcarvers who created wood sculptures using stone tools.

Europeans began to arrive in the mid-1600s. A Dutch sea captain, Abel Tasman, stopped in the islands in 1642. Cartographers named the area *Nieuw Zeeland* after a territory in the Netherlands. English Captain James Cook came to the islands more than a hundred years later, in 1769. He was the first European to explore New Zealand and claimed it for Great Britain.

A British Colony Soon, other Europeans—sailors, explorers, and traders—stopped at the islands. In the late 1700s, about 100,000 Maori lived there. In 1814, British Protestant missionaries established the first European settlement on North Island. Because there was no official government, there was much lawlessness in New Zealand. The Maori and the settlers fought each other. The British colonists asked their government to come and establish law and order.

In 1840, Maori chiefs signed the **Treaty of Waitangi** (WY•TAHNG•gee) with Britain. It stated that the Maori accepted the British queen as their ruler in exchange for British protection. The British believed the treaty allowed them to control the land while protecting Maori rights. But the Maori thought that they still had control of the land and that the British were only minor officials. The difference in the understanding of the treaty terms caused problems. British settlers began buying land without Maori permission. Between 1845 and 1872, land wars broke out. When they were over, the British had control of the islands, and the Maori had suffered great losses. Today, Maoris make up about 15 percent of the population.

After gold was discovered in 1861, more settlers arrived. Like the Australian gold rush, many unsuccessful gold seekers did not have the funds to return home. So, they settled on the land and began farming or raising sheep or cattle.

DIFFERENTIATING INSTRUCTION

Struggling Readers

Create a Time Line

Have students scan this textbook page and point out dates. On the board or chart paper, make a column for each date, in chronological order. For each date, review the related text with students and discuss what is important about it. Write their ideas in the columns. Then have each student create a time line from the information.

Gifted & Talented

Write from Different Points of View

Have students form groups and discuss the settlement of the British in New Zealand, from both the settlers' and Maori's point of view. Have them discuss key issues, such as why they would fight each other and how language barriers might have affected the situation. Have groups share their ideas. Then have each student write a brief journal entry from a settler's and a Maori's perspective.

Even though they were a British colony, the New Zealanders thought of themselves as a separate group. In 1907, they asked Britain to allow them to be a self-governing country.

Self-Government Today, New Zealand has a government similar to that of Australia. New Zealand is a constitutional monarchy. The monarch of the United Kingdom is head of state, but has little control. The monarch appoints a governor general, who also has little power. A parliamentary-style government is in place. The one-house parliament has 120 members. Seven seats in parliament are reserved for Maori members, but Maori may also be elected to other seats.

New Zealanders have always been concerned about equal rights for all citizens. The country was one of the first to guarantee women the vote, in 1893. (The United States did not guarantee this right to women until 1920.) The Maori continued to make efforts to get provisions of the Treaty of Waitangi to be enforced. In the 1990s, the Maori finally gained a settlement that provided land and money to the Maori people. New Zealand also provides many social services and old-age pensions for all its people.

ANALYZE CAUSE AND EFFECT Explain how the British gained control of New Zealand.

by signing the Treaty of Waitangi with the Maori in 1840 and driving the Maori from their lands

ANALYZING Primary Sources

The Treaty of Waitangi was signed by Maori leaders and the British government on February 6, 1840. Under the treaty, Britain annexed New Zealand. Because the treaty was written in both Maori and English, each party interpreted the treaty terms differently. Below is a part of the treaty.

> Her Majesty the Queen of England confirms and guarantees to the Chiefs and Tribes of New Zealand and to the respective families and individuals thereof the full exclusive and undisturbed possession of their Lands and Estates Forests Fisheries and other properties which they may collectively or individually possess so long as it is their wish and desire to retain the same in their possession.
>
> Source: Government of New Zealand

DOCUMENT-BASED QUESTION

What right does the British government appear to be granting to the Maori?

◄ *Present-day rally for Maori rights*

903

INTERDISCIPLINARY ACTIVITIES

CONNECT to Language Arts

Write a Newspaper Caption

Have students write a caption for the photograph of the rally for Maori rights on this page, answering as many of the six journalist's questions as possible: *who, what, where, when, how, why.* Have students share their captions. To extend this activity, have them write a headline and lead paragraph for an accompanying newspaper article.

CONNECT to Civics

Parliamentary Debate Role Play

Have students take on the role of members of New Zealand's parliament, with one or more students acting as Maori members to approximate the minimum representation of seven members out of 120. Then have students debate the hypothetical issue of allocating government college grant money so that some money is set aside only for Maori students.

ANALYZING Primary Sources

Provide students with the following information: New Zealand's government passed a law in 1975 to establish the Waitangi Tribunal. The Tribunal inquires into claims brought by Maori individuals or groups regarding government breaches of the treaty, and makes recommendations to the government. The Tribunal, which can have up to 16 members, has an equal representation of Maori and non-Maori. Since the law was passed, there have been over 1,300 claims.

- Why would the Tribunal be made up of equal numbers of Maori and non-Maori? *(Possible Answer: Both perspectives would get considered; the original treaty was a partnership between both groups.)*
- What might the process of inquiring into a claim involve? *(Possible Answer: hearing the claim, researching local and government records, talking to others who might know some history related to it)*

DOCUMENT–BASED QUESTION ANSWER

Possible Answer: the right to their lands as long as they want to hold those rights

Unit Resource Book: Oceania & Antarctica

- Skillbuilder Practice: Analyzing Primary Sources, p. 28
- Primary and Secondary Sources: The Treaty of Waitangi, p. 41

Comparing Regions

Waitangi Day and Columbus Day

Waitangi Day in New Zealand and Columbus Day in the United States were both originally meant to serve as joyous commemorations of each nation's history. Today, both days have now also become forums for native peoples and other concerned citizens to raise issues regarding native peoples' rights.

Teach
Culture

Reader, Recorder, Reporter

- What are New Zealand's official languages? *(English, Maori)*
- How are the Maori preserving their culture? *(through web sites, cultural schools, newspapers, body tattooing, land claims, traditional religious beliefs)*
- **Find Main Ideas** Why has the film industry become big business in New Zealand? *(because of the country's scenic physical geography)*

Unit Resource Book: Oceania & Antarctica

- History Makers: Moana Jackson, p. 34

More About . . .

Maori Tattooing

Traditional Maori tattoos are called *moko,* and *ta moko* is the process of applying them. *Moko* designs and where they are placed convey a range of information about the wearer, such as a person's genealogy, family and tribal affiliation, occupation, social status, personal achievements, and passage to adulthood (as in women's chin *moko).*

The traditional *ta moko* method involves making groove-like lines by tapping an *uhi* (a chisel often made of bone) with a mallet-like stick. Some *ta moko* artists still use this method today.

More About . . .

***King Kong* Comes to New Zealand**

Another highly publicized movie filmed in New Zealand was the 2005 remake of *King Kong,* directed by New Zealand native Peter Jackson. Filming sites included studios and outdoor sites, many of which were in the Wellington area. The movie's scenes on the ocean steamer *Venture* were filmed around Kapiti Island, a nature reserve several miles off the Wellington coast near the entrance to Cook Strait.

Traditional Maori Costumes and Tattoos
Facial tattoos are common for both men and women.

Culture

KEY QUESTION What have been the main influences on New Zealand's culture?

Most New Zealanders are descendants of early British colonists. All New Zealanders of European ancestry are called ***Pakehas*** (PAH•kee•HAHS), or white people, by the Maori. New Zealand's culture reflects both British and Maori influences. For example, English and Maori are official languages. New Zealand has an urban culture. Almost 85 percent of the people live in coastal cities, mainly on the North Island. In rural areas, most people live in small villages or on farms or ranches.

Maori Culture The Maori have worked hard to preserve their culture. They also have a good relationship with the *Pakehas.* But Maori leaders would like more opportunities for their people. The Maori have set up Web sites, cultural schools, and newspapers to help retain their culture. In recent years, they have filed claims in court to try to recover their traditional lands. One of the most distinctive aspects of Maori culture is body tattooing. The tattoos are signs of power and prestige. Most Maori follow traditional religious beliefs that honor ancestors and the natural environment.

Sports and Recreation Since most New Zealanders live no more than 80 miles from the sea, water sports such as swimming, sailing, surfing, and fishing are popular. Many families own a small beach house where they go on weekends. New Zealand's mild climate and its mountains, especially those on the South Island, make hiking, climbing, and camping year-round activities. Skiing and snowboarding are also popular. Team sports like rugby and cricket, borrowed from the British, are enjoyed by many.

Wellywood New Zealand's scenic mountains, beaches, forests, and pasturelands have become popular locations for international movie productions, as well as the country's own. In fact, the New Zealand film industry has become big business. It is called Wellywood because it has its headquarters in the capital city of Wellington, and also because it is compared to America's Hollywood. Large studios and technical-effects companies work together to produce a variety of films. One of the biggest projects in motion-picture history, the *Lord of the Rings* trilogy, was filmed in New Zealand.

SUMMARIZE Explain the influence of geography on life in New Zealand. The land has influenced sports, recreation, and the economy.

DIFFERENTIATING INSTRUCTION

Struggling Readers

Create Question & Answer Sheets

Read (or have volunteers read) this page out loud. Then assign student pairs one of the subsections. Have them review the section together, and write three questions and answers based on it. Invite pairs to ask their questions for the whole class to answer.

English Learners

Comprehension: Connection

Have students locate sentences in which the word *most* is followed directly by a noun. Explain what a generalization is, if needed, and ask them what generalization each sentence is making. For each sentence, also have them create a new one using *some.* To extend this activity, you may want to point out the sentence on this page containing the phrase *most distinctive,* and discuss the different use of *most* there.

CONNECT Geography & Culture

Film Production in New Zealand

You probably know that J.R.R. Tolkien's *Lord of the Rings* takes place in Middle Earth. But did you know that Middle Earth was New Zealand? Director Peter Jackson selected 150 locations in his native country of New Zealand for the three films. New Zealand also supplied about 2,000 production people and thousands of minor performers.

CRITICAL THINKING

Make Inferences Why did the *Lord of the Rings* director choose to film in New Zealand?

a scene from Lord of the Rings ▶

An Economy Based on Land Resources

KEY QUESTION What are New Zealand's major economic activities?

New Zealand has a few mineral resources, and it also has good cropland and pastureland, abundant forests, and beautiful scenery. These land resources are the basis for New Zealand's economy. Many of New Zealand's goods are exported to Asia, Australia, Great Britain, and the United States.

Agriculture, Forestry, and Fishing Did you know that New Zealand has about 25 times as many sheep and cattle as people? Raised on rich pasturelands, these animals produce wool, meat, and dairy products in great quantities that are exported around the world. North Island forests provide timber for export. The fishing industry also supplies huge catches of fish and rock lobster.

It has good pastureland, and sheep do well in mountainous terrain. ▼

Sheep New Zealand's economy relies heavily on the raising of sheep. **Why is New Zealand suited for sheep raising?**

DIFFERENTIATING INSTRUCTION: TIERED ACTIVITIES

OBJECTIVE Narrate a film tour of New Zealand.

Basic	On Level	Challenge
Have students work in small groups and explain that they will write a brief narration as if creating a film tour of New Zealand's physical and cultural geography. Have groups pick three images from this section and write a brief paragraph that describes each one.	Adapt the Basic activity, so that students choose five images (from their textbooks or other sources). After writing the accompanying paragraph for each image, they should connect the paragraphs with transitional statements (for example, "Next it was on to . . .").	In addition to the On Level activity, have students include an itinerary and a map for their tour. To extend this activity, have students use a video camera to film the tour as they narrate it, or create a multimedia presentation of their tour.

CONNECT Geography & Culture

Film Production in New Zealand

The film industry has also helped New Zealand's tourism industry. One way to tour New Zealand is through themed tours, such as to *Lord of the Rings* filming locations. Many companies offer these—by land, air, and water, for a half day, full day, or more.

CRITICAL THINKING ANSWER

Make Inferences He knew the land well, and its geography fit with the story told by the film.

Teach

An Economy Based on Land Resources

Talk About It

- What are New Zealand's main agricultural products? *(wool, meat, and dairy products from sheep and cattle)*
- What is the benefit of having more manufacturing businesses in the country? *(It would reduce import needs.)*
- **Find Main Ideas** How does physical geography affect New Zealand's energy production? *(It uses its mountains, waterfalls, and geysers to make hydroelectric and geothermal steam power since it has very limited petroleum supplies.)*

CONNECT to the Essential Question

How has the human geography of Australia and New Zealand changed since European settlement?

Ask students what they have learned so far that can help them answer this question. Students might mention:

- After gaining independence, New Zealanders formed a parliamentary-style government, emphasizing equal rights for all its citizens.
- New Zealand culture includes British and Maori influences.
- New Zealand's economy is based on its farmland, forests, sea resources, and beautiful scenery.

4 Assess & Reteach

Assess Have students complete the Section Assessment.

Unit Resource Book: Oceania & Antarctica
- Section Quiz, p. 44

Interactive Review @ClassZone.com
- Section Quiz

Power Presentations

Test Generator

Reteach Have student pairs design a Web home page, giving an overview of New Zealand's history, government, culture, and economy.

Unit Resource Book: Oceania & Antarctica
- Reteaching Activity, p. 47

Hydroelectric Plant New Zealand relies heavily on this form of power. **What aspect of New Zealand's geography makes this possible?**

a mountainous terrain with lots of falling water

Manufacturing There has been a growing emphasis on manufacturing in New Zealand since the end of World War II. Since that time New Zealand has worked hard to build businesses in the country so that they would not need to import so many products. Today food products and manufactured goods, including automobiles, electronics, furniture, and metal products, are produced in New Zealand. Some textiles are produced because of the great quantities of wool available here. About 40 percent of the workforce is employed in manufacturing. Auckland, New Zealand's largest city, is the country's largest manufacturing center.

New Zealand has a very limited supply of petroleum. Its industrial production depends on the use of **hydroelectric power**, electricity produced by moving water. The many waterfalls in mountainous areas are some of the sources of hydroelectric power. About three-fourths of all power in New Zealand is created by moving water.

Geothermal energy, power produced by the internal heat of the Earth, is another source of power. Geothermal energy is generated by harnessing the steam or hot water in New Zealand's **geysers**. In some areas, engineers have drilled deep holes to release the underground steam that is then piped to stations to be converted to electricity.

SUMMARIZE Identify the major economic activities of New Zealand.
agriculture, forestry, fishing, manufacturing, and film production

ONLINE QUIZ For test practice, go to Interactive Review @ClassZone.com

Section 3 Assessment

TERMS & NAMES

1. **Explain the importance of**
   - Maori
   - Treaty of Waitangi
   - *Pakeha*

USE YOUR READING NOTES

2. **Find Main Ideas** Use your completed web diagram to answer the following question:
   In what ways have the Maori preserved their culture?

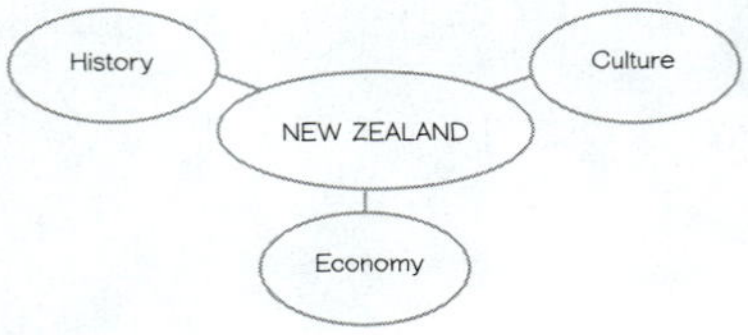

KEY IDEAS

3. Who were the first settlers of New Zealand and when did they arrive?
4. What was the purpose of the Treaty of Waitangi?
5. What are New Zealand's three major agricultural products?

CRITICAL THINKING

6. **Analyze Cause and Effect** What type of government developed in New Zealand? Why?
7. **Evaluate** How did New Zealand's physical geography contribute to its economic development?
8. **CONNECT to Today** Why might the government of New Zealand want to encourage the New Zealand film industry?
9. **TECHNOLOGY** **Plan a Power Presentation** Plan a power presentation about the cultural life of New Zealand. Include pictures and descriptions in your presentation.

SECTION 3 ASSESSMENT ANSWERS

Terms & Names

1. Maori: the first settlers of New Zealand; Treaty of Waitangi: a treaty between the British and the Maori; *Pakeha*: a Maori word for white people

Use Your Reading Notes

2. Maori is one of the official languages; they have Web sites, cultural centers, and newspapers.

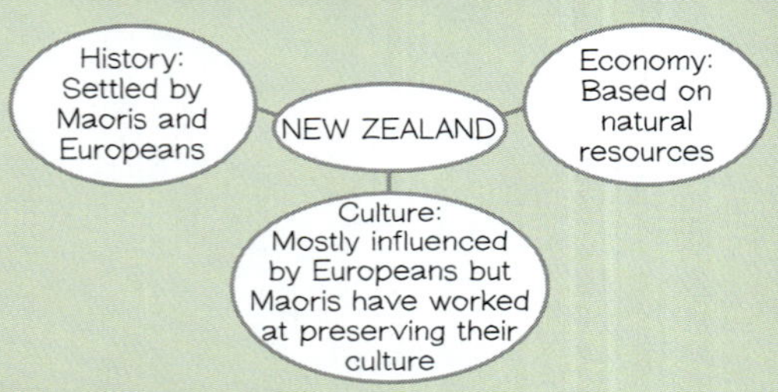

Key Ideas

3. The Maori people arrived sometime between A.D. 950 and 1150.
4. The treaty allowed Britain to gain control of the land of New Zealand.
5. wool, meat, and dairy products

Critical Thinking

6. New Zealand is a constitutional monarchy with a parliamentary government because it adopted the government of its colonizer, Great Britain.
7. Because New Zealand did not have many mineral resources, it used its land resources to provide cropland, pastureland, and energy sources.
8. The film industry brings in millions of dollars and provides many jobs for the economy.
9. Use the rubric to score presentations.

Presentation Rubric

	Text	Illustration	Use of Technology
4	clear and informative	very helpful illustrations	excellent use of technical tools
3	adequate	adequate illustration	good
2	confusing	poor	fair
1	unclear	unclear or missing	poor

Interactive Review

Click here to complete these and other activities online @ClassZone.com

CHAPTER SUMMARY

Key Idea 1
The nations of Australia and New Zealand have plant and animal life found nowhere else on earth.

Key Idea 2
Australia's history, culture, and economic life have been shaped by both aborigines and European immigrants.

Key Idea 3
New Zealand's history, culture, and economy have been influenced by the Maori and European immigrants.

For **Review and Study Notes**, go to **Interactive Review** @ClassZone.com

NAME GAME

Use the Terms & Names list to complete each sentence on paper or online.

1. I am the interior desert region of Australia. Outback
2. I am an animal with an abdominal pouch. ______
3. I am descended from a Polynesian people that settled New Zealand. ______
4. I worked to ensure that aborigines gained the rights they deserve. ______
5. I am a mountain chain in New Zealand. ______
6. I am a British explorer who was one of the first Europeans to explore Australia. ______
7. I am the islands of the central and southern Pacific. ______
8. I am an agreement between the British and the Maori. ______
9. I am an original inhabitant of Australia. ______
10. I am a settlement that serves as a prison. ______

aborigine
James Cook
geyser
Great Barrier Reef
Great Dividing Range
Maori
marsupial
Outback
Pakeha
penal colony
Polynesia
Charles Perkins
Polynesia
Southern Alps
Treaty of Waitangi

Activities

GeoGame

Use this online map to reinforce your understanding of the geography of Australia and New Zealand. Click and drag each place name to its location on the map.

To play the complete game, go to **Interactive Review** @ClassZone.com

Crossword Puzzle

Complete an online crossword puzzle to test your knowledge of Australia and New Zealand.

Across
1. an animal with an abdominal pouch to carry its young

907

ONLINE AND ON-SCREEN ACTIVITIES

GeoGame

Use the interactive map on **Power Presentations** DVD-ROM or the **eEdition** to help students review Australia's and New Zealand's location, geographic features, and important places.

- Demonstrate how to drag and drop the map labels listed onto the appropriate locations on the map.
- Advance to the scorecard to show how students have a chance to improve their scores.

Crossword Puzzle

Students can review important information about Australia and New Zealand by completing this crossword puzzle online at **ClassZone.com.**

- As an extension of this activity, have students use the *Activity Maker* at **ClassZone.com** to make their own crossword puzzle for one section of the chapter.
- Then have them trade puzzles with classmates to solve each other's puzzles.

Chapter Summary

As a follow-up to the chapter, encourage students to pair up and go over the **Review and Study Notes** at **ClassZone.com.**

Alternatively, students pairs can review their Reading Study Guide notes, making sure that they can add important details to each section's "Key Idea."

Name Game

Teaching Tips

ON PAPER Have students work in pairs to complete this activity. As an extension, they can write questions for the terms that are not covered in the items shown here.

ONLINE Make sure that students understand how to drag and drop terms to fill in blanks. Ask them to print out their results and compare them with those of a classmate.

ANSWERS

1. Outback
2. marsupial
3. Maori
4. Charles Perkins
5. Southern Alps
6. James Cook
7. Polynesia
8. Treaty of Waitangi
9. aborigine
10. penal colony

Teacher-Tested Activities

Jim Easton, Roscoe Middle School, Roscoe, Illinois

As a follow up activity or as an alternate to the Connect to Science activity in Section 1, students can follow the same directions to complete an illustrated book on the animal life of the Great Barrier Reef. They should include fish, marine animals, and plants. If technology permits, have student teams use presentation software to create presentations for the class. Use a class period to view the results.

Vocabulary ANSWERS

1. Geysers are found in New Zealand and are used as an energy source.
2. Marsupials are animals unique to Australia.
3. James Cook was the first European to explore Australia.
4. Australia was made into a penal colony by the British.
5. Both Australia and New Zealand have a constitutional monarchy form of government.
6. All are physical features of Australia.
7. Both are mountain ranges in the region.
8. These groups were the first to reach the islands of Australia and New Zealand.
9. All are terms associated with the early settlement of New Zealand.
10. Charles Perkins worked to gain rights for aborigines in Australia.

Key Ideas ANSWERS

11. because both were so far away from inhabited landmasses
12. coastlands, mountains, and deserts
13. North Island and South Island
14. because it is on the Pacific's volcanic "Ring of Fire"
15. The aborigines came from Southeast Asia and arrived in Australia at least 40,000 years ago.
16. gold
17. the aborigines and the British settlers
18. wheat, wool, meat, minerals, and gemstones
19. The Maori were the first settlers and they came from other islands in Polynesia.
20. Because there were English and Maori versions of the treaty, it was interpreted differently by the British and the Maori.
21. Three of the following: living in tribes, woodcarving, body tattooing, and worshiping ancestors and the natural environment
22. its land resources

CHAPTER 29 Assessment

VOCABULARY

Explain the significance of each of the following.

1. geyser
2. marsupial
3. James Cook
4. penal colony
5. constitutional monarchy

Explain how the terms and names in each group are related.

6. Outback, Uluru, and Great Barrier Reef
7. Great Dividing Range and Southern Alps
8. aborigine and Maori
9. Maori, Treaty of Waitangi, and *Pakeha*
10. aborigine and Charles Perkins

KEY IDEAS

1 Unique Lands

11. Why did Australia and New Zealand develop unique plant and animal life?
12. What are the three distinct landforms of Australia?
13. What are the two main islands of New Zealand?
14. Why does New Zealand have so many mountains, geysers, and volcanoes?

2 Australia

15. When did the aborigines arrive in Australia and from what locations did they come?
16. What mineral drew Europeans to Australia in the 1800s?
17. What two peoples influenced Australia's culture?
18. What are five of Australia's major exports?

3 New Zealand

19. Who were the first settlers of New Zealand and from what locations did they come?
20. Why did the Treaty of Waitangi cause problems between the British and the Maori?
21. What are three elements of Maori culture?
22. What are New Zealand's most important natural resources?

CRITICAL THINKING

23. **Compare and Contrast** Create a table to compare and contrast Australia with New Zealand.

FEATURE	AUSTRALIA	NEW ZEALAND
geography		
first settlers		
European colonizer		

24. **Draw Conclusions** How do the landforms of Australia and New Zealand affect where people live?
25. **Connect to Economics** How important is trade to Australia and New Zealand?
26. **Five Themes: Movement** Why did it take so long for Europeans to settle Australia and New Zealand?
27. **Connect Geography & Culture** How did the arrival of British settlers affect the native peoples of Australia and New Zealand?

Answer the ESSENTIAL QUESTION

How has the human geography of Australia and New Zealand changed since European settlement?

Written Response Write a two- or three-paragraph response to the Essential Question. Be sure to consider the key ideas of each section as well as specific ideas about how life in Australia and New Zealand has changed from their earliest settlement. Use the rubric below to guide your thinking.

Response Rubric

A strong response will:

- discuss the first settlers in the region
- identify changes that occurred in culture and economics because of European settlement

ESSENTIAL QUESTION REVIEW

Before students begin their essays, have them review the key chapter elements that will help them answer the Essential Question. Direct them to significant features in the chapter such as

- Physical Geography of Australia and New Zealand, Section 1
- Animated Geography: The Great Barrier Reef, Section 1
- History Makers: Charles Perkins, Section 2
- Worldster: A Day in Mick's Life, Section 2
- Comparing Economic Activities, Section 2
- Analyzing Primary Sources, Section 3
- Connect Geography and Culture, Section 3

Written Response Rubric

	Settlement Summary	Changes Identified
4	clear and concise	5 changes listed
3	complete	3–4 changes listed
2	parts missing	1–2 changes listed
1	missing or minimal	no changes listed

STANDARDS-BASED ASSESSMENT

Test Practice
- Online Test Practice @ClassZone.com
- Test-Taking Strategies and Practice at the front of this book

TABLES

Use the chart to answer questions 1 and 2 on your paper.

World Production of Wool

Country	Metric Tons
Australia	395,000
New Zealand	199,000
China	152,000
United Kingdom	42,000
South Africa	32,000
United States	10,000

Source: Commodity Research Bureau, *CRB Commodity Yearbook 2004*

1. Which country is the largest producer of wool in the world?

A. Australia
B. China
C. New Zealand
D. South Africa

2. How does U.S. wool production compare to the wool production of New Zealand and Australia?

A. They are about the same.
B. The United States produces more than New Zealand.
C. Australia produces less than the United States.
D. Australian and New Zealand produce more than the United States.

PRIMARY SOURCE

Read the primary source below and use context clues to answer questions 3 and 4 on your paper.

In 1812, the schooner *Parramatta* put into the Bay of Islands . . . for provisions and water. She was supplied by the natives [Maori] with potatoes, pork, and fish . . . and when they [Maori] required payment they were thrown overboard, fired at, and wounded. The schooner immediately [left] but was . . . driven on shore in a storm and the islanders revenged themselves by putting the crew to death.

Source: Committee of the Church of England Missionary Society, *Memorandum to the Earl of Bathhurst, Secretary of State of the Colonies,* 1817

3. The subject of this primary source is

A. a list of typical foods eaten by the Maori
B. a story about an early explorer of New Zealand
C. the relationship between European sailors and the Maori
D. missionaries in New Zealand converting Maoris

4. Does this quotation show evidence of a bias? Explain your answer.

GeoActivity

1. INTERDISCIPLINARY ACTIVITY–SCIENCE

With a small group, review New Zealand's energy sources. Choose either hydroelectric power or geothermal energy to research and create a museum-style display. Be sure that your display shows where the energy sources are located in New Zealand and how they work.

2. WRITING FOR SOCIAL STUDIES

Review information on the Maori attempts to regain their traditional lands. Form an opinion on the issue. Then write a paragraph supporting your opinion that could be posted on a Web log (blog).

3. MENTAL MAPPING

Create an outline map of Australia and New Zealand and label the following:

Australia
- Great Dividing Range
- Outback
- Tasmania

New Zealand
- North Island
- South Island
- Southern Alps

Critical Thinking ANSWERS

23. Possible answers: Australia—flat, dry continent; aborigines; Great Britain; New Zealand—two main islands, many mountains and bodies of water; Maori; Great Britain

24. Possible answer: Avoiding deserts and mountainous areas, most people in both countries live on the coastlands where the land is flat, the climate is mild, and there is access to the sea.

25. Possible answer: Both countries are major exporters of goods; trade partners include Asian countries, Great Britain, and the United States.

26. Possible answer: Europeans believed that there were unexplored lands in the southern hemisphere but the distances were so great that it took a long time to locate, explore, and then settle these lands.

27. Possible answer: The arrival of British settlers in Australia and New Zealand displaced the culture of the native peoples and also led to the loss of much of their traditional lands.

Standards-Based Assessment ANSWERS

1. A
2. D
3. C
4. Bias seems to be toward the Maori over the sailors.

GeoActivity

1. INTERDISCIPLINARY ACTIVITY RUBRIC

The museum-style display should:
- show examples of energy sources
- include illustrations, images, or other visuals
- explain how the energy sources work

2. WRITING FOR SOCIAL STUDIES RUBRIC

The paragraph should:
- state an opinion
- support the opinion with logical arguments and facts

3. MENTAL MAPPING RUBRIC

The map should:
- represent the approximate shapes of Australia and New Zealand
- correctly locate and label the required locations

DIFFERENTIATING CHAPTER ASSESSMENT

Unit Resource Book
- **Chapter Test A,** pp. 48–52. *Use for Struggling Readers, Beginning-to-Intermediate-Level English Learners, and for Inclusion of students whose IEP suggests supported test-taking.*
- **Chapter Test B,** pp. 53–57. *Use for On-Level Learners and ESOL students with Advanced-Level English.*
- **Chapter Test C,** pp. 58–62. *Use for Gifted and Talented students or students assigned Pre-AP work.*

Test Generator
- Chapter Tests A, B, and C, in English and Spanish

McDougal Littell Assessment System

TEST **Test Generator** Use three levels of tests in English and Spanish tied to state standards. Plus, choose questions to customize tests for all populations.

SCORE Scan and Score plain paper student tests in minutes.

REPORT Create instant progress reports, sortable by student, by standard, or other criteria.

RETEACH Respond immediately to students' needs by using a wealth of reteaching resources.

CHAPTER 30

Oceania and Antarctica PLANNING GUIDE

Complete print resources all on one USB drive allow you to customize lesson plans for all students.

Section Titles & Key Ideas	Unit Resource Book Oceania and Antarctica	Assessment Resources
1 Oceania: The Pacific World **KEY IDEA** Oceania is home to hundreds of cultural groups living on islands spread across thousands of miles of the Pacific Ocean.	• **Reading Study Guide,** pp. 65–66 • **RSG with Additional Support,** pp. 69–70 • **Vocabulary Practice,** p. 81 • **Map and Graph Skills,** pp. 83–84 • **Connect to Science,** p. 93	**Unit Resource Book** • Section Quiz, p. 95 • Reteaching Activity, p. 97 **Test Generator CD-ROM** **Daily Test Practice Transparency,** TT99
2 Antarctica: A Frozen Desert **KEY IDEA** Antarctica has no permanent residents because it has such a harsh environment.	• **Reading Study Guide,** pp. 67–68 • **RSG with Additional Support,** pp. 71–72 • **Outline Map Activities,** p. 85 • **History Makers,** pp. 87–88 • **Primary Source,** p. 94	**Unit Resource Book** • Section Quiz, p. 96 • Reteaching Activity, p. 98 • Chapter Test (Forms A, B, & C), pp. 99–110 **Test Generator CD-ROM** **Daily Test Practice Transparency,** TT100

INTEGRATED TECHNOLOGY

No Time? To focus students on important content in the chapter, use the **"Reading Study Guide with Additional Support"** in ***Resources2Go.***

Teacher Presentation Options

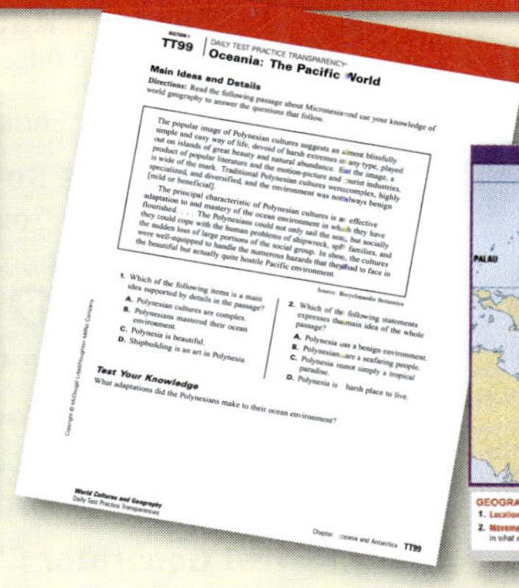

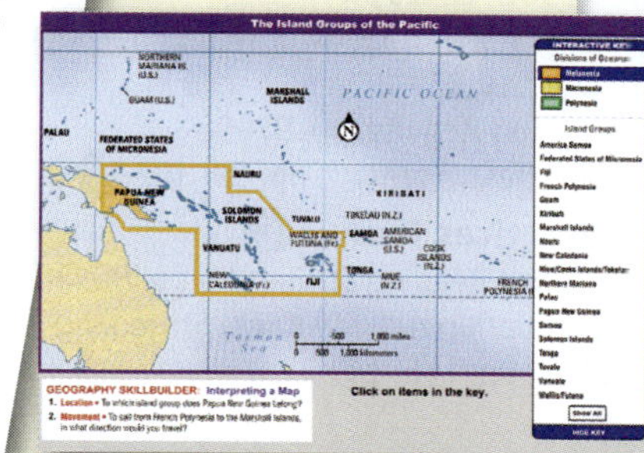

Electronic Books

eEdition DVD-ROM

eEdition Online

Presentation Toolkit

Video/DVD

- World Cultures and Geography

Power Presentations DVD-ROM

- Discussion Starters
- Media Gallery
- Review Game
- Interactive Review
- Animated Features

Unit Transparency Book

- 3-Minute Warm-Ups
- Essential Question
- Critical Thinking
- Maps
- Cultures

Daily Test Practice Transparencies

Animated Geography

- Present-Day Oceania Map
- Subregions of Oceania Map

OnlineActivities at ClassZone.com

Animated Geography

- **Videos and Photos**
- **3-D Animations**
- **Interactive Maps**

Interactive Review

- **Section Quizzes**
- **Review and Study Notes**
- **Vocabulary Flip Cards**
- **Graphic Organizers**
- **Name Game and Crossword Puzzles**
- **GeoGame**

Online Test Practice

Map Center

- **Atlas Maps**
- **Political Maps with Data**
- **Links to Mapping Web Sites**

Activity Center

- **Activity Maker—Review Games, Puzzles, Flip Cards, and Graphic Organizers**
- **Worldster**

Research & Writing Center

- **Web Sites**
- **Data Updates**
- **Primary Sources**
- **Writing Models**

Current Events

Kenneth Metz
Glacier Creek Middle School
Cross Plains, Wisconsin

Make It Fun — Teacher-Tested Activities

As you teach this chapter, look for my classroom-proven ideas for:

- making the lesson fun
- focusing students' attention and comprehension
- integrating technology with traditional teaching

Teaching Bell to Bell When you see this timing guide, have students discuss in pairs, in groups of 3–6, or as a class.

- **Think, Pair, Share** (pairs)
- **Reader, Recorder, Reporter** (3–6)
- **Talk About It** (any size)

CHAPTER 30

RESOURCES FOR DIFFERENTIATED INSTRUCTION

Struggling Readers

eEdition with Audio DVD-ROM

Lesson Plans
- Strategies for Inclusion

Unit Resource Book
- Reading Study Guide with Additional Support, pp. 69–72 B
- Reteaching Activity, pp. 97–98
- Chapter Test (Form A), pp. 99–102 A

Best Practices Toolkit C
- Interactive Reading Strategies, pp. 90–145
- Differentiation Activity Bank: Struggling Readers, pp. 146–149

Teacher's Edition Activities
- Tiered Activities, Sec. 1–2

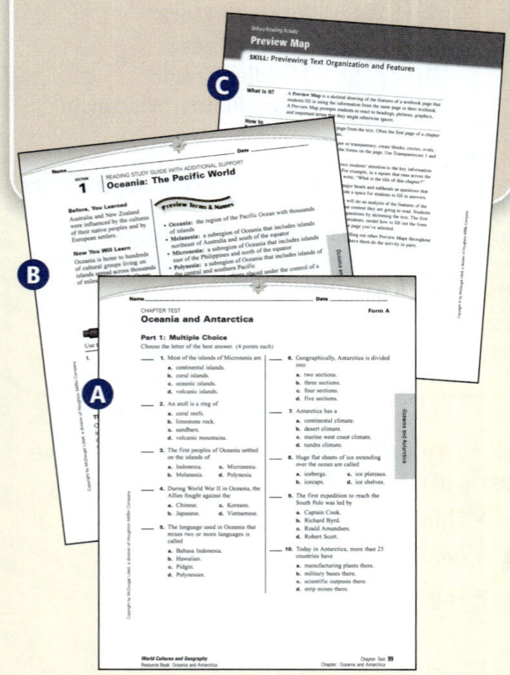

English Learners

Pupil's Edition in Spanish

eEdition with Audio DVD-ROM

Lesson Plans
- Strategies for English Learners

Unit Resource Book
- Reading Study Guide (Spanish), pp. 73–76
- RSG with Additional Support (Spanish), pp. 77–80 A B

Multi-Language Glossary C

Best Practices Toolkit
- Differentiation Activity Bank: English Learners, pp. 165–194

Test Generator CD-ROM
- Section Quizzes in Spanish
- Chapter Test in Spanish

Teacher's Edition Activities
- Pronounce and Preview, Sec. 1–2
- Vocabulary: Prefixes, Sec. 1
- Vocabulary: Key Academic Vocabulary, Sec. 2
- Tiered Activities, Sec. 1–2

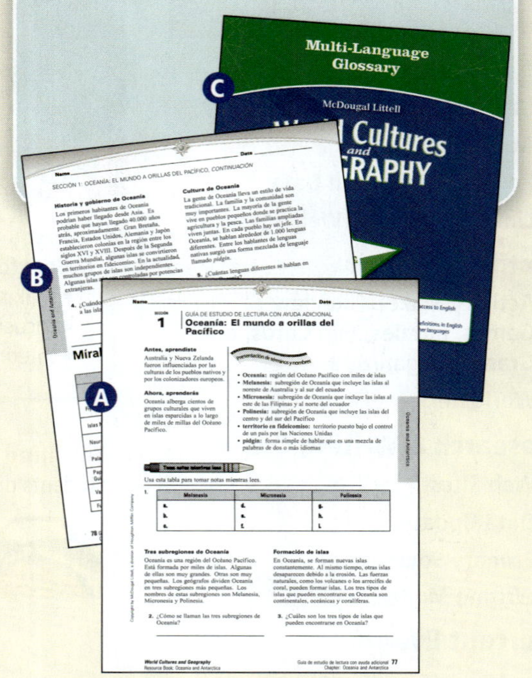

Inclusion

eEdition with Audio DVD-ROM

Lesson Plans
- Strategies for Inclusion

Unit Resource Book
- Reading Study Guide with Additional Support, pp. 69–72 A B
- Reteaching Activity, pp. 97–98
- Chapter Test (Form A), pp. 99–102

Best Practices Toolkit C
- Interactive Reading Strategies, pp. 90–145
- Differentiation Activity Bank: Struggling Readers, pp. 146–149

Teacher's Edition Activities
- Vocabulary: Magnet Words, Sec. 1
- Vocabulary Quiz, Sec. 2
- Tiered Activities, Sec. 1–2

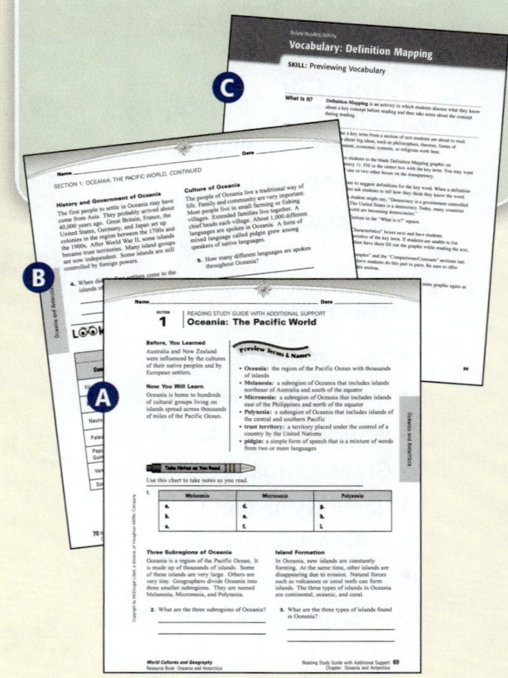

Interdisciplinary Activities

Gifted & Talented Students

Lesson Plans
- Core Lesson Plan

Unit Resource Book
- Outline Map Activities, p. 85
- History Makers, pp. 87–88 **A**
- Interdisciplinary Projects, pp. 89–91 **B**
- Connect to Science, p. 93 **C**
- Chapter Test (Form C), pp. 107–110

Best Practices Toolkit
- Student Products Rubrics, pp. 1–14
- Writing Templates, pp. 78–89

Teacher's Edition Activities
- Tiered Activities, Sec. 1–2

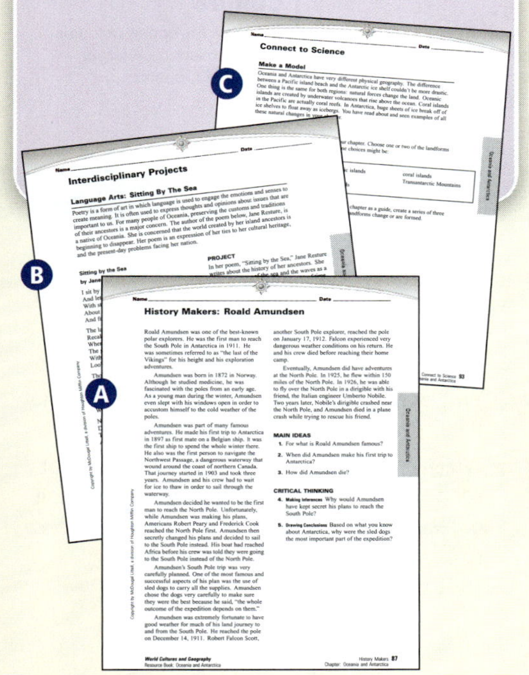

Pre-AP*

Lesson Plans
- Core Lesson Plan

Unit Resource Book
- Map and Graph Skills, pp. 83–84
- Economics in the World, p. 92 **A**
- Primary Source, p. 94 **B**
- Chapter Test (Form C), pp. 107–110
- World Literature, p. 111 **C**

Best Practices Toolkit
- Writing Templates, pp. 78–89
- Differentiation Activity Bank: Pre-AP Students, pp. 195–200

* Pre-AP is a registered trademark of the College Entrance Examination Board, which was not involved in the production of and does not endorse this product.

Teacher's Edition Activities
- Critique Like a Geographer, Sec. 1
- Explore the Poles and Antarctica, Sec. 2
- Tiered Activities, Sec. 1–2

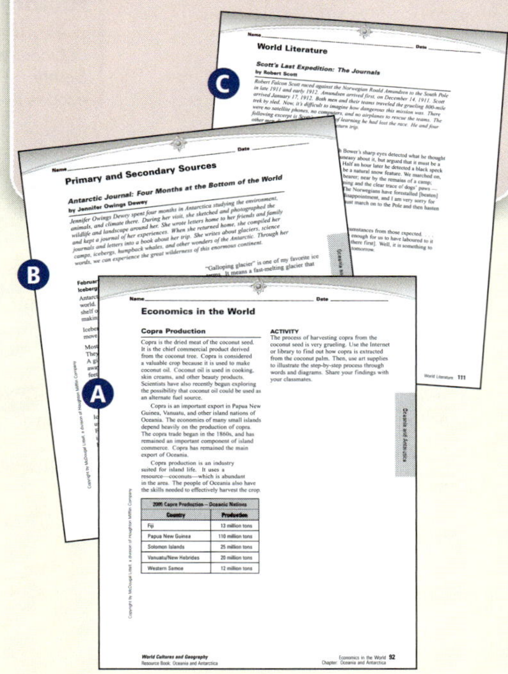

CONNECT to Science

Diagram on Culturing Pearls

CLASS TIME 30 minutes

TASK Prepare a diagram about the culturing of pearls.

SUPPLIES
- information and images about oysters and pearls
- posterboard, colored markers or pencils

ACTIVITY Draw students' attention to the Polynesian page of the Comparing Regions of Oceania feature. Discuss growing mollusks for pearl culturing. Then have students work in pairs to prepare a diagram poster about pearl culturing. Students should begin with a statement about mollusks in their natural habitat. They should then explain pearl culturing, illustrating each step in the process with a drawing or other image. Have students display and compare their diagrams.

Good for Visual Learners

CONNECT to Art

Antarctic Research in Miniature

CLASS TIME 30 minutes

TASK Build a three-dimensional model of an Antarctic research community.

SUPPLIES
- images of McMurdo Station or another Antarctic research facility
- small boxes (such as shoe boxes)
- construction paper, colored markers or pencils, rulers

ACTIVITY Have students work in small groups to create a model of an Antarctic research community. Draw students' attention to descriptions of the Antarctic landscape and scientific research in Section 2 of this chapter. Supply images of McMurdo and other Antarctic research communities. If convenient, supply additional information about Antarctic research. Encourage groups to assign tasks, such as scene designer, artist, caption writer, and presenter. Have students present their finished models with a statement about the research done at their facility.

Good for Visual and Kinesthetic Learners

Introduce the ESSENTIAL QUESTION

- Identify the subregions of Oceania and describe their physical features.
- Identify the groups who settled and governed Oceania, and describe Oceanic cultures.
- Describe the physical features of East and West Antarctica.
- Summarize the exploration of Antarctica and its international role as a scientific outpost.

Key Ideas

Section 1: Oceania is home to hundreds of cultural groups living on islands spread across thousands of miles of the Pacific Ocean.

Section 2: Antarctica has no permanent residents because it has such a harsh environment.

CONNECT to the Essential Question

You may wish to revisit the Essential Question periodically with students as they work through the chapter.

Suggestions for reconnecting are provided at an appropriate point in the *Teacher's Edition* for each section.

Emperor penguins in Antarctica

CHAPTER 30

Oceania and Antarctica

1 FOCUS ON Oceania: The Pacific World

2 FOCUS ON Antarctica: A Frozen Desert

ESSENTIAL QUESTION

How do the regions of Oceania demonstrate cultural diffusion?

CONNECT Geography & History

Use the map and the time line to answer the following questions.

1. Which state of the United States is found in Oceania?
2. Which European nations who staked claims to islands in the Pacific still control them?

40,000 B.C.

Geography
1521 Sailing under the Spanish flag, explorer Ferdinand Magellan reaches Guam. ►

History
c. 40,000 B.C. Settlers of Oceania come from Southeast Asia in outrigger canoes like the one above.

Culture
1800s European colonies are established on larger islands of Oceania.

TIME LINE DISCUSSION

Point out that Oceania includes the islands of the south, west, and central Pacific Ocean. Although separated by thousands of miles of ocean, cultural similarities exist among many Pacific islanders. Explain that Antarctica to the south is sometimes called a white desert, and that it is a vital frontier for scientific study.

- Which event on the time line marks the beginning of a European presence in Oceania? *(1521—Explorer Ferdinand Magellan reaches Guam.)*
- Who was the first explorer to reach the South Pole? *(Norwegian Roald Amundsen in 1911)*
- Which event on the time line is an example of cooperation among nations toward a common goal? *(1959—Antarctic Treaty opens Antarctica to international scientific research.)*

Animated GEOGRAPHY Present-Day Oceania

Click here to explore Oceania and Antarctica @ClassZone.com

Geography
◄ **1911** Norwegian explorer Roald Amundsen is the first to reach the South Pole.

History
1941–1945 Islands of the Pacific are battlegrounds during World War II.

Culture
▼ **1959** Antarctic Treaty opens Antarctica to international scientific research.

Today

911

TEACHER MEDIA FAVORITES

Books for the Teacher

Arnold, Caroline. *Easter Island.* Houghton Mifflin, 2000. Exploration of the history of Easter Island.

Karner, Julie. *Roald Amundsen.* Crabtree Publishing, 2006. An account of the exploration of the South Pole.

Lansing, Alfred. *Endurance: Shackleton's Incredible Voyage.* Audio Partners Publishing Corp.: 2000 (cassette), 2002 (CD). Audio editions of a classic about Antarctic survival.

Webster, Christine. *Polynesians.* Weigl Publishers, 2004. The history and cultures of Polynesia.

Video / DVD

NOVA: *Shackleton's Voyage of Endurance.* (DVD) WGBH Boston, 2005. Gripping survival story of Ernest Shackleton's 1914–16 Antarctic expedition, told with vintage photos and on-site film footage.

Wayfinders: A Pacific Odyssey. (DVD) Maiden Voyage Productions, 1998. Acclaimed PBS documentary about sailing the Pacific in traditional canoes using ancient navigational methods.

Internet

Go to the *Research and Writing Center* at **ClassZone.com** to link to:

- data on the region
- a variety of chapter-specific, content-reviewed sites
- "History Makers" biographies
- primary sources

Animated GEOGRAPHY Present-Day Oceania

CLASSZONE.COM

Point out that this is a political map of Oceania, showing the islands or island groups, their capitals, and selected other cities. Note that control by Britain is indicated by *U.K.*, France by *Fr.*, and New Zealand by *N.Z.*

- Which islands besides Hawaii are controlled by the United States? *(Northern Mariana Islands, Guam, Wake Island, Johnston Atoll, Midway Island, American Samoa)*
- Which country now controls the island that was reached by the Spanish explorer Magellan? *(The United States now controls Guam.)*
- Which countries control islands that lie south of the equator? *(United States, Britain, France, and New Zealand)*
- What is the capital of Fiji? *(Suva)* Of the Solomon Islands? *(Honiara)*

CONNECT GEOGRAPHY & HISTORY ANSWERS

1. Hawaii
2. France, United Kingdom

Presentation Options

Use the **Power Presentations** DVD-ROM to project the animated map.

- Prompt students to use the pop-up images and audio from the map to discover what they already know about the geography, culture, history, economics, or governments of Oceania.
- Challenge students to answer the map questions individually, in small groups, or as a class activity. Use all of the interactive material to help preview chapter content.

1 Plan & Prepare

Objectives

- Identify and describe the three subregions of Oceania
- Summarize the settlement history and governments in Oceania
- Describe culture in Oceania today

Read for the Essential Question

Help students read for a purpose by reminding them of the Essential Question: "How do the regions of Oceania demonstrate cultural diffusion?" Point out that they can use text features such as Key Questions, special features, images, and maps to form answers to the question.

Vocabulary

Best Practices Toolkit

Use the ***Best Practices Toolkit*** to model strategies for vocabulary notetaking. Vary strategies throughout the year. Choose from: *Knowledge Rating, Predicting ABCs, Definition Mapping, Word Sort, Frayer Model (Word Squares), Magnet Words, Word Wheel,* and *Student VOC.*

Vocabulary Strategies, TT9–TT16

Reading Strategy

Best Practices Toolkit

Display the Three-Column Chart transparency.

- Explain to students that comparing and contrasting means finding both similarities and differences.
- Model filling in a few details using the "Three Subregions of Oceania" subsection.
- Have students suggest entries.
- Display the completed chart for students to review.

Three-Column Chart, TT21

SECTION 1 Reading for Understanding

Key Ideas

BEFORE, YOU LEARNED
Australia and New Zealand were influenced by the cultures of their native peoples and by European settlers.

NOW YOU WILL LEARN
Oceania is home to hundreds of cultural groups living on islands spread across thousands of miles of the Pacific Ocean.

Vocabulary

TERMS & NAMES

Oceania the region of the Pacific Ocean with thousands of islands

Melanesia a subregion of Oceania that includes islands northeast of Australia and south of the equator

Micronesia a subregion of Oceania that includes islands east of the Philippines and north of the equator

Polynesia a subregion of Oceania that includes islands of the central and southern Pacific

oceanic island an island rising from the deep sea floor on a volcanic foundation

coral island a type of oceanic island that is a coral reef

atoll a ring-shaped coral reef encircling a central lagoon with no inner island

trust territory a territory placed under the control of a country by the United Nations

pidgin a simple form of speech that is a mixture of words from two or more languages

Visual Vocabulary atoll

Reading Strategy

Re-create the chart shown at right. As you read and respond to the **KEY QUESTIONS**, use the chart to compare and contrast details about the subregions of Oceania.

Skillbuilder Handbook, page R9

COMPARE AND CONTRAST

MELANESIA	MICRONESIA	POLYNESIA
1.	1.	1.
2.	2.	2.
3.	3.	3.

GRAPHIC ORGANIZERS
Go to Interactive Review @ ClassZone.com

912 Chapter 30

PRETEACHING VOCABULARY

English Learners

Pronounce and Preview

Pronounce each term for students. Review the meanings of words within definitions, such as *subregion, equator,* and *lagoon.*

- To modify vocabulary learning, have students complete worksheets as they read, instead of afterward.

Unit Resource Book: Oceania & Antarctica
- Vocabulary Practice, p. 79

Inclusion

Vocabulary: Magnet Words

Have students work in pairs to fill out the graphic organizer for the terms in the section.

- Use the transparency to model the strategy.
- Have volunteers suggest entries.
- Then distribute copies of the organizer for students to complete with a partner.

Best Practices Toolkit
- Vocabulary: Magnet Words, TT15

Cook Island

SECTION 1 FOCUS ON

Emperor penguins in Antarctica

Oceania: The Pacific World

Connecting to Your World

What do you think is the ideal climate? Many people believe that the climate and weather of the Pacific islands are ideal. The Pacific islands have a mild tropical climate with warm, but not too hot, temperatures. Gentle ocean breezes cool off the evenings, and sandy beaches are everywhere. In this section, you will study life in the Pacific islands and decide for yourself if they are your idea of a perfect place.

Three Subregions of Oceania

KEY QUESTION What are the three subregions of Oceania?

Oceania is a region of the Pacific Ocean with thousands of islands. It stretches from Hawaii on the northeast, to Easter Island on the east, south to New Zealand, and west to the island of New Guinea. Geographers estimate there are between 20,000 and 30,000 islands in the Pacific Ocean. Some are just tiny specks in the vast ocean. Others are very large, such as Australia, a continent; New Zealand; and New Guinea, the second largest island in the world. Islands located close to Asia, such as the Philippines, Indonesia, and Japan, are not considered part of the region.

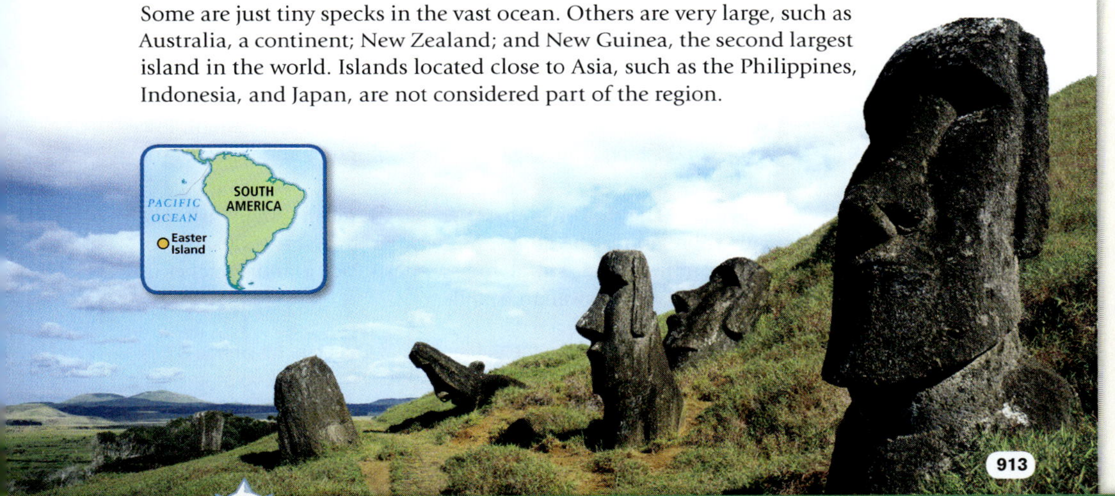

Easter Island Statues Hundreds of carved stone heads on this island look out over the Pacific Ocean.

2 Focus & Motivate

3-Minute Warm-Up

Write on the board or display the transparency:

- Imagine you are living thousands of years ago. How would you find your way to distant islands without technology? *(Possible Answer: Navigate by the stars and currents.)*

Unit Transparency Book: Oceania & Antarctica
- 3-Minute Warm-Ups, TT10

Connecting to Your World

More About . . .

Pacific Island Statues

Stone carvings have been found on many Pacific islands, including statues on Easter Island. These statues, called *moai*, were carved from volcanic stone. The shortest rise up to 20 feet, the tallest almost 40 feet.

3 Teach

Three Subregions of Oceania

Think, Pair, Share

- What is the estimated number of islands in Oceania? *(20,000 to 30,000)*
- Identify the world's second largest island. *(New Guinea)* In which subregion is the island located? *(Melanesia)*
- **Compare and Contrast** How does Polynesia compare in size to other subregions of Oceania? *(It's the largest.)* How do the islands of Micronesia differ in size from islands in other subregions? *(They're smaller.)*

SECTION 1 PROGRAM RESOURCES

Unit Resource Book: Oceania and Antarctica

ON LEVEL

Unit Resource Book
- Reading Study Guide, p. 63
- Map and Graph Skills, p. 81
- Section Quiz, p. 93

STRUGGLING READERS

Unit Resource Book
- Reading Study Guide with Additional Support, p. 67
- Skillbuilder Practice: Making Inferences, p. 80
- Section Quiz, p. 93
- Reteaching Activity, p. 95

eEdition with Audio DVD-ROM

ENGLISH LEARNERS

Pupil Edition in Spanish
eEdition with Audio DVD-ROM
Unit Resource Book
- RSG (Spanish), p. 71
- RSG with Additional Support (Spanish), p. 75

Multi-Language Glossary
Test Generator
- Section Quiz in Spanish

INCLUSION

Unit Resource Book
- RSG with Additional Support, p. 67
- Section Quiz, p. 93
- Reteaching Activity, p. 95

GIFTED & TALENTED

Unit Resource Book
- Connect to Science, p. 91
- Interdisciplinary Project, pp. 87, 89
- Section Quiz, p. 93

PRE-AP

Unit Resource Book
- Economics in the World, p. 90

TECHNOLOGY

Unit Transparency Book
- 3-Minute Warm-Ups, TT10
- Essential Question, TT11
- Critical Thinking, TT12
- Map: East Asia, Australia, and the Pacific Islands: Political TT14
- Culture, TT16

Daily Test Practice Transparencies
- Section 1, TT99

Power Presentations
ClassZone.com
World Cultures and Geography Video on DVD

Animated GEOGRAPHY

Subregions of Oceania

CLASSZONE.COM

CONNECT Geography & Culture

Point out that this map is similar to the map that begins the chapter, but this map also includes lines that show the division of Oceania into subregions.

- Which subregions include islands controlled by the United States? *(Micronesia and Polynesia)* Which subregion includes the state of Hawaii? *(Polynesia)*
- In which subregion are the most trust territories (territories placed under control of other countries)? *(Polynesia)* Identify the countries that control these trust territories. *(the United States, France, New Zealand, the United Kingdom, Chile)*

ANSWER

Movement Melanesia

Unit Resource Book: Oceania & Antarctica

- Map and Graph Skills: Reading a Political Map: Oceania, pp. 81–82

Presentation Options

Use the **Power Presentations** DVD-ROM to project the animated map.

- Click on each region of Oceania.
- Listen to the audio to learn more about the three subregions of Oceania.

Comparing Regions

Shared Islands: New Guinea and Hispaniola

A province of Indonesia occupies the western part of the island of New Guinea. Papua New Guinea, an independent country, lies on the east. Many people in both states are Micronesians. In contrast, peoples of different cultures share the island of Hispaniola, situated southeast of Florida in the Caribbean Sea. Haiti on the western part of Hispaniola reflects African and French culture, while the Dominican Republic on the east is culturally Hispanic.

Animated GEOGRAPHY

Subregions of Oceania

Click here to view the three subregions of Oceania @ ClassZone.com

CONNECT Geography & History

Movement Which of the three subregions was most likely first settled by people from the mainland of Asia?

Geographers divide Oceania into three subregions. These divisions are based on the physical geography of the islands and the culture and ethnic backgrounds of the peoples living there. The subregions are Melanesia, Micronesia, and Polynesia.

Melanesia The islands of **Melanesia** Ⓐ are close to Southeast Asia, northeast of Australia, and south of the equator. The world's second largest island—New Guinea—is located here.

Micronesia As you might have guessed, **Micronesia** Ⓑ means "small islands." Most of the islands of Micronesia are small, low-lying coral islands found east of the Philippines and north of the equator.

Polynesia The third region, **Polynesia** Ⓒ, means "many islands." This subregion is the largest in Oceania. It covers thousands of square miles of the central and southern Pacific and stretches as far north as the Hawaiian Islands.

SUMMARIZE Describe the subregions of Oceania. Melanesia is closest to Southeast Asia and Australia. Micronesia is made up of small coral islands. Polynesia is the largest region.

DIFFERENTIATING INSTRUCTION

English Learners

Vocabulary: Prefixes

Explain that *sub-* when used as a prefix means "below or under" or "division of a larger unit," such as is used here in *subregion*. Ask students to suggest examples of other words with *sub*. *(submarine, subway)* Clarify the prefixes, *micro-* meaning "small" and *poly-* meaning "many." Have students write sentences using words with these prefixes.

Pre-AP

Critique Like a Geographer

Have students reread the first paragraph of text below the map on this page. Then remind them that change is an ever-present fact. Ask students to think about the future of the current divisions of Oceania. For example: Might physical geography, such as Oceania's climate, change? If so, how might that affect Oceania? After the discussion, have students write a paragraph on how Oceania may change in the future.

CONNECT to Science

In Oceania, islands are constantly building and eroding. Islands can be created by natural forces. They are also subject to erosion and, thus, might disappear.

There are three types of islands in Oceania—continental islands, **oceanic islands** rising from the deep sea floor on a volcanic foundation, and **coral islands**, which are coral reefs. An **atoll** is a ring-shaped coral reef encircling a lagoon with no inner island. Below, see oceanic and coral islands.

Activity

Make a Model

Materials

- modeling materials
- colored pencils or markers, paper

1. Study the diagrams below. Choose one of the two types of islands to create in three dimensions.
2. Using the illustrations below as a guide, create a series of 3–5 small models showing in more detail how the type of island you selected is formed.
3. Write a caption for each model.

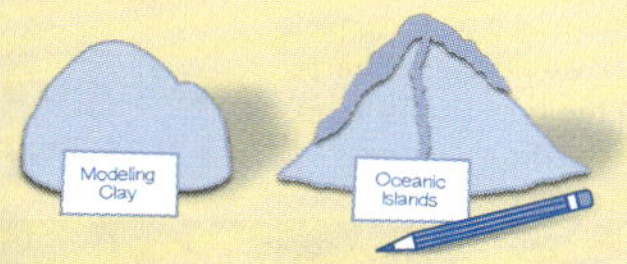

Oceanic Islands

1. Magma sometimes erupts through cracks in the ocean floor.
2. Over time, layers of lava can build up to form a volcanic cone.
3. Some volcanic cones rise above sea level and become islands.

Coral Islands

1. Some corals form reefs on the sides of volcanic islands.
2. As the island erodes, the reef continues to grow upward.
3. In time, only the low islands of the reef remain.

CONNECT to Science

Point out that this page highlights the formation of the different types of islands in Oceania. Have volunteers read aloud the text at the top of the page. Then direct attention to the images and captions.

- Identify the type of island formed by a buildup of lava. *(oceanic island)*
- Why might islands disappear? *(Erosion may occur.)*

Unit Resource Book: Oceania & Antarctica

- Connect to Science: Make a Model, p. 91

More About . . .

Current Undersea Volcanic Activity

Hawaii consists of eight main islands and 114 smaller islands. Scientists now know another island is being formed nearby. They are studying the undersea activity of a volcano named Loihi, located off the southeast coast of the big island of Hawaii. These scientists will continue to monitor the earthquakes and lava flows that are causing Loihi to rise—slowly—from the ocean floor.

Make It Fun

Teacher-Tested Activities

Kenneth Metz, Glacier Creek Middle School, Cross Plains, Wisconsin

To help students become aware of the vocabulary words used during the course of a lesson, I have my students play "How many times?" Have the students look at the highlighted vocabulary words on a page. Have the students write the words on a sheet of paper and check the meaning of the word in the glossary. Then as the lesson proceeds, they should keep track of how many times the word is spoken by the teacher or other students. At the end of the lesson have students compare the numbers to determine the actual times the word was spoken.

INTERDISCIPLINARY ACTIVITIES

CONNECT to Science

Make a Vocabulary Chart

To help students better understand scientific concepts, ask volunteers to reread parts of the text and summarize it. Then help students fill in a chart like the one at right. Each box should have just one term and its definition.

erosion	magma	lava
volcanic cone		
coral reefs		

Unit Resource Book: Oceania & Antarctica

- Interdisciplinary Projects: Sitting by the Sea, pp. 87, 89

Teach

History and Government

Talk About It

- How did the earliest settlers reach the islands of Oceania? *(by crossing land bridges or by large canoes from the Asian mainland and from Indonesia)* How did the arrival of Europeans affect the cultures of Oceania? *(The Europeans passed along their language and religions.)*
- **Compare and Contrast** How do governments of a colony and a trust territory compare? *(The original colonizing country controls a colony; the United Nations gives another country control of a trust territory to prepare it for self-rule.)*

Unit Resource Book: Oceania & Antarctica

- Skillbuilder Practice: Sequencing Events, p. 80
- Economics in the World: Copra Production, p. 90

More About . . .

Ancient Navigation

In 1975, members of the Polynesian Voyaging Society built a traditional Pacific canoe and chose Mau Piailug, a Micronesian, to pilot it. Piailug navigated as the ancient islanders did, using the stars, sun, moon, and ocean swells. In 1976, Piailug piloted the canoe, *Hokule'a*, from Hawaii to Tahiti and back. Since then the *Hokule'a* has made many voyages, proving that ancient Polynesians could skillfully navigate over thousands of miles of ocean.

CONNECT to the Essential Question

How do the regions of Oceania demonstrate cultural diffusion?

Ask students what they have learned so far that can help them answer this question. Students might mention:

- Southeast Asians and Indonesians settled Melanesia and Micronesia, and then moved east into Polynesia.
- Europeans, Japanese, and Americans colonized Oceania.
- Oceania's peoples developed pidgin to communicate with each other.

History and Government

KEY QUESTION Who settled the islands of Oceania?

The islands of Oceania were settled in three waves. The first peoples settled around 40,000 years ago on the islands of Melanesia and, later, those of Micronesia. Thousands of years later, groups from the settled islands moved east as far as Polynesia. In the 1800s, Europeans arrived in Oceania to establish colonies.

Oceania's First Settlers The earliest settlers of Oceania probably moved across land bridges or came by large canoes from the mainland of Asia and from Indonesia. These people may have left crowded conditions where they lived to look for more land. As they moved, they took animals and plants with them to start their new lives. In this way, the environment of the Pacific islands was changed to accommodate the culture of the newcomers.

For those migrating, it would show them a safe way to reach certain locations. ▼

Navigation Chart Pacific Islanders used charts like this to record the patterns of waves and ocean currents. **How would these charts aid the migrations across the Pacific?**

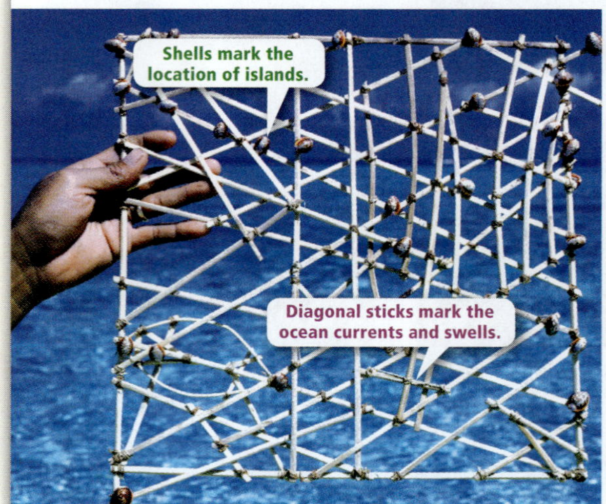

Colonization and Independence The first European contact with Oceania was made by explorer Ferdinand Magellan, who reached Guam in 1521. Later, British and French explorers came to the islands, bringing traders and Christian missionaries and colonists. In time, the United States, Germany, and Japan also set up colonies. The colonizers brought their language and religions with them.

During World War II, the Allies fought the Japanese across the region of Oceania for control of the Pacific. After the war, people living on islands that were European colonies began to pressure the colonizers for self-rule. Some islands became trust territories administered by larger nations, such as the United States, New Zealand, Australia, and Britain. A **trust territory** is a territory placed under the control of another country by the United Nations. The countries prepared the territories to achieve self-government.

From 1962 to 1994, 12 islands or island groups were given independence. Some islands, such as French Polynesia, are still controlled by the original colonizers.

FIND MAIN IDEAS Identify the groups that settled the islands of Oceania. people from mainland Asia, Indonesia, Europe, Japan, and the United States

DIFFERENTIATING INSTRUCTION: TIERED ACTIVITIES

OBJECTIVE Create a chart about colonization and its effects on Oceania.

Basic

Ask students, "What might a Pacific island gain and lose as a colony?" List responses on the board. Then have pairs of students create a two-column chart to display the information. Invite pairs to present or post their charts.

On Level

Adapt the Basic activity so that students work independently to create a two-column chart. In addition, ask students to include a brief summary below their charts. Invite students to present or post their charts.

Challenge

Have students complete the On-Level activity. In addition, have students do research to identify former and current trust territories as an additional column of chart information. Invite students to present or post their charts.

Culture

KEY QUESTION What are the cultures of Oceania like?

The many peoples of Oceania's small islands live a traditional way of life. Family and village are the focus of their lives. Most live in small farming or fishing villages, just as their ancestors did. Households include parents, grandparents, uncles, aunts, and cousins. They may occupy one house or a cluster of houses. A chief heads each village.

About one-third of the world's languages are spoken in Oceania. Scholars estimate about 1,000 languages are found there. About 740 are spoken in Papua New Guinea alone. With so many languages, communication between peoples in Oceania was difficult. So, a form of speech called pidgin developed. **Pidgin** is a mixture of two or more languages and is spoken by people with different native languages.

CONNECT Geography & Culture

Pidgin

Melanesian pidgin is spoken widely in Oceania. Pidgin languages often are a combination of a Western language, such as English, French, or German, and the local language.

The language is quite colorful. Here is how a piano is described in Melanesian pidgin: "bigfala bokis garem plande tit, iu hitim hemi kraeout" (a big box with many teeth, when you hit it, it cries).

MAKE GENERALIZATIONS Describe the cultures of Oceania.
Oceania has many cultures, but most of the people live a traditional way of life in small rural villages.

Section 1 Assessment

ONLINE QUIZ For test practice, go to **Interactive Review @ClassZone.com**

TERMS & NAMES

1. **Explain the importance of**
   - Oceania
   - Melanesia
   - Micronesia
   - Polynesia

USE YOUR READING NOTES

2. **Compare and Contrast** Use your completed chart to answer the following question:
   In which of the subregions of Oceania would you find people who had traveled the farthest from Southeast Asia?

MELANESIA	MICRONESIA	POLYNESIA
1.	1.	1.
2.	2.	2.
3.	3.	3.

KEY IDEAS

3. What are the three subregions of Oceania?
4. How were the Pacific islands first settled?
5. What is the traditional way of life for Pacific Islanders?

CRITICAL THINKING

6. **Draw Conclusions** How did contact with Europeans and Americans affect the peoples of Oceania?
7. **Understand Causes** Why was it necessary to set up trust territories in Oceania?
8. **CONNECT to Today** Why might tourism be an important industry for the islands of Oceania to develop?
9. **WRITING** **Prepare a Travel Brochure** Select one of the nations of Oceania and prepare a travel brochure for tourists. Describe the nation's physical setting, climate, and attractions, and include pictures.

Teach

Culture

Reader, Recorder, Reporter

- How many languages are spoken in Oceania? *(about 1,000 languages)*
- **Compare and Contrast** Compare ways of living between modern and ancient peoples of Oceania. *(Like their ancestors, most people of Oceania today live in farming or fishing villages, but also have some modern conveniences.)*

4 Assess & Reteach

Assess Have students complete the Section Assessment.

Unit Resource Book: Oceania & Antarctica
- Section Quiz, p. 93

Interactive Review @ClassZone.com
- Section Quiz

Power Presentations

Test Generator

Reteach Assign student pairs a subregion of Oceania. Have them write a summary about it.

Unit Resource Book: Oceania & Antarctica
- Reteaching Activity, p. 95

SECTION 1 ASSESSMENT ANSWERS

Terms & Names

1. Oceania: the region of the Pacific with many islands; Melanesia: a subregion of Oceania including islands northeast of Australia and south of the equator; Micronesia: a subregion of Oceania including islands east of the Philippines and north of the equator; Polynesia: a subregion of Oceania including islands in the central and southern Pacific

Use Your Reading Notes

2. Polynesia

MELANESIA	MICRONESIA	POLYNESIA
1. Closest to mainland Asia 2. Earliest settled 3. northeast of Australia and south of the equator	1. tiny islands 2. east of the philippines and north of the equator	1. huge subregion covers thousands of miles in Pacific 2. stretches as far north as Hawaii

Key Ideas

3. Melanesia, Micronesia, Polynesia
4. by peoples from Southeast Asia about 40,000 years ago
5. They live in small rural farming and fishing villages with their extended families.

Critical Thinking

6. Their lands were controlled by the Europe and some people adopted some of the culture of the colonizers.
7. The peoples of Oceania had little experience with Western-style government and required assistance from larger developed nations.
8. Possible Answer: Since the islands of Oceania are mainly agricultural, tourism would be a good source of income for Oceania.
9. Use the rubric to score brochures.

Brochure Rubric

	Description	Illustrations
4	clear and precise	attractive
3	covers adequately	adequate
2	barely explains	sloppy
1	unclear or missing	missing

1 Focus & Motivate

Objective

Gain a deeper appreciation of Oceania by exploring its regional differences.

Springboard Invite students to respond to images in the feature.

- Are any images striking or unusual? Why do you think so?
- Why do you think Oceania might appeal to tourists?
- If you had to choose an image from your own life to place beside each image on these pages, what would they be?

2 Teach

Regions of Oceania

Talk About It

- In which regions of Oceania is farming a major activity? *(Melanesia and Micronesia)*
- What evidence can you find that the economies of Oceania are adapting to meet modern demands? *(Possible Answer: Polynesia has a cultured pearl industry, forestry industry uses modern equipment, and some of its economies are based on tourism.)*
- **Compare and Contrast** In what ways do the cultures of Polynesia differ from those of Melanesia? *(Polynesian cultures are similar—in language, housing, and art; Melanesian cultures vary—with hundreds of languages spoken.)*

More About . . .

Cultured Pearls

A pearl forms when a foreign object enters the shell of a mollusk. Layers of a substance called *nacre* wrap around the object, forming a pearl.

Finding oysters or other mollusks with natural pearls is a costly process. Around 1900, the Japanese developed a new way to help mollusks produce pearls. Young mollusks are placed in barrels to grow for three years. They are then taken to pearl farms where trained workers place tiny beads of *nacre* inside the shells. The mollusks are placed in wire nets and kept in protected waters until their shells can be opened to remove the pearls.

COMPARING Regions of Oceania

The entire population of Oceania, excluding Australia and New Zealand, is over 12 million. The population is spread out over thousands of square miles of the Pacific Ocean. Geographers divide the areas into three regions based on ethnic groups and on language.

Polynesia

Ancestors of Polynesians sailed across thousands of miles of the Pacific to settle these islands. Although the lands are widespread, the cultures of the subregion are similar in language, housing, and art forms. Tourism forms the basis of some economies.

POLYNESIA

▲ *A Polynesian man displays traditional tattoos from head to toe.*

French Polynesia ▼

Polynesian Work

A Polynesian works with an oyster to create a pearl for the pearl industry.

Micronesian Work

Ifalik Island fishermen bring in a catch. Many in Micronesia work in the fishing industry.

918 Chapter 30

DIFFERENTIATING INSTRUCTION: TIERED ACTIVITIES

OBJECTIVE Gain a deeper appreciation of Oceania by exploring its regional differences.

Basic

Ask students to assume the role of travel agents, trying to persuade tourists to visit each region of Oceania. Have them list three reasons to travel to the region. Then create a print ad for a magazine based on their reasons to get people to travel to the region. Invite groups to display their ads.

On Level

In addition to the Basic level activity, have students work in small groups to create copy to accompany the print ads. Invite groups to compare and display their listings.

Challenge

In addition to the Basic level activity, have each group create a script for tour guides to use as they take groups of visitors around each region. Ask the groups to perform their scripts by roleplaying the parts of tourists and tour guides.

MICRONESIA

A Micronesian child in traditional headdress and grass skirt gets ready for a special dance.

Micronesia

The oceanic islands in this subregion depend on agriculture for their economy. The coral islands have coconuts and fishing industries. The region was mostly under control of the United States as a group of trust territories.

Islands of Palau

Melanesia

Hundreds of languages are spoken in this subregion, and the cultures are quite varied. Many of the people living here are subsistence farmers. The rugged lands are rich in minerals, which attracted European settlement.

MELANESIA

A Melanesian tribesman wears a traditional headdress and boar's tusks for a ritual ceremony.

Melanesian Work

New Guinea lumbermen take advantage of the forests on the island.

CRITICAL THINKING

1. **Draw Conclusions** How do the islanders of Oceania use their environment to meet their needs?
2. **Form and Support Opinions** What challenges do you think the people of Oceania face?

Comparing Regions

Tattoos as Cultural Expression

Tattooing has been practiced worldwide for centuries. Europeans who explored America saw tattooed Native Americans. In the early 1770s, British explorer James Cook brought the word *tattoo* to Europe after his contact with tattooed peoples in Polynesia and other parts of the Pacific. *Tattoo* derives from the Tahitian (Polynesian) word tatau, meaning "to strike."

In French Polynesia, tattoos traditionally mark a rite of passage into adulthood for boys and girls. Adult men use elaborate tattoos for spiritual protection and to signal social rank and spiritual power.

More About . . .

Palau Islands

These islands were part of a trust territory administered by the United States from 1947 until 1994. Palau is now an independent state, in free association with the United States. Many people in Palau work in government service. Other economic activities are subsistence farming and fishing. Shellfish and tuna are major exports, but tourism is a major part of the Palau economy.

3 Wrap Up

Encourage students to review their three-column compare-and-contrast charts for Section 1 and add details based on their reading of the Comparing feature.

CRITICAL THINKING ANSWERS

1. **Draw Conclusions** The islanders use the sea for food and transportation, and the natural resources as a base for their economy.
2. **Form and Support Opinions** They are limited by the amount of land they have; they are somewhat isolated from other parts of the world; their economies are small and developing.

INTERDISCIPLINARY ACTIVITIES

CONNECT to Language Arts

Write an Encyclopedia Article on Oceania

Have students write a brief entry about Oceania for an encyclopedia. The entry should include a sentence or two about each of the following topics: landforms, history, governments, cultures, and economies.

CONNECT to Art

Create a New National Flag

After looking at the images in the Comparing feature, have students look at the flags shown in the Country Almanac. Then have them imagine that another new island nation has been created and they need to design a flag for that nation. Along with their design they should provide a short explanation of the elements of the design. Display all the flags for the class.

1 Plan & Prepare

Objectives

- Describe the icy landscape of Antarctica
- Summarize the exploration of Antarctica and explain its importance to scientific research

Read for the Essential Question

Help students read for a purpose by reminding them of the Essential Question: "How do the regions of Oceania demonstrate cultural diffusion?" Point out that they can use text features, such as Key Questions, special features, images, and maps to form answers to the question.

Vocabulary

Best Practices Toolkit

Use the ***Best Practices Toolkit*** to model strategies for vocabulary notetaking. Vary strategies throughout the year. Choose from: *Knowledge Rating, Predicting ABCs, Definition Mapping, Word Sort, Frayer Model (Word Squares), Magnet Words, Word Wheel,* and *Student VOC.*

Vocabulary Strategies, TT9–TT16

Reading Strategy

Best Practices Toolkit

Display the Venn Diagram transparency.

- Remind students that drawing conclusions means finding important details and then putting those details together.
- Model filling in a few details using the "An Icy Landscape" subsection, and explain how a conclusion can be drawn from these details.
- Have students suggest entries and conclusions. Remind students that drawing conclusions will help them answer the Key Questions.
- Display the completed diagram for students to review.

Venn Diagram, TT22

SECTION 2 Reading for Understanding

Key Ideas

BEFORE, YOU LEARNED

Oceania has three distinct subregions, Melanesia, Micronesia, and Polynesia.

NOW YOU WILL LEARN

Antarctica has no permanent residents because it has such a harsh environment.

Vocabulary

TERMS & NAMES

Transantarctic Mountains mountains that divide Antarctica into two parts

geographic South Pole the point where all lines of longitude meet

icecap a glacier that covers a large area

Antarctic Peninsula an S-shaped peninsula that is part of the Andes Mountain chain

ice shelf huge flat sheet of ice extending over the ocean, but attached to the land

iceberg a large piece of floating ice

Robert Scott a British polar explorer

Roald Amundsen Norwegian polar explorer who was the first to reach the South Pole

Antarctic Treaty an agreement signed in 1959 calling for the peaceful use of Antarctica

BACKGROUND VOCABULARY

ozone hole the seasonal thinning of a protective layer of atmospheric gas high above Antarctica

Visual Vocabulary iceberg

Reading Strategy

Re-create the Venn diagram shown at right. As you read and respond to the **KEY QUESTIONS**, use the diagram to organize important details and draw conclusions about the physical geography of Antarctica.

Skillbuilder Handbook, page R13

DRAW CONCLUSIONS

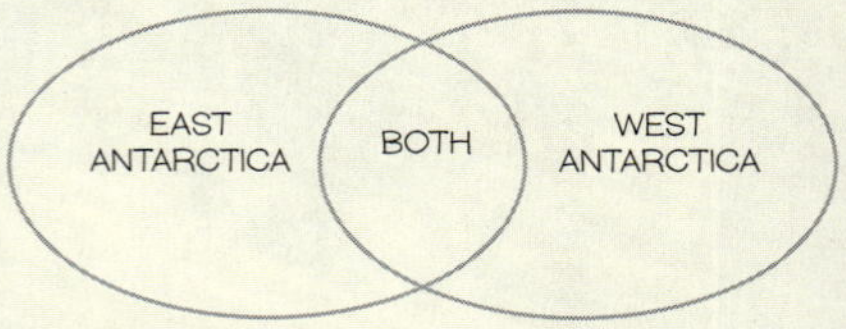

GRAPHIC ORGANIZERS
Go to **Interactive Review** @ClassZone.com

PRETEACHING VOCABULARY

English Learners

Pronounce and Preview

Pronounce each term for students. Review the meanings of words within definitions, such as *longitude*, *glacier*, and *seasonal*.

- To modify vocabulary learning, have students complete worksheets as they read, instead of afterward.

Unit Resource Book: Oceania & Antarctica
- Vocabulary Practice, p. 79

Inclusion

Vocabulary Quiz

Have volunteers read aloud each term and definition. Guide students in correct pronunciation and discuss lengthy or difficult definitions. Then ask pairs of students to quiz each other. One member of the pair should read a term and the partner should supply the definition. Students should then switch roles.

SECTION 2 FOCUS ON

Cook Island

Emperor penguins in Antarctica

Antarctica: A Frozen Desert

Connecting to Your World

You just learned about lands that have a climate many people believe to be ideal. Now try to imagine a place where the summer temperatures range from –30°F to 59°F. In the winter, it is so cold (–40°F to –100°F) that even a short exposure to the elements could mean certain death. This place is Antarctica—the land at the "bottom of the Earth."

Leopard Seal Herds of leopard seals roam the Antarctic waters.

An Icy Landscape

KEY QUESTION What are the physical features of East and West Antarctica?

Like Australia, Antarctica is an island continent. It is the world's fifth largest continent. It is the coldest, windiest, and most isolated continent, and one of the driest. Unlike other continents, Antarctica is buried under thousands of feet of ice. At its thickest point, the ice is around two miles deep. This ice-cover holds as much as 70 percent of the Earth's fresh water. Poking up through the ice are mountains. But just 2 percent of the land is ice-free. Even more floating ice surrounds Antarctica. Geographically, it is divided into sections, East Antarctica and West Antarctica.

Gentoo Penguins An adult feeds a baby penguin at an Antarctic seashore.

921

2 Focus & Motivate

3-Minute Warm-Up

Write on the board or display the transparency:

- What challenges do you think Antarctica presents to wildlife and to people? *(surviving in a cold climate; finding food and water)*

Unit Transparency Book: Oceania & Antarctica
- 3-Minute Warm-Ups, TT10

Connecting to Your World

More About . . .

Antarctic Wildlife

Many different kinds of penguins nest on Antarctica's coasts and nearby islands: Adélie, emperor, chinstrap, rockhopper, gentoo, king, and macaroni.

3 Teach

An Icy Landscape

Reader, Recorder, Reporter

- What are the sections of Antarctica? *(East and West)* Describe the landscape of East Antarctica. *(huge plateau and mountain peaks with ice-free valleys and frozen lakes)*
- **Draw Conclusions** In which section of Antarctica would year-round life be easier? *(Possible Answer: West Antarctica, because it includes the warmest, most ice-free region.)*

SECTION 2 PROGRAM RESOURCES

Unit Resource Book: Oceania and Antarctica

ON LEVEL

Unit Resource Book
- Reading Study Guide, p. 65
- Map and Graph Skills, p. 83
- Section Quiz, p. 94

STRUGGLING READERS

Unit Resource Book
- Reading Study Guide with Additional Support, p. 69
- Skillbuilder Practice: Sequencing Events, p. 80
- Section Quiz, p. 94
- Reteaching Activity, p. 96

eEdition with Audio DVD-ROM

ENGLISH LEARNERS

Pupil Edition in Spanish

eEdition with Audio DVD-ROM

Unit Resource Book
- RSG (Spanish), p. 73
- RSG with Additional Support (Spanish), p. 77

Multi-Language Glossary

Test Generator
- Section Quiz in Spanish

INCLUSION

Unit Resource Book
- RSG with Additional Support, p. 69
- Section Quiz, p. 94
- Reteaching Activity, p. 96

GIFTED & TALENTED

Unit Resource Book
- History Makers, pp. 85, 86
- Interdisciplinary Project, pp. 88–89

PRE-AP

Unit Resource Book
- Primary Source, p. 92
- Geography Project, p. 111

TECHNOLOGY

Unit Transparency Book
- 3-Minute Warm-Ups, TT10
- Essential Question, TT11
- Critical Thinking, TT13
- Map: Exploration and Land Claims in Antarctica, TT15
- Cultures, TT13

Daily Test Practice Transparencies
- Section 2, TT100

Power Presentations

ClassZone.com

World Cultures and Geography Video on DVD

Antarctica: Physical

CONNECT Geography & Culture

Tell students that this map shows the landforms of Antarctica. Have students compare the map to the information in the text. Ask them to identify the Transantarctic Mountains, the ice shelves, and the seas shown on the map.

- What is the highest peak on Antarctica? *(Vinson Massif, 16,066 ft.)*
- Identify Antarctica's major ice shelves. *(Ronne Ice Shelf, Ross Ice Shelf)* In which section of Antarctica are they located? *(West Antarctica)*
- Which sea is located off the Ronne Ice Shelf? *(Weddell Sea)*

ANSWERS

1. **Location** South Pole
2. **Place** Southern Ocean

Unit Resource Book: Oceania & Antarctica
- Outline Map Activities: Antarctica, pp. 83–84
- Interdisciplinary Projects: Animals of Antarctica, pp. 88, 89

COMPARING

Climate: Antarctica and Sahara

Remind students that Algeria is located in North Africa. Tell them that McMurdo is the largest station on Antarctica. Make sure students understand that the graphs show inches of precipitation and temperatures in Fahrenheit, and that the scale for precipitation differs on each chart.

- Why does the temperature drop in McMurdo while the temperatures are the hottest during July and August in Tindouf? *(The seasons are reversed.)*

CRITICAL THINKING ANSWER

Compare and Contrast Tindouf has greater precipitation in every month compared to McMurdo.

Unit Resource Book: Oceania & Antarctica
- Primary and Secondary Sources: Antarctica Journal: *Four Months at the Bottom of the World,* p. 92
- Geography Project: Ozone Hole Over Antarctica, pp. 111–112

Antarctica: Physical

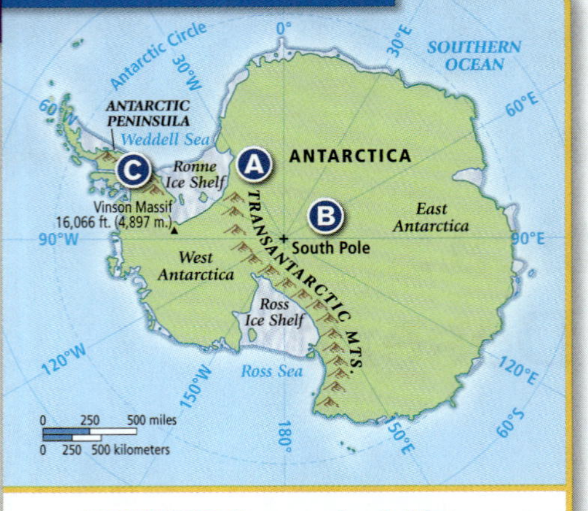

CONNECT Geography & History

1. **Location** At what point do the lines of longitude come together?
2. **Place** Which ocean surrounds the continent of Antarctica?

East Antarctica The **Transantarctic Mountains** (A) run across the continent from the Ross Sea to the Weddell Sea. They divide Antarctica into two sections. Some mountain peaks in East Antarctica are higher than 14,000 feet. Dry, ice-free valleys are located within these mountains where glaciers were once found. What little snow falls is blown out of the valleys by the wind. Some of the valleys have frozen lakes. The eastern side of the continent is covered by a huge plateau, mountains, and the **geographic South Pole** (B). That is the point where all lines of longitude meet. The entire region is a cold, high, and dry land.

It is hard to believe that Antarctica, with its snow and ice, has a dry climate. However, desert climate is one in which precipitation is under 10 inches per year. Only a little more than six inches of snow falls in Antarctica each year. That makes it a polar desert. The yearly amount of precipitation in Antarctica is somewhat larger than the amount of rain that falls in Africa's Sahara desert. See the graphs below.

COMPARING Climate: Antarctica and Sahara

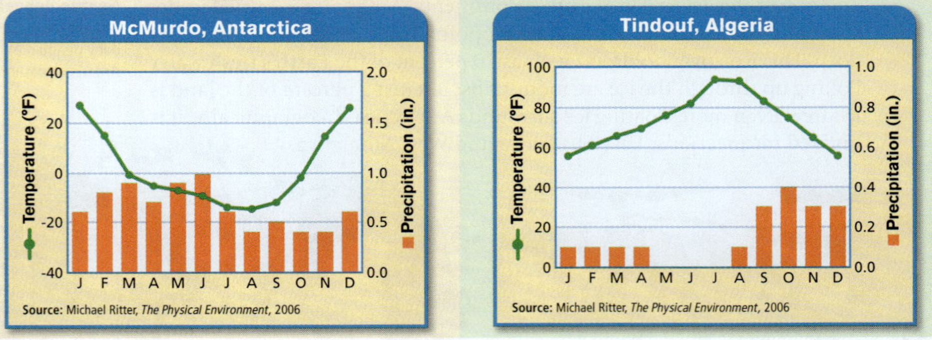

CRITICAL THINKING

Compare and Contrast How do the two regions compare in the amount of precipitation received each year?

DIFFERENTIATING INSTRUCTION

English Learners

Key Academic Vocabulary

Point out the map title, "Antarctica: Physical." Explain that here *physical* means "landforms and other geographical features." Also refer students to *region* in the first paragraph of text. Remind them that *region* is a word used by geographers to define a particular place, so that it can be compared to other places or regions. Have students write sentences using each of these words.

Pre-AP

Explore the Poles and Antarctica

Explain that the text specifies "geographic South Pole" because there is more than one South Pole. Explain that a "magnetic South Pole" exists, as does a magnetic North Pole. Ask students to think about the text on Antarctica. What would they like to know more about or in greater depth? Have students list two questions, and then research and present the answers to their questions.

West Antarctica The western side of Antarctica has a huge **icecap**, a glacier that covers a large area. The icecap links five off-shore islands into a frozen land of mountains, volcanoes, and the **Antarctic Peninsula** C. This S-shaped peninsula points toward South America and actually is a part of the Andes Mountain chain. The peninsula is the warmest, most ice-free region of Antarctica.

Ice Shelves and Icebergs The rugged coastline of Antarctica is marked by bays and channels. **Ice shelves** are huge, flat sheets of ice extending over the ocean, but attached to the land. In the summer, parts of these shelves break off and float away. These large floating pieces of ice are called **icebergs**. Antarctic icebergs are enormous, sometimes covering thousands of square miles. Their smooth, flat shape and massive size make them different from the usually smaller, peaked icebergs found in the Arctic.

COMPARE AND CONTRAST Describe how West Antarctica differs from East Antarctica. East Antarctica is mountainous, and has dry ice-free valleys and the South Pole. The west has a huge ice cap linking the continent to mountains, volcanoes, and the Antarctic Peninsula.

Fun Facts!

WORLD'S LARGEST ICEBERG

In March 2000, an iceberg, called B-15, broke off of Antarctica's Ross Ice Shelf. About the same area as Connecticut (about 5,000 square miles), the iceberg later broke into smaller pieces. These pieces of floating ice disrupted shipping lanes and affected the breeding grounds of several penguin colonies.

Exploration of Antarctica

KEY QUESTION Why do scientists continue to explore Antarctica?

Geographers had long thought there might be a seventh continent, but it was not until 1820 that a Russian explorer probably made what was the first sighting of Antarctica. Soon, European explorers were trying to map the coastline and to sail completely around the island continent. By 1895, explorers thought about crossing Antarctica to reach the geographic South Pole.

The Race to the South Pole Early expeditions failed to reach the South Pole because of the extremely harsh conditions. Then, in 1911, two explorers—**Robert Scott** of Great Britain and **Roald Amundsen** of Norway—organized polar expeditions. Using dog sleds, Amundsen and his team reached the South Pole on December 14, 1911. Scott's team, which used ponies and motor-powered sleds, arrived on January 17, 1912. Unfortunately, all members of Scott's expedition froze to death on the return journey. An American admiral, Richard Byrd, led five expeditions to the continent between the late 1920s and the 1950s. Later expeditions used aircraft and satellites to photograph the continent's features.

Robert F. Scott The British explorer perished on his return trip from the South Pole.

DIFFERENTIATING INSTRUCTION: TIERED ACTIVITIES

OBJECTIVE Create a time line of exploration in Antarctica.

Basic	On Level	Challenge
Have students work in pairs to complete time lines for Antarctic exploration, using the Section 2 text and features. Time lines should list each event in chronological order and include some visuals and captions. Invite students to display their time lines.	Adapt the Basic activity, so that students work individually to complete their time lines. In addition, have them include a descriptive paragraph about the landscape that confronted explorers in Antarctica. Invite students to display their work.	Adapt the On Level activity, so that students write a diary entry from the point of view of an explorer seeing Antarctica's land and waters for the first time. Ask students to display their work and to give an oral reading of their diary entry.

Fun Facts!

World's Largest Iceberg

If B-15 had blocked shipping lanes entirely, it would have kept supplies from reaching McMurdo Station, the largest U.S. research base on Antarctica. Historically, icebergs have had disastrous results: the oceanliner *Titanic* struck an iceberg on its maiden voyage across the North Atlantic in April, 1912, killing about 1,500 people.

Teach

Exploration of Antarctica

Talk About It

- When did Antarctic exploration begin? *(in the 1800s)* What year was the Antarctic Treaty concluded? *(1959)* What uses for Antarctica are specified in the treaty? *(peaceful scientific research and protection of the environment)*
- **Draw Conclusions** Why do you think nations continue to cooperate with the Antarctic Treaty? *(Possible Answers: The benefits help everyone; nations still have common problems, so it is in their best interests to solve them together.)*

Unit Resource Book: Oceania & Antarctica
- Skillbuilder Practice: Sequencing Events, p. 80
- History Makers: Roald Amundson; Admiral Richard Byrd, pp. 85–86
- World Literature: *Scott's Last Expedition: The Journal* by Robert Scott, pp. 109–110

CONNECT to the Essential Question

How do the regions of Oceania demonstrate cultural diffusion?

Ask students what they have learned so far that can help them answer this question. Students might mention:

- People from different countries have explored Antarctica.
- Many countries are cooperating to preserve Antarctica for peaceful scientific research.

Antarctica Territorial Claims

CONNECT Geography & History

Draw students' attention to the fact that the map includes no compass showing four directions. Explain that every line from the South Pole runs north.

- Which country has the largest claim in Antarctica? *(Australia)*
- Why do you think Australia, Argentina, and Chile are interested in claims to Antarctica? *(Possible Answer: These countries lie close to Antarctica.)*

ANSWERS

1. **Place** United Kingdom, Argentina, Chile
2. **Place** United Kingdom, Norway, France

4 Assess & Reteach

Assess Have students complete the Section Assessment.

Unit Resource Book: Oceania & Antarctica
- Section Quiz, p. 94

Interactive Review @ClassZone.com
- Section Quiz

Power Presentations

Test Generator

Reteach Have student pairs create four main idea questions about Antarctica. Collect the questions and answer them as a class.

Unit Resource Book: Oceania & Antarctica
- Reteaching Activity, p. 96

Antarctica Territorial Claims

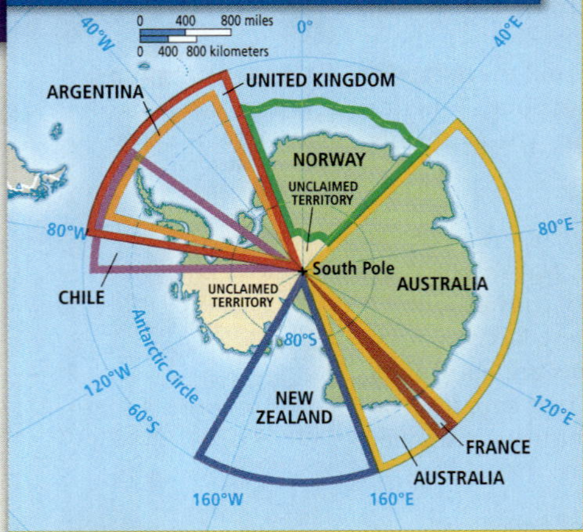

CONNECT Geography & History

1. **Place** Which three countries have overlapping claims?
2. **Place** Which of the nations making claims are not Antarctica's neighbors?

Scientific Outpost In 1957, a worldwide program of scientific investigation called the International Geophysical Year began. Sixty-seven countries agreed to conduct scientific experiments around the world and at science stations in Antarctica. When it ended, 7 of the 12 countries that had set up research stations in Antarctica claimed a part of the continent. Fearing attempts to use the land for military purposes, the nations later agreed to the **Antarctic Treaty** (1959). The treaty stated that Antarctica should be used for peaceful scientific research and the environment should be protected. Today, more than 25 countries have scientific outposts there. They study climate, animal behavior, earthquakes, solar radiation, and changes in the **ozone hole** (the seasonal thinning of a protective layer of atmospheric gas above Antarctica).

MAKE INFERENCES Identify reasons why Antarctica continues to be explored.
We still know very little about the continent.

ONLINE QUIZ For test practice, go to **Interactive Review @ClassZone.com**

Section 2 Assessment

TERMS & NAMES

1. **Explain the importance of**
   - Transantarctic Mountains
   - geographic South Pole
   - Antarctic Peninsula
   - Antarctic Treaty

USE YOUR READING NOTES

2. **Draw Conclusions** Use your completed Venn diagram to answer the following question:

   Which physical features do both East and West Antarctica share?

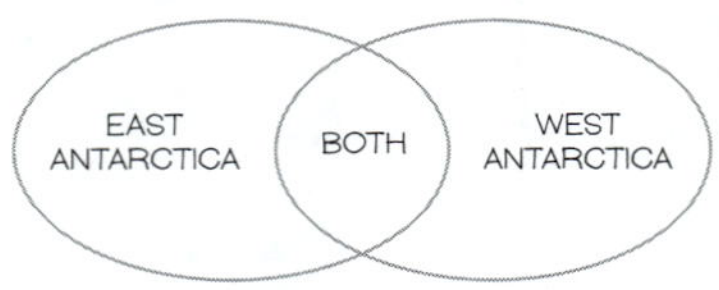

KEY IDEAS

3. What physical feature divides Antarctica?
4. What is the warmest, most ice-free region of Antarctica?
5. Which explorer was the first to lead a successful expedition to the South Pole?

CRITICAL THINKING

6. **Find Main Ideas** What was the purpose of the Antarctic Treaty?
7. **Understand Effects** What might happen to Antarctica if global warming increases?
8. **CONNECT to Today** What might Antarctica be like today if the Antarctic Treaty had not prevented more land claims there?
9. **SCIENCE** **Create a Plant and Animal Life Poster** Use the Internet to research the plants and animals found on Antarctica and collect images. Then label and display them on a poster.

SECTION 2 ASSESSMENT ANSWERS

Terms & Names

1. Transantarctic Mountains: mountains that divide Antarctica into two parts; geographic South Pole: the point where all lines of longitude come together; Antarctic Peninsula: an S-shaped peninsula that is a part of the Andes Mountain chain; Antarctic Treaty: an agreement that guarantees that Antarctica will be used for peaceful purposes

Use Your Reading Notes

2. Both share coastlines with bays, channels, and ice shelves, and a similar climate.

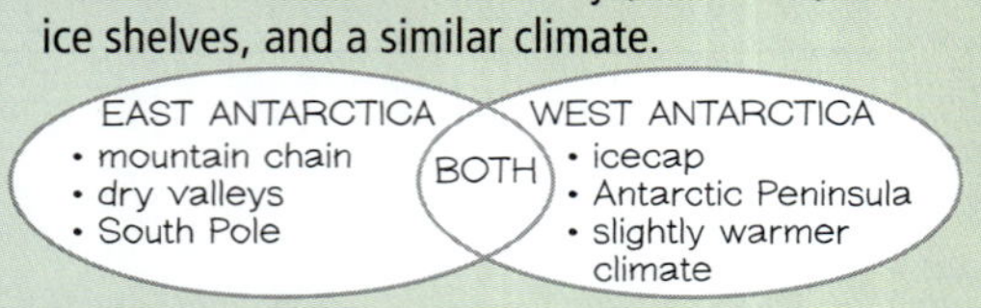

Key Ideas

3. Transantarctic Mountains
4. Antarctic Peninsula
5. Roald Amundsen of Norway

Critical Thinking

6. No country owns Antarctica; it is governed by treaty among countries carrying out scientific research there.
7. Possible Answer: Antarctica's ice cover would begin to melt, sea levels would rise, and more icebergs would cause problems for shipping and animal life.
8. Possible Answers: Antarctica could have various colonies, or there might be little cooperation between the colonizing nations and scientific research would suffer.
9. Use the rubric to evaluate the posters.

Poster Rubric

	Text	Illustration
4	clear and informative	very helpful illustrations
3	adequate to explain the ideas	adequate illustration
2	confusing	poor illustration
1	unclear	unclear or missing

Interactive Review

Click here to complete these and other activities online @ClassZone.com

CHAPTER SUMMARY

Key Idea 1
Oceania is home to hundreds of cultural groups living on islands spread across thousands of miles of the Pacific Ocean.

Key Idea 2
Antarctica has no permanent residents because it has such a harsh environment.

For **Review and Study Notes**, go to **Interactive Review** @ClassZone.com

NAME GAME

Use the Terms & Names list to complete each sentence on paper or online.

1. I am the region in the Pacific Ocean with thousands of islands. **Oceania**
2. I am a ring of reefs around a lagoon. ________
3. I am the point where all lines of longitude meet. ________
4. I am a subregion of Oceania that includes islands of the central and southern Pacific and stretches north. ________
5. I am a glacier that covers a large land area. ________
6. I am a Norwegian polar explorer who was the first to reach the South Pole. ________
7. I am the mountains that divide Antarctica into two parts. ________
8. I am an agreement signed in 1959 calling for the peaceful use of Antarctica. ________
9. I am a subregion of Oceania that includes islands northeast of Australia and south of the equator. ________
10. I am a form of speech that is a mixture of words from two or more languages. ________

atoll
Roald Amundsen
Antarctic Peninsula
Antarctic Treaty
geographic South Pole
icecap
ice shelf
Melanesia
Micronesia
Oceania
pidgin
Polynesia
Robert Scott
Transantarctic Mountains
trust territory

Activities

Flip Cards

Use the online flip cards to quiz yourself on the terms and names introduced in this chapter.

Crossword Puzzle

Complete an online crossword puzzle to test your knowledge of Oceania and Antarctica.

ACROSS
1. a large piece of floating ice

925

ONLINE AND ON-SCREEN ACTIVITIES

Flip Cards

Vocabulary Flip Cards are virtual flash cards at **ClassZone.com.**

Students can quiz each other or work independently to study the flip cards for the chapter "Oceania and Antarctica."

- They can quiz themselves on key terms and names by reviewing definitions.
- Demonstrate how to advance and shuffle the deck for another challenge.

Crossword Puzzle

Students can review important information about Oceania and Antarctica by completing this crossword puzzle online at **ClassZone.com.**

- As an extension of this activity, have students use the *Activity Maker* at **ClassZone.com** to make their own crossword puzzle for one section of the chapter.
- Then have students trade puzzles with classmates to solve answers for each other's puzzles.

Chapter Summary

As a follow-up to the chapter, encourage students to pair up and go over the **Review and Study Notes** at **ClassZone.com.**

Alternatively, students pairs can review their Reading Study Guide notes, making sure that they can add important details to each section's "Key Idea."

Name Game

Teaching Tips

ON PAPER Have students work in pairs to complete this activity. As an extension, they can write questions for the terms that are not covered in the items shown.

ONLINE Make sure that students understand how to drag and drop terms to fill in blanks. Ask them to print out their results and compare them with those of a classmate.

ANSWERS

1. Oceania
2. atoll
3. geographic South Pole
4. Polynesia
5. icecap
6. Roald Amundsen
7. Transantarctic Mountains
8. Antarctic Treaty
9. Melanesia
10. pidgin

Make It Fun

Teacher-Tested Activities

Kenneth Metz, Glacier Creek Middle School, Cross Plains, Wisconsin

For review of vocabulary before a test on the chapter, try classroom scramble.

- Create a list of vocabulary words and make an index card for each letter in each word.
- Each time a student participates in class, he/she earns a letter, one per student until the letters are used up. (No one gets doubles until all have one letter.)
- Place the vocabulary words on the board.
- Have students get together to form one of the words on the board.
- After the word is formed, the group should give a definition of the word.

Vocabulary ANSWERS

1. Pidgin language is used for communication among many Pacific Islanders.
2. Transantarctic Mountains divide east and west Antarctica.
3. The Antarctic Peninsula is an extension of the Andes Mountains.
4. All longitude lines join at the geographic South Pole.
5. The Antarctic Treaty opened Antarctica for scientific study by many nations.
6. These are types of landforms found in Oceania.
7. They are the subregions of Oceania.
8. Many nations of Oceania were once trust territories.
9. All are types of ice formations found in Antarctica.
10. Both explored Antarctica.

Key Ideas ANSWERS

11. Melanesia, Micronesia, Polynesia
12. Early peoples came about 40,000 years ago from Southeast Asia; a second wave of people from the settled islands came thousands of years later; and Europeans arrived in the 1800s.
13. to trade, to do missionary work, and to establish colonies
14. farming and fishing
15. a territory that is put under the control of a country by the United Nations to be prepared for independence
16. Because so many languages are spoken in the islands, pidgin is a way to communicate with others through a second language.
17. because the environment is too harsh
18. East Antarctica and West Antarctica
19. Icebergs are formed when parts of the ice shelf break off, usually when the temperature gets warmer.
20. Roald Amundsen from Norway
21. scientific research
22. to keep Antarctica open to peaceful international scientific research

CHAPTER 30 Assessment

VOCABULARY

Explain the significance of each of the following.

1. pidgin
2. Transantarctic Mountains
3. Antarctic Peninsula
4. geographic South Pole
5. Antarctic Treaty

Explain how the terms and names in each group are related.

6. oceanic island, coral island, and atoll
7. Melanesia, Micronesia, and Polynesia
8. Oceania and trust territory
9. icecap, ice shelf, and iceberg
10. Roald Amundsen and Robert Scott

KEY IDEAS

1 Oceania: The Pacific World

11. How is Oceania divided?
12. What were the three waves of settlement of the Pacific islands?
13. Why did Europeans come to Oceania?
14. What are the two traditional economic activities of the Pacific Islanders?
15. What is a trust territory?
16. Why did pidgin language develop?

2 Antarctica: A Frozen Desert

17. Why are their no permanent residents in Antarctica?
18. What are the two geographical divisions of Antarctica?
19. How are Antarctic icebergs formed?
20. Which European explorer was the first to reach the geographic South Pole?
21. What is the major activity conducted in Antarctica?
22. What is the purpose of the Antarctic Treaty?

CRITICAL THINKING

23. **Compare and Contrast** Create a table to compare and contrast the physical description, climate, population, and political status of Oceania and Antarctica.

CHARACTERISTICS	OCEANIA	ANTARCTICA
Physical Description		
Climate		
Population		
Political Status		

24. **Five Themes: Movement** Why was Antarctica the last continent to be discovered and explored?
25. **Make Generalizations** How has the world benefited from international cooperation in Antarctica?
26. **Connect to Economics** Why is the growth of the tourist industry important for the economies of the Pacific islands?
27. **Connect Geography & History** What happened to the colonies of Oceania after World War II?
28. **Connect Geography & Culture** What is distinctive about Oceania in terms of its languages?

Answer the ESSENTIAL QUESTION

How do the regions of Oceania demonstrate cultural diffusion?

Written Response Write a two- or three-paragraph response to the Essential Question. Be sure to consider the key ideas of each section as well as basic ideas related to cultural diffusion. Use the rubric below to guide your thinking.

Response Rubric
A strong response will:
- point out the waves of settlers and colonizers in the region
- identify aspects of the culture that show changes

ESSENTIAL QUESTION REVIEW

Before students begin their essays, have them review the key chapter elements that will help them answer the Essential Question. Direct them to significant features in the chapter, such as:

- Animated Geography: Subregions of Oceania, Section 1
- Connect Geography & Culture, Section 1
- Comparing Regions of Oceania, Section 1
- Comparing Climate: Antarctica and Sahara, Section 2
- Antarctica Territorial Claims, Section 2

Written Response Rubric

	Explanation	Paragraph Construction
4	Clear and very complete	Clear topic sentence and good support
3	Clear	Clear topic sentence and some support
2	Unclear missing points	Poor topic sentence, some fragments
1	Very unclear or missing	no topic sentence, fragments

STANDARDS-BASED ASSESSMENT

Test Practice
- Online Test Practice @ClassZone.com
- Test-Taking Strategies and Practice at the front of this book

CHART

Use the chart to answer questions 1 and 2 on your paper.

Countries of Oceania

Country	Date of Independence	Population (2005 Estimate)
Fiji	1970	905,949
Marshall Islands	1986	60,422
Nauru	1968	13,287
Palau	1994	20,579
Papua New Guinea	1975	5,670,544
Samoa	1962	176,908
Vanuatu	1980	208,754

Source: *CIA World Fact Book*, 2006

1. Which country was the first to gain independence?

A. Fiji
B. Nauru
C. Papua New Guinea
D. Samoa

2. Which is the smallest nation to gain independence?

A. Marshall Islands
B. Nauru
C. Palau
D. Vanuatu

CLIMATE GRAPH

Examine the climate graph below. Use the information in the graph to answer questions 3 and 4 on your paper.

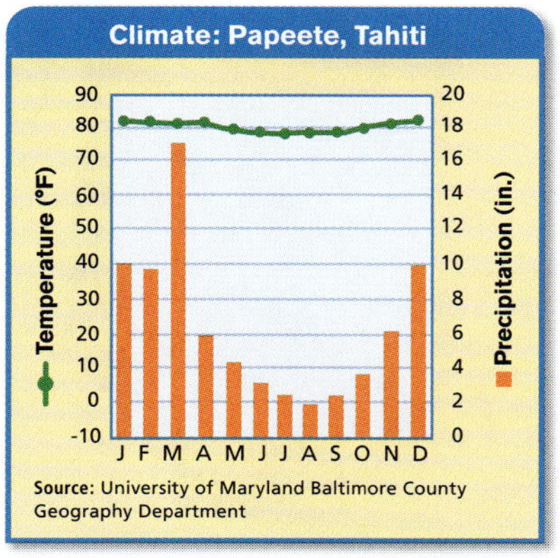

3. In which month does it rain the most?

4. Why would the month of August be a good time to visit Papeete?

GeoActivity

1. INTERDISCIPLINARY ACTIVITY–SCIENCE

With a small group, research the various types of ice formations found in Antarctica. Create a museum-style display of these formations. Be sure your display includes images of the formations, their names, and descriptions.

2. WRITING FOR SOCIAL STUDIES

Unit Writing Project After you have completed your illustrated time line showing the arrival of different groups in the region, add a series of newspaper headlines to the time line that also tell the story of migration in this region.

3. MENTAL MAPPING

Create an outline map of the southern Pacific and label the following:

- Melanesia
- Micronesia
- Polynesia
- Antarctica
- Australia
- New Zealand

Critical Thinking ANSWERS

23. Possible answers: Oceania: 20,000 to 30,000 islands in the Pacific Ocean; mild tropical climate; population not given; mostly independent nations, some colonies. Antarctica: island continent surrounding the South Pole; cold desert climate; no permanent residents; governed by international treaty

24. because it was so far away from other lands and because its climate was so harsh

25. Possible answers: The world has been spared conflicts over land claims, the environment of Antarctica has been protected, and scientific research has been shared.

26. Possible answer: Because the Pacific islands have few natural resources, the growth of the tourism will help their economies expand.

27. Many of the colonies asked for self-government. Eventually, 12 islands and island groups were granted independence.

28. More than 1,000 different languages are spoken and pidgin languages developed so that they could talk with each other.

Standards-Based Assessment ANSWERS

1. D
2. B
3. March
4. It rains very little during the month.

GeoActivity

1. INTERDISCIPLINARY ACTIVITY RUBRIC

The museum display should
- be attractive and easy to navigate
- have a clear and complete explanation
- correctly label and describe images

2. WRITING FOR SOCIAL STUDIES RUBRIC

The illustrated time line should
- include 5 or 6 entries
- have clear newspaper headlines for each event

3. MENTAL MAPPING RUBRIC

The outline map should
- include all geographic features listed in their correct location
- clearly show the 3 subregions

DIFFERENTIATING CHAPTER ASSESSMENT

Unit Resource Book
- **Chapter Test A,** pp. 97–100. *Use for Struggling Readers, Beginning-to-Intermediate-Level English Learners, and for Inclusion of students whose IEP suggests supported test-taking.*
- **Chapter Test B,** pp. 101–104. *Use for On-Level Learners and ESOL students with Advanced-Level English.*
- **Chapter Test C,** pp. 105–108. *Use for Gifted and Talented students or students assigned Pre-AP work.*

Test Generator
- Chapter Tests A, B, and C, in English and Spanish

McDougal Littell Assessment System

TEST **Test Generator** Use three levels of tests in English and Spanish tied to state standards. Plus, choose questions to customize tests for all populations.

SCORE Scan and Score plain paper student tests in minutes.

REPORT Create instant progress reports, sortable by student, by standard, or other criteria.

RETEACH Respond immediately to students' needs by using a wealth of reteaching resources.

N
E
W
S

Reference Section

World Cultures and GEOGRAPHY

Contents

Reading and Critical Thinking Skills

Reading Maps, Graphs, and Other Visuals

Research, Writing, and Presentation Skills

Using the Internet

1.1 Taking Notes

Defining the Skill

When you **take notes,** you write down the important ideas and details of a passage. A chart or an outline can help you organize your notes to use in the future.

Applying the Skill

The following passage describes several different types of bodies of water. Use the strategies listed below to help you take notes on the passage.

How to Take and Organize Notes

Strategy 1 Look at the title to find the main topic of the passage.

Strategy 2 Identify the main ideas and details of the passage. Then summarize the main ideas and details in your notes.

Strategy 3 Identify key terms and define them. The term *hydrosphere* is shown in boldface type and highlighted; both techniques signal that it is a key term.

Strategy 4 In your notes, use abbreviations to save time and space. You can abbreviate words such as *gulf (G.), river (R.),* and *lake (L.)*, as long as you write the proper name of the body of water with the abbreviation.

1 BODIES OF WATER

2 All the bodies of water on Earth form what is called 3 the **hydrosphere**. The world's 2 oceans make up the largest part of the hydrosphere. Oceans have smaller regions. 2 Gulfs, such as the Gulf of Tonkin, and seas, such as the Sea of Japan, are extensions of oceans. Land partially encloses these waters.

2 Oceans and seas contain salt water, but most lakes and rivers contain fresh water. The water in rivers, such as the 2 Nile River in Africa, flows down a channel in one direction. This movement is the current. Lake water, such as that found in 2 Lake Victoria in Africa, can have currents too, even though the water is surrounded by land. Some lakes feed into rivers, and some rivers supply water to lakes.

Make a Chart

Making a chart can help you take notes on a passage. The chart below contains notes from the passage you just read.

2 Item	Notes
1. 3 hydrosphere	all water on Earth
a. oceans	salt water; largest part of hydrosphere
b. gulfs and seas	4 G. of Tonkin; 4 S. of Japan; part of ocean
c. lakes and rivers	usually fresh water; 4 Nile R. flows; 4 L. Victoria surrounded by land

Practicing the Skill

Turn to Chapter 1, Section 2, "Technology Tools for Geographers." Read "The Science of Mapmaking," and use a chart to take notes on the passage.

1.2 Finding Main Ideas

Defining the Skill

A **main idea** is a statement that summarizes the subject of a speech, an article, a section of a book, or a paragraph. Main ideas can be stated or unstated. The main idea of a paragraph is often stated in the first or last sentence. If it is the first sentence, it is followed by sentences that support that main idea. If it is the last sentence, the details build up to the main idea. To find an unstated idea, use the details of the paragraph as clues.

Applying the Skill

The following paragraph provides reasons why Japan and Australia make such good trading partners. Use the strategies listed below to help you identify the main idea.

How to Find the Main Idea

Strategy ❶ Identify what you think may be the stated main idea. Check the first and last sentences of the paragraph to see if either could be the stated main idea.

Strategy ❷ Identify details that support the main idea. Some details explain that idea. Others give examples of what is stated in the main idea.

TRADING PARTNERS

Australia is an island. Japan is also an island, though much smaller than Australia. Australia is not as densely populated as Japan. ❷ In Japan, an average of 881 people live in each square mile. Australia averages 7 people per square mile. ❷ Australia has wide-open lands available for agriculture, ranching, and mining. ❷ Japan buys wool from Australian ranches, wheat from Australian farms, and iron ore from Australian mines. To provide jobs for its many workers, ❷ Japan developed industries. Those industries ❷ sell electronics and cars to Australia. ❶ Australia and Japan are trading partners because each has something the other needs.

Make a Chart

Making a chart can help you identify the main idea and details in a passage or paragraph. The chart below identifies the main idea and details in the paragraph you just read.

Main Idea: Australia and Japan are good trading partners because each supplies something the other needs.
Detail: Japan's population density is 881 people per square mile; Australia's is 7 per square mile.
Detail: Australia has land, natural resources, and agricultural products.
Detail: Japan buys wool, wheat, and iron ore from Australia.
Detail: Japan has many industries.
Detail: Australia buys electronics and cars from Japanese industries.

Practicing the Skill

Turn to Chapter 15, Section 1, "Sweeping Across Eurasia." Read "Human-Environment Interaction." Create a chart that identifies the main idea and the supporting details.

1.3 Summarizing

Defining the Skill

When you **summarize,** you restate a paragraph, passage, or chapter in fewer words. You include only the main ideas and most important details. It is important to use your own words when summarizing.

Applying the Skill

The passage below describes the origins of several state names in the United States. Use the strategies that follow to help you summarize the passage.

How to Summarize

Strategy 1 Look for topic sentences stating the main idea or ideas. These are often at the beginning of a section or paragraph. Briefly restate each main idea in your own words.

Strategy 2 Include key facts and any names, dates, numbers, amounts, or percentages from the text.

Strategy 3 After writing your summary, review it to see that you have included only the most important details.

STATES' NAMES

1 The name of a state often comes from that state's geography. For example, 2 the name for Montana comes from a Spanish word that means "mountainous."

1 Other states are named after people who were in power at the time the area was explored or settled. 2 A French explorer named Louisiana after the French king at the time, Louis XIV. The state of 2 Georgia was named after King George II of England, who granted the right to start the colony.

1 Still other states get their names from Native American tribes living in the area when Europeans arrived. 2 Arkansas, Alabama, and Massachusetts were named for the Native American tribes living there.

Write a Summary

You should be able to write your summary in a short paragraph. The paragraph at right summarizes the passage you just read.

3 The names of states in the United States often came from geographical features, the names of people in power, or the names of Native American tribes.

Practicing the Skill

Turn to Chapter 6, Section 1, "Mountains, Prairies, and Coastlands." Read "Climates of Canada," and write a short paragraph summarizing the passage.

Skillbuilder Handbook

1.4 Sequencing Events

Defining the Skill

Sequence is the order in which events occur. Learning to follow the sequence of events through history will help you to better understand how events relate to one another.

Applying the Skill

The following passage explains the sequence of events that led to the U.S. Civil War. Use the strategies listed below to help you follow the sequence of events.

How to Find the Sequence of Events

Strategy ① Look for specific dates provided in the text. The dates may not always read from earliest to latest, so be sure to match an event with the date.

Strategy ② Look for clues about time that allow you to order events according to sequence. Words and phrases, such as *day, week, month,* or *year,* may help to sequence the events.

THE CIVIL WAR BEGINS

On ① December 20, 1860, South Carolina became the first state to secede, or leave the Union. Many others followed. In ② February of the next year, these states united to form the Confederate States of America, also called the Confederacy. Soon afterward, Confederate soldiers began seizing federal forts. By the time that Lincoln was sworn in as president on ① March 4, 1861, few southern forts remained in Union hands. The most important was Fort Sumter.

On the morning of ① April 12, Confederate cannons began firing on Fort Sumter. The war between North and South had begun.

Make a Time Line

Making a time line can help you sequence events. The time line below shows the sequence of events in the passage you just read.

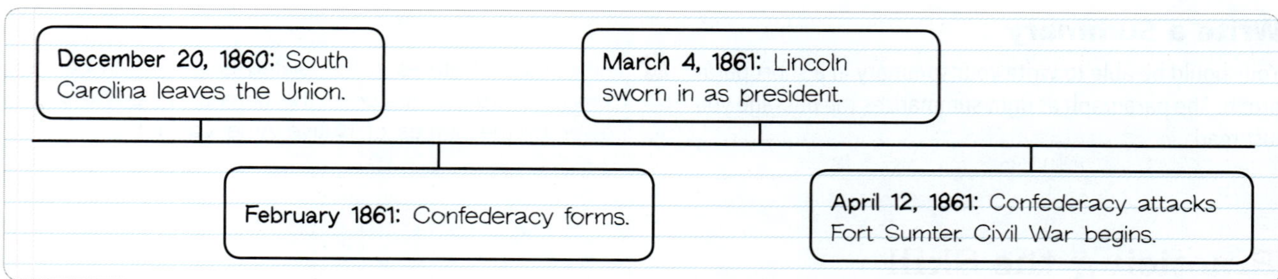

Practicing the Skill

Turn to Chapter 7, Section 4, "From Colonization to Independence." Read "Mexico's Path to Independence," and make a time line showing the sequence of events in the passage.

1.5 Categorizing

Defining the Skill

To **categorize** is to sort people, objects, ideas, or other information into groups, called categories. Geographers categorize information to help them identify and understand patterns in geographical data.

Applying the Skill

The following passage discusses the involvement of various countries in World War II. Use the strategies listed below to help you categorize information.

How to Categorize

Strategy 1 First, decide what kind of information needs to be categorized. Decide what the passage is about and how that information can be sorted into categories. For example, look at the different ways countries reacted to World War II.

Strategy 2 Then find out what the categories will be. To learn how countries reacted to the war, look for clue words such as *in response*, *some*, *other*, and *both*.

Strategy 3 Once you have chosen the categories, sort information into them. Which countries were Axis Powers? Which were Allies? What about the ones who were conquered and those who never fought at all?

WORLD WAR II ALLIANCES

1 During World War II, most countries around the world had to choose which side to take in the conflict. Some countries were conquered so quickly that they could not join either side. 2 *Others* chose to remain neutral. 2 Italy and Japan joined with Germany to form the Axis Powers. In 1939, Germany began the war by invading Poland. 2 *In response,* Great Britain and France (the first two Allied Powers) declared war on Germany. Germany quickly conquered many countries, including Denmark, Norway, the Netherlands, Belgium, and France. 2 During 1941, *both* the Soviet Union and the United States entered the war on the Allied side. 2 Sweden and Switzerland remained neutral during the war.

Make a Chart

Making a chart can help you categorize information. The chart below shows how the information from the passage you just read can be categorized.

3

Name of Alliance	Axis Powers	Allied Powers	Neutral
Countries	• Germany • Japan • Italy	• Great Britain • France • Soviet Union • United States	• Sweden • Switzerland

Practicing the Skill

Turn to Chapter 2, Section 1, "The Earth and Its Forces." Read "External Forces Shaping the Earth." Make a chart in which you categorize the external forces shaping Earth.

1.6 Analyzing Causes and Effects

Defining the Skill

A **cause** is an action that makes something happen. An **effect** is the event that is the result of the cause. A single event may have several causes. It is also possible for one cause to result in several effects. Geographers identify cause-and-effect relationships to help them understand patterns in human movement and interaction.

Applying the Skill

The following paragraph describes events that caused changes in the way of life of the ancient Maya people of Central America. Use the strategies below to help you identify the cause-and-effect relationships.

How to Analyze Causes and Effects

Strategy ❶ Ask why an action took place. Ask yourself a question about the title and topic sentence, such as, "What caused Mayan civilization to decline?"

Strategy ❷ Look for effects. Ask yourself, "What happened?" (the effect). Then ask, "Why did it happen?" (the cause). For example, "What caused the Maya to abandon their cities?"

Strategy ❸ Look for clue words that signal causes, such as *caused, contributed to,* and *led to.*

❶ **DECLINE OF MAYAN CIVILIZATION**

❶ The civilization of the Maya went into a mysterious decline around A.D. 900. ❷ Mayan cities in the southern lowlands were abandoned, trade ceased, and the huge stone pyramids of the Maya fell into ruin. No one really understands what happened to the Maya, but there are many theories.

❸ Some believe that a change in climate *caused* the decline of Mayan civilization. Three long droughts between 810 and 910 meant that there was not enough water for Mayan crops. ❷ As a result, the Maya abandoned their cities. ❸ Other researchers believe that additional problems *contributed to* the crisis. They include overpopulation and warfare among the Mayan nobility.

Make a Diagram

Making a diagram can help you analyze causes and effects. The diagram below shows two causes and an effect for the passage you just read.

Cause: Three long droughts occurred between 810 and 910.

Cause: Other problems such as overpopulation and warfare among the Mayan nobility developed.

→ **Effect:** Mayan civilization declined.

Practicing the Skill

Turn to Chapter 29, Section 1, "Unique Lands." Read "Non-Native Animals." Then make a diagram about the causes and effects of Australia's rabbit population.

1.7 Comparing and Contrasting

Defining the Skill

Comparing means looking at the similarities and differences between two or more things. **Contrasting** means examining only the differences between them. Geographers compare and contrast physical features, regions, cultures, beliefs, and situations in order to understand them.

Applying the Skill

The following passage describes the Dead Sea, in southwest Asia, and the Red Sea, located between northeast Africa and southwest Asia. Use the strategies below to help you compare and contrast these two bodies of water.

How to Compare and Contrast

Strategy (1) Look for two aspects of the subject that can be compared and contrasted. This passage compares the Red Sea and the Dead Sea, two salty bodies of water close to one another.

Strategy (2) To find similarities, look for clue words indicating that two things are alike. Clue words include *both, together,* and *similarly.*

Strategy (3) To contrast, look for clue words that show how two things differ. Clue words include *however, but, on the other hand,* and *yet.*

SALTY SEAS

(1) According to the Bible, God parted the Red Sea so Moses could lead his people across it. The Bible also mentions the Dead Sea, calling it the Salt Sea. The Dead Sea is one of the saltiest bodies of water in the world. (2) *Similarly,* the Red Sea also has a high salt content. Many observers argue that the Dead Sea is not really a sea at all. It is more of a lake since it is fed by the Jordan River and is surrounded by land on all four sides. (3) *On the other hand,* the Indian Ocean feeds the Red Sea. The Red Sea is also the larger of the two seas. It is 170,000 square miles, while the Dead Sea is under 400 square miles. (2) *Both* bodies of water provide minerals for commercial use, especially salt.

Make a Venn Diagram

Making a Venn diagram can help you identify similarities and differences between two things. In the overlapping area, list characteristics shared by both subjects. Then, in the separate ovals, list the characteristics of each subject not shared by the other. This Venn diagram compares and contrasts the Red Sea and the Dead Sea.

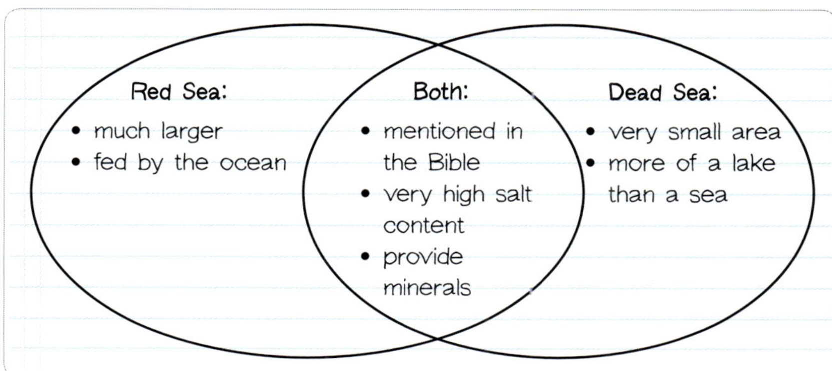

Practicing the Skill

Turn to Chapter 3, Section 3, "Resources and Economics." Read "Measuring Economic Development." Then make a Venn diagram showing similarities and differences between developed and developing nations.

1.8 Identifying Problems and Solutions

Defining the Skill

Identifying problems means finding and understanding the difficulties faced by a particular group of people during a certain time. **Identifying solutions** means understanding how people tried to remedy those problems. By studying how people solved problems in the past, you can learn ways to solve problems today.

Applying the Skill

The following passage describes the problems of floods and droughts in early Mesopotamia. Use the strategies listed below to find and understand these problems.

How to Identify Problems and Solutions

Strategy 1 Look for the difficulties or problems faced by a group of people.

Strategy 2 Consider how situations that existed at that time and place contributed to these problems.

Strategy 3 Look for the solutions that people or groups employed to deal with the problems. Think about whether the solutions were good ones.

FLOOD AND DROUGHT IN MESOPOTAMIA

1 In ancient Mesopotamia, farmers had to deal with both floods and droughts. They often had either too much rain or too little. 2 If too much rain fell, the rivers might overflow and wash everything away. Too little rain also created difficulties. 2 During a drought, the river levels dropped, making it hard to water crops.

3 To combat the lack of rain, farmers in Mesopotamia eventually built canals to carry water from the river to the fields. Such a system is called irrigation. 3 Farmers also built dams to hold back excess water during floods.

Make a Chart

Making a chart can help you identify information about problems and solutions. The chart below shows the problem, contributing factors, and solutions in the passage you just read.

1 Problem	2 Contributing Factors	3 Solutions
Floods and droughts made farming difficult in Mesopotamia.	Too much rain caused floods that washed everything away. Droughts caused the river level to drop, making it hard to water crops.	Farmers built canals to carry water from the river to the fields. Farmers built dams to hold back excess water during floods.

Practicing the Skill

Turn to Chapter 2, Section 4, "Environmental Challenges." Read "Desertification." Then make a chart that summarizes the problem of desertification faced by many arid and semiarid areas and the solutions that people who live in these areas are applying.

1.9 Making Inferences

Defining the Skill

Inferences are ideas that the author has not directly stated. **Making inferences** involves reading between the lines to interpret the information you read. You can make inferences by studying what is stated and using your common sense and previous knowledge.

Applying the Skill

The passage below examines the great pyramids of ancient Egypt. Use the strategies below to help you make inferences from the passage.

How to Make Inferences

Strategy ❶ Read to find statements of facts and ideas. Knowing the facts will give you a good basis for making inferences.

Strategy ❷ Use your knowledge, logic, and common sense to make inferences that are based on facts. Ask yourself, "What does the author want me to understand?" For example, from the facts about the pyramids' purpose, you can make the inference that the Egyptians believed in life after death. See other inferences in the chart below.

THE PYRAMIDS OF EGYPT

One reason that ancient Egypt is famous is for its giant pyramids. ❶ The Egyptians built these magnificent monuments for their kings, or pharaohs. Each pyramid was a resting place where an Egyptian pharaoh planned to spend the afterlife. ❶ Overseers, who ran the construction of each pyramid for the pharaoh, managed the workers and tracked the supplies. The largest pyramid, built for Khufu, required workers to place 6.5 million tons of stone. ❶ Workers built these long-lasting structures with none of the modern cutting tools and machines that we have today.

Make a Chart

Making a chart can help you organize information and make logical inferences. The chart below organizes information from the passage you just read.

❶ Stated Facts and Ideas	❷ Inferences
Egyptians built their pyramids as resting places for their pharaohs in the afterlife.	Egyptians believed in life after death.
A staff oversaw construction of the project.	Pyramid building was a complicated task that required organization.
The Egyptians built the pyramids without modern equipment.	The Egyptians were skilled engineers and hard workers.

Practicing the Skill

Turn to Chapter 26, Section 3, "Contrasting Countries." Read "Economy" under "North Korea." Then use a chart like the one above to make inferences about the effect of North Korea's economy on its citizens.

1.10 Making Generalizations

Defining the Skill

To **make generalizations** means to make broad judgments based on information. When you make generalizations, you should gather information from several sources.

Applying the Skill

The following three passages contain views on the United States as a melting pot of different cultures. Use the strategies below to make a generalization about these ideas.

How to Make Generalizations

Strategy 1 Look for information that the sources have in common. These three sources all look at the relationship of immigrants to the United States.

Strategy 2 Form a generalization that describes this relationship in a way that all three sources would agree with. State your generalization in a sentence.

THE MELTING POT

At the turn of the century, 1 many native-born Americans thought of their country as a melting pot, a mixture of people of different cultures and races who blended together by abandoning their native languages and customs.

—*The Americans*

1 A nation, like a tree, does not thrive well till it is engrafted with a foreign stock.

—*Journals* (Emerson, 1823)

1 The United States has often been called a *melting pot.* But in other ways, U.S. society is an example of *cultural pluralism.* . . . That is, large numbers of its people have retained features of the cultures of their ancestors.

—*World Book Encyclopedia*

Make a Diagram

Using a diagram can help you make generalizations. The diagram below shows how the information you just read can be used to generalize about immigrants' contributions.

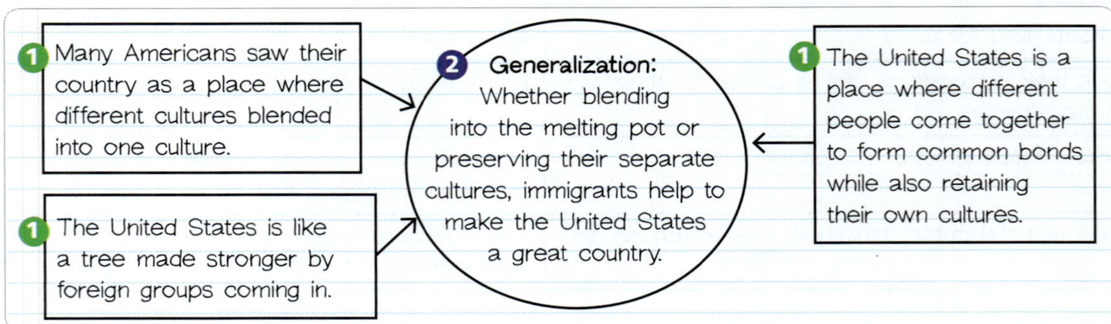

Practicing the Skill

Turn to Chapter 19, Section 1, "Three Trading Empires." Read "European Imperialism," and study the chart on page 584. Also examine the map on page 579. Use a diagram to make a generalization about imperialism in Africa.

1.11 Drawing Conclusions

Defining the Skill

Drawing conclusions means analyzing what you have read and forming an opinion about its meaning. To draw conclusions, look at the facts and then use your own common sense and experience to decide what the facts mean.

Applying the Skill

The following passage presents information about Europe's mountain ranges. Use the strategies listed below to help you draw conclusions about their effects.

How to Draw Conclusions

Strategy 1 Read carefully to identify and understand all the facts.

Strategy 2 List the facts in a diagram and review them. Use your own experiences and common sense to understand how the facts relate to each other.

Strategy 3 After reviewing the facts, write down the conclusions you have drawn about them.

EUROPE'S MOUNTAIN RANGES

Mountains are a key landform in Europe. 1 The Alps stretch across eight countries in southern Europe. The mountain range affects travel within those countries. 1 The Pyrenees once formed a natural barrier between France and Spain. Europe also has several smaller mountain ranges.

Europe's mountain ranges separated the groups of people who settled the land thousands of years ago. 1 With the mountains in between them, many different cultures were able to develop in Europe. 1 Mountain ranges contain natural resources, such as timber. These resources also influenced where people chose to settle.

Make a Diagram

Making a diagram can help you draw conclusions. The diagram below shows how to organize facts to draw a conclusion about the passage you just read.

2 FACTS

- Mountain ranges, such as the Alps and the Pyrenees, formed natural barriers and affected travel between European nations.
- Mountains separated groups of settlers, allowing different cultures to develop.
- Mountain ranges contain natural resources that also influenced settlement.

3 CONCLUSION

Mountain ranges had a significant impact on the settlement of and the development of cultures in Europe.

Practicing the Skill

Turn to Chapter 10, Section 2, "A Multicultural Society." Read "Africans." Then make a diagram to help you draw conclusions about African influences on Brazilian culture.

R13

1.12 Making Decisions

Defining the Skill

Making decisions involves choosing between two or more options, or courses of action. In most cases, decisions have consequences, or results. Sometimes decisions may lead to new problems. By understanding how historical figures made decisions, you can learn how to improve your own decision-making skills.

Applying the Skill

The passage below explains the decision British Prime Minister Chamberlain faced when Germany threatened aggression in 1938. Use the strategies below to analyze his decision.

How to Make Decisions

Strategy ❶ Identify a decision that needs to be made. Think about what factors make the decision difficult.

Strategy ❷ Identify possible consequences of the decision. Remember that there can be more than one consequence to a decision.

Strategy ❸ Identify the decision that was made.

Strategy ❹ Identify actual consequences that resulted from the decision.

PEACE OR WAR?

In 1938, German leader Adolf Hitler demanded a part of Czechoslovakia. ❶ British Prime Minister Neville Chamberlain had to decide how to respond to that aggression. ❷ He could threaten to go to war, but he feared that Britain was not ready. ❷ If he gave Germany the region, he might avoid war, but he would be setting a bad example by giving in to a dictator. ❸ Chamberlain decided to give Germany what it demanded. In exchange, Hitler promised not to take any more land in Europe. ❹ Six months later, Germany took the rest of Czechoslovakia and later invaded Poland.

Make a Flow Chart

Making a flow chart can help you identify the process of making a decision. The flow chart below shows the decision-making process in the passage you just read.

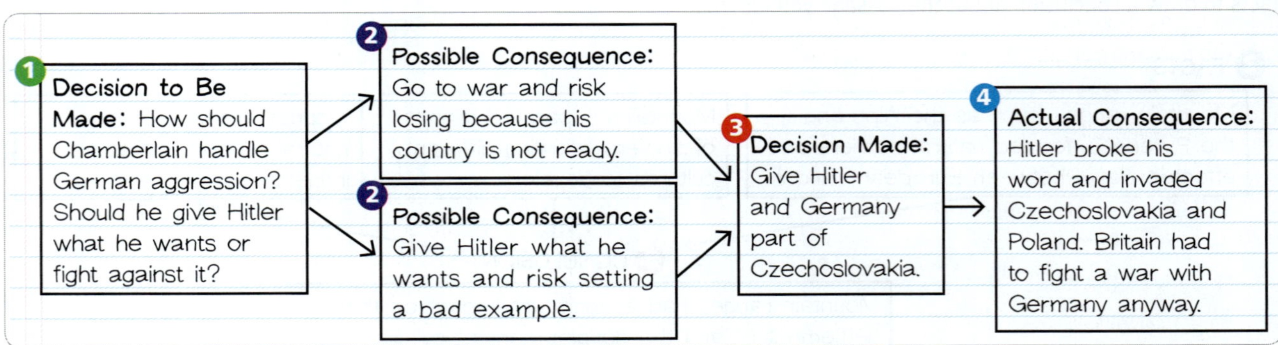

Practicing the Skill

Turn to Chapter 15, Section 2, "Governing a Vast Land." Read "Superpower and Collapse." Then make a flow chart to identify Gorbachev's policy decision and its consequences.

1.13 Evaluating

Defining the Skill

To **evaluate** is to make a judgment about something. Geographers evaluate the actions of people in history. One way to do this is to examine both the positives and the negatives of an action, and then decide which is stronger—the positive or the negative.

Applying the Skill

The following passage examines the rule of Mustafa Kemal of Turkey. Use the strategies listed below to evaluate the success of his reforms.

How to Evaluate

Strategy 1 Before you evaluate a person's actions, first determine what that person was trying to do. In this case, think about what Kemal wanted to accomplish.

Strategy 2 Look for statements that show the positive, or successful, results of his actions. For example, did he achieve his goals?

Strategy 3 Also look for statements that show the negative, or unsuccessful, results of his actions. Did he fail to achieve something he tried to do?

Strategy 4 Write an overall evaluation of the person's actions.

KEMAL RULES TURKEY

In 1923, Turkey became an independent republic. Mustafa Kemal became Turkey's first president. (1) Kemal wanted his nation to resemble European countries rather than its Islamic neighbors. Kemal quickly replaced Turkey's Islamic government with a nonreligious system. He also replaced the Arabic alphabet and the Islamic calendar with Western versions of each. (3) Kemal's actions drew protests from traditionalists. (2) However, Kemal's moves helped to modernize Turkey. (2) They also benefited women, who had lived with many restrictions under Islamic law. Women now had greater social freedom. They also could vote and run for political office.

Make a Diagram

Making a diagram can help you evaluate. List the positives and negatives of the person's actions and decisions. Then make an overall judgment. The diagram below shows how the information from the passage you just read can be evaluated.

(2) Positive Results:
- Turkey becomes more modern.
- Women attain greater freedoms.

(3) Negative Results:
- Traditionalists oppose and protest Kemal's reforms.

(4) Evaluation:
While traditionalists criticized Kemal's actions, the reforms helped modernize Turkey and improve life for women. Therefore, he succeeded at his goal.

Practicing the Skill

Turn to Chapter 11, Section 3, "The Middle Ages and Renaissance." Read "Forces of Change," and make a diagram to evaluate the decision to take part in the Crusades.

1.14 Distinguishing Fact from Opinion

Defining the Skill

Facts are events, dates, statistics, or statements that can be proved to be true. **Opinions** are judgments, beliefs, and feelings. By identifying facts and opinions, you will be able to think critically when a person is trying to influence your own opinion.

Applying the Skill

The following passage describes the church Hagia Sophia. It was one of the more notable structures in the development of Constantinople as the capital of the Byzantine Empire. Use the strategies listed below to distinguish facts from opinions.

How to Distinguish Fact from Opinion

Strategy 1 Look for specific information that can be proved or checked for accuracy.

Strategy 2 Look for assertions, claims, and judgments that express opinions. In this case, one speaker's opinion is addressed in a direct quote.

Strategy 3 Think about whether statements can be checked for accuracy. Then, identify the facts and opinions in a chart.

HAGIA SOPHIA

1 Hagia Sophia was built between A.D. 532 and 537. The Byzantine emperor Justinian ordered that the cathedral be built on the site where an earlier church had burned down. 1 The church is noted for its large central dome, which rises 185 feet above the ground. With numerous windows, arches, and vaults, the structure is hailed by many as an architectural masterpiece. 2 Said one observer, "You would declare that the place is not lighted by the sun from without, but that the rays are produced from within."

Make a Chart

Making a chart can help you distinguish fact from opinion. The chart below analyzes the facts and opinions in the passage above.

Statement	3 Can It Be Proved?	3 Fact or Opinion
Hagia Sophia was built between A.D. 532 and 537.	Yes. Check historical documents.	fact
Hagia Sophia's dome is 185 feet high.	Yes. Check measurement records.	fact
The inside of Hagia Sophia is so beautiful that it appears to make its own sunlight.	No. This cannot be proved. It is what one speaker believes.	opinion

Practicing the Skill

Turn to Chapter 28, Section 2, "The Mainland Countries." Read "Singapore." Then make a chart like the one above in which you analyze key statements about Singapore to determine whether they are facts or opinions.

1.15 Analyzing Points of View

Defining the Skill

Analyzing points of view means looking closely at a person's arguments to understand the reasons behind that person's beliefs. The goal of analyzing a point of view is to understand different thoughts, opinions, and beliefs about a topic.

Applying the Skill

The following passage describes the difference between Native American and European attitudes about land use. Use the strategies below to help you analyze the points of view.

How to Analyze Points of View

Strategy 1 Look for statements that show you a particular point of view on an issue. For example, Native Americans believed land should be preserved for the future. European colonists believed land could be owned and changed as desired.

Strategy 2 Think about why different people or groups held a particular point of view. Ask yourself what they valued. What were they trying to gain or to protect? What were they willing to sacrifice?

Strategy 3 Write a summary that explains why different groups of people might have taken different positions on this issue.

LAND USE

Native Americans and Europeans had many conflicts because of differing ideas about land use. 1 Most Native Americans believed that land must be preserved for future generations. They believed the present generation had the right to use land for hunting and farming. However, no one had the right to buy or sell land.

In contrast, Europeans had a long history of taming wilderness and owning land. As a result, 1 Europeans believed that they could buy land, sell it, and alter it. For example, if they wished to mine for gold underground, they could destroy landscape as they dug the mine. 1 Europeans used land to make money.

Make a Diagram

Making a diagram can help you analyze points of view. The diagram below analyzes the different points of view of the Native Americans and the European colonists in the passage you just read.

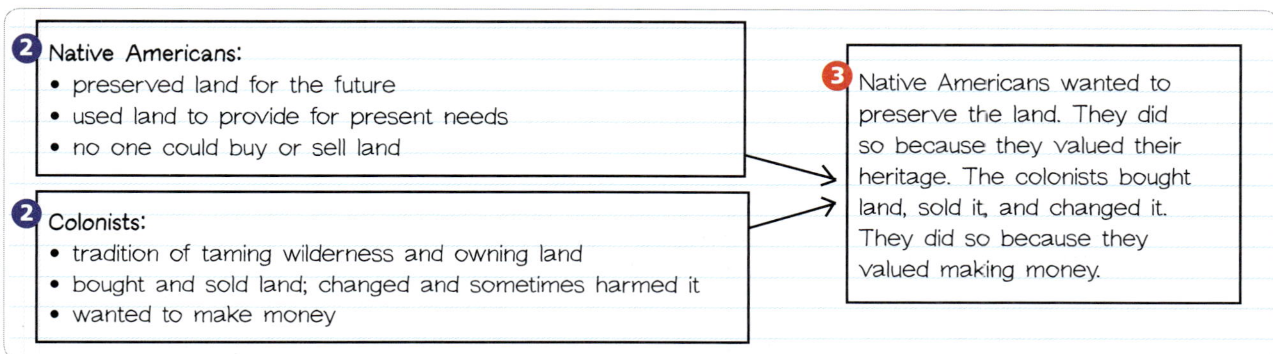

Practicing the Skill

Turn to Chapter 22, Section 2, "Israel and the Palestinian Territories." Read "A New Israel" and "The Palestinian Territories." Make a diagram to analyze the different points of view of the Israelis and the Palestinians.

1.16 Recognizing Bias

Defining the Skill

Bias is a prejudiced point of view. Accounts that are biased tend to present only one side of an issue and reflect the opinions of the writer. Recognizing bias will help you understand how much to trust the primary and secondary sources that you find.

Applying the Skill

The following passage is from an 1883 speech in which Jules Ferry supported the French policy of imperialism. Use the strategies listed below to help you recognize bias in the passage.

How to Recognize Bias

Strategy 1 Identify the author and examine any information about him or her. Does the author belong to a group, social class, or political party that might lead to a one-sided view of the subject?

Strategy 2 Think about the opinions the author is presenting. Look for words, phrases, statements, or images that might convey a positive or negative slant.

Strategy 3 Examine the evidence provided to support the author's opinions. Is the opinion correct? Would the same information appear in another account of the same event?

1 **JULES FERRY ARGUES FOR FRENCH IMPERIALISM**

Nations are great in our times only by means of the activities which they develop; . . . Something else is needed for France: . . . that 2 she must also be a great country exercising all of her rightful influence over the destiny of Europe, that 2 she ought to propagate [spread] this influence throughout the world and carry everywhere that she can 3 her language, her customs, her flag, her arms, and her genius.

—Jules Ferry, Speech Before the French National Assembly, July 28, 1883

Make a Chart

Making a chart can help you recognize bias in primary and secondary sources. The chart below analyzes the bias in the passage you just read.

Author	Jules Ferry
Occasion and Purpose	speech in front of the French National Assembly, arguing for France to practice imperialism
Tone	urging, superior
Slanted Language	"exercising all of her rightful influence;" "ought to propagate [spread] this influence"
Description of Bias	Ferry argues that France had the right and the obligation to practice imperialism and spread French control and culture to colonies around the world.

Practicing the Skill

Look through newspapers and news magazines to find an article related to geography. Then use a chart like the one above to analyze the article for bias.

1.17 Synthesizing

Defining the Skill

Synthesizing involves putting together clues, information, and ideas to form an overall picture. Geographers synthesize information in order to develop interpretations of important facts.

Applying the Skill

The following passage describes the agricultural revolution. The highlighting indicates the different kinds of information that will help you synthesize.

How to Synthesize

Strategy 1 Look carefully for facts that will help you base your interpretations on evidence.

Strategy 2 Look for explanations that link the facts together. This statement is based on the evidence of ancient tools mentioned in the next sentence.

Strategy 3 Consider what you already know that could apply. Your knowledge will probably lead you to accept this statement.

Strategy 4 Bring together the information you have about the subject. This interpretation brings together different kinds of information to arrive at a new understanding.

THE AGRICULTURAL REVOLUTION

1 Flaked arrowheads found with mammoth bones at ancient sites suggest that some early people lived as big game hunters. 2 After the big game became extinct, people shifted to hunting smaller game. They made tools, such as bows and arrows for hunting small game as well as baskets for collecting nuts.

About 10,000 years ago, people began to farm. 3 The rise of agriculture brought tremendous change. More people lived in settled villages, and the storage of surplus food became more important. As their surplus increased, people had time to develop specialized skills and think about the world. 4 From this agricultural base rose larger, more stable, and more complex societies.

Make a Diagram

Making a diagram can help you organize the facts, examples, and interpretations that you need to synthesize. The diagram below synthesizes the passage you just read.

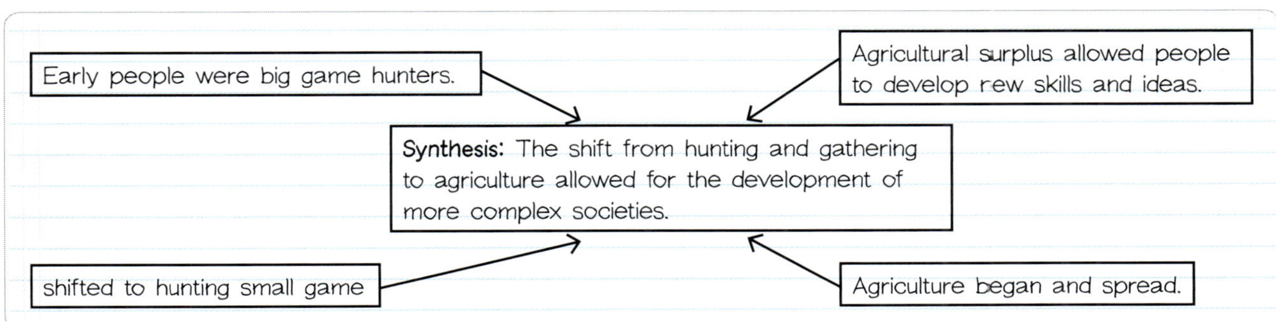

Practicing the Skill

Turn to Chapter 3, Section 2, "Why People Move." Read "Causes of Migration." Then use a diagram like the one above to form a synthesis about why people migrate.

2.1 Reading Maps: Physical

Defining the Skill

Physical maps represent the landforms, or physical features, on Earth's surface, such as mountains or bodies of water. By learning about the different elements of maps, it will be easier for you to read them.

Applying the Skill

A physical map of Peru is shown below. Use the strategies listed below to help you read the map.

How to Read a Physical Map

Strategy 1 Read the title. This will tell you what place or region is represented by the map.

Strategy 2 Look for the grid-lines that form a pattern of squares over the map. These numbered lines represent latitude (horizontal) and longitude (vertical). They indicate the location of the area on Earth.

Strategy 3 Read the map legend. On a physical map, the legend often includes a scale to help you determine elevation as well as information about physical features.

Strategy 4 Use the scale and the pointer or compass rose to determine distance and direction.

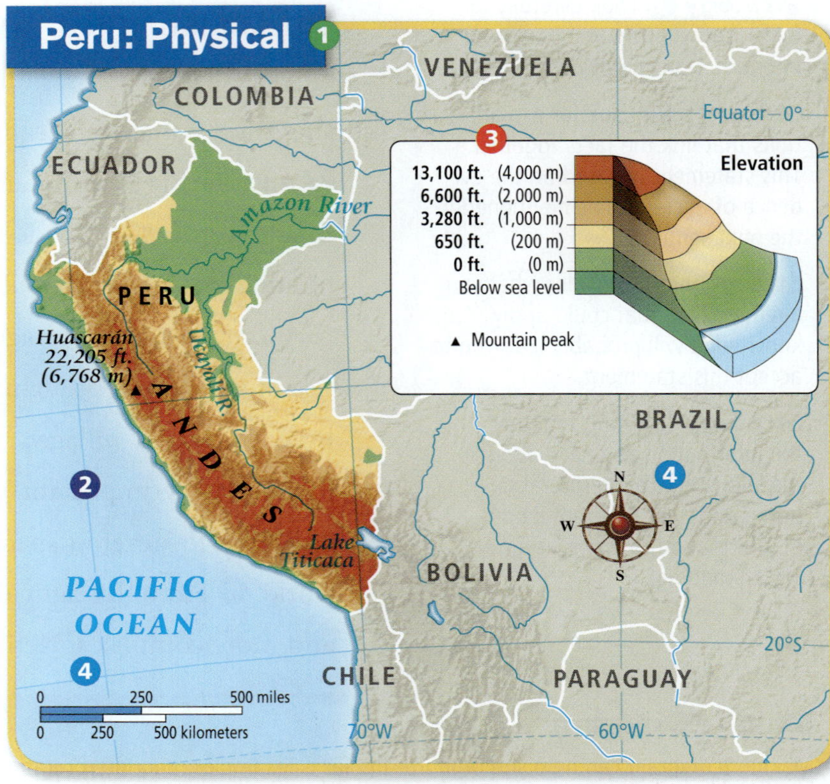

Make a Chart

Making a chart can help you organize the information given on a map. The chart below summarizes information about the physical map you just studied.

Title	Peru: Physical
Location	the Pacific coast of South America, near the equator and latitude 70° W
Legend Information	colors = elevation
Scale	1 in. = 500 miles, 5/8 in. = 500 km
Summary	Peru is located on the Pacific Ocean. The Andes run along the west coast, and the Amazon River and Ucayali River flow through the country.

Practicing the Skill

Turn to Chapter 21, Section 2, "Physical Geography of South Asia." Study the physical map of South Asia on page 661. Make a chart to identify the information on the map.

2.2 Reading Maps: Political

Defining the Skill

Political maps represent political features of various places, such as national borders and capital cities. The countries on political maps are often shaded in different colors to make them easier to distinguish from each other.

Applying the Skill

A political map of Peru is shown below. Use the strategies listed below to help you read the map.

How to Read a Political Map

Strategy 1 Read the title. This identifies the main idea of the map and will tell you what place or region the map represents.

Strategy 2 Look for the grid-lines that form a pattern of squares over the map. These numbered lines represent latitude (horizontal) and longitude (vertical). They indicate the location of the area on Earth.

Strategy 3 Read the map legend. The legend will help you to interpret the symbols on the map. On a political map, these symbols often include icons for major and capital cities.

Strategy 4 Use the scale and the pointer or compass rose to determine distance and direction.

Make a Chart

Making a chart can help you organize the information given on a map. The chart below summarizes information about the political map you just studied.

Title	Peru: Political
Location	the Pacific coast of South America, near the equator and latitude 70° W
Legend Information	star icon = national capital; dot = other city
Scale	1 in. = 500 miles, 5/8 in. = 500 km
Summary	Peru, located on the Pacific Ocean, borders Ecuador, Colombia, Brazil, Bolivia, and Chile. Its capital is Lima.

Practicing the Skill

Turn to Chapter 26, "The Korean Peninsula," and study the political map of North and South Korea on page 795. Make a chart to identify the information on the map.

2.3 Reading Maps: Thematic

Defining the Skill

Thematic maps represent specific themes, or ideas. Climate, language, migration, population density, and economic activity maps are all different kinds of thematic maps.

Applying the Skill

The following map shows selected products of Mexico. Use the strategies listed below to help you read the map.

How to Read a Thematic Map

Strategy 1 Read the title. This identifies the theme of the map and will tell you what place or region is represented by the map.

Strategy 2 Some thematic maps have latitude (horizontal) and longitude (vertical) lines. They can help indicate the location of the area on Earth.

Strategy 3 Read the map legend. On a thematic map, the legend is very important because the colors and symbols on the map are specific to the theme. The legend will help you interpret them.

Strategy 4 Use the pointer or compass rose to determine direction. If the map has a scale, it can help you determine distance.

Make a Chart

Making a chart can help you organize the information given on a map. The chart below summarizes information about the thematic map you just studied.

Title	Selected Products of Mexico
Location	Mexico, which borders the Pacific Ocean and the Gulf of Mexico
Legend Information	Icons represent specific products.
Summary	Mexico's different regions produce a variety of goods and crops, including coffee, corn, fish, petroleum, and silver.

Practicing the Skill

Turn to Chapter 12, Section 2, "Spain and Portugal." Study the map of the Iberian Peninsula on page 372, and make a chart to identify the information on the map.

2.4 Reading Maps: Cartograms

Defining the Skill

A **cartogram** takes statistical data and represents it visually in the form of a map. Rather than representing land area, a cartogram shows a different set of data about a country, such as population or gross domestic product. The size of the nations on the cartogram are adjusted to reflect the selected data for each one.

Applying the Skill

The following cartogram shows world population. Use the strategies listed below to help you read the cartogram.

How to Read a Cartogram

Strategy 1 Read the title. This identifies the data and region represented by the cartogram.

Strategy 2 Look at the sizes of the countries shown to see how they compare to each other. Cartograms adjust the sizes of countries to convey relative data. Because of this, the countries' shapes are altered.

Strategy 3 Compare the cartogram to a conventional map such as the world political map on pp. A16–A17. This will help you determine how much more or less of the selected data a particular country has by comparing its relative size on the cartogram to its size on the map.

World Population 1

Make a Chart

Making a chart can help you organize the information given on a cartogram. The chart below summarizes information about the cartogram you just studied.

Title	World Population
Location	the world
Summary	Countries are shown either smaller or larger than their actual area depending on their population. For instance, India and Japan are shown larger than normal because they have high populations. Canada and Russia, which are more sparsely populated, are smaller than normal.

R23

2.5 Reading Graphs and Charts

Defining the Skill

Graphs and charts translate information into a visual form. Graphs take numerical information and present it using pictures and symbols instead of words. Different kinds of graphs include bar graphs, line graphs, and pie graphs. Charts are created by simplifying, summarizing, and organizing information. This information is then presented in a format that is easy to understand. Tables and diagrams are examples of commonly used charts.

Applying the Skill

Use the strategies listed to read the graphs and chart on these pages.

How to Read Graphs

Strategy 1 Read the title to identify the main idea of the graph. Ask yourself what kind of information the graph shows. Check the graph's source line to make sure the data is reliable.

Strategy 2 Read the labels on the graph. Bar and line graphs have labels on their vertical axis and horizontal axis. Pie graphs have labels for each wedge of the pie. These labels tell you what data the graph represents.

Strategy 3 Look at the graph. Try to find patterns, such as similarities and differences or increases and decreases.

Strategy 4 Summarize the information shown on the graph. Use the title to help you focus on what information the graph is presenting.

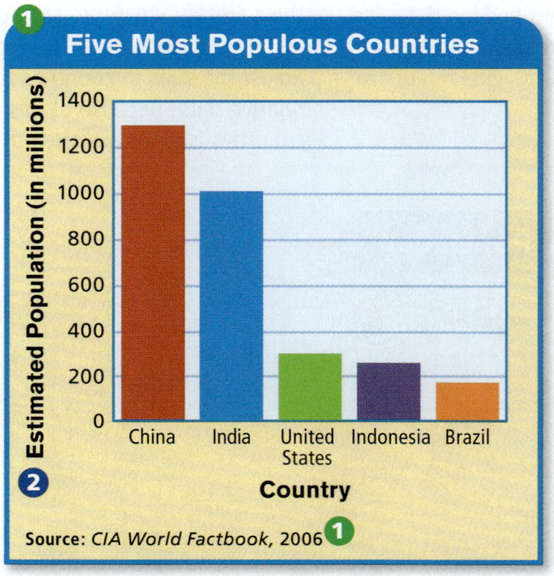

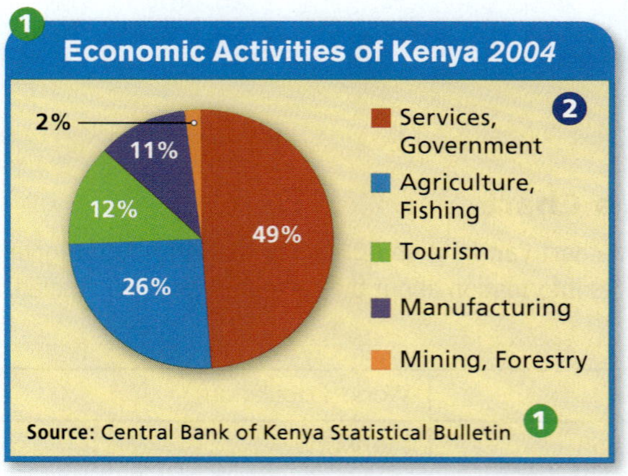

How to Read Charts

Strategy 1 Read the title to find out what the chart is about. Ask yourself what kind of information the chart shows.

Strategy 2 Read the headings to see how the chart is organized.

Strategy 3 Study the data in the chart to understand the facts that it was designed to show.

Strategy 4 Summarize the information shown in each part of the chart. Use the title to help you focus on what information the chart is presenting.

1 **World Economic Development**

2 Country	Status	GDP in US Dollars*	GDP/ Person in US Dollars*	Infant Mortality per 1,000 Live Births	Life Expectancy at Birth	Literacy Rate (Percent)
3 Burkina Faso	developing	5.4 billion	1,300	91.35	48.9	26.6
India	developing	719.8 billion	3,400	54.63	64.7	59.5
Uruguay	developing	13.2 billion	9,600	11.61	76.3	98.0
Germany	developed	2.8 trillion	30,400	4.12	78.8	99.0
Japan	developed	4.9 trillion	31,500	3.24	81.3	99.0
United States	developed	12.5 trillion	42,000	6.43	77.9	99.0

Source: *CIA World Factbook*, 2006

* official exchange rate

Write a Summary

Writing a summary can help you understand the information contained in graphs and charts. The paragraph below summarizes information from the chart above.

4 The chart compares several indicators of economic development for three developing and three developed nations: Burkina Faso, India, Uruguay, Germany, Japan, and the United States. The developed nations have higher GDPs and literacy rates, longer life expectancies, and lower infant mortality than the developing nations. Of the countries shown, the United States has the highest GDP per person, Burkina Faso the lowest.

Practicing the Skill

Turn to Chapter 5, Section 3, "A Diverse Culture." Look at "Comparing Languages, Religion, and Ethnicity," and study the language chart and religion pie graph. Then write a paragraph each summarizing what you learned from the chart and the graph.

2.6 Interpreting Political Cartoons

Defining the Skill

Political cartoons express an opinion about a serious subject. A political cartoonist uses symbols, familiar objects, and people to make his or her point quickly and visually. Sometimes words used in the cartoon will help to clarify its meaning.

Applying the Skill

The cartoon below is about the competition between the Russian Empire and Great Britain over territory in Central Asia. Use the strategies listed below to help you interpret this and other political cartoons.

How to Interpret Political Cartoons

Strategy 1 Identify the subject of the cartoon. Look at the cartoon as a whole. If the cartoon has them, read the title and caption.

Strategy 2 Look for symbols that the cartoonist used to communicate ideas. The bear in this cartoon represents Russia. The lion is a symbol of Great Britain.

Strategy 3 Analyze the other visual details of the cartoon. The man in the middle, wearing Central Asian dress, is crouching in fear as the bear and lion stare at him hungrily. This suggests that the Russian Empire and Great Britain are interested in devouring Central Asian territory.

1 "SAVE ME FROM MY FRIENDS!"

Make a Chart

Making a chart can help you interpret a political cartoon. The chart below summarizes the information from the cartoon above.

Subject	competition between Russia and Great Britain over Central Asia
Symbols and Details	The bear, representing the Russian Empire, and the lion, symbolizing Great Britain, are circling their prey, a man representing Central Asia.
Message	Central Asia is afraid that its "friends," the Russian Empire and Great Britain, are going to take control of its territory.

Practicing the Skill

Turn to Chapter 20, "East, Central, and Southern Africa." Study the political cartoon on page 639. Use a chart like the one above to summarize information from the cartoon to help you interpret its meaning.

2.7 Creating a Map

Defining the Skill

When you **create a map,** you can choose what geographical information to include. You can show political information, such as the area covered by empires or countries. Your map can also be a thematic map, showing data on climates, population, or resources.

Applying the Skill

Below is a map that a student created that shows the vegetation zones of Africa. Use the strategies listed below to learn how to create your own map.

How to Create a Map

Strategy 1 Select a title that identifies the map's geographical area and purpose.

Strategy 2 Use dashes to draw latitude and longitude lines. Using these as a guide, draw the area you are representing.

Strategy 3 Create a legend that identifies the map's colors or icons.

Strategy 4 Draw the colors or icons on the map to show information.

Strategy 5 Draw a compass rose and scale.

Practicing the Skill

Turn to Chapter 14, Section 2, "Hungary and the Czech Republic." Read "Government and Economics" under "Hungary." Using the map on page 325 as a guide, create a map of Europe showing Hungary's trading partners.

Skillbuilder Handbook

R27

2.8 Creating a Model

Defining the Skill

When you **create a model,** you use information and ideas to show an event, situation, or place in a visual way. A model might be a poster or diagram that explains something. Or, it might be a three-dimensional model, such as a diorama, that depicts an important scene or situation.

Applying the Skill

The following sketch shows the early stages of a model of a feudal manor in Europe. Use the strategies listed below to help you create your own model.

How to Create a Model

Strategy 1 Gather the information you need to understand the situation or event. In this case, you need to be able to show the parts of a manor and their uses.

Strategy 2 Visualize an idea for your model. Once you have created a picture in your mind, make an actual sketch to plan how it might look.

Strategy 3 Think of symbols you may want to use. Since the model should give information in a visual way, think about ways you can use color, pictures, or other visuals to tell the story.

Strategy 4 Gather the supplies you will need. Then create the model. For example, you would need pictures of manors and art supplies to make this model.

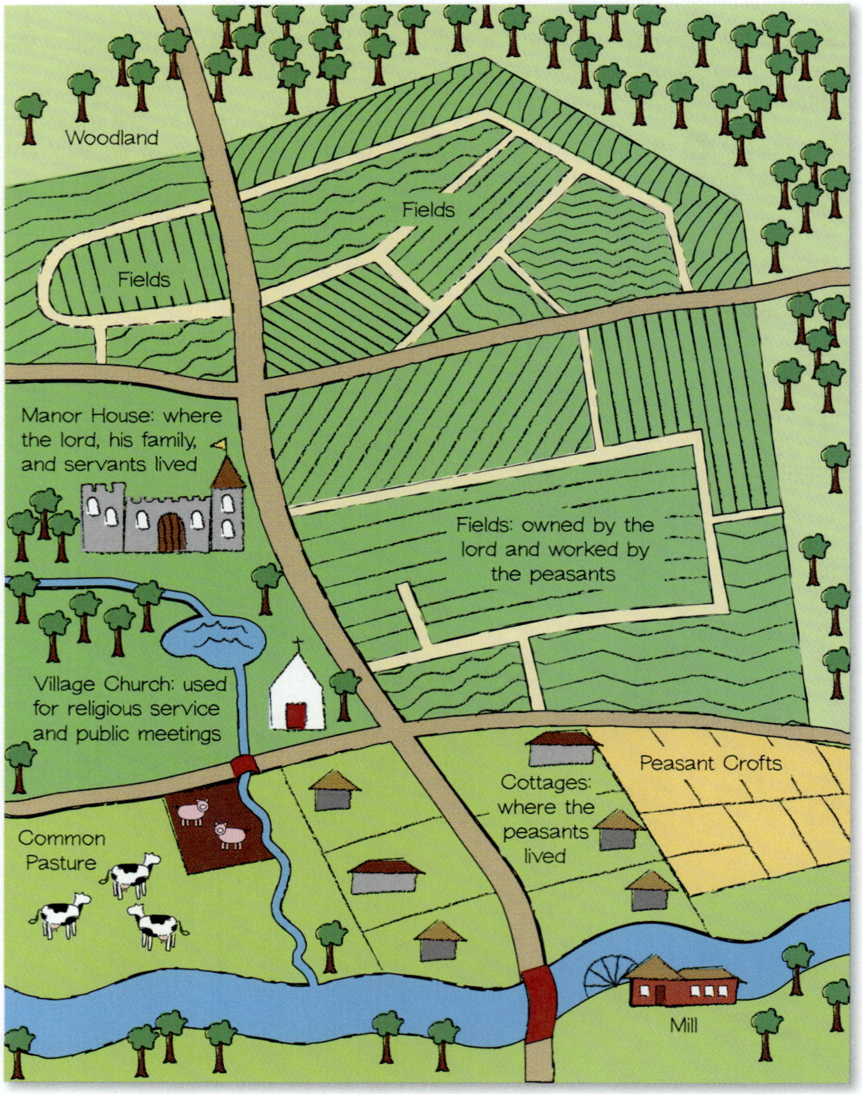

Practicing the Skill

Turn to Chapter 28, Section 3, "Indonesia and the Philippines." Read "Culture and Economy" under "Indonesia" and the text about shadow puppets in the "Fun Facts," and look at the images on page 866. Then create a model of a shadow puppet theater.

3.1 Outlining

Defining the Skill

Outlining is a method of organizing information. Once you have settled on a topic for a report or paper, you must do research. When you have all the information you need, then you have to organize it. An outline lists your report's main ideas in the order in which they will appear. It also organizes main ideas and supporting details according to their importance.

Applying the Skill

The outline at right is for a research report about the causes and effects of earthquakes. Use the strategies listed below to help you make an outline.

How to Outline

Strategy 1 Read the main ideas of this report. They are listed on the left and labeled with capital Roman numerals. Each main idea will need at least one paragraph in your report.

Strategy 2 Read the supporting ideas for each main idea. These are indented and labeled with capital letters. Notice that some main ideas require more supporting ideas than others.

Strategy 3 Read the supporting details in the outline. They are indented farther and labeled with numerals. Some supporting ideas will not have supporting details. It is not necessary to include every piece of information that you have. An outline is merely a guide to follow as you write your report.

Strategy 4 Reports can be organized in different ways. This report is arranged to show the causes and effects of earthquakes. The outline should clearly reflect the way the report is organized.

(1) I. Causes of Earthquakes
- A. Shifting tectonic plates
- (2) B. Volcanoes
- C. Human activities
    - (3) 1. Large underground nuclear explosions, deep mine excavations, and filling large reservoirs can all cause earthquakes.

II. Effects of Earthquakes
- A. Earthquakes usually do not kill people directly.
- B. Damage to natural features
    - 1. shifting of large blocks of Earth's crust
        - a. can cause landslides
        - b. can break down the barks of rivers and lakes and cause flooding
- C. Damage to man-made structures
    - 1. Ground shaking causes structures to sway, bounce, and possibly collapse.
- D. Cause other catastrophic events
    - 1. Fires
        - a. San Francisco, 1906
    - 2. Tsunamis
        - a. Indian Ocean, 2004

Practicing the Skill

Turn to Chapter 17, Section 2, "Human Beginnings in Africa." Read the section and gather information from it about hominid finds in Africa. Write an outline for a report about the topic using the correct outline form.

R29

3.2 Writing for Social Studies

Defining the Skill

Writing is an important skill to learn. It allows you to communicate your thoughts and ideas. There are many different kinds of writing used in social studies.

Narrative Writing

Narrative writing tells a story. Some narratives are fictional, but historians and geographers write narratives about events that really happened. A narrative has three basic parts. The beginning sets the scene. The body presents a conflict. The resolution settles the conflict and ends the narrative. The sample below describes the volcano Krakatoa.

One of the most destructive volcanic eruptions in recorded history happened in 1883, when two-thirds of the Indonesian island of Krakatoa was blown apart. The explosion sent volcanic ash 50 miles into the air. The surrounding area was plunged into darkness for two days. Krakatoa was uninhabited by humans, but the eruption buried all life on the island under a layer of ash. It took five years for plants and animals to thrive again on Krakatoa.

Persuasive Writing

Persuasive writing is writing whose purpose is to convince another person to adopt your opinion or position. People use persuasive writing for many reasons, such as convincing a government to adopt a proposal. Geographers and historians use persuasive writing to propose interpretations of facts. The sample below presents arguments for establishing a tsunami warning system in the Indian Ocean. The system began operations in 2006.

In December 2004, a catastrophic tsunami struck Southeast Asia. Massive losses underline the need for a tsunami early warning system in the Indian Ocean. Such a system would predict and detect tsunamis and communicate warnings to the region's inhabitants. It will cost millions of dollars and require the cooperation of nations already monitoring seismic activity. However, many feel it is worth the effort to reduce destruction from tsunamis.

Expository Writing

Expository writing is meant to explain. It might examine a process or event, compare and contrast two different regions or civilizations, explain causes and effects, or explore problems and solutions. Expository writing has three main parts. The introduction states the main idea. The body supports the main idea with facts and examples. The conclusion summarizes the information and restates the main idea. In the sample below, the author explains some of the major causes of earthquakes.

Earthquakes, the sudden movement of Earth's crust followed by a series of vibrations, have several causes. Most are caused by shifts in tectonic plates. Stress builds up along fault lines until it is finally released, making the ground shake. Earthquakes can also be triggered by the explosive action of a volcano. Human activities, such as deep mining, can prompt earthquakes as well. No matter the cause, earthquakes can have dramatic effects.

Applying the Skill

No matter what kind of writing you are doing—whether it is narrative, persuasive, or expository writing or something more complicated such as a research report—it needs to be clear, concise, and factually accurate. The strategies listed below will help you achieve your goal. The following passage is part of a larger research report about earthquakes. Notice how using these strategies helped the author convey information about the effects of earthquakes.

How to Write for Social Studies

Strategy 1 Focus on your topic. Be sure that you clearly state the main idea of your piece so that your readers know what you intend to say.

Strategy 2 Collect and organize facts and supporting details. Gather accurate information about your topic to support the narrative you are crafting or the point you are trying to make. Creating an outline can help you organize your thoughts and information.

Strategy 3 Write a first draft, using standard grammar, spelling, sentence structure, and punctuation. Proofread your writing. Make sure it is well organized and grammatically correct. Make any necessary revisions in your final draft.

EFFECTS OF EARTHQUAKES

1 Earthquakes usually do not kill people directly. Instead, most deaths and injuries occur as a result of damage to natural features and man-made structures. Earthquakes can also cause other catastrophes that harm people.

2 During an earthquake, large blocks of Earth's crust shift. These shifts can cause landslides or break down the banks of rivers and lakes and cause flooding. 2 The ground shaking that occurs during an earthquake can make buildings sway and bounce. The stress causes some buildings to collapse. All of these situations are dangerous for people.

Sometimes, earthquakes can cause other disasters, such as tsunamis or fires. 2 A 1906 quake in San Francisco started a fire that burned for three days. About 3,000 people died in this event. 2 An undersea earthquake caused the Indian Ocean tsunami in 2004, which claimed at least 280,000 lives.

Practicing the Skill

Turn to Chapter 11, Section 1, "Europe's Dramatic Landscape." Read "Europe's Resources." Then write a persuasive paragraph convincing European government leaders about the importance of developing renewable energy sources. Use the strategies listed above to help you make your paragraph clear and informative.

R31

3.3 Forming and Supporting Opinions

Defining the Skill

When you **form opinions,** you interpret and judge the importance of events and people in history. You should always **support your opinions** with facts, examples, and quotations.

Applying the Skill

The following passage describes the impact of the policy of apartheid on South Africa. Use the strategies listed below to form and support an opinion about the policy.

How to Form and Support Opinions

Strategy 1 Look for important information about the subject. Information can include facts, quotations, and examples.

Strategy 2 Form an opinion about the subject by asking yourself questions about the information. For example, how important was the subject? How does it relate to similar subjects in your own experience?

Strategy 3 Support your opinions with facts, quotations, and examples. If the facts do not support the opinion, then rewrite your opinion so that it is supported by the facts.

THE POLICY OF APARTHEID

From the mid- to late-twentieth century, South Africa had a policy called apartheid, which separated whites and nonwhites. 1 Nonwhites faced discrimination concerning where they could live, what jobs they could hold, and what they were taught in school. Many were forced to leave their homes and relocate in less desirable areas. Nonwhites' protests of these policies led to years of conflict. A number of nations also criticized apartheid. 1 In 1985, the United States and Great Britain restricted trade with South Africa. In 1989, Frederik Willem de Klerk became president of South Africa. He opposed apartheid. His efforts helped to end the country's policy of segregation.

Make a Chart

Making a chart can help you organize your opinions and supporting facts. The following chart summarizes one possible opinion about the policy of apartheid in South Africa.

2 Opinion	The policy of apartheid oppressed many South Africans and hurt the nation's economy.
3 Facts	Nonwhites faced discrimination concerning where they could live, what jobs they could hold, and what they were taught in school.
	Many nonwhites were forced to relocate in less desirable regions.
	Nations opposed to apartheid restricted trade with South Africa.

Practicing the Skill

Turn to Chapter 16, Section 1, "Center of a Landmass." Read "Human-Environment Interaction," and form your own opinion about Soviet projects in Central Asia. Make a chart like the one above to summarize your opinion and the supporting facts and examples.

3.4 Using Primary and Secondary Sources

Defining the Skill

Primary sources are materials written or created by people who lived during historical events. The writers might have been participants or witnesses. Primary sources include letters, journals, articles, and artwork. **Secondary sources** are materials that teach about an event, such as textbooks. When you research, you will use both kinds of sources.

Applying the Skill

The following passage uses primary and secondary sources to describe Kublai Khan's lifestyle. Use the strategies listed below to help you learn how to use these sources.

How to Use Primary and Secondary Sources

Strategy 1 Distinguish secondary sources from primary sources. Most of the paragraph is a secondary source. The observation by Marco Polo is a primary source. The primary source supports the point the secondary source is making.

Strategy 2 Determine the main idea of the secondary source. Look for details that support it.

Strategy 3 Identify the author of the primary source and consider why the author produced it. Consider what the document was supposed to achieve. Is it credible? Is it promoting a particular viewpoint? In this case, Polo seems to be an objective observer.

LIFESTYLE OF KUBLAI KHAN

1 Unlike his Mongol ancestors, Kublai Khan spent almost his entire life in China. 2 He enjoyed living in the luxurious manner of a Chinese emperor. He maintained a beautiful summer palace at Shangdu, on the border between Mongolia and China. He also built a new square-walled capital at the site of modern Beijing. The size of Kublai's palace in Beijing greatly impressed the European traveler 3 Marco Polo. "The whole building is at once so immense and so well constructed that no man in the world . . . could imagine any improvement in design or execution," Polo observed.

Make a Chart

Making a chart can help you assess information from primary and secondary sources. The chart below summarizes information from the passage above.

Questions	Answers
What is the main idea?	Kublai Khan enjoyed a life of luxury.
What are the supporting details?	He had a beautiful summer palace. He built a new capital in Beijing.
Who wrote the primary source?	Marco Polo
What can you tell about the primary source?	Polo appears to be an objective observer. The source seems credible.

Practicing the Skill

Turn to Chapter 21, Section 2, "Physical Geography of South Asia." Read "Monsoons" and the primary source on page 662. Make a diagram to summarize the information found in the sources and to help you read them.

R33

3.5 Using a Database

Defining the Skill

A **database** is a collection of data, or information, that is organized so that you can find and retrieve information on a specific topic quickly and easily. Once a computerized database is set up, you can search it to find specific information without going through the entire database. The database will provide a list of all information in the database related to your topic. Learning how to use a database will help you learn how to create one.

Applying the Skill

The chart below is a database for famous mountains in the Eastern Hemisphere. Use the strategies listed below to help you understand and use the database.

How to Use a Database

Strategy 1 Identify the topic of the database. The most important words in this title are *Mountains* and *Eastern Hemisphere*. These words were used to begin the research for this database.

Strategy 2 Identify the kinds of data you need to enter in your database. These will be the column headings of your database. The keywords *Mountain*, *Location*, *Height*, and *Interesting Facts* were chosen to focus the research.

Strategy 3 Identify the entries included under each heading.

Strategy 4 Use the database to help you find information quickly. For example, in this database, you could search for "Mountains over 28,000 feet" to find a list of famous mountains that are more than 28,000 feet tall.

1 FAMOUS MOUNTAINS OF THE EASTERN HEMISPHERE

2 MOUNTAIN	LOCATION	HEIGHT ABOVE SEA LEVEL (FEET)	INTERESTING FACTS
3 Dhaulagiri	Nepal	26,810	name means "White Mountain"
Everest	border of Nepal and China	4 29,035	tallest mountain in the world
Fuji	Japan	12,388	considered sacred by many Japanese
K2	Pakistan	4 28,250	second tallest mountain in the world
Khan-Tengri	border of China, Kazakhstan, and Kyrgyzstan	22,949	pyramid-shaped mountain is highest point in Kazakhstan
Xixabangma Feng (formerly Gosainthan)	China	26,291	Gosainthan means "place of God" in Sanskrit.

Practicing the Skill

Create a database of countries in Oceania that shows the name of each country, its capital, its land area, and its population. Use the information on pages 880 and 881 in the Unit 9 Atlas "Country Almanac" to provide the data. Use a format like the one above for your database.

3.6 Creating a Multimedia Presentation

Defining the Skill

Audio and video recordings, photographs, CD-ROMs, television, and computer software are all different kinds of media. To **create a multimedia presentation,** you need to collect information in different media and organize them into one presentation so that your audience watches, listens, and learns.

Applying the Skill

The photographs below show students using computers to create multimedia presentations. A multimedia presentation can incorporate computers, but it does not have to. Use the strategies listed below to help you create your own multimedia presentation.

How to Create a Multimedia Presentation

Strategy 1 Identify the topic of your presentation and decide which media are best for an effective presentation. For example, you may want to use video or photographic images to show the dry character of a desert. Or, you may want to use CDs or audio tapes to provide music or to make sounds that go with your presentation, such as the sounds of a camel.

Strategy 2 Research the topic in a variety of sources. Images, text, props, and background music should reflect the region and the historical period of your topic.

Strategy 3 Write the script for the presentation. You could use a narrator and characters' voices to tell the story. Primary sources are an excellent source for script material.

Strategy 4 Put together your presentation. Supplement your script with the images, props, music, and other media that you selected from your research. If you create your presentation on your computer, you can save the file for future viewing.

Practicing the Skill

Create a multimedia presentation about Red Square in Moscow, Russia. Turn to Chapter 15, Section 3, "Blending Europe and Asia," and look at the images on page 479 for ideas about what your presentation might include.

4.1 Using a Search Engine

Defining the Skill

Using a **search engine** can be a useful way to do research on the Internet. A search engine is a tool designed to find information on the Internet. There are billions of Web sites and documents available on the Internet. Using a search engine properly can help you sort through the vast amount of information.

Applying the Skill

The screen below shows a popular search engine. Use the strategies listed below to help you search the Internet.

How to Use a Search Engine

Strategy 1 Brainstorm keywords to enter into the search engine. Make a list of possible search terms. It often helps to try to be specific. If your research topic is what careers geographers can have, try entering *careers in geography* instead of just *geography*. This will narrow the results to relevant information.

Strategy 2 Many search engines will provide an excerpt of a Web site below the link to the site. Read these to help you determine if the Web site is relevant to your topic and worth visiting.

Strategy 3 Click on the link to go to the Web site you are interested in exploring in depth. If you can, open the new Web site in a separate window, so that it is easier to come back to your original search.

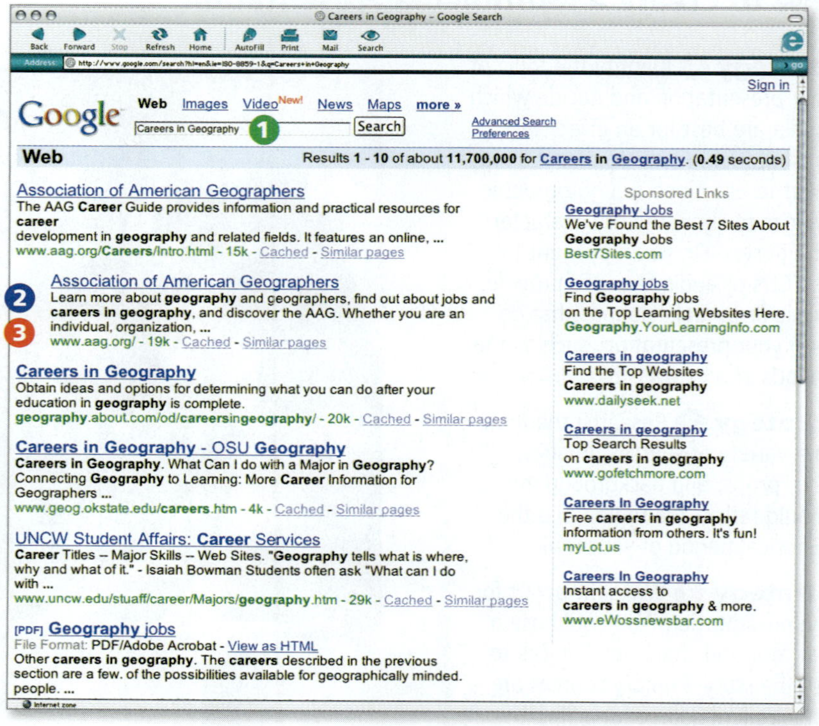

Practicing the Skill

Choose a topic from Unit 1 that you would like to learn more about. Develop a list of keywords to help you search the Internet for information about that topic. Visit a search engine, and enter your keywords. Take a look at your search results, and consider which keywords were the most useful and why.

4.2 Evaluating Internet Sources

Defining the Skill

By **evaluating Internet sources** for credibility, you can make sure that you are only using the most accurate, reliable information as a resource.

Applying the Skill

The screen below shows a Web site about geography. Use the strategies listed below to learn how to assess its credibility.

How to Evaluate Internet Sources

Strategy 1 Look at the Web site's Internet address. The three-letter code in it will help you determine who created the site. Almost anyone can set up a Web site with a ".com," or commercial, address. School Web sites use ".edu," or education, addresses. Addresses that end with ".org" are used by nonprofit organizations. Official government Web sites end in ".gov." These last three will often be more reliable than commercial Web sites.

Strategy 2 Try to identify the author of the Web site and when it was last updated. This information does not always appear. Some sites are anonymous, or created by an unidentified author. You should not use these as sources because the information could be outdated, or the author might not know much about the Web site's topic.

Strategy 3 Use another source to verify the information you find on the Internet. Online encyclopedias contain accurate information. Sources with .gov and .edu addresses are usually reliable. So is information from newspaper, magazine, and television news channel Web sites. Search several sites, and try to find two or three sources with the same information.

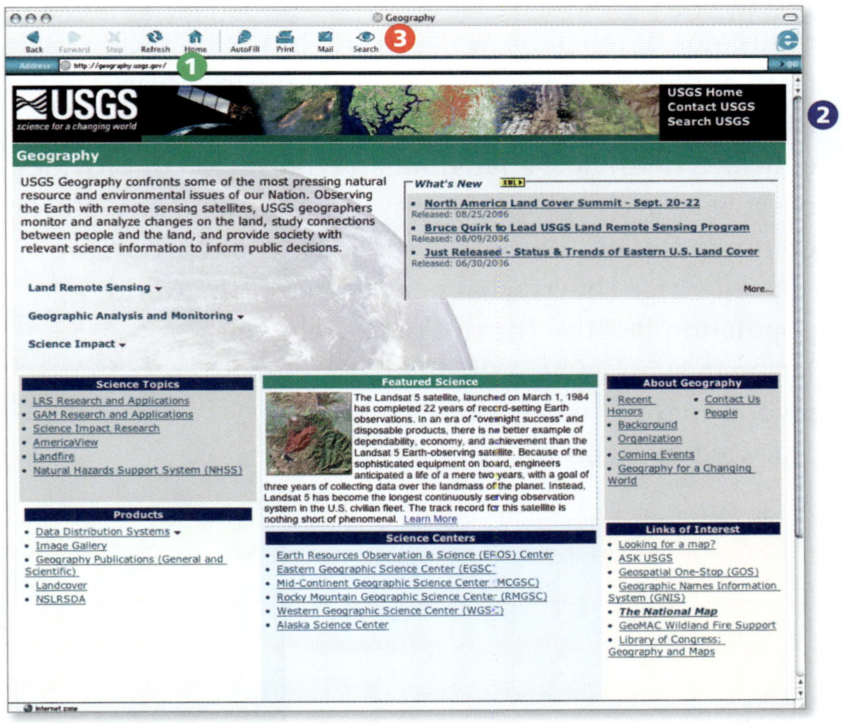

Practicing the Skill

Choose one of the Web sites that you found in your search from "Practicing the Skill" for "Using a Search Engine" on page R36 of this handbook. Evaluate the Web site to assess its credibility and its usefulness as a resource. Use the strategies listed above to help you.

A Global View

A religion is an organized system of beliefs and practices, often centered on one or more gods. In this book, you have learned about many different religions and their impact on the world. Religions have guided people's beliefs and actions for thousands of years. They have brought people together. But they have also torn them apart.

Religions are powerful forces today as well. They affect everything from what people wear to how they behave. There are thousands of religions in the world. In the following pages, you will learn about five major religions: Buddhism, Christianity, Hinduism, Islam, and Judaism. You will also learn about Confucianism, an ethical system. Like a religion, an ethical system provides guidance on how to live your life. However, unlike religions, ethical systems do not center on the worship of gods. The chart on the opposite page shows what percentages of the world population practice the five major religions. The map shows where these religions are predominant or where they are practiced by significant numbers.

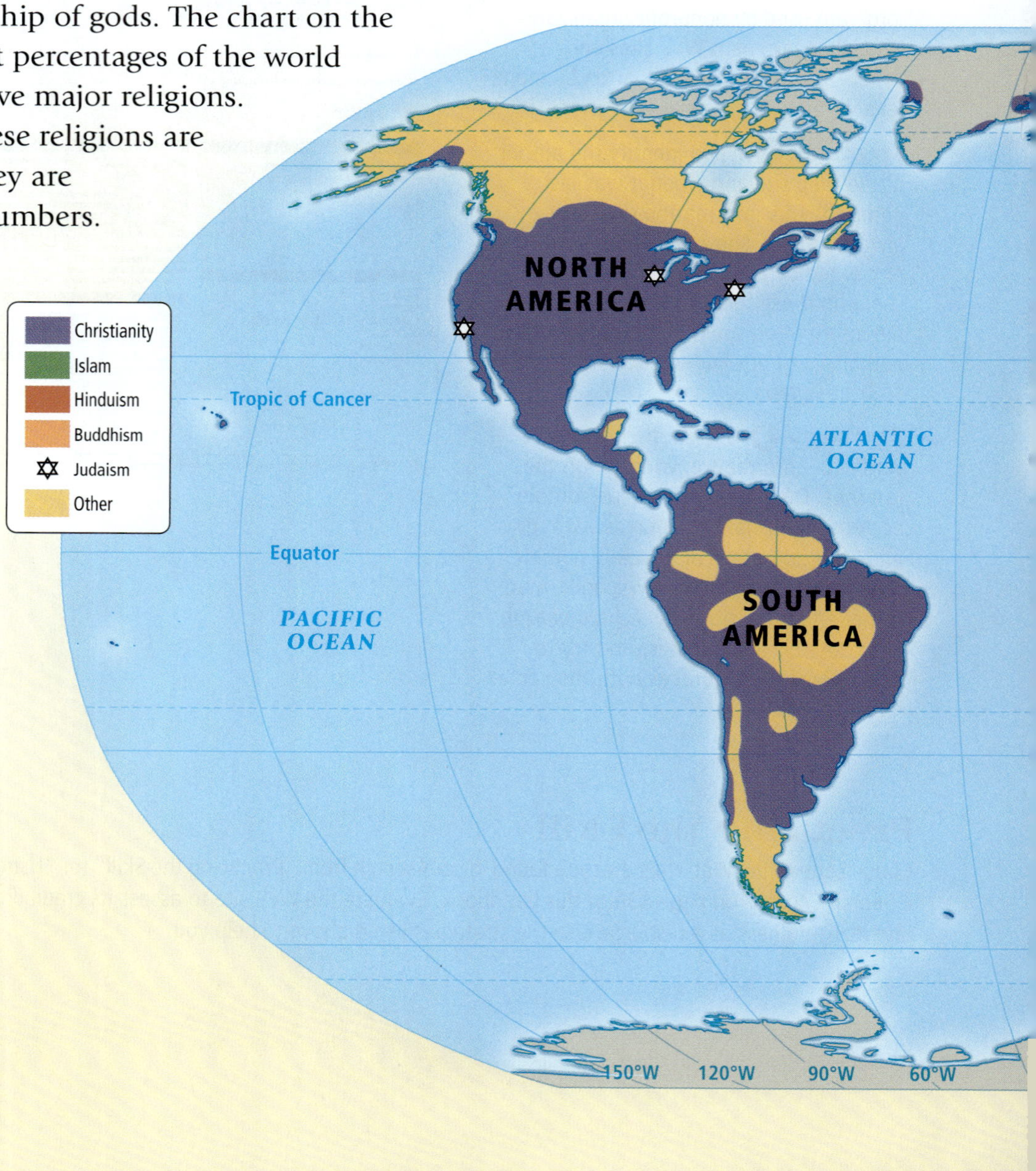

R38

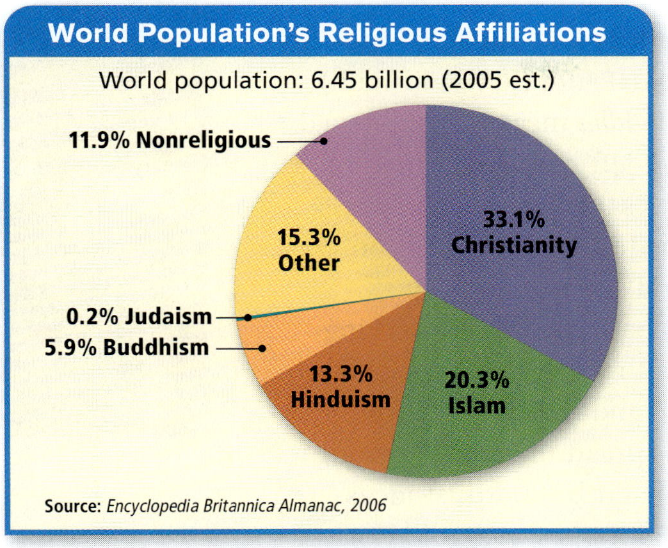
World Population's Religious Affiliations
World population: 6.45 billion (2005 est.)
11.9% Nonreligious
15.3% Other
33.1% Christianity
0.2% Judaism
5.9% Buddhism
13.3% Hinduism
20.3% Islam
Source: Encyclopedia Britannica Almanac, 2006

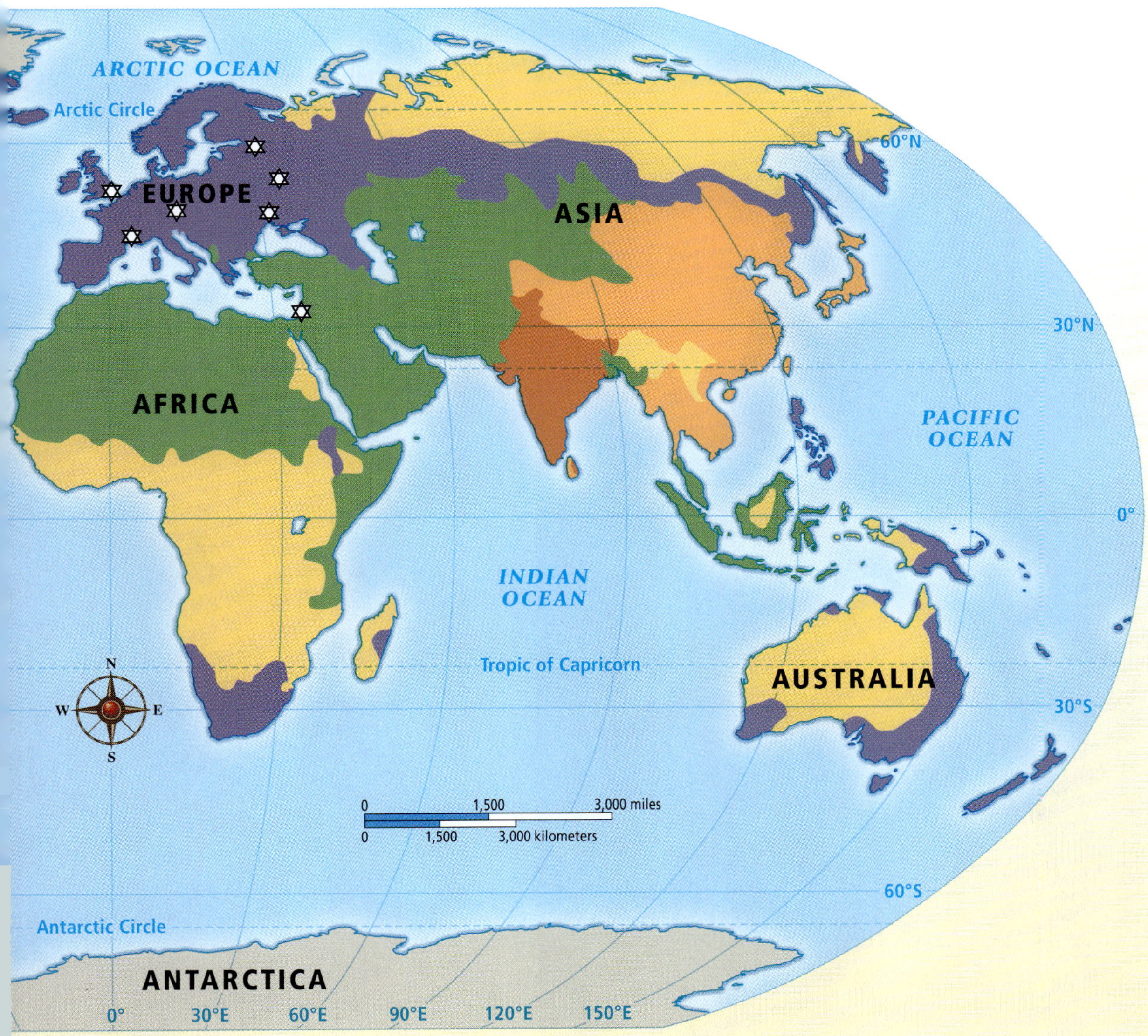
ARCTIC OCEAN
Arctic Circle
EUROPE
ASIA
AFRICA
PACIFIC OCEAN
INDIAN OCEAN
Tropic of Capricorn
AUSTRALIA
ANTARCTICA
Antarctic Circle
60°N
30°N
0°
30°S
60°S
0°
30°E
60°E
90°E
120°E
150°E
N
S
E
W
0
1,500
3,000 miles
0
1,500
3,000 kilometers

R39

Buddhism

Buddhism began in India in the sixth century B.C. The religion was founded by Siddhartha Gautama (sihd•DAHR•tuh GOW•tuh•muh), who came to be known as the Buddha. *Buddha* means "enlightened one." He was born into a noble family but left home to search for enlightenment, or wisdom. The Buddha is said to have achieved enlightenment after long study. According to Buddhist tradition, he taught his followers that the way to end suffering was by practicing the Noble Eightfold Path. This path involved observing the following: right opinions, right desires, right speech, right action, right job, right effort, right concentration, and right meditation.

After the Buddha's death, Buddhism spread in India, Ceylon, and Central Asia. Missionaries spread the faith. Buddhist ideas also traveled along trade routes. The religion, however, did not survive on Indian soil. Today, most Buddhists live in Sri Lanka (formerly Ceylon), East Asia, Southeast Asia, and Japan.

▼ Buddha
Statues of the Buddha, such as this one in Japan, appear in shrines throughout Asia. Buddhists try to follow the Buddha's teachings by meditating, a way of emptying the mind of thought. They also make offerings at shrines, temples, and monasteries.

▼ Monks
Buddhist monks dedicate their entire lives to the teachings of the Buddha. They live together in religious communities called monasteries. There, the monks lead lives of poverty, meditation, and study. In this photograph, Buddhist monks in Myanmar hold their begging bowls.

R40

▲ **Pilgrimage**
For centuries, Buddhists have come to visit places in India and Nepal associated with the Buddha's life. These sites include the Buddha's birthplace and the fig tree where he achieved his enlightenment. Worshipers also visit the Dhamekha Stupa in Sarnath, India, the site of the Buddha's first sermon, shown here.

Learn More About Buddhism

Symbol The Buddha's teaching, known as the dharma, is often symbolized by a wheel because his teaching was intended to end the cycle of births and deaths. The Buddha is said to have "set in motion the wheel of the dharma" during his first sermon.

Primary Source

The Buddha called his insight into the nature of suffering the Four Noble Truths. In the following selection, the Buddha tells his followers how they can end suffering and find enlightenment. The path involves understanding that life on Earth is brief and full of sadness. It also involves giving up selfish desire.

All created things are transitory [short-lived]; those who realize this are freed from suffering. This is the path that leads to pure wisdom.

All created beings are involved in sorrow; those who realize this are freed from suffering. This is the path that leads to pure wisdom.

All states are without self; those who realize this are freed from suffering. This is the path that leads to pure wisdom.

from the Dhammapada
Translated by Eknath Easwaran

R41

Christianity

Christianity is the largest religion in the world, with about 2 billion followers. It is based on the life and teachings of Jesus, as described in the Bible's New Testament. Jesus, a Jew, taught many ideas from the Jewish tradition. Some biblical prophets had spoken of a day when a promised figure would come to save all of humankind. By the end of the first century A.D., many Jews and non-Jews had come to believe that Jesus was the one who would make this happen. Now called "Christians," they spread their faith throughout the Roman Empire.

Christians regard Jesus as the Son of God. They believe that Jesus entered the world and died to save humanity.

▼ Easter and Palm Sunday

On Easter, Christians celebrate their belief in Jesus' resurrection, or his being raised to heavenly life after he was put to death. The Sunday before Easter, Christians observe Palm Sunday. This day celebrates Jesus' triumphal entry into Jerusalem. Palm branches, like those carried in this procession in El Salvador, were spread before him.

R42

▲ **Jesus and the Disciples**
Jesus' followers included 12 disciples, or pupils. Jesus passed on his teachings to his disciples. This painting from the 1400s shows Jesus with his disciples.

▼ **St. Paul's Cathedral**
Paul was a missionary who spread Christian beliefs throughout the Roman Empire. He started churches almost everywhere he went. Many churches today, such as this great cathedral in London, are named for Paul.

Learn More About Christianity

Symbol According to the New Testament, Jesus was crucified, or put to death on a cross. As a result, the cross became an important symbol of Christianity. It represents the belief that Jesus died to save humanity.

Primary Source

One of Jesus' most famous sermons is the Sermon on the Mount. In this talk, Jesus provided guidance to his followers. His words were written down in the New Testament, the part of the Bible that describes the teachings of Jesus. In the following verses, Jesus explains that people can be saved by opening their hearts to God and by treating others as they would like to be treated.

Ask, and it will be given you; seek, and you will find; knock, and it will be opened to you. For every one who asks receives, and he who seeks finds, and to him who knocks it will be opened. Or what man of you, if his son asks him for a loaf, will give him a stone? Or if he asks for a fish, will give him a serpent? If you then, who are evil, know how to give good gifts to your children, how much more will your Father who is in heaven give good things to those who ask him? So whatever you wish that men would do to you, do so to them; for this is the law and the prophets.

Matthew 7:7–12

Hinduism

Hinduism is a way of life guided by religious beliefs and practices that developed over thousands of years. Hindus believe that a supreme being called Brahman is the soul of the universe. The same presence, they believe, can also be found within each person. People can be freed from suffering and desires once they understand the nature of Brahman. The religious practices of Hindus include prayer, meditation, selfless acts, and worship of the various Hindu deities.

Today, Hinduism is the major religion of India and Nepal. It also has followers in Indonesia, Africa, Europe, and the Western Hemisphere.

▼ Festival of Diwali

Diwali, the Festival of Lights, is the most important festival in India. Diwali may have begun as a harvest festival in ancient India. Today, it marks the beginning of the year for many Hindus. They celebrate the festival by lighting candles and lamps, as shown in this photograph.

R44

▲ **Deities**
Brahman often takes the form of three deities in Hinduism. Brahma is the creator of the universe. Vishnu is its protector. Shiva is its destroyer. All three deities are represented in this sculpture.

▼ **Brahmin Priest**
Brahmin priests, like the one shown here, are among Hinduism's religious leaders. These priests take care of the holy images in temples and read from the religion's sacred books.

Learn More About Hinduism

Symbol The syllable *Om* (or *Aum*) is often recited at the beginning of Hindu prayers. *Om* is the most sacred sound in Hinduism because it is believed to contain all other sounds. The syllable is represented by the symbol shown below.

Primary Source

Hinduism has many sacred texts. The Vedas, four collections of prayers, rituals, and other sacred texts, are the oldest Hindu scriptures. They are believed to contain all knowledge, past and future.

The *Bhagavad-Gita* is another sacred Hindu text. In this work, Vishnu takes on the personality of a chariot driver named Krishna. Krishna and the warrior Arjuna discuss the meaning of life and religious faith. In this selection, Krishna explains that Brahman cannot be destroyed.

Weapons do not cut it,
fire does not burn it,
waters do not wet it,
wind does not wither it.

It cannot be cut or burned;
it cannot be wet or withered;
it is enduring, all-pervasive,
fixed, immovable, and timeless.

Bhagavad-Gita 2:23–24

World Religions Handbook

R45

Islam

Islam is a religion based on the teachings of the Qur'an, the religion's holy book. Followers of Islam, known as Muslims, believe that God revealed these teachings to the prophet Muhammad through the angel Gabriel around A.D. 610. Islam teaches that there is only one God—the same God that is worshiped in Christianity and Judaism. In Arabic, God is called Allah. Muslims also believe in the prophets of Judaism and Christianity. In fact, Muslims traditionally refer to Christians and Jews as "people of the book." That is because Christians and Jews have received divine revelations from scriptures in the Bible.

Today, most Muslims live in southwestern and central Asia and parts of Africa. Islam also has many followers in Southeast Asia. Muslims show their devotion by performing acts of worship known as the Five Pillars of Islam. These include faith, prayer, charity, fasting, and a pilgrimage to Mecca.

▲ **Muslim Prayer**
Five times a day—dawn, noon, mid-afternoon, sunset, and evening—Muslims face toward Mecca to pray. Like the people in this photograph, Muslims stop what they are doing when they hear the call to prayer. Everything comes to a halt—even traffic.

▼ **The Dome of the Rock**
The Dome of the Rock in Jerusalem is one of Islam's holiest sites. The rock on the site is the spot from which Muslims say Muhammad rose to heaven to learn Allah's will. With Allah's blessing, Muhammad returned to Earth to bring God's message to all people.

R46

Ramadan

During the holy month of Ramadan, Muslims fast, or do not eat or drink, from dawn to sunset. The family shown here is ending their fast. The most important night of Ramadan is called the Night of Power. This is believed to be the night the angel Gabriel first spoke to Muhammad.

Learn More About Islam

Symbol The crescent moon has become a symbol of Islam. The symbol may be related to the new moon that begins each month in the Islamic lunar calendar.

Primary Source

The Qur'an is the spiritual guide for Muslims. It also contains teachings for Muslim daily life. The following chapter is called the Exordium (introduction). It is also called Al-Fatihah. Muslims recite this short chapter, as well as other passages from the Qur'an, when they pray.

In the Name of God, the Compassionate,
the Merciful

Praise be to God, Lord of the Universe,
The Compassionate, the Merciful,
Sovereign of the Day of Judgment!
You alone we worship, and to You alone
we turn for help.
Guide us to the straight path,
The path of those whom You have favored,
Not of those who have incurred Your wrath,
Nor of those who have gone astray.

Qur'an 1:1–6

Judaism

Judaism was the first major monotheistic religion—that is, based on the concept of one God. The basic teachings of Judaism come from the Torah, the first five books of the Hebrew Bible. Judaism teaches that a person serves God by studying the Torah and living by its teachings. The Torah became the basis for the civil and religious laws of Judaism. The followers of Judaism, or Jews, also believe that God set down many moral laws for all of humanity with the Ten Commandments.

Today, there are more than 15 million Jews throughout the world. Many live in Israel, where a Jewish state was created in 1948.

▼ Abraham

According to the Torah, God chose a Hebrew shepherd named Abraham to be the "father" of the Hebrew people. In the 19th century B.C., Abraham led his family to a land that he believed God had promised them. This painting illustrates their journey.

R48

▲ **Rabbi**
Rabbis are the Jewish people's spiritual leaders and teachers. A rabbi often conducts the services in a synagogue, or Jewish house of worship. Like the rabbi shown here, he or she may also conduct the ceremony that marks Jewish children's entrance into the religious community.

▼ **Western Wall**
Many Jews make the pilgrimage to the Western Wall, shown here. The sacred wall formed the western wall of the courtyard of the Second Temple of Jerusalem. The temple was built in the second century B.C. The Romans destroyed it in A.D. 70.

Learn More About Judaism

Symbol The Star of David, also called the Shield of David, is a very important symbol of Judaism. The symbol honors King David, who ruled the kingdom of Israel about 1000–962 B.C.

Primary Source

The Book of Genesis is the first book of the Hebrew Bible and of the Torah. Genesis tells the history of the Hebrew people. It focuses on the individuals with whom God had a special relationship. In the following verses, God speaks to Abraham. His words express a promise of land and a special pledge to the Hebrew people.

Now the Lord said to Abram [Abraham], "Go from your country and your kindred and your father's house to the land that I will show you. And I will make of you a great nation, and I will bless you, and make your name great, so that you will be a blessing. I will bless those who bless you, and him who curses you I will curse; and by you all the families of the earth will bless themselves."

Genesis 12:1–3

Confucianism

Confucianism is an ethical system based on the teachings of the Chinese scholar Confucius. It stresses social and civic responsibility. Confucius was born in 551 B.C., during a time of crisis in China. He hoped his ideas and teachings would restore the order of earlier times to his society. But although Confucius was active in politics, he never had enough political power to put his ideas into practice. After his death, Confucius' students spread his teachings. As a result, his ideas became the foundation of Chinese thought for more than 2,000 years.

Today, Confucianism guides the actions of millions of Chinese people and other peoples of the East. It has also greatly influenced people's spiritual beliefs. While East Asians declare themselves to follow a number of religions, many also claim to be Confucians.

▼ **Temple**
Although Confucianism has no clergy or gods to worship, temples, like this one in Taiwan, have been built to honor Confucius. In ancient times, the temples provided schools of higher education. Today, many have been turned into museums.

R50

◀ Confucius

Confucius believed that society should be organized around five basic relationships. These are the relationships between (1) ruler and subject, (2) father and son, (3) husband and wife, (4) elder brother and junior brother, and (5) friend and friend.

▲ Confucius' Birthday

Historians do not know for certain the day when Confucius was born, but people in East Asia celebrate his birthday on September 28. In Taiwan and China, it is an official holiday known as Teachers' Day. The holiday pays tribute to teachers because Confucius himself was a teacher. Here, students in Beijing take part in a ceremony honoring their teachers.

Learn More About Confucianism

Symbol The yin-and-yang symbol represents opposite forces in the world working together. Yin represents all that is cold, dark, soft, and mysterious. Yang represents everything that is warm, bright, hard, and clear. The yin-and-yang symbol represents the harmony that Confucius hoped to restore to society.

Primary Source

Confucius' teachings were collected by his students in a book called the *Analects*. In the following selections from the *Analects*, Confucius (called the Master) instructs his students about living a moral and thoughtful life.

The Master said: "Even in the midst of eating coarse rice and drinking water and using a bent arm for a pillow happiness is surely to be found; riches and honors acquired by unrighteous means are to me like the floating clouds." (7.16)

The Master said: "When I walk with two others, I always receive instruction from them. I select their good qualities and copy them, and improve on their bad qualities." (7.22)

The Master said: "The people may be made to follow something, but may not be made to understand it." (8.9)

from the ***Analects***
Translated by Raymond Dawson

World Religions Handbook

Other Important Religions

You have learned about the five major world religions. Now find out about some other important religions: Bahaism, Shinto, Sikhism, and Zoroastrianism. These religions are important both historically and because they have many followers today.

▼ Shinto

Shinto, meaning "way of the gods," is Japan's oldest and only native religion. Shintoists worship many gods, called *kami*. They believe that kami are spirits found in mountains, rivers, rocks, trees, and other parts of nature. Shintoists often worship the kami at shrines in their homes. They also celebrate the gods during special festivals, such as the one shown here. Today, there are about 3 million Shintoists, mostly in Japan.

▲ Bahaism

Bahaism (buh•HAH•IHZ•uhm) is a young religion, with more than 7 million followers throughout the world. It was founded in 1863 in Persia (modern-day Iran) by a man known as Bahaullah, which means "splendor of God" in Arabic. Followers believe that, in time, God will break down barriers of race, class, and nation. When this happens, people will form a single, united society. All of the Baha'i houses of worship have nine sides and a central dome, symbolizing this unity. The Baha'i house of worship shown here is located in Illinois.

R52

◄ Sikhism

Sikhism was founded in India over 500 years ago by Guru Nanak. The religion's 24 million followers, called Sikhs, believe in one God. Like Buddhists and Hindus, Sikhs believe that the soul goes through repeated cycles of life and death. However, Sikhs do not believe that they have to live outside the world to end the cycle. Rather, they can achieve salvation by living a good and simple life. Uncut hair symbolizes this simple life. Many Sikh men cover their long hair with a turban, like the one worn by this man.

▲ Zoroastrianism

Zoroastrianism (ZAWR•oh•AS•tree•uh•NIHZ•uhm) was founded in ancient Persia around 600 B.C. by a prophet named Zoroaster. This prophet taught that Earth is a battleground where a great struggle is fought between the forces of good and the forces of evil. Each person is expected to take part in this struggle. At death, the Zoroastrian god, called Ahura Mazda (ah•HUR•uh MAZ•duh), will judge the person on how well he or she fought. This stone relief shows Ahura Mazda (*right*) giving the crown to a Persian king. Today, there are about 2.5 million Zoroastrians throughout the world.

Comparing World Religions and Ethical Systems

	Buddhism	Christianity	Hinduism	Islam	Judaism	Confucianism
Followers worldwide (estimated 2005 figures)	379 million	2.1 billion	860 million	1.3 billion	15.1 million	6.5 million
Name of god	no god	God	Brahman	Allah	God	no god
Founder	the Buddha	Jesus	no founder	no founder but spread by Muhammad	Abraham	Confucius
Holy book	many sacred books, including the Dhammapada	Bible, including Old Testament and New Testament	many sacred texts, including the Upanishads	Qur'an	Hebrew Bible, including the Torah	*Analects*
Clergy	Buddhist monks	priests, ministers, monks, and nuns	Brahmin priests, monks, and gurus	no clergy but a scholar class, called the ulama, and imams, who may lead prayers	rabbis	no clergy
Basic beliefs	• Followers can achieve enlightenment by understanding the Four Noble Truths and by following the Noble Eightfold Path of right opinions, right desires, right speech, right action, right jobs, right effort, right concentration, and right meditation.	• There is only one God, who watches over and cares for his people. • Jesus is the Son of God. He died to save humanity. His death and resurrection made eternal life possible for others.	• The soul never dies but is continually reborn until it becomes divinely enlightened. • Persons achieve happiness and divine enlightenment after they free themselves from their earthly desires. • Freedom from earthly desires comes from many lifetimes of worship, knowledge, and virtuous acts.	• There is only one God, who watches over and cares for his people. • Persons achieve salvation by following the Five Pillars of Islam and living a just life. The pillars are faith, prayer, charity, fasting, and pilgrimage to Mecca.	• There is only one God, who watches over and cares for his people. • God loves and protects his people but also holds people accountable for their sins and shortcomings. • Persons serve God by studying the Torah and living by its teachings.	• Social order, harmony, and good government should be based on strong family relationships. • Respect for parents and elders is important to a well-ordered society. • Education is important for the welfare of both the individual and society.

Source: *Encyclopedia Britannica Almanac, 2006*

R54

Review

KEY IDEAS

Buddhism (pages R40–R41)

1. How did the Buddha believe that his followers could end their suffering? by practicing the Noble Eightfold Path
2. How did Buddhism spread? through missionaries and along trade routes

Christianity (pages R42–R43)

3. Why is Jesus important to the Christian religion? People can be saved by following Jesus' teachings.
4. What are some Christian beliefs? Jesus died for people's sins, then came back to life.

Hinduism (pages R44–R45)

5. What is the importance of Brahman in Hinduism? People must understand the nature of Brahman to be freed from suffering.
6. What three deities does Brahman often take the form of? Brahma, the creator of the universe; Vishnu, its protector; Shiva, its destroyer

Islam (pages R46–R47)

7. How do Muslims believe the teachings of the Qur'an were revealed? through an angel to the prophet Muhammad
8. Why do Muslims traditionally refer to Christians and Jews as "people of the book"? because they follow the teachings of the Bible

Judaism (pages R48–R49)

9. What does it mean to say that Judaism is a monotheistic religion? It is based on the concept of one God.
10. What are the Ten Commandments? moral laws Jews believe that God gave to humanity

Confucianism (pages R50–R51)

11. What did Confucius hope to restore? order, harmony, good government
12. What five relationships are important in Confucianism? ruler–subject, father–son, husband–wife, older brother–younger brother, friend–friend

Other Important Religions (pages R52–R53)

13. How does Shinto differ from Bahaism, Sikhism, and Zoroastrianism? Shintoists believe in spirits found in nature.
14. How is Sikhism similar to Buddhism and Hinduism? Sikhs believe that the soul goes through cycles of life and death.

CRITICAL THINKING

15. **Compare and Contrast** What goal do Buddhists and Hindus share? to free themselves from suffering
16. **Draw Conclusions** How does Islam affect the everyday lives of its followers? Muslims are required to pray five times a day.

STANDARDS-BASED ASSESSMENT

INTERPRETING A PIE CHART

The pie chart below shows what percentages of the population of India practice the major religions. Use the pie chart to answer the following questions.

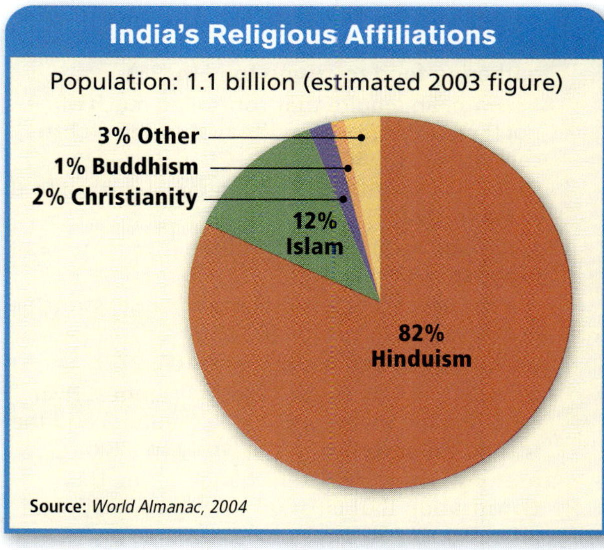

1. **What percentage of the people in India practice Hinduism?** D
   - **A.** 1 percent
   - **B.** 2 percent
   - **C.** 12 percent
   - **D.** 82 percent
2. **Which religion is practiced by 12 percent of the population?** D
   - **A.** Buddhism
   - **B.** Christianity
   - **C.** Hinduism
   - **D.** Islam

Glossary

A

aborigine (AB•uh•RIHJ•uh•nee) *n.* an original inhabitant of Australia (p. 892)

African National Congress (ANC) *n.* South African political party opposing apartheid (p. 624)

Afrikaner *n.* a South African of European ancestry who speaks a language called Afrikaans (p. 624)

agricultural revolution *n.* the shift from gathering food to raising food (p. 96)

Aksum *n.* an empire that controlled much of northern Ethiopia from the first to the eighth century A.D. (p. 606)

alpaca *n.* a South American mammal related to the llama (p. 272)

Alpine *adj.* having to do with the Alps mountain range (p. 382)

Alps *n.* Europe's tallest mountain range, stretching across southern Europe (p. 326)

altiplano *n.* a high plateau (pp. 204, 272)

Amazon River *n.* South America's longest river (about 4,000 miles, or 6,400 kilometers) and the second-longest river in the world (p. 204)

amendment *n.* a written change to the U.S. Constitution that must go through an approval process (p. 146)

Amundsen, Roald *n.* Norwegian polar explorer first to reach the South Pole (p. 920)

Anatolian peninsula *n.* in Southwest Asia between the Black Sea and Mediterranean Sea (p. 652)

ancestor *n.* an early relative (p. 544)

Andes Mountains *n.* a mountain range located on South America's west coast and extending the full length of the continent (p. 204)

animism *n.* the belief that spirits exist in animals, plants, natural forces, and ancestors (p. 544)

annex *v.* to add to an existing territory (p. 230)

Antarctic Peninsula *n.* an S-shaped peninsula that is part of the Andes Mountain chain (p. 920)

Antarctic Treaty *n.* an agreement signed in 1959 calling for the peaceful use of Antarctica (p. 920)

anthropologist (AN•thruh•PAHL•uh•jihst) *n.* a scientist who studies culture (p. 88)

apartheid (uh•PAHRT•hyt) *n.* the official policy of racial segregation practiced in South Africa from 1950 to 1991 (p. 624)

Appalachian (AP•uh•LAY•chee•uhn) **Mountains** *n.* a mountain chain in the Eastern United States, running parallel to the Atlantic Ocean (p. 122)

Aquino (uh•KEE•noh), **Corazon** *n.* politician who opposed Marcos and was president of the Philippines between 1986 and 1992 (p. 864)

Arabian Peninsula *n.* peninsula of Southwest Asia between the Red Sea and Persian Gulf (p. 652)

Aral Sea *n.* an inland body of water in Central Asia that has been steadily shrinking (p. 496)

archipelago (AHR•kuh•PEHL•uh•GOH) *n.* a set of closely grouped islands, which sometimes form a curved arc (pp. 198, 736, 852, 864)

artisan *n.* a worker skilled in making products or art with his or her hands (p. 254)

Aryan *n.* one of a group of Indo-European nomadic herders who are believed to have migrated to the Indian subcontinent (p. 698)

Ashanti (uh•SHAN•tee) *n.* a people of Ghana (pp. 580, 586)

Aswan High Dam *n.* a structure to control the floodwaters of the Nile River and provide electricity (p. 554)

Atlas Mountains *n.* series of mountain ranges stretching from Morocco to Tunisia (p. 568)

atmosphere *n.* the layer of gases that surround the Earth (p. 34)

atoll *n.* a ring-shaped coral reef encircling a central lagoon with no inner island (p. 912)

Aung San Suu Kyi (awng sahn soo chee) *n.* leader of the democracy movement in Myanmar (p. 858)

autonomy *n.* self-governance or independence (p. 376)

Aztec *n.* an early civilization in the Valley of Mexico (p. 212)

B

Bahasa Indonesia (bah•HAH•suh in•duh•NEE•zhuh) *n.* the common language of Indonesia (p. 864)

Baltic States *n.* Latvia, Lithuania, and Estonia, three countries that border the Baltic Sea (p. 424)

Bangalore *n.* south-central Indian city that is home to many high-tech industries (p. 716)

basin *n.* a region drained by a river system (p. 536)

Basque (bask) *n.* an ethnic group living in the western Pyrenees and along the Bay of Biscay in Spain and France; also the name of their language (p. 370)

Bechuanaland (bech•WAH•nuh•LAND) *n.* colonial name for Botswana (p. 630)

Bedouin *n.* an Arabic-speaking, traditionally nomadic people of Southwest Asia (p. 676)

Beirut *n.* capital of Lebanon and a center of banking and finance (p. 688)

Benelux (BEHN•uh•LUHKS) *n.* name for the economic union formed by Belgium, the Netherlands, and Luxembourg to ensure the fast and efficient movement of people, goods, and services within these nations (p. 376)

Berbers (BUR•buhrz) *n.* the original inhabitants of North Africa and the Sahara (p. 568)

Berlin Wall *n.* a wall built by East Germany's Communist government to close off East Berlin from West Berlin; torn down in 1991 (p. 382)

Bhagavad-Gita *n.* one of the sacred writings of Hinduism (p. 706)

Biafra (bee•AF•ruh) *n.* a largely Igbo region of Nigeria that declared its independence in 1967, but surrendered after losing a civil war with Nigeria in 1970 (p. 586)

bilingual *adj.* using or able to use two languages equally well (p. 164)

bipedal *adj.* walking on two feet (p. 544)

birth rate *n.* the number of births per 1,000 people per year (p. 58)

blues *n.* a type of music with lyrics that express sorrow, usually about problems in love or the hardships of life (p. 138)

Boer *n.* a Dutch colonist or descendant in South Africa (p. 624)

Bogotá *n.* capital city of Colombia (p. 266)

Bolívar, Simón (boh•LEE•vahr, see•MOHN) *n.* leader for independence in northern South America (pp. 220, 266)

Bolsheviks (BOHL•shuh•VIHKS) *n.* a group of revolutionaries who took control of the Russian government in 1917 (p. 468)

Bosniac *n.* an ethnic Muslim from Bosnia and Herzegovnia (p. 440)

bossa nova (BAHS•uh NOH•vuh) *n.* a jazz version of the samba, a Brazilian dance (p. 296)

Botero, Fernando *n.* Colombian artist known for portraits of people with exaggerated forms (p. 266)

boycott *v.* to stop buying and using products from certain sources as a way of protest (p. 304)

brain drain *n.* the loss of skilled workers who move in search of better opportunities (p. 424)

Brasília *n.* Brazil's current capital city (p. 290)

bread basket *n.* an abundant grain-producing region (p. 424)

Briton *n.* a British person (p. 408)

Buddha *n.* a name, meaning the "enlightened one," used for Siddhartha Gautama, the founder of Buddhism (pp. 698, 706, 774)

Buddhism *n.* a major world religion that began in India and is based on the teachings of Siddhartha Gautama (pp. 476, 698, 706, 774, 804)

Byzantine Empire *n.* the eastern half of the Roman Empire that survived for a thousand years after the fall of Rome (p. 362)

C

cabinet system *n.* a system of government that links the executive and legislative branches by choosing the prime minister and major department heads from the legislature (p. 176)

Cabral, Pedro Álvares *n.* Portuguese explorer who in 1500 claimed land that is now part of Brazil for Portugal (p. 290)

camel *n.* a humped mammal used in the desert as a pack animal and provider of meat, milk, and wool (p. 580)

Camp David Accords *n.* peace treaty negotiated by U.S. President Jimmy Carter, Egyptian President Anwar el-Sadat, and Israeli Prime Minister Menachem Begin between Egypt and Israel, signed at the U.S. presidential retreat in Maryland in 1979 (p. 562)

Canadian Shield *n.* a horseshoe-shaped, rocky plateau that covers much of eastern central Canada (p. 156)

Candomblé (kahn•duhm•BLEH) *n.* African religious practices that are mixed with Roman Catholic beliefs to produce a unique belief system (p. 296)

canopy *n.* the highest layer in a forest, formed by the treetops (p. 536)

Cape Colony *n.* South African land settled by Dutch colonists called Boers in the 1600s and 1700s (p. 624)

capoeira (KAP•oh•AY•ruh) *n.* a Brazilian dance combined with martial arts (p. 296)

Caracas *n.* capital city of Venezuela (p. 266)

carbon dioxide *n.* a gas composed of carbon and oxygen (p. 48)

Caribbean Community and Common Market (CARICOM) *n.* a trade organization of several Caribbean nations (p. 260)

Carthage (KAR•thihj) *n.* an ancient North African city-state (p. 568)

cartographer (kahr•TAHG•ruh•fur) *n.* a geographer who creates maps (p. 10)

Caspian Sea *n.* a large body of water bordering Russia to the south and situated between Transcaucasia and Central Asia (pp. 460, 496)

caste *n.* in India, a social class that a person belongs to by birth (p. 698)

cataract *n.* a high waterfall or rapids (p. 554)

Caucasus (KAW•kuh•suhs) **Mountains** *n.* a mountain range that runs between the Black Sea and the Caspian Sea (p. 496)

celadon (SEHL•uh•DAHN) *n.* Korean ceramic pottery with a thin blue or green glaze (p. 804)

Central America Free Trade Agreement (CAFTA) *n.* an agreement to promote trade between the United States and countries of Central America (pp. 254, 260)

Central Asia *n.* a region that contains the republics of Kazakhstan (KAH•zahk•STAN), Kyrgyzstan (KEER•gee•STAN), Tajikistan (tah•JIHK•ih•STAN), Turkmenistan (TURK•mehn•ih•STAN), and Uzbekistan (uz•BEHK•ih•STAN) (p. 496)

cession *n.* surrendered territory (p. 230)

Glossary

chaebol (JEH•buhl) *n.* in Korea, a family-owned conglomerate made up of related businesses (p. 812)

Chang Jiang (chahng jyahng) *n.* the longest river in Asia, flowing through eastern China (p. 756)

Channel Tunnel *n.* the underwater railroad tunnel between France and England under the English Channel (p. 414)

Chechnya (CHEHCH•nee•uh) *n.* a largely Islamic republic in southwestern Russia that continues to fight for its independence (p. 486)

Cheju (CHEH•JOO) **Island** *n.* largest island off the southern coast of the Korean Peninsula (p. 796)

chinampas *n.* in Mexico, artificial islands used for farming (p. 212)

Christianity *n.* a monotheistic religion based on the life and teachings of Jesus of Nazareth and on the writings of the Christian Bible (p. 682)

Church of England *n.* the official church of England headed by the Archbishop of Canterbury (p. 408)

citizen *n.* a person who owes loyalty to a country and receives its protection (p. 78)

city-state *n.* a political unit made up of a city and its surrounding lands (pp. 334, 666)

Civil War (1861–1865) *n.* a conflict between the U.S. North and South over the issues of states' rights and slavery (p. 130)

civilization *n.* an advanced form of culture that developed in cities (p. 666)

clan *n.* a group of families related through a common ancestor (p. 830)

climate *n.* the typical weather conditions of a region over a long period of time (p. 40)

climatologist *n.* a geographer who studies climates (p. 16)

coalition *n.* an alliance or partnership, often a temporary one (p. 362)

Cold War *n.* a conflict between the United States and the Soviet Union after World War II that never developed into open warfare (p. 468)

colonia *n.* a neighborhood in Mexico (p. 236)

colonialism *n.* the control by one power over a dependent area or people (p. 858)

colony *n.* a group of people who settle a distant land but are ruled by their homeland, such as the United States; overseas territory ruled by a nation, such as Mexico from the 1500s to 1821 (pp. 130, 220)

Columbian Exchange *n.* the movement of plants and animals between Latin America and Europe after Columbus' voyage to the Americas in A.D. 1492 (p. 220)

Columbus, Christopher *n.* Italian navigator and explorer who sailed for Spain and explored the Caribbean and the coast of Central and South America (p. 370)

command economy *n.* an economic system in which the production of goods and services is decided by a central government (pp. 72, 424, 486, 512, 784)

commonwealth *n.* a self-governing political unit that is associated with another country (p. 260)

Commonwealth of Nations *n.* an association made up of the United Kingdom and many former British colonies (pp. 164, 400, 624)

communal *adj.* referring to shared ownership of property (p. 774)

communism *n.* a type of government in which the Communist Party holds all political power and controls the economy (pp. 78, 260, 424, 468, 764)

compulsory *adj.* required (p. 362)

Confucianism (kuhn•FYOO•shuh•nihz•uhm) *n.* a belief system based on the teachings of Confucius, a Chinese scholar (pp. 764, 774, 804)

conglomerate *n.* an organization made up of several companies in different businesses (p. 812)

Congo River *n.* Africa's second-longest river (p. 618)

conquistador (kahn•KWIHS•tuh•DAWR) *n.* Spanish word for "conqueror" (p. 220)

Constitution *n.* the document that is the basis for the U.S. government (p. 130)

constitution *n.* the set of laws and principles that defines the nature of a government (pp. 176, 230)

Constitution of 1917 *n.* Mexican constitution written during the revolution that is still in effect today (p. 230)

constitutional monarchy *n.* a government in which the powers of the king or queen are limited by the constitution (pp. 176, 414, 736, 858, 892)

continent *n.* one of seven large landmasses on the Earth's surface (pp. 26, 122)

Continental Divide *n.* a high ridgeline in the Rocky Mountains that divides east-flowing from west-flowing rivers (p. 122)

continental shelf *n.* the submerged land at the edge of a continent (p. 34)

Cook, James *n.* a British explorer who claimed land in Australia (p. 892)

coral island *n.* a type of oceanic island that is a coral reef (p. 912)

cottage industry *n.* a small business in which the workers manufacture items in their homes (pp. 716, 728)

coup (koo) *n.* military takeover (p. 586)

cuíca (KWEE•kuh) *n.* a friction drum used in the samba, a Brazilian dance (p. 296)

cultural blending *n.* something new created from combining the elements of two or more cultures (p. 138)

cultural hearth *n.* the heartland of a major culture; an area where a culture originated and spread to other areas (pp. 96, 212)

culture *n.* the shared attitudes, knowledge, and behaviors of a group (pp. 66, 88, 138, 212, 544)

cuneiform (KYOO•nee•uh•form) *n.* the first known writing system, which used wedge-shaped symbols and was developed in Sumer (p. 666)

cyclone *n.* a violent storm with fierce winds that rotate in a circular pattern like a hurricane (p. 658)

Cyrillic (suh•RIHL•ihk) **alphabet** *n.* the official writing system of Russia (p. 476)

czar (zahr) *n.* title for the rulers of Russia from the mid-1500s to the early 1900s (p. 468)

Czechoslovakia (CHEHK•uh•sluh•VAH•kee•uh) *n.* former country in Eastern Europe that existed from 1918 until 1993, when it split into the Czech Republic and Slovakia (p. 432)

D

daimyo (DY•mee•OH) *n.* in Japan, powerful samurai who became warrior-chieftains (p. 830)

Dao (DOW) *n.* according to the belief system of Daoism, a force that guides the whole universe (p. 774)

Daoism (DOW•IHZ•uhm) *n.* a Chinese belief system based on the idea of natural order in the world (p. 774)

database *n.* a collection of information that can be analyzed (p. 10)

Day of the Dead *n.* holiday to remember loved ones who have died (p. 236)

death rate *n.* the number of deaths per 1,000 people per year (p. 58)

debris (duh•BREE) *n.* the scattered remains of something broken or destroyed (p. 10)

Deccan (DEHK•uhn) **Plateau** *n.* a high area of land at the center of the Indian subcontinent (p. 658)

deforestation *n.* the cutting and clearing away of trees (p. 304)

degraded *adj.* of a lower quality than previously existed (p. 48)

delta *n.* the triangle-shaped deposit of rich soil at a river's mouth (p. 554)

demilitarized zone *n.* a buffer zone between North Korea and South Korea (p. 804)

democracy *n.* a government in which the citizens make political decisions, either directly or through elected representatives (p. 334)

demographer *n.* a geographer who studies the characteristics of human populations (p. 58)

demonstration *n.* a public gathering to protest a policy or action (p. 784)

dependency *n.* a place governed by a country that it is not officially a part of (p. 260)

deport *v.* to expel from a country (p. 424)

desert *n.* a region with plants specially adapted to dry conditions (p. 40)

desertification *n.* the change of fertile land to desert; the process in which farmland becomes less productive because the land is degraded (pp. 48, 536)

devolution *n.* the process of shifting some power from national to regional government (p. 414)

dictator *n.* a person with complete control over a country's government (pp. 254, 260, 266, 812)

dictatorship *n.* a state or government ruled by a leader with absolute power (pp. 78, 568)

Diet (DY•iht) *n.* Japan's law-making body, which consists of a House of Representatives and a House of Councillors (p. 840)

diffusion *n.* the spread of ideas, inventions, and patterns of behavior from one group to another (p. 96)

discrimination *n.* actions that might be hurtful to an individual or a group (p. 66)

diversity *n.* having many different ways to think or to do something, or a variety of people (p. 66)

Dom Pedro I *n.* Brazil's first Portuguese emperor; declared Brazil's independence from Portugal in 1822 (p. 290)

Dom Pedro II *n.* second emperor of Brazil, under whose rule slavery was abolished in Brazil in 1888 (p. 290)

domesticate *v.* to adapt or train plants or animals for human use (p. 544)

domestication *n.* the raising and tending of a plant or an animal to be of use to humans (p. 96)

dominion *n.* in the British Empire, a nation that is allowed to govern its domestic affairs (p. 164)

draft animal *n.* an animal used to pull a heavy load (p. 884)

drainage basin *n.* the area drained by a major river (p. 34)

Dravidian (druh•VIHD•ee•uhn) *n.* speakers of a language group found in India since the earliest times; mostly spoken in South India today (p. 698)

drought *n.* a long period of little or no rainfall (p. 594)

duchy (DUHCH•ee) *n.* the territory ruled by a duke or duchess (p. 376)

Dust Bowl *n.* a region of the U.S. Great Plains that suffered drought and suffocating dust storms in the 1930s (p. 130)

dynastic (dy•NAS•tihk) **cycle** *n.* in China, the pattern of the rise and fall of dynasties (p. 764)

dynasty (DY•nuh•stee) *n.* a family or group that rules for several generations, such as in China (pp. 554, 606, 764)

R59

E

earthquake *n.* a sudden movement of the Earth's crust followed by a series of shocks (p. 26)

eclectic *adj.* made up of parts from a variety of sources (p. 432)

economic system *n.* a way people use resources to make and exchange goods (p. 72)

economy *n.* a system for producing and exchanging goods and services among a group of people (pp. 72, 304)

ecotourism *n.* travel to unique environments by people who take care to preserve them in their natural state (pp. 254, 736)

edible *adj.* fit for eating (p. 272)

emissions *n.* substances discharged into the air (p. 48)

emperor *n.* a ruler of an empire, which is a group of different territories or cultures led by a single all-powerful authority (pp. 764, 830)

empire *n.* a political system in which people or lands are controlled by one ruler (pp. 212, 220)

Enlightenment *n.* a philosophical movement in the 1600s and 1700s that was characterized by the use of reason and scientific methods (p. 350)

environment *n.* the physical surroundings of a location (p. 4)

equinox *n.* one of the two times a year when the sun's rays are over the equator and days and nights around the world are equal in length (p. 40)

Eritrea *n.* a former region of Ethiopia that became an independent country in 1993 (p. 606)

erosion *n.* the wearing away and movement of weathered materials by water, wind, or ice (p. 26)

escarpment *n.* a steep slope separating two flat areas of different heights (p. 536)

estancia (eh•STAHN•see•ah) *n.* large farm or ranch in Argentina (p. 280)

ethanol (EHTH•uh•NAWL) *n.* a liquid made by using a chemical process to convert sugar cane to a kind of alcohol that can be used for fuel (p. 304)

ethnic cleansing *n.* removing an ethnic or religious group from an area by force or the mass killing of members of such a group (p. 440)

ethnic group *n.* a group of people who share language, customs, and a common heritage (pp. 88, 476, 504)

Euphrates (yoo•FRAY•teez) **River** *n.* a river of Southwest Asia that flows from a plateau on the Anatolian peninsula to the Persian Gulf (p. 652)

Eurasia *n.* a term used to refer to a single continent made up of Europe and Asia (pp. 460, 496)

European Union (EU) *n.* an organization of European nations whose members cooperate on economic, social, and political issues (pp. 350, 362, 688)

Everglades *n.* a huge wetlands area of southern Florida (p. 122)

export *n.* a product or resource sold to another country (pp. 72, 304)

export *v.* to send something from one country to another (p. 146)

extended family *n.* the family unit that contains more relatives than just parents and children, including grandparents and other close family (p. 688)

F

famine *n.* a severe shortage of food that causes widespread hunger (pp. 594, 658)

fascism (FASH•IHZ•uhm) *n.* a political philosophy that promotes blind loyalty to the state and a strong central government controlled by a powerful dictator (p. 362)

fast *v.* to avoid eating, often for religious reasons (p. 728)

favelas (fuh•VEH•lahs) *n.* Brazilian name for the poor neighborhoods that surround the cities (p. 290)

federal republic *n.* form of government in which power is divided between a national government and state governments (p. 266)

federal system *n.* a government system in which powers are divided between the national and state or provincial governments (pp. 146, 176)

federation *n.* colonies, states, or countries that have joined together (p. 594)

Fertile Crescent *n.* a region stretching from the Persian Gulf northwest up the Tigris and Euphrates rivers and west over to the Mediterranean Sea (p. 666)

feudalism *n.* a political system in which lords gave land to vassals in exchange for services (p. 342)

fiesta *n.* a holiday celebrated with parades, games, and food (p. 236)

Filipino *n.* an inhabitant of the Philippines (p. 864)

First Nations *n.* organized cultural groups of Canada's native peoples (p. 164)

foreign investment *n.* money put into a business by people from another country (p. 512)

fossil *n.* remains of early life preserved in the ground (p. 544)

fossil fuels *n.* sources of energy from ancient plant and animal remains; includes fuels such as coal, oil, and natural gas (pp. 48, 326)

Fox, Vicente *n.* Mexican president from the National Action Party who was elected in 2000 (p. 230)

free enterprise *n.* an economic system in which businesses are free to operate without much government involvement (pp. 146, 176, 784)

French Revolution *n.* a conflict in France between 1789 and 1799 that ended the monarchy and led to changes in the way France was governed (p. 350)

G

Gaelic *n.* any of the Celtic family of languages spoken in Ireland or Scotland (p. 408)

Gandhi, Mohandas (GAHN•dee, MOH•hehn•DAHS) *n.* a 20th-century Indian who helped lead his country to independence by using nonviolent resistance to colonial rule (p. 698)

Ganges (GAN•JEEZ) **River** *n.* a river of South Asia that flows southeast from the Himalayas to the Bay of Bengal (p. 658)

García Márquez, Gabriel *n.* Colombian author and Nobel Prize winner (p. 266)

gaucho (GAOW•choh) *n.* Argentinian cowboy (p. 280)

Gautama, Siddhartha (GAW•tuh•muh, sihd•DAHR•tuh) *n.* the founder of Buddhism also known as the Buddha, or "enlightened one" (p. 774)

Gaza Strip *n.* a narrow territory along the Mediterranean Sea (p. 682)

Geographic Information Systems (GIS) *n.* a computer- or Internet-based mapping technology (pp. 10, 16)

geographic South Pole *n.* the point where all lines of longitude meet (p. 920)

geography *n.* the study of people, places, and environments (p. 4)

geomorphology (JEE•oh•mawr•FAHL•uh•jee) *n.* the study of how the shape of the Earth changes (p. 16)

geothermal energy *n.* power produced by the internal heat of the Earth (p. 900)

geyser (GY•zuhr) *n.* a hot spring that shoots sprays of steam and boiling water into the air (pp. 884, 900)

Ghana *n.* a West African kingdom between the A.D. 700s and the A.D. 1000s in what is now Mauritania and Mali; a modern country in West Africa (p. 580)

glacier *n.* a large, slow-moving mass of ice (p. 26)

global economy *n.* economy in which buying and selling occurs across national borders (pp. 146, 244)

Global Positioning System (GPS) *n.* a system that uses a network of Earth-orbiting satellites to pinpoint location (p. 10)

global warming *n.* an increase in the average temperature of the Earth's atmosphere (pp. 48, 304)

globe *n.* a model of the Earth in the shape of a sphere (p. 10)

glyph *n.* a carved or engraved symbol that stands for a syllable or a word (p. 212)

Gobi (GOH•bee) **Desert** *n.* a high desert in China and Mongolia (p. 756)

Golan Heights *n.* a region that was formerly part of southwest Syria that Israel has occupied since the 1967 war (p. 688)

golden age *n.* a period during which a society attains prosperity and cultural achievements (p. 698)

Good Friday Agreement *n.* an agreement between Northern Ireland's unionists and nationalists that set up a new government (p. 414)

Gorbachev (GAWR•buh•CHAWF), **Mikhail** *n.* Soviet leader from 1985 to 1991 who increased freedoms for the Russian people (p. 468)

government *n.* an organization set up to make and enforce rules for a society (p. 78)

Grand Canyon *n.* a deep gorge cut through northern Arizona by the Colorado River (p. 122)

Great Barrier Reef *n.* a 1,250-mile chain of coral reefs off the coast of Australia (p. 884)

Great Depression *n.* a period of severe economic decline from 1929 into the early 1940s (p. 130)

Great Dividing Range *n.* a range of low mountains in Australia (p. 884)

Great Lakes *n.* five lakes forming the largest group of freshwater lakes in the world (pp. 122, 156)

Great Plains *n.* a vast grassland in central North America (p. 122)

Great Wall *n.* a long, huge wall built over several centuries to keep nomads out of China (p. 764)

Great Zimbabwe *n.* an empire built by the Shona people from the first to the eighth century A.D. (p. 630)

Greater Antilles *n.* the northern largest Caribbean islands that include Cuba, Jamaica, Hispaniola (which includes Haiti and the Dominican Republic), and Puerto Rico (p. 198)

Green Revolution *n.* the use of special seeds, irrigation, fertilizers, and pesticides to produce high crop yields and food production (pp. 716, 728)

greenhouse effect *n.* the trapping of the sun's heat by gases in the Earth's atmosphere (p. 48)

greenhouse gas *n.* any gas in the atmosphere that contributes to the greenhouse effect (p. 48)

grid *n.* a network of horizontal and vertical lines that create squares or rectangles (p. 666)

R61

griot (gree•OH) *n.* a West African storyteller (p. 544)

Gross Domestic Product (GDP) *n.* the total value of all the goods and services produced in a country in a year (pp. 72, 280)

ground water *n.* water found beneath the Earth's surface (p. 34)

guerrilla *n.* a member of an irregular army that operates in small bands (p. 630)

Gulf of Mexico *n.* an arm of the Atlantic Ocean that lies south of the United States (p. 122)

H

habitable lands *n.* lands suitable for human living (p. 58)

Hammurabi's Code *n.* one of the world's first law codes, compiled by the ruler Hammurabi (p. 666)

hanbok *n.* a traditional Korean costume (p. 804)

Harappan civilization *n.* an ancient civilization that developed along the Indus River (p. 666)

Hausa (HOW•suh) *n.* a people of northern Nigeria and southern Niger (p. 586)

Heian (hay•ahn) **Period** *n.* an era in Japanese history from A.D. 794 to 1185; arts and writing flourished during this time (p. 830)

Hidalgo, Miguel ("Father Hidalgo") *n.* father of Mexican independence (p. 220)

hieroglyphics *n.* an Egyptian writing system in which picture symbols stand for meanings or sounds (p. 554)

highlands *n.* mountainous or hilly sections of a country (pp. 198, 204)

Himalayas (HIHM•uh•LAY•uhz) *n.* the highest mountains in the world, which stretch along northern India, separating India from China and the rest of Asia (pp. 658, 756)

Hinduism *n.* the modern name for the major religion of India, which developed from Brahmanism (pp. 698, 706)

Hitler, Adolf *n.* German head of state from 1933 to 1945 (p. 382)

Hokkaido (hah•KY•doh) *n.* Japan's northernmost and second largest island (p. 824)

Holocaust *n.* the systematic murder of Jews and other minorities by the Nazis during World War II (pp. 350, 382, 682)

hominid *n.* a human primate; an early human (p. 544)

Honshu *n.* Japan's largest and most populous island (p. 824)

Horn of Africa *n.* a horn-shaped extension of land on the east coast of Africa (p. 606)

Huang He (hwahng huh) *n.* a river that flows from the Kunlun Mountains to the Yellow Sea (p. 756)

Hudson Bay *n.* a large inland sea to the north of the Canadian province of Ontario (p. 156)

hunter-gatherer *n.* a person who hunts animals and gathers plants for food (pp. 618, 892)

hydroelectric *adj.* having to do with electricity created by water-powered engines (p. 156)

hydroelectricity *n.* electric power generated by water (p. 304)

hydroelectric power *n.* electricity made by water-powered engines (pp. 326, 756, 900)

hydrologic cycle *n.* the circulation of water between the Earth, the oceans, and the atmosphere (p. 34)

I

ice shelf *n.* huge flat sheet of ice extending over a body of water, but attached to the land (p. 920)

iceberg *n.* a large piece of floating ice (p. 920)

icecap *n.* a glacier that covers a large area (p. 920)

icon *n.* religious image used by Orthodox Christians in their worship (p. 476)

Igbo (IHG•bo) *n.* a people of southeastern Nigeria (p. 586)

igloo *n.* a house made of snow (p. 170)

immigrant *n.* a person who leaves one area to settle in another (pp. 66, 130, 296, 408)

immigration *n.* process of coming to another country to live (p. 244)

imperialism *n.* the practice of one country controlling the government and economy of another country or territory (pp. 350, 400, 580)

import *n.* a product or resource that comes into a country (p. 72)

import *v.* to bring something from one country into another (p. 146)

Inca *n.* an early civilization in the Andes Mountains of Peru (p. 212)

indigenous *n.* native to a region (p. 272)

Indochina *n.* the name for a former French colony in Southeast Asia made up of Cambodia, Laos, and Vietnam (p. 858)

Indochinese Peninsula *n.* one of two peninsulas on which mainland Southeast Asia lies; the peninsula is located south of China (p. 852)

Indus River *n.* a river in South Asia that flows from the Himalayas to the Arabian Sea (p. 658)

Industrial Revolution *n.* the shift that began in Great Britain in the 1760s from making goods by hand to making them by machine (pp. 350, 400)

inflation *n.* a price increase or a fall in the purchasing power of money (p. 594)

infrastructure *n.* basic services a community needs, such as roads, railroads, water and power lines, schools, and post offices (p. 594)

innovation *n.* something new that is introduced for the first time (p. 96)

Interstate Highway System *n.* a network of more than 45,000 miles of roads that links every major U.S. city (p. 146)

Inuit (IHN•yoo•iht) *n.* the native peoples who inhabit the Arctic region of North America (p. 164)

Inuktitut (ih•NOOK•tih•tut) *n.* the language of the Inuit (p. 170)

Islam *n.* a monotheistic religion based on the teachings of Muhammad and the writings of the Qur'an, the Muslim holy book (pp. 476, 504, 562, 676)

isolate *v.* to cut off or set apart from a group (p. 334)

Istanbul *n.* the largest city of Turkey, formerly called Constantinople (capital of the Byzantine, or Eastern Roman, Empire) and originally called Byzantium in ancient times (p. 688)

isthmus *n.* strip of land that connects two landmasses (p. 198)

ivory *n.* the hard white substance that makes up an elephant's tusk (p. 618)

J

jaguar *n.* a large cat mainly found in Central and South America (p. 212)

jazz *n.* a type of music that developed from a blending of African rhythms, American band music, and the musical styles of African Americans and Europeans (p. 138)

jellaba *n.* a traditional Berber garment; a robe with full sleeves (p. 568)

Jerusalem (juh•ROO•suh•luhm) *n.* the current capital of Israel and an ancient city that is holy to Judaism, Christianity, and Islam (p. 682)

joropo (huh•ROH•poh) *n.* Venezuelan national folk dance (p. 266)

Juárez, Benito (1806–1872) *n.* Indian who became president and a Mexican national hero (p. 230)

Judaism (JOO•day•IHZ•uhm) *n.* the monotheistic religion of the Jews, based on the writings of the Hebrew Bible (pp. 476, 682)

K

Kalahari *n.* a desert in southern Africa (p. 630)

Kanto Plain *n.* the largest lowland of Japan, which extends from the Japanese Alps east to the Pacific (p. 824)

Kara Kum *n.* a huge desert that covers most of Turkmenistan (p. 496)

karma *n.* in Hinduism, the consequences of a person's actions in this life, which determine his or her fate in the next life (p. 706)

Kashmir *n.* a region in the northwest part of the Indian subcontinent; India and Pakistan dispute control of Kashmir (p. 716)

Kente cloth *n.* a brightly colored, woven cloth made by the Ashanti and Ewe peoples (p. 586)

Kenyatta, Jomo *n.* independent Kenya's first president (p. 612)

Khmer (kmair) **Empire** *n.* an empire that began in the 500s and had gained control of much of mainland Southeast Asia by the 800s (p. 858)

kibbutz (kih•BOOTS) *n.* a type of settlement in Israel in which the community shares all wealth and property (p. 682)

Kim Jong Il *n.* the Communist leader of North Korea (p. 812)

kimchi (KIHM•chee) *n.* a Korean dish made of pickled vegetables (p. 804)

King, William Lyon Mackenzie *n.* Canadian prime minister who led his country to independence (p. 164)

knight *n.* a vassal trained in combat who fought on behalf of lords (p. 342)

Koizumi, Junichiro (KOH•EE•zoo•mee, JU•NEE•chee•roh) *n.* Japan's prime minister from 2001 to 2006 (p. 840)

Kongo *n.* a member of a people living along the lower Congo River; a Bantu-speaking kingdom of the Congo River region arising in the 1300s (p. 618)

Korean Peninsula *n.* a body of land bordered by the Yellow Sea to the west, the East Sea (or Sea of Japan) to the east, the Korea Strait to the south, and China and Russia to the north (p. 796)

Korean War (1950–1953) *n.* a conflict between North Korea and the Soviet Union on one side and South Korea, the United States, and the UN on the other (p. 804)

Korean Workers' Party (KWP) *n.* a Communist political group that controls the government of North Korea (p. 812)

Kosovo *n.* a self-governing province within Serbia (p. 440)

Kurd *n.* a member of an ethnic group that does not have its own country but whose homeland lies in parts of Turkey, Iraq, and Iran (p. 676)

Kush *n.* an ancient Nubian kingdom in northern Sudan (pp. 554, 606)

Kyushu (kee•OO•shoo) *n.* Japan's southernmost island and home to several large cities (p. 824)

L

lacrosse *n.* a team sport invented by Native Americans (p. 170)

Lake Balkhash *n.* a lake in eastern Kazakhstan (p. 496)

R63

land bridge *n.* dry land that appeared between Asia and Alaska during the Ice Age (p. 130)

land reform *n.* dividing up large plots of land to distribute land more evenly (p. 630)

landform *n.* a feature on the Earth's surface formed by physical force (p. 34)

landlocked *adj.* surrounded by land with no access to an ocean or sea (pp. 272, 280, 496, 728)

Landsat *n.* a series of information-gathering satellites that orbit above Earth (p. 10)

language *n.* human communication, either written, spoken, or signed (p. 88)

language family *n.* a group of languages that have a common origin (p. 88)

Latin America *n.* the region that includes Mexico, Central America, the Caribbean Islands, and South America (p. 198)

legislature *n.* the law-making body of a government (p. 716)

Leopold II *n.* King of Belgium, who was forced to give up his private ownership of the Congo Free State (p. 618)

Lesser Antilles *n.* the smaller Caribbean islands southeast of the Greater Antilles (p. 198)

Lewis and Clark Expedition *n.* a journey made by Lewis and Clark to explore the Louisiana Territory (p. 130)

lineage *n.* descent from an ancestor (p. 544)

literacy *n.* the ability to read and write (p. 138)

llama *n.* a South American mammal related to the camel (p. 272)

llanos (YAH•nohs) *n.* grasslands of South America's Central Plains (pp. 204, 266)

location analyst *n.* a person who studies an area to find the best location for a client (p. 16)

location *n.* an exact position using latitude and longitude, or a description of a place in relation to places around it (p. 4)

lord *n.* a powerful landowner (p. 342)

Louisiana Purchase *n.* the action by which President Thomas Jefferson bought the Louisiana Territory from France in 1803 (p. 130)

Loyalist *n.* a person who supported Great Britain during the American Revolutionary War (p. 164)

M

Maghrib (MUHG•ruhb) *n.* the Arab name for North Africa, which means "the West" (p. 568)

magma *n.* molten rock (p. 26)

Magna Carta *n.* a charter, or document, signed by England's King John in 1215 that limited the power of the monarch and guaranteed nobles basic rights (p. 400)

Magyar *n.* an ethnic Hungarian person or the Hungarian language (p. 432)

mainland *n.* the primary landmass of a continent or territory rather than its islands or peninsulas (p. 400)

Malay Peninsula *n.* the second peninsula of mainland Southeast Asia; the narrow strip of land that serves as a bridge between the mainland and islands (p. 852)

Mali *n.* a West African empire between the A.D. 1200s and the A.D. 1500s; a modern West African country (p. 580)

Mandela, Nelson *n.* South Africa's first African president (p. 624)

manifest destiny *n.* the idea that the United States should own all the land between, and even beyond, the Atlantic and Pacific Oceans (p. 130)

manor *n.* a noble's house and the villages on his land where the peasants lived (p. 342)

Mao Zedong (MOW dzuh•dahng) *n.* Communist leader of China from 1949 to 1976 (p. 764)

Maori (MOW•ree) *n.* a Polynesian people who were the first inhabitants of New Zealand (p. 900)

map *n.* a representation of a part of the Earth (p. 10)

maquiladora (mah•kee•la•DOHR•uh) *n.* factory in which materials are imported and assembled into products for export (pp. 244, 254)

Marcos, Ferdinand *n.* dictatorial president of the Philippines between 1965 and 1986 (p. 864)

market economy *n.* an economic system in which the production of goods and services is decided by supply and the demand of consumers (pp. 72, 146, 382, 424, 486, 512)

marsupial (mahr•SOO•pee•uhl) *n.* an animal with an abdominal pouch to carry its young (p. 884)

Marxist *n.* a person who supports the philosophy behind communism, including government ownership of the land and the means of production (p. 630)

Mau Mau *n.* a Kenyan independence movement that began in the 1940s (p. 612)

Maya *n.* an early civilization located in what is now the Yucatán Peninsula, Guatemala, and northern Belize (p. 212)

mechanized *adj.* equipped with machinery (pp. 304, 892)

medieval *adj.* from the Middle Ages (p. 342)

Mekong River *n.* a major river that runs south from southern China through Laos, Cambodia, and Vietnam (p. 852)

Melanesia *n.* a subregion of Oceania that includes islands northeast of Australia and south of the equator (p. 912)

Mercosur *n.* association of several South American countries to promote trade among the countries (p. 280)

Mesopotamia *n.* the land between the Tigris and Euphrates rivers (p. 666)

mestizo (mehs•TEE•zoh) *n.* person with mixed European and Native American ancestry (pp. 220, 254)

Mexican Revolution *n.* a fight for reforms in Mexico from 1910 to 1920 (p. 230)

Micronesia *n.* a subregion of Oceania that includes islands east of the Philippines and north of the equator (p. 912)

microstate *n.* an independent country that is very small in area and population (p. 858)

Middle Ages *n.* the period between the fall of the Roman Empire and the modern era, from about A.D. 476 to 1453 (p. 342)

Middle Kingdom *n.* an ancient Chinese name for China (p. 764)

migration *n.* the process of relocating to a new region (p. 66)

militarism *n.* a government's aggressive use of its armed forces (p. 840)

millet *n.* a grain raised in West Africa, often made into porridge (p. 594)

Milosevic, Slobodan (muh•LOH•suh•VICH, SLAW•baw•dahn) *n.* president of Serbia from 1989 to 1997 and of Yugoslavia from 1997 to 2000; a key figure in the ethnic conflicts in the Balkans in the 1990s (p. 440)

missionary *n.* a person sent to do religious work in another place (p. 88)

Mississippi River *n.* the largest river and chief waterway of the United States (p. 122)

Mobuto Sese Seko (moh•BOO•too SAY•say SAY•koh) *n.* dictator of the Democratic Republic of the Congo from 1965 to 1997 (p. 618)

modernize *v.* to accept new ways or ideas (pp. 562, 606)

monarch *n.* a king or queen who rules a country or territory (p. 716)

monarchy *n.* a type of government in which a ruling family headed by a king or queen holds political power (pp. 78, 334)

monotheism *n.* a belief in one god (p. 676)

monsoon *n.* a seasonal wind system that produces a wet or dry season in a region, sometimes with heavy rainfall (pp. 612, 658, 796, 852)

Moors *n.* the group of Muslims from North Africa who conquered Spain in the eighth century (p. 370)

mosaic *n.* a picture made by placing small, colored pieces of stone, tile, or glass on a surface (p. 266)

Mount Everest *n.* the world's tallest mountain, located in the Himalayas (p. 658)

Mount Fuji *n.* Japan's highest mountain peak (p. 824)

Mount Kilimanjaro *n.* the highest mountain in Africa, in northeast Tanzania (p. 536)

Mount McKinley (also called Denali) *n.* the tallest mountain of North America, located in Alaska (p. 122)

Mount Paektu (PAHK•too) *n.* the highest mountain (9,003 feet) on the Korean Peninsula (p. 796)

Muhammad (also spelled Mohammed) *n.* the founder and major prophet of Islam (p. 676)

multicultural *adj.* relating to or including many different cultures (p. 408)

multilingual *n.* able to speak many languages (p. 376)

multinational corporation *n.* a company that operates in more than one country (p. 146)

mummy *n.* a body prepared for burial according to ancient Egyptian practice (p. 554)

mural *n.* a wall painting (pp. 236, 296)

Musa, Mansa (moo•SAH, MAHN•sah) *n.* leader of the Mali empire from 1312 to about 1337 (p. 580)

Muslim *n.* a follower of Islam (pp. 504, 562, 676)

N

Nagorno-Karabakh (nuh•GAWR•noh-KARH•uh•BAHK) *n.* a province in Azerbaijan that Armenians believe should be a part of their country (p. 512)

National People's Congress (NPC) *n.* China's national law-making body (p. 784)

nationalism *n.* pride in and loyalty to one's nation (pp. 350, 362)

natural resource *n.* something that is found in nature that is necessary or useful to humans (p. 72)

neutrality *n.* a policy of not taking part in war (p. 382)

Nile River *n.* the longest river in the world, which empties into the Mediterranean Sea in northeast Egypt (pp. 536, 554)

nirvana *n.* in Buddhism, a state of wisdom that breaks the cycle of reincarnation (p. 706)

Nkrumah, Kwame (uhn•KROO•muh, KWAHM•eh) *n.* modern Ghana's first president (p. 586)

nomad *n.* member of a group who makes a living by herding animals and moving from place to place as the seasons change (pp. 96, 504)

nonviolent resistance *n.* a method of protest that draws attention to a problem without using violence (p. 698)

Nordic *adj.* relating to Scandinavia (p. 388)

North American Free Trade Agreement (NAFTA) *n.* agreement that reduced trade barriers among Mexico, Canada, and the United States (pp. 244, 254)

North Atlantic Drift *n.* warm ocean current that helps keep Europe's climate mild (p. 326)

R65

North China Plain *n.* the fertile region between the Huang He and Chang Jiang rivers (p. 764)

Northern European Plain *n.* vast area of flat or gently rolling land from France to Russia (pp. 326, 376)

Nubia *n.* in ancient times, the Nile River valley of southern Egypt and northern Sudan (p. 554)

Nunavut (NOO•nuh•VOOT) *n.* a territory created in 1999 from the eastern part of the Northwest Territories and home to many Inuit (p. 156)

O

oasis *n.* fertile or green spot in a desert created by underground water coming to the surface (p. 652)

obelisk *n.* a pillar-shaped stone monument (p. 606)

Oceania *n.* the region of the Pacific Ocean with thousands of islands (p. 912)

oceanic island n. an island rising from the deep sea floor on a volcanic foundation (p. 912)

official language *n.* a language that by law must be used for government, business, and education (p. 138)

oligarchy (AHL•ih•GAHR•kee) *n.* a government ruled by a few powerful individuals (pp. 78, 334)

Olmec *n.* an early civilization along the Gulf Coast of what is now southern Mexico (p. 212)

ombudsman *n.* an official who investigates citizens' complaints against the government (p. 388)

one-crop economy *n.* an economy that depends on a single crop for income (pp. 260, 512)

Organization of the Petroleum Exporting Countries (OPEC) *n.* a group of oil-producing countries founded in 1960 (p. 676)

Orthodox Christianity *n.* the Eastern branch of Christianity that spread to Russia in the 900s and became the state religion (p. 476)

Ottoman Empire *n.* a Muslim empire based in Turkey that lasted from the 1300s to 1922 (pp. 676, 688)

Outback *n.* the interior region of Australia (p. 884)

outcropping *n.* bedrock that rises above the soil (p. 884)

ozone hole *n.* the seasonal thinning of a protective layer of atmospheric gas high above Antarctica (p. 920)

P

Pakeha (PAH•kee•HAH) *n.* a Maori word that means "white people" (p. 900)

Palestine Liberation Organization (PLO) *n.* a political group that claims to represent all Palestinians and to be working toward gaining an independent Palestinian nation (pp. 682, 688)

Palestinian territories *n.* the West Bank and Gaza Strip combined (p. 682)

Pampas *n.* grassy plains in south-central South America (pp. 204, 280)

Pan-Africanism *n.* a movement calling for the unity of all people of African descent (p. 580)

papyrus *n.* a paperlike material made from the stems of the papyrus reed (p. 554)

Parliament *n.* the national legislature of the United Kingdom (pp. 176, 400)

parliamentary *adj.* relating to a government with the parliament form of legislature (p. 716)

patrician (puh•TRIHSH•uhn) *n.* a wealthy landowner who held a high government position in ancient Rome (p. 334)

patrilineal *adj.* descent from father to father (p. 544)

patron *n.* a wealthy or powerful person who provides money, support, and encouragement to an artist or a cause (p. 342)

peacekeeping *n.* referring to military forces whose purpose is to prevent or end war (p. 688)

Peloponnesus (PEHL•uh•puh•NEE•suhs) *n.* the peninsula in southern Greece where Sparta was located (p. 334)

penal colony *n.* a settlement that serves as a prison (p. 892)

peninsula *n.* a body of land nearly surrounded by water (pp. 326, 796)

Perkins, Charles *n.* in Australia, an aboriginal rights leader (p. 892)

permafrost *n.* ground that is frozen throughout the year (p. 460)

persecution *n.* cruel treatment on the basis of religion, race, ethnic group, nationality, political views, gender, or class (p. 66)

Persia *n.* a historic region of Southwest Asia located mostly in what is now Iran (p. 676)

perspective *n.* a technique used by artists to give the appearance of depth and distance (p. 342)

Peter the Great *n.* a czar who ruled Russia from 1682 to 1725 and tried to make Russia more European (p. 468)

pharaoh *n.* a ruler of ancient Egypt (p. 554)

pidgin *n.* a simple form of speech that is a mixture of words from two or more languages (p. 912)

Piedmont (PEED•MAHNT) *n.* a hilly, upland region between the Appalachian Mountains and the coastal plain of the South (p. 122)

place *n.* a geographical term that describes the physical and human characteristics of a location (p. 4)

plantation agriculture *n.* the use of large farms to raise cash crops (p. 736)

plateau *n.* a high area of flat land (pp. 34, 198, 652, 756)

Plateau of Iran *n.* a high plateau in central and eastern Iran, surrounded by mountains (p. 652)

Plaza of the Three Cultures *n.* plaza in Mexico City that shows parts of Aztec, Spanish, and modern influences in Mexico (p. 236)

plebeian (plih•BEE•uhn) *n.* a commoner who was allowed to vote but not to hold government office in ancient Rome (p. 334)

polders (POHL•durz) *n.* land reclaimed by draining it of water with the use of a dam and pump (p. 376)

Polynesia *n.* a subregion of Oceania that includes islands of the central and southern Pacific (pp. 900, 912)

polytheistic *adj.* believing in many gods (p. 666)

population *n.* the number of people who live in a specified area (p. 58)

population density *n.* the average number of people who live in a certain area (p. 58)

porridge *n.* a soft food, made by boiling cereal grain in milk or water (p. 612)

precipitation *n.* falling water droplets in the form of rain, snow, sleet, or hail (p. 40)

prime minister *n.* the head of the cabinet and chief executive of government in a parliamentary system; for example, the head of the Canadian government (pp. 176, 716)

privatization (PRY•vuh•tih•ZAY•shuhn) *n.* the process of selling government-owned businesses to private individuals (p. 486)

privatize *v.* to sell government-owned businesses to private individuals (p. 512)

propaganda *n.* distorted or inaccurate information that is deliberately used to influence opinion (p. 476)

protectorate *n.* a weaker country or area controlled by a stronger country (p. 562)

Protestant *n.* a member of a Christian Church founded on the principles of the Reformation (p. 342)

province *n.* a governmental division like a state (pp. 156, 784)

Prussia *n.* the most powerful German state in the Holy Roman Empire (p. 382)

public education *n.* schooling that is paid for by the government (p. 138)

pull factor *n.* a reason that attracts people to another area (p. 66)

Punic (PYOO•nik) **Wars** *n.* wars fought between Rome and Carthage during the third and second centuries B.C. (p. 568)

push factor *n.* a reason that causes people to leave an area (pp. 66, 236)

Putin (POOT•ihn), **Vladimir** *n.* president of Russia who increased executive power in the 2000s (p. 486)

Pygmy *n.* a member of a people with an average height of less than five feet (p. 618)

pyramid *n.* an ancient Egyptian structure, built over or around a tomb (p. 554)

Q

quilombos (kih•LOHM•buhs) *n.* communities created by escaped African slaves (p. 296)

quinoa (kih•NOH•uh) *n.* a kind of weed from the Andean region that produces a small grain (p. 272)

quota *n.* a target amount for production (p. 618)

R

rain forest *n.* a broadleaf tree region in a tropical climate (pp. 204, 304)

Ramadan *n.* the ninth month of the Muslim calendar, during which Muslims fast from sunrise to sunset each day (p. 728)

rate of natural increase *n.* the death rate subtracted from the birth rate (p. 58)

raw material *n.* an unprocessed natural resource that will be converted to a finished product (p. 72)

Reconquista *n.* the successful effort by the Spanish to drive the Moors out of Spain (p. 370)

Red Sea *n.* a narrow sea that divides northeast Africa from the Arabian Peninsula (p. 606)

reform *n.* action to improve social or economic conditions (p. 784)

Reformation *n.* a movement in the 1500s to change practices in the Catholic Church (p. 342)

refugee *n.* a person who flees a place to find safety (pp. 66, 440)

region *n.* an area that has one or more common characteristics that unite or connect it with other areas (p. 4)

reincarnation *n.* the rebirth of a soul in another body (p. 706)

relief *n.* the difference in the elevation of a landform from its lowest point to its highest point (p. 34)

religion *n.* an organized system of beliefs and practices, often centered on one or more gods (p. 88)

remote sensing *n.* obtaining information about a site by using an instrument that is not physically in contact with the site (p. 10)

Renaissance (REHN•ih•SAHNS) *n.* a rebirth of creativity, literature, and learning in Europe from about 1300 to 1600 (p. 342)

renewable energy sources *n.* sources of energy able to be replaced through ongoing natural processes (p. 326)

R67

representative democracy *n.* a type of government in which citizens hold political power through elected representatives (pp. 78, 146)

representative government *n.* a system of government with a legislature that is at least partly elected by the people (p. 400)

republic *n.* a government in which citizens elect representatives to rule in their name; another term for representative democracy (pp. 146, 334)

Republic of Texas *n.* constitutional government of Texas after independence from Mexico (p. 230)

reunify *v.* to bring something that has been separated back together (p. 382)

revolution *n.* the overthrow of a ruler or government; a major change in ideas (p. 230)

Rhodesia *n.* the colonial name for the countries that are today Zambia and Zimbabwe (p. 630)

rig *v.* to influence or fix the results of an election (p. 594)

Ring of Fire *n.* a geographic zone that extends along the rim of the Pacific Ocean and has numerous volcanoes and earthquakes (pp. 26, 824)

Rio de Janeiro (REE•oh day zhuh•NAYR•OH) *n.* Brazil's capital city from 1763 to 1960 (p. 290)

ritual *n.* a set of regularly followed ceremonies (p. 774)

Rivera, Diego *n.* famous muralist who painted the history of Mexico on the walls of the National Palace (p. 236)

Rocky Mountains *n.* a mountain range that extends about 3,000 miles from New Mexico to Alaska (p. 122)

Romance language *n.* any of the languages that developed from the Roman language, Latin, such as Spanish, Portuguese, French, Italian, and Romanian (p. 362)

Rose Revolution *n.* a peaceful uprising in Soviet Georgia that helped force a corrupt president to resign (p. 512)

Rosetta Stone *n.* an ancient Egyptian stone that provided a key to decipher hieroglyphics (p. 562)

Royal Canadian Mounted Police *n.* an organization of officers who were responsible for law enforcement in the Canadian West (p. 170)

rural *adj.* having to do with the countryside (pp. 58, 236)

Russification *n.* the effort to make countries occupied by the Soviet Union more Russian by replacing local languages and customs (pp. 424, 504)

S

Sadat, Anwar el- *n.* president of Egypt from 1970 to 1981 (p. 562)

safari *n.* an overland expedition for exploring, hunting, or photography (p. 612)

Sahara *n.* the world's largest desert (p. 536)

Sahel (suh•HAYL) *n.* a semiarid region just south of the Sahara (p. 536)

samba (SAM•buh) *n.* music and dance, with roots in African rhythms; the most famous form of Brazilian music worldwide (p. 296)

Sami *n.* people of northern Scandinavia who traditionally herd reindeer; also the name of their language (p. 388)

samurai (SAM•uh•RY) *n.* professional soldiers of Japan (p. 830)

San Martín, José de *n.* leader for independence in southern South America (pp. 220, 280)

sanction *n.* a penalty or pressure applied to a country for not obeying an order or law (p. 624)

Sanskrit *n.* an ancient language of India, first spoken by the Aryans (p. 698)

Santa Anna, Antonio López de (1794–1876) *n.* Mexican general, president, and leader of Mexican independence from Spain (p. 230)

savanna *n.* a vegetation region in the tropics or subtropics with a mix of grassland and scattered trees (pp. 40, 536)

scholar-official *n.* an educated person who worked in China's government (p. 764)

Scientific Revolution *n.* a major change in European thinking in the mid-1500s that led to the questioning of old theories (p. 350)

Scott, Robert *n.* a British polar explorer (p. 920)

seafaring *adj.* using the sea for transportation (p. 326)

secular *adj.* worldly or not related to religion (p. 342)

sediment *n.* pieces of rock in the form of sand, stone, or silt deposited by wind, water, or ice (p. 26)

segregation *n.* the policy of separating people of different races (p. 624)

seismic (SYZ•mihk) *adj.* having to do with earthquakes (p. 756)

Selassie, Haile (suh•LAS•ee, HY•lee) *n.* emperor of Ethiopia from 1930 to 1974 (p. 606)

self-rule *n.* independent government (p. 580)

self-sufficient *adj.* able to provide for one's own needs without outside help (p. 304)

selva *n.* Spanish name for the eastern Peruvian regions that contain rain forests (p. 272)

separatist *n.* a person who wants a region to break away from the nation it is a part of (pp. 164, 370)

serf *n.* in Europe, a person who lived and worked on the manor of a lord or vassal; in Russia, a peasant who is bound to the land (pp. 342, 468)

service industry *n.* an industry that provides services rather than objects (p. 254)

Seven Years' War (1756–1763) *n.* a conflict between Britain and France that was fought in North America, Europe, and India (p. 164)

shadow puppet *n.* traditional Indonesian art, in which puppets are handled from behind a screen lit by a bright light (p. 864)

shah *n.* title for the king of Persia or Iran (p. 676)

Shakespeare, William *n.* English playwright and poet during the late 16th and early 17th centuries (p. 408)

shamanism (SHAH•muh•NIHZ•uhm) *n.* a belief system in which a person called a shaman is believed to be able to communicate with spirits and heal the sick (p. 804)

Shi Huangdi (shee hwahng•dee) *n.* the first emperor to unify much of China (p. 764)

Shikoku (shee•KAW•koo) *n.* Japan's smallest and least populated major island (p. 824)

Shinto *n.* Japan's original religion; involves worshiping gods believed to be found in nature (p. 830)

shock therapy *n.* in Russia, term applied to Yeltsin's economic plan, which called for an abrupt shift from a command economy to a market economy (p. 486)

shogun *n.* the leader of the military government of Japan beginning in 1192 (p. 830)

Siam *n.* the former name of Thailand (p. 858)

Siberia *n.* a huge region in Asian Russia (p. 460)

Sierra Madre Occidental *n.* mountain range that runs from north to south down the western part of Mexico (p. 198)

Sierra Madre Oriental *n.* mountain range that runs from north to south down the eastern part of Mexico (p. 198)

Silk Roads *n.* major trade routes that ran from China to the rest of East Asia, through Central Asia, and into Europe (pp. 504, 774)

silt *n.* the fine, fertile soil carried by rivers and deposited on nearby lands (pp. 554, 666)

Solidarity *n.* Poland's first independent labor union, cofounded by Lech Walesa (p. 424)

solstice *n.* the time during the year when the sun reaches the farthest northern or southern point in the sky (p. 40)

Songhai (sawng•hy) *n.* an ancient West African empire, which was most powerful about A.D. 1500 (p. 580)

souk (sook) *n.* open-air market common in North Africa (p. 568)

Southern Alps *n.* a mountain chain in New Zealand (p. 884)

Southern Cone *n.* South American nations located in the cone-shaped southernmost part of South America (p. 280)

Soviet Union *n.* a Communist nation, consisting of Russia and 14 other states, that existed from 1922 to 1991 (p. 728)

spatial *adj.* referring to where a place is located and its physical relationship to other places, people, or environments (p. 4)

specialization *n.* a focus on producing a limited number of a specific product (p. 72)

squatter *n.* person who settles on unoccupied land without having legal claim to it (p. 236)

St. Lawrence Seaway *n.* a waterway made up of the St. Lawrence River, the Great Lakes, and several canals (p. 156)

stability *n.* durability; resistance to change (p. 586)

Stalin, Joseph *n.* a cruel 20th-century dictator who had millions of his political enemies executed (pp. 468, 512)

standard of living *n.* an economic measure relating to the quality and amount of goods available to a group of people and how those goods are distributed across the group (p. 432)

stateless societies *n.* groups that have no single authority, but make decisions based on compromise (p. 544)

steppe *n.* a grassy plain; in Russia, this area lies south of the taiga (p. 460)

Stone Age *n.* the earliest period of human culture, in which people used stone tools (p. 544)

strategic *adj.* relating to a plan of action designed to achieve a specific goal; in relation to the military, helpful in the fighting of a war (pp. 432, 728)

subcontinent *n.* a large landmass that is part of a continent but is considered a separate region (p. 658)

subsistence farming *n.* farming that produces just enough to feed the farmer's family, with little or nothing left over to sell (pp. 254, 586, 728, 736)

Sudan *n.* a country of northeast Africa, located south of Egypt (p. 562)

Suez Canal *n.* a waterway linking the Red Sea with the Mediterranean Sea (p. 562)

Sumer *n.* a region of city-states in Mesopotamia that was home to the first civilization (p. 666)

surveyor *n.* a person who maps and measures the land (pp. 10, 16)

sustainable *adj.* refers to using natural resources in such a way that they exist for future generations (p. 48)

Swahili *n.* an African language, commonly spoken in East Africa (p. 612)

R69

T

taiga *n.* (TY•guh) a vast forest in Russia that lies south of the tundra (p. 460)

Taino (TY•noh) *n.* Indian group in the Caribbean islands (p. 260)

Taj Mahal *n.* a beautiful tomb in India built by the Mughal ruler Shah Jahan to honor his wife, who died in 1631 (p. 698)

Taklimakan (TAH•kluh•muh•KAHN) **Desert** *n.* the second largest sand desert in the world (p. 756)

Taliban *n.* a conservative Islamic group that took control of Afghanistan in the mid-1990s (p. 728)

technology *n.* people's application of knowledge, tools, and inventions to meet their needs (p. 96)

tectonic plate *n.* a large rigid section of the Earth's crust that is in constant motion (pp. 26, 198, 536)

territory *n.* a Canadian political unit that doesn't have enough people to be a province (p. 156)

Tet *n.* the Vietnamese New Year and most important holiday in Viet Nam (p. 858)

theocracy *n.* a government run by religious leaders (p. 676)

Three Gorges Dam *n.* a barrier built on the Chang Jiang to control floods (p. 756)

Three Kingdoms *n.* the three kingdoms that had formed on the Korean Peninsula by A.D. 300 (p. 804)

three-dimensional *adj.* describing an image in which there is a sense of depth and perspective (p. 4)

Tiananmen (tyahn•ahn•mehn) **Square** *n.* the location in Beijing, China, of a huge demonstration for democratic rights that occurred in 1989 (p. 784)

Tian Shan (TYAHN SHAHN) *n.* a mountain range that runs across Kyrgyzstan and Tajikistan (p. 496)

Tigris River *n.* a river of Southwest Asia that flows from a plateau on the Anatolian peninsula to the Persian Gulf (p. 652)

Tito, Josip Broz *n.* the Communist leader of Yugoslavia from 1953 to 1980 (p. 440)

Tokugawa Ieyasu (TOH•koo•GAH•wah EE•yeh•YAH•soo) *n.* shogun who unified Japan (p. 830)

Tolstoy (TOHL•stoy), **Leo** *n.* a 19th-century writer who is one of Russia's greatest novelists (p. 476)

totalitarian (toh•tal•ih•TAIR•ee•uhn) *adj.* referring to a type of government that controls every aspect of public and private life (pp. 468, 812)

tourism *n.* industry that provides services for travelers (p. 260)

Transantarctic Mountains *n.* mountains that divide Antarctica into two parts (p. 920)

Transcaucasia (TRANS•kaw•KAY•zhyuh) *n.* a region between the Black and Caspian seas that contains the republics of Armenia, Azerbaijan (AZ•uhr•by•ZHYAHN), and Georgia (p. 496)

transcontinental railroad *n.* a railroad that crosses a continent (p. 130)

trans-Saharan *adj.* across the Sahara (p. 580)

Treaty of Tordesillas (TAWR•day•SEEL•yahs) *n.* 1494 treaty between Spain and Portugal that gave Portugal control over land that is now Brazil (p. 290)

Treaty of Waitangi (WY•TAHNG•gee) *n.* a treaty between the British and the Maori signed in 1840 (p. 900)

trust territory *n.* a territory placed under the control of a country by the United Nations (p. 912)

tsunami (tsu•NAH•mee) *n.* a series of giant, destructive ocean waves caused by underwater earthquakes, volcanoes, or landslides (pp. 658, 736, 852)

tundra *n.* a cold, dry climate and vegetation adapted to the climate in the Arctic Circle; in Russia, located in the far north (pp. 156, 460)

typhoon *n.* a tropical storm that starts in the western Pacific or Indian oceans (p. 796)

tyrant *n.* someone who takes power illegally (p. 334)

U

Uluru (oo•LOO•roo) *n.* a spectacular rock outcropping in the Australian Outback (p. 884)

Union of Soviet Socialist Republics (USSR), or **Soviet Union** *n.* a nation with a Communist government created by the Bolsheviks in 1922; dissolved in 1991 (p. 468)

unwritten constitution *n.* a framework for government that is not a single written document but includes many different laws, court decisions, and political customs (p. 414)

Ural Mountains *n.* a north-south mountain range that forms the border between European and Asian Russia (p. 460)

urban *adj.* having to do with a city (pp. 16, 58, 236)

urban planner *n.* a person who creates plans for developing and improving parts of a city (p. 16)

urbanization *n.* growth in the number of people living in urban areas; the process of city development (pp. 58, 290)

V

vassal *n.* a less wealthy noble who paid taxes to and served a lord in exchange for land (p. 342)

Vatican *n.* the official residence of the pope in Vatican City, and the political and religious center of the Roman Catholic Church (p. 362)

Vedas *n.* ancient Sanskrit writings that are the earliest sacred texts of Hinduism (p. 698)

vegetation region *n.* an area that has similar plants (p. 40)

Velvet Revolution *n.* the peaceful protest by the Czech people that led to the smooth end of communism in Czechoslovakia (p. 432)

Victoria Falls *n.* a 355-foot waterfall on the Zambezi River in South Central Africa (p. 536)

Vikings *n.* a seafaring Scandinavian people who raided northern and western Europe from the 9th to the 11th century (p. 388)

volcano *n.* an opening in the Earth's crust from which molten rock, ash, and hot gases flow or are thrown out (p. 26)

Volga River *n.* the longest river in Europe, flowing through European Russia (p. 460)

W

Walesa, Lech (wah•LEHN•suh, lehk) *n.* Polish leader who cofounded Solidarity and served as president of Poland from 1990 to 1995 (p. 424)

weather *n.* the condition of the Earth's atmosphere at a given time and place (p. 40)

weathering *n.* the gradual physical and chemical breakdown of rocks on the surface of the Earth (p. 26)

welfare state *n.* a social system in which the government provides for many of its citizens' needs (p. 388)

West Bank *n.* a region west of the Jordan River (p. 682)

wind farm *n.* a power plant that uses windmills to generate electricity (p. 414)

X

Xi Jiang (she jyahng) *n.* a river in southeast China (p. 756)

Y

Yalu River *n.* the longest river on the Korean Peninsula; forms part of the border between China and North Korea; known as the Amnok River in Korean (p. 796)

Yeltsin (YEHLT•sihn), **Boris** *n.* president who abruptly transformed the Russian economy after the Soviet Union collapsed (p. 486)

Yoruba (YAWR•uh•buh) *n.* a people of southwestern Nigeria, parts of Benin, and Togo (p. 586)

Yugoslavia *n.* a country on the Balkan Peninsula from 1918 to 1991 (p. 440)

yurt *n.* a portable, tentlike structure used by the nomads of Central Asia (p. 504)

Z

zaibatsu (ZY•baht•SOO) *n.* large family-controlled banking and industrial groups that owned many companies in Japan before World War II (p. 840)

Zaire *n.* the Democratic Republic of the Congo's name between 1971 and 1997 (p. 618)

ziggurat (ZIG•oo•rat) *n.* a temple built atop a series of increasingly smaller platforms (p. 666)

R71

Spanish Glossary

A

aborigine [aborigen] *s.* habitante originario de Australia (pág. 892)

African National Congress (ANC) [Congreso Nacional Africano] *s.* partido político sudafricano que se oponía al *apartheid* (pág. 624)

Afrikaner [afrikáner] *s.* sudafricano descendiente de europeos que habla un idioma denominado afrikáans (pág. 624)

agricultural revolution [revolución agrícola] *s.* el cambio de recolectar alimentos a cultivarlos (pág. 96)

Aksum *s.* imperio que controló la mayor parte del norte de Etiopía desde el siglo I hasta el siglo VIII a.C. (pág. 606)

alpaca *s.* mamífero suramericano de la misma familia que la llama (pág. 272)

Alpine [alpino] *adj.* relativo a la cadena montañosa de los Alpes (pág. 382)

Alps [Alpes, los] *s.* la cadena montañosa más alta de Europa, que se extiende por el sur de Europa (pág. 326)

altiplano *s.* meseta situada a gran altitud (págs. 204, 272)

Amazon River [río Amazonas] *s.* el río más extenso de Suramérica (aproximadamente 4,000 millas o 6,400 kilómetros) y el segundo más extenso del mundo (pág. 204)

amendment [enmienda] *s.* cambio realizado por escrito a la Constitución de EUA, que debe pasar un proceso de aprobación (pág. 146)

Amundsen, Roald *s.* explorador polar noruego, primero en llegar al Polo Sur (pág. 920)

Anatolian peninsula [península de Anatolia] *s.* en el sudoeste de Asia, entre el mar Negro y el mar Mediterráneo (pág. 652)

ancestor [ancestro] *s.* antepasado (pág. 544)

Andes Mountains [Cordillera de los Andes] *s.* cadena montañosa ubicada en la costa oeste de Suramérica, que se extiende a lo largo de todo el continente (pág. 204)

animism [animismo] *s.* creencia en la existencia de espíritus en los animales, plantas, fuerzas naturales y ancestros (pág. 544)

annex [anexar] *v.* añadir a un territorio existente (pág. 230)

Antarctic Peninsula [Península Antártica] *s.* península en forma de S que forma parte de la cadena de la cordillera de los Andes (pág. 920)

Antarctic Treaty [Tratado Antártico] *s.* acuerdo firmado en 1959 que acordó el uso pacífico de la Antártida (pág. 920)

anthropologist [antropólogo] *s.* científico que estudia la cultura (pág. 88)

apartheid [*apartheid*] *s.* política oficial de segregación racial establecida en Sudáfrica desde 1950 hasta 1991 (pág. 624)

Appalachian Mountains [montes Apalaches] *s.* cadena montañosa en el este de los Estados Unidos, que corre paralela al océano Atlántico (pág. 122)

Aquino, Corazon [Corazón Aquino] *s.* política que se opuso a Marcos y fue presidenta de las Filipinas entre 1986 y 1992 (pág. 864)

Arabian Peninsula [Península Arábiga] *s.* península del sudoeste de Asia, entre el mar Rojo y el golfo Pérsico (pág. 652)

Aral Sea [Mar de Aral] *s.* cuerpo de agua interior ubicado en Asia central que se está reduciendo de manera continua (pág. 496)

archipelago [archipiélago] *s.* conjunto de islas agrupadas, que a veces toma la forma de un arco curvo (págs. 198, 736, 852, 864)

artisan [artesano] *s.* trabajador hábil en fabricar con sus manos productos o arte (pág. 254)

Aryan [ario] *s.* uno de los grupos de pastores nómadas indoeuropeos, que se cree emigraron al subcontinente indio (pág. 698)

Ashanti [asante o ashanti] *s.* un pueblo de Ghana (págs. 580, 586)

Aswan High Dam [presa de Asuán] *s.* estructura para controlar las inundaciones del río Nilo y para proporcionar electricidad (pág. 554)

Atlas Mountains [montes Atlas] *s.* grupo de cadenas montañosas que se extiende desde Marruecos hasta Túnez (pág. 568)

atmosphere [atmósfera] *s.* capa de gases que rodea la Tierra (pág. 34)

atoll [atolón] *s.* arrecife de coral en forma de anillo que contiene una laguna central sin islas en su interior (pág. 912)

Aung, San Suu Kyi *s.* líder del movimiento democrático de Myanmar (pág. 858)

autonomy [autonomía] *s.* gobernabilidad propia o independencia (pág. 376)

Aztec [azteca] *s.* antigua civilización del valle de México (pág. 212)

B

Bahasa Indonesia [indonesio] *s.* lengua común de Indonesia (pág. 864)

Baltic States [Países Bálticos] *s.* Letonia, Lituania y Estonia, tres países cuyas costas dan al mar Báltico (pág. 424)

Bangalore [Bengaluru] *s.* ciudad en el sur de la región central de la India que alberga muchas industrias de tecnología de punta (pág. 716)

basin [cuenca] *s.* región drenada por un sistema fluvial (pág. 536)

Basque [vasco] *s.* grupo étnico que vive en los Pirineos occidentales y a lo largo del golfo de Vizcaya, en España y Francia; también el nombre de su idioma (pág. 370)

Bechuanaland [Bechuanalandia] *s.* nombre colonial de Botsuana (pág. 630)

Bedouin [beduino] *s.* pueblo de habla árabe y tradicionalmente nómada del sudoeste de Asia (pág. 676)

Beirut *s.* capital del Líbano y centro bancario y financiero (pág. 688)

Benelux *s.* nombre de la unión económica formada por Bélgica, los Países Bajos y Luxemburgo para asegurar la rapidez y la eficiencia en el movimiento de personas, bienes y servicios entre los tres países (pág. 376)

Berbers [bereberes] *s.* habitantes originarios del norte de África y el Sahara (pág. 568)

Berlin Wall [Muro de Berlín] *s.* muro construido por el gobierno comunista de Alemania Oriental para separar Berlín Oriental de Berlín Occidental; derribado en 1991 (pág. 382)

Bhagavad-Gita *s.* uno de los escritos sagrados del hinduismo (pág. 706)

Biafra *s.* región de Nigeria, habitada mayormente por igbos (o ibos), que declaró su independencia en 1967 pero se rindió después de perder una guerra civil con Nigeria en 1970 (pág. 586)

bilingual [bilingüe] *adj.* que usa o es capaz de usar correctamente dos idiomas (pág. 164)

bipedal [bípedo] *adj.* que camina en dos pies (pág. 544)

birth rate [tasa de natalidad] *s.* cantidad de nacimientos por año, por cada 1,000 personas (pág. 58)

blues [*blues*] *s.* tipo de música donde las letras de las canciones expresan congoja, habitualmente debido a problemas amorosos o a las penurias de la vida (pág. 138)

Boer [bóer] *s.* colono o descendiente de colonos holandeses de Sudáfrica (pág. 624)

Bogotá *s.* capital de Colombia (pág. 266)

Bolívar, Simón *s.* líder independentista del norte de Suramérica (págs. 220, 266)

Bolsheviks [bolcheviques] *s.* grupo de revolucionarios que tomó el control del gobierno de Rusia en 1917 (pág. 468)

Bosniac [bosnio] *s.* grupo étnico musulmán de Bosnia-Herzegovina (pág. 440)

bossa nova *s.* versión estilo *jazz* de la samba, un baile brasileño (pág. 296)

Botero, Fernando *s.* artista colombiano famoso por sus retratos de personas con figuras exageradas (pág. 266)

boycott [boicotear] *v.* dejar de comprar y de usar productos de ciertos orígenes, a modo de protesta (pág. 304)

brain drain [fuga de cerebros] *s.* pérdida de trabajadores calificados que emigran en búsqueda de mejores oportunidades laborales (pág. 424)

Brasília [Brasilia] *s.* capital actual de Brasil (pág. 290)

bread basket [granero] *s.* región de producción abundante de granos (pág. 424)

Briton [británico] *s.* natural de Gran Bretaña (pág. 408)

Buddha [Buda] *s.* palabra que significa "el iluminado", usada para denominar a Siddhartha Gautama, el fundador del budismo (págs. 698, 706, 774)

Buddhism [budismo] *s.* una de las principales religiones del mundo, que se inició en la India y se basa en las enseñanzas de Siddhartha Gautama (págs. 476, 698, 706, 774, 804)

Byzantine Empire [Imperio Bizantino] *s.* la mitad oriental del Imperio Romano, que sobrevivió más de mil años después de la caída de Roma (pág. 362)

C

cabinet system [sistema parlamentario] *s.* sistema de gobierno que vincula el poder ejecutivo y el legislativo a través de la elección de un primer ministro y varios ministros de la legislatura (pág. 176)

Cabral, Pedro Álvares *s.* explorador portugués que en el año 1500 reclamó como posesión portuguesa parte del territorio actual de Brasil (pág. 290)

camel [camello] *s.* mamífero con gibas en su dorso utilizado en el desierto como animal de transporte y para la producción de carne, leche y algodón (pág. 580)

Camp David Accords [Acuerdos de Camp David] *s.* tratado de paz entre Egipto e Israel, negociado por el presidente de EUA, Jimmy Carter, el presidente de Egipto, Anwar el-Sadat, y el primer ministro israelí, Menachem Begin, firmado en la residencia presidencial de Maryland, en 1979 (pág. 562)

Canadian Shield [escudo canadiense] *s.* planicie rocosa con forma de herradura que cubre gran parte del este de la región central de Canadá (pág. 156)

candomblé *s.* prácticas religiosas africanas que se mezclan con creencias católicas para crear un sistema único de creencias (pág. 296)

canopy [cubierta forestal] *s.* la capa superior de un bosque, formada por las copas de los árboles (pág. 536)

Cape Colony [Colonia del Cabo] *s.* territorio sudafricano donde se establecieron los colonos holandeses denominados bóers a fines del siglo XVII y principios del siglo XVIII (pág. 624)

R73

capoeira *s.* danza brasileña combinada con artes marciales (pág. 296)

Caracas *s.* capital de Venezuela (pág. 266)

carbon dioxide [dióxido de carbono] *s.* gas compuesto de carbono y oxígeno (pág. 48)

Caribbean Community and Common Market (CARICOM) [Comunidad del Caribe] *s.* organización comercial de varias naciones del Caribe (pág. 260)

Carthage [Cartago] *s.* antigua ciudad estado del norte de África (pág. 568)

cartographer [cartógrafo] *s.* geógrafo que traza mapas (pág. 10)

Caspian Sea [Mar Caspio] *s.* gran cuerpo de agua en el límite sur de Rusia, ubicado entre Transcaucasia y Asia Central (págs. 460, 496)

caste [casta] *s.* en la India, clase social a la cual se pertenece desde el nacimiento (pág. 698)

cataract [catarata] *s.* salto grande de agua o rápidos (pág. 554)

Caucasus Mountains [cordillera del Cáucaso] *s.* cadena montañosa que se extiende entre el mar Negro y el mar Caspio (pág. 496)

celadon [porcelana *celadón*] *s.* cerámica vidriada de Corea con un barniz de brillo azul tenue o verde (pág. 804)

Central America Free Trade Agreement (CAFTA) [Tratado de Libre Comercio entre Estados Unidos y América Central] *s.* acuerdo para fomentar el comercio entre los Estados Unidos y los países de América Central (págs. 254, 260)

Central Asia [Asia Central] *s.* región que comprende las repúblicas de Kazajistán, Kirguistán, Tayikistán, Turkmenistán y Uzbekistán (pág. 496)

cession [cesión] *s.* territorio entregado (pág. 230)

chaebol [*chaebol*] *s.* en Corea, un conglomerado familiar que se compone de empresas afines (pág. 812)

Chang Jiang [Yangsté] *s.* el río más extenso de Asia, que fluye a través del este de China (pág. 756)

Channel Tunnel [*Eurotúnel*] *s.* túnel ferroviario entre Francia e Inglaterra que corre bajo el Canal de la Mancha (pág. 414)

Chechnya [Chechenia] *s.* república mayormente islámica del sudoeste de Rusia que continúa su lucha por la independencia (pág. 486)

Cheju Island [Isla de Jeju] *s.* la isla más grande de la costa sur de la península de Corea (pág. 796)

chinampas *s.* en México, islas artificiales utilizadas para la agricultura (pág. 212)

Christianity [cristianismo] *s.* religión monoteísta basada en la vida y las enseñanzas de Jesús de Nazaret y en las escrituras de la Biblia cristiana (pág. 682)

Church of England [Iglesia de Inglaterra o Anglicana] *s.* iglesia oficial de Inglaterra, cuya cabeza es el Arzobispo de Canterbury (pág. 408)

citizen [ciudadano] *s.* persona que debe lealtad a un país y recibe su protección (pág. 78)

city-state [ciudad estado] *s.* unidad política compuesta de una ciudad y las tierras a su alrededor (págs. 334, 666)

Civil War [Guerra Civil o de Secesión] (1861–1865) *s.* conflicto entre el Norte y el Sur de EUA sobre las cuestiones de los derechos de los estados y la esclavitud (pág. 130)

civilization [civilización] *s.* una forma de cultura avanzada que se desarrolló en las ciudades (pág. 666)

clan *s.* grupo de familias que tienen en común los mismos ancestros (pág. 830)

climate [clima] *s.* condiciones habituales del tiempo de una región durante un largo período de tiempo (pág. 40)

climatologist [climatólogo] *s.* geógrafo que estudia los climas (pág. 16)

coalition [coalición] *s.* alianza o asociación, habitualmente transitoria (pág. 362)

Cold War [Guerra Fría] *s.* conflicto entre los Estados Unidos y la Unión Soviética tras la Segunda Guerra Mundial, que nunca se desarrolló en forma de combate armado explícito (pág. 468)

colonia *s.* un vecindario de México (pág. 236)

colonialism [colonialismo] *s.* control de una potencia sobre un área o sobre personas que dependen de ella (pág. 858)

colony [colonia] *s.* grupo de personas que se asientan en una tierra lejana pero están gobernadas por su madre patria, como los Estados Unidos; territorio de ultramar gobernado por una nación, como México desde el siglo XVI hasta 1821 (págs. 130, 220)

Columbian Exchange [intercambio colombino] *s.* traslado de plantas y animales entre América Latina y Europa, después del viaje de Colón a América en 1492 d.C. (pág. 220)

Columbus, Christopher [Cristobal Colón] *s.* navegante y explorador italiano que navegaba al servicio de España y exploró el Caribe y las costas de América Central y de Suramérica (pág. 370)

command economy [economía planificada] *s.* sistema económico en donde la producción de bienes y servicios es decidida por un gobierno central (págs. 72, 424, 486, 512, 784)

commonwealth [mancomunidad] *s.* unidad política de gobierno autónomo que está asociada a otro país (pág. 260)

Commonwealth of Nations [Mancomunidad Británica de Naciones] *s.* asociación compuesta por el Reino Unido y muchas antiguas colonias británicas (págs. 164, 400, 624)

communal [comunal] *adj.* relativo a la posesión compartida de los bienes (pág. 774)

communism [comunismo] *s.* tipo de gobierno en donde el Partido Comunista posee la totalidad del poder político y controla la economía (págs. 78, 260, 424, 468, 764)

compulsory [obligatorio] *adj.* exigido (pág. 362)

Confucianism [confucianismo] *s.* sistema de creencias basado en las enseñanzas de Confucio, un sabio chino (págs. 764, 774, 804)

conglomerate [conglomerado] *s.* organización compuesta por varias empresas de diferentes industrias (pág. 812)

Congo River [río Congo] *s.* el segundo río más extenso de África (pág. 618)

conquistador *s.* explorador y guerrero español que llegó al continente americano en el siglo XVI (pág. 220)

Constitution [Constitución] *s.* documento que es la base del gobierno de los EUA (pág. 130)

constitution [constitución] *s.* conjunto de leyes y principios que definen la naturaleza de un gobierno (págs. 176, 230)

Constitution of 1917 [Constitución de 1917] *s.* constitución mexicana redactada durante la revolución que aún está en vigencia (pág. 230)

constitutional monarchy [monarquía constitucional] *s.* gobierno donde los poderes del rey o la reina están limitados por la constitución (págs. 176, 414, 736, 858, 892)

continent [continente] *s.* una de las siete enormes extensiones de tierra de la superficie de la Tierra (págs. 26, 122)

Continental Divide [divisoria continental] *s.* formación lineal a lo largo de las montañas Rocosas que divide las aguas que fluyen hacia el oeste de las que fluyen hacia el este (pág. 122)

continental shelf [plataforma continental] *s.* tierra sumergida al borde de un continente (pág. 34)

Cook, James *s.* explorador británico que reclamó tierras en Australia (pág. 892)

coral island [isla de coral] *s.* tipo de isla oceánica que es un arrecife de coral (pág. 912)

cottage industry [industria artesanal] *s.* pequeña empresa donde los trabajadores fabrican ítems en sus hogares (págs. 716, 728)

coup [golpe de Estado] *s.* toma del poder por parte de los militares (pág. 586)

cuíca *s.* zambomba utilizada en la samba, una danza brasileña (pág. 296)

cultural blending [mezcla cultural] *s.* algo nuevo que se crea al combinar los elementos de dos o más culturas (pág. 138)

cultural hearth [núcleo cultural] *s.* la zona central de una cultura importante; el área en la cual una cultura se originó y se expandió a otras áreas (págs. 96, 212)

culture [cultura] *s.* actitudes, conocimientos y comportamientos compartidos por un grupo de personas (págs. 66, 88, 138, 212, 544)

cuneiform [cuneiforme] *s.* el primer sistema de escritura conocido, que usaba símbolos en forma de cuña y fue desarrollado en Sumeria (pág. 666)

cyclone [ciclón] *s.* tormenta violenta de vientos feroces que rotan en una pauta circular como un huracán (pág. 658)

Cyrillic alphabet [alfabeto cirílico] *s.* el sistema oficial de escritura de Rusia (pág. 476)

czar [zar] *s.* título de los soberanos de Rusia desde mediados del siglo XVI hasta comienzos del siglo XX (pág. 468)

Czechoslovakia [Checoslovaquia] *s.* antiguo país de Europa del Este que existió desde 1918 hasta 1993, cuando se dividió en la República Checa y Eslovaquia (pág. 432)

D

daimyo [daimio] *s.* en Japón, samuráis poderosos que se convirtieron en caciques guerreros (pág. 830)

Dao [Tao] *s.* de acuerdo al sistema de creencias del taoísmo, una fuerza que guía todo el universo (pág. 774)

Daoism [taoísmo] *s.* sistema de creencias chino basado en la idea del orden natural del mundo (pág. 774)

database [base de datos] *s.* recopilación de información que puede analizarse (pág. 10)

Day of the Dead [Día de Muertos] *s.* día feriado para recordar a los seres queridos que han fallecido (pág. 236)

death rate [tasa de mortalidad] *s.* cantidad de muertes por año, por cada 1,000 personas (pág. 58)

debris [escombros] *s.* restos dispersos de algo roto o destruido (pág. 10)

Deccan Plateau [meseta de Decán] *s.* área elevada a gran altura en el centro del subcontinente indio (pág. 658)

deforestation [deforestación] *s.* tala y eliminación de árboles (pág. 304)

degraded [degradado] *adj.* de calidad menor a la que existía previamente (pág. 48)

delta *s.* depósito de tierra de forma triangular en la boca de un río (pág. 554)

demilitarized zone [zona desmilitarizada] *s.* zona de separación entre Corea del Norte y Corea del Sur (pág. 804)

democracy [democracia] *s.* gobierno en el cual los ciudadanos toman decisiones políticas, ya sea de manera directa o a través de los representantes que han elegido (pág. 334)

R75

demographer [demógrafo] *s.* geógrafo que estudia las características de las poblaciones humanas (pág. 58)

demonstration [manifestación] *s.* reunión pública para protestar por una política o una decisión (pág. 784)

dependency [dependencia] *s.* lugar gobernado por un país del cual oficialmente no forma parte (pág. 260)

deport [deportar] *v.* expulsar de un país (pág. 424)

desert [desierto] *s.* región con plantas especialmente adaptadas a condiciones secas (pág. 40)

desertification [desertificación] *s.* transformación de tierra fértil en tierra desértica; proceso por el cual la tierra de cultivo pierde su productividad debido a que la tierra se degrada (págs. 48, 536)

devolution [traspaso] *s.* proceso por el cual se transfiere algo del poder del gobierno nacional al gobierno regional (pág. 414)

dictator [dictador] *s.* persona que tiene el control absoluto sobre el gobierno de un país (págs. 254, 260, 266, 812)

dictatorship [dictadura] *s.* estado o gobierno regido por un líder con poder absoluto (págs. 78, 568)

Diet [Dieta] *s.* organismo legislativo del Japón, que se compone de una Cámara de Representantes y una Cámara de Consejeros (pág. 840)

diffusion [difusión] *s.* propagación de ideas, inventos y pautas de conducta de un grupo a otro (pág. 96)

discrimination [discriminación] *s.* acciones que pueden herir a un individuo o a un grupo (pág. 66)

diversity [diversidad] *s.* tener muchas maneras diferentes de pensar o de hacer algo, o una variedad de personas (pág. 66)

Dom Pedro I [Pedro I] *s.* primer emperador portugués de Brasil; proclamó Brasil independiente de Portugal en 1822 (pág. 290)

Dom Pedro II [Pedro II] *s.* segundo emperador de Brasil; durante su reinado, en 1888, se abolió la esclavitud en Brasil (pág. 290)

domesticate [domesticar] *v.* adaptar o entrenar plantas o animales para uso de los seres humanos (pág. 544)

domestication [domesticación] *s.* cría y cuidado de una planta o un animal para ser de utilidad a los seres humanos (pág. 96)

dominion [dominio] *s.* en el Imperio Británico, nación a la que se le permite administrar sus asuntos internos (pág. 164)

draft animal [animal de carga] *s.* animal usado para jalar una carga pesada (pág. 884)

drainage basin [cuenca de drenaje] *s.* área drenada por un río principal (pág. 34)

Dravidian [lenguas drávidas] *s.* grupo de lenguas que se hablan en la India desde tiempos muy remotos; en la actualidad, se las habla principalmente en el sur de la India (pág. 698)

drought [sequía] *s.* largo período de escasa precipitación o carencia total de lluvia (pág. 594)

duchy [ducado] *s.* territorio regido por un duque o una duquesa (pág. 376)

Dust Bowl [Cuenca de Polvo] *s.* región de las Grandes Planicies de los EUA que padeció sequías y tormentas de polvo sofocantes durante la década de 1930 (pág. 130)

dynastic cycle [ciclo dinástico] *s.* en China, el patrón de caída y ascenso de las dinastías (pág. 764)

dynasty [dinastía] *s.* familia o grupo que gobierna durante varias generaciones, como en China (págs. 554, 606, 764)

E

earthquake [terremoto] *s.* movimiento repentino de la corteza terrestre seguido de una serie de sacudidas (pág. 26)

eclectic [ecléctico] *adj.* compuesto a partir de partes de diversos orígenes (pág. 432)

economic system [sistema económico] *s.* manera en que las personas usan los recursos para fabricar e intercambiar bienes (pág. 72)

economy [economía] *s.* sistema de producción e intercambio de bienes y servicios entre un grupo de personas (págs. 72, 304)

ecotourism [ecoturismo] *s.* viajes que algunas personas realizan a medio ambientes únicos, en los que se preocupan por preservarlos en su estado natural (págs. 254, 736)

edible [comestible] *adj.* apto para ser comido (pág. 272)

emissions [emisiones] *s.* sustancias liberadas en el aire (pág. 48)

emperor [emperador] *s.* líder de un imperio, que es un grupo de territorios o de culturas diferentes regido por una autoridad suprema (págs. 764, 830)

empire [imperio] *s.* sistema político en donde las personas o las tierras son controladas por un soberano único (págs. 212, 220)

Enlightenment [Ilustración] *s.* movimiento filosófico de los siglos XVII y XVIII que se caracterizó por el uso de la razón y los métodos científicos (pág. 350)

environment [medio ambiente] *s.* el entorno físico de una ubicación (pág. 4)

equinox [equinoccio] *s.* una de las dos ocasiones en el año en que los rayos del sol caen directamente sobre el ecuador y los días y las noches tienen la misma duración en todo el mundo (pág. 40)

Eritrea *s.* antigua región de Etiopía, que se convirtió en un país independiente en 1993 (pág. 606)

erosion [erosión] *s.* desgaste y movimiento de materiales desgastados por el agua, el viento o el hielo (pág. 26)

escarpment [escarpa] *s.* pendiente empinada que separa dos áreas llanas de diferente altura (pág. 536)

estancia *s.* hacienda o finca agrícola de gran tamaño de la Argentina (pág. 280)

ethanol [etanol] *s.* líquido que se obtiene mediante un proceso químico que convierte la caña de azúcar en un tipo de alcohol que se puede usar como combustible (pág. 304)

ethnic cleansing [limpieza étnica] *s.* eliminar un grupo étnico o religioso de un área a través de la fuerza o del asesinato en masa de los miembros del grupo (pág. 440)

ethnic group [grupo étnico] *s.* grupo de personas que comparten un idioma, costumbres y un patrimonio cultural común (págs. 88, 476, 504)

Euphrates River [río Éufrates] *s.* río del sudoeste de Asia que fluye desde una planicie de la península de Anatolia hasta el golfo Pérsico (pág. 652)

Eurasia *s.* término usado para designar a un único continente compuesto por Europa y Asia (págs. 460, 496)

European Union (EU) [Unión Europea (UE)] *s.* organización de naciones europeas cuyos miembros cooperan en asuntos económicos, sociales y políticos (págs. 350, 362, 688)

Everglades [los *Everglades*] *s.* área inmensa de terrenos pantanosos al sur de Florida (pág. 122)

export [exportación] *s.* producto o recurso vendido a otro país (págs. 72, 304)

export [exportar] *v.* enviar algo de un país a otro (pág. 146)

extended family [familia ampliada] *s.* unidad familiar que agrupa a otros familiares además de los padres y los hijos, incluyendo a los abuelos y a otros parientes cercanos (pág. 688)

F

famine [hambruna] *s.* escasez severa de alimentos que causa hambre generalizada (págs. 594, 658)

fascism [fascismo] *s.* filosofía política que fomenta la lealtad ciega hacia el estado y un fuerte gobierno central controlado por un dictador poderoso (pág. 362)

fast [ayunar] *v.* no comer, habitualmente por motivos religiosos (pág. 728)

favelas [*favelas*] *s.* término brasileño para designar a los vecindarios pobres que rodean las ciudades (pág. 290)

federal republic [república federal] *s.* forma de gobierno donde el poder está dividido entre un gobierno nacional y los gobiernos estatales (pág. 266)

federal system [sistema federal] *s.* sistema de gobierno donde los poderes están divididos entre el Estado nacional y los gobiernos provinciales o estatales (págs. 146, 176)

federation [federación] *s.* colonias, estados o países que se han unido (pág. 594)

Fertile Crescent [Creciente Fértil] *s.* región que se extiende desde el golfo Pérsico hacia el noroeste, hasta los ríos Tigris y Éufrates, y hacia el oeste, hasta el mar Mediterráneo (pág. 666)

feudalism [feudalismo] *s.* sistema político en el cual los señores feudales daban tierras a sus vasallos a cambio de servicios (pág. 342)

fiesta *s.* día feriado que se celebra con desfiles, juegos y comida (pág. 236)

Filipino [filipino] *s.* habitante de las Filipinas (pág. 864)

First Nations [Primeras Naciones] *s.* pueblos indígenas de Canadá organizados en grupos culturales (pág. 164)

foreign investment [inversión extranjera] *s.* dinero aportado a una empresa por personas de otro país (pág. 512)

fossil [fósil] *s.* restos de vida primitiva preservados bajo tierra (pág. 544)

fossil fuels [combustibles fósiles] *s.* fuentes de energía a partir de restos antiguos de animales y plantas; se incluyen los combustibles como el carbón, el petróleo y el gas natural (págs. 48, 326)

Fox, Vicente *s.* presidente mexicano perteneciente al Partido de Acción Nacional, elegido en 2000 (pág. 230)

free enterprise [libre empresa] *s.* sistema económico en el cual las empresas son libres de operar, con poca interferencia del gobierno (págs. 146, 176, 784)

French Revolution [Revolución Francesa] *s.* conflicto ocurrido en Francia, entre 1789 y 1799, que puso fin a la monarquía y produjo cambios en la manera de gobernar Francia (pág. 350)

G

Gaelic [gaélico] *s.* cualquiera de los dialectos de la familia de lenguas celtas habladas en Irlanda o Escocia (pág. 408)

Gandhi, Mohandas *s.* hindú del siglo XX que ayudó a obtener la independencia de su país usando la resistencia pacífica frente al dominio colonial (pág. 698)

Ganges River [río Ganges] *s.* río de Asia del Sur que fluye al sudeste desde los Himalayas hasta el Golfo de Bengala (pág. 658)

R77

García Márquez, Gabriel *s.* autor colombiano ganador del Premio Nobel (pág. 266)

gaucho *s.* vaquero argentino (pág. 280)

Gautama, Siddhartha *s.* fundador del budismo, también conocido como Buda, o "el iluminado" (pág. 774)

Gaza Strip [Franja de Gaza] *s.* territorio estrecho que se extiende a lo largo del Mar Mediterráneo (pág. 682)

Geographic Information Systems (GIS) [Sistemas de Información Geográfica] *s.* tecnología de mapeo basada en las computadoras o Internet (págs. 10, 16)

geographic South Pole [Polo Sur geográfico] *s.* punto de encuentro de todas las líneas de longitud (pág. 920)

geography [geografía] *s.* estudio de las personas, los lugares y los medios ambientes (pág. 4)

geomorphology [geomorfología] *s.* estudio de cómo cambia la forma de la Tierra (pág. 16)

geothermal energy [energía geotérmica] *s.* energía producida por el calor interno de la Tierra (pág. 900)

geyser [géiser] *s.* manantial caliente que dispara chorros de vapor y agua hirviendo al aire (págs. 884, 900)

Ghana *s.* reino de África Occidental entre el siglo VIII y XI d.C. ubicado en lo que hoy es Mauritania y Malí; país moderno de África Occidental (pág. 580)

glacier [glaciar] *s.* gran masa de hielo de movimiento lento (pág. 26)

global economy [economía global] *s.* economía en la cual se compra y vende atravesando las fronteras nacionales (págs. 146, 244)

Global Positioning System (GPS) [Sistema de Posicionamiento Global] *s.* sistema que usa una red de satélites que orbitan la Tierra para dar una ubicación exacta (pág. 10)

global warming [calentamiento global] *s.* aumento de la temperatura promedio de la atmósfera terrestre (págs. 48, 304)

globe [globo] *s.* modelo de la Tierra con forma esférica (pág. 10)

glyph [glifo] *s.* símbolo tallado o grabado que representa una sílaba o una palabra (pág. 212)

Gobi Desert [Desierto de Gobi] *s.* desierto elevado en China y Mongolia (pág. 756)

Golan Heights [Altos de Golán] *s.* región que formaba parte del suroeste de Siria y que Israel ha ocupado desde la guerra de 1967 (pág. 688)

golden age [edad dorada] *s.* período durante el cual una sociedad alcanza prosperidad y realiza logros culturales (pág. 698)

Good Friday Agreement [Acuerdo de Viernes Santo] *s.* acuerdo entre los unionistas y nacionalistas de Irlanda del Norte que estableció un nuevo gobierno (pág. 414)

Gorbachev, Mikhail *s.* líder soviético de 1985 a 1991 que incrementó las libertades del pueblo ruso (pág. 468)

government [gobierno] *s.* organización establecida para crear y hacer cumplir las reglas de una sociedad (pág. 78)

Grand Canyon [Gran Cañón] *s.* barranco profundo que atraviesa el norte de Arizona excavado por el río Colorado (pág. 122)

Great Barrier Reef [Gran Arrecife de Coral] *s.* cadena de arrecifes de coral de 1,250 millas, cerca de la costa de Australia (pág. 884)

Great Depression [Gran Depresión] *s.* período de severo declive económico, desde 1929 hasta comienzos de la década de 1940 (pág. 130)

Great Dividing Range [Gran Cordillera Divisoria] *s.* cordillera de montañas bajas en Australia (pág. 884)

Great Lakes [Grandes Lagos] *s.* cinco lagos que forman el mayor grupo de lagos de agua dulce del mundo (págs. 122, 156)

Great Plains [Grandes Llanuras] *s.* grandes prados del centro de América del Norte (pág. 122)

Great Wall [Gran Muralla] *s.* muralla enorme y extensa construida en el transcurso de varios siglos para mantener a los nómadas fuera de China (pág. 764)

Great Zimbabwe [Gran Zimbabue] *s.* imperio erigido por el pueblo shona desde el siglo I hasta el siglo VIII d.C. (pág. 630)

Greater Antilles [Antillas Mayores] *s.* las islas más grandes del norte del Caribe, entre las que se incluyen Cuba, Jamaica, La Española (la cual incluye Haití y la República Dominicana) y Puerto Rico (pág. 198)

Green Revolution [Revolución Verde] *s.* uso de semillas especiales, irrigación, fertilizantes y pesticidas para lograr un elevado rendimiento de los cultivos y una gran producción de alimentos (págs. 716, 728)

greenhouse effect [efecto invernadero] *s.* retención del calor del sol producido por los gases de la atmósfera terrestre (pág. 48)

greenhouse gas [gas invernadero] *s.* cualquier gas de la atmósfera que contribuye al efecto invernadero (pág. 48)

grid [cuadrícula] *s.* red de líneas horizontales y verticales que crean cuadrados o rectángulos (pág. 666)

griot *s.* narrador de África Occidental (pág. 544)

Gross Domestic Product (GDP) [Producto Bruto Interno (PBI)] *s.* valor total de todos los bienes y servicios producidos por un país en un año (págs. 72, 280)

ground water [agua subterránea] *s.* agua que se encuentra debajo de la superficie terrestre (pág. 34)

guerrilla [guerrillero] *s.* miembro de un ejército irregular que opera en unidades militares pequeña (pág. 630)

Gulf of Mexico [Golfo de México] *s.* ensenada del océano Atlántico que se ubica al sur de los Estados Unidos (pág. 122)

H

habitable lands [tierras habitables] *s.* tierras apropiadas para ser habitadas por los seres humanos (pág. 58)

Hammurabi's Code [Código de Hammurabi] *s.* uno de los primeros códigos legales del mundo, compilado por el gobernante Hammurabi (pág. 666)

hanbok [*hanbok*] *s.* traje tradicional coreano (pág. 804)

Harappan civilization [civilización Harappa] *s.* civilización antigua que se desarrolló a lo largo del río Indo (pág. 666)

Hausa [hausa] *s.* un pueblo del norte de Nigeria y sur de Níger (pág. 586)

Heian Period [Período Heian] *s.* era de la historia japonesa desde 794 d.C. hasta 1185 d.C.; las artes y la escritura florecieron durante esta época (pág. 830)

Hidalgo, Miguel ("Padre Hidalgo") *s.* padre de la independencia mexicana (pág. 220)

hieroglyphics [jeroglíficos] *s.* sistema de escritura egipcio en el cual los símbolos pictográficos representan significados o sonidos (pág. 554)

highlands [tierras altas] *s.* regiones montañosas o accidentadas de un país (págs. 198, 204)

Himalayas [Himalayas] *s.* las montañas más altas del mundo, las cuales se extienden a través del norte de India, separando la India de China y del resto de Asia (págs. 658, 756)

Hinduism [Hinduismo] *s.* nombre actual para la religión principal de la India, la cual se desarrolló a partir del bramanismo (págs. 698, 706)

Hitler, Adolf *s.* jefe de estado alemán desde 1933 a 1945 (pág. 382)

Hokkaido *s.* isla más septentrional de Japón y segunda en tamaño (pág. 824)

Holocaust [Holocausto] *s.* asesinato sistemático de judíos y otras minorías por parte de los nazis durante la Segunda Guerra Mundial (págs. 350, 382, 682)

hominid [homínido] *s.* primate humano; ser humano primitivo (pág. 544)

Honshu *s.* la isla más grande y más poblada de Japón (pág. 824)

Horn of Africa [Cuerno de África] *s.* extensión de tierra con forma de cuerno en la costa este de África (pág. 606)

Huang He *s.* río que fluye desde las montañas Kunlun hasta el Mar Amarillo (pág. 756)

Hudson Bay [Bahía de Hudson] *s.* gran mar en el interior al norte de Ontario, provincia canadiense (pág. 156)

hunter-gatherer [cazador-recolector] *s.* persona que caza animales y recolecta plantas para obtener alimento (págs. 618, 892)

hydroelectric [hidroeléctrico] *adj.* relativo a la electricidad generada por motores que funcionan con agua (pág. 156)

hydroelectricity [hidroelectricidad] *s.* energía eléctrica generada por el agua (pág. 304)

hydroelectric power [energía hidroeléctrica] *s.* electricidad generada por motores que funcionan gracias a la fuerza del agua (págs. 326, 756, 900)

hydrologic cycle [ciclo hidrológico] *s.* circulación del agua entre la Tierra, los océanos y la atmósfera (pág. 34)

I

ice shelf [barrera de hielo] *s.* inmensa capa plana de hielo que se extiende sobre un cuerpo de agua, pero permanece unida a la tierra (pág. 920)

iceberg [glaciar] *s.* gran pedazo de hielo flotante (pág. 920)

icecap [casquete glaciar] *s.* glaciar que cubre un área extensa (pág. 920)

icon [ícono] *s.* imagen religiosa usada por los cristianos ortodoxos en su culto (pág. 476)

Igbo [igbo] *s.* pueblo del sureste de Nigeria (pág. 586)

igloo [iglú] *s.* vivienda hecha de nieve (pág. 170)

immigrant [inmigrante] *s.* persona que deja un área para habitar en otra (págs. 66, 130, 296, 408)

immigration [inmigración] *s.* proceso de ir a vivir a otro país (pág. 244)

imperialism [imperialismo] *s.* la práctica de un país de controlar el gobierno y la economía de otro país o territorio (págs. 350, 400, 580)

import [importación] *s.* producto o recurso que llega a un país (pág. 72)

import [importar] *v.* llevar algo de un país a otro (pág. 146)

Inca [inca] *s.* civilización antigua que se desarrolló en la Cordillera de los Andes, en Perú (pág. 212)

indigenous [indígena] *s.* nativo de una región (pág. 272)

Indochina [Indochina] *s.* nombre de una antigua colonia francesa del sureste de Asia constituida por Camboya, Laos y Vietnam (pág. 858)

Indochinese Peninsula [península de Indochina] *s.* una de las dos penínsulas en las que se ubica el sureste de Asia continental; la península está ubicada al sur de China (pág. 852)

R79

Indus River [río Indo] *s.* río de Asia del sur que fluye desde los Himalayas hasta el Mar Arábigo (pág. 658)

Industrial Revolution [Revolución Industrial] *s.* cambio que comenzó en Gran Bretaña en la década de 1760 de producir bienes a mano a hacerlo mediante máquinas (págs. 350, 400)

inflation [inflación] *s.* aumento de los precios o caída del poder adquisitivo del dinero (pág. 594)

infrastructure [infraestructura] *s.* servicios básicos que necesita una comunidad, tales como carreteras, ferrocarriles, agua y líneas eléctricas, escuelas y oficinas de correos (pág. 594)

innovation [innovación] *s.* algo nuevo que es presentado por primera vez (pág. 96)

Interstate Highway System [Sistema de Carreteras Interestatales] *s.* red de más de 45,000 millas de carreteras que une a todas las ciudades principales de EUA (pág. 146)

Inuit [inuit] *s.* pueblos nativos que habitan la región ártica de América del Norte (pág. 164)

Inuktitut [inuktitut] *s.* idioma de los inuit (pág. 170)

Islam [islam] *s.* religión monoteísta basada en las enseñanzas de Mahoma y los escritos del Corán, el libro sagrado de los musulmanes (págs. 476, 504, 562, 676)

isolate [aislar] *v.* separar o apartar de un grupo (pág. 334)

Istanbul [Estambul] *s.* la ciudad más grande de Turquía, anteriormente llamada Constantinopla (capital del Imperio Bizantino, o Imperio Romano Oriental) y originariamente llamada Bizancio en la antigüedad (pág. 688)

isthmus [istmo] *s.* franja de tierra que conecta dos territorios (pág. 198)

ivory [marfil] *s.* sustancia dura y blanca de la que están hechos los colmillos de los elefantes (pág. 618)

J

jaguar *s.* felino de gran tamaño que habita principalmente en América Central y del Sur (pág. 212)

jazz [*jazz*] *s.* tipo de música que se desarrolló a partir de una mezcla de ritmos africanos, música estadounidense de orquesta y estilos musicales afroamericanos y europeos (pág. 138)

jellaba [chilaba] *s.* vestimenta tradicional bereber; túnica de mangas largas (pág. 568)

Jerusalem [Jerusalén] *s.* capital actual de Israel y ciudad antigua considerada sagrada por el judaísmo, el cristianismo y el islam (pág. 682)

joropo *s.* baile folklórico nacional de Venezuela (pág. 266)

Juárez, Benito (1806–1872) *s.* indio que se convirtió en presidente y héroe nacional mexicano (pág. 230)

Judaism [judaísmo] *s.* religión monoteísta de los judíos, basada principalmente en los escritos de la Biblia hebrea (págs. 476, 682)

K

Kalahari *s.* desierto del sur de África (pág. 630)

Kanto Plain [Planicie de Kanto] *s.* la llanura de Japón de mayor extensión, que se extiende desde los Alpes japoneses al este, hasta el Pacífico (pág. 824)

Kara Kum *s.* desierto inmenso que cubre la mayor parte de Turkmenistán (pág. 496)

karma *s.* en el hinduismo, las consecuencias de las acciones de una persona en esta vida, las cuales determinan su destino en la próxima vida (pág. 706)

Kashmir [Cachemira] *s.* región en la parte noroeste del subcontinente indio; India y Pakistán se disputan el control de Cachemira (pág. 716)

Kente cloth [tela *kente*] *s.* tela tejida de colores vivos, hecha por los pueblos ashanti y ewe (pág. 586)

Kenyatta, Jomo *s.* primer presidente después de la independencia de Kenia (pág. 612)

Khmer Empire [Imperio Khmer] *s.* imperio que comenzó en el siglo VI y había ganado el control de gran parte del sureste de Asia continental hacia el siglo IX (pág. 858)

kibbutz [kibutz] *s.* tipo de colonia agrícola de Israel en el que la comunidad comparte la riqueza y todos los bienes (pág. 682)

Kim Jong Il *s.* líder comunista de Corea del Norte (pág. 812)

kimchi [kimchi] *s.* plato coreano hecho de vegetales encurtidos (pág. 804)

King, William Lyon Mackenzie *s.* primer ministro canadiense que llevó a su país a la independencia (pág. 164)

knight [caballero] *s.* vasallo entrenado en el combate que peleaba en nombre de un señor feudal (pág. 342)

Koizumi, Junichiro *s.* primer ministro japonés de 2001 a 2006 (pág. 840)

Kongo [Congo] *s.* pueblo que habitaba a lo largo del río Congo; reino de habla bantú de la región del río Congo surgido en el siglo XIV (pág. 618)

Korean Peninsula [península de Corea] *s.* masa de tierra bordeada por el Mar Amarillo al oeste, el Mar del Este (o Mar del Japón) al este, el Estrecho de Corea al sur y China y Rusia al norte (pág. 796)

Korean War [Guerra de Corea] (1950–1953) *s.* conflicto entre Corea del Norte y la Unión Soviética por un lado, y Corea del Sur, los Estados Unidos y las Naciones Unidas por el otro (pág. 804)

Korean Workers' Party (KWP) [Partido Coreano de los Trabajadores] *s.* grupo político comunista que controla el gobierno de Corea del Norte (pág. 812)

Kosovo *s.* provincia autónoma dentro de Serbia (pág. 440)

Kurd [kurdo] *s.* miembro de un grupo étnico que no tiene su propio país pero cuya tierra natal se ubica en partes de Turquía, Irak e Irán (pág. 676)

Kush *s.* antiguo reino de Nubia ubicado en el norte de Sudán (págs. 554, 606)

Kyushu *s.* isla del extremo sur de Japón donde se hallan muchas ciudades grandes (pág. 824)

L

lacrosse [*lacrosse*] *s.* deporte de equipo inventado por los americanos nativos (pág. 170)

Lake Balkhash [lago Balkhash] *s.* lago del este de Kazakstán (pág. 496)

land bridge [puente terrestre] *s.* tierra firme que surgió entre Asia y Alaska durante la Edad de Hielo (pág. 130)

land reform [reforma agraria] *s.* acción de dividir grandes porciones de terreno para distribuir la tierra más equitativamente (pág. 630)

landform [accidente geográfico] *s.* rasgo de la superficie terrestre formado por la fuerza física (pág. 34)

landlocked [mediterráneo] *adj.* rodeado por tierra sin acceso a un océano o mar (págs. 272, 280, 496, 728)

Landsat [*Landsat*] *s.* serie de satélites que reúnen información y orbitan sobre la Tierra (pág. 10)

language [lenguaje] *s.* comunicación humana ya sea escrita, hablada o por señas (pág. 88)

language family [familia de lenguas] *s.* conjuntos de lenguas que derivan de una misma lengua originaria (pág. 88)

Latin America [Latinoamérica] *s.* región que incluye México, América Central, las islas del Caribe y Suramérica (pág. 198)

legislature [legislatura] *s.* organismo legislativo de un gobierno (pág. 716)

Leopold II [Leopoldo II] *s.* rey de Bélgica, obligado a entregar su propiedad privada del Estado Libre del Congo (pág. 618)

Lesser Antilles [Antillas Menores] *s.* las islas del Caribe más pequeñas ubicadas al sureste de las Antillas Mayores (pág. 198)

Lewis and Clark Expedition [Expedición de Lewis y Clark] *s.* viaje realizado por Lewis y Clark para explorar el Territorio de Luisiana (pág. 130)

lineage [linaje] *s.* descendencia de un ancestro (pág. 544)

literacy [alfabetización] *s.* la capacidad de leer y escribir (pág. 138)

llama *s.* mamífero suramericano emparentado con el camello (pág. 272)

llanos *s.* prados de las planicies centrales de Suramérica (págs. 204, 266)

location analyst [analista de ubicación] *s.* persona que estudia un área para encontrar la mejor ubicación para un cliente (pág. 16)

location [ubicación] *s.* posición exacta según la latitud y longitud, o la descripción de un lugar en relación a los lugares que lo rodean (pág. 4)

lord [señor] *s.* terrateniente poderoso (pág. 342)

Louisiana Purchase [Adquisición de Luisiana] *s.* acción por la cual el presidente Thomas Jefferson compró el Territorio de Luisiana a Francia en 1803 (pág. 130)

Loyalist [Leal] *s.* persona que apoyó a Gran Bretaña durante la Guerra Revolucionaria Norteamericana (pág. 164)

M

Maghrib [Magreb] *s.* nombre árabe de África del norte, que significa "el oeste" (pág. 568)

magma *s.* roca fundida (pág. 26)

Magna Carta [Carta Magna] *s.* carta, o documento, firmado por el rey Juan de Inglaterra en 1215 que limitaba el poder del monarca y le garantizaba derechos básicos a los nobles (pág. 400)

Magyar [magiar] *s.* persona de la etnia húngara o el idioma húngaro (pág. 432)

mainland [continente] *s.* masa terrestre principal de un continente o territorio, excluyendo a sus islas o penínsulas (pág. 400)

Malay Peninsula [península de Malaya] *s.* la segunda península del sureste de Asia continental; la franja estrecha de tierra que sirve como puente entre el continente y las islas (pág. 852)

Mali [Malí] *s.* imperio de África Occidental entre el siglo XIII d.C. y el siglo XVI d.C.; país actual de África Occidental (pág. 580)

Mandela, Nelson *s.* primer presidente africano de Sudáfrica (pág. 624)

manifest destiny [destino manifiesto] *s.* idea de que Estados Unidos debía poseer toda la tierra entre, e inclusive más allá de, los océanos Atlántico y Pacífico (pág. 130)

manor [finca solariega] *s.* casa del noble y las aldeas ubicadas en sus tierras donde vivían los campesinos (pág. 342)

Mao Zedong [Mao Tsé Tung] *s.* líder comunista de China entre 1949 y 1976 (pág. 764)

Maori [maorí] *s.* pueblo de Polinesia compuesto por los primeros habitantes de Nueva Zelanda (pág. 900)

map [mapa] *s.* representación de una parte de la Tierra (pág. 10)

maquiladora *s.* fábrica en la que los materiales son importados y ensamblados para producir bienes de exportación (págs. 244, 254)

Marcos, Ferdinand *s.* presidente dictatorial de las Filipinas entre 1965 y 1986 (pág. 864)

market economy [economía de mercado] *s.* sistema económico en el que la producción de bienes y servicios se determina según la oferta y la demanda de los consumidores (págs. 72, 146, 382, 424, 486, 512)

marsupial *s.* animal provisto de una bolsa abdominal para llevar a sus crías (pág. 884)

Marxist [marxista] *s.* persona que apoya la filosofía subyacente al comunismo, incluyendo la posesión de la tierra y de los medios de producción por parte del gobierno (pág. 630)

Mau Mau *s.* movimiento por la independencia de Kenia que comenzó en la década de 1940 (pág. 612)

Maya [maya] *s.* civilización antigua ubicada en lo que hoy es la Península de Yucatán, Guatemala y el norte de Belice (pág. 212)

mechanized [mecanizado] *adj.* equipado con maquinaria (págs. 304, 892)

medieval *adj.* relativo a la Edad Media (pág. 342)

Mekong River [río Mekong] *s.* río principal que corre hacia el sur, desde el sur de China a través de Laos, Camboya y Vietnam (pág. 852)

Melanesia *s.* subregión de Oceanía que incluye las islas al noreste de Australia y al sur del ecuador (pág. 912)

Mercosur *s.* asociación de varios países de Suramérica con el fin de promover el comercio entre los países (pág. 280)

Mesopotamia *s.* tierra entre los ríos Tigris y Éufrates (pág. 666)

mestizo *s.* persona con mezcla de ancestros europeos y americanos nativos (págs. 220, 254)

Mexican Revolution [Revolución Mexicana] *s.* lucha por lograr reformas en México entre 1910 y 1920 (pág. 230)

Micronesia *s.* subregión de Oceanía que incluye las islas al este de las Filipinas y al norte del ecuador (pág. 912)

microstate [microestado] *s.* país independiente con un área y una población muy pequeñas (pág. 858)

Middle Ages [Edad Media] *s.* período entre la caída del Imperio Romano y la Edad Moderna, desde aproximadamente el 476 d.C. hasta 1453 (pág. 342)

Middle Kingdom [Reino Medio] *s.* antiguo nombre chino de China (pág. 764)

migration [migración] *s.* proceso de trasladarse a una nueva región (pág. 66)

militarism [militarismo] *s.* uso agresivo de las fuerzas armadas por parte de un gobierno (pág. 840)

millet [mijo] *s.* grano cultivado en África Occidental, a menudo convertido en gachas (pág. 594)

Milosevic, Slobodan *s.* presidente de Serbia desde 1989 a 1997, y de Yugoslavia desde 1997 a 2000; una figura clave en los conflictos étnicos de los países balcánicos en la década de 1990 (pág. 440)

missionary [misionero] *s.* persona enviada para realizar trabajo religioso en otro lugar (pág. 88)

Mississippi River [río Misisipi] *s.* río más largo y vía fluvial principal de los Estados Unidos (pág. 122)

Mobuto Sese Seko *s.* dictador de la República Democrática del Congo desde 1965 a 1997 (pág. 618)

modernize [modernizar] *v.* aceptar nuevas costumbres o ideas (págs. 562, 606)

monarch [monarca] *s.* rey o reina que gobierna un país o territorio (pág. 716)

monarchy [monarquía] *s.* tipo de gobierno en el cual una familia gobernante encabezada por un rey o una reina ejerce el poder político (págs. 78, 334)

monotheism [monoteísmo] *s.* creencia en un solo dios (pág. 676)

monsoon [monzón] *s.* sistema de vientos estacionales que produce una estación húmeda o seca en una región, en ocasiones con lluvias intensas (págs. 612, 658, 796, 852)

Moors [moros] *s.* grupo de musulmanes de África del Norte que conquistó España en el siglo VIII (pág. 370)

mosaic [mosaico] *s.* imagen creada al colocar pequeñas piezas coloreadas de piedra, azulejo o vidrio sobre una superficie (pág. 266)

Mount Everest [Monte Everest] *s.* la montaña más alta del mundo, ubicada en los Himalayas (pág. 658)

Mount Fuji [Monte Fuji] *s.* la cumbre más alta de Japón (pág. 824)

Mount Kilimanjaro [Monte Kilimanjaro] *s.* la montaña más alta de África, al noreste de Tanzania (pág. 536)

Mount McKinley [Monte McKinley (también llamado Denali)] *s.* montaña más alta de América del Norte, ubicada en Alaska (pág. 122)

Mount Paektu [Monte Paektu] *s.* la montaña más alta de la Península Coreana (9,003 pies) (pág. 796)

Muhammad (Mohammed) [Mahoma] *s.* fundador y profeta principal del islam (pág. 676)

multicultural [multicultural] *adj.* relativo a muchas culturas diferentes o que las incluye (pág. 408)

multilingual [multilingüe] *s.* capaz de hablar muchos idiomas (pág. 376)

multinational corporation [corporación multinacional] *s.* compañía que opera en más de un país (pág. 146)

mummy [momia] *s.* cuerpo preparado para el entierro de conformidad con la antigua práctica egipcia (pág. 554)

mural *s.* pintura hecha sobre una pared (págs. 236, 296)

Musa, Mansa *s.* líder del Imperio Malí desde 1312 a 1337 aproximadamente (pág. 580)

Muslim [musulmán] *s.* seguidor del islam (págs. 504, 562, 676)

N

Nagorno-Karabakh [Nagorno Karabaj] *s.* provincia de Azerbaiyán sobre la que los armenios tienen reclamos territoriales, pues creen que debe ser parte de su país (pág. 512)

National People's Congress (NPC) [Congreso Nacional del Pueblo] *s.* organismo legislativo nacional de China (pág. 784)

nationalism [nacionalismo] *s.* orgullo por la propia nación y lealtad a ella (págs. 350, 362)

natural resource [recurso natural] *s.* algo que se halla en la naturaleza que es necesario o útil para los seres humanos (pág. 72)

neutrality [neutralidad] *s.* política de no tomar parte en la guerra (pág. 382)

Nile River [río Nilo] *s.* río más largo del mundo, que desemboca en el Mar Mediterráneo al noreste de Egipto (págs. 536, 554)

nirvana *s.* en el budismo, estado de sabiduría que rompe con el ciclo de la reencarnación (pág. 706)

Nkrumah, Kwame *s.* primer presidente de la actual Ghana (pág. 586)

nomad [nómada] *s.* miembro de un grupo que vive del pastoreo de animales y que se desplaza de un lugar a otro con el cambio de estación (págs. 96, 504)

nonviolent resistance [resistencia pacífica] *s.* método de protesta que atrae la atención hacia un problema sin usar la violencia (pág. 698)

Nordic [nórdico] *adj.* relativo a Escandinavia (pág. 388)

North American Free Trade Agreement (NAFTA) [Tratado de Libre Comercio de América del Norte] *s.* acuerdo que redujo las barreras comerciales aduaneras entre México, Canadá y los Estados Unidos (págs. 244, 254)

North Atlantic Drift [Corriente del Atlántico Norte] *s.* corriente oceánica cálida que ayuda a mantener moderado el clima de Europa (pág. 326)

North China Plain [Planicie del Norte de China] *s.* región fértil entre los ríos Huang He y Chang Jiang (pág. 764)

Northern European Plain [Planicie del Norte de Europa] *s.* área vasta de tierras llanas o de relieve leve que se extiende desde Francia hasta Rusia (págs. 326, 376)

Nubia *s.* en la antigüedad, el valle del río Nilo al sur de Egipto y al norte de Sudán (pág. 554)

Nunavut *s.* territorio creado en 1999 con la parte este de los Territorios del Noroeste, hogar de muchos inuit (pág. 156)

O

oasis *s.* sitio verde o fértil de un desierto, creado por aguas subterráneas que llegan a la superficie (pág. 652)

obelisk [obelisco] *s.* monumento de piedra con forma de columna (pág. 605)

Oceania [Oceanía] *s.* región del Océano Pacífico con miles de islas (pág. 912)

oceanic island [isla oceánica] *s.* isla que emerge del fondo del mar sobre una base volcánica (pág. 912)

official language [idioma oficial] *s.* idioma que de acuerdo con la ley debe ser usado en los asuntos del gobierno, los negocios y la educación (pág. 138)

oligarchy [oligarquía] *s.* gobierno regido por unos pocos individuos poderosos (págs. 78, 334)

Olmec [olmeca] *s.* civilización antigua a lo largo de la Costa del Golfo en lo que hoy es el sur de México (pág. 212)

ombudsman [defensor del pueblo] *s.* funcionario que investiga las quejas de los ciudadanos contra el gobierno (pág. 388)

one-crop economy [economía de monocultivo] *s.* economía que depende de un único cultivo para obtener ingresos (págs. 260, 512)

Organization of the Petroleum Exporting Countries (OPEC) [Organización de Países Exportadores de Petróleo (OPEP)] *s.* grupo de países productores de petróleo fundado en 1960 (pág. 676)

Orthodox Christianity [cristianismo ortodoxo] *s.* rama oriental del cristianismo que se difundió en Rusia en el siglo X y se convirtió en su religión oficial (pág. 476)

Ottoman Empire [Imperio Otomano] *s.* imperio musulmán basado en Turquía que duró desde el siglo XIV hasta 1922 (págs. 676, 688)

Outback [*outback*, desierto australiano] *s.* región interior de Australia (pág. 884)

outcropping [filón] *s.* afloramiento de una masa metalífera o pétrea que asoma a la superficie del terreno (pág. 884)

ozone hole [agujero de ozono] *s.* disminución estacional de una capa protectora de gas atmosférico que se encuentra por encima de la Antártida (pág. 920)

P

Pakeha *s.* palabra maorí que significa "gente blanca" (pág. 900)

Palestine Liberation Organization (PLO) [Organización para la Liberación de Palestina (OLP)] *s.* grupo político que dice representar a todos los palestinos y trabajar para la obtención de una nación palestina independiente (págs. 682, 688)

Palestinian territories [territorios palestinos] *s.* Cisjordania y la Franja de Gaza juntas (pág. 682)

Pampas *s.* llanuras cubiertas de pasto en la zona central del sur de Suramérica (págs. 204, 280)

Pan-Africanism [panafricanismo] *s.* movimiento que reclama la unidad de todas las personas de ascendencia africana (pág. 580)

papyrus [papiro] *s.* material semejante al papel hecho de los tallos de la caña de papiro (pág. 554)

Parliament [Parlamento] *s.* legislatura nacional del Reino Unido (págs. 176, 400)

parliamentary [parlamentario] *adj.* relativo a un gobierno que tiene un parlamento como poder legislativo (pág. 716)

patrician [patricio] *s.* terrateniente acaudalado que ocupaba un alto puesto gubernamental en la antigua Roma (pág. 334)

patrilineal *adj.* descendencia de padre a padre, o predominio de la línea paterna en una organización social (pág. 544)

patron [mecenas] *s.* persona acaudalada o poderosa que provee dinero, apoyo y ánimo a un artista o a una causa (pág. 342)

peacekeeping [fuerza de paz] *s.* relativo a las fuerzas armadas cuyo propósito es prevenir, evitar o poner fin a la guerra (pág. 688)

Peloponnesus [Peloponeso] *s.* la península del sur de Grecia donde se encontraba Esparta (pág. 334)

penal colony [colonia penal] *s.* asentamiento que sirve como prisión (pág. 892)

peninsula [península] *s.* una masa de tierra rodeada casi completamente por agua (págs. 326, 796)

Perkins, Charles *s.* en Australia, un líder de los derechos de los aborígenes (pág. 892)

permafrost [permahielo] *s.* suelo que se mantiene congelado a lo largo del año (pág. 460)

persecution [persecución] *s.* trato cruel por motivo de religión, raza, grupo étnico, nacionalidad, opiniones políticas, género o clase (pág. 66)

Persia *s.* región histórica del suroeste de Asia ubicada principalmente en lo que hoy es Irán (pág. 676)

perspective [perspectiva] *s.* técnica usada por los artistas para dar la impresión de profundidad y distancia (pág. 342)

Peter the Great [Pedro el Grande] *s.* zar que gobernó Rusia desde 1682 a 1725 y trató de hacerla más europea (pág. 468)

pharaoh [faraón] *s.* gobernante del antiguo Egipto (pág. 554)

pidgin [pidgin] *s.* forma simple de hablar que es una mezcla de palabras de dos o más idiomas (pág. 912)

Piedmont [piedemonte] *s.* región alta de montaña entre los montes Apalaches y la llanura costera del sur (pág. 122)

place [lugar] *s.* término geográfico que describe las características físicas y humanas de una ubicación (pág. 4)

plantation agriculture [agricultura de plantación] *s.* uso de grandes granjas para obtener cultivos comerciales (pág. 736)

plateau [meseta] *s.* área plana situada a considerable altura (págs. 34, 198, 652, 756)

Plateau of Iran [Meseta de Irán] *s.* meseta de altura ubicada en el este y en el centro de Irán, rodeada por montañas (pág. 652)

Plaza of the Three Cultures [Plaza de las Tres Culturas] *s.* plaza de la Ciudad de México que muestra partes de las influencias azteca, española y moderna en México (pág. 236)

plebeian [plebeyo] *s.* persona humilde a la que se le permitía votar pero no podía ocupar un cargo gubernamental en la antigua Roma (pág. 334)

polders [pólder] *s.* terreno pantanoso ganado al mar drenándole el agua con el uso de una presa y una bomba y que una vez desecado se dedica al cultivo (pág. 376)

Polynesia [Polinesia] *s.* subregión de Oceanía que incluye las islas del centro y del sur del Pacífico (págs. 900, 912)

polytheistic [politeísta] *adj.* que cree en muchos dioses (pág. 666)

population [población] *s.* número promedio de personas que viven en una área determinada (pág. 58)

population density [densidad de población] *s.* número promedio de personas que vive en una determinada área (pág. 58)

porridge [gacha] *s.* comida blanda que se prepara hirviendo granos de cereal en leche o agua (pág. 612)

precipitation [precipitación] *s.* agua que cae en forma de lluvia, nieve, aguanieve o granizo (pág. 40)

prime minister [primer ministro] *s.* jefe del gabinete y el funcionario principal del poder ejecutivo en un gobierno de sistema parlamentario; por ejemplo, el jefe del gobierno canadiense (págs. 176, 716)

privatization [privatización] *s.* el proceso de vender empresas que son propiedad del gobierno a individuos (pág. 486)

privatize [privatizar] *v.* vender empresas que son propiedad del gobierno a individuos (pág. 512)

propaganda *s.* información tergiversada o imprecisa que se usa deliberadamente para influenciar la opinión pública (pág. 476)

protectorate [protectorado] *s.* área o país más débil controlado por un país más fuerte (pág. 562)

Protestant [protestante] *s.* miembro de una iglesia cristiana basada en los principios de la Reforma (pág. 342)

province [provincia] *s.* división gubernamental similar a un estado (págs. 156, 784)

Prussia [Prusia] *s.* el estado alemán más poderoso del Sacro Imperio Romano (pág. 382)

public education [educación pública] *s.* educación que es financiada por el gobierno (pág. 138)

pull factor [factor de atracción] *s.* motivo que atrae a la gente a otra área (pág. 66)

Punic Wars [Guerras Púnicas] *s.* guerras entre Roma y Cartago durante los siglos III y II a.C. (pág. 568)

push factor [factor de exclusión] *s.* motivo que hace que la gente abandone un área (págs. 66, 236)

Putin, Vladimir *s.* presidente de Rusia que incrementó el poder ejecutivo en la década de 2000 (pág. 486)

Pygmy [pigmeo] *s.* miembro de un pueblo con una altura media aproximada de menos de cinco pies (pág. 618)

pyramid [pirámide] *s.* antigua estructura egipcia, construida sobre una tumba o cerca de ella (pág. 554)

Q

quilombos *s.* comunidades creadas por esclavos africanos fugitivos (pág. 296)

quinoa [quinua] *s.* tipo de planta de la región andina que produce semillas pequeñas comestibles (pág. 272)

quota [cupo] *s.* meta de la cantidad de producción (pág. 618)

R

rain forest [bosque tropical] *s.* región poblada de árboles en un clima tropical (págs. 204, 304)

Ramadan [ramadán] *s.* noveno mes del calendario musulmán, durante el cual los musulmanes ayunan todos los días desde el amanecer hasta la puesta del sol (pág. 728)

rate of natural increase [tasa de crecimiento natural] *s.* tasa de muertes restada de la tasa de nacimientos (pág. 58)

raw material [materia prima] *s.* recurso natural no procesado que será convertido en un producto acabado (pág. 72)

Reconquista *s.* esfuerzo exitoso realizado por los españoles para expulsar de España a los moros (pág. 370)

Red Sea [Mar Rojo] *s.* mar estrecho que separa el noreste de África de la Península Arábiga (pág. 606)

reform [reforma] *s.* acción de mejorar las condiciones sociales o económicas (pág. 784)

Reformation [Reforma] *s.* movimiento del siglo XVI que buscó cambiar las prácticas de la Iglesia Católica (pág. 342)

refugee [refugiado] *s.* persona que huye de un lugar para encontrar seguridad (págs. 66, 440)

region [región] *s.* área que tiene una o más características en común que la unen o conectan con otras áreas (pág. 4)

reincarnation [reencarnación] *s.* el renacimiento de un alma en otro cuerpo (pág. 706)

relief [relieve] *s.* diferencia en la elevación de un accidente geográfico desde su punto más bajo a su punto más alto (pág. 34)

religion [religión] *s.* sistema organizado de creencias y prácticas, a menudo centrado en uno o más dioses (pág. 88)

remote sensing [detección remota] *s.* obtención de información sobre un sitio mediante el uso de un instrumento que no está en contacto físico con el mismo (pág. 10)

Renaissance [Renacimiento] *s.* renacimiento de la creatividad, la literatura y el aprendizaje en Europa desde aproximadamente el siglo XIV hasta el siglo XVII (pág. 342)

renewable energy sources [fuentes de energía renovables] *s.* fuentes de energía que pueden ser reemplazadas a través de procesos naturales en curso (pág. 326)

representative democracy [democracia representativa] *s.* sistema de gobierno en el cual los ciudadanos ejercen el poder político a través de representantes elegidos (págs. 78, 146)

representative government [gobierno representativo] *s.* sistema de gobierno con una legislatura que es al menos en parte elegida por el pueblo (pág. 400)

R85

republic [república] *s.* gobierno en el que los ciudadanos eligen representantes para que gobiernen en su nombre; otro término para la democracia representativa (págs. 146, 334)

Republic of Texas [República de Texas] *s.* gobierno constitucional de Texas luego de la independencia de México (pág. 230)

reunify [reunificar] *v.* volver a unir algo que había sido separado (pág. 382)

revolution [revolución] *s.* el derrocamiento de un gobernante o gobierno; un cambio importante en las ideas (pág. 230)

Rhodesia [Rodesia] *s.* nombre colonial de los países que hoy son Zambia y Zimbabwe (pág. 630)

rig [manipular] *v.* influenciar o arreglar los resultados de una elección (pág. 594)

Ring of Fire [Anillo de Fuego] *s.* zona geográfica que se extiende a lo largo del borde del Océano Pacífico y tiene numerosos volcanes y terremotos (págs. 26, 824)

Rio de Janeiro [Río de Janeiro] *s.* capital de Brasil desde 1763 a 1960 (pág. 290)

ritual *s.* conjunto de ceremonias llevadas a cabo regularmente (pág. 774)

Rivera, Diego *s.* muralista famoso que pintó la historia de México sobre las paredes del Palacio Nacional (pág. 236)

Rocky Mountains [Montañas Rocosas] *s.* cordillera que se extiende aproximadamente 3,000 millas desde Nuevo México hasta Alaska (pág. 122)

Romance language [lengua romance] *s.* cualquiera de las lenguas que se desarrollaron a partir de la lengua romana, el latín, tales como el español, el portugués, el francés, el italiano y el rumano (pág. 362)

Rose Revolution [Revolución de las Rosas] *s.* levantamiento pacífico en la Georgia soviética que contribuyó a forzar la renuncia de un presidente corrupto (pág. 512)

Rosetta Stone [Piedra Roseta] *s.* antigua piedra egipcia que proporcionó un código para descifrar los jeroglíficos (pág. 562)

Royal Canadian Mounted Police [Real Policía Montada de Canadá] *s.* organización de oficiales responsables del cumplimiento de la ley en el oeste canadiense (pág. 170)

rural *adj.* relativo al campo (págs. 58, 236)

Russification [rusificación] *s.* esfuerzo en hacer que los países ocupados por la Unión Soviética se volvieran similares a Rusia mediante la sustitución de los idiomas y costumbres locales (págs. 424, 504)

S

Sadat, Anwar el- *s.* presidente de Egipto desde 1970 a 1981 (pág. 562)

safari *s.* una expedición por tierra en la que se practica la exploración, la caza o la fotografía (pág. 612)

Sahara *s.* el desierto más grande del mundo (pág. 536)

Sahel *s.* región semiárida al sur del Sahara (pág. 536)

samba *s.* música y baile, con raíces en los ritmos africanos; el estilo de música brasileña más famosa en el mundo (pág. 296)

Sami [sami] *s.* pueblo del norte de Escandinavia que tradicionalmente arrea renos; también es el nombre de su idioma (pág. 388)

samurai [samuráis] *s.* soldados profesionales de Japón (pág. 830)

San Martín, José de *s.* líder de la independencia en el sur de Suramérica (págs. 220, 280)

sanction [sanción] *s.* castigo o presión aplicados a un país por no obedecer una orden o ley (pág. 624)

Sanskrit [sánscrito] *s.* idioma antiguo de la India, hablado por primera vez por los arios (pág. 698)

Santa Anna, Antonio López de (1794–1876) *s.* general mexicano, presidente y líder de la independencia mexicana de España (pág. 230)

savanna [sabana] *s.* región de vegetación en los trópicos o subtrópicos con una mezcla de pastizales y árboles dispersos (págs. 40, 536)

scholar-official [académico-funcionario] *s.* persona educada que trabajaba en el gobierno chino (pág. 764)

Scientific Revolution [Revolución Científica] *s.* cambio rotundo en el pensamiento europeo a mediados del siglo XVI que llevó al cuestionamiento de viejas teorías (pág. 350)

Scott, Robert *s.* explorador británico que recorrió el polo (pág. 920)

seafaring [marinero] *adj.* que usa el mar para transportarse (pág. 326)

secular *adj.* mundano, no relacionado con la religión (pág. 342)

sediment [sedimento] *s.* pedazos de roca en forma de arena, piedra o cieno depositados por el viento, el agua o el hielo (pág. 26)

segregation [segregación] *s.* política de separar a la gente de razas diferentes (pág. 624)

seismic [sísmico] *adj.* relativo a los terremotos (pág. 756)

Selassie, Haile *s.* emperador de Etiopía de 1930 a 1974 (pág. 606)

self-rule [gobierno autónomo] *s.* gobierno independiente (pág. 580)

self-sufficient [autosuficiente] *adj.* capaz de satisfacer las propias necesidades sin ayuda externa (pág. 304)

selva *s.* nombre que se le da en español a la región al este de Perú que contiene bosques tropicales (pág. 272)

separatist [separatista] *s.* persona que quiere que una región se separe de la nación de la que forma parte (págs. 164, 370)

serf [siervo] *s.* en Europa, persona que vivía y trabajaba en la finca de un señor feudal o de un vasallo; en Rusia, campesino atado a la tierra (págs. 342, 468)

service industry [industria de servicios] *s.* industria que provee servicios en lugar de objetos (pág. 254)

Seven Years' War [Guerra de los Siete Años] (1756–1763) *s.* conflicto entre Gran Bretaña y Francia cuyos combates tuvieron lugar en América del Norte, Europa y la India (pág. 164)

shadow puppet [títeres de sombras] *s.* arte tradicional indonesio, en el que se manejan títeres desde detrás de una pantalla iluminada por una luz brillante (pág. 864)

shah [sah] *s.* título del rey de Persia o Irán (pág. 676)

Shakespeare, William *s.* dramaturgo y poeta inglés de finales del siglo XVI y principios del siglo XVII (pág. 408)

shamanism [chamanismo] *s.* sistema de creencias en el cual una persona llamada chamán se supone dotada de poderes para comunicarse con los espíritus y sanar a los enfermos (pág. 804)

Shi Huangdi *s.* primer emperador en unificar gran parte de China (pág. 764)

Shikoku *s.* isla principal más pequeña y menos poblada de Japón (pág. 824)

Shinto [sintoísmo] *s.* religión original de Japón; involucra la adoración de dioses que se cree se hallan en la naturaleza (pág. 830)

shock therapy [terapia de choque] *s.* en Rusia, término dado al plan económico de Yeltsin, el cual exigía un cambio abrupto desde una economía dirigida a una economía de mercado (pág. 486)

shogun [sogún] *s.* líder del gobierno militar japonés que comenzó en 1192 (pág. 830)

Siam *s.* antiguo nombre de Tailandia (pág. 858)

Siberia [Siberia] *s.* inmensa región de Rusia asiática (pág. 460)

Sierra Madre Occidental *s.* cordillera que se extiende de norte a sur por la parte oeste de México (pág. 198)

Sierra Madre Oriental *s.* cordillera que se extiende de norte a sur por la parte este de México (pág. 198)

Silk Roads [Rutas de la seda] *s.* principales rutas de comercio que iban desde China hasta el resto del este de Asia, a través de Asia Central y hasta Europa (págs. 504, 774)

silt [cieno] *s.* suelo fértil y blando transportado por los ríos y depositado en los terrenos circundantes (págs. 554, 666)

Solidarity [Solidaridad] *s.* primer sindicato independiente de Polonia, cuyo cofundador fue Lech Walesa (pág. 424)

solstice [solsticio] *s.* época del año en la que el Sol alcanza en el cielo su punto extremo al norte o al sur (pág. 40)

Songhai *s.* antiguo imperio de África Occidental, que tuvo su mayor poderío alrededor del siglo XVI d.C. (pág. 580)

souk [zoco] *s.* mercado al aire libre común en África del Norte (pág. 568)

Southern Alps [Alpes del Sur] *s.* cadena montañosa de Nueva Zelanda (pág. 884)

Southern Cone [Cono Sur] *s.* naciones de Suramérica ubicadas en la parte con forma de cono del extremo sur de Suramérica (pág. 280)

Soviet Union [Unión Soviética] *s.* nación comunista, compuesta por Rusia y otros 14 Estados, que existió entre 1922 y 1991 (pág. 728)

spatial [espacial] *adj.* relativo a la ubicación de un lugar y a su relación física con otros lugares, personas o ambientes (pág. 4)

specialization [especialización] *s.* énfasis en producir una cantidad limitada de un producto específico (pág. 72)

squatter [ocupante ilegal] *s.* persona que se asienta en tierras sin ocupar sin tener un derecho legal sobre ellas (pág. 236)

St. Lawrence Seaway [vía marítima de San Lorenzo] *s.* vía fluvial compuesta por el río San Lorenzo, los Grandes Lagos y diversos canales (pág. 156)

stability [estabilidad] *s.* durabilidad; resistencia al cambio (pág. 586)

Stalin, Joseph *s.* cruel dictador del siglo XX que hizo ejecutar a millones de sus enemigos políticos (págs. 468, 512)

standard of living [estándar de vida] *s.* medida económica relativa a la calidad y cantidad de bienes disponibles para un grupo de personas y cómo se distribuyen esos bienes en el grupo (pág. 432)

stateless societies [sociedades sin estado] *s.* grupos que no tienen una autoridad única, sino que toman decisiones basadas en el acuerdo (pág. 544)

steppe [estepa] *s.* planicie cubierta de hierba; en Rusia, esta área se ubica al sur de la taiga (pág. 460)

Stone Age [Edad de Piedra] *s.* primer período de la cultura humana, en el que las personas usaban herramientas de piedra (pág. 544)

R87

strategic [estratégico] *adj.* relativo a un plan de acción diseñado para alcanzar una meta específica; con relación a lo militar, de ayuda en la lucha de una guerra (págs. 432, 728)

subcontinent [subcontinente] *s.* gran masa de tierra que es parte de un continente pero se considera como una región separada (pág. 658)

subsistence farming [agricultura de subsistencia] *s.* agricultura que produce tan sólo lo suficiente para alimentar a la familia del agricultor, quedando poco excedente o nada para vender (págs. 254, 586, 728, 736)

Sudan [Sudán] *s.* país del noreste de África, ubicado al sur de Egipto (pág. 562)

Suez Canal [Canal de Suez] *s.* vía marítima que une el Mar rojo con el Mar Mediterráneo (pág. 562)

Sumer [Sumeria] *s.* región de ciudades-estado de la Mesopotamia que albergó a la primera civilización (pág. 666)

surveyor [agrimensor] *s.* persona que traza un mapa del terreno y lo mide (págs. 10, 16)

sustainable [sostenible] *adj.* relativo al uso de recursos naturales de tal manera que existan para las generaciones futuras (pág. 48)

Swahili [swahili] *s.* idioma africano, comúnmente hablado en África Oriental (pág. 612)

T

taiga *s.* extenso bosque de Rusia que se ubica al sur de la tundra (pág. 460)

Taino [*taíno*] *s.* pueblo indígena de las islas del Caribe (pág. 260)

Taj Mahal *s.* hermosa tumba que se halla en India, construida por el gobernante mogol Shah Jahan para honrar a su esposa, que murió en 1631 (pág. 698)

Taklimakan Desert [Desierto de Taklimakán] *s.* el segundo desierto de arena más grande del mundo (pág. 756)

Taliban [talibán] *s.* grupo islámico conservador que tomó el control de Afganistán a mediados de la década de 1990 (pág. 728)

technology [tecnología] *s.* aplicación del conocimiento, herramientas e inventos de las personas para satisfacer sus necesidades (pág. 96)

tectonic plate [placa tectónica] *s.* sección rígida de gran tamaño en la corteza terrestre que se encuentra en movimiento constante (págs. 26, 198, 536)

territory [territorio] *s.* unidad política canadiense que no tiene suficientes personas como para ser una provincia (pág. 156)

Tet *s.* año nuevo vietnamita y el feriado más importante en Vietnam (pág. 858)

theocracy [teocracia] *s.* gobierno regido por líderes religiosos (pág. 676)

Three Gorges Dam [Presa de las Tres Gargantas] *s.* barrera construida en el río Chang Jiang para controlar las inundaciones (pág. 756)

Three Kingdoms [Tres Reinos] *s.* los tres reinos que se habían formado en la Península de Corea hacia el año 300 d.C. (pág. 804)

three-dimensional [tridimensional] *adj.* que describe una imagen en la que hay un sentido de profundidad y perspectiva (pág. 4)

Tiananmen Square [Plaza Tiananmen] *s.* la ubicación en Pekín, China, de una inmensa manifestación a favor de los derechos democráticos que se realizó en 1989 (pág. 784)

Tian Shan *s.* cordillera que se extiende a través de Kirguistán y Tayikistán (pág. 496)

Tigris River [río Tigris] *s.* río del suroeste de Asia que fluye desde una meseta en la Península de Anatolia hasta el Golfo Pérsico (pág. 652)

Tito, Josip Broz *s.* líder comunista de Yugoslavia desde 1953 a 1980 (pág. 440)

Tokugawa Ieyasu *s.* sogún que unificó Japón (pág. 830)

Tolstoy, Leo *s.* escritor del siglo XIX que es uno de los más grandes novelistas rusos (pág. 476)

totalitarian [totalitario] *adj.* relativo a un tipo de gobierno que controla todos los aspectos de la vida pública y privada (págs. 468, 812)

tourism [turismo] *s.* industria que provee servicios para los viajeros (pág. 260)

Transantarctic Mountains [Montañas Transantárticas] *s.* montañas que dividen a la Antártida en dos partes (pág. 920)

Transcaucasia *s.* región entre los mares Negro y Caspio que comprende las repúblicas de Armenia, Azerbaiyán y Georgia (pág. 496)

transcontinental railroad [ferrocarril transcontinental] *s.* ferrocarril que cruza un continente (pág. 130)

trans-Saharan [trans-sahariano] *adj.* que atraviesa el Sahara (pág. 580)

Treaty of Tordesillas [Tratado de Tordesillas] *s.* tratado de 1494 entre España y Portugal que le dio a Portugal el control sobre la tierra que hoy es Brasil (pág. 290)

Treaty of Waitangi [Tratado de Waitangi] *s.* tratado entre los británicos y los maoríes firmado en 1840 (pág. 900)

trust territory [territorio en fideicomiso] *s.* territorio puesto bajo el control de un país por las Naciones Unidas (pág. 912)

tsunami [tsunami] *s.* serie de olas oceánicas gigantes y destructivas causadas por terremotos, volcanes o desplazamientos de tierras submarinos (págs. 658, 736, 852)

tundra *s.* clima seco y frío y vegetación adaptada a tal clima que se halla en el Círculo Polar Ártico; en Rusia, ubicada en el extremo norte (págs. 156, 460)

typhoon [tifón] *s.* tormenta tropical que comienza en los océanos Pacífico o Índico (pág. 796)

tyrant [tirano] *s.* alguien que toma el poder ilegalmente (pág. 334)

U

Uluru *s.* espectacular afloramiento rocoso en el desierto interior australiano (pág. 884)

Union of Soviet Socialist Republics (USSR), o **Soviet Union** [Unión de Repúblicas Socialistas Soviéticas (URSS), o Unión Soviética] *s.* nación con un gobierno comunista creada por los bolcheviques en 1922; disuelta en 1991 (pág. 468)

unwritten constitution [constitución no escrita] *s.* marco de gobierno que no es un único documento escrito sino que incluye muchas leyes diferentes, decisiones judiciales y costumbres políticas (pág. 414)

Ural Mountains [montes Urales] *s.* cordillera que se extiende de norte a sur y forma la frontera entre la Rusia europea y la asiática (pág. 460)

urban [urbano] *adj.* relativo a la ciudad (págs. 16, 58, 236)

urban planner [urbanista] *s.* persona que crea planes para desarrollar y mejorar partes de una ciudad (pág. 16)

urbanization [urbanización] *s.* crecimiento en el número de personas que viven en áreas urbanas; el proceso de desarrollo de las ciudades (págs. 58, 290)

V

vassal [vasallo] *s.* noble menos acaudalado que pagaba impuestos y servía a un señor feudal a cambio de tierras (pág. 342)

Vatican [Vaticano] *s.* la residencia oficial del Papa en la Ciudad de Vaticano y el centro político y religioso de la Iglesia Católica Romana (pág. 362)

Vedas *s.* antiguos escritos en sánscrito que son los primeros textos sagrados del hinduismo (pág. 698)

vegetation region [región de vegetación] *s.* un área que tiene plantas similares (pág. 40)

Velvet Revolution [Revolución de Terciopelo] *s.* protesta pacífica del pueblo checo que condujo al fin paulatino del comunismo en Checoslovaquia (pág. 432)

Victoria Falls [cataratas Victoria] *s.* cataratas de 355 pies de alto en el río Zambezi, en el sur de África Central (pág. 536)

Vikings [vikingos] *s.* pueblo marinero escandinavo que realizó incursiones en Europa occidental y del norte desde el siglo IX al siglo XI (pág. 388)

volcano [volcán] *s.* abertura en la corteza terrestre de la que fluyen o salen expelidos roca fundida, ceniza y gases calientes (pág. 26)

Volga River [río Volga] *s.* río más largo de Europa, que fluye a través de la Rusia europea (pág. 460)

W

Walesa, Lech *s.* líder polaco, cofundador de Solidaridad, que fue presidente de Polonia desde 1990 a 1995 (pág. 424)

weather [tiempo] *s.* la condición de la atmósfera terrestre en un momento y lugar dados (pág. 40)

weathering [desgaste] *s.* erosión física y química gradual de las rocas en la superficie de la Tierra (pág. 26)

welfare state [estado de bienestar] *s.* sistema social en el que el gobierno satisface muchas de las necesidades de sus ciudadanos (pág. 388)

West Bank [Cisjordania] *s.* región al oeste del río Jordán (pág. 682)

wind farm [parque eólico] *s.* central de energía que usa molinos de viento para generar electricidad (pág. 414)

X

Xi Jiang *s.* río en el sureste de China (pág. 756)

Y

Yalu River [río Yalu] *s.* el río más largo de la Península de Corea; forma parte de la frontera entre China y Corea del Norte; conocido como el río Amnok en coreano (pág. 796)

Yeltsin, Boris *s.* presidente que transformó abruptamente la economía rusa después del colapso de la Unión Soviética (pág. 486)

Yoruba [yoruba] *s.* pueblo del suroeste de Nigeria, parte de Benín y Togo (pág. 586)

Yugoslavia *s.* país en la Península de los Balcanes desde 1918 a 1991 (pág. 440)

yurt [yurta] *s.* estructura portátil semejante a una tienda usada por los nómadas de Asia Central (pág. 504)

Z

zaibatsu *s.* grandes grupos financieros e industriales controlados por familias que poseían muchas compañías en Japón antes de la Segunda Guerra Mundial (pág. 840)

Zaire *s.* nombre de la República Democrática del Congo entre 1971 y 1997 (pág. 618)

ziggurat [zigurat] *s.* templo construido encima de una serie de plataformas progresivamente más pequeñas (pág. 666)

R89

Index

Page references in **boldface** indicate Key Terms & Names and Background Vocabulary that are highlighted in the main text.

Page references in *italics* indicate illustrations, charts, and maps.

A

B

Index

C

R93

E

Index

F

G

H

Index

I

J

K

L

R99

M

N

R101

Q

R

R103

S

R105

T

U

Index

R107

V

W

X

Y

Z

Acknowledgments

Text Acknowledgments

Front Matter

Houghton Mifflin: Adaption of line graph from "Exports of English Manufactured Goods, 1700-1774," from A History of World Societies, Fifth Edition by John P. McKay, Bennett D. Hill, John Buckler, and Patricia Buckley Ebrey. Copyright © 2000 by Houghton Mifflin Company. All rights reserved. Used by permission.

Unit 1

Christian Science Monitor: Excerpt from "As English Spreads, Speakers Morph It Into World Tongue," by David Rohde from Christian Science Monitor, May 17, 1995. Copyright © 1995 by the Christian Science Monitor. Reprinted by permission of the Christian Science Monitor (www.csmonitor.com). All rights reserved.

Unit 3

John Bierhorst: Excerpt from "The Eagle On the Prickly Pear," from The Hungry Woman by John Bierhorst. Copyright © 1984 by John Bierhorst. Reprinted by permission of the author.

Rhett A. Butler: Excerpt from Interview with William F. Laurance by Rhett A. Butler. Copyright © 2006 by Rhett A. Butler. Reprinted by permission of Rhett A. Butler, mongabay.com.

Unit 4

Andreas Ramos: Excerpt from "A Personal Account of The Fall of the Berlin Wall," by Andreas Ramos from www.andreas.com/berlin.html. Copyright © by Andreas Ramos. Reprinted by permission of the author.

Unit 6

Atlantic Monthly: Excerpt from "An African Voice" interview with Chinua Achebe by Katie Bacon from Atlantic Online, August 2, 2000. Copyright © 2000 by Atlantic Monthly. Reprinted with permission of Atlantic Monthly Online.

Unit 7

Diablo Press: Excerpt from India Emerges: A Concise History by Steven Warshaw. Copyright © 1987, 1988 Diablo Press Inc. Reprinted by permission of Diablo Press Inc.

Newsweek: Excerpt from "A City That Echoes Eternity" by Kenneth Woodward, from Newsweek, July 24, 2000. Copyright © 2000 by Newsweek. Reprinted by permission of Newsweek, Inc. All rights reserved.

Unit 8

Shambhala Publications: "Cool crescent moon," by Bashō, translated from the Japanese by Sam Hamill. Copyright © 1998 by Sam Hamill. Reprinted by permission of Shambhala Publications, Inc.

Photography

Front Matter

cover *front to back* © Dennis Cox/Alamy; © Theo Allofs/zefa/Corbis; © Robert Fried/Alamy; © David Sanger Photography/Alamy; © Bruce Coleman Inc./Alamy; *background* © Adam Crowley/Getty Images; **Senior Consultants** *top* Photograph by Robert S. Bednarz; *center* Photograph by Marci Smith Deal; *bottom* Photograph by Inés M. Miyares; **Senior Consultants** *top* Photograph by Rob Caron; *center* Photograph by Erica Houskeeper; *bottom* Photograph by Joe Wieszczyk; **Student Panel** Photographs by Bac To Trong; **Contents in Brief** *top left* © Keren Su/Getty Images; *top center* © Corbis; *top right* © Beth Dixson/Getty Images; **Introduction to World Geography** *top left* © Joeseph Sohm-Visions of America/Getty Images; *top right* © Tibor Bognar/Alamy; **The United States and Canada** *top left* © Fridmar Damm/zefa/Corbis; *top center* © Getty Images; *second from top right* © Irwin Barrett/Getty Images; *top right* © Art Wolfe/Getty Images; **Latin America** *top left* © Getty Images; *top, reef* © Royalty-Free/Corbis; *top, small dolphin* © David Tipling/Alamy; *top, big dolphin* © Stuart Westmorland/Corbis; **Europe** *top left* © Simon Reddy/Alamy; *top right* © Bill Bachmann/Index Stock Imagery/Jupiter Images; **Russia and The Eurasian Republics** *top left* © Pavel Filatov/Alamy; *top center* © Pavel Filatov/Alamy; *top right* © Pavel Filatov/Alamy; *top right inset* © Royalty-Free/Corbis; **Africa** *top left* © Sylvia Cordaiy Photo Library Ltd./Alamy; *top center* © Joshua D. Polman/Alamy; *top right* © Stan Osolinski/Oxford Scientific/Jupiter Images; **Southwest Asia and South Asia** *top left* © Design Pics Inc./Alamy; *top center* © Dinodia Images/Alamy; *top right* © Michelangelo Gratton/Getty Images; **East Asia and Southeast Asia** *top left* © Jose Fuste Raga/Corbis; *top right* © Angelo Cavalli/Getty Images; **Oceania and Antarctica** *top left* © Chris Cheadle/Getty Images; *top center* © Peter Scoones/Getty Images; *top right* © David Tipling/Getty Images; **Features** *top* Confucius (551-479 B.C.). Scroll painting. Bibliotheque Nationale, Paris. Photo © Snark/Art Resource, New York; *bottom right* © Stockbyte Silver/Alamy; **Features** *top* © Rob Melnychuk/Getty Images; **Graphs, Tables, and Charts** © Gavriel Jecan/Corbis; **Reading for Understanding** *top* © Comstock Images/Alamy; **Geography Handbook** NASA Goddard Space Flight Center; **A11** *bottom left* © Getty Images; *top right* © Bruce Roberts/Photo Researchers, Inc.; *second from top right* © Frans Lanting/Corbis; *third from top right* © David R. Frazier/Photo Researchers, Inc.; **A12–13** Image created by Reto Stockli with the help of Alan Nelson, under the leadership of Fritz Hasler/NASA.

Unit 1 Opener

1 *top right* © Getty Images; *top left* © Digital Vision/Getty Images; *bottom right* © Peter Horree/Alamy; *bottom left* © Philip and Karen Smith/Getty Images.

Chapter 1

2-3 NASA Goddard Space Flight Center; **2** *bottom left* Strabo (1584). Engraving. The Granger Collection, New York; *bottom right* © Science Museum/Science and Society Picture Library; *top left* © Peter Horree/Alamy; **3** *bottom left* © Science Museum/Science and Society Picture Library; *bottom right* © Photolibrary/Alamy; *center right* © Peter Horree/Alamy; *center left* © Photo Researchers, Inc.; **5** *bottom* NASA; *top left* © Keren Su/Getty Images; *top center* © Corbis; *top right* © Beth Dixson/Getty Images; **6** *top* © Silvestre Machado/Getty Images; *second from top* © Gavin Hellier/Getty Images; *center* © Danita Delimont/Alamy; *second from bottom* © Lester Lefkowitz/Corbis; *bottom* © Phil Schermeister/Peter Arnold, Inc.; **7** *second from top* © David R. Frazier/Photo Researchers, Inc.; *center left* © Craig Lovell/Eagle Visions Photography/Alamy; *second from bottom* © Frans Lanting/Corbis; *bottom* © Bruce Roberts/Photo Researchers, Inc.; **8** © Mark Segal/Panoramic Images/NGSImages.com; **11** *center right* © The Image Works, Inc.; *bottom* © Rob Matheson/Corbis; *top left* © Keren Su/Getty Images; *top center* © Corbis; *top right* © Beth Dixson/Getty Images; **12** *bottom left* Strabo (1584). Engraving. The Granger Collection, New York; *bottom right* Horace L. Jones. trans., *The Geography of Strabo*, 8 vols. (New York: G.P. Putnam's Sons, (1917), vol. 1, pp. 451-455, 501-503, 277-385, passim; **13** *bottom center* © NASA/Angela King/Geology.com; *bottom right* © Max Dannenbaum/Getty Images; **14** *center* © David R. Frazier Photolibrary, Inc./Alamy; **15** © Craig Lovell/Eagle Visions Photography/Alamy; **16** © David Vaughan/Photo Researchers, Inc.; **17** *center* © Mark Gibson/Alamy; *bottom* © Mark Edwards/Peter Arnold, Inc.; *top left* © Keren Su/Getty Images; *top center* © Corbis; *top right* © Beth Dixson/Getty Images; **18** © Johnathan Smith; Cordaiy Photo Library Ltd./Corbis; **19** *top* © age fotostock/SuperStock; *bottom* © Jeff Greenberg/Alamy; **20** © Erin Cigliano; **21** *bottom left* © NASA/Angela King/Geology.com; *bottom right* © Scott Rothstein/ShutterStock; **22** © Scott Rothstein/ShutterStock.

Chapter 2

24 *bottom* The Granger Collection, New York; *top left* © Peter Horree/Alamy; **25** © Gary Braasch/Corbis; **26** © Gregory G. Dimijian, M.D./Photo Researchers, Inc.; **27** *bottom* © Roger Ressmeyer/Corbis; *top left* © Robert Glusic/Getty Images; *top center* © Getty Images; *top right* © Royalty-Free/Corbis; *top right inset* © Peter Arnold, Inc./Alamy; **31** *background* © George Steinmetz/Corbis; *bottom* © Roger Ressmeyer/Corbis; **32-33** © James Randklev/Getty Images; **33** © Corbis; **34** D. Brown/PanStock/Panoramic Images/National Geographic Image Collection; **35** *center right* NASA; *bottom* © David Lyons/Alamy; *top left* © Robert Glusic/Getty Images; *top center* © Getty Images; *top right* © Royalty-Free/Corbis; *top right inset* © Peter Arnold, Inc./Alamy; **36** © Scott T. Smith/Corbis; **38** Sitki Tarlan/Panoramic Images/National Geographic Image Collection; **39** © James D. Watt/Image Quest Marine; **40** © Jeff Lepore/Photo Researchers, Inc.; **41** *center right* © Jeff Schultz/AlaskaStock.com; *bottom* © Arend/Soucek/AlaskaStock.com; *top left* © Robert Glusic/Getty Images; *top center* © Getty Images; *top right* © Royalty-Free/Corbis; *top right inset* © Peter Arnold, Inc./Alamy; **43** *all* © Getty Images; © Getty Images; © Getty Images; **46** *center left* © Ed George/NGSImages.com; *bottom left* © Mark Moffett/Minden Pictures; **48** © Frank Kroenke/Peter Arnold, Inc.; **49** *bottom* © Knut Mueller/Peter Arnold, Inc.; *top left* © Robert Glusic/Getty Images; *top center* © Getty Images; *top right* © Royalty-Free/Corbis; *top right inset* © Peter Arnold, Inc./Alamy; **51** Fritz Reiss/AP/Wide World Photos; **52** © William Campbell/Corbis; **53** *bottom left* © Douglas Peebles/Corbis; *bottom right* Sitki Tarlan/Panoramic Images/National Geographic Image Collection; **54** NASA.

Chapter 3

56 *top left* © Peter Adams Photography/Alamy; *bottom left* Kuba mask. Kuba related peoples, Zaire. Wood, pigment. Private collection. Photo © Aldo Tutino/Art Resource, New York; *bottom right* © The British Museum/HIP/The Image Works, Inc.; **57** *center* NASA; *bottom left* © Gary Randall/Getty Images; *bottom center*

R109

© First/zefa/Corbis; **58** © Royalty-Free/Corbis; **59** *bottom* © David Lawrence; *top left* © Joseph Sohm-Visions of America/Getty Images; *top right* © Tibor Bognar/Alamy; **60** © Art Kowalsky/Alamy; **61** © Peter Adams Photography/Alamy; **62** © PCL/Alamy; **64** *center left* © Dave G. Houser/Post-Houserstock/Corbis; *center right* © Australian Scenics; *bottom* © John Hay/Lonely Planet Images; **65** *center left* © Robin Moyer/OnAsia; *center right* © Dan Bigelow/Getty Images; *bottom* © Robert Harding Picture Library Ltd/Alamy; **66** © Yellow Dog Productions/Getty Images; **67** *center right* © Alexander Walter/ Getty Images; *bottom* The Granger Collection, New York; *top left* © Joeseph Sohm-Visions of America/Getty Images; *top right* © Tibor Bognar/Alamy; **68** The Granger Collection, New York; **69** Kuba mask. Kuba related peoples, Zaire. Wood, pigment. Private collection. Photo © Aldo Tutino/Art Resource, New York; **70** Brennan Linsley/ AP/Wide World Photos; **71** © Beatrice Mategwa/Reuters/Corbis; **73** *bottom* © Photo Researchers, Inc.; **74** *top left* © Joeseph Sohm-Visions of America/Getty Images; *top right* © Tibor Bognar/Alamy; **75** Steve Raymer/National Geographic Image Collection; **76** © Sue Cunningham Photographic/Alamy; **77** © Authors Image/Alamy; **78** © Le Segretain Pascal/Corbis; **79** *top left* © Joeseph Sohm-Visions of America/Getty Images; *top right* © Tibor Bognar/Alamy; *center right* © Denise Kappa/ShutterStock; *bottom* © Marc Muench/Getty Images; **81** © Majid/Getty Images; **82** Gregory Bull/AP/Wide World Photos; **83** *bottom left* © Le Segretain Pascal/Corbis; *bottom right* © Dan Lamont/Corbis; **84** © Bettmann/Corbis.

Chapter 4

86 *bottom center* © The British Museum/Topham-HIP/The Image Works, Inc.; *bottom right* © Jane Sweeney/Lonely Planet Images; *top left* © Philip and Karen Smith/Getty Images; *bottom left* © Louis Fox/Getty Images; **87** *bottom left* Limestone tablet with Sumerian pictographic script. Symbol of the hand indicates the proprietor. End of 4th millenium. Mesopotamia. 5 x 4.2 cm. A) 19936. Louvre, Paris. Photo © Erich Lessing/Art Resource, New York; *bottom right* © Peter Horree/Alamy; *top left* © Nathan Benn/Corbis; *center* Benin aquamanile in the shape of a leopard (1700s). British Museum, London. Photo © Art Resource, New York; *top right* © Lowell Georgia/Corbis; **88** © Pascal Goetgheluck/Photo Researchers, Inc.; **89** *bottom* © David L. Brown/PanStock/Jupiter Images; *top left* © Joe Sohm/Pan America/Jupiter Images; *top center* © David Jones/Alamy; *top right* © Kevin Schafer/Alamy; **90** © Sergio Gaudenti/Kipa/Corbis; **91** *bottom left* © Rob Melnychuk/Getty Images; *top left* © Yellow Dog Productions/Getty Images; *top center* © Jeff Greenberg/ PhotoEdit; *second from right* © Michael Newman/PhotoEdit; *bottom right* © Getty Images; **92** Boddhisatva Kuan-Yin (Avalokitesvara) (1271-1368 B.C.). Porcelain Ch'ing Pai ware with blue glaze. National Museum, Beijing. Photo © Erich Lessing/Art Resource, New York; **96** © Images&Stories/Alamy; **97** *center right* © Bill Bachman/Danita Delimont; *bottom* © Robert Azzi/Woodfin Camp and Associates, Inc.; *top left* © Joe Sohm/Pan America/Jupiter Images; *top center* © David Jones/Alamy; *top right* © Kevin Schafer/Alamy; **98** *left* Pot with handles, ribbon-ornaments and animal scenes. Baked clay with white incrustations. The Louvre, Paris. Photo © Erich Lessing/ Art Resource, New York; *center* Basket woven with a killer whale crest. Spruce root. Museum of Anthropology, University of British Columbia, Vancouver, British Columbia, Canada. Photo © Werner Forman/Art Resource, New York; *right* Pouch (1800s). Mixed media. 30cm. Peabody Essex Museum, Salem, Massachusetts. Photo © Bridgeman Art Library; **100** © David Noble Photography/Alamy; **101** *bottom right* Limestone tablet with Sumerian pictographic script. Symbol of the hand indicates the proprietor. End of 4th millenium. Mesopotamia. 5 x 4.2 cm. A) 19936. Louvre, Paris. Photo © Erich Lessing/Art Resource, New York; *bottom left* © Pascal Goetgheluck/ Photo Researchers, Inc.; *center left* © Dynamic Graphics Group/IT Stock Free/Alamy; **102** © J. Marshall - Tribaleye Images/Alamy.

Unit 2 Opener

105 *bottom left* © Galen Rowell/Corbis; *top right* © Royalty-Free/ Corbis; *bottom right* © Robert Harding/Getty Images; **110** © Lee Snider/Corbis; **111** *top right* © Paul A. Souders/Corbis; *bottom* © JupiterMedia/Alamy; **113** © Royalty-Free/Corbis; **114** © Tom Walker/ Getty Images; **115** © Buddy Mays/Corbis; **116** © Getty Images; **117** © franzfoto.com/Alamy; **118** © Joseph Van Os/Riser/Getty Images; **119** © Visual&Written SL/Alamy.

Chapter 5

120 *bottom left* © Lee Snider/Photo Images/Corbis; *bottom right* *Battle of Lexington* (about 1850), Alonzo Chappel. Photo © Bettmann/Corbis; *top left* © Bill Ross/Corbis; **121** *bottom left* The Granger Collection, New York; *bottom right* AP/Wide World Photos; **123** *center left* © Sean Daveys/Australian Picture Library/Corbis; *center right* © bilderlounge/Alamy; *bottom* © Christopher Talbot Frank/Photex/zefa/Corbis; *top left* © Fridmar Damm/zefa/Corbis; *top center* © Getty Images; *second from top right* © Irwin Barrett/ Getty Images; *top right* © Art Wolfe/Getty Images; **124** © Joseph Sohm/Corbis; **126** *top left* © David Muench/Corbis; *bottom right* © Tom Bean/Corbis; **127** *top right* © Royalty-Free/Corbis; *bottom left* © Tom Bean/Corbis; **128-129** *top* © Andrew Sacks/Corbis; *bottom* © Reinhard Eisele/Corbis; **128** *top inset* © Galen Rowell/Corbis; *center* © Luis Castaneda Inc./Getty Images; *bottom left* © Arthur Morris/Corbis; **129** *top* © Tom Bean/Corbis; *second from right* © William Manning/Corbis; *right* © Arthur Morris/Corbis; *bottom* © Marvin E. Newman/Getty Images; **131** *center right* © Veer; *bottom* © Robert Holmes/Corbis; *top left* © Fridmar Damm/zefa/Corbis; *top center* © Getty Images; *second from top right* © Irwin Barrett/Getty Images; *top right* © Art Wolfe/Getty Images; **133** *The Drafting of the Declaration of Independence* (1909), Jean Leon Germome Ferris. Painting. Photograph © PoodlesRock/Corbis; **135** © Francis G. Mayer/Corbis; **136** *top left* The Granger Collection, New York; *bottom* © Bettmann/Corbis; **137** © Corbis; **138** © Richard T. Nowitz/Corbis; **139** *center right* © David Katzenstein/Corbis; *bottom* © Alan Schein Photography/Corbis; *top left* © Fridmar Damm/zefa/Corbis; *top center* © Getty Images; *second from top right* © Irwin Barrett/Getty Images; *top right* © Art Wolfe/Getty Images; **140** © Veer; **141** © Rob Lewine/Corbis; **142** © Getty Images; **143** © Pictorial Press/Alamy; **144** © Natalie Fobes/Corbis; **145** © Paul. J. Sutton/Duomo/Corbis; **147** *bottom* © Roger Ressmeyer/Corbis; *top left* © Fridmar Damm/ zefa/Corbis; *top center* © Getty Images; *second from top right* © Irwin Barrett/Getty Images; *top right* © Art Wolfe/Getty Images; **148** *top left* © Martin H. Simon/Corbis; *top center* © Corbis; *top right* © Brooks Kraft/Corbis; **150** © Chad Ehlers/Alamy; **151** © Francis G. Mayer/Corbis; **152** The Granger Collection, New York; **153** © Bettmann/Corbis.

Chapter 6

154 *bottom left* © David Sanger photography/Alamy; *bottom right* © Philip Gould/Corbis; *top left* © David Noton Photography/Alamy; **155** © Corbis; **156** © Kike Calvo/V&W/The Image Works, Inc.; **157** *bottom* © Ray Juno/Corbis; *top left* © Grant Taylor/Getty Images; *second from top left* © Corbis; *top center* © John Wang/Getty Images; *second from top right* © David Tipling/Alamy; *top right* © Bruce Heinemann/ Getty Images; **158** *center left* © Ron Watts/Corbis; *bottom left* © Terrance Klassen/Alamy; *center right* © Ed Simpson/Getty Images; *bottom right* © Wolfgang Kaehler/Corbis; **160** © Rob Howard/Corbis; **161** © Terry W. Eggers/Corbis; **162** *bottom left* © Richard T. Nowitz/ Corbis; **164** © Robert McGouey/Alamy; **165** *bottom* © Ron Watts/ Corbis; *center right* © David Sanger photography/Alamy; *top left* © Grant Taylor/Getty Images; *second from top left* © Corbis; *top center* © John Wang/Getty Images; *second from top right* © David Tipling/ Alamy; *top right* © Bruce Heinemann/Getty Images; **166** *bottom left* © John Farmer/Cordaiy Photo Library Ltd./Corbis; *John Cabot* (about 1450). Color print. Photo © Image Select/Art Resource, New York; *Jacques Cartier, Samuel de Champlain, Henry Hudson* © North Wind Picture Archives/Alamy; **167** © Bettmann/Corbis; **168** *bottom right* © Royalty-Free/Corbis; **169** © Ted Spiegel/Corbis; **170** *left* © Michael S. Yamashita/Corbis; *right* © Peter Arnold, Inc./Alamy; **171** *center left* © Paul A. Souders/Corbis; *bottom* © Wayne R. Bilenduke/Getty Images; *top left* © Grant Taylor/Getty Images; *second from top left* © Corbis; *top center* © John Wang/Getty Images; *second from top right* © David Tipling/Alamy; *top right* © Bruce Heinemann/Getty Images; **172** © Natalie Fobes/Corbis; **173** *top* © Bryan & Cherry Alexander Photography/Alamy; *bottom* © Corbis; **174** *top left* © Bryan & Cherry Alexander Photography/Alamy; *top center* © Alissa Crandall/Corbis; *second from right* © Wolfgang Kaehler/Corbis; *bottom right* © Bryan & Cherry Alexander Photography/Alamy; *bottom left* © Galen Rowell/Corbis; **175** © Wally McNamee/Corbis; **177** *bottom* © Andre Jenny/Alamy; *top left* © Grant Taylor/Getty Images; *second from top left* © Corbis; *top center* © John Wang/Getty Images; *second from top right* © David Tipling/Alamy; *top right* © Bruce Heinemann/Getty Images; **178** Jeff McIntosh/AP/Wide World Photos; **179** © Getty Images; **181** *bottom right* © David Sanger photography/Alamy; *bottom left* © Paul A. Souders/Corbis; **182** © Robert McGouey/Alamy.

Unit 3 Opener

185 *top right* © Peter M. Wilson/Alamy; *top left* © Ian Cumming/ Axiom; *bottom left* © Mauritius/SuperStock; *bottom right* © Rachael Bowes/Alamy; **190-191** © Wide Group/Iconica/Getty Images; **191** © eStock Photo/Alamy; **193** © Larry Fisher/Masterfile; **194** © Andoni Canela/ASA/IPNstock; **195** © Carol Barrington/IPNstock.

Chapter 7

196 *bottom left* © Werner Forman/Corbis; *bottom right* Chalchihuitlicue. Sculpture. British Museum, London. Photo © Werner Forman/Art Resource, New York; *top left* © Dynamic Graphics Group/IT Stock Free/Alamy; **197** *bottom left* The Granger Collection, New York; *bottom right* The Granger Collection, New York; **199** *top left* © Royalty-Free/Corbis; *top center* © Getty Images; *bottom* © Toby Maudsley/Getty Images; *top waterfall* © imagebroker/Alamy;

top macaw © Getty Images; *top leaves* © Getty Images; **200** © Jeff Hunter/Getty Images; **201** © Danny Lehman/Corbis; **202** *left* © Getty Images; *center* © eStock Photo/Alamy; *right* © Robert Everts/Getty Images; **203** © Gail Shumway/Getty Images; **204** © Gary Cook/Alamy; **205** *top left* © Royalty-Free/Corbis; *top center* © Getty Images; *bottom* © Peter Adams Photography/Alamy; *top waterfall* © imagebroker/Alamy; *top macaw* © Getty Images; *top leaves* © Getty Images; **206** © Jay Dickman/Corbis; **208** © Keith Dannemiller/Alamy; **209** *top left* © Simon Littlejohn/Alamy; *top right* © Derrick Francis Furlong/Alamy; *bottom left* © Pixonnet.com/Alamy; *bottom right* © Jeremy Horner/Corbis; **210** © Hemis/Alamy; **211** © H. Sitton/zefa/Corbis; **213** *top left* © Royalty-Free/Corbis; *top center* © Getty Images; *center right* © SEF/Art Resource, New York; *bottom* © Vanni/Art Resource, New York; *top waterfall* © imagebroker/Alamy; *top macaw* © Getty Images; *top leaves* © Getty Images; **214** *center left* Photo by Fernando Rochaix, University of Texas/Courtesy of the Naachtun Archaeological Project; *bottom left* Drawing by Peter L. Mathews, La Trobe University, Australia/Courtesy of the Naachtun Archaeological Project; *bottom center* Photo by Fernando Rochaix, University of Texas/Courtesy of the Naachtun Archaeological Project; **215** *center right* © Loren McIntyre-Woodfin Camp/IPNstock; © Getty Images; *center left* © Werner Forman/Art Resource, New York; The Granger Collection, New York; *left* Detail of a page from the Codex Becker depicting two war-chiefs. Museum fuer Voelkerkunde, Vienna, Austria. Photo © Werner Forman/Art Resource, New York; **216** © Dagli Orti/Museo Ciudad Mexico/The Art Archive; **217** © Wolfgang Kaehler/Corbis; **218** *bottom left* © Danny Lehman/Corbis; **220** The Granger Collection, New York; **221** *top left* © Royalty-Free/Corbis; *top center* © Getty Images; *top waterfall* © imagebroker/Alamy; *top macaw* © Getty Images; *top leaves* © Getty Images; **222** © James Quine/Alamy; **223** Father Hidalgo at the Battle of Monte de las Cruces, 1810 (1865), Joaquin Ramirez. National Palace, Mexico City. Photo © Art Resource, New York; **224** © J. Marshall - Tribaleye Images/Alamy.

Chapter 8

228 *bottom left* Portrait of Emperor Maximilian of Mexico (1865), Albert Graefle. Museo Nacional de Historia, Castillo de Chapultepec, Mexico City. Photo © Schalkwijk/Art Resource, New York; *bottom right* The Granger Collection, New York; *top left* © Tom Bean/Getty Images; **229** *bottom right* © Reuters/Corbis; **231** *center right* Santa Anna (1849). Lithograph. Photo The Granger Collection, New York; *bottom* Encounter of the Armies (1964), David Alfaro Siqueiros. Mural. Museo Nacional de Historia, Castillo de Chapultepec, Mexico City. Photo © Schalkwijk/Art Resource, New York; *top left* © Macduff Everton/Getty Images; *top center* © blickwinkel/Alamy; *pine* © Getty Images; *top right, large butterfly* © Getty Images; *top right, group of butterflies* © Elvele Images/Alamy; **233** Benito Juarez (1870). Steel engraving. Photo The Granger Collection, New York; **234** Francisco "Pancho" Villa (1914). Oil over a photograph. Photo The Granger Collection, New York; **235** © Janet Schwartz/Getty Images; **236** © Tony Anderson/Getty Images; **237** *bottom* © Royalty-Free/Corbis; *top left* © Macduff Everton/Getty Images; *top center* © blickwinkel/Alamy; *pine* © Getty Images; *top right, large butterfly* © Getty Images; *top right, group of butterflies* © Elvele Images/Alamy; **238** *left* © John Mitchell Stock Photography, photographersdirect.com; *right* © Viviane Moos/Corbis; **239** Self Portrait with Changuito (1945), Frida Kahlo. Fundacion Dolores Olmedo, Mexico City. © Banco de Mexico Trust. Photo © Schalkwijk/Art Resource, New York; **240** *top left* © JupiterMedia/Alamy; *center right* © Danita Delimont/Alamy; *bottom left* © Danita Delimont/Alamy; *bottom right* © Spike Mafford/Getty Images; **241** © Henry Romero/Reuters/Corbis; **242** *top right* © Kathleen Finlay/Masterfile; *center left* © David Hiser/Getty Images; *bottom right* © Kayte Deioma Photography, photographersdirect.com; *background* © Irfan Parvez, http://pg.photos.yahoo.com/ph/iparvezfareast/my_photos; **243** *top left* © Steve Hamblin/Alamy; *bottom* © Russell Monk/Masterfile; *center right* © Rommel/Masterfile; *background* © Chris Barton Travel Photography/photographersdirect.com; **244** © Keith Dannemiller/Corbis; **245** *bottom* © Lynsey Addario/Corbis; *top left* © Macduff Everton/Getty Images; *top center* © blickwinkel/Alamy; *pine* © Getty Images; *top right, large butterfly* © Getty Images; *top right, group of butterflies* © Elvele Images/Alamy; **246** © Reuters/Corbis; **247** © Oxford Scientific/Jupiter Images; **248** © Tom Tracy Photography/Alamy; **249** *bottom left* Benito Juarez (1870). Steel engraving. Photo The Granger Collection, New York; *bottom right* © Tony Anderson/Getty Images; *center left* © Blend Images/Veer; **250** © Danita Delimont/Alamy.

Chapter 9

252 *top left* © Stuart Westmorland/Getty Images; *bottom left* © Royalty-Free/Corbis; *bottom right* © Bettmann/Corbis; **253** *bottom left* © Brian A. Vikander/Corbis; *bottom right* © Bettmann/Corbis; **255** *center right* © Stuart Westmorland/Getty Images; *bottom* © Hemis/Alamy; *top left* © Getty Images; *top, reef* © Royalty-Free/Corbis; *top, small dolphin* © David Tipling/Alamy; *top, big dolphin* © Stuart Westmorland/Corbis; **257** © Matthew Johnston/Alamy; **258** © Mauritius/SuperStock; **259** *top right* © Justine Evans, photographersdirect.com; *top* © Kevin Schafer/Getty Images; **261** *bottom* © Look GMBH/eStock Photo; *top left* © Getty Images; *top, reef* © Royalty-Free/Corbis; *top, small dolphin* © David Tipling/Alamy; *top, big dolphin* © Stuart Westmorland/Corbis; **263** © Alejandro Ernesto/epa/Corbis; **265** © Bob Krist/Corbis; **267** *bottom* © Hisham Ibrahim/Getty Images; *center left two* © Pablo Corral V/Corbis; *center right* © age fotostock/SuperStock; *top left* © Getty Images; *top, reef* © Royalty-Free/Corbis; *top, small dolphin* © David Tipling/Alamy; *top, big dolphin* © Stuart Westmorland/Corbis; **268** *Simon Bolivar*. Painting. The Granger Collection, New York; **269** © Charles W Luzier/Corbis; **270** © Reuters/Corbis; **271** © Carl & Ann Purcell/Corbis; **273** *center right* © Michael J. P. Scott/Getty Images; *bottom* © Eye Ubiquitous/Corbis; *top left* © Getty Images; *top, reef* © Royalty-Free/Corbis; *top, small dolphin* © David Tipling/Alamy; *top, big dolphin* © Stuart Westmorland/Corbis; **274** © David Noton Photography/Alamy; **275** *top left* © Steve Vidler/eStock Photo; *bottom right* Victor R. Caivano/AP/Wide World Photos; **276** *top left* © Ed Kashi/Corbis; *top center* © Alistair Berg/Alamy; *top right* © Owen Franken/Corbis; **277** © Steve Vidler/eStock Photo; **278** *background* © Nik Wheeler/Corbis; *top right* © Stephen Frink/Corbis; *center left* © Ted Spiegel/Corbis; *bottom right* © Danita Delimont/Alamy; **279** *top left* © Robert Harding World Imagery/Corbis; *center right* © Pablo Corral V/Corbis; *bottom left* © Juniors Bildarchiv/Alamy; *background* © Fridmar Damm/zefa/Corbis; **281** *center right* © Hemis/Alamy; *bottom* © Jon Crwys-Williams/Alamy; *top left* © Getty Images; *top, reef* © Royalty-Free/Corbis; *top, small dolphin* © David Tipling/Alamy; *top, big dolphin* © Stuart Westmorland/Corbis; **282** *bottom left* © Chad Ehlers/Alamy; *top left* Jesus Inostroza/AP/Wide World Photos; **283** *right* © Danita Delimont/Alamy; *top left* © Claudia Raschke-Robinson/Just One Productions/The Kobal Collection; **284** © Richard Wareham Fotografie/Alamy; **285** *bottom right* © Michael J. P. Scott/Getty Images; **286** © Hemis/Alamy.

Chapter 10

288 *top left* © Theo Allofs/Getty Images; *bottom left* © Newberry Library/SuperStock; *bottom center* © Roger Day/Alamy; **289** *bottom left* © Luis Pacheco/Getty Images; *bottom right* © Christian Liewig/Tempsport/Corbis; **290** Dom Pedro I of Brazil (1800s). Lithograph. The Granger Collection, New York; **291** *top left* © Panoramic Images/Getty Images; *top right* © Luiz C. Marigo/Peter Arnold, Inc./Alamy; *top, ants* © Digital Archive Japan/Alamy; *top center* © Getty Images; *bottom* © Ary Diesendruck/Getty Images; **292** © The Granger Collection, New York; **294** *top left* © ImageState/Jupiter Images; *top right* © Eduardo Garcia/SuperStock; **295** © Adrian Lascom/Alamy; **296** © Terry Winn Photography, photographersdirect.com; **297** *top left* © Panoramic Images/Getty Images; *top right* © Luiz C. Marigo/Peter Arnold, Inc./Alamy; *top, ants* © Digital Archive Japan/Alamy; *top center* © Getty Images; *bottom* © Edward Parker/EASI-Images/CFWiages.com/photographersdirect.com; **298** *top right* Cover Illustration from *Tales of the Amazon: How the Munduruku Indians Live*. Text © 1996 by Daniel Munduruku. Illustrations © 1996 by Laurabeatriz. First published in Canada by Groundwood Books Ltd. Reprinted by permission of the publisher.; *bottom* © Images&Stories, photographersdirect.com; **299** © James Davis Photography/Alamy; **300** *top inset* © Paulo Fridman/Corbis; *top* © Cassio Vasconcellos/Getty Images; **301** *bottom left* © Rick Gomez/Masterfile; *second from top left* © Julia Waterlow/Eye Ubiquitous/Hutchison; *second from right* © Martina Urban/Getty Images; *bottom right* © Rainer Raffalski Photography, photographersdirect.com; *top left* © David R. Frazier Photolibrary, Inc./Alamy; **302** *left* © Sue Cunningham Photographic/Alamy; *inset* Photograph by Ed-Imaging; **303** © Tom Cockrem Photography, photographersdirect.com; **305** *top left* © Panoramic Images/Getty Images; *top right* © Luiz C. Marigo/Peter Arnold, Inc./Alamy; *top, ants* © Digital Archive Japan/Alamy; *top center* © Getty Images; *bottom* © mediacolor's/Alamy; **306-307** *bottom* © Will and Deni McIntyre/Getty Images; **306** *bottom inset* © Stephen Ferry/Getty Images; *top left* © Reuters News Picture Archive; **307** *top right* © Paulo Backes Fotografia e Paisagismo, photographersdirect.com; *bottom left* © Ricardo Beliel/BrazilPhotos/Alamy; *bottom right* © James Davis Photography/Alamy; **308** © Paulo Backes Fotografia e Paisagismo, photographersdirect.com; **309** *bottom left* © Roger Day/Alamy; *bottom right* © Lebrecht Music and Arts Photo Library/Alamy; *center left* © Creatas; **310** © The Granger Collection, New York.

Unit 4 Opener

313 *top left* © BL Images Ltd/Alamy; *top right* © David Sailors/Corbis; *bottom left* © Kieran Doherty/Reuters/Corbis; *bottom right* © William Manning/Corbis; **318** © Getty Images; **320** Peter Dejong/AP/Wide World Photos; **323** © Peter Adams Photography/Alamy.

R111

Chapter 11

324 *top left* © nagelestock.com/Alamy; *bottom left* The Granger Collection, New York; *bottom right* © Metropolitan Museum of Art; **325** *bottom left Napoleon Crossing the Alps* (1801), Jacques Louis David. Oil on canvas. 259 cm x 221 cm. Chateaux de Malmaison et Bois-Preau, Rueil-Malmaison, France. Photo © Réunion des Musées Nationaux/Art Resource, New York; *bottom right* © First/zefa/Corbis; **326** © K. Yamashita/PanStock/Panoramic Images/NGSImages.com; **327** *center right* © Jon Arnold Images/Alamy; *bottom* © Nik Wheeler/Corbis; *top left* © Josef Beck/Getty Images; *top right* © Su Davies/Life File/Getty Images; *top swan* © David Martyn Hughes/Alamy; **328** © Paul Hardy/Corbis; **330** *left* © Hortus/Alamy; *second from left* © images-of-france/Alamy; *second from right* © Stan Kujawa/Alamy; *right* © Jorma Jaemsen/zefa/Corbis; **331** © Doug Houghton/Alamy; **332** © nagelestock.com/Alamy; **335** *center right* © Ted Russell/Getty Images; *bottom* © Frank Chmura/Panoramic Images/National Geographic Image Collection; *top left* © Josef Beck/Getty Images; *top right* © Su Davies/Life File/Getty Images; *top, swan* © David Martyn Hughes/Alamy; **336** *bottom* © Peter Grumann/Alamy; *bottom right* The Granger Collection, New York; **337** Detail from the Alexander battle. Mosaic. Museo Archeologico Nazionale, Naples, Italy. Photo © Erich Lessing/Art Resource, New York; **339** © Hideo Kurihara/Alamy; **340** *top right* Kleophrades Painter (500-490 B.C.). Attic red figure amphora. 37 cm. Antikensammlung, Staatliche Museen zu Berlin, Berlin, Germany. Photo © Bildarchiv Preussischer Kulturbesitz/Art Resource, New York; *center left* Greek hydria from Vulci (530 B.C.). Museo Nazionale di Villa Giulia, Rome. Photo © Scala/Art Resource, New York; *left mask* © Ancient Art and Architecture; *right mask* © The Lowe Art Museum, The University of Miami/Superstock; *bottom right* © Buddy Mays/Corbis; *background* © John Elk III/Lonely Planet Images; **341** *top left* Neptune Calming the Waves (1757), Lambert-Sigisbert Adam. Marble sculpture. Louvre, Paris. Photo © Giraudon/Bridgeman Art Library; *top right Breadseller in Public Square*, from *House of the Baker*. Fresco. Museo Archeologico Nazionale, Naples, Italy. Photo © Scala/Art Resource, New York; *bottom left* © Gianni Dagli Orti/Corbis; *background* © Art Kowalsky/Alamy; **343** *center right* © Royalty-Free/Corbis; *bottom* Detail from the Bayeux Tapestry (1000s). Wool embroidery on linen. Musee de la Tapisserie, Bayeux, France. Photo © Bridgeman Art Library; *top left* © Josef Beck/Getty Images; *top right* © Su Davies/Life File/Getty Images; *top, swan* © David Martyn Hughes/Alamy; **344** © Metropolitan Museum of Art; **346** © Jason Hawkes/Corbis; **347** *top right Mona Lisa* (1503-1506), Leonardo da Vinci. Oil on wood. 77 x 53 cm. Inv. 779. Louvre, Paris. Photo © Erich Lessing/Art Resource, New York; *bottom left* Maesta (Madonna Enthroned), Duccio di Buoninsegna. Museo dell'Opera Metropolitana, Siena, Italy. Photo © Scala/Art Resource, New York; *bottom right The Peasants' Wedding* (1568), Pieter Brueghel, the Elder. Oil on oakwood. 114 cm x 164 cm. Kunsthistorisches Museum, Vienna, Austria. Photo © Erich Lessing/Art Resource, New York; **349** © akg-images; **351** *center right* © akg-images; *bottom An Experiment at the Accademia del Cimento*, Gaspero Martellini. Tribuna di Galileo, Museo della Scenza, Florence, Italy. Photo © Scala/Art Resource, New York; *top left* © Josef Beck/Getty Images; *top right* © Su Davies/Life File/Getty Images; *top, swan* © David Martyn Hughes/Alamy; **352** *top left* Execution of King Louis XVI of France (1793). Colored etching. Photo The Granger Collection, New York; *bottom right* Napoleon in Coronation Robe (1810), François Gerard. Oil on canvas 205 cm x 150.5 cm. Chateaux de Malmaison et Bois-Preau, Rueil-Malmaison, France. Photo © Réunion des Musées Nationaux/Art Resource, New York; **354** *left column, top* © Bettmann/Corbis; *left column, second from top* © Swim Ink 2, LLC/Corbis; *left column, second from bottom* © Corbis; *left column, bottom* © Bettmann/Corbis; *second column, top* © Hulton-Deutsch Collection/Corbis; *second column, second from top* © Swim Ink 2, LLC/Corbis; *second column, second from bottom* © Bettmann/Corbis; *second column, bottom* © Corbis; **356** © Photograph by Sharon Hoogstraten; **357** © Metropolitan Museum of Art; **358** © Photograph by Sharon Hoogstraten.

Chapter 12

360 *top left* © Rohan/Getty Images; *bottom left* Detail from Theodora's Court, bust of Theodora. S. Vitale, Ravenna, Italy. Photo © Scala/Art Resource, New York; *bottom right* The Taking of the Bastille, July 14, 1789 Anonymous French painter. Musée National du Chateau, Versailles, France. Photo © Erich Lessing/Art Resource, New York; **361** *bottom left* © Tom Stoddart Archive/Getty Images; *bottom right* © Suomen Kuvapalveluoy/Corbis; **363** *top left* © Adrian Neal/Getty Images; *top right* © Westend61/Alamy; *center right* Polynices and Eriphyle (about 450-440 B.C.) Red figure oinochoe. Louvre, Paris. Photo © Réunion des Musées Nationaux/Art Resource, New York; *bottom* © Roger Wood/Corbis; **364** © Sergio Pitamitz/Corbis; **365** © New York Times Co./Getty Images; **367** Luca Bruno/AP/Wide World Photos; **368** *center left* © Franco Origlia/Getty Images; *top right* © Gareth McCormack/Lonely Planet Images; *bottom* © Jose Fuste Raga/Corbis; **369** *top left* © MedioImages/Getty Images; *bottom left* © Robert Harding Picture Library Ltd/Alamy; *bottom* © Atlantide Phototravel/Corbis; **370** The Granger Collection, New York; **371** *top left* © Adrian Neal/Getty Images; *top right* © Westend61/Alamy; *center right* © Iraida Icaza/Getty Images; *bottom* © John Heseltine/Corbis; **372** © Pete Saloutos/Corbis; **373** Jon Dimis/AP/Wide World Photos; **374** Henry the Navigator (1400s), Nuno Goncalves. Museu Nacional de Arte Antiga, Lisbon, Portugal. Photo © Bridgeman-Giraudon/Art Resource, New York; **375** © Buddy Mays/Corbis; **376** © K. M. Westermann/Corbis; **377** *top left* © Adrian Neal/Getty Images; *top right* © Westend61/Alamy; *center right* © Wolfgang Rattay/Reuters/Corbis; *bottom* © Stock Connection/Jupiter Images; **378** Jacques Brinon/AP/Wide World Photos; **379** *right* © Angelo Cavalli/zefa/Corbis; *inset* © char abumansoor/Alamy; **380** *bottom center* © Brand X Pictures/Alamy; *bottom left* © Greg Stott/Masterfile; **381** © David Noble Photography/Alamy; **383** *top left* © Adrian Neal/Getty Images; *top right* © Westend61/Alamy; *bottom* © Josef Beck/Getty Images; **384** © Robert Maass/Corbis; **385** *bottom* © ImageState/Alamy; *center right* © David Bathgate/Corbis; **386-387** © Stefan Matzke/NewSport/Corbis; **387** © Denis Balibouse/Reuters/Corbis; **388** © Corbis; **389** *top left* © Adrian Neal/Getty Images; *top right* © Westend61/Alamy; *center right* © Nik Wheeler/Corbis; *bottom* © Chris Lisle/Corbis; **391** *bottom left* © Royalty-Free/Corbis; *top left* © Maskot; *second from top left* © P. Manner/zefa/Corbis; *second from right* © Jonathan Blair/Corbis; *bottom right* © FoodPix/Jupiter Images; **392** © Royalty-Free/Corbis; **393** © Nik Wheeler/Corbis; **394** © Rohan/Getty Images; **395** *bottom right* © Nik Wheeler/Corbis; **396** Detail from Theodora's Court, bust of Theodora. S. Vitale, Ravenna, Italy. Photo © Scala/Art Resource, New York.

Chapter 13

398 *bottom right* Signing the Magna Carta (1800s). Engraving. The Granger Collection, New York; *top left* © Kim Sayer/Corbis; *bottom left* © David Noble Photography/Alamy; **399** *bottom left* Factories: England (about 1850). Engraving. The Granger Collection, New York; *bottom right* © Tim Hawkins/Eye Ubiquitous/Corbis; **401** *bottom* © Panoramic Images/Getty Images; *top left* © Hideo Kurihara/Alamy; *top center* © images-of-france/Alamy; *top right* © Getty Images; *center right* © Peter Casolino/Alamy; **402** © PanStock/Jupiter Images; **403** *top* King William III and Queen Mary II. The Granger Collection; *bottom* © Swim Ink 2, LLC/Corbis; **404** © Hulton-Deutsch Collection/Corbis; **408** © Nathan Benn/Corbis; **409** *center right* © Daemmrich Bob/Corbis; *top left* © Hideo Kurihara/Alamy; *top center* © images-of-france/Alamy; *top right* © Getty Images; *bottom* © Panoramic Images/Getty Images; **410** © Grant Pritchard/britainonview; **411** *top* © Scott Barbour/Getty Images; *bottom* © Simon Reddy/Alamy; **412** *left* © Christian Liewig/Corbis; *right* © Steve C. Michell/epa/Corbis; **413** © Ashley Cooper/Corbis; **414** © Photo Researchers, Inc.; **415** *center right* © Tim Graham/Corbis; *bottom* © Julian Calder/Corbis; *top left* © Hideo Kurihara/Alamy; *top center* © images-of-france/Alamy; *top right* © Getty Images; **416** *second from right* © Sergio Pitamitz/zefa/Corbis; *left* © Free Agents Limited/Corbis; *second from left* © Photo 24/Brand X Pictures/Getty Images; *right* © Tim Graham/Corbis; **418** © qaphotos.com/Alamy; **419** © Robert Brook/Photo Researchers, Inc.; **420** © Visual Arts Library (London)/Alamy.

Chapter 14

422 *bottom left* Boar (400-300 B.C.). Bronze statuette. Hungarian National Museum, Budapest, Hungary. Photo © Erich Lessing/Art Resource, New York; *bottom right* © Bettmann/Corbis; *top left* © ML Sinibaldi/Corbis; **423** *bottom right* © James Reeve/Corbis; **425** *bottom* © Momatiuk-Eastcott/Corbis; *top left* © Simon Reddy/Alamy; *top right* © Bill Bachmann/Index Stock Imagery/Jupiter Images; **426** © Konrad Zelazowski/Alamy; **427** *bottom right* John Maniaci/Wisconsin State Journal/AP/Wide World Photos; *bottom* © Bernard Bisson/Corbis; **428** © Reuters/Corbis; **429** *hand with egg* © Jim Sugar/Corbis; *right egg* © Ingram Publishing/Alamy; *bottom* © Peter Turnley/Corbis; *left egg* © Hemera Technologies/Alamy; **430** © Tiit Veermae/Alamy; **431** © Peter Turnley/Corbis; **432** © Peter Turnley/Corbis; **433** *center right* © Stock Connection Distribution/Alamy; *bottom* © Panoramic Images/Getty Images; *top left* © Simon Reddy/Alamy; *top right* © Bill Bachmann/Index Stock Imagery/Jupiter Images; **434** *pepper* © imagebroker/Alamy; *can* © Danita Delimont/Alamy; *bottom left* © Catherine Karnow/Corbis; *bottom right* © Ivan Zupic/Alamy; **436** © Reuters/Corbis; **437** *top* © ML Sinibaldi/Corbis; *bottom* © Robert Harding World Imagery/Corbis; **438** *bottom left* © Jose Luis Pelaez Inc/Getty Images; *top left* © Janine Wiedel Photolibrary/Alamy; *top center* © David Young-Wolff/PhotoEdit; *top right* © Simon Reddy/Alamy; *bottom right* © Mario Ponta/Alamy; **439** *top* © Justin Leighton/Alamy; *top inset* © Motoring Picture Library/Alamy; **440** © Bettmann/Corbis; **441** *center right* AP/Wide World Photos; *bottom* © FAN travelstock/Alamy; *top left* © Simon Reddy/Alamy; *top right* © Bill Bachmann/Index Stock Imagery/Jupiter Images; **442** © Jon Hicks/Corbis; **443** © WoodyStock/Alamy; **444** © CapitalCity Images/Alamy; **445** © Bettmann/Corbis; **446** John Maniaci/Wisconsin State Journal/AP/Wide World Photos.

Unit 5 Opener

449 *bottom left* © 2004 Adrienne McGrath; *top right* © John Lamb/Getty Images; *bottom right* © Eitan Simanor/Alamy; **455** © Dean Conger/Corbis; **456** © Klaus Nigge/National Geographic/Getty Images; **457** © Sovfoto/Eastfoto, photographersdirect.com.

Chapter 15

458 *top left* © Ellen Rooney/Getty Images; *bottom left* © Diego Lezama Orezzoli/Corbis; *bottom right* © SIME s.a.s./eStock Photo; **459** *bottom right* The Granger Collection, New York; *bottom center* © Steve Allen Travel Photography/Alamy; *bottom left* © Corbis; **461** *bottom* © Ralph White/Corbis; *top left* © Pavel Filatov/Alamy; *top center* © Pavel Filatov/Alamy; *top right* © Pavel Filatov/Alamy; *top right inset* © Royalty-Free/Corbis; **462** © Boyd Norton; **463** *center left* © Michel Setboun/Corbis; *bottom left* © Carsten Peter/Getty Images; **464** © Bryan & Cherry Alexander Photography; **465** *top left* © Bryan & Cherry Alexander Photography/Alamy; *top right* © Pavel Filatov/Alamy; *bottom left* © Boyd Norton/Evergreen Photo Alliance; *bottom right* © Pat O'Hara/Corbis; **466** *bottom left* © Bryan & Cherry Alexander Photography/Alamy; *bottom center* © Peter Arnold, Inc./Alamy; *bottom right* © Bryan & Cherry Alexander Photography/Alamy; *top left* © Royalty-Free/Corbis; **469** *top left* © Pavel Filatov/Alamy; *top center* © Pavel Filatov/Alamy; *top right* © Pavel Filatov/Alamy; *top right inset* © Royalty-Free/Corbis; *bottom* © Gavin Hellier/Jon Arnold Images/Alamy; **470** © SuperStock; **471** © By courtesy of Sotheby's Picture Library, London; **473** © Bernard Bisson & Thierry Orban/Sygma/Corbis; **474** *top left* © JTB Photo/Alamy; **476** © Mark Sykes/Alamy; **477** *bottom* © Picture Contact/Alamy; *top left* © Pavel Filatov/Alamy; *top center* © Pavel Filatov/Alamy; *top right* © Pavel Filatov/Alamy; *top right inset* © Royalty-Free/Corbis; **478** © Reuters/Corbis; **479** *top right* © Mark Sykes/Alamy; *center inset* © Reuters/Corbis; *center left* © Douglas Armand/Getty Images; *background* © José Fuste Raga/zefa/Corbis; *bottom right* © AFP/Getty Images; **480** *bottom inset* © Franck Fife/AFP/Getty Images; **481** *bottom inset* © Lesegretain/Corbis; *bottom* © Pierre Perrin/Corbis; **482** *top* © 2006 Artists Rights Society (ARS), New York/ADAGP, Paris. Photo © Albright-Knox Art Gallery/Corbis; *bottom* © Kurov Alexander/Itar-Tass/Corbis; **483** © Thomas Johnson/Corbis; **484** *top left* © Bryan & Cherry Alexander; *background* © Marc Garanger/Corbis; *bottom right* © Nick Haslam/Alamy; *top right* © Jack Sullivan/Alamy; **485** *top left* © Iain Masterton/Alamy; *top right* © Peter Turnley/Corbis; *background* © Bryan & Cherry Alexander Photography/Alamy; *bottom left* © Steve Raymer/Corbis; *second from top left* © Iain Masterton/Alamy; **487** *center right* © Panov Alexei/ITAR-TASS/Corbis; *bottom* © Jeremy Nicholl/Alamy; *top left* © Pavel Filatov/Alamy; *top center* © Pavel Filatov/Alamy; *top right* © Pavel Filatov/Alamy; *top right inset* © Royalty-Free/Corbis; **488** *right* © Reuters/Corbis; **489** © Jeremy Nicholl/Alamy; **491** © SuperStock; **492** © Mark Sykes/Alamy.

Chapter 16

494 *top left* © Keren Su/China Span/Alamy; *bottom left* © Paul H. Kuiper/Corbis; *bottom right* Portrait of Timur (1780), Pierre Duflos. © Stapleton Collection/Corbis; **495** *left* © WorldSat International, Inc.; *right* © WorldSat International, Inc.; **497** *bottom* © Marc Garanger/Corbis; *top left* © Upperhall/Robert Harding World Imagery/Getty Images; *top right* © Imageshop/Alamy; *top right inset* © Creatas/Dynamic Graphics Group/Alamy; **498** © Jerry Kobalenko/Getty Images; **499** © Charles and Josette Lenars/Corbis; **500** *left* © Zurab Kurtsikidze/epa/Corbis; *right* © Tony Allen Photography, photographersdirect.com; **501** © Reuters/Corbis; **502** © AFP/Getty Images; **503** © Reuters/Corbis; **504** © Charles and Josette Lenars/Corbis; **505** *bottom* Temp. Credit: © Ludovic Maisant/Corbis; *top left* © Upperhall/Robert Harding World Imagery/Getty Images; *top right* © Imageshop/Alamy; *top right inset* © Creatas/Dynamic Graphics Group/Alamy; *center right* © Nevada Wier/Corbis; **506** © Robert Harding Picture Library Ltd./Alamy; **507** *bottom left* © Titus Moser/Eye Ubiquitous/Hutchison; *center right* © Marco Brivio/Alamy; *bottom right* Robert Harding Picture Library; **508** © HIP-Archive/Topham/The Image Works, Inc.; **509** *bottom* © Simon Richmond/Lonely Planet Images; *top right* © Vladimir Sidoropolev Photography, photographersdirect.com; *center right* © Vladimir Sidoropolev Photography, photographersdirect.com; *bottom right* © James Strachan/Getty Images; **510** *bottom left* © Photo © 2004 Adrienne McGrath; *top left* © Eliane Farray-Sulle/Alamy; *second from top left* © Antoine Gyori/Corbis; *second from right* © Antoine Gyori/Corbis; *right* © Index Stock Imagery/Jupiter Images; **511** © Reuters/Corbis; **513** *bottom* © Alexey Panov/AFP/Getty Images; *top left* © Upperhall/Robert Harding World Imagery/Getty Images; *top right* © Imageshop/Alamy; *top right inset* © Creatas/Dynamic Graphics Group/Alamy; **514** Joseph Stalin (1935), Nikolaj Vasilevic Tomskij. Oil on canvas. © SuperStock; **515** © Sergei Karpukhin/Reuters/Corbis; **516** © Staton R. Winter/Getty Images; **517** *bottom left* © Marc Garanger/Corbis; *bottom right* Robert Harding Picture Library; **518** © Nevada Wier/Corbis; **519** © David Samuel Robbins/Corbis.

Unit 6 Opener

521 *top left* © Renee Lynn/Getty Images; *top right* © Royalty-Free/Corbis; *bottom left* © Remi Benali/Corbis; *bottom right* © W. Robert Moore/National Geographic/Getty Images; **522** © Getty Images; **527** © Peter Turnley/Corbis; **529** © blickwinkel/Alamy; **530-531** © Heinrich van den Berg/Gallo Images/Getty Images; **533** © Peter Johnson/Corbis.

Chapter 17

534 *top left* © Renee Lynn/Getty Images; *bottom left* © Gabriela Staebler/zefa/Corbis; *bottom center* © Bettmann/Corbis; *bottom right* © Javier Trueba/MSF/Photo Researchers, Inc.; **535** *bottom left* © Gianni Dagli Orti/Corbis; *bottom right* © Wildlife Art Ltd.; **536** © Kazuyoshi Nomachi/Corbis; **537** *center right* © Joe McDonald/Corbis; *bottom* © Index Stock/Alamy; *top left* © Getty Images; *top center* © John Conrad/Corbis; *top right* © James Warwick/Getty Images; **538** © ISS-NASA/Getty Images; **539** *left* © Images of Africa Photobank/Alamy; *right* © Torleif Svensson/Corbis; **540** *top left* © Stockbyte Silver/Alamy; **543** © George Steinmetz/Corbis; **545** *center right* © Tom McHugh/Photo Researchers, Inc.; *bottom* © BennettPhoto/Alamy; *top left* © Getty Images; *center left* © John Conrad/Corbis; *top right* © James Warwick/Getty Images; **546** *top left* © Patrick Robert/Corbis; *bottom left* © John Reader/Photo Researchers, Inc.; *bottom right* © Bettmann/Corbis; **547** © John Reader/Photo Researchers, Inc.; **548** © Carla Signorini/Images of Africa Photobank/Alamy; **549** *center left* © Michael Hart/Getty Images; *bottom left* © Bettmann/Corbis; *bottom right* © Kazuyoshi Nomachi/Corbis; **550** *top left* © Torleif Svensson/Corbis; *bottom left* © Stockbyte Silver/Alamy.

Chapter 18

552 *top left* © Royalty-Free/Corbis; *bottom left* © Charles & Josette Lenars/Corbis; *bottom right Selim I* (1808), John Young. Watercolour on paper. Photo © Bridgeman Art Library; **553** *bottom left* © The Art Archive/Corbis; *bottom right* © Jonathan Blair/Corbis; **555** *bottom* © G.J. Owen; *top left* © Sami Sarkis/Getty Images; *top center* © Getty Images; *top right* © Sandi Ford/Alamy; **556** © Lloyd Cluff/Corbis; **557** *bottom left* © Historical Picture Archive/Corbis; *bottom* Photograph by Sharon Hoogstraten **558** © Frank Trapper/Corbis; **559** © Sandro Vannini/Corbis; **560** *top left* © Jean-Pierre Lescourret/Corbis; **562** © Jonathan Blair/Corbis; **563** *center right* © The Art Archive/Corbis; *bottom* © Jean Dominique DALLET/Alamy, *top left* © Sami Sarkis/Getty Images; *top center* © Getty Images; *top right* © Sandi Ford/Alamy; **564** *Mehemet Ali* (1800s), Louis Charles Couder. Oil on canvas. Chateau de Versailles, France. Photo © Bridgeman Art Library; **565** © Premium Stock/Corbis; **566** © Actual Films/Principe Productions; **567** © Wally Nell/epa/Corbis; **568** © Chris Lisle/Corbis; **569** *bottom* © Antoine Gyori/Corbis; *top left* © Sami Sarkis/Getty Images; *top center* © Getty Images; *top right* © Sandi Ford/Alamy; **570** *left* © Chris Lisle/Corbis; *right* © Roger Antrobus/Corbis; **571** *bottom* © Patrick Robert/Corbis; *top* © Digital Vision/Getty Images; *center* © Comstock Images/Jupiter Images; **572** © Peter Adams Photography/Alamy; **574** © K. M. Westermann/Corbis; **575** © Giraudou Laurent/Corbis; **576** © Lloyd Cluff/Corbis.

Chapter 19

578 *top left* © SuperStock; *bottom right* © David Jones/Alamy; *bottom left* © Art Wolfe/Getty Images; **579** *bottom left* © Mary Evans Picture Library/Alamy; *bottom right* © Corbis; **581** *top left* © Sylvia Cordaiy Photo Library Ltd./Alamy; *top center* © Joshua D. Polman/Alamy; *top right* © Stan Osolinski/Oxford Scientific/Jupiter Images; *bottom* © Frans Lemmens/Getty Images; **583** Askia Muhammed. © Leo and Diane Dillon; **585** © Alain DeJean/Corbis; **586** © Robert Estall photo agency/Alamy; **587** *top left* © Sylvia Cordaiy Photo Library Ltd./Alamy; *top center* © Joshua D. Polman/Alamy; *top right* © Stan Osolinski/Oxford Scientific/Jupiter Images; *bottom* © Paul Almasy/Corbis; **589** *top left* © Greenpeace/Corbis; *bottom right* Saurabh Das/AP/Wide World Photos; **590** *bottom left* © Paul Almasy/Corbis; *bottom* © Digital Vision Ltd./SuperStock; **591** © Margaret Courtney-Clarke/Corbis; **592** *top left* © Robert Estall photo agency/Alamy; *top right* © Michael S. Lewis/Corbis; *center* © Yann Arthus-Bertrand/Corbis; *bottom right* © Gary Cook/Alamy; **593** *top left* © Julian Nieman/Alamy; *top right* © Gary Cook/Alamy; *center* © Yann Arthus-Bertrand/Corbis; *bottom left* © Stephanie Colasanti/Corbis; **594** © Peter Bowater/Alamy; **595** *top left* © Sylvia Cordaiy Photo Library Ltd./Alamy; *top center* © Joshua D. Polman/Alamy; *top right* © Stan Osolinski/Oxford Scientific/Jupiter Images; *center right* © Nik Wheeler/Corbis; *bottom* © Karsten Wrobel/Alamy; **596** © David C. Conrad; **597** *top left* © Mike Goldwater/Alamy; *top center* © Finbarr O'Reilly/Reuters/Corbis; *second from right* © Michael Dwyer/Alamy; *bottom right* © Daniel Berehulak/Getty Images; **598** *center left* © Luc Gnago/Reuters/Corbis; *bottom* © Patrick Robert/Corbis; **599** © Christian Liewig/Corbis; **600** © Charles and Josette Lenars/Corbis; **601** © Margaret Courtney-Clarke/Corbis; **602** © Robert Estall photo agency/Alamy.

R113

Chapter 20

604 *top left* © Brian A. Vikander/Corbis; *bottom left* © Werner Forman/Corbis; *bottom right* © David Reed/Corbis; **605** *bottom left* © Marion Kaplan/Alamy; **606** © Pixtal/SuperStock; **607** *top left* © Mike Copeland/Getty Images; *top center* © Paul A. Souders/Corbis; *top right* © Art Wolfe/Getty Images; *center right* © J. P. De Manne/Robert Harding World Imagery/Getty Images; *bottom* © Robert Preston/Alamy; **609** © Werner Forman/Corbis; **610** *bottom left* © imagebroker/Alamy; *bottom right* © Stock Connection/Jupiter Images; **611** © Earl and Nazima Kowall/Corbis; **612** © age fotostock/SuperStock; **613** *top left* © Mike Copeland/Getty Images; *top center* © Paul A. Souders/Corbis; *top right* © Art Wolfe/Getty Images; *center right* © Martin Harvey/Corbis; *bottom* © Images of Africa Photobank/Alamy; **614** *top left* © Charles O. Cecil/Alamy; *top right* © foodfolio/Alamy; **615** © Dylan Martinez/Reuters/Corbis; **616** © Marion Kaplan/Alamy; **617** © Gavriel Jecan/Corbis; **619** *top left* © Mike Copeland/Getty Images; *top center* © Paul A. Souders/Corbis; *top right* © Art Wolfe/Getty Images; *bottom* © David Wall/Alamy; **621** © Jacques Jangoux/Alamy; **622** *left* © Francisco Cruz/SuperStock; *center* © J. Marshall - Tribaleye Images/Alamy; *right* © Robert Estall photo agency/Alamy; **623** Lionel Healing/AFP/Getty Images; **625** *top left* © Mike Copeland/Getty Images; *top center* © Paul A. Souders/Corbis; *top right* © Art Wolfe/Getty Images; *center right* Naashon Zalk/AP/Wide World Photos; *bottom* © Eric Nathan/Alamy; **626** © David Turnley/Corbis; **627** AP/Wide World Photos; **628** *left* © Patrick Durand/Corbis; *right* © Charles O'Rear/Corbis; **630** © David Reed/Corbis; **631** *top left* © Mike Copeland/Getty Images; *top center* © Paul A. Souders/Corbis; *top right* © Art Wolfe/Getty Images; *bottom* © Patrick Ward/Corbis; **632** *bottom* © Robert Harding Picture Library Ltd/Alamy; *center right* Conical tower on the southern side of the great enclosure. The purpose is still shrouded in mystery. Marius Loots, South Africa; **633** © Anthony Bannister/Gallo Images/Corbis; **634** © Brian Seed/Alamy; **636** © Expuesto - Nicolas Randall/Alamy; **637** © Bernard Bisson/Corbis; **638** © Robert Harding Picture Library Ltd/Alamy; **639** © /politicalcartoons.com.

Unit 7 Opener

641 *top left* © Frans Lemmens/zefa/Corbis; *top right* © Medioimages/Alamy; *bottom left* © Sheldan Collins/Corbis; *bottom right* © Jiri Rezac/Alamy; **642** © Getty Images; **646-647** © Jayanta Shaw/Reuters/Corbis; **649** © Massimo Listri/Corbis.

Chapter 21

650-651 © Michael S. Yamashita/Corbis; **650** *top left* © Alistair Baird/Alamy; *bottom left* © Zen Icknow/Corbis; *bottom right* © Visual Arts Library (London)/Alamy; **651** *bottom right* © Gianni Dagli Orti/Corbis; **652** © Jeremy Horner/Corbis; **653** *center right* © Roger Wood/Corbis; *bottom* © Profimedia.CZ s.r.o./Alamy; *top left* © Design Pics Inc./Alamy; *top center* © Dinodia Images/Alamy; *top right* © Michelangelo Gratton/Getty Images; **654** © Ricki Rosen/Corbis; **656** *bottom* © Sindre Ellingsen/Alamy; **657** Burhan Ozbilici/AP/Wide World Photos; **659** *center right* © Arko Datta/Reuters/Corbis; *top left* © Design Pics Inc./Alamy; *top center* © Dinodia Images/Alamy; *top right* © Michelangelo Gratton/Getty Images; **660** *left* © Alistair Baird/Alamy; *right* © David Samuel Robbins/Corbis; **662** © Rafiqur Rahman/Reuters/Corbis; **663** © Anuruddha Lokuhapuarachchi/Reuters/Corbis; **664** *center left* © Reuters/Corbis; *bottom left* © Rafiqur Rahman/Reuters/Corbis; **667** *second from center right* © Science Museum/SSPL/The Image Works, Inc.; *center right* © Science Museum/SSPL/The Image Works, Inc.; *bottom* © Tor Eigeland/Alamy; *top left* © Design Pics Inc./Alamy; *top center* © Dinodia Images/Alamy; *top right* © Michelangelo Gratton/Getty Images; **668** *bottom* © Ali Jarekji/Reuters/Corbis; *bottom left* © Peter Horree/Alamy; **669** © Gianni Dagli Orti/Corbis; **670** © JTB Photo/Alamy; **671** *bottom left* © Michael S. Yamashita/Corbis; *bottom right* © Visual Arts Library (London)/Alamy; **672** © Michael S. Yamashita/Corbis.

Chapter 22

674 *top left* © Medioimages/Alamy; *bottom left* © Bojan Brecelj/Corbis; *bottom right* © Peter Horree/Alamy; **675** *bottom left* © Trip/Alamy; *bottom right* © W. Robert Moore/National Geographic/Getty Images; **676** © Popperfoto/Alamy; **677** *top left* © Tibor Bognár/Corbis; *top right* © Mike Hill/Oxford Scientific/Jupiter Images; *bottom* © George Steinmetz/Corbis; **678** © Dbimages/Alamy; **680** © Robert Patrick/Corbis; **681** © Paul Almasy/Corbis; **683** *top left* © Tibor Bognár/Corbis; *top right* © Mike Hill/Oxford Scientific/Jupiter Images; *center right* © Reinhard Krause/Reuters/Corbis; *bottom* © Annie Griffiths Belt/Corbis; **684** © Nina Large/ArenaPal/Topham/The Image Works, Inc.; **685** *left* © Royalty-Free/Corbis; *center* © Hanan Isachar/Israel images/Alamy; *right* © Sandro Vannini/Corbis; **686** © Hanan Isachar/Israel images/Alamy; **687** © Douglas Kirkland/Corbis; **689** *top left* © Tibor Bognár/Corbis; *top right* © Mike Hill/Oxford Scientific/Jupiter Images; *bottom* © Herbert Spichtinger/zefa/Corbis; **690** © Murat Taner/zefa/Corbis; **692** © Hemis/Alamy; **693** © Tim Graham/Alamy; **694** © Herbert Spichtinger/zefa/Corbis.

Chapter 23

696 *top left* © Wolfgang Kaehler/Corbis; *bottom left* © Borromeo/Art Resource, New York; *bottom right* © The British Library/HIP/The Image Works, Inc.; **697** *bottom left* © Pallava Bagla/Corbis; *bottom right* © Dinodia Images/Alamy; **699** *top left* © Gale Beery/Index Stock Imagery/Jupiterimages Corporation; *top center* © Getty Images; *top right* © blickwinkel/Alamy; *center right* © Maciej Wojtkowiak/Alamy; *bottom* © Grant Dixon/Lonely Planet Images/Getty Images; **700-701** © Historical Picture Archive/Corbis; **701** © The British Library/HIP/The Image Works, Inc.; **702** © Pallava Bagla/Corbis; **703** © Bettmann/Corbis; **704** © Bettmann/Corbis; **705** AP/Wide World Photos; **706** © Burstein Collection/Corbis; **707** *top left* © Gale Beery/Index Stock Imagery/Jupiterimages Corporation; *top center* © Getty Images; *top right* © blickwinkel/Alamy; *center right* © Dinodia Images/Alamy; *bottom* © Michael S. Lewis/Corbis; **709** *bottom* © CMCD, Inc.; *inset top* © Jeremy Horner/Corbis; *inset bottom* © Lindsay Hebberd/Corbis; **710** *left* © Earl and Nazima Kowall/Corbis; *right* © Robert Wallis/Corbis; **711** *center right* © DAMFX/The Kobal Collection; *bottom* © Arko Datta/Reuters/Corbis; **712** *bottom left* © Photosindia.com LLC/Alamy; *top left* © David H. Wells/Corbis; *top center* © Jonah Calinawan/Alamy; *second from right* © Dinodia Images/Alamy; *right* © Dinodia Photo Library; **713** © Kamal Kishore/Reuters/Corbis; **714** *center left* © David Samuel Robbins/Corbis; *top right* © Christine Osborne/Corbis; *background* © Stephanie Colasanti/Corbis; *bottom right* © Chris Lisle/Corbis; **715** *top left* © Kapoor Baldev/Corbis; *top right* © Robert Wallis/Corbis; *background* © Dinodia Images/Alamy; *bottom left* © Brian Lee/Corbis; **716** © Charles and Josette Lenars/Corbis; **717** *top left* © Gale Beery/Index Stock Imagery/Jupiterimages Corporation; *top center* © Getty Images; *top right* © blickwinkel/Alamy; *center right* © Stringer/India/Reuters/Corbis; *bottom* © John Peter Photography/Alamy; **718** © Getty Images; **720** © Pallava Bagla/Corbis; **722** © Reuters/Corbis; **723** *bottom left* © Pallava Bagla/Corbis; *bottom right* © Dinodia Images/Alamy; **724** © Burstein Collection/Corbis.

Chapter 24

726 *top left* © Casa Productions/Corbis; *bottom left* © Joe McDonald/Corbis; *bottom center* © Reuters/Corbis; *bottom right* © Time Life Pictures/Getty Images; **727** *bottom left* © Dennis M. Sabangan/epa/Corbis; *bottom right* © Gopal Chitrakar/Reuters/Corbis; **729** *top left* © Colin Monteath/Oxford Scientific/Jupiter Images; *top center* © Andy Rouse/Getty Images; *center right* © Michael Edwards/Getty Images; *bottom* © Michael S. Yamashita/Corbis; **730** © Reuters/Corbis; **732** *left* © Galen Rowell/Corbis; *right* © Tiziana and Gianni Baldizzone/Corbis; **733** © Liba Taylor/Corbis; **734** *top left* © Zohra Bensemra/Reuters/Corbis; *top right* © Faisal Mahmood/Reuters/Corbis; *center* © Mian Khursheed/Reuters/Corbis; *bottom right* © Goran Tomasevic/Reuters/Corbis; **735** *top left* © Jayanta Shaw/Reuters/Corbis; *top right* © Rafiqur Rahman/Reuters/Corbis; *center* © Rafiqur Rahman/Reuters/Corbis; *bottom left* © Rafiqur Rahman/Reuters/Corbis; **736** © Chad Ehlers/Alamy; **737** *top left* © Colin Monteath/Oxford Scientific/Jupiter Images; *top center* © Andy Rouse/Getty Images; *center right* © Michael Edwards/Getty Images; *center right* © Sergio Pitamitz/zefa/Corbis; *bottom* © Eye Ubiquitous/Corbis; **738** *top* © Stock Connection/Alamy; *bottom* © Keren Su/China Span/Alamy; **740** © Jeremy Horner/Corbis; **741** *center left* © Image Source/Getty Images; *bottom right* © Reuters/Corbis; **742** © Chad Ehlers/Alamy; *bottom* © Keren Su/China Span/Alamy.

Unit 8 Opener

745 *top left* © Panorama Media (Beijing) Ltd./Alamy; *top right* © Jeremy Woodhouse/Getty Images; *bottom left* © Darby Sawchuck Photography, photographersdirect.com; *bottom right* © Tibor Bognár/Corbis; **751** © Arco Images/Alamy; **753** © Ingo Jezierski/Getty Images.

Chapter 25

754 *top left* © Getty Images; *bottom left* Confucius (551-479 B.C.). Scroll painting. Bibliotheque Nationale, Paris. Photo © Snark/Art Resource, New York; *bottom right* Qin Shi Huang (about 1900). Photo © HIP/Art Resource, New York; **755** *bottom left* © Luis Castaneda Inc/Getty Images; *bottom right* © Peter Adams/Getty Images; **756** © John Foster/Masterfile; **757** *top left* © Jose Fuste Raga/Corbis; *top right* © Angelo Cavalli/Getty Images; *bottom* © epa/Corbis; **758** *top left* © Royalty-Free/Corbis; *bottom* © ImagineChina; **760** The Granger Collection, New York; **761** *left* © Keren Su/China Span; *right* © Michael S. Yamashita/Corbis; **762** © China Photos/Reuters/Corbis; **763** Science Museum/Science and Society Picture Library; **765** *top left* © Jose Fuste Raga/Corbis; *top right* © Angelo Cavalli/Getty Images; *center right Qin Shi Huang* (about 1900). Photo © HIP/Art Resource, New York; *bottom* © Juliet Coombe/Getty Images; **766** © Harvey Lloyd/Getty Images; **767** The Granger Collection, New York; **768** *left* Chinese coins with holes for stringing on cords. (tenth century B.C.) © Dagli Orti/The Art Archive; *top right* Marco Polo (1820) Line engraving. Italian. The Granger Collection; *top left* Banknote from Kublai Khan's first issue of banknotes. Private Collection/Photo © The Bridgeman Art Library; **769** *compass* © Science Museum/Science and Society Picture Library; *bottom left* Chinese map of the Eastern hemisphere (1790). Staatsbibliothek zu Berlin, Berlin, Germany. Photo © Bildarchiv Preussischer Kulturbesitz/Art Resource, New York; **771** © Yann Layma/Getty Images; **772** *top left* © Yann Layma/Getty Images; **775** *top left* © Jose Fuste Raga/Corbis; *top right* © Angelo Cavalli/Getty Images; *bottom* © ImagineChina; *center right* © Inga Spence/Index Stock Imagery, Inc.; **776** © ImagineChina; **779** © Bruce Connolly/Corbis; **780** *bottom left* © Jack Hollingsworth/Corbis; *top left* © Rick Lew/FoodPix/Jupiter Images; *second from top left* © Bruno Barbey/Magnum Photos; *second from right* © Royalty-Free/Corbis; *bottom right* © ImagineChina; **782** *bottom* © Keren Su/Getty Images; *center left* © David Butow/Corbis; *top right* Photo © 2004 Adrienne McGrath; *bottom right* © Tom Nebbia/Corbis; **783** *bottom* © Didier Bauweraerts/Van Parys Media/Corbis; *top left* © Yann Layma/Getty Images; *top right* © ImagineChina; *bottom left* © Keren Su/China Span; **784** © Robin Moyer/OnAsia; **785** *top left* © Jose Fuste Raga/Corbis; *top right* © Angelo Cavalli/Getty Images; *center right* © Frederic J. Brown/AFP/Getty Images; *bottom* © Digital Vision/Getty Images; **786** *left* © Joseph Sohm/ChromoSohm Inc./Corbis; *bottom center* © Free Agents Limited/Corbis; *bottom right* © Photo 24/Brand X Pictures/Getty Images; **788** © Jon Hicks/Corbis; **789** © ImagineChina; **790** © China Daily Information Corp/Reuters News Picture Archive; **791** *bottom right* *Qin Shi Huang* (about 1900). Photo © HIP/Art Resource, New York; **792** *top left* Confucius (551-479 B.C.). Scroll painting. Bibliotheque Nationale, Paris. Photo © Snark/Art Resource, New York.

Chapter 26

794 *bottom left* © JTB Photo/Alamy; *top left* © WizData,inc./Alamy; *bottom right* Gold crown from Ch'onmachong. Silla kingdom. National Treasure No. 188. National Kyongiu Museum, Republic of Korea; **795** *bottom left* © Bobby Yip/Reuters/Corbis; *bottom center* © Richard Hamilton Smith/Corbis; **796** NASA; **797** *center right* © Wolfgang Kaehler/Corbis; *bottom* © Trip/Alamy; *top left* © Andre Seale/Alamy; *top right* © blickwinkel/Alamy; *top right inset* © Gary Stones/Alamy; **798** © Craig J. Brown; **799** *top right* © JTB Photo/Alamy; *center right* © FAN travelstock/Alamy; **800** *left* © Jörén Gerhard/OnAsia; *right* © TongRo Image Stock/Alamy; **801** © Steve Kaufman/Corbis; **802** *bottom right* © David White/Alamy; *bottom* © Digital Archive Japan/Alamy; **804** © John Borthwick/Lonely Planet Images; **805** *bottom* © Werner Forman/Corbis; *top left* © Andre Seale/Alamy; *top right* © blickwinkel/Alamy; *top right inset* © Gary Stones/Alamy; **806** © Leonard de Selva/Corbis; **808** *bottom right* © Photojourneys, photographersdirect.com; *inset* © Claro Cortes IV/Reuters/Corbis; **809** *bottom left* © LWA/Dann Tardif/Getty Images; *top left* © Craig J. Brown; *top center* © Fotosearch Stock Photography; *second from right* © Jean Chung/OnAsia; *bottom right* © Chad Ehlers/Alamy; **810-811** Pear-shaped vase (1100s to 1200s). Korea. Celadon ceramic, with incrustations under "sanggam" glaze. Koryo period. Height 32.5 cm. Inv.: MG 18279. Musée des Arts Asiatiques-Guimet, Paris. Photo © Pascal Pleynet/Réunion des Musées Nationaux/Art Resource, New York; **810** *top left* © Chris Lisle/Corbis; *top right* © Chris Lisle/Corbis; **811** © KCNA/epa/Corbis; **812** © Christopher Farina/Corbis; **813** *center right* NASA; *bottom* © KCNA/epa/Corbis; *top left* © Andre Seale/Alamy; *top right* © blickwinkel/Alamy; *top right inset* © Gary Stones/Alamy; **814** © nichola west/Alamy; **815** *top right* © Tom Wagner/Corbis; *bottom* © Jeon Heon-Kyun/epa/Corbis; **816** © Chung Sung-Jun/Getty Images; **818** © Reuters/Corbis; **819** © John Borthwick/Lonely Planet Images; **820** © David White/Alamy.

Chapter 27

822 *top left* © Toyofumi Mori/Getty Images; *bottom left* Shotoku Taishi. Private Collection, Paris. Photo © Dagli Orti/The Art Archive; *bottom right* © Burstein Collection/Corbis; *bottom* © Archivo Iconografico, S.A./Corbis; **823** *bottom left* © Tibor Bognar/Corbis; *bottom right* © Martin Plomer/Dorling Kindersley/Getty Images; **825** *top right* © Seth Rutledge/Alamy; *top center* © Yoshio Sawaragi/Dex Image/Jupiter Images; *top left* © C. Bowman/Robert Harding World Imagery/Getty Images; *center right* © Tim McGuire/Corbis; *bottom* © Free Agents Limited/Corbis; **826** © Reuters/Corbis; **828** © Roger Ressmeyer/Corbis; **829** © DLILLC/Corbis; **830** © Chris Lisle/Corbis; **831** *top right* © Seth Rutledge/Alamy; *top center* © Yoshio Sawaragi/Dex Image/Jupiter Images; *top left* © C. Bowman/Robert Harding World Imagery/Getty Images; *bottom* © Shin Terada, Sweetish Fotostage ST; *center right* © Royalty-Free/Corbis; **832** *left* Eisei-Bunko Museum; *right* Detail of Knight with green and white shields on red horse: Capodilista Codex. Manuscripts. Civic Library of Padua, Italy. Photo © SuperStock; **833** *top* © Ric Ergenbright/Corbis; *bottom* © Iain Masterton/Alamy; **835** *top* © Jon Hicks/Corbis; *bottom left* Photograph by Bac To Trong; **836** *top left* © Christie's Images/Corbis; *bottom* © R. Ian Lloyd/Masterfile; **837** © Andy Rain/epa/Corbis; **838** *top right* © Royalty-Free/Corbis; *top left* © Michael Kim/Corbis; *background* © Catherine Karnow/Corbis; *bottom right* © Jack Fields/Corbis; **839** *top right* © Walt Disney Pictures/ZUMA/Corbis; *top left* © Masterfile; *background* © Douglas Armand/Alamy; *bottom left* © Barry Lewis/Corbis; **840** © Bettmann/Corbis; **841** *top right* © Seth Rutledge/Alamy; *top center* © Yoshio Sawaragi/Dex Image/Jupiter Images; *top left* © C. Bowman/Robert Harding World Imagery/Getty Images; *bottom* © Royalty-Free/Corbis; *bottom right* © Anonymous/Corbis; **842** *bottom left* © Free Agents Limited/Corbis; *bottom center* © Photo 24/Brand X Pictures/Getty Images; *bottom right* © Royalty-Free/Corbis; **843** *bottom left* © Tom Wagner/Corbis; *bottom center* © Reuters/Corbis; *bottom right* © Issei Kato/Reuters/Corbis; **844** © Issei Kato/Reuters/Corbis; **846** © Yuriko Nakao/Reuters/Corbis; **847** Eisei-Bunko Museum; **848** © Chris Lisle/Corbis.

Chapter 28

850 *top left* © Adam Deschamps/Alamy; *bottom left* © Jack Fields/Corbis; *bottom right* © Mary Evans Picture Library/The Image Works, Inc.; **851** *bottom left* © Neil McAllister/Alamy; *bottom right* © David Hodges/Alamy; **852** © Image Makers/Getty Images; **853** *top left* © Bruno Morandi/Robert Harding World Imagery/Getty Images; *top center* © Getty Images; *top right* © JTP Photo/Alamy; *bottom* © Getty Images; **854** © Hemis/Alamy; **855** © Mark Pearson/Alamy; **856** *top left* © Frans Lanting/Corbis; *bottom* © Tim Graham/Corbis; **857** © Terry Whittaker/Alamy; **858** © Chris Lisle/Corbis; **859** *top left* © Bruno Morandi/Robert Harding World Imagery/Getty Images; *top center* © Getty Images; *top right* © JTP Photo/Alamy; *bottom* © Gavin Hellier/Getty Images; **861** © Luca I Tettoni/Corbis; **862** Richard Vogel/AP/Wide World Photos; **863** © DigitalVision/eStock Photo; **864** © Ted Spiegel/Corbis; **865** *top left* © Bruno Morandi/Robert Harding World Imagery/Getty Images; *top center* © Getty Images; *top right* © JTP Photo/Alamy; *bottom* © Chris Tan/Getty Images; **866** *bottom right* © Free Agents Limited/Corbis; *bottom left* © David Hanson/Getty Images; **867** © Reuters/Corbis; **868** © Paul A. Souders/Corbis; **869** *bottom left* © Hemis/Alamy; *bottom right* © Chris Lisle/Corbis; **870** Richard Vogel/AP/Wide World Photos.

Unit 9 Opener

873 *bottom left* © Royalty-Free/Corbis; *top right* © Oliver Strewe/Lonely Planet Images; *bottom right* © Robert Harding Picture Library Ltd./Alamy; **878** The Granger Collection, New York; **879** © Bill Curtsinger/NGSImages.com; **880** © MedioImages/Getty Images; **881** © Owen Franken/Corbis.

Chapter 29

882 *top left* © Image Source/Alamy; *bottom left* Birem, Iguana of the Rocks, Midjau-Midjawu, group Gunwinggu. Aboriginal art, painted on Eucalyptus bark. Musée du Quai Branly, Paris. Photo © J. G. Berizzi/Réunion des Musées Nationaux/Art Resource, New York; *bottom right* The Granger Collection, New York; **883** © Duomo/Corbis; **884** © Nik Wheeler/Corbis; **885** *top left* © Jose Fuste Raga/Corbis; *center* © Matthew Noble/Alamy; *top right* © Dallas and John Heaton/Stock Connection; *bottom* © Charles and Josette Lenars/Corbis; *center right* © Paul A. Souders/Corbis; **886** *top left* © Nik Wheeler/Corbis; *bottom* © Jeff Hunter/The Image Bank/Getty Images; **888** *top* © Frans Lanting/Corbis; *left* © Juergen and Christine Sohns/Animals Animals - Earth Scenes; *center* © Gary Bell/zefa/Corbis; *right* © Dave Watts/Alamy; **889** © Hulton-Deutsch Collection/Corbis; **890** *center left* © Leonard Douglas Zell/Lonely Planet Images; *bottom left* © AM Corporation/Alamy; **892** © John Van Hasselt/Corbis; **893** *bottom* © Suzanne Long/Alamy; *top left* © Jose Fuste Raga/Corbis; *center* © Matthew Noble/Alamy; *top right* © Dallas and John Heaton/Stock Connection; **894** © John Van Hasselt/Corbis; **895** *left* © Free Agents Limited/Corbis; *right* © Photo 24/Brand X Pictures/Getty Images; **896** *top* © Norman Chan/ShutterStock; *bottom* © SCPhotos/Alamy; **897** *bottom left* © Royalty-Free/Corbis; *top left* © Royalty-Free/Corbis; *second from top left* © Mark Richards/PhotoEdit; *third from top left* © Phototake Inc./Alamy; *bottom right* © Paul A. Souders/Corbis; **898** *top left* © Paul A. Souders/Corbis; *left* © Dave G. House/Post-Houserstock/Corbis; *center* © Paul A. Souders/Corbis; *right* © Bill Bachman/Alamy; **900** © Mike Powell/Getty Images; **901** *top left* © Jose Fuste Raga/Corbis; *center* © Matthew Noble/Alamy; *top right* © Dallas and John Heaton/Stock Connection; *center right* © Laureen Middley/Getty Images; *bottom* © Panoramic Images/Getty Images; **902** © Werner Forman/Corbis; **903** © Neil Rabinowitz/Corbis; **904** © Chris Ballentine/Paul Thompson Images/Alamy; **905** *top right* © Pierre Vinet/New Line/Saul Zaentz/Wing Nut Films/The Kobal Collection; *bottom* © Paul Thompson/Eye Ubiquitous/Corbis; **906** © Anthony Cooper/Ecoscene/Corbis; **907** © Frans Lanting/Corbis; **908** © John Warden/Getty Images.

R115

Chapter 30

910 © Bill Curtsinger/NGSImages.com; *bottom left* Owen Franken/Corbis; *bottom* The Granger Collection, New York; **911** *bottom left* The Granger Collection, New York; *bottom right* © Danita Delimont/Alamy; **912** © Paul Chesley/Getty Images; **913** *top left* © Chris Cheadle/Getty Images; *top center* © Peter Scoones/Getty Images; *top right* © David Tipling/Getty Images; *bottom* © Kevin Schafer/Getty Images; **915** © Jodi Cobb/NGSImages.com; **916** © NGSImages.com; **917** © Hippocrene Books; **918** *top right* © Gary Steer/Lonely Planet Images; *bottom left* © Jean-Bernard Carillet/Lonely Planet Images; *bottom right* © Wolfgang Kaehler/Corbis; *center* © Jean-Bernard Carillet/Lonely Planet Images; **919** *top left* © Anders Ryman/Corbis; *center right* © Jodi Cobb/NGSImages.com; *bottom left* © Klaus Nigge/NGSImages.com; *center* © Tim Laman/NGSImages.com; **920** © Photo Researchers, Inc.; **921** *top left* © Chris Cheadle/Getty Images; *top center* © Peter Scoones/Getty Images; *top right* © David Tipling/Getty Images; *center right* © Getty Images; *bottom* © Maria Stenzel/NGSImages.com; **923** *center right* © Photo Researchers, Inc.; *bottom right* The Granger Collection, New York; **925** *top left* © Kevin Cooley/Getty Images; *bottom left* © Jodi Cobb/NGSImages.com; **926** © Paul Chesley/Getty Images.

End Matter

R1 Photograph by Ed-Imaging; **R23** © Mark Newman/University of Michigan; **R26** © HIP-Archive/Topham/The Image Works, Inc.; **R35** *top* © Michael Newman/PhotoEdit; *bottom* © Bob Daemmrich/PhotoEdit; **R36** Screenshot © Google Inc. and is used with permission; **R37** Courtesy of USGS; **R40-R41** *top* © Lindsay Hebberd/Corbis; *bottom* © Dave Bartruff/Corbis; **R40** *bottom left* © Comstock; **R41** *top right* © David Samuel Robbins/Corbis; **R42-R43** *top* The Tribute Money, Masaccio (Maso di San Giovanni). S. Maria del Carmine, Florence, Italy. Photo © Scala/Art Resource, New York; **R42** *bottom left* Luis Romero/AP/Wide World Photos; **R43** *bottom left* © Angelo Hornak/Corbis; *top right* © Comstock Images/Jupiter Images; **R44** © Reuters/Corbis; **R45** *top left* Heads of Brahma, Vichnu and Shiva (late 800's). Phnom Bok (Siemreap). Musée des Arts Asiatiques-Guimet, Paris. Photo © Erich Lessing/Art Resource, New York; *bottom left* © Bennett Dean; Eye Ubiquitous/Corbis; *top right* © ArkReligion.com/Alamy; **R46-R47** © Richard T. Nowitz/Corbis; **R46** *bottom left* Muslim men praying in the street. Photo © Arthur Thévenart/Corbis; **R47** *top left* Jassim Mohammed/AP/Wide World Photos; *top right* © Michael S. Yamashita/Corbis; **R48-R49** *top* Photograph by Bill Aron; **R48** *bottom left* © The Art Archive/Corbis; **R49** *bottom left* © Christie's Images/Corbis; *top right* © Nathan Benn/Corbis; **R50** *bottom* © Yoshio Tomii/SuperStock Inc.; *top* © SuperStock Inc.; **R51** *left* © ImagineChina; *top right* Chinese ying and yang (200's). Roman. Mosaic. Archaeological Museum, Sousse, Tunisia. Photo © Dagli Orti/Art Archive; **R52-R53** © Bob Krist/Corbis; **R52** © Royalty-Free/Corbis; **R53** *top* © Chris Lisle/Corbis; *center right* © Paul Almasy/Corbis.

Illustrations

Maps by Rand McNally **A14-A37**

All other maps, locators and globe locators by GeoNova LLC

Illustration by Nick Rotondo/Bizzy Productions **162-163**

Illustration by Peter Bull **218-219, 218** *both,* **474-475, 474, 560-561, 560** *center left,* **664-665, 890-891, 890** *top left*

Illustration by Peter Dennis **348**

Illustration by Ken Goldammer **A8-A9**

Illustration by Martin Hargraves **345**

Illustration by Michael Jaroszko **Graphs, Charts, and Tables, 558**

Illustration by Bob Kayganich **46-47, 46** *top left*

Illustration by Precision Graphics **28, 29, 37, 42, 50, 655, 659, 915** *both*

Illustration by Sebastian Quigley **772-773, 772** *both*

Illustration by Robert Sikora **406-407, 406** *both*

Illustration by Stephen R. Wagner **540**

All other illustrations by McDougal Littell/Houghton Mifflin Co.

The editors have made every effort to trace the ownership of all copyrighted material found in this book and to make full acknowledgment for its use. Omissions brought to our attention will be corrected in a subsequent edition.

McDougal Littell
World Cultures and Geography

correlated to the

Tennessee Social Studies Curriculum Standards: Process Standards
and
World Geography Seventh Grade (Standards, Learning Expectations, and Performance Indicators)

PROCESS STANDARDS
To be used with ALL grades

Acquiring Information: Acquiring Information involves locating, gathering, observing, comprehending, organizing, and processing information from a variety of primary and secondary sources. These sources include printed materials, maps, graphic representations, and artifacts, physical and human environmental elements, media and technology sources.

"Acquiring Information" emphasizes the learners' use of a broad base of strategies to:

Process Standards	Core Instruction	Application
P1. Read to gain literal meaning	**PE/TE:** 5-9, 11-15, 17-20, 27-33, 35-39, 41-45, 49-52, 59-63, 64-65, 67-71, 73-77, 79-82, 89-95, 97-100, 110-111, 123-127, 131-137, 139-145, 147-150, 157-161, 165-169, 171-175, 177-180, 190-191, 199-203, 205-211, 213-217, 221-224, 231-235, 237-241, 245-248, 255-259, 261-265, 267-271, 273-277, 281-284, 291-295, 297-303, 305-308, 318, 327-333, 335-339, 343-349, 351-356, 363-367, 371-375, 377-381, 383-387, 389-394, 401-405, 409-413, 415-418, 425-431, 433-439, 441-444, 455-456, 461-467, 469-473, 477-483, 487-490, 497-503, 505-511, 513-516,	**PE/TE:** S4, S5, S21, 5, 9, 10, 11, 13, 14, 15, 16, 17, 18, 19, 20, 21, 22, 26, 30, 31, 33, 34, 35, 36, 39, 40, 41, 42, 43, 45, 48, 49, 50, 51, 52, 53, 54, 58, 59, 60, 61, 63, 65, 66, 67, 68, 69, 70, 71, 72, 73, 74, 75, 76, 77, 78, 79, 81, 82, 83, 84, 88, 89, 90, 92, 93, 94, 95, 96, 97, 99, 100, 101, 102, 122, 123, 124, 125, 126, 127, 130, 131, 132, 133, 134, 135, 136, 137, 138, 139, 142, 145, 146, 147, 148, 149, 150, 151, 152, 156, 157, 159, 160, 161, 164, 165, 166, 167, 168, 169, 170, 171, 172, 173, 175, 176, 177, 178, 180, 181, 182, 183, 198, 199, 200, 202, 203, 204, 205, 208, 211, 212, 213, 214, 216, 217, 220, 221, 222, 223, 224, 225, 226, 230, 231, 233, 235, 236, 237, 239, 240, 241, 244, 245, 246, 247, 248, 249, 250, 254, 255, 257, 258, 259, 260, 261, 263, 264, 265, 266, 267, 269, 270, 271, 272, 273, 274, 275, 277, 280, 281, 283, 284, 285, 286, 290, 291, 292, 293, 295, 296, 297, 299, 300, 303, 304, 305, 307, 308, 309, 310, 326, 327, 328, 330, 331, 332, 333, 334, 335, 337, 338, 339, 342, 343, 345, 346, 348, 349, 350, 351, 352, 353, 355,

Process Standards	Core Instruction	Application
(continued) **P1.** Read to gain literal meaning	**PE/TE:** 527, 537-543, 545-548, 555-559, 563-567, 569-574, 581-585, 587-591, 595-600, 607-611, 613-617, 619-623, 625-629, 631-636, 646, 653-657, 659-663, 667-670, 677-681, 683-687, 689-692, 699-705, 707-713, 717-722, 729-733, 737-740, 751, 757-763, 765-771, 775-781, 785-790, 797-803, 805-811, 813-818, 825-829, 831-837, 841-846, 853-857, 859-863, 865-868, 878-879, 885-889, 893-899, 901-906, 913-917, 921-924, R38-R53	**PE/TE:** 356, 357, 358, 362, 363, 364, 365, 367, 370, 371, 373, 374, 375, 376, 377, 379, 380, 381, 382, 383, 385, 386, 387, 388, 389, 392, 393, 394, 395, 396, 400, 401, 402, 404, 405, 408, 409, 411, 412, 413, 414, 415, 417, 418, 419, 420, 424, 425, 427, 428, 429, 430, 431, 432, 433, 435, 436, 439, 440, 441, 442, 444, 445, 446, 460, 461, 462, 464, 466, 467, 468, 469, 471, 472, 473, 476, 477, 480, 481, 482, 483, 486, 487, 489, 490, 491, 492, 496, 497, 498, 500, 501, 502, 503, 504, 505, 508, 509, 511, 512, 513, 514, 515, 516, 517, 518, 536, 537, 539, 540, 541, 542, 543, 544, 545, 546, 547, 548, 549, 550, 554, 555, 556, 557, 558, 559, 562, 563, 564, 565, 566, 567, 568, 569, 572, 573, 574, 575, 576, 580, 581, 583, 584, 585, 586, 587, 589, 590, 591, 594, 595, 596, 598, 600, 601, 602, 606, 607, 608, 609, 610, 611, 612, 613, 615, 616, 617, 618, 619, 620, 621, 622, 623, 624, 625, 626, 627, 628, 629, 630, 631, 633, 634, 635, 636, 637, 638, 652, 653, 654, 656, 657, 658, 659, 662, 663, 666, 667, 668, 669, 670, 671, 672, 676, 677, 679, 680, 681, 682, 683, 684, 686, 687, 688, 689, 690, 691, 692, 693, 694, 698, 699, 700, 701, 703, 705, 706, 707, 709, 710, 711, 713, 716, 717, 719, 720, 722, 723, 724, 728, 729, 731, 732, 733, 736, 737, 739, 740, 741, 742, 756, 757, 758, 759, 761, 762, 763, 764, 765, 767, 769, 771, 774, 775, 777, 778, 779, 781, 784, 785, 786, 788, 789, 790, 791, 792, 796, 797, 798, 800, 801, 803, 804, 805, 808, 810, 811, 812, 813, 815, 816, 817, 818, 819, 820, 824, 825, 828, 829, 830, 831, 834, 837, 840, 841, 842, 843, 844, 845, 846, 847, 848, 852, 853, 854, 856, 857, 858, 859, 860, 861, 862, 863, 864, 865, 866, 867, 868, 869, 884, 885, 886, 888, 889, 892, 893, 894, 895, 896, 898, 899, 900, 901, 903, 904, 905, 906, 907, 908, 912, 913, 914, 916, 917, 920, 921, 923, 924, 925, 926, R3, R4, R5
P2. Use chapter and section headings, and topic sentences to select main ideas	**PE/TE:** S24, R3, R4, R5	**PE/TE:** S26, 7, 8, 11, 12, 13, 14, 17, 18, 20, 24, 27, 28, 29, 30, 31, 32, 33, 35, 36, 38, 39, 41, 42, 43, 45, 49, 50, 51, 52, 56, 59, 60, 61, 63, 67, 68, 69, 70, 71, 73, 75, 76, 77, 79, 81, 82, 89, 90, 92, 93, 94, 95, 97, 98, 99, 100, 123, 124, 125, 126, 131, 132, 133, 134, 135, 136, 137, 139, 140, 142, 143, 144, 145, 147, 148, 149, 150, 154, 157, 158, 159, 160, 161, 165, 166, 167, 168, 169, 171, 172, 173, 175, 177, 178, 179, 180, 196, 199, 200, 202, 203, 205, 206, 208, 210, 211, 213, 214, 216, 217, 221, 222, 223, 224, 228, 231, 232, 233, 234, 235, 237, 238, 239, 240, 241, 245, 246, 247, 248, 252, 255, 256, 257, 258, 259, 261, 262, 263, 264, 267, 268, 269, 270, 271, 273, 274, 275, 276, 277, 281, 282, 283, 284, 288, 291, 292, 293, 295, 297, 298, 299, 300, 303, 305, 306, 307, 308, 324, 327, 328, 330, 331, 332, 333, 335, 336, 337, 338, 339, 343, 344, 345, 346, 347, 348, 349, 351, 352, 353, 355, 356, 360, 363, 364, 365, 366, 367, 371, 372, 374, 375, 377, 378, 379, 380, 383, 384, 385, 386, 387, 389, 390, 392, 393, 394, 398, 401, 402, 403, 404, 405, 409, 410, 411, 412, 413, 415, 416, 417, 418, 422, 425, 426, 427, 428, 429, 430, 433, 434, 435, 436, 439, 441, 442, 444, 458, 461, 462, 464, 466, 467, 469, 470, 471, 472, 473, 477, 478, 480, 481, 482, 483, 487, 488, 489, 490, 494, 497, 498, 500, 501, 502, 503, 505, 506, 507, 508, 509, 511, 513, 514, 515, 516, 534, 537, 538, 539, 540, 542, 543, 545, 546, 547, 548, 552, 555, 556, 557, 558, 559, 563, 564, 565, 566, 567, 569, 570, 571, 573, 574, 578, 581, 582, 583, 584, 585, 587, 588, 589, 590, 591, 595, 596, 598, 599, 600, 604, 607, 608, 609, 610, 611

Process Standards	Core Instruction	Application
(continued) **P2.** Use chapter and section headings, and topic sentences to select main ideas		**PE/TE:** 613, 614, 615, 616, 619, 620, 621, 622, 625, 626, 627, 628, 631, 632, 633, 634, 635, 636, 650, 653, 654, 656, 657, 659, 660, 662, 663, 667, 668, 669, 674, 677, 678, 679, 680, 681, 683, 684, 686, 687, 689, 690, 691, 692, 696, 699, 700, 701, 703, 704, 705, 707, 708, 709, 710, 711, 713, 717, 718, 719, 720, 721, 726, 729, 730, 731, 732, 733, 737, 738, 739, 740, 754, 757, 758, 760, 762, 763, 765, 766, 767, 768, 770, 771, 775, 776, 777, 778, 779, 781, 785, 786, 788, 789, 790, 794, 797, 798, 80,0, 801, 802, 803, 805, 806, 807, 808, 810, 811, 813, 814, 815, 816, 817, 818, 822, 825, 826, 828, 829, 831, 832, 833, 834, 835, 836, 841, 842, 843, 844, 845, 846, 850, 853, 854, 857, 859, 860, 861, 862, 863, 865, 866, 867, 868, 882, 885, 886, 888, 889, 893, 894, 895, 896, 898, 901, 902, 903, 904, 905, 906, 910, 913, 916, 921, 922, 923, 924, R3, R4, R38, R40, R42, R44, R46, R48, R50, R52, R53, R54
P3. Detect cause and effect relationships	**PE/TE:** S22, S25, 29, 31, 32, 33, 36, 37, 41-42, 43, 49, 50, 51-52, 53, 67-68, 69, 70-71, 93, 94-95, 96, 97-100, 133-137, 142, 160, 166-169, 172-173, 202-203, 211, 212, 216-217, 221-222, 223, 230, 231-235, 236, 238-239, 245-246, 247-248, 257, 262-263, 280, 282-283, 292, 293-295, 296, 297, 299, 307-308, 330-331, 336-337, 338, 342, 343-349, 350, 351-352, 353, 354, 355-356, 365-366, 374-375, 376, 377-379, 380-381, 382, 383-385, 386-387, 390, 392, 393-394, 400, 401-405, 408, 417-418, 426, 427, 428-429, 433-434, 435, 436-437, 439, 440, 441-444, 467, 469-471, 472-473, 476, 478, 483, 487-490, 502-503, 505-508, 515-516, 539-540, 542, 543, 555-557, 559, 563-564, 566-567, 569-572, 581-585, 587-589, 590-591, 596, 598-599, 607-608, 609, 610-611, 613-615, 616-617, 620, 621-623, 625-626, 627, 628-629, 631-634, 635-636, 654-656, 657, 663, 667-668, 669-670, 677-681, 683-686, 687, 689-690, 691-692, 699-700, 701-705, 706, 719, 720-722, 729-731, 732-733, 734-735, 738-740, 760-761, 762-763, 766-769, 770-771, 776-777, 778-779, 781, 785-786, 788, 789-790, 796, 798, 800, 801-803, 806-808, 810-811, 813-815, 816-818, 819, 826, 828-829, 831-837, 840, 841-842, 843-844, 845-846, 853-854, 856-857, 859-860, 861, 862, 863, 865-866, 867-868, 886, 889, 892, 893-894, 895-896, 898-899, 900, 901-903, 904, 906, 916, 917, 920, 922, 923-924, R8	**PE/TE:** S21, S23, S27, 33, 36, 37, 39, 40, 45, 48, 50, 52, 54, 56, 59, 60, 63, 66, 70, 71, 84, 96, 99, 100, 102, 120, 127, 137, 139, 141, 145, 152, 160, 161, 167, 169, 170, 175, 182, 196, 202, 203, 211, 221, 222, 224, 226, 230, 233, 235, 243, 245, 246, 248, 250, 257, 259, 260, 261, 265, 271, 275, 277, 281, 283, 286, 291, 295, 299, 307, 308, 310, 330, 335, 339, 343, 345, 346, 348, 349, 351, 353, 354, 355, 356, 358, 367, 374, 375, 377, 380, 381, 383, 386, 392, 394, 396, 398, 401, 402, 405, 411, 418, 420, 428, 431, 433, 436, 439, 441, 444, 446, 467, 468, 469, 472, 483, 487, 489, 492, 493, 494, 502, 503, 505, 508, 511, 516, 518, 519, 534, 539, 540, 542, 543, 550, 552, 554, 559, 562, 567, 568, 569, 574, 576, 577, 578, 581, 584, 585, 591, 600, 602, 603, 604, 607, 609, 610, 611, 613, 614, 616, 617, 618, 622, 623, 625, 627, 629, 631, 632, 633, 635, 636, 638, 639, 650, 654, 657, 661, 663, 666, 672, 674, 676, 677, 681, 684, 687, 689, 692, 694, 696, 696, 699, 703, 705, 713, 716, 722, 724, 726, 729, 732, 733, 735, 740, 742, 743, 754, 760, 761, 762, 763, 767, 769, 770, 771, 778, 781, 788, 790, 792, 793, 799, 800, 803, 808, 810, 811, 818, 820, 821, 824, 826, 827, 828, 829, 830, 831, 837, 841, 843, 846, 848, 850, 853, 855, 856, 857, 859, 861, 863, 865, 867, 868, 870, 882, 889, 893, 895, 899, 901, 904, 906, 908, 917, 923, 924, 926, R8

Process Standards	Core Instruction	Application
P4. Distinguish between fact and opinion to recognize propaganda	**PE/TE:** S4, S6, S25, 818, R16, R18, R33, R37	**PE/TE:** S27, 51, 152, 168, 216, 227, 246, 268, 311, 336, 427, 493, 515, 546, 589, 616, 617, 686, 704, 730, 806, 818, 844, 867, 903, 909, R16, R18, R33, R37, R41, R43, R45, R47, R49, R51
P5. Recognize author bias	**PE/TE:** S4, S6, S25, 818, 909, R16, R17, R18, R33, R37	**PE/TE:** S5, S21, S23, S26, S27, 51, 82, 90, 216, 227, 246, 268, 298, 311, 336, 359, 384, 427, 481, 515, 546, 589, 603, 616, 617, 639, 662, 686, 821, R16, R17, R18, R33, R37, R41, R43, R45, R47, R49, R51
P6. Use picture clues and picture captions to aid comprehension	**PE/TE:** S6, S20, S24	**PE/TE:** S7, S23, S26, 2, 5, 6-7, 8, 11, 13, 14, 15, 16, 17, 18, 19, 20, 25, 26, 27, 28, 29, 30, 31, 32, 33, 34, 35, 36, 37, 38, 39, 41, 42, 44, 46-47, 49, 50, 51, 52, 57, 58, 59, 60, 61, 62, 63, 65, 67, 68, 71, 73, 74, 75, 77, 78, 79, 81, 82, 86, 87, 89, 90, 91, 92, 96, 97, 98, 100, 105, 110, 111, 113, 114, 115, 116, 117, 118, 119, 120, 123, 124, 126, 127, 128, 129, 133, 134, 135, 136, 137, 142, 143, 145, 147, 148, 150, 154, 155, 156, 157, 158, 160, 161, 162, 163, 165, 166, 168, 169, 170, 171, 172, 173, 174, 175, 177, 178, 179, 190, 191, 193, 194, 195, 196, 197, 199, 200, 201, 202, 203, 204, 205, 206, 208, 209, 210, 211, 213, 214, 215, 216, 217, 218, 219, 222, 223, 224, 228, 229, 231, 233, 234, 235, 237, 238, 239, 240, 242, 243, 244, 245, 246, 247, 248, 252, 253, 257, 258, 259, 261, 263, 265, 267, 268, 269, 270, 271, 273, 274, 275, 276, 277, 278, 279, 281, 282, 283, 284, 288, 289, 290, 291, 292, 294, 295, 296, 297, 298, 299, 300, 301, 302, 303, 305, 306, 307, 313, 324, 325, 326, 327, 328, 330, 331, 332, 335, 336, 337, 339, 340, 341, 343, 344, 345, 346, 347, 348, 349, 351, 352, 354, 356, 360, 361, 363, 364, 365, 367, 368, 369, 370, 371, 372, 373, 374, 375, 376, 377, 378, 379, 380, 383, 384, 385, 386, 388, 389, 391, 392, 393, 394, 395, 398, 399, 401, 402, 403, 404, 406, 407, 408, 409, 410, 411, 412, 413, 414, 415, 418, 422, 423, 425, 426, 427, 428, 429, 430, 431, 433, 434, 436, 437, 438, 439, 440, 441, 442, 443, 449, 455, 456, 457, 458, 459, 461, 462, 463, 464, 465, 466, 469, 470, 471, 473, 474, 475, 477, 478, 479, 480, 481, 482, 483, 484, 485, 487, 489, 494, 495, 497, 498, 499, 500, 501, 502, 503, 504, 505, 506, 507, 509, 510, 511, 513, 514, 515, 516, 517, 521, 527, 531, 533, 534, 535, 536, 537, 538, 539, 542, 543, 545, 546, 547, 548, 550, 552, 553, 555, 556, 557, 558, 559, 560, 561, 562, 563, 564, 565, 566, 567, 568, 569, 570, 571, 572, 574, 575, 578, 579, 581, 583, 585, 587, 589, 590, 591, 592, 593, 594, 595, 596, 597, 598, 599, 600, 601, 604, 605, 606, 607, 609, 610, 611, 612, 613, 615, 616, 617, 619, 621, 622, 623, 625, 626, 627, 628, 630, 631, 632, 633, 634, 636, 637, 641, 642, 646, 649, 650, 651, 652, 653, 654, 655, 656, 657, 659, 660, 662, 663, 664, 665, 667, 668, 669, 670, 671, 674, 675, 676, 677, 678, 680, 681, 683, 684, 685, 686, 687, 689, 690, 692, 693, 694, 696, 697, 699, 700, 701, 702, 703, 704, 705, 706, 707, 709, 710, 711, 712, 713, 714, 715, 716, 717, 720, 722, 723, 726, 727, 729, 730, 732, 733, 734, 735, 736, 737, 738, 740, 741, 745, 751, 753, 754, 755, 756, 757, 758, 760, 761, 762, 763, 765, 767, 768, 769, 771, 772, 773, 775, 776, 779, 780, 782, 783, 784, 785, 788, 789, 790, 792, 794, 795, 796, 797, 798, 799, 800, 801, 802, 804, 805, 806, 808, 809, 810, 811, 812, 813, 814, 815, 816, 818, 819, 822, 823, 825, 826, 828, 829, 830, 831, 832, 833,

Process Standards	Core Instruction	Application
(continued) **P6.** Use picture clues and picture captions to aid comprehension		**PE/TE:** 835, 836, 837, 838, 839, 840, 841, 843, 844, 846, 847, 848, 850, 851, 852, 853, 854, 855, 856, 857, 858, 859, 861, 862, 863, 864, 865, 866, 867, 868, 869, 873, 878, 879, 880, 881, 882, 883, 884, 885, 886, 888, 889, 890, 891, 892, 893, 894, 896, 897, 898, 900, 901, 902, 903, 904, 905, 906, 907, 908, 910, 911, 912, 913, 915, 916, 917, 918, 919, 920, 921, 923, 925, R40, R41, R42, R43, R44, R45, R46, R47, R48, R49, R50, R51, R52, R53
P7. Read from a variety of sources	**PE/TE:** 5-9, 11-15, 17-20, 27-33, 35-39, 41-45, 49-52, 59-63, 64-65, 67-71, 73-77, 79-82, 89-95, 97-100, 110-111, 123-127, 131-137, 139-145, 147-150, 157-161, 165-169, 171-175, 177-180, 190-191, 199-203, 205-211, 213-217, 221-224, 231-235, 237-241, 245-248, 255-259, 261-265, 267-271, 273-277, 281-284, 291-295, 297-303, 305-308, 318, 327-333, 335-339, 343-349, 351-356, 363-367, 371-375, 377-381, 383-387, 389-394, 401-405, 409-413, 415-418, 425-431, 433-439, 441-444, 455-456, 461-467, 469-473, 477-483, 487-490, 497-503, 505-511, 513-516, 527, 537-543, 545-548, 555-559, 563-567, 569-574, 581-585, 587-591, 595-600, 607-611, 613-617, 619-623, 625-629, 631-636, 646, 653-657, 659-663, 667-670, 677-681, 683-687, 689-692, 699-705, 707-713, 717-722, 729-733, 737-740, 751, 757-763, 765-771, 775-781, 785-790, 797-803, 805-811, 813-818, 825-829, 831-837, 841-846, 853-857, 859-863, 865-868, 878-879, 885-889, 893-899, 901-906, 913-917, 921-924, R38-R53	**PE/TE:** S4, S5, S21, 5, 9, 10, 11, 13, 14, 15, 16, 17, 18, 19, 20, 21, 22, 26, 30, 31, 33, 34, 35, 36, 39, 40, 41, 42, 43, 45, 48, 49, 50, 51, 52, 53, 54, 58, 59, 60, 61, 63, 65, 66, 67, 68, 69, 70, 71, 72, 73, 74, 75, 76, 77, 78, 79, 81, 82, 83, 84, 88, 89, 90, 92, 93, 94, 95, 96, 97, 99, 100, 101, 102, 122, 123, 124, 125, 126, 127, 130, 131, 132, 133, 134, 135, 136, 137, 138, 139, 142, 145, 146, 147, 148, 149, 150, 151, 152, 156, 157, 159, 160, 161, 1634, 165, 166, 167, 168, 169, 170, 171, 172, 173, 175, 176, 177, 178, 180, 181, 182, 183, 198, 199, 200, 202, 203, 204, 205, 208, 211, 212, 213, 214, 216, 217, 220, 221, 222, 223, 224, 225, 226, 230, 231, 233, 235, 236, 237, 239, 240, 241, 244, 245, 246, 247, 248, 249, 250, 254, 255, 257, 258, 259, 260, 261, 263, 264, 265, 266, 267, 269, 270, 271, 272, 273, 274, 275, 277, 280, 281, 283, 284, 285, 286, 290, 291, 292, 293, 295, 296, 297, 299, 300, 303, 304, 305, 307, 308, 309, 310, 326, 327, 328, 330, 331, 332, 333, 334, 335, 337, 338, 339, 342, 343, 345, 346, 348, 349, 350, 351, 352, 353, 355, 356, 357, 358, 362, 363, 364, 365, 357, 370, 371, 373, 374, 375, 376, 377, 379, 380, 381, 382, 383, 385, 386, 387, 388, 389, 392, 393, 394, 395, 396, 400, 401, 402, 404, 405, 408, 409, 411, 412, 413, 414, 415, 417, 418, 419, 420, 424, 425, 427, 428, 429, 430, 431, 432, 433, 435, 436, 439, 440, 441, 442, 444, 445, 446, 460, 461, 462, 464, 466, 467, 468, 469, 471, 472, 473, 476, 477, 480, 481, 482, 483, 486, 487, 489, 490, 491, 492, 496, 497, 498, 500, 501, 502, 503, 504, 505, 508, 509, 511, 512, 513, 514, 515, 516, 517, 518, 536, 537, 539, 540, 541, 542, 543, 544, 545, 546, 547, 548, 549, 550, 554, 555, 556, 557, 558, 559, 562, 563, 564, 565, 566, 567, 568, 569, 572, 573, 574, 575, 576, 580, 581, 583, 584, 585, 586, 587, 589, 590, 591, 594, 595, 596, 598, 600, 601, 602, 606, 607, 608, 609, 610, 611, 612, 613, 615, 616, 617, 618, 619, 620, 621, 622, 623, 624, 625, 626, 627, 628, 629, 630, 631, 633, 634, 635, 636, 637, 638, 652, 653, 654, 656, 657, 658, 659, 662, 663, 666, 667, 668, 669, 670, 671, 672, 676, 677, 679, 680, 681, 682, 683, 684, 686, 687, 688, 689, 690, 691, 692, 693, 694, 698, 699, 700, 701, 703, 705, 706, 707, 709, 710, 711, 713, 716, 717, 719, 720, 722, 723, 724, 728, 729, 731, 732, 733, 736, 737, 739, 740, 741, 742, 756, 757, 758, 759, 761, 762, 763, 764, 765, 767, 769, 771, 774, 775, 777, 778, 779, 781, 784, 785, 786, 788, 789, 790, 791, 792, 796, 797, 798, 800, 801, 803, 804, 805, 808, 810, 811, 812, 813, 815, 816, 817, 818, 819, 820, 824, 825, 828, 829, 830, 831, 834, 837, 840, 841, 842, 843, 844, 845, 846, 847, 848, 852, 853, 854, 856, 857, 858, 859, 860, 861, 862, 863, 864, 865, 866, 867, 868, 869, 884, 885, 886, 888, 889, 892, 893, 894, 895, 896, 898, 899, 900, 901, 903, 904, 905, 906, 907, 908, 912, 913, 914, 916, 917, 920, 921, 923, 924, 925, 926, R3, R4, R5

WORLD CULTURES AND GEOGRAPHY CORRELATED TO THE TENNESSEE SOCIAL STUDIES CURRICULUM STANDARDS

Process Standards	Core Instruction	Application
P8. Use maps, graphs, globes, media and technology sources	**PE/TE:** S10, S12, S14, S16, lvixx, A2-A3, A4-A7, A10-A11, 13, 14, 57, 60, 62, 64, 132, 141, 215, 428, 443, 463, 464, R20-R25, R34, R35, R36	**PE/TE:** S11, S13, S15, S17, A1-A3, A4-A7, A8-A9, A10, A11, A13-A37, 2-3, 4, 9, 10, 11, 12, 13, 15, 16, 20, 21, 23, 24-25, 26, 30, 33, 34, 39, 40, 44, 45, 46-47, 48, 51, 52, 53, 55, 56, 57, 58, 62, 63, 65, 66, 69, 71, 72, 77, 78, 82, 83, 85, 86-87, 91, 93, 94, 95, 96, 99, 100, 101, 103, 105, 107, 108, 109, 112, 118, 120, 121, 122, 125, 127, 134, 137, 141, 142, 145, 146, 149, 150, 151, 153, 154, 155, 156, 159, 161, 162-163, 164, 167, 168, 169, 174, 175, 176, 180, 181, 183, 186, 187, 188, 189, 192, 196, 197, 198, 203, 204, 207, 209, 211, 212, 216, 217, 218-219, 220, 221, 223, 224, 224, 225, 227, 228, 230, 232, 233, 235, 236, 241, 241, 244, 245, 246, 247, 248, 249, 251, 252, 253, 254, 256, 259, 260, 263, 265, 266, 268, 271, 277, 280, 284, 285, 287, 288, 289, 292, 294, 295, 296, 300, 301, 303, 304, 308, 309, 311, 314, 315, 316, 317, 319, 324, 325, 326, 239, 331, 333, 337, 338, 339, 342, 345, 346, 347, 349, 350, 353, 356, 357, 359, 360, 361, 362, 366, 367, 370, 372, 374, 375, 376, 381, 382, 384, 387, 388, 390, 691, 394, 395, 397, 398, 399, 400, 404, 405, 406-407, 413, 414, 417, 418, 419, 421, 422, 423, 424, 427, 431, 432, 435, 436, 438, 439, 440, 443, 444, 445, 447, 450, 451, 452, 453, 454, 456, 458, 459, 460, 463, 464, 466, 467, 468, 470, 472, 473, 474-475, 476, 483, 486, 490, 491, 493, 494, 195, 496, 499, 503, 504, 506, 507, 510, 511, 512, 514, 516, 517, 519, 522, 523, 524, 525, 526, 528, 534, 535, 536, 538, 541, 543, 544, 545, 546, 548, 549, 551, 552, 553, 554, 555, 557, 558, 559, 560-561, 562, 564, 567, 568, 574, 575, 577, 578, 579, 580, 582, 583, 585, 586, 588, 589, 591, 594, 600, 601, 603, 604, 605, 605, 606, 608, 611, 612, 616, 617, 620, 623, 624, 629, 630, 634, 636, 637, 639, 642, 643, 644, 645, 650, 651, 652, 655, 657, 661, 663, 664-665, 666, 667, 669, 670, 671, 673, 674, 675, 676, 679, 681, 682, 683, 685, 686, 687, 691, 692, 693, 695, 696, 697, 698, 702, 703, 704, 705, 706, 708, 712, 713, 716, 719, 720, 721, 723, 725, 726, 727, 728, 730, 731, 733, 736, 738, 7.9, 740, 741, 743, 746, 747, 748, 749, 750, 754, 755, 756, 757, 759, 763, 764, 767, 768, 770, 771, 772-773, 774, 778, 780, 781, 790, 791, 793, 794, 795, 796, 799, 803, 804, 807, 809, 811, 812, 814, 815, 817, 818, 821, 822, 823, 824, 827, 829, 830, 833, 834, 837, 840, 845, 846, 847, 849, 850, 851, 852, 855, 857, 858, 860, 862, 863, 864, 867, 868, 869, 871, 874, 875, 876, 877, 882, 883, 884, 887, 889, 890-891, 892, 894, 897, 899, 900, 906, 907, 909, 910, 911, 912, 914, 917, 922, 924, 925 927, R20-R25, R34, R35, R36, R38-R39
P9. Discover resources available from museums, historical sites, presidential libraries, and local and state preservation societies	**PE/TE:** 287, 349	**PE/TE:** 183, 241, 251, 287, 349, 375, 670

Analysis of Data and Problem Solving: Problem solving involves the comprehension, analysis, and interpretations of data leading to the development of a solution or conclusion. Students will develop problem-solving skills through comprehension, analysis, interpretations, synthesis, summary, and evaluation.

"Problem Solving and Analysis of Data" emphasizes the learners' use of a broad base of strategies to:

Process Standards	Core Instruction	Application
P10. Identify relevant factual material	**PE/TE:** S4, S6, S20, S22, S24-S25, R32, R33, R37	**PE/TE:** S5, S7, S21, S23, S26-S27, A3, A7, 2, 20, 23, 52, 63, 71, 77, 85, 103, 127, 153, 161, 183, 224, 235, 241, 248, 251, 259, 277, 284, 287, 295, 303, 311, 333, 349, 359, 387, 394, 397, 405, 418, 431, 447, 490, 503, 511, 516, 551, 559, 567, 574, 577, 585, 591, 600, 603, 629, 636, 657, 670, 673, 687, 681, 695, 713, 722, 725, 743, 781, 793, 803, 811, 818, 821, 837, 846, 849, 857, 863, 868, 871, 899, 909, 924, 927, R32, R33, R37
P11. Classify information by source, chronology, and importance	**PE/TE:** S4, S6, S20, S22, S24-S25, R32, R33, R37	**PE/TE:** S5, S7, S21, S23, S26-S27, A7, 13, 85, 103, 153, 235, 251, 259, 387, 394, 405, 418, 447, 503, 567, 603, 687, 743, 821, 846, 909, R32, R33, R37
P12. Critically examine data from a variety of sources	**PE/TE:** S4, S6, S8, S10, S12, S16, S20, S22, S24-S25, R32, R33, R37	**PE/TE:** S5, S7, S9, S11, S13, S17, S21, S23, S26-S27, A3, 12, 13, 153, 248, 251, 259, 287, 387, 394, 405, 418, 447, 490, 503, 567, 577, 603, 636, 670, 687, 722, 743, 803, 821, 846, 909, R32, R33, R37
P13. Detect bias in data presented in a variety of forms	**PE/TE:** S4, S6, S25, 818, 909, R16, R17, R18, R33, R37	**PE/TE:** S5, S21, S23, S26, S27, 51, 82, 90, 216, 227, 246, 268, 298, 311, 336, 359, 384, 427, 481, 515, 546, 589, 603, 616, 617, 639, 662, 686, 821, R16, R17, R18, R33, R37, R41, R43, R45, R47, R49, R51
P14. Compare and contrast data	**PE/TE:** 48, 58, 66, 80, 106, 128-129, 141, 148, 158, 186, 202, 209, 212, 230, 238, 242-243, 278-279, 293, 314, 330, 331, 334, 340-341, 347, 354, 365, 368-369, 370, 410, 416, 432, 443, 450, 465, 484-485, 488, 496, 507, 522, 568, 570, 584, 592-593, 594, 610, 614, 630, 635, 642, 658, 660, 710, 714-715, 718, 732, 734-735, 746, 777, 782-783, 786, 800, 812, 816, 832, 838-839, 842, 856, 874, 895, 912, 918-919, 920, 922, R9, R38-R39	**PE/TE:** 9, 36, 39, 42, 43, 48, 56, 58, 61, 63, 65, 66, 69, 71, 77, 84, 93, 106, 107, 124, 129, 135, 137, 141, 150, 152, 153, 158, 161, 169, 173, 174, 183, 186, 187, 199, 200, 211, 212, 213, 214, 217, 226, 227, 230, 237, 239, 243, 265, 271, 276, 279, 284, 301, 314, 315, 316, 331, 334, 338, 339, 341, 347, 349, 356, 359, 369, 370, 396, 397, 416, 418, 420, 421, 422, 431, 432, 438, 439, 443, 446, 447, 450, 452, 461, 462, 465, 473, 483, 485, 488, 492, 493, 496, 500, 503, 507, 511, 522, 526, 548, 550, 566, 567, 568, 570, 573, 574, 577, 578, 584, 585, 593, 594, 597, 600, 602, 603, 610, 622, 626, 628, 629, 630, 635, 636, 639, 642, 643, 644, 645, 657, 658, 660, 663, 670, 672, 673, 674, 679, 681, 687, 691, 692, 694, 695, 696, 701, 705, 710, 715, 718, 722, 724, 726, 731, 732, 735, 740, 742, 743, 746, 747, 748, 749, 750, 760, 761, 763, 777, 779, 780, 783, 786, 792, 801, 803, 807, 809, 812, 816, 817, 820, 821, 832, 839, 842, 846, 848, 849, 856, 862, 868, 871, 874, 876, 877, 886, 895, 897, 908, 909, 912, 919, 920, 922, 923, 924, 926, 927, R9, R55
P15. Note cause/effect relationship and draw inferences from a variety of data	**PE/TE:** S22, S25, 29, 31, 32, 33, 36, 37, 41-42, 43, 49, 50, 51-52, 53, 67-68, 69, 70-71, 93, 94-95, 96, 97-100, 133-137, 142, 160, 166-169, 172-173, 202-203, 211, 212, 216-217, 221-222, 223, 230, 231-235, 236, 238-239, 245-246, 247-248, 257, 262-263, 280, 282-283, 292, 293-295,	**PE/TE:** S21, S23, S27, 33, 36, 37, 39, 40, 45, 48, 50, 52, 54, 56, 59, 60, 63, 66, 70, 71, 84, 96, 99, 100, 102, 120, 127, 137, 139, 141, 145, 152, 160, 161, 167, 169, 170, 175, 182, 196, 202, 203, 211, 221, 222, 224, 226, 230, 233, 235, 243, 245, 246, 248, 250, 257, 259, 260, 261, 265, 271, 275, 277, 281, 283, 286, 291, 295, 299, 307, 308, 310, 330, 335, 339, 343, 345, 346, 348, 349, 351, 353, 354, 355, 356, 358, 367, 374, 375, 377, 380, 381, 383, 386, 392, 394, 396, 398, 401, 402, 405, 411, 418, 420, 428, 431, 433, 436, 439, 441, 444, 446, 467, 468, 469, 472, 483, 487, 489, 492, 493, 494, 502, 503, 505, 508, 511, 516, 518, 519, 534, 539, 540, 542, 543, 550, 552, 554, 559, 562, 567, 568, 569, 574,

Process Standards	Core Instruction	Application
(continued) **P15.** Note cause/effect relationship and draw inferences from a variety of data	**PE/TE:** 296, 297, 299, 307-308, 330-331, 336-337, 338, 342, 343-349, 350, 351-352, 353, 354, 355-356, 365-366, 374-375, 376, 377-379, 380-381, 382, 383-385, 386-387, 390, 392, 393-394, 400, 401-405, 408, 417-418, 426, 427, 428-429, 433-434, 435, 436-437, 439, 440, 441-444, 467, 469-471, 472-473, 476, 478, 483, 487-490, 502-503, 505-508, 515-516, 539-540, 542, 543, 555-557, 559, 563-564, 566-567, 569-572, 581-585, 587-589, 590-591, 596, 598-599, 607-608, 609, 610-611, 613-615, 616-617, 620, 621-623, 625-626, 627, 628-629, 631-634, 635-636, 654-656, 657, 663, 667-668, 669-670, 677-681, 683-686, 687, 689-690, 691-692, 699-700, 701-705, 706, 719, 720-722, 729-731, 732-733, 734-735, 738-740, 760-761, 762-763, 766-769, 770-771, 776-777, 778-779, 781, 785-786, 788, 789-790, 796, 798, 800, 801-803, 806-808, 810-811, 813-815, 816-818, 819, 826, 828-829, 831-837, 840, 841-842, 843-844, 845-846, 853-854, 856-857, 859-860, 861, 862, 863, 865-866, 867-868, 886, 889, 892, 893-894, 895-896, 898-899, 900, 901-903, 904, 906, 916, 917, 920, 922, 923-924, R8	**PE/TE:** 576, 577, 578, 581, 584, 585, 591, 600, 602, 603, 604, 607, 609, 610, 611, 613, 614, 616, 617, 618, 622, 623, 625, 627, 629, 631, 632, 633, 635, 636, 638, 639, 650, 654, 657, 661, 663, 666, 672, 674, 676, 677, 681, 684, 687, 689, 692, 694, 696, 696, 699, 703, 705, 713, 716, 722, 724, 726, 729, 732, 733, 735, 740, 742, 743, 754, 760, 761, 762, 763, 767, 769, 770, 771, 778, 781, 788, 790, 792, 793, 799, 800, 803, 808, 810, 811, 818, 820, 821, 824, 826, 827, 828, 829, 830, 831, 837, 841, 843, 846, 848, 850, 853, 855, 856, 857, 859, 861, 863, 865, 867, 868, 870, 882, 889, 893, 895, 899, 901, 904, 906, 908, 917, 923, 924, 926, R8
P16. Predict likely outcomes and recognize cases in which more than one interpretation of data is valid	**PE/TE:** R32	**PE/TE:** 15, 22, 39, 52, 63, 84, 127, 137, 150, 152, 182, 217, 235, 248, 265, 277, 284, 295, 308, 310, 333, 339, 356, 367, 381, 387, 394, 398, 416, 418, 420, 431, 467, 493, 494, 516, 559, 591, 600, 602, 604, 617, 623, 638, 657, 663, 674, 681, 694, 713, 715, 733, 742, 743, 783, 790, 794, 818, 820, 821, 850, 857, 868, 870, 889, 924, R32
P17. Reinterpret data to develop alternative outcomes and their likely effects on subsequent events/issues	**PE/TE:** R32	**PE/TE:** 52, 63, 127, 226, 248, 265, 271, 277, 308, 310, 339, 356, 381, 394, 444, 493, 494, 503, 516, 602, 663, 694, 713, 715, 743, 790, 794, 818, 850, 857, 868, 870, 924, R32

Process Standards	Core Instruction	Application
P18. Use available data to devise new situations and outcomes	**PE/TE:** R32	**PE/TE:** 52, 63, 127, 248, 265, 308, 339, 394, 503, 516, 602, 636, 663, 694, 743, 790, 850, 857, 868, 870, R32
P19. Demonstrate an understanding of the data through written, visual, or oral methods	**PE/TE:** S20, S22, S24-S25, 4, 10, 16, 26, 34, 40, 48, 58, 66, 72, 78, 88, 96, 122, 130, 138, 146, 156, 164, 170, 176, 198, 204, 212, 220, 230, 236, 244, 254, 260, 266, 272, 280, 290, 304, 326, 334, 342, 350, 362, 370, 376, 382, 388, 400, 408, 414, 424, 432, 440, 460, 468, 476, 486, 496, 504, 512, 536, 544, 554, 562, 568, 580, 586, 594, 606, 612, 618, 624, 630, 652, 658, 666, 676, 682, 688, 698, 706, 716, 728, 736, 756, 764, 774, 784, 796, 804, 812, 824, 830, 840, 852, 858, 864, 884, 892, 900, 912, 920, R3, R4, R5, R6, R7, R8, R9, R10, R11, R12, R13, R15, R16, R17, R18, R19, R20, R21, R22, R23, R24-R25, R26, R27, R28, R29, R30-R31, R32, R33, R34, R35	**PE/TE:** S21, S23, S26-S27, A3, A7, A11, 2, 4, 8, 9, 10, 12, 13, 15, 16, 18, 19, 20, 21, 22, 23, 24, 26, 30, 31, 33, 34, 36, 37, 39, 40, 42, 43, 44, 45, 47, 48, 50, 51, 52, 53, 54, 55, 58, 60, 61, 62, 63, 65, 66, 68, 69, 71, 72, 74, 75, 76, 77, 78, 80, 81, 82, 83, 84, 85, 86, 88, 90, 91, 93, 95, 96, 98, 99, 100, 101, 102, 103, 106, 107, 108, 109, 120, 122, 124, 125, 126, 127, 129, 130, 132, 133, 134, 135, 137, 138, 141, 142, 143, 145, 146, 148, 149, 150, 151, 152, 153, 154, 156, 158, 159, 160, 161, 163, 164, 166, 167, 168, 169, 170, 171, 172, 174, 175, 176, 178, 179, 180, 181, 182, 183, 186, 187, 188, 189, 196, 198, 200, 201, 202, 203, 204, 206, 207, 208, 209, 211, 212, 214, 215, 216, 217, 219, 220, 221, 222, 224, 225, 226, 227, 228, 230, 232, 233, 235, 236, 238, 239, 240, 241, 243, 244, 246, 247, 248, 249, 250, 251, 252, 254, 256, 257, 259, 260, 262, 263, 264, 265, 266, 268, 269, 271, 272, 274, 275, 276, 277, 279, 280, 282, 283, 284, 285, 286, 287, 288, 290, 292, 293, 294, 295, 296, 298, 299, 301, 302, 303, 304, 306, 307, 308, 309, 310, 311, 314, 315, 316, 317, 324, 326, 328, 329, 330, 331, 332, 333, 334, 336, 337, 338, 339, 341, 342, 344, 345, 346, 347, 348, 349, 350, 352, 353, 354, 356, 357, 358, 359, 360, 362, 364, 365, 366, 367, 369, 370, 372, 373, 375, 376, 378, 379, 381, 382, 384, 385, 386, 387, 388, 390, 391, 392, 393, 394, 395, 396, 397, 398, 400, 402, 403, 404, 405, 407, 408, 411, 412, 413, 414, 416, 417, 418, 419, 420, 421, 422, 424, 427, 428, 429, 431, 432, 435, 437, 438, 439, 440, 442, 443, 444, 445, 446, 447, 450, 451, 452, 453, 454, 458, 460, 461, 462, 463, 464, 465, 466, 467, 468, 469, 471, 472, 473, 475, 476, 480, 481, 482, 483, 485, 486, 488, 489, 490, 491, 492, 493, 494, 496, 498, 499, 500, 501, 502, 503, 504, 506, 508, 509, 510, 511, 512, 514, 515, 516, 517, 518, 519, 522, 523, 524, 525, 526, 534, 536, 538, 539, 540, 541, 542, 543, 544, 546, 547, 548, 549, 550, 551, 552, 554, 556, 557, 558, 559, 561, 562, 564, 566, 567, 568, 570, 572, 574, 575, 576, 577, 578, 580, 582, 583, 584, 585, 586, 588, 589, 590, 591, 593, 594, 596, 597, 600, 601, 602, 603, 604, 606, 608, 609, 610, 611, 612, 614, 615, 616, 617, 618, 620, 622, 623, 624, 626, 627, 628, 629, 630, 632, 633, 634, 635, 636, 637, 638, 639, 642, 643, 644, 645, 650, 652, 654, 655, 656, 657, 658, 660, 661, 662, 663, 665, 666, 668, 669, 670, 671, 672, 673, 674, 676, 677, 679, 680, 681, 682, 684, 685, 686, 687, 688, 689, 690, 691, 692, 693, 694, 695, 696, 700, 701, 702, 703, 704, 705, 706, 707, 708, 709, 710, 712, 713, 715, 716, 718, 719, 720, 721, 722, 723, 724, 725, 726, 728, 729, 730, 731, 732, 733, 735, 736, 738, 739, 740, 741, 742, 743, 746, 747, 748, 749, 750, 754, 756, 757, 758, 759, 761, 762, 763, 764, 766, 767, 768, 769, 771, 773, 774, 776, 777, 778, 779, 780, 781, 783, 784, 786, 788, 789, 790, 791, 792, 793, 794, 796, 798, 799, 800, 801, 802, 803, 804, 806, 807, 808, 809, 810, 811, 812, 815, 816, 817, 818, 819, 820, 821, 822, 824, 826, 827, 828, 829, 830, 832, 834, 835, 837, 839, 840, 842, 843, 844, 845, 846, 847, 848, 849, 850, 852, 854, 855, 856, 857, 858, 859, 860, 861, 862, 863, 864, 866, 867, 868, 869, 870, 871, 874, 875, 876, 877, 882, 884, 886, 887, 888, 889, 891, 892, 894, 895, 896, 897, 899, 900, 903, 904, 905,

Process Standards	Core Instruction	Application
(continued) **P19.** Demonstrate an understanding of the data through written, visual, or oral methods		**PE/TE:** 906, 907, 908, 909, 910, 912, 914, 915, 916, 917, 919, 920, 922, 923, 924, 925, 926, 927, R3, R4, R5, R6, R7, R8, R9, R10, R11, R12, R13, R15, R16, R17, R18, R19, R20, R21, R22, R23, R24-R25, R26, R27, R28, R29, R30-R31, R32, R33, R34, R35, R55
P20. Extract significant ideas from supporting details	**PE/TE:** S4, S6, S20, S22, S24-S25, 4, 10, 16, 26, 34, 40, 48, 58, 66, 72, 78, 88, 96, 122, 130, 138, 146, 156, 164, 170, 176, 198, 204, 212, 220, 230, 236, 244, 254, 260, 266, 272, 280, 290, 304, 326, 334, 342, 350, 362, 370, 376, 382, 388, 400, 408, 414, 424, 432, 440, 460, 468, 476, 486, 496, 504, 512, 536, 544, 554, 562, 568, 580, 586, 594, 606, 612, 618, 624, 630, 652, 658, 666, 676, 682, 688, 698, 706, 716, 728, 736, 756, 764, 774, 784, 796, 804, 812, 824, 830, 840, 852, 858, 864, 884, 892, 900, 912, 920, R4	**PE/TE:** S5, S7, S21, S23, S26-S27, 2, 4, 5, 8, 9, 10, 11, 13, 14, 15, 16, 17, 18, 19, 20, 22, 23, 24, 26, 27, 30, 31, 32, 33, 34, 35, 36, 37, 38, 39, 40, 41, 42, 43, 44, 45, 47, 48, 49, 50, 51, 52, 54, 55, 56, 58, 59, 61, 63, 65, 66, 67, 68, 69, 70, 71, 72, 73, 74, 75, 76, 77, 78, 79, 80, 81, 82, 84, 85, 86, 88, 89, 90, 91, 92, 93, 94, 95, 96, 97, 99, 100, 102, 103, 120, 122, 123, 124, 125, 126, 127, 129, 130, , 131, 132, 133, 134, 135, 136, 137, 138, 139, 141, 142, 143, 145, 146, 147, 148, 149, 150, 152, 153, 154, 156, 157, 159, 160, 161, 164, 165, 166, 167, 168, 169, 170, 171, 172, 173, 175, 176, 177, 178, 180, 182, 183, 196, 198, 199, 200, 201, 202, 203, 204, 205, 207, 208, 209, 211, 212, 213, 214, 215, 216, 217, 220, 221, 222, 223, 224, 226, 227, 228, 230, 231, 232, 233, 235, 236, 237, 238, 239, 240, 241, 243, 244, 245, 246, 247, 248, 250, 251, 252, 254, 255, 256, 257, 258, 259, 260, 261, 262, 263, 264, 265, 266, 267, 268, 269, 270, 271, 272, 273, 274, 275, 276, 277, 279, 280, 281, 283, 284, 286, 287, 288, 290, 291, 292, 293, 294, 295, 296, 297, 298, 299, 300, 301, 303, 304, 305, 306, 307, 308, 310, 311, 324, 326, 327, 328, 329, 330, 331, 332, 333, 334, 335, 336, 337, 338, 339, 341, 342, 343, 345, 346, 347, 348, 349, 350, 351, 352, 353, 354, 355, 356, 358, 359, 360, 362, 363, 364, 365, 366, 367, 369, 370, 371, 372, 373, 374, 375, 376, 377, 379, 380, 381, 382, 383, 384, 385, 386, 387, 388, 389, 390, 392, 393, 394, 396, 397, 398, 400, 401, 402, 403, 404, 405, 408, 409, 411, 412, 413, 414, 415, 416, 417, 418, 420, 421, 422, 424, 425, 427, 428, 429, 430, 431, 432, 433, 435, 436, 437, 439, 440, 441, 442, 443, 444, 446, 447, 458, 460, 461, 462, 463, 464, 465, 466, 467, 468, 469, 471, 472, 473, 476, 477, 480, 481, 482, 483, 485, 486, 487, 488, 489, 490, 492, 493, 494, 496, 497, 498, 499, 500, 501, 502, 503, 504, 505, 506, 508, 509, 511, 512, 513, 514, 515, 516, 517, 518, 534, 536, 537, 539, 540, 541, 542, 543, 544, 545, 546, 547, 548, 550, 551, 552, 554, 555, 556, 557, 558, 559, 562, 563, 564, 565, 566, 567, 568, 569, 572, 573, 574, 576, 577, 578, 580, 581, 582, 583, 584, 585, 587, 589, 590, 591, 593, 594, 595, 596, 598, 600, 602, 603, 604, 606, 607, 608, 609, 610, 611, 612, 613, 614, 615, 616, 617, 618, 619, 620, 621, 622, 623, 624, 625, 626, 627, 628, 629, 630, 631, 632, 633, 634, 635, 636, 638, 639, 650, 652, 654, 655, 656, 657, 658, 659, 660, 661, 662, 663, 666, 667, 668, 669, 670, 672, 673, 674, 676, 677, 679, 680, 681, 682, 683, 684, 685, 686, 687, 688, 689, 690, 691, 692, 693, 694, 695, 696, 698, 699, 700, 701, 702, 703, 704, 705, 706, 707, 708, 709, 710, 711, 713, 715, 716, 717, 718, 719, 720, 721, 722, 723, 724, 726, 728, 729, 730, 731, 732, 733, 735, 736, 737, 739, 740, 742, 743, 754, 756, 757, 758, 759, 760, 761, 762, 763, 764, 765, 766, 767, 768, 769, 770, 771, 773, 774, 775, 777, 778, 779, 781, 783, 784, 785, 786, 788, 789, 790, 792, 793, 794, 796, 797, 798, 799, 800, 801, 802, 803, 804, 805, 806, 807, 808,

Process Standards	Core Instruction	Application
(continued) **P20.** Extract significant ideas from supporting details		**PE/TE:** 810, 811, 812, 813, 815, 816, 817, 818, 820, 821, 822, 824, 825, 826, 827, 828, 829, 830, 831, 832, 834, 837, 839, 840, 841, 842, 843, 844, 845, 846, 848, 849, 850, 852, 853, 854, 855, 856, 857, 858, 859, 860, 861, 862, 863, 864, 865, 866, 867, 868, 870, 871, 882, 884, 885, 886, 887, 888, 889, 891, 892, 893, 894, 895, 896, 898, 899, 900, 901, 903, 904, 905, 906, 908, 909, 910, 912, 913, 914, 916, 917, 919, 920, 921, 922, 923, 924, 926, 927, R4, R55
P21. Combine critical concepts in a statement of conclusion based on information	**PE/TE:** S20, S22, S24-S25, R13	**PE/TE:** S21, S22, S26-S27, 5, 9, 11, 13, 15, 18, 19, 20, 22, 23, 30, 31, 33, 39, 41, 42, 45, 49, 52, 54, 56, 59, 65, 68, 70, 74, 80, 81, 82, 84, 86, 90, 92, 93, 94, 95, 100, 102, 120, 127, 133, 135, 137, 143, 145, 149, 150, 152, 158, 169, 175, 178, 182, 203, 221, 222, 224, 226, 228, 233, 235, 241, 246, 247, 248, 250, 259, 261, 263, 264, 265, 271, 277, 279, 281, 283, 284, 286, 295, 303, 307, 308, 310, 329, 332, 333, 341, 346, 347, 348, 349, 351, 352, 353, 355, 356, 358, 359, 363, 365, 367, 373, 374, 375, 380, 381, 382, 385, 386, 387, 389, 392, 393, 394, 396, 398, 402, 404, 405, 409, 411, 413, 416, 418, 420, 422, 431, 433, 435, 436, 441, 443, 444, 446, 465, 466, 467, 471, 473, 477, 480, 483, 485, 487, 489, 492, 493, 494, 502, 503, 511, 515, 516, 518, 539, 540, 543, 546, 550, 555, 556, 567, 569, 572, 573, 574, 582, 583, 584, 585, 591, 600, 602, 603, 604, 607, 608, 610, 611, 614, 617, 620, 623, 625, 626, 629, 631, 632, 633, 635, 636, 638, 639, 655, 657, 663, 670, 672, 674, 679, 681, 687, 693, 695, 702, 705, 710, 713, 715, 718, 721, 722, 724, 726, 729, 730, 731, 733, 740, 742, 743, 763, 766, 770, 771, 778, 779, 781, 783, 788, 789, 790, 792, 794, 802, 803, 811, 816, 818, 820, 821, 827, 828, 829, 832, 837, 846, 848, 850, 855, 857, 861, 863, 866, 868, 870, 889, 893, 894, 899, 901, 903, 906, 908, 909, 914, 917, 919, 924, 926, R13, R55
P22. Determine whether information is pertinent to the topic	**PE/TE:** S4, S6, S20, S22, S24-S25, R32, R33, R37	**PE/TE:** S5, S7, S21, S23, S26-S27, A3, A7, 2, 20, 23, 52, 63, 71, 77, 85, 103, 127, 153, 161, 183, 224, 235, 241, 248, 251, 259, 277, 284, 287, 295, 303, 311, 333, 349, 359, 387, 394, 397, 405, 418, 431, 447, 490, 503, 511, 516, 551, 559, 567, 574, 577, 585, 591, 600, 603, 629, 636, 657, 670, 673, 687, 681, 695, 713, 722, 725, 743, 781, 793, 803, 811, 818, 821, 837, 846, 849, 857, 863, 868, 871, 899, 909, 924, 927, R32, R33, R37
P23. Test the validity of the information using such criteria as source, objectivity, technical correctness, and currency	**PE/TE:** S4, S8, S10, S12, S20, S22, S24-S25, R16, R17, R18, R33	**PE/TE:** S5, S9, S11, S13, S21, S23, S26-S27, 12, 23, 51, 55, 61, 76, 85, 90, 103, 133, 168, 216, 227, 246, 251, 268, 287, 298, 311, 336, 365, 384, 397, 403, 421, 427, 447, 481, 493, 508, 515, 546, 551, 574, 577, 589, 603, 616, 639, 656, 662, 673, 686, 695, 704, 725, 730, 743, 768, 793, 806, 818, 821, 844, 849, 867, 871, 903, 909, 922, 927, R16, R17, R18, R33, R55

Communication: Communication is the conveyance of ideas, value judgments, beliefs, and emotions through individual expression, group dialogue, cultural communities, and global networks by oral, written, symbolic, visual, and technological means.

"Communication" emphasizes the learners' use of a broad base of strategies to:

Process Standards	Core Instruction	Application
P24. Transmit ideas through speeches	**PE/TE:** 183, 349, 356, 387, 394, 421, 431, 473, 493, 603, 629, 681, 771, 868	**PE/TE:** 183, 349, 356, 387, 394, 421, 431, 473, 493, 603, 629, 681, 771, 868
P25. Demonstrate conflicting ideas through debate	**PE/TE:** 153, 394	**PE/TE:** 153, 394
P26. Summarize judgments through essays	**PE/TE:** S22, S25, R31, R32	**PE/TE:** S23, S27, 20, 23, 54, 82, 84, 102, 137, 152, 182, 183, 226, 227, 250, 286, 287, 310, 311, 333, 358, 359, 375, 396, 418, 420, 439, 446, 492, 493, 503, 518, 550, 576, 577, 602, 603, 638, 663, 672, 673, 694, 705, 724, 733, 742, 743, 792, 803, 820, 821, 846, 848, 870, 871, 908, 909, 926, R31, R32
P27. Demonstrate ideas through dramatizations	**PE/TE:** 183, 227, 421, 475, 493, 551, 629, 681, 837	**PE/TE:** 183, 227, 421, 475, 493, 551, 629, 681, 837
P28. Transmit ideas through discussions	**PE/TE:** 47, 561	**PE/TE:** 47, 561
P29. Demonstrate emotions through the creation of visuals	**PE/TE:** 359, 405, 600	**PE/TE:** 359, 405, 600
P30. Demonstrate beliefs through multimedia projects	**PE/TE:** R35	**PE/TE:** 670, 687, 811, 909
P31. Recognize beliefs through simulation and role play	**PE/TE:** 308, 359, 381, 475, 493, 629, 663	**PE/TE:** 308, 359, 381, 475, 493, 629, 663

Historical Awareness: Historical Awareness, integral to all of the Social Studies disciplines, includes an understanding of chronological placement, historical trends, and historical decision-making. Students will be able to comprehend the significance of historical data using a variety of analytical skills. Such understanding enables students to prioritize events, identify bias, recognize perspectives, interpret trends, and predict outcomes.

"Historical Awareness" emphasizes the learners' use of a broad base of strategies to:

Process Standards	Core Instruction	Application
P32. Read critically a variety of materials including textbooks, historical documents, newspapers, magazines, and other reference sources	**PE/TE:** S4, S8, S10, S12, S20, S22, S24-S25, R16, R17, R18, R33	**PE/TE:** S5, S9, S11, S13, S21, S23, S26-S27, 12, 23, 51, 55, 61, 76, 85, 90, 103, 133, 168, 216, 227, 246, 251, 268, 287, 298, 311, 336, 365, 384, 397, 403, 421, 427, 447, 481, 493, 508, 515, 546, 551, 574, 577, 589, 603, 616, 639, 656, 662, 673, 686, 695, 704, 725, 730, 743, 768, 793, 806, 818, 821, 844, 849, 867, 871, 903, 909, 922, 927, R16, R17, R18, R33, R55
P33. Prepare and analyze maps, charts, and graphs	**PE/TE:** S8, S10, S12, S14, S16, lviii, A2-A3, A4-A7, A11, 282, 428, 656, 720, 738, 756, 764, 815, 824, 827, 840, 852, 864	**PE/TE:** S9, S11, S13, S15, S17, A3, A7, 2-3, 12, 23, 24-25, 26, 30, 33, 44, 47, 48, 52, 55, 56-57, 61, 62, 63, 65, 66, 69, 71, 72, 74, 76, 77, 78, 82, 85, 86-87, 93, 94, 99, 103, 106, 107, 108, 109, 120-121, 122, 125, 127, 134, 138, 141, 142, 145, 146, 149, 150, 153, 154-155, 156, 159, 161, 170, 172, 175, 176, 179, 182, 183, 186, 187, 188, 189, 196-197, 198, 201, 203, 207, 215, 227, 228-229, 230, 232, 235, 247, 251, 252-253, 256, 257, 260, 262, 265, 272, 277, 282, 284, 285, 286, 287, 288-289, 292, 293, 294, 296, 303, 311, 314, 315, 316, 317, 324-325, 326, 329, 331, 333, 334, 338, 339, 353, 357, 359, 360-361, 362, 365, 366, 367, 367, 372, 388, 390, 394, 395, 397, 398-399, 417, 419, 421, 422-423, 424, 428, 431, 435, 443, 445, 447, 450, 451, 452, 453, 454, 458-459, 460, 463, 464, 467, 468, 472, 473, 490, 491, 492, 493, 494-495, 499, 519, 522, 523, 524, 525, 526, 534-535, 536, 541, 543, 551, 552-553, 554, 557, 559, 567, 575, 577, 578-579, 582, 586, 588, 591, 601, 602, 603, 604-605, 608, 612, 617, 620, 623, 629, 636, 637, 638, 639, 642, 643, 644, 645, 650-651, 655, 656, 658, 661, 663, 673, 674-675, 676, 679, 681, 685, 688, 691, 692, 696-697, 698, 702, 705, 708, 716, 720, 721, 722, 725, 726-727, 728, 731, 738, 739, 743, 746, 747, 748, 749, 750, 754-755, 756, 764, 770, 778, 779, 781, 790, 794-795, 799, 807, 815, 817, 821, 822-823, 824, 827, 829, 834, 840, 845, 846, 847, 849, 850-851, 852, 855, 856, 857, 860, 864, 868, 871, 874, 875, 876, 877, 882-883, 887, 889, 910-911, 912, 914, 917, 922, 924, 926, 927, R55
P34. Construct and analyze timelines	**PE/TE:** S18, 166, 220, 290, 350, 356, 378, 382, 547, 562, 624, 690, 830, R20, R21, R22, R23, R24, R25, R26, R27	**PE/TE:** S19, S27, 2-3, 24-25, 56-57, 86-87, 120-121, 154-155, 166, 196-197, 220, 224, 228-229, 236, 252-253, 288-289, 290, 295, 324-325, 350, 356, 360-361, 378, 382, 387, 398-399, 422-423, 458-459, 494-495, 534-535, 547, 548, 552-553, 562, 567, 578-579, 604-605, 624, 629, 650-651, 674-675, 690, 696-697, 724, 726-727, 754-755, 794-795, 822-823, 830, 837, 850-851, 882-883, 908, 909, 910-911, 927, R20, R21, R22, R23, R24, R25, R26, R27

Process Standards	Core Instruction	Application
P35. Utilize community resources such as field trips, guest speakers, and museums	**PE/TE:** 287, 349, 670	**PE/TE:** 287, 349, 670
P36. Incorporate the use of technological resources	**PE/TE:** R34, R35, R36, R37	**PE/TE:** 3, 4, 9, 10, 12, 15, 16, 20, 25, 26, 30, 33, 34, 39, 40, 45, 46-47, 48, 51, 52, 57, 58, 63, 66, 69, 71, 72, 77, 78, 82, 87, 91, 95, 96, 100, 103, 105, 112, 121, 122, 127, 130, 133, 135, 137, 138, 145, 146, 150, 156, 161, 162-163, 164, 169, 170, 174, 175, 176, 180, 183, 192, 197, 198, 203, 204, 211, 212, 215, 216, 218-219, 220, 224, 229, 230, 233, 235, 236, 241, 244, 245, 246, 248, 251, 253, 256, 259, 263, 266, 271, 272, 277, 280, 285, 289, 290, 296, 300, 301, 303, 304, 308, 319, 325, 326, 333, 334, 337, 339, 342, 345, 349, 350, 356, 359, 361, 362, 370, 374, 375, 376, 381, 382, 384, 388, 391, 394, 397, 399, 400, 404, 408, 413, 414, 421, 423, 424, 427, 431, 432, 436, 438, 439, 440, 444, 447, 456, 459, 460, 466, 467, 468, 470, 473, 476, 483, 486, 490, 493, 495, 496, 503, 504, 506, 510, 511, 512, 514, 516, 517, 519, 519, 528, 535, 536, 543, 544, 545, 546, 548, 551, 553, 554, 558, 559, 560-561, 564, 567, 568, 574, 575, 577, 579, 580, 583, 585, 586, 589, 591, 594, 596, 597, 600, 603, 605, 606, 611, 612, 616, 617, 618, 623, 624, 626, 629, 630, 636, 639, 652, 655, 657, 657, 658, 665, 666, 667, 669, 670, 671, 673, 673, 675, 676, 681, 682, 683, 686, 687, 688, 697, 698, 703, 704, 706, 708, 712, 713, 716, 723, 725, 727, 728, 730, 731, 733, 736, 740, 741, 743, 752, 756, 757, 759, 763, 764, 767, 768, 771, 772-773, 774, 781, 784, 790, 795, 796, 799, 803, 804, 809, 811, 812, 814, 818, 819, 821, 823, 824, 829, 830, 833, 837, 840, 846, 847, 849, 851, 852, 857, 858, 862, 863, 864, 867, 868, 869, 871, 880, 882, 884, 889, 890-891, 892, 894, 897, 899, 900, 906, 907, 909, 911, 912, 914, 917, 920, 924, 925, 927, R34, R35, R36, R37
P37. Utilize primary and secondary source material such as biographies and autobiographies; novels; speeches and letters; and poetry, songs, and artwork	**PE/TE:** S4, S20, S22, S24-S25, R33	**PE/TE:** S5, S21, S23, S26-S27, 12, 20, 51, 52, 63, 77, 85, 90, 103, 127, 133, 137, 145, 153, 161, 168, 180, 183, 216, 224, 227, 235, 241, 246, 251, 259, 268, 277, 282, 284, 287, 295, 298, 303, 311, 336, 339, 347, 349, 359, 384, 387, 394, 397, 403, 405, 418, 421, 427, 431, 481, 483, 490, 503, 511, 515, 546, 551, 559, 574, 577, 589, 591, 600, 603, 611, 616, 617, 629, 636, 639, 657, 662, 665, 670, 673, 681, 686, 687, 704, 725, 730, 740, 743, 768, 781, 803, 806, 811, 818, 821, 837, 844, 846, 849, 857, 867, 871, 891, 899, 903, 909, 924, 927, R33

WORLD GEOGRAPHY

Standards, Learning Expectations, and Performance Indicators

CULTURE

Content Standard: 1.0

Culture encompasses similarities and differences among people including their beliefs, knowledge, changes, values, and traditions. Students will explore these elements of society to develop an appreciation and respect for the variety of human cultures.

Learning Expectations	Core Instruction	Application
1.01 Understand the complex nature of culture.	**PE/TE:** 70, 86, 89-95, 97-99, 100, 110, 139-145, 168-169, 171-175, 191,237-241, 242-243, 258-259, 264-265, 269, 271, 274, 277, 281, 283, 284, 286, 287, 297-303, 364, 366, 371, 372-373, 374-375, 379, 380-381, 384, 385, 386, 387, 390, 392, 394, 401-402, 406-407, 409-413, 426, 429, 430, 434, 436-437, 438, 441-443, 477-483, 484-485, 506-511, 542, 547-548, 558-559, 564, 565, 566, 569-572, 588, 590-591, 596, 597, 599-600, 610-611, 613,-614, 615, 619, 621-622, 625, 627, 631, 632, 633, 634, 636, 678, 680, 681, 684-686, 690, 692, 707-713, 714-715, 731, 733, 739-740, 767-768, 775-781, 782-783, 788, 798, 810-811, 828, 834-837, 838-839, 860, 861, 862, 866, 868, 894, 895-896, 897, 904, 917, 918-919, R38-R54	**PE/TE:** 71, 84, 86, 88, 90, 91, 93, 95, 96, 99, 100, 101, 102, 103, 138, 141, 142, 145, 168, 169, 170, 173, 175, 182, 236, 239, 240, 241, 243, 249, 250, 251, 259, 265, 277, 284, 296, 298, 299, 301, 303, 310, 311, 366, 367, 373, 375, 379, 381, 384, 387, 392, 393, 394, 396, 397, 402, 405, 407, 408, 411, 413, 420, 421, 429, 431, 435, 437, 438, 439, 442, 444, 446, 447, 476, 480, 482, 483, 485, 492, 493, 508, 509, 511, 518, 519, 542, 548, 550, 551, 559, 566, 567, 572, 574, 577, 603, 604, 611, 615, 617, 622, 627, 629, 633, 638, 639, 681, 686, 687, 692, 694, 695, 696, 708, 709, 710, 713, 715, 724, 725, 733, 740, 742, 754, 769, 771, 774, 777, 779, 781, 783, 788, 792, 793, 803, 811, 820, 821, 822, 829, 837, 839, 848, 860, 863, 870, 871, 899, 904, 908, 910, 912, 917, 919, 926, R55
1.02 Recognize the role of major religions.	**PE/TE:** 92-93, 140, 141, 237, 271, 297, 364, 366, 372, 374, 379, 380, 385, 386, 387, 390, 394, 411, 426, 430, 434, 436, 442, 443, 478-480, 509, 564, 566-567, 570, 571, 588, 591, 596, 599, 610, 611, 615, 621, 627, 634, 636, 678, 680, 681, 685-686, 707-709, 731, 733, 777, 810, 835, 860, 862, R38-R54	**PE/TE:** 93, 95, 101, 102, 103, 145, 396, 442, 483, 567, 572, 574, 591, 602, 611, 679, 685, 686, 687, 694, 708, 709, 713, 733, 777, 811, 820, 835, R55
1.03 Appreciate the relationship between physical environments and culture.	**PE/TE:** A11, 69, 99-100, 165, 172-173, 218-219, 238, 242-243, 250, 262-263, 275, 281, 283, 303, 306, 307-308, 374, 386, 387, 390, 392, 403, 418, 461-462, 464, 465, 474-475, 505-506, 540, 555-556, 581, 592-593, 607, 613-614, 619, 620, 656, 678-679, 736, 758, 797-798, 896, 904	**PE/TE:** 22, 56, 69, 100, 101, 175, 250, 259, 286, 306, 375, 387, 392, 393, 394, 396, 398, 405, 420, 467, 492, 518, 534, 550, 552, 554, 556, 559, 576, 577, 593, 608, 614, 617, 620, 623, 656, 694, 803, 824, 848, 896, 904, 908
1.04 Understand how cultural perspective impacts perceptions of places and regions.	**PE/TE:** A11, 11-12, 61-63, 68, 172-173, 255, 256-257, 263, 299, 355-356, 403, 405, 467, 583, 684-686, 835	**PE/TE:** 63, 259, 356, 396, 467, 685, 686, 687, 848

SEVENTH GRADE BENCHMARKS

CULTURE

Performance Indicators State:
As documented through state assessment -

Performance Indicators State	Core Instruction	Application
at Level I, the student is able to		
7.1.spi.1. recognize cultural definitions (i.e., language, religion, customs, political system, economic system).	**PE/TE:** 80, 89-90, 92-93, 94-95, 110, 111, 148, 191, 192-195, 234, 237, 260, 264, 293, 294, 318, 319-323, 410, 416, 435, 443, 455, 456-457, 488, 507, 522, 527, 528-533, 573, 584, 614, 627, 635, 646, 647-649, 678, 679, 680, 685, 714-715, 718, 739, 751, 752-753, 777, 779, 780, 786, 809, 816, 832, 838-839, 842, 843, 878-879, 880-881, 895, 897, R38-R54	**PE/TE:** 90, 93, 95, 102, 103, 138, 148, 170, 175, 176, 180, 183, 264, 265, 266, 271, 272, 276, 280, 284, 287, 293, 294, 295, 296, 303, 311, 362, 388, 394, 408, 416, 421, 424, 431, 443, 447, 476, 488, 493, 519, 522, 551, 567, 584, 614, 635, 638, 673, 676, 679, 685, 688, 692, 706, 713, 715, 716, 718, 725, 728, 739, 743, 777, 779, 786, 816, 832, 842, 871, 895, 897, R55
at Level 2, the student is able to		
7.1.spi.2. locate cultural information on a thematic map (i.e., languages, political systems, economic systems, religions).	**PE/TE:** A5, A6-A7, 57, 62, 87, 93, 94, 107, 141, 187, 294, 315, 366, 435, 443, 452, 506-507, 524, 525, 526, 579, 620, 643, 679, 685, 697, 708, 739, 750, 778, 875, R38-R39	**PE/TE:** A7, 56, 86, 93, 94, 103, 107, 141, 187, 287, 294, 315, 366, 397, 435, 443, 452, 506, 524, 525, 526, 577, 579, 620, 643, 679, 685, 696, 708, 739, 750, 778, 875
at Level 3, the student is able to		
7.1.spi.3. compare and contrast the tenets of the five major world religions (i.e., Christianity, Buddhism, Islam, Hinduism, and Judaism).	**PE/TE:** 92, 708, 777, R38-R54	**PE/TE:** 694, 709, 777, R55

Performance Indicators Teacher:

As documented through teacher observation -

Performance Indicators Teacher	Core Instruction	Application
at Level I, the student is able to		
7.1.tpi.1. research a country for a fictional visit.	**PE/TE:** 251, 287, 311, 375, 381, 447, 519, 692, 763, 793, 849, 871, 917	**PE/TE:** 251, 287, 311, 375, 381, 447, 519, 692, 763, 793, 849, 871, 917
7.1.tpi.2. host a cultural food fair bringing in recipes and/or food items from varying cultures.	**PE/TE:** An opportunity to address this standard may be found on page 411	**PE/TE:** An opportunity to address this standard may be found on page 611
7.1.tpi.3. decide what is needed to define the character of a place.	**PE/TE:** 80, 89-90, 92-93, 94-95, 110, 111, 148, 191, 192-195, 234, 237, 260, 264, 293, 294, 318, 319-323, 410, 416, 435, 443, 455, 456-457, 488, 507, 522, 527, 528-533, 573, 584, 614, 627, 635, 646, 647-649, 678, 679, 680, 685, 714-715, 718, 739, 751, 752-753, 777, 779, 780, 786, 809, 816, 832, 838-839, 842, 843, 878-879, 880-881, 895, 897, R38-R54	**PE/TE:** 90, 93, 95, 102, 103, 138, 148, 170, 175, 176, 180, 183, 264, 265, 266, 271, 272, 276, 280, 284, 287, 293, 294, 295, 296, 303, 311, 362, 388, 394, 408, 416, 421, 424, 431, 443, 447, 476, 488, 493, 519, 522, 551, 567, 584, 614, 635, 638, 673, 676, 679, 685, 688, 692, 706, 713, 715, 716, 718, 725, 728, 739, 743, 777, 779, 786, 816, 832, 842, 871, 895, 897, R55
7.1.tpi.4. draw a thematic map.	**PE/TE:** S16	**PE/TE:** S17, R22
at Level 2, the student is able to		
7.1.tpi.5. use the Internet to explore cultural differences among groups of people by analyzing data updates, viewing pictures, and researching cultural facts.	**PE/TE:** R36	**PE/TE** 103, 180, 183, 251, 259, 265, 277, 431, 490, 600, 695, 811, 849, 906, R36
7.1.tpi.6. design a cartouche (a signature in ancient Egyptian times) using global images describing the student's interests. Include only the student's name as the only English word on the page. The student will describe to the class why he/she chose the symbols to include in his/her cartouche.	**PE/TE:** An opportunity to address this standard may be found on page 559	**PE/TE:** An opportunity to address this standard may be found on page 577
at Level 3, the student is able to		
7.1.tpi.7. use a chart, Venn diagram, graphic organizer to compare/contrast tenets of the five major world religions.	**PE/TE:** R54	**PE/TE:** An opportunity to address this standard may be found on pages 694, R55

ECONOMICS

Content Standard: 2.0

Globalization of the economy, the explosion of population growth, technological changes and international competition compel students to understand, both personally and globally, production, distribution, and consumption of goods and services. Students will examine and analyze economic concepts such as basic needs versus wants, using versus saving money, and policy-making versus decision-making.

Learning Expectations	Core Instruction	Application
2.01 Understand fundamental economic concepts and their application to a variety of economic systems.	**PE/TE:** 75, 76-77, 111, 149, 178, 191, 192-195, 282, 356, 365, 375, 385, 392, 394, 427, 431, 435, 439, 489, 515, 566, 788, 817, 843, 862	**PE/TE:** 72, 75, 77, 85, 150, 180, 183, 244, 282, 427, 447, 518, 742, 817, 820, 862
2.02 Understand global economic connections, conflicts, and interdependence.	**PE/TE:** 77, 111, 150, 180, 246, 248, 257, 283, 284, 306, 356, 364, 367, 375, 379, 386, 387, 392, 393, 417-418, 427, 429, 431, 435, 439, 515-516, 589, 591, 609, 614, 617, 623, 626, 633, 634, 636, 678-679, 680, 681, 721-722, 738, 740, 788, 844, 845, 860, 862, 863, 898-899	**PE/TE:** 77, 84, 150, 182, 246, 248, 250, 277, 284, 286, 418, 420, 431, 444, 516, 518, 591, 602, 614, 617, 623, 629, 638, 724, 754, 788, 790, 792, 844, 846, 849, 870, 908, 926
2.03 Understand the changes that occur in the nature, use, distribution, and importance of resources.	**PE/TE:** 73-74, 76, 178-179, 189, 246, 256, 258, 262, 270, 274, 276, 282-283, 305-307, 367, 368-369, 373, 379, 392, 393, 417, 429, 466, 501, 516, 540, 542-543, 573-574, 589, 592-593, 596, 599, 617, 629, 633, 634, 636, 654-656, 678, 681, 687, 721, 733, 803, 815, 829, 844, 857, 866, 898-899, 905	**PE/TE:** 74, 76, 77, 84, 85, 150, 152, 179, 182, 189, 246, 248, 250, 251, 256, 259, 269, 271, 274, 277, 284, 286, 288, 306, 307, 310, 369, 417, 418, 420, 421, 501, 516, 518, 540, 543, 577, 593, 602, 629, 638, 655, 657, 663, 694, 695, 742, 792, 803, 848, 906, 908

SEVENTH GRADE BENCHMARKS

ECONOMICS

Performance Indicators State:
As documented through state assessment -

Performance Indicators State	Core Instruction	Application
at Level I, the student is able to		
7.2.spi.1. recognize basic economic concepts (i.e. imports, exports, barter system, tariffs, closed and emerging markets, supply and demand, inflation, recession, depression).	**PE/TE:** 76-77, 150, 244, 246, 260, 262, 280, 282, 326, 333, 382, 414, 416, 424, 427, 486, 489, 512, 515, 594, 596, 784, 814, 817, 844	**PE/TE:** 150, 248, 250, 265, 420, 490, 492, 516, 600, 695, 790
7.2.spi.2. define renewable and nonrenewable resources.	**PE/TE:** 73-74, 76, 178-179, 189, 246, 256, 258, 262, 270, 274, 276, 282-283, 305-307, 333, 367, 368-369, 373, 379, 392, 393, 417, 429, 466, 501, 516, 540, 542-543, 573-574, 589, 592-593, 596, 599, 617, 629, 633, 634, 636, 654-656, 678, 681, 687, 721, 733, 803, 815, 829, 844, 857, 866, 898-899, 905	**PE/TE:** 74, 76, 77, 84, 85, 150, 152, 179, 182, 189, 246, 248, 250, 251, 256, 259, 269, 271, 274, 277, 284, 286, 288, 306, 307, 310, 369, 417, 418, 420, 421, 501, 516, 518, 540, 543, 577, 593, 602, 629, 638, 655, 657, 663, 694, 695, 742, 792, 803, 848, 906, 908
7.2.spi.3. define demographic concepts. (i.e. population, population distribution, population density, growth rate).	**PE/TE:** 59-63, 64-65	**PE/TE:** 60, 62, 63, 65, 83, 84, 85
at Level 2, the student is able to		
7.2.spi.4. interpret economic issues as expressed with maps, tables, diagrams, and charts.	**PE/TE:** R22, R24, R25	**PE/TE:** 76, 85, 107, 109, 149, 179, 187, 189, 251, 315, 317, 417, 447, 452, 454, 519, 524, 525, 603, 629, 639, 643, 645, 675, 695, 721, 743, 749, 750, 793, 821, 849, 871, 875, 877, 909
7.2.spi.5. select the major resources, industrial, and agricultural products for the three grand divisions from a map of Tennessee.	**PE/TE:** 109, 149	**PE/TE:** 77

Performance Indicators Teacher:
As documented through teacher observation -

Performance Indicators Teacher	Core Instruction	Application
at Level I, the student is able to		
7.2.tpi.1. cite examples of renewable and nonrenewable resources from the community.	**PE/TE:** 74, 109, 149	**PE/TE:** 77

Performance Indicators Teacher	Core Instruction	Application
at Level 1, the student is able to		
7.2.tpi.2. identify through a graphic organizer various economic systems used throughout the world.	**PE/TE:** 75	**PE/TE:** 72, 77
7.2.tpi.3. select a photograph from a global community and cite examples of how this photograph illustrates the local economy.	**PE/TE:** 56, 59, 65, 75, 185, 201, 248, 258, 270, 276, 278, 281, 306-307, 328, 368-369, 375, 399, 439, 466, 506, 507, 516, 567, 568, 572, 592-593, 595, 628, 664, 681, 686, 687, 714-715, 761, 782-783, 795, 802, 815, 843, 898, 918-919	**PE/TE:** 65, 248, 258, 281, 369, 375, 516, 577, 593, 715, 761, 783, 802, 919
at Level 2, the student is able to		
7.2.tpi.4. simulate the basic concepts of economics by classroom activities.	**PE/TE:** 623	**PE/TE:** 623
7.2.tpi.5. use symbolic space through simulations to show real world facts and figures.	**PE/TE:** 180, 265, 311, 493, 611, 663, 681, 713, 722, 740, 793, 909	**PE/TE:** 180, 265, 311, 493, 611, 663, 681, 713, 722, 740, 793, 909
7.2.tpi.6. compare imports and exports on a bar graph.	**PE/TE:** 77, 567, 636	**PE/TE:** 77, 567, 636
7.2.tpi.7. know the major resources, industrial and agricultural products from Tennessee.	**PE/TE:** 109, 149	**PE/TE:** 77
7.2.tpi.8. use a table of facts to analyze differences among countries and communities in economic terms (e.g. exports, imports.)	**PE/TE:** 76, 276, 282, 365, 573, 899	**PE/TE:** 76, 272, 276, 282, 365, 367, 424, 431, 447, 695, 743, 909
at Level 3, the student is able to		
7.2.tpi.9. use a population pyramid to show population and gender of students in the class, school, community, state, and various countries or continents.	**PE/TE:** An opportunity to address this standard may be found on pages 345, 558	**PE/TE:** An opportunity to address this standard may be found on pages 345, 558
7.2.tpi.10. illustrate global trading patterns.	**PE/TE:** 77, 111, 150, 180, 246, 248, 257, 283, 284, 306, 356, 364, 367, 375, 379, 386, 387, 392, 393, 417-418, 427, 429, 431, 435, 439, 515-516, 589, 591, 609, 614, 617, 623, 626, 633, 634, 636, 678-679, 680, 681, 721-722, 738, 740, 788, 844, 845, 860, 862, 863, 898-899	**PE/TE:** 77, 84, 150, 182, 246, 248, 250, 277, 284, 286, 418, 420, 431, 444, 516, 518, 591, 602, 614, 617, 623, 629, 638, 724, 754, 788, 790, 792, 844, 846, 849, 870, 908, 926
7.2.tpi.11. evaluate the domestic and international impact of various economic agreements (e.g., NAFTA, European Union).	**PE/TE:** 169, 180, 246, 254, 257, 260, 350, 355-356, 364, 373, 375, 379, 418, 429, 444, 676, 678-679, 690	**PE/TE:** 248, 259, 356, 375, 420, 431, 692

GEOGRAPHY

Content Standard: 3.0

Geography enables the students to see, understand and appreciate the web of relationships between people, places, and environments. Students will use the knowledge, skills, and understanding of concepts within the six essential elements of geography: world in spatial terms, places and regions, physical systems, human systems, environment and society, and the uses of geography.

Learning Expectations	Core Instruction	Application
3.01 Understand the characteristics and uses of maps, globes, and other geographic tools and technologies.	**PE/TE:** S14, S16, lviii, A2-A3, A4-A7	**PE/TE:** S15, S17, A3, A7
3.02 Know the location of places and geographic features, both physical and human, locally, regionally and globally.	**PE/TE:** lviii, A2, A4, A14-A37, 6, 25, 121, 155, 157, 159, 163, 196, 201, 207, 229, 232, 253, 289, 325, 329, 361, 399, 423, 459, 463, 472, 495, 499, 535, 541, 605, 608, 651, 661, 685, 691, 755, 759, 795, 823, 851, 860, 883, 901, 911, R20	**PE/TE:** A3, A7, 23, 24, 55, 120, 153, 154, 159, 183, 196, 207, 227, 228, 232, 251, 252, 287, 288, 311, 324, 353, 359, 360, 397, 398, 421, 422, 445, 447, 458, 463, 472, 491, 493, 494, 499, 519, 534, 541, 551, 575, 577, 601, 603, 604, 608, 637, 639, 650, 661, 673, 685, 691, 693, 695, 725, 741, 743, 759, 791, 793, 794, 819, 821, 847, 849, 850, 860, 871, 882, 907, 909, 910, 927, R20
3.03 Understand the characteristics and uses of spatial organization of Earth's surface.	**PE/TE:** S14, A5, A14-A37, 5-9, 6, 25, 121, 155, 157, 159, 163, 196, 201, 207, 229, 232, 253, 289, 325, 329, 361, 399, 423, 459, 463, 472, 495, 499, 535, 541, 605, 608, 651, 661, 685, 691, 755, 759, 795, 823, 851, 860, 883, 901, 911, R20	**PE/TE:** S15, 9, 21, 22, 23, 24, 55, 120, 153, 154, 159, 183, 196, 207, 227, 228, 232, 251, 252, 287, 288, 311, 324, 353, 359, 360, 397, 398, 421, 422, 445, 447, 458, 463, 472, 491, 493, 494, 499, 519, 534, 541, 551, 575, 577, 601, 603, 604, 608, 637, 639, 650, 661, 673, 685, 691, 693, 695, 725, 741, 743, 759, 791, 793, 794, 819, 821, 847, 849, 850, 860, 871, 882, 907, 909, 910, 927, R20
3.04 Understand the physical and human characteristics of place.	**PE/TE:** A4-A5, A11, 6, 42-43	**PE/TE:** 22, 43, 45, 55, 103
3.05 Understand that common physical and cultural characteristics create regions.	**PE/TE:** A5, A11, 6, 42-45, 94-95	**PE/TE:** 22, 43, 44, 45, 55, 95, 102, 103
3.06 Understand how physical processes shape the Earth's natural landscapes and affect environments.	**PE/TE:** A11, 9, 25, 26, 27-33, 40, 41-45, 124, 125, 126-127, 160, 188, 202-203, 211, 278-279, 316, 327-328, 330, 331, 453, 500, 539-540, 541, 644, 659-663, 664-665, 748, 826, 853-854, 856-857, 876, 915	**PE/TE:** 24, 26, 30, 31, 33, 42, 43, 45, 54, 55, 160, 198, 203, 211, 226, 279, 316, 330, 333, 358, 453, 500, 541, 543, 549, 644, 657, 658, 663, 665, 672, 748, 849, 857, 876, 915

Learning Expectations	Core Instruction	Application
3.07 Understand how physical systems and the physical environment affect human systems.	**PE/TE:** A11, 7, 9, 30-31, 73-74, 109, 132, 136, 149, 160, 161, 175, 178-180, 189, 200, 209, 256, 274, 275, 278-279, 283, 305-307, 317, 328, 329, 332-333, 368-369, 379, 393, 417-418, 429, 454, 464-466, 480, 498, 501, 516, 525, 536, 539, 541, 542-543, 555-556, 581, 589, 591, 592-593, 617, 622-623, 629, 634, 636, 645, 654, 657, 658, 662-663, 664-665, 667, 678, 692, 720-721, 727, 729, 734-735, 736, 738, 740, 749, 758, 760-761, 762-763, 796, 798, 799, 801-803, 825, 826, 828, 829, 845-846, 853-854, 855, 856-857, 859, 868, 877, 886, 898-899, 905-906	**PE/TE:** 55, 56, 74, 77, 83, 84, 85, 137, 149, 152, 153, 156, 161, 175, 179, 180, 182, 183, 209, 251, 277, 279, 284, 286, 307, 308, 310, 317, 329, 333, 394, 396, 398, 405, 417, 421, 465, 466, 467, 483, 490, 492, 501, 516, 518, 519, 525, 534, 541, 543, 550, 552, 556, 559, 567, 576, 593, 603, 623, 629, 634, 636, 638, 639, 645, 651, 652, 657, 663, 665, 672, 673, 695, 722, 724, 735, 740, 749, 761, 763, 792, 793, 799, 803, 820, 826, 829, 846, 848, 849, 854, 855, 857, 870, 877, 906, 908, 909
3.08 Understand how human activities impact and modify the physical environment.	**PE/TE:** A11, 7, 9, 47, 48, 49-52, 136, 169, 201, 218-219, 221, 257, 270, 275, 306, 307-308, 339, 376, 380, 428, 467, 495, 498, 502-503, 536, 540, 556, 562, 564, 581, 589, 654-656, 668, 687, 721, 733, 740, 789, 801, 802, 808, 889	**PE/TE:** 48, 50, 51, 52, 54, 127, 203, 219, 226, 306, 308, 309, 310, 311, 381, 428, 465, 467, 503, 519, 540, 543, 549, 550, 554, 559, 576, 577, 629, 636, 639, 655, 656, 657, 672, 722, 724, 741, 743, 790, 792, 801, 803, 808, 820, 889
3.09 Understand the nature, distribution and migration of human populations on Earth's surfaces.	**PE/TE:** A11, 6, 7, 9, 61-63, 64-65, 66, 67-71, 106, 107, 124, 127, 132, 135, 150, 166, 172-173, 187, 221-222, 238, 262, 265, 274, 283, 293-295, 299, 315, 348, 371, 374, 390, 401-402, 403, 409-411, 413, 426, 427, 452, 474-475, 524, 539, 542, 556, 557, 571, 572, 579, 583, 599, 610, 620, 621, 625, 634, 643, 653-654, 678, 697, 700, 703, 708-709, 710, 714-715, 729, 750, 767, 768, 778-779, 782-783, 788, 797, 836, 860, 861, 865, 875, 895-896, 902, 904, 916	**PE/TE:** 56, 62, 63, 65, 66, 68, 69, 71, 83, 84, 85, 106, 107, 127, 132, 150, 152, 175, 187, 241, 251, 265, 283, 286, 294, 295, 299, 303, 310, 311, 315, 329, 333, 358, 375, 390, 404, 405, 411, 413, 418, 420, 421, 447, 452, 492, 503, 524, 542, 550, 559, 576, 577, 578, 602, 620, 638, 643, 679, 696, 710, 724, 750, 773, 778, 779, 783, 788, 792, 821, 870, 871, 875, 882, 896, 899, 908, 917, 926

SEVENTH GRADE BENCHMARKS

GEOGRAPHY

Performance Indicators State:

As documented through state assessment -

Performance Indicators State	Core Instruction	Application
at Level 1, the student is able to		
7.3.spi.1. identify and use the basic elements of maps and mapping.	**PE/TE:** lviii,-A3, A4-A7	**PE/TE:** A3, A7, A13-A37, 23, 55, 85, 103, 153, 183 227, 251, 287, 311, 359, 397, 421, 447 493, 519, 551, 577, 603, 639, 673, 695, 725, 743, 793, 821, 849, 871, 909, 927
7.3.spi.2. locate the Earth's major physical characteristics (i.e., 7 continents, 4 oceans).	**PE/TE:** A13-A17, A22-A26, A30, A31, A36-A37, 25, 26, 28	**PE/TE:** 24, 55
7.3.spi.3. identify the major river systems of Tennessee.	**PE/TE:** A19	**PE/TE:** A19
7.3.spi.4. distinguish the differences among rural, suburban, and urban communities.	**PE/TE:** 58, 63, 69, 627	**PE/TE:** 63, 71, 83, 84
7.3.spi.5. select the natural resources found in the 3 grand divisions of Tennessee (coal, copper, timber, plants, animals.)	**PE/TE:** 109, 149	**PE/TE:** 77
7.3.spi.6. Locate on a map specific lines of longitude and latitude. (i.e., Prime Meridian, International Date Line, Equator, North and South Poles, Tropics of Cancer and Capricorn, Arctic /Antarctic circles).	**PE/TE:** lviii, A2, 6, 13, 42, 43-45	**PE/TE:** A3, 23, 44, 882, 922
7.3.spi.7. compare the five largest cities of Tennessee using a bar graph.	**PE/TE:** An opportunity to address this standard may be found on pages A21, 61	**PE/TE:** An opportunity to address this standard may be found on page A10
7.3.spi.8. define demographic concepts (i.e., population, population distribution, population density, growth rate, family size, and infant mortality).	**PE/TE:** 58, 59-60, 61-63, 64-65	**PE/TE:** 58, 63, 83, 84, 85
at Level 2, the student is able to		
7.3.spi.9. identify the location of Earth's major landforms and bodies of water (i.e., Rockies, Andes, Himalayas, Alps, Urals, Sahara desert, Nile River Valley, Great Plains, Mississippi River, Amazon River, Thames River, Seine River, Rhine River, Danube River, Tigris River, Euphrates River, Ganges River, Volga River, Yellow River).	**PE/TE:** A14-A15, A18-A19, A22, A24, A26-A27, A30, A32-A33, A36, A37, 34, 35-39, 122, 124, 125, 126-127, 128-129, 156, 157-159, 198, 199-201, 204, 205-208, 326, 327-329, 459, 460, 461-463, 496, 497-499, 535, 536, 537-539, 652, 653-654, 655, 658, 659-662, 755, 756, 757-759, 795, 796, 797-799, 824, 825-828, 851, 852, 853-854, 883, 884, 885-887, 912, 913-914, 920, 921-923	**PE/TE:** 55, 122, 125, 127, 129, 152, 153, 156, 161, 182, 183, 198, 203, 204, 207, 208, 225, 226, 227, 326, 329, 357, 359, 460, 491, 492, 493, 498, 499, 517, 518, 519, 534, 539, 549, 550, 551, 658, 661, 662, 671, 672, 673, 754, 756, 758, 759, 791, 792, 793, 798, 799, 803, 819, 820, 821, 827, 828, 829, 847, 848, 849, 850, 852, 857, 859, 870, 871, 882, 885, 886, 887, 889, 907, 908, 909, 915, 917, 922, 924, 925, 926, 927

Performance Indicators State	Core Instruction	Application
at Level 2, the student is able to		
7.3.spi.10. identify the characteristics that define a region geographically.	**PE/TE:** A5, A11, 6, 42-45, 94-95	**PE/TE:** 22, 43, 44, 45, 55, 95, 102, 103
7.3.spi.11. recognize specific physical processes that operate on the Earth's surface (i.e., erosion, volcanoes, earthquakes, wind and water currents, plate tectonics, and weathering).	**PE/TE:** A11, 9, 25, 26, 27-33, 40, 41-45, 124, 125, 126-127, 160, 188, 202-203, 211, 278-279, 316, 327-328, 330, 331, 453, 500, 539-540, 541, 644, 659-663, 664-665, 748, 826, 853-854, 856-857, 876, 915	**PE/TE:** 24, 26, 30, 31, 33, 42, 43, 45, 54, 55, 160, 198, 203, 211, 226, 279, 316, 330, 333, 358, 453, 500, 541, 543, 549, 644, 657, 658, 663, 665, 672, 748, 849, 857, 876, 915
7.3.spi.12. identify the six physical regions of Tennessee (i.e., Unaka Mountains, Valley and Ridge, Cumberland Plateau, Highland Rim, Central Basin, Gulf Coastal Plain).	**PE/TE:** 43-45, 126, 128-129	**PE/TE:** 44, 45
7.3.spi.13. recognize the definitions of modifications on the physical environment (i.e. global warming, deforestation, desert, urbanization).	**PE/TE:** A11, 7, 9, 47, 48, 49-52, 136, 169, 201, 218-219, 221, 257, 270, 275, 306, 307-308, 339, 376, 380, 428, 467, 495, 498, 502-503, 536, 540, 556, 562, 564, 581, 589, 654-656, 668, 687, 721, 733, 740, 789, 801, 802, 808, 889	**PE/TE:** 48, 50, 51, 52, 54, 127, 203, 219, 226, 306, 308, 309, 310, 311, 381, 428, 465, 467, 503, 519, 540, 543, 549, 550, 554, 559, 576, 577, 629, 636, 639, 655, 656, 657, 672, 722, 724, 741, 743, 790, 792, 801, 803, 808, 820, 889
7.3.spi.14. distinguish between types of maps. (i.e. political, physical, climatic, land-use resource, contour, elevation, topographic.)	**PE/TE:** A4-A7	**PE/TE:** A7
7.3.spi.15 interpret a map indicating scale, distance and direction.	**PE/TE:** lviii, A2, A13-A37, 125, 534, 661, 887	**PE/TE:** A3, 23, 55, 69, 125, 159, 201, 338, 359, 390, 397, 534, 608, 620, 661, 679, 743, 770, 807, 887
7.3.spi.16. view and discuss a show which shows a diverse global culture.	**PE/TE:** 100, 183, 431, 483, 603, 673, 687, 713, 793, 811, 871	**PE/TE:** 100, 183, 431, 483, 603, 673, 687, 713, 793, 811, 871
at Level 3, the student is able to		
7.3.spi.17. read and interpret a time zone map.	**PE/TE:** An opportunity to address this standard may be found on pages 41-42	**PE/TE:** An opportunity to address this standard may be found on page 54
7.3.spi.18. analyze the environmental consequences of humans changing their physical environment (i.e., air and water pollution, mining, deforestation, global warming).	**PE/TE:** A11, 7, 9, 47, 48, 49-52, 136, 169, 201, 218-219, 221, 257, 270, 275, 306, 307-308, 339, 376, 380, 428, 467, 495, 498, 502-503, 536, 540, 556, 562, 564, 581, 589, 654-656, 668, 687, 721, 733, 740, 789, 801, 802, 808, 889	**PE/TE:** 48, 50, 51, 52, 54, 127, 203, 219, 226, 306, 308, 309, 310, 311, 381, 428, 465, 467, 503, 519, 540, 543, 549, 550, 554, 559, 576, 577, 629, 636, 639, 655, 656, 657, 672, 722, 724, 741, 743, 790, 792, 801, 803, 808, 820, 889
at Level 3, the student is able to		

Performance Indicators State	Core Instruction	Application
7.3.spi.19. examine reasons and patterns of human migration through the use of maps, charts, diagrams (i.e., famine, natural disasters, political and religious oppression, wars).	**PE/TE:** A11, 6, 7, 9, 61-63, 64-65, 66, 67-71, 106, 107, 124, 127, 132, 135, 150, 166, 172-173, 187, 221-222, 238, 262, 265, 274, 283, 293-295, 299, 315, 348, 371, 374, 390, 401-402, 403, 409-411, 413, 426, 427, 452, 474-475, 524, 539, 542, 556, 557, 571, 572, 579, 583, 599, 610, 620, 621, 625, 634, 643, 653-654, 678, 697, 700, 703, 708-709, 710, 714-715, 729, 750, 767, 768, 778-779, 782-783, 788, 797, 836, 860, 861, 865, 875, 895-896, 902, 904, 916	**PE/TE:** 56, 62, 63, 65, 66, 68, 69, 71, 83, 84, 85, 106, 107, 127, 132, 150, 152, 175, 187, 241, 251, 265, 283, 286, 294, 295, 299, 303, 310, 311, 315, 329, 333, 358, 375, 390, 404, 405, 411, 413, 418, 420, 421, 447, 452, 492, 503, 524, 542, 550, 559, 576, 577, 578, 602, 620, 638, 643, 679, 696, 710, 724, 750, 773, 778, 779, 783, 788, 792, 821, 870, 871, 875, 882, 896, 899, 908, 917, 926
7.3.spi.20. predict the consequences of population changes on the Earth's physical and cultural environments.	**PE/TE:** A11, 6, 7, 9, 61-63, 64-65, 66, 67-71, 106, 107, 124, 127, 132, 135, 150, 166, 172-173, 187, 221-222, 238, 262, 265, 274, 283, 293-295, 299, 315, 348, 371, 374, 390, 401-402, 403, 409-411, 413, 426, 427, 452, 474-475, 524, 539, 542, 556, 557, 571, 572, 579, 583, 599, 610, 620, 621, 625, 634, 643, 653-654, 678, 697, 700, 703, 708-709, 710, 714-715, 729, 750, 767, 768, 778-779, 782-783, 788, 797, 836, 860, 861, 865, 875, 895-896, 902, 904, 916	**PE/TE:** 56, 62, 63, 65, 66, 68, 69, 71, 83, 84, 85, 106, 107, 127, 132, 150, 152, 175, 187, 241, 251, 265, 283, 286, 294, 295, 299, 303, 310, 311, 315, 329, 333, 358, 375, 390, 404, 405, 411, 413, 418, 420, 421, 447, 452, 492, 503, 524, 542, 550, 559, 576, 577, 578, 602, 620, 638, 643, 679, 696, 710, 724, 750, 773, 778, 779, 783, 788, 792, 821, 870, 871, 875, 882, 896, 899, 908, 917, 926
7.3.spi.21. interpret a population pyramid.	**PE/TE:** 345, 558	**PE/TE:** 558

Performance Indicators Teacher:

As documented through teacher observation -

Performance Indicators Teacher	Core Instruction	Application
at Level I, the student is able to		
7.3.tpi.1 construct a papier of a world globe showing the seven continents and four oceans.	**PE/TE:** An opportunity to address this standard may be found on pages A13-A17, A22-A26, A30, A31, A36-A37, 25, 26, 28	**PE/TE:** An opportunity to address this standard may be found on pages 24, 55
7.3.tpi.2. color country and regional outline maps.	**PE/TE:** 23, 55, 85, 103, 153, 183, 227, 251, 287, 311, 359, 397, 421, 447, 493, 519, 551, 577, 603, 639, 673, 695, 725, 743, 793, 821, 849, 871, 909, 927	**PE/TE:** 23, 55, 85, 103, 153, 183, 227, 251, 287, 311, 359, 397, 421, 447, 493, 519, 551, 577, 603, 639, 673, 695, 725, 743, 793, 821, 849, 871, 909, 927

Performance Indicators Teacher	Core Instruction	Application
at Level 1, the student is able to		
7.3.tpi.3. apply geographic symbols to outline maps.	**PE/TE:** 23, 55, 85, 103, 153, 183, 227, 251, 287, 311, 359, 397, 421, 447, 493, 519, 551, 577, 603, 639, 673, 695, 725, 743, 793, 821, 849, 871, 909, 927	**PE/TE:** 23, 55, 85, 103, 153, 183, 227, 251, 287, 311, 359, 397, 421, 447, 493, 519, 551, 577, 603, 639, 673, 695, 725, 743, 793, 821, 849, 871, 909, 927
7.3.tpi.4. create student maps illustrating geographic terminology (e.g. archipelago, atoll, bay, canyon, cape, continental divide, delta, estuary, fjord, headwaters, glacier, gulf, island, isthmus, lagoon, mesa, ocean currents, peninsula, plateau, reef, strait.)	**PE/TE:** lviii,-A3, A4-A7	**PE/TE:** A3, A7, A13-A37, 23, 55, 85, 103, 153, 183, 227, 251, 287, 311, 359, 397, 421, 447, 493, 519, 551, 577, 603, 639, 673, 695, 725, 743, 793, 821, 849, 871, 909, 927
7.3.tpi.5. list physical characteristics that comprise a place (e.g., landforms, water systems, climate patterns, and plant and animal life).	**PE/TE:** A4-A5, A11, 6, 42-43	**PE/TE:** 22, 43, 45, 55, 103
7.3.tpi.6. participate in a class-wide analysis of ecosystems through the use of posters and illustrations.	**PE/TE:** An opportunity to address this standard may be found on pages 46-47, 208-211, 331, 464-466, 500, 539-540, 760-761, 798, 825-826, 828-829, 856-857, 886	**PE/TE:** An opportunity to address this standard may be found on pages 47, 183, 211, 465, 540, 761, 886
7.3.tpi.7. explain how geographers use technology to study the Earth.	**PE/TE:** 10, 11-13, 14-15, 16, 17-20	**PE/TE:** 13, 14, 15, 19, 20, 21, 22, 23
at Level 2, the student is able to		
7.3.tpi.8. construct a salt map illustrating the six regions of Tennessee.	**PE/TE:** An opportunity to address this standard may be found on pages 43-45, 126, 128-129	**PE/TE:** An opportunity to address this standard may be found on pages 44, 45
7.3.tpi.9. create a bar graph showing the various physical processes.	**PE/TE:** A10, 331, 522	**PE/TE:** 186, 656, 856, 909
7.3.tpi.10. create a cause and effect chart which shows relationships between geographic issues. (water pollution, immigration, settlement.)	**PE/TE:** 506-507, 540, 555-556, 662, 664-665, 733, 740, 762, 789, 825, 826, 828, 829	**PE/TE:** 54, 310, 467, 503, 519, 559, 577, 655, 662, 663, 665, 733, 740, 743, 821, 824, 909, R8
7.3.tpi.11. construct a cartogram showing a community population variations.	**PE/TE:** An opportunity to address this standard may be found on pages 107, 187, 315, 452, 524, 643, 750, 778, 875, R23	**PE/TE:** An opportunity to address this standard may be found on pages 107, 187, 315, 452, 524, 643, 750, 778, 875, R23
7.3.tpi.12. research the industrial development of a country giving the causes and consequences of urbanization.	**PE/TE:** 180, 247, 293-295, 353, 373, 404, 410, 427, 439, 488, 502-503, 509, 572, 599, 610, 627, 709-710, 778-779, 781, 782-783, 828, 836-837, 838-839	**PE/TE:** 23, 247, 290, 295, 356, 375, 420, 503, 511, 576, 710, 715, 783, 848

Performance Indicators Teacher	Core Instruction	Application
7.3.tpi.13. compare the four major families of map projections; conformal, equal-area, azimuthal, and equidistant. Explain the biases of the Mercator's versus the Peter's projections.	**PE/TE:** A3	**PE/TE:** A3
at Level 3, the student is able to		
7.3.tpi.14. debate the consequences of deforestation.	**PE/TE:** 51-52, 304, 306, 307-308, 467, 801	**PE/TE:** 306, 307, 308, 310, 311, 467, 803
7.3.tpi.15. create a power-point presentation to illustrate and facilitate information on any given topic.	**PE/TE:** R35	**PE/TE:** 259, 483, 559, 585, 670, 673, 740, 771, 811, 837, 857, 871, 906, R35
7.3.tpi.16. construct a volcano demonstrating physical processes.	**PE/TE:** 31	**PE/TE:** 31
7.3.tpi.17. use population data sheets to further understand a community's demographics.	**PE/TE:** 61, 141, 142, 172, 294, 410, 443, 507, 540, 679, 720, 779, 865	**PE/TE:** 61, 85, 141, 142, 227, 251, 287, 294, 311, 421, 447, 493, 540, 679, 720, 779, 871, 927
7.3.tpi.18. conduct a classroom geo-bee.	**PE/TE:** 19	**PE/TE:** 19

GOVERNANCE AND CIVICS

Content Standard: 4.0

Governance establishes structures of power and authority in order to provide order and stability. Civic efficacy requires understanding rights and responsibilities, ethical behavior, and the role of citizens within their community, nation, and world.

Learning Expectations	Core Instruction	Application
4.01 Understand different systems of governance.	**PE/TE:** 79-81, 147-148, 177-178, 234-235, 264, 293, 336, 337, 344, 402-403, 415-417, 587, 591, 628, 635, 678, 680, 717-718, 765-766, 770, 785-786, 814, 816, 841-842, 861, 862, 894, 895, 903	**PE/TE:** 78, 80, 81, 82, 83, 84, 148, 152, 178, 181, 182, 183, 264, 265, 293, 295, 334, 338, 339, 358, 403, 405, 416, 492, 635, 673, 718, 722, 724, 767, 786, 790, 792, 816, 820, 842, 895, 908, 909
4.02 Understand how cooperation and conflict among people influence the division and control of resources, rights, and privileges.	**PE/TE:** 60, 73-74, 77, 132, 134, 135, 150, 166, 180, 221-222, 232, 257, 291-292, 337, 353, 355-356, 404, 428, 429, 431, 439, 470, 472, 488, 489, 501, 502, 506, 516, 542-543, 564, 571, 582, 583-584, 614, 625-626, 629, 636, 654-656, 678-679, 703, 894, 898	**PE/TE:** 63, 77, 84, 134, 137, 152, 180, 182, 232, 235, 292, 295, 356, 404, 405, 420, 431, 439, 446, 488, 492, 501, 503, 516, 518, 519, 543, 574, 576, 585, 614, 629, 636, 655, 656, 657, 681, 694, 695, 894, 899, 908

SEVENTH GRADE BENCHMARKS

GOVERNANCE AND CIVICS

Performance Indicators State:
As documented through state assessment -

Performance Indicators State	Core Instruction	Application
at Level 1, the student is able to		
7.4.spi.1. define the different types of governments (i.e., democracy, autocracy, oligarchy, monarchy, dictatorship).	**PE/TE:** 79-81, 147-148, 177-178, 234-235, 264, 293, 336, 337, 344, 402-403, 415-417, 587, 591, 628, 635, 678, 680, 717-718, 765-766, 770, 785-786, 814, 816, 841-842, 861, 862, 894, 895, 903	**PE/TE:** 78, 80, 81, 82, 83, 84, 148, 152, 178, 181, 182, 183, 264, 265, 293, 295, 334, 338, 339, 358, 403, 405, 416, 492, 635, 673, 718, 722, 724, 767, 786, 790, 792, 816, 820, 842, 895, 908, 909
7.4.spi.2. using a map key, locate various governance systems.	**PE/TE:** Opportunities to address this standard may be found on pages 79-81, 147-148, 177-178, 234-235, 264, 293, 336, 337, 344, 402-403, 415-417, 587, 591, 628, 635, 678, 680, 717-718, 765-766, 770, 785-786, 814, 816, 841-842, 861, 862, 894, 895, 903	**PE/TE:** Opportunities to address this standard may be found on pages 78, 80, 81, 82, 83, 84, 148, 152, 178, 181, 182, 183, 264, 265, 293, 295, 334, 338, 339, 358, 403, 405, 416, 492, 635, 673, 718, 722, 724, 767, 786, 790, 792, 816, 820, 842, 895, 908, 909
at Level 2, the student is able to		
7.4.spi.3. recognize how the boundaries of Congressional districts change in the state of Tennessee. (i.e. statutory requirements, population shifts, political power shifts).	**PE/TE:** 85	**PE/TE:** 85
7.4.spi.4. identify political leaders from selected contemporary settings (i.e., United States, India, Canada, Mexico, Great Britain, Russia, China).	**PE/TE:** 78, 178, 235, 246, 263, 275, 282, 385, 415, 487-488, 513, 620, 628, 633, 814, 841	**PE/TE:** 235, 246, 249, 284, 295, 418, 623, 633

Performance Indicators Teacher:
As documented through teacher observation -

Performance Indicators Teacher	Core Instruction	Application
at Level 1, the student is able to		
7.4.tpi.1 compare different political systems with those of the United States.	**PE/TE:** 80, 293, 416, 488, 623, 717-718, 786, 842, 895	**PE/TE:** 78, 82, 183, 293, 295, 416, 420, 488, 718, 786, 842, 895
7.4.tpi.2 use a graphic organizer to outline information according to importance on how communities govern themselves.	**PE/TE:** 293, 416, 488, 635, 678, 718, 786, 816, 842, 895	**PE/TE:** 183, 293

Performance Indicators Teacher	Core Instruction	Application
at Level 2, the student is able to		
7.4.tpi.3 describe components of various governments.	**PE/TE:** 79-81, 147-148, 177-178, 234-235, 264, 293, 336, 337, 344, 402-403, 415-417, 587, 591, 628, 635, 678, 680, 717-718, 765-766, 770, 785-786, 814, 816, 841-842, 861, 862, 894, 895, 903	**PE/TE:** 78, 80, 81, 82, 83, 84, 148, 152, 178, 181, 182, 183, 264, 265, 293, 295, 334, 338, 339, 358, 403, 405, 416, 492, 635, 673, 718, 722, 724, 767, 786, 790, 792, 816, 820, 842, 895, 908, 909
7.4.tpi.4. write an essay comparing different political systems with that of the United States.	**PE/TE:** 82	**PE/TE:** 82
7.4.tpi.5. debate various governments' laws and compare them to the United States (e.g., China's one child rule, women's rights, children's rights).	**PE/TE:** 903	**PE/TE:** 356, 394, 493, 790, 818, 821, 846
at Level 3, the student is able to		
7.4.tpi.6. research contemporary issues from Internet sites pertaining to different perspectives.	**PE/TE:** R36	**PE/TE:** 145, 180, 235, 295, 503, 519, 577, 603, 639, 743, 790, 821, 857, R36

HISTORY

Content Standard: 5.0

History involves people, events, and issues. Students will evaluate evidence to develop comparative and casual analyses, and to interpret primary sources. They will construct sound historical arguments and perspectives on which informed decisions in contemporary life can be based.

Learning Expectations	Core Instruction	Application
5.1 Understand the patterns of human settlement.	**PE/TE:** 57, 61-63, 64-65, 67-69, 107, 141, 165-166, 172, 173, 187, 221-222, 238, 248, 258, 261-262, 264-265, 269, 271, 274, 275, 282, 283, 291-292, 293-295, 299, 315, 353, 365, 366, 371-372, 374, 378, 380, 381, 385, 386, 387, 389-390, 393, 401-402, 403, 409-411, 427, 429, 430, 433, 434, 441, 444, 452, 469-470, 474-475, 477-478, 505-507, 524, 542, 565, 570, 572, 573, 581-582, 583-584, 595, 599, 610, 613, 614, 621, 625-626, 632, 633, 635, 643, 667-668, 669-670, 678, 680, 683-684, 691, 699-700, 703, 708, 710, 719, 729, 730, 739, 750, 757, 758, 761, 769, 777, 779, 782-783, 797, 808-809, 828, 836-837, 859, 860, 863, 865, 866, 875, 893-894, 902-903, 916, 917	**PE/TE:** 61, 62, 63, 65, 66, 68, 69, 71, 84, 85, 107, 141, 152, 166, 172, 175, 182, 187, 227, 238, 248, 250, 262, 264, 270, 284, 286, 292, 294, 295, 299, 310, 311, 315, 353, 356, 366, 375, 379, 390, 392, 396, 397, 404, 411, 413, 420, 421, 431, 444, 446, 452, 473, 492, 508, 511, 518, 524, 542, 559, 574, 576, 577, 602, 629, 638, 643, 687, 694, 700, 708, 713, 724, 725, 731, 742, 750, 763, 779, 781, 783, 792, 829, 849, 863, 870, 871, 875, 894, 899, 908, 916, 917, 926, 927
5.2 Recognize that places change over time.	**PE/TE:** 49-52, 132, 134, 135, 136, 137, 140, 141, 162-163, 173, 201, 221-222, 232, 257, 262-263, 268, 270, 282, 292, 293-294, 305-308, 332-333, 353, 355-356, 363, 365, 367, 371, 372, 373, 374-375, 377, 378-379, 381, 384-385, 386, 387, 390, 392, 403-405, 409-411, 417-418, 425-426, 427, 428-429, 430-431, 433-435, 436, 437, 439, 442, 469-471, 472-473, 479, 484-485, 487-490, 502-503, 506-507, 509, 514, 515-516, 540, 556, 565, 566-567, 570-571, 573, 581, 583-585, 587, 588, 590-591, 595-596, 599-600, 608, 610-611, 613-615, 621, 623, 625-626, 628-629, 632-634, 635-636, 668, 669, 678-681, 683-684, 687, 691, 700, 703-705, 710, 711-713, 714-715, 720-722, 730, 733, 738-740, 757, 765-767, 766-769, 770-771, 775-777, 778-779, 781, 786-788, 789-790, 805-808, 814-815, 817, 818, 832-834, 836-837, 838-839, 841-842, 843-844, 845-846, 860, 861, 862, 863, 866, 867-868, 889, 894, 895-896, 901-903, 904, 916	**PE/TE:** 52, 134, 137, 141, 152, 153, 163, 173, 182, 201, 222, 226, 232, 250, 259, 263, 271, 286, 295, 299, 306, 308, 310, 311, 324, 333, 353, 356, 358, 364, 367, 375, 379, 381, 387, 396, 404, 405, 411, 413, 418, 420, 421, 422, 427, 429, 431, 439, 444, 446, 447, 458, 471, 473, 483, 485, 489, 490, 492, 503, 508, 511, 516, 518, 519, 540, 556, 567, 576, 577, 578, 585, 589, 591, 602, 603, 611, 623, 626, 629, 634, 636, 638, 639, 672, 674, 681, 684, 687, 692, 694, 696, 700, 705, 710, 713, 715, 722, 724, 725, 733, 742, 767, 769, 771, 777, 779, 781, 788, 790, 792, 808, 818, 820, 834, 837, 839, 842, 844, 846, 848, 863, 868, 870, 882, 894, 899, 908, 909, 911, 917, 926

SEVENTH GRADE BENCHMARKS

HISTORY

Performance Indicators State:

As documented through state assessment -

Performance Indicators State	Core Instruction	Application
at Level 1, the student is able to		
7.5.spi.1. identify the causes and consequences of urbanization (i.e. industrial development.)	**PE/TE:** 180, 247, 293-295, 353, 373, 404, 410, 427, 439, 488, 502-503, 509, 572, 599, 610, 627, 709-710, 778-779, 781, 782-783, 828, 836-837, 838-839	**PE/TE:** 23, 247, 290, 295, 356, 375, 420, 503, 511, 576, 710, 715, 783, 848
at Level 2, the student is able to		
7.5.spi.2. identify reasons why people choose to settle in different places (i.e., occupation, family, climate, natural resources).	**PE/TE:** A11, 6, 7, 9, 61-63, 64-65, 66, 67-71, 106, 107, 124, 127, 132, 135, 150, 166, 172-173, 187, 221-222, 238, 262, 265, 274, 283, 293-295, 299, 315, 348, 371, 374, 390, 401-402, 403, 409-411, 413, 426, 427, 452, 474-475, 524, 539, 542, 556, 557, 571, 572, 579, 583, 599, 610, 620, 621, 625, 634, 643, 653-654, 678, 697, 700, 703, 708-709, 710, 714-715, 729, 750, 767, 768, 778-779, 782-783, 788, 797, 836, 860, 861, 865, 875, 895-896, 902, 904, 916	**PE/TE:** 56, 62, 63, 65, 66, 68, 69, 71, 83, 84, 85, 106, 107, 127, 132, 150, 152, 175, 187, 241, 251, 265, 283, 286, 294, 295, 299, 303, 310, 311, 315, 329, 333, 358, 375, 390, 404, 405, 411, 413, 418, 420, 421, 447, 452, 492, 503, 524, 542, 550, 559, 576, 577, 578, 602, 620, 638, 643, 679, 696, 710, 724, 750, 773, 778, 779, 783, 788, 792, 821, 870, 871, 875, 882, 896, 899, 908, 917, 926
7.5.spi.3. map large civilizations to discover the impact of water as a main reason behind a society's founding.	**PE/TE:** 557, 608, 650, 667, 669-670	**PE/TE:** 577, 650
at Level 3, the student is able to		
7.5.spi.4. analyze the causes and effects of change in a place over time from a written passage.	**PE/TE:** 49-52, 132, 134, 135, 136, 137, 140, 141, 162-163, 173, 201, 221-222, 232, 257, 262-263, 268, 270, 282, 292, 293-294, 305-308, 332-333, 353, 355-356, 363, 365, 367, 371, 372, 373, 374-375, 377, 378-379, 381, 384-385, 386, 387, 390, 392, 403-405, 409-411, 417-418, 425-426, 427, 428-429, 430-431, 433-435, 436, 437, 439, 442, 469-471, 472-473, 479, 484-485, 487-490, 502-503, 506-507, 509, 514, 515-516, 540, 556, 565, 566-567, 570-571, 573, 581, 583-585, 587, 588, 590-591, 595-596, 599-600, 608, 610-611, 613-615, 621, 623, 625-626, 628-629, 632-634, 635-636, 668, 669, 678-681, 683-684, 687, 691, 700, 703-705, 710, 711-713, 714-715, 720-722, 730, 733, 738-740, 757, 765-767, 766-769, 770-771, 775-777, 778-779, 781, 786-788, 789-790, 805-808, 814-815, 817, 818, 832-834, 836-837, 838-839, 841-842, 843-844, 845-846, 860, 861, 862, 863, 866, 867-868, 889, 894, 895-896, 901-903, 904, 916	**PE/TE:** 52, 134, 137, 141, 152, 153, 163, 173, 182, 201, 222, 226, 232, 250, 259, 263, 271, 286, 295, 299, 306, 308, 310, 311, 324, 333, 353, 356, 358, 364, 367, 375, 379, 381, 387, 396, 404, 405, 411, 413, 418, 420, 421, 422, 427, 429, 431, 439, 444, 446, 447, 458, 471, 473, 483, 485, 489, 490, 492, 503, 508, 511, 516, 518, 519, 540, 556, 567, 576, 577, 578, 585, 589, 591, 602, 603, 611, 623, 626, 629, 634, 636, 638, 639, 672, 674, 681, 684, 687, 692, 694, 696, 700, 705, 710, 713, 715, 722, 724, 725, 733, 742, 767, 769, 771, 777, 779, 781, 788, 790, 792, 808, 818, 820, 834, 837, 839, 842, 844, 846, 848, 863, 868, 870, 882, 894, 899, 908, 909, 911, 917, 926

Performance Indicators Teacher:

As documented through teacher observation -

Performance Indicators Teacher	Core Instruction	Application
at Level I, the student is able to		
7.5.tpi.1 diagram or reconstruct the movement of continents and tectonic plates.	**PE/TE:** 29-30	**PE/TE:** 33
at Level 2, the student is able to		
7.5.tpi.2. research past geographic phenomena and their impact on societies (e.g., volcanoes, earthquakes, hurricanes, flooding, typhoons).	**PE/TE:** 27, 30-31, 136, 199, 200, 380, 471, 540, 555-556, 596, 608, 651, 657, 662-663, 664-665, 667, 669, 734-735, 740, 757, 758, 762-763, 800, 826, 827, 853-854, 855, 856, 915	**PE/TE:** 33, 54, 137, 540, 556, 576, 657, 662, 663, 665, 672, 673, 735, 740, 742, 763, 792, 820, 821, 824, 826, 827, 829, 848, 849, 855, 856, 857, 869, 870, 871, 915
at Level 3, the student is able to		
7.5.tpi.3. analyze how geography affected historic events in order to predict future problems and decisions.	**PE/TE:** A11, 7, 9, 30-31, 73-74, 109, 132, 136, 149, 160, 161, 175, 178-180, 189, 200, 209, 256, 274, 275, 278-279, 283, 305-307, 317, 328, 329, 332-333, 368-369, 379, 393, 417-418, 429, 454, 464-466, 480, 498, 501, 516, 525, 536, 539, 541, 542-543, 555-556, 581, 589, 591, 592-593, 617, 622-623, 629, 634, 636, 645, 654, 657, 658, 662-663, 664-665, 667, 678, 692, 720-721, 727, 729, 734-735, 736, 738, 740, 749, 758, 760-761, 762-763, 796, 798, 799, 801-803, 825, 826, 828, 829, 845-846, 853-854, 855, 856-857, 859, 868, 877, 886, 898-899, 905-906	**PE/TE:** 55, 56, 74, 77, 83, 84, 85, 137, 149, 152, 153, 156, 161, 175, 179, 180, 182, 183, 209, 251, 277, 279, 284, 286, 307, 308, 310, 317, 329, 333, 394, 396, 398, 405, 417, 421, 465, 466, 467, 483, 490, 492, 501, 516, 518, 519, 525, 534, 541, 543, 550, 552, 556, 559, 567, 576, 593, 603, 623, 629, 634, 636, 638, 639, 645, 651, 652, 657, 663, 665, 672, 673, 695, 722, 724, 735, 740, 749, 761, 763, 792, 793, 799, 803, 820, 826, 829, 846, 848, 849, 854, 855, 857, 870, 877, 906, 908, 909
7.5.tpi.4. explore how geographic places change throughout time through pictures and illustrations.	**PE/TE:** 29, 30, 33, 50, 201, 257, 294, 305, 306-307, 495, 502, 540, 556, 562, 655, 659, 757, 758, 773-774, 855, 915	**PE/TE:** 33, 55, 127, 161, 519, 857

INDIVIDUALS, GROUPS, AND INTERACTIONS

Content Standard: 6.0

Personal development and identity are shaped by factors including culture, groups, and institutions. Central to this development are exploration, identification, and analysis of how individuals and groups work independently and cooperatively.

Learning Expectations	Core Instruction	Application
6.01 Understand the impact of individual and group decisions on citizens and communities.	**PE/TE:** 49-52, 67-71, 81, 82, 91, 93-95, 97-99, 131-137, 165-169, 213-217, 221-224, 231-235, 255-257, 261-264, 267-268, 270, 274, 275, 282, 283-284, 291-292, 305-308, 332-333, 336-339, 343-349, 351-356, 363-364, 365-367, 371-372, 374, 377-379, 380-381, 383-385, 386-387, 389-390, 392, 393, 401-405, 425-426, 428, 429, 430, 433-434, 436, 441-442, 443, 469-473, 474-475, 502-503, 505-508, 513-514, 516, 540, 556-557, 563-567, 569-571, 573-574, 582-585, 587-588, 590-591, 595-596, 598-599, 607-608, 613-617, 620, 625-627, 631-636, 654-656, 668, 669-670, 677-681, 683-684, 687, 689-692, 699-705, 729-731, 732-733, 765-771, 789-790, 805-808, 813-818, 831-834, 842, 859-860, 861, 862, 863, 866, 867, 889, 890-891, 893-894, 895-896, 901-903, 916	**PE/TE:** 48, 51, 52, 54, 55, 67, 69, 71, 81, 82, 84, 85, 86, 95, 100, 102, 103, 120, 132, 134, 135, 137, 151, 152, 166, 167, 169, 181, 182, 217, 222, 224, 225, 226, 227, 232, 235, 250, 252, 257, 259, 263, 264, 265, 271, 277, 284, 286, 288, 295, 306, 308, 310, 311, 333, 339, 345, 348, 349, 352, 353, 354, 356, 358, 359, 364, 367, 375, 379, 381, 385, 387, 393, 394, 396, 397, 402, 403, 404, 405, 420, 427, 428, 429, 431, 439, 442, 443, 444, 446, 447, 471, 473, 475, 492, 493, 503, 508, 514, 516, 518, 540, 559, 564, 566, 567, 572, 574, 576, 577, 584, 585, 591, 600, 602, 603, 611, 615, 617, 620, 622, 623, 626, 627, 629, 633, 634, 636, 638, 639, 655, 657, 670, 672, 673, 679, 681, 684, 687, 692, 694, 695, 700, 703, 705, 722, 724, 725, 726, 730, 731, 733, 742, 743, 769, 770, 771, 790, 792, 793, 806, 808, 815, 818, 820, 821, 832, 834, 837, 846, 848, 860, 861, 863, 868, 870, 882, 889, 891, 894, 899, 903, 908, 909, 917, 926
6.02 Understand how groups can impact change at local, regional, and global levels.	**PE/TE:** 49-52, 67-71, 81, 82, 91, 93-95, 97-99, 131-137, 165-169, 213-217, 221-224, 231-235, 255-257, 261-264, 267-268, 270, 274, 275, 282, 283-284, 291-292, 305-308, 332-333, 336-339, 343-349, 351-356, 363-364, 365-367, 371-372, 374, 377-379, 380-381, 383-385, 386-387, 389-390, 392, 393, 401-405, 425-426, 428, 429, 430, 433-434, 436, 441-442, 443, 469-473, 474-475, 502-503, 505-508, 513-514, 516, 540, 556-557, 563-567, 569-571, 573-574, 582-585, 587-588, 590-591, 595-596, 598-599, 607-608, 613-617, 620, 625-627, 631-636, 654-656, 668, 669-670, 677-681, 683-684, 687, 689-692, 699-705, 729-731, 732-733, 765-771, 789-790, 805-808, 813-818, 831-834, 842, 859-860, 861, 862, 863, 866, 867, 889, 890-891, 893-894, 895-896, 901-903, 916	**PE/TE:** 48, 51, 52, 54, 55, 67, 69, 71, 81, 82, 84, 85, 86, 95, 100, 102, 103, 120, 132, 134, 135, 137, 151, 152, 166, 167, 169, 181, 182, 217, 222, 224, 225, 226, 227, 232, 235, 250, 252, 257, 259, 263, 264, 265, 271, 277, 284, 286, 288, 295, 306, 308, 310, 311, 333, 339, 345, 348, 349, 352, 353, 354, 356, 358, 359, 364, 367, 375, 379, 381, 385, 387, 393, 394, 396, 397, 402, 403, 404, 405, 420, 427, 428, 429, 431, 439, 442, 443, 444, 446, 447, 471, 473, 475, 492, 493, 503, 508, 514, 516, 518, 540, 559, 564, 566, 567, 572, 574, 576, 577, 584, 585, 591, 600, 602, 603, 611, 615, 617, 620, 622, 623, 626, 627, 629, 633, 634, 636, 638, 639, 655, 657, 670, 672, 673, 679, 681, 684, 687, 692, 694, 695, 700, 703, 705, 722, 724, 725, 726, 730, 731, 733, 742, 743, 769, 770, 771, 790, 792, 793, 806, 808, 815, 818, 820, 821, 832, 834, 837, 846, 848, 860, 861, 863, 868, 870, 882, 889, 891, 894, 899, 903, 908, 909, 917, 926
6.03 Understand how a geographic view is a significant tool in interpreting the present and planning for the future.	**PE/TE:** 5-9	**PE/TE:** 9, 22, 23

SEVENTH GRADE BENCHMARKS

INDIVIDUALS, GROUPS, AND INTERACTIONS

Performance Indicators State:
As documented through state assessment -

Performance Indicators State	Core Instruction	Application
at Level I, the student is able to		
7.6.spi.1. identify ways family, groups, and community influence daily life and personal choices.	**PE/TE:** 49-52, 67-71, 81, 82, 91, 93-95, 97-99, 131-137, 165-169, 213-217, 221-224, 231-235, 255-257, 261-264, 267-268, 270, 274, 275, 282, 283-284, 291-292, 305-308, 332-333, 336-339, 343-349, 351-356, 363-364, 365-367, 371-372, 374, 377-379, 380-381, 383-385, 386-387, 389-390, 392, 393, 401-405, 425-426, 428, 429, 430, 433-434, 436, 441-442, 443, 469-473, 474-475, 502-503, 505-508, 513-514, 516, 540, 556-557, 563-567, 569-571, 573-574, 582-585, 587-588, 590-591, 595-596, 598-599, 607-608, 613-617, 620, 625-627, 631-636, 654-656, 668, 669-670, 677-681, 683-684, 687, 689-692, 699-705, 729-731, 732-733, 765-771, 789-790, 805-808, 813-818, 831-834, 842, 859-860, 861, 862, 863, 866, 867, 889, 890-891, 893-894, 895-896, 901-903, 916	**PE/TE:** 48, 51, 52, 54, 55, 67, 69, 71, 81, 82, 84, 85, 86, 95, 100, 102, 103, 120, 132, 134, 135, 137, 151, 152, 166, 167, 169, 181, 182, 217, 222, 224, 225, 226, 227, 232, 235, 250, 252, 257, 259, 263, 264, 265, 271, 277, 284, 286, 288, 295, 306, 308, 310, 311, 333, 339, 345, 348, 349, 352, 353, 354, 356, 358, 359, 364, 367, 375, 379, 381, 385, 387, 393, 394, 396, 397, 402, 403, 404, 405, 420, 427, 428, 429, 431, 439, 442, 443, 444, 446, 447, 471, 473, 475, 492, 493, 503, 508, 514, 516, 518, 540, 559, 564, 566, 567, 572, 574, 576, 577, 584, 585, 591, 600, 602, 603, 611, 615, 617, 620, 622, 623, 626, 627, 629, 633, 634, 636, 638, 639, 655, 657, 670, 672, 673, 679, 681, 684, 687, 692, 694, 695, 700, 703, 705, 722, 724, 725, 726, 730, 731, 733, 742, 743, 769, 770, 771, 790, 792, 793, 806, 808, 815, 818, 820, 821, 832, 834, 837, 846, 848, 860, 861, 863, 868, 870, 882, 889, 891, 894, 899, 903, 908, 909, 917, 926
at Level 2, the student is able to		
7.6.spi.2. differentiate between the rights, roles and state of the individual in relation to the general welfare in various regions of the world.	**PE/TE:** 49-52, 67-71, 81, 82, 91, 93-95, 97-99, 131-137, 165-169, 213-217, 221-224, 231-235, 255-257, 261-264, 267-268, 270, 274, 275, 282, 283-284, 291-292, 305-308, 332-333, 336-339, 343-349, 351-356, 363-364, 365-367, 371-372, 374, 377-379, 380-381, 383-385, 386-387, 389-390, 392, 393, 401-405, 425-426, 428, 429, 430, 433-434, 436, 441-442, 443, 469-473, 474-475, 502-503, 505-508, 513-514, 516, 540, 556-557, 563-567, 569-571, 573-574, 582-585, 587-588, 590-591, 595-596, 598-599, 607-608, 613-617, 620, 625-627, 631-636, 654-656, 668, 669-670, 677-681, 683-684, 687, 689-692, 699-705, 729-731, 732-733, 765-771, 789-790, 805-808, 813-818, 831-834, 842, 859-860, 861, 862, 863, 866, 867, 889, 890-891, 893-894, 895-896, 901-903, 916	**PE/TE:** 48, 51, 52, 54, 55, 67, 69, 71, 81, 82, 84, 85, 86, 95, 100, 102, 103, 120, 132, 134, 135, 137, 151, 152, 166, 167, 169, 181, 182, 217, 222, 224, 225, 226, 227, 232, 235, 250, 252, 257, 259, 263, 264, 265, 271, 277, 284, 286, 288, 295, 306, 308, 310, 311, 333, 339, 345, 348, 349, 352, 353, 354, 356, 358, 359, 364, 367, 375, 379, 381, 385, 387, 393, 394, 396, 397, 402, 403, 404, 405, 420, 427, 428, 429, 431, 439, 442, 443, 444, 446, 447, 471, 473, 475, 492, 493, 503, 508, 514, 516, 518, 540, 559, 564, 566, 567, 572, 574, 576, 577, 584, 585, 591, 600, 602, 603, 611, 615, 617, 620, 622, 623, 626, 627, 629, 633, 634, 636, 638, 639, 655, 657, 670, 672, 673, 679, 681, 684, 687, 692, 694, 695, 700, 703, 705, 722, 724, 725, 726, 730, 731, 733, 742, 743, 769, 770, 771, 790, 792, 793, 806, 808, 815, 818, 820, 821, 832, 834, 837, 846, 848, 860, 861, 863, 868, 870, 882, 889, 891, 894, 899, 903, 908, 909, 917, 926

Performance Indicators State	Core Instruction	Application
at Level 3, the student is able to		
7.6.spi.3. recognize the causes, consequences and possible solutions applied by governing bodies to persistent global issues using a narrative (i.e., health, security, resource allocation, economic development, environmental quality.)	**PE/TE:** 49-52, 67-71, 81, 82, 91, 93-95, 97-99, 131-137, 165-169, 213-217, 221-224, 231-235, 255-257, 261-264, 267-268, 270, 274, 275, 282, 283-284, 291-292, 305-308, 332-333, 336-339, 343-349, 351-356, 363-364, 365-367, 371-372, 374, 377-379, 380-381, 383-385, 386-387, 389-390, 392, 393, 401-405, 425-426, 428, 429, 430, 433-434, 436, 441-442, 443, 469-473, 474-475, 502-503, 505-508, 513-514, 516, 540, 556-557, 563-567, 569-571, 573-574, 582-585, 587-588, 590-591, 595-596, 598-599, 607-608, 613-617, 620, 625-627, 631-636, 654-656, 668, 669-670, 677-681, 683-684, 687, 689-692, 699-705, 729-731, 732-733, 765-771, 789-790, 805-808, 813-818, 831-834, 842, 859-860, 861, 862, 863, 866, 867, 889, 890-891, 893-894, 895-896, 901-903, 916	**PE/TE:** 48, 51, 52, 54, 55, 67, 69, 71, 81, 82, 84, 85, 86, 95, 100, 102, 103, 120, 132, 134, 135, 137, 151, 152, 166, 167, 169, 181, 182, 217, 222, 224, 225, 226, 227, 232, 235, 250, 252, 257, 259, 263, 264, 265, 271, 277, 284, 286, 288, 295, 306, 308, 310, 311, 333, 339, 345, 348, 349, 352, 353, 354, 356, 358, 359, 364, 367, 375, 379, 381, 385, 387, 393, 394, 396, 397, 402, 403, 404, 405, 420, 427, 428, 429, 431, 439, 442, 443, 444, 446, 447, 471, 473, 475, 492, 493, 503, 508, 514, 516, 518, 540, 559, 564, 566, 567, 572, 574, 576, 577, 584, 585, 591, 600, 602, 603, 611, 615, 617, 620, 622, 623, 626, 627, 629, 633, 634, 636, 638, 639, 655, 657, 670, 672, 673, 679, 681, 684, 687, 692, 694, 695, 700, 703, 705, 722, 724, 725, 726, 730, 731, 733, 742, 743, 769, 770, 771, 790, 792, 793, 806, 808, 815, 818, 820, 821, 832, 834, 837, 846, 848, 860, 861, 863, 868, 870, 882, 889, 891, 894, 899, 903, 908, 909, 917, 926

Performance Indicators Teacher:
As documented through teacher observation -

Performance Indicators Teacher	Core Instruction	Application
at Level I, the student is able to		
7.6.tpi.1. diagram the home, school, or path to school in order to achieve a personal sense of geographic perspective.	**PE/TE:** 103	**PE/TE:** 103
7.6.tpi.2. conduct a discussion about the culture of the school and community to show how regional, ethnic and national culture affects the student's daily life.	**PE/TE:** 103	**PE/TE:** 103
7.6.tpi.3. exhibit model behavior during a guest speaker.	**PE/TE:** An opportunity to address this standard may be found on pages 287, 349, 670	**PE/TE:** An opportunity to address this standard may be found on pages 287, 349, 670

Performance Indicators Teacher	Core Instruction	Application
at Level 2, the student is able to		
7.6.tpi.4. create a web diagram analyzing multiple points of view about geographic issues. (e.g., land use, urban planning, conservation, consumption, production, population.)	**PE/TE:** 554	**PE/TE:** 52, 54, 77, 103, 519, 554, 743
7.6.tpi.5. discuss how individuals exhibit how he/she belongs to a larger culture and, yet, retain their individual identities.	**PE/TE:** 89-90, 91, 92-93, 94-95, 100	**PE/TE:** 86, 91, 93, 95, 100, 102, 103
7.6.tpi.6. compare an average student's calorie consumption both in terms of calorie intake and a total daily energy consumption to that of an average third world student's average consumption.	**PE/TE:** An opportunity to address this standard may be found on page 411	**PE/TE:** An opportunity to address this standard may be found on page 611
7.6.tpi.7. analyze how individuals within the global culture can influence a country's conduct. (e.g., Ghandi, Castro, Mandela, Hussein).	**PE/TE:** 90, 135, 223, 233, 239, 263, 336, 337, 365, 384, 404, 427, 436, 470, 471, 473, 514, 515, 564, 583, 589, 614-615, 616, 620, 626, 634, 669, 680, 703-704, 730, 771, 814, 862, 894	**PE/TE:** 90, 137, 224, 446, 473, 492, 493, 515, 589, 616, 617, 629, 636, 670, 673, 680, 694, 695, 704, 725, 771, 792
at Level 3, the student is able to		
7.6.tpi.8. participate in a simulation in which students assume roles of an international organization in which they attempt to resolve global issues.	**PE/TE:** 271, 308, 503, 663, 681, 687, 743	**PE/TE:** 271, 308, 503, 663, 681, 687, 743
7.6.tpi.9. compare women's rights in the United States, Europe, Asia, Africa, and the Middle East.	**PE/TE:** 361, 385, 572, 621, 709, 776, 832, 867	**PE/TE:** 713, 777, 832
7.6.tpi.10. debate how role, status, and social class affect interactions of individuals and social groups.	**PE/TE:** 140, 142, 171-172, 173, 237-239, 258, 264-265, 269, 271, 274, 277, 283, 297-299, 372, 385, 390, 411, 426, 434, 436-437, 477-478, 507-508, 511, 565-567, 572, 583, 584, 599-600, 610-611, 615, 621-622, 627, 636, 678, 680, 692, 709-710, 731, 733, 739, 775-776, 778-779, 788, 811, 814, 832, 861, 862, 867-868, 864, 895-896, 903, 917	**PE/TE:** 145, 152, 239, 259, 277, 420, 431, 492, 508, 511, 567, 576, 600, 603, 638, 692, 694, 742, 777, 788, 792, 832, 894, 906
7.6.tpi.11 create a montage showing cooperation and conflict throughout the world.	**PE/TE:** 241; An opportunity to address this standard may be found on pages 77, 356, 381, 418, 439, 591, 617, 629, 705, 733	**PE/TE** 241; An opportunity to address this standard may be found on pages 77, 356, 381, 418, 439, 591, 617, 629, 705, 733